Encyclopedic Dictionary of Mathematics

Second Edition

by the
Mathematical Society
of Japan

edited by
Kiyosi Itô

Volume II
O–Z
Appendices and Indexes

The MIT Press
Cambridge, Massachusetts,
and London, England

Third printing, 1996
First MIT Press paperback edition, 1993

This book was set in Monolaser Times Roman by Asco Trade Typesetting Ltd., Hong Kong, and printed and bound by Arcata Graphics, Kingsport in the United States of America.

Library of Congress Cataloging-in-Publication Data

Iwanami sūgaku jiten (ziten). English.
Encyclopedic dictionary of mathematics.
Translation of: Iwanami sūgaku jiten.
Includes bibliographies and indexes.
1. Mathematics—Dictionaries. I. Itô, Kiyosi, 1915– . II. Nihon Sūgakkai. III. Title.
QA5.I8313 1986 510′.3′21 86-21092
ISBN 0-262-09026-0 (HB), 0-262-59020-4 (PB)

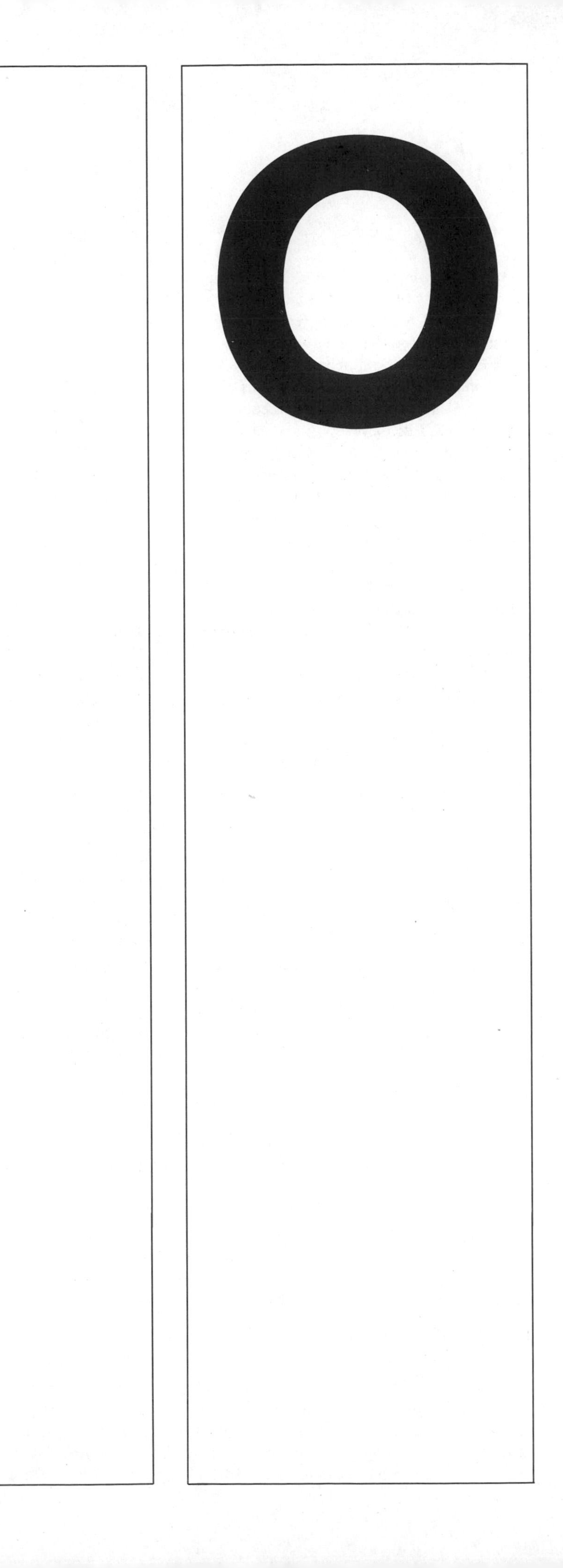

305 (IX.11) Obstructions

A. History

The theory of **obstructions** aims at measuring the extensibility of mappings by means of algebraic tools. Such classical results as the †Brouwer mapping theorem and Hopf's extension and †classification theorems in homotopy theory might be regarded as the origins of this theory. A systematic study of the theory was initiated by S. Eilenberg [1] in connection with the notions of †homotopy and †cohomology groups, which were introduced at the same time. A. Komatu and P. Olum [2] extended the theory to mappings into spaces not necessarily †n-simple. For mappings of polyhedra into certain special spaces, the †homotopy classification problem, closely related to the theory of obstructions, was solved in the following cases (K^m denotes an m-dimensional polyhedron): $K^{n+1} \to S^n$ (N. Steenrod [5]), $K^{n+2} \to S^n$ (J. Adem), $K^{n+k} \to Y$, where $\pi_i(Y)=0$ for $i<n$ and $n<i<n+k$ (M. Nakaoka). There are similar results by L. S. Pontryagin, M. Postnikov, and S. Eilenberg and S. MacLane. Except for the special cases already noted, it is extremely difficult to discuss higher obstructions in general since they involve many complexities. Nevertheless, it is significant that the idea of obstructions has given rise to various important notions in modern algebraic topology, including cohomology operations (→ 64 Cohomology Operations) and characteristic classes (→ 56 Characteristic Classes).

The notion of obstruction is also very useful in the treatment of cross sections of fiber bundles (→ 147 Fiber Bundles), †diffeomorphisms of differentiable manifolds, etc.

B. General Theory for an n-Simple Space Y

The question of whether two (continuous) mappings of a topological space X into another space Y are †homotopic to each other can be reduced to the extensibility of the given mapping: $(X \times \{0\}) \cup (X \times \{1\}) \to Y$ to a mapping of the product $X \times I$ of X and the unit interval $I=[0,1]$ into Y. Therefore the problem of classifying mappings can be treated in the same way as that of the extension of mappings.

Let K be a †polyhedron, L a subpolyhedron of K, and $\hat{K}^n = L \cup K^n$ the union of L and the †n-skeleton K^n of K. Let Y be an †arcwise connected n-simple space, and f' be a mapping of L into Y. Denote by $\Phi^n(f')$ the set of mappings of $\hat{K}^n$ into Y that are extensions of f', and by $\Phi^n(f')$ the set of †homotopy classes of mappings in $\bar{\Phi}^n(f')$ relative to L. The set $\bar{\Phi}^0(f')$ consists of a single element because of the arcwise connectedness of Y, $\bar{\Phi}^1(f')$ is nonempty, and $\bar{\Phi}^n(f')$ $(n \geqslant 2)$ may be empty. Let f^n be an element of $\Phi^n(f')$. If we consider the restriction of f^n to the boundary $\dot{\sigma}^{n+1}$ of an oriented $(n+1)$-cell σ^{n+1} of K, then $f^n: \dot{\sigma}^{n+1} \to Y$ determines an element $c(f^n, \sigma^{n+1})$ of the †homotopy group $\pi_n(Y)$ (→ 202 Homotopy Theory). This element gives a measure of obstruction for extending f^n to the interior of σ^{n+1}. We obtain an $(n+1)$-†cocycle $c^{n+1}(f^n)$ of the †simplicial pair (K, L) with coefficients in $\pi_n(Y)$, called the **obstruction cocycle** of f^n, by assigning $c(f^n, \sigma^{n+1})$ to each $(n+1)$-cell σ^{n+1}. This obstruction cocycle $c^{n+1}(f^n)$ is the measure of obstruction for extending f^n to $\hat{K}^{n+1}$. A necessary and sufficient condition for the extensibility is given by $c^{n+1}(f^n)=0$. Clearly, $c^{n+1}(f^n)$ is uniquely determined for each element f of $\bar{\Phi}^n(f')$. The set of all $c^{n+1}(f^n)$ with $f^n \in \Phi^n(f')$ forms a subset $\mathbf{o}^{n+1}(f')$ of the group of cocycles $Z^{n+1}(K, L; \pi_n(Y))$. $\Phi^{n+1}(f')$ is nonempty if and only if $\mathbf{o}^{n+1}(f')$ contains the zero element 0.

Let $K^\square = K \times I$, $L^\square = (K \times 0) \cup (L \times I) \cup (K \times 1)$. Given two mappings $f_0, f_1: K \to Y$ satisfying $f_0 | L = f_1 | L$, we can define a natural mapping $F': L^\square \to Y$ such that an element F^n of $\Phi^n(F')$ corresponds to a †homotopy h^{n-1} relative to L connecting $f_0 | K^{n-1}$ with $f_1 | K^{n-1}$. Given an element $F^n \in \Phi^n(F')$, we have the element $c^{n+1}(F^n)$ of $Z^{n+1}(K^\square, L^\square; \pi_n(Y))$, which we identify with $Z^n(K, L; \pi_n(Y))$ through the natural isomorphism of chain groups of the pair $(K^\square, L^\square)$ to those of the pair (K, L). Thus we can regard $c^{n+1}(F^n)$ as an element of $Z^n(K, L; \pi_n(Y))$, which is denoted by $d^n(f_0, h^{n-1}, f_1)$, and call it the **separation** (or **difference**) **cocycle**. If $f_0 | \hat{K}^{n-1} = f_1 | \hat{K}^{n-1}$, we have the canonical mapping $F^n: L^\square \cup (K^\square)^n \to Y$, and the separation cocycle is denoted simply by $d^n(f_0, f_1)$. The set of separation cocycles corresponding to elements of $\Phi^n(F')$ is considered to be a subset of $Z^n(K, L; \pi_n(Y))$ and is denoted by $\mathbf{o}^n(f_0, f_1)$. A necessary and sufficient condition for h^{n-1} to be extensible to a homotopy on $\hat{K}^n$ is $d^n(f_0, h^{n-1}, f_1)=0$. Therefore a necessary and sufficient condition for $f_0 | \hat{K}^n \simeq f_1 | \hat{K}^n$ (rel L) (i.e., relative to L) is $0 \in \mathbf{o}^n(f_0, f_1)$. Given $f_0^n, f_1^n: \hat{K}^n \to Y$ with $f_0^n | L = f_1^n | L$, then $d^n(f_0^n, h^{n-1}, f_1^n)$ $(\in \mathbf{o}^n(f_0^n, f_1^n))$ is an element of $Z^n(\hat{K}^n, L; \pi_n(Y))$, which is also considered to be a cochain of the pair (K, L). In this sense, we call $d^n(f_0^n, h^{n-1}, f_1^n)$ the **separation** (or **deformation**) **cochain** over (K, L). The coboundary of the separation cochain $d^n(f_0^n, h^{n-1}, f_1^n)$ coincides (except possibly for sign) with $c^{n+1}(f_0^n) - c^{n+1}(f_1^n)$.

For a fixed $f_0^n \in \Phi^n(f')$, any n-cochain d^n

of the pair (K, L) with coefficients in $\pi_n(Y)$ is expressible as a separation cochain $d^n = d^n(f_0^n, f_1^n)$ where $f_1^n \in \Phi^n(f')$ is a suitable mapping such that $f_0^n | \hat{K}^{n-1} = f_1^n | \hat{K}^{n-1}$ (existence theorem).

Therefore if we take an element f^{n-1} of $\bar{\Phi}^{n-1}(f')$ whose obstruction cocycle $c^n(f^{n-1})$ is zero, the set of all obstruction cocycles $c^{n+1}(f^n)$ of all such $f^n \in \Phi^n(f')$ that are extensions of f^{n-1} forms a subset of $\mathbf{o}^{n+1}(f')$ and coincides with a coset of $Z^{n+1}(K, L; \pi_n(Y))$ factored by $B^{n+1}(K, L; \pi_n(Y))$. Thus a cohomology class $\bar{c}^{n+1}(f^{n-1}) \in H^{n+1}(K, L; \pi_n(Y))$ corresponds to an $f^{n-1} \in \Phi^{n-1}(f')$ such that $c^n(f^{n-1}) = 0$, and $\bar{c}^{n+1}(f^{n-1}) = 0$ is a necessary and sufficient condition for f^{n-1} to be extensible to $\hat{K}^{n+1}$ (**first extension theorem**).

For the separation cocycle, $\bar{d}^n(f_0, h^{n-2}, f_1) \in H^n(K, L; \pi_n(Y))$ corresponds to each homotopy h^{n-2} on $\hat{K}^{n-2}$ such that $d^{n-1}(f_0, h^{n-2}, f_1) = 0$, and $\bar{d}^n(f_0, h^{n-2}, f_1) = 0$ is a necessary and sufficient condition for h^{n-2} to be extensible to a homotopy on $\hat{K}^n$ (**first homotopy theorem**).

The subset of $H^{n+1}(K, L; \pi_n(Y))$ corresponding to $\mathbf{o}^{n+1}(f')$ is denoted by $\mathbf{O}^{n+1}(f')$ and is called the **obstruction to an** $(n+1)$**-dimensional extension** of f'. Similarly, the subset $\mathbf{O}^n(f_0, f_1)$ of $H^n(K, L; \pi_n(Y))$ corresponding to $\mathbf{o}^n(f_0, f_1)$ is called the **obstruction to an** n**-dimensional homotopy** connecting f_0 with f_1. Clearly, a condition for f' to be extensible to $\hat{K}^{n+1}$ is given by $0 \in \mathbf{O}^{n+1}(f')$, and a necessary and sufficient condition for $f_0 | K^n \simeq f_1 | K^n$ (rel L) is given by $0 \in \mathbf{O}^n(f_0, f_1)$.

A continuous mapping $\varphi: (K', L') \to (K, L)$ induces homomorphisms of cohomology groups $\varphi^*: H^{n+1}(K, L; \pi_n(Y)) \to H^{n+1}(K', L'; \pi_n(Y))$, $H^n(K, L; \pi_n(Y)) \to H^n(K', L'; \pi_n(Y))$. Then for $f': L \leftarrow Y$, $\mathbf{O}^{n+1}(f' \circ \varphi) \supset \varphi^* \mathbf{O}^{n+1}(f')$, and for $f_0, f_1: K \to Y$ such that $f_0 | L = f_1 | L$, $\mathbf{O}^n(f_0 \circ \varphi, f_1 \circ \varphi) \supset \varphi^* \mathbf{O}^n(f_0, f_1)$. Therefore we also find that the obstruction to an extension and the obstruction to a homotopy are independent of the choice of subdivisions of K, L, and consequently are topological invariants.

Let f_0, f_1, and f_2 be mappings $K \to Y$ such that $f_0 | L = f_1 | L = f_2 | L$. Given homotopies $h_{01}^{n-1}: f_0 | \hat{K}^{n-1} \simeq f_1 | \hat{K}^{n-1}$ (rel L), $h_{12}^{n-1}: f_1 | \hat{K}^{n-1} \simeq f_2 | \hat{K}^{n-1}$ (rel L), then for the composite $h_{02}^{n-1} = h_{12}^{n-1} \circ h_{01}^{n-1}$, we have

$$d^n(f_0, h_{01}^{n-1}, f_1) + d^n(f_1, h_{12}^{n-1}, f_2)$$

$$= d^n(f_0, h_{02}^{n-1}, f_2),$$

and for the inverse homotopy $h_{10}^{n-1}: f_1 | \hat{K}^{n-1} \simeq f_0 | \hat{K}^{n-1}$ of h_{01}^{n-1}, clearly

$$d^n(f_1, h_{10}^{n-1}, f_0) = -d^n(f_0, h_{01}^{n-1}, f_1).$$

Therefore $\mathbf{O}^n(f_0, f_0)$ forms a subgroup of $H^n(K, L; \pi_n(Y))$ that is determined by the homotopy class of $f_0 | \hat{K}^{n-1}$ relative to L. In general, if $\mathbf{O}^n(f_0, f_1)$ is nonempty, it is a coset of $H^n(K, L; \pi_n(Y))$ factored by the subgroup $\mathbf{O}^n(f_0, f_0)$. Combined with the existence theorem on separation cochains, this can be utilized to show the following theorem.

Assume that $\Phi^n(f')$ is nonempty. The set of all elements $\bar{\Phi}^n(f')$ that are extensions of an element of $\bar{\Phi}^{n-1}(f')$ is put in one-to-one correspondence with the quotient group of $H^n(\hat{K}^n, L; \pi_n(Y))$ modulo $\mathbf{O}^n(f_0^n, f_0^n)$ by pairing the obstruction $\mathbf{O}^n(f_0^n, f^n)$ with each f^n for a fixed f_0^n. Among such elements of $\bar{\Phi}^n(f')$, the set of f^n that are extensible to $\hat{K}^{n+1}$ is in one-to-one correspondence with the quotient group of $H^n(\hat{K}^{n+1}, L; \pi_n(Y)) = H^n(K, L; \pi_n(Y))$ modulo the subgroup $\mathbf{O}^n(f_0^{n+1}, f_0^{n+1})$, assuming that f_0^n is extended to f_0^{n+1} (**first classification theorem**).

C. Primary Obstructions

Assume that $H^{i+1}(K, L; \pi_i(Y)) = H^i(K, L; \pi_i(Y)) = 0$, where $0 < i < p$ (e.g., $\pi_i(Y) = 0$, $0 < i < p$). In this case, by consecutive use of the first extension theorem and the first homotopy theorem, we can show that each $\bar{\Phi}^i(f')$ $(i < p)$ consists of a single element and $\mathbf{O}^{p+1}(f')$ also consists of a single element $\bar{c}^{p+1}(f') \in H^{p+1}(K, L; \pi_p(Y))$. The element $\bar{c}^{p+1}(f')$, called the **primary obstruction** of f', vanishes if and only if f' can be extended to $\hat{K}^{p+1}$ (**second extension theorem**). When $H^{i+1}(K, L; \pi_i(Y)) = 0$ for $i > p$ (for example, when $\pi_i(Y) - 0$ for $p < i < \dim(K - L)$), f' is extendable to K if and only if the first obstruction of f' vanishes (**third extension theorem**).

Correspondingly, if $H^i(K, L; \pi_i(Y)) = H^{i-1}(K, L; \pi_i(Y)) = 0$ $(0 < i < p)$, then for any two mappings $f_0, f_1: K \to Y$, $f_0 | L = f_1 | L$, $\mathbf{O}^p(f_0, f_1)$ consists of a single element $\bar{d}^p(f_0, f_1) \in H^p(K, L; \pi_p(Y))$, which we call the **primary difference** of f_0 and f_1. This element vanishes if and only if $f_0 | \hat{K}^p \simeq f_1 | \hat{K}^p$ (rel L) (**second homotopy theorem**). Moreover, when $H^i(K, L; \pi_i(Y)) = 0$ $(i > p)$, the primary difference is zero if and only if $f_0 \simeq f_1$ (rel L) (**third homotopy theorem**).

Assume that the hypotheses of the second extension theorem and second homotopy theorem are satisfied. If we assign to each element f^p of $\bar{\Phi}^p(f')$ the primary difference of f^p and the fixed element f_0^p, then $\bar{\Phi}^p(f')$ is in one-to-one correspondence with $H^p(\hat{K}^p, L; \pi_p(Y))$ by the first classification theorem (**second classification theorem**). Similarly, assume that the hypotheses of the third extension theorem and third homotopy theorem are satisfied. If $f_0: K \to Y$, $f' = f_0 | L$, then homotopy classes relative to L of extensions f of f' are put in one-to-one correspondence with the elements of $H^p(K, L; \pi_p(Y))$ by pairing $d^p(f, f_0)$ with f (**third classification theorem**).

D. Secondary Obstructions

For simplicity, assume that $\pi_i(Y)=0$ ($i<p$ and $p<i<q$). If the primary obstruction $\bar{c}^{p+1}(f')\in H^{p+1}(K, L; \pi_p(Y))$ of $f': L\rightarrow Y$ vanishes, we can define $\mathbf{O}^{q+1}(f')\subset H^{q+1}(K, L; \pi_q(Y))$, which we call the **secondary obstruction** of f'. When $Y=S^p$, $q=p+1$, $p>2$, the secondary obstruction $\mathbf{O}^{p+2}(f')$ coincides with a coset of $H^{p+2}(K, L; Z_2)$ modulo the subgroup $Sq^2(H^p(K, L; Z))$, where Sq^2 denotes the †Steenrod square operation [5]. In this case, if $L=K^p$, then $\mathbf{O}^{p+2}(f')$ reduces to a cohomology class, $Sq^2(i^*)^{-1}f'^*(\sigma)$ with $i: L\rightarrow K$, where σ is a generator of $H^p(S^p, Z)$ (in this case $(i^*)^{-1}f'^*(\sigma)\neq\varnothing$ is equivalent to $\bar{c}^{p+1}(f')=0$) [5]. Moreover, if $Sq^2f'^*(\sigma)=0$, then there exists a suitable extension $f^{p+2}: \hat{K}^{p+2}\rightarrow Y=S^p$ of f'. The set of obstruction cocycles of all such f^{p+2} defines the **tertiary obstruction** $\mathbf{O}^{p+3}(f')$, which coincides with a coset of $H^{p+3}(K, L; Z_2)$ modulo the subgroup $Sq^2(H^{p+1}(K, L; Z_2))$. By using the †secondary cohomology operation Φ of J. Adem, it can be expressed as $\Phi((i^*)^{-1}f'^*(\sigma))$ (→ 64 Cohomology Operations).

All the propositions in this article remain true if we take †CW complexes instead of polyhedra K.

References

[1] S. Eilenberg, Cohomology and continuous mappings, Ann. Math., (2) 41 (1940), 231–251.
[2] P. Olum, Obstructions to extensions and homotopies, Ann. Math., (2) 52 (1950), 1–50.
[3] P. Olum, On mappings into spaces in which certain homotopy groups vanish, Ann. Math., (2) 57 (1953), 561–574.
[4] N. E. Steenrod, The topology of fibre bundles, Princeton Univ. Press, 1951.
[5] N. E. Steenrod, Products of cocycles and extensions of mappings, Ann. Math., (2) 48 (1947), 290–320.
[6] E. H. Spanier, Algebraic topology, McGraw-Hill, 1966.

306 (XII.20) Operational Calculus

A. General Remarks

The term "operational calculus" in the usual sense means a method for solving †linear differential equations by reducing the operations of differentiation and integration into algebraic ones in a symbolic manner. The idea was initiated by P. S. Laplace in his *Théorie analytique des probabilités* (1812), but the method has acquired popularity since O. Heaviside used it systematically in the late 19th century to solve electric-circuit problems. The method is therefore also called **Heaviside calculus**, but Heaviside gave only a formal method of calculus without bothering with rigorous arguments. The mathematical foundations were given in later years, first in terms of †Laplace transforms, then by applying the theory of †distributions. One of the motivations behind L. Schwartz's creation of this latter theory in the 1940s was to give a sound foundation for the formal method, but the theory obtained has had a much larger range of applications. Schwartz's theory was based on the newly developed theory of †topological linear spaces. On the other hand, J. Mikusiński gave another foundation, based only on elementary algebraic notions and on Titchmarsh's theorem, whose proof has recently been much simplified.

In this article, we first explain the simple theory established by Mikusiński [2] and later discuss its relation to the classical Laplace transform method.

B. The Operational Calculus of Mikusiński

The set $\mathscr{C}$ of all continuous complex-valued functions $a=\{a(t)\}$ defined on $t\geqslant 0$ is a †linear space with the usual addition and scalar multiplication. $\mathscr{C}$ is a †commutative algebra with multiplication $a\cdot b$ defined by the †convolution $\{\int_0^t a(t-s)b(s)\,ds\}$. The ring $\mathscr{C}$ has no †zero divisors (**Titchmarsh's theorem**). (There have been several interesting proofs of Titchmarsh's theorem since the first demonstration given by Titchmarsh himself [3]. For example, a simple proof has been published by C. Ryll-Nardzewski (1952).) Hence we can construct the †quotient field $\mathscr{Q}$ of the ring $\mathscr{C}$. An element of $\mathscr{Q}$ is called a **Mikusiński operator**, or simply an **operator**. If we define $a(t)=0$ for $t<0$ for the elements $\{a(t)\}$ in $\mathscr{C}$, then $\mathscr{C}$ is a subalgebra of $\mathscr{U}$, which is the set of all locally integrable (locally L_1) functions in $(-\infty, \infty)$ whose †support is bounded below. Here we identify two functions that coincide almost everywhere. The algebra $\mathscr{U}$ has no zero divisor, and its quotient field is also $\mathscr{Q}$.

The unity element for multiplication in $\mathscr{Q}$, denoted by $\delta=b/b$ ($b\neq 0$), plays the role of the †Dirac δ-function. It is sometimes called the **impulse function**. The operator $l=\{1\}\in\mathscr{C}$ is the function that takes the values 0 and 1 according as $t<0$ or $t>0$. This operator is **Heaviside's function** and is sometimes called the **unit function**. Usually it is denoted by $\mathbf{1}(t)$ or simply **1**. The value $\mathbf{1}(0)$ may be arbitrary, but usually it is defined as 1/2, the mean of the limit values from both sides. The operator l is an **integral operator**, because, as an operator carrying a into $l \cdot a$, it yields

$$l \cdot a = \left\{ \int_0^t a(s)ds \right\} = \text{the integral of } a \text{ over } [0,t].$$

More generally, the operator $\{t^{\lambda-1}/\Gamma(\lambda)\}$ ($\mathrm{Re}\,\lambda > 0$) gives the λth-order integral. The operator $s=\delta/l$, which is the inverse operator of l, is a **differential operator**. If $a \in \mathscr{C}$ is of †class C^1, then we have

$$s \cdot a = a' + a(+0)\delta = a' + \{a(+0)\}/\{1\}. \qquad (1)$$

Similarly, if $a \in \mathscr{C}$ is of class C^n, we have

$$s^n \cdot a = a^{(n)} + a^{(n-1)}(0)\delta + a^{(n-2)}(0)s + \ldots + a(0)s^{n-1}. \qquad (2)$$

The operator $a \to s \cdot a$ can be applied to functions a that are not differentiable in the ordinary sense, and considering the application of s to be the operation of differentiation, we can treat the differential operator algebraically in the field $\mathscr{Q}$. In particular, we have $s \cdot \mathbf{1} = \delta$, and this relation is frequently represented by the formula

$$d\mathbf{1}(t)/dt = \delta(t). \qquad (3)$$

A rational function of s whose numerator is of lower degree than its denominator is an †elementary function of t. For example, we have the relations

$$1/(s-\alpha)^n = \{t^{n-1}e^{\alpha t}/(n-1)!\},$$
$$1/(s^2+\beta^2) = \{\beta^{-1}\sin\beta t\},$$
$$s/(s^2+\beta^2) = \{\cos\beta t\}. \qquad (4)$$

The solution of an ordinary linear differential equation with constant coefficients $\sum_{r=0}^n \alpha_r \varphi^{(r)}(t) = f(t)$ under the †initial condition $\varphi^{(i)}(0) = \gamma_i$ $(0 \leqslant i \leqslant n-1)$ is thus reduced to an equation in s by using formulas (1) and (2), and is computed by decomposing the following operator into partial fractions:

$$\{\varphi(t)\} = \frac{\beta_{n-1}s^{n-1} + \ldots + \beta_0 + f}{\alpha_n s^n + \ldots + \alpha_1 s + \alpha_0}, \qquad (5)$$

where $\beta_r = \alpha_{r+1}\gamma_0 + \alpha_{r+2}\gamma_1 + \ldots + \alpha_n\gamma_{n-r-1}$, $0 \leqslant r \leqslant n-1$. The general solution is represented by (5) if we consider the constants $\gamma_0, \ldots, \gamma_{n-1}$ or $\beta_0, \ldots, \beta_{n-1}$ as arbitrary parameters. If the rational function in the right-hand side of (5) is $M(s)/D(s)$ and the degree of the numerator is less than that of the denominator, then the right-hand side of (5) is explicitly represented by

$$\sum_{r=0}^{l-1} \frac{t^{l-r-1}}{(l-r-1)!r!} \left[\frac{d^r}{d\lambda^r}\left(\frac{M(\lambda)}{D(\lambda)/(\lambda-\lambda_0)^l} \right) \right]_{\lambda=\lambda_0} e^{\lambda_0 t} + \sum_{i=1}^m \frac{M(\lambda_i)}{D(\lambda_i)} e^{\lambda_i t},$$

where we assume that $\lambda_0, \lambda_1, \ldots, \lambda_m$ exhaust the roots of the equation $D(\lambda)=0$, λ_0 is a multiple root of degree l, and all other roots are simple $(m=n-l)$. The above formula is called the **expansion theorem**.

C. Limits of Operators

A sequence a_n of operators is said to **converge** to the **limit** $a=b/q$ if there exists an operator $q(\neq 0)$ such that $q \cdot a_n \in \mathscr{C}$ and the sequence of functions $q \cdot a_n$ converges †uniformly to b on compact sets. The limit a is determined uniquely without depending on the operator q. Based on this notion of limits of operators, we can construct the theory of series of operators and differential and integral calculus of functions of an independent variable λ whose values are operators. They are completely parallel to the usual theories of elementary calculus (→ 106 Differential Calculus; 216 Integral Calculus; 379 Series). A linear partial differential equation in the function $\varphi(x,t)$ of two variables, in particular its initial value problem, reduces to a linear ordinary differential equation of an operational-valued function of an independent variable x.

For a given operator w, the solution (if it exists) of the differential equation $\varphi'(\lambda) = w \cdot \varphi(\lambda)$ with the initial condition $\varphi(0)=\delta$ is unique, is called the **exponential function of an operator** w, and is denoted by $\varphi(\lambda)=e^{\lambda w}$. If the power series

$$\sum_{n=0}^{\infty} \lambda^n w^n/n! \qquad (6)$$

converges, the limit is identical to the exponential function $e^{\lambda w}$. However, there are several cases in which $e^{\lambda w}$ exists even when the series (6) of operators does not converge.

For example, for $w=-\sqrt{s}$, we have

$$e^{-\lambda\sqrt{s}} = \{(\lambda/2\sqrt{\pi t^3})\exp(-\lambda^2/4t)\}, \qquad (7)$$

and for $w=-s$, we have

$$e^{-\lambda s} = h^\lambda = s \cdot H_\lambda(t), \qquad (8)$$

where the function $H_\lambda(t)$ takes the values 0 and 1 according as $t<\lambda$ or $t>\lambda$. $H_\lambda(t)$ belongs to the ring $\mathscr{U}$ and is called the **jump function** at λ. For $f(t) \in \mathscr{U}$, we have

$$h^\lambda \cdot \{f(t)\} = \{f(t-\lambda)\},$$

and hence we call (8) the **translation operator**

(or **shift operator**). For $w=-s$, the series (6) does not converge, but if we apply the formal relation $e^{-\lambda s}=\sum_{n=0}^{\infty}(-\lambda s)^n/n!$ to $f(t)$, we have a formal Taylor expansion

$$f(t-\lambda)=\sum_{n=0}^{\infty} f^{(n)}(t)(-\lambda)^n/n!.$$

The solution of linear †difference equations are represented by rational functions of h^{λ}. The power series $\sum \alpha_n h^n$ of operators always converges. This fact gives an explicit example of a representation by formal power series. Note that the operators $e^{-\lambda s}$ and $e^{-\lambda\sqrt{s}}$ play an essential role in the solution of the †wave equation

$$\frac{\partial^2\varphi}{\partial x^2}=\frac{\alpha^2\partial^2\varphi}{\partial t^2}$$

and the †heat equation

$$\frac{\partial^2\varphi}{\partial x^2}=\frac{\alpha^2\partial\varphi}{\partial t}.$$

The operator (7) converges to δ for $\lambda\to 0$, and this gives a †regularization of the Dirac δ-function.

D. Laplace Transform

For every function $\{f(t)\}\in\mathscr{C}$, the limit

$$\lim_{\beta\to+\infty}\int_0^{\beta} e^{-\lambda s}f(\lambda)\,d\lambda=\int_0^{\infty} e^{-\lambda s}f(\lambda)\,d\lambda \qquad (9)$$

always exists (in the sense of the limit of operators), and as an operator coincides with the original function $\{f(t)\}$. Therefore, if the usual Laplace transform (→ 240 Laplace Transform) of the function $f(\lambda)$ exists and (9) is a function $g(s)$, then as a function of the differential operator s, $g(s)$ is the operator that is given by the inverse Laplace transform $f(t)$ of $g(s)$. Formulas (4) and (7) are indeed typical examples of this relation, where the left-hand side is the usual Laplace transform of the right-hand side. In the practical computation of (5), we can compute the Laplace inverse transform of the right-hand side. However, if we took the Laplace transform as the foundation of the theory, it would not only be complicated but also be subject to the artificial restriction caused by the convergence condition on Laplace transforms.

In the theory of operational calculus, the transform

$$g(p)=p\int_0^{\infty} e^{-pt}f(t)\,dt \qquad (10)$$

is sometimes used instead of the Laplace transform itself. But the difference is not essential; we obtain the latter transform merely by changing the variable from s to p and multiplying the former transform by p.

E. Relation to Distributions

For $f\in\mathscr{C}$, an operator of the form $h^{\lambda}\cdot s^k\cdot f$ is identified with a distribution of L. Schwartz with support bounded from below. We can identify with a Schwartz distribution the limit of a sequence f_n (or a suitable equivalence class of sequences) of operators of the form $h^{\lambda}\cdot s^k\cdot f$ such that $f_n, f_{n+1}, \ldots$ are identical in the interval $(-n, n)$. The notions of Schwartz distributions and of Mikusiński operators do not include each other, but both are generalizations of the notions of functions and their derivatives. For formulas and examples → Appendix A, Table 12.

References

[1] O. Heaviside, On operators in physical mathematics I, II, Proc. Roy. Soc. London, ser. A, 52 (1893), 504–529; 54 (1893), 105–442.
[2] J. Mikusiński, Operational calculus, Pergamon, 1959. (Original in Polish, 1953.)
[3] E. C. Titchmarsh, The zeros of certain integral functions, Proc. London Math. Soc., 25 (1926), 283–302.
[4] K. Yosida, Operational calculus: A theory of hyperfunctions, Springer, 1984.

307 (XIX.13) Operations Research

A. General Remarks

Operations research in the most general sense can be characterized as the application of scientific methods, techniques, and tools to the operations of systems so as to provide those in control with optimum solutions to problems. This definition is due to Churchman, Ackoff, and Arnoff [1]. Operational research began in a military context in the United Kingdom during World War II, and it was quickly taken up under the name operations research (OR) in the United States. After the war it evolved in connection with industrial organization, and its many techniques have found expanding areas of application in the United States, the United Kingdom, and in other industrial countries. Nowadays OR is used widely in industry for solving practical problems, such as planning, scheduling, inventory, transportation, and marketing. It also has various important applications in the fields of agriculture, commerce, economics, education, public service, etc., and some techniques developed in OR have influenced other fields of science and technology.

B. Applications

Applications of OR to practical problems are often carried out by project teams because knowledge of disparate aspects of the problems are required, and interdisciplinary cooperation is indispensable. The following are the major phases of an OR project: (i) formulating the problem, (ii) constructing a mathematical model to represent the system under study, (iii) deriving a solution from the model, (iv) testing the model and the solution derived from it, (v) establishing controls over the solution and putting it to work (implementation).

When the mathematical model that has been constructed in phase (ii) is complicated and/or the amount of data to be handled is large, a digital †computer is often utilized in phases (iii) and (iv).

C. Mathematical Models [2]

Typical mathematical models and tools that appear frequently in OR are:

(1) **Optimization model** (→ 264 Mathematical Programming). This model is characterized by one or more real-valued functions, which are called **objective functions**, to be minimized (or maximized) under some constraints. According to the number of objective functions, the types of objective functions, and the types of constraints, this model is classified roughly as follows: (i) **Single-objective model**, which includes linear, quadratic, convex, nonlinear and integer programming models (→ 215 Integer Programming, 255 Linear Programming, 292 Nonlinear Programming, 349 Quadratic Programming); (ii) **multi-objective model**; (iii) **stochastic programming model** (→ 408 Stochastic Programming); (iv) **dynamic programming model** (→ 127 Dynamic Programming); (v) **network flow model** (→ 281 Network Flow Problems).

(2) **Game-theoretic model** (→ 173 Game Theory). Game theory is a powerful tool for deriving a solution to practical problems in which more than one person is involved, with each player having different objectives.

(3) **Inventory model** (→ 227 Inventory Control). It is necessary for most firms to control stocks of resources, products, etc., in order to carry out their activities smoothly; various inventory models have been developed for such problems. Mathematically, optimization techniques (→ 264 Mathematical Programming), Markovian decision processes (→ 127 Dynamic Programming, 261 Markov Processes), and basic †probability theory are used to construct models for these problems.

(4) **Queuing model** (→ 260 Markov Chains H). In a telephone system, calls made when all the lines of the system are busy are lost. The problem of computing the **probability of loss** involved was first solved in the pioneering article on †queuing theory by A. K. Erlang in 1917. For systems in which calls can be put on hold when all lines are busy, one deals with the †waiting time distribution instead of the probability of loss. In the 1930s, F. Pollaczeck and A. Ya. Khinchin gave explicit formulas for the characteristic function of the waiting time distribution. In many situations, such as in telephone systems, air and surface traffic, production lines, and computer systems, various congestion phenomena are often observed, and many kinds of queuing models are utilized to analyze the congestion. Mathematically, almost all such models are formulated by using Markov processes. For practical uses, approximation and computational methods are important as well as theoretical results.

(5) **Scheduling model** (→ 376 Scheduling and Production Planning). **Network scheduling** is used to schedule complicated projects (for example, construction of buildings) that consist of a large number of jobs related to each other in some natural order. **PERT** (program evaluation and review technique) and **CPM** (critical path method) are popular computational methods for this model (→ 281 Network Flow Problems). **Job shop scheduling** is used when we have m jobs and n machines and each job requires some of the machines in a given order. The model allows us to find an optimal order (in some certain sense) of jobs to be implemented on each machine.

(6) **Replacement model**. There are two typical cases. One is the **preventive maintenance model**, which is suitable when replacements are done under a routine policy because a replacement or a repair before a failure is more effective than after a failure. Probabilistic treatments are mainly used, and this model resembles those for queues and †Markov processes. The other is a model for deciding whether to replace a piece of equipment in use. In this case, we need to compare costs of both used and new equipment, and evaluations of various types of present cost are important.

(7) †**Simulation**. This is a numerical experiment in a simulated model of a phenomenon which we want to analyze. Simulation is one of the most popular techniques in OR.

(8) **Other models**. Besides the models listed above, many problems are formulated by way of various other models in OR. In modeling,

†probability theory, and mathematical †statistics, especially, †Markov chain, †multivariate analysis, †design of experiments, †regression analysis, †time series analysis, etc. often play important roles.

References

[1] C. W. Churchman, R. L. Ackoff, and E. L. Arnoff, Introduction to operations research, Wiley, 1957.
[2] H. M. Wagner, Principles of operations research, Prentice-Hall, 1975.

308 (XII.19) Operator Algebras

A. Preliminaries

Let $\mathfrak{H}$ be a †Hilbert space. The set of †bounded linear operators on $\mathfrak{H}$ is denoted by $\mathscr{B}(\mathfrak{H})=\mathscr{B}$. It contains the identity operator I. The notions of operator sum $A+B$, operator product AB, and †adjoint A^* are defined on it. A subalgebra of $\mathscr{B}(\mathfrak{H})$ is called an **operator algebra**. In this article we consider mainly von Neumann algebras. For C^*-algebras → 36 Banach Algebras G–K.

Any †Hermitian operator A (i.e., an operator such that $A=A^*$) has the property that (Ax,x) is always real for any $x\in\mathfrak{H}$. If $(Ax,x)\geqslant 0$ for any x, A is called **positive**, and we write $A\geqslant 0$. When Hermitian operators A and B satisfy $A-B\geqslant 0$, we write $A\geqslant B$. Thus we introduce an ordering $A\geqslant B$ between Hermitian operators. A set $\{A_\alpha\}$ of positive Hermitian operators is said to be an **increasing directed set** if any two of them A_α, A_β always have a common majorant A_γ, that is, $A_\alpha\leqslant A_\gamma$ and $A_\beta\leqslant A_\gamma$. If a Hermitian operator A satisfies $(Ax,x)=\sup(A_\alpha x,x)$ for such a set, it is called the **supremum** and is denoted by $\sup A_\alpha$. The supremum $\sup A_\alpha$ exists if and only if the $\sup(A_\alpha x,x)$ is finite for any x, and then A_α converges to A with respect to the weak and strong operator topologies.

B. Topologies in $\mathscr{B}$

Various topologies are introduced in $\mathscr{B}=\mathscr{B}(\mathfrak{H})$: the †uniform operator topology, the †strong operator topology, and the †weak operator topology (→ 251 Linear Operators). These topologies are listed above in order of decreasing fineness. The operation in $\mathscr{B}$ of taking the adjoint, $A\to A^*$, is continuous with respect to the uniform operator topology and weak operator topology, but not with respect to the strong operator topology. The operation from $\mathscr{B}\times\mathscr{B}$ to $\mathscr{B}$ of taking the product, $(A,B)\to AB$, is continuous with respect to the uniform operator topology, is continuous with respect to the strong operator topology when the first factor is restricted to a set bounded in the operator norm, but is not continuous on $\mathscr{B}\times\mathscr{B}$. It is continuous with respect to the weak operator topology when one of the factors is fixed (i.e., it is separately continuous). The set $\mathscr{B}$ is a †Banach space with respect to the operator norm, or, more precisely, a †C^*-algebra. It is a †locally convex topological linear space with respect to the strong or weak operator topology.

The Banach space $\mathscr{B}$ is the †dual of the Banach space $\mathscr{B}_*$ of all †nuclear operators in $\mathfrak{H}$ (→ 68 Compact and Nuclear Operators I). The weak* topology in $\mathscr{B}$ as the dual of $\mathscr{B}_*$ is called the **σ-weak topology**.

C. Von Neumann Algebras

A subset $\mathscr{M}$ of $\mathscr{B}$ is called a **$*$-subalgebra** if it is a subalgebra (i.e., A, $B\in\mathscr{M}$ implies $\lambda A+\mu B$, $AB\in\mathscr{M}$) and contains the adjoint A^* of any $A\in\mathscr{M}$. The **commutant** $\mathscr{A}'$ of a subset $\mathscr{A}$ of $\mathscr{B}$ is the set of operators that commute with both A and A^* for $A\in\mathscr{A}$. The commutant is a $*$-subalgebra, and $\mathscr{A}'=\mathscr{A}'''$.

A **von Neumann algebra** $\mathscr{M}$ is a $*$-subalgebra of $\mathscr{B}$ that is defined by one of the following four equivalent conditions: (i) $\mathscr{M}$ is a $*$-subalgebra of $\mathscr{B}$ containing I, closed under the weak operator topology; (ii) $\mathscr{M}$ is a $*$-subalgebra of $\mathscr{B}$ containing I, closed under the strong operator topology; (iii) $\mathscr{M}$ is the commutant of a subset of $\mathscr{B}$ (or, equivalently, $\mathscr{M}=\mathscr{M}''$); (iv) $\mathscr{M}$ is a $*$-subalgebra of $\mathscr{B}$ containing I, closed under the uniform operator topology, and, as a Banach space, coinciding with the conjugate space of some Banach space. Note that a $*$-subalgebra of $\mathscr{B}$, closed under the uniform operator topology, is a C^*-algebra. Von Neumann algebras are also called **rings of operators** or **W^*-algebras**. The latter term is usually used for a C^*-algebra $*$-isomorphic to a von Neumann algebra in contrast to a concrete von Neumann algebra.

The study of these algebras was started by J. von Neumann in 1929. He showed the equivalence of conditions (i)–(iii) (**von Neumann's density theorem**), and established a foundation for the theory named after him [1]. The notion of von Neumann algebras can be regarded as a natural extension of the notion of matrix algebras in a finite-dimensional space, and therein lies the importance of the theory. The fourth condition of the definition was given by S. Sakai (*Pacific J. Math.*, 1956).

The following theorem is of use in the theory of von Neumann algebras: Given a $*$-subalgebra $\mathcal{A}$ of $\mathcal{B}$ containing I, its closure $\mathcal{M}$ with respect to the weak or strong operator topology is von Neumann algebra; and when we denote the set of elements of operator norm $\leqslant 1$ in $\mathcal{A}$ (resp. $\mathcal{M}$) by $\mathcal{A}_1$ (resp. $\mathcal{M}_1$), $\mathcal{M}_1$ is likewise the closure of $\mathcal{A}_1$ with respect to the weak or strong operator topology (**Kaplansky's density theorem** (*Ann. Math.*, 1951).

If E is a projection operator in a von Neumann algebra $\mathcal{M}$, then $E\mathcal{M}E=\{EAE \mid A\in\mathcal{M}\}$ is a $*$-subalgebra of $\mathcal{B}$ closed with respect to the weak operator topology. It is not a von Neumann algebra because it does not contain I, but since its elements operate exclusively in the closed subspace $E\mathfrak{H}$, we can regard it as an algebra of operators on $E\mathfrak{H}$. In this sense, $E\mathcal{M}E$ can be regarded as a von Neumann algebra on $E\mathfrak{H}$, which we call the **reduced von Neumann algebra** of $\mathcal{M}$ on $E\mathfrak{H}$ and write $\mathcal{M}_E$. If E is a projection operator in $\mathcal{M}'$, $E\mathcal{M}E=E\mathcal{M}$ restricted to the subspace $E\mathfrak{H}$ is called the **induced von Neumann algebra** of $\mathcal{M}$ on $E\mathfrak{H}$ and is denoted also by $\mathcal{M}_E$. In the latter case, the mapping $A\in\mathcal{M}\rightarrow EA\in\mathcal{M}_E$ is a $*$-homomorphism and is called the **induction** of $\mathcal{M}$ onto $\mathcal{M}_E$.

The **tensor product** $\mathfrak{H}_1\otimes\mathfrak{H}_2$ of two Hilbert spaces $\mathfrak{H}_i$ $(i=1,2)$ is the †completion of their †tensor product as a complex linear space equipped with the unique †inner product satisfying $(f_1\otimes f_2, g_1\otimes g_2)=(f_1,g_1)_1(f_2,g_2)_2$ for all $f_i, g_i\in\mathfrak{H}_i$. For any $A_i\in\mathcal{B}(\mathfrak{H}_i)$, there exists a unique operator in $\mathcal{B}(\mathfrak{H}_1\otimes\mathfrak{H}_2)$ denoted by $A_1\otimes A_2$ satisfying $(A_1\otimes A_2)(f_1\otimes f_2)=A_1f_1\otimes A_2f_2$ for all $f_i\in\mathfrak{H}_i$. For von Neumann algebras $\mathcal{M}_i\subset\mathcal{B}(\mathfrak{H}_i)$, the von Neumann algebra generated by $A_1\otimes A_2$ with $A_i\in\mathcal{M}_i$ is denoted by $\mathcal{M}_1\otimes\mathcal{M}_2$ and is called the **tensor product** of $\mathcal{M}_1$ and $\mathcal{M}_2$. The $*$-isomorphism $A\in\mathcal{M}\rightarrow A\otimes I\in\mathcal{M}\otimes\mathbf{I}$, where $\mathbf{I}$ is the trivial von Neumann algebra consisting solely of complex multiples of the identity operator, is called an **amplification**.

For two von Neumann algebras $\mathcal{M}_i\subset\mathcal{B}(\mathfrak{H}_i)$ $(i=1,2)$, a $*$-isomorphism π from $\mathcal{M}_1$ into $\mathcal{M}_2$ is called **spatial** if there exists a unitary (i.e., a bijective isometric linear) mapping U from $\mathfrak{H}_i$ to $\mathfrak{H}_2$ such that $UAU^*=\pi A$ for all $A\in\mathcal{M}_1$, and a $*$-homomorphism π is called **normal** if $\sup_\alpha \pi A_\alpha=\pi(\sup_\alpha A_\alpha)$ whenever A_α is a bounded increasing net in $\mathcal{M}$. Any normal $*$-homomorphism is continuous in the strong and weak operator topologies and is a product of an amplification, a spatial $*$-isomorphism, and an induction. Its kernel is of the form $E\mathcal{M}$, where E is a projection operator belonging to the **center** $\mathcal{Z}=\mathcal{M}\cap\mathcal{M}'$ of $\mathcal{M}$. This gives a complete description of all possible **normal representations** (i.e., a normal $*$-homomorphism into some $\mathcal{B}$) of a von Neumann algebra.

D. States, Weights, and Traces

A **state** φ of a C^*-algebra $\mathcal{A}$ is a complex-valued function on $\mathcal{A}$ that is (1) complex linear: $\varphi(A+B)=\varphi(A)+\varphi(B)$, $\varphi(cA)=c\varphi(A)$ for $A, B\in\mathcal{A}$, $c\in\mathbf{C}$, (2) **positive**: $\varphi(A^*A)\geqslant 0$ for $A\in\mathcal{A}$, and (3) normalized: $\|\varphi\|=1$ (equivalent to $\varphi(I)=1$ if $I\in\mathcal{A}$). For any positive linear functional φ on $\mathcal{A}$, there exists a triplet $(\mathfrak{H}_\varphi, \pi_\varphi, \xi_\varphi)$ (unique up to the unitary equivalence) of a Hilbert space $\mathfrak{H}_\varphi$, a representation π_φ (i.e., a $*$-homomorphism into $\mathcal{B}(\mathfrak{H}_\varphi)$) of $\mathcal{A}$, and a vector ξ_φ in $\mathfrak{H}_\varphi$ such that $\varphi(A)=(\pi_\varphi(A)\xi_\varphi, \xi_\varphi)$ and $\mathfrak{H}_\varphi$ is the closure of $\pi_\varphi(\mathcal{A})\xi_\varphi$. The space $\mathfrak{H}_\varphi$ is constructed by defining the inner product $(\eta(A),\eta(B))=\varphi(B^*A)$ in the quotient of $\mathcal{A}$ by its left ideal $\{A \mid \varphi(A^*A)=0\}$, where η is the quotient mapping, and by completion. Then π_φ is defined by $\pi_\varphi(A)\eta(B)=\eta(AB)$. This is called the **GNS construction** after its originators I. M. Gel'fand, M. A. Naĭmark, and I. E. Siegel.

A **weight** φ on a von Neumann algebra $\mathcal{M}$ is a function defined on the positive elements of $\mathcal{M}$, with positive real or infinite values, which is additive and homogeneous ($\varphi(A+B)=\varphi(A)+\varphi(B)$ and $\varphi(\lambda A)=\lambda\varphi(A)$ for all $A, B\in\mathcal{M}$ and $\lambda\geqslant 0$ with the convention $0\cdot\infty=0$ and $\infty+a=\infty$). It is said to be **faithful** if it does not vanish except for $\varphi(0)=0$, **normal** if $\varphi(A)=\sup\varphi(A_\alpha)$ whenever A_α is an increasing net of positive elements of $\mathcal{M}$ and $A=\sup A_\alpha$, and **semifinite** if the left ideal $\mathfrak{N}_\varphi$, consisting of all elements $A\in\mathcal{M}$ for which $\varphi(A^*A)$ is finite, has the property that the linear span $\mathfrak{M}_\varphi$ of $\mathfrak{N}_\varphi^*\mathfrak{N}_\varphi$ is dense in $\mathcal{M}$. The restriction of φ to positive elements of $\mathfrak{M}_\varphi$ has a unique extension to a linear functional on $\mathfrak{M}_\varphi$, which we denote by the same letter φ. Canonically associated with a normal semifinite weight φ, there exists a Hilbert space $\mathfrak{H}_\varphi$, a normal $*$-representation π_φ of $\mathcal{M}$, and a complex linear mapping η from $\mathfrak{N}_\varphi$ into a dense subset of $\mathfrak{H}_\varphi$ such that $(\eta(B'),\eta(B))=\varphi(B^*B)$, and $\pi_\varphi(A)\eta(B)=\eta(AB)$, where $B, B'\in\mathfrak{N}_\varphi$ and $A\in\mathcal{M}$. If φ is **finite** (i.e., $\mathfrak{N}_\varphi=\mathcal{M}$), then its extension to $\mathcal{M}$ is a positive linear functional for which the triplet $(\mathfrak{H}_\varphi, \pi_\varphi, \eta(1))$ is given by the GNS construction.

The linear span of all normal states of a von Neumann algebra $\mathcal{M}$ is a norm-closed subspace of its dual $\mathcal{M}^*$, called its **predual** and denoted by $\mathcal{M}_*$, because $\mathcal{M}$ turns out to be the dual of $\mathcal{M}_*$.

A **trace** t on a von Neumann algebra $\mathcal{M}$ is a weight satisfying $t(UAU^*)=t(A)$ for U unitary in $\mathcal{M}$ and for all positive A in $\mathcal{M}$ (equivalently, $t(A^*A)=t(AA^*)$ for all $A\in\mathcal{M}$).

E. The von Neumann Classification

A von Neumann algebra for which a semifinite normal trace does not exist is called a **purely infinite** von Neumann algebra or von Neumann algebra of **type III**. In contrast to this, a von Neumann algebra $\mathcal{M}$ is called **semifinite** (resp. **finite**) if for each positive Hermitian operator A ($\neq 0$) in $\mathcal{M}$ there is a semifinite (resp. finite) normal trace t such that $t(A)\neq 0$. Every Abelian von Neumann algebra is finite. If there are no central projection operators $E\neq 0$ such that $\mathcal{M}_E$ is finite, $\mathcal{M}$ is called **properly infinite**. Purely infinite $\mathcal{M}$ and $\mathcal{B}(\mathfrak{H})$ for infinite $\mathfrak{H}$, for example, are properly infinite. A nonzero projection operator E in $\mathcal{M}$ is called **Abelian** when $\mathcal{M}_E$ is Abelian. We call $\mathcal{M}$ a von Neumann algebra of **type I** (or **discrete**) when it contains an Abelian projection E for which I is the only central projection P covering E (i.e., $E\leqslant P$). A von Neumann algebra is of **type II** if it is semifinite and contains no Abelian projection. A von Neumann algebra of type II is called of **type II$_1$** if it is finite and of **type II$_\infty$** if it is properly infinite. A finite von Neumann algebra is also characterized as a von Neumann algebra in which the operation of taking the adjoint is continuous with respect to the strong operator topology on bounded spheres (Sakai, *Proc. Japan Acad.*, 1957). A properly infinite von Neumann algebra $\mathcal{M}$ is characterized by the property that $\mathcal{M}$ and $\mathcal{M}\otimes\mathcal{B}(\mathfrak{H})$ for any separable $\mathfrak{H}$ is $*$-isomorphic.

Given a von Neumann algebra $\mathcal{M}$, there exist mutually orthogonal projections E_{I}, E_{II}, E_{III} in the center $\mathcal{Z}$ of $\mathcal{M}$ such that $E_{\mathrm{I}}+E_{\mathrm{II}}+E_{\mathrm{III}}=I$, and $\mathcal{M}_{E_{\mathrm{I}}}$, $\mathcal{M}_{E_{\mathrm{II}}}$, $\mathcal{M}_{E_{\mathrm{III}}}$ are von Neumann algebras of type I, type II, type III, respectively. This decomposition is unique. There also exist unique central projections E_f and E_i such that $\mathcal{M}_{E_f}$ is finite, $\mathcal{M}_{E_i}$ is properly infinite, and $E_f+E_i=1$. The two decompositions can be combined. (If some of the projections E are 0, the condition on the corresponding $\mathcal{M}_E$ is to be waived.)

F. Factors

A von Neumann algebra whose center consists exclusively of scalar multiples of the identity operator is called a **factor**. Von Neumann's reduction theory (→ Section G) reduces the study of arbitrary von Neumann algebras on a separable Hilbert space more or less to the study of factors. Factors are classified into **types I$_n$** ($n=\infty,1,2,\ldots$), **II$_1$** (i.e., of type II and finite), **II$_\infty$** (i.e., of type II and not finite), and III. A factor of type I$_n$ is isomorphic to the algebra $\mathcal{B}(\mathfrak{H})$ of all bounded operators on an n-dimensional Hilbert space $\mathfrak{H}$. Since the discovery of two nonisomorphic examples of factors of type II$_1$ by F. J. Murray and von Neumann (1943), classification of factors has been a central problem in the theory of von Neumann algebras. After 1967, great progress was made in the investigation of isomorphism classes of factors, and we have uncountably many nonisomorphic examples of factors of types II$_1$, II$_\infty$, and III. After the discovery of the third to ninth nonisomorphic examples of factors of type II$_1$ by J. Schwartz (1963) (the third example, 1963), W. Ching (the fourth), Sakai (the fifth), J. Dixmier and E. C. Lance (the sixth and seventh), and G. Zeller-Meier (the eighth and ninth), D. McDuff showed that there exist countably many nonisomorphic examples of factors of type II$_1$, and finally McDuff (*Ann. Math.*, 1969) and Sakai (*J. Functional Anal.*, 1970) showed the existence of uncountably many nonisomorphic examples of factors of type II$_1$. For type III factors → Section I.

G. The Integral Direct Sum and Decomposition into Factors

The Hilbert spaces considered in this section are all †separable. Let $(\mathfrak{M},\mathcal{E},\mu)$ be a †measure space; with each $\zeta\in\mathfrak{M}$ we associate a Hilbert space $\mathfrak{H}(\zeta)$. We consider functions $x(\zeta)$ on $\mathfrak{M}$ whose values are in $\mathfrak{H}(\zeta)$ for each ζ. Let $\mathbf{K}$ be a set of these functions having the following properties: (i) $\|x(\zeta)\|$ is measurable for $x(\zeta)\in\mathbf{K}$; (ii) if for a function $y(\zeta)$, the numerical function $(x(\zeta),y(\zeta))$ is measurable any $x(\zeta)\in\mathbf{K}$, then $y(\zeta)\in\mathbf{K}$; (iii) there is a countable family $\{x_1(\zeta),x_2(\zeta),\ldots\}$ of functions in $\mathbf{K}$ such that for each fixed $\zeta\in\mathfrak{M}$, the set $\{x_1(\zeta),x_2(\zeta),\ldots\}$ is dense in $\mathfrak{H}(\zeta)$. Then $\mathbf{K}$ is a linear space. We call each function in $\mathbf{K}$ a **measurable vector function**. We introduce in the set of measurable vector functions $x(\zeta)$ with

$$\int\|x(\zeta)\|^2\,d\mu(\zeta)<\infty$$

an equivalence relation by defining $x(\zeta)$ and $y(\zeta)$ as equivalent when

$$\int\|x(\zeta)-y(\zeta)\|^2\,d\mu(\zeta)=0.$$

Thus we obtain a space of equivalence classes which we denote by $\mathfrak{H}$. $\mathfrak{H}$ is a Hilbert space with the inner product

$$(x,y)=\int(x(\zeta),y(\zeta))\,d\mu(\zeta),$$

which is called the **integral direct sum** (or **direct integral**) of $\mathfrak{H}(\zeta)$. An operator function $A(\zeta)$ that associates with each $\zeta \in \mathfrak{M}$ a bounded linear operator $A(\zeta)$ on $\mathfrak{H}(\zeta)$ is called **measurable** if for any measurable $x(\zeta)$, $A(\zeta)x(\zeta)$ is also measurable. If, moreover, $\|A(\zeta)\|$ is bounded, $A(\zeta)$ transforms a function in $\mathfrak{H}$ to a function in $\mathfrak{H}$ and thus defines a bounded linear operator on $\mathfrak{H}$. This operator is called the **integral direct sum** (or **direct integral**) of $A(\zeta)$, and an operator on $\mathfrak{H}$ that can be reduced to this form is called **decomposable**.

Generally, let $\mathfrak{H}$ be a Hilbert space, and consider an Abelian von Neumann algebra $\mathscr{A}$ on $\mathfrak{H}$. Then we construct a measure space $(\mathfrak{M}, \mathscr{E}, \mu)$ and represent $\mathscr{A}$ as the set of bounded measurable functions on $\mathfrak{M}$. (This is possible in different ways. The Gel'fand representation is an example.) Then a Hilbert space $\mathfrak{H}(\zeta)$ can be constructed so that $\mathfrak{H}$ is represented as the integral direct sum of $\mathfrak{H}(\zeta)$. Operators in $\mathscr{A}$ are all decomposable and are called **diagonalizable**. A von Neumann algebra $\mathscr{M}$ on $\mathfrak{H}$ whose elements are all decomposable is characterized by $\mathscr{M} \subset \mathscr{A}'$. The $A(\zeta)$ yielded by the decomposition of operators A in $\mathscr{M}$ generate a von Neumann algebra $\mathscr{M}(\zeta)$ on $\mathfrak{H}(\zeta)$. If we take as $\mathscr{A}$ the center $\mathscr{Z}$ of $\mathscr{M}$, then almost all the $\mathscr{M}(\zeta)$ are factors (**von Neumann's reduction theory**), and if we take as $\mathscr{A}$ a maximal Abelian von Neumann algebra contained in $\mathscr{M}'$, then almost all the $\mathscr{M}(\zeta)$ are type I factors (F. I. Mautner, *Ann. Math.*, 1950).

H. Tomita-Takesaki Theory

Motivated by the problem of proving the commutant theorem for tensor products (i.e., $(\mathscr{M}_1 \otimes \mathscr{M}_2)' = \mathscr{M}_1' \otimes \mathscr{M}_2'$), which remained unsolved for algebras of type III up until that time, Tomita succeeded in 1967, after years of effort, in generalizing the theory of Hilbert algebras, previously developed only for semifinite von Neumann algebras. The most important ingredient of this theory lies in certain one-parameter groups of $*$-automorphisms of a von Neumann algebra, called modular automorphisms (see below), each one-parameter group of modular automorphisms being intrinsically associated with a faithful semifinite normal weight of the algebra. Tomita's theory was perfected by Takesaki [13], who also showed that modular automorphisms satisfy (and are characterized by) a condition originally introduced in statistical mechanics by the physicists R. Kubo, P. C. Martin, and J. Schwinger and accordingly known as the KMS condition. In the mathematical foundations of statistical mechanics, this condition characterizes equilibrium states of a physical system for a given one-parameter group of automorphisms (of a C^*-algebra) describing the time-development of the system. It was a fortunate coincidence that this condition was formulated in a so-called C^*-algebra approach to statistical mechanics by R. Haag, N. M. Hugenholtz, and M. Winnink [14] at about the same time that Tomita's work appeared in 1967. The original proofs of the Tomita-Takesaki theory have been simplified considerably by the work of M. Rieffel [16] and A. Van Daele [15]. Deeper insight into the significance of modular automorphisms is also provided by the work of A. Connes [19], showing that the group of modular automorphisms (up to inner automorphisms) is intrinsic to the von Neumann algebra (i.e., independent of the weight) and belongs to the center of Out $\mathscr{M}$ (the group Aut $\mathscr{M}$ of all $*$-automorphisms of the von Neumann algebra $\mathscr{M}$ modulo the subgroup Int $\mathscr{M}$ of all inner $*$-automorphisms).

Some of the basic definitions and results of the **Tomita-Takesaki theory** are as follows. If φ is a normal semifinite faithful weight on $\mathscr{M}$, the antilinear operator S_φ, defined on a dense subset $\eta(\mathfrak{N}_\varphi \cap \mathfrak{N}_\varphi^*)$ of $\mathfrak{H}_\varphi$ (→ Section D) by the relation $S_\varphi \eta(A) = \eta(A^*)$, is †closable and the polar decomposition $\overline{S_\varphi} = J_\varphi \Delta_\varphi^{1/2}$ of its closure defines a positive self-adjoint operator Δ_φ, called a **modular operator**, and an antiunitary involution J_φ. The principal results of the Tomita-Takesaki theory are (1) if $x \in \mathscr{M}$, then $\sigma_t^\varphi(A) \equiv \Delta_\varphi^{it} A \Delta_\varphi^{-it} \in \mathscr{M}$ for all real t, and this defines a continuous one-parameter group of $*$-automorphisms σ_t^φ of $\mathscr{M}$, called **modular automorphisms**, and (2) if $A \in \mathscr{M}$, then $j_\varphi(A) \equiv J_\varphi A J_\varphi \in \mathscr{M}'$, and j_φ is a conjugate-linear isomorphism of $\mathscr{M}$ onto $\mathscr{M}'$. A weight φ on $\mathscr{M}$ is said to satisfy the **KMS condition** at β (a real number) relative to a one-parameter group of $*$-automorphisms σ_t of $\mathscr{M}$ if, for every pair $A, B \in \mathfrak{N}_\varphi \cap \mathfrak{N}_\varphi^*$, there exists a bounded continuous function $F(z)$ (depending on A, B), on $0 \leqslant \operatorname{Im} z \leqslant \beta$ holomorphic in $0 < \operatorname{Im} z < \beta$ and such that $F(t) = \varphi(A\sigma_t(B))$ and $F(t+i\beta) = \varphi(\sigma_t(B)A)$. A given one-parameter group σ_t coincides with a group of modular automorphisms σ_t^φ if and only if φ satisfies the KMS condition at $\beta = -1$ relative to σ_t. (In statistical mechanics, $\beta = (kT)^{-1}$, where k is the Boltzmann constant and T the absolute temperature of the system.)

I. Structure and Classification of Factors of Type III

At the Baton Rouge Conference in 1967, R. T. Powers reported his results [17] on nonisomorphism of the one-parameter family of factors of type III (now called **Powers's factors**), which had been constructed by von Neumann in 1938 in terms of an infinite tensor product of factors of type I (abbreviated as **ITPFI**). Prior to Powers's work only three different kinds of factors of type III, along with the same number of factors of type II_1, had been distinguished. A systematic classification of ITPFIs was subsequently given by H. Araki and E. J. Woods [18] in terms of two invariants, i.e., the **asymptotic ratio set** $r_\infty(\mathscr{M})$ and the **ρ-set** $\rho(\mathscr{M})$ of the von Neumann algebra $\mathscr{M}$. Using the Tomita-Takesaki theory, Connes [19] introduced two invariants, namely the **S-set** $S(\mathscr{M})$ (the intersection of the spectra of all modular operators) and the **T-set** $T(\mathscr{M})$ (the set of all real t for which the modular automorphism σ_t^φ is inner), and, when $\mathscr{M}$ is an ITPFI, proved the equality $S(\mathscr{M})=r_\infty(\mathscr{M})$ and the relation $T(\mathscr{M})=2\pi|\log\rho(\mathscr{M})|^{-1}$. The S-set $S(\mathscr{M})$ of a factor of type III on a separable Hilbert space is either the set of all nonnegative reals (**type III_1**), the set of all integral powers of λ (where $0<\lambda<1$) and 0 (**type III_λ**), or the set $\{0,1\}$ (**type III_0**). The work of Araki and Woods shows that there exists only one ITPFI of type III_λ for each $\lambda\in(0,1)$ (the examples of Powers) as well as for $\lambda=1$, while there exist continuously many ITPFIs of type III_0. Woods [20] has shown that the classification of ITPFIs of type III_0 is not smooth.

A structural analysis of factors of type III, given independently by Connes [19], Takesaki (*Acta Math.*, 1973), and Araki (*Publ. Res. Inst. Math. Sci.*, 1973) expressed independently a certain class of factors of type III as a kind of crossed product of semifinite von Neumann algebras with their injective endomorphisms (automorphism in the case of Connes). These analyses led Takesaki [21] to the discovery of a duality theorem for crossed products of von Neumann algebras with locally compact groups of their $*$-automorphisms (→ Section J) and its application to the following **structure theorem for von Neumann algebras of type III**. The crossed product of a von Neumann algebra $\mathscr{M}$ with the group of modular automorphisms σ_t^φ is a von Neumann algebra $\mathscr{N}$ of type II_∞, with a canonical action θ_t of the dual group as a one-parameter group of $*$-automorphisms which is trace-scaling, i.e., $\tau\circ\theta_t=e^{-t}\tau$ for some faithful normal trace τ. If $\mathscr{M}$ is properly infinite, the crossed product of $\mathscr{N}$ with θ_t is isomorphic to the original von Neumann algebra $\mathscr{M}$. In particular, any von Neumann algebra $\mathscr{M}$ of type III can be written as the crossed product of a von Neumann algebra $\mathscr{N}$ of type II_∞ with a one-parameter group of trace-scaling $*$-automorphisms θ_t. The isomorphism class of $\mathscr{M}$ is determined by the isomorphism class of $\mathscr{N}$ together with the conjugacy class of θ_t modulo inner automorphisms. The restriction $\tilde{\theta}_t$ of θ_t to the center $\mathscr{Z}$ of $\mathscr{N}$ is of special importance. $\mathscr{M}$ is a factor if and only if $\tilde{\theta}_t$ is ergodic. In that case, $\mathscr{M}$ is of type III_1 if $\mathscr{N}$ is a factor, $\mathscr{M}$ is of type III_λ, $0<\lambda<1$, if $\tilde{\theta}_t$ is periodic with period $-\log\lambda$, and $\mathscr{M}$ is of type III_0 if $\tilde{\theta}_t$ is aperiodic and not isomorphic to the one-parameter group of $*$-automorphisms of $L_\infty(\mathbf{R})$ induced by the translations of the real line $\mathbf{R}$. (The excluded case does not occur for $\mathscr{M}$ of type III.)

A von Neumann algebra on a separable Hilbert space is called **approximately finite-dimensional** (or **approximately finite** or **hyperfinite**) if it is generated by an increasing sequence of finite-dimensional $*$-subalgebras. This class of von Neumann algebras includes many important examples, such as ITPFI and the von Neumann algebra generated by any representation of canonical commutation (or anticommutation) relations on a separable Hilbert space. The classification of approximately finite-dimensional factors is almost complete. In fact the uniqueness of an approximately finite-dimensional factor of type II_1 has been known since the work of von Neumann. It is called the **hyperfinite factor**. The uniqueness of approximately finite-dimensional factors of type II_∞ (which is then the tensor product of the hyperfinite factor with $\mathscr{B}(\mathfrak{H})$) and of type III_λ, $0<\lambda<1$, (which are then Powers's factors) has been demonstrated by Connes [22]. Approximately finite-dimensional factors of type III_0 are classified exactly by the isomorphism classes of the ergodic groups $\tilde{\theta}_t$ of $*$-automorphisms of commutative von Neumann algebras $\mathscr{Z}$. Any such factor is a **Krieger's factor**, i.e., a crossed product of a commutative von Neumann algebra with a single $*$-automorphism. Examples of such factors have been extensively studied by Krieger, who has also shown [23] that isomorphism of a Krieger's factor is equivalent to weak equivalence of the associated nonsingular transformation of the standard measure space.

A von Neumann algebra on a separable Hilbert space is approximately finite-dimensional if and only if it is injective (→ 36 Banach Algebras H).

J. Crossed Products

The **crossed product** $\mathscr{M}\otimes_\alpha G$ of a von Neumann algebra $\mathscr{M}$ (acting on a Hilbert space $\mathfrak{H}$) and a locally compact Abelian group G relative to a continuous action α of G on $\mathscr{M}$ (by $*$-automorphisms α_g, $g\in G$) is the von Neumann algebra $\mathscr{N}$ generated by the operators $\pi(A)$, $A\in\mathscr{M}$ and $\lambda(h)$, $h\in G$, defined on the Hilbert space $L_2(G,\mathfrak{H})$ of all $\mathfrak{H}$-valued L_2-

functions on G (relative to the Haar measure) by

$$[\pi(A)\xi](g)=\alpha_g^{-1}(A)\xi(g),$$

$$[\lambda(h)\xi](g)=\xi(h^{-1}g),$$

where $\xi\in L_2(G,\mathfrak{H})$. The canonical action $\hat{\alpha}$ of the dual $\hat{G}$ on $\mathcal{N}$ is defined by $\hat{\alpha}_p(B)=\mu(p)B\mu(p)^*$ for $B\in\mathcal{N}$ and $p\in\hat{G}$, where $\mu(p)$ is defined by $[\mu(p)\xi](g)=\langle g,p\rangle\xi(g)$. The **duality theorem of Takesaki** [21] asserts that $[\mathcal{M}\otimes_\alpha G]\otimes_{\hat{\alpha}}\hat{G}$ is isomorphic to $\mathcal{M}\otimes\mathcal{B}(L_2(G))$, where the second factor $\mathcal{B}(L_2(G))$ is the algebra of all bounded linear operators on $L_2(G)$.

K. Natural Positive Cone

The closure V^α of the set of vectors $\Delta_\varphi^\alpha\eta(A)$ for all positive A in $\mathfrak{N}_\varphi\cap\mathfrak{N}_\varphi^*$ reflects certain properties of the von Neumann algebra $\mathcal{M}$ for $0\leqslant\alpha\leqslant1/2$ [24, 25]. In particular, $V^{1/4}$ is called the **natural positive cone**. It is a self-dual closed convex cone, and is intrinsic to the von Neumann algebra $\mathcal{M}$ (i.e., independent of the weight φ). Every normal positive linear functional ψ on $\mathcal{M}$ has a unique representative $\xi(\psi)$ in this cone (i.e., $\psi(A)=(\pi_\varphi(A)\xi(\psi),\xi(\psi))$), and the mapping ξ is a concave, monotone, bijective homeomorphism, homogeneous of degree 1/2. The group of all $*$-automorphisms of $\mathcal{M}$ has a natural unitary representation $U(g)$, $g\in\operatorname{Aut}\mathcal{M}$, satisfying the relations $U(g)AU(g)^*=g(A)$, $U(g)\xi(\varphi)=\xi(\varphi\circ g^{-1})$.

L. C^*-Algebras and von Neumann Algebras

Let a C^*-algebra $\mathcal{A}$ be given. A $*$-representation $x\to T_x$ gives rise to a von Neumann algebra $\mathcal{M}$, generated by T_x, $x\in\mathcal{A}$. The type of this representation is defined according to the type of $\mathcal{M}$. A C^*-algebra is called a **C^*-algebra of type I** if its $*$-representations are always of type I. It is known that this class is exactly the class of GCR algebras (→ 36 Banach Algebras H). It is also known that a separable non–type I C^*-algebra has a representation of type II and a general non–type I C^*-algebra has a representation of type III (J. Glimm, *Ann. Math.*, 1961; Sakai [8]).

For a C^*-algebra $\mathcal{A}$, all its representations generate †injective von Neumann algebras if and only if $\mathcal{A}$ is †nuclear [26, 27].

M. Topological Groups and von Neumann Algebras

Consider a unitary representation $g\to U_g$ of a locally compact Hausdorff group G (→ 423 Topological Groups). If this representation is a †factor representation, the type of this representation is defined according to the type of the von Neumann algebra $\mathcal{M}$ generated by U_g, $g\in G$. A †group of type I is a group whose factor representations are all of type I. For example, connected semisimple Lie groups and connected nilpotent Lie groups are of type I (Harish-Chandra, *Trans. Amer. Math. Soc.*, 1953; J. M. G. Fell, *Proc. Amer. Math. Soc.*, 1962). Examples of groups that are not of type I are known (→ 437 Unitary Representation E).

References

[1] J. von Neumann, Collected works II, III, Pergamon, 1961.
[2] W. Arveson, An invitation to C^*-algebras, Springer, 1976.
[3] O. Bratteli and D. W. Robinson, Operator algebras and quantum statistical mechanics, Springer, 1979.
[4] J. Dixmier, Von Neumann algebras, North-Holland, 1981.
[5] J. Dixmier, Les C^*-algèbres et leurs représentations, Gauthier-Villars, 1964.
[6] I. Kaplansky, Rings of operators, Benjamin, 1968.
[7] G. K. Pedersen, C^*-algebras and their automorphism groups, Academic Press, 1979.
[8] S. Sakai, C^*-algebras and W^*-algebras, Springer, 1971.
[9] J. T. Schwartz, W^*-algebras, Gordon & Breach, 1967.
[10] S. Stratila and L. Zsido, Lectures on von Neumann algebras, Editura Academiei/Abacus Press, 1975.
[11] M. Takesaki, Theory of operator algebras I, Springer, 1979.
[12] Y. Nakagami and M. Takesaki, Duality for crossed products of von Neumann algebras, Lecture notes in math. 731, Springer, 1979.
[13] M. Takesaki, Tomita's theory of modular Hilbert algebras and its applications, Lecture notes in math. 128, Springer, 1970.
[14] R. Haag, N. M. Hugenholtz, and M. Winnink, On the equilibrium states in quantum statistical mechanics, Comm. Math. Phys., 5 (1967), 215–236.
[15] A. Van Daele, A new approach to the Tomita-Takesaki theory of generalized Hilbert algebras, J. Functional Anal., 15 (1974), 378–393.
[16] M. Rieffel and A. van Daele, A bounded approach to Tomita-Takesaki theory, Pacific J. Math., 69 (1977), 187–221.
[17] R. T. Powers, Representations of uniformly hyperfinite algebras and their asso-

ciated von Neumann rings, Ann. Math., (2) 86 (1967), 138–171.
[18] H. Araki and E. J. Woods, A classification of factors, Publ. Res. Inst. Math. Sci., (A) 4 (1968), 51–130.
[19] A. Connes, Une classification des facteurs de type III, Ann. Sci. Ecole Norm. Sup., (4) 6 (1973), 133–252.
[20] E. J. Woods, The classification of factors is not smooth, Canad. J. Math., 25 (1973), 96–102.
[21] M. Takesaki, Duality for crossed products and the structure of von Neumann algebras of type III, Acta Math., 131 (1973), 249–310.
[22] A. Connes, Classification of injective factors, Ann. Math., 104 (1976), 73–115.
[23] W. Krieger, On ergodic flows and the isomorphism of factors, Math. Ann., 223 (1976), 19–70.
[24] H. Araki, Some properties of modular conjugation operators of von Neumann algebras and a non-commutative Radon-Nikodym theorem with a chain rule. Pacific J. Math., 50 (1974), 309–354.
[25] A. Connes, Caractérisation des espaces vectoriels ordonnés sous-jacents aux algèbres de von Neumann, Ann. Inst. Fourier, 24, Fasc. 4 (1974), 127–156.
[26] M. Choi and E. Effros, Separable nuclear C^*-algebras and injectivity, Duke Math. J., 43 (1976), 309–322.
[27] M. Choi and E. Effros, Nuclear C^*-algebras and injectivity: The general case, Indiana Univ. Math. J., 26 (1977), 443–446.
[28] R. V. Kadison and J. R. Ringrose, Fundamentals of the theory of operators I, Academic Press, 1983.
[29] S. Stratila, Modular theory in operator algebras, Editura Academiei/Abacus Press, 1981.

309 (XX.7) Orbit Determination

A. General Remarks

The purposes of the theory of **orbit determination** are (1) to study properties of orbits of celestial bodies, (2) to determine orbital elements from observed positions of the celestial bodies, and (3) to compute their predicted positions utilizing the orbital elements. Celestial bodies to which the theory is applied are mainly planets, asteroids, comets, satellites, and artificial satellites in the solar system, although orbits of meteors and visual, photometric, and eclipsing binaries can be determined by similar methods.

B. Kepler's Orbital Elements

Consider, for example, an asteroid moving on an ellipse with one focus at the sum. The elliptic orbit is fixed by the initial conditions of the motion or the integration constants of the †Hamilton-Jacobi equation (→ 55 Celestial Mechanics) and is described by **Kepler's orbital elements** a, e, ω, i, Ω, and t_0 (Fig. 1).

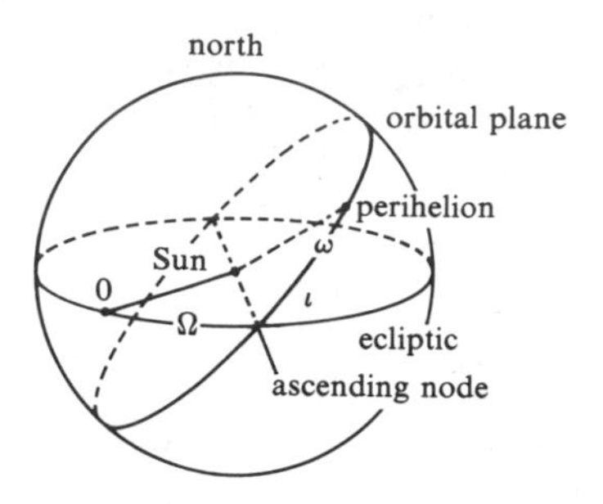

Fig. 1
Orbital elements

The size and shape of the ellipse are determined by the semimajor axis (half the †major axis) a and the †eccentricity e, while the argument ω of perihelion, measured from the ascending node to the perihelion, shows the direction of the major axis. (Sometimes, we adopt as one of the main parameters the **perihelion distance** $q = a(1-e)$ instead of the semimajor axis a.) The position of the orbital plane is determined by the inclination angle i to the ecliptic and the longitude Ω of the ascending node, and then the position of the asteroid on the orbit is determined by the time t_0 of the perihelion passage. The period T of one revolution, or **mean motion** $n = 2\pi/T$, which is the mean angular velocity, is computed by Kepler's third law $n^2a^3 = \mu$, with μ a constant depending on the mass of the asteroid. The mean motion is a fundamental frequency in the solution of the Hamilton-Jacobi equation and is obtained by differentiating the energy constant $-\mu/2a$ with respect to an action variable $\sqrt{\mu a}$.

To express the position of the asteroid on the ellipse as a function of time, we use the **true anomaly** v, which is the angular distance of the asteroid from the perihelion, the **eccentric anomaly** E, and the **mean anomaly** $M = n(t-t_0)$. Of these three anomalies the mean anomaly can be derived directly from Kepler's elements, although it must be transformed to the true anomaly or to the eccentric anomaly when we compute the coordinates of the asteroid. **Kepler's equation**

$$E-e\sin E=M \tag{1}$$

holds between E and M. Solving this equation, we obtain an expression for E as a function of M:

$$E=M+\sum_{n=1}^{\infty}\frac{2}{n}J_n(ne)\sin nM, \tag{2}$$

where J_n is the †Bessel function of order n. However, in practical computations, we often solve equation (1) directly by numerical methods or by using tables.

C. Orbit Determination

An astrometric observation of a celestial body usually consists of measurements of two coordinates (right ascension and declination) on the celestial sphere. Therefore, to derive six orbital elements, three sets of observations should be made at three moments separated by appropriate time intervals. If the topocentric distance of the celestial body is known, the orbital elements can be computed directly from observations. However, since the distance is not usually known, special methods have had to be developed. A method for orbit determination was worked out by C. F. †Gauss at the beginning of the 19th century to find the orbit of Ceres, the first asteroid to be discovered. Although the topocentric distances are not known, we know that orbits of asteroids are planar, and Kepler's second law, the law of conservation of areal velocity, holds approximately. Therefore we can assume that the area of the triangle made by the sun and the two positions of the asteroid observed at different moments is proportional to the corresponding time interval. Using this property of the orbit we can derive the topocentric distance and then the orbital elements. This method is called the indirect method, and similar methods can be developed for parabolic and hyperbolic orbits.

D. Osculating Elements and Orbit Improvement

For the †two-body problem the orbit is a fixed and invariable ellipse, and therefore Kepler's orbital elements are constants. On the other hand, when gravitational interactions from other bodies cannot be disregarded, the orbital elements are found to be variable by computing the †perturbations by the †method of variation of constants. The perturbations are expressed as sums of periodic, †secular, and long-periodic terms.

Because of the perturbations, the orbit deviates from the fixed ellipse, although at every moment the instantaneous velocity and position of the asteroid determines an ellipse. The orbital elements of the ellipse thus defined at each moment, called **osculating elements**, are variable with time. To compute perturbations that cause this change of osculating elements, it is necessary to observe the initial conditions of motion, i.e., the osculating elements at the initial moment. During a time interval shorter than the period of one revolution, the variations of the osculating elements are usually very small. Therefore, by three sets of observations made at three moments at short intervals, it is possible to determine the orbital elements that can be identified with the osculating elements observed at the mean moment. However, if the intervals are very short, errors in the determined values often are very large, and it becomes necessary to carry out observations at distant moments also. When such additional observations are made, those data are compared with the respective values that follow from the initial observations, and the perturbations computed from them; then the method of least squares is used to improve the estimation of the orbital elements.

E. Artificial Satellites

Since the periods of revolution of asteroids are of the order of a few years, the osculating elements change very little in a few weeks. On the other hand, for artificial satellites moving around the earth, the periodic as well as secular perturbations become very large after a few hours because the period of revolution may be as short as two hours. Therefore, to determine orbital elements for artificial satellites, observed positions should be corrected by subtracting the effects of periodic perturbations computed from approximate orbital elements already known. By using the observations thus corrected, mean orbital elements are derived by the method of orbit improvement. The approximate orbital elements can be computed if the launching conditions of the satellites are known. In this manner, mean orbital elements can be derived every day, and variations of the mean orbital elements, or amounts of secular perturbations, for a certain period (say, for 100 days) are found. From them, information on atmospheric density and the gravitational potential of the earth are derived. It should be remarked that for artificial satellites distance measurements have been made by radar, and velocity determinations have been made by measuring the Doppler effect.

For satellites of other planets, measurements of two coordinates with respect to the centers of the planets are made. Masses of planets can be computed by Kepler's third law when the orbital elements of satellites are known, and gravitational potentials of the planets can be determined from their secular perturbations.

F. Binaries

In the study of visual binaries, methods similar to those for satellites can be applied, although the exact estimation of the distances to binaries is often impossible. For photometric binaries radial components of velocities are derived by measuring the Doppler effect; and for eclipsing binaries important information, such as their masses, densities, and sizes, as well as data regarding their internal constitutions, can be derived from the observed orbital elements.

References

[1] G. Stracke, Bahnbestimmung der Planeten und Kometen, Springer, 1929.
[2] A. D. Dubyago, The determination of orbits, Macmillan, 1961.
[3] C. F. Gauss, Theoria motus, Dover, 1963.
[4] P. R. Escobal, Methods of orbit determination, Wiley, 1965.

310 (XII.4) Ordered Linear Spaces

A. History

Many spaces used in functional analysis, such as †Hilbert spaces, †Banach spaces, and †topological linear spaces, are generalizations of Euclidean spaces, where the leading idea has been to generalize the distance in Euclidean spaces in various ways. On the other hand, generalizing the order concept for real numbers has led to spaces of another kind: ordered linear spaces and vector lattices. The theory of vector lattices was presented in a lecture by F. Riesz at the International Congress of Mathematicians in 1928 [1] and has been developed by many authors. Among them we cite H. Freudenthal, L. V. Kantorovitch (*Mat. Sb.*, 2 (44) (1937)), Riesz (*Ann. Math.*, (2) 41 (1940)), S. Kakutani, F. Bohnenblust, G. Birkhoff [2], H. Nakano [3], B. C. Vulikh [5], and H. H. Schaefer [8]. Vector lattices have been used in lattice-theoretic treatments of integration (→ Section I), †spectral resolution, and †ergodic theory. The central notion is Banach lattices (→ Section F), but the theory has been extended to the case where E is a †locally convex topological linear space with the structure of a vector lattice [6–8].

B. Definitions

A real †linear space E is said to be an **ordered linear space** if E is supplied with an †order relation $\geqslant$ with the following two properties: (i) $x \geqslant y \Rightarrow x+z \geqslant y+z$; (ii) $x \geqslant y$ and $\lambda \geqslant 0$ (λ is a real number) $\Rightarrow \lambda x \geqslant \lambda y$.

If, in addition, E forms a †lattice under this order $\geqslant$, we call E a **vector lattice (Riesz space** or **lattice ordered linear space**).

For Sections B through E, we assume E to be a vector lattice. For any $x, y \in E$, the †join and †meet of x, y are denoted by $x \vee y$ and $x \wedge y$ respectively. The following relations are obvious:

$$(x+z) \vee (y+z) = (x \vee y) + z,$$

$$(x+z) \wedge (y+z) = (x \wedge y) + z,$$

$$\lambda x \vee \lambda y = \lambda(x \vee y), \quad \lambda x \wedge \lambda y = \lambda(x \wedge y) \quad (\lambda \geqslant 0),$$

$$\lambda x \vee \lambda y = \lambda(x \wedge y), \quad \lambda x \wedge \lambda y = \lambda(x \vee y) \quad (\lambda \leqslant 0)$$

and

$$(x \vee y) \wedge z = (x \wedge z) \vee (y \wedge z),$$

$$(x \wedge y) \vee z = (x \vee z) \wedge (y \vee z).$$

The last relation means that E is a †distributive lattice.

For $x \in E$, the elements $x \vee 0$, $(-x) \vee 0$, and $x \vee (-x)$ are called respectively the **positive part**, **negative part**, and **absolute value** of the element x, and are denoted respectively by x^+, x^-, and $|x|$. The following identities hold: $x = x^+ - x^-$ **(Jordan decomposition)**, $|x| = x^+ + x^-$, $x^+ \wedge x^- = 0$, $x \vee y + x \wedge y = x + y$, $|\lambda x| = |\lambda||x|$, and $|x-y| = x \vee y - x \wedge y$.

For $a, b \in E$ with $a \leqslant b$, the set $\{x \mid a \leqslant x \leqslant b\}$ is called an **interval** and is denoted by $[a, b]$. A subset of E is called **(order) bounded** if it is included in an interval. An element e of E is said to be a **unit** or an **Archimedean unit** if for any $x \in E$ there exists a natural number n such that $x \leqslant ne$. A linear subspace I of E is called an **ideal** (or **order ideal**) of E if $x \in I$ and $|y| \leqslant |x|$ imply $y \in I$.

C. Order Limits

Given a subset $\{x_\alpha\}$ of E, if an element x of E is an upper bound of $\{x_\alpha\}$ and any upper bound y of $\{x_\alpha\}$ satisfies the relation $y \geqslant x$, then it is called the **least upper bound** (or **su-**

premum) of $\{x_\alpha\}$ and is denoted by $\sup_\alpha x_\alpha$ or $\bigvee_\alpha x_\alpha$. The **greatest lower bound** (or **infimum**) of $\{x_\alpha\}$, denoted by $\inf_\alpha x_\alpha$ or $\bigwedge_\alpha x_\alpha$, is defined dually.

A sequence $\{x_n\}$ $(x_n \in E)$ is said to be **order convergent** to x if there exists a nonincreasing sequence $\{u_n\}$ $(u_n \in E)$ such that $\bigwedge_n u_n = 0$ and $|x_n - x| \leqslant u_n$. In this case x is called the **order limit** of $\{x_n\}$ and is denoted by $x = \text{o-lim}\, x_n$. For order convergent sequences $\{x_n\}$ and $\{y_n\}$, we can show the following relations: $\text{o-lim}(\lambda x_n + \mu y_n) = \lambda(\text{o-lim}\, x_n) + \mu(\text{o-lim}\, y_n)$ and $\text{o-lim}(x_n \underset{\wedge}{\vee} y_n) = (\text{o-lim}\, x_n) \underset{\wedge}{\vee} (\text{o-lim}\, y_n)$. We say that E is **Archimedean** if the relations $0 \leqslant nx \leqslant y$ $(n = 1, 2, 3, \ldots)$ imply $x = 0$. If E is Archimedean, then the relations $x = \text{o-lim}\, x_n$ and $\lambda = \lim \lambda_n$ imply $\lambda x = \text{o-lim}\, \lambda_n x_n$. We say that E is **complete** (σ-**complete**) if any (countable) subset of E that is bounded above has a least upper bound. A σ-complete vector lattice is always Archimedean. If E is σ-complete, for any sequence $\{x_n\}$ bounded above, $\text{o-lim}\sup x_n$ is defined to be $\bigwedge_n \bigvee_{m \geqslant n} x_m$; we define $\text{o-lim}\inf x_n$ similarly. With these definitions, $x = \text{o-lim}\, x_n$ is equivalent to $x = \text{o-lim}\sup x_n = \text{o-lim}\inf x_n$. Any Archimedean vector lattice can be extended to a complete vector lattice in the same way as the real numbers are constructed from the rational numbers by Dedekind cuts (→ 294 Numbers).

D. Examples of Vector Lattices

†Sequence spaces, such as c, m, and l_p, and †function spaces, such as C, M, and L_p, form vector lattices under pointwise ordering (→ 168 Function Spaces). Among these spaces c and C are not σ-complete, but the others are complete. We give two examples. First, let Σ be a σ-algebra of subsets of a space Ω, and let $A(\Omega, \Sigma)$ be the set of all finite σ-additive †set functions defined on Σ. Then $A(\Omega, \Sigma)$ is a complete vector lattice if we define $\mu_1 \geqslant \mu_2$ to mean $\mu_1(S) \geqslant \mu_2(S)$ for any $S \in \Sigma$. The second example is an ordered space consisting of all †bounded symmetric operators T on a Hilbert space H, where we define $T_1 \geqslant T_2$ to mean $(T_1 x, x) \geqslant (T_2 x, x)$ for any $x \in H$. In general, this space is not a vector lattice. However, if A is a commutative †W^*-algebra of operators on H and S is the set of †symmetric operators belonging to A, then S is a complete vector lattice under the ordering just defined. We can replace the conditions of finiteness in $A(\Omega, \Sigma)$ and boundedness in S with weaker ones and still obtain the same situation. The †Radon-Nikodym theorem in $A(\Omega, \Sigma)$ and the †spectral resolution theorem of symmetric operators in S can be extended to theorems of †spectral representations in general vector lattices.

Let E_1 be a linear space of functions defined on a set Ω ordered pointwise. If there exists a bijective mapping defined on a vector lattice E onto E_1 that is linear and order isomorphic, we call E_1 a **representation** of E. If E has an Archimedean unit and is simple (which means that E and $\{0\}$ are the only ideals of E), then E can be represented as the set of real numbers such that the Archimedean unit of E is represented by the number 1 (H. Freudenthal, *Proc. Akad. Amsterdam*, 39 (1936)).

E. Dual Spaces

Let $\mathfrak{L}(E, F)$ be the set of **order bounded** linear mappings of a vector lattice E into a vector lattice F, where order boundedness means that any bounded (in the sense of the order) subset of E is mapped into a bounded set of F. For any $\varphi_1, \varphi_2 \in \mathfrak{L}(E, F)$, define $\varphi_1 \geqslant \varphi_2$ to mean $\varphi_1(x) \geqslant \varphi_2(x)$ $(x \geqslant 0, x \in E)$. If F is complete, then $\mathfrak{L}(E, F)$ is a complete vector lattice. An element $\varphi \in \mathfrak{L}(E, F)$ is called a **positive operator** if $\varphi \geqslant 0$. If F is the set of real numbers $\mathbf{R}$, then $\mathfrak{L}(E, F)$ is the set of all (order) bounded †linear functionals on E. This space, called the **dual lattice** of the vector lattice E and denoted by E^b, is a complete vector lattice. For $f \in E^b$ and $x \geqslant 0$, $x \in E$, we have

$$f^+(x) = \sup_{0 \leqslant y \leqslant x} f(y), \quad f^-(x) = -\inf_{0 \leqslant y \leqslant x} f(y),$$

$$|f|(x) = \sup_{|y| \leqslant x} f(y).$$

F. Banach Lattices

A linear space E is called a **normed vector lattice** if E is a vector lattice having the structure of a †normed space satisfying $|x| \leqslant |y| \Rightarrow \|x\| \leqslant \|y\|$. Furthermore if a normed vector lattice E is complete relative to the norm, we call E a **Banach lattice**. The examples in Section D are Banach lattices (for $\mu \in A(\Omega, \Sigma)$ we define $\|\mu\| = |\mu|(\Omega)$).

In Banach lattices, $\|x_n - x\| \to 0$ and $\|y_n - y\| \to 0$ imply $\|x_n \underset{\wedge}{\vee} y_n - x \underset{\wedge}{\vee} y\| \to 0$. Among relations between order convergence and norm convergence in Banach lattices, the following is one of the most fundamental: In a Banach lattice E, norm convergence of a sequence $\{x_n\}$ to x is equivalent to **relative uniform star convergence** of $\{x_n\}$ to x, i.e., for any subsequence $\{x_{n(m)}\}$ of $\{x_n\}$, there exists a subsequence $\{x_{n(m(l))}\}$ of $\{x_{n(m)}\}$ and an element y of E satisfying the relations $|x_{n(m(l))} - x| \leqslant y/l$ $(l = 1, 2, \ldots)$.

Any set bounded relative to the order is bounded relative to the norm, but the converse does not hold in general. For a linear functional, however, these two concepts of boundedness coincide, and the order dual of

E is the same as the norm dual of E. Moreover, the dual (in any sense) of a Banach lattice is also a Banach lattice.

G. Abstract M Spaces and Abstract L Spaces

For a Banach lattice E, we consider the following three conditions: (M) $x, y \geqslant 0 \Rightarrow \|x \vee y\| = \max(\|x\|, \|y\|)$. (L) $x, y \geqslant 0 \Rightarrow \|x+y\| = \|x\| + \|y\|$. ($L_p$) $x \wedge y = 0 \Rightarrow \|x+y\|^p = \|x\|^p + \|y\|^p$ $(1 < p < \infty)$. If E satisfies one of the conditions (M), (L), or (L_p), we say that E is an **abstract M space**, and **abstract L space**, or an **abstract L_p space**, written AM, AL, and AL_p, respectively. If the unit ball of an AM space has a †greatest element, it is called the **Kakutani unit** of E. The duals of AM spaces, AL spaces, and AL_p spaces are AL spaces, AM spaces with the Kakutani units, and AL_q spaces $(1/p + 1/q = 1)$, respectively. An AM space with a Kakutani unit is represented by $C(\Omega)$, i.e., the set of all real-valued continuous functions defined on a compact Hausdorff space Ω. The AL spaces and AL_p spaces are represented by L_1 and L_p, respectively, on a †measure space. Here the representation of a Banach lattice means a representation of a vector lattice preserving the norm (Kakutani, *Ann. Math.*, (2) 42 (1941); Bohnenblust, *Duke Math. J.*, 6 (1940)).

H. Spectral Properties of Positive Operators

The n-dimensional real vector space E_n is a vector lattice under pointwise order (→ Section D). An element x in E_n is called **strictly positive** if $x_i > 0$ for all i. A square matrix $A = (a_{i,j})$ of order n is called **positive** if $a_{i,j} \geqslant 0$ for all i and j. It corresponds to a positive operator in E_n (→ Section E). A is called **irreducible** if there exists no permutation matrix P such that $P^{-1}AP = \begin{pmatrix} A_1 & A_2 \\ 0 & A_3 \end{pmatrix}$, where A_1 and A_3 are square matrices of order n_i $(1 \leqslant n_i < n)$. We denote by $\sigma(A)$ the †spectrum of A and by $r(A)$ the †spectral radius of A, i.e., $\sup\{|\alpha| \mid \alpha \in \sigma(A)\}$. The spectral circle of A is the circle of radius $r(A)$ having the origin as its center. O. Perron (*Math. Ann.*, 64 (1907)) and G. Frobenius (*S. B. Preuss. Akad. Wiss.*, 1908 and 1912) established the following remarkable result on the spectral properties of positive matrices.

Theorem (Perron-Frobenius). Let A be a positive square matrix. Then its spectral radius $r(A)$ belongs to $\sigma(A)$, and for this spectrum $r(A)$ there exists an eigenvector $x \geqslant 0$. Assume further that A is irreducible and the order of A is greater than 1. Then $r(A) > 0$ and the eigenspace of A for $r(A)$ is a 1-dimentional subspace spanned by a strictly positive element. In this case the eigenvalues of A on the spectral circle are the kth roots of unity for some k multiplied by $r(A)$, each of which is a simple root of the eigenequation of A.

Since a positive matrix of order n corresponds to a positive operator in E_n, extensions of this theorem to positive operators in ordered linear spaces have been studied by many mathematicians. For these extensions, see the following articles: M. G. Kreĭn and M. A. Rutman (*Amer. Math. Soc. Transl.*, 26 (1950); original in Russian, 1948), F. F. Bonsall (*J. London Math. Soc.*, 30 (1955)), S. Karlin (*J. Math. Mech.*, 8 (1959)), T. Ando (*J. Fac. Sci. Hokkaido Univ.*, ser. 1, 13 (1957)), H. H. Schaefer [8], H. P. Lotz (*Math. Z.*, 108 (1968)), F. Niiro and I. Sawashima (*Sci. Pap. Coll. Gen. Educ., Univ. Tokyo*, 1966), I. Sawashima and F. Niiro (*Nat. Rep. Ochanomizu Univ.*, 30 (1979)), and S. Miyajima (*J. Fac. Sci. Univ. Tokyo*, 27 (1980)).

I. Integrals Based on Ordering

As applications of ordered linear space theory, we state the integrals of Daniel-Stone and of Banach. Let us begin with a set $\mathfrak{E}$ of real-valued functions defined on an abstract space S and assume that $\mathfrak{E}$ is a vector lattice with respect to the usual order relation, addition, and scalar multiplication. Assume further that a functional $E(f)$ defined on $\mathfrak{E}$ satisfies the following conditions: (i) additivity, i.e., $E(f+g) = E(f) + E(g)$; (ii) positivity, i.e., $f \geqslant 0$ implies $E(f) \geqslant 0$; (iii) $f, f_n \in \mathfrak{E}$ $(n = 1, 2, \ldots)$ and $|f| \leqslant \sum_{n=1}^{\infty} |f_n|$ imply $E(|f|) \leqslant \sum_{n=1}^{\infty} E(|f_n|)$, where $|f|$ means $f \vee (-f)$. A functional on $\mathfrak{E}$ satisfying both conditions (i) and (ii) is called a positive linear functional. A positive linear functional on $\mathfrak{E}$ satisfies M. H. Stone's condition (iii) if and only if it satisfies P. J. Daniell's condition (iii)′: $f_1 \geqslant f_2 \geqslant \ldots$ and $\lim_{n\to\infty} f_n = 0$ imply $\lim_{n\to\infty} E(f_n) = 0$. Next, we define, for every function φ on S admitting $\pm\infty$ as values, a functional $N(\varphi)$ as follows:

$$N(\varphi) = \inf\left\{ \sum_{n=1}^{\infty} E(|f_n|) \,\middle|\, f_n \in \mathfrak{E}, \quad |\varphi| \leqslant \sum_{n=1}^{\infty} |f_n| \right\}.$$

Here we put $N(\varphi) = +\infty$ when for a function φ there are no functions $\{f_n\}$ such that $|\varphi| \leqslant \sum_{n=1}^{\infty} |f_n|$. A function φ is, by definition, a **null function** if $N(\varphi) = 0$ holds, and a set A is a **null set** if its †characteristic function is a null function. Since each function of $\mathfrak{F}_0 = \{\varphi \mid N(\varphi) < +\infty\}$ takes finite values except on a null set, we can define addition and the scalar multiplication for such functions except on a null set. Let $\mathfrak{F}$ be the set of equivalence classes of $\mathfrak{F}_0$

with respect to the relation $\varphi \sim \psi$ defined as $N(\varphi-\psi)=0$. Then $\mathfrak{F}$ is a Banach lattice with the norm N, and $\mathfrak{E}$ is included in $\mathfrak{F}$ (by identifying f and g of $\mathfrak{E}$ when $E(|f-g|)=0$). Let us denote now by $\mathfrak{L}$ the closure of $\mathfrak{E}$ in $\mathfrak{F}$. Then any function φ belonging to $\mathfrak{L}$ is said to be **Daniell-Stone integrable**, and $L(\varphi)=N(\varphi^+)-N(\varphi^-)(\varphi^+=\frac{1}{2}(|\varphi|+\varphi), \varphi^-=\frac{1}{2}(|\varphi|-\varphi))$ is called the **Daniell-Stone integral** of φ. The integral L thus defined is, as a functional on $\mathfrak{L}$, an extension of the functional E on $\mathfrak{E}$. For this integral, Lebesgue's convergence theorem is easily proved, and a result corresponding to Fubini's theorem has been obtained [9]. Furthermore, the concepts of measurable functions, measurable sets, and measure can be defined by using L and $\mathfrak{L}$. Also, the relation between L and the Lebesgue integral with respect to this measure is known [9]. Since $\mathfrak{L}$ is an †abstract L-space, the Daniell-Stone integral $L(\varphi)$ is represented by the Lebesgue integral of φ on a certain measure space. The Daniell-Stone integral introduced above is due to M. H. Stone [9]. Daniell (*Ann. Math.*, (2) 19 (1917–1918)) originally defined the upper integral $\bar{I}(\varphi)$ by using $E(f)$ on $\mathfrak{E}$ satisfying conditions (i), (ii), and (iii′). Also, he defined the set $\mathfrak{L}$ of Daniell-Stone integrable functions by $L=\{\varphi \mid I(\varphi)=-I(-\varphi)\}$. S. Banach defined an integral by using methods similar to Daniell's, replacing condition (iii′) for a positive linear functional $E(f)$ on $\mathfrak{E}$ by condition (iii″): $\lim_{n\to\infty} f_n=0$, $|f_n|\leqslant g$, and $f_n, g\in\mathfrak{E}$ imply $\lim_{n\to\infty} E(f_n)=0$ [10]. Furthermore, N. Bourbaki [11] and E. J. McShane (*Proc. Nat. Acad. Sci. US*, 1946) have defined a more general integral than the Daniell-Stone with a condition analogous to (iii′), replacing the sequence in it by a †directed family of functions $\{f_\alpha\}$.

Specifically if, in the Daniell-Stone integral, S is a locally compact Hausdorff space and $\mathfrak{E}$ is the set of continuous functions with compact supports, then a functional $E(f)$ on $\mathfrak{E}$ satisfying conditions (i) and (ii) is proved to satisfy the condition (iii′), and the Daniell-Stone integral $L(\varphi)$ can be constructed from $E(f)$ [11].

Banach also defined another integral for all real-valued bounded functions on $[0,1)$ by using the †Hahn-Banach extension theorem [12]. His definition is as follows: Let $\mathfrak{F}$ be the set of all real-valued bounded functions on $[0,1)$ and $\mathfrak{A}$ be the family of all finite sets of real numbers $\alpha=(\alpha_1,\alpha_2,\ldots,\alpha_n)$. Furthermore, we define, for $x(s)\in\mathfrak{F}$ and $\alpha\in\mathfrak{A}$,

$$M(x,\alpha)=\sup_{-\infty<s<+\infty}\frac{1}{n}\sum_{k=1}^{n}x(s+\alpha_k),$$

where $x(s)$ is considered as the periodic extension to $(-\infty,+\infty)$ of the function defined originally on $[0,1)$ and

$$p(x)=\inf_{\alpha\in\mathfrak{A}} M(x,\alpha).$$

Then, by the Hahn-Banach extension theorem, there exists a linear functional F on $\mathfrak{F}$ satisfying $F(x)\leqslant p(x)$. If we write $\int x(s)\,ds$ for $F(x)$, then we can prove immediately that $\int x(s)\,ds$ has the following properties: (1) $\int\{ax(s)+by(x)\}\,ds=a\int x(s)\,ds+b\int y(s)\,ds$, where a and b are real constants. (2) $x(s)\geqslant 0$ implies $\int x(s)\,ds\geqslant 0$. (3) $\int x(s+s_0)\,ds=\int x(s)\,ds$, where s_0 is an arbitrary real number. (4) $\int 1\,ds=1$. If necessary, we can add the property (5) $\int x(1-s)\,ds=\int x(s)\,ds$ by defining

$$\int x(s)\,ds=\frac{1}{2}\{F(x(s))+F(x(1-s))\}.$$

Then

$$F(x)=\int x(s)\,ds$$

or

$$\frac{1}{2}\{F(x(s))+F(x(1-s))\}=\int x(s)\,ds$$

is called the **Banach integral** of $x(s)$.

The construction of the Daniell-Stone integral and the Banach integral opened avenues to several other abstract integrals based on the order relation, such as an integral for more general functions with values in a vector lattice, or an integral considered as a mapping from a vector lattice into another vector lattice (or from an ordered set into another ordered set). Indeed, if the function takes values in a complete vector lattice, then almost all results in this section (e.g., the Hahn-Banach extension theorem) hold trivially. For discussions of these and other abstract integrals → [13–15].

References

[1] F. Riesz, Sur la décomposition des opérations fonctionnelles linéaires, Atti del Congresso Internazionale dei Matematici, 1928, Bologna, 3, 143–148.
[2] G. Birkhoff, Lattice theory, Amer. Math. Soc. Colloq. Publ., revised edition, 1948.
[3] H. Nakano, Modulared semi-ordered linear spaces, Maruzen, 1950.
[4] M. M. Day, Normed linear spaces, Erg. Math., Springer, 1958.
[5] B. C. Vulikh, Introduction to the theory of partially ordered vector spaces, Gronigen, 1967.
[6] I. Namioka, Partially ordered linear topological spaces, Mem. Amer. Math. Soc., 1957.
[7] A. L. Peressini, Ordered topological vector spaces, Harper & Row, 1967.

[8] H. H. Schaefer, Banach lattices and positive operators, Springer, 1974.
[9] M. H. Stone, Notes on integration, Proc. Nat. Acad. Sci. US I–III, 34 (1948), 336–342, 447–455, 483–490; IV, 35 (1949), 50–58.
[10] S. Banach, The Lebesgue integral in abstract spaces, Theory of the Integral, S. Saks (ed.), Stechert, 1937, 320–330 (Dover, 1964).
[11] N. Bourbaki, Eléments de mathématique, Integration, ch. 1–9, Actualités Sci. Ind., Hermann: ch. 1–4, 117a, second edition, 1965; ch. 5, 1244b, second edition, 1967; ch. 6, 1281a, 1959; ch. 7, 8, 1306, 1963; ch. 9, 1343, 1969.
[12] S. Banach, Théorie des opérations linéaires, Warsaw, 1932, 30–32. (Chelsea, 1963).
[13] S. Izumi, N. Matuyama, M. Nakamura, M. Orihara, and G. Sunouchi, An abstract integral, Proc. Imp. Acad. Tokyo, I–III, 16 (1940), 21–25, 87–89, 518–523; IV, 17 (1941), 1–4; V–X, 18 (1942), 45–49, 50–52, 53–56, 535–538, 539–542, 543–547.
[14] E. J. McShane, Order-preserving maps and integration processes, Ann. Math. Studies, Princeton Univ. Press, 1953.
[15] T. H. Hildebrandt, Integration in abstract spaces, Bull. Amer. Math. Soc., 59 (1953), 111–139.
[16] N. Dunford and J. T. Schwartz, Linear operators, Wiley, I (1958), III (1971).
[17] K. Yosida, Functional analysis, Springer, 1965.

311 (II.12) Ordering

A. Ordering

The concept of ordering is abstracted from various relations, such as the inequality relation between real numbers and the inclusion relation between sets. Suppose that we are given a set $X=\{x, y, z, \dots\}$; the relation between the elements of X, denoted by $\leqslant$ or other symbols, is called an **ordering** (**partial ordering**, **semiordering**, **order relation**, or simply **order**) if the following three laws hold: (i) the **reflexive law**, $x \leqslant x$; (ii) the **antisymmetric law**, $x \leqslant y$ and $y \leqslant x$ imply $x=y$; and (iii) the **transitive law**, $x \leqslant y$ and $y \leqslant z$ imply $x \leqslant z$.

A set X with an ordering between its elements is called an **ordered set** (**partially ordered set** or **semiordered set**). A subset of an ordered set X is also an ordered set with respect to the same ordering as in X. If for an arbitrary pair of elements x, y of an ordered set A either $x \leqslant y$ or $y \leqslant x$ must hold, then the ordering $\leqslant$ is called a **total** (or **linear**) **ordering**, and A is called a **totally ordered** (or **linearly ordered**) **set**.

We sometimes write $x \leqslant y$ as $y \geqslant x$. The binary relation $\geqslant$ is called the **dual ordering** of $\leqslant$; it is also an ordering. More generally, the duals of concepts, conditions, and propositions concerning an ordering are defined by replacing the ordering with its dual. For example, $x<y$ means that $x \leqslant y$ and $x \neq y$, while $x>y$ means that $x \geqslant y$ and $x \neq y$; and $>$ is the dual of $<$. If a universal proposition concerning the ordering is true, then its dual is also true; this principle is called the **duality principle for ordering**. Incidentally, $x \leqslant y$ is equivalent to the statement $x<y$ or $x=y$, according to the definition of $<$.

B. Definitions

A subset of an ordered set X of the form $\{x \mid a<x<b\}$ is denoted by (a, b), and a set of the form (a, b), $\{x \mid x<a\}$, or $\{x \mid x>a\}$ is called an **interval**. In particular, $S(c)=\{x \mid x<c\}$ is called the **segment** of X determined by c. A pair of elements a, b satisfying $a \leqslant b$ is called a **quotient** of X and is denoted by b/a.

When $a<c<b$ or $b<c<a$, c is said to lie **between** a and b. A totally ordered set A is said to be **dense** if for any pair of distinct elements a and b in A there exists a third element c lying between a and b. When $a<b$ and there is no element lying between a and b, then a is called a **predecessor** of b, and b a **successor** of a. In this manner, most of the terminology associated with the inequality of numbers is carried over to general ordering.

In an ordered set A, an element a is called an **upper bound** of a subset X if $x \leqslant a$ for every element x of X. When an upper bound exists, X is said to be **bounded from above** (or **bounded above**). The dual concept of an upper bound is a **lower bound** of the subset; and if the subset has a lower bound, it is said to be **bounded from below** (or **bounded below**). A set bounded both from above and from below is simply said to be **bounded**. When a is an upper bound of X and $a \in X$, then a is called the **greatest element** (or **maximum element**) of X. Such an element a (if it exists) is unique and is denoted by $\max X$; its dual is the **least element** (or **minimum element**) and is denoted by $\min X$. If there is a least element in the set of upper bounds of X, it is called the **least upper bound** (or **supremum**) of X and is denoted by **l.u.b.** X or **sup** X. Its dual is the **greatest lower bound** (or **infimum**) and is denoted by **g.l.b.** X or **inf** X.

If the ordered set X is the image $\varphi(\Lambda)$ of a set Λ under a †mapping φ, where Λ is of the form $\{\lambda \mid C(\lambda)\}$, then $\sup X$ is also written

$\sup_{C(\lambda)} \varphi(\lambda)$ and is called the supremum of $\varphi(\lambda)$ for all λ that satisfy $C(\lambda)$. When there is no danger of misunderstanding, it may be written as $\sup_\lambda \varphi(\lambda)$ or $\sup \varphi(\lambda)$ and called simply the supremum of $\varphi(\lambda)$; similar conventions hold for inf, max, min, etc.

An element a of a set X is called a **maximal element** if $a < x$ never holds for any element x of X; its dual is a **minimal element**. If the greatest (least) element exists, it is the only maximal (minimal) element. But in general, a maximal (minimal) element is not necessarily unique.

C. Chain Conditions

An ordered set X is said to satisfy the **minimal condition** if every nonempty subset of X has a minimal element. The dual condition is called the **maximal condition**. An infinite sequence $\{a_1, a_2, \dots, a_n, \dots\}$ of elements of an ordered set X such that $a_1 < a_2 < \dots < a_n < \dots$ is called an **ascending chain**, and the condition that X has no ascending chain is called the **ascending chain condition**. The notions dual to those of ascending chain and ascending chain condition are **descending chain** and **descending chain condition**, respectively. By the **chain condition**, we mean either the ascending or the descending chain condition. Under the †axiom of choice, the maximal condition is equivalent to the ascending chain condition, and the minimal condition to the descending chain condition.

If a totally ordered set X satisfies the minimal condition or, equivalently, if every nonempty subset of X has a least element, then the set X is called a **well-ordered set**, and its ordering is called a **well-ordering**.

The following theorem is called the principle of **transfinite induction**: Let $P(x)$ be a proposition concerning an element x of a well-ordered set X such that (i) $P(x_0)$ is true for the least element x_0 of X, and (ii) $P(x)$ is true if $P(y)$ is true for all y satisfying $y < x$. Then $P(x)$ is true for all x in X. †Mathematical induction is a special case of this principle, where X is the set of all natural numbers. To define a mapping F from a well-ordered set X into a set Y, we may use the following principle: Suppose that $F(x_0)$ is defined for the least element x_0 of X, and for each element x of X there is given a method to associate an element $G(f)$ of Y uniquely with each mapping $f: S(x) \rightarrow Y$ with domain $S(x)$, where $S(x)$ is the segment of X determined by x. Then there exists a unique mapping $F: X \rightarrow Y$ satisfying $F(x) = G(F \mid S(x))$ for all x. The definition of the mapping F by this principle is called a **definition by transfinite induction**. The principles of induction are often used for proving propositions or giving definitions concerning ordinal numbers (→ 312 Ordinal Numbers).

D. Directed Sets

An ordered set (or in general a preordered set (→ Section H)) in which every finite subset is bounded from above is called a **directed set**. Let B be a subset of a directed set A. If $\{b \mid b \geqslant a\} \cap B \neq \varnothing$ for every element a of A, then B is said to be **cofinal** in A; such a subset B is itself a directed set. If $\{b \mid b \geqslant a\} \subset B$ for some element a of A, then B is said to be **residual** in A; such a subset B is also cofinal in A. The condition that B is cofinal in A is equivalent to the condition that $A - B$ is not residual in A.

E. Order-Preserving Mappings

A mapping $\varphi: A \rightarrow A'$ of an ordered set A into an ordered set A' is called an **order-preserving mapping** (**monotone mapping** or **order homomorphism**) if $a \leqslant b$ always implies $\varphi(a) \leqslant \varphi(b)$. Moreover, if φ is bijective and φ^{-1} is also an order-preserving mapping from A' onto A, then φ is called an **order isomorphism**. A' is said to be **order homomorphic** (**order isomorphic**) to A when there exists an order homomorphism (order isomorphism) φ such that $A' = \varphi(A)$. If a mapping $\varphi: A \rightarrow A'$ gives an order isomorphism of A to the dual of A', φ is called a **dual isomorphism** (or **anti-isomorphism**).

F. Direct Sum and Direct Product

Let S be a set that is the †disjoint union of a family $\{A_\lambda\}_{\lambda \in \Lambda}$ of its subsets, and suppose that each A_λ is an ordered set. For $a, b \in S$, define $a \leqslant b$ to mean that $a, b \in A_\lambda$ for some $\lambda \in \Lambda$ and $a \leqslant b$ with respect to the ordering in A_λ. The ordered set S obtained in this way is called the **direct sum** (or **cardinal sum**) of the family $\{A_\lambda\}_{\lambda \in \Lambda}$ of ordered sets. When $(a_\lambda)_{\lambda \in \Lambda}$ and $(b_\lambda)_{\lambda \in \Lambda}$ are elements of the †Cartesian product $P = \prod_{\lambda \in \Lambda} A_\lambda$ of a family $\{A_\lambda\}_{\lambda \in \Lambda}$ of ordered sets, we define $(a_\lambda)_{\lambda \in \Lambda} \leqslant (b_\lambda)_{\lambda \in \Lambda}$ to mean that $a_\lambda \leqslant b_\lambda$ holds for all $\lambda \in \Lambda$. The ordered set P obtained in this way is called the **direct product** (or **cardinal product**) of the family $\{A_\lambda\}_{\lambda \in \Lambda}$ of ordered sets.

G. Ordinal Sum and Ordinal Product

Suppose that $\mathfrak{A} = \{A, B, \dots\}$ is a family of mutually disjoint ordered sets and is itself an ordered set. Then an ordering $\leqslant$ can be defined in the disjoint union $S=\bigcup X$ $(X\in\mathfrak{A})$ as follows: $x\leqslant y$ in S means that either (i) there exists an A satisfying $x, y\in A\in\mathfrak{A}$ and $x\leqslant y$ holds with respect to the ordering in A; or (ii) for A and B satisfying $x\in A\in\mathfrak{A}$, $y\in B\in\mathfrak{A}$, we have $A<B$. The ordered set S obtained in this way is called the **ordinal sum** obtained from $\mathfrak{A}$ and is denoted by $\Sigma_{\mathfrak{A}}X$. In particular, if $\mathfrak{A}=\{A, B\}$ and $A<B$, the ordinal sum is denoted by $A+B$.

Suppose that X is a subset of the Cartesian product $\prod_\lambda X_\lambda$ of a family of ordered sets indexed by an ordered set Λ, and the subset $\{\lambda\mid x_\lambda\neq y_\lambda\}$ of Λ has a least element whenever $x=(x_\lambda)_{\lambda\in\Lambda}$ and $y=(y_\lambda)_{\lambda\in\Lambda}$ are two distinct elements of X. The ordering in X defined by setting $x<y$ when $x_\mu<y_\mu$ for the least element μ of $\{\lambda\mid x_\lambda\neq y_\lambda\}$ is called the **lexicographic ordering** in X. It can be applied to $X=\prod_\lambda X_\lambda$ if Λ is well-ordered; X is then called the **ordinal product**. When $A, B, \ldots$ are ordered sets, $AB\ldots$ is often used to denote the ordinal product obtained from $X_1=A$, $X_2=B, \ldots$ with the ordering $1<2<\ldots$ of indices; the ordering in this ordinal product is called the **lexicographic ordering** in the Cartesian product $A\times B\times\ldots$.

H. Preordering

A relation R between elements of a set X is called a **preordering** (or **pseudoordering**) if it satisfies the reflexive law and the transitive law, but not necessarily the antisymmetric law. By defining $(x_1, y_1)R\,(x_2, y_2)\Leftrightarrow x_1\leqslant x_2$, for example, a preordering of pairs (x, y) of real numbers is obtained. From a preordering R an equivalence relation $\sim$ can be defined in X by $x\sim y\Leftrightarrow(xRy$ and $yRx)$. Let $[X]=X/\sim$ be the †quotient set of set X by this equivalence relation, and let $[x]$, $[y]$ be the equivalence classes determined by $x, y\in X$; then an ordering $\leqslant$ can be defined in $[X]$ by $[x]\leqslant[y]\Leftrightarrow xRy$. (For further topics → 52 Categories and Functors; 409 Structures.)

References

[1] N. Bourbaki, Eléments de mathématique, I. Théorie des ensembles, ch. 3, Actualités Sci. Ind., 1243b, Hermann, second edition, 1967; English translation, Theory of sets, Addison-Wesley, 1968.
[2] G. Birkhoff, Lattice theory, Amer. Math. Soc. Colloq. Publ., 1940, revised edition, 1948.
Also → references to 381 Sets.

312 (II.13) Ordinal Numbers

A. General Remarks

Let $A\cong B$ mean that two †ordered sets A, B are †order isomorphic; then the relation $\cong$ is an †equivalence relation. An equivalence class under this relation is called an **order type**, and the class to which an ordered set A belongs is called the **order type of** A. Historically, an ordinal number was first defined as the order type of a †well-ordered set (Cantor [2]). However, it was found that a contradiction occurs if order type defined in this way are considered to form a set. Hence, another definition was given by J. von Neumann [3], which is stated in Section B. A similar situation was found concerning the definition of †cardinal numbers, which led to a new definition of cardinal numbers using ordinal numbers, which is given in Section D.

B. Definitions

A set α is called an **ordinal number** if it satisfies the following two conditions: (i) α is a well-ordered set with the †binary relation $\in$ as its ordering; and (ii) $\beta\in\alpha$ implies $\beta\subset\alpha$. According to this definition, the empty set is an ordinal number, which is denoted by 0. Also, $1=\{0\}$, $2=\{0,1\}$, $3=\{0,1,2\},\ldots$ are ordinal numbers. These ordinal numbers, which are finite sets, are called **finite ordinal numbers**. The finite ordinal numbers are identified with the natural numbers (including 0). The set $\omega=\{0,1,2,\ldots\}$ of all natural numbers is also an ordinal number. An ordinal number that is an infinite set, like ω, is called a **transfinite ordinal number**.

For every well-ordered set A, there exists one and only one ordinal number order isomorphic to A. This ordinal number is called the **ordinal number** of A. (Throughout this article, lower-case Greek letters denote ordinal numbers.) We also write $\alpha\in\beta$ as $\alpha<\beta$, which defines an †ordering of the ordinal numbers. The least ordinal number is 0, and the ordering of the finite ordinal numbers coincides with the usual ordering of the natural numbers. The least transfinite ordinal number is ω. The ordering $\leqslant$, introduced by defining $\alpha\leqslant\beta$ to mean either $\alpha<\beta$ or $\alpha=\beta$, is a †linear ordering and, in fact, a †well-ordering of the ordinal numbers. Therefore †transfinite induction can be applied to ordinal numbers.

For any ordinal number α, the set $\alpha'=\{\xi\mid\xi\leqslant\alpha\}$ is also an ordinal number, and is the †successor of α. There exists at most one ordinal number that is the †predecessor of α. A transfinite ordinal number without a predecessor is called a **limit ordinal number**, and all the other ordinal numbers are called **isolated ordinal numbers**. The first limit ordinal number is ω. For any set A of ordinal numbers, $\{\xi \mid \exists\eta\,(\xi<\eta\in A)\}$ is an ordinal number and is $\sup A$, the †supremum of A.

C. Sum, Product, and Power

The **sum** $\alpha+\beta$, the **product** $\alpha\cdot\beta$ (or $\alpha\beta$), and the **power** α^{β} of ordinal numbers α, β are defined by transfinite induction on β and have the following properties:

$$\alpha+0=\alpha, \quad \alpha+\beta'=(\alpha+\beta)',$$

$$\alpha+\gamma=\sup\{\alpha+\xi \mid \xi<\gamma\};$$

$$\alpha\cdot 0=0, \quad \alpha\cdot\beta'=\alpha\cdot\beta+\alpha,$$

$$\alpha\cdot\gamma=\sup\{\alpha\cdot\xi \mid \xi<\gamma\};$$

$$\alpha^0=1, \quad \alpha^{\beta'}=\alpha^{\beta}\cdot\alpha, \quad \alpha^{\gamma}=\sup\{\alpha^{\xi} \mid \xi<\gamma\}.$$

Here γ is a limit ordinal number, and for the power we assume that $\alpha>0$. The sum and product thus defined satisfy the associative laws $(\alpha+\beta)+\gamma=\alpha+(\beta+\gamma)$, $(\alpha\cdot\beta)\cdot\gamma=\alpha\cdot(\beta\cdot\gamma)$ and the left distributive law $\alpha\cdot(\beta+\gamma)=\alpha\cdot\beta+\alpha\cdot\gamma$; the power satisfies the laws $\alpha^{\beta+\gamma}=\alpha^{\beta}\cdot\alpha^{\gamma}$, $\alpha^{\beta\gamma}=(\alpha^{\beta})^{\gamma}$. If α and β are the ordinal numbers of the well-ordered sets A and B, respectively, then $\alpha+\beta$ is the ordinal number of the †ordinal sum $A+B$, and $\alpha\cdot\beta$ is the ordinal number of the †ordinal product BA.

When $\pi>1$, any ordinal number α can be written uniquely in the form

$$\alpha=\pi^{\beta_1}\cdot\gamma_1+\pi^{\beta_2}\cdot\gamma_2+\ldots+\pi^{\beta_n}\cdot\gamma_n;$$

$$\beta_1>\beta_2>\ldots\beta_n>0, \quad 0<\gamma_i<\pi, \quad 1\leqslant i\leqslant n,$$

which is called the **π-adic normal form** for α; when $\pi=\omega$, it is called **Cantor's normal form**.

Let f be an ordinal number-valued function of ordinal numbers. We say that f is **strictly monotone** when $\alpha<\beta$ implies $f(\alpha)<f(\beta)$. If f is strictly monotone, then $\alpha\leqslant f(\alpha)$. We say that f is **continuous** when $f(\gamma)=\sup\{f(\xi)\mid\xi<\gamma\}$ for each limit ordinal number γ. A strictly monotone continuous function is called a **normal function**. If f is a normal function, then for any α there exists a β that satisfies $f(\beta)=\beta>\alpha$. In fact, it suffices to define $\beta_n (n<\omega)$ by $\beta_0=f(\alpha+1)$, $\beta_{n+1}=f(\beta_n)$ and put $\beta=\sup\{\beta_n \mid n<\omega\}$. Since $f(\alpha)=\omega^{\alpha}$ is a normal function, there exists an ε that satisfies $\omega^{\varepsilon}=\varepsilon$. Such an ordinal number ε is called an **ε-number**. We say that β is **cofinal** to α when there exists a monotone function f that satisfies $\alpha=\sup\{f(\xi)' \mid \xi<\beta\}$. The first ordinal number that is cofinal to α is called the **cofinality** of α and is denoted by $\mathrm{cf}(\alpha)$.

D. Cardinal Numbers

Let $M\sim N$ mean that a one-to-one correspondence exists between the two sets M and N. An ordinal number α with the property that $\alpha\sim\xi$ implies $\alpha\leqslant\xi$ is called an **initial number** or a **cardinal number**.

With the †axiom of choice, it can be shown that for each set M there exists one and only one cardinal number α satisfying $M\sim\alpha$. This unique α is called the **cardinality** (or **cardinal number**) **of the set** M and is denoted by $\overline{\overline{M}}$.

All finite ordinal numbers are cardinal numbers, and ω is the least transfinite cardinal number. There exists one and only one monotone function that maps the class of ordinal numbers onto the class of transfinite cardinal numbers, and it is a normal function. The value of this function corresponding to α is denoted by $\aleph_{\alpha}$ **(aleph alpha)** or ω_{α}. In particular, $\aleph_0=\omega$, and $\aleph_1$ is both the smallest uncountable cardinal number and the smallest uncountable ordinal number. A finite ordinal number is called an **ordinal number of the first number class**, and an ordinal number α satisfying $\aleph_0\leqslant\alpha<\aleph_1$ is called an **ordinal number of the second number class**. The concept of **ordinal number of the third** (or **higher**) **number class** is defined similarly.

E. Inaccessible Ordinal Numbers

The cofinality $\mathrm{cf}(\alpha)$ of α always satisfies $\mathrm{cf}(\alpha)\leqslant\alpha$. An ordinal number is said to be **regular** when $\mathrm{cf}(\alpha)=\alpha$ and singular when $\mathrm{cf}(\alpha)<\alpha$. For any ordinal number α, $\mathrm{cf}(\alpha)$ is a regular cardinal number; therefore any regular ordinal number is a cardinal number. When $\alpha=\omega_{\beta}$ is regular and β is a limit ordinal number, α is said to be **weakly inaccessible**. Let R be the set-valued function of ordinal numbers, defined by $R(0)=\varnothing$ and $R(\alpha)=\bigcup\{\mathfrak{P}(R(\xi))\mid\xi<\alpha\}$ (by †transfinite induction), where $\mathfrak{P}(M)$ denotes the †power set of M. A regular ordinal number α is said to be **strongly inaccessible** when $\alpha>\omega$ and the following condition is satisfied: If x, y are a pair of sets such that $x\in R(\alpha)$, $y\subset R(\alpha)$, and there exists a mapping of x onto y, then $y\in R(\alpha)$. If a regular ordinal number α is strongly inaccessible, it is weakly inaccessible. A strongly inaccessible ordinal number is usually defined as a regular number $\alpha>\omega$ such that $\beta<\alpha$ implies $\overline{\overline{\mathfrak{P}(\beta)}}<\alpha$. Under the axiom of choice, this definition is equivalent to the one given here. Moreover, under the †generalized continuum hypothesis, strong inaccessibility and weak inaccessibility are equivalent.

References

[1] N. Bourbaki, Eléments de mathématique, I. Théorie des ensembles, ch. 3, Actualités Sci. Ind., 1243b, second edition, Hermann, 1967; English translation, Theory of sets, Addison-Wesley, 1968.
[2] G. Cantor, Beiträge zur Begründung der transfiniten Mengenlehre II, Math. Ann., 49 (1897), 207–246. (Gesammelte Abhandlungen, Springer, 1932; English translation, Contributions to the founding of the theory of transfinite numbers, Open Court, 1915.)
[3] J. von Neumann, Zur Einführung der transfiniten Zahlen, Acts Sci. Math. Szeged., 1 (1923), 199–208. (Collected works I, Pergamon, 1961.)
Also → references to 381 Sets.

313 (XIII.2) Ordinary Differential Equations

A. General Remarks

Let x be a real (complex) variable and y a real (complex) function of x. Assume that $y=F(x)$ is a differentiable function of class C^n if x, y are real, and a holomorphic function if x, y are complex. We write $y', y'', \ldots, y^{(n)}$ for the first n derivatives of y. A relation among $x, y, y', \ldots, y^{(n)}$,

$$f(x, y, y', \ldots, y^{(n)})=0 \tag{1}$$

(which holds identically with respect to x), is called an **ordinary differential equation** for the function $y=F(x)$. Here we assume that the function f in the left-hand side of (1) is a real (complex) function of the $n+2$ variables $x, y, y', \ldots, y^{(n)}$ and is defined in a given domain of $\mathbf{R}^{n+2}$ ($\mathbf{C}^{n+2}$). Usually we assume further that f has a certain regularity, such as being of class C^r ($r=0, 1, \ldots, \infty$), †real analytic, or †complex analytic. A function $y=F(x)$ that satisfies (1) is called a **solution** of (1). To find a solution of (1) is to **solve** or **integrate** it. Ordinary differential equations may be contrasted to **partial differential equations**, which are equations similar to (1) but in which y is a function of two or more variables $x_1, x_2, \ldots$ and which contain the partial derivatives $\partial y/\partial x_1, \partial y/\partial x_2, \ldots$ (— 320 Partial Differential Equations). Ordinarily, the term **differential equation** refers to an ordinary or partial differential equation. In this article, since we are concerned only with ordinary differential equations, we omit the word "ordinary." If the left-hand side f of (1) contains $y^{(n)}$ explicitly or $\partial f/\partial y^{(n)} \neq 0$, then we say that the **order** of (1) is n, and if further f is a polynomial in $y, y', \ldots, y^{(n)}$ that is of degree m with respect to $y^{(n)}$, we say that the **degree** of (1) is m. In particular, if f is a linear form in $y, y', \ldots, y^{(n)}$, then (1) is said to be **linear**. A differential equation that is not linear is said to be **nonlinear** (→ 252 Linear Ordinary Differential Equations; 291 Nonlinear Problems).

Let $\varphi(x, y, c_1, \ldots, c_n)$ be a function of the $n+2$ variables $x, y, c_1, \ldots, c_n$ of class C^r in a domain D, and let $(x_0, y_0, c_1^0, \ldots, c_n^0) \in D$, $\varphi(x_0, y_0, c_1^0, \ldots, c_n^0)=0$, and $\varphi_y(x_0, y_0, c_1^0, \ldots, c_n^0) \neq 0$. Then the equation $\varphi(x, y, c_1^0, \ldots, c_n^0)=0$ defines an †implicit function $y(x)$ of class C^r satisfying the condition $y(x_0)=y_0$. Consider $c_1, \ldots, c_n$ to be constants in $\varphi(x, y, c_1, \ldots, c_n)=0$ and differentiate φ n times with respect to x. Then we obtain a system of n equations in the variables $x, y, y', \ldots, y^{(n)}, c_1, \ldots, c_n$. If we can eliminate $c_1, \ldots, c_n$ from these n equations and $\varphi=0$, then we obtain an nth-order differential equation of the form (1). Conversely, a solution of an nth-order differential equation can usually be written in the form

$$\varphi(x, y, c_1, \ldots, c_n)=0, \tag{2}$$

which contains n **arbitrary constants** $c_1, \ldots, c_n$ (sometimes called **integration constants**). A solution containing n arbitrary constants of the form (2) of an nth-order differential equation is called a **general solution**, and a solution $\varphi(x, y, c_1^0, \ldots, c_n^0)=0$ obtained from a general solution $\varphi=0$ by giving particular values $c_1^0, \ldots, c_n^0$ to the arbitrary constants is called a **particular solution**. Some equations admit solutions that are not particular solutions. They are called **singular solutions** (for example, †Clairaut differential equations; → Appendix A, Table 14.I).

B. Systems of Differential Equations

A set of n differential equations containing n unknown functions $y_1, \ldots, y_n$ of a variable x is called a **system of ordinary differential equations**. Here each equation of the system has a form similar to (1), but each left-hand side contains $y_1, \ldots, y_n$ and their derivatives. A set of n functions $y_1, \ldots, y_n$ of x is called a **solution** if the functions satisfy the given system of differential equations. The highest order of derivatives in the left-hand sides is called the **order** of the system of differential equations.

We consider most frequently a first-order system of the form

$$y_i'=f_i(x, y_1, \ldots, y_n), \qquad i=1, 2, \ldots, n. \tag{3}$$

If we put $y=y_1, y'=y_2, \ldots, y^{(n-1)}=y_n$ and solve (1) with respect to $y^{(n)}$ to get $y^{(n)}=f_n(x, y_1, \ldots, y_n)$, then (1) is equivalent to a system of equations of the form (3), where $f_1=y_2, f_2=y_3, \ldots, f_n=f$. In an analogous way, a general system of equations can be transformed to a system of the form (3). Therefore (3) is called the **normal form** of differential equations.

C. The Geometric Interpretation

When $x, y_1, \ldots, y_n$ are real, (3) can be interpreted as follows: Let $I=(a,b)$ be an open interval and D a domain of $\mathbf{R}^n$. Let

$$y_i=\varphi_i(x, c_1, \ldots, c_n), \qquad i=1,2,\ldots,n, \tag{4}$$

be functions of class C^1 defined for $(x, c_1, \ldots, c_n) \in I \times D$, and let $\mathfrak{D}(x_0)$ be the image in the $y_1, \ldots, y_n$-space of D under the mapping $y_i=\varphi_i(x_0, c_1, \ldots, c_n)$ $(i=1,2,\ldots,n)$ for a fixed $x_0 \in I$. We assume that for each fixed $x_0 \in I$ we have $\partial(\varphi_1, \ldots, \varphi_n)/\partial(c_1, \ldots, c_n) \neq 0$ in D. Then for every $x=x_0 \in I$, $c_1, \ldots, c_n$ are considered to be functions of $(y_1, \ldots, y_n)$ defined in a neighborhood of every point $(y_1^0, \ldots, y_n^0)$ of $\mathfrak{D}(x_0)$, and we have $y_i'=\varphi_i'(x, c_1, \ldots, c_n)=f_i(x, y_1, \ldots, y_n)$ $(i=1,2,\ldots,n)$, i.e., $y_1, \ldots, y_n$ satisfy a system of differential equations of the form (3). On the other hand, (4) represents a family of curves of class C^1 in the $x, y_1, \ldots, y_n$ space $\mathbf{R}^{n+1}$ containing n parameters $c_1, \ldots, c_n$, for which $(y_1', \ldots, y_n')$ is the tangent vector (in the terminology of physics, $(y_i', \ldots, y_n')$ gives the speed and the direction of a stationary flow in $\mathbf{R}^{n+1}$ at each point). By solving (3) we find the family of curves of class C^1 in $\mathbf{R}^{n+1}$ (in the terminology of physics, we find a stationary flow of which the speed and the direction are given at each point). A solution containing n parameters analogous to (4) is called a **general solution** of (3), and a solution obtained from a general solution by giving particular values to the n parameters is called a **particular solution**.

As may be imagined by the interpretation in this section, there exists in general one and only one particular solution passing through the point $(x_0, y_1^0, \ldots, y_n^0)$ for $x_0 \in I$, $(y_1^0, \ldots, y_n^0) \in \mathfrak{D}(x_0)$. The problem of finding this solution, i.e., the solution of (3) for which $y_i(x_0)=y_i^0$ for $x=x_0$, is called the **initial value problem** (→ 316 Ordinary Differential Equations (Initial Value Problems)).

D. Methods of Integration

We have different methods of solving differential equations. To solve differential equations by a finite number of integrations is called the **method of quadrature**. This method is useful for some special types of differential equations (→ Appendix A, Table 14.I). S. Lie gave theoretical foundations for this method by using Lie transformation groups (→ 431 Transformation Groups; Appendix A, Table 14.III). There are many other methods, for example, power series methods (assuming that the solution can be expanded in a power series $\sum a_\nu (x-a)^\nu$, substituting the series for y in (1), and finding its coefficients); methods of successive approximation; methods using †Laplace transforms or †Fourier transforms; †perturbation methods; numerical methods; etc. (→ 303 Numerical Solution of Ordinary Differential Equations).

Historically, finding explicit solutions of various kinds of differential equations has been the main object of the theory. Recently, however, the importance of qualitative studies, in particular theorems on the existence and uniqueness of solutions, has been recognized. For example, if a solution with a property A is given, and if the uniqueness of the solution having the property A and the existence of solutions having the properties A and B can be shown, then the given solution necessarily has the property B. In this way, topological and analytic studies of differential equations are applied to find their solutions (→ 314 Ordinary Differential Equations (Asymptotic Behavior of Solutions); 315 (Boundary Value Problems); 316 (Initial Value Problems); 126 Dynamical Systems).

References

[1] W. Kaplan, Ordinary differential equations, Addison-Wesley, 1958.
[2] P. Hartman, Ordinary differential equations, Wiley, 1964.
[3] E. Hille, Lectures on ordinary differential equations, Addition-Wesley, 1968.
Also → references to 316 Ordinary Differential Equations (Initial Value Problems).

314 (XIII.5) Ordinary Differential Equations (Asymptotic Behavior of Solutions)

A. Linear Differential Equations

A system of linear ordinary differential equations can be written as

$$\mathbf{x}'=A(t)\mathbf{x}, \tag{1}$$

where t is a real independent variable, $\mathbf{x}=(x_1, \ldots, x_n)$ is an n-dimensional complex vector function of t, and $A(t)$ is an $n \times n$ matrix whose elements are complex-valued functions of t. If $A(t)$ is a continuous function of t defined on an open interval I, any solution of (1) is continuously differentiable for $t \in I$. The question naturally arises as to how the solutions behave as t approaches either one of the endpoints of I; that is, the question of the **asymptotic properties** of the solutions. The interval I can always be taken to be $0 \leqslant t < \infty$, by applying a suitable transformation of the independent variable if necessary.

The study of the †asymptotic expansions of solutions when the coefficient $A(t)$ is an analytic function of t was initiated by H. Poincaré in 1880. This work was continued by J. Horn, J. C. C. A. Kneser, and others in the direction of removing assumptions on the structure of $A(t)$ and extending the domain where the expansions are valid. The theory has been almost completed by W. J. Trjitzinsky, J. Malmquist, and M. Hukuhara (→ 254 Linear Ordinary Differential Equations (Local Theory)). On the other hand, O. Perron initiated a new direction of research by weakening the regularity conditions on the coefficients. His work was continued by F. Lettenmeyer, R. A. Späth, Hukuhara, and others. The methods used in these two lines of investigation were originally distinct, but Hukuhara established a unified method of treating the problems arising in these two different types of investigations. Furthermore, he succeeded in sharpening those results previously obtained.

Here we assume that $A(t)$ need not be analytic. The following asymptotic properties of a solution $\mathbf{x}(t)$ as $t \to \infty$ are considered: (i) boundedness of $\limsup t^{-1} \log|\mathbf{x}(t)|$; (ii) boundedness of solution: $\limsup|\mathbf{x}(t)| < \infty$; (iii) convergence of solution: $\lim \mathbf{x}(t)$; (iv) integrability: $\int^{\infty} |\mathbf{x}(s)|^p ds < \infty$, etc. We call $\chi(\mathbf{x}) = \limsup t^{-1} \log|\mathbf{x}(t)|$ the **type number** (or **Lyapunov characteristic number**) of the solution $\mathbf{x}(t)$. The fact that all solutions of (1) are bounded is equivalent to the †stability of the solution $\mathbf{x}=0$, and the fact that all solutions of (1) tend to zero as $t \to \infty$ is equivalent to the †asymptotic stability of the solution $\mathbf{x}=0$.

B. Constant Coefficients and Periodic Coefficients

We begin with the particular case of (1), where $A(t)$ is a constant matrix:

$$\mathbf{x}' = A\mathbf{x}. \tag{2}$$

To study the asymptotic properties of the solutions of (2), it suffices to transform the matrix A into a †Jordan canonical form, since the structure of the solution space of (2) is completely determined by the Jordan canonical form of A. Thus all solutions of (2) are bounded if and only if every eigenvalue of A has a real part not greater than zero, and those with zero real parts are of simple type, that is, the corresponding blocks in the Jordan canonical form are all 1×1 matrices; all solutions of (2) tend to zero as $t \to \infty$ if and only if every eigenvalue of A has negative real part.

Consider the linear system

$$\mathbf{x}' = A_0(t)\mathbf{x}, \tag{3}$$

where $A_0(t)$ is a periodic matrix function of period ω. According to †Floquet's theorem, (3) is transformed into a system with constant coefficients by means of a suitable transformation $\mathbf{x} = P(t)\mathbf{y}$, where $P(t)$ is a nonsingular periodic matrix of period ω. Thus, at least theoretically, the information on the asymptotic behavior of the solutions of the periodic system (3) can be derived from the corresponding theory for the system with constant coefficients (2).

C. Asymptotic Integration

Suppose that $A(t)$ is bounded. Then the type number $\chi(\mathbf{x})$ is finite for any nontrivial solution $\mathbf{x}(t)$ of (1), and the number of distinct type numbers does not exceed n.

Consider the linear system

$$\mathbf{x}' = [A + B(t)]\mathbf{x}, \tag{4}$$

where A is a constant matrix and $B(t)$ is a matrix function such that $\int_t^{t+1} \|B(s)\| ds \to 0$ as $t \to \infty$. For any nontrivial solution $\mathbf{x}(t)$ of (4), the limit $\mu = \lim t^{-1} \log|\mathbf{x}(t)|$ exists and is equal to the real part of one of the eigenvalues of A. Conversely, if at least one eigenvalue of A has real part μ, then there exists a nontrivial solution $\mathbf{x}(t)$ of (4) satisfying $\lim t^{-1} \log|\mathbf{x}(t)| = \mu$. Suppose in addition that $B(t) \to 0$ as $t \to \infty$. Let $\mu_1 \leqslant \mu_2 \leqslant \ldots \leqslant \mu_n$ be the real parts of the eigenvalues of A. Then there exists a †fundamental system of solutions of (4), $\{\mathbf{x}_1(t), \ldots, \mathbf{x}_n(t)\}$, such that for any c_i, $c_k \neq 0$,

$$\log|c_1\mathbf{x}_1(t) + \ldots + c_k\mathbf{x}_k(t)| = \mu_k t + o(t).$$

A sharp estimate of the term $o(t)$ was given by Hukuhara.

Next consider the linear system

$$\mathbf{x}' = [A(t) + B(t)]\mathbf{x}, \tag{5}$$

where the matrices $A(t)$ and $B(t)$ satisfy $\int^{\infty} \|A'(s)\| ds < \infty$ and $\int^{\infty} \|B(s)\| ds < \infty$.

Let $\lambda_1(t), \ldots, \lambda_n(t)$ and $\lambda_1, \ldots, \lambda_n$, $\lambda_k = \lim \lambda_k(t)$, be the eigenvalues of $A(t)$ and $A=\lim A(t)$, respectively. N. Levinson proved the following theorem: Assume that $\lambda_1, \ldots, \lambda_n$ are mutually distinct and $M_{jk}(t)=\operatorname{Re}\int_0^t[\lambda_j(s)-\lambda_k(s)]\,ds$ satisfy either $M_{jk}(t)\to\infty$ as $t\to\infty$ and for each pair (j,k), $M_{jk}(t_2)-M_{jk}(t_1)\geqslant -K$ for $t_1<t_2$, or $M_{jk}(t)\to-\infty$ as $t\to\infty$ and $M_{jk}(t_2)-M_{jk}(t_1)\leqslant K$ for $t_1<t_2$, or $|M_{jk}(t_2)-M_{jk}(t_1)|\leqslant K$ for all t_1, t_2, where K is a positive constant. Then (5) has a fundamental system of solutions $\{\mathbf{x}_1(t), \ldots, \mathbf{x}_n(t)\}$ such that

$$\mathbf{x}_j(t)=\exp\left(\int_0^t \lambda_j(s)\,ds\right)[\xi_j+o(1)], \qquad j=1,\ldots,n,$$

where ξ_j is an eigenvector of A corresponding to λ_j.

D. Boundedness and Convergence of Solutions

Consider again the linear system (5) satisfying $\int^\infty\|A'(s)\|\,ds<\infty$ and $\int^\infty\|B(s)\|\,ds<\infty$. Suppose that all eigenvalues of $A(t)$ have nonpositive real parts and that the eigenvalues of $A=\lim A(t)$ whose real parts vanish are simple. Then all solutions of (5) are bounded. This result is a generalization, due to L. Cesari, of the so-called **Dini-Hukuhara theorem**.

In the case of general $A(t)$, it is known that not all solutions of (5) are bounded even if all solutions of (1) tend to zero as $t\to\infty$ and if the matrix $B(t)$ is such that $\int^\infty\|B(s)\|\,ds<\infty$ and $B(t)\to 0$ as $t\to\infty$. However, if $A(t)$ is periodic or satisfies $\liminf\operatorname{Re}\int^t\operatorname{tr}A(s)\,ds>-\infty$, then under the assumption that $\int^\infty\|B(s)\|\,ds<\infty$, the boundedness of all solutions of (1) implies the boundedness of all solutions of (5).

The following inequalities often provide useful information about the asymptotic behavior of solutions of (1):

$$|\mathbf{x}(0)|\exp\left(-\int_0^t \mu[-A(s)]\,ds\right)\leqslant|\mathbf{x}(t)|$$
$$\leqslant|\mathbf{x}(0)|\exp\left(\int_0^t \mu[A(s)]\,ds\right), \qquad t\geqslant 0,$$

where $\mu[A(t)]=\lim_{h\to+0}[\|I+hA(t)\|-1]/h$. ($\mu[A(t)]$ was introduced by Lozinskiĭ.) If $\limsup\int^t\mu[A(s)]\,ds<\infty$, then all solutions of (1) are bounded; if $\lim\int^t\mu[A(s)]\,ds$ exists, then for every solution $\mathbf{x}(t)$ of (1), $|\mathbf{x}(t)|$ tends to a finite limit as $t\to\infty$; and if $\lim\int^t\mu[A(s)]\,ds=-\infty$, then all solutions of (1) tend to zero as $t\to\infty$. It can be shown that every solution of (1) tends to a finite limit as $t\to\infty$, provided that $\int^\infty\|A(s)\|\,ds<\infty$.

If all solutions of (1) are bounded, then $\limsup\operatorname{Re}\int^t\operatorname{tr}A(s)\,ds<\infty$. If $\liminf\operatorname{Re}\int^t\operatorname{tr}A(s)\,ds>-\infty$, then (1) has a solution $\mathbf{x}(t)$ with the property that $\limsup|\mathbf{x}(t)|>0$. When $\lim|\mathbf{x}(t)|$ exists for every solution $x(t)$ of (1), if there exists a nontrivial solution $\mathbf{x}(t)$ of (1) such that $\lim x(t)=0$, then $\lim\operatorname{Re}\int^t\operatorname{tr}A(s)\,ds=-\infty$, but if there is no such solution, then $\operatorname{Re}\int^t\operatorname{tr}A(s)\,ds$ is bounded.

E. Nonlinear Differential Equations

Consider a system of nonlinear differential equations of the form

$$\mathbf{x}'=A\mathbf{x}+\mathbf{f}(t,\mathbf{x}), \tag{6}$$

where A is an $n\times n$ constant matrix and $\mathbf{f}(t,\mathbf{x})$ is an n-vector function that is continuous for $t\geqslant 0$, $|\mathbf{x}|<\Delta$, and that satisfies $\mathbf{f}(t,\mathbf{0})=\mathbf{0}$. Suppose that $\mathbf{f}(t,\mathbf{x})/|\mathbf{x}|\to\mathbf{0}$ as $|\mathbf{x}|\to 0$ and $t\to\infty$. Then for every eventually nontrivial solution $\mathbf{x}(t)$ of (6) that tends to zero as $t\to\infty$, $\mu=\lim t^{-1}\log|\mathbf{x}(t)|$ exists and equals the real part of one of the eigenvalues of A. Conversely, if at least one eigenvalue of A has real part $\mu<0$, then there exists a solution $\mathbf{x}(t)$ of (6) such that $\lim t^{-1}\log|\mathbf{x}(t)|=\mu$. Suppose that $\mathbf{f}(t,\mathbf{x})/|\mathbf{x}|\to 0$ as $|\mathbf{x}|\to 0$ uniformly with respect to t. Then if all eigenvalues of A have negative real parts, the zero solution $\mathbf{x}(t)=\mathbf{0}$ of (6) is asymptotically stable, and if A has an eigenvalue whose real part is positive, then the zero solution of (6) is †unstable. Suppose that $\mathbf{f}_\mathbf{x}(t,\mathbf{x})=(\partial f_j(t,x)/\partial x_k)\to 0$ as $|\mathbf{x}|\to 0$ uniformly with respect to t. In this case, if A is a matrix such that its k eigenvalues have negative real parts and the other $n-k$ eigenvalues have positive real parts, then there exists a k-dimensional manifold S containing the origin with the following property: For t_0 sufficiently large, any solution $\mathbf{x}(t)$ of (6) tends to zero as $t\to\infty$, provided that $\mathbf{x}(t_0)\in S$, and if $\mathbf{x}(t_0)\notin S$, $\mathbf{x}(t)$ cannot remain in the vicinity of the origin no matter how close $\mathbf{x}(t_0)$ is to the origin.

In the nonlinear system

$$\mathbf{x}'=\mathbf{F}(t,\mathbf{x}), \tag{7}$$

suppose that $\mathbf{F}(t,\mathbf{x})$ is of period ω with respect to t and has continuous partial derivatives with respect to $\mathbf{x}$. Suppose, moreover, that (7) has a solution $\mathbf{p}(t)$ of period ω. If all the †characteristic exponents of the †variational system of (7) with respect to $\mathbf{p}(t)$, $\mathbf{y}'=\mathbf{F}_\mathbf{x}(t,\mathbf{p}(t))\mathbf{y}$ with $\mathbf{F}_\mathbf{x}(t,\mathbf{x})=(\partial F_j(t,x)/\partial x_k)$, have negative real parts, then the periodic solution $\mathbf{p}(t)$ is asymptotically stable. If an autonomous system $\mathbf{x}'=F(\mathbf{x})$ has a periodic solution $\mathbf{p}(t)$ and the corresponding variational system $\mathbf{y}'=\mathbf{F}_\mathbf{x}(\mathbf{p}(t))\mathbf{y}$ has $n-1$ characteristic exponents with negative eigenvalues, then there exists an $\varepsilon>0$ such that for any solution $\mathbf{x}(t)$ satisfying $|\mathbf{x}(t_1)-\mathbf{p}(t_0)|<\varepsilon$ for some t_0 and t_1, we have $|\mathbf{x}(t)-\mathbf{p}(t+c)|\to 0$ as $t\to\infty$ for a suitable choice of c (asymptotic phase).

F. Scalar Differential Equations

The aforementioned results can be specialized to the case of higher-order scalar (or single) ordinary differential equations. Much sharper results can often be derived through direct analysis of scalar equations themselves. In particular, detailed and deep results have been obtained for second-order linear differential equations of the form

$$x''+q(t)x=0, \qquad (8)$$

e.g., †Mathieu's equation.

If $\int^{\infty} s|q(s)|\,ds<\infty$, then (8) has a fundamental system of solutions $\{x_1(t), x_2(t)\}$ satisfying, for $t\to\infty$,

$$x_1(t)=1+o(1), \qquad x_2(t)=t[1+o(1)],$$

$$x_1'(t)=t^{-1}o(1), \qquad x_2'(t)=1+o(1);$$

if $\int^{\infty}|q(s)+1|\,ds<\infty$, then (8) has a fundamental system of solutions satisfying

$$x_1(t)=e^{t}[1+o(1)], \qquad x_2(t)=e^{-t}[1+o(1)],$$

$$x_1'(t)=e^{t}[1+o(1)], \qquad x_2'(t)=-e^{-t}[1+o(1)];$$

and if $\int^{\infty}|q(s)-1|\,ds<\infty$, then (8) has a fundamental system of solutions

$$x_1(t)=e^{it}[1+o(1)], \qquad x_2(t)=e^{-it}[1+o(1)],$$

$$x_1'(t)=ie^{it}[1+o(1)], \qquad x_2'(t)=-ie^{-it}[1+o(1)].$$

Suppose that $q(t)\to c>0$ as $t\to\infty$ and $\int^{\infty}|q'(s)|\,ds<\infty$. Then $x(t)$ and $x'(t)$ are bounded for every solution $x(t)$ of (8). The same is true if $q(t)$ is a positive periodic function of period ω such that $\omega\int_0^{\omega}q(s)\,ds\leqslant 4$. If $q(t)$ is negative, then (8) always has both bounded and unbounded monotone solutions.

The number of linearly independent solutions $x(t)$ of (8) satisfying $\int^{\infty}|x(s)|^2\,ds<\infty$ plays an important role in †eigenvalue problems. It is known that the ordinary differential operator $l[x]=x''+q(t)x$ is of †limit point type at infinity if there exist a positive function $M(t)$ and positive constants k_1, k_2 such that $q(t)\leqslant k_1 M(t)$, $|M'(t)M^{-3/2}(t)|\leqslant k_2$, and $\int^{\infty}M^{-1/2}(s)\,ds=\infty$, and that $l[x]$ is of †limit circle type at infinity if $q(t)>0$, $\int^{\infty}q^{-1/2}(s)\,ds=\infty$, and $\int^{\infty}|[q^{-3/2}(s)q'(s)]'+(1/4)q^{-5/2}(s)q'^2(s)|\,ds<\infty$.

Finally consider the nonlinear equation

$$x''+q(t)|x|^{\gamma}\operatorname{sgn}x=0, \qquad (9)$$

where γ is a positive constant and $q(t)$ is a positive function. If $q'(t)\geqslant 0$, then all solutions of (9) are bounded; if either $q'(t)\geqslant 0$ and $\lim q(t)<\infty$ or $q'(t)\leqslant 0$ and $\lim q(t)>0$, then all solutions $x(t)$ of (9) are bounded together with their derivatives $x'(t)$; and if $q'(t)\geqslant 0$, $\lim q(t)=\infty$ and either $q''(t)\geqslant 0$ or $q''(t)\leqslant 0$, then all solutions of (9) converge to zero as $t\to\infty$.

Equation (9) is said to be **oscillatory** if every solution of (9) that is continuable to $t=\infty$ has arbitrarily large zeros. If (9) is oscillatory and if $q_1(t)\geqslant q(t)$, then the equation $x''+q_1(t)|x|^{\gamma}\operatorname{sgn}x=0$ is also oscillatory. When $\gamma=1$, (9) is oscillatory if $q(t)\geqslant(1+\varepsilon)/4t^2$ for some $\varepsilon>0$, and is not oscillatory if $q(t)\leqslant 1/4t^2$. A necessary and sufficient condition for equation (9) with $\gamma\neq 1$ to be oscillatory is as follows: $\int^{\infty}sq(s)\,ds=\infty$ if $\gamma>1$; $\int^{\infty}s^{\gamma}q(s)\,ds=\infty$ if $0<\gamma<1$.

References

[1] M. Hukuhara, Sur les points singuliers des équations différentielles linéaires; domaine réel, J. Fac. Sci. Hokkaido Univ., (I) 2 (1934), 13–88.

[2] R. Bellman, Stability theory of differential equations, McGraw-Hill, 1953.

[3] L. Cesari, Asymptotic behavior and stability problems in ordinary differential equations, Springer, third edition, 1970.

[4] W. A. Coppel, Stability and asymptotic behavior of differential equations, Heath, 1965.

[5] P. Hartman, Ordinary differential equations, Wiley, 1964.

[6] E. A. Coddington and N. Levinson, Theory of ordinary differential equations, McGraw-Hill, 1955.

[7] J. S. W. Wong, On the generalized Emden-Fowler equation, SIAM Rev., 17 (1975), 339–360.

315 (XIII.4) Ordinary Differential Equations (Boundary Value Problems)

A. General Remarks

Consider the differential equation in the real variable x

$$f(x, y, y', \ldots, y^{(n)})=0. \qquad (1)$$

Let $a_1, \ldots, a_k$ be points in an interval $I\subset\mathbf{R}$ and consider several relations between nk values $y(a_i), y'(a_i), \ldots, y^{(n-1)}(a_i)$, $i=1, \ldots, k$. The problem of finding solutions of (1) satisfying these relations is called a **boundary value problem** of (1), and the relations considered are called **boundary conditions**. When $k=2$ and a_1, a_2 are the endpoints of I, the problem, called a **two-point boundary value problem**, has been a main subject of study. We can consider boundary value problems in the same way for systems of differential equations.

B. Linear Differential Equations

Consider a linear ordinary differential operator L defined by

$$L[y]=p_0(x)y^{(n)}+p_1(x)y^{(n-1)}+\ldots+p_n(x)y,$$

where $p_k(x)$ is a complex-valued function of class C^{n-k} defined on a compact interval $a\leqslant x\leqslant b$ and $p_0(x)\neq 0$ for any $x\in[a,b]$. We define a system of **linear boundary operators** $U_1,\ldots,U_m$ by

$$U_i[y]=\sum_{j=1}^{n} M_{ij}y^{(j-1)}(a)+\sum_{j=1}^{n} N_{ij}y^{(j-1)}(b).$$

Given a function $f(x)$ and complex constants $\gamma_1,\ldots,\gamma_m$, the linear boundary value problem defined by

$$L[y]=f(x),\qquad U_i[y]=\gamma_i,\qquad i=1,\ldots,n, \tag{2}$$

is a two-point boundary value problem. When $f(x)\equiv 0$, $\gamma_i=0$, $i=1,\ldots,n$, the problem is called **homogeneous**; otherwise it is called **inhomogeneous.** Let $L^*[y]$ be a formally †adjoint differential operator of $L[y]$. A set of m^* linear boundary conditions $U_i^*[y]=0$, $i=1,\ldots,m^*$, is said to be an **adjoint boundary condition** of $U_i[y]=0$, $i=1,\ldots,m$, if for any function y of class C^n satisfying $U_i[y]=0$, $i=1,\ldots,m$, and any function y^* of class C^n satisfying $U_i^*[y^*]=0$, $i=1,\ldots,m^*$, we have $\int_a^b L[y]\overline{y^*}\,dx=\int_b^a y\overline{L^*[y^*]}\,dx$. The boundary value problem

$$L^*[y]=0,\qquad U_i^*[y]=0,\qquad i=1,\ldots,m^*, \tag{3}$$

is said to be an **adjoint boundary value problem** of

$$L[y]=0,\qquad U_i[y]=0,\qquad i=1,\ldots,m. \tag{4}$$

We say that the problem (4) is **self-adjoint** if $L[y]=L^*[y]$ and the conditions $U_i[y]=0$, $i=1,\ldots,m$, are equivalent to the conditions $U_i^*[y]=0$, $i=1,\ldots,m^*$.

The boundary value problem containing a parameter λ

$$L[y]=\lambda y,\qquad U_i[y]=0,\qquad i=1,\ldots,n, \tag{5}$$

admits nontrivial solutions only for special values of λ. Such values of λ are called the **eigenvalues** (or **proper values**) of (5), and the corresponding solutions $\not\equiv 0$ are called the **eigenfunctions** (or **proper functions**) of (5). For any value of λ that is not an eigenvalue, there exists a unique function $G(x,\xi,\lambda)$ such that the conditions $L[y]=\lambda y+f$, $U_i[y]=0$ are equivalent to $y=\int_a^b G(x,\xi,\lambda)f(\xi)\,d\xi$. The function $G(x,\xi,\lambda)$ is called the **Green's function** of (5). If $\lambda=0$ is not an eigenvalue, then (5) is equivalent to

$$y(x)=\lambda\int_a^b G(x,\xi)y(\xi)\,d\xi,$$

where $G(x,\xi)=G(x,\xi,0)$. For the Green's function $G(x,\xi,\lambda)$ of (5) and the Green's function $G^*(x,\xi,\lambda)$ of $L^*[y]=\lambda y$, $U_i^*[y]=0$, we have the relation $G(x,\xi,\lambda)=\overline{G}^*(\xi,x,\overline{\lambda})$. Under the assumption that (5) is self-adjoint, we have the following four propositions: (i) Problem (5) has only real eigenvalues which form a finite or countably infinite discrete set; (ii) two eigenfunctions corresponding to two different eigenvalues are orthogonal to each other; (iii) if $\{\varphi_n\}$ is an †orthonormal set of eigenfunctions such that no eigenfunction is linearly independent of $\{\varphi_n\}$, then the system $\{\varphi_n\}$ is a complete orthonormal set in the Hilbert space $L_2(a,b)$ consisting of functions that are square integrable on (a,b), and hence for the †Fourier expansion $f=a_1\varphi_1+a_2\varphi_2+\ldots$ of any $f\in L_2(a,b)$ the †Parseval equality holds; and (iv) if f is a function of class C^n satisfying $U_i[f]=0$, then the Fourier expansion of f converges uniformly to f on $[a,b]$.

The boundary value problem for a second-order equation

$$(p(x)y')'+(q(x)+\lambda r(x))y=0,$$

$$\alpha y(a)+\beta y'(a)=0,\qquad \gamma y(b)+\delta y'(b)=0$$

is called a **Sturm-Liouville problem**, where p, q, r are real-valued functions defined on $[a,b]$ and α, β, γ, are real constants. Suppose that p, q, r are continuous and $p(x)>0$, $r(x)>0$ on $[a,b]$. Then (i) the eigenvalues form an increasing sequence tending to $+\infty$; (ii) the eigenfunction $\varphi_n(x)$ associated with λ_n has precisely n zeros in $a<x<b$, and there exists between two adjacent zeros of $\varphi_n(x)$ a zero of $\varphi_{n-1}(x)$; and (iii) the set of eigenfunctions is an orthogonal set on $[a,b]$ with weight function $r(x)$:

$$\int_a^b r(x)\varphi_m(x)\varphi_n(x)\,dx=0,\qquad m\neq n.$$

When the coefficients $p_0,\ldots,p_n$ of L are defined in an open interval $-\infty\leqslant a<x<b\leqslant\infty$ and p_k is of class C^{n-k}, L defines in a natural way operators in the Hilbert space consisting of functions that are square integrable in $a<x<b$, and the general theory is based on operator theory in Hilbert space (→ 390 Spectral Analysis of Operators).

C. Nonlinear Differential Equations

Boundary value problems for nonlinear differential equations are very difficult, and results are obtained only for equations of special form.

Consider, for example, the second-order equation

$$y''=f(x,y,y') \tag{6}$$

and boundary conditions $y(a)=A$, $y(b)=B$. The following theorem has been proved: Suppose that $f(x,y,y')$ is continuous for $a\leqslant x\leqslant b$, $\omega(x)\leqslant y\leqslant\bar{\omega}(x)$, $-\infty<y'<+\infty$, and $|f(x,y,y')|\leqslant M(1+y'^2)$; $\omega''(x)>f(x,\omega(x),\omega'(x))$ and $\bar{\omega}''<f(x,\bar{\omega}(x),\bar{\omega}'(x))$ for $a\leqslant x\leqslant b$; and $\omega(a)\leqslant A\leqslant\bar{\omega}(a)$ and $\omega(b)\leqslant B\leqslant\bar{\omega}(b)$. Then (6) admits a solution $y(x)$ such that $y(a)=A$, $y(b)=B$, and $\omega(x)\leqslant y(x)\leqslant\bar{\omega}(x)$ for $a\leqslant x\leqslant b$. If in addition f is an increasing function with respect to y, the solution is unique. Moreover, under suitable conditions, the solution is obtainable by the method of successive approximations.

The boundary value problem

$$y'''+2yy''+2\lambda(k^2-y'^2)=0,$$

$$y(0)=y'(0)=0, \qquad y'(x)\to k\ (x\to\infty),$$

where λ and k are constants, appears in the theory of fluid dynamics. It is known that if $\lambda\geqslant 0$, the problem has a solution, and that if $0\leqslant\lambda\leqslant 1$, the solution is unique.

Consider the system of differential equations

$$y_j'=f_j(x,y_1,\ldots,y_n), \qquad j=1,\ldots,n.$$

The problem of finding a solution such that $y_j(a_j)=b_j$, $j=1,\ldots,n$, called **Hukuhara's problem**, reduces to the initial value problem when the a_j coincide. The problem of solving

$$y^{(n)}=f(x,y,y',\ldots,y^{(n-1)}),$$

$$y(a_j)=b_j, \qquad j=1,\ldots,n,$$

is reduced to Hukuhara's problem by a suitable change of variables. The following result is a generalization of †Perron's theorem: Let $\omega_j(x)$, $\bar{\omega}_j(x)$, $j=1,\ldots,n$, be continuous and right and left differentiable functions and $\omega_j(x)\leqslant\bar{\omega}_j(x)$ for $a\leqslant x\leqslant b$. Suppose that the $f_j(x,y_1,\ldots,y_n)$ are continuous for $\alpha\leqslant x\leqslant\beta$ and $\omega_k(x)\leqslant y_k\leqslant\bar{\omega}_k(x)$, $k=1,\ldots,n$; satisfy $(x-a_j)(D^{\pm}\bar{\omega}_j(x)-f_j(x,y_1,\ldots,y_n))\geqslant 0$ for $y_j=\bar{\omega}_j(x)$ and $\omega_k(x)\leqslant y_k\leqslant\bar{\omega}_k(x)$, $k\neq j$; satisfy $(x-a_j)(D^{\pm}\omega_j(x)-f_j(x,y_1,\ldots,y_n))\leqslant 0$ for $y_j=\omega_j(x)$ and $\omega_k(x)\leqslant y_k\leqslant\bar{\omega}_k(x)$, $k\neq j$; and satisfy $\omega_j(a_j)\leqslant b_j\leqslant\bar{\omega}_j(a_j)$. Then there exists a solution $y(x)$ such that $y_j(a_j)=b_j$ and $\omega_j(x)\leqslant y(x)\leqslant\bar{\omega}_j(x)$. This theory was applied by M. Hukuhara to the study of singular points of ordinary differential equations.

References

[1] E. A. Coddington and N. Levinson, Theory of ordinary differential equations, McGraw-Hill, 1955.
[2] P. Hartman, Ordinary differential equations, Wiley, 1964.
[3] E. Hille, Lectures on ordinary differential equations, Addison-Wesley, 1969.
[4] M. A. Naĭmark (Neumark), Linear differential operators I, II, Ungar, 1967, 1968. (Original in Russian, 1954.)
[5] N. Dunford and J. T. Schwartz, Linear operators II, Interscience, 1963.
[6] B. M. Levitan and I. S. Sargsyan (Sargsjan), Introduction to spectral theory: Self-adjoint ordinary differential operators, Amer. Math. Soc. Transl. of Math. Monographs, 1975. (Original in Russian, 1970.)

316 (XIII.3) Ordinary Differential Equations (Initial Value Problems)

A. General Remarks

Consider a system of ordinary differential equations

$$dy_i/dx=f_i(x,y_1,\ldots,y_n), \qquad i=1,\ldots,n. \tag{1}$$

A. L. Cauchy first gave a rigorous proof for the existence and uniqueness of solutions: If f_i, $i=1,\ldots,n$, and their derivatives $\partial f_i/\partial y_k$ are continuous in a neighborhood of a point $(a,b_1,\ldots,b_n)$, then there exists a unique solution of (1) satisfying the conditions $y_i(a)=b_i$, $i=1,\ldots,n$. These conditions are called **initial conditions**, and the values $a,b_1,\ldots,b_n$ **initial values**. The problem of finding solutions that satisfy initial conditions is called an **initial value problem** (or **Cauchy problem**). If we consider $(x,y_1,\ldots,y_n)$ as the coordinates of a point in the $(n+1)$-dimensional space $\mathbf{R}^{n+1}$, then a solution of (1) represents a curve in this space called a **solution curve** (or **integral curve**). The statement that a solution satisfies initial conditions $y_i(a)=b_i$, $i=1,\ldots,n$, means that the integral curve represented by it passes through the point $(a,b_1,\ldots,b_n)$.

Since, in general, we can transform a differential equation of higher order into a system of differential equations of the form (1) by introducing new dependent variables, all definitions and theorems concerning the system (1) can be interpreted as applying to a higher-order equation. For example, for the equation $y^{(n)}=f(x,y,y',\ldots,y^{(n-1)})$ the conditions $y(a)=b$, $y'(a)=b',\ldots,y^{(n-1)}(a)=b^{(n-1)}$ constitute initial conditions, and the values $a,b,b',\ldots,b^{(n-1)}$ are initial values. If f and its derivatives $\partial f/\partial y^{(i)}$ are continuous, then there exists a unique solution satisfying given initial conditions.

Suppose that the f_i are continuous. Then a system of functions $(y_1(x),\ldots,y_n(x))$ is a solution of (1) if and only if

$$y_i(x)=b_i+\int_a^x f_i(x,y_1(x),\ldots,y_n(x))\,dx,\qquad i=1,\ldots,n.$$

When the f_i are not continuous, we define $(y_1(x),\ldots,y_n(x))$ to be a solution of (1) for the initial value problem $y_i(a)=b_i$ if $(y_1(x),\ldots,y_n(x))$ satisfied the integral equation just given.

We use the vectorial notation: $\mathbf{y}=(y_1,\ldots y_n)$, $\mathbf{f}=(f_1,\ldots,f_n)$ together with $\|\mathbf{y}\|^2=y_1^2+\ldots+y_n^2$. The equations (1) are then written as the single equation

$$\mathbf{y}'=\mathbf{f}(x,\mathbf{y}).$$

B. Equations in the Real and Complex Domains

We state main theorems for differential equations in the real domain in Sections C–F and in the complex domain in Section G.

C. Existence Theorems

Suppose that $\mathbf{f}(x,\mathbf{y})$ is continuous for $|x-a|\leqslant r$ and $\|\mathbf{y}-\mathbf{b}\|\leqslant\rho$, and that $\|\mathbf{f}(x,\mathbf{y})\|\leqslant M$ there. Then equation (1) admits a solution satisfying $\mathbf{y}(a)=\mathbf{b}$ and defined in an interval $|x-a|\leqslant\min(r,\rho/M)$ (**existence theorem**). There are two methods of proving this theorem, one using **Cauchy polygons** and one using †fixed-point theorems for function spaces. From this theorem we deduce that if $\mathbf{f}(x,\mathbf{y})$ is continuous in a domain D of $\mathbf{R}^{n+1}$, then there exists a solution curve passing through any point of D. Let $\mathbf{y}=\varphi_1(x)$ and $\mathbf{y}=\varphi_2(x)$ be solutions of (1) defined in the intervals I_1 and I_2, respectively. If $I_1\subset I_2$ and $\varphi_1(x)=\varphi_2(x)$ for $x\in I_1$, we say that φ_2 is a **prolongation** or **extension** of φ_1. Given a solution of (1), there exists a nonextendable solution that is an extension of the solution. The solution curve of a nonextendable solution tends to the boundary of D as x tends to any one of the ends of its interval of definition.

D. Uniqueness Theorems

Continuity does not imply uniqueness of the solution. If (1) admits at most one solution satisfying a given condition, we call this condition a **uniqueness condition**. Various kinds of **uniqueness theorems**, which state uniqueness conditions, are known.

The **Lipschitz condition**:

$$\|\mathbf{f}(x,\mathbf{y})-\mathbf{f}(x,\mathbf{z})\|\leqslant L\|\mathbf{y}-\mathbf{z}\|,\quad L>0 \text{ a constant,}$$

is one of the simplest. When $\mathbf{f}$ is continuous and satisfies the Lipschitz condition, the **method of successive approximation**, initiated by C. E. Picard, is often used to prove the existence of solutions. This method is as follows: We choose a suitable function, for example $\mathbf{y}_0(x)\equiv\mathbf{b}$, and then define $\mathbf{y}_k(x)$, $k=1,2,\ldots$, recursively by $\mathbf{y}_k(x)=b+\int_a^x\mathbf{f}(x,\mathbf{y}_{k-1}(x))dx$. Then $\{\mathbf{y}_k(x)\}$ is uniformly convergent, and its limit is a solution of (1) satisfying $\mathbf{y}(a)=\mathbf{b}$.

Assuming the continuity of $\mathbf{f}$, H. Okamura gave a necessary and sufficient condition for uniqueness: Suppose that $\mathbf{f}$ is continuous in D. Then a necessary and sufficient condition for there to exist a unique solution curve of (1) going from any point of D to the right is that there exist a C^1-function $\varphi(x,\mathbf{y},\mathbf{z})$ defined for $(x,\mathbf{y},\mathbf{z})$ such that $(x,\mathbf{y})$ and $(x,\mathbf{z})\in D$ and satisfying the conditions $\varphi(x,\mathbf{y},\mathbf{z})=0$ for $\mathbf{y}=\mathbf{z}$, $\varphi(x,\mathbf{y},\mathbf{z})>0$ for $\mathbf{y}\neq\mathbf{z}$, and

$$\frac{\partial\varphi}{\partial x}+\sum_i\frac{\partial\varphi}{\partial y_i}f_i(x,\mathbf{y})+\sum_i\frac{\partial\varphi}{\partial z_i}f_i(x,\mathbf{z})\leqslant 0.$$

E. Perron's Theorem

Consider the scalar equation $y'=f(x,y)$. We have **Perron's theorem**: Let $\underline{\omega}(x)$ and $\overline{\omega}(x)$ be continuous functions that are right differentiable in $\alpha\leqslant x<\beta$ and satisfy $\underline{\omega}(x)\leqslant\overline{\omega}(x)$, and let f be a continuous function defined on $D:\alpha\leqslant x<\beta$, $\underline{\omega}(x)\leqslant y\leqslant\overline{\omega}(x)$. Suppose that $D^+\underline{\omega}(x)\leqslant f(x,\underline{\omega}(x))$ and $D^+\overline{\omega}(x)\geqslant f(x,\overline{\omega}(x))$ ($D^+\omega$ denotes the †right derivative of ω.) Then for any $(a,b)\in D$ there exists a solution defined on $a\leqslant x<\beta$ and satisfying $y(a)=b$. The fact that the interval of definition is $a\leqslant x<\beta$ can be expressed by saying that if we denote the set $\infty<x<\beta$, $|y|<\infty$ by Ω, then D is closed in Ω and there exists, among solution curves going from a point in D to the right, a curve that reaches the boundary of Ω.

Perron's theorem was generalized by M. Hukuhara and M. Nagumo. Let Ω be an open set in $\mathbf{R}^{n+1}$, D a closed set in Ω, and $\mathbf{f}$ a continuous function in D. A necessary and sufficient condition for (1) to admit a solution curve going from any point $(a,\mathbf{b})$ in D to the right is that there exist a sequence of points in D, $\{(a_k,\mathbf{b}_k)\}$, such that $a_k\downarrow a$ and $(\mathbf{b}_k-\mathbf{b})/(a_k-a)\to\mathbf{f}(a,\mathbf{b})$. Moreover, every solution curve is prolonged to the right to the boundary of Ω. Let $S(\mathbf{y})$ be a continuous †subadditive and positively homogeneous function and $\omega(x)$ a function continuous and right differentiable on $\alpha\leqslant x<\beta$. A sufficient condition for $D:\alpha\leqslant x<\beta$, $S(\mathbf{y})\leqslant\omega(x)$ to possess the property in the statement of Perron's theorem is given by $D^+\omega(x)\geqslant S(\mathbf{f}(x,\mathbf{y}))$ for $\|\mathbf{y}\|=\omega(x)$.

A continuous function $\omega(x)$ is said to be a

right majorizing function of (1) with respect to $S(\mathbf{y})$ if for any solution $\varphi(x)$, $S(\varphi(a))\leqslant\omega(a)$ implies $S(\varphi(x))\leqslant\omega(x)$ for $x\geqslant a$ if both $S(\varphi(x))$ and $\omega(x)$ are defined. In order for $\omega(x)$ to be a right majorizing function, it suffices that $D^+\omega(x)>S(\mathbf{f}(x,\mathbf{y}))$ for $\|\mathbf{y}\|=\omega(x)$. A function satisfying this inequality is called a **right superior function** of (1) with respect to $S(\mathbf{y})$. If $F(x,S(\mathbf{y}))>S(\mathbf{f}(x,\mathbf{y}))$, then any solution of $y'=F(x,y)$ is a right superior function of (1). Theorems stating such facts are called **comparison theorems**.

If (1) has a unique solution, the condition $D^+\omega(x)\geqslant S(\mathbf{f}(x,\mathbf{y}))$ for $S(\mathbf{y})=\omega(x)$ implies that $\omega(x)$ is a right majorizing function of (1). Conversely, we can derive from comparison theorems general uniqueness theorems, one of which we state. Suppose that $G(x,y)$ is continuous for $\alpha<x<\beta$ and $0\leqslant y<r(x)$; $G(x,0)\equiv 0$; a solution of $y'=G(x,y)$ such that $y=o(r(x))$ as $x\to\alpha+0$ vanishes identically; and finally that $S(\mathbf{f}(x,\mathbf{y}_1)-\mathbf{f}(x,\mathbf{y}_2))\leqslant G(x,S(\mathbf{y}_1-\mathbf{y}_2))$. Then for two solutions φ_1,φ_2 of (1) such that $S(\varphi_1-\varphi_2)=o(r(x))$ as $x\to\alpha+0$, we have $\varphi_1\equiv\varphi_2$. Assuming that $\mathbf{f}$ is continuous at $(a,\mathbf{b})$ and taking $y/(x-a)$ as G, we obtain Nagumo's condition $(x-a)S(\mathbf{f}(x,\mathbf{y}_1)-\mathbf{f}(x,\mathbf{y}_2))\leqslant S(\mathbf{y}_1-\mathbf{y}_2)$.

G. Peano proved the following theorem: With the same notation and assumption as in Perron's theorem, there exist a **maximum solution** $\bar{\varphi}$ and a **minimum solution** $\underline{\varphi}$ of $y'=f(x,y)$ such that $y(\alpha)=b$ for $\underline{\omega}(\alpha)\leqslant b\leqslant\bar{\omega}(\alpha)$, and such that there exists a solution curve passing through any point in $\alpha\leqslant x<\beta$, $\underline{\varphi}(x)\leqslant y\leqslant\bar{\varphi}(x)$. This theorem was extended by Hukuhara as follows. Suppose that $\mathbf{f}(x,\mathbf{y})$ is continuous and bounded in $D:\alpha\leqslant x\leqslant\beta$, $\|\mathbf{y}\|<\infty$. Let C be a †continuum in D, and let $\mathfrak{F}(C)$ denote the set of solutions intersecting C. Then $\mathfrak{F}(C)$ is a continuum of the †function space $C([\alpha,\beta])$. From this theorem we can deduce the **Kneser-Nagumo theorem**, which says that the intersection of the set of points belonging to the members of $\mathfrak{F}(C)$ and a hyperplane $x=\xi$ is a continuum. It was proved by Hukuhara that if C is in the hyperplane $x=\alpha$, then (1) admits a solution connecting the two hyperplanes $x=\alpha$ and $x=\beta$ and passing through the boundary of the set of points belonging to the members of $\mathfrak{F}(C)$.

F. Equations Containing Parameters

We assume uniqueness of the solution of

$$\mathbf{y}'=\mathbf{f}(x,\mathbf{y},\lambda),\qquad \lambda=(\lambda_1,\ldots,\lambda_m),\tag{2}$$

where $\mathbf{f}$ is a continuous function of $(x,\mathbf{y},\lambda)$. Let $\varphi(x,a,\mathbf{b},\lambda)$ denote the solution of (2) satisfying $\mathbf{y}(a)=\mathbf{b}$. Then $\varphi(x,a,\mathbf{b},\lambda)$ is continuous with respect to $(x,a,\mathbf{b},\lambda)$ in its region of definition. If the derivatives $\partial\mathbf{f}/\partial y_k$ are also continuous, then $\varphi(x,a,\mathbf{b},\lambda)$ is a continuously differentiable function of $(x,\mathbf{b})$; $z_{jk}=\partial\varphi_j/\partial b_k$, $j=1,\ldots,n$, satisfy the system of linear ordinary differential equations and the initial condition

$$\frac{d}{dx}z_{jk}=\sum_{l=1}^{n}\left(\frac{\partial f_j}{\partial y_l}\right)z_{lk},\qquad z_{jk}(a)=\delta_{jk};$$

and $z_j=\partial\varphi_j/\partial a$, $j=1,\ldots,n$, satisfy the same system with the initial condition $z_j(a)=-f_j(a,\mathbf{b},\lambda)$, where $(\partial f_j/\partial y_l)$ means $(\partial f_j/\partial y_l)(x,\varphi(x,a,\mathbf{b},\lambda),\lambda)$. If $\mathbf{f}$ further admits continuous derivatives $\partial\mathbf{f}/\partial\lambda_l$, then $\varphi(x,a,\mathbf{b},\lambda)$ is continuously differentiable with respect to λ_l, and moreover, $w_{jl}=\partial\varphi_j/\partial\lambda_l$, $j=1,\ldots,n$, satisfy the system

$$\frac{d}{dx}w_{jl}=\sum_{k=1}^{n}\left(\frac{\partial f_j}{\partial y_k}\right)w_{kl}+\left(\frac{\partial f_j}{\partial\lambda_l}\right),\qquad w_{jl}(a)=0.$$

These differential systems are called the **variational equations** of (1).

C. Carathéodory proved the existence of solutions of (1) under the less restrictive assumption that $\mathbf{f}$ is continuous with respect to $\mathbf{y}$ for any fixed x and measurable with respect to x for any fixed $\mathbf{y}$.

Suppose that $\mathbf{f}$ is continuous and satisfies a Lipschitz condition. Let $\mathbf{z}(x)$ be a function such that $\mathbf{z}(a)=\mathbf{b}$ and $\|z_i'(x)-f_i(x,\mathbf{z}(x))\|\leqslant\varepsilon(x)$, and let $\mathbf{y}(x)$ be a solution of (1) such that $\mathbf{y}(a)=\mathbf{b}$. Then we obtain

$$\|z_i(x)-y_i(x)\|\leqslant e^{L|x-a|}\left|\int_a^x\varepsilon(x)e^{-L|x-a|}dx\right|,$$

which gives approximate solutions of (1).

G. Equations in the Complex Domain

We assume that the variables $x,y_1,\ldots,y_n$ all have complex values. We have the following theorem: If $\mathbf{f}$ is holomorphic at $(a,\mathbf{b})$, then (1) has a unique solution that is holomorphic at $x=a$ and takes the value $\mathbf{b}$ at $x=a$. This theorem can be proved by utilizing the method of successive approximations and fixed-point theorems. Cauchy proved the theorem by using **majorant series**. This method, called the **method of majorants**, proceeds for the scalar equations as follows: Let $f(x,y)=\sum a_{jk}(x-a)^j(y-b)^k$ and $y=\sum c_n(x-a)^n$. Substituting the latter series into both members, we can successively determine the coefficients c_n by the method of undetermined coefficients. Assuming that $|f|<M$ for $|x-a|<r$ and $|y-b|<\rho$, consider the solution $Y=\sum C_n(x-a)^n$ of

$$\frac{dY}{dx}=\frac{M}{(1-(x-a)/r)(1-(Y-b)/\rho)}$$

satisfying $Y(a)=b$. We have $C_n\geqslant|c_n|$ for any n, which shows that $\sum C_n(x-a)^n$ is a majorant series of $\sum c_n(x-a)^n$.

We have the following **uniqueness theorem**: Suppose that $\mathbf{f}(x, \mathbf{y})$ is holomorphic at $(a, \mathbf{b})$. Let C be a curve having the point a as one of its ends and $\varphi(x)$ be a solution with the following properties: φ is holomorphic on C except possibly at $x=a$, and there exists a sequence of points on C, $\{a_k\}$, such that $a_k \to a$ and $\varphi(a_k) \to \mathbf{b}$. Then $\varphi(x)$ is holomorphic at a. By a †theorem of identity, the analytic continuation of a solution continues to be a solution if it does not encounter any singularity of $\mathbf{f}$. If a solution $\varphi(x)$ is holomorphic on a smooth curve $x=\chi(t)$, with $0 \leqslant t \leqslant t_0$ and $\chi(0)=a$, and $\varphi(a)=\mathbf{b}$, then $\mathbf{y}=\varphi(\chi(t))$ is a solution for $0 \leqslant t \leqslant t_0$ of

$$\mathbf{y}'=\chi'(x)\mathbf{f}(\chi(t), \mathbf{y}) \tag{3}$$

satisfying $\mathbf{y}(0)=\mathbf{b}$. Conversely, if (3) has a solution $\mathbf{y}=\psi(t)$ defined in $0 \leqslant t \leqslant t_0$ and satisfying $\mathbf{y}(0)=\mathbf{b}$, and if $\mathbf{f}(x, \mathbf{y})$ is holomorphic at $(\chi(t), \psi(t))$ for $0 \leqslant t \leqslant t_0$, then (1) has a solution $\varphi(x)$ holomorphic on C and $\psi(t)=\varphi(\chi(t))$ for $0 \leqslant t \leqslant t_0$.

Suppose that $f=f_1/f_2$, where f_1 and f_2 are holomorphic at (a, b). If $f_1(a, b) \neq 0$, $f_2(a, y) \not\equiv 0$, and $f_2(a, b)=0$, then the equation $y'=f(x, y)$ admits a unique solution such that $y \to b$ as $x \to a$, and this solution can be expanded into a †Puiseux series:

$$y=\sum_{n=0}^{\infty} c_n(x-a)^{n/r}.$$

If $\mathbf{f}$ is holomorphic at $(a, \mathbf{b})$, then the solution $\mathbf{y}=\varphi(x, x_0, \mathbf{y}_0)$ of (1) satisfying $\mathbf{y}(x_0)=\mathbf{y}_0$ is holomorphic with respect to $(x, x_0, \mathbf{y}_0)$ at $(a, a, \mathbf{b})$. If $\mathbf{f}(x, \mathbf{y}, \lambda)$ is holomorphic at $(a, \mathbf{b}, \lambda_0)$, then the solution of (2), $\mathbf{y}=\varphi(x, x_0, \mathbf{y}_0, \lambda)$, satisfying $\mathbf{y}(x_0)=\mathbf{y}_0$ is holomorphic at $(a, a, \mathbf{b}, \lambda_0)$. Suppose that x is a real variable and $\mathbf{y}$ is a complex vector. If $\mathbf{f}$ is continuous with respect to $(x, \mathbf{y})$ and holomorphic with respect to $\mathbf{y}$, then the solution $\varphi(x, x_0, \mathbf{y}_0)$ is holomorphic with respect to $\mathbf{y}_0$. If $\mathbf{f}(x, \mathbf{y}, \lambda)$ is continuous with respect to $(x, \mathbf{y}, \lambda)$ and holomorphic with respect to $(\mathbf{y}, \lambda)$, then $\varphi(x, x_0, \mathbf{y}_0, \lambda)$ is holomorphic with respect to $(\mathbf{y}_0, \lambda)$.

References

[1] E. Kamke, Differentialgleichungen reeller Funktionen, Akademische Verlag, 1930.
[2] E. A. Coddington and N. Levinson, Theory of ordinary differential equations, McGraw-Hill, 1955.
[3] P. Hartman, Ordinary differential equations, Wiley, 1964.
[4] H. Okamura, Condition nécessaire et suffisante remplie par les équations différentielles ordinaires sans points de Peano, Mem. Coll. Sci. Univ. Kyoto, (A) 24 (1941).
[5] O. Perron, Ein neuer Existenz Beweis für die Integrale der Differentialgleichung $y'=f(x, y)$, Math. Ann., 15 (1945).
[6] M. Hukuhara, Sur la théorie des équations différentielles ordinaires, J. Fac. Sci. Univ. Tokyo, sec. I, vol. VII, pt. 5, 1958.
[7] M. Nagumo, Über die Lage der Integralkurven gewöhnlicher Differentialgleichungen, Proc. Phys.-Math. Soc. Japan, 24 (1943).
[8] E. Hille, Ordinary differential equations in the complex domain, Wiley, 1976.
[9] R. Abraham and J. E. Marsden, Foundations of mechanics, Benjamin-Cummings, second edition, 1978.

317 (X.21) Orthogonal Functions

A. Orthogonal Systems

Let (X, μ) be a †measure space. For complex-valued functions f, g on X belonging to the †function space $L_2(X)$, we define the inner product $(f, g)=\int_X f(x)\overline{g(x)}\,d\mu(x)$ and the norm $\|f\|=(f, f)^{1/2}$. If $(f, g)=0$, then we say that f and g are **orthogonal** on X with respect to the measure μ. If X is a subset of a Euclidean space and μ is the †Lebesgue measure m, then we simply say that they are orthogonal. If the measure has a †density function $\varphi(x)$ with respect to the Lebesgue measure and $(f, g)=\int_X f(x)\overline{g(x)}\varphi(x)\,dm(x)=0$, we say that they are orthogonal with respect to the **weight function** $\varphi(x)$. If $\|f\|^2=1$, then f is said to be **normalized**. A set of functions $\{f_n(x)\}$ $(n=1, 2, \ldots)$ is said to be an **orthogonal system** (or **orthogonal set**), and we write $\{f_n\} \in O(X)$, if any pair of functions in the set are orthogonal. The orthogonal set $\{f_n(x)\}$ is said to be **orthonormal**, and we write $\{f_n\} \in ON(X)$ if each f_n is normalized.

Let $\{f_n\}$ be a set of linearly independent functions in $L_2(X)$, and let R be a subset of $L_2(X)$. If we can approximate any function $f \in R$ arbitrarily closely by a finite linear combination of the $f_n(x)$ with respect to the norm in $L_2(X)$, we say that $\{f_n\}$ is **total** in R. Let $\{f_n\} \in O(X)$. If $(\varphi, f_n)=0$ for all n implies $\varphi(x)=0$ almost everywhere (a.e.), then $\{f_n\}$ is said to be **complete** in $L_2(X)$. An orthogonal system $\{f_n\}$ is complete in $L_2(X)$ if and only if the system is total in $L_2(X)$.

If $\{f_n\} \in O(X)$, then the series $\sum_{n=1}^{\infty} c_n f_n(x)$ is called an **orthogonal series**. If the series converges to $\varphi(x)$ †in the mean of order 2, then $c_n=(\varphi, f_n)/\|f_n\|^2$. We call the c_n $(n=1, 2, \ldots)$ the

expansion coefficients or **Fourier coefficients** of $\varphi(x)$ with respect to $\{f_n\}$.

If $\{g_n\} \subset L_2(X)$ are linearly independent, we can construct an orthonormal system $\{f_n\}$ by forming suitable linear combinations of the g_n; $\{f_n\}$ spans the same subspace as $\{g_n\}$. For this purpose we set $f_1(x)=g_1(x)/\|g_1\|$, $f_n(x)=\varphi_n(x)/\|\varphi_n\|$, where $\varphi_n(x)=g_n(x)-\sum_{v=1}^{n-1}(g_n, f_v)f_v(x)$, $n \geqslant 2$. This procedure is called **Schmidt orthogonalization** or **Gram-Schmidt orthogonalization**.

If the c_n are Fourier coefficients of $\varphi(x) \in L_2(X)$ with respect to $\{f_n(x)\} \in ON(X)$, then we have the †Bessel inequality $\sum_{n=1}^{\infty}|c_n|^2 \leqslant \|\varphi\|^2$. Equality in the Bessel inequality for all $\varphi \in L_2(X)$ (the †Parseval identity) is equivalent to completeness of $\{f_n\}$ in $L_2(X)$. In this case $\sum_{n=1}^{\infty} c_n f_n(x)$ is called the **orthogonal expansion** of φ with respect to $\{f_n\}$, and conversely, for any sequence $\{c_n\}$ such that $\sum|c_n|^2 < \infty$, there is a function $\varphi \in L_2(X)$ that has the c_n as its Fourier coefficients, and

$$\sum_{n=1}^{\infty}|c_n|^2 = \|\varphi\|^2,$$

$$\varphi(x) = \underset{n\to\infty}{\mathrm{l.i.m.}}\left(\sum_{m=1}^{n} c_m f_m(x)\right).$$

This is called the †**Riesz-Fischer theorem**.

B. Orthogonal Systems on the Real Line

We assume that X is a finite interval (a,b) and that functions on X are real-valued. We write $O(a,b)$ or $ON(a,b)$ instead of $O(X)$, $ON(X)$.
(1) If $\{f_n\} \in ON(a,b)$, $|f_n(x)| \leqslant M$ (const.), and $\sum c_n f_n(x)$ converges †a.e., then $c_n \to 0$ as $n \to \infty$.
(2) We can construct a complete orthonormal system $\{f_n(x)\}$ and a function $\varphi(x) \in L_1(a,b)$ such that its orthogonal expansion $\sum c_n f_n(x)$ diverges everywhere. (3) If $\{f_n(x)\} \in ON(a,b)$ and $\sum c_n^2 \log^2 n < \infty$, then $\sum c_n f_n(x)$ converges a.e. The factor $\log^2 n$ cannot be replaced by any other monotone increasing factor $\omega(n)$ satisfying $0 \leqslant \omega(n) = o(\log^2 n)$ (**Rademacher-Men'shov theorem**). K. Tandori proved that if $c_n \downarrow 0$ and $\sum c_n \varphi_n$ converges a.e. for any orthonormal system $\{\varphi_n\}$, then $\sum |c_n|^2 \log^2 n < \infty$.
(4) If the orthogonal expansion of a function $\varphi \in L_2(a,b)$ is †summable by Abel's method on a set E, then it is †$(C,1)$-summable a.e. on E. $(C,1)$-summability a.e. of the orthogonal expansion of a function of $\varphi \in L_2(a,b)$ is equivalent to convergence a.e. of the partial sums $s_{2^n}(x)$ $(n=1,2,\ldots)$ of its expansion. (5) Suppose that $\{f_n(x)\} \in ON(a,b)$, $|f_n(x)| \leqslant M$. Then: (i) If the a_n are the Fourier coefficients of $\varphi(x)$ with respect to $\{f_n(x)\}$, then

$$\left(\sum_{n=1}^{\infty}|a_n|^{p'}\right)^{1/p'} \leqslant M^{(2-p)/p}\left(\int_a^b |\varphi(x)|^p dx\right)^{1/p},$$

where $1 < p \leqslant 2$, $1/p + 1/p' = 1$. Conversely, if $(\sum |a_n|^p)^{1/p} < \infty$ $(1 < p \leqslant 2)$, there exists a function $\varphi(x)$ which has the a_n as its Fourier coefficients and such that

$$\left(\int_a^b |\varphi(x)|^{p'} dx\right)^{1/p'} \leqslant M^{(2-p)/p}\left(\sum_{n=1}^{\infty}|a_n|^p\right)^{1/p}$$

(**F. Riesz's theorem**). When the orthonormal system is the trigonometric system, this is called the **Hausdorff-Young theorem**. (ii) Let $\{a_n^*\}$ be the decreasing rearrangement of $\{|a_n|\}$; then

$$\sum_{n=1}^{\infty} a_n^{*p} n^{p-2} \leqslant A_p \int_a^b |\varphi(x)|^p dx \qquad (1 < p \leqslant 2).$$

If $q \geqslant 2$ and $\sum a_n^{*q} n^{q-2} < \infty$, then there exists a function $\varphi(x)$ which has the a_n as its Fourier coefficients and such that

$$\int_a^b |\varphi(x)|^q dx \leqslant A_q \sum_{n=1}^{\infty} a_n^{*q} n^{q-2}$$

(**Paley's theorem**). When the system is trigonometric, this is called the **Hardy-Littlewood theorem**. (6) If for some positive ε we have $\sum |c_n|^{2-\varepsilon} < \infty$, then $\sum c_n f_n(x)$ converges a.e.
(7) If we set $s^*(x) = \sup_n |\sum_{v=1}^{n} c_v f_v(x)|$, then $\|s^*\|_q \leqslant A_q (\sum c_v^{*q} v^{q-2})^{1/q}$ $(q > 2)$, where $\{c_n^*\}$ is the decreasing rearrangement of $\{|c_n|\}$.

C. Examples of Orthogonal Systems

(1) $\{\cos nx\} \in O(0,\pi)$, $\{\sin nx\} \in O(0,\pi)$. (2) $\{1, \cos nx, \sin nx\} \in O(0, 2\pi)$ (→ 159 Fourier Series). (3) Suppose that $A(x)$ is positive and continuous, and let $y_n(x)$ be solutions of $y''(x) + \lambda_n A(x) y(x) = 0$ satisfying the condition $y_n(a) = y_n(b) = 0$, where λ_n is any †eigenvalue. Then $\{\sqrt{A(x)}\, y_n(x)\} \in O(a,b)$ (for orthogonality of eigenfunctions → 315 Ordinary Differential Equations (Boundary Value Problems) B). (4) Set $r_n(x) = -1$ or 1 according as the nth digit of the binary expansion of x $(0 < x \leqslant 1)$ is 1 or 0, and $r_n(x) = 0$ if x is expandable in two ways. Then $\{r_n(x)\} \in O(0,1)$. This is called **Rademacher's system of orthogonal functions**. The system is not complete, but it is interpreted as a †sample space of coin tossing. Rademacher's system is useful for constructing various counter-examples. (5) Let the binary expansion of n be $n = 2^{v_1} + 2^{v_1} + \ldots + 2^{v_p}$ $(v_1 < v_2 < \ldots < v_p)$, and set $w_n(x) = r_{v_1+1}(x) r_{v_2+1}(x) \ldots r_{v_p+1}(x)$. Then $\{w_n(x)\}$ is a complete orthonormal system called **Walsh's system of orthogonal functions**. This system is interpreted as a system of characters of the group of binary numbers, and there are many theorems for this system analogous to those for the trigono-

metric system. (6) In the interval [0, 1], set

$$\chi_m^k(x)=\sqrt{2^m}, \quad x\in((k-1)/2^m,\ (k-1/2)/2^m)$$
$$=-\sqrt{2^m}, \quad x\in((k-1/2)/2^m,\ k/2^m)$$
$$=0, \quad x\in((l-1)/2^m,\ l/2^m),$$
$$l\neq k,\ 1\leqslant l\leqslant 2^m.$$

The orthonormal system $\chi_m^k(x)$ $(1\leqslant k\leqslant 2^m, 1\leqslant m)$ is called **Haar's system of orthogonal functions**. The Haar expansion of the continuous function $f(x)$ converges to $f(x)$ uniformly.

D. Orthogonal Polynomials (→ Appendix A, Table 20)

Suppose that we are given a weight function $\varphi(x)\geqslant 0$ ($\varphi(x)>0$ a.e.) defined on (a,b) and that the inner product of functions f, g on (a,b) is defined by $(f,g)=\int_a^b f(x)g(x)\varphi(x)dx$. We orthogonalize $\{x^n\}$ by Schmidt orthogonalization and obtain polynomials $p_n(x)$ of degree n. Here the sign of $p_n(x)$ can be determined so that the sign of the coefficient of the highest power of x is positive. We call $\{p_n(x)\}$ the **system of orthogonal polynomials** belonging to the weight function $\varphi(x)$. This system is complete in $L_2^{(\varphi)}(a,b)$, which is defined to be the space of functions f such that $\int_a^b|f(x)|^2\varphi(x)dx<\infty$. In other words, the system $\{\sqrt{\varphi(x)}\,p_n(x)\}$ is a complete orthonormal system in the ordinary $L_2(a,b)$ space. Concerning the convergence problem of the orthogonal expansion by $\{p_n(x)\}$, the **Christoffel-Darboux formula**

$$\sum_{k=0}^{n} p_k(t)p_k(x)=\frac{\alpha_n}{\alpha_{n+1}}\cdot\frac{p_n(x)p_{n+1}(t)-p_n(t)p_{n+1}(x)}{t-x}$$

plays an important role.

Several important special functions in classical mathematical physics are given by orthogonal polynomials:

(1) Setting $\varphi(x)=(1-x)^\alpha(1+x)^\beta$ $(\alpha>-1,\beta>-1)$ in $[-1,1]$, we get the **Jacobi polynomials**, although they are sometimes defined in [0, 1] with respect to $\varphi(x)=x^\alpha(1-x)^\beta$ (→ Appendix A, Table 20.V). If we set $\alpha=\beta$ in the Jacobi polynomials, we get the **ultraspherical polynomials** (or **Gegenbauer polynomials**) (→ Appendix A, Table 20.I). Furthermore, if $\alpha=\beta=0$, then we get the †Legendre polynomials, and if $\alpha=\beta=-1/2$, we get the **Chebyshev polynomials** $T_n(x)=\cos(n\arccos x)$. The $T_n(x)$ also appear in the best approximation problem (→ Appendix A, Table 20.II; 336 Polynomial Approximation).

(2) If we set $\varphi(x)=x^\alpha e^{-x}$ in $(0,\infty)$, we get the **Sonine polynomials** (or **associated Laguerre polynomials**) with appropriate constant factors. If α is a positive integer m, we get

$$S_n^{(m)}(x)=x^{-m}e^x\left(\frac{d^n}{dx^n}(x^{n+m}e^{-x})\right).$$

The particular case $m=0$ gives the **Laguerre polynomials**. In this case, however, it is customary to normalize them as $L_n(x)=(e^x/n!)(d^n/dx^n)(x^ne^{-x})$ (→ Appendix A, Table 20.VI). Laguerre polynomials are used in †numerical integrations of a Gaussian type in $(0,\infty)$. Furthermore, associated Laguerre polynomials appear in the solutions of the Schrödinger equation for the behavior of hydrogen atoms. This system of orthogonal polynomials is useful in the expansions of approximate eigenfunctions of atoms analogous to hydrogen, velocity distribution functions of molecules in gas theory, and so on.

(3) Setting $\varphi(x)=e^{-x^2}$ (or $e^{-x^2/2}$) in $(-\infty,\infty)$, we get **Hermite polynomials** $H_n(x)=(-1)^ne^{x^2}(d^ne^{-x^2}/dx^n)$, modulo constant factors (→ Appendix A, Table 20.VI). Hermite polynomials are special cases of parabolic cylinder functions (→ 167 Functions of Confluent Type). These polynomials appear as eigenfunctions of the Schrödinger equation for harmonic oscillators. They are also connected with probability integrals and are used in mathematical statistics.

(4) Replacing the integral by a finite sum $\sum_{m=0}^n f(m)g(m)$ in the definition of inner product, we get so-called **orthogonality for a finite sum**. (Regarding orthogonal polynomials with respect to a finite sum (→ Appendix A, Table 20.VII) and their application to the mean square approximation → 19 Analog Computation F.) Since orthogonal polynomials with respect to a finite sum are often called simply orthogonal polynomials by engineers, one must be careful not to confuse these with the ordinary ones.

References

[1] S. Kaczmarz and H. Steinhaus, Theorie der Orthogonalreihen, Warsaw, 1935 (Chelsea, 1951).
[2] G. Szegö, Orthogonal polynomials, Amer. Math. Soc. Colloq. Publ., 1939.
[3] R. Courant and D. Hilbert, Methods of mathematical physics I, Interscience, 1953.
[4] F. G. Tricomi, Vorlesungen über Orthogonalreihen, Springer, 1955.
[5] G. Alexits, Convergence problems of orthogonal series, Pergamon, 1961, revised translation of the German edition of 1960.
[6] G. Sansone, Orthogonal functions, Interscience, revised English edition, 1959.

318 (XX.14) Oscillations

A. General Remarks

A **vibration** or **oscillation** is a phenomenon that repeats periodically, either exactly or approximately. Exactly periodic oscillations are studied in the theory of †periodic solutions of differential equations. The †period of a solution $f(t)$ is called the **period** of the oscillation, and its reciprocal the **frequency**. The difference between the greatest and least values of $f(t)$ (globally or in an interval) is the **amplitude**. The theory of vibrations has its origin in the study of mechanical vibrations, but its nomenclature has been used also for electric circuits. As examples of practical applications of the theory of oscillations, we mention, in engineering, the prevention of vibrations and the generation of stable sustained oscillations, and in geophysics, investigations concerning the free oscillation of the earth, the existence of which has recently been confirmed.

B. Linear Oscillation

Periodic solutions of †linear differential equations have been studied in detail for a long time. Perhaps the simplest case of such an oscillation is represented by the differential equation

$$d^2x/dt^2+n^2x=0, \tag{1}$$

where the **restitutive force** is proportional to the displacement from the equilibrium position. Typical examples are the free vibration of a simple pendulum with small amplitude and an electric circuit composed of a self-inductance and capacity (without resistance). The solution is given by $x=A\cos(nt+\alpha)$. This is called **harmonic oscillation** or **simple harmonic motion**. Here the amplitude is A, the period is $2\pi/n$, n is the **circular frequency**, and α is the **initial phase**.

A system of m †degrees of freedom $(x_1, \ldots, x_m)$ is said to be in free harmonic oscillation if the coordinates can be expressed as

$$x_i=\sum_{k=1}^{m} A_{ik}\cos(n_k t+\alpha_k), \qquad i=1,2,\ldots,m.$$

Each of these simple harmonic oscillations is called a **normal vibration**. As a limiting case, where the number of degrees of freedom is infinite, we have the vibration of a string:

$$\frac{\partial^2 u(x,t)}{\partial t^2}=n^2\frac{\partial^2 u(x,t)}{\partial x^2},$$

$$u(0,t)=u(l,t)=0.$$

The solution is given by a series

$$\sum_k A_k\sin(k\pi x/l)\cos(k\pi nt/l),$$

which is just the superposition of the fundamental vibration (corresponding to $k=1$) and simply harmonic motions of frequencies equal to multiples of the fundamental frequency.

If a resisting force proportional to the velocity is acting, the equation becomes

$$d^2x/dt^2+2\varepsilon\, dx/dt+n^2x=0, \qquad n>\varepsilon, \tag{2}$$

whose solution

$$x=Ae^{-\varepsilon t}\cos(\sigma t+\alpha), \qquad \sigma=\sqrt{n^2-\varepsilon^2} \tag{3}$$

is not periodic. However, x becomes zero at a fixed interval π/σ, and the extremal values in the intervals decrease to zero in a geometrical progression with the common ratio $v=\exp(-\pi\varepsilon/\sigma)$. This phenomenon is called **damped oscillation** with **damping ratio** v and **logarithmic decrement** $\log v=-\pi\varepsilon/\sigma$. In this case, too, $2\pi/\sigma$ is called the period.

When a driving force term $\varphi(t)$ is present in the right-hand side of (2), the solution takes on the additional term

$$\frac{e^{-\varepsilon t}}{\sigma}\left(\sin\sigma t\int\varphi(t)e^{\varepsilon t}\cos\sigma t\,dt - \cos\sigma t\int\varphi(t)e^{\varepsilon t}\sin\sigma t\,dt\right),$$

which represents the **forced oscillation** due to $\varphi(t)$.

If $\varepsilon<0$ in (2) (**negative resistance**), the solution (3) increases in amplitude, so that a small disturbance is amplified, resulting in an automatic generation of oscillation. This phenomenon is called **self-excited vibration**. Besides being caused by some special kinds of circuit elements (e.g., tunnel diodes), such a situation often occurs when the vibrating system has time delay characteristics (→ 163 Functional-Differential Equations).

Among sustained vibrations, other than forced oscillations and self-excited vibrations, are the **parametrically sustained vibrations** caused by periodic variation of a parameter of the vibrating system. Electric wires and pantographs for use in high-speed electric railways must be designed to prevent unwanted parametrically sustained vibrations. On the other hand, a parametron is an electric element utilizing parametrically sustained vibration.

C. Nonlinear Oscillation

Actual vibrating systems contain more or less nonlinear elements, which give rise to various kinds of oscillations different from those described by the linear theory (→ 290 Nonlinear Oscillation). For example, $d^2x/dt^2 - \varepsilon(1-x^2)dx/dt + x = 0$ $(\varepsilon > 0)$ represents a stable sustained oscillation such that for large values of ε, two nearly stationary states occur alternately, the transition from one to the other taking place abruptly. This is called **relaxation oscillation**.

References

[1] Lord Rayleigh, The theory of sound, Macmillan, second revised edition, I, 1894; II, 1896 (Dover, 1945).
[2] A. A. Andronov and C. E. Chaĭkin, Theory of oscillations, Princeton Univ. Press, 1949. (Original in Russian, 1937.)
[3] L. S. Pontryagin, Ordinary differential equations, Addison-Wesley, 1962. (Original in Russian, 1961.)
Also → references to 287 Nonlinear Lattice Dynamics, 290 Nonlinear Oscillation, 291 Nonlinear Problems.

P

319 (I.3) Paradoxes

A. General Remarks

A statement that is apparently absurd but not easily disproved is called a **paradox**. A contradiction between a proposition and its negation is called an **antinomy** if both statements can be supported by logically equivalent reasoning. In practical use, however, "paradox" and "antinomy" often mean the same thing.

B. Paradoxes in Set Theory

(1) The Russell Paradox (1903). We classify sets into two kinds as follows: Any set that does not contain itself as an element is called a **set of the first kind**, and any set that contains itself as an element is called a **set of the second kind**. Every set is either a set of the first kind or of the second kind. Denote the set of all sets of the first kind by M. If M is a set of the first kind, M cannot be an element of M. But if M is of the first kind, then M must be an element of M, by definition. This is contradictory. On the other hand, if M is a set of the second kind, M must be an element of M; but since M is an element of M, M is a set of the first kind, so M cannot be an element of M, by definition. This is contradictory.

Since the kind of reasoning employed in this paradox is very simple and is often utilized in mathematics, it became popular in set theory. To remove this paradox from set theory, Russell suggested †ramified type theory. If we adopt this theory, however, it becomes very hard to develop even an ordinary theory of real numbers (→ 156 Foundations of Mathematics). On the other hand, this paradox, together with the Burali-Forti paradox, indicates that the definition of a set should be restrictive. This realization led to the development of †axiomatic set theory.

(2) The Burali-Forti Paradox (1897). Let $W=\{0,1,2,\ldots,\omega,\ldots\}$ be the †well-ordered set (→ 312 Ordinal Numbers A) of all †ordinal numbers. Let Ω be the ordinal number of W. Then every ordinal number, being an element of W, is less than Ω. But Ω is an ordinal number. Hence, $\Omega<\Omega$. This is contradictory.

(3) The Richard Paradox (1905). The expressions in the English language can be enumerated by the device that is applied to the usual enumeration of the algebraic equations with integral coefficients. From the specified enumeration of all expressions in the English language, by striking out those which do not define a real number in the interval $(0,1]$, we obtain an enumeration of those which do. Consider the following expression: "The greatest real number represented by a proper nonterminating decimal fraction whose nth digit, for any natural number n, is not equal to the nth digit of the nonterminating decimal fraction representing the real number defined by the nth expression in the last-described enumeration." Then we have before us a definition of a real number in the interval $(0,1]$ by means of an expression in the English language. This real number, by its definition, must differ from every real number definable by an expression in the English language. This is contradictory.

The following paradox was given by Berry (1906): "The least natural number not nameable in fewer than twenty-two syllables" is actually named by this expression, which has twenty-one syllables. The Epimenides paradox is a traditional ancient Greek paradox of this kind: Epimenides (a Cretan) said, "Cretans are always liars. ..."

C. Paradoxes of the Continuum

The problem of the continuum is important in both mathematics and philosophy. There are several **paradoxes of Zeno** concerning the continuum, among which the following two are best known:

(1) Assume that Achilles and a tortoise start simultaneously from the points A and B, respectively, Achilles running after the tortoise. When Achilles reaches the point B, the tortoise advances to a point B_1. When Achilles reaches the point B_1, the tortoise advances further to a point B_2. Thus Achilles can never overtake the tortoise.

(2) A flying arrow occupies a certain point at each moment. In other words, at each moment the arrow stands still. Therefore the arrow can never move.

References

[1] A. N. Whitehead and B. Russell, Principia mathematica, second edition, Cambridge Univ. Press, 1925.
[2] S. C. Kleene, Introduction to metamathematics, Van Nostrand, 1952.
[3] E. Mendelson, Introduction to mathematical logic, Van Nostrand, 1966.

320 (XIII.19) Partial Differential Equations

A. General Remarks

A partial differential equation is a functional equation

$$F\left(x_1, \ldots, x_n, z, \frac{\partial z}{\partial x_1}, \ldots, \frac{\partial^2 z}{\partial x_1^2}, \frac{\partial^2 z}{\partial x_1 \partial x_2}, \ldots\right)=0$$

that involves a function z of independent variables $x_1, x_2, \ldots, x_n$, its †partial derivatives, and the independent variables $x_1, \ldots, x_n$. The definition of a system of partial differential equations is similar to that of a system of †ordinary differential equations. (The partial differential equation becomes an ordinary differential equation if the number of independent variables is one.)

The †order of the highest derivative appearing in a partial differential equation is called the **order** of the partial differential equation.

Usually we write p_i for $\partial z/\partial x_i$, x for x_1, and y for x_2 when $n=2$, and $p=\partial z/\partial x$, $q=\partial z/\partial y$, $r=\partial^2 z/\partial x^2$, $s=\partial^2 z/\partial x \partial y$, $t=\partial^2 z/\partial y^2$ when z is a function of x and y.

A partial differential equation is called **linear** if it is a linear relation with respect to z and its partial derivatives. For example, the equation

$$A(x, y)r+B(x, y)s+C(x, y)t+D(x, y)p$$
$$+E(x, y)q+F(x, y)z=G(x, y)$$

is a linear partial differential equation. A partial differential equation is **quasilinear** if it is a linear relation with respect to the highest-order partial derivatives. A partial differential equation is called **nonlinear** if it is not linear (→ 291 Nonlinear Problems). A function $z=\varphi(x_1, x_2, \ldots, x_n)$ that satisfies the given partial differential equation is called a **solution** of the partial differential equation. Obtaining such a solution for a given partial differential equation is called **solving** this equation, and by analogy to the case $n=2$, the **integral hypersurface** of the equation is

$$z-\varphi(x_1, x_2, \ldots, x_n)=0.$$

For a system of partial differential equations, we define solutions in the same manner.

Example 1. $X_1, X_2, \ldots, X_n$ are functions of n independent variables $x_1, x_2, \ldots, x_n$. Then solving the partial differential equation

$$X_1\frac{\partial z}{\partial x_1}+X_2\frac{\partial z}{\partial x_2}+\ldots+X_n\frac{\partial z}{\partial x_n}=0 \tag{1}$$

is equivalent to solving the system of ordinary differential equations

$$\frac{dx_1}{X_1}=\frac{dx_2}{X_2}=\ldots=\frac{dx_n}{X_n}. \tag{2}$$

In other words, if $f_1, f_2, \ldots, f_{n-1}$ are $n-1$ independent †integrals of (2), then for an arbitrary function Φ, $z=\Phi(f_1, \ldots, f_{n-1})$ is a general solution of (1) (→ Section C).

Example 2. If $P_1, P_2, \ldots, P_n$, R are functions of independent variables $x_1, \ldots, x_n$ and the dependent variable z, and if the quasilinear partial differential equation (**Lagrange's differential equation**)

$$P_1\frac{\partial z}{\partial x_1}+P_2\frac{\partial z}{\partial x_2}+\ldots+P_n\frac{\partial z}{\partial x_n}=R \tag{3}$$

has an integral hypersurface $V(z, x_1, \ldots, x_n)=0$, then we have

$$P_1\frac{\partial V}{\partial x_1}+P_2\frac{\partial V}{\partial x_2}+\ldots+P_n\frac{\partial V}{\partial x_n}+R\frac{\partial V}{\partial z}=0, \tag{4}$$

which is an equation of type (1). From this we can obtain a general solution by the method of example 1. The same procedure is applicable to solving other systems of partial differential equations.

B. Characteristic Manifolds

We consider a partial differential equation of the nth order of two independent variables x, y:

$$F(x, y, z, p_{10}, p_{01}, \ldots, P_{n0}, \ldots, p_{0n})=0,$$
$$\text{where } p_{jk}=\frac{\partial^{j+k} z}{\partial x^j \partial y^k}. \tag{5}$$

With this equation, we associate a manifold defined by a real parameter λ,

$$x=x(\lambda), \quad y=y(\lambda), \quad p_{jk}=p_{jk}(\lambda),$$
$$j, k=0, 1, \ldots, n-1; \quad j+k\leqslant n-1, \tag{6}$$

and consider the following problem: Find a solution $\varphi(x, y)$ of (5) that satisfies

$$\frac{\partial^{j+k}\varphi(x, y)}{\partial x^j \partial y^k}=p_{jk}(\lambda),$$
$$j, k=0, 1, 2, \ldots, n-1; \quad j+k\leqslant n-1$$

on the curve $x=x(\lambda)$, $y=y(\lambda)$. We call this problem the **Cauchy problem** for equation (5).

If F vanishes for a system of values x^0, y^0, p_{jk}^0 ($j, k=0, 1, 2, \ldots, n$; $p_{00}^0=z^0$; $j+k\leqslant n$) and is †holomorphic in a neighborhood of this system of values, if x, y, p_{jk} are holomorphic functions of λ in a neighborhood of $\lambda=0$ and take the respective values x^0, y^0, p_{jk}^0 at $\lambda=0$, and if

$$P_{n0}\,dy^n-P_{n-1,1}\,dx\,dy^{n-1}+\ldots$$
$$+(-1)^n P_{0n}\,dx^n\neq 0 \tag{7}$$

for (x^0, y^0, p^0_{jk}), then we have a unique holomorphic solution of this Cauchy problem in a neighborhood of (x^0, y^0, p^0_{jk}). (Here we use the notation $P_{jk}=\partial F/\partial p_{jk}$.) This is **Cauchy's existence theorem**. We cannot apply this theorem when the left-hand side of (7) vanishes at (x^0, y^0, p^0_{jk}). At such a point, uniqueness of the solution fails, and there may be several solutions through the point (x^0, y^0, p^0_{jk}). There are n curves on which the left-hand side of (7) vanishes on the integral surface $z=\varphi(x, y)$. These curves are called **characteristic curves** of (5). We associate the values

$$p_{jk}(\lambda)=\left.\frac{\partial^{j+k}\varphi(x,y)}{\partial x^j \partial y^k}\right|_{x=x(\lambda),\, y=y(\lambda)},$$

$$j, k=0, 1, \ldots, n; \qquad j+k\leqslant n,$$

with each point (x, y) on these curves. The manifold $\{x(\lambda), y(\lambda), p(\lambda)\}$ $(p(\lambda)=\{p_{jk}(\lambda)\})$ of the parameter λ is called a **characteristic manifold** of equation (5). Cauchy's existence theorem cannot be applied on the characteristic manifold.

The foregoing considerations can be extended, to some extent, to the space of higher dimensions $\mathbf{R}^n$ or $\mathbf{C}^n$. Let P be a linear partial differential operator of order m:

$$P(x, D)=\sum_{|\alpha|\leqslant m} a_\alpha(x)D^\alpha,$$

$$D^\alpha=\left(\frac{\partial}{\partial x_i}\right)^{\alpha_1}\cdots\left(\frac{\partial}{\partial x_n}\right)^{\alpha_n}, \qquad |\alpha|=\alpha_1+\ldots+\alpha_n.$$

The coefficients are assumed smooth and real in $\mathbf{R}^n$ or holomorphic in $\mathbf{C}^n$. Its homogeneous part of order m, denoted by $P_m(x, D)$, is called the **principal part** of P. Let S be a hypersurface defined by $\varphi(x)=0$ with $\operatorname{grad}\varphi(x)\neq 0$. S is called a **characteristic surface** if $P_m(x, \operatorname{grad}\varphi(x))=0$ holds for x in a neighborhood of S or merely for $x\in S$, and $\varphi(x)$ is called a phase function. A real vector $\xi^0(\neq 0)$, or a complex vector $\xi^0\neq 0$ is called a characteristic direction at the point x_0 if $P_m(x_0, \xi^0)=0$. The zeros (x, ξ) $(\xi\neq 0)$ of $P_m(x, \xi)$ are called the **characteristic set**. Furthermore (x_0, ξ^0) is called simple if $\operatorname{grad}_\xi P_m(x_0, \xi^0)\neq 0$. Suppose that (x_0, ξ^0) is simple. The integral curve $(x(t), \xi(t))$ that satisfies

$$\dot{x}=\operatorname{grad}_\xi P_m(x, \xi),$$

$$\dot{\xi}=-\operatorname{grad}_x P_m(x, \xi)$$

is called the **bicharacteristic strip** of P_m. Evidently $P_m(x, \xi)$ is constant along the bicharacteristic strip. In particular, if this constant is zero, it is said to be **null-bicharacteristic**. The phase function $\varphi(x)$ can be obtained at least locally by using the bicharacteristic strip.

C. Classification of Solutions

First, we consider a partial differential equation of the first order of two independent variables:

$$F(x, y, z, p, q)=0. \tag{8}$$

A solution of (8) that contains two arbitrary constants is called a **complete solution**. If we get one complete solution of (8), then we can obtain all the solutions by differentiations and eliminations. Let (9) be a complete solution:

$$V(x, y, z, a, b)=0. \tag{9}$$

Differentiating this, we get

$$\frac{\partial V}{\partial x}+p\frac{\partial V}{\partial z}=0, \qquad \frac{\partial V}{\partial y}+q\frac{\partial V}{\partial z}=0. \tag{10}$$

Eliminating a and b from (9) and (10), we get the original equation (8). Furthermore, solving equation (8) is equivalent to getting three functions a, b, z of x, y from (8), (9), and (10). If we regard a, b as functions of x, y, in (9), we get

$$\frac{\partial V}{\partial a}\frac{\partial a}{\partial x}+\frac{\partial V}{\partial b}\frac{\partial b}{\partial x}=0, \qquad \frac{\partial V}{\partial a}\frac{\partial a}{\partial y}+\frac{\partial V}{\partial b}\frac{\partial b}{\partial y}=0. \tag{11}$$

Therefore we can replace (9) and (10) by (9) and (11). Hence we have the following three cases: (i) When a, b are constants, we get a complete solution. (ii) When $V=0$, $\partial V/\partial a=0$, and $\partial V/\partial b=0$, we get a solution that does not contain arbitrary constants, because z, a, b are all functions of x and y. We call this solution a **singular solution** of (8). (iii) When $\partial V/\partial a$, $\partial V/\partial b$ do not vanish simultaneously, the Jacobian $D(a,b)/D(x,y)$ vanishes because of (11). This means that there exists a functional relation between a and b. If there are two such relations, a and b are constants and the solution z becomes a complete solution. Therefore we assume that there is only one such relation between a and b, whose form is assumed to be $b=\varphi(a)$. Then we get

$$V(x, y, z, a, \varphi(a))=0, \qquad \frac{\partial V}{\partial a}+\frac{\partial V}{\partial b}\varphi'(a)=0. \tag{12}$$

If we solve (12) for the unknowns a and z, we get a solution z of (8) that contains an arbitrary function φ instead of arbitrary constants. Such a solution is called a **general solution** of the partial differential equation (8) of the first order. By specializing this function φ, we obtain a **particular solution** of (8). Thus (i)–(iii) exhaust all the cases, and by obtaining a complete solution of (8) we can get all the solutions of (8). The number of complete solutions may be more than 1, or it may be infinite. These complete solutions can be transformed into each other by †contact transformations.

Moreover, they are contained in the general solutions.

Now we consider the case where the number of independent variables is n. Take an equation that contains $n-r+1$ arbitrary constants $a_1, a_2, \ldots, a_{n-r+1}$:

$$V(x_1, x_2, \ldots, x_n : a_1, a_2, \ldots, a_{n-r+1}) = 0. \qquad (13)$$

Differentiate this equation assuming that $a_1, \ldots, a_{n-r+1}$ take fixed values. Then we get

$$\frac{\partial V}{\partial x_i} + p_i \frac{\partial V}{\partial z} = 0, \qquad p_i = \frac{\partial z}{\partial x_i}, \qquad i = 1, 2, \ldots, n. \qquad (14)$$

If we eliminate $a_1, a_2, \ldots, a_{n-r+1}$ from (13) and (14), we obtain r partial differential equations of the first order:

$$F_j(x_1, x_2, \ldots, x_n, z, p_1, p_2, \ldots, p_n) = 0, \qquad j = 1, 2, \ldots, r. \qquad (15)$$

We call (13) a **complete solution** of (15). In this case, as in the case when $n=2$, we get all the solutions of (15) from a complete solution (13) of (15). We have the same classification as in the case $n=2$: (i) When we regard $a_1, \ldots, a_{n-r+1}$ as constants, then we have a complete solution of (15). (ii) When we can eliminate the constants $a_1, \ldots, a_{n-r+1}$ from equations

$$V = 0, \qquad \frac{\partial V}{\partial a_1} = 0, \ldots, \frac{\partial V}{\partial a_{n-r+1}} = 0,$$

we get a solution that does not contain an arbitrary constant. Such a solution is called a **singular solution** of (15). (iii) If not all of the $\partial V / \partial a_i$ vanish, there exists at least one functional relation among $a_1, a_2, \ldots, a_{n-r+1}$. We assume that there exist exactly k $(\leqslant n-r)$ relations among $a_1, a_2, \ldots, a_{n-r+1}$:

$$f_j(a_1, a_2, \ldots, a_{n-r+1}) = 0, \qquad j = 1, \ldots, k. \qquad (16)$$

Then there exist numbers $\lambda_1, \lambda_2, \ldots, \lambda_k$ such that

$$\frac{\partial V}{\partial a_l} = \lambda_1 \frac{\partial f_1}{\partial a_l} + \ldots + \lambda_k \frac{\partial f_k}{\partial a_l}, \qquad l = 1, \ldots, n-r+1. \qquad (17)$$

Hence, by eliminating $a_1, a_2, \ldots, a_{n-r+1}, \lambda_1, \lambda_2, \ldots, \lambda_k$ from (13), (16), and (17), we generally obtain exactly one relation between $x_1, \ldots, x_n$ and z. This is a solution of (15) that contains exactly k arbitrary functions $f_1, f_2, \ldots, f_k$. Such a solution is called a **general solution** of (15). In particular, if $k = n-r+1$, then it is a complete solution. We might think that there are $n-r$ general solutions corresponding to $k = 1, \ldots, n-r$. But these general solutions are not essentially different. For the partial equation of the second order $F(x, y, z, p, q, r, s, t) = 0$, this definition of a general solution is not applicable since we cannot successfully define a general solution by using the number of arbitrary functions contained in a solution.

Instead, we now use the following definition, due to J. G. Darboux: A solution of a general partial differential equation is called a **general solution** if by specializing its arbitrary functions and constants appropriately we obtain a solution whose existence is established by Cauchy's existence theorem. A solution $z = \varphi(x, y)$ of a general partial differential equation is called a **singular solution** if Cauchy's existence theorem cannot be applied on any curves on the manifold formed by $z = \varphi(x, y)$, $p = \partial\varphi/\partial x$, $q = \partial\varphi/\partial y$.

D. Cauchy's Method

We can regard equation (8) as a relation between the point (x, y, z) on the integral surface S and the direction cosines of a tangent plane at that point. Therefore the tangent planes at all points of the surface form a one-parameter family. They envelop a cone (T) whose vertex is (x, y, z) on S. The tangent plane at a point M on the integral surface S is tangent to this cone (T) along one generating line G of (T).

A curve on S whose tangents are all generating lines of (T) is a characteristic curve. If we write

$$X = \frac{\partial F}{\partial x}, \quad Y = \frac{\partial F}{\partial y}, \quad Z = \frac{\partial F}{\partial z}, \quad P = \frac{\partial F}{\partial p}, \quad Q = \frac{\partial F}{\partial q},$$

then the characteristic curve is given by the system of ordinary differential equations:

$$\frac{dx}{P} = \frac{dy}{Q} = \frac{dz}{Pp + Qq} = \frac{-dp}{X + pZ} = \frac{-dp}{Y + qZ}. \qquad (18)$$

We call this system the **characteristic differential equation** or **Charpit subsidiary (auxiliary) equation** of the partial differential equation (8) of the first order. System (18) determines not only x, y, z but also p and q. The set of these †surface elements (x, y, z, p, q) is the characteristic manifold. This characteristic manifold is considered as a part of the integral surface with infinitesimal width, and in this case we call it the **characteristic strip**. The characteristic strip is represented by the equations $x = x(\lambda)$, $y = y(\lambda)$, $z = z(\lambda)$, $p = p(\lambda)$, $q = q(\lambda)$ containing a parameter. On the integral surface $z = z(x, y)$, we have

$$\frac{dz}{d\lambda} = \frac{\partial z}{\partial x}\frac{dx}{d\lambda} + \frac{\partial z}{\partial y}\frac{dy}{d\lambda}$$

and

$$dz = p\,dx + q\,dy. \qquad (19)$$

Equation (19) is called the **strip condition**. The equations (18) evidently satisfy this condition.

The solution $u(t,x)=u(t,x_1,\ldots,x_n)$ of

$$u_t+f(t,x_1,\ldots,x_n,u_{x_1},\ldots,u_{x_n})=0,$$

$$u(0,x)=\varphi(x)$$

is obtained (at least locally) as follows. Let the solution of the differential equations

$$\frac{dx_i}{dt}=f_{\xi_i}(t,x_1,\ldots,x_n,\xi_1,\ldots,\xi_n),$$

$$\frac{d\xi_i}{dt}=-f_{x_i}(t,x_1,\ldots,x_n,\xi_1,\ldots,\xi_n),\quad 1\leqslant i\leqslant n,$$

issuing from (x_0,ξ^0) at $t=0$ be $(x(t;x_0,\xi^0)$, $\xi(t;x_0,\xi^0))$. Then specializing $\xi_j^0=\varphi_{x_j}(x_0)$ $(1\leqslant j\leqslant n)$, the solution $u(t,x)$ is obtained by quadrature along these curves (characteristic strips) from the relation

$$\frac{du}{dt}=-f(t,x,\xi)+\sum_{j=1}^{n}\xi_j f_{\xi_j}(t,x,\xi).$$

In particular, when $f(t,x,\xi)$ is homogeneous of degree 1 in ξ, by Euler's identity the right-hand side is identically zero. This means that u is constant along the characteristic strip.

E. Homogeneous Partial Differential Equations

Assume that $f(\xi_1,\xi_2,\ldots,\xi_m)$ is a †homogeneous polynomial of m independent variables $\xi_1,\xi_2,\ldots,\xi_m$. We denote the differential operator $\partial/\partial x_s$ by D_s. Then consider a homogeneous partial differential equation

$$f(D_1,D_2,\ldots,D_m)u=0. \tag{20}$$

We can obtain a homogeneous equation from an inhomogeneous partial differential equation by transformation of the dependent variable. For example, $D_1^2w=D_2w$ becomes the homogeneous partial differential equation $(D_1^2-D_2D_3)u=0$ by the transformation of the dependent variable $u=e^{x_3}w$. The equation $(D_1^2-D_2D_3)u=0$ corresponds to the homogeneous polynomial $f(\xi_1,\xi_2,\xi_3)=\xi_1^2-\xi_2\xi_3$. Generally, for equation (20), we consider the solution

$$u=F(\theta_1,\theta_2,\ldots,\theta_s), \tag{21}$$

where $\theta_1,\theta_2,\ldots,\theta_s$ are functions of $x_1,\ldots,x_m$ and F is an arbitrary function of θ_i. Such a solution is called a **primary solution** of (20). For example, for the equation $(D_1^2-D_2^2)u=0$, there are two primary solutions, $u=F(x_1+x_2)$ and $u=F(x_1-x_2)$. For †Laplace's equation

$$\Delta u=\frac{\partial^2u}{\partial x^2}+\frac{\partial^2u}{\partial y^2}+\frac{\partial^2u}{\partial z^2}=0,$$

we have a primary solution $u=F(z+ix\cos\alpha+iy\sin\alpha)$, where α is a parameter.

Second, for the †wave equation

$$\frac{\partial^2u}{\partial x^2}+\frac{\partial^2u}{\partial y^2}+\frac{\partial^2u}{\partial z^2}-\frac{1}{c^2}\frac{\partial^2u}{\partial t^2}=0, \tag{22}$$

we have a solution

$$u=\gamma f(\alpha,\beta),\quad \gamma=\frac{1}{r},\quad \alpha=t-\frac{r}{c},\quad \beta=\frac{z-r}{x+iy}, \tag{23}$$

where $r^2=x^2+y^2+z^2$, f is an arbitrary function, and γ is a particular solution of (22). Furthermore, $v=f(\alpha,\beta)$ satisfies the equation of a characteristic curve of (22):

$$\left(\frac{\partial v}{\partial x}\right)^2+\left(\frac{\partial v}{\partial y}\right)^2+\left(\frac{\partial v}{\partial z}\right)^2=\frac{1}{c^2}\left(\frac{\partial v}{\partial t}\right)^2.$$

Such a solution, which is the product of particular solutions and some function that contains an arbitrary function, is called a **primitive solution** of the original equation. Equation (22) has another primitive solution of the type

$$u=\frac{1}{r}g\left(t+\frac{r}{c},\ \frac{z-r}{x+iy}\right),$$

where g is an arbitrary function.

Laplace's equation has a primitive solution

$$u=\frac{1}{r}f\left(\frac{z-r}{x+iy}\right).$$

A **basic equation** is an equation, such as Laplace's equation, that has a primary solution and a primitive solution. A solution of an equation that has the same characteristic curves as a basic equation can be obtained from a particular solution of the basic equation by integrations and additions. For example, if we choose a particular solution $u=r^{-1}$ of Laplace's equation, which is a specialization of the primitive solution $u=r^{-1}f((z-r)/(x+iy))$, then

$$u=\iiint[(x-\xi)^2+(y-\eta)^2+(z-\zeta)^2]^{-1/2}\times F(\xi,\eta,\zeta)\,d\xi\,d\eta\,d\zeta$$

is a solution of

$$\Delta u+4\pi F(x,y,z)=0$$

in the interior of the domain of integration.

F. Determined Systems

The general form of a system of partial differential equations in two independent variables is

$$F_i(x,y,u^{(1)},u^{(2)},\ldots,u^{(m)},u_x^{(1)},u_y^{(1)},\ldots,u_x^{(m)},u_y^{(m)},u_{xx}^{(1)},\ldots)=0,\quad i=1,2,\ldots,h, \tag{24}$$

i.e., a system of h equations for m functions

$u^{(1)}, u^{(2)}, \ldots, u^{(m)}$ of the independent variables x and y. The system is called a **determined system** if $h=m$, an **overdetermined system** if $h>m$, and an **underdetermined system** if $h<m$.

An example of a determined system is the †Cauchy-Riemann equation

$$u_x - v_y = 0, \qquad u_y + v_x = 0,$$

for $u(x, y)$, $v(x, y)$, which can be further reduced to two determined equations $\Delta u = 0$ and $\Delta v = 0$.

A simple example of an overdetermined system is

$$u_x = f(x, y), \qquad u_y = f(x, y).$$

A necessary and sufficient condition for the existence of a solution of this system is $f_y = f_x$. The **Cauchy-Riemann differential equations** for a holomorphic function $f(z_1, z_2) = u + iv$ of two complex variables $z_1 = x_1 + iy_1$, $z_2 = x_2 + iy_2$ are

$$u_{x_1} = v_{y_1}, \qquad u_{x_2} = v_{y_2}, \qquad u_{y_1} = -v_{x_1},$$

$$u_{y_2} = -v_{x_2},$$

which can be reduced to

$$u_{x_1x_1} + u_{y_1y_1} = 0, \qquad u_{x_1x_2} + u_{y_1y_2} = 0,$$

$$u_{x_2x_2} + u_{y_2y_2} = 0, \qquad u_{x_1y_2} - u_{x_2y_1} = 0,$$

which is also an overdetermined system.

An example of an underdetermined system is

$$\frac{\partial(u, v)}{\partial(x, y)} = u_x v_y - u_y v_x = 0.$$

If this equation holds, there exists a functional relation $w(u, v) = 0$ that can be regarded as a solution of this underdetermined system.

G. General Theory of Differential Operators

In recent developments of the theory of partial differential equations, there is a trend to construct a general theory for †differential operators regardless of the classical types of differential equations (→ 112 Differential Operators). For example, we take a property that is satisfied by some equation of classical type (e.g., an †elliptic differential equation) and proceed to characterize all equations that have this property. (For example, †hypoellipticity is a property of classical †parabolic and elliptic equations.) There are several basic problems in this general theory: the existence of a fundamental solution, the existence of a local solution, unique continuation of solutions, the differentiability and analyticity of solutions, and the propagation of smoothness. We explain here only two of them: the fundamental solution and the local existence of solutions.

H. Fundamental Solutions

L is assumed to be a linear partial differential operator with constant coefficients. If a †distribution E satisfies the equation

$$LE(x) = \delta(x),$$

where $\delta(x)$ is the †Dirac δ-function, then we call $E(x)$ a **fundamental solution** (or **elementary solution**) of L. Also, if L is a linear differential operator and E satisfies the equation

$$L_x E(x, y) = \delta(x - y),$$

then we call this distribution $E(x, y)$ a **fundamental kernel** (or **elementary kernel**) of L.

Let L be a differential operator with constant coefficients and $E(x)$ be a fundamental solution of L. Then $E(x, y)$ is a fundamental kernel of L. Sometimes $E(x, y)$ itself is called a fundamental solution. L. Ehrenpreis and B. Malgrange proved that any linear differential operator with constant coefficients has a fundamental solution [4].

If we take a fundamental solution (fundamental kernel) E and add to it an arbitrary solution of the equation $Lu = 0$, then we get another fundamental solution (fundamental kernel) of L. This freedom of the fundamental solution (fundamental kernel) can be used to construct †Green's functions of the boundary value problem of elliptic equations and of the mixed initial-boundary value problem for parabolic equations. A Green's function is a fundamental solution (fundamental kernel) that satisfies given boundary conditions (→ 188 Green's Functions; 189 Green's Operator). The fundamental solutions (fundamental kernels) relative to the Cauchy problem are also defined as in this section. For example, consider a fundamental solution of the Cauchy problem concerning the future behavior of a differential operator $L = \partial/\partial t - P(\partial/\partial x)$, namely, a distribution $E(t, x)$ that satisfies $LE = 0$ $(t > 0)$ and $E(t, x)|_{t=0} = \delta(x)$. If we put $\tilde{E}(t, x) = E(t, x)$ $(t \geqslant 0)$ and $\tilde{E}(t, x) = 0$ $(t < 0)$, then $\tilde{E}(t, x)$ is a fundamental solution (or kernel) of the differential operator L, that is, $L\tilde{E} = \delta(t, x)$. Sometimes a fundamental solution of the Cauchy problem for a parabolic equation is called a Green's function. On the other hand, a fundamental solution (or kernel) of the Cauchy problem for a hyperbolic equation is called a †Riemann function. A Riemann function actually is not always a function; in general it is a distribution.

Example 1. A fundamental solution of the 3-dimensional Laplacian

$$\Delta = \frac{\partial^2}{\partial x_1^2} + \frac{\partial^2}{\partial x_2^2} + \frac{\partial^2}{\partial x_3^2}$$

is $E(x) = -1/4\pi r$, where $r = \sqrt{x_1^2 + x_2^2 + x_3^2}$.

Example 2. A fundamental solution of the Cauchy problem for the future of the 3-dimensional wave operator

$$L=\frac{\partial^2}{\partial t^2}-\sum_{i=1}^{3}\frac{\partial^2}{\partial x_i^2},$$

i.e., a distribution $E(t,x)$ $(t \geqslant 0)$ that satisfies $LE=0$ $(t>0)$, $E(0,x)=0$, and $(\partial/\partial t)E(0,x)=\delta(x)$, is given by $E(t,x)=(1/4\pi t)\delta(r-t)$ $(t>0)$, $r=\sqrt{x_1^2+x_2^2+x_3^2}$. A fundamental solution for L is given by $\tilde{E}(t,x)=E(t,x)$ $(t \geqslant 0)$; $=0$ $(t<0)$ (→ Appendix A, Table 15.V).

I. Existence of Local Solutions

Given a linear differential operator L and the equation $Lu=f$, we have the problem of determining whether this equation always has a solution in some neighborhood of a given point. If the coefficients of L and f are holomorphic in a neighborhood of this point and if the homogeneous part of highest order does not vanish, then there exists a solution that is holomorphic in a neighborhood of the given point (†Cauchy-Kovalevskaya existence theorem).

If L is a linear differential operator with constant coefficients, E is a fundamental solution of L, and f is a function (or distribution) that is zero outside of a compact set, then we have a solution u that is the †convolution of E and $f: u=E*f$. On the other hand, H. Lewy proposed the following example [3]:

$$Lu \equiv \frac{\partial u}{\partial x_1}+i\frac{\partial u}{\partial x_2}+2i(x_1+ix_2)\frac{\partial u}{\partial x_3}=f(x_3),$$

where f is a real function of x_3. He showed that if this equation has a solution that is of class C^1, then f must be real analytic. Therefore, if f is of class C^∞ but not real analytic, then this equation has no C^1-solution. Actually, no solution exists even in the distribution sense. (Note that, since the coefficients of L are now complex-valued, the results mentioned at the beginning of this section are no longer applicable.)

For linear differential operators L, a study by L. Hörmander gives some necessary conditions and also some sufficient conditions for the local existence of a solution of the equation $Lu=f$ for sufficiently general f [4]. This result has been developed and completed by L. Nirenberg and F. Treves [18] and by R. Beals and C. Fefferman [19]. The operator considered by S. Mizohata (*J. Math. Kyoto Univ.*, 1 (1962)),

$$L \equiv \frac{\partial}{\partial x_i}+ix_1^k\frac{\partial}{\partial x_2},$$

serves as a standard model in this problem. In the neighborhood of the origin, if k is even, L is locally solvable, and if k is odd, it is not locally solvable. However, for linear partial differential operators with multiple characteristics, the problem of local solvability becomes extremely difficult.

References

[1] E. Goursat, Cours d'analyse mathématique, Gauthier-Villars, II, second edition, 1911; III, fourth edition, 1927.
[2] R. Courant and D. Hilbert, Methods of mathematical physics II, Interscience, 1962.
[3] H. Lewy, An example of a smooth linear partial differential equation without solution, Ann. Math., (2) 66 (1957), 155–158.
[4] L. Hörmander, Linear partial differential operators, Springer, 1963.
[5] S. Mizohata, Theory of partial differential equations, Cambridge Univ. Press, 1973. (Original in Japanese, 1965.)
[6] C. Carathéodory, Calculus of variations and partial differential equations of the first order, Holden-Day, 1967.
[7] I. G. Petrovskiĭ, Über das Cauchysche Problem für ein System linearer partieller Differentialgleichungen im Gebiete der nichtanalytischen Funktionen, Bull. Univ. Moscou, ser. internat., sec. A, vol. 1, fasc. 7 (1938), 1–74.
[8] J. Hadamard, Le problème de Cauchy et les équations aux dérivées partielles linéaires hyperboliques, Hermann, 1932.
[9] I. G. Petrovskiĭ, Lectures on partial differential equations, Interscience, 1954. (Original in Russian, 1953.)
[10] S. L. Sobolev, Partial differential equations of mathematical physics, Pergamon, 1964. (Original in Russian, 1954.)
[11] J. L. Lions, Equations différentielles opérationelles et problèmes aux limites, Springer, 1961.
[12] J. L. Lions and E. Magenes, Problèmes aux limites non homogènes et applications, Dunod, I, 1968; II, 1968; III, 1969.
[13] A. Friedman, Partial differential equations, Holt, Rinehart and Winston, 1969.
[14] F. Treves, Locally convex spaces and linear partial differential equations, Springer, 1967.
[15] F. Treves, Linear partial differential equations with constant coefficients, Gordon & Breach, 1966.
[16] F. John, Partial differential equations, Springer, 1971.
[17] L. Bers, F. John, and M. Schechter, Partial differential equations, Interscience, 1964.
[18] L. Nirenberg and F. Treves, On local solvability of linear partial differential equations I, II, Comm. Pure Appl. Math., 23 (1970), 1–38, 459–510.

[19] R. Beals and C. Fefferman, On local solvability of linear partial differential equations, Ann. Math., 97 (1973), 482–498.
[20] F. Cardoso and F. Treves, A necessary condition of local solvability for pseudodifferential equations with double characteristics, Ann. Inst. Fourier, 24 (1974), 225–292.

321 (XIII.21) Partial Differential Equations (Initial Value Problems)

A. General Remarks

First, we give two examples of initial value problems for †partial differential equations.

(I) Consider the partial differential equation $u_x - u_y = 0$ of two independent variables x and y. If the function $\varphi(y)$ is of class C^1, then $u = \varphi(x+y)$ is a solution of this equation that satisfies $u(0,y) = \varphi(y)$.

(II) We denote a point of $\mathbf{R}^{n+1}$ (or $\mathbf{C}^{n+1}$) by (t,x), $x = (x_1, \ldots, x_n)$. Let L be a linear partial differential operator of order m:

$$L = \frac{\partial^m}{\partial t^m} + \sum_{|\nu| \leqslant m} a_{\nu_0 \nu_1 \ldots \nu_n}(t,x) \frac{\partial^{|\nu|}}{\partial t^{\nu_0} \partial x_1^{\nu_1} \ldots \partial x_n^{\nu_n}},$$

$$|\nu| = \nu_0 + \nu_1 + \ldots + \nu_n, \quad \nu_0 < m,$$

where the coefficients $a_{\nu_0 \nu_1 \ldots \nu_n}(t,x)$ are functions of class C^ω (i.e., †real analytic functions or holomorphic functions) in a neighborhood of $(t,x) = (0,0)$. If the functions $v(t,x)$ and $w_k(x)$ $(0 \leqslant k \leqslant m-1)$ are of class C^ω in a neighborhood of $(t,x) = (0,0)$, then there exists a unique solution $u(t,x)$ of class C^ω in a neighborhood of $(t,x) = (0,0)$ that satisfies

$$L[u] = v(t,x),$$

$$\frac{\partial^k u}{\partial t^k}(0,x) = w_k(x), \quad 0 \leqslant k \leqslant m-1. \tag{1}$$

This is called the **Cauchy-Kovalevskaya existence theorem** (for linear partial differential equations).

As in (II) we choose one of the independent variables as the principal variable and regard the others as parameters. When we assign a value a to the principal variable t, the values of the dependent variables (unknown functions) and their derivatives are called **initial values**, **initial data**, or **Cauchy data**. Conditions to determine initial values are called **initial conditions**. The problem of finding a solution of (1) under given initial conditions is called an **initial value problem** or **Cauchy problem**. We may consider initial value problems not only for initial conditions on a hyperplane $t = a$, but also for initial conditions on a hypersurface, called an **initial surface** (→ Section C).

Let $a(x,D)$ be a linear partial differential operator of order m:

$$a(x,D) = \sum_{|\alpha| \leqslant m} a_\alpha(x) D^\alpha, \qquad D^\alpha = \frac{\partial^{|\alpha|}}{\partial x_1^{\alpha_1} \ldots \partial x_n^{\alpha_n}},$$

$$|\alpha| = \alpha_1 + \ldots + \alpha_n,$$

where the coefficients $a_\alpha(x)$ are of class C^ω in a neighborhood of $x = 0$. Its **characteristic polynomial** is

$$h(x,\xi) = \sum_{|\alpha| = m} a_\alpha(x) \xi^\alpha, \qquad \xi^\alpha = \xi_1^{\alpha_1} \ldots \xi_n^{\alpha_n}.$$

Let $S: s(x) = 0$ be a regular surface (i.e., $s_x = (\partial s/\partial x_1, \ldots, \partial s/\partial x_n) \neq (0)$) of codimension 1. We suppose that S is a †noncharacteristic surface, that is, $h(x, s_x) \neq 0$ on S. Let $v(x)$ and $w_k(x)$ $(0 \leqslant k \leqslant m-1)$ be the functions of class C^ω in a neighborhood of $x = 0$ and on S, respectively. We consider the Cauchy problem

$$a(x,D)u(x) = v(x), \qquad \frac{\partial^k u}{\partial n^k(x)} = w_k(x) \quad \text{on } S,$$

$$0 \leqslant k \leqslant m-1, \tag{1}$$

where n is the outward normal direction of S. S is thus the initial surface. Then there exists a unique solution $u(x)$ of class C^ω in a neighborhood of $x = 0$. In fact, by the change of variables $X_1 = s(x)$, $X_i = x_i$ $(2 \leqslant i \leqslant n)$ if $\partial s/\partial x_1 \neq 0$ on S, this problem can be reduced to the Cauchy problem (1) by taking account of the fact that $h(x, s_x) \neq 0$ on S.

The Cauchy-Kovalevskaya theorem asserts the local existence of solution when the initial values are of class C^ω. Indeed, J. Hadamard noted that if the initial values are not of class C^ω, the initial value problem does not always have a solution. For example, consider the initial value problem

$$\frac{\partial^2 u}{\partial x^2} + \frac{\partial^2 u}{\partial y^2} + \frac{\partial^2 u}{\partial z^2} = 0$$

with the initial values

$$u(0,y,z) = w(y,z), \qquad \frac{\partial u}{\partial x(0,y,z)} = 0.$$

If the function $w(y,z)$ is not of class C^ω in any neighborhood of $y = z = 0$, the solution of this problem can never exist in (or even on one side $x \geqslant 0$ of) any neighborhood of $x = y = z = 0$.

B. The Cauchy-Kovalevskaya Existence Theorem for a System of Partial Differential Equations in the Normal Form

The Cauchy-Kovalevskaya existence theorem (1) is extended for more general systems of partial differential equations in the normal form studied by Kovalevskaya. Consider

$$\frac{\partial^{p_i} u_i}{\partial t^{p_i}} = F_i\left(t, x, u_1, \ldots, u_m, \ldots, \frac{\partial^{|\nu|} u_j}{\partial t^{\nu_0} \partial x_1^{\nu_1} \ldots \partial x_n^{\nu_n}}, \ldots\right),$$

where $1 \leqslant i, j \leqslant m$, $|\nu| = \nu_0 + \nu_1 + \ldots + \nu_n \leqslant p_j$, $\nu_0 < p_j$, and $x = (x_1, \ldots, x_n)$. We assume that F_i ($1 \leqslant i \leqslant m$) are functions of class C^ω with respect to arguments t, x, $u_1, \ldots, u_m, \ldots, \partial^{|\nu|} u_j / \partial t^{\nu_0} \partial x_1^{\nu_1} \ldots \partial x_n^{\nu_n}, \ldots$, in a neighborhood of $(0, 0, \ldots, 0)$. If the functions $w_{ik}(x)$ $(1 \leqslant i \leqslant m, 0 \leqslant k \leqslant p_i - 1)$ are of class C^ω in a neighborhood of $x = 0$, then there exists a unique solution $u_1(t, x), \ldots, u_m(t, x)$ of class C^ω in a neighborhood of $(t, x) = (0, 0)$ that satisfies the equations and the initial values

$$\frac{\partial^k u_i}{\partial t^k}(0, x) = w_{ik}(x), \qquad 1 \leqslant i \leqslant m, \quad 0 \leqslant k \leqslant p_i - 1$$

[2, 4].

C. Single Equations of the First Order

For a single partial differential equation given in the normal form

$$\frac{\partial u}{\partial x} = F\left(x, y_1, \ldots, y_k, u, \frac{\partial u}{\partial y_1}, \ldots, \frac{\partial u}{\partial y_k}\right), \tag{2}$$

we assume that $F(x, y_1, \ldots, y_k, u, q_1, \ldots, q_k)$ is a real-valued function of class C^2 in a neighborhood of $x = a$, $y_i = b_i$, $u = c$, $q_i = d_i$, and that $\varphi(y_1, \ldots, y_k)$ is also a function of class C^2 such that $\varphi = c$, $\partial\varphi/\partial y_i = d_i$ at $y_i = b_i$. Then there is a solution u of (2) in a neighborhood of $x = a$ and $y_i = b_i$ that satisfies $u(a, y_1, \ldots, y_k) = \varphi(y_1, \ldots, y_k)$ and is of class C^2.

For the uniqueness of solutions of this Cauchy problem, A. Haar (*Atti del Congresso Internazionale dei Matematici*, 1928, Bologna, vol. 3) showed that if F satisfies the †Lipschitz condition

$$|F(x, y, u', q') - F(x, y, u, q)| \leqslant A \sum_{i=1}^{k} |q_i' - q_i| + B|u' - u|,$$

then the solution of the initial value problem for (2) is unique. To obtain this result he studied the partial differential inequality

$$\left|\frac{\partial u}{\partial x}\right| \leqslant A \sum_{i=1}^{k} \left|\frac{\partial u}{\partial y_i}\right| + B|u| + C. \tag{3}$$

Next, we consider more general equations of the first order

$$F\left(x_1, \ldots, x_k, u, \frac{\partial u}{\partial x_1}, \ldots, \frac{\partial u}{\partial x_k}\right) = 0. \tag{4}$$

Suppose that $F(x_1, \ldots, x_k, u, p_1, \ldots, p_k)$ is a real-valued function of class C^2 in a neighborhood of $x_i = a_i$, $u = b$, $p_i = c_i$; $\varphi(x_1, \ldots, x_k)$, $S(x_1, \ldots, x_k)$ are functions of class C^2 in a neighborhood of $x_i = a_i$ that satisfy $b = \varphi(a_1, \ldots, a_k)$, $c_i = (\partial\varphi/\partial x_i)_{x=a}$, $S(a_1, \ldots, a_k) = 0$; and

$$\left[\sum_{\nu=1}^{k} \frac{\partial F}{\partial p_\nu} \frac{\partial S}{\partial x_\nu}\right]_{x=a, u=b, p=c} \neq 0. \tag{5}$$

Then there exists a solution u of (4) of class C^2 in a neighborhood of $x = a$ that satisfies $u = \varphi(x)$ on the hypersurface $S(x) = 0$.

Furthermore, if F, φ, S are of class C^1 and satisfy (5), then there is at most one solution u of (4) of class C^1 in a neighborhood of a that satisfies $u = \varphi(x)$ on the hypersurface $S(x) = 0$. These facts can be proved in the following way. By choosing $S_1(x), \ldots, S_{k-1}(x)$ and then $S(x)$ so that the †Jacobian $\partial(S, S_1, \ldots, S_{k-1})/\partial(x_1, \ldots, x_k)$ does not vanish and by changing variables from x to $S, S_1, \ldots, S_{k-1}$, we obtain a normal form solved for $\partial u/\partial S$ by condition (5) (this condition means that the hypersurface $S(x) = 0$ is not tangent to the †characteristic curves).

D. Quasilinear Equations of the Second Order

Consider the equation

$$\sum_{i,j=1}^{k} a_{ij}(x, u, p) \frac{\partial^2 u}{\partial x_i \partial x_j} + b(x, u, p) = 0,$$

where $x = (x_1, \ldots, x_k)$, $p = (p_1, \ldots, p_k)$, and $p_i = \partial u/\partial x_i$. We assume that the initial conditions are $u = \varphi(x)$ on $S(x) = 0$ and

$$\sum_{i=1}^{k} \frac{\partial S}{\partial x_i} \frac{\partial u}{\partial x_i} = \psi(x)$$

on the same hypersurface. Taking the other functions $S_1(x), \ldots, S_{k-1}(x)$ and then $S(x)$ so that the Jacobian $\partial(S, S_1, \ldots, S_{k-1})/\partial(x_1, \ldots, x_k)$ does not vanish, we change the variables x to $s_0, s_1, \ldots, s_{k-1}$ $(s_0 = S, s_i = S_i)$. Then we get

$$\sum_{\mu,\nu=0}^{k-1} Q(S_\mu, S_\nu) \frac{\partial^2 u}{\partial s_\mu \partial s_\nu} + b^* = 0, \tag{6}$$

where

$$Q(S_\mu, S_\nu) = \sum_{i,j=1}^{k} a_{ij} \frac{\partial S_\mu}{\partial x_i} \frac{\partial S_\nu}{\partial x_j},$$

$$p_i = \sum_{\mu=0}^{k-1} \frac{\partial S_\mu}{\partial x_i} \frac{\partial u}{\partial s_\mu},$$

$$b^* = b + \sum_{\lambda=0}^{k-1} \sum_{i,j=1}^{k} a_{ij} \frac{\partial^2 S_\lambda}{\partial x_i \partial x_j} \frac{\partial u}{\partial s_\lambda}.$$

When $s_0 = 0$, the initial conditions are transformed to

$$u=\varphi^*(s_1,\dots,s_{k-1})=\varphi(x(0,s_1,\dots,s_{k-1})),$$

$$\frac{\partial u}{\partial s_0}=\psi^*(s_1,\dots,s_{k-1})$$

$$=\left[\sum_{i=1}^{k-1}\left(\frac{\partial S}{\partial x_i}\right)^2\right]^{-1}\left[\psi(x(0,s_1,\dots,s_{k-1}))-\sum_{\lambda=1}^{k-1}\frac{\partial\psi^*}{\partial s_\lambda}\sum_{i=1}^{k}\frac{\partial S_\lambda}{\partial x_i}\frac{\partial S}{\partial x_i}\right].$$

Suppose that $Q(S,S)\neq 0$, and set $\partial u/\partial s_i=q_i$. Then equation (6) added to the initial conditions $u=\varphi^*(s_1,\dots,s_{k-1})$, $q_0=\psi^*(s_1,\dots,s_{k-1})$ at $s_0=0$ is equivalent to the system in the normal form of partial differential equations of the first order

$$\frac{\partial u}{\partial s_0}=q_0,\qquad \frac{\partial q_\mu}{\partial s_0}=\frac{\partial q_0}{\partial s_\mu},\qquad \mu=1,\dots,k-1,$$

$$\frac{\partial q_0}{\partial s_0}=-\frac{1}{Q(S,S)}\left[\sum_{\mu=1}^{k-1}\left(\sum_{\nu=1}^{k-1}Q(S_\mu,S_\nu)\frac{\partial q_\mu}{\partial s_\nu}-2Q(S_\mu,S)\frac{\partial q_0}{\partial s_\mu}\right)-b^*\right]$$

with initial conditions

$$u=\varphi^*(s_1,\dots,s_{k-1}),\qquad q_0=\psi^*(s_1,\dots,s_{k-1}),$$

$$q_\mu=\partial\varphi^*/\partial s_\mu$$

when $s_0=0$. Thus, if the coefficients are of class C^ω, the preceding theory applies to this equation.

E. Continuous Dependence of Solutions on the Initial Values

First, we consider the following simple example of a linear equation. The wave equation

$$\frac{\partial^2 v}{\partial t^2}=c^2\frac{\partial^2 v}{\partial x^2}$$

for $v(x,t)$ is the simplest †hyperbolic equation. Its solution satisfying the initial conditions $v(x,0)=f(x)$, $(\partial v/\partial t)(x,0)=g(x)$ is given by

$$v(x,t)=\frac{1}{2}(f(x+ct)+f(x-ct))+\frac{1}{2}\int_{x-ct}^{x+ct}g(\lambda)\,d\lambda.$$

It is obvious from this expression that if we regard $f(x)$ and $g(x)$ as elements of the †function space $C(\mathbf{R})$ of continuous functions of $x\in\mathbf{R}$ with the topology of †uniform convergence on compact sets, then $v(x,t)$ is determined as the value of a linear operator from $C(\mathbf{R}^2)$ to $C(\mathbf{R}^2)$ on xt-space. If such continuous dependence on the initial values is established, or more precisely, if there is a unique solution for sufficiently smooth initial values that depends on the initial values continuously in a suitable sense, we say that the initial value problem is **well posed (properly posed** or **correctly posed)**.

Systematic research on the well-posedness of Cauchy problems was initiated by I. G. Petrovskiĭ, who considered the following system of partial differential equations, which is more general than differential equations of the normal form. (The coefficients are all assumed to be functions of t only.)

$$\frac{\partial^{n_j}u_j}{\partial t^{n_j}}=\sum_{k=1}^{N}\sum_{|\nu|\leqslant m}a_{jk\nu_0\nu_1\dots\nu_n}(t)\frac{\partial^{\nu_0+\nu_1+\dots+\nu_n}u_k(x,t)}{\partial t^{\nu_0}\partial x_1^{\nu_1}\dots\partial x_n^{\nu_n}}+B_j(x_1,\dots,x_n,t),\qquad j=1,\dots,N,$$

$$|\nu|=\nu_0+\nu_1+\dots+\nu_n,\qquad \nu_0<n_k. \tag{7}$$

This has a normal form if $n_k=m$ $(k=1,\dots,N)$. Now, taking the derivatives

$$\left(\frac{\partial}{\partial t}\right)^{n_j-1}u_j(x,t),\quad \left(\frac{\partial}{\partial t}\right)^{n_j-2}u_j(x,t),\dots,\quad \left(\frac{\partial}{\partial t}\right)u_j(x,t)$$

as new unknowns, we get another system:

$$\frac{\partial u_j}{\partial t}=\sum_{k=1}^{N'}\sum_{|\nu|\leqslant m}a_{jk\nu_1\dots\nu_n}(t)\frac{\partial^{\nu_1+\dots+\nu_n}}{\partial x_1^{\nu_1}\dots\partial x_n^{\nu_n}}u_k(x,t)+C_j(x,t),\qquad j=1,\dots,N'. \tag{8}$$

We take as the space of initial values the †topological linear space composed of all functions whose derivatives up to a sufficiently large order are bounded on the whole space $\mathbf{R}^n$ and equipped with the topology determined by the †seminorms that are the maximums of derivatives up to a given order on the whole space, and we take as the range space a similar space on the xt-space, where $x\in\mathbf{R}^n$ and $0\leqslant t\leqslant T$. Then we can formulate a necessary and sufficient condition for the well-posedness of the initial value problem for the future (the problem is regarded as specifying a mapping that assigns to the initial values on $t=0$ the values $u(x,t)$ for $t>0$). To give such a condition we consider the following system of ordinary differential equations, which are given by a †Fourier transformation on the x-space of system (8):

$$\frac{d\hat{u}_j}{dt}=\sum_{k=1}^{N'}\sum_{|\nu|\leqslant m}a_{jk\nu_1\dots\nu_n}(t)\times(2\pi i\xi_1)^{\nu_1}\dots(2\pi i\xi_n)^{\nu_n}\hat{u}_k(\xi,t)+C_j(\xi,t),\qquad j=1,\dots,N'. \tag{9}$$

If the †fundamental system of solutions of (9) is $v_i^{(j)}(\xi,t)$ $(i=1,\dots,N',j=1,\dots,N')$, then the condition is that these functions satisfy the inequalities

$$|v_i^{(j)}(\xi,t)|\leqslant C(1+|\xi|)^L,\qquad 0\leqslant t\leqslant T, \tag{10}$$

where C and L are constants independent of ξ. This is the condition obtained by Petrovskiĭ.

If system (7) is of normal form and is well posed for the future, it is also well posed for the past. In this case, equation (7) is said to be of **hyperbolic type** (→ 325 Partial Differential Equations of Hyperbolic Type). An example that is not of a normal form and is well posed is a parabolic equation (→ 327 Partial Differential Equations of Parabolic Type). In Petrovskiĭ's theory it is sufficient to assume that the coefficients in (7) are continuous and bounded, and we can take T arbitrarily large provided that (10) is satisfied. Hence Petrovskiĭ's theory guarantees global existence of solutions.

F. Uniqueness of Solutions

If a linear partial differential equation of the first order of normal form has coefficients of class C^ω, then the solution of class C^1 satisfying the prescribed initial conditions is unique (**Holmgren's uniqueness theorem**, 1901). Every system of partial differential equations of higher order of normal form can be reduced to a system of the first order of normal form. Therefore, if the coefficients are of class C^ω, there is only one solution for the original initial value problem with continuous partial derivatives up to the order of the equation. Moreover, if an analytic manifold S of dimension $n-1$ (n is the number of independent variables) is not †characteristic for the given equation of order m, there is at most one solution whose derivatives of order up to $m-1$ coincide with given functions on the manifold S. The proof of this fact relies on the Cauchy-Kovalevskaya existence theorem.

The uniqueness problem for the initial value problem is in general very difficult even for linear equations when the coefficients are not of class C^ω.

In particular, if the number of independent variables is 2 and the coefficients are all real, then we have a result of T. Carleman (1939) about the system:

$$\frac{\partial u_\mu}{\partial x}=\sum_{\nu=1}^{m} a_{\mu\nu}(x,y)\frac{\partial u_\nu}{\partial y}+\sum_{\nu=1}^{m} b_{\mu\nu}(x,y)u_\nu, \tag{11}$$

where the $a_{\mu\nu}$ are of class C^2 and the $b_{\mu\nu}$ are continuous. He proved that if the eigenvalues of the matrix $(a_{\mu\nu})$ are all distinct, then even when some of the eigenvalues are complex, there is at most one solution of class C^1 for the initial value problem. If we omit the assumption about the eigenvalues, however, the theorem does not hold in general, because we have a counterexample due to A. Pliś (1954) where all coefficients are of class C^∞, $m=2$, $b_{\mu\nu}\equiv 0$.

A. P. Calderón showed that Carleman's result can be extended to the case $n>2$ (*Amer. J. Math.*, 80 (1958)). Consider the following linear partial differential equation of the kth order:

$$\frac{\partial^k u}{\partial x^k}+\sum_{j=1}^{k} P_j\left(x,y,\frac{\partial}{\partial y}\right)\left[\frac{\partial^{k-j}u}{\partial x^{k-j}}\right]+B[u]=0, \tag{12}$$

where $P_j(x,y,\xi)$ is a homogeneous polynomial of degree j of $\xi=(\xi_1,\ldots,\xi_n)$ with real coefficients and B is a differential operation in (x,y) of order at most $k-1$. We assume that the coefficients of $P_j(x,y,\xi)$ are functions of (x,y) of class C^1, their derivatives are Hölder continuous, and the coefficients of B are bounded and continuous. If the characteristic equation of (12),

$$\lambda^k+\sum_{j=1}^{k} P_j(x,y,\xi)\lambda^{k-j}=0,$$

has only distinct roots for $\xi\neq 0$, then the C^k-solution of the Cauchy problem is unique in a neighborhood of $x=a$. Calderón proved this except for the cases $k\geqslant 4$, $n=2$, where a certain topological difficulty arises. S. Mizohata (*J. Math. Soc. Japan*, 11 (1959)) succeeded in obtaining the proof for the exceptional cases.

This result can be extended to systems of equations under similar assumptions. See L. Hörmander [4] for an extension to the complex coefficient case. S. Mizohata, T. Shirota, and H. Kumanogo discuss the uniqueness theorem for equations of double characteristics or of parabolic type.

For nonlinear equations there are, in general, very few results about the global existence of solutions. For example, if the function F in equation (2) in the normal form satisfies the Lipschitz condition with constants A and B that are independent of x, then we get a global existence theorem. The method of proof of this theorem is as follows: First, we prove the existence for $|x|\leqslant\varepsilon_1$ and sufficiently small ε_1 by Picard's †successive iteration method. Then we regard $x=\varepsilon_1$ as the hyperplane on which the initial values are assigned and proved the existence of a solution on $\varepsilon_1\leqslant|x_1|\leqslant\varepsilon_2$, and so on. The same method can be applied to nonlinear systems of the first order if they are special types of quasilinear systems.

G. Construction of Solutions by Asymptotic Expansion

Let $a(x,D)=\sum_{|\alpha|\leqslant m}a_\alpha(x)D^\alpha$ be a linear partial differential operator of order m, with coefficients of class C^∞. We write $a(x,\xi)=\sum_{|\alpha|\leqslant m}a_\alpha(x)\xi^\alpha=h(x,\xi)+h'(x,\xi)+\ldots$ with h and h' homogeneous in ξ of degree m and $m-1$, respectively. Let $K:\varphi(x)=0$ be a regular surface of codimension 1. We assume that K is a simple characteristic, i.e., $h(x, \varphi_x)=0$ and $(\partial h/\partial \xi_i(x, \varphi_x)) \neq (0)$ on K. Let $f_j(t)$ $(j=0, 1, \dots)$ be a sequence of functions satisfying $df_j/dt(t)=f_{j-1}(t), j=1,2,\dots$. Then the equation $a(x, D)u(x)=0$ has a formal solution in the form

$$u(x)=\sum_{j=0}^{\infty} f_j(\varphi(x))u_j(x).$$

The coefficients $u_j(x)$ are obtained by solving successively the equations

$$L_1[u_j]=\sum_{k=2}^{m} L_k[u_{j-k+1}], \tag{13}$$

where

$$L_1=\sum_{i=1}^{n}\frac{\partial h}{\partial \xi_i}(x, \varphi_x)D_i+\frac{1}{2}\sum_{i,j=1}^{n}\frac{\partial^2 h}{\partial \xi_i \partial \xi_j}(x, \varphi_x)\varphi_{x_i x_j} + h'(x, \varphi_x)$$

and L_k $(2 \leqslant k \leqslant m)$ are differential operators of order k depending only on a and φ. By this method of asymptotic expansion of solution, fundamental solutions and singularities of solutions for hyperbolic equations have been studied (Hadamard [5, 6]; R. Courant and P. Lax, *Proc. Nat. Acad. Sci. US*, 42 (1956); Lax, *Duke Math. J.*, 24 (1957); D. Ludwig, *Comm. Pure Appl. Math.*, 22 (1960); S. Mizohata, *J. Math. Kyoto Univ.*, 1 (1962); → 325 Partial Differential Equation of Hyperbolic Type).

If $a(x, D)$ is an operator with analytic coefficients, this formal solution is convergent. By using this fact, Mizohata constructed †null solutions. In fact, put $f_j(t)=t^{p+m+j}/\Gamma(p+m+j)$ $(p \geqslant 0), j=0,1,\dots$. Define $u(x)$ by $u(x)=0$ for $\varphi(x) \leqslant 0$ and $u(x)=\sum_{j=0}^{\infty} f_j(\varphi(x))u_j(x)$ for $\varphi(x) \geqslant 0$ obtained by the preceding process. Then $u(x)$ is a null solution of $a(x, D)u(x)=0$ (Mizohata, *J. Math. Kyoto Univ.*, 1 (1962)).

The Cauchy problem in the case when the initial surface has characteristic points had been studied by J. Leray and by L. Gårding, T. Kotake, and Leray. Let $a(x, D)$ be a differential operator with holomorphic coefficients in a complex domain. Let $S: s(x)=0$ be a regular surface and T be a subvariety of codimension 1 of S. Suppose that S is noncharacteristic on $S-T$, but characteristic on T. Consider the Cauchy problem (1) of Section A. Then the solution $u(x)$ is ramified around a characteristic surface K that is tangent to S on T, and it can be uniformized (Leray, *Bull. Soc. Math. France*, 85 (1957); Gårding, Kotake, and Leray, *Bull. Soc. Math. France*, 92 (1964)).

Next we consider the Cauchy problem (1) when the initial surface S $(x_1=0)$ is noncharacteristic, but the initial values $w_k(x)$ $(0 \leqslant k \leqslant m-1)$ have singularities on a regular subvariety T $(x_1=x_2=0)$ of S. We assume that $h(0, \xi_1, 1, 0, \dots, 0)=0$ has m distinct roots. Then there exist m characteristic surfaces $K_i: \varphi_i(x)=0$ $(1 \leqslant i \leqslant m)$ originating from T. $\varphi_i(x)$ is obtained by solving the †Hamilton-Jacobi equation $h(x, \varphi_x)=0$, $\varphi(0, x_2, \dots, x_n)=x_2$. Now, if $w_k(x)$ $(0 \leqslant k \leqslant m-1)$ has a pole along T, the Cauchy problem (1) has a unique solution in the form

$$u(x)=\sum_{i=1}^{m}\left\{\frac{F_i(x)}{[\varphi_i(x)]^{p_i}}+G_i(x)\log\varphi_i(x)\right\}+H(x),$$

where $F_i(x)$, $G_i(x)$ $(1 \leqslant i \leqslant m)$ and $H(x)$ are holomorphic functions in a neighborhood of $x=0$ and p_i $(1 \leqslant i \leqslant m)$ are integers >0. In order to obtain this solution, we set $u(x)$ in the form $u(x)=\sum_{i=1}^{m}\sum_{j=0}^{\infty} f_j(\varphi_i(x))u_{i,j}(x)$, where $f_0(t)$ is a function with a pole or a logarithmic singularity at $t=0$ chosen so that $u(x)$ satisfies the initial conditions. Thus we can determine the coefficients $u_{i,j}(x)$ by solving successively the equations (13) on each K_i $(1 \leqslant i \leqslant m)$ (Y. Hamada, *Publ. Res. Inst. Math. Sci.*, 5 (1969); C. Wagschal, *J. Math. Pures Appl.*, 51 (1972)). When the multiplicity of characteristic roots is more than 1, the situation is not the same. For example, consider the Cauchy problem

$$D_1^2 u - D_2 u=0, \qquad u(0, x_2)=\frac{1}{x^2},$$

$$D_1 u(0, x_2)=0.$$

The solution is

$$u(x)=\sum_{k=0}^{\infty}(-1)^n n!\, x_1^{2n}/(2n)!\, x_2^{n+1}$$

with essential singularities along $x_2=0$. This situation occurs quite generally. We factor $h(x, \xi)=h_1(x, \xi)^{r_1}\dots h_s(x, \xi)^{r_s}$, where h_i $(1 \leqslant i \leqslant s)$ are irreducible polynomials of degree m_i in ξ with holomorphic coefficients. We assume that the equation $\prod_{i=1}^{s} h_i(0, \xi_1, 1, 0, \dots, 0)=0$ has p distinct roots $(p=m_1+\dots+m_s)$. Then there exist p characteristic surfaces K_i $(1 \leqslant i \leqslant p)$ originating from T. We suppose more generally that the initial values $w_k(x)$ are multivalued functions ramified around T. Then the Cauchy problem (1) has a unique holomorphic solution on the universal covering space of $V - \bigcup_{i=1}^{p} K_i$, where V is a neighborhood of $x=0$. In fact, this is solved by transforming this problem into †integrodifferential equations. Such a method of solution is closely related to the method discussed in this section. See Hamada, Leray, and Wagschal (*J. Math. Pures Appl.*, 55 (1976)). In this case, even if the initial values have only poles, the solution in general may have essential singularities along $\bigcup_{i=1}^{p} K_i$, but if $a(x, D)$ satisfies Levi's condition ($a(x, D)$ is well decomposable), the solution does not yield essential singularities along $\bigcup_{i=1}^{p} K_i$. See J. De Paris (*J. Math. Pures Appl.*, 51 (1972)).

References

[1] R. Courant and D. Hilbert, Methods of mathematical physics II, Interscience, 1962.
[2] I. G. Petrovskiĭ, Lectures on partial differential equations, Interscience, 1954. (Original in Russian, second edition, 1953.)
[3] L. Hörmander, Linear partial differential operators, Springer, 1963.
[4] E. Goursat, Cours d'analyse mathématique, Gauthier-Villars, II, second edition, 1911; III, fourth edition, 1927.
[5] J. Hadamard, Leçons sur la propagation des ones et les équations de l'hydrodynamique, Hermann, 1903 (Chelsea, 1949).
[6] J. Hadamard, Le problème de Cauchy et les équations aux dérivées partielles linéaires hyperboliques, Hermann, 1932.
[7] C. Carathéodory, Calculus of variations and partial differential equations of first order, Holden-Day, 1965. (Original in German, 1935.)
[8] S. Mizohata, Theory of partial differential equations, Cambridge University Press, 1973. (Original in Japanese, 1965.)
[9] F. Treves, Linear partial differential equations with constant coefficients, Gordon & Breach, 1966.
[10] F. John, Partial differential equations, Springer, 1971.
[11] J. Dieudonné, Eléments d'analyse, 4, 7, 8, Gauthier-Villars, 1971–1978.

322 (XIII.20) Partial Differential Equations (Methods of Integration)

A. General Remarks

The methods of integrating partial differential equations are not as simple as those for †ordinary differential equations. For ordinary differential equations, we often obtain the desired solution by first finding the general solution containing several arbitrary constants and then specializing those constants to satisfy prescribed additional conditions. The situation is more complicated, however, for partial differential equations. In the formal general solution of a partial differential equation we have arbitrary functions instead of arbitrary constants, and there are cases where it is impossible, or very difficult, to find a suitable specialization of these functions so that the given additional conditions are fulfilled. For this reason, methods of solution are rather specific and are classified according to the types of additional conditions. In many cases, we assume that a problem is mathematically well posed, guess the possible solutions, and verify it directly.

A problem is said to be mathematically **well posed (properly posed** or **correctly posed)** if, under assigned additional conditions, the solution (i) exists, (ii) is uniquely determined, and (iii) depends continuously on the assigned data. By carefully examining problems in physics and engineering, we usually obtain many well-posed and important problems. In these problems, usually the data are sufficiently smooth functions. In these cases, part (iii) of the above definition (the continuous dependence of the solutions on the data) follows often from assumptions (i) and (ii). For example, †elliptic equations like †Laplace's equation $u_{xx}+u_{yy}=0$ for $u(x, y)$ describe laws of static or stationary phenomena such as the field of universal gravitation, the electrostatic field, the magnetostatic field, the steady flow of incompressible fluids without vortices, and the steady flow of electricity or heat. For this equation, †boundary value problems are well posed, but †initial value problems are not (→ 323 Partial Differential Equations of Elliptic Type). By contrast, for †hyperbolic equations like $u_{tt}-u_{xx}=0$ for $u(x,t)$ and †parabolic equations like $u_t-u_{xx}=0$ for $u(x,t)$ which control the change (in reference to time) of the various stationary phenomena, initial value problems or mixed problems with both boundary conditions and initial conditions are well posed (→ 325 Partial Differential Equations of Parabolic Type).

For the rest of this article we explain fundamental and typical methods of integration (→ Appendix A, Table 15).

B. The Lagrange-Charpit Method

For the partial differential equation of the first order

$$F(x,y,u,p,q)=0, \qquad p=\frac{\partial u}{\partial x}, \qquad q=\frac{\partial u}{\partial y}, \tag{1}$$

we consider a system of ordinary differential equations called †characteristic differential equations:

$$\frac{dx}{F_p}=\frac{dy}{F_q}=\frac{du}{pF_p+qF_q}=\frac{-dp}{F_x+pF_u}=\frac{-dq}{F_y+qF_u}. \tag{2}$$

If we obtain at least one †integral of this system containing p, q, and an arbitrary constant a in the form

$$G(x,y,u,p,q)=a, \tag{3}$$

and if we find p and q from (1) and (3), then $du=p\,dx+q\,dy$ is an †exact differential form, and by integrating it we get a solution $\Phi(x,y,u,a,b)$

$=0$ of (1). A solution of (1) containing two arbitrary constants is called a †complete solution. Here, setting $b=g(a)$ and eliminating a from the two equations $\Phi(x,y,u,a,g(a))=0$ and $\Phi_a+\Phi_b g'(a)=0$, we obtain a solution involving an arbitrary function. Such a solution is called a †general solution of (1). For example, consider $pq=1$. Since the characteristic differential equations are

$$\frac{dx}{q}=\frac{dy}{p}=\frac{du}{2pq}=\frac{-dp}{0}=\frac{-dq}{0},$$

we have $p=a$ (constant), and hence from the original equation we get $q=a^{-1}$. Then $du=a\,dx+a^{-1}\,dy$ is an exact differential form, and by integrating it a complete solution $u=ax+a^{-1}y+b$ is obtained. A general solution is found by eliminating a from $u=ax+a^{-1}y+g(a)$ and $0=x-a^{-2}y+g'(a)$.

Likewise, when we have n independent variables $x_1,\ldots,x_n$ in the equation

$$F(x_1,\ldots,x_n,u,p_1,\ldots,p_n)=0, \qquad p_i=\frac{\partial u}{\partial x_i}, \qquad (1')$$

we can use the †characteristic differential equations to find a complete solution and a general solution (the **Lagrange-Charpit method**; → 82 Contact Transformations).

C. Separation of Variables and the Principle of Superposition

The simplest and most useful method is the **separation of variables**. Concerning the equation $u_x^2+u_y^2=1$ in $u(x,y)$, for example, by setting $u=\varphi(x)+\psi(y)$ we obtain $\varphi'(x)^2+\psi'(y)^2=1$ or $\varphi'(x)^2=1-\psi'(y)^2$. Since the right-hand side is independent of x and the left-hand side is independent of y, both sides must be equal to the same constant α^2. From this we get a (complete) solution involving two arbitrary constants α, β:

$$u=\alpha x+\sqrt{1-\alpha^2}\;y+\beta.$$

For †linear equations, it is often effective to write the solution as a product $u=\varphi(x)\psi(y)$. From the †heat equation $u_y=u_{xx}$, we obtain by this process a relation $\psi'(y)/\psi(y)=\varphi''(x)/\varphi(x)$, from which we get a particular solution $u=ae^{-v^2y}\sin v(x-\alpha)$ containing a parameter v.

Next, when the equation is linear and homogeneous, by forming a linear combination of particular solutions that correspond to various values of a parameter v, we obtain a new solution (the **principle of superposition**). For example, by integrating the solution $e^{-v^2y}\cos vx$ with respect to the parameter v between the limits $-\infty$ and ∞ (namely, by a superposition consisting of a linear combination and a limiting process), we obtain a new solution

$$u=\int_{-\infty}^{\infty}e^{-v^2y}\cos vx\,dv=\sqrt{\frac{\pi}{y}}\exp\left(-\frac{x^2}{4y}\right) \qquad (y>0), \qquad (4)$$

which is the †fundamental solution of the heat equation. This name refers to the fact that the function (4) can be used to obtain solutions of the heat equation under some initial conditions. More exactly, a solution of $u_y-u_{xx}=0$ that is a function of class C^2 in $y>0$, is continuous in $y\geqslant 0$, and coincides with a bounded continuous function $\varphi(x)$ on $y=0$ can be obtained by a superposition of the solution (4) such as

$$u(x,y)=\frac{1}{2\sqrt{\pi y}}\int_{-\infty}^{\infty}\varphi(\xi)\exp\left(-\frac{(x-\xi)^2}{4y}\right)d\xi.$$

The method of separation of variables applied after a suitable transformation of variables is often successful. In particular, by using orthogonal coordinates, †polar coordinates, or †cylindrical coordinates according to the form of boundary, we often obtain satisfactory results. For example, concerning the boundary value problem for Laplace's equation $\Delta u=u_{xx}+u_{yy}=0$, which is smooth in the circle $r^2=x^2+y^2<1$ and takes the value of a given continuous function $g(\theta)$ on the circumference $r=1$, we can use polar coordinates to rewrite the equation in the form

$$\Delta u=u_{rr}+\frac{u_r}{r}+\frac{1}{r^2}u_{\theta\theta}=0$$

and apply the method of separation of variables to obtain particular solutions $r^n\cos n\theta$, $r^n\sin n\theta$. Hence it is reasonable to suspect that by a superposition of these particular solutions we can obtain the desired solution:

$$u(x,y)=\frac{a_0}{2}+\sum_{n=1}^{\infty}(a_n\cos n\theta+b_n\sin n\theta)r^n.$$

In fact, this series is a desired solution if the coefficients a_n, b_n can be chosen so that the series converges uniformly for $0\leqslant r\leqslant 1$ and can be differentiated twice term by term for $0\leqslant r<1$, and if we have

$$g(\theta)=\frac{a_0}{2}+\sum_{n=1}^{\infty}(a_n\cos n\theta+b_n\sin n\theta).$$

By virtue of the uniqueness of the solution of an elliptic equation, this is the unique desired solution. The boundary value problem in the preceeding paragraph is well posed.

D. Mixed Problems

For linear homogeneous equations of hyperbolic or parabolic type, **mixed problems** frequently appear, i.e., problems in which both initial and boundary conditions are assigned. These problems are, furthermore, classified into two types.

For the first type, homogeneous boundary conditions are assigned. For example, in vibration problems of a nonhomogeneous string between $0 \leqslant x \leqslant l$, we must find the solution of the equation

$$(T(x)u_x)_x = \rho u_{tt}$$

satisfying an initial condition $u = f(x)$, $u_t = g(x)$ for $t = 0$, under homogeneous boundary conditions: (i) $u = 0$ for $x = 0$ and $x = l$ in the case of two fixed ends; (ii) $u_x = 0$ for $x = 0$ and $x = l$ in the case of two free ends; (iii) $-u_x + \sigma_0 u = 0$ for $x = 0$, and $u_x + \sigma_1 u = 0$ for $x = l$ ($\sigma_0 > 0$, $\sigma_1 > 0$), where the two ends are tied elastically to the fixed points; (iv) $T(0)u(0) = T(l)u(l)$, $T(0)u'(0) = T(l)u'(l)$ (periodicity condition); (v) u, u' are finite at $x = 0$ and $x = l$ (regularity condition).

The method of separation of variables is applicable to problems of this type also. For example, suppose that we have two fixed ends. Disregarding the initial condition for a while, we find a particular solution fulfilling only the boundary condition by setting $u(x,t) = y(x)\exp ivt$. Here, $y(x)$ must satisfy

$$(T(x)y')' + v^2 \rho(x) y = 0, \quad y(0) = y(l) = 0. \tag{5}$$

Except when the solution is trivial (i.e., $u(x,t) = 0$), $y(x) \not\equiv 0$ must hold. But in the special case $T(x) \equiv 1$, $\rho(x) \equiv 1$, $l = \pi$, the values of v for which functions of this kind exist are only $1, 2, 3, \ldots$ (the corresponding $y(x)$ is $\sin vx$).

Also, in more general cases, nontrivial solutions $y(x)$ exist only for some discrete values of v. These v are called †eigenvalues of (5), and the corresponding solutions $y(x)$ are called †eigenfunctions for the eigenvalues v. That is, the desired particular solution can be obtained by solving the †eigenvalue problem (5). Again, when $T(x) \equiv \rho(x) \equiv 1$, $l = \pi$, particular solutions are $\sin nx \exp int$, $\sin nx \exp(-int)$ ($n = 1, 2, \ldots$). We consider

$$u(x,t) = \sum_{n=1}^{\infty} (a_n \cos nt + b_n \sin nt) \sin nx, \tag{6}$$

which is obtained by a superposition of these particular solutions. If we can determine the coefficients a_n, b_n so that this series converges uniformly and is twice differentiable term by term, and

$$f(x) = \sum_{n=1}^{\infty} a_n \sin nx, \quad g(x) = \sum_{n=1}^{\infty} n b_n \sin nx,$$

then the series (6) is a solution of the mixed problem in question. If the uniqueness of the solution of the mixed problem is proved, it is not necessary to look for any solution other than the one obtained by combining the method of separation of variables and the principle of superposition.

Furthermore, the method described in this section is applicable to solving the following nonhomogeneous equation, which characterizes the motion of a string under the influence of an external force $f(x,t)$:

$$u_{tt} - u_{xx} = f(x,t) \tag{7}$$

under the boundary condition for the first type of mixed problem and the initial condition $u = u_t = 0$ for $t = 0$. In this case, we expand the unknown u and the function $f(x,t)$ in terms of the system of eigenfunctions $\{\sin nx\}$, and by substituting

$$u = \sum_{n=1}^{\infty} a_n(t) \sin nx, \qquad f(x,t) = \sum_{n=1}^{\infty} A_n(t) \sin nx$$

into (7), we reduce the problem to determining $a_n(t)$ ($n = 1, 2, \ldots$). When the external force varies with a harmonic oscillation over time as in $f(x,t) = -\varphi(x)\exp(-i\omega t)$, a similar method can be applied by setting $u = v(x)\exp(-i\omega t)$.

For mixed problems of the second type, the homogeneous initial condition $u = 0$, $u_t = 0$ for $t = 0$ is assigned, but the boundary condition is nonhomogeneous. For example, when an oscillating string is at rest until $t = 0$, and for $t > 0$ its right end is fixed and its left end moves subject to an assigned rule, the behavior of the string is described by the solution of $u_{tt} - u_{xx} = 0$ under the boundary condition $u(0,t) = f(t)$, $u(l,t) = 0$ ($t > 0$). This is called a **transient problem**. If we now choose an arbitrary function $B(x,t)$ that fulfills all the boundary and initial conditions, and if we set $u - B(x,t) = v(x,t)$ and $B_{xx} - B_{tt} = f(x,t)$, then v satisfies

$$v_{tt} - v_{xx} = f(x,t) \qquad \text{for } t > 0$$

with $v = v_t = 0$ for $t = 0$; $v = 0$ for $x = 0$ and $x = l$. Then $v(x,t)$ describes the oscillation of a string that is at rest until $t = 0$ and moves under the effect of an external force represented by $f(x,t)$ for $t > 0$. Problems of this kind often appear in electrical engineering.

Such problems can be reduced to problems of the first type in the manner described in the previous paragraph, but there are some direct methods that are more effective, the first being **Duhamel's method**. Consider the case where $f(t)$ is the unit impulse function:

$$f(t) = \begin{cases} 1, & t > 0, \\ 0, & t < 0, \end{cases}$$

and let $U(x,t)$ be a solution corresponding to this case and vanishing for $t \leqslant 0$, $x > 0$. Then a solution corresponding to the general case is given by

$$u = \int_0^t \frac{\partial}{\partial t} U(x, t - \tau) f(\tau)\, d\tau.$$

The second method is based on application of the †Laplace transformation. Denoting the Laplace transform of a solution $u(x,t)$ by

$$\int_0^\infty ue^{-pt}\,dt=\frac{v(x,p)}{p},$$

we multiply both sides of $u_{tt}-u_{xx}=0$ by e^{-pt} and integrate the result with respect to t from 0 to ∞. Then, taking account of the initial condition, we have

$$v_{xx}-p^2v=0$$

with the boundary condition

$$v(0,p)=p\int_0^\infty f(t)e^{-pt}\,dt, \qquad v(l,p)=0.$$

If for $p=\alpha+i\beta$ $(\alpha>\alpha_0)$ we can find a solution $v(x,p)$ of this boundary value problem for the ordinary differential equation, then the desired solution is given by

$$u(x,t)=\frac{1}{2\pi i}\int_L \frac{v(x,p)}{p}e^{pt}\,dp,$$

where L is a path in $\alpha>\alpha_0$ parallel to the imaginary axis. This is called the **Bromwich integral**.

E. Green's Formula and Application of Fundamental Solutions

Given a linear partial differential operator

$$L[u]\equiv\sum_p a_p(x)D^p u, \quad p=(p_1,\ldots,p_n),$$

$$D^p=\partial^{p_1+\ldots+p_n}/\partial x_1^{p_1}\ldots dx_n^{p_n},$$

we call the operator

$$L^*[v]\equiv\sum_p(-1)^{p_1+\ldots+p_n}D^p(\bar{a}_p(x)v)$$

the **adjoint operator** of L, and the operator

$${}^tL[v]\equiv\sum_p(-1)^{p_1+\ldots+p_n}D^p(a_p(x)v)$$

the **transposed operator** of L. Sometimes ${}^tL$ is also called the adjoint operator of L. In the complex Hilbert space, the adjoint operators are more appropriate than the transposed operators. In this case, the operator L^* defined above is usually called the **formal adjoint operator** to distinguish it from the one defined by

$$(L[u],v)=(u,L^*[v]) \text{ for all } u\in D(L),$$

$D(L)$ being the domain of the definition of L (→ 251 Linear Operators). These transposed or adjoint operators are often used to represent (at least locally) the solutions u of $L[u]=f$.

We explain in more detail the specific case where the number of independent variables is 2. For the linear partial differential operator

$$L(u)=A(x,y)\frac{\partial^2 u}{\partial x^2}+2B(x,y)\frac{\partial^2 u}{\partial x\,\partial y}+C(x,y)\frac{\partial^2 u}{\partial y^2}$$

$$+D(x,y)\frac{\partial u}{\partial x}+E(x,y)\frac{\partial u}{\partial y}+F(x,y)u$$

and its adjoint partial differential operator

$$M(v)=\frac{\partial^2(Av)}{\partial x^2}+2\frac{\partial^2(Bv)}{\partial x\,\partial y}+\frac{\partial^2(Cv)}{\partial y^2}$$

$$-\frac{\partial(Dv)}{\partial x}-\frac{\partial(Ev)}{\partial y}+Fv,$$

we have †Green's formula:

$$\iint_D (vL(u)-uM(v))\,dx\,dy$$

$$=-\int_{\partial D}\left(P\frac{\partial x}{\partial n}+Q\frac{\partial y}{\partial n}\right)ds, \tag{8}$$

where

$$P=v\{Au_x+Bu_y\}-u\{(Av)_x+(Bv)_y\}+Duv,$$

$$Q=v\{Bu_x+Cu_y\}-u\{(Bv)_x+(Cv)_y\}+Euv, \tag{9}$$

and ∂D denotes the boundary curve of the domain D, n the internal normal of ∂D at a point of ∂D, and s the arc length.

We can apply this formula for solving a nonhomogeneous equation $L(u)=f$ as follows. Assume that there exists a solution of $L(u)=f$ satisfying the assigned additional condition, and choose a †fundamental solution of $M(v)=0$ having an adequate singularity at a point (x_0,y_0) of D and fulfilling a suitable boundary condition. Then, if we substitute these solutions u and v into (8), we obtain an explicit representation of $u(x_0,y_0)$. If we can verify that the function $u(x,y)$ thus obtained is a solution of $L(u)=f$ fulfilling the assigned additional conditions, then we see, under the assumption of uniqueness of solutions, that this and only this function $u(x,y)$ is the desired solution. For example, consider a boundary problem for $L(u)=\frac{\partial^2 u}{\partial x^2}+\frac{\partial^2 u}{\partial y^2}$, for which $M(v)=\frac{\partial^2 v}{\partial x^2}+\frac{\partial^2 v}{\partial y^2}$. The problem is, for a circle of radius r with center at the origin, to find a function $u(x,y)$ that is continuous in the interior and on the circumference C_r of the circle, and that satisfies $L(u)=f$ in the interior of the circle (f is bounded and continuous in the interior of the circle) and is equal to a given continuous function g on the circumference C_r. In this case, we consider the circumference K_ε of sufficiently small radius ε with center (x_0,y_0) contained in the interior of the first circle, and set

$$v(x,y)=(1/2\pi)\log 1/\rho+h(x,y),$$

$$\rho=((x-x_0)^2+(y-y_0)^2)^{1/2}. \tag{10}$$

We apply formula (8) to the domain D_ε enclosed by the circumferences K_ε and C_r. If, in (10), h satisfies $M(v)=0$ in the interior of the circle enclosed by C_r and vanishes on C_r, then we get

$$\iint_{D_\varepsilon} vf\,dx\,dy=-\int_{K_\varepsilon}(vu_n-uv_n)\,ds+\int_{C_r} gv\,ds.$$

By letting ε tend to zero, the first term on the right-hand side yields

$$-u(x_0,y_0)\int_0^{2\pi}\frac{1}{2\pi}\frac{\partial\log\rho}{\partial\rho}\rho\,d\theta=-u(x_0,y_0),$$

by the logarithmic singularity of v at the point (x_0,y_0). Therefore we have an explicit representation of u,

$$u(x_0,y_0)=\int_{C_r} gv_n\,ds-\iint_{x^2+y^2\leqslant r^2} vf\,ds\,dy, \qquad (11)$$

and it is easily verified that this is the desired solution.

As stated in the previous paragraph, to apply Green's formula it is necessary to find a solution v of $M(v)=0$ possessing a fundamental singularity as the logarithmic singularity, i.e., a so-called fundamental solution. As fundamental singularities for

$$M(v)=\frac{\partial v}{\partial y}-\frac{\partial^2 v}{\partial x^2} \quad\text{and}\quad M(v)=\frac{\partial^2 v}{\partial y^2}-\frac{\partial^2 v}{\partial x^2}$$

of parabolic type and of hyperbolic type, we must take those given respectively by

$$v(x,y)=\begin{cases}\dfrac{1}{2\sqrt{\pi(y-y_0)}}\exp\left(-\dfrac{(x-x_0)^2}{4(y-y_0)}\right), & y>y_0,\\ 0, \quad y\leqslant y_0,\end{cases}$$

and

$$v(x,y)=\begin{cases}1/2, & |x-x_0|<y-y_0,\\ 0, & |x-x_0|\geqslant y-y_0\end{cases}$$

(→ 323 Partial Differential Equations of Elliptic Type; 325 Partial Differential Equations of Hyperbolic Type; 327 Partial Differential Equations of Parabolic Type).

References

[1] R. Courant and D. Hilbert, Methods of mathematical physics, Interscience, I, 1953; II, 1962.
[2] E. Goursat, Cours d'analyse mathématique, Gauthier-Villars, III, 1927.
[3] A. G. Webster and G. Szegö, Partielle Differentialgleichungen der mathematischen Physik, Teubner, 1930.
[4] L. Schwartz, Mathematics for the physical sciences, Hermann, revised edition, 1966.
[5] J. L. Lions, Equations différentielles opérationelles et problèmes aux limites, Springer, 1961.
[6] A. Friedman, Partial differential equations, Holt, Rinehart and Winston, 1969.

323 (XIII.24) Partial Differential Equations of Elliptic Type

A. General Remarks

Suppose that we are given a †linear second-order partial differential equation

$$L[u]\equiv\sum_{i,j=1}^{n}a_{ij}(x)\frac{\partial^2 u}{\partial x_i\partial x_j}+\sum_{i=1}^{n}b_i(x)\frac{\partial u}{\partial x_i}+c(x)u$$
$$=f(x), \qquad (1)$$

where $x=(x_1,\ldots,x_n)$, $a_{ji}=a_{ij}$. If the quadratic form $\sum a_{ij}\xi_i\xi_j$ in ξ is †positive definite at every point x of a domain G, this equation (or the operator L) is said to be **elliptic** (or of **elliptic type**) in G. For $n=2$, $a_{11}(x)a_{22}(x)-(a_{12}(x))^2>0$ is the condition of ellipticity. In this case, by a change of independent variables, the equation is transformed locally into the canonical form

$$\frac{\partial^2 u}{\partial x^2}+\frac{\partial^2 u}{\partial y^2}+b_1(x,y)\frac{\partial u}{\partial x}+b_2(x,y)\frac{\partial u}{\partial y}+c(x,y)u$$
$$=f(x,y).$$

The operator $\sum_{i=1}^n\partial^2/\partial x_i^2$, denoted by Δ, is called **Laplace's operator** (or the **Laplacian**). The simplest examples of elliptic equations are $\Delta u=0$ (**Laplace's differential equation**) and $\Delta u=f(x)$ (**Poisson's differential equation**) (→ Appendix A, Table 15).

B. Fundamental Solutions

Let K be the n-dimensional ball with radius R, center x_0, boundary Ω (an $(n-1)$-dimensional sphere), and area S_n, and let r be the distance from x_0 to x. Then for any function $u(x)$ of class C^2 we have

$$u(x_0)=\frac{1}{S_n}\int_\Omega u\,dS-\frac{R^{n-1}}{(n-2)S_n}$$
$$\times\int_K(r^{2-n}-R^{2-n})\Delta u\,dx \qquad \text{for } n>2,$$

$$u(x_0)=\frac{1}{2\pi}\int_\Omega u\,dS-\frac{1}{2\pi}\int_K\log\frac{R}{r}\Delta u\,dx$$
$$\text{for } n=2.$$

Thus, if u is a solution of Poisson's equation $\Delta u = f(x)$, we have a representation of $u(x_0)$ by replacing Δu by f in the integrals just given. Next, concerning the solutions of the equation $\Delta u + cu = 0$ ($c > 0$, constant), the following relation holds:

$$\frac{1}{S_n}\int_{\Omega} u\,dS = \frac{\Gamma(n/2)J_{n'}(R\sqrt{c})}{(R\sqrt{c}/2)^{n'}}u(x_0),$$

where Γ is the †gamma function, $n' = (n-2)/2$, and J_ν is the †Bessel function of order ν.

Now, if we put

$$V(x,\xi) = \begin{cases} \left(\sum_{i=1}^{n}(x_i-\xi_i)^2\right)^{1-n/2} = r^{2-n}, & n \geqslant 3, \\ \log\left(\sum_{i=1}^{2}(x_i-\xi_i)^2\right)^{-1/2} = \log\frac{1}{r}, & n = 2, \end{cases}$$

then the function $u(x)$ defined by

$$u(x) = -\frac{1}{\omega_n}\int_G V(x,\xi)f(\xi)\,d\xi,$$

$$\omega_n = 2\pi^{n/2}\frac{n-2}{\Gamma(n/2)}$$

represents a †particular solution of $\Delta u = f(x)$, where $V(x,\xi)$ as a function of the variables x satisfies $\Delta V(x,\xi) = 0$ except at $x = \xi$.

Consider now the more general case (1). A function $E(x,\xi)$ is called a **fundamental solution** (or **elementary solution**) of (1) or of L if

$$u(x) = \int_G E(x,\xi)f(\xi)\,d\xi$$

provides a solution of (1) for any $f(x) \in C^\infty$ with compact support. A fundamental solution $E(x,\xi)$ is a solution of the equation $L[E] = 0$ having a singularity at $x = \xi$ of the form $-\omega_n^{-1}\sqrt{a(\xi)}\,r^{2-n}$ ($n \geqslant 3$) or $\omega_2^{-1}\sqrt{a(\xi)}\log r$ ($n = 2$), where $a(\xi) = \det(a^{ij}(\xi))$ and $r = (\sum a^{ij}(\xi)(x_i-\xi_i)(x_j-\xi_j))^{1/2}$ and (a^{ij}) is the inverse matrix of (a_{ij}).

Roughly, there are three different methods of constructing fundamental solutions. The first and most general is to use pseudodifferential operators (→ 345 Pseudodifferential Operators). The second is that of J. Hadamard [1], which uses the geodesic distance between two points x and ξ with respect to the Riemannian metric $\sum a^{ij}(x)\,dx_i\,dx_j$. This is important in applications of elliptic equations to geometry. The third method is due to E. E. Levi [2] and is as follows: Let

$$V(x,\xi) = \begin{cases} r^{2-n}, & n \geqslant 3, \\ -\log r, & n = 2. \end{cases}$$

To obtain a solution of $L[u] = f(x)$ we set

$$u = \int_G V(x,\xi)\varphi(\xi)\sqrt{a(\xi)}\,d\xi.$$

Writing $L[V(x,\xi)] = \chi(x,\xi)$, we obtain the following †integral equation of Fredholm type in $\varphi(x)$:

$$\varphi(x) - \frac{1}{\omega_n}\int_G \chi(x,\xi)\varphi(\xi)\sqrt{a(\xi)}\,d\xi = \frac{1}{\omega_n}f(x).$$

If we denote the †resolvent of this equation by $\bar{\chi}(x,\xi)$ and put

$$\gamma(x,\xi) = V(x,\xi) - \omega_n\int_G V(x,\xi')\bar{\chi}(\xi',\xi)\sqrt{a(\xi')}\,d\xi',$$

then $E(x,\xi) = -\gamma(x,\xi)\sqrt{a(\xi)}/\omega_n$ is seen to be a fundamental solution of L. Thus Levi's method enables us to construct a fundamental solution locally by successive approximation, because the integral equation in φ as above has a unique solution expressible by Neumann series if the domain G is small enough (→ 189 Green's Operator).

C. The Dirichlet Problem

Let G be a bounded domain with boundary Γ. We call the problem of finding a solution u of the given elliptic equation in G that is continuous on $G \cup \Gamma$ and takes the assigned continuous boundary values on T the **first boundary value problem** (or **Dirichlet problem**). In particular, the Dirichlet problem for $\Delta u = 0$ has been studied in detail (→ 120 Dirichlet Problem).

If $c(x) < 0$ and $f(x) \leqslant 0$ or if $c(x) \leqslant 0$ and $f(x) < 0$ in (1), a solution u does not attain its local negative minimum in G. This is called the **strong maximum principle** (→ [3, 4] for **Hopf's maximum principle** and **Giraud's theorem**). The maximum principle is one of the most powerful tools available for the treatment of elliptic equations of the second order with real coefficients. From this it follows that the solution of the Dirichlet problem for (1) is unique if $c(x) < 0$. Furthermore, concerning the uniqueness of the solution, we have the following criterion: If there exists a function $\omega(x) > 0$ of class C^2 in G and continuous on $G \cup \Gamma$ such that $L[\omega(x)] < 0$, then the solution of the Dirichlet problem is unique.

Let G be a †regular domain in the plane. If we denote †Green's function of Δ relative to the Dirichlet problem in G by $K(x,y;\xi,\eta)$, then the solution u of $\Delta u = f(x,y)$ vanishing on Γ is given by

$$u(x,y) = -\frac{1}{2\pi}\iint_G K(x,y;\xi,\eta)f(\xi,\eta)\,d\xi\,d\eta,$$

where $f(x,y)$ satisfies a †Hölder condition.

The Dirichlet problem for

$$L[u] \equiv \Delta u + a(x,y)\frac{\partial u}{\partial x} + b(x,y)\frac{\partial u}{\partial y} + c(x,y)u = f(x,y)$$

with prescribed boundary values reduces to the problem in the previous paragraph. In fact, let $h(x, y)$ be a function that coincides with the prescribed boundary value on Γ. Then if we put $u=h(x, y)+v$, the problem is reduced to finding a solution v satisfying an equation similar to the one for $L[u]$ (replacing f by $f-L[h]$) and vanishing on Γ. Now suppose that Γ consists of a finite number of †Jordan curves whose †curvatures vary continuously, a and b are continuously differentiable, and c and f satisfy Hölder conditions. Let $K(x, y; \xi, \eta)$ be Green's function for the Dirichlet problem relative to Δ. To find a solution of $L[u]=f$ vanishing on Γ we set

$$u=-\frac{1}{2\pi}\iint_G K(x, y; \xi, \eta)\rho(\xi, \eta)\, d\xi\, d\eta.$$

Then ρ is a solution of the integral equation

$$\rho(x, y)+\iint_G H(x, y, \xi, \eta)\rho(\xi, \eta)\, d\xi\, d\eta=f(x, y),$$

where $H(x, y, \xi, \eta)=(-1/2\pi)\{a(x, y)K_x(x, y; \xi, \eta)+b(x, y)K_y(x, y; \xi, \eta)+c(x, y)K(x, y; \xi, \eta)\}$. Therefore we have the following alternatives: Either $L[u]=f$ has a unique solution for any f and any given boundary values on Γ, or $L[u]=0$ has nontrivial solutions vanishing on Γ (in this case the number of linearly independent solutions is finite).

More general equations of type (1) have been studied by J. Schauder and others. Schauder proved, first, that when the $b_i(x)$ and $c(x)$ are zero, $a_{ij}(x)$ and $f(x)$ satisfy Hölder conditions, and the boundary Γ of G is of class C^2, then there exists a unique solution $u(x)$ vanishing on Γ. Next, when the $b_i(x)$ and $c(x)$ are continuous and a_{ij} and f satisfy Hölder conditions, he showed the following alternatives: Either $L[u]=f$ admits a unique solution vanishing on Γ for every f, or $L[u]=0$ has nontrivial solutions vanishing on Γ (in this case, the number of linearly independent solutions is finite).

In (1), suppose that a_{ij}, b_i, and c are Hölder continuous of exponent α $(0<\alpha<1)$ uniformly on $\bar{G}$ and that Γ is of class $C^{2+\alpha}$. Then we have the following inequality (**Schauder's estimate**) for any $u\in C^{2+\alpha}(\bar{G})$:

$$\|u\|_{2+\alpha,\bar{G}}\leqslant K_1(\|L[u]\|_{\alpha,\bar{G}}+\|u\|_{2+\alpha,\Gamma})+K_2\|u\|_{0,\bar{G}}, \qquad (2)$$

where $\|\cdot\|_{k,S}$ stands for the norm in the function space $C^k(S)$ (→ 168 Function Spaces). K_1 and K_2 are positive constants depending on L, G, and α but independent of u. More precisely, K_1 depends only on the ellipticity constant λ of L defined as the smallest number $\geqslant 1$ such that

$$\lambda^{-1}|\xi|^2\leqslant\sum a_{ij}(x)\xi_i\xi_j\leqslant\lambda|\xi|^2 \qquad (3)$$

for any $(x, \xi)\in\bar{G}\times\mathbf{R}^n$. The inequality (2) is one of the most important **a priori estimates** in the theory of elliptic equations [5, 6] (→ inequality (9) in Section H).

D. Quasilinear Partial Differential Equations

Consider the second-order partial differential equation in $u(x_1, \ldots, x_n)$:

$$F(x_1, \ldots, x_n, u, p_1, \ldots, p_n, p_{11}, \ldots, p_{ij}, \ldots, p_{nn})=0,$$

where $p_i=\partial u/\partial x_i$, $p_{ij}=\partial^2 u/\partial x_i\partial x_j$. If, for a solution $u(x)$, $\sum_{i,j=1}^n(\partial F/\partial p_{ij})\xi_i\xi_j$ is a positive definite form, we say that the equation is **elliptic** at $u(x)$. Moreover, if for any values of u, p_i, p_{ij} this quadratic form is positive definite, we say simply that the equation is of **elliptic type**. The equation is called **quasilinear** if F is linear in p_{ij}. For example, the equation of †minimal surfaces (→ 334 Plateau's Problem)

$$(1+p_2^2)p_{11}-2p_1p_2p_{12}+(1+p_1^2)p_{22}=0$$

is a quasilinear elliptic equation. Furthermore,

$$\Delta u=f(x, y, u, u_x, u_y)$$

is also elliptic. E. Picard solved this equation by the **method of successive approximation** [7]. Specifically, let $h(x, y)$ be the †harmonic function taking the assigned boundary values on Γ. Starting from $u_0(x, y)=h(x, y)$, we define the functions $u_n(x, y)$ successively as the solutions of

$$\Delta u_n=f(x, y, u_{n-1}, \partial u_{n-1}/\partial x, \partial u_{n-1}/\partial y),$$

coinciding with h on Γ. Let $K(x, y; \xi, \eta)$ be Green's function in G and in the region defined by $|u-h(x, y)|\leqslant A$, $|p-h_x|\leqslant B$, $|q-h_y|\leqslant B$, $(x, y)\in G$, and let the supremum of $|f(x, y, u, p, q)|$ be C. Assume now that

$$\frac{C}{2\pi}\iint_G K\, d\xi\, d\eta\leqslant A, \qquad \frac{C}{2\pi}\iint_G |K_x|\, d\xi\, d\eta\leqslant B,$$

$$\frac{C}{2\pi}\iint_G |K_y|\, d\xi\, d\eta\leqslant B,$$

and that f satisfies a Hölder condition in (x, y). Moreover, let

$$|f(x, y, u', p', q')-f(x, y, u, p, q)|$$

$$\leqslant L|u-u'|+L'(|p'-p|+|q'-q|).$$

Assume finally that

$$\iint_G (LK+L'(|K_x|+|K_y|))\, d\xi\, d\eta\leqslant\gamma<1.$$

Under these assumptions, $\{u_n(x, y)\}$ is uniformly convergent, and the limit $u(x, y)$ coinciding with $h(x, y)$ on Γ satisfies

$$\Delta u=f(x, y, \partial u/\partial x, \partial u/\partial y).$$

Furthermore, it is known that the solution is unique within the region mentioned above. This method can be applied when G is small and the values of h_x and h_y are limited.

When f does not contain p and q, the following method is known. Let $\underline{\omega}(x, y)$ and $\overline{\omega}(x, y)$ both be continuous on $G \cup \Gamma$ and of class C^2 in G. Suppose that $f(x, y, u)$ satisfies a Hölder condition in $\underline{\omega}(x, y) \leqslant u \leqslant \overline{\omega}(x, y)$, and that

$$\Delta\underline{\omega} \geqslant f(x, y, \underline{\omega}(x, y)), \qquad \Delta\overline{\omega} \leqslant f(x, y, \overline{\omega}(x, y)).$$

Then, given a continuous function φ on Γ such that $\underline{\omega} \leqslant \varphi \leqslant \overline{\omega}$, there exists a unique solution u of $\Delta u = f(x, y, u)$ such that

$$\underline{\omega}(x, y) \leqslant u(x, y) \leqslant \overline{\omega}(x, y) \text{ in } G,$$

$$u(x, y) = \varphi \text{ on } \Gamma.$$

Finally, consider the equation

$$A\frac{\partial^2 u}{\partial x^2} + 2B\frac{\partial^2 u}{\partial x \partial y} + C\frac{\partial^2 u}{\partial y^2} = 0,$$

where A, B, and C are functions of x, y, u, p, q and $AC - B^2 > 0$. Under the following conditions, there exists a solution of the Dirichlet problem: A, B, C are of class C^2, and their derivatives of order 2 always satisfy Hölder conditions; G is †convex; and the boundary value φ along Γ considered as a curve in xyu-space represented by the parameter of arc length is of class $C^{3+\alpha}$ $(\alpha > 0)$. Moreover, any plane having 3 common points with this curve has slope less than a fixed number Λ. The proof of this theorem is carried out in the following way: For any function u satisfying

$$|u(x, y)| \leqslant \max|\varphi|, \qquad |u_x| \leqslant \Lambda, \qquad |u_y| \leqslant \Lambda,$$

replacing u, p, q by $u(x, y)$, $u_x(x, y)$, $u_y(x, y)$, respectively, in A, B, and C, we have the linear equation in v:

$$A[u]\frac{\partial^2 v}{\partial x^2} + 2B[u]\frac{\partial^2 v}{\partial x \partial y} + C[u]\frac{\partial^2 v}{\partial y^2} = 0.$$

We can obtain the solution v taking the boundary value φ on Γ. Thus we have a mapping $u \to v$. Applying the †fixed-point theorem in function space to this mapping, we have the desired solution $v(x, y) = u(x, y)$.

Concerning the Dirichlet problem for the second-order **semilinear** elliptic partial differential equation

$$\sum_{i,j=1}^{n} a_{ij}(x)\frac{\partial^2 u}{\partial x_i \partial x_j} = f\left(x, u, \frac{\partial u}{\partial x_1}, \ldots, \frac{\partial u}{\partial x_n}\right),$$

a work by M. Nagumo (*Osaka Math. J.*, 6 (1954)) establishes a general existence theorem.

For the general nonlinear equation

$$F(x_1, \ldots, x_n, u, p_1, \ldots, p_n, p_{11}, \ldots, p_{ij}, \ldots, p_{nn}) = 0,$$

if $\sum(\partial F/\partial p_{ij})\xi_i\xi_j > 0$, $F_u \leqslant 0$, the solution of the Dirichlet problem for this equation is unique. Even when $F_u \leqslant 0$ does not hold, the conclusion remains the same if we can reduce the equation to this case by a suitable change of variables.

Quasilinear equations in **divergence form**

$$\sum_{i=1}^{n} \frac{\partial}{\partial x_i} a_i\left(x, u, \frac{\partial u}{\partial x_1}, \ldots, \frac{\partial u}{\partial x_n}\right)$$
$$= f\left(x, u, \frac{\partial u}{\partial x_1}, \ldots, \frac{\partial u}{\partial x_n}\right), \tag{4}$$

or more generally, any quasilinear elliptic equation

$$\sum_{i,j=1}^{n} a_{ij}\left(x, u, \frac{\partial u}{\partial x_1}, \ldots, \frac{\partial u}{\partial x_n}\right)\frac{\partial^2 u}{\partial x_i \partial x_j}$$
$$= f\left(x, u, \frac{\partial u}{\partial x_1}, \ldots, \frac{\partial u}{\partial x_n}\right), \tag{5}$$

and even quasilinear elliptic systems have been treated in detail in several recent works [8, 9]. J. Serrin [10] treated the Dirichlet problem and established the existence and the uniqueness of solutions for some classes of equations of type (5) containing the minimal surface equation. His method is to estimate the maximum norms of u and of its first derivatives, to apply a result of O. A. Ladyzhenskaya and N. N. Ural'tseva [9] and the Schauder estimate (2), and finally to use the **Leray-Schauder fixed-point theorem** [11].

E. Relation to the Calculus of Variations

Consider the bilinear form

$$J = \int_G \left(\sum a_{ij}(x)\frac{\partial u}{\partial x_i}\frac{\partial u}{\partial x_j}\right.$$
$$\left. + 2\sum b_i(x)\frac{\partial u}{\partial x_i}u + c(x)u^2 + 2f(x)u\right)dx,$$

where we assume $\sum a_{ij}(x)\xi_i\xi_j > 0$. Under the boundary conditions imposed on u, if there exists a function u that makes J minimum, then assuming some differentiability condition on $a_{ij}(x)$, $b_i(x)$, $c(x)$, we have the †Euler-Lagrange equation

$$\sum_j \frac{\partial}{\partial x_j}\left(\sum_i a_{ij}(x)\frac{\partial u}{\partial x_i}\right)$$
$$-\left(c(x) - \sum_i \frac{\partial}{\partial x_i} b_i(x)\right)u - f(x) = 0,$$

which is a linear second-order self-adjoint elliptic equation.

B. Riemann treated the simplest case, where $a_{ij}(x) = \delta_i^j$, $b_i(x) = c(x) = 0$, i.e., the case $\Delta u = 0$. He proved, assuming the existence of the minimum of J, the existence of the solution of $\Delta u = 0$ with assigned boundary values. This result, called **Dirichlet's principle**, was used by D. Hilbert, R. Courant, H. Weyl, O. Nikodym, and others to show the existence of solutions for linear self-adjoint elliptic equations.

If $F(x_1, \ldots, x_n, u, p_1, \ldots, p_n)$ satisfies $\Sigma(\partial^2 F/\partial p_i \partial p_j)\xi_i\xi_j > 0$ (and F has some regularity), then the function that minimizes the integral

$$J = \int_G F\left(x_1, \ldots, x_n, u, \frac{\partial u}{\partial x_1}, \ldots, \frac{\partial u}{\partial x_n}\right) dx_1, \ldots, dx_n$$

with the given boundary condition satisfies the Euler-Lagrange equation (of type (4))

$$\sum_{i,j=1}^n F_{p_i p_j} \frac{\partial^2 u}{\partial x_i \partial x_j} + \sum_{i=1}^n F_{u p_i} \frac{\partial u}{\partial x_i} + \sum_{i=1}^n F_{x_i p_i} - F_u = 0$$

($p_i = \partial u/\partial x_i$) and the boundary condition as well. This is also an elliptic partial differential equation in u. The case where F is a function of p alone (in particular, the case of †minimal surfaces) has been studied by A. Haar, T. Radó, J. Serrin, and others, particularly for $F = (1 + p_1^2 + \ldots + p_n^2)^{1/2}$ in the case of the minimal surface equation.

F. The Second and Third Boundary Value Problems

Let G be a domain in $\mathbf{R}^n$ with a smooth boundary consisting of a finite number of hypersurfaces. Also, let $B[u]$ be the boundary operator defined by

$$B[u] = a\frac{\partial u}{\partial \nu} + \beta u, \tag{6}$$

where $a = (\sum_{i=1}^n (\sum_{j=1}^n a_{ij} \cos(\nu_0 x_j))^2)^{1/2}$, ν_0 is the outer normal of unit length at the point $x \in S$, and ν is the **conormal** defined by

$$\cos(\nu x_i) = \sum_{j=1}^n a_{ij} \cos(\nu_0 x_j)/a, \qquad i = 1, \ldots, n.$$

The problem of finding the solution $u(x)$ of the equation $L[u] = f$ continuous on the closed domain $\bar{G}$ and satisfying $B[u] = \varphi$ on the boundary S of G is called the **second boundary value problem** (or **Neumann problem**) when $\beta \equiv 0$, and the **third boundary value problem** (or **Robin problem**) when $\beta \not\equiv 0$. In general, in boundary value problems, the condition that the solutions must satisfy at the boundary is called the **boundary condition**. We assume that the boundary S of G is expressed locally by a function with †Hölder continuous first derivatives (G is then called a **domain of class** $C^{1,\lambda}$). Assume that G is such a domain, $c \leqslant 0$, $\beta \geqslant 0$, and at least one of c and β is not identically 0. Then the second and third boundary value problems admit one and only one solution. When $c \equiv 0$ and $\beta \equiv 0$, the solutions of the second boundary value problem are determined uniquely up to additive constants. Furthermore, let M be the †adjoint operator of L, and let

$$B'[v] = a\frac{\partial v}{\partial \nu} + (\beta - b)v,$$

where

$$b = \sum_{i=1}^n \cos(\nu_0 x_i)\left[b_i - \sum_{j=1}^n \frac{\partial a_{ij}}{\partial x_j}\right].$$

Then if the boundary S of G is of class C^1 and f, φ are continuous, in order that there exist at least one solution u of the second or third boundary value problem relative to $L[u] = f$, it is necessary and sufficient that

$$\int_G fv\, dx - \int_S \varphi v\, dS = 0,$$

where v is any solution of $M[v] = 0$ with the boundary condition $B'[v] = 0$. Here the necessity is easily derived from Green's formula. G. Giraud used the notion of fundamental solution to reduce the second and third boundary value problems relative to $L[u] = f$ to a problem of integral equations, under the assumptions that G is a domain of class C^1, the coefficients of L and f satisfy Hölder conditions, and φ and β are continuous [3; also 12].

G. Method of Orthogonal Projection

The theory of †Hilbert spaces is applicable to the boundary value problems in Section F. In general, let $H^m(G)$ be the space of functions in $L_2(G)$ whose partial derivatives in the sense of †distributions up to order m belong to $L_2(G)$. For elements f and g in $H^m(G)$, we define the following inner product:

$$(f, g)_m = \sum_{|\alpha| \leqslant m} \int_G D^\alpha f(x) \overline{D^\alpha g(x)}\, dx,$$

where

$$D^\alpha = \frac{\partial^{|\alpha|}}{\partial x_1^{\alpha_1} \ldots \partial x_n^{\alpha_n}}, \qquad |\alpha| = \alpha_1 + \ldots + \alpha_n$$

(→ 168 Function Spaces). With respect to this inner product, $H^m(G)$ is a Hilbert space. When u satisfies $L[u] = f$ ($f \in L_2(G)$) and $\varphi(x)$ is an arbitrary element of $H^1(G)$, Green's formula yields

$$-\sum_{i,j=1}^n \left(a_{ij}\frac{\partial u}{\partial x_i}, \frac{\partial \varphi}{\partial x_j}\right) + \sum_{i=1}^n \left(b_i'(x)\frac{\partial u}{\partial x_i}, \varphi\right)$$

$$+ (c(x)u, \varphi) + \int_S a\frac{\partial u}{\partial \nu} \cdot \bar{\varphi}\, dS = (f, \varphi),$$

where $b_i'(x) = b_i(x) = -\Sigma_j(\partial a_{ij}/\partial x_j)$. Taking account of the boundary condition on u:

$a\,\partial u/\partial v+\beta u=0$, we get

$$\sum_{i,j}\left(a_{ij}\frac{\partial u}{\partial x_i},\frac{\partial \varphi}{\partial x_j}\right)-\sum_i\left(b_i'\frac{\partial u}{\partial x_i},\varphi\right)-(cu,\varphi)$$

$$+\int_S \beta u\bar{\varphi}\,dS=-(f,\varphi).$$

Thus the problem is reduced to finding $u(x)\in H^1(G)$ satisfying this equation for all $\varphi\in H^1(G)$. This equation can be regarded as an equation in $H^1(G)$. If necessary, by replacing $c(x)$ by $c(x)-t$ for a large t, we can show that for any $f(x)\in L_2(G)$, there exists a unique solution $u(x)\in H^1(G)$ [12, 13]. Now, for a solution of this functional equation, if we take $\varphi(x)\in \mathcal{D}(G)$, we have $(u, L^*[\varphi])=(f,\varphi)$, where L^* is the †adjoint operator of L. This means that $u(x)$ is a solution of $L[u]=f$ in the sense of distributions, and we call such a $u(x)$ a **weak solution**. Such a treatment may be called the **method of orthogonal projection**, following Weyl. In this case, it can be shown that if we assume smoothness of the coefficients, the boundary S, and β, then the solution $u(x)\in H^1(G)$ belongs to $H^{s+2}(G)$ when $f(x)\in H^s(G)$ $(s=0,1,\ldots)$. Thus, if we apply Green's formula, we can see that u satisfies the boundary condition $a\,\partial u/\partial v+\beta u=0$. In particular, when $s>n/2$, we see that $u(x)\in C^2(\bar{G})$ by †Sobolev's theorem [12]. In other words, $u(x)$ is a **genuine solution**.

Next, we introduce the complex parameter λ and consider the boundary value problem $(L+\lambda I)[u]=f$, $f\in L_2(G)$, $a\,\partial u/\partial v+\beta u=0$. If t is large, $(L-tI)$ is a one-to-one mapping from the domain $\mathcal{D}(L)=\{u\in H^2(G)\mid a\,\partial u/\partial v+\beta u=0\}$ onto $L_2(G)$. Thus, denoting its inverse, which acts on the equation from the left, by G_t, we have

$$(I+(\lambda+t)G_t)[u]=G_t f.$$

Conversely, since the solution $u(x)$ contained in $L_2(G)$ (hence also contained in $\mathcal{D}(L)$) satisfies the equation and the boundary condition, the problem is reduced to the displayed equation in $L_2(G)$ considered above. Now, since G is bounded and G_t is a continuous mapping from $L_2(G)$ into $H^2(G)$, **Rellich's theorem** yields that G_t is a †compact operator when it is regarded as an operator in $L_2(G)$. So we can apply the †Riesz-Schauder theorem (→ 189 Green's Operator).

H. Elliptic Equations of Higher Order

The differential operator of order m:

$$L=\sum_{|\alpha|\leqslant m}a_\alpha(x)D^\alpha,\qquad D^\alpha=\frac{\partial^{|\alpha|}}{\partial x_1^{\alpha_1}\ldots\partial x_n^{\alpha_n}},$$
$$|\alpha|=\alpha_1+\ldots+\alpha_n,$$

is called an **elliptic operator** if $\sum_{|\alpha|=m}a_\alpha(x)\xi^\alpha\neq 0$ $(\xi\neq 0)$. In particular, if

$$\operatorname{Re}\sum_{|\alpha|=m}a_\alpha(x)\xi^\alpha\geqslant c|\xi|^m,\qquad c>0,$$

L is called a **strongly elliptic operator**. In this case, m is even. L. Gårding studied the Dirichlet problem for strongly elliptic operators [14]. If we put $m=2b$, the boundary value condition is stated as $\partial^j u/\partial v^j=f_j(x)$ $(j=0,1,\ldots,b-1)$, where v is the normal of unit length at the boundary. Using the notion of function space, this boundary condition means that the solutions belong to the closure $\mathring{H}^b(G)$ of $\mathcal{D}(G)$ in $H^b(G)$ (→ 168 Function Spaces). In this treatment, **Gårding's inequality**

$$(-1)^b\operatorname{Re}(L[u],u)\geqslant\delta\|u\|_b^2-c\|u\|^2,\qquad u\in\mathring{H}^b(G),\quad(7)$$

where δ and c are positive constants, plays an important role.

In general, for an elliptic operator L defined in an open set G, if $u(x)$ satisfies $L[u]=f(x)$ and $f(x)$ belongs to H^s on any compact set in G, then $u(x)$ belongs to H^{m+s} on every compact set in G (**Friedrich's theorem** [15]).

General boundary value problems for elliptic equations of higher order have been considered by S. Agmon, A. Douglis, and L. Nirenberg [16], M. Schechter [17], and others. These problems are formulated as follows:

$$L[u]=f(x),\qquad B_j(x,D)u(x)=\varphi_j(x),$$
$$x\in S,\qquad j=1,2,\ldots,b(=m/2),\quad(8)$$

where the $B_j(x,D)$ are differential operators at the boundary and f and $\{\varphi_j\}$ are given functions. Under certain algebraic conditions (**Shapiro-Lopatinskiĭ conditions**) on $(L,\{B_j\})$, the problems are treated also in $H^m(G)$; hence the L^2 **a priori estimates** play a fundamental role: If $u\in H^m(G)$,

$$\|u\|_m\leqslant K\left(\|Lu\|+\sum_{j=1}^b\|B_ju\|_{m-m_j-(1/2),S}+\|u\|\right),\quad(9)$$

where K is a constant determined by (L,B_j,G), $\|\cdot\|_{k,S}$ is the norm in $H^k(S)$, and m_j are the orders of B_j (compare (9) with (2)). Under these estimates the boundary value problem is said to be **coercive**. In applications, the theory of interpolation of function spaces are also used [18] (→ 168 Function Spaces). Variational general boundary value problems have been treated by D. Fujiwara and N. Shimakura (*J. Math. Pures Appl.*, 49 (1970)) and others. For systems of such equations, there are works by F. E. Browder (*Ann. math. studies* 33, Princeton Univ. Press, 1954, 15–51) and others.

I. Analyticity of Solutions

In a linear elliptic equation $Lu=f$, suppose that all the coefficients and f are of class C^∞ (resp. real analytic) in an open set G and that u is a distribution solution in G. Then u is also of class C^∞ (resp. real analytic) in G [19]. Hence the linear elliptic operators are **hypoelliptic** (resp. **analytically hypoelliptic**) (→ 112 Differential Operators). In particular, †harmonic functions (i.e., solutions of $\Delta u=0$) are (real) analytic in the domain of existence, whatever the boundary values may be.

Hilbert conjectured that when $F(x, y, u, p, q, r, s, t)$ $(p=p_1, q=p_2, r=p_{11}, s=p_{12}, t=p_{22})$ is analytic in the arguments, then any solution u of the elliptic equation $F=0$ is analytic on the domain of existence (1900, Hilbert's 19th problem; → 196 Hilbert). This conjecture was proved by S. Bernshteĭn, Radó, and others, and then H. Lewy proposed a method of extending this equation to a complex domain so that it can be regarded as a †hyperbolic equation (*Math. Ann.*, 101 (1929)). This result was further extended by I. G. Petrovskiĭ to a general system of nonlinear differential equations of elliptic type (*Mat. Sb.*, 5 (47), (1939)).

J. The Unique Continuation Theorem

Since all the solutions of Laplace's equation $\Delta u=0$ are analytic, it follows that if $u(x)$ vanishes on an open set in a domain, then $u(x)$ vanishes identically in this domain. This **unique continuation theorem** can be extended to linear elliptic partial differential equations with analytic coefficients in view of the analyticity of solutions. This fact is also proved by applying †Holmgren's uniqueness theorem (→ 321 Partial Differential Equations (Initial Value Problems)). The unique continuation theorem, first established by T. Carleman for second-order elliptic partial differential equations $L[u]=0$ with C^1-coefficients in the case of two independent variables, was extended to second-order linear elliptic equations with C^2-coefficients in the case of any number of independent variables by C. Müller, E. Heinz, and finally by N. Aronszajn [20]. This research was extended by A. P. Calderón [21] and others in the direction of establishing the uniqueness of the Cauchy problem. However, it is to be remarked that even if we assume that the coefficients are of class C^∞, we cannot affirm the unique continuation property for general elliptic equations. A counterexample was given by A. Pliś (*Comm. Pure Appl. Math.*, 14 (1961)). See also the work of K. Watanabe (*Tohoku Math J.*, 23 (1971)).

K. Elliptic Pseudodifferential Operators and the Index

A pseudodifferential operator $P(x, D)$ with symbol $p(x,\xi)\in S^m_{0,1}$ (→ 345 Pseudodifferential Operators) is said to be **elliptic**, provided there exists a positive constant c such that $|p(x,\xi)|\geqslant c(1+|\xi|)^m$ for all $x\in\mathbf{R}^n$ and $|\xi|\geqslant c^{-1}$. The notion of ellipticity can be extended to operators on a manifold. The theory of elliptic pseudodifferential operators has been widely applied to the study of elliptic differential equations, and is particularly useful in the calculation of the †index of elliptic operators. B. R. Vaĭnberg and V. V. Grushin [22] calculated the index i of the †coercive boundary value problem for an elliptic operator by showing that i is equal to the index of some elliptic pseudodifferential operator on the boundary.

Example [23]: Given a real vector field (v_1, v_2) on the unit circle $x_1^2+x_2^2=1$, suppose the vector $(v_1(x), v_2(x))$ rotates l times around the origin as the point $x=(x_1, x_2)$ moves once around the unit circle in the positive direction. Then the index of the boundary value problem

$$\left(\frac{\partial^2}{\partial x_1^2}+\frac{\partial^2}{\partial x_2^2}\right)u(x)=f(x), \qquad x_1^2+x_2^2<1,$$

$$v_1(x)\frac{\partial u}{\partial x_1}+v_2(x)\frac{\partial u}{\partial x_2}=g(x), \qquad x_1^2+x_2^2=1,$$

is equal to $2-2l$.

M. F. Atiyah and I. M. Singer determined the index of a general elliptic operator on a manifold in terms of certain topological invariants of the manifold (→ 237 *K*-Theory H). The index of noncoercive boundary value problems has also been studied by Vaĭnberg and Grushin, R. Seeley (*Topics in pseudo-differential operators*, C.I.M.E. 1968, 335–375), and others.

L. The Giorgi-Nash-Moser Result

Let us state the following result (J. Moser [24]): Let L be of the form

$$Lu=\sum_{i,j=1}^{n}\frac{\partial}{\partial x_j}\left\{a_{ij}(x)\frac{\partial u}{\partial x_i}\right\},$$

where $a_{ij}=a_{ji}$ are real-valued, of class $L^\infty(G)$, and such that the ellipticity condition (3) holds at almost everywhere in G with some $\lambda\geqslant 1$. Also, let G' be any subdomain of G whose distance from ∂G is not smaller than $\delta>0$. Then, for any weak solution $u\in H^1(G)$ of the equation $Lu=0$ and for any two points x and y in G', we have the inequality

$$|u(x)-u(y)|\leqslant A|x-y|^\alpha\|u\|_{L^2(G)},$$

where A and α ($A>0, 0<\alpha\leqslant 1$) depend only on (n, λ, δ) and are independent of the particular choice of (L, G, G', u). Moser proved that the above inequality is a corollary to a Harnack-type inequality (→ 327 Partial Differential Equations of Parabolic Type G).

M. Asymptotic Distribution of Eigenvalues

Let L be an elliptic operator on $\bar{G}$ of order m with smooth coefficients realized as a self-adjoint operator in $L^2(G)$ under a nice boundary condition, where G is a bounded domain in $\mathbf{R}^n$ with smooth boundary. Let $N(T)(T>0)$ be the number of eigenvalues of L smaller than T. Then, it holds that

$$N(T)=CT^{n/m}+\text{an error term, as } T\to+\infty,$$

$$C=(2\pi)^{-n}\int_G dx\int_{S^{n-1}} a(x,\xi)^{-n/m}\,dS_\xi, \qquad (10)$$

where $a(x,\xi)$ is the †principal symbol of L and S^{n-1} is the unit sphere on $\mathbf{R}^n$. C is independent of the boundary condition and, if L is of constant coefficients, of the shape of G.

Formula (10) was at first established by H. Weyl [25] for the case of $L=$ Laplacian, and hence it is often called **Weyl's formula**. Weyl's method is based on the minimax principle [26]. T. Carleman (*Ber. Math.-Phys. Klasse der Sächs. Akad. Wiss. Leipzig*, 88 (1936)) studied the behavior of the trace of the Green's function of zI-L as $|z|\to\infty$ in the complex plane (→ [27]). S. Minakshisundaram (*Canad. J. Math.*, 1 (1947)) discussed this formula in connection with the heat equation; see also S. Mizohata and R. Arima (*J. Math. Kyoto Univ.*, 4 (1964)). L. Hörmander (*Acta Math.*, 121 (1968)) treated the case of compact manifolds without boundary and obtained the best possible error estimate. H. P. McKean and I. M. Singer (*J. Differential Geometry*, 1 (1967)) treated the case of manifolds and discussed the geometric meaning of this formula. In general, $N(T)$ is no more than $O(T^{n/m})$ if L is of degenerate elliptic type (C. Nordin, *Ark. Mat.*, 10 (1972)).

N. Equations of Degenerate Elliptic Type

An operator L of the form (1) is said to be degenerate at $x^0\in\bar{G}$ in the direction $\xi\in\mathbf{R}^n$ if ξ is a null vector of the matrix $(a_{ij}(x^0))$. L is said to be of degenerate elliptic type if $(a_{ij}(x))$ is nonnegative definite at any $x\in\bar{G}$ and if L is degenerate at some point of $\bar{G}$ in some direction.

Suppose that the coefficients of L and the boundary Γ of G are smooth enough. At $x\in\Gamma$, denote by $\nu(x)=(\nu_1(x),\dots,\nu_n(x))$ the unit outer normal vector to G. Let Σ_3 be the set of $x\in\Gamma$ at which L is not degenerate in the normal direction. Also, let Σ_1, Σ_2, and Σ_0 be the sets of $x\in\Gamma\setminus\Sigma_3$ at which $b(x)>0$, <0, and $=0$, respectively, where $b(x)$ is defined by

$$b(x)=-\sum_{i=1}^n \nu_i(x)\left\{b_i(x)-\sum_{j=1}^n\frac{\partial a_{ij}(x)}{\partial x_j}\right\}. \qquad (11)$$

Then the Dirichlet problem for equation (1) is to find a function $u(x)$ defined on $G\cup\Sigma_2\cup\Sigma_3$ satisfying

$$L[u]=f \quad \text{in } G, \qquad (1)$$

$$u=g \quad \text{on } \Sigma_2\cup\Sigma_3, \qquad (12)$$

where f and g are given functions.

Let $1<p<\infty$. We set $q=p/(p-1)$. We also put

$$c^*(x)=c(x)-\sum_{i=1}^n\frac{\partial b_i(x)}{\partial x_i}+\sum_{i,j=1}^n\frac{\partial^2 a_{ij}(x)}{\partial x_i\partial x_j}. \qquad (13)$$

We have the following existence theorem [28]: If (i) either $c<0$ on $\bar{G}$ or $c^*<0$ on $\bar{G}$ and if (ii) $pc+qc^*<0$ on $\bar{G}$, the Dirichlet problem (1) and (12) (with $g=0$) has a weak solution $u\in L^p(G)$ for any $f\in L^p(G)$. The regularity of solutions is also discussed in [28]. The value of $b(x)$ is closely related to the regularity near the point $x\in\Gamma$ if L is degenerate at x in the normal direction (→ also M. S. Baouendi and C. Goulaouic, *Arch. Rational Mech. Anal.*, 34 (1969)).

Degenerate elliptic equations of type (1) have also been investigated from the probabilistic viewpoint (→ 115 Diffusion Processes). The general boundary value problems for degenerate elliptic equations of higher order have been treated by M. I. Vishik and V. V. Grushin [29], N. Shimakura (*J. Math. Kyoto Univ.*, 9 (1969)), P. Bolley and J. Camus (*Ann. Scuola Norm. Sup. Pisa*, IV-1 (1974)), and others.

References

[1] J. Hadamard, Lectures on Cauchy's problem in linear partial differential equations, Yale Univ. Press and Oxford Univ. Press, 1923.
[2] E. E. Levi, Sulle equazioni lineari totalmente elliptiche alle derivate parziali, Rend. Circ. Mat. Palermo, 24 (1907), 275–317.
[3] C. Miranda, Partial differential equations of elliptic type, Springer, 1970.
[4] M. H. Protter and H. F. Weinberger, Maximum principles in differential equations, Prentice-Hall, 1967.
[5] J. Schauder, Über lineare elliptische Differentialgleichungen zweiter Ordnung, Math. Z., 38 (1934), 257–282.
[6] J. Schauder, Numerische Abschätzungen in

elliptischen linearen Differentialgleichungen, Studia Math., 5 (1934), 34–42.
[7] E. Picard, Leçon sur quelques problèmes aux limites de la théorie des équations différentielles, Gauthier-Villars, 1930.
[8] C. B. Morrey, Multiple integrals in the calculus of variations, Springer, 1966.
[9] O. A. Ladyzhenskaya and N. N. Ural'tseva, Linear and quasilinear elliptic equations, Academic Press, 1968. (Original in Russian, 1964.)
[10] J. Serrin, The problem of Dirichlet for quasilinear elliptic equations, Philos. Trans. Roy. Soc. London, (A) 264 (1969), 413–493.
[11] J. Leray and J. Schauder, Topologie et équations fonctionnelles, Ann. Ecole Norm. Sup., 51 (1934), 45–78.
[12] S. Mizohata, The theory of partial differential equations, Cambridge Univ. Press, 1973. (Original in Japanese, 1965.)
[13] L. Nirenberg, Remarks on strongly elliptic partial differential equations, Comm. Pure Appl. Math., 8 (1955), 649–675.
[14] L. Gårding, Dirichlet's problem for linear elliptic partial differential equations, Math. Scand., 1 (1953), 55–72.
[15] K. O. Friedrichs, On the differentiability of solutions of linear elliptic equations, Comm. Pure Appl. Math., 6 (1953), 299–325.
[16] S. Agmon, A. Douglis, and L. Nirenberg, Estimates near the boundary for solutions of elliptic partial differential equations satisfying general boundary conditions I, II, Comm. Pure Appl. Math., 12 (1959), 623–727; 17 (1964), 35–92.
[17] M. Schechter, General boundary value problems for elliptic equations, Comm. Pure Appl. Math., 12 (1959), 457–486.
[18] J. L. Lions and E. Magenes, Problèmes aux limites non homogènes et applications I–III, Dunod, 1968.
[19] F. John, Plane waves and spherical means applied to partial differential equations, Interscience, 1955.
[20] N. Aronszajn, A unique continuation theorem for solutions of elliptic equations or inequalities of second order, J. Math. Pure Appl., 36 (1957), 235–249.
[21] A. P. Calderón, Uniqueness in the Cauchy problem for partial differential equations, Amer. J. Math., 80 (1958), 16–36.
[22] B. R. Vaĭnberg and V. V. Grushin, Uniformly nonelliptic problems I, II, Math. USSR-Sb., 1 (1967), 543–568; 2 (1967), 111–133. (Original in Russian, 1967.)
[23] I. N. Vekua, System von Differentialgleichungen erster Ordnung vom elliptischen Typus und Randwertaufgaben, Berlin, 1956.
[24] J. Moser, On Harnack's theorem for elliptic differential equations, Comm. Pure Appl. Math., 14 (1961), 577–591.
[25] H. Weyl, Das asymptotische Verteilungsgesetze der Eigenverte linearer partieller Differentialgleichungen, Math. Ann., 71 (1912), 441–479.
[26] R. Courant and D. Hilbert, Methods of mathematical physics I, II, Interscience, 1962.
[27] S. Agmon, Lectures on elliptic boundary value problems, Van Nostrand, 1965.
[28] O. A. Oleĭnik and E. V. Radkevich, Second-order equations with non-negative characteristic form, Plenum, 1973. (Original in Russian, 1971.)
[29] M. I. Vishik and V. V. Grushin, Boundary value problems for elliptic equations degenerate on the boundary of a domain, Math. USSR-Sb., 9 (1969), 423–454. (Original in Russian, 1969.)

324 (XIII.22) Partial Differential Equations of First Order

A. Quasilinear Partial Differential Equations and Their Characteristic Curves

Suppose that we are given a †quasilinear partial differential equation

$$\sum_{i=1}^{n} P_i(x,u)\frac{\partial u}{\partial x_i}=Q(x,u), \qquad x=(x_1,\dots,x_n). \qquad (1)$$

A curve defined by a solution $x_i=x_i(t)$, $u=u(t)$ of the system of ordinary differential equations

$$\frac{dx_i}{dt}=P_i(x,u), \qquad i=1,\dots,n; \qquad \frac{du}{dt}=Q(x,u)$$

is called a **characteristic curve** of (1) (→ 320 Partial Differential Equations; 322 Partial Differential Equations (Methods of Integration)). A necessary and sufficient condition for $u=u(x)$ to be a solution of (1) is that the characteristic curve passing through any point on the hypersurface $u=u(x)$ (in the $(n+1)$-dimensional xu-space) always be contained in this hypersurface. For example, the characteristic curve of $\sum_{i=1}^{n} x_i\partial u/\partial x_i=ku$ is $x_i=x_i^0 e^t$, $u=u^0 e^{kt}$ (a solution of $x_i'=x_i, u'=ku$). Therefore the solution $u=u(x)$ is a function such that $u(\lambda x_1,\dots,\lambda x_n)=\lambda^k u(x_1,\dots,x_n)$ ($\lambda=e^t>0$), i.e., a homogeneous function of degree k.

B. Nonlinear Partial Differential Equations and Their Characteristic Strips

We denote the value of $\partial u/\partial x_i$ by p_i and define the **surface element** (or **hypersurface element**) by the $(2n+1)$-dimensional vector $(x, u, p) = (x_1, \ldots, x_n, u, p_1, \ldots, p_n)$. Consider the partial differential equation

$$F(x_1, \ldots, x_n, u, p_1, \ldots, p_n) = 0, \qquad p_i = \partial u/\partial x_i. \tag{2}$$

A set $(x(t), u(t), p(t))$ of surface elements depending on a parameter t and satisfying the system of ordinary differential equations

$$\frac{dx_i}{dt} = F_{p_i}, \quad \frac{du}{dt} = \sum_{i=1}^{n} p_i F_{p_i}, \quad \frac{dp_i}{dt} = -(F_{x_i} + p_i F_u)$$

is called a **characteristic strip** of equation (2), and the curve $x = x(t)$, $u = u(t)$ is called a **characteristic curve** of (2). For quasilinear equations, this definition of characteristic curve coincides with the one mentioned in Section A. Furthermore, an r-dimensional †differentiable manifold consisting of surface elements satisfying the relation

$$du - \sum_{i=1}^{n} p_i \, dx_i = 0$$

is called an r-dimensional **union of surface elements**. A solution of the partial differential equation (2) is, in general, formed by the set of all characteristic strips possessing, as initial values, surface elements belonging to an $(n-1)$-dimensional union of surface elements satisfying $F(x, u, p) = 0$.

An example of a nonlinear partial differential equation of first order is $pq - z = 0$ (where $x_1 = x$, $x_2 = y$, $u = z$, $p_1 = p$, $p_2 = q$). The equations of the characteristic strip are $x' = q$, $y' = p$, $z' = 2pq$, $p' = p$, $q' = q$, and therefore the characteristic strip is given by $y = y_0 + (p_0/q_0)x$, $z = z_0 + 2p_0 x + (p_0/q_0)x^2$, $p = p_0 + (p_0/q_0)x$, $q = q_0 + x$ (if we take x as an independent variable and impose an initial condition $y = y_0$, $z = z_0$, $p = p_0$, $q = q_0$ at $x = 0$). Putting $z_0 = W(y_0)$ (an arbitrary function) for $x = 0$, we have, furthermore, $y = y_0 + W(y_0)/(W'(y_0))^2$, $z = W(y_0) + 2W(y_0)/W'(y_0) + W(y_0)x^2/(W'(y_0))^2$. The elimination of y_0 from these expressions yields a general solution $z = z(x, y)$. In this case, a †complete solution is $4az = (x + ay + b)^2$ (where a, b are constants), and a †singular solution is $z = 0$.

C. Complete Systems of Linear Partial Differential Equations

For functions $P_i(x)$ $(i = 1, \ldots, x = (x_1, \ldots, x_n))$ of class C^∞, define a differential operator X by

$$X = \sum_{\nu=1}^{n} P_\nu(x) \frac{\partial}{\partial x_\nu}.$$

We call k differential operators $X_i = \sum_{\nu=1}^{n} P_\nu^i(x) \partial/\partial x_\nu$ $(i = 1, \ldots, k)$ mutually **independent** when the rank of the matrix (P_ν^i) is equal to k. If a system of k independent linear partial differential equations involving one unknown function $f(x)$,

$$X_1 f = 0, \ldots, X_k f = 0, \tag{3}$$

has the maximum number $n-k$ of independent †integrals, then the system (3) is called a **complete system**. A necessary and sufficient condition for the system (3) to be a complete system is that there exist k^3 functions $\lambda_{ij}^\nu(x)$ of class C^∞ such that

$$[X_j, X_i] \equiv X_j X_i - X_i X_j = \sum_{\nu=1}^{k} \lambda_{ij}^\nu(x) X_\nu,$$

that is,

$$\sum_{h=1}^{n} \left(P_h^j \frac{\partial P_\mu^i}{\partial x_h} - P_h^i \frac{\partial P_\mu^j}{\partial x_h} \right) = \sum_{\nu=1}^{k} \lambda_{ij}^\nu P_\mu^\nu.$$

Here, $[X_j, X_i]$ is a differential operator of first order, called the **commutator** of the differential operators X_i, X_j or the **Poisson bracket**.

D. Involutory Systems

For two functions $F(x, u, p)$, $G(x, u, p)$ of x, u, p of class C^∞, we define the **Lagrange bracket** $[F, G]$ by

$$[F, G] = \sum_{\nu=1}^{n} \left(\frac{\partial F}{\partial p_\nu} \left(\frac{\partial G}{\partial x_\nu} + p_\nu \frac{\partial G}{\partial u} \right) - \frac{\partial G}{\partial p_\nu} \left(\frac{\partial F}{\partial x_\nu} + p_\nu \frac{\partial F}{\partial u} \right) \right).$$

If F, G do not contain u and are homogeneous linear forms with respect to p, then F and G are differential operators $F = X_1 u$ and $G = X_2 u$ with respect to u (for $p_\nu = \partial u/\partial x_\nu$), and we see that $[F, G] = [X_1, X_2]_u$. This bracket has the following properties:

$$[F, G] = -[G, F],$$

$$[F, \varphi(G_1, \ldots, G_k)] = \sum_{i=1}^{k} \frac{\partial \varphi}{\partial G_i} [F, G_i],$$

$$[[F, G], H] + [[G, H], F] + [[H, F], G] = \frac{\partial F}{\partial u}[G, H] + \frac{\partial G}{\partial u}[H, F] + \frac{\partial H}{\partial u}[F, G].$$

When F, G are functions of x and p only, we usually use the notation (F, G) and call it also the **Poisson bracket**. In this case, the right-hand side of the third relation vanishes.

Consider k partial differential equations involving one unknown function $u(x_1, \ldots, x_n)$,

$$F_i(x, u, p) = 0, \quad i = 1, \ldots, k; \qquad p_\nu = \frac{\partial u}{\partial x_\nu}. \tag{4}$$

If a common solution $u(x)$ of these equations exists, it is also a solution of $[F_i, F_j] = 0$ $(i, j =$

$1, \ldots, k$). Therefore, from the equations thus obtained, we take independent equations and add them to the original system. If we then have more than $n+1$ equations, the original system has no solution. Otherwise, we obtain a system $F_j=0$ ($j=1, \ldots, r$) for which the F_j are independent (i.e., the rank of $(\partial F_i/\partial p_\nu)$ is equal to r), and all $[F_i, F_j]=0$ can be derived from $F_j=0$. A system (4) such that $[F_i, F_j]\equiv 0$ for all i, j is called an **involutive** (or **involutory**) **system**. We always treat a system by extending it to an involutory system. When $k=1$, we regard the equation itself as an involutory system.

When the equations (4) are mutually independent, a necessary and sufficient condition for them to have in common a solution with $n-k$ degrees of freedom (a solution that coincides with an arbitrary function on an adequate manifold of dimension $n-k$) is that the system (4) be a system of equations involving unknowns p and equivalent to an involutory system.

An involutory system (4) can be extended to an involutory system consisting of n independent equations by adding $n-k$ suitable equations

$$F_{k+1}(x,u,p)=a_{k+1}, \ldots, F_n(x,u,p)=a_n. \qquad (4')$$

That is, we can find successively $f=F_l$ ($l=k+1, \ldots, n$) such that F_l satisfies the system of equations

$$[F_i, f]=0, \qquad i=1, \ldots, l-1,$$

which is a complete system of linear partial differential equations for f, and the F_i ($i=1, \ldots, n$) are mutually independent. Then, if we find that p_ν as functions of (x,u,a) ($a=(a_{k+1}, \ldots, a_n)$) from (4) and (4′), the system of †total differential equations

$$\frac{\partial u}{\partial x_\nu}=\rho_\nu(x,u,a), \quad \nu=1, \ldots, n,$$

is †completely integrable, and we can find u as a solution containing essentially $n-k+1$ parameters $c, a_{k+1}, \ldots, a_n$, that is, a complete solution of (4). Moreover, if we have an involutory system of $n+1$ independent equations $F_1=0, \ldots, F_k=0, F_{k+1}=a_{k+1}, \ldots, F_{n+1}=a_{n+1}$, then we find a complete solution by eliminating $p_1, \ldots, p_n$ between the equations. This method of integrating an involutory system is called **Jacobi's second method of integration**.

E. Relation to the Calculus of Variations

Consider a partial differential equation of first order $F(x_1, \ldots, x_n, u, p_1, \ldots, p_n)=0$. If a solution $u(x)$ of this equation is given as an implicit function by $\varphi(x_1, \ldots, x_n, u)=0$, we have

$$F\left(x_1, \ldots, x_n, u, -\frac{\partial\varphi/\partial x_1}{\partial\varphi/\partial u}, \ldots, -\frac{\partial\varphi/\partial x_n}{\partial\varphi/\partial u}\right)=0.$$

This gives formally a partial differential equation with independent variables $u, x_1, \ldots, x_n$ and a dependent variable φ. It does not contain φ explicitly. That is, this equation has the form

$$F(x_1, \ldots, x_{n+1}, \partial\varphi/\partial x_1, \ldots, \partial\varphi/\partial x_{n+1})=0.$$

Then, by finding a partial derivative, say $\partial\varphi/\partial x_{n+1}$, as a function of the remaining ones from the displayed equation, we get a partial differential equation of the form

$$\partial\varphi/\partial t+H(t,x,\partial\varphi/\partial x)=0,$$

which is called the **normal form** of the partial differential equation of first order. Setting $p_i=\partial\varphi/\partial x_i$, the equations of the characteristic curve of this equation are

$$\frac{dx_i}{dt}=H_{p_i}(t,x,p), \qquad \frac{dp_i}{dt}=-H_{x_i}(t,x,p), \qquad i=1, \ldots, n,$$

which are called **Hamilton's differential equations**.

Now, consider the †Euler-Lagrange differential equations

$$dF_{x_i'}/dt-F_{x_i}=0, \quad i=1, \ldots, n,$$

for the integral

$$J=\int_{t_0}^{t_1} F(t,x,x')\,dt, \qquad x'=\frac{dx}{dt}.$$

Under the assumption that $\det(F_{x_i'x_j'})\neq 0$, we put $F_{x_i'}=p_i$, and solve these relations with respect to x_i' in the form, say, $x_i'=\varphi_i(t,x,p)$. Furthermore, if we put

$$\left(\sum_{\nu=1}^{n} x_\nu' F_{x_\nu'}-F\right)_{x'=\varphi(t,x,p)}=H(t,x,p),$$

then the Euler-Lagrange equations are equivalent to Hamilton's differential equations

$$dx_i/dt=H_{p_i}, \quad dp_i/dt=-H_{x_i}, \qquad i=1, \ldots, n,$$

since $F(t,x,x')=\sum_{i=1}^{n} p_i H_{p_i}-H$.

A curve represented by a solution of the Euler-Lagrange equations is called a **stationary curve**. Now consider a family of stationary curves in a domain G of the $(n+1)$-dimensional tx-space such that passing through every point of G there is one and only one curve in this family, and suppose that the family is †transversal to an r-dimensional manifold $\mathfrak{A}$ ($r\leqslant n$) (that is, $F\delta t-\sum F_{x_i'}\delta x_i=0$ for the differentials $\delta t, \delta x_i$ along $\mathfrak{A}$; in particular, if $\mathfrak{A}$ consists of only one point ($r=0$), a stationary curve passing through this point is transversal to $\mathfrak{A}$). In this case, if we denote by $V(t,x)$ the value of the integral J along the stationary curve from $\mathfrak{A}$ to any point (t,x) of G, then the equation

$$\partial V/\partial t+H(t,x,\partial V/\partial x)=0$$

holds. This equation is called the **Hamilton-Jacobi differential equation** or the **canonical** or **eikonal equation**. Conversely, a solution of this equation is equal to the value of the integral J for a family of stationary curves transversal to an adequate $\mathfrak{A}$.

F. The Monge Differential Equation

Consider the partial differential equation (2). By eliminating p and t between

$$\frac{dx_i}{dt}=F_{p_i}, \quad \frac{du}{dt}=\sum_{i=1}^{n} p_i F_{p_i}, \quad \text{and} \quad F(x,u,p)=0$$

(for example, by eliminating p between $dx_i/dx_1 = F_{p_i}/F_{p_1}$ and $F=0$ when $F_{p_1}\neq 0$), we obtain

$$M(x_1,\dots,x_n,u,\partial x_2/\partial x_1,\dots,\partial x_n/\partial x_1)=0.$$

This equation is called the **Monge differential equation**, and the curve represented by its solution is called an **integral curve** of the equation. In the $(n+1)$-dimensional tx-space, a curve that is an envelope of a 1-parameter family of characteristic curves of the partial differential equation (2) is a solution of the Monge equation. A characteristic curve is also an integral curve. When $n=2$, an integral curve that is not a characteristic curve is a †line of regression of the surface generated by the family of characteristic curves tangent to the integral curve under consideration (which is an integral surface of $F(x,u,p)=0$). If F is linear in p_i, i.e., quasilinear, all integral curves coincide with characteristic curves.

References

[1] R. Courant and D. Hilbert, Methods of mathematical physics II, Interscience, 1962.
[2] I. G. Petrovskiĭ, Lectures on partial differential equations, Interscience, 1954. (Original in Russian, 1953.)
[3] E. Goursat, Cours d'analyse mathématique, Gauthier-Villars, II, second edition, 1911; III, fourth edition, 1927.
[4] F. John, Partial differential equations, Springer, 1971.
[5] C. Carathéodory, Variationsrechnung und Partielle Differentialgleichungen Erste Ordnung, Teubner, 1935; English translation, Calculus of variations and partial differential equations of the first order, Holden-Day, 1967.

325 (XIII.25) Partial Differential Equations of Hyperbolic Type

A. Second-Order Linear Hyperbolic Equations

A †linear partial differential equation in $n+1$ variables t, $x=(x_1,\dots,x_n)$ of the second order,

$$L[u]=\frac{\partial^2 u}{\partial t^2}-\sum_{i=1}^{n} a_{0i}\frac{\partial^2 u}{\partial t\partial x_i}-\sum_{i,j=1}^{n} a_{ij}\frac{\partial^2 u}{\partial x_i\partial x_j}$$

$$-a_0\frac{\partial u}{\partial t}-\sum_{i=1}^{n} a_i\frac{\partial u}{\partial x_i}-au=0, \tag{1}$$

with coefficients $a_{0i},\dots,a$ that are functions in (t,x) is said to be **hyperbolic** (or of **hyperbolic type**) (with respect to the t-direction) in tx-space if the **characteristic equation** of equation (1) considered at each point of tx-space,

$$H(t,x;\lambda,\xi)=\lambda^2-\sum_{i=1}^{n} a_{0i}\xi_i\lambda-\sum_{i,j=1}^{n} a_{ij}\xi_i\xi_j=0, \tag{2}$$

has two distinct real roots $\lambda=\lambda_1(t,x,\xi)$, $\lambda_2(t,x,\xi)$ for any n-tuple of real numbers $\xi=(\xi_1,\dots,\xi_n)\neq(0,\dots,0)$. In particular, (1) is called **regularly hyperbolic** if these two roots are separated uniformly, that is,

$$\lim_{(t,x),|\xi|=1} |\lambda_1(t,x,\xi)-\lambda_2(t,x,\xi)|=c>0.$$

A typical example of hyperbolic equations is the **wave equation**

$$\Box u=\frac{\partial^2 u}{\partial t^2}-\frac{\partial^2 u}{\partial x_1^2}-\dots-\frac{\partial^2 u}{\partial x_n^2}=0. \tag{3}$$

Equation (3) is also called the **equation of a vibrating string**, the **equation of a vibrating membrane**, or the **equation of sound propagation** according as $n=1$, 2, or 3. Another example is

$$\frac{\partial^2 u}{\partial t^2}-c^2\frac{\partial^2 u}{\partial x^2}-2\frac{\partial u}{\partial t}=0,$$

which describes the propagation of electric current in a conducting wire with leakage and is called the **telegraph equation** (→ Appendix A, Table 15).

A hyperplane $\lambda(t-t^0)+\dots+\xi_n(x_n-x_n^0)=0$ passing through a point $p^0=(t^0,x^0)$ in tx-space and having normal direction (λ,ξ) is called a **characteristic hyperplane** of (1) at p^0 if the direction (λ,ξ) satisfies the characteristic equation at p^0: $H(t^0,x^0;\lambda,\xi)=0$. A hypersurface S: $s(t,x)=0$ in tx-space is a **characteristic hypersurface** of (1) if at each point of S the tangent hyperplane of S is a characteristic hyperplane of (1), that is, $H(t,x;s_t,s_x)=0$ everywhere on S. According to the theory of first-order partial differential equations, a characteristic hypersurface of S is generated by so-called **bicharacteristic curves**, i.e., solution curves $t=t(\tau)$, $x(\tau)$ of a system of ordinary differential equations

$$\frac{dt}{d\tau}=H_\lambda, \quad \ldots, \quad \frac{dx_n}{d\tau}=H_{\xi_n},$$

$$\frac{d\lambda}{d\tau}=-H_t, \quad \ldots, \quad \frac{d\xi_n}{d\tau}=-H_{x_n},$$

$$H(t,x;\lambda,\xi)=0.$$

Now if (1) is hyperbolic, the set of all characteristic hyperplanes at $p^0:\{\lambda(t-t^0)+\ldots+\xi_n(x_n-x_n^0)=0 \mid H(t^0,x^0;\lambda,\xi)=0\}$ has as its envelope a cone $C(p^0)$ with the vertex p^0. Moreover, since the intersection of any hyperplane $t=$ constant and the cone $C(p^0)$ is an $(n-1)$-dimensional ellipsoid or two points for $n=1$, a conical body $D_+(p^0)$ $(D_-(p^0))$ is determined, whose boundary consists of the part of $C(p^0)$ with $t\geqslant t_0$ $(t\leqslant t_0)$ and the interior of the ellipsoid on the hyperplane $t=$constant. A †smooth curve γ in tx-space is called **timelike** if the tangent vector of γ at each point p on γ belongs to $D_+(p)$ or $D_-(p)$. Consider the set of points that can be connected with the point p^0 by a timelike curve. We call its closure an **emission**, and a subset $\mathscr{D}_+(p^0)$ $(\mathscr{D}_-(p^0))$ of the closure for which $t\geqslant t_0$ $(t\leqslant t_0)$ a **forward (backward) emission**. An emission is a conical body surrounded by characteristic hyperplanes in some neighborhood of the vertex p^0. If the coefficients of (1) are bounded functions, the emissions $\mathscr{D}_\pm(p^0)$ are contained in a conical body

$$K=\left\{(t,x) \mid \lambda_{\max}^2(t-t^0)^2\geqslant\sum_{i=1}^n(x_i-x_i^0)^2\right\}$$

independently of the situation of p^0, where $\lambda_{\max}=\max_{(t,x),|\xi|=1}(|\lambda_1(t,x,\xi)|,|\lambda_2(t,x,\xi)|)$.

B. The Cauchy Problem

Important for the hyperbolic equation (1) is the †Cauchy problem, i.e., the problem of finding a function $u=u(t,x)$ that satisfies (1) in $t>0$ and the initial conditions

$$u(0,x)=u_0(x), \qquad \partial u/\partial t(0,x)=u_1(x), \tag{4}$$

where the functions $u_0(x)$ and $u_1(x)$ are given on the initial hyperplane $t=0$.

Suppose that (1) is regularly hyperbolic and the coefficients are bounded and sufficiently smooth (i.e., of class C^ν with ν sufficiently large). Then for the Cauchy problem the following theorem holds. Theorem (C): There exists a positive integer $l\,(=[n/2]+3)$, depending on the dimension $n+1$ of the tx-space, such that if the functions $u_0(x)$ and $u_1(x)$ in (4) are of class C^l, then there exists a unique solution $u=u(t,x)$ of class C^2 in the domain $0\leqslant t<\infty$, $-\infty<x_i<\infty$ $(1\leqslant i\leqslant n)$. Moreover, this correspondence $\{u_0(x),u_1(x)\}\rightarrow u(t,x)$ is continuous in the following sense: If a sequence of initial functions $\{u_{0k}(x),u_{1k}(x)\}$ $(k=1,2,\ldots)$ and their derivatives up to the lth order tend to 0 uniformly on every compact set in the hyperplane $t=0$, then the sequence of corresponding solutions $u_k(t,x)$ also tends to 0 uniformly on every compact set in each hyperplane $t=$constant. In other words, the Cauchy problem for regularly hyperbolic equations is †well posed in the sense of Hadamard [2].

For dependence of the solution on initial data, the following proposition is valid: The values of the solution u at a point $p^0=(t^0,x^0)$ depend only on the initial data on a domain G_0 (**domain of dependence**) of the initial hyperplane, which is determined as the intersection of the backward emission $\mathscr{D}_-(p^0)$ and the initial hyperplane. We have the following dual proposition: A change in the initial conditions in a neighborhood of a point Q_0 of the initial hyperplane induces a change of values of the solution only in some neighborhood of the forward emission $\mathscr{D}_+(Q_0)$ (**domain of influence**). If the coefficients of the equation are bounded, the intersection of emissions $\mathscr{D}_\pm(p^0)$ and the hyperplane $t=$constant is always compact. It follows that the domain of dependence and the domain of influence are bounded. In some special cases, there exists a proper subdomain of G_0 such that the solution depends only on the initial data on the subdomain. For example, for the wave equation (3) with $n=3$, the solution for the Cauchy problem at a point $p^0=(t^0,x^0)$ (→ Section D) is determined, as can be seen from the solution formula (12), by the initial data in a neighborhood of the cone with vertex $p^0:(t-t^0)^2=\sum_{i=1}^n(x_i-x_i^0)^2$, namely, in a neighborhood of the intersection of bicharacteristic curves (lines, in this case) passing through p^0 and the initial hyperplane. If the solution of the Cauchy problem has such a property, it is said that **Huygens's principle** is valid, or that diffusion of waves does not occur. For the wave equation (3), Huygens's principle is valid only for odd $n>1$.

C. The Energy Inequality

The energy conservation law for the wave equation (3),

$$E(t)$$

$$=\int_{-\infty}^{\infty}\cdots\int_{-\infty}^{\infty}\left(\left(\frac{\partial u}{\partial t}\right)^2+\sum_{i=1}^n\left(\frac{\partial u}{\partial x_i}\right)^2\right)dx_1\ldots dx_n$$

$$=\text{constant},$$

is generalized to the so-called energy inequality for hyperbolic equations, which plays an essential role in deducing the well-posedness of the Cauchy problem and the properties of the domain of dependence of the solution. Let the coefficients of a hyperbolic equation (1) be bounded functions, and let $G(\tau)$ be the intersection of the conical body $K=\{(t,x)\,|\,\lambda_{\max}^2(t-t^1)^2 \geqslant \sum_{i=1}^n (x_i - x_i^1)^2\}$ with the hyperplane $t=\tau$ $(\tau<t^1)$. The $k(\geqslant 1)$th-order energy integral of the solution $u(t,x)$ of (1) on $G(\tau)$ is defined by

$$E_\tau^{(k)}(u, G(\tau))$$

$$=\int_{G(\tau)} \sum_{\alpha_0+\ldots+\alpha_n\leqslant k} \left|\frac{\partial^{|\alpha|}u}{\partial t^{\alpha_0}\ldots\partial x_n^{\alpha_n}}(\tau,x)\right|^2 dx.$$

Then the following inequality holds:

$$E_\tau^{(k)}(u, G(\tau)) \leqslant C E_{t^0}^{(k)}(u, G(t^0)), \qquad t^0\leqslant\tau<t^1, \tag{5}$$

where the constant C is independent of u. We call (5) the **energy inequality** (J. Schauder [5]). For the wave equation (3), the hypothesis $l=[n/2]+3$ in theorem (C) can be replaced by a weaker condition $l=[n/2]+2$, but if we take $l=[n/2]+1$, there is an example for which no global solution of class C^1 exists. In general, even though the initial functions are of class C^1, the solution in the Cauchy problem for hyperbolic equations may not be of class C^1, while if the energy of the initial functions is bounded, the energy of the solution is also bounded.

D. Representation Formulas for Solutions of the Cauchy Problem

We consider solution formulas that represent solutions of the Cauchy problem explicitly as functionals of the initial functions. The problem of solving (1) under the condition (4), or more generally, the problem of solving an equation $L[u]=f(t,x)$ under (4), can be reduced, by transforming the unknown function u and applying †Duhamel's method, to the Cauchy problem with initial conditions on the hyperplane $t=\tau$:

$$u(\tau,x)=0, \qquad \partial u/\partial t(\tau,x)=\varphi(x) \tag{4'}$$

for arbitrary τ. We define a function $\chi_q(s)$ of a real variable s by

$$\chi_q(s)=|s|^q/4(2\pi i)^{n-1}q! \qquad (n \text{ odd}),$$

$$=-s^q\log|s|/(2\pi i)^n q! \qquad (n \text{ even}), \tag{6}$$

where n is the dimension of the x-space and q is a positive integer such that $q+n$ is even. Then for a function $\varphi(x)$ of class C^ν (with ν sufficiently large) with compact support, the following equality holds [7]:

$$\varphi(x)=\int_{-\infty}^{\infty}\Delta_y^{(n+q)/2}\varphi(y)\,dy\int_{|\omega|=1}\chi_q((x-y)\omega)\,d\omega, \tag{7}$$

where Δ_y is the †Laplacian with respect to the variables $y=(y_1,\ldots,y_n)$ and $d\omega$ is the surface element of the unit sphere $|\omega|=1$ in x-space. Now, since the †principle of superposition is valid because (1) is linear, we can infer from formula (7) that the Cauchy problem (1) with initial condition (4′) can be reduced to the one for initial conditions with parameters y, ω:

$$u(\tau,x)=0, \quad \partial u/\partial t(\tau,x)=\chi_q((x-y)\omega). \tag{4''}$$

In fact, since $\chi_q(s)$ is $(q-1)$-times differentiable by definition, the Cauchy problem (1) with initial condition (4″) has a unique solution $R_q(t,x;\tau,y;\omega)$ for q chosen large enough so that theorem (C) can be applied to (1) with initial condition (4″). Moreover, $R_q(t,x;\tau,y;\omega)$ is a function of (t,x,τ,y,ω) of class C^ν, and ν increases with q. Now, let $\varphi(x)$ be a function of class C^ν with sufficiently large ν and with compact support. Then, by (7) and the definition of R_q, the integral

$$\int_{-\infty}^{\infty}\Delta_y^{(n+q)/2}\varphi(y)\,dy\int_{|\omega|=1}R_q(t,x;\tau,y;\omega)\,d\omega \tag{8}$$

is a solution of the Cauchy problem (1) with initial condition (4′). Therefore, when R_q is found explicitly, (8) yields a solution formula of the Cauchy problem (1) with initial condition (4′) as a functional of the initial functions. Since in (8), the integral $\int_{|\omega|=1}R_q\,d\omega$ is not necessarily of class C^{n+q} as a function of $(t,x;\tau,y)$, $\Delta_y^{(n+q)/2}\int_{|\omega|=1}R_q\,d\omega$ is in general not a function in the ordinary sense. But we denote it by $R(t,x;\tau,y)$ formally, and understand that a linear operator

$$u(t,x)=\int R(t,x;\tau,y)\varphi(y)\,dy \tag{9}$$

is defined by (8). The kernel $R(t,x;\tau,y)$ in this sense is called a **fundamental solution** or **Riemann function** of the Cauchy problem. If we extend the function $\int_{|\omega|=1}R_q(t,x;\tau,y;\omega)\,d\omega$ defined for $t\geqslant\tau$ to $t<\tau$, assigning it the value 0 there, then $R(t,x;\tau,y)=\Delta_y^{(n+q)/2}\int_{|\omega|=1}R_q(t,x;\tau,y;\omega)\,d\omega$ can be considered a †distribution on $(t,x;\tau,y)$-space, and the equality $L(t,x,\partial/\partial t,\partial/\partial x)R(t,x;\tau,y)=L^*(\tau,y,\partial/\partial\tau,\partial/\partial y)R(t,x;\tau,y)=\delta(t-\tau)\delta(x-y)$ is valid, where L^* is the †adjoint operator of L and δ is †Dirac's δ-function. In other words, $R(t,x;\tau,y)$ is a fundamental solution of L in the sense of distribution theory.

The fundamental solution $R(t,x;\tau,y)$ can be analyzed using the asymptotic expansion with respect to the sequence of functions $\{\chi_q(s)\}$, and we have the following important result: If the coefficients of (1) are of class C^∞ (resp. real analytic), then the fundamental solution $R(t,x;\tau,y)$ is of class C^∞ (real analytic) in (t,x) except for points that are on bicharacteristic curves of (1) passing through the point (τ,y). In the language of the Cauchy problem, the smoothness of the solution u at a point $p=(t,x)$ depends only on the smoothness of the initial conditions on a neighborhood of the intersection of the initial hyperplane and all bicharacteristic curves passing through p. This fact is called **Huygens's principle in the wider sense**. Behavior of the fundamental solution $R(t,x;\tau,y)$ near discontinuous points has also been investigated [2].

For the wave equation (3), the fundamental solution can be constructed, and therefore we can write the solution formula explicitly. The solution formula for (3) with initial condition (4′) for $n\geqslant 3$ is

$$u(t,x)$$
$$=\frac{1}{(n-2)!}\frac{\partial^{n-2}}{\partial t^{n-2}}\int_0^t(t^2-\tau^2)^{(n-3)/2}\tau Q(x,\tau)\,d\tau,$$

$$Q(x,\tau)=\frac{1}{\omega_n}\int_{|\omega|=1}\varphi(x+\tau\omega)\,d\omega,$$

where $\omega_n=2\sqrt{\pi^n}/\Gamma(n/2)$ is the surface area of the unit sphere of n-dimensional space. Solutions of the Cauchy problem (3) with initial condition (4) for $n=1$, 2, and 3 are, respectively,

$$u(t,x)=\frac{u_0(x+t)+u_0(x-t)}{2}+\frac{1}{2}\int_{x-t}^{x+t}u_1(\xi)\,d\xi \tag{10}$$

(d'Alembert's solution),

$$u(t,x_1,x_2)$$
$$=\frac{1}{2\pi}\frac{\partial}{\partial t}\int_{C_t}\frac{u_0(\xi_1,\xi_2)\,d\xi_1\,d\xi_2}{\sqrt{t^2-(x_1-\xi_1)^2-(x_2-\xi_2)^2}}$$
$$+\frac{1}{2\pi}\int_{C_t}\frac{u_1(\xi_1,\xi_2)\,d\xi_1\,d\xi_2}{\sqrt{t^2-(x_1-\xi_1)^2-(x_2-\xi_2)^2}} \tag{11}$$

(Poisson's solution), and

$$u(t,x_1,x_2,x_3)=\frac{1}{4\pi}\frac{\partial}{\partial t}\int_{D_t}\frac{u_0(\xi_1,\xi_2,\xi_3)}{t}\,d\omega_t$$
$$+\frac{1}{4\pi}\int_{D_t}\frac{u_1(\xi_1,\xi_2,\xi_3)}{t}\,d\omega_t \tag{12}$$

(Kirchhoff's solution), where C_t is a disk in the $\xi_1\xi_2$-plane with center (x_1,x_2) and radius t, D_t is a sphere in the $\xi_1\xi_2\xi_3$-space with center (x_1,x_2,x_3) and radius t, and $d\omega_t$ is the surface element of D_t.

E. Second-Order Nonlinear Hyperbolic Equations

A second-order nonlinear differential equation

$$\frac{\partial^2 u}{\partial t^2}=A\left(t,x,u,\frac{\partial u}{\partial t},\frac{\partial u}{\partial x_i},\frac{\partial^2 u}{\partial t\partial x_i},\frac{\partial^2 u}{\partial x_i\partial x_j}\right),$$
$$1\leqslant i,\quad j\leqslant n, \tag{13}$$

is called **hyperbolic** in a neighborhood of a function $U(t,x)$ if the linear equation of the form (1) obtained from (13) is hyperbolic, where $a_{0i}(t,x)$ and $a_{ij}(t,x)$ are determined by substituting u by $U(t,x)$ in the partial derivatives of A with respect to $\partial^2u/\partial t\partial x_i$ and $\partial^2u/\partial x_i\partial x_j$, respectively. If the functions A in (13) and $U=x_0(x)+tu_1(x)$ determined by (4) are sufficiently smooth with respect to $t,x,u,\ldots,\partial^2u/\partial x_i\partial x_j$, the Cauchy problem for (13) with initial condition (4) has a unique solution in some neighborhood of the initial hyperplane under the condition that equation (13) is hyperbolic in a neighborhood of U. In general, initial value problems for nonlinear equations have only local solutions.

F. Higher-Order Hyperbolic Equations

An Nth-order linear differential equation in $n+1$ variables t, $x=(x_1,\ldots,x_n)$ with constant coefficients

$$L\left(\frac{\partial}{\partial t},\frac{\partial}{\partial x}\right)u=\sum_{\alpha_0+|\alpha|\leqslant N}a_{\alpha_0\alpha}\left(\frac{\partial}{\partial t}\right)^{\alpha_0}\left(\frac{\partial}{\partial x}\right)^{\alpha}u$$
$$=0, \tag{14}$$

where $\alpha=(\alpha_1,\ldots,\alpha_n)$, $|\alpha|=\alpha_1+\ldots+\alpha_n$, and

$$\left(\frac{\partial}{\partial x}\right)^{\alpha}=\frac{\partial^{|\alpha|}}{\partial x_1^{\alpha_1}\ldots\partial x_n^{\alpha_n}},$$

is **hyperbolic in the sense of Gårding** if the following two conditions are satisfied: (i) the partial derivative $\partial^N/\partial t^N$ appears in L; (ii) the real parts of the roots $\lambda=\lambda_1(\xi),\ldots,\lambda_N(\xi)$ of the **characteristic equation** $L(\lambda,i\xi)=0$ are bounded functions of real variables $\xi=(\xi_1,\ldots,\xi_n)$. When L is a homogeneous equation of the Nth order, condition (ii) is equivalent to the following condition: (ii′) $\lambda_1(\xi),\ldots,\lambda_N(\xi)$ are purely imaginary for all real $\xi=(\xi_1,\ldots,\xi_n)\neq(0,\ldots,0)$. The principal part (consisting of the highest-order terms) of a hyperbolic equation is also hyperbolic. If (14) is hyperbolic, a theorem analogous to theorem (C) holds for the Cauchy problem for (14) with initial conditions

$$\partial^k u/\partial t^k(0,x)=u_k(x),\qquad 0\leqslant k\leqslant n-1; \tag{15}$$

that is, the Cauchy problem for (14) with initial condition (15) is well posed in the sense of Hadamard. Conversely, if the Cauchy problem for (14) with initial condition (15) is well posed, then (14) must satisfy the hyperbolicity conditions (i) and (ii) in the sense of Gårding. In other words, well-posedness of the Cauchy problem is equivalent to hyperbolicity in the sense of L. Gårding [14]. Gårding's conditions for hyperbolicity cannot be generalized to the case of variable coefficients, since the influence of the lower-order terms in the equation is taken into account in the definition of hyperbolicity. However, in the case of constant coefficients, an Nth-order homogeneous equation remains hyperbolic for any addition of lower-order terms if and only if the characteristic equation has N distinct purely imaginary roots for any real $\xi=(\xi_1,\ldots,\xi_n)\neq(0,\ldots,0)$. In this case the equation is called **hyperbolic in the strict sense**. Thus a linear equation with variable coefficients

$$L[u]=\frac{\partial^N u}{\partial t^N}+\sum_{\substack{\alpha_0+|\alpha|\leqslant N\\ \alpha_0<N}} a_{\alpha_0\alpha}(t,x)\left(\frac{\partial}{\partial t}\right)^{\alpha_0}\left(\frac{\partial}{\partial x}\right)^{\alpha}u$$
$$=0 \qquad (16)$$

is called **hyperbolic in the sense of Petrovskiĭ** if the characteristic equation

$$\lambda^N+\sum_{\alpha_0+|\alpha|=N} a_{\alpha_0\alpha}(t,x)\lambda^{\alpha_0}(i\xi)^{\alpha}=0$$

has N distinct purely imaginary roots (called **characteristic roots**) $\lambda_1(t,x,\xi),\ldots,\lambda_N(t,x,\xi)$ for each point $p=(t,x)$ and each $\xi\neq 0$. Moreover, if the characteristic roots $\lambda_1,\ldots,\lambda_N$ are separated uniformly, i.e., the inequality

$$\lim_{(t,x),|\xi|=1,j\neq k}|\lambda_j(t,x,\xi)-\lambda_k(t,x,\xi)|=c>0$$

holds, (16) is said to be **regularly hyperbolic**. In the second-order case, this definition is equivalent to the previous one. Theorem (C) holds for the Cauchy problem for a regularly hyperbolic equation (16) with initial conditions (15). For the domain of dependence of the solution, a result analogous to the case of the second-order equation can be obtained using an energy inequality [9, 10]. If the coefficients are of class C^∞, Huygens's principle in the wider sense is valid, that is, discontinuity of the solution is carried over only along bicharacteristic curves.

G. Systems of Hyperbolic Equations

For systems of equations

$$\sum_{j=1}^{l} L_{ij}[u_j]=0, \qquad 1\leqslant i\leqslant l,$$

where the L_{ij} are higher-order linear differential operators of the form (16), several types of hyperbolicity are formulated in connection with the well-posedness of the Cauchy problem. We take up two important types, hyperbolicity in the sense of Petrovskiĭ and symmetric hyperbolicity due to Friedrichs.

We call a system of linear differential equations

$$\sum_{j=1}^{l}\sum_{\alpha_0+|\alpha|\leqslant n_j} a^{ij}_{\alpha_0\alpha}(t,x)\left(\frac{\partial}{\partial t}\right)^{\alpha_0}\left(\frac{\partial}{\partial x}\right)^{\alpha}u_j=0,$$
$$1\leqslant i\leqslant l, \qquad (17)$$

a **system of hyperbolic differential equations** (in the sense of Petrovskiĭ) if the determinant

$$\det\left(\sum_{\alpha_0+|\alpha|\leqslant n_j} a^{ij}_{\alpha_0\alpha}(t,x)\left(\frac{\partial}{\partial t}\right)^{\alpha_0}\left(\frac{\partial}{\partial x}\right)^{\alpha}\right), \qquad (18)$$

calculated formally using the matrix of differential operators in the system, is hyperbolic in the sense of Petrovskiĭ as a single equation of $N(=\sum_{j=1}^{l} n_j)$th order. Petrovskiĭ showed that the Cauchy problem for a system that is hyperbolic in this sense is well posed [10]. There were some imperfections in his argument, which have been corrected by others (→ S. Mizohata [12]). In the case of constant coefficients, the Cauchy problem for (17) is well posed if and only if (18) is hyperbolic in the sense of Gårding.

K. O. Friedrichs, observing that the energy inequality played an essential role in Petrovskiĭ's research, studied symmetric hyperbolic systems of equations, since for them the energy inequalities are valid most naturally. A system of first-order linear differential equations

$$A_0(t,x)\frac{\partial u}{\partial t}=\sum_{i=1}^{n} A_i(t,x)\frac{\partial u}{\partial x_i}+Bu \qquad (19)$$

is called **symmetric hyperbolic** (in the sense of Friedrichs) if the matrices $A_i(t,x)$ $(0\leqslant i\leqslant n)$ are symmetric and $A_0(t,x)$ is positive definite. A typical example is provided by Maxwell's equations. For this system it has been shown that the Cauchy problem is well posed and the domain of dependence of the solution is bounded [13].

H. Weakly Hyperbolic Operators

We adopt the following definition of hyperbolicity: a linear differential operator of Nth order

$$L=\frac{\partial^N}{\partial t^N}+\sum_{\substack{\alpha_0+|\alpha|\leqslant N\\ \alpha_0\leqslant N-1}} a_{\alpha_0\alpha}(t,x)\left(\frac{\partial}{\partial t}\right)^{\alpha_0}\left(\frac{\partial}{\partial x}\right)^{\alpha}$$

is called **hyperbolic** if the Cauchy problem for $L[u]=0$ with initial condition (15) is well posed in Hadamard's sense. A hyperbolic operator L is called **strongly hyperbolic** if L remains hyperbolic for any addition of lower-order terms, and **weakly hyperbolic** otherwise.

A necessary condition for hyperbolicity is that all characteristic roots of $L_N(t,x,\lambda,\xi)=0$ be real for any (t,x,ξ) (P. D. Lax [15], Mizohata [16]). In the case of constant coefficients, strongly hyperbolic operators have been characterized by K. Kasahara and M. Yamaguti (*Mem. Coll. Sci. Kyoto Univ.* (1960)).

As for operators with variable coefficients, it is known that not only regularly hyperbolic operators but also some special classes of not regularly hyperbolic operators are strongly hyperbolic (V. Ya. Ivriĭ, *Moscow Math. Soc.*, 1976). Let us consider the hyperbolicity of an operator L which is not regularly hyperbolic under the assumption that the multiplicities of the characteristic roots are constant and at most 2; namely, the characteristic polynomial

$$L_N(t,x,\lambda,\xi)=\lambda^N+\sum_{\substack{\alpha_0+|\alpha|=N\\ \alpha_0\leqslant N-1}} a_{\alpha_0\alpha}(t,x)\lambda^{\alpha_0}\xi^\alpha$$

is decomposed in the following way:

$$L_N(t,x,\lambda,\xi)=\prod_{j=1}^{s}(\lambda-\lambda_j(t,x,\xi))^2\prod_{j=s+1}^{N-s}(\lambda-\lambda_j(t,x,\xi)),$$

$$\lim_{\substack{(t,x)\\ |\xi|=1,\, j\neq k}} |\lambda_j(t,x,\xi)-\lambda_k(t,x,\xi)|=c>0.$$

Assume that the $\lambda_j(t,x,\xi)$, $j=1,2,\ldots,N-s$, are real. Then L is hyperbolic if and only if it satisfies **E. E. Levi's condition**, i.e.,

$$l_j(t,x,\xi)=\left[L_{N-1}+\frac{1}{2}\left(\frac{\partial^2 L_N}{\partial\lambda^2}\frac{\partial\lambda_j}{\partial t}+\sum_{\alpha=1}^{n}\frac{\partial^2 L_N}{\partial\lambda\partial\xi_\alpha}\frac{\partial\lambda_j}{\partial x_\alpha}\right)\right]_{\lambda=\lambda_j}=0$$

$$\text{for all } (t,x,\xi), \quad j=1,2,\ldots,s,$$

where L_{N-1} denotes the homogeneous part of $(N-1)$th order among lower-order terms of L (Mizohata and Y. Ohya [17]). Thus, for a **weakly hyperbolic** operator with variable coefficients, even the principal part is not necessarily hyperbolic. J. Chazarain [18] has studied weakly hyperbolic operators with characteristic roots of arbitrary constant multiplicity. O. A. Oleĭnik [19] studied the Cauchy problem with nonconstant multiplicities for second-order equations. For higher-order equations, if the multiplicity of characteristic roots at $(\hat{t},\hat{x})$ is at most p, or more precisely, if there exist positive rational numbers q and r $(q\geqslant r)$ such that

$$\left(\frac{\partial}{\partial\lambda}\right)^p L_N(\hat{t},\hat{x},0,\xi)\neq 0$$

and

$$\left(\frac{\partial}{\partial\lambda}\right)^{\alpha_0}\left(\frac{\partial}{\partial\xi}\right)^{\alpha}\left(\frac{\partial}{\partial t}\right)^{\beta_0}\left(\frac{\partial}{\partial x}\right)^{\beta} L_N(\hat{t},\hat{x},0,\xi)=0 \qquad ({}^\forall\xi\in\mathbf{R}^n\setminus 0)$$

for $\alpha_0+|\alpha|+q\beta_0+r|\beta|<p$, then it is necessary for the well posedness of the Cauchy problem that

$$\left(\frac{\partial}{\partial\lambda}\right)^{\alpha_0}\left(\frac{\partial}{\partial\xi}\right)^{\alpha}\left(\frac{\partial}{\partial t}\right)^{\beta_0}\left(\frac{\partial}{\partial x}\right)^{\beta} L_{N-s}(\hat{t},\hat{x},0,\xi)=0 \qquad ({}^\forall\xi\in\mathbf{R}^n\setminus 0)$$

for $\alpha_0+|\alpha|+q\beta_0+r|\beta|+s(1+q)<p$ are satisfied, where L_{N-s} $(1\leqslant s\leqslant N)$ are the homogeneous lower-order terms of order $N-s$ of L (Ivriĭ and V. M. Petkov [20]).

On the other hand, the sufficient condition in the case of multiplicity 2 is given by some conditions related to the subprincipal symbol

$$L_{N-1}-\frac{1}{2}\left(\frac{\partial^2 L_N}{\partial\lambda\partial t}+\sum_{\alpha=1}^{n}\frac{\partial^2 L_N}{\partial\xi_\alpha\partial x_\alpha}\right),$$

which corresponds to Levi's condition in the case of constant multiplicity, and to the †Poisson brackets (A. Menikoff, *Amer. J. Math.*, 1976; Ohya, *Ann. Scuola Norm. Sup. Pisa*, 1977; L. Hörmander, *J. Anal. Math.*, 1977).

The Cauchy problem for a weakly hyperbolic system of equations is more complicated, because of the essential difficulty that the matrix structure

$$\left(\sum_{\alpha_0+|\alpha|=n_j} a^{ij}_{\alpha_0\alpha}(t,x)\lambda^{\alpha_0}\xi^\alpha\right)$$

associated with (17) is not clear in general (→ references in [20]).

I. Gevrey Classes

Classically, the functions of †class s $(s\geqslant 1)$ of Gevrey (→ 58 C^∞-Functions and Quasi-Analytic Functions G; 168 Function Spaces B (14) were introduced into the studies of the fundamental solution for the heat equation:

$\gamma^{(s)}_{\mathrm{loc}}(\mathbf{R}^n)=\{\varphi(x)\in C^\infty(\mathbf{R}^n)\,|$ for any compact subset K of $\mathbf{R}^n$ and any multi-indices α, there exist constants C_K and A such that $\sup_K|(\partial/\partial x)^\alpha\varphi(x)|\leqslant C_K A^{|\alpha|}|\alpha|!^s\}$.

This class of functions was used efficiently in the studies of the Cauchy problem for †weakly hyperbolic partial differential equations:

$$L\left(t,x,\frac{\partial}{\partial t},\frac{\partial}{\partial x}\right)u(t,x)=f(t,x) \qquad \text{in} \quad [0,T]\times\mathbf{R}^n=\Omega,$$

$$\left(\frac{\partial}{\partial t}\right)^j u(0,x)=u_j(x), \qquad j=0,1,\ldots,N-1.$$

We assume that the multiplicities of the characteristic roots are constant, i.e.,

$$L_N(t,x,\lambda,\xi)=\prod_{i=1}^{k}(\lambda-\lambda_i(t,x,\xi))^{\nu_i},$$

where ν_i is constant for any $(t, x, \xi) \in \Omega \times \mathbf{R}^n$, $\lambda_i(t, x, \xi)$ is real and distinct, and $\sum_{i=1}^{k} \nu_i = N$. Let $\max_{1 \leqslant i \leqslant k} \nu_i = p$. If we suppose that $L(t, x, \partial/\partial t, \partial/\partial x) - \tilde{L}(t, x, \partial/\partial t, \partial/\partial x)$ is a (†pseudo) differential operator of order at most q, where

$$\tilde{L} = \prod_{i=1}^{p} a_i\left(t, x, \frac{\partial}{\partial t}, \frac{\partial}{\partial x}\right)$$

is a (pseudo)differential operator with $a_i(t, x, \partial/\partial t, \partial/\partial x)$ being strictly hyperbolic (pseudo) differential operators associated with L_N, then for any s such that $1 \leqslant s < p/q$, the Cauchy problem is well posed in $\gamma^{(s)}_{\text{loc}}(\Omega)$, provided that all $a_{\alpha_0\alpha}(t, x)$ of L and $f(t, x)$ belong to $\gamma^{(s)}_{\text{loc}}(\Omega)$, and that $\{u_j(x)\}_{0 \leqslant j \leqslant N-1}$ are given in $\gamma^{(s)}_{\text{loc}}(\mathbf{R}^n)$ (Ohya [21], J. Leray and Ohya [22]). This result was proved even for the case of arbitrary nonconstant multiplicity of characteristic roots by M. D. Bronshteĭn [23].

J. Lacunas for Hyperbolic Operators

The theory of **lacunas** of fundamental solutions of hyperbolic operators, initiated by Petrovskiĭ [24], has been developed further in a paper [25] by M. F. Atiyah, R. Bott, and L. Gårding.

Let $L(\xi) = L_N(\xi) + M(\xi)$ be a hyperbolic polynomial with respect to the vector $\theta \in \mathbf{R}^n - 0$, where $L_N(\xi)$ is the principal part of L; this means that $L_N(\theta) \neq 0$ and $L(\xi + t\theta) \neq 0$ when $|\mathrm{Im}\, t|$ is sufficiently large. Then L has a fundamental solution $E = E(L, \theta, x)$ in the form

$$E(L, \theta, x) = (2\pi)^{-n} \int L(\xi - ic\theta)^{-1} e^{ix(\xi - ic\theta)} d\xi,$$

where c is sufficiently large and the integral is taken in the sense of †distribution. The convex hull of the support of E, denoted by $K = K(L, \theta) = K(A, \theta)$, is a cone depending only on the real part $\mathrm{Re}\, A$ of the complex hypersurface $A: L(\xi) = 0$, and contained in the union of the origin and the half-space $x \cdot \theta > 0$. Let A_ξ be the †tangent conoid of A at ξ, transported to the origin, and define the wavefront surface $W(A, \theta)$ by the union of all $K(A_\xi, \theta)$ for $\xi \neq 0$. Then it can be shown that the singular supports of $E(L, \theta)$ and all the $E(L_N^k, \theta)$ are contained in $W(A, \theta)$ and, moreover, that they are locally holomorphic outside W. In [25], the Herglotz-Petrovskiĭ-Leray formulas are generalized to any nonstrict $L(\xi)$. Thus we have

$$D^\beta E(L_N^k, \theta, x) = \mathrm{const} \int_{\alpha^*} (x \cdot \xi)^q \xi^\beta L_N^k(\xi)^{-k} \omega(\xi), \quad q > 0, \tag{20}$$

$$D^\beta E(L_N^k, \theta, x) = \mathrm{const} \int_{t_x \cdot \partial\alpha^*} (x \cdot \xi)^q \xi^\beta L_N^k(\xi)^{-k} \omega(\xi), \quad q \leqslant 0 \tag{20'}$$

when $x \in K(A, \theta) \setminus W(A, \theta)$. Here, $q = mk - |\beta| - n$ is the degree of homogeneity of the left-hand side, and $\omega = \sum_{j=1}^{n} (-1)^{j-1} \xi_j d\xi_1 \wedge \ldots \wedge \widehat{d\xi_j} \wedge \ldots \wedge d\xi_n$. The integrands are closed †rational $(n-1)$-forms on $(n-1)$-dimensional complex †projective space and are integrated over certain †homology classes $\alpha^* = \alpha(A, \theta, x)^*$ and $t_x \cdot \partial\alpha^*$. These formulas provide means of obtaining topological criteria for lacunas. Let $\mathscr{L} \subset \mathring{K}$ be a maximal connected open set, where $E(L, \theta, x)$ is holomorphic. $\mathscr{L}$ is said to be a **weak (strong) lacuna** of L if $E(L, \theta, x)$ is the restriction of an entire function to $\mathscr{L}$ ($E(L, \theta, x) = 0$ in $\mathscr{L}$). In [25] it is shown that x belongs to a weak lacuna for all $E(L_N^k, \theta, \cdot)$ if and only if $\partial\alpha^* = 0$. The sufficiency directly follows from (20′) and the necessity follows from a theorem of A. Grothendieck (*Publ. Math. Inst. HES*, 1966) which implies that the rational forms which appear in (20′) span all the †cohomology classes in question.

K. Mixed Initial-Boundary Value Problems

Let Ω be a domain in $\mathbf{R}^n$ with a sufficiently smooth boundary Γ, let $L(t, x, \partial/\partial t, \partial/\partial x)$ be a linear hyperbolic operator of Nth order defined in $[0, \infty) \times \bar{\Omega} = \{(t, x) \mid t \in [0, \infty), x \in \bar{\Omega}\}$, and let $B_j(t, x, \partial/\partial t, \partial/\partial x), j - 1, 2, \ldots, b$, be linear differential operators of N_jth order defined in a neighborhood of $[0, \infty) \times \Gamma$.

The problem of finding a function $u(t, x)$ satisfying the conditions

$$L[u] = 0 \quad \text{in} \quad (0, \infty) \times \Omega,$$

$$B_j[u] = 0 \quad \text{on} \quad (0, \infty) \times \Gamma, \quad j = 1, 2, \ldots, b,$$

$$\partial^k u/\partial t^k(0, x) = u_k(x), \quad 0 \leqslant k \leqslant N - 1, \tag{21}$$

is called a **mixed initial-boundary value problem**. A typical example of such a mixed problem is provided by the case $L = \square$ $(n = 2)$ and $B[u] = u(t, x)$, which describes the vibration of membranes with a fixed boundary.

The mixed problem (21) is said to be well posed if for any initial data $u_k(x) \in C^\infty(\bar{\Omega})$, $0 \leqslant k \leqslant N - 1$, which are compatible with the boundary conditions, there exists a unique solution $u(t, x) \in C^\infty([0, \infty) \times \bar{\Omega})$.

Mixed problems for second-order hyperbolic equations are considered in [6]. In regard to mixed problems for hyperbolic equations of higher order, we make the following four assumptions: (i) $\Omega = \mathbf{R}^n_+ = \{(x', x_n) \mid x' \in \mathbf{R}^{n-1}, x_n > 0\}$; (ii) $\Gamma = \{x \mid x_n = 0\}$ is not characteristic for L or B_j; (iii) L is regularly hyperbolic; (iv) $N_j \leqslant N - 1$ and $N_j \neq N_k$ if $j \neq k$.

We denote the †principal parts of L and B_j by $L_N(t, x', x_n, \partial/\partial t, \partial/\partial x', \partial/\partial x_n)$ and $B_{j0}(t, x', \partial/\partial t, \partial/\partial x', \partial/\partial x_n)$, respectively. By the hyperbolicity of L, an equation in κ

$$L_N(t,x',0,\lambda,\xi',\kappa)=0 \quad \text{for} \quad \operatorname{Im}\lambda<0, \quad \xi'\in\mathbf{R}^{n-1}$$

has μ roots κ_j^+ with $\operatorname{Im}\kappa_j^+>0$, and $N-\mu$ roots κ_j^- with $\operatorname{Im}\kappa_j^-<0$, and the number μ is independent of (t,x') and (λ,ξ'). A necessary condition for the well-posedness of the mixed problem (21) is that the number of boundary conditions coincide with this integer μ. The function R defined by

$$R(t,x',\lambda,\xi')$$
$$=\det\left[\oint_C \frac{B_{j0}(t,x',\lambda,\xi',\kappa)\kappa^{l-1}}{L^+(t,x',\lambda,\xi',\kappa)}d\kappa\right]_{j,l=1,2,\ldots,\mu},$$

where $L^+=\prod_{j=1}^{\mu}(\kappa-\kappa_j^+(t,x',\lambda,\xi'))$ and C is a contour enclosing all κ_j^+, is called a **Lopatinski determinant**.

We say that L and B_j satisfy the uniform Lopatinski condition if

$$\inf_{\substack{(t,x'),\operatorname{Im}\lambda<0\\ |\xi'|+|\lambda|=1}} |R(t,x',\lambda,\xi')|=c>0.$$

When the uniform Lopatinski condition is satisfied, the mixed problem (21) is well posed, and (21) represents a phenomenon with a finite propagation speed, which is the same as that of the Cauchy problem for $L[u]=0$ (T. Balaban [26], H. O. Kreiss [27], and R. Sakamoto [28]). An analogous result holds in the case of a domain Ω with a compact boundary Γ, provided that L and B_j satisfy the uniform Lopatinski condition at every point of Γ.

In the treatment of mixed problems for L and B_j not satisfying the uniform Lopatinski condition, the well-posed problems have been characterized for operators with constant coefficients when $\Omega=\mathbf{R}^n_+$ (Sakamoto [29]). For general domains, however, the well-posedness of mixed problems depends not only on the properties of the Lopatinski determinant but also on the shape of the domain (M. Ikawa [30]).

L. Asymptotic Solutions

In order to explain some properties of phenomenon governed by hyperbolic equations, asymptotic solutions play an important role. Consider, for example, the **acoustic problem**

$$\square u=0 \quad \text{in} \quad (0,\infty)\times\Omega,$$
$$u=0 \quad \text{on} \quad (0,\infty)\times\Gamma.$$

Let $w(t,x;k)$ be a function defined in $(0,\infty)\times\Omega$ with parameter $k\geqslant 1$ of the form

$$w(t,x;k)=e^{ik(\varphi(x)-t)}\sum_{j=0}^{N} v_j(t,x)k^{-j}, \qquad (22)$$

where φ is a smooth function satisfying the **eikonal equation** $|\nabla\varphi|^2=1$. If v_j, $j=0,1,\ldots,N$, satisfy the **transport equations**

$$2\frac{\partial v_j}{\partial t}+2\nabla\varphi\cdot\nabla v_j+\Delta\varphi v_j=-i\square v_{j-1}, \quad v_{-1}=0,$$

we have

$$\square w=O(k^{-N}).$$

Then w of (22) is an approximate solution of $\square u=0$ for large k, and it represents a wave propagating in the direction $\nabla\varphi$.

When $\operatorname{supp} w\cap(0,\infty)\times\Gamma\neq\varphi$, if w hits the boundary Γ transversally, we can construct

$$w^+(t,x;k)=e^{ik(\varphi^+(x)-t)}\sum_{j=0}^{N} v_j^+(t,x)k^{-j}$$

such that

$$|\nabla\varphi^+|^2=1 \ \text{in}\ \Omega, \quad \varphi^+=\varphi, \quad \text{and}$$
$$\frac{\partial\varphi}{\partial\nu}>0 \ \text{on}\ \Gamma,$$

and v_j^+ satisfy the transport equations and $v_j^+=-v_j$ on $(0,\infty)\times\Gamma$, where ν is the unit inner normal of Γ. Then $w+w^+$ is an approximate solution, and w^+ represents a **reflected wave** propagating in the direction

$$\nabla\varphi^+=\nabla\varphi-2(\nabla\varphi\cdot\nu)\nu.$$

These asymptotic solutions show that the high-frequency waves propagate approximately according to the laws of †geometric optics.

If asymptotic solution (22) has a **caustic**, i.e., $\{\{x+l\nabla\varphi(x)\,|\,l\in R\}\,|\,x\in\operatorname{supp} w\}$ has an envelope, w of the form (22) cannot be an asymptotic solution near the caustic. The asymptotic behavior of high-frequency solutions near the caustic was first considered by G. B. Airy (*Trans. Cambridge Philos. Soc.*, 1838). Under the condition that the principal curvatures of the caustic are positive, w in the form (22) can be prolonged to a domain containing the caustic satisfying the asymptotic solution

$$w(t,x;k)=e^{ik(\theta(x)-t)}\{\operatorname{Ai}(-k^{2/3}\rho(x))g_0(t,x;k)$$
$$+ik^{-1/3}\operatorname{Ai}'(-k^{2/3}\rho(x))g_1(t,x;k)\}, \qquad (23)$$

where Ai is the Airy function

$$\operatorname{Ai}(z)=\frac{1}{2\pi}\int_{-\infty}^{\infty} e^{i(zt+t^3/3)}dt$$

and

$$g_l(t,x;k)=\sum_j g_{lj}(t,x)k^{1/6-j}$$

(D. Ludwig [31]).

Concerning the reflection of **grazing rays** by

strictly convex obstacles, the reflected wave can be constructed by the superposition of asymptotic solutions of the type (23).

The methods of construction of asymptotic solutions of the forms (22) and (23) are also applicable to Maxwell equations or more general hyperbolic systems (R. K. Luneberg [32]; Ludwig and Granoff, *J. Math. Anal. Appl.*, 1968; Guillemin and Sternberg, *Amer. Math. Soc. Math. Surveys*, 14 (1977)).

M. Propagation of Singularities

Let L be a hyperbolic operator with C^∞ coefficients and consider the Cauchy problem $L[u]=0$ with initial condition (15). When the initial data have singularities, the solution also has singularities for $t>0$, which is a property of hyperbolic equations quite different from the properties of parabolic ones. It should be noted that the †propagation of singularities cannot be derived from the Huygens principle in the wider sense, i.e., even for regularly hyperbolic operators of second order we cannot determine the location and the type of singularities of the solutions for initial data with singularities directly from the singularities of the fundamental solution $R(t, x; \tau, y)$.

Suppose that the multiplicities of characteristic roots of L are constant. Assume that u_k, $k=0, 1, \ldots, N-1$, have, on either side of a sufficiently smooth $(n-1)$-dimensional manifold Γ, continuous derivatives of sufficiently high order to suffer jump discontinuities across Γ. Then the solution u has continuous partial derivatives of sufficiently high order everywhere except on the characteristic surfaces of L issuing from Γ, and across these the partial derivatives of u have jump discontinuities (Courant and Hilbert [1]).

For more general singularities of initial data, it is known that the †wavefront propagates along the †bicharacteristic strips satisfying $\varphi_t - \lambda_l(t, x, \nabla\varphi)=0$, $l=1, 2, \ldots, N-s$, that is, $\mathrm{WF}u(\cdot, t)$ is contained in $\{(x(t), \xi(t))\in T^*(\mathbf{R}^n) \mid (dx_j/dt)(s)=(\partial\lambda_l/\partial\xi_j)(s, x, \xi), (d\xi_j/dt)(s)=-(\partial\lambda_l/\partial x_j)(s, x, \xi), (x(0), \xi(0))\in\bigcup_k \mathrm{WF}(u_k)\}$ (J. Chazarain [18]).

The propagation of singularities is more complicated in mixed problems because of the reflections of singularities at the boundary. For the †acoustic problem, R. B. Melrose [33] showed the following: Suppose $\mathcal{O}=C\Omega\subset\{x \mid |x|<R\}$ for some $R>0$, and all the broken rays according to the geometric optics starting from $\Omega_R=\Omega\cap\{|x|<R\}$ go out of Ω_R in a fixed time. Then for initial data with singularities in Ω_R, the solution becomes smooth in Ω_R for sufficiently large t.

References

[1] R. Courant and D. Hilbert, Methods of mathematical physics II, Interscience, 1962.
[2] J. Hadamard, Le problème de Cauchy et les équations aux dérivées partielles linéaires hyperboliques, Hermann, 1932.
[3] J. Hadamard, Leçons sur la propagation des ondes et les équations de l'hydrodynamique, Hermann, 1903.
[4] R. Courant and K. O. Friedrichs, Supersonic flow and shock waves, Interscience, 1948.
[5] J. Schauder, Das Anfangswertproblem einer quasilinearen hyperbolischen Differentialgleichung zweiter Ordnung, Fund. Math., 24 (1935), 213–246.
[6] M. Krzyżański and J. Schauder, Quasilineare Differentialgleichungen zweiter Ordnung vom hyperbolischen Typus, gemischte Randwertaufgaben, Studia Math., 6 (1936), 162–189.
[7] F. John, Plane waves and spherical means applied to partial differential equations, Interscience, 1955.
[8] S. L. Sobolev, Applications of functional analysis in mathematical physics, Amer. Math. Soc., 1963. (Original in Russian, 1950.)
[9] J. Leray, Hyperbolic differential equations, Lecture notes, Institute for Advanced Study, Princeton, 1952.
[10] I. G. Petrovskiĭ, Über das Cauchysche Problem für Systeme von partiellen Differentialgleichungen, Mat. Sb. (N.S.), 2 (44) (1937), 815–870.
[11] I. G. Petrovskiĭ, Lectures on partial differential equations, Interscience, 1954. (Original in Russian, 1953.)
[12] S. Mizohata, The theory of partial differential equations, Cambridge, 1973. (Original in Japanese, 1965.)
[13] K. O. Friedrichs, Symmetric hyperbolic linear differential equations, Comm. Pure Appl. Math., 7 (1954), 345–392.
[14] L. Gårding, Linear hyperbolic equations with constant coefficients, Acta Math., 85 (1951), 1–62.
[15] P. D. Lax, Asymptotic solutions of oscillatory initial value problems, Duke Math. J., 24 (1957), 627–646.
[16] S. Mizohata, Some remarks on the Cauchy problem, J. Math. Kyoto Univ., 1 (1961), 109–127.
[17] S. Mizohata and Y. Ohya, Sur la condition de E. E. Levi concernant des équations hyperboliques, Publ. Res. Inst. Math. Sci., 4 (1968), 511–526; Sur la condition d'hyper-

bolicité pour les équations à caractéristiques multiples II, Japan. J. Math., 40 (1971), 63–104.
[18] J. Chazarain, Opérateurs hyperboliques à caractéristiques de multiplicité constante, Ann. Inst. Fourier, 24 (1974), 173–202; Propagation des singularités pour une classe d'opérateurs à caractéristiques multiples et résolubilité locale, Ann. Inst. Fourier, 24 (1974), 203–223.
[19] O. A. Oleĭnik, On the Cauchy problem for weakly hyperbolic equations, Comm. Pure Appl. Math., 23 (1970), 569–586.
[20] V. Ya. Ivriĭ and V. M. Petkov, Necessary conditions for the correctness of the Cauchy problem for nonstrictly hyperbolic equations, Russian Math. Surveys, 29 (1974), 1–70. (Original in Russian, 1974.)
[21] Y. Ohya, Le problème de Cauchy pour les équations hyperboliques à caractéristique multiple, J. Math. Soc. Japan, 16 (1964), 268–286.
[22] J. Leray and Y. Ohya, Systèmes linéaires, hyperboliques non stricts, 2ième Colloque sur l'Analyse Fonctionnelle, CBRM (1964), 105–144.
[23] M. D. Bronshteĭn, The parametrix of the Cauchy problem for hyperbolic operators with characteristics of variable multiplicity, Functional Anal. Appl., 10 (1976) 4, 83–84. (Original in Russian, 1976.)
[24] I. G. Petrovskiĭ, On the diffusion of waves and the lacunas for hyperbolic equations, Mat. Sb. (N.S.), 17 (59) (1945), 289–370.
[25] M. F. Atiyah, R. Bott, and L. Gårding, Lacunas for hyperbolic differential operators with constant coefficients I, II, Acta Math., 124 (1970), 109–189; 134 (1973), 145–206.
[26] T. Balaban, On the mixed problem for a hyperbolic equation, Mem. Amer. Math. Soc., 112 (1971).
[27] H. O. Kreiss, Initial boundary value problems for hyperbolic systems, Comm. Pure Appl. Math., 23 (1970), 277–298.
[28] R. Sakamoto, Mixed problems for hyperbolic equations I, II, J. Math. Kyoto Univ., 10 (1970), 349–373, 403–417.
[29] R. Sakamoto, $\mathscr{E}$-well posedness for hyperbolic mixed problems with constant coefficients, J. Math. Kyoto Univ., 14 (1974), 93–118.
[30] M. Ikawa, On the mixed problems for the wave equation in an interior domain II, Osaka J. Math., 17 (1980), 253–279.
[31] D. Ludwig, Uniform asymptotic expansions at a caustic, Comm. Pure Appl. Math., 19 (1966), 215–250.
[32] R. K. Luneberg, Mathematical theory of optics, Brown Univ. Press, 1944.
[33] R. B. Melrose, Singularities and energy decay in acoustical scattering, Duke Math. J., 45 (1979), 43–59.

326 (XIII.27) Partial Differential Equations of Mixed Type

A. General Remarks

Let $A[u(x)]=0$ be a †quasilinear second-order partial differential equation. The type (†elliptic, †hyperbolic, or †parabolic) of the equation depends on the location of the point x. If the type varies as the point x moves, the equation is said to be of **mixed type**. An example is the equation

$$\left(1-\frac{u^2}{c^2}\right)\frac{\partial^2\varphi}{\partial x^2}-\frac{2uv}{c^2}\frac{\partial^2\varphi}{\partial x\,\partial y}+\left(1-\frac{v^2}{c^2}\right)\frac{\partial^2\varphi}{\partial y^2}=0 \tag{1}$$

of 2-dimensional stationary flow without rotation of a compressible fluid without viscosity, where φ is the velocity potential, $u=\partial\varphi/\partial x$ and $v=\partial\varphi/\partial y$ are the velocity components, and c is the local speed of sound, which is a known function of the speed $q=(u^2+v^2)^{1/2}$ of the flow. Equation (1) is of elliptic type if $q<c$, i.e., the flow is **subsonic**, and of hyperbolic type if $q>c$, i.e., the flow is **supersonic**. If there exist points where the flow is subsonic as well as points where it is supersonic, (1) is of mixed type. The study of equations of mixed type has become important with the development of high-speed jet planes.

B. Chaplygin's Differential Equation

It is difficult to solve equation (1) directly since it is nonlinear. However, we can linearize it by taking q and $\theta=\arctan(v/u)$ as independent variables (the hodograph transformation). The linearized equation takes the form

$$\frac{\partial^2 z}{\partial x^2}-K(x)\frac{\partial^2 z}{\partial y^2}=0,\qquad xK(x)\geqslant 0, \tag{2}$$

which is called **Chaplygin's differential equation**. Equation (2) is hyperbolic for $x>0$ and elliptic for $x<0$. The study of general equations of mixed type, even when they are linear, is much more difficult and less developed than the study of equations of nonmixed type. Almost all research so far has been on equation (2) or slight modifications of it.

C. Tricomi's Differential Equation

The simplest equation of the form (2) is

$$\frac{\partial^2 z}{\partial x^2}-x\frac{\partial^2 z}{\partial y^2}=0, \tag{3}$$

which is called **Tricomi's differential equation**. F. G. Tricomi considered the following boundary value problem for (3): In Fig. 1, AC and BC are two †characteristic curves of (3) and σ is a Jordan curve connecting A and B. We seek a solution of (3) in the domain D bounded by AC, BC, and σ that takes given values on σ and on one of the two characteristic curves, say on AC. This boundary value problem is called the **Tricomi problem**. Tricomi proved the existence and uniqueness of the solution of his problem under some conditions on the shape of σ and the smoothness of the boundary values. After Tricomi, much research has been done on his and similar problems for equations of form (2) [2]. We can also consider problems such as finding a solution of (3) (or of (2)) satisfying the initial conditions

$$z(0,y)=z_1(y), \qquad (\partial z/\partial x)(0,y)=z_2(y)$$

on the common boundary $x=0$ of the elliptic domain and the hyperbolic domain of the equation. This is called the **singular initial value problem**. S. Bergman [3] obtained an integral formula for the solution under the condition that $z_1(y)$ and $z_2(y)$ are real analytic.

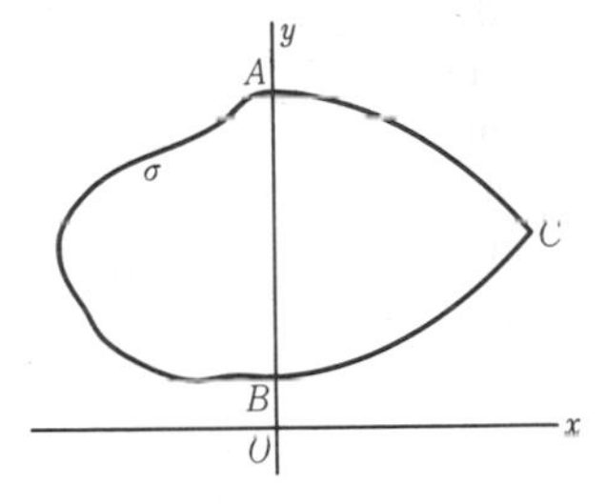

Fig. 1

D. Friedrichs's Theory

For the study of equations of mixed type it would of course be most convenient if there existed a general theory of boundary value problems independent of the type of the equation. However, constructing such a general theory is considered very difficult, because the †well-posedness of boundary conditions as well as the analytic properties of solutions are quite different according to the type. The first contributor to the solution of this difficult problem was K. O. Friedrichs [4], who noticed that although the methods of solving the †Cauchy problem and the †Dirichlet problem are quite different, both methods utilize energy integrals in the proof of the uniqueness of solutions. Using this observation, he succeeded in constructing a unified theory that enables us to treat various types (including the mixed type) of linear equations in a single scheme—an admissible boundary value problem for a **symmetric positive system** of first-order linear partial differential equations. There is, however, a difficulty in Friedrichs's theory since it does not give a unified procedure for reducing a given boundary value problem for a given equation to an admissible boundary value problem for a symmetric positive system of partial differential equations. The study of equations of mixed type that are of more general form than (2) by means of Friedrichs's theory is an open problem.

E. Further Studies

Work on equations of more general type than (2) or (3) has appeared (not all depending on Friedrichs's theory). For example, the following equations are treated in [5, 6, 7], respectively:

$$\sum_{i=1}^{n}\frac{\partial^2 z}{\partial x_i^2}+\frac{t}{|t|}\frac{\partial z}{\partial t}=0,$$

$$\left(\frac{\partial^2}{\partial x^2}+\frac{y}{|y|}\frac{\partial^2}{\partial y^2}\right)^n z=0,$$

$$G(y)\frac{\partial^2 z}{\partial x^2}+\frac{\partial^2 z}{\partial y^2}-K(y)z=h,$$

where $z=(z_1,\ldots,z_n)$ and $G(y)$ and $K(y)$ are symmetric matrices.

References

[1] F. Tricomi, Sulle equazioni lineari alle derivate parziali di 2° ordine di typo misto, Atti Acad. Naz. Lincei, (5) 14 (1923), 133–247.
[2] L. Bers, Mathematical aspects of subsonic and transonic gas dynamics, Wiley, 1958.
[3] S. Bergman, An initial value problem for a class of equations of mixed type, Bull. Amer. Math. Soc., 55 (1949), 165–174.
[4] K. O. Friedrichs, Symmetric positive linear differential equations, Comm. Pure Appl. Math., 11 (1958), 333–418.
[5] A. V. Bitsadze, On the problem of Equations of mixed type in multi-dimensional domains (in Russian), Dokl. Akad. Nauk SSSR, 110 (1956), 901–902.
[6] V. I. Zhegalov, Boundary value problem for higher-order equations of mixed type (in Russian), Dokl. Akad. Nauk SSSR, 136 (1961), 274–276.
[7] V. P. Didenko, Some systems of differential equations of mixed type (in Russian), Dokl. Akad. Nauk SSSR, 144 (1962), 709–712.
[8] A. V. Bitsadze, Equations of the mixed type, Pergamon, 1964. (Original in Russian, 1959.)
[9] M. M. Smirnov, Equations of mixed type, Amer. Math. Soc. Transl., 1978. (Original in Russian, 1970.)

327 (XIII.26) Partial Differential Equations of Parabolic Type

A. General Remarks

Consider a second-order linear partial differential equation

$$\sum_{i,j=1}^{n}\left[a_{ij}(x,t)\frac{\partial^2 u}{\partial x_i \partial x_j}+\sum_{i=1}^{n} b_i(x,t)\frac{\partial u}{\partial x_i}\right] + c(x,t)u-\frac{\partial u}{\partial t}=f \quad (1)$$

for an unknown function u of $(n+1)$ independent variables $(x,t)=(x_1,\ldots,x_n,t)$, where $a_{ij}=a_{ji}$. This equation is said to be **parabolic** (or of **parabolic type**) if and only if the quadratic form $\sum a_{ij}\xi_i\xi_j$ in ξ is positive definite at each point (x,t) of the region under consideration. x and t are sometimes called the variables of space and of time, respectively.

The most widely studied of the parabolic equations is the **equation of heat conduction** (or the **heat equation**):

$$L[u]\equiv\Delta u-\frac{\partial u}{\partial t}=0, \quad (2)$$

where $\Delta=\sum_{i=1}^{n}\partial^2/\partial x_i^2$ is the Laplacian taken over the space variables.

B. The Equation of Heat Conduction

The 1-dimensional case of the heat equation is

$$L[u]\equiv\frac{\partial^2 u}{\partial x^2}-\frac{\partial u}{\partial t}=0 \quad (3)$$

for the temperature $u(x,t)$ in a rod, considered as a function of the distance x measured along the rod and the time t. Equation (3) was one of the first treated in the theory of partial differential equations. Consider a finite rod with constant temperature 0 at its ends $x=a$ and $x=b$. Thermodynamics suggests that the initial temperature $\varphi(x)$ $(\varphi(a)=\varphi(b)=0)$ prescribed at $t=0$ is sufficient to determine the distribution of heat $u(x,t)$ in the rod at all later times $t>0$. On such physical grounds, we can expect that a solution to the following problem exists: Find a continuous function $u(x,t)$ that satisfies equation (3) for $a\leqslant x\leqslant b$, $t>0$, and the boundary conditions

$$u(a,t)=u(b,t)=0,$$

$$\lim_{t\downarrow 0}u(x,t)=\varphi(x), \qquad a\leqslant x\leqslant b. \quad (4)$$

According to J. Fourier the answer to this problem is expressed in a series $\sum_{n=1}^{\infty}c_n u_n$ constructed by superposition of the particular solutions $u_n=\sin\sqrt{\lambda_n}\,(x-a)\exp(-\lambda_n t)$. Here the λ_n are the roots of $\sin\sqrt{\lambda_n}\,(b-a)=0$, and the c_n are chosen so that $\sum_{n=1}^{\infty}c_n u_n(x,0)=\varphi(x)$. In fact, if $\varphi(x)$ is continuously differentiable, then $\sum_{n=1}^{\infty}c_n u_n(x,t)$ is the required solution. Thus we are led to the problem of expanding a given function $\varphi(x)$ in a †Fourier series.

The temperature distribution in an infinite rod is given by a continuous function $u(x,t)$ that satisfies equation (3) for $t>0$ and that, for $t=0$, takes values given by $\varphi(x)$, where

$$\lim_{t\downarrow 0}u(x,t)=\varphi(x). \quad (5)$$

If $\varphi(x)$ is bounded, then it can also be represented by superposition of particular solutions $e^{-(\alpha-x)^2/4t}$ of (3) as

$$u(x,t)=\begin{cases}\dfrac{1}{2\sqrt{\pi t}}\displaystyle\int_{-\infty}^{\infty}\varphi(\alpha)e^{-(\alpha-x)^2/4t}\,d\alpha, & t>0,\\ \varphi(x), & t=0.\end{cases} \quad (6)$$

Partial differential equations of parabolic type are important because of their connection with various phenomena in the physical world; they include not only equations that govern the flow of heat but also those that describe diffusion processes (→ 115 Diffusion Processes).

C. Partial Differential Equations of Parabolic Type in Two Variables

We are concerned mainly with the partial differential equation of parabolic type in two variables:

$$a(x,t)\frac{\partial^2 u}{\partial x^2}+2b(x,t)\frac{\partial^2 u}{\partial x\,\partial t}+c(x,t)\frac{\partial^2 u}{\partial t^2} + d(x,t)\frac{\partial u}{\partial x}+e(x,t)\frac{\partial u}{\partial t}+f(x,t)u=g, \quad (7)$$

with $ac=b^2$. In the region where $|a|+|c|>0$, equation (7) can be reduced to the form

$$\frac{\partial^2\omega}{\partial\xi^2}+d'(\xi,\tau)\frac{\partial\omega}{\partial\xi}+e'(\xi,\tau)\frac{\partial\omega}{\partial\tau}+f'(\xi,\tau)\omega=g' \quad (7')$$

by an appropriate change of variables $\xi=U(x,t)$, $\tau=V(x,t)$. If $e'<0$ in this region, we can assume without loss of generality that our equation takes the canonical form

$$a(x,t)\frac{\partial^2 u}{\partial x^2}+b(x,t)\frac{\partial u}{\partial x}+c(x,t)u-\frac{\partial u}{\partial t}=g, \quad (8)$$

with $a>0$, from the outset. It has the single family of †characteristics

$$t=\text{constant}. \quad (9)$$

There are four typical problems to be posed with regard to equation (8).

The first consists of determining, in some neighborhood of a given curve C nowhere tangent to a characteristic, a solution u that possesses prescribed values of u and $\partial u/\partial n$, or of a linear combination of u and $\partial u/\partial n$, along C. For instance, the problem of finding a solution $u(x,t)$ such that $u(x_0,t)=g(t)$ and $u_x(x_0,t)=h(t)$ for given functions $g(t)$ and $h(t)$ is a problem of this type. Consider the equation

$$\frac{\partial^2 u}{\partial x^2}-\frac{\partial u}{\partial t}=0 \tag{10}$$

in the region $a<t<b$, $x_0 \leqslant x$. According to E. Holmgren, a solution $u(x,t)$ satisfying the conditions

$$\lim_{x\downarrow x_0} u(x,t)=g(t), \qquad \lim_{x\downarrow x_0} u_x(x,t)=h(t), \tag{11}$$

where $g'(t)$ is bounded and continuous, exists if and only if

$$h(t)+\frac{1}{\sqrt{\pi}}\int_a^t \frac{g'(\tau)}{\sqrt{t-\tau}}d\tau=k(t) \tag{12}$$

is of class C^∞ and satisfies

$$|k^{(n)}(t)|\leqslant M(n!)^2/r^n \tag{13}$$

for positive constants M and r.

In the second type of problem we are required to find in a region of the form

$$\varphi_1(t)\leqslant x\leqslant \varphi_2(t), \qquad t_1<t<t_2, \tag{14}$$

a solution of (8) that takes prescribed values on part of the boundary of that region. Here we impose the hypothesis on the curves $x=\varphi_1(t)$ and $x=\varphi_2(t)$ that they are nowhere tangent to a characteristic (9). M. Gevrey [3] showed that if such a solution does exist, the functions $\varphi_1(t)$ and $\varphi_2(t)$ must satisfy the †Hölder condition with exponent $\alpha>1/2$:

$$|\varphi_i(t+h)-\varphi_i(t)|\leqslant c|h|^\alpha, \qquad c=\text{constant}, \tag{15}$$

for sufficiently small h. The problem of heat conduction mentioned in Section B corresponds to the particular case $\varphi_1(t)=\text{constant}$, $\varphi_2(t)=\text{constant}$, for which condition (15) is automatically fulfilled.

The third type of problem is to find in a region of the form

$$a\leqslant x\leqslant b, \qquad t>0, \tag{14'}$$

a solution of (8) that satisfies the conditions

$$\lim_{t\downarrow 0} u(x,t)=\varphi(x);$$

$$\frac{\partial u}{\partial x}-hu=0 \quad \text{for} \quad x=a, \quad h=\text{constant}>0;$$

$$\frac{\partial u}{\partial x}+Hu=0 \quad \text{for} \quad x=b, \quad H=\text{constant}>0. \tag{16}$$

This problem, posed for equation (10), is also a mathematical formulation of the problem of heat conduction in a rod [4]. If $\varphi(x)$ is of class C^1, then the solution to this problem can be expanded as

$$u(x,t)=\sum_{n=1}^{\infty} c_n e^{-\lambda_n t}\varphi_n(x),$$

$$c_n=\int_a^b \varphi(x)\varphi_n(x)\,dx, \tag{17}$$

where $\varphi_n(x)$ is a normalized function ($\int_a^b \varphi_n^2(x)dx=1$) that satisfies the boundary conditions

$$\varphi_n'(a)-h\varphi_n(a)=0, \qquad \varphi_n'(b)+H\varphi_n(b)=0 \tag{18}$$

and the equation

$$\frac{d^2\varphi_n(x)}{dx^2}=-\lambda_n\varphi_n(x). \tag{19}$$

The fourth type of problem is to find a function $u(x,t)$ that satisfies (8) for $t>0$ and the initial condition $\lim_{t\downarrow 0}u(x,t)=\varphi(x)$. It corresponds to the problem of heat conduction in an infinite rod.

D. Green's Formula

The †adjoint of the differential operator $L[u]$ in (3) is given by

$$M[v]\equiv\frac{\partial^2 v}{\partial x^2}+\frac{\partial v}{\partial t}. \tag{20}$$

Integration by parts yields the identity

$$\iint_G (\psi L[\varphi]-\varphi M[\psi])\,dx\,dt$$

$$=\int_C \varphi\psi\,dx+\int_C\left(\psi\frac{\partial\varphi}{\partial x}-\varphi\frac{\partial\psi}{\partial x}\right)dt, \tag{21}$$

where G is the region bounded by the closed curve C, and the line integral on the right is evaluated in the counterclockwise direction over C. We call (21) **Green's formula** for the partial differential equation of parabolic type (3). As in the case of partial differential equations of elliptic type (→ 323 Partial Differential Equations of Elliptic Type), this formula is used to establish the uniqueness of the solution of (3) and to derive an integral representation for it.

For example, the uniqueness of the solution is established in the following way: Let the curve $\widehat{AD}$ and $\widehat{BE}$ in Fig. 1 be such that no characteristic meets either of them in more than one point. If $u(x,t)$ is continuous in the closed region $(ABED)$, vanishes on $\widehat{AD}$, $\widehat{BE}$, and the segment $\overline{AB}$, and satisfies equation (3) in $(ABED)$ except on $\overline{AB}$, then it vanishes identically. Green's formula (21), applied to the region $G=(ABQP)$ and the functions $\psi\equiv 1$,

$\varphi \equiv u^2$, yields

$$2\iint_G \left(\frac{\partial u}{\partial x}\right)^2 dx\,dt + \int_{\overline{PQ}} u^2\,dx = 0.$$

In a similar way, by extending Green's formula suitably, we are able to prove uniqueness theorems for the four problems stated in Section C for more general linear parabolic equations.

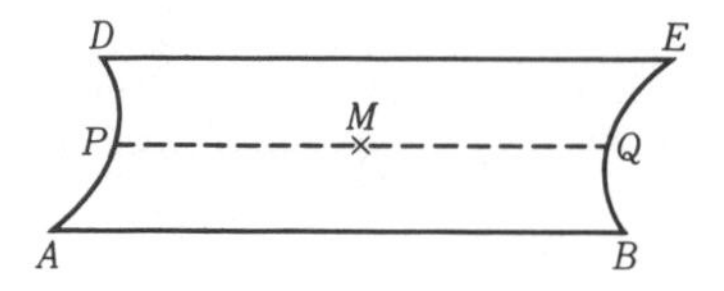

Fig. 1

To obtain a representation for solutions we proceed as follows: Let $u(x,t)$ be a solution of (3) and Let

$$U(\alpha,\beta,x,t) = \frac{1}{\sqrt{4\pi(\beta-t)}} e^{-(\alpha-x)^2/4(\beta-t)} \tag{22}$$

be a particular solution. Applying Green's formula to the region $(PABQMP)$ and the functions $\varphi = u(x,t)$, $\psi = U(x_0, t_0+h, x, t)$, where h is a positive number and M is a point with coordinates (x_0, t_0), we obtain

$$\int_{\overline{PQ}} u(x,t_0) e^{-(x_0-x)^2/4h} \frac{dx}{\sqrt{4\pi h}}$$

$$= \int_{\overline{PABQ}} u(x,t)U(x_0, t_0+h, x, t)\,dx + \left(U\frac{\partial u}{\partial x} - u\frac{\partial U}{\partial x}\right)dt.$$

Since the integral in the left-hand side of this equality approaches $u(x_0,t_0)$ as $h \downarrow 0$, we can establish the basic representation formula

$$u(x_0,t_0) = \int_{\overline{PABQ}} u(x,t)U(x_0,t_0,x,t)\,dx + \left(U\frac{\partial u}{\partial x} - u\frac{\partial U}{\partial x}\right)dt \tag{23}$$

for solutions u of (3). Formula (23) shows that $u(x_0,t_0)$ is determined in terms of the particular solution (22) if we know the values of u and $\partial u/\partial x$ on the part $\overline{PABQ}$ of the boundary of the region $(ABED)$. The function (22) is called the **fundamental solution** of (3) because it plays the same role as the fundamental solution $\log r$ $(r = ((\alpha-x)^2+(\beta-y)^2)^{1/2})$ of Laplace's equation $\partial^2 u/\partial x^2 + \partial^2 u/\partial y^2 = 0$.

Similarly, the following function E (called the **Gauss kernel**) is a fundamental solution of equation (2):

$$E(\alpha,\beta;x,t) = (4\pi(\beta-t))^{-n/2} \exp\left\{-\frac{|\alpha-x|^2}{4(\beta-t)}\right\}, \tag{24}$$

where $(x,\alpha) \in \mathbf{R}^n \times \mathbf{R}^n$ and $t < \beta$. For equation (3), the following **maximum principle** holds: In the region $(ABED)$ of Fig. 1, suppose that $L[u] \geqslant 0$ and that u takes its maximum value K at an interior point M. Then u is identically equal to K on the segment $\overline{QP}$ and in the region $(ABQP)$. More generally, various versions of the maximum principle are known for equation (1) [5].

E. The Laplace Transform Method

Let $u(x,t)$ be a solution of (3) for $t > 0$ and

$$v(x,\lambda) = \int_0^\infty e^{-\lambda t} u(x,t)\,dt, \qquad \lambda > 0, \tag{25}$$

be its †Laplace transform with respect to t. Utilizing integration by parts, we have

$$\int_0^\infty e^{-\lambda t} u_t(x,t)\,dt = [e^{-\lambda t}u(x,t)]_{t=0}^{t=\infty} + \lambda\int_0^\infty e^{-\lambda t}u(x,t)\,dt$$

$$= -\varphi(x) + \lambda v(x,\lambda), \tag{26}$$

provided that $\lim_{t\to\infty} e^{-\lambda t}u(x,t) = 0$ and $\lim_{t\downarrow 0} u(x,t) = \varphi(x)$. We find in view of (3):

$$\frac{\partial^2 v}{\partial x^2} = \lambda v(x,\lambda) - \varphi(x). \tag{26$'$}$$

Once the solution of (26′) has been found, the desired solution of (3) can be derived by inverting the Laplace transform (25). This idea can also be applied to the solution of parabolic equations with constant coefficients in $(n+1)$ variables, such as (1).

F. General Second-Order Equations of Parabolic Type

Consider the equation (1) with $f = 0$, which can be written as

$$\frac{\partial u}{\partial t} = A(t)u, \tag{27}$$

where $A(t)$ is a second-order †elliptic operator with parameter t. Let D be a region (bounded or unbounded) of points x whose boundary is a smooth hypersurface S. We pose the following initial boundary value problem for (27): Find a function $u(x,t)$ that satisfies in $D \times (0,\infty)$ equation (27) together with the conditions

$$\lim_{t\downarrow 0} u(x,t) = \varphi(x), \quad x \in D,$$

$$\partial u(x,t)/\partial n + h(x,t)u(x,t) = f(x,t), \quad x \in S, \tag{16$'$}$$

where $\partial/\partial n$ is the directional derivative in the outward †conormal direction at (x,t) $(x\in S)$, and $h(x,t)\geqslant 0$.

The Laplace transform is not suitable for solving problems (27) and (16′). Instead, the theory of 1-parameter semigroups of linear operators (→ 378 Semigroups of Operators and Evolution Equations) can be applied to establish similar fundamental results. Let m be a large positive integer. For $t>0$, put $t_k=k\delta$ for $k=0,1,\dots,m-1$ with $\delta=t/m$. By the Laplace transform method as described above, we can associate with ψ a unique solution v of $A(t_k)v=\lambda v-\psi$ with $\lambda=1/\delta$. We put $R_k\psi=\lambda v$. By iterating this procedure m times, we have a function $u_m(x,t)=R_{m-1}R_{m-2}\dots R_0\varphi$ starting from the initial value φ at $t=0$. Then we obtain a solution $u(x,t)$ of (27) and (16′) as the limit of $u_m(x,t)$ as $m\to\infty$. The following results are known [6,7]: (i) There exists a $u=U(\xi,\tau,x,t)$ $(x,\xi\in D, t>\tau\geqslant 0)$ that, as a function of x and t, satisfies equation (27) and the homogeneous boundary conditions (16′) with $\varphi=0$, $f=0$. (ii) The function $u(x,t)$ defined by

$$u(x,t)=\int_D\varphi(\xi)U(\xi,0,x,t)\,d\xi+\int_0^t d\tau\int_S f(\xi,\tau)\left(U(\xi,\tau,x,t)-\frac{\partial U(\xi,\tau,x,t)}{\partial n_\xi}\right)dS_\xi \quad (23')$$

is a solution of (27) and satisfies (16′). Thus $U(\xi,\tau,x,t)$ is a generalization of the function (24), called the **fundamental solution** of the linear parabolic equation (27) with boundary conditions (16′). Besides the properties (i) and (ii), the fundamental solution satisfies $U(\xi,\tau,x,t)\geqslant 0$, $\int_D U(\xi,\tau,z,w)U(z,w,x,t)\,dz=U(\xi,\tau,x,t)$ $(\tau<w<t)$, and further $\int_D U(\xi,\tau,x,t)\,dx=1$ under some additional assumptions. Therefore this theory is of considerable significance from the point of view of the theory of probability (→ 115 Diffusion Processes).

It can be shown that a weak solution of the parabolic equation (27) is a genuine solution. That is, if $u(x,t)$ is locally summable and

$$\int_0^\infty\int_D u(x,t)\left(\frac{\partial\varphi(x,t)}{\partial t}+A(t)^*\varphi(x,t)\right)dx\,dt=0$$

for any function $\varphi(x,t)$ of class C^2 in $D\times(0,\infty)$ and vanishing outside a compact subset of D ($A(t)^*$ is the adjoint of the partial differential operator $A(t)$ and $dx=dx_1\dots dx_{n-1}$), then $u(x,t)$ satisfies (27) in $D\times(0,\infty)$ in the usual sense. In particular, when the coefficients of $A(t)$ are infinitely differentiable, any solution $u(x,t)$ of (27) in the distribution sense is a genuine solution (→ 125 Distributions and Hyperfunctions).

If the function h in the boundary conditions (16′) and the coefficients of A are independent of t, then the fundamental solution $U(\xi,\tau,x,t)$ depends only on ξ, x, and $t-\tau$ and is written as $U(\xi,x;t-\tau)$ $(t>\tau)$. Furthermore, if A is †self-adjoint, then there exist a sequence of †eigenfunctions $\{\psi_p(x;\lambda)\mid p=1,2,\dots\}$ $(A\psi_p+\lambda\psi_p=0)$ and a sequence $\{\rho_p(\Lambda)\}$ of measures on the real line for which the following hold: (1) The fundamental solution $U(\xi,x;t)$ is expressed in the form

$$U(\xi,x;t)=\sum_{p=1}^\infty\int_{-\infty}^\infty e^{-\lambda t}\psi_p(\xi;\lambda)\psi_p(x;\lambda)\,d\rho_p(\lambda).$$

(2) The solution $u(x,t)$ of (27) satisfying (16′) with $f(x,t)\equiv 0$ is expanded as

$$u(x,t)=\sum_{p=1}^\infty\int_{-\infty}^\infty e^{-\lambda t}\psi_p(x;\lambda)\varphi_p(\lambda)\,d\rho_p(\lambda),$$

where

$$\varphi_p(\lambda)=\int_D\psi_p(x;\lambda)\varphi(x)\,dx$$

($\varphi(x)$ is the function given in (16′).

G. Nash's Results

Let us consider a parabolic equation

$$\frac{\partial u}{\partial t}=\sum_{i,j=1}^n\frac{\partial}{\partial x_j}\left\{a_{ij}(x,t)\frac{\partial u}{\partial x_i}\right\}, \quad (28)$$

where $a_{ij}=a_{ji}$ are real-valued functions of class C^∞ and equal to constants outside a fixed compact set of $\mathbf{R}^n$ for all $t\geqslant t_0$ (this regularity assumption can be relaxed). Suppose that there exists a constant $\lambda\geqslant 1$ such that $\lambda^{-1}|\xi|^2\leqslant\sum a_{ij}(x,t)\xi_i\xi_j\leqslant\lambda|\xi|^2$ for all $(t,x,\xi)\in(t_0,\infty)\times\mathbf{R}^n\times\mathbf{R}^n$. Then, for any bounded solution $u(x,t)$ of (28) in $(t_0,\infty)\times\mathbf{R}^n$ and for any (x,y,t,s) such that $x\in\mathbf{R}^n$, $y\in\mathbf{R}^n$, and $t_0<t\leqslant s$, the inequality

$$|u(x,t)-u(y,s)|\leqslant AB\left\{\left(\frac{x-y}{\sqrt{t-t_0}}\right)^\alpha+\left(\frac{s-t}{t-t_0}\right)^\beta\right\}, \quad (29)$$

holds, where $B=\sup\{|u(x,t)|\mid t\geqslant t_0, x\in\mathbf{R}^n\}$ and $\beta=\alpha/(2\alpha+2)$. In this inequality, the constants α and A are positive, depending on (n,λ) but independent of the particular choice of (a_{ij}), t_0 and of u [9].

As a corollary to this theorem, J. Nash proved that, if the a_{ij} do not contain t and if $v(x)$ is a bounded solution in $\mathbf{R}^n$ of the elliptic equation obtained by replacing $\partial u/\partial t$ by 0 in (28), the inequality

$$|v(x)-v(y)|\leqslant A'B'|x-y|^\sigma \quad (30)$$

holds for any $(x,y)\in\mathbf{R}^n\times\mathbf{R}^n$, where $\sigma=\alpha/(\alpha+1)$ with α in (29) and $B'=\sup|v(x)|$. The constant

A' depends only on (n, λ) (→ 323 Partial Differential Equations of Elliptic Type L).

H. Partial Differential Equations of p-Parabolic Type

Let p and m be given positive integers. Let us consider an equation for an unknown function u of $(n+1)$ independent variables (x, t) of the type

$$\frac{\partial^m u}{\partial t^m}+\sum_{\alpha,j} a_{\alpha,j}(x,t)\frac{\partial^\alpha}{\partial x^\alpha}\frac{\partial^j u}{\partial t^j}=f, \qquad (31)$$

where $\alpha=(\alpha_1,\ldots,\alpha_n)$ and $\partial^\alpha/\partial x^\alpha=(\partial/\partial x_1)^{\alpha_1}\ldots(\partial/\partial x_n)^{\alpha_n}$. We write also $|\alpha|=\alpha_1+\ldots+\alpha_n$. In (31), $\Sigma_{\alpha,j}$ is the summation taken over the (α,j) such that $pj+|\alpha|\leqslant pm$ and $0\leqslant j<m$. Let us denote by $\{\lambda_k(x,t,\xi)\}_{k=1}^m$ the roots λ of the equation

$$\lambda^m+\sum_{\alpha,j}{}' a_{\alpha,j}(x,t)(i\xi)^\alpha\lambda^j=0, \qquad (32)$$

where $\Sigma'_{\alpha,j}$ is the summation over the (α,j)'s such that $pj+|\alpha|=pm$ and that $0\leqslant j<m$. We say that the equation (31) is ***p*-parabolic** (or of ***p*-parabolic type**) in the sense of I. G. Petrovskiĭ if and only if there exists a positive number δ such that

$$\operatorname{Re}\lambda_k(x,t,\xi)\leqslant-\delta|\xi|^p, \quad 1\leqslant k\leqslant m, \qquad (33)$$

for any (x,t) in the region under consideration and for any $\xi\in\mathbf{R}^n$. The integer p is then seen to be even. Equation (27) is p-parabolic if $-A(t)$ is strongly elliptic of order p. The heat equation (2) is 2-parabolic in this sense. Similarly, we can define the p-parabolic systems of equations [10].

p-parabolic equations are known to be †hypoelliptic if the coefficients are of class C^∞ [11]. S. D. Eĭdel'man obtained precise estimates of the fundamental solutions and of their derivatives for p-parabolic equations [10]. The mixed initial boundary value problems are investigated in detail also by Eĭdel'man [10] and by R. Arima (*J. Math. Kyoto Univ.*, 4 (1964)).

References

[1] L. Bers, S. Bochner, and F. John, Contributions to the theory of partial differential equations, Ann. Math. Studies, Princeton Univ. Press, 1954.
[2] A. Friedman, Partial differential equations of parabolic type, Prentice-Hall, 1964.
[3] M. Gevrey, Les équations aux dérivées partielles du type paraboliques, J. Math. Pures Appl., 9 (1913), 305–471; 10 (1914), 105–147.
[4] R. Courant and D. Hilbert, Methods of mathematical physics, Interscience, I, 1953, II, 1962.
[5] M. H. Protter and H. F. Weinberger, Maximum principles in differential equations, Prentice-Hall, 1967.
[6] S. Ito, Fundamental solutions of parabolic differential equations and boundary value problems, Japan, J. Math., 27 (1957), 55–102.
[7] A. M. Il'in, A. S. Kalashnikov, and O. A. Oleĭnik, Linear equations of the second order of parabolic type, Russian Math. Surveys, 17-3 (1962), 1–143. (Original in Russian, 1962.)
[8] J. L. Lions, Equations différentielles opérationnelles et problèmes aux limites, Springer, 1961.
[9] J. Nash, Continuity of solutions of parabolic and elliptic equations, Amer. J. Math., 80 (1958), 931–954.
[10] S. D. Eĭdel'man, Parabolic systems, North-Holland, 1969. (Original in Russian, 1963.)
[11] S. Mizohata, Hypoellipticité des équations paraboliques, Bull. Soc. Math. France, 85 (1957), 15–50.

328 (V.6) Partitions of Numbers

A **partition** of a positive integer n is an expression of n as the sum of positive integers. The number of partitions of n, where the order of the summands is ignored and repetition is permitted, is denoted by $p(n)$ and is called the **number of partitions** of n. For example $p(5)=7$ since $5=4+1=3+2=3+1+1=2+2+1=2+1+1+1=1+1+1+1+1$. Therefore, $p(n)$ equals the number of †conjugate classes of the †symmetric group of order n and is closely related to the †representation theory of this group.

The †generating function of $p(n)$ is

$$f(x)=1+\sum_{n=1}^{\infty}p(n)x^n=\left(\sum_{n=1}^{\infty}(1-x^n)\right)^{-1}.$$

The unit circle $|x|=1$ is the †natural boundary of $f(x)$, which is holomorphic in $|x|<1$. The **Dedekind eta function**, which is closely related to $f(x)$, is defined by the following formula for the complex variable τ taking values in the upper half-plane:

$$\eta(\tau)=\exp(\pi i\tau/12)\prod_{n=1}^{\infty}(1-\exp(2\pi i n\tau)).$$

Hence $\eta(\tau+1)=\exp(\pi i/12)\eta(\tau)$. L. Euler (1748) obtained the following formula (called the **pentagonal number theorem** because $n(3n-1)/2$

is a †pentagonal number):

$$\prod_{n=1}^{\infty}(1-x^n)$$

$$=1+\sum_{n=1}^{\infty}(-1)^n(x^{n(3n-1)/2}+x^{n(3n+1)/2}).$$

This follows easily from the **Jacobi-Biehler equality**

$$\prod_{n=1}^{\infty}\{(1-q^{2n})(1+q^{2n-1}z)(1+q^{2n-1}z^{-1})\}$$

$$=1+\sum_{n=1}^{\infty}q^{n^2}(z^n+z^{-n}) \qquad (|q|<1, \quad z\neq 0).$$

By using the †transformation formula for ϑ-functions, we can infer from the pentagonal number theorem that $\eta(-1/\tau)=\sqrt{\tau/i}\,\eta(\tau)$. Hence, if a, b, c, d are integers satisfying $ad-bc=1$ and $c>0$, then

$$\eta\left(\frac{a\tau+b}{c\tau+d}\right)=\varepsilon\sqrt{\frac{c\tau+d}{i}}\,\eta(\tau),$$

where ε is a 24th root of unity. It is known that $\eta(\tau)$ is a †cusp form of weight $-1/2$ with respect to the full †modular group $\Gamma(1)$ [9, 11]. C. L. Siegel (1955) gave a simple proof of the formula $\eta(-1/\tau)=\sqrt{2/i}\,\eta(\tau)$. Later S. Iseki (1957) gave another proof by using a new method, known as the $\alpha-\beta$ formula [12].

The size of $p(n)$ increases rapidly with n; for instance $p(10)=42$ and $p(100)=190{,}569{,}292$. By making use of a remarkable identity, G. H. Hardy and S. Ramanujan (1918) proved the following inequalities, where A and B are suitable constants:

$$(A/n)e^{2\sqrt{n}}<p(n)<(B/n)e^{2\sqrt{2}\sqrt{n}}.$$

Subsequently they obtained

$$p(n)\sim(1/4\sqrt{3}\,n)\exp(\pi\sqrt{2n/3}).$$

After, P. Erdös (1942) and D. J. Newman (1951), A. G. Postnikov [8] succeeded in proving

$$p(n)=\frac{\exp(\pi\sqrt{2n/3})}{4\sqrt{3}\,n}\left(1+O\left(\frac{\log n}{n^{1/4}}\right)\right)$$

by means of an elementary function-theoretic method. Multiplying both sides of Euler's formula by the generating function of $p(n)$ and comparing the coefficients, we obtain

$$\sum_{0\leqslant\omega_k\leqslant n}(-1)^{k-1}p(n-\omega_k)=0,$$

where $\omega_k=k(3k-1)/2$ $(k=0,\pm1,\pm2,\ldots)$ is a pentagonal number. From this formula we can calculate $p(n)$ successively; in fact P. A. MacMahon obtained in this way the values of $p(n)$ for n up to 200.

Hardy and Ramanujan proved the following **transformation formula** for the generating function $f(x)$ of $p(n)$:

$$f\left(\exp\left(\frac{2\pi ih}{k}-\frac{2\pi z}{k}\right)\right)$$

$$=W_{h,k}\sqrt{z}\exp\left(\frac{\pi}{12kz}-\frac{\pi z}{12k}\right)$$

$$\times f\left(\exp\left(\frac{2\pi ih'}{k}-\frac{2\pi}{kz}\right)\right),$$

$$(h,k)=1, \quad hh'\equiv-1\ (\mathrm{mod}\ k),$$

where $W_{h,k}$ is defined by

$$W_{h,k}=\exp(\pi is(h,k))$$

and the value of $s(h,k)$ was given by Rademacher in the form

$$s(h,k)=\sum_{m=1}^{k-1}\frac{m}{k}\left(\left(\frac{hm}{k}\right)\right).$$

Here the symbol $((t))$ in the sum denotes the function that is 0 for integral t and $t-[t]-1/2$ otherwise ([ ] is the †Gauss symbol). With regard to the **Dedekind sum** $s(h,k)$, we have the **reciprocity law for Dedekind sums**:

$$s(h,k)+s(k,h)=-\frac{1}{4}+\frac{1}{12}\left(\frac{h}{k}+\frac{k}{h}+\frac{1}{hk}\right).$$

If we make substitutions $a=h'$, $b=(hh'+1)/k$, $c=k$, and $d=-h$ in the Hardy-Ramanujan transformation formula, then the ε appearing in the transformation formula of $\eta(\tau)$ is seen to be equal to $\exp(-\pi is(c,d)+\pi i(a+d)/12c)$. A direct proof of the transformation formula and the reciprocity law was given by K. Iseki (1952).

According to Cauchy's integral formula, $p(n)$ can be represented as an integral:

$$\frac{1}{2\pi i}\int_{\Gamma}\frac{f(x)}{x^{n+1}}dx,$$

where the contour Γ is taken inside the unit circle around the origin. The generating function $f(x)$ varies greatly: namely, letting $r\to1-0$ in $x=r\exp(2\pi ip/q)$, where p and q are fixed integers, it follows from the transformation formula that $f(x)\sim\exp(\pi^2/6q^2(1-r))$. Nevertheless, we can deal with the integral by the †circle method, introduced by Hardy and Ramanujan, which threw light on recent additive number theory. Hardy and Ramanujan thus obtained

$$p(n)=\frac{1}{2\pi\sqrt{2}}\frac{d}{dn}\left(\frac{\exp(C\lambda_n)}{\lambda_n}\right)+O(\exp(D\sqrt{n})),$$

where

$$\lambda_n=\sqrt{n-1/24}, \quad C=\pi\sqrt{2/3}, \quad D>C/2.$$

The theory was improved by Rademacher (1937, 1943), who expanded $p(n)$ into the series

$$p(n)=\frac{1}{\pi\sqrt{2}}\sum_{k=1}^{\infty} A_k(n)k^{1/2}\frac{d}{dn}\left(\frac{\sinh(C\lambda_n/k)}{\lambda_n}\right),$$

where

$$A_k(n)=\sum_{h(\operatorname{mod} k),(h,k)=1} W_{h,k}\exp(-2\pi i h n/k).$$

Rademacher (1954) proved further that

$$p(n)=2\pi\left(\frac{\pi}{12}\right)^{3/2} \times\sum_{k=1}^{\infty} A_k(n)k^{-5/2}L_{3/2}\left(\left(\frac{\pi}{12k}\right)^2(24n-1)\right),$$

where

$$L_\nu(z)=\sum_{n=0}^{\infty}\frac{z^n}{n!\,\Gamma(\nu+n+1)}.$$

Rademacher (1943) had developed an ingenious proof by taking "Ford's circle" as the contour Γ.

Ramanujan observed that $p(5m+4)\equiv 0$ (mod 5), $p(7m+5)\equiv 0$ (mod 7), and $p(11m+6)\equiv 0$ (mod 11). Rademacher (1942) and Newman studied these cases by using $\eta(\tau)$. More generally, A. O. L. Atkin proved that if $24n\equiv 1$ (mod $5^a7^b11^c$) then $p(n)\equiv 0$ (mod $5^a7^{1+[b/2]}11^c$) (*Glasgow Math. J.*, 8 (1967)). At present, this is the best result.

Let $n=n_1+n_2+\ldots+n_s$ be a partition of n. Many problems arise when we put additional conditions on the n_j. For instance, we may require that the n_j satisfy certain congruence relations (L. K. Hua (1942), S. Iseki (1959)) or are powers of integers (E. M. J. Wright (1934), L. Schoenfeld (1944), S. Iseki (1959)) or are powers of primes (T. Mitsui (1957)). The partition problem can also be extended to the case of an algebraic number field of finite degree (Rademacher, G. Meinardus (1953), Mitsui (1978)).

References

[1] R. G. Ayoub, An introduction to the analytic theory of numbers, Amer. Math. Soc. Math. Surveys, 1963.
[2] G. H. Hardy and S. Ramanujan, Asymptotic formulae in combinatory analysis, Proc. London Math. Soc., (2) 17 (1918), 75–115.
[3] G. H. Hardy and E. M. Wright, An introduction to the theory of numbers, Clarendon Press, fourth edition, 1965.
[4] K. Iseki, A proof of a transformation formula in the theory of partitions, J. Math. Soc. Japan, 4 (1952), 14–26.
[5] H. A. Rademacher, Lectures on analytic number theory, Tata Inst. Fund. Res., 1954–1955.
[6] S. Ramanujan, Collected papers of Srinivasa Ramanujan, Cambridge Univ. Press, 1927 (Chelsea, 1962).
[7] H. Petersson, Über Modulfunktionen und Partitionenprobleme, Abh. Deutsch. Akad. Wiss., Berlin Kl. Math. Allg. Nat., 1954, no. 2.
[8] A. G. Postnikov, Introduction to the analytical theory of numbers (in Russian), Moscow, 1971.
[9] M. I. Knopp, Modular functions in analytic number theory, Markham, 1970.
[10] T. Mitsui, On the partition problem in an algebraic number field, Tokyo J. Math., 1 (1978).
[11] H. A. Rademacher, Topics in analytic number theory, Springer, 1973.
[12] S. Iseki, The transformation formula for the Dedekind modular function and related functional equations, Duke Math. J., 24 (1957), 653–662.

329 (XXI.38) Pascal, Blaise

Blaise Pascal (June 19, 1623–August 19, 1662) was born in Clermont-Ferrand in southern France. He lost his mother when still an infant and was brought up by his father, Etienne Pascal (discoverer of the curve called Pascal's †limaçon). As a youth, he demonstrated a remarkable ability for mathematics. In 1640, under the influence of Desargues, he discovered †Pascal's theorem on conic sections, and in 1642 invented an adding machine. After hearing of Toricelli's experiments in 1646, he became interested in the theory of fluids and on his own began to conduct experiments; this research put to rest the prevailing opinion that nature abhors a vacuum and that, therefore, a vacuum cannot exist. Pascal formulated the principle stating that pressure, when applied at any point within a contained liquid, is transmitted throughout the fluid. By means of this principle, he explained various phenomena concerning fluids such as the atmosphere and laid the foundations for hydrostatics.

Between 1652 and 1654, Pascal was preoccupied with social affairs, but subsequently he began to devote himself to religion. He entered the Abbey Port-Royal of the Jansenist sect, where he remained until his death. Immediately before his entry, however, he and Fermat exchanged correspondence about games of chance, and these letters proved to be the beginning of the theory of †probability. Concerning games of chance, Pascal had conducted research on †Pascal's triangle, and in this study he formulated and used †mathematical induction. He also indicated a way to obtain the sum of the *m*th powers of the consecutive terms of an arithmetic progression, and with an intuitive idea of limits obtained the formula $\int_0^a x^m dx = a^{m+1}/(m+1)$. While in Port-Royal, he published *Lettres provinciales* (1657), in which he carried on a dispute with the Jesuits. His book *Pensées* shows his deep involvement with religion; however, he did not abandon mathematics. In 1658, he determined the area enclosed by a †cycloid and its base, the barycenter and area of the figure enclosed by a cycloid and straight lines parallel to its base, and the volume of the figure obtained by rotating it around these lines. The study of the methods used by Pascal to obtain these results, which were forerunners of differential and integral calculus, led †Leibniz to discover the fundamental theorem of calculus. Pascal also formulated clear ideas about axioms.

References

[1] L. Brunschvicg (ed.), B. Pascal, Oeuvres I–XIV, Hachette, 1904–1914.
[2] J. Chevalier (ed.), B. Pascal, Oeuvres complètes, Bibliothèque de Pléiades, Gallimard, 1954.
[3] Kôkiti Hara, L'Oeuvre mathématique de Pascal, Mém. de la Faculté des Lettres de l'Univ. d'Osaka, no. 21, 1981.

330 (II.8) Permutations and Combinations

Let there be given a set Ω of n elements. If we choose k distinct elements of Ω and arrange them in a row, we have a ***k*-array** or ***k*-permutation** of elements of Ω. The number of such arrays is $(n)_k = n(n-1)\dots(n-k+1)$. The polynomial $(x)_k = x(x-1)\dots(x-k+1)$ in x of degree k is called the **Jordan factorial** of degree k. In particular, $(n)_n = n!$, ***n* factorial**, is the number of permutations of Ω. A subset of Ω is called a ***k*-subset** if it contains exactly k elements. The number of k-subsets (or ***k*-combinations**) of Ω is $\binom{n}{k} = (n)_k/k!$. The **binomial coefficients** $\binom{x}{n}$ are defined by the generating function $(1+z)^x = \sum_{n=0}^{\infty} \binom{x}{n} z^n$. For any complex number x, the series is convergent for $|z|<1$, and it is verified that $\binom{x}{n} = (x)_n/n!$ in terms of the Jordan factorial $(x)_n$. The same results hold in a †complete field with †valuation, in particular in a †p-adic number field. In any case, we have the recursive relation

$$\binom{x}{n} = \binom{x-1}{n} + \binom{x-1}{n-1} \qquad (n \geqslant 1),$$

$$\binom{x}{0} = 1,$$

and in general

$$\sum_{k=0}^{n} \binom{x}{k}\binom{y}{n-k} = \binom{x+y}{n},$$

which leads to many identities involving binomial coefficients. The recursive relation allows us to compute the values of $\binom{n}{k}$ easily for small integers n, k, as was noticed by Pascal. The arrangement of these values in a triangular form:

```
        1
      1   1
    1   2   1
  1   3   3   1
1   4   6   4   1
```

is called **Pascal's triangle**. For integral values x, $(1+z)^x$ are polynomials, and we have $(a+b)^n = \sum_{k=0}^{n} \binom{n}{k} a^{n-k} b^k$ (**binomial theorem**). As a generalization, we have

$$(a_1 + \dots + a_m)^n = \sum \frac{n!}{p_1! \dots p_m!} a_1^{p_1} \dots a_m^{p_m}$$

(**multinomial theorem**), where the sum is extended over all nonnegative p_i with $\sum p_i = n$.

The number of ways of choosing k elements, allowing repetition, from a set of n elements is $(-1)^k \binom{-n}{k} = \binom{n+k-1}{k}$. This is also the number of nonnegative integral solutions of $\sum_{i=1}^{n} x_i = k$. As an example of binomial coefficients with noninteger arguments, we have

$$\binom{-1/2}{n} = (-1)^n 2^{-2n} \binom{2n}{n}.$$

References

[1] E. Netto, Lehrbuch der Combinatorik, Teubner, second edition, 1927 (Chelsea, 1958).
[2] P. A. MacMahon, Combinatory analysis I, II, Cambridge Univ. Press, 1915–1916 (Chelsea, 1960).
[3] J. Riordan, Combinatorial identities, Wiley, 1968.

331 (XII.13) Perturbation of Linear Operators

A. General Remarks

Historically, the perturbation method was developed as an approximation device in classical and quantum mechanics. In the perturbation theory of eigenvalues and eigenfunctions, created by L. Rayleigh and E. Schrödinger, the main concern was to find solutions as power series in a parameter κ that could be regarded as small. In the perturbation theory for linear operators, however, we are concerned more generally with the behavior of spectral properties of linear operators when the operators undergo small change. The foundation of the mathematical theory, including a complete convergence proof of perturbation series, was laid down by F. Rellich [1] and T. Kato [2, 3]. Another major topic in the perturbation theory for linear operators is the perturbation of continuous spectra, which was initiated by K. O. Friedrichs (*Math. Ann.*, 115 (1938); [4]). It is closely related to scattering theory and is discussed more fully in 375 Scattering Theory. A standard reference in this field is [5] (also → [6, 7]). Most of the material presented in this article is taken from [5].

For problems in Hilbert spaces there are two general frameworks in which to formulate perturbation situations: the operator formulation and the form formulation. In the former we deal with a family of operators $T(\kappa)$ directly, while in the latter, we deal with associated semibounded Hermitian (or, more generally, sectorial) forms $\mathbf{t}(\kappa)$. The latter is applicable only when there is semiboundedness (or a sectorial property) inherent in the problem, but is usually more general than the former in such problems, since the latter (resp. the former) requires roughly the constancy of the domain of the "square root" of $T(\kappa)$ (resp. the domain of $T(\kappa)$). In this article we discuss problems in the operator formulation. For the form formulation → [5] and [7].

In this article $X, Y, \ldots$, are complex Banach spaces and $T, A, \ldots$ are linear operators unless other specifications are made. The following notations defined in 251 Linear Operators are used without further explanation: $D(T)$, $R(T)$, $\mathbf{B}(X, Y)$, $\mathbf{B}(X)$, $\Gamma(T)$ (the graph of T), $\sigma(T)$ (the spectrum of T), $\rho(T)$ (the resolvent set of T), and $R(\zeta; T)$ (the resolvent of T). We also use $C(X, Y)$ (resp. $A(X, Y)$) to denote the set of all †closed linear operators (resp. all †linear operators) from X to Y and $C(X)=C(X, X)$.

B. Stability of Basic Properties

(1) Let $T\in C(X, Y)$. Important notions for characterizing the smallness of $A\in A(X, Z)$ relative to T are the following. (i) A is said to be **relatively bounded** with respect to T (or simply **T-bounded**) if $D(A)\supset D(T)$ and there exist $a, b\geqslant 0$ such that

(*) $\quad \|Au\|_Z\leqslant a\|u\|_X+b\|Tu\|_Y$ for all $u\in D(T)$.

The infimum, denoted by $\|A\|_T$, of b for which (*) holds with some a is called the T-bound of A. (ii) A is said to be **relatively compact** with respect to T (or **T-compact**) if $D(A)\supset D(T)$ and A is compact from $D(T)$ with the graph norm of T to Z (→ 68 Compact and Nuclear operators F). T-compactness of A implies T-boundedness (and in Hilbert spaces $\|A\|_T=0$).

(2) Let $T\in C(X, Y)$, and let $A\in A(X, Y)$ be T-bounded. (i) If $\|A\|_T<1$ (or if A is T-compact), then $T+A\in C(X, Y)$. (ii) If, in addition, $X=Y$ is a Hilbert space, T is †self-adjoint, and A is †symmetric, then $T+A$ is self-adjoint (**Rellich-Kato theorem**) [1, 8]. (iii) Suppose that T is a †Fredholm operator. If either A is T-compact or the inequality (*) holds with constants a, b satisfying $b\rho+a<\rho$ for a certain positive number ρ determined by T, then $T+A$ is a Fredholm operator and $\operatorname{ind}(T+A)=\operatorname{ind} T$ (for ind T, nul T, and def T → 251 Linear Operators). In the latter case where $b\rho+a<\rho$, we also have $\operatorname{nul}(T+A)\leqslant\operatorname{nul} T$ and $\operatorname{def}(T+A)\leqslant\operatorname{def} T$.

C. Continuity and Analyticity of Families of Closed Operators

In order to handle unbounded operators, which are important in applications, it is necessary to introduce generalized notions of convergence and analyticity of families of closed operators.

(1) $C(X, Y)$ becomes a †metric space by a distance function $\hat{d}(S, T)$ having the property that $\hat{\delta}(\Gamma(S), \Gamma(T))\leqslant\hat{d}(S, T)\leqslant 2\hat{\delta}(\Gamma(S), \Gamma(T))$, where for closed subspaces M and N we put

$$\hat{\delta}(M, N)=\max[\delta(M, N), \delta(N, M)],$$

$$\delta(M, N)=\sup_{u\in M,\ \|u\|=1}\operatorname{dist}(u, N);\ \delta(0, N)=0.$$

$\hat{\delta}(M, N)$ is called the **gap** between M and N [5]. When $T_n\to T$ in this metric, T_n is said to converge to T in the generalized sense. This **generalized convergence** coincides with the norm convergence if $T_n, T\in\mathbf{B}(X, Y)$. If $X=Y$ and $\rho(T)\neq\varnothing$, then $T_n\to T$ in the generalized sense if and only if for some (or equivalently all) $\zeta\in\rho(T)$ we have $\zeta\in\rho(T_n)$ for sufficiently large n and $\|R(\zeta; T_n)-R(\zeta; T)\|\to 0$, $n\to\infty$. This is called **norm resolvent convergence**.

(2) When $X=Y$, there is also the notion of strong convergence in the generalized sense [5], which is roughly the strong convergence of resolvents. In particular, when T_n and T are self-adjoint operators in a Hilbert space, $T_n \to T$ strongly in the generalized sense if $R(\zeta; T_n) \to R(\zeta; T)$ strongly for some (or equivalently all) ζ with $\mathrm{Im}\,\zeta \neq 0$. This is called **strong resolvent convergence**.

(3) Let $D \subset \mathbf{C}$ be a domain. The notion of analyticity (holomorphic property) of a family $T(\kappa) \in \mathbf{B}(X, Y)$, $\kappa \in D$, of bounded operators is well known (→ 37 Banach Spaces K). This notion is generalized to a family $T(\kappa) \in C(X, Y)$, $\kappa \in D$, of closed operators [1, 5]. Namely, $T(\kappa)$ is said to be **holomorphic** in D if at each $\kappa_0 \in D$ there exist a Banach space Z and $U(\kappa) \in \mathbf{B}(Z, X)$, $V(\kappa) \in \mathbf{B}(Z, Y)$, defined near κ_0, such that (i) $U(\kappa)$ and $V(\kappa)$ are holomorphic at κ_0 as families of bounded operators; (ii) $U(\kappa)$ is one to one and onto $D(T(\kappa))$; (iii) $T(\kappa)U(\kappa) = V(\kappa)$. Let us mention several special cases. (I) if $X=Y$ and if $\zeta \in \rho(T(\kappa))$ for all $\kappa \in D$, then $T(\kappa)$ is holomorphic in D if and only if $R(\zeta; T(\kappa))$ is holomorphic in D. (II) If $D(T(\kappa))$ is independent of κ and if $T(\kappa)u$ is holomorphic in D for every $u \in D(T(\kappa))$ then $T(\kappa)$ is holomorphic in D (holomorphic family of type (A) [5]). (III) Let $T \in C(X, Y)$, and let $T^{(n)} \in A(X, Y)$ such that $D(T^{(n)}) \supset D(T)$ and $\|T^{(n)}u\| \leqslant c^{n-1}(a\|u\| + b\|Tu\|)$, $u \in D(T)$, where a, b, $c \geqslant 0$. Then $T(\kappa)u = Tu + \kappa T^{(1)}u + \ldots + \kappa^n T^{(n)} + \ldots$, $u \in D(T)$ defines a holomorphic family of type (A) in $D = \{\kappa \mid |\kappa| < (b+c)^{-1}\}$. (IV) If $X=Y$ is a Hilbert space and if $T(\kappa)$ is self-adjoint for real κ, $T(\kappa)$ is said to be a self-adjoint family. In particular, the family discussed in (III) is a self-adjoint holomorphic family if T is self-adjoint and $T^{(n)}$ is symmetric.

D. Perturbation of Isolated Eigenvalues

(1) Separation of the spectrum. Let $T \in C(X)$. Suppose that a bounded subset Δ of $\sigma(T)$ is separated from the rest of $\sigma(T)$ by a simple closed contour Γ (i.e., $\Gamma \subset \rho(T)$ and $\Delta(\sigma(T) \setminus \Delta)$ lies inside (outside) of Γ). Then the operator

$$P = \frac{1}{2\pi i}\int_\Gamma R(\zeta; T)\,d\zeta,$$

which is independent of Γ, is a projection (i.e., $P \in \mathbf{B}(X)$ and $P^2 = P$). The closed subspaces $X_1 \equiv PX$ and $X_2 \equiv (I-P)X$ †reduce T and give rise to the decomposition $T = T|_{X_1} \oplus T|_{X_2} \equiv T_1 \oplus T_2$. In particular, $\sigma(T_1) = \sigma(T) \cap \{\text{inside of } \Gamma\}$ and $\sigma(T_2) = \sigma(T) \cap \{\text{outside of } \Gamma\}$.

(2) Let $T(\kappa)$ be holomorphic in D. We assume that $0 \in D$ and regard $T^{(0)} = T(0)$ as the unperturbed operator. Suppose that Δ and Γ are as in (1) with T replaced by $T^{(0)}$. Then there exists $\delta > 0$ such that $|\kappa| < \delta$ implies $\Gamma \subset \rho(T(\kappa))$. This follows from the upper continuity of compact components of the spectrum with respect to the metric $\hat{d}$ of $C(X)$ [5]. Thus the separation of the spectrum discussed in (1) is applicable to $T(\kappa)$. In particular, corresponding to the projection

$$P(\kappa) = \frac{1}{2\pi i}\int_\Gamma R(\zeta; T(\kappa))\,d\zeta, \qquad |\kappa| < \delta,$$

$T(\kappa)$ is decomposed as $T(\kappa) = T_1(\kappa) \oplus T_2(\kappa)$; and the problem of determining the spectrum of $T(\kappa)$ inside Γ is reduced to the problem of determining the spectrum of $T_1(\kappa)$ ($|\kappa| < \delta$). Suppose now that $\Delta = \{\lambda_0\}$ is an isolated eigenvalue of $T^{(0)}$ and that $m \equiv \dim P(0)X < \infty$. Then $\dim P(\kappa)X = m$, $|\kappa| < \delta$. Moreover, a base $\{\varphi_1(\kappa), \ldots, \varphi_m(\kappa)\}$ of $P(\kappa)X$ can be constructed in such a way that the $\varphi_j(\kappa)$ are holomorphic in $\{|\kappa| < \delta' \leqslant \delta\}$ [3, 5]. Thus the problem for $T_1(\kappa)$ in this case is just the finite-dimensional eigenvalue problem $\det\{\lambda\delta_{jk} - (T(\kappa)\varphi_j(\kappa), \varphi_k(\kappa))\} = 0$. The totality $\{\lambda_j(\kappa)\}$ of solutions of this equation, i.e., the totality of eigenvalues of $T(\kappa)$ near λ_0, is expressed by one or several power series of $\kappa^{1/p}$ with a suitable integer $p > 0$. If $T(\kappa)$ is a self-adjoint family, we can take $p=1$ so that the eigenvalues are holomorphic near λ_0. If $H(\kappa) = H^{(0)} + \kappa H^{(1)} + \ldots$ is a self-adjoint holomorphic family described in example (IV) in (3) of Section C and if $m=1$, the power series $\lambda(\kappa) = \sum \lambda_j \kappa^j$ can be explicitly computed as $\lambda_1 = (H^{(1)}u_0, u_0)$, $\lambda_2 = (H^{(2)}u_0, u_0) + (SH^{(1)}u_0, H^{(1)}u_0), \ldots$, where $H^{(0)}u_0 = \lambda_0 u_0$ with $\|u_0\| = 1$ and where $S = \lim_{\zeta \to \lambda_0} R(\zeta; H^{(0)})(I - P(0))$ is the reduced resolvent. This series is known as the **Rayleigh-Schrödinger series**. The power series for the associated eigenvectors $u(\kappa) = \sum \kappa^j u_j$ can also be computed. For details, including the case of a degenerate λ_0 ($m > 1$), in which the situation becomes more complicated due to the splitting of eigenvalues, → [5]. The perturbation theory discussed in this subsection is called **analytic** (or **regular**) **perturbation theory**.

(3) Even when a problem cannot be handled by means of analytic perturbation, it may happen that the coefficients λ_j and u_j of formal power series can be computed up to a certain j. In many such cases it can be shown under general assumptions that an asymptotic expansion such as $\lambda(\kappa) = \lambda_0 + \lambda_1\kappa + o(\kappa)$ is valid as long as the coefficients involved can be computed legitimately [2, 5]. Estimates for $o(\kappa)$ can also be given. This provides a rigorous foundation for the perturbation method in many important practical problems. The case of degenerate λ_0 can be treated similarly. The strong convergence in the generalized sense mentioned in (2) of Section C is used here. This theory is called **asymptotic perturbation theory**.

E. Perturbation of Continuous Spectra

For continuous spectra, studying the mode of change under perturbation is not usually a tractable problem. Rather, certain parts of continuous spectra tend to be stable under perturbation; and the study of this stability has been a major topic in perturbation theory (also → 375 Scattering Theory). In this section we discuss only self-adjoint operators and let $H=\int\lambda dE(\lambda)$, $H_0, \ldots$, be self-adjoint operators in a Hilbert space X. For $\mathbf{B}_p$ and $\| \ \|_p$ to be used below → 68 Compact and Nuclear Operators.

(1) The essential spectrum (→ 390 Spectral Analysis of Operators E) is stable under compact perturbation. Namely, if $H=H_0+K$ with compact K, then $\sigma_e(H)=\sigma_e(H_0)$ (H. Weyl, *Rend. Circ. Mat. Palermo*, 27 (1909)). More generally, it suffices to assume that $R(\zeta;H)-R(\zeta;H_0)$ is compact for some (or equivalently all) $\zeta\in\rho(H)\cap\rho(H_0)$. Conversely, if X is separable and if $\sigma_e(H)=\sigma_e(H_0)$, then there exist a unitary operator U and a compact operator K such that $H=UH_0U^{-1}+K$ (J. von Neumann, *Actualités Sci. Ind.*, 229 (1935)). Moreover, any self-adjoint operator H in a separable Hilbert space can be changed into $H+K$ with a pure point spectrum by adding a $K\in\mathbf{B}_p$ with $\|K\|_p<\varepsilon$ for any $p>1$ and $\varepsilon>0$ (S. T. Kuroda, *Proc. Japan Acad.*, 34 (1958)). I. D. Berg (1971), W. Sikonia (1971), J. Voigt (1977), and D. Voiculescu (1979) have extended these results to normal operators and m-tuples of commutative self-adjoint operators. Also → 390 Spectral Analysis of Operators I, J.

(2) The absolutely continuous spectrum (→ 390 Spectral Analysis of Operators E) is stable under perturbation by the †trace class. Namely, if $H=H_0+K$, with $K\in\mathbf{B}_1$, then the absolutely continuous parts of H_0 and H are †unitarily equivalent, and in particular $\sigma_{ac}(H)=\sigma_{ac}(H_0)$ (M. Rosenblum, *Pacific J. Math.*, 7 (1957); T. Kato, *Proc. Japan Acad.*, 33 (1957)). Among generalizations we mention the following two. (i) If $R(\zeta;H)-R(\zeta;H_0)\in\mathbf{B}_1$ for some $\zeta\in\rho(H)\cap\rho(H_0)$, then the absolutely continuous parts of $\varphi(H_0)$ and $\varphi(H)$ are unitarily equivalent for any smooth strictly increasing real function φ (M. Sh. Birman, *Izv. Akad. Nauk SSSR*, ser. mat., 27, (1963); T. Kato, *Pacific J. Math.*, 15 (1965)). (ii) If H_0 and H act in different Hilbert spaces X_0 and X, respectively, and if there exists $J\in\mathbf{B}(X_0,X)$ such that $JD(H_0)\subset D(H)$ and such that the closure of $HJ-JH_0$ belongs to $\mathbf{B}_1(X_0,X)$, then the same conclusion as in (i) holds (D. Pearson, *J. Functional Anal.*, 28 (1978)). (i) can be derived from (ii). Perturbation theory for absolutely continuous spectra is closely related to the study of generalized wave operators in scattering theory. In fact, the existence and the completeness of the latter implies the unitary equivalence of absolutely continuous parts. All the results mentioned above are proved by scattering-theoretic methods, either by the wave operator approach or by the abstract stationary approach (→ 375 Scattering Theory, esp. B, C).

F. Some Other Topics

(1) For the perturbation theory for semigroups of operators and evolution equations, not discussed in this article, → [5, 7, 9].

(2) The detailed structure of continuous spectra is hard to analyze. An eigenvalue λ_0 of H_0 which is embedded in the continuous spectrum may diffuse into the continuous spectrum in the presence of a perturbation. In such a case, $H(\kappa)$, $\kappa\neq 0$, has no eigenvalues near λ_0 but may have a continuous spectrum highly concentrated around λ_0. This phenomenon of **spectral concentration** is studied, especially for some concrete problems, in relation to **resonance poles** (or poles of the holomorphic continuation of the resolvent or the scattering matrix). In some problems, it is proved that the first few terms of the perturbation series for $\lambda(\kappa)$ that are still computable are related to the real part of the resonance. Some problems of resonance can be treated by the technique of dilation analyticity, a technique which is also effective in other problems of spectral analysis (J. Aguilar and J. M. Combes, *Comm. Math. Phys.*, 22 (1971)).

(3) A vast quantity of results in the spectral theory of the Schrödinger operators appearing in the †Schrödinger equation in quantum mechanics can be obtained by perturbation methods.

For the topics mentioned in (2) and (3) → [7].

References

[1] F. Rellich, Störungstheorie der Spektralzerlegung I–V, Math. Ann., 113 (1937), 600–619; 113 (1937), 677–685; 116 (1939), 555–570; 117 (1940), 356–382; 118 (1942), 462–484.
[2] T. Kato, On the convergence of the perturbation method, J. Fac. Sci. Univ. Tokyo, sec. I, 6 (1951), 145–226.
[3] T. Kato, On the perturbation theory of closed linear operators, J. Math. Soc. Japan, 4 (1952), 323–337.
[4] K. O. Friedrichs, On the perturbation of continuous spectra, Comm. Pure Appl. Math., 1 (1948), 361–406.
[5] T. Kato, Perturbation theory for linear operators, Springer, 1966.

[6] N. Dunford and J. T. Schwartz, Linear operators I–III, Wiley-Interscience, 1958, 1963, 1971.
[7] M. Reed and B. Simon, Methods of modern mathematical physics I–IV, Academic Press, 1972, 1975, 1979, 1978.
[8] T. Kato, Fundamental properties of Hamiltonian operators of Schrödinger type, Trans. Amer. Math. Soc., 70 (1951), 195–211.
[9] H. Tanabe, Equation of evolution, Pitman, 1979. (Original in Japanese, 1975.)

332 (VI.7) Pi (π)

The **ratio of the circumference of a circle to its diameter** in a Euclidean plane is denoted by π, the initial letter of $\pi\varepsilon\rho\iota\mu\varepsilon\tau\rho o\varsigma$ (perimeter). Thus π can be defined as

$$2\int_0^1 dx/\sqrt{1-x^2}.$$

The symbol π has been used since W. Jones (1675–1749) and L. Euler. The fact that this ratio is a constant is stated in Euclid's *Elements*; however, Euclid gave no statement about the numerical value of π. As an approximate value of π, 3 has been used from antiquity. According to the Rhind Papyrus, $(4/3)^4$ was used in ancient Egypt. Let $L_n(l_n)$ be the perimeter of a regular n-gon circumscribed about (inscribed in) a circle of radius 1. Then the relations

$$L_n > \pi > l_n, \qquad \frac{2}{L_{2n}} = \frac{1}{L_n} + \frac{1}{l_n},$$

$$l_{2n} = \sqrt{l_n L_{2n}}$$

hold. Archimedes obtained $3\frac{10}{71} < \pi < 3\frac{1}{7}$ by calculating L_{96} and l_{96}. In 3rd-century China Liu Hui used $\pi \fallingdotseq 3.14$. In 5th-century China, Tsu Chung-Chih mentioned 22/7 as an inaccurate approximate value and 355/113 as an accurate approximate value of π. These values were obtained by methods similar to those of Archimedes. In 5th-century India, Aryabhatta obtained $\pi \fallingdotseq 3.1416$, and in 16th-century Europe, Adriaen van Roomen obtained $\pi \fallingdotseq 355/113$.

F. Viète represented $2/\pi$ in the following infinite product:

$$\prod_{n=2}^{\infty} \cos\frac{\pi}{2^n} = \sqrt{\frac{1}{2}}\sqrt{\frac{1}{2}+\frac{1}{2}\sqrt{\frac{1}{2}}} \times \sqrt{\frac{1}{2}+\frac{1}{2}\sqrt{\frac{1}{2}+\frac{1}{2}\sqrt{\frac{1}{2}}}}\cdots.$$

Using this formula, L. van Ceulen (1540–1610) calculated π to 35 decimals. In the 17th and 18th centuries, the Japanese mathematicians T. Seki, K. Takebe, and Y. Matunaga computed π to 50 decimals. Since the 17th century, many formulas that represent π as a sum of infinite series or as a limit have been used to obtain more accurate approximate values. The following are representations of π known in those days:

$$\frac{\pi}{2} = \frac{2\cdot2\cdot4\cdot4\cdot6\cdot6\ldots}{1\cdot3\cdot3\cdot5\cdot5\cdot7\ldots} \qquad \text{(J. Wallis)}$$

$$\frac{\pi}{4} = \frac{1}{1+}\,\frac{1^2}{2+}\,\frac{3^2}{2+}\,\frac{5^2}{2+}\ldots \qquad \text{(W. Brouncker; for the notation} \rightarrow \text{83 Continued Fractions)}$$

$$= 1 - 1/3 + 1/5 - 1/7 + \ldots \qquad \text{(J. Gregory, G. W. F. Leibniz)}$$

$$= 4\,\mathrm{Arc\,tan}\,1/5 - \mathrm{Arc\,tan}\,1/239 \qquad \text{(J. Machin).}$$

A formula combining Machin's representation of π and the power series $\mathrm{Arc\,tan}\,x = x - (1/3)x^3 + (1/5)x^5 - \ldots$ is called **Machin's formula** and was often used for calculating an approximate value of π. By utilizing this formula, in 1873 W. Shanks obtained an approximate value of π up to 707 decimals. No improvement of his approximation was obtained until 1946 when D. F. Ferguson calculated 710 digits of π and found that Shanks's value was correct only up to the 527th digit. The computation of an accurate approximate value of π has been made easier by the recent development of computing machines, and an approximate value up to 1,000,000 decimals has been obtained. P. Beckmann [2] gives a detailed and humorous historical account of the calculation of π from ancient times up to the present computer age. Various numerical results obtained by electronic computers are not formally published, some being deposited in the UMT repository of the editorial office of the journal *Mathematics of computation*. Choong et al. [3], using information in [1], obtained the first 21,230 partial denominators of the regular continued fraction representation of π and described how their numerical evidence tallies with theoretical results, obtained by the metrical theory of continued fractions, which is valid for almost all irrational numbers (e.g. → [4]).

In 1761, J. H. Lambert used Brouncker's expression of π in a continued fraction to prove that π is irrational. In 1882, C. L. F. Lindemann proved that π is a †transcendental number using Euler's formula $e^{\pi i} = -1$. The approximate value of π up to 50 decimals is 3.14159265358979323846264338327950288419716939937510... (→ Appendix B, Table 6).

References

[1] D. Shanks and J. W. Wrench, Calculation of π to 100,000 decimals, Math. Comp., 16 (1962), 76–99.
[2] P. Beckmann, A history of pi, second edition, Golem Press, 1971.
[3] K. Y. Choong, D. E. Daykin, and C. R. Rathbone, Rational approximations to π, Math. Comp., 25 (1971), 387–392.
[4] A. Khinchin, Continued fractions, Noordhoff, 1963.

333 (II.18) Plane Domains

A. Domains in the Complex Plane

A †domain (i.e., a †connected open set) in the †complex plane or on the †complex sphere is called a **plane domain**. The †closure of such a domain is called a **closed plane domain**. In this article, we consider only subsets of the complex plane (or sphere), and a plain domain is called simply a domain. The †interior of a †Jordan curve J in the complex plane is a domain called a **Jordan domain**. In a domain D, a Jordan arc whose two endpoints lie on the boundary of D is called a **cross cut** of D.

For a domain D, each of the following three conditions is equivalent to the condition that D is †simply connected: (1) For every cross cut Q of D, $D-Q$ has exactly two †connected components. (2) Every Jordan curve in D is †homotopic to one point, that is, it can always be continuously deformed to a point. (3) The †monodromy theorem holds in D.

If D is a domain on a complex sphere, each of the following three conditions is equivalent to the condition that D is simply connected: (4) The boundary of D consists of a single †continuum or a single point. (5) For every Jordan curve C in D, either the interior or the exterior of C is contained in D. (6) The complement of D with respect to the complex sphere is a connected (not necessarily arcwise connected) closed set. Jordan domains are simply connected.

Let $n \geqslant 2$ be an integer and D a plane domain. The †homology group $H_1(D, \mathbf{Z})$ is identical to $\mathbf{Z}^{n-1}$ if and only if the complement of D in the complex sphere has n connected components. Then D is said to be ***n*-ply connected** or **multiply connected** without specifying n. If D is an n-ply connected domain, there exist $n-1$ suitable mutually disjoint cross cuts $Q_1, \ldots, Q_{n-1}$ such that $D-(Q_1 \cup \ldots \cup Q_{n-1})$ is simply connected.

Some typical examples of domains are as follows: (1) **Circular domain**: $|z-c|<r$. (2) **Half-plane**: $\operatorname{Re} z>0$, or $\operatorname{Im} z>0$. (3) **Angular domain**: $\alpha<\arg(z-c)<\beta$. (4) **Annular domain**: $r<|z-c|<R$. (5) **Slit domain**: a domain obtained by excluding a Jordan arc Γ from a domain D, where all points on Γ (except an endpoint lying on the boundary of D) are contained in D. In this case, the Jordan arc Γ is called the **slit** of the domain.

B. Boundary Elements

A boundary point P of a domain D is called **accessible** if there exists a sequence of points P_ν tending to P such that the line segments $P_1 P_2, \ldots$ lie completely in D. For example, for the domain obtained by removing $x=1/(n+1)$, $0<y\leqslant 1/2$ $(n=1,2,\ldots)$ from the square $0<x<1$, $0<y<1$, the boundary points with $x=0$, $0\leqslant y<1/2$ are all inaccessible (Fig. 1).

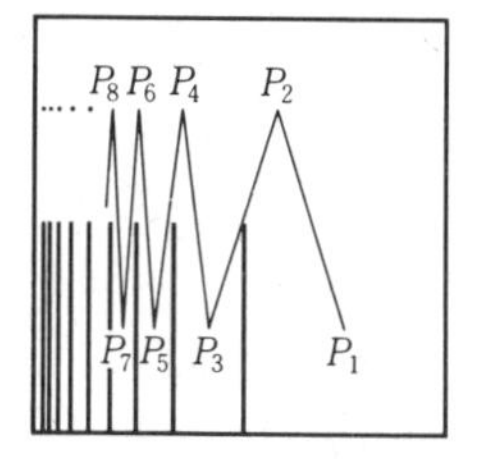

Fig. 1

Let the domain D be bounded by a smooth Jordan curve, and let P be a boundary point of D. Take an angular domain D' with vertex at the point P and the initial parts of the two sides of D' lying in D. A curve in D converging to the point P from the interior of the angular domain D' is called a **Stolz's path** or a **nontangential path** ending at the point P.

Let D be a simply connected domain. A sequence $\{q_\nu\}$ of cross cuts mutually disjoint except for their endpoints is called a **fundamental sequence of cross cuts** if it satisfies the following two conditions (Fig. 2): (1) Every q_ν separates $q_{\nu-1}$ and $q_{\nu+1}$ on D. (2) For $\nu \rightarrow \infty$, the sequence q_ν tends to a point on the boundary. Let $\{q_\nu\}$ be a fundamental sequence of cross cuts, and denote by D_ν the subdomain of D separated by q_ν that contains $q_{\nu+1}$. The intersection $\bigcap \bar{D}_\nu$ consists only of the boundary points of D. Two fundamental sequences $\{q_\nu\}$, $\{q'_\nu\}$ of cross cuts are equivalent if every D_μ contains all q'_ν except for a finite number of ν, and every D'_μ contains all q_ν except for a finite number of ν. Here D_μ, D'_μ are the subdomains constructed from q_μ and q'_μ as above. This condition determines an equivalence relation, under which the equivalence class of fundamental sequences of cross cuts is called a **boundary element**. This notion is due to C. Carathéodory [2]. The boundary element of a multiply connected domain is defined similarly for each isolated component of the boundary. For example, each point of a slit domain, except for the endpoint of Γ lying on the boundary of the domain, determines two distinct boundary elements on each side. A closed domain is usually considered to be the union of a domain and the set of all its boundary elements. Various notions of †ideal boundary come from considering suitable boundary elements for various purposes (→ 207 Ideal Boundaries).

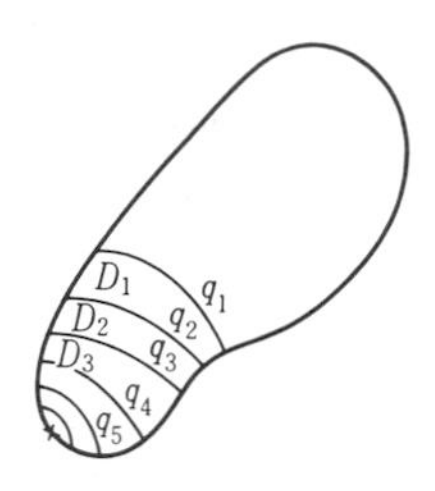

Fig. 2

C. Domain Kernels

Let $\{G_\nu\}$ be a sequence of domains containing the origin 0. If a suitable neighborhood of the origin is contained in G_ν for all ν, there exists a domain G such that every closed domain containing the origin and contained in G is contained in G_ν except for a finite number of ν. The union K of such domains G is called the **domain kernel** of the sequence $\{G_\nu\}$ (Carathéodory). If there is no neighborhood of the origin contained in G_ν for all ν, we put $K=\{0\}$.

If every infinite subsequence of $\{G_\nu\}$ has the same domain kernel K, then we say that the sequence $\{G_\nu\}$ converges to K. The notion of domain kernel is important in considering the limits of a sequence of conformal mappings (→ 77 Conformal Mappings).

References

[1] M. H. A. Newman, Elements of the topology of plane sets of points, Cambridge Univ. Press, second edition, 1951.
[2] C. Carathéodory, Über die Begrenzung einfach zusammenhängender Gebiete, Math. Ann., 73 (1913), 323–370.

334 (X.33) Plateau's Problem

A. Origin

Because of surface tension, a soap membrane bounded by a given closed space curve takes the shape of a minimal surface, i.e., a surface of the least area. This experiment was performed by the Belgian scientist J. A. Plateau (1873) to realize minimal surfaces; hence **Plateau's problem** is that of determining the minimal surfaces bounded by given closed space curves. It is a problem of the †calculus of variations.

B. Formulation

Let Γ be a †simple closed curve in xyz-space such that its projection C on the xy-plane is also a simple closed curve. Let D be the finite domain bounded by C. We consider surfaces $z=z(x,y)$ having common boundary Γ. Then under suitable assumptions on the smoothness of $z(x,y)$, the problem is to minimize the †functional

$$J[z]=\iint_D\sqrt{1+p^2+q^2}\,dx\,dy;$$

$$p=\frac{\partial z}{\partial x},\qquad q=\frac{\partial z}{\partial y},$$

with the condition that $z=z(x,y)$ has Γ as its boundary. The †Euler-Lagrange differential equation for the functional $J[z]$ is

$$\frac{\partial}{\partial x}\frac{p}{\sqrt{1+p^2+q^2}}+\frac{\partial}{\partial y}\frac{q}{\sqrt{1+p^2+q^2}}=0,$$

or $(1+q^2)r-2pqs+(1+p^2)t=0$, $r=\partial^2 z/\partial x^2$, $s=\partial^2 z/\partial x\partial y$, $t=\partial^2 z/\partial y^2$, which is a second-order †quasilinear partial differential equation of elliptic type and whose geometric interpretation had already been given by M. C. Meusnier (1776).

To formulate the problem more generally, let a surface be expressed in vector form $\mathfrak{x}=\mathfrak{x}(u,v)$ by means of parameters u, v. Let its †first fundamental form be $d\mathfrak{x}^2=E\,du^2+2F\,du\,dv+G\,dv^2$ and its †second fundamental form be $-d\mathfrak{x}\,d\mathfrak{n}=L\,du^2+2M\,du\,dv+N\,dv^2$, with $\mathfrak{n}=\mathfrak{n}(u,v)$ the unit normal vector. By equating to zero the †first variation of the **areal functional**

$$\iint\sqrt{EG-F^2}\,du\,dv$$

based on infinitesimal displacement in the normal direction, we obtain the Euler-Lagrange equation in the form

$$2H\equiv(NE-2MF+LG)/(EG-F^2)=0,$$

where $H=(R_1^{-1}+R_2^{-1})/2$ is the †mean curvature of the surface and R_1 and R_2 are the †radii of principal curvature. Since †Beltrami's second differential form satisfies $\Delta_2\mathfrak{x}=H\mathfrak{n}$, the condition for a minimal surface becomes $\Delta\mathfrak{x}=0$ (with Δ the †Laplace operator) provided that **isothermal parameters** u, v satisfying $E=$

G, $F=0$ are chosen, i.e., the vector $\mathfrak{x}(u, v)$ representing a minimal surface is harmonic (the components of this vector are †harmonic functions of u, v). Let $\mathfrak{y}(u, v)$ be a harmonic vector conjugate to $\mathfrak{x}(u, v)$. Then isothermality is expressed by the condition that the analytic vector $\mathfrak{F}(w)=\mathfrak{x}(u, v)+i\mathfrak{y}(u, v)$ $(w=u+iv, i=\sqrt{-1})$ satisfies $\mathfrak{F}'(w)^2=0$ (Weierstrass). In general, a **minimal surface** is defined as a surface with everywhere vanishing mean curvature, and Plateau's problem is to determine the minimal surface with a preassigned boundary. In this formulation, the problem can be easily generalized to an n-dimensional Euclidean space $\mathbf{R}^n$ (→ 275 Minimal Submanifolds).

C. Existence of a Solution

The existence of a solution of Plateau's problem was discussed by S. N. Bernshteĭn (1910) from the viewpoint of a †boundary value problem of the first kind for the elliptic partial differential equation in the previous section. A. Haar (1927) dealt with the minimal problem for the functional $J[z]$ by a †direct method in the calculus of variations. Previously, Riemann, Weierstrass, Schwarz, and others discussed the case where the given space curve Γ is a polygon, in connection with the †monodromy group concerning a second-order linear ordinary differential equation. Subsequently, R. Garnier (1928) investigated the existence of a solution by the limit process when Γ is a simple closed curve with bounded curvature. However, when Γ is assumed merely to be †rectifiable, the existence of a solution was first shown by the limit process by T. Radó (1930). He further discussed the general case where Γ can bound a surface with finite area. On the other hand, by introducing a new functional depending on boundary values instead of the areal functional, J. Douglas (1931) succeeded in giving a satisfactory result for the existence problem. R. Courant (1937) gave another existence proof by reducing Plateau's problem to the †Dirichlet principle [3].

At present, the methods of discussing the existence of solutions of Plateau's problem can be classified into the following three sorts (represented, respectively, by Radó, Courant, and Douglas):

(1) The first method is to minimize directly the areal functional $\iint\sqrt{EG-F^2}\,du\,dv$. The variational equation of the areal functional becomes $H=0$.

(2) **Dirichlet's functional** for a scalar function $f(u, v)$ is defined by $D[f]=\iint(f_u^2+f_v^2)\,du\,dv$, and for an n-dimensional vector function $\mathfrak{f}(u, v)$ with components $f_j(u, v)$ $(j=1, \ldots, n)$ by $D[\mathfrak{f}]=\sum_{j=1}^n D[f_j]$. The existence of a solution of Plateau's problem can be discussed by starting from the variational problem of minimizing $D[\mathfrak{f}]$. The variational equation of $D[\mathfrak{f}]$ is $\Delta\mathfrak{f}=0$.

(3) An analytic vector $\mathfrak{F}(w)$ is representable in terms of the boundary values of its real part. For instance, if the domain of w is the unit disk $|w|<1$, then Poisson's integral formula

$$\mathfrak{F}(w)=\frac{1}{2\pi}\int_0^{2\pi}\mathfrak{h}(\theta)\frac{e^{i\theta}+w}{e^{i\theta}-w}d\theta+i\,\mathrm{Im}\,\mathfrak{F}(w)$$

with the boundary function $\mathfrak{h}(\theta)=\mathrm{Re}\,\mathfrak{F}(e^{i\theta})$ can be used. On the other hand, the vector function that minimizes the Dirichlet integral among functions with fixed boundary values is harmonic. Based on these facts, Douglas transformed Dirichlet's functional with harmonic argument functions into a functional whose arguments are boundary functions. Specifically, by starting from the problem of minimizing **Douglas's functional**

$$A[\mathfrak{h}]=\frac{1}{4\pi}\int_0^{2\pi}\int_0^{2\pi}\frac{(\mathfrak{h}(\theta)-\mathfrak{h}(\varphi))^2}{4\sin^2(\theta-\varphi)/2}d\theta\,d\varphi,$$

we can prove the existence of solution of Plateau's problem satisfactorily.

D. The Generalized Case

Up to now we have been concerned with Plateau's problem in the case of a single simple closed curve. Douglas, Courant, and others treated the generalized case of a finite number of boundary curves, where †genus and orientability are assigned as the topological structure of the surface to be found. The existence of a solution has been shown in this case also. The problem is further generalized from the case of fixed boundary to the case where the boundary is merely restricted to lie on a given manifold [3]. On the other hand, C. B. Morrey (1948) generalized the problem by replacing the ambient space $\mathbf{R}^n$ by an n-dimensional †Riemannian manifold and gave the existence proof in considerable generality [6].

E. Relation to Conformal Mappings

There is a notable relation between Plateau's problem and conformal mapping when the dimension of the space is 2. Namely, the existence proof of the solution of the former for a Jordan domain implies †Riemann's mapping theorem together with W. F. Osgood and C. Carathéodory's result on boundary correspondence (→ 77 Conformal Mappings).

F. New Developments

Among recent contributions to the study of Plateau's problem, the following remarkable results have emerged. One of them is connected with the final result of Douglas (1939) on the existence of solution surfaces. The mapping of a 2-dimensional manifold with boundary into $\mathbf{R}^n$ defining Douglas's solution of the Plateau problem for a finite number of simple closed curves is a †minimal immersion with the possible exception of isolated points where it fails to be an immersion. These points are called branch points. It was then proved by R. Osserman (1970) and R. D. Gulliver (1973) that for $n=3$ the mapping of Douglas's theorem, which is a surface of least area, is free of branch points, i.e., is an immersion. Osserman also gave examples of generalized minimal surfaces in $\mathbf{R}^n$ $(n>3)$ with true branch points. In this connection, Gulliver also dealt with an analogous problem for surfaces of prescribed mean curvature.

Next, we mention the question of boundary regularity. H. Lewy (1951) proved that if the boundary of a minimal surface is analytic, then the surface is analytic up to the boundary. Subsequently, S. Hildebrandt (1969) and others proved that if the boundary is of class $C^{m,\alpha}$, $m \geqslant 1$, the surface is also of class $C^{m,\alpha}$ up to the boundary. There are also some recent results on the number of solutions of Plateau's problem. For instance, J. C. C. Nitsche (1973) proved the uniqueness of solutions for analytic boundaries of †total curvature at most 4π.

Further developments in connection with Plateau's problem have emerged in the work of E. R. Reifenberg (1960) and others, who sought to minimize the †Hausdorff measure among general classes of geometric objects, not as parametrized manifolds, but as subsets of $\mathbf{R}^n$. The existence and regularity of solutions of Plateau's problem from this point of view have been discussed by H. Federer (1969), W. H. Fleming, F. J. Almgren, and others [10].

References

[1] T. Radó, On the problem of Plateau, Erg. Math., Springer, 1933.
[2] J. Douglas, Solution of the problem of Plateau, Trans. Amer. Math. Soc., 33 (1931), 263–321.
[3] R. Courant, Dirichlet's principle, conformal mapping and minimal surfaces, Interscience, 1950.
[4] J. C. C. Nitsche, Vorlesungen über Minimalflächen, Springer, 1975.
[5] R. Osserman, A survey of minimal surfaces, Van Nostrand, 1969.
[6] C. B. Morrey, Multiple integrals in the calculus of variations, Springer, 1966.
[7] R. Osserman, A proof of the regularity everywhere of the classical solution to Plateau's problem, Ann. Math., (2) 91 (1970), 550–569.
[8] S. Hildebrandt, Boundary behavior of minimal surfaces, Arch. Rational Mech. Anal., 35 (1969), 47–82.
[9] J. C. C. Nitsche, A new uniqueness theorem for minimal surfaces, Arch. Rational Mech. Anal., 52 (1973), 319–329.
[10] H. Federer, Geometric measure theory, Springer, 1969.
[11] F. J. Almgren, Plateau's problem. An invitation to varifold geometry, Benjamin, 1966.

335 (XXI.39) Poincaré, Henri

Henri Poincaré (April 29, 1854–July 17, 1912) was born in Nancy, France. After graduating from the Ecole Polytechnique, he taught at the University of Caen in 1879, then at the University of Paris in 1881. He was made a member of the Académie des Sciences in 1887 and of the Académie Française in 1908. He died in Paris.

His achievements center on analysis and applications to theoretical physics and astronomy. However, his work covered many fields of mathematics, including arithmetic, algebraic geometry, spectral theory, and topology. His †uniformization of analytic functions by means of the theory of †automorphic functions in 1880 is especially notable. His paper on the †three-body problem won the prize offered by the king of Sweden in 1889.

The methods he developed in his three-volume *Mécanique céleste* (1892–1899) began a new epoch in celestial mechanics. In addition, Poincaré opened the road to †algebraic topology and made suggestive contributions to the †theory of relativity and †quantum theory. He asserted that science is for science's own sake [4], and his popular philosophical works concerning the foundations of natural science and mathematics exhibit a lucid style.

References

[1] H. Poincaré, Oeuvres I–XI, Gauthier-Villars, 1916–1956.
[2] H. Poincaré, Les méthodes nouvelles de la mécanique céleste 1–III, Gauthier-Villars, 1892–1899.
[3] H. Poincaré, Science et hypothèse, Flammarion, 1903.
[4] H. Poincaré, La valeur de la science, Flammarion, 1914.

[5] H. Poincaré, Science et méthode, Flammarion, 1908.

336 (X.20) Polynomial Approximation

A. General Remarks

On the existence of polynomial approximations, we have **Weierstrass's approximation theorem**, which is formulated in the following two forms: (i) If $f(x)$ is a function that is continuous in the finite interval $[a,b]$, then for every $\varepsilon>0$ there exists a polynomial $P_n(x)$ of degree $n=n(\varepsilon)$ such that the inequality $|f(x)-P_n(x)|\leqslant\varepsilon$ holds throughout the interval $[a,b]$. (ii) If $f(\theta)$ is a continuous function of period 2π, then corresponding to every positive number ε there exists a trigonometric polynomial of degree $n=n(\varepsilon)$,

$$P_n(\theta)=a_0+\sum_{k=1}^{n}(a_k\cos k\theta+b_k\sin k\theta), \tag{1}$$

such that the inequality $|f(\theta)-P_n(\theta)|\leqslant\varepsilon$ holds for all values of θ. The second form of Weierstrass's theorem follows from the first, and conversely. M. H. Stone obtained a theorem that generalizes Weierstrass's theorem to the case of functions of several variables. Of the many direct proofs now available for Weierstrass's theorem, we mention two simple ones. To prove version (i) of the theorem, we can assume that the given function $f(x)$ is defined in the segment $[0,1]$. Consider the **Bernshteĭn polynomial**

$$B_n(x)=\sum_{k=0}^{n}f\left(\frac{k}{n}\right)\binom{n}{k}x^k(1-x)^{n-k}.$$

Then $B_n(x)$ converges to $f(x)$ †uniformly. To prove (ii), we can apply †Fejér's theorem on †Fourier series. We have the following generalization of (i): Let $p_1, p_2, \ldots$ be a sequence of positive numbers such that $\lim p_n=\infty$. Then linear combinations of $x^0=1, x^{p_1}, x^{p_2}, \ldots$ can uniformly approximate each continuous function on $[0,1]$ with arbitrary precision if and only if $\sum p_n^{-1}=-\infty$ (**Müntz's theorem**).

B. Best Approximations

Let $\varphi_0(x), \varphi_1(x), \ldots$ be a sequence of linearly independent continuous functions on a bounded closed domain A in $\mathbf{R}^n$. For any given continuous function $f(x)$, a function $P_n(x)=\sum_{k=0}^{n}c_k\varphi_k(x)$ attains

$$\inf_{c_0,\ldots,c_n}\max_{x\in A}|f(x)-P_n(x)|$$

is called the **best approximation** of $f(x)$ by a linear combination of $\{\varphi_k(x)\}$. For any given n there is a best approximation of $f(x)$ by a linear combination of $\varphi_0(x),\ldots,\varphi_n(x)$, but such an approximation is not always unique. For such an approximation to be unique it is necessary and sufficient that the determinant of the matrix $(\varphi_k(x_i))$ $(k,i=0,1,2,\ldots,n)$ is not zero, where $x_0,x_1,\ldots,x_n$ are $n+1$ arbitrary distinct points of A (**Haar's condition**) (*Math. Ann.*, 78 (1918)). If $\{\varphi_k(x)\}$ satisfies this condition, the system of functions $\{\varphi_k\}_{k=0,\ldots,n}$ is called a **Chebyshev system** (or **unisolvent system**). The sets $\{1,x,x^2,\ldots,x^n\}$ on $[a,b]$, $\{1,\cos x,\ldots,\cos nx\}$ on $[0,\pi]$ and $\{\sin x,\ldots,\sin nx\}$ on $[0,\pi]$ are Chebyshev systems. For a Chebyshev system $\{\varphi_k(x)\}$ on $[a,b]$, let $P_n(x)$ be a linear combination of $\varphi_0(x),\ldots,\varphi_n(x)$ that is not identical to the function $f\in C[a,b]$. Then $P_n(x)$ is the best approximation for $f(x)$ if and only if there are at least $n+2$ distinct points $x_1<\ldots<x_{n+2}$ of $[a,b]$, where $|f(x)-P_n(x)|$ attains its maximum (these points are called **deviation points**) and $(f(x_i)-P_n(x_i))(f(x_{i+1})-P_n(x_{i+1}))<0$ $(i=1,\ldots,n+1)$ (**Chebyshev's theorem**).

For example, consider the polynomial $P_n(x)=a_{n-1}x^{n-1}+\ldots+a_1x+a_0$ with real coefficients such that

$$\max_{-1\leqslant x\leqslant 1}|x^n-a_{n-1}x^{n-1}-\ldots-a_0|$$

takes its smallest value. Then $x^n-P_n(x)=2^{-(n-1)}T_n(x)$, where $T_n(x)=\cos(n\arccos x)$ is the Chebyshev polynomial of degree n.

Since the best approximation is desired for numerical computation, several methods have been developed to find it (→ 300 Numerical Methods). However, when the set $A\subset\mathbf{R}^n$ $(n\geqslant 2)$ contains three nonintersecting arcs emanating from a common point, A admits no Chebyshev system. Thus we do not always have a unique best-approximation polynomial [16].

C. Degrees of Approximation and Moduli of Continuity

For a continuous function $f(x)$ defined on $[a,b]$, the **modulus of continuity of kth order** is defined by

$$\omega_k(f;t)=\sup_{\substack{|h|\leqslant t\\ a\leqslant x\leqslant b\\ a\leqslant x+kh\leqslant b}}\left|\sum_{v=0}^{k}(-1)^{k-v}\binom{k}{v}f(x+vh)\right|$$

for $t\leqslant(b-a)/k$. In particular, ω_1 is the ordinary modulus of continuity. Put $E_n^*(f)=\inf_{a_0,\ldots,a_n,b_1,\ldots,b_n}\max_{a\leqslant x\leqslant b}|f(x)-P_n(x)|$, where f is a continuous periodic function of period 2π and $P_n(x)$ is a trigonometric polynomial of the form (1). Then $E_n^*(f)\leqslant c_k\omega_k(f;1/(n+1))$ (**Jackson's theorem** [1]), where c_k is independent of f. The best possible coefficient c_k has been determined by J. Favard [2]. Further investigations on the relation between $E_n^*(f)$ and $\omega_k(f;t)$ have been carried out by S. N. Bernshteĭn [3] and A. Zygmund [4]. S. B. Stechkin obtained the following results:

$$\omega_k\left(f;\frac{1}{n}\right)\leqslant\frac{c_k}{n^k}\sum_{v=0}^{n}(v+1)^{k-1}E_v^*(f)$$

[5, 6]. For the approximation of $f\in C^r([-1,1])$ by polynomials, there exists a polynomial $P_n(x)$ of degree at most n such that for any $x\in[-1,1]$,

$$|f(x)-P_n(x)|\leqslant M_r(t(x))^r\omega_1(f^{(r)};t(x)),$$

where M_r is a constant not depending on f, x, and n, $f^{(r)}(x)$ is the rth derivative of $f(x)$, and $t(x)=(1/n)(\sqrt{1-x^2}+(|x|/n))$. We also have theorems evaluating $\omega_k(f^{(r)};t)$ in terms of $|f(x)-P_n(x)|$. For the proof of these theorems, estimation of the magnitude of the derivative of the polynomial of degree n plays an essential role. For example, we have the **Bernshteĭn inequality** $\max_x|T_n'(x)|\leqslant n\max_x|T_n(x)|$ for any trigonometric polynomial $T_n(x)$ of degree n and the **Markov inequality**

$$|P_n'(x)|\leqslant n\min[1/\sqrt{1-x^2},n]\max_{-1\leqslant x\leqslant 1}|P_n(x)|$$

for $x\in[-1,1]$ and any polynomial $P_n(x)$ of degree n.

D. Approximation by Fourier Expansions

If $\{\varphi_n(x)\}$ is an †orthonormal system of functions in $L_2(a,b)$ and f is any function in $L_2(a,b)$, then among all linear combinations of $\varphi_0(x),\ldots,\varphi_n(x)$ the one that gives the best mean square approximation to f (i.e., the one for which the integral

$$\int_a^b\left(f(x)-\sum_{k=0}^{n}c_k\varphi_k(x)\right)^2dx$$

attains its minimum) is the Fourier polynomial $\sum_{k=0}^{n}a_k\varphi_k(x)$, where $a_k=\int_a^b f(x)\varphi_k(x)dx$. Consequently, the **least square approximation** (or best approximation with respect to the L_2-norm) by trigonometric polynomials is given by the partial sum $s_n(x)$ of the Fourier series of $f(x)$. For L_p $(1<p<\infty)$, $s_n(x)$ also gives the best approximation up to a constant factor, but in the case of uniform approximation we have $|f(x)-s_n(x)|\leqslant A(\log n)\omega_k(f;n^{-1})$, and this result cannot be improved in general. There is no linear operation that gives the best trigonometric approximation. In approximation with a linear combination of $\varphi_0(x),\ldots,\varphi_n(x)$, the saturation phenomenon of approximation often appears. For example, observe the arithmetic means of $s_n(x)$ (i.e., †Fejér means $\sigma_n(x)$). If $f\in\operatorname{Lip}\alpha$ $(0<\alpha<1)$ (→ 84 Continuous Functions A), then $|f(x)-\sigma_n(x)|=O(n^{-\alpha})$. However, $|f(x)-\sigma_n(x)|=O(n^{-1})$ if and only if the †conjugate function $\tilde{f}(x)\in\operatorname{Lip}1$; $|f(x)-\sigma_n(x)|=o(n^{-1})$ if and only if $f(x)$ is constant (M. Zamansky [7], G. Sunouchi and C. Watari [8]).

E. Trigonometric Interpolation Polynomials

Since the trigonometric system is a Chebyshev system, given $2n+1$ distinct points $x_0,x_1,\ldots,x_{2n}$ and arbitrary numbers $c_0,c_1,\ldots,c_{2n}$, there is always a unique trigonometric polynomial of degree n with prescribed values c_k at the points x_k. Given any continuous function $f(x)$ with period 2π, the trigonometric polynomial that coincides with $f(x)$ at the points x_k is called the **trigonometric interpolation polynomial** with nodes at x_k. If $x_k=2\pi k/(2n+1)$ $(k=0,1,\ldots,2n)$, then the interpolating trigonometric polynomial is given by

$$U_n(f,x)=\frac{1}{2n+1}\sum_{j=0}^{2n}f(x_j)\frac{\sin((n+1/2)(x-x_j))}{\sin((x-x_j)/2)}$$

$$=\frac{1}{2\pi}\int_0^{2\pi}f(t)\frac{\sin((n+1/2)(x-t))}{\sin((x-t)/2)}d\varphi_n(t),$$

where $\varphi_n(t)$ is a step function that has the value $2\pi j/(2n+1)$ in $[2\pi j/(2n+1), 2\pi(j+1)/(2n+1)]$. $U_n(f,x)$ resembles the partial sum $s_n(x)$ of the Fourier series of $f(x)$. If $f(x)$ is continuous and of †bounded variation, then $U_n(f,x)$ converges uniformly to $f(x)$ (D. Jackson [1]). Although the partial sum $s_n(x)$ of the Fourier series of a continuous function $f(x)$ converges almost everywhere to $f(x)$, there is a continuous function for which $U_n(f,x)$ diverges everywhere (J. Marcinkiewicz [9]). Moreover, there exists a continuous function for which $(1/n)(\sum_{k=1}^{n}U_k(f,x))$ diverges everywhere (P. Erdös [10], G. Grünwald [11]). Restating these facts for the algebraic polynomial case, we can conclude that there is a continuous function defined in $[-1,1]$ for which the †Lagrange interpolation polynomial and its arithmetic mean are both divergent everywhere if we take as nodes the roots of the Chebyshev polynomial of degree n.

F. The Case of a Complex Domain

If a given function $f(z)$ is holomorphic in a bounded †simply connected domain E in the complex plane and continuous in $\bar{E}$, then $f(z)$ is approximated uniformly by polynomials on any compact set in E (**Runge's theorem**). This theorem was first studied by C. Runge, and his results were developed by J. L. Walsh and M. V. Keldysh (e.g., [12]). When E contains no interior point, the polynomial approximation of a continuous function defined in E was given by M. A. Lavrent'ev. Unifying these two results, S. N. Mergelyan obtained the following theorem [13]: A necessary and sufficient condition for an arbitrary function continuous on a compact set E and holomorphic inside E to be approximated on E uniformly by polynomials is that the set E does not divide the complex plane.

On the degree of approximation of polynomials to $f(z)$ on a simply connected domain, there are the following results: Let D be a closed bounded set whose complement K is connected and regular in the sense that K possesses a †Green's function $G(x, y)$ with a pole at infinity. Let D_R be the locus $G(x, y) = \log R > 0$. When $f(z)$ is holomorphic on D, there exists the largest number ρ with the following property: $f(z)$ is single-valued and holomorphic at every interior point of D_ρ. If $R < \rho$, there exist polynomials $P_n(z)$ of degree n ($n = 1, 2, \ldots$) such that $|f(z) - P_n(z)| \leqslant M/R^n$ for $z \in D$, where M is a constant independent of n and z. On the other hand, there exist no such polynomials $P_n(z)$ on D for $R > \rho$ (Bernshteĭn and Walsh [12]).

G. Lagrange's Interpolation Formula

For each n ($n = 0, 1, \ldots$), let $z_1^{(n)}, z_2^{(n)}, \ldots, z_{n+1}^{(n)}$ be a given set of real or complex numbers, and let $f(z)$ be an arbitrary function. Then there is a unique polynomial of degree n that coincides with $f(z)$ at each point $z_k^{(n)}$ ($k = 1, \ldots, n+1$). This is called **Lagrange's interpolation polynomial** and is given by

$$P_n(z) = \sum_{k=1}^{n+1} f(z_k^{(n)}) \frac{\omega(z)}{(z - z_k^{(n)})\omega'(z_k^{(n)})},$$

$$\omega(z) = (z - z_1^{(n)}) \ldots (z - z_{n+1}^{(n)}).$$

The sequence $P_n(z)$ does not always converge to $f(z)$. For example, if we take $f(z) = 1/z$ and the $(n+1)$st roots of 1 as $z_k^{(n)}$, then $P_n(z) = z^n$ and $P_n(z)$ converges to $f(z)$ only at the point 1. For real variables also, there are examples of divergent $P_n(z)$. However, if $f(z)$ is holomorphic in $|z| < \rho$ ($\rho > 1$), then $P_n(z)$ with the $(n+1)$st roots of 1 as nodes converges to $f(z)$ uniformly in $|z| \leqslant 1$.

When $z_k^{(n)}$ is independent of the choice of n, $P_n(z)$ coincides with the sum of the first n terms of †Newton's interpolation formula. In this case, $P_n(z)$ is called **Newton's interpolation polynomial** and is given by

$$P_n(z) = a_0 + a_1(z - z_1) + a_2(z - z_1)(z - z_2) + \ldots$$
$$+ a_n(z - z_1) \ldots (z - z_n),$$

where $a_0 = f(z_1)$; $a_1 = (f(z_2) - f(z_1))/(z_2 - z_1)$ ($z_2 \neq z_1$), $a_1 = f'(z_1)$ ($z_2 = z_1$); and so on. Successive coefficients of the polynomial $P_n(z)$ can be calculated by †finite differences. Convergence of Newton's interpolation polynomial is closely connected to convergence of †Dirichlet series.

H. Chebyshev Approximation

Let D be a bounded closed subset of the complex plane, and $f(z)$ a continuous function on D. Then there exists a polynomial $\pi_n(z)$ of degree n such that $\max_{z \in D} |f(z) - \pi_n(z)|$ attains the infimum $E_n(f)$. The polynomial $\pi_n(z)$ is unique and is called the **best approximation polynomial (in the sense of Chebyshev)**. If D is simply connected and $f(z)$ is single-valued and holomorphic on D, then $\pi_n(z)$ converges to $f(z)$ uniformly on D. Moreover, in this case there exist a number M that does not depend on n and a number $R > 1$ such that $|f(z) - \pi_n(z)| \leqslant M/R^n$. Assuming that $f(z)$ satisfies certain additional conditions, W. E. Sewell [14] proved the existence of a constant r such that $|f(z) - \pi_n(z)| \leqslant M/n^r R^n$. Furthermore, by approximating $f(z) = z^n$ by polynomials of degree $n-1$, we can show that there exists a polynomial $T_n(z)$ of degree n such that

$$\min \left\{ \max_{z \in D} |z^n + a_1 z^{n-1} + \ldots + a_n| \right\} = |T_n(z)|.$$

$T_n(z)$ is called a **Chebyshev polynomial** of degree n with respect to the domain D. Similar statements are valid for functions of a real variable. In particular, when $D = [-1, 1]$, we have

$$T_n(x) = \cos(n \arccos x)/2^{n-1},$$

which is the ordinary (real) Chebyshev polynomial. Generally, the limit

$$\lim_{n \to \infty} \left(\max_D |T_n(z)| \right)^{1/n} = \rho(D)$$

exists, and $\rho(D)$ coincides with the †capacity and †transfinite diameter of D [15]. For new results and applications of Chebyshev polynomials → [17].

If the method of evaluating the degree of approximation using the absolute value $|f(z) - \pi_n(z)|$ is replaced by methods using a †curvilinear integral or †surface integral, as explained below, we still obtain similar results. Let D be a closed domain in the complex plane with a boundary C that is a rectifiable Jordan curve. If $f(z)$ is single-valued and holomorphic on D, then there exists a polynomial $\pi_n(z)$ of degree n that minimizes the integral $\int_C u(z) |f(z) - \pi_n(z)|^p |dz|$ ($p > 0$), where $u(z)$ is a given positive continuous function on C. Moreover, $|f(z) - \pi_n(z)| \leqslant M/R^n$ for some $R > 1$ (actually $\{\pi_n(z)\}$ is †overconvergent). If D is a closed

Jordan domain and if $f(z)$ is single-valued and holomorphic in D, then there exists a polynomial $\pi_n(z)$ of degree n that minimizes the integral $\int\int_D u(z)|f(z)-\pi_n(z)|^p dS$, where $u(z)$ is a given positive continuous function on D. Moreover, $|f(z)-\pi_n(z)| \leqslant M/R^n$ for some $R>1$ on D.

I. Approximation by Orthogonal Polynomials on a Curve

Let C be a rectifiable Jordan curve in the complex plane, and let $p_k(z) \in {}^\dagger L_2(C)$. If $\int_C p_k(z)\overline{p_n(z)}|dz| = \delta_{kn}$, then $\{p_k(z)\}$ is called an orthonormal system on C. Given a holomorphic function $f(z)$ on D, we set $a_k = \int_C f(z)p_k(z)|dz|$ and consider the formal series $\sum_{k=0}^{\infty} a_k p_k(z)$. If we denote the nth partial sum of this series by $s_n(z)$, then $s_n(z)$ is the least square approximation by a linear combination of $p_0(z), \ldots, p_n(z)$. This and other results, such as †Bessel's inequality, the †Riesz-Fischer theorem, etc., are all valid here as in the theory of general †orthogonal systems. In particular, if we take $|z|=R$ as C, then $\{1, z, z^2, \ldots\}$ is an orthogonal system. Since in this case

$$a_k = \frac{1}{2\pi R^{2k+1}} \int_C f(z)\bar{z}^k |dz| = \frac{1}{2\pi i} \int_C \frac{f(z)}{z^{k+1}} |dz|$$

and $s_n(z) = a_0 + a_1 z + \ldots + a_n z^n$, the †Taylor expansion of $f(z)$ coincides with the orthogonal expansion of $f(z)$.

Given a compact domain D and a holomorphic function on D, if there exist orthogonal polynomials $p_n(z)$ such that the orthogonal expansion of $f(z)$ with respect to $p_n(z)$ converges to $f(z)$ uniformly on D, we say that $\{p_n(z)\}$ belongs to the domain D. The problem of existence and determination of such polynomials for any given domain was proposed and first solved by G. Faber. Generalizations were given by G. Szegö, T. Carleman, and Walsh. Roughly speaking, $p_n(z)$ is given by the orthogonalization of the system $\{1, z, z^2, \ldots\}$ with respect to the curvilinear integral on $C = \partial D$ or the surface integral on D.

J. Numerical Approximation of Functions

The accuracy of the approximation of a given function $f(z)$ by the partial sums of its †Taylor expansion $\sum a_n(x-x_0)^n$ decreases rapidly as the distance $|x-x_0|$ increases. The accuracy of the approximation of $f(x)$ defined on a compact interval $[A, B]$ by a (polynomial) function $\varphi(x)$ can be evaluated by means of the least square approximation, the best approximation with respect to the uniform norm, and so on. The second method is best suited to numerical calculation of functions. To get the best approximating polynomial $\varphi(x) = P_n(x) = \sum_{k=0}^{n} c_k \varphi_k(x)$ of $f(x)$ (→ Section B), we must determine coefficients c_k that satisfy the conditions of Chebyshev's theorem. The first step in this process is the orthogonal development of $f(x)$ by Chebyshev polynomials $\{T_n(x)\}$: $\varphi_n(x) = \sum_{k=0}^{n} a_k T_k(u)$, $u = (x-(A+B)/2)/((B-A)/2)$. The error $|f(x)-\varphi_n(x)|$ is estimated by a constant multiple of $T_{n+1}(u)$: $|f(x)-\varphi_n(x)| \leqslant K|T_{n+1}(u)|$. This **Chebyshev interpolation** is actually given by $a_0 = N^{-1}\sum_{i=1}^{N} f(x_i)$, $a_k = 2N^{-1}\sum_{i=1}^{N} f(x_i)T_k(u_i)$ $(k=1, \ldots, n)$, where $N = n+1$ and the $u_i = (x_i-(A+B)/2)/((B-A)/2)$ $(i=1, \ldots, N)$ are the roots of $T_N(u)$. Let M be the extremum of the error $|f(x)-\varphi_n(x)|$ of such an approximation, and set $f(x_i) - \varphi_n(x_i) = \pm M_i$ $(i=1,2,\ldots,N)$. Consider a function $\bar{\varphi}_n(x) = \sum \bar{a}_k T_k(u)$ satisfying $f(x_i) - \bar{\varphi}(x_i) = \pm M$. Then solve the linear equation $\bar{\varphi}_n(x_i) - \varphi_n(x_i) = \pm(M - M_i)$ with respect to $\Delta a_k = \bar{a}_k - a_k$ and M. Repeat this process until Δa_k becomes sufficiently small.

A computer can perform the division very quickly, and the rational approximation of a function, for example by its †continued fraction expansion, is often useful.

References

[1] D. Jackson, The theory of approximation, Amer. Math. Soc. Colloq. Publ., 1930.

[2] J. Favard, Sur les meilleurs procédés d'approximation des certains classes des fonctions par des polynomes trigonometrique, Bull. Sci. Math., 61 (1937), 209–224, 243–265.

[3] S. N. Bernshteĭn (Bernstein), Sur l'ordre de meilleure approximation des fonctions continues par des polynomes de degré donné, Mém. Acad. Roy. Belgique, ser. 2, 4 (1912), 1–104.

[4] A. Zygmund, Smooth functions, Duke Math. J., 12 (1945), 47–75.

[5] N. I. Achieser, Theory of approximation, Unger, 1956. (Original in Russian, 1947.)

[6] A. Zygmund, Trigonometric series, Cambridge, 1959.

[7] M. Zamansky, Class de saturation de certains procédés d'approximation des séries de Fourier des fonctions continues et applications à quelques problèmes d'approximation, Ann. Sci. Ecole Norm. Sup., sér. 3, 66 (1949), 19–93.

[8] G. Sunouchi and C. Watari, On the determination of the class of saturation in the theory of approximation of functions II, Tôhoku Math. J., 11 (1959), 480–488.

[9] J. Marcinkiewicz, Sur la divergence des polynomes d'interpolation, Acta Sci. Math. Szeged., 8 (1937), 131–135.

[10] P. Erdös, Some theorems and remarks on interpolations, Acta Sci. Math. Szeged., 12 (1950), 11–17.
[11] G. Grünwald, Über Divergenzerscheinungen der Lagrangeschen Interpolationspolynome der stetigen Functionen, Ann. Math., (2) 37 (1936), 908–918.
[12] J. L. Walsh, Interpolation and approximation by rational functions in the complex domain, Amer. Math. Soc. Colloq. Publ. 1935; fifth edition, 1981.
[13] S. N. Mergelyan (Mergeljan), On the representation of functions by series of polynomials on closed sets, Amer. Math. Soc. Transl., ser. I, 3 (1962), 287–293. (Original in Russian, 1951.)
[14] W. E. Sewell, Degree of approximation by polynomials in the complex domain, Ann. math. studies 9, Princeton Univ. Press, 1942.
[15] E. Hille, Analytic function theory II, Ginn, 1962.
[16] J. R. Rice, The approximation of functions, Addison-Wesley, I, 1964; II, 1969.
[17] S. J. Karlin and W. J. Sneddon, Tchebysheff systems with applications in analysis and statistics, Interscience, 1966.

337 (III.4) Polynomials

A. Polynomials in One Variable

Let R be a commutative †ring and $a_0, a_1, \ldots, a_n$ elements of R. An expression $f(X)$ of the form

$$f(X)=a_0+a_1X+\ldots+a_nX^n \tag{1}$$

is called a **polynomial** in a variable X over R; if $a_n \neq 0$, the number n is called the **degree** of the polynomial $f(X)$ and is denoted by $\deg f$. If $a_n = 1$, the polynomial (1) is called a **monic polynomial**. The totality of polynomials in X over R forms a commutative ring with respect to ordinary addition and multiplication (whose definition will be given later). It is called the **ring of polynomials** (or the **polynomial ring**) of X over R and is denoted by $R[X]$. We say that we **adjoin** X to R to obtain $R[X]$.

B. Polynomials in Several Variables

Let $R[X, Y]$ denote the ring $R[X][Y]$, namely, the ring obtained by adjoining Y to $R[X]$. An element of $R[X, Y]$ can then be expressed as $\sum a_{\mu\nu} X^\mu Y^\nu$. This expression is called a polynomial in X and Y over R. Generally, $R[X_1, \ldots, X_m] = R[X_1, \ldots, X_{m-1}][X_m]$ is called the **polynomial ring in** m **variables** (on m **indeterminates**) $X_1, \ldots, X_m$ over R, and its element

$$F(X_1, X_2, \ldots, X_m) = \sum a_{\nu_1\nu_2\ldots\nu_m} X_1^{\nu_1} X_2^{\nu_2} \ldots X_m^{\nu_m} \tag{2}$$

($\sum$ denotes a finite sum for nonnegative integral ν_i beginning with $\nu_1=\nu_2=\ldots=\nu_m=0$) is called a **polynomial in** m **variables** $X_1, \ldots, X_m$ over R. We call each summand a **term** of the polynomial, $a_{\nu_1\nu_2\ldots\nu_m}$ the **coefficient** and $\nu_1+\nu_2+\ldots+\nu_m$ the **degree** of this term. The greatest degree of terms is called the **degree** of the polynomial F. The term $a_{0\ldots0}$ of degree 0 is called the **constant term** of F. If a polynomial F in $X_1, \ldots, X_m$ is composed of terms of the same degree n, then F is called a **homogeneous polynomial** (or **form**) of degree n; a polynomial consisting of a single term, such as $aX_1^{\nu_1}X_2^{\nu_2}\ldots X_m^{\nu_m}$, is called a **monomial**. Now let $\alpha_1, \alpha_2, \ldots, \alpha_m$ be elements of R (or a commutative ring S containing R), and let $F(\alpha_1, \alpha_2, \ldots, \alpha_m)$ denote the element of R (or S) obtained by substitution of $\alpha_1, \alpha_2, \ldots, \alpha_m$ for $X_1, X_2, \ldots, X_m$ in $F(X_1, X_2, \ldots, X_m)$. It is also called a polynomial in $\alpha_1, \alpha_2, \ldots, \alpha_m$. If $F(\alpha_1, \alpha_2, \ldots, \alpha_m)=0$, then $(\alpha_1, \alpha_2, \ldots, \alpha_m)$ is called a **zero point** (in S) of the polynomial $F(X_1, X_2, \ldots, X_m)$ (or a solution of the algebraic equation $F(X_1, \ldots, X_m)=0$). In the case of one variable, it is called a **root** of $F(X_1)$ (or of $F(X_1)=0$).

C. Polynomial Rings

Addition and multiplication in $R[X]$ are defined by

$(\sum a_i X^i)+(\sum b_j X^j)=\sum(a_i+b_i)X^i$,

$(\sum a_i X^i)(\sum b_j X^j)=\sum_k(\sum_{i+j=k} a_i b_j)X^k$.

A polynomial $f(X)\in R[X]$ can be regarded as a function of a commutative ring R' containing R into itself such that $c \mapsto f(c)$. In this sense, $f(X)+g(X)$ and $f(X)g(X)$ are the functions such that $c \mapsto f(c)+g(c)$ and $c \mapsto f(c)g(c)$, respectively.

It holds that $\deg(f(X)+g(X)) \leqslant \max\{\deg f(X), \deg g(X)\}$, $\deg f(X)g(X) \leqslant \deg f(X)+\deg g(X)$. If R is an †integral domain, then the latter inequality is an equality, and therefore $R[X]$ is an integral domain. For these inequalities and for convenience elsewhere, we define the degree of 0 to be indefinite.

Assume that R is a field. For given $f, g \in R[X]$ ($\deg g \geqslant 1$), we can find unique $q, r \in R[X]$ such that $f=gq+r$ and $\deg r<\deg g$ or $r=0$ (**division algorithm**). This q is called the **integral quotient** of f by g, and r is called the **remainder** of f divided by g. The same fact remains true in the general $R[X]$ if $g(X)$ is monic. (→ 369 Rings of Polynomials).

D. Factorization into Primes

Let k be an integral domain. Since $k[X]$ and hence $k[X_1, \ldots, X_m]$ are integral domains, we can define the concepts concerning divisibility (such as a divisor, a multiple, etc.) (→ 67 Commutative Rings). If k is a †unique factorization domain, then so are $k[X]$ and $k[X_1, \ldots, X_m]$. A polynomial over k is said to be **primitive** if the greatest common divisor of all the coefficients is equal to 1. Every polynomial over k can be uniquely expressed as a product of some primitive polynomials and an element of k; a product of primitive polynomials is primitive (**Gauss's theorem**).

If k is a field, then $k[X]$ and $k[X_1, \ldots, X_m]$ are unique factorization domains. Furthermore, to find the greatest common divisor (f, g) of $f, g \in k[X]$, we can use the **Euclidean algorithm**, that is, apply the division algorithm repeatedly to obtain

$$f = gq_1 + r_1, \quad g = r_1 q_2 + r_2, \quad r_1 = r_2 q_3 + r_3, \ldots,$$

$$\deg g > \deg r_1 > \deg r_2 > \ldots,$$

so that after a finite number of steps we attain $r_{\nu-1} = r_\nu q_{\nu+1}$ $(r_{\nu+1} = 0)$. Then $r_\nu = (f, g)$. Accordingly, $k[X]$ is a †principal ideal domain. This algorithm is applied to Euclid rings (→ 67 Commutative Rings L).

E. The Remainder Theorem

Let k be an integral domain, $f(X) \in k[X]$, and let $g(X) = X - \alpha$ $(\alpha \in k)$. Then using the division algorithm, we get

$$f(X) = (X - \alpha) q(X) + r,$$

$$q(X) \in k[X], \quad r \in k.$$

Therefore, $f(\alpha) = r$; that is, the remainder of $f(X)$ divided by $X - \alpha$ is equal to $f(\alpha)$. This is called the **remainder theorem**. If $f(\alpha) = 0$, then $f(X)$ is divisible by $X - \alpha$ in $k[X]$.

F. Irreducible Polynomials

Let k be a field. A polynomial $f(X) \in k[X]$ of degree n is said to be **reducible** over k if f is divisible by a polynomial of degree $\nu < n$ in $k[X]$ $(\nu \neq 0)$; otherwise, it is said to be **irreducible** over k. Any polynomial of degree 1 is irreducible. A polynomial f is a †prime element of $k[X]$ if and only if f is irreducible over k. Let I be a unique factorization domain. If $f(X)$ is a polynomial (1) in $I[X]$ such that for a prime element p in I, $a_n \not\equiv 0 \pmod p$, $a_{n-1} \equiv a_{n-2} \equiv \ldots \equiv a_0 \equiv 0 \pmod p$ but $a_0 \not\equiv 0$ $(\bmod\, p^2)$, then $f(X)$ is irreducible over the field of quotients of I (**Eisenstein's theorem**). If a polynomial (2) in m variables over an †algebraic number field k is irreducible, we can obtain an irreducible polynomial in $X_1, \ldots, X_\mu$ $(0 < \mu < m)$ from the polynomial $F(X_1, \ldots, X_m)$ by assigning appropriate values in k to $X_{\mu+1}, \ldots, X_m$ (**Hilbert's irreducibility theorem**, *J. Reine Angew. Math.*, 110 (1892)). These two theorems have been generalized in many ways and given precise formulations. In Hilbert's irreducibility theorem, the algebraic number field may be replaced, for example, by any infinite field that is †finitely generated over its †prime field (K. Dörge, W. Franz, E. Inaba).

G. Derivatives

Given a polynomial

$$f(X_1, \ldots, X_m) = \sum a_{\nu_1 \nu_2 \ldots \nu_m} X_1^{\nu_1} X_2^{\nu_2} \ldots X_m^{\nu_m}$$

over a field k, we define the (formal) **derivative** of f with respect to X_i as $f_i(X_1, \ldots, X_m) = \sum \nu_i a_{\nu_1 \nu_2 \ldots \nu_m} X_1^{\nu_1} \ldots X_i^{\nu_i - 1} \ldots X_m^{\nu_m}$ and denote it by $\partial f / \partial X_i$. The map $f \mapsto \partial f / \partial X_i$ is called the (formal) **derivative** with respect to X_i. In particular, if $m = 1$, then $\partial f / \partial X$ is denoted by df/dX. The usual rules of †derivatives also hold for the formal derivative. If $df/dX = 0$ for an irreducible polynomial $f(X)$ in $k[X]$, then $f(X)$ is said to be **inseparable**; otherwise, $f(X)$ is **separable**. If the †characteristic of the field k is 0, then every irreducible polynomial $f(X)$ $(\neq 0)$ is separable. When k is of characteristic $p \neq 0$, an irreducible polynomial $f(X)$ is inseparable if and only if we can write $f(X) = g(X^p)$.

H. Rational Expressions

The †field of quotients of the polynomial ring $k[X_1, \ldots, X_n]$ over a field k is denoted by $k(X_1, \ldots, X_n)$ and is called the **field of rational expressions** (or **field of rational functions**) in variables $X_1, \ldots, X_n$ over k. Its element is called a **rational expression** in $X_1, \ldots, X_n$. It can be written as a quotient of one polynomial $f(X_1, \ldots, X_n)$ by another polynomial $g(X_1, \ldots, X_n) \neq 0$. Also, an expression $f(\alpha_1, \ldots, \alpha_n)/g(\alpha_1, \ldots, \alpha_n)$ obtained by replacing $X_1, \ldots, X_n$ with elements $\alpha_1, \ldots, \alpha_n$ of k in the above expression is called a rational expression in $\alpha_1, \ldots, \alpha_n$ (provided that $g(\alpha_1, \ldots, \alpha_n) \neq 0$).

I. Symmetric Polynomials and Alternating Polynomials

Let $f(X_1, \ldots, X_n)$ be a polynomial in variables $X_1, \ldots, X_n$ over an integral domain I. If

$f(X_1, \ldots, X_n)$ is invariant under every permutation of $X_1, \ldots, X_n$, it is called a **symmetric polynomial** (or **symmetric function**) of $X_1, \ldots, X_n$. If $f(X_1, \ldots, X_n)$ is transformed into $-f(X_1, \ldots, X_n)$ by every †odd permutation of $X_1, \ldots, X_n$, it is called an **alternating polynomial** (or **alternating function**). Also, an expression $f(\alpha_1, \ldots, \alpha_n)$ obtained from $f(X_1, \ldots, X_n)$ by replacing $X_1, \ldots, X_n$ with elements $\alpha_1, \ldots, \alpha_n$ of I is called a symmetric (alternating) function of $\alpha_1, \ldots, \alpha_n$ if $f(X_1, \ldots, X_n)$ is symmetric (alternating).

Let the coefficient of X^{n-k} in the expansion of $(X-X_1)\ldots(X-X_n)$ be denoted by $(-1)^k\sigma_k$. Then we have $\sigma_1 = \sum X_i = X_1 + \ldots + X_n$, $\sigma_2 = \sum X_i X_j = X_1X_2 + X_1X_3 + \ldots + X_{n-1}X_n, \ldots, \sigma_n = X_1 X_2, \ldots, X_n$. Obviously, these are symmetric polynomials of $X_1, \ldots, X_n$. Moreover, for every element φ of the polynomial ring $I[Y_1, Y_2, \ldots, Y_n]$, $\varphi(\sigma_1, \sigma_2, \ldots, \sigma_n)$ is a symmetric polynomial of $X_1, \ldots, X_n$. Conversely, every symmetric polynomial of $X_1, \ldots, X_n$ can be uniquely expressed as a polynomial $\varphi(\sigma_1, \sigma_2, \ldots, \sigma_n)$. Thus the totality of symmetric polynomials of $X_1, \ldots, X_n$ is identical with the ring $I[\sigma_1, \sigma_2, \ldots, \sigma_n]$. This is called the **fundamental theorem on symmetric polynomials**, and $\sigma_1, \sigma_2, \ldots, \sigma_n$ are called **elementary symmetric polynomials** (or **elementary symmetric functions**). For example, for $s_\nu = \sum X_i^\nu$ $(\nu = 1, 2, \ldots)$, we have $s_1 = \sigma_1$, $s_2 = \sigma_1^2 - 2\sigma_2$, $s_3 = \sigma_1^3 - 3\sigma_1\sigma_2 + 3\sigma_3$, $s_4 = \sigma_1^4 - 4\sigma_1^2\sigma_2 + 2\sigma_2^2 + 4\sigma_1\sigma_3 - 4\sigma_4$. Concerning the elementary symmetric polynomials and the s_ν, we have the relations $s_\nu - \sigma_1 s_{\nu-1} + \sigma_2 s_{\nu-2} - \ldots + (-1)^{\nu-1}\sigma_{\nu-1}s_1 + (-1)^\nu \nu\sigma_\nu = 0$ $(\nu = 1, 2, \ldots)$, and $s_\mu - \sigma_1 s_{\mu-1} + \ldots + (-1)^n \sigma_n s_{\mu-n} = 0$ $(\mu = n+1, n+2, \ldots)$ (**Newton's formulas**).

Let $p(X_1, \ldots, X_n) = (X_2 - X_1)(X_3 - X_1)\ldots(X_n - X_1)(X_3 - X_2)\ldots(X_n - X_{n-1})$ be the product of $n(n-1)/2$ differences between $X_1, \ldots, X_n$. Then the polynomial p is invariant under even permutations of $X_1, \ldots, X_n$, and p becomes $-p$ under odd permutations. Hence p is an alternating polynomial of $X_1, \ldots, X_n$. It is called the **simplest alternating polynomial** of these variables. Because of its particular expression, p is also called the **difference product** of $X_1, \ldots, X_n$. If the characteristic of I is different from 2, an alternating polynomial f is divisible by the simplest alternating polynomial p; it can be written as $f = ps$, where s stands for a symmetric polynomial.

J. Discriminants

The square $D(X_1, \ldots, X_n) = p^2(X_1, \ldots, X_n)$ of the simplest alternating polynomial p is a symmetric polynomial, and it is therefore a polynomial in $\sigma_1, \ldots, \sigma_n$. $D(\alpha_1, \alpha_2, \ldots, \alpha_n) = 0$ gives a criterion for the condition that some of $\alpha_1, \ldots, \alpha_n$ are equal. If $\alpha_1, \ldots, \alpha_n$ are the roots of an †algebraic equation $a_0X^n + a_1X^{n-1} + \ldots + a_n = 0$ of degree n, then $D(\alpha_1, \ldots, \alpha_n)$ is called the **discriminant of the equation**. It can be expressed in terms of coefficients $a_0, a_1, \ldots, a_n$ of the equation. For instance, if $n=2$, we have $a_0^2 D = a_1^2 - 4a_0a_2$; if $n=3$, we have $a_0^4 D = a_1^2a_2^2 + 18a_0a_1a_2a_3 - 4a_0a_2^3 - 4a_1^3a_3 - 27a_0^2a_3^2$.

References

[1] B. L. van der Waerden, Algebra I, Springer, seventh edition, 1966.
[2] N. Bourbaki, Eléments de mathématique, Algèbre, ch. 4, Actualités Sci. Ind., 1102b, Hermann, second edition, 1959.
[3] A. G. Kurosh, Lectures on general algebra, Chelsea, 1970. (Original in Russian, 1955.)
[4] R. Godement, Cours d'algèbre, Hermann, 1963.
For Hilbert's irreducibility theorem,
[5] S. Lang, Diophantine geometry, Interscience, 1962, ch. 8.

338 (X.28)
Potential Theory

A. Newtonian Potential

In dynamics, a **potential** means a function u of n variables $x_1, \ldots, x_n$ such that $-\operatorname{grad} u = -(\partial u/\partial x_1, \ldots, \partial u/\partial x_n)$ gives a field of force in the n-dimensional $(n \geqslant 2)$ Euclidean space $\mathbf{R}^n$. Given a point P in $\mathbf{R}^n$ and a measure μ, the functions $u(P)$ given by the integrals

$$u(P) = -\int \log \overline{PQ}\, d\mu(Q), \qquad n=2,$$

$$u(P) = \int \overline{PQ}^{2-n}\, d\mu(Q), \qquad n \geqslant 3,$$

are typical examples of potential functions. They are called the **logarithmic potential** and **Newtonian potential**, respectively. However, some authors mean by Newtonian potential the function $u(P) = \int \overline{PQ}^{-1}\, d\mu(Q)$ in $\mathbf{R}^3$. Usually, the measure μ is taken to be a nonnegative †Radon measure with compact †support. These potentials are †superharmonic in $\mathbf{R}^n$ and harmonic outside the support of μ. Conversely, any harmonic function defined on a domain in $\mathbf{R}^n$ can be expressed as the sum of a potential of a single layer and a potential of a double layer (defined in the next paragraph). Because of this close relation between potentials and harmonic functions, sometimes potential theory means the study of harmonic functions (→ 193 Harmonic Functions and Subharmonic Functions). (For the representation by potentials of superharmonic functions → 193 Harmonic Functions and Subharmonic Functions S.)

Suppose that the measure μ of $\mathbf{R}^3$ satisfies $d\mu=\rho\, d\tau$ with sufficiently smooth density ρ and volume element $d\tau$. Then the Newtonian potential u of μ satisfies **Poisson's equation** $\Delta u=-4\pi\rho$. If the support of μ is contained in a surface S and $d\mu=\rho\, d\sigma$ with density ρ and surface element $d\sigma$, then the potential u of μ is called the potential of a **single layer** (or **simple distribution**). If ρ is continuous on S, then u is continuous in the whole space, and the †directional derivative of u at a point P in the direction of the normal line to S at P_0 tends to

$$-2\pi\rho(P_0)+\int_S \rho \left.\frac{\partial}{\partial n_P}\frac{1}{\overline{PQ}}\right|_{P=P_0} d\sigma(Q)$$

as P approaches P_0 along the normal line. Therefore, as P moves on the line, the directional derivative jumps by $-4\pi\rho(P_0)$ at P_0. If ρ satisfies the †Hölder condition at $P_0\in S$, then the derivative at P in the direction of any fixed tangent line at P_0 has a finite limit as P tends to P_0 along the normal line. The integral

$$u(P)=\int_S \rho\frac{\partial}{\partial n_Q}\frac{1}{\overline{PQ}}\, d\sigma(Q)$$

is called the potential of a **double layer** (or **double distribution**). If ρ is continuous on S, then the limits at P_0 of u from the two directions along the normal line at P_0 exist and are $2\pi\rho(P_0)+u(P_0)$ and $-2\pi\rho(P_0)+u(P_0)$. If, further, ρ is of class C^2 on S, then each partial derivative of u has a finite limit as P tends to a point of S.

B. Generalized Potential

The classical notion of potentials is generalized as follows: Let Ω be a space supplied with a measure μ ($\geqslant 0$) and $\Phi(P,Q)$ a real-valued function on the product space $\Omega\times\Omega$. When the integral $\int\Phi(P,Q)\,d\mu(Q)$ is well defined at each point $P\in\Omega$, it is called the **potential** of μ with **kernel** Φ and is denoted by $\Phi(P,\mu)$ or $\Phi\mu(P)$. The function $\check{\Phi}(P,Q)=\Phi(Q,P)$ is called the **adjoint kernel** of Φ. When

$$(\mu,\nu)=\iint\Phi(P,Q)\,d\mu(Q)\,d\nu(P)$$

$$=\int\Phi(P,\mu)\,d\nu(P)$$

exists for measures $\mu,\nu\geqslant 0$, the value is called the **mutual energy** of μ and ν. In particular, (μ,μ) is called the **energy** of μ. The definition of potential given above may be too general, and some restrictions are called for. We assume that Ω is a †locally compact Hausdorff space; Φ is a †lower semicontinuous function on $\Omega\times\Omega$ satisfying $-\infty<\Phi\leqslant\infty$; and μ, ν, and λ are nonnegative Radon measures with compact support in Ω. In particular, when $\Omega=\mathbf{R}^n$, the potential with the kernel $\Phi(P,Q)=\overline{PQ}^{-\alpha}$ $(0<\alpha<n)$ is called a **potential of order** α (sometimes of order $n-\alpha$) or a **Riesz potential**.

C. The Maximum Principle and the Continuity Principle

Let $\Omega=\mathbf{R}^n$, and let $\Phi(P,\mu)$ be the kernel of the Newtonian potential. Then $\Phi(P,\mu)$ satisfies the following principles: (1) **Frostman's maximum principle (first maximum principle)**: $\sup_{P\in\Omega}\Phi(P,\mu)\leqslant\sup_{P\in S_\mu}\Phi(P,\mu)$ for any μ, where S_μ is the support of μ. (2) **Ugaheri's maximum principle (dilated maximum principle)**: There is a constant $c\geqslant 0$ such that $\sup_{P\in\Omega}\Phi(P,\mu)\leqslant c\sup_{P\in S_\mu}\Phi(P,\mu)$ for any μ. (3) A variation of Ugaheri's maximum principle: Given any compact set $K\subset\Omega$, there exists a constant c which may depend on K such that $\sup_{P\in K}\Phi(P,\mu)\leqslant c\sup_{P\in S_\mu}\Phi(P,\mu)$ for any μ with $S_\mu\subset K$. (4) **Upper boundedness principle**: If $\Phi(P,\mu)$ is bounded from above on S_μ, then it is bounded on Ω also. (5) For any compact set $K\subset\Omega$ and any μ with $S_\mu\subset K$, if $\Phi(P,\mu)$ is bounded above on S_μ, then it is bounded on K also. (6) **Continuity principle**: If $\Phi(P,\mu)$ is continuous as a function on S_μ, then it is also continuous in Ω. Generally, the relations shown in Fig. 1 hold among principles (1)–(6), where (a)→(b) means that (a) implies (b), and (c)↛(d) means the negation of (c)→(d).

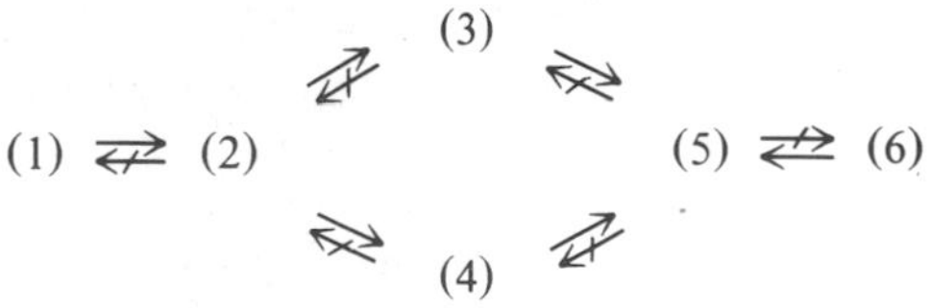

(3)

(1) ⇄ (2) (5) ⇄ (6)

(4)

Fig. 1

If the continuity principle holds for a general kernel $\Phi(p,\mu)$ of a potential and if for any μ there is a sequence $\{P_k\}$ of points in $\Omega-S_\mu$ that has an accumulation point on S_μ and along which $\Phi(P_k,\mu)\to\sup_{\Omega-S_\mu}\Phi(P,\mu)$, then (1) holds also. The second condition is valid, for instance, when $\Omega=\mathbf{R}^n$, $\Phi(P,\mu)$ is †subharmonic in $\mathbf{R}^n-S_\mu$, and $\limsup\Phi(P,\mu)<\sup_{Q\in\mathbf{R}^n}\Phi(Q,\mu)$ as P tends to the point at infinity. T. Ugaheri, G. Choquet, and N. Ninomiya studied (2) and (6). Ugaheri showed that for any nonnegative decreasing function $\varphi(r)$ defined in $[0,\infty)$ and satisfying $\varphi(0)=\infty$, the kernel $\Phi(P,Q)=\varphi(\overline{PQ})$ satisfies (2) in $\mathbf{R}^n$. M. Ohtsuka established that (6)→(5) and (5)↛(6) in general and proved (5)→(6) in the special case where Φ is continuous on $\Omega \times \Omega$ in the wider sense (i.e., Φ may have ∞ as its value) and Φ is finite outside the diagonal set of $\Omega \times \Omega$. The examples in Sections F, I, and J show that the potentials with kernels satisfying a weak condition such as (6) possess a number of important properties. We note that (6) does not necessarily hold in general. (For literature on (1)–(6) and other related principles → [18].)

D. The Energy Principle

Denote by E the class of all measures of finite energy. A symmetric kernel is called **positive definite** (or of **positive type**) if $(\mu-\nu, \mu-\nu)= (\mu, \mu)+(\nu, \nu)-2(\mu, \nu) \geqslant 0$ for any $\mu, \nu \in E$. If the equality $(\mu-\nu, \mu-\nu)=0$ always implies $\mu=\nu$, then the kernel is said to satisfy the **energy principle**. Some characterizations for a kernel to be of positive type or to satisfy the energy principle were given by Ninomiya. Using them, he showed that a symmetric kernel that satisfies Frostman's maximum principle or the domination principle (→ Section L) and a certain additional condition is of positive type. Choquet and Ohtsuka generalized these definitions and results [18].

E. Topologies on Classes of Measures

Let C_0 be the space of continuous functions with compact support in Ω and M_0^+ be the class of measures in Ω. The †seminorms $\mu-\nu \rightarrow |\int f d\mu - \int f d\nu|$ $(f \in C_0)$ define the **vague topology** on M_0^+. The class of unit distributions can be topologized by the vague topology, which induces a topology on Ω itself. This topology coincides with the original topology in Ω. A subclass M of M_0^+ is relatively compact with respect to the vague topology if on every compact set in Ω, the values of the measures of M are bounded. Denote by L the class of measures λ such that (λ, μ) is finite for all $\mu \in M_0^+$, and define the **fine topology** on M_0^+ by the seminorms $\mu-\nu \rightarrow |(\lambda, \mu)-(\lambda, \nu)|$ $(\lambda \in L)$. This topology was introduced by H. Cartan [4] for the Newtonian kernel. It is the weakest topology that makes each $\check{\Phi}(P, \lambda)$, $\lambda \in L$, continuous. Further, when the kernel is of positive type, the **weak topology** is defined on E by the seminorms $\mu-\nu \rightarrow |(\lambda, \mu)-(\lambda, \nu)|$ $(\lambda \in E)$. The **strong topology** is defined on E by the seminorm $\sqrt{(\mu-\nu, \mu-\nu)}$.

For the Newtonian kernel, Cartan [5] showed that vague≺fine≺weak≺strong on E, where vague≺fine, for instance, means that the fine topology is stronger than the vague topology. He proved that the fine, weak, and strong convergences are equivalent for any sequence $\{\mu_n\}$ of measures with bounded energies. B. Fuglede [16] called a kernel of positive type **consistent** when any †Cauchy net with respect to the strong topology that converges vaguely to a measure converges strongly to the same measure. This notion is used to give conditions for E to be †complete with respect to the strong topology [11, 18]. Moreover, Fuglede called a consistent kernel satisfying the energy principle **perfect**, and studied the cases where †convolution kernels on a locally compact topological group are consistent or perfect [11]. For instance, $\overline{PQ}^{-\alpha}$ $(0<\alpha<n)$ in $\mathbf{R}^n$ and the Bessel kernel, which was studied by N. Aronszajn and K. T. Smith, are perfect.

F. Convergence of Sequences of Potentials

We are concerned with determining when a family of potentials corresponding to a class of measures $\{\mu_\omega\}$ with indices ω in a directed set converges. If all S_{μ_ω} are contained in a fixed compact set and μ_ω converges vaguely to μ_0, then $\liminf \Phi(P, \mu_\omega) \geqslant \Phi(P, \mu_0)$ in Ω. If, moreover, Φ is continuous in the wider sense and both Φ and $\check{\Phi}$ satisfy the continuity principle, then equality holds quasi-everywhere (q.e.) in Ω in this inequality [18]; we now define the notion q.e. First, for a nonempty compact set K in Ω, define $W(K)$ to be $\inf(\mu, \mu)$, where $S_\mu \subset K$ and $\mu(K)=1$, and $W(\varnothing)$ to be ∞ for the empty set $\varnothing$. Next, set $W_i(X)=\inf_{K \subset X} W(K)$ for an arbitrary set X in Ω and $W_e(X)=\sup_{X \subset G} W_i(G)$, where G is an open set in Ω. When a property holds except on a set X such that $W_e(X)=\infty$ (resp. $W_i(X)=\infty$), we say that the property holds **quasi-everywhere** (q.e.) (resp. **nearly everywhere** (n.e.)). The terms q.e. and n.e. are also used in the theory of †capacity, although their meaning is not the same as here. (For results on the convergence of sequences of potentials → [18].)

G. Thin Sets

A set $X \subset \Omega$ is called **thin** at P_0 when either P_0 is an isolated point of the set $X \cup \{P_0\}$ with respect to the original topology of Ω or there exists a measure μ such that $\liminf \Phi(P, \mu) > \Phi(P_0, \mu)$ as $P \in X-\{P_0\}$ tends to P_0. If P_0 is an isolated point of $X \cup \{P_0\}$ with respect to the topology weakest among those stronger than both the original topology and the fine topology, then X is thin with respect to the adjoint kernel $\check{\Phi}$. The converse is true in a special case [5]. The notion of thinness was introduced in 1940 by M. Brelot and investigated in detail in [3]. Let $\varphi(r)$ be a positive decreasing function. Suppose there exist positive numbers r_0, δ, a such that $\varphi(r) \leqslant a\varphi((1+\delta)r)$ in $0<r<r_0$. Assume that Ω is a metric space with distance ρ, and take $\varphi(\rho(P,Q))$ as the kernel $\Phi(P,Q)$. Then a necessary and sufficient condition for X to be thin at P_0 is $\sum_{j=1}^{\infty} s^j/W_e(X_j)<\infty$ $(s>1)$ [19], where $X_j=\{P\in X \mid s^j \leqslant \varphi(\rho(P,P_0)) \leqslant s^{j+1}\}$. This criterion was obtained by N. Wiener in 1924 and utilized to give a condition for a boundary point P_0 of a domain D in $\mathbf{R}^3$ to be †regular with respect to the †Dirichlet problem for D. In this situation, P_0 is regular if and only if the complement of D is thin at P_0. When every compact subset of X is thin at P_0, X is called **internally thin** at P_0. A necessary and sufficient condition for X to be internally thin at P_0 is $\sum s^j/W_i(X_j)<\infty$.

H. Polar Sets

Brelot (1941) called a set A **polar** when there is a measure μ for which $\Phi(P,\mu)=\infty$ on A. Consider $\check{U}(\Omega,K)$ $(=V(K,\Omega))$ as defined in 48 Capacity C, and define $\check{U}_e(\Omega,X)$ by $\sup_{G\supset X}\inf_{K\subset G}\check{U}(\Omega,K)$ for an arbitrary set X, where G is an open set. Then for any μ, $X=\{P \mid \Phi(P,\mu)=\infty\}$ is a G_δ set for which the value of $\check{U}_e(\Omega,X)$ is infinite. Conversely, given a G_δ set A of †Newtonian outer capacity zero in $\mathbf{R}^n$ $(n\geqslant 3)$, there is a measure μ such that the set of points where the Newtonian potential of μ is equal to ∞ coincides with A and $\mu(\mathbf{R}^n-A)=0$ (Choquet [7]). This result is called **Evans's theorem** (or the **Evans-Selberg theorem**) (→ 48 Capacity) in the special case where A is compact.

I. Quasicontinuity

A function f in Ω is called **quasicontinuous** if there is an open subset G in Ω of arbitrarily small capacity such that the restriction of f to $\Omega-G$ is continuous. Naturally, quasicontinuity depends on the definition of capacity. Suppose that whenever the potential of a measure μ with kernel Φ is continuous as a function on S_μ it is quasicontinuous in Ω; Φ is then said to satisfy the **quasicontinuity principle**. Assuming in addition that Φ is positive symmetric and taking $1/U(\Omega,G)$ as the capacity of G, we find that every potential is quasicontinuous in Ω (M. Kishi). A similar result is valid for a nonsymmetric kernel if the continuity principle is assumed (Choquet [6]).

J. The Gauss Variational Problem

Given a compact set K and a function f on K, the problem of minimizing the **Gauss integral** $(\mu,\mu)-2\int f\,d\mu$ for a measure μ such that $S_\mu\subset K$ is called the **Gauss variational problem**. When an additional condition is imposed on μ, the problem is called conditional. Among many results obtained for this problem [18], the following is typical: If Φ is symmetric, K supports a nonzero measure of finite energy, and f is finite upper semicontinuous on K, then there exists a μ such that $f(P)\leqslant\Phi(P,\mu)$ n.e. on K and $f(P)\geqslant\Phi(P,\mu)$ on S_μ. When Φ is not symmetric, the same relations hold for some μ if Φ is positive and $\check{\Phi}$ satisfies the continuity principle (Kishi [15]), although the method using Gauss variation is not applicable. When Φ is symmetric and of positive type, the Gauss integral with $f(P)=\Phi(P,\nu)$ $(\nu\in E)$ is equal to $\|\mu-\nu\|^2-\|\nu\|^2$, and the minimizing problem is equivalent to finding the projection of ν to $\{\mu\in E \mid S_\mu\subset K\}$. In some cases, this projection is equal to the measure obtained by the balayage of ν to K (→ Section L).

K. Equilibrium Mass Distributions

A unit measure μ supported by a compact set K is called an **equilibrium mass distribution** on K if $\Phi(P,\mu)$ is equal to a constant a n.e. on K and $\Phi(P,\mu)\leqslant a$ in Ω. The kernel is said to satisfy the **equilibrium principle** if there exists an equilibrium mass distribution on every compact set. If $a>0$, $1/a$ can be regarded as a kind of capacity, and μ/a is called a **capacitary mass distribution**. Corresponding to †inner and outer capacities, inner and outer capacitary mass distributions and their coincidence can be discussed [11]. When Φ is symmetric, Frostman's maximum principle is equivalent to the equilibrium principle.

L. The Sweeping-Out Principle

A kernel is said to satisfy the **sweeping-out principle** (or **balayage principle**) if for any compact set K and measure μ there exists a measure ν supported by K such that $\Phi(P,\nu)=\Phi(P,\mu)$ n.e. on K and $\Phi(P,\nu)\leqslant\Phi(P,\mu)$ in Ω. When we find such a ν, we say that we **sweep out** μ to K, and finding ν is called a **sweeping-out process** (or **balayage**). This terminology originated in the classical process for the Newtonian potential in which the exterior of K is covered by a countable number of balls and the masses inside the balls are repeatedly swept out onto the spherical surfaces. For any general kernel, the balayage principle implies the **domination principle** (also called **Cartan's maximum principle**), which asserts that if the inequality $\Phi(P,\mu)\leqslant\Phi(P,\nu)$ is valid on S_μ for $\mu\in E$ and any ν, then the same inequality holds in Ω. The converse is true if Φ is positive, symmetric, continuous in the wider sense, and finite outside the diagonal set. In contrast to the domination principle, Φ is said to satisfy the **inverse domination principle** if the inequality $\Phi(P,\mu)\leqslant\Phi(P,\nu)$ on S_ν for $\mu\in E$ and any ν implies the same inequality in Ω. In a special case, the domination principle implies Frostman's maximum principle [17]. For the equilibrium and domination principles for non-symmetric kernels → [14].

Corresponding to inner and outer capacitary mass distributions, we can examine inner and outer balayage mass distributions and inner and outer Gauss variational problems and their coincidences [5, 18]. With respect to the Newtonian potential, a point P is called an **internally (externally) irregular point** of X if the inner (outer) balayage mass distribution to X of the unit measure ε_P at P is different from ε_P. X is thin (internally thin) at P if and only if P is an externally (internally) irregular point of X (Cartan [5]).

M. Other Principles

A kernel Φ is said to satisfy the **uniqueness principle** if $\mu\equiv\nu$ follows from the equality $\Phi(P,\mu)=\Phi(P,\nu)$, which is valid n.e. in Ω. Ninomiya and Kishi studied this principle. A kernel Φ is said to satisfy the **lower envelope principle** if given μ and ν, there is a λ such that $\Phi(P,\lambda)=\min(\Phi(P,\mu),\Phi(P,\nu))$. If Φ satisfies the domination principle and $\check{\Phi}$ satisfies the continuity principle, then Φ satisfies the lower envelope principle on every compact set considered as a space. Conversely, if Φ satisfies the lower envelope principle on every compact set considered as a space, then Φ satisfies the domination principle or the inverse domination principle under some additional conditions (Kishi). A kernel is said to satisfy the **complete maximum principle** if the inequality $\Phi(P,\mu)\leqslant\Phi(P,\nu)+a$ on S_μ implies the same inequality in Ω for $\mu\in E$, any ν, and $a\geqslant 0$ (Cartan and J. Deny, 1950). This principle implies both Frostman's maximum principle and the domination principle. Potentials of order α $(n-2\leqslant\alpha<n)$ in $\mathbf{R}^n$ and the **Yukawa potential** with kernel $a\overline{PQ}^{-1}\exp(-\lambda\overline{PQ})$ in $\mathbf{R}^3$ satisfy the complete maximum principle. Relations between this principle and some other principles were studied by Kishi [14].

While all principles discussed so far are global, Choquet and Ohtsuka made a local study.

N. Diffusion Kernels

By means of the bilinear form $\int f\,d\mu$ $(f\in C,$ $\mu\in M_0)$, we now introduce weak topologies in the space C of continuous functions in Ω and in the class M_0 of Radon measures of general sign with compact support. Similarly, we introduce weak topologies in C_0 and in the class M of measures of general sign with not necessarily compact support. A positive linear mapping G of M_0 into M that is continuous with respect to these two weak topologies is called a **diffusion kernel**. The linear mapping G^* of C_0 into C that is determined by $\int f\,dG\mu=\int G^*f\,d\mu$ is called a **transposed mapping**. We can define the balayage principle for G and the domination principle for G^* as in the case where kernels are functions. Then G satisfies the balayage principle if and only if G^* satisfies the domination principle (Choquet and Deny). The complete maximum principle has been defined and studied for G^* (Deny). G. A. Hunt obtained a relation between this principle and the representation of G^*f in the form $\int_0^\infty P_t f\,dt$ with a †semigroup P_t. His result is important in the theory of †stochastic processes (→ 261 Markov Processes).

O. Convolution Kernels

A diffusion kernel G on a locally compact Abelian group induces a convolution kernel κ if G is translation-invariant. It is called a **Hunt kernel** when there exists a vaguely continuous semigroup $\{\alpha_t\}_{t\geqslant 0}$ such that $\kappa=\int_0^\infty\alpha_t\,dt$ and $\alpha_0=\varepsilon_0$, the Dirac measure at the origin. A Hunt kernel satisfies the domination principle and the balayage principle to all open sets, and it satisfies the complete maximum principle if and only if $\{\alpha_t\}_{t\geqslant 0}$ is sub-Markovian, $\int d\alpha_t\leqslant 1$. The Fourier transform of such a semigroup has a closed connection with a negative-definite function [1].

For a convolution kernel κ satisfying the domination principle, or, equivalently, the balayage principle, the inequality

$$\kappa(\omega_1)\kappa(\omega_2)\leqslant\kappa(\omega_1-\omega_1)\kappa(\omega_1+\omega_2)$$

is valid for all relatively compact open sets ω_1 and ω_2 [8]. It has a unique decomposition $\kappa=\varphi\cdot(\kappa_0+\kappa_1)$, where φ is a continuous exponential function, κ_0 is equal to 0 or a Hunt kernel satisfying the complete maximum principle, and κ_1 is a singular kernel satisfying the domination principle such that $\kappa_1*\varepsilon_x=\kappa_1$ for every $x\in S_{\kappa_0}$ [12].

P. Potentials with Distribution Kernels

A function f in $\mathbf{R}^n$ is called **slowly increasing in the sense of Deny** if there exists a positive integer q such that $\int f(P)(1+\overline{OP}^2)^{-q}d\tau(P)<\infty$. A †distribution K is called a **distribution kernel** if the †Fourier transform FK of K is a function $k\geqslant 0$ and both k and $1/k$ are slowly increasing in the sense of Deny. Given such a distribution K, the family W of distributions T for which FT are functions and $\|T\|^2=\int k(FT)^2\,d\tau<\infty$ (this is called the energy of T) is a †Hilbert space with inner product $(T_1, T_2)=\int kFT_1FT_2\,d\tau$. However, the family of Newtonian potentials of measures with finite energy is not a Hilbert space (Cartan). The family of functions of class C^∞ with compact support is †dense in W. For every $T\in W$ the function $FK\times FT$ is slowly increasing in the sense of Deny. There exists a distribution $U=U^T$ that satisfies $FU=FK\times FT$, called the K-potential of T. Since W is complete, the method of projection is applicable, and problems of equilibrium, balayage, and capacity can be examined. For instance, if †Dirac's distribution δ is taken as a distribution kernel, then the corresponding capacity is the Lebesgue measure. In the case of the Newtonian kernel, $\partial U^T/\partial x_j=f_j$ is defined a.e. in $\mathbf{R}^n$ for any $T\in W$. These f_j are square integrable, $T=-c_n\sum_{j=1}^n \partial f_j/\partial x_j$, and $\|T\|^2=c_n\sum_{j=1}^n\int|f_j|^2\,d\tau$, where $1/c_n=2(n-2)\pi^{n/2}/\Gamma(n/2)$. Furthermore,

$$U^T=(n-2)\int\sum_{j=1}^n (x_j-y_j)f_j(Q)\overline{PQ}^{-n}\,d\tau(Q),$$

where x_j, y_j are components of P, Q, respectively. Every ordinary potential of a double layer is a special case of U^T. Conversely, let f be a function in $\mathbf{R}^n$ that is absolutely continuous along almost every line parallel to each coordinate axis and whose partial derivatives are square integrable. Then f is equal to the potential of some $T\in W$ with Newtonian kernel up to an additive constant. These results are due to Deny [9].

Q. Dirichlet Spaces

In this section, functions are assumed to be complex-valued. Let Ω be a locally compact Hausdorff space, $\xi\geqslant 0$ a Radon measure in Ω, and C_0 the space of continuous functions with compact support. A Hilbert space D consisting of locally ξ-integrable functions is called a **Dirichlet space** if $C_0\cap D$ is dense in both C_0 and D, the relations $|v(P)-v(Q)|\leqslant|u(P)-u(Q)|$ and $|v(P)|\leqslant|u(P)|$ for $u\in D$ and a function v always imply $v\in D$ and $\|v\|\leqslant\|u\|$, and for any compact set $K\subset\Omega$, there exists a constant $A(K)$ such that $\int_K|u|\,d\xi\leqslant A(K)\|u\|$ for every $u\in D$. The notion of Dirichlet space was introduced by A. Beurling. A function $u\in D$ is called a potential if there exists a Radon measure μ such that $(u,\varphi)=\int\bar{\varphi}\,d\mu$ holds for every $\varphi\in C_0\cap D$. If in addition $\mu\geqslant 0$, then u is called a pure potential. For any pure potential, the lower envelope principle, the equilibrium principle, the balayage principle, and the complete maximum principle hold [2]. Suppose that Ω is a locally compact Abelian group and D is a Dirichlet space such that $U_yu(x)=u(x-y)\in D$ and $\|U_yu\|=\|u\|$ for every $u\in D$ and $y\in\Omega$. Then we call D special and characterize it in terms of a real-valued continuous function on Ω [2]. (For axiomatic potential theory → 193 Harmonic Functions and Subharmonic Functions.)

References

[1] C. Berg and G. Forst, Potential theory on locally compact Abelian groups, Springer, 1975.
[2] A. Beurling and J. Deny, Dirichlet spaces, Proc. Nat. Acad. Sci. US, 45 (1959), 208–215.
[3] M. Brelot, Sur les ensembles effilés, Bull. Sci. Math., (2) 68 (1944), 12–36.
[4] M. Brelot, Eléments de la théorie classique du potentiel, Centre de Documentation Universitaire, third edition, 1965.
[5] H. Cartan, Théorie généale du balayage en potentiel newtonien, Ann. Univ. Grenoble (N.S.), 22 (1946), 221–280.
[6] G. Choquet, Sur les fondements de la théorie fine du potentiel, C. R. Acad. Sci. Paris, 244 (1957), 1606–1609.
[7] G. Choquet, Potentiels sur un ensemble de capacité nulle, Suites de potentiels, C. R. Acad. Sci. Paris, 244 (1957), 1707–1710.
[8] G. Choquet and J. Deny, Noyaux de convolution et balayage sur tout ouvert, Lecture notes in math. 404, Springer, 1974, 60–112.
[9] J. Deny, Les potentiels d'énergie finie, Acta Math., 82 (1950), 107–183.
[10] O. Frostman, Potentiel d'équilibre et capacité des ensembles avec quelques applications à la théorie des fonctions, Medd. Lunds Univ. Mat. Sem., 3 (1935).
[11] B. Fuglede, On the theory of potentials in locally compact spaces, Acta Math., 103 (1960), 139–215.
[12] M. Itô, Caractérisation du principe de domination pour les noyaux de convolution non-bornés, Nagoya Math. J., 57 (1975), 167–197.
[13] O. D. Kellogg, Foundations of potential theory, Springer, 1929.
[14] M. Kishi, Maximum principles in the potential theory, Nagoya Math. J., 23 (1963), 165–187.
[15] M. Kishi, An existence theorem in potential theory, Nagoya Math. J., 27 (1966), 133–137.

[16] N. S. Landkof, Foundations of modern potential theory, Springer, 1972.
[17] N. Ninomiya, Etude sur la théorie du potentiel pris par rapport au noyau symétrique, J. Inst. Polytech. Osaka City Univ., (A) 8 (1957), 147–179.
[18] M. Ohtsuka, On potentials in locally compact spaces, J. Sci. Hiroshima Univ., 25 (1961), 135–352.
[19] M. Ohtsuka, On thin sets in potential theory, Sem. Anal. Funtions, Inst. for Adv. Study, Princeton, 1957, vol. 1, p. 302–313.

339 (XI.2) Power Series

A. General Remarks

Let a and $c_0, c_1, c_2, \ldots$ be elements of a †field K and z be a variable. A series of the form $P=\sum_{n=0}^{\infty} c_n(z-a)^n$ is called a **power series** (in one variable). We assume that K is the field of complex numbers. For a given power series P, we can determine a unique real number R $(0 \leqslant R \leqslant \infty)$ such that P converges if $|z-a|<R$ and diverges if $R<|z-a|$. We call R the **radius of convergence** and the circle $|z-a|=R$ (sometimes $|z-a|<R$) the **circle of convergence** of P. The value of R is given by $R=1/\limsup \sqrt[n]{|c_n|}$ (**Cauchy-Hadamard formula**) with the conventions $0=1/\infty$, $\infty=1/0$. Also $R=\lim |c_n/c_{n+1}|$, provided that the limit on the right-hand side exists.

A power series †converges absolutely and uniformly on every compact subset inside its circle of convergence and defines there a single-valued complex function. Since the series is †termwise differentiable, the function is actually a holomorphic function of a complex variable. Conversely, any function $f(z)$ holomorphic in a domain can be represented by a power series in a neighborhood of each point a of the domain. Such a representation is called the **Taylor expansion** of $f(z)$ at a (or in the neighborhood of a). A power series that represents a holomorphic function is called a holomorphic **function element**. K. Weierstrass defined an analytic function as the set of all elements that can be obtained by †analytic continuations starting from a given function element (→ 198 Holomorphic Functions).

Besides the series $\sum_{n=0}^{\infty} c_n(z-a)^n$, a series of the form $Q=\sum_{n=0}^{\infty} c_n z^{-n}$ is called a **power series with center at the point at infinity**, and its value at ∞ is defined to be c_0. By putting $z-a=t$ when its center a is a finite point and $z^{-1}=t$ when its center is ∞, every power series can be written in the form $\sum_{n=0}^{\infty} c_n t^n$, and such a t is called a **local canonical parameter**.

When t is a local canonical parameter, a series of the form $\sum_{n=-\infty}^{\infty} c_n t^n$ is called a **Laurent series** and a series $\sum_{n=-\infty}^{\infty} c_n t^{n/k}$ (k a fixed natural number) is called a **Puiseux series**, after the French mathematicians A. Laurent and V. A. Puiseux. Power series are sometimes called **Taylor series**.

If we perform †analytic continuations of a power series from its center along radii of its circle of convergence, we encounter a †singularity on the circumference for at least one radius. For a power series with the circle of convergence $|z|=R$, the argument α of the singularity on $|z|=R$ nearest $z=R$ is given in the following way. Suppose, for simplicity, that the radius of convergence R of $\sum c_n z^n$ equals 1, and put

$$\rho_n(h)=\sum_{\nu=0}^{n}\binom{n}{\nu} c_{n-\nu} h^{\nu},$$

$$P(h)=\limsup_{n\to\infty} \sqrt[n]{|\rho_n(h)|}.$$

Then α is obtained from

$$\cos\alpha = P'_{+}(0)=\lim_{h\to+0}(P(h)-1)/h$$

(S. Mandelbrojt, 1937). In particular, if all the c_n are real and nonnegative, $z=R$ is a singularity (Vivanti's theorem).

B. Abel's Continuity Theorem

As a property of the power series on the circle of convergence, we have **Abel's continuity theorem**: If the radius of convergence of $f(z)=\sum_{n=0}^{\infty} a_n z^n$ is equal to 1 and $\sum_{n=0}^{\infty} a_n$ coverges (or is †(C,k)-summable $(k>-1)$) to A, then $f(z)\to A$ when z approaches 1 in any sector $\{z \mid |z|<1, |\arg(1-z)|<(\pi/2)-\delta\}$, $\delta>0$ (†Stolz's path).

The converse of this theorem is not always true. The existence of $\lim_{z\to1} f(z)$ does not necessarily lead to the convergence of $\sum_{n=0}^{\infty} a_n$. Even Cesàro summability of $\sum_{n=0}^{\infty} a_n$ does not always follow from the existence of $\lim_{z\to1} f(z)$. If $a_n=o(1/n)$ and $f(z)\to A$ when z approaches 1 along a curve ending at $z=1$, then $\sum a_n$ converges to A (**Tauber's theorem**, 1897). The theorems concerning additional sufficient conditions for the validity of the converse of Abel's theorem are called **theorems of Tauberian type** (or **Tauberian theorems**). In Tauber's theorem, the hypothesis on the a_n may be replaced by $a_n=O(1/n)$ or $n\operatorname{Re}a_n$, $n\operatorname{Im}a_n$ may be bounded from above but not necessarily from below (G. H. Hardy and J. E. Littlewood). Here, the condition $a_n=O(1/n)$ cannot be weakened (Littlewood). Sufficient conditions for $\sum a_n$ to be †summable for various summation methods are also known.

C. Lambert Series

A series of the form

$$\sum_{n=1}^{\infty} a_n \frac{z^n}{1-z^n}, \qquad |z| \neq 1, \tag{1}$$

is called a **Lambert series**. If $\sum a_n$ is convergent, (1) converges for any z with $|z| \neq 1$, and moreover, it converges uniformly on any compact set contained in $|z|<1$ or $|z|>1$. If $\sum a_n$ is divergent, (1) and the power series $\sum a_n z^n$ converge or diverge simultaneously for $|z| \neq 1$.

There is a detailed study of Lambert series by K. Knopp (1913). If R is the radius of convergence of $\sum a_n z^n$ and $\sum_{d|n} a_d = A_n$ is the sum extending over all divisors of n, then we have the reciprocity relation

$$\sum_{n=1}^{\infty} a_n \frac{z^n}{1-z^n} = \sum_{n=1}^{\infty} A_n z^n,$$

which holds for $|z| < \min(R, 1)$. As special cases of this relation, we have

$$z = \sum_{n=1}^{\infty} \mu(n) \frac{z^n}{1-z^n},$$

$$\frac{z}{(1-z)^2} = \sum_{n=1}^{\infty} \varphi(n) \frac{z^n}{1-z^n},$$

where μ and φ are the †Möbius function and †Euler function, respectively.

If the na_n are real and bounded from below,

$$\lim_{z \to 1-0} \sum_{n=1}^{\infty} (1-z) n a_n \frac{z^n}{1-z^n} = s$$

implies $\sum a_n = s$. Hardy and Littlewood (1921) showed that this theorem of Tauberian type is equivalent to the †prime number theorem.

D. Singularities of Power Series

Given a power series $P = \sum a_n z^n$, if the †branch in $|z| < R^*$ of the analytic function $f(z)$ determined by P is single-valued meromorphic but the branch in $|z| < R'$ with $R' > R^*$ has singularities other than poles, then R^* is called the **radius of meromorphy** of P and $|z| = R^*$ (sometimes $|z| < R^*$) is called the **circle of meromorphy** of P. R^* can be computed in the following way. Put

$$l_p = \limsup_{n \to \infty} |\sqrt[n]{D_n^{(p)}}|,$$

$$D_n^{(p)} = \begin{vmatrix} a_n & a_{n+1} & \cdots & a_{n+p} \\ a_{n+1} & a_{n+2} & \cdots & a_{n+p+1} \\ \cdots & \cdots & \cdots & \cdots \\ a_{n+p} & a_{n+p+1} & \cdots & a_{n+2p} \end{vmatrix}.$$

Then the sequence of numbers l_{p-1}/l_p is increasing, and $\lim l_{p-1}/l_p = R^*$. If $R_1, R_2, \ldots$ are different valued of l_{p-1}/l_p, then $f(z)$ is holomorphic at points not lying on $|z| = R_n$ (**Hadamard's theorem**, 1892).

When a point a and a set A of points in the complex plane are given, the set of points that can be joined with a by a segment disjoint from A is called the **star region** determined by a and A. Take any half-line starting at a; the point of A lying on it and nearest to a is called a **vertex**. For a power series $\sum c_n z^n$, the set of centers of the function elements obtained by analytic continuations along half-lines starting at the origin is called the star region of $\sum c_n z^n$ with respect to the origin. Let $\{\alpha\}$ and $\{\beta\}$ be the set of vertices of the star regions with respect to the origin of $\sum a_n z^n$ and $\sum b_n z^n$, respectively. Then the star region determined by the origin and the set $\{\alpha\beta\}$ is contained in the star region of $\sum a_n b_n z^n$ (**Hadamard's multiplication theorem**, 1892).

The following are some results concerning conditions for the coincidence of the circle of convergence of a power series and its †natural boundary. Let the a_n be positive numbers and b a natural number greater than 1. If the radius of convergence of $\sum_{n=0}^{\infty} a_n z^{b^n}$ is equal to 1, then $|z|=1$ is its natural boundary (Weierstrass, E. I. Fredholm). If the radius of convergence of $\sum_{n=0}^{\infty} a_n z^{\lambda_n}$ (with $\{\lambda_n\}$ an increasing sequence of natural numbers) is 1 and $\liminf_{n \to \infty} (\lambda_{n+1} - \lambda_n)/\lambda_n > 0$, then $|z|=1$ is the natural boundary (**Hadamard's gap theorem**). The latter condition was weakened to $\liminf_{n \to \infty} (\lambda_{n+1} - \lambda_n)/\sqrt{\lambda_n} > 0$ by E. Borel (1896). E. Fabry (1896) showed that with the radius of convergence of $\sum_{n=0}^{\infty} a_n z^n$ being 1, if there exist a suitable sequence of natural numbers $m_1 < m_2 < \ldots$ and a number θ $(0 < \theta < 1)$ such that $\lim s_i / m_i = 0$, where the s_i are the number of nonzero a_n contained in the interval $(m_i(1-\theta), m_i(1+\theta))$, then $|z|=1$ is the natural boundary of $\sum a_n z^n$. By applying this theorem to $\sum a_n z^{\lambda_n}$ with radius of convergence 1, it can be shown that if $\lim_{n \to \infty} \lambda_n / n = \infty$, then $|z|=1$ is its natural boundary. These theorems are called **gap theorems** because they concern power series with gaps in their exponents. A generalization of Fabry's theorem was obtained by G. Pólya. It is known that Fabry's last condition above is in a sense the best possible (Pólya, 1942).

Regarding the natural boundary of a power series, we also have the following result: When the radius of convergence of $\sum a_n z^n$ is 1, by a suitable choice of the sequence $\{\varepsilon_n\}$ $(\varepsilon_n = \pm 1)$, the series $\sum \varepsilon_n a_n z^n$ has $|z|=1$ as its natural boundary (A. Hurwitz, P. Fatou, Pólya).

E. Overconvergence

If the radius of convergence of $f(z)=\sum a_n z^n$ is 1, the sequence of partial sums $S(1,z)$, $S(2,z)$, $\ldots$, where $S(n,z)=\sum_{\nu=0}^{n} a_\nu z^\nu$, is naturally divergent for $|z|>1$, but a suitable subsequence $S(n_k,z)$ $(k=1,2,\ldots)$ may still be convergent. A. Ostrowski [7, 8] called this phenomenon **overconvergence** and proved the following result: By definition, $f(z)=\sum_{n=0}^{\infty} a_n z^{\lambda_n}$ has a **lacunary structure** if the sequence $\{\lambda_n\}$ has a subsequence $\{\lambda_{n_k}\}$ such that $\lambda_{n_k+1}>\lambda_{n_k}(1+\theta)$ $(\theta>0)$. If this situation occurs, then in a sufficiently small neighborhood of a point on $|z|=1$ where $f(z)$ is holomorphic, $S(\lambda_{n_k},z)$ $(k=1,2,\ldots)$ converges uniformly. This result includes Hadamard's gap theorem as a special case. Conversely, any power series for which overconvergence takes place can be represented as the sum of a power series having a lacunary structure and a power series whose radius of convergence is greater than 1. G. Bourion [9] gave a unified theory of these results using †superharmonic functions.

R. Jentsch [10] showed that all singularities of a power series on its circle of convergence are accumulation points of the zeros of the partial sums. On the other hand, if the zeros of a subsequence $S(n_k,z)$ $(k=1,2,\ldots)$ has no accumulation point on $|z|=1$, then the power series has a lacunary structure and overconvergence takes place for $S(n_k,z)$ $(k=1,2,\ldots)$. If $\log n_{k+1}=O(n_k)$ and $S(n_k,z)$ $(k=1,2,\ldots)$ is overconvergent, then all boundary points of the domain of overconvergence are accumulation points of the zeros of $S(n_k,z)$ (Ostrowski [8]).

A power series is completely determined by its coefficients, but little is known about the relations between the arithmetical properties of its coefficients and the function-theoretic properties of the function represented by the series. A known result is that if the power series $\sum c_n z^n$ with rational coefficients represents a branch of an †algebraic function, then we can find an integer γ such that the $c_n\gamma^n$ $(n\geqslant 1)$ are all integers (Eisenstein's theorem, 1852).

For power series of several variables → 21 Analytic Functions of Several Complex Variables; for formal power series → 370 Rings of Power Series. For power series expansions → Appendix A, Table 10.IV.

References

For the general theory of power series,
[1] E. G. H. Landau, Darstellung und Begründung einiger neuerer Ergebnisse der Funktionentheorie, Springer, second edition, 1929.
[2] P. Dienes, The Taylor series, Clarendon Press, 1931.
[3] L. Bieberbach, Analytische Fortsetzung, Springer, 1955.
Also → references to 198 Holomorphic Functions.
For singularities,
[4] S. Mandelbrojt, Les singularités des fonctions analytiques représentées par une série de Taylor, Gauthier-Villars, 1932.
For theorems of Tauberian type,
[5] N. Wiener, Tauberian theorems, Ann. Math., (2) 33 (1932), 1–100.
[6] H. R. Pitt, Tauberian theorems, Oxford Univ. Press, 1958.
For overconvergence,
[7] A. Ostrowski, Über eine Eigenschaft gewisser Potenzreihen mit unendlichvielen verschwinden der Koeffizienten, S.-B. Preuss. Akad. Wiss., (1921), 557–565.
[8] A. Ostrowski, Über vollständige Gebiete gleichmässiger Konvergenz von Folgen analytischer Funktionen, Abh. Math. Sem. Univ. Hamburg, 1 (1922), 327–350.
[9] G. Bourion, L'ultraconvergence dans les séries de Taylor, Actualités Sci. Ind., Hermann, 1937.
[10] R. Jentsch, Fortgesetzte Untersuchungen über die Abschnitte von Potenzreihen, Acta Math., 41 (1918), 253–270.

340 (XVII.17) Probabilistic Methods in Statistical Mechanics

A. Introduction

Probabilistic methods are often very useful in the rigorous treatment of the mathematical foundations of statistical mechanics and also in some other problems related to statistical mechanics. As examples of such methods, we explain here (1) Ising models, (2) Markov statistical mechanics, (3) percolation processes, (4) random Schrödinger equations, and (5) the †Boltzmann equation.

B. Ising Models

The **Ising model** was proposed by E. Ising [1] to explain the **phenomena of phase transitions** of a ferromagnet, in which either a + or − spin is put on each site of a crystal lattice, and interaction between nearest neighboring sites is taken into consideration.

Let V be a cube in the d-dimensional integer lattice space $\mathbf{Z}^d$ and $X_V=\{+1,-1\}^V$ be the totality of spin configurations in V. Each element of X_V is denoted by $\sigma=\{\sigma_i\}_{i\in V}$ ($\sigma_i=+1$ or -1). We suppose that a spin configuration σ has a potential of the type

$$U_V(\sigma)=-\sum_{\langle i,j\rangle\subset V}\sigma_i\sigma_j+h\sum_{i\in V}\sigma_i,$$

where $\langle i,j\rangle$ means that (i,j) is a nearest neighboring pair of sites and h stands for the parameter of an external field. A †probability measure on X_V is called a **state** on V. For each state μ, the **free energy** is defined by

$$F_V(\mu)=\sum_{\sigma\in X_V}\mu(\sigma)U_V(\sigma)+\frac{1}{\beta}\sum_{\sigma\in X_V}\mu(\sigma)\log\mu(\sigma),$$

where $\beta=1/kT$ (k is the †Boltzmann constant, T is the †absolute temperature). Then there exists a unique state

$$g_V^{\beta,h}(\sigma)=\frac{\exp(-\beta U_V(\sigma))}{\sum_{\gamma\in X_V}\exp(-\beta U_V(\gamma))},\qquad \sigma\in X_V,$$

on V which minimizes the free energy F_V (**variational principle**). $g_V^{\beta,h}$ is called a **Gibbs state** on V of the Ising model with parameter (β,h). Physically, a Gibbs state is an **equilibrium state**, in which various physical quantities are calculated.

Since for each $\mathbf{Z}^d$-homogeneous (i.e., translationally invariant with respect to $\mathbf{Z}^d$) state μ the **mean free energy**

$$f(\mu)=\lim_{V\to\mathbf{Z}^d}\frac{1}{|V|}F_V(\mu)$$

is well defined, a (limiting) Gibbs state can also be defined for the infinite domain $V=\mathbf{Z}^d$ by the above mentioned variational principle. However, at present the probabilistic definition of Gibbs states given by Dobrushin [2] and Lanford and Ruelle [3] is prevalent.

It is known that if an external field is present (i.e., $h\neq 0$) there is only one Gibbs state for any β. On the other hand when an external field is absent ($h=0$) and $d\geqslant 2$, there are at least two Gibbs states, i.e., a phase transition occurs, for a sufficiently low temperature.

Finally, we mention some known facts in this field. In the following we assume $h=0$. (1) In the 1-dimensional case, the phase transition never occurs. (2) For $d\geqslant 2$, there exists a **critical value** $\beta_c(d)$ such that the phase transition does not occur for any $\beta<\beta_c(d)$ but it occurs for every $\beta>\beta_c(d)$. The calculation of $\beta_c(2)$ has been carried out by Onsager [4]. (3) In the 2-dimensional case, every Gibbs state is $\mathbf{Z}^2$-homogeneous [5, 6]. (4) For $d\geqslant 3$, there is a $\mathbf{Z}^d$-inhomogeneous Gibbs state for sufficiently large β [7].

C. Markov Statistical Mechanics

Stochastic Ising models, infinite interacting particle systems, and many models occurring in physics, biology, and sociology are formulated as a class of infinite-dimensional †Markov processes. The field of investigation of stationary states and statistical or ergodic properties of these processes is called **Markov statistical mechanics**, which has made rapid progress during the last decade. We explain this field by looking at a typical class of processes.

Let $\mathbf{Z}^d$ be the d-dimensional integer lattice space. Putting $+$ or $-$ spin on each site on $\mathbf{Z}^d$, let us consider a random motion of spins which evolves while interacting with neighboring spins. Let $X=\{+1,-1\}^{\mathbf{Z}^d}$ be the totality of spin configurations and an element of X be denoted by $\eta=\{\eta(i)\}_{i\in\mathbf{Z}^d}$ ($\eta(i)=+1$ or -1). The process is described in terms of a collection of nonnegative functions $c_i(\eta)$ defined for $i\in\mathbf{Z}^d$ and $\eta\in X$. For the configuration η_t at time t, $\eta_t(i)$ changes to $-\eta_t(i)$ in the time interval $[t,t+\Delta t]$ with probability $c_i(\eta_t)\Delta t+o(\Delta t)$. This process on X is called a spin-flip model. For an initial distribution μ we denote by μ_t the distribution at time t. If $\mu_t=\mu$ for all $t\geqslant 0$, μ is called a **stationary state**.

Example 1. Stochastic Ising models. A **stochastic Ising model** was proposed by Glauber [8] to describe the random motion in a ferromagnet upon contact with a heat bath. Then the flip rate $\{c_i(\eta)\}$ is defined by the potential of the Ising model. It is known that any Gibbs state of the Ising model is a reversible stationary state of the stochastic Ising model, and the converse is also valid. Free energy plays an important role in the study of the ergodic properties of these models. In particular, the mean free energy is a nondecreasing functional along the distributions μ_t, $0\leqslant t<\infty$.

Example 2. Contact processes. A **contact process** was introduced by Harris [11] to investigate the spread of infection. The flip rate of the contact process is given by

$$c_i(\eta)=\begin{cases}1 & \text{if}\quad \eta(i)=+1,\\ k\lambda & \text{if}\quad \eta(i)=-1\quad\text{and}\end{cases}$$
$$\#\{j\,|\,|j-i|=1,\quad \eta(j)=+1\}=k.$$

Here $+1$ denotes an infected individual and -1 denotes a healthy one. Denote by $-\mathbf{1}$ the configuration at which all sites are healthy and by $\delta_{-\mathbf{1}}$ the unit point mass at $-\mathbf{1}$; then $\delta_{-\mathbf{1}}$ is a stationary state. The most important result is the following: There exists a **critical value** λ_d ($0<\lambda_d<\infty$) such that if $\lambda<\lambda_d$ $\delta_{-\mathbf{1}}$ is a unique stationary state, and if $\lambda>\lambda_d$ there is another stationary state μ satisfying

$$\mu[\eta\in X;\ \eta(i)=+1 \text{ for infinitely many } i]=1.$$

D. Percolation Processes

A **percolation process** is a mathematical model which describes the random spread of a fluid through a medium. It can be used to describe phenomena such as the penetration through a porous solid by a liquid or the spread of an infectious disease [14]. Usually, the process is identified as a **site percolation process** or a **bond percolation process**. Here we describe the latter only.

Let $L=\{S,B\}$ be a countable connected †graph with a set of sites (†vertices) S and a set of bonds (†edges) B. Each bond b is open with probability p and closed with probability $1-p$ independently of all other bonds. Suppose that a fixed point o is the source of a fluid which flows from o along open bonds only.

Let $\theta(p)$ be the probability that the fluid spreads infinitely far, and define the **critical percolation probability** $p_H=\inf\{p\,|\,\theta(p)>0\}$. Then it is known that (1) $p_H=1/2$ for the square lattice [16], (2) $p_H=2\sin(\pi/18)$ for the triangular lattice, and (3) $p_H=1-2\sin(\pi/18)$ for the honeycomb lattice.

E. Random Schrödinger Equations

Random Schrödinger equations are †Schrödinger equations in $\mathbf{R}^n$ with random potentials $U(x,\omega)$; therefore the corresponding operators are of the form

$$A(\omega)=-\Delta+U(\cdot,\omega),$$

where ω denotes a random parameter in a probability space $(\Omega,\mathfrak{F},P)$ and Δ denotes the †Laplacian in $\mathbf{R}^n$. It is assumed that this system of potentials forms a spatially homogeneous random field with the †ergodic property. This system of equations is considered to be a model describing the motion of quantum-mechanical particles in a random medium. Mathematically, the problem is to investigate various spectral properties of the self-adjoint operators $A(\omega)$. Since $A(\omega)$ and its shifted operator $A(\omega(\cdot+x))$ are †unitarily equivalent, the above assumption on the potentials $U(\cdot,\omega)$ implies that the spectral structures are independent of each sample ω a.s. if their structures of $A(\omega)$ are measurable with respect to $(\Omega,\mathfrak{F})$.

Several rigorous results have been obtained. In the 1-dimensional case, if the potentials $U(\cdot,\omega)$ are functionals of a strongly ergodic †Markov process, then it is proved that $A(\omega)$ has only a pure point spectrum [17], and each eigenfunction decays exponentially fast [18]. In multidimensional cases, asymptotic behavior at the left edge of the mean of the resolutions of the identity for a certain $A(\omega)$ have been investigated [19]. It is assumed that the random potential for this $A(\omega)$ takes the form

$$U(x,\omega)=\int_{\mathbf{R}^n}\varphi(x-y)\pi(dy,\omega),$$

where $\{\pi(dy,\omega)\}$ is a †Poisson random measure with mean measure dy and $\varphi(x)$ is a nonnegative measurable function satisfying $\varphi(x)=o(|x|^{-n-2})$ as $|x|\to\infty$. Let $E_\lambda(x,y,\omega)$ be the continuous kernel for the resolution of the identity for $A(\omega)$, and denote by $N(\lambda)$ the mean of $E_\lambda(0,0,\omega)$. Then $N(\lambda)$ is a nondecreasing function vanishing on $(-\infty,0)$ and has the following asymptotic form at $\lambda=0$:

$$\lambda^{n/2}\log N(\lambda)\to-\gamma_1^{n/2}\quad\text{as}\quad\lambda\to0,$$

where γ_1 is the first eigenvalue of $-\Delta$ with a Dirichlet boundary condition on the ball in $\mathbf{R}^n$ with unit volume. The quantity $N(\lambda)$ can be identified with a limiting state density of $A(\omega)$, namely, the limit function of

$$N_V(\lambda,\omega)=\#\{j\,|\,\lambda_j^V(\omega)\leqslant\lambda\}/|V|$$

as V tends to $\mathbf{R}^n$ regularly, where $\{\lambda_j^V(\omega)\}$ is the set of eigenvalues of $A(\omega)$ in a smooth bounded domain V in $\mathbf{R}^n$ with a Dirichlet boundary condition and $|V|$ is the volume of V. To obtain the above asymptotic behavior of $N(\lambda)$, the theory of †large deviation for Markov processes plays a crucial role [20].

F. Boltzmann Equation

In the kinetic theory of gases the †Boltzmann equation is derived from the †Liouville equation by considering the BBGKY hierarchy of particle distribution functions for N particles and then by taking the limit $N\to\infty$ under certain conditions (→ 402 Statistical Mechanics). Mathematically rigorous discussions were given by O. E. Lanford [21] for a gas of hard spheres; he showed that solutions of the BBGKY hierarchy converge to those of the Boltzmann hierarchy for small time under the Boltzmann-Grad limit ($N\to\infty$, $Nd^2\to1$, $d=$ the diameter of the hard spheres).

An approach to the Boltzmann equation in the spatially homogeneous case can also be based on a †master equation. M. Kac [22] considered a Poisson-like process describing the random time evolution of the n-tuple of the velocities of n particles. For a gas of hard spheres this is determined by the master equation

$$\frac{\partial}{\partial t}u(t,x_1,\dots,x_n)$$
$$=\frac{1}{n}\sum_{1\leqslant i<j\leqslant n}\int_{S^2}\{u(t,x_1,\dots,x_i',\dots,x_j',\dots,x_n)-u(t,x_1,\dots,x_n)\}\,|(x_i-x_j,l)|\,dl,$$
$$t>0,\quad x_1,\dots,x_n\in\mathbf{R}^3,\qquad(1)$$

where S^2 is the 2-dimensional unit sphere, dl is

the uniform distribution on S^2 and

$$x_1' = x_i + (x_j - x_i, l)l, \qquad x_j' = x_j - (x_j - x_i, l)l.$$

Let σ be a positive constant and $S(\sqrt{n\sigma})$ denote the $(3n-1)$-dimensional sphere with center 0 and radius $\sqrt{n\sigma}$. Given a symmetric probability density u_n on $S(\sqrt{n\sigma})$ for each $n \geqslant 1$, a sequence $\{u_n\}$ is said to have **Boltzmann's property** or to be **chaotic** (or u-chaotic to stress u), if there exists a probability density u on $\mathbf{R}^3$ such that

$$\lim_{n\to\infty} \int_{S(\sqrt{n\sigma})} \varphi_1(x_1) \ldots \varphi_m(x_m) u_n(x_1, \ldots, x_n) dx_1 \ldots dx_n = \prod_{k=1}^{m} \int_{\mathbf{R}^3} \varphi_k(x) u(x) dx$$

for each $m \geqslant 1$ and $\varphi_k \in C_0(\mathbf{R}^3)$, $1 \leqslant k \leqslant m$. Kac's assertion is that the Boltzmann equation is to be derived from the master equation via the **propagation of chaos**; more precisely, if $\{u_n\}$ is a u-chaotic sequence, then $\{u_n(t)\}$ is also $u(t)$-chaotic, where $u_n(t)$ is the solution of the master equation (1) with $u_n(0) = u_n$ and $u(t)$ is the solution of the following Boltzmann equation with $u(0) = u$:

$$\frac{\partial}{\partial t} u(t,x) = \int_{S^2 \times \mathbf{R}^3} \{u(t,x')u(t,y') - u(t,x)u(t,y)\} \times |(x-y,l)| \, dl \, dy,$$

$$x' = x + (y-x,l)l, \quad y' = y - (y-x,l)l.$$

The propagation of chaos was verified by Kac [22], H. P. McKean [24], and others for a considerably wide class of nonlinear equations of Boltzmann type (with cutoff). The propagation of chaos is the stage corresponding to the †law of large numbers. The next stage is the †central limit theorem or fluctuation theory, which was also discussed by Kac [23] and McKean [25]. Moreover, based on Kac's work [22], McKean [26] introduced a class of †Markov processes associated with certain nonlinear evolution equations including the Boltzmann equation; a process of this type describes the time evolution of the velocity of a particle interacting with other similar particles. In the case of the spatially homogeneous Boltzmann equation of Maxwellian molecules without cutoff, such a Markov process was constructed by solving a certain †stochastic differential equation [27]. This implies the existence of probability measure-valued solutions of the equation.

References

[1] E. Ising, Beitrag zur Theorie des Ferromagnetismus, Z. Phys., 31 (1925), 253–264.

[2] R. L. Dobrushin, The description of a random field by means of conditional probabilities and conditions of its regularity, Theory Prob. Appl., 13 (1968), 197–224. (Original in Russian, 1968.)

[3] O. E. Lanford III and D. Ruelle, Observables at infinity and states with short range correlations in statistical mechanics, Comm. Math. Phys., 13 (1969), 194–215.

[4] L. Onsager, Crystal statistics I: A two-dimensional model with an order-disorder transition, Phys. Rev., 65 (1944), 117–149.

[5] M. Aizenman, Translation invariance and instability of phase coexistence in the two dimensional Ising system, Comm. Math. Phys., 73 (1980), 83–94.

[6] Y. Higuchi, On the absence of non-translationally invariant Gibbs states for the two-dimensional Ising model, Proc. Conf. on Random Fields, Esztergom, North-Holland, 1981.

[7] R. L. Dobrushin, Gibbs state describing coexistence of phases for a three-dimensional Ising model, Theory Prob. Appl., 17 (1972), 582–600. (Original in Russian, 1972.)

[8] R. J. Glauber, Time-dependent statistics of the Ising model, J. Math. Phys., 4 (1963), 294–307.

[9] R. L. Dobrushin, Markov processes with a large number of locally interacting components: Existence of a limit process and its ergodicity, Problems of Information Transmission, 7 (1971), 149–164; Markov processes with many locally interacting components: The reversible case and some generalizations, ibid., 235–241. (Originals in Russian, 1971.)

[10] F. Spitzer, Interaction of Markov processes, Advances in Math., 5 (1970), 246–290.

[11] T. E. Harris, Contact interactions on a lattice, Ann. Probability, 2 (1974), 969–988.

[12] R. Holley and T. M. Liggett, The survival of contact processes, Ann. Probability, 6 (1978), 198–206.

[13] T. M. Liggett, The stochastic evolution of infinite systems of interacting particles, Lecture notes in math. 598, Springer, 1977, 187–248.

[14] S. R. Broadbent and J. M. Hammersley, Percolation processes I: Crystals and mazes, Proc. Cambridge Philos. Soc., 53 (1957), 629–641.

[15] R. T. Smythe and J. C. Wierman, First-passage percolation on the square lattice, Lecture notes in math. 671, Springer, 1978.

[16] H. Kesten, The critical probability of bond percolation on the square lattice equals 1/2, Comm. Math. Phys., 74 (1980), 41–59.

[17] I. J. Goldseid, S. A. Molchanov, and L. A. Pastur, One-dimensional random Schrödinger operator with purely point spectrum, Functional Anal. Appl., 11 (1977), 1–10. (Original in Russian, 1977.)

[18] S. A. Molchanov, The structure of eigenfunctions of one-dimensional unordered structures, Math. USSR-Izv., 12 (1978), 69–101. (Original in Russian, 1978.)
[19] S. Nakao, On the spectral distribution of the Schrödinger operator with random potential, Japan. J. Math., 3 (1977), 111–139.
[20] M. D. Donsker and S. R. S. Varadhan, Asymptotics for the Wiener sausage, Comm. Pure Appl. Math., 28 (1975), 525–565.
[21] O. E. Lanford III, Time evolution of large classical systems, Lecture notes in physics 38, Springer, 1975.
[22] M. Kac, Probability and related topics in physical sciences, Interscience, 1959.
[23] M. Kac, Some probabilistic aspects of the Boltzmann-equation, Acta Phys. Austrica Supp. 10, Springer, 1973, 379–400.
[24] H. P. McKean, Speed of approach to equilibrium for Kac's caricature of a Maxwellian gas, Arch. Rational Mech. Anal., 21 (1966), 347–367.
[25] H. P. McKean, Fluctuations in the kinetic theory of gases, Comm. Pure Appl. Math., 28 (1975), 435–455.
[26] H. P. McKean, A class of Markov processes associated with nonlinear parabolic equations, Proc. Nat. Acad. Sci. US, 56 (1966), 1907–1911.
[27] H. Tanaka, Probabilistic treatment of the Boltzmann equation of Maxwellian molecules, Z. Wahrscheinlichkeitstheorie und Verw. Gebiete, 46 (1978), 67–105.

341 (XVII.2) Probability Measures

A. General Remarks

A **probability measure** Φ on a †measurable space $(S, \mathfrak{S})$ is defined to be a †measure on $(S, \mathfrak{S})$ with $\Phi(S)=1$ (→ 270 Measure Theory). In probability theory, probability measure appears usually as the †probability distribution of a †random variable (→ 342 Probability Theory). Unless stated otherwise, we regard a †topological space T as a measurable space endowed with the topological σ-algebra $\mathfrak{B}(T)$ on T, i.e., the †σ-algebra generated by the †open subsets of T. Hence the distribution of an $\mathbf{R}^n$-valued random variable is a probability measure on $(\mathbf{R}^n, \mathfrak{B}^n=\mathfrak{B}(\mathbf{R}^n))$. From this probabilistic background we often call a probability measure on $(\mathbf{R}^n, \mathfrak{B}^n)$ an n-dimensional (probability) distribution. For probability measures on topological spaces → 270 Measure Theory.

B. Quantities Characterizing Probability Distributions

Several different quantities characterize the properties of probability distributions in one dimension: the **mean** (or **mathematical expectation**) $m=\int_{-\infty}^{\infty} x d\Phi(x)$, the **variance** $\sigma^2=\int_{-\infty}^{\infty}|x-m|^2 d\Phi(x)$, the **standard deviation** σ, the kth **moment** $\alpha_k=\int_{-\infty}^{\infty} x^k d\Phi(x)$, the kth **absolute moment** $\beta_k=\int_{-\infty}^{\infty}|x|^k d\Phi(x)$, the kth **moment about the mean** $\mu_k=\int_{-\infty}^{\infty}(x-m)^k d\Phi(x)$, etc.

A one-to-one correspondence exists between a 1-dimensional distribution Φ and its **(cumulative) distribution function** F defined by $F(x)=\Phi((-\infty, x])$. A distribution function is characterized by the following properties: (1) It is monotone nondecreasing; (2) it is right continuous; (3) $\lim_{x\to-\infty} F(x)=0$ and $\lim_{x\to+\infty} F(x)=1$. Similar statements hold for the multidimensional case.

Let $X(\omega)$ be a real random variable on a †probability space $(\Omega, \mathfrak{F}, P)$. Then the distribution of X is given by $\Phi(E)=P(\{\omega \mid X(\omega)\in E\})$, $E\in\mathfrak{B}^1$, and the characteristic quantities of Φ defined above are given in terms of $X(\omega)$ as follows: $m=E(X)$, $\sigma^2=E(X-m)^2$, $F(x)=P(\{\omega \mid X(\omega)\leqslant x\})$, etc. The moments and the moments about the mean are connected by the relation $\mu_r=\sum_{k=0}^{r}\binom{r}{k}\alpha_{r-k}(-m)^k$ $(r=1, 2, \ldots)$. When Φ is an n-dimensional distribution, the following quantities are frequently used: the **mean vector**, which is an n-dimensional vector whose ith component is given by $m_i=\int x_i d\Phi(x)$; the **covariance matrix**, which is an $n\times n$ matrix whose (i,j)-element is $\sigma_{ij}=\int(x_i-m_i)(x_j-m_j)d\Phi(x)$; the **moment matrix**, which is an $n\times n$ matrix whose (i,j)-element is $m_{ij}=\int x_i x_j d\Phi(x)$. (The covariance matrix is also called the **variance matrix** or the **variance-covariance matrix**.) The covariance matrix and the moment matrix are †positive definite and symmetric. The quantities listed above are defined only under some integrability conditions.

C. Characteristic Functions

Consider a probability measure Φ defined on a measurable space $(\mathbf{R}^n, \mathfrak{B}^n)$, where $\mathfrak{B}^n$ is the σ-algebra of all †Borel sets in $\mathbf{R}^n$. The **characteristic function** of Φ is the †Fourier transform φ defined by

$$\varphi(z)=\int_{\mathbf{R}^n} e^{i(z,x)} d\Phi(x), \qquad z\in\mathbf{R}^n, \tag{1}$$

where (z, x) denotes the †scalar product of z and x $(z, x\in\mathbf{R}^n)$. Let X be an n-dimensional random variable with probability distribution Φ defined on a †probability space $(\Omega, \mathfrak{B}, P)$.

Then the Fourier transform of Φ is also called the **characteristic function** of X, which can also be written as $E(e^{i(z,X)})$ (→ 342 Probability Theory).

The following properties play a fundamental role in the study of the relationship between probability distributions and characteristic functions: (i) the correspondence defined by (1) between the n-dimensional probability distribution Φ and its characteristic function φ is one-to-one. (ii) For any $a_p, b_p \in \mathbf{R}$, $a_p < b_p$ $(p = 1, 2, \ldots, n)$, we have

$$\int_{\mathbf{R}^n} \prod_{p=1}^{n} f(x_p; a_p, b_p)\, d\Phi(x)$$
$$= \lim_{c \to \infty} \left(\frac{1}{2\pi}\right)^n \int_{-c}^{c} \cdots \int_{-c}^{c} \prod_{p=1}^{n} \frac{e^{-ib_p z_p} - e^{-ia_p z_p}}{-iz_p} \times \varphi(z_1, \ldots, z_n)\, dz_1, \ldots, dz_n, \quad (2)$$

where $f(t; a, b)$ denotes the **modified indicator function** of $[a, b]$ defined by

$$f(t; a, b) = \begin{cases} 1, & t \in (a, b), \\ 1/2, & t = a \text{ or } b, \\ 0, & t \notin [a, b], \end{cases}$$

and $x = (x_1, \ldots, x_n) \in \mathbf{R}^n$. If an n-dimensional interval $I = [a_1, \ldots, a_n; b_1, \ldots, b_n]$ defined by $a_i \leqslant x_i \leqslant b_i$ $(i = 1, 2, \ldots, n)$ is an **interval of continuity** for the probability distribution Φ, i.e., $\Phi(\partial I) = 0$, where ∂I denotes the boundary of I, then the left-hand side of (2) is equal to $\Phi(I)$. Equation (2) is called the **inversion formula** for the characteristic function φ.

The characteristic function φ of an n-dimensional probability distribution has the following properties: (i) For any points $z^{(1)}, \ldots, z^{(p)}$ of the n-dimensional space $\mathbf{R}^n$ and any complex numbers $a_1, \ldots, a_p$, we have

$$\sum_{j,k=1}^{p} \varphi(z^{(j)} - z^{(k)}) a_j \bar{a}_k \geqslant 0.$$

(ii) $\varphi(z^{(k)})$ converges to $\varphi(0)$ as $z^{(k)} \to 0$. (iii) $\varphi(0) = 1$. A complex-valued function φ of $z \in \mathbf{R}^n$ is called †positive definite if it satisfies the inequality in (i). Any continuous positive definite function φ on $\mathbf{R}^n$ such that $\varphi(0) = 1$ is the characteristic function of an n-dimensional probability distribution (†Bochner's theorem) (→ 192 Harmonic Analysis). A counterpart to Bochner's theorem holds for any positive definite sequence as well (†Herglotz's theorem).

The characteristic function is often useful for giving probability distributions explicitly. (For characteristic functions of typical probability distributions → Appendix A, Table 22. For general information about criteria that can be used to decide whether a given function is a characteristic function → [8].)

The **moment generating function** defined by $f(z) = \int \exp(-(z, x))\, d\Phi(x)$ $(z \in \mathbf{R}^n)$ does not necessarily exist for all n-dimensional distributions but does exist for a number of useful probability distributions Φ, and then $f(z)$ uniquely determines Φ.

Given a 1-dimensional distribution Φ with $\beta_k < +\infty$, we denote by γ_k the coefficient of $(iz)^k/k!$ in the †Maclaurin expansion of $\log \varphi(z)$. We call γ_k the (kth order) **semi-invariant** of Φ. The moments and semi-invariants are connected by the relations $\gamma_1 = \alpha_1$, $\gamma_2 = \alpha_2 - \alpha_1^2 = \sigma^2$, $\gamma_3 = \alpha_3 - 3\alpha_1\alpha_2 + 2\alpha_1^3$, $\gamma_4 = \alpha_4 - 3\alpha_2^2 - 4\alpha_1\alpha_3 + 12\alpha_1^2\alpha_2 - 6\alpha_1^4, \ldots$.

D. Specific Distributions

Given an n-dimensional distribution Φ, a point a with $\Phi(\{a\}) > 0$ is called a discontinuity point of Φ. The set D of all discontinuity points of Φ is at most countable. When $\Phi(D) = 1$, Φ is called a **purely discontinuous distribution**. In particular, if D is a lattice, Φ is called a **lattice distribution**. If the distribution function of Φ is a continuous function, Φ is called a **continuous distribution**. By virtue of the †Lebesgue decomposition theorem, every probability distribution can be expressed in the form

$$\Phi = a_1\Phi_1 + a_2\Phi_2 + a_3\Phi_3,$$

$$a_1, a_2, a_3 \geqslant 0, \qquad a_1 + a_2 + a_3 = 1,$$

where Φ_1 is purely discontinuous, Φ_2 is †absolutely continuous with respect to †Lebesgue measure, and Φ_3 is continuous and †singular. Let Φ be an absolutely continuous distribution. Then there exists a unique (up to Lebesgue measure zero) measurable nonnegative function $f(x)$ $(x \in \mathbf{R}^n)$ such that $\Phi(E) = \int_E f(x)\, dx$. This function $f(x)$ is called the **probability density** of Φ.

We now list some frequently used 1-dimensional lattice distributions (for explicit data → Appendix A, Table 22): the **unit distribution** with $\Phi(\{0\}) = 1$; the **binomial distribution** $Bin(n, p)$ with parameters n and p; the **Poisson distribution** $P(\lambda)$ with parameter λ; the **geometric distribution** $G(p)$ with parameter p; the **hypergeometric distribution** $H(N, n, p)$ with parameters N, n, and p; and the **negative binomial distribution** $NB(m, q)$ with parameters m and q. The following k-dimensional lattice distributions are used frequently: the **multinomial distribution** $M(n, p)$ with parameters n and p; the **multiple hypergeometric distribution**; the **negative multinomial distribution**; etc.

The following 1-dimensional distributions are absolutely continuous: the **normal distribution** (or **Gaussian distribution**) $N(\mu, \sigma^2)$ with mean μ and variance σ^2 (sometimes $N(0, 1)$ is called the **standard normal distribution**); the

Cauchy distribution $C(\mu, \sigma)$ with parameters μ (†median) and σ; the **uniform distribution** $U(\alpha, \beta)$ on an interval $[\alpha, \beta]$; the **exponential distribution** $e(\sigma)$ with parameter σ; the **gamma distribution** $\Gamma(p, \sigma)$; the †χ^2 distribution $\chi^2(n)$; the **beta distribution** $B(p, q)$; the **F-distribution** $F(m, n)$; the **Z-distribution** $Z(m, n)$; the **t-distribution** $t(n)$; etc. Furthermore, there are several k-dimensional absolutely continuous distributions, such as the **k-dimensional normal distribution** $N(\mu, \Sigma)$ with mean vector $\mu = (\mu_1, \mu_2, \ldots, \mu_k)$ and covariance matrix $\Sigma = (\sigma_{ij})$, the **Dirichlet distribution**, etc.

E. Convolution

Given any two n-dimensional distributions Φ_1, Φ_2, the n-dimensional distribution $\Phi(E) = \int_{\mathbf{R}^{2n}} \chi_E(x+y)\, d\Phi_1(x)\, d\Phi_2(y)$ is called the **composition** (or **convolution**) of Φ_1 and Φ_2 and is denoted by $\Phi_1 * \Phi_2$, where χ_E is the indicator function of the set E. Let X_1 and X_2 be †independent random variables with distributions Φ_1 and Φ_2. Then the distribution of $X_1 + X_2$ is $\Phi_1 * \Phi_2$. When F_i is the distribution function of Φ_i $(i = 1, 2)$, the distribution function $F_1 * F_2$ of $\Phi_1 * \Phi_2$ is expressed in the form $F_1 * F_2(x) = \int_{\mathbf{R}^n} F_1(x-y)\, dF_2(y)$. If Φ_1 has a density $f_1(x)$, then $\Phi_1 * \Phi_2$ has a density $f(x) = \int_{\mathbf{R}^n} f_1(x-y)\, dF_2(y)$. If $\varphi(z)$ is the characteristic function of the convolution of two probability distributions Φ_1 and Φ_2 with characteristic functions φ_1 and φ_2, then φ is the product of φ_1 and φ_2: $\varphi(z) = \varphi_1(z) \cdot \varphi_2(z)$. Therefore, for every k, the kth order semi-invariant of the convolution of two distributions is equal to the sum of their kth order semi-invariants. Suppose that we are given a family of distributions $\Phi = \{\Phi(\alpha, \beta, \ldots)\}$ indexed with parameters $\alpha, \beta, \ldots$. If for $(\alpha_1, \beta_1, \ldots)$ and $(\alpha_2, \beta_2, \ldots)$ there exists $(\alpha_3, \beta_3, \ldots)$ such that $\Phi(\alpha_1, \beta_1, \ldots) * \Phi(\alpha_2, \beta_2, \ldots) = \Phi(\alpha_3, \beta_3, \ldots)$, then we say that Φ has a **reproducing property**. Some of the distributions listed above have the reproducing property: $P(\lambda_1) * P(\lambda_2) = P(\lambda_1 + \lambda_2)$, $Bin(n_1, p) * Bin(n_2, p) = Bin(n_1 + n_2, p)$, $NB(m_1, q) * NB(m_2, q) = NB(m_1 + m_2, q)$, $N(\mu_1, \sigma_1^2) * N(\mu_2, \sigma_2^2) = N(\mu_1 + \mu_2, \sigma_1^2 + \sigma_2^2)$, $\Gamma(p_1, \sigma) * \Gamma(p_2, \sigma) = \Gamma(p_1 + p_2, \sigma)$, $C(\mu_1, \sigma_1) * C(\mu_2, \sigma_2) = C(\mu_1 + \mu_2, \sigma_1 + \sigma_2)$, etc.

Given a 1-dimensional distribution function $F(x)$,

$$Q_F(l) = \max_{-\infty < x < \infty} (F(x+l) - F(x-l)), \quad l > 0,$$

is called the **maximal concentration function** of F (P. Lévy [6]). Since it satisfies the relation $Q_{F_1 * F_2}(l) \leqslant Q_{F_i}(l)$ $(i = 1, 2)$, we can use it to study the properties of sums of independent random variables. The **mean concentration function** defined by

$$C_F(l) = \frac{1}{2l} \int_{-\infty}^{\infty} (F(x+l) - F(x-l))^2\, dx$$

is also useful for similar purposes.

Let $N(m, v)$ be the 1-dimensional normal distribution with mean m and variance v, and let $P(\lambda, a)$ be the distribution obtained through translation by a of the Poisson distribution with parameter λ. If 1-dimensional distributions Φ_k, Ψ_k $(k = 1, 2)$ exist such that $N(m, v) = \Phi_1 * \Phi_2$, $P(\lambda, a) = \Psi_1 * \Psi_2$, we have $\Phi_k = N(m_k, v_k)$, $\Psi_k = P(\lambda_k, a_k)$ $(k = 1, 2)$ for some m_k, v_k, λ_k, a_k $(k = 1, 2)$. These are known, respectively, as Cramér's theorem and Raikov's theorem. Yu. V. Linnik proved a similar fact (the decomposition theorem) for a more general family with reproducing property by using the theory of analytic functions [9].

F. Convergence of Probability Distributions

The concept of convergence of distributions plays an important role in limit theorems and other fields of probability theory. When Ω is a topological space, we consider **convergence** of probability measures on Ω with respect to the †weak topology introduced in the space of measures on Ω (→ 37 Banach Spaces). Such convergence is called **weak convergence** in probability theory. For a sequence of n-dimensional distributions Φ_k $(k = 1, 2, \ldots)$ to converge to Φ weakly, each of the following conditions is necessary and sufficient. (1) For every continuous function with compact support, $\lim_{k \to \infty} \int_{\mathbf{R}^n} f(x)\, d\Phi_k(x) = \int_{\mathbf{R}^n} f(x)\, d\Phi(x)$. (2) At every continuity point of the distribution function $F(x_1, \ldots, x_n)$ of Φ, $\lim_{k \to \infty} F_k(x_1, \ldots, x_n) = F(x_1, \ldots, x_n)$ (F_k is the distribution function of Φ_k). (3) For every continuity set E of Φ (namely, a set such that $\Phi(\bar{E} - E^0) = 0$), $\lim_{k \to \infty} \Phi_k(E) = \Phi(E)$. (4) For all open $G \subset \mathbf{R}^n$, $\liminf_{k \to \infty} \Phi_k(G) \geqslant \Phi(G)$. (5) For all closed $F \subset \mathbf{R}^n$, $\limsup_{k \to \infty} \Phi_k(F) \leqslant \Phi(F)$. (6) $\lim_{k \to \infty} \rho(\Phi_k, \Phi) = 0$, where ρ is a metric defined in the following way: Given any n-dimensional distributions Φ_1, Φ_2, we put $\varepsilon_{ij} = \inf\{\varepsilon \mid \Phi_j(F) < \Phi_i(F^\varepsilon) + \varepsilon$ for every closed $F\}$ (F^ε is the ε-neighborhood of F) and define $\rho(\Phi_1, \Phi_2) = \max(\varepsilon_{12}, \varepsilon_{21})$. The metric ρ, called the **Lévy distance**, was introduced by Lévy [6] in one dimension and by Yu. V. Prokhorov in metric spaces [10]. Each of these conditions except (2) is still necessary and sufficient for Φ_n to converge weakly to Φ when Φ_n and Φ are probability measures on a †complete separable metric space. It should also be noted that the probability measures on a complete separable metric space constitute a complete separable metric space with respect to the Lévy distance.

A family $\Phi_\alpha (\alpha\in\Lambda)$ of probability measures on a complete separable metric space is said to be **tight** if for every $\varepsilon>0$ there exists a †compact set $K=K(\varepsilon)$ such that $\Phi_\alpha(K^c)<\varepsilon$ for all $\alpha\in\Lambda$. A family Φ_α $(\alpha\in\Lambda)$ is tight if and only if it is †totally bounded with respect to the topology induced by the Lévy distance. Hence a tight family Φ_α $(\alpha\in\Lambda)$ has a weakly convergent subsequence.

We can give a criterion for the convergence of probability measures in terms of their characteristic functions. Suppose that Φ_k and Φ are n-dimensional probability measures with characteristic functions φ_k and φ. Then Φ_k converges weakly to Φ if and only if for every z, $\lim_k \varphi_k(z)=\varphi(z)$. Let φ_k be the characteristic function of an n-dimensional probability distribution Φ_k. If the sequence $\{\varphi_k\}$ converges pointwise to a limit function φ and the convergence of φ_k is uniform in some neighborhood of the origin, then φ is also the characteristic function of an n-dimensional probability distribution Φ and the sequence $\{\Phi_k\}$ converges weakly to Φ [7] (**Lévy's continuity theorem**).

For any probability distribution concentrated on $[0,\infty)$, the use of †Laplace transforms as a substitute for Fourier transforms provides a powerful tool. The method of **probability generating functions** is available for the study of arbitrary probability distribution concentrated on the nonnegative integers [14]. The method of moment-generating functions is also useful. There are many results on the relation between these functions, probability distributions, and their convergence [14–16].

Let $\Phi_1, \Phi_2, \ldots$ and Φ be 1-dimensional distributions. If all absolute moments exist, $\sum_{j=1}^{\infty}\beta_j^{-1/j}=\infty$ for $\beta_j=\int_{-\infty}^{\infty}|x|^j d\Phi(x)<\infty$, and $\lim_{k\to\infty}\int_{-\infty}^{\infty}x^j d\Phi_k(x)=\int_{-\infty}^{\infty}x^j d\Phi(x)$ $(j=0,1,2,\ldots)$, then Φ_k weakly converges to Φ. This condition is sufficient but not necessary.

G. Infinitely Divisible Distributions

An n-dimensional probability distribution Φ is called **infinitely divisible** if for every positive interger k, there exists a probability distribution Φ_k such that $\Phi=\Phi_k*\Phi_k*\ldots*\Phi_k$ $(=\Phi_k^{*k})$. Both normal distributions and Poisson distributions are infinitely divisible. If an n-dimensional distribution Φ satisfies the condition $\int_{|x|>\varepsilon}\Phi(dx)<\varepsilon$, we say that $\Phi\in v(\varepsilon)$. Then Φ is an infinitely divisible distribution if and only if for every $\varepsilon>0$ we can find $\Phi_1, \Phi_2, \ldots, \Phi_k\in v(\varepsilon)$ such that $\Phi=\Phi_1*\Phi_2*\ldots*\Phi_k$.

Let X_{ki}, $i=1,2,\ldots,n(k)$, be independent random variables for every k, and assume that the distribution of X_{ki} belongs to $v(\varepsilon_k)$, $i=1,2,\ldots,n(k)$, where $\varepsilon_k\to 0$ as $k\to\infty$. If the probability distributions of the sums $X_k=\sum_{i=1}^{n(k)}X_{ki}$ converge to a probability distribution as $k\to\infty$, then the limit distribution is infinitely divisible.

The characteristic function of a 1-dimensional infinitely divisible distribution can be written in the form

$$\varphi(z)=\exp\left(irz-\frac{v}{2}z^2+\int_{-\infty}^{\infty}A(u,z)\frac{1+u^2}{u^2}dG(u)\right), \quad (3)$$

where γ is a constant, v is a nonnegative constant, $G(u)$ is a nondecreasing bounded function with $G(-\infty)=0$, $A(u,z)=\exp(iuz)-1-izu/(1+u^2)$, and the value of $A(u,z)(1+u^2)/u^2$ at $u=0$ is defined to be $-z^2/2$. Formula (3) is called **Khinchin's canonical form**. For the characteristic function of an infinitely divisible n-dimensional distribution, the canonical form is as follows:

$$\varphi(z)=\exp\left(i(m,z)-\sum_{p,q=1}^{n}c_{pq}z_pz_q+\int_{\mathbf{R}^n}\left(e^{i(z,x)}-1-\frac{i(z,x)}{1+|x|^2}\right)n(dx)\right),$$
$$z=(z_1,\ldots,z_n)\in\mathbf{R}^n, \quad (4)$$

where $m\in\mathbf{R}^n$, (c_{pq}) is a positive semidefinite matrix, and $n(dx)$ is a measure on $\mathbf{R}^n$ such that $n(\{0\})=0$ and

$$\int_{\mathbf{R}^n}\frac{|x|^2}{1+|x|^2}n(dx)<\infty.$$

Formula (4) is called **Lévy's canonical form**. If a 1-dimensional infinitely divisible distribution Φ satisfies $\int_{\mathbf{R}^1}x^2\,d\Phi(x)<\infty$, then its characteristic function is given by

$$\varphi(z)=\exp\left(imz-\frac{v}{2}z^2+\int_{-\infty}^{\infty}(e^{izu}-1-izu)\frac{1}{u^2}dK(u)\right), \quad (5)$$

where m is a real constant, v is a nonnegative constant, and $K(u)$ is a nondecreasing bounded function such that $K(-\infty)=0$. It is called **Kolmogorov's canonical form**. (For infinitely divisible distributions on a homogeneous space → 5 Additive Processes.)

Let Φ and Ψ be n-dimensional distributions. If for some $\lambda>0$, $\Psi(E)=\Phi(\lambda E)$ $(\lambda E=\{\lambda\xi\mid\xi\in E\})$ for every set E, we say that Φ and Ψ are equivalent. Let Φ and Ψ be probability distributions with distribution functions F and G

and characteristic functions φ and ψ. Then the following three statements are equivalent: (1) Φ and Ψ are equivalent; (2) $G(x)=F(\lambda x)$ for every x; and (3) $\psi(\lambda z)=\varphi(z)$ for every z. We call Φ a **stable distribution** if for every pair of distributions Φ_1, Φ_2 equivalent to Φ, the convolution $\Phi_1 * \Phi_2$ is equivalent to Φ. If Φ is stable, every distribution equivalent to Φ is also stable. We can characterize stable distributions in terms of their characteristic functions $\varphi(z)$ as follows: For every pair $\lambda_1, \lambda_2>0$, there exists a $\lambda=\lambda(\lambda_1,\lambda_2)>0$ such that $\varphi(\lambda z)=\varphi(\lambda_1 z)\varphi(\lambda_2 z)$. We can restate this characterization as follows: Φ is stable if and only if for every pair of independent random variables X_1 and X_2 with identical distribution Φ and for any positive numbers λ_1 and λ_2, there exists a positive number λ such that $(\lambda_1 X_1+\lambda_2 X_2)/\lambda$ has the distribution Φ. By the definition we see that all stable distributions are infinitely divisible.

In the 1-dimensional case, putting $\varphi(z)=\exp\psi(z)$, we have $\psi(\lambda z)=\psi(\lambda_1 z)+\psi(\lambda_2 z)$, which implies $\psi(z)=(-c_0+i(z/|z|)c_1)|z|^\alpha$, where $c_0\geqslant 0$, $-\infty<c_1<\infty$, $0<\alpha\leqslant 2$. The parameter α is called the **exponent** (or **index**) of the stable distribution. The stable distributions with exponent $\alpha=2$ are the normal distributions, and the stable distributions with exponent $\alpha=1$ are the Cauchy distributions. We have $\psi(z)=-c_0|z|^\alpha$ for a symmetric stable distribution. (For the stable distribution with exponent 1/2 → Appendix A, Table 22).

Generalizing stable distributions, we can define **quasistable distributions**, which B. V. Gnedenko and A. N. Kolmogorov [17] called stable distributions also. Let F be the distribution function of a 1-dimensional distribution Φ. Φ is said to be quasistable if to every $b_1>0$, $b_2>0$ and real λ_1, λ_2 there correspond a positive number b and a real number λ such that we have the relation $F((x-\lambda_1)/b_1)*F((x-\lambda_2)/b_2)=F((x-\lambda)/b)$.

Let $\{X_i\}$ be a sequence of independent random variables with identical distribution. If for suitably chosen constants A_n and B_n the distributions of the sums $B_n^{-1}(\sum_{i=1}^n X_i)-A_n$ converge to a distribution, the limit distribution is a quasistable distribution (Lévy). A necessary and sufficient condition for a distribution to be quasistable is that its characteristic function $\varphi(z)$ satisfy the relation $\varphi(b_1 z)\varphi(b_2 z)=\varphi(bz)e^{i\gamma z}$ $(\gamma=\lambda-\lambda_1-\lambda_2)$. The characteristic function of a quasistable distribution has the canonical representation

$$\varphi(z)=\exp\psi(z),$$

$$\psi(z)=imz-c|z|^\alpha(1+i\beta(z/|z|)\omega(z,\alpha)),$$

where m is a real number, $c\geqslant 0$, $0<\alpha\leqslant 2$, $|\beta|\leqslant 1$, and $\omega(z,\alpha)=\tan(\pi\alpha/2)$ $(\alpha\neq 1)$, $\omega(z,\alpha)=(2/\pi)\log|z|$ $(\alpha=1)$. The parameter α is called the exponent of the quasistable distribution. A quasistable distribution with $\alpha\neq 1$ is obtained from a stable distribution by translation, but quasistable distributions with $\alpha=1$ are not.

Semistable distributions are another generalization of stable distributions. A distribution is called semistable if its characteristic function $\varphi(z)$ satisfies the relation $\psi(qz)=q^\alpha\psi(z)$ for a positive number q $(\neq 1)$, where $\varphi(z)=\exp(\psi(z))$. Also in this case, the general form was obtained by Lévy [6].

A 1-dimensional probability distribution Φ is called an ***L*-distribution** if the distribution function F of Φ is the convolution of $F(x/a)$ and some other distribution function $F_a(x)$ for every $0<a<1$. Φ is an L-distribution if and only if there exists a sequence of independent random variables $\{X_k\}$ such that for suitably chosen constants $B_n>0$ and A_n the distributions of the sums $B_n^{-1}(\sum_{k=1}^n X_k)-A_n$ converge to Φ and $\sup_{1\leqslant k\leqslant n}P(|X_k/B_n|>\varepsilon)\to 0$ as $n\to\infty$ for every $\varepsilon>0$. Quasistable distributions are L-distributions.

H. The Shape of Distributions

Let $F(x)$ be a 1-dimensional distribution function. The quantity ζ_p such that $F(\zeta_p-0)\leqslant p\leqslant F(\zeta_p)$ $(0<p<1)$ is called the **quantile of order** p of F. In particular, the quantity $\zeta_{1/2}$ is called a **median**. If F satisfies the relation $1-F(m+x)=F(m-x)$, it is called **symmetric**. In any 1-dimensional symmetric distribution, every moment of odd order about the mean (if it exists) is equal to zero.

The ratio $\gamma_1=\mu_3/\sigma^3$ is used as a measure of departure from symmetry of a distribution and is called the **coefficient of skewness**. Furthermore, the ratio $\gamma_2=\mu_3/\sigma^4-3$ is called the **coefficient of excess**. For the normal distribution, we have $\gamma_1=\gamma_2=0$. If $\gamma_2\neq 0$, γ_2 expresses the degree of deviation from the normal distribution.

A distribution function $F(x)$ is called **unimodal** if there exists one value $x=a$ such that $F(x)$ is convex for $x<a$ and concave for $x>a$. All L-distributions (and hence quasistable distributions) are unimodal [18].

I. Kolmogorov's Extension Theorem

Let $\Omega=\mathbf{R}^T$, where T is an arbitrary index set. We associate with $\mathbf{R}^T$ the σ-algebra $\mathfrak{B}^T$ generated by the cylinder sets, i.e., $\{\omega\in\Omega\,|\,\pi_{t_1}(\omega)\in E_1,\ldots,\pi_{t_n}(\omega)\in E_n\}$, where $\pi_t(\omega)$ denotes the tth coordinate of ω, $E_k\in\mathfrak{B}(\mathbf{R}^1)$, $1\leqslant k\leqslant n$, $t_1<t_2<\ldots<t_n$, and $n=1,2,\ldots$. Given a probability

measure Φ on $(\mathbf{R}^T, \mathfrak{B}^T)$, we can define a finite-dimensional †marginal distribution Φ_S for any finite subset S of T by $\Phi_S(E)=\Phi(\pi_S^{-1}(E))$, $E\in\mathfrak{B}^S$, where π_S is the natural †projection π_S: $\mathbf{R}^T\to\mathbf{R}^S$. The measures $\{\Phi_S\}$ satisfy the following **consistency condition**: If $S_1\subset S_2$ $(\subset T)$ are finite and if $E\in\mathfrak{B}^{S_1}$, then

$$\Phi_{S_1}(E)=\Phi_{S_2}(\pi_{S_1S_2}^{-1}(E)), \tag{6}$$

where $\pi_{S_1S_2}:\mathbf{R}^{S_2}\to\mathbf{R}^{S_1}$ is the natural projection.

Conversely, if we are given a family of finite-dimensional probability measures $\{\Phi_S\}$ which satisfies the consistency condition (6), then **Kolmogorov's extension theorem** [1] asserts that there exists a unique probability measure Φ on $(\mathbf{R}^T, \mathfrak{B}^T)$ such that $\Phi_S(E)=\Phi(\pi_S^{-1}(E))$, $E\in\mathfrak{B}^S$, for any finite $S\subset T$.

This theorem is useful in constructing †stochastic processes. For example, let $\Phi_{n_1,n_2,\ldots,n_k}$ be the †product measure of k copies of a given probability measure Φ on $\mathbf{R}^1$. Then the family $\{\Phi_{n_1,n_2,\ldots,n_k}; n_1,n_2,\ldots,n_k\in\mathbf{Z}, k\in\mathbf{N}\}$ satisfies the consistency condition and hence, by Kolmogorov's extension theorem, determines a probability measure on $\mathbf{R}^{\mathbf{Z}}$, which is denoted by $\Phi^{\mathbf{Z}}$. Thus $X_n(\omega)=\pi_n(\omega)$, $n\in\mathbf{Z}$, $(\omega\in(\mathbf{R}^{\mathbf{Z}}, \mathfrak{B}^{\mathbf{Z}}, \Phi^{\mathbf{Z}}))$, are independent identically Φ-distributed random variables.

Kolmogorov's extension theorem is generalized to the case where the component spaces are †standard measurable spaces (→ 270 Measure Theory) instead of $\mathbf{R}^1$, and also to the case where product spaces are replaced by †projective systems [19].

J. Characteristic Functionals on Infinite-Dimensional Spaces

Contrary to the finite-dimensional case, Bochner's theorem does not necessarily hold in infinite-dimensional spaces. For example, let $(T, \|\cdot\|)$ be an infinite-dimensional †Hilbert space and $\varphi(t)=\exp(-\|t\|^2)$. φ is continuous and positive definite, and $\varphi(0)=1$. But it is known that there is no probability measure on $T=T^*$ (topological dual of T) which corresponds to φ. Bochner's theorem is generalized to infinite-dimensional spaces as follows.

Let T be a real †vector space endowed with the topology τ defined by a system of †Hilbertian seminorms $\{\|\cdot\|_\alpha, \alpha\in A\}$. Define a new topology $I(\tau)$ of T by all Hilbertian seminorms $\|\cdot\|$ each of which is HS-dominated by some $\|\cdot\|_\alpha$, $\alpha\in A$, i.e., $\sup\{(\sum_i\|e_i\|^2)^{1/2}; \{e_i\}:\alpha\text{-orthonormal}\}<\infty$. If $I(\tau)=\tau$, then (T,τ) is called a †nuclear space.

Let T_τ^* be the topological dual of (T,τ) (i.e., the set of all τ-continuous real valued linear functionals on T). Define a Borel structure $\mathfrak{B}(T_\tau^*)$ of T_τ^* as the σ-algebra generated by the system of half-spaces $\{x\in T_\tau^*; x(t)<a\}$, $t\in T$, $a\in\mathbf{R}^1$. For a probability measure Φ on $(T_\tau^*, \mathfrak{B}(T_\tau^*))$, define

$$\varphi(t)=\int_{T_\tau^*}\exp(i\langle x,t\rangle)\,d\Phi(x), \quad t\in T,$$

which is called the characteristic functional of Φ. A functional φ on T is the characteristic functional of a probability measure Φ on $(T_\tau^*, \mathfrak{B}(T_\tau^*))$ such that $\Phi^*(\bigcup_n T_{\alpha_n}^*)=1$ for some sequence $\{\alpha_n\}\subset A$, where Φ^* is the †outer measure (→ 270 Measure Theory) and T_α^* is the topological dual of $(T, \|\cdot\|_\alpha)$, if and only if φ is positive definite, $\varphi(0)=1$, and continuous with respect to the topology $I(\tau)$ [23].

As special cases of the foregoing theorem, we have the following. If (T,τ) is a nuclear space, then every positive definite τ-continuous functional φ with $\varphi(0)=1$ is the characteristic functional of a probability measure on T_τ^* (Minlos [24]). Schwartz's spaces $\mathcal{S}(\mathbf{R}^n)$ and $\mathcal{D}(\mathbf{R}^n)$ are nuclear. Let $(T,\tau=\|\cdot\|)$ be a Hilbert space. A †Hilbert-Schmidt operator U is, by definition, a †bounded linear operator on T such that $\sum_i\|Ue_i\|^2<\infty$, by any †complete orthonormal system $\{e_i\}$ (this quantity does not depend on the choice of $\{e_i\}$). Define a seminorm $\|\cdot\|_U$ by $\|t\|_U=\|Ut\|$, $t\in T$, for a Hilbert-Schmidt operator U. Then the topology $I(\tau)$ coincides with the topology induced by the system of seminorms $\|\cdot\|_U$, where U are Hilbert-Schmidt operators, which is called the **Sazonov topology**. Thus every functional φ on T, which is positive definite, $\varphi(0)=1$, and continuous with respect to the Sazonov topology, is the characteristic functional of a probability measure on T_τ^* (Sazonov [25]).

The probability measure on $\mathcal{S}'$ with the characteristic functional $\exp(-\int_{-\infty}^{\infty}|f(s)|^2\,ds)$, $f\in\mathcal{S}$, is the probability measure of a Gaussian †white noise on $\mathcal{S}'$.

References

[1] A. N. Kolmogorov, Grundbegriffe der Wahrscheinlichkeitsrechnung, Springer, 1933; English translation, Foundations of the theory of probability, Chelsea, 1950.
[2] H. Cramér, Mathematical methods of statistics, Princeton Univ. Press, 1946.
[3] J. L. Doob, Stochastic processes, Wiley, 1953.
[4] M. M. Loève, Probability theory, Van Nostrand, third edition, 1963.
[5] W. Feller, An introduction to probability theory and its applications II, Wiley, 1966.
[6] P. Lévy, Théorie de l'addition des variables aléatoires, Gauthier-Villars, 1937.

[7] P. Lévy, Calcul des probabilités, Gauthier-Villars, 1925.
[8] E. Lukacs, Characteristic functions, Hafner, 1960.
[9] Yu. V. Linnik, Decomposition of probability distributions, Oliver & Boyd, 1964. (Original in Russian, 1960.)
[10] Yu. V. Prokhorov, Convergence of random processes and limit theorems in probability theory, Theory of Prob. Appl., 1 (1956), 155–214. (Original in Russian, 1956.)
[11] L. LeCam, Convergence in distribution of stochastic processes, Univ. California Publ. Statist., 2 (1957), 207–236.
[12] K. R. Parthasarathy, Probability measures on metric spaces, Academic Press, 1967.
[13] P. Billingsley, Convergence of probability measures, Wiley, 1968.
[14] W. Feller, An introduction to probability theory and its applications I, Wiley, second edition, 1957.
[15] J. A. Shohat and J. D. Tamarkin, The problem of moments, Amer. Math. Soc. Math. Surveys, 1943.
[16] D. V. Widder, Laplace transform, Princeton Univ. Press, 1941.
[17] B. V. Gnedenko and A. N. Kolmogorov, Limit distributions for sums of independent random variables, Addison-Wesley, 1954. (Original in Russian, 1949.)
[18] M. Yamazato, Unimodality of infinitely divisible distribution functions of class L, Ann. Prob., 6 (1978), 523–531.
[19] S. Bochner, Harmonic analysis and the theory of probability, Univ. of California Press, 1955.
[20] E. Nelson, Regular probability measures on function space, Ann. Math., (2) 69 (1959), 630–643.
[21] I. M. Gel'fand and N. Ya. Vilenkin, Generalized functions IV, Applications of harmonic analysis, Academic Press, 1964. (Original in Russian, 1961.)
[22] Yu. V. Prokhorov, The method of characteristic functionals, Proc. 4th Berkeley Symp. Math. Stat. Prob. II, Univ. of California Press, 1961, 403–419.
[23] A. N. Kolmogorov, A note on the papers of R. A. Minlos and V. Sazonov, Theory of Prob. Appl., 4 (1959), 221–223. (Original in Russian, 1959.)
[24] R. A. Minlos, Generalized random processes and their extension to a measure, Selected Transl. Math. Stat. and Prob., Amer. Math. Soc., 3 (1962), 291–313. (Original in Russian, 1959.)
[25] V. Sazonov, A remark on characteristic functionals, Theory of Prob. Appl., 3 (1958), 188–192. (Original in Russian, 1958.)
[26] M. Kac, Probability and related topics in physical sciences, Interscience, 1959.

342 (XVII.1) Probability Theory

A. History

The origin of the **theory of probability** goes back to the mathematical problems connected with dice throwing that were discussed in letters exchanged by B. Pascal and P. de Fermat in the 17th century. These problems were concerned primarily with concepts such as †permutations, †combinations, and †binomial coefficients, whose theory was established at about the same time [1]. This elementary theory of probability was later enriched by the work of scholars such as Jakob Bernoulli [2], A. de Moivre [3], T. Bayes, L. de Buffon, Daniel Bernoulli, A. M. Legendre, and J. L. Lagrange. Finally, P. S. Laplace completed the classical theory of probability in his book *Théorie analytique des probabilités* (1812). In this work, Laplace not only systematized but also greatly extended previous important results by introducing new methods, such as the use of †difference equations and †generating functions. Since the 19th century, the theory of probability has been extensively applied to the natural sciences and even to the social sciences.

The definition of a priori probability due to Laplace provoked a great deal of argument when it was applied. For example, R. von Mises advocated an empirical theory of probability based on the notion of **Kollektiv (collective)**, which is a mathematical model of mass phenomena [5]. However, these arguments are concerned with philosophical rather than mathematical aspects. Nowadays, the main concern of mathematicians lies not in the intuitive or practical meaning of probability but in the logical setup governing probability. From this viewpoint the mathematical model of a random phenomenon is given by a probability measure space $(\Omega, \mathfrak{B}, P)$, where Ω is the set of all possible outcomes of the phenomenon, $P(E)$ represents the probability that an outcome belonging to E be realized, and $\mathfrak{B}$ is a σ-algebra consisting of all sets E for which $P(E)$ is defined. All probabilistic concepts, such as random variables, independence, etc., are defined on $(\Omega, \mathfrak{B}, P)$ in terms of measure theory. Such a measure-theoretic basis of probability theory is due to A. Kolmogorov [6], though similar considerations had been made before him for special problems, for example, in the work of E. Borel concerning the strong law of large numbers [7] and in the rigorous definition of Brownian motion by N. Wiener [8].

Ever since probability theory was given

solid foundations by Kolmogorov, it has made tremendous progress. The most important concept in today's probability theory is that of †**stochastic processes**, which correspond to functions in analysis. In applications a stochastic process is used as the mathematical model of a random phenomenon varying with time. The following types of stochastic processes have been investigated extensively: †additive processes, †Markov processes and †Markov chains, †martingales, †stationary processes, and †Gaussian processes. †Brownian motion and †branching processes are important special stochastic processes. In the same way as functions are often defined by differential equations, there are stochastic processes which can be defined by †stochastic differential equations. The theory of stochastic processes and stochastic differential equations can be applied to †stochastic control, †stochastic filtering, and †statistical mechanics. The †ergodic theory that originated in statistical mechanics is now regarded as an important branch of probability theory closely related to the theory of stationary processes.

B. Probability Spaces

Let Ω be an †abstract space and $\mathfrak{B}$ be a †σ-algebra of subsets of Ω. A **probability measure** (or **probability distribution**) over $\Omega(\mathfrak{B})$ is a set function $P(E)$ defined for $E \in \mathfrak{B}$ and satisfying the following conditions: (P1) $P(E) \geqslant 0$; (P2) for every sequence $\{E_n\}$ $(n=1,2,\ldots)$ of pairwise disjoint sets in $\mathfrak{B}$,

$$P\left(\bigcup_n E_n\right) = \sum_n P(E_n);$$

(P3) $P(\Omega)=1$. The triple $(\Omega, \mathfrak{B}, P)$ is called a **probability space**. The space Ω (resp. each element ω of Ω) is called the **basic space**, **space of elementary events**, or **sample space** (resp. **sample point** or **elementary event**). We say that a condition $\varepsilon(\omega)$ involving a generic sample point ω is an **event**; in particular, it is called a **measurable event** or **random event** if the set E of all sample points satisfying $\varepsilon(\omega)$ belongs to $\mathfrak{B}$. We assume that an event is always a measurable one, since we encounter only measurable events in the theory of probability. Because of the obvious one-to-one correspondence between measurable events and $\mathfrak{B}$-measurable sets (i.e., the correspondence of each event ε with the set E of all sample points ω satisfying ε), a $\mathfrak{B}$-measurable set itself is frequently called an event. If ε is an event and E is the $\mathfrak{B}$-measurable set corresponding to ε, we call $P(E)$ or $\Pr(\varepsilon)$ the **probability that the event** ε **occurs**, i.e., the **probability of the event** E. The **complementary event** (resp. **impossible event**, **sure event**) is the complementary set E^c (empty set $\varnothing$, whole space Ω). For a finite or infinite family $\{E_\lambda\}$ $(\lambda \in \Lambda)$, the **sum event** (resp. **intersection** or **product event**) of E_λ is the set $\bigcup_\lambda E_\lambda$ $(\bigcap_\lambda E_\lambda)$. If $E \cap F = \varnothing$, then we say that E and F are mutually exclusive or that they are **exclusive events**.

By the definition of P, we have $0 \leqslant P(E) \leqslant 1$ for any event E, $P(\varnothing)=0$, and $P(\Omega)=1$. Moreover, if $\{E_n\}$ $(n=1,2,\ldots)$ is a sequence of pairwise exclusive events and E is the sum event of E_n, we have

$$P(E) = \sum_{n=1}^{\infty} P(E_n).$$

This property is called the **additivity of probability**. If $P(E)=1$, the event E is said to occur **almost certainly** (**almost surely** (abbrev. a.s.), **for almost all** ω, or **with probability** 1).

Given a finite sequence $\{E_n\}$ $(n=1,2,\ldots,N)$ of events, we say that the events E_n $(n=1,2,\ldots,N)$ are mutually **independent** or that the sequence $\{E_n\}$ $(n=1,2,\ldots,N)$ is **independent** if every subsequence satisfies

$$P(E_{i_1} \cap E_{i_2} \cap \ldots \cap E_{i_k}) = \prod_{j=1}^{k} P(E_{i_j}).$$

Given an infinite family $\{E_\lambda\}$ $(\lambda \in \Lambda)$ of events, we say that the events E_λ $(\lambda \in \Lambda)$ are mutually **independent** or that the family $\{E_\lambda\}$ $(\lambda \in \Lambda)$ is **independent** if every finite subfamily is independent. The concept of independence of events can be generalized to a family $\{\mathfrak{B}_\lambda\}$ $(\lambda \in \Lambda)$ of σ-subalgebras of $\mathfrak{B}$ as follows. A family $\{\mathfrak{B}_\lambda\}$ $(\lambda \in \Lambda)$ of σ-subalgebras of events is said to be independent if for every choice of $E_\lambda \in \mathfrak{B}_\lambda$, the family $\{E_\lambda\}$ $(\lambda \in \Lambda)$ of events is independent.

For a sequence $\{E_n\}$ $(n=1,2,\ldots)$ of events, the sets $\limsup_n E_n$ and $\liminf_n E_n$ are called the **superior limit event** and **inferior limit event**, respectively. The superior limit event (inferior limit event) is the set of all ω for which infinitely many events among E_n (all events except finitely many E_n) occur. Therefore $P(\liminf_n E_n)$ is the probability that infinitely many events among E_n occur, and $P(\limsup_n E_n)$ is the probability that the events E_n occur for all n after some number n_0, where n_0 depends on ω in general. The **Borel-Cantelli lemma**, which is concerned with the evaluation of $P(\limsup_n E_n)$, reads as follows: Given a sequence $\{E_n\}$ $(n=1,2,\ldots)$ of events, we have (i) whether the events E_n $(n=1,2,\ldots)$ are mutually independent or not, $\sum_n P(E_n) < \infty$ implies that $P(\limsup_n E_n)=0$; and (ii) if the events E_n $(n=1,2,\ldots)$ are mutually independent, $\sum_n P(E_n)=\infty$ imples that $P(\limsup_n E_n)=1$. Frequently, applications of part (ii) are greatly hampered by the requirement of independence; a number of sufficient conditions for dependent events to have the same conclusion as (ii) have been discovered. The **Chung-Erdös theorem** [9] is quite useful in this connection.

C. Random Variables

Let $(\Omega, \mathfrak{B}, P)$ be a probability space. A **random variable** is a real-valued function X defined on Ω that is $\mathfrak{B}$-measurable (i.e., for every real number a, the set $\{\omega \mid X(\omega)<a\}$ is in $\mathfrak{B}$). If $X_1, X_2, \ldots, X_n$ are random variables, the mapping $X=(X_1, X_2, \ldots, X_n)$ from Ω into $\mathbf{R}^n$ is said to be an **n-dimensional** (or **$\mathbf{R}^n$-valued**) **random variable**. More generally, a mapping X from $(\Omega, \mathfrak{B})$ into another †measurable space $(S, \mathfrak{E})$ is called an **$(S, \mathfrak{E})$-valued random variable** if it is measurable, that is, for every set A of $\mathfrak{E}$, the set $\{\omega \mid X(\omega) \in A\}$ belongs to $\mathfrak{B}$.

Let $\mathfrak{B}^1$ be the σ-algebra of all †Borel subsets of the real line $\mathbf{R}$. Then each random variable X induces a probability measure Φ on $(\mathbf{R}, \mathfrak{B}^1)$ such that

$$\Phi(A)=P(\{\omega \mid X(\omega) \in A\}), \qquad A \in \mathfrak{B}^1.$$

The measure Φ is called the (**1-dimensional**) **probability distribution of the random variable** X or simply the **distribution of** X. The point function F defined by

$$F(a)=P(\{\omega \mid X(\omega) \leqslant a\}), \qquad a \in \mathbf{R},$$

is a monotone nondecreasing and right continuous function such that $\lim_{a \to \infty} F(a)=1$, $\lim_{a \to -\infty} F(a)=0$. The function F is called the **cumulative distribution function** (or simply the **distribution function**) of the random variable X. Similarly, an n-dimensional random variable $X=(X_1, \ldots, X_n)$ induces its **n-dimensional probability distribution** (or simply **n-dimensional distribution**) and its **n-dimensional distribution function** $F(a_1, \ldots, a_n)=P(\{\omega \mid X_1(\omega) \leqslant a_1, \ldots, X_n(\omega) \leqslant a_n\})$. If the X_n $(n=1,2, \ldots, N)$ are k_n-dimensional random variables $(n=1,2, \ldots, N)$, we say that the $l(=\sum_{n=1}^{N} k_n)$-dimensional random variable $X=(X_1, X_2, \ldots, X_N)$ is the **joint random variable** of X_n $(n=1,2, \ldots, N)$ and that the (l-dimensional) distribution Φ of X is the **joint distribution** (or **simultaneous distribution**) of X_n $(n=1,2, \ldots, N)$. On the other hand, the k_n-dimensional distribution Φ_n of X_n is called the **marginal distribution** of the l-dimensional distribution Φ.

Given a finite sequence $\{X_n\}$ $(n=1,2, \ldots, N)$ of random variables, if the relation

$$P(\{\omega \mid X_n(\omega) \in A_n\ (n=1,2, \ldots, N)\}) = \prod_{n=1}^{N} P(\{\omega \mid X_n(\omega) \in A_n\}) \tag{1}$$

holds for every choice of 1-dimensional †Borel sets A_n $(n=1,2, \ldots, N)$, we say that the random variables X_n $(n=1,2, \ldots, N)$ are mutually **independent** or that the sequence $\{X_n\}$ $(n=1,2, \ldots, N)$ is **independent**. Given an infinite family $\{X_\lambda\}$ $(\lambda \in \Lambda)$ of random variables, we say that the random variables are mutually **independent** or that the family is independent if every finite subfamily is independent. The latter definition of independence of random variables is compatible with the previous definition of independence of σ-subalgebras of $\mathfrak{B}$; i.e., if $\mathfrak{B}[X_\lambda]$ denotes the σ-subalgebras of $\mathfrak{B}$ generated by the sets $\{\omega \mid X_\lambda(\omega) \in A_\lambda\}$, with A_λ an arbitrary 1-dimensional Borel set, the independence of the family $\{X_\lambda\}$ $(\lambda \in \Lambda)$ in the latter sense is equivalent to the independence of the family $\{\mathfrak{B}[X_\lambda]\}$ $(\lambda \in \Lambda)$ in the previous sense. If the X_n $(n=1,2, \ldots, N)$ $(X_\lambda$ $(\lambda \in \Lambda))$ are k_n- (k_λ-) dimensional random variables, then the independence of the family $\{X_n\}$ $(n=1,2, \ldots, N)$ ($\{X_\lambda\}$ $(\lambda \in \Lambda)$) is defined similarly; it is enough to take k_n- (k_λ-) dimensional Borel sets A_n (A_λ) for 1-dimensional Borel sets in equation (1).

Given a family $\{X_\lambda\}$ $(\lambda \in \Lambda)$ of random variables, the smallest σ-algebra with respect to which every X_λ is measurable is called the σ-algebra generated by $\{X_\lambda\}$ $(\lambda \in \Lambda)$ and is denoted by $\mathfrak{B}[X_\lambda \mid \lambda \in \Lambda]$. Each element of this class is said to be **measurable with respect to the family** $\{X_\lambda\}$ $(\lambda \in \Lambda)$ **of random variables**.

Since a random variable X is a $\mathfrak{B}$-measurable function, we can speak of the †integral of X relative to the measure P on $\mathfrak{B}$. If X is integrable relative to P, the integral of X over A is denoted by $E(X; A)$. $E(X; \Omega)$, usually denoted by $E(X)$, is called the **mean**, **expectation**, or **expected value** of X, denoted also by $M(X)$ or m_X. If $(X-E(X))^2$ is integrable,

$$V(X)=E((X-E(X))^2)$$

is called the **variance** of X, denoted by $\sigma^2(X)$. The **standard deviation** of X is the nonnegative square root $\sigma(X)$ of the variance. If X and Y are two random variables for which $E((X-E(X))(Y-E(Y)))$ exists, the value $E((X-E(X))(Y-E(Y)))$ is called the **covariance** of X and Y. When X and Y have finite variances, the **correlation coefficient** of X and Y is defined by

$$\rho(X, Y)=\frac{E((X-E(X))(Y-E(Y)))}{\{E((X-E(X))^2) E((Y-E(Y))^2)\}^{1/2}}.$$

It follows that $E(aX+bY)=aE(X)+bE(Y)$, $V(aX+b)=a^2 V(X)$ for any real numbers a, b, and that, in particular, $E(XY)=E(X)E(Y)$, $V(X+Y)=V(X)+V(Y)$ for mutually independent random variables X and Y. It also follows from the definition that $-1 \leqslant \rho(X, Y) \leqslant 1$ in all cases. The independence of X and Y implies that $\rho(X, Y)=0$, but the converse is false in general. The variance is important because of the well-known **Chebyshev inequal-**

ity: If X is a random variable with finite variance σ^2,

$$P(|X-E(X)|\geqslant c)\leqslant\sigma^2/c^2$$

for every positive number c.

D. Convergence of Random Variables

If $P(\lim_{n\to\infty}X_n=X_\infty)=1$, the sequence $\{X_n\}$ is said to **converge almost everywhere (almost certainly, almost surely** (a.s.), or **with probability** 1) to X_∞. If $\lim_{n\to\infty}P(|X_n-X_\infty|>\varepsilon)=0$ for every positive number ε, the sequence $\{X_n\}$ is said to **converge in probability** to X_∞. For a given positive number p the sequence $\{X_n\}$ is said to **converge in the mean of order** p to X_∞ if $\lim_{n\to\infty}E(|X_n-X_\infty|^p)=0$. Finally, if the random variables X_n $(n=1,2,\ldots,\infty)$ have distributions Φ_n $(n=1,2,\ldots,\infty)$, respectively, and if

$$\lim_{n\to\infty}\int_{-\infty}^{\infty}f(x)\,d\Phi_n(x)=\int_{-\infty}^{\infty}f(x)\,d\Phi_\infty(x)$$

for every continuous function f with compact support, the sequence $\{X_n\}$ is said to **converge in distribution** to X_∞. Note that the sequence of random variables converging in distribution may not converge in any ordinary sense. For example, random variables converging in distribution may even be defined on different probability spaces. On one hand, almost sure convergence does not in general imply convergence in the mean. On the other hand, either almost sure convergence or convergence in the mean implies convergence in probability, and convergence in probability implies convergence in distribution. However, P. Lévy [10] proved that if the X_n $(n=1,2,\ldots)$ are mutually independent, the sequence

$$Y_k=\sum_{n=1}^{k}X_n,\qquad k=1,2,\ldots,$$

is convergent almost everywhere if and only if it is convergent in distribution (or in probability). The famous **three-series theorem** of Khinchin and Kolmogorov [11] claims that the series $\sum_n X_n$ with $X_1,X_2,\ldots$ independent is convergent almost surely if and only if there exists a sequence of independent random variables $X_1',X_2',\ldots$ such that each of the three series

$$\sum_n P(X_n\neq X_n'),\quad \sum_n E(X_n'),\quad \sum_n V(X_n')$$

is convergent.

E. Conditional Probability and Conditional Expectation

Let $(\Omega,\mathfrak{B},P)$ be a probability space and $\mathfrak{F}$ a σ-subalgebra of $\mathfrak{B}$. If X is a random variable with finite mean, the function

$$\mu(E)=\int_E X(\omega)\,dP,\qquad E\in\mathfrak{F},$$

defines on $(\Omega,\mathfrak{F})$ a completely additive set function which is †absolutely continuous with respect to P. Therefore, by the †Radon-Nikodym theorem, there is an $\mathfrak{F}$-measurable function f such that

$$\mu(E)=\int_E f(\omega)\,dP\qquad\text{for every } E\in\mathfrak{F}.$$

This function is unique up to a set of P-measure zero and is called the **conditional expectation** (or **conditional mean**) of X relative to $\mathfrak{F}$, denoted by $E(X\,|\,\mathfrak{F})$. When $\mathfrak{F}$ is generated by a random variable Y, we also write $E(X\,|\,Y)$ for $E(X\,|\,\mathfrak{F})$ and call it the conditional expectation of X relative to Y. In this case, there is a †Borel measurable function f such that $E(X\,|\,Y)=f(Y(\omega))$, and we write $E(X\,|\,Y=y)$ for $f(y)$. The same fact holds when Y is a multidimensional random variable. It follows from the definition that the conditional expectation has the following properties, up to a set of P-measure zero: (i) if $X\geqslant 0$, then $E(X\,|\,\mathfrak{F})\geqslant 0$; (ii) $E(aX+bY\,|\,\mathfrak{F})=aE(X\,|\,\mathfrak{F})+bE(Y\,|\,\mathfrak{F})$; (iii) $E(E(X\,|\,\mathfrak{F}))=E(X)$; (iv) if X and $\mathfrak{F}$ are mutually independent, i.e., $\mathfrak{B}[X]$ and $\mathfrak{F}$ are mutually indepent, then $E(X\,|\,\mathfrak{F})=E(X)$; (v) if X is $\mathfrak{F}$-measurable, then $E(X\,|\,\mathfrak{F})=X$ and $E(XY\,|\,\mathfrak{F})=XE(Y\,|\,\mathfrak{F})$; (vi) if $\lim_{n\to\infty}X_n=X_\infty$ with $|X_n|\leqslant Y$ and Y is an integrable random variable, then $\lim_{n\to\infty}E(X_n\,|\,\mathfrak{F})=E(X_\infty\,|\,\mathfrak{F})$; (vii) if $\mathfrak{G}$ is a σ-subalgebra of $\mathfrak{F}$, then $E(E(X\,|\,\mathfrak{F})\,|\,\mathfrak{G})=E(X\,|\,\mathfrak{G})$; (viii) if X^2 is integrable and Y is any $\mathfrak{F}$-measurable random variable, then $E((X-E(X\,|\,\mathfrak{F}))^2)\leqslant E((X-Y)^2)$.

When X is the **indicator function** (i.e., the †characteristic function) χ_E of a set E in $\mathfrak{B}$, $E(\chi_E\,|\,\mathfrak{F})$ is called the **conditional probability** of E relative to $\mathfrak{F}$ and is denoted by $P(E\,|\,\mathfrak{F})$. In particular, if $\mathfrak{F}=\{F,F^c,\varnothing,\Omega\}$ with $1>P(F)>0$, $P(E\,|\,\mathfrak{F})$ is the simple function which takes the values $P(E\cap F)/P(F)$ on F and $P(E\cap F^c)/P(F^c)$ on F^c. These values are denoted respectively by $P(E\,|\,F)$ and $P(E\,|\,F^c)$. The definition of $P(E\,|\,Y)$ or $P(E\,|\,Y=y)$ is also the same as in the case of the conditional expectation.

Let $\mathfrak{F}$ be a σ-subalgebra of $\mathfrak{B}$ and Y a real random variable. According to the foregoing definition, $P(Y\in E\,|\,\mathfrak{F})$ or $P(Y^{-1}(E)\,|\,\mathfrak{F})$ is the conditional probability of the occurrence of the event $Y\in E$ under $\mathfrak{F}$. Since $P(Y\in E\,|\,\mathfrak{F})$ is determined except on a P-null set depending on E, an arbitrary version of $P(Y\in E\,|\,\mathfrak{F})$, viewed as a function of E, does not always satisfy the conditions of a probability measure. However, we can prove that there exists a nice version of $P(Y\in E\,|\,\mathfrak{F})$ which is a probability measure in $E \in \mathfrak{B}^1$ for every $\omega \in \Omega$ and that such a version is unique almost surely. This version is called a **regular conditional probability** of $Y \in E$ under $\mathfrak{F}$ or the **conditional probability distribution** of Y under $\mathfrak{F}$; this is written as $P_Y(E|\mathfrak{F})$. $P(Y \in E|X)$, $P(Y \in E|X=x)$, $P_Y(E|X)$, and $P_Y(E|X=x)$ are interpreted similarly. The conditional probability distribution can be defined not only for real random variables but also for every random variable which takes values in an †analytic measurable space.

F. Bayes's Formula

Let $E_1, E_2, \ldots, E_n$ be pairwise exclusive events, and assume that one of them must occur. If E is another random event, we have

$$P(E_i|E) = \frac{P(E_i)P(E|E_i)}{P(E_1)P(E|E_1) + \ldots + P(E_n)P(E|E_n)},$$

where $P(E_i)$ is the probability of the event E_i and $P(E|E_i)$ is the conditional probability of E under the assumption that the event E_i has occurred. This is called **Bayes's formula**. In practical applications $E_1, \ldots, E_n$ usually represent n unknown hypotheses. Suppose that the probabilities on the right-hand side of the formula are given. We then apply Bayes's formula to reevaluate the probability of each hypothesis E_i knowing that some event E has occurred as the result of a trial. This is why $P(E_i)$ $(P(E_i|E))$ is called the **a priori (a posteriori) probability**. However, the determination of the values of a priori probabilities is sometimes difficult, and we often set $P(E_i) = 1/n$ in practical applications, although this has caused a great deal of criticism.

When X is a random variable subject to the distribution with continuous probability density $f(x)$, Bayes's formula is extended to the following form:

$$f(x_0|E) = \frac{f(x_0)P(E|X=x_0)}{\int_{-\infty}^{\infty} P(E|X=x)f(x)dx},$$

where $f(x|E)$ is the conditional probability density of the random variable X under the assumption that the event E has occurred, and $P(E|X=x_0)$ is the conditional probability of E relative to X.

G. Zero-One Laws

In probability theory there are many theorems claiming that an event with certain properties has probability 0 or 1. Such theorems are called **zero-one laws**. Here, we mention two famous examples, Kolmogorov's zero-one law [6] and the Hewitt-Savage zero-one law. Let $\alpha = \alpha(X_1, X_2, \ldots)$ be an event concerning a sequence of random variables $\{X_n\}$. α is called a **tail event** concerning $\{X_n\}$ if for every n, occurrence or nonoccurrence of α depends only on $\{X_n, X_{n+1}, \ldots\}$. For example, $\{\lim_{n \to \infty} X_n = 0\}$ is a tail event. α is called a **symmetric event** concerning $\{X_n\}$ if occurrence or nonoccurrence of α is invariant under every finite permutation of $X_1, X_2 \ldots$. For example, the event that $\sum_{k=1}^{n} X_k > 0$ for infinitely many n's is a symmetric event. **Kolmogorov's zero-one law**: Every tail event concerning a sequence of independent random variables has probability 0 or 1. **Hewitt-Savage zero-one law**: Every symmetric event concerning a sequence of independent and identically distributed random variables has probability 0 or 1.

Kolmogorov's zero-one law can be extended as follows: Let $\mathfrak{F}_n$, $n = 1, 2, \ldots$, be a sequence of independent σ-subalgebras of $\mathfrak{B}$. Then the σ-algebra $\mathfrak{T} = \bigcap_k \bigcup_{n>k} \mathfrak{F}_n$, called the **tail σ-algebra** of $\{\mathfrak{F}_n\}$, is trivial, i.e., $P(A) = 0$ or 1 for every $A \in \mathfrak{T}$. Kolmogorov's zero-one law is a special case where $\mathfrak{F}_n$ is the σ-algebra generated by X_n for every n.

References

[1] I. Todhunter, A history of the mathematical theory of probability from the time of Pascal to that of Laplace, Macmillan, 1865 (Chelsea, 1949).
[2] J. Bernoulli, Ars conjectandi, 1713.
[3] A. de Moivre, The doctrine of chances, 1718.
[4] P. S. Laplace, Théorie analytique des probabilités, Paris, 1812.
[5] R. von Mises, Vorlesungen aus dem Gebiete der angewandten Mathematik I, Wahrscheinlichkeitsrechnung und ihre Anwendung in der Statistik und theoretischen Physik, Franz Deuticke, 1931.
[6] A. Kolmogorov (Kolmogoroff), Grundbefriffe der Wahrscheinlichkeitsrechnung, Erg. Math., Springer, 1933; English translation, Foundations of the theory of probability, Chelsea, 1950.
[7] E. Borel, Les probabilités dénombrables et leur application arithmetiques, Rend. Circ. Mat. Palermo, 27 (1909), 247–271.
[8] N. Wiener, Differential space, J. Math. Phys., 2 (1923), 131–174.
[9] K. L. Chung and P. Erdös, On the application of the Borel-Cantelli lemma, Trans. Amer. Math. Soc., 72 (1952), 179–186.
[10] P. Lévy, Théorie de l'addition des variables aléatoires, Gauthier-Villars, 1937.
[11] J. L. Doob, Stochastic processes, Wiley, 1953.

[12] M. M. Loève, Probability theory, Van Nostrand, third edition, 1963.
[13] W. Feller, An introduction to probability theory and its applications, Wiley, I, second edition, 1957; II, 1966.
[14] L. Breiman, Probability, Addison-Wesley, 1968.
[15] J. Lamperti, Probability, Benjamin, 1966.

343 (VI.14) Projective Geometry

A. Introduction

Projective geometry is the most fundamental of classical geometries and one of the first examples of axiomatized mathematics.

B. Construction of Projective Geometry

We construct projective geometry axiomatically [4]. Given two sets P, Q and a †relation $\Gamma \subset P \times Q$, consider the triple $\mathfrak{P} = \{P, Q, \Gamma\}$. We call each element of P a **point** and each element of Q a **line**. If $(p, l) \in \Gamma$ holds for a point p and a line l, then we say that the line l contains the point p. When two lines l_1 and l_2 contain a point p, we say that they intersect at p. When several points are contained in the same line, these points are said to be **collinear**, and when several lines contain the same point, these lines are said to be **concurrent**. For $\mathfrak{P}$ we impose the following axioms:

(I) There exists one and only one line that contains two given distinct points.

(II) Suppose that we are given noncollinear points p_0, p_1, and p_2, and distinct points q_1, q_2. Now suppose that $\{p_0, p_1, q_1\}$ and $\{p_0, p_2, q_2\}$ are collinear triples. Then the line containing p_1, p_2 and the line containing q_1, q_2 necessarily intersect (Fig. 1).

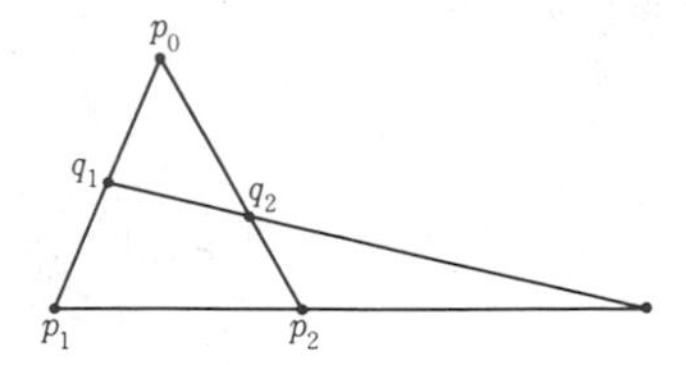

Fig. 1

(III) Every line contains at least three distinct points.

The $\mathfrak{P}$ that satisfy axioms (I) and (II) and axioms (I), (II), and (III) are called **general projective geometry** and **projective geometry**, respectively. The set of all points that are contained in a line is called the **point range** with the line as its **base**. In projective geometry, there exists a one-to-one correspondence between the set of lines and the set of point ranges, so we can identify every line with a point range. In this case a line $l \in Q$ is represented as a subset of P, and the relation $(p, l) \in \Gamma$ means that the point p belongs to the set l.

Let S be a subset of P and p_1, p_2 be any two distinct points of S. If the line that contains p_1 and p_2 is always contained in S, then S is called a **subspace**. Points and lines are subspaces. Now we impose the following axiom:

(IV) There exist a finite number of points such that any subspace that contains all of them contains P.

We call a projective geometry satisfying axiom (IV) a **finite-dimensional projective geometry**, which from this point on will be the sole object of our consideration. We call P a **projective space**. Consider sequences of subspaces of the type $P \supsetneqq P_{n-1} \supsetneqq \ldots \supsetneqq P_1 \supsetneqq P_0 \neq \varnothing$, where $\varnothing$ is the empty set. The number n of the longest sequence is called the **dimension** of P. If P is of dimension n, we write P^n instead of P. We call P^1 a **projective line** and P^2 a **projective plane**. Each subspace S of P, together with the set of lines of P contained in S, gives a finite-dimensional projective geometry, and so S is a projective space. Lines and points are projective spaces of dimensions 1 and 0, respectively. By convention, the empty set is a (-1)-dimensional projective space. We call each 2-dimensional subspace a **plane** and each $(n-1)$-dimensional subspace in P^n a **hyperplane**.

Let M, N be subspaces of P, and for a pair of points $p \in M$, $q \in N$ consider the set $p \cup q$ of all points on the line that contains p and q. The set $\{p \cup q \mid p \in M, q \in N\}$ is denoted by $M \cup N$, and we call it the set spanned by M and N. By convention, we put $\varnothing \cup M = M$ and $p \cup p = p$. Then $P^r \cup P^s$ is the projective space of the lowest dimension which contains P^r and P^s. On the other hand, if we denote the intersection of P^r and P^s by $P^r \cap P^s$, then it is the projective space of highest dimension that is contained in both of them. We call $P^r \cup P^s$ and $P^r \cap P^s$ the **join** and the **intersection** of P^r and P^s, respectively. When the dimension of the space spanned by $r+1$ points is r, we say that these points are **independent**; otherwise they are **dependent**. If any $r+1$ points of a given subset M of P^n are independent for each $r \leqslant n$, we say that points of M lie in a **general position**. The space P^r necessarily contains $r+1$ independent points, and there necessarily exists a P^r that contains $r+1$ arbitrary given points in a projective space; it is unique if the points are independent. If $P^r \cup P^s = P^t$ and $P^r \cap P^s = P^u$, then $r+s=t+u$. We call the latter the **dimension theorem** (or **intersection theorem**) of projective geometry.

The set Σ_1 of all hyperplanes that contain a P_0^{n-2} in P^n is called a **pencil of hyperplanes**, and the P_0^{n-2} common to them is called the **center** of Σ_1. If a pencil of hyperplanes contains two distinct hyperplanes of P^n, then the pencil is determined uniquely by these two. When $n=2$ and 3, it is called a **pencil of lines** and a **pencil of planes**, respectively. Each pencil of hyperplanes of P^n, or more generally, each pencil of hyperplanes of a subspace of an arbitrary dimension in P^n, is called a **linear fundamental figure** of P^n or simply a **fundamental figure**. In P^n, the set Σ_r of all $P^{n-r}, P^{n-r+1}, \ldots, P^{n-1}$ that contain the same P_0^{n-r-1} is called the **star** with center P_0^{n-r-1}. Each set that consists of the totality of subspaces of an arbitrary demension in the same P^r or a subset of it is called a **P^r-figure**.

Under the assumption that P^r and P^s do not have points in common, the operation of constructing $P^r \cup P^s$ from P^r and P^s is called **projecting** P^s from P^r. Assuming that P^r and P^s have points in common the operation of constructing $P^r \cap P^s$ from P^r and P^s is called **cutting** P^s by P^r. Suppose that we are given spaces P_0, P_1, and P_2, and a fundamental figure Σ in the space P_1. By projecting Σ from P_0 and then cutting it by P_2, we can construct a fundamental figure Σ' on P_2. This operation is called **projection** of Σ from P_0 onto P_2, and we call P_0 the **center of projection** (Fig. 2). In this case, we say that Σ and Σ' are **in perspective** and denote the relation by $\Sigma \overline{\wedge} \Sigma'$. If for two fundamental figures Σ and Σ' there exist a finite number of fundamental figures F_i $(1 \leqslant i \leqslant l)$ such that $\Sigma \overline{\wedge} F_1 \overline{\wedge} \ldots \overline{\wedge} F_l \overline{\wedge} \Sigma'$, then we say that Σ and Σ' are **projectively related** to each other and denote this by $\Sigma \barwedge \Sigma'$ (Fig. 3). Now for arbitrary subspaces P_1^r, P_2^r $(0 \leqslant r \leqslant n)$, we take P_0^{n-r-1} that have no points in common with them and project each point of P_1^r onto P_2^r from P_0^{n-r-1}. The one-to-one correspondence $P_1^r \rightarrow P_2^r$ thus obtained is called a **perspective mapping**. If a one-to-one correspondence $P_1^r \rightarrow P_2^r$ is represented as the composite of a finite number of perspective mappings, then we call it a **projective mapping**. These mappings are extended to those of fundamental figures, too.

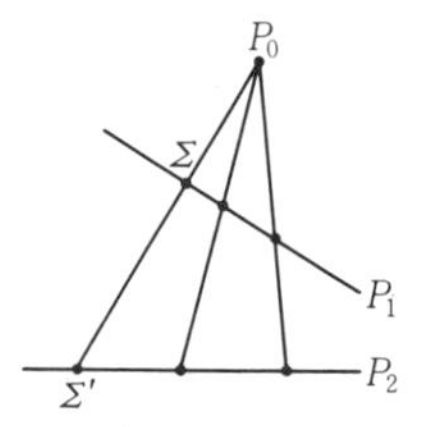

Fig. 2

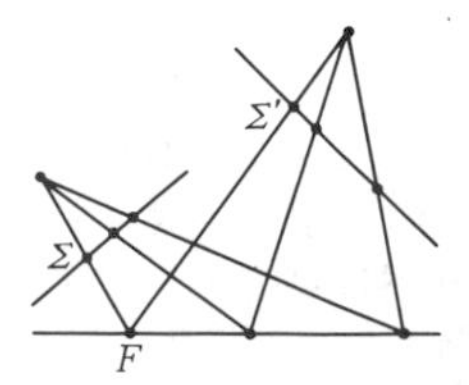

Fig. 3

Suppose that in a proposition or a figure in P^n, we interchange P^r and P^{n-r-1} $(0 \leqslant r \leqslant n)$ and also interchange *contains* and *is contained* (and related terms). The proposition or the figure thus obtained is said to be **dual** to the original one. In projective geometry, if a proposition is true, then its dual proposition is also true (**duality principle**). This is assured because propositions dual to axioms (I)–(IV) hold; and P^r and Σ_r are dual to each other. The projective space P_*^n obtained by the principle of duality, by regarding the hyperplanes of P^n as points of P_*^n, is called the **dual space** of P^n.

C. Projective Coordinates

Here we introduce **projective coordinates** in P^n. Consider **Desargues's theorem**: Suppose that p_1, p_2, p_3 and q_1, q_2, q_3, are two sets of points in P^n, each of which is independent and satisfies $p_i \neq q_i$ $(i=1,2,3)$. If the three lines $p_i \cup q_i$ $(i=1,2,3)$ are concurrent, then the three points $(p_2 \cup p_3) \cap (q_2 \cup q_3)$, $(p_3 \cup p_1) \cap (q_3 \cup q_1)$, $(p_1 \cup p_2) \cap (q_1 \cup q_2)$ are collinear. The converse is also true. This theorem holds for $n \geqslant 3$ generally. However, when $n=2$, there exist projective geometries for which it does not hold; we call these **non-Desarguesian geometries**. In such cases it is impossible to introduce coordinates, so we assume Desargues's theorem for $n=2$.

When four points p_i $(1 \leqslant i \leqslant 4)$ in P^n lie on the same plane and in general position, we call the figure that consists of these four points and the six straight lines $g_{ij} = p_i \cup p_j$ $(1 \leqslant i \leqslant j \leqslant 4)$ a **complete quadrangle** $p_1 p_2 p_3 p_4$; each p_i is called a **vertex**, and each g_{ij} is called a **side**. If six points q_i $(1 \leqslant i \leqslant 6)$ on a line l are points of intersection of six sides $g_{12}, g_{13}, g_{14}, g_{34}, g_{24}, g_{23}$ of a complete quadrangle with l, we call

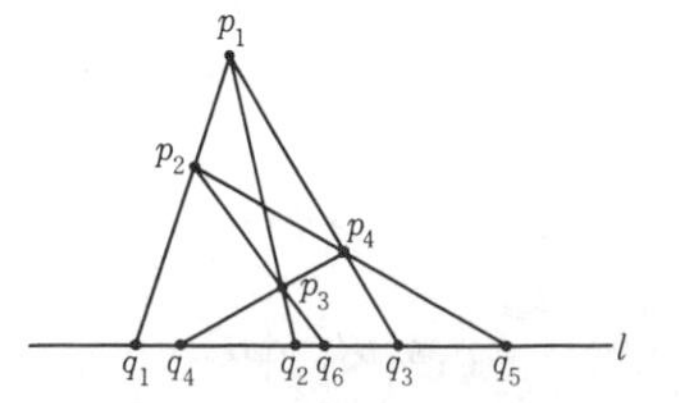

Fig. 4

them a **quadrangular set of six points** (Fig. 4). By Desargues's theorem, we can show that if there are given three fixed distinct points on a line l, then any pair of distinct points on l determines uniquely a point on l such that the six points thus obtained constitute a quadrangular set. The quadrangular property is invariant under projective mappings. On a line l we fix three mutually distinct points p_0, p_1, p_∞. For any two points p_x, p_y different from p_∞ on l, we take the point s such that $p_\infty, p_x, p_0, p_\infty, p_y, s$ constitute a quadrangular set of six points and call s the **sum** of p_x and p_y with respect to $[p_0, p_\infty, p_1]$ (Fig. 5). On the other hand, the point t such that $p_0, p_x, p_1, p_\infty, p_y, t$ constitute a quadrangular set of six points is called the **product** of p_x and p_y with respect to $[p_0, p_\infty, p_1]$ (Fig. 6). When we are given a fixed triple $[p_0, p_\infty, p_1]$ on a line l, as before, the set of points on l not equal to p_∞ is called a **point range of the number system**, provided that we exclude p_∞ from the point range. We call the set of three points $[p_0, p_\infty, p_1]$ a **frame** (or **projective frame**) of l, and we call p_0 the **origin**, p_1 the **unit point**, and p_∞ the **supporting point**.

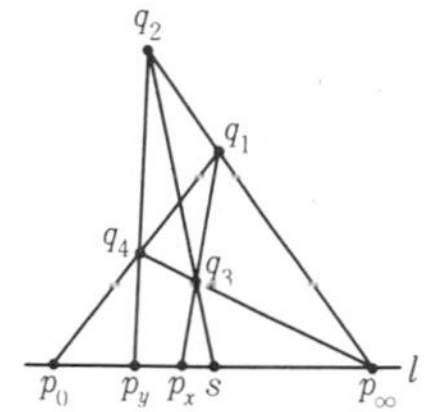

Fig. 5

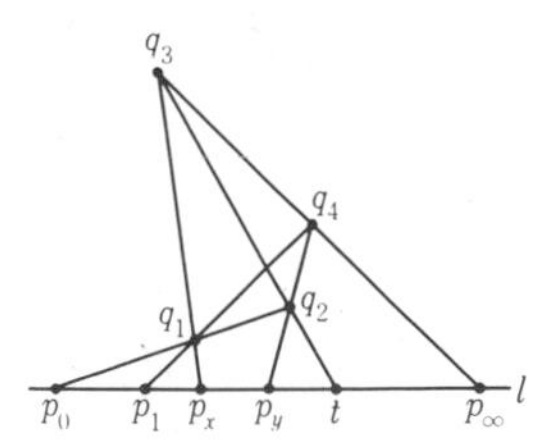

Fig. 6

A point range of the number system constitutes a †field (which may be noncommutative) with respect to the previously defined sum and product. We call the field a **Staudt algebra**, and an abstract algebra isomorphic to it is called a **coefficient field** of P^n. We denote by $K(p_0, p_\infty, p_1)$ the Staudt algebra that is determined by a frame $[p_0, p_\infty, p_1]$. A projective mapping of l onto itself that leaves invariant each of three distinct points p_0, p_∞, p_1 on l is necessarily an †inner automorphism of the field $K(p_0, p_\infty, p_1)$. Denoting by $[p_0, p_\infty, p_1]$ a frame on a line l of P^n with coefficient field K, we call each isomorphism $\theta: K(p_0, p_\infty, p_1) \rightarrow K$ a **coordinate system** of l. For each point p on l we call the element $\xi = \theta(p)$ of K the **inhomogeneous coordinate** of p with respect to this frame. Also, we call the pair (x^0, x^1) such that $x^0, x^1 \in K$ and $x^1(x^0)^{-1} = \xi$ **homogeneous coordinates** of p. Since the supporting point p_∞ is excluded from $K(p_0, p_\infty, p_1)$, we fix $(0, x^1)$ such that $x^1 \neq 0$ as the homogeneous coordinates of P_∞. In order for (x^0, x^1) and (y^0, y^1) to be homogeneous coordinates of the same point, it is necessary and sufficient that there exist an element $\lambda \neq 0$ of K such that $y^\alpha = x^\alpha \lambda$ $(\alpha = 0, 1)$.

In conformity with these results, we now introduce coordinates in P^n. A set $\mathfrak{F} = [a_0, a_1, \ldots, a_n, u]$ of ordered $n+2$ points in a general position is called a **frame** (or **projective frame**) of P^n; each of a_α $(0 \leqslant \alpha \leqslant n)$ is called a **fundamental point**, and u is called a **unit point**. For $A = a_{\alpha_0} \cup \ldots \cup a_{\alpha_r}$ $(0 \leqslant \alpha_0 < \ldots < \alpha_r \leqslant n)$, we denote by A^* the space spanned by the remaining fundamental points. For any point p that is not contained in A^*, we put $p_A = A \cap (p \cup A^*)$ and call it the **component** of p on A. Then $\mathfrak{F}_A = [a_{\alpha_0}, \ldots, a_{\alpha_r}, u_A]$ is a frame of A. Hereafter, we shall omit u_A for brevity. Suppose that isomorphisms $\theta_{\alpha\beta}: K(a_\alpha, a_\beta) \rightarrow K$ are assigned for each pair α, β $(0 \leqslant \alpha < \beta \leqslant n)$. Under a certain condition, the system $\{\theta_{\alpha\beta}\}$ is determined by one of the $\theta_{\alpha\beta}$. In this case we denote $\{\theta_{\alpha\beta}\}$ by θ and call $\{\mathfrak{F}, \theta\}$ a **projective coordinate system** of P^n. For any point p of P^n not contained in $A_0 = a_1 \cup \ldots \cup a_n$, we denote by p_i the component of p on $a_0 \cup a_i$ $(1 \leqslant i \leqslant n)$, and we put $\xi^i = \theta(p_i)$. The elements of the ordered set $(\xi^1, \xi^2, \ldots, \xi^n)$ are called the **inhomogeneous coordinates** of p with respect to $\mathfrak{F}$, and those of the set $(x^0, x^1, \ldots, x^n)$ such that $x^i(x^0)^{-1} = \xi^i$ are called the **homogeneous coordinates** of p. When p is contained in A_0, we define $(0, x^1, \ldots, x^n)$ as homogeneous coordinates of p with respect to $\mathfrak{F}$, provided that $(x^1, \ldots, x^n)$ are homogeneous coordinates of p with respect to $\mathfrak{F}_{A_0}$.

Now we represent the point whose coordinates are $(x^0, x^1, \ldots, x^n)$ simply by x. In P^n, when coordinates are introduced, a necessary and sufficient condition for points z to be on the line that passes through two distinct points x and y is that $z^\alpha = x^\alpha \lambda + y^\alpha \mu$ $(0 \leqslant \alpha \leqslant n)$, when $\lambda, \mu \in K$ are parameters. More generally, a point z is contained in the space spanned by $r+1$ independent points x_β $(0 \leqslant \beta \leqslant r)$ in P^n if and only if $z^\alpha = \sum_{\beta=0}^{r} x_\beta^\alpha \lambda^\beta$ $(0 \leqslant \alpha \leqslant n, \lambda^\beta \in K)$. In particular, the equation of a hyperplane is represented in the form $\sum_{\alpha=0}^{n} X_\alpha z^\alpha = 0$ $(X_\alpha \in K)$ with respect to variable coordinates z^α. Therefore each hyperplane is uniquely determined by the ratio of $X_0, X_1, \ldots, X_n$. We call $X_0, X_1, \ldots, X_n$ **hyperplane coordinates** of the hyperplane. If $n=2$, they are called **coordinates** of a line, and if $n=3$, **plane coordinates** of a plane. (For coordinates of P^r in P^n → 90 Coordinates B.)

D. Projective Transformations

A one-to-one correspondence φ between the point sets of two projective spaces P^n and $\bar{P}^n$ is called a **collineation in the wider sense** if for any three points p_1, p_2, p_3 that are collinear, $\varphi(p_i)$ $(i=1,2,3)$ are also collinear and vice versa. If $\bar{P}^n = P^n_*$, we call φ a **correlation**; if $\bar{P}^n = P^n$, we call φ a **collineation**. If we denote a correlation by τ_0, any other correlation is obtained as a composite of τ_0 and a collineation. If τ is a correlation, it naturally induces a mapping $P^n_* \to P^n$, which we also denote by τ. Then $\tau \circ \tau$ is a collineation. If $\tau \circ \tau$ is an identity, we call τ an **involutive correlation**. Suppose that $\varphi: P^n \to \bar{P}^n$ is a collineation in the wider sense and $0 \leqslant r \leqslant n-1$. Then φ induces a one-to-one correspondence between the set of r-dimensional subspaces of P^n and the set of r-dimensional subspaces of $\bar{P}^n$; and if $P^r \supset P^s$ in P^n, then $\varphi P^r \supset \varphi P^s$.

Next, suppose that we are given two projective spaces P^n and $\bar{P}^n$ that are subspaces of a space P^N $(n<N)$. (When Desargues's theorem holds, any two projective spaces of the same dimension can be identified with subspaces of a projective space of higher dimension.) In this case, when a collineation in the wider sense $\varphi: P^n \to \bar{P}^n$ is a projective mapping, we call it a **projective collineation in the wider sense**. A **projective collineation** is also called a **projective transformation**. The totality of collineations of P^n constitutes a †transformation group and is called the **group of collineations** of P^n; we denote it by $\mathfrak{C}(P^n)$. The totality of projective transformations of P^n constitutes a †normal subgroup of $\mathfrak{C}(P^n)$; we denote it by $\mathfrak{G}(P^n)$ and call it the **group of projective transformations**. The totality of projective transformations that leave invariant a frame $\mathfrak{F}$ of P^n constitutes a subgroup $\mathfrak{G}_0^{n+1}(\mathfrak{F})$. It is isomorphic to the group of †inner automorphisms $\mathfrak{J}(K)$ of the coefficient field K of P^n. A collineation is not necessarily a projective transformation. The former is obtained as a composite of a projective transformation and an automorphism of the coefficient field. Specifically, if we denote the group of †automorphisms of K by $\mathfrak{A}(K)$, then $\mathfrak{C}(P^n)/\mathfrak{G}(P^n) \cong \mathfrak{A}(K)/\mathfrak{J}(K)$. Hence in order for all collineations to be projective transformations, it is necessary and sufficient that all automorphisms of the coefficient field be inner automorphisms. If the coefficient field is the real number field, then collineations are always projective transformations. For the complex number field, however, this is not necessarily true.

Now, we consider the following three propositions: (1) The coefficient field of P^n is commutative. (2) Given frames $\mathfrak{F}$ and $\mathfrak{F}'$ of P^n, there exists a unique projective transformation sending $\mathfrak{F}$ onto $\mathfrak{F}'$. (3) Given two distinct lines l_1 and l_2 contained in a plane in P^n and two sets of three distinct points p_i $(i=1,2,3)$ and q_i $(i=1,2,3)$ that lie on l_1 and l_2 respectively, then the three points $(p_2 \cup q_3) \cap (p_3 \cup q_2)$, $(p_3 \cup q_1) \cap (p_1 \cup q_3)$, and $(p_1 \cup q_2) \cap (p_2 \cup q_1)$ are collinear. These three propositions are mutually equivalent. We call proposition (2) the **fundamental theorem of projective geometry** and proposition (3) the **theorem of Pappus**. If the coefficient field is the real (complex) number field, we call the projective space a **real (complex) projective space**. In classical geometry, only these cases were studied.

Suppose that the coefficient field is commutative. Then, if we assign an isomorphism $\theta_0: K(p_0, p_\infty, p_1) \to K$ for the Staudt algebra $K(p_0, p_\infty, p_1)$ on a line in a space, then the isomorphism θ of the Staudt algebra $K(q_0, q_\infty, q_1)$ on an arbitrary line onto K can be uniquely determined so that $\theta^{-1} \circ \theta_0$ is a projective mapping. Utilizing such isomorphisms, we can determine homogeneous coordinates in an arbitrary subspace of P^n by a frame on it.

Suppose that the coefficient field is a commutative field whose characteristic is not 2. For four collinear points p_i $(1 \leqslant i \leqslant 4)$ in P^n, where p_1, p_2, p_3 are distinct and $p_4 \neq p_1$, we consider a frame such that p_1, p_2, and p_3 are, respectively, the supporting point, the origin, and the unit point. The inhomogeneous coordinate λ of p_4 with respect to this frame is called the **anharmonic ratio (cross ratio** or **double ratio**) of these four points and is denoted by $[p_1, p_2; p_3, p_4]$. If we denote the inhomogeneous coordinates of p_i with respect to a general frame by (x_i^0, x_i^1) $(i=1,2,3,4)$, then λ can be expressed as

$$\lambda = [p_1, p_2; p_3, p_4] = \frac{(x_1^0 x_3^1 - x_3^0 x_1^1)(x_2^0 x_4^1 - x_4^0 x_2^1)}{(x_1^0 x_4^1 - x_4^0 x_1^1)(x_2^0 x_3^1 - x_3^0 x_2^1)}.$$

Moreover, if we interchange the order of the four points, then we have

$$\lambda = [p_2, p_1; p_4, p_3] = [p_3, p_4; p_1, p_2] = [p_4, p_3; p_2, p_1],$$

$$[p_1, p_2; p_4, p_3] = \frac{1}{\lambda},$$

$$[p_1, p_3; p_2, p_4] = 1 - \lambda,$$

$$[p_1, p_3; p_4, p_2] = \frac{1}{1-\lambda},$$

$$[p_1, p_4; p_2, p_3] = \frac{\lambda - 1}{\lambda},$$

$$[p_1, p_4; p_3, p_2] = \frac{\lambda}{\lambda - 1}.$$

In general, these six values are different; however, there are the following two exceptions: when $\lambda = -1$, $1/2$, and 2; and when λ is a root of $\lambda^2 - \lambda + 1 = 0$. When $\lambda = -1$, these four points are called a **harmonic range of points**, and the points p_3, p_4 are called **harmonic conjugates** with respect to p_1, p_2; or p_1, p_2 and p_3, p_4 are said to be **harmonically separated** from each other. When $\lambda^2 - \lambda + 1 = 0$, these four points are said to be an **equianharmonic range of points**. For the dual of these, we can consider the anharmonic ratio of four hyperplanes of a pencil of hyperplanes. The concept of the anharmonic ratio can be extended further to the case of four elements of fundamental figures in general. The anharmonic ratio is a quantity that is invariant under projective transformations.

Each projective transformation $x \to \bar{x}$ is expressed with respect to homogeneous coordinates x^α $(0 \leqslant \alpha \leqslant n)$ of P^n as

$$\rho \bar{x}^\alpha = \sum_{\beta=0}^{n} t^\alpha_\beta x^\beta, \qquad \rho, t^\alpha_\beta \in K,$$

$$\rho \neq 0, \qquad \det(t^\alpha_\beta) \neq 0. \tag{1}$$

Conversely, if $T = (t^\alpha_\beta)$ is a †regular matrix $(t^\alpha_\beta \in K)$, then (1) determines a projective transformation. So there is a one-to-one correspondence between projective transformations and †equivalence classes of the regular matrices $T = (t^\alpha_\beta)$ with the †equivalence relation $T \sim \lambda T$ $(\lambda \in K \setminus \{0\})$. Therefore, when K is commutative, the group of projective transformations $\mathfrak{G}(P^n)$ of P^n is isomorphic to the factor group $PGL(n+1, K)$ of the †general linear group $GL(n+1, K)$ with the coefficient field K by its center $\{\rho I \mid \rho \in K \setminus \{0\}\}$; that is, $\mathfrak{G}(P^n) \cong PGL(n+1, K)$.

Extending the definition of projective transformations, we call the transformation represented by (1), with an arbitrary square matrix that is not necessarily regular, a projective transformation. When T is regular, it is called a **regular projective transformation**, and when T is not regular, it is called a **singular projective transformation**. In particular, if the †rank of T is $n+1-h$, then we say that the projective transformation is **singular of the hth species**. If (1) is singular of the hth species, $n+1$ hyperplanes $\sum_{\beta=0}^{n} t^\alpha_\beta x^\beta = 0$ $(0 \leqslant \alpha \leqslant n)$ have a space P^{h-1} in common. We call P^{h-1} the **singular subspace** of this transformation. A projective transformation is not defined on its singular subspace. A singular projective transformation of the hth species is the composite of the projection of P^n onto some P^{n-h} with the singular subspace as its center and a regular projective transformation of P^{n-h}.

If the coordinates of a point are denoted by (x^α) and hyperplane coordinates with respect to some frame by (X_α), then the linear transformation

$$\tau_* : \rho X_\alpha = \sum_{\beta=0}^{n} t^*_{\alpha\beta} x^\beta, \qquad \rho, t^*_{\alpha\beta} \in K, \tag{2}$$

is a correlation. (Here also, we extend the definition of correlation and include the case where $T_* = (t^*_{\alpha\beta})$ is not regular.) The condition that τ_* is an involutive correlation is given by $T_* = \pm {}^t T_*$. When $T_* = -{}^t T_*$, the involutive correlation τ_* is called a **null system**. The correlation τ_* is a null system if and only if any point x of P^n is contained in the hyperplane $\tau_*(x)$. When $T_* = {}^t T_*$, we call the involutive correlation τ_* a **polar system**. For a polar system τ_*, the set of points x that are contained in hyperplanes $\tau_*(x)$ constitutes a **quadric hypersurface** (or **hyperquadric**).

E. Quadric Hypersurfaces

Let τ_* be a polar system, and let Q_2^{n-1} be the totality of points x contained in $\tau_*(x)$. Then the equation of the quadric hypersurface Q_2^{n-1} is given by

$$\sum_{\alpha,\beta=0}^{n} t^*_{\alpha\beta} x^\alpha x^\beta = 0. \tag{3}$$

For such a correlation τ_* we call a relation between the set of points x of P^n and the set of hyperplanes $\tau_*(x)$ a **polarity** with respect to Q_2^{n-1}. We call $\tau_*(x)$ the **polar** of x with respect to Q_2^{n-1}, and x the **pole** of $\tau_*(x)$ with respect to Q_2^{n-1}. If the points of intersection of a line passing through a point x with Q_2^{n-1} and $\tau_*(x)$ are denoted by z_1, z_2; y, then x, y; z_1, z_2 is a harmonic range of points. When a point x lies on the polar of a point y, we say that x and y are mutually **conjugate**. Each point on Q_2^{n-1} is conjugate with itself, and the converse is also true. We call the polar of a point on Q_2^{n-1} the **tangent hyperplane** of Q_2^{n-1} at that point.

If τ_* is regular or singular of the hth species, we call the corresponding quadric hypersurface **regular** or **singular of the hth species.** If τ_* is singular of the hth species, its singular subspace is contained in Q_2^{n-1}. We call points on this singular subspace **singular points** of Q_2^{n-1}. Q_2^{n-1}, which is singular of the first species (i.e., Q_2^{n-1} with just one singular point), is called a **cone**.

We call a subspace contained in Q_2^{n-1} a **generating space**. If it is a line we call it a **generating line**. We put $q = (n-2)/2$ or $(n-1)/2$ according as n is even or odd. Then, if the coefficient field is an †algebraically closed field, for each regular Q_2^{n-1} there necessarily exist q-dimensional generating spaces. Also, Q_2^2 is a †ruled surface covered by two families of generating lines, and Q_2^{2k} is covered by two families of k-dimensional generating spaces.

If p_i $(1 \leqslant i \leqslant 5)$ are five points in a general position in a plane, then there exists one and only one Q_2^1 passing through these points; we call Q_2^1 a **conic**. In order that six points p_i $(1 \leqslant i \leqslant 6)$ in a plane lie on Q_2^1, it is necessary and sufficient that the three points $(p_1 \cup p_2) \cap (p_4 \cup p_5)$, $(p_2 \cup p_3) \cap (p_5 \cup p_6)$, and $(p_3 \cup p_4) \cap (p_6 \cup p_1)$ be collinear (**Pascal's theorem**). The dual of the last theorem is called **Brianchon's theorem**.

Given two hypersurfaces Q_2^{n-1}, $\bar{Q}_2^{n-1}$ in P^n, we consider another $\bar{\bar{Q}}_2^{n-1}$ such that the polar of an arbitrary point x with respect to $\bar{\bar{Q}}_2^{n-1}$ belongs to the pencil of hyperplanes determined by polars of x with respect to Q_2^{n-1} and $\bar{Q}_2^{n-1}$. The set of all such $\bar{\bar{Q}}_2^{n-1}$ is called a **pencil of quadric hypersurfaces**. It is the set of all $\bar{\bar{Q}}_2^{n-1}$ that pass through the intersection of Q_2^{n-1} and $\bar{Q}_2^{n-1}$. In the cases $n=2$ and 3, we call it a **pencil of conics** and a **pencil of quadrics**, respectively.

Denoting by l^{ij} and $\bar{l}^{ij}$ $(0 \leqslant i \leqslant j \leqslant 3)$ the †Plücker coordinates of two straight lines l and $\bar{l}$ in P^3, we put

$$(l,\bar{l}) = l^{01}\bar{l}^{23} - l^{02}\bar{l}^{13} + l^{03}\bar{l}^{12} + \bar{l}^{01}l^{23} - \bar{l}^{02}l^{13} + \bar{l}^{03}l^{12}. \qquad (4)$$

Then $(l,l)=0$ holds. If we regard these l^{ij} as homogeneous coordinates of P^5, then there exists a one-to-one correspondence between the points on the regular quadric hypersurface Q_2^4 defined by $(l,l)=0$ and the lines in P^3. We say that each point of Q_2^4 is the **image** of the line corresponding to it in P^3. Two lines l and $\bar{l}$ intersect if and only if $(l,\bar{l})=0$. Geometrically, this means that the images of l and $\bar{l}$ are conjugate with respect to Q_2^4. Therefore the line passing through the images of l and $\bar{l}$ is a generating line of Q_2^4. The image of a pencil of lines in P^3 is a generating line of Q_2^4. Quadric hypersurfaces and sets of lines in P^3 are important objects of study in both projective and algebraic geometry. In particular, **linear line congruences (linear line complexes)** that are families of lines dependent upon two (three) parameters are of great interest. In these theories, quadric hypersurfaces play a fundamental role. When the coefficient field is noncommutative, the above theory has to be greatly modified.

F. Projective Geometry and the Erlangen Program

From the standpoint of the †Erlangen program of F. Klein, the aim of projective geometry is to study properties that are invariant under the group of projective transformations. Utilizing various subgroups of this group, we can reconstruct various classical geometries. For example, consider the projective space P^n whose coefficient field is the real number field, and fix a hyperplane Π_∞. Let $\mathfrak{S}(P^n)$ be the subgroup of projective transformations formed by all projective transformations that leave Π_∞ invariant. Then the geometry that belongs to this group is †affine geometry. Similarly, fix an imaginary regular quadric hypersurface Q_2^{n-2} in Π_∞ and consider the geometry that belongs to the subgroup of $\mathfrak{S}(P^n)$ leaving this Q_2^{n-2} invariant. We thus obtain Euclidean geometry. Moreover, if we assign some regular quadric hypersurface Q_2^{n-1}, then the geometry belonging to the subgroup of $\mathfrak{G}(P^n)$ that leaves the Q_2^{n-1} invariant is a †non-Euclidean or †conformal geometry according as the transformation space is the set of inner points of Q_2^{n-1} or the whole Q_2^{n-1}.

G. Projective Geometry and Modular Lattices

†Lattices (lattice-ordered sets) and projective geometry are intimately related. The totality of subspaces of each dimension in general projective geometry $\mathfrak{P}$ constitutes a †complete †modular lattice $L(\mathfrak{P})$ with respect to the inclusion relation. If $\mathfrak{P}$ is a finite-dimensional projective geometry, then it is an †irreducible complemented modular lattice of finite †height. Conversely, suppose that L is a modular lattice with †minimum element Φ, and denote by P the totality of elements p †prime over Φ (i.e., †atomic elements) and by Q the totality of elements l prime over atomic elements. Then, if $p<l$ and $(p,l) \in \Gamma$, $\mathfrak{P}(L)=\{P,Q,\Gamma\}$ is a general projective geometry. If L is an irreducible complemented modular lattice of finite height, then $\mathfrak{P}(L)$ is a finite-dimensional projective geometry; in this case we have $\mathfrak{P} \approx \mathfrak{P}(L(\mathfrak{P}))$ and $L \approx L(\mathfrak{P}(L))$. So we may consider projective geometry and irreducible complemented modular lattices as having the same mathematical structure. If a lattice L is an n-dimensional projective geometry, its †dual lattice is also an n-dimensional projective geometry, and this is the principle of duality.

H. Analytic Representations of Projective Geometry

Let K be an arbitrary field, commutative or noncommutative. For an arbitrary natural number n, we consider an $(n+1)$-dimensional (for the noncommutative case, right or left) linear space $V^{n+1}(K)$ over K. The totality of linear subspaces in it constitutes an irreducible complemented modular lattice $P^n(K)$ with respect to the inclusion relation, and $P^n(K)$ gives rise to an n-dimensional projective geometry. We call it a **right** or **left projective space**. Points of $P^n(K)$ correspond to (right or left) 1-dimensional linear subspaces. Conversely, it can be shown that an n-dimensional projective geometry over K is isomorphic to $P^n(K)$. Therefore projective geometries can be completely classified by means of the natural number n and the field K except when $n=2$ and the geometry is non-Desarguesian. We may restate this fact as follows: We consider a space $\tilde{P}=V^{n+1}(K)-\{0\}$. If we fix a basis of $V^{n+1}(K)$, then we can represent $\tilde{P}=\{x=(x^0, x^1, \ldots, x^n) \mid x^\alpha \in K, 0 \leqslant \alpha \leqslant n\}$, $x \neq (0,0,\ldots,0)$. If there exists a nonzero element λ of K such that $y=x\lambda$, then the elements x and y are called equivalent; we write $x \sim y$. We denote by $\mathbf{P}^n(K)$ the factor set of $\tilde{P}$ under the foregoing equivalence relation, and by $[x]$ the equivalence class that contains x. We put $l([x],[y])=\{[z] \mid z^\alpha = x^\alpha\lambda + y^\alpha\mu, \forall\lambda, \mu \in K\}$ and $Q=\{l([x],[y]) \mid [x], [y] \in \mathbf{P}^n(K)\}$. We call each element of $\mathbf{P}^n(K)$ a point and each element of Q a line. Then these points and lines and the natural inclusion relation satisfy axioms (I)–(IV) and give an n-dimensional projective geometry. When K is a †topological field (e.g., the real number field, the complex number field, or the †quaternion field), we may define the topology of $\mathbf{P}^n(K)$ as the factor space $\mathbf{P}^n(K)=\tilde{P}/\sim$. In particular, if K is the real number field $\mathbf{R}$, then $\mathbf{P}^n(\mathbf{R})$ is homeomorphic to the factor space obtained from the n-dimensional hypersphere $S^n:(x^0)^2+\ldots+(x^n)^2=1$ in the $(n+1)$-dimensional Euclidean space E^{n+1} by identifying the end points of each diameter. Hence $\mathbf{P}^n(\mathbf{R})$ is compact. Similar facts hold for the cases of the complex and quaternion number fields. Since the group of projective transformations $\mathfrak{G}(\mathbf{P}^n(K))$ acts †transitively on $\mathbf{P}^n(K)$, if K is a topological field we can regard $\mathbf{P}^n(K)$ as a †homogeneous space of the topological group $\mathfrak{G}(\mathbf{P}^n(K))$. Moreover, the totality of r-dimensional subspaces in $\mathbf{P}^n(K)$ constitutes a †Grassmann manifold. In algebraic geometry the †direct product of two projective spaces is important; we call it a **biprojective space**.

I. Tits's Theory of Buildings (Generalization of Projective Geometry)

In a situation when a triple (G, B, N) consisting of a group G and its subgroups B, N satisfies the axioms of a BN**-pair** or **Tits system** (→ 13 Algebraic Groups R), a new geometric object, called a "building," was introduced by J. Tits [9]. His theory contains projective geometry as a particular case. The theory of buildings has deep connection with algebraic groups. The Tits system corresponds to a projective geometry in the following case. Let k be any commutative field, and let G be the general linear group of degree n over k, i.e., G consists of all nonsingular square matrices of degree n with entries in k. Let B be the subgroup of G consisting of all upper triangular matrices (i.e., matrices whose entries below the principal diagonal are all zero). Let N be the subgroup of G consisting of all monomial matrices (i.e., matrices such that each column and each row contain just one nonzero entry). Then (G, B, N) forms a Tits system called type (A_{n-1}). The corresponding **theory of buildings** of the type above is nothing but the projective geometry. Thus by means of Tits's theory of buildings the relationships among projective geometry and other geometries have been clarified [9].

References

[1] G. Birkhoff, Lattice theory, Amer. Math. Soc. Colloq. Publ., revised edition, 1967.
[2] W. V. D. Hodge and D. Pedoe, Methods of algebraic geometry I, Cambridge, 1947.
[3] O. Schreier and E. Sperner, Einführung in die analytische Geometrie und Algebra, Vandenhoeck & Ruprecht, II, 1951; English translation, Projective geometry of n dimensions, Chelsea, 1961.
[4] O. Veblen and J. W. Young, Projective geometry I, II, Ginn, 1910–1938.
[5] E. Artin, Geometric algebra, Interscience, 1957.
[6] S. Iyanaga and K. Matsuzaka, Affine geometry and projective geometry, J. Fac. Sci. Univ. Tokyo, 14 (1967), 171–196.
[7] A. Seidenberg, Lectures in projective geometry, Van Nostrand, 1962.
[8] R. Hartshorne, Foundations of projective geometry, Benjamin, 1967.
[9] J. Tits, Buildings of spherical type and finite BN-pairs, Lecture notes in math. 386, Springer, 1974.

344 (VII.22) Pseudoconformal Geometry

A. Definitions

Let A and A' be subsets (with relative topology) of †complex manifolds X and X' of dimension n, respectively. A homeomorphism f of A onto A' is called a **pseudoconformal transformation** if there exists a †biholomorphic mapping $\tilde{f}$ of an open neighborhood of A in X onto an open neighborhood of A' in X' such that $\tilde{f}(x)=f(x)$ for $x \in A$. If there exists such a mapping f, A is said to be **pseudoconformally equivalent** to A'. **Pseudoconformal geometry** is a geometry that studies geometric properties invariant under the pseudoconformal equivalence. However, most studies in pseudoconformal geometry so far have concentrated mainly on the investigation of smooth hypersurfaces in a complex manifold—more specifically, the smooth (or real analytic) boundaries of bounded domains in $\mathbf{C}^n$. In fact, to pseudoconformal geometry on hypersurfaces we can apply the methods of differential geometry as well as those of the theory of functions of several complex variables.

H. Poincaré [1] studied perturbations of the boundary of the unit ball in $\mathbf{C}^2$ that are pseudoconformally equivalent. E. Cartan [2] studied the equivalence problem of hypersurfaces in $\mathbf{C}^2$ and gave the complete list of all simply connected hypersurfaces on which the group of pseudoconformal automorphisms acts transitively. Such a hypersurface is called **homogeneous**.

Let M be a smooth hypersurface in a complex manifold X with the †almost complex structure tensor J, i.e., $J_x: T_xX \rightarrow T_xX$ is an involutive linear automorphism of the tangent space T_xX of X at x induced by the complex structure of X. Put $H_xM = T_xM \cap J_xT_xM$ for $x \in M$. The union of all H_xM is called the bundle of holomorphic tangent vectors of M and is denoted by $H(M)$. $H(M)$ is also called the **CR** (Cauchy-Riemann) **structure** of M. Let M' be a smooth hypersurface in a complex manifold X'. A diffeomorphism $f: M \rightarrow M'$ is called a **CR-equivalence** if the †differential mapping $Tf: TM \rightarrow TM'$ of f preserves the CR-structures, where TM denotes the †tangent bundle of M. If $f: M \rightarrow M'$ is a pseudoconformal transformation, then f is clearly a CR-equivalence. Let E_x be the annihilator of H_xM in $T_x^*(M)$. Then the union of E_x $(x \in M)$ defines a †line bundle E over M. The **Levi form** L_x at $x \in M$, defined only up to a multiplier, is the quadratic form on H_xM defined by $L_x(u,v) = d\theta(u,v)$ for $u, v \in H_xM$, where θ is a nonvanishing section of E in a neighborhood of x. If the Levi form is nondegenerate at every point of M, M is called a **nondegenerate hypersurface**. In particular, if the Levi form is definite, then M is called **strictly pseudoconvex**.

B. Equivalence Problem

Cartan studied the equivalence problem for the case $n=2$, and obtained a criterion for two hypersurfaces in $\mathbf{C}^2$ to be pseudoconformally equivalent. N. Tanaka (1965) generalized the method of Cartan for the case $n \geqslant 3$ and obtained a criterion in terms of †Cartan connections in some fiber bundle over the hypersurfaces. However, he did not publish the proof of his result until S. S. Chern and J. Moser [4], independently of Tanaka [3], obtained a similar result and gave the first proof of this result. Let M be a real analytic hypersurface in $\mathbf{C}^{n+1}$ $(n \geqslant 1)$ whose Levi form has p positive and q negative eigenvalues $(p+q=n)$. Let H be the subgroup of $SU(p+1, q+1)$ leaving the point $(1, 0, \ldots, 0) \in \mathbf{C}^{n+2}$ fixed. According to the Cartan-Tanaka-Chern-Moser result, we can construct functorially a principal fiber bundle Y over M with structure group H and a Cartan connection ω on Y with values in the Lie algebra of $SU(p+1, q+1)$ such that if M and M' are pseudoconformally equivalent, then there is a bundle isomorphism φ of Y to Y' preserving the Cartan connections: $\varphi^*\omega' = \omega$, where Y' is the corresponding principal fiber bundle over M' and ω' is the Cartan connection on Y'. Conversely, if there is a bundle isomorphism φ of Y to Y' such that $\varphi^*\omega' = \omega$, then M and M' are pseudoconformally equivalent. By using this solution of the equivalence problem, we can prove that the group $A(M)$ of all pseudoconformal automorphisms of a nondegenerate real analytic hypersurface M in a complex manifold X of dimension n is a Lie transformation group of dimension not exceeding n^2+2n. H. Jacobowitz [5] constructed a similar bundle B over M and a Cartan connection on B in a different way from that of Chern and Moser. We do not know whether B and Y actually coincide.

C. Classification

Cartan (1932) classified all simply connected homogeneous hypersurfaces in $\mathbf{C}^2$. In particular, he proved that if M is a compact homogeneous strictly pseudoconvex hypersurface with $\dim M = 3$, then M is pseudoconformally equivalent to either (1) S^3 or its quotient by the action of a root of unity or (2) the hypersurface given in the 2-dimensional projective space by the equation in homogeneous coordinates: $(z_1\bar{z}_1 + z_2\bar{z}_2 + z_3\bar{z}_3)^2 = m^2|z_1^2 + z_2^2 + z_3^2|^2$ $(m > 1)$ or the double covering of such a surface. A. Morimoto and T. Nagano [6] and later H. Rossi [7] tried to generalize this result and obtained a partial classification of simply connected compact homogeneous hypersurfaces with dimension $\geqslant 5$. D. Burns and S. Shnider [8] classified all simply connected compact homogeneous strictly pseudoconvex hypersurfaces M with $\dim M = 2n+1 \geqslant 5$. They proved that M is pseudoconformally equivalent to S^{2n+1} or the tangent sphere bundle of a rank one †symmetric space or the unit circle bundle of a homogeneous negative line bundle over a homogeneous algebraic manifold.

On the other hand, a real hypersurface M in a complex manifold X of complex dimension $n+1$ is called **spherical** if at every point $p \in M$, there is a neighborhood of p in X such that $U \cap M$ is pseudoconformally equivalent to an open submanifold of S^{2n+1}. The hyperquadric $Q^{n+1} = \partial U_{n+1}$ is spherical, where $U_{n+1} = \{(z_1, \ldots, z_{n+1}) \in \mathbf{C}^{n+1} \mid \mathrm{Im}(z_{n+1}) > |z_1|^2 + \ldots + |z_n|^2\}$. If M is spherical, then the universal covering space $\tilde{M}$ of M is also spherical. If M is a homogeneous spherical hypersurface, then there is a covering into mapping $f: \tilde{M} \to S^{2n+1}$, and $f(\tilde{M})$ is a homogeneous domain in S^{2n+1}. We know that the only compact simply connected spherical M is S^{2n+1}. Burns and Shnider classified all homogeneous domains M in S^{2n+1}: M is pseudoconformally equivalent to (I) or (II) of the following: (Ia) $S^{2n+1} - V \cap S^{2n+1}$, where V is a complex vector subspace of $\mathbf{C}^{n+1}$ with $0 \leqslant \dim_{\mathbf{C}} V \leqslant n$. (Ib) $Q^{n+1} - L_m \cdot 0$, $0 \leqslant m \leqslant n$, where L_m is a certain subgroup of $SU(n+1,1)/(\text{center})$. (II) $S^{2n+1} - S^{2n+1} \cap \mathbf{R}^{n+1}$. At present, it seems difficult to extend Cartan's classification of all simply connected homogeneous hypersurfaces to higher dimensions. K. Yamaguchi (1976) treated a hypersurface M in a complex manifold of dimension n with a large automorphism group $A(M)$. He showed that if $\dim A(M) = n^2 + 2n$, then M is a real hyperquadric in the n-dimensional complex projective space $P_n\mathbf{C}$ (→ Section B). He then showed that the second largest dimension for $A(M)$ is equal to $n^2 + 1$ except when $n = 3$ and the index $r = 1$, for which $\dim A(M) = 11 (= n^2 + 1)$. Under the additional assumption that M is homogeneous, he showed that if $\dim A(M) = n^2 + 1$, then M is the affine part of a real hyperquadric in $P_n\mathbf{C}$ (except when $n = 5$ and $r = 2$). He also obtained a similar result in the nonhomogeneous case.

D. Relations to Other Equivalences

Let D_1 and D_2 be bounded domains in $\mathbf{C}^n$ with smooth boundary $\partial D_i = M_i$ $(i = 1, 2)$ for which we denote by $H(M_i)$ the CR structures. We consider the following propositions (A)–(D) for these domains: (A) D_1 is biholomorphically equivalent to D_2. (B) M_1 is CR equivalent to M_2. (C) M_1 is pseudoconformally equivalent to M_2. (D) There is a diffeomorphism $f: \bar{D}_1 \to \bar{D}_2$ such that $f|_{D_1}: D_1 \to D_2$ is biholomorphic.

It is clear that (C) implies (D) and that (D) implies (A). On the other hand, we can prove that (B) is equivalent to (D). When does (A) imply (B) and when does (B) imply (C)?

C. Fefferman [10] proved that (A) implies (D) when D_1 and D_2 are strictly pseudoconvex. S. Bell generalized the result of Fefferman in the case when one of D_1 and D_2 is strictly pseudoconvex. Bell and E. Ligocka [11] proved that if M_1 and M_2 are real analytic and if D_1 and D_2 are pseudoconvex, then (A) implies (D). When D_1 and D_2 are not strictly pseudoconvex and M_i is not real analytic, we do not know whether (A) implies (D) or not. As remarked by Burns, Shnider, and Wells (1978), by using the theorem of Fefferman, we can prove that (A) implies (C) when M_1 and M_2 are real analytic and if D_1 and D_2 are strictly pseudoconvex. I. Naruki [12] obtained the same result. We do not know whether (A) implies (C) when M_1 and M_2 are real analytic and D_1 and D_2 are pseudoconvex, though we know that (A) implies (B). We do not know whether (B) implies (C) in general.

S. I. Pinchuk [13] proved the following: Let D, D' be strictly pseudoconvex domains in $\mathbf{C}^n$ with simply connected real analytic boundaries ∂D, $\partial D'$. Let $f: U \to \mathbf{C}^n$ be a nonconstant holomorphic mapping from a connected neighborhood U of a point $p \in \partial D$ in $\mathbf{C}^n$ into $\mathbf{C}^n$ such that $f(U \cap \partial D) \subset \partial D'$. Then we can find a holomorphic mapping $\tilde{f}: D \to D'$ such that $\tilde{f}(x) = f(x)$ for $x \in D \cap U$. Combining this theorem with Fefferman's result we see that for two domains as above, D is biholomorphically equivalent to D' if and only if ∂D is locally pseudoconformally equivalent to $\partial D'$, i.e, there are neighborhoods U and V of a point $p \in \partial D$ and $q \in \partial D'$, respectively, such that $U \cap \partial D$ and $V \cap \partial D'$ are pseudoconformally equivalent.

Concerning the †proper holomorphic mappings rather than diffeomorphisms, Burns and Shnider (1979) proved the following theorem: Let $M_i = \partial D_i$ $(i = 1, 2)$ be strictly pseudoconvex, and let $f: D_1 \to D_2$ be a proper holomorphic mapping. (a) If $D_1 = D_2$, then f extends smoothly up to the boundary D_1. (b) If ∂D_i is real analytic for $i = 1, 2$, then f extends holomorphically past the boundary.

E. Deformations of Domains

Let M be a compact connected strictly pseudoconvex real hypersurface in a complex manifold X of dimension $n+1$. Let φ be a smooth strictly †plurisubharmonic function defined on a neighborhood V of M such that $M = \{x \in V \mid \varphi(x) = 0\}$ and $d\varphi \neq 0$ on M. Let $U = \{x \in V \mid -\varepsilon < \varphi(x) < \varepsilon\}$ for small $\varepsilon > 0$ such that $\bar{U}$ is compact and ∂U is smooth. Let $\mathscr{P}(\bar{U})$ be the open set in $\mathbf{C}^\infty(\bar{U})$ of strictly plurisubharmonic functions ψ with $d\psi \wedge \bar{d}\psi \wedge (d\bar{d}\psi)^n \neq 0$ on $\bar{U}$. Let $B \subset \mathbf{R}^k$ be a small open ball around 0. We denote by $\mathscr{P}(\bar{U} \times \bar{B})$ the set of $\psi \in \mathbf{C}^\infty(\bar{U}$

$\times \bar{B})$ such that $\psi(x,t)=\psi_t(x)\in\mathscr{P}(\bar{U})$ for all $t\in\bar{B}$. For $\psi\in\mathscr{P}(\bar{U}\times\bar{B})$ we set $M_{t,\delta}=\{x\in U \mid \psi_t(x)=\delta\}$. After introducing these notations, Burns, Shnider, and Wells (1978) proved the following theorem. There exists an open dense set $\mathscr{V}\subset\mathscr{P}(\bar{U}\times\bar{B})$ with $\varphi\in\mathscr{V}$ and a set of †second category $\mathscr{R}\subset\mathscr{V}$ such that for every $\psi\in\mathscr{R}$, $t_i\in\bar{B}$ and $\delta_i\in\mathbf{R}$ small enough, (i) M_{t_1,δ_1} is CR-equivalent to M_{t_2,δ_2} if and only if $t_1=t_2$, $\delta_1=\delta_2$. (ii) The group of CR-automorphisms of M_{t_1,δ_1} reduces to the identity only. For $\psi\in\mathscr{P}(\bar{U}\times\bar{B})$, taking $t\in B$, $\delta\in\mathbf{R}$ small enough, $M_{t,\delta}$ is a compact connected strictly pseudoconvex hypersurface in X. If M bounds the relatively compact region D in X then $M_{t,\delta}$ also bounds a relatively compact region $D_{t,\delta}$. In particular, there exist smooth families of deformations of the unit ball in $\mathbf{C}^{n+1}$ of arbitrary high dimension. There are arbitrary small perturbations of the unit sphere in $\mathbf{C}^{n+1}$ that admit no pseudoconformal transformations other than the identity.

F. Topics Related to Pseudoconformal Geometry

(1) Pinchuk (1975) proved the following: Let D_1, D_2 be strictly pseudoconvex domains in $\mathbf{C}^n$ with C^ω boundary ∂D_1, ∂D_2. Let U be a neighborhood of a point $p\in\partial D_1$ in $\mathbf{C}^n$. If there is a C^1-mapping $f: U\cap\bar{D}_1\to C^n$ such that f is holomorphic on $U\cap D_1$ and $f(U\cap\partial D_1)\subset\partial D_2$, then there is a holomorphic mapping $\tilde{f}: U'\to\mathbf{C}^n$ of a neighborhood U' of $U\cap\partial D_1$ into C^n such that $\tilde{f}(x)=f(x)$ for $x\in U\cap\bar{D}_1\cap U'$. This result is related to the implication (B)$\Rightarrow$(C) in Section D.

(2) H. Alexander [14] proved the following: Let U be a connected neighborhood of a point $p\in S^{2n-1}$ in $\mathbf{C}^n$ and $f: U\to\mathbf{C}^n$ a holomorphic mapping such that $f(U\cap S^{2n-1})\subset S^{2n-1}$. Then either f is a constant mapping or there is a biholomorphic automorphism $\tilde{f}: B_n\to B_n$ of the unit open ball B_n such that $\tilde{f}(x)=f(x)$ for $x\in U\cap B_n$. He also proved that every proper holomorphic mapping $f: B_n\to B_n$ is necessarily an automorphism of B_n if $n>1$. G. M. Henkin (1973) proved that every proper holomorphic mapping $f: D_1\to D_2$ of a strictly pseudoconvex domain D_1 into a strictly pseudoconvex D_2 can be extended continuously to a function $\tilde{f}: \bar{D}_1\to\bar{D}_2$. More precisely, there is a constant $c>0$ such that $|f(z_1)-f(z_2)|\leqslant c|z_1-z_2|^{1/2}$ for every $z_1, z_2\in D_1$.

(3) Let M be a real hypersurface in $\mathbf{C}^{n+1}$ with $H(M)$ the bundle of holomorphic tangent vectors to M. We take a real nonvanishing 1-form θ that annihilates $H(M)$. S. M. Webster (1978) called the pair (M,θ) a **pseudo-Hermitian manifold**. He considered the equivalence problem of pseudo-Hermitian manifolds by applying Cartan's method of equivalence. He proves, among other things, that the group of all pseudo-Hermitian transformations of the nondegenerate pseudo-Hermitian manifold (M,θ) of dimension $2n+1$ is a Lie transformation group of dimension not exceeding $(n+1)^2$. Webster considered the relation between pseudo-Hermitian manifolds and pseudoconformal geometry and proved that for $n\geqslant 2$ the ellipsoid E given by the equation $A_1x_1^2+B_1y_1^2+\ldots+A_{n+1}x_{n+1}^2+B_{n+1}y_{n+1}^2=1$, where $z_k=x_k+iy_k$ $(k=1,\ldots,n+1)$ is pseudoconformally equivalent to the hypersphere S^{2n+1} if and only if $A_k=B_k$ $(k=1,\ldots,n+1)$. This result gives, by virtue of Fefferman's theorem, a necessary and sufficient condition for an ellipsoidal domain to be biholomorphically equivalent to the unit ball.

References

[1] H. Poincaré, Les fonctions analytiques de deux variables et la représentation conforme, Rend. Circ. Mat. Palermo, 23 (1907), 185–220.

[2] E. Cartan, Sur la géométrie pseudo-conforme des hypersurfaces de l'espace de deux variables complexes I, II, Ann. Mat. Pura Appl., 11 (1932), 17–90; Ann. Scuola Norm. Sup. Pisa, 1 (1932), 333–354.

[3] N. Tanaka, On nondegenerate real hypersurfaces, graded Lie algebras and Cartan connections, Japan. J. Math., 2 (1976), 131–190.

[4] S. S. Chern and J. Moser, Real hypersurfaces in complex manifolds, Acta Math., 133 (1974), 219–271.

[5] H. Jacobowitz, Induced connections on hypersurfaces in $\mathbf{C}^n$, Inventiones Math., 43 (1977), 109–123.

[6] A. Morimoto and T. Nagano, On pseudo-conformal transformations of hypersurfaces, J. Math. Soc. Japan, 14 (1963), 289–300.

[7] H. Rossi, Homogeneous strongly pseudo-convex hypersurfaces, Rice Univ. Studies, 59 (1973), 131–145.

[8] D. Burns and S. Shnider, Spherical hypersurfaces in complex manifolds, Inventiones Math., 33 (1976), 223–246.

[9] K. Yamaguchi, Non-degenerate hypersurfaces in complex manifolds admitting large groups of pseudoconformal transformations I, Nagoya Math. J., 62 (1976), 55–96.

[10] C. Fefferman, The Bergman kernel and biholomorphic mappings of pseudoconvex domains, Inventiones Math., 26 (1974), 1–65.

[11] S. Bell and E. Ligocka, A simplification and extension of Fefferman's theorem on biholomorphic mappings, Inventiones Math., 57 (1980), 283–289.

[12] I. Naruki, On extendability of isomorphisms of Cartan connections and biholomorphic mappings of bounded domains, Tôhoku Math. J., 28 (1976), 117–122.
[13] S. I. Pinchuk, On holomorphic mappings of real analytic hypersurfaces, Math. USSR-Sb., 34 (1978), 503–519.
[14] H. Alexander, Proper holomorphic mappings in $\mathbf{C}^n$, Indiana Math. J., 26 (1977), 137–146.
[15] S. M. Webster, Pseudo-Hermitian structure on a real hypersurface, J. Differential Geometry, 13 (1978), 25–41.

345 (XIII.33) Pseudodifferential Operators

A. Pseudodifferential Operators

Pseudodifferential operators are a natural extension of linear partial differential operators. The theory of pseudodifferential operators grew out of the study of singular integral operators, and developed rapidly after 1965 with the systematic studies by J. J. Kohn and L. Nirenberg [1], L. Hörmander [2], and others. The term "pseudodifferential operator" first appeared in Kohn and Nirenberg [1].

Let P be a †linear partial differential operator of the form

$$P=p(x,D_x)=\sum_{|\alpha|\leqslant m}a_\alpha(x)D_x^\alpha, \qquad (1)$$

and let $u(x)$ be a function of class $C_0^\infty(\Omega)$ $(\subset C_0^\infty(\mathbf{R}^n))$. Then by means of the †Fourier inversion formula, $Pu(x)$ can be written in the form

$$Pu(x)=(2\pi)^{-n/2}\int_{\mathbf{R}^n}\exp(ix\cdot\xi)p(x,\xi)\hat{u}(\xi)d\xi, \qquad (2)$$

where $\hat{u}(\xi)$ denotes the †Fourier transform of $u(x)$ (→ 160 Fourier Transform H). But this representation of $Pu(x)$ has a meaning even if $p(x,\xi)$ is not a polynomial in ξ. Thus, for a general function $p(x,\xi)$, the **pseudodifferential operator** $P=p(x,D_x)$ with the **symbol** $p(x,\xi)$ is defined by (2). A symbol class is determined in accordance with various purposes, but it is always required that the corresponding operators have essential properties in common with partial differential operators. Hörmander [3] defined a symbol class $S^m_{\rho,\delta}(\Omega)$ for real numbers m, ρ, and δ with $\rho\geqslant 0$ and $\delta\geqslant 0$ in the following way: Let $p(x,\xi)$ be a C^∞-function defined in $\Omega\times\mathbf{R}^n$. If for any pair of multi-indices α, β and any compact set $K\subset\mathbf{R}^n$, there exists a constant $C_{\alpha,\beta,K}$ such that

$$|D_x^\alpha D_\xi^\beta p(x,\xi)|\leqslant C_{\alpha,\beta,K}(1+|\xi|)^{m+\delta|\alpha|-\rho|\beta|}, \qquad x\in K, \quad \xi\in\mathbf{R}^n,$$

then $p(x,\xi)$ is said to be of class $S^m_{\rho,\delta}(\Omega)$. The operator P defined by (2) is called a pseudodifferential operator (of order m) of class $S^m_{\rho,\delta}(\Omega)$ and is often denoted by $P=p(x,D_x)\in S^m_{\rho,\delta}(\Omega)$. When $\Omega=\mathbf{R}^n$ and constants $C_{\alpha,\beta,K}=C_{\alpha,\beta}$ are independent of K, we denote $S^m_{\rho,\delta}(\mathbf{R}^n)$ simply by $S^m_{\rho,\delta}$, and set

$$S^{-\infty}=\bigcap_{-\infty<m<\infty}S^m_{1,0}\left(=\bigcap_{-\infty<m<\infty}S^m_{\rho,\delta}\right),$$

$$S^m_{\rho,\delta}=\bigcup_{-\infty<m<\infty}S^m_{\rho,\delta}.$$

Differential operators (1) with coefficients of class $\mathscr{B}$ (→ 168 Function Spaces B(13)) belong to $S^m_{1,0}$. The complex power $(1-\Delta)^{z/2}$ of $1-\Delta=1-\sum_{j=1}^n\partial^2/\partial x_j^2$ is defined as a pseudodifferential operator of class $S^{\mathrm{Re}\,z}_{1,0}$ by the symbol $(1+|\xi|^2)^{z/2}$. Operators of class $S^m_{\rho,\delta}$ are continuous mappings of $\mathscr{S}$ into $\mathscr{S}$. Therefore, for any real s, the operator $(1-\Delta)^{s/2}$ can be uniquely extended to be a mapping of $\mathscr{S}'$ into $\mathscr{S}'$ by the relation

$$\langle(1-\Delta)^{s/2}u,v\rangle=\langle u,(1-\Delta)^{s/2}v\rangle, \qquad u\in\mathscr{S}', \quad v\in\mathscr{S}.$$

For any $1\leqslant r\leqslant\infty$ and real s, the †Sobolev space $H^{s,r}$ is defined by

$$H^{s,r}=\{u\in\mathscr{S}'\,|\,(1-\Delta)^{s/2}u\in L_r(\mathbf{R}^n)\},$$

which is a Banach space provided with the norm $\|u\|_{s,r}=\|(1-\Delta)^{s/2}u\|_{L_r}$. In particular, $H^s=H^{s,2}$ is a Hilbert space with the norm $\|u\|_s=\|u\|_{s,2}$. Set

$$H^{-\infty,r}=\bigcup_{-\infty<s<\infty}H^{s,r}, \quad H^{-\infty}=H^{-\infty,2},$$

$$H^{\infty,r}=\bigcap_{-\infty<s<\infty}H^{s,r}, \quad H^\infty=H^{\infty,2}.$$

Then

$$\mathscr{S}'\supset H^{-\infty}\supset\mathscr{E}', \quad H^{-\infty}\supset L_2(\mathbf{R}^n)\supset H^\infty(\subset\mathscr{B}).$$

Choosing the Hörmander class $S^m_{\rho,\delta}$ in the case $0\leqslant\delta\leqslant\rho\leqslant 1$ and $\delta<1$ as a model class, we here list the main results of the theory of pseudodifferential operators:

(i) Pseudolocal property. The operator P of class $S^m_{\rho,\delta}$ in general does not have the **local property** $u\in\mathscr{S}'\Rightarrow\operatorname{supp}Pu\subset\operatorname{supp}u$, but if $\rho>0$, then P has the **pseudolocal property** $u\in\mathscr{S}'\Rightarrow\operatorname{sing\,supp}Pu\subset\operatorname{sing\,supp}u$ [3].

(ii) Algebra of pseudodifferential operators. Let $P=p(x,D_x)\in S^m_{\rho,\delta}$ and $P_j=p_j(x,D_x)\in S^{m_j}_{\rho,\delta}$, $j=1,2$. Then there exist $P^*=p^*(x,D_x)\in S^m_{\rho,\delta}$ and $Q=q(x,D_x)\in S^{m_1+m_2}_{\rho,\delta}$ such that $(Pu,v)=(u,P^*v)$ for $u,v\in\mathscr{S}$, i.e., P^* is the formal adjoint of P, and $Q=P_1P_2$. Furthermore, if we set $p^*_\alpha(x,\xi)=D^\alpha_x(iD_\xi)^\alpha\overline{p(x,\xi)}$ and $q_\alpha(x,\xi)=(iD_\xi)^\alpha p_1(x,\xi)D^\alpha_x p_2(x,\xi)$, then for any integer N we have

$$p^*(x,\xi)-\sum_{|\alpha|<N}\frac{1}{\alpha!}p^*_\alpha(x,\xi)\in S^{m-(\rho-\delta)N}_{\rho,\delta} \quad (3)$$

and

$$q(x,\xi)-\sum_{|\alpha|<N}\frac{1}{\alpha!}q_\alpha(x,\xi)\in S^{m_1+m_2-(\rho-\delta)N}_{\rho,\delta}. \quad (4)$$

Hence the operator class $S^m_{\rho,\delta}$ is an algebra in the sense

$$P\in S^m_{\rho,\delta},\quad P_j\in S^{m_j}_{\rho,\delta},\quad j=1,2,$$

$$\Rightarrow\quad P^*\in S^m_{\rho,\delta},\quad P_1+P_2\in S^{m_0}_{\rho,\delta},\quad P_1P_2\in S^{m_1+m_2}_{\rho,\delta},$$

where $m_0=\max(m_1,m_2)$. In particular, if $0\leqslant\delta<\rho\leqslant 1$, we have $m-(\rho-\delta)N\to-\infty$ and $m_1+m_2-(\rho-\delta)N\to-\infty$ as $N\to\infty$. Then, we say that $p^*(x,\xi)$ and $q(x,\xi)$ have **asymptotic expansions** in the sense of (3) and (4), respectively:

$$p^*(x,\xi)\sim\sum_\alpha\frac{1}{\alpha!}p^*_\alpha(x,\xi)$$

and

$$q(x,\xi)\sim\sum_\alpha\frac{1}{\alpha!}q_\alpha(x,\xi)$$

[3, 4].

(iii) H^s-boundedness. For $P\in S^m_{\rho,\delta}$ and any real s there exists a constant C_s such that

$$\|Pu\|_s\leqslant C_s\|u\|_{s+m},\qquad u\in H^{s+m} \quad (5)$$

[4].

(iv) A sharp form of Gårding's inequality. Let $p(x,\xi)=(p_{jk}(x,\xi);j,k=1,\dots,l)$ be a Hermitian symmetric and nonnegative matrix of $p_{j,k}(x,\xi)\in S^m_{\rho,\delta}$. Then for $P=p(x,D_x)$ there exists a constant C such that

$$\operatorname{Re}(Pu,u)\geqslant -C\|u\|^2_{(m-(\rho-\delta))/2}, \quad (6)$$

where $u=(u_1,\dots,u_l)$ with $u_j\in H^{m/2}$, $j=1,\dots,l$, and $\|u\|^2_{m/2}=\sum^l_{j=1}\|u_j\|^2_{m/2}$.

(v) Invariance under coordinate transformations. Assume that $0\leqslant 1-\rho\leqslant\delta\leqslant\rho\leqslant 1$. Let $x(y)=(x_1(y),\dots,x_n(y))$ be a C^∞-coordinate transformation from $\mathbf{R}^n_y$ onto $\mathbf{R}^n_x$ such that $\partial x_k(y)/\partial y_j\in\mathscr{B}$, $j,k=1,\dots,n$, and $C^{-1}\leqslant|\det(\partial_y x(y))|\leqslant C$ for a constant $C>0$, where $\det(\partial_y x(y))$ denotes the determinant of the Jacobian matrix $(\partial_y x(y))=(\partial x_k(y)/\partial y_j)$. Then for any $P=p(x,D_x)\in S^m_{\rho,\delta}$ in $\mathbf{R}^n_x$, there exists an operator $Q=q(x,D_x)\in S^m_{\rho,\delta}$ in $\mathbf{R}^n_y$ such that $(Qw)(y)=(Pu)(x(y))$ for $w(y)=u(x(y))\in\mathscr{S}$. This fact enables us to define pseudodifferential operators on C^∞-manifolds [3, 4].

(vi) Parametrix. For a given $P\in S^m_{\rho,\delta}$, an operator $E\in S^\infty_{\rho,\delta}$ is called a **left** (resp. **right**) **parametrix** of P if $EP-I$ (resp. $PE-I$) is of class $S^{-\infty}$. If E is a left and right parametrix of P, we call it a **parametrix** of P. For a differential operator P, the existence of a left (resp. right) parametrix is a sufficient condition for P to be †hypoelliptic if $\rho>0$, (resp. the equation $Pu=f\in\mathscr{D}'$ is locally solvable).

The estimate (5), in particular, when $m=s=0$, has been obtained by Hörmander [3], V. V. Grushin (*Functional Anal. Appl.*, 4 (1970)), H. Kumano-go (*J. Fac. Sci. Univ. Tokyo*, 17 (1970)) when $0\leqslant\delta<\rho\leqslant 1$, and A. P. Calderón and R. Vaillancourt [5], H. O. Cordes (*J. Functional Anal.* 18 (1975)), T. Kato (*Osaka J. Math.*, 13 (1976)), Kumano-go [4], and others when $0\leqslant\delta\leqslant\rho\leqslant 1$ and $\delta<1$. A sharp form of Gårding's inequality has been proved by Hörmander [6], P. D. Lax and L. Nirenberg [7], and sharpened by A. Melin (*Ark. Mat.*, 9 (1971)), C. Fefferman and D. H. Phong (*Proc. Nat. Acad. Sci. US*, 76 (1979)), and Hörmander [8]. A general sufficient condition for the existence of a parametrix for an operator of class $S^m_{\rho,\delta}$ was obtained by Hörmander [3]. Let $P=p(x,D_x)$ belong to $S^m_{\rho,\delta}$ with $0\leqslant\delta<\rho\leqslant 1$. Assume that the symbol $p(x,\xi)$ satisfies the following conditions: (i) for some $C_0>0$, real m' ($\leqslant m$), and $R>0$, we have $|p(x,\xi)|\geqslant C_0|\xi|^{m'}$ ($|\xi|\geqslant R$); (ii) for any α, β there exists a constant $C_{0,\alpha,\beta}$ such that

$$|D^\beta_x D^\alpha_\xi p(x,\xi)/p(x,\xi)|\leqslant C_{0,\alpha,\beta}|\xi|^{\delta|\beta|-\rho|\alpha|},\qquad |\xi|\geqslant R.$$

Then there exists a parametrix $Q=q(x,D_x)$ of P in the class $S^{-m'}_{\rho,\delta}$.

By means of operators of class $S^m_{1,0}(\Omega)$ we can define the wave front set of $u\in\mathscr{D}'(\Omega)$, which enables us to resolve sing supp u on $T^*(\Omega)\setminus 0$, the cotangent bundle of Ω minus its zero section. An operator $P=p(x,D_x)\in S^m_{1,0}(\Omega)$ is said to be **microlocally elliptic** at $(x^0,\xi^0)\in T^*(\Omega)\setminus 0$ if $\lim_{\tau\to\infty}|p(x^0,\tau\xi^0)|/|\tau\xi^0|^m>0$. For a distribution $u\in\mathscr{D}'(\Omega)$, we say that a point (x^0,ξ^0) of $T^*(\Omega)\setminus 0$ does not belong to the **wave front set** (or the **singular spectrum**) of u, denoted by $\mathrm{WF}(u)$, if there exist $a(x)$, $b(x)\in C^\infty_0(\Omega)$, $a(x^0)\neq 0$, $b(x^0)\neq 0$, and $P\in S^m_{1,0}(\Omega)$, which is microlocally elliptic at (x^0,ξ^0), such that $aPbu\in C^\infty_0(\Omega)$. Then we easily see that $\mathrm{WF}(u)$ is a closed conic subset of $T^*(\Omega)\setminus 0$. An important fact is that the relation sing supp $u=\mathrm{Proj}_x\mathrm{WF}(u)$ (the projection of $\mathrm{WF}(u)$ on Ω) holds, from which we can perform a so-called **microlocal analysis**, the analysis on $T^*(\Omega)\setminus 0$, of sing supp u. As the sharp form of the pseudolocal property of an operator $P\in S^m_{\rho,\delta}$, if $0\leqslant\delta<\rho\leqslant 1$, P has the **micro-pseudolocal property**: $u\in\mathscr{S}'\Rightarrow\mathrm{WF}(Pu)\subset\mathrm{WF}(u)$.

Pseudodifferential operators of multiple symbol have been defined by K. O. Friedrichs (*Courant Inst.*, 1968) and Kumano-go [4]. More refined and useful classes of pseudodifferential operators have been defined by R. Beals (*Duke Math. J.*, 42 (1975)) Hörmander [8], and others.

The theory of pseudodifferential operators has found many fields of application, such as M. F. Atiyah and R. Bott (*Ann. Math.*, 86 (1967)) on the †Lefschetz fixed-point formula; Friedrichs and P. D. Lax (*Comm. Pure Appl.*, 18 (1965)) on symmetrizable systems; Hörmander [6], Yu. V. Egorov (*Russian Math. Surveys*, 30 (1975)) on subelliptic operators; Kumano-go (*Comm. Pure Appl. Math.*, 22 (1969)), F. Treves (*Amer. J. Math.*, 94 (1972)), S. J. Alinhac and M. S. Bouendi (*Amer. J. Math.*, 102 (1980)) on uniqueness of the Cauchy problem; S. Mizohata and Y. Ohya (*Publ. Res. Inst. Math. Sci.*, 4 (1968)), Hörmander (*J. Analyse Math.*, 32 (1977)) on †weakly hyperbolic equations; C. Morawetz, J. V. Ralston, and W. A. Strauss (*Comm. Pure Appl. Math.*, 30 (1977)), M. Ikawa (*Pub. Res. Inst. Math. Sci.*, 14 (1978)) on the exponential decay of solutions; and Nirenberg and Treves [16], Beals and Fefferman [17] on local solvability theory.

For recent developments in the theory of pseudodifferential operators and its applications → Kumano-go [4], M. Taylor (Princeton Univ. Press, 1981), Treves (Plenum, 1981), and others.

B. Fourier Integral Operators

A **Fourier integral operator** $A: C_0^\infty(\mathbf{R}^n)\to\mathcal{D}'(\mathbf{R}^n)$ is a locally finite sum of linear operators of the type

$$Af(x)=(2\pi)^{-(n+N)/2}\int_{\mathbf{R}^{N+n}}\exp(i\varphi(x,\theta,y))\times a(x,\theta,y)f(y)\,dy\,d\theta. \quad (7)$$

Here $a(x,\theta,y)$ is a C^∞-function satisfying the inequality

$$|D_x^\alpha D_\theta^\beta D_y^\gamma a(x,\theta,y)|\leqslant C(1+|\theta|)^{m-\rho|\beta|+(1-\rho)(|\alpha|+|\gamma|)}$$

for some fixed m and ρ, $1/2\leqslant\rho\leqslant 1$, and any triple of multi-indices α,β,γ, and for $\varphi(x,\theta,y)$ a real-valued function of class C^∞ for $\theta\neq 0$ and homogeneous of degree 1 in θ there. The function φ is called the **phase function** and a the **amplitude function**.

Let $C_\varphi=\{(x,\theta,y)\mid d_\theta\varphi(x,\theta,y)=0,\ \theta\neq 0\}$ and $W=\{(x,y)\in\mathbf{R}^n\times\mathbf{R}^n\mid \exists\theta\neq 0$ such that $(x,\theta,y)\in C_\varphi\}$. If $d_{x,\theta,y}\varphi(x,\theta,y)\neq 0$ for $\theta\neq 0$, then the kernel distribution $k(x,y)$ of A is of class C^∞ outside W. There have been detailed studies of the case where the $d_{x,\theta,y}(\partial\varphi(x,\theta,y)/\partial\theta_j)$, $j=1,2,\ldots,N$, are linearly independent at every point of C_φ. In this case, C_φ is a smooth manifold in $\mathbf{R}^n\times(\mathbf{R}^N\setminus 0)\times\mathbf{R}^n$, and the mapping $\Phi: C_\varphi\ni(x,\theta,y)\to(x,y,\xi,\eta)$, $\xi=d_x\varphi(x,\theta,y)$, $\eta=d_y\varphi(x,\theta,y)$, is an immersion of C_φ to $T^*(\mathbf{R}^n\times\mathbf{R}^n)\setminus 0$, the cotangent bundle of $\mathbf{R}^n\times\mathbf{R}^n$ minus its zero section. The image $\Phi C_\varphi=\Lambda_\varphi$ is a **conic Lagrange manifold**, i.e., the canonical 2-form $\sigma=\sum_j d\xi_j\wedge dx_j-\sum_j d\eta_j\wedge dy_j$ vanishes on Λ_φ and the multiplicative group of positive numbers acts on Λ_φ. Let $\lambda_1,\ldots,\lambda_{2n}$ be a system of local coordinates in Λ_φ. These, together with $\partial\varphi/\partial\theta_1,\ldots,\partial\varphi/\partial\theta_N$, constitute a system of local coordinate functions of $\mathbf{R}^n\times(\mathbf{R}^N\setminus 0)\times\mathbf{R}^n$ in a conic neighborhood of C_φ. Let J denote the Jacobian determinant

$$\frac{D\left(\lambda_1,\ldots,\lambda_{2n},\dfrac{\partial\varphi}{\partial\theta_1},\ldots,\dfrac{\partial\varphi}{\partial\theta_N}\right)}{D(x,\theta,y)}.$$

The function $a_{\Lambda_\varphi}=\sqrt{J}\,a|_{C_\varphi}\Phi^{-1}$ is called the local **symbol** of A. Here $a|_{C_\varphi}$ is the restriction of a to C_φ. The conic Lagrange manifold $\Lambda_\varphi=\Lambda_\varphi(A)$ and the symbol $a_{\Lambda_\varphi}=a_{\Lambda_\varphi}(A)$ essentially determine the singularity of the kernel distribution $k(x,y)$ of the Fourier integral operator A. Conversely, given a conic Lagrange manifold Λ in $T^*(\mathbf{R}^n\times\mathbf{R}^n)\setminus 0$ and a function a_Λ on it, one can construct a Fourier integral operator A such that $\Lambda_\varphi(A)=\Lambda$ and $a_{\Lambda_\varphi}(A)=a_\Lambda$. Those Fourier integral operators whose associated conic Lagrange manifolds are the graphs of homogeneous †canonical transformations of $T^*(\mathbf{R}^n)\setminus 0$ are most frequently used in the theory of linear partial differential equations. Let A be a Fourier integral operator such that $\Lambda_\varphi(A)$ is the graph of a homogeneous canonical transformation χ. Then the adjoint of A is a Fourier integral operator such that the associated conic Lagrange manifold is the graph of the inverse transformation χ^{-1}. Let A_1 be another such operator; if the associated conic Lagrange manifold is the graph of χ_1, then the composed operator A_1A is also a Fourier integral operator and the associated conic Lagrange manifold is the graph of the composed homogeneous canonical transformation $\chi_1\chi$.

A pseudodifferential operator of class $S^m_{\rho,1-\rho}(\mathbf{R}^n)$ is a particular type of Fourier integral operator. In fact, a Fourier integral operator A is a pseudodifferential operator of class $S^m_{\rho,1-\rho}(\mathbf{R}^n)$ if and only if $\Lambda_\varphi(A)$ is the graph of the identity mapping of $T^*(\mathbf{R}^n)\setminus 0$. Hence for any Fourier integral operator A, A^*A and AA^* are pseudodifferential operators.

The following theorem is due to Egorov [11]: Let P and Q be pseudodifferential operators of class $S^m_{\rho,1-\rho}(\mathbf{R}^n)$ with the symbols $p(x,\xi)$ and $q(x,\xi)$, respectively, and let A be a Fourier integral operator such that the associated conic Lagrange manifold $\Lambda_\varphi(A)$ is the graph of a homogeneous canonical transformation χ of $T^*(\mathbf{R}^n)\setminus 0$. If the equality $PA=AQ$

holds, then $q(x,\xi)-p(\chi(x,\xi))$ belongs to the class $S^{m+1-2\rho}_{\rho,1-\rho}(\mathbf{R}^n)$.

Assume that $m=1$, $\rho=1$, and that $p_1(x,\xi)$ is a real-valued C^∞-function, homogeneous of degree 1 in ξ for $|\xi|>1$, such that $p(x,\xi)-p_1(x,\xi)\in S^0_{1,0}(\mathbf{R}^n)$ and $d_\xi p_1(x^0,\xi^0)\neq 0$ at (x^0,ξ^0), $|\xi^0|>1$, where $p_1(x^0,\xi^0)=0$. Then one can find a Fourier integral operator A such that the function $q(x,\xi)$ of Egorov's theorem satisfies the relation $q(x,\xi)-\xi_1\in S^0_{1,0}(\mathbf{R}^n)$.

The theory of Fourier integral operators has its origin in the asymptotic representation of solutions of the wave equation (→ 325 Partial Differential Equations of Hyperbolic Type L; also, e.g., [12, 13, 14]). G. I. Eskin (*Math. USSR-Sb.*, 3 (1976)) used a type of Fourier integral operator in deriving the energy estimates and constructing the fundamental solutions for strict hyperbolic operators. Hörmander (*Acta Math.*, 121 (1968)) introduced the term "Fourier integral operators," and applied these operators to the derivation of highly accurate asymptotic formulas for spectral functions of elliptic operators. Egorov (*Math. USSR-Sb.*, 11 (1970)) applied his theorem and the corollary stated above to the study of hypoellipticity and local solvability for pseudodifferential operators of principal type. Using Egorov's theorem and the same corollary, Nirenberg and Treves [16] obtained decisive results concerning local solvability for linear partial differential operators of principal type; these results were completed by Beals and Fefferman [17]. Hörmander and J. J. Duistermaat [9, 15] constructed a general global theory of Fourier integral operators making use of Maslov's theory [14], which was originally published in 1965. By virtue of this research, the Fourier integral operator has come to be recognized as a powerful tool in the theory of linear partial differential operators. An interesting application of the global theory of Fourier integral operators appeared in J. Chazarain (*Inventiones Math.*, 24 (1974)). The boundedness of Fourier integral operators in the spaces $L^2(\mathbf{R}^n)$ (or the space H^s) has been studied in several cases. Some sufficient conditions for boundedness have been obtained by Eskin (*Math. USSR-Sb.*, 3 (1967)), Hörmander [9], D. Fujiwara [18], Kumano-go (*Comm. Partial Diff. Eq.*, 1 (1976)), K. Asada and Fujiwara (*Japan. J. Math.*, 4 (1978)), and others. A calculus of Fourier integral operators in $\mathbf{R}^n$ was given in Kumano-go [4].

The propagation of wave front sets by means of a Fourier integral operator is described as follows. Let us consider a phase function of the form $\varphi(x,\xi,y)=S(x,\xi)-y\cdot\xi$ in $\mathbf{R}^n_x\times\mathbf{R}^n_\xi\times\mathbf{R}^n_y$, and let $a(x,\xi)$ be an amplitude function independent of y of class $S^m_{1,0}$. Then by (7) the Fourier integral operator $A=A_S$ is defined by

$$A_S u(x)=(2\pi)^{-n/2}\int_{\mathbf{R}^n}\exp(iS(x,\xi))a(x,\xi)\hat{u}(\xi)\,d\xi. \tag{8}$$

Let T be the canonical transformation with the †generating function $S(x,\xi)$, i.e., T is defined by $y=\nabla_\xi S(x,\eta)$, $\xi=\nabla_x S(x,\eta)$. Then for the Fourier integral operator A_S we have

$$\mathrm{WF}(A_S u)\subset\{(x,\xi)=T(y,\eta)\,|\,(y,\eta)\in\mathrm{WF}(u)\},\quad u\in\mathscr{S}'. \tag{9}$$

Next consider a hyperbolic operator $L=D_t+p(t,x,D_x)$ for a real-valued symbol $p(t,x,\xi)\in\mathscr{B}^0([0,T_0];S^1_{1,0})$ with some $T_0>0$. For a small $0<T\leqslant T_0$ the solution $S(t,x,\xi)$ of the eikonal equation $\partial_t S+p(t,x,\nabla_x S)=0$ on $[0,T]$ with the initial condition $S|_{t=0}=x\cdot\xi$ exists in $\mathscr{B}^1([0,T];S^1_{1,0})$. Consider the Cauchy problem $Lu=0$ on $[0,T]$, $u|_{t=0}=u_0$. Then there exists an amplitude function $e(t,x,\xi)\in\mathscr{B}^1([0,T];S^0_{1,0})$ such that the solution $u(t)$ is found in the form $u(t)=E_S(t)u_0$. On the other hand, let $(x,\xi)=(X(t,y,\eta),\Xi(t,y,\eta))$ be the bicharacteristic strip defined by †Hamilton's canonical equation $dx/dt=\nabla_\xi p(t,x,\xi)$, $d\xi/dt=-\nabla_x p(t,x,\xi)$ with $(x,\xi)_{|t=0}=(y,\eta)$. Then $(X(t,y,\eta),\Xi(t,y,\eta))$ can be solved by means of the relations $y=\nabla_\xi S(t,X,\eta)$, $\Xi=\nabla_x S(t,X,\eta)$, as a family of canonical transformations with a parameter $t\in[0,T]$. Thus by means of (9) we have

$$\mathrm{WF}(u(t))\subset\{(x,\xi)=(X(t,y,\eta),\Xi(t,y,\eta))\,|\,(y,\eta)\in\mathrm{WF}(u_0)\}, \tag{10}$$

which is the fundamental result in the study of the propagation of wave front sets as solutions of general hyperbolic equations (→ 325 Partial Differential Equations of Hyperbolic Type M).

The works of Egorov, Nirenberg and Treves, and Hörmander motivated the theory of hyperfunctions developed by M. Sato and gave rise to the concept of †quantized contact transformations, which correspond to Fourier integral operators in the theory of distributions. The above-stated transformation theorem of Egorov has been studied in detail with reference to systems of pseudodifferential equations with analytic coefficients [19] (→ 274 Microlocal Analysis).

References

[1] J. J. Kohn and L. Nirenberg, An algebra of pseudo-differential operators, Comm. Pure Appl. Math., 18 (1965), 269–305.

[2] L. Hörmander, Pseudo-differential operators, Comm. Pure Appl. Math., 18 (1965), 501–517.
[3] L. Hörmander, Pseudo-differential operators and hypoelliptic equations, Amer. Math. Soc. Proc. Symp. Pure Math., 10 (1967), 138–183.
[4] H. Kumano-go, Pseudo-differential operators, MIT Press, 1981. (Original in Japanese, 1974.)
[5] A. P. Calderón and R. Vaillancourt, A class of bounded pseudo-differential operators, Proc. Nat. Acad. Sci. US, 69 (1972), 1185–1187.
[6] L. Hörmander, Pseudo-differential operators and non-elliptic boundary problems, Ann. Math., 83 (1966), 129–209.
[7] P. D. Lax and L. Nirenberg, On stability for difference schemes, a sharp form of Gårding's inequality, Comm. Pure Appl. Math., 19 (1966), 473–492.
[8] L. Hörmander, The Weyl calculus of pseudo-differential operators, Comm. Pure Appl. Math., 32 (1979), 359–443.
[9] L. Hörmander, Fourier integral operators I, Acta Math., 128 (1971), 79–183.
[10] J. J. Duistermaat, Fourier integral operators, Lecture notes, Courant Institute, 1973.
[11] Yu. V. Egorov, On canonical transformations of pseudo-differential operators (in Russian), Uspekhi Mat. Nauk, 25 (1969), 235–236.
[12] P. D. Lax, Asymptotic solutions of oscillatory initial value problems, Duke Math. J., 24 (1957), 627–646.
[13] D. Ludwig, Exact and asymptotic solutions of the Cauchy problem, Comm. Pure Appl. Math., 13 (1960), 473–508.
[14] V. P. Maslov, Théorie des perturbations et méthodes asymptotiques, Gauthier-Villars, 1972.
[15] J. J. Duistermaat and L. Hörmander, Fourier integral operators II, Acta Math., 128 (1972), 183–269.
[16] L. Nirenberg and F. Treves, On local solvability of linear partial differential equations I, II, Comm. Pure Appl. Math., 23 (1970), 1–38, 459–510.
[17] R. Beals and C. Fefferman, On local solvability of linear partial differential equations, Ann. Math., (2) 97 (1973), 482–498.
[18] D. Fujiwara, On the boundedness of integral transformations with highly oscillatory kernels, Proc. Japan Acad., 51 (1975), 96–99.
[19] H. Komatsu (ed.), Hyperfunctions and pseudo-differential equations, Lecture notes in math. 287, Springer, 1973.

346 (XVIII.17) Psychometrics

A. General Remarks

Psychometrics is a collection of methods for drawing statistical conclusions from various psychological phenomena which are expressed numerically or quantitatively. It consists chiefly of statistical methods to deal with psychological measurements and of theories dealing with mathematical models concerning learning processes, social attitudes, and mental abilities.

B. Sensory Tests

A measurement wherein human senses are taken as the gauge is called a **sensory test**. The panel of judges must be composed appropriately, and the examining circumstances must be controlled. Various methods of psychological measurements are applied. In the following sections we describe the basic statistical procedures used in sensory testing.

C. Paired Comparison

When there are t objects (treatments or stimuli in some cases) $O_1, O_2, \ldots, O_t$, the method of comparing them two at a time in every possible way is called **paired comparison**. The following are typical mathematical models of this method.

(1) Thurstone-Mosteller Model. Suppose that the probability that O_i is preferred to O_j for a pair (O_i, O_j) is p_{ij}. Of the n judges who compare this pair, the number who prefer O_i is n_{ij}, and the number who prefer O_j is $n_{ji}=n-n_{ij}$. In this comparison it is assumed that the strengths of the stimuli O_i, O_j to the senses are random variables X_i, X_j, and O_i is preferred when $X_i > X_j$. Furthermore, it is assumed that the joint probability distribution of X_i and X_j is the 2-dimensional †normal distribution with μ_i and σ^2 as mean and variance of X_i, and ρ as correlation coefficient of X_i and X_j. There is no loss of generality in assuming that $2\sigma^2(1-\rho)=1$ and $\sum_{i=1}^{t}\mu_i=0$. Let $\Phi(x)$ be the standardized normal distribution function and $p_{ij}=\Phi(\mu_i-\mu_j)$. Using $p'_{ij}=n_{ij}/n$ as estimates of the true p_{ij}, we can obtain the estimates $\hat{\mu}_i$. Using $p''_{ij}=\Phi(\hat{\mu}_i-\hat{\mu}_j)$ and p'_{ij} we can test the hypothesis that $\mu_1=\mu_2=\ldots=\mu_t$.

(2) The Bradley-Terry Model. The experimental method in the Bradley-Terry model is the same as in the Thurstone-Mosteller model. It is postulated that, associated with $O_1, O_2, \ldots, O_t$, there exist parameters π_i for O_i ($\pi_i \geqslant 0$, $\Sigma_{i=1}^{t} \pi_i = 1$) such that $p_{ij} = \pi_i/(\pi_i + \pi_j)$. Obtaining the †maximum likelihood estimator of π_i, we can test the appropriateness of the models.

(3) Scheffé's Model. Each pair O_i, O_j is presented to $2n$ judges; n of them examine O_i first and O_j next, and the remaining n examine the pair in the opposite order. A judgment is recorded on a 7-point (or 9-, 5-, or 3-point) scale. In the 7-point scaling system a judge presented with the ordered pair (O_i, O_j) marks one of the seven points $3, 2, 1, 0, -1, -2, -3$, meaning, respectively, that O_i is strongly preferable to O_j, O_i is moderately preferable to O_j, O_i is slightly preferable to O_j, no preference, O_j is slightly preferable to O_i, etc. The mark given by the kth judge on his preference of O_i to O_j is denoted by X_{ijk}, which can be regarded as the sum of a main effect, deviation of subtractivity, order effect, and error. Significance tests for these effects and estimates of various parameters are given by using statistical †linear models. †BIBD, †PBIBD, etc., can also be applied to paired comparisons.

D. The Pair Test, Triangle Test, and Duo-Trio Test

The pair test, triangle test, and duo-trio test are sensory difference tests. The methods are as follows. **Pair test**: A judge is requested to designate a preference between the paired samples A and B. **Triangle test**: A judge is requested to select two samples of the same kind out of A, A, B. **Duo-trio test**: A judge is first acquainted with a sample A and then is requested to choose from A and B the one he has seen in the previous step. In all the above cases, the hypothesis that A and B are different and that the judge has no ability to determine the difference between them is tested by using the †binomial distribution.

E. Scaling

(1) One-Dimensional Case. Psychometric **scaling methods** are procedures for constructing scales for psychological phenomena. Some of them require judgments concerning a particular attitude that is considered unidimensional. Under the assumption that a psychological phenomenon is a random variable with some distribution law and the parameters of the distribution law determine psychological scales, a psychological scaling is given by estimating the parameters. The Thurstone-Mosteller model is a method for scaling a set of stimuli by means of observable proportions.

(2) Multidimensional Scaling (MDS). Multidimensional scaling is a collection of methods to deal with data consisting of many measurements on many objects and to characterize the mutual distance (dissimilarity), or closeness (affinity), by representing those objects by a small number of indices or by points in a small-dimensional Euclidean space. It has seen useful applications in the analysis of people's attitude and perception and their characterizations by means of a few numbers or points in a space of low dimension.

Historically, MDS was first developed by Torgerson (1958) and refined further by Shepard (1962) and Kruskal (1964). The method developed by Torgerson and also the INDSCAL method by Carrol and Chang (1970) are called **metric multidimensional scaling**, while the method by Shepard and Kruskal is called the **nonmetric MDS**. The former is applied when the data are represented in continuous scales and the latter when the data are in discrete nominal or ordinal scales. Techniques of multidimensional scaling are closely related and sometimes actually equivalent to various methods of multivariate analysis, especially principal component analysis, canonical correlation analysis, and discriminant analysis (→ 280 Multivariate Analysis).

F. Factor Analysis

Though factor analysis can be considered to be a method to deal with multivariate data in general (→ 280 Multivariate Analysis), it has had close connections with psychometric studies, in both theoretical developments and applications. Historically, it was initiated by Spearman (1927) and developed further by Thurston (1945) in order to measure human abilities from test scores. Mathematically, the model of factor analysis is formulated as follows: Let z_{jk} be the standardized score of the jth test achieved by the kth subject, $j = 1, \ldots, p$; $k = 1, \ldots, N$; then it is assumed that it can be represented as a linear combination of r common factors and one specific factor as

$$z_{jk} = a_{j1} f_{1k} + a_{j2} f_{2k} + \ldots + a_{jr} f_{rk} + u_j v_{jk}, \qquad (1)$$

where f_{ik}, $i = 1, \ldots, r$, $k = 1, \ldots, N$, represents the magnitude of the ith common factor (ability) in the kth subject and a_{ji} is the size of contribution of the ith factor to the score of the jth test.

Usually it is assumed that (i) $V(z_j) = 1$, (ii) $V(f_i) = 1$ and $\mathrm{Cov}(f_i, f_{i'}) = 0$ for $i \neq i'$, (iii) $V(v_j) = 1$ and $\mathrm{Cov}(v_j, v_{j'}) = 0$, $j \neq j'$, (iv) $\mathrm{Cov}(f_i, v_j) = 0$.

Then it follows that

$$r_{jj'}=\mathrm{Cor}(z_j,z_{j'})=a_{j1}a_{j'1}+a_{j2}a_{j'2}+\ldots+a_{jr}a_{j'r},$$

and

$$1=a_{j1}^2+a_{j2}^2+\ldots+a_{jr}^2+u_j{}^2=h_j^2+u_j^2,$$

where h_j^2 is called the **communality** of the jth variable z_j and u_j^2 is called the **specificity**. It follows easily from (1) that any orthogonal transformation of the scores does not affect the model.

Problems of factor analysis are classified into three types:
(1) Estimation of communality: There are several methods of determining communality or initial estimates of it when some iterative procedure is used.
(2) Determination of **factor loadings**, which is the estimation of the a_{ji}: A number of methods have been proposed, among which those often used are the MINRES by Harman (1967), the varimax by Kaiser (1958), and the †maximum likelihood based on the normal model by Lawley and Maxwell.
(3) Estimation of **factor scores** f_{ik}: Usually factor scores are estimated after factor loadings have been determined. Thurstone proposed $\hat{F}=ZR^{-1}A$ and Harman $\hat{F}=ZA(A'A)^{-1}$, where F, Z, A are the matrices of factor scores, test scores, factor loadings, respectively, and R is the correlation matrix of the z's.

G. Learning Theory

(1) General Description. Assume that a sequence of trials is done in order to study some given behavior and that on each trial particular events occur (stimuli, responses, reinforcements, etc.) that influence the ensuing behavior. Then the behavior itself is modified by such a sequence of trials. **Learning models** refer to such processes of behavior modification, and they are frequently represented by recursive formulas for response probabilities.

Assume that two mutually exclusive response alternatives A_1 and A_2 occur on the nth trial $(n=1,2,\ldots)$ with respective probabilities P_n and $1-P_n$, and that an event E_i occurs on the nth trial with $\Pr(\mathscr{E}_n=E_i)=\pi_i$ $(i=1,2,\ldots,t;$ $\sum_{i=1}^t\pi_i=1)$. Then the recursive formula for P_n is of the form $P_{n+1}=f(P_n;\mathscr{E}_n,\mathscr{E}_{n-1},\ldots,\mathscr{E}_1)$. If the formula can be written as $f=f(P_n;\mathscr{E}_n,$ $\mathscr{E}_{n-1},\ldots,\mathscr{E}_{n-d})$, then the response probability is called **d-trial path dependent**. In the special case $d=0$, it is called **path independent**. For simplicity we write $f(P_n;\mathscr{E}_n=E_i,\mathscr{E}_{n-1}=E_j,\ldots,$ $\mathscr{E}_1=E_k)=f_{ij\ldots k}(P_n)$ $(i,j,\ldots,k=1,2,\ldots,t)$. If the response probability is path independent, then $f_{ij\ldots k}(P_n)=f_if_j\cdots f_k(P_1)$, where $f_v(P_n)=f(P_n;\mathscr{E}_n=E_v)$ $(v=i,j,\ldots,k)$. When the recursive formula can be expressed as $f=f(P_n;\mathscr{E}_n,n)$, the response probability is said to be **quasi-independent of path** [11]. In the recursive formula f, two events E_i and E_j $(i\neq j)$ are said to be commutative if $f_{\ldots i\ldots j\ldots}(P_n)=f_{\ldots j\ldots i\ldots}(P_n)$. If any two events are commutative, the condition of **event commutativity** is satisfied. By making f explicit with respect to n, we write $P_n=F(n;\mathscr{E}_{n-1},\ldots,\mathscr{E}_1;P_1)$. Under the condition of event commutativity, the explicit formula can be written $P_n=F(N_1,N_2,\ldots,N_t;P_1)$, where N_i is the frequency of occurrence of E_i in the first $(n-1)$ trials $(\sum_{i=1}^t N_i=n-1)$. If both event commutativity and path independence of response probability are satisfied, the explicit formula $P_n=f_1^{N_1}f_2^{N_2}\ldots f_t^{N_t}(P_1)$ can be obtained.

(2) Linear Models. In a linear model, the recursive formula is written as a linear function of P_n.

Example (1). Bush-Mosteller model [6]. The Bush-Mosteller model assumes the response probability to be path independent. The recursive formula is expressed as $f_i(P_n)=\alpha_iP_n+(1-\alpha_i)\lambda_i$, $\mathscr{E}_n=E_i$. Here α_i $(0\leqslant\alpha_i\leqslant 1)$ represents the degree of ineffectiveness of E_i for learning and λ_i $(0\leqslant\lambda_i\leqslant 1)$ is the †fixed point of f_i. A necessary and sufficient condition for E_i and E_j $(i\neq j)$ to be commutative is that either f_i or f_j be an †identity operator or $\lambda_i=\lambda_j$.

Example (2). Estes's stimulus-sampling model [7]. We can consider the stimulus as a set composed of m elements, each of which corresponds to either response A_1 or A_2; the manner of their correspondences depends on each trial. If J_n elements correspond to A_1 on the nth trial, then we have $P_n=J_n/m$. Suppose that on the nth trial s $(\leqslant m)$ elements are sampled, among which X_n elements correspond to A_2, and the remaining X_n' $(=s-X_n)$ elements correspond to A_1. As a result of the nth trial, if A_1 is reinforced, we set $Y_n=1$; otherwise, we set $Y_n=0$. Furthermore, assume that $J_{n+1}=J_n+X_nY_n-X_n'(1-Y_n)$. Hence, we obtain the recursive formula $P_{n+1}=P_n+\{X_nY_n-X_n'(1-Y_n)\}/m$. In this model, the response probability is path independent and $\mathscr{E}_n=(X_n,Y_n)$. Other linear models have been proposed in which the response probability is either quasi-independent of path [8] or path dependent [10].

(3) Nonlinear Models. In nonlinear models the recursive formula cannot be written as a linear function of P_n.

Example (3). Luce's β-model [9]. Let the response strengths of A_1 and A_2 on the nth trial be v_n and v_n', respectively (both positive), and assume that P_n, the response probability of A_1, is expressed as $v_n/(v_n+v_n')$. The response strengths v_n and v'_n depend on each trial. Under the assumption that the response strength is path independent and that v_{n+1} changes independently from v'_n, the recursive formula of v_n is written as $v_{n+1} = \varphi_i(v_n)$, $\mathscr{E}_n = E_i$. Here, if we assume $\varphi_i(v) > 0$ for $v > 0$ and $\varphi_i(cv) = c\varphi_i(v)$ for $v > 0$ and $c > 0$, then $\varphi_i(v_n) = \beta_i v_n$ with $\beta_i > 0$. In a similar way, the recursive formula for v'_n can be expressed as $\varphi'_i(v'_n) = \beta'_i v'_n$ $(\beta'_i > 0)$. Therefore we have $P_{n+1} = P_n / \{P_n + b_i(1 - P_n)\}$ $(b_i = \beta'_i / \beta_i)$, $\mathscr{E}_n = E_i$. This model is nonlinear, and the response probability is path independent. By making the recursive formula explicit, we obtain $P_n = P_1 / \{P_1 + (1 - P_1)\exp(\sum_{i=1}^{t} N_i \log b_i)\}$. Hence it is clear that the events are commutative. Other nonlinear models in which the response probability is either quasi-independent of path [5] or path dependent [9] have also been proposed.

Here we have taken up only the case in which the number of response alternatives is 2, but we can generalize to the case of more than two alternatives. For fitting a model and experimental data, expected response probabilities and various other statistics deduced from the model (total error, trial number of first success or last error, and sequential statistics such as length of response †run or †autocorrelation between responses) are used. Estimation methods have also been devised for the parameters involved.

References

For sensory tests,

[1] H. A. David, The method of paired comparisons, Hafner, 1963.

[2] J. P. Guilford, Psychometric methods, McGraw-Hill, second edition, 1954.

[3] M. G. Kendall, Rank correlation methods, Griffin, third edition, 1962.

[4] W. S. Torgerson, Theory and method of scaling, Wiley, 1958.

For learning theory,

[5] R. J. Audley and A. R. Jonckheere, The statistical analysis of the learning process, Brit. J. Statist. Psychol., 9 (1956), 87–94.

[6] R. R. Buch and F. Mosteller, Stochastic models for learning, Wiley, 1955.

[7] W. K. Estes, Component and pattern models with Markovian interpretations, Studies in Mathematical Learning Theory, R. R. Bush and W. K. Estes (eds.), Stanford Univ. Press, 1959, 9–52.

[8] M. I. Hanania, A generalization of the Bush-Mosteller model with some significance tests, Psychometrika, 24 (1959), 53–68.

[9] R. D. Luce, Individual choice behavior: A theoretical analysis, Wiley, 1959.

[10] S. H. Sternberg, A path-dependent linear model, Studies in Mathematical Learning Theory, R. R. Bush and W. K. Estes (eds.), Stanford Univ. Press, 1959, 308–339.

[11] S. H. Sternberg, Stochastic learning theory, Handbook of Mathematical Psychology II, R. D. Luce, R. R. Bush, and E. Galanter (eds.), Wiley, 1963, 1–120.

For MDS,

[12] J. B. Kruskal, Nonmetric multidimensional scaling: A numerical method, Psychometrika, 29 (1964), 115–129.

[13] R. N. Shepard, The analysis of proximities: Multidimensional scaling with an unknown distance function, Psychometrika, 27 (1962), 125–219.

[14] L. Guttman, Multiple rectinear prediction and the resolution into components, Psychometrika, 5 (1940), 75–99.

[15] J. B. Carrol and J. J. Chang, Analysis of individual difference in multidimensional scaling via an N-way generalization of Eckart-Young decomposition, Psychometrika, 35 (1970), 283–319.

For factor analysis,

[16] C. Spearman, The abilities of man, Macmillan, 1927.

[17] H. H. Harman, Modern factor analysis, Univ. of Chicago Press, 1967.

[18] H. F. Kaiser, The varimax criterion for analytic rotation in factor analysis, Psychometrika, 23 (1958), 187–240.

[19] D. N. Lawley and A. E. Maxwell, Factor analysis as a statistical criterion, Butterworth, 1963.

[20] K. Takeuchi, H. Yanai, and B. N. Mukherjee, Foundations of multivariate analysis, Wiley, 1982.

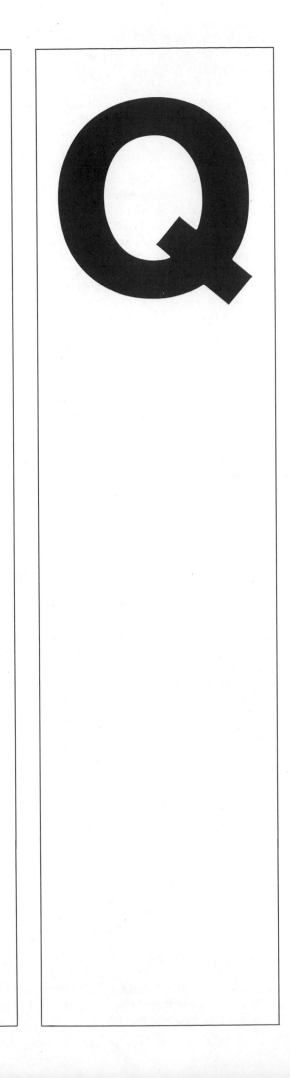

Q

347 (V.12) Quadratic Fields

A. General Remarks

Any †extension field of the rational number field $\mathbf{Q}$ of degree 2 is called a **quadratic field**. Any quadratic field k is obtained from $\mathbf{Q}$ by adjoining a square root of a **square-free integer** (i.e., an integer $\neq 0, 1$ with no square factor) $m: k=\mathbf{Q}(\sqrt{m})$. If m is positive (negative), k is called a **real (imaginary** or **complex) quadratic field** (→ 14 Algebraic Number Fields). Let

$$\omega=\begin{cases}(1+\sqrt{m})/2 & \text{for } m\equiv 1 \pmod 4,\\ \sqrt{m} & \text{for } m\equiv 2,3 \pmod 4.\end{cases}$$

Then $(1,\omega)$ is a †minimal basis of k. That is, any †algebraic integer α of k has the unique expression $\alpha=a+b\omega$ with $a,b\in\mathbf{Z}$. The †discriminant d of k is given by $d=m$ in case $m\equiv 1 \pmod 4$ and $d=4m$ in case $m\equiv 2,3 \pmod 4$. The conjugate element of an element $\alpha=a+b\sqrt{m}$ $(a,b\in\mathbf{Q})$ of k over $\mathbf{Q}$ is given by $\alpha'=a-b\sqrt{m}$. The mapping $\sigma:\alpha\rightarrow\alpha'$ is an †automorphism of the field k.

B. Units

Let k be an imaginary quadratic field. The †units of k are ± 1, $\pm i$ in case $m=-1$; ± 1, $\pm\omega_0$, $\pm\omega_0^2$ $(\omega_0=(1+\sqrt{-3})/2)$ in case $m=-3$; and ± 1 in all other cases.

Let k be a real quadratic field. There exists a unit ε_0 that is the smallest one among the units (>1) of k. Any unit ε of k can be uniquely expressed in the form $\varepsilon=\pm\varepsilon_0^n$ $(n\in\mathbf{Z})$. That is, $\pm\varepsilon_0^{\pm 1}$ is a †fundamental unit of k. The fundamental unit $\varepsilon_0=(x+y\sqrt{d})/2$ (>1) can be calculated by finding a minimal positive integral solution (x,y) of †Pell's equation $x^2-dy^2=\pm 4$ by using continued fractions (→ 83 Continued Fractions; for a table of the fundamental unit of k for $m<100$ → [1]).

C. Prime Ideals

The decomposition of a prime number p in the †principal order $\mathfrak{o}$ of k is given as follows: (i) Let $p|d$, where d is the discriminant of k. Then p is decomposed in $\mathfrak{o}$ in the form $(p)=\mathfrak{p}^2$. (ii) Let $p\neq 2$ and $(p,m)=1$. If $(m/p)=1$, then $(p)=\mathfrak{p}\mathfrak{p}'$ in $\mathfrak{o}$ $(\mathfrak{p}\neq\mathfrak{p}')$ and $N(\mathfrak{p})=N(\mathfrak{p}')=p$. If $(m/p)=-1$, then $(p)=\mathfrak{p}$ in $\mathfrak{o}$ and $N(\mathfrak{p})=p^2$. Here $\mathfrak{p}$, $\mathfrak{p}'$ are prime ideals of $\mathfrak{o}$, N means the †norm, and (m/p) is the †Legendre symbol. (iii) Let $2\nmid d$, that is, $m\equiv 1 \pmod 4$. If $m\equiv 1 \pmod 8$, then $(2)=\mathfrak{p}\mathfrak{p}'(\mathfrak{p}\neq\mathfrak{p}')$ and $N(\mathfrak{p})=N(\mathfrak{p}')=2$. If $m\equiv 5 \pmod 8$, then $(2)=\mathfrak{p}$ and $N(\mathfrak{p})=4$.

D. The Kronecker Symbol

Let $k=\mathbf{Q}(\sqrt{m})$, and let d be the discriminant of k. We define the symbol χ for k as follows: (i) $\chi(p)=1$ if $(p)=\mathfrak{p}\mathfrak{p}'$ $(\mathfrak{p}\neq\mathfrak{p}')$ in $\mathfrak{o}$; (ii) $\chi(p)=-1$ if $(p)=\mathfrak{p}$ in $\mathfrak{o}$; (iii) $\chi(p)=0$ if $(p)=\mathfrak{p}^2$ in $\mathfrak{o}$; and (iv) $\chi(n)=\prod_i\chi(p_i)^{e_i}$ for $n=\prod_i p_i^{e_i}>0$. In particular, we define $\chi(1)=1$. If $(n,d)=1$, the symbol χ can also be defined using the †Jacobi symbol as follows: If $m\equiv 1 \pmod 4$, then $\chi(n)=(n/|m|)$; and if $m\equiv 2,3 \pmod 4$, then $\chi(n)=\chi_2^*(n)(n/|m'|)$ for $d=2^e m'$, where (i) $\chi_2^*(n)=(-1)^{(n-1)/2}$ for $e=2$, $m'\equiv 3 \pmod 4$; (ii) $\chi_2^*(n)=(-1)^{(n^2-1)/8}$ for $e=3$, $m'\equiv 1 \pmod 4$; and (iii) $\chi_2^*(n)=(-1)^{(n^2-1)/8+(n-1)/2}$ for $e=3$, $m'\equiv 3 \pmod 4$. If $(n,d)\neq 1$, then $\chi(n)=0$. For a negative integer $-n$, we define $\chi(-n)=(\operatorname{sgn} d)\chi(n)$. The symbol $\chi(n)$ for $n\in\mathbf{Z}$ is called the **Kronecker symbol** for k.

The Kronecker symbol for k has the following four properties: (1) $\chi(n)=0$ if $(n,d)\neq 1$, and $\chi(n)=\pm 1$ if $(n,d)=1$; (2) $\chi(m)=\chi(n)$ if $m\equiv n \pmod d$; (3) $\chi(mn)=\chi(m)\chi(n)$; (4) $\chi(n)=1$ if and only if $n\equiv N(\mathfrak{a}) \pmod d$ for some integral ideal $\mathfrak{a}$ of k such that $(\mathfrak{a},(d))=1$. (Property (4) shows that a quadratic field provides a class field; → 59 Class Field Theory.)

E. Ideal Classes

The †class number h of k was calculated by P. G. L. †Dirichlet (1840) by analytical methods as follows:

$$h\log\varepsilon_0=-\frac{1}{2}\sum_{r=1}^{d-1}\chi(r)\log\sin\frac{r\pi}{d} \quad \text{for } m>0,$$

$$h=\frac{-w}{2|d|}\sum_{r=1}^{|d|-1}\chi(r)r \quad \text{for } m<0,$$

where d is the discriminant of k, ε_0 is the positive fundamental unit (>1) of k, w is the number of roots of unity in k, and χ is the Kronecker symbol.

Denote by $h(d)$ the class number of the imaginary quadratic field with discriminant d. It was conjectured from the time of C. F. †Gauss that $h(d)\rightarrow\infty$ as $|d|\rightarrow\infty$. This conjecture was proved by H. Heilbronn (1934). More precisely, C. L. Siegel (*Acta Arith.*, 1 (1935)) proved

$$\lim_{|d|\rightarrow\infty}(\log h(d))/(\log\sqrt{|d|})=1.$$

$h(d)=1$ holds for $|d|=3,4,7,8,11,19,43,67,163$. In 1934, Heilbronn and E. H. Linfoot proved that there can be at most one more such d, and finally A. Baker and H. M. Stark independently proved that these nine numbers are the only ones for which $h(d)=1$ (Baker, *Mathematica*, 13 (1966); Stark, *Michigan Math*

J., 14 (1967)). Also, Baker and Stark proved independently (*Ann. Math.*, (2) 94 (1971); *Math. Comp.*, 29 (1975)) that $h(d)=2$ holds only for $|d|=15, 20, 24, 35, 40, 51, 52, 88, 91, 115, 123, 148, 187, 232, 235, 267, 403$, and 427.

For real quadratic fields, we have

$$\lim_{d\to\infty}(\log(h(d)\log\varepsilon_d))/(\log\sqrt{d})=1,$$

where ε_d is the fundamental unit of $k=\mathbf{Q}(\sqrt{d})$ $(\varepsilon_d>1)$. However, it is not yet determined whether there exist infinitely many d with $h(d)=1$ (→ Appendix B, Table 4).

F. Genera

Let G be the group of all (†fractional) ideals of k, and let H be the group of all †principal ideals (α) of k such that $N(\alpha)>0$. Each coset of G modulo H is called an **ideal class in the narrow sense**. (This notion is a special case of the notion of ideal classes in the narrow sense of algebraic number fields; → 14 Algebraic Number Fields G.) In the cases (i) $m<0$ and (ii) $m>0$, $N(\varepsilon_0)=-1$, the usual classification of ideals and classification of ideals in the narrow sense are identical. When $m>0$, $N(\varepsilon_0)=1$, each ideal class is divided into two ideal classes in the narrow sense. We call an ideal class in the narrow sense simply an ideal class.

Let $p_1, \ldots, p_t$ be the set of all prime numbers dividing d. For $n\in\mathbf{Z}$ with $(n,d)=1$, we define $\chi_1(n), \ldots, \chi_t(n)$ as follows: For $p_i\neq 2$, we define $\chi_i(n)=(n/p_i)$; for $p_i=2$, we identify χ_i with χ_2^* in the definition of the Kronecker symbol. In order that $\chi_1(n)=\ldots=\chi_t(n)=1$ for $n\in\mathbf{Z}$ with $(n,d)=1$, it is necessary and sufficient that $n\equiv N(\alpha)\ (\mathrm{mod}\, d)$ for an integer α of k (where $N(\alpha)=\alpha\alpha'$). Since $\chi_1\chi_2\ldots\chi_t$ is equal to the Kronecker symbol, it follows that $n\equiv N(\mathfrak{a})\ (\mathrm{mod}\, d)$ for an integral ideal $\mathfrak{a}$ of k is a necessary and sufficient condition for $\chi_1(n)\ldots\chi_t(n)=1$ to hold for $n\in\mathbf{Z}$ with $(n,d)=1$. Put $\varepsilon_i=\chi_i(N(\mathfrak{a}))$ $(i=1,\ldots,t)$, where $\mathfrak{a}$ is an integral ideal with $(\mathfrak{a}, (d))=1$. Then $(\varepsilon_1,\ldots,\varepsilon_t)$ is uniquely determined for the ideal class C containing $\mathfrak{a}$ and does not depend on the choice of $\mathfrak{a}$. The set $\mathfrak{H}$ of all ideal classes of k such that $(\varepsilon_1,\ldots,\varepsilon_t)=(1,\ldots,1)$ is called the **principal genus** of k, and $\mathfrak{H}$ is a subgroup of the ideal class group $\mathfrak{C}$ of k. Each coset of $\mathfrak{C}$ modulo $\mathfrak{H}$ is called a **genus** of k. For each genus, the values $\varepsilon_i=\chi_i(N(\mathfrak{a}))$ $(i=1,\ldots,t)$ are uniquely determined. We call $(\varepsilon_1,\ldots,\varepsilon_t)$ the **character system** of this genus, and each genus is uniquely determined by its character system. A necessary and sufficient condition for $(\varepsilon_1,\ldots,\varepsilon_t)$ to be a character system for some genus is that $\varepsilon_i=\pm1$ and $\varepsilon_1\varepsilon_2\ldots\varepsilon_t=1$. Hence there are 2^{t-1} genera of k, and $\mathfrak{C}/\mathfrak{H}$ is an Abelian group of type $(2,2,\ldots,2)$.

In order that an ideal class C belong to the principal genus, it is necessary and sufficient that $C=C_1^{1-\sigma}=C_1^2$ hold for some ideal class C_1. From this it follows that there are $t-1$ †invariants of the ideal class group $\mathfrak{C}$ of k that are powers of 2. Each ideal class C such that $C^\sigma=C$ is called an **ambig class** of k. There are 2^{t-1} ambig classes, and they form an Abelian group of type $(2,2,\ldots,2)$. Each ideal $\mathfrak{a}$ of k with $\mathfrak{a}^\sigma=\mathfrak{a}$ is called an **ambig ideal** of k. Let $(p_i)=\mathfrak{p}_i^2$ $(i=1,\ldots,t)$. Then each ambig ideal is uniquely expressed in the form $\mathfrak{a}=\mathfrak{p}_1^{\nu_1}\ldots\mathfrak{p}_t^{\nu_t}(a)$ by some $a\in\mathbf{Q}$ and $\nu_i=0, 1$. Each ambig class contains exactly two ambig ideals of the form $\mathfrak{p}_1^{\nu_1}\ldots\mathfrak{p}_t^{\nu_t}$. For example, for $k=Q(\sqrt{-65})$ we have $d=-2^2\cdot5\cdot13$, $t=3$, $h=8$, $\mathfrak{C}=\{E, A, A^2, A^3, B, AB, A^2B, A^3B\}$, where $A^4=E$, $B^2=E$, $A^\sigma=A^3$, and $B^\sigma=B$; the principal genus is $\mathfrak{H}=\{E, A^2\}$, and the ambig classes are $\{E, A^2, B, A^2B\}$.

G. Norm Residues

A quadratic field k is the †class field over $\mathbf{Q}$ for an ideal group H. The †conductor of H is said to be the **conductor** of $k/\mathbf{Q}$ or simply the conductor of k. The conductor $f=\prod_p f_p$ of $k=\mathbf{Q}(\sqrt{m})$ is given by $f=d$ for $m>0$ and $f=dp_\infty$ for $m<0$. That is, the p-conductor $f_p=p$ for $p\,|\,d$, $p\neq2$; and $f_2=2^e$ for $2\,|\,d$, $d=2^em'$ $((2,m')=1)$. By means of the †Hilbert norm-residue symbol, the Kronecker symbol χ is expressed by

$$\chi(a)=\begin{cases}\prod_{p|d}\left(\dfrac{a,d}{p}\right) & \text{for } (a,d)=1,\\ 0 & \text{for } (a,d)\neq1.\end{cases}$$

H. History

The arithmetic of quadratic fields was originally developed in terms of the theory of binary quadratic forms with rational integral coefficients by Gauss and Dirichlet [2]. The theory was then translated into the terms of ideal theory by J. W. R. Dedekind [2] (→ 348 Quadratic Forms M). For example, the theory of genera for quadratic fields explained in Section F was first developed by Gauss in terms of binary quadratic forms, and the class number formula was obtained by Dirichlet as a formula for binary quadratic forms. Hilbert [4, ch. 2] developed the arithmetic of quadratic fields systematically by introducing the Hilbert norm-residue symbol (→ [1, 5, 6]). Later the arithmetic of quadratic fields assumed the aspect of a simple example of class field theory (→ 59 Class Field Theory).

References

[1] Z. I. Borevich and I. R. Shafarevich, Number theory, Academic Press, 1966. (Original in Russian, 1964.)
[2] P. G. L. Dirichlet, Vorlesungen über Zahlentheorie, herausgegeben und mit Zusätzen versehen von R. Dedekind, Viewig, fourth edition, 1894 (Chelsea, 1969).
[3] H. Hasse, Vorlesungen über Zahlentheorie, Springer, 1950.
[4] D. Hilbert, Die Theorie der algebraischen Zahlkörper, Jber. Deutsch. Math. Verein. 4, I–XVIII (1897), 175–546. (Gesammelte Abhandlungen I, Springer, 1932, 63–363; Chelsea, 1967.)
[5] B. W. Jones, The arithmetic theory of quadratic forms, Math. Assoc. of America and Wiley, 1950.
[6] E. G. H. Landau, Vorlesungen über Zahlentheorie I, II, III, Hirzel, 1927 (Chelsea, 1969).

348 (III.15) Quadratic Forms

A. General Remarks

A **quadratic form** Q is a quadratic homogeneous polynomial with coefficients in a †field K, written

$$Q(x_1, \ldots, x_n) = \sum_{1 \leqslant i \leqslant k \leqslant n} c_{ik} x_i x_k.$$

If the coefficients c_{ik} belong to the field of real (complex) numbers, we call Q a **real (complex) quadratic form**. Let V be an n-dimensional vector space over K. For a vector x in V whose coordinates are $x_1, \ldots, x_n$, we put $Q(x) = Q(x_1, \ldots, x_n)$. This gives rise to a mapping $x \rightarrow Q(x)$ of V into K. Such a mapping satisfies the following two conditions: (i) $Q(ax) = a^2 Q(x)$ ($a \in K$); and (ii) $Q(x+y) - Q(x) - Q(y) = B(x, y)$ is a †symmetric bilinear form on V. Conversely, if a mapping $Q: V \rightarrow K$ satisfies these two conditions, then Q must come from a quadratic form (→ 256 Linear Spaces). $B(x, y)$ is called the **symmetric bilinear form associated with** Q.

We assume that the †characteristic of K is not 2. Putting $a_{ik} = a_{ki} = c_{ik}/2$ $(i < k)$, $a_{ii} = c_{ii}$ $(i = 1, \ldots, n)$, we have $Q(x) = \sum_{i,k=1}^{n} a_{ik} x_i x_k$. The matrix $A = (a_{ik})$ is the **matrix of the quadratic form** Q, and the determinant $|A|$ is the **discriminant** of Q, denoted by $\Delta(Q)$. (Sometimes, instead of $|A|$, we call $(-1)^{n(n-1)/2} 2^n |A|$, the discriminant of Q.) The rank of the matrix A is called the **rank** of Q. Using the notation for the †inner product of vectors, we can write $Q(x) = (x, Ax) = {}^t x A x$; $2^{-1} B(x, y) = (x, Ay) = {}^t x A y$. We say that Q is **nondegenerate** if $B(x, y)$ is nondegenerate (i.e., if $|A| \neq 0$).

Consider a linear substitution $x_i = \sum_{j=1}^{m} P_{ij} x'_j$ (i.e., $x = Px'$, with an $n \times m$ matrix P). Then we get a new quadratic form $Q'(x')$ with the matrix ${}^t PAP$. If each p_{ij} belongs to the field K (to a subring R of K that contains the unit element of K), we say that Q represents Q' over K (resp. R). A basic problem in the theory of quadratic forms is to determine the exact conditions under which a given quadratic form Q represents another quadratic form Q'. The problem of representing numbers by a quadratic form (representation problem) is the particular case corresponding to $m = 1$. Any quadratic form Q represents 0 by taking the zero matrix as P. Hence, by the expression "Q represents 0 over K," we usually mean the nontrivial representation of 0 over K by Q, i.e., $Q(x) = 0$ for some nonzero vector x. If Q is nondegenerate and represents 0, then it represents any element of K^*. Given an element μ in K^*, we consider the quadratic form Q' defined by $Q'(x_1, \ldots, x_{n+1}) = Q(x) - \mu x_{n+1}^2$. Then Q represents μ if and only if Q' represents 0.

Another important special case is that of $n = m$. Then the discriminant of Q' is given by $|P|^2 |A|$. In particular, if $|P| \neq 0$ ($|P|$ is an invertible element of R), we say that Q is **equivalent** to Q' over K (resp. R). This gives rise to an equivalence relation. Equivalent forms have the same rank. On the other hand, if the rank of Q is r, then Q is equivalent to a form $\sum_{i=1}^{r} a_i x_i^2$ over K ($a_i \neq 0$, $i = 1, \ldots, r$). Generally, for elements a and b in $K^* = K - \{0\}$, we write $a \sim b$ if $a \cdot b^{-1} \in (K^*)^2$. Then if Q is equivalent to Q', we have $\Delta(Q) \sim \Delta(Q')$.

When we specify a field K, we assume that the coefficients of the quadratic forms and the coordinates of linear transformations are all contained in the field K. In particular, the equivalence of the forms is equivalence over K.

B. Complex Quadratic Forms

If K is the field of complex numbers $\mathbf{C}$, then a form of rank r is equivalent to the form $\sum_{i=1}^{r} x_i^2$; hence over $\mathbf{C}$ two forms of the same dimension are equivalent if and only if they are of the same rank.

C. Real Quadratic Forms

Now let K be the field of real numbers $\mathbf{R}$. If Q is of rank r, then it is equivalent to the form $\sum_{i=1}^{p} x_i^2 - \sum_{j=1}^{q} x_{p+j}^2$ $(p + q = r)$. Here, p and q are uniquely determined by Q (**Sylvester's law of inertia**). We call (p, q) the **signature** of Q. Two quadratic forms of the same dimension are equivalent if and only if they have the same signature. A quadratic form with the signature $(n, 0)$ (resp. $(0, n)$) is called a **positive (negative) definite quadratic form**. Q is called a **definite quadratic form** if it is either positive or negative definite; otherwise it is an **indefinite quadratic form**. Each of the following conditions is necessary and sufficient for a form Q to be positive definite: (i) for any nonzero real vector x we have $Q(x)>0$; (ii) all the †principal minors of the matrix of Q are positive. Q is negative definite if and only if $-Q$ is positive definite. A form with n variables is called a **positive (negative) semidefinite quadratic form** if its signature is $(r, 0)$ (resp. $(0, r)$), where $1 \leqslant r<n$.

A linear transformation $x' \rightarrow x=Px'$ that leaves invariant the unit form $\sum_{i=1}^{n} x_i^2$ is an **orthogonal transformation**. Then P is an †orthogonal matrix. Any quadratic form can be transformed into a diagonal form $\sum_{i=1}^{n} a_i x_i^2$ via an orthogonal transformation. Here $a_1, \ldots, a_n$ are the †eigenvalues of the matrix of the form. Two forms Q and Q' are equivalent with respect to an orthogonal transformation if and only if the corresponding matrices have the same eigenvalues.

D. Quadratic Forms over Finite Fields and p-adic Number Fields

Let Q and Q' be nondegenerate quadratic forms over the †finite field $\mathbf{F}_q$. They are equivalent if and only if they have the same rank and $\Delta(Q) \sim \Delta(Q')$. Moreover, if the rank of Q is not less than 3, then Q represents 0.

Next, suppose that Q and Q' are nondegenerate quadratic forms over the †p-adic number field K. They are equivalent if and only if they have the same rank, $\Delta(Q) \sim \Delta(Q')$, and they have the same **Minkowski-Hasse character** χ, where χ is defined as follows: Let $C(Q)$ be the †Clifford algebra of Q, and let $C^*(Q)$ denote $C(Q)$ if n is even and $C^+(Q)$ if n is odd. Then $\chi(Q)=1$ or -1 according as $C^*(Q) \cong M_t(K)$ or $M_t(K) \otimes D(K)$. (Here M_t is the total matrix algebra of degree t over K and $D(K)$ is the unique †quaternion algebra over K.) Also, if Q has rank not less than 5, then Q represents 0 in K.

E. Quadratic Forms over a General Field K

The following facts are valid on any field K whose characteristic is not 2. Given a quadratic form Q_1 with variables $x_1, \ldots, x_n$ and another form Q_2 with variables $x_{n+1}, \ldots, x_{n+m}$, we get a new quadratic form $Q_1 \oplus Q_2$ or Q_1+Q_2 defined by $Q_1 \oplus Q_2(x_1, \ldots, x_{n+m})= Q_1(x_1, \ldots, x_n)+Q_2(x_{n+1}, \ldots, x_{n+m})$. $Q_1 \oplus Q_2$ is called the **direct sum** of Q_1 and Q_2. The matrix of $Q_1 \oplus Q_2$ is the direct sum of the matrices of Q_1 and Q_2. If Q_1 and Q_1' are equivalent and $Q_1 \oplus Q_2$ and $Q_1' \oplus Q_2'$ are also equivalent, then Q_2 and Q_2' are equivalent (**Witt's theorem**).

The quadratic form $x_1x_2+x_3x_4+\ldots+x_{2r-1}x_{2r}$ is called the **kernel form** and is denoted by N_r. Any nondegenerate quadratic form $Q(x_1, \ldots, x_n)$ is equivalent to the direct sum of a kernel form $N_r(x_1, \ldots, x_{2r})$ and a form $Q_0(x_{2r+1}, \ldots, x_n)$, where if $Q_0 \neq 0$, i.e., $n>2r$, we have $Q_0(x_{2r+1}, \ldots, x_n)=0$ only if $x_{2r+1}=\ldots=x_n=0$. N_r and Q_0 are uniquely determined by Q up to equivalence. The decomposition $N_r \oplus Q_0$ is called the **Witt decomposition** of Q (E. Witt [5]). The number r is called the **index** of Q. An element x in V is said to be **singular** with respect to Q if $Q(x)=0$. A subspace W of V is said to be **totally singular** if all the elements in W are singular. Let B be the symmetric bilinear form associated with Q. Then x is singular with respect to Q if and only if $B(x, x)=0$ (characteristic of $K \neq 2$). We say that x is **isotropic** if $B(x, x)=0$. Thus a subspace W is totally singular if and only if it is **totally isotropic** (i.e., $B(x, y)=0$ for all $x, y \in W$). The index r of Q is the dimension of a maximal totally singular subspace of V. In particular, if $K=\mathbf{R}$ and (p, q) is the signature of Q, then the index $r=\min(p, q)$. Here we must be careful, since some authors call the number $p-q$ or p or q the index of Q. To make the distinction clear, we also call our r the **index of total isotropy**, and the number $p-q$ the **index of inertia**.

Necessary and sufficient conditions for a nondegenerate Q to be a kernel form are: $n=$ the rank of $Q \equiv 0 \pmod 2$ when $K=\mathbf{C}$; $n \equiv 0 \pmod 2$ and $p-q=0$ when $K=\mathbf{R}$; $n \equiv 0 \pmod 2$ and $\Delta(Q) \sim 1$ when $K=\mathbf{F}_q$; $n \equiv 0 \pmod 2$, $\Delta(Q) \sim 1$, and $\chi(Q)=1$ when K is a p-adic number field.

Let $N_r \oplus Q_0$, $N_s \oplus Q_0'$ be Witt decompositions of Q and Q', respectively. We say that Q and Q' belong to the same **type** if Q_0 and Q_0' are equivalent and denote the set of types of nondegenerate quadratic forms over K by $\mathbf{W}$. We define the sum of the types of Q and Q' as the type of $Q \oplus Q'$, and this gives $\mathbf{W}$ the structure of a commutative group. The type of a kernel form is the identity element of this group $\mathbf{W}$, called the **Witt group**. The structure of the Witt group depends on K. If $K=\mathbf{C}$, then $\mathbf{W} \cong \mathbf{Z}/2\mathbf{Z}$; if $K=\mathbf{R}$, then $\mathbf{W} \cong \mathbf{Z}$; if K is a †local field with a †non-Archimedean valuation, then $\mathbf{W}$ is a finite group; if $K=\mathbf{F}_q$, then $\mathbf{W} \cong (\mathbf{Z}/2\mathbf{Z})+(\mathbf{Z}/2\mathbf{Z})$ if $q \equiv 1 \pmod 4$, $\mathbf{W} \cong \mathbf{Z}/4\mathbf{Z}$ if $q \equiv 3 \pmod 4$, and $\mathbf{W} \cong \mathbf{Z}/2\mathbf{Z}$ if q is a power of 2.

F. Hermitian Forms

An expression $H(x)=\sum_{i,k=1}^{n} a_{ik}\bar{x}_i x_k$ is called a **Hermitian form** if $a_{ik}\in\mathbf{C}$, $a_{ik}=\bar{a}_{ki}$. (Here $\bar{a}_{ik}$, $\bar{x}_i$ are the complex conjugates of a_{ik}, x_i, respectively.) The value $H(x)$ is a real number. As for quadratic forms, we define the notions of the matrix of H and the discriminant, rank, and †sesquilinear form associated with H. The matrix A of H is a †Hermitian matrix whose principal minors are real numbers. If we apply a linear transformation $P(x')=x$, we obtain a Hermitian form with respect to x' whose matrix is given by ${}^t\bar{P}AP$. Any Hermitian form is equivalent to a form $\sum_{i=1}^{p}\bar{x}_i x_i - \sum_{j=1}^{q}\bar{x}_{p+j}x_{p+j}$; (p,q) is called the **signature** of H. We define the notions of **positive definite**, **negative definite**, and **indefinite Hermitian forms** as we did for quadratic forms over the field of real numbers. Each of the following conditions is necessary and sufficient for H to be positive definite: (i) $H(x)>0$ for any nonzero complex vector x; (ii) all the principal minors of the matrix of H are positive. The definition of a **semidefinite Hermitian form** is given in the same manner as for a quadratic form.

A linear transformation that leaves the Hermitian form $\sum_{i=1}^{n}\bar{x}_i x_i$ invariant is called a **unitary transformation**, and its matrix is a †unitary matrix. Any Hermitian form can be transformed via a unitary transformation into a diagonal form $\sum_{i=1}^{n} a_i\bar{x}_i x_i$, where $a_1,\ldots,a_n$ are the eigenvalues of the matrix of the Hermitian form.

The notion of Hermitian forms can be generalized as follows: Suppose that K is a †division ring with an **involution** $a\to\bar{a}$ $(a\in K)$ (i.e., $a\to\bar{a}$ is a linear mapping of K onto itself such that $\bar{\bar{a}}=a$, $\overline{ab}=\bar{b}\bar{a}$). Then a Hermitian form H over K is defined by

$$H(x)=\sum_{i,k=1}^{n}\bar{x}_i a_{ik} x_k,$$

where $x_i\in K$, $a_{ik}\in K$, $a_{ik}=\bar{a}_{ki}$. In particular, if we have, for any given vector x whose coordinates belong to K, an element a in K such that $H(x)=a+\bar{a}$, then we have a Witt decomposition for H. Two examples of such K having involutions that differ from the identity mapping are a separable quadratic extension K of a field L and a †quaternion algebra K over a field L.

G. Quadratic Forms over Algebraic Number Fields

Let K be an algebraic number field of finite degree, $\mathfrak{p}$ be an †Archimedean or non-Archimedean †prime divisor of K, $K_\mathfrak{p}$ be the $\mathfrak{p}$-completion of K, and Q and Q' be nondegenerate quadratic forms over K. Then Q represents Q' over K if and only if Q represents Q' over $K_\mathfrak{p}$ for all $\mathfrak{p}$, and Q represents 0 in K (i.e., there exists a nonzero vector x whose coordinates belong to K such that $Q(x)=0$) if and only if it represents 0 in $K_\mathfrak{p}$ for all $\mathfrak{p}$ [3, 4]. In particular, Q and Q' are equivalent over K if and only if they are equivalent over $K_\mathfrak{p}$ for all $\mathfrak{p}$. Hence the invariants with respect to equivalence over K of a nondegenerate quadratic form Q over K are $n=$ the rank of Q, Δ $=$ the discriminant of Q, the Minkowski-Hasse character $\chi_\mathfrak{p}$ for †prime divisors $\mathfrak{p}$ of K, and the index of inertia j_λ of Q over $K_{\mathfrak{p}_{\infty,\lambda}}$ for each †real infinite prime divisor $\mathfrak{p}_{\infty,\lambda}$ $(\lambda=1,\ldots,r_1)$ of K. Here the following properties hold: (i) $\chi_\mathfrak{p}=1$ for all but finitely many $\mathfrak{p}$; (ii) $\prod_\mathfrak{p}\chi_\mathfrak{p}=1$ (this is equivalent to the †product formula of norm-residue symbols); (iii) $\Delta\sim(-1)^{(n^2+j_\lambda)/2}$ in $K_{\mathfrak{p}_{\infty,\lambda}}$; and (iv) $\chi_{\mathfrak{p}_{\infty,\lambda}}=1$ if $j_\lambda\equiv 0, 1, 2, 7 \pmod 8$, $=-1$ if $j_\lambda\equiv 3, 4, 5, 6 \pmod 8$ [3, 4]. Conversely, if the system $\{n,\chi_\mathfrak{p},\chi_{\mathfrak{p}_{\infty,\lambda}}, j_\lambda,\Delta\}$ satisfies conditions (i)–(iv), then it is the set of invariants of a quadratic form over K (**Minkowski-Hasse theorem**). In general, if a property concerning K holds if and only if it holds for all $K_\mathfrak{p}$, we say that the **Hasse principle** holds for the property.

H. Class and Genus of a Quadratic Form

Let K be an algebraic number field of finite degree. Quadratic forms Q and Q' over K are said to be of the same **class** if they are equivalent over the †principal order $\mathfrak{o}$ in K. On the other hand, Q and Q' are said to be of the same **genus** if (i) they are equivalent over the principle order $\mathfrak{o}_\mathfrak{p}$ in $K_\mathfrak{p}$ for all non-Archimedean prime divisors $\mathfrak{p}$ of K and (ii) they are equivalent over $K_\mathfrak{p}$ for all the Archimedean prime divisors $\mathfrak{p}$ of K. A genus is decomposed into a finite number of classes. For example, if K is the field of rational numbers, the number of classes in the genus of $\sum_{i=1}^{m}x_i^2$ is 1 for $m\leqslant 8$, while it is $\geqslant 2$ for $m>8$.

I. Reduction of Real Quadratic Forms

Let A be an $m\times m$ matrix and X an $m\times n$ matrix. We put $A[X]={}^tXAX$. Then we can write $Q(x)=S[x]={}^txSx$, where S is the matrix of the quadratic form Q. In this section we put $K=\mathbf{R}$ and define two forms to belong to the same **class** if they are equivalent over the ring of rational integers. We identify the form Q with its matrix $S=(s_{ij})$. Let S be a positive definite form in m variables. Then S is said to be a **reduced quadratic form** if $S[\mathfrak{g}_k]\geqslant s_{kk}$ and $s_{l\,l+1}\geqslant 0$ $(1\leqslant k\leqslant m,\ 1\leqslant l\leqslant m-1)$, where $\mathfrak{g}_k$ is an arbitrary vector whose coordinates $g_1, \ldots, g_m$ are integers such that $(g_k, \ldots, g_m)=1$. Any class of positive definite quadratic forms contains at least one (and generally only one) reduced form. For a reduced form $R=(r_{kl})$, the following inequalities hold: $0<r_{11}\leqslant r_{22}\leqslant \ldots \leqslant r_{mm}$; $\pm 2r_{kl}\leqslant r_{ll}$, $k<l$; $r_{11}r_{22}\ldots r_{mm}<c(m)|R|$, where $c(m)$ depends only on m. The set of all symmetric matrices of degree m forms a linear space of dimension $m(m+1)/2$ in which the subset $\mathfrak{P}$ formed by the positive definite symmetric matrices is a convex open subset. Moreover, the subset $\mathfrak{R}$ formed by all reduced positive definite symmetric matrices is a convex cone whose boundary consists of finitely many hypersurfaces and whose vertex is the origin. Let S be an indefinite quadratic form whose signature is $(n, m-n)$. The set of positive definite quadratic forms H such that $S^{-1}[H]=S$ forms a variety of dimension $n(m-n)$, which is denoted by $H(S)$. We say that S is reduced if $H(S)\cap\mathfrak{R}\neq\varnothing$. Given a natural number D, there are only a finite number of definite or indefinite reduced quadratic forms with rational integral coefficients whose discriminant is $\pm D$. Hence the number of classes of quadratic forms with rational integral coefficients and discriminant $\pm D$ is also finite.

J. Units

Let S be a symmetric matrix of degree m with rational coordinates. Let $O(S)$ be the set of all real $m\times m$ matrices W for which $S[W]=S$, and $\Gamma(S)$ be the subset of $O(S)$ consisting of the integral matrices. An element of $\Gamma(S)$ is called a **unit** of S. $\Gamma(S)$ is a finite group if S is definite, but otherwise it is infinite (except for the case $m=2$, $-|S|=r^2$, with rational r). $O(S)$ is a †Lie group, and $\Gamma(S)$ is a discrete subgroup with a finite number of generators. The †homogeneous space $O(S)/\Gamma(S)$ is of finite measure with respect to a †Haar measure defined on the space.

K. Minkowski-Siegel-Tamagawa Theory

Let S and T be rational integral positive definite symmetric matrices of degree m and n, respectively $(m\geqslant n)$. Let $A(S,T)$ be the number of rational integral solutions for the equation $S[X]=T$, and $E(S)$ be the order of the group of units $\Gamma(S)$. We put

$$M(S,T)=\frac{A(S_1,T)}{E(S_1)}+\frac{A(S_2,T)}{E(S_2)}+\ldots,$$

$$M(S)=\frac{1}{E(S_1)}+\frac{1}{E(S_2)}+\ldots,$$

$$A_0(S,T)=\frac{M(S,T)}{M(S)},$$

where $S_1, S_2, \ldots$ is a complete system of representatives of the classes in the genus of S. $M(S)$ is called the **measure of genus** of S. On the other hand, for a natural number q, we denote by $A_q(S,T)$ the number of the solutions of the congruence equation $S[X]\equiv T \pmod q$. If q is a prime power p^a, then the ratio $\varepsilon_{m,n}q^{-mn+n(n+1)/2}A_q(S,T)$ takes a constant value $\alpha_p(S,T)$ for sufficiently large a (where $\varepsilon_{m,n}=1/2$ if $m=n\geqslant 2$; $=1$ otherwise). Furthermore, let us consider a domain B in the Euclidean space of dimension $n(n+1)/2$ formed by the set of $n\times n$ symmetric matrices containing T, and let B_1 be the domain formed by the matrices X such that $S[X]\in B$. Then B_1 is a domain in the space of dimension mn formed by all $m\times n$ matrices. Let $\alpha_\infty(S,T)$ be the limit of the ratio $v(B_1)/v(B)$ of the volumes of B and B_1 as the domain B shrinks toward the point T. Then **Siegel's theorem** states: $\alpha_\infty(S,T)\prod_p \alpha_p(S,T)=\varepsilon A_0(S,T)$, where $\varepsilon=2$ if $m=n+1$ or $m=n\geqslant 2$ and $\varepsilon=1$ otherwise. The infinite product of the left-hand side of this equation does not converge absolutely if either $m=n=2$ or $m=n+2$, and in those cases the order of the product $\prod_p$ is considered to be the natural order of the primes p.

A special case of Siegel's theorem was proved by H. Minkowski, but it was C. L. Siegel [6] who proved it in its general form. Except for a finite number of p, the numbers $\alpha_p(S,T)$ have been calculated. The explicit form of $\alpha_\infty(S,T)$ is also known. In particular, if we take the identity matrix $E^{(m)}$ of degree m as S, then the formula in Siegel's theorem is related to the problem of expressing natural numbers as sums of m squares. For $m=2,3,\ldots,8$, the genus of $E^{(m)}$ contains only one class. Hence, putting $n=1$, $T=t$ $(=$a natural number), we obtain from Siegel's theorem the number of ways in which we can express t as the sum of m squares [6, pt. I]. Siegel's result was generalized by Siegel himself to the case where the form S is indefinite [6, pt. II] and where the coefficients of the forms are elements of an algebraic number field of finite degree [6, pt. III]. Also, regarding the number of possible ways to express a natural number t as a sum of m squares, the following formula was obtained by C. G. J. Jacobi for the case where $m=4$, $n=1$:

$$A(E^{(4)},t)=8\left(\sum_{d|t} d-\sum_{4d|t} d\right).$$

For the case $m=3$, $n=1$, it is known that if t is odd and $A(E^{(3)},t)>0$, then $t\not\equiv 7 \pmod 8$ (for details → P. T. Batemann, *Trans. Amer. Math. Soc.*, 71 (1951)).

T. Tamagawa used the theory of †adelized algebraic groups and proved that the †Tamagawa number $\tau(SO(n,S))$ of the special orthogonal group is 2. He also showed that from this fact, Siegel's theory in this section can be deduced (→ 13 Algebraic Groups P) [10].

L. Theta Series

Let $Q(x_1, \ldots, x_m)$ be a positive definite form with integral coefficients. For a complex number z, we put

$$F(z,Q)=\sum_{x_1,\ldots,x_m}\exp(2\pi iQ(x_1,\ldots,x_m)z),$$

where $x_1, \ldots, x_m$ run over all the integers. If $\mathrm{Im}\, z>0$, the series converges and represents an †entire function of z. These series are called **theta series**. If we denote by $A(n)$ the number of integral solutions of the equation $Q(x_1, \ldots, x_m)=n$, we have

$$F(z,Q)=\sum_{n=0}^{\infty}A(n)e^{2\pi inz}.$$

Moreover, if $m=2k$, we have the following **transformation formula**:

$$F\left(\frac{az+b}{cz+d},Q\right)=\varepsilon(d)(cz+d)^kF(z,Q),$$

where a, b, c, d are integers such that $ad-bc=1$, $c\equiv 0 \pmod N$, N is a natural number determined by Q, and ε is a character mod N. In other words, $F(z,Q)$ is a †modular form with respect to the †congruence subgroup of level N. Using the theory of modular forms, E. Hecks showed that $A(n)=A_0(n)+O(n^{k/2})$, where $A_0(n)$ is a number-theoretic function of n determined by the genus of Q.

M. Binary Quadratic Forms with Integral Coefficients

Now put $m=2$. Given a form $Q(x,y)=ax^2+bxy+cy^2$, we put $D(Q)=b^2-4ac$ and call it the **discriminant** of Q (i.e., $D(Q)=-4\Delta(Q)$). Q is said to be **primitive** if $(a,b,c)=1$. When $D(Q)$ is not a square, the theory concerning Q is closely related to the arithmetic theory of the †quadratic field $\mathbf{Q}(\sqrt{D})=k$. Let d be the †discriminant of k and put $D=df^2$. When $f=1$, there is a one-to-one correspondence between the †ideal classes of k and the classes of quadratic forms with discriminant d (when $D<0$, we consider the classes of positive definite forms). The correspondence is given in the following manner: If $\mathfrak{a}$ is an ideal in k with a basis α_1, α_2, then the corresponding form is given by $Q(x,y)=N(\mathfrak{a})^{-1}N(\alpha_1x+\alpha_2y)$, where N is the absolute †norm. If $f>1$, we must replace the ring of integers $\mathfrak{o}$ by the †order of the †conductor f. That is, if we consider the ring formed by the elements $x+fy\omega$ (x,y are rational integers; the meaning of ω is explained shortly), we again have a one-to-one correspondence between the classes of ideals of this order and the classes of quadratic forms [3]. We let $\omega=\sqrt{d}/2$ if $d\equiv 0 \pmod 4$; $=(1+\sqrt{d})/2$ if $d\equiv 1 \pmod 4$. When $D>0$, we can introduce the notion of proper equivalence as follows: Q and Q' are **properly equivalent** if the matrix of Q is transformed to the matrix of Q' by a linear transformation P whose determinant is 1. Then, in the correspondence for the case $f=1$, if we take $\alpha_1=r>0$, $\alpha_2=s+t\omega$, $t>0$ ($r,s,t\in\mathbf{Q}$), then we get a relation between the classification of the forms and the classification in the finer sense of the ideals in k.

Suppose that $D>0$ is not a square. Let (t,u) be an integral solution of †Pell's equation $t^2-Du^2=\pm 4$. Then the units of the form $Q(x,y)=ax^2+bxy+cy^2$ with discriminant D are given by

$$\pm\begin{pmatrix}(t-bu)/2 & -cu\\ au & (t+bu)/2\end{pmatrix}.$$

Let (t_0,u_0) be the smallest positive integral solution of $t^2-Du^2=4$, put $\varepsilon_D=(t_0+u_0\sqrt{D})/2$, and let h_D be the class number in the finer sense of the forms of discriminant D. Then the following formula holds (Dirichlet):

$$\frac{1}{\sqrt{D}}h_D\log\varepsilon_D=\sum_{n=1}^{\infty}\frac{1}{n}\left(\frac{D}{n}\right),$$

where (D/n) is the †Kronecker symbol (here $D=f^2d$; we put $(D/n)=0$ if $(f,n)\neq 1$, and $(D/n)=(d/n)$ if $(f,n)=1$). For $D<0$, the order w_D of the units is known: it is 6 if $D=-3$; 4 if $D=-4$; and 2 otherwise. We also have

$$\frac{2\pi}{\sqrt{|D|}}\frac{h_D}{w_D}=\sum_{n=1}^{\infty}\frac{1}{n}\left(\frac{D}{n}\right).$$

With respect to the numbers h_D and ε_D, little else is known.

References

[1] B. W. Jones, The arithmetic theory of quadratic forms, Wiley, 1950.
[2] N. Bourbaki, Eléments de mathématique, Algèbre, ch. 9, Actualités Sci. Ind., 1272a, Hermann, 1959.
[3] M. Eichler, Quadratische Formen und orthogonale Gruppen, Springer, 1952.
[4] O. T. O'Meara, Introduction to quadratic forms, Springer, 1963.
[5] E. Witt, Theorie der quadratischen Formen in beliebigen Körpern, J. Reine Angew. Math., 176 (1937), 31–44.
[6] C. L. Siegel, Über die analytische Theorie der quadratischen Formen I, II, III, Ann. Math., (2) 36 (1935), 527–606; 37 (1936), 230–263; 38 (1937), 212–291 (Gesammelte Abhand-

lungen, Springer, 1966, vol. 1, 326–405, 410–443, 469–548).
[7] C. L. Siegel, On the theory of indefinite quadratic forms, Ann. Math., (2) 45 (1944), 577–622. (Gesammelte Abhandlungen, Springer, 1966, vol. 2, 421–466.)
[8] E. Hecke, Analytische Arithmetik der positiven quadratischen Formen, Kgl. Danske Videnskab. Selskab, Mat.-Fys. Medd., XIII, 12 (1940). (Mathematische Werke, Vandenhoeck & Ruprecht, 1959, 789–918.)
[9] G. L. Watson, Integral quadratic forms, Cambridge Univ. Press, 1960.
[10] T. Tamagawa, Adèles, Amer. Math. Soc. Proc. Symp. Pure Math., 9 (1966), 113–121.
For binary quadratic forms,
[11] P. G. L. Dirichlet, Vorlesungen über Zahlentheorie, Vieweg, fourth edition, 1894 (Chelsea, 1969).

349 (XIX.3) Quadratic Programming

A. Problems

A **quadratic programming problem** is a special type of mathematical programming (→ 264 Mathematical Programming) where the objective function is quadratic while the constraints are linear. A typical formulation of the problem is as follows.

(Q) Maximize $z=\mathbf{c}'\mathbf{x}-\frac{1}{2}\mathbf{x}'D\mathbf{x}$ under the condition $A\mathbf{x}\leqslant\mathbf{b}$ and $\mathbf{x}\geqslant 0$, $\mathbf{x}\in\mathbf{R}^n$.

Let the Lagrangian form for this problem be

$$\varphi(\mathbf{x},\boldsymbol{\lambda})=\mathbf{c}'\mathbf{x}-\tfrac{1}{2}\mathbf{x}'D\mathbf{x}+\boldsymbol{\lambda}'(\mathbf{b}-A\mathbf{x}).$$

Then, from the general properties of Lagrangian forms, the following theorem can be proved.

Theorem: If $\mathbf{x}=\mathbf{x}^*$ is an optimal solution of the problem (Q), there exists a vector λ^* satisfying the conditions

$$-D\mathbf{x}^*+\mathbf{c}\leqslant\boldsymbol{\lambda}^*, \qquad \boldsymbol{\lambda}^*\geqslant 0;$$

$$\mathbf{b}'\boldsymbol{\lambda}^*=(\mathbf{c}-D\mathbf{x}^*)'\mathbf{x}^*.$$

Moreover, if the matrix D is nonnegative definite, the above conditions are also sufficient for $\mathbf{x}=\mathbf{x}^*$ to be optimal. The second condition can be shown to be equivalent to

$$\boldsymbol{\lambda}^{*\prime}(\mathbf{b}-A\mathbf{x}^*)=0 \quad\text{and}\quad \mathbf{x}^{*\prime}(A'\boldsymbol{\lambda}^*+D\mathbf{x}^*-\mathbf{c})=\mathbf{0}.$$

By introducing the slack vectors $\mathbf{u}\geqslant\mathbf{0}$ and $\mathbf{v}\geqslant\mathbf{0}$, the conditions can be expressed as

$$\mathbf{x}\geqslant\mathbf{0},\quad \mathbf{y}\geqslant\mathbf{0},\quad \mathbf{u}\geqslant\mathbf{0},\quad \mathbf{v}\geqslant\mathbf{0};$$

$$\text{(C)}\quad A\mathbf{x}+\mathbf{u}=\mathbf{b},\quad A'\mathbf{y}+D\mathbf{x}-\mathbf{v}=\mathbf{c};$$

$$\mathbf{y}'\mathbf{u}=0,\quad \mathbf{x}'\mathbf{v}=0.$$

When D is nonnegative definite, any feasible solution of the above system of equalities gives an optimal solution of the primary problem (Q), and when D is positive definite, the solution is unique. When D is not nonnegative definite, the optimal solution of (Q), if it exists, is one of the feasible solutions of (C). The last line of (C) implies that the solution must be a basic solution of the linear system of equalities, and it also restricts the possible combinations of the basic variables. Since there exist only a finite number of possible combinations of the basic variables, the quadratic programming problem can be solved in a finite number of steps, if it has an optimal solution.

B. Duality

The **dual problem** of (Q) is the following.

(QD) Minimize $w=\mathbf{b}'\mathbf{y}+\frac{1}{2}\mathbf{x}'D\mathbf{x}$ under the condition $A'\mathbf{y}+D\mathbf{x}\geqslant\mathbf{c}$ and $\mathbf{x}\geqslant 0$, $\mathbf{y}\geqslant 0$.

If D is nonnegative definite, the following theorem holds.

Theorem: If the primary problem (Q) has a solution $\mathbf{x}=\mathbf{x}^*$, then the dual problem has a solution $\mathbf{x}=\mathbf{x}^*$ and $\mathbf{y}=\mathbf{y}^*$, and $\max z=\min w$.

A more general form of the quadratic programming problem can be given as follows.

(Q) Maximize $z=\mathbf{c}'\mathbf{x}-\frac{1}{2}\mathbf{x}'D\mathbf{x}$ under the condition $\mathbf{x}\in V$ and $\mathbf{b}-A\mathbf{x}\in W$, where V and W are closed convex cones in $\mathbf{R}^n$ and $\mathbf{R}^m$, respectively.

Then the dual problem is expressed as follows.

(QD) Minimize $w=\mathbf{b}'\mathbf{y}-\frac{1}{2}\mathbf{x}'D\mathbf{x}$ under the condition $\mathbf{x}\in V$, $\mathbf{y}\in W^*$ and $A'\mathbf{y}+D\mathbf{x}-\mathbf{c}\in V^*$, where V^* and W^* are the dual cones of V and W.

The above theorem holds for both (Q) and (QD).

C. Algorithms

Various algorithms have been proposed for quadratic programming [1,2,4], most of which are based on condition (C). Wolfe [4] proposed a method based on the simplex method for linear programming. If we introduce the artificial vectors ξ and η, we can find a feasible solution of (C) by solving the following linear programming problem.

(LQ) Maximize $z=-1'\boldsymbol{\xi}-1'\boldsymbol{\eta}$ under the condition that

$$A\mathbf{x}+\mathbf{u}-\boldsymbol{\xi}=\mathbf{b},\quad A'\mathbf{y}+D\mathbf{x}-\mathbf{v}+\boldsymbol{\eta}=\mathbf{c};$$

$$\mathbf{x}\geqslant\mathbf{0},\quad \mathbf{y}\geqslant\mathbf{0},\quad \mathbf{u}\geqslant\mathbf{0},\quad \mathbf{v}\geqslant\mathbf{0},\quad \boldsymbol{\xi}\geqslant\mathbf{0},\quad \boldsymbol{\eta}\geqslant\mathbf{0};$$

$$\mathbf{y}'\mathbf{u}=0,\quad \mathbf{x}'\mathbf{v}=0.$$

(LQ) can be solved by applying the simplex algorithm with the only modification being that the last line of the condition restricts the possible changes in the basic variables. When D is positive definite, we can always obtain a solution if there is a feasible solution of the original problem, and Wolfe proposed a modification of the foregoing algorithm for the case when D is nonnegative definite that tells whether or not it has an optimal solution, and gives it if it has one. Some other algorithms are also effective when D is positive or nonnegative definite, but when D is not nonnegative definite, no simple effective method has been found to reach the optimal solution even when its existence has been established.

References

[1] E. M. L. Beale, On quadratic programming, Naval Res. Logistic Quart., 6 (1959), 227–243.
[2] C. Hildreth, A quadratic programming procedure, Naval Res. Logistic Quart., 4 (1957), 79–85.
[3] W. S. Dorn, Duality in quadratic programming, Quart. Appl. Mat., 18 (1960), 155–162.
[4] P. Wolfe, The simplex method for quadratic programming, Econometrica, 27 (1959), 382–398.

350 (VI.10) Quadric Surfaces

A. Introduction

A subset F of a 3-dimensional Euclidean space E^3 is called a **quadric surface** (**surface of the second order** or simply **quadric**) if F is the set of zeros of a quadratic equation $G(x, y, z)=0$, where the coefficients of G are real numbers. The equation $G(x, y, z)=0$ is written as

$$ax^2+by^2+cx^2+d+2fyz+2gzx+2hxy+2f'x+2g'y+2h'z=0. \tag{1}$$

In general, a straight line intersects a quadric surface at two points. If it intersects the surface at more than two points, then the whole straight line lies on the surface. Suppose that we are given a quadric surface and a point O. Suppose further that we are given a straight line passing through the point O and intersecting the quadric surface at two points A and A'. If $AO=OA'$ for all such straight lines, then the point O is called the **center** of the quadric surface.

B. Classification

The subset defined by equation (1) may be empty; for example, $x^2+y^2+z^2+1=0$. In this article, we consider only quadric surfaces that are not the empty set. When a quadric surface F without †singular points has a center or centers, we say that F is **central**.

If we choose a suitable rectangular coordinate system, the equation of a central quadric surface is written in one of the following forms:

$$\frac{x^2}{a^2}+\frac{y^2}{b^2}+\frac{z^2}{c^2}=1, \tag{2}$$

$$\frac{x^2}{a^2}+\frac{y^2}{b^2}-\frac{z^2}{c^2}=1, \tag{3}$$

$$-\frac{x^2}{a^2}-\frac{y^2}{b^2}+\frac{z^2}{c^2}=1, \tag{4}$$

$$\frac{x^2}{a^2}+\frac{y^2}{b^2}=1, \tag{5}$$

$$\frac{x^2}{a^2}-\frac{y^2}{b^2}=1, \tag{6}$$

$$\frac{x^2}{a^2}=1. \tag{7}$$

When the equation takes the form (2), (3), (4), (5), or (6), we call the quadric surface an **ellipsoid**, **hyperboloid of one sheet**, **hyperboloid of two sheets**, **elliptic cylinder** (or **elliptic cylindrical surface**), or **hyperbolic cylinder** (**hyperbolic cylindrical surface**), respectively. When the equation takes the form (7), the surface coincides with a pair of parallel planes. If $a=b$ in (2), (3), (4), or (5), the surface is a †surface of revolution with the z-axis as the axis of revolution. In this case, we call the surface an **ellipsoid of revolution**, **hyperboloid of revolution of one sheet**, **hyperboloid of revolution of two sheets**, or **circular cylinder** (or **circular cylindrical surface**), respectively. If $a=b=c$ for an ellipsoid of revolution, then the surface is a sphere with radius a.

If we choose a suitable rectangular coordinate system, the equation of a noncentral surface of the second order is written in one of the following forms:

$$2z=\frac{x^2}{a^2}+\frac{y^2}{b^2}, \tag{8}$$

$$2z=\frac{x^2}{a^2}-\frac{y^2}{b^2}, \tag{9}$$

$$2z=\frac{x^2}{a^2}. \tag{10}$$

When the equation takes the form (8), (9), or (10), we call the surface an **elliptic paraboloid**, **hyperbolic paraboloid**, or **parabolic cylinder** (or **parabolic cylindrical surface**), respectively. If a

$=b$ in (8), the surface is called an **elliptic paraboloid of revolution**.

Among these, (2), (3), (4), (8), and (9) are sometimes called **proper quadric surfaces**, and the others **degenerate quadric surfaces**.

Equations (2)–(10) are called the **canonical forms of the equations** of these surfaces (a, b, c in canonical forms should not be confused with a, b, c in (1)). The planes $x=0$, $y=0$, and $z=0$ in surfaces (2), (3), and (4) and the planes $x=0$ and $y=0$ in surfaces (8) and (9) are called **principal planes** of the respective surfaces; and lines of intersection of principal planes are called **principal axes**. For a surface of revolution, positions of principal planes and principal axes are indeterminate. We call a, b, c in equations of canonical form the lengths of the principal axes, or simply the principal axes. If F is a hyperboloid of one sheet or a hyperbolic paraboloid, there are two systems of straight lines lying on F; two straight lines belonging to the same system never meet (and are not parallel), and two straight lines belonging to different systems always meet (or are parallel). If F satisfies (3), these systems of straight lines are given by

$$\begin{cases} \dfrac{x}{a}-\dfrac{z}{c}=\lambda\left(1-\dfrac{y}{b}\right) \\ \dfrac{x}{a}+\dfrac{z}{c}=\dfrac{1}{\lambda}\left(1+\dfrac{y}{b}\right) \end{cases} \quad \begin{cases} \dfrac{x}{a}-\dfrac{z}{c}=\mu\left(1+\dfrac{y}{b}\right) \\ \dfrac{x}{a}+\dfrac{z}{c}=\dfrac{1}{\mu}\left(1-\dfrac{y}{b}\right). \end{cases}$$

If F satisfies (9), then two such systems are given by

$$\begin{cases} \dfrac{x}{a}-\dfrac{y}{b}=\lambda \\ \dfrac{x}{a}+\dfrac{y}{b}=\dfrac{2z}{\lambda} \end{cases} \quad \begin{cases} \dfrac{x}{a}+\dfrac{y}{b}=\mu \\ \dfrac{x}{a}-\dfrac{y}{b}=\dfrac{2z}{\mu} \end{cases}$$

(λ and μ are parameters). We call these straight lines **generating lines** of the respective surfaces. A hyperboloid of one sheet and a hyperbolic paraboloid are †ruled surfaces described by these generating lines.

When a quadric surface has singular points, they are double points. The set of double points of a quadric surface F is either a single point O, a straight line l, or a plane π. In the second case, F consists of two planes passing through l or l itself, and in the third case, F coincides with π. In the first case, we say that F is a **quadric conical surface** (or **quadric cone**) with vertex O. Its equation is written in the form $Ax^2+By^2+Cz^2=0$ ($ABC\neq 0$). When A, B, C are of the same sign, F consists of only one point O. Otherwise, we can assume that A, $B>0$, $C=-1$. In this case, if $A=B$, F is called a **right circular cone**, and if $A\neq B$, F is called an **oblique circular cone**.

Given hyperboloids (3) and (4), we call the quadric cones

$$\frac{x^2}{a^2}+\frac{y^2}{b^2}-\frac{z^2}{c^2}=0 \qquad (3')$$

and

$$-\frac{x^2}{a^2}-\frac{y^2}{b^2}+\frac{z^2}{c^2}=0 \qquad (4')$$

asymptotic cones of (3) and (4), respectively.

C. Poles and Polar Planes

Suppose that we are given a straight line S passing through a fixed point P not contained in a quadric surface F, and S intersects the surface at two points X, Y. The locus of the point Q that is the †harmonic conjugate of P with respect to X and Y is a plane. We call this plane π the **polar plane** of P with respect to the quadric surface F, and P the **pole** of the plane π. If the polar plane of a point P contains a point Q, then the polar plane of Q contains P. In this case, we say that the two points P and Q are **conjugate** to each other with respect to the quadric surface. When the point P is on the quadric surface, the tangent plane at P is regarded as the polar plane of P. If the polar plane (with respect to a quadric surface) of each vertex of a tetrahedron is the face corresponding to that vertex, we call this tetrahedron a **self-polar tetrahedron**. If the polar planes (with respect to a quadric surface) of four vertices of a tetrahedron A are four faces of a tetrahedron B, the same property holds when we interchange A and B. We say that such tetrahedrons are **polar tetrahedrons** with respect to the quadric surface. Suppose that we are given a quadric surface and two planes. If the pole (with respect to the quadric surface) of one plane is on the other plane, these two planes are said to be **conjugate** with respect to the surface.

When we are given two †pencils of planes in †projective correspondence, the locus of lines of intersection of two corresponding planes is generally a hyperboloid of one sheet or a hyperbolic paraboloid. In particular, if the axes of these pencils of planes intersect, the locus is a quadric conical surface, and if the axes are parallel, the locus is a quadric cylindrical surface (i.e., an elliptic or hyperbolic cylinder). When there exists a projective correspondence between two straight lines not on a plane, the locus of lines joining corresponding points is a quadric surface (M. Chasles).

D. Surfaces of the Second Class

A surface F in $\mathbf{E}^3$ is called a **surface of the second class** if it admits two tangent planes passing through an arbitrary straight line L provided that $F \cap L = \varnothing$. This surface can be represented as the set of zeros of a homogeneous equation of the second order in †plane coordinates u_1, u_2, u_3, u_4. It is possible that a surface of the second class degenerates into a conic or two points. In general, quadrics are surfaces of the second class, and vice versa.

As in the case of quadrics, we can define poles, polar planes, and polar tetrahedrons with reference to surfaces of the second class. Four straight lines joining corresponding vertices of two tetrahedrons polar with respect to a surface of the second class are on the same quadric. We say that two such tetrahedrons are in **hyperboloid position**.

E. Confocal Quadrics

A family of central quadrics represented by the following equations is called a **family of confocal quadrics**:

$$\frac{x^2}{a+k}+\frac{y^2}{b+k}+\frac{z^2}{c+k}=1, \qquad a>b>c>0, \tag{11}$$

where k is a parameter. For a quadric belonging to this family, any point on the ellipse $x^2/(a-c)+y^2/(b-c)=1$, $z=0$ or the hyperbola $x^2/(a-b)-z^2/(b-c)=1$, $y=0$ is called a **focus**. This ellipse and hyperbola are called **focal conics** of the quadric.

Given an ellipsoid F and a point $X(x,y,z)$ not contained in the principal plane, we can draw three quadrics F', F'', F''' passing through X and confocal with F. These surfaces F', F'', F''' intersect each other and are mutually perpendicular. One of them is an ellipsoid, another one a hyperboloid of one sheet, and the third a hyperboloid of two sheets. Let k_1, k_2, k_3 be the values of the parameter k in (11) corresponding to these three surfaces. Then the coordinates x, y, z of the point X are given by

$$x=\sqrt{\frac{(a+k_1)(a+k_2)(a+k_3)}{(b-a)(c-a)}},$$

$$y=\sqrt{\frac{(b+k_1)(b+k_2)(b+k_3)}{(a-b)(c-b)}},$$

$$z=\sqrt{\frac{(c+k_1)(c+k_2)(c+k_3)}{(a-c)(b-c)}}.$$

We call k_1, k_2, k_3 the **elliptic coordinates** of the point X.

Two points (x,y,z), (x',y',z') are called **corresponding points** if they belong to confocal quadrics of the same kind,

$$\frac{x^2}{a}+\frac{y^2}{b}+\frac{z^2}{c}=1,$$

$$\frac{(x')^2}{a+k}+\frac{(y')^2}{b+k}+\frac{(z')^2}{c+k}=1,$$

and satisfy

$$\frac{x}{\sqrt{a}}=\frac{x'}{\sqrt{a+k}}, \qquad \frac{y}{\sqrt{b}}=\frac{y'}{\sqrt{b+k}},$$

$$\frac{z}{\sqrt{c}}=\frac{z'}{\sqrt{c+k}}.$$

If $P_1, P_2; Q_1, Q_2$ are corresponding points, then $P_1Q_2=P_2Q_1$ (J. Ivory).

F. Circular Sections

When the intersection of a plane and a quadric is a circle, the intersection is called a **circular section**. In general, circular sections are cut off by two systems of parallel planes through a quadric. The point of contact on the tangent plane parallel to these is an †umbilical point of the quadric.

G. Quadric Hypersurfaces

A subset F of an n-dimensional Euclidean space E^n is called a **quadric hypersurface** (or simply **hyperquadric**) if it is the set of points $(x_1,\dots,x_n)$ satisfying the following equation of the second degree:

$$\sum_{i,k=1}^{n} a_{ik}x_ix_k+2\sum_{i=1}^{n} b_ix_i+c=0, \tag{12}$$

where a_{ik}, b_i, c are all real numbers. We can assume without loss of generality that the matrix $A=(a_{ik})$ is symmetric. Assume that A is not a zero matrix. In the case $n=2$, F is a conic, and in the case $n=3$, it is a quadric surface. The theory of classification of quadric surfaces can be generalized to the n-dimensional case as follows: Let $r(A^*)=r^*$ be the rank of the $(n+1)\times(n+1)$ matrix

$$A^*=\begin{bmatrix} a_{11} & \dots & a_{1n} & b_1 \\ & \dots & & \\ a_{n1} & \dots & a_{nn} & b_n \\ b_1 & \dots & b_n & c \end{bmatrix} = \begin{bmatrix} & & & b_1 \\ & A & & \vdots \\ & & & b_n \\ b_1 & \dots & b_n & c \end{bmatrix},$$

and put $r(A)=r$. Then we have the following three cases: (I) $r=r^*$; (II) $r+1=r^*$; and (III) $r+2=r^*$. Corresponding to each case, equation (12) can be simplified (by a coordinate transformation in E^n) to the following canonical forms, respectively:

(I) $\sum_{i=1}^{r} \lambda_i x_i^2=0$,

(II) $\sum_{i=1}^{r} \lambda_i x_i^2+1=0$,

(III) $\sum_{i=1}^{r} \lambda_i x_i^2+2x_{r+1}=0$,

where $(\lambda_1, \ldots, \lambda_r, 0, \ldots, 0)$ (with $n-r$ zeros) is proportional to the †eigenvalues of the matrix A. In general, we have $1 \leqslant r \leqslant n$. In the cases where $r=n$ in forms (I) and (II) and $r+1=n$ in (III), the hypersurface is called a **properly $(n-1)$-dimensional quadric hypersurface**, and in other cases, a **quadric cylindrical hypersurface**. In cases (I) and (II), the quadric cylindrical hypersurface is the locus of $(n-r)$-dimensional subspaces passing through each point of a properly $(r-1)$-dimensional quadric hypersurface and parallel to a fixed $(n-r)$-dimensional subspace. In case (III), the quadric cylindrical hypersurface is the locus of $(n-r-1)$-dimensional subspaces passing through each point of a properly r-dimensional quadric hypersurface and being parallel to a fixed $(n-r-1)$-dimensional subspace. For form (I) with $\lambda_i>0$ $(i=1, \ldots, n)$, a properly $(n-1)$-dimensional quadric hypersurface reduces to a point in E^n; for form (II) with $\lambda_i>0$ $(i=1, \ldots, n)$ it becomes the empty set. Suppose that we are given a quadric hypersurface F that is neither a point nor empty. Then the system $\{\lambda_1, \ldots, \lambda_r\}$ associated with F in its canonical equation is unique up to order (and signature in form (III)). Suppose that we are given a quadric surface F and a point P on F. Suppose further that if a point X other than P is on F, then the whole straight line PX lies on F. In this case, the hypersurface is called a **quadric conical hypersurface** (or simply **quadric cone**). For example, for case (I), we can take $P=O$ (the origin), and the hypersurface is a quadric cone. In cases (I) and (II), the hypersurface is symmetric with respect to the origin. In these cases, a hypersurface is called a **central quadric hypersurface**, and in case (III), it is called a **noncentral quadric hypersurface** or **parabolic quadric hypersurface**. When we cut a parabolic quadric hypersurface by a (2-dimensional) plane containing the x_{n+1}-axis, the section is a parabola. If $\lambda_i<0$ $(i=1, \ldots, r)$ in form (II), then the surface is called an **elliptic quadric hypersurface**, and if there are both positive and negative numbers among the λ_i, the surface is called a **hyperbolic quadric hypersurface**. The section of an elliptic quadric hypersurface by a plane is always an ellipse. The section of a hyperbolic quadric hypersurface by a plane is an ellipse, a hyperbola, or two straight lines. In general, the section of a quadric hypersurface by a subspace is a quadric hypersurface on that subspace.

H. Quadric Hypersurfaces in an Affine Space

In Section G we considered a quadric hypersurface defined by (12) in an n-dimensional Euclidean space E^n and transformed the equation to canonical form by an orthogonal transformation of coordinates in E^n. If we regard E^n as an n-dimensional †affine space over the real number field and reduce (12) to the simplest form by a coordinate transformation in the affine space, we have the following canonical forms corresponding to cases (I), (II), and (III) discussed in Section G:

$$\text{(I)} \quad (s,t): \sum_{i=1}^{s} x_i^2 - \sum_{j=s+1}^{r} x_j^2 = 0,$$

$$\text{(II)} \quad (s,t): \sum_{i=1}^{s} x_i^2 - \sum_{j=s+1}^{r} x_j^2 + 1 = 0,$$

$$\text{(III)} \quad (s,t): \sum_{i=1}^{s} x_i^2 - \sum_{j=s+1}^{r} x_j^2 + 2x_{r+1} = 0,$$

where $0 \leqslant s \leqslant r$ and $r-s=t$. The terms *properly $(n-1)$-dimensional, cylindrical, conical, parabolic, elliptic*, and *hyperbolic* can be defined in terms of this affine classification. For example, a cone is of type (I), a parabolic hypersurface is of type (III), an elliptic hypersurface is of type (II) $(0,r)$, a hyperbolic hypersurface is of type (II) (s,t) $(s,t>0)$, and type (II) $(s,0)$ represents the empty set. A necessary and sufficient condition for a (nonempty) hypersurface to be represented by two canonical forms $N(s,t)$, $N'(s',t')$ is that (i) $N=N'$ and (ii) if $N=$(I) or $N=$(III), then $s=s'$, $t=t'$ or $s=t'$, $t=s'$, and if $N=$(II), then $s=s'$, $t=t'$.

I. Quadric Hypersurfaces in a Projective Space

Suppose that we are given a field K of characteristic not equal to 2 and an n-dimensional †projective space $\mathbf{P}^n$ over K. A subset F of $\mathbf{P}^n$ is called a **quadric hypersurface** (or simply **hyperquadric**) if F is represented by a homogeneous equation of the second degree $\sum_{i,k=0}^{n} a_{ik} x_i x_k = 0$, where $(x_0, x_1, \ldots, x_n)$ are homogeneous coordinates in $\mathbf{P}^n$ and $a_{ik} \in K$; $A=(a_{ik})$ is a nonzero symmetric matrix. The problem of classifying such surfaces is reduced to that of †quadratic forms or, equivalently, to that of symmetrix matrices in K. Two symmetric matrices A and B are equivalent if there exists a regular matrix T such that $B={}^tTAT$ (→ 348 Quadratic Forms). In particular, when K is an †algebraically closed field or a †real closed field, a simple result is obtained. If K is an algebraically closed field, then the equation of the quadric hypersurface is reduced to the canonical form $\sum_{i=0}^{r} x_i^2 = 0$, where $r=r(A)=$ the rank of A. When K is a real closed field, then the canonical form is $\sum_{i=0}^{s} x_i^2 - \sum_{j=s+1}^{r} x_j^2 = 0$.

References

[1] G. Salmon, A treatise on the analytic geometry of three dimensions, Hodges, Figgis & Co., seventh edition, 1928 (Chelsea, 1958).

[2] G. Salmon and W. Fiedler, Analytische Geometrie des Raumes I, Teubner, fourth edition, 1898.
[3] H. F. Baker, Principles of geometry III, Solid geometry, Cambridge Univ. Press, 1922.
[4] O. Schreier and E. Sperner, Einführung in die analytische Geometrie und Algebra II, Vandenhoeck & Ruprecht, 1951; English translation, Projective geometry of n-dimensions, Chelsea, 1961.
[5] M. Protter and C. Morrey, Analytic geometry, Addison-Wesley, 1975.

351 (XX.23) Quantum Mechanics

A. Historical Remarks

†Newtonian mechanics (classical mechanics) successfully explained the motion of mechanical objects, both celestial and terrestrial, on a macroscopic scale. It failed, however, to explain blackbody radiation, which was discovered in the last decade of the 19th century. M. Planck introduced a hypothesis of discrete energy quanta, each of which contains an amount of energy E equal to the frequency of the radiation v multiplied by a universal constant h (called **Planck's constant**). He applied this hypothesis to derive a new formula for radiation that gives predictions in good agreement with observations. A. Einstein proposed the hypothesis of the photon as a particlelike discrete unit of light rays. Assuming that many physical quantities, including energy, have only discrete values, N. H. Bohr explained the stability of electronic states in atoms. As illustrated in these examples, **quantum mechanics** is applied to study the motion of microscopic objects, including molecules, atoms, nuclei, and elementary particles.

B. Quantum-Mechanical Measurement

Fundamental differences between the new mechanics and classical mechanics are due to the facts that many physical quantities, for example, energy, can take only discrete values in the microscopic world, and that states of microscopic objects are disturbed by observation.

A (pure) **state** at a certain time is expressed by a unit vector ψ in a †Hilbert space $\mathscr{H}$, and **observables**, or physical quantities, are expressed by †self-adjoint operators in such a space.

Let a_n $(n=1,2,\dots)$ be †eigenvalues of an observable A, and let P_n be a †projection operator onto the eigenspace spanned by †eigenvectors belonging to the eigenvalue a_n. Suppose that $A=\sum_n a_n P_n$. Then the hypothesis on measurement in quantum mechanics is given as follows.

When an observable A is observed in a state ψ, one of the eigenvalues a_n is found with probability proportional to $(\psi, P_n\psi)^2$. When an eigenvalue a_n is once observed, a state jumps from ψ to an eigenstate $P_n\psi$ which belongs to the eigenvalue a_n. Quantum mechanics predicts only a probability p_n with which a certain value a_n is found when an observable A is observed. This probability, given by $(\psi, P_n\psi)$, is not changed even if ψ is replaced by $e^{i\theta}\psi$, $0\leqslant\theta\leqslant 2\pi$. Therefore $e^{i\theta}\psi$ represents the same state as ψ. The set of $e^{i\theta}\psi$, $0\leqslant\theta\leqslant 2\pi$, for a fixed ψ $(\|\psi\|=1)$ is called a **unit ray**.

If P_n is 1-dimensional and $P_n\varphi=\varphi$, $\|\varphi\|=1$, then $(\psi, P_n\psi)=|(\psi,\varphi)|^2$ is called the **transition probability** between the two states.

The **expectation value** (or **expectation**) of an operator A in a state ψ, usually normalized to $(\psi,\psi)=1$, is defined to be $\langle A\rangle=(\psi, A\psi)=\sum_n a_n p_n$.

A general self-adjoint operator A can be written as $A=\int\lambda\, dP(\lambda)$. When A is observed in a state ψ, the probability for a value to be found between λ_1 and λ_2 ($\lambda_2>\lambda_1$; λ_2 included and λ_1 excluded if $P(\lambda)$ is right continuous) is $(\psi(P(\lambda_2)-P(\lambda_1)),\psi)$ (→ 390 Spectral Analysis of Operators).

The quantity $(\varphi, A\psi)$ is called the **matrix element** of A between the two states φ and ψ. A state ψ can be viewed as a functional $\psi(A)=\langle A\rangle$ on the set of all observables A (its value being the expectation), which is linear in A, positive in the sense $\psi(A^*A)\geqslant 0$ for any operator A, and normalized: $\psi(1)=1$. If $0\leqslant\lambda\leqslant 1$ and $\psi(A)=\lambda\psi_1(A)+(1-\lambda)\psi_2(A)$ for all observables A, then the state ψ is called a **mixture** of states ψ_1 and ψ_2 with weights λ and $(1-\lambda)$. If a state is not a nontrivial (i.e., $\lambda\neq 0, 1, \psi_1\neq\psi_2$) mixture, it is called **pure**. The state $\langle A\rangle=(\psi, A\psi)$ on the set of all self-adjoint operators A given by a vector ψ is pure in this sense. If $\sup_\alpha\psi(A_\alpha)=\psi(A)$ whenever A_α is an increasing net of positive operators with A as its limit, then ψ is called **normal**. Any normal positive linear functional on the set of all self-adjoint operators can be described by a trace-class positive operator ρ, called the **density matrix**, as $\langle A\rangle=\mathrm{tr}(A\rho)$. If $\{\psi_n\}$ is a complete orthonormal set, where each ψ_n is an eigenvector of ρ belonging to the eigenvalue λ_n, then $\langle A\rangle=\sum_n\lambda_n(\psi_n, A\psi_n)$.

C. Canonical Commutation Relations

In quantum mechanics, canonical variables are represented by the self-adjoint operators

Q_k (coordinates) and P_k (momenta), $k=1, \ldots, N$, which satisfy the **canonical commutation relations**

$$[Q_k, P_l]=i\hbar\delta_{kl}\mathbf{1},$$

where $\mathbf{1}$ is the identity operator, $[A, B]$ denotes the commutator $AB-BA$, $\hbar=h/(2\pi)$, and the relation is supposed to hold on a certain dense domain of vectors. Self-adjoint operators Q_k and P_k satisfying the above relations are unique up to quasi-equivalence under a suitable domain assumption, e.g., if $\Sigma_k Q_k^2+\Sigma_k P_k^2$ is essentially self-adjoint on a dense domain invariant under multiplication of the Q's and P's and on which the above relations are satisfied. Under such an assumption, Q_k and P_k are unitarily equivalent to a direct sum of the **Schrödinger representation** on $L_2(\mathbf{R}^N, dx_1 \ldots dx_N)$, where Q_k is multiplication by the kth coordinate x_k and P_k is the differentiation $-i\hbar(\partial/\partial x_k)$ (**Rellich-Dixmier theorem**).

The above Schrödinger representation is called the **position representation** (or q-representation). The formulation using the function space L_2 of real variables p_k, $k=1, 2, \ldots, N$, on which the operators P_k act as multiplications by p_k, is called the **momentum representation** (or p-representation).

If Hermitian operators A and B satisfy the canonical commutation relations in the form $(A\psi, B\psi)-(B\psi, A\psi)=i\hbar(\psi, \psi)$, then the following **Heisenberg uncertainty relation** holds for the expectation:

$$\langle(A-\langle A\rangle)^2\rangle\langle(B-\langle B\rangle)^2\rangle\geqslant\frac{\hbar^2}{4}.$$

This gives the **uncertainty** in observations, which means that two observables A and B cannot simultaneously be observed with accuracy. This is another important property of microscopic motion that cannot be found in macroscopic motion.

In a direct sum of the Schrödinger representation of the canonical commutation relations, the unitary operators

$$U(a)=\exp i\sum_k a_k Q_k, \qquad V(b)=\exp i\sum_k b_k P_k$$

with real parameters a_k and b_k, $k=1, \ldots, N$, satisfy the following **Weyl form** of the canonical commutation relations:

$$U(a)U(a')=U(a+a'), \qquad V(b)V(b')=V(b+b'),$$

$$U(a)V(b)=V(b)U(a)\exp\left(-i\sum_k a_k b_k\right).$$

Conversely, any pair of families of unitary operators $U(a)$ and $V(b)$ satisfying these relations and depending continuously on parameters a and b are unitarily equivalent to those obtained as above (**von Neumann uniqueness theorem**).

D. Time Evolution and the Schrödinger Equation

The time t of an observation is fixed in the foregoing discussion. A state changes, however, as the time t changes, in such a way that the transition probability between states is preserved. By Wigner's theorem (→ Section H), this time evolution of states can be implemented by unitary operators U_t defined by the transformation of vectors $\psi\rightarrow U_t\psi=\psi_t$. Furthermore, under some continuity assumption, such as that of $(\varphi, U_t\psi)$, U_t can be made a continuous one-parameter group. By Stone's theorem, $U_t=e^{-iHt/\hbar}$ for a self-adjoint operator H. This operator is called the **Hamiltonian operator** (or simply **Hamiltonian**) determined by the structure of a system. An infinitesimal change in ψ corresponding to an infinitesimal change in t can be generated by this operator H as follows:

$$i\hbar\frac{\partial\psi_t}{\partial t}=H\psi_t.$$

This equation is called the **time-dependent Schrödinger equation**.

A state ψ changes but observables do not change with time in the **Schrödinger picture** above. The other picture, known as the **Heisenberg picture**, is equally possible. In this picture, the state is expressed by a time-independent vector, while operators A vary with time as follows: $A\rightarrow U_t^*AU_t=A(t)$. Rates of change of operators $A(t)$ can be calculated by means of the equation

$$\frac{dA(t)}{dt}=\frac{i}{\hbar}[H, A(t)],$$

which is called the **Heisenberg equation of motion**. When time t changes, the expectation value of an operator A in a state ψ changes in both pictures according to

$$d\langle A\rangle/dt=i\langle[H, A]\rangle/\hbar.$$

According to classical analytical dynamics, a change of a dynamical quantity that is a function of †canonical variables q_i (positions) and p_i (momenta) is given by

$$dA/dt=-(H, A),$$

where H is a †Hamiltonian function and the parentheses (,) denote the †Poisson bracket. A replacement of the Poisson bracket (A, B) by $[A, B]/i\hbar$ transforms this classical equation into the quantum-mechanical equation above. It should be noticed that the mathematical structure of the Poisson bracket is similar to that of commutator. In this transition from classical to quantum mechanics the **correspondence principle** can be used. This requires that the laws of quantum mechanics must lead to the equations of classical mechanics in the classical situation, where many quanta are involved and $\hbar$ can be regarded as infinitesimally small in the commutation relation. The correspondence principle suggests that Hamiltonian operators in quantum mechanics can be obtained from Hamiltonian functions $H(p_k, q_k)$ of canonical variables p_k and q_k in classical mechanics after replacing p_k and q_k by the operators P_k and Q_k in the Schrödinger representation (up to uncertainty of about the order of operators). This process of moving from canonical variables and the Hamiltonian function in classical mechanics to canonical operators and the Hamiltonian in quantum mechanics is called **quantization**. Taking a system of s particles and letting x_k, y_k, and z_k be the Cartesian coordinates of the kth particle, we usually write the equation of motion as

$$i\hbar\frac{\partial\psi}{\partial t}=\left(-\sum_{k=1}^{s}\frac{\hbar^2}{2m_k}\Delta_k + V(x_1,y_1,z_1,\ldots,x_s,y_s,z_s)\right)\psi,$$

which is a second-order partial differential equation. Here m_k is the mass of the kth particle; Δ_k is the †Laplacian of x_k, y_k, and z_k; and V is a real function called the potential energy. This equation is the time-dependent **Schrödinger equation**. The partial differential operator on the right-hand side is called the (s-body) **Schrödinger operator** and $\psi(x_1, y_1, z_1, \ldots, x_s, y_s, z_s, t)$ is called a **wave function**. The probability of finding a particle in the volume $dx_k dy_k dz_k$ bounded by x_k, x_k+dx_k, y_k, y_k+dy_k, and z_k, z_k+dz_k is proportional to $|\psi(x_1, y_1, z_1, \ldots, x_s, y_s, z_s, t)|^2$. Usually $|\psi|^2$ is normalized so that its integral over the whole space is 1. We sometimes call ψ the **probability amplitude**. When ψ is given by $e^{-iEt/\hbar}\varphi(x_1, \ldots, z_s)$, the expectation value of an operator A in a state ψ, $\langle A\rangle=\int\psi^*A\psi\,dx_1\ldots dz_s$, does not depend on time. When this is the case, ψ is called a **stationary state**.

A real value E and a function $\varphi(x_1, \ldots, z_s)$ are found by solving an †eigenvalue problem $H\varphi=E\varphi$. This equation is the **time-independent Schrödinger equation**, which is a second-order partial differential equation. Since the Hamiltonian H stands for the energy of this system, the eigenvalues E are the energy values that this system can take.

When a potential function V is given, it is a nontrivial matter to prove that the (s-body) Schrödinger operator with the given V is essentially self-adjoint on the set of, for example, all C^∞-functions with compact supports so that its closure H defines mathematically the continuous one-parameter group of unitaries $U_t=e^{-iHt/\hbar}$ for the time evolution of the quantum system of s particles with the given interaction potential V. If V satisfies an estimate $\|V\psi\|\leqslant\lambda\|H_0\psi\|+\mu\|\psi\|$ (called the **Kato perturbation** on H_0) for some nonnegative $\lambda<1$ and $\mu\geqslant 0$ and for all ψ in a dense domain on which H_0 is essentially self-adjoint, where H_0 denotes the Schrödinger operator with $V=0$ (called the **free Hamiltonian** or the **kinetic energy** term), then H_0+V is essentially self-adjoint on the same domain. For the case where V consists of Coulomb interactions between electrons and Coulomb potentials on electrons by fixed nuclei, for example, such an estimate and hence the essential self-adjointness of Hamiltonians for atoms and molecules were established first by T. Kato (*Trans. Amer. Math. Soc.*, 70 (1951)).

For a 1-body Schrödinger operator (or 2-body Schrödinger operator after the center of mass motion has been separated out), the point spectrum is that of the particle trapped by the potential, and the state represented by its eigenvector is called a **bound state**. The eigenvalue is nonpositive for a reasonable class of potentials V (for example, if $V(x)$ $(x\in\mathbf{R}^3)$ is continuous and $\mathcal{O}(|x|^{-1-\varepsilon})$ as $|x|\to\infty$ for some $\varepsilon>0$), and its absolute value is called the **binding energy**. The eigensolutions of the Schrödinger equation are what have been called stationary states above.

There are also stationary solutions that do not correspond to the point spectrum and hence are not square integrable. They are used in the stationary methods of scattering theory (→ 375 Scattering Theory).

E. Some Exact Solutions for the 1-Body Schrödinger Equation

(1) **Harmonic oscillator**. First consider the case in which the space is of 1 dimension, so that the Laplacian Δ is $(d/dx)^2$. Let m be the mass of the particle and $V(x)=m\omega^2x^2/2$ for a positive constant ω (called the **angular frequency**). The Hamiltonian

$$H=-(\hbar^2/2m)(d/dx)^2+m\omega^2x^2/2$$

has simple eigenvalues

$$E_n=\hbar\omega(n+(1/2)),\qquad n=0,1,2,\ldots,$$

with a complete orthonormal set of eigenfunctions

$$\psi_n(x)=c_nH_n(q)e^{-q^2/2},\qquad q=(m\omega/\hbar)^{1/2}x,$$

where $H_n(q)$ is a †Hermite polynomial and c_n is the normalization constant:

$$H_n(q)=\sum_{k=0}^{[n/2]}(-1)^k n!(2q)^{n-2k}/\{(n-2k)!k!\},$$

$$c_n=\{2^{2n}(n!)^2\pi\hbar/(m\omega)\}^{-1/4}.$$

When the space is of r dimensions, n in E_n is replaced by $n_1+\ldots+n_r$ with nonnegative integers $n_1,\ldots,n_r$, and the corresponding eigenfunction is $\prod_{j=1}^{r}\psi_{n_j}(x_j)$.

(2) **One-dimensional square-well potential.** Let $V(x)=V$ for $|x|\leqslant a/2$ and $V(x)=0$ for $|x|>a/2$ $(x\in\mathbf{R})$. If $V\geqslant 0$, there are no point spectra. If $V<0$ and

$$N-1<a(-2mV)^{1/2}/(\pi\hbar)\leqslant N,$$

then there are N eigenvalues $(N=1,2,\ldots)$ obtained as the roots $E<0$ of one of the following equations:

$$\{(V/E)-1\}^{1/2}=\tan\{a(-2mE)^{1/2}/(2\hbar)\},$$

$$\{(V/E)-1\}^{1/2}=-\cot\{a(-2mE)^{1/2}/(2\hbar)\}.$$

(3) **Separation of angular dependence for central potential.** If $V(x)$ $(x\in\mathbf{R}^3)$ depends only on $r=|x|=(\sum_{j=1}^{3}(x_j)^2)^{1/2}$ (called a **central potential**), then all eigenvalues E and a basis for eigenfunctions $\psi(x)$ can be obtained in terms of the polar coordinate r, θ, φ $(x_1=r\sin\theta\cos\varphi$, $x_2=r\sin\theta\sin\varphi$, $x_3=r\cos\theta)$ as

$$\psi(x)=Y_{lm}(\theta,\varphi)r^{-1}u(r),$$

$$-(\hbar^2/(2m))u''(r)+\{\hbar^2 l(l+1)/(2mr) + V(r)-E\}u(r)=0,$$

$$\|\psi\|=\int_0^\infty |u(r)|^2\,dr<\infty,$$

where the angular function Y_{lm} is an eigenfunction of the square $\mathbf{L}^2$ of the **orbital angular momentum** $\mathbf{L}=-ix\times\nabla$:

$$Y_{lm}=c_{lm}P_l^m(\cos\theta)e^{im\varphi},$$

$$c_{lm}=(-1)^m\{(2l+1)(l-m)!/(4\pi(l+m)!)\}^{1/2},$$

$$m=l,\,l-1,\,\ldots,\,-l+1,\,-l,\qquad l=0,\,1,\,2,\,\ldots.$$

Here $P_l^m(x)$ is an †associated Legendre polynomial:

$$P_l^m(x)=(1-x^2)^{m/2}(d^{l+m}(x^2-1)^l/dx^{l+m})/(2^l l!).$$

The above equation for $u(r)$ is called the **radial equation**. The nonnegative integer l is the **azimuthal quantum number**, and the integer m is the **orbital magnetic quantum number**. The wave function $\psi(x)$ with the angular dependence $Y_{lm}(\theta,\varphi)$ is called the ***S*-wave, *P*-wave, *D*-wave,** ... according as $l=0,1,2,\ldots$.

(4) **Hydrogen-type atom.** Let $V(r)=-Ze^2/r$ $(Z>0)$. For each l and m, there are eigenvalues $-e^2Z^2/(2an^2)$ with eigenfunctions $\psi_{nlm}=r^{-1}u_{nl}(r)Y_{lm}(\theta,\varphi)$, where $n=l+1,\,l+2,\ldots$ is the **principal quantum number**,

$$u_{nl}(r)=c_{nl}L_{n+l}^{(2l+1)}(s)s^{l+1}e^{-s/2},\qquad s=2Zr/(na),$$

$$c_{nl}=-\{(n-l-1)![2Z/(na(n+l)!)]^3/(2n)\}^{1/2},$$

and $L_N^{\mu}(x)$ is the μth derivative of the †Laguerre polynomial

$$L_N(x)=\sum_{\nu=0}^{N}(N!)^2(-x)^\nu/\{(N-\nu)!(\nu!)^2\}.$$

The eigenvalue is determined by n, and its multiplicity is n^2, corresponding to the different possible values of l and m.

F. Path Integrals

R. P. Feynman (*Rev. Mod. Phys.*, 20 (1948)) has given the solution of the Schrödinger equation as an integral of $e^{iL/\hbar}$ over all possible paths $q(t)$, where $L=L(q,\dot q)$ $(\dot q=(d/dt)q(t))$ is the classical Lagrangian for the Hamiltonian system. This integral is called the Feynman path integral. Mathematical reformulation of the formula in terms of the Wiener measure has been given by M. Kac (*Proc. 2nd Berkeley Symp. Math. Statist. Probability*, 1950; *Probability and related topics in the physical sciences*, Wiley, 1959).

Consider the 1-body Schrödinger operator $H=H_0+V$ (form sum), where V is the sum of a locally integrable function bounded below and a Kato perturbation on H_0. Let $b(t)$ $(t\geqslant 0)$ be the Wiener process and $q(t)=\hbar b(t)/(2m)^{1/2}$. For any L_2 functions f,

$$(e^{-tH/\hbar}f)(x)$$

$$=E\left(f(x+q(t))\exp\left\{-\int_0^t V(x+q(s))\,ds/\hbar\right\}\right)$$

for almost all x, where E denotes the expectation for the Wiener process. If V is a sum of L_2 and L_∞ functions (for spatial dimension $\leqslant 3$), then the right-hand side is continuous in x for $t>0$. This is called the **Feynman-Kac formula**.

Let L_0 be the Hamiltonian for a 1-dimensional harmonic oscillator with $m=\omega=\hbar=1$ and ψ_0 be the eigenfunction $\psi_0(x)=\pi^{-1/4}\exp(-x^2/2)$. Consider $H=L_0+V$ (form sum), where V is a sum of a locally integrable function bounded below and a Kato perturbation on L_0. Let $q(t)$ $(t\in\mathbf{R})$ be Gaussian random variables with mean 0 and covariance $E(q(t)q(s))=2^{-1}\exp(-|t-s|)$ (called the **oscillator process**). For any $f_j(x)$ in $L_2(\mathbf{R},\psi_0^2\,dx)$ $(j=1,\ldots,n)$ and $t_0\leqslant t_1\ldots\leqslant t_n\leqslant t_{n+1}$,

$$(\psi_0,e^{-(t_1-t_0)H}f_1e^{-(t_2-t_1)H}f_2\ldots f_ne^{-(t_{n+1}-t_n)H}\psi_0)$$

$$=E\left(\left\{\prod_{j=1}^{n}f_j(q(t_j))\right\}\exp\left\{-\int_{t_0}^{t_{n+1}}V(q(s))\,ds\right\}\right).$$

The above path integral formulas are closely related to the **Trotter product formula**

$$e^{-t(A+B)}=\lim_{n\to\infty}(e^{-tA/n}e^{-tB/n})^n\qquad(t\geqslant 0),$$

where A and B are self-adjoint operators

bounded below and $A+B$ is essentially self-adjoint. (The same formula holds without the boundedness assumption when $t \in i\mathbf{R}$.)

G. The Dirac Equation

The Schrödinger equation is not relativistically invariant. The **Klein-Gordon equation**

$$(\Box - \kappa^2)\psi = 0, \qquad \Box \equiv \Delta - \frac{1}{c^2}\frac{\partial^2}{\partial t^2}, \qquad \kappa = \frac{mc}{\hbar}$$

is obtained by replacing p_k by $(\hbar/i)\partial/\partial x_k$ $(k=1,2,3)$ and E by $i\hbar\partial/\partial t$ in the relativistic identity

$$E^2 = m^2c^4 + \mathbf{p}^2c^2,$$

where c is the speed of light. Wave functions of free particles are believed to satisfy this equation. P. A. M. Dirac assumed that the ψ of a free electron is expressed in terms of a †spinor with four components satisfying a linear differential equation that automatically implies the Klein-Gordon equation. Relativity requires the equal handling of space and time. The **Dirac equation**

$$\sum_{\mu=0}^{3} \gamma^\mu \frac{\partial\psi}{\partial x^\mu} + i\kappa\psi = 0 \tag{8}$$

$(x^0 = ct)$ satisfies these requirements. The coefficients γ^μ can be so determined that every component of ψ also satisfies the Klein-Gordon equation. Thus the γ^μ are found to be 4×4 matrices satisfying the commutation relations $\gamma^\mu\gamma^\nu + \gamma^\nu\gamma^\mu = 2g^{\mu\nu}$ $(\mu, \nu = 0, 1, 2, 3)$, where $g^{\mu\nu} = 0$ for $\mu \neq \nu$ and $g^{00} = -g^{kk} = 1$ $(k = 1, 2, 3)$. Sixteen linearly independent matrices are obtained by repeated multiplication of five matrices, which include the four matrices γ^0, $\gamma^1, \gamma^2, \gamma^3$ and the identity matrix. Any 4×4 matrix can be expressed as a linear combination of these sixteen matrices.

The Dirac equation has †plane wave solutions

$$\psi(\mathbf{x}, t) = u\exp((i/\hbar)\mathbf{p}\cdot\mathbf{x} - (i/\hbar)Et),$$

where the energy eigenvalues E are

$$\pm\sqrt{m^2c^4 + p^2c^2}.$$

There are four independent eigensolutions $u^{(1)}$, $u^{(2)}$, $u^{(3)}$, and $u^{(4)}$, because u has four components. Two of them are of positive energy and the other two are of negative energy. Although the negative energy case is physically undesirable, it has to be taken into account in order to obtain a mathematically complete set. To solve this difficulty, Dirac proposed the hypothesis (**Dirac's hole theory**) that all the negative energy states are filled up by an infinite number of electrons in the normal state of the vacuum. The absence from the vacuum of a negatively charged electron in a negative energy state could then be expected to manifest itself as a positively charged particle (positron) with positive mass and kinetic energy. If gamma rays are absorbed to excite an electron from a negative energy state into a positive energy state, an electron-position pair must be created. Y. Nishina and O. Klein calculated the cross section of Compton scattering (the **Klein-Nishina formula**) by use of the Dirac equation and found good agreement with observations, thus providing evidence that the Dirac equation is correct. The existence of negative energy states, however, forces us to give up considering the Dirac equation as an equation of one electron. The positron theory is introduced, and the Dirac equation is considered as the classical field of electron waves and is second-quantized (→ 377 Second Quantization).

The Klein-Gordon equation can also be considered to be the classical wave equation of matter and can be second-quantized. Motions of particles with zero spin, pi mesons (π) for example, obey this equation.

We can rewrite the Dirac equation as $i\hbar\partial\psi/\partial t = H\psi$, $H = c\alpha\cdot\mathbf{p} + mc^2\beta$, where $\gamma^k = \beta\alpha_k$ $(k=1,2,3)$, and $\gamma^0 = \beta$. Since H cannot commute with the orbital angular momentum of an electron $\mathbf{L} = \mathbf{r}\times\mathbf{p} = -i\hbar\mathbf{r}\times\nabla$, $\mathbf{L}$ is not conserved. However, the total angular momentum $\mathbf{J} = \mathbf{L} + (\hbar/2)\boldsymbol{\sigma}$ can be conserved when $\boldsymbol{\sigma}$ is a vector whose components can be given as $\begin{pmatrix}\sigma_k & 0\\ 0 & \sigma_k\end{pmatrix}$, where $\beta = \begin{pmatrix}1 & 0\\ 0 & -1\end{pmatrix}$, $\alpha_k = \begin{pmatrix}0 & \sigma_k\\ \sigma_k & 0\end{pmatrix}$, and the σ_k, called **Pauli spin matrices**, are given by $\sigma_x = \begin{pmatrix}0 & 1\\ 1 & 0\end{pmatrix}$, $\sigma_y = \begin{pmatrix}0 & -i\\ i & 0\end{pmatrix}$, $\sigma_z = \begin{pmatrix}1 & 0\\ 0 & -1\end{pmatrix}$. (The γ^μ are called **Dirac's γ matrices**.) The quantity $\mathbf{S} = \frac{\hbar}{2}\boldsymbol{\sigma}$ is the **intrinsic angular momentum** of the electron, also called the **spin**. Many particles besides the electron, the neutron for example have spin. The matrix $\mathbf{S}^2 = S_x^2 + S_y^2 + S_z^2$ is diagonal and is equal to $s(s+1)\hbar^2 I$ (I is the identity matrix). For the electron $s = 1/2$, and $\hbar/2$ is called the absolute value of the spin. Therefore we say that electrons are particles of spin $\hbar/2$. This was predicted in the theory of light spectra.

When the speed of an electron is very small, so that $(v/c)^2$ can be neglected, states of the electron can be expressed in terms of two-component wave functions. This approximation is called the **Pauli approximation**. If the spin-orbit term that appears in the Pauli approximation is also neglected, these two components become independent of each other and individually satisfy the Schrödinger equation.

H. Application of Representation Theory of Lie Groups

A **symmetry** (with active interpretation) is a bijective mapping of pure states (represented by unit rays in a Hilbert space) preserving transition probabilities between them. **Wigner's theorem** says that any symmetry can be implemented by either a unitary or an antiunitary mapping of the underlying Hilbert space as a mapping of unit rays onto themselves. Furthermore a connected Lie group of symmetries is implemented by unitaries, which form a projective representation. Eigenstates ψ of the Schrödinger equation are functions of the coordinates of each particle. Let these coordinates be denoted together as x.

Suppose that an operator T operates on the x, as, for example, a rotation of the coordinate system or a permutation of the labels of the particles. If T commutes with H or H is invariant under the transformation $x \to x' = Tx$, then $T\psi(x) = \psi'(x) = \psi(T^{-1}x)$ satisfies the same Schrödinger equation as ψ, where the transformation of the function is defined by $\psi'(x') = \psi(x)$. The set of transformations $x \to x' = Tx$ forms a †group $\{T\}$, and the corresponding transformations $\psi \to T\psi$ give a (generally infinite-dimensional) representation of this group, which should be †unitary on L_2-space relative to the Lebesgue measure dx if T leaves the measure invariant. There are **degeneracies** of the energy eigenvalues, each of which is equal to the dimension of the corresponding representation of the group $\{T\}$. If the representation for each eigenvalue is decomposed into irreducible ones, then the decomposed stationary state can be labeled by an †irreducible representation.

When H is spherically symmetric, i.e., H is invariant under the 3-dimensional rotation group, states are classified by the irreducible representations D_L of the †rotation group (→ 258 Lorentz Group). The square of the sum $\mathbf{L}$ of all orbital angular momenta has eigenvalues $L(L+1)\hbar^2$, where L must be 0 or a positive integer. There are $2L+1$ degenerate states, each of which belongs to a different M, the z-component of $\mathbf{L}$, where M ranges from L to $-L$ by unit steps. Even when there is an interaction between the orbital angular momentum and the spin angular momentum, states are labeled by the irreducible representations D_J of the rotation group, where $\mathbf{J}$ is the sum of the orbital angular momentum $\mathbf{L}$ and the spin angular momentum $\mathbf{S}$ ($\mathbf{J} = \mathbf{L} + \mathbf{S}$) and eigenvalues of $\mathbf{J}^2$ are given by $J(J+1)\hbar^2$. Each J must be zero, a positive integer, or a half-integer. Adding inversions to the pure rotations, we obtain the 3-dimensional orthogonal group (→ 60 Classical Groups I). Irreducible representations of this group are written as $D_J^{\pm}$, where $\pm$ corresponds to the characters of the inversion relative to the origin. States with + are called **even states** and those with −, **odd states**. For example, energy levels of atoms and nuclei can be classified by $D_J^{\pm}$.

To obtain matrix elements of observables between two stationary states, group representation theory is useful. The transformation of every observable obeys a certain rule under the transformation of coordinates. The scalar is transformed according to D_0^+, the vector according to D_1^-, the pseudovector according to D_1^+, and the traceless tensor according to D_2^+. If the transformation of an observable is given by D_J, then a matrix element of this observable between the states belonging to $D_{J'}$ and $D_{J''}$ vanishes unless the tensor product representation $D_J \otimes D_{J'}$ contains as a factor a representation equivalent to $D_{J''}$. In electromagnetic transitions in atoms or nuclei, $D_1 \otimes D_{J'} = D_{J'+1} + D_{J'} + D_{J'-1}$ $(J' \geqslant 1)$ if the electric dipole transition dominates $(J=1)$. This implies the **selection rule** $J' \to J'+1, J', J'-1$. When $J'=0$, only the transition $0 \to 1$ is possible. More general selection rules can be obtained in the same way for general multipole transitions. Representation theory is useful in determining general formulas of transition strengths.

There is a class of particles, many of which can occupy the same state, called **bosons**. There is another class of particles, of which only one can occupy a given state, called **fermions**. For example, the electron, neutron, and proton are fermions, while the photon and pi meson are bosons. Two identical particles, both of which are either fermions of the same kind or bosons of the same kind, cannot be distinguished. Therefore the Hamiltonian should be invariant under permutations between identical fermions, or between identical bosons. A system consisting of N identical particles can be classified by the irreducible representations of the †symmetric group S_N of N elements. When the particles are fermions, two of them cannot occupy the same state (this law is called the **Pauli principle**), so that only totally antisymmetric states are permissible for fermions. When a system consists only of fermions of the same kind with spin $\hbar/2$ and a Hamiltonian of this system does not include terms depending on spins, then the wave functions are just products of spin and orbital parts. In order to make wave functions totally antisymmetric, orbital wave functions with a total spin $v/2$ should be limited to those corresponding to the †Young diagram $[2^{N-v}, 1^v] = T(2, 2, \ldots, 1, 1)$.

I. Polyatomic Molecules

The group generated by the 2-dimensional rotations about the axis connecting two atoms and the reflections with respect to the planes containing this axis is used to classify states of diatomic molecules. Stationary states can be classified by the absolute value Λ of the angular momentum of the diatomic system around this axis, which can be zero or a positive integer. When $\Lambda \geqslant 1$, the corresponding state has twofold degeneracy, whereas when $\Lambda = 0$, two states (labeled by $\pm$) arise, depending on the character of the reflections. If these two atoms are identical, the molecular states can be classified further as even and odd according to the character of reflections with respect to the plane containing the center of mass and perpendicular to the axis. The classification of the spectral terms of a polyatomic molecule is related to its symmetry, described by the set of all transformations that interchange identical atoms. For example, stationary states of methane molecules CH_4 are classified by the irreducible representations of the group T_d (which is generated by adjoining reflection symmetry to the †tetrahedral group T). Level structures of a crystal are classified by the irreducible representations of its symmetry groups.

In the approximation of many-body problems by means of independent particles, the wave function of the total system is constructed by multiplying the wave functions of the individual particles. To construct such a wave function, it is useful to consider the reduction to irreducible parts of the †tensor products of representations of the groups attached to the individual particles. For example, an atom with two electrons carrying the angular momenta J_1 and J_2 has $2J'+1$ different angular momentum states, where $J' = \min(J_1, J_2)$ corresponding to the decomposition $D_{J_1} \otimes D_{J_2} = D_{J_1+J_2} + D_{J_1+J_2-1} + \ldots + D_{|J_1-J_2|}$. The right-hand side of this equation gives all possible states of the atom.

J. Charge Symmetry

The proton and the neutron can be considered to be different states of the same particle, called the nucleon, because these two particles have very similar natures except for their charges, masses, etc. As an approximation, the Hamiltonian of a system consisting of protons and neutrons may be taken to be invariant under the interchange of protons and neutrons. This invariance is called **charge symmetry**. Analogous to ordinary spin, **isospin** can be introduced to describe the two states of nucleons. The up state of the isospin corresponds to the proton and the down state corresponds to the neutron. Consider transformations belonging to the †special unitary group $SU(2)$ in the 2-dimensional space spanned by the proton state and the neutron state. If a Hamiltonian of N nucleons is invariant under any transformation belonging to $SU(2)$, then the eigenstate of these N nucleons is classified by its irreducible representations D_T, where T stands for the total isospin of each state. This invariance, called **isospin invariance**, holds in the nucleus and elementary particles if electromagnetic and weak interactions, and possibly the interaction responsible for the proton-neutron mass difference, are neglected. When a state of N nucleons has an isospin $T = v/2$, the orbital-spin wave function of this state must correspond to the Young diagram $[2^{N-v}, 1^v]$, since the isospin wave function multiplied by the orbital-spin wave function is a totally antisymmetric wave function. If this Hamiltonian is also independent of spin, it is invariant under unitary transformations in the 4-dimensional space spanned by the four internal states of the nucleon: up and down spins, up and down isospins. Therefore the states of N nucleons can be classified by the irreducible representations of the group $U(4)$ (Wigner's **supermultiplet theory**).

K. The C^*-Algebra Approach

The uniqueness of operators satisfying the canonical commutation relations (representations of CCR's) up to quasi-equivalence (→ Section C) no longer holds if the number of canonical variables become infinite (a so-called system of infinitely many degrees of freedom), a point first emphasized by K. O. Friedrichs (*Mathematical aspects of the quantum theory of fields*, Interscience, 1953), and physical examples illustrating this point were given by L. van Hove (*Physica*, 18 (1952)) and R. Haag (*Mat. Fys. Medd. Danske Vid. Selsk.*, 29 (1955)). The use of C^*-algebras in physics was first advocated by I. E. Segal (*Ann. Math.*, 48 (1947)), and the physical relationship among all inequivalent representations of a C^*-algebra was first discussed by R. Haag and D. Kastler (*J. Math. Phys.*, 5 (1964)).

In C^*-algebra approach, a physical observable is an element of a C^*-algebra $\mathfrak{A}$ and a state is a functional φ on $\mathfrak{A}$ (its value $\varphi(A)$ is the expectation value of the observable $A \in \mathfrak{A}$ when measured in that state) that is linear, positive in the sense $\varphi(A^*A) \geqslant 0$ for any $A \in \mathfrak{A}$, and normalized, i.e., $\|\varphi\| = 1$ or equivalently $\varphi(1) = 1$ if $1 \in \mathfrak{A}$. The †GNS construction associates with every state φ a Hilbert space H_φ, a representation $\pi_\varphi(A)$, $A \in \mathfrak{A}$, of $\mathfrak{A}$ by bounded linear operators on H_φ, and a unit cyclic vector Ω_φ in H_φ such that $\varphi(A) = (\Omega_\varphi, \pi_\varphi(A)\Omega_\varphi)$. Two states φ and ψ (or rather π_φ and π_ψ) are called disjoint if there is no nonzero mapping T from H_φ to H_ψ such that $T\pi_\varphi(A) = \pi_\psi(A)T$ for all $A \in \mathfrak{A}$. Abundant disjoint states occur for a system of infinitely many degrees of freedom, e.g., equilibrium states of a infinitely extended system with different temperatures (→ 402 Statistical Mechanics), superselection sectors explained below (→ 150 Field Theory) and equilibrium or ground states with broken symmetry.

Because actual measurement can be performed only on a finite number of observables (though chosen at will from an infinite number of possibilities) and only with nonzero experimental errors, information on any state φ can be obtained by measurements only up to a neighborhood in the weak topology: $|\varphi(A_i) - a_i| < \varepsilon_i$, $i = 1, \ldots, n$. A set K_1 of states can describe measured information on states at least equally well as another set K_2 (K_1 **physically contains** K_2) if the closure of K_1 in the weak topology contains K_2. From another viewpoint, all states of K_2 are weak limits of states in K_1 and are physically relevant if states in K_1 are physically relevant. The set of all mixtures of vector states $(\psi, \pi(A)\psi)$ for any fixed faithful representation π of $\mathfrak{A}$ is weakly dense in the set of all states of $\mathfrak{A}$, a point emphasized by Haag and Kastler as a foundation of the algebraic viewpoint in the formulation of quantum theory.

Under 360° rotation a vector representing a state of a particle with spin $\hbar/2$ acquires a factor -1 (→ 258 Lorentz Group), while the vacuum vector would be unchanged. A nontrivial linear combination (superposition) of these two would then be changed to a vector in a different ray. If the 360° rotation is not to produce a physically observable effect, then we should either forbid nontrivial superpositions of states of the two classes or, equivalently, restrict observables to those leaving the subspace spanned by vectors in each class invariant so that the relevant linear combinations of vectors, when considered as states on the algebra of observables in the form of expectation functionals $(\psi, A\psi)$, are actually mixtures (rather than superpositions) of states in two classes and are invariant under the 360° rotation. This is called the **univalence superselection rule** and has been pointed out by A. S. Wightman, G. C. Wick, and E. P. Wigner (*Phys. Rev.*, 88 (1952)).

In quantum field theory, the vacuum state can be taken to be pure (by central decomposition if necessary) and in the associated GNS representation (called the vacuum sector) all vectors can be assumed to be physically relevant pure states. In principle, all physically relevant information is in the vacuum representation; for example, a particle with spin $\hbar/2$ can also be discussed in the vacuum sector if we consider a state of this particle in the presence of its antiparticle at a far distance, such as behind the moon (the **behind-the-moon argument**). However, it is mathematically more convenient to consider the states of the particle without any compensating object (in the same way that an infinitely extended gas is more convenient for some purposes than a finitely extended gas surrounded by walls), which can be obtained as weak limits of states in the vacuum sector by removing the compensating particle to spatial infinity and which produce inequivalent representations called **superselection sectors**.

L. Foundation of Quantum Mechanics

Hilbert spaces and the underlying field of complex numbers, which constitute a mathematical background for quantum mechanics, are not immediately discernible from physical observations, and hence there are various attempts to find axioms for quantum mechanics that imply the usual mathematical structure and at the same time allow direct physical interpretation.

One approach of this kind focuses attention on the set of all observables that have only two possible measured values 1 (yes) and 0 (no), called **questions**, together with their order structure (logical implications) and associated lattice structure (join, meet, and orthocomplementation as logical sum, product, and negation). This is called **quantum logic** in contradistinction to the situation in classical physics, where it would form a Boolean lattice. The lattice of all orthogonal projections (corresponding to all closed subspaces of a Hilbert space) in quantum mechanics is a †complete, †orthocomplemented, **weakly modular** (also called **orthomodular**) †atomic lattice satisfying the **covering law**, where weak modularity means $c \wedge (c' \vee b) = b$ and $b \vee (b' \wedge c) = c$ whenever $b \leqslant c$, and the covering law means that every $b \neq 0$ possesses an atom p under it ($p \leqslant b$) and that if an atom q satisfies $q \wedge b = 0$, then any c between $q \vee b$ and b ($q \vee b \geqslant c \geqslant b$) is b or $q \vee b$. Conversely any such lattice is a direct sum of irreducible ones, each of which, if of dimension (length of longest chain) > 3, can be obtained as the lattice of subspaces V satisfying $(V^{\perp})^{\perp} = V$ in a vector space over a (generally noncommutative) field with an anti-automorphic involution $*$, equipped with a nondegenerate Hermitian form. In this approach, an additional requirement is needed to restrict the underlying field and its $*$ to be more familiar ones, such as real, complex, or quaternion fields and their usual conjugations $*$. If that is done, then the set of all probability measures on the lattice (i.e., assignment of expectation values $0 \leqslant \mu(a) \leqslant 1$ for all elements a in the lattice such that $\mu(\bigvee_i a_i) = \sum_i \mu(a_i)$ if $a_i \perp a_j$ for all pairs $i \neq j$, $\mu(a) \geqslant 0$, $\mu(1) = 1$) is exactly the restriction to questions of states $\rho(A) = \mathrm{tr}(\rho A)$ given by the density matrices ρ (**Gleason's theorem**).

It is also possible to characterize the set of all states equipped with the convex structure (mixtures) geometrically. The set of all states (without the normalization condition) of a finite-dimensional, formally real, irreducible †Jordan algebra over the field of reals (the positivity of a state φ is defined by $\varphi(a^2) \geqslant 0$) has been characterized as a transitively homogeneous self-dual cone in a finite-dimensional real vector space (a cone V is transitively homogeneous if the group of all nonsingular linear transformations leaving V invariant is transitive on the topological interior of V) by E. B. Vinberg (*Trans. Moscow Math. Soc.*, 12 (1963); 13 (1965)), where the relevant Jordan algebras were completely classified earlier by P. Jordan, J. von Neumann, and E. P. Wigner (*Ann. Math.*, 36 (1934)) as direct sums of the following irreducible ones: the Jordan algebra (with the product $A \circ B = (AB + BA)/2$) of all Hermitian $n \times n$ matrices over the real, complex, or quaternion field, all 3×3 Hermitian matrices over octanions, or the so-called **spin balls** (the set of all normalized states being a ball) linearly generated by the identity and γ_j $(j = 1, \ldots, n)$ satisfying $\gamma_j \circ \gamma_k = 0$ if $j \neq k$ and $\gamma_j^2 = 1$.

In infinite-dimensional cases, this type of characterization extends to the "natural" positive cones of vectors (A. Connes, *Ann. Inst. Fourier*, 24 (1974); J. Bellissard and B. Iochum, *Ann. Inst. Fourier*, 28 (1978)); while the convex cone of all states (without normalization) of Jordan algebras and C^*-algebras have been characterized in terms of a certain class of projections associated with faces of the cone, called P-projections, by E. M. Alfsen, F. W. Shultz, and others (*Acta Math.*, 140 (1978); 144 (1980)). In finite-dimensional cases, Araki (*Commun. Math. Phys.*, 75 (1980)) has given a characterization allowing direct physical interpretation by replacing P-projection with a notion of filtering corresponding to quantum-mechanical measurement.

Due to some features of quantum-mechanical measurement not in conformity with common sense, there have arisen **hidden variable theories** that are deterministic and reproduce the quantum-mechanical prediction. For a situation where a pair of (correlated) particles in states a and b are created and their spins (1 or -1, i.e., up or down spin) measured at positions distant from each other, the expectation value $E(a, b)$ for the product would be given in a hidden variables theory by $E(a, b) = \int A_a(\lambda) B_b(\lambda) d\rho(\lambda)$ for a probability measure ρ and the functions A_a and B_b of hidden variables λ, representing spins and hence satisfying $|A_a| \leqslant 1$ and $|B_b| \leqslant 1$. Then the following **Bell's inequality** holds:

$$|E(a, b) - E(a, b') + E(a', b) + E(a', b')| \leqslant 2.$$

This contradicts both quantum-mechanical predictions and experimental results, so that hidden variable theories of this type have been rejected.

References

[1] P. A. M. Dirac, Principles of quantum mechanics, Clarendon Press, fourth edition, 1958.
[2] A. Messiah, Quantum mechanics I, II, North-Holland, 1961, 1962. (Original in French, 1959.)
[3] W. Pauli, General principles of quantum mechanics, Springer, 1980. (Original in German, 1958.)
[4] S. Tomonaga, Quantum mechanics I, II, North-Holland, 1962–1966.
[5] J. von Neumann, Mathematical foundations of quantum mechanics, Princeton Univ. Press, 1955. (Original in German, 1932.)
[6] B. L. van der Waerden, Sources of quantum mechanics, Dover, 1967.
[7] B. L. van der Waerden, Group theory and quantum mechanics, Springer, 1974.
[8] H. Weyl, The theory of groups and quantum mechanics, Dover, 1949. (Original in German, 1928.)
[9] E. P. Wigner, Group theory and its application to the quantum mechanics of atomic spectra, Academic Press, 1959.
[10] B. Simon, Functional integration and quantum mechanics, Academic Press, 1979.
[11] G. G. Emch, Algebraic methods in statistical mechanics and quantum field theory, Wiley, 1972.
[12] T. Bastin (ed.), Quantum theory and beyond, Cambridge Univ. Press, 1971.

[13] B. d'Espagnat, Conceptual foundations of quantum mechanics, Benjamin, second edition, 1976.
Also → references to 150 Field Theory, 375 Scattering Theory, 377 Second Quantization, and 386 S-Matrices.

352 (XI.15) Quasiconformal Mappings

A. History

H. Grötzsch (1928) introduced quasiconformal mappings as a generalization of conformal mappings. Let $f(z)$ be a continuously differentiable homeomorphism with positive Jacobian between plane domains. The image of an infinitesimal circle $|dz|=$constant is an infinitesimal ellipse with major axis of length $(|f_z|+|f_{\bar{z}}|)|dz|$ and minor axis of length $(|f_z|-|f_{\bar{z}}|)|dz|$. When the ratio $K(z)=(|f_z|+|f_{\bar{z}}|)/(|f_z|-|f_{\bar{z}}|)$ is bounded, f is called **quasiconformal**. If $K\equiv 1$, then f is conformal. Grötzsch noticed that Picard's theorem still holds under the weaker condition; he determined the quasiconformal mappings between two given domains, which are not conformally equivalent to each other, providing the smallest $\sup K$, that is, those closest to conformality [1].

We cannot speak of the history of quasiconformal mappings without mentioning the discovery of extremal length by A. Beurling and L. V. Ahlfors (→ 143 Extremal Length), which has led to the precise definition for quasiconformality itself.

Quasiconformal mappings have less rigidity than conformal mappings, and for this reason they have been utilized for the type problem or the classification of open Riemann surfaces (Ahlfors, S. Kakutani, O. Teichmüller, K. I. Virtanen, Y. Tôki; → 367 Riemann Surfaces). Quasiconformal mappings have important applications in other fields of mathematics, e.g., in the theory of †partial differential equations of elliptic type (M. A. Lavrent'ev [2]) and especially in the problem of moduli of Riemann surfaces, including the theory of Teichmüller spaces (→ 416 Teichmüller Spaces). These applications are explained in Sections C and D.

B. Definitions

The current definitions of quasiconformality, which dispense with continuous differentiability, are due to Ahlfors [3], A. Mori [4], and L. Bers [5] (→ C. B. Morrey [6]). Consider an orientation-preserving topological mapping f of a domain D on the $z(=x+iy)$-plane. The quasiconformality of f is defined as follows. (1) (the geometric defintion) Let Q be a curvilinear quadrilateral, i.e., a closed Jordan domain with four specified points on the boundary, and let the interior of Q be mapped conformally onto a rectangular domain I. The ratio $(\geqslant 1)$ of the sides of I, called the modulus of Q and denoted by $\operatorname{mod} Q$, is uniquely determined. If $\operatorname{mod} f(Q)\leqslant K \operatorname{mod} Q$ for any curvilinear quadrilateral Q in D, then f is called a K-**quasiconformal mapping** of D. This is equivalent to: (2) (the analytic definition) f is absolutely continuous on almost every line segment parallel to the coordinate axes contained in D (this condition is often referred to as ACL in D) and satisfies the inequality $|f_{\bar{z}}|\leqslant\dfrac{K-1}{K+1}|f_z|$ almost everywhere in D with some constant $K\geqslant 1$. When the value of K is irrelevant to the problem considered, K-quasiconformal mappings are simply said to be quasiconformal.

The K-quasiconformal mapping f satisfies the so-called **Beltrami differential equation**

$$f_{\bar{z}}=\mu f_z$$

almost everywhere in D with the measurable coefficient μ. The **maximal dilatation** $(1+\|\mu\|_\infty)/(1-\|\mu\|_\infty)$ does not exceed K. Sometimes f is called, for short a μ-conformal mapping. These notions are also defined for mappings between †Riemann surfaces, where the $(-1,1)$-form $\mu\, d\bar{z}\, dz^{-1}$ is independent of the choice of the local parameter z.

If in the above statements f is not necessarily topological but merely a continuous function satisfying the same requirements, we call it a μ-**conformal function**. (If in addition $\|\mu\|_\infty<(K-1)/(K+1)$, we call it a K-**quasiregular function** or K-**pseudoanalytic function**.) A μ-conformal function is represented as the composite $g\circ h$ of an analytic function g with a μ-conformal mapping h.

C. Principal Properties and Results

The inverse mapping of a K-quasiconformal mapping is also K-quasiconformal. The composite mapping $f_2\circ f_1$ of a K_1-quasiconformal mapping f_1 with a K_2-quasiconformal mapping f_2, if it can be defined, is K_1K_2-quasiconformal. A 1-quasiconformal mapping is conformal. Every quasiconformal mapping is †totally differentiable a.e. (almost everywhere), its Jacobian is positive a.e., and $(|f_z|+|f_{\bar{z}}|)/(|f_z|-|_{\bar{z}}|)\leqslant K$ a.e.

Let f be a K-quasiconformal mapping of $|z|<1$ onto $|w|<1$. Then f extends to a homeomorphism of $|z|\leqslant 1$ onto $|w|\leqslant 1$. If, furthermore, $f(0)=0$, then the Hölder condition

$$\left(\frac{|z_1-z_2|}{16}\right)^K \leqslant |f(z_1)-f(z_2)| \leqslant 16|z_1-z_2|^{1/K}$$

holds for $|z_1|\leqslant 1$, $|z_2|\leqslant 1$, and 16 is the best coefficient obtainable independently of K (Mori). This shows that any family of K-quasiconformal mappings of $|z|<1$ onto $|w|<1$ is †normal. For further properties and bibliography → O. Lehto and Virtanen [7] and Ahlfors [8].

(a) Boundary Correspondences and Extensions. Ahlfors and Beurling characterized the correspondence between $|z|=1$ and $|w|=1$ induced by f [9]. What amounts to the same thing, the following theorem holds: Let $\mu(x)$ be a real-valued monotone increasing continuous function on R such that $\lim_{x\to\pm\infty}\mu(x)=\pm\infty$. Then there exists a quasiconformal mapping of the upper half-plane $y>0$ onto itself with boundary correspondence $x\mapsto\mu(x)$ if and only if

$$\frac{1}{\rho}\leqslant\frac{\mu(x+t)-\mu(x)}{\mu(x)-\mu(x-t)}\leqslant\rho$$

for some constant $\rho\geqslant 1$ and for all $x, t\in\mathbf{R}$.

Theorem of quasiconformal reflection (Ahlfors [10]). Let L denote a curve which passes through ∞ and divides $\mathbf{C}\cup\{\infty\}$ into two domains Ω, Ω^* such that $\Omega\cup L\cup\Omega^*=\mathbf{C}\cup\{\infty\}$. Then there exists an orientation-reversing quasiconformal mapping of Ω onto Ω^* which keeps every point of L fixed if and only if some constant C exists satisfying $|\zeta_3-\zeta_1|/|\zeta_2-\zeta_1|\leqslant C$ for any three points ζ_1, ζ_2, ζ_3 on L such that $\zeta_3\in\widehat{\delta_1\xi_2}$.

(b) Mapping Problem. Given a measurable function μ in a simply connected domain D with $\|\mu\|_\infty<1$, there exists a μ-conformal mapping of D onto a plane domain Δ which is unique up to conformal mappings of Δ [8]. When μ is real analytic and the derivatives of functions are defined in the usual manner, a classical result concerning the †conformal mapping of surfaces asserts the existence of a solution of Beltrami's differential equation $f_{\bar z}=\mu f_z$.

Concerning the dependence of μ-conformal mapping on μ, Ahlfors and Bers [11] obtained the following important result: Denote by f^μ a μ-conformal mapping of the whole finite plane onto itself that preserves 0 and 1. The space of functions μ has the structure of a Banach space with L_∞-norm, and the space of mappings f^μ also has the structure of a Banach space with respect to a suitable norm. If $\{\mu(t)=\mu(z;t)\}$ is a family of μ depending on the parameter t with $\|\mu(t)\|_\infty\leqslant k<1$ and $\mu(t)$ is continuous (resp. continuously differentiable, real analytic, complex analytic) in t, then $f^{\mu(t)}$ is also continuous (continuously differentiable, real analytic, complex analytic). For the proofs of these important results, which have opened up a new way to study †Teichmüller space, the extension and reflection of quasiconformal mappings are made essential use of.

(c) Extremal Quasiconformal Mappings. Let $K(f)$ denote the maximal dilatation of a quasiconformal mapping f. Suppose that a family $\mathscr{F}=\{f\}$ of quasiconformal mappings is given. If some $f_0\in\mathscr{F}$ exists such that $K(f_0)$ attains the infimum of $K(f)$ for all $f\in\mathscr{F}$, then f_0 is called an **extremal quasiconformal mapping** in $\mathscr{F}$

Let $R=\{(x,y)\,|\,0<x<a, 0<y<b\}$, $R'=\{(x',y')\,|\,0<x'<a', 0<y'<b'\}$ be a pair of rectangular domains. Let $\mathscr{F}$ be the family of all quasiconformal mappings of R onto R' which map each vertex to a vertex with $(0,0)\mapsto(0,0)$. Then the unique extremal quasiconformal mapping for $\mathscr{F}$ is the affine mapping $x'=(a'/a)x$, $y'=(b'/b)y$ (Grötzsch [1]).

Next suppose that we are given two homeomorphic closed Riemann surfaces R, S and a †homotopy class $\mathscr{F}$ of orientation-preserving homeomorphisms of R onto S. Then $\mathscr{F}$ contains a unique extremal quasiconformal mapping. More precisely, either R and S are conformally equivalent to each other or else R admits an essentially unique analytic $(2,0)$-form Φ such that the respective local coordinates z, w of R, S satisfy the differential equation

$$(\partial w/\partial\bar z)/(\partial w/\partial z)=[(K-1)/(K+1)]\bar\Phi/|\Phi| \qquad (1)$$

with some constant $K>1$ everywhere on R, at which $\Phi\neq 0$ (Teichmüller [12], Ahlfors [3]). This turns out to be a generalization of Grötzsch's extremal affine mapping. The extremal mapping f satisfying equation (1) is sometimes referred to as the **Teichmüller mapping**.

Consider again a μ-conformal mapping g of the unit disk $D:|z|<1$ onto itself which induces a topological automorphism of the boundary $|z|=1$. If we define $\mathscr{F}$ as the family of all quasiconformal automorphisms f of D satisfying $f(e^{i\theta})=g(e^{i\theta})$, then the extremal quasiconformal mapping in $\mathscr{F}$ exists but is not always determined uniquely (K. Strebel [13]). As to the Teichmüller mapping, the uniqueness theorem is as follows: If the norm $\|\Phi\|=\iint_D|\Phi(z)|\,dx\,dy$ of Φ in (1) is finite, the Teichmüller mapping is the unique extremal quasiconformal mapping in $\mathscr{F}$. Otherwise, the uniqueness does not hold in general (Strebel [13]). On the other hand, a necessary and sufficient condition is proved for the Beltrami coefficient μ of a quasiconformal mapping of $\mathscr{F}$ to be extremal (R. S. Hamilton [14], E. Reich and Strebel in [15]). Moreover, this last result can be extended to the extremal quasiconformal mapping between arbitrary Riemann surfaces.

D. Applications

In the earlier stage of development of this theory, quasiconformal mappings were applied only to the †type problem of simply connected Riemann surfaces and to the classification of Riemann surfaces of infinite genus (→ 367 Riemann Surfaces). This application is based on the fact that it is often possible to find a quasiconformal mapping with the prescribed boundary correspondence even when no equivalent conformal mapping exists and the fact that the classes O_G and O_{HD} (→ 367 Riemann Surfaces) of Riemann surfaces are invariant under quasiconformal mappings, as they are under conformal mappings.

It is worth remarking that the investigation of quasiconformal mappings is intimately connected with the recent development of the theory of †Kleinian groups via Teichmüller spaces.

The theory of quasiconformal mappings was also applied by Lavrent'ev [16] and Bers [2] to partial differential equations, particularly to those concerning the behavior of fluids. They utilized the fact that if the density and its reciprocal are bounded in a steady flow of a 2-dimensional †compressible fluid, then the mapping of the physical plane to the potential plane (the plane on which the values of the †velocity potential and the †stream function are taken as coordinates) is quasiconformal, and that if in addition the supremum of the †Mach number is smaller than 1, then the mapping from the physical plane to the †hodograph plane is pseudoanalytic.

E. Similar Notions

The term *quasiconformal* was used differently by Lavrent'ev, as follows: A topological mapping $f=u+iv$ is called **quasiconformal** with respect to a certain system of linear partial differential equations when u and v satisfy the system. This is a generalized definition because the system may not be equivalent to a Beltrami equation. However, it is reduced to a quasiconformal mapping if the system is uniformly elliptic. Bers used the term **pseudoanalytic** to describe a certain function related to linear partial differential equations of elliptic type. This function is pseudoanalytic in the sense of Section B on every relatively compact subset and has properties similar to those of analytic functions.

Analytic transformations in the theory of functions of several variables are called pseudoconformal by some mathematicians, and there is a similar term *quasi-analytic*. The latter is an entirely different notion from the one discussed in this article.

F. Generalization to Higher Dimensions

Let f be a continuous ACL-mapping of a subdomain G of $\mathbf{R}^n$ into $\mathbf{R}^n$ whose Jacobian matrix is denoted by $f'(x)$. Furthermore, the operator norm and the determinant of f' are denoted by $\|f'\|$ and $\det f'$, respectively. Then f is said to be quasiregular if all the partial derivatives of f are locally of class L^n on G and if there exists a constant $K\geqslant 1$ such that $(\|f'\|(x))^n\leqslant K\cdot\det f'(x)$ almost everywhere in G. The smallest $K\geqslant 1$ for which this inequality is true is called the outer dilatation of f and is denoted by $K_0(f)$. If f is quasiregular, then the smallest $K\geqslant 1$ for which the inequality $\det f'(x)\leqslant K\cdot[\min_{|y|=1}|f'(x+y)|^n]$ holds almost everywhere in G is called the inner dilatation of f and is denoted by $K_I(f)$. If $\max(K_I(f),K_0(f))\leqslant K'$, then f is said to be K'-quasiregular. An orientation-preserving mapping is called K-quasiconformal (J. Väisälä [17]) if it is a K-quasiregular homeomorphism. When $n=2$, these definitions agree with those given in Section B.

For $n\geqslant 3$ the following properties also still hold: A quasiregular mapping is discrete, open, totally differentiable a.e. and is absolutely continuous (O. Martio, S. Rickman, and Väisälä [18]). Quasiconformal extension of higher-dimensional half-spaces have been studied by Ahlfors and L. Carleson [15].

References

[1] H. Grötzsch, Über möglichst konforme Abbildungen von schlichten Bereichen, Ber. Verh. Sächs. Akad. Wiss. Leipzig, 84 (1932), 114–120.
[2] M. A. Lavrent'ev, Varational methods for boundary value problems for systems of elliptic equations, Noordhoff, 1963. (Original in Russian, 1962.)
[3] L. V. Ahlfors, On quasiconformal mappings, J. Analyse Math., 3 (1954), 1–58, 207–208.
[4] A. Mori, On quasi-conformality and pseudo-analyticity, Trans. Amer. Math. Soc., 84 (1957), 56–77.
[5] L. Bers, On a theorem of Mori and the definition of quasi-conformality, Trans. Amer. Math. Soc., 84 (1957), 78–84.

[6] C. B. Morrey, On the solutions of quasilinear elliptic partial differential equations, Trans. Amer. Math. Soc., 43 (1938), 126–166.
[7] O. Lehto and K. I. Virtanen, Quasikonforme Abbildungen, Springer, 1965; English translation, Quasiconformal mappings in the plane, Springer, second edition, 1973.
[8] L. V. Ahlfors, Lectures on quasiconformal mappings, Van Nostrand, 1966.
[9] A. Beurling and L. V. Ahlfors, The boundary correspondence under quasiconformal mappings, Acta Math., 96 (1956), 125–142.
[10] L. V. Ahlfors, Quasiconformal reflections, Acta Math., 109 (1963), 291–301.
[11] L. V. Ahlfors and L. Bers, Riemann's mapping theorem for variable metrics, Ann. Math., (2) 72 (1960), 385–404.
[12] O. Teichmüller, Extremale quasikonforme Abbildungen und quadratische Differentiale, Abh. Preuss. Akad. Wiss. Math.-Nat. Kl., 22 (1939), 1–197.
[13] K. Strebel, Zur Frage der Eindeutigkeit extremaler quasikonformer Abbildungen des Einheitskreises, Comment. Math. Helv., 36 (1962), 306–323; 39 (1964), 77–89.
[14] R. S. Hamilton, Extremal quasiconformal mappings with prescribed boundary values, Trans. Amer. Math. Soc., 138 (1969), 399–406.
[15] L. V. Ahlfors et al., Contributions to analysis, Academic Press, 1974.
[16] L. Bers, Mathematical aspects of subsonic and transonic gas dynamics, Wiley, 1958.
[17] J. Väisälä, Lectures on n-dimensional quasiconformal mappings, Lecture notes in math. 229, Springer, 1971.
[18] O. Martio, S. Rickman, and J. Väisälä, Definitions for quasiregular mappings, Ann. Acad. Sci. Fenn., 448 (1969), 1–40.
[19] L. Bers, A new proof of a fundamental inequality for quasiconformal mappings, J. Analyse Math., 36 (1979), 15–30.

R

353 (XX.25) Racah Algebra

A. General Remarks

Racah algebra is a systematic method of calculating the †matrix element $(\psi, A\psi')$ in †quantum mechanics, where A is a dynamical quantity and ψ and ψ' are irreducible components of the state obtained by combining n †angular momenta. The angular momentum **j** has x-, y-, z-components j_x, j_y, j_z, respectively. Each component is i times the infinitesimal rotation around the respective axis and is the generator of the infinitesimal rotation for every irreducible component ψ. The addition of two angular momenta leads to a †tensor representation $D(j_1)\otimes D(j_2)$ of two †irreducible representations of the 3-dimensional rotation group. The problem is to decompose this tensor representation into the direct sum of irreducible representations.

B. Irreducible Representations of the Three-Dimensional Rotation Group

Irreducible representations of the group $SO(3)$ of 3-dimensional rotations can be obtained from irreducible representations $D(j)$ ($j=0,1,2,\ldots$) of its †universal covering group $SU(2)$ of 2×2 matrices with determinant 1, through the 2-fold covering isomorphism $SO(3)\cong SU(2)/\{\pm I\}$ (→ 60 Classical Groups I). The representation $D(j)$ ($j=0,1/2,1,3/2,\ldots$) of $SU(2)$ is the $2j$-fold tensor product $A\otimes\ldots\otimes A$ of $A\in SU(2)$ restricted to the totally symmetric part of the $2j$-fold tensor product space. Let $u=\binom{1}{0}$ and $v=\binom{0}{1}$ be a basis for the complex 2-dimensional space on which $SU(2)$ operates. The symmetrized tensor product of $(j+m)$-fold u and $(j-m)$-fold v multiplied by a positive normalization constant ($m=j,j-1,\ldots,-j$) defines an orthonormal basis of the representation space of $D(j)$, which we shall denote by $\psi(jm)$.

Decomposition of the tensor product of two irreducible representations $D(j_1)$ and $D(j_2)$ into irreducible components leads to

$$D(j_1)\otimes D(j_2)=\sum D(j),$$

$$j=j_1+j_2,\quad j_1+j_2-1,\ldots,|j_1-j_2|.$$

For the basis we can write

$$\psi(jm)=\sum_{m_1,m_2}\psi(j_1m_1)\psi(j_2m_2)(j_1m_1j_2m_2\,|\,j_1j_2jm),$$

and the coefficients are called the **Clebsch-Gordan coefficients** or **Wigner coefficients**. The vectors $\psi(jm)$ in each irreducible representation space are determined only up to an overall phase factor (a complex number of modulus 1). By a suitable choice of the resulting arbitrary phase (which may depend on j_1, j_2, j), the coefficients are given by

$$j_1+j_2\geqslant j\geqslant|j_1-j_2|,$$

$$(j_1m_1j_2m_2\,|\,j_1j_2jm)=\delta(m_1+m_2,m)$$
$$\times\sqrt{\frac{(2j+1)(j_1+j_2-j)!(j+j_1-j_2)!(j+j_2-j_1)!}{(j_1+j_2+j+1)!}}$$
$$\times\sum_\nu\left((-1)^\nu\frac{\sqrt{(j_1+m_1)!(j_1-m_1)!(j_2+m_2)!}}{\nu!(j_1+j_2-j-\nu)!(j_1-m_1-\nu)!}\right.$$
$$\left.\times\frac{\sqrt{(j_2-m_2)!(j+m)!(j-m)!}}{(j_2+m_2-\nu)!(j-j_2+m_1+\nu)!(j-j_1-m_2+\nu)!}\right).$$

They satisfy †orthogonality relations. Another concrete expression for the same coefficients, but of a different appearance, was obtained earlier by Wigner. Wigner introduced the **3j-symbol**, given by

$$\begin{pmatrix} j_1j_2j_3\\ m_1m_2m_3\end{pmatrix}=(-1)^{j_1-j_2-m_3}(2j_3+1)^{-1/2}$$
$$\times(j_1m_1j_2m_2\,|\,j_1j_2j_3\ -m_3)$$

for $m_1+m_2+m_3=0$ and zero otherwise. This is invariant under cyclic permutations of 1, 2, 3 and is multiplied by $(-1)^{j_1+j_2+j_3}$ under transpositions of indices as well as under the simultaneous sign change of all the m's. The 3j-symbol multiplied by $(-1)^{j_2+j_3-j_1}$ is the V-coefficient of Racah.

There are two ways, $(D(j_1)\otimes D(j_2))\otimes D(j_3)$ and $D(j_1)\otimes(D(j_2)\otimes D(j_3))$, to reduce the tensor product of three irreducible representations, and two corresponding sets of basis vectors. The transformation coefficient for the two ways of reduction is written in the form

$$\langle j_1j_2(j_{12})j_3;j\,|\,j_1,j_2j_3(j_{23});j\rangle$$
$$=\sqrt{(2j_{12}+1)(2j_{23}+1)}\,W(j_1j_2j\ j_3;j_{12}j_{23}).$$

Here $W(abcd;ef)$, called the **Racah coefficient**, can be written as the sum of products of four Wigner coefficients. W has the following symmetry properties:

$$W(abcd;ef)=W(badc;ef)$$
$$=W(cdab;ef)$$
$$=W(acbd;fe)$$
$$=(-1)^{e+f-a-d}W(ebcf;ad)$$
$$=(-1)^{e+f-b-c}W(aefd;bc)$$

and satisfies an orthogonality relation. The **6j-symbol** $\left\{\begin{smallmatrix}abe\\cdf\end{smallmatrix}\right\}$ is related to the Racah coefficient by

$$W(abdc;ef)=(-1)^{a+b+c+d}\begin{Bmatrix}abe\\cdf\end{Bmatrix}.$$

C. Irreducible Tensors

A dynamical quantity T_q^k $(q=k, k-1, \ldots, -k)$ that transforms in the same way as the basis of $D(k)$ under rotation of coordinates is called an **irreducible tensor of rank** k. That is, it satisfies

$$[j_x \pm ij_y, T_q^k] = \sqrt{(k \mp q)(k \pm q+1)}\ T_{q\mp 1}^k,$$
$$[j_z, T_q^k] = qT_q^k.$$

Here $[a,b]=ab-ba$. The matrix element of this quantity between two irreducible components can be written in the form

$$(\alpha jm|T_q^k|\alpha' j'm')$$
$$=(1/\sqrt{2j+1})(\alpha j\|T^{(k)}\|\alpha' j')(j'm'kq|j'kjm),$$

where α is a parameter to distinguish multiple components with the same j, and components of different α are assumed to be orthogonal. In this formula the Clebsch-Gordan coefficients are determined from group theory, while $(\alpha j\|T^{(k)}\|\alpha' j')$ depends on the dynamics of the system.

When $T^{(k)}$ and $U^{(k)}$ operate only on the state vectors in the subspaces H_1 and H_2, respectively, of the total space (†Hilbert space) $H=H_1\times H_2$, their scalar product $(T^{(k)}, U^{(k)})=\Sigma_q(-1)^q T_q^k U_q^k$ has the matrix element

$$(\alpha_1\alpha_2 j_1 j_2 jm|(T^{(k)}, U^{(k)})|\alpha_1'\alpha_2' j_1' j_2' jm)$$
$$=(-1)^{j_1+j_2'-j}W(j_1 j_2 j_1' j_2'; jk)$$
$$\times(\alpha_1 j_1\|T^{(k)}\|\alpha_1' j_1')(\alpha_2 j_2\|U^{(k)}\|\alpha_2' j_2').$$

For an irreducible component of the tensor product of two irreducible tensors,

$$[T^{(k_1)}\otimes U^{(k_2)}]_q^k$$
$$=\sum_{q_1+q_2=q} T_{q_1}^{k_2}U_{q_2}^{k_2}(k_1q_1k_2q_2|k_1k_2kq),$$

the matrix can be written as

$$(\alpha j_1 j_2 j\|[T^{(k_1)}\otimes U^{(k_2)}]^{(k)}\|\alpha' j_1' j_2' j')$$
$$=\sqrt{(2k+1)(2j+1)(2j'+1)}$$
$$\times\sum_{\alpha''}(\alpha j_1\|T^{(k_1)}\|\alpha'' j_1')(\alpha'' j_2\|U^{(k_2)}\|\alpha' j_2')$$
$$\times\begin{Bmatrix} j_1 & j_2 & j \\ j_1' & j_2' & j' \\ k_1 & k_2 & k \end{Bmatrix}.$$

The last factor, the **9*j*-symbol**, is defined as the matrix element between basis vectors of $[D(j_1)\times D(j_2)]\times[D(j_3)\times D(j_4)]$ and $[D(j_1)\times D(j_3)]\times[D(j_2)\times D(j_4)]$:

$$\langle j_1 j_2(j_{12}) j_3 j_4(j_{34}) jm| j_1 j_3(j_{13}) j_2 j_4(j_{24}) jm\rangle$$
$$=\sqrt{(2j_{12}+1)(2j_{34}+1)(2j_{13}+1)(2j_{24}+1)}$$
$$\times\begin{Bmatrix} j_1 & j_2 & j_{12} \\ j_3 & j_4 & j_{34} \\ j_{13} & j_{24} & j \end{Bmatrix}.$$

The 9*j*-symbol can be written as a weighted sum of the products of the three W's.

See [6] and [7] for explicit formulas of Clebsch-Gordan coefficients and [8] for Racah coefficients.

References

[1] E. P. Wigner, Group theory and its application to the quantum mechanics of atomic spectra, Academic Press, 1959.
[2] M. E. Rose, Elementary theory of angular momentum, Wiley, 1957.
[3] U. Fano and G. Racah, Irreducible tensorial sets, Academic Press, 1959.
[4] A. R. Edmonds, Angular momentum in quantum mechanics, Princeton Univ. Press, 1959.
[5] F. Bloch, S. G. Cohen, A. De-Shalit, S. Sambutsky, and I. Talmi, Spectroscopic and group-theoretical methods in physics (Racah memorial volume), North-Holland, 1968.
[6] E. U. Condon and G. H. Shortley, Theory of atomic spectra, Cambridge Univ. Press, 1935 (reprinted with corrections, 1951).
[7] M. Morita, R. Morita, T. Tsukada, and M. Yamada, Clebsch-Gordan coefficients for $j_2=5/2$, 3, and 7/2, Prog. Theoret. Phys., Suppl., 26 (1963), 64–74.
[8] L. C. Biedenharn, J. M. Blatt, and M. E. Rose, Some properties of the Racah and associated coefficients, Rev. Mod. Phys., 24 (1952), 249–257.

354 (XVI.5) Random Numbers

A. General Remarks

A sequence of numbers that can be regarded as realizations of independent and identically distributed †random variables is called a sequence or table of **random numbers**. It is a basic tool for the †Monte Carlo method, †simulation of stochastic phenomena in nature or in society, and †sampling or †randomization techniques in statistics. Random numbers used in practice are pseudorandom numbers (→ Section B); theoretically, the definition of random numbers leads to an algorithmic approach to the foundations of probability [1, 2].

B. Pseudorandom Numbers

Tables of numbers generated by random mechanisms have been statistically tested and published. To generate random numbers on a large scale, electronic devices based on stochastic physical phenomena, such as thermoelectron noise or radioactivity, can be used. For digital computers, however, numbers generated by certain simple algorithms can be viewed practically as a sequence of random numbers; this is called a sequence of **pseudorandom numbers**.

Distribution of random numbers that are easily generated and suitable for general use is the continuous uniform distribution on the interval (0, 1), which is approximated by the discrete distribution on $\{0, 1, \ldots, N-1\}$ $(N \gg 1)$. Random numbers with distribution function $F(\cdot)$ are obtained by transforming uniform distributions by $F^{-1}(\cdot)$. For typical distributions, computation tricks avoiding the direct computation of $F^{-1}(\cdot)$ have been devised. Among them the use of †order statistics and acceptance-rejection techniques have wide applicability.

For the generation of uniform pseudorandom numbers on $\{0, 1, \ldots, N-1\}$, $N = n^s$ (n = a computer word length), the following algorithms are used. Each of them is written in terms of simple computer instructions. (1) The **middle-square method** was proposed by von Neumann. We square an integer of s digits of radix (or base) n and take out the middle s digits as the next term. We repeat this process and obtain a sequence of pseudorandom numbers. The sequence thus generated might be cyclic with a short period, possibly after many repetitions. The lengths from initial values to the terminal cycles are empirically checked. (2) The †Fibonacci sequence $\{u_n\}$ defined by $u_{k+1} \equiv u_{k-1} + u_k \pmod{n^s}$ is apparently regular, but it is uniformly distributed. (3) The **congruence method** [3]: Define a sequence by $u_{k+1} \equiv au_k + c \pmod{n^s}$ or $(\mathrm{mod}\, n^s \pm 1)$. If $c = 0$, the procedure is called the multiplicative congruence method, otherwise the mixed congruence method. The cycle, that is, the minimum k such that $u_k = u_0$, and the constants a, c, and u_0 that make the cycle maximum for given n and s are determined by number theory. The points $(u_{kl}, u_{kl+1}, \ldots, u_{kl+l-1})$, $k = 0, 1, 2, \ldots$, lie on a small number of parallel hyperplanes in the l-dimensional cube. Good choice of the constant a makes the sequence quite satisfactory. (4) H. Weyl considered sequences $f(k) = k\alpha \pmod 1$, where α is an irrational number and $k = 1, 2, \ldots$, whose values are uniformly distributed on the interval (0, 1). They are not independent, though they can be used for some special purposes. A modified sequence $x_k = k^2\alpha \pmod 1$ is known to be random for any irrational α in the sense that the †serial correlation $N^{-1}\sum_{k=1}^{N} x_k x_{k+l} - 1/4$ converges to 0 uniformly in l as $N \to \infty$.

C. Statistical Tests

To check uniform random numbers on (0, 1) the following tests are used: (1) Divide (0, 1) into subintervals; then the frequency of random numbers falling into these is a multinomial sample. †Goodness of fit can be tested by the †chi-square test; independence can be tested by observing the frequency of transitions of subintervals in which a pair of consecutive numbers falls, as well as by observing the overall properties, such as uniformity of the frequency of patterns of subintervals in which a set of random numbers falls. (2) For a set of random numbers, the distance of the empirical distribution function from that of the theoretical one is tested by the †Kolmogorov-Smirnov test. (3) Observe the rank orders of a set of random numbers, and test the randomness of their permutations (test the number of runs up and down).

D. Kolmogorov-Chaitin Complexity and Finite Random Sequences

As Shannon's entropy is a quantity for measuring the randomness of random variables, the Kolmogorov-Chaitin complexity [4, 5] is that of individual objects based on logic instead of probability. For constructive objects $x \in X$, $y \in Y$ and a partial recursive function $A: Y \times \{1, 2, \ldots\} \to X$, define

$$K_A(x \mid y) = \begin{cases} \min(\log_2 n \mid A(y, n) = x), \\ \infty \quad (\text{if } A(y, n) = x \text{ for no } n). \end{cases}$$

The function A is said to be **asymptotically optimal** if for any B there exists a constant C such that $K_A(x \mid y) \leqslant K_B(x \mid y) + C$ for any $x \in X$ and $y \in Y$. For an asymptotically optimal A, which is known to exist, $K_A(x \mid y)$ is simply denoted by $K(x \mid y)$ and is called the **Kolmogorov-Chaitin complexity** of x given y.

P. Martin-Löf [6] discussed a relation between complexity and randomness. Consider any statistical test for the randomness on the set of (say) finite decimal sequences which is effective in the sense that it has a finite algorithm. Then there exists a constant C independent of L and M such that

$$K(\xi_1, \ldots, \xi_L \mid L) \geqslant L \log_2 10 - M$$

implies the acceptance of the decimal sequence $\xi_1, \ldots, \xi_L$ by the test at the level $1 - 2^{-M-C}$. This condition on the complexity is satisfied by at least $(1 - 2^{-M})10^L$ sequences among the decimal sequences of length L.

E. Collective and Infinite Random Sequences

For finite sequences, the notion of randomness is obscure by nature. For infinite sequences, however, clearer definition is possible. Based on the notion of collectives by R. von Mises, a definition of infinite random sequences has been given by A. Church [7]. A **selection function** is a $\{0,1\}$-valued function on the set of (say) finite decimal sequences such that $\{n, \varphi(\xi_1,\xi_2,\ldots,\xi_{n-1})=1\}$ is an infinite set for any infinite sequence $\xi_1, \xi_2, \ldots$. For a selection function φ and an infinite sequence $\xi_1, \xi_2, \ldots$, the **φ-subsequence** is defined as $\xi_{n_1}, \xi_{n_2}, \ldots$, where $\{n_1<n_2<\ldots\}=\{n, \varphi(\xi_1,\ldots,\xi_{n-1})=1\}$. For a class ψ of selection functions, an infinite decimal sequence is called a **ψ-collective** if each of the numbers $0, 1, \ldots, 9$ appears in it with a limiting relative frequency of $1/10$, and the same thing holds for any φ-subsequence with $\varphi\in\psi$. By definition, a **random sequence** is a ψ-collective for the class ψ of recursive selection functions. Almost all real numbers are random in their decimal expansions.

F. Normal Numbers

Let $x-[x]=\sum x_n r^{-n}$ be the r-adic expansion of the fractional part of a real number x. For any ordered set $B_k=(b_1,\ldots,b_k)$ of numbers $0, 1, \ldots, r-1$, let $N_n(x, B_k)$ be the number of occurrences of the block B_k in the sequence $x_1, \ldots, x_n$. If $N_n(x, B_k)/n\rightarrow r^{-k}$ as $n\rightarrow\infty$ for every k and every B_k, then x is said to be **normal** to base r. Almost all real numbers are normal to any r. D. G. Champernowne [8] constructed a normal number given by the decimal expansion 0.1, 2, 3, 4, 5, 6, 7, 8, 9, 10, 11, 12, 13, ... No one has so far been able to prove or disprove the normality of such irrational numbers as π, e, $\sqrt{2}$, $\sqrt{3}$, ... W. Schmidt [9] proved that the normality to base r implies the normality to base p if and only if $\log r/\log p$ is rational. A real number whose decimal expansion is random in the above sense is normal to base 10. For the converse, a necessary and sufficient condition for a selection function φ (for which $\varphi(\xi_1,\ldots,\xi_L)$ depends only on L) to have the property that the normality implies the $\{\varphi\}$-collectiveness has been obtained in [10].

References

[1] D. E. Knuth, The art of computer programming II, second edition, Addison-Wesley, 1981, ch. 3.
[2] E. R. Sowey, A second classified bibliography on random number generation and testing, Int. Statist. Rev., 46 (1978), 89–102.
[3] D. H. Lehmer, Mathematical methods in large-scale computing units, Proc. 2nd Symp. on Large-Scale Digital Calculating Machinery, Harvard Univ. Press, 1951, 141–146.
[4] A. N. Kolmogrov, Logical basis for information theory and probability theory, IEEE Trans. Information Theory, IT-14 (1968), 662–664.
[5] G. J. Chaitin, Algorithmic information theory, IBM J. Res. Develop., 21 (1977), 350–359.
[6] P. Martin-Löf, The definition of random sequences, Information and Control, 9 (1966), 602–619.
[7] A. Church, On the concept of a random sequence, Bull. Amer. Math. Soc., 47 (1940), 130–135.
[8] L. Kuipers and H. Niederreiter, Uniform distribution of sequences, Wiley-Interscience, 1974.
[9] W. Schmidt, On normal numbers, Pacific J. Math., 10 (1960), 661–672.
[10] T. Kamae, Subsequences of normal sequences, Israel J. Math., 16 (1973), 121–149.

355 (II.10) Real Numbers

A. Axioms for the Real Numbers

The set $\mathbf{R}$ of all **real numbers** has the following properties:

(1) Arithmetical properties: (i) For each pair of numbers $x, y\in\mathbf{R}$, there exists one and only one number $w\in\mathbf{R}$, called their **sum** and denoted by $x+y$, for which $x+y=y+x$ (commutative law) and $(x+y)+z=x+(y+z)$ (associative law) hold. Furthermore, there exists a unique number 0 (**zero**) such that $x+0=x$ for every $x\in\mathbf{R}$ (existence of †zero element). Also, for each x, there exists one and only one number $-x\in\mathbf{R}$ for which $x+(-x)=0$. (ii) For each pair of numbers $x, y\in\mathbf{R}$, there exists one and only one number $w\in\mathbf{R}$, called their **product** and denoted by xy, for which $xy=yx$ (commutative law), $(xy)z=x(yz)$ (associative law), and $(x+y)z=xz+yz$ (distributive law) hold. Furthermore, there exists a unique number 1 (**unity**)$\in\mathbf{R}$ such that $1x=x$ for every $x\in\mathbf{R}$ (existence of †unity element). Also, for each $x\neq0$ $(x\in\mathbf{R})$ there exists one and only one number $x^{-1}\in\mathbf{R}$ for which $xx^{-1}=1$. Owing to properties (i) and (ii), all †four arithmetic operations obey the usual laws (with the single exception of division by zero); in other words, $\mathbf{R}$ is a †field.

(2) Order properties: (i) For each $x, y\in\mathbf{R}$, one and only one of the following three relations holds: $x<y$, $x=y$, or $x>y$. With $x\leqslant y$ meaning $x<y$ or $x=y$, the relation $\leqslant$ obeys the transitive law: $x\leqslant y$ and $y\leqslant z$ imply $x\leqslant z$, which makes $\mathbf{R}$ †totally ordered. (ii) Order and

arithmetical properties are related by: $x \leqslant y$ implies $x+z \leqslant y+z$ for any $z \in \mathbf{R}$, and $x \leqslant y$ and $0 \leqslant z$ imply $xz \leqslant yz$; in other words, **R** is an †ordered field.

In particular, $x \in \mathbf{R}$ with $x>0$ is called a **positive number**, and $x \in \mathbf{R}$ with $x<0$ a **negative number**. We write $|x|=x$ if $x \geqslant 0$ and $|x|=-x$ if $x<0$, and call $|x|$ the **absolute value** of x.

(3) Continuity property: If nonempty subsets A and B of **R**, with $a<b$ for each pair $a \in A$ and $b \in B$, satisfy $\mathbf{R}=A \cup B$ and $A \cap B=\varnothing$ (empty set), then the pair (A, B) of sets is called a **cut** of **R**. For each cut (A, B) of **R**, there exists a number $x \in \mathbf{R}$ (necessarily unique) such that for every $a \in A$, $a \leqslant x$, and for every $b \in B$, $b \geqslant x$ (i.e., $x=\sup A=\inf B$). This property of **R** is called **Dedekind's axiom of continuity** (→ 294 Numbers).

The set **R** of all real numbers is determined uniquely up to an isomorphism, with respect to arithmetic operations and ordering, by properties (1)–(3). **R** forms an additive Abelian group; its subgroup $\{0, \pm 1, \pm 2, \ldots, \pm n, \ldots\}$ generated by 1 can be identified with the group **Z** of integers. The subset of all positive integers $\{1, 2, \ldots, n, \ldots\}$ may be identified with the set **N** of all natural numbers. The subset $\{m/n \mid m, n \in \mathbf{Z}, n \neq 0\}$ of **R** forms the subfield of **R** generated by 1. It can be identified with the field **Q** of all rational numbers. A real number that is not rational is called an **irrational number**.

B. Properties of Real Numbers

(1) For each pair of positive numbers a and $b>a$, there exists a natural number n with $a<nb$ (**Archimedes' axiom**).

(2) For each pair of positive real numbers a and b with $a<b$, there exists a rational number x such that $a<x<b$ (**denseness of rational numbers**).

(3) For any subset A in **R** †bounded from above (below), the †least upper bound of A: $a=\sup A$ (†greatest lower bound of A: $b=\inf A$) exists.

Let $\{a_n\}$ be a †sequence of real numbers. Assume that for each arbitrary positive number ε there exists a number n_0 such that $|a_n-b|<\varepsilon$ for all $n>n_0$. Then we write $\lim_{n\to\infty} a_n=b$ (or $a_n \to b$) and call b the **limit** of $\{a_n\}$. We also say that $\{a_n\}$ is a **convergent sequence** or that a_n **converges** to b.

(4) If for two sequences $\{a_n\}$, $\{b_n\}$, we have $a_1 \leqslant a_2 \leqslant \ldots \leqslant a_n \leqslant \ldots \leqslant b_n \leqslant \ldots \leqslant b_2 \leqslant b_1$ and $\lim(b_n-a_n)=0$, then there exists one and only one number $c \in \mathbf{R}$ with $\lim a_n=\lim b_n=c$ (**principle of nested intervals**).

(5) Let $\{a_n\}$ be a sequence of real numbers. If for any arbitrary positive ε there exists a natural number n_0 satisfying $|a_m-a_n|<\varepsilon$ for all $m, n>n_0$, then $\{a_n\}$ is called a **fundamental sequence** or **Cauchy sequence**. Any fundamental sequence of real numbers is convergent (**completeness of real numbers**).

For a set with properties 1 and 2 of Section A, it can be proved that property 3 of Section A is equivalent to property 3, or properties 1 and 4, or properties 1 and 5 of this section.

C. Intervals

For two numbers $a, b \in \mathbf{R}$ with $a<b$, we write

$(a, b)=\{x \mid a<x<b\}$,

$(a, b]=\{x \mid a<x \leqslant b\}$,

$[a, b)=\{x \mid a \leqslant x<b\}$,

$[a, b]=\{x \mid a \leqslant x \leqslant b\}$,

and call them (**finite**) **intervals**, of which a and b are their **left** and **right endpoints**, respectively. Specifically, (a, b) is called an **open interval** and $[a, b]$ a **closed interval**. The symbols ∞ and $-\infty$ are introduced as satisfying $\infty>x$, $x>-\infty$, $\infty>-\infty$ for all $x \in \mathbf{R}$. Writing $+\infty$ for ∞, we call $+\infty$ and $-\infty$ **positive infinity** and **negative infinity**, respectively. To extend the concept of intervals, we define $(-\infty, b)=\{x \mid x<b, x \in \mathbf{R}\}$, $(-\infty, b]=\{x \mid x \leqslant b, x \in \mathbf{R}\}$, $(a, \infty)=\{x \mid a<x, x \in \mathbf{R}\}$, $[a, \infty)=\{x \mid a \leqslant x, x \in \mathbf{R}\}$, and $(-\infty, \infty)=\mathbf{R}$, and call them **infinite intervals**.

Let $\{a_n\}$ be a sequence of real numbers. If for each infinite interval (a, ∞) ($(-\infty, a)$) there exists a number n_0 such that $a_n \in (a, \infty)$ ($a_n \in (-\infty, a)$) for all $n>n_0$, then we write $a_n \to +\infty$ ($a_n \to -\infty$) and call ∞ ($-\infty$) the limit of a_n, denoted as before by $\lim a_n$.

D. Topology of R

With the collection of all its open intervals (a, b) as an †open base, **R** is a †topological space (†order topology) that satisfies the †separation axioms $\mathbf{T}_2$, $\mathbf{T}_3$, $\mathbf{T}_4$. In **R** every (finite or infinite) interval (including **R** itself) is †connected, and the set **Q** of rational numbers is dense. A necessary and sufficient condition for a subset F of **R** to be †compact is that F be bounded and closed (**Weierstrass's theorem**). In particular, any finite closed interval is compact. **R** is a locally compact space satisfying the second †countability axiom. Further, any (finite or infinite) open interval is homeomorphic to **R**. The topology of **R** may also be defined by the notion of convergence (→ 87 Convergence).

Arithmetic operations in **R** are all continuous: If $a_n \to a$ and $b_n \to b$, then $a_n + b_n \to a+b$, $a_n - b_n \to a-b$, $a_n b_n \to ab$, and $a_n/b_n \to a/b$ (where $b \neq 0, b_n \neq 0$). Hence **R** is a †topological field (regarding the characterization of **R** as a topological group or a topological field → 422 Topological Abelian Groups).

R as a topological Abelian group (with respect to addition) is isomorphic to the topological Abelian group $\mathbf{R}^+$ of all positive real numbers with respect to multiplication. To be precise, there exist topological mappings $f: \mathbf{R} \to \mathbf{R}^+$ with $f(x+y) = f(x)f(y)$ and $g: \mathbf{R}^+ \to \mathbf{R}$ with $g(xy) = g(x) + g(y)$. If $f(1) = a$, $g(b) = 1$, then f, g are uniquely determined and are written $f(x) = a^x$, $g(x) = \log_b x$.

Regarding **R** as a topological Abelian group (with respect to addition), any proper closed subgroup Γ of **R** is discrete and isomorphic to the additive group **Z** of integers. That is, for some $\varepsilon > 0$ we have $\Gamma = \{n\varepsilon \mid n \in \mathbf{Z}\}$. In particular, the quotient group $\mathbf{R}/\mathbf{Z}$ as a topological group is isomorphic to the rotation group of a circle (1-dimensional †torus group). Elements of $\mathbf{R}/\mathbf{Z}$ are called **real numbers** mod 1.

E. The Real Line

Let l be a Euclidean straight line considered to lie horizontally, say from left to right. Let p_0, p_1 be two distinct points on l, with p_0 situated to the left of p_1. Then there exists one and only one bijection φ from the set L of all points of l to **R** satisfying (i) $\varphi(p_0) = 0$, $\varphi(p_1) = 1$; (ii) if p lies to the left of q, then $\varphi(p) < \varphi(q)$; and (iii) for two line segments pq and $p'q'$ (where p and p' are to the left of q and q', respectively), $pq \equiv p'q'$ (pq and $p'q'$ are †congruent) $\Leftrightarrow \varphi(q) - \varphi(p) = \varphi(q') - \varphi(p')$. Then $\varphi(p)$ is called the **coordinate** of the point p, and (p_0, p_1) the **frame** of the line l. A Euclidean straight line with a fixed frame is called a **real line** (identified with **R** by the mapping φ) and is usually denoted by the same notation **R** or $\mathbf{R}^1$.

References

[1] N. Bourbaki, Eléments de mathématique III. Topologie générale, ch. 4. Nombres réels, Actualités Sci. Ind., 1143c, Hermann, third edition, 1960; English translation, General topology pt. 1, Addison-Wesley, 1966.
[2] G. Cantor, Gesammelte Abhandlungen, Springer, 1932.
[3] R. Dedekind, Gesammelte mathematische Werke I–III, Braunschweig, 1930–1932.
[4] J. Dieudonné, Foundations of modern analysis, Academic Press, 1960, enlarged and corrected printing, 1969.
[5] K. Weierstrass, Gesammelte Abhandlungen 1–7, Mayer & Miler, Akademische Verlag., 1894–1927.
Also → references to 381 Sets.

356 (I.9) Recursive Functions

A. General Remarks

A function whose †domain and †range are both the set of natural numbers $\{0, 1, 2, \dots\}$ is called a **number-theoretic function**. In this article, the term *natural number* is used to mean a nonnegative integer. Hilbert (1926) and K. Gödel [1] considered certain number-theoretic functions, called recursive functions by them and now called **primitive recursive functions** after S. C. Kleene [2] (the definition is given in Section B). Gödel introduced an efficient method of arithmetizing metamathematics based on representing certain finitary procedures in metamathematics by primitive recursive functions. Then the following problem naturally arises: How shall we define a finitary method? In other words, how shall we characterize a number-theoretic function that is effectively computable, or provided with an algorithm of computation? Gödel defined the notion of general recursive function by introducing a formal system for the elementary calculation of functions, following the suggestion given by J. Herbrand. Kleene later improved Gödel's definition and developed the theory of general recursive functions [2]. Furthermore, A. Church and Kleene defined λ-calculable functions using the λ-notation (Church [4]), and E. L. Post and A. M. Turing defined the notion of computable functions by introducing the concept of Turing machines. These notions, introduced independently and almost simultaneously, were found to be equivalent. Hence such functions are now simply called recursive functions. Here, instead of giving the definition of recursive functions in the original style (the Herbrand-Gödel-Kleene definition), we give it by utilizing the idea of introducing schemata, a natural extension of the notion of primitive recursive functions. We employ the letters x, y, z, x_1, x_2, ... for variables ranging over the natural numbers.

B. Primitive Recursive Functions

Consider the following five definition schemata:

(I) $\varphi(x)=x' \quad (=x+1)$,

(II) $\varphi(x_1, \dots, x_n)=q$

(q a given natural number),

(III) $\varphi(x_1, \dots, x_n)=x_i \quad (1 \leqslant i \leqslant n)$,

(IV) $\varphi(x_1, \dots, x_n)=\psi(\chi_1(x_1, \dots, x_n), \dots, \chi_m(x_1, \dots, x_n))$,

(V) $\varphi(0, x_2, \dots, x_n)=\psi(x_2, \dots, x_n)$,

$\varphi(y', x_2, \dots, x_n)=\chi(y, \varphi(y, x_2, \dots, x_n), x_2, \dots, x_n)$,

where $\psi(\ \ )$ is a constant natural number if $n=1$. A function is called **primitive recursive** if it is definable by a finite series of applications of the operations (IV) and (V) ($\psi, \chi, \chi_1, \dots, \chi_m$ are already-introduced functions) starting from functions each of which is given by (I), (II), or (III). Given the functions $\psi_1, \dots, \psi_l$, we define the **relativization** (with respect to $\psi_1, \dots, \psi_l$) of the definition of primitive recursive functions as follows: A function is called **primitive recursive in** $\psi_1, \dots, \psi_l$ if it is definable by a finite series of applications of (IV) and (V) starting from $\psi_1, \dots, \psi_l$ and from functions each of which is given by (I), (II), or (III).

We say that a function $\varphi(x_1, \dots, x_n)$ is the **representing function** of a †predicate $P(x_1, \dots, x_n)$ if φ takes only 0 and 1 as values and satisfies

$$P(x_1, \dots, x_n) \Leftrightarrow \varphi(x_1, \dots, x_n)=0.$$

Then we call P a **primitive recursive predicate** if its representing function φ is primitive recursive. The following functions and predicates are examples of primitive recursive ones: $a+b$, $a \cdot b$, a^b, $a!$, $\min(a,b)$, $\max(a,b)$, $|a-b|$, $a=b$, $a<b$, $a|b$ (a divides b), $\mathrm{Pr}(a)$ (a is a prime number), p_i (the $(i+1)$st prime number, $p_0=2$, $p_1=3, \dots$), $(a)_i$ (the exponent of p_i of the unique factorization of a into prime numbers if $a \neq 0$; otherwise, 0).

Whenever we are given a concept or a theorem, we always transform it by replacing the predicates contained in it (if any) by corresponding representing functions. Then an operation Ω is called **primitive recursive** if the function or the predicate $\Omega(\psi_1, \dots, \psi_l, Q_1, \dots, Q_m)$ that results from the application of Ω to functions $\psi_1, \dots, \psi_l$ and predicates $Q_1, \dots, Q_m$ ($l, m \geqslant 0, l+m>0$) is primitive recursive in $\psi_1, \dots, Q_m$. Put $\varphi(x_1, \dots, x_n, z) = \sum_{y<z} \psi(x_1, \dots, x_n, y)$. Then φ is primitive recursive in ψ, and the finite sum $\sum_{y<z}$ is a primitive recursive operation in this sense.

Similarly, the following operations are primitive recursive: the finite product $\prod_{y<z}$, the logical connectives $\neg, \vee, \wedge, \rightarrow$ (→ 411 Symbolic Logic), definitions by cases, the **bounded quantifiers** $\exists y_{y<z}$, $\forall y_{y<z}$, and the **bounded μ-operator** $\mu y_{y<z}$ defined as follows: $\mu y_{y<z} R(x,y)$ is the least y such that $y<z$ and $R(x,y)$ holds, if there exists such a number y; otherwise, it equals z. The following operation is also primitive recursive:

$$\varphi(y, x_2, \dots, x_n)$$

$$=\chi(y, \tilde{\varphi}(y; x_2, \dots, x_n), x_2, \dots, x_n),$$

where

$$\tilde{\varphi}(y; x_2, \dots, x_n)=\prod_{i<y} p_i^{\varphi(i, x_2, \dots, x_n)}.$$

A function φ is said to be primitive recursive **uniformly** in $\psi_1, \dots, \psi_l$ when φ is definable by applying a primitive recursive operation to $\psi_1, \dots, \psi_l$.

Almost all results mentioned in this section were given by Gödel [1]. There are further investigations on primitive recursive functions by R. Péter (1934), R. Robinson (1947), and others. Note that a function definable by a †double recursion is not necessarily primitive recursive. Péter (1935, 1936) investigated in detail functions that are definable, in general, by k-fold recursions for every positive interger k [8].

C. General Recursive Functions

The following μ-operator is used to define general recursive functions by extending primitive recursive functions. For a predicate $R(y)$ on the natural numbers, $\mu y R(y)$ is the least y such that $R(y)$, if $\exists y R(y)$; otherwise, $\mu y R(y)$ is undefined. Generally, $\mu y(\psi(x_1, \dots, x_n, y)=0)$ is not necessarily defined for each n-tuple $(x_1, \dots, x_n)$ of natural numbers. Now, a function is called a **general recursive function** (or simply **recursive function**) if it is definable by a series of applications of schemata including a new schema

(VI) $\varphi(x_1, \dots, x_n)=\mu y(\psi(x_1, \dots, x_n, y)=0)$

for the definition of φ from any function ψ that satisfies

$$\forall x_1 \dots \forall x_n \exists y(\psi(x_1, \dots, x_n, y)=0),$$

in addition to those used to define the primitive recursive functions. Thus, by definition, a primitive recursive function is general recursive. **A general recursive predicate** is a predicate such that its representing function is general recursive. The facts, including the ones concerning relativization, that are valid for primitive recursive functions are also valid for general recursive functions.

Kleene's Normal Form Theorem. For each n, we can construct a primitive recursive predicate $T_n(z, x_1, \dots, x_n, y)$ and a primitive recursive function $U(y)$ such that given any general recursive function $\varphi(x_1, \dots, x_n)$, a natural number e can be found such that

$$\forall x_1 \dots \forall x_n \exists y T_n(e, x_1, \dots, x_n, y), \tag{1}$$

$$\varphi(x_1, \dots, x_n) = U(\mu y T_n(e, x_1, \dots, x_n, y)). \tag{2}$$

Any natural number e for which (1) and (2) hold is said to **define φ recursively** or to be a **Gödel number** of a recursive function φ. Let $\psi_1, \dots, \psi_l$ (abbreviated Ψ) be any given functions. We can relativize Kleene's normal form theorem with respect to them as follows: For each n, we can construct a predicate $T_n^\Psi(z, x_1, \dots, x_n, y)$ that is primitive recursive in Ψ such that given any function φ that is general recursive in Ψ, a natural number e can be found such that

$$\forall x_1 \dots \forall x_n \exists y T_n^\Psi(e, x_1, \dots, x_n, y), \tag{3}$$

$$\varphi(x_1, \dots, x_n) = U(\mu y T_n^\Psi(e, x_1, \dots, x_n, y)), \tag{4}$$

where $U(y)$ is the primitive recursive function mentioned in Kleene's normal form theorem. A natural number e for which (3) and (4) hold is said to **define φ recursively in Ψ** or to be a **Gödel number** of φ from Ψ. In particular, a Gödel number e of φ from Ψ can be found independently of Ψ (except for l and the respective numbers of variables of $\psi_1, \dots, \psi_l$) when φ is general recursive uniformly in Ψ.

Now let S be a †formal system containing ordinary number theory. A number-theoretic predicate $P(x_1, \dots, x_n)$ is said to be **decidable** within S if there is a formula $\mathrm{P}(\mathrm{a}_1, \dots, \mathrm{a}_n)$ (with no †free variables other than the distinct variables $\mathrm{a}_1, \dots, \mathrm{a}_n$) of S such that for each n-tuple $(x_1, \dots, x_n)$ of natural numbers (the symbol $\vdash$ means provable in S),

(i) $\vdash \mathrm{P}(\mathbf{x}_1, \dots, \mathbf{x}_n)$ or $\vdash \neg \mathrm{P}(\mathbf{x}_1, \dots, \mathbf{x}_n)$

and

(ii) $P(x_1, \dots, x_n) \Leftrightarrow \vdash \mathrm{P}(\mathbf{x}_1, \dots, \mathbf{x}_n)$,

where $\mathbf{x}_1, \dots, \mathbf{x}_n$ designate the numerals corresponding to $x_1, \dots, x_n$ in S. If S is a consistent system such that primitive recursive predicates are decidable within S and the predicates Pf_A (for any formula A, $Pf_A(x_1, \dots, x_n, y)$ means that y is the Gödel number of a proof of $\mathrm{A}(\mathbf{x}_1, \dots, \mathbf{x}_n)$) are primitive recursive, then a necessary and sufficient condition for P to be decidable within S is that P is a general recursive predicate (A. Mostowski, 1947).

Church (1936) proposed the following statement: Every **effectively calculable function** is a general recursive function. The converse of this is evidently true by the definition of recursiveness. So **Church's thesis** and its converse provide the exact definition of the notion of effectively computable functions. Though this notion is somewhat vague and intuitive, the definition seems to be satisfactory, as mentioned at the beginning of this article. Therefore, any function with a computation procedure or **algorithm** can be assumed to be general recursive. Utilizing this, various decision problems have been negatively solved (→ 97 Decision Problem). Furthermore, traditional descriptive set theory can be reinvestigated from this point of view, and the concept of effectiveness used in †semi-intuitionism is clarified using general recursive functions (→ 22 Analytic Sets).

D. Recursive Enumerability

A set $\{\varphi(0), \varphi(1), \varphi(2), \dots\}$ enumerated by a general recursive function φ (allowing repetitions) is called a **recursively enumerable set**. The empty set is also considered recursively enumerable. It is known that in this definition "general recursive" can be replaced by "primitive recursive" (J. B. Rosser, 1936). A set E of natural numbers is recursively enumerable if and only if there is a primitive recursive predicate $R(x, y)$ such that $x \in E \Leftrightarrow \exists y R(x, y)$ (Kleene [2]).

Generally, a predicate $E(x_1, \dots, x_m)$ is called a **recursively enumerable predicate** if there is a general recursive predicate $R(x_1, \dots, x_m, y_1, \dots, y_n)$ such that $E(x_1, \dots, x_m) \Leftrightarrow \exists y_1 \dots \exists y_n R(x_1, \dots, x_m, y_1, \dots, y_n)$. (Here "general recursive" can be replaced by "primitive recursive.")

We call a set E a **recursive set** if the predicate $x \in E$ is general recursive. The set $C = \{x \mid \exists y T_1(x, x, y)\}$ is an example of a set that is recursively enumerable but not recursive, and it has the following remarkable property: For every recursively enumerable set E, there is a primitive recursive function such that $x \in E \Leftrightarrow \varphi(x) \in C$. In this sense, the set C is said to be **complete** for the class of recursively enumerable sets. Post's problem, which asked whether the sets that are recursively enumerable but not recursive have the same †degree of (recursive) unsolvability as that of C, was negatively solved simultaneously by R. M. Friedberg (1957) and A. A. Muchnik (1956–1958). A recursively enumerable set E is general recursive if and only if there is a general recursive predicate $R(x, y)$ such that $x \notin E \Leftrightarrow \exists y R(x, y)$ (Kleene [5], Post [6]).

E. Partial Recursive Functions

A function $\varphi(x_1, \ldots, x_n)$ is called a **partial function** if it is not necessarily defined for all n-tuples $(x_1, \ldots, x_n)$ of natural numbers. For two partial functions $\psi(x_1, \ldots, x_n)$ and $\chi(x_1, \ldots, x_n)$, $\psi(x_1, \ldots, x_n) \simeq \chi(x_1, \ldots, x_n)$ means that if either $\psi(x_1, \ldots, x_n)$ or $\chi(x_1, \ldots, x_n)$ is defined for $x_1, \ldots, x_n$, so is the other, and the values are the same. For any given natural number e, $\varphi(x_1, \ldots, x_n) \simeq U(\mu y T_n(e, x_1, \ldots, x_n, y))$ (or $\simeq U(\mu y T_n^{\Psi}(e, x_1, \ldots, x_n, y)))$ is a partial function, in general. We say that such a function is **partial recursive (partial recursive (uniformly) in** Ψ) and that a natural number e defines φ **recursively ((uniformly) in** Ψ) or is a **Gödel number** of a partial recursive function φ (from Ψ). When a natural number e is a Gödel number of $\varphi(x_1, \ldots, x_n)$ (a Gödel number of φ from Ψ), $\varphi(x_1, \ldots, x_n)$ is sometimes written as $\{e\}(x_1, \ldots, x_n)(\{e\}^{\Psi}(x_1, \ldots, x_n))$. If a predicate $R(x_1, \ldots, x_n, y)$ is general recursive, then $\mu y R(x_1, \ldots, x_n, y)$ is partial recursive. Therefore, $\{z\}(x_1, \ldots, x_n)$ is a partial recursive function of the variables of z and of $x_1, \ldots, x_n$.

On the partial recursive functions, the following two theorems, given by Kleene [3], are most useful. (1) For natural numbers m, n, a primitive recursive function $S_n^m(z, y_1, \ldots, y_m)$ can be found such that, for any natural number e, $\{e\}(y_1, \ldots, y_m, x_1, \ldots, x_n) \simeq \{S_n^m(e, y_1, \ldots, y_m)\}(x_1, \ldots, x_n)$. (2) For any partial recursive function $\psi(z, x_1, \ldots, x_n)$, a natural number e can be found such that $\{e\}(x_1, \ldots, x_n) \simeq \psi(e, x_1, \ldots, x_n)$.

The notion of partial recursive functions appeared first in the theory of †constructive ordinal numbers of Church and Kleene (1963). Partial recursive functions can be defined in the Herbrand-Gödel-Kleene style as a natural extension of general recursive functions, and they are also definable by a finite series of applications of the schemata (IV), (V), and (VI) (in each schema, $=$ used for the definition of φ should be replaced by $\simeq$) starting from functions given by (I), (II), and (III).

F. Extension of Recursive Functions to Number-Theoretic Functionals

Let $\alpha_1, \ldots, \alpha_m$ be number-theoretic functions of one variable. If $\varphi(x_1, \ldots, x_n)$ is (partial) recursive uniformly in $\alpha_1, \ldots, \alpha_m$, then a Gödel number e of φ is found independently of $\alpha_1, \ldots, \alpha_m$, and $\varphi(x_1, \ldots, x_n)$ is expressed as $U(\mu y T_n^{\alpha_1 \cdots \alpha_m}(e, x_1, \ldots, x_n, y))$. Now suppose that $\alpha_1, \ldots, \alpha_m$ range over the set N^N of all number-theoretic functions of one variable, and put

$$\varphi(\alpha_1, \ldots, \alpha_m, x_1, \ldots, x_n)$$
$$\simeq U(\mu y T_n^{\alpha_1 \cdots \alpha_m}(e, x_1, \ldots, x_n, y)).$$

We call such a functional $\varphi(\alpha_1, \ldots, \alpha_m, x_1, \ldots, x_n)$ (partial) recursive, and with it we can develop a theory of recursive functions of variables of two types.

Extending the notion of recursive functionals, Kleene introduced and investigated the recursive functionals of variables of arbitrary (finite) types [10, 11]. The natural numbers are the **objects of type** 0, and the one-place functions from type-j objects to natural numbers are **objects of type** $j+1$. Denote variables ranging over the type-j objects by α^j, β^j, $\gamma^j, \ldots$, or $\alpha_1^j, \alpha_2^j, \alpha_3^j, \ldots$, etc. Consider a functional (simply called a function) of a given finite number of such variables of types taking natural numbers as values. A function φ is called a **primitive recursive function** if it is definable by a finite series of applications of the following schemata (I)–(VIII), where a is a variable of type 0, $\mathfrak{b}$ is any list (possibly empty) of variables that are mutually distinct and different from the other variables of the schema, and ψ, χ are given functions of the indicated variables. (I) $\varphi(a, \mathfrak{b}) = a'$; (II) $\varphi(\mathfrak{b}) = q$ (q is a natural number); (III) $\varphi(a, \mathfrak{b}) = a$; (IV) $\varphi(\mathfrak{b}) = \psi(\chi(\mathfrak{b}), \mathfrak{b})$; (V) $\varphi(0, \mathfrak{b}) = \psi(\mathfrak{b})$, $\varphi(a', \mathfrak{b}) = \chi(a, \varphi(a, \mathfrak{b}), \mathfrak{b})$; (VI) $\varphi(\mathfrak{a}) = \psi(\mathfrak{a}_1)$ ($\mathfrak{a}_1$ is a list of variables from which $\mathfrak{a}$ is obtained by changing the order of two variables of the same type); (VII) $\varphi(\alpha^1, a, \mathfrak{b}) = \alpha^1(a)$; (VIII) $\varphi(\alpha^j, \mathfrak{b}) = \alpha^j(\lambda\alpha^{j-2}\chi(\alpha^j, \alpha^{j-2}, \mathfrak{b}))$ ($\lambda\alpha^{j-2}$ designates that χ is a function of the variables α^{j-2}).

We assign to each function $\varphi(\mathfrak{a})$ a natural number called an **index** (which plays the same role as a Gödel number) in such a way that it reflects the manner of application of the schemata used to introduce $\varphi(\mathfrak{a})$. Now, we write $\varphi(\mathfrak{a})$ with an index e as $\{e\}(\mathfrak{a})$. We call a function $\varphi(\mathfrak{a})$ **partial recursive** if it is definable by a finite series of applications of the schemata (I)–(VIII) ($\simeq$ is employed instead of $=$ in (IV)–(VI) and (VIII)), and (IX) $\varphi(a, \mathfrak{b}, \mathfrak{c}) \simeq \{a\}(\mathfrak{b}, \mathfrak{c})$ ($\mathfrak{c}$ is a finite list of variables of arbitrary types). In particular, $\varphi(\mathfrak{a})$ is called a **general recursive function** if it is defined for all values of the argument $\mathfrak{a}$. These notions can be **relativized** also with respect to any given functions. Note that for the case of types $\leqslant 1$, primitive recursive functions, partial recursive functions, and also general recursive functions in the present sense are equivalent to the corresponding notions (introduced via relativization with uniformity) in the ordinary sense already described. The following theorem is important: Let r be the maximum type of $\mathfrak{a}$. Then there are two primitive recursive predicates M, N such that

$$\{e\}(\mathfrak{a}) \simeq$$

$$w \Leftrightarrow \forall \xi^{r-1} \exists \eta^{r-2} M(e, \mathfrak{a}, w, \xi^{r-1}, \eta^{r-2}), \quad r \geqslant 2,$$

$$w \Leftrightarrow \exists \xi^{r-1} \forall \eta^{r-2} N(e, \mathfrak{a}, w, \xi^{r-1}, \eta^{r-2}), \quad r > 2.$$

Every function definable using (IX′) $\psi(\mathfrak{a}) \simeq \mu x(\psi(\mathfrak{a}, x) = 0)$ instead of (IX) is partial recursive. However, not all the partial recursive functions of variables of types $\geqslant 2$ can be obtained by applying schemata (I)–(VIII) and (IX′).

Further developments have been pursued by J. E. Fenstad, J. Moldestad, and others in abstract computation theory [20–23].

G. Recursive Functions of Ordinal Numbers

G. Takeuti introduced a notion of primitive recursiveness for functions from a segment of the ordinal numbers to ordinal numbers. Using this, he constructed a model of set theory in ordinal number theory. In connection with recursive functions of ordinal numbers, there are also investigations by A. Lévy, M. Machover, Takeuti and A. Kino, T. Tugué, S. Kripke, and others.

Early treatments of recursive functions of ordinal numbers dealt only with functions on infinite cardinals. For example, Takeuti considered functions with a fixed infinite cardinal κ as a domain and a range, and defined κ-recursive functions using schemata similar to the abovementioned (I)–(VI). Subsequently Kripke observed that the assumption that κ is a cardinal is not necessary, and introduced the notion of **admissible ordinals**. An admissible ordinal κ has the closure properties required for the construction of the calculus, and whenever $\alpha, \beta < \kappa$ and $\beta = f(\alpha)$ is computable, then $\beta = f(\alpha)$ is computable in fewer than κ stages. Given an admissible ordinal κ, κ-**recursiveness** can be defined, as in the case of general recursiveness, by various equivalent methods, e.g., schemata, the equation calculus, and definability in both quantifier forms. Most of the elementary properties of general recursive functions (e.g., the normal form theorem, parametrization theorem, enumeration theorem, etc.) are also valid for κ-recursive functions. The notions of degrees of unsolvability and recursive enumerability can also be generalized, yielding the notions of κ-degrees and κ-recursive enumerability, respectively. The fine structures of these properties are currently the objects of intensive research.

Every infinite cardinal is admissible. The least admissible ordinal is ω, and the next is the ordinal ω_1 of Church and Kleene, i.e., the first nonconstructive ordinal. In fact, for every $n \geqslant 1$, the first ordinal not expressible as the order type of a Δ_n^1 predicate is admissible (→ Section H). For each infinite cardinal κ there are κ^+ admissible ordinals of power κ. Platek investigated recursion theory in a still wider setting. He dealt with functions defined, not on a segment of ordinal numbers, but on a set, and introduced the notion of admissible sets, i.e., sets on which a well-behaved recursion theory can be developed. An admissible set is a transitive ε-model of a certain weak set theory, and an ordinal κ is admissible if and only if there exists an admissible set A such that $A \cap O_n = \kappa$, where O_n is the class of all ordinal numbers.

Recent developments have shown that generalized recursion theory, set theory, and infinitary logic are closely related. In addition to the abovementioned, there are some investigations by Y. N. Moschovakis and others [14–27].

H. Hierarchies

Utilizing the theory of †recursive functions, S. C. Kleene succeeded in establishing a theory of hierarchies that essentially contains classical descriptive set theory as an extreme case [5, 10, 31, 32]. Although research following a similar line had also been done by M. Davis, A. Mostowskiĭ, and others, it was Kleene who succeeded in bringing the theory to its present form.

Sets or functions are described by †predicates, which we classify as follows. Let $a, b, \ldots, a_1, a_2, \ldots, x, y, \ldots$ be variables ranging over the set $\mathbf{N}$ of natural numbers, and $\alpha, \beta, \ldots, \alpha_1, \alpha_2, \ldots, \xi, \eta, \ldots$ be variables ranging over the set $\mathbf{N}^{\mathbf{N}}$ of all †number-theoretic functions with one argument. Let $\psi_1, \ldots, \psi_l (l \geqslant 0)$ be number-theoretic functions. A predicate $P(\alpha_1, \ldots, \alpha_m, a_1, \ldots, a_n)$ $(m, n \geqslant 0, m+n>0)$ with variables of two †types is called **analytic** in $\psi_1, \ldots, \psi_l$ $(l \geqslant 0)$ if it is expressible syntactically by applying a finite number of logical symbols: $\rightarrow$, $\vee$, $\wedge$, $\neg$, $\exists x$, $\forall x$, $\exists \xi$, $\forall \xi$, to †general recursive predicates in $\psi_1, \ldots, \psi_l$. In particular, when P is expressible without function quantifiers $\exists \xi$, $\forall \xi$, it is called **arithmetical** in $\psi_1, \ldots, \psi_l$ $(l \geqslant 0)$. When $l = 0$, they are called simply analytic and arithmetical, respectively.

For brevity, consider the case $l = 0$, and denote by $\mathfrak{a}$ a finite list of variables $(\alpha_1, \ldots, \alpha_m, a_1, \ldots, a_n)$. Every arithmetical predicate $P(\mathfrak{a})$ is expressible in a form contained in the following table (a):

(a) $R(\mathfrak{a})$;

$\exists x R(\mathfrak{a}, x), \quad \forall x \exists y R(\mathfrak{a}, x, y), \ldots,$

$\forall x R(\mathfrak{a}, x), \quad \exists x \forall y R(\mathfrak{a}, x, y), \ldots,$

where each R is †general recursive. In order to obtain such an expression we first transform the given predicate into its †prenex normal form and then contract successive quantifiers of the same kind by the formula

$$\exists x_1 \ldots \exists x_n A(x_1, \ldots, x_n) \Leftrightarrow \exists x A((x)_0, \ldots, (x)_{n-1}) \quad (1)$$

and its "dual form." Each form in (a) (or the class of all predicates with that form) is denoted by Σ_k^0 or Π_k^0, where the suffix k refers to the number of quantifiers prefixed, and Σ or Π shows that the outermost quantifier is existential or universal, respectively. A predicate that is expressible in both forms Σ_k^0 and Π_k^0 (or the class of such predicates) is denoted by Δ_k^0. A predicate belongs to Δ_1^0 if and only if it is general recursive (an analog of †Suslin's theorem).

For $k \geqslant 1$, there exists in Σ_k^0 (or Π_k^0) an **enumerating predicate** that specifies every predicate in Σ_k^0 (Π_k^0). For example, for Π_2^0 and $m=n=1$, there is a †primitive recursive predicate $S(\alpha, z, a, x, y)$ such that, given a general recursive predicate $R(\alpha, a, x, y)$, we have a natural number e such that

$$\forall x \exists y R(\alpha, a, x, y) \Leftrightarrow \forall x \exists y S(\alpha, e, a, x, y)$$

(**enumeration theorem**). In this theorem, we can take $T_2^{\alpha}(z, a, x, y)$ (→ Section F) as $S(\alpha, z, a, x, y)$. For each $k \geqslant 0$, there exists a $\Sigma_{k+1}^0(\Pi_{k+1}^0)$ predicate that is not expressible in its dual form Π_{k+1}^0 (Σ_{k+1}^0) (hence, of course, in neither Σ_k^0 not Π_k^0) (**hierarchy theorem**). Therefore, table (a) gives the classification of the arithmetical predicates in a hierarchy. This hierarchy is called the **arithmetical hierarchy**. For each $k \geqslant 1$, there exists a **complete** predicate with respect to $\Sigma_k^0(\Pi_k^0)$, that is, a $\Sigma_k^0(\Pi_k^0)$ predicate with only one variable such that any $\Sigma_k^0(\Pi_k^0)$ predicate is expressible by substituting a suitable general (or more strictly, primitive) recursive function for its variable (**theorem on complete form**). When $m=0$, all the general recursive predicates in Σ_k^0 exhaust Δ_{k+1}^0 (**Post's theorem**).

Concerning the function quantifiers, we have

$$\exists \alpha_1 \ldots \exists \alpha_m A(\alpha_1, \ldots, \alpha_m) \Leftrightarrow \exists \alpha A(\lambda t(\alpha(t))_0, \ldots, \lambda t(\alpha(t))_{m-1}) \quad (2)$$

$$\exists x A(x) \Leftrightarrow \exists \alpha A(\alpha(0)), \quad (3)$$

$$\forall x \exists \alpha A(x, \alpha) \Leftrightarrow \exists \alpha \forall x A(x, \lambda t \alpha(2^x \cdot 3^t)), \quad (4)$$

and their dual forms. For any general recursive predicate R, there is a primitive recursive predicate S such that

$$\exists \alpha R(\alpha, \mathfrak{a}) \Leftrightarrow \exists \alpha \exists x S(\bar{\alpha}(x), \mathfrak{a}) \Leftrightarrow \exists y S(y, \mathfrak{a}) \quad (5)$$

and its dual hold. Using these facts, we can classify the forms of all analytic predicates by the table (b):

(b) $A(\mathfrak{a})$;

$\forall \alpha \exists x R(\mathfrak{a}, \alpha, x), \quad \exists \alpha \forall \beta \exists x R(\mathfrak{a}, \alpha, \beta, x), \ldots,$

$\exists \alpha \forall x R(\mathfrak{a}, \alpha, x), \quad \forall \alpha \exists \beta \forall x R(\mathfrak{a}, \alpha, \beta, x), \ldots,$

where A is arithmetical and each R is general recursive. Similarly, denote by Σ_k^1, Π_k^1 each form of predicate in (b) (or the class of all predicates reducible to that form), where k is the number of function quantifiers prefixed; also, denote by Δ_k^1 the (class of) predicates expressible in both forms Σ_k^1 and Π_k^1. For Σ_k^1, Π_k^1 ($k \geqslant 1$), we have the enumeration theorem, the hierarchy theorem, and the theorem on complete form. The hierarchy given by table (b) is called the **analytic hierarchy**.

For $l > 0$ (namely, when predicates are arithmetical or analytic in $\psi_1, \ldots, \psi_l$), we can †uniformly †relativize the above results with respect to $\psi_1, \ldots, \psi_l$. Now let $\{\Sigma_k^{r, \psi_1, \ldots, \psi_l}, \Pi_k^{r, \psi_1, \ldots, \psi_l}\}_k$ ($r=0, 1$) be the corresponding hierarchy relative to $\psi_1, \ldots, \psi_l$. Given a set C ($\subset \mathbf{N}^{\mathbf{N}}$) of functions with one argument, we can consider hierarchies of predicates which are arithmetical or analytic in a finite number of functions in C. Such a hierarchy is called a C-**arithmetical** or C-**analytic hierarchy** and denoted by $\{\Sigma_k^0[C], \Pi_k^0[C]\}_k$ or $\{\Sigma_k^1[C], \Pi_k^1[C]\}_k$, respectively. That is, when we regard $\Sigma_k^r[C]$ as a class of predicates (or sets) P, it is the family $\{P \mid P \in \Sigma_k^{r, \xi_1, \ldots, \xi_l}, \xi_1, \ldots \xi_l \in C, l=0, 1, 2, \ldots\}$. These notations have been given by J. W. Addison [28, 29]. The $\mathbf{N}^{\mathbf{N}}$-arithmetical hierarchy and the $\mathbf{N}^{\mathbf{N}}$-analytic hierarchy for sets correspond respectively to the finite Borel hierarchy and the projective hierarchy in the †space of irrational numbers. Addison called the theory of those hierarchies **classical descriptive set theory**, and in contrast to this, the theory of arithmetical and analytic hierarchies for sets ($C = \varnothing$) **effective descriptive set theory** [28].

We now restrict our consideration to predicates for natural numbers (i.e., to the case $m=0$). Define the predicates L_k by $L_0(a) \Leftrightarrow a=a$, $L_{k+1}(a) \Leftrightarrow \exists x T_1^{L_k}(a, a, x)$. For each $k \geqslant 0$, $L_{k+1}(a)$ is a Σ_{k+1}^0 predicate which is of the highest †degree of recursive unsolvability among the Σ_{k+1}^0 predicates, and its degree is properly higher than that of $L_k(a)$. Thus L_k, $k=0, 1, 2, \ldots$, determine the **arithmetical hierarchy of degrees of recursive unsolvability**. Kleene has extended the series of L_k by using the system S_3 (→ 81 Constructive Ordinal Numbers) of notations for the constructive ordinal numbers as follows [6]: $H_1(a) \Leftrightarrow a=a$; for $y \in O$, $H_{2^y}(a) \Leftrightarrow \exists x T_1^{H_y}(a,a,x)$; for $3 \cdot 5^y \in O$, $H_{3 \cdot 5^y}(a) \Leftrightarrow H_{y_{(a)_1}}((a)_0)$, where $y_n = \{y\}(n_0)$. This H_y is defined for each $y \in O$, and it is of a properly higher degree than that of H_z when $z <_O y$. If $|y|=|z|$ ($|y|$ is the ordinal number represented by y), then H_y and H_z are of the same degree (C. Spector [34]). Thus a hierarchy of degrees is uniquely determined by constructive ordinal numbers. This hierarchy is called the **hyperarithmetical hierarchy of degrees of recursive unsolvability**. A function or predicate is said to be **hyperarithmetical** if it is recursive in H_y for some $y \in O$. These concepts and the results mentioned below can be relativized to any given functions or predicates.

A necessary (Kleene [31]) and sufficient (Kleene [32]) condition for a predicate to be hyperarithmetical is that it be expressible in both one-function quantifier forms Δ_1^1 (an effective version of Suslin's theorem). Denote by Hyp the set ($\subset \mathbf{N}^{\mathbf{N}}$) of all hyperarithmetical functions α. For an arithmetical predicate $A(\alpha, a)$, $\exists \alpha_{\alpha \in \mathrm{Hyp}} A(\alpha, a)$ is always a Π_1^1 predicate (Kleene [33]). Conversely, for any Π_1^1 predicate P, there is a general recursive predicate R such that $P(a) \Leftrightarrow \exists \alpha_{\alpha \in \mathrm{Hyp}} \forall x R(a, \alpha, x)$ (Spector [35]). As to †uniformization, for a Π_1^1 predicate $P(a,b)$, we have $\forall x \exists y P(x,y) \Rightarrow \exists \alpha_{\alpha \in \mathrm{Hyp}} \forall x P(x, \alpha(x))$ (G. Kreisel, 1962). Let $\mathbf{E}$ be an object of type 2 defined by: $\mathbf{E}(\alpha)=0$ if $\exists x(\alpha(x)=0)$, otherwise $\mathbf{E}(\alpha)=1$. A function $\varphi(a_1, \ldots, a_n)$ is hyperarithmetical if and only if it is general recursive in $\mathbf{E}$ (Kleene [10]). A predicate that is hyperarithmetical relative to Π_k^1 predicates ($k \geqslant 0$) is of Δ_{k+1}^l (Kleene [32]), but the converse does not hold in general (Addison and Kleene, 1957).

Kleene extended his theory of hierarchy to the case of predicates of variables of any type by utilizing the theory of general recursive functions with variables of finite types $0, 1, 2, \ldots$ [10]. Let $\mathfrak{a}^t$ be a list of variables of types $\leqslant t$. We say a predicate $P(\mathfrak{a}^t)$ is of order r in completely defined functions $\psi_1, \ldots, \psi_l$ ($l \geqslant 0$) (for brevity, denote them by Ψ) if P is syntactically expressible in terms of variables of finite types, predicates that are general recursive in Ψ, and symbols of the †predicate calculus with quantification consisting only of variables of types $<r$. The predicates of order 0 in Ψ are exactly the general recursive ones in Ψ. When $t \geqslant 1$, and Ψ are functions of variables of type 0, a predicate $P(\mathfrak{a}^t)$ is of order 1 (of order 2) in Ψ if and only if P is arithmetical (analytic).

We have theorems similar to (2)–(4) and the following theorem and its dual for $r \geqslant 2$: For any given general recursive predicate $P(\mathfrak{a}^r, \sigma^r, \xi^{r-2})$, there is a primitive recursive predicate $R(\mathfrak{a}^r, \eta^{r-1}, \xi^{r-2})$ such that

$$\exists \sigma^r \forall \xi^{r-2} P(\mathfrak{a}^r, \sigma^r, \xi^{r-2}) \Leftrightarrow \exists \eta^{r-1} \forall \xi^{r-2} R(\mathfrak{a}^r, \eta^{r-1}, \xi^{r-2}). \quad (6)$$

Using these equivalences, each predicate $P(\mathfrak{a}^t)$ of order $r+1$ ($r \geqslant 0$) is expressible in one of the following forms:

(c) $B(\mathfrak{a})$;

$\forall \alpha^r \exists \xi^{r-1} R(\mathfrak{a}, \alpha^r, \xi^{r-1})$,

$\exists \alpha^r \forall \beta^r \exists \xi^{r-1} R(\mathfrak{a}, \alpha^r, \beta^r, \xi^{r-1}), \ldots,$

$\exists \alpha^r \forall \xi^{r-1} R(\mathfrak{a}, \alpha^r, \xi^{r-1})$,

$\forall \alpha^r \exists \beta^r \forall \xi^{r-1} R(\mathfrak{a}, \alpha^r, \beta^r, \xi^{r-1}), \ldots,$

where B is of order r and each R is general recursive. When $t=r+1$, table (c) gives the classification of the predicates of order $r+1$ into the hierarchy. In fact, for the predicates $P(\mathfrak{a}^{r+1})$ in each form, we have the enumeration theorem, the hierarchy theorem, and the theorem on complete form (Kleene [10]). D. A. Clarke [30] has published a detailed review of the general theory of hierarchies.

References

[1] K. Gödel, Über formal unentscheidbare Sätze der Principia Mathematica und verwandter Systeme I, Monatsh. Math. Phys., 38 (1931), 173–198.
[2] S. C. Kleene, General recursive functions of natural numbers, Math. Ann., 112 (1936), 727–742.
[3] S. C. Kleene, On notation for ordinal numbers, J. Symbolic Logic, 3 (1938), 150–155.
[4] A. Church, The calculi of lambda-conversion, Ann. Math. Studies, Princeton Univ. Press, 1941.
[5] S. C. Kleene, Recursive predicates and quantifiers, Trans. Amer. Math. Soc., 53 (1943), 41–73.
[6] E. L. Post, Recursively enumerable sets of positive integers and their decision problems, Bull. Amer. Math. Soc., 50 (1944), 284–316.
[7] S. C. Kleene, Introduction to metamathematics, Van Nostrand, 1952.
[8] R. Péter, Rekursive Funktionen, Akademische Verlag., 1951.
[9] A. A. Markov, Theory of algorithms (in Russian), Trudy Mat. Inst. Steklov., 42 (1954).
[10] S. C. Kleene, Recursive functionals and quantifiers of finite types I, Trans. Amer. Math. Soc., 91 (1959), 1–52.
[11] S. C. Kleene, Recursive functionals and quantifiers of finite types II, Trans. Amer. Math. Soc., 108 (1963), 106–142.
[12] H. Hermes, Enumerability, decidability, computability, Springer, 1965.

[13] H. Rogers, Theory of recursive functions and effective computability, McGraw-Hill, 1967.
[14] J. Barwise, Infinitary logic and admissible sets, J. Symbolic Logic, 34 (1969), 226–252.
[15] R. B. Jensen and C. Karp, Primitive recursive set functions, Proc. Symposia in Pure Math., XIII (1971), 143–176.
[16] S. Kripke, Transfinite recursions on admissible ordinals, J. Symbolic Logic, 29 (1964), 161–162.
[17] M. Machover, The theory of transfinite recursion, Bull. Amer. Math. Soc., 67 (1961), 575–578.
[18] G. Takeuti, A formalization of the theory of ordinal numbers, J. Symbolic Logic, 30 (1965), 295–317.
[19] T. Tugué, On the partial recursive functions of ordinal numbers, J. Math. Soc. Japan, 16 (1964), 1–31.
[20] J. Moldestad, Computation in higher types, Springer, 1977.
[21] J. E. Fenstad, R. O. Gandy, and G. E. Sacks (eds.), Generalized recursion theory, North-Holland, 1974.
[22] J. E. Fenstad, R. O. Gandy, and G. E. Sacks (eds.), Generalized recursion theory II, North-Holland, 1978.
[23] J. E. Fenstad, General recursion theory, Springer, 1980.
[24] Y. N. Moschovakis, Elementary induction on abstract structures, North-Holland, 1974.
[25] Y. N. Moschovakis, Descriptive set theory, North-Holland, 1980.
[26] J. Barwise, Admissible sets and structures, Springer, 1975.
[27] P. G. Hinman, Recursion theoretic hierarchies, Springer, 1978.
[28] J. W. Addison, Separation principles in the hierarchies of classical and effective descriptive set theory, Fund. Math., 46 (1958–1959), 123–135.
[29] J. W. Addison, Some consequences of the axiom of constructibility, Fund. Math., 46 (1958–1959), 337–357.
[30] D. A. Clarke, Hierarchies of predicates of finite types, Mem. Amer. Math. Soc., 1964.
[31] S. C. Kleene, Arithmetical predicates and function quantifiers, Trans. Amer. Math. Soc., 79 (1955), 312–340.
[32] S. C. Kleene, Hierarchies of number-theoretic predicates, Bull. Amer. Math. Soc., 61 (1955), 193–213.
[33] S. C. Kleene, Quantification of number-theoretic functions, Compositio Math., 14 (1959), 23–40.
[34] C. Spector, Recursive well-orderings, J. Symbolic Logic, 20 (1955), 151–163.
[35] C. Spector, Hyperarithmetical quantifiers, Fund. Math., 48 (1959–1960), 313–320.

357 (VI.6) Regular Polyhedra

A. Regular Polygons

A †polygon in a Euclidean plane bounding a †convex cell whose sides and interior angles are all respectively congruent is called a **regular polygon**. When the number of vertices (which equals the number of sides) is n, it is called a **regular n-gon**. There exist a circle (**circumscribed circle**) passing through all the vertices of a regular n-gon and a concentric circle (**inscribed circle**) tangent to all the sides. We call the center of these circles the **center** of the regular n-gon. The n vertices of a regular n-gon are obtained by dividing a circle into n equal parts. (When a polygon in a Euclidean plane bounds a †convex cell, this 2-cell is sometimes called a convex polygon. Thus *regular polygon* sometimes means the convex cell bounded by a regular polygon as described above.) A necessary and sufficient condition for a regular n-gon to be geometrically constructible is that n be decomposable into the product of prime numbers $n = 2^m p_1 \dots p_k$ $(m \geqslant 0)$, where the p_i $(i = 1, 2, \dots)$ are different †Fermat numbers (→ 179 Geometric Construction).

B. Regular Polyhedra

Consider a regular polygon on a plane, and take a point on the line perpendicular to the plane at the center of the polygon. The set of points on all half-lines joining this point and points on the polygon (considered as a convex cell) is called a **regular polyhedral angle** having this point as vertex (Fig. 1).

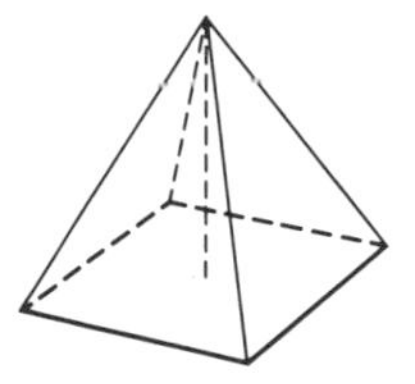

Fig. 1

When a †convex polyhedron $\mathfrak{F}$ in E^3 satisfies the following two conditions, we call it a **regular polyhedron**: (1) Each face of $\mathfrak{F}$, which is a 2-dimensional cell, is a regular polygon, and all faces of $\mathfrak{F}$ are congruent to each other. (2) Its vertices are all surrounded alike. That is, by the projection of $\mathfrak{F}$ from each vertex of $\mathfrak{F}$, we obtain a regular polyhedral angle; these regular polyhedral angles are all congruent to each other. From (2) we see that the number of edges emanating from each vertex of $\mathfrak{F}$ is independent of the vertex. It has been known since Plato's time that there are only five kinds of regular polyhedra: **tetrahedrons** (Fig. 2), **octahedrons** (Fig. 3), **icosahedrons** (Fig. 4), **cubes** or **hexahedrons** (Fig. 5), and **dodecahedrons** (Fig. 6) (see also see Table 1).

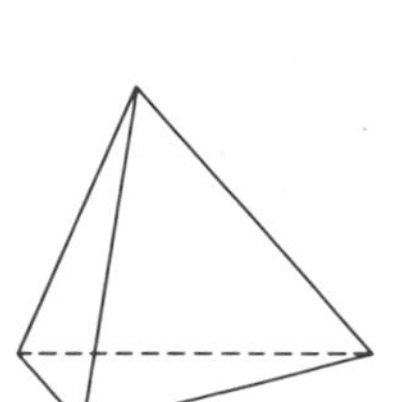

Fig. 2
Regular tetrahedron.

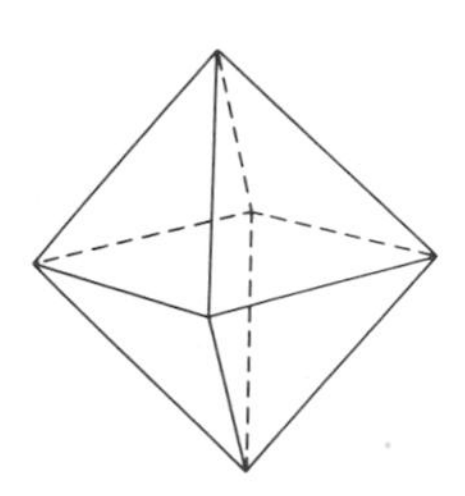

Fig. 3
Regular octahedron.

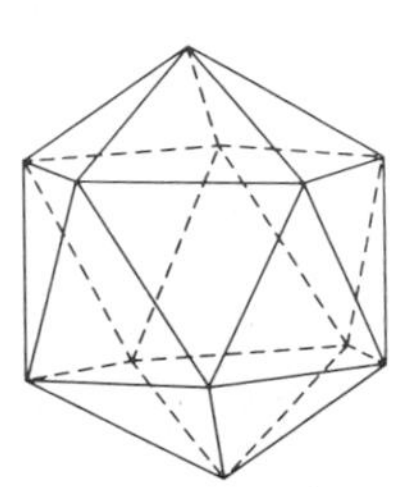

Fig. 4
Regular icosahedron.

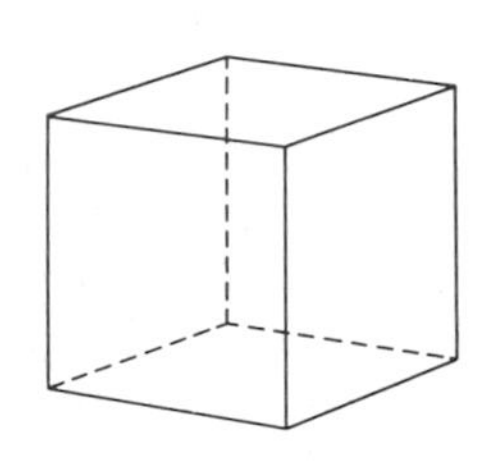

Fig. 5
Regular hexahedron (or cube).

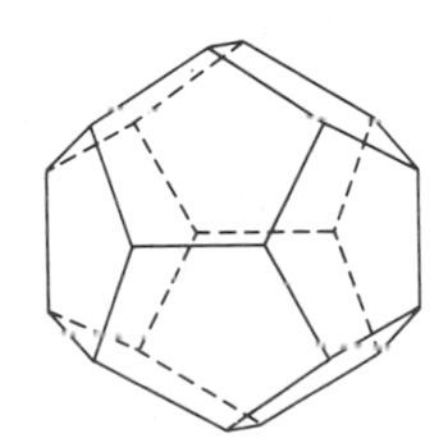

Fig. 6
Regular dodecahedron.

From a given regular polyhedron, we can obtain another one by taking as vertices the centers of all the faces of the given polyhedron (Fig. 7). We say that the given regular polyhedron and the one obtained in this way are **dual** to each other. The octahedron and hexahedron are dual to each other, as are the icosahedron and dodecahedron. The tetrahedron is dual to itself. For a regular polyhedron $\mathfrak{F}$, there exist concentric circumscribed and inscribed spheres whose center is the center of symmetry of $\mathfrak{F}$ and is called the **center** of $\mathfrak{F}$. Drawing tangent planes to the circumscribed sphere at each vertex, we can obtain a regular polyhedron dual to the given one (Fig. 8).

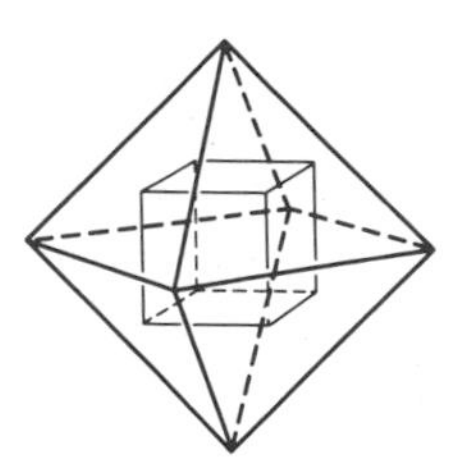

Fig. 7

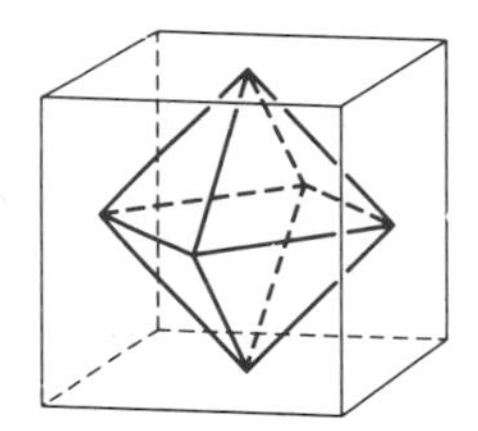

Fig. 8

In a regular polyhedron, let a be the length of an edge, θ the magnitude of the dihedral angle at each edge, and R and r the radii of circumscribed and inscribed spheres, respectively. Then the following relations hold (we assume that each face is a regular p-gon and q faces meet at each vertex):

$$\sin\frac{\theta}{2}=\cos\frac{\pi}{q}\Big/\sin\frac{\pi}{p},$$

$$R=\frac{a}{2}\sin\frac{\pi}{q}\Big/\sin\frac{\pi}{p}\cos\frac{\theta}{2}, \tag{1}$$

$$r=\frac{a}{2}\cot\frac{\pi}{p}\tan\frac{\theta}{2}, \qquad \frac{R}{r}=\tan\frac{\pi}{p}\tan\frac{\pi}{q}$$

(see Table 2). Corresponding to these polyhedra, we have finite subgroups of $O(3)$ called regular polyhedral groups (→ 151 Finite Groups).

C. Higher-Dimensional Cases

It is possible to generalize these considerations to higher dimensions to define inductively

Table 1. Regular Polyhedra in 3-Dimensional Euclidean Space E^3

Figure	Face	Number of Vertices	Number of Edges	Number of Faces	Number of Faces around a Vertex
Regular tetrahedron	Equilateral triangle	4	6	4	3
Regular octahedron	Equilateral triangle	6	12	8	4
Regular icosahedron	Equilateral triangle	12	30	20	5
Regular hexahedron	Square	8	12	6	3
Regular dodecahedron	Regular pentagon	20	30	12	3

Table 2. Numerical Values for Eqs. (1)

Number of Faces	$\sin\theta$	θ	R/a	r/a
4	$2\sqrt{2}/3$	70°31′43.6″	$\sqrt{6}/4$	$\sqrt{6}/12$
6	1	90°	$\sqrt{3}/2$	$1/2$
8	$2\sqrt{2}/3$	109°28′16.4″	$1/\sqrt{2}$	$1/\sqrt{6}$
12	$2/\sqrt{5}$	116°33′54.2″	$\dfrac{\sqrt{3}(\sqrt{5}+1)}{4}$	$\dfrac{\sqrt{25+11\sqrt{5}}}{2\sqrt{10}}$
20	$2/3$	138°11′22.8″	$\dfrac{\sqrt{(5+\sqrt{5})}}{2\sqrt{2}}$	$\dfrac{3+\sqrt{5}}{4\sqrt{3}}$

Table 3. Regular Polyhedra in 4-Dimensional Euclidean Space E^4

Figure	3-Dimensional Regular Polyhedra		Number of Vertices	Duality
	Kind	Number		
Regular 5-hedron	Tetrahedron	5	5	a
Regular 8-hedron	Cube	8	16	b
Regular 16-hedron	Tetrahedron	16	8	b
Regular 24-hedron	Octahedron	24	24	a
Regular 120-hedron	Dodecahedron	120	600	b
Regular 600-hedron	Tetrahedron	600	120	b

a: dual to itself; b: dual to each other

Table 4. Regular Polyhedra in n-Dimensional Euclidean Space ($n \geqslant 5$)

Figure	Regular Polyhedron in $\mathbf{R}^{n-1}$		Number of Vertices	Duality
	Kind	Number		
Regular $(n+1)$-hedron	Regular n-hedron	$n+1$	$n+1$	a
Regular $2n$-hedron	Regular $(2n-2)$-hedron	$2n$	2^n	b
Regular 2^n-hedron	Regular n-hedron	2^n	$2n$	b

a: dual to itself; b: dual to each other

regular polyhedra in E^n, $n \geqslant 4$. When $n=4$ we have 6 kinds of regular polyhedra (Table 3). For $n \geqslant 5$ we have only 3 kinds (Table 4) (→ 70 Complexes).

References

[1] J. S. Hadamard, Leçons de géométrie élémentaire II, Armand Colin, second edition, 1906, 425–427.
[2] D. Hilbert and S. Cohn-Vossen, Anschauliche Geometrie, Springer, 1932; English translation, Geometry and the imagination, Chelsea, 1952.
[3]E. Steinitz and H. Rademacher, Vorlesungen über die Theorie der Polyeder, Springer, 1934.
[4] H. S. M. Coxeter, Regular polytopes, Methuen, 1948; third edition, Dower, 1973.
[5] H. S. M. Coxeter, Regular complex polytopes, Cambridge Univ. Press, 1974.

358 (II.2) Relations

A. General Remarks

In its wider sense the term *relation* means †n-ary relation ($n=1,2,3,\ldots$) (→ 411 Symbolic Logic G), but in this article we restrict ourselves to its most ordinary meaning, i.e., to the case $n=2$. Let X, Y be two sets and x, y be two variables taking their values in X, Y, respectively. A proposition $R(x,y)$ containing x, y is called a **relation** or a **binary relation** if it can be determined whether $R(a,b)$ is true or false for each pair (a,b) in the †Cartesian product $X \times Y$. For example, if both X and Y are the set of rational integers, then the following propositions are relations: $x \leqslant y$, $x-y$ is even, x divides y. A relation $R(x,y)$ is sometimes written as xRy.

For a given relation R, we define its **inverse relation** R^{-1} by $yR^{-1}x \Leftrightarrow xRy$. Then R is the inverse relation of R^{-1}. In the example above, the inverse relation of $x \leqslant y$ is $y \geqslant x$, and the inverse relation of x is a divisor of y is y is a multiple of x. A relation R is called **reflexive** if xRx holds. R is called **symmetric** if $xRy \Leftrightarrow yRx$ (namely, if R and R^{-1} are identical). R is called **transitive** if xRy and yRz imply xRz. R is called **antisymmetric** if xRy and yRx imply $x = y$. A reflexive, symmetric, and transitive relation is called an **equivalence relation** (→ 135 Equivalence Relations). A reflexive and transitive relation is called a †preordering. A reflexive, transitive, and antisymmetric relation is called an †ordering (→ 311 Ordering).

Suppose that we are given a relation xRy $(x \in X, y \in Y)$. Then the set $G = \{(x, y) \mid xRy\}$, which consists of elements (x, y) of the Cartesian product $X \times Y$ satisfying xRy, is called the **graph** of the relation R. Conversely, for any subset G of $X \times Y$, there exists a unique relation R with the graph G given by $xRy \Leftrightarrow (x, y) \in G$.

B. Correspondences

For a subset G of the Cartesian product $X \times Y$, the triple $\Gamma = (G, X, Y)$ is called a **correspondence** from X to Y. The set X is called the **initial set** of the correspondence Γ, and Y the **final set** of Γ. A relation xRy $(x \in X, y \in Y)$ determines a correspondence $\Gamma = (G, X, Y)$ by its graph G, and conversely, a correspondence Γ determines a relation R. Given a correspondence $\Gamma = (G, X, Y)$, the sets $A = \mathrm{pr}_X G$ and $B = \mathrm{pr}_Y G$, where $\mathrm{pr}_X : X \times Y \to X$ and $\mathrm{pr}_Y : X \times Y \to Y$ are the †canonical projections, are called the **domain** and **range** of the correspondence Γ, respectively. For $x \in X$, the set $\{y \in Y \mid (x, y) \in G\}$ is denoted by $G(x)$ or $\Gamma(x)$, and we say that any element y of $G(x)$ **corresponds** to x by Γ.

For a subset G of $X \times Y$, we define a subset G^{-1} of $Y \times X$ by $(x, y) \in G \Leftrightarrow (y, x) \in G^{-1}$. Given a correspondence $\Gamma = (G, X, Y)$, the correspondence (G^{-1}, Y, X) is denoted by Γ^{-1} and is called the **inverse correspondence** of Γ. If G is the graph of a relation R, then G^{-1} is the graph of the inverse relation R^{-1}. The domain of the correspondence Γ is the range of Γ^{-1}, and vice versa. We have $(\Gamma^{-1})^{-1} = \Gamma$.

Suppose that we are given correspondences $\Gamma_1 = (G_1, X, Y)$ and $\Gamma_2 = (G_2, Y, Z)$. We define a subset G of $X \times Z$ by: $(x, z) \in G \Leftrightarrow$ there exists $y \in Y$ satisfying $(x, y) \in G_1$ and $(y, z) \in G_2$. Then the correspondence $\Gamma = (G, X, Z)$ is denoted by $\Gamma_2 \circ \Gamma_1$ and is called the **composite** of Γ_1 and Γ_2. We have the **associative law** $(\Gamma_3 \circ \Gamma_2) \circ \Gamma_1 = \Gamma_3 \circ (\Gamma_2 \circ \Gamma_1)$ and the law $(\Gamma_2 \circ \Gamma_1)^{-1} = \Gamma_1^{-1} \circ \Gamma_2^{-1}$.

Let Γ be a correspondence from X to Y, and assume that to any x belonging to the domain A of Γ there corresponds one and only one $y \in Y$, namely, $\Gamma(x) = \{y\}$ for any $x \in A$. Then Γ is called a **univalent correspondence**. If Γ and Γ^{-1} are both univalent correspondences, Γ is called a **one-to-one correspondence**. For given sets A and B, a univalent correspondence with domain A and range B is called a †mapping (or †function) with domain A and range B (→ 381 Sets C).

References

See references to 381 Sets.

359 (XX.21) Relativity

A. History

The theory of relativity is a system of theoretical physics established by A. Einstein and is composed of special relativity and general relativity. Toward the end of the 19th century, it was believed that electromagnetic waves propagate through the ether, a hypothetical medium. A number of experimenters tried to find the motion of the earth relative to the ether, but all these attempts were unsuccessful (A. A. Michelson, E. W. Morley). Studying these results, in 1905 Einstein proposed the **special theory of relativity**, which extended Galileo's relativity principle of †Newtonian mechanics to †electromagnetism and radically revised the concepts of space and time. Almost all the conclusions of special relativity theory are now confirmed by experiments, and this theory has even become a guiding principle for developing new theories in physics. By extending special relativity, Einstein established (1915) the **general theory of relativity**. Its principal part is a new theory of gravitation containing Newton's theory as a special case. Its conclusions about the solar system are compatible with observed facts that are regarded as experimental support for the theory. Effects due to general relativity other than those just described have been studied to a considerable extent, but it is hard at present to test theoretical results experimentally, and there are some doubts about the limit of its applicability.

B. Special Relativity

In Newtonian mechanics, natural phenomena are described in a 3-dimensional Euclidean space considered independent of time. In special relativity, however, it is postulated that space and time cannot be separated but are unified into a 4-dimensional pseudo-Euclidean space with the †fundamental form

$$ds^2 = \sum_{a,b} g_{ab}\, dx^a dx^b \equiv c^2 dt^2 - dx^2 - dy^2 - dz^2,$$

where $a, b = 0, 1, 2, 3$ and

$$(x^0, x^1, x^2, x^3) = (ct, x, y, z).$$

Here (x, y, z) are spatial Cartesian coordinates, t is time, and c is the speed of light. This space was introduced by H. Minkowski and is called **Minkowski space-time**. By means of it Minkowski gave an ingenious geometric interpretation to special relativity.

A nonzero vector V is called timelike, null (or lightlike), or spacelike according as $V^2 > 0$, $= 0$, or < 0, where $V^2 = \sum_{a,b} g_{ab} V^a V^b$.

The group of motions in Minkowski space-time is called the **inhomogeneous Lorentz group**. Its elements can be written as

$$x^{i'} = \sum_a c_a^i x^a + c^i, \qquad \sum_{a,b} g_{ab} c_i^a c_j^b = g_{ij},$$

where c_j^i and c^i are constants. The transformations with $c^i = 0$ are usually called **Lorentz transformations**, and the group G composed of these transformations is called the **homogeneous Lorentz group** or simply the **Lorentz group**. These are important concepts in special relativity. If G_0 denotes the †connected component of the identity element of G, the factor group G/G_0 is an Abelian group of type (2, 2) and of order 4. We call G_0 the **proper Lorentz group**. A frequently used element of G_0 is

$$L_v : x' = \frac{x - vt}{\sqrt{1 - v^2/c^2}}, \qquad t' = \frac{t - (v/c^2)x}{\sqrt{1 - v^2/c^2}},$$

$$y' = y, \qquad z' = z; \qquad |v| < c. \tag{1}$$

Such transformations form a 1-dimensional subgroup of G_0 with v as a parameter, and the composition law of the subgroup is given by

$$L_u \cdot L_v = L_w, \qquad w = \frac{u + v}{1 + uv/c^2}.$$

Elements of G not belonging to G_0 are

$$T : x^{0'} = -x^0, \quad x^{i'} = x^i; \qquad i = 1, 2, 3,$$

$$S : x^{0'} = x^0, \quad x^{i'} = -x^i; \qquad i = 1, 2, 3.$$

Both T (**time reversal**) and S (**space reflection** or **parity transformation**) have aroused much interest among physicists.

Historically, the transformation formula (1) was first obtained by H. A. Lorentz, under the assumption of contraction of a rod in the direction of its movement in order to overcome the difficulties of the ether hypothesis, but his theoretical grounds were not satisfactory. On the other hand, Einstein started with the following two postulates: (i) **Special principle of relativity**: A physical law should be expressed in the same form in all **inertial systems**, namely, in all coordinate systems that move relative to each other with uniform velocity. (ii) **Principle of invariance of the speed of light**: The speed of light in a vacuum is the same in all inertial systems and in all directions, irrespective of the motion of the light source. From these assumptions Einstein derived (1) as the transformation formula between inertial systems $\mathbf{x} = (ct, x, y, z)$ and $\mathbf{x}' = (ct', x', y', z')$ that move relative to each other with uniform velocity v along the common x-axis. This was the first step in special relativity, and along this line of thought, Einstein solved successively the problems of the Lorentz-Fitzgerald contraction, the dilation of time as measured by moving clocks, the aberration of light, the Doppler effect, and Fresnel's dragging coefficient.

C. Relativity and Electromagnetism

In special relativity, a physical quantity is represented by a †tensor (or a scalar or a vector) in Minkowski space-time, and physical laws are written in tensor form and are invariant under Lorentz transformations of coordinates. This is the mathematical expression of the special principle of relativity. Since the transformation (1) tends to a **Galileo transformation** in Newtonian mechanics as $c \to \infty$, the special principle of relativity is a generalization of the Newton-Galileo principle of relativity. To summarize mathematically, it may safely be said that special relativity is a theory of invariants with respect to the Lorentz group. To illustrate this conclusion we consider electromagnetic theory.

The electric field $\mathbf{E}$ is usually represented by a "polar vector" and the magnetic field $\mathbf{H}$ by an "axial vector" in a 3-dimensional Euclidean space. Even if the magnetic field does not exist in one inertial system, the field can arise in another system that moves uniformly relative to the original system. In view of this, electric and magnetic fields are considered in relativity to form one physical quantity with components

$$(F_{ij}) = \begin{pmatrix} 0 & -E_x & -E_y & -E_z \\ E_x & 0 & H_z & -H_y \\ E_y & -H_z & 0 & H_x \\ E_z & H_y & -H_x & 0 \end{pmatrix}.$$

This quantity transforms as an †alternating tensor of degree 2 under Lorentz transformations. In like manner, the electric charge density ρ and the electric current density $\mathbf{J}$ are unified into a †contravariant vector with respect to Lorentz transformations:

$$s=(s^0,s^1,s^2,s^3)$$
$$=(\rho,J_x/c,J_y/c,J_z/c).$$

Such a vector **s** in Minkowski space-time is sometimes called a **four-vector** as distinguished from an ordinary vector such as **J**. If the electromagnetic field F_{ij} and the current four-vector s are thus defined, the †Maxwell equations, the basic equations of electromagnetism, can be written in tensor form:

$$\sum_a \frac{\partial F^{ia}}{\partial x^a}=s^i, \qquad \frac{\partial F_{jk}}{\partial x^i}+\frac{\partial F_{ki}}{\partial x^j}+\frac{\partial F_{ij}}{\partial x^k}=0,$$

where

$$F^{ij}=\sum_{a,b} g^{ia}g^{jb}F_{ab}, \qquad \sum_a g^{ia}g_{ja}=\delta^i_j.$$

In the same way, the equation of motion for a charged particle in an electromagnetic field can be expressed as

$$\frac{d^2x^i}{ds^2}=\frac{e}{mc}\sum_a F^i_a\frac{dx^a}{ds}, \qquad F^i_j=\sum_a g^{ia}F_{ja},$$

where e and m are the charge and mass of the particle, respectively, and s is the arc length along the particle trajectory (→ 130 Electromagnetism).

Though special relativity originated in studies of electromagnetic phenomena, it has gradually become clear that the theory is valid also for other phenomena. One interesting result is that the energy of a particle moving with uniform velocity v is given by

$$E=mc^2/\sqrt{1-v^2/c^2},$$

and accordingly even a particle at rest has energy mc^2 (**rest energy**). This shows the equivalence of mass and energy, with the conversion formula given by $E=mc^2$. This conclusion has been verified experimentally by studies of nuclear reactions and has become the basis of the development of nuclear power. The special principle of relativity also showed its validity in the electron theory of P. A. M. Dirac (1928) and the quantum electrodynamics of S. Tomonaga (1943) and others. It has, however, been shown that the invariance for space reflection (namely for the coset SG_0 of the Lorentz group G) is violated in the decay of elementary particles (T. D. Lee and C. N. Yang, 1956; C. S. Wu et al., 1957). Similar results have been obtained for time reversal (J. H. Christenson, J. W. Cromin, V. L. Fitch, and R. Turlay, 1964).

D. General Relativity

Special relativity has its origin in studies of electromagnetic phenomena, while the central part of general relativity is a theory of gravitation founded on the **general principle of relativity** and the **principle of equivalence**. The first principle is an extension of the special principle of relativity to accelerated systems in general. It requires that a physical law should be independent of the choice of local coordinates in a 4-dimensional †differentiable manifold representing space and time (**space-time manifold**). Since a physical quantity is represented by a tensor on the space-time manifold, physical laws are expressed in tensor form, in agreement with the first principle. The second principle claims that gravitational and inertial mass are equal, and accordingly fictitious forces due to acceleration (such as centrifugal force) cannot be distinguished from gravitational force. This had been shown with high accuracy by the experiments of R. von Eötvös (1890) and others.

Starting from these two principles, Einstein was led to the following conclusion. If a gravitational field is produced by matter, the corresponding space-time structure is altered; namely, flat Minkowski space-time is changed into a curved 4-dimensional manifold with †pseudo-Riemannian metric of †signature (1, 3). The †fundamental tensor g_{ij} of this manifold represents the gravitational potential, and the gravitational equation satisfied by g_{ij} can be expressed as a geometric law of the manifold. Gravitational phenomena are thus reduced to properties of the geometric structure of the space-time manifold. This idea, which was not seen in the older physics, became the motif in the development of †unified field theories.

Now the gravitational law proposed by Einstein is an analog of the †Poisson equation in Newtonian mechanics. Let R_{ij} and R be the †Ricci tensor and the †scalar curvature formed from g_{ij}. Then outside the source of a gravitational field, g_{ij} must satisfy

$$G_{ij}\equiv R_{ij}-g_{ij}R/2=0, \quad \text{that is,} \quad R_{ij}=0, \tag{2}$$

and inside the source,

$$G_{ij}=\kappa T_{ij}, \tag{3}$$

where κ is the gravitational constant. Here the **energy-momentum tensor** T_{ij} is a †symmetric tensor representing the dynamical state of matter (energy, momentum, and stress). Usually (2) and (3) are called the **exterior** and **interior field equations**, respectively.

Next, the **equation of motion** of a particle in a gravitational field is given by

$$\frac{\delta^2x^i}{\delta s^2}=0, \qquad \sum_{a,b} g_{ab}\frac{dx^a}{ds}\frac{dx^b}{ds}=+1, \tag{4}$$

if the particle mass is so small that its effect on the field is negligible. Here $\delta/\delta s$ stands for †covariant differentiation with respect to the arc length s along the particle trajectory. In other words, a particle in a gravitational field moves along a timelike †geodesic in the space-time manifold. Similarly, the path of light is represented by a **null geodesic**, whose equation is formally obtained from (4) by replacing the right-hand side of the second equation by zero.

Experimental verification of the theory of general relativity has been obtained by detection of the following effects: the shift of spectral lines due to the gravity of the earth and of white dwarfs, the deflection of light or radio waves passing near the sun, the time delay of radar echo signals passing near the sun, and the advance of the perihelion of Mercury. All the observational data are compatible with the theoretical results. Time delay and the advance of the perihelion have been observed in a binary system of neutron stars. It is generally accepted that these results are experimental verifications of general relativity.

It should be noted that (2) has wave solutions, which have no counterpart in Newton's gravitational theory. This fact implies that gravitational effects propagate with the velocity of light. The gravitational waves transport energy and momentum, and the gravitational mass is decreased by the emission of the waves. Experiments to detect gravitational waves generated in the universe have been planned, and the decrease of the orbital period of a binary star system due to the emission of gravitational waves has been observed.

The concept of gravitational waves suggests the existence of a quantum of the gravitational field (graviton); however, the detection of the graviton is far from feasible.

In the interior equation (3), the matter producing a gravitational field is represented by a tensor T_{ij} of class C^0. But there is also a way of representing it by singularities of a solution of the exterior equation (2). From this point of view, the equation of motion of a material particle (i.e., singularity) is not assumed a priori as in (4), but is derived as a result of (2) (A. Einstein, L. Infeld, and B. Hoffman, 1938).

In the static and weak field limits, the fundamental form is approximately given by Newton's gravitational potential φ as

$$ds^2 \doteq c^2(1-\varphi/c^2)\,dt^2 - (1+\varphi/c^2)(dx^2+dy^2+dz^2),$$

and (2) and (3) reduce in this limit to Laplace's and Poisson's equations, respectively. Newton's theory of gravity is valid in the limit $\varphi/c^2 \ll 1$.

Stimulated by the discoveries of neutron stars and black holes and by the big-bang theory of the universe in the 1960s, numerous studies of general relativity have been carried out on such problems as the gravitational field of a spinning mass, the dynamical process of gravitational collapse, the space-time structure of black holes, the generation of gravitational waves, the global structure and dynamics of the universe, and so on. A comparison of the theoretical predictions and the observations is generally favorable, but the phenomena in the universe are so complex that the effects of general relativity cannot always be isolated.

E. Solutions of Einstein's Equations

The isometric symmetry of space-time is described by †Killing vectors. The stationary metric is characterized by a timelike Killing vector, in which case equation (2) reduces to an †elliptic partial differential equation on a 3-dimensional manifold. If the space-time is axially symmetric as well as stationary, (2) reduces to the **Ernst equation**:

$$(\varepsilon+\varepsilon^*)\nabla^2\varepsilon = 2\nabla\varepsilon\cdot\nabla\varepsilon, \tag{5}$$

where ∇ represents divergence in a flat space. The metric tensors are derived from the complex potential ε. The solutions of (5) can be obtained using techniques developed for the soliton problem.

One example of stationary and axially symmetric solutions is the **Kerr metric**, which is written as

$$ds^2 = c^2\,dt^2 - \frac{2mr}{\rho^2}(a\sin^2\theta\,d\varphi - c\,dt)^2 - \rho^2\left(\frac{dr^2}{\Delta}+d\theta^2\right)^2 - (r^2+a^2)\sin^2\theta\,d\varphi^2, \tag{6}$$

with $\rho^2 = r^2 + a^2\cos^2\theta$ and $\Delta = r^2 - 2mr + a^2$. This metric solution represents a gravitational field around a spinning mass with mass $M = mc^2/G$ and angular momentum $J = Mac$. When $a = 0$, this metric reduces to the Schwarzschild metric.

Applying a Bäcklund transformation to the Kerr metric, an infinite series of stationary and axially symmetric solutions can be derived. All these solutions belong to the space-time metric with, in general, two Killing vectors.

The dynamical evolution of space-time structure has been studied by means of the †Cauchy problem of general relativity. Choosing appropriate dynamical variables, equation (2) or (3) is divided into constraint equations in terms of the initial data and evolution equations in terms of the dynamical variables. The latter †hyperbolic equations may also be written in Hamiltonian form.

A typical example of such a problem is the dynamics of a spatially homogeneous 3-

dimensional manifold; this has been studied as a cosmological model. The space-time with a constant scalar curvature R is called **de Sitter space**, and reduces to the Minkowski space if $R=0$. If the 3-dimensional space is isotropic as well as homogeneous, the metric takes the form

$$ds^2=c^2\,dt^2-a(t)^2\{dx^2+f(x)^2 \times(d\theta^2+\sin^2\theta\,d\varphi^2)\},$$

where $f(x)=\sin x$, x, or $\sinh x$. These are called **Robertson-Walker metrics** and are considered to describe a realistic expanding universe.

F. Global Structure of Space-Time

Following the advances of modern differential geometry, manifestly coordinate-independent techniques to analyze space-time properties have been applied to general relativity. The mathematical model of space-time is a connected 4-dimensional †Hausdorff C^∞-manifold endowed with a metric of signature (1,3). The metric allows the physical description of local causality and of local conservation of energy and momentum. The metric functions obey the Einstein field equation (2) or (3).

In order to clarify the global structure of the solutions of Einstein's equations, maximally analytic extension of the solutions has been studied. The maximal extension of the Schwarzschild metric is given as

$$ds^2=\frac{32m^3}{r}e^{-r/2m}(dv^2-du^2)-r^2(d\theta^2+\sin^2\theta\,d\varphi^2),$$

using the **Kruskal coordinates**, which are related to the coordinates of (6) by

$$\left(\frac{r}{2m}-1\right)e^{r/2m}=u^2-v^2,$$

$$\tanh t/4m=u/v \quad \text{or} \quad v/u.$$

To study the global structure at infinity, a †conformal mapping of the metric is used. For example, the Minkowski metric is written in the form $ds^2=\Omega^2\,d\bar{s}^2$, where

$$d\bar{s}^2=dp\,dq-\tfrac{1}{4}\sin^2(p-q)(d\theta^2+\sin^2\theta\,d\varphi^2)$$

and $\Omega=\sec p\sec q$, $\tan p=t+r$, $\tan q=t-r$. By means of this mapping, all points, including infinity, are assigned finite p, q coordinate values in $-\pi/2\leqslant q\leqslant p\leqslant\pi/2$.

Singularities in space-time are one of the major problems concerning the global structure of the manifold. For some Cauchy problems relevant to cosmology and gravitational collapse, the inevitable occurrence of a singularity has been proved (**singularity theorem**). A sufficient condition for occurrence of a singularity is that there be some point p such that all the null geodesics starting from p converge to p again. In addition to this condition, for the proof of the singularity theorem it is presumed that the space-time is free of closed nonspacelike curves, that a Cauchy surface exists, and that the energy-momentum tensor satisfies the condition

$$\sum_{i,j}\left(T_{ij}-\frac{1}{2}Tg_{ij}\right)V^iV^j\geqslant 0$$

for any timelike vector V^i. The singularity whose existence is implied by this theorem means that the space-time manifold is geodesically incomplete (the space-time is complete if every geodesic can be extended to arbitrary values of its affine parameter).

The causal structure of space-time is also related to the global structure of the manifold. In this regard, black holes have been introduced as the final state of gravitational collapse. In the black-hole structure of space-time, there exists a closed surface called an **event horizon** in an asymptotically flat space. The event horizon is the boundary (the set of points) in space-time from which one can escape to infinity, or the boundary of the set of points that one can see from the infinite future. Then the black hole is a region from which no signal can escape to the exterior of the event horizon.

If we assume that singularities do not exist in the exterior of the event horizon, a stationary black-hole structure is uniquely described by the Kerr metric [6]. In the case of spherically symmetric collapse, this assumption is verified and the final metric is given by the Schwarzschild metric. However, it is not known whether this assumption is true in more general gravitational collapse.

References

[1] A. Einstein, The meaning of relativity, Princeton Univ. Press, fifth edition, 1956.
[2] H. Weyl, Space, time, matter, Dover, 1950. (Original in German, 1923.)
[3] A. Lichnerowicz, Théories relativistes de la gravitation et de l'électromagnétisme, Masson, 1955.
[4] C. W. Misner, K. S. Thorne, and J. A. Wheeler, Gravitation, Freeman, 1973.
[5] L. Witten (ed.), Gravitation, an introduction to current research, Wiley, 1962.
[6] C. DeWitt and B. S. DeWitt (eds.), Black holes, Gordon & Breach, 1972.
[7] S. W. Hawking and G. F. R. Ellis, The large scale structure of space-time, Cambridge Univ. Press, 1973.

[8] M. Carmeli, Group theory and general relativity, McGraw-Hill, 1977.
[9] S. W. Hawking and W. Israel (eds.), General relativity, an Einstein centenary survey, Cambridge Univ. Press, 1979.
[10] A. Held (ed.), General relativity and gravitation I, II, Plenum, 1980.

360 (XXI.8) Renaissance Mathematics

Toward the middle of the 13th century, scholastic theology and philosophy were at their height with the *Summa theologiae* of Thomas Aquinas (1225?–1274); but in the latter half of the century, the English philosopher Roger Bacon (1214–1294) attacked Aquinian philosophy in his *Opus majus*, insisted on the importance of experimental methods in science, and strongly urged the study of mathematics. The Renaissance flourished first in Italy, then in other European countries in the 15th and 16th centuries, in the domains of the arts and literature. Newer ideas in mathematics and the natural sciences dominated the 17th century. However, it was the invention of printing in the 15th century, the translation of the Greek texts of Euclid and Archimedes into European languages, and the importation of Arabian science into Europe during the Renaissance that prepared for this development.

In the 15th century, the German priest Nicolaus Cusanus (1401–1464) discussed infinity, the convergence of infinite series, and some problems of quadrature. During the same period, the German scholar Regiomontanus (1436–1476) wrote the first systematic treatise on trigonometry independent of astronomy. Leonardo da Vinci (1452–1519), the all-encompassing genius born in the same century, left manuscripts in which he wrote about mechanics, geometric optics, and perspective. Da Vinci's contemporary, the German painter A. Dürer (1471–1528), wrote a textbook on perspective. In 1494, L. Pacioli (1445?–1514) published *Summa de arithmetica*, one of the first printed books on mathematics. Its content, influenced by Arabian mathematics, includes practical arithmetic and double-entry bookkeeping. The book enjoyed wide popularity.

The best known result of 16th-century mathematics is the solution of algebraic equations of degrees 3 and 4 by the Italian mathematicians Scipione del Ferro, N. Tartaglia (1506–1557), G. Cardano (1501–1576), and L. Ferrari (1522–1565). Cardano published the solution of equations of the third degree in his book *Ars magna* (1545). The solution was due to Tartaglia, to whom acknowledgment was made, although publication of the method was against his will. This constitutes a famous episode in the history of mathematics, but what is historically more important is the fact that essential progress beyond Greek mathematics was made by mathematicians of this period, since the Greeks were able to solve equations only of degrees 1 and 2. Algebra was subsequently systematized by the French mathematician F. Viète (1540–1603).

By the end of the 15th century, practical mathematics (influenced by the Arabians) had become popular in Europe, and more advanced mathematics began to be studied in European universities, especially in Italy. In 1543, N. Copernicus (1473–1543) published his heliocentric theory (1543); G. Galilei (called Galileo) (1564–1642), the indomitable proponent of this theory, was also born in the 16th century. Copernicus studied at the Universities of Bologna, Padua, and Ferrara; Galileo studied at the University of Pisa and taught at the Universities of Pisa, Padua, and Florence. A system of numeration was imported from Arabia to Europe in the 13th century; by the time of S. Stevin (1548?–1620?) it took the definite form of a decimal system, and with the development and acceptance of printing, the forms of the numerals became fixed.

References

[1] M. Cantor, Vorlesungen über Geschichte der Mathematik II, Teubner, second edition, 1900.
[2] J. Cardan (G. Cardano), The book of my life (trans. J. Stoner), Dover, 1963.
[3] G. Cardano, Opera, 10 vols., Lyon, 1663.
[4] N. Bourbaki, Eléments d'histoire des mathématiques, Hermann, 1960.

361 (XX.32) Renormalization Group

A. Introduction

The concept of renormalization was introduced by S. Tomonaga, J. S. Schwinger, M. Gell-Mann, and F. E. Low in order to overcome the difficulty of divergence in field theory. If the upper bound of the momentum is limited to a finite cutoff value Λ, then physical quantities, for example, the mass m of an electron, can be obtained as finite quantities by letting

Λ go to infinity after summing all divergent terms. This is called the **renormalization method** using **subtraction**. Since the cutoff Λ is arbitrary insofar as it is finite, the Green's functions are indefinite because they depend on Λ. This dependence on the cutoff Λ corresponds to the response for the scale transformation of length, and this transformation is a certain (semi) group, called a **renormalization group**. Several kinds of renormalization group have been used in field theory, as well as in the statistical mechanics of phase transition.

B. Renormalization Group in Field Theory [1–3]

A typical method to resolve the ultraviolet divergence is to add **subtraction terms** in the Lagrangian so that they cancel the divergence. This cancellation is usually performed in each order of the perturbation expansion. When the addition of a finite number of subtraction terms cancels the divergence, the relevant Hamiltonian is said to be **renormalizable**. Otherwise it is called **unrenormalizable**. The Lagrangian density

$$\mathscr{L}_0=\frac{1}{2}\partial^\mu\varphi_0\partial_\mu\varphi_0-\frac{1}{2}m_0^2\varphi_0^2-\frac{1}{4!}g_0\varphi_0^4,$$

for example, can be renormalized by the transformation $\varphi_0=Z_3^{1/2}\varphi$, $g_0=Z_1Z_3^{-2}g$, and $m_0^2=m^2+\delta m^2$. All the divergences are taken into the renormalization constants Z_3, Z_1, and δm^2, so that the renormalized quantities φ, g, and m are finite. These renormalization constants can be calculated by means of a perturbation method, but the requirement of circumventing the divergences alone is not sufficient to determine them explicitly. This indeterminacy is usually expressed as $Z_3(\mu)$ and $Z_1(\mu)$, i.e., in terms of a parameter μ, called the renormalization point. The μ-dependence of these functions can be determined by means of the following renormalization conditions:

$$\lim_{p^2+\mu^2\to 0}(p^2+m^2)G^{(2)}(p,\mu)=1;$$

$$\Gamma(p_i,\mu)=1, \qquad p_ip_j=\frac{\mu^2}{3}(1-4\delta_{ij}).$$

Since the renormalization constants depend on the continuous parameter μ, the renormalized Green's function and coupling constant g are also functions of μ, and consequently the quantity $d^{(2N)}(\{p_i\},\mu,g(\mu))$ defined by

$$d^{(2N)}(\{p_i\},\mu,g(\mu))$$

$$\equiv\left[\prod_{i=1}^{2N}(p_i^2+m^2)\right]G^{(2N)}(p_i,\mu,g(\mu))$$

satisfies the **renormalization equation**

$$\left(\mu\frac{\partial}{\partial\mu}+\beta\frac{\partial}{\partial g}-2N\gamma\right)d^{(2N)}(\{p_i\},\mu,g)=0,$$

where $\beta=\mu\,dg/d\mu$ and $\gamma=\frac{1}{2}\mu\,d\log Z_3(\mu)/d\mu$. If the coefficients β and γ are calculated perturbationally up to a certain order, the renormalized Green's function is obtained up to the same order by solving the foregoing renormalization equation. This is the first kind of renormalization group. A second kind expresses the response of the renormalized Green's function to the change in the mass and coupling constant, and is expressed by the **Callan-Symanzik equation** [4, 5]

$$\left(m\frac{\partial}{\partial m}+\beta(g)\frac{\partial}{\partial g}+2N\gamma_\varphi(g)\right)G_R^{(2N)}=\Delta G_G^{(2N)},$$

where $\beta(g)=Zm_0\partial g/\partial m_0$ and $\gamma_\varphi(g)=\frac{1}{2}Zm_0\cdot\partial\log Z_3/\partial m_0$, and where Z is determined by $Zm_0\partial m/\partial m_0=m$. The inhomogeneous term is defined by

$$\Delta G_R^{(2N)}=Zm_0\frac{\partial}{\partial m_0}G_R^{(2N)}$$

$$+Zm_0\frac{\partial}{\partial m_0}(N\log Z_3)G_R^{(2N)}.$$

The irreducible Green's function $\Gamma_R^{(2N)}$ satisfies

$$\left(m\frac{\partial}{\partial m}+\beta(g)\frac{\partial}{\partial g}-2N\gamma_\varphi(g)\right)\Gamma_R^{(2N)}=i\Delta\Gamma_R^{(2N)}.$$

Since the inhomogeneous term can be neglected in the high-energy region, the foregoing equation becomes homogeneous, and consequently its solution is

$$\Gamma_R^{(2N)}\left(p_i,\frac{m}{\lambda},g\right)$$

$$=\exp\left[-2N\int_g^{g(\lambda)}\gamma(g')\beta^{-1}(g')dg'\right]$$

$$\times\Gamma_R^{(2N)}(p_i,m,g(\lambda)).$$

Here $g(\lambda)$ is the solution of the equation $\int_g^{g(\lambda)}\beta^{-1}(g')dg'=\log\lambda$. Since dimensional analysis yields $\Gamma_R^{(2N)}(\lambda p_i,m,g)=\lambda^{4-2N}\Gamma_R(p_i,m/\lambda,g)$, the foregoing solution shows that high-energy phenomena can be described by the low-energy phenomena whose coupling constant is given by $g_\infty\equiv g(\infty)$. In particular, when $g_\infty=0$, high-energy phenomena are described by the asymptotically free field. This circumstance is called **asymptotic freedom**.

C. Renormalization Group Theory in Statistical Physics

The renormalization group technique has proved to be powerful in statistical physics, particularly in studies of phase transitions and critical phenomena [6–10]. The correlation length ξ diverges like $\xi(T)\sim(T-T_c)^{-\nu}$ near the critical point T_c, where ν is called the **critical exponent** of ξ. Similarly, the correlation function $C(R)$ for the distance R behaves like $C(R)\sim R^{-(d-2+\eta)}\times\exp(-\kappa R)$, $\kappa=\xi^{-1}$, where d is the dimensionality of the system and η is the exponent describing the deviation of the singular behavior of $C(R)$ from classical theory. Renormalization is useful in evaluating these critical exponents systematically. The fundamental idea is to eliminate some degrees of freedom, to find recursion formulas for interaction parameters, and then to evaluate critical exponents from their asymptotic behavior near the fixed point. There are many different ways of carrying out this idea explicitly. Roughly classifying these into two groups, we have (i) momentum-space renormalization group theories [6–10] and (ii) real-space renormalization group theories [10, 11].

The common fundamental structure of these renormalization group techniques is explained as follows. First the momentum space or real space is divided into cells and rapidly fluctuating parts, namely, small-momentum parts inside each cell are integrated or eliminated, and consequently the remaining slowly fluctuating parts, namely, long-wave parts, are renormalized. The original Hamiltonian $\mathscr{H}_0$ is transformed into $\mathscr{H}_1$ by means of this elimination process and by some scale transformation that preserves the phase space volume. This renormalization operation is written as $\mathbf{R}_b$: i.e., $\mathscr{H}_1=\mathbf{R}_b\mathscr{H}_0$. Similarly we have $\mathscr{H}_2=\mathbf{R}_b\mathscr{H}_1=\mathbf{R}_b^2\mathscr{H}_0,\ldots,\mathscr{H}_n=\mathbf{R}_b\mathscr{H}_{n-1}=\mathbf{R}_b^n\mathscr{H}_0,\ldots$. This transformation $\mathbf{R}_b$ has the (semi) group property $\mathbf{R}_{bb'}=\mathbf{R}_b\mathbf{R}_{b'}$. A generator $\mathbf{G}$ is defined by $\mathbf{G}=\lim_{b\to1+0}(\mathbf{R}_b-1)/(b-1)$. That is, $\mathbf{R}_b=\exp(l\mathbf{G})$, $e^l=b$. The transformation of $\mathscr{H}$ is expressed as $d\mathscr{H}/dl=\mathbf{G}[\mathscr{H}]$. The fixed point $\mathscr{H}^*=\mathbf{R}_b\mathscr{H}^*$ is the solution of $\mathbf{G}[\mathscr{H}^*]=0$. In order to find critical exponents from the asymptotic behavior of $\mathbf{G}$ near $\mathscr{H}^*$, we consider a Hamiltonian of the form $\mathscr{H}=\mathscr{H}^*+wQ$ and expand $\mathbf{G}[\mathscr{H}]$ as $\mathbf{G}[\mathscr{H}^*+wQ]=w\mathbf{K}Q+O(w^2)$. If the operator $\mathbf{K}$ thus defined has a negative eigenvalue λ_i, the corresponding physical quantity Q_i becomes irrelevant after repeating the renormalization procedure, and the physical quantity Q_j corresponding to a positive eigenvalue $\lambda_j>0$ becomes relevant. Thus, by introducing a field h_j conjugate to the relevant operator Q_j, we study the Hamiltonian $\mathscr{H}=\mathscr{H}^*+\sum_j h_jQ_j$. The free energy per unit volume $f[\mathscr{H}]\equiv f(h_1,h_2,\ldots)$ is found to have the scaling property

$$f(h_1,\ldots,h_j,\ldots)\simeq b^{-d}f(b^{\lambda_1}h_1,\ldots,b^{\lambda_j}h_j,\ldots).$$

By taking Q_1 as the energy operator, we have $h_1\sim T-T_c\equiv t$. By the normalization $b^{\lambda_1}h_1=1$, we obtain the scaling law

$$f(t,\ldots,h_j,\ldots)\simeq t^{d/\lambda_1}f(1,\ldots,h_j/t^{\varphi_j},\ldots).$$

The critical exponent of the specific heat defined by $C\sim t^{-\alpha}$ is given by the formula $\alpha=2-d/\lambda_1$. Other scaling exponents $\{\varphi_j\}$ can be obtained via the formula $\varphi_j=\lambda_j/\lambda_1$ from the eigenvalues of $\mathbf{K}$. The simplest example of R_b is the case where a single interaction parameter K is transformed into a new parameter K' by $K'=f_b(K)$. The fixed point K^* is given by the solution of $K^*=f_b(K^*)$. The correlation exponent ν defined by $\xi\sim(K-K^*)^{-\nu}$ is given by the Wilson formula $\nu=\log b/\log\Lambda$, $\Lambda=(df_b/dK)_{K=K^*}$. In most cases, $\mathbf{R}_b$ is constructed perturbationally, and critical exponents are usually calculated in power series of $\varepsilon\equiv d_c-d$, as $\varphi=\varphi_0+\varphi_1\varepsilon+\varphi_2\varepsilon^2+\ldots$, where d_c denotes the critical dimension. This is called the **ε-expansion**. The first few terms are calculated explicitly for specific models, such as the φ^4-model. By applying the Borel sum method to these ε-expansions, one can estimate critical exponents [10, 11].

The renormalization group method can be applied to other many-body problems, such as the Kondo effect [12].

References

[1] E. C. G. Stueckelberg and A. Petermann, La normalisation des constantes dans la théorie des quanta, Helv. Phys. Acta, 26 (1953), 499–520.

[2] M. Gell-Mann and F. E. Low, Quantum electrodynamics at small distances, Phys. Rev., 95 (1954), 1300–1312.

[3] N. N. Bogolyubov and D. V. Shirkov, Introduction to the theory of quantized fields, Interscience, 1959.

[4] C. G. Callan, Jr., Broken scale invariance in scalar field theory, Phys. Rev., D2 (1970), 1541–1547.

[5] K. Symanzik, Small distance behaviour in field theory and power counting, Comm. Math. Phys., 18 (1970), 227–246; Small distance-behaviour analysis and Wilson expansions, Comm. Math. Phys., 23 (1971), 49–86.

[6] C. DiCastro and G. Jona-Lasinio, On the microscopic foundation of scaling laws, Phys. Lett., 29A (1969), 322–323; C. DiCastro, The multiplicative renormalization group and the critical behavior in $d=4-\varepsilon$ dimensions, Lett. Nuovo Cimento, 5 (1972), 69–74.

[7] K. G. Wilson, Renormalization group and critical phenomena. I, Renormalization group and the Kadanoff scaling picture, Phys. Rev.,

B4 (1971), 3174–3183; II, Phase-space cell analysis of critical behavior, Phys. Rev., B4 (1971), 3184–3205.
[8] K. G. Wilson and M. E. Fisher, Critical exponents in 3.99 dimensions, Phys. Rev. Lett., 28 (1972), 240–243.
[9] K. G. Wilson and J. Kogut, The renormalization group and the ε expansion, Phys. Rep., 12C (1974), 75–200.
[10] C. Domb and M. S. Green (eds.), Phase transitions and critical phenomena VI, Academic Press, 1976.
[11] A. A. Migdal, Recursion equations in gauge field theories, Sov. Phys. JETP, 42 (1975), 413–418; Phase transitions in gauge and sublattice systems, Sov. Phys. JETP, 42 (1975), 743–746. (Original in Russian, 1975.)
[12] K. G. Wilson, The renormalization group: Critical phenomena and the Kondo problem, Rev. Mod. Phys., 47 (1975), 773–840.
[13] L. P. Kadanoff, Notes on Migdal's recursion formulas, Ann. Phys., 100 (1976), 359–394.

362 (IV.16) Representations

A. General Remarks

For a mathematical system A, a mapping from A to a similar (but in general "more concrete") system preserving the structure of A is called a **representation** of A. In this article, we consider the representations of †groups and †associative algebras. For representations of other algebraic systems → 42 Boolean Algebras; 231 Jordan Algebras; 248 Lie Algebras. For topological, analytic, and algebraic groups → 13 Algebraic Groups; 69 Compact Groups; 249 Lie Groups; 422 Topological Abelian Groups; 423 Topological Groups; 437 Unitary Representations. For specific groups → 60 Classical Groups; 61 Clifford Algebras.

B. Permutation Representations of Groups

We denote by $\mathfrak{S}_M$ the group of all †permutations of a set M (→ 190 Groups B). A **permutation representation** of a group G in M is a homomorphism $G \rightarrow \mathfrak{S}_M$. We denote by a_M the permutation of M corresponding to $a \in G$ and write $a_M(x) = ax$ $(x \in M)$. Then we have a condition $(ab)x = a(bx)$, $1x = x$ ($a, b \in G$, 1 is the identity element, $x \in M$). In general, if the product $ax \in M$ of $a \in G$ and $x \in M$ is defined and satisfies this condition, then G is said to **operate on** M **from the left**, and M is called a **left** G**-set**. Giving a permutation representation of G in M is equivalent to giving the structure of a left G-set to M. A **reciprocal permutation representation** of G in M is an †antihomomorphism $G \rightarrow \mathfrak{S}_M$, which becomes a homomorphism if we define the multiplication in $\mathfrak{S}_M$ by the right notation $x(fg) = (xf)g$. If the product $xa \in M$ of $a \in G$ and $x \in M$ is defined and satisfies the conditions $x(ab) = (xa)b$ and $x1 = x$, then, as before, G is said to **operate on** M **from the right**, and M is called a **right** G**-set**. Giving a reciprocal permutation representation of G in M is equivalent to giving the structure of a right G-set to M.

A (reciprocal) permutation representation is said to be **faithful** if it is injective; the corresponding G-set is also said to be faithful. In particular, we can take G itself as M and define the left (right) operation by the multiplication from the left (right). Then we have a faithful permutation representation (reciprocal permutation representation), which is called the **left (right) regular representation** of G. For $a \in G$, the induced permutation $a_G: x \rightarrow ax$ (xa) is called the **left (right) translation** by a.

We call a left G-set simply a G-set. If a subset N of a G-set M satisfies the condition that $a \in G$, $x \in N$ implies $ax \in N$, then N forms a G-set, which is called a G**-subset** of M. If a G-set M has no proper G-subset (i.e., one different from M itself and the empty subset), then for any two elements $x, y \in M$ there exists an element $a \in G$ satisfying $ax = y$. In this case, the operation of G on M is said to be **transitive**, and the corresponding permutation representation is also said to be transitive. If an equivalence relation R in a G-set M is compatible with the operation of G (i.e., R satisfies the condition that $a \in G$, $R(x, y)$ implies $R(ax, ay)$), then the quotient set M/R forms a G-set in the natural way, called the **quotient** G**-set** of M by R. If a G-set M has no nontrivial quotient G-set, i.e., if the only equivalence relations compatible with the operation are

$$R(x, x') \quad \text{for any} \quad x, x' \in M$$

and

$$R(x, x') \quad \text{if and only if} \quad x = x',$$

then the operation of G on M and the corresponding permutation representation are said to be **primitive**.

A mapping $f: M \rightarrow M'$ of G-sets is called a G**-mapping** (G**-map**) if the condition $f(ax) = af(x)$ $(a \in G, x \in M)$ is satisfied. G-injection, G-surjection, and G-bijection are defined naturally. The inverse mapping of a G-bijection is also a G-bijection. Two permutation representations are said to be **similar** if there exists a G-bijection between the corresponding G-sets.

Let M be a transitive G-set, and fix any element $x \in M$. If we view G as a G-set, the mapping $f: G \rightarrow M$ defined by $f(a) = ax$ is a G-surjection and induces a G-bijection $\bar{f}: G/R \rightarrow M$. Here an equivalence class of R is precisely a left coset of the **stabilizer (stability group** or **isotropy group)** $H_x = \{a \in G \mid ax = x\}$. Hence we have a G-bijection $G/H_x \rightarrow M$. Conversely, for any subgroup H of G, G/H is a transitive left G-set. A transitive G-set is called a **homogeneous space** of G.

For a family $\{M_\lambda\}_{\lambda \in \Lambda}$ of G-sets, the Cartesian product $\prod_{\lambda \in \Lambda} M_\lambda$ and the †direct sum $\sum_{\lambda \in \Lambda} M_\lambda$ become G-sets in the natural way; they are called the **direct product** of G-sets and the **direct sum** (i.e., disjoint union) of G-sets, respectively. Every G-set M is the direct sum of a family $\{M_\lambda\}$ of transitive G-subsets, and each M_λ is called an **orbit** (or **system of transitivity**). For a G-set M, the direct product G-set $M^k = M \times \ldots \times M$ (k times) contains a G-subset $M^{(k)} = \{(x_1, \ldots, x_k) \mid i \neq j \text{ implies } x_i \neq x_j\}$. If $M^{(k)}$ is transitive, M is said to be k**-ply transitive**. If M is transitive and the stabilizer of each point of M consists of the identity element alone, M is said to be **simply transitive**.

If M has n elements, a permutation representation of a group G in M is said to be of **degree** n. When G is a group of permutations of M, the canonical injection $G \rightarrow \mathfrak{S}_M$ is a faithful permutation representation; this case has been studied in detail (→ 151 Finite Groups G).

C. Linear Representations of Groups and Associative Algebras

Let K be a †commutative ring with unity element and M be a K-module. Though we shall mainly treat the case where K is a field and M is a finite-dimensional †linear space over K, the case where K is an †integral domain and M is a †free module over K of finite rank is also important. Since K is commutative, we can write $\lambda x = x\lambda$ ($\lambda \in K, x \in M$). Let $\mathscr{E}_K(M)$ be the †associative algebra over K consisting of all K-endomorphisms of M, and let $GL(M)$ be the group of all †invertible elements in $\mathscr{E}_K(M)$, where we assume $M \neq \{0\}$. Let A be an associative algebra over K. A **linear representation of the algebra** A in M is an algebra homomorphism $A \rightarrow \mathscr{E}_K(M)$. We always assume that A has a unity element and the homomorphisms are unitary. For convenience, we can also consider a linear representation in the trivial space $M = \{0\}$, which is called the **zero representation**. A **reciprocal linear representation** is an antihomomorphism $A \rightarrow \mathscr{E}_K(M)$. A **linear representation of a group** G in M is a group homomorphism $G \rightarrow GL(M)$. This can be extended uniquely to a linear representation of the †group ring $K[G]$ in M, and conversely, the restriction of a linear representation of $K[G]$ in M to G is a linear representation of G; and similarly for reciprocal linear representations. Thus the study of (reciprocal) linear representations of a group G in M can be reduced to the study of (reciprocal) linear representations of the group ring $K[G]$ in M.

We now consider the linear representation of associative algebras, which we call simply "algebras." (Note that a group ring has a canonical basis—the group itself—and allows a more detailed investigation; → Sections G, I.)

Given a commutative ring K with unity and a linear representation ρ of a K-algebra A in a K-module M, we introduce the structure of a left A-module into M by defining $ax = \rho(a)x$ ($a \in A, x \in M$); the structure of a K-module in M obtained by the canonical homomorphism $K \rightarrow A$ coincides with the original one. This A-module is called the **representation module** of ρ. Conversely, for any left A-module M we can define a linear representation ρ of A in M (with M viewed as a K-module via $K \rightarrow A$) by putting $\rho(a)x = ax$; the representation module of ρ coincides with the original one. This representation ρ is called the **linear representation associated with** M. A reciprocal linear representation of A corresponds to a right A-module. Thus the study of (reciprocal) linear representations of A is equivalent to the study of left (right) A-modules. For instance, if the operation of a group G on M is trivial: $\sigma x = x$ ($\sigma \in G, x \in M$), the corresponding representation of G in M assigns the identity mapping I_M to every $\sigma \in G$. Furthermore, if $M = K$, this representation is called the **unit representation** of G (over K).

Let ρ, ρ' be linear representations of A in K-modules M, M', respectively. Then an A-homomorphism $M \rightarrow M'$ is precisely a K-homomorphism $f: M \rightarrow M'$ satisfying the condition $f \circ \rho(a) = \rho'(a) \circ f$ ($a \in A$); this is sometimes called a **homomorphism** from ρ to ρ'. In particular, an A-isomorphism is a K-isomorphism $f: M \rightarrow M'$ satisfying the condition $f \circ \rho(a) \circ f^{-1} = \rho'(a)$ ($a \in A$); in this case we say that ρ and ρ' are **similar (isomorphic** or **equivalent)** and write $\rho \cong \rho'$.

Let M be the representation module of a linear representation ρ of an algebra A. If ρ is injective, ρ and the corresponding M are said to be **faithful**. For example, the linear representation associated with the left A-module A is faithful; this is called the **(left) regular representation** of A. If M is †simple as an A-module, ρ is said to be **irreducible** (or **simple**). A homomorphism from an irreducible representation ρ to ρ must be an isomorphism or the zero homomorphism (**Schur's lemma**). In particular, if K is an †algebraically closed field and M is finite-dimensional, then such a homomorphism is a scalar multiplication. A linear representation is said to be **reducible** if it is not irreducible. If M is †semisimple as an A-module, ρ is said to be **completely reducible** (or **semisimple**). If A is a semisimple ring, any linear representation of A is completely reducible. The converse also holds (→ 368 Rings G).

The linear representations associated with a submodule and a quotient module of M as an A-module are called a **subrepresentation** and a **quotient representation**, respectively. The linear representation associated with the direct sum $M_1+\ldots+M_r$ of the representation modules $M_1,\ldots,M_r$ of linear representations $\rho_1,\ldots,\rho_r$ is written $\rho_1+\ldots+\rho_r$ and called the **direct sum** of representations. If ρ is never similar to the direct sum of two nonzero linear representations, then ρ is said to be **indecomposable**; this means that M is †indecomposable as an A-module.

For linear representations ρ, ρ' of a group G in M, M', we define the linear representation $\rho\otimes\rho'$ in $M\otimes M'$ by $(\rho\otimes\rho')(g)=\rho(g)\otimes\rho'(g)$ $(g\in G)$; this is called the **tensor product** of representations ρ and ρ'.

D. Matrix Representations

Let K^n be the K-module consisting of all n-tuples (ξ_i) of elements in a commutative ring K. $\mathscr{E}_K(K^n)$ is identified with the K-algebra $M_n(K)$ of all $n\times n$ matrices (λ_{ij}) over $K:(\lambda_{ij})(\xi_j)=(\sum_{j=1}^n\lambda_{ij}\xi_j)$. Thus a linear representation of A in K^n, i.e., a homomorphism $A\rightarrow M_n(K)$, is called a **matrix representation** of A over K, and n is called its **degree**. A matrix representation of a group G over K of degree n is a homomorphism $G\rightarrow GL(n,K)$, where $GL(n,K)$ is the group of all $n\times n$ invertible matrices. If $(e_1,\ldots,e_n)$ is a †basis of a K-module M, then by the K-isomorphism $K^n\rightarrow M$ given by the assignment $(\xi_i)\rightarrow\sum_{i=1}^n e_i\xi_i$, we have a bijective correspondence between the matrix representations of A of degree n and the linear representations of A in M, and the corresponding representations are similar. Explicitly, the linear representation ρ corresponding to a matrix representation $a\rightarrow(\lambda_{ij}(a))$ is given by

$$\rho(a)e_j=\sum_{i=1}^n e_i\lambda_{ij}(a),\quad a\in A.$$

Hence giving the finite-dimensional linear representations over a field K is equivalent to giving the matrix representations over K. Let T, T' be matrix representations of degree n, n'. Then a homomorphism from T to T' is an $n'\times n$ matrix P satisfying $PT(a)=T'(a)P$ $(a\in A)$. Therefore T and T' are similar if and only if $n=n'$ and $PT(a)P^{-1}=T'(a)$ $(a\in A)$ for some $n\times n$ invertible matrix P. For a representation of a group G, it suffices that this equation is satisfied by all $a\in G$.

We always assume that K is a field. Then a K-module is a linear space over K. A linear representation ρ over K of a K-algebra A is said to be of **degree** n if its representation module M is of dimension n over K. Suppose that a sequence $\{0\}=M_0\subset M_1\subset\ldots\subset M_r=M$ of A-submodules of M is given. We take a basis $(e_1,\ldots,e_n)$ of M over K such that $(e_1,\ldots,e_{m_i})$ forms a basis of M_i over K $(1\leqslant i\leqslant r)$. Then the matrix representation corresponding to ρ relative to the basis $(e_1,\ldots,e_n)$ has the form

$$a\rightarrow T(a)=\begin{bmatrix} T_{11}(a) & T_{12}(a) & \ldots & T_{1r}(a)\\ & T_{22}(a) & \ldots & T_{2r}(a)\\ & & \ddots & \vdots\\ 0 & & & T_{rr}(a)\end{bmatrix}$$

where, if we put $n_i=\dim M_i/M_{i-1}=m_i-m_{i-1}$, $T_{ij}(a)$ is an $n_i\times n_j$ matrix and $T_{ij}(a)=0$ for $i>j$. The residue classes of $e_{m_{i-1}+1},\ldots,e_{m_i}$ form a basis of the quotient space M_i/M_{i-1} over K, and the matrix representation corresponding to the linear representation ρ_i associated with M_i/M_{i-1} relative to this basis is given by T_{ii}. The sequence $\{M_i\}$ is a †composition series if and only if each ρ_i (hence T_{ii}) is irreducible. In this case, $\rho_1,\ldots,\rho_r$ are uniquely determined by ρ up to their order and similarity (**Jordan-Hölder theorem**). An irreducible representation ρ' similar to some ρ_i is called an **irreducible component** of ρ. The number $p>0$ of ρ_i similar to ρ' is called the **multiplicity** of ρ' as an irreducible component of ρ. We also say that ρ contains ρ' p times or ρ' appears p times as an irreducible component of ρ. The representation ρ is completely reducible if and only if it is similar to the direct sum of its irreducible components (admitting repetition). In this case, ρ is similar to the matrix representation

$$a\rightarrow\begin{bmatrix} T_{11}(a) & & 0\\ & \ddots & \\ 0 & & T_{rr}(a)\end{bmatrix}.$$

E. Coefficients and Characters of Linear Representations

We consider the linear representations of an algebra over a field K. A right (left) A-module M is regarded as a linear space over K. In its dual space M^*, we introduce the structure of a left (right) A-module using the inner product $\langle\ ,\ \rangle$ as follows: $\langle x,ax^*\rangle=\langle xa,x^*\rangle$ $(\langle x,x^*a\rangle$

$=\langle ax, x^*\rangle$), where $a \in A$, $x \in M$, $x^* \in M^*$. If ρ is the representation associated with M, the representation associated with M^* is called the **transposed representation (dual representation** or **adjoint representation**) of ρ, and is denoted by ${}^t\rho$. The linear mapping ${}^t\rho(a)$ is the †transposed mapping of $\rho(a)$. If M is finite-dimensional over K, we have $(M^*)^* = M$ as an A-module. For a linear representation ρ of a group G, the mapping $g \to {}^t\rho(g)^{-1}$ $(g \in G)$ is called the **contragredient representation** of ρ. The reciprocal linear representation associated with the right A-module A is called the **right regular representation** of A, and its transposed representation (i.e., the representation associated with the left A-module A^*) is called the **coregular representation** of A. For any finite-dimensional semisimple algebra and group ring of a finite group, the regular representation and the coregular representation are similar (→ 29 Associative Algebras H).

Let ρ be a linear representation of A over K and M be its representation module. For any $x \in M$, $x^* \in M^*$, we define a †linear form $\rho_{x,x^*} \in A^*$ on A by $\rho_{x,x^*}(a) = \langle ax, x^*\rangle$ $(a \in A)$. This is called the **coefficient** of ρ relative to x, x^* and is determined by its values at generators of A as a linear space. In particular, a coefficient of a linear representation ρ of a group G can be regarded as a function on G taking values in K. For a fixed $x^* \in M^*$ the assignment $x \to \rho_{x,x^*}$ gives an A-homomorphism $M \to A^*$, where A^* is considered as a left A-module. Therefore any nonzero coefficient ρ_{x,x^*} of an irreducible representation ρ generates an A-submodule of A^* isomorphic to M. In other words, any irreducible representation of A is similar to some subrepresentation of the coregular representation of A. In particular, any irreducible representation of a finite-dimensional semisimple algebra or a finite group is an irreducible component of the regular representation. Let A^*_ρ be the subspace of A^* generated by all coefficients ρ_{x,x^*} $(x \in M$, $x^* \in M^*)$ for a given linear representation ρ. Then $\rho \cong \rho'$ implies $A^*_\rho = A^*_{\rho'}$. If $\rho_1, \ldots, \rho_r$ are irreducible representations of A such that ρ_i and ρ_j are not similar unless $i=j$, the sum $A^*_{\rho_1} + \ldots + A^*_{\rho_r}$ in A^* is direct. In particular, for a semisimple algebra A, let the ρ_i $(1 \leqslant i \leqslant r)$ be the irreducible representations associated with the minimal left ideals of the †simple components A_i of A. Then any irreducible representation is similar to one and only one of $\rho_1, \ldots, \rho_r$, and A^* can be decomposed into the direct sum of $A^*_{\rho_1}, \ldots, A^*_{\rho_r}$. In addition, each $A^*_{\rho_i}$ is canonically identified with A^*_i.

We shall treat finite-dimensional representations exclusively. Let $(e_1, \ldots, e_n)$ be a basis of the representation module M of ρ over K, and let $a \to T(a) = (\lambda_{ij}(a))$ be the matrix representation that corresponds to ρ with respect to this basis. Then $\lambda_{ij} = \rho_{e_j, e_i^*}$ $(1 \leqslant i, j \leqslant n)$, where $(e_1^*, \ldots, e_n^*)$ is the dual basis. If K is algebraically closed and ρ is irreducible (or more generally, †absolutely irreducible), then $\{\lambda_{ij}\}$ is linearly independent; therefore we have $\dim A^*_\rho = n^2$ (G. Frobenius and I. Schur). We take a matrix representation T corresponding to ρ and put $\chi_\rho(a) = \operatorname{tr} T(a)$ $(a \in A)$. Then χ_ρ is a function on A that is uniquely determined by ρ and belongs to A^*_ρ; χ_ρ is called the **character** of ρ. For a linear representation ρ of a group G, the character of ρ can be regarded as a function on G. Moreover, it can be viewed as a function on the set of all †conjugate classes of G. The character of ρ is equal to the sum of the characters of the irreducible components of ρ taken with their multiplicities. The character of an irreducible representation is called an **irreducible character** (or **simple character**). If K is of characteristic 0, then $\rho \cong \rho'$ is equivalent to $\chi_\rho = \chi_{\rho'}$, and the different irreducible characters are linearly independent. The character of an absolutely irreducible representation (→ Section F) is called an **absolutely irreducible character**. If we consider absolutely irreducible characters only, the statement holds irrespective of the characteristic of K.

The sum of all absolutely irreducible characters of A is called the **reduced character** (or **reduced trace**) of A. The direct sum of all absolutely irreducible representations of A is called the **reduced representation** of A, and its character is equal to the reduced character. The determinant of the reduced representation is called the **reduced norm** of A.

F. Scalar Extension of Linear Representations

Let K, L be commutative rings with unity element, and fix a homomorphism $\sigma: K \to L$. We denote by M^σ the scalar extension $\sigma^*(M) = M \otimes_K L$ of a K-module M relative to $\sigma: x\lambda \otimes \mu = x \otimes \lambda^\sigma \mu$ $(x \in M; \lambda \in K, \mu \in L)$ (→ 277 Modules L). For an algebra A over K, the scalar extension A^σ of the K-module A has the natural structure of an algebra over $L: (a \otimes \lambda)(b \otimes \mu) = ab \otimes \lambda\mu$ $(a, b \in A; \lambda, \mu \in L)$. For a group G, we can regard $(K[G])^\sigma = L[G]: g \otimes \lambda = g\lambda$ $(g \in G, \lambda \in L)$. If M is a left A-module, then M^σ has the natural structure of a left A^σ-module; $(a \otimes \lambda)(x \otimes \mu) = ax \otimes \lambda\mu$ $(a \in A, x \in M; \lambda, \mu \in L)$. For the linear representation ρ associated with M, the linear representation ρ^σ over L associated with M^σ is called the **scalar extension** of ρ relative to $\sigma: \rho^\sigma(a \otimes 1) = \rho(a) \otimes 1_L$. Let $(e_1, \ldots, e_n)$ be a basis of M over K. If the matrix representation $a \to (\lambda_{ij}(a))$ corresponds to M relative to this basis, then the matrix representation corresponding to M^σ relative to the basis ($e_1 \otimes 1, \ldots, e_n \otimes 1$) over L is given by $a \otimes 1 \rightarrow (\lambda_{ij}(a)^\sigma)$. A linear representation over L is said to be **realizable** in K if it is similar to the scalar extension ρ^σ of some linear representation ρ over K.

In particular, if $\sigma: K \rightarrow L$ is an isomorphism, ρ^σ is called the **conjugate representation** of ρ relative to σ. The conjugate representation relative to the automorphism $\sigma: \lambda \rightarrow \bar{\lambda}$ (complex conjugation) of the complex number field is called the **complex conjugate representation**. If $\mathfrak{m}$ is an ideal of K and $\sigma: K \rightarrow K/\mathfrak{m}$ (†residue class ring) is the canonical homomorphism, then the construction of ρ^σ from ρ is called the **reduction modulo** $\mathfrak{m}$ (→ Section I). If $\mathfrak{p}$ is a prime ideal of K and $\sigma: K \rightarrow K_\mathfrak{p}$ (†local ring) is the canonical homomorphism, then the construction of ρ^σ from ρ is called the **localization** relative to $\mathfrak{p}$. If K is an integral domain and $\mathfrak{p} = K - \{0\}$, then $K_\mathfrak{p}$ is the †field of quotients of K. We can also consider the "completion of representation" with respect to $\mathfrak{p}$.

Let K be a field, L a field extension, and $\sigma: K \rightarrow L$ the canonical injection. Then for a linear representation ρ of A and its representation module M, the scalar extensions ρ^σ, M^σ are written ρ^L, M^L, respectively. In view of $M \subset M^L$, $A \subset A^L$, $\mathscr{E}_K(M) \subset \mathscr{E}_L(M^L)$ by the natural injections, we can regard ρ^L as an extension of the mapping ρ. We shall consider finite-dimensional representations exclusively. For linear representations ρ_1, ρ_2 over K, $\rho_1 \cong \rho_2$ is equivalent to $\rho_1^L \cong \rho_2^L$. An irreducible representation ρ over K is said to be **absolutely irreducible** if its scalar extension ρ^L to any field extension L is irreducible; an equivalent condition is that the scalar extension $\rho^{\bar{K}}$ to the †algebraic closure $\bar{K}$ is irreducible. Another equivalent condition is that every endomorphism of the representation module M of ρ must be a scalar multiplication. If every irreducible representation of A over K is absolutely irreducible, K is called a **splitting field** for A. For a group G, if the field K is a splitting field for the group ring $K[G]$, then K is called a splitting field for G. Let A be finite-dimensional over K. If K is a splitting field for A, any irreducible representation of A^L is realizable in K for any field extension L of K. For an arbitrary field K, the scalar extension ρ^L of an irreducible representation ρ to a †separable algebraic extension L of K is completely reducible. For simplicity, we assume that K is †perfect and $L = \bar{K}$. Then the multiplicities of all irreducible components of ρ^L are the same; this multiplicity is called the **Schur index** of ρ.

The set $S(K)$ of †algebra classes over K, each of which is represented by a (central) simple component of the group algebra $K[G]$ of some finite group G, is a subgroup of the †Brauer group $B(K)$ of K, known as the **Schur subgroup** of $B(K)$. Recent research has clarified considerably the structure of this group [19].

G. Linear Representations of Finite Groups

Let G be a finite group of order g. The linear representation of G over K is equivalent to the linear representation of the group ring $K[G]$, concerning which we have already stated the general facts. If K is the ring $\mathbf{Z}$ of rational integers, a linear representation over K is sometimes called an **integral representation**. We assume that K is a field. If the characteristic of K is zero or more generally not a divisor of g, every linear representation of G over K is completely reducible (H. Maschke). Such a representation is called an **ordinary representation**. If g is divisible by the characteristic of K, we have a **modular representation** (→ Section I).

The **exponent** of G is the smallest positive integer n satisfying $a^n = 1$ for every element $a \in G$. A field containing all the nth roots of unity is a splitting field for G (R. Brauer, 1945). Consequently, for such a field K, any scalar extension of an irreducible representation over K is irreducible, and any irreducible representation over any field extension of K is realizable in K. We fix a splitting field K for G and assume that K is of characteristic 0; for example, we can assume $K = \mathbf{C}$.

The number of nonsimilar irreducible representations of G is equal to the number of conjugate classes in G. Each irreducible representation appears as an irreducible component of the regular representation with multiplicity equal to the degree. In addition, each degree is a divisor of the order g of G. Let ρ be a linear representation of a subgroup H of G and M be its representation module. Then the linear representation of G associated with the $K[G]$-module $K[G] \otimes_{K[H]} M$ is called the **induced representation** and is denoted by ρ^G. If the matrix representation T corresponds to ρ, then using the partition of G into the cosets $G = a_1 H \cup \ldots \cup a_r H$ we can write the matrix representation corresponding to ρ^G as

$$a \rightarrow \begin{pmatrix} T(a_1^{-1}aa_1) & \ldots & T(a_1^{-1}aa_r) \\ \ldots & \ldots & \ldots \\ T(a_r^{-1}aa_1) & \ldots & T(a_r^{-1}aa_r) \end{pmatrix},$$

where we define $T(b) = 0$ for $b \notin H$. The induced representation from a representation of degree 1 of a subgroup is called a **monomial representation**. To such a representation corresponds a matrix representation T such that $T(a)$ has exactly one nonzero entry in each row and column for every $a \in G$. For the trivial subgroup $H=\{e\}$, we obtain the regular representation of G. In general, for an irreducible representation σ of G and an irreducible representation ρ of a subgroup H, the multiplicity of σ in ρ^G coincides with that of ρ in the restriction σ_H of σ to H (the **Frobenius theorem**). The following **orthogonality relations** hold for irreducible characters χ and ψ of G:

$$\sum_{a\in G}\chi(a)\psi(a^{-1})=\begin{cases} g, & \chi=\psi, \\ 0, & \chi\neq\psi, \end{cases}$$

$$\sum_{\chi}\chi(a)\chi(b^{-1})=\begin{cases} g/g_a, & C(a)=C(b), \\ 0, & C(a)\neq C(b). \end{cases}$$

In the second formula, χ ranges over all the irreducible characters of G, $C(a)$ denotes the conjugate class of G containing a, and g_a is the number of elements in $C(a)$.

H. Linear Representations of Symmetric Groups

All irreducible representations of a †symmetric group $\mathfrak{S}_n$ over the field $\mathbf{Q}$ of rational numbers are absolutely irreducible. Hence the representation theory of $\mathfrak{S}_n$ over a field of characteristic zero reduces to that over $\mathbf{Q}$. Since the group algebra $A=\mathbf{Q}[\mathfrak{S}_n]$ is semisimple, to obtain an irreducible representation of $\mathfrak{S}_n$ it is sufficient to find a †primitive idempotent (i.e., an idempotent that is not the sum of two orthogonal nonzero idempotents) of A. Such an idempotent can be obtained in the following way. As in Fig. 1, we draw a diagram T consisting of n squares arranged in rows of decreasing lengths, the left ends of which are arranged in a single column. Such a diagram T is called a **Young diagram**; if it has k rows of lengths $f_1\geqslant f_2\geqslant\ldots\geqslant f_k>0$, $f_1+f_2+\ldots+f_k=n$, then it is written $T=T(f_1,\ldots,f_k)$. We put the numerals 1 to n in any order into the n squares of $T=T(f_1,f_2,\ldots,f_k)$, as in Fig. 1, for example. We then denote by σ any permutation of $\mathfrak{S}_n$ that preserves each row and construct an element of $A: s=\sum\sigma$. Similarly, we denote by τ any permutation of $\mathfrak{S}_n$ that preserves each column and set $t=\sum(\operatorname{sgn}\tau)\tau$. If we set $u=t\cdot s=\sum\pm\tau\cdot\sigma$, then u is a primitive idempotent of A except for a numerical factor. This implies that u yields an irreducible representation of $\mathfrak{S}_n$. The element u of A is called the **Young symmetrizer** associated with T.

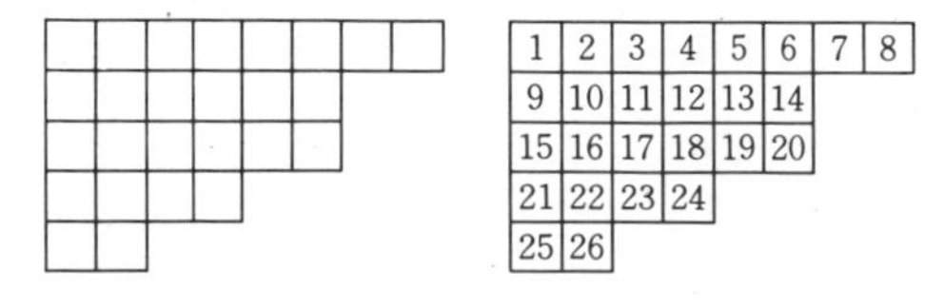

Fig. 1
Young diagram. $n=26$; $f_1=8$, $f_2=6$, $f_3=6$, $f_4=4$, $f_5=2$.

If we put the numerals 1 to n into the n squares of T in a different order, we obtain another symmetrizer u' associated with T. However, these two irreducible representations associated with u and u' are similar. Hence there corresponds to T a fixed class of irreducible representations of $\mathfrak{S}_n$, i.e., a fixed irreducible character of $\mathfrak{S}_n$. Moreover, any two different Young diagrams yield different irreducible characters, and any irreducible character is obtained by a suitable Young diagram. Thus there exists a one-to-one correspondence between the Young diagrams and the irreducible characters of $\mathfrak{S}_n$.

The method of determining the character associated with a given diagram was found by Schur and H. Weyl (→ 60 Classical Groups).

I. Modular Representations of Finite Groups

Let G be a finite group of order g, and let K be a splitting field of G of characteristic $p\neq 0$. If p is a divisor of g, we have the case of modular representation, in which the situation is quite different from the case of ordinary representation. The theory of modular representations of a finite group was developed mainly by Brauer after 1935.

The elements of G whose orders are prime to p are called **p-regular**. Let k be the number of p-regular classes of G, i.e., conjugate classes of G containing the p-regular elements. Then there exist exactly k nonsimilar absolutely irreducible modular representations F_1, F_2, $\ldots$, F_k. The number of nonsimilar indecomposable components of the regular representation R of G is also equal to k, and we denote these representations by $U_1, U_2, \ldots, U_k$. We can number them in such a way that F_κ appears in U_κ as both its top and bottom component. If the degree of F_κ is f_κ and that of U_κ is u_κ, then U_κ appears f_κ times in R and F_κ appears u_κ times in R. The multiplicities $c_{\kappa\lambda}$ of F_λ in U_κ are called the **Cartan invariants** of G.

Take an algebraic number field Ω that is a splitting field of G. Let $\mathfrak{p}$ be a prime ideal in Ω dividing p, and let $\mathfrak{o}$ be the domain of †$\mathfrak{p}$-integers of Ω. Then the residue class field $\mathfrak{o}/\mathfrak{p}$ is a finite field of characteristic p and a splitting field of G. Hence we can assume that $\mathfrak{o}/\mathfrak{p}=K$, where K is the field considered at the beginning of this section. Let $Z_1, Z_2, \ldots, Z_n$ be the nonsimilar irreducible representations of G in Ω. We can assume that all the coefficients of Z_i are contained in $\mathfrak{o}$. Replacing every coefficient in Z_i by its residue class mod $\mathfrak{p}$, we obtain a modular representation $\bar{Z}_i$. The modular representations $\bar{Z}_1, \ldots, \bar{Z}_n$ thus obtained may be reducible. The multiplicities $d_{i\kappa}$ of F_κ in $\bar{Z}_i$ are called the **decomposition numbers** of G.

They are related to the Cartan invariants by the fundamental relations

$$c_{\kappa\lambda}=\sum_{i=1}^{n} d_{i\kappa}d_{i\lambda}.$$

The determinant $|c_{\kappa\lambda}|$ of degree k is a power of p. We set $g=p^e g'$, $(p,g')=1$. Then we may assume that Ω contains a primitive g'th root of unity $\delta(\in\mathfrak{o})$. Since $(p,g')=1$, the residue class $\bar{\delta}(\in K)$ of δ is a primitive g'th root of unity. Let M be a modular representation of G. The characteristic roots of $M(a)$ for a p-regular element a are powers $\bar{\delta}^{\nu}$ of $\bar{\delta}$. We replace each $\bar{\delta}^{\nu}$ by δ^{ν} and obtain an element $\xi(a)$ of Ω as the sum of these δ^{ν}. In this manner we define a complex-valued function ξ on the set of p-regular elements of G. We call ξ the **modular character** (or **Brauer character**) of M. Two modular representations have the same irreducible components if and only if their modular characters coincide. Denoting by φ_κ the modular character of F_κ and by η_κ that of U_κ, we have the following orthogonality relations for the modular characters:

$$\sum_a \varphi_\kappa(a)\eta_\lambda(a^{-1})=\begin{cases} g, & \kappa=\lambda,\\ 0, & \kappa\neq\lambda,\end{cases}$$

$$\sum_\kappa \varphi_\kappa(a)\eta_\kappa(b^{-1})=\begin{cases} g/g_a, & C(a)=C(b),\\ 0, & C(a)\neq C(b).\end{cases}$$

In the first sum, a ranges over all p-regular elements of G.

We say that F_κ and F_λ belong to the same **block** if there exists a sequence of indices κ, α, $\beta, \ldots, \gamma$, λ such that $c_{\kappa\alpha}\neq 0$, $c_{\alpha\beta}\neq 0, \ldots, c_{\gamma\lambda}\neq 0$. This is obviously an equivalence relation, and $F_1, F_2, \ldots, F_k$ are classified into a finite number, say s, of blocks $B_1, B_2, \ldots, B_s$. If F_κ belongs to a block B_τ, we say by a stretch of language that the corresponding U_κ also belongs to B_τ. All the irreducible components of $\bar{Z}_i$ belong to the same block since $c_{\kappa\lambda}\neq 0$ if $d_{i\kappa}\neq 0$ and $d_{i\lambda}\neq 0$. If the irreducible components of $\bar{Z}_i$ belong to B_τ, we say that Z_i belongs to B_τ. Let x_τ be the number of Z_i belonging to B_τ and y_τ the number of F_κ belonging to B_τ. Then $x_\tau\geqslant y_\tau$. If χ_i is the ordinary character of Z_i, then χ_i can be considered as the modular character of $\bar{Z}_i$. If we denote the degree of Z_i by z_i, then $g_a\chi_i(a)/z_i$ for $a\in G$ is an algebraic integer and hence belongs to $\mathfrak{o}$. Now Z_i and Z_j belong to the same block if and only if $g_a\chi_i(a)/z_i\equiv g_a\chi_j(a)/z_j \pmod{\mathfrak{p}}$ for all p-regular elements a of G.

If p^α is the highest power of p that divides all the degrees z_i of Z_i belonging to B_τ, then it is also the highest power of p dividing all the degrees f_κ of F_κ belonging to B_τ. We call $d=e-\alpha$ the **defect** of B_τ; obviously $0\leqslant d\leqslant e$. If Z_i belongs to a block of defect d, then the power of p dividing z_i is p^{e-d+h_i} $(h_i\geqslant 0)$. A block of defect 0 contains exactly one ordinary representation Z_i, hence also exactly one modular representation F_κ $(x_\tau=y_\tau=1)$. Moreover, we have $Z_i=F_\kappa=U_\kappa$. It follows that all the degrees z_i of Z_i belonging to a block of defect 1 are exactly divisible by p^{e-1}; the converse is also true. Z_i belongs to a block of defect 0 if and only if $\chi_i(a)=0$ for any element a of G whose order is divisible by p.

Let D be any p-Sylow subgroup of the †centralizer $C_G(a)$ of an element a of G, and let $(D:1)=p^d$. Then d is called the **defect** of the class $C(a)$, and D is called a **defect group** of $C(a)$. The number of blocks of defect e is equal to the number of p-regular classes of defect e. Let B_τ be a block of defect d. Then there exists a p-regular class of defect d containing an element a such that $g_a\chi_i(a)/z_i\not\equiv 0 \pmod{\mathfrak{p}}$ for any Z_i in B_τ. The defect group D of $C(a)$ is called the **defect group** of B_τ, and D is uniquely determined up to conjugacy in G. The number of blocks of G with defect group D is equal to the number of blocks of the †normalizer $N_G(D)$ with defect group D.

An arbitrary element x of G can be written uniquely as a product $x=sr=rs$, where s, called the ***p*-factor** of x, is an element whose order is a power of p, and r is a p-regular element. We say that two elements of G belong to the same **section** if and only if their p-factors are conjugate in G. This is an equivalence relation. Obviously, each section is the union of conjugate classes of G. If the p-factor of x is not conjugate to any element of the defect group D of B_τ, then $\chi_i(x)=0$ for all Z_i in B_τ. Let $\varphi_1^s, \varphi_2^s, \ldots, \varphi_{k(s)}^s$ be the absolutely irreducible modular characters of $C_G(s)$, and let χ_l' be the absolutely irreducible ordinary characters of $C_G(s)$. Since

$$\chi_l'(sr)=\varepsilon_l\chi_l'(r)=\varepsilon_l\sum_\sigma d_{l\sigma}'\varphi_\sigma^s(r), \quad r\in C_G(s),$$

we have

$$\chi_i(sr)=\sum_l r_{il}\chi_l'(sr)=\sum_\sigma d_{i\sigma}^s\varphi_\sigma^s(r).$$

The $d_{i\sigma}^s$ are called the **generalized decomposition numbers** of G. If the order of s is p^l, then the $d_{i\sigma}^s$ are algebraic integers of the field of the p^lth roots of unity. Let s be conjugate to an element of D. There corresponds to B_τ a union $\tilde{B}_\tau$ of blocks of $C_G(s)$, and if $\sigma\neq\rho$, then $\tilde{B}_\tau$ and $\tilde{B}_\rho$ contain no irreducible modular representations in common. We have $d_{i\sigma}^s=0$ for any Z_i in B_τ (i.e., $\varphi_\sigma^s\notin\tilde{B}_\tau$). Brauer's original proof of this result was considerably complicated; simpler proofs were given independently by K. Iizuka and H. Nagao. From these relations we get the following refinement of the orthogonality relations for group characters. If Z_i and Z_j belong to different blocks of G, then $\sum_{a\in S}\chi_i(a)\chi_j(a^{-1})=0$, where a ranges over all the elements belonging to a fixed section S of G. If elements a and b of G belong to different sections, then $\sum_{\chi_i}\chi_i(a)\chi_i(b^{-1})=0$, where χ_i ranges over all the characters of G belonging to a fixed block B_τ.

J. Projective Representations of Finite Groups

Let V be a finite-dimensional linear space over a field K, and let $P(V)$ be the †projective space associated with V (→ 343 Projective Geometry). The set of all projective transformations of $P(V)$ forms the group $PGL(V)$, which can be identified with the quotient group $GL(V)/K^*1_V$. Here $K^*=K-\{0\}$ and K^*1_V is the set of all scalar multiples of the identity transformation 1_V of V and is the center of $GL(V)$. A homomorphism $G\rightarrow PGL(V)$ is called a **projective representation** of G in V or simply a projective representation of G over K. Two projective representations (ρ, V) and (ρ', V') of G are said to be **similar** if there exists an isomorphism $\varphi: PGL(V)\rightarrow PGL(V')$ induced by a suitable isomorphism $V\rightarrow V'$ such that $\varphi\circ\rho(a)\circ\varphi^{-1}=\rho'(a)$ $(a\in G)$. Let $V_1\neq\{0\}$ be a subspace of V. We can assume that $P(V_1)\subset P(V)$. If (ρ, V) is a projective representation of G such that each $\rho(a)$ $(a\in G)$ leaves $P(V_1)$ invariant, we get a projective representation (ρ_1, V_1) by restricting the $\rho(a)$ to $P(V_1)$. In this case (ρ_1, V_1) is called a **subrepresentation** of ρ. A projective representation is said to be **irreducible** if there exists no proper subrepresentation of ρ.

A mapping $\sigma: G\rightarrow GL(V)$ is called a section for (ρ, V) if $\pi(\sigma(a))=\rho(a)$ for each $a\in G$, where π is the natural projection of $GL(V)$ onto $PGL(V)$. Any section σ defines a mapping $f: G\times G\rightarrow K^*$ satisfying $\sigma(a)\sigma(b)=f(a,b)\sigma(ab)$ $(a,b\in G)$. The set $\{f(a,b)\}_{a,b\in G}$ is called the **factor set** of ρ with respect to σ. The mapping f is a †2-cocycle of G with values in K^*. The 2-cohomology class $c_\rho\in H^2(G,K^*)$ of f is determined by ρ and is independent of the choice of sections for ρ. A projective representation ρ has a section σ which is a linear representation of G in V if and only if $c_\rho=1$. If G is a finite group, for any $c\in H^2(G,K^*)$ there exists an irreducible projective representation ρ of G over K which belongs to c, i.e., $c_\rho=c$. If ρ and ρ' are similar, then $c_\rho=c_{\rho'}$. The tensor product $\rho\otimes\rho'$ of two projective representations ρ and ρ' can be defined as in the case of linear representations, and we have $c_{\rho\otimes\rho'}=c_\rho\cdot c_{\rho'}$. If K is algebraically closed, then $H^2(G,K^*)$ is determined by the characteristic of K. When K is the complex field $\mathbf{C}$, the group $H^2(G,\mathbf{C}^*)=\mathfrak{M}(G)$ is called the **multiplier** of G. If $\mathfrak{M}(G)=1$, then G is called a **closed group**, and any projective representation of G is induced by a linear representation of G. In general, if ρ is a projective representation of G over $\mathbf{C}$, then the order of c_ρ is a divisor of the degree of ρ (dimension of V). Moreover, if ρ is irreducible, then both the degree of ρ and the square of the order of c_ρ are divisors of the order of G. K. Yamazaki, among others, studied the projective representations of finite groups in detail.

K. Integral Representations

Every complex matrix representation of G is equivalent to a matrix representation in the ring of algebraic integers. If an algebraic number field K is specified, every $K[G]$-module V contains G-invariant R-†lattices (briefly, G-lattices), where R is the ring of integers in K. A G-lattice L is characterized as an $R[G]$-module, which is finitely generated and †torsion free (hence †projective) as an R-module. It provides an **integral representation** of G as an automorphism group of the R-projective module L.

$R[G]$-modules L and M need not be isomorphic even when the $K[G]$-modules $K\otimes L$ and $K\otimes M$ are isomorphic. The set of G-lattices in a fixed $K[G]$-module V is divided into a finite number of $R[G]$-isomorphism classes (**Jordan-Zassenhaus theorem**). Let $\mathfrak{p}$ be a prime ideal of R and $R_\mathfrak{p}$ be the localization of R at $\mathfrak{p}$. The study of $R_\mathfrak{p}$-representations is intimately related with modular representation theory. For any $R[G]$-module L there is an associated family of $R_\mathfrak{p}[G]$-modules $L_\mathfrak{p}=R_\mathfrak{p}\otimes L$, where $\mathfrak{p}$ ranges over all primes of R. G-lattices L and M in a $K[G]$-module V are said to be of the same **genus** if $L_\mathfrak{p}\cong M_\mathfrak{p}$ for every $\mathfrak{p}$. The number of genera of G-lattices in V is given by $\prod_{\mathfrak{p}|g}h_\mathfrak{p}$ ($g=$ order G), where $h_\mathfrak{p}$ denotes the number of $R_\mathfrak{p}[G]$-equivalence classes of $R_\mathfrak{p}[G]$-lattices in V. When V is absolutely irreducible, the number of $R[G]$-equivalence classes in a genus equals the (ideal) †class number of K (J. M. Maranda and S. Takahashi).

The †Krull-Schmidt theorem, asserting the uniqueness of a direct sum decomposition into indecomposable $R[G]$-modules, holds if R is a complete discrete valuation ring or if R is a discrete valuation ring and K is a splitting field of G. The condition for the finiteness of the number of nonisomorphic indecomposable G-lattices is known. In particular, for $R=\mathbf{Z}$ it reduces to the requirement that the Sylow p-subgroup of G be cyclic of order p or p^2 for every $p|g$. Regarding projective $\mathbf{Z}[G]$-modules → 200 Homological Algebra G.

The **isomorphism problem**, i.e., the question of whether the isomorphism $\mathbf{Z}[G]\cong\mathbf{Z}[H]$ of integral group algebras implies the isomorphism $G \cong H$ of groups, has been answered affirmatively for certain special cases such as †meta-Abelian groups.

$R[G]$ is an R-†order in $K[G]$, and in this context, a considerable portion of the integral representation theory has been extended to more general orders in separable algebras [14–16].

References

[1] B. L. van der Waerden, Gruppen von linearen Transformationen, Springer, 1935 (Chelsea, 1948).
[2] B. L. van der Waerden, Algebra II, Springer, fifth edition, 1967.
[3] H. Boerner, Darstellungen von Gruppen, Springer, 1955.
[4] N. Bourbaki, Eléments de mathématique, Algèbre, ch. 8, Actualités Sci. Ind., 1261a, Hermann, 1958.
[5] C. W. Curtis and I. Reiner, Representation theory of finite groups and associative algebras, Interscience, 1962.
[6] R. Brauer and C. Nesbitt, On the modular characters of groups, Ann. Math., (2) 42 (1941), 556–590.
[7] R. Brauer, Zur Darstellungstheorie der Gruppen endlicher Ordnung I, II, Math Z., 63 (1956), 406–444; 72 (1959–1960), 25–46.
[8] M. Osima, Notes of blocks of group characters, Math. J. Okayama Univ., 4 (1955), 175–188.
[9] I. Schur, Über die Darstellung der endlichen Gruppen durch gebrochene lineare Substitutionen, J. Reine Angew. Math., 127 (1904), 20–50.
[10] I. Schur, Untersuchungen über die Darstellung der endlichen Gruppen durch gebrochene lineare Substitutionen, J. Reine Angew. Math., 132 (1907), 85–137.
[11] I. Schur, Über die Darstellung der symmetrischen und der alternierenden Gruppe durch gebrochene lineare Substitutionen, J. Reine Angew. Math., 139 (1911), 155–250.
[12] R. Brauer, Representations of finite groups, Lectures on modern mathematics, T. L. Saaty (ed.), Wiley, 1963, vol. 1, 133–175.
[13] W. Feit, Characters of finite groups, Benjamin, 1967.
[14] I. Reiner, A survey of integral representation theory, Bull. Amer. Math. Soc., 76 (1970), 159–227.
[15] R. Swan, K-theory of finite groups and orders, Lecture notes in math. 149, Springer, 1970.
[16] K. W. Roggenkamp (with V. Huber-Dyson), Lattices over orders I, II, Lecture notes in math. 115, 142, Springer, 1970.
[17] L. Dornhoff, Group representation theory, Dekker, A, 1971; B, 1972.
[18] J.-P. Serre, Représentations linéaires des groupes finis, second edition, Hermann, 1972.
[19] T. Yamada, The Schur subgroup of the Brauer group, Lecture notes in math. 397, Springer, 1974.

363 (XXI.40) Riemann, Georg Friedrich Bernhard

Georg Friedrich Bernhard Riemann (September 17, 1826–July 20, 1866) was born the son of a minister in Breselenz, Hanover, Germany. He attended the universities of Göttingen and Berlin. In 1851 he received his doctorate at the University of Göttingen and in 1854 became a lecturer there. In 1857 he rose to assistant professor, and in 1859 succeeded P. G. L. †Dirichlet as full professor. In 1862 he contracted tuberculosis, and he died at age 40. Despite his short life, his contributions encompassed all aspects of mathematics.

His doctoral thesis (1851) stated the basic theorem on †conformal mapping and became the foundation for the geometric theory of functions. In his paper presented for the position of lecturer (1854), he defined the †Riemann integral and gave the conditions for convergence of trigonometric series. In his inaugural lecture in the same year, he discussed the foundations of geometry, introduced n-dimensional manifolds, formulated the concept of †Riemannian manifolds, and defined their curvature. In his paper of 1857 on †Abelian functions, he systematized the theory of †Abelian integrals and Abelian functions. In his paper of 1858 on the distribution of prime numbers, he considered the †Riemann zeta function as a function of a complex variable and stated †Riemann's hypothesis concerning the distribution of its zeros. It remains for modern mathematics to investigate whether this hypothesis is correct. In his later years, influenced by W. Weber, Riemann became interested in theoretical physics. He gave lectures on the uses of partial differential equations in physics that were edited and published by H. Weber.

References

[1] G. F. B. Riemann, Gesammelte mathematische Werke und wissenschaftlicher Nachlass, R. Dedekind and H. Weber, (eds.), Teubner, second edition, 1892 (Nachträge, M. Noether and W. Wirtinger (eds.), 1902) (Dover, 1953).

[2] F. Klein, Vorlesungen über die Entwicklung der Mathematik im 19. Jahrhundert I, Springer, 1926 (Chelsea, 1956).

364 (VII.3) Riemannian Manifolds

A. Riemannian Metrics

Let M be a †differentiable manifold of class C^r $(1 \leqslant r \leqslant \omega)$, and g be a †Riemannian metric of class C^{r-1} on M. Then (M, g) or simply M is called a **Riemannian manifold** (or **Riemannian space**) of class C^r (→ 105 Differentiable Manifolds). The metric g is a †covariant tensor field of order 2 and of class C^{r-1}; it is called the **fundamental tensor** of M. Using the value g_p of g at each point $p \in M$, a positive definite inner product $g_p(X, Y)$, X, $Y \in T_p$, is introduced on the †tangent vector space T_p to M at p, and hence T_p can be considered as a †vector space over $\mathbf{R}$ with inner product that can be identified with the Euclidean space E^n of dimension $n = \dim M$. Utilizing the properties of the space E^n, we can introduce various notions on T_p and M. (For example, given a tangent vector $L \in T_p$, we define the length $\|L\| = \|L\|_{g,p}$ of L to be the quantity $g_p(L, L)^{1/2}$. A **normal vector** at a point p of a submanifold N of M is well defined as an element of the orthogonal complement of the subspace $T_p(N)$ of $T_p(M)$ with respect to g_p; a differential form of degree 1 is identified with a tangent vector field.) A necessary and sufficient condition for a differentiable manifold M of class C^r to have a Riemannian metric is that M be †paracompact. A Euclidean space E^n has a Riemannian metric expressed by $\sum_{i=0}^n dx^i \otimes dx^i$ in terms of an orthogonal coordinate system (x^i).

We assume that M is connected and of class C^∞. A curve $x: [a, b] \to M$ is called **piecewise smooth** or of **class** D^∞ if x is continuous and there exists a partition of $[a, b]$ into finite subintervals $[t_{i-1}, t_i]$ such that the restrictions $x | [t_{i-1}, t_i]$ are †immersions of class C^∞. The length $\|x\|$ of such a curve x is defined to be $\int_a^b \|x'(t)\| \, dt$, where $x'(t)$ is the tangent vector of x defined for almost all values of t. As in a Euclidean space, the length $\|x\|$ is independent of the choice of parameter t, and the concepts of †canonical parameter and orientation of x can be defined (→ 111 Differential Geometry of Curves and Surfaces). A function $d: M \times M \to [0, \infty)$ is defined so that the value $d(p, q)$, p, $q \in M$, is the infimum of the lengths of curves of class D^∞ joining p and q. The function d is a †distance function on M, and the topology of M defined by d coincides with the original topology of M. There exists an essentially unique structure of a Riemannian manifold on (real or complex) †elliptic or †hyperbolic space (→ 285 Non-Euclidean Geometry), and d is the distance function of these spaces.

If there exists an immersion φ of a differentiable manifold N in a Riemannian manifold (M, g), then a Riemannian metric $\varphi^* g$ is defined on N by the †pullback process ($\|L\|_{\varphi^* g} = \|d\varphi(L)\|_g$). (For example, a submanifold and a †covering manifold of M have Riemannian manifold structures induced by the natural mappings (→ 365 Riemannian Submanifolds).) If $M = E^3$ and N is a 2-dimensional submanifold of M, then $\varphi^* g$ is the †first fundamental form of N. Assume further that φ is a diffeomorphism and N has a Riemannian metric h. If $\varphi^* g = h$, then (N, h) is said to be **isometric** to (M, g), and φ is called an **isometry**. The set $I(M)$ of all isometries (isometric transformations) of M onto M is a group. A necessary and sufficient condition for a mapping ψ: $N \to M$ to be an isometry is that $d_N(p, q) = d_M(\psi(p), \psi(q))$, p, $q \in N$. In particular, $I(E^n)$ is the †congruent transformation group.

If a differentiable manifold M is the product manifold of Riemannian manifolds (M_1, g_1) and (M_2, g_2), then $(M, \pi_1^* g_1 + \pi_2^* g_2)$ is called the **Riemannian product** of M_1 and M_2, where π_α, $\alpha = 1, 2$, are projections from M to M_α.

Let F be the †tangent n-frame bundle over M and $B = B_g(M)$ be the subset of F consisting of all orthonormal frames with respect to g. Then B is an $O(n)$-subbundle of F of class C^∞, called the **tangent orthogonal** n**-frame bundle** (or **orthogonal frame bundle**). In this way we get a one-to-one correspondence between the set of all $O(n)$-subbundles of F and the set of all Riemannian metrics of M.

B. Riemannian Connections

There exists a unique †affine connection in the orthogonal frame bundle B whose †torsion tensor is zero. This connection is called the **Riemannian connection** (or **Levi-Civita connection**; → 80 Connections K). Let ∇ denote the †covariant differential operator defined by this connection (→ 80 Connections, 417 Tensor Calculus). (For a vector field X, the covariant differential operator ∇_X acts on any tensor field T defined on a submanifold having X as a tangent vector field.) The covariant differential ∇g of the fundamental tensor g vanishes identically. The †connection form of the Riemannian connection is expressed by n^2 differential 1-forms $(\omega_j^i)_{1 \leqslant i,j \leqslant n}$ on B, and we have $\omega_j^i + \omega_i^j = 0$. Let $(\omega^i)_{1 \leqslant i \leqslant n}$ be the †canonical 1-forms on B. Then $(\omega_j^i)_{1 \leqslant i \leqslant j \leqslant n}$ together with (ω^i) give rise to an absolute parallelism on B (that is, they are linearly independent at every point). Let (θ^i) and (θ_i^j) be the corresponding set of differential 1-forms on the orthogonal frame bundle B_N over another Riemannian manifold N with $\dim N = \dim M$. If there exists an isometry $\psi: M \rightarrow N$, then the differential $d\psi$ is a diffeomorphism from $B = B_M$ to B_N, and we get $(d\psi)^*(\theta^i) = \omega^i$, $(d\psi)^*(\theta_i^j) = \omega_i^j$. Conversely, if there exists a diffeomorphism $\Psi: B_M \rightarrow B_N$ satisfying $\Psi^*(\theta^i) = \omega^i$, $\Psi^*(\theta_i^j) = \omega_i^j$ and M is †orientable, then there exists an isometry $\psi: M \rightarrow N$ such that $d\psi = \Psi$ holds on a connected component B_0 of B. Moreover, ψ is uniquely determined if we choose one B_0. In this way the problem of the existence of an isometry from M may be reduced to one of the existence of a diffeomorphism from B preserving absolute parallelism (as well as the order of the basis (ω^i, ω_i^j)).

According to the general theory of affine connections, the Riemannian connection on M determines a †Cartan connection uniquely with $E^n = I(E^n)/O(n)$ as †fiber, which is called the **Euclidean connection**. As a consequence, every tangent vector space $T_p(M)$ is regarded as a Euclidean space E_p^n, and for a given curve $x: [a, b] \rightarrow M$ of class D^∞ and for $t \in [a, b]$ there exists an isometry $I_{x,t}: E_{x(t)}^n \rightarrow E_{x(a)}^n$ satisfying the following three conditions (we denote $I_{x,b}$ by I_x): (1) If x is a composite of two curves y and z, then $I_x = I_y \cdot I_z$. (2) Differentiability: If x is of class C^∞ at t_0, then $t \rightarrow I_{x,t}$ is of class C^∞ at t_0. (3) I_x depends on the orientation of x but not on the choice of its parameter. The **development** $\bar{x}$ of x is the curve in $E_{x(a)}^n$ defined by $\bar{x}(t) = I_{x,t}(x(t))$, and we get $\|\bar{x}\| = \|x\|$. (I_x is sometimes called the **development along** x.) Utilizing the concept of development, the theory of curves in E^n can be used to study curves on M (→ 111 Differential Geometry of Curves and Surfaces). For example, if $\bar{x}$ is a segment, then x or $x([a, b])$ is called the **geodesic arc** (→ 80 Connections L); the †Frenet formula is automatically formulated and proved. The rotation part I_x^R of I_x (the composite of I_x and the parallel displacement of E_p^n translating $I_x(x(b))$ to $x(a)$) is regarded as an isomorphism of the inner product space $T_{x(b)}$ to $T_{x(a)}$. I_x^R is extended to an isomorphism of the †tensor algebra $\mathcal{T}(T_{x(b)})$ to $\mathcal{T}(T_{x(a)})$, which is denoted by the same symbol I_x^R and called the **parallel displacement** or **parallel translation along** x. Given a tensor field K on M, we have $\nabla_{x'(a)} K = [dI_{x,t}^R(K(x(t)))/dt]_{t=a}$. In particular, a necessary and sufficient condition for $\nabla K = 0$ is that $I_x^R(K(x(b))) = K(x(a))$ for any x, in which case K is said to be **parallel**.

C. Exponential Mapping (→ 178 Geodesics)

A curve x on M or the image of x is called a **geodesic** if any subarc $x \mid [a, b]$ of x is a geodesic arc. Let $N(S)$ be the **normal bundle** of a submanifold S of M, that is, the differentiable vector bundle over S consisting of all normal vectors at all points of S. Then S is contained in $N(S)$ as the set of zero vectors at all points of S. There exist a neighborhood U of S in $N(S)$ and a mapping $\mathrm{Exp}_S: U \rightarrow M$ of class C^∞ with the following property: There exists a geodesic arc x with the initial tangent vector $L \in U$, length $\|x\| = \|L\|$, and final point $\mathrm{Exp}_S(L)$. Let U_S be the largest U with this property. Then $\mathrm{Exp}_S: U_S \rightarrow M$ is determined uniquely by S. The mapping Exp_S is called the **exponential mapping** on S. If the rank of the Jacobian matrix of Exp_S is less than n at $L \in U_S$, then L or $\mathrm{Exp}_S(L)$ is called the **focal point** of S on the geodesic $s \rightarrow \mathrm{Exp}_S(sL)$ ($0 \leqslant s$, $sL \in U_S$). If S is compact, then S has an open neighborhood V_S in $N(S)$ satisfying the following three conditions: (i) $V_S \subset U_S$; (ii) $\|L\| = d(\mathrm{Exp}_S(L), S)$ for $L \in V_S$, where the right-hand member expresses the infimum of the distance between the point $\mathrm{Exp}_S(L)$ and points of S; (iii) the restriction $\mathrm{Exp}_S \mid V_S$ is an embedding. The image $\mathrm{Exp}_S(V_S)$ is the **tubular neighborhood** of S. In the special case where S consists of only one point p, $N(\{p\})$ coincides with the tangent vector space $T_p(M)$, and the focal point of p is called the **conjugate point** of p, given as the zero point of the †Jacobi field (→ 178 Geodesics, 279 Morse Theory). In this case, V_S is denoted by V_p. If T_p is identified with $\mathbf{R}^n$ (or E^n) by means of an orthonormal basis of T_p, then $(\mathrm{Exp}_p)^{-1}$ defined on $\mathrm{Exp}_p(V_p)$ is a coordinate mapping, called the **normal coordinate mapping**. Furthermore, $\mathrm{Exp}_p(V_p)$ contains a neighborhood W_p of p such that there exists a unique geodesic arc x joining any two points q and r of W_p with $\|x\| = d(q, r)$ and contained in W_p. W_p is called a **convex neighborhood** of p.

D. Curvature

The set of differential 1-forms (ω^i, ω_i^j), by means of which absolute parallelism is given in the orthogonal frame bundle B of M, satisfies the †structure equation $d\omega^i = -\sum_j \omega_j^i \wedge \omega^j$, $d\omega_j^i = -\sum_k \omega_k^i \wedge \omega_j^k + \Omega_j^i$, and (Ω_j^i) is called the **curvature form** of the Riemannian connection of M. This form is expressed by a tensor field R (→ 80 Connections; 417 Tensor Calculus) of type $(1, 3)$ on M, called the **curvature tensor**; if R_{jkl}^i are the components of R with respect to an orthonormal frame $b \in B$ of the tangent vector space T_p of M, then $\Omega_j^i = (1/2) \sum R_{jkl}^i \omega^k \wedge \omega^l$ at b. Let (X, Y) be an orthonormal basis of a 2-dimensional subspace P of T_p. Then the inner product $K_p(P)$ of X and $R(X, Y)Y$ is determined by P independently of the choice of the basis (X, Y), where the i-component of

$R(X, Y)Z$ with respect to the basis b of T_p is given by $\sum R^i_{jkl} Z^j X^k Y^l$. $K_p(P)$ is the †Gaussian curvature of the surface $\mathrm{Exp}_p(V_p \cap P)$ and is called the **sectional curvature** (or **Riemannian curvature**) of P. The curvature tensor R is uniquely determined by the function $K_p(P)$ of p and P. If $\dim M \geqslant 3$ and if at every point p of M, $K_p(P)$ has a constant value M_p independent of the choice of P, then M_p is a constant independent of the choice of p (F. Schur). If $K_p(P)$ is constant, then M is called a **space of constant curvature**. If $\nabla R = 0$, then M is called a **locally symmetric space** (→ 412 Symmetric Riemannian Spaces and Real Forms; 413 Symmetric Spaces). In a local sense, Riemannian metrics of these spaces are uniquely determined by the curvature tensor R up to a constant factor. If M is of constant curvature K, complete, and simply connected, then M is isometric to E^n, the sphere (which is the universal covering Riemannian manifold of a real †elliptic space), or a real †hyperbolic space according as K is 0, positive, or negative. The compact spaces of positive constant curvature, that is, the Riemannian manifolds having the sphere as the universal covering Riemannian manifold, were completely classfied by J. A. Wolf [1]. A complete, simply connected, and locally symmetric space is a †symmetric Riemannian space. The **Ricci tensor** (R_{ij}) is defined by $R_{ij} = -\sum_k R^k_{ijk}$. Let Q be the quadratic form on T_p given by (R_{ij}). Then the value $Q(L)$ for a unit vector $L \in T_p$ is the mean of $K_p(P)$ for all sections P (2-dimensional subspaces of T_p) containing L and is called the **Ricci curvature** (or **mean curvature**) of the direction L at p. The mean R of $Q(L)$ for all the unit vectors L at p is called the **scalar curvature** at p (→ 417 Tensor Calculus). $Q(L)$ and R are expressed by $Q(L) = \sum_i g_p(R(X_i, L)L, X_i)$ and $R = \sum_i Q(X_i)$, up to positive constant factors, in terms of an orthonormal basis (X_i) of T_p. If the Ricci tensor of M is a scalar multiple of the fundamental tensor, then M is called an **Einstein space**. (When $\dim M \geqslant 3$, this scalar is constant.) If M is a †Kähler manifold and P is restricted to a complex plane (invariant under the almost complex structure), then $K(P)$ is called the **holomorphic sectional curvature**. A Kähler manifold M of constant holomorphic sectional curvature is locally isometric to a complex Euclidean space, elliptic space, or hyperbolic space.

The properties of the sectional curvature and the Ricci curvature are closely related to the behavior of geodesics of Riemannian manifolds, and these properties reflect those of the topological structures of the manifolds (→ 178 Geodesics). The compact simply connected homogeneous Riemannian manifolds of strictly positive sectional curvature have been classified [2–4]. Related to algebraic geometry, as the solution of the Frankel conjecture, the following holds: If a compact Kähler manifold has strictly positive sectional curvature, then it is biholomorphic to the complex projective space [5, 6] (→ 232 Kähler Manifolds). Furthermore, curvature tensors are related to †characteristic classes. For example, we have the **Gauss-Bonnet formula**: If M is an even-dimensional compact and oriented Riemannian manifold, the integral of $a_n K_{(n)} \omega$ on M is equal to the †Euler-Poincaré characteristic, where

$$a_n = n!/(2^n \pi^{n/2} (n/2)!),$$

ω is the volume element of M, and $K_{(n)}$ is defined as follows: For a positive even number s, $K_{(s)}$ is a real-valued function of the s-dimensional subspaces P of the tangent vector spaces T_p of M, which is given by

$$K_{(s)}(P) = b_s \sum \varepsilon_{i_1 \dots i_s} \varepsilon_{j_1 \dots j_s} \langle R_p(X_{i_1}, X_{i_2}) X_{j_1}, X_{j_2} \rangle \dots \langle R_p(X_{i_{s-1}}, X_{i_s}) X_{j_{s-1}}, X_{j_s} \rangle$$

in terms of an orthonormal basis $(X_1, \dots, X_s)$ of P, where $b_s = (-1)^{s/2}/(2^{s/2} s!)$, $\sum$ is summation over all pairs of s-tuples satisfying $\{i_1, \dots, i_s\}$, $\{j_1, \dots, j_s\} \subset \{1, 2, \dots, n\}$, $\varepsilon_{i_1 \dots i_s}$ is the sign of $(i_1, \dots, i_s)$, $\langle\ ,\ \rangle$ is the inner product in T_p with respect to g_p, R_p is the value of the tensor R at p, and $R_p(X_i, X_j)X_k$ is as already defined at the beginning of this section. In particular, $K_{(2)} = K$. If $K_{(s)}$ of a compact and orientable M is constant for a certain s, then the kth †Pontryagin class of M (with real coefficients) vanishes for all $k \geqslant s/2$.

E. Holonomy Groups

Let p be a fixed point of M, and let Ω_p be the set of all closed oriented curves of class D^∞ with initial and final points p and with parameters neglected. The set $H = \{I_x \mid x \in \Omega_p\}$, called the **holonomy group** of M, is a subgroup of $I(T_p)$ (T_p is identified with E^n) independent of the choice of p (→ 80 Connections), and $x \to I_x$ is a homomorphism from Ω_p to H. The restriction H_0 of this homomorphism to all closed curves homotopic to zero is called the **restricted holonomy group**. The rotation part h of H, called the **homogeneous holonomy group**, is a subgroup of the orthogonal group $O(n)$ of T_p. The rotation part h_0 of H_0, called the **restricted homogeneous holonomy group**, is a connected component of h and a †compact Lie group. The †Lie algebra of h_0 is spanned by $\{I_x(R_{x(b)}(X, Y)) \mid x : [a, b] \to M$ is of class D^∞, $x(a) = p$, and $X, Y \in T_{x(b)}\}$, where $R_{x(b)}(X, Y)$ is the endomorphism of the linear space $T_{x(b)}$ defined in Section D.

If $M=E^n$, then $H=\{e\}$, where e is the identity element. If $h=\{e\}$ ($h_0=\{e\}$), then M is called **flat (locally flat)** (→ 80 Connections E). Local flatness is equivalent to M being locally isometric to E^n. If M is complete and H (regarded as a transformation group of E^n) has a fixed point, then M is isometric to E^n. Any finite rotation group h is the homogeneous holonomy group of some locally flat and compact Riemannian manifold.

With respect to the linear group h of T_p, we get a unique decomposition $T_p=V_{(0)}\oplus V_{(1)}\oplus\ldots\oplus V_{(r)}$ of mutually orthogonal subspaces, where $V_{(0)}$(dim $V_{(0)}\geqslant 0$) consists of all h-invariant vectors and $V_{(i)}$, $i=1,\ldots,r$, are irreducible h-invariant subspaces. If h or h_0 is irreducible (reducible) on T_p, then M is called **irreducible (reducible)**. If M is complete and simply connected (hence $h=h_0$), then M is the Riemannian product of closed submanifolds $M_{(\alpha)}$, $\alpha=0,1,\ldots,r$, satisfying $V_{(\alpha)}=T(M_{(\alpha)})$. This decomposition $M=\prod M_{(\alpha)}$ is determined uniquely by M and called the **de Rham decomposition** of M [7]. In this case h is the direct product of closed subgroups $h_{(\alpha)}$, where every $h_{(\alpha)}$ acts on $V_{(\beta)}$, $\beta\neq\alpha$ as the identity, and can be regarded as the homogeneous holonomy group of $M_{(\alpha)}$. If h_0 is irreducible and M is not locally symmetric, then h_0 acts †transitively on the unit sphere of T_p. The classification of possible candidates for such h_0 has been made [8, 9]. For example, if n is even and h_0 is the †unitary group $U(n/2)$, then h acts transitively on the unit sphere. A necessary and sufficient condition for h to be contained in $U(n/2)$ is that M have a †complex structure and the structure of a Kähler manifold.

The group h acts naturally on the †tensor algebra $\mathscr{T}(T_p)$ of T_p. If a tensor field A on M is parallel, then A_p is invariant under h. Conversely, if $A_0\in\mathscr{T}(T_p)$ is invariant under h, there exists a unique parallel tensor field A satisfying $A_p=A_0$. The orthogonal frame bundle B is †reducible to the h-bundle.

F. Transformation Groups

The group $I(M)$ consisting of all isometries of M with the †compact-open topology is a †Lie transformation group. The isotropy subgroup at any point is compact. In particular, if M is compact, so is $I(M)$. The differential $d\varphi$ of $\varphi\in I(M)$ is a transformation of the orthogonal frame bundle B. If b_0 is a fixed point of B, then the mapping β defined by $\varphi\rightarrow d\varphi(b_0)$ embeds $I(M)$ as a closed submanifold of B, and the differentiable structure of $I(M)$ is thus determined. If β is surjective, it follows from the structure equation that M is of constant curvature and equals E^n, a real †hyperbolic space, or a real †elliptic space (or a sphere). A necessary and sufficient condition for the image of β to be a subbundle of B is that $I(M)$ be transitive. If the image of β contains the h-bundle, then M is a symmetric space. If M is compact and $I(M)$ is transitive, then the image of β is contained in the h-bundle (→ 191 G-Structures). If $I(M)$ is transitive on M, then M is complete and is the †homogeneous space of $I(M)$. Conversely, a homogeneous space $M=G/K$ of a Lie group G by a compact subgroup K has a Riemannian metric invariant under G. In general, an element of $I(M)$ preserves quantities uniquely determined by the Riemannian metric g, such as the Riemannian connection, its curvature, the set of all geodesics, etc. Furthermore, any element of $I(M)$ commutes with ∇ and the †Laplace-Beltrami operator. If M is compact and oriented, then the connected component $I_0(M)$ of $I(M)$ preserves any †harmonic differential form. If M is complete and simply connected, then $I_0(M)$ is clearly decomposed into a direct product by the de Rham decomposition of M. An element of the Lie algebra of $I(M)$ is regarded as a vector field X on M, called the **infinitesimal motion**, which satisfies the equation $L_X g=0$; that is, $\nabla_j\xi_i+\nabla_i\xi_j=0$, where L_X denotes †Lie derivation and the ξ_i are †covariant components of X with respect to a natural frame $(\partial/\partial X_i)$, $i=1,\ldots,n$ (→ 417 Tensor Calculus). This equation is called **Killing's differential equation**, and a solution X of this equation is called a **Killing vector field**. The set of all Killing vector fields is a Lie algebra of finite dimension ($\leqslant\dim B$). If M is complete, then this Lie algebra coincides with that of $I(M)$. If M is compact and the Ricci tensor is negative definite, then $I(M)$ is discrete. If, furthermore, the sectional curvature is nonpositive, then an isometry of M homotopic to the identity transformation is the identity transformation itself.

It is known that $\dim I(M)\leqslant n(n+1)/2$ if $\dim M=n$, and the maximum dimension is attained only when M is a space of constant curvature. For Riemannian manifolds with large $I(M)$, extensive work on the structures of M and $I(M)$ has been done by I. P. Egorov, S. Ishihara, N. H. Kuiper, L. N. Mann, Y. Muto, T. Nagano, M. Obata, H. Wakakuwa, K. Yano, and others [10, 11].

The fixed point set of a family of isometries has interesting differential geometric properties [10]. For example, let G be any subset of $I(M)$ and F the set of points of M which are left fixed by all the elements of G. Then each connected component of F is a closed †totally geodesic submanifold of M. If M is compact and f is an isometry of M, then $\Lambda_f=\chi(F)$, where Λ_f denotes the †Lefschetz number and

$\chi(F)$ the †Euler characteristic of the fixed point set F of f. As for the existence of fixed points of an isometry, the following are known: Let f be an isometry of a compact, orientable Riemannian manifold M with positive sectional curvature. If $\dim M$ is even and f is orientation preserving, or if $\dim M$ is odd and f is orientation reversing, then f has a fixed point. In the case of nonpositive curvature, the following is basic: Every compact group of isometries of a complete, simply connected Riemannian manifold with nonpositive sectional curvature has a fixed point (E. Cartan). If a compact, orientable Riemannian manifold admits a fixed-point-free 1-parameter group of isometries, then its †Pontryagin numbers vanish.

On a Riemannian manifold M, a transformation of M which preserves the Riemannian connection, or equivalently which commutes with covariant differentiation ∇ is called an **affine transformation**. Let $A(M)$ denote the group of all affine transformations of M. A transformation preserving the set of all geodesics is called a **projective transformation**. Let $P(M)$ denote the group of all projective transformations of M. A transformation preserving the angle between tangent vectors is called a **conformal transformation**. Let $C(M)$ denote the group of all conformal transformations. They are Lie transformation groups with respect to suitable topologies. Clearly, $I(M) \subset A(M) \subset P(M)$, $I(M) \subset C(M)$ (→ 191 G-Structures).

$A_0(M)$, the connected component of $A(M)$, is decomposed into a direct product according to the de Rham decomposition of M when M is complete and simply connected (J. Hano). If M is complete and irreducible, then $A(M) = I(M)$ except when M is a 1-dimensional Euclidean space. If M is complete and its restricted homogeneous holonomy group h_0 leaves no nonzero vectors, then $A_0(M) = I_0(M)$. If M is compact, then $A_0(M) = I_0(M)$ always.

If M is complete and has a parallel Ricci tensor, then the connected component $P_0(M) = A_0(M)$, unless M is a space of positive constant sectional curvature ($n > 2$) (Nagano, N. Tanaka, Y. Tashiro). If M is compact, simply connected, and has constant scalar curvature, then $P_0(M) = I_0(M)$, unless M is a sphere ($n > 2$) (K. Yamauchi).

Similarly to the case of $P(M)$, it is known that if M is complete and has a parallel Ricci tensor, then the connected component $C_0(M) = I_0(M)$, unless M is a sphere ($n > 2$) (Nagano). A conformal transformation remains conformal if the Riemannian metric g is changed conformally, namely, to $e^{2f}g$, f being any smooth function on M. A subset of $C(M, g)$ is called **essential** if it cannot be reduced to a subset of $I(M, \bar{g})$ for any metric $\bar{g}$ conformal to g. When M is compact, $C(M)$ or $C_0(M)$ is essential if and only if it is not compact. If $C_0(M)$ is essential, then M is conformally diffeomorphic to a sphere or a Euclidean space ($n > 2$) [12–15]. When M is compact and has constant scalar curvature and $C_0(M) \neq I_0(M)$, sufficient conditions for M to be isometric to a sphere have been obtained by S. I. Goldberg and S. Kobayashi, C. C. Hsiung, S. Ishihara, A. Lichnerowicz, Obata, S. Tanno, Tashiro, K. Yano, and others. For example, if $C_0(M)$ is essential, then M is a sphere [14]. In general, however, there are compact Riemannian manifolds with constant scalar curvature for which $C_0(M) \neq I_0(M)$ (N. Ejiri).

G. Spheres as Riemannian Manifolds

A Euclidean n-sphere S^n ($n \geqslant 2$) has the properties of a Riemannian manifold. It is a space of positive constant sectional curvature $1/r^2$ (r = radius) with respect to the natural Riemannian metric as a hypersurface of the Euclidean $(n+1)$-space $\mathbf{E}^{n+1}$. A sphere is characterized by the existence of solutions of certain differential equations on a Riemannian manifold. On a unit sphere S^n in $\mathbf{E}^{n+1}$, the eigenvalues of the †Laplace-Beltrami operator Δ on smooth functions are given by $0 < \lambda_1 < \ldots < \lambda_k < \ldots$, $\lambda_k = k(n+k-1)$. It is known that eigenfunctions f corresponding to λ_k, $\Delta f = \lambda_k f$, are the restrictions to S^n of harmonic homogeneous polynomial functions F of degree k on $\mathbf{E}^{n+1}$. On a compact Riemannian manifold M, if the Ricci curvature of M is not less than that of S^n, then the first eigenvalue $\bar{\lambda}_1$ of Δ on M satisfies $\bar{\lambda}_1 \geqslant \lambda_1 = n$ [16]. Conversely, under the same assumption on the Ricci curvature, if $\bar{\lambda}_1 = n$, then M is a sphere (Obata). On the other hand, if g is the standard metric on S^n, then $\Delta f = nf$ is equivalent to the system of differential equations

$$\nabla_j \nabla_i f + f g_{ji} = 0. \qquad (\mathrm{E}_1)$$

A complete Riemannian manifold M ($n \geqslant 2$) admits a nontrivial solution of (E_1) if and only if M is a sphere (Obata, Tashiro). In general, the restriction f to S^n of a harmonic homogeneous polynomial of degree k satisfies $\Delta f = k(n+k-1)f$ as well as a certain system (E_k) of differential equations of degree $k+1$ involving the Riemannian metric. For example,

$$\nabla_k \nabla_j \nabla_i f + 2 g_{ji} \nabla_k f + g_{ik} \nabla_j f + g_{kj} \nabla_i f = 0. \qquad (\mathrm{E}_2)$$

If a complete Riemannian manifold M admits a nontrivial solution of (E_k) for some integer $k \geqslant 2$, then M is locally isometric to a sphere (Obata, Tanno, S. Gallot [17]). The gradient of a solution of (E_1) is an infinitesimal conformal transformation and that of (E_2) is an infinitesimal projective transformation.

As the Kähler or quaternion Kähler version of (E_2), there is a system of differential equations characterizing the complex projective space or the quaternion projective space as a Kähler manifold (Obata, Tanno, D. E. Blair, Y. Maeda).

On a sphere, a Riemannian metric which is conformal to the standard metric and has the same scalar curvature as the standard one is always standard, namely, it has a positive constant sectional curvature [14].

H. Scalar Curvature

On a 2-dimensional Riemannian manifold M, the sectional curvature, the Ricci curvature, and the scalar curvature all coincide with the †Gaussian curvature, which is a function on M. If M is compact, by the †Gauss-Bonnet theorem the Gaussian curvature K of M must satisfy the following sign condition in terms of the †Euler characteristic $\chi(M)$:
if $\chi(M)>0$, then K is positive somewhere;
if $\chi(M)=0$, then K changes sign unless it is identically zero;
if $\chi(M)<0$, then K is negative somewhere.
This sign condition is also sufficient for a given function K to be the Gaussian curvature of some metric on M. More precisely, starting with a Riemannian metric with constant Gaussian curvature, one can say that a smooth function K is the Gaussian curvature of some metric conformally equivalent to the original metric if and only if K satisfies the foregoing sign condition [18].

H. Yamabe [19] announced that on every compact Riemannian manifold (M, g) of dimension $n \geqslant 3$, there exists a strictly positive function u such that the Riemannian metric $\bar{g}=u^{4/(n-2)}g$ has constant scalar curvature. N. S. Trudinger, however, pointed out that his original proof contains a gap in some cases. The problem reduces to the following nonlinear partial differential equation on a compact manifold M:

$$\bar{R}u^{(n+2)/(n-2)}=4\frac{n-1}{n-2}\Delta u+Ru,$$

where R is the scalar curvature of g and $\bar{R}$ a constant which should be the scalar curvature of $\bar{g}=u^{4/(n-2)}g$ (→ 183 Global Analysis). Nevertheless, Yamabe's original proof can be pushed to cover a large class of metrics with $\int_M R dM<0$. Furthermore, it has since been solved for a wider class: namely, if M is not conformally flat and $n \geqslant 6$, or if it is conformally flat and its fundamental group is finite, then the problem has been solved affirmatively [20].

On the other hand, any smooth function on a compact manifold M of dimension $n \geqslant 3$ that is negative somewhere is the scalar curvature of some metric on M. In particular, on a compact manifold ($n \geqslant 3$) there always exists a Riemannian metric with constant negative scalar curvature [18]. Any smooth function can be the scalar curvature if and only if M admits a metric of constant positive scalar curvature. The foregoing results show that there is no topological obstruction to the existence of metrics with negative scalar curvature of a compact manifold of dimension $n \geqslant 3$.

For positive scalar curvature, there is a topological obstruction. A compact †spin structure (spin manifold) having nonvanishing †$\hat{A}$-genus cannot carry a Riemannian metric of positive scalar curvature. The existence of such a manifold has been shown. If a compact, simply connected manifold M of dimension $n \geqslant 5$ is not a spin manifold, then there exists a Riemannian metric of positive scalar curvature. Furthermore, if M is a spin manifold and spin †cobordant to M' with positive scalar curvature, then M carries a Riemannian metric of positive scalar curvature [22]. A torus T^n cannot carry a metric of positive scalar curvature. In fact, any metric of nonnegative scalar curvature on T^n must be flat [22].

Let K_g and R_g denote the sectional curvature and the scalar curvature, respectively, of a Riemannian metric g. Then the following are known for a compact manifold M of dimension $\geqslant 3$: If M carries a metric g with $K_g \leqslant 0$, then it carries no metric with $R>0$. If M carries a metric g with $K_g<0$, then it carries no metric with $R \geqslant 0$. If M carries metrics g_1, g_2 with $K_{g_1} \leqslant 0$ and $R_{g_2} \geqslant 0$, then both metrics are flat [22].

If the assignment of the scalar curvature to a Riemannian metric is viewed as a mapping of a space of Riemannian metrics into a space of functions on a manifold M, then locally it is almost always surjective when M is compact (A. E. Fischer and J. E. Marsden, O. Kobayashi, J. Lafontaine).

I. Ricci Curvature and Einstein Metrics

In this paragraph the manifolds under consideration are assumed to be of dimension $n \geqslant 3$. The Ricci tensor (R_{ij}) is a symmetric tensor field of type (0, 2) on a Riemannian manifold. The problem of finding a Riemannian metric g which realizes a given Ricci tensor reduces to the one of solving a system of nonlinear second-order partial differential equations for g. The Bianchi identity (→ 417

Tensor Calculus)

$$g^{kj}\nabla_k R_{ji} - \frac{1}{2}\nabla_i R = 0$$

must be satisfied. There is a symmetric (0, 2)-tensor on $\mathbf{R}^n$ which cannot be the Ricci tensor for any Riemannian metric in a neighborhood of $0 \in \mathbf{R}^n$. However, if a C^∞ (or C^ω) symmetric tensor field (R_{ij}) of type (0, 2) is invertible at a point p, then in a neighborhood of p there exists a C^∞ (or C^ω) Riemannian metric g such that (R_{ij}) is the Ricci tensor of g [24].

The positivity of the Ricci curvature on a Riemannian manifold puts rather strong restrictions on the topology of the manifold (→ 178 Geodesics). However, nonnegative Ricci curvature and positive Ricci curvature are not too far from each other. If, on a complete Riemannian manifold M with nonnegative Ricci curvature, there is a point at which the Ricci curvature is positive, then there exists a complete metric on M with positive Ricci curvature [25–27].

If a Riemannian manifold (M, g) is an Einstein space, then g is called an **Einstein metric** on the manifold M. Let v_g denote the volume element determined by g. When M is compact, $\mathscr{M}$ denotes the space of Riemannian metrics on M with total volume 1. The integral of the scalar curvature $\mathscr{G}(g) = \int_M R_g v_g$ is a functional on $\mathscr{M}$. The critical points of $\mathscr{G}$ are Einstein metrics (D. Hilbert). Let $\mathscr{M}_1 (\subset \mathscr{M})$ denote the space of metrics with constant scalar curvature. Then if $\mathscr{G}$ is restricted to $\mathscr{M}_1$, then the †nullity and †coindex at the critical point are finite [28, 29].

An Einstein metric is always real analytic in some coordinate system. In particular, if two simply connected Einstein spaces have neighborhoods on which metrics are isometric, then they are isometric [30]. Though S^n with standard Riemannian metric is a typical example of an Einstein space, S^{4k+3} $(k \geqslant 1)$ carries an Einstein metric that is not standard [31].

References

[1] J. A. Wolf, Spaces of constant curvature, Publish or Perish, 1977.
[2] M. Berger, Les variétés riemanniennes homogènes simplement connexes à courbure strictment positive, Ann. Scuola Norm. Sup. Pisa, (3) 15 (1961), 179–246.
[3] N. R. Wallach, Compact homogeneous Riemannian manifolds with strictly positive curvature, Ann. Math., (2) 96 (1972), 277–295.
[4] L. Berard-Bergery, Les variétés riemanniennes homogènes simplement connexes de dimension impaire à courbure strictment positive, J. Math. Pures Appl., (9) 55 (1976), 47–67.
[5] S. Mori, Projective manifolds with ample tangent bundles, Ann. Math., (2) 110 (1979), 593–606.
[6] Y. T. Siu and S. T. Yau, Compact Kähler manifolds of positive bisectional curvature, Inventiones Math., 59 (1980), 189–204.
[7] G. de Rham, Sur la reductibilité d'un espace de Riemann, Comment. Math. Helv., 26 (1952), 328–344.
[8] M. Berger, Sur les groupes d'holonomie homogène des variétés à connexion affine et des variétés riemanniennes, Bull. Soc. Math. France, 83 (1955), 279–330.
[9] J. Simons, On the transitivity of holonomy systems, Ann. Math., (2) 76 (1962), 213–234.
[10] S. Kobayashi, Transformation groups in differential geometry, Erg. Math., 70, 1972.
[11] K. Yano, The theory of Lie derivatives and its applications, North-Holland, 1957.
[12] J. Lelong-Ferrand, Transformations conformes et quasi conformes des variétés riemanniennes compactes (démonstration de la conjecture de Lichnerowicz), Acad. Roy. Belg., Cl. Sc. Mém. Colloq. 39, no. 5 (1971), 1–44.
[13] A. J. Ledger and M. Obata, Compact Riemannian manifolds with essential group of conformorphisms, Trans. Amer. Math. Soc., 150 (1970), 645–651.
[14] M. Obata, The conjectures on conformal transformations of Riemannian manifolds, J. Differential Geometry, 6 (1971), 247–258.
[15] D. V. Alekseeviskiĭ, Groups of conformal transformations of Riemann spaces, Math. USSR-Sb., 18 (1972), 285–301.
[16] A. Lichnerowicz, Géométrie des groupes de transformations, Dunod, 1958.
[17] S. Gallot, Équations différentielles caractéristiques de la sphère, Ann. Sci. Ecole Norm. Sup., 12 (1979), 235–267.
[18] J. L. Kazdan and F. W. Warner, Existence and conformal deformation of metrics with prescribed Gaussian and scalar curvature, Ann. Math., (2) 101 (1975), 317–331.
[19] H. Yamabe, On a deformation of Riemannian structures on compact manifolds, Osaka Math. J., 12 (1960), 21–37.
[20] T. Aubin, Equations différentielles non linéaires et problème de Yamabe concernant la courbure scalaire, J. Math. Pure Appl., 55 (1976), 269–296.
[21] N. Hitchin, Harmonic spinors, Advances in Math., 14 (1974), 1–55.
[22] M. Gromov and L. B. Lawson, The classification of simply connected manifold of positive scalar curvature, Ann. Math., (2) 111 (1980), 423–434.
[23] R. Schoen and S. T. Yau, The structure

of manifolds with positive scalar curvature, Manuscripta Math., 28 (1979), 159–183.
[24] D. DeTurck, Existence of metrics with prescribed Ricci curvature: Local theory, Inventiones Math., 65 (1981), 179–207.
[25] T. Aubin, Metriques riemanniennes et courbure, J. Differential Geometry, 4 (1970), 383–424.
[26] P. Ehrlich, Metric deformations of curvature I: local convex deformations, Geometriae Dedicata, 5 (1976), 1–23.
[27] J. P. Bourguignon, Ricci curvature and Einstein metrics, Lecture notes in math. 838, Springer, 1981, 42–63.
[28] Y. Muto, On Einstein metrics, J. Differential Geometry, 9 (1974), 521–530.
[29] N. Koiso, On the second derivative of the total scalar curvature, Osaka J. Math., 16 (1979), 413–421.
[30] D. M. DeTurck and J. L. Kazdan, Some regularity theorems in Riemannian geometry, Ann. Sci. Ecole Norm. Sup., 14 (1981), 249–260.
[31] J. E. D'Atri and W. Ziller, Naturally reductive metrics and Einstein metrics on compact Lie groups, Mem. Amer. Math. Soc., 18 (1979).
See also references to 80 Connections, 105 Differentiable Manifolds, 109 Differential Geometry, 111 Differential Geometry of Curves and Surfaces, 178 Geodesics, 191 G-Structures, 365 Riemannian Submanifolds, 417 Tensor Calculus.

365 (VII.13) Riemannian Submanifolds

A. Introduction

If an †immersion (or an †embedding) f of a †Riemannian manifold (M, g) into a Riemannian manifold $(\tilde{M}, \tilde{g})$ satisfies the condition $f^*\tilde{g} = g$, then f is called an **isometric immersion** (or embedding) and M is called a **Riemannian submanifold** of $\tilde{M}$. In this article, $f(M)$ will be identified with M except where there is danger of confusion. Suppose $\dim M = n$ and $\dim \tilde{M} = n+p$. Then the †bundle $F(M)$ of orthonormal tangent frames of M, the bundle $F_\nu(M)$ of orthonormal normal frames of M, and their †Whitney sum $F(M) \oplus F_\nu(M)$ are †principal fiber bundles over M with †structure groups $O(n)$, $O(p)$, and $O(n) \times O(p)$, respectively. These are subbundles of the restriction to M of the bundle $F(\tilde{M})$ of orthonormal frames of $\tilde{M}$. The vector bundles associated with $F(M)$, $F_\nu(M)$, and $F(M) \oplus F_\nu(M)$ are, respectively, the †tangent bundle $T(M)$, the †normal bundle $\nu(M)$, and their Whitney sum $T(M) \oplus \nu(M)$.

B. General Results for Immersibility

An n-dimensional real analytic Riemannian manifold can be locally isometrically embedded into any real analytic Riemannian manifold of dimension $n(n+1)/2$ (M. Janet (1926), E. Cartan (1927)). The generalization to the C^∞ case is an open question even when the ambient space is Euclidean.

An n-dimensional compact C^r Riemannian manifold $(3 \leqslant r \leqslant \infty)$ can be isometrically embedded into an $(n(3n+11)/2)$-dimensional Euclidean space (J. F. Nash (1956)). An n-dimensional noncompact C^r Riemannian manifold $(3 \leqslant r \leqslant \infty)$ can be isometrically embedded into a $2(2n+1)(3n+7)$-dimensional Euclidean space (Nash (1956), R. E. Greene (1970)).

Let M be an n-dimensional Riemannian manifold with †sectional curvature K_M and $\tilde{M}$ an $(n+p)$-dimensional Riemannian manifold with sectional curvature $K_{\tilde{M}}$. Then M cannot be isometrically immersed into $\tilde{M}$ in the following cases:
(1) $p \leqslant n-2$ and $K_M < K_{\tilde{M}}$ (T. Otsuki (1954));
(2) $p \leqslant n-1$, $K_M \leqslant K_{\tilde{M}} \leqslant 0$, M is compact, and $\tilde{M}$ is complete and simply connected (C. Tompkins (1939), S. S. Chern and N. H. Kuiper (1952), B. O'Neill (1960));
(3) $p \leqslant n-1$, $K_M \leqslant 0$, $K_{\tilde{M}}$ is constant $(\leqslant 0)$, M is compact, and $\tilde{M}$ is complete and simply connected [2].

C. Fundamental Equations

Let $f:(M, g) \to (\tilde{M}, \tilde{g})$ be an isometric immersion. Let ∇ and $\tilde{\nabla}$ denote the †covariant differentiations with respect to the †Riemannian connections of M and $\tilde{M}$, respectively. For vector fields X and Y on M, the tangential component of $\tilde{\nabla}_X Y$ is equal to $\nabla_X Y$. Put

$$\sigma(X, Y) = \tilde{\nabla}_X Y - \nabla_X Y. \tag{1}$$

Then σ is a $\nu(M)$-valued symmetric $(0,2)$ tensor field on M, which is called the **second fundamental form** of M (or of f). For a normal vector ξ at $x \in M$, put $g(A_\xi X, Y) = \tilde{g}(\sigma(X, Y), \xi)$. Then A_ξ defines a symmetric linear transformation on $T_x(M)$, which is called the second fundamental form in the direction of ξ. The eigenvalues of A_ξ are called the **principal curvatures** in the direction of ξ. The connection on $\nu(M)$ induced from the Riemannian connection of $\tilde{M}$ is called the **normal connection** of M (or of f). Let $\nabla^\perp$ denote the covariant differentiation with respect to the normal connection. For a tangent vector field X and a normal vector field ξ on M, the tangential (resp. normal) component of $\tilde{\nabla}_X\xi$ is equal to $-A_\xi X$ (resp. $\nabla_X^\perp\xi$), that is to say, the relation

$$\tilde{\nabla}_X\xi=-A_\xi X+\nabla_X^\perp\xi \tag{2}$$

holds. (1) is called the **Gauss formula**, and (2) is called the **Weingarten formula**.

Let R, $\tilde{R}$, and $R^\perp$ be the †curvature tensors of ∇, $\tilde{\nabla}$, and $\nabla^\perp$, respectively. Then the †integrability condition for (1) and (2) implies

$$\tilde{R}(X,Y)Z=R(X,Y)Z+A_{\sigma(X,Z)}Y-A_{\sigma(Y,Z)}X$$
$$+(\nabla'_X\sigma)(Y,Z)-(\nabla'_Y\sigma)(X,Z) \tag{3}$$

for vector fields X, Y, Z tangent to M, where ∇' denotes covariant differentiation with respect to the connection in $T(M)\oplus\nu(M)$. (3) is called the equation of Gauss and Codazzi. More precisely, the tangential component of (3) is given by the **equation of Gauss** and the normal component of (3) is given by the **equation of Codazzi**. Similarly, for vector fields ξ and η normal to M, the relation

$$\tilde{g}(\tilde{R}(X,Y)\xi,\eta)=\tilde{g}(R^\perp(X,Y)\xi,\eta)$$
$$-g([A_\xi,A_\eta]X,Y) \tag{4}$$

holds, which is called the **equation of Ricci**. Formulas (1)–(4) are the fundamental equations for the isometric immersion $f:M\to\tilde{M}$.

As a particular case, suppose $\tilde{M}$ is a †space form of constant curvature c. Then the equations of Gauss, Codazzi, and Ricci reduce respectively to

$$R(X,Y)Z=c(g(Y,Z)X-g(X,Z)Y)+A_{\sigma(Y,Z)}X$$
$$-A_{\sigma(X,Z)}Y, \tag{3_c)$_t$}$$

$$(\nabla'_X\sigma)(Y,Z)-(\nabla'_Y\sigma)(X,Z)=0, \tag{3_c)$_n$}$$

$$\tilde{g}(R^\perp(X,Y)\xi,\eta)=g([A_\xi,A_\eta]X,Y). \tag{4_c}$$

Conversely, let (M,g) be an n-dimensional simply connected Riemannian manifold, and suppose there is given a p dimensional †Riemannian vector bundle $\nu(M)$ over M with curvature tensor $R^\perp$ and a $\nu(M)$-valued symmetric $(0,2)$ tensor field σ on M. For a †cross section ξ of $\nu(M)$, define A_ξ by $g(A_\xi X,Y)=\langle\sigma(X,Y),\xi\rangle$, where $\langle\ ,\ \rangle$ is the fiber metric of $\nu(M)$. If they satisfy $(3_c)_t$, $(3_c)_n$, and (4_c), then M can be immersed isometrically into an $(n+p)$-dimensional complete and simply connected space form $M^{n+p}(c)$ of curvature c in such a way that $\nu(M)$ is the normal bundle and σ is the second fundamental form. Moreover, such an immersion is unique up to an †isometry of $M^{n+p}(c)$.

Let $(e_A)_{1\leqslant A\leqslant n+p}$ be a local cross section of $F(\tilde{M})$ such that its restriction to M gives a local cross section of $F(M)\oplus F_\nu(M)$, and let (ω^A) be its dual. Then $f^*\omega^\alpha=0$ for $n+1\leqslant\alpha\leqslant n+p$. Let $(\tilde{\omega}_B^A)_{1\leqslant A,B\leqslant n+p}$ and $(\tilde{\Phi}_B^A)_{1\leqslant A,B\leqslant n+p}$ be the †connection form and the †curvature form of $\tilde{M}$ with respect to (e_A), and put $\omega_B^A=f^*\tilde{\omega}_B^A$. Then $(\omega_j^i)_{1\leqslant i,j\leqslant n}$ is the connection form of M with respect to $(e_i)_{1\leqslant i\leqslant n}$. $(\omega_i^\alpha)_{1\leqslant i\leqslant n<\alpha\leqslant n+p}$ gives the second fundamental form, that is,

$$\sigma(e_i,e_j)=\sum\omega_i^\alpha(e_j)e_\alpha. \tag{1'}$$

Put $\omega_i^\alpha=\sum h_{ij}^\alpha\omega^j$. Then (h_{ij}^α) is the matrix representing the symmetric linear transformation A_{e_α} with respect to (e_i), that is,

$$A_{e_\alpha}e_i=\sum h_{ij}^\alpha e_j. \tag{2'}$$

Moreover, $(\omega_\beta^\alpha)_{n+1\leqslant\alpha,\beta\leqslant n+p}$ is the connection form of the normal connection with respect to $(e_\alpha)_{n+1\leqslant\alpha\leqslant n+p}$. Let $(\Phi_j^i)_{1\leqslant i,j\leqslant n}$ and $(\Phi_\beta^\alpha)_{n+1\leqslant\alpha,\beta\leqslant n+p}$ be the curvature forms of (ω_j^i) and (ω_β^α), respectively. Then the equations of Gauss, Codazzi, and Ricci are given respectively by

$$f^*\tilde{\Phi}_j^i=\Phi_j^i+\sum_{\alpha=n+1}^{n+p}\omega_\alpha^i\wedge\omega_j^\alpha\quad(1\leqslant i,j\leqslant n), \tag{3')$_t$}$$

$$f^*\tilde{\Phi}_i^\alpha=d\omega_i^\alpha+\sum_{A=1}^{n+p}\omega_A^\alpha\wedge\omega_i^A\quad(1\leqslant i\leqslant n<\alpha\leqslant n+p), \tag{3')$_n$}$$

$$f^*\tilde{\Phi}_\beta^\alpha=\Phi_\beta^\alpha+\sum_{k=1}^{n}\omega_k^\alpha\wedge\omega_\beta^k\quad(n+1\leqslant\alpha,\beta\leqslant n+p). \tag{4'}$$

D. Basic Notions

Let M be a Riemannian submanifold of $\tilde{M}$. A point $x\in M$ is called a **geodesic point** if $\sigma=0$ at x. If every point of M is a geodesic point, then M is called a **totally geodesic submanifold** of $\tilde{M}$. M is a totally geodesic submanifold of $\tilde{M}$ if and only if every geodesic of M is a geodesic of $\tilde{M}$.

A mapping $\mathfrak{h}:M\to\nu(M)$ defined by $x\to\frac{1}{n}\sum_{i=1}^n\sigma(e_i,e_i)$ is independent of the choice of an orthonormal basis (e_i). $\mathfrak{h}$ is called the **mean curvature vector** and $\|\mathfrak{h}\|$ is called the **mean curvature**. M is called a **minimal submanifold** of $\tilde{M}$ if $\mathfrak{h}\equiv0$ (→ 275 Minimal Submanifolds).

A point $x\in M$ is called an **umbilical point** if $\sigma=g\otimes\mathfrak{h}$ at x. $x\in M$ is an umbilical point if and only if A_ξ is proportional to the identity transformation for all $\xi\in\nu_x(M)$. If every point of M is an umbilical point, then M is called a **totally umbilical submanifold** of $\tilde{M}$.

A point $x\in M$ is called an **isotropic point** if $\|\sigma(X,X)\|/\|X\|^2$ does not depend on $X\in T_x(M)$. If every point of M is an isotropic point, then M is called an **isotropic submanifold** of $\tilde{M}$. It is clear that an umbilical point is an isotropic point.

$\mu(x)=\dim\bigcap_{\xi\in\nu_x(M)}\ker A_\xi$ is called the **index of relative nullity** at $x\in M$.

E. Rigidity

An isometric immersion $f: M \to \tilde{M}$ is said to be **rigid** if it is unique up to an isometry of $\tilde{M}$, that is, if $f': M \to \tilde{M}$ is another isometric immersion, then there exists an isometry φ of $\tilde{M}$ such that $f' = \varphi \circ f$. If $f: M \to \tilde{M}$ is rigid, then every isometry of M can be extended to an isometry of $\tilde{M}$.

An isometric immersion $f: M \to M^{n+1}(c)$ of an n-dimensional Riemannian manifold into an $(n+1)$-dimensional complete and simply connected space form is rigid in each of the following cases:
(1) $n=2$, $c=0$, and M is compact and of positive curvature (S. Cohn-Vossen (1929)).
(2) The index of relative nullity is $\leqslant n-3$ at each point (R. Beez (1876); → [8]).
(3) $n \geqslant 5$, $c>0$, M is complete, and the index of relative nullity is $\leqslant n-2$ at each point (D. Ferus (1970)).
(4) $n \geqslant 4$, $c \neq 0$, and the †scalar curvature of M is constant ($\neq n(n-1)c$) (C. Harle (1971)).

A generalization of (2) for the case of higher codimension was obtained by C. Allendoerfer (1939). Various rigidity conditions have been studied by S. Dolbeault-Lemoine, R. Sacksteder, E. Kaneda and N. Tanaka, and others.

F. Totally Geodesic and Totally Umbilical Submanifolds

A totally geodesic submanifold of a space form is also a space form of the same curvature. Totally geodesic submanifolds of compact †symmetric spaces of rank 1 were completely classified by J. A. Wolf (1963), and totally geodesic submanifolds of symmetric spaces of rank $\geqslant 2$ were studied by Wolf and B. Y. Chen and T. Nagano [4].

Let $f: M \to M^{n+p}(\tilde{c})$ be a totally umbilical immersion of an n-dimensional Riemannian manifold M into an $(n+p)$-dimensional space form. Then M is a space form $M^n(c)$ with $c \geqslant \tilde{c}$, and $f(M)$ is contained in a certain $(n+1)$-dimensional totally geodesic submanifold $M^{n+1}(\tilde{c})$ of $M^{n+p}(\tilde{c})$. If $\tilde{c}>0$, then $f(M)$ is locally a hypersphere; if $\tilde{c}=0$, then $f(M)$ is locally a hyperplane or a hypersphere; if $\tilde{c}<0$, then $f(M)$ is locally a geodesic sphere, a horosphere, or a parallel hypersurface of a totally geodesic hypersurface [2].

G. Minimal Submanifolds

For general properties of minimal submanifolds → 275 Minimal Submanifolds.

There is no compact minimal submanifold in a simply connected Riemannian manifold with nonpositive sectional curvature (O'Neill (1960)). On the contrary, a sphere has plenty of compact minimal submanifolds.

For each positive integer s, an n-dimensional sphere of curvature $\dfrac{n}{s(s+n-1)}$ can be minimally immersed into a $\left\{(2s+n-1)\dfrac{(s+n-2)!}{s!(n-1)!}-1\right\}$-dimensional unit sphere and the immersion is rigid if $n=2$ or $s \leqslant 3$ (E. Calabi (1967), M. do Carmo and N. Wallach (1971)).

Among all n-dimensional compact minimal submanifolds of an $(n+p)$-dimensional unit sphere, the totally geodesic submanifold is isolated in the sense that it is characterized by each of the following conditions:
(1) sectional curvature $> \dfrac{n}{2(n+1)}$ (T. Itoh (1978)),
(2) Ricci curvature $> n-2$ (N. Ejiri (1979)),
(3) scalar curvature $> n(n-1) - \dfrac{n}{2-1/p}$ (J. Simons (1968)).

H. Submanifolds of Constant Mean Curvature

A manifold of constant mean curvature is a solution to a variational problem. In particular, with respect to any volume-preserving variation of a domain D in a Euclidean space, the mean curvature of $M = \partial D$ is constant if and only if the volume of M is critical.

The interesting question "If the mean curvature of an isometric immersion $f: M \to M^{n+1}(c)$ of an n-dimensional compact Riemannian manifold M into an $(n+1)$-dimensional space form $M^{n+1}(c)$ is constant, is M a sphere?" has not yet been completely solved, where $M^{n+1}(c)$ denotes a Euclidean space, a hyperbolic space, or an open hemisphere according as $c=0$, <0, or >0. The answer is affirmative in the following cases: (1) $\dim M = 2$, and the †genus of M is zero (H. Hopf (1951), Chern (1955)). (2) f is an embedding (A. D. Alexandrov (1958); → [8]).

These results remain true even if the assumption "the mean curvature is constant" is replaced by the weaker condition "the principal curvatures $k_1 \geqslant \ldots \geqslant k_n$ satisfy a relation $\varphi(k_1, \ldots, k_n)=0$ such that $\partial\varphi/\partial k_i > 0$."

Unlike an open hemisphere, a sphere S^{n+1} admits many compact hypersurfaces of constant mean curvature, among which totally umbilical hypersurfaces and the product of two spheres are the only ones with nonnegative sectional curvature (B. Smyth and K. Nomizu (1969)).

A nonnegatively or nonpositively curved complete surface of nonzero constant mean curvature in a 3-dimensional space form $M^3(c)$ is either a sphere or a †Clifford torus if $c>0$ and is either a sphere or a right circular cylinder if $c \leqslant 0$ (T. Klotz and R. Osserman (1966–1967), D. Hoffman (1973)).

I. Isoparametric Hypersurfaces

A hypersurface M of $\tilde{M}$ is said to be **isoparametric** if M is locally defined as the †level set of a function f on (an open set of) $\tilde{M}$ with property

$$df \wedge d\|df\|^2 = 0 \quad \text{and} \quad df \wedge d(\Delta f) = 0.$$

A hypersurface M of a complete and simply connected space form $M^{n+1}(c)$ is isoparametric if and only if M has locally constant principal curvatures (Cartan). If $c \leqslant 0$, M has at most two distinct principal curvatures (Cartan). If $c > 0$, the number of distinct principal curvatures of M is 1, 2, 3, 4, or 6 (H. Münzner (1980)). If $c = 0$, then M is locally $S^k \times E^{n-k}$, and if $c < 0$, then M is locally E^n or $S^k \times H^{n-k}$ (Cartan). Isoparametric hypersurfaces of S^{n+1} having at most three distinct principal curvatures were completely classified by Cartan. R. Takagi, T. Takahashi, and H. Ozeki and M. Takeuchi obtained several results for isoparametric hypersurfaces of S^{n+1} with four or six distinct principal curvatures [7].

If a subgroup of the isometry group of $M^{n+1}(c)$ acts transitively on M, then M is isoparametric. The converse is true if $c \leqslant 0$, or if $c > 0$ and M has at most three distinct principal curvatures (Cartan), but not true in general (Ozeki and Takeuchi [7]).

J. Isometric Immersions between Space Forms

Let $f: M^n(c) \to M^{n+p}(\tilde{c})$ be an isometric immersion of an n-dimensional space form into an $(n+p)$-dimensional space form.

(1) If $n=2$, $p=1$, $c>0$, $c \geqslant \tilde{c}$, $M^2(c)$ is complete, and $M^3(\tilde{c})$ is complete and simply connected, then f is totally umbilical (H. Liebmann (1901); → [2]).

(2) If $n=2$, $p=1$, $c<0$, $c<\tilde{c}$, $M^2(c)$ is complete, and $M^3(\tilde{c})$ is complete and simply connected, then f does not exist (D. Hilbert (1901); → [2]).

(3) If $n=2$, $p=1$, $c=0<\tilde{c}$, $M^2(0)$ is complete, and $M^3(\tilde{c})$ is complete and simply connected, then there exist infinitely many f (L. Bianchi (1896); → [2]).

(4) If $n=2$, $p=1$, $c=0>\tilde{c}$, $M^2(0)$ is complete, and $M^3(\tilde{c})$ is complete and simply connected, then $f(M^2(0))$ is either a horosphere or a set of points at a fixed distance from a geodesic (J. Volkovand S. Vladimirova, S. Sasaki; → [2]).

(5) If $n \geqslant 3$, $p=1$, and $c>\tilde{c}$, then f is totally umbilical.

(6) If $p=1$, $c=\tilde{c}=0$, $M^n(0)$, is complete, and $M^{n+1}(0)$ is complete and simply connected, then f is cylindrical (A. Pogorelov (1956), P. Hartman and L. Nirenberg (1959), and others).

(7) If $p \leqslant n-1$, $c=\tilde{c}>0$, and both $M^n(c)$ and $M^{n+p}(\tilde{c})$ are complete, then f is totally geodesic (D. Ferus (1975)).

(8) If $p \leqslant n-1$, $\tilde{c}>c>0$, $M^n(c)$ is complete, and $M^{n+p}(\tilde{c})$ is complete and simply connected, then f does not exist (J. Moore (1972)).

K. Homogeneous Hypersurfaces

Let M be an n-dimensional †homogeneous Riemannian manifold which is isometrically immersed into an $(n+1)$-dimensional complete and simply connected space form $M^{n+1}(c)$.

(1) If $c=0$, then M is isometric to $S^k \times E^{n-k}$ (S. Kobayashi (1958), Nagano (1960), Takahashi [9]).

(2) If $c>0$, then M is isometric to E^2 or else is given as an orbit of a subgroup of the isometry group of $M^{n+1}(c)$ (W. Y. Hsiang, H. B. Lawson, Takagi; → [7]).

(3) If $c<0$, then M is isometric to E^n, $S^k \times H^{n-k}$, or a 3-dimensional group manifold

$$B = \left\{ \begin{pmatrix} e^t & 0 & x \\ 0 & e^{-t} & y \\ 0 & 0 & 1 \end{pmatrix} \middle| x, y, t \in \mathbf{R} \right\}$$

with the metric $ds^2 = e^{-2t}dx^2 + e^{2t}dy^2 + dt^2$ (Takahashi (1971)). Each of the hypersurfaces above except E^2 in (2) and B in (3) is given as an orbit of a certain subgroup of the isometry group of $M^{n+1}(c)$.

L. Kähler Submanifolds

A †complex submanifold of a †Kähler manifold is a Kähler manifold with respect to the induced Riemannian metric. A complex analytic and isometric immersion of a Kähler manifold (M, J, g) into a Kähler manifold $(\tilde{M}, \tilde{J}, \tilde{g})$ is called a **Kähler immersion**, and M is called a **Kähler submanifold** of $\tilde{M}$. A Kähler submanifold is a minimal submanifold. A compact Kähler submanifold M of a Kähler manifold $\tilde{M}$ can never be homologous to 0, that is, there exists no submanifold M' of $\tilde{M}$ such that $M = \partial M'$. If $[M] \in H_*(\tilde{M}, Z)$ denotes the †homology class represented by a Kähler submanifold M of $\tilde{M}$, then $\mathrm{vol}(M) \leqslant \mathrm{vol}(M')$ holds for any submanifold $M' \in [M]$ with equality if and only if M' is a Kähler submanifold (W. Wirtinger (1936)).

A Kähler manifold of constant †holomorphic sectional curvature is called a **complex space form**. An n-dimensional complete and simply connected complex space form is either $P_n(C)$, C^n, or D_n. Every Kähler submanifold of a complex space form is rigid (Calabi [10]).

Kähler immersions of complex space forms

into complex space forms were completely determined by Calabi [10] and by H. Nakagawa and K. Ogiue (1976).

C^n (resp. D_n) is the only †Hermitian symmetric space that can be immersed in C^m (resp. D_m) as a Kähler submanifold (Nakagawa and Takagi [11]), and Kähler immersions of Hermitian symmetric spaces into $P_m(C)$ were precisely studied by Nakagawa, Y. Sakane, Takagi, Takeuchi, and others. More generally, Kähler immersions of homogeneous Kähler manifolds into $P_m(C)$ were determined by Takeuchi (1978).

$Q_n=\{[z_i]\in P_{n+1}(C)\mid \sum z_i^2=0\}$ in $P_{n+1}(C)$ is the only Einstein-Kähler hypersurface of a complex space form that is not totally geodesic (B. Smyth (1967), S. S. Chern (1967)). The result remains true even if "Einstein" is replaced by "parallel Ricci tensor" (Takahashi (1967)). Besides linear subspaces, Q_n is the only Einstein-Kähler submanifold of $P_m(C)$ that is a complete intersection (J. Hano (1975)).

Integral theorems and pinching problems with respect to various curvatures for compact Kähler submanifolds of $P_m(C)$ have been studied by K. Ogiue, S. Tanno (1973), S. T. Yau (1975), and others [12]. For example, if the holomorphic sectional curvature of $P_{n+p}(C)$ is 1, then each of the following is sufficient for an n-dimensional compact Kähler submanifold to be totally geodesic:
(1) holomorphic sectional curvature $>1/2$ (A. Ros (1985)),
(2) sectional curvature $>1/8$ (A. Ros and L. Verstraelen (1984)),
(3) Ricci curvature $>n/2$ [12],
(4) embedded and scalar curvature $>n^2$ (J. H. Cheng (1981)).

The index of relative nullity $\mu(x)$ of an n-dimensional complete Kähler submanifold M of $P_m(C)$ satisfies $\mathrm{Min}_{x\in M}\mu(x)=0$ or $2n$ (K. Abe (1973)).

M. Totally Real Submanifolds

An isometric immersion of a Riemannian manifold (M,g) into a Kähler manifold $(\tilde{M},\tilde{J},\tilde{g})$ satisfying $\tilde{J}T_x(M)\subset\nu_x(M)$ at each point $x\in M$ is called a **totally real immersion**, and M is called a **totally real submanifold** of $\tilde{M}$. A totally geodesic submanifold $P_n(R)$ in $P_n(C)$, $S^1\times S^1$ in $P_2(C)$ and an immersion $P_n(C)\to P_{n(n+2)}(C)$ defined by $[z_i]\to[z_i\bar{z}_j]$ give typical examples of totally real submanifolds.

N. Submanifolds with Planar Geodesics

A surface in E^3 whose geodesics are all plane curves is (a part of) a plane or sphere. More generally, let $f: M\to M^m(c)$ be an isometric immersion of M into a complete and simply connected space form $M^m(c)$. If the image of each geodesic of M is contained in a 2-dimensional totally geodesic submanifold of $M^m(c)$, then f is either a totally geodesic immersion, a totally umbilical immersion or a minimal immersion of a compact symmetric space of rank 1 by harmonic functions of degree 2; the last case occurs only when $c>0$ (S. L. Hong (1973), J. Little (1976), K. Sakamoto [14]).

A Kähler submanifold of a complete and simply connected complex space form with the same property as above is either a totally geodesic submanifold or a Veronese submanifold (a Kähler immersion of $P_n(C)$ into $P_{n(n+3)/2}(C)$) (K. Nomizu (1976)).

Submanifolds with the above property are closely related to isotropic submanifolds with $\nabla'\sigma=0$. Submanifolds with $\nabla'\sigma=0$ in symmetric spaces have been studied by Ferus, H. Naito, Takeuchi, K. Tsukada, and others.

O. Total Curvature

Let $f: M\to E^m$ be an isometric immersion of an n-dimensional compact Riemannian manifold M into a Euclidean space. Let $\nu_1(M)$ be the unit normal bundle, S^{m-1} the unit sphere centered at $O\in E^m$, and let $f_\perp:\nu_1(M)\to S^{m-1}$ be the parallel translation. Let ω and Ω be the †volume elements of $\nu_1(M)$ and S^{m-1}, respectively. Then for each $\xi\in\nu_1(M)$, $f_\perp^*\Omega=(\det A_\xi)\omega$ holds. As a generalization of the †total curvature for a space curve, the **total curvature** of the immersion f is defined as

$$\tau(f)=\frac{1}{\mathrm{vol}(S^{m-1})}\int_{\nu_1(M)}|f_\perp^*\Omega|$$

$$=\frac{1}{\mathrm{vol}(S^{m-1})}\int_{\nu_1(M)}|\det A_\xi|\omega.$$

If $\beta(M)$ is the least number of critical points of a †Morse function on M, then

$$\inf_f \tau(f)=\beta(M)\geqslant 2$$

holds (Chern and R. Lashof (1957, 1958), Kuiper (1958)). $\tau(f)=2$ if and only if f is an embedding and $f(M)$ is a convex hypersurface of some E^{n+1} in E^m (Chern and Lashof (1958)). If $\tau(f)<3$, then M is homeomorphic to S^n (Chern and Lashof (1958)). These results generalize theorems for space curves by W. Fenchel (1929), I. Fary (1949), J. Milnor (1950), and others.

An isometric immersion f which attains $\inf(f)$ is called a **minimum immersion** or a **tight immersion**. †Exotic spheres do not have minimum immersions (Kuiper (1958)). †R-

spaces have minimum immersions and, in particular, a minimum immersion of a symmetric R-space is a †minimal immersion into a hypersphere (Kobayashi and Takeuchi (1968)).

The **total mean curvature** of an isometric immersion $f: M \to E_m$ of an n-dimensional compact Riemannian manifold into a Euclidean space is, by definition,

$$\int_M \|\mathfrak{h}\|^n * 1.$$

It satisfies

$$\int_M \|\mathfrak{h}\|^n * 1 \geqslant \operatorname{vol}(S^n),$$

where S^n is the n-dimensional unit sphere. The equality holds if and only if f is totally umbilical (T. J. Willmore (1968), Chen [3]).

References

[1] S. Kobayashi and K. Nomizu, Foundations of differential geometry II, Interscience, 1969.
[2] M. Spivak, A comprehensive introduction to differential geometry III, IV, V, Publish or Perish, 1975.
[3] B. Y. Chen, Geometry of submanifolds, Dekker, 1973.
[4] B. Y. Chen and T. Nagano, Totally geodesic submanifolds of symmetric spaces I, II, III, Duke Math. J., 44 (1977), 745–755; 45 (1978), 405–425.
[5] S. S. Chern, M. doCarmo, and S. Kobayashi, Minimal submanifolds of a sphere with second fundamental form of constant length, Functional Analysis and Related Fields, Springer, 1970.
[6] A. D. Alexandrov, A characteristic property of spheres, Ann. Mat. Pura Appl., 58 (1962), 303–315.
[7] H. Ozeki and M. Takeuchi, On some types of isoparametric hypersurfaces I, II, Tôhoku Math. J., 27 (1975), 515–559; 28 (1976), 7–55.
[8] P. Ryan, Homogeneity and some curvature conditions for hypersurfaces, Tôhoku Math. J., 21 (1969), 363–388.
[9] T. Takahashi, Homogeneous hypersurfaces in spaces of constant curvature, J. Math. Soc. Japan, 22 (1970), 395–410.
[10] E. Calabi, Isometric imbedding of complex manifolds, Ann. Math., (2) 58 (1953), 1–23.
[11] H. Nakagawa and R. Takagi, On locally symmetric Kähler submanifolds in a complex projective space, J. Math. Soc. Japan, 28 (1976), 638–667.
[12] K. Ogiue, Differential geometry of Kähler submanifolds, Advances in Math., 13 (1974), 73–114.
[13] B. Y. Chen and K. Ogiue, On totally real submanifolds, Trans. Amer. Math. Soc., 193 (1974), 257–266.
[14] K. Sakamoto, Planar geodesic immersions, Tôhoku Math. J., 29 (1977), 25–56.
[15] D. Ferus, Totale Absolutkrümmung in Differentialgeometrie und Topologie, Springer, 1968.

366 (VIII.6) Riemann-Roch Theorems

A. General Remarks

The †Riemann-Roch theorem (abbreviation: R. R. theorem) is one of the most significant results in the classical theory of †algebraic functions of one variable. Let X be a compact †Riemann surface of †genus g, and let $D = \sum m_i P_i$ be a †divisor on X. We denote by $\deg D$ the degree of D, which is defined to be $\sum m_i$. The divisor D is said to be positive if $D \neq 0$ and $m_i \geqslant 0$ for all i. A nonzero †meromorphic function f on X determines a divisor $(f) = \sum a_i Q_i - \sum b_j R_j (a_i, b_j > 0)$, where the Q_i are the zeros of order a_i and the R_j are the poles of order b_j. The set of meromorphic functions f such that $(f) + D$ is positive, together with the constant $f = 0$, forms a finite-dimensional linear space $L(D)$ over $\mathbf{C}$. The R. R. theorem asserts that $\dim L(D) = \deg D - g + 1 + r(D)$, where $r(D)$ is a nonnegative integer determined by D. If K is the †canonical divisor of X, then $r(D) = \dim L(K - D)$ (→ 9 Algebraic Curves C; 11 Algebraic Functions D). (For the R. R. theorem for algebraic surfaces → 15 Algebraic Surfaces D.)

Generalizations of this important theorem to the case of higher-dimensional compact †complex manifolds were obtained by K. Kodaira, F. Hirzebruch, A. Grothendieck, M. F. Atiyah and I. M. Singer, and others. Let X be a compact complex manifold, B be a †complex line bundle over X, and $\mathcal{O}(B)$ be the †sheaf of germs of holomorphic cross sections of B. When B is determined by a divisor D of X, we have $H^0(X, \mathcal{O}(B)) \cong L(D)$. Hence a desirable generalization of the R. R. theorem will provide a description of $\dim_{\mathbf{C}} H^0(X, \mathcal{O}(B))$ in terms of quantities relating to the properties of X and B. Following this idea, various theorems of Riemann-Roch type have been obtained.

B. Hirzebruch's Theorem of R. R. Type

Keeping the notation given in Section A, we put $\chi(X, \mathcal{O}(B)) = \sum_q (-1)^q \dim H^q(X, \mathcal{O}(B))$.

Generally, if $\mathscr{F}$ is an arbitrary †coherent analytic sheaf on X, we can define $\chi(X, \mathscr{F})$ using the same formula (replacing $\mathcal{O}(B)$ by $\mathscr{F}$). The quantity $\chi(X, \mathscr{F})$ has simple properties in various respects. For example, if the sequence $0 \to \mathscr{F}' \to \mathscr{F} \to \mathscr{F}'' \to 0$ is exact, we have $\chi(X, \mathscr{F}) = \chi(X, \mathscr{F}') + \chi(X, \mathscr{F}'')$. If an †analytic vector bundle F depends continuously on auxiliary parameters, then $\chi(X, \mathcal{O}(F))$ remains constant. Let X be a projective algebraic manifold of complex dimension n. We consider the †Chern class $c = 1 + c_1 + \ldots + c_n$ of X and express it formally as the product $\prod_{i=1}^{n}(1+\gamma_i)$. Thus the ith Chern class c_i is expressed as the ith elementary symmetric function of $\gamma_1, \ldots, \gamma_n$. Consider the formal expression $T(X) = \prod_{i=1}^{n} \gamma_i / (1 - e^{-\gamma_i})$. $T(X)$ can be expanded as a formal power series in the γ_i, and each homogeneous term, being symmetric in the γ_i, can be expressed as a polynomial in $c_1, \ldots, c_n$, and thus determines a cohomology class of X. Similarly, we consider the formal expression of the Chern class of the vector bundle F as $1 + d_1 + \ldots + d_q = \prod_{j=1}^{q}(1+\delta_j)$, where q is the dimension of the fiber of F. We put $ch(F) = \sum_{j=1}^{q} e^{\delta_j}$. The formal series $ch(F)$ is also an element of the cohomology ring of X whose $(\nu+1)$st term consists of a 2ν-dimensional cohomology class. We call $ch(F)$ the †Chern character of F (→ 237 *K*-Theory B), and define $T(X, F)$ to be the value of $ch(F)\,T(X)$ at the †fundamental cycle X. (The multiplication $ch(F) \cdot T(X)$ is formal. $T(X, F)$ is determined by the term of dimension $2n$ alone.) $T(X, F)$ is called the **Todd characteristic** with respect to F. **Hirzebruch's theorem of R. R. type** asserts that $\chi(X, \mathcal{O}(F)) = T(X, F)$. In particular, when $n = 1$, $F = [D]$ (the line bundle determined by the divisor D), Hirzebruch's formula yields the classical R. R. theorem. If F satisfies the conditions for the vanishing theorem of cohomologies, the formula gives an estimate for $\dim H^0(X, \mathcal{O}(F))$ [11].

In 1963, Atiyah and Singer developed a theory on indices of elliptic differential operators on a compact orientable differentiable manifold and obtained a general result that includes the proof of Hirzebruch's theorem for an arbitrary compact complex manifold [4, 5] (→ 237 *K*-Theory H).

C. R. R. Theorem for Surfaces

If X is a compact complex surface, i.e., a compact complex manifold of dimension 2, then for complex line bundles F_1 and F_2, the intersection number $(F_1 F_2)$ is defined to be $c_1(F_1) \cup c_1(F_2)[X]$. The **R. R. theorem for a line bundle** F on X is stated as follows: $\chi(X, \mathcal{O}(F)) = (F^2)/2 - (KF)/2 + ((K^2) + c_2(X))/12$, where K is the canonical line bundle of X and $c_2(X)$ denotes the value at X of the 2nd Chern class of X, that is, the Euler number of X.

The Noether formula $\chi_X = ((K^2) + c_2(X))/12$ follows from the above identity. The R. R. theorem for surfaces is a powerful tool for the study of compact complex surfaces.

D. Grothendieck's Theorem of R. R. Type

Grothendieck took an entirely new point of view in generalizing Hirzebruch's theorem. The following is a description of his idea as reformulated by A. Borel and J.-P. Serre [8]. We consider a nonsingular quasiprojective algebraic variety X (→ 16 Algebraic Varieties) over a ground field of arbitrary characteristic. Namely, X is a closed subvariety of an open set in a projective space (over an algebraically closed ground field). Consider the group $K(X)$, which is the quotient of the free Abelian group generated by the equivalence classes of algebraic vector bundles over X modulo the subgroup generated by the elements of the form $F - F' - F''$, where F, F', F'' are classes of bundles such that there exists an exact sequence $0 \to F' \to F \to F'' \to 0$. A similar construction for the (equivalence classes of the) †coherent algebraic sheaves instead of the vector bundles yields another Abelian group $K'(X)$. It can be shown that $K(X)$ is isomorphic to $K'(X)$ by the correspondence $F \to \mathcal{O}(F)$ (= the sheaf of germs of regular cross sections of F). Addition in $K(X)$ is induced by the †Whitney sum of the bundles, and $K(X)$ has the structure of a ring with multiplication induced by the tensor product. For a vector bundle F, its Chern class $c(F) = 1 + c_1(F) + \ldots + c_q(F)$ (q = the dimension of the fiber) is defined as an element of the †Chow ring $A(X)$ with appropriate properties. ($A(X)$ is the ring of the rational equivalence classes of algebraic cycles on X, and $c_i(F)$ is the class of a cycles of codimension i.) We define $ch(F)$ as before. It can be shown that $c(F)$ and $ch(F)$ are determined by the image of F in $K(X)$, and we have $c(\xi + \eta) = c(\xi)c(\eta)$, $ch(\xi + \eta) = ch(\xi) + ch(\eta)$, $ch(\xi\eta) = ch(\xi)ch(\eta)$ ($\xi, \eta \in K(X)$). If we have a †proper morphism $f: Y \to X$ between quasiprojective algebraic varieties Y and X, we have homomorphisms $f^!: K(X) \to K(Y)$ and $f_!: K(Y) \to K(X)$. The former is defined by taking the induced vector bundle and the latter by the correspondence

$$\mathscr{F} \to \sum (-1)^q (\mathscr{R}^q f) \mathscr{F},$$

where $\mathscr{F}$ is a coherent algebraic sheaf on Y and $(\mathscr{R}^q f)\mathscr{F}$ is the qth †direct image of $\mathscr{F}$ under f. (Since f is proper, $(\mathscr{R}^q f)\mathscr{F}$ is coherent.) Between Chow rings we have homomorphisms $f^*: A(X) \to A(Y)$ and $f_*: A(Y) \to A(X)$, defined by taking inverse and direct images of cycles. With this notation, the theorem asserts that if X and Y are quasiprojective and $f: Y \to X$ is a proper morphism, then $f_*(ch(\xi)T(Y)) = ch(f_!(\xi))T(X)$. This is called **Grothendieck's theorem of R. R. type**. If X consists of a single point, the theorem gives Hirzebruch's theorem for algebraic bundles. Since algebraic and analytic theories of coherent sheaves on a complex projective space are isomorphic, this result covers Hirzebruch's theorem (→ 237 *K*-Theory).

The subgroup $R(Y)$ of $A(Y)$ given by $R(Y) = \{ch(\xi) \cdot T(Y) \mid \xi \in K(Y)\}$ is called the **Riemann-Roch group** of Y. Thus, using the notions developed by Grothendieck, the R. R. theorem can be expressed as follows: $R(Y)$ is mapped into $R(X)$ by a proper morphism $f: Y \to X$. Generalizations to †almost complex manifolds and †differentiable manifolds were made by Atiyah and Hirzebruch in this latter form [1]. One of the remarkable results is that an element of $R(Y)$ takes an integral value at the fundamental cycle. This theorem is obtained by taking X to be a single point.

E. R. R. Theorem for Singular Varieties

Let X be a projective variety over $\mathbf{C}$ and let $H_.(X)$ (resp. $H^{\cdot}(X)$) denote the singular homology (resp. cohomology) group with rational coefficients. Note that $K(X)$ may not agree with $K'(X)$ when X is singular. The R. R. theorem for X formulated by P. Baum, W. Fulton, and R. MacPherson [6] says that there exists a unique natural transformation $\tau: K'(X) \to H_.(X)$ such that (1) if $\xi \in K(X)$ and $\eta \in K'(X)$, then $\xi \otimes \eta \in K'(X)$ and $\tau(\xi \otimes \eta) = ch(\xi)(\tau(\eta))$; (2) whenever X is nonsingular, $\tau(\mathcal{O}_X) = T(X)(X)$. Note that the naturality of τ means that for any $f: X \to Y$ and any $\eta \in K'(X)$, $f_* \tau(\eta) = \tau(f_* \eta)$.

References

[1] M. F. Atiyah and F. Hirzebruch, Riemann-Roch theorems for differentiable manifolds, Bull. Amer. Math. Soc., 65 (1959), 276–281.
[2] M. F. Atiyah, R. Bott, and V. K. Patodi, On the heat equation and the index theorem, Inventiones Math., 19 (1973), 279–330.
[3] M. F. Atiyah, V. K. Patodi, and I. M. Singer, Spectral asymmetry and Riemann geometry, Bull. Lond. Math. Soc., 5 (1973), 229–234.
[4] M. F. Atiyah, G. B. Segal, and I. M. Singer, The index of elliptic operators, Ann. Math., (2) 87 (1968), 484–604 (I by Atiyah and Singer, II by Atiyah and Segal, III by Atiyah and Singer).
[5] M. F. Atiyah and I. M. Singer, The index of elliptic operators on compact manifolds, Bull. Amer. Math. Soc., 69 (1963), 422–433.
[6] P. Baum, W. Fulton, and R. MacPherson, Riemann-Roch for singular varieties, Publ. Math. Inst. HES, no. 45 (1975), 101–145.
[7] P. Berthelot, A. Grothendieck, L. Illusie, et al., Théorie des intersections et théorème de Riemann-Roch, SGA 6, Lecture notes in math. 225, Springer, 1971.
[8] A. Borel and J.-P. Serre, La théorème de Riemann-Roch, Bull Soc. Math. France, 86 (1958), 97–136.
[9] W. Fulton, Rational equivalence on singular varieties, Publ. Math. Inst. HES, no. 45 (1975), 147–167.
[10] W. Fulton and R. MacPherson, Intersecting cycles on an algebraic variety, Real and Complex Singularities, Sijthoff & Noordhoff, 1976, 179–197.
[11] F. Hirzebruch, Topological methods in algebraic geometry, Springer, third edition, 1966.
[12] K. Kodaira, On compact analytic surfaces I, Ann. Math., (2) 71 (1961), 111–152.
[13] R. MacPherson, Chern classes of singular varieties, Ann. Math., (2) 100 (1974), 423–432.
[14] B. Segre, Nuovi metodi e risultati nella geometria sulle varieta algebriche, Ann. Mat. (4) 35 (1953), 1–128.
[15] J. Todd, The geometrical invariants of algebraic loci, Proc. London Math. Soc. (2) 43 (1937), 127–138.

367 (XI.12) Riemann Surfaces

A. General Remarks

Riemann considered certain surfaces, now named after him, obtained by modifying in a suitable manner the domains of definition of multiple-valued †analytic functions on the complex plane in order to obtain single-valued functions defined on the surfaces. For example, consider the function $z = f(w) = w^2$, where w varies in the complex plane, and its inverse function $w = g(z) = \sqrt{z}$. Then $g(0) = 0$ and $g(\infty) = \infty$, whereas if $z \neq 0, \infty$, there exist two values of w satisfying $g(z) = w$. By setting $z = re^{i\theta}$ $(r > 0, 0 \leqslant \theta < 2\pi)$, the corresponding two values of w are $w_1 = \sqrt{r}\, e^{(\theta/2)i}$ and $w_2 = \sqrt{r}\, e^{(\theta/2 + \pi)i}$. Now consider how we should modify the complex z-plane so that we can obtain a single-valued function on the modified surface representing the same relationship between z and w. Let π_1 and π_2 be two copies of the complex plane. Delete the nonnegative real axes from π_1 and π_2 and patch them crosswise along the slits (Fig. 1). The surface R thus obtained is locally homeomorphic to the complex plane except for the origin and the point at infinity, and situations at the origin and the point at infinity are as indicated in Fig. 1. For $z \neq 0, \infty$, there are two points z_1 and z_2 in π_1 and π_2, respectively, with the same coordinate z. Let w_1 and w_2 correspond to z_1 and z_2, respectively. Then the function $w = \sqrt{z}$ becomes single-valued on R, and w_1 and w_2 are holomorphic functions of z_1 and z_2, respectively. The surface R is called the Riemann surface determined by $w = \sqrt{z}$.

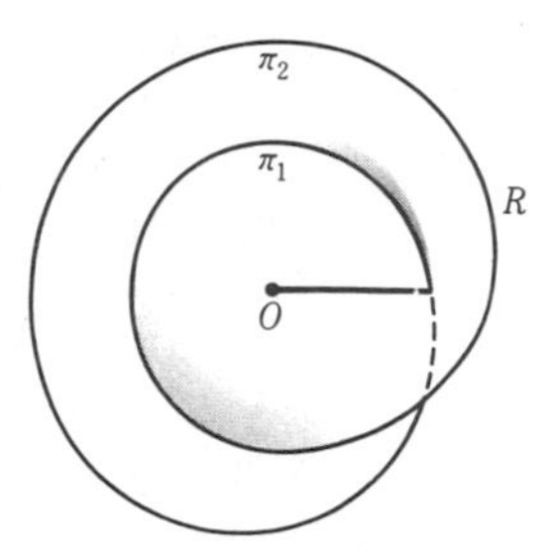

Fig. 1

Working from the idea illustrated by this example, H. Weyl and T. Radó gave rigorous definitions of Riemann surfaces. The usual definition nowadays is as follows: Let $\mathfrak{A}$ be a set of pairs (U, ψ_U) of open sets U in a †connected †Hausdorff space R and topological mappings ψ_U of U onto plane regions satisfying the following two conditions: (i) $R = \bigcup_{(U, \psi_U) \in \mathfrak{A}} U$; (ii) for each $(U_1, \psi_{U_1}), (U_2, \psi_{U_2}) \in \mathfrak{A}$ with $V = U_1 \cap U_2 \neq \varnothing$, $\psi_{U_1} \circ \psi_{U_2}^{-1}$ gives an (orientation-preserving) †conformal mapping of each connected component of $\psi_{U_2}(V)$ onto a corresponding one of $\psi_{U_1}(V)$. Two such sets $\mathfrak{A}_1$ and $\mathfrak{A}_2$ are equivalent, by definition, if $\mathfrak{A}_1 \cup \mathfrak{A}_2$ also satisfies conditions (i) and (ii). The equivalence class $(\mathfrak{A})$ of such $\mathfrak{A}$ is called a **conformal structure** (**analytic structure** or **complex structure**) on R (→ 72 Complex Manifolds). A pair $(R, (\mathfrak{A}))$ consisting of a connected Hausdorff space R and a conformal structure $(\mathfrak{A})$ is called a **Riemann surface**, with R its **base space** and $(\mathfrak{A})$ its **conformal structure**. (A Riemann surface in this sense is sometimes called an **abstract Riemann surface**.) It is a complex manifold of †complex dimension 1. For (U, ψ_U) in $\mathfrak{A} \in (\mathfrak{A})$, (U, ψ_U) (or sometimes U itself) is called an **analytic neighborhood**, and ψ_U is called a **local uniformizing parameter** (or simply a **local parameter**). In the remainder of this article we call R itself a Riemann surface.

From condition (i) it follows that a Riemann surface R is a real 2-dimensional †topological manifold. Moreover, by condition (ii) it can be deduced that R satisfies the †second countability axiom and consequently is a †surface and admits a †simplicial decomposition (T. Radó, 1925). It is also orientable (→ 410 Surfaces). Therefore R is a †locally compact metric space. It is not possible to define curve lengths on R compatible with the conformal structure, but since angles can be defined, R is considered to be a real 2-dimensional space with a †conformal connection. It is customary in the theory of functions to call R **closed** if it is compact and **open** otherwise. A plane region D is considered an open Riemann surface with the conformal structure $\mathfrak{A} = (D, 1: D \rightarrow D)$. A Riemann sphere is also considered to be a closed Riemann surface whose analytic neighborhoods are given by $\{U_1, \varphi\}$ and $\{U_2, 1/\varphi\}$, where $U_1 (U_2)$ is the domain corresponding to $\{|z| < 2\} (\{|z| > 1/2\} \cup \{\infty\})$ under the stereographic projection φ.

A function f on a Riemann surface is said to be meromorphic, holomorphic, or harmonic on R if $f \circ \psi_U^{-1}$ is †meromorphic, †holomorphic, or †harmonic in the usual sense on $\psi_U(U)$ for every analytic neighborhood (U, ψ_U). More generally, suppose that for mappings between plane regions we are given a property $\mathfrak{P}$ that is invariant under conformal mappings. A mapping T of a Riemann surface R_1 onto another Riemann surface R_2 is said to have the property $\mathfrak{P}$ if the mapping $\psi_{U_2} \circ T \circ \psi_{U_1}^{-1}$ of $\psi_{U_1}(U_1)$ into $\psi_{U_2}(U_2)$ has the property $\mathfrak{P}$ for every pair of analytic neighborhoods (U_1, ψ_{U_1}) nd (U_2, ψ_{U_2}) in R_1 and R_2, respectively. Thus such a mapping T may be conformal, †analytic, †quasiconformal, harmonic, etc. If there exists a one-to-one conformal mapping of a Riemann surface R_1 onto another Riemann surface R_2, then R_1 and R_2 are said to be **conformally equivalent**. Two such Riemann surfaces are sometimes identified with each other.

B. Covering Surfaces

One of the main themes of the theory of functions is the study of analytic mappings of a Riemann surface R into another Riemann surface R_0, i.e., the theory of covering surfaces of Riemann surfaces.

Suppose, in general, that there are two surfaces R and R_0 and a continuous open mapping T of R into R_0 such that the inverse image of a point in R_0 under T is an †isolated set in R. Then T is called an **inner transformation** in the sense of Stoïlow and $(R, R_0; T)$ a **covering surface** with R_0 its **basic surface** and T its **projection**. Often R is called simply a covering surface of R_0. A point p_0 with $p_0 =$

$T(p)$ is called the projection of p, and p is said to lie above p_0. In this case, there exist surface coordinates (U, ψ) and (U_0, ψ_0) at p and p_0, respectively, such that $\psi(U)=\{z \mid |z|<1\}$, $\psi(p)=0$, $\psi_0(U_0)=\{w \mid |w|<1\}$, $\psi_0(p_0)=0$, and $w=(\psi_0 \circ T \circ \psi^{-1})(z)=z^n$, with the positive integer n independent of the choice of coordinate neighborhoods. If $n>1$, then p is called a **branch point**, n the **multiplicity**, and $n-1$ the **degree of ramification**. The set of all branch points forms an at most countably infinite set of isolated points. If there is no branch point, then the covering surface $(R, R_0; T)$ and the projection T are said to be **unramified**. For a given curve C_0 in R_0 and a point p in R lying above the initial point of C_0, a curve C in R with p its initial point satisfying $T(C)=C_0$ is called the **prolongation** along C_0 (or the **lift** of C_0) starting from p. If any proper subarc of C_0 sharing the initial point with C_0 admits a prolongation along itself starting from p but C_0 does not admit a prolongation along itself starting from p, then R is said to have a **relative boundary** above the endpoint of C_0. A covering surface without a relative boundary is called **unbounded**. A †simply connected surface R^∞ that is an unramified unbounded covering surface of R_0 is said to be a **universal covering surface** of R_0. The universal covering surface exists for every R.

Suppose that R is an unbounded covering surface of R_0. Then the number of points on R that lie above each point of R_0 is always constant, say n ($\leqslant +\infty$), where the branch points of R are counted with their multiplicities. n is called the **number of sheets** of R over R_0, or R is said to be ***n*-sheeted** over R_0. If R and R_0 are compact surfaces with †Euler characteristic χ and χ_0, respectively, and if R is an n-sheeted covering surface of R_0, then we have the **Riemann-Hurwitz relation**: $\chi=n\chi_0-V$, V being the sum of the degrees of ramification. A topological mapping S of an unramified unbounded covering surface R of R_0 onto itself such that $T \circ S=T$ is called a **covering transformation**. The group of all covering transformations of a universal covering surface of R_0 is isomorphic to the †fundamental group (i.e., the 1-dimensional homotopy group) of R_0.

In a covering surface $(R, R_0; T)$ whose basic surface R_0 is a Riemann surface, T can be regarded as an analytic mapping of R onto R_0 by giving R a conformal structure in a natural manner. In particular, if R_0 is the sphere, then its covering surface is a Riemann surface. Conversely, any Riemann surface can be regarded as a covering surface of the sphere. This fact had long been known for closed Riemann surfaces; for open Riemann surfaces, it can be deduced from the existence theorem of analytic functions proved by H. Behnke and K. Stein, which states that an open Riemann surface is a †Stein manifold. Historically, by a Riemann surface mathematicians meant either the abstract Riemann surface or a covering surface of the sphere, until the two notions were proved identical.

Suppose that R is a covering surface of the z-sphere R_0 with the projection $T: R \rightarrow R_0$, and denote by R_a the region that lies above $0<|z-a|<r_0$. If there exists a topological mapping ψ of R_a onto $\{(r, \theta) \mid 0<r<r_0, -\infty<\theta<\infty\}$ such that $a+re^{i\theta}=T(\psi^{-1}(r, \theta))$, then R is said to have a **logarithmic branch point** above a; in contrast, a branch point of multiplicity n of the type defined previously is sometimes called an **algebraic branch point**.

Ahlfors's theory of covering surfaces, which treats covering surfaces $(R, R_0; T)$ not only from the topological viewpoint but also from the metrical one, is particularly important. Let R and R_0 be either compact surfaces with simplicial decompositions or their closed subregions with boundaries consisting of 1-simplexes and vertices such that T preserves simplicial decompositions. Here we call the part of the boundary of R whose projection is in the interior of R_0 the relative boundary. With respect to a suitable metric on R_0, let S be the ratio of areas of R and R_0, L the length of the relative boundary, and $-\rho$ and $-\rho_0$ the †Euler characteristics of R and R_0, respectively. Then **Ahlfors's principal theorem** asserts that $\max(0, \rho) \geqslant \rho_0 S-hL$, where $h>0$ is a constant determined only by R_0. This has been applied widely in various branches of mathematics, including the theory of distribution of values of analytic mappings between Riemann surfaces.

C. Uniformization

Suppose that we are given a correspondence between the z-plane and w-plane determined by a †function element $p_0=(z_0^*, w_0^*)$ ($z_0^*=P_0(t)$, $w_0^*=Q_0(t)$). This correspondence generally gives rise to a multiple-valued analytic function $w=f(z)$. We show how to construct a Riemann surface so that the function $w=f(z)$ can be considered a single-valued function on it. We use f again to mean the connected component of the set of function elements $p=(z^*, w^*)$ ($z^*=P(t)$, $w^*=Q(t)$) in the wider sense containing p_0, where the analytic neighborhood of each point p is defined to be the set of elements that are direct analytic continuations of p. Then f is a Riemann surface. For a point $p=(z^*, w^*)$ ($z^*=P(t)$, $w^*=Q(t)$) in f, set $z=F(p)=P(0)$, $w=G(p)=Q(0)$. Then two meromorphic functions $z=F(p)$ and $w=G(p)$

are defined on f, and f can be considered as a covering surface of the z-sphere and the w-sphere. We call f an †analytic function in the wider sense. Thus we obtain a single-valued function $w=G(p)$ on the Riemann surface f that can be regarded as a modification of the original function $w=f(z)$. Suppose that there exist two meromorphic functions $z=\varphi(\zeta)$ and $w=\psi(\zeta)$ on a region D in the ζ-plane, and let $z=P(\zeta-\zeta_0)$ and $w=Q(\zeta-\zeta_0)$ be †Laurent expansions of φ and ψ at each point ζ_0 in D. If the function element $p_{\zeta_0}=(z^*, w^*)$ $(z^*=P(t), w^*=Q(t))$ belongs to the Riemann surface f, then the correspondence $w=f(z)$ determined by the function element p_{ζ_0} is said to be **locally uniformized** on D by $z=\varphi(\zeta)$ and $w=\psi(\zeta)$. In particular, if $\{p_\zeta \mid \zeta\in D\}=f$, then f is said to be **uniformized** by $z=\varphi(\zeta)$ and $w=\psi(\zeta)$. If an analytic function f in the wider sense, considered as a Riemann surface, is conformally equivalent to a region D in the ζ-plane, then f can be uniformized by $z=F(p)$ and $w=G(p)$. In general, f is not conformally equivalent to a plane region, but if an unramified unbounded covering surface $(\hat{f}, f; T)$ of f is conformally equivalent to a region D in the ζ-plane, then f is uniformized by $z=F\circ T(\zeta)$ and $w=G\circ T(\zeta)$ (**Schottky's uniformization**). In particular, since the universal covering surface $(f^\infty, f; T)$ of f is simply connected, it is conformally equivalent to the sphere $|\zeta|\leqslant\infty$, the finite plane $|\zeta|<\infty$, or the unit disk $|\zeta|<1$. Consequently, f is uniformized by $z=F\circ T(\zeta)$ and $w=G\circ T(\zeta)$. Therefore analytic functions in the wider sense are always uniformizable.

For example, an †algebraic function f considered as a Riemann surface is always closed. If its †genus $g=0$, then f is the sphere and is thus uniformized by rational functions $z=F(\zeta)$ and $w=G(\zeta)$. If $g>0$, then $(f^\infty, f; T)$ is conformally equivalent to $|\zeta|<1$ or $|\zeta|<\infty$, and hence f is uniformized by $z=F\circ T(\zeta)$ and $w=G\circ T(\zeta)$. When $|\zeta|<1$, z and w are †automorphic functions with respect to the group of linear transformations preserving $|\zeta|<1$, while if $|\zeta|<\infty$, they are †elliptic functions.

D. The Type Problem

A simply connected Riemann surface R is conformally equivalent to the sphere, the finite plane, or the unit disk. Then R is said to be **elliptic**, **parabolic**, or **hyperbolic**, respectively. The problem of determining the types of simply connected covering surfaces of the sphere by their structures, such as the distributions of their branch points, is called the **type problem** for Riemann surfaces. For example, if a simply connected covering surface does not cover three points on the sphere, it must be hyperbolic (Picard's theorem). The Nevanlinna theory of meromorphic functions stimulated this type problem. However, it is difficult to measure the ramifications of covering surfaces, and many detailed reuslts of the type problem obtained in the 1930s are limited mainly to the case where all branch points lie above a finite number of points on the sphere. A sufficient condition for R to be of parabolic type, given by Z. Kobayashi (using the so-called Kobayashi net, and a sufficient condition for R to be of hyperbolic type, given by S. Kakutani (using quasiconformal mappings), are significant results on the type problem. The type problem had by then been extensively generalized to the following classification theory.

E. Classification Theory of Riemann Surfaces

Riemann surfaces are, as pointed out by Weyl, "not merely devices for visualizing the many-valuedness of analytic functions, but rather an essential component of the theory ... the only soil in which the functions grow and thrive." So the problem naturally arises of how to extend various results in the theory of analytic functions of a complex variable to the theory of analytic mappings between Riemann surfaces. In general, open Riemann surfaces can have infinite genus and are quite complicated. So to obtain fruitful results and systematic development, one usually sets certain restrictions on the properties of the Riemann surfaces. In connection with this, R. Nevanlinna, L. Sario, and others initiated the **classification theory of Riemann surfaces**, which classifies Riemann surfaces by the existence (or nonexistence) of functions with certain properties.

Denote by $\mathfrak{X}(R)$ the totality of functions on a Riemann surface R with a certain property $\mathfrak{X}$. The set of all Riemann surfaces R for which $\mathfrak{X}(R)$ does not contain any function other than constants is denoted by $O_{\mathfrak{X}}$. The family of analytic functions and the family of harmonic functions are denoted by $A(R)$ and $H(R)$, respectively. The family of positive functions, that of bounded functions, and that of functions with finite Dirichlet integrals are denoted by $P(R)$, $B(R)$, and $D(R)$, respectively. From these families, various new families are created, e.g., $ABD(R)=A(R)\cap B(R)\cap D(R)$. Usually O_{HB}, O_{HD}, O_{HBD}, O_{AB}, O_{AD}, O_{ABD}, and also O_G, the family of Riemann surfaces on which there are no Green's functions, are considered. P. J. Myrberg found an example of a Riemann surface of infinite genus which has a large boundary but belongs to O_{AB}. The idea behind Myberg's example is often used to construct examples in classification theory. From works of Y. Tôki, L. Sario, K. I. Virtanen, H. L. Royden, M. Parreau, M. Sakai, and others, it can be seen that there are inclusion relations, as indicated in Fig. 2, among the classes just mentioned. There is no inclusion relation between O_{AB} and O_{HD}. For Riemann surfaces of finite genus, $O_G = O_{HD}$. Closed Riemann surfaces are all in O_G.

Open Riemann surfaces in O_G are also said to be **parabolic** (or **of null boundary**), and those not in O_G, **hyperbolic** (or **of positive boundary**). Several characterizations for parabolic Riemann surfaces are known.

$$O_G \subsetneqq O_{HP} \subsetneqq O_{HB} \subsetneqq \begin{matrix} O_{HD} = O_{HBD} \\ O_{AB} \end{matrix} \subsetneqq O_{AD} = O_{ABD}$$

Fig. 2

From a similar point of view, the classification problem for subregions was studied in detail by Parreau, A. Mori, T. Kuroda, and others. We call a noncompact region Ω which is the complement of a compact subset of a Riemann surface a **Heins's end**. M. Heins called the minimal number ($\leqslant \infty$) of generators of the semigroup of the additive class of HP-functions that vanish continuously at the relative boundary of a Heins's end Ω the **harmonic dimension** of Ω. Its properties were investigated by Z. Kuramochi, M. Ozawa, and others. Generally, a function f is said to be **$\mathfrak{X}$-minimal** if f is positive and contained in $\mathfrak{X}(R)$ and there exists a constant C_g for every g in $\mathfrak{X}(R)$ with $f \geqslant g \geqslant 0$ such that $g \equiv C_g f$. The family of Riemann surfaces R not belonging to O_G and admitting $\mathfrak{X}$-minimal functions on R is denoted by $U_{\mathfrak{X}}$. C. Constantinescu and A. Cornea and others studied Riemann surfaces in U_{HB} and $U_{\underline{HD}}$ where $\underline{HD}$ is the class of positive functions in HD or limits of monotone decreasing sequences of such functions. There are inclusion relations $U_{HB} \supsetneqq O_{HB} - O_G$, $U_{\underline{HD}} \supsetneqq O_{HD} - O_G$, and $U_{\underline{HD}} \subsetneqq U_{UB}$. One of the interesting results in classification theory is the fact, discovered by Kuramochi, that $U_{HB} \cup O_G \subsetneqq O_{AB}$ and $U_{\underline{HD}} \cup O_G \subsetneqq O_{AD}$.

Classification theory has a very deep connection with the theory of †ideal boundaries. A. Pfluger and Royden showed that the classes O_G and O_{HD} are invariant under quasiconformal mappings. However, it is still an open problem whether O_{HB} is invariant in this sense.

F. Prolongations of Riemann Surfaces

As classification theory shows, pathological phenomena occur for Riemann surfaces from the viewpoint of function theory in the plane. These are caused by the complexity of ideal boundaries of Riemann surfaces, and in particular by the complexity of the set of ideal boundary points at which handles of Riemann surfaces (i.e., parts with cycles not homologous to 0) accumulate. Hence it is desirable to find larger Riemann surfaces so that ideal boundary points of original surfaces that are not accumulating points of handles become interior points. Suppose that a Riemann surface R is conformally equivalent to a proper subregion R' of another Riemann surface R_1. Then R_1 is said to be a **prolongation** of R, and R is **prolongable**. A nonprolongable Riemann surface is said to be **maximal**. Closed Riemann surfaces are always maximal, but there also exist maximal open Riemann surfaces (Radó). However, every open Riemann surface is homeomorphic to a prolongable Riemann surface (S. Bochner). Characterizations of prolongable Riemann surfaces and relationships between the various null classes mentioned in Section E and prolongations were investigated from several viewpoints by R. de Possel, J. Tamura, and others.

G. Analytic Mappings of Riemann Surfaces

Apart from the development of the classification theory of Riemann surfaces, efforts have been made to extend various results in the theory of analytic functions of a complex variable to the case of analytic mappings between Riemann surfaces. L. Sario studied the method of **normal operators**, which is utilized to construct harmonic functions on Riemann surfaces with given singularities at their ideal boundaries; and he extended the main theorems of Nevanlinna to analytic mappings between arbitrary Riemann surfaces (→ 124 Distributions of Values of Functions of a Complex Variable). M. Heins introduced the notions of Lindelöf type, Blaschke type, and others which are special classes of analytic mappings. Utilizing these notions, Constantinescu and Cornea, Kuramochi, and others extended various results in the theory of cluster sets (→ 62 Cluster Sets) by studying the behavior of analytic mappings at ideal boundaries. The theory of †capacities on ideal boundaries has also been developed.

On every open Riemann surface R there exists a nonconstant holomorphic function (Behnke and Stein). Furthermore, Gunning and Narasimhan proved that there exists a holomorphic function on R whose derivative never vanishes. In other words, R is conformally equivalent to an unramified covering surface of the sphere. Such a locally homeomorphic analytic mapping is called the **immersion** of R. The proof is based on the following deep result (S. Mergelyan's theorem): Suppose that K is a compact set on R such that $R-K$ has no relatively compact connected component. Then every continuous function which is holomorphic on the interior of K can be approximated uniformly on K by a holomorphic function on R [22].

A surface of genus 0 is said to be **planar** (or **of planar character** or **schlichtartig**). A simply connected surface is planar. Using the Dirichlet principle, P. Koebe proved the following **general uniformization theorem**: Every planar Riemann surface R can be mapped conformally to the canonical slit regions on the extended complex plane **C**. More precisely, given a point p on R, there exists the **extremal horizontal (vertical) slit mapping** $F_1(F_2)$ such that (i) $F_1(F_2)$ maps R conformally to a region on **C** whose boundary consists of horizontal (vertical) slits and possibly points; (ii) F_1 and F_2 have a simple pole at p with residue 1; (iii) the total area of the slits and points is zero. Suppose that $F_i=1/z+a_i+c_iz+\ldots$ $(i=1,2)$ in terms of local parameter z at p. Then $s=(c_1-c_2)/2$ is called the **span** of R. It is known that $s\ (=\|d(F_1-F_2)/2\|^2)\geqslant 0$, where the equality holds if and only if R belongs to O_{AD}. In the case of finite genus g, there also exist the conformal mappings of R onto the parallel slit regions on the $(g+1)$-sheeted covering surface of **C** (Z. Nehari, Y. Kusunoki, and others). L. Ahlfors proved that a Riemann surface of genus g bounded by m contours can be mapped conformally to an at most $(2g+m)$-sheeted unbounded covering surface of the unit disk.

The structures of closed Riemann surfaces are determined by the algebraic structures of meromorphic function fields on them. H. Iss'sa obtained a noteworthy result which established that open Riemann surfaces are also determined by their meromorphic function fields [24]. It is known too that open Riemann surfaces are determined by their rings of holomorphic functions.

H. Differential Forms on Riemann Surfaces

Since Riemann surfaces are considered real 2-dimensional differentiable manifolds of class C^∞, differential 1-forms $\omega=u\,dx+v\,dy$ and differential 2-forms $\Omega=c\,dx\wedge dy$ are defined on them, and operations such as the exterior derivative $d\omega=(\partial v/\partial x-\partial u/\partial y)dx\wedge dy$ and exterior product can be defined (→ 105 Differentiable Manifolds). Since coordinate transformations satisfy the Cauchy-Riemann differential equation, the **conjugate differential** $*\omega=-v\,dx+u\,dy$ of ω can be defined. A differential form ω satisfying $d\omega=d*\omega=0$ is called a **harmonic differential**, and one with $*\omega=-i\omega$ is said to be **pure**. A pure differential is expressed as $\omega=f\,dz$. Here, if f is a holomorphic function, then ω is called a **holomorphic** (or **analytic**) **differential**, and if f is a meromorphic function, then ω is called a **meromorphic** (or **Abelian**) **differential**. The differential form ω is a holomorphic differential if and only if it is **closed** (i.e., $d\omega=0$) and pure. The differential ω is called **exact** (or **total**) if ω is written as $dF=F_x\,dx+F_y\,dy$ with a globally single-valued function F.

Next, the set of all measurable differentials ω with $\|\omega\|^2=\iint_R\omega\wedge\overline{*\omega}<\infty$ forms a †Hilbert space with respect to the norm $\|\omega\|$. The method of orthogonal decomposition in the theory of Hilbert spaces is the main device to study this space and also its suitable subspaces and to obtain the existence theorem of harmonic and holomorphic differentials with various properties (→ 194 Harmonic Integrals). However, in contrast to differentiable manifolds, it should be noted that finer orthogonal decompositions into subspaces with specific properties hold for open Riemann surfaces. For instance, let $\Gamma_a(\Gamma_h)$ be the Hilbert space of analytic (harmonic) differentials with finite norm, and set $\Gamma_{ae}=\{\omega\in\Gamma_a\mid\omega$ is exact$\}$, $\Gamma_{hse}=\{\omega\in\Gamma_h\mid\omega$ is †semiexact$\}$; then we have the orthogonal decompositions

$$\Gamma_a=\Gamma_{ae}\dot{+}\Gamma_{aS},$$

$$\Gamma_h=*\Gamma_{hse}\dot{+}\Gamma_{hm}=\Gamma_{hse}\dot{+}*\Gamma_{hm},\quad\text{etc.},$$

where $*\Gamma_x$ stands for the space $\{\omega\mid*\omega\in\Gamma_x\}$ and Γ_{as} and Γ_{hm} are known as the space of analytic Schottky differentials and the space of differentials of harmonic measures, respectively. Both spaces Γ_{aS} and Γ_{hm} have remarkable properties [9].

I. Abelian Integrals on Open Riemann Surfaces

The systematic effort to extend the theory of Abelian integrals on closed Riemann surfaces to open Riemann surfaces was initiated by Nevanlinna in 1940. At the first stage of the development, only those Riemann surfaces with small boundaries (i.e., ones in O_G or O_{HD}) were treated; later a more general treatment was made possible by the discovery of the notion of semiexact differentials (K. Virtanen, Ahlfors).

Let R be an arbitrary open Riemann surface. A 1-dimensional cycle C is called a **dividing cycle** if for any compact set in R there exists a cycle outside the compact set homologous to C. A differential is said to be **semiexact** if its period along every dividing cycle vanishes.

Ahlfors defined the distinguished (complex harmonic) differentials with polar singularities and obtained in terms of them a generalization of Abel's theorem. Independently, Y. Kusunoki defined the semiexact canonical (meromorphic) differentials and gave in terms of them a formulation of Abel's theorem and the Riemann-Roch theorem on R [27]. It was proved that a meromorphic differential $df = du + idv$ is semiexact canonical if and only if du is (real) distinguished, and then u is almost constant on every ideal boundary component of R (in appropriate †compactification of R). Hence every meromorphic function f for which df is distinguished is almost constant on every ideal boundary component, and therefore f reduces to a constant by the boundary theorem of Riesz type if R has a large boundary. Whereas by the Riemann-Roch theorem above a nonconstant meromorphic function f such that df is (exact) canonical exists on any open Riemann surface R with finite genus, and f gives a canonical parallel slit mapping of R (→ Section G). H. L. Royden [28] and B. Rodin also gave generalizations of the Riemann-Roch theorem. M. Yoshida, H. Mizumoto, M. Shiba, and others further generalized the Kusunoki type theorems. The Riemann-Roch theorem for a closed Riemann surface can be deduced from that for open Riemann surfaces by considering an open Riemann surface obtained from a closed surface by deleting a point.

Riemann's period relation on R has been studied for various classes of differentials, but for the case of infinitely many nonvanishing periods, no definitive result has been obtained. The same is true for the theory of Abelian differentials with infinitely many singularities. On the other hand, the analogy to the classical theory is lost completely if no restriction is posed on the differentials on R. In this context the following results due to Behnke and Stein [26] are outstanding: (1) There exists an Abelian differential of the first kind on R with infinitely many given periods. (2) For two discrete sequences $\{p_n\}$ and $\{q_n\}$ of points in R, there exists a single-valued meromorphic function with zeros at p_n and poles at q_m. It is further proved that there exists an Abelian differential with prescribed divisor and periods (Kusunoki and Sainouchi). This generalizes the results above and the Gunning-Narasimhan theorem.

References

[1] H. Weyl, Die Idee der Riemannschen Fläche, Teubner, 1913, third edition, 1955; English translation, The concept of a Riemann surface, Addison-Wesley, 1964.
[2] S. Stoïlow, Leçons sur les principes topologiques de la théorie des fonctions analytiques, Gauthier-Villars, 1938.
[3] L. Ahlfors, Zur Theorie der Überlagerungsflächen, Acta Math., 65 (1935), 157–194.
[4] R. Nevanlinna, Uniformisierung, Springer, 1953.
[5] H. Behnke and F. Sommer, Theorie der analytischen Funktionen einer komplexen Veränderlichen, Springer, second revised edition, 1962.
[6] A. Pfluger, Theorie der Riemannschen Flächen, Springer, 1957.
[7] G. Springer, Introduction to Riemann surfaces, Addison-Wesley, 1957.
[8] L. V. Ahlfors (ed.), Contribution to the theory of Riemann surfaces, Ann. Math. Studies, Princeton Univ. Press, 1953.
[9] L. V. Ahlfors and L. Sario, Riemann surfaces, Princeton Univ. Press, 1960.
[10] M. Schiffer and D. C. Spencer, Functionals of finite Riemann surfaces, Princeton Univ. Press, 1954.
[11] L. Bers, Riemann surfaces, Lecture notes, New York Univ., 1957.
[12] M. Tsuji, Potential theory in modern function theory, Maruzen, 1959 (Chelsea, 1975).
[13] R. C. Gunning, Lectures on Riemann surfaces, Princeton Univ. Press, 1966.
For the classification of Riemann surfaces → [6, 9] and the following:
[14] C. Constantinescu and A. Cornea, Ideale Ränder Riemannscher Flächen, Springer, 1963.
[15] L. Sario and M. Nakai, Classification theory of Riemann surfaces, Springer, 1970.
[16] M. Sakai, Analytic functions with finite Dirichlet integrals on Riemann surfaces, Acta Math., 142 (1979), 199–220.
For analytic mappings of Riemann surfaces → [4, 6, 9] and the following:
[17] B. Rodin and L. Sario, Principal functions, Van Nostrand, 1968.
[18] L. Sario and K. Noshiro, Value distribution theory, Van Nostrand, 1966.
[19] L. Sario and K. Oikawa, Capacity functions, Springer, 1969.
[20] L. Ahlfors, Open Riemann surfaces and extremal problems on compact subregions, Comm. Math. Helv., 24 (1950), 100–134.
[21] R. Gunning and R. Narasimhan, Immersion of open Riemann surfaces, Math. Ann., 174 (1967), 103–108.
[22] E. Bishop, Subalgebras of functions on a Riemann surface, Pacif. J. Math., 8 (1958), 29–50.
[23] M. H. Heins, Algebraic structure and conformal mapping, Trans. Amer. Math. Soc., 89 (1958), 267–276.

[24] H. Iss'sa, On the meromorphic function field of a Stein variety, Ann. Math., (2) 83 (1966), 34–46.
For the generalization of algebraic functions and Abelian integrals → [4, 6, 9] and the following:
[25] R. Nevanlinna, Quadratisch integrierbare Differentiale auf einer Riemannschen Mannigfaltigkeit, Ann. Acad. Sci. Fenn., 1 (1941), 1–34.
[26] H. Behnke and K. Stein, Entwicklung analytischer Functionen auf Riemannschen Flächen, Math. Ann., 120 (1949), 430–461.
[27] Y. Kusunoki, Theory of Abelian integrals and its applications to conformal mappings, Mem. Coll. Sci. Univ. Kyôto, (A, Math.) 32 (1959), 235–258; 33 (1961), 429–433.
[28] H. L. Royden, The Riemann-Roch theorem, Comment. Math. Helv., 34 (1960), 37–51.
Also → reference to 11 Algebraic Functions, 416 Teichmüller Spaces.

368 (III.9) Rings

A. Definition

A nonempty set A is called a **ring** if the following conditions are satisfied.

(1) Two †operations, called **addition** and **multiplication** (the **ring operations**), are defined, which send an arbitrary pair of elements a, b of A to elements $a+b$ and ab of A.

(2) For arbitrary elements a, b, c of A, these operations satisfy the following four laws: (i) $a+b=b+a$ (**commutative law of addition**); (ii) $(a+b)+c=a+(b+c)$, $(ab)c=a(bc)$ (**associative laws**); (iii) $a(b+c)=ab+ac$, $(a+b)c=ac+bc$ (**distributive laws**); and (iv) for every pair a, b of elements of A, there exists a unique element c of A such that $a+c=b$. Thus a ring A is an †Abelian group under addition. Each element a of a ring A determines operations L_a and R_a defined by $L_a(x)=ax$, $R_a(x)=xa$ $(x \in A)$. Thus a ring A has the structure of a †left A-module and a †right A-module. Since the operations L_a and R_b commute for every pair a, b of elements of A, the ring A is also an A-A-bimodule (→ 277 Modules).

The identity element of A under addition is called the **zero element** and is denoted by 0. It satisfies the equation $a0=0a=0$ $(a \in A)$. An element $e \in A$ is called a **unity element** (**identity element** or **unit element**) of A if it satisfies $ae=ea=a$ $(a \in A)$. If A has such a unity element, it is unique and is often denoted by 1. A ring with unity element is called a **unitary ring**. Most of the important examples of rings are unitary. Hence we often call a unitary ring simply a ring. If a ring has only one member (namely, 0), then 0 is the unity element of the ring. Such a ring is called a **zero ring**. However, if a ring has more than one element, the unity element is distinct from the zero element. A ring is called a **commutative ring** if it satisfies (v) $ab=ba$ $(a, b \in A)$ (**commutative law for multiplication**) (→ 67 Commutative Rings).

In this article we shall discuss associative rings. Certain nonassociative rings are important; an example is †Lie algebra. (An algebra is a ring having a †ground ring.)

B. Further Definitions

An element $a \neq 0$ of a ring A is called a **zero divisor** if there exists an element $b \neq 0$ such that $ab=0$ or $ba=0$. A commutative unitary ring having more than one element is called an **integral domain** if it has no zero divisors (→ 67 Commutative Rings). Elements a and b of a ring are said to be **orthogonal** if $ab=ba=0$. An element a satisfying $a^n=0$ for some positive integer n is called a **nilpotent element**, and a nonzero element a satisfying $a^2=a$ is called an **idempotent element**. An idempotent element is said to be **primitive** if it cannot be represented as the sum of two orthogonal idempotent elements. For any subsets S and T of a ring A, let $S+T(ST)$ denote the set of elements $s+t(st)$ $(s \in S, t \in T)$. In particular, SS is denoted by S^2 (similarly for S^3, S^4, etc.), and furthermore, $\{a\}+S(\{a\}S)$ is denoted by $a+S(aS)$. If $ST=TS=\{0\}$, then subsets S and T are said to be **orthogonal**. A subset S of a ring is said to be **nilpotent** if $S^n=0$ for some positive integer n, and **idempotent** if $S^2=S$.

For an element a of a unitary ring A, an element a' such that $a'a=1$ $(aa'=1)$ is called a **left (right) inverse element** of a. There exists a left (right) inverse element of a if and only if A is generated by a as a left (right) A-module. If there exist both a right inverse and a left inverse of a, then they coincide and are uniquely determined by a. This element is called the **inverse element** of a and is denoted by a^{-1}. An element that has an inverse element is called an **invertible element** (**regular element** or **unit**). The set of all invertible elements of a unitary ring forms a group under multiplication. A nonzero unitary ring is called a †**skew field** (or †**division ring**) if every nonzero element is invertible. Furthermore, a skew field that satisfies the commutative law is called a **commutative field** or simply a field (→ 149 Fields). In a general ring A, if we define a new operation $(a, b) \rightarrow a \circ b$ by setting $a \circ b = a+b-ab$, then A is a †semigroup with the identity element 0 with respect to this operation. The inverse element under this operation is called the **quasi-inverse element**, and an element that has a quasi-inverse element is called a **quasi-invertible element** (or **quasiregular element**). An element a of a unitary ring is quasi-invertible if and only if the element $1-a$ is invertible.

C. Examples

(1) Rings of numbers. The ring $\mathbf{Z}$ of rational integers, the rational number field $\mathbf{Q}$, the real number field $\mathbf{R}$, and the complex number field $\mathbf{C}$ are familiar examples (→ 14 Algebraic Number Fields, 257 Local Fields).

(2) Rings of functions. The set K^I of functions defined on a set I and taking values in a ring K forms a commutative ring under pointwise addition and multiplication. In particular, let $K=\mathbf{R}$, and let I be an interval of $\mathbf{R}$. Then the set $C^0(I)$ of continuous functions, the set $C^r(I)$ of functions that are r-times continuously differentiable, and the set $C^\omega(I)$ of analytic functions on I are subrings (→ Section E) of $\mathbf{R}^I$.

(3) Rings of expressions. The set $K[X_1, \ldots, X_n]$ of polynomials and the set $K[[X_1, \ldots, X_n]]$ of †formal power series in indeterminates $X_1, \ldots, X_n$ with coefficients in a commutative ring K are commutative rings (→ 369 Rings of Polynomials, 370 Rings of Power Series).

(4) **Endomorphism rings** of modules. The set $\mathscr{E}_K(M)$ of †endomorphisms of a †module M over a ring K is in general a noncommutative ring. In particular, if M is a finite-dimensional †linear space over a field K, then $\mathscr{E}_K(M)$ can be identified with a †full matrix ring (→ 256 Linear Spaces, 277 Modules).

(5) For other examples → 29 Associative Algebras, 36 Banach Algebras, 67 Commutative Rings, 284 Noetherian Rings, and 439 Valuations.

D. Homomorphisms

A mapping $f: A \rightarrow B$ of a ring A into a ring B satisfying conditions (i) $f(a+b)=f(a)+f(b)$ and (ii) $f(ab)=f(a)f(b)$ $(a, b \in A)$ is called a **homomorphism**. If a homomorphism f is †bijective, then the inverse mapping $f^{-1}: B \rightarrow A$ is also a homomorphism, and in this case f is called an **isomorphism**. More precisely, a homomorphism (isomorphism) of rings is often called a **ring homomorphism (ring isomorphism)**. There exists only one homomorphism of any ring onto the zero ring. For unitary rings A and B, a homomorphism $f: A \rightarrow B$ is said to be **unitary** if it maps the unity element of A to the unity element of B. By a homomorphism, a unitary homomorphism is usually meant. In this sense there exists a unique homomorphism of the ring $\mathbf{Z}$ of rational integers into an arbitrary unitary ring. A composite of homomorphisms is also a homomorphism. The identity mapping 1_A of a ring A is an isomorphism. A homomorphism of a ring A into itself is called an **endomorphism**, and an isomorphism of A onto itself is called an **automorphism** of A. If a is an invertible element of a unitary ring A, then the mapping $x \rightarrow axa^{-1}$ $(x \in A)$ is an automorphism of A, called an **inner automorphism**.

When condition (ii) for a homomorphism is replaced by (ii') $f(ab)=f(b)f(a)$ $(a, b \in A)$, a mapping satisfying (i) and (ii') is called an **antihomomorphism**. In particular, if an antihomomorphism f is bijective, then the inverse mapping f^{-1} is also an antihomomorphism, and f is called an **anti-isomorphism. Antiendomorphisms** and **antiautomorphisms** are defined similarly.

E. Subrings, Factor Rings, and Direct Products

A subset S of a ring A is called a **subring** of A if a ring structure is given on S and the canonical injection $S \rightarrow A$ is a ring homomorphism. Thus the ring operations of S are the restrictions of those of A. If we deal only with unitary rings and unitary homomorphisms, then a subring S necessarily contains the unity element of A. The smallest subring containing a subset T of a ring A is called the subring **generated** by T. The set of elements that commute with every element of T forms a subring and is called the **commutor** (or **centralizer**) of T. In particular, the commutor of A itself is called the **center** of A.

A †quotient set A/R of a ring A by an equivalence relation R is called a factor ring (quotient ring) of A if a ring structure is given on A/R and the canonical surjection $A \rightarrow A/R$ is a ring homomorphism. This is the case if and only if the equivalence relation R is compatible with the ring operations (i.e., aRa' and bRb' imply $(a+b)R(a'+b')$ and $(ab)R(a'b')$). Let α and β be elements of the factor ring A/R, represented by a and b, respectively. Then the definition of factor ring implies that $\alpha+\beta(\alpha\beta)$ is the equivalence class represented by $a+b(ab)$. Every ring A has two trivial factor rings, namely, A itself and the zero ring 0. If A has no nontrivial factor rings, then A is called a **quasisimple ring** (→ Section G). If $f: A \rightarrow B$ is a ring homomorphism, then the image $f(A)$ is a subring of B, and the equivalence relation R in A defined by $f(aRb \Leftrightarrow f(a)=f(b))$ is compatible with the ring operations of A. Thus f induces an isomorphism $A/R \rightarrow f(A)$ (→ Section F).

If $\{A_i\}_{i \in I}$ is a family of rings, the Cartesian

product $A=\prod_{i\in I}A_i$ forms a ring under the componentwise operations $(a_i)+(b_i)=(a_i+b_i)$ and $(a_i)(b_i)=(a_ib_i)$. This ring is called the **direct product** of the family of rings $\{A_i\}_{i\in I}$. The mapping $p_i: A\rightarrow A_i$ that assigns to each (a_i) its ith component a_i is called a **canonical homomorphism**. For any set of homomorphisms $f_i: B\rightarrow A_i$ $(i\in I)$, there exists a unique homomorphism $f: B\rightarrow A$ such that $f_i=p_i\circ f$ for each i.

F. Ideals

A subset of a ring A is called a **left (right) ideal** of A if it is a submodule of the left (right) A-module of A (→ 277 Modules). In other words, a left (right) ideal J of A is an additive subgroup of A such that $AJ\subset J$ $(JA\subset J)$. Under the operations induced from A, J is a ring (however, J is not necessarily unitary). A subset of A is called a **two-sided ideal** or simply an **ideal** of A if it is a left and right ideal.

For an ideal J of a ring A, we define a relation R in A by $aRb\Leftrightarrow a-b\in J$. Then R is an equivalence relation that is compatible with the operations of A. Each equivalence class is called a **residue class** modulo J, and the quotient ring A/R is denoted by A/J and called the **residue (class) ring** (or **factor ring**) **modulo** J. If it is a field, it is called a **residue (class) field**. Conversely, given an equivalence relation R that is compatible with the operations of A, the equivalence class of 0 forms an ideal J of A, and the equivalence relation defined by J coincides with R.

If $f: A\rightarrow B$ is a ring homomorphism, then the †kernel of f as a homomorphism of additive groups forms an ideal J of A, and f induces an isomorphism $A/J\rightarrow f(A)$. It S is a subring and J is an ideal of a ring A, then $S+J$ is a subring of A and $S\cap J$ is an ideal of S. Furthermore, the natural homomorphism $S\rightarrow(S+J)/J$ induces an isomorphism $S/S\cap J\rightarrow(S+J)/J$ (**isomorphism theorem**).

A left (right) ideal of a ring A is said to be **maximal** if it is not equal to A and is properly contained in no left (right) ideal of A other than A. Similarly, a left (right) ideal of A is said to be **minimal** if it is nonzero and properly contains no nonzero left (right) ideal of A.

If e is an idempotent element of a unitary ring A, then $1-e$ and e are orthogonal idempotent elements, and $A=Ae+A(1-e)$ is the direct sum of left ideals. This is called **Peirce's left decomposition. Peirce's right decomposition** is defined similarly. A left ideal J of A can be expressed as $J=Ae$ for some idempotent element e if and only if there exists a left ideal J' such that $A=J+J'$ is a direct sum decomposition. More generally, if $e_1,\ldots,e_n$ are orthogonal idempotent elements whose sum is equal to 1, then $A=Ae_1+\ldots+Ae_n$ is the direct sum of left ideals. Conversely, if $A=J_1+\ldots+J_n$ is the direct sum of left ideals and $1=e_1+\ldots+e_n$ $(e_i\in J_i)$ is the corresponding decomposition of the unity element, then $e_1,\ldots,e_n$ are orthogonal idempotent elements. In particular, if $J_1,\ldots,J_n$ are two-sided ideals, then each J_i is a ring with unity element e_i, and by a natural correspondence, the ring A is isomorphic to the direct product $\prod_{i=1}^n J_i$. In this case, A is called the **direct sum** of ideals $J_1,\ldots,J_n$ and is denoted by $A=\bigoplus_{i=1}^n J_i$, or $A=\sum_{i=1}^n J_i$.

A ring A is called a **left (right) Artinian ring** if it is †Artinian as a left (right) A-module (i.e., if A satisfies the †minimal condition for left (right) ideals of A). A ring A is called a **left (right) Noetherian ring** if it is †Noetherian as a left (right) A-module (i.e., if A satisfies the †maximal condition for left (right) ideals of A). If A is commutative, *left* and *right* are omitted in these definitions. The property of being Artinian or Noetherian is inherited by quotient rings and the direct product of a finite family of rings, but not necessarily by subrings. For general rings, the maximal and minimal conditions for left (right) ideals are independent, but for unitary rings, left (right) Artinian rings are necessarily left (right) Noetherian (Y. Akizuki, C. Hopkins).

G. Semisimple Rings

The statement that a unitary ring A is †semisimple as a left A-module is equivalent to the statement that A is semisimple as a right A-module; in this case A is called a **semisimple ring** (→ Section H). Every module over a semisimple ring is also semisimple. A semisimple ring is left (right) Artinian and Noetherian. A semisimple ring is called a **simple ring** if it is nonzero and has no proper ideals except $\{0\}$, that is, if A is a quasisimple ring. Thus A is a simple ring if and only if A is a nonzero, unitary, quasisimple, left (right) Artinian ring. If A is a semisimple ring, then it has only a finite number of minimal ideals $A_1,\ldots,A_n$, and A is expressible as the direct sum $A=A_1+\ldots+A_n$, where each A_i is a simple ring, called a **simple component** of A. Any ideal of A is the direct sum of a finite number of simple components of A. Quotient rings of a semisimple ring and the direct product of a finite number of semisimple rings are also semisimple.

Any left (right) ideal of a semisimple ring A is expressible as Ae (eA) for some idempotent element e, and Ae (eA) is minimal if and only if e is primitive. In particular, a minimal left (right) ideal is a simple left (right) A-module that is contained in a certain simple component. Two minimal left (right) ideals are isomorphic as A-modules if and only if they are contained in the same simple component. If A_i, $1 \leqslant i \leqslant n$, are the simple components of A, then for each simple left A-module M there exists a unique i such that $A_i M \neq \{0\}$, and M is isomorphic to a minimal left ideal contained in A_i.

If M is a finite-dimensional linear space over a (skew) field D, then the endomorphism ring $A = \mathscr{E}_D(M)$ of M over D is a simple ring. Conversely, for any ring A, the endomorphism ring $D = \mathscr{E}_A(M)$ of a simple A-module M is a (skew) field (**Schur's lemma**). If A is a simple ring, then any simple A-module M can be considered as a finite-dimensional linear space over $D = \mathscr{E}_A(M)$, and A is isomorphic to $\mathscr{E}_D(M)$ (**Wedderburn's theorem**). Furthermore, if r is the dimension of M over D, then $\mathscr{E}_D(M)$ is isomorphic to the full matrix ring $M_r(D^\circ)$ of degree r over the field D°, which is anti-isomorphic to D. The dimension r is also equal to the †length of A as an A-module. The center of $A = \mathscr{E}_D(M)$ is isomorphic to the center of D, which is a commutative field. Thus a simple ring is an associative algebra over a commutative field (→ 29 Associative Algebras).

H. Radicals

Let A be a ring. Then among ideals consisting only of quasi-invertible elements of A, there exists a largest one, which is called the **radical** of A and denoted by $\mathfrak{R}(A)$. The radical of the residue ring $A/\mathfrak{R}(A)$ is $\{0\}$. A ring A is called a **semiprimitive ring** if $\mathfrak{R}(A)$ is $\{0\}$. On the other hand, A is called a left (right) **primitive ring** if it has a †faithful simple left (right) A-module. The radical $\mathfrak{R}(A)$ is equal to the intersection of all ideals J such that A/J is a left (right) primitive ring. In a unitary ring A, $\mathfrak{R}(A)$ coincides with the intersection of all maximal left (right) ideals of A. Furthermore, in a left (right) Artinian ring A, $\mathfrak{R}(A)$ is the largest nilpotent ideal of A, and the condition $\mathfrak{R}(A) = \{0\}$ is equivalent to the condition that A is a semisimple ring.

Among ideals consisting only of nilpotent elements of A, there exists a largest one, which is called the **nilradical** (or simply the radical) and denoted by $\mathfrak{N}(A)$ (→ 67 Commutative Rings). The nilradical of $A/\mathfrak{N}(A)$ is $\{0\}$. In general, $\mathfrak{N}(A)$ is contained in $\mathfrak{R}(A)$, and if A is left (right) Artinian, we have $\mathfrak{N}(A) = \mathfrak{R}(A)$. A ring A is called a **semiprimary ring** if $A/\mathfrak{N}(A)$ is left (right) Artinian and therefore semisimple. Furthermore, a ring A is called a **primary ring** (**completely primary ring**) if $A/\mathfrak{N}(A)$ is a simple ring (skew field). A primary ring is isomorphic to a full matrix ring over a completely primary ring.

References

[1] B. L. van der Waerden, Algebra, Springer, I, 1966; II, 1967.
[2] N. Jacobson, The theory of rings, Amer. Math. Soc. Math. Surveys, 1943.
[3] E. Artin, C. J. Nesbitt, and R. M. Thrall, Rings with minimum condition, Univ. of Michigan Press, 1944.
[4] N. Jacobson, Structure of rings, Amer. Math. Soc. Colloq. Publ., 1956.
[5] N. Bourbaki, Eléments de mathématique, Algèbre, ch. 1, 8, Actualités Sci. Ind., 1144b, 1261a, Hermann, 1964, 1958.
[6] C. Chevalley, Fundamental concepts of algebra, Academic Press, 1956.
[7] N. Jacobson, Lectures in abstract algebra I, Van Nostrand, 1951 (Springer, 1976).
[8] R. Godement, Cours d'algèbre, Hermann, 1963.
[9] J. P. Jans, Rings and homology, Holt, Rinehart and Winston, 1964.
[10] S. Lang, Algebra, Addison-Wesley, 1965.
[11] N. H. McCoy, The theory of rings, Macmillan, 1964.

369 (III.13) Rings of Polynomials

A. General Remarks

In this article, we mean by a *ring* a †commutative ring with †unity element. Let R be a ring, and let $X_1, \ldots, X_n$ be **variables** (**letters**, **indeterminates**, or **symbols**). Then the set of †polynomials in $X_1, \ldots, X_n$ with coefficients in R is called the **ring of polynomials** (or **polynomial ring**) in n variables $X_1, \ldots, X_n$ over R and is denoted by $R[X_1, \ldots, X_n]$ (→ 67 Commutative Rings; 284 Noetherian Rings; 337 Polynomials; 368 Rings). On the other hand, when R and R' are rings with common unity element and $R \subset R'$, then for a subset S of R' we denote the subring of R', generated by S over R, by $R[S]$. When $S = \{x_1, \ldots, x_n\}$, then there is a homomorphism φ of the polynomial ring $R[X_1, \ldots, X_n]$ onto $R[S]$ defined by

$$\varphi(\Sigma a_{i_1 \ldots i_n} X_1^{i_1} \ldots X_n^{i_n}) = \Sigma a_{i_1 \ldots i_n} x_1^{i_1} \ldots x_n^{i_n} \qquad (a_{i_1 \ldots i_n} \in R).$$

If φ is an isomorphism, then $x_1, \ldots, x_n$ are said to be **algebraically independent** over R; and otherwise, **algebraically dependent** over R. Thus the ring of polynomials in n variables over R may be regarded as a ring $R[x_1, \ldots, x_n]$ generated by a finite system of algebraically independent elements $x_1, \ldots, x_n$ over R.

B. Ideals, Homogeneous Rings, and Graded Rings

Consider the polynomial ring $R[X]=R[X_1, \ldots, X_n]$ in n variables over a ring R. A polynomial $f \in R[X]$ is a †zero divisor if and only if there is a nonzero member a of R for which $af=0$. If $\mathfrak{a}$ is an †ideal of R, then $R[X]/\mathfrak{a}R[X] \cong (R/\mathfrak{a})[X_1, \ldots, X_n]$. Therefore, if $\mathfrak{p}$ is a †prime ideal of R, then $\mathfrak{p}R[X]$ is a prime ideal of $R[X]$. If R is a †unique factorization domain (u.f.d.), then $R[X]$ is also a u.f.d. If R is a normal ring, then so is $R[X]$. By the †Hilbert basis theorem, if R is †Noetherian, then $R[X]$ is also Noetherian. If m is the †Krull dimension of R, then Krull dim $R[X] \geqslant n+m$; the equality holds if R is Noetherian. If R is a field, then $R[X]$ is not only a u.f.d. but also a †Macaulay ring.

A **homogeneous ideal** of $R[X]$ is an ideal generated by a set of †homogeneous polynomials f_λ (the degree of f_λ may depend on λ). When $\mathfrak{a}$ is a homogeneous ideal, an element in $R[X]/\mathfrak{a}$ is defined to be a **homogeneous element** of degree d if it is the class of a homogeneous polynomial of degree d modulo $\mathfrak{a}$, and the quotient ring $R[X]/\mathfrak{a}$ is called a **homogeneous ring**. More generally, assume that a ring R is, as a †module, the †direct sum $\sum_{i=0}^{\infty} R_i$ of its submodules R_i $(i=0,1,2,\ldots)$ and that $R_iR_j \subset R_{i+j}$ for every pair (i,j). Then we call R a **graded ring**, and an element in R_d a **homogeneous element** of degree d. (In some literature the term graded ring is used in a wider sense; see below) In a graded ring R, if an ideal is generated by homogeneous elements, then the ideal is called a **homogeneous ideal** (or **graded ideal**). In a graded ring $R=\sum_{i=0}^{\infty} R_i$, if the ideal $\sum_{i=1}^{\infty} R_i$ has a finite basis, then R is generated (as a ring) by a finite number of elements over its subring R_0. Therefore, the graded ring $R=\sum R_i$ is Noetherian if and only if R_0 is Noetherian and R is generated by a finite number of elements over R_0. In this case, every homogeneous ideal is the intersection of a finite number of homogeneous †primary ideals, and every prime divisor of a homogeneous ideal is a homogeneous prime ideal.

The notion of a graded ring is generalized further as follows: A ring R is **graded** by an additive semigroup I (containing 0) if R is $\sum_{i \in I} R_i$ (direct sum) and if $R_iR_j \subset R_{i+j}$.

C. Zero Points

(1) Zero Points in an Affine Space. We consider the polynomial ring $K[X]=K[X_1, \ldots, X_n]$ in n variables over a field K and a field Ω containing K. A point $(a_1, \ldots, a_n)$ of an n-dimensional †affine space $\Omega^n=\{(a_1, \ldots, a_n) \mid a_i \in \Omega\}$ is called a **zero point** of a subset S of $K[X]$ if $f(a_1, \ldots, a_n)=0$ for every $f(X_1, \ldots, X_n) \in S$. A point $(a_1, \ldots, a_n)$ is called **algebraic** over K (**K-rational**) if every a_i is algebraic over K (is an element of K). In this way we define algebraic zero points and rational zero points. Zero points of S are zero points of the †ideals generated by S. Therefore, in order to investigate the set of zero points of S, we may restrict ourselves to the case where S is an ideal. Denote by $V(S)$ the set of zero points of S. If $\mathfrak{a}_1$, $\mathfrak{a}_2$ are ideals of $K[X]$, then (i) $V(\mathfrak{a}_1 \cap \mathfrak{a}_2)=V(\mathfrak{a}_1\mathfrak{a}_2)=V(\mathfrak{a}_1) \cup V(\mathfrak{a}_2)$; (ii) $V(\mathfrak{a}_1+\mathfrak{a}_2)=V(\mathfrak{a}_1) \cap V(\mathfrak{a}_2)$; and (iii) if $\mathfrak{a}_1$ and $\mathfrak{a}_2$ have a common †radical, then $V(\mathfrak{a}_1)=V(\mathfrak{a}_2)$.

(2) Zero Points in a Projective Space. A point $(\lambda a_1, \ldots, \lambda a_n)$ of an $(n-1)$-dimensional †projective space over Ω (with $a_i \in \Omega$, some $a_i \neq 0$, $\lambda \in \Omega$, $\lambda \neq 0$) is called a **zero point** of a polynomial $f(X_1, \ldots, X_n)$ if, f being expressed as $\sum f_i$ with homogeneous polynomials f_i of degree i, $f_i(a_1, \ldots, a_n)=0$ for every i (this condition holds if and only if $f(\lambda a_1, \ldots, \lambda a_n)=0$ for any element λ in Ω, provided that Ω contains infinitely many elements). Therefore, zero points of a subset S of $K[X]$ are zero points of the smallest homogeneous ideal containing S. Thus, in order to study the sets of zero points, it is sufficient to consider sets of zero points of homogeneous ideals, and propositions similar to (i), (ii), and (iii) of part 1 of this section hold for homogeneous ideals $\mathfrak{a}_1$, $\mathfrak{a}_2$.

D. The Normalization Theorem

Let $\mathfrak{a}$ be an ideal of †height h in the polynomial ring $K[X]=K[X_1, \ldots, X_n]$ in n variables over a field K. Then there exist elements $Y_1, \ldots, Y_n$ of $K[X]$ such that (i) $K[X]$ is †integral over $K[Y]=K[Y_1, \ldots, Y_n]$ and (ii) $Y_1, \ldots, Y_h$ generate $\mathfrak{a} \cap K[Y]$ (**normalization theorem for polynomial rings**).

Using this theorem, we obtain the following important theorems on finitely generated rings.

(1) **Normalization theorem for finitely generated rings**. If a ring R is finitely generated over an integral domain I, then there exist an element a $(\neq 0)$ of I and algebraically independent elements $z_1, \ldots, z_t$ of R over I such that the †ring of quotients R_S (where $S=\{a^n \mid n=1,2,\ldots\}$) is integral over $I[a^{-1}, z_1, \ldots, z_t]$.

(2) If $\mathfrak{p}$ is a prime ideal of an integral domain R that is finitely generated over a field K, then (height of $\mathfrak{p}$)+(†depth of $\mathfrak{p}$)=(†transcendence degree of R over K), and the depth of $\mathfrak{p}$ coincides with the transcendence degree of $R/\mathfrak{p}$ over K. In particular, if $\mathfrak{m}$ is a maximal ideal of R, then $R/\mathfrak{m}$ is algebraic over K.

(3) **Hilbert's zero point theorem (Hilbert Nullstellensatz).** Let $\mathfrak{a}$ be an ideal of the polynomial ring $K[X]=K[X_1, \ldots, X_n]$ over the field K, and assume that the field Ω containing K is †algebraically closed. If $f \in K[X]$ satisfies the condition that every algebraic zero point (→ Section C) of $\mathfrak{a}$ is a zero point of f, then some power of f is contained in $\mathfrak{a}$.

E. Elimination Theory

Let $f_1, \ldots, f_N$ be elements of the polynomial ring $R=I[X_1, \ldots, X_m, Y_1, \ldots, Y_n]$ in $m+n$ variables over an integral domain I. For each maximal ideal $\mathfrak{m}$ of I, let $\varphi_{\mathfrak{m}}$ be the canonical homomorphism with modulus $\mathfrak{m}$, and let $\Omega_{\mathfrak{m}}$ be an algebraically closed field containing $I/\mathfrak{m}$. Let $W_{\mathfrak{m}}$ be the set of points $(a_1, \ldots, a_n)$ of the n-dimensional affine space $\Omega_{\mathfrak{m}}^n$ over $\Omega_{\mathfrak{m}}$ such that the system of equations $\varphi_{\mathfrak{m}}(f_i)(X_1, \ldots, X_m, a_1, \ldots, a_n)=0$ $(i=1,2, \ldots, N)$ has a solution in $\Omega_{\mathfrak{m}}^m$. To **eliminate** $X_1, \ldots, X_m$ from $f_1, \ldots, f_N$ is to obtain $g(Y_1, \ldots, Y_n) \in I[Y_1, \ldots, Y_n]$ such that every point of $W_{\mathfrak{m}}$ is a zero point of $\varphi_{\mathfrak{m}}(g)$ for every $\mathfrak{m}$; such a g (or an equation $g=0$) is called a **resultant** of $f_1, \ldots, f_N$. The set $\mathfrak{a}$ of resultants forms an ideal of $I[Y_1, \ldots, Y_n]$, and $\{g_1, \ldots, g_M\}$ is called a **system of resultants** if the radical of the ideal generated by it coincides with $\mathfrak{a}$. If I is finitely generated over a field, then, denoting by $\mathfrak{b}$ the radical of the ideal generated by $f_1, \ldots, f_N$, we have $\mathfrak{a}=\mathfrak{b} \cap I[Y_1, \ldots, Y_n]$. In particular, let I be a field. It is obvious that $W_{(0)}$ is contained in the set V of zero points of $\mathfrak{a}$. However, it is not necessarily true that $V=W_{(0)}$. If every f_i is homogeneous in $X_1, \ldots, X_m$ and also in $Y_1, \ldots, Y_n$, then we have $V=W_{(0)}$.

If we wish to write a system of resultants explicitly, we can proceed as follows: Regard the f_i as polynomials in X_1 with coefficients in $I[X_2, \ldots, X_m, Y_1, \ldots, Y_n]$, and obtain resultants $R(f_i, f_j)$ by eliminating X_1 from the pairs f_i, f_j. Then eliminate X_2 from these resultants, and so forth. To obtain $R(f_i, f_j)$, we may use **Sylvester's elimination method**. Namely, let f and g be polynomials in x with coefficients in $I: f=a_0 x^m+a_1 x^{m-1}+\ldots+a_m$, $g=b_0 x^n+b_1 x^{n-1}+\ldots+b_n$. Let $D(f, g)$ be the following determinant of degree $m+n$:

$$\begin{array}{l} n\left\{\begin{array}{l} \\ \\ \\ \\ \\ \end{array}\right. \\ m\left\{\begin{array}{l} \\ \\ \\ \\ \end{array}\right. \end{array}\left|\begin{array}{cccccccc} a_0 & a_1 & \cdots & a_m & 0 & & \cdots & 0 \\ 0 & a_0 & \cdots & a_{m-1} & a_m & 0 & \cdots & 0 \\ & & & \cdots & & & & \\ & & & \cdots & & & & \\ 0 & & \cdots & 0 & a_0 & a_1 & \cdots \; a_{m-1} & a_m \\ b_0 & b_1 & \cdots & & b_{n-1} & b_n & 0 \; \cdots & 0 \\ & & & & \cdots & & & \\ & & & & \cdots & & & \\ 0 & \cdots & 0 & b_0 & b_1 & & \cdots \; b_{n-1} & b_n \end{array}\right| .$$

Then $D(f, g)=0$ if and only if either f and g have a common root or $a_0=b_0=0$. Therefore, if I is a u.f.d. and a_0 and b_0 have no common factor, then $D(f, g)$ is the required resultant $R(f, g)$. (For other methods of elimination see B. L. van der Waerden, *Algebra*, vol. II. For criteria on whether a finitely generated ring over a field is †regular → 370 Rings of Power Series B.)

F. Syzygy Theory

(1) Classical Case. The notion of **syzygy** was introduced by Sylvester (*Phil. Trans.*, 143 (1853)), then generalized and clarified by Hilbert [3], whose definition can be formulated as follows: Let $R=k[X_1, \ldots, X_n]$ be a polynomial ring of n variables over a field k. R has the natural gradation (i.e., R is a graded ring in which each X_i $(1 \leqslant i \leqslant n)$ is of degree 1 and elements of k are of degree 0). Let M be a finitely generated graded R-module. If $f_1, \ldots, f_m$ form a minimal basis of M over R consisting of homogeneous elements, we introduce m indeterminates $u_1, \ldots, u_m$ and put $F=\sum_{1 \leqslant j \leqslant m} R u_j$, the free R-module generated by $u_1, \ldots, u_m$. Set $\operatorname{deg}(u_j)=\operatorname{deg}(f_j)$ $(1 \leqslant j \leqslant m)$ and supply F with the structure of a graded R-module. Let φ be the graded R-homomorphism of F onto M defined by $\varphi(u_j)=f_j$. Then $N=\operatorname{Ker}(\varphi)$ is a graded R-module uniquely determined by M up to isomorphism (of graded R-modules); N is called the **first syzygy** of M. For a positive integer r, the r**th syzygy** of M is inductively defined as the first syzygy of the $(r-1)$st syzygy of M. The **Hilbert syzygy theorem** states that for any finitely generated graded R-module M, the nth syzygy of M is free. In other words, M admits a **free resolution** of length $\leqslant n$, i.e., an exact sequence of the form

$$0 \rightarrow F^{(\nu)} \rightarrow \ldots \rightarrow F^{(1)} \rightarrow F^{(0)} \rightarrow M \rightarrow 0,$$

where $\nu \leqslant n$ and each $F^{(i)}$ $(0 \leqslant i \leqslant \nu)$ is a finitely generated free graded R-module. It follows that if M_d denotes the homogeneous part of degree d in M, there exists a polynomial $P(X)$ of degree $\leqslant n-1$ such that $\operatorname{dim}_k(M_d)=P(d)$ for sufficiently large d; $P(X)$ is called the **Hilbert polynomial** (or **characteristic function**) of the graded R-module M.

(2) Generalization by Serre. The syzygy theory was generalized by J.-P. Serre [2] as follows: Let R be a Noetherian ring and M a finitely generated R-module. Then we can find a finitely generated free R-module F and an R-homomorphism φ of F onto M. The kernel of φ, called a **first syzygy** of M, is not uniquely determined by M. However, if N_1 and N_2 are first syzygies of M, then there exist finitely generated †projective R-modules P_1 and P_2 such that $N_1 \oplus P_1 \cong N_2 \oplus P_2$ (→ 277 Modules K). For a positive integer r, an **rth syzygy** is defined inductively as in (1) of this section. An important result of Serre is that R is a †regular ring of †Krull dimension at most n if and only if an nth syzygy of every finitely generated R-module is †projective.

(3) Serre Conjecture. D. Quillen (*Invertiones Math.*, 36 (1976)) and A. Suslin (Dokl. Akad. Nauk SSSR (26 Feb. 1976)) solved the Serre conjecture by proving that every projective module over a ring of polynomials over a field is free.

(4) Special Cases. In the following special cases, we can define the first syzygy of M uniquely up to isomorphism: (i) R is a Noetherian †local ring and M is a finitely generated R-module; and (ii) R is a graded Noetherian ring $\sum_{d \geqslant 0} R_d$, where R_0 is a field and M is a finitely generated graded R-module [4].

References

[1] M. Nagata, Local rings, Wiley, 1962 (Krieger, 1975).
[2] J.-P. Serre, Algèbre locale, multiplicités, Lecture notes in math., Springer, 1965.
[3] D. Hilbert, Über die Theorie der algebraischen Formen, Math. Ann., 36 (1890), 473–534.
[4] J.-P. Serre, Sur la dimension homologique des anneaux et des modules noethériens, Proc. Intern. Symp. Alg. Number Theory, Tokyo and Nikko (1955), 175–189.
Also → references to 284 Noetherian Rings.

370 (III.14) Rings of Power Series

A. Rings of Formal Power Series (→ 67 Commutative Rings; 284 Noetherian Rings; 368 Rings)

Let R be a commutative ring with unity element 1. Let F_d be the module of †homogeneous polynomials of degree d in $X_1, \ldots, X_n$ with coefficients in R. A formal infinite sum $\sum_{d=0}^{\infty} a_d = a_0 + a_1 + \ldots + a_n + \ldots +$ of elements $a_d \in F_d$ is called a **formal power series** or simply **power series** in n variables $X_1, \ldots, X_n$ with coefficients in R, and a_d is called the **homogeneous part** of degree d of the power series. The homogeneous part a_0 of degree zero is called the **constant term**. Addition and multiplication are defined by $(\sum a_d)+(\sum b_d)=\sum(a_d+b_d)$, $(\sum a_d)(\sum b_d)=\sum_d(\sum_{i+j=d} a_i b_j)$. By these operations, the set of power series forms a commutative ring, which is called the **ring of (formal) power series** (or **(formal) power series ring**) in $X_1, \ldots, X_n$ over R and is denoted by $R[[X_1, \ldots, X_n]]$ or $R\{X_1, \ldots, X_n\}$. If there is a natural number N such that $a_d = 0$ for every $d > N$, then the power series $\sum a_d$ is identified with the polynomial $a_0 + a_1 + \ldots + a_N$. Thus $R[X_1, \ldots, X_n] \subset R[[X_1, \ldots, X_n]]$. Set $\mathfrak{X} = \sum_i X_i R[[X_1, \ldots, X_n]]$. Then $R[[X_1, \ldots, X_n]]$ is †complete under the $\mathfrak{X}$-adic topology (→ 284 Noetherian Rings B).

Assume that R' is a commutative ring containing R and having a unity element in common with R, $\mathfrak{a}'$ is an ideal of R' such that R' is complete under the $\mathfrak{a}'$-adic topology, and $x_1, \ldots, x_n$ are elements of $\mathfrak{a}'$. Then an infinite sum $\sum c_{i_1 \ldots i_n} x_1^{i_1} \ldots x_n^{i_n}$ (each i_j ranges over nonnegative rational integers and $c_{i_1 \ldots i_n} \in R$) has a well-defined meaning in R' (namely, if S_d is a finite sum of these terms such that $\sum i_j \leqslant d$, then the infinite sum is defined to be $\lim_{d \to \infty} S_d$). This element $\sum c_{i_1 \ldots i_n} x_1^{i_1} \ldots x_n^{i_n}$ is also called a **power series** in $x_1, \ldots, x_n$ with coefficients in R. The set of such power series in $x_1, \ldots, x_n$ is a subring of R', called the **power series ring** in $x_1, \ldots, x_n$ over R and denoted by $R[[x_1, \ldots, x_n]]$ or $R\{x_1, \ldots, x_n\}$. Defining φ by $\varphi(\sum c_{i_1 \ldots i_n} X_1^{i_1} \ldots X_n^{i_n}) = \sum c_{i_1 \ldots i_n} x_1^{i_1} \ldots x_n^{i_n}$, we obtain a ring homomorphism $\varphi: R[[X_1, \ldots, X_n]] \to R[[x_1, \ldots, x_n]]$. If φ is an isomorphism, then we say that $x_1, \ldots, x_n$ are **analytically independent** over R.

If $\overline{\mathfrak{m}}$ is a †maximal ideal of the formal power series ring $R[[X_1, \ldots, X_n]]$, then $\mathfrak{m} = \overline{\mathfrak{m}} \cap R$ is a maximal ideal of R and $\overline{\mathfrak{m}}$ is generated by $\mathfrak{m}$ and $X_1, \ldots, X_n$. An element f of the power series ring is †invertible if and only if its constant term f_0 is an invertible element of R, and in this case $f^{-1} = \sum_{d=0}^{\infty} f_0^{-d-1} \cdot (f_0 - f)^d$. If R is one of the following, then $R[[X_1, \ldots, X_n]]$ is also of the same kind: (i) †Noetherian ring, (ii) †local ring, (iii) †semilocal ring, (iv) †integral domain, (v) †regular local ring, (vi) Noetherian †normal ring. But even if R is a †unique factorization domain (u.f.d.), $R[[X_1, \ldots, X_n]]$ need not be a u.f.d. (If R is a field, or more generally, if R is a regular semilocal ring, then $R[[X_1, \ldots, X_n]]$ is a u.f.d.). In particular, a formal power series ring $k[[X]]$ in one variable X over a field k is an integral domain whose field of quotients is called the **field of (formal) power series** (or **(formal) power series field) in one variable** X over k and is denoted by $k((X))$; an element of $k((X))$ is expressed uniquely in the form $\sum_{n=r}^{\infty} a_n X^n$ $(a_n \in k, r \in \mathbf{Z})$.

B. Rings of Convergent Power Series

Let K be a field with multiplicative †valuation v (for instance, $K=\mathbf{C}$, the complex number field, and $v(\alpha)=|\alpha|$ for $\alpha\in\mathbf{C}$). A formal power series $f(X_1,\ldots,X_n)=\sum c_{i_1\ldots i_n}X_1^{i_1}\ldots X_n^{i_n}$ is said to be a **convergent power series** if there are positive numbers $r_1,\ldots,r_n$, M such that $v(c_{i_1\ldots i_n})r_1^{i_1}\ldots r_n^{i_n}\leqslant M$ for every $(i_1,\ldots,i_n)$. In this case, if $a_i\in K$ and $v(a_i)<r_i$, then $\sum c_{i_1\ldots i_n}a_1^{i_1}\ldots a_n^{i_n}$ has its sum in the †completion of K. The set of convergent power series is a subring of $K[[X_1,\ldots,X_n]]$. It is called the **ring of convergent power series** (or **convergent power series ring**) in n variables over K and is denoted by $K\langle\langle X_1,\ldots,X_n\rangle\rangle$ or $K\{X_1,\ldots,X_n\}$. It is a regular local ring of †Krull dimension n. Hence it is a u.f.d. and its completion is $K[[X_1,\ldots,X_n]]$. If v is a †trivial valuation, then $K\langle\langle X_1,\ldots,X_n\rangle\rangle=K[[X_1,\ldots,X_n]]$.

Weierstrass's Preparation Theorem. For an element $f=\sum c_{i_1\ldots i_n}X_1^{i_1}\ldots X_n^{i_n}\in K\langle\langle X_1,\ldots,X_n\rangle\rangle$, assume that $c_{0\ldots 0i}=0$ for $i=0,1,\ldots,r-1$ and $c_{0\ldots 0r}\neq 0$. Then for an arbitrary element g of $K\langle\langle X_1,\ldots,X_n\rangle\rangle$, there exists a unique $q\in K\langle\langle X_1,\ldots,X_n\rangle\rangle$ such that $g-qf\in\sum_{i=0}^{r-1}X_n^iK\langle\langle X_1,\ldots,X_{n-1}\rangle\rangle$. In particular (considering the case where $g=X_n^r$), there is an invertible element u of $K\langle\langle X_1,\ldots,X_n\rangle\rangle$ such that $fu=f_0+f_1X_n+\ldots+f_{r-1}X_n^{r-1}+X_n^r$ $(f_i\in K\langle\langle X_1,\ldots,X_{n-1}\rangle\rangle)$.

By this theorem, we see easily that if $\mathfrak{a}$ is an ideal of †height h of $K\langle\langle X_1,\ldots,X_n\rangle\rangle$, then $K\langle\langle X_1,\ldots,X_n\rangle\rangle/\mathfrak{a}$ is isomorphic to a ring that is a finite module over $K\langle\langle X_1,\ldots,X_{n-h}\rangle\rangle$.

If $\mathfrak{p}$ is a †prime ideal of $K\langle\langle X_1,\ldots,X_n\rangle\rangle$, then $\mathfrak{p}K[[X_1,\ldots,X_n]]$ is a prime ideal.

The Jacobian Criterion. Let K be a field, and let R be the ring of polynomials $K[X_1,\ldots,X_n]$, the ring of formal power series $K[[X_1,\ldots,X_n]]$, or the ring of convergent power series $K\langle\langle X_1,\ldots,X_n\rangle\rangle$ in n variables $X_1,\ldots,X_n$ over K. †Partial derivatives $\partial/\partial X_i$ are well defined in R. For $f_1,\ldots,f_t\in R$, a †Jacobian matrix $J(f_1,\ldots,f_t)$ is defined to be the $t\times n$ matrix whose (i,j)-entry is $\partial f_i/\partial X_j$. Let $\mathfrak{p}$ be a †prime divisor of the ideal $\sum_i f_iR$, and let $\mathfrak{q}$ be a prime ideal containing $\mathfrak{p}$. If the †rank of ($J(f_1,\ldots,f_t)$ modulo $\mathfrak{q}$) is equal to the height of $\mathfrak{p}$, then the ring $R_\mathfrak{q}/\sum f_iR_\mathfrak{q}$ is a regular local ring. The converse is also true if K is a †perfect field (if K is not a perfect field, then, modifying $J(f_1,\ldots,f_t)$, we can have a similar criterion [1]).

C. Hensel Rings

A **Hensel ring** (or **Henselian ring**) is a commutative ring with unity element satisfying the following two conditions: (i) R has only one maximal ideal $\mathfrak{m}$ (i.e., R is a †quasilocal ring); and (ii) if f, g_0, h_0 are monic polynomials in one variable x (here, a polynomial in x is called monic if the coefficient of the term of the highest degree is 1) such that $f-g_0h_0\in\mathfrak{m}R[x]$, $g_0R[x]+h_0R[x]+\mathfrak{m}R[x]=R[x]$, then there are monic polynomials g, $h\in R[x]$ such that $f=gh$ and $g\equiv g_0$, $h\equiv h_0$ modulo $\mathfrak{m}$.

Important examples of Hensel rings are complete local rings, rings of convergent power series, and †complete valuation rings. When R is a Hensel ring, a commutative ring R' with unity element such that R' is a finite R-module is the direct sum of a finite number of Hensel rings. For any quasilocal ring Q, there exists a Hensel ring $\tilde{Q}$, called the **Henselization** of Q (for details → [1]), for which the following statements hold: (i) $\tilde{Q}$ is a †faithfully flat Q-module; (ii) if $\mathfrak{m}$ is the maximal ideal of Q, then the maximal ideal of $\tilde{Q}$ is $\mathfrak{m}\tilde{Q}$, and $\tilde{Q}/\mathfrak{m}\tilde{Q}=Q/\mathfrak{m}Q$; (iii) if R is a Hensel ring that contains Q and has a maximal ideal $\mathfrak{n}$, and $\mathfrak{n}\cap Q=\mathfrak{m}$, then there is one and only one Q-homomorphism φ of $\tilde{Q}$ into R; (iv) if Q is a †normal ring, then the Q-homomorphism φ is an injection; (v) if Q is a local ring, then $\tilde{Q}$ is also a local ring, and Q is dense in $\tilde{Q}$.

References

[1] M. Nagata, Local rings, Interscience, 1962 (Krieger, 1975).
[2] O. Zariski and P. Samuel, Commutative algebra II, Van Nostrand, 1960; new edition, Springer, 1975.

371 (XVIII.10) Robust and Nonparametric Methods

A. General Remarks

Robust and **nonparametric** or **distribution-free methods** are statistical procedures specifically devised to deal with broad families of probability distributions.

In the theory of statistical inference it is usual to assume that the probability distribution of the population from which the observed values are chosen at random is specified exactly except for a small number of unknown parameters (→ 401 Statistical Inference). In practical applications, however, it often happens that the assumptions made for the model, especially those about the shape of the distribution, may not hold for the actual data. In such cases robust and/or nonparametric procedures that do not require exact knowledge of the shape of the distribution and yet prove to be relatively efficient or valid are required. The term **nonparametric** or **distribution-free** is used for problems of testing hypotheses, and the term **robust** is mainly used for problems of point estimation (→ 396 Statistic; 399 Statistical Estimation; 400 Statistical Hypothesis Testing).

Although the idea of the sign test appears in the work of J. Arbuthnot (1710), the theoretical foundation for nonparametric tests was first given in the proposals for the permutation test by R. A. Fisher (1935), the rank test by F. Wilcoxon (1945), and the test based on U-statistics by H. B. Mann and D. R. Whitney (1947). In the years that followed two important ideas appeared: the concept of asymptotic relative efficiency by E. J. G. Pitman (1948) and the development of the theory of U-statistics by W. Hoeffding (1948). H. Chernoff and I. R. Savage (1958) showed, in studying the asymptotic distribution of a class of rank statistics, that the asymptotic efficiencies of nonparametric tests are incredibly high. These findings accelerated the studies of nonparametric tests; recent progress is summarized in the books by J. Hájek and Z. Šidák [1], M. L. Puri and P. K. Sen [2], R. H. Randles and D. A. Wolfe [3], and P. J. Huber [6].

On the other hand, G. E. P. Box (1953) first coined the term robustness in his sensitivity studies, in which he investigated how the standard statistical procedures obtained under certain assumptions are influenced when such assumptions are violated. Two papers by J. W. Tukey (1960, 1962) provided the initial foundation for robust estimation. J. L. Hodges and E. L. Lehmann (1963) noticed that estimators of location could be derived from nonparametric tests and that these estimators have sometimes much higher efficiency than the sample mean. A similar study for scale was made by S. Kakeshita and T. Yanagawa (1967). Huber (1964) proposed an estimator of location by generalizing the method of least squares. Along with the idea of the influence curve introduced by F. R. Hampel (1974) the estimator proposed by Huber has become a core of subsequent studies of robust estimation. K. Takeuchi (1971) proposed an adaptive estimate that is asymptotically fully efficient for a wide class of underlying distribution functions. The developments of the theory of robust estimation are reviewed by Huber [4–6] and R. V. Hogg [7]. Various proposed estimators are compared in the book by D. F. Andrews et al. [8].

B. The One-Sample Problem

Let $F(x)$ be a †distribution function of a †random variable X, $(X_1, \ldots, X_n)$ be an independent †random sample of size n from $F(x)$, and $(x_1, \ldots, x_n)$ be an observed sample value. The $100p$ percentile of F is denoted by ξ_p, i.e., $F(\xi_p)=p$. For testing the †hypothesis $H:\xi_p \leqslant \xi^0$ against the †alternative hypothesis A: $\xi_p > \xi^0$, the following procedure is proposed. Let $i(x_1, \ldots, x_n)$ be the number of x_i that are greater than ξ^0. A test procedure by the following †test function φ is †uniformly most powerful in some neighborhood of $\xi_p = \xi^0$ for the double exponential distribution, where $\varphi(x_1, \ldots, x_n)$ is defined by the equations

$$\varphi(x_1, \ldots, x_n) = \begin{cases} 1 & \text{when } i(x_1, \ldots, x_n) > c, \\ a & \text{when } i(x_1, \ldots, x_n) = c, \\ 0 & \text{when } i(x_1, \ldots, x_n) < c \end{cases}$$

$(0 < a < 1, 0 \leqslant c \leqslant n)$. This procedure is called the **sign test**.

Suppose that $F(x)$ is symmetric about $x = \xi_{1/2}$. Let R_i^+ be the rank of $|X_i - \xi^0|$ among $|X_1 - \xi^0|, \ldots, |X_n - \xi^0|$, and let $\Psi(t)=1, 0$ according as $t > 0$, $\leqslant 0$. Set

$$S_n(X_1, \ldots, X_n) = \sum_{i=1}^{n} a_n(R_i^+)\Psi(X_i - \xi^0)$$

for some weights $a_n(1), \ldots, a_n(n)$. The following procedure φ, called the **signed rank test**, is also used for testing the hypothesis $H:\xi_{1/2} \leqslant \xi^0$ against the alternative hypothesis $A:\xi_{1/2} > \xi^0$:

$$\varphi(x_1, \ldots, x_n) = \begin{cases} 1 & \text{when } S_n(x_1, \ldots, x_n) > c, \\ a & \text{when } S_n(x_1, \ldots, x_n) = c, \\ 0 & \text{when } S_n(x_1, \ldots, x_n) < c. \end{cases}$$

The procedure with $a_n(i) = i$ is frequently used and is called the **Wilcoxon signed rank test,** which is the uniformly most powerful rank test in a neighborhood of $\xi_{1/2} = \xi^0$ for $F(x) = 1/(1+e^{-x})$, the logistic distribution.

C. The Two-Sample Problem

Let F and G be continuous distribution functions of random variables X and Y, respectively, $(X_1, \ldots, X_m)$ and $(Y_1, \ldots, Y_n)$ be the corresponding random samples, and $(x_1, \ldots, x_m)$ and $(y_1, \ldots, y_n)$ be the respective sample values. Consider the problem of testing the hypothesis $H: F(x) \equiv G(x)$ against the alternative hypothesis $A_1: F(x) \not\equiv G(x)$ or $A_2: F(x) \geqslant G(x)$ for all x and $F(x) \not\equiv G(x)$. When the alternative hypothesis A_2 is true, we say that the random variable Y is **stochastically larger** than X and write $F > G$. A frequently used example of such an alternative hypothesis A_2 is $G(x) = F(x - \theta)$, θ

>0. Let $\mathscr{H}$ be the family of all strictly increasing continuous functions. Then the hypothesis H and the alternative hypothesis A_2 are invariant under the group of transformations of the form $x_i'=h(x_i)$, $y_j'=h(y_j)$ $(i=1,\ldots,m; j=1,\ldots,n; h\in\mathscr{H})$. The †maximal invariant statistic in this case is the rank $(R_1,\ldots,R_m)$ of $(X_1,\ldots,X_m)$ or the rank $(S_1,\ldots,S_n)$ of $(Y_1,\ldots,Y_n)$ when the combined sample $(X_1,\ldots,X_m;Y_1,\ldots,Y_n)$ is ordered in an ascending order. If a test function $\varphi(x_1,\ldots,x_m;y_1,\ldots,y_n)$ satisfies $P_\varphi(F,F)\leqslant\alpha$ and $P_\varphi(F,G)\geqslant\alpha$ for any $F>G$, then φ is considered a desirable test, where

$$P_\varphi(F,G)=\int\ldots\int\varphi(x_1,\ldots,x_m;y_1,\ldots,y_n)\times\prod_i dF(x_i)\prod_j dG(y_j).$$

Lehmann's Theorem. If φ satisfies the conditions stating that $y_j^*>y_j$ $(j=1,\ldots,n)$ yield $\varphi(x_1,\ldots,x_m;y_1^*,\ldots,y_n^*)\geqslant\varphi(x_1,\ldots,x_m;y_1,\ldots,y_n)$, then $P_\varphi(F,G)\geqslant P_\varphi(F,F)$. If in addition φ is a †similar test, then φ is unbiased (→ 400 Statistical Hypothesis Testing C).

The **Wilcoxon test** (or the **Mann-Whitney U-test**) is described by a test function $\varphi=1$ when $U\geqslant c$ and $\varphi=0$ when $U<c$, where U is a †U-statistic defined by

$$U=\frac{1}{mn}\sum_{i=1}^{m}\sum_{j=1}^{n}\psi(x_i,y_j)$$

with $\psi(x,y)=1$ when $x\leqslant y$ and $\psi(x,y)=0$ when $x>y$. This test is similar and unbiased.

Testing the hypothesis $H:F=G$ against the alternative hypothesis $A:F\neq G=F(x/\sigma)$, $\sigma>1$, is another two-sample problem, for which the following test was proposed by T. Tamura. The test function is given by $\varphi=1$ for $U\geqslant c$ and $\varphi=0$ for $U<c$, where $U=\binom{m}{2}^{-1}\binom{n}{2}^{-1}\sum_{i<i'}\sum_{j<j'}\psi(x_i,x_{i'};y_j,y_{j'})$ with $\psi(u,u';v,v')=1$ when $v<u<v'$, $v<u'<v'$ or $v'<u<v$, $v'<u'<v$ and $\psi(u,u';v,v')=0$ otherwise.

The following statistic T_N is used frequently in nonparametric problems. Let $x_1,\ldots,x_m;y_1,\ldots,y_n$ be arranged in order of magnitude. Set $z_k=+1$ or 0 when the kth value $(k=1,2,\ldots,n+m)$ in the arrangement is an x_i or y_j, respectively. For a given set of $N=n+m$ reals $\{e_k\}$, T_N is defined by $T_N=m^{-1}\sum_{k=1}^{N}e_kz_k$. Set $H_N(x)=\lambda_NF_m(x)+(1-\lambda_N)G_n(x)$, where $F_m(x)$ and $G_n(x)$ are the †empirical distribution functions based on $(x_1,\ldots,x_m)$ and $(y_1,\ldots,y_n)$, respectively, and $0<\lambda_0\leqslant\lambda_N=m/N\leqslant1-\lambda_0<1$. Then T_N is represented by the integral

$$\int J_N(H_N(x))\,dF_m(x)$$

with $e_k=J_N(k/N)$. Chernoff and Savage [9] proved that under some regularity conditions the asymptotic distribution of T_N is normal and that the asymptotic mean μ and the variance σ^2 of T_N are given by

$$\mu=\int J(H(x))\,dF(x),$$

$$N\sigma^2=2(1-\lambda)\Bigg\{\iint_{x<y}G(x)(1-G(y))\times J'(H(x))J'(H(y))\,dF(x)\,dF(y)+\frac{1-\lambda}{\lambda}\iint_{x<y}F(x)(1-F(y))\times J'(H(x))J'(H(y))\,dG(x)\,dG(y)\Bigg\},$$

where $J(H)=\lim_{N\to\infty}J_N(H)$, $H(x)=\lambda F(x)+(1-\lambda)G(x)$, and $\lambda=\lim\lambda_N$. When $e_k=k/N$, the statistic T_N is equivalent to the †U-statistic in the Wilcoxon test. When e_k is the mean $E(Z_k)$ of the kth order statistic Z_k in an independent sample of size N from $N(0,1)$, then the test by T_N is called the **Fisher-Yates-Terry normal score test**. When J is the inverse function of the distribution function of $N(0,1)$, then the test is called the **van der Waerden test**.

D. The k-Sample Problem

Let $(X_{ij}, j=1,\ldots,n_i)$ be a random sample of size n_i from the population with a distribution function $F_i(x)$ for each $i=1,\ldots,k$. The **k-sample problem** is concerned with testing the hypothesis $H:F_1(x)=\ldots=F_k(x)$ against an alternative hypothesis A_1: not all the $F_i(x)$ are equal, $A_2:F_i(x)=F(x-\theta_i)$ with $\theta_i\not\equiv\theta$, or $A_3:F_i(x)=F(x/\sigma_i)$ with $\sigma_i\not\equiv\sigma$. Several tests have been proposed for this problem, using quadratic forms of the vector-valued U-statistic $\mathbf{U}=(U^1,\ldots,U^k)$ whose coordinates U^i are defined by means of a function

$$\psi^i(x_{11},\ldots,x_{1m_{1i}};\ldots;x_{k1},\ldots,x_{km_{ki}}),\qquad i=1,\ldots,k.$$

When $N=\sum n_i\to\infty$ with $n_i=\rho_iN$, $0<\rho_i<1$, and $\sum\rho_i=1$, then

$$\mathbf{V}=(\sqrt{N}(U^1-E(U^1)),\ldots,\sqrt{N}(U^k-E(U^k)))$$

has asymptotically a †multivariate normal distribution $N(0,\Sigma)$. Let B be the projection matrix corresponding to the eigenspace for the zero eigenvalues of the matrix Σ, and let Λ be a matrix such that $\Lambda B=0$, $\Sigma\Lambda=I-B$. The statistic $\mathbf{V}\Lambda^t\mathbf{V}$ has asymptotically a †noncentral chi-square distribution with degrees of freedom $=$ rank Σ. Several kinds of test represented by a critical region of the form $\mathbf{V}\Lambda\mathbf{V}'>c$ are proposed, among which the **Kruskal-Wallis test** is a particular one having

$$\psi^i(x_1;\ldots;x_k)=\sum_\alpha\frac{1}{n_in_\alpha}\delta(x_\alpha,x_i),\qquad i=1,\ldots,k,$$

as basic functions, where $\delta(x, y)=1$ when $x<y$ and $\delta(x, y)=0$ otherwise.

E. Asymptotic Relative Efficiency of Tests

If there is more than one test procedure for a given testing problem, then one may wish to compare these procedures. Let $\{\varphi_n\}$ and $\{\psi_n\}$ be two sequences of level α tests, where φ_n and ψ_n are test functions based on a sample of size n. The †power functions of φ_n and ψ_n are denoted by $\beta(\theta \mid \varphi_n)$ and $\beta(\theta \mid \psi_n)$, respectively. Let θ be a real parameter and $\{\theta_i\}$ be such that $\theta_i \to \theta_0$ as $i \to \infty$. Consider a hypothesis $\theta=\theta_0$ and a sequence $\{\theta_i\}$ of alternative hypotheses. If, for any increasing sequences $\{n_i\}$ and $\{n_i^*\}$ of positive integers satisfying $\alpha<\lim \beta(\theta_i \mid \varphi_{n_i})= \lim \beta(\theta_i \mid \psi_{n_i^*})<1$, $\lim n_i^*/n_i$ $(=e(\{\varphi_n\},\{\psi_n\})$, say) exists and is independent of α and $\lim \beta(\theta_i \mid \varphi_{n_i})$, then e is called Pitman's asymptotic relative efficiency of $\{\varphi_n\}$ against $\{\psi_n\}$. Suppose further that the tests $\{\varphi_n\}$ and $\{\psi_n\}$ are based on statistics $T_n=t_n(\mathbf{X})$ and $T_n^*=t_n^*(\mathbf{X})$, respectively, in the following manner:

$$\varphi_n(\mathbf{x})= \begin{cases} 1 & \text{when } t_n(\mathbf{x})>c, \\ a & \text{when } t_n(\mathbf{x})=c, \\ 0 & \text{when } t_n(\mathbf{x})<c, \end{cases}$$

$$\psi_n(\mathbf{x})= \begin{cases} 1 & \text{when } t_n^*(\mathbf{x})>c^*, \\ b & \text{when } t_n^*(\mathbf{x})=c^*, \\ 0 & \text{when } t_n^*(\mathbf{x})<c^*, \end{cases}$$

where $\mathbf{X}=(X_1, \ldots, X_n)$ and $\mathbf{x}=(x_1, \ldots, x_n)$. Put $\theta_0=0$ and $\theta_n=k/\sqrt{n}$ ($k=$constant) for simplicity. If T_n *and* T_n^* are asymptotically normal, then under some conditions e is given by the formula

$$e=\lim \frac{(dE_\theta(T_n)/d\theta \mid_{\theta=0})^2 \sigma_0^2(T_n^*)}{(dE_\theta(T_n^*)/d\theta \mid_{\theta=0})^2 \sigma_0^2(T_n)}.$$

As an example, consider a two-sample problem on a †location parameter. If the population distribution is normal and the Wilcoxon test is used to test the hypothesis of equality of means, then its asymptotic relative efficiency against Student's test is $3/\pi$. For the same problem, the asymptotic relative efficiencies of the Fisher-Yates-Terry normal score test and the van der Waerden test against Student's test are both unity. For the hypothesis of equality of means in the k-sample problem, the asymptotic relative efficiency of the Kruskal-Wallis test against the F-test is $3/\pi$, provided that the sample is distributed normally.

F. Kolmogorov-Smirnov Tests

Let $F_n(x)$ be the empirical distribution function based on a random sample of size n from $F_0(x)$. Set

$$d_n=\sup |F_n(x)-F_0(x)|,$$

$$D_n=\sup (F_n(x)-F_0(x)),$$

$$s_n=\sup_{a \leqslant F_0(x)} \left|\frac{F_n(x)-F_0(x)}{F_0(x)}\right|,$$

$$S_n=\sup_{a \leqslant F_0(x)} \frac{F_n(x)-F_0(x)}{F_0(x)}.$$

Then

$$\lim_{n \to \infty} P_r(d_n<z/\sqrt{n})=L(z) = \sum_{k=-\infty}^{\infty}(-1)^k e^{-2k^2 z^2},$$

$$\lim_{n \to \infty} P_r(D_n<z/\sqrt{n})=K(z)=1-e^{-2z^2},$$

$$P_r(D_n<D)=1-D \sum_{j=0}^{[n(1-D)]}\binom{n}{j} \times\left(1-D-\frac{j}{n}\right)^{n-j}\left(D+\frac{j}{n}\right)^{j-1},$$

$$\lim_{n \to \infty} P_r(s_n<z/\sqrt{n})=\frac{4}{\pi} \sum_{k=0}^{\infty}(-1)^k \times(2k+1)^{-1} e^{-((2k+1)^2 \pi^2/8)((1-a)/a) z^2},$$

$$\lim_{n \to \infty} P_r(S_n<z/\sqrt{n})=\sqrt{\frac{2}{\pi}} \int_0^{z\sqrt{a/(1-a)}} e^{-t^2/2}\, dt.$$

The statistics d_n, D_n, s_n, S_n are frequently used to test the hypothesis $F(x)=F_0(x)$. (This problem is called testing goodness of fit.)

In a two-sample problem, let $F_m(x)$ and $G_n(x)$ be two empirical distribution functions based on samples of sizes m and n from $F(x)$ and $G(x)$, respectively. Set

$$d_{m,n}=\sup |F_m(x)-G_n(x)|,$$

$$D_{m,n}=\sup (F_m(x)-G_n(x)).$$

If the hypothesis $F=G$ is true, we have

$$\lim P_r(d_{m,n}<z/\sqrt{N})=L(z),$$

$$\lim P_r(D_{m,n}<z/\sqrt{N})=K(z),$$

provided that m and n tend to ∞ so that $N=mn/(m+n) \to \infty$ and m/n is constant. Taking account of these facts, $d_{m,n}$ and $D_{m,n}$ are used to test the hypothesis $F=G$. The tests using the statistics d_n, D_n, $d_{m,n}$, $D_{m,n}$ are called **Kolmogorov-Smirnov tests.**

G. Interval Estimation

Let $(X_1, \ldots, X_n)$ be an independent random sample from the population with a distribution function $F(x-\theta)$, where θ is an unknown location parameter, and $(x_1, \ldots, x_n)$ be its observed value. Suppose that $F(x)$ is continuous and symmetric about the origin.

Using the statistic $S_n = S(x_1, \ldots, x_n)$ for the one-sample nonparametric test for testing the hypothesis $H: \theta = 0$, a †confidence interval of θ is constructed as follows. For an appropriately given γ $(0 < \gamma < 1)$, select constants d_1 and d_2 in the range of $S(x_1, \ldots, x_n)$ that satisfy

$$P_0\{d_1 < S(X_1, \ldots, X_n) < d_2\} = 1 - \gamma,$$

where P_0 means the probability under the hypothesis $H: \theta = 0$. If there exist statistics $L_s(x_1, \ldots, x_n)$ and $U_s(x_1, \ldots, x_n)$ such that $L_s(x_1, \ldots, x_n) \leqslant \theta < U_s(x_1, \ldots, x_n)$ if and only if $d_1 < S(x_1 - \theta, \ldots, x_n - \theta) < d_2$ for all θ, then the confidence interval of θ with $100(1-\gamma)\%$ confidence coefficient is given by $(L_s(x_1, \ldots, x_n), U_s(x_1, \ldots, x_n))$. This interval is distribution-free, i.e., it holds that

$$P\{L_s(X_1, \ldots, X_n) < \theta < U_s(X_1, \ldots, X_n)\} = 1 - \gamma \quad \text{for all } F.$$

When S_n is the statistic for the Wilcoxon signed rank test, L_s and U_s are given by $L_s = W_{(M+1-d_2)}$ and $U_s = W_{(M-d_1)}$, where $M = n(n+1)/2$ and $W_{(1)} \leqslant \ldots \leqslant W_{(M)}$ are ordered values for M averages $(x_i + x_j)/2$ $(i \leqslant j = 1, 2, \ldots, n)$.

H. Point Estimation

Let $(X_1, \ldots, X_n)$ be an independent random sample from the population with a distribution function $F(x - \theta)$, where θ is an unknown location parameter, and let $(x_1, \ldots, x_n)$ be its observed value.

There are four methods of constructing robust estimators of θ. Let $X_{(1)} < \ldots < X_{(n)}$ be ordered values of $X_1, \ldots, X_n$. The first method is to use $T_n = a_1 X_{(1)} + \ldots + a_n X_{(n)}$ for some given constants $a_1, \ldots, a_n$ such that $\sum_{i=1}^n a_i = 1$. T_n is called the **L-estimator**. An example is

$$T_n(\alpha) = (pX_{([\alpha n]+1)} + X_{([\alpha n]+2)} + \ldots + pX_{(n-[\alpha n])})/n(1-2\alpha),$$

where $p = 1 + [\alpha n] - \alpha n$. This estimator is called the **α-trimmed mean**. Let J be a real-valued function such that $\int_0^1 J(t)\,dt = 1$, and set $a_k = \int_{(k-1)/n}^{k/n} J(t)\,dt$; then as $n \to \infty$, T_n converges to $T(F) = \int_0^1 J(t) F^{-1}(t)\,dt$ in probability. Suppose that F is a distribution function having an †absolutely continuous density function f. Denote the derivative of f by f', and let $I(F)$ be the †Fisher information on θ. Set $\psi(t) = -f'(t)/f(t)$ and $J(t) = \psi'(F^{-1}(t))/I(F)$. Then Chernoff, J. L. Gastwirth, and M. V. Johns [10] proved that under some regularity conditions T_n is an †asymptotically efficient estimator of θ for F.

Let ρ be a real-valued (usually convex) function of a real parameter with derivative $\psi = \rho'$. The second method is to estimate θ by T_n by minimizing $\sum_{i=1}^n \rho(x_i - T_n)$ or by satisfying

$$\sum_{i=1}^n \psi(x_i - T_n) = 0.$$

T_n is called the **M-estimator**. When $\rho(t) = t^2$, it agrees with the least squares estimator. Let $\Phi(x)$ be the distribution function of the standard normal distribution, $H(x)$ be an arbitrary continuous distribution function which is symmetric about the origin, $\mathscr{F}$ be a class of distribution functions of the form $F(x) = (1-\varepsilon)\Phi(x) + \varepsilon H(x)$ for a given ε $(0 < \varepsilon < 1)$, and $V(\rho, F)$ be the asymptotic variance of T_n. Huber [11] proved that ρ minimizing $\sup_{F \in \mathscr{F}} V(\rho, F)$ is given by $\rho_K(t) = t^2/2$, $K|t| - K^2/2$ defined for $|t| \leqslant K$, $> K$, respectively, for some constant K. Under quite general conditions, the M-estimator converges as $n \to \infty$ to $T(F)$ in probability, which is defined by $\int \psi(x - T(F))\,dF(x) = 0$. If $\psi(t) = -f'(t)/f(t)$ is chosen for $\psi(x)$, then T_n is the †maximum likelihood estimator of θ for F and is asymptotically efficient under some regularity conditions. Generally, the M-estimator defined above is not scale invariant. A scale invariant version of the M-estimator is obtained by replacing the defining equation by

$$\sum_{i=1}^n \psi\left(\frac{x_i - T_n}{S_n}\right) = 0, \tag{1}$$

where S_n is any robust estimate of scale, e.g., the median of $\{|x_i - M|/0.6745\}_{i=1,2,\ldots,n}$ where M is the sample median, or by solving the simultaneous equations (1) and

$$\sum_{i=1}^n \chi\left(\frac{x_i - T_n}{S_n}\right) = 0$$

with respect to T_n and S_n. In the context of the maximum likelihood estimation, χ is chosen to be $\chi(t) = t\psi(t) - 1$ (→ 399 Statistical Estimation P).

The third method employs nonparametric tests for testing the hypothesis $H: \theta = 0$ against the alternative hypothesis $A: \theta > 0$. Let J be a real-valued and nondecreasing function such that $\int_0^1 J(t)\,dt = 0$ and $R_k^+(\theta)$ be the rank of $|X_k - \theta|$ among $|X_1 - \theta|, \ldots, |X_n - \theta|$. Set $\Psi(t) = 1, 0$ according as $t > 0$, $\leqslant 0$ and $S(X_1 - \theta, \ldots, X_n - \theta) = \sum_{k=1}^n J((R_k^+(\theta) + n)/(2n+1))\Psi(X_k - \theta)$. Let

$$\theta^* = \sup\{\theta; S(X_1 - \theta, \ldots, X_n - \theta) > \mu\},$$

$$\theta^{**} = \inf\{\theta; S(X_1 - \theta, \ldots, X_n - \theta) < \mu\},$$

where μ is the expected value of $S(X_1, \ldots, X_n)$ under the hypothesis $H: \theta = 0$. Then an estimator of θ is defined by $T_n = (\theta^* + \theta^{**})/2$. Hodges and Lehmann [12] first proposed this technique, and this estimator is called the **R-estimator**. When $F(x)$ is symmetric about the origin and $J(t) = t - \frac{1}{2}$, S tends to be the statistic for the Wilcoxon signed rank test, and the

R-estimator reduces to the median of $n(n+1)/2$ averages $(X_i+X_j)/2$ $(1 \leqslant i \leqslant j \leqslant n)$. As $n \to \infty$, the R-estimator converges in probability to $T(F)$, defined by

$$\int J\left(\frac{F(x)+1-F(2T(F)-x)}{2}\right) dF(x)=0.$$

For symmetric F, the R-estimator defined by the statistic S with $J(t)=-f'(F^{-1}(t))/f(F^{-1}(t))$ is asymptotically efficient under some regularity conditions.

Although the above three methods provide robust estimators, which are seldom affected by outlying observations or contamination by gross errors, their behavior still depends on F. The last method of constructing robust estimators consists of estimating θ adaptively by utilizing information on the shape of F. Among these, a striking one is the asymptotically fully efficient estimator for a wide class of F proposed by Takeuchi [13]. The estimator is constructed by using subsamples of size K $(K<n)$ drawn from the original sample, estimating the elements of the †covariance matrix of the order statistics by U-statistics, and selecting the best weights of the L-estimator. L. A. Jaeckel [14] made an α-trimmed mean adaptive estimator by selecting an α that minimizes the estimated asymptotic variance.

I. The Influence Curve

Let $T(F)$ be a functional of a distribution function F, and let an estimator T_n of θ calculated from an empirical distribution function F_n converge to $T(F)$ in probability as $n \to \infty$. A real-valued function $\mathrm{IC}(x; F, T)$ defined by

$$\mathrm{IC}(x; F, T)=\lim_{\varepsilon \to 0} \frac{T((1-\varepsilon)F+\varepsilon\delta_x)-T(F)}{\varepsilon} \quad \text{for all } x$$

is called the **influence curve**, where δ_x is the distribution function of a point mass 1 at x. The curve was first introduced by Hampel [15] to study the stability aspect of estimators against a small change of F. As an example, when F is symmetric about the origin, the influence curve for the α-trimmed mean $\mathrm{IC}(x; F, T)$ is given by

$$\begin{aligned} &F^{-1}(\alpha)/(1-2\alpha) && \text{when } x<F^{-1}(\alpha),\\ &x/(1-2\alpha) && \text{when } F^{-1}(\alpha) \leqslant x < F^{-1}(1-\alpha),\\ &F^{-1}(1-\alpha)/(1-2\alpha) && \text{when } x>F^{-1}(1-\alpha). \end{aligned}$$

By substituting the empirical distribution function F_n for F in $T(F)$, we can represent the robust estimators discussed in Section H, at least approximately, by $T_n=T(F_n)$. Under some conditions, it can be proved that as $n \to \infty$,

$$n^{1/2}\left(T(F_n)-T(F)-\frac{1}{n}\sum_{i=1}^{n} \mathrm{IC}(X_i; F, T)\right) \to 0$$

in probability. Thus it follows that $n^{1/2}(T_n - T(F))$ is asymptotically normally distributed with asymptotic variance $\int (\mathrm{IC}(x; F, T))^2\, dF(x)$.

J. The Regression Problem

Consider the linear regression problem (→ 403 Statistical Models D)

$$X_i=\sum_{j=1}^{p} \theta_j a_{ij}+\varepsilon_i, \qquad i=1, 2, \ldots, n,$$

where the X_i are observable variables, the θ_j are regression coefficients to be estimated, the a_{ij} are given constants, and $\varepsilon_1, \varepsilon_2, \ldots, \varepsilon_n$ are identically and independently distributed random errors whose distribution function is given by $F(x)$. The idea of the M-estimator is a direct generalization of the method of least squares; namely, to adopt $(\hat{\theta}_1, \ldots, \hat{\theta}_p)$ as an estimator for $(\theta_1, \ldots, \theta_p)$ that minimizes $\sum_{i=1}^{n} \rho(X_i-\sum_j \theta_j a_{ij})$ for some function ρ such as the one described above.

R-estimators of the regression coefficients are obtained by minimizing $\sum_{i=1}^{n} a_n(R_i)\Delta_i$, where $\Delta_i=X_i-\sum_j \theta_j a_{ij}$, R_i is the rank of Δ_i among $\Delta_1, \ldots, \Delta_n$, and $a_n(\cdot)$ is some monotone function satisfying $\sum_{i=1}^{n} a_n(i)=0$. It has been proved that minimizing $\sum_{i=1}^{n} a_n(R_i)\Delta_i$ is asymptotically equivalent to minimizing

$$\sum_{j=1}^{p} \left| \sum_{i=1}^{n} a_n(R_i) a_{ij} \right|,$$

the properties of which were first studied by J. Jurečková [16].

K. Dependence

Let $(X_1, Y_1), \ldots, (X_n, Y_n)$ be random samples from a population with a bivariate distribution function, R_i be the rank of X_i among $X_1, \ldots, X_n$ when they are rearranged in an ascending order, and S_j be the rank of Y_j among $Y_1, \ldots, Y_n$ defined similarly as R_i. Various quantities are devised to measure the degree of dependence between X and Y.

(1) Spearman's Rank Correlation. Set $d_i=R_i-S_i$. Then $r_s=1-6\sum_i d_i^2/(n^3-n)$ is called **Spearman's rank correlation**. If there is no dependence between X and Y, i.e., if the X_i and Y_j are independent random variables, then $E(r_s)=0$ and $V(r_s)=(n-1)^{-1}$.

(2) Kendall's Rank Correlation. Take pairs (R_i, S_i) and (R_j, S_j). If $(R_j-R_i)(S_j-S_i)>0$, set

$\varphi_{ij}=1$; otherwise, $\varphi_{ij}=-1$. The statistic $r_k=\binom{n}{2}^{-1}\sum\varphi_{ij}$ is called **Kendall's rank correlation**, where $\sum$ runs over all possible pairs chosen from $(R_1, S_1), \ldots, (R_n, S_n)$.

If there is no dependence between X and Y, then $E(r_k)=0$ and $V(r_k)=2(2n+5)/(9n(n-1))$.

(3) Rankit Correlation. R_i and S_i, $i=1, \ldots, n$, are replaced by the corresponding normal scores, i.e., the means of order statistics in an independent sample of size n from $N(0,1)$; then the usual †sample correlation coefficient is calculated from these scores. This correlation coefficient r_R is called rankit correlation; and if there is no dependence between X and Y, then $E(z)=0$, $V(z)=(n-3)^{-1}$, asymptotically, where $z=\frac{1}{2}\log\frac{1+r_R}{1-r_R}$ (†Fisher's z-transformation).

These statistics are used to test the hypothesis of independence.

References

[1] J. Hájek and Z. Šidák, Theory of rank tests, Academic Press, 1967.
[2] M. L. Puri and P. K. Sen, Nonparametric methods in multivariate analysis, Wiley, 1971.
[3] R. H. Randles and D. A. Wolfe, Introduction to the theory of nonparametric statistics, Wiley, 1979.
[4] P. J. Huber, Robust statistics: A review, Ann. Math. Statist., 43 (1972), 1041–1067.
[5] P. J. Huber, Robust regression: Asymptotics, conjectures and Monte Carlo, Ann. Statist., 1 (1973), 799–821.
[6] P. J. Huber, Robust statistics, Wiley, 1981.
[7] R. V. Hogg, Adaptive robust procedures: A partial review and some suggestions for future applications and theory, J. Amer. Statist. Assoc., 69 (1974), 909–923.
[8] D. F. Andrews, P. J. Bickel, F. R. Hampel, P. J. Huber, W. H. Rogers, and J. W. Tukey, Robust estimates of location: Survey and advances, Princeton Univ. Press, 1972.
[9] H. Chernoff and I. R. Savage, Asymptotic normality and efficiency of certain nonparametric test statistics, Ann. Math. Statist., 29 (1958), 972–994.
[10] H. Chernoff, J. L. Gastwirth, and M. V. Johns, Asymptotic distribution of linear combinations of functions of order statistics with application to estimation, Ann. Math. Statist., 38 (1967), 52–72.
[11] P. J. Huber, Robust estimation of a location parameter, Ann. Math. Statist., 35 (1964), 73–101.
[12] J. L. Hodges and E. L. Lehmann, Estimates of location based on rank tests, Ann. Math. Statist., 34 (1963), 598–611.
[13] K. Takeuchi, A uniformly asymptotically efficient estimator of a location parameter, J. Amer. Statist. Assoc., 66 (1971), 292–301.
[14] L. A. Jaeckel, Some flexible estimates of location, Ann. Math. Statist., 42 (1971), 1540–1552.
[15] F. R. Hampel, The influence curve and its role in robust estimation, J. Amer. Statist. Assoc., 69 (1974), 383–393.
[16] J. Jurečková, Nonparametric estimate of regression coefficients, Ann. Math. Statist., 42 (1971), 1328–1338.

372 (XXI.3) Roman and Medieval Mathematics

The Romans were interested in mathematics for everyday use; their arithmetic consisted of computation (by means of the **abacus**), weights and measures, and money. For their monetary system, they developed a computational method using duodecimal fractions. Julius Caesar (102?–44 B.C.), known for his calendar reform in 46 B.C., also undertook to measure his territory, which aroused a demand for land surveying techniques. Books on practical geometry which provided this knowledge were called *gromatics* (a "groma" was a land surveying instrument). Toward the end of the Western Roman Empire (476), Greek mathematics was studied; during this period Boethius (c. 480–524) wrote his two books on arithmetic and geometry. The former was a summarized translation of Nichomachus' book, and the latter included propositions from the first three books of Euclid's *Elements* (without proof) and practical geometry.

Music, astronomy, geometry, and arithmetic, which constituted the "mathemata" of Plato's Academy (closed in 529), were treated as the "quadrivium" (the four major subjects) in the *Encyclopedia* of Martianus Capella, Cassiodorus, Isidorus, Hispalensis, and others. After the establishment of the Roman Church in the 5th century, the quadrivium was to be studied for the glory of God. Books on mathematics from this period laid emphasis on the computation of an ecclesiastical calendar and mystical interpretations of integers, as seen in books by Bede Venerabilis, Alcuin, and Maurus from the 7th through 9th centuries.

Arabian science was imported first through Spain—under Moorish influence beginning in 711, the year of the fall of the Visigoths—and then through the Crusades (1096–1270). Computation with figures, originating in India, replaced the abacus in the 12th century, when

Arabian books on arithmetic and algebra, along with Greek books on geometry and astronomy (such as books by Euclid and Ptolemy), were translated into Latin. Italian merchants, whose occupation necessitated computation, rapidly adopted the new system. Representative books of this period are *Liber abaci* (1202) and *Practica geometrica* (1220) by Leonardo da Pisa (also known as Fibonacci, c. 1170–1250). The former includes the four arithmetic operations, showing Indian influence, commercial arithmetic, and algebra. The new methods were not limited to merchants. The French bishop Nicole Oresme (c. 1323–1382), who influenced Leonardo da Vinci (1452–1519), introduced fractional exponents and conceived the graphic representation of temperature, a precursor of coordinates and functions.

From the 11th century, universities developed from theological seminaries, first in Italian cities such as Bologna and Palermo, and later in Paris, Oxford, and Cambridge. Mathematics was taught in these universities, although no remarkable creative contributions were made. However, theologians such as Albertus Magnus (c. 1193–1280) and Thomas Aquinas (1225?–1274) discussed infinity in a way that went beyond Greek thought and thus helped to lay a basis for the modern philosophy of mathematics.

References

[1] M. Cantor, Vorlesungen über Geschichte der Mathematik, Teubner, I, 1880; II, 1892.
[2] M. Clagett, Greek science in antiquity, Abelard-Schuman, 1955.
[3] M. Clagett, Archimedes in the Middle Ages I, II, Univ. of Wisconsin Press, 1964.
[4] M. Clagett, The science of mechanics in the Middle Ages, Univ. of Wisconsin Press, 1959.
[5] G. Sarton, Introduction to the history of science I, From Homer to Omar Khayyám, Carnegie Institution, 1927.

373 (XVIII.13) Sample Survey

A. General Remarks

The sample survey is a means of getting statistical information about a certain aggregate from the observation of some but not all of it. The aggregate concerned is usually called the **finite population**, and the observed part is called the **sample**. Introducing a random mechanism, J. Neyman established a method of ascertaining objectively the reliability of such information. This method is mainly applied to demographic statistical surveys and opinion polls, but it is also applicable to random samples of physical materials. We briefly sketch the mathematical structure of this method without going into detail about technical problems that arise in the practical survey situation.

Suppose that the population consists of N units, where N is called its **size**. Each unit has some characteristic α, which is an element of some set Ω. The set of all characteristics of the units in the population is designated by $\theta = \{\alpha_1, \ldots, \alpha_N\}$, which we regard as a parameter. The set of all possible θ is denoted by $\Theta \subset \Omega^N$.

Suppose that one unit is chosen and observed according to some procedure, the index number of the unit in the population is J, and the observed characteristic is X. It is assumed that the observation is without error, hence $X = \alpha_J$.

Denote the whole sample by $(\mathbf{J}, \mathbf{X}) = (J_1, \ldots, J_n, X_1, \ldots, X_n)$, where n is the **sample size** and $X_i = \alpha_{J_i}$. The sample size n may be a random variable, and among $J_1, \ldots, J_n$, duplication may be allowed. The probabilistic scheme for $\mathbf{J}$ is called the **sampling procedure**, and if it satisfies the condition

(c) $\Pr\{J_i = j\}$ is independent of α_{J_i} and of $X_{i+1}, \ldots, X_n$ (but may depend on $X_1, \ldots, X_{i-1}$),

it is called a **random sampling procedure**. Moreover, if the †joint distribution of $\mathbf{J}$ is independent of θ, it is called **regular**. Specifically, if n is constant and $\Pr\{\mathbf{J}\}$ is symmetric in $\mathbf{J}$, it is called **uniform**.

The two main mathematical problems of sample surveys are to determine a random sampling procedure and to provide methods whereby statistical inferences can be made concerning θ (→ 401 Statistical Inference).

B. The Problem of Inference

Condition (c) is assumed. The probability of $(\mathbf{J}, \mathbf{X})$ is given by

$$\Pr\{(\mathbf{J}, \mathbf{X}) = (j_1, \ldots, j_n, X_1, \ldots, X_n)\}$$
$$= \Pr\{J_1 = j_1\} \chi_\theta(X_1) \Pr\{J_2 = j_2 \mid J_1, X_1\} \chi_\theta(X_2)$$
$$\ldots \Pr\{J_n = j_n \mid J_1, X_1, \ldots, J_{n-1}, X_{n-1}\} \chi_\theta(X_n),$$

where $\chi_\theta(X_i)$ is defined as 1 if $X_i = \alpha_{J_i}$ and 0 otherwise. This formula can be shortened to the form

$$\Pr\{(\mathbf{J}, \mathbf{X})\} = P(\mathbf{J}, \mathbf{X}) \chi_\theta(\mathbf{X}, \mathbf{J}), \tag{1}$$

where

$$\chi_\theta(\mathbf{X}, \mathbf{J}) = \begin{cases} 1 & \text{if } X_i = \alpha_{J_i},\ i = 1, \ldots, n, \\ 0 & \text{otherwise.} \end{cases}$$

Expression (1) is the fundamental model for the sample survey problem. Note that $P(\mathbf{J}, \mathbf{X})$ is independent of the parameter θ. Therefore if we let $\mathbf{I} = (I_1, \ldots, I_m)$ $(I_1 < \ldots < I_m)$ be the set of numbers in $(J_1, \ldots, J_n)$ after deleting duplications, and let $\mathbf{Y} = (Y_1, \ldots, Y_m)$ be the corresponding X values, then the joint distribution of $\mathbf{I}$ and $\mathbf{Y}$ can also be expressed as

$$\Pr\{(\mathbf{I}, \mathbf{Y})\} = P(\mathbf{I}, \mathbf{Y}) \chi_\theta(\mathbf{Y}, \mathbf{I}),$$

where

$$\chi_\theta(\mathbf{Y}, \mathbf{I}) = \begin{cases} 1 & \text{if } Y_j = \alpha_{I_j},\ j = 1, \ldots, m, \\ 0 & \text{otherwise.} \end{cases}$$

Since $\chi_\theta(\mathbf{Y}, \mathbf{I}) = \chi_\theta(\mathbf{X}, \mathbf{J})$ for all θ, the †conditional probability distribution of $(\mathbf{J}, \mathbf{X})$ for given $(\mathbf{I}, \mathbf{Y})$ is independent of $\chi_\theta(\mathbf{X}, \mathbf{J})$, and hence $(\mathbf{J}, \mathbf{X})$ is a †sufficient statistic. According to the general theory of sufficient statistics, we can restrict ourselves to the class of procedures depending only on $(\mathbf{I}, \mathbf{Y})$.

C. Estimation

Suppose that $g(\theta) = g(\alpha_1, \ldots, \alpha_N)$ is a real parameter whose unbiased estimators are under discussion.

Theorem. There exists an unbiased estimator of $g(\theta)$ if and only if there exists a decomposition

$$g(\theta) = \sum h_\nu(\alpha_{j_1(\nu)}, \ldots, \alpha_{j_n(\nu)}),$$
$$\Pr\{\mathbf{I} = (j_1(\nu), \ldots, j_n(\nu))\} > 0, \qquad \nu = 1, 2, \ldots. \tag{2}$$

If the sampling procedure is regular, the second condition can be replaced by $\Pr\{\mathbf{I} \supset (j_1(\nu), \ldots, j_n(\nu))\} > 0$, $\nu = 1, 2, \ldots$. Hence, for example, if α_i is real and the sampling procedure is regular, $\bar{\alpha} = \sum \alpha_i / N$ is †estimable if and only if $\Pr\{\mathbf{I} \ni i\} > 0$ for all i, and $\sigma_\alpha^2 = \sum(\alpha_i - \bar{\alpha})^2/(N-1) = \sum\sum(\alpha_i - \alpha_j)^2/N(N-1)$ is estimable if and only if $\Pr\{\mathbf{I} \ni i, j\} > 0$ for all i and j. Also, $\prod_{i=1}^N \alpha_i$ is not estimable unless $\Pr\{\mathbf{I} = (1, \ldots, N)\} > 0$. The decomposition (2) is not unique, and corresponding to different decompositions, different unbiased estimators are derived. Also, for the case of regular sampling procedures, for any $\theta=\theta_0$ it is always possible to construct an unbiased estimator $g(\theta)$ such that $\Pr\{\hat{g}(\theta)\equiv g(\theta_0)\mid\theta=\theta_0\}=1$ if $g(\theta)$ is estimable. Hence the variance of the locally best unbiased estimator is always 0, and the †uniformly minimum variance unbiased estimators exist only in the trival case.

If some kind of symmetry exists among the population units as well as the sampling procedure and the parameter, it would be natural to require the same kind of symmetry for the estimators. Let G be a group of permutations among N numbers. Assume that for any $\theta\in\Theta$ and $\gamma\in G$, we have $\gamma\theta\in\Theta$ and $g(\gamma\theta)=g(\theta)$. If $\Pr\{\gamma\mathbf{J}\}=\Pr\{\mathbf{J}\}$ for all γ, then the sampling procedure is said to be **invariant** with respect to G. An estimator is also called invariant if its value does not change under any permutation $\gamma\in G$ of the numbers of sample units. Thus if G is the set of all permutations (i.e., the †symmetric group), then the invariant estimator is a function of $\mathbf{Y}$ (or $\mathbf{X}$) only. Moreover, if the dimension m of $\mathbf{Y}$ is constant, $\mathbf{Y}$ is complete (under some mild conditions); hence there exists a unique minimum variance invariant unbiased estimator.

When there is some additional information, it can be represented by **auxiliary variables** $\beta_1,\ldots,\beta_N$, which are known and assumed to have some relation to $\alpha_1,\ldots,\alpha_N$. Assume that the α_i are real numbers and that the parameter to be estimated is $\theta=\bar{\alpha}=(\sum\alpha_i)/N$. If we can assume that α_i and β_i are approximately proportional, we can estimate the unknown population mean $\bar{\alpha}$ by $\hat{\bar{\alpha}}^*=(\bar{X}/\bar{Z})\times\bar{\beta}$ where $\bar{X}$ is the sample mean of the α's and $\bar{Z}$ the sample mean of the β's. Although $\hat{\bar{\alpha}}^*$ is not unbiased, we may expect that it has small error if the relation between two variables is close. $\hat{\bar{\alpha}}^*$ is usually called the ratio estimator.

In practical research, as an estimator of the population total $A=\sum\alpha_i$, we usually use $\hat{A}=\sum X_i/P_i$, where $P_i=\Pr\{\mathbf{J}\ni i\}$. Its variance is $V(\hat{A})=\sum\sum(P_iP_j-P_{ij})(\alpha_i/P_i-\alpha_j/P_j)^2$ and is estimated by $v(\hat{A})=\sum\sum\{(P_iP_j-P_{ij})/P_{ij}\}(X_i/P_i-X_j/P_j)^2$, where $P_{ij}=\Pr\{\mathbf{J}\supset(i,j)\}$. When N is unknown, it can be estimated by the same procedure as $\hat{A}$ (say $\hat{N}$), and the population mean $\bar{\alpha}$ can be estimated by $\hat{\bar{\alpha}}=\hat{A}/\hat{N}$, which is called a ratio estimator. $\hat{\bar{\alpha}}$ is biased except when N is known.

D. Asymptotic Confidence Intervals

It is usually impossible to obtain any meaningful †confidence interval based on exact small-sample theory. But when the α_i are real and the sampling procedure is uniform and without replacement, the sample mean $\bar{X}$ is asymptotically normal with mean $\bar{\alpha}$ and variance $\frac{1}{n}\left(1-\frac{n}{N}\right)\sigma_\alpha^2$, where $\sigma_\alpha^2=\frac{1}{N-1}\sum(\alpha_i-\bar{\alpha})^2$, as N and $n\to\infty$, and $\limsup n/N<1$, provided that

$$\limsup_{N\to\infty}\frac{\max_{1<i\leqslant N}(\alpha_i-\bar{\alpha})^2}{\sum(\alpha_i-\bar{\alpha})^2}=0.$$

Also, the sample variance converges to σ_α^2 as $n\to\infty$. From these results we can construct asymptotic confidence intervals for $\bar{\alpha}$.

E. The Problem of Sampling Procedures

In determining the sampling procedures, both the technical aspects of sampling and the accuracy of the estimators should be considered. The most commonly used methods are **multistage sampling** and **stratified sampling**, or some combination of the two. For example, the population is partitioned into several clusters. First we select some of them according to a probability scheme and then choose units from the selected clusters. This procedure is called **two-stage sampling**. The probabilities for the selection of clusters may be uniform or proportional to the size of the clusters. **Stratified sampling** is the method of dividing the population into several subpopulations, called **strata**, and selecting the sample units within each stratum. If the size of the ith stratum is N_i, the size of the sample chosen from this stratum is n_i, and the probability is uniform within each stratum, then the most common estimator for the population mean $\bar{\alpha}$ is given by

$$\hat{\bar{\alpha}}=\sum N_i\bar{X}_i/N,$$

where $\bar{X}_i$ is the mean of the sample values in the ith stratum. The variance of $\hat{\bar{\alpha}}$ is given by

$$V(\hat{\bar{\alpha}})=\left(\frac{1}{N}\right)^2\sum\frac{N_i^2}{n_i}\left(1-\frac{n_i}{N_i}\right)\sigma_i^2,$$

where σ_i^2 is the population variance within the ith stratum.

If the cost of drawing one sample unit in the ith stratum is equal to c_i, then for fixed cost, the variance of the estimator is minimized when

$$n_i/N_i\propto\sigma_i/\sqrt{c_i},$$

which is called the condition of **optimum allocation**.

F. Replicated Sampling Plan

W. E. Deming proposed an effective method in practical sample surveys, called a replicated sampling plan, following J. W. Tukey's hint. It enables us to easily evaluate variances of estimates for any estimator and any sampling procedure. Let the sample be composed of k subordinate samples which are selected by the same random sampling procedure from the same population, and let $\hat{\theta}_i(\mathbf{J}_i, X_i)$ be the estimate from the ith subordinate sample by the same estimator and $\hat{\theta}$ be the estimate from the whole sample by the same estimator. Then, provided $\hat{\theta} = \sum \hat{\theta}_i / k$ approximately, the variance of $\hat{\theta}$ can be estimated by $v(\hat{\theta}) = \sum(\hat{\theta}_i - \hat{\theta})^2 / k(k-1)$. If the sample is selected by the simple random sampling procedure and is of large scale, $\hat{\theta}_i$ and $\hat{\theta}$ are approximately normal variates, and $v(\theta)$ is evaluated by using the sample range of the $\hat{\theta}_i$. In large-scale sample surveys, even when the random sampling procedure is not simple, the theory related to the normal distribution can be applied to the $\hat{\theta}_i$ and $\hat{\theta}$. It has been shown that $v(\hat{\theta})$ evaluated by the above method includes not only the sampling error but also the random part of the nonsampling errors.

G. Conceptual Problems

Although it has been established that the sample survey method is useful in large-scale social or economic surveys, there are difficult conceptual problems about the foundations of the method (especially when auxiliary information exists) that are still far from being settled.

References

[1] W. G. Cochran, Sampling techniques, Wiley, third edition, 1977.
[2] W. E. Deming, On simplifications of sampling design through replication with equal probabilities and without stages, J. Amer. Statist. Assoc., 51 (1956), 24–53.
[3] R. J. Jessen, Statistical survey techniques, Wiley, 1978.
[4] J. Neyman, Contribution to the theory of sampling human populations, J. Amer. Statist. Assoc., 33 (1938), 101–116.
[5] P. V. Sukhatme and B. V. Sukhatme, Sampling theory of surveys with applications, Asia Publishing House, second edition, 1970.
[6] K. Takeuchi, Some remarks on general theory for unbiased estimation of a finite population, Japan. J. Math., 35 (1966), 73–84.

374 (XVIII.4) Sampling Distributions

A. General Remarks

To perform statistical inference, it is necessary to find the †probability distribution of a †statistic involved in the situation (→ 396 Statistic, 401 Statistical Inference). In general, the probability distribution of a statistic is called the **sampling distribution**. A set $\{X_1, X_2, \ldots, X_n\}$ of †random variables that are independently and identically distributed according to a distribution F is called a **random sample** from F. A common sampling distribution described in this article is that of the statistic $Y = f(X_1, \ldots, X_n)$, where the set $\{X_1, \ldots, X_n\}$ is a random sample from a †normal distribution. Examples of such a statistic Y of dimension 1 are the †sample mean, †sample variance, linear or quadratic forms of $\{X_1, \ldots, X_n\}$, their ratios, and †order statistic, while examples of Y of higher dimensions are the sample mean vector and the sample covariance matrix and its eigenvalues. The †normal distribution with †mean μ and †variance σ^2 is denoted by $N(\mu, \sigma^2)$, while the †p-dimensional (p-variate) normal distribution with mean vector $\boldsymbol{\mu}$ and covariance matrix Σ is denoted by $N(\boldsymbol{\mu}, \Sigma)$ (→ Appendix A, Table 22).

B. Samples from Univariate Normal Distributions

If random variables $X_1, \ldots, X_n$ are distributed independently according to $N(\mu_1, \sigma_1^2), \ldots, N(\mu_n, \sigma_n^2)$, then a linear form $\sum_i a_i X_i$ has the distribution $N(\sum_i a_i \mu_i, \sum_i a_i^2 \sigma_i^2)$. In particular, if $\{X_1, \ldots, X_n\}$ is a random sample from $N(\mu, \sigma^2)$, then the sample mean $\bar{X} = \sum_i X_i / n$ has the distribution $N(\mu, \sigma^2/n)$.

Let $\{X_1, \ldots, X_n\}$ be a random sample from the distribution $N(0,1)$. The sampling distribution of the statistic $Y = \sum_i X_i^2$ is called the **chi-square distribution** with n **degrees of freedom** and is denoted by $\chi^2(n)$. It has the †probability density

$$f_n(y) = 2^{-n/2} \left(\Gamma\left(\frac{n}{2}\right) \right)^{-1} y^{-1+n/2} e^{-y/2}$$

for $0 < y < \infty$, $f_n(y) = 0$ elsewhere, where Γ is the †gamma function. The distribution of $Y = \sum_{i=1}^n (X_i + \mu_i)^2$ depends only on n and $\lambda = \sum_i \mu_i^2$, and is called the **noncentral chi-square distribution** with n **degrees of freedom** and **noncentrality** λ and denoted by $\chi^2(n, \lambda)$. It has the probability density

$$f_{n,\lambda}(y) = \sum_{k=0}^{\infty} e^{-\lambda/2} \left(\frac{\lambda}{2}\right)^k \frac{1}{k!} f_{n+2k}(y)$$
$$= e^{-\lambda/2} {}_0F_1\left(\frac{n}{2}; \frac{\lambda y}{4}\right) f_n(y)$$

for $0 < y < \infty$, where f_{n+2k} and f_n are the probability densities of chi-square distributions and ${}_0F_1$ is an extended hypergeometric function. Noncentral chi-square distributions have the following †reproducing property: If $Y_1, \ldots, Y_k$ are distributed independently according to $\chi^2(n_1, \lambda_1), \ldots, \chi^2(n_k, \lambda_k)$, then $\sum_i Y_i$ has the distribution $\chi^2(\sum_i n_i, \sum_i \lambda_i)$. Also, we have **Cochran's theorem** (*Proc. Cambridge Philos. Soc.*, 30 (1934)): If $X_1, \ldots, X_n$ are distributed independently according to $N(\mu_1, 1)$, $\ldots, N(\mu_n, 1)$ and if for quadratic forms $Q_m = \sum_i \sum_j a_{ij}^{(m)} X_i X_j$ for $m = 1, \ldots, k$ the matrices $A_m = (a_{ij}^{(m)})$ with †rank r_m satisfy the condition $A_1 + \ldots + A_k = I$ (unit matrix), then a necessary and sufficient condition for $Q_1, \ldots, Q_k$ to have independent noncentral chi-square distributions with $r_1, \ldots, r_k$ degrees of freedom, respectively, is that $r_1 + \ldots + r_k = n$. In particular, when $\mu_i = 0$ for all i, they have (central) chi-square distributions, and the theorem implies their reproducing property.

Let X and Y be independent random variables having distributions $N(\delta, 1)$ and $\chi^2(n)$, respectively. Then the sampling distribution of $T = X/\sqrt{Y/n}$ is called the **noncentral *t*-distribution** with n **degrees of freedom** and **noncentrality** δ and is denoted by $t(n, \delta)$. Its probability density is given by

$$f_{n,\delta}(t) = \sum_{k=0}^{\infty} e^{-\delta^2/2} \delta^k \frac{2^{k/2}}{k!} \frac{\Gamma((n+k+1)/2)}{\sqrt{\pi n}\,\Gamma(n/2)} \times \frac{(t/\sqrt{n})^k}{(1+t^2/n)^{(n+k+1)/2}}$$

for $-\infty < t < \infty$. In particular, when $\delta = 0$, the distribution is called the ***t*-distribution** with n **degrees of freedom** and is denoted by $t(n)$. Its probability density is simplified to

$$f_n(t) = \frac{\Gamma((n+1)/2)}{\sqrt{\pi n}\,\Gamma(n/2)} \left(1 + \frac{t^2}{n}\right)^{-(n+1)/2}$$

for $-\infty < t < \infty$.

Let $\{X_1, \ldots, X_n\}$ be a random sample from $N(\mu, \sigma^2)$. Exact sampling distributions of the †sample variance $S^2 = \sum (X_i - \bar{X})^2/(n-1)$ and of the ***t*-statistic** $T = \sqrt{n}(\bar{X} - \mu_0)/\sqrt{S^2}$, where $\mu = \mu_0$ is a given number, were essentially obtained by Student [5]: $(n-1)S^2/\sigma^2$ and T are distributed according to $\chi^2(n-1)$ and $t(n-1)$, respectively. His proof was made rigorous by R. A. Fisher (*Metron*, 5 (1925)), who proved in particular that $\bar{X}$ and S^2 are independent. If $\mu \neq \mu_0$, then T follows the distribution $t(n-1, \sqrt{n}(\mu - \mu_0)/\sigma)$.

Let X and Y be distributed independently according to $\chi^2(m, \lambda)$ and $\chi^2(n)$, respectively. The distribution of $Z = (X/m)/(Y/n)$ is called the **noncentral *F*-distribution** with m and n **degrees of freedom** and **noncentrality** λ. In the special case when $\lambda = 0$ it is called the ***F*-distribution** with m and n **degrees of freedom** and is denoted by $F(m, n)$. The probability densities $f_{m,n,\lambda}$ and $f_{m,n}$ of these distributions are given by

$$f_{m,n}(z) = \frac{(m/n)^{m/2}}{B(m/2, n/2)} \frac{z^{(m/2)-1}}{(1+mz/n)^{(m+n)/2}},$$

$$f_{m,n,\lambda}(z) = \sum_{k=0}^{\infty} e^{-\lambda/2} \left(\frac{\lambda}{2}\right)^k \frac{1}{k!} \frac{(m/n)^{(m/2)+k}}{B((m/2)+k, n/2)} \times \frac{z^{(m/2)+k-1}}{(1+mz/n)^{((m+n)/2)+k}}$$

$$= e^{-\lambda/2}\, {}_1F_1\left(\frac{m+n}{2}; \frac{m}{2}; \frac{\lambda}{2} \frac{mz/n}{1+mz/n}\right) f_{m,n}(z)$$

for $0 < z < \infty$, where B and ${}_1F_1$ are the †beta function and the confluent hypergeometric function (→ 167 Functions of Confluent Type), respectively.

Let X be a random variable having the distribution $F(m, n)$. The distribution of $Z = \frac{1}{2}\log X$ is called the ***z*-distribution** with m and n **degrees of freedom** and is denoted by $z(m, n)$. Its probability density is given by

$$\frac{2(m/n)^{m/2}}{B(m/2, n/2)} \frac{e^{mz}}{(1+me^{2z}/n)^{(m+n)/2}}$$

for $-\infty < z < \infty$. If $S_1^2 = \sum_{i=1}^m (X_i - \bar{X})^2/(m-1)$ and $S_2^2 = \sum_{i=1}^n (Y_i - \bar{Y})^2/(n-1)$ are sample variances based on independent samples of sizes m and n taken from $N(\mu, \sigma^2)$ and $N(\nu, \tau^2)$, respectively, then the statistic $z = \frac{1}{2}\log(S_1^2/S_2^2)$, which was introduced by Fisher (*Proc. Int. Math. Congress*, 1924), is distributed according to $z(m-1, n-1)$ under the hypothesis $\sigma^2 = \tau^2$. Fisher [6] tabulated percent points of $z(m, n)$.

C. Samples from Multivariate Normal Distributions

Let $\mathbf{X}$ be a p-dimensional random vector, namely, a vector having real random variables as its components. $\mathbf{X}$ has the p-variate normal distribution $N(\boldsymbol{\mu}, \Sigma)$ if and only if for any real vector $\mathbf{a} = (a_1, \ldots, a_p)'$, the random variable $\mathbf{a}'\mathbf{X}$ has the normal distribution $N(\mathbf{a}'\boldsymbol{\mu}, \mathbf{a}'\Sigma\mathbf{a})$. If $\mathbf{X}_1, \ldots, \mathbf{X}_n$ are independent and have p-variate normal distributions $N(\boldsymbol{\mu}_1, \Sigma_1), \ldots, N(\boldsymbol{\mu}_n, \Sigma_n)$, respectively, and if $A_1, \ldots, A_n$ are $m \times p$ real matrices, then the random vector $A_1\mathbf{X}_1 + \ldots + A_n\mathbf{X}_n$ has the m-variate normal distribution $N(\boldsymbol{\mu}, \Sigma)$, where $\boldsymbol{\mu} = \sum_{j=1}^n A_j \boldsymbol{\mu}_j$ and $\Sigma = \sum_{j=1}^n A_j \Sigma_j A_j'$.

Suppose that $\{\mathbf{X}_1, \ldots, \mathbf{X}_n\}$ is a random sample from the p-variate normal distribution $N(\mathbf{0}, \Sigma)$, and let $\mathbf{X} = (\mathbf{X}_1, \ldots, \mathbf{X}_n)$ be a $p \times n$ matrix. Then the probability distribution of $W = \mathbf{X}\mathbf{X}'$ is called the **Wishart distribution** with **scale matrix** Σ and n **degrees of freedom** and is denoted by $W_p(\Sigma, n)$ or simply $W(\Sigma, n)$. If $n > p - 1$, the joint probability density function of $p(p+1)/2$ arguments of $W=(W_{ij})$ is

$$f_{\Sigma,n}(W)=(\Gamma_p(n/2)|2\Sigma|^{n/2})^{-1} \times \mathrm{etr}(-\Sigma^{-1}W/2)|W|^{(n-p-1)/2}$$

for $W>0$, where $W>0$ means that W is †positive definite, $\mathrm{etr}(A)=\exp(\mathrm{tr}\,A)$, and Γ_p, a **multidimensional gamma function**, is defined as

$$\Gamma_p(a)=\pi^{p(p-1)/4}\prod_{i=1}^{p}\Gamma\left(a-\frac{1}{2}(i-1)\right)$$

for $a>(p-1)/2$. When $n\leqslant p-1$ the distribution is **singular** and has no probability density.

Suppose that $\mathbf{X}_1,\ldots,\mathbf{X}_n$ are independent and obey normal distributions $N(\boldsymbol{\mu}_1,\Sigma)$, $\ldots,N(\boldsymbol{\mu}_n,\Sigma)$, and let $\mathbf{X}=(\mathbf{X}_1,\ldots,\mathbf{X}_n)$, $M=(\boldsymbol{\mu}_1,\ldots,\boldsymbol{\mu}_n)$. Then the distribution of $W=\mathbf{XX}'$ is called the p-**dimensional noncentral Wishart distribution** with **scale matrix** Σ, n **degrees of freedom**, and **noncentrality matrix** $\Omega=\Sigma^{-1}MM'$ and is denoted by $W(\Sigma,n,\Omega)$. If $n>p-1$, the probability density function is

$$\mathrm{etr}(-\Omega/2)\,{}_0F_1(n/2;\Omega\Sigma^{-1}W/4)f_{\Sigma,n}(W)$$

for $W>0$. ${}_\alpha F_\beta$ is a hypergeometric function with matrix argument, which is defined by ${}_\alpha F_\beta(a_1,\ldots,a_\alpha;b_1,\ldots,b_\beta;S)={}_\alpha F_\beta^{(p)}(a_1,\ldots,a_\alpha;b_1,\ldots,b_\beta;S,I)$, where

$${}_\alpha F_\beta^{(p)}(a_1,\ldots,a_\alpha;b_1,\ldots,b_\beta;S,T) =\sum_{k=0}^{\infty}\sum_{\kappa}\left(\frac{\prod_{i=1}^{\alpha}(a_i)_\kappa}{\prod_{j=1}^{\beta}(b_j)_\kappa}\right)\frac{C_\kappa(S)C_\kappa(T)}{k!\,C_\kappa(I_p)},$$

$\kappa=(k_1,\ldots,k_p)$ is an ordered set of integers such that $k_1+\ldots+k_p=k$ and $k_1\geqslant\ldots\geqslant k_p\geqslant 0$, and where $C_\kappa(S)$ is a **zonal polynomial** (→ [8]) of a symmetric matrix S. The multivariate hypergeometric coefficient $(a)_\kappa$ is given by

$$(a)_\kappa=\prod_{i=1}^{p}\left(a-\frac{1}{2}(i-1)\right)_{k_i},$$

$$(a)_k=a(a+1)\ldots(a+k-1).$$

The noncentral Wishart distribution is singular when $n\leqslant p-1$. Similarly to the noncentral chi-square distribution, the noncentral Wishart distribution has the reproducing property with respect to both the number of degrees of freedom and the noncentrality matrix. Also, Cochran's theorem can be extended to the multivariate case: Let $\mathbf{X}_1,\ldots,\mathbf{X}_n$ be p-variate random vectors independently distributed according to $N(\boldsymbol{\mu}_1,\Sigma),\ldots,N(\boldsymbol{\mu}_n,\Sigma)$, respectively, and let $A_m=(a_{ij}^{(m)})$, $m=1,\ldots,k$, be $p\times p$ real matrices of †rank r_m and such that $A_1+A_2+\ldots+A_k=I$ (unit matrix). A necessary and sufficient condition for random matrices $\mathbf{Q}_m=\sum_{i,j}a_{ij}^{(m)}\mathbf{X}_i\mathbf{X}_j'$, $m=1,\ldots,k$, to be independently distributed according to noncentral Wishart distributions with $r_1,\ldots,r_k$ degrees of freedom, respectively, is that $r_1+\ldots+r_k=n$. If, in particular, $\boldsymbol{\mu}_1=\ldots=\boldsymbol{\mu}_n=0$ and if $r_m\geqslant p$, then $\mathbf{Q}_m$ is distributed according to $W(\Sigma,r_m)$.

If $\mathbf{W}$ has the distribution $W(\Sigma,n)$ with $n>p-1$, then the eigenvalues $\lambda_1,\ldots,\lambda_p$ ($\lambda_1\geqslant\ldots\geqslant\lambda_p>0$) of W have the joint probability density function $C_{p,n}|\Sigma|^{-n/2}\,{}_0F_0^{(p)}(-\Sigma^{-1}/2,\boldsymbol{\Lambda})|\boldsymbol{\Lambda}|^{(n-p-1)/2}\prod_{i<j}(\lambda_i-\lambda_j)$, where $\boldsymbol{\Lambda}$ is a diagonal matrix with diagonal elements $\lambda_1,\ldots,\lambda_p$ and $C_{p,n}=\pi^{p^2/2}(2^{pn/2}\Gamma_p(p/2)\Gamma_p(n/2))^{-1}$. If $\Sigma=I$, then the joint probability density function becomes

$$C_{p,n}\mathrm{etr}(-\boldsymbol{\Lambda}/2)|\boldsymbol{\Lambda}|^{(n-p-1)/2}\prod_{i<j}(\lambda_i-\lambda_j).$$

Suppose that $\mathbf{S}_1$ and $\mathbf{S}_2$ have independent Wishart distributions $W(\Sigma,n_1)$ and $W(\Sigma,n_2)$, respectively. The random matrix $\mathbf{B}=(\mathbf{S}_1+\mathbf{S}_2)^{-1/2}\mathbf{S}_1(\mathbf{S}_1+\mathbf{S}_2)^{-1/2}$ is called the beta matrix, and its distribution is denoted by $B(n_1/2,n_2/2)$. Its probability density function is

$$\left(\Gamma_p\left(\frac{n_1+n_2}{2}\right)\Big/\Gamma_p\left(\frac{n_1}{2}\right)\Gamma_p\left(\frac{n_2}{2}\right)\right)|\mathbf{B}|^{(n_1-p-1)/2} \times|I-\mathbf{B}|^{(n_2-p-1)/2}$$

for $0<\mathbf{B}<I$.

Suppose that $\mathbf{S}$ has the distribution $W(\Sigma,n)$ and $\mathbf{B}$ has $B(n_1/2,n_2/2)$; then, for any nonsingular symmetric matrix $\boldsymbol{\Omega}$,

$$P\{\mathbf{S}<\Omega\}=\left\{|2\Sigma\Omega^{-1}|^{-n/2}\Gamma_p\left(\frac{p+1}{2}\right)\Big/\Gamma_p\left(\frac{n+p+1}{2}\right)\right\}{}_1F_1\left(\frac{n}{2};\frac{n+p+1}{2};-\frac{1}{2}\Sigma^{-1}\Omega\right)$$

and

$$P\{\mathbf{B}<\Omega\}=\left\{\Gamma_p\left(\frac{n_1+n_2}{2}\right)\Gamma_p\left(\frac{p+1}{2}\right)\Big/\Gamma_p\left(\frac{n_1+p+1}{2}\right)\Gamma_p\left(\frac{n_2}{2}\right)\right\}|\Omega|^{n_1/2} \times{}_2F_1\left(\frac{n_1}{2},-\frac{n_2}{2}+\frac{p+1}{2};\frac{n_1+p+1}{2};\Omega\right).$$

If $\{\mathbf{X}_1,\ldots,\mathbf{X}_n\}$ is a random sample from $N(\boldsymbol{\mu},\Sigma)$, then the sample mean $\bar{\mathbf{X}}=\sum_{\alpha=1}^{n}\mathbf{X}_\alpha/n$ and the sample covariance matrix $\mathbf{S}=\sum_{\alpha=1}^{n}(\mathbf{X}_\alpha-\bar{\mathbf{X}})(\mathbf{X}_\alpha-\bar{\mathbf{X}})'/(n-1)$ are distributed independently according to the respective distributions $N(\boldsymbol{\mu},\Sigma/n)$ and $W(\Sigma,n-1)$. If $n>p-1$, $T^2=n(\bar{\mathbf{X}}-\boldsymbol{\mu}_0)'\mathbf{S}^{-1}(\bar{\mathbf{X}}-\boldsymbol{\mu}_0)$ is called the **noncentral Hotelling T^2 statistic** with $n-1$ **degrees of freedom** and **noncentrality** $\lambda=n(\boldsymbol{\mu}-\boldsymbol{\mu}_0)'\Sigma^{-1}(\boldsymbol{\mu}-\boldsymbol{\mu}_0)$. $(n-p)T^2/p(n-1)$ has a †noncentral F-distribution with p and $n-p$ degrees of freedom and noncentrality λ.

Let $\mathbf{X}=(X_1,\ldots,X_p)'$ and $\mathbf{Y}=(Y_1,\ldots,Y_q)'$, $p\leqslant q$ denote two random vectors, Σ_{11} and Σ_{22} their respective covariance matrices, and Σ_{12} the $p\times q$ matrix of covariances between the components of $\mathbf{X}$ and $\mathbf{Y}$. Each of the nonnegative roots $\rho_1, \ldots, \rho_p$ of the equation $|\Sigma_{12}\Sigma_{22}^{-1}\Sigma_{21} - \rho^2\Sigma_{11}| = 0$ is called the **canonical correlation coefficient**. Let $(\mathbf{X}_1, \mathbf{Y}_1), \ldots, (\mathbf{X}_n, \mathbf{Y}_n)$ be a random sample from the $(p+q)$-variate normal distribution $N(\mathbf{0}, \Sigma)$, and let $\mathbf{S}_{11} = \sum_{\alpha=1}^{n} \mathbf{X}_\alpha \mathbf{X}'_\alpha / n$, $\mathbf{S}_{12} = \mathbf{S}'_{21} = \sum_{\alpha=1}^{n} \mathbf{X}_\alpha \mathbf{Y}'_\alpha / n$, $\mathbf{S}_{22} = \sum_{\alpha=1}^{n} \mathbf{Y}_\alpha \mathbf{Y}'_\alpha / n$, and $\mathbf{S} = (\mathbf{S}_{ij})_{i,j=1,2}$. The sample canonical correlation coefficients are the nonnegative roots $r_1, \ldots, r_p$ of $|\mathbf{S}_{12}\mathbf{S}_{22}^{-1}\mathbf{S}_{21} - r^2\mathbf{S}_{11}| = 0$ and for $n \geqslant p+q$ the probability density function of $r_1^2, \ldots, r_p^2$ is

$$C_{p,q,n}|I-P^2|^{n/2}|R^2|^{(q-p-1)/2}|I-R^2|^{(n-q-p-1)/2} \times \prod_{i<j}(r_i^2 - r_j^2)\,{}_2F_1^{(p)}\left(\frac{n}{2}, \frac{n}{2}; \frac{q}{2}; R^2, P^2\right),$$

where

$$C_{p,q,n} = \pi^{p^2/2}\Gamma_p\left(\frac{n}{2}\right) \Big/ \left\{\Gamma_p\left(\frac{n-q}{2}\right)\Gamma_p\left(\frac{p}{2}\right)\Gamma_p\left(\frac{q}{2}\right)\right\},$$

and R^2 and P^2 are diagonal matrices with elements r_i^2 and ρ_j^2, respectively. If, in particular, $p=1$, then ρ_1 and r_1 are, respectively, the population and the sample multiple correlation coefficients, and $(n-q)r_1^2/q(1-r_1^2)$ follows the distribution $F(q, n-q)$ whenever $\rho_1 = 0$.

Let $\{(X_1, Y_1), \ldots, (X_n, Y_n)\}$ be a random sample from the 2-dimensional normal distribution with †correlation coefficient ρ. Then the sample correlation coefficient

$$R = \frac{\Sigma_i(X_i - \bar{X})(Y_i - \bar{Y})}{(\Sigma_i(X_i - \bar{X})^2 \Sigma_i(Y_i - \bar{Y})^2)^{1/2}}$$

has probability density

$$f_n(r; \rho) = (2^{n-3}/\pi(n-3)!)(1-\rho^2)^{(n-1)/2}(1-r^2)^{(n-4)/2} \times \sum_{k=0}^{\infty} \Gamma^2\left(\frac{1}{2}(n+k-1)\right)\frac{(2\rho r)^k}{k!}$$

for $-1 < r < 1$. For the special case $\rho = 0$, the probability density becomes

$$f_n(r) = \frac{1}{\sqrt{\pi}} \frac{\Gamma\left(\frac{1}{2}(n-1)\right)}{\Gamma\left(\frac{1}{2}(n-2)\right)}(1-r^2)^{(n-4)/2},$$

which implies that $T = \sqrt{n-2}R/\sqrt{1-R^2}$ has the †t-distribution with $n-2$ degrees of freedom.

Given a random sample from a p-variate normal distribution, the probability density of the †sample partial correlation coefficient $R_{12\cdot3\ldots p}$ between the first and the second components with the remaining components fixed is given by $f_{n-p+2}(r; \rho_{12\cdot3\ldots p})$, where f is the density of R mentioned in the previous paragraph and $\rho_{12\cdot3\ldots p}$ is the population partial correlation.

D. Large-Sample Theory

So far we have dealt with random samples $\{X_1, \ldots, X_n\}$ composed of finitely many random variables (or vectors). The theory dealing with such finite cases is called small-sample theory, which is not always suitable for numerical applications. In comparison with this, in large-sample theory, where the sample size is assumed to be sufficiently large, an approximation of the sampling distribution can often be obtained easily by means of the †central limit theorem.

If for three sequences X_n, μ_n, and σ_n, $n = 1, 2, \ldots$, of random variables, real numbers, and positive numbers, respectively, the sequence $(X_n - \mu_n)/\sigma_n$ †converges in distribution to $N(0,1)$ as $n \to \infty$, then the sequence X_n is said to be **asymptotically distributed** according to $N(\mu_n, \sigma_n^2)$. The definition can be extended to higher dimensions. We write $X_n = o_p(r_n)$ for a sequence r_n of positive numbers if and only if X_n/r_n †converges in probability to zero as $n \to \infty$. The following theorem is useful: If $X_n = a + o_p(r_n)$, where a is a constant and $r_n = o(1)$, and if a real-valued function $f(x)$ is of class C^s in a neighborhood of $x = a$, then

$$f(X_n) - \sum_{k=0}^{s} \frac{1}{k!} f^{(k)}(a)(X_n - a)^k + o_p(r_n^s).$$

If X_n is asymptotically distributed according to $N(\mu, \sigma^2/n)$ and $f(x)$ is differentiable at $x = \mu$ with the derivative $f'(\mu) \neq 0$, then $f(X_n)$ is asymptotically distributed according to $N(f(\mu), (f'(\mu))^2\sigma^2/n)$. In higher-dimensional cases, if $\mathbf{X}_n$ is asymptotically distributed according to $N(\boldsymbol{\mu}, \Sigma/n)$ and $f(\mathbf{x})$ is continuously differentiable in a neighborhood of $\mathbf{x} = \boldsymbol{\mu}$ with nonzero vector $\mathbf{c} = (\partial f/\partial x_1, \ldots, \partial f/\partial x_p)_{\mathbf{x}=\boldsymbol{\mu}}$, then $f(\mathbf{X}_n)$ is asymptotically distributed according to $N(f(\boldsymbol{\mu}), \mathbf{c}\Sigma\mathbf{c}'/n)$.

Let $\{X_1, \ldots, X_n\}$ be a random sample from a univariate distribution with finite †moments $v_i = E(X^i)$ for $i = 1, \ldots, k$, and let $a_i = \Sigma_\alpha X_\alpha^i/n$ be its ith sample moment. Then the random vector $(a_1, \ldots, a_k)$ asymptotically follows the k-variate normal distribution as $n \to \infty$ with mean vector $(v_1, \ldots, v_k)$ and covariance matrix $n^{-1}(\sigma_{ij})$, where $\sigma_{ij} = v_{i+j} - v_i v_j$. Let $M_i = \Sigma_\alpha(X_\alpha - \bar{X})^i/n$ and $\mu_i = E(X - v_1)^i$ for $i = 2, \ldots, k$ be the sample central moment of order i and population central moment of order i, respectively. Then the random vector $(\bar{X}, M_2, \ldots, M_k)$ obeys the k-variate normal distribution asymptotically as $n \to \infty$ with mean vector $(v_1, \mu_2, \ldots, \mu_k)$

and covariance matrix $n^{-1}(\sigma_{ij})$, where $\sigma_{11}=\mu_2$, $\sigma_{1i}=\mu_{i+1}-i\mu_2\mu_{i-1}$, $\sigma_{ij}=\mu_{i+j}-i\mu_{i-1}\mu_{j+1}-j\mu_{i+1}\mu_{j-1}-\mu_i\mu_j+ij\mu_2\mu_{i-1}\mu_{j-1}$ for $i,j\geqslant 2$.

A random variable χ_n^2 that has a †chi-square distribution with n degrees of freedom obeys the distribution $N(n,2n)$ asymptotically as $n\to\infty$. Also, $\sqrt{2\chi_n^2}-\sqrt{2n-1}$ obeys $N(0,1)$ asymptotically. The latter distribution approximates χ_n^2 indirectly better than $N(n,2n)$ approximates χ_n^2 directly. The t-distribution with n degrees of freedom obeys $N(0,1)$ asymptotically as $n\to\infty$. If X_n obeys an F-distribution with m and n degrees of freedom, then mX_n obeys asymptotically the distribution $\chi^2(m)$ as $n\to\infty$. If X_n obeys a †binomial distribution $Bin(n,p)$, then X_n obeys asymptotically the distribution $N(np,np(1-p))$, and $\operatorname{Arcsin}\sqrt{X_n/n}$ obeys asymptotically $N(\operatorname{Arcsin}\sqrt{p},1/4n)$ as $n\to\infty$. This transformation is called the **arcsin** (or **angular**) **transformation**. If $(X_1,X_2,\ldots,X_k)$ obeys the multinomial distribution $Mu(n;p_1,p_2,\ldots,p_k)$, then it is asymptotically distributed according to the normal distribution $N(\boldsymbol{\mu}_n,\Sigma_n)$, where $\boldsymbol{\mu}_n=(np_1,\ldots,np_k)$, $\Sigma_n=(\sigma_{ij}^{(n)})$, $\sigma_{ii}^{(n)}=np_i(1-p_i)$, and $\sigma_{ij}^{(n)}=-np_ip_j$, $(i\neq j)$, and the random variable $\sum_{j=1}^{k+1}(X_j-np_j)^2/np_j$, where $X_{k+1}=n-(X_1+\ldots+X_k)$ and $p_{k+1}=1-(p_1+\ldots+p_k)$, obeys asymptotically the distribution $\chi^2(k)$ [11].

If X_n has the †Poisson distribution with mean λ_n. where $\lambda_n\to\infty$ as $n\to\infty$, then X_n and $\sqrt{X_n}$ obey the respective distributions $N(\lambda_n,\lambda_n)$ and $N(\sqrt{\lambda_n},1/4)$ asymptotically. If R is the sample correlation coefficient based on a random sample of size n from a 2-dimensional (bivariate) normal distribution with population correlation coefficient ρ, then R is asymptotically distributed according to $N(\rho,(1-\rho^2)^2/n)$ as $n\to\infty$, and therefore $z=\frac{1}{2}\log((1+R)/(1-R))$ obeys asymptotically the distribution $N(\frac{1}{2}\log((1+\rho)/(1-\rho)),1/n)$ asymptotically. This transformation is called **Fisher's z-transformation**. The distribution

$$N\left(\frac{1}{2}\log\frac{1+\rho}{1-\rho}+\frac{\rho}{2(n-1)},\frac{1}{n-3}\right)$$

gives a better approximation.

E. Empirical Distribution Function

Let $\{X_1,\ldots,X_n\}$ be a random sample from a distribution F. The random function

$$F_n(x)=\frac{1}{n}\{\text{number of } X\text{'s that are }\leqslant x\}$$

is called the **empirical distribution function**. For any collection of fixed x's $(-\infty=x_0<x_1<\ldots<x_k<\infty)$, the random vector $(nF_n(x_1), n(F_n(x_2)-F_n(x_1)),\ldots,n(F_n(x_k)-F_n(x_{k-1})))$ obeys the †multinomial distribution $Mu(n;p_1,\ldots,p_k)$, where $p_j=F(x_j)-F(x_{j-1})$, $j=1,\ldots,k$, provided that the p's are positive. In particular, the vector is asymptotically distributed according to the k-variate normal. The result is substantially strengthened as follows: the **Glivenko-Cantelli theorem** states that $\sup_x|F_n(x)-F(x)|$ converges to zero with probability 1 as n tends to infinity. If $F(x)$ is continuous, then the random function $\sqrt{n}(F_n(t)-F(t))$ converges in distribution to a †Gaussian process $X(t)$ such that $E(X(t))=0$ and $E(X(s)X(t))=F(s)(1-F(t))$ for $s\leqslant t$. A Gaussian process $X(t)$, $0\leqslant t\leqslant 1$, with this moment condition is called a **Brownian bridge** if $F(t)=t$, for $0\leqslant t\leqslant 1$. If $F(x)$ is continuous, then the distributions of the random variables $C_n=\sqrt{n}\sup_x(F_n(x)-F(x))$ and $D_n=\sqrt{n}\sup|F_n(x)-F(x)|$ do not depend on F. Asymptotically, they have identical distributions with $\sup_t B(t)$ and $\sup_t|B(t)|$, respectively, where $B(t)$ is a Brownian bridge. We have

$$P(\sup_t B(t)\leqslant x)=1-e^{-2x^2},$$

$$P(\sup_t|B(t)|\leqslant x)=1+2\sum_{k=1}^{\infty}(-1)^k e^{-2k^2x^2},\qquad x>0.$$

Let $\{X_1,\ldots,X_m\}$ and $\{Y_1,\ldots,Y_n\}$ be random samples from continuous distributions F and G, respectively, and let $F_m(x)$ and $G_n(x)$ be their empirical distribution functions. Under the hypothesis $H_0:F\equiv G$, the distribution of the **Kolmogorov-Smirnov test statistic**

$$D_{m,n}=\sup_x|F_m(x)-G_n(x)|$$

does not depend on F (or G), and asymptotically, as $m\to\infty$ and $m/n\to\lambda\leqslant 1$, the random function $\sqrt{m}(F_m(t)-G_n(t))$ converges in distribution to a Gaussian process $X(t)$ such that $E(X(t))=0$ and $E(X(s)X(t))=(1+\lambda)F(s)(1-F(t))$, $s\leqslant t$.

F. Edgeworth and Cornish-Fisher Expansions

Let $\{X_1,X_2,\ldots,X_n\}$ be a sample from a distribution with mean μ and variance σ^2. The random variable $(X_1+X_2+\ldots+X_n-n\mu)/\sqrt{n}\sigma$ is called the normalized sum of the sample. The distribution function $F_n(x)$ of the normalized sum of a sample from an absolutely continuous distribution F with higher-order moments admits the **Edgeworth expansion** [15]:

$$F_n(x)=\Phi(x)+\sum_{k=1}^{\nu-2}R_k(x)\left(\frac{1}{\sqrt{n}}\right)^k\phi(x)+B\left(\frac{1}{\sqrt{n}}\right)^{\nu-1},$$

where Φ and ϕ are the †cumulative distribution and the †probability density functions, respectively, of $N(0,1)$, and B is a quantity bounded by a constant depending on F and ν. $R_k(x)$ is the polynomial given by

$$R_k(x) = -\Sigma H_{k+2l-1}(x) \prod_{j=1}^{k} \frac{1}{m_j!}\left(\frac{\gamma_{j+2}}{(j+2)!}\right)^{m_j},$$

where $H_k(x)$ is the †Hermite polynomial of degree k, γ_k is the †cumulant (†semi-invariant) of order k of the distribution of $(X_1-\mu)/\sigma$, the summation extends over all nonnegative m's such that $m_1+2m_2+\ldots+km_k=k$, and $l=m_1+m_2+\ldots+m_k$. In particular, we have $R_1(x) = -\gamma_3(x^2-1)/6$ and $R_2(x) = -\gamma_4(x^3-3x)/24 - \gamma_3^2(x^5-10x^3+15x)/72$. For a †lattice distribution F concentrated on $0, \pm 1, \pm 2, \ldots$ but not on $0, \pm\rho, \pm 2\rho, \ldots$ for any $\rho > 1$, the following expansion is valid for $x = 0, \pm 1, \pm 2, \ldots$:

$$F_n\left(\frac{x-n\mu}{\sqrt{n}\sigma}\right) = \Phi(z) + \sum_{k=1}^{\nu-2} Q_k(z)\left(\frac{1}{\sqrt{n}}\right)^k \phi(z) + B\left(\frac{1}{\sqrt{n}}\right)^{\nu-1},$$

where $z = (x - n\mu + 1/2)/\sqrt{n}\sigma$ and the Q's are suitable polynomials; $Q_1(z) = R_1(z)$ and $Q_2(z) = R_2(z) + z/24\sigma^2$.

The Edgeworth expansion makes it possible to derive asymptotic formulas for the relation between those u and v such that $F_n(v) = \Phi(u)$. If F is an absolutely continuous distribution with moments of order ν $(\geqslant 3)$, then we have the **Cornish-Fisher expansions** [16]:

$$u = v + \sum_{k=1}^{\nu-2} A_k(v)\left(\frac{1}{\sqrt{n}}\right)^k + O(n^{-(\nu-1)/2})$$

and

$$v = u + \sum_{k=1}^{\nu-2} B_k(u)\left(\frac{1}{\sqrt{n}}\right)^k + O(n^{-(\nu-1)/2}),$$

where the A's and B's are polynomials derived from the R's of the Edgeworth expansion; $A_1(v) = -\gamma_3(v^2-1)/6$, $A_2(v) = -\gamma_4(v^3-3v)/24 + \gamma_3^2(4v^3-7v)/36$, $B_1(u) = \gamma_3(u^2-1)/6$, $B_2(u) = \gamma_4(u^3-3u)/24 - \gamma_3^2(2u^3-5u)/36$.

The expansions imply, in particular, that the random variable $v + \Sigma_{k=1}^{\nu-2} A_k(v) n^{-k/2}$ with $v = (X_1 + X_2 + \ldots + X_n - n\mu)/\sqrt{n}\sigma$ is asymptotically distributed according to $N(0,1)$ and that the $100\alpha\%$ point v_α of F_n is approximated by $u_\alpha + \Sigma_{k=1}^{\nu-2} B_k(u_\alpha) n^{-k/2}$,where u_α is the $100\alpha\%$ point of $N(0,1)$. These approximations can be improved further in some cases by a suitable transformation of the sum $X_1 + X_2 + \ldots + X_n$. Thus, for example, if X is distributed according to $\chi^2(n)$, then the Cornish-Fisher expansions with $\nu = 3$ are

$$u = v - \frac{1}{3}\sqrt{\frac{2}{n}}(v^2-1) + O\left(\frac{1}{n}\right)$$

and

$$v = u + \frac{1}{3}\sqrt{\frac{2}{n}}(u^2-1) + O\left(\frac{1}{n}\right),$$

where $v = (X-n)/\sqrt{2n}$. However, the distribution of the random variable

$$\sqrt{\frac{9n}{2}}\left\{\left(\frac{X}{n}\right)^{1/3} - 1 + \frac{2}{9n}\right\}$$

is much better approximated by $N(0,1)$, and

$$n\left(1 - \frac{2}{9n} + \sqrt{\frac{2}{9n}}\, u_\alpha\right)^3$$

gives a more accurate approximation to the $100\alpha\%$ point of the distribution $\chi^2(n)$. These are called the **Wilson-Hilferty approximations** (*Proc. Nat. Acad. Sci. US*, 17 (1931)).

The Edgeworth expansion was shown to be valid in more general situations by R. N. Bhattacharya and J. K. Ghosh [17]. In particular, they obtained the following: Let $\{\mathbf{X}_1, \mathbf{X}_2, \ldots, \mathbf{X}_n\}$ be a random sample from a p-variate distribution with a nonzero †absolutely continuous component w.r.t. †Lebesgue measure on $\mathbf{R}^p$. Let $f_0\ (\equiv 1), f_1, \ldots, f_k$ be linearly independent, real-valued, and continuously differentiable functions. For $i = 1, \ldots, n$, put $\mathbf{Z}_i = (f_1(\mathbf{X}_i), f_2(\mathbf{X}_i), \ldots, f_k(\mathbf{X}_i))$, and assume that the distribution of $\mathbf{Z}_1$ has moments up to the order ν $(\geqslant 3)$. Let H be a real-valued function on $\mathbf{R}^k$ such that the νth order derivatives are continuous in a †neighborhood of $\boldsymbol{\mu} \equiv E(\mathbf{Z}_1)$. Let $V = (v_{ij})$, $i, j = 1, \ldots, k$, be the covariance matrix of the random vector $\mathbf{Z}_1$, and put $\sigma^2 = \Sigma v_{ij} l_i l_j$, where $l_i = \partial H(\mathbf{z})/\partial \mathbf{z}_i|_{\mathbf{z}=\boldsymbol{\mu}}$, and $\mathbf{z} = (z_1, \ldots, z_k)$. Then

$$\sup_B \left| \Pr\{\sqrt{n}(H(\overline{\mathbf{Z}}) - H(\boldsymbol{\mu})) \in B\} - \int_B \psi_{\nu,n}(x)\,dx \right| = O(n^{-(\nu-2)/2}),$$

where $\overline{\mathbf{Z}} = \Sigma_{i=1}^n \mathbf{Z}_i/n$, the supremum is taken over all Borel measurable sets B,

$$\psi_{\nu,n}(x) = \left(1 + \sum_{k=1}^{\nu-2}\left(\frac{1}{\sqrt{n}}\right)^k P_k\left(-\frac{d}{dx}\right)\right)\phi_\sigma(x),$$

$\phi_\sigma(x)$ is the probability density function of the normal distribution $N(0,\sigma^2)$, and the P's are polynomials whose coefficients are independent of n.

G. Order Statistics

Let $\{X_1, \ldots, X_n\}$ be a random sample from a univariate distribution with continuous probability density $f(x)$ and distribution function $F(x)$, and let $X_{(1)} \leqslant \ldots \leqslant X_{(n)}$ be †order statistics. The joint probability density of $Y_1 = X_{(\alpha)}$, $Y_2 =$

$X_{(\beta)}, \ldots, Y_{p-1}=X_{(\varepsilon)}$, and $Y_p=X_{(\eta)}$ is given by

$$\frac{n!}{(\alpha-1)!(\beta-\alpha-1)!\ldots(\eta-\varepsilon-1)!(n-\eta)!}$$
$$\times(F(y_1))^{\alpha-1}(F(y_2)-F(y_1))^{\beta-\alpha-1}\ldots$$
$$\times(F(y_p)-F(y_{p-1}))^{\eta-\varepsilon-1}(1-F(y_p))^{n-\eta}\ldots$$
$$\times f(y_1)f(y_2)\ldots f(y_p)$$

for $-\infty<y_1<\ldots<y_p<\infty$, where $\alpha<\beta<\ldots<\varepsilon<\eta$. If for given constants $0<\lambda_1<\ldots<\lambda_p<1$, each of the subscripts $\alpha, \ldots, \eta$ tends to infinity as $n\to\infty$ under the conditions $r_i=n\lambda_i+o(\sqrt{n})$, where $\alpha=r_1, \ldots, \eta=r_p$, then the random vector $(Y_1, \ldots, Y_p)$ asymptotically obeys the p-dimensional normal distribution with mean vector $(\xi_1, \ldots, \xi_p)$ and covariance matrix $n^{-1}(\sigma_{ij})$, where $\sigma_{ij}=\sigma_{ji}=\lambda_i(1-\lambda_j)/f(\xi_i)f(\xi_j)$ for $i\leqslant j$ and ξ_i is the λ_i-quantile of the population, defined by $\lambda_i=F(\xi_i)$.

Suppose that there exist two sequences a_n and b_n of real and positive numbers, respectively, such that as $n\to\infty$ the sequence $(X_{(n)}-a_n)/b_n$ converges in distribution to a nondegenerate distribution G. The underlying distribution F is said to belong to the **domain of attraction** of the limiting distribution G (DA(G), for short). Except for the change of location and scale, only the following three distributions have nonempty domains of attraction:

$$G_1(x)=\begin{cases}0 & x<0\\ e^{-x^{-\gamma}} & x\geqslant 0\end{cases},$$

$$G_2=\begin{cases}e^{-(-x)^{\gamma}} & x<0\\ 1, & x\geqslant 0\end{cases},$$

$$G_3(x)=e^{-e^{-x}}.$$

Writing $\bar{F}(x)=1-F(x)$, $\alpha(F)=\inf\{x\mid F(x)>0\}$ and $\omega(F)=\sup\{x\mid F(x)<1\}$, we have the following theorem (B. V. Gnedenko, *Ann. Math.*, 44 (1943)):
$F\in\mathrm{DA}(G_1)$ iff $\omega(F)=\infty$ and there exists an a such that

$$\lim_{u\to\infty}\bar{F}(a+ux)/\bar{F}(a+u)=x^{-\gamma}\quad\text{for all } x;$$

$F\in\mathrm{DA}(G_2)$ iff $a\equiv\omega(F)<\infty$ and

$$\lim_{u\to 0}\bar{F}(a-ux)/\bar{F}(a-u)=x^{\gamma}\quad\text{for all } x>0;$$

$F\in\mathrm{DA}(G_3)$ iff there exists a positive function $R(t)$ such that

$$\lim_{t\to\omega(F)}\bar{F}(t+xR(t))/\bar{F}(t)=e^{-x}\quad\text{for all } x.$$

If $F(x)$ is twice differentiable, $f(x)=F'(x)$ is positive for sufficiently large x, and $\lim_{x\to\omega(F)}d\{(1-F(x))/f(x)\}/dx=0$, then $F\in\mathrm{DA}(G_3)$. Noticing the relation $X_{(1)}=-\max\{-X_1, -X_2, \ldots, -X_n\}$, we can also derive the possible limiting distributions for the sequence $(X_{(1)}-a_n)/b_n$ and their domains of attraction. The statistics $R_n=X_{(n)}-X_{(1)}$ and $M_n=(X_{(1)}+X_{(n)})/2$ are called the range and the **midrange** of the sample, respectively. If, for some a_n, a_n', and b_n, both sequences $(X_{(n)}-a_n)/b_n$ and $(X_{(1)}-a_n')/b_n$ converge to nondegenerate distributions G and H, respectively, then they are asymptotically independent, and we have

$$\lim \Pr\{(R_n-a_n+a_n')/b_n\leqslant x\}$$
$$=\int_{-\infty}^{\infty}(1-H(y-x))\,dG(y),$$
$$\lim \Pr\{(2M_n-a_n-a_n')/b_n\leqslant x\}$$
$$=\int_{-\infty}^{\infty}H(x-y)\,dG(y).$$

H. Characterization of the Distribution by means of a Property of the Sampling Distribution

A distribution or a family of distributions can be characterized by a property of the sampling distribution of a suitable statistic. Let $\{X_1, X_2, \ldots, X_n\}$ be a random sample from a nondegenerate distribution F, and let $X_{(1)}\leqslant\ldots\leqslant X_{(n)}$ be the †order statistics. The †sample mean $\bar{X}$ is independent of the †sample variance $S^2=\Sigma(X_j-\bar{X})^2/(n-1)$ iff F is normal $N(\mu,\sigma^2)$ (Kawata and Sakamoto, *J. Math. Soc. Japan*, 1 (1949)). Let a_{ij}, $i,j=1, \ldots, n$, be real numbers such that $\Sigma a_{ij}=0$ and $\Sigma a_{ii}\neq 0$. If F has a finite second moment, then the condition $E(\Sigma a_{ij}X_iX_j\mid\bar{X})=\text{const.}$ implies that F is normal. Two linear statistics $L_1=a_1X_1+\ldots+a_nX_n$ and $L_2=b_1X_1+\ldots+b_nX_n$ are independent only if F is normal, provided that $a_jb_j\neq 0$ for some j. In fact, the X's need not be identically distributed: If L_1 and L_2 are independent and $a_jb_j\neq 0$, then the distribution of X_j is normal (**Skitovich-Darmois theorem**). Yu. V. Linnik [23] gave a necessary and sufficient condition for the normality of F to be equivalent to the identity of the distributions of L_1 and L_2. The condition is stated in terms of the zeros of the entire function $\sigma(z)=|a_1|^z+\ldots+|a_n|^z-|b_1|^z-\ldots-|b_n|^z$. The result contains as a special case the following characterization theorem for the normal distribution: If $\Sigma a_j^2=1$ and L_1 has a distribution identical to that of X_1, then F is normal $N(\mu,\sigma^2)$ with $\mu(\Sigma a_j-1)=0$. R. Shimizu gave a complete description of the characteristic function of the distribution for which L_1 has the same distribution as X_1. In particular, it was proved that if $\log|a_1|/\log|a_2|$ is an irrational number, α is the positive number given by $\Sigma|a_j|^\alpha=1$, and if L_1 has a distribution identical to that of X_1, then F is the †stable distribution with †characteristic exponent α. The result was extended in [24] to the cases where the a's are random variables independent of the X's. If $E(X_i)=0$ and $E(\bar{X} \mid X_1-\bar{X}, X_2-\bar{X}, \ldots, X_n-\bar{X})=0$, then F is normal [26]. Let $\mu_1, \mu_2, \ldots, \mu_n$ be a set of real numbers. If the sampling distribution of the statistic $\sum(X_i+\mu_i)^2$ depends on the μ's only through $\sum \mu_i^2$, then F is normal. If X_i is positive and has finite mean, then the condition $E(\bar{X} \mid X_2/X_1, X_3/X_1, \ldots, X_n/X_1)=\text{const.}$ implies that F is the gamma distribution. If the distribution F is not concentrated on a lattice 0, ρ, $2\rho, \ldots$, then $E(X_{(k+1)}-X_{(k)} \mid X_{(k)})=\text{const.}$ for some k implies that F is the †exponential distribution: $F(x)=1-e^{-\lambda x}$, $x>0$. $X_{(k+1)}-X_{(k)}$ has the identical distribution for some k with $\min\{X_1, X_2, \ldots, X_{n-k}\}$ iff F is exponential. If $a_1, a_2, \ldots, a_n$ are positive numbers such that $a_1+a_2+\ldots+a_n=1$ and such that $\log a_1/\log a_2$ is irrational, then the sampling distribution of $\min\{X_1/a_1, X_2/a_2, \ldots, X_n/a_n\}$ is the same as that of X_1 iff F is exponential. $X_{(1)}$ is independent of $X_{(1)}-\bar{X}$ iff F is exponential.

Suppose that the distribution F has a bounded density function and that the integral $\int_{-\infty}^{\infty} e^{tx} dF(x)$ is finite on a neighborhood of $t=0$. Let $\{X_1, X_2, \ldots, X_n\}$ be a random sample from a distribution of the family $\mathscr{P}=\{F((x-\mu)/\sigma) \mid -\infty<\mu<\infty, \sigma>0\}$. For $n \geqslant 9$, the sampling distribution of the statistics

$$\left\{\frac{(X_3-X_1)}{(X_9-X_7)}, \frac{(X_2-X_1)}{(X_8-X_7)}, \frac{(X_6-X_4)}{(X_9-X_7)}, \frac{(X_5-X_4)}{(X_8-X_7)}, \operatorname{sgn}(X_3-X_1), \operatorname{sgn}(X_2-X_1), \operatorname{sgn}(X_6-X_4), \operatorname{sgn}(X_5-X_4)\right\}$$

uniquely determines the family $\mathscr{P}$. If F is symmetric, then for $n \geqslant 6$ the distribution of $\{|(X_4-X_3)/(X_2-X_1)|, |(X_6-X_5)/(X_2-X_1)|\}$ uniquely determines $\mathscr{P}$ [25].

I. U-Statistics

Let $\{X_1, \ldots, X_n\}$ be a random sample from a certain distribution, and let $\varphi(x_1, \ldots, x_m)$ be a real-valued function that is symmetric with respect to the arguments $x_1, \ldots, x_m$. The statistic

$$U=\binom{n}{m}^{-1} \sum \varphi(X_{\alpha_1}, \ldots, X_{\alpha_m}),$$

where the summation extends over all combinations $(\alpha_1, \ldots, \alpha_m)$ taken from $(1, 2, \ldots, n)$, is called a **U-statistic**. Let $\theta=E(\varphi(X_1, \ldots, X_m))$, and assume that $E(\varphi(X_1, \ldots, X_m)^2)$ is finite. Then the mean and variance of U are given by

$$E(U)=\theta, \; V(U)=\binom{n}{m}^{-1} \sum_{i=1}^{m} \binom{m}{i}\binom{n-m}{m-i} \zeta_i,$$

where ζ_i is the covariance between $\varphi(X_1, \ldots, X_m)$ and $\varphi(X_1, \ldots, X_i, X'_{i+1}, \ldots, X'_m)$, with $X'_{i+1}, \ldots, X'_m$ an additional independent random sample of size $m-i$ from the same distribution. If $\zeta_1 \neq 0$, then U obeys the distribution $N(\theta, m^2\zeta_1/n)$ asymptotically as $n \rightarrow \infty$ (W. Hoeffding, *Ann. Math. Statist.*, 19 (1948)).

These results can be generalized to the case of several populations and samples. Let $X_{11}, \ldots, X_{1n_1}; \ldots; X_{c1}, \ldots, X_{cn_c}$ be c independent random samples, each drawn from one of c populations. Assume that a real-valued function $\varphi(x_{11}, \ldots, x_{1m_1}; \ldots; x_{c1}, \ldots, x_{cm_c})$ is symmetric with respect to the arguments $x_{i1}, \ldots, x_{im_i}$ for each $i=1, \ldots, c$. Then

$$U=\prod_{i=1}^{c}\binom{n_i}{m_i}^{-1} \cdot \sum \varphi(X_{1\alpha(11)}, \ldots, X_{1\alpha(1m_1)}; \ldots; X_{c\alpha(c1)}, \ldots, X_{c\alpha(cm_c)}),$$

where the summation extends over all combinations $(\alpha(i1), \ldots, \alpha(im_i))$ of m_i numbers taken from $(1, 2, \ldots, n_i)$ for each $i=1, \ldots, c$, is called a **U-statistic**. The mean and variance of U can be obtained as before, while U is asymptotically normally distributed as the sample sizes $n_1, \ldots, n_c$ tend to infinity in fixed proportion. Furthermore, if there are given several U-statistics, their †joint distribution is asymptotically normal.

J. Distributions Having Monotone Likelihood Ratio, and Pólya-Type Distributions

Let $(\mathscr{X}, \mathfrak{B})$ be a sample space and

$\mathscr{P}=\{p_\theta(x) \mid \theta \in \Omega\}$

be a family of probability densities with respect to a fixed †σ-finite measure. The function $p_\theta(x)$ regarded as a function of θ with a fixed observed value of x is called the **likelihood function**, and its value at a particular point θ is called the **likelihood** of that point. The family $\mathscr{P}$ with $\Omega \subset \mathbf{R}$ is said to have **monotone likelihood ratio** with respect to a real-valued function $T(x)$ if and only if for any $\theta<\theta'$ such that θ and θ' belong to Ω the ratio $p_{\theta'}(x)/p_\theta(x)$ is a nondecreasing function of $T(x)$. Under the assumption that $\mathscr{X} \subset \mathbf{R}$ and $\partial^2 \log p_\theta(x)/\partial x \partial\theta$ exists, a necessary and sufficient condition for $\mathscr{P}$ to have monotone likelihood ratio with respect to $T(x) \equiv x$ is that $\partial^2 \log p_\theta(x)/\partial x \partial\theta \geqslant 0$ for any x and θ. If $\{X_1, \ldots, X_n\}$ is a random sample from a distribution that has a monotone likelihood ratio and if a real-valued function $\psi(x_1, \ldots, x_n)$ is nondecreasing in each of its arguments, then the expectation $E_\theta(\psi(X_1, \ldots, X_n))$ is a nondecreasing function of θ.

The family $\mathscr{P}$ is said to be of **Pólya type** n if and only if for any $m=1, 2, \ldots, n$ and any real numbers $x_1 < \ldots < x_m$ and $\theta_1 < \ldots < \theta_m$, the determinant of the matrix $(p_{\theta_i}(x_j))$, $i, j = 1, \ldots, m$, is nonnegative, and $\mathscr{P}$ is said to be **strictly of Pólya type** n if the determinant is positive. Being of Pólya type 2 is equivalent to having a monotone likelihood ratio. If $\mathscr{P}$ is (strictly) of Pólya type n for any positive integer n, then it is said to be (strictly) of **Pólya type**. An †exponential family of distributions with probability density $p_\theta(x) = \exp(\theta x + \alpha(\theta) + s(x))$ for $x \in \mathscr{X} \subset \mathbf{R}$ and $\theta \in \Omega \subset \mathbf{R}$ is strictly of Pólya type. Each of the noncentral chi-square distribution, noncentral t-distribution, and noncentral F-distribution is strictly of Pólya type with respect to the noncentrality parameter.

References

[1] H. Cramér, Mathematical methods of statistics, Princeton Univ. Press, 1946.
[2] S. S. Wilks, Mathematical statistics, Wiley, 1962.
[3] M. G. Kendall and A. Stuart, The advanced theory of statistics I, Griffin, fifth edition, 1952.
[4] N. L. Johnson and S. Kotz, Distributions in statistics, 4 vols., Wiley, 1969–1972.
[5] Student (W. G. Gosset), The probable error of a mean, Biometrika, 6 (1908), 1–25.
[6] R. A. Fisher, Statistical methods for research workers, Oliver & Boyd, 1925.
[7] T. W. Anderson, An introduction to multivariate statistical analysis, Wiley, 1958.
[8] A. T. James, Distributions of matrix variates and latent roots derived from normal samples, Ann. Math. Statist., 35 (1964), 475–501.
[9] A. G. Constantine, Some non-central distribution problems in multivariate analysis, Ann. Math. Statist., 34 (1963), 1270–1285.
[10] H. Chernoff, Large-sample theory: Parametric case, Ann. Math. Statist., 27 (1956), 1–22.
[11] K. Pearson, On the criterion that a given system of deviations from the probable in the case of a correlated system of variables is such that it can be reasonably supposed to have arisen from random sampling, Philos. Mag., (5) 50 (1900). Reproduced in K. Pearson, Early statistical papers, Cambridge Univ. Press, 1948.
[12] K. R. Parthasarathy, Probability measures on metric spaces, Academic Press, 1967.
[13] P. Billingsley, Convergence of probability measures, Wiley, 1968.
[14] R. H. Randles and D. A. Wolfe, Introduction to the theory of nonparametric statistics, Wiley, 1979.
[15] F. Y. Edgeworth, The law of errors, Trans. Cambridge Philos. Soc., 20 (1904).
[16] E. A. Cornish and R. A. Fisher, Moments and cumulants in the specification of distributions, Rev. Int. Statist. Inst., 5 (1937). Reproduced in R. A. Fisher, Contributions to mathematical statistics, Wiley, 1950.
[17] R. N. Bhattacharya and J. K. Ghosh, On the validity of the formal Edgeworth expansion, Ann. Statist., 6 (1978), 434–451.
[18] R. N. Bhattacharya and R. Ranga Rao, Normal approximation and asymptotic expansions, Wiley, 1976.
[19] E. J. Gumbel, Statistics of extremes, Columbia Univ. Press, 1958.
[20] A. E. Sarhan and B. G. Greenberg (eds.), Contributions to order statistics, Wiley, 1962.
[21] H. A. David, Order statistics, Wiley, second edition, 1981.
[22] J. Galambos, The asymptotic theory of extreme order statistics, Wiley, 1978.
[23] Yu. V. Linnik, Linear forms and statistical criteria, Selected Transl. Math. Statist. Prob., 3 (1962), 1–90. (Original in Russian, 1953.)
[24] R. Shimizu and L. Davies, General characterization theorems for the Weibull and the stable distributions, Sankhyā, ser. A, 43 (1981), 282–310.
[25] Yu. V. Prokhorov, On a characterization of a class of probability distributions by distributions of some statistics, Theory of Prob. Appl., 10 (1965), 438–445. (Original in Russian, 1965.)
[26] A. M. Kagan, Yu. V. Linnik, and C. R. Rao, Characterization problems in mathematical statistics, Wiley, 1973.
[27] J. Galambos and S. Kotz, Characterizations of probability distributions, Springer, 1978.
[28] G. P. Patil, S. Kotz, and J. K. Ord (eds.), Statistical distributions in scientific work III, Characterizations and applications, Reidel, 1975.
[29] S. Karlin, Decision theory for Pólya type distributions, Case of two actions I, Proc. 3rd Berkeley Symp. Math. Stat. Prob. I, Univ. of California Press, 1956, 115–128.

375 (XX.26) Scattering Theory

A. General Remarks

The path of a moving (incident) particle is distorted when it interacts with another (target) particle, such as an atom or a molecule. Phenomena of this sort are generally called

scattering. Scattering is called **elastic** when the internal properties of the incident particle and the target remain unchanged after the collision, and **inelastic** when the internal properties change, other particles are emitted, or the two particles form a bound state.

The extent of scattering depends on the sizes of the incident and target particles. The **scattering cross section** is defined as the probability that the incident beam will be scattered per unit time (normalized to one particle per unit time crossing unit area perpendicular to the direction of incidence). In †classical mechanics the scattering cross section is equal to the cross section of the target perpendicular to the incoming beam, hence the term "cross section." The probability of scattering into a unit solid angle in a particular direction is called the **differential cross section**. The probability that the incoming particle is absorbed by the target, called the **absorption cross section**, is intimately connected with the scattering cross section. Analyses of scattering give information on the structure and interactions of atoms, molecules, and elementary particles. One can also study the scattering of acoustic and electromagnetic waves by inhomogeneous media and obstacles, by considering notions similar to the above.

Scattering theory may be dated back to Lord Rayleigh. Since the advent of quantum mechanics in the mid-1920s, scattering problems, mainly for central (spherically symmetric) potentials, have been investigated strenuously by physicists. It may be said, however, that a scattering theory having mathematically rigorous foundations began around the 1950s, when the pioneering work of K. Friedrichs (*Comm. Pure Appl. Math.*, 1 (1948)), A. Ya. Povzner (*Mat. Sb.*, 32 (1953)), T. Kato (*J. Math. Soc. Japan*, 9 (1957)), J. M. Cook (*J. Math. Phys.*, 36 (1957)), and J. M. Jauch (*Helv. Phys. Acta*, 31 (1958)), among others, appeared, and scattering theory has now grown into a branch of mathematical physics.

General references for mathematical scattering theory are, e.g., [1–5].

B. Wave and Scattering Operators

In †quantum mechanics the dynamics of an interacting system is given by a †one-parameter group of unitary operators e^{-itH}, where t denotes the time and H, called the Hamiltonian of the system, is a †self-adjoint operator acting in a †Hilbert space $\mathscr{H}$. Elements of $\mathscr{H}$ represent (pure) states of the system. Let H_0 be the "free" Hamiltonian of the corresponding "noninteracting" system. (There are at present no generally accepted definite criteria for "free" and "noninteracting.") H_0 is assumed to be †absolutely continuous, which is the case in most practical situations. Then the **outgoing** and **incoming wave operators** $W_\pm \equiv W_\pm(H, H_0)$ are defined, if they exist, by

$$W_\pm = \operatorname*{s-lim}_{t\to\pm\infty} e^{itH}e^{-itH_0} \qquad (\text{s-lim} = {}^\dagger\text{strong limit}).$$

This means that given any free motion $e^{-itH_0}u$ there is an initial $(t=0)$ state $u_\pm$ $(=W_\pm u)$ such that $e^{-itH}u_\pm$ and $e^{-itH_0}u$ are asymptotically equal at $t=\pm\infty$. $W_\pm$ are †isometric, intertwine the two dynamics: $e^{-itH}W_\pm = W_\pm e^{-itH_0}$, and map $\mathscr{H}$ (which is nothing but the †absolutely continuous subspace $\mathscr{H}_{\mathrm{ac}}(H_0)$ for H_0) onto a closed subspace of $\mathscr{H}_{\mathrm{ac}}(H)$. The **scattering operator** S is defined as $S = W_+^* W_-$ (A^* is the Hilbert-space †adjoint of A). S commutes with H_0 and maps states in the remote past into states in the distant future. One of the most important problems in scattering theory is to prove the †unitarity of S, or equivalently, $\operatorname{Ran} W_+ = \operatorname{Ran} W_-$ (Ran = †range = †image). $W_\pm$ is called **complete** if $\operatorname{Ran} W_\pm = \mathscr{H}_{\mathrm{ac}}(H)$. The completeness of $W_\pm$ implies that S is unitary.

As a typical example we consider the 1-body problem. Note that the 2-body problem reduces to the 1-body problem by separating out the center-of-mass motion, which is free. Then $H_0 = -\Delta$ (the negative †Laplacian in $\mathbf{R}^3$), $H = -\Delta + V$, the operator V being multiplication by a real-valued function $V(x)$, called the **potential**, and $\mathscr{H} = L_2(\mathbf{R}^3)$. If V is **short range**, i.e., if, roughly speaking, $V(x) = O(|x|^{-1-\varepsilon})$ $(\varepsilon > 0)$ at ∞, the wave operators are known to exist and to be complete (S. Agmon, *Ann. Scuola Norm. Sup. Pisa*, (4) 2 (1975)). If the potential $V(x)$ is **long range**, i.e., if, roughly speaking, $V(x) = O(|x|^{-\varepsilon})$ $(\varepsilon > 0)$ at ∞, then the foregoing definition of the wave operators has to be modified. For the Coulomb potential $V(x) = c/x$, for instance, one can adopt the following definition of **modified wave operators**:

$$\widetilde{W}_\pm = \operatorname*{s-lim}_{t\to\pm\infty} e^{itH}\exp(-itH_0 - i(c/2)H_0^{-1/2}\log t).$$

It can be shown that the $\widetilde{W}_\pm$ exist (which implies that the ordinary wave operators do not exist) and are complete: $\operatorname{Ran}\widetilde{W}_\pm = \mathscr{H}_{\mathrm{ac}}(H)$ (J. D. Dollard, in [6]). The same result obtains for more general long-range potentials (H. Kitada, *J. Math. Soc. Japan*, 30 (1978); T. Ikebe and H. Isozaki, *Integral Equations and Operator Theory*, 5 (1982)).

If the wave operators exist and are complete, they give †unitary equivalence between the †absolutely continuous parts of H_0 and H (→ T. Kato [7]; 331 Perturbation of Linear Operators).

In the foregoing discussion it was tacitly assumed that in dealing with scattering phenomena we adhere to states in $\mathscr{H}_{\mathrm{ac}}(H_0)$ and

$\mathscr{H}_{ac}(H)$. A more physically intuitive definition in the case of potential scattering ($H_0 = -\Delta$, $H = H_0 + V$) of **scattering states** $\Sigma_{\pm}(H)$ is the following: $f \in \Sigma_{\pm}(H)$ if and only if for any $r > 0$,

$$\lim_{T \to \pm\infty} \frac{1}{T} \int_0^T \| F_r e^{-itH} f \|^2 dt = 0$$

$$(\| \ \| = L_2\text{-norm}),$$

where F_r is the (projection) operator of multiplication by the characteristic function of $\{x \in \mathbf{R}^3 \,|\, |x| < r\}$. In general no inclusion relations between $\Sigma_{\pm}(H)$ and $\mathscr{H}_{ac}(H)$ are known. But for a wide class of potentials it is known that $\mathscr{H}_{ac}(H)$ as well as the †continuous subspace of H coincides with $\Sigma_{\pm}(H)$ (in this case there is no †singular continuous spectrum) (W. O. Amrein and V. Georgescu, *Helv. Phys. Acta*, 46 (1973); C. Wilcox, *J. Functional Anal.*, 12 (1973); Amrein, in [8]).

A purely abstract result in scattering theory may be noted. Let H_0 and H be self-adjoint operators in an abstract Hilbert space $\mathscr{H}$ such that $V = H - H_0$ is a †trace-class (†nuclear) operator. Then the **generalized wave operators** $W_{\pm}(H, H_0) = \text{s-lim}_{t \to \pm\infty} \exp(itH) \exp(-itH_0) P_{ac}(H_0)$ exist, where $P_{ac}(H_0)$ is the †projection onto $\mathscr{H}_{ac}(H_0)$. Since this statement is symmetric in H_0 and H, the "inverse" generalized wave operators $W_{\pm}(H_0, H)$ also exist, from which one can conclude that $W_{\pm}(H, H_0)$ are complete. Moreover, the **invariance principle** holds, which means roughly the following: If ϕ is a strictly increasing function on R, then $W_{\pm}(\phi(H), \phi(H_0))$ exists and is equal to $W_{\pm}(H, H_0)$. This result can be applied to potential scattering when $V(x) \in L_2(\mathbf{R}^3) \cap L_1(\mathbf{R}^3)$ (Kato [7]; D. B. Pearson, *J. Functional Anal.*, 28 (1978)).

C. Stationary (Time-Independent) Approach

We again consider the 1-body problem as in Section B. $V(x)$ is assumed to verify certain appropriate decay conditions at ∞ as the case may be. Consider the †resolvents $R_0(z) = (H_0 - z)^{-1}$ and $R(z) = (H - z)^{-1}$ for $z \in C - R$, which are well-defined bounded integral operators on $\mathscr{H} = L_2(\mathbf{R}^3)$. Here we note the following: $[0, \infty)$ is the †continuous spectrum of H_0 and H, $(-\infty, 0)$ is contained in the †resolvent set of H_0, and H has possibly †discrete †eigenvalues in $[-a, 0)$ with $(-\infty, -a)$ contained in the resolvent set, where a is a positive number. When z approaches a positive real value, $R_0(z)$ and $R(z)$ do not have limits as †bounded operators on $\mathscr{H}$. But if we regard them as operators from $L_{2,\gamma}$ to $L_{2,-\gamma}$ ($L_{2,\alpha} = \{u \,|\, (1+|x|)^{\alpha} u(x) \in L_2(\mathbf{R}^3)\}$), $\gamma > 1/2$, they can be shown to have boundary values $R_0(\lambda \pm i0)$ and $R(\lambda \pm i0)$, $\lambda > 0$ (**limiting absorption principle**; → Agmon, *loc. cit.*, for short-range potentials; for long-range potentials → R. Lavine, *J. Functional Anal.*, 12 (1973); T. Ikebe and Y. Saitō, *J. Math. Kyoto Univ.*, 12 (1972); and Saitō, *Publ. Res. Inst. Math. Sci.*, 9 (1974). With these boundary values we can define "stationary" wave operators whose range is easily proved to coincide with $\mathscr{H}_{ac}(H)$, and show their equality to the time-dependent wave operators discussed in Section B, thus obtaining the completeness of the latter.

H_0 is known to have **generalized** (improper) **eigenfunctions** $\varphi_0(x, \xi) = e^{ix \cdot \xi}$ with **generalized** (improper) **eigenvalues** $|\xi|^2$. The associated eigenfunction expansion is nothing but the Fourier integral expansion. An eigenfunction expansion, similar to the Fourier expansion, that diagonalizes H can be obtained by using generalized eigenfunctions

$$\varphi_{\pm}(x, \xi)$$
$$= \varphi_0(x, \xi) - (R(|\xi|^2 \pm i0) V \varphi_0(\cdot, \xi))(x),$$

which are the solutions to the **Lippmann-Schwinger equation**

$$\varphi_{\pm}(x, \xi) = \varphi_0(x, \xi)$$
$$- \frac{1}{4\pi} \int_{\mathbf{R}^3} \frac{e^{\pm i|\xi||x-y|}}{|x-y|} V(y) \varphi_{\pm}(y, \xi) dy.$$

A rough statement of this is the following: Let $\tilde{u}_{\pm}(\xi) = (2\pi)^{-3/2} \int \overline{\varphi_{\pm}(x, \xi)} u(x) dx$. Then $\|\tilde{u}\| = \|u\|$, $(Hu)^{\sim}(\xi) = |\xi|^2 \tilde{u}(\xi)$, and $u(x) = (2\pi)^{-3/2} \int \varphi_{\pm}(x, \xi) \tilde{u}(\xi) d\xi$ (→ e.g. [4, XI.6] for a more precise statement).

In view of the fact that S commutes with H_0, we can show that S admits the following representation: Let $\hat{u}(\xi) = (2\pi)^{-3/2} \int \overline{\varphi_0(x, \xi)} u(x) dx$ be the †Fourier transform of u. Then

$$(Su)^{\wedge}(\xi) \equiv (\hat{S}\hat{u})(\xi)$$
$$= \hat{u}(\xi) - \pi i \int_{S^2} |\xi| T(\xi, |\xi| \omega') \hat{u}(|\xi| \omega') d\omega',$$

where

$$T(\xi, \xi') = (2\pi)^{-3} \int \overline{\varphi_0(x, \xi)} V(x) \varphi(x, \xi') dx$$

is the kernel of the so-called ***T*-operator**, which is a †compact operator on $L_2(S^2)$ under suitable conditions on $V(x)$. $T(\xi, \xi')$ is related to the experimentally measurable total cross section (for incident momentum ξ) $\sigma(\xi)$:

$$\sigma(\xi) = 2\pi \int_{S^2} |f(|\xi|; \omega, \omega')|^2 d\omega' \qquad (\omega = \xi / |\xi|),$$

$$f(\lambda; \omega, \omega') = -2\pi^2 T(\lambda\omega, \lambda\omega') \qquad (\lambda > 0).$$

The quantity $f(\lambda; \omega, \omega')$ is called the **scattering amplitude** and appears in the asymptotic expansion of $\varphi_\pm(x,\xi)$ as

$$\varphi_\pm(x,\xi)\sim\varphi_0(x,\xi)+\frac{e^{\pm i|\xi||x|}}{|x|}f(|\xi|;\hat{x},\hat{\xi}),$$

where $\hat{a}=a/|a|$ for $a\in\mathbf{R}^3$.

An abstract version of the limiting absorption principle and eigenfunction expansion is known as the Kato-Kuroda theory, for which the reader is referred to Kato and S. T. Kuroda in [6] and in *Functional analysis and related fields*, Springer, 1970, and to Kuroda (*J. Math. Soc. Japan*, 25 (1973)).

D. Time-Dependent Approach

Consider the same situation as in Section C. Since scattering is a time-dependent phenomenon, it seems natural to develop scattering theory in a time-dependent fashion. Indeed, there is an approach to the completeness of wave operators that does not resort to any eigenfunction expansion results, but instead follows the temporal development of the wave packet $e^{-itH}u$. The completeness of $W_\pm$ will be established if one can show that any $u\in\mathscr{H}_{\mathrm{ac}}(H)$ orthogonal to Ran $W_\pm$ is 0. A crucial step to prove this is to find a clever decomposition of a wave packet into an outgoing and an incoming one, or to find projections $P_\pm$ such that $P_+ + P_- = I$ and $P_\pm e^{-itH_0}$ goes to 0 as $t\to\mp\infty$. Some compactness arguments are also needed.

To construct such a decomposition or projections one looks at the scalar product $x\cdot\xi$, x and ξ being the position and momentum (operators), respectively. The main idea is that if this is positive, the particle will be outgoing (to infinity), and if negative, incoming (from infinity). But since we work in the framework of quantum mechanics, this classical-mechanical intuition should be properly modified.

Besides the completeness of wave operators it can also be shown through the above approach that the singular continuous spectrum of H is absent. For details → [4, XI.17] and V. Enss, *Comm. Math. Phys.*, 61 (1978).

E. Partial Wave Expansion

In this section we assume that the potential $V(x)$ is central, i.e., $V(x)$ is a function of $|x|$ alone. Then the scattering operator S turns out to commute not only with H_0 but also with the †angular momentum operator $L=x\times i^{-1}\nabla$ (vector product). The eigenvalues of $L^2=L\cdot L$ (scalar product) are $l(l+1)$ $(l=0,1,2,\dots)$, and those of L_z, the third component of L, are $m=-l,-(l-1),\dots,l-1,l$ if L^2 has eigenvalue $l(l+1)$, while simultaneous eigenfunctions are given by suitably normalized †spherical harmonics $Y_{lm}(\omega)$, $\omega\in S^2$. Let E_l be the projection onto the subspace spanned by functions of the form $\sum_{m=-l}^{l}u_{lm}(r)Y_{lm}(\omega)$ $(r=|x|>0,\ \omega=x/r)$. $\mathscr{H}$ becomes an orthogonal sum of $\mathscr{H}_l=E_l\mathscr{H}$. The aforementioned commutation property claims that $E_lS=SE_l\equiv S_l$, and the operator $\hat{S}_l$ reduces to multiplication by a scalar function $e^{2i\delta_l(\lambda)}$ in $\hat{\mathscr{H}}_l$ (→ Section C). $\delta_l(\lambda)$ is called the **phase shift**. Defining the **partial wave scattering amplitude** $f_l(\lambda)=(2i\lambda)^{-1}(e^{2i\delta_l(\lambda)}-1)$, one obtains the **partial wave expansion** of the scattering amplitude:

$$f(\lambda;\omega,\omega')=\sum_{l=0}^{\infty}(2l+1)f_l(\lambda)P_l(\cos\theta),$$

where θ is the angle between ω and ω', and P_l is a †Legendre polynomial. The total cross section is $\sigma(\lambda)=4\pi\lambda^{-2}\sum_{l=0}^{\infty}(2l+1)|f_l(\lambda)|^2$ [1–5, 9].

F. Many-Body Problem (Multichannel Scattering)

We consider only the 3-body case, which is complicated enough compared with the 2-(essentially 1-) body case. The complications are both kinematical and dynamical. The configuration of a 3-body system is given by a point in $\mathbf{R}^9$. Once we choose the center-of-mass coordinates, there is no kinematically natural way to choose the remaining six coordinates. In the 2-body case a freely moving particle in the remote past will be freely moving in the distant future. But in the 3-body case there come into play various other dynamical processes, such as capture, breakup, rearrangement, and excitation.

The 3-body Hamiltonian is a self-adjoint operator in $L_2(\mathbf{R}^9)$ of the form

$$\tilde{H}=-\sum_{i=1}^{3}\frac{1}{2m_i}\Delta_i+\sum_{i<j}V_{ij}(x_i-x_j),$$

where Δ_i is the 3-dimensional Laplacian associated with particle i, m_i is the mass of particle i, and each $V_{ij}(x)$ $(=V_{ji}(x)=V_{ij}(-x))$ is a real-valued function decaying at ∞ in $\mathbf{R}^3$ (not in $\mathbf{R}^9$). If we remove the center-of-mass motion, $\tilde{H}$ can be written in the form

$$\tilde{H}=H_0\otimes I+I\otimes H\quad(\otimes=\text{†tensor product}),$$

where H_0 is the center-of-mass Hamiltonian in $L_2(\mathbf{R}^3)$ representing the uniform free motion of the center of mass, and H is the Hamiltonian of relative motion in $L_2(\mathbf{R}^6)$. One should note as mentioned above that there is no unique natural way of choosing coordinates in $\mathbf{R}^6$ and representing H, but there are many equivalent representations. Suppose, for instance, that particles 1 and 2 and particles 1 and 3 form †bound states (12) and (13) and that there are no bound states between 2 and 3. We partition the whole system into **clusters**: (1) (2) (3), (12) (3), and (13) (2) (( ) represents a cluster and figures in ( ) are the particles forming the cluster). A **channel** is a partition into clusters together with a specified bound-state eigenfunction. Take, for instance, channel (12) (3), and suppose $\psi \in L_2(\mathbf{R}^3)$ is the eigenfunction in question. If we take $x = x_2 - x_1$ and $y = x_3 - (m_1 + m_2)^{-1}(m_1 x_1 + m_2 x_2)$, then

$$H = -\frac{1}{2m}\Delta_y - \frac{1}{2n}\Delta_x + V_{12}(x)$$

$$+ V_{23}\left(\frac{m_1 x}{m_1 m_2} - y\right) + V_{31}\left(\frac{m_2 x}{m_1 + m_2} + y\right),$$

where $m^{-1} = m_3^{-1} + (m_1 + m_2)^{-1}$, $n^{-1} = m_1^{-1} + m_2^{-1}$. Let us neglect the interactions between (12) and (3), i.e., set $V_{23} = V_{31} = 0$ to define the **cluster decomposition Hamiltonian**

$$H_{(12)(3)} = -\frac{1}{2m}\Delta_y - \frac{1}{2n}\Delta_x + V_{12}(x).$$

Let $\mathscr{H}_{(12)(3)} = L_2(\mathbf{R}^3)$ (called the **channel Hilbert space** consisting of functions of y), and define a mapping $\tau: \mathscr{H}_{(12)(3)} \to \mathscr{H} = L_2(\mathbf{R}^6)$ (functions of x and y) by $(\tau f)(x, y) = \psi(x) f(y)$. The **channel wave operators** $W_{(12)(3)\pm}$ are defined by

$$W_{(12)(3)\pm} = \operatorname*{s-lim}_{t \to \pm\infty} e^{itH} e^{-itH_{(12)(3)}} \tau$$

as isometries from $\mathscr{H}_{(12)(3)}$ into $\mathscr{H}$. Their ranges are not expected to coincide as in the 2-body case. $W_{(1)(2)(3)\pm}$ and $W_{(13)(2)\pm}$ are similarly defined. Note that $\mathscr{H}_{(1)(2)(3)} = \mathscr{H}$. If α and β are distinct channels, we have $\operatorname{Ran} W_{\alpha+} \perp \operatorname{Ran} W_{\beta+}$ and $\operatorname{Ran} W_{\alpha-} \perp \operatorname{Ran} W_{\beta-}$, but no such relations exist between $\operatorname{Ran} W_{\alpha+}$ and $\operatorname{Ran} W_{\beta-}$ or $\operatorname{Ran} W_{\alpha-}$ and $\operatorname{Ran} W_{\beta+}$. Define, for channels α and β, $S_{\alpha\beta}: \mathscr{H}_\beta \to \mathscr{H}_\alpha$ by $S_{\alpha\beta} = (W_{\alpha+})^* W_{\beta-}$. Now the **scattering operator** S for the 3-body system is defined as the †direct sum of $S_{\alpha\beta}$ acting in the Hilbert space $\sum_\alpha \oplus \mathscr{H}_\alpha: S = \sum_{\alpha,\beta} \oplus S_{\alpha\beta}$. Naturally the question arises: Is S unitary? The first affirmative answer was made by L. D. Faddeev (*Israel Program for Scientific Translations*, 1965 (in English; original in Russian, 1963)), and later the work of J. Ginibre and M. Moulin (*Ann. Inst. H Poincaré*, A21 (1974)) and L. Thomas (*Ann. Phys.*, 90 (1975)) came out. The method of these authors is stationary. There have also been some attempts using time-dependent methods.

G. Inverse Problem

The **inverse problem** in potential scattering may be formulated at least mathematically as follows: Given the scattering operator or scattering amplitude, determine the potential(s) giving rise to the operator or amplitude. We consider here only the 2-body case (as to the many-body case, almost nothing is known). The central-potential case can be reduced to 1-dimensional problems on $(0, \infty)$. In the 1-dimensional case the celebrated Gel'fand-Levitan theory (→ 112 Differential Operators O) has long been known and has been successfully applied even to nonlinear problems such as the †Korteweg–de Vries equation. In the 3-dimensional case, however, the problem becomes difficult; so far there has not been any satisfactory theory comparable to that for the 1-dimensional case. The potential $V(x)$ is a function $\mathbf{R}^3 \to \mathbf{R}$. The scattering amplitude $f(\lambda; \omega, \omega')$ is a function $f: R \times S^2 \times S^2 \to C$. Let M be the mapping that takes V into f. The inverse problem deals with M^{-1}. Several questions may be posed (in order of increasing difficulty): (1) Is M one-to-one? (2) When it is known that M is one-to-one and f is in the image of M, how does one (re-)construct the V that yields f? (3) What conditions characterize the image of M? Question (1) has been rather satisfactorily answered insofar as short-range potentials are concerned. Concerning questions (2) and (3), attempts have been and are being made to generalize the Gel'fand-Levitan theory, but it may be said that we are still at the beginning stage. References are [2, 9, 10] and R. G. Newton (*J. Math. Phys.*, 20 (1980); 21 (1981); 22 (1982)).

H. Scattering for the Wave Equation

Consider the †wave equation $u_{tt} - \Delta u = 0$ in $\mathbf{R}^3$. The solution $u(t) = u(t, x)$ is uniquely determined by the initial data $\{f, g\} = \{u(0), u_t(0)\}$, and $U_0(t)\{f, g\} = \{u(t), u_t(t)\}$ defines the solution operator $U_0(t)$. The set of data $\{f, g\}$ with finite energy: $\int (|\nabla f|^2 + |g|^2) dx < \infty$ forms a Hilbert space $\mathscr{H}_0$. $U_0(t)$ is a unitary group on $\mathscr{H}_0$. A similar description is possible for solutions of the wave equation in an exterior domain Ω outside an obstacle with zero boundary condition. Denote the resulting Hilbert space and solution operator by $\mathscr{H}$ and $U(t)$, respectively. Let $J: \mathscr{H}_0 \to \mathscr{H}$ be the identification operator defined by $(J\{f, g\})(x) = \{f, g\}(x)$, $x \in \Omega$. The **wave operators** are defined by $W_\pm = \text{s-lim}_{t \to \pm\infty} U(-t) J U_0(t)$. The existence of $W_\pm$ is shown rather easily by using †Huygens's principle. As in Section B, we define the **scattering operator** $S = W_+^* W_-$ and say that $W_\pm$ is **complete** if $\operatorname{Ran} W_\pm = \mathscr{H}_{\text{ac}}(H)$, where H is the self-adjoint †infinitesimal generator of $U(t)$: $U(t) = e^{-itH}$. The completeness of $W_\pm$ and the unitarity of S are proved on the basis of the abstract **translation representation**

theorem: Let $U(t)$ be a unitary group on a Hilbert space $\mathscr{H}$. Suppose there exist subspaces D_+ and D_-, called **outgoing** and **incoming subspaces**, such that $U(t)D_\pm \subset D_\pm$ for $\pm t \geqslant 0$, $\bigcap_{t\in R} U(t)D_\pm = \{0\}$, and $\bigcup_{t\in R} U(t)D_\pm$ is dense in $\mathscr{H}$. Then we have two unitary operators $\mathscr{R}_\pm : \mathscr{H} \to L_2(R; N)$, where N is an auxiliary Hilbert space, such that $\mathscr{R}_\pm U(t) \cdot \mathscr{R}_\pm^{-1}$ is right translation by t, and D_+ (D_-) is mapped onto $L_2(0,\infty; N)$ ($L_2(-\infty, 0; N)$) by $\mathscr{R}_+$ ($\mathscr{R}_-$).

Turning to the concrete situation, one can study the detailed properties of S. The uniqueness theorem in the inverse problem is also obtained, to the effect that S determines the obstacle uniquely. The foregoing treatment of scattering is known as the Lax-Phillips theory (P. D. Lax and R. S. Phillips, in [6, 8, 11].

References

[1] W. O. Amrein, J. M. Jauch, and K. B. Sinha, Scattering theory in quantum mechanics, Benjamin, 1977.
[2] R. G. Newton, Scattering theory of waves and particles, McGraw-Hill, 1966.
[3] E. Prugovečki, Quantum mechanics in Hilbert space, Academic Press, 1971.
[4] M. Reed and B. Simon, Methods of modern mathematical physics III, Scattering theory, Academic Press, 1979.
[5] J. R. Taylor, Scattering theory, Wiley, 1972.
[6] Rocky Mountain J. Math., 1 (1) (1971) (special issue on scattering theory).
[7] T. Kato, Perturbation theory for linear operators, Springer, 1976.
[8] J. A. Lavita and J.-P. Marchand (eds.), Scattering theory in mathematical physics, Proc. NATO Adv. Study Inst., Denver, Colo., 1973, Reidel, 1974.
[9] V. De Alfaro and T. Regge, Potential scattering, North-Holland, 1965.
[10] K. Chadan and P. C. Sabatier, Inverse problems in quantum scattering theory, Springer, 1977.
[11] P. D. Lax and R. S. Phillips, Scattering theory, Academic Press, 1967.

376 (XIX.15) Scheduling and Production Planning

Production planning emerges in many situations. Models of economic planning can be classified as (1) fiscal policy-oriented, (2) final demand-oriented, (3) structure-oriented, (4) expenditure-oriented, and (5) industrialization-oriented. Types (2), (3), and (5) belong to production planning in a broad sense, as emphasis is placed on production in these models.

A typical **production planning** theory of primary importance is **activity analysis**, which has made remarkable progress since its initiation by T. C. Koopmans [1]. Its principal theoretical content consists of †linear programming. Most applications of linear programming are more or less concerned with production activities. Because of the additivity and divisibility of production as well as the limitation of production intensities, problems of production planning can be formulated as problems of linear programming. The methods of †linear algebra are used to obtain an optimal production plan and are very important in modern economic analysis, because these methods not only provide practical algorithms but also clarify the role of price, especially in †dual linear programming problems.

The originators of general equilibrium theory (→ 128 Econometrics) failed to give an analytical demonstration of the existence of solutions of certain systems of equations of economic relevance. The existence of a determinate equilibrium was established first by A. Wald for a system of equations of the Walras-Cassel type. On the other hand, J. von Neumann [3] proved the existence of nonnegative solutions α, β, x_i, y_j for a system of inequalities

$$\alpha \sum_{i=1}^{m} a_{ij} x_i \leqslant \sum_{i=1}^{m} b_{ij} x_i, \qquad j = 1, 2, \dots, n,$$

$$\beta \sum_{j=1}^{n} a_{ij} y_j \geqslant \sum_{j=1}^{n} b_{ij} y_j, \qquad i = 1, 2, \dots, m,$$

by reducing the problem to the proof of the existence of a †saddle point (→ 292 Nonlinear Programming A) of the function

$$\Phi(X, Y) \equiv \sum_{i=1}^{m} \sum_{j=1}^{n} b_{ij} x_i y_j \Big/ \sum_{i=1}^{m} \sum_{j=1}^{n} a_{ij} x_i y_j$$

by means of Brouwer's fixed-point theorem. In this result an equilibrium is defined in the broad sense that demand for goods does not exceed their supply, rather than requiring exact balance.

A second important kind of production planning is related to both †inventory control (→ 227 Inventory Control) and sales planning. An example is the minimization of

$$\int_0^T (z(t) + \beta \max(dz/dt, 0))\, dt$$

subject to the condition $z(t) \geqslant r(t)$, where $z(t)$ and $r(t)$ are the output and demand, respectively, at time t. Stabilization of employment and production can also be classified as a production planning problem of this kind, in which inventory holding is considered as a means for lessening the change of employment level. This is related to the problem of smoothing production by inventory control. Dynamic programming (→ 127 Dynamic Programming) is very useful in dealing with problems of smoothing production.

Production planning as a production management tool is often embodied in **scheduling**. Consider a project that consists of n indivisible tasks (jobs or activities) J_i, $i=1,2,\ldots,n$, each requiring p_i units of time for processing, where p_i is given either deterministically or probabilistically. A precedence constraint (generally a †partial ordering) partially specifying the order in which these tasks are to be processed is also imposed by technical considerations: One attempts to find a schedule (i.e., a specification of the time to process the tasks J_i) consistent with the given precedence constraint.

Well-known techniques developed for this purpose are **PERT** (program evaluation and review technique) [4] and **CPM** (critical path method) [5], in which the precedence constraint is represented by an acyclic †directed graph, called an arrow diagram, a project network, or a PERT network, such that each task J_i corresponds to an arc of length equal to p_i. An arrow diagram is illustrated in Fig. 1. The longest path from the start node to the end node in the network is called the **critical path**, and gives the minimum time necessary to complete the project. Following computation of the critical path by means of dynamic programming, computations are also made for the earliest (latest) start time, the earliest (latest) finish time, and the floats (i.e., the allowances for such start and finish times) of each task to be satisfied in order to complete the project within the indicated minimum time. These are used to review and control the progress of the project. In PERT the processing time of each task is probabilistically treated on the basis of three estimates: most likely, optimistic, and pessimistic. From these data other parameters,

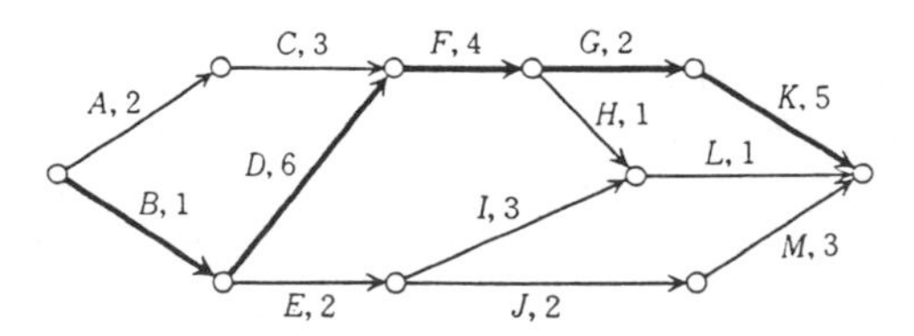

Fig. 1
$A, B, \ldots, M$ denote tasks, while the associated integers are their processing times. Bold arrows indicate the critical path. The start node is on the left, the end node is on the right.

such as the probability of completing the project before the specified due date, are computed. In CPM, on the other hand, a minimum cost schedule to attain the given due date is obtained by utilizing †network flow algorithms (J. E. Kelley [6], D. R. Fulkerson [7]), in which the processing time of a task is determined by linear interpolation between the normal time (achieved with low cost) and the crash time (high cost).

PERT and CPM are used in various areas of application, e.g., civil engineering and the construction industry, shipbuilding, production of automobiles, machines, and electric apparatus, and management of research and development programs. PERT was originally developed by the US Navy to monitor and control the development of the Polaris fleet ballistic missile program, while CPM was developed by the RAND Corporation and the Du Pont Corporation, both in the late 1950s. Computers have been essential from the beginning, to handle the large amount of associated data. A number of application program packages, each with some additional features, are currently available, e.g., PERT/TIME, PERT/COST, CPM, and RAMPS.

The **machine sequencing (scheduling) problem** arises when the resources, instruments, workers, and so forth, required to process a task are abstractly formulated as machines and if the restriction on the number of available machines is taken into consideration (i.e., the conflict between tasks competing for the same machine at the same time must be resolved). Usually one machine is assigned to each task. Such a machine is either (a) uniquely determined for each task or (b) chosen from a given set of machines; in the latter case, there might be (i) parallel machines with the same capability or (ii) machines with different capabilities. The precedence constraints are also ramified into independent (i.e., no constraint), in-tree, out-tree, series-parallel, and general partial ordering constraints. Each task has a ready time (release time) r_i such that J_i cannot be processed before it, and a due time d_i. One is asked to find a schedule satisfying the above machine constraints, precedence constraints, and ready time constraints, while considering an optimality criterion that is a function of the completion time C_i of J_i ($i=1,2,\ldots,n$). Typical criteria for minimization are: (1) maximum completion time (makespan) $C_{\max}=\max_i C_i$; (2) flowtime (total completion time) $F=\sum C_i$; weighted flowtime $\sum w_i C_i$, where $w_i \geqslant 0$ are weights representing the relative importance of J_i; (3) maximum lateness $L_{\max}=\max_i L_i$, $L_i=C_i-d_i$; (4) total tardiness $T=\sum T_i$, $T_i=\max(0, L_i)$, and weighted total tardiness $\sum w_i T_i$; (5) number of tardy tasks $U = \sum U_i$, $U_i = 1$ (if $C_i > d_i$), 0 (otherwise), and weighted number of tardy tasks $\sum w_i U_i$.

Numerous problems can be defined by combining the above conditions. Typical ones might be: the **job-shop scheduling problem**, in which n tasks are scheduled on m machines of type (a), and where the maximum completion time is minimized; the **flow-shop scheduling problem**, which is the same as the job-shop scheduling problem except that $n = n'm$ tasks are divided into n' groups of m tasks processed on machine 1, machine 2, ..., machine m, respectively, in this order; the **multiprocessor scheduling problem**, in which the maximum completion time of n independent tasks on m parallel machines is minimized; the **one-machine sequencing problem**, assuming only one machine (with various types of precedence constraints and optimality criteria), and others [8, 9].

These machine sequencing problems are examples of the combinatorial optimization problem (→ 281 Network Flow Problems E), as the processing time p_i is usually considered to be a given constant. Their computational complexity (→ 71 Complexity of Computations) has been extensively studied with an emphasis on the classification between those problems solvable in polynomial time and those that are †NP-complete, as summarized in [10]. Table 1 lists representative results for one-machine sequencing problems with $r_i = 0$ ($i = 1, 2, \ldots, n$). The improvement of the algorithm efficiency is pursued for both polynomially solvable problems and NP-complete problems. †Branch and bound (→ 215 Integral Programming D) is a common approach used to solve NP-complete problems such as the job-shop and flow-shop scheduling problems [8, 9]. Many approximation algorithms to obtain good suboptimal schedules in reasonable computation time are also known, and their worst-case and average accuracies have been analyzed [10], as these are important in practical applications.

In more realistic scheduling situations, other factors, such as the set-up cost, balancing of production lines, frequent modifications and updatings of project data, capacity of factories, manpower planning including the possibility of overtime and part-time employment, should be taken into account. Both deterministic and probabilistic models have been proposed for these cases. Mathematical tools used to compute adequate schedules include †mathematical programming, †queuing theory, and †simulation techniques.

References

[1] T. C. Koopmans, Activity analysis of production and allocation, Wiley, 1951.

[2] O. Morgenstern, Economic activity analysis, Wiley, 1954.

[3] J. von Neumann, Über ein ökonomisches Gleichungssystem und eine Verallgemeinerung des Brouwerischen Fixpunktsatzes, Ergebnisse eines Mathematischen Kolloquiums, 8 (1937), 73–83; English translation, A model of general economic equilibrium, Collected Works, Pergamon, 1963, vol. 6, 29–37.

[4] D. G. Malcolm, J. H. Reseboom, C. E. Clark, and W. Fazar, Application of a technique for research and development program evaluation, Operations Res., 7 (1959), 646–669.

Table 1. One-Machine Sequencing Problems with $r_i = 0$

Optimality Criterion	Precedence Constraint	Other Constraints	Complexity
$C_{\max}$	Partial order	None	$O(n^2)$
$\sum C_i$	Partial order	None	NP-complete
$\sum w_i C_i$	Series-parallel	None	$O(n \log n)$
	Partial order	$p_i = 1$ ($i = 1, 2, \ldots, n$)	NP-complete
$L_{\max}$	Partial order	None	$O(n^2)$
$\sum T_i$	Independent	None	Not known[a]
	Partial order	None	NP-complete
$\sum w_i T_i$	Independent	None	NP-complete
$\sum U_i$	Independent	None	$O(n \log n)$
	In-tree, out-tree	$p_i = 1$ ($i = 1, 2, \ldots, n$)	NP-complete
$\sum w_i U_i$	Independent	$p_i < p_j \Rightarrow w_i \geqslant w_j$	$O(n \log n)$
	Independent	None	NP-complete[b]

a. An algorithm with $O(n^5 p_{\max})$ running time is known, where $p_{\max} = \max_i p_i$.
b. An algorithm with $O(n \sum p_i)$ running time is known.

[5] J. E. Kelley, Jr., and M. R. Walker, Critical path planning and scheduling, Proc. Eastern Joint Computer Conference, 1959, 160–173.
[6] J. E. Kelley, Jr., Critical-path planning and scheduling: Mathematical basis, Operations Res., 9 (1961), 296–320.
[7] D. R. Fulkerson, A network flow computation for project cost curves, Management Sci., 7 (1961), 167–178.
[8] R. W. Conway, W. L. Maxwell, and L. W. Miller, Theory of scheduling, Addison-Wesley, 1967.
[9] K. R. Baker, Introduction to sequencing and scheduling, Wiley, 1974.
[10] R. L. Graham, E. L. Lawler, J. K. Lenstra, and A. H. G. Rinnooy Kan, Optimization and approximation in deterministic sequencing and scheduling: A survey, Ann. Discrete Math., 5 (1979), 287–326.

377 (XX.27) Second Quantization

A. Fock Space

For a complex Hilbert space K with $\dim K \geqslant 1$, $K^{\otimes n}$ denotes the n-fold tensor product of K with itself (where the vectors $f_1 \otimes \ldots \otimes f_n$ with $f_j \in K$ are total). Let $E_\pm^{(n)}$ be the projection operators on totally symmetric and antisymmetric parts of $K^{\otimes n}$:

$$E_\pm^{(n)}(f_1 \otimes \ldots \otimes f_n)$$
$$=(n!)^{-1} \sum \varepsilon_\pm(P) f_{P(1)} \otimes \ldots \otimes f_{P(n)},$$

where the sum is over all permutations P, $\varepsilon_+(P)=1$ and $\varepsilon_-(P)$ is the signature of P ($+1$ for even permutations and -1 for odd permutations). The following orthogonal direct sum is called a **Fock space** (**symmetric** for E_+, and **antisymmetric** for E_-):

$$\mathscr{F}_\pm(K)=\sum_{n=0}^{\infty}{}^{\oplus} E_\pm^{(n)} K^{\otimes n}.$$

Here the term for $n=0$ is the 1-dimensional space $\mathbf{C}$, and a vector Ω represented by $1 \in \mathbf{C}$ is called the **vacuum vector** in $\mathscr{F}_\pm(K)$. The subspace $\mathscr{F}_\pm(K)_n = E_\pm^{(n)} K^{\otimes n}$ is called the **n-particle subspace**. The operator $N=\sum^{\oplus\infty}_{n=0} n$, which takes the value n on $\mathscr{F}_\pm(K)_n$, is called the **number operator**.

On the algebraic sum

$$D_\pm = \bigcup_N \sum_{n=0}^{N}{}^{\oplus} \mathscr{F}_\pm(K)_n \subset \mathscr{F}_\pm(K),$$

the **creation operator** $a^+(f)$ for $f \in K$ is defined as the unique linear operator with domain $D_\pm$ satisfying

$$a^+(f) E_\pm^{(n)} f_1 \otimes \ldots \otimes f_n$$
$$=(n+1)^{1/2} E_\pm^{(n+1)} f \otimes f_1 \otimes \ldots \otimes f_n.$$

For $f \in K$, $\bar{f}$ denotes the element of the dual K^* satisfying $\bar{f}(g)=(g,f)$ for $g \in K$ (the inner product is linear in the first entry). The **annihilation operator** $a(\bar{f})$ is defined by

$$a(\bar{f}) E_\pm^{(n)} f_1 \otimes \ldots \otimes f_n$$
$$=n^{-1/2} \sum_{j=1}^{n} \varepsilon^{j-1}(f_j, f) E_\pm^{(n-1)} f_1 \otimes \ldots \hat{j} \ldots \otimes f_n,$$

where the tensor product of f_k, $k \neq j$, appears in the jth term and $\varepsilon = \pm 1$ depending on which of $\pm$ is taken in $\mathscr{F}_\pm(K)$. For $n=0$, $a(\bar{f})\Omega$ is defined to be 0. The adjoint of $a^+(f)$ coincides with $a(\bar{f})$ on $D_\pm$.

The creation and annihilation operators map $D_\pm$ into itself and satisfy the following commutation relations on $D_\pm$:

$$[a^+(f_1), a^+(f_2)]_\mp = [a(\bar{f}_1), a(\bar{f}_2)]_\mp = 0,$$
$$[a(\bar{f}_1), a^+(f_2)]_\mp = (f_2, f_1),$$

where $[A,B]_\mp = AB \mp BA$ and $\mp$ is used depending on the choice of $\pm$ in $\mathscr{F}_\pm(K)$. These relations are often called **canonical commutation relations** for $[\ ,\ ]_-$ (**CCRs**) and **canonical anticommutation relations** for $[\ ,\ ]_+$ (**CARs**).

On $\mathscr{F}_-(K)$, $a^+(f)$ and $a(\bar{f})$ are bounded with $\|a^+(f)\| = \|a(\bar{f})\| = \|f\|$. On $\mathscr{F}_+(K)$, both $a^+(f)$ and $a(\bar{f})$ are not bounded, though $a^+(f)N^{-1/2}$ and $a(\bar{f})N^{-1/2}$ are bounded.

On $\mathscr{F}_+(K)$, $2^{-1/2}(a^+(f)+a(\bar{f}))$ is essentially self-adjoint. Let $\psi(f)$ be its closure. The operator $W(f)=e^{i\psi(f)}$ is unitary and satisfies the identity

$$W(f_1)W(f_2)=W(f_1+f_2)\exp(-i\,\mathrm{Im}(f_1,f_2)/2).$$

It also satisfies $(W(f)\Omega,\Omega)=\exp(-4^{-1}\|f\|^2)$.

Let K_φ be a real subspace of K such that the inner product (f,g) in K is real for any f and g in K_φ and $K=K_\varphi+iK_\varphi$. (K_φ is then a real Hilbert space.) The unitary operators $U(f)=W(f)$ and $V(f)=W(if)$ for $f \in K_\varphi$ satisfy the following **Weyl form of the CCRs**:

$$U(f_1)U(f_2)=U(f_1+f_2),$$
$$V(f_1)V(f_2)=V(f_1+f_2),$$
$$U(f_1)V(f_2)=V(f_2)U(f_1)\exp(-i(f_1,f_2)).$$

The infinitesimal generators of the continuous one-parameter groups of unitaries $U(tf)$ and $V(tg)$ ($t \in \mathbf{R}$) are denoted by $\varphi(f)$ and $\pi(g)$ and satisfy the following CCRs:

$$[\varphi(f_1),\varphi(f_2)]\Psi=[\pi(g_1),\pi(g_2)]\Psi=0,$$
$$[\varphi(f),\pi(g)]\Psi=i(f,g)\Psi,$$

where $[A,B]=AB-BA$ and $\Psi \in D_+$.

If Q is a linear operator on K, $\Gamma(Q)$ denotes the linear operator on $\mathscr{F}_{\pm}(K)$ defined as the closure of $\sum^{\oplus\infty}_{n=0} Q^{\otimes n} | D_{\pm}$. It is bounded if $\|Q\| \leqslant 1$. $\Gamma(Q_1)\Gamma(Q_2)=\Gamma(Q_1 Q_2)$ on $D_{\pm}$. If H is a self-adjoint operator on K, then $\Gamma(e^{itH})= \exp it\, d\Gamma(H)$ defines a self-adjoint operator $d\Gamma(H)$ on $\mathscr{F}_{\pm}(K)$, usually called a **bilinear Hamiltonian** and denoted (a^+, Ha). More explicitly,

$$d\Gamma(H)E^{(n)}_{\pm} f_1 \otimes \ldots \otimes f_n$$

$$= \sum_{j=1}^{n} E^{(n)}_{\pm} f_1 \otimes \ldots \otimes Hf_j \otimes \ldots \otimes f_n.$$

If U is a unitary operator on K, then

$$\Gamma(U) W(f) \Gamma(U)^{-1} = W(Uf).$$

B. Second Quantization

A single (scalar) particle in quantum mechanics is described by a wave function $\Psi(x)$, $x \in \mathbf{R}^3$, considered as a unit vector in a Hilbert space $K=L_2(\mathbf{R}^3)$. The system consisting of n such identical particles is described by a totally symmetric function $\Psi(x_1, \ldots, x_n)$, $x_j \in \mathbf{R}^3$, considered as a unit vector in the totally symmetric part $E^{(n)}_{+} K^{\otimes n}$ of the n-fold tensor product of the one-particle Hilbert space K, where the restriction to totally symmetric wave functions is referred to as **Bose statistics**. In a nonrelativistic system, the Hamiltonian operator on a 1-particle space is

$$T = -\hbar^2 (2m)^{-1} \Delta_{\mathbf{x}},$$

called the kinetic energy ($\Delta_{\mathbf{x}}$ denotes the Laplacian); on an n-particle space it is typically given by

$$H_n = -\hbar^2 (2m)^{-1} \sum_{j=1}^{n} \Delta_{\mathbf{x}_j} + \sum_{i<j}^{n} V(\mathbf{x}_i - \mathbf{x}_j),$$

where V is a 2-body potential.

The totality of multiparticle spaces $E^{(n)}_{+} K^{\otimes n}$ can be described in terms of the Fock space $\mathscr{F}_{+}(K)$, the vacuum vector Ω (no-particle state), and the annihilation and creation operators, denoted by

$$a(\bar{f}) = \int \Psi(\mathbf{x}) \overline{f(\mathbf{x})} d^3\mathbf{x},$$

$$a^*(f) = \int \Psi^+(\mathbf{x}) f(\mathbf{x}) d^3 x.$$

Since the CCRs (for operator-valued distributions)

$$[\Psi(\mathbf{x}), \Psi(\mathbf{y})]_- = [\Psi^+(\mathbf{x}), \Psi^+(\mathbf{y})]_- = 0,$$

$$[\Psi(\mathbf{x}), \Psi^+(\mathbf{y})]_- = \delta^3(\mathbf{x}-\mathbf{y})$$

are a continuous generalization of CCRs for canonical variables in quantum mechanics and since $\Psi(\mathbf{x})$ comes from the wave function by way of quantization, the above formalism is called **second quantization**. The Hamiltonians H_n for all n can now be combined into the expression

$$H = \int \Psi^+(\mathbf{x}) T \Psi(\mathbf{x}) d^3\mathbf{x}$$

$$+ \frac{1}{2} \int \Psi^+(\mathbf{x}) \Psi^+(\mathbf{y}) V(\mathbf{x}-\mathbf{y})$$

$$\times \Psi(\mathbf{x}) \Psi(\mathbf{y}) d^3\mathbf{x}\, d^3\mathbf{y},$$

where the first term is $d\Gamma(T)$.

For particles such as electrons the system of n identical particles is described by a totally antisymmetric wave function, the total antisymmetry being referred to as **Fermi statistics**. Then the antisymmetric Fock space $\mathscr{F}_{-}(K)$ can be used in exactly the same manner as $\mathscr{F}_{+}(K)$ in the preceding case.

The method of second quantization was introduced by P. Dirac [2] for the case of bosons and extended by P. Jordan and E. Wigner [3] to fermions. Electromagnetic waves, when quantized in this way, represent a system of **photons**, and the quantization of electron waves leads to the particle picture of the **electron**. The method of second quantization is intimately connected with the notion of fields, as shown below for free fields, and is the basis of the perturbation approach in field theory (→ 150 Field Theory).

C. Free Fields

Let σ_j $(j=1,2,3)$ be †Pauli spin matrices and $\sigma_0 = \left(\begin{smallmatrix} 1 & 0 \\ 0 & 1 \end{smallmatrix}\right)$. Let $\tilde{p} = \sum_{\mu=0}^{3} \sigma_\mu p^\mu$ for a 4-vector $p = (p^0, \mathbf{p})$ with $\mathbf{p} = (p^1, p^2, p^3)$ and $p^0 = (m^2 + \mathbf{p}^2)^{1/2}$. Let $u_j(a, A)$ be the irreducible unitary representation $[m_+, j]$ of $\mathscr{P}^{\uparrow}_{+}$ on a Hilbert space $K_j = L_2(\mathbf{R}^3, \mathbf{C}^{2j}, (2p^0)^{-1} (m/\tilde{p})^{\otimes 2j} d^3\mathbf{p})$ (→ 258 Lorentz Group C (3)).

Consider first the symmetric Fock space $\mathscr{F}_{+}(K_0)$. For any complex-valued rapidly decreasing C^∞-function f ($f \in \mathscr{S}(\mathbf{R}^4)$), let

$$\tilde{f}(\mathbf{p}) = (2\pi)^{-3/2} \int e^{ip\cdot x} f(x) d^4 x$$

$$(p = (p^0, \mathbf{p}), \quad p^0 = (\mathbf{p}^2 + m^2)^{1/2}),$$

$$A(f) = a^+(\tilde{f}) + a(\bar{\tilde{\bar{f}}}) \quad \left(= \int A(x) f(x) d^4 x \right),$$

where $p \cdot x = p^0 x^0 - \sum_{j=1}^{3} p^j x^j$ and the bar denotes the complex conjugate. Then $A(x)$ as an operator-valued distribution satisfies the †Wightman axiom and is called the **free scalar field** of mass m. It satisfies the **Klein-Gordon equation**

$$(\square_x + m^2) A(x) \Psi = 0$$

$$\left(\square_x = (\partial/\partial x^0)^2 - \sum_{j=1}^{3} (\partial/\partial x^j)^2 \right),$$

it has the 4-dimensional scalar commutator

$$i[A(x),A(y)]_-\Psi=\Delta_m(x-y)\Psi,$$

and it has the two-point function

$$i(A(x)A(y)\Omega,\Omega)=\Delta_m^+(x-y).$$

Here Ψ is any vector in D_+; for example, Ω is the vacuum vector, and the **invariant distributions** Δ_m and Δ_m^+ are defined by

$$\Delta_m^+(x)=i(2\pi)^{-3}\int e^{-ip\cdot x}(2p^0)^{-1}d^3\mathbf{p},$$

$$\Delta_m(x)=(2\pi)^{-3}\int(\sin p\cdot x)(p^0)^{-1}d^3\mathbf{p}.$$

If we define $U(a,\Lambda(A))=\Gamma(u_0(a,A))$, then

$$U(a,\Lambda)A(x)U(a,\Lambda)^*=A(\Lambda x+a).$$

If $g\in\mathscr{S}(\mathbf{R}^3)$ and $h\in\mathscr{S}(\mathbf{R}^3)$, $\tilde{g}(\mathbf{p})$ and $\tilde{h}(\mathbf{p})$ obtained by substituting $g(\mathbf{x})\delta(x^0)$ and $-h(\mathbf{x})\delta'(x^0)$ into f in the defining equation of $\tilde{f}$ above are in K_0. We then define

$$\varphi(g)=a^+(\tilde{g})+a(\overline{\tilde{\tilde{g}}})\qquad\left(=\int\varphi(\mathbf{x})g(\mathbf{x})d^3\mathbf{x}\right),$$

$$\pi(h)=a^+(\tilde{h})+a(\overline{\tilde{\tilde{h}}})\qquad\left(=\int\pi(\mathbf{x})h(\mathbf{x})d^3\mathbf{x}\right).$$

The operator-valued distributions $\varphi(\mathbf{x})$ and $\pi(\mathbf{x})$ are the **canonical field** and its **conjugate field** at time 0 for the free scalar field and satisfy the following **canonical commutation relations**:

$$[\varphi(\mathbf{x}),\varphi(\mathbf{y})]_-\Psi=[\pi(\mathbf{x}),\pi(\mathbf{y})]_-\Psi=0,$$

$$[\varphi(\mathbf{x}),\pi(\mathbf{y})]_-\Psi=i\delta^3(\mathbf{x}-\mathbf{y})\Psi.$$

If we set $T(t)\equiv U(te_0,1)$ for $e_0=(1,0,0,0)$ and $(\alpha\otimes g)(x)=\alpha(x^0)g(\mathbf{x})$ for $\alpha\in\mathscr{S}(\mathbf{R})$ and $g\in\mathscr{S}(\mathbf{R}^3)$, then for $\Psi\in D_+$

$$A(\alpha\otimes g)\Psi=\int_{-\infty}^{\infty}T(t)\varphi(g)T(t)^*\Psi\alpha(t)\,dt,$$

$$-A(\alpha'\otimes g)\Psi=\int_{-\infty}^{\infty}T(t)\pi(g)T(t)^*\Psi\alpha(t)\,dt,$$

or, equivalently,

$$A(x)=T(x^0)\varphi(\mathbf{x})T(x^0)^*,$$

$$\partial A(x)/\partial x^0=T(x^0)\pi(\mathbf{x})T(x^0)^*.$$

If $A(x)$ is a classical field, then $\varphi(\mathbf{x})$ and $\pi(\mathbf{x})$ are the value of $A(x)$ and its time derivative at $x^0=0$, and they serve as initial data for the Klein-Gordon equation

$$(\square_x+m^2)A(x)=0.$$

Consider next the antisymmetric Fock space $\mathscr{F}_-(K_{1/2}\otimes K_{1/2})$. For $f_\pm\in\mathscr{S}(\mathbf{R}^4,\mathbf{C}^2)$ ($\mathbf{C}^2$-valued rapidly decreasing C^∞-functions), write $f=(f_+,f_-)$, $f_+=(f_1,f_2)$, $f_-=(f_3,f_4)$, define $\tilde{f}_\pm$ as before, and let

$$\psi(f)=m^{1/2}\{a(\overline{(\tilde{\tilde{f}}_-+(\tilde{p}/m)\tilde{\tilde{f}}_+)}\oplus 0)+a^+(0\oplus(\sigma_2\tilde{f}_+-(\tilde{p}/m)\sigma_2\tilde{f}_-))\}$$

$$=\sum_{\alpha=1}^{4}\int\psi_\alpha(x)f_\alpha(x)\,d^4x,$$

where $\sigma_2=\begin{pmatrix}0&-i\\ i&0\end{pmatrix}=-i\varepsilon$. Then $\psi_\alpha(x)$ as an operator-valued distribution satisfies the †Wightman axiom and is called the **free Dirac field** of mass m. In the present formulation, $\{\psi_1,\psi_2\}$ is a contravariant spinor of rank $(1,0)$ and $\{\psi_3,\psi_4\}$ is a covariant spinor of rank $(0,1)$. This field satisfies the **Dirac equation**

$$\left(\sum_\mu\gamma^\mu(\partial/\partial x^\mu)+im\right)\psi(x)=0,$$

as well as the relations

$$[\psi_\alpha(x),\psi_\beta(y)]_+\Psi=0,$$

$$[\psi_\alpha(x),\psi_\beta(y)^*]_+\Psi$$

$$=\left\{\left(\sum_\mu\gamma^\mu(\partial/\partial x^\mu)-im\right)\gamma^0\right\}_{\alpha\beta}\Delta_m(x-y),$$

$$(\psi_\alpha(x)\psi_\beta(y)^*\Omega,\Omega)$$

$$=\left\{\left(\sum_\mu\gamma^\mu(\partial/\partial x^\mu)-im\right)\gamma^0\right\}_{\alpha\beta}\Delta_m^+(x-y).$$

Here $\Psi\in D_+$, Δ_m and Δ_m^+ are as described above, and the γ's are **Dirac matrices** in the following form (somewhat different from but equivalent to the usual form; → 351 Quantum Mechanics):

$$\gamma^0=\begin{pmatrix}0&1\\1&0\end{pmatrix},\qquad\gamma^j=\begin{pmatrix}0&-\sigma_j\\ \sigma_j&0\end{pmatrix},$$

and the σ's are †Pauli spin matrices.

D. Coherent Vectors and Exponential Hilbert Space

In the symmetric Fock space $\mathscr{F}_+(K)$, a vector of the form

$$\exp f\equiv\Omega\oplus\sum_{n=1}^{\infty}{}^{\oplus}(n!)^{-1/2}f^{\otimes n}\in\mathscr{F}_+(K)$$

for $f\in K$ is called a **coherent vector**. The set of $\exp f$ is linearly independent (in the algebraic sense) and total. The inner product is given by

$$(\exp f_1,\exp f_2)=\exp(f_1,f_2).$$

Conversely, we can define $\mathscr{F}_+(K)$ abstractly by introducing this inner product into the formal linear combinations of $\exp f$, $f\in K$, and by completion. In this sense, $\mathscr{F}_+(K)$ is also denoted as $\exp K$ and is called an **exponential Hilbert space** [5]. Then

$$\exp\sum{}^{\oplus}K_j=\otimes\exp K_j,$$

where $\exp \sum^{\oplus} f_j$ is identified with $\otimes \exp f_j$. If the number of indices is infinite, the right-hand side is the incomplete infinite tensor product containing the product of the vacuum vector Ω.

If $K = \int_{\Xi} K_\lambda d\mu(\lambda)$ and the measure μ is nonatomic, then for any measurable set S in Ξ, there corresponds a decomposition $\exp K = (\exp K(S)) \otimes (\exp K(S^c))$, where $K(S) = \int_S K_\lambda d\mu(\lambda)$ and S^c is the complement of S in Ξ, and an associated von Neumann algebra $R(S) = B(K(S)) \otimes 1$, where $B(K(S))$ is the set of all bounded linear operators on $K(S)$. The system $\{R(S)\}$ forms a complete Boolean lattice of type I factors on $\exp K$. Coherent vectors are characterized by the property of being a product vector for $\{R(S)\}$ in the sense that for any S, $A \in R(S)$ and $A' \in R(S^c)$, the vector $\Psi = \exp f$ satisfies

$$(AA'\Psi, \Psi)\|\Psi\|^2 = (A\Psi, \Psi)(A'\Psi, \Psi).$$

In this sense, we can interpret $\exp K$ as a **continuous tensor product** of $\exp K_\lambda$ and also, if $\Psi = \int \Psi_\lambda d\mu(\lambda)$, $\exp \Psi$ as a continuous tensor product of $\exp \Psi_\lambda$ [5].

References

[1] V. Fock, Konfigrationsraum und zweite Quantelung, Z. Phys., 75 (1932), 622–647.
[2] P. A. M. Dirac, The quantum theory of the emission and absorption of radiation, Proc. Roy. Soc. London, 114 (1927), 243–265.
[3] P. Jordan and E. P. Wigner, Über das Paulische Äquivalenzverbot, Z. Phys., 47 (1928), 631–651.
[4] F. A. Berezin, The method of second quantization, Academic Press, 1966. (Original in Russian, 1965.)
[5] H. Araki and E. J. Woods, Complete Boolean algebras of type I factors, Publ. Res. Inst. Math. Sci., A2 (1966), 157–242, 451–452.
[6] J. Avery, Creation and annihilation operators, McGraw-Hill, 1976.

378 (XII.14) Semigroups of Operators and Evolution Equations

A. Introduction

The analytical theory of semigroups was inaugurated around 1948 in order to define exponential functions in infinite-dimensional function spaces. Then it was generalized to the theory of evolution equations as ordinary differential equations in infinite-dimensional linear spaces.

B. The Hille-Yosida Theorem

Let X be a †locally convex topological linear space, and denote by $L(X)$ the totality of continuous linear operators defined on X with values in X. A family $\{T_t \mid t \geqslant 0\}$ of operators $T_t \in L(X)$ is called a **(one-parameter) semigroup of class** (C^0) or a **strongly continuous semigroup** if it satisfies the following two conditions: (i) $T_t T_s = T_{t+s}$ (the semigroup property), $T_0 = I$ (the identity operator); and (ii) $\lim_{t \to t_0} T_t x = T_{t_0} x$ ($\forall x \in X, \forall t_0 \geqslant 0$). When X is a Banach space, (ii) is implied by w-$\lim_{t \downarrow 0} T_t x = x$ ($\forall x \in X$), as proved by N. Dunford in 1938. In this case there exist constants $M > 0$ and $\beta \geqslant 0$ such that (iii') $\|T_t\| \leqslant Me^{\beta t}$ ($\forall t \geqslant 0$). Hence, considering $e^{-\beta t} T_t$ in place of T_t, we can assume the equicontinuity: (iii) For any continuous seminorm p on X, there exists a continuous seminorm q on X such that $p(T_t x) \leqslant q(x)$ ($\forall x \in X, \forall t \geqslant 0$). Such semigroups are called **equicontinuous semigroups of class** (C^0) (abbreviated e.c.s.g. (C^0)).

Example 1. $X = L_p(0, \infty)$ with $\infty > p \geqslant 1$.

$$(T_t x)(s) = x(t+s).$$

Example 2. $X = L_p(-\infty, \infty)$ with $\infty > p \geqslant 1$.

$$(T_t x)(s) = (2\pi t)^{-1/2} \int_{-\infty}^{\infty} \exp\left(\frac{(s-u)^2}{2t}\right) x(u)\, du, \quad t > 0,$$

$$= x(s), \quad t = 0.$$

Example 3. $X = BC(-\infty, \infty)$.

$$(T_t x)(s) = e^{-\lambda t} \sum_{k=0}^{\infty} \frac{(\lambda t)^k}{k!} x(s - k\mu), \quad t \geqslant 0.$$

Here λ and μ are positive constants. (For these examples, we have $\|T_t\| \leqslant 1$; hence (iii) is satisfied.) For L_p and BC → 168 Function Spaces.

We assume in the remainder of the article that X is sequentially complete, that is, if a sequence $\{x_n\}$ of X satisfies $\lim_{n,m \to \infty} p(x_n - x_m) = 0$ for every continuous seminorm p on X, then there exists a unique $x \in X$ such that $\lim_{n \to \infty} p(x - x_n) = 0$.

The **infinitesimal generator** A of an e.c.s.g. (C^0) $\{T_t \mid t \geqslant 0\}$ is defined by

$$Ax = \lim_{t \downarrow 0} t^{-1}(T_t - I)x. \tag{1}$$

(This is also called the **generator** of T_t.) Then we have the following results.

(I) Differentiability theorem. For every complex number λ with $\mathrm{Re}\,\lambda>0$, the resolvent $(\lambda I-A)^{-1}\in L(X)$ exists and

$$(\lambda I-A)^{-1}x=\int_0^\infty e^{-\lambda t}T_t x\,dt \qquad (\forall x\in X), \tag{2}$$

where the integration is Riemannian. Hence the domain $D(A)$ of A is dense in X and coincides with the range $R((\lambda I-A)^{-1})$, and A is a closed linear operator such that the family

$$\{(\lambda(\lambda I-A)^{-1})^n \mid \lambda>0,\ n=0,1,2,\dots\} \tag{3}$$

is equicontinuous.

(II) Representation theorem. Let $J_n=(I-n^{-1}A)^{-1}$ and consider the approximations to T_t:

$$T_t^{(n)}x=e^{-nt}\sum_{m=0}^{\infty}(m!)^{-1}(ntJ_n)^m x,$$

$$\hat{T}_t^{(n)}x=(J_{nt^{-1}})^n x.$$

Then

$$T_t x=\lim_{n\to\infty}T_t^{(n)}x=\lim_{n\to\infty}\hat{T}_t^{(n)}x \tag{5}$$

uniformly on every compact set of t.

(III) Converse theorem. Let a linear operator A with both dense domain $D(A)$ and range $R(A)$ in X satisfy the condition $(nI-A)^{-1}\in L(X)$ for $n=1,2,\dots$. Then a necessary and sufficient condition for A to be the infinitesimal generator of an e.c.s.g. (C^0) is that the family of operators

$$\{(I-n^{-1}A)^{-m} \mid n=1,2,\dots;\ m=0,1,\dots\} \tag{3'}$$

be equicontinuous. Since such a semigroup $\{T_t \mid t\geqslant 0\}$ is uniquely determined by A, we can write $T_t=\exp(tA)$.

These three theorems together are called the **Hille-Yosida theorem** or the Hille-Yosida-Feller-Phillips-Miyadera theorem.

Examples of Infinitesimal Generators. $A=d/ds$ for example 1 above, $A=2^{-1}d^2/ds^2$ for example 2, and $(Ax)(s)=\lambda(x(s-\mu)-x(s))$ for example 3.

C. Groups

An operator A in a Hilbert space X generates a group $\{T_t \mid -\infty<t<\infty\}$ of †unitary operators of class (C^0) satisfying $T_tT_s=T_{t+s}$ for $-\infty<t,\ s<\infty$ if and only if A is equal to iH for some †self-adjoint operator H (**M. H. Stone's theorem**, 1932). In a locally convex space, a necessary and sufficient condition for a given e.c.s.g. (C^0) $\{T_t \mid t\geqslant 0\}$ to be extended to an **equicontinuous group of class** (C^0) $\{T_t \mid -\infty<t<\infty\}$ is that the family (3′) be equicontinuous also for $n=\pm1,\pm2,\dots$.

D. Holomorphic Semigroups

For an e.c.s.g. (C^0) $\{T_t \mid t\geqslant 0\}$, the following three conditions are equivalent (K. Yosida, 1963; the equivalence between (ii) and (iii) for Banach spaces was proved earlier by E. Hille, 1948): (i) When $t>0$,

$$T_t'x=\lim_{h\to 0}h^{-1}(T_{t+h}-T_t)x$$

exists for all $x\in X$ and $\{(CtT_t')^n \mid n=1,2,\dots$ and $0<t\leqslant 1\}$ is equicontinuous for a certain constant $C>0$. (ii) When $t>0$, T_t admits a convergent expansion T_λ given locally by $T_\lambda x=\sum_{n=0}^{\infty}(\lambda-t)^n T_t^{(n)}x/n!$. The extension exists for $|\arg\lambda|\leqslant\arctan(Ce^{-1})$, and the family of operators $\{e^{-\lambda}T_\lambda\}$ is equicontinuous in λ for $|\arg\lambda|\leqslant\arctan(2^{-k}Ce^{-1})$ with some positive constant k. (iii) For the infinitesimal generator A of T_t, there exists a positive constant C_1 such that $\{(C_1\lambda(\lambda I-A)^{-1})^n\}$ is equicontinuous in $n=1,2,\dots$ and in λ with $\mathrm{Re}(\lambda)\geqslant 1+\varepsilon$, $\varepsilon>0$. An e.c.s.g. (C^0) $\{T_t\}$ satisfying the above conditions is called a **holomorphic semigroup**.

For example, introduce

$$\begin{aligned} f_{t,\alpha}(\lambda)&=(2\pi i)^{-1}\int_{\sigma-i\infty}^{\sigma+i\infty}e^{z\lambda-tz^\alpha}dz,\\ &\qquad \lambda\geqslant 0,\quad t>0,\quad \sigma>0,\quad 0<\alpha<1,\\ &=0,\quad \lambda<0, \end{aligned} \tag{6}$$

where the branch of z^α is taken so that $\mathrm{Re}\,z^\alpha>0$ for $\mathrm{Re}\,z>0$. Following S. Bochner (1949), we define

$$\begin{aligned} \hat{T}_{t,\alpha}x=\hat{T}_t x&=\int_0^\infty f_{t,\alpha}(s)T_s x\,ds,\qquad t>0,\\ &=x,\quad t=0, \end{aligned} \tag{7}$$

from a given e.c.s.g. (C^0) $\{T_t \mid t\geqslant 0\}$. Then $\{\hat{T}_{t,\alpha} \mid t\geqslant 0\}$ is a holomorphic semigroup (Yosida, T. Kato, and A. V. Balakrishnan, 1960). Its infinitesimal generator $\hat{A}_\alpha$ can be considered as the **fractional power** $(-A)^\alpha$ of $-A$, multiplied by -1.

Fractional powers $(-A)^\alpha$, $\alpha\in\mathbf{C}$, of operators have also been defined for operators A satisfying the weaker condition than (3′) that $\{\lambda(\lambda-A)^{-1} \mid \lambda>0\}$ is equicontinuous (Balakrishnan, H. Komatsu). If A is such an operator, $-\sqrt{-A}$ generates a holomorphic semigroup and the unique uniformly bounded solution of the "elliptic" equation

$$x_t''=-Ax_t,\qquad t>0,\qquad \lim_{t\to 0}x_t=x_0 \tag{8}$$

is the solution of

$$x_t'=-\sqrt{-A}\,x_t,\qquad t>0,\qquad \lim_{t\to 0}x_t=x_0,$$

and therefore $x_t=\exp(-t\sqrt{-A})x_0$ (Balakrishnan). Equation (8) has also been discussed by

M. Sova and H. O. Fattorini from a different point of view.

E. Convergence of Semigroups

Let a sequence $\{\exp(tA_n) \mid n=1,2,\dots\}$ of e.c.s.g. (C^0) be equicontinuous as a family of operators $\in L(X)$. Then a necessary and sufficient condition for there to exist an e.c.s.g. (C^0) $\exp(tA)$ such that $\lim_{n\to\infty}(\exp(tA_n))x=(\exp(tA))x$ uniformly on every compact interval of t is that $\lim_{n\to\infty}(\lambda_0 I-A_n)^{-1}x=J_{\lambda_0}x$ exist (for some λ_0 with $\mathrm{Re}\,\lambda_0>0$ and for all $x\in X$) and be such that $R(J_{\lambda_0})$ is dense in X (H. F. Trotter, Kato).

F. Miscellaneous Semigroups

(i) **Distribution semigroups**. The semigroup of translations $(T_tx)(s)=x(t+s)$ in $X=L_\infty(-\infty,\infty)$ is not continuous and hence not measurable in t. However, T_tx is an X-valued distribution. For semigroups $\{T_t\}$ such that T_tx is an X-valued distribution for $x\in X$, an analog of the Hille-Yosida theorem is known (J.-L. Lions, 1960). It has been generalized to ultradistribution semigroups by J. Chazarain and to hyperfunction semigroups by S. Ōuchi.

(ii) **Dual semigroups**. The above semigroup $\{T_t\}$ of translations in $L_\infty(-\infty,\infty)$ is obtained as $T_t=S_t^*$ from the e.c.s.g. (C^0) $\{S_t\}$ defined by $(S_tx)(s)=x(s-t)$ in $L_1(-\infty,\infty)$. Let $B=d/ds$ be the infinitesimal generator of $\{S_t\}$. The restriction of $\{S_t^*\}$ to the space of uniformly continuous functions, which is the closure of the domain $D(B^*)$ of the dual B^* in $L_\infty(-\infty,\infty)$, is an e.c.s.g. (C^0). This fact holds for the semigroup $\{S_t^*\}$ of an e.c.s.g. (C^0) $\{S_t\}$ in a Banach space X in general (R. Phillips, 1955) and also in a locally convex space.

(iii) **Locally equicontinuous semigroups**. The infinitesimal generator A of the semigroup of translations $(T_tx)(s)=x(t+s)$ in $X=C(-\infty,\infty)$ is d/ds. A has no resolvent since all complex numbers are eigenvalues of A. $\{T_t\}$ is not equicontinuous but locally equicontinuous, i.e., $\{T_t \mid 0\leqslant t\leqslant t_1\}$ is equicontinuous for any $t_1>0$. For locally equicontinuous (C^0) semigroups an analog of the Hille-Yosida theorem is obtained by using the notion of generalized resolvents (T. Kômura, 1968; Ōuchi, 1973).

(iv) **Differentiable semigroups**. The notion of the holomorphy of semigroups in Section D is weakened to the differentiability. A characterization of a semigroup $\{T_t\}$ such that T_tx is infinitely differentiable in $t>0$ is given by using the resolvent of the infinitesimal generator (A. Pazy, 1968).

(v) **Nonlinear semigroups**. For a (C^0) semigroup $\{T_t\}$ of contractions (i.e., $\|T_tx-T_ty\|\leqslant\|x-y\|$ for $x, y\in X$) in a Hilbert space X, an analog of the Hille-Yosida theorem is known (Y. Kômura, 1969). This result has been partially extended to Banach spaces (→ 286 Nonlinear Functional Analysis X).

G. The Evolution Equation

Let $T_t=\exp(tA)$ be an e.c.s.g. (C^0). Then for $x\in D(A)$,

$$T_t'x=AT_tx(=T_tAx). \qquad (9)$$

Considered in suitable function spaces, the †equation of heat conduction ($A=\Delta=$ the †Laplacian), the †Schrödinger equation ($A=\sqrt{-1}\,(\Delta-V(x))$), and the †wave equation given in matrix form

$$\begin{pmatrix}u\\v\end{pmatrix}'=\begin{pmatrix}u'\\v'\end{pmatrix}=\begin{pmatrix}0&I\\\Delta&0\end{pmatrix}\begin{pmatrix}u\\v\end{pmatrix}$$

are all of the form (9). For a linear operator A_t in X depending on t, the ordinary differential equation in X

$$x_t'=A_tx_t+f(t), \qquad t\geqslant0, \qquad (10)$$

is called the **evolution equation**. A family of operators $\{V(r,s) \mid r\geqslant s\geqslant0\}$ in X which gives general solutions to the homogeneous evolution equation

$$x_t'=A_tx_t \qquad (11)$$

(i.e., for any $s\geqslant0$, $a\in D(A_s)$, $x_t=V(t,s)a$ is a solution to (11) for $x_s=a$) is called the **evolution operator** associated with the generators $\{A_t\}$. An evolution operator $\{V(r,s)\}$ satisfies (i) $V(r,r)=I$, (ii) $V(r,s)V(s,t)=V(r,t)$. The solution to (10) is formally expressed by

$$x_t=V(t,0)x_0+\int_0^t V(t,s)f(s)\,ds. \qquad (12)$$

H. Integration of the Evolution Equation

For equation (11) we have the following result (Kato, 1953; Yosida, 1966). Assume the following four conditions: (i) $D(A_t)$ is independent of t and dense in the Banach space X, and for all $\alpha>0$, $(I-\alpha A_t)^{-1}\in L(X)$ with the estimate $\|(I-\alpha A_t)^{-1}\|\leqslant1$; (ii) $B_{t,s}=(I-A_t)(I-A_s)^{-1}$ is uniformly bounded in the norm for $0\leqslant s,t\leqslant l$; (iii) $\sum_{j=0}^{n-1}\|B_{t_{j+1},t_0}-B_{t_j,t_0}\|\leqslant N$, where N is independent of the partition ($0=t_0<t_1<\ldots<t_n=l$); (iv) $B_{t,0}$ is weakly differentiable with respect to t such that the differentiated operator $\partial B_{t,0}/\partial t$ is strongly continuous in t. Under these assumptions, we can prove that for $x_0\in D(A_0)$, the limit $V(t,0)x_0=\text{s-lim}_{k\to\infty}V_k(t,0)x_0$,

with

$$V_k(t,s) = \left(I - \left(t - \frac{[kt]}{k}\right) A\left(\frac{[kt]}{k}\right)\right)^{-1}$$
$$\times \left(I - \frac{1}{k} A\left(\frac{[kt]-1}{k}\right)\right)^{-1}$$
$$\times \ldots \times \left(I - \frac{1}{k} A\left(\frac{[ks]+1}{k}\right)\right)^{-1}$$
$$\times \left(I - \left(\frac{[ks]+1}{k} - s\right) A\left(\frac{[ks]}{k}\right)\right)^{-1}$$

$(0 \leqslant s \leqslant t \leqslant l)$, exists and gives the unique solution of (11). If $f(t)$ is continuously differentiable, the right-hand side of (12) exists and gives a unique solution to the inhomogeneous equation (10).

I. The Evolution Equation of Parabolic Type

Equation (10), for which every A_t is the infinitesimal generator of a holomorphic semigroup, is said to be of **parabolic type** by analogy to parabolic partial differential equations. Under weaker conditions, especially without the condition that $D(A_t)$ is independent of t, the existence of solutions of an equation of this type is obtained. Moreover, differentiability or analyticity of solutions follow from some natural assumptions.

(i) Existence of weak solutions. Let X be a Hilbert space. For $t \in [0, l]$, let V_t be a subspace and at the same time a Hilbert space with respect to a norm $|||\cdot|||_t$ stronger than $\|\cdot\|$. Since the form $(A_t x, y)$ is †sesquilinear (linear in x and antilinear in y), we get a sesquilinear functional $a(t, \cdot, \cdot)$ on $V_t \times V_t$ such that

$$a(t, x, y) = -(A_t x, y), \qquad x, y \in D(A_t),$$

if $D(A_t)$ is dense in V_t with respect to $|||\cdot|||_t$ and if

$$|(A_t x, y)| \leqslant C |||x|||_t |||y|||_t, \qquad x, y \in D(A_t).$$

V_t, $a(t, \cdot, \cdot)$ should be measurable in a certain sense. A solution x_t of the equation (10) in $[0, l]$ satisfies

$$\int_0^l a(t, x_t, v_t)\, dt - \int_0^l (x_t, v_t')\, dt$$
$$= \int_0^l (f(t), v_t)\, dt + (x_0, v_0) \qquad (13)$$

for any differentiable X-valued function v_t such that $v_t \in V_t$, $v_l = 0$, $\int_0^l \|v_t\|^2 dt < \infty$, and $\int_0^l \|v_t'\|^2 dt < \infty$. A solution x_t of (13) is called a **weak solution** of equation (10), though it does not necessarily satisfy (10). If the relation

$$a(t, x, x) + \lambda \|x\|^2 \geqslant \alpha |||x|||_t, \qquad x \in V_t,$$

holds for some λ, $\alpha > 0$, a weak solution of (10) in the sense of (13) exists for a given $x_0 \in X$ (Lions, 1961). In order to obtain the uniqueness or the differentiability of weak solutions, we need some additional conditions.

(ii) Some properties of strong solutions. Let X be a Banach space. Let every semigroup $\{T_\lambda^{(t)}\}$ generated by A_t be holomorphic in a complex sector $|\arg \lambda| \leqslant \theta$, $\theta > 0$, independent of t. Suppose one of the following conditions holds: (1) For some α, $0 < \alpha < 1$, $D(A_t^\alpha)$ is independent of t and for $1 - \alpha < \beta < 1$,

$$\|(A_t^\alpha - A_s^\alpha) A_0^{-\alpha}\| \leqslant C' |t-s|^\beta, \qquad t, s \in [0, 1]$$

(P. E. Sobolevskiĭ, 1958–1961; Kato, 1961); (2) A_t^{-1} is differentiable in t,

$$\|dA_t^{-1}/dt - dA_s^{-1}/ds\| \leqslant C' |t-s|^\beta$$

for some $C' > 0$, $0 < \beta \leqslant 1$, and

$$\left\| \frac{\partial}{\partial t} (A_t - \lambda)^{-1} \right\| \leqslant \frac{N}{|\lambda|^\alpha}$$

for every $\lambda : |\arg \lambda| \geqslant \pi/2 - \theta$ for some N, $0 < \alpha < 1$ (Kato and H. Tanabe, 1962). Then a differentiable evolution operator $\{V(t, s)\}$ associated with (11) exists.

The most interesting property of evolution equations of parabolic type is the analyticity of solutions. If A_t is holomorphic in t in a certain sense, then the solutions are holomorphic (Tanabe, 1967; first noted by Komatsu, 1961). Furthermore, a characterization of evolution operators $\{V(t, s)\}$ holomorphic in some complex neighborhood of $[0, l]$ (called **holomorphic evolution operators**) is obtained by using the resolvent of A_t (Kato and Tanabe, 1967; K. Masuda, 1972; → [8]).

J. Application to Semilinear Evolution Equations

The evolution equation with a nonlinear additive term $f(t, x_t)$: $x_t' = A_t x_t + f(t, x_t)$ can be written as an inhomogeneous integral equation $x_t = V(t, 0) x_0 + \int_0^t V(t, s) f(s, x_s)\, ds$ in the Banach space X, by means of the evolution operator $\{V(t, s)\}$ introduced in Section G. The existence, differentiability (Kato, H. Fujita, and Sobolevskiĭ, 1963–1966), and analyticity (Masuda, 1967) of solutions of the Navier-Stokes equation has been obtained by reducing it to an integral equation of this type.

Concerning quasilinear equations in which A_t depends on x_t, the existence, differentiability, and analyticity of their solutions have been discussed.

References

[1] E. Hille and R. S. Phillips. Functional analysis and semi-groups, Amer. Math. Soc. Colloq. Publ., 1957.

[2] N. Dunford and J. T. Schwartz, Linear operators I, Interscience 1958.
[3] J. L. Lions, Equations différentielles opérationnelles et problèmes aux limites, Springer, 1961.
[4] K. Yosida, Functional analysis, Springer, 1980.
[5] S. G. Kreĭn, Linear differential equations in a Banach space, Amer. Math. Soc., 1971. (Original in Russian, 1967.)
[6] P. L. Butzer and H. Berens, Semigroups of operators and applications, Springer, 1967.
[7] R. W. Carroll, Abstract methods in partial differential equations, Harper & Row, 1969.
[8] H. Tanabe, Equations of evolution, Pitman, 1979. (Original in Japanese, 1975.)

379 (X.18) Series

A. Convergence and Divergence of Infinite Series

Let $\{a_n\}$ $(n=1,2,3,\ldots)$ be a sequence of real or complex numbers. Then the formal infinite sum $a_1+a_2+\ldots$ is called an **infinite series** (or **series**) and is denoted by $\sum_{n=1}^{\infty}a_n$ or $\sum a_n$. The number a_n is the nth **term** of the series $\sum a_n$, and $s_n=a_1+a_2+\ldots+a_n$ is the nth **partial sum** of $\sum a_n$. Also, for a finite sequence $a_1,a_2,\ldots,a_n$, the sum $a_1+a_2+\ldots+a_n$ is called a series. To distinguish these two series, the latter is called a **finite series**. In this article, *series* means an infinite series. If the sequence of partial sums $\{s_n\}$ †converges to s, we say that the series $\sum a_n$ **converges** or is **convergent** to the **sum** s and write $\sum_{n=1}^{\infty}a_n=s$ or $\sum a_n=s$. If the sequence $\{s_n\}$ is not convergent, we say that the series **diverges** or is **divergent**. In particular, if $\{s_n\}$ is divergent to $+\infty$ $(-\infty)$ or †oscillating, we say that the series is **properly divergent** to $+\infty$ $(-\infty)$ or **oscillating**, respectively.

The notation $\sum a_n$ is customarily used for both the sum s of the convergent series and the formal series, which may or may not be convergent. When the series is convergent, the sum is sometimes called the **Cauchy sum** in contrast to the "summations" of series, which are not necessarily convergent (→ Sections K ff.).

Applying the †Cauchy criterion for the convergence of a sequence, we see that a necessary and sufficient condition for $\sum a_n$ to be convergent is that for any given $\varepsilon>0$, we can take N sufficiently large so that

$$|s_m-s_n|=|a_{n+1}+\ldots+a_m|<\varepsilon$$

for all m, n such that $m>n>N$. Hence if $\sum a_n$ converges, the $a_n\to0$ as $n\to\infty$, but the converse is not always true.

Elementary properties of the convergence of series are: (1) If $\sum a_n$ and $\sum b_n$ converge to a, b, respectively, then $\sum(a_n+b_n)$ converges to $a+b$. (2) If $\sum a_n$ converges to a, then $\sum ca_n$ converges to ca for any constant c. (3) Suppose that $\{b_m\}$ is a subsequence of $\{a_n\}$ obtained by deleting a finite number of terms a_n from $\{a_n\}$. Then $\sum b_m$ is convergent if and only if $\sum a_n$ is convergent. (4) When a series $\sum a_n$ converges to a and $\{b_m\}$ is a sequence such that $b_1=a_1+a_2+\ldots+a_{r_1}$, $b_2=a_{r_1+1}+a_{r_1+2}+\ldots+a_{r_2}$, $b_3=a_{r_2+1}+\ldots+a_{r_3},\ldots$, then $\sum b_m$ also converges to a. The converse, however, is not always true. For example, $1-1+1-1+\ldots$ is oscillating, but $(1-1)+(1-1)+\ldots=0$.

B. Series of Positive Terms

Suppose that $\sum a_n$ is a **series of positive** (or **nonnegative**) **terms**. Since its partial sums s_n form a †monotonically increasing sequence, the series is convergent if and only if $\{s_n\}$ is bounded. For example, the series $\sum_{n=1}^{\infty}n^{-p}$ $(p>0)$ converges if $p>1$ because $s_n<2^{p-1}/(2^{p-1}-1)$, whereas it diverges if $p\leqslant1$ because $s_{2^{m+1}}>1+(m+1)/2$. The **geometric series** $\sum_{n=1}^{\infty}a^{n-1}$ $(a>0)$ converges for $a<1$ because $s_n=(1-a^n)/(1-a)$, and diverges for $a\geqslant1$ because $s_n\geqslant n$.

Some criteria for the convergence of a series $\sum a_n$ of nonnegative terms are: (1) If $\{a_n\}$ is monotone decreasing, then the series $\sum a_n$ and $\sum 2^{\nu}a_{2^{\nu}}$ have the same convergence behavior (**Cauchy's condensation test**). (2) Suppose that $f(x)$ is a positive monotone decreasing function defined for $x\geqslant1$ such that $f(n)=a_n$ $(n=1,2,\ldots)$. Then the series $\sum a_n$ and the integral $\int_1^{\infty}f(x)\,dx$ have the same convergence behavior (**Cauchy's integral test**), for example, $\sum n^{-p}$ $(p>0)$ and $\int_1^{\infty}x^{-p}\,dx$. (3) If for any positive constant k we have $a_n\leqslant kb_n$ except for a finite number of n, then the convergence of $\sum b_n$ implies the convergence of $\sum a_n$. If $kb_n\leqslant a_n$ and $\sum b_n$ diverges, then $\sum a_n$ also diverges (**comparison test**). (4) Let $a_n>0$ and $b_n>0$. If $a_{n+1}/a_n\leqslant b_{n+1}/b_n$ except for a finite number of values of n and $\sum b_n$ converges, then $\sum a_n$ also converges; if $a_{n+1}/a_n\geqslant b_{n+1}/b_n$ and $\sum b_n$ diverges, then $\sum a_n$ also diverges (→ Appendix A, Table 10).

C. Absolute Convergence and Conditional Convergence

A series $\sum a_n$ (with real or complex terms a_n) is called **absolutely convergent** if the series $\sum|a_n|$ is convergent. If a convergent series is not absolutely convergent, then it is called **conditionally convergent**. An absolutely convergent series is convergent. A real series $\sum a_n$ whose terms have alternating signs is called an **alternating series**. An alternating series $\sum a_n$ is convergent if the absolute values of terms $|a_n|$ form a monotone decreasing sequence which converges to zero (**Leibniz's test**). An absolutely convergent series remains absolutely convergent under every rearrangement of terms and retains its sum under the rearrangement (**Dirichlet's theorem**). If a series with real terms is conditionally convergent, then it is possible to rearrange its terms so that the rearranged series converges to any given number, diverges to $+\infty$ (or $-\infty$), or is oscillating (**Riemann's theorem**). A convergent series whose convergence behavior is unaffected by rearrangement and whose sum remains unchanged is called **unconditionally convergent** (or **commutatively convergent**). A real or complex series is unconditionally convergent if and only if it is absolutely convergent. The notion of infinite series can be extended to any complete †normed linear space, and absolute convergence can be defined by replacing the absolute values of the terms by the norm of the terms. However, in general, unconditional convergence is not always equivalent to absolute convergence.

D. Abel's Partial Summation

Let $\{a_0, a_1, a_2, \dots\}$ and $\{b_0, b_1, b_2, \dots\}$ be arbitrary sequences, and put $A_n = a_0 + a_1 + \dots + a_n$ for $n \geqslant 0$. Then the following formula of **Abel's partial summation** holds:

$$\sum_{\nu=n+1}^{n+k} a_\nu b_\nu = \sum_{\nu=n+1}^{n+k} A_\nu (b_\nu - b_{\nu+1}) - A_n b_{n+1} + A_{n+k} b_{n+k-1}$$

for any $n \geqslant 0$ and any $k \geqslant 1$; this formula also holds for $n = -1$ if we put $A_{-1} = 0$.

Abel's partial summation enables us to deduce a number of tests of convergence for series of the form $\sum a_n b_n$. In particular, the following criteria are easy to apply:
(1) $\sum a_n b_n$ is convergent if $\sum a_n$ is convergent and if the sequence $\{b_n\}$ is monotone and bounded (**Abel's test**).
(2) $\sum a_n b_n$ is convergent if the sequence $\{s_n\}$ of partial sums of $\sum a_n$ is bounded and if $\{b_n\}$ is monotone and converges to zero (**Dirichlet's test**).
(3) $\sum a_n b_n$ is convergent if $\sum (b_n - b_{n+1})$ is absolutely convergent and if $\sum a_n$ is (at least conditionally) convergent (**test of du Bois-Reymond and Dedekind**).

For example, criterion (2) implies that if $\{b_n\}$ is monotone and converges to zero, then the power series $\sum b_n z^n$ of a complex variable z is convergent on the unit circle $|z| = 1$ except at most for $z = 1$; the case $z = -1$ gives Leibniz's test for alternating series (→ Section C).

E. Double Series

A sequence with two indices, i.e., a mapping from the Cartesian product $\mathbf{N} \times \mathbf{N}$ of two copies of the set of natural numbers $\mathbf{N}$ to a subset of the real or complex numbers, is called a **double sequence** and is denoted by $\{a_{mn}\}$ or $\{a_{m,n}\}$. If there exists a number l such that for any positive ε there is a natural number $N(\varepsilon)$ satisfying $|a_{mn} - l| < \varepsilon$ for all $m > N(\varepsilon)$ and $n > N(\varepsilon)$, then we say that the sequence $\{a_{mn}\}$ has a limit l and write $\lim_{m \to \infty, n \to \infty} a_{mn} = l$. This limit should not be confused with repeated limits such as $\lim_{n \to \infty} (\lim_{m \to \infty} a_{mn})$. If $\lim_{m \to \infty} a_{mn} = \alpha_n$ uniformly in n and $\lim_{n \to \infty} \alpha_n = l$, then $\lim_{m \to \infty, n \to \infty} a_{mn} = l$. For a given double sequence $\{a_{mn}\}$, the formal series $\sum_{m,n=1}^{\infty} a_{mn}$ is called a **double series** and is sometimes denoted by $\sum a_{mn}$. In contrast with double series, the ordinary series discussed previously is called a **simple series**.

Given a double series $\sum a_{mn}$, when the double sequence of partial sums $s_{mn} = \sum_{k=1}^{m} \sum_{l=1}^{n} a_{kl}$ is convergent to s, then $\sum a_{mn}$ is said to be **convergent** to the **sum** s. On the other hand, if s_{mn} is not convergent, $\sum a_{mn}$ is said to be **divergent**. If $\sum_{n=1}^{\infty} a_{mn}$ converges to b_m for each m, then $\sum_{m=1}^{\infty} b_m = \sum_{m=1}^{\infty} (\sum_{n=1}^{\infty} a_{mn})$ is called the **repeated** (or **iterated**) **series by rows**. If $\sum_{m=1}^{\infty} a_{mn}$ converges to c_n for each n, then $\sum_{n=1}^{\infty} c_n = \sum_{n=1}^{\infty} (\sum_{m=1}^{\infty} a_{mn})$ is called the **repeated** (or **iterated**) **series by columns**. Even if two repeated series by rows and columns are convergent, the two sums are not always identical, and $\sum a_{mn}$ is not always convergent. However, if the double series $\sum a_{mn}$ is convergent and $\sum_n a_{mn}$ is convergent for each m, then the repeated series by rows is convergent to the same sum. A similar statement is valid for the repeated series by columns.

Suppose that we are given a double series $\sum a_{mn}$ of nonnegative terms. If any one of $\sum_{m,n} a_{mn}$, $\sum_m \sum_n a_{mn}$, and $\sum_n \sum_m a_{mn}$ is convergent, the other two converge to the same sum. If the **diagonal partial sum** $s_{mm} = \sum_{k=1}^{m} \sum_{l=1}^{m} a_{kl}$ converges to a, then the double series $\sum a_{mn}$ also converges to a.

If $\sum |a_{mn}|$ converges, the double series $\sum a_{mn}$ is called **absolutely convergent**, whereas if $\sum a_{mn}$ converges but not absolutely, then $\sum a_{mn}$ is called **conditionally convergent**. If $\sum a_{mn}$ is absolutely convergent, then any series obtained from $\sum a_{mn}$ by arranging the terms in an arbitrary order is convergent to the same sum.

F. Multiplication of Series

The series $\sum_{n=1}^{\infty} c_n$, where $c_n = a_1 b_n + a_2 b_{n-1} + \ldots + a_n b_1$, is called the **Cauchy product** of two series $\sum_{n=1}^{\infty} a_n$ and $\sum_{n=1}^{\infty} b_n$. (1) Let $\sum a_n$ and $\sum b_n$ be two convergent series and A, B be the sums of these series. If their Cauchy product $\sum c_n$ is also convergent, then it has the sum $C = AB$ (**Abel's theorem**). (2) If at least one of the two convergent series $\sum a_n$ and $\sum b_n$ with the sums A, B, respectively, is absolutely convergent, then their Cauchy product $\sum c_n$ is also convergent and has the sum $C = AB$ (**Mertens's theorem**). (3) If $\sum a_n$ and $\sum b_n$ are absolutely convergent, then their Cauchy product $\sum c_n$ is absolutely convergent (**Cauchy's theorem**). (4) Let $\sum a_n$ and $\sum b_n$ be two convergent series with the sums A, B, respectively. If $\{na_n\}$ and $\{nb_n\}$ are bounded from below, then $\sum c_n$ is convergent and has the sum $C = AB$ (**Hardy's theorem**).

G. Infinite Product

Let $\{a_n\}$ be a given sequence with terms $a_n \neq 0$ $(n = 1, 2, \ldots)$. The formal **infinite product** $a_1 \cdot a_2 \cdot a_3 \cdot \ldots$ is denoted by $\prod_{n=1}^{\infty} a_n$. We call $p_n = a_1 \cdot a_2 \cdot \ldots \cdot a_n$ its nth **partial product**. If the sequence $\{p_n\}$ is convergent to a nonzero limit p, then this infinite product is said to **converge** to p, and p is called the **value** of the infinite product. We write $\prod a_n = p$. If $\{p_n\}$ is not convergent or is convergent to 0, then the infinite product is called **divergent**. Sometimes we consider the infinite product $\prod a_n$ with $a_n = 0$ for a finite number of n's; and then by convergence or divergence of $\prod a_n$ we mean that of the infinite product $\prod a'_n$, where the sequence $\{a'_n\}$ is obtained by deleting zero terms from $\{a_n\}$. Usually we do not treat an infinite product with $a_n = 0$ for an infinite number of n's.

A necessary and sufficient condition for $\prod_{n=1}^{\infty} a_n$ to be convergent is that for any positive ε there is a number N such that $|p_m/p_n - 1| < \varepsilon$ for all $n, m > N$. If $\prod a_n$ converges, then $a_n \rightarrow 1$, but the converse is not always true.

It is often convenient to write an infinite product as $\prod (1 + a_n)$. Then $\prod (1 + a_n)$ and $\sum \log(1 + a_n)$ have the same convergence behavior, where the imaginary part $i\theta$ of the logarithm is assumed to satisfy $0 \leqslant |\theta| < \pi$. If $a_n \geqslant 0$, convergence of $\prod (1 + a_n)$ implies convergence of $\sum a_n$, and vice versa.

If $\prod (1 + |a_n|)$ converges, then $\prod (1 + a_n)$ is said to be **absolutely convergent**. An absolutely convergent infinite product is also convergent, and the value of the infinite product is unchanged by the alteration of the order of terms.

H. Termwise Differentiation of Infinite Series with Function Terms

Uniform convergence of an infinite series $\sum f_n(x)$ is defined by uniform convergence of the sequence of the partial sums $\sum_{k=1}^{n} f_k(x)$ ($\rightarrow$ 435 Uniform Convergence). If the infinite series $\sum f_n(x)$ defined on an interval I of the real line is convergent at least at one point of I and $\sum f'_n(x)$ is convergent uniformly in I when the derivatives $f'_n(x)$ exist, then $\sum f_n(x)$ is convergent to $f(x)$ uniformly in I, and $\sum f_n(x)$ is **termwise differentiable**, that is, $f'(x) = \sum f'_n(x)$. If the $\varphi_n(z)$ are holomorphic in a complex domain D and $\sum \varphi_n(z)$ converges to $\varphi(z)$ uniformly on every compact subset of D, then $\sum \varphi'_n(z)$ also converges to $\varphi'(z)$ uniformly on every compact subset of D (**Weierstrass's theorem of double series**). (For termwise integration $\rightarrow$ 216 Integral Calculus.)

I. Numerical Evaluation of Series

In some special cases, we can express the nth partial sum s_n of a series $\sum a_n$ as a well-known function of n. Specifically, if $\sum a_n$ is an **arithmetic progression** $\sum_{k=1}^{n} (a + (k-1)d)$ or a **geometric progression** $\sum_{k=1}^{n} aq^{k-1}$, we have

$$s_n = \frac{n}{2}(2a + (n-1)d), \qquad s_n = \frac{a(q^n - 1)}{q - 1},$$

respectively. If $|q| < 1$, then $\sum_{n=0}^{\infty} aq^n$ converges to $a/(1-q)$. If $B_{r+1}(x)$ is the $(r+1)$st †Bernoulli polynomial, then $s_n = 1^r + 2^r + \ldots + n^r = [B_{r+1}(x)]_1^{n+1}/(r+1)$. This sum was studied by J. Bernoulli, who gave formulas up to $r = 10$ in his *Ars conjectandi*.

In the series $\sum u_n$, if we can find another sequence $\{v_n\}$ such that $u_n = v_n - v_{n-1}$, then $s_n = u_1 + u_2 + \ldots + u_n = v_n - v_0$. For example, if $u_n = n(n+1)(n+2)$, then $v_n = n(n+1)(n+2)(n+3)/4$ and $s_n = v_n$ because $v_0 = 0$ ($\rightarrow$ 104 Difference Equations). Series with trigonometric function terms are calculated analogously.

There are cases where the sum $\sum a_n$ itself can be expressed in a satisfactory form although we cannot find an appropriately simple expression for each partial sum s_n. For example, $\zeta(r) = \sum_{m=1}^{\infty} 1/m^r$ can be represented by †Bernoulli numbers if r is even. In particular, $\zeta(2) = \pi^2/6$, $\zeta(4) = \pi^4/90$ ($\rightarrow$ Appendix A, Table 10).

If an infinite series converges rapidly, we can get a good approximation by taking a suitable partial sum. On the other hand, if the series converges less rapidly, an effective means for evaluating series is afforded by transformation of series. If the kth †difference is exactly zero, then

$$s_n = \sum_{i=1}^{k} \binom{n+1}{i} \Delta^{i-1} u_1.$$

Since the absolute value of finite differences often decreases rapidly, it is sometimes convenient to consider the series whose terms are the differences of the original series. One finite difference method is **Euler's transformation** of infinite series. In particular, the formula

$$\sum_{k=0}^{\infty}(-1)^k a_k=\sum_{n=0}^{\infty}\frac{\Delta^n a_0}{2^{n+1}}$$

is useful for numerical calculation of sums of slowly converging alternating series. In numerical calculation of the series, we usually start calculating the numerical values of the first few terms; we then apply such transformations as Euler's to the remainder, and calculate the partial sums of the transformed series.

When we calculate the sum of an infinite series approximately, we must estimate the error, i.e., the remainder that must be added to yield the sum of the series itself. We can estimate the maximum error by derivatives or differences of higher orders. We also have the transformations of Markov and Kummer. In the former, every term of the series is represented by convergent series, and in the latter, the given series is reduced by subtracting another convergent series, which has a known sum and similar terms [1].

J. Infinite Series and Integrals

In numerical calculation of functions, we sometimes use the **Euler-Maclaurin formula** [5]:

$$f(x+\xi\omega)=\frac{1}{\omega}\int_x^{x+\omega}f(z)dz+\sum_{r=1}^{m}\left(\frac{\omega^r}{r!}B_r(\xi)\Delta_\omega f^{(r-1)}(x)\right)+R_m,$$

$$R_m=-\omega\int_0^1\frac{\bar{B}_m(\xi-z)}{m!}f^{(m)}(x+\omega z)dz,$$

where

$$\bar{B}_m(\xi-z)=\begin{cases}B_m(\xi-z+1), & \xi<z,\\ B_m(\xi-z), & \xi\geqslant z.\end{cases}$$

The speed of the convergence for this formula is greater than that for Taylor's expansion when $\xi\omega$ is large, since the terms of the formula are †Bernoulli polynomials $B_r(\xi)$ in $0\leqslant\xi\leqslant 1$. We also have Boole's formula, with †Euler polynomials as its terms [4]. The formulas discussed in this section are also used to calculated approximately the partial sums of infinite series.

Another method of evaluating sums of infinite series analytically entails transforming infinite series to definite integrals using the †residue theorem. If an analytic function $f(z)$ is holomorphic except at poles a_n $(n=1,2,\ldots,k)$ in a domain bounded by the closed curve C and containing the points $z=m$ $(m=1,2,\ldots,N)$, then

$$\sum_{m=1}^{N}f(m)=\frac{1}{2\pi i}\int_C\pi(\cot\pi z)f(z)dz-\sum_{n=1}^{k}\mathrm{Res}[\pi(\cot\pi z)f(z)]_{z=a_n}.$$

When the left-hand side of this equation is replaced by

$$\sum_{m=1}^{N}(-1)^m f(m),$$

we replace $\cot\pi z$ by $\operatorname{cosec}\pi z$. $\mathrm{Res}[F(z)]_{z=a}$ is the †residue of $F(z)$ at $z=a$. The line integral along C is often calculated easily by choosing a suitable deformation of C. Sometimes it can be shown immediately that the integral along C is zero, or its asymptotic value can be evaluated by the †method of steepest descent.

K. History of the Study of Divergent Series

Mathematicians in the 18th century did not concern themselves with the question of whether †series were †convergent or †divergent. This indiscriminateness led to various contradictions. In 1821 an exact definition of the notion of convergence of series (Section A) was given by A. L. Cauchy; since then, mathematicians have mostly concerned themselves with convergent series. However, since divergent series appeared in many problems in analysis, the study of such series could not be neglected, and it became desirable to give a suitable definition of their sum. Although some results were given by L. Euler, N. Abel, and others, it was during the latter part of the 19th century that methods of summation of divergent series were studied systematically. This study constituted a new branch of mathematics.

In the following sections, some important methods of summation of divergent series are mentioned. Cesàro's method (→ Section M) was the forerunner of the theory whose general foundation is now the theory of linear transformations.

L. Linear Transformations

For a sequence $\{s_n\}$ $(n=0,1,2,\ldots)$ of real or complex numbers, assume that $\sigma_n=\sum_{i=0}^{\infty}a_{ni}s_i$ converges for $n=0,1,2,\ldots$, where (a_{ik}) is a given matrix $(i,k=0,1,2,\ldots)$. The mapping $T:\{s_n\}\rightarrow\{\sigma_n\}$ is called a **linear transformation**, and $\{\sigma_n\}$ is called the **transform** of $\{s_n\}$ under T. If the matrix satisfies $a_{ik}=0$ $(k>i)$, then T is defined for any sequence $\{s_n\}$ and T is said to be **triangular**. If the transform $\{\sigma_n\}$ under T is defined and convergent whenever $\{s_n\}$ converges, then T is called a **semiregular transformation**. If in addition $\{\sigma_n\}$ has the same limit as $\{s_n\}$, then T is called a **regular transformation**. If for any bounded sequence $\{s_n\}$ the transform $\{\sigma_n\}$ is defined and convergent, then T is called a **normal transformation**. If T is triangular and the transform $\{\sigma_n\}$ of $\{s_n\}$ under T is divergent to ∞ whenever $s_n \to \infty$, then T is called a **totally regular transformation**.

Let T be a regular transformation. If for at least one divergent sequence $\{s_n\}$ the transform $\{\sigma_n\}$ of $\{s_n\}$ under T converges, then T is called a **method of summation**. The limit s of $\{\sigma_n\}$ is called the **sum** of $\{s_n\}$ under the method T of summation, and $\{s_n\}$ is said to be **T-summable** to s. For a given method of summation T_1, let $D(T_1)$ be the set of sequences that are T_1-summable. If $D(T_1)=D(T_2)$, then the methods T_1 and T_2 are called **equivalent**. If $D(T_1)\subset D(T_2)$, then we say that T_1 is **weaker** than T_2 and T_2 is **stronger** than T_1. If $D(T_1)\not\subset D(T_2)$ and $D(T_1)\not\supset D(T_2)$, then T_1 and T_2 are called **mutually noncomparable**. The following theorems on linear transformations of sequences are important:

(1) **Kojima-Schur theorem**. In order that T be semiregular it is necessary and sufficient that (i) $\lim_{n\to\infty} a_{nk}$ exist for each k; (ii) $t_n = \sum_{k=0}^{\infty}|a_{nk}|$ exist and $\{t_n\}$ be bounded; and (iii) $\lim_{n\to\infty}\sum_{k=0}^{\infty} a_{nk}$ exist. In that case, we have

$$\sum_{k=0}^{\infty} |\lim_{n\to\infty} a_{nk}| < \infty$$

and

$$\lim_{n\to\infty}\sigma_n = \lim_{n\to\infty}\sum_{k=0}^{\infty} a_{nk}s_k$$

$$= \left(\lim_{n\to\infty} s_n\right)\left(\lim_{n\to\infty}\sum_{k=0}^{\infty} a_{nk}\right) + \sum_{k=0}^{\infty}\left(\lim_{n\to\infty} a_{nk}\right)\left(s_k - \lim_{n\to\infty} s_n\right).$$

In particular, in order that $\{\sigma_n\}$ converge whenever $s_n\to 0$ it is necessary and sufficient that conditions (i) and (ii) be satisfied. In that case,

$$\lim_{n\to\infty}\sigma_n = \lim_{n\to\infty}\sum_{k=0}^{\infty} a_{nk}s_k = \sum_{k=0}^{\infty}\left(\lim_{n\to\infty} a_{nk}\right)s_k$$

(I. Schur, *J. Reine Angew, Math.*, 151 (1921); T. Kojima, *Tôhoku Math. J.*, 12 (1917)).

(2) **Toeplitz's theorem**. In order that T be regular it is necessary and sufficient that (i′) $\lim_{n\to\infty} a_{nk}=0$ for each k; (ii); and (iii′) $\lim_{n\to\infty}\sum_{k=0}^{\infty} a_{nk}=1$ (O. Toeplitz, *Prace Mat.-Fiz.*, 22 (1914)). In particular, (ii) and (i″) $\lim_{n\to\infty}\sum_{k=K}^{\infty} a_{nk}=1$ for each K imply that T is regular (**Perron's theorem**).

(3) **Schur's theorem**. In order that T be normal it is necessary and sufficient that (i); (ii); (iii); and (iv) for any $\varepsilon>0$ there exist a $K>0$ such that $\sum_{k=K+1}^{\infty}|a_{nk}|<\varepsilon$ for each n.

(4) In order that the regular triangular transformation T be totally regular, it is necessary and sufficient that $a_{nk}\geqslant 0$ except for a finite number of k.

M. Cesàro's Method of Summation

We write

$$(1-x)^{-\alpha-1} = \sum_{n=0}^{\infty} A_n^{\alpha} x^n,$$

$$A_n^{\alpha} = \binom{n+\alpha}{n} \sim \frac{n^{\alpha}}{\Gamma(\alpha+1)},$$

$$(1-x)^{-\alpha-1}\sum_{n=0}^{\infty} u_n x^n = (1-x)^{-\alpha}\sum_{n=0}^{\infty} s_n x^n = \sum_{n=0}^{\infty} s_n^{\alpha} x^n,$$

where $s_n=\sum_{i=0}^{n}u_i$. Thus the series $\sum u_i$ is associated with the sequence $\{s_n^{\alpha}\}$. If $\sigma_n^{(\alpha)}=s_n^{\alpha}/A_n^{\alpha}$ converges to s as $n\to\infty$, then we say that $\sum u_n$ is **summable by Cesàro's method of order** α (or simply (C, α)-summable) to s and write $\sum_{n=0}^{\infty}u_n = s$ (C, α). This method of summation is called **Cesàro's method of summation of order** α (or simply **(C, α)-summation**).

It is natural to consider (C, α)-summation for $\alpha>-1$. We say that $\sum u_n$ is (C, -1)-summable if $\sum u_n$ converges and $nu_n=o(1)$. Here $\sum_{n=0}^{\infty}u_n=s$ (C, 0) means $\lim_{n\to\infty}s_n=s$, and $\sum_{n=1}^{\infty}u_n=s$ (C, 1) means $s=\lim_{n\to\infty}(s_0+s_1+\ldots+s_{n-1})/n$. Generally, we have the following results:

(1) A_n^{α} is increasing if $\alpha>0$ and decreasing if $0>\alpha>-1$. $A_n^0=1$ and $A_n^{\alpha}>0$ if $\alpha>-1$.

(2) $s_n^{\alpha+\beta+1}=\sum_{k=0}^{n}A_{n-k}^{\beta}s_k^{\alpha}$; $A_n^{\alpha}-A_{n-1}^{\alpha}=A_n^{\alpha-1}$, $s_n^{\alpha}-s_{n-1}^{\alpha}=s_n^{\alpha-1}$.

(3) (C,α) $(\alpha\geqslant 0)$ is regular, and $D(C,\alpha)\supset D(C,\beta)$ if $\alpha>\beta>-1$.

(4) If $\sum u_n=s$ (C, α), then $u_n=o(n^{\alpha})$. Moreover, if $\sum u_n'=s'$ (C, α), then $\sum(u_n+u_n')=s+s'$ (C, α) and $\sum\lambda u_n=\lambda s$ (C, α) for any number λ.

(5) If $\sum u_n=s$ (C, α) and $\sum u_n'=s'$ (C, β), then their †Cauchy product $\sum v_n=ss'$ (C, $\alpha+\beta+1$) (**Chapman's theorem**). Moreover, if $\sum_{k=0}^{n}A_k^{\alpha}|s_{n-k}^{\beta-\alpha-1}(u_{n-k}')|/A_n^{\beta}=O(1)$, then $\sum v_n=ss'$ (C, β) (T. Kojima). If $\alpha', \beta'>-1$, $s_n^{(\alpha')}(u_n)=O(n^{\alpha'})$, and $s_n^{(\beta')}(u_n')=O(n^{\beta'})$, then $\sum v_n=ss'$ (C, $\alpha'+\beta'+2$) (G. Doetsch).

(6) For any integer $\alpha>0$, in order that $\sum u_n=s$ (C, α) it is necessary and sufficient that there exist $\{v_n\}$ such that $u_n=(n+1)(v_n-v_{n+1})$ and $\sum v_n=s$ (C, $\alpha-1$) (G. H. Hardy, *Proc. London Math. Soc.*, (2) 8 (1910)). This condi-

tion is equivalent to (i)–(iii) together: (i) the series $\sum_{k=n}^{\infty} s_n^{\alpha-2}/(k+1)\ldots(k+\alpha)$ converges to the limit b_n; (ii) $b_n=o(1)$ as $n\to\infty$; and (iii) $(s_n^{\alpha-1}/A_n^{\alpha-1})+(n+\alpha)\Gamma(\alpha)b_{n+1}\to s$ as $n\to\infty$.

(7) If $\sum u_n=s$ (C, α) ($\alpha>0$), one of the following five conditions is sufficient for the convergence of $\sum u_n$ (this is a kind of †Tauberian theorem): (i) $nu_n=o(1)$; (ii) $t_n=\sum_{\nu=1}^{n}\nu u_\nu=o(n)$; (iii) $\sum n^p|u_n|^{p+1}<\infty$ ($p\geqslant 0$); (iv) $nu_n>-K$ (K is independent of n); (v) $\liminf(s_m-s_n)\geqslant 0$ as $m>n\to\infty$, $m/n\to 1$ (**R. Schmidt's condition**).

(8) If $\alpha'>\alpha>-1$ and $\sum u_n=s(C,\alpha')$, then $\sum u_n=s$ (C, $\alpha+\varepsilon$) for any $\varepsilon>0$.

For a given series $\sum u_n$, we write H_n^0 for s_n, H_n^1 for the arithmetic mean of $\{H_n^0\}$, and H_n^2 for the arithmetic mean of $\{H_n^1\}$. Similarly, we can define $\{H_n^p\}$ for any integer p. If $H_n^p\to s$ as $n\to\infty$, then $\sum u_n$ is said to be **summable by Hölder's method of order** p (or **(H,** p**)-summable**) to s, and we write $\sum u_n=s(\mathrm{H},p)$. For any integer $p\geqslant 0$, (H, p)-summability is equivalent to (C, p)-summability (**Knopp-Schnee theorem**).

N. Abel's Method of Summation

If the radius of convergence of the power series $\sum_{n=0}^{\infty}u_nr^n$ is $\geqslant 1$ and $\sum_{n=0}^{\infty}u_nr^n\to s$ as $r\to 1$, then $\sum u_n$ is said to be **summable by Abel's method** (or **A-summable**) to s, and we write $\sum u_n=s$ (A). The transformation matrix is denoted by A, and the transformation is called **Abel's method of summation**.

(1) If $\sum u_n=s$ (A), then $\limsup_{n\to\infty}|u_n|^{1/n}\leqslant 1$. (2) If $\sum u_n=s$ (A) and $\sum u_n'=s'$ (A), then $\sum(u_n+u_n')s+s'$ (A) and $\sum\lambda u_n=\lambda s$ (A) for any constant λ. Moreover, $\sum_{n=k+1}^{\infty}u_n=s-u_0-u_1-\ldots-u_k$ (A). (3) If $\sum u_n=s$ (A) and $\sum u_n'=s'$ (A), then the Cauchy product is $\sum u_n=ss'$ (A). (4) If $\sum u_n=s$ (A) and one of the following five conditions is satisfied, then $\sum u_n=s$: (i) $nu_n=o(1)$; (ii) $t_n=\sum_{\nu=1}^{n}\nu u_\nu=o(n)$; (iii) $nu_n=O(1)$; (iv) $nu_n>-M$; (v) $\liminf(s_m-s_n)\geqslant 0$ as $m>n\to\infty$ and $m/n\to 1$. These theorems are †Tauberian, in the original form proved by Tauber (*Monatsh. Math.*, 8 (1897)). (5) The matrix A is regular, and $D(C,\alpha)\subset D(A)$ for any $\alpha>-1$. (6) If $\sum u_n=s$ (A) and $s_n\geqslant 0$, then $\sum u_n=s$ (C, 1). Moreover, if $\sigma_n^{(\alpha)}=O(1)$, then $\sum u_n=s$ (C, $\alpha+\varepsilon$) for $\alpha>-1$ and $\varepsilon>0$.

O. Borel's Method of Summation

If for a given series $\sum u_n$,

$$u(x)=\sum_{n=0}^{\infty}\frac{s_nx^n}{n!}$$

is convergent for all x, and $u(x)/e^x\to s$ as $x\to\infty$, then $\sum u_n$ is said to be **summable by Borel's exponential method** to the sum s, and we write $\sum u_n=s$ (B). The transformation thus determined is denoted by B and is called **Borel's method of summation.** If

$$\int_0^\infty u(x)e^{-x}dx=s,$$

then $\sum u_n$ is said to be **summable by Borel's integral method** (or **$\mathfrak{B}$-summable**) to s, and we write $\sum u_n=s$ ($\mathfrak{B}$). Then we have: (1) B is regular and $D(C,\alpha)\subset D(B)$ ($\alpha>-1$), while $D(C,\alpha)$ and $D(\mathfrak{B})$ are noncomparable. (2) If the radius of convergence of $\sum_{n=0}^{\infty}u_nx^n$ is $\geqslant 1$ and $\sum u_n=s$ (B), then $\sum u_n=s$ (A). (3) If $\sum_{n=k+1}^{\infty}u_n=s$ (B) (resp. ($\mathfrak{B}$)), then $\sum u_n=s+u_0+u_1+\ldots+u_k$(B) (resp. ($\mathfrak{B}$)), but the converse is not always true. (4) $\sum u_n=s$ (B) implies $|u_n|^{1/n}=o(n)$. (5) If $\sum u_n=s$ (B) and $\sum u_n'=s'$ (B), then $\sum(u_n+u_n')=s+s'$ (B), and $\sum\lambda u_n=\lambda s$ (B) for any constant λ. The same is true for summation ($\mathfrak{B}$). (6) If $\sum u_n=s$ (B) and if one of the following two conditions is satisfied, then $\sum u_n=s$: (i) $\sqrt{n}\,u_n=o(1)$; (ii) $\liminf(s_m-s_n)\geqslant 0$ as $m>n\to\infty$ and $(m-n)\sqrt{n}\to 0$. (7) if $\sum u_n=s$ (B) and $s_{n-1}^{\alpha-1}=o(n^{\alpha-1/2})$, then $\sum u_n=s$ (C, α) ($\alpha\geqslant 0$). (8) If $\sum u_n=s$ (A) and $u(t)>-Mt^{-1}\exp t$, then $\sum u_n=s$ (B). (9) If the sequences $\{n_k\}$ and $\{n_{k'}\}$ satisfy $n_{k+1}>n_{k'}$, $n_{k'}/n_k>1+\varepsilon$ ($k=1,2,\ldots;\varepsilon>0$), $u_\nu=0$ ($n_k<\nu\leqslant n_{k'}$), and $\sum u_n=s$ (B), then $s_{n_k}\to s$ as $k\to\infty$.

If $\sum u_n=s$ ($\mathfrak{B}$) and

$$\int_0^\infty\left|\frac{d^\lambda u(x)}{dx^\lambda}\right|e^{-x}dx$$

converges for all $\lambda=0,1,2,\ldots$, then $\sum u_n$ is said to be **absolute Borel summable** (or **$|\mathfrak{B}|$-summable**). Concerning this we have: (1) If $\sum|u_n|$ converges, then $\sum u_n$ is $|\mathfrak{B}|$-summable, but even if $\sum u_n$ converges, $\sum u_n$ is not always $|\mathfrak{B}|$-summable. If $\sum u_n$ is $|\mathfrak{B}|$-summable, then $\sum u_n$ is $\mathfrak{B}$-summable to s. In this case, we write $\sum u_n=s$ ($|\mathfrak{B}|$). (2) $\sum_{n=0}^{\infty}u_n=s$ ($|\mathfrak{B}|$) implies $\sum_{n=k+1}^{\infty}u_n=s-(u_0+u_i+\ldots+u_k)$ ($|\mathfrak{B}|$). (3) If $\sum u_n=s$ (B), $\sum u_n'=s'$ (B), and if at least one of them is $|\mathfrak{B}|$-summable, then their Cauchy product is $\sum v_n=ss'$ ($|\mathfrak{B}|$).

P. Euler's Method of Summation

In the series $\sum u_n$, if

$$\left\{\binom{k+1}{1}s_0+\binom{k+1}{2}s_1+\ldots+\binom{k+1}{k+1}s_k\right\}2^{-(k+1)}$$

($s_n=\sum_{k=0}^{n}u_k$) converges to s as $k\to\infty$, then $\sum u_n$ is said to be **summable by Euler's method**, and we write $\sum u_n=s$ (E). The transformation thus obtained is called **Euler's method of summation**. A necessary and sufficient condition for $\sum u_n = s$ (E) is that $\sum v_n = s$, where $v_n = 2^{-(n+1)}\sum_{k=0}^{n}\binom{n}{k}u_k$. This summation method is regular. We have $\sum u_n = s$ if $\sum u_n = s$ (E) and if one of the following two conditions is satisfied: (i) $\sqrt{n}\,u_n = O(1)$; (ii) $\liminf(s_m - s_n) \geqslant 0$ for $m > n \to \infty$, $(m-n)/\sqrt{n} \to 1$. Cesàro and Euler summations are noncomparable. As an extension of Euler's method, the Euler method of summation of pth order is also defined (e.g., → [6]).

Q. Nörlund's Method of Summation

For a positive sequence $\{p_n\}$, let $P_n = \sum_{\nu=0}^{n} p_\nu \to \infty$ as $n \to \infty$ and $p_n/P_n \to 0$ as $n \to \infty$. If

$$\left(\sum_{k=0}^{n} s_k p_{n-k}\right)\Big/ P_n = \left(\sum_{k=0}^{n} u_k P_{n-k}\right)\Big/ P_n$$

converges to s as $n \to \infty$, then $\sum u_n$ is said to be **summable by Nörlund's method** of type $\{p_n\}$, and we write $\sum u_n = s\ (\mathrm{N}, \{p_n\})$. The transformation thus obtained is also regular and is called **Nörlund's method of summation**. If $\sum u_n = s$ (C, 1) and $0 \leqslant p_0 \leqslant p_1 \leqslant \ldots$, then $\sum u_n = s\ (\mathrm{N}, \{p_n\})$. Cesàro's method is actually a special case of this method.

R. M. Riesz's Method of Summation

Let $\{\lambda_n\}$ be a sequence with increasing terms and tending to $+\infty$ as $n \to \infty$. If

$$R(\lambda_n, k, \tau) = \left(\sum_{\lambda_n < \tau} (\tau - \lambda_n)^k u_n\right)\Big/ \tau^k$$

converges to s as $\tau \to \infty$, then $\sum u_n$ is said to be **summable by Riesz's method of order** k and type λ_n, and we write $\sum u_n = s\ (\mathrm{R}, \lambda_n, k)$. The transformation thus obtained is regular and is called **Riesz's method of summation of the** k**th order**. In particular, if $\lambda_n = n$, then $D(R, \lambda_n, k) = D(C, k)$.

S. Riemann's Method of Summation

If

$$\sum_{n=1}^{\infty} u_n \left(\frac{\sin nh}{nh}\right)^k, \quad u_0 = 0,$$

converges for $h > 0$ and tends to s as $h \to 0$, then $\sum u_n$ is said to be **(R,** k**)-summable** to s. When $k = 1$, this method, often called **Lebesgue's method of summation**, is not regular. When $k = 2$, it is ordinarily called **Riemann's method of summation** and is regular. Corresponding to these cases, if

$$\frac{2}{\pi}\sum s_n \frac{\sin nh}{n} \to s \quad \text{or} \quad \frac{2}{\pi h}\sum s_n \left(\frac{\sin nh}{n}\right)^2 \to s$$

as $h \to 0$, then $\sum u_n$ is called (R_1)-summable or (R_2)-summable to s, respectively. The summation method (R_1) is not regular, while (R_2) is regular. If $\sum u_n$ is (R_1)-summable, then it is also (R_2)-summable, but (R, 2) and (R_2) are noncomparable.

Other methods of summation were developed by G. H. Hardy and J. E. Littlewood, E. Le Roy, C. J. de La Vallée Poussin, and others (e.g., → [6]).

For related topics → 121 Dirichlet Series, 159 Fourier Series, 339 Power Series.

References

[1] K. Knopp, Theorie und Anwendung der unendlichen Reihen, Springer, 1921; fifth edition, 1964; English translation from the second edition, Theory and application of infinite series, Blackie, 1928.
[2] T. J. I'A. Bromwich, An introduction to the theory of infinite series, Macmillan, second edition, 1926.
For numerical computation of series,
[3] L. B. W. Jolley, Summation of series, Chapman & Hall, 1925.
[4] N. E. Nörlund, Vorlesungen über Differenzenrechnung, Springer, 1924.
For the Euler-Maclaurin formula,
[5] N. Bourbaki, Eléments de mathématique, Fonctions d'une variable réele, ch. 6, Actualités Sci. Ind., 1132a, Hermann, second edition, 1961.
For the summation of divergent series and Tauberian theorems,
[6] G. H. Hardy, Divergent series, Clarendon Press, 1949.
[7] R. G. Cooke, Infinite matrices and sequence spaces, Macmillan, 1950.
[8] K. Zeller, Theorie der Limitierungsverfahren, Erg. Math., Springer, 1958; second edition, 1970.
[9] K. Chandrasekharan and S. Minakshisundaram, Typical means, Oxford Univ. Press, 1952.
[10] H. R. Pitt, Tauberian theorems, Oxford Univ. Press, 1958.

380 (X.15) Set Functions

A. General Remarks

A †function whose domain is a †family of sets is called a **set function**. Usually we consider set functions that take real values or $\pm\infty$. For example, if $f(x)$ is a real-valued function defined on a set X, and if we assign to each subset A of X values such as $\sup_A f$, $\inf_A f$, or $\sup_A f - \inf_A f$, then we obtain a corresponding set function. In particular, a set function whose domain is the family of left open intervals in $\mathbf{R}^m$ is called an **interval function**. To distinguish between set functions and ordinary functions defined at each point of a set, we call the latter **point functions**. For example, if $f(x)$ is an †integrable (point) function with $\mathbf{R}$ $(=\mathbf{R}^1)$ as its domain and we put $F(I)=\int_a^b f(x)\,dx$ for $I=(a,b]$, then we obtain an interval function F on $\mathbf{R}$.

B. Finitely Additive Set Functions

Let $\Phi(E)$ be a real-valued set function defined on a †finitely additive class $\mathfrak{B}$ in a space X. If Φ satisfies the finite additivity condition:

$$E_1, E_2 \in \mathfrak{B}, \quad E_1 \cap E_2 = \varnothing \quad \text{imply}$$

$$\Phi(E_1 \cup E_2) = \Phi(E_1) + \Phi(E_2),$$

then $\Phi(E)$ is called a **finitely additive set function** on $\mathfrak{B}$. For each $E \in \mathfrak{B}$ we denote $\sup\{\Phi(A) \mid A \subset E, A \in \mathfrak{B}\}$ $(\inf\{\Phi(A) \mid A \subset E, A \in \mathfrak{B}\})$ by $\overline{V}(\Phi; E)$ $(\underline{V}(\Phi; E))$, the **upper (lower) variation** of Φ. Since $\Phi(\varnothing)=0$, we have $\underline{V}(\Phi; E) \leqslant 0 \leqslant \overline{V}(\Phi; E)$. $V(\Phi; E) = \overline{V}(\Phi; E) + |\underline{V}(\Phi; E)|$ is called the **total variation** of Φ on E. When we deal with a fixed Φ, instead of $\overline{V}(\Phi; E)$, $\underline{V}(\Phi; E)$, $V(\Phi; E)$ we write simply $\overline{V}(E)$, $\underline{V}(E)$, $V(E)$. If $V(\Phi; E)$ is bounded, then Φ is said to be **of bounded variation**. If $\Phi(E) \geqslant 0$ $(\leqslant 0)$ for every $E \in \mathfrak{B}$, i.e., $E \subset E'$ implies $\Phi(E) \leqslant \Phi(E')$ $(\Phi(E) \geqslant \Phi(E'))$, then Φ is said to be **monotone increasing (decreasing)**. Every finitely additive set function of bounded variation can be represented as the difference of two monotone increasing finitely additive set functions.

Let I_0 be a fixed interval in $\mathbf{R}^m$ and $F(I)$ be an interval function defined for left open intervals $I \subset I_0$, where $\varnothing$ is considered as a degenerate left open interval. If, for any two left open intervals I_1, I_2 such that $I_1 \cup I_2$ is an interval and $I_1 \cap I_2 = \varnothing$, we have $F(I_1 \cup I_2) = F(I_1) + F(I_2)$, then we call $F(I)$ an **additive interval function** in I_0. Specifically, if $f(x)$ is a real-valued bounded function on $\mathbf{R}$ and D is an interval function determined by $D(I)=f(b)-f(a)$, where $I=[a,b)$ (i.e., $D(I)$ is the increment of $f(x)$), then D is an additive interval function on $\mathbf{R}$, called the **increment function** of f. For a given f the increment function is determined uniquely. Conversely, for a given D, a function f such that D is its increment function is determined uniquely up to an additive constant. In this sense an additive interval function in $\mathbf{R}$ may be identified with the corresponding point function on $\mathbf{R}$.

Let $\mathfrak{R}(I_0)$ be the finitely additive class of all finite unions R of left open intervals in I_0. Then any additive interval function $F(I)$ can be extended to a finitely additive set function $F(R)$ defined on $\mathfrak{R}(I_0)$. For the rest of this article, it is understood that an additive interval function means this extended set function. If for any $\varepsilon>0$ there exists a $\delta>0$ such that $|I|<\delta$ implies $F(I)<\varepsilon$ (where $|I|$ is the volume of the interval I), then we say that F is **continuous**.

C. Completely Additive Set Functions

Let $\Phi(E)$ be a real-valued set function defined on a †completely additive class $\mathfrak{B}$ in a space X. If Φ satisfies the complete additivity condition:

$$E_j, E_k \in \mathfrak{B}, \quad E_j \cap E_k = \varnothing \quad (j \neq k)$$

$$\text{imply} \quad \Phi\left(\sum_{j=1}^{\infty} E_j\right) = \sum_{j=1}^{\infty} \Phi(E_j),$$

then $\Phi(E)$ is called a **completely additive set function** (or simply **additive set function**) on $\mathfrak{B}$. In this case the corresponding upper variation $\overline{V}(E)$, lower variation $\underline{V}(E)$, and total variation $V(E)$ are all completely additive set functions, and for every $E \in \mathfrak{B}$ we have $\Phi(E) = \overline{V}(E) + \underline{V}(E)$ (**Jordan decomposition**). Furthermore, $V(E) = \sup \sum_{j=1}^{n} |\Phi(E_j)|$, where the supremum is taken over all decompositions of E such that $E = \bigcup_{j=1}^{n} E_j$ $(E_j \in \mathfrak{B}, E_j \cap E_k = \varnothing, j \neq k)$. The completely additive nonnegative set functions are the same as the finite measures. Hence the Jordan decomposition implies that every completely additive set function is represented as the difference of two finite measures. A completely additive set function is also called a **signed measure**.

Any continuous additive interval function of bounded variation can be extended to a completely additive set function. The notion of additive interval function of bounded variation is a generalization of that of function of bounded variation (→ 166 Functions of Bounded Variation).

Let Φ be a completely additive set function and μ a finite or σ-finite measure, both defined on $\mathfrak{B}$. If $\mu(E)=0$ implies $\Phi(E)=0$, then Φ is said to be **absolutely continuous** with respect to μ or **μ-absolutely continuous**. Then Φ is μ-absolutely continuous if and only if for any $\varepsilon>0$, there exists a $\delta>0$ such that $\mu(E)<\delta$ implies $|\Phi(E)|<\varepsilon$. If for given Φ and μ there exists an $E_0 \in \mathfrak{B}$ such that $\mu(E_0)=0$ and $\Phi(E) = \Phi(E \cap E_0)$ for every $E \in \mathfrak{B}$, then Φ is said to be **singular** with respect to μ or **μ-singular**.

In a †σ-finite measure space $(X, \mathfrak{B}, \mu)$, every completely additive set function $\Phi(E)$ defined on $\mathfrak{B}$ can be represented uniquely as the sum of a μ-absolutely continuous set function and a

μ-singular set function (**Lebesgue decomposition theorem**). Also, $\Phi(E)$ is μ-absolutely continuous if and only if $\Phi(E)$ can be represented as the indefinite integral $\int_E f(x)\,d\mu$ of a function f that is integrable on X with respect to μ (**Radon-Nikodym theorem**). This function $f(x)$ is called the **Radon-Nikodym derivative**, $d\Phi/d\mu$, of Φ with respect to μ (→ 270 Measure Theory L (iii)).

D. Differentiation of Set Functions

Let m be the Lebesgue measure in $\mathbf{R}^m$ and E a Lebesgue measurable set. We denote $\sup(m(E)/m(Q))$ for all cubes Q such that $E \subset Q$ by $r(E)$ and call it the **parameter of regularity** of E. If for a sequence of sets $\{E_n\}$ there exists an α such that $r(E_n) > \alpha > 0$, then $\{E_n\}$ is called a **regular sequence**. If all the E_n contain a point P and the †diameter of E_n tends to 0 as $n \to \infty$, then we say that $\{E_n\}$ converges to the point P. Let Φ be a set function in $\mathbf{R}^m$. For a regular sequence $\{E_n\}$ of closed sets converging to a point P, we put $l = \limsup(\Phi(E_n)/m(E_n))$ and define the **general upper derivative** of Φ at P to be the least upper bound of l for all such sequences $\{E_n\}$, denoted by $\bar{D}\Phi(P)$. Similarly, the **general lower derivative** $\underline{D}\Phi(P)$ of Φ at P is defined to be the greatest lower bound of $\liminf(\Phi(E_n)/m(E_n))$ for all regular sequences $\{E_n\}$ of closed sets converging to P. The **ordinary upper (lower) derivative**, denoted by $\bar{\Phi}(E)(\underline{\Phi}(E))$, is defined in the same way by taking regular sequences of closed intervals instead of closed sets. $\bar{D}\Phi$, $\underline{D}\Phi$, $\bar{\Phi}$, $\underline{\Phi}$ are point functions derived from Φ. Clearly, $\underline{D}\Phi(P) \leqslant \underline{\Phi}(P) \leqslant \bar{\Phi}(P) \leqslant \bar{D}\Phi(P)$. If $\bar{D}\Phi(P) = \underline{D}\Phi(P)$, then we write it simply as $D\Phi(P)$. If $D\Phi(P)$ is finite, then we call it the **general derivative** of Φ at P and say that Φ is **derivable in the general sense** at P. If $\bar{\Phi}(P) = \underline{\Phi}(P)$, then we write it as $\Phi(P)$. If $\Phi(P)$ is finite, then we call it the **ordinary derivative** of Φ at P and say that Φ is **derivable in the ordinary sense**. We have the following theorems: (1) A completely additive set function is derivable in the general sense †almost everywhere (Lebesgue). The Radon-Nikodym derivative of a set function absolutely continuous with respect to the Lebesgue measure is equal almost everywhere to the generalized derivative of the set function. (2) An additive interval function of bounded variation is derivable in the ordinary sense almost everywhere (Lebesgue). (3) An additive interval function Φ is derivable in the ordinary sense at almost all points such that $\bar{\Phi}(P) < +\infty$ or $\underline{\Phi}(P) > -\infty$.

For the proof of these theorems, **Vitali's covering theorem** is essential: Let A be a given set and $\mathfrak{F}$ a family of measurable sets in Euclidean space. If for each $x \in A$ there is a regular sequence of sets belonging to $\mathfrak{F}$ that converges to x, then there exists a finite or countable set of disjoint $E_n \in \mathfrak{F}$ such that $m(A \setminus \bigcup_{j=1}^{\infty} E_j) = 0$.

References

[1] H. Lebesgue, Leçons sur l'intégration et la recherche des fonctions primitives, Gauthier-Villars, 1904, second edition, 1928.
[2] S. Saks, Theory of the integral, Stechert, 1937 (Dover, 1964).
[3] W. Rudin, Real and complex analysis, McGraw-Hill, second edition, 1974.
[4] H. L. Royden, Real analysis, Macmillan, second edition, 1963.

381 (II.1) Sets

A. Definitions and Symbols

G. Cantor defined a **set** as a collection of objects of our intuition or thought, within a certain realm, taken as a whole. Each object in the collection is called an **element** (or **member**) of the set. The notation $a \in A$ ($A \ni a$) means that a is an element of the set A. In this case we say that a is a member of A or a **belongs** to A. The negation of $a \in A$ ($A \ni a$) is written $a \notin A$ or $a \bar{\in} A$ ($A \not\ni a$ or $A \bar{\ni} a$). The set having no element, namely the set A such that $a \notin A$ for every object a, is called the **empty set** (or **null set**) and is usually denoted by $\varnothing$. Two sets A and B are identical, i.e., $A = B$, if every element of A belongs to B, and vice versa. The set containing $a, b, c, \ldots$ as its elements is said to consist of $a, b, c, \ldots$ and is denoted by $\{a, b, c, \ldots\}$. The symbol $\{x \mid C(x)\}$ (or $\{x; C(x)\}$, sometimes $E_x[C(x)]$) denotes the set of all objects that have the property $C(x)$. Thus $\{a\}$ is the set whose only element is a, and $\{a, b\}$ is the set with two elements a and b, provided that $a \neq b$. A set is called a **finite set** or an **infinite set** according as the number of its elements is finite or infinite.

A set A is a **subset** of a set B if each element of A is an element of B. In this case we also say that A is contained in B or that B **contains** A, and we write $A \subset B$ and $B \supset A$. The negation of $A \subset B$ ($B \supset A$) is $A \not\subset B$ ($B \not\supset A$). For every set A, $\varnothing \subset A$. $A \subset B$ and $B \subset C$ imply $A \subset C$. If $A \subset B$ and $B \subset A$, then $A = B$. A is a **proper subset** of B (in symbols: $A \subsetneqq B$, $B \supsetneqq A$) if $A \subset B$ and $A \neq B$. Some authors use $\subseteqq$ ($\supseteqq$) for $\subset$ ($\supset$), and $\subset$ ($\supset$) for $\subsetneqq$ ($\supsetneqq$).

B. Algebra of Sets

The **union** (**join** or **sum**) of sets A and B, written $A \cup B$, is the set of all elements which belong either to A or to B or to both. The **intersection** (**meet** or **product**) of sets A and B, written $A \cap B$, is the set of all elements which belong to both A and B. In other words, $x \in A \cup B$ if and only if $x \in A$ or $x \in B$ or both, and $x \in A \cap B$ if and only if $x \in A$ and $x \in B$. Given sets A, B, and C, $A \cup B = B \cup A$, $A \cap B = B \cap A$ (**commutative law**); $(A \cup B) \cup C = A \cup (B \cup C)$, $(A \cap B) \cap C = A \cap (B \cap C)$ (**associative law**); $A \cup (B \cap C) = (A \cup B) \cap (A \cup C)$, $A \cap (B \cup C) = (A \cap B) \cup (A \cap C)$ (**distributive law**); $A \cup (A \cap B) = A$, $A \cap (A \cup B) = A$ (**absorption law**).

Two sets A and B are **disjoint** if $A \cap B = \varnothing$. In this case the set $C = A \cup B$ is said to be the **disjoint union** (or **sum**) of A and B, and is written sometimes as $C = A + B$. The set of elements of A which are not members of B is denoted by $A - B$, and is called the **difference** of A and B (or **relative complement** of B in A). If $A \supset B$, $A - B$ is called the **complement** (or **complementary set**) of B with respect to A.

We often consider a theory in which we restrict our attention to elements and subsets of a certain fixed set Ω, and call it the **universal set** of the theory. In geometic terms, Ω is also called the **space** or the **abstract space**, elements of Ω are called **points**, and subsets of Ω **point sets**. If A is a subset of Ω, $\Omega - A$ is simply called the **complement** of A and is denoted A^c. For $A \subset \Omega$ and $B \subset \Omega$, $A \supset B$ and $A^c \subset B^c$ are equivalent. Furthermore, we have $A \cup A^c = \Omega$, $A \cap A^c = \varnothing$, $A^{cc} = A$; and $(A \cap B)^c = A^c \cup B^c$, $(A \cup B)^c = A^c \cap B^c$ (**de Morgan's law**).

The **power set** of a set X, written $\mathfrak{P}(X)$, is the set of all subsets of X. A set whose elements are sets is often called a **family of sets**.

The **pair** consisting of objects a and b is denoted by (a, b). Two pairs (a, b), (c, d) are defined to be equal if and only if $a = c$ and $b = d$. A pair (a, b) is called an **ordered pair**, while the set $\{a, b\}$ is sometimes called an **unordered pair**. Generally the n-**tuple** $(a, b, c, \dots, d)$ of n given objects a, b, c, $\dots$, d is defined to be $((\dots((a, b), c), \dots), d)$, so that $(a, b, c, \dots, d) = (a', b', c', \dots, d')$ if and only if $a = a'$, $b = b'$, $\dots$, $d = d'$. The **Cartesian product** (or **direct product**) of sets A and B, written $A \times B$, is the set of all pairs (a, b) such that $a \in A$ and $b \in B$. $A \times B = \varnothing$ if and only if either $A = \varnothing$ or $B = \varnothing$; $A \times B \subset C \times D$ if and only if $A \subset C$ and $B \subset D$, provided that neither A nor B is empty. Furthermore,

$$(A \times B) \cup (A' \times B) = (A \cup A') \times B,$$

$$(A \times B) \cap (C \times D) = (A \cap C) \times (B \cap D).$$

The subset $\{(a, a) \mid a \in A\}$ of $A \times A$ is called the **diagonal** of $A \times A$ and is denoted by Δ_A. Generally the **Cartesian product** (or **direct product**) of sets A, B, $\dots$, D, written $A \times B \times \dots \times D$, is defined as the set $\{(a, b, \dots, d) \mid a \in A, b \in B, \dots, d \in D\}$.

C. Mappings

If there exists a rule which assigns to each element of a set A an element of a set B, this rule is said to define a **mapping** (or simply **map**), **function**, or **transformation** from A into B. The term *transformation* is sometimes restricted to the case where $A = B$. Usually letters f, g, φ, ψ, $\dots$ stand for mappings. The expression $f: A \to B$ ($A \xrightarrow{f} B$) means that f is a function which maps A into B. If $f: A \to B$ and $a \in A$, then $f(a)$ denotes the element of B which is assigned to a by f. We call $f(a)$ the **image** of a under f. The notation $f: a \mapsto b$ (or $f: a \to b$) is often used to mean $f(a) = b$ (but not in the present volumes). The **domain** of a mapping $f: A \to B$ is the set A, and its **range** (or **codomain**), written $f(A)$, is the subset $\{f(a) \mid a \in A\}$ of B. Two functions f and g are equal ($f = g$) if their domains coincide and $f(a) = g(a)$ for each a in the common domain.

For a mapping $f: A \to B$ and a set $C \in \mathfrak{P}(A)$, $f(C)$ is defined to be the set $\{f(x) \mid x \in C\}$. This definition induces the mapping from $\mathfrak{P}(A)$ to $\mathfrak{P}(B)$ which is usually also denoted by f. If $A_i \in \mathfrak{P}(A)$ $(i = 1, 2)$, then $f(A_1 \cup A_2) = f(A_1) \cup f(A_2)$ and $f(A_1 \cap A_2) \subset f(A_1) \cap f(A_2)$. The **inverse image** of $D \in \mathfrak{P}(B)$, denoted by $f^{-1}(D)$, is defined to be the set $\{x \mid x \in A, f(x) \in D\}$; thus the mapping $f^{-1}: \mathfrak{P}(B) \to \mathfrak{P}(A)$ is defined. If $B_i \in \mathfrak{P}(B)$ $(i = 1, 2)$, then $f^{-1}(B_1 \cup B_2) = f^{-1}(B_1) \cup f^{-1}(B_2)$; $f^{-1}(B_1 \cap B_2) = f^{-1}(B_1) \cap f^{-1}(B_2)$; $f^{-1}(B - B_1) = A - f^{-1}(B_1)$. Furthermore, $A_1 \subset f^{-1} \circ f(A_1)$ and $f \circ f^{-1}(B_1) \subset B_1$.

A mapping g is an **extension** of a mapping f to a set A' if A' is the domain of g and contains the domain A of f, and if $g(a) = f(a)$ for each a in A. In this case f is called a **contraction** (or **restriction**) of g to A or simply a **partial mapping** of g, and is denoted by $g \mid A$. A mapping f is the **constant mapping** (or **constant function**) with the value b_0 if $f(a) = b_0$ for every a in the domain of f. The **identity mapping** (or **identity function**) on A, often denoted by 1_A, is the mapping with the domain A such that $f(a) = a$ for every a in A. Given two mappings $f: A \to B$ and $g: B \to C$, the mapping from A to C which assigns $g(f(a))$ to each $a \in A$ is called the **composite** of f and g and is denoted by $g \circ f$. If $f: A \to B$, $g: B \to C$, and $h: C \to D$, then $(h \circ g) \circ f = h \circ (g \circ f)$ (associative law for composition of mappings).

A mapping $f: A \to B$ is from A **onto** B if $f(A) = B$. In this case f is also called a **surjection** (or a **surjective mapping**). A mapping $f: A \to B$

is **one-to-one** (**1-1**, or **injective**) if $a \neq a'$ implies $f(a) \neq f(a')$ for every pair of elements a and a' in A, that is, if for each b in the range of A, there exists only one element a of A such that $f(a)=b$. Such an f is also called an **injection**. In particular, given a subset B of a set A, the injection $f: B \rightarrow A$ defined by the condition $f(b)=b$ for each $b \in B$ is called the **inclusion mapping** (**inclusion** or **canonical injection**). A necessary and sufficient condition for $f: A \rightarrow B$ to be a surjection is that $g \circ f = h \circ f$ imply $g=h$ for every pair of mappings $g: B \rightarrow C$ and $h: B \rightarrow C$. For $f: A \rightarrow B$ to be an injection it is necessary and sufficient that $f \circ g = f \circ h$ imply $g=h$ for every pair of mappings $g: C \rightarrow A$ and $h: C \rightarrow A$. A mapping which is both a surjection and an injection is called a **bijection** (or **bijective mapping**). If $f: A \rightarrow B$ is a bijection, then the mapping from B to A which assigns to each element b of B the unique element a of A such that $f(a)=b$ is called the **inverse mapping** (**inverse function** or simply **inverse**) of f, and is denoted by f^{-1}. We have $f \circ f^{-1} = 1_B$ and $f^{-1} \circ f = 1_A$ for every bijection $f: A \rightarrow B$.

If the domain A of a mapping $f: A \rightarrow B$ is the Cartesian product of A_1 and A_2, $f(a)=b$ (where $a=(a_1, a_2)$) is written as $f(a_1, a_2)=b$. Given $A = A_1 \times A_2$, $B = B_1 \times B_2$, and $f_i: A_i \rightarrow B_i$ $(i=1,2)$, the mapping $f: A \rightarrow B$ defined by the condition $f(a_1, a_2) = (f_1(a_1), f_2(a_2))$ is called the **Cartesian product** (or **direct product**) of the mappings of f_1 and f_2, and is denoted by $f_1 \times f_2$.

For a mapping $f: A \rightarrow B$, the subset $G = \{(a, f(a)) \mid a \in A\}$ of $A \times B$ is called the **graph** of f. The basic properties of the graph G of f are: (1) For every $a \in A$ there exists a $b \in B$ such that $(a, b) \in G$. (2) $(a, b) \in G$ and $(a, b') \in G$ imply $b=b'$. Conversely, a subset G of $A \times B$ with these two properties determines a mapping $f: A \rightarrow B$ such that $(a, b) \in G$ if and only if $f(a)=b$. All notions concerning a mapping $f: A \rightarrow B$ can be transferred by means of its graph to those concerning a subset of a Cartesian product $A \times B$.

Given sets A and B, we denote by B^A the set of mappings from A to B. If a mapping is identified with its graph, B^A is considered to be a subset of $\mathfrak{P}(A \times B)$. For $X \in \mathfrak{P}(A)$, the mapping $c_X: A \rightarrow \{0,1\}$ such that $c_X(x)=1$ if $x \in X$ and $c_X(x)=0$ if $x \notin X$ is called the **characteristic function** (or **representing function**) of X. By assigning to each $X \in \mathfrak{P}(A)$ its characteristic function $c_X \in \{0,1\}^A$, we obtain a one-to-one correspondence between $\mathfrak{P}(A)$ and $\{0,1\}^A$; hence $\mathfrak{P}(A)$ is sometimes denoted by 2^A.

D. Families of Sets

A mapping from a set Λ to a set A is also called a family of elements of A indexed by Λ. Λ is its **index set** (or **indexing set**). In this case, the image $f(\lambda)$ of $\lambda \in \Lambda$ is denoted by a_λ, and the mapping itself is denoted by $\{a_\lambda\}_{\lambda \in \Lambda}$ ($\{a_\lambda\}$ $(\lambda \in \Lambda)$, or simply $\{a_\lambda\}$). In particular, if the set A is the power set of a set, the family $\{a_\lambda\}_{\lambda \in \Lambda}$ is called a **family of sets indexed by** Λ, or simply a **family of sets**. (Moreover, if Λ is chosen to be a subset of the power set $\mathfrak{P}(X)$ of a set X and f to be the identity mapping on Λ, then the family of sets resulting from f can be identified with the set of subsets Λ itself.)

The union $\bigcup_{\lambda \in \Lambda} A_\lambda$ of a family of sets $\{A_\lambda\}_{\lambda \in \Lambda}$ is the set of all elements a such that $a \in A_\lambda$ for at least one λ in Λ. Their intersection $\bigcap_{\lambda \in \Lambda} A_\lambda$ is the set of all elements a such that $a \in A$ for all λ in Λ. A family of sets $\{A_\lambda\}_{\lambda \in \Lambda}$ is **mutually disjoint** if $\lambda \neq \mu$ implies $A_\lambda \cap A_\mu = \varnothing$. In this case $A = \bigcup_{\lambda \in \Lambda} A_\lambda$ is called the **disjoint union** (or **direct sum**) of the sets of the family, and $\{A_\lambda\}_{\lambda \in \Lambda}$ is called a **partition** (or **decomposition**) of A. For families of sets, the following hold: $\bigcup A_{\lambda\mu} = \bigcup_\lambda (\bigcup_\mu A_{\lambda\mu})$, $\bigcap A_{\lambda\mu} = \bigcap_\lambda (\bigcap_\mu A_{\lambda\mu})$ $(\lambda \in \Lambda, \mu \in M_\lambda)$ (associative law); $(\bigcup_{\lambda \in \Lambda} A_\lambda) \cap (\bigcup_{\mu \in M} B_\mu) = \bigcup_{(\lambda,\mu) \in \Lambda \times M} (A_\lambda \cap B_\mu)$, $(\bigcap_{\lambda \in \Lambda} A_\lambda) \cup (\bigcap_{\mu \in M} B_\mu) = \bigcap_{(\lambda,\mu) \in \Lambda \times M} (A_\lambda \cup B_\mu)$ (distributive law); $(\bigcup_{\lambda \in \Lambda} A_\lambda)^c = \bigcap_{\lambda \in \Lambda} A_\lambda^c$, $(\bigcap_{\lambda \in \Lambda} A_\lambda)^c = \bigcup_{\lambda \in \Lambda} A_\lambda^c$ (de Morgan's law).

A family $\{A_\lambda\}_{\lambda \in \Lambda}$ of sets is a **covering** of a set A, or **covers** A, if $A \subset \bigcup_{\lambda \in \Lambda} A_\lambda$.

Given $f: X \rightarrow Y$, $\{A_\lambda\}_{\lambda \in \Lambda}$ and $\{B_\lambda\}_{\lambda \in \Lambda}$ (where $A_\lambda \subset X$ and $B_\lambda \subset Y$), then $f(\bigcup_{\lambda \in \Lambda} A_\lambda) = \bigcup_{\lambda \in \Lambda} f(A_\lambda)$, $f(\bigcap_{\lambda \in \Lambda} A_\lambda) \subset \bigcap_{\lambda \in \Lambda} f(A_\lambda)$; and $f^{-1}(\bigcup_{\lambda \in \Lambda} B_\lambda) = \bigcup_{\lambda \in \Lambda} f^{-1}(B_\lambda)$, $f^{-1}(\bigcap_{\lambda \in \Lambda} B_\lambda) = \bigcap_{\lambda \in \Lambda} f^{-1}(B_\lambda)$.

E. Direct Sum and Direct Product of Families of Sets

Given a family $\{A_\lambda\}_{\lambda \in \Lambda}$ of sets indexed by Λ, a set S, and a family of injections $\{i_\lambda: A_\lambda \rightarrow S\}_{\lambda \in \Lambda}$, then the pair $(S, \{i_\lambda\}_{\lambda \in \Lambda})$ is called the **direct sum** of $\{A_\lambda\}_{\lambda \in \Lambda}$ if $\{i_\lambda(A_\lambda)\}_{\lambda \in \Lambda}$ is a partition of S. In this case, S is written $\sum_{\lambda \in \Lambda} A_\lambda$ (or $\sum_\lambda A_\lambda$ or $\coprod_\lambda A_\lambda$). Each A_λ is called a **direct summand** of S, and each i_λ is called a **canonical injection**.

The **Cartesian product** (or **direct product**) $\prod_{\lambda \in \Lambda} A_\lambda$ ($\prod_\lambda A_\lambda$) of $\{A_\lambda\}_{\lambda \in \Lambda}$ (where $A_\lambda \subset X$) is the set of all mappings from Λ to X such that $f(\lambda) \in A_\lambda$ for every $\lambda \in \Lambda$. The sets A_λ are the **direct factors** of $\prod_{\lambda \in \Lambda} A_\lambda$. Each element f of $\prod_{\lambda \in \Lambda} A_\lambda$ is denoted by $\{x_\lambda\}_{\lambda \in \Lambda}$ or $(\ldots, x_\lambda, \ldots)$ (where $x_\lambda = f(\lambda)$). The element x_λ is the λth **component** (or **coordinate**) of f. The mapping $\mathrm{pr}_\lambda: \prod_{\lambda \in \Lambda} A_\lambda \rightarrow A_\lambda$ which assigns x_λ to each $\{x_\lambda\}_{\lambda \in \Lambda} \in \prod_{\lambda \in \Lambda} A_\lambda$ is called the **projection** of $\prod_{\lambda \in \Lambda} A_\lambda$ onto its λth component. If $\Lambda = \{1,2\}$, $\prod_{\lambda \in \Lambda} A_\lambda$ can be identified with $A_1 \times A_2$.

F. Set Theory

It was G. †Cantor who introduced the concept of the set as an object of mathematical study.

Cantor stated: "A set is a collection of definite, well-distinguished objects of our intuition or thought. These objects are called the elements of the set" (G. Cantor, *Math. Ann.*, 46 (1895)). Cantor introduced the notions of †cardinal number and †ordinal number and developed what is now known as **set theory**. He proved that the cardinal number of the set of transcendental numbers is greater than that of algebraic numbers, and that all Euclidean spaces have the same cardinal number regardless of their dimension. He stated the †continuum hypothesis and also conjectured the †well-ordering theorem (G. Cantor, *Math. Ann.*, 21 (1883)), which was proved by E. Zermelo [2]. In this proof Zermelo stated the †axiom of choice explicitly for the first time, and used it in an essential way.

Meanwhile it was pointed out that Cantor's naive set concept leads to various logical †paradoxes (→ 319 Paradoxes). Since the set concept plays a fundamental role in every branch of mathematics, the discovery of the paradoxes had a serious impact upon mathematics, and led to a systematic investigation of the †foundations of mathematics. In the course of attempts to avoid paradoxes, set theory was reconstructed as †axiomatic set theory (→ 33 Axiomatic Set Theory), in which Cantor's theory of cardinal numbers and ordinal numbers was restored. Also, the theory of the algebra of sets, which forms a basis for various branches of mathematics, was reconstructed. Axiomatic set theory is considered to be free from paradoxes.

G. Classes

A set in the naive sense is a collection $\{x \mid C(x)\}$ of all x which satisfy a certain condition $C(x)$. The only principle for generating sets in naive set theory is the **axiom of comprehension**, which asserts the existence of the set $\{x \mid C(x)\}$ for any condition $C(x)$. However, this principle leads to paradoxes if the notion of an **arbitrary set** is considered to be well defined; for example, the †Russell paradox is caused by the set $\{x \mid x \notin x\}$. This situation necessitates some restrictions on the axiom of comprehension. The simplest way to overcome the paradoxes is to adopt Zermelo's **axiom of subsets**: Given a set M and a condition $C(x)$, there exists a set $\{x \mid x \in M, C(x)\}$. But this axiom cannot produce any sets other than subsets of sets whose existence is preassumed. Hence further generating principles of sets had to be introduced. The following axioms are usually chosen as generating principles.

Axiom of pairing: For any two objects (possibly sets) a and b, there exists a set $\{a, b\}$.

Axiom of power set: Given a set A, its power set $\mathfrak{P}(A)$ exists.

Axiom of union: For any family of sets the union exists.

Axiom of substitution (or **replacement**): For any set A and any mapping f from A, there exists a set of all images $f(x)$ with $x \in A$.

In ordinary theories of mathematics the set of natural numbers, the set of real numbers, etc., are assumed to exist, in addition to sets generated by the axioms in this section. In pure set theory the **axiom of infinity** is needed to secure the existence of infinite sets.

The concept of the "set" $\{x \mid x \notin x\}$ does not automatically lead to Russell's paradox. The trouble arises when this "set" is regarded as a member of a collection represented by x. This leads to a narrower concept of sets. Consider a fixed collection V consisting of sets in the naive sense and closed under the set-theoretic operations mentioned in the axioms. Call a member of V a set in the narrow sense. Then set theory becomes free from the known paradoxes if the qualification for being a set is restricted in this narrow sense. When sets in the narrow sense are called simply sets, sets in the naive sense are called **classes**. The object $\{x \mid x \notin x\}$ (where x ranges over sets in the narrow sense) is a class which is not a set. Those classes which are not sets are called **proper classes**; for example, the class V of all sets and the class of all ordinal numbers are both proper classes. For classes, unrestricted use of the comprehension axiom again leads to paradoxes, but other set-theoretic operations are justifiably applicable to classes.

The notion of classes was first introduced in connection with the construction of an axiomatic set theory. The term *class* was used originally to denote certain subclasses of the class V of all sets. In these volumes the term *set* is mostly used to mean a set in the naive sense, and most of the notions defined for sets are applicable to classes.

References

[1] G. Cantor, Gesammelte Abhandlungen, Springer, 1932.
[2] E. Zermelo, Beweis, dass jede Menge wohlgeordnet werden kann, Math. Ann., 59 (1904), 511–516.
[3] A. Schoenflies, Entwicklung der Mengenlehre und ihrer Anwendungen, Teubner, 1913.
[4] F. Hausdorff, Grundzüge der Mengenlehre, Veit, 1914 (Chelsea, 1949).
[5] A. Fraenkel, Einleitung in die Mengenlehre, Springer, third edition, 1928.
[6] F. Hausdorff, Mengenlehre, Teubner, 1927; English translation, Set theory, Chelsea 1962.

[7] A. Fraenkel, Abstract set theory, North-Holland, 1953.
[8] J. L. Kelley, General topology, Van Nostrand, 1955, appendix.
[9] N. Bourbaki, Eléments de mathématique, I. Théorie des ensembles, ch. 2, Actualités Sci. Ind., 1212c, Hermann, second edition, 1960; English translation, Theory of sets, Addison-Wesley, 1968.
[10] P. R. Halmos, Naive set theory, Van Nostrand, 1960.

382 (IX.23) Shape Theory

A. General Remarks

In 1968 K. Borsuk introduced the notion of shape as a modification of the notion of †homotopy type. His idea was to take into account the global properties of topological spaces and neglect the local ones. It is a classification of spaces that is coarser than the homotopy type but that coincides with it on †ANR-spaces.

Let $\mathscr{P}$ be the category whose objects are all †polyhedra and whose morphisms are homotopy classes of continuous mappings between them. For spaces X and Y, denote the set of all †homotopy classes of continuous mappings of X to Y by $[X, Y]$ and the homotopy class of a mapping f by $[f]$. For a space X, let Π_X be the functor from $\mathscr{P}$ to the †category of sets and functions that assigns to a polyhedron P the set $\Pi_X(P)=[X,P]$. A morphism $\varphi:P\rightarrow Q$ of $\mathscr{P}$ induces the function $\varphi_{\#}:[X,P]\rightarrow[X,Q]$ defined by $\varphi_{\#}([f])=\varphi\cdot[f]$ for $[f]:X\rightarrow P$. A †natural transformation from Π_Y to Π_X is a **shape morphism** from X to Y. A continuous mapping $f:X\rightarrow Y$ defines the shape morphism $f^{\#}$ of X to Y as follows: For $[g]:Y\rightarrow P$ in $\Pi_Y(P)$, the composition $[g\cdot f]$ is an element of $\Pi_X(P)$. The correspondence: $[g]\rightarrow[g\cdot f]$ defines a natural transformation from Π_Y to Π_X and hence determines the shape morphism $f^{\#}:X\rightarrow Y$. The identity mapping 1_X on X defines the identity shape morphism $1_X^{\#}$ on X. Given spaces X and Y, X **shape dominates** Y if there are shape morphisms $\xi:Y\rightarrow X$ and $\eta:X\rightarrow Y$ such that $\xi\eta=1_X^{\#}$, and we write $\mathrm{Sh}(X)\leqslant\mathrm{Sh}(Y)$. If, in addition, $\eta\xi=1_Y^{\#}$, then X and Y are of the **same shape**, and we write $\mathrm{Sh}(X)=\mathrm{Sh}(Y)$. A **shape category** $\mathscr{S}$ is the category whose objects are all topological spaces and whose morphisms are shape morphisms between them. If we replace topological spaces by pointed ones, the **pointed shape** category is obtained. In what follows, for simplicity, we assume that all spaces are metrizable and the mappings are continous. General references are Borsuk [1], J. Dydak and J. Segal [3], R. H. Fox [4], S. Mardešić [7].

B. Chapman's Complement Theorem

Let X be a compactum. A closed set A of X is a **Z-set** in X if for any $\varepsilon>0$ there is a mapping $f:X\rightarrow X-A$ such that $d(x,f(x))<\varepsilon$ for $x\in X$, where d is a metric on X. The **Hilbert cube** Q is the countable product $\prod_{i=1}^{\infty}I_i$, where I_i is the closed interval $[0,1]$. The subset $s=\prod_{i=1}^{\infty}I_i^0$ ($I_i^0=(0,1)$) is called the **pseudointerior** of Q. The following facts are known: (1) If a compact metric space X is contained in s or $Q-s$, then X is a Z-set in Q. (2) For any continuous mapping f of a compact metric space X into Q, there exists an embedding g of X into Q such that g is arbitrarily close to f and the image $g(X)$ in Q is a Z-set. The **complement theorem** (T. A. Chapman [2]) states: Let X and Y be Z-sets in Q. Then $\mathrm{Sh}(X)=\mathrm{Sh}(Y)$ iff $Q-X$ and $Q-Y$ are homeomorphic.

C. FAR, FANR, Movability, and Shape Group: Shape Invariants

A closed set A of a compactum X is a **fundamental retract** of X if there is a shape morphism $r:X\rightarrow A$ such that $r\cdot i^{\#}=1_A^{\#}$, where i is the inclusion of A into X. A compactum X is a **fundamental absolute retract (FAR)** (resp. **fundamental absolute neighborhood retract (FANR)**) if for any compactum Y containing X, X is a fundamental retract of Y (resp. of some closed neighborhood of X in Y). A compactum X is **movable** if for any embedding $X\subset Q$ and for any neighborhood U of X in Q there is a neighborhood V of X satisfying the following condition: For any mapping f of a compactum Y to V and for any neighborhood W of X, there is a homotopy $H:Y\times I\rightarrow U$ such that $H(y,0)=f(y)$ and $H(y,1)\in W$ for $y\in Y$. In this definition, if Y is replaced by a compactum with dimension $\leqslant k$, then X is said to be **k-movable**. Pointed FAR, FANR, movability, and k-movability are defined similarly in the pointed shape category. The following facts are known (Borsuk [1], Dydak and Segal [3], J. Keeslings [5], J. Krasinkiewicz [8]). A compactum X is a FAR if and only if X is a pointed FAR iff $\mathrm{Sh}(X)=\mathrm{Sh}(\mathit{point})$, i.e., X has the same shape as a one-point space. A pointed compactum (X,x) is a pointed FANR iff $\mathrm{Sh}(X,x)\leqslant\mathrm{Sh}(K,k)$ for some pointed polyhedron (K,k). An FANR is movable. A compact connected Abelian topological group is movable if and only if it is locally connected.

A continuous image of a pointed 1-movable compactum is pointed 1-movable. It is unknown whether (i) an FANR is a pointed FANR and (ii) movability means pointed movability. For a pointed compactum (X, x), let $\{(K_i, k_i) | i=1,2,\ldots\}$ be a countable †inverse system consisting of pointed finite polyhedra whose limit is (X, x). The limit group $\varprojlim \pi_n(K_i, k_i)$ is the kth **shape group** of (X, x), where $\pi_n(K, k)$ is the kth homotopy group of (K, k). It is known that the shape groups for movable compacta behave like homotopy groups for ANR. A property P of spaces is a **shape invariant** if whenever X has P and $\mathrm{Sh}(X)=\mathrm{Sh}(Y)$, then Y has P. FAR, FANR, movability, k-movability, and shape groups are shape invariants.

D. CE Mappings

A mapping f of a space X onto a space Y is a **cell-like (CE) mapping** if it is proper and $\mathrm{Sh}(f^{-1}(y))=\mathrm{Sh}(\mathit{point})$ for each point y of Y. It is known (R. B. Sher [9], Y. Kodama [6]) that if there is a CE mapping of X to Y with finite dimension, then $\mathrm{Sh}(X)=\mathrm{Sh}(Y)$. Here the finite-dimensionality of Y is essential. A **Q-manifold** is a space, each point of which has a closed neighborhood homeomorphic to Q. The following are known (Chapman [2], J. E. West [10]): (1) If f is a CE mapping of a Q-manifold M to an ANR X, then the mapping $g: M \times Q \rightarrow X \times Q$ defined by $g(m, x)=(f(m), x)$ for $(m, x) \in M \times Q$ is approximated by homeomorphisms. As a consequence, if X is a locally compact ANR, then $X \times Q$ is a Q-manifold. (2) Every compact ANR is a CE image of a compact Q-manifold. (3) Every compact ANR has the same homotopy type as that of a compact polyhedron. The following problem raised by R. H. Bing is open: Is a CE image of a finite-dimensional compactum finite-dimensional?

References

[1] K. Borsuk, Theory of shape, Monograf. mat. 59, Polish Scientific Publishers, 1975.
[2] T. A. Chapman, Lectures on Hilbert cube manifolds, CBMS 28 regional conf. ser. in math., Amer. Math. Soc., 1976.
[3] J. Dydak and J. Segal, Shape theory, Lecture notes in math. 688, Springer, 1978.
[4] R. H. Fox, On shape, Fund. Math., 74 (1972), 47–71.
[5] J. Keesling, Shape theory and compact connected Abelian topological groups, Trans. Amer. Math. Soc., 194 (1974), 349–358.
[6] Y. Kodama, Decomposition spaces and shape in the sense of Fox, Fund. Math., 97 (1977), 199–208.
[7] S. Mardešić, Shapes for topological spaces, General Topology and Appl., 3 (1973), 265–282.
[8] J. Krasinkiewicz, Continuous images of continua and 1-movability, Fund. Math., 98 (1978), 141–164.
[9] R. B. Sher. Realizing cell-like maps in Euclidean spaces, General Topology and Appl., 2 (1972), 75–89.
[10] J. E. West, Mapping Hilbert cube manifolds to ANR's: A solution of a conjecture of Borsuk, Ann. Math., (2) 106 (1977), 1–8.

383 (II.26) Sheaves

A. Presheaves

Let X be a †topological space. Suppose that the following conditions are satisfied: (i) There exists an (additive) Abelian group $\mathscr{F}(U)$ for each open set U of X, and $\mathscr{F}(\varnothing)=\{0\}$; and (ii) there exists a homomorphism $r_{UV}: \mathscr{F}(V) \rightarrow \mathscr{F}(U)$ for each pair $U \subset V$, such that $r_{UU}=1$ (identity) and $r_{UW}=r_{UV} \circ r_{VW}$ for $U \subset V \subset W$. We call $\mathscr{F}$, consisting of a family $\{\mathscr{F}(U)\}$ of Abelian groups and a family of mappings $\{r_{UV}\}$, a **presheaf** (of Abelian groups) on X. If $a \in \mathscr{F}(V)$ and $U \subset V$, we write $r_{UV}(a)=a | U$ and call it the **restriction** of a to U. A **homomorphism** φ between two presheaves $\mathscr{F}$ and $\mathscr{G}$ on X is a family $\{\varphi(U)\}$ of group homomorphisms $\varphi(U): \mathscr{F}(U) \rightarrow \mathscr{G}(U)$ satisfying $r_{UV} \circ \varphi(V)=\varphi(U) \circ r_{UV}$ whenever $U \subset V$. The presheaves on X and their homomorphisms form a †category.

B. Axioms for Sheaves

A presheaf $\mathscr{F}$ is called a **sheaf** (of Abelian groups) if it satisfies the following condition: If U is open in X and $(U_i)_{i \in I}$ is an †open covering of U, and if for each $i \in I$ an element s_i of $\mathscr{F}(U_i)$ is given such that $s_i | U_i \cap U_j = s_j | U_i \cap U_j$ for all i and j, then there exists a unique $s \in \mathscr{F}(U)$ such that $s | U_i = s_i$ for all i. By definition, a **homomorphism** between two sheaves is a homomorphism of the presheaves. The sheaves on X also form a category.

Let $\mathscr{F}$ be a presheaf, x a point of X, and $\mathfrak{N}_x$ the †directed set of open neighborhoods of x, with the order opposite to that of inclusion. Then $\{\mathscr{F}(U) | U \in \mathfrak{N}_x\}$ is an inductive system of groups. The †inductive limit $\varinjlim_{x \in U} \mathscr{F}(U)$ of groups $\{\mathscr{F}(U)\}$ is denoted by $\mathscr{F}_x$ and called the **stalk** of $\mathscr{F}$ over x. The image of $s \in \mathscr{F}(U)$ in $\mathscr{F}_x$ is called the **germ** of s at x and is written s_x.

A homomorphism $\varphi:\mathscr{F}\rightarrow\mathscr{G}$ of presheaves induces a homomorphism $\varphi_x:\mathscr{F}_x\rightarrow\mathscr{G}_x$ of stalks.

C. Sheaf Spaces

We introduce a topology on the †direct sum $\mathscr{F}'=\coprod_{x\in X}\mathscr{F}_x$ in the following way: For each open set U of X and each $s\in\mathscr{F}(U)$, consider the set $M_{U,s}=\{s_x \mid x\in U\}$ of the germs defined by s at the points of U, and take the set of all such $M_{U,s}$ as a †base of open sets of the topology. If $p:\mathscr{F}'\rightarrow X$ is the mapping that maps the points of $\mathscr{F}_x$ to x, then p is continuous, and each $p^{-1}(x)$ $(=\mathscr{F}_x)$ has the structure of an Abelian group. Moreover, the following conditions are satisfied: (i) p is a †local homeomorphism, and (ii) the group operations on $p^{-1}(x)$ are continuous in the sense that $(a,b)\rightarrow a+b$ is a continuous mapping from the †fiber product $\mathscr{F}'\times_X\mathscr{F}'$ (i.e., the subspace $\{(a,b)\mid p(a)=p(b)\}$ of the product space $\mathscr{F}'\times\mathscr{F}'$) to $\mathscr{F}'$ and $a\rightarrow -a$ is a continuous mapping from $\mathscr{F}'$ to itself. In general, a topological space $\mathscr{F}'$ with a structure satisfying these conditions is called a **sheaf space** over X.

When $\mathscr{F}'$ is a sheaf space, a continuous mapping s from a subspace A of X to $\mathscr{F}'$ such that $p\circ s=1_A$ is called a **section** of $\mathscr{F}'$ over A. The **set of sections** over A, denoted by $\Gamma(A,\mathscr{F}')$, is an Abelian group in the obvious way. If we associate $\Gamma(U,\mathscr{F}')$ with each open set U and define r_{UV} by the restriction of sections ($r_{UV}(s)=s\mid U$), then we get a sheaf $\mathscr{F}''$ on X. If we start from a presheaf $\mathscr{F}$ and get $\mathscr{F}''$ via $\mathscr{F}'$, the correspondence $\mathscr{F}\rightarrow\mathscr{F}''$ is a †covariant functor from the category of presheaves to the category of sheaves, and $\mathscr{F}''$ is called the **sheaf associated with the presheaf** $\mathscr{F}$. If $\mathscr{F}$ is a sheaf, we can prove $\mathscr{F}''\cong\mathscr{F}$. Conversely, if we start from a sheaf space $\mathscr{F}'$ and construct the sheaf $\mathscr{F}''$ and then the sheaf space $\mathscr{F}'''$, then $\mathscr{F}'''$ is canonically isomorphic to $\mathscr{F}'$. Since we can identify a sheaf and the corresponding sheaf space, both are usually denoted by the same letter. In particular, when $\mathscr{F}$ is a sheaf, $\mathscr{F}(U)$ is usually written $\Gamma(U,\mathscr{F})$.

Given a section $s\in\Gamma(X,\mathscr{F})$ of a sheaf, the points $x\in X$ for which $s_x\neq 0$ in $\mathscr{F}_x$ from a closed set (the sheaf space $\mathscr{F}$ is not necessarily Hausdorff even if X is so). This set is called the **support** of s and is denoted by $\operatorname{supp} s$.

In the theory of this and the previous two sections, we can replace Abelian groups by groups, rings, etc. Then $\mathscr{F}(U)$ is a group or ring accordingly, and $\mathscr{F}(\varnothing)$ the group consisting of the identity element or the ring consisting of the zero element, respectively. We thus obtain the theories of **sheaves of groups**, **sheaves of rings**, etc. In general, a presheaf $\mathscr{F}$ on X with values in a category $\mathscr{C}$ is a †contravariant functor from the category of open sets of X to $\mathscr{C}$, and a homomorphism between presheaves $\mathscr{F}$ and $\mathscr{G}$ on X is a †natural transformation between the functors $\mathscr{F}$ and $\mathscr{G}$.

The presheaves (sheaves) of Abelian groups on a space X form an †Abelian category, denoted by $\mathscr{P}^X$ ($\mathscr{C}^X$). For a homomorphism f: $\mathscr{F}\rightarrow\mathscr{G}$ of presheaves, the **image**, **coimage**, **kernel**, and **cokernel** of f in $\mathscr{P}^X$ are given by

$(\operatorname{Im} f)(U)=\operatorname{Im} f(U)$,

$(\operatorname{Coim} f)(U)=\operatorname{Coim} f(U)$,

$(\operatorname{Ker} f)(U)=\operatorname{Ker} f(U)$,

$(\operatorname{Coker} f)(U)=\operatorname{Coker} f(U)$.

When $\mathscr{F}$ and $\mathscr{G}$ are sheaves, the kernel of f in $\mathscr{C}^X$ coincides with the kernel in $\mathscr{P}^X$, while the image and cokernel of f in $\mathscr{C}^X$ are the associated sheaves of the image and the cokernel in $\mathscr{P}^X$, respectively. Thus, $f:\mathscr{F}\rightarrow\mathscr{G}$ induces $f_x:\mathscr{F}_x\rightarrow\mathscr{G}_x$ at each $x\in X$, $(\operatorname{Ker} f)_x=\operatorname{Ker} f_x$, $(\operatorname{Im} f)_x=\operatorname{Im} f_x$, $(\operatorname{Coker} f)_x=\operatorname{Coker} f_x$, and a sequence of sheaves $0\rightarrow\mathscr{F}\xrightarrow{f}\mathscr{G}\xrightarrow{g}\mathscr{H}\rightarrow 0$ is **exact** if and only if $0\rightarrow\mathscr{F}_x\xrightarrow{f_x}\mathscr{G}_x\xrightarrow{g_x}\mathscr{H}_x\rightarrow 0$ is exact at each $x\in X$.

D. Examples

(1) Let G be an Abelian group (or some other †algebraic system) with †discrete topology. The Cartesian product $X\times G$ gives rise to a sheaf on X, called a **constant sheaf** (or **trivial sheaf**).

(2) Let X be a topological space and Y be a topological Abelian group (e.g., the real or complex numbers). We obtain a sheaf $\mathscr{F}$ on X by putting $\mathscr{F}(U)=$ the set of all continuous mappings $U\rightarrow Y$ and $r_{UV}=$ the natural restriction. The stalk over $x\in X$ is the set of germs at x of continuous functions into Y. This sheaf is called the **sheaf of germs of continuous functions** with values in Y.

(3) When X is an †analytic manifold and Y is a commutative †Lie group, we define the **sheaf of germs of analytic mappings** with values in Y in the same way. If Y is the complex number field $\mathbf{C}$, this sheaf is the **sheaf $\mathscr{O}$ of germs of analytic** (or **holomorphic**) **functions**. A †connected component of the sheaf space $\mathscr{O}$ can be identified with the †analytic function determined by the function element corresponding to a point on that component. The **sheaf of germs of functions of class C^r** on a C^s-manifold $(r\leqslant s)$ is similarly defined.

(4) Given a †vector bundle B over a topological space X, we define a sheaf on X by $\mathscr{F}(U)$ $=\Gamma(U)$ (= the module of sections of B over U) and $r_{UV}=$ the natural restriction. Here the stalk over $x\in X$ consists of the germs at x of sections of B, and is called the **sheaf of germs of sec-**

tions of the vector bundle B. We have similar definition for the **sheaf of germs of differentiable (analytic) sections** when X is a †differentiable (complex) manifold. The case where B is a †tensor bundle (e.g., the †cotangent bundle $\mathfrak{T}^*(X)$) is important. The sheaf $\mathfrak{A}^r(X)$ of germs of C^∞-sections of the r-fold †exterior power of $\mathfrak{T}^*(X)$ is called the **sheaf of germs of differential forms of degree** r $(0 \leqslant r \leqslant \dim X)$.

E. Sheaf Cohomology

The category $\mathscr{C}^X$ of sheaves of Abelian groups on X has sufficiently many †injective objects. A sheaf $\mathscr{F}$ with the property that $r_{UX}:\Gamma(X,\mathscr{F}) \to \Gamma(U,\mathscr{F})$ is surjective for any open set U is said to be **flabby** (or **scattered**). An injective sheaf is flabby.

Fix a nonempty family Φ of closed subsets of X satisfying the following two conditions: (i) $A, B \in \Phi \Rightarrow A \cup B \in \Phi$; (ii) any closed set contained in an element of Φ belongs to Φ. Putting $\Gamma_\Phi(\mathscr{F}) = \{s \mid s \in \Gamma(X,\mathscr{F}), \operatorname{supp} s \in \Phi\}$ for each $\mathscr{F} \in \mathscr{C}^X$, we obtain a †left-exact †covariant functor Γ_Φ from $\mathscr{C}^X$ to the category (Ab) of Abelian groups. Therefore, by the general theory of homological algebra, we can define the †right derived functors $R^q\Gamma_\Phi:\mathscr{C}^X \to (\mathrm{Ab})$ $(q=0,1,2,\ldots)$. We put $R^q\Gamma_\Phi(\mathscr{F}) = H^q_\Phi(X,\mathscr{F})$ and call the $H^q_\Phi(X,\mathscr{F})$ $(q=0,1,\ldots)$ the **cohomology groups with coefficient sheaf** $\mathscr{F}$ and family of supports Φ (→ 200 Homological Algebra I). When Φ is the family of all closed subsets of X, we write $H^q(X,\mathscr{F})$ instead of $H^q_\Phi(X,\mathscr{F})$.

Thus the cohomology group $H^q_\Phi(X,\mathscr{F})$ is the qth cohomology of the complex $\Gamma_\Phi(\mathfrak{L}^0) \xrightarrow{d^0} \Gamma_\Phi(\mathfrak{L}^1) \xrightarrow{d^1} \Gamma_\Phi(\mathfrak{L}^2) \xrightarrow{d^2} \ldots$ induced by an †injective resolution $0 \to \mathscr{F} \to \mathfrak{L}^0 \to \mathfrak{L}^1 \to \ldots$ of the sheaf $\mathscr{F}: H^q_\Phi(X,\mathscr{F}) = \operatorname{Ker} d^q/\operatorname{Im} d^{q-1}$ $(q=0,1,\ldots; d^{-1}=0)$.

$H^0_\Phi(X,\mathscr{F}) = \Gamma_\Phi(\mathscr{F})$, and from an exact sequence of sheaves $0 \to \mathscr{F} \to \mathscr{G} \to \mathscr{H} \to 0$ we get an exact sequence $0 \to H^0_\Phi(X,\mathscr{F}) \to H^0_\Phi(X,\mathscr{G}) \to H^0_\Phi(X,\mathscr{H}) \to H^1_\Phi(X,\mathscr{F}) \to H^1_\Phi(X,\mathscr{G}) \to H^1_\Phi(X,\mathscr{H}) \to H^2_\Phi(X,\mathscr{F}) \to \ldots$.

Similarly, the cohomology groups $H^q_\Phi(X,\mathscr{F})$ can also be calculated with an exact sequence $0 \to \mathscr{F} \to \mathfrak{L}^0 \to \mathfrak{L}^1 \to \ldots$, where each $\mathfrak{L}^i$ is assumed to be Γ_Φ-acyclic (i.e., $H^q_\Phi(X,\mathfrak{L}^i)=0$ for $q>0$) instead of injective. The flabby sheaves, for instance, are Γ_Φ-acyclic, so we can compute $H^q_\Phi(X,\mathscr{F})$ by a flabby resolution of $\mathscr{F}$ (R. Godement). For example, let X be an n-dimensional †paracompact C^∞-manifold and $\mathscr{F} = \mathbf{R}$. Then $0 \to \mathbf{R} \to \mathfrak{A}^0(X) \xrightarrow{d^0} \mathfrak{A}^1(X) \xrightarrow{d^1} \ldots$ is exact, where $\mathfrak{A}^q(X)$ is the †sheaf of germs of C^∞-differential forms of degree q, d^q is †exterior differentiation (**Poincaré's theorem**), and we have $H^p(X,\mathfrak{A}^q)=0$ for $p>0$. Therefore $H^q(X,\mathbf{R})$ is the qth cohomology of the complex $0 \to D^0(X) \xrightarrow{d^0} D^1(X) \xrightarrow{d^1} \ldots$, where $D^i(X) = \Gamma(X,\mathfrak{A}^i(X))$ = the group of C^∞-differential forms of degree i on X. This proves the de Rham theorem, which says that the †de Rham cohomology group is isomorphic to the †(singular) cohomology group of X with real coefficients (→ 105 Differentiable Manifolds R). For a sheaf $\mathscr{F}$ of noncommutative groups, we can define the first cohomology $H^1(X,\mathscr{F})$ [2].

F. The Čech Cohomology Group

Let $\mathfrak{U} = \{U_i\}$ be an open covering of X, and write $U_i \cap U_j = U_{ij}$, etc. Put

$$C^p(\mathscr{F}) = \prod_{i_0,\ldots,i_p} \Gamma(U_{i_0\ldots i_p},\mathscr{F}), \qquad p=0,1,2,\ldots.$$

An element of $C^p(\mathscr{F})$ is called a cochain of degree p. Define $d:C^p(\mathscr{F}) \to C^{p+1}(\mathscr{F})$ by $(df)_{i_0\ldots i_{p+1}} = \sum_{r=0}^{p+1}(-1)^r (f_{i_0\ldots\hat{i}_r\ldots i_{p+1}} \mid U_{i_0\ldots i_{p+1}})$, and denote the qth cohomology of the complex $(C^p(\mathscr{F}),d)$ thus obtained by $H^q(\mathfrak{U},\mathscr{F})$. When an open covering $\mathfrak{B}$ is a refinement of $\mathfrak{U}$, there is a canonical homomorphism $H^q(\mathfrak{U},\mathscr{F}) \to H^q(\mathfrak{B},\mathscr{F})$. So we can take the inductive limit of the groups $H^q(\mathfrak{U},\mathscr{F})$ with respect to the refinement of open coverings. This limit group is denoted by $\check{H}^q(X,\mathscr{F})$ and is called the **Čech cohomology group with coefficient sheaf** $\mathscr{F}$. It coincides with $H^q(X,\mathscr{F})$ for $q \leqslant 1$, and if X is paracompact, for all q.

G. Relation to Continuous Mappings

Let X and Y be topological spaces and $f:X \to Y$ be a continuous mapping. If $\mathscr{F}$ is a sheaf on Y, the fiber product $X \times_Y \mathscr{F}$ (where $\mathscr{F}$ is viewed as a sheaf space over Y) is a sheaf on X. It is denoted by $f^*(\mathscr{F})$ or $f^{-1}(\mathscr{F})$ and is called the **inverse image** of $\mathscr{F}$. The correspondence $\mathscr{F} \to f^*(\mathscr{F})$ is an exact functor from $\mathscr{C}^Y$ to $\mathscr{C}^X$. Next, let $\mathscr{G}$ be a sheaf on X. Associating $\Gamma(f^{-1}(U),\mathscr{G})$ with each open set U of Y, we obtain a sheaf on Y, which we denote by $f_*(\mathscr{G})$ and call the **direct image** of $\mathscr{G}$. The correspondence f_* is a left-exact functor $\mathscr{C}^X \to \mathscr{C}^Y$, and we can consider its right derived functors R^qf_*. The sheaf $R^qf_*(\mathscr{G})$ is the sheaf associated with the presheaf that associates $H^q(f^{-1}(U),\mathscr{G})$ with each open set U.

A homomorphism ψ from $\mathscr{F}$ to $f_*(\mathscr{G})$ is also called an f-homomorphism from $\mathscr{F}$ to $\mathscr{G}$. To give such a ψ is equivalent to giving a family of homomorphisms of the stalks $\psi_x:\mathscr{F}_{f(x)} \to \mathscr{G}_x$ $(x \in X)$ satisfying the continuity condition: For any open set U of Y and any section $s \in \Gamma(U,F)$ over U, the mapping φ from $f^{-1}(U)$ to $\mathscr{G}$ defined by $\varphi(x) = \psi_x(s(f(x)))$ is continuous.

The functors f^* and f_* are related by

$\mathrm{Hom}(f^*(\mathscr{F}),\mathscr{G})\cong\mathrm{Hom}(\mathscr{F},f_*(\mathscr{G}))$. The Leray †spectral sequence

$$E_2^{p,q}=H^p(Y,R^qf_*(\mathscr{G}))\Rightarrow H^n(X,\mathscr{G})$$

exists and connects the cohomologies of X and of Y.

H. Ringed Spaces

Let X be a topological space and $\mathscr{O}$ be a sheaf on X of commutative rings with unity element such that $\mathscr{O}_x\neq\{0\}$ for any $x\in X$. Then the pair $(X,\mathscr{O})$ is called a **ringed space**, and $\mathscr{O}$ is called its **structure sheaf**. A morphism $(X,\mathscr{O})\to(X',\mathscr{O}')$ is by definition a pair (f,θ) consisting of a continuous mapping $f:X\to X'$ and an f-homomorphism $\theta:\mathscr{O}\to\mathscr{O}'$. When each $\mathscr{O}_x$ is a †local ring, $(X,\mathscr{O})$ is called a **local ringed space**. A morphism of local ringed spaces is defined to be a pair $(f,\theta):(X,\mathscr{O})\to(X',\mathscr{O}')$ as before, satisfying the additional condition that θ is local (i.e., $\theta_x:\mathscr{O}'_{f(x)}\to\mathscr{O}_x$ maps the maximal ideal into the maximal ideal for each $x\in X$). These concepts are important in algebraic geometry and the theory of functions of several complex variables.

I. Direct Products and Tensor Products

Let $\mathscr{F}_\lambda$ $(\lambda\in\Lambda)$ be sheaves of Abelian groups on a topological space X. The sheaf $\mathscr{F}$ on X defined by $\mathscr{F}(U)=\prod_\lambda\mathscr{F}_\lambda(U)$ and $r_{UV}=\prod_\lambda r_{UV}^{\mathscr{F}_\lambda}$ is denoted by $\mathscr{F}=\prod_\lambda\mathscr{F}_\lambda$ and called the **direct product** of sheaves $\{\mathscr{F}_\lambda\}$. For each $x\in X$ there is a natural mapping $\mathscr{F}_x\to\prod_\lambda(\mathscr{F}_\lambda)_x$, which is in general neither injective nor surjective. When Λ is a finite set, $\prod\mathscr{F}_i$ is also written $\mathscr{F}=\mathscr{F}_1+\ldots+\mathscr{F}_n$ and is called the **direct sum** of the sheaves. The **inductive limit** $\mathscr{F}=\mathrm{ind\,lim}\,\mathscr{F}_\lambda$ of an inductive system of sheaves on X also exists, and $\mathscr{F}_x=\mathrm{ind\,lim}\,\mathscr{F}_{\lambda,x}$.

Let $(X,\mathscr{O})$ be a ringed space. A sheaf of Abelian groups $\mathscr{F}$ on X is called a **sheaf of $\mathscr{O}$-modules** (or simply an **$\mathscr{O}$-module**) if $\mathscr{F}(U)$ is an $\mathscr{O}(U)$-module for each U and $r_{UV}:\mathscr{F}(V)\to\mathscr{F}(U)$ is a module homomorphism compatible with $\mathscr{O}(V)\to\mathscr{O}(U)$ for each $U\subset V$. Then $\mathscr{F}_x$ is an $\mathscr{O}_x$-module for each $x\in X$. For a fixed $(X,\mathscr{O})$, the $\mathscr{O}$-modules form an Abelian category. When $\mathscr{F}$ and $\mathscr{G}$ are $\mathscr{O}$-modules, the **tensor product** $\mathscr{H}=\mathscr{F}\otimes_{\mathscr{O}}\mathscr{G}$ of $\mathscr{F}$ and $\mathscr{G}$ as sheaves over $\mathscr{O}$ is defined as follows: Define a presheaf by $U\to\mathscr{F}(U)\otimes_{\mathscr{O}(U)}\mathscr{G}(U)$ and $r_{UV}=r_{UV}^{\mathscr{F}}\otimes r_{UV}^{\mathscr{G}}$, and let $\mathscr{H}$ be the associated sheaf of this presheaf. Then we have $\mathscr{H}_x=\mathscr{F}_x\otimes_{\mathscr{O}_x}\mathscr{G}_x$.

The notion of coherent sheaves is important in the theory of $\mathscr{O}$-modules (→ 16 Algebraic Varieties E).

J. History

About 1945, J. Leray established the theory of sheaf coefficient cohomology groups (in a form slightly different from that in Sections E and F) and the theory of spectral sequences to study the relation between the local properties of a continuous mapping and the global cohomologies. In the theory of functions of several complex variables, K. Oka conceived the idea of "ideals of indefinite domain." These two ideas were unified by H. Cartan into the present form of sheaf theory. As a link between local properties and global properties, sheaf theory has been applied in many branches of mathematics (→ 16 Algebraic Varieties; 21 Analytic Functions of Several Complex Variables; 23 Analytic Spaces; 72 Complex Manifolds).

References

[1] F. Hirzebruch, Neue topologische Methoden in der algebraischen Geometrie, Erg. Math., Springer, second edition, 1962; English translation, Topological methods in algebraic geometry, Springer, third enlarged edition, 1966.
[2] A. Grothendieck, Sur quelques points d'algèbre homologique, Tôhoku Math. J., (2) 9 (1957), 119–221.
[3] R. Godement, Topologie algébrique et théorie des faisceaux, Actualités Sci. Ind., 1252, Hermann, 1958.
[4] R. G. Swan, The theory of sheaves, Univ. of Chicago Press, 1964.
[5] G. E. Bredon, Sheaf theory, McGraw-Hill, 1967.
[6] J. Leray, L'anneau spectral et l'anneau filtré d'homologie d'un espace localement compact et d'une application continue, J. Math. Pure Appl., 29 (1950), 1–139.
[7] H. Cartan, Sém. de topologie algébrique, Paris, 1948–1949; 1950–1951.
[8] A. Borel, Sém. de topologie algébrique de l'Ecole polytechnique fédérale, 1951. (Cohomologie des espaces localement compacts d'après J. Leray, Lecture notes in math. 2, Springer, 1964.)

384 (VII.20) Siegel Domains

A. Siegel Domains

Let D be a bounded domain in $\mathbf{C}^n$ and $G_h(D)$ the full †holomorphic automorphism group of

D, which is a †Lie transformation group with respect to the †compact-open topology. If $G_h(D)$ acts transitively on D, then D is called a †**homogeneous bounded domain**. The study of homogeneous bounded domains was initiated systematically by E. Cartan in 1936, while the notion of **Siegel domains**, which was introduced by I. I. Pyatetskiĭ-Shapiro, has made remarkable contributions to the study of homogeneous bounded domains.

Let V be a convex domain in an n-dimensional real vector space R. V is called a **regular cone** if for every $x \in V$ and $\lambda > 0$, $\lambda x \in V$ and if V contains no entirely straight lines. Let W be a complex vector space. A mapping F: $W \times W \rightarrow R^{\mathbf{C}}$ (the †complexification of R) is a V-**Hermitian form** if the following conditions are satisfied: (Fi) $F(u, v)$ is $\mathbf{C}$-linear in u, (Fii) $\overline{F(u, v)} = F(v, u)$, where the bar denotes the †conjugation with respect to R, (Fiii) $F(u, u) \in \bar{V}$ (the closure of V), and (Fiv) $F(u, u) = 0$ implies $u = 0$. Given a regular cone $V \subset R$ and a V-Hermitian form F on W, one can define a **Siegel domain** $D(V, F)$ (**of the second kind**) by putting $D(V, F) = \{(x+iy, u) \in R^{\mathbf{C}} \times W \mid y - F(u, u) \in V\}$, which is holomorphically equivalent to a bounded domain in $R^{\mathbf{C}} \times W$. When $W = (0)$, $D(V, F)$ is reduced to $D(V) = \{x + iy \in R^{\mathbf{C}} \mid y \in V\}$, which is called a **Siegel domain of the first kind**. A mapping $L: W \times W \rightarrow R^{\mathbf{C}}$ is a nondegenerate semi-Hermitian form if L can be written as $L = L_1 + L_2$, where L_1 and L_2 are $R^{\mathbf{C}}$-valued functions satisfying the conditions: (Li) L_1 satisfies (Fi) and (Fii); (Lii) L_2 is a symmetric $\mathbf{C}$-bilinear form; (Liii) $L(u, v) = 0$ for all $u \in W$ implies $v = 0$. Let B be a bounded domain in a complex vector space X and L_p ($p \in B$) an $R^{\mathbf{C}}$-valued nondegenerate semi-Hermitian form on W depending differentiably on $p \in B$. Consider a domain $D(V, L, B)$ in $R^{\mathbf{C}} \times W \times X$ defined by putting $D(V, L, B) = \{(x + iy, u, p) \in R^{\mathbf{C}} \times W \times X \mid y - \operatorname{Re} L_p(u, u) \in V, p \in B\}$. The domain $D(V, L, B)$ is called a **Siegel domain of the third kind** over B if it is holomorphically equivalent to a bounded domain. $D(V, L, B)$ is a fiber space over B.

By the affine automorphism group G_a of a Siegel domain $D(V, F) \subset R^{\mathbf{C}} \times W$ we mean the group consisting of all elements in the complex affine transformation group of $R^{\mathbf{C}} \times W$ leaving $D(V, F)$ stable. The full holomorphic automorphism group G_h of $D(V, F)$ contains G_a as a closed subgroup. If G_h acts transitively on $D(V, F)$, then $D(V, F)$ is said to be homogeneous. A homogeneous Siegel domain is necessarily affinely homogeneous, i.e., G_a acts transitively on $D(V, F)$ [2]. The †Bergman metric of $D(V, F)$ which is a G_h-invariant †Kähler metric, is †complete [3], and so $D(V, F)$ is a †domain of holomorphy.

Examples. $H(n, \mathbf{R})$ denotes the vector space of all real symmetric matrices of degree n, and $H^+(n, \mathbf{R})$ the regular cone consisting of all positive definite matrices in $H(n, \mathbf{R})$.

(i) The Siegel domain of the first kind $D(H^+(n, \mathbf{R})) = \{X + iY \mid X \in H(n, \mathbf{R}), Y \in H^+(n, \mathbf{R})\}$ is called the †Siegel upper half-plane, which is holomorphically equivalent to the classical †symmetric domain of type III.

(ii) Let $u, v \in \mathbf{C}$ and $F(u, v)$ be the 2×2 diagonal matrix $\operatorname{diag}(u\bar{v}, 0)$, which is an $H^+(2, \mathbf{R})$-Hermitian form on $\mathbf{C}$. The resulting Siegel domain is

$$D(H^+(2, \mathbf{R}), F) = \left\{ (z_1, z_2, z_3, u) \in \mathbf{C}^4 \;\middle|\; \begin{pmatrix} \operatorname{Im} z_1 - |u|^2 & \operatorname{Im} z_3 \\ \operatorname{Im} z_3 & \operatorname{Im} z_2 \end{pmatrix} \in H^+(2, R) \right\}.$$

(iii) Let $B = \{t \in \mathbf{C} \mid |t| < 1\}$, and let $u, v \in \mathbf{C}$. Put $L_t(u, v) = (1 - |t|^2)^{-1}(u\bar{v} + \bar{t}uv)$. Then $L_t(u, v)$ is a nondegenerate semi-Hermitian form, and we have $D(H^+(2, \mathbf{R}), L, B) = \{(z, u, t) \in \mathbf{C}^3 \mid \operatorname{Im} z - (1 - |t|^2)^{-1} \operatorname{Re}(|u|^2 + \bar{t}u^2) > 0, |t| < 1\}$, which is a Siegel domain of the third kind and is holomorphically equivalent to the Siegel upper half-plane of dimension 3.

The domains in (i) and (ii) are both affinely homogeneous; the latter was originally found by Pyateteskiĭ-Shapiro in 1959 [1] and provides the least-dimensional example of non-symmetric homogeneous bounded domains, which answered affirmatively Cartan's conjecture (1936): Are there non-†symmetric homogeneous bounded domains in $\mathbf{C}^n$ ($n \geqslant 4$)?

B. Infinitesimal Automorphisms of Siegel Domains

For a Siegel domain $D(V, F) \subset R^{\mathbf{C}} \times W$, the Lie algebra $\mathfrak{g}_h$ of G_h can be identified with the Lie algebra of infinitesimal automorphisms, i.e., all complete holomorphic vector fields on $D(V, F)$. Let $G(V)$ be the group consisting of all the linear automorphisms of R leaving V stable. Let us fix a base in R, and let $(z_1, z_2, \ldots, z_n)$ be the complex linear coordinate system in $R^{\mathbf{C}}$ corresponding to it. Choose a complex linear coordinate system $(u_1, u_2, \ldots, u_m)$ in W. We write $F(u, v)$ as $F(u, v) = (F_1(u, v), \ldots, F_n(u, v))$. Consider the following two vector fields in the Lie algebra $\mathfrak{g}_a$ of G_a:

$$E = 2\sum_k z_k \frac{\partial}{\partial z_k} + \sum_\alpha u_\alpha \frac{\partial}{\partial u_\alpha}, \qquad I = \sum_\alpha i u_\alpha \frac{\partial}{\partial u_\alpha},$$

and put $\mathfrak{g}_a^\lambda = \{X \in \mathfrak{g}_a \mid [E, X] = \lambda X\}$, $\lambda \in \mathbf{Z}$. Then $\mathfrak{g}_a$ can be written as a †graded Lie algebra in the following way: $\mathfrak{g}_a = \mathfrak{g}_a^{-2} + \mathfrak{g}_a^{-1} + \mathfrak{g}_a^0$. Here we have

$$\mathfrak{g}_a^{-2} = \left\{ \sum_k a_k \frac{\partial}{\partial z_k} \,\middle|\, a_k \in \mathbf{R} \right\},$$

$$\mathfrak{g}_a^{-1} = \left\{ 2i \sum_k F_k(u,c) \frac{\partial}{\partial z_k} + \sum_\alpha c_\alpha \frac{\partial}{\partial u_\alpha} \,\middle|\, c = (c_\alpha), \quad c_\alpha \in \mathbf{C} \right\},$$

and $\mathfrak{g}_a^0$ consists of all vector fields $X_{(A,B)}$ of the form

$$\sum_{k,j} a_{kj} z_j \frac{\partial}{\partial z_k} + \sum_{\alpha,\beta} b_{\alpha\beta} u_\beta \frac{\partial}{\partial u_\alpha},$$

where the matrices $A = (a_{kj})$, $B = (b_{\alpha\beta})$ satisfy the conditions: $\exp tA \in G(V)$, $t \in \mathbf{R}$; $AF(u,v) = F(Bu,v) + F(u,Bv)$. For $X_{(A,B)} \in \mathfrak{g}_a^0$ we define $\operatorname{tr} X_{(A,B)}$ to be the sum of the trace of A and that of B. Let $\mathfrak{g}^\lambda$ be the λ-eigenspace of $\operatorname{ad} E$ in $\mathfrak{g}_h$ ($\lambda \in \mathbf{Z}$). Then $\mathfrak{g}_h$ can be written also in the form of a graded Lie algebra: $\mathfrak{g}_h = \mathfrak{g}^{-2} + \mathfrak{g}^{-1} + \mathfrak{g}^0 + \mathfrak{g}^1 + \mathfrak{g}^2$, and $\mathfrak{g}^\lambda = \mathfrak{g}_a^\lambda$ is valid for $\lambda = -2, -1, 0$. Furthermore $\mathfrak{g}_h$ can be nicely determined by $\mathfrak{g}_a$ in the following manner. $p_{\mu\lambda}$ denotes a polynomial on $R^{\mathbf{C}} \times W$ homogeneous of degree μ in $z_1, \ldots, z_n$, and homogeneous of degree λ in $u_1, \ldots, u_m$. Let $\hat{\mathfrak{g}}^1$ (resp. $\hat{\mathfrak{g}}^2$) be the set of all polynomial vector fields of the form

$$\sum_k p_{1,1}^k \frac{\partial}{\partial z_k} + \sum_\alpha (p_{1,0}^\alpha + p_{0,2}^\alpha) \frac{\partial}{\partial u_\alpha}$$

$$\left(\text{resp. } \sum_k p_{2,0}^k \frac{\partial}{\partial z_k} + \sum_\alpha p_{1,1}^\alpha \frac{\partial}{\partial u_\alpha} \right).$$

Then we have $\mathfrak{g}^1 = \{ X \in \hat{\mathfrak{g}}^1 \mid [X, \mathfrak{g}^{-1}] \subset \mathfrak{g}^0 \}$, and $\mathfrak{g}^2 = \{ X \in \hat{\mathfrak{g}}^2 \mid [X, \mathfrak{g}^{-2}] \subset \mathfrak{g}^0, [X, \mathfrak{g}^{-1}] \subset \mathfrak{g}^1, \operatorname{Im} \operatorname{Tr}[X, Y] = 0 \text{ for } Y \in \mathfrak{g}^{-2} \}$. Another description of $\mathfrak{g}^1$ and $\mathfrak{g}^2$ has been given in terms of Jordan triple systems [4]. The explicit descriptions of $\mathfrak{g}^1$ and $\mathfrak{g}^2$ have been given for most homogeneous Siegel domains $D(V,F)$ over irreducible self-dual cones V (→ Section D; T. Tsuji, *Nagoya Math. J.*, 55 (1974)). $\mathfrak{g}_h = \mathfrak{g}_a$ is valid for the Siegel domains which are irreducible quasisymmetric but not symmetric (→ Section D). Main references for this section are [2–6].

C. *j*-Algebras and Homogeneous Bounded Domains

The notion of ***j*-algebra** was introduced by Pyatetskiĭ-Shapiro [1], which reduces the study of homogeneous bounded domains to purely algebraic problems. Let $\mathfrak{g}$ be a Lie algebra over $\mathbf{R}$, and $\mathfrak{k}$ a subalgebra of $\mathfrak{g}$, (j) a collection of linear endomorphisms of $\mathfrak{g}$, and ω be a linear form on $\mathfrak{g}$. Then the quadruple $\{\mathfrak{g}, \mathfrak{k}, (j), \omega\}$ (sometimes abbreviated $\mathfrak{g}$) is called a j-algebra if the following conditions are satisfied: (i) $j\mathfrak{k} \subset \mathfrak{k}$ for $j \in (j)$ and $j \equiv j' \pmod{\mathfrak{k}}$ for $j, j' \in (j)$; (ii) $j^2 \equiv -id \pmod{\mathfrak{k}}$; (iii) $j[k,x] \equiv [k, jx] \pmod{\mathfrak{k}}$ for $k \in \mathfrak{k}$, $x \in \mathfrak{g}$; (iv) $[jx, jy] \equiv j[jx, y] + j[x, jy] + [x, y] \pmod{\mathfrak{k}}$ for $x, y \in \mathfrak{g}$; (v) $\omega([k, x]) = 0$ for $k \in \mathfrak{k}$; (vi) $\omega([jx, jy]) = \omega([x, y])$; (vii) $\omega([jx, x]) > 0$ for $x \notin \mathfrak{k}$. Let $\mathfrak{g}'$ be a subalgebra of $\mathfrak{g}$ such that $j\mathfrak{g}' \subset \mathfrak{g}' + \mathfrak{k}$. Then, putting $\mathfrak{k}' = \mathfrak{g}' \cap \mathfrak{k}$, one can naturally induce a j-algebra structure on the pair $\{\mathfrak{g}', \mathfrak{k}'\}$. The j-algebra thus obtained is called a j-subalgebra of $\{\mathfrak{g}, \mathfrak{k}, (j), \omega\}$. A j-algebra $\{\mathfrak{g}, \mathfrak{k}, (j), \omega\}$ is called proper (resp. effective), if, for any j-subalgebra $\{\mathfrak{g}', \mathfrak{k}'\}$ with $\mathfrak{g}'$ compact semisimple, $\mathfrak{g}'$ is contained in $\mathfrak{k}$ (resp. if $\{\mathfrak{g}, \mathfrak{k}\}$ is an effective pair).

Now let D be a homogeneous bounded domain in $\mathbf{C}^n$, G a connected †Lie subgroup of $G_h(D)$ acting †transitively on D, and K the †isotropy subgroup at a point in D. The Lie algebras of G and K are denoted by $\mathfrak{g}$ and $\mathfrak{k}$, respectively. Then the pair $\{\mathfrak{g}, \mathfrak{k}\}$ becomes an effective proper j-algebra. Conversely, to every effective proper j-algebra there corresponds a homogeneous bounded domain. The identity component of $G_h(D)$ is isomorphic to the identity component of a †real algebraic group via the †adjoint representation. Let $\{\mathfrak{g}, \mathfrak{k}, (j), \omega\}$ be a j-algebra. Suppose that $\mathfrak{g}$ satisfies the following conditions: (i) $\mathfrak{g} = \mathfrak{g}^{-2} + \mathfrak{g}^{-1} + \mathfrak{g}^0$ as a graded Lie algebra; (ii) $\mathfrak{g}^0 = \mathfrak{k} + j\mathfrak{g}^{-2}$; (iii) there exists a $j \in (j)$ such that $j\mathfrak{g}^{-1} = \mathfrak{g}^{-1}$; and (iv) there exists an $r \in \mathfrak{g}^{-2}$ such that $[jx, r] = x$ for $x \in \mathfrak{g}^{-2}$. Such a decomposition is called a Siegel decomposition of $\mathfrak{g}$. To an effective j-algebra admitting a Siegel decomposition there corresponds a unique Siegel domain up to affine equivalence. Vinberg, Gindikin, and Pyatetskiĭ-Shapiro (Appendix in [1] or *Trans. Moscow Math. Soc.*, 12 (1963)) proved that the Lie algebra $\mathfrak{g}_h(D)$ of $G_h(D)$ contains a j-subalgebra admitting a Siegel docomposition and corresponding to the same domain D, and obtained the realization theorem: Every homogeneous bounded domain D is holomorphically equivalent to a Siegel domain. In consequence, D is diffeomorphic to a Euclidean space, and the isotropy subgroup $K_h(D)$ is a maximal compact subgroup of $G_h(D)$. We have the decomposition $G_h(D) = K_h(D) \cdot T$ (semidirect), where T is an $\mathbf{R}$-splittable solvable subgroup of $G_h(D)$ acting simply transitively on D. T is uniquely determined up to conjugacy in $G_h^0(D)$ (= the identity component of $G_h(D)$), and is called the **Iwasawa group** of D.

The j-algebra structure of the Lie algebra $\mathfrak{t}$ of the Iwasawa group T is characterized by the following properties: (i) for every $t \in \mathfrak{t}$, the eigenvalues of $\operatorname{ad} t$ are all real; (ii) there exists a †complex structure j such that $[jx, jy] =$

$j[jx, y]+j[x, jy]+[x, y]$ for $x, y \in \mathfrak{t}$; and (iii) there exists a linear form ω on $\mathfrak{t}$ such that $\omega([jx, jy])=\omega([x, y])$ and that $\omega([jx, x])>0$ for $x \neq 0$. A Lie algebra satisfying (i)–(iii) is called a **normal *j*-algebra**. There exists a one-to-one correspondence between the set of holomorphic equivalence classes of homogeneous bounded domains and the set of j-isomorphism classes of normal j-algebras; by a j-isomorphism here we mean an isomorphism which commutes with j. Let $\{\mathfrak{g}, j, \omega\}$ be a normal j-algebra and define an inner product $\langle\ ,\ \rangle$ on $\mathfrak{g}$ by $\langle x, y\rangle=\omega([jx, y])$. The orthogonal complement $\mathfrak{h}$ with respect to $\langle\ ,\ \rangle$ of the †commutator subalgebra $\mathfrak{g}_1$ of $\mathfrak{g}$ is an Abelian subalgebra of $\mathfrak{g}$, and the adjoint representation of $\mathfrak{h}$ on $\mathfrak{g}_1$ is fully reducible. One has $\mathfrak{g}=\sum_\alpha \mathfrak{k}_\alpha$, $\mathfrak{h}=\mathfrak{k}_0$, and $\mathfrak{g}_1=\sum_{\alpha \neq 0} \mathfrak{k}_\alpha$, where $\mathfrak{k}_\alpha=\{x \in \mathfrak{g} \mid [h, x]=\alpha(h)x, h \in \mathfrak{h}\}$. The linear form α on $\mathfrak{h}$ is called a root of $\mathfrak{g}$. There exist l roots $\alpha_1, \ldots, \alpha_l$ ($l=\dim \mathfrak{h}$) such that $\mathfrak{h}$ can be written in the form $\mathfrak{h}=j\mathfrak{k}_{\alpha_1}+\ldots+j\mathfrak{k}_{\alpha_l}$, l being the **rank** of $\mathfrak{g}$. Then, after a suitable change of the numbering of the α_i's, any root α will be seen to be of the form $(\alpha_i+\alpha_k)/2$, $(\alpha_i-\alpha_k)/2$ or $\alpha_i/2$, where $1 \leqslant i \leqslant k \leqslant l$. A normal j-algebra admits a unique Siegel decomposition which can be constructed by using root spaces.

D. Equivariant Holomorphic Embedding

We retain the notation of Section B. Let $D(V, F) \subset R^{\mathbf{C}} \times W$ be a Siegel domain and $\mathfrak{g}_{\mathbf{C}}$ be the †complexification of the Lie algebra $\mathfrak{g}_h$. $\mathfrak{g}^{-1}$ has the complex structure defined by the endomorphism $\operatorname{ad} I$. Let $\mathfrak{g}_{\pm}^{-1}$ be the $\pm i$-eigenspaces in the complexification $g_{\mathbf{C}}^{-1}$ of g^{-1} under $\operatorname{ad} I$. Let us consider the complex subalgebras $\mathfrak{b}=\mathfrak{g}_{-}^{-1}+\mathfrak{g}_{\mathbf{C}}^0+\mathfrak{g}_{\mathbf{C}}^1+\mathfrak{g}_{\mathbf{C}}^2$ and $\mathfrak{n}=\mathfrak{g}_{\mathbf{C}}^{-2}+\mathfrak{g}_{+}^{-1}$ of $\mathfrak{g}_{\mathbf{C}}$, where the subscripts $\mathbf{C}$ denote the complexification of the respective space. Let $G_{\mathbf{C}}$ be the connected †complex Lie group generated by $\mathfrak{g}_{\mathbf{C}}$ and containing G_h^0 (—the identity component of G_h) as a subgroup. The Lie algebra of the normalizer B of $\mathfrak{b}$ in $G_{\mathbf{C}}$ coincides with $\mathfrak{b}$. Identifying $R^{\mathbf{C}} \times W$ with $\mathfrak{n}$ as a complex vector space, and denoting by π the natural projection of $G_{\mathbf{C}}$ onto the complex coset space $G_{\mathbf{C}}/B$, the composite mapping $\tau=\pi \exp$ is a holomorphic embedding of $\mathfrak{n}$ into $G_{\mathbf{C}}/B$, which induces a holomorphic G_h^0-†equivariant embedding of $D(V, F)$ into $G_{\mathbf{C}}/B$ as an open submanifold. This embedding is called the **Tanaka embedding**. By the **(Shilov) boundary** S of $D(V, F)$ we mean the real submanifold $S=\{(x+iy, u) \in R^{\mathbf{C}} \times W \mid y=F(u, u)\}$ of $R^{\mathbf{C}} \times W$, which is a subset of the boundary of $D(V, F)$. S has the natural †CR-structure induced from the complex structure of $R^{\mathbf{C}} \times W$. Every element of $\mathfrak{g}_h$ can be extended to a unique holomorphic vector field on $R^{\mathbf{C}} \times W$ which is tangent to S, and its restriction to S is an infinitesimal †CR-automorphism on S, i.e., a complete vector field generating a 1-parameter group of †CR-equivalences of S onto itself. Conversely, every infinitesimal CR-automorphism on S can be extended uniquely to a holomorphic vector field on $R^{\mathbf{C}} \times W$, and an element of $\mathfrak{g}_h$ is characterized as an infinitesimal CR-automorphism on S whose extension leaves the †Bergman kernel form of $D(V, F)$ invariant [6]. Let $n=\dim_{\mathbf{C}} \mathfrak{g}_{\mathbf{C}}$, $m=\dim_{\mathbf{C}} \mathfrak{b}$, and let $k=\binom{n}{m}-1$. Then $G_{\mathbf{C}}/B$, and consequently $D(V, F)$, is embedded holomorphically into the complex †Grassmann manifold of m-dimensional subspaces in $\mathfrak{g}_{\mathbf{C}}$ and so into the †complex projective space $P_k(\mathbf{C})$. Any element of G_h^0 is induced from a projective transformation and hence is a birational transformation on $D(V, F)$.

Let D be a homogeneous bounded domain in $\mathbf{C}^n$, $\mathfrak{g}_h$ the Lie algebra of $G_h(D)$, and $\mathfrak{k}_h$ the isotropy subalgebra of $\mathfrak{g}_h$; and let $\mathfrak{g}_{\mathbf{C}}$ be the complexification of $\mathfrak{g}_h$. $\mathfrak{g}_h$ is a j-algebra. Let us define the complex subalgebra $\mathfrak{g}_-$ of $\mathfrak{g}_{\mathbf{C}}$ by putting $\mathfrak{g}_-=\{x+ijx \mid x \in \mathfrak{g}_h, j \in (j)\}$. Then we have $\mathfrak{g}_{\mathbf{C}}=\mathfrak{g}_h+\mathfrak{g}_-$, $\mathfrak{g}_h \cap \mathfrak{g}_-=\mathfrak{k}_h$. Let $G_{\mathbf{C}}$ be the connected Lie group generated by $\mathfrak{g}_{\mathbf{C}}$ and containing $G_h^0(D)$ as a subgroup. Let G_- be the connected (closed) Lie subgroup of $G_{\mathbf{C}}$ generated by $\mathfrak{g}_-$. Then D can be holomorphically embedded in $G_{\mathbf{C}}/G_-$ as the open $G_h^0(D)$-orbit of the origin of $G_{\mathbf{C}}/G_-$ [8]. This embedding is called the **generalized Borel embedding**. $G_{\mathbf{C}}/G_-$ is compact if and only if D is symmetric, and in this case $G_{\mathbf{C}}/G_-$ coincides with the compact dual [9].

Let $\{\mathfrak{t}_0, j, \omega\}$ be a normal j-algebra of rank l corresponding to a homogeneous bounded domain D_0, and define the Hermitian inner product h by $h(x, y)=\omega([jx, y])+i\omega([x, y])$ for $x, y \in \mathfrak{t}_0$. $\mathfrak{t}_0$ has $l-1$ (normal) nontrivial j-ideals (i.e., j-invariant ideals) up to j-isomorphisms. Take a j-ideal $\mathfrak{t}_1$ of $\mathfrak{t}_0$. Then we have $\mathfrak{t}_0=\mathfrak{t}_1+\mathfrak{t}_2$, $\mathfrak{t}_2$ being a (normal) j-subalgebra of $\mathfrak{t}_0$ defined as the orthogonal complement of $\mathfrak{t}_1$ in $\mathfrak{t}_0$ with respect to h. The geometric version of this is that D_0 is represented as a holomorphic fiber space over the homogeneous bounded domain D_2 corresponding to $\mathfrak{t}_2$, with fibers holomorphically equivalent to the homogeneous bounded domain D_1 corresponding to $\mathfrak{t}_1$. For this fibering there exists a universal fiber space D over the product $\hat{D}_2$ of certain classical symmetric domains, with the same fibers, which plays the same role as that of a †universal fiber bundle in topology. Here, D is again a homogeneous bounded domain. The fiber space $D_0 \to D_2$ is induced from the fiber space $D \to \hat{D}_2$ by the classifying mapping λ of D_2 to $\hat{D}_2$. Let β be the generalized Borel embedding of D into $G_{\mathbf{C}}/G_-$. Then there exists a complex Abelian subalgebra $\mathfrak{m}$ of $\mathfrak{g}_{\mathbf{C}}$ satisfying $\mathfrak{g}_{\mathbf{C}}=\mathfrak{g}_{-}+\mathfrak{m}$ (semidirect), and one can construct a biholomorphic mapping f of $\beta(D)$ onto a certain Siegel domain of the third kind over D_2 in the vector space $\mathfrak{m}$ [8]. The fiber space $D_0\rightarrow D_2$ coincides with the one induced from the aforementioned Siegel domain of the third kind by the composite mapping of λ and $f\cdot\beta$. Every realization of a homogeneous bounded domain as a Siegel domain of the third kind is obtained by this method.

E. Classification of Homogeneous Bounded Domains

The main concern is to classify all homogeneous bounded domains in $\mathbf{C}^n$ up to holomorphic equivalence. Since the realization as Siegel domains has been set up, the second step is to get the uniqueness theorem: The holomorphic equivalence of two homogeneous Siegel domains implies that they are linearly equivalent, that is, there exists a (complex) linear isomorphism between the ambient vector spaces which carries the one domain to the other. The uniqueness theorem was first stated in 1963 (Appendix in [1]), rigorously proved in 1967 [10], and in 1970 the homogeneity assumption was removed [2]. A homogeneous Siegel domain is called **irreducible** if it is not holomorphically equivalent to a product of two homogeneous Siegel domains. Every homogeneous Siegel domain is linearly equivalent to a product of irreducible homogeneous Siegel domains [3, 10]. A homogeneous Siegel domain $D(V,F)$ is irreducible if and only if the regular cone V is irreducible, i.e., if V cannot be written as a direct sum of two regular cones. So the problem is to classify irreducible homogeneous Siegel domains up to linear equivalence. This reduces to classifying two kinds of nonassociative algebras with bigradation, called T-algebras and S-algebras [11]. Nonsymmetric homogeneous Siegel domains appear in dimension 4. The numbers of such domains are finite up to dimension 6, but in every dimension $\geqslant 7$ there is at least one continuous family of nonsymmetric irreducible homogeneous Siegel domains, which are not mutually holomorphically equivalent.

There is a remarkable class of homogeneous Siegel domains, called **quasisymmetric** [12], which contains the class of symmetric bounded domains. A regular cone $V\subset R$ is called **self-dual** if there exists an inner product $(\ ,\ )$ on R such that $V=\{x\in R\,|\,(x,y)>0$ for $y\in\bar{V}-(0)\}$, $\bar{V}$ denoting the closure of V. V is called homogeneous if the group $G(V)$ is transitive on V. Suppose that V is homogeneous self-dual. Then the group $G(V)$ is †self-adjoint with respect to the inner product $(\ ,\ )$ on R. Let $\mathfrak{g}(V)$ be the Lie algebra of $G(V)$. Then the totality $\mathfrak{k}(V)$ of skew-symmetric operators in $\mathfrak{g}(V)$ with respect to $(\ ,\ )$ is a †maximal compact subalgebra of $\mathfrak{g}(V)$ and is the isotropy subalgebra of $\mathfrak{g}(V)$ at a point $e\in V$. Consider the associated †Cartan decomposition $\mathfrak{g}(V)=\mathfrak{k}(V)+\mathfrak{p}(V)$. For each $x\in R$ there exists a unique element $T(x)\in\mathfrak{p}(V)$ such that $T(x)e=x$. Let F be a V-Hermitian form on a complex vector space W. Define a Hermitian inner product $\langle\ ,\ \rangle$ on W by $\langle u,v\rangle=(e,F(u,v))$ for u, $v\in W$, and let $H(W)$ be the set of Hermitian operators on W with respect to $\langle\ ,\ \rangle$. A (homogeneous) Siegel domain $D(V,F)\subset R^{\mathbf{C}}\times W$ is called quasisymmetric if V is homogeneous self-dual and if for each $x\in R$ there exists $R(x)\in H(W)$ such that $F(R(x)u,v)+F(u,R(x)v)=T(x)F(u,v)$ for u, $v\in W$. The normal j-algebra $\mathfrak{t}$ of an irreducible quasisymmetric Siegel domain is characterized by the following conditions: $\dim\mathfrak{k}_{(\alpha_i+\alpha_k)/2}=a$ $(1\leqslant i\leqslant k\leqslant l)$; and $\dim k_{\alpha_i/2}=b$ $(1\leqslant i\leqslant l)$, where a, b are some constants and l is the rank of $\mathfrak{t}$ (D'Atri and de Miatello). Quasisymmetric Siegel domains have been completely classified (M. Takeuchi, *Nagoya Math. J.*, 59 (1975), also [12]).

F. Generalized Siegel Domains and Further Results

Let Ω be a domain in $\mathbf{C}^n\times\mathbf{C}^m$ $(n,m\geqslant 0)$ which is holomorphically equivalent to a bounded domain and contains a point of the form $(z,0)$, $z\in\mathbf{C}^n$. Ω is called a **generalized Siegel domain** with exponent c $(c\in\mathbf{R})$, if Ω is invariant under holomorphic transformations of $\mathbf{C}^n\times\mathbf{C}^m$ of the types

$$(z,u)\mapsto(z+a,u)\quad\text{for all}\quad a\in\mathbf{R}^n,$$

$$(z,u)\mapsto(z,e^{it}u)\quad\text{for all}\quad t\in\mathbf{R},$$

$$(z,u)\mapsto(e^{t}z,e^{ct}u)\quad\text{for all}\quad t\in\mathbf{R}.$$

Let D be a bounded domain in $\mathbf{C}^n$, and Γ a subgroup of $G_h(D)$. Γ is said to **sweep** D if there exists a compact set $K\subset D$ such that $\Gamma K=D$. Γ is said to **divide** D if Γ, provided with discrete topology, acts properly on D and sweeps D. D is called **sweepable** (resp. **divisible**) if there exists a subgroup Γ of $G_h(D)$ which sweeps (resp. divides) D. A divisible generalized Siegel domain is symmetric. A sweepable generalized Siegel domain with exponent $c\neq 0$ (resp. $c=0$) is a Siegel domain (resp. a product of a Siegel domain of the first kind and of a homogeneous bounded circular domain) ([13]; also A. Kodama, *J. Math. Soc. Japan*, 33 (1981)).

Some results have been obtained concerning geometry of bounded domains, homogeneous bounded domains, and Siegel domains in complex Banach spaces [14], and also concerning the unitary representations of the generalized Heisenberg group on the square-integrable cohomology spaces of $\bar{\partial}_b$-complexes on the Shilov boundary of a Siegel domain [15].

References

[1] I. I. Pyatetskiĭ-Shapiro, Automorphic functions and the geometry of classical domains, Gordon & Breach, 1969.
[2] W. Kaup, Y. Matsushima, and T. Ochiai, On the automorphisms and equivalences of generalized Siegel domains, Amer. J. Math., 92 (1970), 267–290.
[3] K. Nakajima, Some studies on Siegel domains, J. Math. Soc. Japan, 27 (1975), 54–75.
[4] J. Dorfmeister, Homogene Siegel-Gebiete, Habilitation, Univ. Münster, 1978.
[5] S. Murakami, On automorphisms of Siegel domains, Lecture notes in math. 286, Springer, 1972.
[6] N. Tanaka, On infinitesimal automorphisms of Siegel domains, J. Math. Soc. Japan, 22 (1970), 180–212.
[7] S. G. Gindikin, E. B. Vinberg, and I. I. Pyatetskiĭ-Shapiro, Homogeneous Kähler manifolds, Geometry of Homogeneous Bounded Domains, C.I.M.E. 1967; Cremonese, 1968, 3–87.
[8] S. Kaneyuki, Homogeneous bounded domains and Siegel domains, Lecture notes in math. 241, Springer, 1971.
[9] K. Nakajima, On Tanaka's embeddings of Siegel domains, J. Math. Kyoto Univ., 14 (1974), 533–548.
[10] S. Kaneyuki, On the automorphism groups of homogeneous bounded domains, J. Fac. Sci. Univ. Tokyo, 14 (1967), 89–130.
[11] M. Takeuchi, Homogeneous Siegel domains, Publ. study group geometry 7, Kyoto, 1973.
[12] I. Satake, Algebraic structure of symmetric domains, Iwanami and Princeton Univ. Press, 1980.
[13] J. Vey, Sur la division des domaines de Siegel, Ann. Sci. Ecole Norm. Sup., 3 (1970), 479–506.
[14] J. P. Vigué, Le groupe des automorphismes analytiques d'un domaine borné d'un espace de Banach complexe, Ann. Sci. Ecole Norm. Sup., 9 (1976), 203–282.
[15] H. Rossi and M. Vergne, Group representations on Hilbert spaces defined in terms of $\bar{\partial}_b$-cohomology on the Shilov boundary of a Siegel domains, Pacific J. Math., 65 (1976), 193–207.

385 (XVI.6) Simulation

A. General Remarks

Simulation, in its widest sense, is a method of utilizing models to study the nature of certain phenomena. This method is employed when experimentation with the actual phenomena in question is difficult because of high cost in time or money. Also, it is sometimes almost impossible to carry out observations when the behavior of the objects can be influenced by their surroundings.

We can classify simulation techniques into the following four types, although simulations in practical use are usually a mixture of them.

The first type is **model experimentation**, which includes model basins and wind tunnels in hydrodynamics and pilot plants in the chemical industry. In advance of construction in a real situation, we perform experiments on a small scale and verify or modify those theories upon which the construction is based.

The second type is **analog simulation** or **experimental analysis**. We investigate the properties of real objects by experiments on alternative phenomena satisfying the same differential equations as those known for or assumed to be satisfied by the real objects. For example, we use an equivalent electric network to study dynamic vibration, and dynamic systems to study heat conduction problems. When theoretical analysis of the actual phenomenon is difficult, we look for other phenomena with similar properties and study them in order to construct mathematical models for them. This type of simulation has come into practical use mainly in engineering problems, but recently it has been utilized for the study of economic phenomena, nervous systems, the circulating system of an artificial heart, etc. Analog simulation was in the past often performed by means of †analog computers. Nowadays, analog simulation is more frequently performed by digital computers than by analog ones. And with the progress of electronics it has become easier to make special-purpose simulators.

The third method of simulation, **simulation in the narrow sense**, has become more important as †digital computers have been developed. In general this method is applied to problems that are more complicated and of larger scale than problems treated by analog simulation. When the mathematical expressions of the phenomenon and the algorithms of its dynamic structure are known, it is easy to simulate it by means of a computer program. In particular, when these techniques are used to study systems such as sets of machines, equipment at factories, or management organization, we call them **system simulations**. Major fields where system simulation techniques have been used are traffic control on highways or at airports, arrangement or operation of machines at factories, balancing problems in chemical processes, production scheduling in connection with demands and stocks, overall management problems, and design of information systems. The method has also been applied in designing plants and highways and in the study of social or biological phenomena. Also, when we investigate instruction systems of computers that are yet to be completed or develop programming systems for such computers, existing computers can be used to simulate the new ones. †Random numbers play an important role where the simulation must include random fluctuations (→ 354 Random Numbers). In such instances, the method is often called the Monte Carlo method (→ Section C).

The fourth method of simulation deals with systems containing human beings. Among them are war games for training in military-operation planning, business games for training in business enterprises, and simulators for training pilots and operators of atomic power plants. The contribution of human decision to simulation processes is characteristic of these cases. For example, the participants in a business game are divided into several enterprise groups. Each group discusses and decides how to invest in plants, equipment, research, and advertising and how to schedule production for each quarter. On the basis of the decisions, a computer outputs the records of sales, stocks, and cash for each quarter, according to hidden rules. From the results, each group decides on the next steps. In this way, the groups compete for development. This type of simulation is important not only for training but also for investigating the mechanism of human decision. Slightly different from this type of simulation, the "perceptron" and EPAM (Elementary Perceiver and Memorizer) are related to artificial intelligence and have been used extensively in the cognitive sciences and in research into the structure and function of the human brain.

The third type of simulation has attracted attention in particular and has been used both in theoretical problems, such as the explication of various phenomena, and in practical problems, such as design or optimum operation of systems or prediction of their behavior. For school education and training of technicians, this is put to use together with simulations of the fourth type. But their use has also induced heated discussions and controversies on the validity of results.

B. Programming Languages for Simulation

We usually describe models by using general purpose language: such as FORTRAN, or list processing languages such as LISP to simulate situations on computers. For system simulation, a number of programming languages have been developed and put to practical use. They can be divided roughly into two categories: those for which systems change continuously and those describing discrete changes. CSMP (Continuous System Modeling Program), CSSI (Continuous System Simulation Language), and DDS (Digital Dynamics Simulator) belong to the former, and hence all involve integration mechanisms; but each has a different way of describing a model. DYNAMO, which has been implemented, or J. W. Forrester's Industrial Dynamics and World Dynamics, are used extensively. To control simulation time, one may use GPSS (General Purpose Simulation System), SIMULA (Simulation Language), or SIMSCRIPT (Simulation Scriptor), each employing a different method to describe state transitions.

C. The Monte Carlo Method

The **Monte Carlo method** was introduced by J. von Neumann and S. M. Ulam around 1945. They defined this as a method of solving deterministic mathematical problems using †random numbers. L. de Buffon's needle experiment, in which the approximate value of π is obtained by dropping needles at random many times, is a classical example of this method.

Another example is the problem of evaluating a definite integral $I=\int_a^b f(x)\,dx$ $(B\geqslant f(x)\geqslant A\geqslant 0)$. First we generate many pairs of (uniform) random numbers (x,y), where $a\leqslant x\leqslant b$ and $A\leqslant y\leqslant B$. The proportion (p) of pairs satisfying $y\leqslant f(x)$ gives an estimate of the integral, i.e., $I\fallingdotseq p(B-A)(b-a)$. The techniques for inverting matrices, solving †boundary value problems of partial differential equations, and so on, are also examples of the Monte Carlo method in this sense. However, direct numerical calculation seems to be more useful in dealing with this sort of problem. At present, *Monte Carlo methods* are usually used when it is difficult to construct (or solve) mathematical equations describing the phenomena in question, for example, when the phenomena involve †stochastic processes such as †random walks. Some methods have been devised so as to get precise results efficiently.

References

[1] B. P. Zeigler, Theory of modeling and simulation, Wiley, 1976.
[2] R. S. Lehman, Computer simulation and modeling, Wiley, 1977.
[3] R. E. Shannon, Systems simulations: The art and science, Prentice-Hall, 1975.
[4] J. M. Hammersley and D. C. Handscomb, Monte Carlo methods, Wiley, 1964.
[5] A. Newell and H. A. Simon, Human problem solving, Prentice-Hall, 1972.
[6] M. Minsky and S. Papert, Perceptrons, MIT Press, 1969; second printing with corrections by the authors, June 1972.
[7] J. W. Forrester, World dynamics, Wright-Allen Press, 1972.

386 (XX.29) *S*-Matrices

A. Basic Notion

It is often useful to focus attention on the relation between a physical system's input and output, without worrying about intermediate processes (the black box), which may be insufficiently understood or too complicated to analyze. For the scattering of particles, this leads to the notion of an ***S*-matrix** that directly relates the state of incoming particles (before scattering processes take place) to that of outgoing (scattered) particles.

In typical cases, the incoming and outgoing particles are described as mutually noninteracting (these are called free particles). This implies particle motions along straight lines at constant speeds (asymptotic to the actual motion at infinite past for incoming particles and at infinite future for outgoing particles) in classical mechanics, and wave functions (or vectors in a Hilbert space) obeying the Schrödinger equation with a free Hamiltonian H_0 in quantum mechanics and more or less the same in quantum field theory.

A wave function for n particles is an L_2-function of their momenta $p_1, \ldots, p_n$ (each p_j being a 3-dimensional vector) with respect to an appropriate measure (normally the Lebesgue measure $d\mu(p)=\prod d^3 p_j$ in quantum mechanics and the Lorentz-invariant measure $d\mu(p)=\prod\{(m^2+p_j^2)^{-1/2}d^3p_j\}$ in quantum field theory, where m is the mass of the particle), and the S-matrix S is described in terms of the S-matrix elements $(p_1, \ldots, p_n|S|p'_1, \ldots, p'_{n'})$ (which is a distribution) as

$$(\Phi, S\Psi)=\int \overline{\Phi(p)}(p|S|p')\Psi(p')\,d\mu(p)\,d\mu(p'),$$

where Φ and Ψ are wave functions for n and n' particles, $p=(p_1, \ldots, p_n)$, and similarly for p'. $(p|S|p')$ gives quantities measured in scattering experiments, as will be explained in Section B (2).

In quantum mechanics, the free and actual (interacting) motion of particles is described in the same Hilbert space with free and interacting Hamiltonians H_0 and H. A vector Φ in interacting motion behaves like a vector φ in free motion at infinite past if

$$\|(\exp[-iHt])\Phi-(\exp[-iH_0t])\varphi\|\to 0 \qquad \text{as} \quad t\to-\infty$$

and hence

$$\Phi=W_-(H;H_0)\varphi,$$

$$W_-(H;H_0)=\lim_{t\to-\infty} e^{iHt}e^{-iH_0t}.$$

Such a Φ is often written as

$$\Phi=\Phi^{\text{in}}(\varphi)=\int \Phi^{\text{in}}(p)\varphi(p)\,d\mu(p),$$

and is called an **in-state**. A definition for an **out-state** Φ^{out} is obtained by changing $t\to-\infty$ to $t\to+\infty$ and W_- to W_+. The S-matrix element is defined (as a distribution) by

$$(p|S|p')=(\Phi^{\text{out}}(p), \Phi^{\text{in}}(p')).$$

The existence and properties of $W_\pm$ (called †wave operators) are central subjects in scattering theory (→ 375 Scattering Theory).

In quantum field theory, the asymptotic description is given in terms of vectors in †Fock space, and the in- and out-states are constructed in terms of †asymptotic fields (→ 150 Field Theory).

The foregoing description actually applies only to a system of one-component particles of the same kind. More generally, additional (discrete) variables, say α, are needed to distinguish different kinds of particles and different spin components of each kind of particle, and the p's appearing in the above formulas should be replaced by (p, α)'s, along with related changes in the measure.

Even in a quantum mechanics of many identical particles a bound state, if it exists, is to be treated as another particle (different from the original one) and should be distinguished by α's in the asymptotic description φ.

If the interaction is of long range (e.g.,

Coulomb interaction), the classical path of a particle does not have an asymptote in general, and correspondingly the wave operators $W_{\pm}$ do not exist for the usual free Hamiltonian. Still, an asymptotic description of scattering is possible in some cases.

In the presence of massless particles, such as photons in quantum field theory, another difficulty, called the infrared problem, can arise in the asymptotic description of scattering because the scattered particle may be accompanied by an infinite number of massless particles (with very small energy). In such a situation, a representation of a free massless field not equivalent to the standard Fock representation is believed to be a possible candidate for the asymptotic description of scattering.

B. Basic Properties

(1) Invariance. Let $\mathscr{H}_0$ be the Hilbert space for the asymptotic description of scattering, such as the space of L_2-functions $\varphi(p_1, \ldots, p_n)$ relative to the measure $d\mu(p)$. The **S-matrix** is an operator on $\mathscr{H}_0$ whose matrix element is as described above. (The corresponding operator **S** in the Hilbert space $\mathscr{H}$ describing the interacting states is defined by $\mathbf{S}\Phi^{\text{out}}(\varphi)=\Phi^{\text{in}}(\varphi)$ and is sometimes called an **S-operator**.)

S is said to be **invariant** under a group G of transformations of the p's (and possibly the α's) if $(U(g)\varphi)(p)=\varphi(g^{-1}p)$ defines a continuous unitary representation $U(g)$ and if $U(g)S=SU(g)$ for all $g\in G$. First, S is usually invariant under time translation. In the quantum mechanics of a particle scattered by a rotationally invariant potential, S is invariant under the group of rotations of 3-dimensional vectors p; in the quantum mechanics of many particles (mutually interacting through central potentials), S is invariant under the 3-dimensional Euclidean group of transformations $p\to Rp+a$ with rotation R; in relativistic field theory, S is assumed to be invariant under the †inhomogeneous Lorentz group of transformations $p\to\Lambda p+a$ with $p=(p^0,\mathbf{p})$ and homogeneous Lorentz transformation Λ.

In all these examples, G is of type I, i.e., there is a direct integral decomposition

$$\mathscr{H}_0=\int\mathscr{H}(k)\otimes\mathscr{L}(k)\,dv(k),$$

$$U(g)=\int U_k(g)\otimes 1_{\mathscr{L}(k)}\,dv(k),$$

into irreducible representations U_k on $\mathscr{H}(k)$, which are mutually inequivalent, where $\mathscr{L}(k)$ is some Hilbert space for each k. The S-matrix is invariant under G if and only if

$$S=\int 1_{\mathscr{H}(k)}\otimes S(k)\,dv(k).$$

When a (scalar) particle is scattered by a central potential, irreducible representations are labeled by the energy $E(p)=p^2/(2m)$ (for time translation) and the angular momentum l (for rotations) with $\dim\mathscr{L}(E(p),l)=1$. Therefore each $S(k)=S_l(|p|)$ is a number. For any given energy, $l=0,1,2,\ldots$ are referred to as the S-wave, P-wave, D-wave, ... or generally as **partial waves**.

For relativistic scalar particles, irreducible representations (with positive energy and nonzero real mass) are labeled by the center-of-mass energy squared s $(=(\Sigma(m^2+p_i^2)^{1/2})^2-(\Sigma p_i)^2)$ and the total angular momentum l. If s is below the threshold $(3m)^2$ of 3-particle scattering, $\dim\mathscr{L}(k)=1$ and $S(k)=S_l(s)$ is a number.

(2) Unitarity. $S\varphi$ is supposed to represent the $t=+\infty$ asymptotic (free) behavior of the state that initially (i.e., at $t=-\infty$) behaves like a free state φ. If there is no loss of probability in the description of scattering, the S-matrix is isometric. If all asymptotic configurations are realized as a result of scattering, the S-matrix must also be unitary. The mappings $\Phi^{\substack{\text{out}\\ \text{in}}}:\varphi\in\mathscr{H}_0\to\Phi^{\substack{\text{out}\\ \text{in}}}(\varphi)\in\mathscr{H}$ ($W_{\pm}$ in quantum mechanics) are proved to be isometric under a general assumption. The unitarity is then proved in potential scattering (under some conditions on the potential) by showing that the two wave operators $W_{\pm}$ have the same range. In fact, a somewhat stronger result—that this range is the same as the absolutely continuous spectral subspace for the interacting Hamiltonian H—is usually proved and is called **completeness (of scattering states** in the absolutely continuous spectral subspace of H). If the mappings Φ^{out} and Φ^{in} (or $W_{\pm}$) are isometric and have the same range, then $\Phi^{\text{in}}(\varphi)=\Phi^{\text{out}}(S\varphi)$, which shows that the state behaving like φ at $t=-\infty$ behaves like $S\varphi$ at $t=+\infty$.

In the simple multiplicity cases such as $S_l(|p|)$ and $S_l(s)$ above (which correspond to the physical situation of purely elastic scattering without any production or change of particles), these numbers must be of the form $e^{2i\delta_l}$ due to unitarity, where the real number δ_l is called the **phase shift**. In terms of the phase shift, the **differential cross section** $d\sigma/d\Omega$, which is the average number of particles scattered per unit time per unit solid angle around the direction forming an angle θ with the incident uniform parallel beam of unit intensity (one particle per unit time per unit area) when viewed in the center-of-mass system, and the **total elastic cross section** $\sigma_{el} = \int (d\sigma/d\Omega) d\Omega$ are given by the following formulas, called the **partial wave expansion**:

$$d\sigma/d\Omega = |f(s,\theta)|^2,$$

$$f(s,\theta) = k^{-1} \sum_{l=0}^{\infty} (2l+1) f_l P_l(\cos\theta), \qquad f_l = \sin\delta_l e^{i\delta_l},$$

$$\sigma_{el} = 4\pi k^{-2} \sum_l (2l+1) \sin^2 \delta_l,$$

where $d\Omega = \sin\theta \, d\theta \, d\varphi$ (invariant measure on a 2-dimensional sphere S^2) and k is the wave number of the particle in the center-of-mass system ($k = \hbar^{-1}|p| = \hbar^{-1}[(s/4) - m^2]^{1/2}$). The function f is called the **scattering amplitude**. The forward scattering amplitude $f(s,0)$ is related to the total cross section σ_{tot} by the **optical theorem**:

$$\sigma_{tot} = 4\pi k^{-1} \operatorname{Im} f(s,0),$$

which follows from the unitarity of the S-matrix. Here the **total cross section** is expressed as the area of the transverse cross section of a classical (impenetrable) scatterer that would scatter the same amount of particles. In a purely elastic region, $\sigma_{tot} = \sigma_{el}$, and the optical theorem is the same as the assertion $\operatorname{Im} f_l = |f_l|^2$.

Even for values of s above the threshold of inelastic scattering, the restriction of S_0 to the subspace of two particles is again described by numbers $e^{2i\delta_l}$, where δ_l is now complex, and the same formulas for the differential and total elastic cross sections hold, except that unitarity of S_0 now implies $\operatorname{Im} f_l \geq |f_l|^2$.

(3) TCP Symmetry. In the framework of either †axiomatic quantum field theory or the †theory of local observables, the TCP theorem (or PCT theorem) shows [1] that to every particle there corresponds another particle with the same mass and spin, which is called the **antiparticle** and can be the original particle itself, such that any particle is the antiparticle of its antiparticle, and the following relation, called **TCP invariance** (or **PCT invariance** [2]) holds for S-matrix elements:

$$(p,\alpha|S|p',\alpha')$$
$$= \eta(\alpha)\eta(\alpha') i^{F(\alpha)-F(\alpha')} (p', \theta_0\alpha'|S|p, \theta_0\alpha).$$

Here $F(\alpha)$ and $F(\alpha')$ are the number of particles with half odd integer spin amongst the incoming and outgoing particles (i.e., in α and α'), $\eta(\alpha)$ and $\eta(\alpha')$ are the product of ± 1 assigned to each particle in α and α', respectively, with the assignment to a particle and its antiparticle the same for bosons and opposite for fermions, (p,α) is an abbreviation for an ordered n-tuple of energy-momentum 4-vectors p_i and other indices α_i (for spin components and particle species), $i = 1, \ldots, n$, and similarly for (p', α'), and θ_0 is the mapping from particles to their antiparticles (without changing spin indices). We may view the mapping $|p,\alpha) \to \eta(\alpha) i^{F(\alpha)} |p, \theta_0\alpha)$ as an antiunitary operator Θ_0 on $\mathscr{H}_0$ satisfying $\Theta_0 S \Theta_0^{-1} = S^*$ so that

$$(\varphi_1, S_0\varphi_2) = (\Theta_0\varphi_2, S_0\Theta_0\varphi_1).$$

This is related to the †TCP operator Θ in quantum field theory by $\Theta\Psi^{out}(\varphi) = \Psi^{in}(\Theta_0\varphi)$ and $\Theta\Psi^{in}(\varphi) = \Psi^{out}(\Theta_0\varphi)$. The name TCP comes from the combination of T for time reversal (incoming $\rightleftarrows$ outgoing, $v \equiv p/m \to -v$), C for charge conjugation (particle $\rightleftarrows$ antiparticle), and P for parity, which is a quantum number for space inversion ($p \to -p$).

TCP symmetry was suggested to G. Lüders (*Dansk. Mat. Fys. Medd.*, 1954) by B. Zumino in the form that P-invariance implies TC symmetry. W. Pauli (*Niels Bohr and the development of physics*, Pergamon 1955, 30) realized that TCP is a symmetry. R. Jost (*Helv. Phys. Acta*, 1957) gave its proof in the framework of axiomatic quantum field theory.

(4) Crossing Symmetry. In the framework of either axiomatic quantum field theory or the theory of local observables, it has been shown by J. Bros, H. Epstein, and V. Glaser (*Comm. Math. Phys.*, 1 (1965); also [3]) that there exist analytic functions $H(k,\alpha)$ of complex variables $k = (k_1, \ldots, k_4)$ in a certain domain D such that each k_j is a complex 4-vector, the variables are on the mass-shell manifold defined by $k_j^2 = m_j^2$ (k_j^2 is in the Minkowski metric) and $\sum k_j = 0$, and the boundary value of $H(k,\alpha)$ as k_j approaches $\varepsilon_j p_j$ from $\operatorname{Im} s > 0$ (s being the square of the sum of two k's for incoming particles) in D is the scattering amplitude for the following processes involving the particles A_j and their antiparticles $\bar{A}_j$ ($j = 1, \ldots, 4$) with 4-momenta p_j, $\varepsilon_j = 1$ for A_j and -1 for $\bar{A}_j$ (some of the A's and $\bar{A}$'s may coincide):

(i) $\quad A_1 + A_2 \to \bar{A}_3 + \bar{A}_4,$

(i′) $\quad A_3 + A_4 \to \bar{A}_1 + \bar{A}_2,$

(ii) $\quad A_1 + A_3 \to \bar{A}_2 + \bar{A}_4,$

(ii′) $\quad A_2 + A_4 \to \bar{A}_1 + \bar{A}_3,$

(iii) $\quad A_1 + A_4 \to \bar{A}_2 + \bar{A}_3,$

(iii′) $\quad A_2 + A_3 \to \bar{A}_1 + \bar{A}_4.$

Any pair of relations taken from (i)–(iii) constitutes a **crossing symmetry**.

(5) High Energy Theorems. M. Froissart (*Phys. Rev.*, 123 (1961)) obtained from the Mandelstam representation the following bound for the forward scattering amplitude (called the

Froissart bound) at large s:

$$|F(s,0)|<(\text{const})s(\log s)^2,$$

$$F(s,t)=kf(s,\theta), \qquad t=2k^2(\cos\theta-1).$$

A. Martin has obtained such a bound in the axiomatic framework. As a consequence we have the **Froissart-Martin bound**:

$$\sigma_{\text{tot}}<4\pi R^{-1}(1+\varepsilon)^2(\log(s/s_0))^2,$$

where $\varepsilon>0$ is arbitrary, R can be taken to be $4m_\pi^2$ (m_π is the mass of a pion) for many cases, such as $\pi\pi$, πK, πN, and $\pi\Lambda$ scattering, and s_0 is an unknown constant.

Many other upper and lower bounds for scattering amplitudes have been obtained under other assumptions [4, 5].

I. Ya. Pomeranchuk (*Sov. Phys. JETP*, 7 (1958)) suggested the asymptotic coincidence of total cross sections at high energy for scattering of AB and $A\bar{B}$, where $\bar{B}$ is the antiparticle of B. This **Pomeranchuk theorem** has been shown to hold by Martin (*Nuovo Cimento*, 39 (1965)) by using the analyticity derived in the axiomatic framework under the following sufficient condition: The existence of $\lim_{s\to\infty}[\sigma(AB)-\sigma(A\bar{B})]$ and $\lim_{s\to\infty}(s\log s)^{-1}f(s,0)=0$.

C. *S*-Matrix Approach

(1) History. All the information needed to understand the experimental elementary-particle scattering data seems to be expressible by *S*-matrix elements. It was therefore natural to try to develop a foundation (and practical methods of calculation) for the theory of elementary particles on the basis of some general properties of the *S*-matrix, especially when other approaches, such as quantum field theory, faced difficulties. W. Heisenberg (*Z. Phys.*, 120 (1943)) first pointed out the possibility of such an approach soon after the introduction of the *S*-matrix by J. A. Wheeler (*Phys. Rev.*, 52 (1937)). Unfortunately, in the early 1940s not much dynamical content could be given to such an *S*-matrix-theoretic approach because only unitarity and some invariance properties, such as †Lorentz invariance, were used to characterize the *S*-matrix. In the late 1950s, through the study of the analyticity of the *S*-matrix in connection with dispersion relations in quantum field theory, it became evident that causality is another important determinant of *S*-matrix structure. In practice, causality in position space is used in the form of the analyticity of the *S*-matrix elements in the energy-momentum space (variables dual to space-time positions in the Fourier transform). Subsequently it was realized that analyticity combined with unitarity gives surprisingly strong control over the structure of the *S*-matrix (G. F. Chew [6]; H. P. Stapp, *Phys. Rev.*, 125 (1962); J. C. Polkinghorne, *Nuovo Cimento*, 23 and 25 (1962); J. Gunson, *J. Math. Phys.*, 6 (1965); D. I. Olive, *Phys. Rev.*, 135B (1964); Chew [7]; R. J. Eden et al. [8]). The study of the *S*-matrix based on its general properties, such as invariance, unitarity, and analyticity, is called ***S*-matrix theory**. In the present form, it is adapted only to massive particles with short-range interactions, and its applicability is believed to be limited to strongly interacting systems.

(2) Landau-Nakanishi Variety. C. Chandler and Stapp (*J. Math. Phys.*, 10 (1969)) and D. Iagolnitzer and Stapp (*Comm. Math. Phys.*, 14 (1969)) clarified the analytic structure of the *S*-matrix in terms of Landau equations (→ 146 Feynman Integrals) based on the important physical idea of **macroscopic causality**, which gives much more precise information in the physical region than a superficial application of †locality (also called microcausality) in axiomatic quantum field theory, though it is possible that a detailed study starting from microcausality and incorporating †asymptotic completeness (the so-called nonlinear program in axiomatic quantum field theory) might eventually entail the macroscopic causality condition (e.g., J. Bros, in [9]; Iagolnitzer [10, ch. IV]; also K. Symanzik, *J. Math. Phys.*, 1 (1960)).

(3) Microlocal Analysis. An important fact is that the **normal analytic structure** of the *S*-matrix discussed by Iagolnitzer and Stapp essentially coincides with the notion of analyticity in microlocal analysis, i.e., with micro-analyticity (→ 274 Microlocal Analysis; F. Pham and Iagolnitzer, *Lecture notes in math. 449*, Springer, 1975; M. Sato, *Lecture notes in phys. 39*, Springer, 1975)). In a word, the †Landau equations have acquired a new interpretation in the description of the micro-analytic structure of the *S*-matrix. In the new developments, the Landau equations define a variety in the cotangent bundle (over the mass-shell manifold in momentum space), and the †singularity spectrum of *S*-matrix elements is assumed to be confined to $\bigcup_G \mathscr{L}^+(G)$ (except for the so-called $\mathscr{M}_0$-points), where G ranges over all possible **Feynman graphs** and $\mathscr{L}^+(G)$ denotes the positive-α Landau-Nakanishi variety associated with G (→ 146 Feynman Integrals). The union $\bigcup_G \mathscr{L}^+(G)$ is known to be locally finite and hence makes sense (Stapp, *J. Math. Phys.*, 8 (1967)). The old interpretation of Landau equations, as defining a variety in energy-momentum space, corresponds now to considering the variety $L^+(G)$

obtained by projecting $\mathscr{L}^+(G)$ from the cotangent bundle to the base manifold (i.e., the mass shell manifold). The new interpretation of Landau equations led Sato (*Lecture notes in phys. 39*, Springer, 1975) to make a further intriguing conjecture that the S-matrix would satisfy a special †overdetermined system (a †holonomic system) of †(micro-) differential equations whose †characteristic variety is given by the complexification of Landau-Nakanishi varieties. This conjecture is closely related to the monodromy-theoretic approach by T. Regge (*Publ. Res. Inst. Math. Sci.*, 12 suppl. (1977) and the references cited therein) and his co-workers.

(4) Discontinuity Formula. It has turned out that the above approach is closely related to the so-called **discontinuity formula** obtained by combining the unitarity and the analyticity of the S-matrix. Actually T. Kawai and Stapp (*Publ. Res. Inst. Math. Sci.*, 12 suppl. (1977)) have shown that Sato's conjecture can be verified at several physically important points on the basis of the discontinuity formula. The discontinuity formula was first found by R. E. Cutkosky (*J. Math. Phys.*, 1 (1960)) for Feynman integrals. It describes the ramification property of the integral around its singularities (→ 146 Feynman Integrals). An analogous formula has been shown to be valid also for the S-matrix, and it demonstrates how strict are the constraints derived from unitarity and analyticity (Eden et al. [8, ch. 4]; M. J. D. Bloxham et al., *J. Math. Phys.*, 10 (1969); J. Coster and Stapp, *J. Math. Phys.*, 10 (1969); also Stapp in [11] and Iagolnitzer [10]). Note, however, that the derivation of the hitherto-known discontinuity uses either some *ad hoc* assumptions or some heuristic reasoning which is not rigorous or sometimes is even erroneous from the mathematical viewpoint. Efforts to give a rigorous proof are still being made, and these present several mathematically interesting problems (e.g., Iagolnitzer in [9] and M. Kashiwara and Kawai in [9]).

(5) Regge Poles. The results stated so far concerning the analyticity of the S-matrix have been primarily derived in the low-energy region. It is commonly hoped that these results can be related to its high-energy behavior through the inner consistency of S-matrix theory, even though it is still unclear to what extent such a relationship can be developed. Such a hope was advocated by Chew, who had been inspired by the results of Regge (*Nuovo Cimento*, 14 (1959); 18 (1960)) for potential scattering. After being adapted to the relativistic case, Regge's idea took the following form: Consider the scattering of two incoming and two outgoing scalar particles with equal mass $m>0$. Let $f_l(s)$ be the partial scattering amplitude defined earlier. Regge introduced the idea of extending the function $f_l(s)$ to an analytic function $f(l,s)$ $(l\in\mathbf{C})$ and of applying the Sommerfeld-Watson transformation in order to replace the partial wave expansion by the integral

$$\sqrt{-1}/2\int_C(2l+1)f(l,s)P_l(-\cos\theta)\,dl/\sin\pi l$$

$$=F(s,t)$$

for a certain contour C in the complex l-plane which encircles $\{0,1,2,\dots\}$. If $f(l,s)$ is meromorphic in $\operatorname{Re} l>-1/2$ and if it tends to zero sufficiently rapidly at infinity, then one can change the contour C so that, with the help of Cauchy's integral formula, $F\sim\text{constant}\cdot t^{\alpha(s)}$, $\alpha(s)=\max\operatorname{Re} l(s)$, where the maximum is taken over all the poles of $f(l,s)$. Thus the poles of the extended function $f(l,s)$ determine the asymptotic behavior of F as $t\to\infty$ (**Regge behavior**) under the assumption that $f(l,s)$ can be chosen to satisfy suitable analyticity and growth order conditions. These poles are called **Regge poles**. Even though meromorphy conditions are found to be satisfied for scattering by a (Yukawa) potential, they do not seem to be satisfied for the full S-matrix in the relativistic case. More general cases than those discussed here, i.e., the cases where more variables are considered, are also being studied but without full success at the moment. For details and references → [7, 12, 13].

(6) Veneziano Model. In connection with Regge-pole theory, we note an interesting observation by G. Veneziano (*Nuovo Cimento*, 57A (1968)) to the effect that $\Gamma(1-\alpha(s))\Gamma(1-\alpha(t))/\Gamma(1-\alpha(s)-\alpha(t))$, with $\alpha(s)$ being linear in s, satisfies a crossing symmetry (in s and t) and shows an exact Regge-pole behavior. Although the many results that have been obtained give rise to a hope of constructing a realistic model of the S-matrix starting from the aforementioned function, no one has yet succeeded [14]. A more promising approach is the topological expansion procedure in which the first term of the expansion apparently shares with the potential-scattering functions the property of having only poles in the complex l-plane, along with several other physically important properties of Veneziano's function.

References

[1] H. Epstein, CPT invariance of the S-matrix in a theory of local observables, J. Math. Phys., 8 (1967), 750–767.

[2] R. F. Streater and A. S. Wightman, PCT, spin and statistics, and all that, Benjamin, 1964.
[3] H. Epstein, in Axiomatic field theory, M. Chretien and S. Deser (eds.), Gordon & Breach, 1966.
[4] A. Martin and F. Cheung, Analyticity property and bounds of the scattering amplitudes, Gordon & Breach, 1970.
[5] A. Martin, Scattering theory in unitarity, analyticity and crossing, Lecture notes in phys. 3, Springer, 1969.
[6] G. F. Chew, S-matrix theory of strong interaction physics, Benjamin, 1962.
[7] G. F. Chew, The analytic S-matrix: A basis for nuclear democracy, Benjamin, 1966.
[8] R. J. Eden, P. V. Landshoff, D. I. Olive, and J. C. Polkinghorne, The analytic S-matrix, Cambridge Univ. Press, 1966.
[9] D. Iagolnitzer (ed.), Proc. Les Houches, Lecture notes in phys. 126, Springer, 1980.
[10] D. Iagolnitzer, The S-Matrix, North-Holland, 1978.
[11] R. Balian, and D. Iagolnitzer (ed.), Structural analysis of collision amplitudes, North-Holland, 1976.
[12] S. Frautschi, Regge poles and S-matrix theory, Benjamin, 1963.
[13] P. O. B. Collins, An introduction to Regge theory and high energy physics, Cambridge Univ. Press, 1977.
[14] M. Jacob, Dual theory, North-Holland, 1974

387 (XX.34) Solitons

A. General Remarks

Solitons are nonlinear waves that preserve their shape under interaction. Mathematically, the theory of solitons continues to develop as a theory of completely integrable mechanical systems. Typical examples are the **Korteweg–de Vries equation** (→ Section B), the **Toda lattice** (→ 287 Nonlinear Lattice Dynamics), and the **Sine-Gordon equation**

$$u_{tt}-u_{xx}=\sin u,$$

studied classically in connection with transformations of surfaces of constant negative curvature.

B. The KdV Equation

In the late 19th century J. Boussinesq and then D. Korteweg and G. de Vries obtained equations describing water waves having traveling-wave solutions. The equation

$$u_t-6uu_x+u_{xxx}=0, \qquad u=u(x,t), \tag{1}$$

derived by de Vries is called the **KdV equation** for short. Putting $u(x,t)=s(x-ct-\delta)$, we find that s is an †elliptic function, and we obtain

$$s=-\frac{c}{2}\operatorname{sech}^2\left(\frac{\sqrt{c}}{2}x\right)$$

as its degenerate form. This solution is called a **solitary wave**.

Around 1965, M. Kruskal and N. Zabusky solved the KdV equation numerically, taking several separated solitary waves as the initial data. They found that the waves interact in a complicated way but that eventually the initial solitary waves reappear. Noting the particle-like character of the waves, they called each of these waves a **soliton**. Subsequently, the KdV equation was found to have an infinite number of constants of motion.

C. Gardner, J. Greene, Kruskal, and R. Miura associated the 1-dimensional †Schrödinger operator $-d^2/dx^2+u(x,t)$ to each solution $u(x,t)$ of the KdV equation and showed that its †eigenvalues are preserved in time. Moreover, they applied inverse scattering theory (→ Section D) and obtained explicit formulas for the solutions.

C. Lax Representation

Let

$$L=-D^2+u(x), \qquad D=d/dx.$$

For

$$M=-4D^3+6uD+3u_x,$$

the commutator $[M,L]=ML-LM$ is the operator of multiplication by the function $6uu_x-u_{xxx}$. So $u=u(x,t)$ is a solution of the KdV equation if and only if

$$L_t=[M,L]. \tag{2}$$

Equation (2) is also the condition that all $L=L(t)$ are †unitarily equivalent to each other and the †spectrum of L is preserved through time.

Most equations having soliton solutions can be represented in the form of (2) for a suitable pair of L and M. This representation is called the **Lax representation**. One sometimes says that an **isospectral deformation** of L is given by (2).

On the other hand, isomonodromic deformations (→ 253 Linear Ordinary Differential Equations (Global Theory)) have been studied extensively by M. Sato and his co-workers, and relations to soliton theory have been discovered (Sato, T. Miwa, and M. Jimbo, *Publ. Res. Inst. Math. Sci.*, 14 (1978), 15 (1979)).

In the present case, the requirement that the commutator $[M, L]$ be a multiplication by a function determines an essentially unique $(2n+1)$th-order ordinary differential operator $M = A_n$, the differential operator part of the †fractional power $L^{n+1/2}$. $[A_n, L]$ is a polynomial in u and its derivatives, denoted by $K_n[u]$. The equation $u_t = K_n[u]$ is called the nth KdV equation. The transformation taking the initial data $u(x)$ to the solution $u_n(x,t)$ of the nth KdV equation is denoted by $T_n(t)$. Then $T_n(t)T_m(s) = T_m(s)T_n(t)$, i.e., the flows defined by these higher-order KdV equations commute. This property and the existence of infinite number of invariant integrals are consequences of the complete integrability of the higher-order KdV equations considered as infinite-dimensional Hamiltonian systems. The KdV equations can be studied group-theoretically as Hamiltonian systems on a certain coadjoint orbit in the †dual space of the †Lie algebra of a certain class of †pseudo-differential operators (M. Adler, *Inventiones Math.*, 50 (1979); also → B. Kostant, *Advances in Math.*, 34 (1979) for the analogous facts for the Toda lattice).

The ordinary differential equation $K_n[u] = 0$ is called a stationary KdV equation. By the commutativity of the flows $T_n(t)$, each KdV flow leaves invariant the space of solutions of $K_n[u] = 0$. The flows restricted to this space form a completely integrable Hamiltonian system with finitely many degrees of freedom (S. P. Novikov, *Functional Anal. Appl.*, 8 (1974)).

$K_n[u] = 0$ is also the condition that there exist an ordinary differential operator M which commutes with L. J. Burchnal and T. Chaundy (*Proc. London Math. Soc.*, (2) 21 (1922)) studied this problem and showed that such L and M are connected by the relation $M^2 = P(L)$, where P is a certain polynomial of degree $2n+1$ and that the potential u is expressed by the †theta function associated with the †hyperelliptic curve $w^2 = P(z)$.

D. Inverse Scattering Method

Let $u(x)$ be a potential such that $u(x) \to 0$ as $x \to \pm\infty$. The equation $Lf = \zeta^2 f$ $(\operatorname{Im}\zeta \geqslant 0)$ has solutions $f_\pm(x, \zeta)$ that can be represented as

$$f_\pm(x,\zeta) = e^{\pm i\zeta x}\left(1 + \int_0^\infty K_\pm(x,t)e^{\pm 2i\zeta t}\,dt\right).$$

Putting $\zeta = \xi + i\eta$ and noting that $f_+(x,\xi)$ and $f_+(x,-\xi)$ are independent solutions of $Lf = \xi^2 f$, one can express f_- as

$$f_-(x,\xi) = a(\xi)f_+(x,-\xi) + b(\xi)f_+(x,\xi).$$

The coefficient $a(\xi)$ can be continued analytically to the upper half-plane, where it has only a finite number of zeros, all of which are simple and lie on the imaginary axis. Denote them by $i\eta_j$ $(j = 1, \ldots, n)$. The †point spectrum of the operator L consists of the numbers $-\eta_j^2$, and the associated eigenfunctions are $f_+(x, i\eta_j)$, which are real-valued. Put

$$c_j = \left(\int f_+(x, i\eta_j)^2\,dx\right)^{-1},$$

and call $t(\xi) = a(\xi)^{-1}$ and $r(\xi) = b(\xi)/a(\xi)$ the **transmission coefficient** and the **reflection coefficient**, respectively. The triplet $r(\xi)$, η_j, c_j $(j = 1, \ldots, n)$ is called the **scattering data**. It is connected with the kernel $K = K_+$ by the **Gel'fand-Levitan-Marchenko equation**

$$K(x,t) + F(x+t) + \int_0^\infty F(x+t+s)K(x,s)\,ds \qquad (t > 0),$$

$$F(x) = \pi^{-1}\int_{-\infty}^{\infty} r(\xi)e^{2i\xi x}\,dx + 2\sum_{j=1}^{n} c_j e^{-2\eta_j x}.$$

The potential is given by

$$u(x) = -(\partial K/\partial x)(x, 0).$$

When the reflection coefficient $r(\xi)$ vanishes identically, the potential is called a **reflectionless potential**. The kernel K then becomes a †degenerate kernel and the potential is expressed by

$$u(x) = -2\frac{d^2}{dx^2}\log D(x),$$

where $D(x)$ is the determinant of the $n \times n$ matrix whose j, k entry is $\delta_{jk} + c_j \exp\{-(\eta_j + \eta_k)x\}/(\eta_j + \eta_k)$.

The authors of [1] showed that if $u(x,t)$ is a solution of the KdV equation and if $u(x,t) \to 0$ $(x \to \pm\infty)$, then the time development of the scattering data of the potential $u(x,t)$ is as follows. n and η_j do not depend on t, and

$$c_j(t) = c_j e^{8\eta_j^3 t}, \qquad r(\xi, t) = r(\xi)e^{8i\xi^3 t}.$$

The solution associated with the reflectionless potentials are obtained by replacing c_j by $c_j(t)$ in the formula for $D(x)$. These are soliton solutions of the KdV equation.

R. Hirota developed a method of treating functions like $D(x)$ directly for most of the equations in the soliton theory (*Lecture notes in math. 515*, Springer, 1976).

A certain geometric method that enables one to obtain solutions of the Sine-Gordon equation from a known solution has been studied in the transformation theory of surfaces of constant negative curvature (G. Darboux, *Leçons sur la théorie générale des surfaces*, Chelsea, 1972, vol. 3, ch. 12), and its generalizations are called Bäcklund transformations. Soliton solutions of the KdV equation can also be constructed by this method. Relations to the inverse scattering method, differential systems, and transformation groups have also been studied (*Lecture notes in math.* *515*, Springer, 1976).

E. Periodic Problem

Let the potential $u(x)$ be of period l, and consider Hill's equation (→ 268 Mathieu Functions E) $Lf=\lambda f$. The real (λ-) axis is divided into intervals of unstable solutions (u-intervals) alternating with intervals of stable solutions. One of the u-intervals is of the form $(-\infty, \lambda_0]$ and the others are finite, possibly degenerating to points. The special potentials that have only a finite number of nondegenerate u-intervals are the periodic analog of the reflectionless potentials, and are called the **finite-gap** (or **-band**) **potentials**.

Consider the eigenvalue problem $Lf=\lambda f$, $f(\tau)=f(\tau+l)=0$ for a fixed real parameter τ. Exactly one of its eigenvalues belongs to each finite u-interval. These eigenvalues move with τ, but those in degenerate intervals cannot move.

Let u be a finite-gap potential, $I_j=[\lambda_{2j-1}, \lambda_{2j}]$ $(j=1,\ldots,g)$ be the nondegenerate (finite) u-intervals, and $\mu_j(\tau)$ be the associated eigenvalues in I_j. The potential is recovered by the formula

$$u(x)=\sum_{j=0}^{2g}\lambda_j-2\sum_{j=1}^{g}\mu_j(x).$$

Put $P(\lambda)=\prod_{j=0}^{2g}(\lambda-\lambda_j)$, realize the †Riemann surface S of the hyperelliptic curve $w^2=P(z)$ as a two-sheeted cover of the †Riemann sphere, and consider the $\mu_j(\tau)$ as points on S. Let ω_j $(j=1,\ldots,g)$ be a basis for the space of the †differentials of the first kind on S. Fix a point P_0 in S and put

$$w_j(\tau)=\sum_{k=1}^{g}\int_{p_0}^{\mu_k(\tau)}\omega_j \qquad (j=1,\ldots,g)$$

(→ 3 Abelian Varieties L). Then the locus $(w_1(\tau),\ldots,w_g(\tau))(-\infty<\tau<\infty)$ on the †Jacobian variety turns out to be a straight line, the direction v_x being determined by the †periods of certain †differentials of the second kind on S. Employing the solution of †Jacobi's inverse problem, we can write the potential in terms of the †Riemann theta function as

$$u(x)=-2\frac{d^2}{dx^2}\log\theta(xv_x+c)+C, \tag{3}$$

where c is a certain constant vector and C is a constant.

Suppose now that $u(x,t)$ is a solution of the KdV equation which is a finite-gap potential for each t. Then g and λ_j are preserved in time. The determination of the direction v_t on the Jacobian variety is similar to that of v_x, and the solutions of the KdV equation are obtained by replacing the vector c by tv_t+c' in (3). The case $g=1$ is the elliptic traveling wave solution (→ Section B). Most of these results have been extended to the general periodic problem (H. P. McKean and E. Trubowitz, *Comm. Pure Appl. Math.*, 29 (1976)).

F. Two-Dimensional KdV Equation

Let S be a compact †Riemann surface of †genus g, and let p_∞ be a fixed point on S. Put $F(\kappa)=a_n\kappa^n+\ldots+a_0$ and $G(\kappa)=b_m\kappa^m+\ldots+b_0$. A function $\psi(x,y,t,p)$ of $p\in S$ and of x, y, t is uniquely determined by the following conditions: (a) it is meromorphic on $S-\{p_\infty\}$, and its pole †divisor is a general divisor of degree g and does not depend on x, y, t; (b) for a †local parameter z at p_∞ $(z(p_\infty)=0)$ and $\kappa=z^{-1}$, $\psi_0=\psi\exp(-\kappa x-F(\kappa)y-G(\kappa)t)$ is holomorphic near p_∞, and $\psi_0(p_\infty)=1$.

Moreover, there is a differential operator $L=a_nD^n+a_{n-1}D^{n-1}+\sum_{j=0}^{n-2}u_j(x,y,t)D^j$ such that $\psi_y=L\psi$. Expanding ψ_0 at p_∞ as $1+\sum_{j=1}\xi_j(x,y,t)z^j$, one can express the coefficients u_j by $\xi_1,\xi_2,\ldots$. Analogously, there is an $M=b_mD^m+b_{m-1}D^{m-1}+\sum_{j=0}^{m-2}v_jD^j$ such that $\psi_t=M\psi$. The operators L and M satisfy the relation

$$L_t-M_y=[L,M], \tag{4}$$

which is a generalization of the Lax representation. The coefficients of L and M satisfy a certain system of nonlinear differential equations (V. E. Zakharov and A. B. Shabat, *Functional Anal. Appl.*, 8 (1974); I. M. Krichever, *Functional Anal. Appl.*, 11 (1977)).

Example. Let $F(\kappa)=\kappa^2$ and $G(\kappa)=\kappa^3+c\kappa$. Then one finds

$$L=D^2+u, \qquad M=D^3+(3u/2+c)D+v,$$

where $u=-2\xi_1'$ and $v=3\xi_1\xi_1'-3\xi_1''-3\xi_2'$. Eliminating v from (4), one has

$$3u_{yy}+(-4u_t+4cu_x+u_{xxx}+6uu_x)_x=0,$$

the so-called **two-dimensional KdV equation (Kadomtsev-Petvyashvili equation)**. If $u(x,y,t)$ does not depend on y, the equation reduces to the KdV equation, and if u does not depend on t, to the **Boussinesq equation**.

The condition for reduction to these special cases can be described in terms of the meromorphic functions admitted by S. Suppose that there is a meromorphic function $E_F(p)$ holomorphic for $p\neq p_\infty$ and of †principal part $F(\kappa)$ at $p=p_\infty$. Then ψ is written as $\varphi(x,t,p)\exp\{E_F(p)y\}$, and the coefficients of L and M do not depend on y. φ satisfies $L\varphi = E\varphi$, and (4) reduces to the Lax representation.

If such an E_F exists, S is hyperelliptic, and p_∞ is one of its †branch points over the Riemann sphere. Thus the result of the previous section is recovered.

G. Solvable Models in Field Theory

The Sine-Gordon equation has been studied extensively as a solvable model in †field theory. It is a special case of a field in two space-time dimensions with values in a †symmetric space; this can also be treated by a variant of the inverse scattering method (V. E. Zakharov and A. V. Mikhailov, *Sov. Phys. JETP*, 47 (1978)).

Much work has been done on the semiclassical †quantization of equations encountered in soliton theory. Recently, a method of exact quantization (called quantum inverse scattering) was developed (see, for example, L. A. Takhtadzhyan and L. D. Faddeev, *Russian Math. Surveys*, 34 (5) (1979)).

References

[1] C. S. Gardner, J. M. Greene, M. D. Kruskal, and R. M. Miura, Method for solving the Korteweg-de Vries equation, Phys. Rev. Lett., 19 (1967), 1095–1097.

[2] P. D. Lax, Integrals of nonlinear equations of evolution and solitary waves, Comm. Pure Appl. Math., 21 (1968), 467–490.

[3] B. A. Dubrovin, V. B. Matveev, and S. P. Novikov, Nonlinear equations of Korteweg-de Vries type, finite-band operators and Abelian varieties, Russian Math. Surveys 31 (1) (1976), 59–146. (Original in Russian, 1976.)

[4] H. P. McKean, Integrable systems and algebraic curves, Lecture notes in math. 755, Springer, 1979, 83–200.

[5] Yu. I. Manin, Algebraic aspects of nonlinear differential equations, J. Sov. Math., 11 (1) (1979), 1–122. (Original in Russian, 1978.)

388 (XIII.32) Special Functional Equations

A. General Remarks

The term **special functional equations** usually means functional equations that do not involve limit operations. Such functional equations appear in various fields, but there is no systematic method for solving them. Usually they are solved by reduction to functional equations of some standard type. In this article functions are all real-valued functions of real variables unless otherwise specified.

B. Additive Functional Equations and Related Equations

Suppose that we are given an equation

$$f(x+y)=f(x)+f(y). \tag{1}$$

Clearly, $f(x)=cx$ (c a constant) is a solution. If $f(x)$ is continuous, (1) has no other solution (Cauchy). The same conclusion holds under any one of the following weaker conditions: (i) $f(x)$ is continuous at a point; (ii) $f(x)$ is bounded in a neighborhood of a point; (iii) $f(x)$ is †measurable in a neighborhood of a point. However, it was shown by G. Hamel and H. Lebesgue by means of †transfinite induction that equation (1) has infinitely many nonmeasurable solutions. On the other hand, it was proved by A. Ostrowski [2] that if a solution $f(x)$ of equation (1) does not take any value between two distinct numbers for x on a set of positive measure, then $f(x)$ is continuous. This result can be extended to the case where x is a point $(x_1, \ldots, x_n)$ of an n-dimensional Euclidean space. In this case, any continuous solution is of the form

$$f(x)=\sum_{j=1}^{n} C_j x_j$$

(C_j constant).

Next, we consider the equation

$$f\left(\frac{x+y}{2}\right)=\frac{f(x)+f(y)}{2}. \tag{2}$$

Any solution of equation (1) satisfies this equation. When x is a point $(x_1, \ldots, x_n)$ of an n-dimensional Euclidean space, any continuous solution of (2) is of the form $f(x)=\sum_{j=1}^{n} C_j x_j + C_0$ (C_j constant). If a solution $f(x)$ of (2) defined on a †convex set K does not take any value between two distinct numbers for x on a set of positive measure, then $f(x)$ is continuous (M. Hukuhara).

Consider the equation

$$g(x+y)=g(x)g(y). \tag{3}$$

If a solution $g(x)$ vanishes at some point ξ, then $g(x)\equiv 0$. Excluding this trivial case, we assume that $g(x)$ never vanishes. Then, putting $y=x$, we see that $g(x)>0$ for all x. The substitution $f(x)=\log g(x)$ then reduces equation (3) to equation (1). Thus we see that any continuous solution of (3) is of the form $g(x)=\exp(cx)$.

Next, we consider the equation

$$g(uv)=g(u)g(v). \tag{4}$$

If a solution $g(u)$ vanishes at some $\xi \neq 0$, then $g(u) \equiv 0$. Excluding this case, we assume that $g(u) \neq 0$ for $u \neq 0$. For $u, v > 0$, by the substitution $x = \log u$, $y = \log v$, we have an equation of the form (3). On the other hand, putting $v = -1$, we have $g(-u) = g(-1)g(u)$. Since $g^2(-1) = g(1) = 1$, we see that any continuous solution of (4) is of the form $|u|^c$ or $(\operatorname{sgn} u)|u|^c$ according as $g(-1) = 1$ or -1.

C. The General Addition Theorem and Related Functional Equations

The **general addition theorem** is: If the equation

$$f(x+y) = F(f(x), f(y)) \tag{5}$$

has a continuous nonconstant solution $f(x)$ on $-\infty < x < +\infty$, then $f(x)$ is strictly monotone, and $F(u, v)$ is strictly monotone increasing and continuous with respect to u and v for $\alpha < u$, $v < \beta$ and satisfies $\alpha < F(u, v) < \beta$ for $\alpha < u$, $v < \beta$. There is also a constant c satisfying $F(c, c) = c$, and the identity $F(F(u, v), w) = F(u, F(v, w))$ holds for any u, v, w in the interval (α, β). Conversely, if $F(u, v)$ is such a function, then (5) has a continuous nonconstant solution on $-\infty < x < +\infty$. Let $f(x)$ be such a solution. Then any other continuous solution is given by $f(cx)$. When $F(u, v)$ is continuously differentiable, a continuous solution $f(x)$ of (5) can be obtained as a solution of the differential equation

$$f'(x) = F_v(f(x), a)c, \qquad c = f'(0),$$

satisfying the initial condition $f(0) = a$.

Consider the equation

$$F(f(x), f(y), f(x+y)) = 0.$$

Suppose that $F(u, v, w)$ is a polynomial in u, v, and w. If this equation has a †meromorphic solution $f(x)$, then $f(x)$ must be a rational function, a rational function of $\exp cx$, or an elliptic function [1, p. 64].

Next, consider the equation

$$f(x+y) + f(x-y) = 2(f(x) + f(y)). \tag{6}$$

Any solution continuous on $-\infty < x < \infty$ is of the form $f(x) = cx^2$. When x is a point $(x_1, \ldots, x_n)$ of an n-dimensional Euclidean space, any continuous solution of (6) is given by a quadratic form $f(x) = \sum_{i,j}^{n} c_{ij} x_i x_j$.

Consider the equation

$$f(x+y) + f(x-y) = 2f(x)f(y). \tag{7}$$

Any solution continuous on $-\infty < x < \infty$ is of the form $f(x) = \cosh cx = (e^{cx} + e^{-cx})/2$ or $f(x) = \cos cx$. If $f(x)$ is allowed to take complex values, then any continuous solution can be written in the form $f(x) = (e^{bx} + e^{-bx})/2$ in terms of a complex number b. A special case is $\cos x$, since b may take purely imaginary values.

D. Schröder's Functional Equation

Schröder's functional equation is

$$f(\theta(x)) = cf(x), \tag{8}$$

where $\theta(x)$ is a given function and c is a constant. A general solution of (8) can be written as $f(x) = f_1(x)\varphi(x)$, where $f_1(x) \neq 0$ is a particular solution of (8) and $\varphi(x)$ is a general solution of the equation $\varphi(\theta(x)) = \varphi(x)$. Suppose that there is a point a such that $\theta(a) = a$, and $\theta(x)$ and $f(x)$ are both differentiable in a neighborhood of $x = a$. Then we have $f'(a) = 0$ or $\theta'(a) = c$. Consider the case where $\theta'(a) = c$, and suppose that $\theta(x)$ is twice differentiable at $x = a$. When $|c| = |\theta'(a)| < 1$, define $\theta_n(x)$ $(n = 0, 1, 2, \ldots)$ by $\theta_0(x) = x$ and $\theta_n(x) = \theta(\theta_{n-1}(x))$. Then the sequence $\{(\theta_n(x-a)c^{-n}\}$ $(n = 0, 1, 2, \ldots)$ converges uniformly in a neighborhood of $x = a$, and its limit function $f(x)$ is a solution of (8). When $|c| = |\theta'(a)| > 1$, put $\theta(x) = u$. Then we have the equation $f(\theta^{-1}(u)) = c^{-1}f(u)$, and the problem reduces to the previous case. The results obtained for equation (8) can be extended to the following system of equations:

$$\begin{aligned} &f_j(\theta_1(x), \theta_2(x), \ldots, \theta_n(x)) \\ &= \lambda_j f_j(x) + \delta_j f_{j-1}(x) + g_j(x), \quad j = 1, 2, \ldots, n, \end{aligned} \tag{9}$$

where the $\theta_j(x)$ are given functions holomorphic in a neighborhood of $x = 0$, $\theta_j(0) = 0$, the coefficients of f in the right-hand side of (9) are numbers such that the matrix $A = (a_{ij})$ ($a_{ij} = 0$ except for $a_{jj} = \lambda_j, a_{j-1,j} = \delta_j$) is the †Jordan canonical form of the matrix formed by $\{(\partial\theta_j/\partial x_k)_{x=0}\}$, and the $g_j(x)$ are polynomials consisting of terms of the form constant $\times$ $x_1^{m_1} x_2^{m_2} \ldots x_n^{m_n}$ with exponents $m_1, m_2, \ldots, m_n$ for which $\lambda_j = \lambda_1^{m_1} \ldots \lambda_n^{m_n}$. If $0 < |\lambda_j| < 1$ $(j = 1, 2, \ldots, n)$, then we can always choose the coefficients of the polynomials $g_j(x)$ so that equation (9) has a solution $\{f_j(x)\}$ holomorphic in a neighborhood of $x = 0$. The same conclusion can be obtained for $|\lambda_j| > 1$ $(j = 1, \ldots, n)$.

Consider **Abel's functional equation**

$$f(\theta(x)) = f(x) + 1. \tag{10}$$

If we put $\exp f(x) = \varphi(x)$, then we have Schröder's functional equation

$$\varphi(\theta(x)) = e\varphi(x).$$

Consider the equation

$$f(x+1) = A(x)f(x). \tag{11}$$

If we put $\varphi(x) = \log f(x)$, then we have a †linear difference equation of the form

$$\varphi(x+1) - \varphi(x) = \log A(x).$$

References

[1] E. C. Picard, Leçons sur quelques équations fonctionnelles, Gauthier-Villars, 1928.
[2] A. Ostrowski, Mathematische Miszellen XIV, Uber die Funktionalgleichung der Exponentialfunktion und verwandte Funktionalgleichungenm, Jber. Deutsch. Math. Verein., 38 (1929), 54–62.
[3] F. Hausdorff, Mengenlehre, de Gruyter, second edition, 1927, 175; English translation, Set theory, Chelsea, 1962.
[4] J. Aczél, Lectures on functional equations and their applications, Academic Press, 1966. (Original in German, 1961.)

389 (XIV.1) Special Functions

A. Special Functions

The term **special functions** usually refers to the classes of functions listed in (1)–(4) (other terms, such as **higher transcendental functions**, are sometimes used). (1) The †gamma function and related functions (→ 174 Gamma Function); (2) †Fresnel's integral, the †error function, the †logarithmic integral, and other functions that can be expressed as indefinite integrals of elementary functions (→ 167 Functions of Confluent Type D); (3) †elliptic functions (→ 134 Elliptic Functions); (4) solutions of †linear ordinary differential equations of the second order derived by the method of separation of variables in certain partial differential equations, e.g., †Laplace's equation, in various †curvilinear coordinates. Recently, new types of special functions, such as †Painlevé's, have been introduced as the solutions of special differential equations.

In this article we discuss class (4); for the other classes, see the articles quoted. Class (4) is further divided into the following three types, according to the character of the †singular points of the differential equations of which the functions are solutions. Equations with a smaller number of singular points than those indicated in (1)–(3) below can be integrated in terms of elementary functions.

(1) **Special functions of hypergeometric type** are solutions of differential equations with three †regular singular points on the Riemann sphere. Examples are the †hypergeometric function and the †Legendre function. Any function of this type reduces to a hypergeometric function through a simple transformation (→ 206 Hypergeometric Functions; 393 Spherical Functions).

(2) **Special functions of confluent type** are solutions of differential equations that are derived from †hypergeometric differential equations by the confluence of two regular singular points, that is, by making one of the regular singular points tend to the other one so that the resulting singularity is an †irregular singular point of class 1 (→ 167 Functions of Confluent Type). Any function of this type can be expressed by means of †Whittaker functions, of which many important special functions, such as †Bessel functions, are special cases (→ 39 Bessel Functions). Also, one can reduce to this type the †parabolic cylindrical functions, that is, the solutions of differential equations with only one singular point which is at infinity and is irregular of class 2.

(3) **Special functions of ellipsoidal type** are solutions of differential equations with four or five regular singular points, some of which may be confluent to become irregular singular points. Examples are †Lamé functions, †Mathieu functions, and †spheroidal wave functions (→ 133 Ellipsoidal Harmonics; 268 Mathieu Functions). In contrast to types (1) and (2), functions of type (3) are difficult to characterize by means of †difference-differential equations and have not been fully explored. Sometimes the term *special function* in the strict sense is not applied to them. To specify the special functions of types (1) and (2), the term *classical special functions* has been proposed.

B. Unified Theories of Special Functions

Though many special functions were introduced separately to solve practical problems, several unified theories have been proposed. The classification in Section A based on differential equations may be regarded also as a kind of unified theory. Other examples are:

(1) Expression by †Barnes's extended hypergeometric function or its extension to the case of several variables by means of a definite integral of the form

$$\int(\zeta-a_1)^{b_1}(\zeta-a_2)^{b_2}\ldots(\zeta-a_m)^{b_m}(\zeta-z)^c\,d\zeta$$

(→ 206 Hypergeometric Functions).

(2) A unified theory [14] that includes the gamma function and is based upon Truesdell's difference-differential equation

$$\partial F(z,\alpha)/\partial z=F(z,\alpha+1).$$

(3) Unification from the standpoint of expansions in terms of †zonal spherical functions of a differential operator (the Laplacian) invariant under a transitive group of motions on a †symmetric Riemannian manifold (→ 437

Unitary Representations). With this approach a great variety of formulas can be derived in a unified way [3, 4].

References

[1] J. Dieudonné, Special functions and linear representations of Lie groups, Expository lectures from the CBMS regional conference held at East Carolina Univ., March 1979, Amer. Math. Soc., 1980.
[2] B. C. Carlson, Special functions of applied mathematics, Academic Press, 1977.
[3] J. D. Talman, Special functions: A group-theoretic approach, Benjamin, 1968.
[4] N. Ya. Vilenkin, Special functions and the theory of group representations, Amer. Math. Soc. Transl. of Math. Monographs 22, 1968. (Original in Russian, 1965.)
[5] W. Miller, Jr., Lie theory and special functions, Academic Press, 1968.
[6] Y. L. Luke, The special functions and their approximations I, II, Academic Press, 1969.
[7] S. Moriguti, K. Udagawa, and S. Hitotumatu, Mathematical formulas, Special functions (in Japanese), Iwanami, 1960.
[8] E. T. Whittaker and G. N. Watson, A course of modern analysis, Cambridge Univ. Press, 1902; fourth edition, 1958.
[9] E. Jahnke and F. Emde, Funktionentafeln mit Formalen und Kurven, Teubner, 1933; bilingual edition, Tables of functions with formulas and curves, Dover, 1945.
[10] W. Magnus and F. Oberhettinger, Formulas and theorems for the special functions of mathematical physics, Springer, third enlarged edition, 1966.
[11] R. Courant and D. Hilbert, Methods of mathematical physics I, II, Interscience, 1953, 1962.
[12] A. Erdélyi (ed.), Higher transcendental functions I–III, McGraw-Hill, 1953–1955.
[13] J. Meixner, Spezielle Funktionen der mathematischen Physik, Handbuch der Physik II, Springer, second edition, 1933.
[14] C. A. Truesdell, An essay toward a unified theory of special functions based upon the functional equation $\partial F(z,\alpha)/\partial z = F(z, \alpha+1)$, Ann. Math. Studies, Princeton Univ. Press, 1948.
[15] I. N. Sneddon, Special functions of mathematical physics and chemistry, Oliver & Boyd, 1956.
[16] F. W. Schäfke, Einführung in die Theorie der spezielle Funktionen der mathematischen Physik, Springer, 1963.
[17] G. Szegö, Orthogonal polynomials, Amer. Math. Soc. Colloq. Publ., revised edition, 1959.
[18] E. D. Rainville, Special functions, Macmillan, 1960; Chelsea, second edition, 1971.

390 (XII.12) Spectral Analysis of Operators

A. General Remarks

Throughout this article, X stands for a †complex linear space and A for a †linear operator in X. Except when X is finite-dimensional, A need not be defined over all X. A linear operator A in X is by definition a linear mapping whose †domain $D(A)$ and †range $R(A)$ are linear subspaces of X. A complex number λ is said to be an **eigenvalue** (**proper value** or **characteristic value**) of A if there exists an $x \in D(A)$ such that $Ax = \lambda x$, $x \neq 0$. Any such x is called an **eigenvector** (**eigenelement**, **proper vector**, **characteristic vector**) associated with λ. When X is a †function space, the word **eigenfunction** is also used. For an eigenvalue λ of A, the subspace $M(\lambda)$ of X given by

$$M(\lambda) = M(\lambda; A) = \{x \mid Ax = \lambda x\},$$

i.e., the subspace consisting of 0 and all eigenvectors associated with λ, is called the **eigenspace** associated with λ, and the number $m(\lambda) = \dim M(\lambda)$ is called the **geometric multiplicity** of λ. The eigenvalue λ is said to be (**geometrically**) **simple** or **degenerate** according as $m(\lambda) = 1$ or $m(\lambda) \geqslant 2$. The problem of seeking eigenvalues and eigenvectors is referred to as the **eigenvalue problem**.

When X is a †topological linear space, the notion of eigenvalues leads to a more general object called the spectrum of A. Let λ be a complex number and put $A_\lambda = \lambda I - A$, where I is the †identity operator in X. Furthermore, put $R_\lambda = (A_\lambda)^{-1} = (\lambda I - A)^{-1}$, if the inverse exists. Then the **resolvent set** $\rho(A)$ of A is defined to be the set of all λ such that R_λ exists, has domain †dense in X, and is continuous. The **spectrum** $\sigma(A)$ of A is, by definition, the complement of $\rho(A)$ in the complex plane, and it is divided into three mutually disjoint sets: the **point spectrum** $\sigma_P(A)$, the **continuous spectrum** $\sigma_C(A)$, and the **residual spectrum** $\sigma_R(A)$. These are defined as follows: $\sigma_P(A) = \{\lambda \mid R_\lambda$ does not exist$\} = \{\lambda \mid \lambda$ is an eigenvalue of $A\}$; $\sigma_C(A) = \{\lambda \mid R_\lambda$ exists and has domain dense in X, but is not continuous$\}$; $\sigma_R(A) = \{\lambda \mid R_\lambda$ exists, but its domain is not dense in $X\}$.

Let X be a †Banach space and $\mathbf{B}(X)$ the set of all †bounded linear operators with domain X. If A is a †closed operator in X, then $\lambda \in \rho(A)$ if and only if $R_\lambda \in \mathbf{B}(X)$. Moreover, $\sigma(A)$ is a closed set. In particular, if $A \in \mathbf{B}(X)$, then $\sigma(A)$ is a nonempty compact set. In this case the **spectral radius** $r_\sigma(A)$ is defined as $r_\sigma(A) = \sup_{\lambda \in \sigma(A)} |\lambda|$. Then $r_\sigma(A) \leqslant \|A^n\|^{1/n}$, $n = 1, 2, \ldots$, and $\|A^n\|^{1/n} \to r_\sigma(A)$, $n \to \infty$.

In many problems of analysis crucial roles have been played by methods involving the spectrum and other related concepts. This branch of analysis is called **spectral analysis**. For an infinite-dimensional X the theory is well developed when X is a †Hilbert space and A is †self-adjoint or †normal.

B. Eigenvalue Problems for Matrices

Throughout this section let X be an N-dimensional complex linear space ($N<\infty$) and A a linear operator in X. (We assume that A is defined over all X.) With respect to a fixed basis $(\psi_1, \ldots, \psi_N)$ of X, the operator A is represented by an $N \times N$ matrix, also denoted by A. Then the eigenvalues of A coincide with the roots of the †characteristic equation $\det(\lambda I - A)=0$. There are no points of the spectrum other than eigenvalues, that is $\sigma(A)=\sigma_P(A)$. Let $\lambda \in \sigma(A)$. The multiplicity $\tilde{m}(\lambda)$ of λ as a root of the characteristic equation is called the **(algebraic) multiplicity** of the eigenvalue λ. The sum of $\tilde{m}(\lambda)$ over all the eigenvalues of A is equal to N. The eigenvalue λ is said to be **(algebraically) simple** or **degenerate** according as $\tilde{m}(\lambda)=1$ or $\tilde{m}(\lambda)\geqslant 2$. Let $v=1,2,\ldots,$ and $N_v(\lambda)=\{x \mid (\lambda I - A)^v x=0\}$. Then $\{N_v(\lambda)\}$ forms a nondecreasing chain of subspaces $M(\lambda)=N_1(\lambda)\subset N_2(\lambda)\subset \ldots$, which ceases to increase after a finite number of steps. When $v\geqslant \tilde{m}(\lambda)$, the space $N_v(\lambda)$ is equal to a fixed subspace $\tilde{M}(\lambda)$, sometimes called the **root subspace** (or **generalized eigenspace** or **principal subspace**) of A associated with λ. A vector in the root subspace is called a **root vector** (or a **generalized eigenvector**). Then $\dim \tilde{M}(\lambda)=\tilde{m}(\lambda)$ and hence $m(\lambda)\leqslant \tilde{m}(\lambda)$. When A is a †normal matrix, $M(\lambda)=\tilde{M}(\lambda)$ and $m(\lambda)=\tilde{m}(\lambda)$.

If two matrices A and B are †similar, i.e., if there exists an invertible matrix P such that $B=P^{-1}AP$, then A and B have the same eigenvalues with the same algebraic (and geometric) multiplicities. The same conclusion holds for A and A', where A' is the †transpose of A. For the adjoint matrix A^* we have $\sigma(A^*)=\overline{\sigma(A)}\equiv\{\bar{\lambda} \mid \lambda\in\sigma(A)\}$. For an arbitrary polynomial f the relation $\sigma(f(A))=f(\sigma(A))\equiv\{f(\lambda) \mid \lambda\in\sigma(A)\}$ holds **(Frobenius's theorem)**. These relations can be extended to operators in a Banach space. In particular, $\sigma(f(A))=f(\sigma(A))$ if A is a bounded operator and f is a function holomorphic in a neighborhood of $\sigma(A)$ (for the †spectral mapping theorem → 251 Linear Operators).

In the next four paragraphs, in which the spectral properties of †normal or †Hermitian matrices is discussed, we introduce into X the Euclidean †inner product (,), regarding X as a space of N-tuples of scalars. Let A be an $N \times N$ normal matrix. Then the eigenspaces associated with different eigenvalues of A are mutually orthogonal. Moreover, the eigenspaces of A as a whole span the entire space X. One can therefore choose a †basis of X formed by a †complete orthonormal set of eigenvectors of A. Specifically, there exists a basis $\{\varphi_j \mid j=1,\ldots,N\}$ of X such that $A\varphi_j=\mu_j\varphi_j$ and $(\varphi_i,\varphi_j)=\delta_{ij}$, where δ_{ij} is the †Kronecker delta. Moreover, $\mu_1,\ldots,\mu_N$ exhaust all the eigenvalues of A. In terms of the basis $\{\varphi_j\}$, an arbitrary $x\in X$ can be expanded as

$$Ax=\sum_{j=1}^{N}\mu_j(x,\varphi_j)\varphi_j=\sum_{\lambda\in\sigma(A)}\lambda P_\lambda x, \tag{1}$$

where P_λ is the orthogonal †projection on the eigenspace associated with the eigenvalue λ.

Of particular importance among normal matrices are Hermitian matrices and †unitary matrices. The eigenvalues of a Hermitian matrix are real, and those of a unitary matrix have the absolute value 1.

Solving the eigenvalue problem of a normal matrix A leads immediately to the diagonalization of A. For instance, let U be the $N \times N$ matrix whose jth column is equal to φ_j. Here the basis $\{\varphi_j\}$ is as before, and each φ_j is regarded as a column vector. Then U is unitary, and the transform U^*AU of A by U is the diagonal matrix whose diagonal entries are the μ_j. The problem of transforming a †Hermitian form to its canonical form can also be solved by means of U. In fact, a Hermitian form $Q=Q(x)$ on X is expressed as $Q(x)=(Ax,x)$ with a Hermitian matrix A. For this A, construct U as before. Then by the transformation $x=Uy$ of the coordinates of X, the form Q is converted to its canonical form $Q=\mu_1|y_1|^2+\ldots+\mu_N|y_N|^2$. When X is a real linear space and A is a real symmetric matrix, U is an orthogonal matrix. By means of the orthogonal transformation $x=Uy$, the surface of the second order $Q(x)=1$ in $\mathbf{R}^N$ is converted to the form $\mu_1 y_1^2+\ldots+\mu_N y_N^2=1$. The orthogonal transformation $x=Uy$ is called the **transformation to principal axes** of the surface $Q(x)=1$.

When A is not normal, it can be transformed into †Jordan's canonical form by a basis φ_j taken from the root subspaces $\tilde{M}(\lambda)$. However, φ_j need not be orthonormal even when A is diagonalizable.

C. Spectral Analysis in Hilbert Spaces

Throughout the rest of this article except for the last section, X is assumed to be a Hilbert space with inner product (,). Furthermore, the most complete discussions will be confined to †normal or †self-adjoint operators. A fundamental theorem in spectral analysis for such operators is the spectral theorem, which asserts that a representation such as (1) holds in a generalized form. When the operator is †compact, we have only to replace the sum by an infinite sum (→ 68 Compact and Nuclear Operators). In the general case we need a kind of integral. This is discussed in detail in Sections D and E. The general theory of spectral analysis for nonnormal operators, however, is rather involved even in Hilbert spaces, but two important developments can be noted. One is the theory of Volterra operators, and the other is the theory of essentially normal operators. The former is discussed in Section H and the latter and its related results in Sections I and J.

D. Spectral Measure

Let $\mathscr{B}$ be a †completely additive class of subsets of a set Ω, that is, $(\Omega, \mathscr{B})$ is a †measurable space. An operator-valued set function $E = E(\cdot)$ defined on $\mathscr{B}$ is said to be a (self-adjoint) **spectral measure** if (i) $E(M)$, $M \in \mathscr{B}$, is an †orthogonal projection in X; (ii) $E(\Omega) = I$; and (iii) E is †countably additive, that is,

$$E\left(\bigcup_{n=1}^{\infty} M_n\right) = \sum_{n=1}^{\infty} E(M_n)$$

(†strong convergence) for a disjoint sequence $\{M_n\}$ of subsets in $\mathscr{B}$. A spectral measure E satisfies $E(M \cap N) = E(M)E(N) = E(N)E(M)$, $M, N \in \mathscr{B}$. Spectral measures which are frequently used in spectral analysis are those defined on the family $\mathscr{B}_r$ ($\mathscr{B}_c$) of all †Borel sets in the field of real (complex) numbers **R** (**C**). A spectral measure on $\mathscr{B}_r$ ($\mathscr{B}_c$) is sometimes referred to as a **real (complex) spectral measure**. For such a spectral measure E the **support** (or the **spectrum**) of E, denoted by $\Lambda(E)$, is defined to be the complement of the largest open set G for which $E(G) = 0$. A complex spectral measure such that $\Lambda(E) \subset \mathbf{R}$ can be identified with a real spectral measure.

Let E be a spectral measure on $\mathscr{B}_r$, and put

$$E_\lambda = E((-\infty, \lambda]), \qquad -\infty < \lambda < \infty. \tag{2}$$

Then E_λ satisfies the relations

$$E_\lambda E_\mu = E_{\min(\lambda,\mu)}, \qquad \operatorname*{s-lim}_{\lambda \to \mu+0} E_\lambda = E_\mu,$$

$$\operatorname*{s-lim}_{\lambda \to -\infty} E_\lambda = 0, \qquad \operatorname*{s-lim}_{\lambda \to +\infty} E_\lambda = I, \tag{3}$$

where s-lim stands for strong convergence. A family $\{E_\lambda\}_{\lambda \in \mathbf{R}}$ of orthogonal projections satisfying the relation (3) is called a **resolution of the identity**. Relation (2) gives a one-to-one correspondence between the resolutions of the identity and the spectral measures on $\mathscr{B}_r$.

Let E be a spectral measure on $\mathscr{B}_r$, and let $x, y \in X$. Then the set function $M \to (E(M)x, x) = \|E(M)x\|^2$ is a bounded regular †measure in the ordinary sense, and the set function $M \to (E(M)x, y)$ is a complex-valued regular †completely additive set function. For every complex Borel †measurable function f on **R**, the operator $S(f)$ in X is defined by the relations

$$D(S(f)) = \left\{ x \,\middle|\, \int_{-\infty}^{\infty} |f(\lambda)|^2 (E(d\lambda)x, x) < \infty \right\},$$

$$(S(f)x, y) = \int_{-\infty}^{\infty} f(\lambda)(E(d\lambda)x, y), \tag{4}$$

$$x \in D(S(f)), \qquad y \in X.$$

$S(f)$ is a densely defined closed operator and is denoted by $S(f) = \int_{-\infty}^{\infty} f(\lambda) E(d\lambda)$. The correspondence $f \mapsto S(f)$ satisfies formulas of the so-called operational calculus (→ 251 Linear Operators). In particular, $S(\bar{f}) = S(f)^*$, and hence $S(f)$ is self-adjoint if f is real-valued. If f is bounded on the support of E, then $S(f)$ is everywhere defined in X and is bounded. $S(f)$ is sometimes called the **spectral integral** of f with respect to E. The operator $S(f)$ can be defined in a similar way for a spectral measure on $\mathscr{B}_c$ (and for a more general spectral measure).

E. Spectral Theorems

For every self-adjoint operator H in a Hilbert space X, there exists a unique real spectral measure E such that

$$H = \int_{-\infty}^{\infty} \lambda E(d\lambda). \tag{5}$$

In other words, H and E correspond to each other by the relations

$$D(H) = \left\{ x \,\middle|\, \int_{-\infty}^{\infty} \lambda^2 (E(d\lambda)x, x) < +\infty \right\},$$

$$(Hx, y) = \int_{-\infty}^{\infty} \lambda (E(d\lambda)x, y), \tag{6}$$

$$x \in D(H), \qquad y \in X.$$

This is the **spectral theorem** for self-adjoint operators. The support of E is equal to the spectrum $\sigma(H)$, so that we can write

$$H = \int_{\sigma(H)} \lambda E(d\lambda) = \int_{-\infty}^{\infty} \lambda \chi_{\sigma(H)}(\lambda) E(d\lambda),$$

where χ_M stands for the †characteristic function of M. Formulas (5) and (6) are called the **spectral resolution** (or **spectral representation**) of the **self-adjoint operator** H. We call E the spectral measure for H, and the $\{E_\lambda\}$ corresponding to E by formula (2) (or sometimes E itself) the resolution of the identity for H. Let λ be a real number. Then $\lambda \in \sigma_P(H)$ if and only if $E(\{\lambda\}) \neq 0$. Also, $\lambda \in \sigma_C(H)$ if and only if $E(\{\lambda\}) = 0$ and $E(V) \neq 0$ for any neighborhood

V of λ. The spectral measure E can be represented in terms of the resolvent $R(\alpha; H) = (\alpha I - H)^{-1}$ of H by the formula

$$E((a,b)) = \lim_{\delta \downarrow 0} \lim_{\varepsilon \downarrow 0} \frac{1}{2\pi i} \int_{a+\delta}^{b-\delta} \{R(\mu - \varepsilon i; H) - R(\mu + \varepsilon i; H)\} d\mu$$

(strong convergence).

For every normal operator A in X, there exists a unique spectral measure E on the family of all complex Borel sets $\mathscr{B}_c$ such that

$$A = \int_{\mathbf{C}} z E(dz).$$

This is called the **complex spectral resolution** (or **complex spectral representation**) of the **normal operator** A. The support $\Lambda(E)$ is equal to $\sigma(A)$. There are characterizations of point and continuous spectra similar to the case of self-adjoint operators. Normal operators have no residual spectra. For a unitary operator U, the support of the associated spectral measure is contained in the unit circle Γ, so that U can be represented as

$$U = \int_{\Gamma} e^{i\theta} F(d\theta) \tag{7}$$

with a spectral measure F defined on Γ. Formula (7) is the **spectral resolution** of the **unitary operator** U.

For a self-adjoint operator $H = \int_{-\infty}^{\infty} \lambda E(d\lambda)$ the following two types of classification of $\sigma(H)$ are often useful.

(i) The **essential spectrum** $\sigma_e(H)$ is by definition the set $\sigma(H)$ minus all the isolated eigenvalues of H with finite multiplicity. When H is bounded this definition of the essential spectrum coincides with that to be given in Section I for a general $A \in B(X)$. $\sigma(H) \setminus \sigma_e(H)$ is called the **discrete spectrum** of H.

(ii) The set $X_{ac}(H)$ (resp. $X_s(H)$) (called the **space of absolute continuity** (resp. **singularity**) with respect to H) of all $u \in X$ such that the measure $(E(d\lambda)u, u)$ is absolutely continuous (resp. singular) with respect to the †Lebesgue measure is a closed subspace of X that reduces H. The restriction of H to X_{ac} (resp. $X_s(H)$) is called the absolutely continuous (resp. singular) part of H, and its spectrum, denoted by $\sigma_{ac}(H)$ (resp. $\sigma_s(H)$), is called the **absolutely continuous** (resp. **singular**) **spectrum** of H. Note that $\sigma_{ac}(H)$ and $\sigma_s(H)$ may not be disjoint.

F. Functions of a Self-Adjoint Operator

Let $H = \int_{-\infty}^{\infty} \lambda E(d\lambda)$ be a self-adjoint operator in X. For a complex-valued Borel measurable function f on $\mathbf{R}$, we define $f(H)$ to be the operator $S(f)$ determined by (4) in reference to the resolution of the identity E associated with H:

$$f(H) = \int_{-\infty}^{\infty} f(\lambda) E(d\lambda).$$

For an arbitrary $a \in X$, let $L_2(a)$ be the L_2-space over the measure $\mu_a = \mu_a(\cdot) = (E(\cdot)a, a)$. In other words, $f \in L_2(a)$ if and only if $a \in D(f(H))$. The correspondence $f \leftrightarrow f(H)a$ gives an isometric isomorphism between $L_2(a)$ and the subspace $M(a) = \{f(H)a \mid f \in L_2(a)\}$ of X. (In particular, $M(a)$ is closed.) H is reduced by $M(a)$, and the part of H in $M(a)$ corresponds to the multiplication $f(\lambda) \to \lambda f(\lambda)$ in $L_2(a)$.

For a given self-adjoint operator H there exists a (not necessarily countable) family $\{a_\theta\}_{\theta \in \Theta}$ of elements a_θ of X such that

$$X = \sum_{\theta \in \Theta} M(a_\theta), \tag{8}$$

where $\sum$ stands for the †direct sum of mutually orthogonal closed subspaces. Consequently, X is represented by the direct sum $\sum_{\theta \in \Theta} L_2(a_\theta)$ of L_2-spaces. If $x \in D(H)$ is represented by $\{f_\theta\}_{\theta \in \Theta}$ in this representation, Hx is represented by $\{\lambda f_\theta\}_{\theta \in \Theta}$.

G. Unitary Equivalence and Spectral Multiplicity

In this section X is assumed to be a †separable Hilbert space. Then (8) can be made more precise. Namely, for a self-adjoint operator H, we can find a countable family $\{a_n\}_{n=1}^{\infty}$ of elements of X such that

$$X = \sum_{n=1}^{\infty} M(a_n) \cong \sum_{n=1}^{\infty} L_2(a_n), \tag{9}$$

$$\mu_{a_{n+1}} \text{ is †absolutely continuous with respect to } \mu_{a_n}, \quad n = 1, 2, \ldots. \tag{10}$$

Furthermore, if $\{a_n'\}$ is another family satisfying (9) and (10), then μ_{a_n} and $\mu_{a_n'}$ are absolutely continuous with respect to each other (**Hellinger-Hahn theorem**). μ_{a_1} is said to be the **maximum spectral measure**.

Two operators H_1 and H_2 are said to be **unitarily equivalent** if there exists a unitary operator U such that $H_2 = U^* H_1 U$. A criterion for unitary equivalence of self-adjoint operators can be given in terms of the spectral representation given previously. Namely, let $\{a_n^{(i)}\}$, $i = 1, 2$, be a sequence satisfying (9) and (10) with respect to H_i. Then H_1 and H_2 are unitarily equivalent if and only if $\mu_{a_n^{(1)}}$ and $\mu_{a_n^{(2)}}$ are absolutely continuous with respect to each other for all $n = 1, 2, \ldots$.

A self-adjoint operator H is said to have a **simple spectrum** if there exists an $a \in X$ such that $M(a) = X$. Such an $a \in X$ is called a **generating element** of X with respect to H.

Self-adjoint operators with simple spectra are closely related to Jacobi matrices. Let H be such an operator with a generating element $a \in X$. Take a complete orthonormal set $\{G_n\}_{n=1}^{\infty}$ in $L_2(a)$ such that $G_n = G_n(\lambda)$ is a polynomial of degree $n-1$ and $\lambda G_n(\lambda) \in L_2(a)$. Then $\{g_n\}_{n=1}^{\infty}$, $g_n = G_n(H)a$, is a complete orthonormal set in X. The matrix representation $\{a_{mn}\}$, $a_{mn} = (Hg_n, g_m)$ of H with respect to the basis $\{g_n\}$ has the following properties: (i) $a_{mn} = 0$ if $|m-n| \geqslant 2$; (ii) $a_{n,n+1} = \overline{a_{n+1,n}} \neq 0$; (iii) a_{nn} is real. Any infinite matrix $\{a_{mn}\}$ satisfying (i), (ii), and (iii) is called a **Jacobi matrix**. A Jacobi matrix determines a †symmetric operator whose †deficiency index is either (0, 0) or (1, 1). Any self-adjoint extension has a simple spectrum. (For more details about Jacobi matrices and their applications → [8].)

H. Triangular Representation of Volterra Operators

A linear operator A in a Hilbert space X is called a †Volterra operator if it is †compact and †quasinilpotent (i.e., 0 is the only spectrum). The name is justified because under very general assumptions such an operator is unitarily equivalent to the integral operator of Volterra type in the vector-valued L_2 space on $[0, 1]$. Let $\mathfrak{P}$ be a maximal †totally ordered family of orthogonal projections in X such that PX is an †invariant subspace of a Volterra operator A for every $P \in \mathfrak{P}$. Such a family $\mathfrak{P}$ always exists and is called a maximal **eigenchain** of A. Then generalizing the triangular representation of nilpotent matrices, we have the integral representation

$$A = 2i \int_{\mathfrak{P}} PA_I \, dP,$$

where $A_I = (A - A^*)/(2i)$ is the imaginary part of A and the integral is the limit in norm of approximating sums of the form $\sum Q_i A_I (P_i - P_{i-1})$ for finite partitions $0 = P_0 < P_1 < \ldots < P_n = I$ of $\mathfrak{P}$ in which Q_i is an arbitrary projection in $\mathfrak{P}$ such that $P_{i-1} \leqslant Q_i \leqslant P_i$ (M. S. Brodskiĭ). Conversely, let $\mathfrak{P}$ be a totally ordered family of orthogonal projections that contains 0 and the identity. If the integral $A = \int_{\mathfrak{P}} PBdP$ converges in norm for a compact linear operator B, then A is a Volterra operator and $\mathfrak{P}$ is an eigenchain of A. If, moreover, B is self-adjoint, we have $B = A_I$ (I. C. Gokhberg and M. G. Kreĭn; Brodskiĭ). Furthermore, assume for simplicity that $\mathfrak{P}$ is continuous in the sense that for every $P_1 < P_2$ in $\mathfrak{P}$ there exists an element P in $\mathfrak{P}$ such that $P_1 < P < P_2$. If B is a †Hilbert-Schmidt operator, then the integral $A = \int_{\mathfrak{P}} PBdP$ converges in the †Hilbert-Schmidt norm and the mapping $B \mapsto A$ is an orthogonal projection to the set of all Volterra operators of Hilbert-Schmidt class possessing $\mathfrak{P}$ as an eigenchain (Gokhberg and Kreĭn). Volterra operators with the imaginary part of the †trace class are especially important for applications. In this case we have the following fundamental theorem on the density of the spectrum of the real part A_R of the Volterra operator A: Let $n_+(r; A_R)$ and $n_-(r; A_R)$ be the numbers of eigenvalues of A_R in the intervals $[1/r, \infty)$ and $(-\infty, -1/r]$, respectively.

Then $\lim_{r \to \infty} n_+(r; A_R)/r = \lim_{r \to \infty} n_-(r; A_R) = h/\pi$.

The number h is given by

$$h = \int_{\mathfrak{P}} \|dP \, A_I \, dP\|_1$$

$$= \inf \sum \|(P_i - P_{i-1}) A_I (P_i - P_{i-1})\|_1,$$

where the norm is the †trace norm and the infimum is taken over all finite partitions P_i of a maximal eigenchain $\mathfrak{P}$ for A. We refer for further details and applications to the books by Gokhberg and Kreĭn [9, 10].

I. Fredholm Operators and Essential Spectra of Operators

Throughout Sections I and J we assume that X is a separable infinite-dimensional complex Hilbert space, and we consider only bounded linear operators in X. The set $\mathbf{B}^{(c)}(X)$ of all †compact linear operators in X is a †maximal two-sided ideal of the †C^*-algebra $\mathbf{B}(X)$ of all bounded linear operators. The simple quotient C^*-algebra $\mathbf{A}(X) = \mathbf{B}(X)/\mathbf{B}^{(c)}(X)$ is called the **Calkin algebra**. We denote the quotient mapping by $\pi: \mathbf{B}(X) \to \mathbf{A}(X)$. Then an operator $A \in \mathbf{B}(X)$ is a †Fredholm operator if and only if its image $\pi(A)$ is an invertible element of $\mathbf{A}(X)$. Let $\mathbf{F}(X)$ be the set of all Fredholm operators in X, and let $\mathbf{F}_n(X)$, $n \in \mathbf{Z}$, be its subset of all operators of †index n. $\mathbf{F}_n(X)$ is a connected set in $\mathbf{B}(X)$, and in particular, $\mathbf{F}_0(X)$ is the inverse image of the connected component of the identity in the multiplicative topological group $\pi(\mathbf{F}(X))$ of all invertible elements in $\mathbf{A}(X)$. The index gives the group isomorphisms $\mathbf{F}(X)/\mathbf{F}_0(X) \cong \pi(\mathbf{F}(X))/\pi(\mathbf{F}_0(X)) \cong \mathbf{Z}$. More generally, we have for any compact topological space Y the group isomorphisms $[Y, \mathbf{F}(X)] \cong [Y, \pi(\mathbf{F}(X))] \cong K(Y)$ of the groups of †homotopy classes of continuous mappings and the K-group in the †K-theory (M. F. Atiyah [12]).

If N is a †normal operator, its **essential spectrum** $\sigma_e(N)$ is defined to be the set of all $\lambda \in \sigma(N)$ that is not an isolated eigenvalue of finite multiplicity. Let H_1 and H_2 be self-adjoint operators. Then $\sigma_e(H_1) = \sigma_e(H_2)$ if and only if $U^*H_1U=H_2+K$ for a unitary operator U and a compact operator K (Weyl-von Neumann theorem).

I. D. Berg and W. Sikonia (1971) extended this result to normal operators N_1 and N_2. Moreover, for any compact subset Y of $\mathbf{C}$ there exists a normal operator N such that $\sigma_e(N)=Y$. Hence it follows that the essential spectrum $\sigma_e(N)$ of a normal operator N coincides with the †spectrum $\sigma(\pi(N))$ of the image in $\mathbf{A}(X)$. Thus we define the **essential spectrum** $\sigma_e(A)$ of an arbitrary operator A to be the spectrum $\sigma(\pi(A))$ in $\mathbf{A}(X)$. An operator $A\in\mathbf{B}(X)$ is said to be **essentially normal** (resp. **essentially self-adjoint, essentially unitary**) if $\pi(A)$ is normal (resp. self-adjoint, unitary) in $\mathbf{A}(X)$. (Note that this definition of essentially self-adjoint operators is completely different from that in 251 Linear Operators E.) Since an essentially self-adjoint operator is the sum of a self-adjoint operator and a compact operator, the Weyl-von Neumann theorem classifies essentially self-adjoint operators up to †unitary equivalence modulo $\mathbf{B}^{(c)}(X)$. An operator A is essentially normal if and only if the commutator $[A, A^*]$ is compact, but it need not be the sum of a normal operator and a compact operator. For example, let V be the **unilateral shift operator** that maps the orthonormal basis $\mathbf{e}_i$ of X into $\mathbf{e}_{i+1}$ for every $i=1,2,\ldots$. Then V is essentially unitary, but it cannot be written as the sum of a normal operator and a compact operator. The essential spectrum $\sigma_e(V)$ is the unit circle, whereas the spectrum $\sigma(V)$ is the unit disk and $\operatorname{ind}(V-\lambda)=-1$ for $|\lambda|<1$.

J. The Brown-Douglas-Fillmore (BDF) Theory

The following is the main theorem for essentially normal operators, due to L. G. Brown, R. G. Douglas, and P. A. Fillmore [14]. Let A_1 and A_2 be essentially normal operators. There are a unitary operator U and a compact operator K such that $U^*A_1U=A_2+K$ if and only if $\sigma_e(A_1)=\sigma_e(A_2)$ and $\operatorname{ind}(A_1-\lambda)=\operatorname{ind}(A_2-\lambda)$ for every λ in the complement of the essential spectrum. An essentially normal operator A is the sum of a normal operator and a compact operator if and only if $\operatorname{ind}(A-\lambda)=0$ for every λ in the complement of $\sigma_e(A)$.

To prove this and many other facts, they developed the theory of extension of $\mathbf{B}^{(c)}(X)$ by the C^*-algebra $C(Y)$ of continuous complex-valued functions on a compact metrizable space Y [14–16]. This revealed deep relations between the theory of operator algebras on Hilbert spaces (→ 36 Banach Algebras, 308 Operator Algebras) and algebraic topology (in particular, K-theory; → 237 K-Theory). Extension theory also gives a natural setting for the †index theory of elliptic differential operators due to Atiyah and I. M. Singer [13, 16].

An **extension** of $\mathbf{B}^{(c)}(X)$ by $C(Y)$ is a †short exact sequence

$$0\to\mathbf{B}^{(c)}(X)\xrightarrow{\varepsilon}E\xrightarrow{\varphi}C(Y)\to 0 \tag{11}$$

of a C^*-subalgebra E of $\mathbf{B}(X)$ and †$*$-homomorphisms, i.e., E is a C^*-subalgebra of $\mathbf{B}(X)$ containing the identity I and including $\mathbf{B}^{(c)}(X)$ as a C^*-subalgebra, and φ is a $*$-homomorphism onto $C(Y)$ whose kernel is equal to $\mathbf{B}^{(c)}(X)$. Or equivalently, an extension is a unital (identity preserving) $*$-monomorphism τ: $C(Y)\to\mathbf{A}(X)$ defined by $\tau=\pi\circ\varphi^{-1}$. (For general extension of C^*-algebras → 36 Banach Algebras.) Two extensions (E_1,φ_1) and (E_2,φ_2) (or τ_1 and τ_2) are said to be equivalent if there exists a $*$-isomorphism $\psi: E_1\to E_2$ such that $\varphi_2\circ\psi=\varphi_1$ (or equivalently there exists a unitary operator U such that $\pi(U^*)\,\tau_1(f)\pi(U)=\tau_2(f)$ for every $f\in C(Y)$). We denote by $\operatorname{Ext}(Y)$ the set of all equivalence classes of extensions of $\mathbf{B}^{(c)}(X)$ by $C(Y)$.

Let A be an essentially normal operator in X with the essential spectrum $\sigma_e(A)=Y$. Then the C^*-algebra E_A generated by $\mathbf{B}^{(c)}(X)$, A and the identity I, and the $*$-homomorphism φ_A of E_A onto $C(Y)$ which sends A to the function $\chi(z)=z$ define an extension (E_A,φ_A). It is easy to see that two essentially normal operators A_1 and A_2 are unitarily equivalent modulo $\mathbf{B}^{(c)}(X)$ if and only if $\sigma_e(A_1)=\sigma_e(A_2)$ and the extensions (E_{A_1},φ_{A_1}) and (E_{A_2},φ_{A_2}) are equivalent. Conversely, if Y is a compact subset of $\mathbf{C}$ and (11) is an extension, then (E,φ) is equivalent to (E_A,φ_A), where A is an essentially normal operator in E such that $\varphi(A)=\chi$.

Extensions of $\mathbf{B}^{(c)}(X)$ by $C(Y)$ appear also in different parts of analysis. Let X be the Hilbert space $L_2(M)$ on a compact differentiable manifold M relative to a fixed smooth measure and let E be the C^*-subalgebra of $\mathbf{B}(X)$ generated by all zeroth-order †pseudodifferential operators together with $\mathbf{B}^{(c)}(X)$. Then E and the †symbol mapping $\varphi: E\to C(S^*(M))$ define an extension of $\mathbf{B}^{(c)}(X)$ by $C(S^*(M))$, where $S^*(M)$ is the †cosphere bundle of M. This extension is closely related to the †Atiyah-Singer index theorem. Let Ω be a †strongly pseudoconvex domain in $\mathbf{C}^n$. Then the C^*-algebra generated by †Toeplitz operators with continuous symbol gives rise to an extension of $\mathbf{B}^{(c)}(H_2(\partial\Omega))$ by $C(\partial\Omega)$.

Let τ_1 and τ_2 be $*$-monomorphisms from $C(Y)$ into $\mathbf{A}(X)$ and $a_1=[\tau_1]$ and $a_2=[\tau_2]$ be corresponding elements in $\operatorname{Ext}(Y)$. Then the sum $a_1+a_2\in\operatorname{Ext}(Y)$ is defined to be the equivalence class of $\tau: C(Y)\to\mathbf{A}(X)$ which sends f to $\tau_1(f)\oplus\tau_2(f)\in\mathbf{A}(X)\oplus\mathbf{A}(X)\subset\mathbf{A}(X\oplus X)\cong\mathbf{A}(X)$. It does not depend on the choice of τ_1, τ_2 and the unitary $X \oplus X \cong X$. An extension $\tau: C(Y) \to \mathbf{A}(X)$ is said to be **trivial** if there exists a unital $*$-monomorphism $\sigma: C(Y) \to \mathbf{B}(X)$ such that $\tau = \pi \circ \sigma$. For each compact metrizable space Y there exists a unique equivalence class of trivial extensions in Ext(Y). Ext(Y) is an Abelian group in which the class of trivial extensions is the identity element. The extension (E_A, φ_A) for an essentially normal operator A is trivial if and only if $A = N + K$ with N normal and K compact. Hence follows the Weyl-von Neumann-Berg-Sikonia theorem. The BDF theorem for essentially normal operators is proved by the pairing $\mathrm{Ext}(Y) \times K^1(Y) \to \mathbf{Z}$ defined by the index, where $K^1(Y) = \tilde{K}(SY^+) = \lim_{n\to\infty} [Y, GL(n, \mathbf{C})]$ (→ 237 K-Theory; [13]). The induced homomorphism $\gamma_\infty: \mathrm{Ext}(Y) \to \mathrm{Hom}(K^1(Y), \mathbf{Z})$ is always surjective, and it is an isomorphism for $Y \subset \mathbf{R}^3$ or $Y = S^n$ but not for $Y \subset \mathbf{R}^4$.

Ext is a †covariant functor from the category of compact metrizable spaces to the category of Abelian groups. It is †homotopy invariant. Define for $n = 0, 1, \ldots$ the group $\mathrm{Ext}_{1-n}(Y)$ by $\mathrm{Ext}(S^n Y)$, where $S^n Y$ is the n-fold †suspension. Then we have the **periodicity** $\mathrm{Ext}_{n+2}(Y) \cong \mathrm{Ext}_n(Y)$. Moreover, for each pair of compact metrizable spaces $Y \supset Z$ we have the long exact sequence

$$\underset{n}{\mathrm{Ext}}(Z) \xrightarrow{i_*} \underset{n}{\mathrm{Ext}}(Y) \xrightarrow{p_*} \underset{n}{\mathrm{Ext}}(Y/Z) \xrightarrow{\partial} \underset{n-1}{\mathrm{Ext}}(Z) \xrightarrow{i_*} \ldots,$$

where Y/Z is the space obtained from Y by collapsing Z to a point and ∂ is $q_* r_*^{\,1}$: $\mathrm{Ext}_n(Y/Z) \to \mathrm{Ext}_n(SZ)$ defined by $q: Y \cup CZ \to (Y \cup CZ)/Y = SZ$ and $r: Y \cup CZ \to (Y \cup CZ)/CZ = Y/Z$, CZ being the †cone over Z. Ext_* satisfies the †Eilenberg-Steenrod axioms for homology theory except for the dimension axiom, which is replaced by $\mathrm{Ext}_n(S^0) = \mathrm{Ext}(S^{n-1}) = \mathbf{Z}$ for n even and $= 0$ for n odd.

The von Neumann algebra $\mathbf{B}(X)$ is classified as a †factor of type I_∞. In the case of a factor M of type II_∞ another index theory has been developed by H. Breuer [17] and others replacing $\mathbf{B}^{(c)}(X)$ by the closed ideal of M generated by finite projections and using the †semifinite trace on M for the dimensions of kernels and cokernels of operators in M.

K. Spectral Analysis in Banach Spaces

Spectral analysis becomes rather involved for general operators in a Banach space as well as for nonnormal operators in Hilbert space.

For a †compact operator A, the nature of the spectrum $\sigma(A)$ and the structure of A in the root subspace associated with a nonzero eigenvalue are well known (→ 68 Compact and Nuclear Operators). However, a full spectral analysis may not be possible without further assumptions.

For a †closed operator with nonempty resolvent set $\rho(A)$, an †operational calculus can be developed by means of a function-theoretic method based on the fact that the resolvent $R_\lambda = (\lambda I - A)^{-1}$ is a $\mathbf{B}(X)$-valued holomorphic function of λ in $\rho(A)$. In particular, the †spectral mapping theorem holds (→ 251 Linear Operators G).

A general class of operators having associated spectral resolution was introduced by N. Dunford. Let X be a Banach space. An operator $E \in \mathbf{B}(X)$ is called a projection if $E^2 = E$. As before we can define a (projection-valued countably additive) **spectral measure** on $\mathscr{B}_c$. An operator $A \in \mathbf{B}(X)$ is said to be a **spectral operator** if there exists a **spectral measure** E on $\mathscr{B}_c$ satisfying the following properties: (i) $E(M)A = AE(M)$, $M \in \mathscr{B}_c$; (ii) $\sigma(A|_{E(M)X}) \subset \overline{M}$, $M \in \mathscr{B}_c$, where $A|_Y$ is the restriction of A to Y and $\overline{M}$ is the closure of M; (iii) there exists a $k \geqslant 0$ such that $\|E(M)\| \leqslant k$ for all $M \in \mathscr{B}_c$. E is unique. A spectral operator A is expressed as $A = S + N$, where $S = \int_{\mathbf{C}} zE(dz)$ and N is †quasi-nilpotent. A is said to be a **scalar operator** if $N = 0$. Unbounded spectral operators are defined similarly, with (i') $E(M)A \subset AE(M)$ in place of (i). However, for unbounded spectral operators A we no longer have the decomposition $A = S + N$. (For more details about spectral operators → [3]. For other topics related to the material discussed in this section → 68 Compact and Nuclear Operators; 251 Linear Operators; and 287 Numerical Computation of Eigenvalues.)

References

[1] N. I. Akhiezer and I. M. Glazman, Theory of linear operators in Hilbert space I, II, Ungar, 1961, 1963. (Original in Russian, 1950.)
[2] F. Riesz and B. Sz.-Nagy, Functional analysis, Ungar, 1955. (Original in French, 1952.)
[3] N. Dunford and J. T. Schwartz, Linear operators, I, II, III, Interscience, 1958, 1963, 1971.
[4] K. Yosida, Functional analysis, Springer, 1965.
[5] T. Kato, Perturbation theory for linear operators, Springer, 1966.
[6] M. Reed and B. Simon, Methods of modern mathematical physics, Academic Press, I, 1972; II, 1975; III, 1979; IV, 1978.
[7] P. R. Halmos, Introduction to Hilbert space and the theory of spectral multiplicity, Chelsea, second edition, 1957.

[8] M. H. Stone, Linear transformations in Hilbert space and their applications to analysis, Amer. Math. Soc. Colloq. Publ., 1932.
[9] I. C. Gokhberg (Gohberg) and M. G. Kreĭn, Introduction to the theory of linear non-self-adjoint operators, Amer. Math. Soc., 1969. (Original in Russian, 1965.)
[10] I. C. Gokhberg (Gohberg) and M. G. Kreĭn, Theory and applications of Volterra operators in Hilbert space, Amer. Math. Soc., 1970. (Original in Russian, 1967.)
[11] R. G. Douglas, Banach algebra techniques in operator theory, Academic Press, 1972.
[12] M. F. Atiyah, K-theory, Benjamin, 1967.
[13] M. F. Atiyah, Global theory of elliptic operators, Proc. Intern. Conf. on Functional Analysis and Related Topics, Univ. Tokyo Press, 1970, 21–30.
[14] L. G. Brown, R. G. Douglas, and P. A. Fillmore, Unitary equivalence modulo the compact operators and extensions of C^*-algebras, Lecture notes in math. 345, Springer, 1973, 58–128.
[15] L. G. Brown, R. G. Douglas, and P. A. Fillmore, Extensions of C^*-algebras and K-homology, Ann. Math., 105 (1977), 265–324.
[16] R. G. Douglas, C^*-algebra extensions and K-homology, Ann. math. studies 95, Princeton Univ. Press, 1980.
[17] M. Breuer, Fredholm theories on von Neumann algebras, Math. Ann., I, 178 (1968), 243–254; II, 180 (1969), 313–325.

391 (VII.21) Spectral Geometry

A. General Remarks

Let E^3 be a Euclidean 3-space with a standard coordinate system (x_1, x_2, x_3) and E^2 be the x_1x_2-plane. Consider a domain D in E^2 as a vibrating membrane with the fixed boundary ∂D. Then the height $x_3 = F(x_1, x_2, t)$ obeys the differential equation of hyperbolic type, $\partial^2 F(x_1, x_2, t)/\partial t^2 = c^2 \Delta F(x_1, x_2, t)$, where $\Delta = \partial^2/\partial x_1^2 + \partial^2/\partial x_2^2$ denotes the †Laplacian in E^2 and c is a constant (we put $c=1$ in the following). For solutions of the form $F(x_1, x_2, t) = U(x_1, x_2)V(t)$, U is a solution of the †Dirichlet problem in $D \subset E^2$; $\Delta U + \lambda U = 0$, where λ is a positive constant called an †eigenvalue of Δ. Solutions of the form $F(x_1, x_2, t) = U(x_1, x_2)\sin\sqrt{\lambda}t$ represent the pure tones that the membrane produces as normal modes. That is, the shape of D is related to the possible sounds or vibrations (i.e., to the eigenvalues of Δ) through the Dirichlet problem. The set of eigenvalues of Δ is called the **spectrum of** D and is denoted by Spec(D). There arises the question of how much information Spec(D) can impart about the geometric properties (i.e., the shape, extent, and connectedness) of D. Generally, **spectral geometry** is the study of the relations between the spectrum of domains D of †Riemannian manifolds or compact Riemannian manifolds (M, g) and the geometric properties of D or (M, g).

B. Spectra

Let $\mathscr{D}^p(M)$ denote the space of smooth †p-forms on a compact m-dimensional C^∞-Riemannian manifold (M, g). Then eigenvalues of the †Laplacian (Laplace-Beltrami operator) Δ acting on $\mathscr{D}^p(M)$ are discretely distributed in $[0, \infty)$, and each multiplicity is finite (→ 68 Compact and Nuclear Operators, 323 Partial Differential Equations of Elliptic Type). The **spectrum for p-forms** $\mathrm{Spec}^p(M, g)$ is $\{\lambda_{p,1} \leqslant \lambda_{p,2} \leqslant \ldots\}$, where each eigenvalue is repeated as many times as its multiplicity indicates. If 0 is an eigenvalue, its multiplicity is equal to the pth †Betti number of M.

In the following, mainly the case $p=0$ is explained. $\mathrm{Spec}^0(M, g)$ is denoted by Spec(M, g). 0 is always in Spec(M, g) and its multiplicity is 1. So we put $\lambda_0 = 0$, and λ_1 is the first nonzero eigenvalue. A geometric meaning of Δf at x for a function f is as follows: If $\{\gamma_h\}_{h=1}^m$ are m geodesics mutually orthogonal at x and parametrized by arc length, then $(\Delta f)(x) = \Sigma_h (f \circ \gamma_h)''(0)$.

Let $\{\varphi_i\}_{i=0}^\infty$ be an orthonormal basis of $\mathscr{D}^0(M)$ consisting of eigenfunctions: $\Delta\varphi_i + \lambda_i\varphi_i = 0$, $\langle\varphi_i, \varphi_j\rangle = \int_M \varphi_i\varphi_j = \delta_{ij}$. Then the †fundamental solution $E(x, y, t)$ of the heat equation $\Delta - \partial/t = 0$ is given by $E(x, y, t) = \Sigma_i e^{-\lambda_i t}\varphi_i(x) \otimes \varphi_i(y)$ as a function on $M \times M \times (0, \infty)$. We put $Z(t) = \int_M E(x, x, t) = \Sigma_i e^{-\lambda_i t}$. $Z(t)$ and Spec(M, g) are equivalent. The Minakshisundaram-Pleijel **asymptotic expansion of** $Z(t)$,

$$Z(t) \sim (1/4\pi t)^{m/2}(a_0 + a_1 t + a_2 t^2 + \ldots), \qquad t \downarrow 0,$$

is the bridge connecting Spec(M, g) and geometric properties of (M, g), because $a_0, a_1, \ldots$ can be expressed as the integrals of functions over M defined by $g = (g_{ij})$, the components R^i_{jkl} of the †Riemannian curvature tensor, and their derivatives of finite order [1]. a_0 is the volume of (M, g) and $a_1 = (1/6)\int_M S$, where S is the †scalar curvature. a_2 was calculated by H. P. Mckean, I. M. Singer, and M. Berger, and a_3 by T. Sakai.

Let D be a bounded domain in E^2 or, more generally, a bounded domain in a Riemannian manifold, and assume that the boundary ∂D is piecewise smooth. For smooth functions which take the value 0 on ∂D, eigenvalues of the Laplacian Δ are discretely distributed in $(0, \infty)$, and each multiplicity is finite. We denote the spectrum of D by $\mathrm{Spec}(D)=\{\lambda_1<\lambda_2\leqslant\ldots\}$. The multiplicity of λ_1 is 1, and an eigenfunction f corresponding to λ_1 takes the same sign in D. The behavior of $Z(t)=\sum_i e^{-\lambda_i t}$ for D is different from that for (M,g) since $Z(t)$ reflects the geometric situation of ∂D in this case.

(M,g) and (N,h), or D_1 and D_2 are called **isospectral** if they have the same spectra.

Examples for which the spectrum is explicitly calculable are as follows: spheres ($S^m, g_0=$ canonical), real projective spaces (RP^m, g_0), complex projective spaces (CP^n, J_0, g_0), (S^{2n+1}, $g_s=$suitably deformed from g_0), flat tori, (and for domains D) unit disks, rectangles, equilateral triangles, etc.

C. Congruence and Characterization

Let D_1 and D_2 be bounded domains in E^2. An open question is whether isospectral D_1, D_2 are congruent. Concerning this, there is M. Kac's paper with the famous title "Can one hear the shape of a drum?" Let D be a bounded domain in E^2 with smooth boundary ∂D. If D has r holes, then

$$Z(t)\sim A(D)/4\pi t-L(\partial D)/4\sqrt{4\pi t}+(1-r)/12, \qquad t\downarrow 0,$$

holds (A. Pleijel, Kac, P. Mckean, I. M. Singer; → [7]), where $A(D)$ denotes the area of D and $L(\partial D)$ denotes the length of ∂D. This theorem implies that the area, the length of ∂D, and the number of holes are determined by $\mathrm{Spec}(D)$. In particular, if D_2 is a unit disk, $\mathrm{Spec}(D_1)=\mathrm{Spec}(D_2)$ implies that D_1 and D_2 are congruent. There are some other results on $Z(t)$ for domains D of surfaces in E^m or for domains D in Riemannian manifolds (M,g).

Two isometric (M,g), (N,h) are isospectral. Concerning the question of whether isospectral (M,g), (N,h) are isometric, there are some counterexamples. The first is the case of two flat tori T^{16}, given by J. Milnor. Examples with nonflat metrics were given by N. Ejiri using warped products and by M. F. Vigneras for surfaces of constant negative curvature. In those examples, M and N are homeomorphic. A. Ikeda showed that there are †lens spaces that are isospectral but not homotopy equivalent.

Examples of affirmative cases are as follows: $\mathrm{Spec}(M,g)=\mathrm{Spec}(S^m,g_0)$, $m\leqslant 6$, implies that (M,g) is isometric to (S^m,g_0) (M. Berger, S. Tanno). The result is the same for (RP^m,g_0), $m\leqslant 6$. For $n\leqslant 6$, (CP^n,J_0,g_0) is characterized by a spectrum among †Kählerian manifolds (M,J,g).

The number of nonisometric flat tori (or more generally, compact flat Riemannian manifolds) with the same spectrum is finite (M. Kneser, T. Sunada).

If one considers spectra for two types of forms, then the situation turns out to be simpler. For example, if $\mathrm{Spec}^p(M,g)=\mathrm{Spec}^p(N,h)$ for $p=0$, 1, then (M,g) is of constant curvature K if and only if (N,h) is also and $K'=K$ (V. K. Patodi [9]).

D. The First Eigenvalue

The first eigenvalue λ_1 for (M,g) or for a domain D in (M,g) reflects the geometric situation of (M,g) or D. A lower bound of λ_1 given by J. Cheeger is $\lambda_1\geqslant h(M)^2/4$, where $h(M)$ is the **isoperimetric constant**, defined by

$$h(M)=\inf\{\mathrm{vol}(S)/\min[\mathrm{vol}(M_1),\mathrm{vol}(M_2)]\},$$

where the inf is taken over all smoothly embedded hypersurfaces S dividing M into two open submanifolds M_1, M_2, $\partial M_1=\partial M_2=S$, and vol means the volume. 1/4 is the best possible estimate.

Let ρ denote the maximum radius of a disk included in a simply connected $D\subset E^2$. Then $\lambda_1\geqslant 1/(4\rho^2)$ (W. K. Hayman, R. Osserman).

If (M,g) has †Ricci curvature $\geqslant k>0$, then $\lambda_1\geqslant mk/(m-1)$ holds, and the equality holds if and only if (M,g) is isometric to $(S^m,(1/k)g_0)$ (A. Lichnerowicz, M. Obata). λ_1 can also be estimated from k and the diameter $d(M)$ of M (P. Li, S. T. Yau).

To obtain upper bounds of λ_1 the **minimum principle of** λ_1 is useful. We state it only for (M,g):

$$\lambda_1=\inf\left\{\int_M(\nabla f,\nabla f)\Big/\int_M f^2\right\},$$

where inf is taken over all piecewise smooth functions f satisfying $\int_M f=0$, $f\neq 0$, and (,) denotes the local inner product with respect to g. An upper bound of λ_1 for (M,g) of nonnegative curvature is $\lambda_1\leqslant c(m)/d(M)^2$, where $c(m)$ is some constant depending on m (Cheeger).

For (M,g) or D a submanifold of another Riemannian manifold, there exist some estimates of λ_1 in terms of second fundamental forms, etc.

$\lambda_1\geqslant j^2/(A(D))$ holds for $D\subset E^2$, and the equality holds if and only if ∂D is a circle (C. Faber, E. Krahn), where j denotes the first zero of the †Bessel function J_0. This estimate is generalized in many directions; for example, for $D\subset(M^2,g)$ in terms of the integral of the Gauss curvature, etc. It is very useful to note that the estimate of λ_1 for D is deeply related to the isoperimetric inequality (→ 228 Isoperimetric Problems).

E. Hersch Type Theorem

With respect to the first eigenvalue $\lambda_1(g)$ and the volume $\mathrm{Vol}(M,g)$ of (M,g), $\lambda_1(g)\cdot \mathrm{vol}(M,g)^{2/m}$ is invariant under a change of metric $g\to c^2 g$ (c is a constant). **Hersch's problem** is stated as follows: Is there a constant $k(M)$ depending on M so that for any Riemannian metric g on M, $\lambda_1(g)\cdot \mathrm{vol}(M,g)^{2/m}\leqslant k(M)$? J. Hersch proved this for a 2-sphere $M=S^2$ with $k(S^2)=8\pi$, and in this case the equality holds if and only if g is proportional to the canonical metric g_0.

The Hersch type theorem holds for an oriented Riemann surface M of genus q with $k(M)=8\pi(q+1)$ (P. C. Yang, S. T. Yau). There is no such constant $k(S^m)$ for an m-sphere S^m, $m\geqslant 3$ (H. Urakawa, H. Muto, S. Tanno).

F. The Multiplicity of λ_i

By a theorem of K. Uhlenbeck each eigenvalue for a Riemannian manifold (M,g) is simple. However, for (S^m, g_0) the first eigenvalue λ_1 is m and its multiplicity is $m+1$. Furthermore, for some g_s deformed from g_0, $\lambda_1(g_s)$ of (S^{2n+1}, g_s) has multiplicity n^2+4n+2, which is larger than $m+1$ $(=2n+2)$.

The multiplicity $m(\lambda_i)$ of the ith eigenvalue λ_i for a Riemann surface of genus q satisfies $m(\lambda_i)\leqslant 4q+2i+1$ (S. Y. Cheng, G. Besson).

G. kth Eigenvalue

The **minimum principle for** λ_k of $\mathrm{Spec}(M,g)$ is stated as follows: Let f_i be an eigenfunction corresponding to λ_i, $0\leqslant i\leqslant k-1$. Define H_{k-1} to be the set of piecewise smooth functions $f\neq 0$ orthogonal to each f_i, i.e., $\int_M ff_i=0$. Then

$$\lambda_k=\inf\left\{\int_M (\nabla f,\nabla f)\Big/\int_M f^2\right\},$$

where inf is taken over $f\in H_{k-1}$. We have the **minimax principle for** λ_k of the first and second type. We state the second type only:

$$\lambda_k=\inf_{L_{k+1}}\ \sup_{0\neq f\in L_{k+1}}\left\{\int_M (\nabla f,\nabla f)\Big/\int_M f^2\right\},$$

where L_k denotes a k-dimensional linear subspace of $\mathcal{D}^0(M)$. From this, for 1-parameter metrics g_u $(a<u<b)$ on M, the continuity of $\lambda_k(g_u)$ with respect to u follows.

H. Courant-Cheng Nodal Domain Theorem

Let f be an eigenfunction on (M,g) or D. The set of all zero points of f is called the **nodal set** of f (or the **nodal curve** of f if $m=2$). Each connected component of the complement of the nodal set in (M,g) or D is called a **nodal domain** of f. The nodal set of f is a smooth submanifold of (M,g) or D except for a set of lower dimension. The number of nodal domains of an eigenfunction corresponding to the ith eigenvalue is $\leqslant i+1$ for (M,g) and $\leqslant i$ for D (**Courant-Cheng nodal domain theorem**).

I. Estimate of $N(\lambda)$

$N(\lambda)$ is defined as the number of eigenvalues of (M,g) or D which are less than or equal to λ. For $D\subset E^2$, H. A. Lorentz conjectured that the behavior of $N(\lambda)$ for $\lambda\to\infty$ does not depend on the shape of D but only on the area $A(D)$ of D, i.e., $\lim_{\lambda\to\infty} N(\lambda)/\lambda=A(D)/4\pi$. This was proved by H. Weyl. Generally, for D or (M,g), the behavior of $N(\lambda)$ for $\lambda\to\infty$ is $\mathrm{vol}(D)\lambda^{m/2}/(4\pi)^{m/2}\Gamma(m/2+1)$, and this is related to the first term $\mathrm{vol}(D)/(4\pi t)^{m/2}$ of the asymptotic expansion of $Z(t)$ by †Tauberian and †Abelian theorems.

J. $\mathrm{Spec}(M,g)$ and Geodesics

Let $T^m=E^m/\Gamma$ be a flat torus, Γ being the lattice for T^m. Let Γ^* be the lattice dual to Γ. Then **Poisson's formula**,

$$\sum_{y\in\Gamma^*} e^{-4\pi^2|y|^2 t}=(\mathrm{vol}(T^m)/(4\pi t)^{m/2})\sum_{x\in\Gamma} e^{-|x|^2/4t},$$

gives a clear relation between $\mathrm{Spec}(T^m)=\{4\pi^2|y|^2, y\in\Gamma^*\}$ and the set $\{|x|, x\in\Gamma\}$ of lengths of closed geodesics on T^m.

If (M,g) satisfies some conditions, then $\mathrm{Spec}(M,g)$ determines the set of lengths of periodic geodesics (Y. Colin de Verdiere), and the spectrum characterizes those Riemannian manifolds whose geodesics are all periodic (J. J. Duistermaat, V. W. Guillemin).

K. $\mathrm{Spec}^p(M,g)$ and the Euler-Poincaré Characteristic

Let (M,g) be oriented and even dimensional. Let $E^p(x,y,t)$ be the †fundamental solution of the †heat equation for p-forms. Corresponding to $Z(t)$ for $\mathrm{Spec}(M,g)$, we get $Z^p(t)=\int_M E^p=\sum_i e^{-\lambda_{p,i}t}$. Then

$$\sum_{p=0}^{m}(-1)^p Z^p(t)=\sum_{p=0}^{m}(-1)^p\int_M \mathrm{tr}\,E^p=\chi(M),$$

where $\chi(M)$ denotes the †Euler-Poincaré characteristic of M (Mckean, Singer). On the other hand, the †Gauss-Bonnet theorem is $\chi(M)=\int_M C$, where C is a function on M expressed as a homogeneous polynomial of components of the Riemannian curvature tensor. Then

Patodi proved

$$\sum_{p=0}^{m}(-1)^p \operatorname{tr} E^p = C + O(t), \qquad t\downarrow 0.$$

L. η-Function

Let (X,g) be a compact oriented $4k$-dimensional Riemannian manifold with boundary $\partial X = Y$ and assume that some neighborhood of Y in (X,g) is isometric to a Riemannian product $Y\times[0,\varepsilon)$. Define an operator B acting on forms of even degree on Y by

$$Bw = (-1)^{k+r+1}(*d - d*)w, \quad w\in\mathscr{D}^{2r}(Y),$$

where $*$ denotes the †Hodge star operator and d denotes exterior differentiation on Y. Then $B^2=\Delta$ holds. Using the spectrum $\{\mu\}$ of B, we define the **η-function** by

$$\eta(s) = \sum_{\mu\neq 0}(\operatorname{sgn}\mu)|\mu|^{-s}.$$

$\eta(s)$ is a spectral invariant, and

$$\operatorname{sgn}(X) = \int_X L_k(p_1,\dots,p_k) - \eta(0)$$

holds (Atiyah, Patodi, and Singer [2]), where $\operatorname{sgn}(X)$ is the †signature of the quadratic form defined by the †cup product on the image of $H^{2k}(X,Y)$ in $H^{2k}(X)$, L_k is the kth †Hirzebruch L-polynomial, and $p_1,\dots,p_k$ are the Pontryagin forms of (X,g).

M. Analytic Torsion

Let χ be a representation of the fundamental group $\pi_1(M)$ of (M,g) by the orthogonal group and E_χ be the associated vector bundle. Let Δ^χ be the Laplacian acting on E_χ-valued p-forms on M and $\{\lambda^\chi_{p,i}\}$ be its spectrum. Then

$$\log T(M,\chi) = \sum_{p=0}^{m}(-1)^p p \log \frac{d}{ds}\left(\sum_i (\lambda^\chi_{p,i})^{-s}\right)\Bigg|_{s=0}$$

is independent of the choice of g. $T(M,\chi)$ is called the **analytic torsion** of M. $T(M,\chi)$ is equal to the †R-torsion $\tau(M,\chi)$ (W. Müller, Cheeger).

N. Concluding Remarks

An †isometry ψ of (M,g) commutes with the Laplacian and induces a linear transformation $\psi^\#_\lambda$ of each eigenspace V_λ. Using the asymptotic expansion of $\sum \operatorname{tr}(\psi^\#_\lambda)e^{-\lambda t}$, the Atiyah-Singer †$G$-signature theorem has been proved (H. Donnelly, Patodi).

The Atiyah-Singer †index theorem has been proved by using Gilkey's theory and the heat equation (Atiyah, R. Bott, Patodi).

Let (N,h) be a complete Riemannian manifold of negative curvature. Then Δ is extended to an unbounded self-adjoint operator for $L_2(N)$. Generally Δ has a continuous spectrum. Some conditions for (N,h) to have pure point spectrum were given in terms of curvature (Donnelly, P. Li).

If D is a minimal submanifold of another Riemannian manifold, estimates of λ_1 are related to the stability of D (→ 275 Minimal Submanifolds).

On the nonexistence of the 1-parameter isospectral deformation $(M,g_0)\to(M,g_t)$, there are results for (i) $m=2$ and g_0 of negative curvature, $m\geqslant 3$ and negatively pinched g_0 (V. Guillemin, D. Kazhdan); (ii) flat metrics g_0 (R. Kuwabara); and (iii) g_0 of constant positive curvature (Tanno).

As for spectral geometry for complex Laplacian on Hermitian manifolds, there are results by P. Gilkey, Tsukada, and others.

References

[1] M. F. Atiyah, R. Bott, and V. K. Patodi, On the heat equation and the index theorem, Inventiones Math., 19 (1973), 279–330.
[2] M. F. Atiyah, V. K. Patodi, and I. M. Singer, Spectral asymmetry and Riemannian geometry I, II, III, Proc. Cambridge Philos. Soc., 77 (1975), 43–69, 405–432; 79 (1976), 71–99.
[3] M. Berger, P. Gauduchon and E. Mazet, Le spectre d'une variété riemannienne, Lecture notes in math. 194, Springer, 1971.
[4] J. Cheeger, A lower bound for the smallest eigenvalue of the Laplacian, Problems in Analysis: Symposium in Honor of Salomon Bochner, R. C. Gunning (ed.), Princeton Univ. Press, 1970, 175–199.
[5] P. B. Gilkey, The index theorem and the heat equation, Publish or Perish, 1974.
[6] M. Kac, Can one hear the shape of a drum ? Amer. Math. Monthly, 73 (April 1966), 1–23.
[7] H. P. Mckean and I. M. Singer, Curvature and eigenvalues of the Laplacian, J. Differential Geometry, 1 (1967), 43–69.
[8] S. Minakshisundarum and A. Pleijel, Some properties of the eigenfunctions of the Laplace operator on Riemannian manifolds, Canad. J. Math., 1 (1949), 242–256.
[9] V. K. Patodi, Curvature and the fundamental solution of the heat operator, J. Indian Math. Soc., 34 (1970), 269–285.

392 (XX.5) Spherical Astronomy

Spherical astronomy is concerned with the apparent positions of celestial bodies and their motions on a celestial sphere with center at an observer on the Earth, while †celestial mechanics is concerned with computing heliocentric true positions of planets and comets and geocentric true positions of satellites. The purpose of spherical astronomy is to find all possible causes of displacement of the apparent positions of celestial bodies on the celestial sphere from their geocentric positions and to study their effects. Atmospheric refraction, geocentric parallax, aberration, annual parallax, precession, nutation, and proper motion are examples of these causes.

When light from a celestial body passes through the Earth's atmosphere, it is refracted since air densities at different heights are different. This phenomenon is called **atmospheric refraction**. The effect of refraction on the apparent direction of the celestial body is a minimum when the body is at its culmination, and vanishes when this coincides with the observer's zenith, while the maximum refraction of 34′.5 occurs when the body is at the horizon.

Topocentric positions differ appreciably from geocentric positions for the Moon and planets, since their geocentric distances are not large compared with the radius of the Earth. The difference is largest when the observer is on the equator and the celestial body is at the horizon, and this largest value is called the **geocentric parallax**. The geocentric parallax of the Moon is between 53′.9 and 60′.2; those of the Sun, Mercury, Venus, Mars, Jupiter, and Saturn are, respectively, 8″.64–8″.94, 6″–16″.5, 5″–32″, 3″.5–23″.5, 1″.4–2″.1, and 0″.8–1″.1. For fixed stars geocentric parallaxes can be regarded as zero since the stars are far from the Earth.

The Earth moves in an orbit around the Sun with period of one year (365.2564 days) and rotates around the polar axis, which is inclined at 66°.5 to the orbital plane (the **ecliptic**), with period of one day (23 hours, 56 minutes, 4.091 seconds). Therefore the observer on the Earth moves with a speed depending on the latitude (0.465 km/sec on the equator) due to the rotation and moves with an average speed of 29.785 km/sec on the ecliptic. Due to these motions of the observer, apparent directions of celestial bodies are displaced from their geometric directions. Displacement due to the rotation is called **diurnal aberration**, and that due to the orbital motion, **annual aberration**. The effect of diurnal aberration is between 0″ and 0″.32 and varies with a period of one day, while that of the annual aberration is between 0″ and 20″.496 and varies with a period of one year. Moreover, to compute the positions of celestial bodies, the travel time of light to the observer should be taken into account.

Annual parallax for a fixed star is half the difference of its apparent directions, which are measured at the ends of a diameter perpendicular to the direction of the star from the orbit of the Earth. The effect of the annual parallax varies with a period of one year. However, except for nearby stars, it is not necessary to take this effect into account when computing apparent positions.

The pole of the Earth on the celestial sphere moves on a circle around the pole of the ecliptic due to the gravitational attraction of the Moon, Sun, and planets, and therefore the equinox moves clockwise on the ecliptic. Because the resultant of the attractive force of the Moon, Sun, and planets changes periodically, the motion of the equinox is not uniform. Therefore the motion is expressed as the sum of secular motion, called **precession**, and periodic motion, called **nutation**. Since the positons of fixed stars on the celestial sphere are measured with respect to the equator and the equinox, their right ascensions and declinations are continuously changing because of precession and nutation.

Since the stars are not fixed in space but themselves have **proper motions**, their positions on the celestial sphere are continuously changing.

Spherical astronomy also includes predictions of solar and lunar eclipses, the theory of †orbit determination to compute apparent positions of celestial bodies in the solar system by use of orbital elements, and the computation of ephemerides for the Sun, Moon, planets, and fixed stars. Practical astronomy, which develops theories and methods of observation by use of meridian circles, transit instruments, zenith telescopes, sextants, theodolites, telescopes with equatorial mountings, and astronomical clocks, and navigational astronomy, which deals with methods for determining the positions of ships and aircraft, are closely connected to spherical astronomy.

It should be noted that recently radar has been used to measure distances to the Moon and planets accurately, a contribution to determining the size of the solar system with precision.

References

[1] W. Chauvenet, A manual of spherical and practical astronomy I, II, Dover, 1960.
[2] S. Newcomb, A compendium of spherical astronomy, Dover, 1960.
[3] W. M. Smart, Textbook on spherical astronomy, Cambridge Univ. Press, fifth edition, 1965.

[4] E. W. Woolard and G. M. Clemence, Spherical astronomy, Academic Press, 1966.

393 (XIV.6) Spherical Functions

A. Spherical Functions

The term *spherical functions* in modern terminology means a certain family of functions on †symmetric Riemannian spaces obtained as simultaneous †eigenfunctions of certain integral operations (→ 437 Unitary Representations). In this article, however, we explain only the classical theory of **Laplace's spherical functions** with respect to the rotation group in 3-dimensional space.

Solutions of †Laplace's equation $\Delta V=0$ that are homogeneous polynomials of degree n with respect to the orthogonal coordinates x, y, z are called **solid harmonics** of degree n. If n is a positive integer, there are $2n+1$ linearly independent solid harmonics of degree n. In †polar coordinates (r,θ,φ) they are of the form $r^n Y_n(\theta,\varphi)$, where $Y_n(\theta,\varphi)$, the **surface harmonic** of degree n, satisfies the differential equation

$$\frac{1}{\sin\theta}\frac{\partial}{\partial\theta}\left(\sin\theta\frac{\partial Y_n}{\partial\theta}\right)+\frac{1}{\sin^2\theta}\frac{\partial^2 Y_n}{\partial\varphi^2}+n(n+1)Y_n=0.$$

Here, if we apply †separation of variables to θ and φ and put $z=\cos\theta$, then the component in φ is represented by trigonometric functions, and the other component in θ reduces to a solution of **Legendre's associated differential equation**

$$(1-z^2)\frac{d^2w}{dz^2}-2z\frac{dw}{dz}+\left(n(n+1)-\frac{m^2}{1-z^2}\right)w=0. \quad (1)$$

B. Legendre Functions

With $m=0$ in (1) and n replaced by an arbitrary complex number ν, the equation is reduced to **Legendre's differential equation**

$$(1-z^2)\frac{d^2w}{dz^2}-2z\frac{dw}{dz}+\nu(\nu+1)w=0, \quad (2)$$

whose fundamental solutions are represented by

$$P_\nu(z)=\frac{1}{2\pi i}\oint^{(1+,z+)}\frac{(\zeta^2-1)^\nu}{2^\nu(z-\zeta)^{\nu+1}}d\zeta, \quad (3)$$

$$Q_\nu(z)=\frac{1}{4i\sin\nu\pi}\oint^{(1+,-1+)}\frac{(\zeta^2-1)^\nu}{2^\nu(z-\zeta)^{\nu+1}}d\zeta, \quad (4)$$

where the contour of integration in (3) is a closed curve with positive direction on the ζ-plane, avoiding the half-line $(-\infty,-1)$, and admitting 1 and z as inner points of the domain it bounds, whereas the contour in (4) is a closed ∞-shaped curve encircling the point 1 once in the negative direction and the point -1 once in the positive direction. The functions $P_\nu(z)$ and $Q_\nu(z)$ are called **Legendre functions of the first and second kind**, respectively. The integral representation (3) is **Schläfli's integral representation**. If $\mathrm{Re}(\nu+1)>0$, we can deform the contour of integration and obtain

$$Q_\nu(z)=\frac{1}{2^{\nu+1}}\int_{-1}^{1}\frac{(1-\zeta^2)^\nu}{(z-\zeta)^{\nu+1}}d\zeta. \quad (5)$$

If ν is an integer, it is convenient to use the representation (5).

From (3)–(5), we can obtain the recurrence formulas for Legendre functions of distinct degrees. The recurrence formulas for $P_\nu(z)$ and $Q_\nu(z)$ have exactly the same form (→ Appendix A, Table 18.II). Expanding the integrand in (3) and (4) with respect to $z-1$ and ζ/z, the following identities are obtained:

$$P_\nu(z)=F\left(\nu+1,-\nu,1,\frac{1-z}{2}\right), \quad |1-z|<2,$$

$$Q_\nu(z)=\frac{\sqrt{\pi}\,\Gamma(\nu+1)}{(2z)^{\nu+1}\Gamma(\nu+3/2)}\times F\left(\frac{\nu+1}{2},\frac{\nu+2}{2},\nu+\frac{3}{2},\frac{1}{z^2}\right),$$

$$|z|>1,\ |\arg z|<\pi,$$

where $F(\alpha,\beta,\gamma,z)$ is the †hypergeometric function. These expansions are the solutions in series of Legendre's differential equation in the neighborhood of the †regular singular points $z=1$ and ∞, respectively (→ Appendix A, Table 18.II).

If ν is a positive integer, since $\zeta=1$ is not a †branch point in (3), $P_n(z)$ is represented by **Rodrigues's formula**

$$P_n(z)=\frac{1}{2\pi i}\oint\frac{(\zeta^2-1)^n}{2^n(z-\zeta)^{n+1}}d\zeta=\frac{1}{2^n n!}\frac{d^n}{dz^n}(z^2-1)^n. \quad (6)$$

In this case, $P_n(z)$ is a polynomial of degree n such that

$$P_n(z)=\sum_{r=0}^{[n/2]}(-1)^r\frac{(2n-2r)!}{2^n r!(n-r)!(n-2r)!}z^{n-2r},$$

$$P_0(z)=1,$$

which is called the **Legendre polynomial** (Legendre, 1784). The †generating function for the

Legendre polynomials is $(1-2\rho\cos\theta+\rho^2)^{-1/2}$, whose expansion with respect to ρ is of the form $\sum_{n=0}^{\infty} P_n(z)\rho^n$, $z=\cos\theta$. Here the †generating function $(1-2\rho\cos\theta+\rho^2)^{-1/2}$ is the inverse of the distance between two points (ρ,θ) and $(1,0)$ in polar coordinates. Hence $P_n(z)$ is also called the **Legendre coefficient**. If z is real, $\{((2n+1)/2)^{1/2}P_n(z)\}_{n=0}^{\infty}$ constitutes an orthonormal system on $[-1,1]$ (→ 317 Orthogonal Functions). The n zeros of $P_n(z)$ are all real, simple, and lie in $(-1,1)$. For sufficiently large n, we have

$$P_n(\cos\theta)$$
$$=\sqrt{\frac{2}{n\pi\sin\theta}}\sin\left(\left(n+\frac{1}{2}\right)\theta+\frac{\pi}{4}\right)+O(1/n^{3/2}),$$
$$Q_n(\cos\theta)$$
$$=\sqrt{\frac{\pi}{2n\sin\theta}}\cos\left(\left(n+\frac{1}{2}\right)\theta+\frac{\pi}{4}\right)+O(1/n^{3/2})$$

as $n\to\infty$.

C. Associated Legendre Functions

For any positive integer m, the functions

$$P_\nu^m(z)=(1-z^2)^{m/2}\,d^mP_\nu(z)/dz^m,$$
$$Q_\nu^m(z)=(1-z^2)^{m/2}\,d^mQ_\nu(z)/dz^m$$

are called the **associated Legendre functions of the first and second kind**, respectively. This definition, due to N. M. Ferrers, is convenient for the case $-1<z<1$. For arbitrary complex z in a domain G obtained by deleting the segment $[-1,1]$ from the complex plane, the following definition, due to H. E. Heine and E. W. Hobson, is used:

$$P_\nu^m(z)=(z^2-1)^{m/2}\,d^mP_\nu(z)/dz^m,$$
$$Q_\nu^m(z)=(z^2-1)^{m/2}\,d^mQ_\nu(z)/dz^m.$$

The associated Legendre functions satisfy the associated Legendre differential equation (1). In particular, for $\nu=n$ (a positive integer) and $z=x$ (real),

$$\{(2n+1)(n-m)!/2(n+m)!\}^{1/2}P_n^m(z),$$
$$n=0,1,\ldots\ m=\text{constant},$$

constitute an orthonormal system on $[-1,1]$.

The **addition theorem** for the Legendre functions is

$$P_n\left(z_1z_2\pm\sqrt{\pm(1-z_1^2)}\sqrt{\pm(1-z_2^2)}\cos\varphi\right)$$
$$=P_n(z_1)P_n(z_2)+2\sum_{m=1}^{n}\frac{(n-m)!}{(n+m)!}P_n^m(z_1)$$
$$\times P_n^m(z_2)\cos m\varphi,$$

where the equality with the plus sign was obtained by Ferrers and that with the minus sign by Heine and Hobson.

D. Surface Harmonics

From the considerations so far for the surface harmonics $Y_n(\theta,\varphi)$, $2n+1$ independent solutions

$$P_n(\cos\theta),\qquad P_n^m(\cos\theta)\sin m\varphi,$$
$$qP_n^m(\cos\theta)\cos m\varphi,\qquad 1\leqslant m\leqslant n,$$

are obtained. Since $P_n(\cos\theta)$ vanishes on n latitudes of the unit sphere, and $P_n^m(\cos\theta)\cdot\cos m\varphi$ and $P_n^m(\cos\theta)\sin m\varphi$ vanish on $n-m$ latitudes and m longitudes of the unit sphere, respectively, the former functions are called **zonal harmonics** and the latter, **tesseral harmonics**. The general form of surface harmonics Y_n of order n is given by a linear combination of zonal and tesseral harmonics:

$$Y_n(\theta,\varphi)=A_{n,0}P_n(\cos\theta)$$
$$+\sum_{m=1}^{n}(A_{n,m}\cos m\varphi+B_{n,m}\sin m\varphi)P_n^m(\cos\theta).\qquad(7)$$

Expressing two surface harmonics $Y_n^{(1)}$ and $Y_l^{(2)}$ in linear combinations such as (7), the following orthogonality relations hold:

$$\int_{-\pi}^{\pi}\int_0^{\pi}Y_n^{(1)}(\theta,\varphi)Y_l^{(2)}(\theta,\varphi)\sin\theta\,d\theta\,d\varphi$$
$$=\delta_{n,l}\frac{4\pi}{2n+1}\left(A_{n,0}^{(1)}A_{n,0}^{(2)}\right.$$
$$\left.+\frac{1}{2}\sum_{m=1}^{n}\frac{(n+m)!}{(n-m)!}(A_{n,m}^{(1)}A_{n,m}^{(2)}+B_{n,m}^{(1)}B_{n,m}^{(2)})\right).$$

Since the family of all zonal and tesseral harmonics constitutes a †complete orthogonal system, it is possible to expand a function $f(\theta,\varphi)$ on the sphere into an orthogonal series:

$$f(\theta,\varphi)-\sum_{n=0}^{\infty}Y_n(\theta,\varphi)$$
$$=\sum_{n=0}^{\infty}\left(A_{n,0}P_n(\cos\theta)+\sum_{m=1}^{n}(A_{n,m}\cos m\varphi\right.$$
$$\left.+B_{n,m}\sin m\varphi)P_n^m(\cos\theta)\right).$$

To obtain surface harmonics, the following method is effective. Let ν be a direction proportional to the direction cosines l, m, n. Then a function

$$\left(l\frac{\partial}{\partial x}+m\frac{\partial}{\partial y}+n\frac{\partial}{\partial z}\right)\frac{1}{r}=\alpha\frac{\partial}{\partial\nu}\left(\frac{1}{r}\right),$$
$$\alpha=\sqrt{l^2+m^2+n^2}$$

is a solution of Laplace's equation. Physically, this corresponds to a †potential of double pole with moment α and direction ν. A more general multipole potential

$$V=c\left(\prod_{i=1}^{n}\alpha_i\frac{\partial}{\partial v_i}\right)\left(\frac{1}{r}\right)$$

also satisfies Laplace's equation. If we put $V=U_n(x,y,z)r^{-2n-1}$, U_n is a spherical function of order n (**Maxwell's theorem**). Various spherical functions correspond to particular directions v_i. For example, if every v_i is equal to z, we have zonal harmonics; and if $n-m$ of the v_i are equal to z and m of the v_i are symmetric on the xy-plane, we obtain tesseral harmonics. Let γ be an angle between two segments connecting the origin to the points (r,θ,φ) and (r',θ',φ') in polar coordinates. Then $\cos\gamma=\cos\theta\cos\theta'+\sin\theta\sin\theta'\cos(\varphi-\varphi')$, and if we choose the line connecting the origin to a point (r',θ',φ') as the axis defining P_n, we have

$$P_n(\cos\gamma)=(-1)^n\frac{r^{n+1}}{n!}\left(\frac{x'}{r'}\frac{\partial}{\partial x}+\frac{y'}{r'}\frac{\partial}{\partial y}+\frac{z'}{r'}\frac{\partial}{\partial z}\right)^n\left(\frac{1}{r}\right).$$

These are called **biaxial spherical harmonics**, which can also be represented (by the addition theorem) by means of spherical harmonics with respect to each axis.

E. Extension of the Legendre Functions

We extend the associated functions with positive integer m to any number m. First, if m is a negative integer, we put

$$P_\nu^{-m}(z)=(1-z^2)^{-m/2}\int_1^z d\zeta_m\int_1^{\zeta_m}d\zeta_{m-1}\cdots\times\int_1^{\zeta_3}d\zeta_2\int_1^{\zeta_2}P_\nu(\zeta_1)d\zeta_1,$$

$$Q_\nu^{-m}(z)=(1-z^2)^{-m/2}\int_\infty^z d\zeta_m\int_\infty^{\zeta_m}d\zeta_{m-1}\cdots\times\int_\infty^{\zeta_3}d\zeta_2\int_\infty^{\zeta_2}Q_\nu(\zeta_1)d\zeta_1,$$

a definition due to Ferrers. Then

$$P_\nu^{-m}(z)=(-1)^m\frac{\Gamma(\nu-m+1)}{\Gamma(\nu+m+1)}P_\nu^m(z),$$

$$Q_\nu^{-m}(z)=(-1)^m\frac{\Gamma(\nu-m+1)}{\Gamma(\nu+m+1)}Q_\nu^m(z).$$

When we use the definition due to Heine and Hobson, the factor $(-1)^m$ in these formulas is excluded. Two fundamental solutions, called **hypergeometric functions** of the **hyperspherical differential equation**

$$(1-z^2)d^2w/dz^2-2(\mu+1)z\,dw/dz+(\nu-\mu)(\nu+\mu+1)w=0,$$

are

$$P_{\nu-\mu}^{(\mu,\mu)}(z)=\frac{e^{-\nu\pi i}}{2^{\nu-\mu}4\pi\sin\nu\pi}\times\oint^{(z+,1+,z-,1-)}\frac{(\zeta^2-1)^\nu}{(\zeta-2)^{\nu+\mu+1}}d\zeta,$$

$$Q_{\nu-\mu}^{(\mu,\mu)}(z)=\frac{e^{(1+\nu)\pi i}}{2^{\nu-\mu}4i\sin\nu\pi}\oint^{(-1+,1-)}\frac{(\zeta^2-1)^\nu}{(\zeta-z)^{\nu+\mu+1}}d\zeta,$$

where the contour of integration for the latter integral is a curve encircling the point -1 once in the positive direction and the point 1 once in the negative direction. Then the associated Legendre functions for an arbitrary number μ are defined as follows:

$$P_\nu^\mu(z)=\frac{\Gamma(\nu+\mu+1)}{2^\mu\Gamma(\nu+1)}(z^2-1)^{\mu/2}P_{\nu-\mu}^{(\mu,\mu)}(z),$$

$$Q_\nu^\mu(z)=\frac{\Gamma(\nu+\mu+1)}{2^\mu\Gamma(\nu+1)}(z^2-1)^{\mu/2}Q_{\nu-\mu}^{(\mu,\mu)}(z).$$

If $\nu-\mu$ is a positive integer, $P_{\nu-\mu}^{(\mu,\mu)}$ is called the **Gegenbauer polynomial**, also denoted by $C_{\nu-\mu}(z)$. The $C_{\nu-\mu}(z)$ are obtained as coefficients of the expansion of the generating function $(1-2hz+z^2)^{-(2\mu+1)/2}$ (→ Appendix A, Table 20.I).

For spherical functions of several variables there is an investigation by P. Appell and J. Kempé de Fériet [2] (→ 206 Hypergeometric Functions D).

References

[1] E. W. Hobson, The theory of spherical and ellipsoidal harmonics, Cambridge Univ. Press, 1931 (Chelsea, 1955).
[2] P. Appell and J. Kempé de Fériet, Fonctions hypergéométriques et hypersphériques, polynomes d'Hermite, Gauthier-Villars, 1926.
[3] C. Müller, Spherical harmonics, Lecture notes in math. 7, Springer, 1967.
[4] T. MacRobert, Spherical harmonics, Snedden, third revised edition, 1967.
Also → references to 206 Hypergeometric Functions, 389 Special Functions.

394 (XIII.13) Stability

A. General Remarks

Stability was originally a concept concerned with stationary physical states. When a state is affected by a small disturbance, this state is said to be **stable** if the disturbance subsequently remains small, and **unstable** if the disturbance gradually increases. For instance, consider a rod placed in the Earth's gravitational field with one end fixed at a point around which the rod can rotate freely. When the rod is placed vertically, this state is stationary. It is stable if the rod is hanging down from the fixed end, and unstable if it is standing on the fixed end. In physical systems only the stable state is practically realizable, so this distinction is important.

The concept of stability is used not only in relation to physical states but also in many other fields of science. We shall restrict ourselves to stability of solutions of differential equations. There, the term *stability* is used in the sense that a small change in the initial values results in a small change in the solution. As long as the solution is considered within a finite interval of the independent variable, this stability is naturally guaranteed by the continuity of the solution with respect to its initial values (→ 316 Ordinary Differential Equations (Initial Value Problems)). The problem arises when an independent variable moves over an unbounded interval.

Let $(x_1, \ldots, x_n) = \mathbf{x}$, $(x_1(t), \ldots, x_n(t)) = \mathbf{x}(t)$, $(x_1'(t), \ldots, x_n'(t)) = \mathbf{x}'(t)$ (the symbol $'$ means differentiation by t), and $|\mathbf{x}| = \sum_{j=1}^{n} |x_j|$. Consider the differential equation

$$\mathbf{x}' = \mathbf{f}(t, \mathbf{x}), \tag{1}$$

for which the existence and uniqueness of the solution of the initial value problem is assumed for $|t| < \infty$, $|\mathbf{x}| < \infty$. Let $\mathbf{x} = \varphi(t)$ be a solution of (1). If for any $\varepsilon > 0$ and t_0, a $\delta > 0$ can be chosen so that $|\mathbf{x}(t_0) - \varphi(t_0)| < \delta$ implies $|\mathbf{x}(t) - \varphi(t)| < \varepsilon$ for $t_0 \leqslant t < \infty$ $(-\infty < t \leqslant t_0)$, where $\mathbf{x}(t)$ is any solution of (1), then $\mathbf{x} = \varphi(t)$ is said to be **(Lyapunov) stable in the positive (negative) direction**. If it is stable both in the positive and negative directions, it is said to be **stable in both directions**. In the remainder of this article we will consider stability in the positive direction only. Corresponding assertions for stability in the negative direction can be obtained by reversing the sign of t.

B. Classification

We denote by $\mathbf{x} = \mathbf{x}(t, t_0, \mathbf{x}_0)$ a solution of (1) such that $\mathbf{x} = \mathbf{x}_0$ at $t = t_0$.

Suppose a solution $\mathbf{x} = \varphi(t)$ is stable. If for any t_0 there exists a $\zeta > 0$ such that $|\mathbf{x}(t, t_0, \mathbf{x}_0) - \varphi(t)| \to 0$ as $t \to \infty$ for any $x(t, t_0, x_0)$ with $|x_0 - \varphi(t_0)| < \zeta$, $\varphi(t)$ is said to be **asymptotically stable**.

If a constant δ in the definition of stability can be chosen independently of t_0, $\varphi(t)$ is said to be **uniformly stable**. When equation (1) is †autonomous, stability implies uniform stability.

If (1) $\varphi(t)$ is uniformly stable and (2) for any t_0 and $\eta > 0$, there exist a $\zeta > 0$ independent of t_0 and a $T > 0$ independent of t_0 such that $|\mathbf{x}_0 - \varphi(t_0)| < \zeta$ and $t > t_0 + T$ imply $|\mathbf{x}(t, t_0, \mathbf{x}_0) - \varphi(t)| < \eta$, then $\varphi(t)$ is said to be **uniformly asymptotically stable**.

Suppose that there exists a positive number λ with the following property: For any t_0 and $\varepsilon > 0$ one can take a $\delta(\varepsilon) > 0$ such that $|\mathbf{x}_0 - \phi(t_0)| < \delta(\varepsilon)$ implies $|\mathbf{x}(t, t_0, \mathbf{x}_0) - \phi(t)| < \varepsilon e^{-\lambda(t - t_0)}$ for $t \geqslant t_0$. Then $\phi(t)$ is said to be **exponentially stable**. Exponential stability implies uniform asymptotic stability.

C. Criteria

To deal with the stability of $\mathbf{x} = \varphi(t)$, we need consider only the case $\varphi(t) \equiv 0$, since the transformation $\mathbf{x} = \mathbf{y} + \varphi(t)$ reduces equation (1) to

$$\mathbf{y}' = \mathbf{f}(t, \mathbf{y} + \varphi(t)) - \mathbf{f}(t, \varphi(t)) = \mathbf{F}(t, \mathbf{y}),$$

$$\mathbf{F}(t, \mathbf{0}) \equiv \mathbf{0}, \tag{2}$$

and thus $\mathbf{x} = \varphi(t)$ is transformed into $\mathbf{y} \equiv \mathbf{0}$. If $\mathbf{F}$ is continuously differentiable with respect to y, (2) can be written in the form

$$\mathbf{y}' = \mathbf{F}_{\mathbf{y}}(t, \mathbf{0})\mathbf{y} + \mathbf{g}(t, \mathbf{y}), \qquad \mathbf{g}(t, \mathbf{y}) = o(|\mathbf{y}|).$$

The linear part of this equation,

$$\mathbf{y}' = \mathbf{F}_{\mathbf{y}}(t, \mathbf{0})\mathbf{y},$$

is called the **variational equation** for (1). So, in this section, we can state several criteria for stability of the null solution $\mathbf{y} \equiv \mathbf{0}$ of the equation

$$\mathbf{y}' = P(t)\mathbf{y} + \mathbf{g}(t, \mathbf{y}), \qquad |\mathbf{g}(t, \mathbf{y})| = o(|\mathbf{y}|). \tag{3}$$

(I) If (3) is linear (i.e., $\mathbf{g}(t, \mathbf{y}) \equiv \mathbf{0}$), then $\mathbf{y} \equiv \mathbf{0}$ is stable if and only if every solution of (3) is bounded as $t \to \infty$.

(II) If (3) is linear, uniform asymptotic stability implies exponential stability.

Let $f(t, \mathbf{y})$ be a function defined for $|\mathbf{y}| < \rho$, $t > \alpha$. If there exists a continuous function $w(\mathbf{y})$ such that $w(\mathbf{0}) = 0$, $w(\mathbf{y}) > 0$ $(\mathbf{y} \neq \mathbf{0})$, $f(t, \mathbf{y}) \geqslant w(\mathbf{y})$ $(|\mathbf{y}| < \rho, t > \alpha)$, then $f(t, \mathbf{y})$ is said to be **positive definite**. If $-f(t, \mathbf{y})$ is positive definite, then $f(t, \mathbf{y})$ is said to be **negative definite**.

(III) The existence of a **Lyapunov function** $V(t, \mathbf{y})$ with the following properties implies the stability of $\mathbf{y} \equiv \mathbf{0}$: (i) $V(t, \mathbf{y})$ is positive definite and differentiable, (ii) $V(t, \mathbf{0}) = 0$, (iii) $\dot{V}(t, \mathbf{y}) = V_t + V_{\mathbf{y}} \cdot (P(t)\mathbf{y} + \mathbf{g}(t, \mathbf{y})) \leqslant 0$.

The existence of $V(t, \mathbf{y})$ with the following properties implies the uniform asymptotic stability of $\mathbf{y} \equiv \mathbf{0}$: (i) same as (i) above, (ii) there exists a continuous function $v(\mathbf{y})$ such that

$v(\mathbf{0})=0$, $v(\mathbf{y})>0$ $(\mathbf{y}\neq\mathbf{0})$, $V(t,\mathbf{y})\leqslant v(\mathbf{y})$, (iii) $\dot{V}(t,\mathbf{y})$ is negative definite.

Hereafter we shall assume that $|\mathbf{g}(t,\mathbf{y})|=o(|\mathbf{y}|)$ as $\mathbf{y}\to\mathbf{0}$ uniformly with respect to t.

(IV) If $P(t)$ is a constant matrix all of whose †eigenvalues have negative real parts, then $\mathbf{y}\equiv\mathbf{0}$ is asymptotically stable [3, 4].

(V) Let $P(t)$ be continuous and periodic with period T and Z be a †fundamental system of solutions of the variational equation

$$\mathbf{z}'=P(t)\mathbf{z}. \tag{4}$$

Then there exists a constant matrix C such that $Z(t+T)=Z(t)C$. Let $\lambda_1,\ldots,\lambda_n$ be the eigenvalues of C. Then the numbers $\mu_k=(\log\lambda_k)/T$ $(k=1,\ldots,n)$ are called the **characteristic exponents** of (4). Obviously they are determined up to integral multiples of $2\pi i/T$. If the real parts of the characteristic exponents are all negative, then $\mathbf{y}\equiv\mathbf{0}$ is asymptotically stable [3, 4].

(VI) If $\mathbf{f}(t,\mathbf{x})$ in (1) is periodic in t with period T and (1) admits a periodic solution $\mathbf{x}=\varphi(t)$ with period T, then (1) can be reduced to (3) by putting $\mathbf{x}=\mathbf{y}+\varphi(t)$, and $P(t)$, $\mathbf{g}(t,\mathbf{y})$ are both periodic in t with period T. Thus criterion (V) can be applied as a stability criterion for the periodic solution of (1). There are many other criteria for various particular forms of the equation (→ 290 Nonlinear Oscillation). For the †autonomous case where $\mathbf{f}(t,\mathbf{x})$ is of the form $\mathbf{p}(\mathbf{x})$ or $\mathbf{p}(\mathbf{x})+\mathbf{q}(t)$ with $\mathbf{q}(t+T)=\mathbf{q}(t)$, many results have been found.

(VII) If the solution $\mathbf{z}\equiv\mathbf{0}$ of

$$\mathbf{z}'=P(t)\mathbf{z}$$

is uniformly asymptotically stable, then the solution $\mathbf{y}\equiv\mathbf{0}$ of (3) is also uniformly asymptotically stable [4].

D. Conditional Stability

Let $\varphi(t)$ be a solution and $\mathfrak{F}$ a family of solutions of (1). If for any $\varepsilon>0$, a $\delta>0$ can be determined so that $|\mathbf{x}(t_0)-\varphi(t_0)|<\delta$ implies $|\mathbf{x}(t)-\varphi(t)|<\varepsilon$ for $t_0\leqslant t<\infty$ for any solution $\mathbf{x}(t)$ in $\mathfrak{F}$, then $\varphi(t)$ is said to be stable with respect to the family $\mathfrak{F}$. If a family $\mathfrak{F}$ can be found so that a solution is stable with respect to $\mathfrak{F}$, the solution is said to be **conditionally stable**. For instance, in equation (3), if $P(t)$ is a constant matrix some of whose eigenvalues have negative real parts, $\mathbf{g}(t,\mathbf{y})$ is differentiable with respect to $\mathbf{y}$, and $\mathbf{g}_y(t,\mathbf{y})=o(1)$ uniformly in t as $\mathbf{y}\to 0$, then $\mathbf{y}\equiv 0$ is conditionally stable.

We now mention a weaker kind of stability called orbital stability. Let $\varphi(t)$ be a solution and ε any positive number. If there can be found a positive number δ such that for any solution $\mathbf{x}(t)$ with $|\mathbf{x}(t_1)-\varphi(t_0)|<\delta$ for some t_0 and t_1, $\bigcup_{t_1\leqslant t<\infty}\mathbf{x}(t)$ belongs to the ε-neighborhood of $\bigcup_{t_0\leqslant t<\infty}\varphi(t)$, then $\varphi(t)$ is said to have **orbital stability**.

When $\mathbf{f}(t,\mathbf{x})$ in equation (1) is independent of t, (1) is often called a †dynamical system. In the theory of dynamical systems, not merely the stability of a solution itself but also the stability of a closed invariant set is of importance (→ 126 Dynamical Systems).

It is also of importance to investigate the change in solution caused by a small change in the right-hand member of the equation. Suppose, for instance, that the right-hand member of the equation depends continuously on a parameter ε. Then the question arises as to how the solution changes if ε changes. The theory of such problems is called †perturbation theory. Suppose that the equation

$$\mathbf{x}'=\mathbf{f}(t,\mathbf{x},\varepsilon)$$

admits a periodic solution $\varphi(t)$ for $\varepsilon=0$. Then $\varphi(t)$ is said to be stable under perturbation if for $\varepsilon\neq 0$ the same equation admits a periodic solution lying near $\varphi(t)$. In †celestial mechanics and †nonlinear oscillation theory this concept plays an important role.

References

[1] R. E. Bellman, Stability theory of differential equations, McGraw-Hill, 1953.
[2] L. Cesari, Asymptotic behavior and stability problems in ordinary differential equations, Springer, third edition, 1971.
[3] E. A. Coddington and N. Levinson, Theory of ordinary differential equations, McGraw-Hill, 1955.
[4] S. Lefschetz, Differential equations, geometric theory, Interscience, 1957.
[5] A. Lyapunov (Ljapunov), Problème général de la stabilité du mouvement, Ann. Math. Studies, Princeton Univ. Press, 1947.
[6] V. V. Nemytskiĭ (Nemyckiĭ) and V. V. Stepanov, Qualitative theory of differential equations, Princeton Univ. Press, 1960. (Original in Russian, 1947.)
[7] O. Perron, Über eine Matrixtransformation, Math. Z., 32 (1930), 465–473.
[8] O. Perron, Die Stabilitätsfrage bei Differentialgleichungen, Math. Z., 32 (1930), 703–728.
[9] A. Halanay, Differential equations, stability, oscillations, time lags, Academic Press, 1966.
[10] T. Yoshizawa, Stability theory by Ljapunov's second method, Publ. Math. Soc. Japan, 1966.

395 (XVII.12) Stationary Processes

A. Definitions

Stationary process is a general name given to all †stochastic processes (→ 407 Stochastic Processes) that have the property of being stationary (to be defined in the next paragraph) under a shift of a time parameter t that extends over T, which is either the set of all real numbers $\mathbf{R}$ (a continuous parameter) or the set of all integers $\mathbf{Z}$ (a discrete parameter).

Let $(\Omega, \mathfrak{B}, P)$ be a †probability space and $\{X_t(\omega)\}$ $(t \in T, \omega \in \Omega)$ a complex-valued †stochastic process. If for every n, every $t_1, t_2, \ldots, t_n \in T$, and every †Borel subset E_n of complex n-dimensional space $\mathbf{C}^n$, the equality

$$P((X_{t_1+t}, \ldots, X_{t_n+t}) \in E_n)$$
$$= P((X_{t_1}, \ldots, X_{t_n}) \in E_n) \qquad (1)$$

holds, then $\{X_t\}$ is called a **strongly** (or **strictly**) **stationary process**; while if $E(|X_t|^2)$ is finite for every t, and if the †moments up to the second order are stationary, i.e., if

$$E(X_{t+h}) = E(X_t),$$
$$E(X_{t+h}\bar{X}_{s+h}) = E(X_t\bar{X}_s), \qquad (2)$$

then $\{X_t\}$ is called a **weakly stationary process** or a **stationary process in the wider sense**. The "stationary" in the latter sense obviously includes the former if $E(|X_t|^2) < \infty$. Condition (2) is equivalent to

$$E(X_t) = m \quad \text{(a constant independent of } t),$$
$$E((X_t - m)\overline{(X_s - m)}) = \rho(t-s)$$
$$\text{(a function of } t-s). \quad (3)$$

We call m and $\rho(t)$ the **mean** and the **covariance function** of $\{X_t\}$.

In the continuous parameter case, we assume †continuity in probability,

$$\lim_{h \to 0} P(|X_{t+h} - X_t| > \varepsilon) = 0, \quad \varepsilon > 0,$$

for a strongly stationary process, and continuity in mean square,

$$\lim_{h \to 0} E(|X_{t+h} - X_t|^2) = 0,$$

for a weakly stationary process. The latter assumption is equivalent to continuity of the covariance function $\rho(t)$.

A †Gaussian process is strongly stationary if and only if it is weakly stationary; and so it is simply called a †stationary Gaussian process. Such processes constitute a typical class of stationary processes (→ 176 Gaussian Processes C).

B. Spectral Decomposition of Weakly Stationary Processes

The covariance function $\rho(t)$ is obviously †positive definite and continuous. Therefore, by †Bochner's theorem, we have the †spectral decomposition of $\rho(t)$:

$$\rho(t) = \int_{T'} e^{i\lambda t} F(d\lambda), \qquad (4)$$

where T' is either $\mathbf{R}$ (when $T = \mathbf{R}$) or $[-\pi, \pi]$ (when $T = \mathbf{Z}$) and F is a bounded measure on T'. The decomposition (4) is called the **Khinchin decomposition** of $\rho(t)$, and $F(d\lambda)$ is called the **spectral measure**. If the process $\{X_t\}$ is real-valued, then the spectral measure $F(d\lambda)$ is symmetric with respect to the origin.

To obtain the spectral decomposition of a weakly stationary process X_t itself, we introduce the †Hilbert space $L_2(\Omega)$ (where $\Omega = \Omega(\mathfrak{B}, P)$ is the basic probability space on which each X_t is regarded as a †square integrable function). Let $\mathfrak{M}(X)$ be the subspace of $L_2(\Omega)$ spanned by the X_t $(t \in T)$ and the constant function 1. Since $\{X_t\}$ is weakly stationary, we can define a one-parameter group of †unitary operators U_t $(t \in T)$ determined by $U_t X_s = X_{t+s}$ and $U_t 1 = 1$. By †Stone's theorem we have the spectral decomposition of U_t:

$$U_t = \int_{T'} e^{i\lambda t} E(d\lambda). \qquad (5)$$

Setting $M(\Lambda) = E(\Lambda)(X_0 - m)$, we obtain the **spectral decomposition** of X_t:

$$X_t = U_t X_0 = m + \int_{T'} e^{i\lambda t} M(d\lambda). \qquad (6)$$

We also have

$$(M(\Lambda), 1) = 0,$$
$$(M(\Lambda_1), M(\Lambda_2)) = F(\Lambda_1 \cap \Lambda_2). \qquad (7)$$

The study of weakly stationary processes is based on the decomposition (6). For example, the **weak law of large numbers** for $\{X_t\}$,

$$\underset{B-A \to \infty}{\text{l.i.m.}} \frac{1}{B-A} \int_A^B X_t \, dt = m + M(\{0\}), \qquad (8)$$

is an immediate consequence of (6). In the discrete parameter case a similar result is obtained by replacing the integral sign in expression (8) by the summation sign. In particular, if F is continuous at the origin, we have $M(\{0\}) = 0$, and only the constant m remains in the right-hand side of (8) [1, 2].

C. Weakly Stationary Random Distributions

Just as we introduce †distributions as generalizations of ordinary functions, we define weakly stationary random distributions as generalizations of weakly stationary processes. Let $\mathscr{D}$ be the space of all functions of class C^∞ on $T=\mathbf{R}$ with compact support. We introduce the same topology on $\mathscr{D}$ as in the theory of distributions. If a random variable $X_\varphi \in L_2(\Omega)$ is defined for every $\varphi \in \mathscr{D}$ and the mapping $\varphi \to X_\varphi$ is continuous in the L_2-sense and linear, then the family $\{X_\varphi\}$ of random variables is called a **random distribution in the wider sense** (→ 407 Stochastic Processes). Furthermore, if

$$(X_{\tau_h\varphi}, 1)=(X_\varphi, 1),$$

$$(X_{\tau_h\varphi}, X_{\tau_h\psi})=(X_\varphi, X_\psi) \tag{9}$$

for every $h \in \mathbf{R}^1$, where (,) stands for the inner product in $L_2(\Omega)$ and

$$\tau_h\varphi(t)=\varphi(t-h),$$

then $\{X_\varphi\}$ is said to be a **weakly stationary random distribution**. With a weakly stationary process we can associate a weakly stationary random distribution by the relation

$$X_\varphi=\int_T X_t\varphi(t)\,dt. \tag{10}$$

This correspondence is one-to-one, and therefore we can identify $\{X_t\}$ with $\{X_\varphi\}$ as we identify an ordinary function with a distribution. From equations (9) it follows that there exist a constant m and a distribution ρ such that $E(X_\varphi)=m\int\varphi(t)\,dt$ and $E(X_\varphi-E(X_\varphi))\overline{(X_\psi-E(X_\psi))}=\rho(\varphi * \check{\bar\psi})$, where $*$ denotes convolution, and $\check{\bar\psi}(t)=\overline{\psi(-t)}$. We call m and ρ the **mean value and covariance distribution** of $\{X_\varphi\}$, respectively. By the generalized Bochner theorem ρ can be expressed in the form

$$\rho(\varphi)=\int\hat\varphi(\lambda)F(d\lambda), \quad \hat\varphi(\lambda)=\int e^{i\lambda t}\varphi(t)\,dt,$$

where $F(d\lambda)$ is a slowly increasing measure, i.e.,

$$\int(1+\lambda^2)^{-k}F(d\lambda)<\infty \tag{11}$$

for some positive integer k. $F(d\lambda)$ is called the **spectral measure**. This expression is the generalization of the Khinchin decomposition. The spectral decomposition corresponding to (6) and the †law of large numbers for X_φ can be discussed in a manner similar to that for weakly stationary processes (K. Itô [3]).

D. Prediction Theory

Let $\{X_t\}$ be a weakly stationary process. Suppose that its values X_s $(s\leqslant t)$ up to time t are observed. **Prediction theory** deals with the problem of forecasting the future value $X_{t+\tau}$ $(\tau>0)$ from the known values X_s $(s\leqslant t)$. If the domain of the admissible predictors is limited to linear functions of X_s $(s\leqslant t)$, the theory is called **linear prediction theory**. We can assume without loss of generality that the mean value m of X_t is zero and that the spectral measure $F(d\lambda)$ of $\{X_t\}$ is not a zero measure.

Let $\mathscr{M}_t(X)$ be the subspace of $L_2(\Omega)$ spanned by the X_s $(s\leqslant t)$, then $\mathscr{M}(X)=\bigvee_t\mathscr{M}_t(X)$. A **linear predictor** for $X_{t+\tau}$ from X_s $(s\leqslant t)$ is an element Y of $\mathscr{M}_t(X)$. If a linear predictor minimizes the prediction error $\sigma^2(\tau)=E(|X_{t+\tau}-Y|^2)$ in $\mathscr{M}_t(X)$, it is called an **optimum linear predictor**, which turns out to be the †(orthogonal) projection of $X_{t+\tau}$ on $\mathscr{M}_t(X)$ and which is denoted by $\hat X_{t,\tau}$. Since $\{X_t\}$ is stationary, the error $\sigma^2(\tau)$ does not depend on t for such a predictor. Corresponding to the spectral decomposition (6) of X_t, the optimum linear predictor is expressed in the form

$$\hat X_{t,\tau}=\int_{T'} e^{i\lambda t}\hat\varphi_\tau(\lambda)M(d\lambda), \tag{12}$$

where $\hat\varphi_\tau(\cdot)$ is square integrable with respect to the spectral measure $F(d\lambda)$.

The subspace $\mathscr{M}_t(X)$ is nondecreasing in t. If $\mathscr{M}_t(X)$ is independent of t, i.e., $\mathscr{M}_t(X)=\mathscr{M}(X)$ for every t, then $\{X_t\}$ is said to be **deterministic**. In this case we have $\hat X_{t,\tau}=X_{t+\tau}$ for every t and $\tau>0$, since $X_{t+\tau}\in\mathscr{M}_t(X)$. This means that the linear predictor enables us to determine the unknown quantities without error, and therefore such a process is of no probabilistic interest. On the other hand, if $\bigcap_t\mathscr{M}_t(X)=\{0\}$, then $\{X_t\}$ is said to be **purely nondeterministic**. A general $\{X_t\}$ is expressed as the sum of the deterministic part $\{X_t^d\}$ and the purely nondeterministic part $\{X_t^n\}$ (**Wold decomposition**). Furthermore, we have $\mathscr{M}(X^d)=\bigcap_t\mathscr{M}_t(X)$, and $\mathscr{M}_t(X)=\mathscr{M}(X^d)+\mathscr{M}_t(X^n)$ (direct sum). Thus $\{X_t^d\}$ and $\{X_t^n\}$ can be dealt with separately.

A weakly stationary process $\{X_t\}$ is purely nondeterministic if and only if the spectral measure $F(d\lambda)$ is absolutely continuous with respect to the Lebesgue measure, and the density $f(\lambda)$ is positive almost everywhere and satisfies

$$\int_{-\pi}^{\pi}\log f(\lambda)\,d\lambda>-\infty$$

(discrete parameter case),

$$\int_{-\infty}^{\infty}\frac{\log f(\lambda)}{1+\lambda^2}\,d\lambda>-\infty$$

(continuous parameter case).

By using $f(\lambda)$ the optimum linear predictor can be obtained.

First, we explain the discrete parameter case. There exists a function $\gamma(z)=\sum_{t=0}^{\infty}a_tz^t$ in the †Hardy class H_2 relative to the unit disk such that its boundary value satisfies

$f(\lambda)=(1/2\pi)|\gamma(e^{-i\lambda})|^2.$

Then we can find a sequence of mutually orthogonal random variables $\{\xi_t\}$ $(t\in\mathbf{Z})$ such that $\{X_t\}$ admits a **backward moving average representation**

$$X_t=\sum_{s=-\infty}^{t} a_{t-s}\xi_s. \qquad (13)$$

There are many pairs $\{a_t\}$ and $\{\xi_t\}$ which give the representation (13), but if $\gamma(z)$ is maximal (optimal), namely, if $\gamma(z)$ is expressed as

$$\gamma(z)=\sqrt{2\pi}\exp\left(\frac{1}{4\pi}\int_{-\pi}^{\pi}\log f(\lambda)\frac{e^{-i\lambda}+z}{e^{-i\lambda}-z}d\lambda\right), \qquad (14)$$

then the representation (13) is **canonical** in the sense that $\mathscr{M}_t(X)=\mathscr{M}_t(\xi)$ for every t. Hence the optimum predictor $\hat{X}_{t,\tau}$ for $X_{t+\tau}$ is given by

$$\hat{X}_{t,\tau}=\sum_{s=-\infty}^{t} a_{t+\tau-s}\xi_s=\int_{-\pi}^{\pi} e^{i\lambda t}\hat{\varphi}_\tau(\lambda)M(d\lambda), \qquad (15)$$

where

$$\hat{\varphi}_\tau(\lambda)=\frac{e^{i\lambda\tau}}{\gamma(e^{-i\lambda})}\left(\gamma(e^{-i\lambda})-\sum_{s=0}^{\tau-1}a_s e^{-i\lambda s}\right).$$

The prediction error $\sigma^2(\tau)$ of this predictor is given by

$$\sigma^2(\tau)=\sum_{s=0}^{\tau-1}|a_s|^2.$$

Example. Let the covariance function of a weakly stationary process $\{X_t\}$ be $e^{-\alpha|t|}$ $(\alpha>0)$. Then we have

$f(\lambda)=(1/2\pi)(1-\beta^2)|1-\beta e^{-i\lambda}|^{-2},$

$\beta=e^{-\alpha}.$

The maximal $\gamma(z)$ is expressed as $\sqrt{1-\beta^2}(1-\beta z)^{-1}$, and

$$\hat{X}_{t,\tau}=\int_{-\pi}^{\pi}e^{i\lambda t}\frac{e^{i\lambda\tau}}{1-\beta e^{-i\lambda}}\sum_{s=\tau}^{\infty}\beta^s e^{-i\lambda s}M(d\lambda)$$

$$=\beta^\tau X_t,$$

$$\sigma^2(\tau)=1-2\beta^\tau e^{-\alpha\tau}+\beta^{2\tau}=1-e^{-2\alpha\tau}.$$

We now come to the continuous parameter case. By replacing the holomorphic function $\gamma(z)$ on the unit disk with the one on the half-plane, we see that almost all results obtained in the discrete parameter case hold similarly in this case. The maximal $\gamma(z)$ is expressed as

$$\gamma(z)=\sqrt{\pi}\exp\left(\frac{1}{2\pi i}\int_{-\infty}^{\infty}\log f(\lambda)\frac{1+\lambda z}{z-\lambda}\frac{d\lambda}{1+\lambda^2}\right).$$

Using the †Fourier transform a_t of the boundary function of $\gamma(z)$ and a process $\{\xi_t\}$ with orthogonal increments, we have the **canonical backward moving average representation** for the process $\{X_t\}$:

$$X_t=\int_{-\infty}^{t} a_{t-s}d\xi_s,$$

which enables us to obtain the optimum predictor and the prediction error in a manner similar to the discrete parameter case [4]. In particular, if the optimal $\gamma(z)$ is of the form $\gamma(z)=c/P(iz)$, where c is a constant and $P(z)$ is a polynomial of degree p, then X_t is $p-1$ times differentiable and $P(d/dt)X_t=(d/dt)\xi_t$ up to a multiplicative constant; therefore $\hat{X}_{t,\tau}$ is obtained explicitly. To obtain the optimum linear predictor for $Y\in\mathscr{M}(X)$, we first establish the expression

$$Y=\sum_{s=-\infty}^{\infty}f(s)\xi_s \quad \text{or} \quad \int_{-\infty}^{\infty}f(s)d\xi_s,$$

and then take $\sum_{s=-\infty}^{0}f(s)\xi_s$ or $\int_{-\infty}^{0}f(s)d\xi_s$ for the optimum linear predictor.

The results stated above can be generalized to multivariate (n-dimensional) stationary processes [6, 7] and to the case where the parameter space T is multidimensional.

N. Wiener observed the individual †sample process $X(t,\omega)$ and discussed a method of finding the optimum predictor for $X(t+\tau,\omega)$ by using a linear functional

$$\int_{0}^{\infty}X(t-s,\omega)dK(s)$$

(K is of †bounded variation) of the values X_s $(s\leqslant t)$ [8]. The spectral measure played an important role in his observation. Calculations in this case are analogous to those mentioned above.

For a weakly stationary random distribution $\{X(\varphi)\}$ $(\varphi\in\mathscr{D})$, the prediction theory is reduced to that for ordinary stationary processes. Assume that the spectral measure $F(d\lambda)$ of $\{X(\varphi)\}$ satisfies (11). Set $e(t)=\exp(t)$ $(t\leqslant0)$, $=0$ $(t>0)$, and let $e_k(t)$ be the k-times convolution of $e(t)$ with itself. Set $Y(\varphi)=X(e_k*\varphi)$. Then $\{Y(\varphi)\}$ is equivalent to a weakly stationary process. It is obvious that $\mathscr{M}_t(X)=\mathscr{M}_t(Y)$ for every t, where $\mathscr{M}_t(X)$ is the linear subspace spanned by $\{X(\varphi)\,|\,\text{support of }\varphi\subset(-\infty,t]\}$. This consideration allows us to reduce the prediction problem for $\{X(\varphi)\}$ to that for the stationary process corresponding to $\{Y(\varphi)\}$.

Nonlinear prediction theory is formulated as follows. Let $\mathfrak{B}_t$ be the smallest σ-algebra with respect to which every X_s $(s\leqslant t)$ is measurable and $H_t(X)$ be the subspace of $L_2(\Omega)$ consisting of all $\mathfrak{B}_t$-measurable elements. The problem is to forecast $X_{t+\tau}$ $(\tau>0)$ by using an element of $H_t(X)$. The optimum predictor is obviously equal to $E(X_{t+\tau}|\mathfrak{B}_t)$. For a stationary Gaussian process it has been proved that the optimum predictor found in $H_t(X)$ belongs to $\mathscr{M}_t(X)$. Therefore the optimum nonlinear predictor coincides with the optimum linear predictor. However, except for stationary Gaussian processes, no systematic approach for nonlinear prediction theory has been established so far. (For a typical case that arises from a stationary Gaussian process → 176 Gaussian Processes H.)

E. Interpolation and Filtering

Interpolation and **filtering** of stationary processes have many similarities with prediction theory, both in the formulation of the problems and in their method of solution.

Let $\{X_t\}$ be a weakly stationary process, all of whose values $\{X_t \mid t \notin T_1\}$. T_1 some interval, are known with the exception of those at $t \in T_1$. The problem of linear interpolation of the unknown value X_t $(t \in T_1)$ is to find the best approximation of this random variable by the limit of linear combinations of the known values. The following example illustrates the problem in the discrete parameter case.

Example. Let $T_1 = \{t_0\}$ and $f(\lambda)\,d\lambda$ be the spectral measure of $\{X_t\}$. The interpolation of X_{t_0} has an error if and only if

$$\int_{-\pi}^{\pi} \frac{1}{f(\lambda)} d\lambda < \infty.$$

Expressing X_t in the form (6) with $m=0$, the best (linear) interpolation $\hat{X}_{t_0}$ of X_{t_0} is given by

$$\int_{-\pi}^{\pi} e^{i\lambda t_0}\left(1 - 2\pi\left(f(\lambda)\int_{-\pi}^{\pi}\frac{1}{f(\mu)}d\mu\right)^{-1}\right)M(d\lambda),$$

and the error of the interpolation is expressed by

$$E(|X_{t_0} - \hat{X}_{t_0}|^2) = 4\pi^2\left(\int_{-\pi}^{\pi}\frac{1}{f(\lambda)}d\lambda\right)^{-1}.$$

The problem of interpolation for multivariate stationary processes has also been discussed [7].

The filtering problem originated in communication theory as a technique to extract the relevant component from a received signal with noise [8, 13]. Suppose that a complex-valued stationary process $\{X_t\}$ with continuous parameter is expressed in the form

$$X_t = \int_{-\infty}^{\infty} e^{i\lambda t} M(d\lambda) = S_t + N_t,$$

where (S_t, N_t) is a (2-dimensional) weakly stationary process with mean vector 0. Here, S_t and N_t indicate the signal and noise, respectively. The filtering problem is to find the element of $\mathscr{M}_t(X)$ that approximates $S_{t+\tau}$ as closely (relative to the $\mathscr{M}(X)$-norm) as possible. The best approximation is the projection of $S_{t+\tau}$ on $\mathscr{M}_t(X)$, but its expression in terms of the spectral measure becomes extremely complicated [8]. This problem is usually discussed under the assumption that S_t and N_t are orthogonal. Let us further assume that their spectral measures are absolutely continuous. The density functions are denoted by $f_S(\lambda)$ and $f_N(\lambda)$, respectively. If $\{X_t \mid t \in T\}$ is observed, then the best (linear) approximation $\hat{S}_t$ of S_t is given by

$$\hat{S}_t = \int_{-\infty}^{\infty} e^{i\lambda t}\varphi_0(\lambda) M(d\lambda),$$

where $\varphi_0(\lambda) = f_S(\lambda)/(f_S(\lambda) + f_N(\lambda))$. The mean square error $E(|S_t - \hat{S}_t|^2)$ of this filtering is

$$\int_{-\infty}^{\infty} f_S(\lambda) f_N(\lambda)/(f_S(\lambda) + f_N(\lambda))\,d\lambda.$$

F. Strongly Stationary Processes and Flows

Let $\{X_t(\omega)\}$ $(t \in T,\ \omega \in \Omega(\mathfrak{B}, P))$ be a strongly stationary process. To study it we take the coordinate representation of $\{X_t\}$ as follows. Let Ω be the complex vector space $\mathbf{C}^T$, $\mathfrak{B}$ the σ-algebra generated by the Borel †cylinder sets, $X_t(\omega)$ the tth coordinate of the function $\omega \in \mathbf{C}^T$, and P the probability distribution of the process $\{X_t\}$ defined on $(\Omega, \mathfrak{B})$. Define the shift transformation S_t of Ω onto itself by $(S_t\omega)(s) = \omega(s+t)$. Then $\{S_t\}$ forms a group of †measure-preserving transformations on $\Omega(\mathfrak{B}, P)$ (→ 136 Ergodic Theory) since $\{X_t\}$ is strongly stationary. Thus we are given a (measure-preserving) †flow $\{S_t\}$ $(t \in T)$ on $\Omega(\mathfrak{B}, P)$. Conversely, if $\{S_t\}$ $(t \in T)$ is a (measure-preserving) flow on a probability space $\Omega(\mathfrak{B}, P)$, then $\{X_t\}$, given by $X_t(\omega) = f(S_t\omega)$, is a strongly stationary process, provided that f is measurable. Many properties of a strongly stationary process are closely related to those of the corresponding flow. For example, the †strong law of large numbers for a strongly stationary process follows from †Birkhoff's individual ergodic theorem for flows. †Ergodicity, several kinds of †mixing properties, and the spectral properties of a strongly stationary process are defined in accordance with the respective notions for the corresponding flow. Now we give some examples of flows corresponding to strongly stationary processes.

(1) If X_t $(t \in \mathbf{Z})$ are mutually independent and have the same probability distribution, then the process $\{X_t\}$ $(t \in \mathbf{Z})$ is strongly stationary and $\bigcap_t \mathfrak{B}_t$ is trivial (the definition of $\mathfrak{B}_t$ → D). Hence the corresponding flow is a †Kolmogorov flow.

(2) Similarly to (1), the flow corresponding to †Gaussian white noise (→ 176 Gaussian Process) is also a Kolmogorov flow.

(3) The mixing properties of the flow corresponding to a stationary Gaussian process is determined by the smoothness of its spectral measure $F(d\lambda)$. The flow is ergodic if and only if F is continuous (i.e., F has no point mass). In this case, the flow is also †weakly mixing. For the flow to be †strongly mixing, it is necessary and sufficient that the covariance function $\rho(t)$ of the process tend to zero as $|t|\to\infty$. In this case the flow is †mixing of all orders (→ 136 Ergodic Theory) [14–16].

G. Analytic Properties of Sample Functions of Stationary Processes

In the continuous parameter case, we always assume that †continuity in probability holds for strongly stationary processes and †mean square continuity holds for weakly stationary processes. Hence the processes discussed here are all continuous in probability, and without loss of generality we can assume that the stationary processes are †separable and †measurable (→ 407 Stochastic Processes).

Let $\{X_t\}$ be a weakly stationary process. Assume that the moments up to order $2n$ of the spectral measure $F(d\lambda)$ are all finite. Then almost all †sample functions of $\{X_t\}$ are $n-1$ times continuously differentiable, and almost all sample functions of $\{X_t^{(n-1)}\}$ are absolutely continuous. Define the spectral distribution function $F(\lambda)=F((-\infty,\lambda])$ for the spectral measure $F(d\lambda)$ of $\{X_t\}$. If F satisfies

$$\sum_{n=-\infty}^{\infty} |n|^r(F(n+1)-F(n))^{1/2}<\infty \tag{16}$$

for a nonnegative integer r, then almost all sample functions of $\{X_t\}$ have continuous rth derivatives. In particular, if the condition (16) is satisfied for $r=0$, then almost all sample functions are continuous. Conditions for Hölder continuity of almost all sample functions of a weakly stationary process have also been obtained [17, 18]. (For sample functions of stationary Gaussian processes → 176 Gaussian Processes F.)

For a strongly stationary process $\{X_t\}$ with $E(X_0)=0$ and finite $E(X_0^2)$, the **sample covariance function**

$$R(t)=\lim_{T\to\infty}\frac{1}{2T}\int_{-T}^{T} X_{t+s}(\omega)\overline{X_s(\omega)}\,ds$$

is determined with probability 1. We can therefore apply the theory of †generalized harmonic analysis, due to Wiener [9]. (Further results on sample function properties are found in [19].)

H. Strongly Stationary Random Distributions

Let $\mathscr{D}$ be the space of all C^∞-functions with compact support and $\mathscr{D}'$ be the space of †distributions. If $X_\varphi(\omega)$ is defined for $\omega\in\Omega(\mathfrak{B},P)$ and $\varphi\in\mathscr{D}$, and if for almost all ω, $X_\varphi(\omega)$ belongs to $\mathscr{D}'$ as a linear functional of φ, then $\{X_\varphi\}$ is called a **random distribution**. Suppose that the joint distribution of the random variables $X_{\tau_h\varphi_1}, X_{\tau_h\varphi_2}, \ldots, X_{\tau_h\varphi_n}$ $(\tau_h\varphi(t)=\varphi(t-h))$ is independent of h. Then we call $\{X_\varphi(\omega)\}$ a **strongly** (or **strictly**) **stationary random distribution**. If we identify random distributions that have the same probability law, then $\{X_\varphi\}$ is determined by the characteristic functional

$$C(\varphi)=E(e^{iX_\varphi}).$$

For $\{X_\varphi\}$ to be strictly stationary it is necessary and sufficient that the equality $C(\tau_h\varphi)=C(\varphi)$ hold. The simplest example of a strictly stationary random distribution is the Gaussian white noise (→ 176 Gaussian Processes, 341 Probability Measures) [20].

I. Generalizations of Stationary Processes

The concept of stationary processes is generalized in many directions.

(1) Let T be a set different from $\mathbf{R}$ or $\mathbf{Z}$, and suppose that there is given a group G of transformations that map T onto itself. If a family $\{X_t\}$ of random variables with parameter $t\in T$ has the property that for every choice of random variables $X_{t_1}, X_{t_2}, \ldots, X_{t_n}$, the joint distribution of $(X_{gt_1}, \ldots, X_{gt_n})$ is always independent of $g\in G$, then $\{X_t\}$ $(t\in T)$ is said to be a **strictly G-stationary** system of random variables. Similarly, a **weakly G-stationary** system of random variables can be defined [21, 22].

(2) Let T be a Riemannian space, and let G be the group of all isometric transformations on T or one of its subgroups. Suppose that a †tensor field $\mathbf{X}_t(\omega)$ of constant rank is associated with any $\omega\in\Omega(\mathfrak{B},P)$ at every point t. Then $\mathbf{X}(\omega)\equiv\{\mathbf{X}_t(\omega)\,|\,t\in T\}$ is called a **random tensor field** over the Riemannian space T. Any $g\in G$ induces an isometric transformation of the tangent vector space at t to that at gt. Hence g maps a tensor field $\mathbf{X}(\omega)$ to another tensor field $g\mathbf{X}(\omega)$ for every ω. If $\mathbf{X}(\omega)$ and $g\mathbf{X}(\omega)$ have the same probability law, then $\mathbf{X}(\omega)$ is said to be **strictly G-stationary**. $X(\omega)$ is defined to be **weakly G-stationary** in a similar way [21, 22].

(3) In the same way as we extended stochastic processes to random distributions, we can generalize random tensor fields to **random currents** and discuss stationary random currents [21].

(4) **Stochastic process with stationary incre-**

ments of order n. Assume that $\{X_t\}$ ($t \in \mathbf{R}$) is not necessarily a stationary process but that the nth-order increment of X_t is stationary. Then by taking the nth derivative $D^n X_t$ in the sense of random distributions, we obtain a stationary random distribution. From the properties of $D^n X_t$ we can investigate the original process itself. Brownian motion is an example of a stochastic process with stationary increments of order 1.

(5) **Weakly stationary processes of degree** k. A weakly stationary process is a process whose moments up to order 2 are stationary. Generalizing this, we can define a weakly stationary process of degree k by requiring the moments up to order k to be stationary. We can obtain more detailed properties of such processes than those of weakly stationary processes [23].

References

[1] J. L. Doob, Stochastic processes, Wiley, 1953.
[2] U. Grenander and M. Rosenblatt, Statistical analysis of stationary time series, Wiley, 1957.
[3] K. Itô, Stationary random distributions, Mem. Coll. Sci. Univ. Kyôto, (A) 28 (1954), 209–223.
[4] K. Karhunen, Über die Struktur stationärer zufälliger Funktionen, Ark. Mat., 1 (1950), 141–160.
[5] H. Cramér, On the linear prediction problem for certain stochastic processes, Ark. Mat., 4 (1963), 45–53.
[6] N. Wiener and P. Masani, The prediction theory of multivariate stochastic processes I, II, Acta Math., 98 (1957), 111–150; 99 (1958), 93–137.
[7] Yu. A. Rozanov, Stationary random processes, Holden-Day, 1967. (Original in Russian, 1963.)
[8] N. Wiener, Extrapolation, interpolation, and smoothing of stationary time series, MIT Press, 1949.
[9] N. Wiener, Generalized harmonic analysis, Acta Math., 55 (1930), 117–258.
[10] N. Wiener, The homogeneous chaos, Amer. J. Math., 60 (1938), 897–936.
[11] N. Wiener, Nonlinear problems in random theory, MIT Press, 1958.
[12] P. Masani and N. Wiener, Non-linear prediction, in Probability and Statistics: The H. Cramér Volume, Wiley, 1959, 190–212.
[13] A. Blanc-Lapierre and R. Fortet, Théorie des fonctions aléatoires, Masson, 1953.
[14] G. Maruyama, The harmonic analysis of stationary stochastic processes, Mem. Fac. Sci. Kyushu Univ., (A) 4 (1949), 45–106.
[15] S. Kakutani, Spectral analysis of stationary Gaussian processes, Proc. 4th Berkeley Symp. Math. Stat. Prob. II, Univ. of California Press (1961), 239–247.
[16] H. Totoki, The mixing property of Gaussian flows, Mem. Fac. Sci. Kyushu Univ., (A) 18 (1964), 136–139.
[17] I. Kubo, On a necessary condition for the sample path continuity of weakly stationary processes, Nagoya Math. J., 38 (1970), 103–111.
[18] T. Kawata and I. Kubo, Sample properties of weakly stationary processes, Nagoya Math. J., 39 (1970), 7–21.
[19] H. Cramér and M. R. Leadbetter, Stationary and related processes, Wiley, 1967.
[20] I. M. Gel'fand and N. Ya. Vilenkin, Generalized functions IV, Academic Press, 1964. (Original in Russian, 1961.)
[21] K. Itô, Isotropic random current, Proc. 3rd Berkeley Symp. Math. Stat. Prob. II, Univ. of California Press (1956), 125–132.
[22] A. M. Yaglom, Second-order homogeneous random fields, Proc. 4th Berkeley Symp. Math. Stat. Prob. II, Univ. of California Press (1961), 593–622.
[23] A. N. Shiryaev, Some problems in the spectral theory of higher-order moments I, Theory Prob. Appl., 5 (1960), 265–284. (Original in Russian, 1960.)

396 (XVIII.3) Statistic

A. General Remarks

A statistic is a function of a value (i.e., a sample value) observed in the process of statistical inference (→ 401 Statistical Inference). A statistic is used for two purposes: (a) to characterize the set of observed values or sample values, and (b) to summarize the information contained in the sample about the unknown parameters of the population from which it is assumed to have been drawn.

B. Samples and Statistics

The basic concepts in statistical inference are †population and †sample. Let $(\Omega, \mathscr{B}, P)$ be a †probability space, where P is a †probability measure on $\mathscr{B}$. A †random variable X defines a 1-dimensional probability distribution $\Phi(A) = P\{\omega \mid X(\omega) \in A\}$, where A is a 1-dimensional †Borel set, which gives rise to a 1-dimensional probability space $(\mathbf{R}, \mathscr{B}^1, \Phi)$. Here $\mathscr{B}^1$ is the family of all 1-dimensional †Borel sets. Let $X_1, X_2, \ldots, X_n$ be †independent random variables with identical 1-dimensional distributions. The †n-dimensional random variable $X=(X_1, X_2, \ldots, X_n)$ is called a **random sample** of **size** n from the population $(\Omega, \mathscr{B}, P)$. In particular, when each of $X_1, \ldots, X_n$ takes only two values (usually 0 and 1), the sample is called a **Bernoulli sample** or a **sequence of Bernoulli trials**. Generally, if Φ_n is the †n-dimensional probability distribution determined by X (i.e., the direct product of n copies of the 1-dimensional probability distribution Φ), then the n-dimensional probability space $(\mathbf{R}^n, \mathscr{B}^n, \Phi_n)$, where $\mathscr{B}^n$ is the family of n-dimensional Borel sets, is called an **n-dimensional sample space**. A point belonging to the set of actually observed values of the sample X, which is a random variable by definition, is called a **sample value** and is denoted by x. Thus the sample value can be expressed as $x=X(\omega)$ $(\omega \in \Omega)$ and regarded as a point in the sample space (**sample point**). The basic underlying structure which determines the probability distribution is the set Ω, which we can view as describing the physical structure of the observed phenomena, but statistical procedures are always carried out through the observations of samples, and Ω itself is often disregarded. The 1-dimensional probability distribution Φ (the n-dimensional probability distribution Φ_n determined by X) is called the **population distribution** in the 1-dimensional (n-dimensional) sample space, since it is induced from the probability measure on $(\Omega, \mathscr{B})$.

A **statistic** Y is a random variable expressed as $Y=f(X)$, where f is a †measurable function from the sample space $(\mathbf{R}^n, \mathscr{B}^n, \Phi_n)$ into a measurable space $(\mathbf{R}, \mathscr{B}^1)$. The value of the statistic Y corresponding to a sample value x of the sample X is denoted by $y=f(x)$.

When we deal with a statistical problem we often have no exact knowledge of the population distribution $\Phi(\Phi_n)$ except that it belongs to a family $\mathscr{P}=\{P_\theta \mid \theta \in \Theta\}$ of probability measures on $\mathscr{B}^1(\mathscr{B}^n)$. We call θ the **parameter** of the probability distribution and Θ the **parameter space**. The typical cases described in this section can be extended as follows: (1) The distribution Φ may be an r-dimensional probability distribution. In this case a sample of size n induces an nr-dimensional sample space. (2) Random variables $X_1, \ldots, X_n$, being mutually independent, may not have identical distributions. (3) Random variables $X_1, \ldots, X_n$ may not be mutually independent. In both cases (2) and (3) the sample space is of the form $(\mathbf{R}^n, \mathscr{B}^n, \Phi_n)$, but n may not be the sample size itself, nor Φ_n be the direct product of n copies of identical 1-dimensional components. (4) The most general sample space is expressed as a certain measurable space $(\mathscr{X}, \mathscr{A})$ and a family $\mathscr{P}=\{P_\theta \mid \theta \in \Theta\}$ of probability measures on $\mathscr{A}$.

A **statistic**, in general, is a random variable expressed as $Y=f(X)$ by a measurable function f defined on a sample space $(\mathscr{X}, \mathscr{A})$ taking values in another meaurable space $(\mathscr{Y}, \mathscr{C})$. When $(\mathscr{Y}, \mathscr{C})$ is $(\mathbf{R}, \mathscr{B}^1)$ or $(\mathbf{R}^n, \mathscr{B}^n)$, $Y=f(X)$ is accordingly called a **1-dimensional** or **n-dimensional statistic**.

C. Population and Sample Characteristics in the 1-Dimensional Case

In a 1-dimensional probability space $(\mathbf{R}, \mathscr{B}^1, P_0)$ the following quantities, called **population characteristics**, are used to characterize the population distribution P_0: Letting $F(z)= P_0((-\infty, z])$ be the †distribution function of P_0, we use the **population mean** $\mu=\int z\,dF(z)$; the **population variance** $\sigma^2=\int (z-\mu)^2\,dF(z)$; the **population standard deviation** $\sigma(\geqslant 0)$; the **population moment of order** k $\mu_k'=\int z^k\,dF(z)$ $(\mu_1'=\mu)$; the **kurtosis** μ_4/σ^4; the **coefficient of excess** $(\mu_4/\sigma^4)-3$; the **skewness** μ_3/σ^3; the **α-quantile** or **$100\alpha\%$-point** m satisfying $F(m-0)\leqslant\alpha\leqslant F(m+0)$; the **median**, which is the 50%-point; the **first** and **third quartiles**, which are the 25%-point and 75%-point, respectively; the **range**, which is the third quartile minus the first quartile; and the **mode**, which is the value or values of z for which $dF(z)/dz$ attains its maximum.

Sometimes the kurtosis and others are called **population kurtosis**, etc. Here the word "population" is used when it is desirable to distinguish population characteristics from the sample characteristics defined in 341 Probability Measures.

Let $x=(x_1, \ldots, x_n)$ be a point of an n-dimensional sample space (a sample value). Corresponding to each 1-dimensional Borel set A, the number of components of x that belong to A is called the **frequency** of A in the sample value $x=(x_1, \ldots, x_n)$, and (frequency)/n is called the **relative frequency** of A. If we take $A=(-\infty, z]$ and regard its relative frequency $F_x(z)$ as a function of z, it becomes a †distribution function for every $x \in \mathbf{R}^n$, called the **empirical distribution function** based on x.

Various characteristics can be defined from the empirical distribution function in exactly the same way as population characteristics are derived from a population distribution function. These are called **sample characteristic values** and can be expressed as functions of $x_1, \ldots, x_n$.

Assuming that $x=(x_1, \ldots, x_n)$ is a sample value of a sample $X=(X_1, \ldots, X_n)$, the statistic obtained by substituting X for x in the function denoting a sample characteristic value is called a **sample characteristic** and given the same name as the corresponding population characteristic, except that the word "population" is replaced by the word "sample." A sample characteristic is a function of random variables. Hence it is also a random variable, and the problem of deriving its probability distribution from the assumed population distribution is called that of sampling distribution (→ 374 Sampling Distributions). Thus we define the **sample mean** $\bar{X}=\sum_{i=1}^{n} X_i/n$; the **sample variance** $\sum_{i=1}^{n}(X_i-\bar{X})^2/n$ (sometimes $\sum_{i=1}^{n}(X_i-\bar{X})^2/(n-1)$ is taken as the sample variance); the **sample standard deviation**

$$\sqrt{\sum_{i=1}^{n}(X_i-\bar{X})^2/n},$$

which is the positive square root of the sample variance; the **sample mode**, which is the value taken by the largest number of X_i; and the **sample moment of order** k $\sum_{i=1}^{n}(X_i-\bar{X})^k/n$.

Among other statistics of frequent use are the **order statistic**, i.e., the set of values of $X_1, \ldots, X_n$ arranged in order of magnitude and usually denoted by $X_{(1)}<X_{(2)}<\ldots<X_{(n)}$. Various other statistics are defined in terms of order statistics: the **sample median** $X_{\text{med}}=X_{((n+1)/2)}$ for odd n and $=(X_{(n/2)}+X_{((n/2)+1)})/2$ for even n, the **sample range** $R=\max X_i-\min X_i=X_{(n)}-X_{(1)}$, and so on. The empirical distribution function $F_X(z)$ or its standardized form $S_x(z)=\sqrt{n}\{F_x(z)-F(z)\}$ can also be considered to be a function of the order statistics, and hence is a statistic taking values in the space of functions of a real variable. So is the **empirical characteristic function**

$$\phi_x(t)=\int \exp(itz)\,dF_x(z)=\sum_j \exp(itx_{(j)})/n.$$

In a sequence of Bernoulli trials, a set of successive components with an identical value is called a **run**. For example, (01100010) has a run of 0 of the length 3 and a run of 1 of the length 2.

Among the statistics listed in the previous paragraphs, the order statistic is an n-dimensional statistic and all others are 1-dimensional.

D. Other Cases

Let $(\mathbf{R}^2, \mathscr{B}^2, P_0)$ be a 2-dimensional probability space with a 2-dimensional population distribution P_0, and let $(X_1, \ldots, X_n)$ $(X_i=(U_i, V_i))$ be a random sample of size n from P_0. In this case also, the population characteristics for the †marginal distributions of U_i and V_i and sample characteristics for $(U_1, \ldots, U_n)$ and $(V_1, \ldots, V_n)$ are defined as in Section C.

As an index for association between U_i and V_i the **population covariance** $\int(u-\mu_{(1)})(v-\mu_{(2)})\,dF(u,v)$ of U_i and V_i and the **population correlation coefficient**, which is equal to (population covariance)/$\sigma_{(1)}\sigma_{(2)}$, are defined. Here $F(u,v)$ is the joint distribution function of U_i and V_i, $\mu_{(1)}$ and $\mu_{(2)}$ are the respective population means of U_i and V_i, and $\sigma_{(1)}$ and $\sigma_{(2)}$ are the respective standard deviations. As corresponding sample characteristics, we have the **sample covariance** $\sum_{i=1}^{n}(U_i-\bar{U})(V_i-\bar{V})/n$ of $(U_1, \ldots, U_n)$ and $(V_1, \ldots, V_n)$ and the **sample correlation coefficient**

$$\frac{\sum_{i=1}^{n}(U_i-\bar{U})(V_i-\bar{V})}{\left(\sum_{i=1}^{n}(U_i-\bar{U})^2\right)^{1/2}\left(\sum_{i=1}^{n}(V_i-\bar{V})^2\right)^{1/2}},$$

where $\bar{U}=\sum_{i=1}^{n} U_i/n$ and $\bar{V}=\sum_{i=1}^{n} V_i/n$.

Similarly, statistics of the samples from a population of k-dimensional distribution $(k\geqslant 3)$ can be defined (→ 280 Multivariate Analysis). More generally, in statistical inference we encounter samples where observed values may not be mutually independent or identically distributed, but have more complicated probability structures. Statistics as functions of such samples are also considered.

E. General Properties of Statistics

The general theory of statistics has been studied in a measure-theoretic framework. $(\mathscr{X}, \mathscr{A}, \mathscr{P})$ is called a **statistical structure**, where $(\mathscr{X}, \mathscr{A})$ is a measurable space and $\mathscr{P}$ is a family of probability measures on $(\mathscr{X}, \mathscr{A})$. A σ-subfield $\mathscr{B}$ of $\mathscr{A}$ (hereafter abbreviated σ-field) is called **sufficient** for $\mathscr{P}$ if for any $A\in\mathscr{A}$ there exists a $\mathscr{B}$-measurable conditional probability of A independent of $P_\theta\in\mathscr{P}$, that is, a $\mathscr{B}$-measurable function $\varphi_A(x)$ satisfying

$$P_\theta(A\cap B)=\int_B \varphi_A(x)\,dP_\theta(x)$$

for all $P_\theta\in\mathscr{P}$ and $B\in\mathscr{B}$.

For any two σ-fields $\mathscr{B}_1$ and $\mathscr{B}_2$, the notation $\mathscr{B}_1\subset\mathscr{B}_2[\mathscr{P}]$ means that to each set A_1 in $\mathscr{B}_1$ there corresponds an A_2 in $\mathscr{B}_2$ satisfying $P_\theta((A_1-A_2)\cup(A_2-A_1))=0$ for all $P_\theta\in\mathscr{P}$. When the reverse relation $\mathscr{B}_2\subset\mathscr{B}_1[\mathscr{P}]$ also holds, we write $\mathscr{B}_1=\mathscr{B}_2[\mathscr{P}]$.

For a statistic t which is a measurable function from $(\mathscr{X}, \mathscr{A})$ to $(\mathscr{Y}, \mathscr{C})$, $\mathscr{B}(t)=\{B \mid B\in\mathscr{A}, t(B)\in\mathscr{C}\}$ is a σ-field and is called the σ-field induced by t. If $\mathscr{B}(t)$ is sufficient for $\mathscr{P}$, t is said to be sufficient for $\mathscr{P}$. Since sufficiency of a statistic means that of a σ-field, we consider only sufficiency of a σ-field.

$\mathscr{B}$ is called **necessary** if for any sufficient $\mathscr{B}_0$ we have $\mathscr{B}\subset\mathscr{B}_0[\mathscr{P}]$. A necessary and sufficient σ-field is called a **minimal sufficient σ-field**. A necessary and sufficient statistic is also called a **minimal sufficient statistic**. Such a statistic does not always exist; $\mathscr{B}_2$ containing a sufficient $\mathscr{B}_1$ is not always sufficient.

$\mathscr{B}$ is said to be **complete** if for every $\mathscr{B}$-measurable integrable function φ, $\int_{\mathscr{X}} \varphi(x) dP_\theta(x) = 0$ for all $P_\theta \in \mathscr{P}$ implies $P_\theta(\{x \mid \varphi(x) \neq 0\}) = 0$ for all $P_\theta \in \mathscr{P}$. $\mathscr{B}$ is said to be **boundedly complete** if for every bounded $\mathscr{B}$-measurable φ, $\int_{\mathscr{X}} \varphi(x) dP_\theta(x) = 0$ for all $P_\theta \in \mathscr{P}$ implies $P_\theta(\{x \mid \varphi(x) \neq 0\}) = 0$ for all $P_\theta \in \mathscr{P}$. When $\mathscr{B}(t)$ is (boundedly) complete, t is called (boundedly) complete. If $\mathscr{B}_1 \subset \mathscr{B}_2$ and $\mathscr{B}_2$ is (boundedly) complete, $\mathscr{B}_1$ is also (boundedly) complete. If $\mathscr{B}_1$ is (boundedly) complete and sufficient and $\mathscr{B}_2$ is minimal sufficient, we have $\mathscr{B}_1 = \mathscr{B}_2[\mathscr{P}]$.

F. Dominated Statistical Structure

When all $P_\theta \in \mathscr{P}$ are absolutely continuous with respect to a σ-finite measure λ on $\mathscr{A}$, then $(\mathscr{X}, \mathscr{A}, \mathscr{P})$ is said to be a **dominated** statistical structure and $\mathscr{P}$ is said to be a dominated family of probability distributions. In this case, P_θ has the density $f_\theta(x) = dP_\theta / d\lambda$ with respect to λ by the Radon-Nikodym theorem. If $\mathscr{A}$ is separable, $\mathscr{P}$ is a separable metric space with respect to the metric $\rho(P_{\theta_1}, P_{\theta_2}) = \sup_{B \in \mathscr{A}} |P_{\theta_1}(B) - P_{\theta_2}(B)|$. There exists a countable subset $\mathscr{P}' = \{P_{\theta_1}, P_{\theta_2}, \dots\}$ of $\mathscr{P}$ such that $P_\theta(N) = 0$ for all $P_\theta \in \mathscr{P}'$ implies $P_\theta(N) = 0$ for all $P_\theta \in \mathscr{P}$. If we put $\lambda_0 = \sum_i c_i P_{\theta_i}$, $c_i > 0$, $\sum_i c_i = 1$, λ_0 dominates $\mathscr{P}$, and if $\mathscr{B}$ is sufficient for $\mathscr{P}$ we can choose a $\mathscr{B}$-measurable version of $dP_\theta / d\lambda_0$. Conversely, if there exists a σ-finite measure λ such that we can choose a $\mathscr{B}$-measurable version of $dP_\theta / d\lambda$ for all $P_\theta \in \mathscr{P}$, then $\mathscr{B}$ is sufficient.

If $\mathscr{P}$ is dominated by a σ-finite λ, $\mathscr{B}$ is sufficient if and only if there exist a $\mathscr{B}$-measurable g_θ and an $\mathscr{A}$-measurable h independent of θ satisfying

$$\frac{dP_\theta}{d\lambda} = g_\theta h \text{ a.e. } (\mathscr{A}, \lambda) \text{ for all } P_\theta \in \mathscr{P}.$$

This is called **Neyman's factorization theorem**.

With a dominated statistical structure, there exists a minimal sufficient σ-field, and a σ-field containing a sufficient σ-field is also sufficient.

We say that a σ-field $\mathscr{B}$ is **pairwise sufficient** for $\mathscr{P}$ if it is sufficient for every pair $\{P_{\theta_1}, P_{\theta_2}\}$ of measures in $\mathscr{P}$. A necessary and sufficient condition for $\mathscr{B}$ to be sufficient for a dominated set $\mathscr{P}$ is that $\mathscr{B}$ be pairwise sufficient for $\mathscr{P}$.

Recently, a more general statistical structure has been studied. Put $\mathscr{A}_e(\mu) = \{A \mid A \in \mathscr{A}, \mu(A) < \infty\}$. A measure μ on $\mathscr{A}$ is said to be a localizable measure if there exists ess-sup $\mathscr{F}(\mu)$ for any subfamily $\mathscr{F} \subset \mathscr{A}_e(\mu)$, that is, if there exists a set $E \in \mathscr{A}$ such that $\mu(A - E) = 0$ holds for all $A \in \mathscr{F}$, and $\mu(A - S) = 0$ for all $A \in \mathscr{F}$ implies $\mu(E - S) = 0$. A σ-finite measure is localizable. A measure μ is said to have the **finite subset property** if for any A satisfying $0 < \mu(A)$, there exists a $B \subset A$ satisfying $0 < \mu(B) < \infty$. A statistical structure $(\mathscr{X}, \mathscr{A}, \mathscr{P})$ is said to be **weakly dominated** if $\mathscr{P}$ is dominated by a localizable measure μ with the finite subset property and a density $dP_\theta / d\mu$ exists for all $P_\theta \in \mathscr{P}$. In this case a minimal sufficient σ-field exists, and a pairwise sufficient σ-field is sufficient. For example, let $\mathscr{A} = 2^{\mathscr{X}}$ and $\mathscr{P}$ be the set of all discrete probability measures on $\mathscr{A}$. $\mathscr{P}$ is weakly dominated by the counting measure μ which is localizable on $\mathscr{A}$.

The order statistic is sufficient if $\mathscr{P}$ is dominated, $X_1, \dots, X_n$ are mutually independent and identically distributed random variables with $\mathscr{X} = \mathbf{R}^n$, and each $P_\theta \in \mathscr{P}$ is invariant under every permutation of the components of the points $x = (x_1, \dots, x_n)$ in $\mathscr{X}$. Moreover, the order statistic is complete if $\mathscr{P}$ is large enough. For example, we have the following theorem: The order statistic is complete if every P_θ^0 (the component of P_θ on $\mathscr{B}_0 \subset \mathscr{B}^1$, $\theta \in \Theta$) is absolutely continuous with respect to the Lebesgue measure l on $\mathbf{R}$ and $\{P_\theta^0 \mid \theta \in \Theta\}$ contains all P_θ^0 for which $g(z) = dP_\theta^0 / dl$ is constant on some finite disjoint intervals in $\mathscr{X}_0 \subset \mathbf{R}$. A similar result holds for discrete distributions.

We call θ a **selection parameter** when $f_\theta(x) = c(\theta) \chi_{E_\theta}(x) h(x)$, where $h(x)$ is a positive $\mathscr{B}$-measurable function, $\chi_{E_\theta}(x)$ is the indicator function of a set $E_\theta \in \mathscr{B}$, and $c(\theta)$ is a constant depending on θ. Here θ determines the carrier E_θ of $f_\theta(x)$ but does not essentially affect the functional form of $f_\theta(x)$. A necessary and sufficient statistic is given by $t^*(x) = \bigcap \{E_\theta \mid E_\theta \ni x, P_\theta \in \mathscr{P}\}$. Here the class of sets of the form given in the right-hand side of this expression is taken as $\mathscr{Y}$, and we set $\mathscr{C} = \{C \mid C \subset \mathscr{Y}, t^{*-1}(C) \in \mathscr{B}\}$. We call $t^*(x)$ the **selection statistic**. Two examples follow.

(I) †Uniform distributions. Let $\Theta = \{(\alpha, \beta) \mid -\infty < \alpha < \beta < \infty\}$, $\mathscr{X}_0 = \mathbf{R}$, P_θ^0 be the uniform distribution on (α, β), and $X = (X_1, \dots, X_n)$ be a random sample of size n having P_θ^0 as its population distribution. Then $E_\theta = \{x \mid \alpha \leqslant \min_i x_i \leqslant \max_i x_i \leqslant \beta\}$ and $f_\theta(x) = (\beta - \alpha)^{-n} \chi_{E_\theta}(x)$. If we put $t(x) = (\min_i x_i, \max_i x_i)$, $\mathscr{Y} = \mathbf{R}^2$, and $\mathscr{C} =$ the set of all Borel sets of $\mathbf{R}^2$, it follows that $t(x) = t^*(x)[\mathscr{P}]$, where $t^*(x)$ is the selection statistic. Hence $t(x)$ itself is necessary and sufficient.

(II) †Exponential distributions. Put $\Theta = (-\infty, \infty)$, $\mathscr{X}_0 = \mathbf{R}$, and

$$g_\theta(z) = \begin{cases} \alpha e^{-\alpha(z-\theta)}, & z \geqslant \theta, \\ 0, & z < \theta, \end{cases}$$

where α is a known constant, and let $X = (X_1, \dots, X_n)$ be a random sample of size n

having a population distribution with density function $g_\theta(z)$. Then

$$E_\theta=\left\{x\mid\theta\leqslant\min_i x_i\right\},$$

$$f_\theta(x)=\alpha^n e^{n\alpha\theta}\chi_{E_\theta}(x)\exp\left(-\alpha\sum_{i=1}^n x_i\right).$$

If we put $t(x)=\min_i x_i$ and let $t^*(x)$ be the selection statistic, it follows that $t(x)=t^*(x)[\mathscr{P}]$, and $t(x)$ is necessary and sufficient. If $\Theta=\{(\alpha,\theta)\mid 0<\alpha<\infty,-\infty<\theta<\infty\}$, then $t(x)=\{\min_i x_i,\sum_i x_i\}$ is a necessary and sufficient statistic.

G. Exponential Families of Distributions

A dominated $\mathscr{P}$ is called an **exponential family of distributions** if and only if $f_\theta(x)=dP_\theta/d\lambda$ can be expressed in the form

$$f_\theta(x)=\exp\left(\sum_{j=1}^k s_j(x)\alpha_j(\theta)+\alpha_0(\theta)+s_0(x)\right),$$

$$x\in\mathscr{X},\quad\theta\in\Theta,\quad(1)$$

where the $s_j(x)$ $(j=0,1,\ldots,k)$ are real-valued $\mathscr{B}$-measurable functions and the $\alpha_j(\theta)$ $(j=0,1,\ldots,k)$ are constants depending on θ. If there exists a sufficient statistic for $\mathscr{P}$ that is not equivalent to but is in a certain sense simpler than the sample itself or the order statistics, then it can be shown under some regularity conditions that $\mathscr{P}$ must be an exponential family. The following theorem provides an instance of the hypotheses that guarantee such a conclusion: Let X be a sample from a 1-dimensional probability space $(\mathscr{X}_0,\mathscr{B}_0,P_\theta^0)$ with P_θ^0 the population distribution, where $\mathscr{X}_0$ is a finite or infinite interval in $\mathbf{R}$ and $\mathscr{B}_0$ is the class of all Borel sets. Let l denote the Lebesgue measure. Assume that $\{P_\theta^0\}$ is dominated by l and $g_\theta(z)=dP_\theta^0/dl$ is greater than a positive constant and continuously differentiable in z on $\mathscr{X}_0$. Assume further that there exists a sufficient statistic $t(x)$ with the property that for each open subset B of $\mathscr{X}$ $(\subset\mathbf{R}^n)$ and λ-null set N there are two points $x\not\equiv x'$ in $B-N$ such that $t(x)\neq t(x')$. Then $\mathscr{P}$ is an exponential family, and the k given in (1) is less than n. Similar results are known also for cases where $X_1,\ldots X_n$ are not identically distributed.

It is evident from the construction of a necessary and sufficient statistic that the statistic $t(x)=(s_1(x),\ldots,s_k(x))$, where the $s_j(x)$ are those appearing in (1), is sufficient for an exponential family and necessary if $\alpha_1(\theta),\ldots,\alpha_k(\theta)$ are linearly independent. If $\{(\alpha_1(\theta),\ldots,\alpha_k(\theta))\mid\theta\in\Theta\}$ contains a k-dimensional interval, $t(x)$ is complete. The distribution of $t(x)$ is of exponential type. When $X_1,\ldots,X_n$ are mutually independent with a common distribution of exponential type, the distribution of $X=(X_1,\ldots,X_n)$ is of exponential type, and vice versa. The family (1) of distributions is a special form of †Pólya-type distributions, and various distributions given in (III)–(VII) below are written in this form. In the following examples, for a sample of size n from the specified distribution, $f_\theta(x)$ is the density with respect to Lebesgue measure in (III), (IV), and (V) and to counting measure in (VI) and (VII) (→ Appendix A, Table 22).

(III) †Normal distributions $N(\mu,\sigma^2)$, $\mathscr{X}_0=(-\infty,\infty)$.

$$f_\theta(x)=\exp\left(-\left(\sum_{i=1}^n x_i^2\right)\frac{1}{2\sigma^2}+\left(\sum_{i=1}^n x_i\right)\frac{\mu}{\sigma^2}-\frac{n\mu^2}{2}-n\log\sigma-\frac{n}{2}\log 2\pi\right).$$

(IV) †Γ-distributions $\Gamma(p,\sigma)$, $\mathscr{X}_0=(0,\infty)$.

$$f_\theta(x)=\exp\left((p-1)\sum_{i=1}^n\log x_i-\frac{1}{\sigma}\sum_{i=1}^n x_i-np\log\sigma-n\log\Gamma(p)\right).$$

(V) †Exponential distributions $e(\mu,\sigma)$, $\mathscr{X}_0=(\mu,\infty)$.

$$f_\theta(x)=\exp\left(-\left(\sum_{i=1}^n x_i\right)\frac{1}{\sigma}-n\log\sigma+n\frac{\mu}{\sigma}\right).$$

(VI) †Binomial distributions $Bin(N,p)$, $\mathscr{X}_0=\{0,1,\ldots,N\}$.

$$f_\theta(x)=\exp\left(\left(\sum_{i=1}^n x_i\right)\log\frac{p}{1-p}+nN\log(1-p)-\sum_{i=1}^n\log\binom{N}{x_i}\right).$$

(VII) †Poisson distributions $P(\lambda)$, $\mathscr{X}_0=\{0,1,2,3,\ldots\}$.

$$f_\theta(x)=\exp\left(\left(\sum_{i=1}^n x_i\right)\log\lambda-\sum_{i=1}^n\log(x_i!)-n\lambda\right).$$

H. Ancillary Statistics

A statistic $t(x)$ is called an **ancillary statistic** when for every element A in $\mathscr{A}(t)$, $P_\theta(A)$ is independent of θ, or in other words, when the distribution of $t(x)$ is independent of θ. A sufficient condition for a statistic to be ancillary is that it is independent of some sufficient statistic. Conversely, an ancillary statistic is independent of all boundedly complete sufficient statistics.

I. Invariant Statistics

Suppose that we are given groups of one-to-one measurable transformations G and $\bar{G}$ on

$\mathscr{X}$ and Θ, respectively. Suppose also that we are given a †homomorphism $g \to \bar{g}$ from G to $\bar{G}$ satisfying $P_\theta(g^{-1}B) = P_{\bar{g}\theta}(B)$. In this case $\mathscr{P} = \{P_\theta \mid \theta \in \Theta\}$ is called **G-invariant**. If $\bar{G}$ is transitive, there exists a fixed element θ_0 of Θ that is sent to an arbitrary θ by an element g of G. In this case, θ is called a **transformation parameter**. In particular, if $\Theta = \mathbf{R}$, X is a random sample from a population distribution, and $P_\theta(B) = P_{\theta_0}(B - \theta)$, where $B - \theta = \{x \mid (x_1 + \theta, \ldots, x_n + \theta) \in B\}$, then θ is called a **location parameter**. When $\Theta = (0, \infty)$ and $P_\theta(B) = P_1(B/\theta)$, where $B/\theta = \{x \mid (\theta x_1, \ldots, \theta x_n) \in B\}$, θ is called a **scale parameter**. Now assume that θ is a combination of these two kinds of parameters such that $\theta = (\alpha, \beta)$ $(-\infty < \alpha < \infty, 0 < \beta < \infty)$ and $P_\theta(B) = P_{\theta_0}((B - \alpha)/\beta)$, where $\theta_0 = (0, 1)$. Then if $\mathscr{P}$ is an exponential family, (1) of Section G can be written as

$$g_e(z) = \frac{dP_\theta^0}{dl} = \frac{1}{\beta}\exp\left(\sum_{j=0}^{m} k_j\left(\frac{z-\alpha}{\beta}\right)^j\right),$$

where the k_j $(j = 0, 1, \ldots, m)$ are constants.

We call $t(x)$ an **invariant statistic** with respect to a general transformation group G when $t(gx) = t(x)$ for all $g \in G$ and $x \in \mathscr{X}$. An invariant statistic is said to be **maximal invariant** with respect to G if, for $t(x) = t(x')$, there exists a $g \in G$ such that $x = gx'$. If t_0 is maximal invariant with respect to G, a statistic t is invariant under G if and only if $t_0(x) = t_0(x')$ implies $t(x) = t(x')$.

When $\mathscr{P}$ is G-invariant, a set A $(\in \mathscr{A})$ is called G-invariant if $gA = A$ for all $g \in G$. We denote by $\mathscr{A}^0$ the set of all G-invariant sets in $\mathscr{A}$. $\mathscr{A}^0$ is clearly a σ-field. A set A $(\in \mathscr{A})$ is called **almost G-invariant** if $gA = A$ $(\mathscr{A}, \mathscr{P})$ for all $g \in G$. We denote by $\mathscr{A}^*$ the σ-field consisting of all almost G-invariant sets. If $\mathscr{P}$ is G-invariant, $\mathscr{B}$ is sufficient for $\mathscr{P}$, $g\mathscr{B} = \mathscr{B}$ for all $g \in G$, and moreover, $\mathscr{B}^0 = \mathscr{B}^*$ $(\mathscr{A}, \mathscr{P})$, then $\mathscr{B}^0$ is a sufficient σ-subfield of $\mathscr{A}^0$, where $\mathscr{B}^0 = \mathscr{B} \cap \mathscr{A}^0$ and $\mathscr{B}^* = \mathscr{B} \cap \mathscr{A}^*$.

J. Various Definitions of Sufficiency

There are many different definitions of sufficiency, and the relations among them have been investigated. A σ-field $\mathscr{B}$ is called **decision-theoretically sufficient** or **D-sufficient** if for a given $\mathscr{A}$-measurable decision function δ there exists a $\mathscr{B}$-measurable decision function δ' such that

$$\int_{\mathscr{X}} \delta(x, E)\, dP_\theta(x) = \int_{\mathscr{X}} \delta'(x, E)\, dP_\theta(x)$$

for all $E \in \mathscr{D}$, $P_\theta \in \mathscr{P}$, where a decision space $(D, \mathscr{D})$ is quite arbitrary. $\mathscr{B}$ is called **test sufficient** if for any given $\mathscr{A}$-measurable test function φ, there exists a $\mathscr{B}$-measurable test function φ' satisfying $E_\theta(\varphi) = E_\theta(\varphi')$ for all $P_\theta \in \mathscr{P}$. Let $(\Theta, \mathscr{C})$ be a measurable space of parameter θ and $\hat{\Theta}$ be the set of all probability measures on $\mathscr{C}$. Moreover we assume that $P_\theta(B)$ is $\mathscr{C}$-measurable as a function of θ for any fixed $B \in \mathscr{A}$. For any $\xi \in \hat{\Theta}$, we define λ_ξ by

$$\lambda_\xi(A \times C) = \int_C P_\theta(A)\, d\xi(\theta), \qquad A \in \mathscr{A},\ C \in \mathscr{C}.$$

We denote by $\tilde{\lambda}_\xi$ the extension of λ_ξ to $\mathscr{A} \times \mathscr{C}$. $\mathscr{B}$ is said to be **Bayes sufficient** if

$$E_{\tilde{\lambda}_\xi}(I_{\mathscr{X} \times C} \mid \mathscr{A} \times \mathscr{C}) = E_{\tilde{\lambda}_\xi}(I_{\mathscr{X} \times C} \mid \mathscr{B} \times \mathscr{C})$$

for all $\xi \in \hat{\Theta}$, $C \in \mathscr{C}$, that is, the a posteriori distribution on Θ given $\mathscr{A}$ coincides with that given $\mathscr{B}$ for any a priori ξ. When $\mathscr{P}$ is dominated, these definitions coincide with the classical definition of sufficiency. Generally, a D-sufficient σ-field contains at least one sufficient σ-field. A σ-field containing a sufficient σ-field is Bayes sufficient. Hence Bayes sufficiency follows from D-sufficiency and from classical sufficiency. If a D-sufficient σ-field is separable it is sufficient.

The notion of **prediction sufficiency** or **adequacy** was defined by Skibinsky [10]. Let (X, Y) be a pair of random variables defined over the probability space $(\mathscr{X} \times \mathscr{Y}, \mathscr{A} \times \mathscr{B}, \mathscr{P}_\theta)$. We suppose that X is the sample to be observed, and Y is (are) the value(s) of future observation(s) about which we are to make prediction(s) based on X. X and Y have joint probability distribution with an unknown parameter. A statistic $T = T(X)$ or a subfield $\mathscr{C}$ of $\mathscr{A}$ is said to be prediction sufficient or adequate if (a) given T, X and Y are conditionally independent (or given $\mathscr{C}$, $\mathscr{A}$ and $\mathscr{B}$ are conditionally independent or Markov) and (b) T is sufficient for X ($\mathscr{C}$ is sufficient for $\mathscr{A}$). It was proved that in any form of prediction on Y, we may restrict ourselves to the class of procedures that are functions of T (or are $\mathscr{C}$-measurable).

References

[1] D. A. S. Fraser, Nonparametric methods in statistics, Wiley, 1957.
[2] E. L. Lehmann, Testing statistical hypotheses, Wiley, 1959.
[3] S. S. Wilks, Mathematical statistics, Wiley, 1962.
[4] R. R. Bahadur, Sufficiency and statistical decision functions, Ann. Math. Statist., 25 (1954), 423–462.
[5] T. S. Pitcher, A more general property than domination for sets of probability measures, Pacific J. Math., 15 (1965), 597–611.
[6] L. Brown, Sufficient statistics in the case of independent random variables, Ann. Math. Statist., 35 (1964), 1456–1474.

[7] E. W. Barankin and A. P. Maitra, Generalization of the Fisher-Darmois-Koopman-Pitman theorem on sufficient statistics, Sankhyā, ser. A, 25 (1963), 217–244.
[8] W. J. Hall, R. A. Wijsman, and J. K. Ghosh, The relationship between sufficiency and invariance with applications in sequential analysis, Ann. Math. Statist., 36 (1965), 575–614.
[9] K. K. Roy and R. V. Ramamoorthi, Relationship between Bayes, classical and decision theoretic sufficiency, Sankhyā, ser. A, 41 (1979), 48–58.
[10] M. Skibinsky, Adequate subfields and sufficiency, Ann. Math. Statist., 38 (1967), 155–161.

397 (XVIII.1) Statistical Data Analysis

A. Statistical Data

Statistical data analysis is comprised of a collection of mathematical methods whereby we can deal with numerical data obtained through observations, measurements, surveys, or experiments on the "objective" world. The purpose of statistical analysis is to extract the relevant information from that numerical data pertinent to the subject under consideration. The nature and the properties of the subject and also the purpose of the analysis may vary greatly. The subject may be physical, biological, chemical, sociological, psychological, economic, etc. in nature, and the purpose of the analysis can be purely scientific, as well as technological, medical, or managerial. Because of the great diversity of statistical data, the methods of statistical data analysis and the manner of application should differ greatly from situation to situation; we cannot expect a single unified system of methods to be applicable to all cases. Nevertheless, we have several formal methods of statistical analysis that are more or less mutually related and have been successfully applied to most, if not all, statistical data.

Statistical data can be classified into several types according to a few criteria: according to the property of each observation or measurement, they can be either **quantitative** or **qualitative**; according to whether only one observation is made on each object under investigation or many observations on the same object, they can be either **univariate** or **multivariate**; and according to whether the observations are made at one time or consecutively in the course of time, they may be either **cross sectional** or **time series**. Each different type of statistical data requires a different type of procedure (→ 280 Multivariate Analysis, 421 Time Series Analysis).

B. Frequency Distributions and Histograms

Statistical data have the simplest structure when they consist of a collection of observations made on an aggregate of objects supposedly of the same kind. Such an aggregate is usually called a **population**, and the number of its members (its size) is denoted by N. When the data are qualitative or **categorical**, each member of the population is classified into several types according to some criteria, the data consist of the numbers of the members of the population classified into each of the categories. Such numbers are usually called **frequencies**, and the set of frequencies is called the **frequency distribution**.

When the data are quantitative and univariate, one quantitative attribute of each member of the population is observed, and the results are given as a set of N real numbers $(x_1, x_2, \ldots, x_N)$. When N is large, as is usually expected, it is necessary to summarize these results in some manner. One common method is to tabulate the frequency distribution: We define a certain number of intervals $(a_{i-1}, a_i]$, $i=1, \ldots, K$, $a_0 < a_1 < \ldots < a_K$, $a_0 \leqslant \min x_i$, $\max x_i \leqslant a_K$; and we count the numbers of those x's falling within each of the intervals and tabulate those numbers or frequencies f_i, $i=1, 2, \ldots, K$. Frequency distribution is often represented in the form of a **histogram**, where the endpoints of the intervals are marked on the horizontal axis, and above each interval a rectangle of area proportional to the frequency for the interval is drawn. It is usually recommended that the widths of the intervals in the frequency distribution be equal, especially when it is to be represented by a histogram. It is, however, often impossible or impractical to do so, and sometimes a logarithmic or other functional scale is used in the abscissa of the histogram; then it is desirable that the intervals of the transformed values are approximately of equal lengths. The number K of the intervals should also be of an appropriate magnitude, neither too large nor too small; K is often constrained by the size N of the population, the shape of the distribution, or other factors. Usually, K is chosen to be between 6 and 20.

From the frequency distribution, we obtain the **cumulative distribution** by associating with each endpoint a_i of the intervals the number F_i of x's not greater than a_i, namely, $F_i = \sum_{j \leqslant i} f_j$. The curve obtained by connecting $K+1$ points of coordinates (a_i, F_i), $i=0, 1, \ldots, K$, by linear segments is called the **cumulative distribution curve** (or **polygon**).

C. Characteristics of the Distribution

In order to summarize univariate quatitative data, various values are calculated from the values $x_1, \ldots, x_N$. Such values are called **statistics** (singular, †statistic) and are used to characterize the distribution of the values. Various types of statistics characterize different aspects of the distribution:

(a) Representative value or **measure of location**: a value which is supposed to give the "representative," "typical," or "most common" value in the population. By far the most commonly used measure is the **mean** $\bar{x}=\sum_{i=1}^N x_i/N$. $\bar{x}$ is sometimes called the **arithmetic mean**, and some other "means" are also calculated: especially when all the values are positive, the **geometrical mean** $\bar{x}_G=(\prod_i x_i{}^{1/N})$ or **harmonic mean** $\bar{x}_H=(\sum_i(1/x_i)/N)^{-1}$ may be calculated; more generally, for some monotone function $f(x)$ we can calculate the f-mean by $\bar{x}_f=f^{-1}(\sum_i f(x_i)/N)$, of which the geometric and the harmonic means are special cases. Another measure of location is the **median**, which is the value in the population located exactly in the middle of the ordering of the magnitudes; more precisely, if $x_{(1)}<x_{(2)}<\ldots<x_{(N)}$ are the values in the population arranged according to their magnitudes, the median $x_{\text{med}}=x_{((N+1)/2)}$ for odd N, and $=\frac{1}{2}(x_{(N/2)}+x_{((N/2)+1)})$ for even N. The **mode** is also sometimes used; this is defined as the value (usually the center) corresponding to the highest frequency.

(b) The **measure of variability** or **dispersion** shows how widely the values in the population vary. The most common measure is the **standard deviation**, which is defined by $s_x=\sqrt{\sum_i(x_i-\bar{x})^2/N}$, and its square is called the **variance** V_x^2. A similar measure is the **mean absolute deviation** $D_x=\sum_i|x_i-\bar{x}|/N$. Another type of a measure of dispersion is the **range** $R_x=\max x_i-\min x_i$, and the **interquartile range** $Q_x=x_{(3N/4)}-x_{(N/4)}$ and more generally the interquantile range $x_{(\alpha N)}-x_{((1-\alpha)N)}$ for some α. The ratio of the standard deviation to the mean is called the **coefficient of variation** (C. V. for short) and is used as a measure of relative variability when all the values in the population are positive.

(c) Characteristics often used to characterize the "shape" of distributions are the **moments** (around the origin) $m_k=N^{-1}\sum_i x_i^k$ and the **central moments** (moments around the mean) $M_k=N^{-1}\sum_i(x_i-\bar{x})^k$ for a positive integer k; for a specific k these are called the kth moments. Central kth moments with odd k are equal to zero when the distribution is symmetric, i.e., when the histogram is symmetrically shaped. Hence the third moment or its ratio to the third power of the standard deviation s_x^3 is used as a measure of the asymmetry of a distribution; this is called the **skewness**. The fourth moment is large if there are some values which are far off from others and small when all values are concentrated; hence it tends to be large when the histogram has a rather sharp peak in the center and has a long tail in either direction or both, and tends to be small when the histogram is flat in the center and drops off sharply at both ends. Accordingly, the ratio M_4/V_x^2 is used as a measure of long-tailedness of the distribution; this ratio minus 3 is called the **kurtosis**.

D. Theoretical Frequency Distribution

When the observed values can be any real number (sometimes in an interval), the size of the population N is increased indefinitely, and the widths of the intervals are decreased to 0, the histogram is expected to approach a smooth curve. And in the limit when N is infinity, we can assume that the distribution is represented by a mathematically well-behaved function $f(x)$ and that the ratio of the numbers of those values in the population within the interval (a, b) to the size of the population approaches $\int_a^b f(x)\,dx$. Such a function $f(x)$ is called the **frequency function** or **density function**. Various types of functions have been proposed and used as "theoretical" frequency functions to approximate the actually observed frequency distributions. The most important is the **normal density function**

$$\varphi(x)=\frac{1}{\sigma\sqrt{2\pi}}\exp\left\{-\frac{1}{2\sigma^2}(x-\mu)^2\right\}.$$

The following density functions most commonly appear in applications: the **gamma density**, $f(x)=x^{p-1}\exp(-x/a)/a^p\Gamma(p)$ for $x>0$ and $=0$ for $x\leqslant 0$; the **beta density**, $f(x)=x^{p-1}(1-x)^{q-1}/B(p,q)$ for $0<x<1$ and $=0$ otherwise.

We can conceive of a population of infinite size with some density function; the term **theoretical distribution** is used to mean such a population with its density function, and more specifically the †normal distribution, etc. Such a population and associated density is often called a continuous distribution. For a theoretical distribution, the mean, variance, and moments are naturally defined by

$$\mu=\int xf(x)\,dx, \qquad \sigma^2=\int(x-\mu)^2 f(x)\,dx,$$

$$\mu_k=\int(x-\mu)^k f(x)\,dx, \qquad \mu_k'=\int x^k f(x)\,dx.$$

It should, however, be noted that the mean, the variance, or the moments may not exist for particular distributions.

K. Pearson introduced a system of density functions defined as solutions of the differential equation

$$d\ln f(x)/dx=(A+Bx)/(C+Dx+Ex^2),$$

where $A, B, \ldots, E$ are constants. A distribution thus obtained is called a **Pearson distribution**. The normal, gamma, and beta distributions together with some other commonly used distributions, such as the t- and F-distributions, are Pearson distributions.

E. Measure of Concentration

When all the observed values are nonnegative in nature, we may sometimes require some measure of inequality or concentration of the distribution. For such a purpose we order the observed values according to their magnitudes and obtain $x_{(1)} \leqslant x_{(2)} \leqslant \ldots \leqslant x_{(N)}$; we define $S_i = \sum_{j \leqslant i} x_{(j)}$ for $i=1, \ldots, N$, and $S_0=0$, plot $N+1$ points $(S_i/S_N, i/N)$, $i=0,1,\ldots,N$, and connect them by line segments. The graph thus obtained is called the **Lorentz curve** or the curve of concentration, and it connects the origin and the point (1,1). It lies below the 45° line, and if all the values are nearly equal the curve comes close to the 45° line, but if values are widely unequal, the curve comes close to the horizontal axis and suddenly jumps to the point (1,1). The area between the curve and the 45° line is called the **area of concentration**, and it is equal to one-fourth of the mean difference δ divided by the mean, where δ is defined by

$$\delta=\frac{1}{N^2}\sum_i\sum_j |x_i-x_j|.$$

$G=\delta/\bar{x}$ is called the **Gini coefficient of concentration** and is used as a measure of concentration or inequality of distribution. Other measures, including the coefficient of variation, are also used to represent the concentration.

F. Discrete Distributions

There are cases where the observed values are taken only from the nonnegative integers, e.g., the number of individual animals of a specific species in an area, of accidents during a specified time, etc. In such cases, when we increase the number of observations, the distribution does not approach one with a continuous density function but rather one with a certain theoretical discrete distribution. Among theoretical discrete distributions, the most commonly used are the **binomial distribution**: $f_j={}_nC_j p^j(1-p)^{n-j}$ for $0 \leqslant j \leqslant n$, and the **Poisson distribution**: $f_j=e^{-\lambda}\lambda^j/j!$ for $j=0,1,\ldots$. The **hypergeometric distribution**: $f_j={}_MC_j \cdot {}_{N-M}C_{n-j}/{}_NC_n$, for $\max(0, M+n-N) \leqslant j \leqslant \min(n, M)$, and the **negative binomial distribution**: $f_j={}_{j+r-1}C_{r-1}p^r(1-p)^j, j=0,1,\ldots$, are also often used.

For discrete distributions we can define moments by $\mu=\sum_j jf_j$ and $\mu_k=\sum_j (j-\mu)^k f_j$, and $\mu'_k=\sum_j j^k f_j$, $k=2,3,\ldots$.

G. Generating Functions and Cumulants

For a theoretical distribution with the density function $f(x)$, the **moment generating function** $M(\theta)$ is defined by $M(\theta)=\int e^{\theta x}f(x)\,dx$. When $M(\theta)$ is well defined in an open interval including the origin, the distribution has all kth moments, and it can be expanded as

$$M(\theta)=1+\mu'_1\theta+\frac{1}{2!}\mu'_2\theta^2+\ldots+\frac{1}{k!}\mu'_k\theta^k,$$

from which the term "moment generating function" is derived. When θ is replaced by it with real t, we have the **characteristic function** $\varphi(t)=M(it)$, which can be expanded as

$$\phi(t)=1+\mu'_1(it)+\frac{1}{2!}\mu'_2(it)^2+\ldots+\frac{1}{k!}\mu'_k(it)^k + o(|t|^k)$$

if the distribution has moments up to the kth. The function $K(\theta)=\ln M(\theta)$ is called the **cumulant generating function**, and the coefficients κ_j in the expansion

$$K(\theta)=\kappa_1\theta+\frac{\kappa_2}{2}\theta^2+\ldots$$

are called the **cumulants**. The kth cumulant κ_k is expressed as a polynomial of the moments of order not exceeding k; thus

$$\kappa_1=\mu'_1, \quad \kappa_2=\mu'_2-\mu'^2_1, \quad \kappa_3=\mu'_3-3\mu'_2\mu'_1+2\mu'^3_1, \quad \text{etc.}$$

For the normal distribution,

$$M(\theta)=\exp\left\{\mu\theta+\frac{\sigma^2\theta^2}{2}\right\};$$

hence

$$K(\theta)=\mu\theta+\frac{\sigma^2\theta^2}{2},$$

and the kth cumulant for $k \geqslant 3$ is equal to 0. Cumulants are used as measures indicating whether the distribution is close to or different from the normal.

For discrete distributions, the moment generating function is defined as $M(\theta)=\sum e^{\theta j}f_j$, but the **probability generating func-**

tion $P(t)=\sum f_j t^j = M(\ln t)$, the **factorial moment generating function** $\tilde{M}(\theta)=P(\theta+1)$, and the **factorial cumulant generating functions** $\tilde{K}(t)=\ln \tilde{M}(t)$ are also of use. The coefficient of t^j in the Maclaurin expansion of $\tilde{M}(t)$ is expressed as $\mu_{[k]}$ and called the kth **factorial moment**; it is equal to $\mu_{[k]}=\sum_j j(j-1)\ldots(j-k+1)f_j$. That in the expansion of $\tilde{K}(t)$ is the **factorial cumulant**. The factorial cumulant $\kappa_{[k]}$ is expressed by the same polynomial in $\mu_{[k]}$ as κ_k is expressed in μ'_k. For the Poisson distribution, $M(\theta)=\exp\lambda(e^\theta-1)$; hence $\tilde{K}(t)=\lambda t$ and it follows that the factorial cumulants $\kappa_{[k]}$ for $k \geqslant 2$ are all equal to zero if and only if the distribution is Poisson.

H. Bivariate Distribution

When two quantitative observations are obtained for each member of a population of size N, the results are given as N pairs of real numbers (x_i, y_i), $i=1,2,\ldots,N$. Such data are called **bivariate data** and the distribution, **bivariate distribution**. Those data can be illustrated as N points in a plane with coordinates (x_i, y_i), and such an illustration is called a **scatter diagram**. In order to characterize a bivariate distribution, we often use **bivariate moments**

$$M_{k,l}=\frac{1}{N}\sum(x_i-\bar{x})^k(y_i-\bar{y})^l,$$

where $\bar{x}$ and $\bar{y}$ are the means of x and y, respectively; especially, the (1,1) moment $M_{1,1}$ is called the **covariance** and is denoted as $\mathrm{Cov}(x, y)$. The most often used measure of the strength of the relation between x and y values is the **correlation coefficient** $r_{x,y}=\mathrm{Cov}(x,y)/s_x s_y$, where s_x and s_y are the standard deviations of x and y. It is easily shown that $-1 \leqslant r_{x,y} \leqslant 1$, and when there exists a nearly linear relationship between x and y values, $r_{x,y}$ is close to either $+1$ or -1 according to whether the x and y values change in the same direction or in opposite directions. When there is no clear relationship between x and y, the correlation coefficient is close to zero, but it may not be a good measure of the relationship when x and y values are related nonlinearly.

A linear function $y=a+bx$ is called the **linear regression function** of y on x, for which the sum of the square distances $\sum_i(y_i-a-bx_i)^2$ is minimized. For the linear regression function the coefficients a and b are determined by $b=\mathrm{Cov}(x,y)/s_x^2$ and $a=\bar{y}-b\bar{x}$, and b is called the **regression coefficient**. We have that $\sum_i(y_i-a-bx_i)^2/\sum(y_i-\bar{y})^2=1-r_{xy}^2$, i.e., that the square of the correlation coefficient is equal to 1 minus the ratio of the variance of the residual y_i-a-bx_i to that of y; hence it is sometimes called the **coefficient of determination**. Similarly, the linear regression function of x on y is defined by $x=c+dy$, where $d=\mathrm{Cov}(x,y)/s_y^2$ and $c=\bar{x}-d\bar{y}$. We have $bd=r_{xy}^2$, and $|1/d|=|b/r_{xy}^2|\geqslant|b|$.

We can tabulate the bivariate frequency distribution by splitting the range of x values into K intervals $(a_{i-1}, a_i]$, $i=1,\ldots,K$, and the range of y values into L intervals $(b_{j-1}, b_j]$, $j=1,\ldots,L$, and counting the number f_{ij} of cases for which $a_{i-1}<x\leqslant a_i$ and $b_{j-1}<y\leqslant b_j$. In contrast to the bivariate frequency distribution, the distributions of x and y values are called the **marginal distributions**.

I. Bivariate Density Function

As we did for the univariate distribution, we can consider the limiting shape of the bivariate frequency distribution when the size N of the population tends to infinity and define a continuous bivariate distribution with density function $f(x,y)$, with which the ratio of those members in the population with values (x,y) in a set S in a plane is given by $\iint_S f(x,y)\,dx\,dy$. The bivariate density function is also called the **joint density**, and then the density functions of x and y are called the **marginal density functions** and are given by $f_1(x)=\int f(x,y)\,dy$ and $f_2(y)=\int f(x,y)\,dx$. The joint moments of a continuous bivariate distribution are defined by $\mu_{k,l}=\iint(x-\mu_1)^k(y-\mu_2)^l f(x,y)\,dx\,dy$, where μ_1 and μ_2 are the means of x and y, respectively. The **joint moment generating function** is defined by $M(t_1,t_2)=\iint e^{t_1x+t_2y}f(x,y)\,dx\,dy$, and the cumulant generating function by $K(t_1,t_2)=\log M(t_1,t_2)$, from which the **joint cumulants**

$$\kappa_{k,l}=\frac{\partial^{k+l}}{\partial t_1^k \partial t_2^l}K(t_1,t_2)|_{t_1=0,t_2=0}$$

are derived.

The **conditional density** of y given x is defined by $f(y|x)=f(x,y)/f_1(x)$, and the distribution with this density function is the **conditional distribution** of y given x; this latter can be interpreted as the distribution of y of those members in the population with x values in the interval $(x, x+dx]$, where dx is small. The conditional density and the conditional distribution of x given y are similarly defined. The mean and the moments of the conditional distribution are called the **conditional mean** and the **conditional moments**. The conditional mean of y given x, considered as a function of x, is called the **regression function** of y on x.

By far the most important theoretical bivariate density is the **bivariate normal density**, which is given by

$$f(x,y)=(2\pi\sigma_1\sigma_2\sqrt{1-\rho^2})^{-1}$$
$$\times\exp\left\{\frac{1}{2(1-\rho^2)}\left[\frac{(x-\mu_1)^2}{\sigma_1^2}+\frac{(y-\mu_2)^2}{\sigma_2^2}-2\rho\frac{(x-\mu_1)(y-\mu_2)}{\sigma_1\sigma_2}\right]\right\},$$

for which the mean of x is μ_1 and that of y is μ_2, the variances of x and y are σ_1^2 and σ_2^2, respectively, and the covariance of x and y is equal to $\rho\sigma_1\sigma_2$. The bivariate normal distribution has several remarkable properties: All the k, l joint cumulants are equal to zero for $k+l\geqslant 3$; the marginal distributions of x and y are normal; the regression functions of y on x and x on y are both linear; the conditional distribution of y given x (and x given y) is normal and the conditional variance is constant; and the contours $f(x,y)=c$ for different values of c are equicentric ellipsoids.

J. Higher-Dimensional Data

When the data are of more than two dimensions, i.e., more than two observations are made on each of the objects, we designate the data by Nk-tuples of real numbers $(x_{i1}, x_{i2}, \dots, x_{ik})$, $i=1,\dots,N$ $(k\geqslant 3)$. Then we can calculate the moments of each of the variates and the joint moments, which are defined by

$$M_{k_1,k_2,\dots,k_k}=\frac{1}{N}\sum(x_{i1}-\bar{x}_1)^{k_1}(x_{i2}-\bar{x}_2)^{k_2}\dots(x_{ik}-\bar{x}_k)^{k_k}.$$

Also, we can arrange the variances and covariances in a symmetric matrix of order k, and we call it the **(variance-) covariance matrix**. A covariance matrix is easily shown to be nonnegative definite. The determinant of the covariance matrix is called the **generalized variance**. The matrix with the (i,j) element equal to the correlation coefficient of the ith and the jth variates r_{ij} (r_{ii} is set equal to 1) is called the **correlation matrix** and is denoted by R. R is also nonnegative definite. If we denote the (i,j) cofactor of R by R_{ij}, the quantity defined by

$$R_{i|1,\dots,(i),\dots,k}=\sqrt{1-|R|/R_{ii}}$$

is called the **multiple correlation coefficient** of the ith variate and all other variates; and

$$r_{ij|1,\dots,(1),\dots,(j),\dots,k}=-R_{ij}/\sqrt{R_{ii}R_{jj}}$$

is called the **partial correlation coefficient** of the ith and the jth variates given all other variates. The meaning of these coefficients will be elucidated below. A linear function $a_0+a_1x_1+a_2x_2+\dots+a_{k-1}x_{k-1}$ is called the **linear regression function** of x_k on $x_1,\dots,x_{k-1}$, when the coefficients $a_1,\dots,a_{k-1}$ are so determined that the sum $Q=\Sigma_i(x_{ik}-a_0-a_1x_{i1}-\dots-a_{k-1}x_{ik-1})^2$ is minimized. They are determined from the equation

$$C_{i1}a_1+C_{i2}a_2+\dots+C_{ik-1}a_{k-1}=C_{ik},\qquad (*)$$
$$i=1,\dots,k-1,$$

with $a_0=\bar{x}_k-a_1\bar{x}_1-\dots-a_{k-1}\bar{x}_{k-1}$, where C_{ij} are the covariances. $a_1,\dots,a_{k-1}$ thus determined are called the **regression coefficients** of x_k on $x_1,\dots,x_{k-1}$, and such a procedure is called the **method of least squares**. The equation (*) is called the **normal equation**. If we write $\hat{x}_{ik}=a_0+a_1x_{i1}+\dots+a_{k-1}x_{i\,k-1}$, we have $Q=\Sigma(x_{ik}-\hat{x}_{ik})^2=\Sigma(x_{ik}-\bar{x}_k)^2-\Sigma(\hat{x}_{ik}-\bar{x}_k)^2=\Sigma(x_{ik}-\bar{x}_k)^2\times(1-R_{k|1,\dots,k-1}{}^2)$, where $R_{k|1,\dots,k-1}$ is the multiple correlation coefficient of x_k and $x_1,\dots,x_{k-1}$, which is also equal to the correlation coefficient of x_k and $\hat{x}_k$. The square of the multiple correlation coefficient is also called the **coefficient of determination**. The quantities $x_{ik}-\hat{x}_{ik}$ are called the residuals. Let $\hat{x}_{i\,k-1}$ and $\hat{x}_{ik}$ be the values of regression functions of x_{k-1} and x_k, respectively, on $x_1,\dots,x_{k-2}$, and let $y_{i\,k-1}=x_{i\,k-1}-\hat{x}_{i\,k-1}$ and $y_{ik}=x_{ik}-\hat{x}_{ik}$ be the residuals; then the correlation coefficient of y_{k-1} and y_k is equal to the partial correlation coefficient of x_{k-1} and x_k given $x_1,\dots,x_{k-2}$. We have the following relation between the multiple and the partial correlation coefficients:

$$1-R_{k|1,\dots,k-1}^2$$
$$=(1-R_{k|1,\dots,k-2}^2)(1-r_{k\ 1,k|1,\dots,k-2}^2).$$

Multiple and partial correlation coefficients are also expressed in terms of the correlation coefficients of the variates. For example, it can be shown that

$$R_{3|12}^2=(r_{13}^2+r_{23}^2-2r_{12}r_{13}r_{23})/(1-r_{12}^2)$$

and that

$$r_{23|1}=(r_{23}-r_{12}r_{13})/\sqrt{(1-r_{12}^2)(1-r_{13}^2)}.$$

For higher dimensions, we can also define the **(joint) density function** $f(x_1,x_2,\dots,x_k)$ and the **(joint) moment generating function** $M(t_1,t_2,\dots,t_k)$

$$=\int\dots\int\exp(t_1x_1+t_2x_2+\dots+t_kx_k)$$
$$\times f(x_1,x_2,\dots,x_k)\,dx_1\dots dx_k.$$

The most important multivariate joint density is that of the **multivariate normal distribution**, which is expressed by

$$f(x_1,\dots,x_k)=(2\pi)^{-R/2}|\Sigma|\exp\left\{\frac{1}{2}\sum\sum\sigma^{ij}(x_i-\mu_i)(x_j-\mu_j)\right\},$$

where Σ is the covariance matrix with elements σ_{ij}, $\Sigma^{-1}=(\sigma^{ij})$, and μ_i is the mean of the ith variate. For the multivariate normal distribution the moment generating function is given by

$$M(t_1,\ldots,t_k)=\exp\left(\sum\mu_i t_i+\frac{1}{2}\sum\sigma_{ij}t_i t_j\right).$$

K. Contingency Tables

When several qualitative observations are made on N objects, each object is classified according to the combination of the categories, and the data are summarized by the numbers $N(i_1, i_2, \ldots, i_k)$ of the objects that fall in the i_1th category according to the first observation, i_2th category in the second observation, etc. A table that shows the results of such observations is called a (k-**way**) **contingency table**. If there are m_1 categories in the first criterion, m_2 categories in the second, etc., the contingency table is also called an m_1 by m_2 by ... by m_k table. The numbers $\tilde{N}(j, i_j)$ of the objects which are classified into the i_jth category according to the jth criterion are called marginal frequencies. If we have

$$N(i_1, i_2, \ldots, i_k)/N=\tilde{N}(1,i_1)\tilde{N}(2,i_2)\ldots\tilde{N}(k,i_k)/N^k$$
$$\text{for all } i_1, i_2, \ldots, i_k,$$

then the k observations or criteria are independent.

The simplest contingency table is a 2 by 2 table, where several measures for the relation of two observations or criteria have been proposed, among which the most commonly used are the **measure of association** defined by

$$Q=\frac{N(1,1)N(2,2)-N(1,2)N(2,1)}{N(1,1)N(2,2)+N(1,2)N(2,1)}$$

and the **odds ratio**

$$\delta=\frac{N(1,1)N(2,2)}{N(1,2)N(2,1)}$$

and also

$$V=\frac{N(1,1)N(2,2)-N(1,2)N(2,1)}{\sqrt{\tilde{N}(1,1)\tilde{N}(1,2)\tilde{N}(2,1)\tilde{N}(2,2)}},$$

where $N(i,j_i)$ are marginal frequencies. The two observations are independent if and only if $Q=0$ or $\sigma=1$ and $V=0$. V is equal to the correlation coefficient of the variables x_1 and x_2, for which $x_i=0$ if the object is classified into the first category according to the ith criterion and $x_i=1$ if it is classified into the second category. In a two-way m_1 by m_2 table a measure of association is defined by

$$\frac{X^2}{N}=\sum\sum\left(\frac{N(i_1,i_2)N}{\tilde{N}(1,i_1)\tilde{N}(2,i_2)}-1\right)^2;$$

it can be shown that $X^2/N=0$ if and only if the two criteria are independent, and that $0\leqslant X^2/N\leqslant\min(m_1-1, m_2-1)$. When we take $x_i=1$ if the object is classified into the ith category according to the first criterion and $x_i=0$ otherwise and take $y_j=1$ if it belongs to the jth category in the second criterion and $y_j=0$ otherwise, it is shown that the sum of squares of the multiple correlation coefficients of x_i and $y_1, \ldots, y_{m_2-1}$ is equal to X^2/N.

L. Decomposition of the Variance

When one observation is qualitative while another is quantitative the objects are classified into several categories according to the first observation, while for each object the value of the second observation is also given. Let x_{ij} be the observed value of the jth object in the group of the ith category; then for each i we can obtain frequency distributions of x_{ij}, and compare these distributions. Let N_i be the number of objects in the ith category, and $\bar{x}_i=\sum x_{ij}/N_i$ be the mean in the ith category. Then the weighted variance of the $\bar{x}_i$ defined by $v_{\mathrm{B}}=\sum_i N_i(\bar{x}_i-\bar{x})^2/N$, where $\bar{x}$ is the mean of all the observations, i.e., $\bar{x}=\sum N_i\bar{x}_i/N$, is called the **between-group variance**, and the weighted mean of the variances of each of the groups defined by $v_{\mathrm{W}}=\sum_i\sum_j(x_{ij}-\bar{x}_i)^2/N$ is called the **within-group variance**. It can be shown that $V_{\mathrm{B}}+V_{\mathrm{W}}=V=\sum\sum(x_{ij}-\bar{x})^2/N$, i.e., the variance of all the observations is decomposed as a sum of the between-group and the within-group variances. The ratio V_{B}/V is called the **correlation ratio**, which is equal to the square of the multiple correlation coefficient of x and $y_1, \ldots, y_m$, where $y_i=1$ if the object is in the ith category and $y_i=0$ otherwise.

M. Ordinal Data

When the observation is not quantitative but there exists a natural ordering among the categories into which the objects are classified, the observation is said to be in an **ordinal scale**, or simply ordinal.

When two ordinal observations are made on the same set of N objects, we can define several measures of association between the two ordinal scales. For each pair of objects we define a variable c_{ij}, $i,j=1,\ldots,N$, $i\neq j$, as $c_{ij}=1$ if the ith object is classified as "better" (or "superior") than the jth object according to both of the measurements, $c_{ij}=-1$ if the orderings are different in the two scales, or $c_{ij}=0$ if they are in the same category according to either or both of the scales. A measure of association is then given by $S=\sum\sum c_{ij}/N(N-1)$, which takes a value between -1 and $+1$ but usually cannot attain ± 1. Other ways of normalizing the sum $\sum\sum c_{ij}$ have been proposed.

Another method of calculating the association is the **scoring method**, i.e., giving a set of ordered real numbers to the categories of each of the scales and calculating the correlation coefficient between the scores. The simplest scores are $0, 1, \ldots, m-1$ when there are m categories, but other methods of scoring are also used. Scores that give the largest possible correlation are called **canonical scores**, which are obtained as the characteristic vectors of the matrices NN' and $N'N$, where N is the matrix of the contingency table.

N. Time Series Data

Time series data can be recorded in a continuous time scale, but usually measurements are made at discrete times, which are most commonly equally spaced. Hence we here denote them as $x(t)$, $t=1,2,\ldots,T$. First we consider the quantitative univariate case. The intertemporal change of $x(t)$ is often decomposed into three parts:

$$x(t)=m(t)+c(t)+e(t),$$

where $m(t)$ is called the **trend**, and represents the secular, systematic change of x; $c(t)$ is called the **cycle**, and represents the recurrent pattern of the change; and $e(t)$ is called the error or random fluctuation, and represents the irregular changes. Such a decomposition cannot be defined rigorously without assuming some probabilistic or stochastic model for $x(t)$, but it is intuitively clear and practically useful in many applications.

There are two ways to estimate the trend. One is to calculate the **moving average** $\hat{x}(t)=(x(t-k)+x(t-k+1)+\ldots+x(t)+\ldots+x(t+k))/(2k+1)$ and use it as an estimate of the trend of $x(t)$; here k should be chosen to substantially eliminate the cyclic and random parts. More generally we can use the weighted moving average defined as $x(t)=\sum_{j=-k}^{k} w(j)x(t+j)$, where $w(-j)=w(j)$ and $\sum w(j)=1$. The second method is to assume some functional form, usually a polynomial in t, for the trend: $m(t)=a_0+a_1t+\ldots+a_kt^k$, and to determine the coefficient by least squares, i.e., to calculate the values of $a_0, a_1, \ldots, a_k$ which minimize $\sum_t (x(t)-a_0-a_1t-\ldots-a_kt^k)^2$.

There are two cases of cyclical changes. One is the case when there is a clearly defined relevant external time period, such as the seasons of the year or the days of the week. In such cases the effects of such external periodical cycles must be eliminated, and the process which does that is called **seasonal adjustment** of the time series data. Various methods of seasonal adjustment have been proposed and applied, but none is definitive. The other case is where the cyclical changes are produced from the observed process itself; here, the length of the period and the pattern of the cyclical change must be estimated.

Now assume that the data do not contain any trend, or that the trend has been effectively eliminated. First we calculate the correlation coefficient between $x(t)$ and $x(t+s)$ by

$$r(s)=\frac{\sum(x(t)-\bar{x}_s')(x(t+s)-\bar{x}_s'')}{\sqrt{\sum(x(t)-\bar{x}_s')^2\sum(x(t+s)-\bar{x}_s'')^2}},$$

where

$$\bar{x}_s'=\sum_{t=1}^{T-s} x(t)/(T-s), \qquad \bar{x}_s''=\sum_{t=s+1}^{T} x(t)/(T-s),$$

or more simply by

$$r(s)=T\sum(x(t)-\bar{x})(x(t+s)-\bar{x})/((T-s)\sqrt{\sum(x(t)-\bar{x})^2}),$$

where $\bar{x}=\sum x(t)/T$. $r(s)$ is called the **serial correlation coefficient** or the **autocorrelation coefficient** of lag s. When there exists a clear and definite cyclical change of period s in the data, $r(s)$ is close to 1. The diagram in which the serial correlation coefficient $r(s)$ is plotted against s is called the **correlogram**. In order to see the cyclical properties of the data more clearly, we calculate the **power spectral density**

$$w(\lambda)=1+2\sum_s r(s)\cos\lambda s.$$

The graph of $w(\lambda)$ is called the **power spectrum**, or simply the **spectrum**. $w(\lambda)$ represents the square of the width of the sine curve of frequency $\lambda/2\pi$ or of period $2\pi/\lambda$ contained in the data. The spectral density is closely related to the intensity, defined by

$$I(\lambda)=\frac{1}{T}\left(\left(\sum_t x(t)\cos\lambda t\right)^2+\left(\sum_t x(t)\sin\lambda t\right)^2\right),$$

which is proportional to the square of the multiple correlation coefficient of $x(t)$ and the functions $\cos\lambda t$ and $\sin\lambda t$ and is large if the data contains a sine curve of frequency $\lambda/2\pi$. It can be shown that $I(\lambda)$ is approximately equal to $w(\lambda)V(x)/\pi$. The spectral density thus obtained usually oscillates irregularly and far from smoothly; hence smoothing by use of a "spectral window" is often applied (→ 421 Times Series Analysis).

When several observations are made in time series data, we speak of multivariate time series. Let $x_i(t)$ be the ith observation in the tth period. The correlation coefficient between $x_i(t)$ and $x_j(t+s)$ is called the **serial cross-correlation coefficient** and is denoted by $r_{ij}(s)$ $(s=0, \pm1, \pm2, \ldots)$. Analogously to the univariate case we define

$$w_{ij}(\lambda)=\sum_s r_{ij}(s)\cos\lambda s+i\sum_s r_{ij}(s)\sin\lambda s$$
$$=p_{ij}(\lambda)+iq_{ij}(\lambda),$$

and we call $p_{ij}(\lambda)$ the **cospectral density** between the ith and the jth variables, and $q_{ij}(\lambda)$ the **quadrature spectral density**. $p_{ij}^2+q_{ij}^2$ is called the **amplitude**, and

$$C_{ij}(\lambda)=\frac{p_{ij}^2(\lambda)+q_{ij}^2(\lambda)}{\sqrt{w_i(\lambda)w_j(\lambda)}},$$

where w_i and w_j are the spectral densities of the ith and the jth variables, is called the **coherence**.

O. Events in Time Scale

Some data give us the time points at which a specific event occurs. Let T_i be the time when the event occurs for the ith time. Then usually the most important information we want to obtain is about the time intervals $d_i=T_{i+1}-T_i$. If there is a periodicity in the occurrences of the event, the d_i will be approximately equal. On the other hand, if the event tends to occur repeatedly after its first appearance, some d_i will be small while others will be large. When there is no periodicity, no tendency to repetition, and no increasing or decreasing trend in the occurrences, we can suppose that the event occurs simply by chance, and this is good reason to suppose that the density function of the distribution of the intervals is exponential, i.e., it can be expressed by $f(d)=(\exp(-d/\alpha))/\alpha$ for $d>0$. Also in such a case the number of occurrences in fixed time intervals are distributed according to the Poisson distribution. Such a sequence of occurrences of an event is called a †Poisson process. More generally, let $f(d)$ be the density function of the time intervals; then

$$h(d)=\frac{f(d)}{1-\int_0^d f(c)\,dc}$$

is called the **hazard rate** or **hazard function**. The hazard function is constant if and only if the process is Poisson.

P. Probabilistic Models

In many applications of statistical data analysis, the data exhibit variabilities and fluctuations that are due to fortuitous or hazardous causes or chance effects and that obscure the information contained in the data. In such cases we assume that the chance variabilities and fluctuations are random variables distributed according to some probability distribution, and the information we require is represented by a set of unknown constants that characterize the probability distribution of the data as unknown parameters. Suppose, to take the simplest case, that we have repeated observations of results of some experiment under a fixed condition and that we obtain the values $x_1, x_2, \ldots, x_n$. Since the experimental condition is fixed, the variations among the x_i values can be considered to be due to chance causes, such as variations in materials, uncontrolled small fluctuations in experimental conditions or instruments, and various other variations usually called the errors. Whatever the true causes of the variations, we can consider them to be random, and we can regard the values $x_1, x_2, \ldots, x_n$ as the results of random experiments or the realizations of random variables $X_1, X_2, \ldots, X_n$ independently and identically distributed according to some probability distribution. Or we may think of a **hypothetical infinite population** of the results of supposedly infinite replications of the experiment under the same fixed condition, and regard the actual observations as n values chosen from this population at random. We may also consider that in this hypothetical infinite population, the frequency distribution is represented by a density function f, which in turn determines the probability distribution of each observation. We may be interested in the "average" values of the result of the experiment as well as the magnitude of the variability; then those values are represented by the mean and the variance of the population distribution. If the form of the population distribution is assumed to be completely specified except for the mean μ and the variance σ^2, the density function f is determined without these two parameters, and is expressed as $f(x;\mu,\sigma)$. The joint density for n repeated observations is $\prod_i f(x_i;\mu,\sigma)$. The set of assumptions that determines the probability distributions of the observations in terms of the unknown parameters is called the **probabilistic model**, and its determination is called the **problem of specification**.

Once the probabilistic model is given, the purpose of statistical data analysis can be formulated as making judgments on the values of the parameters, which may sometimes go wrong but can be relied on with some margin of probability of error that can be mathematically rigorously ascertained. The formal procedure of making such judgments is called statistical inference, and its mathematical theory has been well established over the last hundred years (→ 401 Statistical Inference).

In most cases of statistical inference, the joint density function of the data plays an important role, and when it is regarded as a function of the unknown parameters for given values of observations it is called the †likelihood function.

Q. Exploratory Procedures

In many applications of statistical data analysis, we are not quite sure of the validity of the probabilistic model assumed, or we admit that the models are, at best, approximations to reality and hence cannot be exactly correct; the approximation may not be precise enough for the conclusions drawn from the assumptions to be practically reliable. Therefore we have to check whether the model assumed is at least approximately valid for the data, and if not, we have to look for a better model that reflects more accurately the structure of the actual data. Thus in many practical applications of data analysis, we have to scrutinize the structure of the data and try various models before settling on a model and drawing final conclusions (which are still susceptible to further revisions when more data are obtained). Methods used in such a process are called **exploratory procedures**, which depend partly on the formal procedure of testing hypotheses and partly on intuitive reasoning sometimes combined with graphical presentations of the data, and also on scientific and empirical understanding of the subject matter.

Suppose in the simplest case that n observations $X_1, \ldots, X_n$ are assumed to be independently and identically distributed. Under the condition that all those values are observed under the same well-controlled situation, this assumption is reasonable. But in reality some of the observations may be subject to some unexpected effect due to either a fortuitous outside cause or some "gross error" in the measurement procedure, the process of reporting, etc., and may show much greater variation than others. Such observations can be detected by certain **outlier tests** or simply by looking at the data carefully, and if it is established that some observations are definitely outside of the possible random variability or are subject to some hazardous external effect, those data could be omitted from consideration. Further, the assumed probability density $f(x, \theta)$ may not well approximate the distribution of the actual data even after the "outliers" are omitted. Some test for **goodness of fit** should be applied, and if the hypothesis is rejected, we have to modify our model. Also, if we are provided with several candidates for the model to be adopted, we have to apply some procedure of model selection (→ 400 Statistical Hypothesis Testing, 403 Statistical Models). It could also happen that the supposedly uniform conditions of observation, measurement, or experimentation did not actually prevail but that there has been some heterogeneity among the observations. Bimodality or multimodality of the histogram, i.e., existence of two or more peaks in the histogram, usually strongly suggests such heterogeneity. In such cases, grouping or stratification of the observations is required to make the conditions of observation within each group nearly uniform.

References

[1] R. A. Fisher, Statistical methods for research workers, Oliver & Boyd, first edition, 1925.
[2] M. G. Kendall and A. Stuart, The advanced theory of statistics, 3 vols., Griffin, 1958–1966.
[3] J. W. Tukey, Exploratory data analysis, Addison-Wesley, 1977.

398 (XVIII.6) Statistical Decision Functions

A. General Remarks

The theory of statistical decision functions was established by A. Wald as a mathematically unified theory of statistics (→ 401 Statistical Inference). In this theory, the problems of mathematical statistics, for example, statistical hypothesis testing and †statistical estimation, are formulated in a unified way [1].

A †measurable space $(\mathscr{X}, \mathfrak{B})$ with a fixed †probability measure is called a **sample space**, and an element $x \in \mathscr{X}$ is called a **sample point**. Suppose that we are given a family $\mathscr{P} = \{P_\theta \mid \theta \in \Omega\}$ of probability measures on $(\mathscr{X}, \mathfrak{B})$, where Ω is called the **parameter space**, and a †random variable X takes values in $\mathscr{X}$ according to a true probability distribution P assumed to belong to $\mathscr{P}$. This article deals with the problems involved in making a decision about the parameter θ, called determining the **true value of the parameter**, such that $P = P_\theta$. To describe the procedure for such a decision based on the observation of the behavior of X, we need a triple $(\mathscr{A}, \mathfrak{C}; D)$ consisting of a set $\mathscr{A}$, a †σ-algebra $\mathfrak{C}$ of subsets of $\mathscr{A}$, and a set D of mappings δ from $\mathscr{X}$ into the set of probability measures on $(\mathscr{A}, \mathfrak{C})$, $\delta: x \to \delta(\cdot \mid x)$, such that for a fixed $C \in \mathfrak{C}$, the function $\delta(C \mid x)$ is $\mathfrak{B}$-measurable. We call $\mathscr{A}$ an **action space** or **decision space**, δ a **statistical decision function** (or simply a **decision function**) or **statistical deci-**

sion procedure, and D a **space of decision functions**. In actual decision procedures, $\delta(C|x)$ is the probability that an action belonging to C is taken, based on the observation of sample point x. We further consider a nonnegative function $w:\Omega\times\mathscr{A}\rightarrow\mathbf{R}$, called a **loss function**, such that for a fixed θ, $w(\theta,a)$ is $\mathfrak{C}$-measurable. By averaging the loss, we obtain the **risk function**:

$$r(\theta,\delta)=\int_{\mathscr{X}}\int_{\mathscr{A}}w(\theta,a)\delta(da|x)P_\theta(dx).$$

Two decision functions δ and δ' are identified if $\delta(C|x)=\delta'(C|x)$ for almost every x with respect to P_θ, for all $\theta\in\Omega$ and all $C\in\mathfrak{C}$. When to each $x\in\mathscr{X}$ there corresponds a unique action a_x such that $\delta(\{a_x\}|x)=1$, the decision function δ is said to be **nonrandomized**; otherwise, **randomized**. The system $(\mathscr{X},\mathfrak{B},\mathscr{P},\Omega,\mathscr{A},\mathfrak{C},W,D)$ is called a **statistical decision problem**.

From the point of view of this theory, †point estimation, †interval estimation, and statistical hypothesis testing (→ 400 Statistical Hypothesis Testing) are described as follows.

(1) In point estimation we assume that the action space $\mathscr{A}$ is a subset of $\mathbf{R}$ and that we are given functions $\varphi:\mathscr{X}\rightarrow\mathscr{A}$ and $l:\mathscr{P}\rightarrow\mathbf{R}$. The problem is to estimate the value of $l(P)$ by using the real value $\varphi(x)$ at an observed sample point $x\in\mathscr{X}$. As the loss function, we often set $w(\theta,a)=C(\theta)(a-l(P_\theta))^2$, where $C(\theta)$ is a function of θ, and call it a **quadratic loss function** (→ 399 Statistical Estimation).

(2) In interval estimation we assume that each action is represented as an interval in $\mathbf{R}$. Each interval $[u,v]$ can be represented by a point (u,v) of the half-space $\mathbf{R}^{2\prime}=\{z=(z_1,z_2)|z_1\leqslant z_2\}$, which may be taken as the action space. A weighted sum $\alpha w_1(\theta,z)+\beta w_2(\theta,z)$ $(\alpha,\beta\geqslant 0)$ of two functions,

$$w_1(\theta,z)=\begin{cases}1 & (\theta\notin[z_1,z_2]),\\ 0 & (\theta\in[z_1,z_2]),\end{cases}$$

$$w_2(\theta,z)=z_2-z_1,$$

often supplies the loss function (→ 399 Statistical Estimation).

(3) In testing a hypothesis $\mathbf{H}:\theta\in\omega_0$ versus an †alternative $\mathbf{A}:\theta\in\omega_1$ $(\omega_0\cap\omega_1=\varnothing,\ \omega_0\cup\omega_1=\Omega)$, the action space can be expressed by the set consisting of two points a_1, a_2, where a_1 denotes the decision to reject $\mathbf{H}$ and a_2 the decision to accept $\mathbf{H}$. The loss function is defined as follows:

$$w(\theta,a_1)=\begin{cases}1 & (\theta\in\omega_0),\\ 0 & (\theta\in\omega_1),\end{cases}$$

$$w(\theta,a_2)=\begin{cases}0 & (\theta\in\omega_0),\\ 1 & (\theta\in\omega_1).\end{cases}$$

This is called a **simple loss function**. Whatever testing procedure δ is adopted, the probabilities of the †errors of the first or second kind coincide with the values of $r(\theta,\delta)$ for $\theta\in\omega_0$ or $\theta\in\omega_1$, respectively (→ 400 Statistical Hypothesis Testing).

When Ω is the union of mutually disjoint nonempty sets $\omega_1,\omega_2,\ldots,\omega_n$, $\mathscr{A}=\{a_1,a_2,\ldots,a_n\}$, and $w(\theta,a_j)=c_{ij}$ $(\theta\in\omega_i)$ with $c_{ij}\geqslant 0$ $(i\neq j)$, $c_{ii}=0$, the decision problem is called an ***n*-decision problem**.

B. Optimality of Statistical Decision Functions

Consider the problem of choosing the best decision function δ. When $r(\theta,\delta_1)\leqslant r(\theta,\delta_2)$ for all θ and there exists at least one θ_0 such that $r(\theta_0,\delta_1)<r(\theta_0,\delta_2)$, the decision function δ_1 is said to be **uniformly better** than δ_2. If there exists a decision function δ_0 in D that is uniformly better than any other δ in D, it is the **best** decision function. However, such a function δ_0 does not always exist. A decision function $\tilde{\delta}$ in D is said to be **admissible** if there exists no other decision function δ in D that is uniformly better than $\tilde{\delta}$. In other words, $\tilde{\delta}$ is admissible if and only if the validity of the inequality $r(\theta,\delta)\leqslant r(\theta,\tilde{\delta})$ for some $\delta\in D$ and all $\theta\in\Omega$ implies $r(\theta,\delta)=r(\theta,\tilde{\delta})$ for all $\theta\in\Omega$. When there is no information about P except that it is a member of $\mathscr{P}$, we follow the **minimax principle** and choose a function δ^* for which we have $\inf_{\delta\in D}\sup_{\theta\in\Omega}r(\theta,\delta)=\sup_{\theta\in\Omega}r(\theta,\delta^*)$. This decision function δ^* is called a **minimax decision function** or **minimax solution**.

Let $\mathfrak{F}$ be a σ-algebra of subsets of Ω, and suppose that $r(\theta,\delta)$ is $\mathfrak{F}$-measurable for any fixed δ. If, furthermore, we are given a probability measure ξ, called an **a priori distribution**, on $(\Omega,\mathfrak{F})$, we choose a $\hat{\delta}$ that satisfies

$$\inf_{\delta\in D}\int_\Omega r(\theta,\delta)d\xi(\theta)=\int_\Omega r(\theta,\hat{\delta})d\xi(\theta).$$

Such a $\hat{\delta}$ and the integral of $r(\theta,\hat{\delta})$ are called a **Bayes solution** and the **Bayes risk** relative to ξ, respectively. Let F be a family of a priori distributions on $(\Omega,\mathfrak{F})$. If $\hat{\delta}$ satisfies

$$\inf_{\xi\in F}\left(\int_\Omega r(\theta,\hat{\delta})d\xi(\theta)-\inf_{\delta\in D}\int_\Omega r(\theta,\delta)d\xi(\theta)\right)=0,$$

$\hat{\delta}$ is called a **Bayes solution in the wider sense** relative to F. If $\mathscr{P}$ is †dominated by λ with a $\mathfrak{B}\times\mathfrak{F}$-measurable $f(x,\theta)=dP_\theta/d\lambda$, $w(\theta,a)$ is $\mathfrak{F}\times\mathfrak{C}$-measurable, and

$$A_x=\left\{\tilde{a}\in\mathscr{A}\,\middle|\,\int_\Omega w(\theta,\tilde{a})f(x,\theta)d\xi(\theta)=\inf_{a\in\mathscr{A}}\int_\Omega w(\theta,a)f(x,\theta)d\xi(\theta)\right\}$$

is nonempty and $\mathfrak{C}$-measurable for λ-almost every x, then a Bayes solution δ with respect to ξ satisfies $\delta(A_x|x)=1$ for λ-almost every x. For a sample point x with $\int_\Omega f(x,\theta)d\xi(\theta)\neq 0$, a probability measure $\eta(\cdot|x,\xi)$ on $(\Omega,\mathfrak{F})$ defined by

$$\eta(B|x,\xi)=\frac{\int_B f(x,\theta)d\xi(\theta)}{\int_\Omega f(x,\theta)d\xi(\theta)}$$

is called an **a posteriori distribution**. To get a Bayes solution it is enough to minimize the value of $\int_\Omega w(\theta,a)d\eta(\theta|x,\xi)$ for every observed x.

If ξ in the definition of A_x defined above is a σ-finite measure on $(\Omega,\mathfrak{F})$, a decision function δ satisfying $\delta(A_x|x)=1$ for λ-almost every x is called a **generalized Bayes solution** with respect to ξ.

Let D' be a subset of the space D. If for any $\delta\in D-D'$ there exists a $\delta'\in D'$ that is uniformly better than δ, then D' is called a **complete class**. If for any $\delta\in D$ there exists a $\delta'\in D'$ that is either uniformly better than δ or has the same risk function as δ, then D' is called an **essentially complete class**. If D' is complete and any proper subset of D' is not complete, then D' is called a **minimal complete class**. If a minimal complete class exists, it is unique and coincides with the set of all admissible decision functions.

C. n-Decision Problems

In an n-decision problem where $\mathscr{A}=\{a_1,\ldots,a_n\}$, we set $\delta_i(x)=\delta(a_i|x)$ for a decision function δ, where $\delta_i(x)$ is $\mathfrak{B}$-measurable and satisfies $\delta_i(x)\geqslant 0$, $\delta_1(x)+\ldots+\delta_n(x)=1$. We consider the set $\mathscr{D}$ of vector-valued functions $\Delta(x)=(\delta_1(x),\ldots,\delta_n(x))$ whose components $\delta_i(x)$ satisfy the conditions just given. Such a vector-valued function $\Delta(x)$ can be identified with $\delta(x)$; we write $\delta(x)$ instead of $\Delta(x)$ also. If in addition the parameter space Ω is a finite set $\{\theta_1,\theta_2,\ldots,\theta_k\}$, we can consider a mapping $\psi:\mathscr{D}\rightarrow\mathbf{R}^k$ defined by $\psi(\delta)=(r(\theta_1,\delta),\ldots,r(\theta_k,\delta))$, and then $S=\psi(\mathscr{D})=\{\psi(\delta)|\delta\in\mathscr{D}\}$ is convex and closed in $\mathbf{R}^k$. If δ is nonrandomized, then for each x, one and only one of the $\delta_i(x)$ is 1, and all others are 0. Hence, in this case, $\mathscr{X}$ is the disjoint union of $\mathfrak{B}$-measurable subsets B_i $(i=1,\ldots,n)$ such that $\delta(a_i|x)=1$ if and only if $x\in B_i$. A probability measure m on $(\mathscr{X},\mathfrak{B})$ is said to be **atomless** if for any set $A\in\mathfrak{B}$ with $m(A)>0$ and any b with $0<b<m(A)$, there exists a subset $B\in\mathfrak{B}$ of A such that $m(B)=b$. If Ω is finite and every member of $\mathscr{P}$ is atomless, the image $\psi(\mathscr{D}^0)$ of the set $\mathscr{D}^0$ of all nonrandomized decision functions coincides with $\psi(\mathscr{D})$. This shows that $\mathscr{D}^0$ is an essentially complete class in $\mathscr{D}$. However, when some members of $\mathscr{P}$ are not atomless, $\psi(\mathscr{D})$ is not always equal to $\psi(\mathscr{D}^0)$, but $\psi(\mathscr{D}^0)$ is a closed subset of $\mathbf{R}^k$. In particular, when $n=2$ for given probability measures $P_1,\ldots,P_k$ over $(\mathscr{X},\mathfrak{B})$, the set S of all points $(P_1(B),\ldots,P_k(B))$ $(B\in\mathfrak{B})$ in $\mathbf{R}^k$ is a closed set. If in addition $P_1,\ldots,P_k$ are all atomless, S is convex. These results are known as the **Lyapunov theorem**.

A two-decision problem with $\omega_1=\{1\}$, $\omega_2=\{2\}$, and $c_{ij}=1$ $(i\neq j)$, $=0$ $(i=j)$ is called a **dichotomy**. We discuss this problem in some detail in order to explain the concept of optimality of decision functions. For a dichotomy, $S=\psi(\mathscr{D})$ is a set in $\mathbf{R}^2$ that (i) is convex, (ii) is closed, (iii) is symmetric about $(1/2,1/2)$, (iv) contains the points $(0,1)$, $(1,0)$, and (v) is a subset of interval $[0,1]\times[0,1]$ (Fig. 1). The set of decision functions δ that is mapped under ψ onto the curve $ACDB$ in Fig. 1 constitutes the minimal complete class, and a decision function δ mapped onto the point D is a minimax solution. Let ξ be an a priori distribution such that

$$\xi(1)=\alpha,\quad \xi(2)=\beta\quad(\alpha+\beta=1,\alpha\geqslant 0,\beta\geqslant 0).$$

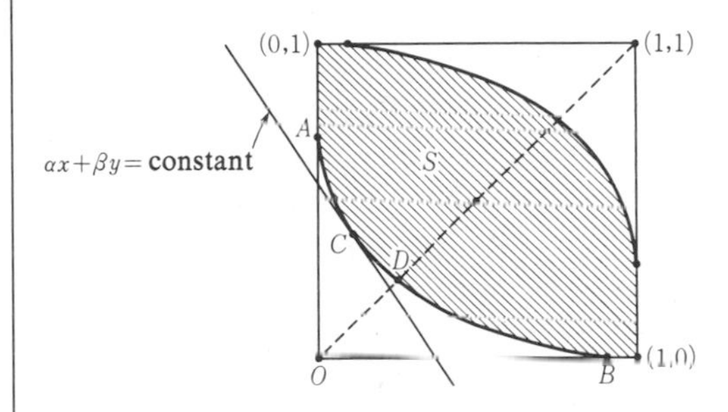

Fig. 1

Then a Bayes solution relative to ξ is mapped under ψ onto a supporting point c with direction ratio $-\alpha/\beta$ and is obtained as the characteristic function of the set $E=\{x|\alpha f(x)<\beta g(x)\}$, where f and g are †Radon-Nikodym derivatives $dP_1/d\lambda$ and $dP_2/d\lambda$ with respect to $\lambda=P_1+P_2$, that is, f, g are measurable functions on $\mathscr{X}$ such that for any measurable set E, we have

$$P_1(E)=\int_E fd\lambda,\qquad P_2(E)=\int_E gd\lambda.$$

This fact implies that the most powerful test constructed in the Neyman-Pearson fundamental lemma (→ 400 Statistical Hypothesis Testing B) is precisely a Bayes solution.

D. Complete Class Theorems

Suppose that $\mathscr{P}$ is †dominated by a σ-finite measure λ, and let $f(x,\theta)$ be the Radon-Nikodym derivative of P_θ with respect to λ.

Consider a subspace L of $L_1(\mathscr{X}, \lambda)$ containing $\{f(x,\theta)|\theta\in\Omega\}$, and the following equivalence relation between bounded $\mathfrak{B}$-measurable functions on $\mathscr{X}$: φ_1 and φ_2 are defined to be equivalent if and only if $\int\varphi_1 f d\lambda = \int\varphi_2 f d\lambda$ holds for every $f\in L$. We further assume that the dual space of L is the linear space $\mathfrak{M}$ of the equivalence classes of bounded measurable functions just defined. Let $C_0(\mathscr{A})$ be the set of all continuous functions on $\mathscr{A}$ with compact support, $\pi\circ\alpha$ the integral of $\alpha\in C_0(\mathscr{A})$ with respect to a probability measure π over $\mathscr{A}$, and $g\circ\delta$ the integral

$$\int\delta(\cdot|x)g(x)d\lambda$$

for $g\in L$, $\delta\in D$. As a base for neighborhoods of the space of decision functions around $\delta_0\in D$, we consider $V(\delta_0:\alpha_1,\ldots,\alpha_n,g_1,\ldots,g_n,\varepsilon)=\{\delta\,|\,|g_i\circ\delta_0\circ\alpha_i - g_i\circ\delta\circ\alpha_i|<\varepsilon,\ i=1,\ldots,n\}$, where $\alpha_i\in C_0(\mathscr{A})$, $g_i\in L$, $\varepsilon>0$.

Let F be the set of all a priori distributions ξ on $(\Omega,\mathfrak{F})$ each of which assigns the total mass 1 to a finite subset of Ω, B the set of all Bayes solutions relative to some $\xi(\in F)$, and W the set of all Bayes solutions in the wide sense relative to F. Suppose that $\mathscr{A}$ is a †locally compact, separable metric space and $w(\theta,a)$ is lower semicontinuous with respect to a for every fixed θ. Then if D is compact and convex, the intersection of W and the closure $\bar{B}$ of B constitutes an essentially complete class [2]. Moreover, if Ω is compact with respect to the metric $d(\theta_1,\theta_2)=\sup_{B\in\mathfrak{B}}|P_{\theta_1}(B)-P_{\theta_2}(B)|$ and $\{w(\cdot,a)|a\in\mathscr{A}\}$ is a uniformly bounded and †equicontinuous family on Ω, then the class of Bayes solutions relative to some a priori distribution is a complete class [1]. These propositions are called **complete class theorems** (for complete classes in specified problems and admissibility of individual procedures → 399 Statistical Estimation; 400 Statistical Hypothesis Testing; and Appendix A, Table 23).

Two measurable spaces $(S,\mathscr{S})$ and $(R,\mathscr{R})$ are said to be **isomorphic** if there exists a correspondence ρ of S and R ($E\in\mathscr{S}$ implies $\rho(E)\in\mathscr{R}$ and, conversely, $\rho(E)\in\mathscr{R}$ implies $E\in\mathscr{S}$) where ρ is one-to-one onto. Let $\mathscr{D}$ be the set of all decision functions associated with a sample space $(\mathscr{X},\mathfrak{B})$ and an action space $(\mathscr{A},\mathfrak{C})$, T be a †statistic on $(\mathscr{X},\mathfrak{B})$ taking values in a measurable space $(\mathscr{Y},\mathfrak{E})$, and $\mathscr{D}^*$ be the set of all decision functions having sample space $(\mathscr{Y},\mathfrak{E})$ and action space $(\mathscr{A},\mathfrak{C})$. The set of all $\delta\in\mathscr{D}$ for which there exists a $\delta^*\in\mathscr{D}^*$ satisfying $\delta(C|x)=\delta^*(C|T(x))$ $(C\in\mathfrak{C})$ is denoted by $\mathscr{D}_T$. If (i) T is a †sufficient statistic and if (ii) $(\mathscr{A},\mathfrak{C})$ is isomorphic to the measurable space $\mathbf{R}^k$ associated with the σ-algebra of its Borel subsets, then $\mathscr{D}_T$ is essentially complete in $\mathscr{D}$. Conversely, if (i) $\mathscr{P}$ is a dominated family, (ii) $f(x,\theta)>0$ always holds on $\mathscr{X}$, and (iii) for any pair $\theta_1,\theta_2(\in\Omega)$, there exists a $\theta_3(\in\Omega)$ such that no element a in $\mathscr{A}$ makes two of the $w(\theta_i,a)$ $(i=1,2,3)$ attain their minimum simultaneously, then essential completeness of $\mathscr{D}_T$ in $\mathscr{D}$ implies that T is sufficient [3] (→ 399 Statistical Estimation).

E. Invariance

Suppose that there exist one-to-one transformation groups G, $\bar{G}$, and $\tilde{G}$ of X, Ω, and $\mathscr{A}$, respectively, onto themselves (transformations belonging to G, $\bar{G}$, and $\tilde{G}$ are †measurable with respect to $\mathfrak{B}$, $\mathfrak{F}$, and $\mathfrak{C}$, respectively) and that there exist homomorphisms $g\to\bar{g}$, $g\to\tilde{g}$ of G to $\bar{G}$ and $\tilde{G}$, respectively, such that for $B\in\mathfrak{B}$ we have $P_\theta(g^{-1}B)=P_{\bar{g}\theta}(B)$ and $w(\bar{g}\theta,\tilde{g}a)=w(\theta,a)$. Then a decision problem $(\mathscr{X},\mathfrak{B},\mathscr{P},\Omega,\mathscr{A},\mathfrak{C},w,D)$ is said to be **invariant** under $(G,\bar{G},\tilde{G})$. In a decision problem invariant under $(G,\bar{G},\tilde{G})$, a decision function satisfying $\delta(\tilde{g}C|gx)=\delta(C|x)$ is called an **invariant decision function**.

Suppose that the transformation group G is locally compact and is the union of a countable family $\{K_n\}$ of compact subsets. Let Γ be the σ-algebra of Borel subsets of G such that the mapping $(g,x)\to(g,gx)$ is measurable in the sense that the inverse image of any set in $\Gamma\times\mathfrak{B}$ is also a set in $\Gamma\times\mathfrak{B}$. For such a Γ the †orbit $G(x)$ of G through x is Γ-measurable. We assume that following conditions: (1) $\mathscr{P}$ is dominated; (2) G operates †effectively on $\mathscr{X}$; (3) $\bar{G}$ operates †transitively on Ω; (4) for any compact subset J of G,

$$\lim_{n\to\infty}\{\mu^r(K_n)/\mu^r(K_n\cdot J^{-1})\}=1,$$

where μ^r is a right-invariant Haar measure (→ 225 Invariant Measures C) on G and $K_n\cdot J^{-1}=\{gh^{-1}|g\in K_n, h\in J\}$; (5) there is a †conditional probability distribution $P_\theta(\cdot:z)$, given $z\in G(x)$, on $\mathscr{X}$; (6) the integral $\int_{G(x)}w(\theta,\tilde{g}a)P_\theta(dgx:z)$ attains its minimum value $b(\theta,z)$ for any θ and $z\in G(x)$; and (7)

$$\lim_{n\to\infty}\left\{\int_{K_nx}w(\theta,\tilde{g}a)P_\theta(dgx:z)-b(\theta,z)\right\}\geqslant 0$$

uniformly in a, where $K_nx=\{gx|g\in K_n\}$. Then the best invariant decision function exists which is also minimax in $\mathscr{D}$ [4] (→ 400 Statistical Hypothesis Testing). It is shown in [9] that some invariant minimax decision functions are not admissible.

F. Sequential Decision Problems

In the general framework of statistical decision theory, not only the decisions to be taken but the number of samples to be observed may be determined based on the previous observations.

A simple formulation is illustrated for the sequential decision problem given below. Suppose that $X_1, X_2, \ldots, X_n$ are independently and identically distributed random variables with probability measure P_θ. We assume that the X's are to be observed one by one, and at the ith stage, when we have observed $X_1, \ldots, X_i$, we are to decide whether to continue sampling and observe X_{i+1} or to stop observation and choose an action or decision in $\mathscr{D}$, utilizing all the observations thus far obtained. Then a decision rule is defined by a sequence of pairs (δ_n, s_n), $n=0,1,\ldots$, where δ_n is a mapping from the space X^n to $\mathscr{D}$ (for the sake of simplicity we here exclude randomized decisions), and $s_n=s_n(x_1,\ldots,x_n)$ is a measurable function from X^n to the interval $[0,1]$. s_n gives the probability of stopping the sampling when the first n of the X's are observed, and δ_n defines the decision taken when the observations are stopped. We include s_0 and δ_0 to denote the probability of taking a decision without making any observation and the decision to be taken then. We call δ_n the **terminal decision rule** and s_n the **stopping rule**. Then for such a decision rule $\tilde{\delta}$ the total expected loss or the risk is given by

$$r(\theta,\tilde{\delta})=\sum_{n=0}^{\infty}\int\prod_{j=0}^{n-1}(1-s_j(x_1,\ldots,x_j))$$

$$\times\{s_n(x_1,\ldots,x_n)w(\theta,\delta_n(x_1,\ldots,x_n))$$

$$+c_n(x_n|x_1,\ldots,x_{n-1})\}\prod_{j=1}^{n}dP_\theta(x_j),$$

where $c_n(x_n|x_1,\ldots,x_{n-1})$ is the **cost of observation** of $X_n=x_n$ when $X_1=x_1,\ldots,X_{n-1}=x_{n-1}$. The rule $\tilde{\delta}$ is called a **sequential decision rule** or a **sequential decision function**, and the whole setup a **sequential decision problem**. In most of the sequential decision problems the cost of observation is assumed to be equal to a constant c per observation, and then the Bayes risk $r^*(\xi_\theta)$ for the prior distribution ξ_θ satisfies the relation

$$r^*(\xi_\theta)=\min\left\{\inf_d\int w(\theta,d)\,d\xi_\theta,\right.$$

$$\left.\iint r^*(\pi_\theta(x_1,\xi_\theta))\,dP_\theta(x_1)\,d\xi_\theta+c\right\},$$

where $\pi_\theta(x_1,\xi_\theta)$ denotes the posterior distribution when $X_1=x_1$ is observed under the prior distribution ξ_θ; and the Bayes decision rule satisfies the condition

$$s_n(x_1,\ldots,x_n)=1\ \ \text{if}\ r^*(\pi_\theta(x_1,\ldots,x_n;\xi_\theta))$$

$$=\inf\int w(\theta,d)\,d\pi_\theta(x_1,\ldots,x_n;\xi_\theta),$$

$$=0\ \ \text{otherwise},$$

and $\delta_n(x_1,\ldots,x_n)=d^*$ if $\int w(\theta,d^*)\,d\pi_\theta(x_1,\ldots,x_n;\xi_\theta)=\inf_d\int w(\theta,d)\,d\pi_\theta(x_1,\ldots,x_n;\xi_\theta)$, where $\pi_\theta(x_1,\ldots,x_n;\xi_\theta)$ is the posterior distribution given $X_1=x_1,\ldots,X_n=x_n$ under the prior ξ_θ.

For the dichotomy problem with the simple loss function discussed above, the Bayes decision rule has the form

$$s_n(x_1,\ldots,x_n)=1\ \ \text{if}\ \pi_0(x_1,\ldots,x_n)\leqslant\pi_1\ \text{or}\ \geqslant\pi_2,$$

$$=0\ \ \text{otherwise},$$

and

$$\delta_n(x_1,\ldots,x_n)=d_1\ \ \text{if}\ \pi_0(x_1,\ldots,x_n)\leqslant\pi_1,$$

$$=d_0\ \ \text{if}\ \pi_0(x_1,\ldots,x_n)\geqslant\pi_2,$$

where $\pi_0(x_1,\ldots,x_n)$ is the posterior probability that $\theta=\theta_0$, given $X_1=x_1$, $X_n=x_n$. This amounts to the following rule: Continue sampling as long as

$$\gamma_1<\lambda(x_1,\ldots,x_n)=\prod_i P_{\theta_0}(x_i)/\prod P_{\theta_1}(x_i)<\gamma_2,$$

and decide on d_0 as soon as $\lambda(x_1,\ldots,x_n)\geqslant\gamma_2$ and on d_1 as soon as $\lambda(x_1,\ldots,x_n)\leqslant\gamma_1$, which is actually equivalent to a †sequential probability ratio test (→ 400 Statistical Hypothesis Testing).

G. Information in Statistical Experiments

The part $\mathscr{E}=(\mathscr{X},\mathfrak{B},\mathscr{P},\Omega)$ of a decision problem $\Delta=(\mathscr{X},\mathfrak{B},\mathscr{P},\Omega,\mathscr{A},\mathfrak{C},w,D)$ is called a **statistical experiment**. In this section, we consider n-decision problems, i.e., those wherein Ω consists of a finite number $\{1,2,\ldots,n\}$ of states, and we denote the set $S=\{r(1,\delta),r(2,\delta),\ldots,r(n,\delta)\,|\,\delta\in\mathscr{D}\}$ by $L(\Delta)$. Let $\mathscr{E}_1$ and $\mathscr{E}_2$ be two experiments having a common parameter space Ω, and let Δ_1 and Δ_2 be two decision problems composed of $\mathscr{E}_1$ and $\mathscr{E}_2$, respectively, and a common $(\mathscr{A},\mathfrak{C},w)$. We say that the experiment $\mathscr{E}_1$ is **more informative** than the experiment $\mathscr{E}_2$ if $L(\Delta_1)\supset L(\Delta_2)$ for any action space $(\mathscr{A},\mathfrak{C})$ and any loss function w [6]; that is, whatever the actions proposed and the loss incurred, the experiment $\mathscr{E}_1$ can offer a decision procedure at least as efficient as the experiment $\mathscr{E}_2$. Thus the set $L(\Delta)$ with $\Delta=(\mathscr{E},\mathscr{A},\mathfrak{C},w,D)$ represents some feature of information that $\mathscr{E}$ can provide about the states Ω. However, comparison of $L(\Delta)$ is not easy to carry out. S. Kullback and R. A. Leibler defined the concept of information for the case of a dichotomy Δ [5]. If the probability distribution induced by a random variable X has a Radon-Nikodym derivative $f_1(x)$ (>0) or $f_2(x)$ (>0) with respect to λ, we define

$$I(X:1,2)=I(f_1,f_2)=\int_{\mathscr{X}}\log\frac{f_1(x)}{f_2(x)}f_1(x)\,d\lambda,$$

calling this the **Kullback-Leibler information number** (or **K-L information number**). This number is uniquely determined by the set $S = L(\Delta)$ in Fig. 1, and the larger S becomes, the larger $I(f_1, f_2)$ becomes. If α, β $(\alpha+\beta=1)$ are a priori probabilities and $\eta(1|x)$ and $\eta(2|x)$ are a posteriori probabilities of 1 and 2, respectively, we have, from the Bayes theorem,

$$\log\frac{f_1(x)}{f_2(x)} = \log\frac{\eta(1|x)}{\eta(2|x)} - \log\frac{\alpha}{\beta}.$$

Here the right-hand side stands for the change in the probability of the occurrence of the state after an observation of x, and the †expectation of the left-hand side under f_1 is $I(f_1, f_2)$. The K-L information number has the following properties: (i) $I(X:1,2) \geqslant 0$, and $I(X:1,2)=0 \Leftrightarrow f=g$; (ii) †independence of X and Y implies $I(X:1,2)+I(Y:1,2)=I(X,Y:1,2)$; (iii) for a statistic $T=t(x)$, $I(X:1,2) \leqslant I(T:1,2)$, where the equality holds if and only if T is sufficient for $\mathscr{P}=\{P_1, P_2\}$, in which $dP_1/d\lambda=f_1$ and $dP_2/d\lambda=f_2$; and, as a result of (ii), (iv) if $X_1, \ldots, X_n$ are distributed independently with the same distribution, $I(X_1, \ldots, X_n:1,2)=nI(X_1:1,2)$.

Suppose next that Ω is the real line and that the Radon-Nikodym derivative $g(t:\theta)$ of the distribution of a statistic T with respect to a measure λ has the following properties: (i) the set of all t at which $g(t:\theta)>0$ is independent of θ; (ii) $g(t:\theta)$ is continuously twice differentiable; and (iii) the order of differentiation with respect to θ and integration with respect to t can be interchanged. Then $I(T:\theta, \theta+d\theta) = I(T:\theta)\,d\theta^2$ for an infinitesimal displacement $d\theta$ of θ, where

$$I(T:\theta) = \int_{\mathscr{X}} \left(\frac{\partial \log g(t:\theta)}{\partial\theta}\right)^2 g(t:\theta)\,d\lambda.$$

Here $I(T:\theta)$ coincides with the Fisher information (→ 399 Statistical Estimation).

Suppose that we are given a sequence $(X_1, X_2, \ldots)$ of independent random variables whose distributions have as their density either $(f_1, f_2, \ldots)$ or $(g_1, g_2, \ldots)$, that is, f_i and g_i are candidates for the density of distribution of X_i $(i=1,2,\ldots)$. A method to determine which of the sequences (f_i), (g_i) actually corresponds to (X_i) is given by the **Kakutani theorem**. Let F be the distribution of $(X_1, X_2, \ldots)$ when each X_i is distributed according to f_i, and let G be that of $(X_1, X_2, \ldots)$ when each X_i is distributed according to g_i. To see how $X_1, X_2, \ldots$ are actually distributed, we assume that the loss incurred by an incorrect decision is 1 and the loss incurred by a correct decision is 0. Denote such a decision problem (dichotomy) by Δ. Then we generally have $L(\Delta) \subset I$, where $I=\{(x,y) \mid 0 \leqslant x, y \leqslant 1\}$. A necessary and sufficient condition for $L(\Delta)=I$ is that $\prod_{n=1}^{\infty} \rho(f_n, g_n)=0$, where

$$\rho(f_n, g_n) = \int_x \sqrt{f_n(x) g_n(x)}\,dx.$$

Consequently, if the X_n $(n=1,2,\ldots)$ have the same distribution, we have $L(\Delta)=I$, and the correct decision can be made with no error based on infinitely many independent observations of $X_1, X_2, \ldots$.

H. Relation to Game Theory

The theory of statistical decision functions is closely related to game theory. From the game-theoretic viewpoint, a statistical decision problem is considered to be a zero-sum two-person game played by the statistician against nature. A strategy of nature is the true distribution P of the variable X or the true value of θ, and a strategy of the statistician is a decision δ. In this setup, the risk function $r(\theta, \delta)$ can be regarded as a †payoff function paid by the statistician to nature. An a priori distribution ξ is a †mixed strategy of nature. A randomized decision function is a mixed strategy of the statistician. A minimax decision function corresponds to a minimax strategy of the statistician. A minimax strategy of nature is called a **least favorable a priori distribution**. If a decision problem is †strictly determined as a game, a minimax solution is a Bayes solution in the wide sense.

If δ_0 is a Bayes solution with respect to ξ_0 and $r(\theta, \delta_0) \leqslant R(\xi_0, \delta_0)$ for all θ, the decision problem is strictly determined, and δ_0 is minimax and ξ_0 is a least favorable a priori distribution, where $R(\xi_0, \delta_0)=\int_\Omega r(\theta, \delta_0)\,d\xi_0(\theta)$. If δ_0 is a Bayes solution in the wide sense and $r(\theta, \delta_0)$ is constant as a function of θ, the decision problem is strictly determined, and δ_0 is minimax. If δ_0 is admissible and $r(\theta, \delta_0) \equiv c < \infty$, δ_0 is minimax. If Ω is a finite set and $\inf_\delta \sup_\theta r(\theta, \delta) < \infty$, the decision problem is strictly determined and there exists a least favorable a priori distribution. We have few general results about generalized Bayes solutions (→ 173 Game Theory; [6]).

References

[1] A. Wald, Statistical decision functions, Wiley, 1950 (Chelsea, 1971).
[2] L. LeCam, An extension of Wald's theory of statistical decision functions, Ann. Math. Statist., 26 (1955), 69–81.
[3] R. R. Bahadur, Sufficiency and statistical decision functions, Ann. Math. Statist., 25 (1954), 423–462.

[4] H. Kudō, On minimax invariant estimates of the transformation parameter, Nat. Sci. Rep. Ochanomizu Univ., 6 (1955), 31–73.
[5] S. Kullback, Information theory and statistics, Wiley, 1959.
[6] D. H. Blackwell and M. A. Girshick, Theory of games and statistical decisions, Wiley, 1954.
[7] L. Weiss, Statistical decision theory, McGraw-Hill, 1961.
[8] H. Raiffa and R. Schlaifer, Applied statistical decision theory, Harvard Univ. Graduate School of Business Administration, 1961.
[9] W. James and C. Stein, Estimation with quadratic loss, Proc. 4th Berkeley Symp. Math. Stat. Prob. I, Univ. of California Press (1961), 361–379.
[10] J. O. Berger, Statistical decision theory, foundations, concepts, and methods, Springer, 1980.
[11] R. H. Farrell, Weak limits of sequences of Bayes procedures in estimation theory, Appendix, Decision theory, Proc. 5th Berkeley Symp. Math. Stat. Prob. I, Univ. of California Press, 1967, 83–111.

399 (XVIII.7) Statistical Estimation

A. General Remarks

Statistical estimation is one of the most important methods of statistical inference (→ 401 Statistical Inference). Its purpose is to estimate the values of †parameters (or their functions) involved in a distribution of a statistical †population by using observations on the population (→ 396 Statistic). Let $\mathscr{P}=\{P_\theta \mid \theta\in\Theta\}$ be a family of †probability distributions, indexed by a parameter θ and defined over a †measurable space (i.e., sample space) $(\mathscr{X},\mathfrak{B})$. Let X be a †random variable taking values in $\mathscr{X}$ and distributed according to a probability distribution P that is a member of $\mathscr{P}$. **Statistical estimation** is a method of estimating the †true value of the parameter θ (i.e., the parameter θ such that $P=P_\theta$) or the (true) value $g(\theta)$ of a given **parametric function** g (i.e., a function defined over Θ) or both, at θ, based on the observed value x of the random variable X. The function g maps the parameter values into $\mathbf{R}$, $\mathbf{R}^k$, or some function space. Statistical estimation methods are classified into two types: point estimation, which deals with individual values of $g(\theta)$, and interval (or region) estimation, by means of which regions that may contain the value $g(\theta)$ are considered. We can also include in statistical estimation the problem of predicting the tolerance region in which the value of a yet unobserved random variable may come out.

B. Point Estimation

In the method of **point estimation** for a given parametric function g, we choose a measurable mapping φ from the sample space $(\mathscr{X},\mathfrak{B})$ into a measurable space $(\mathscr{A},\mathfrak{C})$ and state that "the value of $g(\theta)$ is $\varphi(x)$" for an observed value x, where $\mathscr{A}$ is a set containing the range of g and $\mathfrak{C}$ is a †complete additive class of subsets in $\mathscr{A}$. The mapping φ, or the random variable $\varphi(X)$ taking values in the space $\mathscr{A}$, is called an **estimator** of $g(\theta)$, while the value $\varphi(x)$ determined by the observed value x is called an **estimate** of $g(\theta)$. This estimate is sometimes termed a **nonrandomized estimate** in contrast to the following generalized notion of estimator. A mapping from $\mathscr{X}$ to a set of probability distributions defined over $(\mathscr{A},\mathfrak{C})$ is called a **randomized estimator**, which reduces to a nonrandomized estimator when each image distribution degenerates to a single point. We assume that $\mathscr{A}=\mathbf{R}$ and $\mathfrak{C}=$ the class $\mathfrak{B}$ of all †Borel sets in $\mathbf{R}$, unless stated otherwise. We denote the †expectation and †variance with respect to P_θ by E_θ and V_θ, respectively.

C. Unbiasedness

An estimator $\varphi(X)$ of $g(\theta)$ may not be exactly equal to $g(\theta)$ for any $\theta\in\Theta$ except for trivial cases, but could instead be stochastically distributed around it. An estimator $\varphi(X)$ is said to have **unbiasedness** if it is stochastically balanced around $g(\theta)$ in some sense, such as mean, median, or mode. A statistic $\varphi(X)$ is called a (mean) **unbiased estimator** of $g(\theta)$ if

$$E_\theta(\varphi(X))=g(\theta)$$

for any $\theta\in\Theta$. A parametric function g is said to be **estimable** if it has an unbiased estimator. For example, the sample mean is unbiased for the population mean: $E_\theta(\bar{X})=E_\theta(X)$ for any $\theta\in\Theta$. Unbiasedness usually implies mean unbiasedness, and we assume this unless stated otherwise. The function

$$b(\theta)=E_\theta(\varphi(X))-g(\theta)$$

is called the **bias** of the estimator $\varphi(X)$. If we restrict ourselves to unbiased estimators $\varphi(X)$ only, it is best to choose, if possible, a $\varphi(X)$ whose variance $V_\theta(\varphi(X))$ is minimum uniformly for every $\theta\in\Theta$.

Theorem (Rao-Blackwell). If $T=t(X)$ is a †sufficient statistic, then for any unbiased estimator $\varphi(X)$ of g the †conditional expectation $\psi(t)=E(\varphi(X)|T=t)$ yields another unbiased estimator $\varphi^*(X)=\psi(t(X))$ of g, which satisfies $V_\theta(\varphi^*)\leqslant V_\theta(\varphi)$ for all $\theta\in\Theta$, with the equality holding if and only if $\varphi(x)=\varphi^*(x)$ (a.e.$\mathscr{P}$). The notation a.e.$\mathscr{P}$ means that the statement concerned holds with probability 1 with respect to P_θ for each $\theta\in\Theta$. An estimator φ of $g(\theta)$ is called a **uniformly minimum variance** (or **UMV**) **unbiased estimator** if φ is unbiased for $g(\theta)$ and has a minimum variance uniformly in Θ among the class of unbiased estimators for $g(\theta)$.

Theorem (Lehmann-Scheffé). If T is a sufficient and †complete statistic, then for any estimable parametric function $g(\theta)$, there exists a unique UMV unbiased estimator of $g(\theta)$ that is a function of T. For example, suppose that $X=(X_1,X_2,\ldots,X_n)$ is a random sample from a population with exponential type distribution P_θ with density $p_\theta(x)=\beta(\theta)u(x)\exp(\sum_{i=1}^k\alpha_i(\theta)t_i(x))$ with respect to Lebesgue measure and that the set $\{(\alpha_1(\theta),\ldots,\alpha_k(\theta))\,|\,\theta\in\Theta\}$ contains some open set of $\mathbf{R}^k$. In this case, $T=(t_1(X),\ldots,t_k(X))$ is a sufficient and complete statistic, and hence every real-valued measurable function $\psi(T)$ is the unique UMV unbiased estimator of the parametric function $E_\theta(\psi(T))$. If for any $\theta\in\Theta$ the †median of the distribution of an estimator $\varphi(X)$ equals a real parametric function $g(\theta)$ when X is distributed as P_θ, i.e., if

$$P_\theta\{\varphi(X)<g(\theta)\}\leqslant\tfrac{1}{2}\leqslant P_\theta\{\varphi(X)\leqslant g(\theta)\},$$

then $\varphi(X)$ is called a **median unbiased estimator**. For example, a sample median (suitably defined for the case of an even number of samples) is median unbiased for the population median. If $\varphi(X)$ is a median unbiased estimator of g, then for any real-valued monotone function h, an estimator $h(\varphi(X))$ for $h(g(\theta))$ is median unbiased, that is, median unbiasedness is preserved under monotone transformations, which is not the case with mean unbiasedness.

Restricting our consideration to the class of all median unbiased estimators, we can use the function

$$a(u,\theta,\varphi)=\begin{cases}P_\theta\{\varphi(X)\leqslant u\} & \text{for } u<\theta\\ P_\theta\{\varphi(X)\geqslant u\} & \text{for } u>\theta\end{cases}$$

as an indicator of the behavior pattern of an estimator φ. The estimator φ that minimizes $a(u,\theta,\varphi)$ for all values of u and θ $(u\neq\theta)$ is said to be **uniformly best**. This property is also preserved under monotone transformations.

Theorem (Birnbaum). If a family of distributions $\{P_\theta\,|\,\theta\in\Theta\subset\mathbf{R}\}$ has a monotone †likelihood ratio with respect to a statistic $t(x)$ and the distribution function $F(t,\theta)$ of $T=t(X)$ is continuous both in t for any θ and in θ for any t, then there exists a uniformly best median unbiased estimator of θ. Actually, if $\theta=\hat\theta(t)$ is a solution of $F(t,\theta)=1/2$, then $\varphi(X)=\hat\theta(t(X))$ is such an estimator.

If for any $\theta\in\Theta$ the †mode of the density function or the probability mass function of an estimator $\varphi(X)$ is equal to $g(\theta)$, then $\varphi(X)$ is called a **modal unbiased estimator** of $g(\theta)$.

D. Lower Bounded of the Variance of an Unbiased Estimator

When there does not exist a sufficient and complete statistic, we can still seek to minimize the variance of the (mean) unbiased estimator at every fixed point $\theta=\theta_0$. In the remainder of this section, $\mathscr{P}$ is assumed to be dominated by a measure μ, and $p_\theta(x)$ denotes the density function of P_θ with respect to μ and $\pi_\theta(x)=p_\theta(x)/p_{\theta_0}(x)$. The following theorem guarantees the existence of the locally best unbiased estimator.

Theorem (Barankin). Let $\mathscr{M}$ be the set of all unbiased estimators of a parametric function $g(\theta)$ with finite variance at $\theta=\theta_0$. Assume that $\mathscr{M}$ is not empty and $E_{\theta_0}((\pi_\theta(X))^2)<\infty$. Then there exists an estimator φ_0 in $\mathscr{M}$ that minimizes the variance at θ_0 within $\mathscr{M}$. Actually, $\{\varphi_0\}=\mathscr{M}\cap\mathscr{P}_0$, where $\mathscr{P}_0$ is the linear space generated by $\{\pi_\theta(x)\,|\,\theta\in\Theta\}$. The minimum variance is given as follows:

$$V_{\theta_0}(\varphi_0)$$
$$=\inf\{V_{\theta_0}(\varphi(X));\varphi\in\mathscr{M}\}$$
$$=\sup\nolimits_{\mathrm{I}}\left\{\left(\sum_{i=1}^n a_i h(\theta_i)\right)^2\Big/E_{\theta_0}\left(\sum_{i=1}^n a_i\pi_{\theta_i}(X)\right)^2\right\}$$
$$=\sup\nolimits_{\mathrm{II}}\left\{\sum_{i=1}^n\sum_{j=1}^n h(\theta_i)h(\theta_j)\lambda^{ij}\right\},$$

where $h(\theta)=g(\theta)-g(\theta_0)$ and λ^{ij} is the (i,j)-component of the inverse of the $n\times n$ matrix (λ_{ij}) with $\lambda_{ij}=E_{\theta_0}(\pi_{\theta_i}(X)\pi_{\theta_j}(X))$ and where the supremum $\sup_{\mathrm{I}}$ is taken over all positive integers $n,\theta_1,\ldots,\theta_n\in\Theta$ and $a_1,\ldots,a_n\in R$, and becomes $\sup_{\mathrm{II}}$ when the supremum is taken over n and the θ_j. This theorem leads to the following three theorems with respect to the lower bound of the variance of an unbiased estimator. The first is immediate and the last two are obtained by replacing some order differences of π_θ with the corresponding differentials.

Theorem. For any unbiased estimator $\varphi(X)$ of g and $\theta_0\in\Theta$, we have

$$V_{\theta_0}(\varphi(X))\geqslant\sup_{\theta\in\Theta}\{(g(\theta)-g(\theta_0))^2/E_{\theta_0}(\pi_\theta(X)-\pi_{\theta_0}(X))^2\}$$

(**Chapman-Robbins-Kiefer inequality**).

Theorem. Suppose $\Theta \subset \mathbf{R}$. For any unbiased estimator $\varphi(X)$ of g and under certain regularity conditions, we have

$$V_{\theta_0}(\varphi(X)) \geqslant \frac{(g'(\theta_0))^2}{E_{\theta_0}((\partial \log p_\theta(X)/\partial\theta|_{\theta=\theta_0})^2)}$$

(**Cramér-Rao inequality**), where the equality holds only for the exponential type distribution $p_\theta(x)=\beta(\theta)u(x)\exp(\alpha(\theta)\varphi(x))$. An example of such regularity conditions is (i)–(iii): (i) $E_{\theta_0}((\pi_\theta(X))^2)<\infty$ for all $\theta\in\Theta$; (ii) $p_\theta(x)$ has a partial derivative $p'_\theta(x)$ at $\theta=\theta_0$ (a.e. P_{θ_0}); and (iii)

$$\lim_{\Delta\theta\to 0} E_{\theta_0}\left[\left(\frac{p_{\theta_0+\Delta\theta}(X)-p_{\theta_0}(X)}{p_{\theta_0}(X)\Delta\theta}-\frac{p'_{\theta_0}(X)}{p_{\theta_0}(X)}\right)^2\right]=0.$$

Corollary. Let $X=(X_1,\dots,X_n)$ be a random sample from a distribution with density $f(x,\theta)$ and let

$$I(\theta)=E_\theta((\partial\log f(X_1,\theta)/\partial\theta)^2).$$

Since $E_\theta((\partial\log p_\theta(X)/\partial\theta)^2)=nI(\theta)$, the Cramér-Rao inequality implies

$$V_{\theta_0}(\varphi(X))\geqslant(g'(\theta_0))^2/nI(\theta_0).$$

The number $I(\theta_0)$ is called the **Fisher information** of the distribution $f(x,\theta)$. When the equality holds for an unbiased estimator $\varphi(X)$, $\varphi(X)$ is called an **efficient estimator** of $g(\theta)$. In general, the **efficiency** of an unbiased estimator φ at $\theta=\theta_0$ is defined by

$$\mathrm{Eff}(\varphi)=(g'(\theta_0))^2/(nI(\theta_0)V_{\theta_0}(\varphi)).$$

Theorem. For any unbiased estimator $\varphi(X)$ of $g(\theta)$ and under certain regularity conditions, we have

$$V_{\theta_0}(\varphi(X))\geqslant\sum_{i=1}^k\sum_{j=1}^k g^{(i)}(\theta_0)g^{(j)}(\theta_0)K^{ij}$$

(**Bhattacharyya inequality**), where $g^{(i)}(\theta_0)=d^ig(\theta)/d\theta^i|_{\theta=\theta_0}$ and K^{ij} is the (i,j)-component of the inverse of the matrix (K_{ij}) with

$$K_{ij}=E_{\theta_0}\left(\frac{p^{(i)}_{\theta_0}(X)}{p_{\theta_0}(X)}\frac{p^{(j)}_{\theta_0}(X)}{p_{\theta_0}(X)}\right),\qquad i,j=1,\dots,k.$$

An example of such regularity conditions is (i)–(iii): (i) $E_{\theta_0}((\pi_\theta(X))^2)<\infty$; (ii) $p_\theta(x)$ is k-times differentiable with respect to θ at $\theta=\theta_0$; (iii) the ith partial derivative $p^{(i)}_{\theta_0}(x)$, $i\leqslant k$, satisfies

$$\lim_{\Delta\theta\to 0}E_{\theta_0}\left[\left(\frac{\Delta^i p_\theta(X)|_{\theta=\theta_0}}{p_{\theta_0}(X)(\Delta\theta)^i}-\frac{p^{(i)}_{\theta_0}(X)}{p_{\theta_0}(X)}\right)^2\right]=0,$$

where $\Delta P_\theta(X)|_{\theta=\theta_0}=P_{\theta_0+\Delta\theta}(X)-P_{\theta_0}(X)$ and $\Delta^iP_\theta(X)=\Delta(\Delta^{i-1}P_\theta(X))$ for $i\geqslant 2$. For $k=1$ the Bhattacharyya lower bound is the same as the Cramér-Rao lower bound. In general, the former gives a sharper lower bound than the latter.

If the parameter is multidimensional, $\theta=(\theta_1,\dots,\theta_k)'$, then for any unbiased estimator $\varphi(X)$ of $g(\theta)$ and under similar conditions to those for the 1-dimensional case, we have

$$V_{\theta_0}(\varphi(X))\geqslant\sum_{i=1}^k\sum_{j=1}^k g'_i(\theta_0)g'_j(\theta_0)J^{ij},$$

where $g'_i(\theta_0)=\partial g(\theta)/\partial\theta_i|_{\theta=\theta_0}$ and J^{ij} is the (i,j)-component of the inverse of the matrix (J_{ij}) with

$$J_{ij}=E_{\theta_0}(\partial\log p_\theta(X)/\partial\theta_i|_{\theta=\theta_0}\partial\log p_\theta(X)/\partial\theta_j|_{\theta=\theta_0}).$$

If $\theta^*(X)=(\theta^*_1(X),\dots,\theta^*_k(X))'$ is an unbiased estimator of $\theta=(\theta_1,\dots,\theta_k)'$ (i.e., $E_\theta(\theta^*_i(X))=\theta_i$ for $i=1,2,\dots,k$), then the †covariance matrix $V_{\theta_0}(\theta^*(X))$ of $\theta^*(X)$ at θ_0 satisfies $V_{\theta_0}(\theta^*(X))\geqslant J^{-1}$, that is, the difference $V_\theta(\theta^*(X))-J^{-1}$ is a nonnegative definite matrix. If $X=(X_1,\dots,X_n)$ is a random sample from a distribution having density $f(x_1,\theta)$, then by setting

$$I_{ij}=E_{\theta_0}(\partial\log f(X_1,\theta)/\partial\theta_i|_{\theta=\theta_0}\times\partial\log f(X_1,\theta)/\partial\theta_j|_{\theta=\theta_0}),$$

we have $J_{ij}=nI_{ij}$. The matrix $I=(I_{ij})$ is called the **Fisher information matrix** of the distribution $f(x_1,\theta)$.

E. Decision-Theoretic Formulation

(→ 398 Statistical Decision Functions)

Let $W(\theta,a)$ ($\geqslant 0$) be the loss incurred from an estimate (or action) a of the parameter when the true value of the parameter is θ. The risk function of an estimator $\varphi(X)$ of the parametric function $g(\theta)$ is then defined as

$$r(\theta,\varphi)=E_\theta(W(\theta,\varphi(X))).$$

Statistical decision theory deals with the problem of minimizing, in an appropriate manner, the risk function by a suitable choice of φ. The notions of complete class, Bayes estimator, admissibility, minimax estimator, and invariant estimator, explained here and in Sections F–I, are the most important of the theory. The unbiased estimator explained in Section C may also be considered an important concept of the theory.

A class C of estimators is said to be essentially complete if for any estimator φ there exists an estimator φ_0 in C such that

$$r(\theta,\varphi_0)\leqslant r(\theta,\varphi)$$

for any $\theta\in\Theta$. The following two theorems hold, provided that the action space $\mathscr{A}$ is $\mathbf{R}$ and the loss function $W(\theta,a)$ is convex with respect to $a\in\mathscr{A}$ for any $\theta\in\Theta$.

Theorem (Hodges-Lehmann). If $W(\theta,a)\to\infty$ as $|a|\to\infty$, then the class of all nonrandomized estimators is essentially complete.

Theorem. If $T=t(X)$ is a sufficient statistic, then the class of all functions of T is essentially complete. Actually, given any estimator $\varphi(X)$ of $g(\theta)$, the conditional expectation $\psi(t)=E(\varphi(X)|T=t)$ yields an estimator $\varphi_0(X)=\psi(t(X))$ satisfying $r(\theta,\varphi_0)\leqslant r(\theta,\varphi)$ for any θ, where the equality holds when and only when $\varphi\equiv\varphi_0$, provided that W is convex in the strict sense.

A loss function of the form

$$W(\theta,a)=\lambda(\theta)(a-g(\theta))^2,\qquad \lambda(\theta)>0,$$

is called a **quadratic loss functions**. If, in particular, $\lambda(\theta)\equiv 1$, then

$$r(\theta,\varphi)=E_\theta((\varphi(X)-g(\theta))^2)$$

is called the **mean square error** of the estimator $\varphi(X)$ of $g(\theta)$. This error coincides with the variance when the estimator is unbiased.

F. Bayes Estimators

Let ξ be an a priori distribution over the parameter space Θ associated with a certain †σ-algebra $\mathfrak{F}$, and assume that $r(\theta,\varphi)$ is $\mathfrak{F}$-measurable for every φ. Denote by E^ξ the average operator relative to ξ. The infimum of the average risk $r(\xi,\varphi)=E^\xi(r(\theta,\varphi))=E^\xi E_\theta(W(\theta,\varphi(X)))$ for φ running over its range is called the †Bayes risk relative to ξ, while an estimator $\varphi(X)$ of $g(\theta)$ at which the average risk $r(\xi,\varphi)$ attains the infimum is called a **Bayes estimator** relative to ξ. A Bayes estimator is obtained as follows: Assume that $\mathscr{P}$ is dominated by a measure μ with $\mathfrak{B}\times\mathfrak{F}$-measurable density $p_\theta(x)$, the loss function $W(\theta,a)$ is $\mathfrak{F}\times\mathfrak{C}$-measurable, and

$$\int_\Theta p_\theta(x)\,d\xi(\theta)<\infty.$$

For each observed value of x, the Bayes estimator $\varphi(x)$ takes the value a that minimizes

$$r(a|x)=E^\xi(W(\theta,a)|x)$$
$$=\int_\Theta W(\theta,a)p(\theta|x)\,d\xi(\theta),$$

where $p(\theta|x)$ is the probability density, with given x, of θ. We call $r(a|x)$ the **posterior risk**.

Theorem (Girshick-Savage). Suppose that the loss function is quadratic. For any x the value of the posterior risk is either ∞ (for every value of a) or finite (for all or only one value of a). If the Bayes risk relative to ξ is finite, then a Bayes estimator $\varphi^*(X)$ relative to ξ is determined uniquely as follows: $\varphi^*(x)=a_0$ if $r(a_0|x)<\infty$ for only one value a_0, whereas $\varphi^*(x)=E^\xi(g(\theta)\lambda(\theta)|x)/E^\xi(\lambda(\theta)|x)$ if $r(a|x)<\infty$ for every a. If $E^\xi(\lambda(\theta))<\infty$, then a Bayes estimator is either biased or has average risk zero.

G. Admissibility of Estimators

An estimator $\varphi_0(X)$ of a parametric function $g(\theta)$ is said to be **admissible** if and only if for any estimator $\varphi(X)$ of $g(\theta)$ the inequality $r(\theta,\varphi)\leqslant r(\theta,\varphi_0)$ for all $\theta\in\Theta$ implies that

$$r(\theta,\varphi)=r(\theta,\varphi_0)$$

for all $\theta\in\Theta$. If an estimator is the unique Bayes estimator relative to some a priori distribution, then it is admissible. For example, let $X_1,\ldots,X_n$ be a random sample from $N(\theta,1)$, and let $W(\theta,a)=(a-\theta)^2$. Then $\varphi(X)=(c+n\sigma^2\bar{X})/(1+n\sigma^2)$ is the unique Bayes estimator relative to the prior distribution $N(c,\sigma^2)$ of θ, where $\bar{X}$ is the sample mean. Hence any estimator of the form $\varphi(X)=a\bar{X}+b$ is admissible when $0<a<1$ and $-\infty<b<\infty$.

In the rest of this section and the next section, we restrict ourselves to quadratic loss functions. If a statistic of the form $c\varphi(X)$ with real c is an admissible estimator of $g(\theta)$, then

$$\inf_\theta\left(g(\theta)\frac{E_\theta(\varphi)}{E_\theta(\varphi^2)}\right)\leqslant c\leqslant\sup_\theta\left(g(\theta)\frac{E_\theta(\varphi)}{E_\theta(\varphi^2)}\right).$$

Theorem (Karlin). Let X be a random variable having a 1-dimensional exponential type distribution $dP_\theta(x)=\beta(\theta)e^{\theta x}d\mu(x)$ with a parameter space $\Theta=I(\underline{\theta},\bar{\theta})=\{\theta\,|\,\int_{-\infty}^{\infty}e^{\theta x}d\mu(x)<\infty\}$ a closed or open interval, and let $g(\theta)=E_\theta(X)=-\beta'(\theta)/\beta(\theta)$ be a parametric function to be estimated. Then the estimator $\varphi_\lambda(X)=X/(\lambda+1)$ for real λ is admissible provided that $\int_c^b(\beta(\theta))^{-\lambda}d\theta\to\infty$ as $b\to\bar{\theta}$ and $\int_a^c(\beta(\theta))^{-\lambda}d\theta\to\infty$ as $a\to\underline{\theta}$ for any $c\in(\underline{\theta},\bar{\theta})$.

Corollary. When $\Theta=(-\infty,\infty)$ the estimator $\varphi_0(X)=X$ is admissible.

Corollary. Let $\Theta=(-\infty,\infty)$, and assume that both intervals $(-\infty,0]$ and $[0,\infty)$ have positive measure with respect to μ. Then every estimator of the form $\varphi(X)=aX$ with $0<a\leqslant 1$ is admissible. This theorem can be applied to a random sample $X_1,\ldots,X_n$ drawn from an exponential type distribution because the sufficient statistic $\bar{X}=\sum X_i/n$ has an exponential type distribution and $E_\theta(\bar{X})=E_\theta(X_1)$.

Theorem (Karlin). Let X be a random variable having a distribution $dP_\theta(x)=q(\theta)\cdot r(x)dx$ for $0\leqslant x\leqslant\theta$ and $=0$ otherwise, with the parameter space $\Theta=(0,\infty)$, where $\int_0^\theta r(x)dx<+\infty$ and $\int_0^\infty r(x)dx=\infty$. Among the estimators of the type $\varphi_c(X)=c(q(X))^{-\alpha}$ for $g(\theta)=(q(\theta))^{-\alpha}$ with $\alpha>0$, only one estimator with the value $c=(2\alpha+1)/(\alpha+1)$ is admissible. This theorem is applicable also when the size of the random sample is larger than 1.

Theorem (Stein). Let $X_1,\ldots,X_n$ be a random sample from a univariate distribution $dP_\theta(x)=f(x-\theta)dx$ with a location parameter θ. Define

$$A_k=A_k(x_1,\ldots,x_n)=\int_{-\infty}^{\infty}\theta^k\left(\prod_{i=1}^n f(x_i-\theta)\right)d\theta$$

for $k=0, 1, 2$. If

$$\int_{-\infty}^{\infty} \cdots \int_{-\infty}^{\infty} \left(\frac{A_2}{A_0}-\left(\frac{A_1}{A_0}\right)^2\right)^{3/2} \prod_{i=1}^{n} f(x_i) \times \prod_{i=1}^{n} dx_i < \infty,$$

then the **Pitman estimator** $\varphi_0(X_1, \ldots, X_n)= A_1(X_1, \ldots, X_n)/A_0(X_1, \ldots, X_n)$ of the parameter θ is admissible.

Inadmissibility of the Usual Estimator for Three or More Location Parameters. Let $X=(X_1, \ldots, X_k)'$ be a k-variate normal random variable with mean $\theta=(\theta_1, \ldots, \theta_k)'$ and covariance matrix I, the identity. Then the Pitman estimator of θ is $\hat{\theta}=X$. However, Stein showed that X is inadmissible. It is strictly dominated by the estimator $\theta^*(X)=(1-(k-2)/|X|^2)X$, where $|\cdot|$ denotes the Euclidean norm $|X|^2=\sum_{i=1}^{k} X_i^2$. That is, if $k \geqslant 3$, $E|\theta^*(X)-\theta|^2 < E|X-\theta|^2$ for any θ.

An estimator such as $\theta^*(X)$ is called **Stein's shrinkage estimator** (James and Stein, *Proc. 4th Berkeley Symp.*, 1 (1960)).

H. Minimax Estimation

An estimator $\varphi^*(X)$ is said to be **minimax** if and only if

$$\sup_{\theta} r(\theta, \varphi^*)=\inf_{\varphi} \sup_{\theta} r(\theta, \varphi).$$

If an estimator φ^* is admissible and the risk $r(\theta, \varphi^*)$ is constant with respect to θ, then φ^* is minimax.

Theorem (Hodges-Lehmann). A Bayes estimator φ^* relative to an a priori distribution ξ is minimax if ξ assigns the whole probability to a subset $\omega \subset \Theta$, $r(\theta, \varphi^*)$ is constant (say, c) for $\theta \in \omega$, and $r(\theta, \varphi^*) \leqslant c$ for $\theta \in \Theta$. Let X have a binomial distribution $B(n, \theta)$, $0<\theta<1$; θ is unknown. If the prior distribution of θ is a beta distribution $\beta(\sqrt{n}/2, \sqrt{n}/2)$, then the Bayes estimator is $T^*(X)=(X+\sqrt{n}/2)/(n+\sqrt{n})$, which has constant risk $E_\theta(T^*(X)-\theta)^2=(2(1+\sqrt{n}))^{-2}$ for all θ, $0<\theta<1$. Thus, according to this theorem, $T^*(X)$ is minimax. It is interesting to compare the mean square error of $T^*(X)$ to that of minimum variance unbiased estimator $\hat{\theta}=X/n$, $\theta(1-\theta)/n$.

Theorem (Wald). If there exists a sequence of prior distributions $\{\xi_n\}$ such that

$$\liminf_{n \to \infty}\left(r(\theta, \varphi^*)-\inf_{\varphi} \int_{\Theta} r(\theta, \varphi)\, d\xi_n(\theta)\right) \leqslant 0$$

for any $\theta \in \Theta$, then φ^* is minimax. For example, the last theorem led to the proof of the fact that the Pitman estimator of a location parameter is minimax [4]. In the discussion of robust estimation, Huber (*Ann. Math. Statist.*, 35 (1964)) proved that Huber's minimax robust location estimator minimizes the maximum asymptotic variance over some family of symmetric distributions in a neighborhood of the normal distribution (→ 371 Robust and Nonparametric Methods).

I. Invariant Estimator

For simplicity, assume that the range A of a parametric function $g(\theta)$ coincides with the parameter space, and consider the point estimation of $g(\theta) \equiv 0$ (θ is not necessarily real). Suppose that there exist two groups $G=\{\tau\}$ and $\bar{G}=\{\bar{\tau}\}$ of one-to-one measurable transformations of $\mathscr{X}$ and Θ, respectively, onto themselves such that (i) there exists a homeomorphic mapping $\tau \to \bar{\tau}$ from G to $\bar{G}$; (ii) if X has a distribution P_θ, then τX has the distribution $P_{\bar{\tau}\theta}$; and (iii) $W(\bar{\tau}\theta, \bar{\tau}a)=W(\theta, a)$ for any θ, a, and τ. An estimator φ is said to be **invariant** if it satisfies $\varphi(\tau x)=\bar{\tau}\varphi(x)$ for any $\tau \in G$, a.e. $\mathscr{P}$. An estimator φ is called a **best invariant estimator** if the risk function $r(\theta, \varphi')$, where φ' is an invariant estimator, takes its minimum value when $\varphi'=\varphi$.

If the group $\bar{G}$ is †transitive on Θ, then the risk function of any invariant estimator is independent of the value of θ, and hence any admissible invariant estimator is best invariant. For example, in the point estimation of a location parameter with a quadratic loss function, the Pitman estimator is best invariant.

Theorem. If Θ is a compact topological space, $\bar{G}$ is a group of homeomorphisms of Θ onto itself, and $\bar{G}$ is homeomorphic to Θ under $g \in \bar{G} \to \bar{g}\theta_0 \in \Theta$ with a fixed $\theta_0 \in \Theta$, then any Bayes estimator relative to the †right-invariant Haar measure over $\bar{G}$ is best invariant. This result can be generalized to a locally compact Θ (→ 398 Statistical Decision Functions).

J. Sequential Estimation

Estimation methods based on sequential sampling are not as popular as †sequential tests, because their efficiency is not very large compared to that of nonsequential estimation. A generalization of the Cramér-Rao inequality to any sequential unbiased estimator $\varphi(X)$ of a parametric function $g(\theta)$ is the **Wolfowitz inequality,**

$$V_\theta(\varphi(X)) \geqslant (g'(\theta))^2/(E_\theta(N)I(\theta))$$

for every $\theta \in \Theta$, under regularity conditions similar to those for the fixed-size sample problem, where N is the sample size and $I(\theta)$ the Fisher information.

K. Asymptotic Theory

In practical problems of statistical inference the sample size n is often large enough to give sharp estimates of the parameters involved; then the sample distributions of estimators can be approximated closely by their asymptotic distributions, which are of a simpler nature. Assume that $X=(X_1, X_2, \ldots)$ is a sequence of independent and identically distributed (i.i.d.) random variables with the common distribution P_θ, $\theta \in \Theta$. For each n, let $\varphi_n = \varphi_n(X_1, \ldots, X_n)$ be an estimator of $g(\theta)$ that is a function of θ to $\mathscr{A}$ $(\subset \mathbf{R}^p)$. Thus φ_n is a measurable mapping from $(\mathscr{X}^n, \mathfrak{B}^n)$ to $(\mathscr{A}, \mathscr{C})$. Let us denote the distribution of φ_n by $\mathscr{L}(\varphi_n)$, $\mathscr{L}(\varphi_n | \theta)$ or $\mathscr{L}(\varphi_n | P_\theta)$, the last two emphasizing that the underlying probability distribution is P_θ. For example, if the mean vector $E_\theta(\varphi_n) = m_n(\theta)$ and the covariance matrix $v_n(\theta) = V_\theta(\varphi_n)$ exist for every n, and if $\mathscr{L}[v_n(\theta)^{-1/2}(\varphi_n(X) - m_n(\theta)) | \theta] \rightarrow N_p(0, I)$ as $n \rightarrow \infty$, then $\mathscr{L}(\varphi_n | \theta)$ is approximated by a p-variate normal distribution $N_p(m_n(\theta), v_n(\theta))$ (→ 341 Probability Measures D). φ_n is said to be **asymptotically (mean) unbiased** for $g(\theta)$ if $m_n(\theta) \rightarrow g(\theta)$ for any $\theta \in \Theta$ as $n \rightarrow \infty$. But we often calculate the asymptotic distribution without obtaining the exact mean and covariance matrix of the estimator φ_n for each n. In the asymptotic theory it may be reasonable to regard the sequence of estimators $\{\varphi_n\}$ rather than each estimator φ_n as an "estimator," but we do not bother with the difference between these definitions of an estimator.

Consistency. $\{\varphi_n\}$ is called a **consistent estimator** of $g(\theta)$ if φ_n converges to $g(\theta)$ in probability as $n \rightarrow \infty$:

$$\lim P_\theta\{|\varphi_n - g(\theta)| > \varepsilon\} = 0 \quad \text{for any } \varepsilon > 0 \text{ and every } \theta \in \Theta.$$

If the convergence is almost sure, it is called **a.s. consistent**. For example, if φ_n is asymptotically unbiased with the covariance matrix $v_n(\theta)$ such that $|v_n(\theta)| \rightarrow 0$ as $n \rightarrow \infty$, then φ_n is a consistent estimator of $g(\theta)$. A sufficient condition for existence of a consistent estimator is given by the following result.

Theorem (LeCam). Let $\mathscr{X}$ be a Euclidean space and $\mathfrak{B}$ the σ-algebra of all Borel sets in $\mathscr{X}$. If the parameter space Θ is a locally compact subset of $\mathbf{R}^k$, $P_\theta \neq P_{\theta'}$ for any $\theta \neq \theta'$ (identifiability condition), and $P_{\theta_n} \rightarrow P_\theta$ whenever $\theta_n \rightarrow \theta$, then there exists a consistent estimator of θ. $\varphi_n = g(T_n)$ is a consistent estimator of $g(\theta)$ if $\{T_n\}$ is a consistent estimator of θ and $g(\theta)$ is a continuous function of θ.

Asymptotic Normality. The class of estimators is restricted to what are called consistent estimators. An estimator $\{\varphi_n\}$ is said to be **asymptotically normally distributed** if the asymptotic distribution of $n^{1/2}(\varphi_n - g(\theta))$ is normal:

$$\mathscr{L}[n^{1/2}(\varphi_n - g(\theta)) | \theta] \rightarrow N_p(\mu(\theta), v(\theta)) \quad \text{as } n \rightarrow \infty.$$

$\mu(\theta)$ and $v(\theta)$ are called the **asymptotic bias** and **asymptotic covariance matrix**, respectively. They are not always equal to the limits, if any, of the mean and covariance matrix of $n^{1/2}(\varphi_n - g(\theta))$. $\{\varphi_n\}$ is usually called a **consistent and asymptotically normal (CAN) estimator** if the asymptotic distribution of $n^{1/2}(\varphi_n - g(\theta))$ is normal with the asymptotic bias zero. Then the distribution $\mathscr{L}(\varphi_n | \theta)$ is approximated by $N_p(0, v(\theta)/n)$. For example, the †moment method estimator is a CAN estimator.

Theorem. Let $\{\varphi_n\}$ be a CAN estimator of $g(\theta) \in \mathbf{R}^1$ with asymptotic variance $v(\theta)$. Then

$$\liminf_{n \rightarrow \infty} E_\theta\{n|\varphi_n - g(\theta)|^2\} \geqslant v(\theta) \quad \text{for every } \theta \in \Theta.$$

Theorem. Suppose that $g(\theta)$ is a continuously differentiable function from Θ $(\subset \mathbf{R}^k)$ to $\mathbf{R}^p$ $(p \leqslant k)$. Let $G(\theta) = (\partial g_i(\theta)/\partial \theta_j)$, the †Jacobian matrix. If $\{T_n\}$ is a CAN estimator of θ with asymptotic covariance matrix $v(\theta)$, then $\{g(T_n)\}$ is a CAN estimator of $g(\theta)$:

$$\mathscr{L}[n^{1/2}(g(T_n) - g(\theta)) | \theta] \rightarrow N_P(0, G(\theta)v(\theta)G(\theta)').$$

An estimator $\{T_n\}$ is said to be a **best asymptotically normal (BAN) estimator of** θ if $\{T_n\}$ is a CAN estimator of θ with asymptotic variance $I^{-1}(\theta)$, where $I(\theta)$ is the †Fisher information matrix on θ in a single observation. We can see that the maximum likelihood (ML) estimator (→ Section M) is a BAN estimator.

Functional on Distribution Functions. Let $\varphi(F)$ be a functional on distribution functions to $\mathbf{R}^1$. Let us consider the class of estimators that are defined by $\varphi_n = \varphi(\hat{F}_n)$, where $\hat{F}_n$ is the †empirical distribution function of n samples $X_1, \ldots, X_n$. An estimator $\{\varphi_n\}$ with $\varphi_n = \varphi(\hat{F}_n)$ for each n is said to be **Fisher consistent** for $g(\theta)$ if $\varphi(F_\theta) = g(\theta)$ for every $\theta \in \Theta$ when F_θ is the true distribution function. $\{\varphi_n\}$ is also a.s. consistent for $g(\theta)$ if $\{\varphi_n\}$ is a Fisher consistent estimator of $g(\theta)$ and if φ is a continuous functional. Furthermore, if φ is differentiable, we can see that $\{\varphi_n\}$ is also a CAN estimator by using the fact that $n^{1/2}(\hat{F}_n(F_\theta^{-1}(t)) - t)$, $0 \leqslant t \leqslant 1$, converges weakly to the †Brownian bridge. Let S be a set of distribution functions. S is said to be a star-shaped set of F if $H \in S$ implies $F^{(t)} = (1-t)F + tH \in S$ for any $t \in [0, 1]$.

Theorem (Von Mises). Assume that
(F1) there exists a star-shaped set S_θ at F such that $\lim_n P_\theta\{\hat{F}_n \in S_\theta\} = 1$;
(F2) for any $t \in [0, 1]$ and $H \in S_\theta$, there exist derivatives $(d^i/dt^i)\varphi[(1-t)F + tH]$, $i = 1, 2$;
(F3) there exists $\psi_\theta(y)$ from $\mathbf{R}^1$ to $\mathbf{R}^1$ such that

$$\frac{d}{dt}\varphi[(1-t)F_\theta+tH]|_{t=0}=\int_{-\infty}^{\infty}\psi_\theta(y)d(H(y)$$

$-F_\theta(y))$ for all $H\in S_\theta$; and

(F4) $\lim_{n\to\infty} P_\theta\left\{n^{1-\delta}\sup_{t\in[0,1]}\left|\frac{d^2}{dt^2}\varphi[F_{\theta n}^{(t)}]\right|>\varepsilon\right\}=0$

for any δ and $\varepsilon>0$, where $F_{\theta n}^{(t)}=(1-t)F_\theta+t\hat{F}_n$. Then if φ_n with $\varphi_n=\varphi(\hat{F}_n)$ is Fisher consistent for $g(\theta)$, we have

$$\mathscr{L}[n^{1/2}(\varphi_n-g(\theta))\,|\,\theta]\to N(0,v(\theta)),$$

where

$$v(\theta)=\int_{-\infty}^{\infty}\psi_\theta(y)^2\,dF_\theta(y)-\left\{\int_{-\infty}^{\infty}\psi_\theta(y)\,dF_\theta(y)\right\}^2.$$

$\{c_n\}$-Consistency. For a sequence of positive numbers c_n tending to infinity as $n\to\infty$, an estimator T_n is called **consistent for $\theta\in\Theta$ with order** c_n (or **$\{c_n\}$-consistent** for short) if for every $\varepsilon>0$ and every $\theta\in\Theta$ there exist a sufficiently small positive number δ and a sufficiently large number K satisfying

$$\limsup_{n\to\infty}\sup[P_\tau\{c_n|T_n-\tau|\geqslant K\};|\tau-\theta|<\delta]<\varepsilon.$$

Let $\{c_n\}$ be a maximal order of consistency. This notion was introduced by Takeuchi and Akahira. They studied consistent estimators of location parameters with various orders. Let $\mathscr{X}=\Theta=\mathbf{R}^1$. Suppose that for every $\theta\in\Theta$, P_θ has a density function $f(x-\theta)$ with respect to the Lebesgue measure.

Theorem. Assume that

(OC1) $f(x)>0$ if $a<x<b$ and $f(x)=0$ if $x\leqslant a$ or $x\leqslant b$;

(OC2) there exist positive numbers $0<\alpha\leqslant\beta<\infty$ and $0<A'$, $B'<\infty$ such that

$$\lim_{x\to a+0}(x-a)^{1-\alpha}f(x)=A',$$

$$\lim_{x\to b-0}(b-x)^{1-\beta}f(x)=B';$$

(OC3) $f(x)$ is twice continuously differentiable in the interval (a,b) and there exist positive numbers $0<A''$, $B''<\infty$ such that

$$\lim_{x\to a+0}(x-a)^{2-\alpha}|f'(x)|=A'',$$

$$\lim_{x\to b-0}(b-x)^{2-\beta}|f'(x)|=B''.$$

Assume further that $f''(x)$ is bounded if $\alpha\geqslant 2$. Then for each α there exists a consistent estimator with the order given in Table 1.

Table 1

α	order c_n	$\{c_n\}$-consistent estimator
$0<\alpha<2$	$n^{1/\alpha}$	$\{\min X_i+\max X_i-(a+b)\}/2$
$\alpha=2$	$(n\log n)^{1/2}$	ML estimator
$\alpha>2$	$n^{1/2}$	ML estimator

L. Moment Method

The moment method is also utilized to obtain estimators. Suppose that $\mathscr{X}\subset\mathbf{R}^1$ and $\Theta\subset\mathbf{R}^k$. Denote the †population distribution function by F_θ and the †empirical distribution function of n samples $X_1,\dots,X_n$ by $\hat{F}_n$. The following system of simultaneous equations is derived from letting the jth †population moment

$$\mu_j(\theta)=E_\theta(X^j)=\int x^j F_\theta(dx)$$

be equal to the jth †sample moment

$$m_{nj}=n^{-1}\sum_{i=1}^{n}X_i^j=\int x^j\hat{F}_n(dx).$$

For example, for $j=1,\dots,k$,

$$\mu(\theta)=(\mu_1(\theta),\dots,\mu_k(\theta))'=(m_{n1},\dots,m_{nk})'=m_n.$$

A **moment method estimator** is determined as a solution $\theta=\tilde{\theta}_n(x)\in\Theta$ of k numbers of simultaneous equations.

Theorem. Assume that the function $\mu(\theta)$ from Θ to $\mathbf{R}^k$ is continuously differentiable and that the Jacobian matrix $M(\theta)=\partial\mu(\theta)/\partial\theta=(\partial\mu_i(\theta)/\partial\theta_j)$, $i,j=1,\dots,k$, is nonsingular in a neighborhood of the true parameter. Then the moment method estimator exists and is a CAN estimator:

$$\mathscr{L}[n^{1/2}(\tilde{\theta}_n-\theta)\,|\,\theta]\to N_k(0,M(\theta)^{-1}v(\theta)M(\theta)'^{-1}),$$

where $v(\theta)=(\mathrm{cov}_\theta(X^i,X^j))$, $i,j=1,\dots,k$. In general, a moment method estimator is not a BAN estimator. However, in view of its simple form, a moment method estimator is important and often utilized as a first-step estimator in order to determine the maximum likelihood estimator by the iteration method.

M. Maximum Likelihood Method

Suppose that a distribution P_θ has the density function $f(x,\theta)$, $\theta\in\Theta\subset\mathbf{R}^k$, with respect to a †$\sigma$-finite measure μ, and let $x_1,\dots,x_n$ be observed values of random samples $X_1,\dots,X_n$ from the population $f(x,\theta)$. Then the function L_n of θ defined by

$$L_n(\theta;x_1,\dots,x_n)=\prod_{i=1}^{n}f(x_i,\theta)$$

is called the **likelihood function**. If $\theta=\hat{\theta}_n(x_1,\dots,x_n)$ maximizes the value of $L_n(\theta)$ for fixed $x_1,\dots,x_n$ and if it is a measurable mapping from $(\mathscr{X}^n,\mathfrak{B}^n)$ to $(\Theta,\mathscr{C})$ with $\mathscr{C}$ a †σ-algebra of subsets of Θ, then $\hat{\theta}_n(X)=\hat{\theta}_n(X_1,\dots,X_n)$ is called the **maximum likelihood (ML) estimator of θ**. This method of finding estimators is called the **maximum likelihood method**. If the parameter is transformed into a new parameter $\eta=h(\theta)$ by means of a known one-to-one bimeasurable transformation h and if there exists a unique ML estimator $\hat{\theta}_n$ of θ, then $\hat{\eta}_n(X)=h(\hat{\theta}_n(X))$ is a unique ML estimator of η. In other words, the ML estimator is invariant for every one-to-one transformation.

Many statisticians have investigated and improved known adequate regularity conditions under which the ML estimator exists and is a BAN estimator.

Theorem (Wald). Assume that
(C1) Θ is a closed subset of $\mathbf{R}^k$ with nonempty interior Θ^0;
(C2) for any $x\in\mathscr{X}$, $f(x,\theta)$ is continuous with respect to θ and $\lim_{|\theta|\to\infty} f(x,\theta)=0$ if Θ is not bounded;
(C3) if $\theta_1\neq\theta_2$, then $P_{\theta_1}\neq P_{\theta_2}$ and $\int|f(x,\theta_1)-f(x,\theta_2)\,d\mu(x)>0$;
(C4) $E_{\theta_0}(|\log f(X,\theta_0)|)<\infty$; and
(C5) $E_{\theta_0}(\log^+ f(X,\theta,\rho))<\infty$ and $E_{\theta_0}(\log^+\varphi(X,r))<\infty$, where $f(x,\theta,\rho)=\sup\{f(x,\theta');|\theta'-\theta|\leqslant\rho\}$ and $\varphi(x,r)=\sup\{f(x,\theta);|\theta|>r\}$. (The last two functions are measurable according to assumption (C2).)
Then if a sequence of measurable functions, $\{\bar{\theta}_n(x_1,\ldots,x_n)\}$, satisfies

$$\liminf_{n\to\infty}\left\{\frac{L_n(\bar{\theta}_n;x_1,\ldots,x_n)}{L_n(\theta_0;x_1,\ldots,x_n)}\right\}\geqslant C>0\qquad(\text{a.s. }P_{\theta_0}),$$

then as $n\to\infty$, $\bar{\theta}_n(X_1,\ldots,X_n)$ converges a.s. to the true value θ_0 of the parameter. Hence if the ML estimator exists, it is a.s. consistent. Pfanzagl (*Metrika*, 14 (1969)) and Fu and Gleser (*Ann. Inst. Statist. Math.*, 27 (1975)) gave rigorous proofs for the existence of the ML estimator.

Theorem. Under assumptions (C1) and (C2), there exists a maximum likelihood estimator $\hat{\theta}_n$ for any positive integer n. That is, $\hat{\theta}_n=\hat{\theta}_n(x_1,\ldots,x_n)$ is a measurable function from $(\mathscr{X}^n,\mathfrak{B}^n)$ to $(\Theta,\mathscr{C})$ and satisfies $L(\hat{\theta}_n;x_1,\ldots,x_n)=\sup_\theta L(\theta;x_1,\ldots,x_n)$.

In the remainder of this section we suppose that assumptions (C1)–(C5) are satisfied. We use the notation

$$\frac{\partial f}{\partial\theta}=\left(\frac{\partial}{\partial\theta_i}f(x,\theta)\right),\quad\text{a }k\text{-column vector,}$$

$$\frac{\partial^2 f}{\partial\theta^2}=\left(\frac{\partial^2}{\partial\theta_i\partial\theta_j}f(x,\theta)\right),\quad\text{a }k\times k\text{ matrix,}$$

$$\frac{\partial^3\log f}{\partial\theta^3}=\left(\frac{\partial^3\log f(x,\theta)}{\partial\theta_h\partial\theta_i\partial\theta_j}\right),\qquad h,i,j=1,\ldots,k.$$

Theorem (Cramér). Assume that
(AN1) for a.s. $[\mu]x$, $f(x,\theta)$ is three-times continuously differentiable with respect to each component of $\theta=(\theta_1,\ldots,\theta_k)'\in\Theta^0$;
(AN2) for $\theta\in\Theta^0$, $\displaystyle\int_{\mathscr{X}}\frac{\partial f(x,\theta)}{\partial\theta}d\mu=0$ and $\displaystyle\int_{\mathscr{X}}\frac{\partial^2 f(x,\theta)}{\partial\theta^2}d\mu=0$;
(AN3) $E_\theta\left(\left|\frac{\partial}{\partial\theta}\log f(x,\theta)\right|^3\right)<\infty$ for $\theta\in\Theta^0$;
(AN4) the Fisher information matrix $I(\theta)=E_\theta\left[\left(\frac{\partial}{\partial\theta}\log f(x,\theta)\right)\left(\frac{\partial}{\partial\theta}\log f(x,\theta)\right)'\right]$ exists and is positive definite for $\theta\in\Theta^0$; and
(AN5) there exists an $H(x)$ such that $\left|\frac{\partial^3}{\partial\theta^3}\log f(x,\theta)\right|<H(x)$ and $E_\theta(H(X))<C$, a constant for $\theta\in\Theta^0$.
Then the maximum likelihood estimator is a BAN estimator: $\mathscr{L}[n^{1/2}(\hat{\theta}_n-\theta)\,|\,\theta]\to N(0,I(\theta)^{-1})$ as $n\to\infty$ for $\theta\in\Theta^0$. Note that under assumption (AN1) the likelihood function attains the maximum in Θ^0 with probability tending to 1 as $n\to\infty$ if the true value of the parameter exists in Θ^0. Hence the ML estimator is determined as a root $\theta=\hat{\theta}_n(x_1,\ldots,x_n)$ of the **likelihood equation** with the same probability as above:

$$\frac{\partial}{\partial\theta}\log L_n(\theta)=\left(\sum_{i=1}^n\frac{\partial}{\partial\theta_j}\log f(x_i,\theta)\right)_{j=1,\ldots,k}=0.$$

We also call $n^{-1}\partial\log L_n(\theta)/\partial\theta$ the **likelihood estimating function**. The essential fact used in the proof of the above theorem is the asymptotic equivalence of the ML estimator and the likelihood estimating function:

$$\Delta_n(\theta)-I(\theta)n^{1/2}(\hat{\theta}_n-\theta)\to 0\ \text{ in }\ P_\theta\ \text{ as }\ n\to\infty,$$

where $\Delta_n(\theta)=n^{-1/2}\partial\log L_n(\theta)/\partial\theta$. Note the fact that

$$\mathscr{L}[\Delta_n(\theta)\,|\,\theta]\to N_k(0,I(\theta))\ \text{ as }\ n\to\infty$$

holds according to the central limit theorem (→ 250 Limit Theorems in Probability Theory B (1)).

Contiguity. We now turn to the situation where we need asymptotic distributions under the alternative distribution $P_{\theta+n^{-1/2}h}$ with $\theta+n^{-1/2}h\in\Theta$ in estimation as we do in testing hypotheses (→ 400 Statistical Hypothesis Testing). The notion of contiguity, due to LeCam (1960), is basic for the asymptotic methods of estimation theory. We consider sequences $\{P_n\}$ and $\{P_n'\}$ of probability measures on $(\mathscr{X}^n,\mathfrak{B}^n)$ with the †Radon-Nikodym derivatives p_n and p_n' with respect to a σ-finite measure, such as P_n+P_n'. Denote by $\chi_n=\Lambda[P_n';P_n]$ a generalized log-likelihood ratio that is defined by $\log p_n'/p_n$ on the set $\{p_np_n'>0\}$ and is an arbitrary measurable function on the set $\{p_np_n'=0\}$. Let $\{B_n\}$ be any sequence of $\{\mathfrak{B}^n\}$-measurable sets, and let $\{T_n\}$ be any sequence of $\{\mathfrak{B}^n\}$-measurable functions. A sequence of distributions $\{\mathscr{L}[T_n|P_n]\}$ is said to be **relatively compact** if every subsequence $\{n'\}\subset\{n\}$ contains a further subsequence $\{m\}\subset\{n'\}$ along which it converges to a probability distribution. In the Euclidean space relative compactness is equivalent to **tightness**: that is, for every $\varepsilon>0$ there is a $b(\varepsilon)$ such that $P_n\{|T_n|>b(\varepsilon)\}<\varepsilon$ for every n.

Theorem. The following statements are all equivalent.
(1) For any $\{T_n\}$, $T_n \to 0$ in P_n if and only if $T_n \to 0$ in P'_n.
(2) For any $\{T_n\}$, $\{\mathscr{L}[T_n | P_n]\}$ is relatively compact if and only if $\{\mathscr{L}[T_n | P'_n]\}$ is relatively compact.
(3) For any $\{B_n\}$, $P_n\{B_n\} \to 0$ if and only if $P'_n\{B_n\} \to 0$.
(4) Whatever the choice of χ_n, $\{\mathscr{L}[\chi_n | P_n]\}$ and $\{\mathscr{L}[\chi_n | P'_n]\}$ are relatively compact.
(5) Whatever the choice of χ_n, $\{\mathscr{L}[\chi_n | P_n]\}$ is relatively compact. Furthermore, if $\{m\} \subset \{n\}$ is a subsequence of $\{n\}$ such that $\mathscr{L}[\chi_m | P_m]$ converges to $\mathscr{L}[\chi]$, then $E\{e^{\chi}\}=1$.

Two sequences $\{P_n\}$ and $\{P'_n\}$ satisfying requirements (1)–(5) of the above theorem are said to be **contiguous**.

Theorem. Suppose that $\{P_n\}$ and $\{P'_n\}$ are contiguous. Let $\{m\} \subset \{n\}$ be a subsequence such that $\mathscr{L}[\chi_m, T_m | P_m]$ converges to a limit $\mathscr{L}[\chi, T]$. Then $\mathscr{L}[\chi_m, T_m | P'_m]$ converges to $e^{\chi}\mathscr{L}[\chi, T]$, where $v=e^{\chi}\mathscr{L}[\chi, T]$ is given by $v(A)=\int_A e^{\chi} d\mathscr{L}[\chi, T]$.

Now, set $P_n = P_\theta^n$ and $P'_n = P^n_{\theta+n^{-1/2}h}$ for each n. Under suitable regularity conditions, such as assumptions (C1)–(C5) and (AN1)–(AN5), it is easy to see that $\{P_\theta^n\}$ and $\{P^n_{\theta+n^{-1/2}h}\}$ are contiguous. At the same time, we can see that the asymptotic linearity of $\Lambda_n(\theta+n^{-1/2}h;\theta)=\Lambda[P^n_{\theta+n^{-1/2}h}; P^n_\theta]$ holds (say) in the vicinity of the true parameter as follows:

$$\Lambda_n(\theta+n^{-1/2}h;\theta)-h'\Delta_n(\theta)+\frac{1}{2}h'I(\theta)h \to 0 \quad \text{in} \quad P_\theta.$$

The asymptotic equivalence of the ML estimator and $\Delta_n(\theta)$, and the asymptotic linearity of the log-likelihood function and $\Delta_n(\theta)$, leads to the regularity (→ Section N) of the ML estimator.

Theorem. Under suitable regularity conditions as above, the ML estimator is regular:

$$\mathscr{L}[n^{1/2}(\hat{\theta}_n-\theta)-h \,|\, \theta+n^{-1/2}h] \to N_k(0, I(\theta)^{-1}),$$

for any $h \in \mathbf{R}^k$ with $\theta+n^{-1/2}h \in \Theta$.

N. Asymptotic Efficiency

In Section D we discussed lower bounds of variances of unbiased estimators for finite sample size and defined the efficiency of an unbiased estimator with variance $v_n(\theta)$ by $(v_n(\theta)nI(\theta))^{-1}$. In this section we first discuss the asymptotic efficiency of a CAN estimator in the same vein as in the case of finite sample size. Second, we see a specific approach to the large-sample theory of estimation. Throughout this section we assume (C1)–(C5) and (AN1)–(AN5) stated in Section M.

BAN Estimators. We suppose that the parameter space Θ is a subset of $\mathbf{R}^1$ in this paragraph. We restrict our attention to the class of CAN estimators $\{T_n\}$ of the real-valued parameter θ for which $\mathscr{L}[n^{1/2}(T_n-\theta)|\theta] \to N(0, v(\theta))$ as $n \to \infty$. Fisher's conjecture concerning the lower bound to asymptotic variance $v(\theta)$ of any CAN estimator is

$$v(\theta) \geqslant I(\theta)^{-1},$$

where $I(\theta)$ is the Fisher information on θ in a single observation. The **asymptotic efficiency** of a CAN estimator with asymptotic variance $v(\theta)$ is defined by $(v(\theta)I(\theta))^{-1}$. A CAN estimator with asymptotic variance $I(\theta)^{-1}$ is called a **BAN estimator** or an **asymptotically efficient estimator**. Note that under suitable regularity conditions there always exists an asymptotic efficient estimator, for example a ML estimator, although for a sample of finite size there exists an efficient estimator if and only if the family of density functions is of the exponential type.

Unfortunately, Fisher's conjecture is not true without further conditions on the competing estimators. A counterexample was provided by Hodges and reported by LeCam (1953). Let $\{T_n\}$ be any CAN estimator with the asymptotic variance $v(\theta)$. Consider the estimator

$$T'_n = \begin{cases} \alpha T_n & \text{if } |T_n| < n^{-1/4}, \\ T_n & \text{if } |T_n| \geqslant n^{-1/4}, \end{cases}$$

where $0<\alpha<1$ is a constant. $\{T'_n\}$ is also a CAN estimator with asymptotic variance $v'(\theta)$ such that

$$v'(\theta) = \begin{cases} \alpha^2 v(0) & \text{if } \theta=0, \\ v(\theta) & \text{otherwise.} \end{cases}$$

Let $\{T_n\}$ be a BAN estimator; then T'_n is an estimator with asymptotic variance less than $I(\theta)^{-1}$ and is called a **superefficient estimator**.

Theorem (LeCam). The set of points θ for which the inequality due to Fisher fails is of Lebesgue measure zero. A condition due to Bahadur leads to the continuity of asymptotic variance which implies the validity of the above inequality.

Theorem (Bahadur). Suppose that $\{T_n\}$ is a CAN estimator with asymptotic variance $v(\theta)$ satisfying the condition

$$\liminf_{n\to\infty} P_{\theta_0+n^{-1/2}}\{T_n < \theta_0 + n^{-1/2}\} \leqslant \frac{1}{2}$$

or

$$\liminf_{n\to\infty} P_{\theta_0+n^{-1/2}}\{T_n > \theta_0 + n^{-1/2}\} \leqslant \frac{1}{2}.$$

Then the following inequality, due to Fisher, holds at $\theta=\theta_0$:

$$v(\theta_0)\geqslant I(\theta_0)^{-1}.$$

Regular Estimators. Wolfowitz and Kaufman considered an operationally more justifiable restriction on competing estimators, called the **uniformity property**, stating that, for an estimator $\{T_n\}$ of θ, $\mathscr{L}[n^{1/2}(T_n-\theta)|\theta](y)$ converges to any limit $L_\theta(y)$ uniformly in $(y,\theta)\in \mathbf{R}^k\times C$, where C is any compact subset of the interior Θ^0 of $\Theta\subset\mathbf{R}^k$. The ML estimator $\{\hat{\theta}_n\}$ also has this uniformity property under suitable regularity conditions, such as (C1)–(C5) and (AN1)–(AN5) to which some uniformity properties are added [29]. We note that asymptotic variance is not a good measurement of asymptotic efficiency unless an estimator is a CAN estimator, and that **asymptotic concentration** is in general a more pertinent measurement.

Theorem. For an estimator $\{T_n\}$ with uniformity property above, it holds that the limit $L_\theta(y)$ is a probability distribution function and continuous for either one of the variables y or θ if the other is fixed and furthermore that the probability measure L_θ is absolutely continuous with respect to the Lebesgue measure on $\mathbf{R}^k$.

Theorem. The asymptotic concentration of the ML estimator $\{\hat{\theta}_n\}$ about θ is not less than that of any estimator $\{T_n\}$ with uniformity property: For any convex and symmetric set $S\subset\mathbf{R}^k$ about the origin,

$$\lim_{n\to\infty} P_\theta\{n^{1/2}(\hat{\theta}_n-\theta)\in S\}\geqslant \lim_{n\to\infty} P_\theta\{n^{1/2}(T_n-\theta)\in S\}.$$

Schmetterer (*Research papers in statistics*, F. N. David (ed.), 1966) provided the notion of the continuous convergence of distributions of estimators of θ which is weaker than uniform convergence. An estimator $\{T_n\}$ is said to be **regular** if

$$\mathscr{L}[n^{1/2}(T_n-\theta)-h\,|\,\theta+n^{-1/2}h]\to L_\theta \quad\text{as}\quad n\to\infty,$$

where L_θ is a probability distribution independent of h with $\theta+n^{-1/2}h\in\Theta$. It was shown that the ML estimator $\{\hat{\theta}_n\}$ is regular under ordinary regularity conditions. Hájek obtained the following characterization of the asymptotic distribution of any regular estimator and, independently Inagaki (*Ann. Inst. Statist. Math.*, 22 (1970); 25 (1973)) obtained a similar result.

Theorem. The asymptotic distribution L_θ of any regular estimator $\{T_n\}$ is represented as the †convolution of a normal distribution N_θ and some residual distribution G_θ:

$$L_\theta=N_\theta * G_\theta,$$

where $N_\theta=N_k(0,I(\theta)^{-1})$, the asymptotic distribution of the ML estimator $\{\hat{\theta}_n\}$ in the ordinary regular case. It follows from the characterization $L_\theta=N_\theta * G_\theta$, that the first two theorems in this section hold also for regular estimators. (→ LeCam, *Proc. 6th Berkeley Symp.*, 1972, and Roussas and Soms, *Ann. Inst. Statist. Math.*, 25 (1973).)

O. Higher-Order Asymptotic Efficiency

In Section N it was shown that the ML estimator is a BAN estimator. In general, however, there exist many BAN estimators. For example, consider the case of a †multinomial distribution where probabilities of events are parametrized. That is, let $X=(n_1,\ldots,n_m)'$, $n=n_1+\ldots+n_m$, be distributed as a multinomial distribution $M(n;\pi_1,\ldots,\pi_m)$, $\pi_1+\ldots+\pi_m=1$, and let Π_Θ be a subset of m-vectors, $\Pi_\Theta=\{\pi(\theta)=(\pi_1(\theta),\ldots,\pi_m(\theta))';\theta\in\Theta\}$, $\Theta\subset\mathbf{R}^1$. Define $\hat{p}=(\hat{p}_1,\ldots,\hat{p}_m)'=(n_1/n,\ldots,n_m/n)'$. Then we consider (i) the ML estimator, (ii) the minimum-chi-square estimator, (iii) the minimum modified chi-square estimator, (iv) the minimum Haldane discrepancy (D_k) estimator, (v) the minimum Hellinger distance (HD) estimator, and (vi) the minimum Kullback-Leibler (K-L) information estimator. These are defined as the values of the parameter θ that minimize the following quantities, respectively: (i) $\mathrm{ML}=-\log L_n=-n\sum_{i=1}^m \hat{p}_i\log\pi_i(\theta)$; (ii) $\chi^2=\sum_{i=1}^m(n\hat{p}_i-n\pi_i(\theta))^2/(n\pi_i(\theta))$; (iii) $\mathrm{mod}\,\chi^2=\sum_{i=1}^m(n\hat{p}_i-n\pi_i(\theta))^2/(n\hat{p}_i)$; (iv) $D_k=\sum_{i=1}^m\pi_i(\theta)^{k+1}/\hat{p}_i^k$; (v) $\mathrm{HD}=\cos^{-1}\sum_{i=1}^m(\hat{p}_i\pi_i(\theta))^{1/2}$; and (vi) $\text{K-L}=\sum_{i=1}^m\pi_i(\theta)\log(\pi_i(\theta)/\hat{p}_i)$. Rao [32] showed that under suitable regularity conditions these estimators are Fisher consistent and BAN estimators.

Fisher-Rao Approach to Second-Order Asymptotic Efficiency. For $\theta\in\Theta\subset\mathbf{R}^1$, let $p_{\theta n}$ be the density for n i.i.d. observations $x=(x_1,\ldots,x_n)$, and let $q_{\theta n}$ be the density of estimator T_n. The †Fisher information contained in X and in T_n are defined by $nI(\theta)=E(d\log p_{\theta n}/d\theta)^2$ and $nI_{T_n}(\theta)=E(d\log q_{\theta n}/d\theta)^2$, respectively. Rao defined the **first-order (asymptotic) efficient estimator** T_n satisfying one of the following conditions: (1) $n^{1/2}|d\log p_{\theta n}/d\theta-d\log q_{\theta n}/d\theta|\to 0$ in probability; (2) $I(\theta)-I_{T_n}(\theta)\to 0$ as $n\to\infty$; (3) the asymptotic correlation between $n^{1/2}(T_n-\theta)$ and $n^{1/2}d\log p_{\theta n}/d\theta$ is unity; (4) $|n^{1/2}d\log p_{\theta n}/d\theta-\alpha-\beta n^{1/2}(T_n-\theta)|\to 0$ in probability. We note that the larger the condition number (j), the stronger the condition. A first-order efficient estimator is a BAN estimator. Fisher proposed

$$E_2'=\lim_{n\to\infty}(nI(\theta)-nI_{T_n}(\theta))=\lim_{n\to\infty}V_\theta(d\log p_{\theta n}/d\theta-d\log q_{\theta n}/d\theta)$$

as a measure of **second-order (asymptotic) efficiency** to distinguish different BAN estimators. Fisher stated without any sort of proof that the maximum likelihood estimator minimizes E_2', i.e., maximizes second-order efficiency. Rao proposed

$$E_2=\min_{\lambda} V_\theta(d\log p_{\theta n}/d\theta-\alpha n^{1/2}-\beta n(T_n-\theta)-\lambda n(T_n-\theta)^2)$$

as a measure of second-order efficiency for first-order efficient estimators $\{T_n\}$ satisfying condition (4). He showed that the estimators mentioned above for multinomial distribution are first-order efficient estimators satisfying condition (4) and furthermore calculated second-order efficiencies of these estimators measured in terms of E_2: (i) $E_2(\mathrm{ML})=\gamma^2(\theta)I(\theta)$; (ii) $E_2(\chi^2)=\Delta(\theta)+E_2(\mathrm{ML})$; (iii) $E_2(\mathrm{mod}\,\chi^2)=4\Delta(\theta)+E_2(\mathrm{ML})$; (iv) $E_2(D_k)=(k+1)^2\Delta(\theta)+E_2(\mathrm{ML})$; (v) $E_2(\mathrm{HD})=\Delta(\theta)/4+E_2(\mathrm{ML})$; (vi) $E_2(\mathrm{K\text{-}L})=\Delta(\theta)+E_2(\mathrm{ML})$ with

$$\gamma^2(\theta)=(\mu_{02}-2\mu_{21}+\mu_{40})/I^2(\theta)-1-(\mu_{11}^2+\mu_{30}^2-2\mu_{11}\mu_{30})/I^3(\theta),$$

$$\Delta(\theta)=\sum_{i=1}^{m}(\pi_i'(\theta)/\pi_i(\theta))^2/2-\mu_{40}/I(\theta)+\mu_{30}^2/(2I^2(\theta)),$$

where

$$\mu_{rs}=\mu_{rs}(\theta)=E_\theta\{((df(X,\theta)/d\theta)/f(X,\theta))^r\times((d^2f(X,\theta)/d\theta^2)/f(X,\theta))^s\}$$
$$=\sum_{i=1}^{m}\pi_i(\theta)(\pi_i'(\theta)/\pi_i(\theta))^r(\pi_i''(\theta)/\pi_i(\theta))^s.$$

Rao [33] gave another definition of second-order efficiency based on the expansion of the variance of T_n after correcting for bias: $V_\theta(T_n)=(nI(\theta))^{-1}+\psi(\theta)n^{-2}+o(n^{-2})$. The quantity $\psi(\theta)$ is considered to be a measure of second-order efficiency. The results of Fisher and Rao were confined to multinomial distributions. Efron (*Ann. Statist.*, 3 (1975)) and Ghosh and Subramanyan (*Saṇkhyā*, sec. A, 36 (1974)) extended the results to the so-called curved exponential family of distributions. Efron gave a geometric interpretation to the effect that second-order efficiency is related to the curvature of a statistical problem corresponding to $\gamma(\theta)$ above, and S. Amari recently extended this differential-geometric approach to the discussion of higher-order efficiency of estimators. Rao suggested that E_2' is equal to E_2. Ghosh and Subramanyan gave a sufficient condition for the equality to hold, whereas Efron provided a counterexample to show that $E_2'\neq E_2$ in general.

Pfanzagl and Takeuchi-Akahira Approaches to Higher-Order Asymptotic Efficiency. For each $k=1,2,\ldots$, a †$\{c_n\}$-consistent estimator $\{T_n\}$ is said to be the ***k*th-order asymptotically median unbiased (AMU) estimator** if for any $\theta\in\Theta$ $(\subset\mathbf{R}^1)$ there exists a positive number δ such that

$$\lim_{n\to\infty}\sup_{|\tau-\theta|<\delta}c_n^{k-1}\left|P_\tau\{T_n\leqslant\tau\}-\frac{1}{2}\right|=0$$

or

$$\lim_{n\to\infty}\sup_{|\tau-\theta|<\delta}c_n^{k-1}\left|P_\tau\{T_n\geqslant\tau\}-\frac{1}{2}\right|=0.$$

This notion, which is an extension of the condition due to Bahadur for the asymptotic efficiency, was introduced by Takeuchi and Akahira. For a kth-order AMU estimator $\{T_n\}$, $G_0(t,\theta)+c_n^{-1}G_1(t,\theta)+\ldots+c_n^{-k+1}G_{k-1}(t,\theta)$ is called the ***k*th-order asymptotic distribution** of $c_n(T_n-\theta)$ if

$$\lim_{n\to\infty}c_n^{k-1}|P_\theta\{c_n(T_n-\theta)\leqslant t\}-G_0(t,\theta)-c_n^{-1}G_1(t,\theta)-\ldots-c_n^{-k+1}G_{k+1}(t,\theta)|=0.$$

Pfanzagl and Takeuchi and Akahira obtained the concrete form of the second- or third-order asymptotic distribution of the ML estimator. A kth-order AMU estimator is said to be ***k*th-order asymptotic efficient** if the kth-order asymptotic distribution of it attains uniformly the bound for the kth-order asymptotic distributions of the kth-order AMU estimators. Takeuchi and Akahira showed that under suitable regularity conditions, T_n is second-order asymptotic efficient if

$$P_\theta\{(nI(\theta))^{1/2}(T_n-\theta)\leqslant t\}=\Phi(t)+n^{-1/2}(3\mu_{11}(\theta)+2\mu_{30}(\theta))/(6I(\theta)^{3/2})\times t^2\varphi(t)+o(n^{-1/2}),$$

where $\Phi(t)$ is the standard normal distribution function and $\varphi(t)$ is its density function, and further that the modified ML estimator

$$\hat\theta_n^*=\hat\theta_n+n^{-1}\mu_{30}(\hat\theta_n)/(6I(\hat\theta_n)^2)$$

for the ML estimator $\hat\theta_n$ is second-order asymptotic efficient. Pfanzagl (*Ann. Statist.*, 1 (1973)) obtained a similar result. The formulation due to Takeuchi and Akahira is more extensive since it can be applied to the so-called non-regular cases.

P. Estimating Equations

We often determine an estimator as a solution $\theta=T_n(x_1,\ldots,x_n)$ of an equation $\Psi_n(x_1,\ldots,x_n;\theta)=0$; for example, the ML estimator as a solution of the likelihood equation. In such case, such an equation is called an **estimating equation** and $\Psi_n(\theta)=\Psi_n(X_1,\ldots,X_n;\theta)$, a random function, is called an **estimating function** [3].

Call T_n an **estimator based on an estimating function** $\Psi_n(\theta)$. The following result is a modification of a theorem due to Hodges and Lehmann (*Ann. Math. Statist.*, 34 (1963)).

Theorem. Let Θ be an interval of $\mathbf{R}^1$. Suppose that a real-valued estimating function $\Psi_n(\theta)$ satisfies the following three conditions:
(M1) $\Psi_n(\theta)$ is a nonincreasing function of the real parameter θ;
(M2) for any real number h, $n^{1/2}\Psi_n(\theta_0+n^{-1/2}h)-n^{1/2}\Psi_n(\theta_0)+\gamma h\rightarrow 0$ in probability, where γ is a positive constant; and
(M3) $\mathscr{L}[n^{1/2}\Psi_n(\theta_0)](y)\rightarrow\Phi(y/\sigma)$, where Φ is a continuous distribution function with zero mean and unit variance.
Define an estimator based on Ψ_n by $T_n=(\theta_n^*+\theta_n^{**})/2$, where $\theta_n^*=\sup\{\theta\mid\Psi_n(\theta)>0\}$ and $\theta_n^{**}=\inf\{\theta\mid\Psi_n(\theta)<0\}$, Then we have $\mathscr{L}[n^{1/2}(T_n-\theta_0)](y)\rightarrow\Phi(y\gamma/\sigma)$ as $n\rightarrow\infty$. Huber considered a formulation that guarantees the asymptotic normality of an ***M*-estimator**. An M-estimator T_n is defined by a minimum problem of the form $\sum_{i=1}^n\rho(X_i,T_n)=\min_\theta\sum_{i=1}^n\rho(X_i,\theta)$ or by an implicit equation $\sum_{i=1}^n\psi(X_i,T_n)=0$, where ρ is an arbitrary function and $\psi(x,\theta)=\partial\rho(x,\theta)/\partial\theta$. Note that $\rho(x,\theta)=-\log f(x,\theta)$ gives the ordinary ML estimator. Let Θ be a closed subset of $\mathbf{R}^k$, let $(\mathscr{X},\mathfrak{B},P)$ be a probability space, and let $\psi(x,\theta)$ be some function on $\mathscr{X}\times\Theta$ with value in $\mathbf{R}^k$. Assume that X_1, $X_2,\ldots$ are independent random variables with values in $\mathscr{X}$ having the common probability distribution P that need not be a member of the parametric family. Consider an estimating function

$$\Psi_n(\theta)=\frac{1}{n}\sum_{i=1}^n\psi(X_i,\theta).$$

Assume that
(N1) for each fixed $\theta\in\Theta$, $\psi(x,\theta)$ is $\mathfrak{B}$-measurable and $\psi(x,\theta)$ is †separable;
(N2) the expected value $\lambda(\theta)=E\{\psi(X,\theta)\}$ exists for all $\theta\in\Theta$, and has a unique zero at $\theta=\theta_0\in\Theta^0$;
(N3) there exists a continuous function that is bounded away from zero, $b(\theta)\geqslant b_0>0$, such that $E(\sup_\theta\{|\psi(X,\theta)|/b(\theta)\})<\infty$, $\liminf_{|\theta|\rightarrow\infty}\{|\lambda(\theta)|/b(\theta)\}\geqslant 1$, and $E(\limsup_{|\theta|\rightarrow\infty}\{|\psi(X_i,\theta)-\lambda(\theta)|/b(\theta)\})<1$;
(N4) for $u(x,\theta,d)=\sup\{|\psi(x,\tau)-\psi(x,\theta)|\mid|\tau-\theta|\leqslant d\}$,
(i) as $d\rightarrow 0$, $E(u(X,\theta,d))\rightarrow 0$,
(ii) there exist positive numbers d_0, b_1, and b_2 such that $E(u(X,\theta,d))\leqslant b_1 d$ and $E(u(X,\theta,d)^2)\leqslant b_2 d$ for θ and d satisfying $0<|\theta-\theta_0|+d\leqslant d_0$;
(N5) in some neighborhood of θ_0, $\lambda(\theta)$ is continuously differentiable and the Jacobian matrix at $\theta=\theta_0$, $\Lambda=(\partial\lambda_i(\theta_0)/\partial\theta_j)$, is nonsingular; and
(N6) the covariance matrix $\Sigma=E\{\psi(X,\theta_0)\cdot\psi(X,\theta_0)'\}$ exists and is positive definite.

Theorem. Suppose that an estimator $\{T_n\}$ satisfies $\Psi_n(T_n)\rightarrow 0$ a.s. (or in probability) as $n\rightarrow\infty$. Then, under (N1)–(N4) (i), $T_n\rightarrow\theta_0$ a.s. (or in probability) as $n\rightarrow\infty$.

Lemma. Under (N1)–(N5),

$$\sup\{|\Psi_n(\tau)-\Psi_n(\theta_0)-\lambda(\tau)|/(n^{-1/2}+|\lambda(\tau)|);\quad |\tau-\theta_0|\leqslant d_0\}\rightarrow 0$$

in probability as $n\rightarrow\infty$.

Suppose that $\{\mathscr{L}[n^{1/2}\Psi_n(T_n)]\}$ is †tight. It implies that $\Psi_n(T_n)\rightarrow 0$ in probability, and hence from the above theorem $T_n\rightarrow\theta_0$ in probability. Thus letting $\tau=T_n$ we have

$$|\Psi_n(T_n)-\Psi_n(\theta_0)-\lambda(T_n)|/(n^{-1/2}+|\lambda(T_n)|)\rightarrow 0$$

in probability. That is, for any $\varepsilon>0$ there exists an n_0 such that for $n\geqslant n_0$, $P\{n^{1/2}|\lambda(T_n)|<\varepsilon+|n^{1/2}\Psi_n(T_n)-n^{1/2}\Psi_n(\theta_0)|\}>1-\varepsilon$. This and the tightness of $\{\mathscr{L}[n^{1/2}\Psi_n(T_n)]\}$ and $\{\mathscr{L}[n^{1/2}\Psi_n(\theta_0)]\}$ lead to the tightness of $\{\mathscr{L}[n^{1/2}(T_n-\theta_0)]\}$, and the converse is also true. At the same time we have $n^{1/2}\Psi_n(T_n)-n^{1/2}\Psi_n(\theta_0)-\Lambda n^{1/2}(T_n-\theta_0)\rightarrow 0$ in probability. The following theorem is a straightforward consequence.

Theorem. Suppose that an estimator $\hat{\theta}_n$ satisfies $n^{1/2}\Psi_n(\hat{\theta}_n)\rightarrow 0$ as $n\rightarrow\infty$. Then, under (N1)–(N6), $\mathscr{L}[n^{1/2}(\hat{\theta}_n-\theta_0)\mid P]\rightarrow N_k(0,\Lambda^{-1}\Sigma\Lambda'^{-1})$ as $n\rightarrow\infty$.

Q. Interval Estimation

Interval estimation or **region estimation** is a method of statistical inference utilized to estimate the true value $g(\theta)$ of the given parametric function by stating that $g(\theta)$ belongs to a subset $S(x)$ of $\mathscr{A}$, based on the observed value x of the random variable X. If

$$P_\theta\{g(\theta)\in S(X)\}\geqslant 1-\alpha\qquad\text{for any }\theta\in\Theta$$

for a constant α $(0<\alpha<1)$, then the random region $S(X)$ is called a **confidence region** of $g(\theta)$ of **confidence level** $1-\alpha$, and the infimum of the left-hand side with respect to $\theta\in\Theta$ is called the **confidence coefficient**. In particular, if $\Theta\subset\mathbf{R}$ and a confidence region is an interval, as is often the case, then the region is called a **confidence interval**, and two boundaries of the interval are called **confidence limits**. If a particular subset $S(X)$ among the set of confidence regions of $g(\theta)$ of confidence level $1-\alpha$ minimizes $P_\theta\{g(\theta')\in S(X)\}$ for all pairs θ and θ' $(\neq\theta)$, $S(X)$ is said to be **uniformly most powerful**. If a confidence region $S(X)$ of $g(\theta)$ of confidence level $1-\alpha$ satisfies $P_\theta\{g(\theta')\in S(X)\}\leqslant 1-\alpha$ for all pairs θ and θ' $(\neq\theta)$, then it is said to be **unbiased**. The notion of **invariance of a confidence region** can be defined similarly, and the definition for $S(X)$ being **uniformly most powerful unbiased** (UMPU) or **uniformly most**

powerful invariant (**UMPI**) can be formulated in an obvious manner.

For each $\theta_0 \in \Theta$ let $A(\theta_0)$ be an †acceptance region of a †test of level α of the †hypothesis $\theta = \theta_0$. For each $x \in \mathscr{X}$ let $S(x) = \{\theta \mid x \in A(\theta), \theta \in \Theta\}$. Then $S(X)$ is a confidence region of θ of confidence level $1-\alpha$. If $A(\theta_0)$ is an acceptance region of a UMP test of the hypothesis $\theta = \theta_0$ for each θ_0, then $S(X)$ is a UMP confidence region of θ of confidence level $1-\alpha$. Furthermore, corresponding to an acceptance region $A(\theta_0)$ of a UMP unbiased (invariant) test, a UMP unbiased (invariant) confidence region can be constructed in a similar manner.

R. Tolerance Regions

Let X and Y be distributed according to probability distributions P_θ^X and P_θ^Y labeled by a common $\theta \in \Theta$ over measurable spaces $(\mathscr{X}, \mathfrak{B})$ and $(\mathscr{Y}, \mathfrak{C})$, respectively, and consider the problem of predicting a future value of the random variable Y using the observed value x of the random variable X. If a mapping $S(x)$ sending a point x to a set belonging to $\mathfrak{C}$ is used to predict that the value of Y will lie in the set $S(x)$, then the random region is called a **tolerance region** of Y. In particular, if a tolerance region of a real random variable is an interval, it is called a **tolerance interval** and its boundaries **tolerance limits**.

For simplicity, suppose that X and Y are independent. There are several kinds of tolerance regions. First, if $P_\theta^X(P_\theta^Y\{Y \in S(X)\} \geqslant \beta) \geqslant \gamma$ for any $\theta \in \Theta$, then $S(X)$ is called a tolerance region of Y of **content** β and **level** γ. Second, if $E_\theta^X(P_\theta^Y\{Y \in S(X)\}) \geqslant \beta$ for any $\theta \in \Theta$, then $S(X)$ is called a tolerance region of Y of **mean content** β. Suppose that the random variable $X = (X_1, \ldots, X_n)'$ is a †random sample and that both X_1 and Y obey the same distribution. If further the set $\{P_\theta^Y \mid \theta \in \Theta\}$ forms the totality of 1-dimensional †continuous distributions and the distribution of $P_\theta^Y\{Y \in S(X)\}$ does not depend on the choice of θ, then $S(X)$ is called a distribution-free tolerance region. For example, if $X_{(1)} \leqslant X_{(2)} \leqslant \ldots \leqslant X_{(n)}$ is an †order statistic, then the interval $[X_{(r)}, X_{(s)}]$ (for $r < s$) is a distribution-free tolerance interval for a random variable Y, independent of $X_1, \ldots, X_n$, which has the same distribution as X_1.

References

[1] H. Cramér, Mathematical methods of statistics, Princeton Univ. Press, 1946.
[2] C. R. Rao, Linear statistical inference and its applications, Wiley, second edition, 1973.
[3] S. S. Wilks, Mathematical statistics, Wiley, 1962.
[4] S. Zacks, The theory of statistical inference, Wiley, 1971.
[5] A. Birnbaum, On the foundation of statistical inference, J. Amer. Statist. Assoc., 57 (1962), 269–326.
[6] E. L. Lehmann and H. Scheffé, Completeness, similar regions and unbiased estimation I, II, Sankhyā, sec. A, 10 (1950), 305–340; 15 (1955), 219–236.
[7] H. Morimoto, N. Ikeda, and Y. Washio, Unbiased estimates based on sufficient statistics, Bull. Math. Statist., 6 (1956).
[8] E. W. Barankin, Locally best unbiased estimates, Ann. Math. Statist., 20 (1949), 477–501.
[9] A. Bhattacharya, On some analogues to the amount of information and their uses in statistical estimation, Sankhyā, sec. A, 8 (1946), 1–14.
[10] K. Ishii, Inequalities of the types of Chebyshev and Cramer-Rao and mathematical programming, Ann. Inst. Statist. Math., 16 (1964), 277–293.
[11] C. R. Rao, Information and the accuracy attainable in the estimation of statistical parameters, Bull. Calcutta Math. Soc., 37 (1945), 81–91.
[12] M. A. Girshick and L. G. Savage, Bayes and minimax estimates for quadratic loss function, Proc. 2nd Berkeley Symp. Math. Prob., 1 (1951), 53–74.
[13] S. Karlin, Admissibility for estimation with quadratic loss, Ann. Math. Statist., 29 (1958), 406–436.
[14] E. J. Pitman, The estimation of the location and scale parameters of a continuous population of any given form, Biometrika, 30 (1939), 391–421.
[15] C. Stein, Inadmissibility of the usual estimator for the mean of a multivariate normal distribution, Proc. 3rd Berkeley Symp. Math. Prob., 1 (1956), 197–206.
[16] C. Stein, The admissibility of the Pitman's estimator for a single location parameter, Ann. Math. Statist., 30 (1959), 970–979.
[17] J. L. Hodges and E. L. Lehmann, Some problems in minimax estimation, Ann. Math. Statist., 21 (1950), 182–197.
[18] H. Kudo, On minimax invariant estimates of the transformation parameter, Nat. Sci. Rep. Ochanomizu Univ., 6 (1955), 31–73.
[19] J. Wolfowitz, The efficiency of sequential estimates and Wald's equation for sequential processes, Ann. Math. Statist., 18 (1947), 215–230.
[20] R. A. Fisher, On the mathematical foundations of theoretical statistics, Philos. Trans. Roy. Soc. London, 222 (1922), 309–368.
[21] R. A. Fisher, Theory of statistical estimation, Proc. Cambridge Philos. Soc., 22 (1925), 700–725.

[22] R. von Mises, On the asymptotic distributions of differentiable statistical functions, Ann. Math. Statist., 18 (1947), 309–348.
[23] L. LeCam, On some asymptotic properties of maximum likelihood estimates and related Bayes estimates, Univ. California Publ. Statist., 1 (1953), 277–330.
[24] L. LeCam, Locally asymptotically normal families of distributions, Univ. California Publ. Statist., 3 (1960), 37–98.
[25] G. G. Roussas, Contiguity of probability measures, Cambridge Univ. Press, 1972.
[26] A. Wald, Note on the consistency of the maximum likelihood estimate, Ann. Math. Statist., 20 (1949), 595–601.
[27] R. R. Bahadur, On Fisher's bound for asymptotic variances, Ann. Math. Statist., 35 (1964), 1545–1552.
[28] J. Hájek, A characterization of limiting distributions of regular estimates, Z. Wahrsheinlichkeitstheorie und Verw. Gebiete, 14 (1970), 323–330.
[29] S. Kaufman, Asymptotic efficiency of the maximum likelihood estimator, Ann. Inst. Statist. Math., 18 (1966), 155–178.
[30] J. Wolfowitz, Asymptotic efficiency of the maximum likelihood estimator, Theory Prob. Appl., 10 (1965), 247–260.
[31] M. Akahira and K. Takeuchi, Asymptotic efficiency of statistical estimators: Concepts and higher order asymptotic efficiency, Lecture notes in statist. 7, Springer, 1981.
[32] C. R. Rao, Asymptotic efficiency and limiting information, Proc. 4th Berkeley Symp. Math. Prob., 1 (1961), 531–546.
[33] C. R. Rao, Criteria of estimation in large samples, Sankhyā, sec. A, 25 (1963), 189–206.
[34] P. J. Huber, The behavior of maximum likelihood estimates under nonstandard conditions, Proc. 5th Berkeley Symp. Math. Statist. Prob., 1 (1967), 221–233.

400 (XVIII.8) Statistical Hypothesis Testing

A. General Remarks

A **statistical hypothesis** is a proposition about the †probability distribution of a †sample X. If it is known that the †distribution of X belongs to a family of distributions $\mathscr{P}=\{P_\theta \mid \theta\in\Omega\}$ with a parameter space Ω, the hypothesis can be stated as follows: The value of the parameter θ belongs to ω_H, where ω_H is a nonempty subset of the parameter space Ω. This hypothesis is also written simply as $\mathbf{H}:\theta\in\omega_H$. When ω_H consists of one point, it is called a **simple hypothesis**, otherwise a **composite hypothesis**.

Let $\mathscr{X}$ be a †sample space associated with a †σ-algebra $\mathfrak{B}$ of subsets of $\mathscr{X}$. To test a hypothesis $\mathbf{H}$ is to decide whether $\mathbf{H}$ is false, based on the observation of a sample $X\,(\in\mathscr{X})$. The assertion that $\mathbf{H}$ is not false does not necessarily imply the validity of $\mathbf{H}$. Such an assertion is called the **acceptance** of $\mathbf{H}$, while the opposite assertion, that $\mathbf{H}$ is false, is called the **rejection** of $\mathbf{H}$. In this framework for the testing problem, $\mathbf{H}$ is often called a **null hypothesis** (→ 401 Statistical Inference).

Consider a testing procedure in which $\mathbf{H}$ is rejected with probability $\varphi(x)$ $(0\leqslant\varphi(x)\leqslant 1)$ and accepted with probability $1-\varphi(x)$, when $x\in\mathscr{X}$ is observed. This testing procedure is characterized by the function φ on $\mathscr{X}$ with range in $[0,1]$. Here $\varphi(x)$ is taken as $\mathfrak{B}$-measurable on $\mathscr{X}$, and is called a **test function** or **test**. If $\varphi(x)$ is the indicator function $\chi_B(x)$ of a set $B\,(\in\mathfrak{B})$, then the test is rejecting $\mathbf{H}$ when x belongs to B and accepting $\mathbf{H}$ otherwise. The set B is called a **critical region**, and its complementary set an **acceptance region**. A test is called a **nonrandomized test** if it is the indicator function of a set. Other tests are called **randomized tests**.

Suppose that the †distribution of the sample X is a probability measure P_θ on $(\mathscr{X},\mathfrak{B})$. The probability of rejecting $\mathbf{H}$ when θ is the true value of the parameter is calculated from

$$E_\theta(\varphi)=\int_{\mathscr{X}}\varphi(x)P_\theta(dx).$$

Let α be a given constant in $(0,1)$. If a test $\varphi(x)$ satisfies $E_\theta(\varphi)\leqslant\alpha$ for all $\theta\in\omega_H$, or, in other words, if the probability of rejecting $\mathbf{H}$ when $\mathbf{H}$ is true is not greater than α, α is called the **level** of φ and such a test is called a **level** α **test**. We denote the set of all level α tests by $\Phi(\alpha)$, and $\sup_{\theta\in\omega_H}E_\theta(\varphi)$ is called the **size** of φ. To judge the merit of tests, we introduce a different hypothesis $\mathbf{A}$: The true value of θ belongs to $\omega_A\subset\Omega-\omega_H$. This is called an **alternative hypothesis**, or, for simplicity, an **alternative**. The errors of a test are divided into two kinds: errors owing to the rejection of the hypothesis $\mathbf{H}$ when it is true, and errors owing to the acceptance of $\mathbf{H}$ when it is false. The former are called **errors of the first kind**, and the latter, **errors of the second kind**. The probability $E_\theta(\varphi)$ of rejecting $\mathbf{H}$ when $\theta\in\omega_A$, that is, the probability of the correct decision being made for $\theta\in\omega_A$, is called the **power** of a test or the **power function**. The probability of committing an error of the second kind is $1-E_\theta(\varphi)$ for $\theta\in\omega_A$. A testing problem is indicated by the notation $(\mathscr{X},\mathfrak{B},\mathscr{P},\omega_H,\omega_A)$. A test φ in a class $\mathbf{\Phi}(\alpha)$ of tests is said to be **uniformly most powerful** in $\mathbf{\Phi}(\alpha)$ (or **UMP** in $\mathbf{\Phi}(\alpha)$) if for any

$\psi \in \Phi(\alpha)$, $E_\theta(\varphi) \geqslant E_\theta(\psi)$ for all $\theta \in \omega_A$. When ω_A consist of a single point, it is said to be **most powerful**.

B. The Neyman-Pearson Fundamental Lemma

Let μ be a †σ-finite measure over $(\mathscr{X}, \mathfrak{B})$ and $f_1, \ldots, f_{n+1}$ be μ-integrable real-valued functions. If $c_1, \ldots, c_n$ are constants such that the set $\Phi(c_1, \ldots, c_n)$ of test functions φ satisfying

$$\int \varphi f_i \, d\mu \leqslant c_i, \qquad i = 1, 2, \ldots, n,$$

is not empty, then there exists at least one test φ_0 in $\Phi(c_1, \ldots, c_n)$ that maximizes $\int \varphi f_{n+1} \, d\mu$ among all φ in $\Phi(c_1, \ldots, c_n)$. A test $\tilde{\varphi}$ is one of these tests if it satisfies the following two conditions:
(1) For appropriate constants $k_1, \ldots, k_n \geqslant 0$,

$$\tilde{\varphi}(x) = \begin{cases} 1 & \text{when } f_{n+1}(x) > \sum_{i=1}^{n} k_i f_i(x), \\ 0 & \text{when } f_{n+1}(x) < \sum_{i=1}^{n} k_i f_i(x) \end{cases}$$

almost everywhere with respect to μ, and
(2) the equation

$$\int \tilde{\varphi} f_i \, d\mu = c_i, \qquad i = 1, 2, \ldots, n,$$

holds.

If $(c_1, \ldots, c_n)$ is an interior point of the subset

$$\mathbf{M} = \left\{ \left(\int \varphi f_1 \, d\mu, \ldots, \int \varphi f_n \, d\mu \right) \middle| \varphi \text{ is a test} \right\}$$

of the n-space $\mathbf{R}^n$ and $\tilde{\varphi}$ satisfies (2) and maximizes $\int \varphi f_{n+1} \, d\mu$ among all φ in $\Phi(c_1, \ldots, c_n)$, then $\tilde{\varphi}$ satisfies (1). These statements are called the **Neyman-Pearson lemma**.

As an illustration, suppose that $\Omega = (1, 2, \ldots, n, n+1)$ and that $\{P_\theta \mid \theta \in \Omega)$ is dominated by a σ-finite measure μ. Let $f_i(x)$ be the density of P_i with respect to μ. When ω_H is a finite set $\{1, \ldots, n\}$ and ω_A consists of a single point $n+1$, then $\tilde{\varphi}$ satisfying (1) and (2) with $c_1 = \ldots = c_n = \alpha$ is a uniformly most powerful level α test.

If $\mathfrak{B}$ is generated by a countable number of sets, then there exists a most powerful level α test for any hypothesis against a simple alternative $\mathbf{A}: \theta = \tilde{\theta}$. A method of obtaining such a most powerful test is given in the **Lehman-Stein theorem**: Denote by f_θ the density function of P_θ with respect to a σ-finite measure μ, and define

$$h_\lambda(x) = \int_{\omega_H} f_\theta(x) \, d\lambda(\theta)$$

for a probability measure λ on ω_H. Consider testing the simple hypotheses $\mathbf{H}_\lambda$: The density of distribution of the sample is h_λ, against the alternative $\mathbf{A}: \theta = \tilde{\theta}$, and let φ_λ be a most powerful level α test for this problem $(\mathbf{H}_\lambda : \mathbf{A})$. If $\sup_{\theta \in \omega_H} E_\theta(\varphi_\lambda) \leqslant \alpha$, φ_λ is a most powerful level α test for testing $\mathbf{H}: \theta \in \omega_H$ against $\mathbf{A}: \theta = \tilde{\theta}$. The measure λ satisfies $E_\theta(\varphi_{\lambda'}) \geqslant E_\theta(\varphi_\lambda)$ for any probability measure λ' on ω_H and is called a **least favorable distribution**.

When the alternative hypothesis ω_A consists of more than one point, a uniformly most powerful test does not generally exist. However, if $\Omega = \mathbf{R}$, $\omega_H = (-\infty, \theta_0]$, $\omega_A = (\theta_0, \infty)$, and $f_\theta(x)$ is a density function with †monotone likelihood ratio with respect to a statistic $T(x)$, then a UMP level α test $\varphi(x)$ exists and is defined by $\varphi(x) = 1$ if $T(x) > c$; $=$ a constant if $T(x) = c$; and $= 0$ if $T(x) < c$. For a one-parameter exponential family of distributions, there exists a UMP level α test for testing $\mathbf{H}: \omega_H = (-\infty, \theta_1] \cup [\theta_2, \infty)$ against $\mathbf{A}: \omega_A = (\theta_1, \theta_2)$ $(\theta_1 < \theta_2)$. However, a UMP test does not exist for the problem obtained by interchanging the positions of ω_H and ω_A.

Since hypothesis-testing problems admitting UMP tests are rather rare, alternative ways for judging the merit of tests are needed, and two have been devised. The first is to restrict the class Φ of tests and to find a UMP test in this restricted class. The second is to introduce an alternative criterion of optimality and to select a test accordingly. The first is discussed in detail in Sections C, D, and E, and the second in Section F.

C. Unbiased Tests

The unbiasedness criterion is based on the idea that the probability of rejecting the hypothesis $\mathbf{H}$ when it is true (the probability of an error of the first kind) should preferably be no larger than that of rejecting $\mathbf{H}$ when it is false (the power). If a level α test φ satisfies $E_\theta(\varphi) \geqslant \alpha$ for $\theta \in \omega_A$, then φ is called an **unbiased level** α **test**. Let $\Phi_u(\alpha)$ be the set of all unbiased level α tests. A UMP test in $\Phi_u(\alpha)$ is called a **uniformly most powerful** (or **UMP**) **unbiased level** α **test**.

If P_θ is of the †exponential family whose parameter space Ω is a finite or an infinite open interval of $\mathbf{R}^k$, then there exists a UMP unbiased level α test for the following two problems: (1) $\mathbf{H}: \theta_1 \leqslant a$, $\mathbf{A}: \theta_1 > a$, where θ_1 is the first coordinate of $\theta = (\theta_1, \ldots, \theta_k)$; and (2) $\mathbf{H}: \theta_1 = a$, $\mathbf{A}: \theta_1 \neq a$. For example, when the sample is normally distributed with unknown mean μ and unknown variance σ^2, the Student test (defined in Section G) for a hypothesis $\mathbf{H}: \mu = \mu_0$ against an alternative $\mathbf{A}: \mu \neq \mu_0$ is a UMP unbiased test.

D. Similar Tests and Neyman Structure

If $E_\theta(\varphi)$ is constant for all $\theta \in \omega$ $(\subset \Omega)$, φ is called a **similar test** with respect to ω. If $E_\theta(\varphi)$ is a continuous function of θ, an unbiased test φ is similar with respect to the common boundary $\tilde{\omega}$ of ω_H and ω_A, provided that Ω is a topological space and the density function is continuous in ω. Therefore, in this case, unbiased tests are found in the class of all tests similar with respect to $\tilde{\omega}$. Let a statistic $y = T(x)$ be †sufficient for $\mathscr{P}_\omega = \{P_\theta \mid \theta \in \omega\}$. A test φ is said to have **Neyman structure** with respect to T if the conditional expectation $E(\varphi \mid T(x) = y)$ of φ equals a constant $[\mathscr{P}_\omega]$. (For the notation $[\mathscr{P}_\omega]$ → 396 Statistic.) A test φ having Neyman structure with respect to the statistic $T(x)$ is similar with respect to ω. A test φ similar with respect to ω has Neyman structure with respect to $T(x)$ if and only if the family $\mathscr{Q} = \{Q_\theta = P_\theta T^{-1} \mid \theta \in \omega\}$ of †marginal distributions of T is †boundedly complete.

E. Invariant Tests

Consider groups G and $\bar{G}$ of one-to-one transformations on $\mathscr{X}$ and Ω, respectively. Suppose that each element of G is a measurable transformation of $\mathscr{X}$ onto itself (i.e., $gB \in \mathfrak{B}$ for any $B \in \mathfrak{B}$) and that a homomorphism $g \to \bar{g}$ of G into $\bar{G}$ is defined so that $P_\theta(g^{-1}B) = P_{\bar{g}\theta}(B)$. The hypothesis $\mathbf{H}: \theta \in \omega_H$ and the alternative $\mathbf{A}$: $\theta \in \omega_A$ are said to be **invariant** under G if $\bar{g}\omega_H = \omega_H$ and $\bar{g}\omega_A = \omega_A$ for all $g \in G$, and in this case the testing problem $(\mathscr{X}, \mathfrak{B}, \mathscr{P}, \omega_H, \omega_A)$ is said to be **invariant** under G. A test is called an **invariant test** if $\varphi(gx) = \varphi(x)$ for every $g \in G$, and $E_{\bar{g}\theta}(\varphi) = E_\theta(\varphi)$ holds for any invariant φ. Accordingly, if $\bar{G}$ is †transitive on ω_H, and invariant test is similar with respect to ω_H. If the sample space $\mathscr{X}$ is a subset of $\mathbf{R}^n$ that is invariant under the translation $(x_1, \ldots, x_n) \to (x_1 + a, \ldots, x_n + a)$ with a real a and if there exists a $\theta' = \bar{a} \cdot \theta \in \Omega$ such that $P_{\theta'}(B) = P_\theta(\{(x_1 - a, \ldots, x_n - a) \mid (x_1, \ldots, x_n) \in B\})$ for any $\theta \in \Omega$, then $\bar{a}$ is a transformation on Ω. In this case the real number a is called a **location parameter**. Furthermore, if the sample space $\mathscr{X}$ is a subset invariant under the similarity transformation $(x_1, \ldots, x_n) \to (ax_1, \ldots, ax_n)$ $(a > 0)$ and if there exists a $\theta' = \bar{a} \cdot \theta \in \Omega$ such that $P_{\theta'}(B) = P_\theta(\{(x_1/a, \ldots, x_n/a) \mid (x_1, \ldots, x_n) \in B\})$ for any $\theta \in \Omega$, then the real number a is called a **scale parameter**. The **invariance principle** states that a test for a testing problem invariant under G should preferably be invariant under G. A test $\varphi(x)$ is called an **almost invariant test** if $\varphi(gx) = \varphi(x)$ $[\mathscr{P}]$ for all $g \in G$.

Suppose that the testing problem of a hypothesis under consideration is invariant under a transformation group G on the sample space $\mathscr{X}$. Denote the set of all invariant level α tests by $\mathbf{\Phi}_I(\alpha)$. A test that is uniformly most powerful in $\mathbf{\Phi}_I(\alpha)$ is called a **uniformly most powerful** (in short, **UMP**) **invariant level** α **test**. If there exists a unique UMP unbiased level α test φ^*, then a UMP invariant level α test (if it exists) coincides with $\varphi^*[\mathscr{P}]$. When $T(x)$ is †maximal invariant under G, a necessary and sufficient condition for $\varphi(x)$ to be invariant is that φ be a function of $T(x)$.

For example, suppose that the sample $X = (X_1, \ldots, X_n)$ is taken from $N(\mu, \sigma^2)$ with unknown μ and σ^2. In this situation, $Y = (\bar{X}, S)$ is a sufficient statistic, where $\bar{X} = \sum_i X_i/n$ and $S = \sqrt{\sum (X_i - \bar{X})^2}$. Let G be the group of transformations $(\bar{x}, s) \to (c\bar{x}, cs)$ $(c > 0)$ on the range $\mathscr{Y}$ of Y and $\bar{G}$ be the group of transformations $(\mu, \sigma^2) \to (c\mu, c\sigma^2)$ $(c > 0)$ on the parameter space. Both the hypotheses $\mathbf{H}_1 : \mu/\sigma^2 \leqslant 0$ and $\mathbf{H}_2 : \mu/\sigma^2 = 0$ are invariant under G. Since $t = \sqrt{n} \cdot \bar{x}/(s/\sqrt{n-1})$ is maximal invariant, any invariant level α test is in the class of functions of t. The Student test, defined in Section G, is UMP invariant under G.

F. Minimax Tests and Most Stringent Tests

Minimax tests and most stringent tests are sometimes used as alternatives to UMP tests. Suppose that $\mathscr{P} = \{P_\theta \mid \theta \in \Omega\}$ is a †dominated family and $\mathfrak{B}$ is generated by a countable number of sets. A level α test φ^* is called a **minimax level** α **test** if for any level α test φ,

$$\inf_{\theta \in \omega_A} E_\theta(\varphi^*) \geqslant \inf_{\theta \in \omega_A} E_\theta(\varphi).$$

Such a test exists for any $\alpha \in (0, 1)$. If a group G of measurable transformations on $\mathscr{X}$ leaves a testing problem invariant, then an intimate relation exists between the minimax property and invariance. Concerning this relation, we have the following theorem: For each $\alpha \in (0, 1)$ there is an almost invariant level α test that is minimax if there exists a σ-field $\mathfrak{A}$ of subsets of G and a sequence $\{\nu_n\}$ of probability measures on $(G, \mathfrak{A})$ such that (i) $B \in \mathfrak{B}$ implies $\{(x, g) \mid gx \in B\} \in \mathfrak{B} \times \mathfrak{A}$; (ii) $A \in \mathfrak{A}$, $g \in G$ implies $Ag \in \mathscr{A}$; and (iii) $\lim_{n \to \infty} |\nu_n(Ag) - \nu_n(A)| = 0$ for any $A \in \mathfrak{A}$ and $g \in G$. Fundamental in the invariant testing problem is the **Hunt-Stein lemma**: Under the condition just stated, for any φ there exists an almost invariant test ψ such that

$$\inf_{\bar{g} \in \bar{G}} E_{\bar{g}\theta}(\varphi) \leqslant E_\theta(\psi) \leqslant \sup_{\bar{g} \in \bar{G}} E_{\bar{g}\theta}(\varphi).$$

The following six types of transformation groups satisfy the condition of the theorem: (1) the group of translations on $\mathbf{R}^n$, (2) the group of similarity transformations on $\mathbf{R}^n$, (3) the group of transformations $g=(a,b):(x_1,\dots,x_n)\in\mathbf{R}^n\rightarrow(ax_1+b,\dots,ax_n+b)\in\mathbf{R}^n$ $(0<a<\infty,\ -\infty<b<\infty)$, (4) finite groups, (5) the group of orthogonal transformations on $\mathbf{R}^n$, and (6) the direct product of a finite number of the groups mentioned in (1)–(5).

We call $\beta_\alpha^*(\theta)=\sup_{\varphi\in\Phi(\alpha)}E_\theta(\varphi)$ an **envelope power function**, and φ^* $(\in\Phi(\alpha))$ is called a **most stringent level α test** if

$$\sup_{\theta\in\omega_A}(\beta_\alpha^*(\theta)-E_\theta(\varphi^*))\leqslant\sup_{\theta\in\omega_A}(\beta_\alpha^*(\theta)-E_\theta(\varphi))$$

for any $\varphi\in\Phi(\alpha)$. There exists a most stringent level α test for each $\alpha\in(0,1)$. If a testing problem is invariant under a transformation group G on X and G satisfies the condition in the Hunt-Stein lemma, then a uniformly most powerful invariant level α test is most stringent among the level α tests (→ 398 Statistical Decision Functions).

Admissibility of a test and completeness of a class of tests are defined with respect to the probability of an error of the second kind (→ 398 Statistical Decision Functions). The uniformly most powerful level α test and the uniformly most powerful unbiased level α test are admissible.

G. Useful Tests Concerning Normal Distributions (→ Appendix A, Table 23)

In this section, we treat the rejection regions S that are commonly used in testing problems related to normal distributions. Let α be the level of S, and let $c(\alpha)$ and $d(\alpha)$ be positive numbers determined by α. In (1)–(5) below, the sample consists of n mutually independent random variables $X_1,\dots,X_n$ each of which is assumed to be normally distributed with mean μ and variance σ^2. For any sample point $x=(x_1,\dots,x_n)$, denote $\sum_{i=1}^n x_i/n$ by $\bar{x}$ and $\sum_{i=1}^n(x_i-\bar{x})^2$ by s^2. (1) To test the hypothesis $\mu\leqslant\mu_0$ against an alternative $\mu>\mu_0$, we can use as a critical region $S=\{x\,|\,t(x)>c(\alpha)\}$, where the test statistic $t(x)$ is given by $\sqrt{n}(\bar{x}-\mu_0)/\sqrt{s^2/(n-1)}$. (2) To test the hypothesis $\mu=\mu_0$ against an alternative $\mu\neq\mu_0$, we can use $S=\{x\,|\,|t(x)|\geqslant c(\alpha)\}$ with the same test statistic $t(x)$ as in (1). These tests based on the statistic $t(x)$ are generally called **Student tests** or ***t*-tests**. (3) To test the hypothesis $\sigma^2=\sigma_0^2$ against the alternative $\sigma^2>\sigma_0^2$ with $\sigma_0^2>0$, we can use $S=\{x\,|\,\chi^2(x)\geqslant c(\alpha)\}$, where $\chi^2=s^2/\sigma_0^2$. (4) To test the hypothesis $\sigma^2>\sigma_0^2$ against the alternative $\sigma^2<\sigma_0^2$, we can use $S=\{x\,|\,\chi^2(x)\leqslant c(\alpha)\}$, where χ^2 is the same as in (3). (5) To test the hypothesis $\sigma^2=\sigma_0^2$ against $\sigma^2\neq\sigma_0^2$, we can use $S=\{x\,|\,\chi^2(x)\leqslant c(\alpha)$ or $\geqslant d(\alpha)\}$, where χ^2 is the same as in (3). Each of these tests based on the statistic χ^2 is called a **chi-square test**. Among these tests, (3) is UMP and (1) is UMP when $\alpha\geqslant 1/2$. All tests (1)–(5) are UMP unbiased, and (3)–(5) are UMP invariant under the translations $(x_1,\dots,x_n)\rightarrow(x_1+a,\dots,x_n+a)$ $(-\infty<a<\infty)$. Since (1) and (2) are also UMP invariant under the transformations $(x_1,\dots,x_n)\rightarrow(ax_1,\dots,ax_n)$ $(0<a<\infty)$, they are most stringent tests.

Suppose that $X_1,\dots,X_m$ are independently distributed according to $N(\mu_1,\sigma_1^2)$ and that $Y_1,\dots,Y_n$ are independently distributed according to $N(\mu_2,\sigma_2^2)$, where μ_1, μ_2, σ_1, and σ_2 are assumed unknown unless otherwise stated. Here we give the important tests for μ_1, μ_2, σ_1^2, σ_2^2. Let $x=(x_1,\dots,x_m)$ and $y=(y_1,\dots,y_n)$ be sample points in $\mathbf{R}^m$ and $\mathbf{R}^n$, respectively, and denote $\sum_{i=1}^m x_i/m$, $\sum_{i=1}^n y_i/n$, $\sum_{i=1}^m(x_i-\bar{x})^2$, and $\sum_{i=1}^n(y_i-\bar{y})^2$ by $\bar{x}$, $\bar{y}$, s_x^2, and s_y^2, respectively. (6) Assume that σ_1 and σ_2 are known, and consider a hypothesis $\mu_1=\mu_2$. When an alternative $\mu_1>\mu_2$ $(\mu_1\neq\mu_2)$ is taken, we can use as a critical region $S=\{(x,y)\,|\,T(x,y)\geqslant c(\alpha)\}$ $(S=\{(x,y)\,|\,|T(x,y)|\geqslant c(\alpha)\})$, where $T(x,y)=(\bar{x}-\bar{y})/\sqrt{\sigma_1^2/m+\sigma_2^2/n}$. These tests are both UMP unbiased and invariant under the translations $(x_1,\dots,x_m,y_1,\dots,y_n)\rightarrow(x_1+a,\dots,x_m+a,y_1+a,\dots,y_n+a)$. (7) Assume that $\sigma_1=\sigma_2$, and consider a hypothesis $\mu_1=\mu_2$. When $\mu_1\neq\mu_2$ $(\mu_1>\mu_2)$ is the alternative, $S=\{(x,y)\,|\,|T(x,y)|\geqslant c(\alpha)\}$ $(S=\{(x,y)\,|\,T(x,y)>c(\alpha)\})$ can be used as a critical region, where $T(x,y)=(\bar{x}-\bar{y})\sqrt{m+n-2}/(\sqrt{1/m+1/n}\sqrt{s_x^2+s_y^2})$. Both tests are UMP unbiased and are invariant under $(x_1,\dots,x_m,y_1,\dots,y_n)\rightarrow(ax_1+b,\dots,ax_m+b,ay_1+b,\dots,ay_n+b)$ $(-\infty<b<\infty,0<a<\infty)$. (8) Testing the hypothesis $\mathbf{H}$: $\mu_1=\mu_2$ is called the **Behrens-Fisher problem**. Note that nothing is assumed about the relation of the variances σ_1^2 and σ_2^2 of the two samples X and Y, in contrast to (7). It is difficult to construct a statistic whose distribution is independent of σ_1^2 and σ_2^2 when $\mathbf{H}$ is true. Compare this with (1)–(7), where the proposed statistics have this property and are used to construct similar critical regions S. The critical region

$$\{(x,y)\,|\,(\bar{x}-\bar{y})/\sqrt{s_x^2/m(m-1)+s_y^2/n(n-1)}\geqslant f(s_y^2/s_x^2)\},$$

with an appropriately chosen f, is similar to such a region S. This test is called **Welch's test**. (9) For a hypothesis $\sigma_1=\sigma_2$, we can use as a critical region $S=\{(x,y)\,|\,F(x,y)\leqslant c(\alpha)\}$, $S=\{(x,y)\,|\,F(x,y)\geqslant c(\alpha)\}$, or $S=\{(x,y)\,|\,F(x,y)\geqslant d(\alpha)$ or $<c(\alpha)\}$, when $\sigma_1<\sigma_2$, $\sigma_2<\sigma_1$, or $\sigma_1\neq\sigma_2$, respectively, is taken as alternative, where $F(x,y)=(n-1)s_x^2/(m-1)s_y^2$. All these tests are UMP unbiased and are invariant under the transformations $(x_1,\dots,x_m,y_1,\dots,y_n)\rightarrow(ax_1+b,\dots,ax_m+b,ay_1+b,\dots,ay_n+b)$ $(-\infty<$

$b<\infty, 0<a<\infty$). A test based on $F(x, y)$ is called an F-**test**.

H. Linear Hypotheses

Let $X_1, \ldots, X_n$ be independent and distributed according to $N(\mu_i, \sigma^2)$ $(i=1,2,\ldots,n)$, where $\mu_1, \mu_2, \ldots, \mu_s$ $(s<n)$, σ are assumed to be unknown and $\mu_i=0$ $(s<i\leqslant n)$. The hypothesis is $\mathbf{H}: \mu_1=\mu_2=\ldots=\mu_r=0$ $(r\leqslant s)$, and the alternative hypothesis is that at least one μ_i, $1\leqslant i\leqslant r$, does not vanish. The critical region $S=\{(x_1,\ldots,x_n) | F(x_1,\ldots,x_r; x_{s+1},\ldots,x_n)=\sum_{i=1}^r x_i^2/\sum_{i=s+1}^n x_i^2 \geqslant c(\alpha)\}$ is a UMP unbiased test for this problem, and S is invariant under the group g_1 of translations

$$(x_1,\ldots,x_r,x_{r+1},\ldots,x_s,x_{s+1},\ldots,x_n)$$

$$\rightarrow(x_1,\ldots,x_r,x_{r+1}+a,\ldots,x_s+a,x_{s+1},\ldots,x_n),$$

the group g_2 of similarity transformations $(x_1,\ldots,x_n)\rightarrow(cx_1,\ldots,cx_n)$, the group $g_3=O(r)$ of orthogonal transformations in $\mathbf{R}^r=\{(x_1,\ldots,x_r)\}$, the group $g_4=O(n-s)$ of orthogonal transformations in $\mathbf{R}^{n-s}=\{(x_{s+1},\ldots,x_n)\}$, and finite products of elements of the groups g_1, g_2, g_3, and g_4. This test is also a kind of F-test. More generally, let us denote $\mathbf{X}=(X_1, X_2,\ldots,X_n)'$ and assume that it is expressed as

$$\mathbf{X}=A\xi+\mathbf{W}, \qquad \mathbf{W}=(W_1,W_2,\ldots,W_n)', \tag{1}$$

where $\xi=(\xi_1,\xi_2,\ldots,\xi_s)'$, $s\leqslant n$ is a vector of unknown parameters and A a matrix of known constants, and $W_1, W_2, \ldots, W_n$ are distributed independently according to the normal distribution with mean θ and variance σ^2. Then a **general linear hypothesis** is a hypothesis stating that the vector ξ lies within a linear subspace M of $\mathbf{R}^s$. The set of points $A\xi$ with ξ satisfying a linear hypothesis $\mathbf{H}$ is the linear subspace $L(B)$ of $L(A)$ spanned by the column vectors of an $n\times k_1$ matrix B. Assume, for example, that the dimension of $L(B)$ (=the rank of B) is $s-r$. Let C be an $n\times k_2$ matrix whose column vectors span the orthocomplement $L_A^{\perp}(B)$ of the space $L(B)$ with respect to the space $L(A)$. Then the model (1) can be written as

$$\mathbf{X}=B\boldsymbol{\eta}+C\zeta+\mathbf{W}, \qquad E(\mathbf{W})=\mathbf{0}, \tag{2}$$

with a k_1-vector $\boldsymbol{\eta}$ and a k_2-vector ζ, and hence the hypothesis $\mathbf{H}$ is represented by $\zeta=\mathbf{0}'$. We denote by $\mathbf{Y}=P_A\mathbf{X}$ the projection of $\mathbf{X}$ onto the space $L(A)$ and by $\mathbf{Z}$ the projection $P_B\mathbf{X}$ of $\mathbf{X}$ onto the space $L(B)$. The quantity $Q_H=\mathbf{X}'(P_A-P_B)\mathbf{X}$ equals the square of the length of the vector $\mathbf{Y}-\mathbf{Z}$ and represents the sum of squares of residuals for the hypothesis $\mathbf{H}$. The error mean square $\hat{\sigma}^2=\mathbf{X}'(I-P_A)\mathbf{X}/(n-s)=Q_e/(n-s)$ and also $\hat{\sigma}_H^2=Q_H/r$ under $\mathbf{H}$ are unbiased estimators of σ^2.

Assume that $B'C=0$. Let U be an orthogonal transformation in $\mathbf{R}^n$ such that the first, $\ldots$, rth, $(s+1)$st, $\ldots$, nth rows of UB and the $(s+1)$st, $\ldots$, nth rows of UC are all equal to zero vectors. Using the notation $\tilde{\mathbf{X}}=U\mathbf{X}$, $\tilde{\boldsymbol{\eta}}=UB\boldsymbol{\eta}$, $\tilde{\zeta}=UC\zeta$, and $\tilde{\mathbf{W}}=U\mathbf{W}$, we obtain the **canonical form** $\tilde{\mathbf{X}}=\tilde{\boldsymbol{\eta}}+\tilde{\zeta}+\tilde{\mathbf{W}}$ of the model (2). $\tilde{\mathbf{W}}$ is also a vector of independently and identically distributed normal random variables. The hypothesis $\mathbf{H}$ is expressed as $\tilde{\zeta}=\mathbf{0}$. In this model, we have $E(\tilde{X}_i)=0$ for $i=s+1,\ldots,n$, and moreover, $E(\tilde{X}_i)=0$ for $i=1,\ldots,r,s+1,\ldots,n$ if and only if $\mathbf{H}$ is true.

The least squares estimator $\mathbf{Y}$ of $A\xi$ is the †maximum likelihood estimator, and $\mathbf{X}-\mathbf{Y}$, $\mathbf{Y}-\mathbf{Z}$, $\mathbf{Z}$ are distributed independently according to the n-variate normal distributions $N(\mathbf{0},\sigma^2(I-P_A))$, $N((P_A-P_B)C\zeta, \sigma^2(P_A-P_B))$, and $N(B\boldsymbol{\eta}+P_BC\zeta, \sigma^2P_B)$, respectively. Hence Q/σ^2, Q_H/σ^2, and $\mathbf{X}'P_B\mathbf{X}/\sigma^2$ are distributed independently according to the †noncentral χ^2-distributions with $n-s$, r, and $s-r$ degrees of freedom and †noncentrality parameters 0, $\zeta'C'(P_A-P_B)C\zeta/\sigma^2$, and $(B\boldsymbol{\eta}+P_BC\zeta)'(B\boldsymbol{\eta}+P_BC\zeta)/\sigma^2$, respectively. The †likelihood ratio test of the hypothesis $\mathbf{H}$ has a critical region $\hat{\sigma}_H^2/\hat{\sigma}^2>c$, is a †uniformly most powerful invariant test with respect to the group of linear transformations leaving the hypothesis $\mathbf{H}$ invariant, and is the †most stringent test. This test is also uniformly most powerful among the tests whose †power function has a single variable $\zeta'C'(P_A-P_B)C\zeta/\sigma^2$. Furthermore, for $s-r=1$, this test is a †uniformly most powerful unbiased test. In the decomposition $\mathbf{X}'\mathbf{X}=\mathbf{X}'(P_A-P_B)\mathbf{X}+\mathbf{X}'P_B\mathbf{X}+\mathbf{X}'(I-P_A)\mathbf{X}=Q_H+Q_B+Q_e$ the terms Q_H and Q_e are called the sum of squares due to the hypothesis and due to the error, respectively. Such a process of decomposition is called the analysis of variance and its result is summarized in the analysis of variance table (Table 1).

I. The Likelihood Ratio Test

The likelihood ratio test is comparatively easy to construct. Let $L(x_1,\ldots,x_n;\theta)$ be the †likelihood function. Then

$$\Lambda(x_1,\ldots,x_n)=\frac{\sup_{\theta\in\omega_H}L(x_1,\ldots,x_n;\theta)}{\sup_{\theta\in\omega_H\cup\omega_A}L(x_1,\ldots,x_n;\theta)}$$

is called the **likelihood ratio**, and the test corresponding to the critical region $S=\{(x_1,\ldots,x_n)| \Lambda(x_1,\ldots,x_n)\leqslant c_\alpha\}$ is called the **likelihood ratio test**, where c_α is a positive constant determined by the level α. Let $\hat{\theta}_H(x_1,\ldots,x_n)$ and $\hat{\theta}_{H\vee A}(x_1,\ldots,x_n)$ be the †maximum likelihood estimators for θ in ω_H and in $\omega_H\cup\omega_A$, respectively; that is, $L(x;\hat{\theta}_H(x))=\sup_{\theta\in\omega_H}L(x;\theta)$ and $L(x;\hat{\theta}_{H\vee A}(x))=\sup_{\theta\in\omega_H\cup\omega_A}L(x;\theta)$. Then

Table 1

Factor	Sum of Squares	Degrees of Freedom	Mean Square	Ratio of Variances
H	$Q_H=\mathbf{X}'(P_A-P_B)\mathbf{X}$	r	$\hat{\sigma}_H^2=Q_H/r$	$\hat{\sigma}_H^2/\hat{\sigma}^2$
B	$Q_B=\mathbf{X}'P_B\mathbf{X}$	$s-r$	$\hat{\sigma}_B^2/(s-r)$	
Error	$Q_e=\mathbf{X}'(I-P_A)\mathbf{X}$	$n-s$	$\hat{\sigma}^2=Q_e/(n-s)$	
Total	$\mathbf{X}'\mathbf{X}$	n		

$$\Lambda(x)=\frac{L(x;\hat{\theta}_H)}{L(x;\hat{\theta}_{H\vee A})}.$$

The F-test for a linear hypothesis is a likelihood ratio test, and other examples of the likelihood ratio test are shown in Appendix A, Table 23. However, the likelihood ratio test does not necessarily have the desirable properties stated in the preceding sections.

J. Complete Classes

The set of critical regions of the type $\{x\,|\,T(x)>c\}$ for the problems $\mathbf{H}:\theta\leqslant\theta_0$ and $\mathbf{A}:\theta>\theta_0$, where the distribution family $\mathscr{P}$ of the statistic T is of †Pólya type 2 in the strict sense, and the set of regions of the type $\{x\,|\,c<T(x)<d\}$ for the problem $\mathbf{H}:\theta_1\leqslant\theta\leqslant\theta_2$ and $\mathbf{A}:\theta<\theta_1$ or $\theta_2<\theta$, where the distribution family $\mathscr{P}$ of the statistic T is of †Pólya type 3 in the strict sense, are examples of minimal complete classes [6]. It has been proved under a mild condition that the set of all tests with convex critical regions is essentially complete when the underlying distributions are of exponential type and the null hypothesis is simple.

Let $P_{\theta_i}\in\mathscr{P}$ $(i=0,1)$ and let $\mathfrak{B}_0$ be a σ-subalgebra of $\mathfrak{B}$. $\mathfrak{B}_0$ is sufficient for $\mathfrak{B}$ w.r.t. $\{P_{\theta_0},P_{\theta_1}\}$ if and only if the class of all $\mathfrak{B}_0$-measurable test functions is essentially complete, i.e. iff for every critical region $B\in\mathfrak{B}$ there exists a $\mathfrak{B}_0$-measurable test function φ_0 such that $E_{\theta_0}(\varphi_0)\leqslant E_{\theta_0}(\chi_B)$ and $E_{\theta_1}(\varphi_0)\geqslant E_{\theta_1}(\chi_B)$. Assume that $\mathscr{P}=\{P_\theta\,|\,\theta\in\Theta\}$ is dominated; if for every $\mathfrak{B}$-measurable test function φ there exists a $\mathfrak{B}_0$-measurable test function ψ such that $E_\theta(\psi)=E_\theta(\varphi)$ for all $\theta\in\Theta$, then $\mathfrak{B}_0$ is sufficient for $\mathfrak{B}$ w.r.t. $\mathscr{P}$ [9, 10] (→ 398 Statistical Decision Functions D, 399 Statistical Estimation E).

K. Asymptotic Theory

Let $(\mathscr{X}_v,\mathfrak{B}_v,P_{v\theta})$ $(v=1,2,\ldots,n)$ be a sequence of probability spaces, where the parameter space Ω is common to all v. Let $(\mathscr{X}^{(n)},\mathfrak{B}^{(n)},P_\theta^{(n)})$ be the direct product probability space of $(\mathscr{X}_v,\mathfrak{B}_v,P_{v\theta})$ for $v=1,2,\ldots,n$. For each sample space $(\mathscr{X}^{(n)},\mathfrak{B}^{(n)},P_\theta^{(n)})$, denote a test function for $\mathbf{H}:\theta\in\omega_H(\subset\Omega)$ and $\mathbf{A}:\theta\in\omega_A(\subset\Omega-\omega_H)$ by $\varphi_n(x_1,\ldots,x_n)$. A sequence $\{\varphi_n\}$ $(n=1,2,\ldots)$ is often called a test. For example, a likelihood ratio test is frequently understood as a sequence of tests $S_n=\{(x_1,\ldots,x_n)\,|\,\Lambda_n(x_1,\ldots,x_n)\leqslant\lambda_n\}$, where $\{\lambda_n\}$ is a sequence of constants and $\Lambda_n(x_1,\ldots,x_n)$ is the likelihood ratio defined by $(\mathscr{X}^{(n)},\mathfrak{B}^{(n)},P_\theta^{(n)})$ and ω_H. If a test $\{\varphi_n\}$ satisfies $E_\theta(\varphi_n)\to0$ $(\theta\in\omega_H)$ and $E_\theta(\varphi_n)\to1$ $(\theta\in\omega_A)$ as $n\to\infty$, $\{\varphi_n\}$ is said to be a **consistent test**. If these convergences are uniform with respect to θ, $\{\varphi_n\}$ is said to be a **uniformly consistent test**. When a uniformly consistent test exists, ω_H and ω_A are said to be **finitely distinguishable**. Suppose that the observed values are identically distributed (that is, $(\mathscr{X}_v,\mathfrak{B}_v,P_{v\theta})$ is a copy of a probability space $(\mathscr{X},\mathfrak{B},P_\theta)$) and ω_H and ω_A are both compact with respect to the metric $\rho(\theta,\theta')=\sup_{B\in\mathfrak{B}}|P_\theta(B)-P_{\theta'}(B)|$. In this case, ω_H and ω_A are finitely distinguishable if $E_\theta(\varphi)$ is a continuous function of θ for any φ [4]. Kakutani's theorem (→ 398 Statistical Decision Functions) is regarded as a proposition concerning distinguishability when the null hypothesis and the alternative are both simple.

The following result about the †limit distribution of a likelihood ratio is due to H. Chernoff [3]: Let $\mathscr{X}^{(n)}$ be an n-space and Ω be an open subset of $\mathbf{R}^k$ containing the origin 0. Suppose that the observed random variables are independent and distributed according to a density $f(x,\theta)$; that is, the likelihood function $L(x;\theta)$ is $\prod_{i=1}^n f(x_i,\theta)$. Moreover, assume the following regularity conditions:
(1) $\log f(x,\theta)$ is three-times differentiable with respect to θ at every point of the closure of some neighborhood N of $\theta=0$.
(2) There exist an integrable function F and a measurable function H such that (i) $|\partial f/\partial\theta_i|<F(x)$ for every $\theta\in N$; (ii) $|\partial^2 f/\partial\theta_i\partial\theta_j|<F(x)$ for every $\theta\in N$; (iii) $|\partial^3\log f/\partial\theta_i\partial\theta_j\partial\theta_m|<H(x)$; and (iv) $\sup_\theta E_\theta(H(x))<\infty$.
(3) For every $i,j=1,2,\ldots,k$, we have $J_\theta^{ij}=E_\theta[(\partial\log f/\partial\theta_i)(\partial\log f/\partial\theta_j)]<\infty$, and the matrix $J_\theta=(J_\theta^{ij})$ is positive definite for all $\theta\in N$.
Let $P(x_1,\ldots,x_n;\omega)=\sup_{\theta\in\omega}L(x_1,\ldots,x_n;\theta)$ for a subset ω of Ω. Consider testing a hypothesis $\theta\in\omega_H$ against an alternative $\theta\in\omega_A$, where 0 is an accumulation point of ω_H. If $\lambda^*(x_1,\ldots,x_n)=P(x_1,\ldots,x_n;\omega_H)/P(x_1,\ldots,x_n;\omega_A)$, λ^* plays essentially the same role as the likelihood ratio λ and hence can be used in its place. We call a subset C of Ω a cone if $\theta\in C$ implies $a\theta\in C$ for

any $a>0$. A subset ω of Ω is said to be approximated by C if ω satisfies $\inf_{x\in C}\|x-y\|=o(\|y\|)$ for all $y\in\omega$ and $\inf_{y\in\omega}\|x-y\|=o(\|x\|)$ for all $x\in C$ around the origin, where $\|x\|^2=\sum_{i=1}^n x_i^2$. Suppose that ω_H and ω_A are approximated by two cones C_H and C_A, respectively. Then, setting $z_n=\sqrt{n}\,J_0^{-1}A(x)$ with

$$A(x)=\left[\frac{1}{n}\sum_{\alpha=1}^{n}\frac{\partial\log f(x_\alpha,0)}{\partial\theta_1},\ldots,\frac{1}{n}\sum_{\alpha=1}^{n}\frac{\partial\log f(x_\alpha,0)}{\partial\theta_k}\right]',$$

the limit distribution of $-2\log\lambda^*$, when $\theta=0$, coincides with that of $\inf_{\theta\in C_H}(z_n-\theta)'J_0(z_n-\theta)-\inf_{\theta\in C_A}(z_n-\theta)'J_0(z_n-\theta)$. In particular, when $\Omega=\mathbf{R}^k$ and $\omega_H=\{(\theta_1^0,\ldots,\theta_s^0,\theta_{s+1},\ldots,\theta_k)\mid -\infty<\theta_i<\infty, i=s+1,\ldots,k\}$ and some regularity conditions are assumed, the limit distribution of $-2\log\lambda^*$ is the †χ^2-distribution with s degrees of freedom if the hypothesis is true.

The asymptotic behavior of the chi-square test of goodness of fit is also very important. Suppose that $(X_1,\ldots,X_k)$ has a multinomial distribution $n!(x_1!\ldots x_k!)^{-1}p_1^{x_1}\ldots p_k^{x_k}(\sum_{i=1}^k x_i=n, x_i\geqslant 0)$, and consider testing $\mathbf{H}:p_1=p_1^0,\ldots,p_k=p_k^0$. The **chi-square test of goodness of fit** has a critical region of the type $\{x\mid\chi^2(x_1,\ldots,x_k)\geqslant c\}$, where $\chi^2(x_1,\ldots,x_k)$ is $\sum_{i=1}^k((x_i-np_i^0)^2/np_i^0)$, i.e., the weighted sum of the squares of the differences between the value p_i^0 of p_i and the maximum likelihood estimator x_i/n of p_i. The limit distribution of $\chi^2(x_1,\ldots,x_k)$ when $p_i=p_i^0$, $i=1,2,\ldots,k$, is the chi-square distribution with $k-1$ degrees of freedom.

Suppose moreover that k functions $p_i(\theta)$ $(i=1,\ldots,k;s<k)$ of θ $(\in\mathbf{R}^s)$ are given and that the hypothesis to be tested is that the sample has been drawn from a population having a distribution determined by $\mathbf{H}:p_i=p_i(\theta)$ $(i=1,\ldots,k;s<k)$ for some value of θ. In this case the chi-square test of goodness of fit could be applied after replacing the parameter θ in $p_i=p_i(\theta)$ $(i=1,2,\ldots,n)$ by the solutions $\tilde\theta_n(x_1,\ldots,x_k)$ of the system of equations of the **modified minimum chi-square method**,

$$\sum_{i=1}^{k}\frac{x_i-np_i}{p_i}\frac{\partial p_i}{\partial\theta_j}=0$$

$(j=1,\ldots,s)$. Suppose that (1) $p_i(\theta)>c^2>0$ $(i=1,\ldots,k)$ and $\sum_{i=1}^k p_i(\theta)=1$; (2) $p_i(\theta)$ is twice continuously differentiable with respect to the coordinates of θ; and (3) the rank of the matrix $(\partial p_i/\partial\theta_j)$ is k. Then the system of equations above has a unique solution $\theta=\tilde\theta_n(x_1,\ldots,x_k)$, and $\tilde\theta_n$ converges in probability to θ_0 when $\theta=\theta_0$. The asymptotic distribution of $\chi^2(x)=\sum_{i=1}^k((x_i-np_i(\hat\theta_n))^2/np_i(\hat\theta_n))$ is the chi-square distribution with $n-s-1$ degrees of freedom [5]. For the test of goodness of fit, the empirical distribution function may also be used (→ 371 Robust and Nonparametric Methods).

A test of independence by contingency tables is one application of the chi-square test of goodness of fit. We suppose that n individuals are classified according to two categories A and B, where A has r ranks $A_1,A_2,\ldots,A_r$ and B has s ranks $B_1,B_2,\ldots,B_s$. Let $p_{i\cdot},p_{\cdot j},p_{ij}$ be the probabilities that the observed value of an individual belongs to A_i, B_j, $A_i\cap B_j$, respectively. Let $x_{i\cdot},x_{\cdot j},x_{ij}$ be the numbers of individuals belonging to A_i, B_j, and $A_i\cap B_j$, respectively. Table 2 is called a **contingency table**. To test the null hypothesis $\mathbf{H}$ that the divisions of A and B into their ranks are independent, that is, $\mathbf{H}:p_{ij}=p_{i\cdot}p_{\cdot j}$, the statistic $\chi^2=\sum_{i=1}^r\sum_{j=1}^s(x_{ij}-x_{i\cdot}x_{\cdot j}/n)^2/(x_{i\cdot}x_{\cdot j}/n)$ is applied. When $\mathbf{H}$ is true, χ^2 is asymptotically distributed according to the chi-square distribution with $(r-1)(s-1)$ degrees of freedom as $n\to\infty$.

Likelihood ratio tests and chi-square tests of goodness of fit are consistent tests under conditions stated in their respective descriptions. In general, there are many consistent tests for a problem. Therefore it is necessary to consider another criterion that has to be satisfied by the best test among consistent tests. Pitman's asymptotic relative efficiency is such a criterion. Other notions of efficiency have also been introduced.

A completely specified form of distribution is rather exceptional in applications. More often we encounter cases where distribution of the sample belongs in a large domain. Various tests independent of the functional form of distribution have been proposed, and the asymptotic theory plays an important role in those cases (→ 371 Robust and Nonparametric Methods).

The following concept of asymptotic efficiency is due to R. R. Bahadur [11]: Let $\{T_n\}$ be a sequence of real-valued statistics defined on $\mathscr{X}^{(n)}$. $\{T_n\}$ is said to be a standard sequence (for testing $\mathbf{H}$) if the following three conditions are satisfied.

(I) There exists a continuous probability distribution function F such that for each $\theta\in\omega_H$, $\lim_{n\to\infty}P_\theta^{(n)}\{T_n<t\}=F(t)$ for every $t\in\mathbf{R}^1$.

Table 2. Contingency Table

	B_1	B_2	$\cdots$	B_s	Total
A_1	x_{11}	x_{12}	$\cdots$	x_{1s}	$x_{1\cdot}$
A_2	x_{21}	x_{22}	$\cdots$	x_{2s}	$x_{2\cdot}$
$\vdots$	$\vdots$	$\vdots$		$\vdots$	$\vdots$
A_r	x_{r1}	x_{r2}	$\cdots$	x_{rs}	$x_{r\cdot}$
Total	$x_{\cdot 1}$	$x_{\cdot 2}$	$\cdots$	$x_{\cdot s}$	n

(II) There exists a constant a, $0<a<\infty$, such that $\log\{1-F(t)\}=-(at^2/2)\{1+o(1)\}$ as $t\to\infty$.

(III) There exists a function $b(\theta)$ on $\Omega-\omega_H$ with $0<b(\theta)<\infty$ such that for each $\theta\in\Omega-\omega_H$,

$$\lim_{n\to\infty} P_\theta^{(n)}\{|(T_n/n^{1/2})-b(\theta)|>t\}=0$$

for every $t>0$.

Suppose that $\{T_n\}$ is a standard sequence. Then T_n has the asymptotic distribution F if **H** is satisfied, but otherwise $T_n\to\infty$ in probability. Consequently, large values of T_n are significant when T_n is regarded as a test statistic for **H**. Accordingly, for any given $x\in\mathscr{X}^{(n)}$, $1-F(T_n(x))$ is called the critical level in terms of T_n, and is regarded as a random variable defined on $\mathscr{X}^{(n)}$ [1]. It is convenient to describe the behavior of this random variable as $n\to\infty$ in terms of K_n, where $K_n(x)=-2\log[1-F(T_n(x))]$. Then for each $\theta\in\omega_H$, K_n is asymptotically distributed as a chi-square variable χ_2^2 with 2 degrees of freedom and for $\theta\in\Omega-\omega_H$, $K_n/n\to ab^2(\theta)$ in probability as $n\to\infty$. The asymptotic slope of the test based on $\{T_n\}$ (or simply the slope of $\{T_n\}$) is defined to be $c(\theta)=ab^2(\theta)$. Note that the statistic $K_n^{1/2}$ is equivalent to T_n in the following technical sense: (i) $\{K_n^{1/2}\}$ is a standard sequence; (ii) for each $\theta\in\Omega$, the slope of $\{K_n^{1/2}\}$ equals that of $\{T_n\}$; and (iii) for any given n and x, the critical level in terms of $K_n^{1/2}$ equals the critical level in terms of T_n. Since the critical level of $K_n^{1/2}$ is found by substituting $K_n^{1/2}$ into the function representing the upper tail of a fixed distribution independent of F, $\{K_n^{1/2}\}$ is a normalized version of $\{T_n\}$. Suppose that $\{T_n^{(1)}\}$ and $\{T_n^{(2)}\}$ are two standard sequences defined on $\mathscr{X}^{(n)}$, and let $F^{(i)}(x)$, a_i, and $b_i(\theta)$ be the functions and constants prescribed by conditions (I)–(III) for $i=1, 2$. Consider an arbitrary but fixed θ in $\Omega-\omega_H$, and suppose that x is distributed according to P_θ. The asymptotic efficiency of $\{T_n^{(1)}\}$ relative to $\{T_n^{(2)}\}$ is defined to be $\varphi_{12}(\theta)=c_1(\theta)/c_2(\theta)$, where $c_i(\theta)=a_ib_i^2(\theta)$ is the slope of $\{T_n^{(i)}\}$, $i=1, 2$. The asymptotic efficiency is called **Bahadur efficiency**.

Several comparisons of standard sequences are given in [11]. The relationship between Bahadur efficiency and Pitman efficiency for hypothesis-testing problems has also been studied. Under suitable conditions the two efficiencies coincide.

L. Sequential Tests

Let $X_1, X_2, \ldots$ be a given sequence of random variables. To test a hypothesis concerning the distributions of these variables (sample sizes are not predetermined), we observe first X_1, then X_2, etc. At each stage a decision is made on the basis of the previously obtained data whether the observation should be stopped and a judgment made on the acceptability of the hypothesis. Such a test is called a **sequential test**. Let $X_1, X_2, \ldots$ be independent and identically distributed by $f_\theta(x)$. For testing a simple hypothesis $\mathbf{H}:\theta=0$ against a simple alternative $\mathbf{A}:\theta=1$, we have the **sequential probability ratio test**: Let $G_n(x_1, x_2, \ldots, x_n)=\prod_{i=1}^n f_1(x_i)/\prod_{i=1}^n f_0(x_i)$, and preassign two constants $a_0<a_1$. After the observations of $X_1, \ldots, X_n$ are performed, the next random variable X_{n+1} is observed if $a_0<G_n(x_1, \ldots, x_n)<a_1$. Otherwise the experiment is stopped, and we accept **H** when $G_n\leqslant a_0$ or accept **A** when $a_1\leqslant G_n$. The constants a_0 and a_1 are determined by the desired probabilities α_1 and α_2 of errors of the first and second kind, respectively. It is known that, among the class of sequential tests in which the probability of error of the first (second) kind is not greater than α_0 (α_1), the sequential probability ratio test minimizes the expected number of observations when either **H** or **A** is true (→ 398 Statistical Decision Functions; 404 Statistical Quality Control).

References

[1] E. L. Lehmann, Testing statistical hypotheses, Wiley, 1959.
[2] A. Wald, Sequential analysis, Wiley, 1947.
[3] H. Chernoff, On the distribution of the likelihood ratio, Ann. Math. Statist., 25 (1954), 573–578.
[4] W. Hoeffding and J. Wolfowitz, Distinguishability of sets of distributions, Ann. Math. Statist., 29 (1958), 700–718.
[5] M. G. Kendall and A. Stuart, The advanced theory of statistics II, Griffin, 1961.
[6] S. Karlin, Decision theory for Pólya type distributions, Case of two actions I, Proc. 3rd Berkeley Symp. Math. Stat. Prob. I, Univ. of California Press (1956), 115–128.
[7] A. Birnbaum, Characterizations of complete classes of tests of some multiparametric hypotheses, with applications to likelihood ratio tests, Ann. Math. Statist., 26 (1955), 21–36.
[8] H. Cramér, Mathematical methods of statistics, Princeton Univ. Press, 1946.
[9] K. Takeuchi and M. Akahira, Characterizations of prediction sufficiency (adequacy) in terms of risk functions, Ann. Statist., 3 (1975), 1018–1024.
[10] J. Pfanzagl, A characterization of sufficiency by power functions, Metrika, 21 (1974), 197–199.

[11] R. R. Bahadur, Stochastic comparison of tests, Ann. Math. Statist., 31 (1960), 276–295.

401 (XVIII.2) Statistical Inference

A. The Statistical Model

Broadly and loosely speaking, the term "statistical inference" may imply any procedure for drawing conclusions from statistical data. But now it is usually understood more rigorously to mean those procedures based upon a †probabilistic model of the data to obtain conclusions concerning the unknown parameters of the population that represents the probabilistic model by viewing the observed data as a †random sample extracted from the population.

As the simplest example, suppose that, for some system, we have a number of observations from repeated measurements or experiments under a supposedly uniform condition. If we can assume that there are no systematic trends or tendencies involved, we can suppose that the variations among repeated observations are due to random causes and assume that the observed values $X_1, X_2, \ldots$ are independently and identically distributed random variables. Our purpose in making observations is to draw some information from the data, that is, to make some judgment on an unknown system quantity θ, which together with some other quantity (quantities) η characterizing the measurement or the experiment, determines the distribution of the X_i. We assume that the distribution has a density function $f(x;\theta,\eta)$. This amounts to assuming that the observed values $X_1, X_2, \ldots$ are a sample randomly drawn from a hypothetical population of the results of the measurements or experiments supposedly continued indefinitely. Then the problem of statistical inference is one of making some judgment based on the random sample. The set of hypotheses postulating the distribution of the observed values is called the probabilistic model of the observations, and the problem of determining a model in a specific situation is called that of **specification**.

B. Bayesian and Non-Bayesian Approaches

There are two different ways to make inferences on the population parameters: the **Bayesian approach** and the **non-Bayesian approach**.

In the Bayesian approach it is assumed that we have some probability density $\pi(\theta,\eta)$ for the parameters θ and η. Then, given the observations $X_1=x_1, X_2=x_2, \ldots$, the conditional probability density for θ and η is given by

$$\tilde{P}(\theta,\eta\,|\,x_1,x_2,\ldots)=\frac{\prod f(x_i;\theta,\eta)\pi(\theta,\eta)}{\iint\prod f(x_i;\theta,\eta)\pi(\theta,\eta)\,d\theta\,d\eta}.$$

π is called the **prior density** and $\tilde{p}$ the **posterior density** for the parameters. Then all the information obtained from the sample is considered to be contained in the **posterior distribution** with the density $\tilde{p}(\theta,\eta)$, and conclusions on the parameters can be drawn from it.

The prior density $\pi(\theta,\eta)$ does not necessarily represent a frequency function of a population of which the parameters are a random sample, but in most cases treated by the Bayesian approach it is considered to be a summary of the statistician's judgment over relative possibilities of the different values of the parameters based on all the information obtained before the observations are made. Bayesians claim that it is always possible to determine such a **prior distribution** in a coherent way, specifying the **subjective probability**, representing a person's judgment under uncertainty, as opposed to the **objective probability**, representing the relative frequencies in a population. L. J. Savage [7] succeeded in developing a formal mathematical theory of the subjective probability from a set of postulates about the consistent behavior of a person under uncertainty.

The non-Bayesian statisticians, however, do not accept the Bayesians' viewpoint and insist that statistical inference should be free from any subjective judgment and be based solely on the objective properties of the sample derived from the assumed model. The theory developed by R. A. Fisher, J. Neyman, and E. S. Pearson, and others is based on the non-Bayesian approach.

C. Problems of Non-Bayesian Inference

The most commonly used forms of statistical inference are **point estimation**, used when we want to get a value as the estimate for the parameter; **interval estimation**, when we want to get an interval that contains the true value of the parameter with a probability not smaller than the preassigned level; and **hypothesis testing**, when it is required to determine whether or not some hypothesis about the parameter values is wrong (→ 399 Statistical Estimation, 400 Statistical Hypothesis Testing).

In any type of statistical inference, the problem can be abstractly formulated by determining a procedure that defines a rule, based on the sample observed, for choosing an element from the set of possible conclusions.

Such a procedure is evaluated by the probabilistic properties derived under different values of the parameter from the distribution of the sample, and it is usually required to satisfy some type of validity criteria (such as unbiasedness of an estimator, size of a test, etc.), and among those satisfying them, one which is considered to be best according to some optimality criterion (such as minimum variance or most-powerfulness) is looked for. But in the sense of objective probability, the probabilistic property of a procedure is relevant only for the frequencies in repeated trials when the same procedure is applied to a sequence of samples obtained from the population and has no direct implications for the conclusion obtained by applying the procedure to a specific sample we have in hand. For this reason Neyman argued that in statistical inference there is really no such thing as inductive inference but only inductive behavior. Fisher disagreed strongly with this argument and emphasized that statistical analysis is induction and that its purpose is to allow us to draw the proper conclusions from a particular sample and that the probabilistic properties of the procedure should and could have relevance for a particular conclusion obtained from a specific sample, provided that all the information contained in the sample is used. The arguments between Fisher and Neyman led to a heated controversy between their followers that is still not completely settled. Fisher's arguments lead to the **principle of sufficiency** and the **principle of conditionality**. The principle of sufficiency dictates that all inferences should be based on a sufficient statistic if there is one, and the principle of conditionality requires that any inference should be based on the conditional distribution given the **ancillary statistic**, i.e., a statistic whose distribution is independent of the parameter, if there is such a statistic. These two principles are accepted by many statisticians who do not follow all of Fisher's arguments, though the principle of conditionality sometimes leads to difficulties due to nonuniqueness of the ancillary.

D. Specification Problem

It is often difficult and sometimes impossible to have an exactly correct model for the data, and we must be satisfied with a model that gives a sufficiently close approximation and is mathematically tractable as well. It may also happen, however, that a model first specified may be far from reality and could lead to erroneous conclusions if relied on blindly. Here, the problem of **model selection** arises (→ 403 Statistical Models), i.e., choosing the best of various possible models.

We may also seek procedures that are little affected by the departure of the distribution of the data from the assumed or some other model that satisfies the condition of validity without any assumption about the exact shape of the distributions (→ 371 Robust and Nonparametric Methods). Generally, the problem of determining the model or specification should not be dealt with by mathematical methods alone, and it should be considered by taking into account the properties and nature of the subject under consideration and also the process of measurement or experimentation.

E. History

The first appearance of statistical inference as a method of grasping numerical characteristics of a collective was seen in the study by J. Graunt (1662) of the number of people who died in London. W. Petty applied Graunt's method further to the comparison of communities in his *Political arithmetic* (1690). J. P. Süssmilch, a member of the Graunt school, perceived the regularity in mass observations and stressed the statistical importance of this regularity. The development of the theory of probability inevitably affected the theory of statistical inference. The method of T. Bayes was the first procedure of statistical inference in the current meaning of this expression. We now have a theorem bearing his name (the Bayes theorem), which is stated in current language as follows: If we know the probability $P_C(E)$ that a cause C produces an effect E and if the prior (or †a priori) probability $P(C)$ of the existence of the cause C is also known, then the posterior (or †a posteriori) probability of C, given an effect E, is equal to

$$P_E(C)=\frac{P(C)P_C(E)}{\sum_C P(C)P_C(E)}$$

(→ 342 Probability Theory F). This theorem, easily extendable to the continuous case, suggests the following inference procedure: If we are informed that an effect E has taken place, then we calculate the probabilities $P_E(C)$ for every cause C, compare them, and infer that the C^* with $P_E(C^*)=\max_C P_E(C)$ is the **most probable cause** of E.

Both P. S. Laplace and C. F. Gauss discussed the theory of estimation of parameters (→ 399 Statistical Estimation) as an application of the Bayes theorem. In his research, Laplace considered a monotone function $W(|t-\theta|)$, and $W=|t-\theta|$ in particular, of the distance $|t-\theta|$ between a parameter value θ and its estimate t as a measure of significance of the error of the estimate t. Gauss, following Laplace, used this weight function $W(|t-\theta|)$ of error, and going beyond Laplace, realized that it would be mathematically fruitful to put $W(|t-\theta|)=(t-\theta)^2$. Such considerations led him to the study of the †least squares method, in which the terminology and notation he devised are still in use. He also developed the theory of errors and recognized the importance of the normal distribution and found that the least square estimate is equal to the most probable value if the errors are normally distributed.

F. Galton, a biologist, revealed the usefulness of statistical methods in biological research and explored what we call †regression analysis (→ 403 Statistical Models) by introducing the concepts of regression line and †correlation coefficient. His research on regression analysis originated from the study of the correlation between characteristics of parents and children, but he failed to realize the difference between †population characteristics and †sample characteristics.

Following Galton, K. Pearson developed the theory of regression and correlation, with which he succeeded in establishing the basis of biometrics (→ 40 Biometrics). He arrived at the concept of population in statistics: A statistical **population** is a collective consisting of observable individuals, while a **sample** is a set of individuals drawn out of the population and containing something telling us about characteristics of the population. Thus statistical research is regarded as investigation that focuses not on a sample as such but on a population from which the sample has been drawn. This consideration raised the problem of the **goodness-of-fit test** (→ 400 Statistical Hypothesis Testing), that is, the problem of knowing whether a sample is likely to have been drawn from a population whose distribution was determined by theoretical considerations. K. Pearson characterized some population distributions occurring in practice by a differential equation, and classified them into several types. Using this classification, he discussed goodness-of-fit tests and developed the χ^2-distribution (†chi-square distribution) in relation to the problem of testing hypotheses.

Statisticians in the time of K. Pearson thought of a population as a collective having infinitely many individuals (i.e., an **infinite population**), which led to the idea that the larger the **size** of a sample (i.e., the number of individuals in the sample), the more precisely could the sample give information about the population. They carried out inferences, including the testing of hypotheses (→ 400 Statistical Hypothesis Testing), by approximate methods, which later came to be termed **large sample theory**. Suppose, for instance, that $\{X_1, \ldots, X_n\}$ is an †independent sample of size n from a normal population $N(\mu, \sigma^2)$. The random variable $Z=\sqrt{n}(\bar{X}-\mu_0)/\sigma$ with $\bar{X}=\sum_i X_i/n$ is distributed according to $N(0,1)$ when $\mu=\mu_0$. Therefore, if the size n is sufficiently large ($n\to\infty$), we estimate σ by $\hat{\sigma}=\{\sum_i(X_i-\bar{X})^2/(n-1)\}^{1/2}$ and deal with the random variable $T=\sqrt{n}(\bar{X}-\mu_0)/\hat{\sigma}$ obtained by inserting $\hat{\sigma}$ in place of σ in the expression for Z, as if T itself were distributed according to $N(0,1)$.

F. Development in the 20th Century

W. S. Gosset, writing under the pen name "Student," reported in 1908 the discovery of the exact distribution of T and thereby opened the new epoch of **exact sampling theory** (→ 374 Sampling Distributions). This work of Student made it possible to perform statistical inference by means of **small samples** and consequently changed statistical research from the study of collectives to that of uncertain phenomena; in other words, the concept of population was once again related to a †probability space with a †probability distribution (i.e., a **population distribution**) containing unknown **parameters**. Thus it began to be emphasized that a sample has to be drawn **at random** (i.e., a **random sample**) from the population if we are to make an inference about a parameter based on the sample.

Fisher presented a complete derivation, using the multiple integration method, of the †t-distribution (the sampling distribution of T). In addition, Fisher introduced the concepts of †null hypothesis and significance test, which were the starting points for later progress in the theory of hypothesis testing. He also added the concepts of †consistency, †efficiency, and †sufficiency to the list of possible properties of †estimators, and he studied the connection between the information contained in a sample and the accuracy of an estimator, which led to the idea of amount of information. Fisher also proposed the †maximum likelihood estimator, which is formally equivalent to the most probable value, but he renamed it and gave it a foundation completely independent of any prior information and showed that it leads to the at least asymptotically efficient estimator.

Fisher made efforts to obtain a distribution of the parameter directly from the sample observation, hence independently of the concept of prior probability. He sought in this way to be released from the weakness of the Bayes method. For this purpose he introduced the concept of **fiducial distribution**, which was the subject of bitter controversy in the period that followed. As an example of a fiducial distribution, we consider here the †Behrens-Fisher problem: Let $X_1, \ldots, X_m$ and $Y_1, \ldots, Y_n$ be samples drawn independently from the populations $N(\mu_1, \sigma_1^2)$ and $N(\mu_2, \sigma_2^2)$, respectively, where the parameters μ_1, μ_2, σ_1, and σ_2 are all unknown. The problem raised is to test the hypothesis $\mu_1 = \mu_2$ or to estimate $\delta = \mu_1 - \mu_2$ by an interval. To solve this problem, we put

$$\bar{X} = \sum_i X_i/m, \qquad \bar{Y} = \sum_j Y_j/n,$$

$$S_1^2 = \sum_i (X_i - \bar{X})^2/(m-1),$$

$$S_2^2 = \sum_j (Y_j - \bar{Y})^2/(n-1),$$

and learn that $T_1 = \sqrt{m}(\bar{X} - \mu_1)/S_1$ and $T_2 = \sqrt{n}(\bar{Y} - \mu_2)/S_2$ are mutually independent and distributed according to the t-distribution with degrees of freedom $m-1$ and $n-1$, respectively. From this fact Fisher reasoned as follows: Given observed values $\bar{x}, \bar{y}, s_1, s_2$ of the variables $\bar{X}, \bar{Y}, S_1, S_2$, the distributions of the parameters μ_1 and μ_2 are induced from the distributions of T_1 and T_2 by means of transformations

$$\mu_1 = \bar{x} - \frac{T_1 s_1}{\sqrt{m}}, \qquad \mu_2 = \bar{y} - \frac{T_2 s_2}{\sqrt{n}}.$$

Consequently the distribution of $\delta = \bar{x} - \bar{y} - (T_1 s_1/\sqrt{m} - T_2 s_2/\sqrt{n})$ is obtained. These distributions are called the fiducial distributions of the parameters μ_1, μ_2, and δ. The interval $|\delta - (\bar{x} - \bar{y})| < c$ of δ deduced from the fiducial distribution of δ is called a **fiducial interval** of δ.

Neyman and E. S. Pearson developed a mathematical theory of testing hypotheses, in which they deliberately defined a family of population distributions admissible for formal treatment and considered alternative hypotheses within the family. They proposed to relate a test to its †power function, on the basis of which the test would be judged. Their ideas brought mathematical clarity to the theory of inference. Furthermore, concerning interval estimation, Neyman devised an alternative to the fiducial interval, the †confidence interval, which has full mathematical justification. Unfortunately it was later found that the confidence interval, fiducial interval, and the Bayes posterior interval based on the posterior distribution often gave distinctly different results to the same problem, which became a source of controversy among different schools of thought.

Since the publication of A. Wald's theory of statistical decision functions (→ 398 Statistical Decision Functions) in 1939, there has been a steady increase in its importance. In this theory the totality $\mathscr{D}$ of available statistical procedures, which is considered implicitly in the Neyman-Pearson theory, is put forth explicitly as a set and defined as the space of decision functions. Wald also defined the †risk function of a statistical decision procedure and used it as a basis for judging procedures. In addition, he employed the concept of prior probability and the Bayes procedure for the purpose of proving the †complete class theorem. Wald's idea of bringing the concept of prior probability back into statistical theory carried a great deal of weight, and much literature has now been accumulated on this subject. Prior probability as a technique in statistics was abandoned after Fisher's introduction of the maximum likelihood method independent of prior probability and Neyman's assertion that a probability distribution on the †parameter space made no sense. In addition, Wald linked statistical inference to games (→ 173 Game Theory) and introduced the †minimax principle into statistics. The decision-theoretic setup also enabled him to develop a theory of sequential analysis by comparing the cost of sampling with the risk of erroneous decisions (→ 400 Statistical Hypothesis Testing).

After the publication of Savage's book in 1954, there was a revival of the Bayesian approach, i.e., one based on the concept of subjective probability, and now the group of those statisticians who accept the Bayesian approach are called Bayesians or neo-Bayesians.

G. Applications

Methods of statistical inference are applied in many fields where statistical data are used for scientific, engineering, medical, or managerial purposes. Methods of producing data that are appropriate for statistical inference have also been developed. R. A. Fisher developed the method of statistical †design of experiments (→ 102 Design of Experiments) that when it is impossible or impractical to eliminate completely experimental errors or variabilities, provides the procedures to obtain such data. These data, though subject to random errors, are susceptible to rigorous statistical inference. For this purpose Fisher introduced the principles of †randomization, †local control, and †replication in the design of experiments. W. A. Shewhart defined the †state of statistical control in mass-production processes, where the variabilities of the products can be considered to be due to chance causes alone and hence are statistically analyzable.

Applying the idea of statistical inference to this situation, Shewhart established the method of statistical quality control (→ 404 Statistical Quality Control). Neyman introduced the method of †random sampling into statistical surveys and developed the theory of estimation and allocation based on the theory of statistical inference (→ 373 Sample Survey).

In many applied fields there exist systems of statistical methods which have been developed specifically for the respective fields, and although all of them are based essentially on the same general principles of statistical inference, each has its own special techniques and procedures. Specific names have been invented, such as biometrics (→ 40 Biometrics) econometrics (→ 128 Econometrics), psychometrics (→ 346 Psychometrics), technometrics, sociometrics, etc.

References

[1] R. A. Fisher, Statistical methods for research workers, Oliver & Boyd, 1925; eleventh edition, 1950.
[2] R. A. Fisher, The design of experiments, Oliver & Boyd, 1935; fourth edition, 1947.
[3] R. A. Fisher, Contributions to mathematical statistics, Wiley, 1950.
[4] R. A. Fisher Statistical methods and scientific inference, Oliver & Boyd, 1956.
[5] J. Neyman and E. S. Pearson, Contributions to the theory of testing statistical hypotheses I, II, Statist. Res. Mem. London Univ., 1 (1936), 1–37; 2 (1938), 25–57.
[6] A. Wald, Statistical decision functions, Wiley, 1950 (Chelsea, 1971).
[7] L. J. Savage, The foundation of statistical inference, Methuen, 1962.
[8] C. R. Rao, Linear statistical inference and its applications, Wiley, 1965.
[9] D. A. Fraser, The structure of inference, Wiley, 1968.
[10] D. V. Lindley, Introduction to probability and statistics from a Bayesian viewpoint I, II, Cambridge Univ. Press, 1965.

402 (XX.19) Statistical Mechanics

A. General Remarks

One cubic centimeter of water contains about 3×10^{22} water molecules. A macroscopic system of matter thus consists of an enormous number of particles incessantly moving in accordance with the laws of dynamics (→ 271 Mechanics; 351 Quantum Mechanics). Dynamical description of such microscopic motion in full detail is impossible and even meaningless. A physical process in thermodynamics or †hydrodynamics is described in terms of a relatively small number of macroscopic variables, such as temperature, pressure, and a velocity field. Such a process shows a remarkable simplicity which is a statistical result of the molecular chaos. This is the reason why statistical mechanics is needed as a theoretical model to unify microscopic dynamics and †probability theory. Thus **statistical mechanics** aims at deriving physical laws in the macroscopic world from the atomistic structures of the microscopic world on the basis of microscopic dynamical laws and probabilistic laws. Its function is twofold. First, statistical mechanics should give microscopic proofs of the macroscopic laws of physics, such as those of thermodynamics or the laws of macroscopic electromagnetism. Second, it should also provide us with detailed knowledge of physical properties of a given material system once its microscopic structure is known. In this sense, statistical mechanics is an essential basis of the modern science of materials.

Strictly speaking, the dynamics of the microscopic world obeys †quantum mechanics. However, even before the birth of quantum mechanics, statistical mechanics had progressed on the basis of classical mechanics. This stage of statistical mechanics is often called **classical statistical mechanics**, in contrast to **quantum statistical mechanics** based on quantum mechanics. Statistical mechanics has a fully developed formalism to apply to physical systems in thermal equilibrium. This is sometimes called **statistical thermodynamics** or **equilibrium statistical mechanics**. Until the 1950s the term "statistical mechanics" had often been used in this narrow sense. In a wider sense it is concerned with systems in more general states, for instance, in nonequilibrium states. In the modern literature, a general statistical mechanical theory of nonequilibrium systems is often referred to as the statistical mechanics of **irreversible processes**.

B. History

The early stage of statistical mechanics can be traced back to the **kinetic theory of gases**, which started in the 18th century. In dilute gases, gas molecules fly freely through the whole volume of the vessel and collide only from time to time. In thermal equilibrium, the average energy of each molecule is determined by the temperature of the gas; namely

$m\,\overline{v_x^2}/2=m\,\overline{v_y^2}/2=m\,\overline{v_z^2}/2=kT/2,$

where (v_x, v_y, v_z) is the velocity, an overbar means the average, m is the mass of a molecule, T is the absolute temperature, and k is the **Boltzmann constant** $(=1.38\times 10^{-16}\ \mathrm{erg}\cdot\mathrm{deg}^{-1})$. The velocity of each molecule is only probabilistic, and a †distribution function $f(v_x, v_y, v_z)$ is defined as the †probability density that the velocity of a given molecule is found to be in the neighborhood of (v_x, v_y, v_z). In a dilute gas, this is given by

$$f(v_x, v_y, v_z) = C\exp\left[-\frac{m}{2}(v_x^2+v_y^2+v_z^2)/kT\right], \quad (1)$$

the **Maxwell-Boltzmann distribution law**.

L. Boltzmann viewed the velocity distribution function as changing in time as a result of molecular collisions and gave an equation of the form

$$\frac{\partial f}{\partial t} = A[f] + \Gamma[f], \quad (2)$$

where $A[f]$ is the change of the distribution function f by acceleration due to the presence of external forces and $\Gamma[f]$ is the change caused by molecular collisions. $\Gamma[f]$ is an integral which is nonlinear in f. This type of equation is called a **Boltzmann equation** [1, 2].

Boltzmann introduced the H-**function** by the definition

$$H = \iiint f\log f\, dv_x\, dv_y\, dv_z \quad (3)$$

and proved on the basis of equation (2) that $dH/dt \leqslant 0$. This theorem is known as the **H-theorem** [1–4]. The equilibrium distribution (1) is therefore obtained from equation (2) as the solution that makes H a minimum. In fact the H-function is related to the **entropy** S by

$$S = -kH. \quad (4)$$

Boltzmann further showed (1877) that the distribution function of a system in thermal equilibrium can be obtained on more general grounds without relying on a kinetic equation of the type (2) and that the statistical mechanics of systems in equilibrium can thus be constructed on a basis much more general than that given by a kinetic theory. It was W. Gibbs, however, who clearly established (1902) the complete framework of statistical thermodynamics, although he had to confine himself to classical statistical mechanics [5].

C. The Ergodic Hypothesis

For a given dynamical system with n †degrees of freedom, the **phase space** is defined as a $2n$-dimensional space with †generalized coordinates $q_1, \ldots, q_n$ and †generalized momenta $p_1, \ldots, p_n$. Dynamical states of the system constitute a set of points in this space. At a given time, the state of the system is represented by a point P in the phase space, and hence the motion of the system is represented by the motion of P. If the system is conservative, its energy function is constant. Let $\mathscr{H}$ be the †Hamiltonian function. Then the motion of P is confined to an **energy surface** defined by the condition $\mathscr{H} = E =$ constant. Measure on an energy surface is defined as the limit of the volume element lying between two neighboring energy surfaces corresponding to the energies E and $E + dE$. The motions of P form a †topological group that makes this measure invariant (†Liouville's theorem).

A dynamical quantity $A(p, q)$ of the system changes its value as the phase point P moves on the energy surface. The time average $\bar{A}$ of A is identified with the value of A observed in the equilibrium state of the system, namely, the average of A with respect to the invariant measure. Boltzmann justified this assumption by the following reasoning. If the energy surface has a finite measure and the trajectory of P does not make a closed curve on the energy surface, it can be assumed that the trajectory will move around practically everywhere on the surface. Mathematically formulated, the only measurable subset of the surface that has a nonzero measure and is invariant under the motion is the whole surface. This assumption is the **ergodic hypothesis** [6–9]. The long-time average of A will then equal the average of A over the entire energy surface with weight function equal to the measure previously introduced. The latter average is called the **phase average** and is denoted by $\langle A\rangle$. Boltzmann thus asserted that

$$\bar{A} = \langle A\rangle. \quad (5)$$

Efforts of mathematicians to study the ergodic hypothesis created an important branch of mathematics called ergodic theory (→ 136 Ergodic Theory).

D. Ensembles in Classical Statistical Mechanics

Once we admit the ergodic hypothesis, or more specifically the assumption (5), the calculation of the observed value of a physical quantity A for a system in equilibrium is reduced to finding the phase average of A on an energy surface. The task of statistical mechanics of systems in equilibrium is thus reduced essentially to calculating phase averages and establishing relationships between them [10–13].

For a set (called an **ensemble** in this case) of

identical systems with the same energy, we consider the phase average for the †probability space with the measure mentioned in Section C on the energy surface corresponding to the given energy value. Gibbs called a probability space of this kind a **microcanonical ensemble**. An average in this probability space is defined by

$$\langle A\rangle=\int_{\mathscr{H}=E}\frac{A\,dS}{|\operatorname{grad}\mathscr{H}|}\bigg/\int_{\mathscr{H}=E}\frac{dS}{|\operatorname{grad}\mathscr{H}|},\tag{6}$$

where $\mathscr{H}$ is the Hamiltonian function, grad $\mathscr{H}$ is its gradient in the $2n$-dimensional phase space, and the integration is carried over the energy surface with dS as surface element.

When the observed system is in mechanical contact with a heat reservoir, the composite system consisting of the system and the heat reservoir is regarded as an isolated system with constant energy. Then an ensemble of the composite systems is treated as a microcanonical ensemble. It is more convenient and more physical, however, to consider the heat reservoir simply as providing an environment characterized by its temperature T, and to concentrate only on the system in which we are interested. Then the system is no longer isolated and exchanges energy with its environment. Since the energy of the system is no longer constant, the system will be found in any part of the phase space with a certain probability. To find the probability distribution for an ensemble of this system is a problem of asymptotic evaluation which is solved on the basis of the ergodic hypothesis and the fact that a heat reservoir has an extremely large number of degrees of freedom. This asymptotic evaluation is traditionally done with the help of †Stirling's formula or by using the Fowler-Darwin method [10], but it is essentially based on the †central limit theorem [11].

The probability space of this kind of ensemble of systems in contact with heat reservoirs was called a **canonical ensemble** by Gibbs [5]. If $d\Gamma$ is a volume element of the phase space of the system, the probability of finding a system arbitrarily chosen from the ensemble in a volume element $d\Gamma$ is given by

$$\Pr(d\Gamma)=C\exp(-\mathscr{H}/kT)\,d\Gamma.\tag{7}$$

Accordingly, the average of a dynamical quantity A is given by

$$\langle A\rangle=\int Ae^{-\mathscr{H}/kT}\,d\Gamma\bigg/\int e^{-\mathscr{H}/kT}\,d\Gamma.\tag{8}$$

For example, the average energy is

$$\langle\mathscr{H}\rangle\equiv E=\int\mathscr{H}e^{-\mathscr{H}/kT}\,d\Gamma\bigg/\int e^{-\mathscr{H}/kT}\,d\Gamma.\tag{9}$$

By a traditional convention we introduce the parameter

$$\beta=1/kT$$

and write (9) as

$$E=-\partial\log Z(\beta)/\partial\beta,$$

where $Z(\beta)$ is called the **partition function** or the **sum over states** and is given for a system composed of N identical particles by

$$Z(\beta)=\int e^{-\beta\mathscr{H}}\,d\Gamma/N!.\tag{10}$$

If an exchange of particles with the environment takes place in addition to an exchange of energy, the probability of finding a system with particle number N in the volume element $d\Gamma$ is given by

$$C\exp(-\beta\mathscr{H}+\beta\mu N)\,d\Gamma/N!,$$

where μ is a real parameter called the **chemical potential**; this characterizes the environment with regard to the exchange of particles. This ensemble is called the **grand canonical ensemble**. The average of a dynamical quantity A is then given by

$$\langle A\rangle=\sum_N N!^{-1}\int A_N e^{-\beta\mathscr{H}(N)+\beta\mu N}\,d\Gamma/\Xi(\beta,\mu),\tag{11}$$

where the dependence of A and $\mathscr{H}$ on N is now explicitly written, and where

$$\Xi(\beta,\mu)=\sum_N N!^{-1}\int e^{-\beta\mathscr{H}(N)+\beta\mu N}\,d\Gamma\tag{12}$$

is called the **grand partition function**.

E. Ensembles in Quantum Statistical Mechanics

The quantum counterpart of the classical ergodic hypothesis is that to each of these quantum states an equal probability weight should be assigned [10]. A **microcanonical ensemble** is then defined by this **principle of equal weight**, which yields in turn

$$\langle A\rangle=\sum_l A_l\Big/\sum_l 1\tag{13}$$

instead of (6). Here the index l refers to the quantum states lying in the interval ΔE, and A_l is the quantum-mechanical expectation of a dynamical variable A in the quantum state l. A canonical ensemble is now defined by assigning

$$P_j=e^{-\beta E_j}\Big/\sum_j e^{-\beta E_j}\tag{14}$$

to the jth quantum state as the probability that the system will be found in that state. The expectation value of A must be given by

$$\langle A\rangle=\sum_j A_j e^{-\beta E_j}/\sum_j e^{-\beta E_j}=\operatorname{tr} Ae^{-\beta H}/\operatorname{tr} e^{-\beta H}, \tag{15}$$

where H is the Hamiltonian. The partition function is defined by

$$Z=\sum_j e^{-\beta E_j}=\operatorname{tr} e^{-\beta H}, \tag{16}$$

corresponding to (10).

For a system consisting of identical particles, quantum mechanics requires a particular symmetry of its †wave function; namely, the wave function must be even or odd with respect to permutation of any two particles according as the particles are bosons or fermions. This symmetry requirement is peculiar to quantum mechanics. Thus, even for an ideal gas consisting of noninteracting particles, quantum statistics leads to results characteristically different from those of classical statistical mechanics. This difference becomes more significant when the particle mass is smaller, the density is larger, and the temperature is lower. Quantum effects of this kind are seen in metallic electrons, in liquid helium, in an assembly of photons or phonons, and in high-density stars. The statistical laws obeyed by bosons are called **Bose statistics**, and those obeyed by fermions, **Fermi statistics**.

The expectation value of A in the **grand canonical ensemble** is given in quantum statistics by

$$\langle A\rangle=\Xi(\beta,\mu)^{-1}\operatorname{tr}(Ae^{-\beta H+\beta\mu N}), \tag{17}$$

where H is the †second-quantized Hamiltonian, N is the number operator, the trace tr is taken on the (nonrelativistic) †Fock space (symmetric or antisymmetric according to Bose or Fermi statistics), and $\Xi(\beta,\mu)$ is the **grand partition function** given by

$$\Xi(\beta,\mu)=\operatorname{tr} e^{-\beta H+\beta\mu N}. \tag{18}$$

F. Many-Body Problems in Statistical Mechanics

Since statistical mechanics is primarily concerned with systems with large numbers of particles, problems in statistical mechanics are essentially many-body problems. In practice, however, there are some cases where extreme idealization is possible, as in ideal gases, where the interaction between gas molecules is ignored. In some cases we can proceed by successive approximation, taking the particle interactions as perturbations. Such perturbational treatments are, however, entirely useless for some problems, such as phase transitions, of which an example is the condensation of gases into liquid states, where the interaction of particles plays a critical role. Such problems are clearly many-body problems. There are a number of important and interesting problems in this category, for example, transitions between ferromagnetic and paramagnetic states and those between the superconducting and normal states of metals. Transition from a high-temperature phase to a low-temperature phase is generally regarded as a consequence of the appearance of a certain type of order in thermal motion. This kind of phase change is called an **order-disorder transition** [14–16].

G. Thermodynamic Limit and Characterization of Equilibrium States

Although an actual system is finitely extended, the enormous sizes of the usual macroscopic systems in comparison to the sizes of their constituent particles justifies the idealization to infinitely extended systems. At the same time, there are several mathematical advantages in considering infinitely extended systems, such as the absence of walls (replaced by the boundary condition at infinity, should it be relevant), appearance of phase transitions as mathematical discontinuities rather than mathematically smooth though quantitatively sudden changes, and mathematically clear-cut occurrence of broken symmetries.

Equilibrium states of infinitely extended systems are usually obtained by taking the limit of the equilibrium states of systems in a finite volume V as both V and the number of particles N tend to ∞ with the density $\rho=N/V$ fixed; this is called the **thermodynamic limit**.

It is sometimes possible to formulate the dynamics of infinitely extended systems directly and to characterize their equilibrium states, which more or less coincide with the thermodynamic limit of equilibrium states of finitely extended systems [17–21]. The simplest and most fully investigated case of **lattice spin systems** is explained below in detail [17]. Since classical systems can be viewed as special cases of quantum systems, we start with the latter. To be definite, we take a ν-dimensional cubic lattice $\mathbf{Z}^\nu$ with a lattice site $n=(n_1,\ldots,n_\nu)$ specified by its integer coordinates n_j. (In the lattice case, the thermodynamic limit is simply the limit as $V\to\infty$.)

The C^*-algebra $\mathfrak{A}$ of observables is generated by the subalgebra $\mathfrak{A}_n$ at each lattice site n, which is assumed to be the algebra of all $d\times d$ matrices (for example, linear combinations of †Pauli spin matrices $\boldsymbol{\sigma}^{(n)}=(\sigma_x^{(n)},\sigma_y^{(n)},\sigma_z^{(n)})$ and the identity for $d=2$) and to commute with operators at other lattice sites. The group of lattice translations $n\to n+a$ is represented by automorphisms γ_a of $\mathfrak{A}$, satisfying $\gamma_a\mathfrak{A}_n=\mathfrak{A}_{n+a}$ ($\gamma_a\boldsymbol{\sigma}^{(n)}=\boldsymbol{\sigma}^{(n+a)}$). For any subset Λ of $\mathbf{Z}^\nu$, $\mathfrak{A}(\Lambda)$ denotes the C^*-subalgebra of $\mathfrak{A}$ generated by $\mathfrak{A}_n$, $n\in\Lambda$.

A model is specified by giving a **potential** Φ which assigns to each finite nonempty subset I of $\mathbf{Z}^\nu$ an operator $\Phi(I)=\Phi(I)^*\in\mathfrak{A}(I)$. The Hamiltonian for a finite subset Λ of $\mathbf{Z}^\nu$ is given by $U(\Lambda)=\Sigma_{I\subset\Lambda}\Phi(I)$. In order to control long-range interactions, various assumptions are introduced. Examples are finiteness of either of the following:

$$\|\Phi\|=\sup_n\sum_{I\ni n}N(I)^{-1}\|\Phi(I)\|, \tag{19}$$

$$|||\Phi|||=\sup_n\sum_{I\ni n}\|\Phi(I)\|. \tag{20}$$

Here $N(I)$ is the number of points in I.

Let $\mathfrak{A}_n^{\mathrm{cl}}$ be a maximal Abelian *-subalgebra of $\mathfrak{A}_n$ (such as $\{c_1+c_2\sigma_z^{(n)}\}$) satisfying $\gamma_a\mathfrak{A}_n^{\mathrm{cl}}=\mathfrak{A}_{n+a}^{\mathrm{cl}}$ and $\mathfrak{A}^{\mathrm{cl}}$ be the Abelian C^*-subalgebra of $\mathfrak{A}$ generated by $\mathfrak{A}_n^{\mathrm{cl}}$, $n\in\mathbf{Z}^\nu$. If $\Phi(I)$ is in $\mathfrak{A}^{\mathrm{cl}}$ for all I, we call the potential Φ **Abelian** or **classical**. There exists a conditional expectation π^{cl} which is a positive mapping of norm 1 from $\mathfrak{A}$ onto $\mathfrak{A}^{\mathrm{cl}}$ satisfying $\pi^{\mathrm{cl}}(ABC)=A\pi^{\mathrm{cl}}(B)C$ for A and C in $\mathfrak{A}^{\mathrm{cl}}$ and $\pi^{\mathrm{cl}}(1)=1$. If a state φ on $\mathfrak{A}$ satisfies $\varphi(A)=\varphi(\pi^{\mathrm{cl}}(A))$, we call the state φ **classical**. Classical states are in one-to-one correspondence with the restriction on $\mathfrak{A}^{\mathrm{cl}}$, which can be viewed as a probability measure on the spectrum (also called **configuration space**) of the C^*-algebra $\mathfrak{A}^{\mathrm{cl}}$ of observables for classical spin lattice systems. This correspondence makes it possible to view classical spin lattice systems as quantum spin lattice systems with Abelian interactions.

For a given potential Φ, the time evolution of the infinitely extended system is described by the one-parameter group α_t, $t\in\mathbf{R}$, of *-automorphisms of $\mathfrak{A}$ defined as the following limit:

$$\alpha_t(A)=\lim_{\Lambda\nearrow\mathbf{Z}^\nu}e^{itU(\Lambda)}Ae^{-itU(\Lambda)}\qquad(A\in\mathfrak{A}). \tag{21}$$

The limit exists if $\Phi(I)=0$ for $N(I)>N$ and $|||\Phi|||<\infty$, or if for some $\lambda>0$ $\Sigma_n e^{\lambda n}(\sup_x\Sigma_I\{\|\Phi(I)\|\,|\,I\ni x,\,N(I)=n\})<\infty$, or if $\nu=1$ (1-dimensional lattice) and $\sup_x\Sigma_I\{\|\Phi(I)\|\,|\,I\cap(-\infty,x)\neq\varnothing,\,I\cap[x,\infty)\neq\varnothing\}<\infty$. An alternative way is first to define $\delta_\Phi(A)=\Sigma_I i[\Phi(I),A]$ for $A\in\bigcup_\Lambda\mathfrak{A}(\Lambda)$ (Λ is a finite subset of $\mathbf{Z}^\nu$), which exists if $|||\Phi|||<\infty$, and to prove that the closure $\bar\delta_\Phi$ of δ_Φ is a generator of a one-parameter subgroup α_t ($=\exp t\bar\delta_\Phi$). In the above cases, $\bar\delta_\Phi$ is a generator.

A general canonical ensemble for a system in a finite subset Λ of the lattice $\mathbf{Z}^\nu$, with some boundary condition in the outside $\Lambda^c=\mathbf{Z}^\nu\backslash\Lambda$, is given by $\varphi_\Lambda(A)=(\tau_\Lambda\otimes\psi)(e^{-\beta H(\Lambda)}\times Ae^{-\beta H(\Lambda)/2})$, where $\tau_\Lambda\otimes\psi$ is the product of the unique tracial state τ_Λ on $\mathfrak{A}(\Lambda)$ and a state ψ on $\mathfrak{A}(\Lambda^c)$ (the boundary condition), $H(\Lambda)=U(\Lambda)+W(\Lambda)$ and $W(\Lambda)=\Sigma_I\{\Phi(I)\,|\,I\cap\Lambda\neq\varnothing,\,I\cap\Lambda^c\neq\varnothing\}$ (the **surface energy**). The following conditions on a state φ of $\mathfrak{A}$ are mutually equivalent under the condition that $\bar\delta_\Phi$ is a generator (which holds under any one of the conditions described above) and is satisfied by any limit state of the above φ_Λ as $\Lambda\nearrow\mathbf{Z}^\nu$ (i.e., a state in $\bigcap_{\Lambda'}\overline{\{\varphi_\Lambda\,|\,\Lambda'\subset\Lambda,\psi\}}$, with the bar denoting weak closure).

1. **KMS condition**: $\varphi(A\alpha_{i\beta}(B))=\varphi(BA)$ for any $A,B\in\mathfrak{A}$ such that $\alpha_t(B)$ is an entire function of t. (φ is called a **β-KMS state**.)

2. **Local thermodynamic stability**: For any finite subset Λ of $\mathbf{Z}^\nu$ and for any state ψ having the same restriction to $\mathfrak{A}(\Lambda)$ as the state φ under consideration, $\tilde F_{\Lambda,\beta}(\varphi)\leqslant\tilde F_{\Lambda,\beta}(\psi)$ (the minimality of the **free energy** multiplied by β), where $\tilde F_{\Lambda,\beta}(\varphi)=\beta\varphi(H(\Lambda))-S_\Lambda(\varphi)$, $S_\Lambda(\varphi)=\lim\{S_{\Lambda'}(\varphi)-S_{\Lambda'\backslash\Lambda}(\varphi)\}$ as $\Lambda'\nearrow\mathbf{Z}^\nu$ (the **open system entropy**), $S_\Lambda(\varphi)=-\varphi(\log\rho_\Lambda(\varphi))$ (the **closed system entropy**) and the density matrix $\rho_\Lambda(\varphi)\in\mathfrak{A}(\Lambda)^+$ is defined by $\varphi(A)=\tau_\Lambda(\rho_\Lambda(\varphi)A)$ for all $A\in\mathfrak{A}(\Lambda)$.

3. **Gibbs condition**: For every finite subset Λ of $\mathbf{Z}^\nu$, the perturbed state $\varphi^{\beta W(\Lambda)}$ (not necessarily normalized) is the product $\varphi_\Lambda^G\times\psi$ of the Gibbs state $\varphi_\Lambda^G(A)=\mathrm{tr}(e^{-\beta U(\Lambda)}A)/\mathrm{tr}\,e^{-\beta U(\Lambda)}$ on $A\in\mathfrak{A}(\Lambda)$ and some (unknown) state ψ on $\mathfrak{A}(\Lambda^c)$, where the representative vector Φ for φ for the GNS representation π_φ is assumed to be separating for $\pi_\varphi(\mathfrak{A})''$, and then $\varphi^{\beta W(\Lambda)}(A)=(\Omega,\pi_\varphi(A)\Omega)$ for

$$\Omega=\sum_{n=0}^{\infty}\beta^n\int_0^{1/2}ds_1\int_0^{s_1}ds_2\ldots\int_0^{s_{n-1}}ds_n\Delta_\Phi^{s_n}\pi_\varphi \times(W(\Lambda))\Delta_\Phi^{s_{n-1}-s_n}\ldots\Delta_\Phi^{s_1-s_2}\pi_\varphi(W(\Lambda))\Phi, \tag{22}$$

where Δ_Φ is the †modular operator for Φ and the series converges.

For a classical potential, this condition reduces to the conditions that φ is classical and that the restriction of φ to $\mathfrak{A}^{\mathrm{cl}}$ as a measure on the configuration space $\{1\ldots d\}^{\mathbf{Z}^\nu}$ satisfies the following **DLR equation** due to R. L. Dobrushin, O. E. Lanford, and D. Ruelle: The conditional probability for $\xi(\Lambda)\in\{1\ldots d\}^\Lambda$ knowing $\xi(\Lambda^c)\in\{1\ldots d\}^{\mathbf{Z}^\nu\backslash\Lambda}$ is proportional to $\exp(-\beta H(\Lambda))$, where $H(\Lambda)=U(\Lambda)+W(\Lambda)$ is a function of $\xi(U(\Lambda)$ depending only on $\xi(\Lambda))$.

4. **Roepstorff-Araki-Sewell inequality**: For any $A\in\bigcup_\Lambda\mathfrak{A}(\Lambda)$, $i\varphi(A^*\delta_\Phi(A))$ is real and

$$-i\beta\varphi(A^*\delta_\Phi(A))\geqslant S(\varphi(A^*A),\varphi(AA^*)), \tag{23}$$

where $S(u,v)=u\log(u/v)$ if $u>0$, $v>0$, $S(0,v)=0$ for $v\geqslant0$ and $S(u,0)=+\infty$ for $u>0$.

5. **Roepstorff-Fannes-Verbeure inequality**:

For α-entire $A \in \mathfrak{A}$,

$$\beta^{-1}\int_0^\beta \varphi(A^*\alpha_{i\lambda}(A))\,d\lambda \leqslant F(\varphi(A^*A), \varphi(AA^*)),$$
(24)

where $F(u,v)=(u-v)/\log(u/v)$ for $u>0$, $v>0$, $u \neq v$, $F(u,u)=u$ for $u>0$, $F(u,0)=F(0,v)=0$.

If the interaction is translationally invariant (i.e., $\gamma_a\Phi(I)=\Phi(I+a)$ for all $a \in \mathbf{Z}^\nu$ and I) and if we restrict our attention to translationally invariant states (i.e., $\varphi(\gamma_a(A))=\varphi(A)$ for all $A \in \mathfrak{A}$ and $a \in \mathbf{Z}^\nu$), then the following conditions are also equivalent to the above.

6. **Variational principle**: $\beta e(\varphi)-s(\varphi) \leqslant \beta e(\psi) - s(\psi)$ for all translationally invariant ψ (the minimality of the **mean free energy**), where $e(\varphi)=\lim N(\Lambda)^{-1}\varphi(U(\Lambda))=\lim N(\Lambda)^{-1}\varphi(H(\Lambda))$ (the **mean energy**), $s(\varphi)=\lim N(\Lambda)^{-1}S_\Lambda(\varphi)$ (the **mean entropy**), the infimum value $\beta e(\varphi)-s(\varphi)$ is $-P(\beta\Phi)$ with $P(\beta\Phi)=\lim N(\Lambda)^{-1}\log\tau_\Lambda(e^{-\beta U(\Lambda)})$ (the **pressure**), and the limits exist if $\Lambda \nearrow \mathbf{Z}^\nu$ is taken in the following **van Hove sense**: For any given cube C of lattice points, the minimal number $n_\Lambda^+(C)$ of translations of C that cover Λ and the maximal number $n_\Lambda^-(C)$ of mutually disjoint translations of C in Λ satisfy $n_\Lambda^+(C)/n_\Lambda^-(C) \to 1$ as $\Lambda \nearrow \mathbf{Z}^\nu$.

7. **Tangent to the pressure function**: $P(\Phi)$ is a continuous convex function on the Banach space of translationally invariant Φ with $\|\Phi\|<\infty$. A continuous linear functional α on this Banach space is a **tangent** to P at Φ if $P(\Phi+\Psi) \geqslant P(\Phi)+\alpha(\Psi)$ for all Ψ. For a translationally invariant state ψ, we define $\alpha_\psi(\Psi)=\psi(\sum_{I \ni 0} N(I)^{-1}\Psi(I))$. The condition is that $-\alpha_\varphi$ is a tangent to P at $\beta\Phi$. (Conversely, any tangent α to P at $\beta\Phi$ arises in this manner.)

The set K_β of all (normalized) β-KMS states is nonempty, compact, and convex. A β-KMS state φ is an extremal point of K_β if and only if it is factorial (i.e., the associated von Neumann algebra $\pi_\varphi(\mathfrak{A})''$ has a trivial center). It then has the **clustering property** $\lim_{a\to\infty}\{\varphi(A\gamma_a(B)) - \varphi(A)\varphi(\gamma_a(B))\}=0$ and is interpreted as a **pure phase**. Any β-KMS state has a unique integral decomposition into extremal β-KMS states.

For any Φ, K_β is a one-point set for sufficiently small $|\beta|$. For a 1-dimensional system ($\nu=1$), K_β consists of only one point (uniqueness of equilibrium states usually interpreted as indication of no phase transition) if the surface energy $W([-N,N])$ is uniformly bounded (H. Araki, *Comm. Math. Phys.*, 44 (1975); A. Kishimoto, *Comm. Math. Phys.*, 47 (1976)). For the two-body interaction $\Phi(\{m,n\})=-J|m-n|^{-\alpha}\sigma_z^{(m)}\sigma_z^{(n)}$, this condition is satisfied if $\alpha>2$ while $\|\Phi\|<\infty$ and α_t defined if $\alpha>1$. There is more than one KMS state (with spontaneous magnetization) for $2 \geqslant \alpha > 1$ and large $\beta J>0$, and hence a phase transition exists (F. J. Dyson, *Comm. Math. Phys.*, 12 (1969); J. Fröhlich and T. Spencer, *Comm. Math. Phys.*, 83 (1982)). If a 1-dimensional interaction has a finite range (i.e., $\Phi(I)=0$ if the diameter of I exceeds some number r_0) or if it is classical and $\sum_{I\ni 0} N(I)^{-1}(\operatorname{diam} I+1)\|\Phi(I)\|<\infty$ for $d=2$, then $\varphi(A)$ for $\varphi \in K_\beta$ and $A \in \mathfrak{A}(\Lambda)$ for a finite Λ is real analytic in β and any other analytic parameter in the potential (Araki, *Comm. Math. Phys.*, 14 (1969); [22]; M. Cassandro and E. Olivieri, *Comm. Math. Phys.*, 80 (1981)).

For a 2-dimensional **Ising model** with the nearest-neighbor ferromagnetic interaction [23], K_β consists of only one point for $0 \leqslant \beta \leqslant \beta_c$ while K_β for $\beta>\beta_c$ has exactly two extremal points corresponding to positive and negative magnetizations (M. Aizenman, *Comm. Math. Phys.*, 73 (1980); Y. Higuchi, *Colloquia Math. Soc. János Bolyai*, 27 (1979)). In this case, all KMS states are translationally invariant, while there exist (infinitely many) translationally noninvariant KMS states for sufficiently large β if $\nu=3$ (Dobrushin, *Theory Prob. Appl.*, 17 (1972); H. van Beijeren, *Comm. Math. Phys.*, 40 (1975)).

The accumulation points of β-KMS states as $\beta \to +\infty$ (or $-\infty$) provide examples of **ground** (or **ceiling**) **states** defined by any one of the following mutually equivalent conditions 1_+, 2_+ (or 1_-, 2_-) (O. Bratteli, A. Kishimoto, and D. W. Robinson, *Comm. Math. Phys.*, 64 (1978)):

1_+ (1_-). Positivity (negativity) of energy: For any $A \in \bigcup_\Lambda \mathfrak{A}(\Lambda)$, $-i\varphi(A^*\delta_\Phi(A))$ is real and positive (negative).

2_+ (2_-). Local minimality (maximality) of energy: For any finite subset Λ of $\mathbf{Z}^\nu$ and for any state ψ with the same restriction to $\mathfrak{A}(\Lambda^c)$ as the state φ under consideration, $\varphi(H(\Lambda)) \leqslant \psi(H(\Lambda))$ ($\varphi(H(\Lambda)) \geqslant \psi(H(\Lambda))$).

For translationally invariant potentials and states, the following condition is also equivalent to the above:

3_+ (3_-). Global minimality (maximality) of energy: $e(\varphi) \leqslant e(\psi)$ ($e(\varphi) \geqslant e(\psi)$) for all translationally invariant states ψ.

The totality of KMS, ground, and ceiling states can be characterized by the following formulation of the impossibility of **perpetual motion**: Let $P_t=P_t^* \in \mathfrak{A}$ be a norm-differentiable function of the time $t \in \mathbf{R}$ with a compact support, representing (external) time-dependent perturbations. Then there exists a unique perturbed time evolution α_t^P as a one-parameter family of *-automorphisms of $\mathfrak{A}$ satisfying $(d/dt)\alpha_t^P(A)=\alpha_t^P(\delta_\Phi(A)+i[P_t,A])$ for all $A \in \mathfrak{A}$ in the domain of δ_Φ. A state φ changes with time t as $\varphi_t(A)=\varphi(\tau_t^P(A))$ under the perturbed dynamics α_t^P, and the total energy given to the system (mechanical work performed by the external forces) is given by $L^P(\varphi)=\int_{-\infty}^{\infty}\varphi_t(dP_t/dt)\,dt$. For KMS states at any β, as well as ground and ceiling states, $L^P(\varphi)\geqslant 0$ for any P_t. If φ is a factor state, the converse holds, i.e., $L^P(\varphi)\geqslant 0$ for all P_t implies that φ is either a KMS, ground, or ceiling state. The condition $L^P(\varphi)\geqslant 0$ for all P_t is equivalent to $-i\varphi(U^*\delta_\Phi(U))\geqslant 0$ for all unitary U in the domain of δ_Φ and in the identity component of the group of all unitaries of $\mathfrak{A}$. A state φ satisfying this condition is called **passive**, and a state φ whose n-fold product with itself as a state on $\mathfrak{A}^{\otimes n}$ is passive relative to $\alpha_t^{\otimes n}$ for all n is called **completely passive**. The last property holds if and only if φ is a KMS, ground, or ceiling state (W. Pusz and S. L. Woronowicz, *Comm. Math. Phys.*, 16 (1970)).

The totality of KMS, ground, and ceiling states can be characterized by a certain stability under perturbations (P_t considered above) under some additional condition on α_t (R. Haag, D. Kastler, and E. B. Trych-Pohlmeyer, *Comm. Math. Phys.*, 38 (1974); O. Bratteli, A. Kishimoto, and D. W. Robinson, *Comm. Math. Phys.*, 61 (1978)).

When a lattice spin system is interpreted as a lattice gas, an operator $N_n\in\mathfrak{A}_n^{\mathrm{cl}}$ (such as $(\sigma_z^{(n)}+1)/2$) is interpreted as the particle number at the lattice site n and $N(\Lambda)=\sum_{n\in\Lambda}N_n$ is the particle number in Λ. It defines a representation of a unit circle T by automorphisms τ_θ of $\mathfrak{A}$ defined as $\tau_\theta(A)=\lim e^{iN(\Lambda)\theta}Ae^{-iN(\Lambda)\theta}$ $(\Lambda\nearrow\mathbf{Z}^\nu)$, called **gauge transformations** (of the first kind). The **grand canonical ensemble** can be formulated as a β-KMS state with respect to $\alpha_t\tau_{\mu t}$ (instead of α_t), where the real constant μ is called the **chemical potential**. It can be interpreted as an equilibrium state when the gauge-invariant elements $\{A\in\mathfrak{A}\mid\tau_\theta(A)=A\}$ instead of $\mathfrak{A}$ are taken to be the algebra of observables or as a state stable under those perturbations that do not change the particle number.

H. The Boltzmann Equation

Statistical mechanics of irreversible processes originated from the kinetic theory of gases. Long ago, Maxwell and Boltzmann tried to calculate viscosity and other physical quantities characterizing gaseous flow in nonequilibrium. The †Boltzmann equation is generally a nonlinear †integrodifferential equation. On the basis of this equation mathematical theories were developed by D. Enskog, S. Chapman, and D. Hilbert [2].

Free electrons in a metal can be regarded as forming an electron gas, in which electron scattering by lattice vibrations or by impurities is more important than electron-electron scattering. Following the example of gas theories H. A. Lorentz set forth a simple theory of irreversible processes of metallic electrons. His theory was, however, not quite correct, since metallic electrons are highly quantum-mechanical and classical theories cannot be applied to them. Quantum-mechanical theories of metal electrons were developed by A. Sommerfeld and F. Bloch.

I. Master Equations

The Boltzmann equation gives the velocity distribution function of a single particle in the system. This line of approach can be extended in two directions. The first is the so-called master equation. For example, consider a gaseous system consisting of N particles, and ask for the probability distribution of all the momenta, namely, the distribution function $f_N(p_1,\ldots,p_N;t)$, where $p_1,\ldots,p_N$ are the momenta of the N particles. The equations of motion are deterministic with respect to the complete set of dynamical variables $(x_1,p_1,\ldots,x_N,p_N)$. The equation for $f(p_1,\ldots,p_N,t)$ may not be deterministic, but it may be stochastic because we are concerned only with the variables $p_1,\ldots,p_N$, with all information about the space coordinates $x_1,\ldots,x_N$ disregarded. This situation is essentially the same in both classical and quantum statistical mechanics. If the duration of the observation process is limited to a finite length of time and the precision of the observation to a certain degree of crudeness, the time evolution of the momentum distribution function f_N can be regarded as a †Markov process. In general, an equation describing a Markov process of a certain distribution function is called a **master equation**. Typically it takes the following form for a suitable choice of variables x:

$$(\partial/\partial t)f(x,t)=\int dx'(W(x',x)f(x',t)-W(x,x')f(x,t)),\qquad(25)$$

where $W(x,x')$ is the transition probability from x to x'. By expanding the first integrand into a power series in $x-x'$, with x' fixed and by retaining the first few terms, we obtain the **Fokker-Planck equation**:

$$(\partial/\partial t)f(x,t)=-(\partial/\partial x)(\alpha_1(x)f(x,t))+(\partial^2/\partial x^2)(\alpha_2(x)f(x,t))/2,\qquad(26)$$

$$a_n(x)=\int W(x,x+r)r^n\,dr.\qquad(27)$$

J. The Hierarchy of Particle Distribution Functions

Another way of extending the Boltzmann equation is to consider a set of distribution functions of one particle, of two particles, and generally of n ($<N$) particles selected from the whole system of N particles. For example, a two-particle distribution function is the function $f_2(x_1, v_1, x_2, v_2, t)$ for positions and velocities of two particles at time t. The complete dynamics of the entire system of particles can be projected to the time evolution of this hierarchy of distribution functions. The equation of motion for f_1 then contains the function f_2 if the interaction of particles if pairwise, the equation for f_2 contains f_3, and so on. Thus the equations of motion for the set of distribution functions make a chain of equations. The whole chain is equivalent to the deterministic equations of motion for the dynamics of N particles. However, if the particle number N is made indefinitely large, with the time scale of observation always finite, the chain of equations for the distribution functions asymptotically approaches a stochastic process if certain conditions are satisfied. Approximate methods of solving the hierarchy equations in classical cases have been developed by J. Yvon, J. G. Kirkwood, M. Born, H. S. Green, and others.

In quantum statistics, similar hierarchy equations can be considered. A typical example is the so-called **Green's function method** [27].

K. Irreversible Processes and Stochastic Processes

The statistical mechanics of physical processes evolving in time is a hybrid of dynamics and the mathematical theory of stochastic processes. A typical example is the theory of Brownian motion. A colloidal particle floating in a liquid moves incessantly and irregularly because of thermal agitation from surrounding liquid molecules. For simplicity, an example of 1-dimensional Brownian motion is considered here. Phenomenologically we assume that a colloid particle follows an equation of motion of the form

$$m\dot{u} = -m\gamma u + f(t), \tag{28}$$

called the **Langevin equation**, where m is the mass of the colloid particle and u is the velocity. The first term on the right-hand side is the friction force due to viscous resistance, and the second term represents a random force acting on the particle from surrounding molecules.

If (28) describes the Brownian motion in thermal equilibrium, the friction constant $m\gamma$ and the random force cannot be independent, but are related by a theorem asserting that

$$m\gamma = \int_0^\infty \langle f(t_1) f(t_1 + t)\rangle \, dt. \tag{29}$$

In an electric conductor, the thermal motion of charge carriers necessarily induces irregularities of charge distribution, and so an electromotive force that varies in time in a random manner is created. This random electromotive force is similar to the random motion of a Brownian particle and is called the **thermal noise**. For such a thermal noise there exists a relation similar to (29) between the resistance and the random electromotive force. This relation is known as the **Nyquist theorem**. These theorems are contained in a more general theorem called the **fluctuation-dissipation theorem**.

When an external force is applied to a system in thermal equilibrium, some kind of irreversible flow, an electric current, for example, is induced in the system. The relationship between the flow and the external force is generally represented by an admittance. If the external force is periodic, the admittance is a function of frequency ω and is given by

$$\chi(\omega) = \int_0^\infty e^{-i\omega t} \psi(t) \, dt, \tag{30}$$

where $\psi(t)$ is equal to the correlation function of the flow that appears spontaneously as the fluctuation in thermal equilibrium when no external force is applied. This general expression for an admittance, often called the **Kubo formula**, gives a unified viewpoint from which responses of physical systems to weak external disturbances can be treated without recourse to the traditional kinetic approach.

The static limit ($\omega \to 0$) of the admittance is the **transport coefficient**. The reversibility of dynamics leads to relations among transport coefficients, called **Onsager's reciprocity relations** in the thermodynamics of irreversible processes.

When external disturbances are so large that the system deviates considerably from thermal equilibrium, the responses may show characteristic nonlinearities. Such nonlinear phenomena are important from both experimental and theoretical points of view, and constitute a central subject of modern research (→ 433 Turbulence and Chaos).

References

[1] L. Boltzmann, Lectures on gas theory, Univ. of California Press, 1964. (Original in German, 1912.)

[2] S. Chapman and T. G. Cowling, The mathematical theory of non-uniform gases, Cambridge Univ. Press, third edition, 1970.
[3] R. C. Tolman, The principles of statistical mechanics, Oxford Univ. Press, 1938.
[4] D. ter Haar, Elements of statistical mechanics, Holt, Rinehart and Winston, 1961.
[5] J. W. Gibbs, Elementary principles in statistical mechanics, Yale Univ. Press, 1902 (Dover, 1960).
[6] P. and T. Ehrenfest, The conceptual foundations of the statistical approach in mechanics, Cornell Univ. Press, 1959. (Original in German, 1911.)
[7] I. E. Farquhar, Ergodic theory in statistical mechanics, Interscience, 1964.
[8] V. I. Arnol'd and A. Avez, Ergodic problems of classical mechanics, Benjamin, 1968.
[9] R. Jancel, Foundations of classical and statistical mechanics, Pergamon, 1963.
[10] R. H. Fowler, Statistical mechanics, Cambridge Univ. Press, second edition, 1936.
[11] A. Ya. Khinchin, Mathematical foundation of statistical mechanics, Dover, 1949. (Original in Russian, 1943.)
[12] L. D. Landau and E. M. Lifshits, Statistical physics, Pergamon, 1969. (Original in Russian, 1964.)
[13] R. Kubo, Statistical mechanics, North-Holland, 1965.
[14] H. S. Green and C. A. Hurst, Order-disorder phenomena, Interscience, 1964.
[15] H. E. Stanley, Introduction to critical phenomena, Oxford Univ. Press, 1971.
[16] C. Domb and M. S. Green, Phase transitions and critical phenomena, Academic Press, 1972.
[17] O. Bratteli and D. W. Robinson, Operator algebras and quantum statistical mechanics II, Springer, 1981.
[18] D. Ruelle, Statistical mechanics: Rigorous results, Benjamin, 1969.
[19] D. Ruelle, Thermodynamic formalism, Addison-Wesley, 1978.
[20] D. W. Robinson, The thermodynamic pressure in quantum statistical mechanics, Springer, 1971.
[21] R. B. Israel, Convexity in the theory of lattice gas, Princeton Univ. Press, 1979.
[22] D. H. Mayer, The Ruelle-Araki transfer operator in classical statistical mechanics, Springer, 1980.
[23] B. M. McCoy and T. T. Wu, The two-dimensional Ising model, Harvard Univ. Press, 1973.
[24] N. S. Krylov, Works on the foundation of statistical physics, Princeton Univ. Press, 1979.
[25] K. Huang, Statistical mechanics, Wiley, 1963.
[26] F. Reif, Fundamentals of statistical and thermal physics, McGraw-Hill, 1965.
[27] A. A. Abrikosov, L. P. Gor'kov, and I. E. Dzyaloshinskiĭ, Methods of quantum field theory in statistical physics, Prentice-Hall, 1963. (Original in Russian, 1962.)
[28] E. H. Lieb and D. C. Mattis, Mathematical physics in one dimension, Academic Press, 1966.

403 (XVIII.5) Statistical Models

A. General Remarks

A statistical model is defined by specifying the structure of the probability distributions of the relevant quantities. When a statistical model is used for the analysis of a set of data, its role is to measure the characteristics of a certain configuration of the data points. R. A. Fisher [1] advanced a systematic procedure for the application of statistical models. The process of statistical inference contemplated by Fisher may be characterized by the following three phases: (1) specification of the model, (2) estimation of the unknown parameters, and (3) testing the goodness of fit. The last phase is followed by the first when the result of the testing is negative. Thus the statistical inference contemplated by Fisher is realized through the process of introduction and selection of statistical models.

We always assume that the true distribution of an observation exists in each particular application of statistical inference, even though it may not be precisely known to us. Our partial knowledge of the generating mechanism of the observation suggests various possible constraints on the form of the true distribution. The basic problem of statistical inference is then to generate an approximation to the true distribution by using the available observational data and a model defined by a set of probability distributions satisfying the constraints.

B. The Criterion of Fit

The use of statistical models can best be explained by adopting the predictive point of view, which defines the objective of statistical inference as the determination of the predictive distribution, the probability distribution of a future observation defined as a function of the information available at present. The performance of a statistical inference procedure is then evaluated in terms of the expected discrepancy of the predictive distribution from the true distribution of the future observation.

The probabilistic interpretation of thermodynamic entropy developed by L. Boltzmann [2] provides a natural measure of the discrepancy between two probability distributions. The **entropy of a distribution** specified by the density $f(y)$ with respect to the distribution specified by $g(y)$ is defined by

$$B(f;g) = -\int \frac{f(y)}{g(y)} \log\left[\frac{f(y)}{g(y)}\right] g(y)\,dy,$$

where, as in what follows, the integral is taken with respect to some appropriate measure dy. This definition of entropy is a faithful reproduction of the original probabilistic interpretation of the thermodynamic entropy by Boltzmann and allows the interpretation that the entropy $B(f;g)$ is proportional to the logarithm of the probability of getting a statistical distribution of observations closely approximating $f(y)$ by taking a large number of independent observations from the distribution $g(y)$. (For a detailed discussion → [3].)

Obviously we have

$$B(f;g) = \int f(y)\log g(y)\,dy - \int f(y)\log f(y)\,dy.$$

The second term on the right-hand side is a constant depending only on $f(y)$. The first term is the expected log likelihood of the distribution $g(y)$ with respect to the true distribution $f(y)$. Thus a distribution with a larger value of the expected log likelihood provides a better approximation to the true distribution. Even when $f(y)$ is unknown, $\log g(y)$ provides an unbiased estimate of the expected log likelihood. This fact constitutes the basis of the objectivity of the †likelihood as a criterion for judging the goodness of a distribution as an approximation to the true distribution.

C. Parametrization of Probability Distributions

When we construct a statistical model it is a common practice to represent the uncertain aspect of the true distribution by a family of probability distributions with unknown parameters. This type of family is called a parametric family; the model is called a parametric model. The parameters in a parametric model are the keys to the realization of the information extraction from data by statistical methods. Accordingly, the introduction of mathematically manageable parametric models forms the basis for the advance of statistical methods.

(1) Pearson's System of Distributions. A wide family of distributions can be generated by assuming a rational-function representation of the sensitivity of the density function $f(y)$ given by

$$\frac{d}{dy}\log f(y) = \frac{a_0 + a_1 y + \ldots + a_p y^p}{b_0 + b_1 y + \ldots + b_q y^q}.$$

Pearson's system of distributions is defined by putting $p=1$ and $q=2$ and by assuming various constraints on the parameters a_i and b_j and the support of $f(y)$ [4].

E. Wong [5] discussed the construction of continuous-time stationary Markov processes with the distributions of Pearson's system as their stationary distributions. This allows a structural interpretation of the parameters of a distribution of the system.

(2) Maximum Entropy Principle and the Exponential Family. To develop a formal theory of statistical mechanics E. T. Janes [6] introduced the concept of the maximum entropy estimate of a probability distribution. This concept leads to a natural introduction of the exponential family. Following Kullback [7], we start with a distribution $g(y)$ and try to find $f(y)$ with prescribed expectations of statistics $T_1(y), \ldots, T_k(y)$ and with maximum entropy $B(f;g)$. Such a distribution $f(y)$ is given by the relation

$$f(y) \propto \exp[\tau_1 T_1(y) + \ldots + \tau_k T_k(y)]\,g(y),$$

where it is assumed that the right-hand side is integrable. By varying the parameters $\tau_1, \ldots, \tau_k$ over the allowable range we get the exponential family of distributions. I. J. Good [8] considered the Janes procedure as a principle for the generation of statistical hypotheses and called it the maximum entropy principle.

(3) Parametric Models of Normal Distributions. Of particular interest within the exponential family is the family of normal distributions. This is obtained by assuming the knowledge of the first- and second-order moments of a distribution and applying the maximum entropy principle [9]. Obviously, the parametrization of a normal distribution is concerned only with the mean vector and the variance-covariance matrix.

Let $\mathbf{X} = (X_1, \ldots, X_n)'$ be an n-dimensional normal random variable with mean $E\mathbf{X} = (m_1, \ldots, m_n)'$ and variance-covariance matrix $\Sigma = (\sigma_{ij})$, where $\sigma_{ij} = E(X_i - m_i)(X_j - m_j)$. ($E$ denotes expectation and $'$ denotes the transpose.) A nonrestrictive family of n-dimensional normal distributions is characterized by $n + n(n+1)/2$ parameters, m_i $(i=1, \ldots, n)$, and σ_{ij} $(i=1, \ldots, n; j=1, \ldots, n)$. The prior information on the generating mechanism of $\mathbf{X}$ introduces constraints on these parameters and reduces the number of free parameters.

Reduction of the dimensionality of the

parameter vector $\mathbf{m}=(m_1,\ldots,m_n)'$ is realized by assuming that $\mathbf{m}$ is an element of a k-dimensional subspace of $\mathbf{R}^n$ spanned by the vectors $\mathbf{a}_1=(a_{11},\ldots,a_{1n})',\ldots,\mathbf{a}_k=(a_{k1},\ldots,a_{kn})'$, i.e., by assuming the relation

$$\mathbf{m}=A\mathbf{c},$$

where $A=(\mathbf{a}_1,\ldots,\mathbf{a}_k)$ is an $n\times k$ matrix and $\mathbf{c}=(c_1,\ldots,c_k)'$ is a k-dimensional vector with $k<n$. This parametrization is obtained when for each X_i the observation $(a_{i1},\ldots,a_{ik})$ is made on a set of k factors and the analysis of the linear effect of these factors on the mean of X_i is required. We have the representation $\mathbf{X}=A\mathbf{c}+\mathbf{W}$, where $\mathbf{W}$ is an n-dimensional normal random vector with $E\mathbf{W}=\mathbf{0}$ and variance-covariance matrix Σ (→ Section D).

To complete the model we have to specify the variance-covariance matrix Σ. One of the simplest possible specifications is obtained by assuming that the X_i are mutually independent and of the same variance σ^2. This reduces Σ to $\sigma^2 I$, where I denotes an $n\times n$ identity matrix. With this assumption the number of necessary parameters to represent the variance-covariance matrix reduces from $n(n+1)/2$ to 1. The model obtained with the assumptions $\Sigma=\sigma^2 I$ and $\mathbf{m}=A\mathbf{c}$ is called the general linear model (or regression model) with normal error, or simply the **normal linear model**. The model is called a regression model also when the a_i are random variables (→ e.g., [10, 11]).

A typical example of nontrivial parametrization of the covariance structure of $\mathbf{X}$ is obtained by assuming the representation

$$\mathbf{X}=\mathbf{m}+A\mathbf{F}+\mathbf{W},$$

where $\mathbf{F}=(f_1,\ldots,f_g)'$ denotes the vector of random effects and $\mathbf{W}$ the vector of measurement errors. It is assumed that $\mathbf{F}$ and $\mathbf{W}$ are mutually independent and normally distributed with $E\mathbf{F}=\mathbf{0}$ and $E\mathbf{W}=\mathbf{0}$. Also the components of $\mathbf{W}$ are assumed to be mutually independent. The variance-covariance matrix Σ of $\mathbf{X}$ is then given by

$$\Sigma=A\Phi A'+\Delta,$$

where $\Phi=E\mathbf{F}\mathbf{F}'$ and $\Delta=E\mathbf{W}\mathbf{W}'$, which is diagonal.

When A is a design matrix, the parametrization provides a **components-of-variance model** (or **random-effects model**) of which the main use is the measurement of the variance-covariance matrix Φ of the random effects $f_1,\ldots,f_g$ rather than the measurement of $\mathbf{F}$ itself. If we consider $\mathbf{F}$ to be representing the effects of some latent factors for which A is not uniquely specified, the above representation of Σ gives merely a formal, or noncausal, parametrization of Σ. In this case the model is called the **factor analysis model** and the dimension g of $\mathbf{F}$ is called the number of factors. By keeping the number of factors sufficiently smaller than the dimension of $\mathbf{X}$, we get a parametrization of Σ with a smaller number of free parameters than the unconstrained model. Starting with $g=1$ and successively increasing the number of factors, we can get a hierarchy of models with successively increasing numbers of parameters. (→ [12] for a very general modeling of the variance-covariance matrix.)

(4) Parametrization of Discrete Distribution. Consider the situation where the observation produces one of the events represented by $r=0,1,2,\ldots,k$ with probability $p(r)$, where k may be infinite. Represent by $\mathbf{X}=(X_1,\ldots,X_n)$ the result of n independent observations. The probability $p(\mathbf{X})$ of getting such a result is given by the relation

$$\log p(\mathbf{X})=\sum_{r=0}^{k}\theta_r n_r(\mathbf{X}),$$

where $\theta_r=\log p(r)$ and $n_r(\mathbf{X})$ denotes the number of X_i's which are equal to r. (The term not depending on the θ_r's is omitted in the above and subsequent formulas, since it is immaterial for problems of inference.) Thus a nonrestrictive model is obtained by assuming only the relations $\theta_r\leqslant 0$ and $\sum_{r=0}^{k}e^{\theta_r}=1$. Obviously the model defines an exponential family and various useful parametrizations are realized by introducing some constraints on the parameters θ_r.

When the events r are arranged in a 2-dimensional array (i,j) $(i=1,\ldots,m; j=1,\ldots,n)$ we have

$$\log p(\mathbf{X})=\sum_{i=1}^{m}\sum_{j=1}^{n}\theta_{ij}n_{ij}(\mathbf{X}).$$

One simple parametrization is given by

$$\theta_{ij}=\mu+\alpha_i+\beta_j+\gamma_{ij},$$

where it is assumed that $\sum_{i=1}^{m}\alpha_i=\sum_{j=1}^{n}\beta_j=\sum_{i=1}^{m}\gamma_{ij}=\sum_{j=1}^{n}\gamma_{ij}=0$. Obviously this is a parametrization of θ_{ij} as a linear function of the parameters α_i, β_j, and γ_{ij}, and the model thus obtained is called the **log linear model**. The model shows a formal similarity to the analysis-of-variance model (→ Section D). By introducing successively more restrictive assumptions on the parameters, we can get a hierarchy of models for the analysis of a two-way contingency table. Extension to cases when more than two factors are involved is obvious (→ e.g., [13]).

Here we consider that X_i is a dichotomous variable, i.e., $k=1$, and that the probability of $X_i=1$ may depend on i, i.e., we have

$$\log p(X_i)=n\theta_{0i}+(\theta_{1i}-\theta_{0i})n_1(X_i),$$

where $\theta_{0i} = \log \mathrm{Prob}(X_i = 0)$ and $\theta_{1i} = \log \mathrm{Prob}(X_i = 1)$. We assume that a vector of observations $\mathbf{a}_i = (a_{i1}, \ldots, a_{im})'$ is available simultaneously with X_i and that we are interested in analyzing the relation between $\mathbf{a}_i$ and the probability distribution of X_i. The analysis is realized by exploring the functional relation between $\tau_i = \theta_{1i} - \theta_{0i}$ and $\mathbf{a}_i$. We can assume a linear relation

$$\boldsymbol{\tau} = A\mathbf{c},$$

where $\boldsymbol{\tau} = (\tau_1, \ldots, \tau_n)'$, $A = (\mathbf{a}_1, \ldots, \mathbf{a}_n)'$, and $\mathbf{c} = (c_1, \ldots, c_m)'$. The parameter $\tau_i = \log\{p(X_i = 1)/(1 - p(X_i = 1))\}$ is the log odds ratio or logit of the event $X_i = 1$, and the model is called the **linear logistic model** [14]. A hierarchy of models can be generated by assuming a successively more restrictive linear relations among the components of $\mathbf{C}$.

D. General Linear Models

Another class of models often used in practical applications is composed of general linear models or linear regression models, where the observed value is considered to be the sum of the effects of some fixed causes and the error.

Let $\mathbf{X} = (X_1, \ldots, X_n)'$ be an n-dimensional †random variable, and denote the expectation of $\mathbf{X}$ by $E(\mathbf{X}) = (\mu_1, \ldots, \mu_n)'$. If $E(\mathbf{X})$ is of the form $A\xi$ with an unknown parameter $\xi = (\xi_1, \ldots, \xi_k)'$, and a given $n \times k$ matrix A, then we can express $\mathbf{X}$ as

$$\mathbf{X} = A\xi + \mathbf{W}, \qquad E(\mathbf{W}) = (0, \ldots, 0)', \tag{1}$$

with the **error term** $\mathbf{W} = (W_1, \ldots, W_n)'$. We frequently assume a set of conditions on the distribution of $\mathbf{X}$; for example, (i) $X_1, \ldots, X_n$ are mutually †independent, (ii) $X_1, \ldots, X_n$ have a common unknown †variance σ^2, (iii) $(X_1, \ldots, X_n)$ is distributed according to an n-dimensional †normal distribution. The equations (1) together with conditions on the distribution are called a **linear model**.

Among the methods of statistical analysis of linear models are regression analysis, analysis of variance, and analysis of covariance as explained below, but these are not clearly distinguished from each other. (I) In **design-of-experiment analysis**, i.e., **analysis of variance**, the matrix A and the vector ξ in (1) are called a **design matrix** and an **effect**, respectively. In this case entries of A are assumed to be either 1 or 0. (II) In **regression analysis**, we are first given a linear form $x = \sum_{j=1}^{k} a_j \xi_j$ of a vector $\mathbf{a} = (a_1, \ldots, a_k)'$ with coefficient vector $\xi = (\xi_1, \ldots, \xi_k)'$. Let $X_1, \ldots, X_n$ be the observed values of x at n points $\mathbf{a}_1 = (a_{11}, \ldots, a_{1k})', \ldots, \mathbf{a}_n = (a_{n1}, \ldots, a_{nk})'$, respectively, where $n > k$. If the observations are unbiased, that is, if $E(X_i) = \sum_{j=1}^{k} a_{ij} \xi_j$, $i = 1, \ldots, n$, then the model (1) is obtained with $A = (a_{ij})$. Usually one of the components of the vector $\mathbf{a}$ is taken as unity. In this framework the form $x = \sum_j a_j \xi_j$ is called the **linear regression function** or **regression hyperplane**, and for $k = 2$, the graph of the linear function $x = a_1 \xi_1 + \xi_2$ and its coefficient ξ_1 are called the **regression line** and **regression coefficient**, respectively. The components of the vector $\mathbf{a}$ are called **fixed variates** or **explanatory variables**. Frequently we encounter the case where the vector $\mathbf{a}$, and consequently the matrix A, are random variables. When this is the case, a discussion like that above can be carried out for given A by regarding $A\xi$ in (1) as the conditional expectation of the vector $\mathbf{X}$.

E. The Method of Least Squares

Consider the subspace $L(A)$ of the †sample space $\mathbf{R}^n$ spanned by the column vectors of A. Then the dimension s of $L(A)$ equals the rank of A, and $L(A)$ and its †orthocomplement $L^{\perp}(A)$ are called the **estimation space** and the **error space**, respectively. The †orthogonal projection $\mathbf{y}$ of a point $\mathbf{x}$ to the space $L(A)$ is expressed as $\mathbf{y} = P_A \mathbf{x}$ with a real †projection matrix P_A. The variable $\mathbf{Y} = P_A \mathbf{X}$ is called the **least squares estimator** of $E(\mathbf{X})$, and the routine of getting such an estimator $\mathbf{Y}$, called the **method of least squares**, minimizes the squared error $(\mathbf{X} - A\xi)'(\mathbf{X} - A\xi)$ for a given $\mathbf{X}$. This method consists of two operations solving the **normal equation** $A'A\xi = A'\mathbf{X}$ with respect to ξ, and setting $\mathbf{Y} = A\hat{\xi}$, where $\hat{\xi}$ is a solution of the equation. For $s = k$, we obtain $\mathbf{Y} = A(A'A)^{-1}A'\mathbf{X}$ directly. Even when $s < k$, where the solution of the normal equation is not unique, $\mathbf{Y}$ is uniquely determined. The quantity $Q = \mathbf{X}'(I - P_A)\mathbf{X}$, where I is the unit $n \times n$ matrix, is the squared distance of the point $\mathbf{X}$ from the space $L(A)$ and is called the **error sum of squares with $n - s$ degrees of freedom**.

A linear function $\boldsymbol{\beta}'\xi$ of the parameter ξ with coefficient vector $\boldsymbol{\beta} = (\beta_1, \ldots, \beta_k)'$ is called a **linearly estimable parameter** (or **estimable parameter**) if there is a linear unbiased estimator, that is, an unbiased estimator of the form $\mathbf{b}'\mathbf{X}$, of $\boldsymbol{\beta}'\xi$. In order that $\boldsymbol{\beta}'\xi$ be estimable it is necessary and sufficient that $\boldsymbol{\beta}'$ be a linear combination $\mathbf{u}'A$ of the row vectors of the matrix A. A linear unbiased estimator that has minimum variance among all linear unbiased estimator uniformly in ξ is called the **best linear unbiased estimator (b.l.u.e.)**. If the conditions (i) and (ii) of Section D are satisfied, then for any given n-vector $\mathbf{u}$ the b.l.u.e. of a parameter $\gamma = \mathbf{u}'A\xi$ is given by $\hat{\gamma} = \mathbf{u}'\mathbf{Y}$ with $\mathbf{Y} = P_A \mathbf{X}$, and its variance equals $(\mathbf{u}'P_A\mathbf{u})\sigma^2$, while the expectation of the quantity Q is given by $(n-s)\sigma^2$. This proposition is known as the **Gauss-Markov theorem**. Hence the b.l.u.e. $\hat{\gamma}=\mathbf{u}'\mathbf{Y}$ is frequently cited as the least squares estimator of γ. The quantity $\hat{\sigma}^2=Q/(n-s)$ is an unbiased estimator of σ^2 and is called the **mean square error**. If in addition the condition (iii) of Section D is assumed, then $\hat{\gamma}$ and $\hat{\sigma}^2$ are the †uniformly minimum variance unbiased estimators of γ and σ^2, respectively.

When the error term $\mathbf{W}$ in (1) has covariance matrix $\Sigma=\sigma^2\Sigma_0$ with an unknown real parameter σ^2 and a known matrix Σ_0, the valaue of the parameter ξ minimizing $Q=(\mathbf{X}-A\xi)'\Sigma_0^{-1}(\mathbf{X}-A\xi)$ is called the **generalized least squares estimator** of ξ if it exists. This estimator has properties similar to those of the least squares estimator.

F. Model Selection and the Method of Maximum Likelihood

When a parametric family of distributions $\{f(\cdot\,|\,\theta);\theta\in\Theta\}$ is given and an observation x is made, $\log f(x\,|\,\theta)$ provides an unbiased estimate of the expected log likelihood of the distribution $f(\cdot\,|\,\theta)$ with respect to the true distribution of the observation. The value of θ which maximizes this estimate is the maximum likelihood estimate of the parameter and is denoted by $\theta(x)$ (→ 399 Statistical Estimation).

In a practical application we often have to consider a **multiple model**, defined by a set of component models $\{f_i(\cdot\,|\,\theta_i);\theta_i\in\Theta_i\}$ $(i=1,\ldots,k)$. The problem of model selection is concerned with the selection of a **component model** from a multiple model. The difference of the difficulties of handling a **simple model** defined by a one-component model, and a multiple model is quite significant. For a simple model $\{f(\cdot\,|\,\theta);\theta\in\Theta\}$, each member of the family is a probability distribution. In the case of a multiple model, its member is a model which is simply a collection of distributions and does not uniquely specify a probabilistic structure for the generation of an observation. Thus the likelihood of each component model with respect to the observation x cannot be defined and the direct extension of the method of maximum likelihood to the problem of model selection is impossible. This constitutes a serious difficulty for the handling of multiple models. Apparently, Fisher used the procedure of testing to solve this difficulty.

(1) Analysis of log Likelihood Ratios. The procedure of model selection by testing, which is applicable to a wide class of models, is the method of analysis of log likelihood ratios [15]. Consider the situation where a model is to be determined by using a hierarchy of models $\{f(\cdot\,|\,\theta_i);\theta_i\in\Theta_i\}$ $(i=1,\ldots,k)$ such that $\Theta_1\subset\Theta_2\subset\ldots\subset\Theta_k$. The comparison of models is then realized through the comparison of the maximum likelihoods $f(x\,|\,\theta_i(x))$, where $\theta_i(x)$ denotes the maximum likelihood estimate of θ_i based on the data x. For $\Theta_i\subset\Theta_j$ the log likelihood ratio is defined by

$$\Lambda(\Theta_i/\Theta_j;x)=-2\log\{f(x\,|\,\theta_i(x))/f(x\,|\,\theta_j(x))\}.$$

The analysis of log likelihood ratios is realized by the decomposition

$$\Lambda(\Theta_1/\Theta_k;x)=\Lambda(\Theta_1/\Theta_2;x)+\Lambda(\Theta_2/\Theta_3;x)+\ldots$$
$$+\Lambda(\Theta_{k-1}/\Theta_k;x).$$

The log likelihood ratios $\Lambda(\Theta_{k-1}/\Theta_k;x)$, $\Lambda(\Theta_{k-2}/\Theta_{k-1};x),\ldots,\Lambda(\Theta_1/\Theta_2;x)$ are successively tested by referring to chi-square distributions with the degrees of freedom $d(k)-d(k-1)$, $d(k-1)-d(k-2),\ldots,d(2)-d(1)$, respectively, where $d(i)$ denotes the dimension of the manifold Θ_i. The assumption of the chi-square distributions is only asymptotically valid under the usual regularity conditions (→ 400 Statistical Hypothesis Testing). The model defined with Θ_i for which $\Lambda(\Theta_i/\Theta_{i-1};x)$ first becomes significant is selected and $f(y\,|\,\theta_i(x))$ is accepted as the predictive distribution. The problem of how to choose the levels of significance to make the test procedure a procedure for model selection remains open.

(2) Model selection by AIC. One way out of the difficulty of model selection is to assume a prior distribution over Θ_i for each model $\{f_i(\cdot\,|\,\theta_i);\theta_i\in\Theta_i\}$. This leads to Bayesian modeling, which is discussed in Section G. Another possibility is to replace each component model $\{f_i(\cdot\,|\,\theta_i);\theta_i\in\Theta_i\}$ by a distribution $f_i(\cdot\,|\,\theta_i(x))$ specified by the maximum-likelihood estimate $\theta_i(x)$. The problem here is how to define the likelihood of each distribution $f_i(\cdot\,|\,\theta_i(x))$. An information criterion **AIC** was introduced by H. Akaike [16] for this purpose; it is defined by

$$\mathrm{AIC}=(-2)\log_e(\text{maximum likelihood})$$
$$+2\,(\text{number of estimated parameters}).$$

We may consider -0.5 AIC to be the log "likelihood" of $f(\cdot\,|\,\theta(x))$ which is corrected for its bias as an estimate of $E_xE_y\log f(y\,|\,\theta(x))$, where E_x denotes the expectation with respect to the true distribution of x, and where it is assumed that x and y are independent and identically distributed. The maximum "likelihood" estimate of the model is then defined by the model with minimum AIC. This realizes a procedure of model selection that avoids the ambiguity of the testing procedure. It is applicable, at least formally, even to the case of a nonhierarchical set of models.

G. Bayesian Models

Consider the situation where an observation x is made and it is desired to produce an estimate $p(y|x)$ of the true distribution of a future observation y. $p(y|x)$ is called a **predictive distribution**. Assume that x and y are sampled from one of the distributions within the family $\{g(y|\theta)f(x|\theta); \theta\in\Theta\}$, i.e., x and y are stochastically independent but share common structural information represented by θ. As a design criterion of $p(y|x)$ we assume a probability distribution $\pi(\theta)$ of θ. The model $\{f(\cdot|\theta); \theta\in\Theta\}$ with $\pi(\theta)$ is called a **Bayesian model** and $\pi(\theta)$ is called the **prior distribution**. From the relation

$$E_\theta E_{x|\theta} E_{y|\theta} \log p(y|x) = E_x E_{y|x} \log p(y|x),$$

where $E_{y|x}$ denotes the expectation of y conditional on x and E_x the expectation with respect to the marginal distribution of x, it can be seen that the optimal choice of $p(y|x)$ which maximizes the expected log likelihood is given by the conditional distribution

$$p(y|x) = \int g(y|\theta)p(\theta|x)\,d\theta,$$

where $p(\theta|x)$ is the **posterior distribution** of θ defined by

$$p(\theta|x) = f(x|\theta)\pi(\theta)\left(\int f(x|\theta)\pi(\theta)\,d\theta\right)^{-1}.$$

When a prior distribution $\pi(\theta)$ is specified, the parametric family of distributions $\{f(\cdot|\theta); \theta\in\Theta\}$ is converted into a stochastic structure which specifies a probability distribution of the observations. The likelihood of the structure, or the Bayesian model, with respect to an observation x is defined by

$$\int f(x|\theta)\pi(\theta)\,d\theta.$$

When there is uncertainty about the choice of the prior distribution we can consider a set of possible prior distributions and apply the method of maximum likelihood. Such a procedure is called the method of type II maximum likelihood by I. J. Good [17]. For a multiple model $\{f_i(\cdot|\theta_i); \theta_i\in\Theta_i\}$ $(i=1, \ldots, k)$, if prior distributions $\pi_i(\theta_i)$ are defined, a model selection procedure is realized by selecting the Bayesian model with maximum likelihood.

Bayesian modeling has often been considered as not quite suitable for scientific applications unless the prior distribution is objectively defined. However, even the construction of an ordinary statistical model is always heavily dependent on our subjective judgment. Once the objective nature of the likelihood of a Bayesian model is recognized, the selection or determination of a Bayesian model can proceed completely analogously to Fisher's scheme of statistical inference (→ e.g., [18]). The basic underlying idea of both the minimum AIC procedure and Bayesian modeling is the balancing of the complexity of the model against the amount of information available from the data. This unifying view of the construction of statistical models is obtained by the introduction of entropy as the criterion for judging the goodness of fit of a statistical model (→ [19] for more details).

References

[1] R. A. Fisher, Statistical methods for research workers, Oliver & Boyd, 1925; Hafner, fourteenth edition, 1970.
[2] L. Boltzmann, Über die Beziehung zwischen dem Zweiten Hauptsatze der mechanischen Wärmetheorie und der Wahrscheinlichkeitsrechnung respektive den Sätzen über das Wärmegleichgewicht, Wiener Berichte, 76 (1877), 373–435.
[3] I. N. Sanov, On the probability of large deviations of random variables, IMS and AMS Selected Transl. Math. Statist. Prob., 1 (1961), 213–244. (Original in Russian, 1957.)
[4] K. Pearson, Contributions to the mathematical theory of evolution II, Skew variation in homogeneous material, Philos. Trans. Roy. Soc. London, ser. A, 186 (1895), 343–414. Also included in Karl Pearson's early statistical papers, Cambridge Univ. Press, 1948, 41–112.
[5] E. Wong, The construction of a class of stationary Markoff processes, Proc. Amer. Math. Soc. Symp. Appl. Math., 16 (1963), 264–276. Also included in A. H. Haddad (ed.), Nonlinear systems, Dowden, Hutchinson & Ross, 1975, 33–45.
[6] E. T. Janes, Information theory and statistical mechanics, Phys. Rev., 106 (1957), 620–630.
[7] S. Kullback, Information theory and statistics, Wiley, 1959 (Dover, 1967).
[8] I. J. Good, Maximum entropy for hypothesis formulation, especially for multidimensional contingency tables, Ann. Math. Statist., 34 (1963), 911–934.
[9] C. E. Shannon and W. Weaver, The mathematical theory of communication, Univ. of Illinois Press, 1949.
[10] S. R. Searle, Linear models, Wiley, 1971.
[11] F. A. Graybill, Theory and application of the linear model, Duxbury Press, 1976.
[12] K. G. Jöreskog, A general method for analysis of covariance structures, Biometrika, 57 (1970), 239–251.
[13] Y. M. M. Bishop, S. E. Feinberg, and P. W. Holland, Discrete multivariate analysis: Theory and practice, MIT Press, 1975.

[14] D. R. Cox, The analysis of binary data, Chapman & Hall, 1970.
[15] I. J. Good, Comments on the paper by Professor Anscombe, J. Roy. Statist. Soc., ser. B, 29 (1967), 39–42.
[16] H. Akaike, A new look at the statistical model identification, IEEE Trans. Automatic Control, AC-19 (1974), 716–723.
[17] I. J. Good, The estimation of probabilities, MIT Press, 1965.
[18] G. E. P. Box, Sampling and Bayes' inference in scientific modeling and robustness, J. Roy. Statist. Soc., ser. A, 143 (1980), 383–430.
[19] H. Akaike, A new look at the Bayes procedure, Biometrika, 65 (1978), 53–59.

404 (XVIII.14) Statistical Quality Control

A. General Remarks

According to the Japanese Industrial Standard (JIS) Z 8101, "**Quality Control** (QC) is a system comprising all the methods used in manufacturing products or providing services economically that meet the quality requirements of consumers." To emphasize that modern quality control makes use of statistical methods, it is sometimes referred to as **Statistical Quality Control** (SQC). In order to implement effective QC, statistical concepts and methods must be applied and the "Plan-Do-Check-Action" (PDCA) cycle must be followed in research and development, design, procurement, production, sales, and so on. These QC activities are executed on a company-wide basis from the top management to the production workers. This type of QC is called Company-Wide Quality Control (CWQC) or Total Quality Control (TQC).

The quality Q is an abstract notion of the conformity of a product or service to consumers' requirements; it also refers to the total of the characteristics of a product or service as perceived by consumers. The quality characteristics may include both measurable physical and/or chemical features, such as strength and purity, or features such as color or texture as appreciated by individuals. These latter characteristics could be called "consumer qualities." Furthermore, the concept "quality" has also been used to describe the social impact of a product or service. This might be called "social quality." Examples of social-quality issues are pollution by solid waste or drainage in the production stage; degradation, maintainability, and safety of a product in daily use; and pollution following disposal. For a product or service to conform to this sort of quality it has become necessary to conduct QC activities not only during production but also at early stages of design and development of new products.

The measured characteristics of quality vary from one product to another because of natural variability in the material and production process involved, ability of individual workers, errors in different sorts of measurement, etc. If the variations among the measured values from a process can be attributed to "chance causes" and their distribution expressed by a probability or a probability density function, the process is said to be in a "state of statistical control" according to W. A. Shewhart or in a "stable state" by JIS. In this case the value of a characteristic is deemed to be the realization of a random variable X. Sometimes the variations are attributed to "assignable causes," which must be identified and eliminated.

B. Control Charts

The control chart provides a means of evaluating whether a process is in a stable state.

The control chart is made by plotting points illustrating a statistic of the quality characteristics or manufacturing conditions for an ordered series of samples or subgroups. A sheet of the control chart is provided with a middle line between a pair of lines depicting the upper control limit (UCL) and the lower control limit (LCL). The stable state is assumed to be exhibited by points within the control limits. Points falling outside the control limits suggest some assignable causes, which should be eliminated through corrective measures.

The idea underlying control charts as developed by Shewhart is to apply the statistical principle of significance to the control of production processes. Other types of control charts have also been developed, for example, acceptance control charts and adaptive control charts. These have been successfully applied to many quality control problems.

The foundation of Shewhart's control chart is the division of observations into what are called "rational subgroups." A rational subgroup is the one within which variability is due only to chance causes. Between different subgroups, however, variations due to assignable causes might be detected. In most production processes the rational subgroup comprises the data collected over a short period of time during which essentially the same condition in material, tool setting, environmental factors, etc., prevails.

The limits on the control charts are placed, according to Shewhart, at a 3σ distance from the middle line, where σ is the population standard deviation (or standard error) of the statistic; 3σ expresses limits of variability within the subgroup. Assuming that the population distribution of an observed characteristic is "normal," the range between the limits should include 99.7% of the points plotted so long as the process is "in control" at the middle value. Accordingly, 0.3% of the plotted points from the "in control" process fall outside the limits, and thus give an erroneous "out of control" signal.

To determine that the process is in control for a normally distributed characteristic $N(\mu, \sigma^2)$, we have to investigate the variability between the means μ and the standard deviations σ of different distributions of X for different subgroups. Thus the state of control of a process is determined with control charts for

$$\bar{X} = \sum_{i=1}^{n} X_i/n \quad \text{and} \quad s = \sqrt{\sum_{i=1}^{n} (X_i - \bar{X})/(n-1)},$$

which are the appropriate statistics corresponding to μ and σ. Despite the theoretical drawback of the statistical range $R = \max_i X_i - \min_i X_i$ against s, use of the range is often preferred in QC work because of its simplicity in computation. Hence the $\bar{X}-R$ charts are obtained from the previously collected k rational subgroups each of size n as follows:

$$\mathrm{UCL} = \bar{\bar{X}} + A_2\bar{R}, \qquad \mathrm{UCL} = D_4\bar{R},$$

$$\mathrm{LCL} = \bar{\bar{X}} - A_2\bar{R}. \qquad \mathrm{LCL} = D_3\bar{R},$$

where $\bar{\bar{X}}$ and $\bar{R}$ are the averages of the k values of $\bar{X}$ and R, respectively, and

$$A_2 = \frac{3}{\sqrt{n}d_2}, \qquad D_4 = 1 + 3\frac{d_3}{d_2}, \qquad D_3 = 1 - 3\frac{d_3}{d_2},$$

$$E[R] = d_2\sigma, \qquad V[R] = E[(R - E[R])^2] = {d_3}^2\sigma^2.$$

For $n < 7$, LCL for R cannot be given because D_3 becomes negative.

The other commonly used control charts are the p chart (proportion of nonconformity: binomial distribution) and the c chart (number of defects: Poisson distribution). For those charts the above theory of normal distribution is also used to approximate the binomial and Poisson values.

It is generally sufficient to use the agreed-on decision criterion (3σ limits) and to recognize a relatively small risk ($\alpha = 0.003$) for practical purposes. It should be noted, however, that a shift of the process mean μ by 1σ would not be observed at a ratio of 97.7%, which is the value of the risk β for each plotted point, under the normal distribution of the plotted statistic. One reason why the control chart has been a practical tool in many applications is this lack of sensitivity to a relatively small shift of the level. If greater sensitivity is required, 2σ limits, "warning limits" are used. This results in a greater risk α of erroneously finding a process out of control.

Other decision criteria based on aspects of run theory are also used. Charts using accumulated data from several rational subgroups for each plotted value are sometimes recommended: Moving average and moving range charts and the cusum ("cumulative sum") chart are examples. The statistical theory for these charts is more complicated than that for the simple charts discussed here.

C. Sampling Inspection

Sampling inspection determines whether a lot should be accepted or rejected by drawing a sample from it, observing a quality characteristic of the sample, and comparing the observed value to a prescribed acceptance criterion. Sampling may be conducted in several stages. Definite criteria are required to decide at each stage whether to accept or reject the lot or continue sampling on the basis of sample values observed so far. There also must be some rules to determine the size of the next sample if it is to be taken. These criteria and rules together are called the **sampling inspection plan**. The number of samples eventually drawn and observed and their sizes are generally random variables. In **single sampling inspection** the final decision is always reached after one stage of sampling is completed. **Double sampling inspection** makes the final decision after at most two stages of sampling are completed. **Multiple sampling inspection** makes the final decision after at most N stages of sampling are completed ($N < \infty$). Inspection without a predetermined limit on the number of sampling stages, **sequential sampling inspection**, is usually constructed so that the probability of the indefinite continuation of sampling is 0.

Once a sampling inspection plan is determined, the probability for accepting a lot with given composition can be calculated. This probability as a function of lot composition is called the **operating characteristic** of the plan. In most cases, the quality of a lot is expressed by a real parameter θ (e.g., fraction defective, i.e., percentage of defective products, or the average of some quality characteristic), and we use only inspection plans whose operating characteristics are expressed as a function of θ. The graph of this function is called an **OC-curve**. We impose certain desirable conditions on the OC-curve and design plans to satisfy them. Tables for this purpose, **sampling inspection tables**, are prepared for practical use. The condition most frequently employed is expressed in the following form in terms of four constants $\theta_0, \theta_1, \alpha, \beta$: The probability of rejection is required to be at most α when $\theta \leqslant \theta_0$ (or $\theta \geqslant \theta_0$), and the probability of acceptance at most β when $\theta \geqslant \theta_1$ (or $\theta \leqslant \theta_1$). Here α is called the **producer's risk**, and β the **consumer's risk**.

If the rejection of a lot is identified with the rejection of a statistical hypothesis $\theta \leqslant \theta_0$, the OC-curve is actually the power curve of the test upside down (i.e., the graph of 1 minus the †power function), and the producer's and consumer's risks are precisely the †errors of the first and second kind. The choice of a plan is actually the choice of a test under certain conditions on its power curve. Commonly used plans are mostly based on well-established tests, some of which have certain optimum properties. A few examples, (1)–(4), are given below. Here **sampling inspection by attribute** is an inspection plan that uses a statistic with a discrete distribution, whereas **sampling inspection by variables** uses a statistic with a continuous distribution (→ 400 Statistical Hypothesis Testing).

(1) Single sampling inspection by attribute concerning the fraction of defective items in a lot: Let the defective fraction be denoted by p and identified with θ in the preceding paragraph. Assign two values of p, say p_0 and p_1 $(0<p_0<p_1<1)$, the producer's risk α, and the consumer's risk β. Together they give conditions to be fulfilled by the OC-curve. Draw n items from a lot at random, and suppose that they contain Z defective items. The decision is then made after observing Z, whose distribution is †hypergeometric and approximately †binomial when the size of the lot is large enough. There exists a plan that minimizes n among all plans satisfying the imposed conditions under either of the two assumptions about the distribution of Z. It rejects the lot when Z is greater than a fixed number determined by p_0, p_1, α, and β. This plan is based on the †UMP test of the hypothesis $p \leqslant p_0$ against the alternative $p \geqslant p_1$.

(2) Single sampling inspection by variables concerning the population mean μ in the case where the population distribution is $\mathbf{N}(\mu, \sigma^2)$ with known σ^2: Draw a sample $(X_1, \ldots, X_n)$ of size n from a lot. Assume that the X are independently distributed with the same distribution $\mathbf{N}(\mu, \sigma^2)$. Suppose that smaller values of the quality characteristic stand for a more desirable quality. If two values of μ, say μ_0 and μ_1, α, and β are assigned, a plan can be established to minimize n. It rejects the lot when the sample mean $\bar{X} = \sum_{i=1}^n X_i/n$ exceeds a fixed number determined by μ_0, μ_1, α, and β. This too is based on the UMP test of $\mu \leqslant \mu_0$ against $\mu \geqslant \mu_1$.

(3) Cases where the samples are drawn in more than one stage: As in (2), assign two values of θ, α, and β; there is still liberty to choose $n_1, n_2, \ldots$, which are the sizes of the samples drawn at each stage. Hence there are many possible plans fulfilling the imposed conditions. Among them a plan is sought to minimize the expectation of $n = n_1 + n_2 + \ldots$ (called the **average sample number**). For example, plans based on the sequential probability ratio tests are in common use (→ 400 Statistical Hypothesis Testing).

(4) Among other special plans, **sampling inspection with screening** and **sampling inspection with adjustment** are worthy of mention. In the first plan, all the units in the rejected lots are inspected and defective units replaced by nondefective ones. In this case, fixing p_1 (the **lot tolerance percent defective**) and β, or the average fraction defective after the inspection (**average outgoing quality level**), we attempt to minimize the **expected amount of inspection**, that is, the expected number of inspected units including those in the rejected lots. In the second plan, acceptance criteria are tightened or loosened according to the quality of the lots just inspected.

References

[1] J. M. Juran, Quality control handbook, McGraw-Hill, 1974.
[2] Japanese Standard Association, Terminology, JIS Z 8101, 1963.
[3] Japanese Standard Association, Sampling inspection, JIS Z 9001–9006, 1957.
[4] Japanese Standard Association, Control charts, JIS Z 9021–9023, 1963.

405 (XVII.16) Stochastic Control and Stochastic Filtering

A. General Description of Stochastic Control

Stochastic control is an optimization method for systems subject to random disturbance. Let Γ be a compact convex subset of $\mathbf{R}^k$, called a control region. Let W_t be an †n-dimensional Brownian motion and $\sigma_t(W) = \sigma(W_s; s \leqslant t)$ (say $\mathscr{F}_t$) be the least †σ-field for which W_s, $s \leqslant t$, are measurable. An †$\mathscr{F}_t$-progressible measurable Γ-valued process is called an **admissible control**. For an admissible control U_t the system evolves according to an n-dimensional **controlled stochastic differential equation** (CSDE)

$$dX_t = \alpha(X_t, U_t)\, dW_t + \gamma(X_t, U_t)\, dt,$$

where a symmetric $n \times n$ matrix $\alpha(x, u)$ and n-vector $\gamma(x, u)$ are continuous in $\mathbf{R}^n \times \Gamma$ and Lipschitz continuous in $x \in \mathbf{R}^n$. Hence the CSDE has a unique solution, called the **response for** U_t. The problem is to maximize (or minimize) the performance J:

$$J(\tau, x, \varphi, U) = E_x\left[\int_0^\tau e^{-\int_0^t c(X_s, U_s)\, ds} f(X_t, U_t)\, dt + e^{-\int_0^\tau c(X_s, U_s)\, ds} \varphi(X_\tau)\right],$$

where X_t is the response for U_t with $X_0 = x$ and τ is a constant time or a †hitting time associated with a target set. We put $V(\tau, x, \varphi) = \sup_{U \in \mathrm{adm.\,control}} J(\tau, x, \varphi, U)$ the **value function** as a function of x. If the supremum value is attained at an admissible control $\tilde{U}_t$, then $\tilde{U}_t$ is called an **optimal control**.

B. Bellman Principle

In order to get $V(t+s, x, \varphi)$, R. Bellman applied the following two-stage optimization. After using any U up to time t, a controller changes U to an optimal one. Then at time $t+s$ the performance $J(t, x, V(s, \cdot, \varphi), U)$ is obtained. Taking the supremum with respect to U, one gets $V(t+s, x, \varphi)$. This is called the **Bellman principle**. Let C be the †Banach lattice of the totality of bounded and uniformly continuous functions on $\mathbf{R}^n$. Suppose that α, γ, f and $c (\geqslant 0)$ are bounded and smooth; then for constant time t, the value function $V(t, x, \varphi)$ belongs to C whenever $\varphi \in C$. Moreover, the family of operators $V(t)$ defined by $V(t)\varphi(x) = V(t, x, \varphi)$ becomes a †monotone contraction semigroup on C. The semigroup property $V(t+s, x, \varphi) = V(t, x, V(s, \cdot, \varphi))$ is nothing but the Bellman principle. The †generator G is expressed by

$$G\varphi(x) = \sup_{u \in \Gamma} \{L^u \varphi(x) - c(x, u)\varphi(x) + f(x, u)\}$$

for a smooth function φ, where L^u is the generator of †diffusion of the response for constant control $u (\in \Gamma)$, namely,

$$L^u = \frac{1}{2}\sum_{i,j,p=1}^{n} \alpha_{ip}(x, u)\alpha_{jp}(x, u)\frac{\partial^2}{\partial x_i \partial x_i} + \sum_{i=1}^{n} \gamma_i(x, u)\frac{\partial}{\partial x_i}.$$

Furthermore, assume that α is uniformly positive definite and φ is smooth. Then $V(t)\varphi(x)$ belongs to †$W^{1;2}_{p,\mathrm{loc}}$ for any p and is the unique solution of the **Bellman equation** (= dynamic programming equation)

$$\frac{\partial W}{\partial t} = \sup_{u \in \Gamma} \{L^u W - c(x, u) W + f(x, u)\}$$

a.e. on $(0, \infty) \times \mathbf{R}^n$, $\quad W(0, x) = \varphi(x) \quad$ on $\mathbf{R}^n$.

In addition if $\inf_{x,u} c(x, u) > 0$, then $W = \lim_{t \to \infty} V(t)\varphi$ exists and is the unique solution of the Bellman equation $\sup_{u \in \Gamma}\{L^u W - c(x, u) W + f(x, u)\} = 0$ a.e. on $\mathbf{R}^n$. When τ is the hitting time, the value function is related to the Bellman equation with a boundary condition.

C. Feedback Control

In practical problems we specify the kind of information on which the decision of the controller can be based at each time. We frequently assume that the data obtained up to that time is available. The following situations are possible: (1) The controller knows the complete state of the system. This is called the case of **complete observation**. (2) The controller has partial knowledge of the state of system. This is called the case of **partial observation**. A feedback control (= **policy**) is a function of available information, namely, a Γ-valued progressible measurable function defined on $[0, \infty) \times C^j[0, \infty)$, where j is the dimension of data and $C^j[0, \infty)$ is a metric space of totality of j-vector valued continuous functions on $[0, \infty)$. A policy U is called a **Markovian policy** if $U(t, \xi)$ is a Borel function of t and the tth coordinate of ξ. When a policy U is applied, the system is governed by the †SDE

$$dX_t = \alpha(X_t, U(t, Y))\, dW_t + \gamma(X_t, U(t, Y))\, dt$$

with data process Y. When the SDE has a †weak solution, U is called admissible. For example, when $X = Y$, any Markovian policy is admissible if α is uniformly positive definite. Let X_t be a weak solution for U. Then its performance $J(\tau, x, \varphi, U) \leqslant V(t, x, \varphi)$.

(1) The case of **complete observation**. When α is uniformly positive definite, an optimal Markovian policy can be constructed in the following way. Since Γ is compact, there exists a Borel function $\tilde{U}$ on $[0, \infty) \times \mathbf{R}^n$ which gives the supremum, namely,

$$\begin{aligned}&\sup_{u \in \Gamma} \{L^u V(t)\varphi(x) - c(x, u) V(t)\varphi(x) + f(x, u)\} \\ &= L^{\tilde{U}(t,x)} V(t)\varphi(x) - c(x, \tilde{U}(t, x)) V(t)\varphi(x) \\ &\quad + f(x, \tilde{U}(t, x)).\end{aligned}$$

This relation implies that $V(t)\varphi(x) = J(t, x, \varphi, \tilde{U}(t, X_t))$ for any weak solution X_t. Hence

$\tilde{U}$ is an optimal Markovian policy. Especially when α, c, and f are independent of u and $\gamma(x,u)=R(x)\cdot u$, where $R(x)$ is an $n\times k$ matrix, an optimal Markovian policy $\tilde{U}$ is obtained by a measurable selection of maximum points of (grad. $V(t)\varphi$, $R(x)\cdot u$). Since the supremum of a linear form is attained at the boundary $\partial\Gamma$, one can suppose that $\tilde{U}(t,x)$ belongs to $\partial\Gamma$. This is called **bang-bang control**.

(2) The case of **partial observation**. One useful method is the **separation principle**. This means that the control problem can be split into two parts. The first is the mean square estimate for the system using a †filtering. The second is a stochastic optimal control with complete observation. But generally speaking, the problem of deciding under what conditions the separation principle is valid is difficult. In the case of the following linear regulator the separation principle holds.

Suppose that the system process X_t and the observation process Y_t obey the following SDEs:

$$dX_t=A(t)\,dW_t+(B(t)X_t+b(t,U(t,Y))\,dt,$$

$$dY_t=d\tilde{W}_t+H(t)X_t\,dt,$$

where $A(t)$, $B(t)$, and $H(t)$ are nonrandom matrix-valued functions and $\tilde{W}_t$ is a j-dimensional Brownian motion independent of W_t. The problem is to search for a feedback control which gives the maximum value. Suppose that a feedback control $U(t,\xi)$ is Lipschitz continuous in $\xi\in C^j[0,\infty)$. Put

$$Q(s,x)$$

$$=\sup_{U,\mathrm{Lip}} E_{sx}\left[\int_s^T L(t,X_t,U(t,Y))\,dt+\Phi(X_\tau)\right],$$

where (X_t, Y_t) is the unique solution for U with the initial condition $X_s=x$, $Y_s=0$. By the Lipschitz condition of U, $\sigma_t=\sigma(Y_s; s\leqslant t)$ is independent of U, and the †conditional expectation $\hat{X}_t=E(X_t/\sigma_t)$ is governed by the following SDE, by way of the †Kalman-Bucy filter:

$$d\hat{X}_t=P(t)H'(t)\,dW_t^*+(B(t)\hat{X}_t+b(t,\hat{U}_t))\,dt$$

with some σ_t-progressible measurable control $\hat{U}_t$ and an n-dimensional Brownian motion W_t^* adapted to σ_t. Moreover, $P(t)$ is the †error matrix satisfying the †Riccati equation, and H' is the transpose of H. Let $g(t,x)$ be the probability density of the normal distribution $N(0,P(t))$, and put $\hat{L}(t,\hat{x},u)=\int L(t,x,u)\,g(t,x-\hat{x})\,dx$ and $\hat{\Psi}(\hat{x})=\int\Psi(x)g(T,x-\hat{x})\,dx$; then the problem turns into

$$Q(s,x)=\sup_{\hat{U}} E_{sx}\left[\int_s^T \hat{L}(t,\hat{X}_t,\hat{U}_t)\,dt+\hat{\Psi}(\hat{X}_\tau)\right].$$

Recalling the SDE for $\hat{X}_t$, we can use the Bellman equation for choosing an optimal one.

D. Stochastic Maximum Principle

A stochastic version of †Pontryagin's maximum principle gives a necessary condition for optimality. This means that the instantaneous value of optimal control maximizes the stochastic analog of Pontryagin's Hamiltonian. Suppose that the system evolves according to an n-dimensional CSDE

$$dX_t=\alpha(X_t)\,dW_t+\gamma(X_t,U_t)\,dt.$$

The problem is to seek conditions on admissible control U_t such that $E_x[\int_0^T f(X_t,U_t)\,dt]$ is maximized, where T is a constant time. Assume that α, γ, and f are bounded and smooth. Define a Hamiltonian H on $\mathbf{R}^n\times\Gamma\times\mathbf{R}^n$ by $H(x,u,\Psi)=\gamma(x,u)\cdot\Psi+f(x,u)$. Let $\tilde{U}_t$ be optimal and $\tilde{X}_t$ its response starting at x. Then under some conditions there exist $\lambda\geqslant 0$ and $\mathscr{F}_t$-progressively measurable $q_{t,k}=(q_{t,k_1}, q_{t,k_2},\dots, q_{t,k_n})$ $(k=1,\dots,n)$ and $\Psi_t=(\Psi_{t,1}\dots\Psi_{t,n})$ which satisfy the SDE

$$d\Psi_{t,k}=-\left(\frac{\partial\gamma}{\partial x_k}(\tilde{X}_t,\tilde{U}_t)\cdot\Psi_t+\lambda\frac{\partial f}{\partial x_k}(\tilde{X}_t,\tilde{U}_t)\right)dt$$

$$+q_{t,k}\,dW_t,\quad k=1,\dots,n,$$

and $H(\tilde{X}_t,\tilde{U}_t,\Psi_t)=\max_{u\in\Gamma}H(\tilde{X}_t,u,\Psi_t)$ a.e.

E. Optimal Stopping and Impulse Control

Suppose that X_t is an n-dimensional diffusion whose generator A is an elliptic differential operator. Let τ be a †stopping time. The optimal stopping problem is to seek a stopping time $\tilde{\tau}$ so that $E_x[g(X_\tau)]$ is maximized, where g is nonnegative and continuous. $\tilde{\tau}$ is called optimal. The value function $V(x)=\sup_\tau E_x[g(X_\tau)]$ is characterized as the least †excessive majorant of g. Moreover, under some conditions V belongs to the domain of A and is the unique solution of the †free boundary problem; $V\geqslant g$, $AV\leqslant 0$, and $(V-g)\cdot AV=0$. Therefore, in the Hilbert space framework, the value function is related to the variational inequality. An optimal stopping time is provided by the hitting time for the set $\{x\mid V(x)=g(x)\}$.

Impulse control is a variant of the optimal stopping problem. At some moment (=stopping time) a controller shifts the current state to some other state. But not all shifts are allowed: State x can be shifted to a state of $x+[0,\infty)^n$. Let τ_k, $k=1,2,$ be a sequence of increasing stopping times and ξ_k be a $[0,\infty)^n$-valued $\sigma_{\tau_k}(X)$-measurable random variable. The sequence $U=\{\tau_1,\xi_1,\tau_2,\xi_2,\dots\}$ is called an impulse control. U transfers the process X_t to

$$Y_t^U=X_0+\int_0^t\alpha(Y_s^U)\,dw_s+\int_0^t\gamma(Y_s^U)\,ds+\sum_{\tau_i\leqslant t}\xi_i,$$

if

$$A=\frac{1}{2}\sum_{i,j}\alpha\alpha^*(x)_{ij}\frac{\partial^2}{\partial x_i \partial x_j}+\sum_i \gamma_i(x)\frac{\partial}{\partial x_i},$$

and the problem is to seek U so as to maximize $E_x[\int_0^\tau e^{-\lambda t}f(Y_t^U)dt-\sum_{k=1}^\infty e^{-\lambda\tau_k}K(\xi_k)]$, where $\lambda\,(>0)$ is constant and the function $K\,(\geqslant 0)$ stands for the cost of shifting. The value function is related to a quasivariational inequality.

F. General Description of Stochastic Filtering

The problem of estimating the original signal from data disturbed by noises is called a **stochastic filtering** problem. Let X_t, $t\in[0,T]$, be a continuous stochastic process with values in $\mathbf{R}^n$, called a **signal** (or **system**) process. It is transformed (or coded) to $h(t,X_t)$, where $h(t,x)$ is an m-vector-valued continuous function. Suppose that it is disturbed by a noise $\dot{W}_t$ and we observe $\dot{Y}_t=h(t,X_t)+\dot{W}_t$. Usually $\dot{W}_t$ is assumed to be a †white noise independent of X_t. Since the white noise is a generalized function, the integral of $\dot{Y}_t$, i.e.,

$$Y_t=\int_0^t h(s,X_s)ds+W_t,$$

is called an **observation process**, where W_t is an †m-dimensional Brownian motion independent of X_t. It is assumed for convenience that $|X_t|^2$ and $\int_0^t|h(s,X_s)|^2ds$ are integrable.

Assume that X_t is a 1-dimensional signal process. The least square estimation of X_t by nonlinear functions of observed data Y_s, $s\leqslant t$, is called a **nonlinear filter** of X_t and is denoted by $\hat{X}_t$. Let $\mathscr{F}_t$ or $\sigma(Y_s;s\leqslant t)$ be the least †σ-field for which Y_s, $s\leqslant t$, are measurable. Then the filter $\hat{X}_t$ is equal to an †$\mathscr{F}_t$-measurable random variable such that $E|X_t-\hat{X}_t|^2\leqslant E|X_t-Z|^2$ holds for any $\mathscr{F}_t$-measurable L^2 random variable Z. Hence it coincides with the †conditional expectation $E[X_t|\mathscr{F}_t]$. Now let H_t be the closed linear space spanned by Y_s, $s\leqslant t$. The least square estimation of X_t by elements of H_t, i.e., the orthogonal projection of X_t onto H_t, is called the **linear filter** of X_t and is denoted by $\tilde{X}_t$. Obviously, the mean square error of a nonlinear filter is less than or equal to that of a linear filter, but a linear filter is calculated more easily. If (X_t, Y_t) is a †Gaussian process, both filters coincide.

When X_t is an n-dimensional process $(X_t^1, \ldots, X_t^n)$, the n-vector process $\hat{X}_t=(\hat{X}_t^1,\ldots,\hat{X}_t^n)$ (or $\tilde{X}_t=(\tilde{X}_t^1,\ldots,\tilde{X}_t^n)$) is called the **nonlinear** (or **linear**) **filter** of X_t.

G. Kalman-Bucy Filter

Suppose that the signal process X_t is governed by a †linear stochastic differential equation (LSDE)

$$X_t=X_0+\int_0^t A(s)X_s\,ds+\int_0^t B(s)\,d\tilde{W}_s,$$

where $A(s)$ (or $B(s)$) is an $n\times n$ (or $n\times r$) matrix-valued continuous function, $\tilde{W}_t$ is an r-dimensional Brownian motion independent of the noise W_t, and the initial data X_0 is a Gaussian random variable independent of W_t and $\tilde{W}_t$. Suppose further that $h(t,x)$ is linear, i.e., $h(t,x)=H(t)x$, where $H(t)$ is an $m\times n$-matrix-valued function. Then the joint process (X_t, Y_t) is Gaussian. Hence the nonlinear filter $\hat{X}_t$ coincides with the linear filter and satisfies

$$\hat{X}_t=E[X_0]+\int_0^t(A(s)-P(s)H(s)'H(s))\hat{X}_s\,ds$$
$$+\int_0^t P(s)H(s)'\,dY_s,$$

where $H(s)'$ is the transpose of $H(s)$, and $P(t)=(P_{ij}(t))$ is the **error matrix** defined by $P_{ij}(t)=E(X_t^i-\hat{X}_t^i)(X_t^j-\hat{X}_t^j)$. It satisfies the **matrix Riccati equation**

$$\frac{dP(t)}{dt}=A(t)P(t)+P(t)A(t)'$$
$$-P(t)H(t)'H(t)P(t)+B(t)B(t)',$$

$P(0)=$ covariance of X_0.

Let $\Phi(t,s)$ be the †fundamental solution of the linear differential equation $dx/dt=(A(t)-P(t)H(t)'H(t))x$. Then the solution $\hat{X}_t$ is represented by

$$\hat{X}_t=\Phi(t,0)E[X_0]+\int_0^t\Phi(t,s)P(s)H(s)'\,dY_s.$$

This algorithm is called the **Kalman-Bucy filter** [1]. Analogous results for discrete-time models have been obtained by Kalman.

H. Nonlinear Filter

In the study of nonlinear filters, the †conditional distribution $\pi(dx)=P(X_t\in dx|F_t)$ is considered besides $\hat{X}_t$. Suppose that X_t is governed by the SDE

$$X_t=X_0+\int_0^t a(s,X_s)\,ds+\int_0^t b(s,X_s)\,d\tilde{W}_s,$$

where $a(s,x)$ (or $b(s,x)$) is an n-vector ($n\times r$-matrix) valued Lipschitz continuous function. Then $\pi_t(f)=\int f(x)\pi_t(dx)$ satisfies the SDE

$$\pi_t(f)=E[f(X_0)]+\int_0^t\pi_s(Lf)\,ds$$
$$+\int_0^t(\pi_s(h_sf)-\pi_s(h_s)\pi_s(f))(dY_s-\pi_s(h_s)\,ds),$$

where $h_s(x)=h(s,x)$ and

$$Lf(x)=\sum_i a^i(s,x)\frac{\partial f}{\partial x^i} +\frac{1}{2}\sum_{i,j}\left(\sum_k b_{ik}(s,x)b_{jk}(s,x)\right)\frac{\partial^2 f}{\partial x^i \partial x^j}.$$

Under additional conditions on $a(s,x)$ and $b(s,x)$, $\pi_t(dx)$ has a density function $\pi_t(x)$, and it satisfies

$$\pi_t(x)=\pi_0(x)+\int_0^t L^*\pi_s(x)\,ds+\int_0^t \pi_s(x)(h_s(x) -\int \pi_s(x)h_s(x)\,dx)\left(dy_s-\left(\int \pi_s(x)h_s(x)\,dx\right)ds\right),$$

where L^* is the formal adjoint of L.

The process $I_t \equiv Y_t - \int_0^t \pi_s(h_s)\,ds$ is a Brownian motion such that $\sigma(I_s; s\leqslant t)\subset \mathscr{F}_t$ holds for any t. If $\sigma(I_s; s\leqslant t)=\mathscr{F}_t$ holds for all t, I_t is called the **innovation** of Y_t. The innovation property is not valid in general. A sufficient condition is that (X_t, Y_t) is a Gaussian process or $h(t,x)$ be a bounded function. However, in any case, †$\mathscr{F}_t$-adapted martingales are always represented as †stochastic integrals of the form $\sum_{i=1}^n \int_0^t f_s^i(\omega)\,dI_s^i$, where the f_s^i are $\mathscr{F}_s$-adapted processes.

I. Bayes Formula

Let C (or D) be the space of all continuous mappings x (or y) from $[0,T]$ into $\mathbf{R}^n$ (or $\mathbf{R}^m$) equipped with the uniform topology. x_t (y_t) is the value of x (y) at time t. Let $\mathscr{B}_t(C)$ be the least σ-field of C for which x_s, $s\leqslant t$, are measurable. $\mathscr{B}_t(D)$ is defined similarly. We denote by Φ_X, Φ_W, $\Phi_{X,Y}$ the †laws of processes X_t, W_t, and (X_t, Y_t), respectively. These are defined on $\mathscr{B}_T(C)$, $\mathscr{B}_T(D)$, and $\mathscr{B}_T(C)\otimes\mathscr{B}_T(D)$, respectively. Then $\Phi_{X,Y}$ is equivalent (mutually †absolutely continuous) to the product measure $\Phi_X\otimes\Phi_W$ on each $\mathscr{B}_t(C)\otimes\mathscr{B}_t(D)$. The †Radon-Nikodym density α_t of $\Phi_{X,Y}$ with respect to $\Phi_X\otimes\Phi_W$ is written as

$$\alpha_t(x,y)=\exp\left\{\sum_i\int_0^t h^i(s,x_s)\,dy_s^i -\frac{1}{2}\sum_i\int_0^t h^i(s,x_s)^2\,ds\right\},$$

where h^i and y^i are the corresponding components of vectors and dy_s^i denotes the †Ito integral.

The conditional distribution $\pi_t(dx)$ is computed by the Bayes formula:

$$\pi_t(f)=\frac{\rho_t(f)}{\rho_t(1)},\qquad \rho_t(f)=\int f(x_t)\alpha_t(x,Y)\Phi_X(dx),$$

where $Y=(Y_t; 0\leqslant t\leqslant T)$. Moreover, $\rho_t(f)$ satisfies the LSDE

$$\rho_t(f)=\rho_0(f)+\int_0^t \rho_s(Lf)\,ds+\int_0^t \rho_s(h_s)\rho_s(f)\,dI_s.$$

The density $\rho_t(x)$ (if x exists) satisfies

$$\rho_t(x)=\rho_0(x)+\int_0^t L^*\rho_s(x)\,ds +\int_0^t \rho_s(x)\left(\int \rho_s(y)h_s(y)\,dy\right)dI_s.$$

If $h(t,x)$ is a smooth function, then $\alpha_t(x,y)$ is continuous in y, so that $\pi_t(f)$ or $\rho_t(f)$ is a continuous functional of the observed data $(Y_s; s\leqslant t)$. Thus the filter π_t is a †robust statistic.

Remarks. (i) the signal and noise are not independent if the signal is controlled based on the observed data. In these cases, correction terms are sometimes needed for the SDE of the nonlinear filter. (ii) If the †sample paths of the signal process are not continuous, a similar SDE for a nonlinear filter is valid with L being replaced by some integrodifferential operator. If it is a †Markov chain with finite state, L is the generator of the chain. (iii) Several results are known for the case where the noise W_t is not a Brownian motion but a †Poisson process.

References

[1] W. F. Fleming and R. W. Rishel, Deterministic and stochastic optimal control, Springer, 1975.
[2] R. Bellman, Dynamic programming, Princeton Univ. Press, 1957.
[3] N. V. Krylov, Controlled diffusion processes, Springer, 1980. (Original in Russian, 1977.)
[4] M. Nisio, Stochastic control theory, Indian Statist. Inst. Lect. Notes, Macmillan, 1981.
[5] H. J. Kushner, On the stochastic maximum principle, fixed time of control, J. Math. Anal. Appl., 11 (1965), 78–92.
[6] A. N. Shiryaev, Optimal stopping rules, Springer, 1980. (Original in Russian, 1976.)
[7] A. Bensoussan and J. L. Lions, Sur la théorie du contrôle optimal. I, Temps d'arrêt; II, Contrôle impulsionnel, Hermann, 1977.
[8] R. S. Bucy and R. E. Kalman, New results in linear filtering and prediction theory, J. Basic Eng. ASME, (D) 83 (1961), 95–108.
[9] R. S. Bucy and D. D. Joseph, Filtering for stochastic processes with applications to guidance, Interscience, 1968.
[10] M. Fujisaki, G. Kallianpur, and H. Kunita, Stochastic differential equations for the nonlinear filtering problem, Osaka J. Math., 9 (1972), 19–40.

[11] R. S. Liptzer and A. N. Shiryaev, Statistics of stochastic processes. I, General theory; II, Applications, Springer, 1977, 1978. (Original in Russian, 1974.)
[12] G. Kallianpur, Stochastic filtering theory, Springer, 1980.

406 (XVII.14) Stochastic Differential Equations

A. Introduction

Stochastic differential equations were rigorously formulated by K. Itô [7] in 1942 to construct diffusion processes corresponding to Kolmogorov's differential equations. For this purpose he introduced the notion of stochastic integrals, and thus a differential-integral calculus for sample paths of stochastic processes was established. This theory, often called Itô's stochastic analysis or stochastic calculus, has brought an epoch-making method to the theory of stochastic processes. It provides us with a fundamental tool for describing and analyzing diffusion processes that we can apply effectively to limit theorems and to the probabilistic study of problems in analysis. It also plays an important role in the statistical theory of stochastic processes, such as †stochastic control or †stochastic filtering. Stochastic differential equations on manifolds provide a probabilistic method for differential geometry, sometimes called stochastic differential geometry. Recently, many interesting examples of infinite-dimensional stochastic differential equations have been introduced to describe probabilistic models in physics, biology, etc.

A unified theory of stochastic calculus has been developed in the framework of Doob's martingale theory and this, combined with Stroock and Varadhan's idea of martingale problems, provides an important method in the theory of stochastic processes (→ 262 Martingales).

B. Stochastic Integrals

As is well known, almost all sample paths of a Wiener process are continuous but nowhere differentiable (→ 45 Brownian Motion), and hence integrals with respect to these functions cannot be defined as the usual Stieltjes integrals. But these integrals can be defined by making use of the stochastic nature of Brownian motion. Wiener defined them (the Wiener integrals) for nonrandom integrands, but Itô defined them for a large class of random integrands. Itô's integrals have been extended in the martingale framework by H. Kunita and S. Watanabe, and by others [14, 19], as shown below.

Let $(\Omega, \mathscr{F}, P)$ be a probability space, and let $\mathbf{F}=\{\mathscr{F}_t\}_{t\geqslant 0}$ be an increasing family of σ-subfields of $\mathscr{F}$. Usually we assume that $\{\mathscr{F}_t\}$ is right continuous, i.e., $\mathscr{F}_{t+0}:=\bigcap_{\varepsilon>0}\mathscr{F}_{t+\varepsilon}=\mathscr{F}_t$ for every $t\geqslant 0$. Denote by $\mathscr{M}=\mathscr{M}(\mathbf{F})$ the totality of all continuous square-integrable martingales $X=(X_t)$ relative to $\{\mathscr{F}_t\}$; to be precise, X is an $\{\mathscr{F}_t\}$-martingale such that, with probability 1, $X_0=0$, $t\rightarrow X_t$ is continuous and $E(X_t^2)<\infty$ for every $t\geqslant 0$. We introduce the metric $\|X-Y\|=\sum_{k=1}^{\infty}2^{-n}\min(1,\|X_n-Y_n\|_2)$ on $\mathscr{M}$ where $\|\ \|_2$ stands for the $L_2(\Omega, P)$-norm. We always identify two stochastic processes $X=(X_t)$ and $Y=(Y_t)$ if sample functions $t\rightarrow X_t$ and $t\rightarrow Y_t$ coincide with probability 1. Then, by virtue of Doob's inequality $\|\max_{0\leqslant s\leqslant t}|X_s-Y_s|\|_2\leqslant 2\|X_t-Y_t\|_2$ (→ 262 Martingales), $\mathscr{M}$ becomes a complete metric vector space.

Next, by an **integrable increasing process** we mean a process $A=(A_t)$ with the following properties: (i) A is adapted to $\{\mathscr{F}_t\}$, i.e., A_t is $\mathscr{F}_t$-measurable for every $t\geqslant 0$; (ii) with probability 1, $A_0=0$, $t\rightarrow A_t$ is continuous and nondecreasing; (iii) A_t $(\geqslant 0)$ is integrable for every $t\geqslant 0$, i.e., $E(A_t)<\infty$. We denote by $\mathscr{A}=\mathscr{A}(\mathbf{F})$ the totality of integrable increasing processes. We call a process $V=(V_t)$ an **integrable process of bounded variation** if V is expressed as $V_t=A_t^1-A_t^2$ with $A^1, A^2\in\mathscr{A}$. The totality of integrable processes of bounded variation is denoted by $\mathscr{V}=\mathscr{V}(\mathbf{F})$. It follows from the †Doob-Meyer decomposition theorem that, for every $M, N\in\mathscr{M}$, there exists a unique $V\in\mathscr{V}$ such that $M_tN_t-V_t$ is an $\{\mathscr{F}_t\}$-martingale. We denote this V as $\langle M, N\rangle$. In particular, $\langle M, M\rangle\in\mathscr{A}$, and it is denoted simply by $\langle M\rangle$. $\langle M, N\rangle$ is called the **quadratic variation process** because $\sum_{i=1}^{n}(M_{t_i}-M_{t_{i-1}})(N_{t_i}-N_{t_{i-1}})\rightarrow\langle M, N\rangle_t$ in probability as $|\Delta|\rightarrow 0$, where $\Delta: t_0=0<t_1<\ldots<t_n=t$ is a partition and $|\Delta|=\max_{1\leqslant i\leqslant n}|t_i-t_{i-1}|$. Brownian motion is the most important example of continuous square-integrable martingales, and this is characterized in our framework as follows. Suppose that a d-dimensional continuous $\{\mathscr{F}_t\}$-adapted process $X=(X_t^i)$ satisfies $M_t^i=X_t^i-X_0^i\in\mathscr{M}$ and $\langle M^i, M^j\rangle_t=\delta^{ij}t$, $i, j=1, 2, \ldots, d$. Then X is a d-dimensional Brownian motion such that X_u-X_v, and the $\mathscr{F}_t$ are independent for every $u\geqslant v\geqslant t$. Such a Brownian motion is called an **$\{\mathscr{F}_t\}$-Brownian motion**, and a system of martingales $M^i\in\mathscr{M}$ having this property is often called a system of **$\{\mathscr{F}_t\}$-Wiener martingales**.

Now, we fix $M\in\mathscr{M}$. We denote by $\mathscr{L}_2(M)$ the totality of real, $\{\mathscr{F}_t\}$-adapted, and measurable processes $\Phi=(\Phi(t))$ such that $\|\Phi\|_{t,M}^2=$

$E[\int_0^t \Phi(s)^2 d\langle M\rangle_s] < \infty$ for every $t \geqslant 0$. Two $\Phi_1, \Phi_2 \in \mathscr{L}_2(M)$ are identified if $\|\Phi_1 - \Phi_2\|_{t,M} = 0$ for all $t \geqslant 0$. Since $\|\ \|_{t,M}$ is an L_2-norm on $[0,t] \times \Omega$ with respect to the measure $\mu_M(ds, d\omega) = d\langle M\rangle_s(\omega) P(d\omega)$, it is easy to see that $\mathscr{L}_2(M)$ is a complete metric vector space with the metric $\|\Phi - \Phi'\|_M = \sum_{n=1}^{\infty} 2^{-n} \min(1, \|\Phi - \Phi'\|_{n,M})$, $\Phi, \Phi' \in \mathscr{L}_2(M)$. If $\Phi = (\Phi(t))$ is given, for a partition $0 = t_0 < t_1 < \ldots < t_n \ldots \to \infty$ and $\mathscr{F}_{t_i}$-measurable bounded functions f_i, $i = 0, 1, \ldots$, by

$$\Phi(t, \omega) = \begin{cases} f_0(\omega), & t = 0, \\ f_{i-1}(\omega), & t_{i-1} < t \leqslant t_i, \quad i = 1, 2, \ldots, \end{cases}$$

then $\Phi \in \mathscr{L}_2(M)$ and the totality $\mathscr{L}_0$ of such processes are dense in $\mathscr{L}_2(M)$. If $\Phi \in \mathscr{L}_0$, we define $I^M(\Phi) = (I^M(\Phi)(t))_{t \geqslant 0}$ by

$$I^M(\Phi)(t) = \sum_{i=0}^{n-1} f_i(M_{t_{i+1}} - M_{t_i}) + f_n(M_t - M_{t_n}),$$

$$t_n \leqslant t \leqslant t_{n+1}.$$

Then $I^M(\Phi) \in \mathscr{M}$, and it holds also that $\langle I^M(\Phi), I^N(\Psi)\rangle = \int_0^t \Phi(s)\Psi(s) d\langle M, N\rangle_s$ for $M, N \in \mathscr{M}$ and $\Phi, \Psi \in \mathscr{L}_0$. In particular, $\|I^M(\Phi)(t)\|_2^2 = E[\langle I^M(\Phi)\rangle_t] = \|\Phi\|_{t,M}^2$, and hence $\|I^M(\Phi)\| = \|\Phi\|_M$. This implies that $\Phi \in \mathscr{L}_0 \subset \mathscr{L}_2(M) \to I^M(\Phi) \in \mathscr{M}$ is an isometric linear mapping, and hence it can be extended to $\mathscr{L}_2(M)$ uniquely, preserving the isometric property. $I^M(\Phi) \in \mathscr{M}$ is called the **stochastic integral** of $\Phi \in \mathscr{L}_2(M)$ by $M \in \mathscr{M}$. $I^M(\Phi)(t)$ is often denoted by $\int_0^t \Phi(s) dM_s$, and the random variable $I^M(\Phi)(t)$ obtained by fixing t is also called a stochastic integral.

The definition of stochastic integrals can be extended further by the following localization method. For an $\{\mathscr{F}_t\}$-†progressively measurable process $X = (X_t)$ and $\{\mathscr{F}_t\}$-stopping time σ, the stopped process $X^\sigma = (X_t^\sigma)$ is defined by $X_t^\sigma = X_{t \wedge \sigma}$ $(t \wedge \sigma = \min(t, \sigma))$. It follows from the †optional sampling theorem of Doob that $X^\sigma \in \mathscr{M}$ if $X \in \mathscr{M}$. For $\Phi = (\Phi(t)) \in \mathscr{L}_2(M)$ and an $\{\mathscr{F}_t\}$-stopping time σ, $\Phi_\sigma = (\Phi_\sigma(t))$ defined by $\Phi_\sigma(t) = I_{\{t \leqslant \sigma\}} \Phi(t)$ also belongs to $\mathscr{L}_2(M)$ and it holds that $I^M(\Phi_\sigma) = [I^M(\Phi)]^\sigma$. Keeping these facts in mind, we give the following definition. Let $\mathscr{M}^{\mathrm{loc}} = \{M = (M_t) \mid$ there exists a sequence of $\{\mathscr{F}_t\}$-stopping times σ_n such that $\sigma_n < \infty$, $\sigma_n \uparrow \infty$ as $n \to \infty$ a.s. and $M^{\sigma_n} \in \mathscr{M}$ for every $n = 1, 2, \ldots\}$. $\mathscr{A}^{\mathrm{loc}}$ and $\mathscr{V}^{\mathrm{loc}}$ are defined in a similar way. For $M, N \in \mathscr{M}^{\mathrm{loc}}$, $\langle M, N\rangle \in \mathscr{V}^{\mathrm{loc}}$ is defined to be the unique process in $\mathscr{V}^{\mathrm{loc}}$ such that $\langle M^{\sigma_n}, N^{\sigma_n}\rangle = \langle M, N\rangle^{\sigma_n}$ for a sequence of stopping times $\{\sigma_n\}$ as above, which can be chosen common to M and N. $\langle M, M\rangle$ is denoted by $\langle M\rangle$ as before. We fix $M \in \mathscr{M}^{\mathrm{loc}}$ and set $\mathscr{L}_2^{\mathrm{loc}}(M) = \{\Phi = (\Phi(t)) \mid$ a real, $\{\mathscr{F}_t\}$-adapted and measurable process such that, with probability one, $\int_0^t \Phi(s)^2 d\langle M\rangle_s < \infty$ for every $t \geqslant 0\}$.

For $\Phi \in \mathscr{L}_2^{\mathrm{loc}}(M)$, we can choose a sequence of $\{\mathscr{F}_t\}$-stopping times $\{\sigma_n\}$ such that, $\sigma_n < \infty$, $\sigma_n \uparrow \infty$ a.s. and $M^{\sigma_n} \in \mathscr{M}$, $\Phi_{\sigma_n} \in \mathscr{L}_2(M^{\sigma_n})$, $n = 1, 2, \ldots$. For example, set $\sigma_n = \min[n, \inf\{t \mid \langle M\rangle_t + \int_0^t \Phi(s)^2 d\langle M\rangle_s \geqslant n\}]$. Then there exists an $I^M(\Phi) \in \mathscr{M}^{\mathrm{loc}}$ such that $I^M(\Phi)^{\sigma_n} = I^{M^{\sigma_n}}(\Phi_{\sigma_n})$ for $n = 1, 2, \ldots$, which is unique and independent of a particular choice of $\{\sigma_n\}$. $I^M(\Phi)$ is called the **stochastic integral** of $\Phi \in \mathscr{L}_2^{\mathrm{loc}}(M)$ by $M \in \mathscr{M}^{\mathrm{loc}}$. $I^M(\Phi)(t)$ is often denoted by $\int_0^t \Phi(s) dM_s$, and the random variable $I^M(\Phi)(t)$ obtained by fixing t is also called a stochastic integral.

Some of the basic properties of stochastic integrals are: (i) If $M \in \mathscr{M}^{\mathrm{loc}}$, $\Phi \in \mathscr{L}_2^{\mathrm{loc}}(M)$, and $\Psi \in \mathscr{L}_2^{\mathrm{loc}}(I^M(\Phi))$, then $\Phi\Psi \in \mathscr{L}_2^{\mathrm{loc}}(M)$ and $I^M(\Phi\Psi) = I^{I^M(\Phi)}(\Psi)$. (ii) If $M \in \mathscr{M}^{\mathrm{loc}}$ and $\Phi, \Psi \in \mathscr{L}_2^{\mathrm{loc}}(M)$, then for every $\alpha, \beta \in \mathbf{R}$ we have $\alpha\Phi + \beta\Psi = (\alpha\Phi(t) + \beta\Psi(t)) \in \mathscr{L}_2^{\mathrm{loc}}(M)$ and $I^M(\alpha\Phi + \beta\Psi) = \alpha I^M(\Phi) + \beta I^M(\Psi)$. Also if $M, N \in \mathscr{M}^{\mathrm{loc}}$ and $\Phi \in \mathscr{L}_2^{\mathrm{loc}}(M) \cap \mathscr{L}_2^{\mathrm{loc}}(N)$, then for $\alpha, \beta \in \mathbf{R}$ we have $\Phi \in \mathscr{L}_2^{\mathrm{loc}}(\alpha M + \beta N)$ and $I^{\alpha M + \beta N}(\Phi) = \alpha I^M(\Phi) + \beta I^N(\Phi)$. (iii) If $M, N \in \mathscr{M}^{\mathrm{loc}}$, $\Phi \in \mathscr{L}_2^{\mathrm{loc}}(M)$, and $\Psi \in \mathscr{L}_2^{\mathrm{loc}}(N)$, then $\Phi\Psi \in \mathscr{L}_1^{\mathrm{loc}}(\langle M, N\rangle)$ and $\langle I^M(\Phi), I^N(\Psi)\rangle_t = \int_0^t \Phi_s \Psi_s d\langle M, N\rangle_s$. Here, $\mathscr{L}_p^{\mathrm{loc}}(V)$ for $V \in \mathscr{V}^{\mathrm{loc}}$ is defined as follows: With probability 1, $s \to V_s$ is of bounded variation on every finite interval $[0, t]$, the total variation of which is denoted by $|\mathbf{V}|_t$. Then $|\mathbf{V}| \in \mathscr{A}^{\mathrm{loc}}$, and we define $\mathscr{L}_p^{\mathrm{loc}}(V)$ $(p \geqslant 1)$ to be the totality of real, $\{\mathscr{F}_t\}$-adapted, and measurable processes $\Phi = (\Phi(t))$ such that, with probability1, $\int_0^t |\Phi(s)|^p d|V|_s < \infty$ for every $t \geqslant 0$. In particular, $\mathscr{L}_2^{\mathrm{loc}}(M) = \mathscr{L}_2^{\mathrm{loc}}(\langle M\rangle)$. (iv) If $M \in \mathscr{M}^{\mathrm{loc}}$, $\Phi \in \mathscr{L}_2^{\mathrm{loc}}(M)$, and σ is an $\{\mathscr{F}_t\}$-stopping time, then $I^M(\Phi_\sigma) = I^{M^\sigma}(\Phi) = I^{M^\sigma}(\Phi_\sigma) = [I^M(\Phi)]^\sigma$. (v) If $\Phi(t) = I_{\{\sigma < t\}} \cdot f$, where σ is an $\{\mathscr{F}_t\}$-stopping time and f is a bounded $\mathscr{F}_\sigma$-measurable random variable, then $\Phi \in \mathscr{L}_2^{\mathrm{loc}}(M)$ for every $M \in \mathscr{M}^{\mathrm{loc}}$ and $I^M(\Phi)(t) = f(M_t - M_{t \wedge \sigma})$. (vi) The definition of stochastic integrals is independent of the increasing family of σ-subfields in the following sense: If $\{\mathscr{G}_t\}$ is another family such that M belongs to the class $\mathscr{M}^{\mathrm{loc}}$ for both $\{\mathscr{F}_t\}$ and $\{\mathscr{G}_t\}$ and Φ belongs to the class $\mathscr{L}_2(M)$ for both $\{\mathscr{F}_t\}$ and $\{\mathscr{G}_t\}$, then $I^M(\Phi)$ is the same whether it is defined with respect to $\{\mathscr{F}_t\}$ or $\{\mathscr{G}_t\}$.

In particular, $N = I^M(\Phi)$, $M \in \mathscr{M}^{\mathrm{loc}}$, $\Phi \in \mathscr{L}_2^{\mathrm{loc}}(M)$, satisfies $\langle N, L\rangle_t = \int_0^t \Phi(s) d\langle M, L\rangle_s$ for all $L \in \mathscr{M}^{\mathrm{loc}}$. Conversely, $N \in \mathscr{M}^{\mathrm{loc}}$ having this property is unique, and hence it coincides with $I^M(\Phi)$. $I^M(\Phi) \in \mathscr{M}$ if and only if $\int_0^\cdot \Phi(s)^2 d\langle M\rangle_s \in \mathscr{A}$.

The above definition of stochastic integrals can be extended with a slight technical modification to the case when M_t is not necessarily continuous [19]. Among such general stochastic integrals, a particularly important role is played by stochastic integrals describing point processes, including Poisson point processes as an important special case. These stochastic integrals are important in the study of discontinuous processes including †Lévy processes; even in the study of continuous processes, such as diffusion, they provide an important tool for the treatment of excursions [4, 6, 8].

Let $(\Omega, \mathscr{F}, P)$ and $\{\mathscr{F}_t\}$ be as above. By a **continuous semimartingale** with respect to $\{\mathscr{F}_t\}$, or simply a **semimartingale** when there is no danger of confusion, we mean a process $X=(X(t))$ of the following form: $X(t)=X(0)+M(t)+V(t)$, where $X(0)$ is an $\mathscr{F}_0$-measurable random variable, $M=(M(t))\in\mathscr{M}^{\mathrm{loc}}$ and $V=(V(t))\in\mathscr{V}^{\mathrm{loc}}$. M and V are uniquely determined from X, and this decomposition is called the **semimartingale decomposition** of X. M is called the **martingale part** and V the **drift part** of the semimartingale X. A semimartingale X is often called an **Itô process** if $M(t)=\int_0^t\Phi(s)\,dB(s)$ and $V(t)=\int_0^t\Psi(s)\,ds$, where $B(t)\in\mathscr{M}$ is an $\{\mathscr{F}_t\}$-Brownian motion, $\Phi\in\mathscr{L}_2^{\mathrm{loc}}$, and $\Psi\in\mathscr{L}_1^{\mathrm{loc}}$. (When $V_t=t$, $\mathscr{L}_p^{\mathrm{loc}}(V)$ is denoted simply by $\mathscr{L}_p^{\mathrm{loc}}$.) A d-dimensional process whose components are semimartingales is called a d-dimensional semimartingale. The following formula, originally due to Itô and extended by Kunita and Watanabe, is of fundamental importance in stochastic calculus.

Itô's formula. Let $X(t)=(X^1(t),\dots,X^d(t))$ be a d-dimensional semimartingale and $X^i(t)=X^i(0)+M^i(t)+V^i(t)$ be the semimartingale decomposition of components. Let $F(x)=F(x^1,\dots,x^d)$ be a C^2-function defined on $\mathbf{R}^d$. Then $F(X(t))$ is also a semimartingale, and we have

$$F(X(t))=F(X(0))+\sum_{i=1}^d\int_0^t D_iF(X(s))\,dM^i(s)$$
$$+\sum_{i=1}^d\int_0^t D_iF(X(s))\,dV^i(s)$$
$$+\frac{1}{2}\sum_{i,j=1}^d\int_0^t D_iD_jF(X(s))\,d\langle M^i,M^j\rangle(s)$$

(where $D_i=\partial/\partial x^i$).

In others words, if $Y(t)=Y(0)+M(t)+V(t)$ is the semimartingale decomposition of $Y(t)=F(X(t))$, then $M(t)=\sum_{i=1}^d\int_0^t D_iF(X(s))\,dM^i(s)$ and $V(t)=\sum_{i=1}^d\int_0^t D_iF(X(s))\,dV^i(s)+1/2\sum_{i,j=1}^d\int_0^t D_iD_jF(X(s))\,d\langle M^i,M^j\rangle(s)$.

We now discuss other important transformations on semimartingales.

Time change. Let $A\in\mathscr{A}^{\mathrm{loc}}$, and assume further that with probability 1, $t\to A_t$ is strictly increasing and $\lim_{t\uparrow\infty}A_t=\infty$. Let $u\to C_u$ be the inverse function of $t\to A_t$, i.e., $C_u=\min\{t\mid A_t\geqslant u\}$. Then for every $u\geqslant0$, C_u is an $\{\mathscr{F}_t\}$-stopping time. Set $\tilde{\mathscr{F}}_t=\mathscr{F}_{C_t}$, $t\geqslant0$. For an $\{\mathscr{F}_t\}$-†progressively measurable process $X=(X(t))$, we define $X^A=(X^A(t))$ by $X^A(t)=X(C_t)$ and call it the **time change of** X **determined by** A. Then X^A is progressively measurable with respect to $\{\tilde{\mathscr{F}}_t\}$. If $X:X(t)=X(0)+M(t)+V(t)$ is a semimartingale with respect to $\{\mathscr{F}_t\}$, then X^A is a semimartingale with respect to $\{\tilde{\mathscr{F}}_t\}$, and its semimartingale decomposition is given by $X^A(t)=X(0)+M^A(t)+V^A(t)$. The mappings $M\to M^A$ and $V\to V^A$ are bijections between $\mathscr{M}^{\mathrm{loc}}$ and $\tilde{\mathscr{M}}^{\mathrm{loc}}$ and between $\mathscr{V}^{\mathrm{loc}}$ and $\tilde{\mathscr{V}}^{\mathrm{loc}}$, respectively, where $\tilde{\mathscr{M}}^{\mathrm{loc}}$ and $\tilde{\mathscr{V}}^{\mathrm{loc}}$ are defined relative to $\{\tilde{\mathscr{F}}_t\}$. Furthermore, $\langle M^A,N^A\rangle=\langle M,N\rangle^A$ for every $M,N\in\mathscr{M}^{\mathrm{loc}}$. Noting that $\Phi\in\mathscr{L}_2^{\mathrm{loc}}(M)$ can always be chosen $\{\mathscr{F}_t\}$-progressively measurable (in fact $\{\mathscr{F}_t\}$-†predictable), the mapping $\Phi\to\Phi^A$ defines a bijection between $\mathscr{L}_2^{\mathrm{loc}}(M)$ and $\mathscr{L}_2^{\mathrm{loc}}(M^A)$, and we have $I^{M^A}(\Phi^A)=[I^M(\Phi)]^A$.

Transformation of drift (Girsanov transformation). For $m\in\mathscr{M}^{\mathrm{loc}}$, set $D_m(t)=\exp[m_t-\frac{1}{2}\langle m\rangle_t]$. Then $D_m-1\in\mathscr{M}^{\mathrm{loc}}$, and if m satisfies a certain integrability condition (for example, $E(\exp[\frac{1}{2}\langle m\rangle_t])<\infty$ for every $t\geqslant0$, in particular, $\langle m\rangle_t\leqslant ct$ for all t for some constant $c>0$), then D_m is a martingale, i.e., $E(D_m(t))=1$ for all $t\geqslant0$. If $E(D_m(t))=1$ for all t, then there exists a probability $\tilde{P}$ on $(\Omega,\mathscr{F})$ (if $(\Omega,\mathscr{F})$ is a nice measurable space and $\mathscr{F}=\bigvee_{t\geqslant0}\mathscr{F}_t$, which we can assume without loss of generality) such that $\tilde{P}(A)=E(D_m(t):A)$ for all $A\in\mathscr{F}_t$, $t\geqslant0$. Let X be a semimartingale with the decomposition $X(t)=X(0)+M(t)+V(t)$. On the probability space $(\Omega,\mathscr{F},\tilde{P})$ with the same family $\{\mathscr{F}_t\}$, X is still a semimartingale but its semimartingale decomposition is given by $X(t)=X(0)+\tilde{M}(t)+\tilde{V}(t)$, where $\tilde{M}(t)=M(t)-\langle M,m\rangle(t)$ and $\tilde{V}(t)=V(t)+\langle M,m\rangle(t)$. Furthermore, it holds that $\langle\tilde{M},\tilde{N}\rangle=\langle M,N\rangle$, $M,N\in\mathscr{M}^{\mathrm{loc}}$. This result is known as **Girsanov's theorem**. The transformation of probability spaces given above is called a **transformation of drift** or a **Girsanov transformation** since it produces a change as shown above of the drift part in the semimartingale decomposition.

In the discussion above, the increasing family $\{\mathscr{F}_t\}$ was fixed. It is also important to study how the semimartingale character changes under a changing increasing family [12].

C. Stochastic Differentials

In this section, we introduce stochastic differentials of semimartingales and rewrite the results in the previous section in more convenient form. Let $(\Omega,\mathscr{F},P)$ and $\{\mathscr{F}_t\}$ be as above and $\mathscr{M},\mathscr{A},\mathscr{V},\mathscr{M}^{\mathrm{loc}},\mathscr{A}^{\mathrm{loc}},\mathscr{V}^{\mathrm{loc}}$ be defined as in Section B. By $\mathscr{Q}$ we denote the totality of continuous semimartingales relative to $\{\mathscr{F}_t\}$.

For $X \in \mathcal{Q}$, let $X(t)=X(0)+M_X(t)+V_X(t)$ be the semimartingale decomposition. We write formally $X(t)-X(0)=\int_0^t dX(s)$ and call dX (denoted also by dX_t or $dX(t)$) the **stochastic differential** of X. To be precise, dX can be considered as a random interval function $dX(I)=X(t)-X(s)$, $I=(s,t]$ or the equivalence class containing X under the equivalence relation $X \sim Y$ on $\mathcal{Q}$ defined by $X \sim Y$ if and only if $X(t)-X(0)=Y(t)-Y(0)$, $t \geqslant 0$. For X, $Y \in \mathcal{Q}$ and α, $\beta \in \mathbf{R}$, $\alpha dX+\beta dY$ is defined by $d(\alpha X+\beta Y)$ and $dX \cdot dY$ by $d\langle M_X, M_Y\rangle$. Let $d\mathcal{Q}$ be the totality of stochastic differentials of elements in $\mathcal{Q}$ and $d\mathcal{M}$ and $d\mathcal{V}$ be that of elements in $\mathcal{M}^{\mathrm{loc}}$ and $\mathcal{V}^{\mathrm{loc}}$, respectively. $d\mathcal{Q}$ is a commutative algebra under the operations just introduced. Note that $dX \cdot dY \in d\mathcal{V}$ and that $dX \cdot dY=0$ if either of dX and dY is in $d\mathcal{V}$. In particular, $dX \cdot dY \cdot dZ=0$ for every dX, dY, and dZ. Let $\mathcal{B}$ be the totality of $\{\mathcal{F}_t\}$-progressively measurable processes $\Phi=(\Phi(t))$ such that, with probability 1, $\sup_{0 \leqslant s \leqslant t}|\Phi(s)| < \infty$ for every $t \geqslant 0$. Noting that $\mathcal{B} \subset \mathcal{L}_2^{\mathrm{loc}}(M)$ for any $M \in \mathcal{M}^{\mathrm{loc}}$, we define $\Phi \cdot dX \in d\mathcal{Q}$ for $\Phi \in \mathcal{B}$ and $X \in \mathcal{Q}$ to be the stochastic differential of the semimartingale $\int_0^t \Phi(s)dM_X(s)+\int_0^t \Phi(s)dV_X(s)$. $\Phi \cdot dX$ is uniquely determined by Φ and dX. Itô's formula is stated, in this context, as follows: For $X=(X^1, \dots, X^d)$, $X^i \in \mathcal{Q}$, and $F: \mathbf{R}^d \to \mathbf{R}$, which is of class C^2, $F(X) \in \mathcal{Q}$, and

$$dF(X)=\sum_{i=1}^{d} D_i F(X)\cdot dX^i + \frac{1}{2}\sum_{i,j=1}^{d} D_i D_j F(X)\cdot dX^i \cdot dX^j.$$

We now define another important operation on the space $d\mathcal{Q}$. Noting that $\mathcal{Q} \subset \mathcal{B}$, we define $X \circ dY$ for X, $Y \in \mathcal{Q}$ by

$$X \circ dY = X \cdot dY + \frac{1}{2} dX \cdot dY.$$

This is uniquely determined from X and dY, and is called the **symmetric multiplication** of X and dY. It is also called a **stochastic differential of the Stratonovich type** or **Itô's circle operation** since the notation was introduced by Itô [9]. $\int_0^t X \circ dY$ is called the **stochastic integral of the Stratonovich type**, whereas $\int_0^t X \cdot dY$ is that of the **Itô type**. Under this operation, Itô's formula is rewritten as follows: For $X=(X^1, \dots, X^d)$, $X^i \in \mathcal{Q}$, and $F: \mathbf{R}^d \to \mathbf{R}$, which is of class C^3, $F(X)$, $D_i F(X) \in \mathcal{Q}$, and

$$dF(X)=\sum_{i=1}^{d} D_i F(X) \circ dX^i.$$

This chain rule for stochastic differentials takes the same form as in the ordinary calculus. For this reason symmetric multiplication plays an important role in transferring notions used in ordinary calculus into stochastic calculus and in defining intrinsic (i.e., coordinate-free) notions probabilistically. In particular, it is fundamental to the study of stochastic differential equations on manifolds (→ Section G).

D. Stochastic Differential Equations

Here, we give a general formulation of stochastic differential equations in which the infinitesimal change of the system may depend on the past history of the system; however, equations of **Markovian type**, in which the infinitesimal change of the system depends only on the present state of the system, are considered in most cases. Let W^d be the space of d-dimensional continuous paths: $W^d=C([0,\infty)\to\mathbf{R}^d)$:= the totality of all continuous functions $w:[0,\infty)\to\mathbf{R}^d$, endowed with the topology of the uniform convergence on finite intervals and $\mathcal{B}(W^d)$ be the topological σ-field. For each $t \geqslant 0$, define $\rho_t: W^d \to W^d$ by $(\rho_t w)(s)=w(t \wedge s)$, and let $\mathcal{B}_t(W^d)=\rho_t^{-1}(\mathcal{B}(W^d))$, $t \geqslant 0$. Let $\mathcal{A}^{d,r}$ be the totality of functions $\alpha(t,w)=(\alpha_j^i(t,w))$: $[0,\infty)\times W^d \to \mathbf{R}^d \otimes \mathbf{R}^r$ (:= the totality of $d \times r$ real matrices) such that each component $\alpha_j^i(x,w)$ $(i=1,2,\dots,d; j=1,2,\dots,r)$ is $\mathcal{B}([0,\infty))\times\mathcal{B}(W^d)$-measurable and $\mathcal{B}_t(W^d)$-measurable for each fixed $t \geqslant 0$. In general, $\alpha_j^i(t,w)$ is called **nonanticipative** if it satisfies the second property above. An important case of $\alpha \in \mathcal{A}^{d,r}$ is when it is given as $\alpha(t,w)=\sigma(t,w(t))$ by a Borel function $a:[0,\infty)\times\mathbf{R}^d\to\mathbf{R}^d\otimes\mathbf{R}^r$. In this case, α is called **independent of the past history** or **of Markovian type**. For a given $\alpha \in \mathcal{A}^{d,r}$ and $\beta \in \mathcal{A}^{d,1}$, we consider the following stochastic differential equation:

$$(1) \quad dX^i(t)=\sum_{j=1}^{r}\alpha_j^i(t,X)dB^j(t)+\beta^i(t,X)dt, \qquad i=1,2,\dots,d,$$

also denoted simply as

$$dX(t)=\alpha(t,X)dB(t)+\beta(t,X)dt.$$

Here $X(t)=(X^1(t),\dots,X^d(t))$ is a d-dimensional continuous process. $B(t)=(B^1(t),\dots,B^r(t))$ is a r-dimensional Brownian motion with $B(0)=0$. A precise formulation of equation (1) is as follows. $X=(X(t))$ is called a **solution** of equation (1) if it satisfies the following conditions: (i) X is a d-dimensional, continuous, and $\{\mathcal{F}_t\}$-adapted process defined on a probability space $(\Omega,\mathcal{F},P)$ with an increasing family $\{\mathcal{F}_t\}$, i.e., $X:\Omega\to W^d$ which is $\mathcal{F}_t/\mathcal{B}_t(W^d)$-measurable for every $t \geqslant 0$; (ii) $\alpha_j^i(t,X)\in\mathcal{L}_2^{\mathrm{loc}}$, $\beta^i(t,X)\in\mathcal{L}_1^{\mathrm{loc}}$, $i=1,\dots,d$, $j=1,\dots,r$ (→ Section B for the definition of $\mathcal{L}_p^{\mathrm{loc}}$); (iii) there exists an r-

dimensional $\{\mathcal{F}_t\}$-Brownian motion $B(t)$ with $B(0)=0$ such that the equality

$$X^i(t)-X^i(0)=\sum_{j=1}^{r}\int_0^t \alpha_j^i(s,X)\,dB^j(s)$$
$$+\int_0^t \beta^i(s,X)\,ds, \qquad i=1,2,\ldots,d,$$

holds with probability 1.

Thus a solution X is always accompanied by a Brownian motion B. To emphasize this, we often call X a solution with the Brownian motion B or call the pair (X,B) itself a solution of (1). In the above definition, a solution is given with reference to an increasing family $\{\mathcal{F}_t\}$. The essential point is that σ-fields $\sigma(B(u)-B(v); u\geqslant v\geqslant t)$ and $\sigma(X(s),B(s); 0\leqslant s\leqslant t)$ are independent for every t: If X satisfies the conditions of solutions stated above, then the specified independence is obvious, and conversely, if this independence is satisfied, then by setting $\mathcal{F}_t=\bigcap_{\varepsilon>0}\sigma(X(s),B(s); 0\leqslant s\leqslant t+\varepsilon)$, the conditions of solutions stated above are satisfied. But it is usually convenient to introduce some increasing family $\{\mathcal{F}_t\}$ into the definition of solutions as above. When α and β are of the Markovian type, $\alpha(t,w)=\sigma(t,w(t))$, $\beta(t,w)=b(t,w(t))$, the corresponding equation

(2) $\quad dX(t)=\sigma(t,X(t))\,dB(t)+b(t,X(t))\,dt$

is called a **stochastic differential equation of Markovian type**. Furthermore, if $\sigma(t,x)$ and $b(t,x)$ are independent of t, i.e., $\sigma(t,x)=\sigma(x)$ and $b(t,x)=b(x)$, the equation

(3) $\quad dX(t)=\sigma(X(t))\,dB(t)+b(X(t))\,dt$

is called a stochastic differential equation of **time homogeneous** (or **time-independent**) **Markovian type**.

Next, we define the notions of the uniqueness of solutions. There are two kinds of uniqueness: uniqueness in the sense of law (in distribution) and pathwise uniqueness. When we consider the stochastic differential equations as a means to determine the laws of continuous stochastic processes, uniqueness in the sense of law is sufficient. If, on the other hand, we regard the stochastic differential equation as a means to define the sample paths of solutions as a functional of the accompanying Brownian motion, i.e., if we regard the equation as a machine that produces a solution as an output when we input a Brownian motion, the notion of pathwise uniqueness is more natural and more important. As we shall see, this notion is closely related to the notion of **strong solutions**.

These notions are defined as follows. For a solution $X=(X(t))$ of (1), $X(0)$ is called the **initial value**, its law on $\mathbf{R}^d$ is called the **initial law (distribution)**, and the law of X on W^d is called the **law (distribution) of** X. We say that the **uniqueness in the sense of law of solutions** for (1) holds if the law of any solution X is uniquely determined by its initial law, i.e., if whenever X and X' are two solutions whose initial laws coincide, then the laws of X and X' coincide. In this definition, we restrict ourselves to the solutions whose initial values are nonrandom, i.e., the initial laws are δ-distributions at some points in $\mathbf{R}^d$. Next, we say that the **pathwise uniqueness of solutions** for (1) holds if whenever X and X' are any two solutions defined on the same probability space $(\Omega,\mathcal{F},P)$ with the same increasing family $\{\mathcal{F}_t\}$ and the same r-dimensional $\{\mathcal{F}_t\}$-Brownian motion such that $X(0)=X'(0)$ a.s., then $X(t)=X'(t)$ for all $t\geqslant 0$ a.s. In this definition also, the solutions can be restricted to those having nonrandom initial values.

We say that equation (1) has a **unique strong solution** if there exists a function $F(x,w)$: $\mathbf{R}^d\times W_0^r\to W^d$ ($W_0^r=\{w\in W^r \mid w(0)=0\}$) such that the following are true: (i) For any solution (X,B) of (1), $X=F(X(0),B)$ holds a.s.; (ii) for any $\mathbf{R}^d$-valued random variable $X(0)$ and an r-dimensional Brownian motion $B=(B(t))$ with $B(0)=0$ which are mutually independent, $X=F(X(0),B)$ is a solution of (1) with the Brownian motion B and the initial value $X(0)$. If this is the case, $F(x,w)$ itself is a solution of (1) with the initial value x, and with respect to the canonical Brownian motion $B(t,w)=w(t)$ on the r-dimensional Wiener space $(W_0^r,\mathcal{F},P)$, $\mathcal{F}$ is the completion of $\mathcal{B}(W_0^r)$ with respect to the r-dimensional Wiener measure P. If equation (1) has a unique strong solution, then it is clear that pathwise uniqueness holds. Conversely, if pathwise uniqueness holds for (1) and if a solution exists for any given initial law, then equation (1) has a unique strong solution, [6, 25].

The existence of solutions was discussed by A. V. Skorokhod [20]. If the coefficients α and β are bounded and continuous on $[0,\infty)\times W^d$, a solution of (1) exists for any given initial law. This is shown as follows [6]. We first construct approximate solutions by Cauchy's polygonal method and then show that their probability laws are †tight. A limit process in the sense of probability law can be shown to be a solution. The assumption of boundedness above can be weakened, e.g., to the following condition: For every $T>0$, a constant $K_T>0$ exists such that

(4) $$\|\alpha(t,w)\|+\|\beta(t,w)\|\leqslant K_T(1+\|w\|_t), \qquad t\in[0,T], \quad w\in W^d.$$

Here $\|w\|_t=\max_{0\leqslant s\leqslant t}|w(s)|$. In the case of the

Markovian equation (2), it is sufficient to assume that $\sigma(t,x)$ and $b(t,x)$ are continuous:

(5) $\|\sigma(t,x)\|+\|b(t,x)\|\leqslant K_T(1+|x|),$

$$t\in[0,T],\quad x\in\mathbf{R}^d.$$

If these conditions are violated, a solution $X(t)$ does not exist globally in general but exists up to a certain time e, called the **explosion time**, such that $\lim_{t\uparrow e}|x(t)|=\infty$ if $e<\infty$. To extend the notion of solutions in such cases, we have to replace the path space W^d by the space $\hat{W}^d$ that consists of all continuous functions $w:[0,\infty)\to\hat{\mathbf{R}}^d$ $(=\mathbf{R}^d\cup\{\Delta\}=$ the one-point compactification) satisfying $w(t)=\Delta$ for every $t\geqslant e(w)$ $(=\inf\{t\,|\,w(t)=\Delta\})$.

Now, we list some results on the uniqueness of solutions. First consider the equations of the Markovian type (2), and assume that the coefficients are continuous and satisfy the condition (5). (i) If σ, b are Lipschitz continuous, i.e., for every $N>0$ there exists a constant K_N such that $\|\sigma(t,x)-\sigma(t,y)\|+\|b(t,x)-b(t,y)\|\leqslant K_N|x-y|$, $t\in[0,T]$, x, $y\in B_N:=\{z\in R^d\,|\,|z|\leqslant N\}$, then the pathwise uniqueness of solutions holds for equation (2). Thus the unique strong solution of (2) exists, and this is constructed directly by Picard's successive approximation (Itô [7,8]). (ii) If $d=1$, σ is Hölder continuous with exponent 1/2 and b is Lipschitz continuous, i.e., for every $N>0$, K_N exists such that

$$|\sigma(t,x)-\sigma(t,y)|^2+|b(t,x)-b(t,y)|\leqslant K_N|x-y|,$$

$$t\in[0,N],\quad x,\ y\in B_N,$$

then the pathwise uniqueness of solutions holds for equation (2) (T. Yamada and Watanabe [25]). (iii) If the matrix $a(t,x)=\sigma(t,x)\sigma(t,x)^*$ (i.e., $a^{ij}(t,x)=\sum_{k=1}^r\sigma_k^i(t,x)\sigma_k^j(t,x)$) is strictly positive definite, then the uniqueness in the sense of law of the solution for (2) holds (D. W. Stroock and S. R. S. Varadhan [21]). (iv) An example of stochastic differential equations for which the uniqueness in the sense of law holds but the pathwise uniqueness does not hold was given by H. Tanaka as follows: $d=r=1$, $b(t,x)\equiv 0$ and $\sigma(t,x)=I_{\{x\geqslant 0\}}-I_{\{x<0\}}$. Another example in the non-Markovian cases was given by B. S. Tsirel'son (see below).

Next, consider non-Markovian equations of the following form:

(6) $dX(t)=dB(t)+\beta(t,X)\,dt;$

i.e., the case $d=r$ and $\alpha(t,w)=I$ (identity matrix). Assume further that $\beta\in\mathscr{A}^{d,1}$ is bounded. Then a solution of (6) exists for any given initial distribution, unique in the sense of law, and it can be constructed by the Girsanov transformation of Section B as follows. On a suitable probability space $(\Omega,\mathscr{F},P)$ with an increasing family $\{\mathscr{F}_t\}$ such that $\mathscr{F}=\bigvee_{t>0}\mathscr{F}_t$, we set up an $\mathscr{F}_0$-measurable, d-dimensional random variable $X(0)$ with a given law and a d-dimensional $\{\mathscr{F}_t\}$-Brownian motion $\tilde{B}=(\tilde{B}(t))$ such that $\tilde{B}(0)=0$. Set $X(t)=X(0)+\tilde{B}(t)$ and $M(t)=\exp[\int_0^t\beta(s,X)\,d\tilde{B}(s)-\frac{1}{2}\int_0^t|\beta(s,X)|^2\,ds]$. Then $M(t)$ is an $\{\mathscr{F}_t\}$-martingale, and the probability $\tilde{P}$ on $(\Omega,\mathscr{F})$ is determined by $\tilde{P}(A)=E(M_t;A)$, $A\in\mathscr{F}_t$. By Girsanov's theorem, $B(t)=X(t)-X(0)-\int_0^t\beta(s,X)\,ds$ is a d-dimensional $\{\mathscr{F}_t\}$-Brownian motion on $(\Omega,\mathscr{F},\tilde{P})$, and hence (X,B) is a solution of (6). Any solution is given in this way and hence the uniqueness in the sense of law holds. But the pathwise uniqueness does not hold in general; an example was given by Tsirel'son [1, 6] as follows. Let $\{t_n\}$ be a sequence such that $0<\ldots<t_n<t_{n-1}<t_0=1$ and $\lim_{n\uparrow\infty}t_n=0$. Set

$$\beta(t,w)=\begin{cases}0, & t\geqslant t_0 \text{ and } t=0,\\ \theta\left(\dfrac{w(t_{i+1})-w(t_{i+2})}{t_{i+1}-t_{i+2}}\right), & \\ & t\in[t_{i+1},t_i),\quad i=0,\,1,\,2,\ldots,\end{cases}$$

where $\theta(x)=x-[x]$, $x\in R$, is the decimal part of x.

Time changes (→ Section B) are also used to solve some stochastic differential equations [6].

E. Stochastic Differential Equations and Diffusion Processes

In this section we consider equations of time-independent Markovian type (3) only. The time-dependent case can be reduced to the time-independent case by adding one more component $X^{d+1}(t)$ such that $dX^{d+1}(t)=dt$. Further, we assume that coefficients $\sigma(x)\in\mathbf{R}^d\otimes\mathbf{R}^r$ and $b(x)\in\mathbf{R}^d$ are continuous on $\mathbf{R}^d$ and the uniqueness in the sense of law of solutions holds. Let P_x, $x\in\mathbf{R}^d$, be the law on W^d, or on $\hat{W}^d$ if there is an explosion, of a solution with the initial law δ_x (= the unit measure at x). Then $\{P_x\}$ possesses the †strong Markov property with respect to $\{\mathscr{F}_t\}$, where $\mathscr{F}_t$ is a suitable completion of $\mathscr{B}_t(W^d)$ or $\mathscr{B}_t(\hat{W}^d)$, and hence $(W^d,\{\mathscr{F}_t\},P_x)$ or $(\hat{W}^d,\{\mathscr{F}_t\},P_x)$ is a diffusion process †on $\mathbf{R}^d$ (→ 115 Diffusion Processes, 261 Markov Processes).

Let A be the differential operator

$$A=\frac{1}{2}\sum_{i,j=1}^d a^{ij}(x)D_iD_j+\sum_{i=1}^d b^i(x)D_i\qquad(D_i=\partial/\partial x^i)$$

with the domain $C_0^2(\mathbf{R}^d)$ (= the totality of C^2-functions on $\mathbf{R}^d$ with compact supports), where $a^{ij}(x)=\sum_{k=1}^r\sigma_k^i(x)\sigma_k^j(x)$. By Itô's formula,

(7) $f(w(t))-f(w(0))-\displaystyle\int_0^t(Af)(w(s))\,ds$

is an $\{\mathcal{F}_t\}$-martingale for every $f \in C_0^2(\mathbf{R}^d)$ (we set $f(\Delta)=0$), and this property characterizes the diffusion process. The diffusion is generated by the operator A in this sense. Furthermore, if for some $\lambda>0$, $(\lambda-A)(C_0^2(\mathbf{R}^d))$ is a dense subset of $C_\infty(\mathbf{R}^d)$ (= the totality of continuous functions f on $\mathbf{R}^d$ such that $\lim_{|x|\to\infty} f(x)=0$) then the †transition semigroup of the diffusion is a †Feller semigroup on $C_\infty(\mathbf{R}^d)$, and its infinitesimal generator A is the closure of $(A, C_0^2(\mathbf{R}^d))$. Hence $u(t,x)= E_x[f(w(t))]$, $f \in C_0^2(R^d)$, is the unique solution of the evolution equation $du/dt=Au$, $u|_{t=0}=f$. Generally, if the coefficients σ and b are sufficiently smooth, we can show, by using the stochastic differential equation (3), that $u(t,x)$ is also smooth for a smooth f and satisfies the heat equation $\partial u/\partial t=Au$. Taking the expectations in (7), we have the relation $E_x[f(w(t))]=f(x)+\int_0^t E_x[Af(w(s))]\,ds$, which implies that the transition probability $P(t,x,dy)$ of the diffusion satisfies the equation $\partial p/\partial t=A^*p$ in (t,y) in a weak sense, where A^* is the adjoint operator of A. If $\partial/\partial t-A^*$ is †hypoelliptic, we can conclude that $P(t,x,dy)$ possesses a smooth density $p(t,x,y)$ by appealing to the theory of partial differential equations. Recently, P. Malliavin showed that a probabilistic method based on the stochastic differential equations can also be applied to this problem effectively, [6, 16, 17].

If $c(t,x)$ is continuous and $v(t,x)$ is sufficiently smooth in (t,x) on $[0,\infty)\times\mathbf{R}^d$, then the following fact, more general than (7), holds:

$$(8)\quad v(t,w(t))\exp\left[\int_0^t c(s,w(s))\,ds\right]-v(0,x)$$
$$-\int_0^t \exp\left[\int_0^s c(u,w(u))\,du\right]\times(\partial v/\partial t+(A+c)v)(s,w(s))\,ds$$

is a local martingale (i.e., $\in\mathcal{M}^{\mathrm{loc}}$) with respect to $\{\mathcal{F}_t, P_x\}$. By applying the †optional sampling theorem to (8) for a class of $\{\mathcal{F}_t\}$-stopping times, we can obtain the probabilistic representation in terms of the diffusion of solutions for initial or boundary value problems related to the operator A [3, 4].

F. Stochastic Differential Equations with Boundary Conditions

As we saw in the previous section, diffusion processes generated by differential operators can be constructed by stochastic differential equations. A diffusion process on a domain with boundary is generated by a differential operator that describes the behavior of the process inside the domain, and a boundary condition that describes the behavior of the process on the boundary of the domain. For example, consider a reflecting †Brownian motion on the half-line $[0,\infty)$. This is a diffusion process $X=(X_t)$ on $[0,\infty)$ obtained by setting $X_t=|x_t|$ from a 1-dimensional Brownian motion x_t. The corresponding differential operator is $A=\frac{1}{2}d^2/dx^2$, and the boundary condition is $Lu\equiv du/dx|_{x=0}=0$, that is, the transition expectation $u(t,x)=E_x[f(X_t)]$ is determined by $\partial u/\partial t=Au$, $Lu=0$, and $u|_{t=0}=f$. In constructing such diffusion processes with boundary conditions, stochastic differential equations can be used effectively. In the case of reflecting Brownian motion, it was formulated by Skorokhod in the form

$$(9)\quad dX(t)=dB(t)+d\varphi(t).$$

Here $B(t)$ is a 1-dimensional Brownian motion ($B(0)=0$), $X(t)$ is a continuous process such that $X(t)\geqslant 0$, and $\varphi(t)$ has the following property with probability 1: $\varphi(0)=0$, $t\to\varphi(t)$ is continuous and nondecreasing and increases only on such t that $X(t)=0$, i.e., $\int_0^t I_{\{0\}}(X(s))\,d\varphi(s)=\varphi(t)$. Given a Brownian motion $B(t)$ and a nonnegative random variable $X(0)$ which are mutually independent, $X(t)$ satisfying (9) and with the initial value $X(0)$ is unique and given by $X(t)=X(0)+B(t)$, $t<\sigma_0=\min\{t\,|\,X(0)+B(t)=0\}$ and $X(t)=B(t)-\min_{\sigma_0\leqslant s\leqslant t}B(s)$, $t\geqslant\sigma_0$ (P. Lévy, Skorokhod; → [6, 18]).

In the case of multidimensional processes, possible boundary conditions were determined by A. D. Venttsel' [24]. Stochastic differential equations describing these diffusions were formulated by N. Ikeda [5] in the 2-dimensional case and by Watanabe [23] in the general case as follows. Let D be the upper half-space $\mathbf{R}^d_+=\{x=(x^1,\dots,x^d)\,|\,x^d\geqslant 0\}$, $\partial D=\{x\,|\,x^d=0\}$, and $\mathring{D}=\{x\,|\,x^d>0\}$. The general case can be reduced, at least locally, to this case. Suppose that the following system of functions is given: $\sigma(x): D\to\mathbf{R}^d\times\mathbf{R}^r$, $b(x): D\to\mathbf{R}^d$, $\tau(x):\partial D\to\mathbf{R}^{d-1}\times\mathbf{R}^s$, $\beta(x):\partial D\to\mathbf{R}^{d-1}$, and $\rho(x):\partial D\to[0,\infty)$, which are all bounded and continuous. Consider the following stochastic differential equation:

$$(10)\quad \begin{cases} dX^i(t)=\sum_{j=1}^r \sigma^i_j(X(t))I_{\mathring{D}}(X(t))\,dB^j(t) \\ \quad +b^i(X(t))I_{\mathring{D}}(X(t))\,dt \\ \quad +\sum_{k=1}^s \tau^i_k(X(t))I_{\partial D}(X(t))\,dM^k(t) \\ \quad +\beta^i(X(t))\,d\psi(t), \\ \qquad\qquad i=1,2,\dots,d-1, \\ dX^d(t)=\sum_{j=1}^r \sigma^d_j(X(t))I_{\mathring{D}}(X(t))\,dB^j(t) \\ \quad +b^d(X(t))I_{\mathring{D}}(X(t))\,dt+d\varphi(t), \\ I_{\partial D}(X(t))\,dt=\rho(X(t))\,d\varphi(t). \end{cases}$$

By a solution of this equation, we mean a system of continuous semimartingales $\mathfrak{X}=(X(t), B(t), M(t), \varphi(t))$ over a probability space $(\Omega, \mathscr{F}, P)$ with an increasing family $\{\mathscr{F}_t\}$ satisfying the following conditions: (i) $X(t)=(X^1(t), \ldots, X^d(t))$ is D-valued, i.e., $X^d(t)\geqslant 0$; (ii) with probability 1, $\varphi(0)=0$, $t\rightarrow\varphi(t)$ is nondecreasing, and $\int_0^t I_{\partial D}(X(s))\,d\varphi(s)=\varphi(t)$; (iii) $B(t)$ and $M(t)$ are r-dimensional and s-dimensional systems of elements in $\mathscr{M}^{\mathrm{loc}}$, respectively, such that $\langle B^i, B^j\rangle_t=\delta^{ij}t$, $\langle B^i, M^m\rangle_t=0$, and $\langle M^m, M^n\rangle_t=\delta^{mn}\varphi(t)$, $i, j=1, \ldots, r$, $m, n=1, \ldots, s$; and finally (iv) the stochastic differentials of these semimartingales satisfy (10).

The processes $B(t)$, $M(t)$, and $\varphi(t)$ are subsidiary, and the process $X(t)$ itself is often called a solution. We say that the uniqueness of solution holds if the law of $X=(X(t))$ is uniquely determined from the law of $X(0)$. As before, the existence and the uniqueness of solutions imply that solutions define a diffusion process on D, and these are guaranteed if, for example, $\min_{x\in\partial D} a^{dd}(x)>0$ and σ, b, τ, β are Lipschitz continuous, [6, 23]. Here, we set $a^{ij}(x)=\sum_{k=1}^r \sigma_k^i(x)\sigma_k^j(x)$ and $\alpha^{ij}(x)=\sum_{k=1}^s \tau_k^i(x)\tau_k^j(x)$. It is a diffusion process generated by the differential operator

$$A=\frac{1}{2}\sum_{i,j=1}^{d} a^{ij}(x)D_iD_j+\sum_{i=1}^{d} b^i(x)D_i,$$

and by the Venttsel' boundary condition,

$$Lu(x)\equiv\frac{1}{2}\sum_{i,j=1}^{d-1} \alpha^{ij}(x)D_iD_ju(x)+\sum_{i=1}^{d-1} \beta^i(x)D_iu(x)$$
$$+D_du(x)-\rho(x)(Au)(x)=0 \text{ on } \partial D.$$

G. Stochastic Differential Equations on Manifolds

Let M be a connected σ-compact C^∞-manifold of dimension d, and let $W_M=C([0,\infty)\rightarrow M)$ be the space of all continuous paths in M. If M is not compact, let $\hat{M}=M\cup\{\Delta\}$ be the one-point compactification of M and $\hat{W}_M$ be the space of all continuous paths in $\hat{M}$ with Δ as a †trap. These path spaces are endowed with the σ-fields $\mathscr{B}(W_M)$ and $\mathscr{B}(\hat{W}_M)$, respectively, which are generated by Borel cylinder sets. By a continuous process on M we mean a $(W_M, \mathscr{B}(W_M))$-valued random variable, and by a continuous process on M admitting explosions we mean a $(\hat{W}_M, \mathscr{B}(\hat{W}_M))$-valued random variable. In this section the probability space is taken to be the r-dimensional Wiener space $(W_0^r, \mathscr{F}, P)$ with the increasing family $\{\mathscr{F}_t\}$, where $\mathscr{F}_t$ is generated by $\mathscr{B}_t(W_0^r)$ and P-null sets. Then $w=(w(t))$, $w\in W_0^r$, is an r-dimensional $\{\mathscr{F}_t\}$-Brownian motion.

Suppose that we are given a system of C^∞-vector fields $A_0, A_1, \ldots, A_r$ on M. We consider the following stochastic differential equation on M:

(11) $\quad dX_t=A_k(X_t)\circ dw^k(t)+A_0(X_t)\,dt.$

(Here, the usual convention for the omission of the summation sign is used.) A precise meaning of equation (11) is as follows: We say that $X=(X_t)$ satisfies equation (11) if X is an $\{\mathscr{F}_t\}$-adapted continuous process on M admitting explosions such that, for any C^∞-function f on M with compact support (we set $f(\Delta)=0$), $f(X_t)$ is a continuous semimartingale satisfying

(12) $\quad df(X_t)=(A_kf)(X_t)\circ dw^k(t)+(A_0f)(X_t)\,dt,$

where $\circ$ is Itô's circle operation defined in Section C. This is equivalent to saying that $X_t=(X_t^1, \ldots, X_t^d)$, in each local coordinate, is a d-dimensional semimartingale such that

(13) $$dX_t^i=\sigma_k^i(X_t)\circ dw^k(t)+b^i(X_t)\,dt$$
$$=\sigma_k^i(X_t)\,dw^k(t)$$
$$+\left[\frac{1}{2}\sum_{k=1}^{r} D_j\sigma_k^i\sigma_k^j+b^i\right](X_t)\,dt,$$

where $A_k(x)=\sigma_k^i(x)D_i$, $k=1, 2, \ldots, r$, and $A_0(x)=b^i(x)D_i$. By solving the equation in each local coordinate and then putting these solutions together, we can obtain for each $x\in M$ a unique solution X_t of (11) such that $X_0=x$. We can also embed the manifold M in a higher-dimensional Euclidean space and solve the stochastic differential equation there. We denote the solution by $X(t,x,w)$. The law P_x on $\hat{W}_M$ of $[t\rightarrow X(t,x,w)]$ defines a diffusion process on M which is generated by the differential operator $A=\frac{1}{2}\sum_{k=1}^r A_k^2+A_0$.

Next, if we consider the mapping $x\rightarrow X(t,x,w)$; then, except for w belonging to a set of P-measure 0, the following is valid: For all (t,w) such that $X(t,x_0,w)\in M$, the mapping $x\rightarrow X(t,x,w)$ is a diffeomorphism between a neighborhood of x_0 and a neighborhood of $X(t,x_0,w)$. This is based on the following fact for stochastic differential equations on $\mathbf{R}^d$. If in equation (3) the coefficients σ_k^i and b^i are C^∞-functions with bounded derivatives of all orders α, $|\alpha|\geqslant 1$, then, denoting by $X(t,x,w)$ the solution such that $X(0)=x$, we have that $x\rightarrow X(t,x,w)$ is, with probability 1, a diffeomorphism of $\mathbf{R}^d$ for all t [13].

Example 1: **Stochastic moving frame** [6, 15]. Let M be a Riemannian manifold of dimension d, $O(M)$ be the orthonormal frame bundle over M, and $L_1, L_2, \ldots, L_d$ be the basic vector fields on $O(M)$, that is,

$$(L_if)(x,\mathbf{e})=\lim_{t\rightarrow 0}\frac{1}{t}[f(x_t,\mathbf{e}_t)-f(x,\mathbf{e})],$$

$$i=1, \ldots d,$$

where $\mathbf{e}=(e_1, \ldots, e_d)$ is an orthonormal basis in $T_x(M)$, $x_t=\mathrm{Exp}(te_i)$, i.e., the geodesic such that $x_0=x$ and $\dot{x}=e_i$, and $\mathbf{e}_t$ is the parallel translate of $\mathbf{e}$ along x_t. Let b be a vector field on M and L_b be its horizontal lift on $O(M)$, i.e., L_b is a vector field on $O(M)$ determined by the following two properties: (i) L_b is horizontal and (ii) $d\pi(L_b)=b$, where $\pi: O(M)\rightarrow M$ is the projection. Consider the following stochastic differential equation on $O(M)$:

$$dr(t)=L_i(r(t))\circ dw^i(t)+L_b(r(t))\,dt.$$

Solutions determine a family of (local) diffeomorphisms $r\rightarrow r(t,r,w)=(X(t,r,w), \mathbf{e}(t,r,w))$ on $O(M)$. The law of $[t\rightarrow X(t,r,w)]$ depends only on $x=\pi(r)$, and it defines a diffusion process on M that is generated by the differential operator $\frac{1}{2}\Delta_M+b$ (Δ_M is the Laplace-Beltrami operator). Using this stochastic moving frame $r(t,r,w)$, we can realize a stochastic parallel translation of tensor fields along the paths of Brownian motion on M (a diffusion generated by $\frac{1}{2}\Delta_M$) that was first introduced by Itô [10], and by using it we can treat heat equations for tensor fields by means of a probabilistic method.

Example 2: **Brownian motion on Lie groups**. Let G be a Lie group. A stochastic process $\{g(t)\}$ on G is called a **right-invariant Brownian motion** if it satisfies the following conditions: (i) With probability 1, $g(0)=e$ (the identity), and $t\rightarrow g(t)$ is continuous; (ii) for every $t\geqslant s$, $g(t)g(s)^{-1}$ and $\sigma(g(u); u\leqslant s)$ are independent; and (iii) for every $t\geqslant s$, $g(t)g(s)^{-1}$ and $g(t-s)$ are equally distributed.

Let $A_0, A_1, \ldots, A_r$ be a system of right-invariant vector fields on G, and consider the stochastic differential equation

$$(14)\quad dg_t=A_i(g_t)\circ dw^i(t)+A_0(g_t)\,dt.$$

Then a solution of (14) with $g_0=e$ exists uniquely and globally; we denote this solution by $g^0(t,w)$. It is a right-invariant Brownian motion G, and conversely, every right-invariant Brownian motion can be obtained in this way. The system of diffeomorphisms $g\rightarrow g(t,g,w)$ defined by the solutions of (14) is given by $g\rightarrow g(t,g,w)=g^0(t,w)g$.

Generally, if M is a compact manifold, the system of diffeomorphisms $g_t: x\rightarrow X(t,x,w)$ defined by equation (11) can be considered as a right-invariant Brownian motion on the infinite-dimensional Lie group consisting of all diffeomorphisms of M [2].

References

[1] B. S. Tsirel'son (Cirel'son), An example of stochastic differential equation having no strong solution, Theory Prob. Appl., 20 (1975), 416–418.

[2] K. D. Elworthy, Stochastic dynamical systems and their flows, Stochastic Analysis, A. Friedman and M. Pinsky (eds.), Academic Press, 1978, 79–95.

[3] A. Friedman, Stochastic differential equations and applications I, II, Academic Press, 1975.

[4] I. I. Gikhman and A. V. Skorokhod, Stochastic differential equations, Springer, 1972.

[5] N. Ikeda: On the construction of two-dimensional diffusion processes satisfying Wentell's boundary conditions and its application to boundary value problems, Mem. Coll. Sci. Univ. Kyoto, (A, Math.) 33 (1961), 367–427.

[6] N. Ikeda and S. Watanabe, Stochastic differential equations and diffusion processes, Kodansha and North-Holland, 1981.

[7] K. Itô, Differential equations determining Markov processes (in Japanese), Zenkoku Shijô Sûgaku Danwakai, 1077 (1942), 1352–1400.

[8] K. Itô, On stochastic differential equations, Mem. Amer. Math. Soc., 4 (1951).

[9] K. Itô, Stochastic differentials, Appl. Math. and Optimization, 1 (1975), 347–381.

[10] K. Itô, The Brownian motion and tensor fields on Riemannian manifold, Proc. Intern. Congr. Math., Stockholm 1962, 536–539.

[11] K. Itô and S. Watanabe, Introduction to stochastic differential equations, Proc. Intern. Symp. SDE, Kyoto, 1976, K. Itô (ed.), Kinokuniya, 1978, i–xxx.

[12] T. Jeulin, Semi-martingales et grossissement d'une filtration, Lecture notes in math. 833, Springer, 1980.

[13] H. Kunita, On the decomposition of solutions of stochastic differential equations, Proc. LMS Symp. Stochastic Integrals, Durham, 1980.

[14] H. Kunita and S. Watanabe, On square integrable martingales, Nagoya Math. J., 30 (1967), 209–245.

[15] P. Malliavin, Géometrie différentielle stochastique, Les Presse de l'Université de Montréal, 1978.

[16] P. Malliavin, Stochastic calculus of variation and hypoelliptic operators, Proc. Intern. Symp. SDE, Kyoto, 1976, K. Itô (ed.), Kinokuniya, 1978, 195–263.

[17] P. Malliavin, C^k-hypoellipticity with degeneracy, Stochastic Analysis, A. Friedman and M. Pinsky (eds.), Academic Press, 1978, 199–214, 327–340.

[18] H. P. McKean, Stochastic integrals, Academic Press, 1969.

[19] P. A. Meyer, Un cours sur les intégrales stochastiques, Lecture notes in math. 511, Springer, 1976, 245–400.

[20] A. V. Skorokhod, Studies in the theory of random processes, Addison-Wesley, 1965.
[21] D. W. Stroock and S. R. S. Varadhan, Diffusion processes with continuous coefficients I, II, Comm. Pure Appl. Math., 22 (1969), 345–400, 479–530.
[22] D. W. Stroock and S. R. S. Varadhan, Multidimensional diffusion processes, Springer, 1979.
[23] S. Watanabe, On stochastic differential equations for multi-dimensional diffusion processes with boundary conditions, I, II, J. Math. Kyoto Univ., 11 (1971), 169–180, 545–551.
[24] A. D. Venttsel' (Wentzell), On boundary conditions for multidimensional diffusion processes, Theory Prob. Appl., 4 (1959), 164–177.
[25] T. Yamada and S. Watanabe, On the uniqueness of solutions of stochastic differential equations, J. Math. Kyoto Univ., 13 (1973), 497–512.

407 (XVII.4) Stochastic Processes

A. Definitions

The theory of stochastic processes was originally involved with forming mathematical models of phenomena whose development in time obeys probabilistic laws. Given a basic †probability space $(\Omega, \mathfrak{B}, P)$ and a set T of real numbers, a family $\{X_t\}_{t \in T}$ of real-valued †random variables defined on $(\Omega, \mathfrak{B}, P)$ is called a **stochastic process** (or simply **process**) over $(\Omega, \mathfrak{B}, P)$, where t is usually called the **time parameter** of the process. For each finite t-set $\{t_1, \ldots, t_n\}$, the †joint distribution of $(X_{t_1}, \ldots, X_{t_n})$ is called a **finite-dimensional distribution** of the process $\{X_t\}_{t \in T}$. Stochastic processes are classified into large groups such as †additive processes (or processes with independent increments), †Markov processes, †Markov chains, †diffusion processes, †Gaussian processes, †stationary processes, †martingales, and †branching processes, according to the properties of their finite-dimensional distributions. This classification is possible because of the following fact, a consequence of Kolmogorov's †extension theorem (→ 341 Probability Measures I): Given a system $\mathscr{P}$ of finite-dimensional distributions satisfying certain †consistency conditions, we can construct a suitable probability measure on the space $W = \mathbf{R}^T$ of real-valued functions on T so that the stochastic process $\{X_t\}_{t \in T}$, obtained by setting $X_t(w)$ = the value of $w \in W$ at t, has $\mathscr{P}$ as its system of finite-dimensional distributions. Now, consider two stochastic processes $\mathscr{X} = \{X_t\}_{t \in T}$ and $\mathscr{Y} = \{Y_t\}_{t \in T}$. $\mathscr{Y}$ is called a **modification** of $\mathscr{X}$ if they are defined over a common probability space $(\Omega, \mathfrak{B}, P)$ and $P(X_t = Y_t) = 1$ $(t \in T)$. Regardless of whether $\mathscr{X}$ and $\mathscr{Y}$ are defined over a common probability space or over different probability spaces, X and Y are said to be **equivalent** or each is said to be a **version** of the other if their finite-dimensional distributions are the same. According to Kolmogorov's extension theorem, every stochastic process has a version over the space $W = \mathbf{R}^T$.

The function $X(\omega)$ of t obtained by fixing ω in a stochastic process $\{X_t\}_{t \in T}$ is called the **sample function (sample process** or **path)** corresponding to ω. In applying various operations to stochastic processes and studying detailed properties of stochastic processes, such as continuity of sample functions, the notions of measurability and separability play important roles. We assume that T is an interval in the real line, and (if needed) that the probability measure P is †complete. Denote by $\mathfrak{F}$ the class of all †Borel subsets of T. A stochastic process $\{X_t\}_{t \in T}$ is said to be **measurable** if the function $X_t(\omega)$ of (t, ω) is $\mathfrak{F} \times \mathfrak{B}$-measurable. Continuity in probability defined in the next paragraph gives a sufficient condition for a stochastic process to have a measurable modification. A stochastic process $\{X_t\}_{t \in T}$ is said to be **separable** if there exists a countable subset S of T such that

$$P\left\{\omega \,\middle|\, \liminf_{s \to t, s \in S} X_s(\omega) \leqslant X_t(\omega) \leqslant \limsup_{s \to t, s \in S} X_s(\omega) \quad \text{for any } t \in T\right\} = 1.$$

It was proved by J. L. Doob that every stochastic process has a separable modification [6].

Various types of continuity are considered for stochastic processes. $\{X_t\}_{t \in T}$ is said to be **continuous in probability** at $s \in T$ if $P(|X_t - X_s| > \varepsilon) \to 0$ $(t \to s, t \in T)$ for each $\varepsilon > 0$; it is said to be **continuous in the mean (of order** 1) at $s \in T$ if $E(|X_t \to X_s|) \to 0$ $(t \to s, t \in T)$. **Continuity in the mean of order** p (>1) is defined similarly. Continuity in the mean of any order implies continuity in probability. Suppose that $\{X_t\}_{t \in T}$ is separable. Then

$$D_s = \left\{\omega \,\middle|\, X_s(\omega) = \lim_{t \to s, t \in T} X_t(\omega)\right\}^c \text{ and } \bigcup_{s \in T} D_s$$

are measurable events. If $P(D_s) > 0$, then $s \in T$ is called a **fixed point of discontinuity**. The condition $P(\bigcup_{s \in T} D_s) = 0$ means that almost all sample functions are continuous. Regularity properties of sample functions of processes, such as continuity or right continuity, have been studied by many people. The following theorem is due to A. N. Kolmogorov: Let $T=[0,1]$. If

$$E(|X_t-X_s|^\gamma)\leqslant c|t-s|^{1+\varepsilon}$$

for constants $\gamma>0$, $\varepsilon>0$, and $c>0$, then $\{X_t\}_{t\in T}$ has a modification $\{\tilde{X}_t\}_{t\in T}$ for which almost all sample functions are continuous, and

$$P\left[\lim_{h\downarrow 0} h^{-\delta}\sup_{\substack{|t-s|\leqslant h\\ t,s\in T}}|\tilde{X}_t-\tilde{X}_s|=0\right]=1$$

for any $\delta(0<\delta<\varepsilon/\gamma)$. Each of the following is a sufficient condition for $\mathscr{X}=\{X_t\}_{t\in T}$ to have a modification for which almost all sample paths are right continuous functions with left limits. (i) $\mathscr{X}$ is an additive process which is continuous in probability (P. Lévy [3,4], K. Itô [8]; → 5 Additive Processes B). (ii) $\mathscr{X}$ is a supermartingale that is continuous in probability (Doob [6]; → 262 Martingales C).

B. Increasing Families of σ-Algebras

In the investigation of stochastic processes (especially Markov processes, martingales, and stochastic differential equations), the notion of increasing families of σ-algebras often plays an important role. Let $(\Omega,\mathfrak{B},P)$ be a probability space, and let $T=[0,\infty)$. A family $\{\mathfrak{B}_t\}_{t\in T}$ of σ-subalgebras of $\mathfrak{B}$ is called an **increasing family of σ-algebras** on $(\Omega,\mathfrak{B},P)$ if $\mathfrak{B}_s\subset\mathfrak{B}_t$ for $s<t$. A process $\{X_t\}_{t\in T}$ is said to be **adapted** to $\{\mathfrak{B}_t\}$ if X_t is $\mathfrak{B}_t$-measurable for each $t\in T$. $\{X_t\}$ is said to be **progressively measurable** (or a **progressive process**) with respect to $\{\mathfrak{B}_t\}$ if for every $t\in T$ the mapping $(s,\omega)\mapsto X_s(\omega)$ of $[0,t]\times\Omega$ into $\mathbf{R}$ is measurable with respect to the σ-field $\mathfrak{B}([0,t])\times\mathfrak{B}_t$. A process $\{X_t\}$ with right continuous paths, adapted to $\{\mathfrak{B}_t\}$, is progressively measurable with respect to $\{\mathfrak{B}_t\}$. The same conclusion holds for a process with left continuous paths. A subset A of $[0,\infty)\times\Omega$ is said to be **progressive** if the indicator process $a_t(\omega)=\mathbf{1}_A(t,\omega)$ of A is a progressive process. A random time τ on Ω with values in $[0,\infty]$ is called a **stopping time** (or **Markov time**) if $\{\tau\leqslant t\}\in\mathfrak{B}_t$ for all $t\geqslant 0$. Constants ($\geqslant 0$) are stopping times. If σ and τ are stopping times, then $\min(\sigma,\tau)$ and $\max(\sigma,\tau)$ are also stopping times. The limit of an increasing sequence of stopping times is a stopping time, while the limit of a decreasing sequence of stopping times is a stopping time with respect to $\{\mathfrak{B}_{t+}\}$, where $\mathfrak{B}_{t+}=\bigcap_{s>t}\mathfrak{B}_s$. Let $\mathfrak{B}_\tau$ be the class of $A\in\mathfrak{B}$ such that $A\cap\{\tau\leqslant t\}\in\mathfrak{B}_t$ ($^\forall t\in T$); then it is a σ-algebra if τ is a stopping time. If $\{X_t\}$ is a progressive process and if τ is a stopping time, then $X_\tau\mathbf{1}_{\{\tau<\infty\}}$ is $\mathfrak{B}_\tau$-measurable. An increasing family $\{\mathfrak{B}_t\}$ of σ-algebras on $(\Omega,\mathfrak{B},P)$ is said to be complete if the probability space $(\Omega,\mathfrak{B},P)$ is complete and if all the P-negligible sets belong to $\mathfrak{B}_0$. Fron now on we assume that $\{\mathfrak{B}_t\}$ is complete and right continuous (i.e., $\mathfrak{B}_t=\mathfrak{B}_{t+}$ for all $t\geqslant 0$). Let B be a subset of $\mathbf{R}$ and $\{X_t\}$ be a process. We call $\tau_B=\inf\{t\geqslant 0\mid X_t(\omega)\in B\}$ a **hitting time** for B. Measurability of τ_B is not always guaranteed, that is, τ_B is not always a stopping time. G. A. Hunt showed that for a wide class of Markov processes hitting times for analytic sets are stopping times. This result is based on a theorem of G. Choquet on capacitability and was generalized by P. A. Meyer as follows: (i) For every progressively measurable process, hitting times for analytic sets are stopping times; and (ii) for every progressive set A, $D_A=\inf\{t\geqslant 0\mid(t,\omega)\in A\}$ is a stopping time. The following notions on measurability are also important. The **predictable σ-algebra** on $[0,\infty)\times\Omega$, denoted by $\mathscr{P}$, is defined to be the least σ-algebra on $[0,\infty)\times\Omega$ with respect to which every process $X_t(\omega)$ that is adapted to $\{\mathfrak{B}_t\}$ and has left continuous paths is measurable in (t,ω). The **well-measurable** or **optional σ-algebra** on $[0,\infty)\times\Omega$, denoted by $\mathscr{O}$, is defined to be the least σ-algebra on $[0,\infty)\times\Omega$ with respect to which every process $X_t(\omega)$ that is adapted to $\{\mathfrak{B}_t\}$ and has right continuous paths with left limits is measurable in (t,ω). A process $\{X_t\}$ defined on Ω is said to be **predictable** (resp. **well-measurable** or **optional**) if the function $(t,\omega)\mapsto X_t(\omega)$ on $[0,\infty)\times\Omega$ is measurable with respect to the predictable σ-algebra $\mathscr{P}$ (resp. the optional σ-algebra $\mathscr{O}$). For further information regarding the notions given in this section → [10].

Up to this point it was assumed that the space in which a process $\{X_t\}_{t\in T}$ takes values, namely, the **state space** of $\{X_t\}_{t\in T}$, is a set of real numbers; but in general, topological spaces or merely measurable spaces can be taken for the state spaces of stochastic processes. The general definitions and results already given can be extended to stochastic processes whose state spaces are †locally compact Hausdorff spaces satisfying the second †countability axiom.

Moreover, the time parameter set T of a process $\{X_t\}_{t\in T}$ need not be a set of real numbers. For example, P. Lévy [12] and H. P. McKean [13] investigated stochastic processes with several-dimensional time; such processes are sometimes called **random fields**. The case in which T is $\mathscr{D}$, $\mathscr{S}$, or in general a space of functions (which is nuclear) has also been investigated (→ Section C). A probabilistic formulation of equilibrium states given by R. L. Dobrushin [14] initiated recent probabilistic study of statistical mechanics. For further information concerning processes with general time parameter spaces → 136 Ergodic Theory, 176 Gaussian Processes, 340 Probabilistic Methods in Statistical Mechanics.

C. Random Distributions and Generalized Stochastic Processes

The investigation of random distributions was initiated by I. M. Gel'fand [15] and Itô [17]. Denote by $\mathscr{D}$ the space of functions of t $(-\infty < t < \infty)$ of class C^∞ with compact †support, and by $\mathscr{D}'$ the space of †distributions. At function $X(\varphi, \omega)$ of $\omega \in \Omega$ and $\varphi \in \mathscr{D}$ is called a **random distribution** (or **generalized stochastic process**) if $X(\varphi, \omega)$ is a distribution as a function of φ for almost all ω and is measurable as a function of ω for each fixed φ. Denote by $\mathfrak{B}(\mathscr{D}')$ the smallest σ-algebra containing sets of the form $\{y \in \mathscr{D}' \mid y(\varphi) \in E\}$ ($\varphi \in \mathscr{D}$, E is a Borel set). A random distribution is nothing but a $\mathscr{D}'$-valued random variable. For a random distribution $X(\varphi, \omega)$, a probability measure Φ_X on $\mathfrak{B}(\mathscr{D}')$ is induced by

$$\Phi_X(B) = P\{\omega \mid X(\cdot, \omega) \in B\}, \qquad B \in \mathfrak{B}(\mathscr{D}').$$

The functional

$$c(\varphi) = E(e^{iX(\varphi,\omega)}) = \int e^{iy(\varphi)} \Phi_X(dy), \qquad \varphi \in \mathscr{D},$$

is called the **characteristic functional** of $X(\varphi, \omega)$ or Φ_X. The functional $c(\varphi)$ is continuous positive definite, and $c(0)=1$. Conversely, given a functional $c(\varphi)$ with these properties, a theorem of R. A. Minlos (→ 341 Probability Measures J) states that there exists a unique probability measure Φ on $\mathfrak{B}(\mathscr{D}')$ whose characteristic functional equals $c(\varphi)$. In other words, a random distribution with the characteristic functional $c(\varphi)$ can be constructed over $(\mathscr{D}', \mathfrak{B}(\mathscr{D}'), \Phi)$.

Typical classes of random distributions that have been investigated so far are stationary random distributions and random distributions with independent values at every point. (For stationary ones → 395 Stationary Processes.) A random distribution $X(\varphi, \omega)$ is called a **random distribution with independent values at every point** if $\varphi_1(t)\varphi_2(t) \equiv 0$ implies the independence of $X(\varphi_1, \omega)$ and $X(\varphi_2, \omega)$, that is, $c(\varphi_1 + \varphi_2) = c(\varphi_1)c(\varphi_2)$. A sufficient condition for the functional of the form

$$c(\varphi) = \exp\left(\int_{-\infty}^{\infty} f(\varphi(t), \varphi'(t), \ldots, \varphi^{(k)}(t))\,dt\right)$$

(f is continuous and $f(0, \ldots, 0)=0$) to be the characteristic functional of a stationary random distribution with independent values at every point is that the function $\exp(sf(x_0, x_1, \ldots, x_k))$ of $(x_0, x_1, \ldots, x_k) \in \mathbf{R}^k$ be positive definite for each $s>0$ [16]. Under this condition, a concrete representation of f is known [16]. When $k=0$, a necessary and sufficient condition for $c(\varphi)$ to be the characteristic functional of a stationary random distribution with independent values at every point is that $\exp f(x)$ be the characteristic function of an †infinitely divisible distribution. One such random distribution is the so-called **Guassian white noise**, namely, the distribution derivative of Brownian motion whose characteristic functional is

$$\exp\left(-\frac{1}{2}\int |\varphi(t)|^2\,dt\right).$$

A family $\{X_\varphi\}_{\varphi \in \mathscr{D}}$ of real-valued random variables indexed by $\mathscr{D}$ is called a **random distribution in the wide sense** if X_φ is linear in φ, namely, $X_{a\varphi + b\psi} = aX_\varphi + bX_\psi$ with probability 1 for fixed $\varphi, \psi \in \mathscr{D}$, and real constants a, b, and if $X_\varphi \mapsto 0$ in probability whenever $\varphi \to 0$ in the topology of $\mathscr{D}$. (For a typical class of random distributions in the wide sense → 395 Stationary Processes C.) A random distribution in the wide sense has a modification that is a random distribution.

In the definition of random distributions (in the wide sense) one can replace the space $\mathscr{D}$ by the space $\mathscr{S}$ of rapidly decreasing C^∞-functions or in general by some space Φ of functions that is nuclear. For example, one can define random distributions as $\mathscr{S}'$-valued random variables.

Up to this point random distributions of one variable have been considered. Random distributions of several variables are called generalized random fields and have been investigated by Itô [18], A. M. Yaglom [19], Gel'fand and N. Ya. Vilenkin [16], and others. Moreover, K. Urbanik [20, 21] developed a theory of generalized stochastic processes based on G. Mikusiński's theory instead of L. Schwartz's theory of distributions.

D. Random Measure

Let $(S, \mathfrak{F}, m)$ be any σ-finite measure space, and put $\mathfrak{F}_0 = \{A \in \mathfrak{F} \mid m(A) < \infty\}$. By virtue of †Kolmogorov's extension theorem, there exists a family $\{W(A)\}_{A \in \mathfrak{F}_0}$ of real random variables indexed by $\mathfrak{F}_0$ such that (i) for any mutually disjoint $A_1, \ldots, A_n \in \mathfrak{F}_0$, $\{W(A_1), \ldots, W(A_n)\}$ is independent; (ii) for any $A \in \mathfrak{F}_0$, $W(A)$ is Gaussian distributed with mean 0 and variance $m(A)$; and (iii) for any A, $B \in \mathfrak{F}_0$, $E(W(A)W(B)) = m(A \cap B)$. Similarly, there exists a family $\{N(A)\}_{A \in \mathfrak{F}_0}$ of real-valued random variables indexed by $\mathfrak{F}_0$ such that (i) for any mutually disjoint $A_1, \ldots, A_n \in \mathfrak{F}_0$, $\{N(A_1), \ldots, N(A_n)\}$ is independent; (ii) for any $A \in \mathfrak{F}_0$, $N(A)$ is †Poisson distributed with mean $m(A)$; and (iii) for any $A, B \in \mathfrak{F}_0$, $E(N(A)N(B)) = m(A \cap B)$. $\{W(A)\}_{A \in \mathfrak{F}_0}$ and $\{N(A)\}_{A \in \mathfrak{F}_0}$ are called a **Gaussian random measure** and a **Poisson random measure** associated with the measure space $(S, \mathfrak{F}, m)$, respectively. By using these random measures, the theory of multiple integrals can be developed.

By a **point function** p on S we mean a mapping $p: D_p \rightarrow S$, where the domain D_p is a countable subset of $(0, \infty)$. p defines a counting measure N_p on $(0, \infty) \times S$ such that $N_p((0,t] \times U) = \# \{s \in D_p; s \leqslant t, p(s) \in U\}$ $(t > 0, U \in \mathfrak{F}_0)$. For a point function p and $t > 0$, the shift point function $\Theta_t p$ is defined by $D_{\Theta_t p} = \{s \in (0, \infty); s + t \in D_p\}$ and $(\Theta_t p)(s) = p(s+t)$. Let Π_S be the totality of point functions on S and $\mathscr{B}(\Pi_S)$ be the smallest σ-field on Π_S with respect to which all $p \rightarrow N_p((o,t] \times U, t > 0, U \in \mathfrak{F}_0)$ are measurable. A **point process** on S is a $(\Pi_S, \mathscr{B}(\Pi_S))$-valued random variable. Then there exists a point process p on S such that (i) for any $t > 0$, p and $\Theta_t p$ have the same probability law, and (ii) N_p is a Poisson random measure associated with $((0, \infty) \times S, \mathscr{B}(0, \infty) \times \mathscr{B}(\Pi_S), dt \times m(ds))$. The point process p is called the **stationary Poisson point process** with the **characteristic measure** m.

References

[1] A. N. Kolmogorov, Grundbegriffe der Wahrscheinlichkeitsrechnung, Erg. Math., 1933; English translation, Foundations of the theory of probability, Chelsea, 1950.
[2] N. Wiener, Differential-space, J. Math. Phys., 2 (1923), 131–174.
[3] P. Lévy, Processus stochastiques et mouvement brownien, Gauthier-Villars, 1948.
[4] P. Lévy, Théorie de l'addition des variables aléatoires, Gauthier-Villars, 1937.
[5] M. S. Bartlet, An introduction to stochastic processes, Cambridge Univ. Press, 1955.
[6] J. L. Doob, Stochastic processes, Wiley, 1953.
[7] J. L. Doob, Stochastic processes depending on a continuous parameter, Trans. Amer. Math. Soc., 42 (1937), 107–140.
[8] K. Itô, Stochastic processes, Aarhus Univ. lecture notes 16, 1969.
[9] P. A. Meyer, Une présentation de la théorie des ensembles sousliniens; application aux processus stochastiques, Sém. Théorie du Potentiel, 1962–1963, no. 2, Inst. H. Poincaré, Univ. Paris, 1964.
[10] C. Dellacherie and P. A. Meyer, Probabilités et potentiel, Hermann, 1975, chs. I–IV.
[11] C. Dellacherie, Capacités et processus stochastiques, Springer, 1972.
[12] P. Lévy, Le mouvement brownien fonction d'un ou de plusieurs paramètres, Rend. Sem. Math. Univ. Padova, (5) 22 (1963), 24–101.
[13] H. P. McKean, Jr., Brownian motion with a several-dimensional time, Theory Prob. Appl., 8 (1963), 335–354.
[14] R. L. Dobrushin, The description of the random field by its conditional distributions and its regularity conditions, Theory Prob. Appl., 13–14 (1968), 197–224. (Original in Russian, 1968).
[15] I. M. Gel'fand, Generalized random processes (in Russian), Dokl. Akad. Nauk SSSR (N.S.), 100 (1955), 853–856.
[16] I. M. Gel'fand and I. Ya. Vilenkin, Generalized functions IV, Academic Press, 1964. (Original in Russian, 1961.)
[17] K. Itô, Stationary random distributions, Mem. Coll. Sci. Univ. Kyoto, (A) 28 (1954), 209–223.
[18] K. Itô, Isotropic random current, Proc. 3rd Berkeley Symp. Math. Stat. Prob. II, Univ. of California Press, 1956, 125–132.
[19] A. M. Yaglom (Jaglom), Some classes of random fields in n-dimensional space related to stationary random processes, Theory Prob. Appl., 2 (1957), 273–320. (Original in Russian, 1957.)
[20] K. Urbanik, Stochastic processes whose sample functions are distributions, Theory Prob. Appl., 1 (1956), 132–134. (Original in Russian, 1956.)
[21] K. Urbanik, Generalized stochastic processes, Studia Math., 16 (1958), 268–334.

408 (XIX.7) Stochastic Programming

A. General Remarks

Stochastic programming is a method of finding optimal solutions in mathematical programming in its narrow sense (→ 264 Mathematical Programming), when some or all coefficients are stochastic variables with known probability distributions. There are essentially two different types of models in stochastic programming situations: One is **chance-constrained programming** (CCP), and the other is a **two-stage stochastic programming** (TSSP). The difference between them depends mainly on the informational structure of the sequence of observations and decisions. For simplicity, let us here consider stochastic linear programming, which is the best-known model at present. Let A_0, A be $m \times n$-dimensional matrices and $x, c \in \mathbf{R}^n$ and $b, b_0 \in \mathbf{R}^m$. Suppose further that components of A, b, c are random variables, while those of A_0, b_0 are constants. Consider the following formally defined linear programming problem: $\min_x\{c'x \mid Ax \leqslant b, x \in X_0\}$, $X_0 = \{x \mid A_0 x \leqslant b_0, x \geqslant 0\}$. Let $(\Omega, \mathfrak{B}, P)$ be a probability space (→ 342 Probability Theory) such that $\{A(\omega), b(\omega), c(\omega)\}$ is a measurable transformation on Ω into $\mathbf{R}^{m \times n + m + n}$.

B. Chance-Constrained Programming (CCP)

This method is based on the assumption that a decision x has to be made in advance of the realization of the random variables. Suppose that $A_i(\omega)$ is the ith row of $A(\omega)$, and $b_i(\omega)$ is the ith component of $b(\omega)$. We call $P(\{\omega \mid A_i(\omega)x \leqslant b_i(\omega)\}) \geqslant \alpha_i$ a **chance constraint**, where α_i is a prescribed fractional value determined by the decision maker according to his attitude toward the constraint $A_i(\omega)x \leqslant b_i(\omega)$: if he attaches importance to it, he will take α_i as great as possible. Defining feasible sets $X_i(\alpha_i)$ and X by $X_i(\alpha_i) = \{x \mid P(\{\omega \mid A_i(\omega)x \leqslant b_i(\omega)\}) \geqslant \alpha_i\}$, $X = X_0 \cap \{\bigcap_{i=1}^m X_i(\alpha_i)\}$, we can formulate CCP as follows: $\min_x\{F(x) \mid x \in X\}$, where $F: X \to R$, is the **certainty equivalent** of the stochastic objective function $c'x$. We have four models of CCP corresponding to different types of $F(x)$: (i) E-model: $F(x) = \bar{c}'x$, $\bar{c} = E_\omega c(\omega)'x$. (ii) V-model: $F(x) = \mathrm{Var}(c(\omega)'x) = x'V_c x$, where V_c is a variance-covariance matrix of $c(\omega)$. (iii) P_1-model: $F(x) = f$, $P(\{\omega \mid c(\omega)'x \leqslant f\}) \geqslant \alpha_0$, $1/2 \leqslant \alpha_0 \leqslant 1$. (iv) P_2-model: $F(x) = P(\{\omega \mid c(\omega)'x \geqslant \gamma\})$ for a given constant γ. In particular, if the components of $A(\omega)$, $b(\omega)$, $c(\omega)$ have a multidimensional normal distribution, the certainty equivalent of the ith chance constraint is derived in the following form: $\bar{A}_i x + \Phi^{-1}(\alpha_i)(x'V_i x + 2w_i' x + v_i^2)^{1/2} \leqslant \bar{b}_i$, where $\bar{A}_i$, V_i, w_i, v_i, $\bar{b}_i$ are expectation vectors or a variance-covariance matrix of $A_i(\omega)$ and $b_i(\omega)$ and $\Phi(t) = \int_{-\infty}^t \exp(-z^2/2)\,dz/\sqrt{2\pi}$. The set $X_i(\alpha_i)$ for this constraint can be shown to be convex for $1/2 \leqslant \alpha_i \leqslant 1$, by using the convexity of the function $\sqrt{x'Vx}$ for a positive semidefinite matrix V. Under the same assumption we can obtain the objective functions $F(x) = \bar{c}'x + \Phi^{-1}(\alpha_0)\sqrt{x'V_c x}$ for the P_1-model and $F(x) = (\bar{c}'x - \gamma)/\sqrt{x'V_c x}$ for the P_2-model. These four models have been shown to be computable by applying convex programming techniques, including the conjugate gradient method. Further studies on the convexity of a more general chance constraint $P(\{\omega \mid A(\omega)x \leqslant b(\omega)\}) \geqslant \alpha$, $0 \leqslant \alpha \leqslant 1$, appear in several articles.

C. Two-Stage Stochastic Programming (TSSP)

This method divides the decision process into two stages. First stage: Before the realization of random variables, one makes a decision x, being allowed to compensate for it after the specification of those values. Second stage: One obtains an optimal compensation $y \in \mathbf{R}^r$ for the given x and the realized values of the random variables. Assuming that $q \in \mathbf{R}^r$ is a random vector in addition to A, b, c, we can formulate TSSP as follows. First stage: $\min_x E_\omega\{(c(\omega)'x + Q(x,\omega) \mid x \in X)\}$; second stage: $Q(x,\omega) = \min_y\{q(\omega)'y \mid Wy = A(\omega)x - b(\omega), y \geqslant 0\}$, where $X = X_0 \cap K$, $K = \{x \mid Q(x,\omega) < +\infty$ with probability $1\}$ and $q(\omega)'y$ is a loss function for the deviation $A(\omega)x - b(\omega)$. The $m \times n$ matrix W is called a compensation matrix. Several theorems have been proved: (i) K is a closed convex set; (ii) $Q(x) = E_\omega Q(x,\omega)$ is a convex function on K if the random variables in $A(\omega)$, $b(\omega)$, $q(\omega)$ are square integrable; (iii) if P has a density function, then $Q(x)$ has a continuous gradient on K; (iv) when P has a finite discrete probability distribution, a TSSP problem is reduced to a linear programming problem having a dual decomposition structure.

References

[1] A. Charnes and W. W. Cooper, Chance-constrained programming, Management Sci., 6 (1959), 73–79.
[2] G. B. Dantzig and A. Madansky, On the solution of two-stage linear programs under uncertainty, Proc. 4th Berkeley Symp. Math. Statist. Probab. I, Univ. of California Press, 1961.
[3] S. Kataoka, A stochastic programming model, Econometrica, 31 (1963), 181–196.
[4] I. M. Stancu-Minasian and M. J. Wets, A research bibliography in stochastic programming 1955–1975, Operations Res., 24 (1976), 1078–1119.
[5] S. Vajda, Probabilistic programming, Academic Press, 1972.
[6] P. Kall, Stochastic linear programming, Springer, 1976.
[7] V. V. Kolbin, Stochastic programming, Reidel 1977. (Original in Russian, 1977.)

409 (II.7) Structures

A. Examples of Structure

Structure is a unified description of mathematical objects such as †ordered sets, †rings, †linear spaces, †topological spaces, †probability spaces, †manifolds, etc., using only the concepts †set and †relation. The following are examples.

(1) Order. An †ordering on a set A is a binary relation in A (→ 358 Relations) with a †graph α such that (i) if $a \in A$, then $(a, a) \in \alpha$; (ii) if $(a, b) \in \alpha$ and $(b, a) \in \alpha$, then $a = b$; and (iii) if $(a, b) \in \alpha$ and $(b, c) \in \alpha$, then $(a, c) \in \alpha$, where α is an element of the †power set $\mathfrak{P}(A \times A)$. We say that α determines a structure of ordering on the set A.

(2) Law of composition. A **law of composition** on a set A is a mapping from $A \times A$ (or a subset) to A. This mapping is considered as a ternary relation with a graph $\alpha \in \mathfrak{P}(A \times A \times A)$, satisfying the following two conditions (or possibly only (ii)): (i) if $(a, b) \in A \times A$, then $(a, b, c) \in \alpha$ for some $c \in A$; (ii) if $(a, b, c) \in \alpha$ and $(a, b, c') \in \alpha$, then $c = c'$. We say that α determines the structure of a law of composition in A. The †associative law in the law of composition determined by α is given by: (iii) if $(a, b, x) \in \alpha$, $(b, c, y) \in \alpha$, $(x, c, z) \in \alpha$, and $(a, y, z') \in \alpha$, then $z = z'$. A set with conditions (i), (ii), and (iii) is called a **semigroup**.

(3) Operation. An **operation** of a set A on a set B is a mapping from $A \times B$ (or a subset) to B. It is considered as a ternary relation with the graph $\gamma \in \mathfrak{P}(A \times B \times B)$, satisfying the following two conditions (or possibly only (ii)): (i) if $(a, b) \subset A \times B$, then $(a, b, c) \in \gamma$ for some $c \in B$; (ii) if $(a, b, c) \in \gamma$ and $(a, b, c') \in \gamma$, then $c = c'$. We say that γ determines the structure of an operation of A on B. Each element α of A is called an **operator** on B; A is called a **domain of operators** on B, and B is called an A-**set**. When B is the main object of consideration, an operation of A on B is sometimes called an **external law of composition** of A on B. The law of composition of A as described in (2) is then called an **internal law of composition** of A. When a domain of operators A on B determined by $\gamma \in \mathfrak{P}(A \times B \times B)$ has an internal law of composition determined by $\alpha \in \mathfrak{P}(A \times A \times A)$, it is usually assumed that the following conditions on α, γ are satisfied: (iii) if $(a, b, x) \in \alpha$, $(b, c, y) \in \gamma$, $(x, c, z) \in \gamma$, and $(a, y, z') \in \gamma$, then $z = z'$. If we denote the law of composition by $(a, b) \rightarrow ab$ and the operation by $(a, b) \rightarrow a \cdot b$, then condition (iii) may be written: (iii') $(ab) \cdot c = a \cdot (b \cdot c)$ for $a, b \in A$ and $c \in B$. When an A-set B with an external law of composition determined by $\gamma \in \mathfrak{P}(A \times B \times B)$ has an internal law of composition determined by $\beta \in \mathfrak{P}(B \times B \times B)$, it is usually assumed that the following condition on β, γ is satisfied: (iv) if $(a, b, x) \in \gamma$, $(a, c, y) \in \gamma$, $(b, c, z) \in \beta$, $(x, y, w) \in \beta$, and $(a, z, w') \in \gamma$, then $w = w'$. According to the notation ab and $a \cdot b$, it is described as: (iv') $(a \cdot b)(a \cdot c) = a \cdot (bc)$ for $a \in A$ and $b, c \in B$.

The mapping $A \times B \rightarrow B (B \times A \rightarrow B)$ is called a **left (right) operation** of A on B. To emphasize leftness or rightness, "left-" or "right-" is attached to corresponding concepts.

(4) Topology. A †topology on a set A is determined by a set $\alpha \in \mathfrak{P}\mathfrak{P}(A)$, called the †system of open sets, satisfying the following conditions: (i) $\varnothing \in \alpha$ and $A \in \alpha$; (ii) if $\beta \subset \alpha$, then $\bigcup \beta \in \alpha$; and (iii) if $\beta \subset \alpha$ is finite, then $\bigcap \beta \in \alpha$. We say that α defines the structure of a topology on the set A (→ 425 Topological Spaces).

B. Mathematical Structures

We now explain the concept of mathematical structure for the case of a †linear space (→ 256 Linear Spaces). A linear space has two basic sets, one of which is a set K of elements called scalars and the other, a set V of elements called †vectors, two laws of compositions in K called addition and multiplication, a law of composition in V called addition, and an operation of K on V called scalar multiplication. The laws of composition and the operation are given by elements of power sets: $\alpha_1, \alpha_2 \in \mathfrak{P}(K \times K \times K)$, $\alpha_3 \in \mathfrak{P}(V \times V \times V)$, and $\alpha_4 \in \mathfrak{P}(K \times V \times V)$; and the basic properties of the linear space, such as $\lambda(a+b) = \lambda a + \lambda b$ ($\lambda \in K$, $a, b \in V$), are described as propositions on K, V, $\alpha_1, \ldots, \alpha_4$ and denoted by $P(K, V, \alpha_1, \ldots, \alpha_4), \ldots$.

Up to now, we have been considering a given linear space. To give a description of a linear space in general, we use the symbols $X_1, X_2, \xi_1, \ldots, \xi_4$ instead of the symbols K, V, $\alpha_1, \ldots, \alpha_4$, replace conditions such as $\alpha_1 \in \mathfrak{P}(K \times K \times K), \ldots$ by $\xi_1 \in \mathfrak{P}(X_1 \times X_1 \times X_1), \ldots$, and consider the set Σ of these symbols and formulas:

$$\Sigma : X_1, X_2, \xi_1, \ldots, \xi_4;$$

$$\xi_1 \in \mathfrak{P}(X_1 \times X_1 \times X_1), \ldots,$$

$$\xi_4 \in \mathfrak{P}(X_1 \times X_2 \times X_2).$$

The set Σ is called the **type** of linear space. Similarly, we consider the set Γ of all $P(X_1, X_2, \xi_1, \ldots, \xi_4), \ldots$ corresponding to the basic properties $P(K, V, \alpha_1, \ldots, \alpha_4), \ldots$ of the linear space. The set Γ is called the **axiom system** of the linear space.

In general, let $A_1, \ldots, A_m$ be the **basic sets** (K and V in the preceding example). The **basic concepts** $\alpha_1, \ldots, \alpha_n$ ($\alpha_1, \ldots, \alpha_4$ in the preceding example) are given as elements of finitely generated sets from $A_1, \ldots, A_m$, i.e., elements of sets obtained by a finite number of applications of the operations of forming the †Cartesian product and the †power set from $A_1, \ldots, A_m$. **Basic properties** are given as propositions on $A_1, \ldots, A_m, \alpha_1, \ldots, \alpha_n$. These basic properties and $A_1, \ldots, A_m, \alpha_1, \ldots, \alpha_n$ determine a **mathematical system**. We consider also a type Σ of symbols $X_1, \ldots, X_m$ of basic sets and symbols $\xi_1, \ldots, \xi_n$ of basic concepts, and an axiom system Γ of basic properties. The pair (Σ, Γ) determines a **mathematical structure**. When we substitute sets $A_1, \ldots, A_m$ for $X_1, \ldots, X_m$ and $\alpha_1, \ldots, \alpha_n$ for $\xi_1, \ldots, \xi_n$, where $X_1, \ldots, X_m$ and $\xi_1, \ldots, \xi_n$ satisfy the axiom system Γ, then $(A_1, \ldots, A_m, \alpha_1, \ldots, \alpha_n)$ is called a mathematical system with the mathematical structure (Σ, Γ), or a **model** of the structure (Σ, Γ). Two mathematical systems are called **similar** or **of the same kind** if they have the same mathematical structure. Groups, rings, topological spaces, etc., are mathematical structures. Mathematical systems are sometimes called **algebraic systems in the wider sense**, and when we consider mainly their laws of composition (and operations), we call them **algebraic systems**. We explain this in further detail in Section C.

C. Algebraic Systems

Algebraic systems are sets with laws of composition and operations satisfying certain axiom systems; the laws of composition and operations and the axiom systems they satisfy determine their type (→ 2 Abelian Groups, 29 Associative Algebras, 42 Boolean Algebra, 67 Commutative Rings, 149 Fields, 151 Finite Groups, 190 Groups, 231 Jorden Algebras, 248 Lie Algebras, 368 Rings). Each algebraic system has its own theory, but general properties and related concepts are dealt with from a common standpoint. From this common ground we often get an insight into concepts from which arose a general theory of mathematical systems. We describe here only algebraic systems, but it is possible to describe similar concepts for mathematical systems. The following is a description based mainly on one law of composition $(a, b) \rightarrow ab$; a similar description is possible for the case of two or more laws of composition.

The law of composition ab is sometimes written $a+b$, $a \cdot b$, $[a, b]$, etc. A mapping $f: A \rightarrow A'$ of similar algebraic systems A and A' is called a **homomorphism** provided that $f(ab) = f(a) f(b)$ $(a, b \in A)$. A' is said to be **homomorphic** to A if there is homomorphism from A onto A'. If f is one-to-one, onto, and its inverse mapping $f^{-1}: A' \rightarrow A$ is also a homomorphism, then f is called an **isomorphism**, A' is said to be **isomorphic** to A, and the relation is written $A \cong A'$. The composition of homomorphisms is a homomorphism, and the identity mapping is an isomorphism. A homomorphism of A to A itself is called an **endomorphism**, and if it is also an isomorphism, then it is an **automorphism**. The set of all endomorphisms of A forms a †semigroup under composition, and the set of automorphisms forms a group under composition. The concept of homomorphism is a fundamental concept appearing in all algebraic systems. A homomorphism is sometimes called a **representation**.

An element e of an algebraic system A with a law of composition ab is called an **identity element** if $ae = ea = a$ (for all $a \in A$). If such an element exists, then it is unique. In the case of a ring, two laws of composition, addition and multiplication, are given. In this case the identity element for multiplication (if it exists) is called the identity element (or †unity element) of the ring. In the case of homomorphism between groups, the identity element is mapped to the identity element, but this does not always hold in general algebraic systems. Since the identity element plays an important role, it is frequently added to the basic concepts. Homomorphism between mathematical systems is generally defined to induce a mapping between basic concepts. A semigroup and a ring with a unity element are called a **unitary semigroup (monoid)** and a **unitary ring**, respectively, and homomorphisms between these systems are restricted to mappings that map the identity element to the identity element.

Let A and A' be similar algebraic systems, and let A be a subset of A'. A is called a **subsystem** of A' if the mapping $f: A \rightarrow A'$ defined by $f(a) = a (a \in A)$ is a homomorphism. A subsystem of a group (ring) is called a †subgroup (†subring), and similarly for other algebraic systems.

An †equivalence relation R in an algebraic system A is called **compatible** with the law of composition if $R(a, a')$ and $R(b, b')$ imply $R(ab, a'b')$. Consider the †quotient set A/R. Then the law of composition in A/R is uniquely determined so that the mapping $f: A \rightarrow A/R$ defined by $a \in f(a)$ is a homomorphism. The algebraic system thus obtained is called a **quotient system**. A quotient system of a group (ring) is a group (ring), called a †quotient group (†quotient ring). Other cases, including those where operations are given, are treated similarly (→ 52 Categories and Functors).

References

[1] N. Bourbaki, Eléments de mathématique, I. Théorie des ensembles, ch. 4, Structures, Actualités Sci. Ind., 1258a, Hermann, second edition, 1966; English translation, Theory of sets, Addison-Wesley, 1968.
[2] C. Chevalley, Fundamental concepts of algebra, Academic Press, 1956.

410 (VI.21) Surfaces

A. The Notion of a Surface

The notion of a surface may be roughly expressed by saying that by moving a curve we get a surface or that the boundary of a solid body is a surface. But these propositions cannot be considered mathematical definitions of a surface. We also make a distinction between surfaces and planes in ordinary language, where we mean by surfaces only those that are not planes. In mathematical language, however, planes are usually included among the surfaces.

A surface can be defined as a 2-dimensional †continuum, in accordance with the definition of a curve as a 1-dimensional continuum. However, while we have a theory of curves based on this definition, we do not have a similar theory of surfaces thus defined (→ 93 Curves).

What is called a surface or a curved surface is usually a 2-dimensional †topological manifold, that is, a topological space that satisfies the †second countability axiom and of which every point has a neighborhood †homeomorphic to the interior of a circular disk in a 2-dimensional Euclidean space. In the following sections, we mean by a surface such a 2-dimensional topological manifold.

B. Examples and Classification

The simplest examples of surfaces are the 2-dimensional †simplex and the 2-dimensional †sphere. Surfaces are generally †simplicially decomposable (or triangulable) and hence homeomorphic to 2-dimensional polyhedra (T. Radó, *Acta Sci. Math. Szeged.* (1925)). A †compact surface is called a **closed surface**, and a noncompact surface is called an **open surface**. A closed surface is decomposable into a finite number of 2-simplexes and so can be interpreted as a †combinatorial manifold. A 2-dimensional topological manifold having a boundary is called a **surface with boundary**. A 2-simplex is an example of a surface with boundary, and a sphere is an example of a closed surface without boundary.

Surfaces are classified as †orientable and †nonorientable. In the special case when a surface is †embedded in a 3-dimensional Euclidean space E^3, whether the surface is orientable or not depends on its having two sides (the "surface" and "back") or only one side. Therefore, in this special case, an orientable surface is called **two-sided**, and a nonorientable surface, **one-sided**. A nonorientable closed surface without boundary cannot be embedded in the Euclidean space E^3 (→ 56 Characteristic Classes, 114 Differential Topology).

The first example of a nonorientable surface (with boundary) is the so-called **Möbius strip** or **Möbius band**, constructed as an †identification space from a rectangle by twisting through 180° and identifying the opposite edges with one another (Fig. 1).

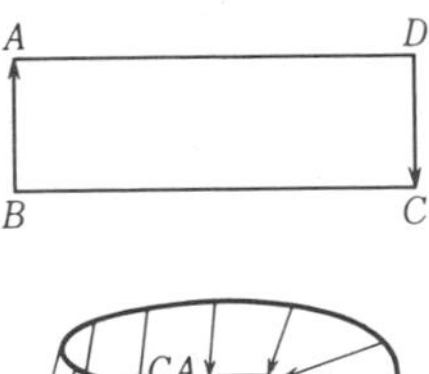

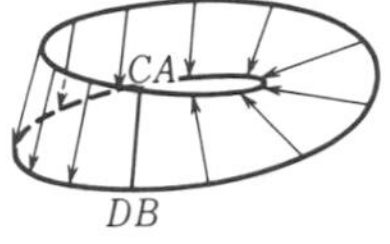

Fig. 1

As illustrated in Fig. 2, from a rectangle $ABCD$ we can obtain a closed surface homeomorphic to the product space $S^1 \times S^1$ by identifying the opposite edges AB with DC and BC with AD. This surface is the so-called 2-dimensional **torus** (or **anchor ring**). In this case, the four vertices A, B, C, D of the rectangle correspond to one point p on the surface, and the pairs of edges AB, DC and BC, AD correspond to closed curves a' and b' on the surface. We use the notation $aba^{-1}b^{-1}$ to represent a torus. This refers to the fact that the torus is obtained from an oriented four-sided polygon by identifying the first side and the third (with reversed orientation), the second side and the fourth (with reversed orientation). Similarly, aa^{-1} represents a sphere (Fig. 3), and $a_1b_1a_1^{-1}b_1^{-1}a_2b_2a_2^{-1}b_2^{-1}$ represents the closed surface shown in Fig. 4.

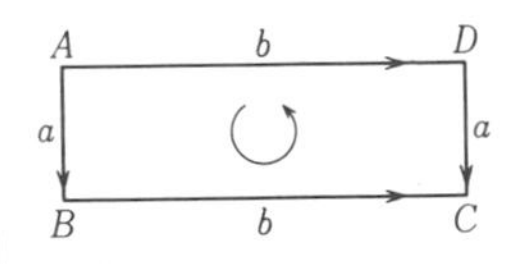

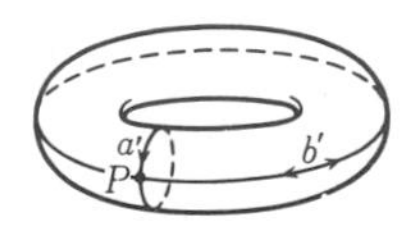

Fig. 2

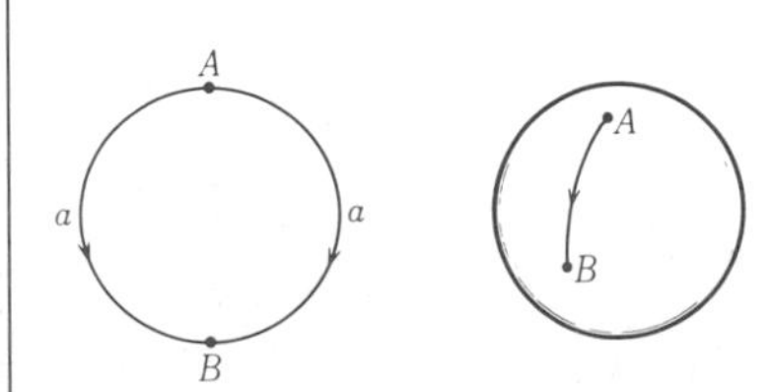

Fig. 3

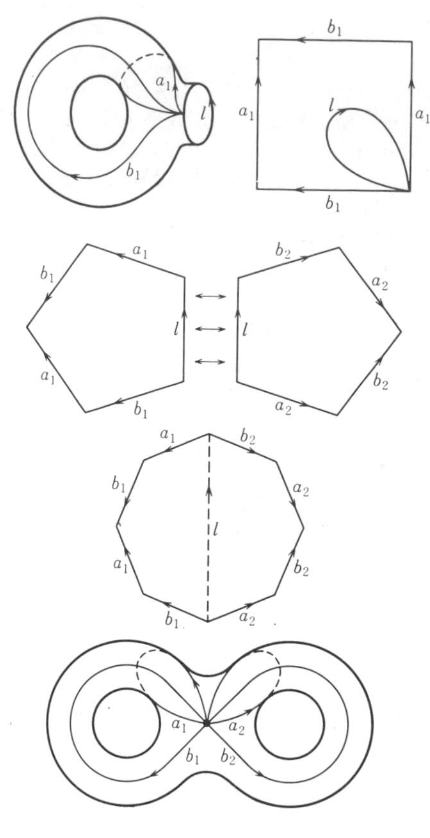

Fig. 4

All closed surfaces without boundary are constructed by identifying suitable pairs of sides of a $2n$-sided polygon in a Euclidean plane E^2. Furthermore, a closed orientable surface without boundary is homeomorphic to the surface represented by aa^{-1} or

$$a_1 b_1 a_1^{-1} b_1^{-1} \dots a_p b_p a_p^{-1} b_p^{-1}. \tag{1}$$

The 1-dimensional †Betti number of this surface is $2p$, the 0-dimensional and 2-dimensional †Betti numbers are 1, the †torsion coefficients are all 0, and p is called the **genus** of the surface. Also, a closed orientable surface of genus p with boundaries $c_1, \dots, c_k$ is represented by

$$w_1 c_1 w_1^{-1} \dots w_k c_k w_k^{-1} a_1 b_1 a_1^{-1} b_1^{-1} \dots a_p b_p a_p^{-1} b_p^{-1} \tag{2}$$

(Fig. 5). A closed nonorientable surface without boundary is represented by

$$a_1 a_1 a_2 a_2 \dots a_q a_q. \tag{3}$$

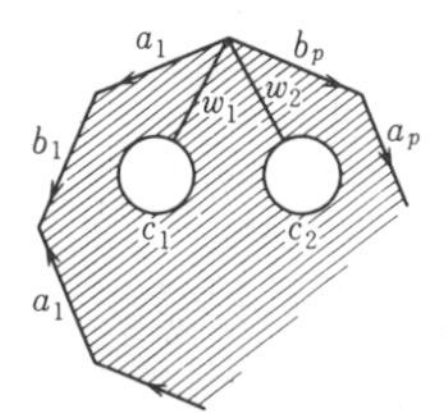

Fig. 5

The 1-dimensional Betti number of this surface is $q-1$, the 0-dimensional and 2-dimensional Betti numbers are 1 and 0, respectively, the 1-dimensional torsion coefficient is 2, the 0-dimensional and 2-dimensional torsion coefficients are 0, and q is called the **genus** of the surface. A closed nonorientable surface of genus q with boundaries $c_1, \dots, c_k$ is represented by

$$w_1 c_1 w_1^{-1} \dots w_k c_k w_k^{-1} a_1 a_1 \dots a_q a_q. \tag{4}$$

Each of forms (1)–(4) is called the **normal form** of the respective surface, and the curves a_i, b_j, w_k are called the **normal sections** of the surface. To explain the notation in (3), we first take the simplest case, aa. In this case, the surface is obtained from a disk by identifying each pair of points on the circumference that are endpoints of a diameter (Fig. 6). The surface aa is then homeomorphic to a †projective plane of which a decomposition into a complex of triangles is illustrated in Fig. 7. On the other hand, $aabb$ represents a surface like that shown in Fig. 8, called the **Klein bottle**. Fig. 9 shows a **handle**, and Fig. 10 shows a **cross cap**.

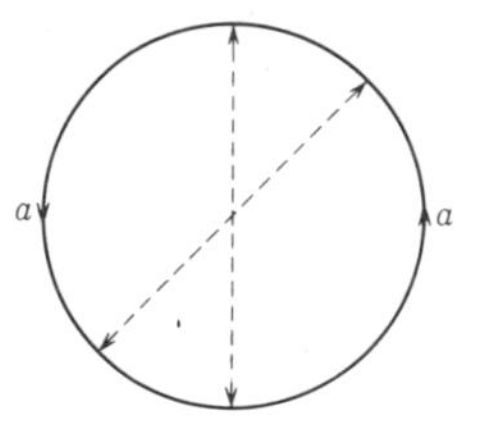

Fig. 6

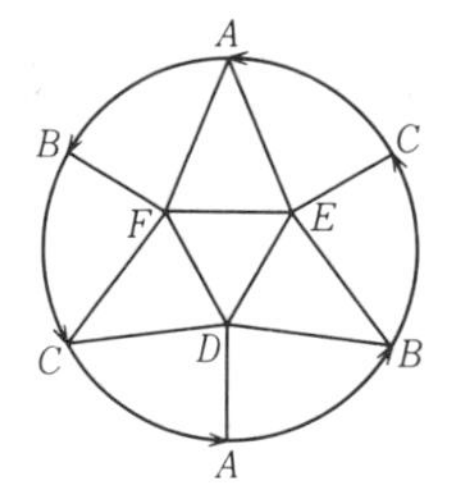

Fig. 7

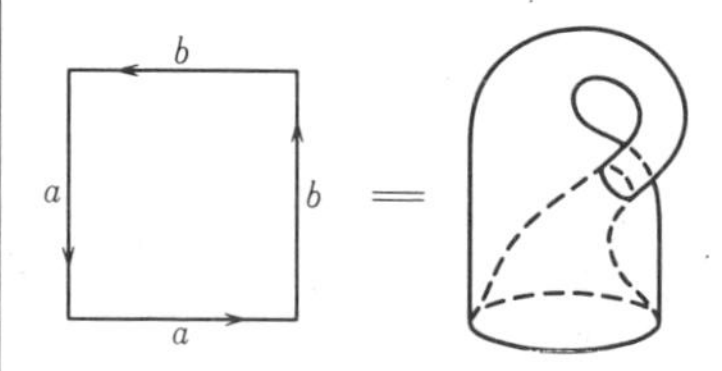

Fig. 8

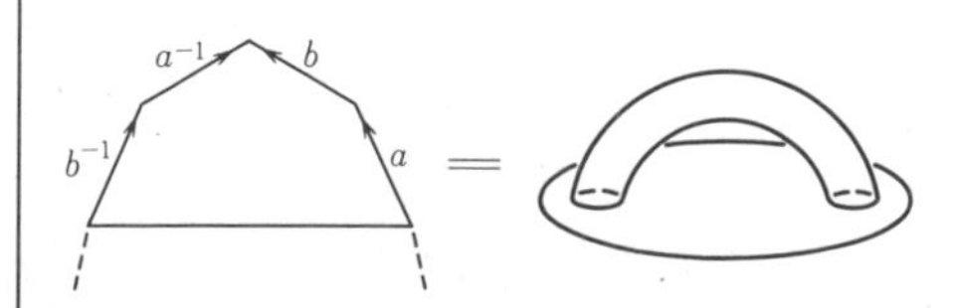

Fig. 9

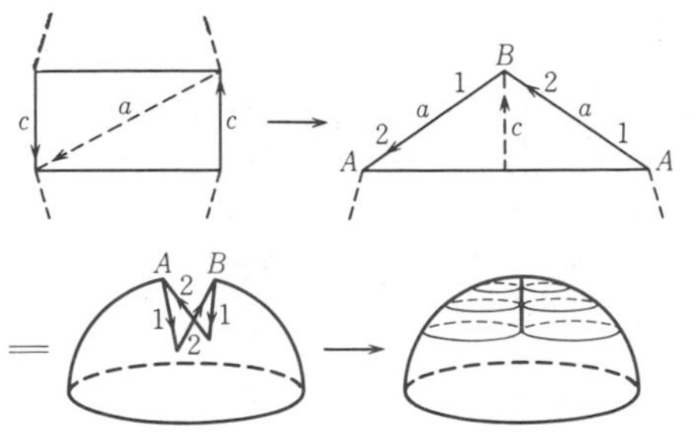

Fig. 10

The last two surfaces have boundaries; a handle is orientable, while a cross cap is nonorientable and homeomorphic to the Möbius strip. If we delete p disks from a sphere and replace them with an equal number of handles, then we obtain a surface homeomorphic to the surface represented in (1), while if we replace the disks by cross caps instead of by handles, then the surface thus obtained is homeomorphic to that represented in (3). Now we decompose the surfaces (1) and (3) into triangles and denote the number of i-dimensional simplexes by α_i $(i=0,1,2)$. Then in view of the †Euler-Poincaré formula, the surfaces (1) and (3) satisfy the respective formulas

$$\alpha_0-\alpha_1+\alpha_2=2(1-p),$$

$$\alpha_0-\alpha_1+\alpha_2=2-q.$$

The †Riemann surfaces of †algebraic functions of one complex variable are always surfaces of type (1), and their genera p coincide with those of algebraic functions.

All closed surfaces are homeomorphic to surfaces of types (1), (2), (3), or (4). A necessary and sufficient condition for two surfaces to be homeomorphic to each other is coincidence of the numbers of their boundaries, their orientability or nonorientability, and their genera (or †Euler characteristic $\alpha^0-\alpha^1+\alpha^2$). This proposition is called the **fundamental theorem of the topology of surfaces**. The †homeomorphism problem of closed surfaces is completely solved by this theorem. The same problem for n $(n\geqslant 3)$ manifolds, even if they are compact, remains open. (For surface area → 246 Length and Area. For the differential geometry of surfaces → 111 Differential Geometry of Curves and Surfaces.)

References

[1] B. Kerékjártó, Vorlesungen über Topologie, Springer, 1923.
[2] H. Seifert and W. Threlfall, Lehrbuch der Topologie, Teubner, 1934 (Chelsea, 1945).
[3] S. Lefschetz, Introduction to topology, Princeton Univ. Press, 1949.
[4] D. Hilbert and S. Cohn-Vossen, Anschauliche Geometrie, Springer, 1932; English translation, Geometry and the imagination, Chelsea, 1952.
[5] W. S. Massey, Algebraic topology: An introduction, Springer, 1967.
[6] E. E. Moise, Geometric topology in dimensions 2 and 3, Springer, 1977.

411 (I.4) Symbolic Logic

A. General Remarks

Symbolic logic (or **mathematical logic**) is a field of logic in which logical inferences commonly used in mathematics are investigated by use of mathematical symbols.

The **algebra of logic** originally set forth by G. Boole [1] and A. de Morgan [2] is actually an algebra of sets or relations; it did not reach the same level as the symbolic logic of today. G. Frege, who dealt not only with the logic of propositions but also with the first-order predicate logic using quantifiers (→ Sections C and K), should be regarded as the real originator of symbolic logic. Frege's work, however, was not recognized for some time. Logical studies by C. S. Peirce, E. Schröder, and G. Peano appeared soon after Frege, but they were limited mostly to propositions and did not develop Frege's work. An essential development of Frege's method was brought about by B. Russell, who, with the collaboration of A. N. Whitehead, summarized his results in *Principia mathematica* [4], which seemed to have completed the theory of symbolic logic at the time of its appearance.

B. Logical Symbols

If A and B are propositions, the propositions (A and B), (A or B), (A implies B), and (not A) are denoted by

$$A\wedge B,\qquad A\vee B,\qquad A\rightarrow B,\qquad \neg A,$$

respectively. We call $\neg A$ the **negation** of A, $A\wedge B$ the **conjunction** (or **logical product**), $A\vee B$ the **disjunction** (or **logical sum**), and $A\rightarrow B$ the **implication** (or B by A). The proposition $(A\rightarrow B)\wedge(B\rightarrow A)$ is denoted by $A\leftrightarrow B$ and is read "A and B are **equivalent**." $A\vee B$ means that at least one of A and B holds. The propositions (For all x, the proposition $F(x)$ holds) and (There exists an x such that $F(x)$ holds) are denoted by $\forall xF(x)$ and $\exists xF(x)$, respectively. A proposition of the form $\forall xF(x)$

is called a **universal proposition**, and one of the form $\exists x F(x)$, an **existential proposition**. The symbols $\wedge$, $\vee$, $\rightarrow$, $\leftrightarrow$, $\neg$, $\forall$, $\exists$ are called **logical symbols**.

There are various other ways to denote logical symbols, including:

$A \wedge B$: $A \& B$, $A \cdot B$,

$A \vee B$: $A + B$,

$A \rightarrow B$: $A \supset B$, $A \Rightarrow B$,

$A \leftrightarrow B$: $A \rightleftarrows B$, $A \equiv B$, $A \sim B$, $A \supset\subset B$, $A \Leftrightarrow B$,

$\neg A$: $\sim A$, $\bar{A}$,

$\forall x F(x)$: $(x) F(x)$, $\prod x F(x)$, $\bigwedge x F(x)$,

$\exists x F(x)$: $(Ex) F(x)$, $\sum x F(x)$, $\bigvee x F(x)$.

C. Free and Bound Variables

Any function whose values are propositions is called a **propositional function**. $\forall x$ and $\exists x$ can be regarded as operators that transform any propositional function $F(x)$ into the propositions $\forall x F(x)$ and $\exists x F(x)$, respectively. $\forall x$ and $\exists x$ are called **quantifiers**; the former is called the **universal quantifier** and the latter the **existential quantifier**. $F(x)$ is transformed into $\forall x F(x)$ or $\exists x F(x)$ just as a function $f(x)$ is transformed into the definite integral $\int_0^1 f(x)\,dx$; the resultant propositions $\forall x F(x)$ and $\exists x F(x)$ are no longer functions of x. The variable x in $\forall x F(x)$ and in $\exists x F(x)$ is called a **bound variable**, and the variable x in $F(x)$, when it is not bound by $\forall x$ or $\exists x$, is called a **free variable**. Some people employ different kinds of symbols for free variables and bound variables to avoid confusion.

D. Formal Expressions of Propositions

A formal expression of a proposition in terms of logical symbols is called a **formula**. More precisely, formulas are constructed by the following **formation rules**: (1) If $\mathfrak{A}$ is a formula, $\neg\mathfrak{A}$ is also a formula. If $\mathfrak{A}$ and $\mathfrak{B}$ are formulas, $\mathfrak{A} \wedge \mathfrak{B}$, $\mathfrak{A} \vee \mathfrak{B}$, $\mathfrak{A} \rightarrow \mathfrak{B}$ are all formulas. (2) If $\mathfrak{F}(a)$ is a formula and a is a free variable, then $\forall x \mathfrak{F}(x)$ and $\exists x \mathfrak{F}(x)$ are formulas, where x is an arbitrary bound variable not contained in $\mathfrak{F}(a)$ and $\mathfrak{F}(x)$ is the result of substituting x for a throughout $\mathfrak{F}(a)$.

We use formulas of various scope according to different purposes. To indicate the scope of formulas, we fix a set of formulas, each element of which is called a **prime formula** (or **atomic formula**). The scope of formulas is the set of formulas obtained from the prime formulas by formation rules (1) and (2).

E. Propositional Logic

Propositional logic is the field in symbolic logic in which we study relations between propositions exclusively in connection with the four logical symbols $\wedge$, $\vee$, $\rightarrow$, and $\neg$, called **propositional connectives**.

In propositional logic, we deal only with operations of **logical operators** denoted by propositional connectives, regarding the variables for denoting propositions, called **proposition variables**, only as prime formulas. We examine problems such as: What kinds of formulas are identically true when their proposition variables are replaced by any propositions, and what kinds of formulas can sometimes be true?

Consider the two symbols $\curlyvee$ and $\curlywedge$, read **true** and **false**, respectively, and let $\mathbf{A} = \{\curlyvee, \curlywedge\}$. A univalent function from $\mathbf{A}$, or more generally from a Cartesian product $\mathbf{A} \times \ldots \times \mathbf{A}$, into $\mathbf{A}$ is called a **truth function**. We can regard $\wedge$, $\vee$, $\rightarrow$, $\neg$ as the following truth functions: (1) $A \wedge B = \curlyvee$ for $A = B = \curlyvee$, and $A \wedge B = \curlywedge$ otherwise; (2) $A \vee B = \curlywedge$ for $A = B = \curlywedge$, and $A \vee B = \curlyvee$ otherwise; (3) $A \rightarrow B = \curlywedge$ for $A = \curlyvee$ and $B = \curlywedge$, and $A \rightarrow B = \curlyvee$ otherwise; (4) $\neg A = \curlywedge$ for $A = \curlyvee$, and $\neg A = \curlyvee$ for $A = \curlywedge$.

If we regard proposition variables as variables whose domain is $\mathbf{A}$, then each formula represents a truth function. Conversely, any truth function (of a finite number of independent variables) can be expressed by an appropriate formula, although such a formula is not uniquely determined. If a formula is regarded as a truth function, the value of the function determined by a combination of values of the independent variables involved in the formula is called the **truth value** of the formula.

A formula corresponding to a truth function that takes only $\curlyvee$ as its value is called a **tautology**. For example, $\mathfrak{A} \vee \neg\mathfrak{A}$ and $((\mathfrak{A} \rightarrow \mathfrak{B}) \rightarrow \mathfrak{A}) \rightarrow \mathfrak{A}$ are tautologies. Since a truth function with n independent variables takes values corresponding to 2^n combinations of truth values of its variables, we can determine in a finite number of steps whether a given formula is a tautology. If $\mathfrak{A} \leftrightarrow \mathfrak{B}$ is a tautology (that is, $\mathfrak{A}$ and $\mathfrak{B}$ correspond to the same truth function), then the formulas $\mathfrak{A}$ and $\mathfrak{B}$ are said to be **equivalent**.

F. Propositional Calculus

It is possible to choose some specific tautologies, designate them as axioms, and derive all tautologies from them by appropriately given rules of inference. Such a system is called a **propositional calculus**. There are many ways to stipulate axioms and rules of inference for a propositional calculus.

The abovementioned propositional calculus corresponds to the so-called classical propositional logic (→ Section L). By choosing appropriate axioms and rules of inference we can also formally construct intuitionistic or other propositional logics. In intuitionistic logic the law of the †excluded middle is not accepted, and hence it is impossible to formalize intuitionistic propositional logic by the notion of tautology. We therefore usually adopt the method of propositional calculus, instead of using the notion of tautology, to formalize intuitionistic propositional logic. For example, V. I. Glivenko's theorem [5], that if a formula $\mathfrak{A}$ can be proved in classical logic, then $\neg\neg\mathfrak{A}$ can be proved in intuitionistic logic, was obtained by such formalistic considerations. A method of extending the classical concepts of truth value and tautology to intuitionistic and other logics has been obtained by S. A. Kripke. There are also studies of logics intermediate between intuitionistic and classical logic (T. Umezawa).

G. Predicate Logic

Predicate logic is the area of symbolic logic in which we take quantifiers in account. Mainly propositional functions are discussed in predicate logic. In the strict sense only single-variable propositional functions are called **predicates**, but the phrase **predicate of** n **arguments** (or n**-ary predicate**) denoting an n-variable propositional function is also employed. Single-variable (or unary) predicates are also called **properties**. We say that a has the property F if the proposition $F(a)$ formed by the property F is true. Predicates of two arguments are called **binary relations**. The proposition $R(a,b)$ formed by the binary relation R is occasionally expressed in the form aRb. Generally, predicates of n arguments are called n**-ary relations**. The domain of definition of a unary predicate is called the **object domain**, elements of the object domain are called **objects**, and any variable running over the object domain is called an **object variable**. We assume here that the object domain is not empty. When we deal with a number of predicates simultaneously (with different numbers of variables), it is usual to arrange things so that all the independent variables have the same object domain by suitably extending their object domains.

Predicate logic in its purest sense deals exclusively with the general properties of quantifiers in connection with propositional connectives. The only objects dealt with in this field are **predicate variables** defined over a certain common domain and object variables running over the domain. Propositional variables are regarded as predicates of no variables. Each expression $F(a_1, \ldots, a_n)$ for any predicate variable F of n variables $a_1, \ldots, a_n$ (object variables designated as free) is regarded as a prime formula ($n = 0, 1, 2, \ldots$), and we deal exclusively with formulas generated by these prime formulas, where bound variables are also restricted to object variables that have a common domain. We give no specification for the range of objects except that it be the common domain of the object variables.

By designating an object domain and substituting a predicate defined over the domain for each predicate variable in a formula, we obtain a proposition. By substituting further an object (object constant) belonging to the object domain for each object variable in a proposition, we obtain a proposition having a definite truth value. When we designate an object domain and further associate with each predicate variable as well as with each object variable a predicate or an object to be substituted for it, we call the pair consisting of the object domain and the association a **model**. Any formula that is true for every model is called an **identically true formula** or **valid formula**. The study of identically true formulas is one of the most important problems in predicate logic.

H. Formal Representations of Mathematical Propositions

To obtain a formal representation of a mathematical theory by predicate logic, we must first specify its object domain, which is a nonempty set whose elements are called **individuals**; accordingly the object domain is called the **individual domain**, and object variables are called **individual variables**. Secondly we must specify **individual symbols**, **function symbols**, and **predicate symbols**, signifying specific individuals, functions, and †predicates, respectively. Here a function of n arguments is a univalent mapping from the Cartesian product $D \times \ldots \times D$ of n copies of the given set to D. Then we define the notion of **term** as in the next paragraph to represent each individual formally. Finally we express propositions formally by formulas.

Definition of terms (formation rule for terms): (1) Each individual symbol is a term. (2) Each free variable is a term. (3) $f(t_1, \ldots, t_n)$ is a term if $t_1, \ldots, t_n$ are terms and f is a function symbol of n arguments. (4) The only terms are those given by (1)–(3).

As a prime formula in this case we use any

formula of the form $F(t_1, \ldots, t_n)$, where F is a predicate symbol of n arguments and $t_1, \ldots, t_n$ are arbitrary terms. To define the notions of term and formula, we need logical symbols, free and bound individual variables, and also a list of individual symbols, function symbols, and predicate symbols.

In pure predicate logic, the individual domain is not concrete, and we study only general forms of propositions. Hence, in this case, predicate or function symbols are not representations of concrete predicates or functions but are **predicate variables** and **function variables**. We also use free individual variables instead of individual symbols. In fact, it is now most common that function variables are dispensed with, and only free individual variables are used as terms.

I. Formulation of Mathematical Theories

To formalize a theory we need **axioms** and **rules of inference**. Axioms constitute a certain specific set of formulas, and a rule of inference is a rule for deducing a formula from other formulas. A formula is said to be **provable** if it can be deduced from the axioms by repeated application of rules of inference. Axioms are divided into two types: **logical axioms**, which are common to all theories, and **mathematical axioms**, which are peculiar to each individual theory. The set of mathematical axioms is called the **axiom system** of the theory.

(I) Logical axioms: (1) A formula that is the result of substituting arbitrary formulas for the proposition variables in a tautology is an axiom. (2) Any formula of the form

$$\forall x \mathfrak{F}(x) \rightarrow \mathfrak{F}(t) \quad \text{or} \quad \mathfrak{F}(t) \rightarrow \exists x \mathfrak{F}(x)$$

is an axiom, where $\mathfrak{F}(t)$ is the result of substituting an arbitrary term t for x in $\mathfrak{F}(x)$.

(II) Rules of inference: (1) We can deduce a formula $\mathfrak{B}$ from two formulas $\mathfrak{A}$ and $\mathfrak{A} \rightarrow \mathfrak{B}$ (**modus ponens**). (2) We can deduce $\mathfrak{A} \rightarrow \forall x \mathfrak{F}(x)$ from a formula $\mathfrak{A} \rightarrow \mathfrak{F}(a)$ and $\exists x \mathfrak{F}(x) \rightarrow \mathfrak{A}$ from $\mathfrak{F}(a) \rightarrow \mathfrak{A}$, where a is a free individual variable contained in neither $\mathfrak{A}$ nor $\mathfrak{F}(x)$ and $\mathfrak{F}(a)$ is the result of substituting a for x in $\mathfrak{F}(x)$.

If an axiom system is added to these logical axioms and rules of inference, we say that a **formal system** is given.

A formal system S or its axiom system is said to be **contradictory** or to contain a **contradiction** if a formula $\mathfrak{A}$ and its negation $\neg\mathfrak{A}$ are provable; otherwise it is said to be **consistent**. Since

$$(\mathfrak{A} \wedge \neg\mathfrak{A}) \rightarrow \mathfrak{B}$$

is a tautology, we can show that any formula is provable in a formal system containing a contradiction. The validity of a proof by **reductio ad absurdum** lies in the fact that

$$(\mathfrak{A} \rightarrow (\mathfrak{B} \wedge \neg\mathfrak{B})) \rightarrow \neg\mathfrak{A}$$

is a tautology. An affirmative proposition (formula) may be obtained by reductio ad absurdum since the formula (of propositional logic) representing the **discharge of double negation**

$$\neg\neg\mathfrak{A} \rightarrow \mathfrak{A}$$

is a tautology.

J. Predicate Calculus

If a formula has no free individual variable, we call it a **closed formula**. Now we consider a formal system S whose mathematical axioms are closed. A formula $\mathfrak{A}$ is provable in S if and only if there exist suitable mathematical axioms $\mathfrak{C}_1, \ldots, \mathfrak{C}_n$ such that the formula

$$(\mathfrak{C}_1 \wedge \ldots \wedge \mathfrak{C}_n) \rightarrow \mathfrak{A}$$

is provable without the use of mathematical axioms. Since any axiom system can be replaced by an equivalent axiom system containing only closed formulas, the study of a formal system can be reduced to the study of pure logic.

In the following we take no individual symbols or function symbols into consideration and we use predicate variables as predicate symbols in accordance with the commonly accepted method of stating properties of the pure predicate logic; but only in the case of **predicate logic with equality** will we use predicate variables and the equality predicate $=$ as a predicate symbol. However, we can safely state that we use function variables as function symbols.

The formal system with no mathematical axioms is called the **predicate calculus**. The formal system whose mathematical axioms are the equality axioms

$$a = a, \quad a = b \rightarrow (\mathfrak{F}(a) \rightarrow \mathfrak{F}(b))$$

is called the **predicate calculus with equality**.

In the following, by being provable we mean being provable in the predicate calculus.

(1) Every provable formula is valid.

(2) Conversely, any valid formula is provable (K. Gödel [6]). This fact is called the **completeness** of the predicate calculus. In fact, by Gödel's proof, a formula $\mathfrak{A}$ is provable if $\mathfrak{A}$ is always true in every interpretation whose individual domain is of †countable cardinality. In another formulation, if $\neg\mathfrak{A}$ is not provable, the formula $\mathfrak{A}$ is a true proposition in some interpretation (and the individual domain in this case is of countable cardinality). We can

extend this result as follows: If an axiom system generated by countably many closed formulas is consistent, then its mathematical axioms can be considered true propositions by a common interpretation. In this sense, **Gödel's completeness theorem** gives another proof of the †Skolem-Löwenheim theorem.

(3) The predicate calculus is consistent. Although this result is obtained from (1) in this section, it is not difficult to show it directly (D. Hilbert and W. Ackermann [7]).

(4) There are many different ways of giving logical axioms and rules of inference for the predicate calculus. G. Gentzen gave two types of systems in [8]; one is a natural deduction system in which it is easy to reproduce formal proofs directly from practical ones in mathematics, and the other has a logically simpler structure. Concerning the latter, Gentzen proved **Gentzen's fundamental theorem**, which shows that a formal proof of a formula may be translated into a "direct" proof. The theorem itself and its idea were powerful tools for obtaining consistency proofs.

(5) If the proposition $\exists x\mathfrak{A}(x)$ is true, we choose one of the individuals x satisfying the condition $\mathfrak{A}(x)$, and denote it by $\varepsilon x\mathfrak{A}(x)$. When $\exists x\mathfrak{A}(x)$ is false, we let $\varepsilon x\mathfrak{A}(x)$ represent an arbitrary individual. Then

$$\exists x\mathfrak{A}(x)\rightarrow\mathfrak{A}(\varepsilon x\mathfrak{A}(x)) \qquad (1)$$

is true. We consider εx to be an operator associating an individual $\varepsilon x\mathfrak{A}(x)$ with a proposition $\mathfrak{A}(x)$ containing the variable x. Hilbert called it the **transfinite logical choice function**; today we call it **Hilbert's ε-operator** (or **ε-quantifier**), and the logical symbol ε used in this sense **Hilbert's ε-symbol**. Using the ε-symbol, $\exists x\mathfrak{A}(x)$ and $\forall x\mathfrak{A}(x)$ are represented by

$$\mathfrak{A}(\varepsilon x\mathfrak{A}(x)), \quad \mathfrak{A}(\varepsilon x \neg \mathfrak{A}(x)),$$

respectively, for any $\mathfrak{A}(x)$. The system of predicate calculus adding formulas of the form (1) as axioms is essentially equivalent to the usual predicate calculus. This result, called the **ε-theorem**, reads as follows: When a formula $\mathfrak{C}$ is provable under the assumption that every formula of the form (1) is an axiom, we can prove $\mathfrak{C}$ using no axioms of the form (1) if $\mathfrak{C}$ contains no logical symbol ε (D. Hilbert and P. Bernays [9]). Moreover, a similar theorem holds when axioms of the form

$$\forall x(\mathfrak{A}(x)\leftrightarrow\mathfrak{B}(x))\rightarrow\varepsilon x\mathfrak{A}(x)=\varepsilon x\mathfrak{B}(x) \qquad (2)$$

are added (S. Maehara [10]).

(6) For a given formula $\mathfrak{A}$, call $\mathfrak{A}'$ a normal form of $\mathfrak{A}$ when the formula

$$\mathfrak{A}\leftrightarrow\mathfrak{A}'$$

is provable and $\mathfrak{A}'$ satisfies a particular condition. For example, for any formula $\mathfrak{A}$ there is a normal form $\mathfrak{A}'$ satisfying the condition: $\mathfrak{A}'$ has the form

$$Q_1x_1\ldots Q_nx_n\mathfrak{B}(x_1,\ldots,x_n),$$

where Qx means a quantifier $\forall x$ or $\exists x$, and $\mathfrak{B}(x_1,\ldots,x_n)$ contains no quantifier and has no predicate variables or free individual variables not contained in $\mathfrak{A}$. A normal form of this kind is called a **prenex normal form**.

(7) We have dealt with the classical first-order predicate logic until now. For other predicate logics (→ Sections K and L) also, we can consider a predicate calculus or a formal system by first defining suitable axioms or rules of inference. Gentzen's fundamental theorem applies to the intuitionistic predicate calculus formulated by V. I. Glivenko, A. Heyting, and others. Since Gentzen's fundamental theorem holds not only in classical logic and intuitionistic logic but also in several systems of first-order predicate logic or propositional logic, it is useful for getting results in modal and other logics (M. Ohnishi, K. Matsumoto). Moreover, Glivenko's theorem in propositional logic [5] is also extended to predicate calculus by using a rather weak representation (S. Kuroda [12]). G. Takeuti expected that a theorem similar to Gentzen's fundamental theorem would hold in higher-order predicate logic also, and showed that the consistency of analysis would follow if that conjecture could be verified [13]. Moreover, in many important cases, he showed constructively that the conjecture holds partially. The conjecture was finally proved by M. Takahashi [14] by a nonconstructive method. Concerning this, there are also contributions by S. Maehara, T. Simauti, M. Yasuhara, and W. Tait.

K. Predicate Logics of Higher Order

In ordinary predicate logic, the bound variables are restricted to individual variables. In this sense, ordinary predicate logic is called **first-order predicate logic**, while predicate logic dealing with quantifiers $\forall P$ or $\exists P$ for a predicate variable P is called **second-order predicate logic**.

Generalizing further, we can introduce the so-called **third-order predicate logic**. First we fix the individual domain D_0. Then, by introducing the whole class D_1^n of predicates of n variables, each running over the object domain D_0, we can introduce predicates that have D_1^n as their object domain. This kind of predicate is called a **second-order predicate** with respect to the individual domain D_0. Even when we restrict second-order predicates to one-variable predicates, they are divided into various types, and the domains of independent variables do not coincide in the case of more than two variables. In contrast, predicates having D_0 as their object domain are called **first-order predicates**. The logic having quantifiers that admit first-order predicate variables is second-order predicate logic, and the logic having quantifiers that admit up to second-order predicate variables is third-order predicate logic. Similarly, we can define further **higher-order predicate logics**.

Higher-order predicate logic is occasionally called **type theory**, because variables arise that are classified into various types. Type theory is divided into **simple type theory** and **ramified type theory**.

We confine ourselves to variables for single-variable predicates, and denote by P such a bound predicate variable. Then for any formula $\mathfrak{F}(a)$ (with a a free individual variable), the formula

$$\exists P \forall x (P(x) \leftrightarrow \mathfrak{F}(x))$$

is considered identically true. This is the point of view in simple type theory.

Russell asserted first that this formula cannot be used reasonably if quantifiers with respect to predicate variables occur in $\mathfrak{F}(x)$. This assertion is based on the point of view that the formula in the previous paragraph asserts that $\mathfrak{F}(x)$ is a first-order predicate, whereas any quantifier with respect to first-order predicate variables, whose definition assumes the totality of the first-order predicates, should not be used to introduce the first-order predicate $\mathfrak{F}(x)$. For this purpose, Russell further classified the class of first-order predicates by their **rank** and adopted the axiom

$$\exists P^k \forall x (P^k(x) \leftrightarrow \mathfrak{F}(x))$$

for the predicate variable P^k of rank k, where the rank i of any free predicate variable occurring in $\mathfrak{F}(x)$ is $\leqslant k$, and the rank j of any bound predicate variable occurring in $\mathfrak{F}(x)$ is $< k$. This is the point of view in ramified type theory, and we still must subdivide the types if we deal with higher-order propositions or propositions of many variables. Even Russell, having started from his ramified type theory, had to introduce the **axiom of reducibility** afterwards and reduce his theory to simple type theory.

L. Systems of Logic

Logic in the ordinary sense, which is based on the **law of the excluded middle** asserting that every proposition is in principle either true or false, is called **classical logic**. Usually, propositional logic, predicate logic, and type theory are developed from the standpoint of classical logic. Occasionally the reasoning of intuitionistic mathematics is investigated using symbolic logic, in which the law of the excluded middle is not admitted (→ 156 Foundations of Mathematics). Such logic is called **intuitionistic logic**. Logic is also subdivided into propositional logic, predicate logic, etc., according to the extent of the propositions (formulas) dealt with.

To express **modal propositions** stating **possibility, necessity**, etc., in symbolic logic, J. Łukaszewicz proposed a propositional logic called **three-valued logic**, having a third truth value, neither true nor false. More generally, **many-valued logics** with any number of truth values have been introduced; classical logic is one of its special cases, **two-valued logic** with two truth values, true and false. Actually, however, many-valued logics with more than three truth values have not been studied much, while various studies in **modal logic** based on classical logic have been successfully carried out. For example, studies of **strict implication** belong to this field.

References

[1] G. Boole, An investigation of the laws of thought, Walton and Maberly, 1854.
[2] A. de Morgan, Formal logic, or the calculus of inference, Taylor and Walton, 1847.
[3] G. Frege, Begriffsschrift, eine der arithmetischen nachgebildete Formalsprache des reinen Denkens, Halle, 1879.
[4] A. N. Whitehead and B. Russell, Principia mathematica I, II, III, Cambridge Univ. Press, 1910–1913; second edition, 1925–1927.
[5] V. Glivenko, Sur quelques points de la logique de M. Brouwer, Acad. Roy. de Belgique, Bulletin de la classe des sciences, (5) 15 (1929), 183–188.
[6] K. Gödel, Die Vollständigkeit der Axiome des logischen Funktionenkalküls, Monatsh. Math. Phys., 37 (1930), 349–360.
[7] D. Hilbert and W. Ackermann, Grundzüge der theoretischen Logik, Springer, 1928, sixth edition, 1972; English translation, Principles of mathematical logic, Chelsea, 1950.
[8] G. Gentzen, Untersuchungen über das logische Schliessen, Math. Z., 39 (1935), 176–210, 405–431.
[9] D. Hilbert and P. Bernays, Grundlagen der Mathematik II, Springer, 1939; second edition, 1970.
[10] S. Maehara, Equality axiom on Hilbert's ε-symbol, J. Fac. Sci. Univ. Tokyo, (I), 7 (1957), 419–435.

[11] A. Heyting, Die formalen Regeln der intuitionistischen Logik I, S.-B. Preuss. Akad. Wiss., 1930, 42–56.
[12] S. Kuroda, Intuitionistische Untersuchungen der formalistischen Logik, Nagoya Math. J., 2 (1951), 35–47.
[13] G. Takeuti, On a generalized logic calculus, Japan. J. Math., 23 (1953), 39–96.
[14] M. Takahashi, A proof of the cut-elimination theorem in simple type-theory, J. Math. Soc. Japan, 19 (1967), 399–410.
[15] S. C. Kleene, Mathematical logic, Wiley, 1967.
[16] J. R. Shoenfield, Mathematical logic, Addison-Wesley, 1967.
[17] R. M. Smullyan, First-order logic, Springer, 1968.

412 (IV.13) Symmetric Riemannian Spaces and Real Forms

A. Symmetric Riemannian Spaces

Let M be a †Riemannian space. For each point p of M we can define a mapping σ_p of a suitable neighborhood U_p of p onto U_p itself so that $\sigma_p(x_t)=x_{-t}$, where x_t $(|t|<\varepsilon, x_0=p)$ is any †geodesic passing through the point p. We call M a **locally symmetric Riemannian space** if for any point p of M we can choose a neighborhood U_p so that σ_p is an †isometry of U_p. In order that a Riemannian space M be locally symmetric it is necessary and sufficient that the †covariant differential (with respect to the †Riemannian connection) of the †curvature tensor of M be 0. A locally symmetric Riemannian space is a †real analytic manifold. We say that a Riemannian space M is a **globally symmetric Riemannian space** (or simply **symmetric Riemannian space**) if M is connected and if for each point p of M there exists an isometry σ_p of M onto M itself that has p as an isolated fixed point (i.e., has no fixed point except p in a certain neighborhood of p) and such that σ_p^2 is the identity transformation on M. In this case σ_p is called the **symmetry** at p. A (globally) symmetric Riemannian space is locally symmetric and is a †complete Riemannian space. Conversely, a †simply connected complete locally symmetric Riemannian space is a (globally) symmetric Riemannian space.

B. Symmetric Riemannian Homogeneous Spaces

A †homogeneous space G/K of a connected †Lie group G is a **symmetric homogeneous space** (with respect to θ) if there exists an **involutive automorphism** (i.e., automorphism of order 2) θ of G satisfying the condition $K_\theta^0 \subset K \subset K_\theta$, where K_θ is the closed subgroup consisting of all elements of G left fixed by θ and K_θ^0 is the connected component of the identity element of K_θ. In this case, the mapping $aK \rightarrow \theta(a)K$ $(a \in G)$ is a transformation of G/K having the point K as an isolated fixed point; more generally, the mapping $\theta_{a_0}: aK \rightarrow a_0\theta(a_0)^{-1}\theta(a)K$ is a transformation of G/K that has an arbitrary given point a_0K of G/K as an isolated fixed point. If there exists a G-invariant Riemannian metric on G/K, then G/K is a symmetric Riemannian space with symmetries $\{\theta_{a_0} \mid a_0 \in G\}$ and is called a **symmetric Riemannian homogeneous space**. A sufficient condition for a symmetric homogeneous space G/K to be a symmetric Riemannian homogeneous space is that K be a compact subgroup. Conversely, given a symmetric Riemannian space M, let G be the connected component of the identity element of the Lie group formed by all the isometries of M; then M is represented as the symmetric Riemannian homogeneous space $M=G/K$ and K is a compact group. In particular, a symmetric Riemannian space can be regarded as a Riemannian space that is realizable as a symmetric Riemannian homogeneous space.

The Riemannian connection of a symmetric Riemannian homogeneous space G/K is uniquely determined (independent of the choice of G-invariant Riemannian metric), and a geodesic $x_t(|t|<\infty, x_0=a_0K)$ passing through a point a_0K of G/K is of the form $x_t=(\exp tX)a_0K$. Here X is any element of the Lie algebra $\mathfrak{g}$ of G such that $\theta(X)=-X$, where θ also denotes the automorphism of $\mathfrak{g}$ induced by the automorphism θ of G and $\exp tX$ is the †one-parameter subgroup of G defined by the element X. The covariant differential of any G-invariant tensor field on G/K is 0, and any G-invariant †differential form on G/K is a closed differential form.

C. Classification of Symmetric Riemannian Spaces

The †simply connected †covering Riemannian space of a symmetric Riemannian space is also a symmetric Riemannian space. Therefore the problem of classifying symmetric Riemannian spaces is reduced to classifying simply connected symmetric Riemannian spaces M and determining †discontinuous groups of isometries of M. When we take the †de Rham decomposition of such a space M and represent M as the product of a real Euclidean space and a number of simply connected irreducible Riemannian spaces, all the factors are symmetric Riemannian spaces. We say that M is an **irreducible symmetric Riemannian space** if it is a symmetric Riemannian space and is irreducible as a Riemannian space.

A simply connected irreducible symmetric Riemannian space is isomorphic to one of the following four types of symmetric Riemannian homogeneous spaces (here Lie groups are always assumed to be connected):

(1) The symmetric Riemannian homogeneous space $(G\times G)/\{(a,a)\mid a\in G\}$ of the direct product $G\times G$, where G is a simply connected compact †simple Lie group and the involutive automorphism of $G\times G$ is given by $(a,b)\rightarrow(b,a)$ $((a,b)\in G\times G)$. This space is isomorphic, as a Riemannian space, to the space G obtained by introducing a two-sided invariant Riemannian metric on the group G; the isomorphism is induced from the mapping $G\times G\ni(a,b)\rightarrow ab^{-1}\in G$.

(2) A symmetric homogeneous space G/K_θ of a simply connected compact simple Lie group G with respect to an involutive automorphism θ of G. In this case, the closed subgroup $K_\theta=\{a\in G\mid\theta(a)=a\}$ of G is connected. We assume here that θ is a member of the given complete system of representatives of the †conjugate classes formed by the elements of order 2 in the automorphism group of the group G.

(3) The homogeneous space $G^{\mathbf{C}}/G$, where $G^{\mathbf{C}}$ is a complex simple Lie group whose †center reduces to the identity element and G is an arbitrary but fixed maximal compact subgroup of $G^{\mathbf{C}}$.

(4) The homogeneous space G_0/K, where G_0 is a noncompact simple Lie group whose center reduces to the identity element and which has no complex Lie group structure, and K is a maximal compact subgroup of G. In Section D we shall see that (3) and (4) are actually symmetric homogeneous spaces. All four types of symmetric Riemannian spaces are actually irreducible symmetric Riemannian spaces, and G-invariant Riemannian metrics on each of them are uniquely determined up to multiplication by a positive number. On the other hand, (1) and (2) are compact, while (3) and (4) are homeomorphic to Euclidean spaces and not compact. For spaces of types (1) and (3) the problem of classifying simply connected irreducible symmetric Riemannian spaces is reduced to classifying †compact real simple Lie algebras and †complex simple Lie algebras, respectively, while for types (2) and (4) it is reduced to the classification of noncompact real simple Lie algebras (→ Section D) (for the result of classification of these types → Appendix A, Table 5.II). On the other hand, any (not necessarily simply connected) irreducible symmetric Riemannian space defines one of (1)–(4) as its †universal covering manifold; if the covering manifold is of type (3) or (4), the original symmetric Riemannian space is necessarily simply connected.

D. Symmetric Riemannian Homogeneous Spaces of Semisimple Lie Groups

In Section C we saw that any irreducible symmetric Riemannian space is representable as a symmetric Riemannian homogeneous space G/K on which a connected semisimple Lie group G acts †almost effectively (→ 249 Lie Groups). Among symmetric Riemannian spaces, such a space $M=G/K$ is characterized as one admitting no nonzero vector field that is †parallel with respect to the Riemannian connection. Furthermore, if G acts effectively on M, G coincides with the connected component $I(M)^0$ of the identity element of the Lie group formed by all the isometries of M.

We let $M=G/K$ be a symmetric Riemannian homogeneous space on which a connected semisimple Lie group G acts almost effectively. Then G is a Lie group that is †locally isomorphic to the group $I(M)^0$, and therefore the Lie algebra of G is determined by M. Let $\mathfrak{g}$ be the Lie algebra of G, $\mathfrak{k}$ be the subalgebra of $\mathfrak{g}$ corresponding to K, and θ be the involutive automorphism of G defining the symmetric homogeneous space G/K. The automorphism of $\mathfrak{g}$ defined by θ is also denoted by θ. Then $\mathfrak{k}=\{X\in\mathfrak{g}\mid\theta(X)=X\}$. Putting $\mathfrak{m}=\{X\in\mathfrak{g}\mid\theta(X)=-X\}$, we have $\mathfrak{g}=\mathfrak{m}+\mathfrak{k}$ (direct sum of linear spaces), and $\mathfrak{m}$ can be identified in a natural way with the tangent space at the point K of G/K. The †adjoint representation of G gives rise to a representation of K in $\mathfrak{g}$, which induces a linear representation $\mathrm{Ad}_{\mathfrak{m}}(k)$ of K in $\mathfrak{m}$. Then $\{\mathrm{Ad}_{\mathfrak{m}}(k)\mid k\in K\}$ coincides with the †restricted homogeneous holonomy group at the point K of the Riemannian space G/K.

Now let φ be the †Killing form of $\mathfrak{g}$. Then $\mathfrak{k}$ and $\mathfrak{m}$ are mutually orthogonal with respect to φ, and denoting by $\varphi_{\mathfrak{k}}$ and $\varphi_{\mathfrak{m}}$ the restrictions of φ to $\mathfrak{k}$ and $\mathfrak{m}$, respectively, $\varphi_{\mathfrak{k}}$ is a negative definite quadratic form on $\mathfrak{k}$. If $\varphi_{\mathfrak{m}}$ is also a negative definite quadratic form on $\mathfrak{m}$, $\mathfrak{g}$ is a compact real semisimple Lie algebra and G/K is a compact symmetric Riemannian space; in this case we say that G/K is of **compact type**. In the opposite case, where $\varphi_{\mathfrak{m}}$ is a †positive definite quadratic form, G/K is said to be of **noncompact type**. In this latter case, G/K is homeomorphic to a Euclidean space, and if the center of G is finite, K is a maximal compact subgroup of G. Furthermore, the group of isometries $I(G/K)$ of G/K is canonically isomorphic to the automorphism group of the Lie algebra $\mathfrak{g}$. When G/K is of compact type (noncompact type), there exists one and only one G-invariant Riemannian metric on G/K, which induces in the tangent space $\mathfrak{m}$ at the point K the positive definite inner product $-\varphi_{\mathfrak{m}}(\varphi_{\mathfrak{m}})$.

A symmetric Riemannian homogeneous space G/K_θ of compact type defined by a simply connected compact semisimple Lie group G with respect to an involutive automorphism θ is simply connected. Let $\mathfrak{g}=\mathfrak{m}+\mathfrak{k}_\theta$ be the decomposition of the Lie algebra $\mathfrak{g}$ of G with respect to the automorphism θ of $\mathfrak{g}$, and let $\mathfrak{g}^{\mathbf{C}}$ be the †complex form of $\mathfrak{g}$. Then the real subspace $\mathfrak{g}_\theta=\sqrt{-1}\,\mathfrak{m}+\mathfrak{k}_\theta$ in $\mathfrak{g}^{\mathbf{C}}$ is a real semisimple Lie algebra and a †real form of $\mathfrak{g}^{\mathbf{C}}$. Let G_θ be the Lie group corresponding to the Lie algebra $\mathfrak{g}_\theta$ with center reduced to the identity element, and let K be the subgroup of G_θ corresponding to $\mathfrak{k}_\theta$. Then we get a (simply connected) symmetric Riemannian homogeneous space of noncompact type G_θ/K.

When we start from a symmetric Riemannian space of noncompact type G/K instead of the symmetric Riemannian space of compact type G/K_θ and apply the same process as in the previous paragraphs, taking a simply connected G_θ as the Lie group corresponding to $\mathfrak{g}_\theta$, we obtain a simply connected symmetric Riemannian homogeneous space of compact type. Indeed, each of these two processes is the reverse of the other, and in this way we get a one-to-one correspondence between simply connected symmetric Riemannian homogeneous spaces of compact type and those of noncompact type. This relationship is called **duality** for symmetric Riemannian spaces; when two symmetric Riemannian spaces are related by duality, each is said to be the **dual** of the other.

If one of the two symmetric Riemannian spaces related by duality is irreducible, the other is also irreducible. The duality holds between spaces of types (1) and (3) and between those of types (2) and (4) described in Section C. This fact is based on the following theorem in the theory of Lie algebras, where we identify isomorphic Lie algebras. (i) Complex simple Lie algebras $\mathfrak{g}^{\mathbf{C}}$ and compact real simple Lie algebras $\mathfrak{g}$ are in one-to-one correspondence by the relation that $\mathfrak{g}^{\mathbf{C}}$ is the complex form of $\mathfrak{g}$. (ii) Form the Lie algebra $\mathfrak{g}_\theta$ in the above way from a compact real simple Lie algebra $\mathfrak{g}$ and an involutive automorphism θ of $\mathfrak{g}$. We assume that θ is a member of the given complete system of representatives of conjugate classes of involutive automorphisms in the automorphism group of $\mathfrak{g}$. Then we get from the pair $(\mathfrak{g}, \theta)$ a noncompact real simple Lie algebra $\mathfrak{g}_\theta$, and any noncompact real simple Lie algebra is obtained by this process in one and only one way.

Consider a Riemannian space given as a symmetric Riemannian homogeneous space $M=G/K$ with a semisimple Lie group G, and let K be the †sectional curvature of M. Then if M is of compact type the value of K is $\geqslant 0$, and if M is of noncompact type it is $\leqslant 0$. On the other hand, the **rank** of M is the (unique) dimension of a commutative subalgebra of $\mathfrak{g}$ that is contained in and maximal in $\mathfrak{m}$. (For results concerning the group of isometries of M, distribution of geodesics on M, etc. → [3].)

E. Symmetric Hermitian Spaces

A connected †complex manifold M with a †Hermitian metric is called a **symmetric Hermitian space** if for each point p of M there exists an isometric and †biholomorphic transformation of M onto M that is of order 2 and has p as an isolated fixed point. As a real analytic manifold, such a space M is a symmetric Riemannian space of even dimension, and the Hermitian metric of M is a †Kähler metric. Let $I(M)$ be the (not necessarily connected) Lie group formed by all isometries of M, and let $A(M)$ be the subgroup consisting of all holomorphic transformations in $I(M)$. Then $A(M)$ is a closed Lie subgroup of $I(M)$. Let G be the connected component $A(M)^0$ of the identity element of $A(M)$. Then G acts transitively on M, and M is expressed as a symmetric Riemannian homogeneous space G/K.

Under the de Rham decomposition of a simply connected symmetric Hermitian space (regarded as a Riemannian space), all the factors are symmetric Hermitian spaces. The factor that is isomorphic to a real Euclidean spaces as a Riemannian space is a symmetric Hermitian space that is isomorphic to the complex Euclidean space $\mathbf{C}^n$. A symmetric Hermitian space defining an irreducible symmetric Riemannian space is called an **irreducible symmetric Hermitian space**. The problem of classifying symmetric Hermitian spaces is thus reduced to classifying irreducible symmetric Hermitian spaces.

In general, if the symmetric Riemannian space defined by a symmetric Hermitian space M is represented as a symmetric Riemannian homogeneous space G/K by a connected semisimple Lie group G acting effectively on M, then M is simply connected, G coincides with the group $A(M)^0$ introduced in the previous paragraph, and the center of K is not a †discrete set. In particular, an irreducible symmetric Hermitian space is simply connected. Moreover, in order for an irreducible symmetric Riemannian homogeneous space G/K to be defined by an irreducible symmetric Hermitian space M, it is necessary and sufficient that the center of K not be a discrete set. If G acts effectively on M, then G is a simple Lie group whose center is reduced to the identity element, and the center of K is of dimension 1. For a space G/K satisfying these conditions, there are two kinds of structures of symmetric Hermitian spaces defining the Riemannian structure of G/K.

As follows from the classification of irreducible symmetric Riemannian spaces, an irreducible Hermitian space defines one of the following symmetric Riemannian homogeneous spaces, and conversely, each of these homogeneous spaces is defined by one of the two kinds of symmetric Hermitian spaces.

(I) The symmetric homogeneous space G/K of a compact simple Lie group G with respect to an involutive automorphism θ such that the center of G reduces to the identity element and the center of K is not a discrete set. Here θ may be assumed to be a representative of a conjugate class of involutive automorphisms in the automorphism group of G.

(II) The homogeneous space G_0/K of a noncompact simple Lie group G_0 by a maximal compact subgroup K such that the center of G_0 reduces to the identity element and the center of K is not a discrete set.

An irreducible symmetric Hermitian space of type (I) is compact and is isomorphic to a †rational algebraic variety. An irreducible symmetric Hermitian space of type (II) is homeomorphic to a Euclidean space and is isomorphic (as a complex manifold) to a bounded domain in $\mathbf{C}^n$ (Section F).

By the same principle as for irreducible symmetric Riemannian spaces, a duality holds for irreducible symmetric Hermitian spaces which establishes a one-to-one correspondence between the spaces of types (I) and (II). Furthermore, an irreducible symmetric Hermitian space M_b of type (II) that is dual to a given irreducible symmetric Hermitian space $M_a = G/K$ of type (I) can be realized as an open complex submanifold of M_a in the following way. Let $G^{\mathbf{C}}$ be the connected component of the identity element in the Lie group formed by all the holomorphic transformations of M_a. Then $G^{\mathbf{C}}$ is a complex simple Lie group containing G as a maximal compact subgroup, and the complex Lie algebra $\mathfrak{g}^{\mathbf{C}}$ of $G^{\mathbf{C}}$ contains the Lie algebra $\mathfrak{g}$ of G as a real form. Let θ be the involutive automorphism of G defining the symmetric homogeneous space G/K, and let $\mathfrak{g} = \mathfrak{m} + \mathfrak{k}$ be the decomposition of $\mathfrak{g}$ determined by θ. We denote by G_0 the real subgroup of $G^{\mathbf{C}}$ corresponding to the real form $\mathfrak{g}_0 = \sqrt{-1}\,\mathfrak{m} + \mathfrak{k}$ of $\mathfrak{g}^{\mathbf{C}}$. Then G_0 (i) is a closed subgroup of $G^{\mathbf{C}}$ whose center reduces to the identity element and (ii) contains K as a maximal compact subgroup. By definition the space M_b is then given by G_0/K. Now the group G_0 acts on M_a as a subgroup of $G^{\mathbf{C}}$, and the orbit of G_0 containing the point K of M_a is an open complex submanifold that is isomorphic to M_b (as a complex manifold). M_a regarded as a complex manifold can be represented as the homogeneous space $G^{\mathbf{C}}/U$ of the complex simple Lie group $G^{\mathbf{C}}$.

F. Symmetric Bounded Domains

We denote by D a bounded domain in the complex Euclidean space $\mathbf{C}^n$ of dimension n. We call D a **symmetric bounded domain** if for each point of D there exists a holomorphic transformation of order 2 of D onto D having the point as an isolated fixed point. On the other hand, the group of all holomorphic transformations of D is a Lie group, and D is called a **homogeneous bounded domain** if this group acts transitively on D. A symmetric bounded domain is a homogeneous bounded domain. The following theorem gives more precise results: On a bounded domain D, †Bergman's kernel function defines a Kähler metric that is invariant under all holomorphic transformations of D. If D is a symmetric bounded domain, D is a symmetric Hermitian space with respect to this metric, and its defining Riemannian space is a symmetric Riemannian homogeneous space of noncompact type G/K with semisimple Lie group G. Conversely, any symmetric Hermitian space of noncompact type is isomorphic (as a complex manifold) to a symmetric bounded domain. When D is isomorphic to an irreducible symmetric Hermitian space, we call D an **irreducible symmetric bounded domain**. A symmetric bounded domain is simply connected and can be decomposed into the direct product of irreducible symmetric bounded domains.

The connected component of the identity element of the group of all holomorphic transformations of a symmetric bounded domain D is a semisimple Lie group that acts transitively on D. Conversely, D is a symmetric bounded domain if a connected semisimple Lie group, or more generally, a connected Lie group admitting a two-sided invariant †Haar measure, acts transitively on D. Homogeneous bounded domains in $\mathbf{C}^n$ are symmetric bounded domains if $n \leqslant 3$ but not necessarily when $n \geqslant 4$.

G. Examples of Irreducible Symmetric Riemannian Spaces

Here we list irreducible symmetric Riemannian spaces of types (2) and (4) (→ Section C) that can be represented as homogeneous spaces of classical groups, using the notation introduced by E. Cartan. We denote by $M_u=G/K$ a simply connected irreducible symmetric Riemannian space of type (2), where G is a group that acts almost effectively on M_u and K is the subgroup given by $K=K_\theta^0$ for an involutive automorphism θ of G. For such an M_u, the space of type (4) that is dual to M_u is denoted by $M_\theta=G_\theta/K$. Clearly $\dim M_u=\dim M_\theta$. (For the dimension and rank of M_u and for those M_u that are represented as homogeneous spaces of simply connected †exceptional compact simple Lie groups → Appendix A, Table 5.III.) In this section (and also in Appendix A, Table 5.III), $O(n)$, $U(n)$, $Sp(n)$, $SL(n,\mathbf{R})$, and $SL(n,\mathbf{C})$ are the †orthogonal group of degree n, the †unitary group of degree n, the †symplectic group of degree $2n$, and the real and complex †special linear groups of degree n, respectively. Let $SO(n)=SL(n,\mathbf{R})\cap O(n)$ and $SU(n)=SL(n,\mathbf{C})\cap U(n)$. We put

$$I_{p,q}=\begin{pmatrix}-I_p & 0\\ 0 & I_q\end{pmatrix},\quad J_n=\begin{pmatrix}0 & I_n\\ -I_n & 0\end{pmatrix},$$

$$K_{p,q}=\begin{pmatrix}-I_p & 0 & 0 & 0\\ 0 & I_q & 0 & 0\\ 0 & 0 & -I_p & 0\\ 0 & 0 & 0 & I_q\end{pmatrix},$$

where I_p is the $p\times p$ unit matrix.

Type AI. $M_u=SU(n)/SO(n)$ $(n>1)$, where $\theta(s)=\bar{s}$ (with $\bar{s}$ the complex conjugate matrix of s). $M_\theta=SL(n,\mathbf{R})/SO(n)$.

Type AII. $M_u=SU(2n)/Sp(n)$ $(n>1)$, where $\theta(s)=J_nsJ_n^{-1}$. $M_\theta=SU^*(2n)/Sp(n)$. Here $SU^*(2n)$ is the subgroup of $SL(2n,\mathbf{C})$ formed by the matrices that commute with the transformation $(z_1,\dots,z_n,z_{n+1},\dots,z_{2n})\to(\bar{z}_{n+1},\dots,\bar{z}_{2n},-\bar{z}_1,\dots,-\bar{z}_n)$ in $\mathbf{C}^n$; $SU^*(2n)$ is called the **quaternion unimodular group** and is isomorphic to the commutator group of the group of all regular transformations in an n-dimensional vector space over the quaternion field $\mathbf{H}$.

Type AIII. $M_u=SU(p+q)/S(U_p\times U_q)$ $(p\geqslant q\geqslant 1)$, where $S(U_p\times U_q)=SU(p+q)\cap(U(p)\times U(q))$, with $U(p)\times U(q)$ being canonically identified with a subgroup of $U(p+q)$, and $\theta(s)=I_{p,q}sI_{p,q}$. This space M_u is a †complex Grassmann manifold. $M_\theta=SU(p,q)/S(U_p\times U_q)$, where $SU(p,q)$ is the subgroup of $SL(p+q,\mathbf{C})$ consisting of matrices that leave invariant the Hermitian form $z_1\bar{z}_1+\dots+z_p\bar{z}_p-z_{p+1}\bar{z}_{p+1}-\dots-z_{p+q}\bar{z}_{p+q}$.

Type AIV. This is the case $q=1$ of type AIII. M_u is the $(n-1)$-dimensional complex projective space, and M_θ is called a **Hermitian hyperbolic space**.

Type BDI. $M_u=SO(p+q)/SO(p)\times SO(q)$ $(p\geqslant q\geqslant 1, p>1, p+q\neq 4)$, where $\theta(s)=I_{p,q}sI_{p,q}$. M_u is the †real Grassmann manifold formed by the oriented p-dimensional subspaces in $\mathbf{R}^{p+q}$. $M_\theta=SO_0(p,q)/SO(p)\times SO(q)$, where $SO(p,q)$ is the subgroup of $SL(n,\mathbf{R})$ consisting of matrices that leave invariant the quadratic form $x_1^2+\dots+x_p^2-x_{p+1}^2-\dots-x_{p+q}^2$, and $SO_0(p,q)$ is the connected component of the identity element.

Type BDII. This is the case $q=1$ of type BDI. M_u is the $(n-1)$-dimensional sphere, and M_θ is called a **real hyperbolic space**.

Type DIII. $M_u=SO(2n)/U(n)$ $(n>2)$, where $U(n)$ is regarded as a subgroup of $SO(2n)$ by identifying $s\in U(n)$ with

$$\begin{pmatrix}\operatorname{Re}s & \operatorname{Im}s\\ -\operatorname{Im}s & \operatorname{Re}s\end{pmatrix}\in SO(2n),$$

and $\theta(s)=J_nsJ_n^{-1}$. $M_\theta=SO^*(2n)/U(n)$. Here $SO^*(2n)$ denotes the group of all complex orthogonal matrices of determinant 1 leaving invariant the skew-Hermitian form $z_1\bar{z}_{n+1}-z_{n+1}\bar{z}_1+z_2\bar{z}_{n+2}-z_{n+2}\bar{z}_2+\dots+z_n\bar{z}_{2n}-z_{2n}\bar{z}_n$; this group is isomorphic to the group of all linear transformations leaving invariant a nondegenerate skew-Hermitian form in an n-dimensional vector space over the quaternion field $\mathbf{H}$.

Type CI. $M_u=Sp(n)/U(n)$ $(n\geqslant 1)$, where $U(n)$ is considered as a subgroup of $Sp(n)$ by the identification $U(n)\subset SO(2n)$ explained in type DIII and $\theta(s)=s(=J_nsJ_n^{-1})$. $M_\theta=Sp(n,\mathbf{R})/U(n)$, where $Sp(n,\mathbf{R})$ is the real symplectic group of degree $2n$.

Type CII. $M_u=Sp(p+q)/Sp(p)\times Sp(q)$ $(p\geqslant q\geqslant 1)$, where $Sp(p)\times Sp(q)$ is identified with a subgroup of $Sp(p+q)$ by the mapping

$$\left(\begin{pmatrix}A_1 & B_1\\ C_1 & D_1\end{pmatrix},\begin{pmatrix}A_2 & B_2\\ C_2 & D_2\end{pmatrix}\right)$$

$$\to\begin{pmatrix}A_1 & 0 & B_1 & 0\\ 0 & A_2 & 0 & B_2\\ C_1 & 0 & D_1 & 0\\ 0 & C_2 & 0 & D_2\end{pmatrix}$$

and $\theta(s)=K_{p,q}sK_{p,q}$. $M_\theta=Sp(p,q)/Sp(p)\times Sp(q)$. Here $Sp(p,q)$ is the group of complex symplectic matrices of degree $2(p+q)$ leaving invariant the Hermitian form $(z_1,\dots,z_{p+q})K_{p,q}\,{}^t(\bar{z}_1,\dots,\bar{z}_{p+q})$; this group is interpreted as the group of all linear transformations leaving invariant a nondegenerate Hermitian form of index p in a $(p+q)$-dimensional vector space over the quaternion field $\mathbf{H}$. For $q=1$, M_u is the quaternion projective space, and M_θ is called the **quaternion hyperbolic space**.

Among the spaces introduced here, there are some with lower p, q, n that coincide (as Riemannian spaces) (→ Appendix A, Table 5.III).

H. Space Forms

A Riemannian manifold of †constant curvature is called a **space form**; it is said to be **spherical**,

Euclidean, or **hyperbolic** according as the constant curvature K is positive, zero, or negative. A space form is a locally symmetric Riemannian space; a simply connected complete space form is a sphere if $K>0$, a real Euclidean space if $K=0$, and a real hyperbolic space if $K<0$. More generally, a complete spherical space form of even dimension is a sphere or a projective space, and one of odd dimension is an orientable manifold. A complete 2-dimensional Euclidean space form is one of the following spaces: Euclidean plane, cylinder, torus, †Möbius strip, †Klein bottle. Except for these five spaces and the 2-dimensional sphere, any †closed surface is a 2-dimensional hyperbolic space form (for details about space forms → [6]).

I. Examples of Irreducible Symmetric Bounded Domains

Among the irreducible symmetric Riemannian spaces described in Section H, those defined by irreducible symmetric Hermitian spaces are of types AIII, DIII, BDI ($q=2$), and CI. We list the irreducible symmetric bounded domains that are isomorphic to the irreducible Hermitian spaces defining these spaces. Positive definiteness of a matrix will be written $\gg 0$.

Type $\mathbf{I}_{m,m'}$ ($m'\geqslant m\geqslant 1$). The set of all $m\times m'$ complex matrices Z satisfying the condition $I_{m'}-{}^t\bar{Z}Z\gg 0$ is a symmetric bounded domain in $\mathbf{C}^{mm'}$, which is isomorphic (as a complex manifold) to the irreducible symmetric Hermitian space defined by M_θ of type AIII ($p=m$, $q=m'$).

Type $\mathbf{II}_m$ ($m\geqslant 2$). The set of all $m\times m$ complex †skew-symmetric matrices Z satisfying the condition $I_m-{}^t\bar{Z}Z\gg 0$ is a symmetric bounded domain in $\mathbf{C}^{m(m-1)/2}$ corresponding to the type DIII ($n=m$).

Type $\mathbf{III}_m$ ($m\geqslant 1$). The set of all $m\times m$ complex symmetric matrices satisfying the condition $I_m-{}^t\bar{Z}Z\gg 0$ is a symmetric bounded domain in $\mathbf{C}^{m(m+1)/2}$ corresponding to the type CI ($n=m$). This bounded domain is holomorphically isomorphic to the †Siegel upper half-space of degree m.

Type $\mathbf{IV}_m$ ($m\geqslant 1, m\neq 2$). This bounded domain in $\mathbf{C}^m$ is formed by the elements $(z_1,\ldots,z_m)$ satisfying the condition $|z_1|^2+\ldots+|z_m|^2<(1+|z_1^2+\ldots+z_m^2|)/2<1$, and corresponds to the type BDI ($p=m, q=2$).

Among these four types of bounded domains, the following complex analytic isomorphisms hold: $\mathrm{I}_{1,1}\cong\mathrm{II}_2\cong\mathrm{III}_1\cong\mathrm{IV}_1$, $\mathrm{II}_3\cong\mathrm{I}_{1,3}$, $\mathrm{IV}_3\cong\mathrm{III}_2$, $\mathrm{IV}_4\cong\mathrm{I}_{2,2}$, $\mathrm{IV}_6\cong\mathrm{II}_4$. (For details about these symmetric bounded domains → [2].) There are two more kinds of irreducible symmetric bounded domains, which are represented as homogeneous spaces of exceptional Lie groups.

J. Weakly Symmetric Riemannian Spaces

A generalization of symmetric Riemannian space is the notion of weakly symmetric Riemannian space introduced by Selberg. Let M be a Riemannian space. M is called a **weakly symmetric Riemannian space** if a Lie subgroup G of the group of isometries $I(M)$ acts transitively on M and there exists an element $\mu\in I(M)$ satisfying the relations (i) $\mu G\mu^{-1}=G$; (ii) $\mu^2\in G$; and (iii) for any two points x, y of M, there exists an element m of G such that $\mu x=my$, $\mu y=mx$. A symmetric Riemannian space M becomes a weakly symmetric Riemannian space if we put $G=I(M)$ and $\mu=$the identity transformation; as the element m in condition (iii) we can take the symmetry σ_p at the midpoint p on the geodesic arc joining x and y. There are, however, weakly symmetric Riemannian spaces that do not have the structure of a symmetric Riemannian space. An example of such a space is given by $M=G=SL(2,\mathbf{R})$ with a suitable Riemannian metric, where μ is the inner automorphism defined by

$$\begin{pmatrix}1 & 0\\ 0 & -1\end{pmatrix}$$

(Selberg [4]). On a weakly symmetric Riemannian space, the ring of all G-invariant differential-integral operators is commutative; this fact is useful in the theory of spherical functions (→ 437 Unitary Representations).

References

[1] S. Helgason, Differential geometry, Lie groups, and symmetric spaces, Academic Press, 1978.
[2] C. L. Siegel, Analytic functions of several complex variables, Princeton Univ. Press, 1950.
[3] J.-L. Koszul, Exposés sur les espaces homogènes symétriques, São Paulo, 1959.
[4] A. Selberg, Harmonic analysis and discontinuous groups in weakly symmetric Riemannian spaces with application to Dirichlet series, J. Indian Math. Soc., 20 (1956), 47–87.
[5] E. Cartan, Sur certaines formes riemanniennes remarquables des géométries à groupe fondamental simple, Ann. Sci. Ecole Norm. Sup., 44 (1927), 345–467 (Oeuvres complètes, Gauthier-Villars, 1952, pt. I, vol. 2, 867–989).
[6] E. Cartan, Sur les domaines bornés homogènes de l'espace de n-variables complexes, Abh. Math. Sem. Univ. Hamburg, 11 (1936), 116–162 (Oeuvres complètes, Gauthier-Villars, 1952, pt. I, vol. 2, 1259–1305).

[7] J. A. Wolf, Spaces of constant curvature, McGraw-Hill, 1967.
[8] S. Kobayashi and K. Nomizu, Foundations of differential geometry II, Interscience, 1969.
[9] O. Loos, Symmetric spaces. I, General theory; II, Compact spaces and classification, Benjamin, 1969.

413 (VII.7) Symmetric Spaces

A †Riemannian manifold M is called a **symmetric Riemannian space** if M is connected and if for each $p \in M$ there exists an involutive †isometry σ_p of M that has p as an isolated fixed point. For the classification and the group-theoretic properties of symmetric Riemannian spaces → 412 Symmetric Riemannian Spaces and Real Forms. We state here the geometrical properties of a symmetric Riemannian space M. Let M be represented by G/K, a †symmetric Riemannian homogeneous space. The †Lie algebras of G and K are denoted by $\mathfrak{g}$ and $\mathfrak{k}$ respectively. Let us denote by τ_a the †left translation of M defined by $a \in G$, and by X^* the vector field on M generated by $X \in \mathfrak{g}$. We denote by θ the differential of the involutive automorphism of G defining G/K and identify the subspace $\mathfrak{m} = \{X \in \mathfrak{g} \mid \theta(X) = -X\}$ of $\mathfrak{g}$ with the tangent space $T_o(M)$ of M at the origin $o = K$ of M. The †representation of $\mathfrak{k}$ on $\mathfrak{m}$ induced from the †adjoint representation of $\mathfrak{g}$ is denoted by $\mathrm{ad}_{\mathfrak{m}}$.

A. Riemannian Connections

M is a complete real analytic †homogeneous Riemannian manifold. If M is a †symmetric Hermitian space, it is a †homogeneous Kählerian manifold. The †Riemannian connection ∇ of M is the †canonical connection of the homogeneous space G/K and satisfies $\nabla_Y X^* = [X, Y]$ $(Y \in \mathfrak{m})$ for each $X \in \mathfrak{k}$ and $\nabla_Y X^* = 0$ $(Y \in \mathfrak{m})$ for each $X \in \mathfrak{m}$. For each $X \in \mathfrak{m}$, the curve γ_X of M defined by $\gamma_X(t) = (\exp tX)o$ $(t \in \mathbf{R})$ is a †geodesic of M such that $\gamma_X(0) = o$ and $\dot{\gamma}_X(0) = X$. In particular, the †exponential mapping Exp_o at o is given by $\mathrm{Exp}_o X = (\exp X)o$ $(X \in \mathfrak{m})$. For each $X \in \mathfrak{m}$, the †parallel translation along the geodesic arc $\gamma_X(t)$ $(0 \leqslant t \leqslant t_0)$ coincides with the differential of $\tau_{\exp t_0 X}$. If M is compact, for each $p \in M$ there exists a smooth simply closed geodesic passing through p. Any G-invariant tensor field on M is †parallel with respect to ∇. Any G-invariant †differential form on M is closed. The Lie algebra $\mathfrak{h}$ of the †restricted homogeneous holonomy group of M at o coincides with $\mathrm{ad}_{\mathfrak{m}}[\mathfrak{m}, \mathfrak{m}]$. If the group $I(M)$ of all isometries of M is †semisimple, one has $\mathfrak{h} = \{A \in \mathfrak{gl}(\mathfrak{m}) \mid A \cdot g_o = 0, A \cdot R_o = 0\} = \mathrm{ad}_{\mathfrak{m}}\mathfrak{k}$. Here, g_o and R_o denote the values at o of the Riemannian metric g and the †Riemannian curvature R of M, respectively, and $A \cdot$ is the natural action of A on the tensors over $\mathfrak{m}$. If, moreover, M is a symmetric Hermitian space, the value J_0 at o of the †almost complex structure J of M belongs to the center of $\mathfrak{h}$. In general, $\mathfrak{h} = \{0\}$ if and only if M is †flat, and $\mathfrak{h}$ has no nonzero invariant on $\mathfrak{m}$ if and only if $I(M)$ is semisimple.

B. Riemannian Curvature Tensors

The Riemannian curvature tensor R of M is parallel and satisfies $R_0(X, Y) = -\mathrm{ad}_{\mathfrak{m}}[X, Y]$ $(X, Y \in \mathfrak{m})$. Assume that $\dim M \geqslant 2$ in the following. Let P be a 2-dimensional subspace of $\mathfrak{m}$, and $\{X, Y\}$ an orthonormal basis of P with respect to g_o. Then the †sectional curvature $K(P)$ of P is given by $K(P) = g_o([[X, Y], X], Y)$. $K = 0$ everywhere if and only if M is flat. If M is of †compact type (resp. of †noncompact type), then $K \geqslant 0$ (resp. $K \leqslant 0$) everywhere. $K > 0$ (resp. $K < 0$) everywhere if and only if the †rank of M is 1 and M is of compact type (resp. of noncompact type). For any four points p, q, p', q' of a manifold M of any of these types satisfying $d(p, q) = d(p', q')$, d being the †Riemannian distance of M, there exists a $\phi \in I(M)$ such that $\phi(p) = p'$ and $\phi(q) = q'$. Other than the aforementioned M's, the only Riemannian manifolds having this property are circles and Euclidean spaces. If $K > 0$ everywhere, any geodesic of M is a smooth simply closed curve and all geodesics are of the same length. For a symmetric Hermitian space M, the †holomorphic sectional curvature H satisfies $H = 0$ (resp. $H > 0$, $H < 0$) everywhere if and only if M is flat (resp. of compact type, of noncompact type).

C. Ricci Tensors

The †Ricci tensor S of M is parallel. If $\varphi_{\mathfrak{m}}$ denotes the restriction to $\mathfrak{m} \times \mathfrak{m}$ of the †Killing form φ of $\mathfrak{g}$, the value S_o of S at o satisfies $S_o = -\frac{1}{2}\varphi_{\mathfrak{m}}$. If M is †irreducible, it is an †Einstein space. $S = 0$ (resp. positive definite, negative definite, nondegenerate) everywhere if and only if M is flat (resp. M is of compact type, M is of noncompact type, $I(M)$ is semisimple). If M is a †symmetric bounded domain and g is the †Bergman metric of M, one has $S = -g$.

D. Symmetric Riemannian Spaces of Noncompact Type

Let M be of noncompact type. For each $p \in M$, p is the only fixed point of the †symmetry σ_p, and the exponential mapping at p is a diffeomorphism from $T_p(M)$ to M. In particular, M is diffeomorphic to a Euclidean space. For each pair $p, q \in M$, a geodesic arc joining p and q is unique up to parametrization. For each $p \in M$ there exists neither a †conjugate point nor a †cut point of p. If M is a symmetric Hermitian space, that is, if it is a symmetric bounded domain, then it is a †Stein manifold and holomorphically homeomorphic to a †Siegel domain.

E. Groups of Isometries

The isotropy subgroup at o in $I(M)$ is denoted by $I_o(M)$. Then the smooth mapping $I_o(M) \times \mathfrak{m} \rightarrow I(M)$ defined by the correspondence $\phi \times X \mapsto \phi\tau_{\exp X}$ is surjective, and it is a diffeomorphism if M is of noncompact type. If M is of noncompact type, $I(M)$ is isomorphic to the group $A(\mathfrak{g})$ of all automorphisms of $\mathfrak{g}$ in a natural way, and $I_o(M)$ is isomorphic to the subgroup $A(\mathfrak{g}, \mathfrak{k}) = \{\phi \in A(\mathfrak{g}) \mid \phi(\mathfrak{k}) = \mathfrak{k}\}$ of $A(\mathfrak{g})$, provided that G acts almost effectively on M. Moreover, in this case the center of the identity component $I(M)^0$ of $I(M)$ reduces to the identity, and the isotropy subgroup at a point in $I(M)^0$ is a maximal compact subgroup of $I(M)^0$. If $I(M)$ is semisimple, any element of $I(M)^0$ may be represented as a product of an even number of symmetries of M. In the following, let M be a symmetric Hermitian space, and denote by $A(M)$ (resp. $H(M)$) the group of all holomorphic isometries (resp. all holomorphic homeomorphisms) of M, and by $A(M)^0$ and $H(M)^0$ their identity components. All these groups act transitively on M. If M is compact or if $I(M)$ is semisimple, one has $A(M)^0 = I(M)^0$. If $I(M)$ is semisimple, M is simply connected and the center of $I(M)^0$ reduces to the identity. If M is of compact type, M is a †rational †projective algebraic manifold, and $H(M)^0$ is a complex semisimple Lie group whose center reduces to the identity, and it is the †complexification of $I(M)^0$. In this case, the isotropy subgroup at a point in $H(M)^0$ is a †parabolic subgroup of $H(M)^0$. If M is of noncompact type, one has $H(M)^0 = I(M)^0$. In the following we assume that G is compact.

F. Cartan Subalgebras

A maximal Abelian †Lie subalgebra in $\mathfrak{m}$ is called a **Cartan subalgebra** for M. Cartan subalgebras are conjugate to each other under the †adjoint action of K. Fix a Cartan subalgebra $\mathfrak{a}$ and introduce an inner product (,) on $\mathfrak{a}$ by the restriction to $\mathfrak{a} \times \mathfrak{a}$ of g_o. For an element α of the dual space $\mathfrak{a}^*$ of $\mathfrak{a}$, we put $\mathfrak{m}_\alpha = \{X \in \mathfrak{m} \mid [H, [H, X]] = -\alpha(H)^2 X$ for any $H \in \mathfrak{a}\}$. The subset $\Sigma = \{\alpha \in \mathfrak{a}^* - \{0\} \mid \mathfrak{m}_\alpha \neq \{0\}\}$ of $\mathfrak{a}^*$ is called the **root system** of M (relative to $\mathfrak{a}$). We write $m_\alpha = \dim \mathfrak{m}_\alpha$ for $\alpha \in \Sigma$. The subset $D = \{H \in \mathfrak{a} \mid \alpha(H) \in \pi\mathbf{Z}$ for some $\alpha \in \Sigma\}$ of $\mathfrak{a}$ is called the **diagram** of M. A connected component of $\mathfrak{a} - D$ is called a **fundamental cell** of M. The quotient group W of the normalizer of $\mathfrak{a}$ in K modulo the centralizer of $\mathfrak{a}$ in K is called the **Weyl group** of M. W is identified with a finite group of orthogonal transformations of $\mathfrak{a}$.

G. Conjugate Points

For a geodesic arc γ with the initial point o, any †Jacobi field along γ that vanishes at o and the end point of γ is obtained as the restriction to γ of the vector field X^* generated by an element $X \in \mathfrak{k}$. For $H \in \mathfrak{a} - \{0\}$, $\operatorname{Exp}_o H$ is a conjugate point to o along the geodesic γ_H if and only if $\alpha(H) \in \pi\mathbf{Z} - \{0\}$ for some $\alpha \in \Sigma$. In this case, the multiplicity of the conjugate point $\operatorname{Exp}_o H$ is equal to $\frac{1}{2}\sum_{\alpha \in \Sigma, \alpha(H) \in \pi\mathbf{Z} - \{0\}} M_\alpha$. From this fact and Morse theory (→ 279 Morse Theory), we get a †cellular decomposition of the †loop space of M. The set of all points conjugate to o coincides with $K \operatorname{Exp}_o D$ and is stratified to a disjoint union of a finite number of connected regular submanifolds with dimension $\leqslant \dim M - 2$.

H. Cut Points

We define a †lattice group Γ of $\mathfrak{a}$ by $\Gamma = \{A \in \mathfrak{a} \mid \operatorname{Exp}_o A = o\}$, and put $C_\mathfrak{a} = \{H \in \mathfrak{a} \mid \operatorname{Max}_{A \in \Gamma - \{0\}} 2(H, A)/(A, A) = 1\}$. Then, for $H \in \mathfrak{a} - \{0\}$, $\operatorname{Exp}_o H$ is a cut point of o along the geodesic γ_H if and only if $H \in C_\mathfrak{a}$. The set C_o of all cut points of o coincides with $K \operatorname{Exp}_o C_\mathfrak{a}$ and is stratified to a disjoint union of a finite number of connected regular submanifolds with dimension $\leqslant \dim M - 1$. The set of all points †first conjugate to o coincides with C_o if and only if M is simply connected.

I. Fundamental Groups

Let Γ_0 denote the subgroup of $\mathfrak{a}$ generated by $\{(2\pi/(\alpha, \alpha))\alpha \mid \alpha \in \Sigma\}$, identifying $\mathfrak{a}^*$ with $\mathfrak{a}$ by means of the inner product (,) of $\mathfrak{a}$. This is a subgroup of Γ. We regard Γ as a subgroup of the group $I(\mathfrak{a})$ of all motions of $\mathfrak{a}$ by parallel translations. The subgroup $\tilde{W}=W\Gamma$ of $I(\mathfrak{a})$ generated by Γ and the Weyl group W is called the **affine Weyl group** of M. $\tilde{W}$ leaves the diagram D invariant and acts transitively on the set of all fundamental cells of M. Take a fundamental cell σ such that its closure $\bar{\sigma}$ contains 0, and put $\tilde{W}_\sigma=\{w\in\tilde{W}\mid w(\sigma)=\sigma\}$. Then the fundamental group $\pi_1(M)$ of M is an †Abelian group isomorphic to the groups $\tilde{W}_\sigma$ and Γ/Γ_0. $\pi_1(M)$ is a finite group if and only if M is of compact type. In this case, the order of $\pi_1(M)$ is equal to the cardinality of the set $\Gamma\cap\bar{\sigma}$ as well as to the index $[\Gamma:\Gamma_0]$. Moreover, if we denote by $\tilde{W}_\sigma^*$ the group $\tilde{W}_\sigma$ for the symmetric Riemannian space $M^*=G^*/K^*$ defined by the †adjoint group G^* of G and $K^*=\{a\in G^*\mid a\theta=\theta a\}$, then $\tilde{W}_\sigma$ is isomorphic to a subgroup of $\tilde{W}_\sigma^*$. If M is irreducible, $\tilde{W}_\sigma^*$ is isomorphic to a subgroup of the group of all automorphisms of the †extended Dynkin diagram of the root system Σ.

J. Cohomology Rings

Let $P(\mathfrak{g})$ (resp. $P(\mathfrak{k})$) be the †graded linear space of all †primitive elements in the †cohomology algebra $H(\mathfrak{g})$ of $\mathfrak{g}$ (resp. $H(\mathfrak{k})$ of $\mathfrak{k}$), and $P(\mathfrak{g},\mathfrak{k})$ the intersection of $P(\mathfrak{g})$ with the image of the natural homomorphism $H(\mathfrak{g},\mathfrak{k})\rightarrow H(\mathfrak{g})$, where $H(\mathfrak{g},\mathfrak{k})$ denotes the relative cohomology algebra for the pair $(\mathfrak{g},\mathfrak{k})$. Then one has $\dim P(\mathfrak{g},\mathfrak{k})+\dim P(\mathfrak{k})=\dim P(\mathfrak{g})$. Denote by $\Lambda P(\mathfrak{g},\mathfrak{k})$ the exterior algebra over $P(\mathfrak{g},\mathfrak{k})$. The †graded algebra of all G-invariant polynomials on $\mathfrak{g}$ (resp. all K-invariant polynomials on $\mathfrak{k}$) is denoted by $I(G)$ (resp. $I(K)$), where the degree of a homogeneous polynomial with degree p is defined to be $2p$. We denote by $I^+(G)$ the ideal of $I(G)$ consisting of all $f\in I(G)$ such that $f(0)=0$, and regard $I(K)$ as an $I^+(G)$-module through the restriction homomorphism. Then the †real cohomology ring $H(M)$ of M is isomorphic to the tensor product $\Lambda P(\mathfrak{g},\mathfrak{k})\otimes(I(K)/I^+(G)I(K))$. If K is connected and the †Poincaré polynomials of $P(\mathfrak{g})$, $P(\mathfrak{k})$, and $P(\mathfrak{g},\mathfrak{k})$ are $\sum_{i=1}^r t^{2m_i-1}$, $\sum_{i=1}^s t^{2n_i-1}$, and $\sum_{i=s+1}^r t^{2m_i-1}$, respectively, then the Poincaré polynomial of $H(M)$ is given by $\prod_{i=s+1}^r(1+t^{2m_i-1})\prod_{i=1}^s(1-t^{2m_i})\prod_{i=1}^s(1-t^{2n_i})^{-1}$.

References

[1] E. Cartan, Sur certaines formes riemanniennes remarquables des géométries à groupe fondamental simple, Ann. Sci. Ecole Norm. Sup., 44 (1927), 345–467.
[2] S. Helgason, Differential geometry, Lie groups, and symmetric spaces, Academic Press, 1978.
[3] S. Kobayashi and K. Nomizu, Foundations of differential geometry II, Interscience, 1969.
[4] H. C. Wang, Two point homogeneous spaces, Ann. Math., (2) 55 (1952), 177–191.
[5] A. Korányi and J. A. Wolf, Realization of Hermitian symmetric spaces as generalized half-planes, Ann. Math., (2) 81 (1965), 265–288.
[6] R. Bott and H. Samelson, Applications of the theory of Morse to symmetric spaces, Amer. J. Math., 80 (1958), 964–1029.
[7] T. Sakai, On cut loci of compact symmetric spaces, Hokkaido Math. J., 6 (1977), 136–161.
[8] M. Takeuchi, On conjugate loci and cut loci of compact symmetric spaces I, Tsukuba J. Math., 2 (1978), 35–68.
[9] R. Crittenden, Minimum and conjugate points in symmetric spaces, Canad. J. Math., 14 (1962), 320–328.
[10] J. L. Koszul, Sur un type d'algèbre différentielles avec la transgression, Colloque de Topologie (Espaces fibrès), Brussels, 1950, 73–81.

414 (XX.1) Systems of Units

A. International System of Units

Units representing various physical quantities can be derived from a certain number of **fundamental (base) units**. By a **system of units** we mean a system of fundamental units. Various systems of units have been used in the course of the development of physics. Today, the standard is set by the **international system of units** (système international d'unités; abbreviated SI) [1], which has been developed in the spirit of the meter-kilogram system. This system consists of the seven fundamental units listed in Table 1, units induced from them, and unit designations with prefixes representing the powers of 10 where necessary. It also contains two **auxiliary units** for plane and solid angles, and a large number of derived units [1].

B. Systems of Units in Mechanics

Units in mechanics are usually derived from length, mass, and time, and SI uses the meter, kilogram, and second as base units. Neither the CGS system, derived from centimeter, gram, and second, nor the **system of gravitational units**, derived from length, force, and time, are recommended for general use by

Table 1

Quantity	SI unit	Symbol	Description
Length	meter	m	The meter is the length equal to 1,650,763.73 wavelengths in vacuum of the radiation corresponding to the transmission between the levels $2p^{10}$ and $5d^5$ of the krypton-86 atom.
Mass	kilogram	kg	The kilogram is equal to the mass of the international prototype of the kilogram.
Time	second	s	The second is the duration of 9,192,631,770 periods of the radiation corresponding to the transmission between the two hyperfine levels of the ground state of the cesium-133 atom.
Intensity of electric current	ampere	A	The ampere is the intensity of the constant current maintained in two parallel, rectilinear conductors of infinite length and of negligible circular section, placed 1 m apart in vacuum, and producing a force between them equal to 2×10^{-7} newton ($m \cdot kg \cdot s^{-2}$) per meter of length.
Temperature	kelvin	K	The kelvin, the unit of thermodynamical temperature, is 1/273.16 of the thermodynamical temperature of the triple point of water.
Amount of substance	mole	mol	The mole is the amount of substance of a system containing as many elementary entities as there are atoms in 0.012 kg of carbon-12.
Luminous intensity	candela	cd	The candela is the luminous intensity in a given direction of a source emitting monochromatic radiation of frequency 540×10^{12} hertz ($= s^{-1}$), the radiant intensity of which in that direction is 1/683 watt per steradian. (This revised definition of candela was adopted in 1980.)

Table 2

Quantity	SI unit	Symbol	Unit in terms of SI base or derived units
Frequency	hertz	Hz	$1\ Hz = 1\ s^{-1}$
Force	newton	N	$1\ N = 1\ kg \cdot m/s^2$
Pressure and stress	pascal	Pa	$1\ Pa = 1\ N/m^2$
Work, energy, quantity of heat	joule	J	$1\ J = 1\ N \cdot m$
Power	watt	W	$1\ W = 1\ J/s$
Quantity of electricity	coulomb	C	$1\ C = 1\ A \cdot s$
Electromotive force, potential difference	volt	V	$1\ V = 1\ W/A$
Electric capacitance	farad	F	$1\ F = 1\ C/V$
Electric resistance	ohm	Ω	$1\ \Omega = 1\ V/A$
Electric conductance	siemens	S	$1\ S = 1\ \Omega^{-1}$
Flux of magnetic induction magnetic flux	weber	Wb	$1\ Wb = 1\ V \cdot s$
Magnetic induction, magnetic flux density	tesla	T	$1\ T = 1\ Wb/m^2$
Inductance	henry	H	$1\ H = 1\ Wb/A$
Luminous flux	lumen	lm	$1\ lm = 1\ cd \cdot sr$
Illuminance	lux	lx	$1\ lx = 1\ lm/m^2$
Activity	becquerel	Bq	$1\ Bq = 1\ s^{-1}$
Adsorbed dose	gray	Gy	$1\ Gy = 1\ J/kg$
Radiation dose	sievert	Sv	$1\ Sv = 1\ J/kg$

the SI Committee. Besides the base units, minute, hour, and day, degree, minute, and second (angle), liter, and ton have been approved by the SI Committee. Units such as the electron volt, atomic mass unit, astronomical unit, and parsec (not SI) are empirically defined and have been approved. Several other units, such as nautical mile, knot, are (area), and bar, have been provisionally approved.

C. System of Units in Thermodynamics

The base unit for temperature is the degree Kelvin (°K; formerly called the absolute temperature). Degree Celsius (°C), defined by $t = T - 273.15$, where T is in °K, is also used. The unit of heat is the joule J, the same as the unit for other forms of energy. Formerly, one calorie was defined as the quantity of heat that must be supplied to one gram of water to raise its temperature from 14.5°C to 15.5°C; now one calorie is defined by 1 cal = 4.1855 J.

D. Systems of Units in Electricity and Magnetism

Three distinct systems of units have been developed in the field of electricity and magnetism: the electrostatic system, which originates from Coulomb's law for the force between two electric charges and defines magnetic quantities by means of the Biot-Savart law; the electromagnetic system, which originates from Coulomb's law for magnetism; and the Gaussian system, in which the dielectric constant and permeability are taken to be nondimensional. At present, however, the rationalized MKSA system of units is adopted as the international standard. It uses the **derived units** listed in Table 2 (taken from [2]), where the derived units with proper names in other fields are also listed.

E. Other Units

In the field of photometry, the following definition was adopted in 1948: One candela (cd) ($\fallingdotseq 0.98$ old candle) is defined as $1/(6 \times 10^5)$ of the luminous intensity in the direction normal to a plane surface of 1 m^2 area of a black body at the temperature of the solidifying point of platinum. The total luminous flux emanating uniformly in all directions from a source of luminous intensity 1 cd is defined as 4π lumen (lm). One lux (lx) is defined as the illuminance on a surface area of 1 m^2 produced by a luminous flux of 1 cd uniformly incident on the surface. In 1980, the definition was revised as shown in Table 1.

For theoretical purposes, a system of units called the absolute system of units is often used, in which units of mass, length, and time are chosen so that the values of universal constants, such as the universal gravitational constant, speed of light, Planck's constant, and Boltzmann's constant, are equal to 1.

References

[1] Bureau International des Points et Mesures, Le système international d'unités, 1970, fourth revised edition, 1981.
[2] R. G. Lerner and G. L. Trigg (eds.), Encyclopedia of physics, Addison-Wesley, 1981.

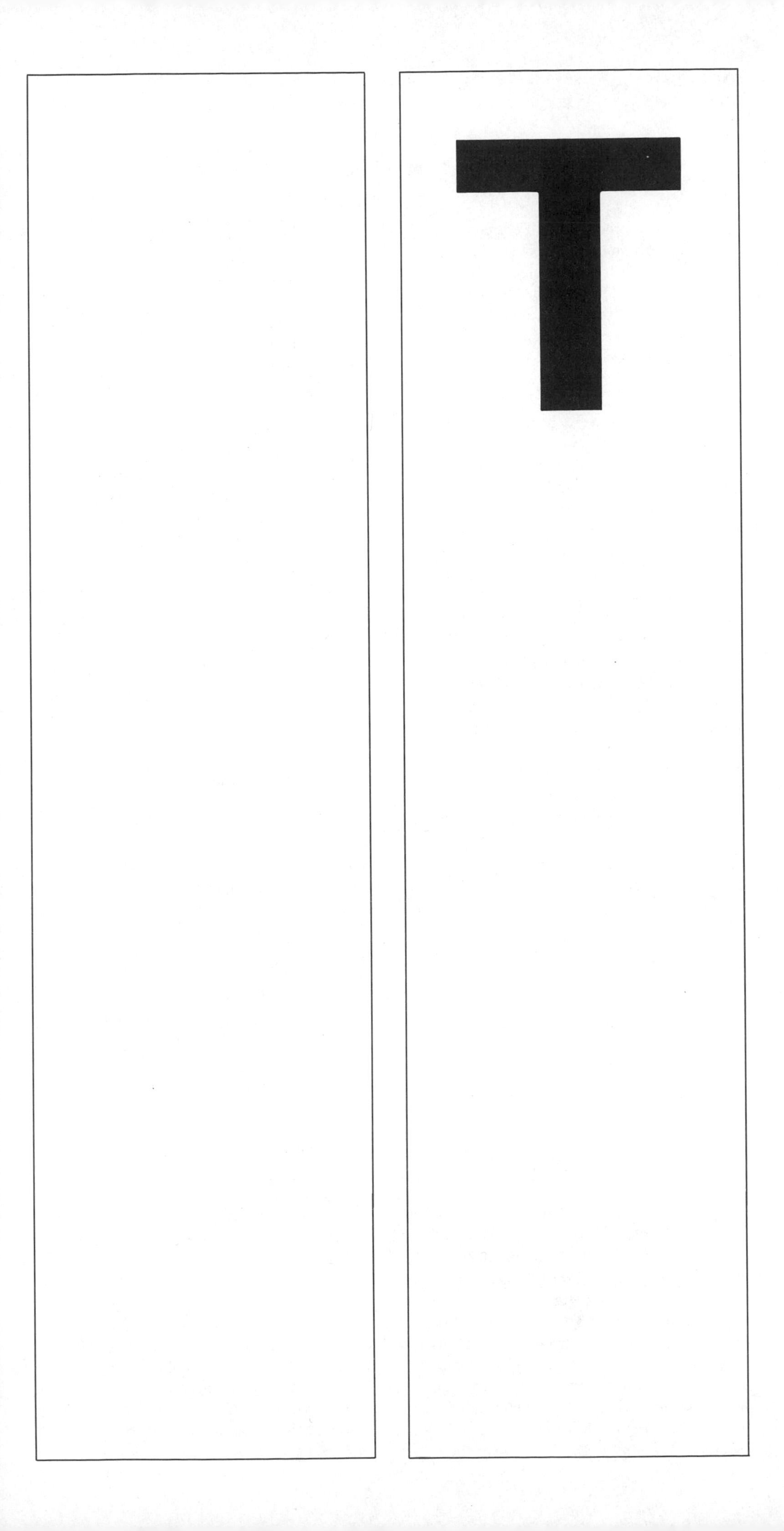

415 (XXI.41) Takagi, Teiji

Teiji Takagi (April 21, 1875–February 28, 1960) was born in Gifu Prefecture, Japan. After graduation from the Imperial University of Tokyo in 1897, he continued his studies in Germany, first with Frobenius in Berlin and then with Hilbert in Göttingen. He returned to Japan in 1901 and taught at the Imperial University of Tokyo until 1936, when he retured. He died in Tokyo of cerebral apoplexy.

Since his student years he had been interested in Kronecker's conjecture on †Abelian extensions of imaginary quadratic number fields. He solved it affirmatively for the case of $\mathbf{Q}(\sqrt{-1})$ while still in Göttingen and presented this result as his doctoral thesis. During World War I, he pursued his research in the theory of numbers in isolation from Western countries. It developed into †class field theory, a beautiful general theory of Abelian extensions of algebraic number fields. This was published in 1920, and was complemented by his 1922 paper on the †reciprocity law of power residues and then by †Artin's general law of reciprocity published in 1927. Besides these arithmetical works, he also published papers on algebraic and analytic subjects and on the foundations of the theories of natural numbers and of real numbers. His book (in Japanese) on the history of mathematics in the 19th century and his *General course of analysis* (also in Japanese) as well as his teaching and research activities at the University exercised great influence on the development of mathematics in Japan.

Reference

[1] S. Kuroda (ed.), The collected papers of Teiji Takagi, Iwanami, 1973.

416 (XI.16) Teichmüller Spaces

Consider the set $\mathbf{M}_g$ consisting of the conformal equivalence classes of closed Riemann surfaces of genus g. In 1859 Riemann stated, without rigorous proof, that $\mathbf{M}_g$ is parametrized by $m(g)$ ($=0$ if $g=0$, $=1$ if $g=1$, $=3g-3$ if $g\geqslant 2$) complex parameters (→ 11 Algebraic Functions). Later, the introduction of a topology and $m(g)$-dimensional complex structure on $\mathbf{M}_g$ were discussed rigorously in various ways. The following explanation of these methods is due to O. Teichmüller [1, 2], L. V. Ahlfors [3, 4], and L. Bers [5–7]. For the algebraic-geometric approach → 9 Algebraic Curves.

The trivial case $g=0$ is excluded, since $\mathbf{M}_0$ consists of a single point. Take a closed Riemann surface $\mathfrak{R}_0$ of genus $g\geqslant 1$, and consider the pairs $(\mathfrak{R}, H)$ consisting of closed Riemann surfaces $\mathfrak{R}$ of the same genus g and the †homotopy classes H of orientation-preserving homeomorphisms of $\mathfrak{R}_0$ into $\mathfrak{R}$. Two pairs $(\mathfrak{R}, H)$ and $(\mathfrak{R}', H')$ are defined to be conformally equivalent if the homotopy class $H'H^{-1}$ contains a conformal mapping. The set $\mathbf{T}_g$ consisting of the conformal equivalence classes $\langle\mathfrak{R}, H\rangle$ is called the **Teichmüller space** (with center at $\mathfrak{R}_0$). Let $\mathfrak{H}_g$ be the group of homotopy classes of orientation-preserving homeomorphisms of $\mathfrak{R}_0$ onto itself. $\mathfrak{H}_g$ is a transformation group acting on $\mathbf{T}_g$ in the sense that each $\eta\in\mathfrak{H}_g$ induces the transformation $\langle\mathfrak{R}, H\rangle\rightarrow\langle\mathfrak{R}, H\eta\rangle$. It satisfies $\mathbf{T}_g/\mathfrak{H}_g=\mathbf{M}_g$. The set $\mathfrak{J}_g$ of elements of $\mathfrak{H}_g$ fixing every point of $\mathbf{T}_g$ consists only of the unity element if $g\geqslant 3$ and is a normal subgroup of order 2 if $g=1, 2$. For the remainder of this article we assume that $g\geqslant 2$. The case $g=1$ can be discussed similarly, and the result coincides with the classical one: $\mathbf{T}_1$ can be identified with the upper half-plane and $\mathfrak{H}_1/\mathfrak{J}_1$ is the †modular group.

Denote by $B(\mathfrak{R}_0)$ the set of measurable invariant forms $\mu\,\overline{dz}\,dz^{-1}$ with $\|\mu\|_\infty<1$. For every $\mu\in B(\mathfrak{R}_0)$ there exists a pair $(\mathfrak{R}, H)$ for which some $h\in H$ satisfies $h_{\bar z}=\mu h_z$ (→ 352 Quasiconformal Mappings). This correspondence determines a surjection $\mu\in B(\mathfrak{R}_0)\mapsto\langle\mathfrak{R}, H\rangle\in\mathbf{T}_g$. Next, if $Q(\mathfrak{R}_0)$ denotes the space of holomorphic quadratic differentials $\varphi\,dz^2$ on $\mathfrak{R}_0$, a mapping $\mu\in B(\mathfrak{R}_0)\mapsto\varphi\in Q(\mathfrak{R}_0)$ is obtained as follows: Consider μ on the universal covering space U ($=$upper half-plane) of $\mathfrak{R}_0$. Extend it to U^* ($=$lower half-plane) by setting $\mu=0$, and let f be a quasiconformal mapping f of the plane onto itself satisfying $f_{\bar z}=\mu f_z$. Take the †Schwarzian derivative $\psi=\{f, z\}$ of the holomorphic function f in U^*. The desired φ is given by $\varphi(z)=\overline{\psi(\bar z)}$ on U. It has been verified that two μ induce the same φ if and only if the same $\langle\mathfrak{R}, H\rangle$ corresponds to μ. Consequently, an injection $\langle\mathfrak{R}, H\rangle\in\mathbf{T}_g\mapsto\varphi\in Q(\mathfrak{R}_0)$ is obtained. Since $Q(\mathfrak{R}_0)=\mathbf{C}^{m(g)}$ by the Riemann-Roch theorem, this injection yields an embedding $\mathbf{T}_g\subset\mathbf{C}^{m(g)}$, where $\mathbf{T}_g$ is shown to be a domain.

As a subdomain of $\mathbf{C}^{m(g)}$, the Teichmüller space is an $m(g)$-dimensional complex analytic manifold. It is topologically equivalent to the unit ball in real $2m(g)$-dimensional space and is a bounded †domain of holomorphy in $\mathbf{C}^{m(g)}$.

Let $\{\alpha_1, \ldots, \alpha_{2g}\}$ be a 1-dimensional homology basis with integral coefficients in $\mathfrak{R}_0$ such that the intersection numbers are $(\alpha_i, \alpha_j)=(\alpha_{g+i}, \alpha_{g+j})=0$, $(\alpha_i, \alpha_{g+j})=\delta_{ij}$, $i, j=1, \ldots, g$.

Given an arbitrary $\langle\mathfrak{R}, H\rangle \in \mathbf{T}_g$, consider the †period matrix Ω of $\mathfrak{R}$ with respect to the homology basis $H\alpha_1, \ldots, H\alpha_{2g}$ and the basis $\omega_1, \ldots, \omega_g$ of †Abelian differentials of the first kind with the property that $\int_{H\alpha_i}\omega_j = \delta_{ij}$. Then Ω is a holomorphic function on $\mathbf{T}_g$. Furthermore, the analytic structure of the Teichmüller space introduced previously is the unique one (with respect to the topology defined above) for which the period matrix is holomorphic.

$\mathfrak{H}_g$ is a properly discontinuous group of analytic transformations, and therefore $\mathbf{M}_g$ is an $m(g)$-dimensional normal †analytic space. $\mathfrak{H}_g$ is known to be the whole group of the holomorphic automorphisms of $\mathbf{T}_g$ (Royden [8]); thus $\mathbf{T}_g$ is not a †symmetric space.

To every point τ of the Teichmüller space, there corresponds a Jordan domain $D(\tau)$ in the complex plane in such a way that the fiber space $\mathbf{F}_g = \{(\tau, z) \mid z \in D(\tau), \tau \in \mathbf{T}_g \subset \mathbf{C}^{m(g)}\}$ has the following properties: $\mathbf{F}_g$ is a bounded domain of holomorphy of $\mathbf{C}^{m(g)+1}$. It carries a properly discontinuous group $\mathfrak{G}_g$ of holomorphic automorphisms, which preserves every fiber $D(\tau)$ and is such that $D(\tau)/\mathfrak{G}_g$ is conformally equivalent to the Riemann surface corresponding to τ. $\mathbf{F}_g$ carries holomorphic functions $F_j(\tau, z)$, $j = 1, \ldots, 5g-5$ such that for every τ the functions $F_j/F_1, j = 2, \ldots, 5g-5$ restricted to $D(\tau)$ generate the meromorphic function field of the Riemann surface $D(\tau)/\mathfrak{G}_g$.

By means of the †extremal quasiconformal mappings, it can be verified that $\mathbf{T}_g$ is a complete metric space. The metric is called the **Teichmüller metric**, and is known to be a Kobayashi metric.

The Teichmüller space also carries a naturally defined Kähler metric, which for $g = 1$ coincides with the †Poincaré metric if $\mathbf{T}_1$ is identified with the upper half-plane. The †Ricci curvature, †holomorphic sectional cruvature, and †scalar curvature are all negative (Ahlfors [9]).

By means of the quasiconformal mapping f, which we considered previously in order to construct the correspondence $\mu \mapsto \varphi$, it is possible to regard the Teichmüller space as a space of quasi-Fuchsian groups (→ 234 Kleinian Groups). To the boundary of $\mathbf{T}_g$, it being a bounded domain in $\mathbf{C}^{m(g)}$, there correspond various interesting Kleinian groups, which are called †boundary groups (Bers [10], Maskit [11]).

The definition of Teichmüller spaces can be extended to open Riemann surfaces $\mathfrak{R}_0$ and, further, to those with signatures. A number of propositions stated above are valid to these cases as well. In particular, the Teichmüller space for the case where $\mathfrak{R}_0$ is the unit disk is called the **universal Teichmüller space**. It is a bounded domain of holomorphy in an infinite-dimensional Banach space and is a symmetric space. Every Teichmüller space is a subspace of the universal Teichmüller space.

References

[1] O. Teichmüller, Extremale quasikonforme Abbildungen und quadratische Differentiale, Abh. Preuss. Akad. Wiss., 1939.

[2] O. Teichmüller, Bestimmung der extremalen quasikonformen Abbildung bei geschlossenen orientierten Riemannschen Flächen, Abh. Preuss. Akad. Wiss., 1943.

[3] L. V. Ahlfors, The complex analytic structure of the space of closed Riemann surfaces, Analytic functions, Princeton Univ. Press, 1960, 45–66.

[4] L. V. Ahlfors, Lectures on quasiconformal mappings, Van Nostrand, 1966.

[5] L. Bers, Spaces of Riemann surfaces. Proc. Intern. Congr. Math., Edinburgh, 1958, 349–361.

[6] L. Bers, On moduli of Riemann surfaces, Lectures at Forschungsinstitut für Mathematik, Eidgenössische Technische Hochschule, Zürich, 1964.

[7] L. Bers, Uniformization, moduli, and Kleinian groups, Bull. London Math. Soc., 4 (1972), 257–300.

[8] H. L. Royden, Automorphisms and isometries of Teichmüller spaces, Advances in the Theory of Riemann Surfaces, Princeton Univ. Press, 1971, 369–383.

[9] L. V. Ahlfors, Curvature properties of Teichmüller's space, J. Analyse Math., 9 (1961), 161–176.

[10] L. Bers, On boundaries of Teichmüller spaces and on Kleinian groups I, Ann. Math., (2) 91 (1970), 570–600.

[11] B. Maskit, On boundaries of Teichmüller spaces and on Kleinian groups II, Ann. Math., (2) 91 (1970), 608–638.

417 (VII.5) Tensor Calculus

A. General Remarks

In a †differentiable manifold with an †affine connection (in particular, in a †Riemannian manifold), we can define an important operator on tensor fields, the operator of covariant differentiation. The **tensor calculus** is a differential calculus on a differentiable manifold that deals with various geometric objects and differential operators in terms of covariant differentiation, and it provides an important tool for studying geometry and analysis on a differentiable manifold.

B. Covariant Differential

Let M be an n-dimensional smooth manifold. We denote by $\mathfrak{F}(M)$ the set of all smooth functions on M and by $\mathfrak{X}^r_s(M)$ the set of all smooth tensor fields of type (r, s) on M. $\mathfrak{X}^1_0(M)$ is the set of all smooth vector fields on M, and we denote it simply by $\mathfrak{X}(M)$.

In the following we assume that an affine connection ∇ is given on M. Then we can define the **covariant differential** of tensor fields on M with respect to the connection (→ 80 Connections). We denote the **covariant derivative** of a tensor field K in the direction of a vector field X by $\nabla_X K$ and the covariant differential of K by ∇K. The operator ∇_X maps $\mathfrak{X}^r_s(M)$ into itself and has the following properties:

(1) $\nabla_{X+Y}=\nabla_X+\nabla_Y$, $\nabla_{fX}=f\nabla_X$,

(2) $\nabla_X(K+K')=\nabla_X K+\nabla_X K'$,

(3) $\nabla_X(K\otimes K')=(\nabla_X K)\otimes K'+K\otimes(\nabla_X K')$,

(4) $\nabla_X f=Xf$,

(5) ∇_X commutes with contraction of tensor fields, where K and K' are tensor fields on M, $X, Y\in\mathfrak{X}(M)$ and $f\in\mathfrak{F}(M)$.

The **torsion tensor** T and the **curvature tensor** R of the affine connection ∇ are defined by

$$T(X,Y)=\nabla_X Y-\nabla_Y X-[X,Y],$$

$$R(X,Y)Z=\nabla_X(\nabla_Y Z)-\nabla_Y(\nabla_X Z)-\nabla_{[X,Y]}Z$$

for vector fields X, Y, and Z. The torsion tensor is of type $(1,2)$, and the curvature tensor is of type $(1,3)$. Some authors define $-R$ as the curvature tensor. We here follow the convention used in [1–6], while in [7, 8] the sign of the curvature tensor is opposite. The torsion tensor and the curvature tensor satisfy the identities

$$T(X,Y)=-T(Y,X),\qquad R(X,Y)=-R(Y,X),$$

$$R(X,Y)Z+R(Y,Z)X+R(Z,X)Y$$
$$=(\nabla_X T)(Y,Z)+(\nabla_Y T)(Z,X)+(\nabla_Z T)(X,Y)$$
$$+T(T(X,Y),Z)+T(T(Y,Z),X)$$
$$+T(T(Z,X),Y),$$

$$(\nabla_X R)(Y,Z)+(\nabla_Y R)(Z,X)+(\nabla_Z R)(X,Y)$$
$$=R(X,T(Y,Z))+R(Y,T(Z,X))$$
$$+R(Z,T(X,Y)).$$

The last two identities are called the **Bianchi identities**.

The operators ∇_X and ∇_Y for two vector fields X and Y are not commutative in general, and they satisfy the following formula, the **Ricci formula**, for a tensor field K:

$$\nabla_X(\nabla_Y K)-\nabla_Y(\nabla_X K)-\nabla_{[X,Y]}K=R(X,Y)\cdot K,$$

where in the right-hand side $R(X,Y)$ is regarded as a derivation of the tensor algebra $\sum_{r,s}\mathfrak{X}^r_s(M)$.

A **moving frame** of M on a neighborhood U is, by definition, an ordered set $(e_1,\ldots,e_n)$ of n vector fields on U such that $e_1(p),\ldots,e_n(p)$ are linearly independent at each point $p\in U$. For a moving frame $(e_1,\ldots,e_n)$ of M on a neighborhood U we define n differential 1-forms $\theta^1,\ldots,\theta^n$ by $\theta^i(e_j)=\delta^i_j$, and we call them the **dual frame** of $(e_1,\ldots,e_n)$. For a tensor field K of type (r,s) on M, we define n^{r+s} functions $K^{i_1\ldots i_r}_{j_1\ldots j_s}$ on U by

$$K^{i_1\ldots i_r}_{j_1\ldots j_s}=K(e_{j_1},\ldots,e_{j_s},\theta^{i_1},\ldots,\theta^{i_r})$$

and call these functions the components of K with respect to the moving frame $(e_1,\ldots,e_n)$.

Since the covariant differentials ∇e_j are tensor fields of type $(1,1)$, n^2 differential 1-forms ω^i_j are defined by

$$\nabla e_j=\omega^i_j\otimes e_i,$$

where in the right-hand side (and throughout the following) we adopt **Einstein's summation convention**: If an index appears twice in a term, once as a superscript and once as a subscript, summation has to be taken on the range of the index. (Some authors write the above equation as $de_j=\omega^i_j e_i$ or $De_j=\omega^i_j e_i$.) We call these 1-forms ω^i_j the **connection forms** of the affine connection with respect to the moving frame $(e_1,\ldots,e_n)$. The torsion forms Θ^i and the curvature forms Ω^i_j are defined by

$$\Theta^i=d\theta^i+\omega^i_j\wedge\theta^j,\qquad \Omega^i_j=d\omega^i_j+\omega^i_k\wedge\omega^k_j.$$

These equations are called the **structure equation** of the affine connection ∇. If we denote the components of the torsion tensor and the curvature tensor with respect to $(e_1,\ldots,e_n)$ by T^i_{jk} and R^i_{jkl} $(=\theta^i(R(e_k,e_l)e_j))$, respectively, then they satisfy the relations

$$\Theta^i=\frac{1}{2}T^i_{jk}\theta^j\wedge\theta^k,\qquad \Omega^i_j=\frac{1}{2}R^i_{jkl}\theta^k\wedge\theta^l.$$

Using these forms, the Bianchi identities are written as

$$d\Theta^i+\omega^i_j\wedge\Theta^j=\Omega^i_j\wedge\theta^j,$$

$$d\Omega^i_j+\omega^i_k\wedge\Omega^k_j-\omega^k_j\wedge\Omega^i_k=0.$$

Let K be a tensor field of type (r,s) on M and $K^{i_1\ldots i_r}_{j_1\ldots j_s}$ be the components of K with respect to $(e_1,\ldots,e_n)$. We define the covariant differential $DK^{i_1\ldots i_r}_{j_1\ldots j_s}$ and the covariant derivative $K^{i_1\ldots i_r}_{j_1\ldots j_s,k}$ by

$$DK^{i_1\ldots i_r}_{j_1\ldots j_s}=K^{i_1\ldots i_r}_{j_1\ldots j_s,k}\theta^k=dK^{i_1\ldots i_r}_{j_1\ldots j_s}+\sum_{v=1}^{r}K^{i_1\ldots a\ldots i_r}_{j_1\ldots j_s}\omega^{i_v}_a$$
$$-\sum_{v=1}^{s}K^{i_1\ldots i_r}_{j_1\ldots a\ldots j_s}\omega^a_{j_v},$$

Then $K^{i_1\dots i_r}_{j_1\dots j_s,k}$ are the components of ∇K with respect to the moving frame $(e_1,\dots,e_n)$. Some authors write $\nabla_k K^{i_1\dots i_r}_{j_1\dots j_s}$ instead of $K^{i_1\dots i_r}_{j_1\dots j_s}$ [5, 6].

Using components, the Bianchi identities are written as

$$R^h_{ijk}+R^h_{jki}+R^h_{kij}=T^h_{ij,k}+T^h_{jk,i}+T^h_{ki,j}$$
$$+T^h_{ai}T^a_{jk}+T^h_{aj}T^a_{ki}+T^h_{ak}T^a_{ij},$$
$$R^h_{ijk,l}+R^h_{ikl,j}+R^h_{ilj,k}=R^h_{iak}T^a_{jl}+R^h_{iaj}T^a_{lk}+R^h_{ial}T^a_{kj}.$$

The Ricci formula is written as

$$K^{i_1\dots i_r}_{j_1\dots j_s,kl}-K^{i_1\dots i_r}_{j_1\dots j_s,lk}=\sum_{v=1}^{r}R^{i_v}_{alk}K^{i_1\dots a\dots i_r}_{j_1\dots\dots\dots j_s}$$
$$-\sum_{v=1}^{s}R^a_{j_vlk}K^{i_1\dots\dots\dots i_r}_{j_1\dots a\dots j_s}$$
$$+T^a_{kl}K^{i_1\dots i_r}_{j_1\dots j_s,a}.$$

Let $(x^1,\dots,x^n)$ be a local coordinate system defined on a neighborhood U of M. Then $(\partial/\partial x^1,\dots,\partial/\partial x^n)$ is a moving frame of M on U, and we call it the **natural moving frame** associated with the coordinate system $(x^1,\dots,x^n)$. Components of a tensor field with respect to the natural moving frame $(\partial/\partial x^1,\dots,\partial/\partial x^n)$ are often called components with respect to the coordinate system $(x^1,\dots,x^n)$. We define an n^3 function Γ^i_{kj} on U by $\omega^i_j=\Gamma^i_{kj}dx^k$, where ω^i_j are the connection forms for the natural moving frame. Γ^i_{kj} are called the coefficients of the affine connection ∇. The components of the torsion tensor and the curvature tensor with respect to $(x^1,\dots,x^n)$ are given by

$$T^i_{jk}=\Gamma^i_{jk}-\Gamma^i_{kj},$$
$$R^h_{ijk}=\partial_j\Gamma^h_{ki}-\partial_k\Gamma^h_{ji}+\Gamma^a_{ki}\Gamma^h_{ja}-\Gamma^a_{ji}\Gamma^h_{ka},$$

where $\partial_i=\partial/\partial x^i$.

With respect to the foregoing coordinate system, the components $K^{i_1\dots i_r}_{j_1\dots j_s,k}$ of the covariant differential ∇K of a tensor field K of type (r,s) are given by

$$K^{i_1\dots i_r}_{j_1\dots j_s,j}=\partial_j K^{i_1\dots i_r}_{j_1\dots j_s}+\sum_{v=1}^{r}\Gamma^{i_v}_{ja}K^{i_1\dots a\dots i_r}_{j_1\dots j_s}$$
$$-\sum_{v=1}^{s}\Gamma^a_{jj_v}K^{i_1\dots i_r}_{j_1\dots a\dots j_s}.$$

C. Covariant Differential of Tensorial Forms

A **tensorial p-form** of type (r,s) on a manifold M is an alternating $\mathfrak{F}(M)$-multilinear mapping of $\overbrace{\mathfrak{X}(M)\times\dots\times\mathfrak{X}(M)}^{p}$ to $\mathfrak{X}^r_s(M)$. A tensorial p-form of type $(0,0)$ is a differential p-form in the usual sense. A tensorial p-form of type $(1,0)$ is often called a **vectorial p-form**.

If an affine connection ∇ is provided on M, we define the covariant differential of tensorial forms. Let α be a tensorial p-form of type (r,s). The covariant differential $D\alpha$ of α is a tensorial $(p+1)$-form of type (r,s) and is defined by

$$(p+1)D\alpha(X_1,\dots,X_{p+1})$$
$$=\sum_{i=1}^{p+1}(-1)^{i-1}\nabla_{X_i}(\alpha(X_1,\dots,\hat{X}_i,\dots,X_{p+1}))$$
$$+\sum_{i<j}(-1)^{i+j}\alpha([X_i,X_j],$$
$$X_1,\dots,\hat{X}_i,\dots,\hat{X}_j,\dots,X_{p+1}),$$

where $\hat{X}_i$ means that X_i is deleted. If α is of type $(0,0)$, $D\alpha$ coincides with the usual exterior differential $d\alpha$.

The simplest example of a tensorial form is the identity mapping of $\mathfrak{X}(M)$, which will be denoted by θ. Some authors write this vectorial form as dp or dx, where p or x expresses an arbitrary point of a manifold. We call θ the **canonical vectorial form** of M. The torsion tensor T can be regarded as a vectorial 2-form, and we have $2D\theta=T$. The curvature tensor R can be regarded as a tensorial 2-form of type $(1,1)$, i.e., $(X,Y)\to R(X,Y)\in\mathfrak{X}^1_1(M)$, and the Bianchi identities are written as $DT=R\wedge\theta$, $DR=0$, where the exterior product $R\wedge\alpha$ of R and a tensorial p-form α is defined by

$$(p+1)(p+2)(R\wedge\alpha)(X_1,\dots,X_{p+2})$$
$$=2\sum_{i<j}(-1)^{i+j-1}R(X_i,X_j)\alpha(X_1,\dots,\hat{X}_i,\dots,\hat{X}_j,$$
$$\dots,X_{p+2}).$$

In general, $2D^2\alpha=R\wedge\alpha$ holds for an arbitrary tensorial form α.

Let $(e_1,\dots,e_n)$ be a moving frame of M on a neighborhood U and $\theta^1,\dots,\theta^n$ be its dual frames. A tensorial p-form α of type (r,s) is written as

$$\alpha=\alpha^{i_1\dots i_r}_{j_1\dots j_s}\otimes e_{i_1}\otimes\dots\otimes e_{i_r}\otimes\theta^{j_1}\otimes\dots\otimes\theta^{j_s},$$

on U, where the $\alpha^{i_1\dots i_r}_{j_1\dots j_s}$ are the usual differential p-forms on U. We call them the components of α with respect to $(e_1,\dots,e_n)$. Then the components of $D\alpha$, which we denote by $D\alpha^{i_1\dots i_r}_{j_1\dots j_s}$, are given by

$$D\alpha^{i_1\dots i_r}_{j_1\dots j_s}=d\alpha^{i_1\dots i_r}_{j_1\dots j_s}+\sum_{v=1}^{r}\omega^{i_v}_a\wedge\alpha^{i_1\dots a\dots i_r}_{j_1\dots j_s}$$
$$-\sum_{v=1}^{s}\omega^a_{j_v}\wedge\alpha^{i_1\dots i_r}_{j_1\dots a\dots j_s}.$$

Then we have

$$D^2\alpha^{i_1\dots i_r}_{j_1\dots j_s}=\sum_{v=1}^{r}\Omega^{i_v}_a\wedge\alpha^{i_1\dots a\dots i_r}_{j_1\dots j_s}-\sum_{v=1}^{s}\Omega^a_{j_v}\wedge\alpha^{i_1\dots i_r}_{j_1\dots a\dots j_s}.$$

This is an expression of $2D^2\alpha=R\wedge\alpha$ in terms of components. The components of the canonical vectorial form θ are the dual forms $\theta^1,\dots,\theta^n$ of $(e_1,\dots,e_n)$, and we have $D\theta^i=\Theta^i$, which means that the components of $D\theta$ are the torsion forms Θ^i.

D. Tensor Fields on a Riemannian Manifold

Let (M, g) be an n-dimensional Riemannian manifold (→ 364 Riemannian Manifolds). The fundamental tensor g defines a one-to-one correspondence between vector fields and differential 1-forms. A differential 1-form α which corresponds to a vector field X is defined by $\alpha(Y)=g(X, Y)$ for any vector field Y. This correspondence is naturally extended to a one-to-one correspondence between $\mathfrak{X}^r_s(M)$ and $\mathfrak{X}^{r'}_{s'}(M)$, where $r+s=r'+s'$. Let $(e_1, \ldots, e_n)$ be a moving frame of M on a neighborhood U and g_{ij} be the components of g with respect to the moving frame. Let (g^{ij}) be the inverse matrix of the matrix (g_{ij}). The g^{ij} are the components of a symmetric contravariant tensor field of order 2. Let X^i be the components of a vector field X and α_i be the components of the differential 1-form α corresponding to X. Then X^i and α_i satisfy the relations $\alpha_i=g_{ij}X^j$ and $X^i=g^{ij}\alpha_j$. If K^h_{ij} are the components of a tensor field K of type (1, 2) (here taken for simplicity), then

$$K_{hij}=K^a_{ij}g_{ah}, \quad K^{hi}_j=K^h_{aj}g^{ai},$$

$$K^{hij}=K^h_{ab}g^{ai}g^{bj}, \ldots,$$

are the components of a tensor field of type (0, 3), (2, 1), (3, 0), ..., respectively, all of which correspond to K. We call this process of obtaining the components of the corresponding tensor fields from the components of a given tensor field **raising the subscripts** and **lowering the superscripts** by means of the fundamental tensor g.

On a Riemannian manifold, we use the †Riemannian connection, unless otherwise stated. The covariant derivative with respect to the Riemannian connection is given by

$$\begin{aligned}2g(\nabla_X Y, Z)&=Xg(Y, X)+Yg(X, Z)-Zg(X, Y)\\&\quad+g([X, Y], Z)-g([X, Z], Y)\\&\quad-g(X, [Y, Z])\end{aligned}$$

for vector fields X, Y, and Z. The coefficients of the Riemannian connection with respect to a local coordinate system $(x^1, \ldots, x^n)$ are usually written as $\{{i \atop kj}\}$, called the **Christoffel symbols**, which are given by $\{{i \atop kj}\}=g^{ia}(\partial_k g_{ja}+\partial_j g_{ka}-\partial_a g_{kj})/2$. The curvature tensor R of the Riemannian connection satisfies the identities

$$R(X, Y)Z+R(Y, Z)X+R(Z, X)Y=0,$$

$$(\nabla_X R)(Y, Z)+(\nabla_Y R)(Z, X)+(\nabla_Z R)(X, Y)=0,$$

$$R(X, Y)=-R(Y, X),$$

$$\begin{aligned}g(R(X, Y)Z, W)&=g(R(Z, W)X, Y)\\&=-g(Z, R(X, Y)W),\end{aligned}$$

$$\begin{aligned}g(R(X, Y)Z, W)&+g(R(X, Z)W, Y)\\&+g(R(X, W)Y, Z)=0.\end{aligned}$$

In terms of the components, these identities are

$$R^h_{ijk}+R^h_{jki}+R^h_{kij}=0,$$

$$R^h_{ijk,l}+R^h_{ikl,j}+R^h_{ilj,k}=0,$$

$$R^h_{ijk}=-R^h_{ikj}, \; R_{hijk}=R_{jkhi}=-R_{ihjk},$$

$$R_{hijk}+R_{hjki}+R_{hkij}=0,$$

where $R_{hijk}=R^a_{ijk}g_{ah}$.

The †Ricci tensor S of the Riemannian manifold is a tensor field of type (0, 2) defined by

$$S(X, Y)=\text{trace of the mapping } Z\to R(Z, X)Y$$

for vector fields X and Y. The components S_{ji} of the Ricci tensor are given by $S_{ji}=R^a_{jai}$. The †scalar curvature k of the Riemannian manifold M is a scalar on M defined by $k=g^{ji}S_{ji}$. The Ricci tensor and the scalar curvature satisfy the identities

$$S(X, Y)=S(Y, X) \quad \text{or} \quad S_{ji}=S_{ij},$$

$$S_{ij,k}-S_{ik,j}=R^a_{ikj,a}, \qquad 2g^{jk}S_{ij,k}=\partial_i k.$$

For a moving frame of a Riemannian manifold, it is convenient to use an **orthonormal moving frame**. A moving frame $(e_1, \ldots, e_n)$ is orthonormal if $e_1, \ldots, e_n$ satisfy $g(e_i, e_j)=\delta_{ij}$. Since the components of the fundamental tensor with respect to an orthonormal moving frame are δ_{ij}, raising or lowering the indices does not change the values of the components. Some authors write all the indices as subscripts. Also they write the dual 1-forms, the connection forms, and the curvature forms as θ_i, ω_{ji}, and Ω_{ji}, respectively, instead of θ^i, ω^i_j, and Ω^i_j. With respect to an orthonormal moving frame, the connection forms ω^i_j and the curvature forms Ω^i_j satisfy

$$\omega^i_j+\omega^j_i=0 \quad \text{and} \quad \Omega^i_j+\Omega^j_i=0.$$

On a Riemannian manifold, the divergence of a vector field and the operators d, δ, and Δ on differential forms (→ 194 Harmonic Integrals) can be expressed by using the covariant derivatives with respect to the Riemannian connection.

If X^i are the components of a vector field X with respect to a local coordinate system $(x^1, \ldots, x^n)$, the divergence $\operatorname{div} X$ of X is given by $\operatorname{div} X=X^i{}_{,i}$.

Let α be a differential p-form on M. α is written locally in the form $\alpha=(1/p!)\alpha_{i_1\ldots i_p}dx^{i_1}\wedge \ldots \wedge dx^{i_p}$, where the coefficients $\alpha_{i_1\ldots i_p}$ are skew-symmetric in all the indices. We call $\alpha_{i_1\ldots i_p}$ the components of α with respect to the coordinate system. Since α is regarded as an alternating tensor field of type $(0, p)$, we can define the covariant differential $\nabla\alpha$ of α. Then the components of $d\alpha$, $\delta\alpha$, and $\Delta\alpha$ are given by

$$(d\alpha)_{i_1\ldots i_{p+1}}=\sum_{v=1}^{p+1}(-1)^{v-1}\alpha_{i_1\ldots\hat{i_v}\ldots i_{p+1},i_v},$$

$$(\delta\alpha)_{i_1\ldots i_{p-1}}=-g^{ab}\alpha_{ai_1\ldots i_{p-1},b},$$

$$(\Delta\alpha)_{i_1\ldots i_p}=-g^{ab}\left[\alpha_{i_1\ldots i_p,ab}-\sum_{v=1}^{p}S_{i_v a}\alpha_{i_1\ldots b\ldots i_p}\right.$$
$$\left.-\sum_{v<w}R^c_{ai_vi_w}\alpha_{i_1\ldots b\ldots c\ldots i_p}\right].$$

For a smooth function f and a differential 1-form β we have

$$\Delta f=-\frac{1}{\sqrt{g}}\partial_i(g^{ij}\sqrt{g}\,\partial_j f),$$

$$(\Delta\beta)_i=-g^{ab}[\beta_{i,ab}-S_{ia}\beta_b],$$

where $g=\det(g_{ij})$.

E. Van der Waerden–Bortolotti Covariant Differential

Let E be a finite dimensional smooth †vector bundle over a smooth manifold M and $\Gamma(E)$ be an $\mathfrak{F}(M)$-module of all smooth sections of E. A connection ∇' in E is a mapping of $\mathfrak{X}(M)\times\Gamma(E)$ to $\Gamma(E)$ such that

(1) $\nabla'_X(\xi+\eta)=\nabla'_X\xi+\nabla'_X\eta$,

(2) $\nabla'_X(f\xi)=Xf\cdot\xi+f\nabla'_X\xi$,

(3) $\nabla'_{X+Y}\xi=\nabla'_X\xi+\nabla'_Y\xi$,

(4) $\nabla'_{fX}\zeta=f\nabla'_X\xi$,

for $X,Y\in\mathfrak{X}(M)$, $\xi,\eta\in\Gamma(E)$, and $f\in\mathfrak{F}(M)$. $\nabla'_X\xi$ is called the covariant derivative of ξ in the direction X.

An element K of $\mathfrak{X}^r_s(M)\otimes\Gamma(E)$ is called a **tensor field of type** (r,s) **with values in** E (or simply an E-valued tensor field of type (r,s)). K can be regarded as an $\mathfrak{F}(M)$-linear mapping of $\mathfrak{X}^s_r(M)$ to $\Gamma(E)$ or an $\mathfrak{F}(M)$-multilinear mapping of $\overbrace{\mathfrak{X}(M)\times\ldots\times\mathfrak{X}(M)}^{s}$ to $\mathfrak{X}^r_0(M)\otimes\Gamma(E)$. For a given $\xi\in\Gamma(E)$, a mapping $X\to\nabla'_X\xi$ defines a tensor field of type $(0,1)$ with values in E which we call the covariant differential of ξ, denoted by $\nabla'\xi$.

The curvature tensor R' of ∇' is a tensor field of type $(0,2)$ with values in $E^*\otimes E$ (E^* is the dual vector bundle of E), and is defined by

$$R'(X,Y)\xi=\nabla'_X(\nabla'_Y\xi)-\nabla'_Y(\nabla'_X\xi)-\nabla'_{[X,Y]}\xi$$

for any vector fields X and Y and any $\xi\in\Gamma(E)$.

If an affine connection ∇ is given on M, we can define the **van der Waerden–Bortolotti covariant derivative** $\overline{\nabla}_X K$ for ∇ and ∇' of a tensor field K of type (r,s) with values in E. It is defined by

$$(\overline{\nabla}_X K)(S)=\nabla'_X(K(S))-K(\nabla_X S)$$

for any $S\in\mathfrak{X}^s_r(M)$. If we regard $\xi\in\Gamma(E)$ as an E-valued tensor field of type $(0,0)$, we have $\overline{\nabla}_X\xi=\nabla'_X\xi$. The covariant derivative $\overline{\nabla}_X R'$ of the curvature tensor R' of ∇' is a tensor field of type $(0,2)$ with values in $E^*\otimes E$ is defined by

$$(\overline{\nabla}_X R')(Y,Z)\xi=\nabla'_X(R'(Y,Z)\xi)-R'(\nabla_X Y,Z)\xi$$
$$-R'(Y,\nabla_X Z)\xi-R'(Y,Z)\nabla'_X\xi.$$

The Bianchi identity is written as

$$(\overline{\nabla}_X R')(Y,Z)+(\overline{\nabla}_Y R')(Z,X)+(\overline{\nabla}_Z R')(X,Y)$$
$$=R'(X,T(Y,Z))+R'(Y,T(Z,X))$$
$$+R'(Z,T(X,Y)),$$

where T is the torsion tensor of ∇. The Ricci formula is given by

$$(\overline{\nabla}_X(\overline{\nabla}_Y K))(S)-(\overline{\nabla}_Y(\overline{\nabla}_X K))(S)-(\overline{\nabla}_{[X,Y]}K)(S)$$
$$=R'(X,Y)\cdot K(S)-K(R(X,Y)\cdot S),$$

where R is the curvature tensor of ∇, $K\in\mathfrak{X}^r_s(M)\otimes\Gamma(E)$ and $S\in\mathfrak{X}^s_r(M)$.

In the following we assume that the fiber of E is of finite dimension m. A moving frame of E on a neighborhood U of M is an ordered set $(\xi_1,\ldots,\xi_m)$ of local sections $\xi_1,\ldots,\xi_m$ on U such that $\xi_1(p),\ldots,\xi_m(p)$ are linearly independent at each point p of U. Let $(e_1,\ldots,e_n)$ be a moving frame of M on U. Then an E-valued tensor field K of type (r,s) is locally written as

$$K^{i_1\ldots i_r\alpha}_{j_1\ldots j_s}e_{i_1}\otimes\ldots\otimes e_{i_r}\otimes\theta^{j_1}\otimes\ldots\otimes\theta^{j_s}\otimes\xi_\alpha,$$

where $\theta^1,\ldots,\theta^n$ are the dual 1-forms of $(e_1,\ldots,e_n)$. The $n^{r+s}m$ functions $K^{i_1\ldots i_r\alpha}_{j_1\ldots j_s}$ on U are called the components of K with respect to $(e_1,\ldots,e_n)$ and $(\xi_1,\ldots,\xi_m)$. We define the connection forms ω'^α_β of the connection ∇' by $\nabla'\xi_\beta=\omega'^\alpha_\beta\otimes\xi_\alpha$. Then the curvature forms Ω'^α_β are defined by

$$\Omega'^\alpha_\beta=d\omega'^\alpha_\beta+\omega'^\alpha_\lambda\wedge\omega'^\lambda_\beta=\tfrac{1}{2}R^\alpha_{\beta ji}\theta^j\wedge\theta^i,$$

where $R^\alpha_{\beta ji}$ are the components of the curvature tensor R', i.e., $R'(e_j,e_i)\xi_\beta=R^\alpha_{\beta ji}\xi_\alpha$.

For a given tensor field K of type (r,s) with values in E, the mapping $X\to\overline{\nabla}_X K$ defines a tensor field $\overline{\nabla}K$ of $(r,s+1)$ with values in E which we call the van der Waerden–Bortollotti covariant differential of K. Then if $K^{i_1\ldots i_r\alpha}_{j_1\ldots j_s}$ are the components of K with respect to $(e_1,\ldots,e_n)$ and $(\xi_1,\ldots,\xi_m)$, the components $K^{i_1\ldots i_r\alpha}_{j_1\ldots j_s,k}$ of $\overline{\nabla}K$ are given by

$$K^{i_1\ldots i_r\alpha}_{j_1\ldots j_s,k}\theta^k=dK^{i_1\ldots i_r\alpha}_{j_1\ldots j_s}+\sum_{v=1}^{r}K^{i_1\ldots a\ldots i_r\alpha}_{j_1\ldots j_s}\omega^{i_v}_a$$
$$-\sum_{v=1}^{s}K^{i_1\ldots i_r\alpha}_{j_1\ldots a\ldots j_s}\omega^a_{j_v}+K^{i_1\ldots i_r\beta}_{j_1\ldots j_s}\omega'^\alpha_\beta.$$

Let f be a smooth mapping of M into a smooth manifold M'. The differential f_*(or df) can be regarded as a tensor field of type $(0,1)$ with values in $f^*T(M')$. Assume that M (resp. M') has a Riemannian metric g (resp. g'). We denote the Riemannian connection of M by ∇.

From the Riemannian connection of M' a connection ∇' in $f^*T(M')$ can be defined. Let $(y^1, \ldots, y^m)$ be a local coordinate system of M' on a neighborhood V and $(x^1, \ldots, x^n)$ be a local coordinate system on a neighborhood U of M such that $f(U) \subset V$. Put $\xi_\alpha(p) = (\partial/\partial y^\alpha)(f(p))$ for a point $p \in U$. Then $(\xi_1, \ldots, \xi_m)$ is a moving frame of $f^*T(M')$. The components of f_* with respect to $(\partial/\partial x^1, \ldots, \partial/\partial x^n)$ and $(\xi_1, \ldots, \xi_m)$ are given by $f^\alpha_i(p) = (\partial y^\alpha/\partial x^i)(p)$. The Laplacian Δf of the mapping f is a tensor field of type $(0,0)$ with values in $f^*T(M')$ and is defined by $(\Delta f)^\alpha = g^{ij} f^\alpha_{i,j}$. If $\Delta f = 0$, the mapping f is called a harmonic mapping (→ 195 Harmonic Mappings).

F. Tensor Fields on a Submanifold

Consider an n-dimensional smooth manifold M immersed in an $(n+m)$-dimensional Riemannian manifold $(\overline{M}, \overline{g})$. If we denote the immersion $M \to \overline{M}$ by f, then $g = f^*\overline{g}$ is a Riemannian metric on M, and we denote its Riemannian connection by ∇. The induced bundle $f^*T(\overline{M})$ splits into the sum of the tangent bundle $T(M)$ of M and the normal bundle $T^\perp(M)$. The Riemannian connection on $\overline{M}$ induces connections in $f^*T(\overline{M})$ and in $T^\perp(M)$ which are denoted by $\overline{\nabla}$ and $\nabla^\perp$, respectively. The van der Waerden-Bortolotti covariant derivative for ∇ and $\nabla^\perp$ is denoted by $\tilde{\nabla}$.

For vector fields X and Y on M, the tangential part of $\overline{\nabla}_X Y$ (here we regard Y as a section of $f^*T(\overline{M})$) is $\nabla_X Y$, and we denote the normal part of $\overline{\nabla}_X Y$ by $h(X, Y)$. Then h is a symmetric tensor field of type $(0,2)$ with values in $T^\perp(M)$, and we call h the **second fundamental tensor** of the immersion f. For $\xi \in \Gamma(T^\perp(M))$, the tangential part of $\overline{\nabla}_X \xi$ (here ξ is also regarded as a section of $f^*T(\overline{M})$) is denoted by $-A_\xi X$ and the normal part of $\overline{\nabla}_X \xi$ is $\nabla^\perp_X \xi$. Thus we have

$$\overline{\nabla}_X Y = \nabla_X Y + h(X, Y), \quad \overline{\nabla}_X \xi = -A_\xi X + \nabla_X \xi.$$

h and A are related by

$$\overline{g}(h(X, Y), \xi) = g(A_\xi X, Y).$$

We have the following formulas, called the equations of Gauss, Codazzi, and Ricci:

$$\begin{aligned}\overline{g}(\overline{R}(X, Y)Z, W) = g(R(X, Y)Z, W) \\ + \overline{g}(h(X, Z), h(Y, W)) \\ - \overline{g}(h(X, W), h(Y, X)),\end{aligned}$$

$$\begin{aligned}\overline{g}(\overline{R}(X, Y)Z, \xi) = \overline{g}((\tilde{\nabla}_X h)(Y, Z), \xi) \\ - \overline{g}((\tilde{\nabla}_Y h)(X, Z), \xi),\end{aligned}$$

$$\begin{aligned}\overline{g}(\overline{R}(X, Y)\xi, \eta) = \overline{g}(R^\perp(X, Y)\xi, \eta) \\ + g([A_\xi, A_\eta]X, Y),\end{aligned}$$

for $X, Y, Z, W \in X(M)$ and $\xi, \eta \in \Gamma(T^\perp(M))$, where R, $\overline{R}$, and $R^\perp$ are the curvature tensors of ∇, $\overline{\nabla}$, and $\nabla^\perp$, respectively.

For the manifold M immersed in $\overline{M}$, we use a moving frame $(e_1, \ldots, e_n, \xi_1, \ldots, \xi_m)$ such that $(e_1, \ldots, e_n)$ is an orthonormal moving frame of M on a neighborhood U and $(\xi_1, \ldots, \xi_m)$ is a moving frame of $T^\perp(M)$ on U with $\overline{g}(\xi_\alpha, \xi_\beta) = \delta_{\alpha\beta}$. Then we can define the connection forms ω^i_j for ∇ and ω^α_β for $\nabla^\perp$. If we extend $(e_1, \ldots, e_n, \xi_1, \ldots, \xi_m)$ to an orthonormal moving frame $(\overline{e}_1, \ldots, \overline{e}_{n+m})$ of $\overline{M}$ such that $\overline{e}_i(p) = e_i(p)$ $(i = 1, \ldots, n)$ and $\overline{e}_{n+\alpha}(p) = \xi_\alpha(p)$ $(\alpha = 1, \ldots, m)$ for $p \in U$, then the restriction $f^*\overline{\theta}^A$ and $f^*\overline{\omega}^A_B$ of the dual 1-forms and the connection forms of $\overline{M}$ with respect to $(\overline{e}_1, \ldots, \overline{e}_{n+m})$ satisfy the relations

$$f^*\overline{\theta}^i = \theta^i, \quad f^*\overline{\theta}^{n+\alpha} = 0, \quad f^*\overline{\omega}^i_j = \omega^i_j,$$

$$f^*\overline{\omega}^{n+\alpha}_{n+\beta} = \omega^\alpha_\beta, \quad f^*\overline{\omega}^{n+\alpha}_i = \sum_j h^\alpha_{ij}\theta^j,$$

where h^α_{ij} are the components of the second fundamental tensor h with respect to $(e_1, \ldots, e_n, \xi_1, \ldots, \xi_m)$.

The components $h^\alpha_{ij,k}$ of the covariant differential $\tilde{\nabla}h$ of h are defined by

$$h^\alpha_{ij,k}\theta^k = dh^\alpha_{ij} - h^\alpha_{aj}\omega^a_i - h^\alpha_{ia}\omega^a_j + h^\beta_{ij}\omega^\alpha_\beta.$$

In terms of the components, the equations of Gauss, Codazzi, and Ricci are given by

$$\overline{R}_{hijk} = R_{hijk} + \sum_\alpha (h^\alpha_{ij}h^\alpha_{hk} - h^\alpha_{ik}h^\alpha_{hj}),$$

$$\overline{R}^\alpha_{ijk} = h^\alpha_{ik,j} - h^\alpha_{ij,k},$$

$$\overline{R}^\alpha_{\beta jk} = R^\perp{}^\alpha{}_{\beta jk} - \sum_a (h^\alpha_{ja}h^\beta_{ak} - h^\beta_{ja}h^\alpha_{ak}).$$

Let $(x^1, \ldots, x^n)$ be a local coordinate system on a neighborhood U of M and $(y^1, \ldots, y^{n+m})$ be a local coordinate system on a neighborhood V of $\overline{M}$ such that $f(U) \subset V$. Regarding the differential f_* of the immersion f as a tensor field of type $(0,1)$ with values in $f^*T(\overline{M})$, we denote the components of f_* with respect to $(x^1, \ldots, x^n)$ and $(y^1, \ldots, y^{n+m})$ by B^A_i $(i = 1, \ldots, n;\ A = 1, \ldots, n+m)$. Then we have $B^A_i = \partial y^A/\partial x^i$. We denote by ∇' the van der Waerden-Bortolotti covariant derivative for ∇ and $\overline{\nabla}$. Then the components $B^A_{i,j}$ of $\nabla' f_*$ are given by

$$B^A_{i,j} = \partial_j B^A_i - \left\{{a \atop ji}\right\} B^A_a + B^C_j B^B_i \overline{\left\{{A \atop CB}\right\}},$$

where $\partial_j = \partial/\partial x^j$, $\left\{{h \atop ji}\right\}$, and $\overline{\left\{{A \atop CB}\right\}}$ are the Christoffel symbols of the Riemannian metrics g and $\overline{g}$, respectively.

Let $(\xi_1, \ldots, \xi_m)$ be an orthonormal moving frame of $T^\perp(M)$ on U and ξ^A_α be the components of ξ_α with respect to $(y^1, \ldots, y^{n+m})$. Then we have

$$B^A_{i,j} = h^\alpha_{ij}\xi^A_\alpha,$$

where h_{ij} are the components of the second

fundamental tensor with respect to $(\partial/\partial x^1, \ldots, \partial/\partial x^n)$ and $(\xi_1, \ldots, \xi_m)$.

A tensor field K with values in $T^{\perp}(M)$ can be regarded as a tensor field with values in $f^*T(M)$, and $\tilde{\nabla}K$ is the normal component of $\nabla' K$. For example, if we regard the second fundamental tensor h as a tensor field with values in $f^*T(\overline{M})$, the components of h with respect to the coordinates $(x^1, \ldots, x^n)$ and $(y^1, \ldots, y^{n+m})$ are equal to $B^A_{i,j}$, and we have

$$h^{\alpha}_{ij,k} = B^A_{i,jk} \xi^B_\alpha \bar{g}_{AB}.$$

References

[1] E. Cartan, Leçon sur la géometrie des espaces de Riemann, Gauthier-Villars, second edition, 1946; English translation, Geometry of Riemannian Spaces, Math-Sci Press, 1983.
[2] B.-Y. Chen, Geometry of submanifolds, Dekker, 1973.
[3] S. Helgason, Differential geometry, Lie groups, and symmetric spaces, Academic Press, 1978.
[4] S. Kobayashi and K. Nomizu, Foundations of differential geometry, Interscience, I, 1963; II, 1969.
[5] J. A. Schouten, Ricci-calculus, Springer, second edition, 1954.
[6] K. Yano, Integral formulas in Riemannian geometry, Dekker, 1970.
[7] G. de Rham, Variétés différentiables, Actualités Sci. Ind., Hermann, second edition, 1960.
[8] L. P. Eisenhart, Riemannian geometry, Princeton Univ. Press, second edition, 1949.
[9] R. L. Bishop and S. I. Goldberg, Tensor analysis on manifolds, Macmillan, 1968.

418 (IX.20) Theory of Singularities

A. Introduction

Let $f_1, f_2, \ldots, f_r$ be †holomorphic functions defined in an open set U of the complex space $\mathbf{C}^n$. Let X be the analytic set $f_1^{-1}(0) \cap \ldots \cap f_r^{-1}(0)$. Let $z_0 \in X$, and let $g_1, \ldots, g_s$ be a system of generators of the ideal $\mathcal{I}(X)_{z_0}$ of the germs of the holomorphic functions which vanish identically on a neighborhood of z_0 in X. z_0 is called a **simple point** of X if the matrix $(\partial g_i/\partial z_j)$ attains its maximal rank, say k, at $z = z_0$. In this case, X is a †complex manifold of dimension $n-k$ near z_0. Otherwise, z_0 is called a **singular point** of X.

B. Resolution of Singularities

Let X be a complex analytic space, and let Y be its singular locus. A **resolution of the singularity** of X is a pair of a complex manifold $\tilde{X}$ and a proper surjective holomorphic mapping $\pi: \tilde{X} \to X$ such that the restriction $\pi|_{\tilde{X}-\pi^{-1}(Y)}$ is biholomorphic and $\tilde{X} - \pi^{-1}(Y)$ is dense in $\tilde{X}$. H. Hironaka proved that there exists a resolution for any X such that $\pi^{-1}(Y)$ is a divisor in $\tilde{X}$ with only †normal crossings [16, 17].

Suppose that a compact connected analytic subset $\tilde{Y}$ of a complex manifold $\tilde{X}$ has a †strongly pseudoconvex neighborhood in $\tilde{X}$. Then the contraction $\tilde{X}/\tilde{Y}$ naturally has a structure of a †normal complex analytic variety such that the projection $\tilde{X} \to \tilde{X}/\tilde{Y}$ is a resolution of $\tilde{X}/\tilde{Y}$ (H. Grauert [14]).

C. Two-Dimensional Singularities

Let X be a normal 2-dimensional analytic space. Then the singular points of X are discrete.

Among the resolutions of X, there exists a unique resolution $\pi: \tilde{X} \to X$ with the following universal property: For any resolution $\pi': \tilde{X}' \to X$, there exists a unique mapping $\rho: \tilde{X}' \to \tilde{X}$ with $\pi' = \pi \circ \rho$. This resolution is called the **minimal resolution**.

Let $\pi: \tilde{X} \to X$ be a resolution of a singular point x of X, and let A_i $(i = 1, \ldots, m)$ be the irreducible components of $\pi^{-1}(x)$. The matrix $(A_i \cdot A_j)$ of the †intersection numbers is known to be negative definite (P. Du Val [12]).

The resolution $\pi: \tilde{X} \to X$ is called **good** if (i) each A_i is nonsingular, (ii) $A_i \cap A_j$ $(i \neq j)$ is at most one point and the intersection is transverse and (iii) no three A_i's meet at a point. For a given good resolution $\pi: \tilde{X} \to X$, we associate a diagram in which the vertices v_i $(i = 1, \ldots, m)$ correspond to A_i $(i = 1, \ldots, m)$ and v_i and v_j are joined by a segment if and only if $A_i \cap A_j \neq \varnothing$.

The **geometric genus** $p_g(X, x)$ of a singular point $x \in X$ is the dimension of the †stalk at x of the first direct image sheaf $R^1\pi_* \mathcal{O}_{\tilde{X}}$, where $\pi: \tilde{X} \to X$ is a resolution of $x \in X$ and $\mathcal{O}_{\tilde{X}}$ is the †structure sheaf of $\tilde{X}$. The definition is independent of the choice of the resolution, and $p_g(X, x)$ is a finite integer.

Among the positive cycles of the form $Z = \sum_{i=1}^n n_i A_i$ (i.e., $n_i \geqslant 0$) such that $Z \cdot A_i < 0$ for each $i = 1, \ldots, m$, there exists a smallest one Z_0, which is called the **fundamental cycle** [3].

(1) **Rational singularities**. A singular point x of X is called **rational** if $p_g(X, x) = 0$. (The singularity (X, x) is also called rational even when $\dim X \geqslant 3$ if the direct image sheaf $R^i \pi_* \mathcal{O}_{\tilde{X}} = 0$ for $i > 0$.)

For a rational singularity $x \in X$, the †multiplicity of X at x equals $-Z_0^2$ and the local embedding dimension of X at x is $-Z_0^2+1$. Hence a rational singularity with multiplicity 2, which is called a **rational double point**, is a hypersurface singularity. The following weighted homogeneous polynomials (→ Section D) give the complete list of the defining equations up to analytic isomorphism:

$A_n: x^{n+1}+y^2+z^2$,

weights $(1/(n+1),\ 1/2,\ 1/2)$, $n \geqslant 1$;

$D_n: x^{n-1}+xy^2+z^2$,

weights $(1/(n-1),\ (n-2)/2(n-1),\ 1/2)$, $n \geqslant 4$;

$E_6: x^4+y^3+z^2$,

weights $(1/4,\ 1/3,\ 1/2)$;

$E_7: x^3y+y^3+z^2$,

weights $(2/9,\ 1/3,\ 1/2)$;

$E_8: x^5+y^3+z^2$,

weights $(1/5,\ 1/3,\ 1/2)$,

where the labels appearing at the left are given according to the coincidence of the diagram of the respective minimal resolutions and the †Dynkin diagrams. Rational double points have many different characterizations [11].

The generic part of the singular locus of the unipotent variety of a †complex simple Lie group G (=the orbit of the subregular †unipotent elements in G) is locally expressed as the product of a rational double point and a polydisk. The †universal deformation of a rational double point and its †simultaneous resolution are constructed by restricting the following diagram on a transverse slice to the subregular unipotent orbit (Brieskorn [7]; [34]):

$$\begin{array}{ccc} Y & \longrightarrow & G \\ \downarrow & & \downarrow \\ T & \longrightarrow & T/W \end{array}$$

where T is a †Cartan subgroup of G with the action of the Weyl group W, $G \rightarrow T/W$ is the quotient mapping by the †adjoint action of G and $Y=\{(x,B) \mid x \in G$ and B is a †Borel subgroup of G with $x \in B\}$, and other morphisms are defined naturally so that the diagram commutes. Here, $Y \rightarrow T$ is the simultaneous resolution of the morphism $G \rightarrow T/W$.

(2) **Quotient singularities**. A singular point $x \in X$ is called a **quotient singularity** if there exists a neighborhood of x which is analytically isomorphic to an orbit space U/G, where U is a neighborhood of 0 in $\mathbf{C}^2$ and G is a finite group of analytic automorphisms of U with the unique fixed point 0. The quotient singularities are rational, and their resolutions have been well studied [6]. U/G has a rational double point at 0 if and only if G is conjugate to a nontrivial finite subgroup of $SU(2)$.

(3) **Elliptic singularities**. The singularity (X,x) is called **minimally elliptic** if $p_g(X,x)=1$ and (X,x) is Gorenstein [23]. The following are examples of minimally elliptic singularities.

A singular point $x \in X$ is called **simply elliptic** if the exceptional set A of the minimal resolution is a smooth †elliptic curve [33]. When $A^2=-1,\ -2,\ -3$, (X,x) is a hypersurface singularity given by the following weighted homogeneous polynomials:

$\tilde{E}_6: x^3+y^3+z^3+axyz$,

weights $(1/3,\ 1/3,\ 1/3)$, $A^2=-3$;

$\tilde{E}_7: x^4+y^4+z^2+axyz$,

weights $(1/4,\ 1/4,\ 1/2)$, $A^2=-2$;

$\tilde{E}_8: x^6+y^3+z^2+axyz$,

weights $(1/6,\ 1/3,\ 1/2)$, $A^2=-1$,

(4) **Cusp singularities**. A singular point $x \in X$ is called a **cusp singularity** if the exceptional set of the minimal resolution is either a single rational curve with a †node or a cycle of smooth rational curves. Cusp singularities appear as the boundary of †Hilbert modular surfaces [18]. The hypersurface cusp singularities are given by the polynomials

$T_{p,q,r}: x^p+y^q+z^r+axyz$,

where $1/p+1/q+1/r<1$ and $a \neq 0$.

D. The Milnor Fibration for Hypersurface Singularities

Let V be an analytic set in $\mathbf{C}^N$, and take a point $z_0 \in V$. Let $S_\varepsilon=S(z_0,\varepsilon)$ be a $(2N-1)$-dimensional sphere in $\mathbf{C}^N$ with center z_0 and radius $\varepsilon>0$, and let $K_\varepsilon=V \cap S_\varepsilon$. If ε is sufficiently small, the topological type of the pair $(S_\varepsilon, K_\varepsilon)$ is independent of ε [27]. By virtue of this fact, the study of singular points constitutes an important aspect of the application of topology to the theory of functions of several complex variables.

A singular point z_0 of V is said to be **isolated** if, for some open neighborhood W of z_0 in $\mathbf{C}^N$, $W \cap V-\{z_0\}$ is a smooth submanifold of $W-\{z_0\}$. In that case, K_ε is a closed smooth submanifold of S_ε, and the diffeomorphism type of $(S_\varepsilon, K_\varepsilon)$ is independent of (sufficiently small) $\varepsilon>0$. So far, the topological study of such singular points has been primarily focused on isolated singularities. When V is a plane curve, that is, $N=2$ and $r=1$, all the singular points of V are isolated, and the submanifold K_ε of the 3-sphere S_ε can be described as an iterated torus link, where type numbers are completely determined by the †Puiseaux expansion of the defining equation f of V at the point z_0 [5]. In 1961, D. Mumford, using a resolution argument, showed that if an algebraic surface V is †normal at z_0 and if the closed 3-manifold K_ε is simply connected, then K_ε is diffeomorphic to the 3-sphere and z_0 is nonsingular [29]. The following theorem in the higher-dimensional case is due to E. Brieskorn [8] (1966):

Every †homotopy $(2n-1)$-sphere $(n \neq 2)$ that is a boundary of a †π-manifold is diffeomorphic to the K_ε of some complex hypersurface defined by an equation of the form $f(z)=z_1^{a_1}+\ldots+z_{n+1}^{a_{n+1}}=0$ at the origin in $\mathbf{C}^{n+1}$, provided that $n \neq 2$. The hypersurface of this type is called the **Brieskorn variety**. Inspired by Brieskorn's method, J. W. Milnor developed topological techniques for the study of hypersurface singularities and obtained results such as the **Milnor fibering theorem**, which can be briefly stated as follows:

Suppose that V is defined by a single equation $f(z)=0$ in the neighborhood of $z_0 \in \mathbf{C}^{n+1}$. Then there is an associated smooth †fiber bundle $\varphi: S_\varepsilon - K_\varepsilon \rightarrow S^1$, where $\varphi(z)=f(z)/|f(z)|$ for $z \in S_\varepsilon - K_\varepsilon$. The fiber $F=\varphi^{-1}(p)$ $(p \in S^1)$ has the homotopy type of a finite CW-complex of dimension n, and K_ε is $(n-2)$-connected.

Suppose that z_0 is an isolated critical point of f. Then F has the homotopy type of a †bouquet of spheres of dimension n [27]. **The Milnor number** $\mu(f)$ of f is defined by the nth Betti number of F, and it is equal to $\dim_{\mathbf{C}} \mathcal{O}_{\mathbf{C}^{n+1}, z_0} / (\partial f/\partial z_1, \ldots, \partial f/\partial z_{n+1})$, where $\mathcal{O}_{\mathbf{C}^{n+1}, z_0}$ is the ring of the germs of analytic functions of $n+1$ variables at $z=z_0$. **The Milnor monodromy** h_* is the automorphism of $H_n(F)$ that is induced by the action of the canonical generator of the fundamental group of the base space S^1. The †Lefschetz number of h_* is zero if z^0 is a singular point of V. Let $\Delta(t)$ be the characteristic polynomial of h_*. Then K_ε is a homology sphere if and only if $\Delta(1)=\pm 1$ [27]. It is known that $\Delta(t)$ is a product of †cyclotomic polynomials.

The diffeomorphism class of $(S_\varepsilon, K_\varepsilon)$ is completely determined by the congruence class of the linking matrix $L(e_i, e_j)$ $(1 \leqslant i, j \leqslant \mu(f)$, where $e_1, \ldots, e_{\mu(f)}$ is an integral basis of $H_n(F)$ and $L(e_i, e_j)$ is the †linking number [21, 10].

The Milnor fibration is also described in the following way. Let $E(\varepsilon, \delta)$ be the intersection of $f^{-1}(D_\delta^*)$ and $B(\varepsilon)$, the open disk of radius ε and center z_0, where D_δ^* is $\{\eta \in \mathbf{C} \mid 0<|\eta|<\delta\}$. The restriction of f to $E(\varepsilon, \delta)$ is a †locally trivial fibration over D_δ^* if δ is sufficiently smaller than ε [27].

Let $f(z)$ be an analytic function; suppose that $f(0)=0$ and let $\sum_{p \in \mathbf{N}^{n+1}} a_p z^p$ be the Taylor expansion of f at $z=0$. Let $\Gamma_+(f)$ be the convex hull of the union of $\{p+(\mathbf{R}^+)^{n+1}\}$ for $p \in \mathbf{N}^{n+1} \subset \mathbf{R}^{n+1}$ with $a_p \neq 0$, where $\mathbf{R}^+=\{x \in \mathbf{R} \mid x \geqslant 0\}$, and let $\Gamma(f)$ be the union of compact faces of $\Gamma_+(f)$. We call $\Gamma(f)$ the **Newton boundary** of f in the **coordinates** $z_1, \ldots, z_{n+1}$. For a closed face Δ of $\Gamma(f)$ of any dimension, let $f_\Delta(z)=\sum_{p \in \Delta} a_p z^p$. We say that f has a **nondegenerate Newton boundary** if $(\partial f_\Delta/\partial z_1, \ldots, \partial f_\Delta/\partial z_{n+1})$ is a nonzero vector for any $z \in (\mathbf{C}^*)^{n+1}$ and any $\Delta \in \Gamma(f)$. Suppose that f has a nondegenerate Newton boundary and 0 is an isolated critical point of f. Then the Milnor fibration of f is determined by $\Gamma(f)$ and $\mu(f)$, and the characteristic polynomial can be explicitly computed by $\Gamma(f)$ [22, 38].

$f(z)$ is called **weighted homogeneous** if there exist positive rational numbers $r_1, \ldots, r_{n+1}$, which are called **weights**, such that $a_p=0$ if $\sum_{i=1}^{n+1} p_i r_i \neq 1$. An analytic function $f(z)$ with an isolated critical point at 0 is weighted homogeneous in suitable coordinates if and only if f belongs to the ideal $(\partial f/\partial z_1, \ldots, \partial f/\partial z_{n+1})$ (K. Saito [32]). Suppose that $f(z)$ is a weighted homogeneous polynomial with an isolated critical point at 0. Then the Milnor fibration of f is uniquely determined by the weights, and $\mu(f)=\prod_{i=1}^{n+1}\left(\frac{1}{r_i}-1\right)$. The surface $f^{-1}(0)$ for $n=2$ is a rational double point if and only if $\sum_{i=1}^{3} r_i>1$.

E. Unfolding Theory

An **unfolding** of a germ of an analytic function $f(z)$ at 0 is a germ of an analytic function $F(z, t)$, where $t \in \mathbf{C}^m$ (m is finite) such that $F(z, 0)=f(z)$. We assume that f has an isolated critical point at 0. Among all the unfoldings of f, there exists a universal one, in a suitable sense, that is unique up to a local analytic isomorphism. It is called the **universal unfolding** of f [36, 37, 26] (→ 51 Catastrophe Theory). Explicitly it can be given by $F(z, t)=f(z)+t_1 \varphi_1(z)+\ldots+t_\mu \varphi_\mu(z)$, where $\varphi_i(z)$ $(i=1, \ldots, \mu)$ are holomorphic functions which form a $\mathbf{C}$-basis of the Jacobi ring $\mathcal{O}_{\mathbf{C}^{n+1}, 0}/(\partial f/\partial z_1, \ldots, \partial f/\partial z_{n+1})$ $(\mu=\mu(f))$.

In the universal unfolding $F(z, t)$ of f, the set of points (z_0, t_0) such that $F(z, t_0)$ has an isolated critical point at z_0 with the Milnor number $\mu(f)$ and $F(z_0, t_0)=0$ forms an analytic set at $(z, t)=0$. **The modulus number** of f is the dimension of this set at 0. This set is sometimes called the **μ-constant stratum**. Let g be a germ of an analytic function. g is said to be **adjacent** to f (denoted by $f \rightarrow g$), if there exists a sequence of points $(z(m), t(m))$ in $\mathbf{C}^{n+1} \times \mathbf{C}^\mu$ that converges to the origin such that the term of $F(z, t(m))$ at $z(m)$ is equivalent to g. Adjacency relations are important for the understanding of the degeneration phenomena of functions. The unfolding theory can be considered in exactly the same way as that for the germ of a real-valued smooth function that is finitely determined [36, 26].

The germs of analytic functions with modulus number 0, 1, and 2 are called **simple**, **unimodular**, and **bimodular**, respectively. They were classified by V. I. Arnold [1] (→ Appendix A, Table 5.V). Simple germs correspond to the equations for the rational double points, and unimodular germs define simply elliptic singularities or cusp singularities. Every unimodular or bimodular germ defines a singularity with $p_g=1$.

F. Picard-Lefschetz Theory

Let $f(z)$ be a holomorphic function such that $f(0)=0$ and 0 is an isolated critical point with the Milnor number μ. Let $F(z,t)$ be a universal unfolding of f at 0. Let $f: E(\varepsilon,\delta)\to D_\delta^*$ be the Milnor fibration of f by the second description in Section D. There exists a positive number r and a codimension 1 analytic subset Δ (called the **bifurcation set**) of $B'(r)$, the open disk of radius r with the center 0 in the parameter space $\mathbf{C}^\mu$, such that for any $t_0\in B'(r)-\Delta$, $f_{t_0}=F|_{B(\varepsilon)\times t_0}$ has μ different nondegenerate critical points in $B(\varepsilon)$. Let $p_1,\dots,p_\mu$ be the critical points of f_{t_0}. For each p_i, one can choose local coordinates $(y_1,\dots,y_{n+1})$ so that $f_{t_0}(y)=f_{t_0}(p_i)+y_1^2+\dots+y_{n+1}^2$. Such an f_{t_0} is called a **Morsification** of f.

Let B_i be a small disk with center p_i in $\mathbf{C}^{n+1}$. Then for any q_i which is near enough to $f_{t_0}(p_i)$, the intersection $f_{t_0}^{-1}(q_i)\cap B_i$ is diffeomorphic to the tangent disk bundle of the sphere S^n. The **vanishing cycle** e_i is the corresponding n-dimensional homology class of $f_{t_0}^{-1}(q_i)\cap B_i$. (We fix q_i.) The self-intersection number of e_i is given by

$$\langle e_i,e_i\rangle=\begin{cases}2(-1)^{n(n-1)/2}, & n \text{ even},\\ 0, & n \text{ odd}.\end{cases}$$

For a sufficiently small $t_0\in B'(r)-\Delta$, one has the following: (i) $|f_{t_0}(p_i)|<\delta$; (ii) the restriction of f_{t_0} to E is a fiber bundle over D', where $D'=\{w\in\mathbf{C}\,|\,|w|\leqslant\delta$, and $w\neq f_{t_0}(p_i)$ for $i=1,\dots,\mu\}$ and $E=f_{t_0}^{-1}(D')\cap B(\varepsilon)$; (iii) the restriction of the above fibration to $\{w\,|\,|w|=\delta\}$ is equivalent to the restriction of the Milnor fibration of f to $\{w\,|\,|w|=\delta\}$. Let w_0 be a fixed point of D', and let $F=f_{t_0}^{-1}(w_0)\cap E$. Then F is diffeomorphic to the Milnor fiber of f. Let l_i be a simple path from w_0 to q_i, and let γ_i be the loop $|w-f_{t_0}(p_i)|=|q_i-f_{t_0}(p_i)|$. We suppose that the union of the l_i is contractible to w_0. By parallel translation of the vanishing cycle e_i along l_i, we consider $e_i\in H_n(F)$. The collection $\{e_i\,|\,i=1,\dots,\mu\}$ is an integral basis of $H_n(F)$, which is called a **strongly distinguished basis** (Fig. 1).

Now let h_i be the linear transformation of $H_n(F)$ that is induced by the parallel translation along $l_i\gamma_i l_i^{-1}$. The **Picard-Lefschetz formula** says that

$$h_i(e)=e-(-1)^{n(n-1)/2}\langle e,e_i\rangle\cdot e_i \quad \text{for } e\in H_n(F).$$

Here $\langle\ ,\ \rangle$ is the intersection number in $H_n(F)$. For n even, h_i is a †reflection.

The Milnor monodromy h_* of f is equal to the composition $h_1\dots h_\mu$ under a suitable ordering of the h_i. The subgroup of the group of linear isomorphisms of $H_n(F)$ generated by $h_1,\dots,h_\mu$ is called the **total monodromy group**.

When f is a simple germ and $n\equiv 2 \bmod 4$, the total monodromy group is isomorphic to the †Weyl group of the corresponding Dynkin diagram. Even-dimensional simple singularities are the only ones for which the monodromy group is finite. These are also characterized as the singularities with definite intersection forms.

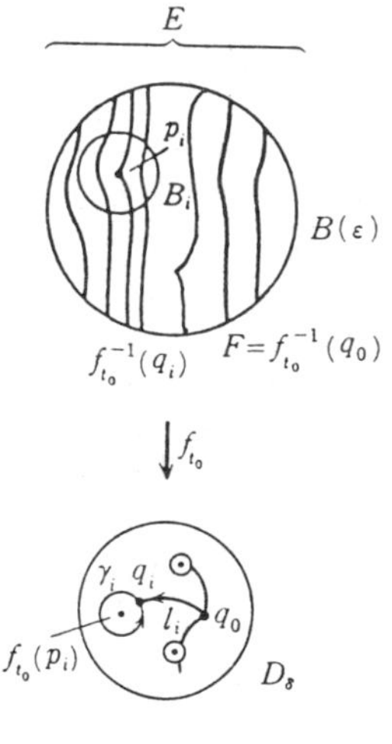

Fig. 1

G. Stratification Theory

The notion of **Whitney stratification** was first introduced by H. Whitney to study the singularities of analytic varieties [39] and was developed by R. Thom for the general case [37].

Let X and Y be submanifolds of the space $\mathbf{R}^n$. We say that the pair (X,Y) satisfies the **Whitney condition (b) at a point** $y\in Y$ if the following holds: Let x_i $(i=1,2,\dots)$ and y_i $(i=1,2,\dots)$ be sequences in X and Y, respectively, that converge to y. Suppose that the tangent space $T_{x_i}X$ converges to a plane T in the corresponding Grassmannian space and the secant $\overline{x_iy_i}$ converges to a line L. Then $L\subset T$. We say that (X,Y) satisfies the **Whitney condition (b)** if it satisfies the Whitney condition (b) at any point $y \in Y$. Let h be a local diffeomorphism of a neighborhood of y. One can see that $(h(X), h(Y))$ satisfies the Whitney condition (b) at $h(y)$ if (X, Y) satisfies it at y. Thus the Whitney condition can be considered for a pair of submanifolds X and Y of a manifold M using a local coordinate system. Let S be a subset of a manifold M, and let $\mathscr{S}$ be a family of submanifolds of M. $\mathscr{S}$ is called a **Whitney prestratification** of S if $\mathscr{S}$ is a locally finite disjoint cover of S satisfying the following: (i) For any $X \in \mathscr{S}$, the frontier $\bar{X} - X$ is a union of $Y \in \mathscr{S}$; (ii) for any pair (X, Y) $(X, Y \in \mathscr{S})$, the Whitney condition (b) is satisfied. A submanifold X in $\mathscr{S}$ is called a **stratum**. There exists a canonical partial order in $\mathscr{S}$ that is defined by $X < Y$ if and only if $X \subset \bar{Y} - Y$.

Let V be an analytic variety, and let $\mathscr{S}$ be an analytic stratification of V that satisfies the frontier condition (i). Then there exists a Whitney prestratification $\mathscr{S}'$ that is finer than $\mathscr{S}$ (Whitney [39]).

For a given Whitney prestratification $\mathscr{S}$, one can construct the following **controlled tubular neighborhood system**: For each $X \in \mathscr{S}$, a †tubular neighborhood $|T_X|$ of X in M and the projection $\pi_X: |T_X| \rightarrow X$ and a tubular function $\rho_X . |T_X| \rightarrow \mathbf{R}^+$ (=the square of a norm under the identification of $|T_X|$ with the †normal disk bundle of X) are given such that the commutation relations

$$\pi_X \cdot \pi_Y(m) - \pi_X(m), \ \rho_X \pi_Y(m) = \rho_X(m) \quad \text{for } m \in M, \ X < Y.$$

are satisfied whenever both sides are defined.

By virtue of this, the notions of vector fields and their integral curves can be defined on a Whitney prestratified set so that several important results on a differentiable manifold can be generalized to the case of stratified sets. For example, the following is **Thom's first isotopy lemma**: Let M and P be differentiable manifolds, and let $(S, \mathscr{S})$ be a Whitney prestratified subset of M. Let $f: S \rightarrow P$ be a continuous mapping that is the restriction of a differentiable mapping from M to P. Suppose that the restriction of f to each stratum X of $\mathscr{S}$ is a proper submersion onto P. Then $f: S \rightarrow P$ is a fiber bundle [37].

H. b-Functions

Let $f(z)$ be a germ of an analytic function in $\mathbf{C}^{n+1}$ with $f(0)=0$. The **b-function** of f at 0 is the monic polynomial $b_f(s)$ of lowest degree among all polynomials $b(s)$ with the following property [4, 20]: There exists a differential operator $P(z, \partial/\partial z, s)$, which is a polynomial in s, such that $b(s) f^s(z) = P(z, \partial/\partial z, s) f^{s+1}(z)$. Since $b_f(s)$ is always divisible by $s+1$, we define $\tilde{b}_f(s) = b_f(s)/(s+1)$. All the roots of $\tilde{b}_f(s) = 0$ are negative rational numbers (M. Kashiwara [20]. When f has an isolated critical point at 0, the set $\{\exp(2\pi i \alpha \mid \alpha$ is a root of $b_f(s)=0\}$ coincides with the set of eigenvalues of the Milnor monodromy [25].

The name "b-function" is due to M. Sato. He first introduced it in the study of †prehomogeneous vector spaces. Some authors call it the **Bernstein (Bernshteĭn) polynomial**.

I. Hyperplane Sections

Let V be an algebraic variety of complex dimension k in the complex projective space $\mathbf{P}^n$. Let L be a hyperplane that contains the singular points of V. Then the †relative homotopy group $\pi_i(V, V \cap L)$ is zero for $i < k$. Thus the same assertion is true for the †relative homology groups (S. Lefschetz [24]; [28]).

Let f be a holomorphic function defined in the neighborhood of $0 \in \mathbf{C}^{n+1}$ and $f(0)=0$. Let H be the hypersurface $f^{-1}(0)$. There exists a †Zariski open subset U of the space $(=\mathbf{P}^n)$ of hyperplanes such that for each $L \in U$, there exists a positive number ε such that $\pi_i(B(r) - H, (B(r) - H) \cap L) = 0$ for $i < n$ and $0 < r \leqslant \varepsilon$, where $B(r)$ is a disk of radius r (D. T. Lê and H. Hamm [15]). This implies the following theorem of Zariski: Let V be a hypersurface of $\mathbf{P}^n$, and let $\mathbf{P}^2$ be a general plane in $\mathbf{P}^n$. Then the fundamental group of $\mathbf{P}^n - V$ is isomorphic to the fundamental group of $\mathbf{P}^2 - C$, where $C = V \cap \mathbf{P}^2$. The fundamental group of $\mathbf{P}^2 - C$ is an Abelian group if C is a nodal curve [9, 13].

Suppose that f has an isolated critical point at 0. Let $\mu^{(n+1)}$ be the Milnor number $\mu(f)$. Take a generic hyperplane L. The Milnor number of $f|_L$ is well defined, and we let $\mu^{(n)} = \mu(f|_L)$. Similarly one can define $\mu^{(i)}$ of f and let $\mu^* = (\mu^{(n+1)}, \mu^{(n)}, \ldots, \mu^{(1)})$. Let $f_t(z)$ be a deformation of f. Each f_t has an isolated critical point at 0, and t is a point of a disk D of the complex plane. Let $W = \{(z, t) \mid f_t(z) = 0\}$ and $D' = \{0\} \times D$. $W - D'$ and D' satisfy the Whitney condition (b) if and only if $\mu^*(f_t)$ is invariant under the deformation [35]. The Whitney condition (b) implies topological triviality of the deformation.

References

[1] V. I. Arnol'd, Local normal forms of functions, Inventiones Math., 35 (1976), 87–109.
[2] V. I. Arnol'd, Normal forms for functions near degenerate critical points, the Weyl group of A_k, D_k, E_k and Lagrangian singularities, Functional Anal. Appl., 6 (1972), 254–272. (Original in Russian, 1972.)

[3] M. Artin, On isolated rational singularities of surfaces, Amer. J. Math., 88 (1966), 129–136.
[4] I. N. Bernshteĭn, The analytic continuation of generalized functions with respect to parameter, Functional Anal. Appl., 6 (1972), 273–285. (Original in Russian, 1972.)
[5] K. Brauer, Zur Geometrie der Funktionen zweier komplexen Veränderlichen III, Abh. Math. Sem. Hamburg, 6 (1928), 8–54.
[6] E. Brieskorn, Rationale Singularitäten Komplexer Flächen, Inventiones Math., 4 (1968), 336–358.
[7] E. Brieskorn, Singular elements of semisimple algebraic groups, Actes Congrès Int. Math., 1970, vol. 2, 279–284.
[8] E. Brieskorn, Beispiele zur Differentialtopologie, Inventiones Math., 2 (1966), 1–14.
[9] P. Deligne, Le groupe fondamental du complément d'une courbe plane n'ayant que des points doubles ordinaires est abelian, Sém. Bourbaki, no. 543, 1979–1980.
[10] H. Durfee, Fibered knots and algebraic singularities, Topology, 13 (1974), 47–59.
[11] A. H. Durfee, Fifteen characterizations of rational double points and simple critical points, Enseignement Math., 25 (1979), 131–163.
[12] P. Du Val, On isolated singularities of surfaces which do not affect the condition of adjunction I–III, Proc. Cambridge Philos. Soc., 30 (1933/34), 453–465, 483–491.
[13] W. Fulton, On the fundamental group of the complement of a node curve, Ann. Math., (2) 111, (1980), 407–409.
[14] H. Grauert, Über Modifikationen und exzeptionelle anaytische Mengen, Math. Ann., 146 (1962), 331–368.
[15] H. Hamm and D. T. Lê, Un théorème de Zariski du type de Lefschetz, Ann. Sci. Ecole Norm. Sup., 3, 1973.
[16] H. Hironaka, Resolution of singularities of an algebraic variety over a field of characteristic zero, Ann. Math., (2) 79 (1964), 109–326.
[17] H. Hironaka, Bimeromorphic smoothing of a complex analytic space, Acta Math. Vietnamica, 2, no. 2 (1977), 106–168.
[18] F. Hirzebruch, Hilbert modular surfaces, Enseignement Math., 19 (1973), 183–281.
[19] P. Holm, Real and complex singularities, Sijthoff and Noordhoff, 1976.
[20] M. Kashiwara, *B*-functions and holonomic systems, Inventiones Math., 38 (1976), 33–54.
[21] M. Kato, A classification of simple spinnable structure on a 1-connected Alexander manifold, J. Math. Soc. Japan, 26 (1974), 454–463.
[22] A. G. Kouchnirenko, Polyèdre de Newton et nombres de Milnor, Inventiones Math., 32 (1976), 1–31.
[23] H. Laufer, On minimally elliptic singularities, Amer. J. Math., 99 (1977), 1257–1295.
[24] S. Lefschetz, L'analysis situs et la géométrie algébrique, Gauthiers-Villars, 1924.
[25] B. Malgrange, Le polynome de Bernstein d'une singularité isolée, Lecture notes in math. 459, Springer, 1976, 98–119.
[26] J. N. Mather, Stability of C^∞-mappings. I, Ann. Math., (2) 87 (1968), 89–104; II, (2) 89 (1969), 254–291; III, Publ. Math. Inst. HES, 35 (1970), 301–336; IV, Lecture notes in math. 192, Springer, 1971, 207–253.
[27] J. W. Milnor, Singular points of complex hypersurfaces, Ann. math. studies 61, Princeton Univ. Press, 1968.
[28] J. W. Milnor, Morse theory, Ann. math. studies 51, Princeton Univ. Press, 1963.
[29] D. Mumford, The topology of normal singularities of an algebraic surface and a criterion for simplicity, Publ. Math. Inst. HES, 9 (1961), 5–22.
[30] F. Pham, Formules de Picard-Lefschetz généralisées et ramification des intégrales, Bull. Soc. Math. France, 93 (1965), 333–367.
[31] E. Picard and S. Simart, Traité des fonctions algébriques de deux variables I, Gauthier-Villars, 1897.
[32] K. Saito, Quasihomogene isolierte Singularitäten von Hyperflächen, Inventiones Math., 14 (1971), 123–142.
[33] K. Saito, Einfach elliptische Singularitäten, Inventiones Math., 23 (1974), 289–325.
[34] P. Slodowy, Simple singularities and simple algebraic groups, Lecture notes in math. 815, Springer, 1980.
[35] B. Teissier, Cycles evanescents, sections planes et conditions de Whitney, Astérisque, 78 (1973), 285–362.
[36] R. Thom, La stabilité topologique des applications polynomiales, Enseignement Math., 8 (1962), 24–33.
[37] R. Thom, Ensembles et morphismes stratifiés, Bull. Amer. Math. Soc., 75 (1969), 240–284.
[38] A. N. Varchenko, Zeta function of the monodromy and Newton's diagram, Inventiones Math., 37 (1976), 253–262.
[39] H. Whitney, Tangents to an analytic variety, Ann. Math., (2) 81 (1964), 496–549.
[40] O. Zariski, Algebraic surfaces, Erg. math. 61, Springer, 1971.

419 (XX.18) Thermodynamics

A. Basic Concepts and Postulates

Thermodynamics traditionally focuses its attention on a particular class of states of a given system called (thermal) equilibrium states, although a more recent extension, called the thermodynamics of irreversible processes, deals with certain nonequilibrium states. In a simple system, an **equilibrium state** is completely specified (up to the shape of the volume it occupies) by the volume V (a positive real number), the **mole numbers** $N_1, \ldots, N_r$ (nonnegative reals) of its chemical components, and the **internal energy** U (real). (More variables might be needed if the system were, e.g., inhomogeneous, anisotropic, electrically charged, magnetized, chemically not inert, or acted on by electric, magnetic, or gravitational fields.) This means that any of the quantities associated with equilibrium states (called **thermodynamical quantities**) of a simple system under consideration is a function of V, $N_1, \ldots, N_r$, and U.

When n copies of the same state are put next to each other and the dividing walls are removed, V, $N_1, \ldots, N_r$, and U for the new state will be n times the old values of these variables under the assumptions that each volume is sufficiently large and that the effects of the boundary walls can be neglected. Thermodynamical quantities behaving in this manner are called **extensive**. Those that are invariant under the foregoing procedure are called **intensive**. More precisely, the thermodynamic variables are defined by homogeneity of degree 1 and 0 as functions of V, $N_1, \ldots, N_r$, and U.

By a shift of the position of the boundary (called an **adiabatic wall** if energy and chemical substances do not move through it) or by transport of energy through the boundary (called a **diathermal wall** if this is allowed) or by transport of chemical components through the boundary (called a **permeable membrane**) (in short, by thermodynamical processes), these variables can change their values. If these shifts or transports are not permitted (especially for a composite system consisting of several simple systems, at its boundary with the outside), the system is called **closed**. Otherwise it is called **open**.

Those equilibrium states that do not undergo any change when brought into contact with each other across an immovable and impermeable diathermal wall (called a **thermal contact**) form an equivalence class. This is sometimes called the **0th law of thermodynamics**. The equivalence class, called the **temperature** of states belonging to it, is an intensive quantity.

The force needed to keep a movable wall at rest, divided by the area of the wall, is called the **pressure**. It is another intensive quantity. For a (slow) change of the volume by an amount dV under a constant pressure P, mechanical work of amount $-P\,dV$ is done on the system. Together with a possible change of the internal energy, say of amount dU, the amount

$$\delta Q = dU - P\,dV \tag{1}$$

of energy is somehow gained (if it is positive) or lost (if it is negative) by the system. This amount of energy is actually transported from or to a neighboring system through diathermal walls so that the total energy for a bigger closed composite system is conserved. This is called the **first law of thermodynamics**, and δQ is called the **heat gain** or **loss** by the system.

If two states of different temperatures T_1 and T_2 are brought into thermal contact, energy is transferred from one, say T_1, to the other (called heat transfer). This defines a binary class relation denoted by $T_1 > T_2$. The Clausius formulation of the **second law of thermodynamics** says that it is impossible to make a positive heat transfer from a state of lower temperature to another state of higher temperature without another change elsewhere. By considering a certain composite system, one reaches the conclusion that there exists a labeling of temperatures by positive real numbers T, called the **absolute temperature**, for which the following is an exact differential:

$$\delta Q/T = (dU - P\,dV)/T = dS. \tag{2}$$

The integral S is an extensive quantity, called the **entropy**. Furthermore, the sum of the entropies of component simple systems in an isolated composite system is nondecreasing during any thermodynamic process, and the following **entropy maximum principle** holds: An isolated composite system reaches an equilibrium at those values of extensive parameters that maximize the sum of the entropies of component simple systems (for constant total energy and volume and within the set of allowed states under a given constraint).

A relation expressing the entropy of a given system as a function of the extensive parameters (specifying equilibrium states) is known as the **fundamental relation** of the system. If it is given as a continuous and differentiable homogeneous function of V, $N_1, \ldots, N_r$, and U and is monotone increasing in U for fixed V, $N_1, \ldots, N_r$, then one can develop the thermodynamics of the system based on the above entropy maximum principle. A relation expressing an intensive parameter as a function of some other independent variables is called an **equation of state**.

Another postulate, which is much less frequently used, is the **Nernst postulate** or the **third law of thermodynamics**, which says that the entropy vanishes at the vanishing absolute temperature.

B. Various Coefficients and Relationships

The partial derivative $\partial/\partial x$ of a function $f(x, y, \ldots)$ with respect to the variable x with the variables $y, \ldots$ fixed is denoted by $(\partial f/\partial x)_{y,\ldots}$. We abbreviate $N_1, \ldots, N_r$ as N in the following.

If the fundamental relation is written as $U = U(V, N_1, \ldots, N_r, S)$ (instead of S being represented as a function of the other quantities), then (2) implies

$$(\partial U/\partial S)_{V,N} = T, \qquad (\partial U/\partial V)_{N,S} = -P.$$

The other first-order partial derivatives of U are

$$\mu_j = (\partial U/\partial N_j)_{V,N_1,\ldots,\hat{j},\ldots,N_r,S},$$

with μ_j called the **chemical potential** (or electrochemical potential) of the jth component.

If a system is surrounded by an adiabatic wall (i.e., the system is thermally isolated) and goes through a gradual reversible change (**quasistatic adiabatic process**), then the entropy has to stay constant. If a system is in thermal contact through a diathermal wall with a large system (called the **heat bath**) whose temperature is assumed to remain unchanged during the thermal contact, then the temperature of the system itself remains constant (an **isothermal process**). The decrease of the volume per unit increase of pressure under the latter circumstance is called the **isothermal compressibility** and is given by

$$\kappa_T = -V^{-1}(\partial V/\partial P)_{T,N}.$$

Under constant pressure, the increase of the volume per unit increase of the temperature is called the **coefficient of thermal expansion** and is given by

$$\alpha = V^{-1}(\partial V/\partial T)_{P,N}.$$

Under constant pressure, the amount of (quasistatic) heat transfer into the system per mole required to produce a unit increase of temperature is called the **specific heat at constant pressure** and is given by

$$c_P = N^{-1}T(\partial S/\partial T)_{P,N},$$

where $N = N_1 + \ldots + N_r$. The same quantity under constant volume is called the **specific heat at constant volume** and is given by

$$c_v = N^{-1}T(\partial S/\partial T)_{V,N}.$$

The positivity of c_v is equivalent to the convexity of energy as a function of entropy for fixed values of V and N.

Because of the first-order homogeneity of an extensive quantity as a function of other extensive variables, one can derive an **Euler relation**, such as

$$U = TS - PV + \mu_1 N_1 + \ldots + \mu_r N_r,$$

for a simple system. Its differential form implies the following **Gibbs-Duhem relation**:

$$S\,dT - V\,dP + N_1\,d\mu_1 + \ldots + N_r\,d\mu_r = 0.$$

Because of the identity

$$\left(\frac{\partial}{\partial x}\right)\left(\frac{\partial f}{\partial y}\right) = \left(\frac{\partial}{\partial y}\right)\left(\frac{\partial f}{\partial x}\right),$$

there arise relationships among second derivatives, known as the **Maxwell relations**:

$$(\partial T/\partial V)_{S,N} = -(\partial P/\partial S)_{V,N},$$

$$(\partial V/\partial S)_{P,N} = (\partial T/\partial P)_{S,N},$$

$$(\partial S/\partial V)_{T,N} = (\partial P/\partial T)_{V,N},$$

$$(\partial S/\partial P)_{T,N} = -(\partial V/\partial T)_{P,N}.$$

By computing the Jacobian of transformations of variables, further relations can be obtained. For example,

$$c_P = c_v + N^{-1}TV\alpha^2/\kappa_T.$$

C. Legendre Transform and Variational Principles

The **Legendre transform** of a function $f(x_1, \ldots, y_1, \ldots)$ relative to the variables x is given by

$$g(p_1, \ldots, y_1, \ldots) = f - \sum_j x_j p_j$$

as a function of the variables $p_j = \partial f/\partial x_j$ and y. The original variables x can be recovered as $-x_j = \partial g/\partial p_j$.

In terms of Legendre transforms, the entropy maximum principle can be reformulated in various forms:

Energy minimum principle: For given values of the total entropy and volume, the equilibrium is reached at those values of unconstrained parameters that minimize the total energy. This principle is applicable in reversible processes where the total entropy stays constant.

Helmholtz free energy minimum principle: For given values of the temperature (equal to that of a heat bath in thermal contact with the system) and the total volume, the equilibrium is reached at those values of the unconstrained parameters that minimize the total Helmholtz free energy, where the **Helmholtz free energy** for a simple system is defined as a function of $T, V, N_1, \ldots, N_r$ by

$$F = U - TS,$$

$$dF = -S\,dT - P\,dV + \mu_1\,dN_1 + \mu_r\,dN_r.$$

Enthalpy minimum principle: For given values of the pressure and the total entropy, the equilibrium is reached at those values of unconstrained parameters that minimize the total enthalpy, where the **enthalpy** for a simple system is defined as a function of S, P, $N_1, \ldots, N_r$ by

$$H = U + PV,$$

$$dH = T\,dS + V\,dP + \mu_1\,dN_1 + \ldots + \mu_r\,dN_r.$$

Gibbs free energy minimum principle: For constant temperature and pressure, the equilibrium is reached at those values of unconstrained parameters that minimize the total Gibbs free energy, where the **Gibbs free energy** for a simple system is given as a function of T, P, $N_1, \ldots, N_r$ by

$$G = U - TS + PV,$$

$$dG = -S\,dT + V\,dP + \mu_1\,dN_1 + \ldots + \mu_r\,dN_r.$$

References

[1] M. W. Zemansky, Heat and thermodynamics, McGraw-Hill, 1951.
[2] H. B. Callen, Thermodynamics, Wiley, 1960.
[3] J. W. Gibbs, The collected works of J. Willard Gibbs, vol. 1, Thermodynamics, Yale Univ. Press, 1948.
[4] M. Planck, Treatise on thermodynamics, Dover, 1945. (Original in German, seventh edition, 1922.)
[5] E. A. Guggenheim, Thermodynamics, North-Holland, 1949.
[6] F. C. Andrews, Thermodynamics: Principles and applications, Wiley, 1971.

420 (XX.8)
Three-Body Problem

A. *n*-Body Problem and Classical Integrals

In the ***n*-body problem**, we study the motions of n particles $P_i(x_i, y_i, z_i)$ $(i = 1, 2, \ldots, n)$ with arbitrary masses $m_i(>0)$ following †Newton's law of motion,

$$m_i \frac{d^2 w_i}{dt^2} = \frac{\partial U}{\partial w_i}, \qquad i = 1, 2, \ldots, n, \tag{1}$$

where w_i is any one of x_i, y_i, or z_i,

$$U = \sum_{i \neq j} k^2 m_i m_j / r_{ij},$$

with k^2 the gravitation constant, and

$$r_{ij} = \sqrt{(x_i - x_j)^2 + (y_i - y_j)^2 + (z_i - z_j)^2}.$$

Although the one-body and two-body problems have been completely solved, the problem has not been solved for $n > 2$. The **three-body problem** is well known and is important both in celestial mechanics and in mathematics. For $n > 3$ the problem is called the **many-body problem.**

The equations (1) have the so-called ten classical integrals, that is, the **energy integral** $\sum_i (m_i/2)((\dot{x}_i)^2 + (\dot{y}_i)^2 + (\dot{z}_i)^2) - U = \text{constant}$ $(\dot{w} = dw/dt)$, six **integrals of the center of mass** $\sum_i m_i \dot{w}_i = \text{constant}$, $\sum_i m_i w_i = (\sum_i m_i \dot{w}_i)t + \text{constant}$, and three **integrals of angular momentum** $\sum_i m_i(u_i \dot{w}_i - w_i \dot{u}_i) = \text{constant}$ $(u \neq w)$. Using these integrals and eliminating the time t and the ascending node by applying Jacobi's method, the order of the equations (1) can be reduced to $6n - 12$. H. Bruns proved that algebraic integrals cannot be found except for the classical integrals, and H. Poincaré showed that there is no other single-valued integral (Bruns, *Acta Math.*, 11 (1887); Poincaré [2, I, ch. 5]). These results are called **Poincaré-Bruns theorems**. Therefore we cannot hope to obtain general solutions for the equations (1) by †quadrature. General solutions for $n \geqslant 3$ have not been discovered except for certain specific cases.

B. Particular Solutions

Let r_i be the position vector of the particle P_i with respect to the center of mass of the n-body system. A configuration $r \equiv \{r_1, \ldots, r_n\}$ of the system is said to form a **central figure** (or **central configuration**) if the resultant force acting on each particle P_i is proportional to $m_i r_i$, where each proportionality constant is independent of i. The proportionality constant is uniquely determined as $-U/\sum_{i=1}^n m_i r_i^2$ by the configuration of the system. A configuration r is a central figure if and only if r is a †critical point of the mapping $r \mapsto U^2(r) \sum_{i=1}^n m_i r_i^2$ [5, 6]. A rotation of the system, in planar central figure, with appropriate angular velocity is a particular solution of the planar n-body problem.

Particular solutions known for the three-body problem are the **equilateral triangle solution** of Lagrange and the **straight line solution** of Euler. They are the only solutions known for the case of arbitrary masses, and their configuration stays in the central figure throughout the motion.

C. Domain of Existence of Solutions

The solutions for the three-body problem are analytic, except for the collison case, i.e., the case where $\min r_{ij} = 0$, in a strip domain enclosing the real axis of the t-plane (Poincaré, P.

Painlevé). K. F. Sundman proved that when two bodies collide at $t=t_0$, the solution is expressed as a power series in $(t-t_0)^{1/3}$ in a neighborhood of t_0, and the solution which is real on the real axis can be uniquely and analytically continued across $t=t_0$ along the real axis. When all three particles collide, the total angular momentum f with respect to the center of mass must vanish (and the motion is planar) (**Sundman's theorem**); so under the assumption $f\neq 0$, introducing $s=\int^t (U+1)\,dt$ as a new independent variable and taking it for granted that any binary collision is analytically continued, we see that the solution of the three-body problem is analytic on a strip domain $|\mathrm{Im}\, s|<\delta$ containing the real axis of the s-plane. The conformal mapping

$$\omega=(\exp(\pi s/2\delta)-1)/(\exp(\pi s/2\delta)+1)$$

maps the strip domain onto the unit disk $|\omega|<1$, where the coordinates of the three particles w_κ, their mutual distances r_{kl}, and the time t are all analytic functions of ω and give a complete description of the motion for all real time (Sundman, *Acta Math.*, 36 (1913); Siegel and Moser [7]).

When a triple collision occurs at $t=t_0$, G. Bisconcini, Sundman, H. Block, and C. L. Siegel showed that as $t\to t_0$, (i) the configuration of the three particles approaches asymptotically the Lagrange equilateral triangle configuration or the Euler straight line configuration, (ii) the collision of the three particles takes place in definite directions, and (iii) in general the triple-collision solution cannot be analytically continued beyond $t=t_0$.

D. Final Behavior of Solutions

Suppose that the center of mass of the three-body system is at rest. The motion of the system was classified by J. Chazy into seven types according to the asymptotic behavior when $t\to+\infty$, provided that the angular momentum f of the system is different from zero. In terms of the †order of the three mutual distances r_{ij} (for large t) these types are defined as follows:

(i) H^+: **Hyperbolic motion**. $r_{ij}\sim t$.
(ii) HP^+: **Hyperbolic-parabolic motion**. r_{13}, $r_{23}\sim t$ and $r_{12}\sim t^{2/3}$.
(iii) HE^+: **Hyperbolic-elliptic motion**. $r_{13}, r_{23}\sim t$ and $r_{12}<a$ (a=finite).
(iv) P^+: **Parabolic motion**. $r_{ij}\sim t^{2/3}$.
(v) PE^+: **Parabolic-elliptic motion**. $r_{13}, r_{23}\sim t^{2/3}$ and $r_{12}<a$.
(vi) L^+: **Lagrange-stable motion** or **bounded motion**. $r_{ij}<a$.
(vii) OS^+: **Oscillating motion**. $\overline{\lim}_{t\to\infty}\sup r_{ij}=\infty$, $\underline{\lim}_{t\to\infty}\sup r_{ij}<\infty$.

Define H^-, HE^-, etc. analogously but with $t\to-\infty$. There are three classes for each of the motions HP, HE, and PE, depending on which of the three bodies separates from the other two bodies and recedes to infinity, denoted by HP_i, HE_i, PE_i ($i=1,2,3$), respectively. The energy constant h is positive for H- and HP-motion, zero for P-motion, and negative for PE-, L-, and OS-motion. For HE-motion, h may be positive, zero, or negative.

We say that a **partial capture** takes place when the motion is H^- for $t\to-\infty$ and HE_i^+ for $t\to+\infty$ (for $h>0$), and a **complete capture** when the motion is HE_i^- for $t\to-\infty$ and L^+ for $t\to+\infty$ (for $h<0$). We say also that an **exchange** takes place when HE_i^- for $t\to-\infty$ and HE_j^+ for $t\to+\infty$ ($t\neq j$). The probability of complete capture in the domain $h<0$ is zero (J. Chazy, G. A. Merman).

E. Perturbation Theories

The radius of convergence in the s-plane for Sundman's solution is too small and the convergence is too slow in the ω-plane to make it possible to compute orbits of celestial bodies, and for that purpose a perturbation method is usually adopted. When the masses $m_2,\ldots,m_n$ are negligibly small compared with m_1 for the n-body problem, the motion of the nth body is derived as the solution of the two-body problem for m_1 and m_n by assuming $m_2=\ldots=m_{n-1}=0$ as a first approximation, and then the deviations of the true orbit from the ellipse are derived as †perturbations. In the **general theory of perturbations** the deviations are derived theoretically by developing a disturbing function, whereas in the **special theory of perturbations** they are computed by numerical integration. In general perturbation theory, problems concerning convergence of the solution are important, and it becomes necessary to simplify the disturbing function in dealing with the actual relations among celestial bodies. Specific techniques have to be developed in order to compute perturbations for lunar motion, motions of characteristic asteroids, and motions of satellites (e.g., the system of the Sun, Jupiter, and Jovian satellites).

F. The Restricted Three-Body Problem

Since the three-body problem is very difficult to handle mathematically, mathematical interest has been concentrated on the **restricted three-body problem** (in particular, the planar problem) since Hill studied lunar theory in the 19th century. For the restricted three-body problem, the third body, of zero mass, cannot have any influence on the motion of the other two bodies, which are of finite masses and which move uniformly on a circle around the center of mass. In the planar case, let us choose units so that the total mass, the angular velocity of the two bodies about their center of mass, and the gravitation constant are all equal to 1, and let (q_1, q_2) be the coordinates of the third body with respect to a rotating coordinate system chosen in such a way that the origin is at the center of mass and the two bodies of finite masses μ and $1-\mu$ are always fixed on the q_1-axis. Then the equations of motion for the third body are given by a Hamiltonian system:

$$\frac{dq_i}{dt}=\frac{\partial H}{\partial p_i}, \quad \frac{dp_i}{dt}=-\frac{\partial H}{\partial q_i}, \quad i=1,2, \tag{2}$$

with

$$H=\frac{1}{2}(p_1^2+p_2^2)+q_2p_1-q_1p_2-U(q_1,q_2),$$

$$U=\frac{1-\mu}{\sqrt{(q_1+\mu)^2+q_2^2}}+\frac{\mu}{\sqrt{(q_1+\mu-1)^2+q_2^2}}.$$

The equations (2) have the energy integral $H(p,q)=$constant, called **Jacobi's integral**. Siegel showed that there is no other algebraic integral, and it can be proved by applying Poincaré's theorem that there is no other single-valued integral. Regularization of the two singular points for the equations (2) and solutions passing through the singular points were studied by T. Levi-Civita, and solutions tending to infinity were studied by B. O. Koopman.

After reducing the number of variables by means of the Jacobi integral, the equations (2) give rise to a flow in a 3-dimensional manifold of which the topological type was clarified by G. D. Birkhoff (*Rend. Circ. Mat. Palermo*, 39 (1915)). Since this flow has an †invariant measure, the equations have been studied topologically, and important results for the restricted three-body problem, particularly on periodic solutions, have been obtained.

G. Stability of Equilateral Triangular Solutions

Suppose that the origin $q_i=p_i=0$ is an †equilibrium point for an autonomous Hamiltonian system with two degrees of freedom:

$$\frac{dq_i}{dt}=\frac{\partial H}{\partial p_i}, \quad \frac{dp_i}{dt}=-\frac{\partial H}{\partial q_i}, \quad i=1,2,$$

with the Hamiltonian H being analytic at the origin. When the †eigenvalues of the corresponding linearized system are purely imaginary and distinct, denoted by $\pm\lambda_1$, $\pm\lambda_2$, and $\lambda_1k_1+\lambda_2k_2\neq 0$ for $0<|k_1|+|k_2|\leqslant 4$ (where k_i is an integer), we can find suitable coordinates ξ_i, η_i so that the Hamiltonian H takes the form

$$H=\lambda_1\zeta_1+\lambda_2\zeta_2+\frac{1}{2}(c_{11}\zeta_1^2+2c_{12}\zeta_1\zeta_2+c_{22}\zeta_2^2)$$
$$+H_5+\dots$$

with $\zeta_i\equiv\xi_i\eta_i$ and real c_{ij}. It is necessary that $\eta_i=\sqrt{-1}\,\bar{\xi}_i$ for the solutions to be real. In addition, if the condition

$$D\equiv c_{11}\lambda_2^2-2c_{12}\lambda_1\lambda_2+c_{22}\lambda_1^2\neq 0$$

is satisfied, then the origin is a †stable equilibrium point of the original system (V. I. Arnol'd, J. Moser) [7].

For Lagrange equilateral triangular solutions of the planar restricted three-body problem, the eigenvalues λ of the linearized system derived from (2) are given as roots of the equation $\lambda^4+\lambda^2+(27/4)\mu(1-\mu)=0$ and are purely imaginary if $\mu(1-\mu)<1/27$. Applying the Arnol'd-Moser result, A. M. Leontovich and A. Deprit and Bartholomé showed that the Lagrange equilibrium points are stable for μ such that $0<\mu<\mu_0$, where μ_0 is the smaller root of $27\mu(1-\mu)=1$, excluding three values: μ_1, μ_2 at which $\lambda_1k_1+\lambda_2k_2=0$ $|k_1|+|k_2|\leqslant 4$ and μ_3 at which $D=0$.

Arnol'd proved that if the masses $m_2,\dots,m_n$ are negligibly small in comparison with m_1, the motion of the n-body system is †quasi-periodic for the majority of initial conditions for which the eccentricities and inclinations of the osculating ellipses are small.

References

[1] Y. Hagihara, Celestial mechanics. I, Dynamical principles and transformation theory, MIT Press, 1970; II, Perturbation theory, pts. 1, 2, MIT 1972; III, Differential equations in celestial mechanics, pts. 1, 2, Japan Society for the Promotion of Science, 1974; IV, Periodic and quasi-periodic solutions, pts. 1, 2, Japan Society, 1975; V, Topology of the three-body problem, pts. 1, 2, Japan Society, 1976.

[2] H. Poincaré, Les méthodes nouvelles de la mécanique céleste I–III, Gauthier-Villars, 1892–1899.

[3] E. T. Whittaker, A treatise on the analytical dynamics of particles and rigid bodies, Cambridge Univ. Press, fourth edition, 1937.

[4] H. Happel, Das Dreikörperproblem, Koehler, 1941.

[5] A. Wintner, The analytical foundations of celestial mechanics, Princeton Univ. Press, 1947.

[6] R. Abraham and J. E. Marsden, Foundations of mechanics, Benjamin, second edition, 1978.

[7] C. L. Siegel and J. Moser, Lectures on celestial mechanics, Springer, 1971.

[8] V. M. Alekseev (Alexeyev), Sur l'allure finale de mouvement dans le problème des trois corps, Actes Congrès Intern. Math., 1970, vol. 2, 893–907.
[9] V. Szebehely, Theory of orbits, Academic Press, 1967.

421 (XVIII.11) Time Series Analysis

A. Time Series

A time series is a sequence of observations ordered in time. Here we assume that measurements are quantitative and the times of measurements are equally spaced. We consider this sequence to be a realization of a stochastic process X_t (→ 407 Stochastic Processes). Usually time series analysis means a statistical analysis based on samples drawn from a stationary process (→ 395 Stationary Processes) or a related process. In what follows we denote the sample by $\mathbf{X}=(X_1, X_2, \ldots, X_T)'$.

B. Statistical Inference of the Autocorrelation

Let us assume X_t (t an integer) to be real-valued and weakly stationary (→ 395 Stationary Processes) and for simplicity $EX_t=0$ and consider the estimation of the **autocorrelation** $\rho_h=R_h/R_0$ of time lag h, where $R_l=EX_tX_{t+l}$. We denote the **sample autocovariance** of time lag h as

$$\tilde{R}_h=\frac{1}{T-|h|}\sum_{t=1}^{t-|h|}X_tX_{t+|h|},$$

and define the **serial correlation coefficient** of time lag h by $\tilde{\rho}_h=\tilde{R}_h/\tilde{R}_0$. It can be shown that the joint distribution of $\{\sqrt{T}(\tilde{\rho}_h-\rho_h) \mid 1\leqslant h\leqslant H\}$ tends to an H-dimensional †normal (Gaussian) distribution with mean vector $\mathbf{0}$, if one assumes that X_t is expressed as $X_t=\sum_{j=-\infty}^{\infty}b_j\xi_{t-j}$, where $\sum_{j=-\infty}^{\infty}|b_j|<+\infty$, $\sum_{j=-\infty}^{\infty}|j|^{1/2}b_j^2<+\infty$, and the ξ_t are independently and identically distributed random variables with $E\xi_t=0$ and $E\xi_t^4<+\infty$.

When X_t is an autoregressive process of order K (→ Section D) and also a †Gaussian process, it can be shown that the asymptotic distribution of $\{\sqrt{T}(\tilde{\rho}_h-\rho_h) \mid 1\leqslant h\leqslant K\}$ as $T\to\infty$ is equal to the asymptotic distribution of $\{\sqrt{T}(\hat{\rho}_h-\rho_h) \mid 1\leqslant h\leqslant K\}$, where $\hat{\rho}_h$ is the †maximum likelihood estimator of ρ_h. In general, it is difficult to obtain the maximum likelihood estimator of ρ_h. The statistical properties of other estimators of ρ_h, e.g., an estimator constructed by using $\operatorname{sgn}(X_t)$ ($\operatorname{sgn}(y)$ means 1 $(y>0)$, 0 $(y=0)$, -1 $(y<0)$) have also been investigated.

Testing hypotheses concerning autocorrelation can be carried out by using the above results. Let us now consider the problem of testing the hypothesis that X_t is a †white noise. Assume that X_t is a Gaussian process and that a white noise with $EX_t^2=\sigma^2$ exists, and define $\tilde{C}_h=\sum_{t=1}^{T}(X_t-\bar{X})(X_{t+h}-\bar{X})$ and $\tilde{\gamma}_h=\tilde{C}_h/\tilde{C}_0$ for $h\geqslant 0$, where $X_{T+j}=X_j$ and $\bar{X}=\sum_{t=1}^{T}X_t/T$. Then the probability density function of $\tilde{\gamma}_1$ can be obtained and it can be shown that

$$P(\tilde{\gamma}_1>\gamma)=\sum_{j=1}^{m}(\lambda_j-\gamma)^{(T-3)/2}\frac{1}{\Lambda_j}, \qquad \lambda_{m+1}\leqslant\gamma\leqslant\lambda_m,$$

where $\lambda_j=\cos 2\pi j/T$ and

$$\Lambda_j=\prod_{\substack{k=1\\(k\neq j)}}^{(T-1)/2}(\lambda_j-\lambda_k), \qquad T=3,5,\ldots,$$

$$\Lambda_j=\prod_{\substack{k=1\\(k\neq j)}}^{T/2-1}(\lambda_j-\lambda_k)\sqrt{1+\lambda_j}, \qquad T=4,6,\ldots,$$

$$1\leqslant m\leqslant(T-3)/2 \qquad \text{if } T \text{ is odd},$$

$$1\leqslant m\leqslant T/2-1 \qquad \text{if } T \text{ is even}.$$

This can be used to obtain a test of significance.

C. Statistical Inference of the Spectrum

To find the periodicities of a real-valued †weakly stationary process X_t with mean 0, the statistic, called the **periodogram**,

$$I_T(\lambda)=\left|\frac{1}{\sqrt{T}}\sum_{t=1}^{T}X_te^{-2\pi it\lambda}\right|^2$$

is used. If X_t is expressed as

$$X_t=\sum_{l=1}^{L}\{m_l\cos 2\pi\lambda_lt+m_l'\sin 2\pi\lambda_lt\}+Y_t,$$

where $\{m_l\}$, $\{m_l'\}$, and $\{Y_t\}$ are mutually independent random variables with $Em_l=Em_l'=0$ and $V(m_l)=V(m_l')=\sigma_l^2$ and $\{Y_t\}$ is independent and identically distributed with means 0 and finite variances σ^2, the distribution of $I_T(\lambda)$ converges to a distribution with finite mean and finite variance at $\lambda\neq\pm\lambda_l$ for $1\leqslant l\leqslant L$ when T tends to infinity. On the other hand, the magnitude of $I_T(\lambda)$ is of the order of T at $\lambda=\pm\lambda_l$, $1\leqslant l\leqslant L$. This means that we can find the periodicities of X_t by using $I_T(\lambda)$. When $X_t=Y_t$, we find that the distribution of $2I_T(\lambda)/\sigma^2$ (when $\lambda\neq 0, \pm 1/2$) or $I_T(\lambda)/\sigma^2$ (when $\lambda=0$ or $\pm 1/2$) tends to the †χ^2 distribution with degrees of freedom 2 or 1, respectively, and $I(\mu_1), I(\mu_2), \ldots, I(\mu_M)$ are asymptotically independent random variables for $0\leqslant|\mu_1|<|\mu_2|<\ldots<|\mu_M|\leqslant 1/2$ when $T\to\infty$. Applying this result, we can test for periods in the data.

Let $f(\lambda)$ be the spectral density function of a real-valued weakly stationary process X_t. In general, the variance of $|\sum_{t=1}^{T} X_t e^{-2\pi it\lambda}/\sqrt{T}|$ does not tend to 0 as T tends to infinity; hence $I_T(\lambda)$ cannot be used as a good estimator for the spectral density. To obtain an estimate of $f(\lambda)$, several estimators defined by using weight functions have been proposed by several authors. Let $W_T(\lambda)$ be a weight function defined on $(-\infty, \infty)$, and construct a statistic $\tilde{f}(\lambda)=\int_{-1/2}^{1/2} I_T(\mu) W_T(\lambda-\mu)\,d\mu$. Let us use $\tilde{f}(\mu)$ for the estimation of $f(\lambda)$. $W_T(\lambda)$ is called a **window**. An important class of $W_T(\lambda)$ is as follows. Let $W(\lambda)$ be continuous, $W(\lambda)=W(-\lambda)$, $W(0)=1$, $|W(\lambda)|<1$, and $\int_{-\infty}^{\infty} W(\lambda)^2\,d\lambda< +\infty$, and let H be a positive integer depending on T such that $H\to\infty$ and $H/T\to 0$ as $T\to\infty$. Putting $w_j=W(j/H)$, we define $W_T(\lambda)$ by $W_T(\lambda)=\sum_{j=-T+1}^{T-1} w_j e^{-2\pi ij\lambda}$. Then $\tilde{f}(\lambda)$ can be expressed as $\tilde{f}(\lambda)=\sum_{h=-T+1}^{T-1} \tilde{\tilde{R}}_h w_h e^{-2\pi ih\lambda}$, where $\tilde{\tilde{R}}_h=\sum_{t=1}^{T-h} X_{t+h}X_t/T$ for $h\geqslant 0$ and $\tilde{\tilde{R}}_h=\sum_{t=|h|+1}^{T} X_{t+h}X_t/T$ for $h<0$. Let X_t be stationary to the fourth order (→ 395 Stationary Processes) and satisfy

$$\sum_{h=-\infty}^{\infty} |R_h|< +\infty,$$

$$\sum_{h,l,p=-\infty}^{\infty} |C_{o,h,l,p}|< +\infty,$$

where $C_{o,h,l,p}$ is the fourth-order joint †cumulant of X_t, X_{t+h}, X_{t+l}, and X_{t+p}. Then we have

$$\lim_{T\to\infty} \frac{T}{H} V(\tilde{f}(0)) - 2f(0)^2 \int_{-\infty}^{\infty} W(\lambda)^2\,d\lambda,$$

$$\lim_{T\to\infty} \frac{T}{H} V(\tilde{f}(\pm 1/2))=2f(1/2)^2 \int_{-\infty}^{\infty} W(\lambda)^2\,d\lambda,$$

$$\lim_{T\to\infty} \frac{T}{H} V(\tilde{f}(\lambda))=f(\lambda)^2 \int_{-\infty}^{\infty} W(\lambda)^2\,d\lambda, \qquad \lambda\neq 0, \quad \pm 1/2,$$

$$\lim_{T\to\infty} \frac{T}{H} \operatorname{Cov}(\tilde{f}(\lambda),\ \tilde{f}(\mu))=0, \quad \lambda\neq\mu. \tag{1}$$

$\{w_h\}$ or $W_T(\lambda)$ should have an optimality, e.g., to minimize the mean square error of $\tilde{f}(\lambda)$. But, generally, it is difficult to obtain such a $\{w_h\}$ or $W_T(\lambda)$.

Several authors have proposed specific types of windows. The following are some examples: (i) (Bartlett) $w_h=(1-|h|/H)$ for $|h|\leqslant H$ and $w_h=0$ for $|h|>H$; (ii) (Tukey) $w_h=\sum_{l=-\infty}^{\infty} a_l\cos(\pi lh/H)$ for $|h|\leqslant H$ and $w_h=0$ for $|h|>H$, where the a_l are constants such that $\sum_{l=-\infty}^{\infty}|a_l|< +\infty$, $\sum_{l=-\infty}^{\infty} a_l=1$ and $a_l=a_{-l}$. The Hanning and Hamming windows are $a_0=0.50$, $a_1=a_{-1}=0.25$, and $a_l=0$ for $|l|\geqslant 2$ and $a_0=0.54$, $a_1=a_{-1}=0.23$, and $a_l=0$ for $|l|\geqslant 2$, respectively [2]. Let $X_t=\sum_{j=-\infty}^{\infty} b_j\varepsilon_{t-j}$, where $\sum_{j=-\infty}^{\infty}|b_j|< +\infty$ and the ε_t are independently and identically distributed random variables with $E\varepsilon_t=0$ and $E\varepsilon_t^4< +\infty$. Let $\{\lambda_j | 1\leqslant j\leqslant M\}$ be arbitrary real numbers such that $0\leqslant\lambda_1<\lambda_2<\ldots<\lambda_M\leqslant 1/2$, where M is an arbitrary positive integer. Then the joint distribution of $\{\sqrt{T/H}(\tilde{f}(\lambda_v)-E\tilde{f}(\lambda_v)) | 1\leqslant v\leqslant M\}$ tends to the normal distribution with means 0 and covariance matrix Σ, which is defined by (1). Let us assume, furthermore, that $\lim_{x\to 0}(1-w(x))/|x|^q = C$ and $\sum_{h=-\infty}^{\infty}|h|^p|R_h|< +\infty$, where C, q, and p are some positive constants satisfying the following conditions: (i) when $p\geqslant q$, $H^q/T\to 0$ $(p\geqslant 1)$ and $H^{q+1-p}/T\to 0$ $(p\geqslant 1)$ as $T\to\infty$ and $\lim_{T\to\infty} T/H^{2q+1}$ is finite; (ii) when $p<q$, $H^p/T\to 0$ $(p\geqslant 1)$ and $H/T\to 0$ $(p\leqslant 1)$ as $T\to\infty$ and $\lim_{T\to\infty} T/H^{2p+1}=0$. Then $\sqrt{T/H}(\tilde{f}(\lambda_v)-E\tilde{f}(\lambda_v))$ in the results above can be replaced by $\sqrt{T/H}(\tilde{f}(\lambda_v)-f(\lambda_v))$.

Estimation of higher-order spectra, particularly the bispectrum, has also been discussed. Let X_t be a weakly stationary process with mean 0, and let its spectral decomposition be given by $X_t=\int_{-1/2}^{1/2} e^{2\pi it\lambda}\,dZ(\lambda)$ (→ 395 Stationary Processes). We assume that X_t is a weakly stationary process of degree 3 and put $R_{h_1,h_2}=EX_tX_{t+h_1}X_{t+h_2}$ for any integers h_1 and h_2. Then we have

$$R_{h_1,h_2}=\int_{-1/2}^{1/2}\int_{-1/2}^{1/2} e^{2\pi i(h_1\lambda_1+h_2\lambda_2)}\,dF(\lambda_1,\lambda_2).$$

Symbolically, $dF(\lambda_1,\lambda_2)=EdZ(\lambda_1)dZ(\lambda_2)dZ(-\lambda_1-\lambda_2)$. If $F(\lambda_1,\lambda_2)$ is absolutely continuous with respect to the Lebesgue measure of $\mathbf{R}^2$ and $\partial F(\lambda_1,\lambda_2)/\partial\lambda_1\partial\lambda_2=f(\lambda_1,\lambda_2)$, we call $f(\lambda_1,\lambda_2)$ the **bispectral density function**. When X_t is Gaussian, $R_{h_1,h_2}=0$ and $f(\lambda_1,\lambda_2)=0$ for any h_1, h_2 and any λ_1, λ_2. $f(\lambda_1,\lambda_2)$ can be considered to give a kind of measure of the departure from a Gaussian process or a kind of nonlinear relationship among waves of different frequencies. We can construct an estimator for $f(\lambda_1,\lambda_2)$ by using windows as in the estimation of a spectral density [3].

D. Statistical Analysis of Parametric Models

When we assume merely that X_t is a stationary process and nothing further, then X_t contains infinite-dimensional unknown parameters. In this case, it may be difficult to develop a satisfactory general theory for statistical inference about X_t. But in most practical applications of time series analysis, we can safely assume at least some of the time dependences to be known. For this reason, we can often use a model with finite-dimensional parameters. This means, mainly, that the moments (usually, second-order moments) or the spectral density are assumed to be expressible in terms of finite-dimensional parameters. As examples of such models, autoregressive models, moving average models, and autoregressive moving average models are widely used.

A process X_t is called an **autoregressive process** of order K if X_t satisfies a difference equation $\sum_{k=0}^{K} a_k X_{t-k} = \xi_t$, where the a_k are constants, $a_0 = 1$, $a_K \neq 0$, and the ξ_t are mutually uncorrelated with $E\xi_t = 0$ and $V(\xi_t) = \sigma_\xi^2 > 0$. We usually assume that X_t is a weakly stationary process with $EX_t = 0$. We sometimes use the notation AR(K) to express a weakly stationary and autoregressive process of order K. Let $\{\xi_t\}$ be as above. If X_t is expressed as $X_t = \sum_{l=0}^{L} b_l \xi_{t-l}$, where the b_l are constants, $b_0 = 1$ and $b_L \neq 0$, X_t is called a **moving average process** of order L (MA(L) process). Furthermore, if X_t is weakly stationary with $EX_t = 0$ and expressed as $\sum_{k=0}^{K} a_k X_{t-k} = \sum_{l=0}^{L} b_l \xi_{t-l}$ with $a_0 = 1$, $b_0 = 1$, and $a_K b_L \neq 0$, then X_t is called an **autoregressive moving average process** of order (K, L) (ARMA(K, L) process). Let $A(Z)$ and $B(Z)$ be two polynomials of Z such that $A(Z) = \sum_{k=0}^{K} a_k Z^{K-k}$ and $B(Z) = \sum_{l=0}^{L} b_l Z^{L-l}$, and let $\{\alpha_k \mid 1 \leqslant k \leqslant K\}$ and $\{\beta_l \mid 1 \leqslant l \leqslant L\}$ be the solutions of the associated polynomial equations $A(Z) = 0$ and $B(Z) = 0$, respectively, we assume that $|\alpha_k| < 1$ for $1 \leqslant k \leqslant K$ and $|\beta_l| < 1$ for $1 \leqslant l \leqslant L$. This condition implies that X_t is purely nondeterministic. Let the observed sample be $\{X_t \mid 1 \leqslant t \leqslant T\}$. If we assume that X_t is Gaussian and an ARMA(K, L) process, we can show that the †maximum likelihood estimators $\{\hat{a}_k\}$ and $\{\hat{b}_l\}$ of $\{a_k\}$ and $\{b_l\}$ are †consistent and asymptotically efficient when $T \to \infty$ ("asymptotically efficient" means that the covariance matrix of the distribution of the estimators is asymptotically equal to the inverse of the information matrix) [5] (→ 399 Statistical Estimation D). Furthermore, if X_t is an AR(K) process, the joint distribution of $\{\sqrt{T}(\hat{a}_k - a_k) \mid 1 \leqslant k \leqslant K\}$ tends to a K-dimensional normal distribution with means 0, and this distribution is the same as the one to which the distribution of the †least-square estimators $\{\hat{a}_k\}$ minimizing $Q = \sum_{t=K+1}^{T} (X_t + \sum_{k=1}^{K} a_k X_{t-k})^2$ tends when $T \to \infty$. If X_t is a MA(L) or ARMA(K, L) process ($L \geqslant 1$), the likelihood equations are complicated and cannot be solved directly. Many approximation methods have been proposed to obtain the estimates.

When X_t is an AR(K) process with $|\alpha_k| < 1$ for $1 \leqslant k \leqslant K$, R_t satisfies $\sum_{k=0}^{K} a_k R_{h-k} = 0$ for $h \geqslant 1$. These are often called the **Yule-Walker equations**. R_h can be expressed as $R_h = \sum_{j=1}^{K} C_j \alpha_j^h$ if the α_k are distinct and $a_K \neq 0$, where $\{C_j\}$ are constants and determined by R_h for $0 \leqslant h \leqslant K-1$. When X_t is an ARMA(K, L) process, $\sum_{k=0}^{K} a_k R_{h-k} = 0$ for $h \geqslant L+1$, and the C_j of $R_h = \sum_{j=1}^{k} C_j \alpha_j^h$ are determined by $\{R_h \mid 0 \leqslant h \leqslant \max(K, L)\}$.

The spectral density is expressed as $f(\lambda) = \sigma_\xi^2 |B(e^{2\pi i\lambda})|^2 / |A(e^{2\pi i\lambda})|^2$. If X_t is Gaussian, the maximum likelihood estimator of $f(\lambda)$ is asymptotically equal to the statistic obtained by replacing σ_ξ^2, $\{b_l\}$, and $\{a_k\}$ in $f(\lambda)$ with $\hat{\sigma}_\xi^2$, $\{\hat{b}_l\}$, and $\{\hat{a}_k\}$, respectively, where $\hat{\sigma}_\xi^2$ is the maximum likelihood estimator of σ_ξ^2, when $T \to \infty$.

When we analyze a time series and intend to fit an ARMA(K, L) model, we have to determine the values of K and L. For AR(K) models, many methods have been proposed to determine the value of K. Some examples are: (i) (Quenouille) Let $(Z^K A(1/Z))^2 = \sum_{j=0}^{2K} A_j Z^j$, and $G_K = \sum_{K-j=0}^{2K} \hat{A}_j (\tilde{R}_j / \tilde{R}_0)$, where $\hat{A}_j$ is obtained by replacing $\{a_k\}$ in A_j by $\{\hat{a}_k\}$, and we construct the statistic $\chi_f^2 = \sum_{l=1}^{f} G_{K+l}$. Then χ_f^2 has a †χ^2 distribution asymptotically with f degrees of freedom under the assumption that $K \geqslant K_0$, where K_0 is the true order, as $T \to \infty$. Using this fact, we can determine the order of an AR model. (ii) (Akaike) We consider choosing an order K satisfying $K_L \leqslant K \leqslant K_M$, where K_L and K_M are minimum order and maximum order, respectively, specified a priori. Then we construct the statistic $\mathrm{AIC}(K) = (T-K)\log \hat{\hat{\sigma}}_\xi^2(K) + 2K$, where

$$\hat{\hat{\sigma}}_\xi^2(K) = \sum_{t=K+1}^{T} (X_t + \hat{a}_1 X_{t-1} + \ldots + \hat{a}_K X_{t-K})^2 / T$$

and $\{\hat{a}_k \mid 1 \leqslant k \leqslant K\}$ are the least square estimators of the autoregressive coefficients of an AR(K) model fitting X_t. Calculate AIC(K) for $K = K_L, K_L + 1, \ldots, K_M$. If AIC($K$) has the minimum value at $K = \hat{K}$, we determine the order to be $\hat{K}$ [6] (→ 403 Statistical Models F). Parzen proposed another method by using the criterion autoregressive transfer function (CAT). Here $\mathrm{CAT}(K) = 1 - \tilde{\sigma}^2(\infty) / \tilde{\sigma}_\xi^2(K) + K/T$, where $\tilde{\sigma}_\xi^2(K) = (T/(T-K)) \hat{\hat{\sigma}}_\xi^2(K)$ and $\tilde{\sigma}^2(\infty)$ is an estimator of $\sigma^2(\infty) = \exp(\int_{-1/2}^{1/2} \log f(\lambda) d\lambda)$ [7]. (iii) We can construct a test statistic for the null hypothesis AR(K) against the alternative hypothesis AR($K+1$) (Jenkins) or use a multiple decision procedure (T. W. Anderson [8]).

Not much is known about the statistical properties of the above methods, and few comparisons have been made among them.

Another parametric model is an exponential model for the spectrum. The spectral density is expressed by $f(\lambda) = C^2 \exp\{2 \sum_{k=1}^{K} \theta_k \cos(2\pi k\lambda)\}$, where the θ_k and C are constants.

We now discuss some general theories of estimation for finite-dimensional-parameter models. Let X_t be a real-valued Gaussian process of mean 0 and of spectral density $f(\lambda)$ which is continous and positive in $[-1/2, 1/2]$, and let the moving average representation of X_t be $X_t = \sum_{l=0}^{\infty} b_l \xi_{t-l}$, where ξ_t is a white noise and $\sigma_\xi^2 = E\xi_t^2$. We assume that $f(\lambda)/\sigma_\xi^2 = g(\lambda)$

depends only on M parameters $\boldsymbol{\theta}=(\theta_1, \theta_2, \ldots, \theta_M)'$ which are independent of σ_ξ^2. Then the logarithm of the †likelihood function can be approximated by $-(1/2)\{T\log 2\pi\sigma_\xi^2 + \mathbf{X}'\Sigma_T^{-1}(\boldsymbol{\theta})\mathbf{X}/\sigma_\xi^2\}$ by ignoring the lower-order terms in T, where $\sigma_\xi^2\Sigma_T(\boldsymbol{\theta})$ is the covariance matrix of $\mathbf{X}$. Usually, it is difficult to find an explicit expression for each element of $\Sigma_T^{-1}(\boldsymbol{\theta})$. Another approximation for the logarithm of the likelihood function is given by

$$-\frac{T}{2}\int_{-1/2}^{1/2}\left[\log f(\lambda)+\frac{I_T(\lambda)}{f(\lambda)}\right]d\lambda.$$

Under mild conditions on the regularity of $g(\lambda)$, the estimators $\hat{\boldsymbol{\theta}}=(\hat\theta_1, \hat\theta_2, \ldots, \hat\theta_M)$ and $\hat\sigma_\xi^2$, obtained as the solutions of the likelihood equations, are †consistent and asymptotically normal as T tends to infinity. This means that the distribution of $\sqrt{T}(\hat\sigma_\xi^2-\sigma_\xi^2)$ is asymptotically normal and $\sqrt{T}(\hat\sigma_\xi^2-\sigma_\xi^2)$ and $\sqrt{T}(\hat{\boldsymbol{\theta}}-\boldsymbol{\theta})$ are asymptotically independent. The asymptotic distribution of $\sqrt{T}(\hat{\boldsymbol{\theta}}-\boldsymbol{\theta})$ is the normal distribution $N(\mathbf{0}, \Gamma^{-1})$, where the (k,l)-component Γ_{kl} of Γ is given by

$$\Gamma_{kl}=\frac{1}{2}\int_{-1/2}^{1/2}\left(\frac{\partial\log g(\lambda)}{\partial\theta_k}\cdot\frac{\partial\log g(\lambda)}{\partial\theta_l}\right)_{\boldsymbol{\theta}}d\lambda.$$

E. Statistical Analysis of Multiple Time Series

Let $\mathbf{X}_t=(X_t^{(1)}, X_t^{(2)}, \ldots, X_t^{(p)})'$ be a complex-valued weakly stationary process with $E\mathbf{X}_t=\mathbf{0}$ and $E\mathbf{X}_t\bar{\mathbf{X}}_s'=R_{t-s}$. R_{t-s} is the $p\times p$ matrix whose (k,l)-component is $R_{t-s}^{(k,l)}=EX_t^{(k)}\bar{X}_s^{(l)}$. We discuss the case when t is an integer. R_h has the spectral representation

$$R_h=\int_{-1/2}^{1/2}e^{2\pi ih\lambda}dF(\lambda),$$

where $F(\lambda)$ is a $p\times p$ matrix and $F(\lambda_1)-F(\lambda_2)$, $\lambda_1\geqslant\lambda_2$, is Hermitian nonnegative. Let $f^{k,l}(\lambda)$ be the (k,l)-component of the spectral density matrix $f(\lambda)$, i.e., $F_a(\lambda)=\int_{-1/2}^{\lambda}f(\mu)d\mu$, of the absolutely continuous part in the Lebesgue decomposition of $F(\lambda)$. The function $f^{k,l}(\lambda)$ for $k\neq l$ is called the **cross spectral density function**. $f^{k,l}(\lambda)$ represents a kind of correlation between the wave of frequency λ included in $X_t^{(k)}$ and the one included in $X_t^{(l)}$.

Let $\mathbf{X}_t=(X_t^{(1)}, X_t^{(2)}, \ldots, X_t^{(p)})'$ and $\mathbf{Y}_t=(Y_t^{(1)}, Y_t^{(2)}, \ldots, Y_t^{(q)})'$ be two complex-valued weakly stationary processes with $E\mathbf{X}_t=0$, $E\mathbf{Y}_t=0$, $E\mathbf{X}_t\bar{\mathbf{X}}_s'=R_{t-s}^{\mathbf{X}}$, $E\mathbf{Y}_t\bar{\mathbf{Y}}_s'=R_{t-s}^{\mathbf{Y}}$ and $E\mathbf{X}_t\bar{\mathbf{Y}}_s'=R_{t-s}^{\mathbf{XY}}$. We assume $\mathbf{Y}_t=\sum_{s=-\infty}^{\infty}A_s\mathbf{X}_{t-s}$, where A_s is a $q\times p$ matrix whose components are constants depending on s. Put $A(\lambda)=\sum_{s=-\infty}^{\infty}A_se^{-2\pi is\lambda}$. $A(\lambda)$ should exist in the sense of mean square convergence with respect to the spectral distribution function F for $\mathbf{X}_t$.

The function $A(\lambda)$ is called the matrix **frequency response function**.

As a measure of the strength of association between $X_t^{(k)}$ and $X_t^{(l)}$ at frequency λ, we introduce the quantity $\gamma^{k,l}(\lambda)=|f^{k,l}(\lambda)|^2/f^{k,k}(\lambda)f^{l,l}(\lambda)$. $\gamma^{k,l}(\lambda)$ is called the **coherence**. Let $X_t^{(k)}=\sum_{s=-\infty}^{\infty}a_s^{k,l}X_{t-s}^{(l)}+\eta_t$, where η_t is a weakly stationary process with mean 0 and uncorrelated with $X_s^{(l)}$, $-\infty<s<\infty$. If $E|\eta_t|^2=0$, $\gamma^{k,l}(\lambda)=1$. If $E|\sum_{-\infty}^{\infty}a_s^{k,l}X_{t-s}^{(l)}|^2=0$, $\gamma^{k,l}(\lambda)=0$. Generally, we have $0\leqslant\gamma^{k,l}(\lambda)\leqslant 1$.

For the estimation of $F(\lambda)$, $A(\lambda)$, and $\gamma^{k,l}(\lambda)$, the theories have been similar to those for the estimation of the spectral density of a scalar time series. For example, an estimator of $f(\lambda)$ is given [11] in the form

$$\hat f(\lambda)=\sum_{h=-(T-1)}^{T-1}\tilde{\tilde{R}}_hw_he^{-2\pi ih\lambda},$$

where

$$\tilde{\tilde{R}}_h=\sum_{t=1}^{T-|h|}\mathbf{X}_{t+|h|}\bar{\mathbf{X}}_t'/T$$

and the w_h are the same as in Section C.

We can define an autoregressive, moving average, or autoregressive moving average process in a similar way as for a scalar time series. The a_k and b_l in Section D should be replaced by $p\times p$ matrices and the associated polynomial equations $A(Z)=0$ and $B(Z)=0$ should be understood in the vector sense [11]. There are problems with determining the coefficients uniquely or identifying an ARMA(K,L) model, and these problems have been discussed to some extent.

F. Statistical Inference of the Mean Function

Let X_t be expressed as $X_t=m_t+Y_t$, where m_t is a real-valued deterministic function of t and Y_t is a real-valued weakly stationary process with mean 0 and spectral distribution function $F(\lambda)$. This means that $EX_t=m_t$. We consider the case when $m_t=\sum_{j=1}^{M}C_j\varphi_t^{(j)}$, where $\mathbf{C}=(C_1, C_2, \ldots, C_M)'$ is a vector of unknown coefficients and $\boldsymbol{\varphi}_t=(\varphi_t^{(1)}, \varphi_t^{(2)}, \ldots, \varphi_t^{(M)})'$ is a set of known (regression) functions.

Let us construct †linear unbiased estimators $\{\tilde C_j=\sum_{t=1}^{T}\gamma_{jt}X_t \mid 1\leqslant j\leqslant M\}$ for the coefficients C_j, where the γ_{jt} are known constants. Put $\Phi=(\boldsymbol{\varphi}_1, \boldsymbol{\varphi}_2, \ldots, \boldsymbol{\varphi}_T)'$. Then the †least squares estimator of $\mathbf{C}$ is given by $\hat{\mathbf{C}}=(\Phi'\Phi)^{-1}\Phi'\mathbf{X}$ when $\Phi'\Phi$ is nonsingular. Let Σ be the covariance matrix of $\mathbf{X}$. Then the †best linear unbiased estimator is $\hat{\mathbf{C}}^*=(\Phi'\Sigma^{-1}\Phi)^{-1}\Phi'\Sigma^{-1}\mathbf{X}$. We put $\|\varphi^{(j)}\|_T^2=\sum_{t=1}^{T}(\varphi_t^{(j)})^2$ and assume that $\lim_{T\to\infty}\|\varphi^{(j)}\|_T^2=\infty$, $\lim_{T\to\infty}\|\varphi^{(j)}\|_{T+h}^2/\|\varphi^{(j)}\|_T^2=1$ for $1\leqslant j\leqslant M$ and any fixed h and assume the existence of $\psi_h^{(j,k)}=\lim_{T\to\infty}\sum_{t=1}^{T}\varphi_{t+h}^{(j)}\varphi_t^{(k)}/\|\varphi^{(j)}\|_T\|\varphi^{(k)}\|_T$ for $1\leqslant j$, $k\leqslant M$. We also assume that $F(\lambda)$ is absolutely continuous and $F'(\lambda)=f(\lambda)$ is positive and piecewise continuous. Let ψ_h be the $M \times M$ matrix whose (j,k)-component is $\psi_h^{(j,k)}$. Then ψ_h can be represented by

$$\psi_h=\int_{-1/2}^{1/2} e^{2\pi i h\lambda}\,dG(\lambda),$$

where $G(\lambda)-G(\mu)$ is a nonnegative definite matrix for $\lambda>\mu$. Assume that $\psi_0=G(1/2)-G(-1/2)$ is nonsingular and put $H(\lambda)=\psi_0^{-1/2}G(\lambda)\psi_0^{-1/2}$, and for any set S, $H(S)=\int_S H(d\lambda)$. Suppose further that $S_1, S_2, \ldots, S_q$ are q sets such that $H(S_j)>0$, $\sum_{j=1}^q H(S_j)=I$, $H(S_j)H(S_k)=0$, $j\neq k$, and for any j there is no subset $S_j'\subset S_j$ such that $H(S_j')>0$, $H(S_j-S_j')>0$ and $H(S_j')H(S_j-S_j')=0$. We have $q\leqslant M$. It can be shown that the spectrum of the regression can be decomposed into such disjoint sets $S_1,\ldots,S_q$. Then we can show that $\hat{\mathbf{C}}$ is asymptotically efficient in the sense that the asymptotic covariance matrix of $\hat{\mathbf{C}}$ is equivalent to that of $\hat{\mathbf{C}}^*$ if and only if $f(\lambda)$ is constant on each of the elements S_j. Especially, if $\psi_t^{(j)}=t^j e^{2\pi i t\mu_j}$, $\hat{\mathbf{C}}$ is asymptotically efficient.

G. Nonstationary Models

It is difficult to develop a statistical theory for a general class of nonstationary time series, but some special types of nonstationary processes have been investigated more or less in detail. Let X_t (t an integer) be a real-valued stochastic process and ∇ be the backward difference operator defined by $\nabla X_t=X_t-X_{t-1}$ and $\nabla^d X_t=\nabla(\nabla^{d-1}X_t)$ for $d\geqslant 2$. We assume that X_t is defined for $t\geqslant t_0$ (t_0 a finite integer), and $EX_t^2<+\infty$. For analyzing a nonstationary time series, Box and Jenkins introduced the following model: For a positive integer d, $Y_t=\nabla^d X_t$, $t\geqslant t_0+d$, is stationary and is an autoregressive moving average process of order (K,L) for $t\geqslant t_0+d+\max(K,L)$. They called such an X_t an **autoregressive integrated moving average process** of order (K,d,L) and denoted it by ARIMA(K,d,L). The word "integrated" means a kind of summation; in fact, X_t can be expressed as a sum of the weakly stationary process Y_t, i.e.,

$$X_t=X_0+(\nabla X_0)t+(\nabla^2 X_0)\left(\sum_{s_2=1}^{t}\sum_{s_1=1}^{s_2}\right)+\ldots$$
$$+(\nabla^{d-1}X_0)\left(\sum_{s_{d-1}=1}^{t}\cdots\sum_{s_1=1}^{s_2}\right)$$
$$+\sum_{s_d=1}^{t}\sum_{s_{d-1}=1}^{s_d}\cdots\sum_{s_1=1}^{s_2}Y_{s_1}$$

when $t_0=-d+1$. Using this model, methods of forecasting and of model identification and estimation can be discussed [13].

Another nonstationary model is based on the concept of evolutionary spectra [14]. In this approach, spectral distribution functions are taken to be time-dependent. Let X_t be a complex-valued stochastic process (t an integer) with $EX_t=0$ and $R_{t,s}=EX_t\bar{X}_s$. In the following, we write simply $\int$ for $\int_{-1/2}^{1/2}$. We now restrict our attention to the class of X_t for which there exist functions $\{u_t(\lambda)\}$ defined on $[-1/2,1/2]$ such that $R_{t,s}$ can be expressed as $R_{t,s}=\int u_t(\lambda)\bar{u}_s(\lambda)\,d\mu(\lambda)$, where $\mu(\lambda)$ is a measure. $u_t(\lambda)$ should satisfy $\int|u_t(\lambda)|^2\,d\mu(\lambda)<+\infty$. Then X_t admits a representation of the form $X_t=\int u_t(\lambda)\,dZ(\lambda)$, where $Z(\lambda)$ is a process with orthogonal increments and $E|dZ(\lambda)|^2=d\mu(\lambda)$. If $u_t(\lambda)$ is expressed as $u_t(\lambda)=\gamma_t(\lambda)e^{2\pi i\theta(\lambda)t}$ and $\gamma_t(\lambda)$ is of the form $\gamma_t(\lambda)=\int e^{2\pi i t w}\,d\Gamma_\lambda(w)$ with $|d\Gamma_\lambda(w)|$ having the absolute maximum at $w=0$, we call $u_t(\lambda)$ an oscillatory function and X_t an oscillatory process. The evolutionary power spectrum $dF_t(\lambda)$ is defined by $dF_t(\lambda)=|\gamma_t(\lambda)|^2\,d\mu(\lambda)$.

Other models, such as an autoregressive model whose coefficients vary with time or whose associated polynomial has roots outside the unit circle, have also been discussed.

References

[1] E. J. Hannan, Time series analysis, Methuen, 1960.
[2] R. B. Blackman and J. W. Tukey, The measurement of power spectra from the point of view of communications engineering, Dover, 1959.
[3] D. R. Brillinger, Time series: Data analysis and theory, Holt, Rinehart and Winston, 1975.
[4] M. S. Bartlett, An introduction to stochastic processes with special reference to methods and applications, Cambridge Univ. Press, second edition, 1966.
[5] P. Whittle, Estimation and information in stationary time series, Ark. Mat., 2 (1953), 423–434.
[6] H. Akaike, Information theory and an extension of the maximum likelihood principle, 2nd Int. Symp. Information Theory, B. N. Petrov and F. Csáki (eds.), Akadémiai Kiadó, 1973, 267–281.
[7] E. Parzen, Some recent advances in time series modeling, IEEE Trans. Automatic Control, AC-19 (1974), 723–730.
[8] T. W. Anderson, The statistical analysis of time series, Wiley, 1971.
[9] P. Whittle, Hypothesis testing in time series analysis, Almqvist and Wiksell, 1951.
[10] H. B. Mann and A. Wald, On the statistical treatment of linear stochastic difference equations, Econometrica, 11 (1943), 173–220.

[11] E. J. Hannan, Multiple time series, Wiley, 1970.
[12] U. Grenander and M. Rosenblatt, Statistical analysis of stationary time series, Wiley, 1957.
[13] G. E. P. Box and G. M. Jenkins, Time series analysis: Forecasting and control, Holden-Day, revised edition, 1976.
[14] M. B. Priestley, Evolutionary spectra and nonstationary processes, J. Roy. Statist. Soc., ser. B, 27 (1965), 204–237.

422 (IV.7) Topological Abelian Groups

A. Introduction

A commutative topological group is called a **topological Abelian group**. Throughout this article, except in Section L, all topological groups under consideration are locally compact Hausdorff topological Abelian groups and are simply called groups (→ 423 Topological Groups).

B. Characters

A **character** of a group is a continuous function $\chi(x)$ ($x \in G$) that takes on as values complex numbers of absolute value 1 and satisfies $\chi(xy)=\chi(x)\chi(y)$. Equivalently, χ is a 1-dimensional and therefore an irreducible †unitary representation of G. Conversely any irreducible unitary representation of G is 1-dimensional. Indeed, for a topological Abelian group, the set of its characters coincides with the set of its irreducible unitary representations. If the product of two characters χ, χ' is defined by $\chi\chi'(x)=\chi(x)\chi'(x)$, then the set of all characters forms the **character group** $C(G)$ of G. With †compact-open topology, $C(G)$ itself becomes a locally compact topological Abelian group.

C. The Duality Theorem

For a fixed element x of G, $\chi(x)$ ($\chi \in C(G)$) is a character of $C(G)$, namely, an element of $CC(G)$. Denote this character of $C(G)$ by $x(\chi)$, and consider the correspondence $G \ni x \rightarrow x(\chi)$. That this correspondence is one-to-one follows from the fact that any locally compact G has †sufficiently many irreducible unitary representations (→ 437 Unitary Representations) and the fact that if G is an Abelian group, then any irreducible unitary representation of G is a character of G. Furthermore, any character of $C(G)$ is given as one of the $x(\chi)$; indeed, by this correspondence, we have $G \cong CC(G)$ (**Pontryagin's duality theorem**).

By the duality theorem, each of G and $C(G)$ is isomorphic to the character group of the other. In this sense, G and $C(G)$ are said to be **dual** to each other.

D. Correspondence between Subgroups

Let $G, G'=C(G)$ be groups that are dual to each other. Given a closed subgroup g of G, the set of all χ' such that $\chi'(x)=1$ for all x in g forms a closed subgroup of G', usually denoted by (G', g). The definition of (G, g') is similar. Then $g \leftrightarrow (G', g)=g'$ gives a one-to-one correspondence between the closed subgroups of G and those of G'. If $g_1 \supset g_2$, then g_1/g_2 and $(G', g_2)/(G', g_1)$ are dual to each other. If the group operations of G, G' are written in additive form, with 0 for the identity, then $x(\chi')=1$ is written as $x(\chi')=0$. In this sense, (G', g) is called the **annihilator** (or **annulator**) of g.

E. The Structure Theorem

Let $\mathfrak{A}$ be the set of all groups (more precisely, of all locally compact Hausdorff topological Abelian groups). If $G_1, G_2 \in \mathfrak{A}$, then the direct product $G_1 \times G_2 \in \mathfrak{A}$, and if $G \in \mathfrak{A}$ and H is a closed subgroup of G, then $H \in \mathfrak{A}$ and $G/H \in \mathfrak{A}$. In addition, if H is a closed subgroup of a group G such that $H \subset \mathfrak{A}$ and $G/H \in \mathfrak{A}$, then $G \in \mathfrak{A}$. In other words, $\mathfrak{A}$ is closed under the operations of forming direct products, closed subgroups, quotient groups, and †extensions by members of $\mathfrak{A}$. Furthermore, the operation C that assigns to each element of $\mathfrak{A}$ its dual element is a reflexive correspondence of $\mathfrak{A}$ onto $\mathfrak{A}$, and if $G \supset H$, the annihilator $(C(G), H)$ of H is a closed subgroup of $C(G)$. Also, $C(G/H) \cong (C(G), H)$, $C(H) \cong C(G)/(C(G), H)$. Furthermore, $C(G_1 \times G_2) \cong C(G_1) \times C(G_2)$. Finally, $H=(G,(C(G), H))$ (**reciprocity of annihilators**).

Typical examples of groups in $\mathfrak{A}$ are the additive group $\mathbf{R}$ of real numbers, the additive group $\mathbf{Z}$ of rational integers, the 1-dimensional †torus group $\mathbf{T}=\mathbf{R}/\mathbf{Z}$, and finite Abelian groups $\mathbf{F}$. The torus group $\mathbf{T}$ is also isomorphic to the multiplicative group $U(1)$ of complex numbers of absolute value 1. The direct product $\mathbf{R}^n$ of n copies of $\mathbf{R}$ is the **vector group** of dimension n, and the direct product $\mathbf{T}^n$ of n copies of $\mathbf{T}$ is the **torus** (or **torus group**) of dimension n (or n-**torus**). Both $\mathbf{T}^n$ and $\mathbf{F}$ are compact, while $\mathbf{R}^n$ and $\mathbf{Z}^n$ are not. We have $C(\mathbf{R})=\mathbf{R}$, $C(\mathbf{T})=\mathbf{Z}$, $C(\mathbf{Z})=\mathbf{T}$. Any finite Abelian group $\mathbf{F}$ is isomorphic to its character group $C(\mathbf{F})$. The direct product of a finite number of copies of **R**, **T**, **Z**, and a finite Abelian group **F**, namely, a group of the form $\mathbf{R}^l \times \mathbf{T}^m \times \mathbf{Z}^n \times \mathbf{F}$, is called an **elementary topological Abelian group**.

Any group in $\mathfrak{A}$ is isomorphic to the direct product of a vector group of some dimension and the extension of a compact group by a discrete group (the **structure theorem**). Hence, if the effect of the operation C is explicitly known, then the problem of finding the structure of groups in $\mathfrak{A}$ is reduced to the problem concerning discrete groups alone. For the structure of groups in $\mathfrak{A}$, the following theorem is known: If $G \in \mathfrak{A}$ is generated by a compact neighborhood of the identity e, then G is isomorphic to the direct product of a compact subgroup K and a group of the form $\mathbf{R}^n \times \mathbf{Z}^m$ (n, m are nonnegative integers). Then any compact subgroup of G is contained in K, which is the unique maximal compact subgroup of G. A group $G \in \mathfrak{A}$ generated by a compact neighborhood of e is the †projective limit of elementary topological Abelian groups. L. S. Pontryagin first proved a structure theorem of this type and then the duality theorem.

F. Compact Elements

An element a of a group $G \in \mathfrak{A}$ is called a **compact element** if the cyclic group $\{a^n | n \in \mathbf{Z}\}$ generated by a is contained in a compact subset of G. The set C_0 of all compact elements of G is a closed subgroup of G, and the quotient group G/C_0 does not contain any compact element other than the identity. In particular, if G is generated by a compact neighborhood of the identity, then C_0 coincides with the maximal compact subgroup K of G. Let C_0 be the set of all compact elements of a group $G \in \mathfrak{A}$. The annihilator $(C(G), C_0)$ is a connected component of the character group $C(G)$ of G. If G is a discrete group, then a compact element of G is an element of G of finite order.

G. Compact Groups and Discrete Groups

Suppose that two groups $G, X \in \mathfrak{A}$ are dual to each other. Then one group is compact if and only if the other group is discrete. By the duality theorem, the properties of a compact Abelian group G can be stated, in principle, through the properties of the discrete Abelian group $C(G)$. The following are a few such examples. Let G be a compact Abelian group. Then its †dimension is equal to the †rank of the discrete Abelian group $C(G)$. A subgroup Y of a discrete Abelian group X is called a **divisible subgroup** if the quotient group X/Y contains no element of finite order other than the identity. A compact Abelian group G is locally connected if and only if any finite subset of the character group $C(G)$ is contained in some divisible subgroup of $C(G)$ generated by a finite number of elements. Hence if a compact locally connected Abelian group G has an †open basis consisting of a countable number of open sets, then G is of the form $\mathbf{T}^a \times \mathbf{F}$, where **F** is a finite Abelian group and $\mathbf{T}^a$ is the direct product of an at most countable number of 1-dimensional torus groups **T**.

H. Dual Decomposition into Direct Products

Let G be a compact or discrete Abelian group, and let $\mathfrak{M} = \{H_\alpha | \alpha \in A\}$ be a family of closed subgroups of G. Let $\Delta(\mathfrak{M}) = \bigcap_{\alpha \in A} H_\alpha$, and denote by $\Sigma(\mathfrak{M})$ the smallest closed subgroup of G containing $\bigcup_{\alpha \in A} H_\alpha$. Then, with $\Omega = \{(C(G), H_\alpha) | \alpha \in A\}$, the relations $\Delta(\Omega) = (C(G), \Sigma(\mathfrak{M}))$ and $\Sigma(\Omega) = (C(G), \Delta(\mathfrak{M}))$ hold. Furthermore, suppose that G is decomposed into the direct product $G = \prod_{\alpha \in A} H_\alpha$, and for each $\alpha \in A$ put $K_\alpha = \Sigma(\mathfrak{M} - \{H_\alpha\})$, $X_\alpha = (C(G), K_\alpha)$. Then X_α is the character group of H_α, and $C(G)$ can be decomposed into the direct product $C(G) = \prod_{\alpha \in A} X_\alpha$. This decomposition of $C(G)$ into a direct product is called the **dual direct product decomposition** corresponding to the decomposition $G = \prod_{\alpha \in A} H_\alpha$.

I. Orthogonal Group Pairs

Suppose that for two groups G, G' there exists a mapping $(x, x') \rightarrow xx'$ of the Cartesian product $G \times G'$ into the set $U(1)$ of all complex numbers of absolute value 1 such that

$$(x_1 x_2)x' = (x_1 x')(x_2 x'),$$

$$x(x_1' x_2') = (xx_1')(xx_2').$$

Then G, G' are said to form a **group pair**. Suppose that G, G' form a group pair, and consider xx' to be a function $x(x')$ in x'. If two functions $x_1(x')$ and $x_2(x')$ coincide only when $x_1 = x_2$ and the same is true when the roles of G and G' are interchanged, then G, G' are said to form an **orthogonal group pair**. If G is a compact Abelian group, G' is a discrete Abelian group, and G, G' form an orthogonal group pair, then G, G' are dual to each other.

J. Commutative Lie Groups

An elementary topological Abelian group $\mathbf{R}^l \times \mathbf{T}^m \times \mathbf{Z}^n \times \mathbf{F}$ is a commutative †Lie group. Conversely, any commutative Lie group G

generated by a compact neighborhood of the identity is isomorphic to an elementary topological Abelian group. In particular, any connected commutative Lie group G is isomorphic to $\mathbf{R}^l \times \mathbf{T}^m$ for some l and m. A closed subgroup H of the vector group $\mathbf{R}^n$ of dimension n is isomorphic to $\mathbf{R}^p \times \mathbf{Z}^q$ $(0 \leqslant p+q \leqslant n)$. More precisely, there exists a basis $a_1, \ldots, a_n$ of the vector group $\mathbf{R}^n$ such that $H=\{\sum_{i=1}^p x_i a_i + \sum_{j=p+1}^r n_j a_j \mid x_i \in \mathbf{R}, n_j \in \mathbf{Z}\}$. Hence the quotient groups of $\mathbf{R}^n$ that are †separated topological groups are all isomorphic to groups of the form $\mathbf{R}^l \times \mathbf{T}^m$ $(0 \leqslant l+m \leqslant n)$. Any closed subgroup of the torus group $\mathbf{T}^n$ of dimension n is isomorphic to a group of the form $\mathbf{T}^p \times \mathbf{F}$ $(0 \leqslant p \leqslant n)$, where $\mathbf{F}$ is a finite Abelian group. Hence the quotient groups of $\mathbf{T}^n$ that are separated topological groups are all isomorphic to $\mathbf{T}^m$ $(0 \leqslant m \leqslant n)$. A †regular linear transformation of the linear space $\mathbf{R}^n$ is a continuous automorphism of the vector group $\mathbf{R}^n$, and in fact, any continuous automorphism of $\mathbf{R}^n$ is given by a regular linear transformation. Indeed, the group of all continuous automorphisms of $\mathbf{R}^n$ is isomorphic to the †general linear group $GL(n, \mathbf{R})$ of degree n. Any continuous automorphism of the torus group $\mathbf{T}^n = \mathbf{R}^n/\mathbf{Z}^n$ of dimension n is given by a regular linear transformation φ of $\mathbf{R}^n$ such that $\varphi(\mathbf{Z}^n) = \mathbf{Z}^n$. Hence the group of continuous automorphisms of $\mathbf{T}^n$ is isomorphic to the multiplicative group of all $n \times n$ matrices, with determinant ± 1 and with entries in the set of rational integers.

K. Kronecker's Approximation Theorem

Let H be a subgroup of a group $G \in \mathfrak{A}$ (not necessarily closed). Then $(G,(C(G),H))$ coincides with the closure $\bar{H}$ of H. In particular, H is †dense in G if and only if the annihilator $(C(G),H)$ consists of the identity alone. Now let $G=\mathbf{R}^n$ and let H be the subgroup of $\mathbf{R}^n$ generated by $\theta=(\theta_1, \ldots, \theta_n) \in \mathbf{R}^n$ and the natural †basis $e_1=(1,0,\ldots,0), \ldots, e_n=(0,\ldots,0,1)$ of $\mathbf{R}^n$. Then H is dense in $\mathbf{R}^n$ if and only if $(\mathbf{R}^n, H)=\{0\}$; that is, $\theta_1, \ldots, \theta_n, 1$ are linearly independent over the rational number field $\mathbf{Q}$ (**Kronecker's approximation theorem**). This theorem implies that the torus group $\mathbf{T}^n$ of dimension n has a cyclic subgroup and a 1-parameter subgroup that are both dense in $\mathbf{T}^n$.

L. Linear Topology

Consider the discrete topology in a field Ω. Suppose that an Ω-module G has a topology that satisfies †Hausdorff's separation axiom and is such that a base for the neighborhood system of the zero element 0 consists of Ω-submodules, and suppose that G together with this topology constitutes a topological Abelian group. Then this topology is called a **linear topology**. If a linear topology is restricted to a Ω-submodule, then it is also a linear topology. If G is of finite rank, then any linear topology is the discrete topology. The discrete topology on G is a linear topology. Let H be a Ω-submodule. Then the subset $V=H+g$ of G obtained by translating H by an element g of G is called a **linear variety** in G. If V is a linear variety, then $\bar{V}$ is also a linear variety. If Ω-modules G, G' have linear topologies, a homomorphism of G into G' is always assumed to be open and continuous with respect to these topologies. A linear variety V in G is said to be **linearly compact** if, for any system $\{V_\alpha\}$ of linear varieties closed in V with the †finite intersection property, we have $\bigcap_\alpha V_\alpha \neq \varnothing$. In this case V is closed in G. If linearly compact Ω-submodules can be chosen as a base for the neighborhood system of the zero element of G, we say that G is **locally linearly compact**. The set $C_\Omega(G)$ of homomorphisms of an Ω-module G with linear topology into Ω is also an Ω-module. For any linearly compact Ω-submodule H of G, let $U(H)=\{\chi \mid \chi(g)= 0, g \in H\}$. Then, with $\{U(H)\}$ as a base for the neighborhood system, a linear topology can be introduced in $C_\Omega(G)$. According as G is discrete, linearly compact, or locally linearly compact, $C_\Omega(G)$ is linearly compact, discrete, or locally linearly compact. Let G, H be Ω-modules each of which has a linear topology, and let $\varphi: G \ni g \rightarrow \varphi_g \in C_\Omega(H)$, $\psi: H \ni h \rightarrow \psi_h \in C_\Omega(G)$ be homomorphisms such that $\varphi_g(h)=\psi_h(g)$. Then if one of φ, ψ is an isomorphism, so is the other. This is an analog of the Pontryagin duality theorem and is called the **duality theorem for Ω-modules**. In particular, a linearly compact Ω-module is the direct sum of 1-dimensional spaces (S. Lefschetz [3]).

References

[1] L. S. Pontryagin, Topological groups, first edition, Princeton Univ. Press, 1939; second edition, Gordon & Breach, 1966 (translation of the 1954 Russian edition).
[2] A. Weil, L'intégration dans les groupes topologiques et ses applications, Actualités Sci. Ind., Hermann, 1940.
[3] S. Lefschetz, Algebraic topology, Amer. Math. Soc. Colloq. Publ., 1942.
[4] W. Rudin, Fourier analysis on groups, Interscience, 1962.
[5] E. Hewitt and K. A. Ross, Abstract harmonic analysis, Springer, I, 1963; II, 1970.

423 (IV.6) Topological Groups

A. Definitions

If a †group G has the structure of a †topological space such that the mapping $(x, y) \rightarrow xy$ (product) of the Cartesian product $G \times G$ into G and the mapping $x \rightarrow x^{-1}$ (inverse) of G into G are both continuous, then G is called a **topological group**. The group G without a topological structure is called the **underlying group** of the topological group G, and the topological space G is called the **underlying topological space** of the topological group G. Let G, G' be topological groups. A mapping f of G into G' is called an **isomorphism** of the topological group G onto the topological group G' if f is an †isomorphism of the underlying group G onto the underlying group G' and also a †homeomorphism of the underlying topological space G onto the underlying topological space G'. Two topological groups are said to be **isomorphic** if there exists an isomorphism of one onto the other.

B. Neighborhood Systems

Let $\mathfrak{N}$ be the †neighborhood system of the identity e of a topological group G. Namely, $\mathfrak{N}$ consists of all subsets of G each of which contains an open set containing the element e. Then $\mathfrak{N}$ satisfies the following six conditions: (i) If $U \in \mathfrak{N}$ and $U \subset V$, then $V \in \mathfrak{N}$. (ii) If U, $V \in \mathfrak{N}$, then $U \cap V \in \mathfrak{N}$. (iii) If $U \in \mathfrak{N}$, then $e \in U$. (iv) For any $U \in \mathfrak{N}$, there exists a $W \in \mathfrak{N}$ such that $WW = \{xy \mid x, y \in W\} \subset U$. (v) If $U \in \mathfrak{N}$, then $U^{-1} \in \mathfrak{N}$. (vi) If $U \in \mathfrak{N}$ and $a \in G$, then $aUa^{-1} \in \mathfrak{N}$. Conversely, if a nonempty family $\mathfrak{N}$ of subsets of a group G satisfies conditions (i)–(vi), then there exists a †topology $\mathfrak{O}$ of G such that $\mathfrak{N}$ is the neighborhood system of e and G is a topological group with this topology. Moreover, such a topology is uniquely determined by $\mathfrak{N}$. †Left translation $x \rightarrow ax$ and †right translation $x \rightarrow xa$ in a topological group G are homeomorphisms of G onto G; thus if $\mathfrak{N}$ is the neighborhood system of the identity e, then $a\mathfrak{N} = \mathfrak{N}a$ is the neighborhood system of a, where $a\mathfrak{N} = \{aU \mid U \in \mathfrak{N}\}$.

If the underlying topological space of a topological group G is a †Hausdorff space, G is called a **T_2-topological group** (**Hausdorff topological group** or **separated topological group**). If the underlying topological space of a topological group G is a †T_0-topological space, then, as is easily seen, it is a †T_1-topological space. If it is a T_1-topological space, then by the fact that the topology may be defined by a †uniformity, it is a †completely regular space, hence, in particular, a Hausdorff space (→ Section G). Thus a topological group whose underlying topological space is a T_0-topological space is a T_2-topological group.

C. Direct Product of Topological Groups

Consider a family $\{G_\alpha\}_{\alpha \in A}$ of topological groups. The Cartesian product $G = \prod_{\alpha \in A} G_\alpha$ of the underlying groups of G_α is a topological group with the †product topology of the underlying topological spaces of G_α. This topological group $G = \prod_{\alpha \in A} G_\alpha$ is called the **direct product** of topological groups G_α $(\alpha \in A)$.

D. Subgroups

Let H be a subgroup of the underlying group of a topological group G. Then H is a topological group with the topology of a †topological subspace of G (†relative topology). This topological group H is called a **subgroup** of G. A subgroup that is a closed (open) set is called a **closed (open) subgroup**. Any open subgroup is also a closed subgroup. For any subgroup H of a topological group G, the closure $\bar{H}$ of H is also a subgroup. If H is a normal subgroup, so is $\bar{H}$. If H is commutative, so is $\bar{H}$. In a T_2-topological group G, the †centralizer $C(M) = \{x \in G \mid xm = mx\ (m \in M)\}$ of a subset M of G is a closed subgroup of G. In particular, the †center $C = C(G)$ of a T_2-topological group is a closed normal subgroup.

E. Quotient Spaces

Given a subgroup H of a topological group G, let $G/H = \{aH \mid a \in G\}$ be the set of †left cosets, and let p be the canonical surjection $p(a) = aH$ of G onto G/H. Consider the †quotient topology on G/H, namely, the strongest topology such that p is a continuous mapping. Since a subset A of G/H is open when $p^{-1}(A)$ is an open set of G, p is also an †open mapping. The set G/H with this topology is called the **left quotient space** (or **left coset space**) of G by H. The **right quotient space** (or **right coset space**) $H \backslash G = \{Ha \mid a \in G\}$ is defined similarly. The quotient space G/H is discrete if and only if H is an open subgroup of G. The quotient space is a Hausdorff space if and only if H is a closed subgroup. If G/H and H are both †connected, then G itself is connected. If G/H and H are both †compact, then G is compact. If H is a closed subgroup of G and G/H, H are both †locally compact, then G is locally compact.

Suppose that H is a normal subgroup of a topological group G. Then the quotient group

G/H is a topological group with the topology of the quotient space G/H. This topological group is called the **quotient group** of the topological group G by the normal subgroup H.

F. Connectivity

The †connected component G_0 containing the identity e of a topological group G is a closed normal subgroup of G. The connected component that contains an element $a \in G$ is the coset $aG_0 = G_0 a$. G_0 is called the **identity component** of G. The quotient group G/G_0 is †totally disconnected. A connected topological group G is generated by any neighborhood U of the identity. Namely, any element of G can be expressed as the product of a finite number of elements in U. Totally disconnected (in particular, discrete) normal subgroups of a connected topological group G are contained in the center of G.

G. Uniformity

Let $\mathfrak{N}_0$ be the neighborhood system of the identity of a topological group G, and let $U_l = \{(x, y) \in G \times G \mid y \in xU\}$ for $U \in \mathfrak{N}_0$. Then a †uniformity having $\{U_l \mid U \in \mathfrak{N}_0\}$ as a base is defined on G. This uniformity is called the **left uniformity** of G. Left translation $x \rightarrow ax$ of G is †uniformly continuous with respect to the left uniformity. The **right uniformity** is defined similarly by $U_r = \{(x, y) \mid y \in Ux\}$. These two uniformities do not necessarily coincide. The mapping $x \rightarrow x^{-1}$ is a †uniform isomorphism of G considered as a uniform space with respect to the left uniformity onto the same group G considered as a uniform space with respect to the right uniformity. A topological group G is thus a †uniform space under a uniformity †compatible with its topology, and hence it is a completely regular space if the underlying topological space is a T_1-space.

H. Completeness

If a topological group G is †complete with respect to the left uniformity, then it is also complete with respect to the right uniformity, and conversely. In this case the topological group G is said to be **complete**. A locally compact T_2-topological group is complete. If a T_2-topological group G is isomorphic to a dense subgroup of a complete T_2-topological group $\hat{G}$, then $\hat{G}$ is called the **completion** of G, and G is said to be **completable**. A T_2-topological group G is not always completable. For a T_2-topological group G to be completable it is necessary and sufficient that any †Cauchy filter of G considered as a uniform space with respect to the left uniformity is mapped to a Cauchy filter of the same uniform space G under the mapping $x \rightarrow x^{-1}$. Then the completion $\hat{G}$ of G is uniquely determined up to isomorphism. A commutative T_2-topological group always has a completion $\hat{G}$, and $\hat{G}$ is also commutative. If each point of a T_2-topological group G has a †totally bounded neighborhood, there exists a completion $\hat{G}$, and $\hat{G}$ is locally compact.

I. Metrization

If a †metric can be introduced in a T_2-topological group G so that the metric gives the topology of G, then G is said to be **metrizable**. For a T_2-topological group G to be metrizable it is necessary and sufficient that G satisfy the †first axiom of countability. Then the metric can be chosen so that it is **left invariant**, i.e., invariant under left translation. Similarly, it can be chosen so that it is right invariant. In particular, the topology of a compact T_2-topological group that satisfies the first axiom of countability can be given by a metric that is both left and right invariant.

J. Isomorphism Theorems

Let G and G' be topological groups. If a homomorphism f of the underlying group of G into the underlying group of G' is a continuous mapping of the underlying topological space of G into that of G', f is called a **continuous homomorphism**. If f is a continuous open mapping, f is called a **strict morphism** (or **open continuous homomorphism**). A continuous homomorphism of a †paracompact locally compact topological group onto a locally compact T_2-topological group is an open continuous homomorphism.

A topological group G' is said to be **homomorphic** to a topological group G if there exists an open continuous homomorphism f of G onto G'. Let N denote the kernel $f^{-1}(e)$ of f. Then the quotient group G/N is isomorphic to G', with G/N and G' both considered as topological groups (**homomorphism theorem**). Let f be an open continuous homomorphism of a topological group G onto a topological group G', and let H' be a subgroup of G'. Then $H = f^{-1}(H')$ is a subgroup of G, and the mapping φ defined by $\varphi(gH) = f(g)H'$ is a homeomorphism of the quotient space G/H onto G'/H'. In particular, if H' is a normal subgroup, then H is also a normal subgroup and φ is an isomorphism of the quotient group G/H onto G'/H' as topological groups (**first isomorphism theorem**). Let H and N be subgroups of a topological group G such that $HN=NH$. Then the canonical mapping $f: h(H\cap N)\to hN$ of the quotient space $H/H\cap N$ to HN/N is a continuous bijection but not necessarily an open mapping. In particular, if N is a normal subgroup of the group HN, then f is a continuous homomorphism. In addition, if f is an open mapping, the quotient groups $H/H\cap N$ and HN/N are isomorphic as topological groups (**second isomorphism theorem**). For example, f is an open mapping (1) if N is compact or (2) if G is locally compact, HN and N are closed subgroups of G, and H is the union of a countable number of compact subsets. Let H be a subgroup of a topological group G and N be a normal subgroup of G such that $H\supset N$. Then the canonical mapping of the quotient space $(G/N)/(H/N)$ onto G/H is a homeomorphism. In particular, if H is also a normal subgroup, the quotient groups $(G/N)/(H/N)$ and G/H are isomorphic as topological groups (**third isomorphism theorem**).

K. The Projective Limit

Let $\{G_\alpha\}_{\alpha\in A}$ be a family of topological groups indexed by a †directed set A, and suppose that if $\alpha\leqslant\beta$, there exists a continuous homomorphism $f_{\alpha\beta}: G_\beta\to G_\alpha$ such that $f_{\alpha\gamma}=f_{\alpha\beta}\circ f_{\beta\gamma}$ if $\alpha\leqslant\beta\leqslant\gamma$. Then the collection $\{G_\alpha, f_{\alpha\beta}\}$ of the family $\{G_\alpha\}_{\alpha\in A}$ of topological groups together with the family $\{f_{\alpha\beta}\}$ of mappings is called a **projective system** of topological groups. Consider the direct product $\prod_{\alpha\in A} G_\alpha$ of topological groups $\{G_\alpha\}$, and denote by G the set of all elements $x=\{x_\alpha\}_{\alpha\in A}$ of $\prod G_\alpha$ that satisfy $x_\alpha=f_{\alpha\beta}(x_\beta)$ for $\alpha\leqslant\beta$. Then G is a subgroup of $\prod G_\alpha$. The topological group G obtained in this way is called the **projective limit** of the projective system $\{G_\alpha, f_{\alpha\beta}\}$ of topological groups and is denoted by $G=\varprojlim G_\alpha$. If each G_α is a T_2-topological (resp. complete) group, then G is also a T_2-topological (complete) group.

Now consider another projective system $\{G'_\alpha, f'_{\alpha\beta}\}$ of topological groups indexed by the same A, and consider continuous homomorphisms $u_\alpha: G_\alpha\to G'_\alpha$ such that $u_\alpha\circ f_{\alpha\beta}=f'_{\alpha\beta}\circ u_\beta$ for $\alpha\leqslant\beta$. Then there exists a unique continuous homomorphism u of $G=\varprojlim G_\alpha$ into $G'=\varprojlim G'_\alpha$ such that for any $\alpha\in A$, $u_\alpha\circ f_\alpha=f'_\alpha\circ u$ holds, where $f_\alpha(f'_\alpha)$ is the restriction to $G(G')$ of the projection of $\prod G_\alpha$ $(\prod G'_\alpha)$ onto $G_\alpha(G'_\alpha)$. The homomorphism u is called the **projective limit** of the family $\{u_\alpha\}$ of continuous homomorphisms and is denoted by $u=\varprojlim u_\alpha$. Let G be a T_2-topological group, and let $\{H_\alpha\}_{\alpha\in A}$ be a decreasing sequence ($H_\alpha\supset H_\beta$ for $\alpha\leqslant\beta$) of closed normal subgroups of G. Consider the quotient group G/H_α, and let $f_{\alpha\beta}$ be the canonical mapping $gH_\beta\to gH_\alpha$ of G_β to G_α for $\alpha\leqslant\beta$. Then $\{G_\alpha, f_{\alpha\beta}\}$ is a projective system of topological groups. Let f_α be the projection of G onto $G_\alpha=G/H_\alpha$, and let $f=\varprojlim f_\alpha$. Now assume that any neighborhood of the identity of G contains some H_α and that some H_α is complete. Then $f=\varprojlim f_\alpha$ is an isomorphism of G onto $\varprojlim G/H_\alpha$ as topological groups. (For a general discussion of the topological groups already discussed → [1,4].)

L. Locally Compact Groups

For the rest of this article, all topological groups under consideration are assumed to be T_2-topological groups. The identity component G_0 of a locally compact group G is the intersection of all open subgroups of G. In particular, any neighborhood of the identity of a totally disconnected locally compact group contains an open subgroup. A totally disconnected compact group is a projective limit of finite groups with discrete topology.

A T_1-topological space L is called a **local Lie group** if it satisfies the following six conditions: (i) There exist a nonempty subset M of $L\times L$ and a continuous mapping $\mu: M\to L$, called **multiplication** ($\mu(a,b)$ is written as ab). (ii) If (a,b), (ab,c), (b,c), (a,bc) are all in M, then $(ab)c=a(bc)$. (iii) L contains an element e, called the **identity**, such that $L\times\{e\}\subset M$ and $ae=a$ for all $a\in L$. (iv) There exists a nonempty open subset N of L and a continuous mapping $v: N\to L$ such that $av(a)=e$ for all $a\in N$. (v) There exist a neighborhood U of e in L and a homeomorphism f of U into a neighborhood V of the origin in the Euclidean space $\mathbf{R}^n$. (vi) Let D be the open subset of $V\times V$ defined by $D=\{(x,y)\in V\times V\,|\,(f^{-1}(x),f^{-1}(y))\in M, f^{-1}(x), f^{-1}(y)\in U\}$. Then the function $F: D\to V$ defined by $F(x,y)=f\mu(f^{-1}(x),f^{-1}(y))$ is of †class C^ω.

For any neighborhood U of the identity e of a connected locally compact group G, there exist a compact normal subgroup K and a subset L that is a local Lie group under the †induced topology and the group operations of G such that the product LK is a neighborhood of e contained in U. Furthermore, under $(l,k)\to lk$, LK is homeomorphic to the product space $L\times K$. Any compact subgroup of a connected locally compact group G is contained in a maximal compact subgroup, and maximal compact subgroups of G are †conjugate. For a maximal compact subgroup K of G, there exists a finite number of subgroups $H_1,\ldots,H_r$ of G, each of which is isomorphic to the additive group of real numbers such that $G=KH_1\ldots H_r$, and the mapping $(k,h_1,\ldots,h_r)\to kh_1\ldots h_r$ is a homeomorphism of the direct product $K\times H_1\times\ldots\times H_r$ onto G. Any locally compact group has a left-invariant positive measure and a right-invariant positive measure, which are uniquely determined up to constant multiples (→ 225 Invariant Measures). Using these measures, the theory of harmonic analysis on the additive group **R** of real numbers can be extended to that on G (→ 69 Compact Groups; 192 Harmonic Analysis; 422 Topological Abelian Groups; 437 Unitary Representations).

M. Locally Euclidean Groups

Suppose that each point of a topological group G has a neighborhood homeomorphic to an open set of a given Euclidean space. Then G is called a **locally Euclidean group**. If the underlying topological space of a topological group has the structure of a †real analytic manifold such that the group operation $(x, y) \rightarrow xy^{-1}$ is a real analytic mapping, then G is called a †**Lie group**. A Lie group is a locally Euclidean group.

N. Hilbert's Fifth Problem

Hilbert's fifth problem asks if every locally Euclidean group is a Lie group (→ 196 Hilbert). This problem was solved affirmatively in 1952; it was proved that any †locally connected finite-dimensional locally compact group is a Lie group (D. Montgomery and L. Zippin [3]). In connection with this, the relation between Lie groups and general locally compact groups has been studied, and the following results have been obtained: A necessary and sufficient condition for a locally compact group to be a Lie group is that there exist a neighborhood of the identity e that does not contain any subgroup (or any normal subgroup) other than $\{e\}$. A locally compact group has an open subgroup that is the projective limit of Lie groups. Hilbert's fifth problem is closely related to the following problem: Find the conditions for a †topological transformation group operating †effectively on a manifold to be a Lie group (→ 431 Transformation Groups).

O. Covering Groups

Let $\mathfrak{G}$ be the collection of all †arcwise connected and †locally arcwise connected T_2-topological groups. Suppose that $G^* \in \mathfrak{G}$ is a †covering space of $G \in \mathfrak{G}$ and the †covering mapping $f: G^* \rightarrow G$ is an open continuous homomorphism, with G^* and G considered as topological groups. Then G^* (or, more precisely, (G^*, f)) is called a **covering group** of G. Then the kernel $f^{-1}(e) = D$ of f is a discrete subgroup contained in the center of G^*, and G^*/D and G, considered as topological groups, are isomorphic to each other. Let $\pi_1(G)$ be the †fundamental group of G. The natural homomorphism $f^*: \pi_1(G^*) \rightarrow \pi_1(G)$ induced by f is an injective homomorphism, and if we identify $\pi_1(G^*)$ with the subgroup $f^*(\pi_1(G^*))$ of $\pi_1(G)$, we have $D \cong \pi_1(G)/\pi_1(G^*)$. Conversely, if D is any discrete subgroup contained in the center of $G^* \in \mathfrak{G}$, then G^* is a covering group of $G = G^*/D$. For any covering space (G^*, f) of $G \in \mathfrak{G}$, a multiplication law can be introduced in G^* so that G^* is a topological group belonging to $\mathfrak{G}$ and (G^*, f) is a covering group of G. In particular, any $G \in \mathfrak{G}$ has a †simply connected covering group $(\tilde{G}, \varphi)$. Then for any covering group (G^*, f) of G, there exists a homomorphism $f^*: \tilde{G} \rightarrow G^*$, and $(\tilde{G}, f^*)$ is a covering group of G^*. Furthermore, $\varphi = f \circ f^*$. Hence, in particular, any simply connected covering group of G is isomorphic to $\tilde{G}$, with G and $\tilde{G}$ considered as topological groups. This simply connected covering group $(\tilde{G}, \varphi)$ is called the **universal covering group**.

Let G and G' be topological groups, and let e and e' be their identities. A homeomorphism f of a neighborhood U of e onto a neighborhood U' of e' is called a **local isomorphism** of G to G' if it satisfies the following two conditions: (i) If a, b, ab are all contained in U, then $f(ab) = f(a)f(b)$. (ii) Let $f^{-1} = g$, then if a', b', $a'b' \in U'$, $g(a'b') = g(a')g(b')$ holds. If there exists a local isomorphism of G to G', we say that G and G' are **locally isomorphic**. If G^* is a covering group of G, then G^* and G are locally isomorphic. For two topological groups G and G' to be locally isomorphic it is necessary and sufficient that the universal covering groups of G and G' be isomorphic. For two connected Lie groups to be locally isomorphic it is necessary and sufficient that their †Lie algebras be isomorphic.

Let f be a mapping of a neighborhood U of the identity of a topological group G into a group H such that if a, b, ab are all contained in U, then $f(ab) = f(a)f(b)$. Then f is called a **local homomorphism** of G into H and U is called its **domain**. A local homomorphism of a simply connected group $G \in \mathfrak{G}$ into a group H can be extended to a homomorphism of G into H if the domain is connected [2, 4].

P. Topological Rings and Fields

If a ring R has the structure of a topological group such that $(x, y) \rightarrow x + y$ (sum) and $(x, y) \rightarrow xy$ (product) are both continuous mappings of $R \times R$ into R, then R is called a **topological ring**. If a topological ring K is a field (not necessarily commutative) such that $x \rightarrow x^{-1}$

(inverse element) is a continuous mapping of $K^* = K - \{0\}$ into K^*, then K is called a **topological field**. Let us assume that K is a topological field that is a locally compact Hausdorff space and is not discrete. If K is connected, then K is a †division algebra of finite rank over the field $\mathbf{R}$ of real numbers; hence it is isomorphic to the field $\mathbf{R}$ of real numbers, the field $\mathbf{C}$ of complex numbers, or the †quaternion field $\mathbf{H}$. If K is not connected, then K is totally disconnected and is isomorphic to a division algebra of finite rank over the †p-adic number field $\mathbf{Q}_p$ or a division algebra of finite rank over the †formal power series field with coefficients in a finite field [4].

For various important classes of topological groups → 69 Compact Groups; 249 Lie Groups; 422 Topological Abelian Groups; 424 Topological Linear Spaces.

References

[1] N. Bourbaki, Eléments de mathématique, Topologie générale, ch. 3, Actualités Sci. Ind., 1143c, Hermann, third edition, 1960; English translation, General topology, pt. 1, Addison-Wesley, 1966.
[2] C. Chevalley, Theory of Lie groups I, Princeton Univ. Press, 1946.
[3] D. Montgomery and L. Zippin, Topological transformation groups, Interscience, 1955.
[4] L. S. Pontryagin, Topological groups, first edition, Princeton Univ. Press, 1939, second edition, Gordon & Breach, 1966. (Second English edition translated from the second Russian edition, 1954.)

424 (XII.5) Topological Linear Spaces

A. Definition

A †linear space E over the real or complex number field K is said to be a **topological linear space**, **topological vector space**, or **linear topological space** if E is a †topological space and the basic operations $x+y$ and αx ($x, y \in E$, $\alpha \in K$) in the linear space are continuous as mappings of $E \times E$ and $K \times E$, respectively, into E. The coefficient field K may be a general †topological field, although it is usually assumed to be the real number field $\mathbf{R}$ or the complex number field $\mathbf{C}$, and accordingly E is called a **real topological linear space** or a **complex topological linear space**. Topological linear spaces are generalizations of †normed linear spaces and play an important role in the study of †function spaces, such as the †space of distributions, that are not †Banach spaces.

Each topological linear space E is equipped with a †uniform topology in which translations of the neighborhoods of zero form a †uniform family of neighborhoods, and the addition $x+y$ and the multiplication αx by a scalar α are uniformly continuous relative to this uniform topology. In particular, if for each $x \neq 0$ there is a neighborhood of the origin that does not contain x, then E satisfies the †separation axiom T_1 and hence is a †completely regular space. The †completion $\hat{E}$ of E is also a topological linear space.

We assume in this article that K is the real or complex number field and E is a topological linear space over K satisfying the axiom of T_1-spaces. Then E is finite-dimensional if and only if E has a †totally bounded neighborhood of zero. The topology of E is †metrizable if and only if it satisfies the †first countability axiom.

B. Linear Functional

A K-valued function $f(x)$ on E is said to be a **linear functional** if it satisfies (i) $f(x+y) = f(x) + f(y)$ and (ii) $f(\alpha x) = \alpha f(x)$. A linear functional that is continuous relative to the topologies of E and K is said to be a continuous linear functional. (Sometimes continuous linear functionals are simply called linear functionals, while abstract linear functionals are called **algebraic linear functionals**.) The following three statements are equivalent for linear functionals $f(x)$: (i) $f(x)$ is continuous; (ii) the half-space $\{x \in E \mid \operatorname{Re} f(x) > 0\}$ is open; (iii) the hyperplane $\{x \in E \mid f(x) = 0\}$ is closed.

C. The Hahn-Banach Theorem

A linear functional $f(x)$ defined on a linear subspace F of E can be extended to a continuous linear functional on E if and only if there exists an open †convex neighborhood V of the origin in E that is disjoint with $\{x \in F \mid f(x) = 1\}$. Furthermore, if $f(x)$ can be extended, at least one extension $f(x)$ never takes the value 1 on V (**Hahn-Banach theorem**).

D. Dual Spaces

The set E' of all continuous linear functionals on E is called the **dual space** of E. It is often denoted by E^* and is also called the **conjugate space** or **adjoint space**. It forms a linear space when $f+g$ and αf ($f, g \in E'$, $\alpha \in K$) are defined by $(f+g)(x) = f(x) + g(x)$ and $(\alpha f)(x) = \alpha(f(x))$ for $x \in E$.

E. Locally Convex Spaces

A topological linear space is said to be **locally convex** if it has a family of convex sets as a †base of the neighborhood system of 0. It follows from the Hahn-Banach theorem that for each $x \neq 0$ in a locally convex space E there is a continuous linear functional f such that $f(x) \neq 0$. A subset M of E is said to be **circled** if M contains $\alpha M = \{\alpha x \mid x \in M\}$ whenever $|\alpha| \leqslant 1$. A set that is both circled and convex is called **absolutely convex**. In a locally convex space, a family of absolutely convex and closed sets can be chosen as a base of the neighborhood system of the origin. Let A and B be subsets of E. A is said to **absorb** B if there is an $\alpha > 0$ such that $\alpha A \supset B$. A set V that absorbs every point $x \in E$ is called **absorbing**. Neighborhoods of 0 are absorbing.

F. Seminorms

A real-valued function $p(x)$ on E is said to be a **seminorm** (or **pseudonorm**) if it satisfies (i) $0 \leqslant p(x) < +\infty$ $(x \in E)$; (ii) $p(x+y) \leqslant p(x) + p(y)$; and (iii) $p(\alpha x) = |\alpha| p(x)$. The relation $V = \{x \mid p(x) \leqslant 1\}$ gives a one-to-one correspondence between seminorms $p(x)$ and absolutely convex absorbing sets V whose intersection with any line through the origin is closed. In terms of seminorms, the Hahn-Banach theorem states: Let E be a linear space on which a seminorm $p(x)$ is given. If a linear functional $f(x)$ defined on a linear subspace F of E satisfies $|f(x)| \leqslant p(x)$ on F, then $f(x)$ can be extended to the whole space E in such a way that the inequality holds on E.

The topology of a locally convex space is determined by the family of continuous seminorms on it. Conversely, if there is a family of seminorms $\{p_\lambda(x)\}$ $(\lambda \in \Lambda)$ on a linear space E over K that satisfies (iv) $p_\lambda(x) = 0$ for all λ implies $x = 0$, then there exists on E the weakest locally convex topology that renders the seminorms continuous. This topology is called the locally convex topology determined by $\{p_\lambda(x)\}$.

We assume that E is a locally convex space whose topology is determined by the family of seminorms $\{p_\lambda(x)\}$ $(\lambda \in \Lambda)$. Then a †net x_ν of E converges to x if and only if $p_\lambda(x_\nu - x) \to 0$ for all $\lambda \in \Lambda$. If F is a locally convex space whose topology is determined by the family of seminorms $\{q_\mu(y)\}$, then a necessary and sufficient condition for a linear mapping $u: E \to F$ to be continuous is that for every $q_\mu(y)$ there exist a finite number of $\lambda_1, \ldots, \lambda_n \in \Lambda$ and a constant C such that $q_\mu(u(x)) \leqslant C(p_{\lambda_1}(x) + \ldots + p_{\lambda_n}(x))$ $(x \in E)$.

A set is said to be **bounded** if it is absorbed by every neighborhood of zero. When the topology of E is determined by the family $\{p_\lambda(x)\}$ of seminorms a set B is bounded if and only if every p_λ is bounded on B. Totally bounded sets are bounded. The unit ball in a normed space is bounded. Conversely, a locally convex space is normable if it has a bounded neighborhood of 0. A locally convex space is called **quasicomplete** if every bounded closed set is complete. Since Cauchy sequences $\{x_n\}$ are totally bounded, all Cauchy sequences converge in a quasicomplete space (i.e., the space is sequentially complete).

G. Pairing of Linear Spaces

Let E and F be linear spaces over the same field K. A K-valued function $B(x, y)$ $(x \in E, y \in F)$ on $E \times F$ is called a **bilinear functional** or **bilinear form** if for each fixed $y \in F$ (resp. $x \in E$), it is a linear functional of x (resp. y). When a bilinear functional $\langle x, y \rangle$ on $E \times F$ is given so that $\langle x, y \rangle = 0$ for all $y \in F$ (all $x \in E$) implies $x = 0$ $(y = 0)$, then E and F are said to form a (separated) **pairing** relative to the **inner product** $\langle x, y \rangle$. A locally convex space E and its dual space E' form a pairing relative to the natural inner product $\langle x, x' \rangle = x'(x)$ $(x \in E, x' \in E')$.

H. Weak Topologies

When E and F form a pairing relative to an inner product $\langle x, y \rangle$, the locally convex topology on E determined by the family of seminorms $\{|\langle x, y \rangle| \mid y \in F\}$ is called the **weak topology (relative to the pairing** $\langle E, F \rangle$) and is denoted by $\sigma(E, F)$. A net x_ν in E is said to **converge weakly** if it converges in the weak topology. When E and E' are a locally convex space and its dual space, $\sigma(E, E')$ is called the **weak topology** of E, and $\sigma(E', E)$ the **weak* topology** of E'. The weak topology on a locally convex space E is weaker than the original topology on E. Consequently, a weakly closed set is closed. If the set is convex, the converse holds, and hence a convex closed set is weakly closed. Also, boundedness is preserved if we replace the original topology by the weak topology. Thus a weakly bounded set is bounded.

Let E and F form a pairing relative to $\langle x, y \rangle$, and let A be a subset of E. Then the set A° of points $y \in F$ satisfying $\mathrm{Re}\langle x, y \rangle \geqslant -1$ for all $x \in A$ is called the **polar** of A (relative to the pairing). If A is absolutely convex, A° is also absolutely convex and is the set of points y such that $|\langle x, y \rangle| \leqslant 1$ for all $x \in A$. If A is a convex set containing zero, its (weak) closure is equal to the **bipolar** $A^{\circ\circ} = (A^\circ)^\circ$ (**bipolar theorem**). In general, let A be a subset of a topological linear space E. We call the smallest closed convex set containing A the **closed convex hull** of A. If E is locally convex, the bipolar $A^{\circ\circ}$ relative to E' coincides with the closed convex hull of $A \cup \{0\}$.

A subset B of the dual space E' is †equicontinuous on E if and only if it is contained in the polar V° of a neighborhood V of 0 in E. Also, V° is weak*- compact in E' (**Banach-Alaoglu theorem**).

I. Barreled Spaces and Bornological Spaces

An absorbing absolutely convex closed set in a locally convex space E is called a **barrel**. In a sequentially complete space (hence in a quasi-complete space also), a barrel absorbs every bounded set. A locally convex space is said to be **barreled** if each barrel is a neighborhood of 0. A locally convex space is said to be **quasi-barreled** (or **evaluable**) if each barrel that absorbs every bounded set is a neighborhood of 0. Furthermore, a locally convex space is said to be **bornological** if each absolutely convex set that absorbs every bounded set is a neighborhood of 0. Bornological spaces are quasi-barreled. However, they are not necessarily barreled. Furthermore, barreled spaces are not necessarily bornological. A metrizable locally convex space, i.e., a space whose topology is determined by a countable number of seminorms, is bornological. A complete metrizable locally convex space is called a **locally convex Fréchet space** (**(F)-space** or simply **Fréchet space**). To distinguish it from Fréchet space as in 37 Banach Spaces, it is sometimes called a **Fréchet space in the sense of Bourbaki**. (F)-spaces are bornological and barreled.

A continuous linear mapping $u: E \to F$ of one locally convex space into another maps each bounded set of E to a bounded set in F. Conversely, if E is bornological, then each linear mapping that maps every bounded sequence to a bounded set is continuous.

J. The Banach-Steinhaus Theorem

In the dual space of a barreled space E, each (weak*-)bounded set is equicontinuous. Thus if a sequence of continuous linear mappings u_n of E into a locally convex space F converges at each point of E, then u_n converges uniformly on each totally bounded set of E, and the limit linear mapping is continuous (**Banach-Steinhaus theorem**).

K. The S-Topology

Let E and F be paired linear spaces relative to the inner product $\langle x, y\rangle$. When a family S of (weakly) bounded sets of F generates a dense subspace of F, the family of seminorms $\{\sup_{y\in B}|\langle x, y\rangle| \mid B\in S\}$ determines a locally convex topology on E. This is called the **S-topology** or **topology of uniform convergence on members of** S, because $x_\nu \to x$ in the S-topology is equivalent to the uniform convergence of $\langle x_\nu, y\rangle \to \langle x, y\rangle$ on each $B\in S$. The space E with the S-topology is denoted by E_S. The weak topology is the same as the topology of pointwise convergence. The S-topology in which S is the family of all bounded sets in F is called the **strong topology** and is denoted by $\beta(E, F)$. The dual space E' of a locally convex space E is usually regarded as a locally convex space with the strong topology $\beta(E', E)$. It is called the **strong dual space**. The topology of a locally convex space E is that of uniform convergence on equicontinuous sets of E'. The topology of a barreled space E coincides with the strong topology $\beta(E, E')$.

L. Grothendieck's Criterion of Completeness

Let E and F be paired spaces, and let S be a family of absolutely convex bounded sets of F such that: (i) the sets of S generate F; (ii) if B_1, $B_2\in S$, then there is a $B_3\in S$ such that $B_3\supset B_1$ and $B_3\supset B_2$. Then E_S is complete if and only if each algebraic linear functional $f(y)$ on F that is weakly continuous on every $B\in S$ is expressed as $f(y)=\langle x, y\rangle$ for some $x\in E$. When E_S is not complete, the space of all linear functionals satisfying this condition gives the completion $\hat{E}_S$ of E_S.

M. Mackey's Theorem

Let E, F, and S satisfy the same conditions as in Section L. Then the dual space of E_S is equal to the union of the weak completions of λB, where $\lambda > 0$ and $B\in S$ (**Mackey's theorem**).

N. The Mackey Topology

When E and F form a pairing, the topology on E of uniform convergence on convex weakly compact sets of F is called the **Mackey topology** and is denoted by $\tau(E, F)$. The dual space of E endowed with a locally convex topology T coincides with F if and only if T is stronger than the weak topology $\sigma(E, F)$ and weaker than the Mackey topology $\tau(E, F)$ (**Mackey-Arens theorem**). A locally convex space is said to be a **Mackey space** if the topology is equal to the Mackey topology $\tau(E, E')$. Every quasi-barreled space is a Mackey space.

O. Reflexivity

Let E be a locally convex space. The dual space E'' of the dual space E' equipped with the strong topology contains the original space E. We call E **semireflexive** if $E''=E$, and **reflexive** if in addition the topology of E coincides with the strong topology $\beta(E, E')$. E is semireflexive if and only if every bounded weakly closed set of E is weakly compact. E is reflexive if and only if E is semireflexive and (quasi)barreled.

A barreled space in which every bounded closed set is compact is called a **Montel space** or **(M)-space**. (M)-spaces are reflexive, and their strong dual spaces are also (M)-spaces.

Many of the function spaces that appear in applications are (F)-spaces or their dual spaces. For these spaces detailed consequences of the countability axiom are known [7, 8]. A convex set C in the dual space E' of an (F)-space E is weak*-closed if and only if for every neighborhood V of 0 in E, $C \cap V^\circ$ is weak*-closed (**Kreĭn-Shmul'yan theorem**). The strong dual space E' of an (F)-space E is (quasi)barreled if and only if it is bornological. In particular, the dual space of a reflexive (F)-space is bornological.

P. (DF)-Spaces

A locally convex space is called a **(DF)-space** if it satisfies: (i) There is a countable base of bounded sets (i.e., every bounded set is included in one of them); (ii) if the intersection V of a countable number of absolutely convex closed neighborhoods of 0 absorbs every bounded set, then V is also a neighborhood of 0. The dual space of an (F)-space is a (DF)-space, and the dual space of a (DF)-space is an (F)-space. A linear mapping of a (DF)-space E into a locally convex space F is continuous if and only if its restriction to every bounded set of E is continuous. A quasicomplete (DF)-space is complete.

Q. Bilinear Mappings

A bilinear mapping $b(x, y)$ on locally convex spaces E and F ($x \in E, y \in F$) to a locally convex space G is said to be **separately continuous** if for each fixed $y \in F$ ($x \in E$) it is continuous as a function of x (y). The linear mappings obtained from $b(x, y)$ by fixing x (y) are denoted by $b_x(y)$ ($b_y(x)$). We call $b(x, y)$ **hypocontinuous** if for each bounded set B of E and B' of F, $\{b_x(y) \mid x \in B\}$ and $\{b_y(x) \mid y \in B'\}$ are equicontinuous. A continuous bilinear mapping is hypocontinuous. However, the converse is not always true. A separately continuous bilinear mapping is not necessarily hypocontinuous. If both E and F are barreled, however, then every separately continuous mapping is hypocontinuous. If E is an (F)-space and F is metrizable, then every separately continuous bilinear mapping is continuous. Similarly, if both E and F are (DF)-spaces, then every hypocontinuous bilinear mapping is continuous.

R. Tensor Products

It is possible to introduce many topologies in the †tensor product $E \otimes F$ of locally convex spaces E and F. The **projective topology** (or **topology** π) is defined to be the strongest topology such that the natural bilinear mapping $E \times F \to E \otimes F$ is continuous. The dual space of $E \otimes_\pi F$ is identified with the space $B(E, F)$ of all continuous bilinear functionals on $E \times F$. The completion of $E \otimes_\pi F$ is denoted by $E \hat{\otimes} F$. The **topology of biequicontinuous convergence** (or **topology** ε) is defined to be the topology of uniform convergence on sets $V^\circ \times U^\circ$, where V and U are neighborhoods of 0 in E and F, respectively, considering the elements of $E \otimes F$ as linear functionals on $E' \otimes F'$ by the natural pairing of $E \otimes F$ and $E' \otimes F'$. The completion of $E \otimes_\varepsilon F$ is denoted by $E \hat{\hat{\otimes}} F$. The dual space of $E \otimes_\varepsilon F$ coincides with the subspace $J(E, F)$ of $B(E, F)$ composed of the union of the absolute convex hulls of the products $V^\circ \otimes U^\circ$ of equicontinuous sets. The elements of $J(E, F)$ are called **integral bilinear functionals**.

Closely related to $E \hat{\hat{\otimes}} F$ is L. Schwartz's ε **tensor product** $E \varepsilon F$ [12]. (They coincide if E and F are complete and if E or F has the †approximation property.) $E \varepsilon F$ can be regarded as (i) a space of vector-valued functions if E is a space of functions and F is an abstract locally convex space, especially a space of functions of two variables if E and F are, respectively, spaces of functions of one variable, and (ii) a space of operators $G \to F$ if E is the dual space G' of a locally convex space G.

S. Nuclear Spaces

Let E be a locally convex space, V be an absolutely convex closed neighborhood of the origin, and $p(x)$ be the seminorm corresponding to V. Then we denote by E_V the normed space with norm $p(x)$ obtained from E by identifying the two elements x and y with $p(x-y)=0$. If $U \subset V$, then a natural linear mapping $\varphi_{U,V}: E_U \to E_V$ is defined.

A locally convex space E is said to be a

nuclear space (resp. **Schwartz space** or simply **(S)-space**) if for each absolutely convex closed neighborhood V of 0 there is another U such that $\varphi_{U,V}$ is a †nuclear operator (resp. †compact operator) as an operator of E_U into the completion of E_V. A nuclear space or (S)-space is an (M)-space if it is quasicomplete and quasibarreled. A locally convex space E is a nuclear space if and only if the topologies π and ε coincide on the tensor product $E \otimes F$ with any locally convex space F. Accordingly, it follows that $B(E, F) = J(E, F)$. This can be regarded as a generalization of Schwartz's **kernel theorem**, which says that every separately continuous bilinear functional on $\mathscr{D}_x \times \mathscr{D}_y$ is represented by an integral with kernel in $\mathscr{D}'_{xy}$. The theory of topological tensor products and nuclear spaces is due to Grothendieck [9].

A locally convex space E is a nuclear (F)-space if and only if E is isomorphic to a closed subspace of $C^\infty(-\infty, \infty)$ (T. Kōmura and Y. Kōmura, 1966). An example of a nuclear (F)-space without basis is known (B. S. Mityagin and N. M. Zobin, 1974).

T. Gel'fand Triplet

Let H and L be Hilbert spaces. If L is a dense subspace of H and the injection $L \rightarrow H$ is a †Hilbert-Schmidt operator, then $H = H'$ is regarded as a dense subspace of L' and the injection $H' \rightarrow L'$ is a Hilbert-Schmidt operator. In this case, (L, H, L') is called a **Gel'fand triplet** (or a **rigged Hilbert space**).

A subset of H is called a cylindrical set if it is expressed in the form $P_F^{-1}(B)$ by the orthogonal projection P_F onto a finite-dimensional subspace F and a Borel subset B of F. If a finitely additive positive measure μ with $\|\mu\|_1 = 1$ defined on the cylindrical sets of H satisfies (i) μ is countably additive on cylindrical sets for a fixed F and (ii) for any $\varepsilon > 0$ there exists a $\delta > 0$ such that $\|x\| < \delta$ implies $\mu\{y \in H \mid |\langle x, y \rangle| \geqslant 1\} < \varepsilon$, then μ is the restriction of a countably additive measure $\tilde{\mu}$ defined on the Borel subsets of L' (**Minlos's theorem**, 1959).

Let T be a self-adjoint operator in H. Then T has a natural extension $\tilde{T}$ in L' and almost every continuous spectrum λ of T has an associated eigenvector x_λ in L': $\tilde{T}x_\lambda = \lambda x_\lambda$, $x_\lambda \in L'$.

U. The Extreme Point Theorem

Let A be a subset of a linear space E. A point $x \in A$ is said to be an **extreme point** if x is an extreme point of any real segment containing x and contained in A. If A is a compact convex subset of a locally convex space E, A is the convex closed hull of (i.e., smallest convex closed set containing) the set of its extreme points (**Kreĭn-Milman theorem**). In applications it is important to know whether every point of A is represented uniquely as an integral of extreme points. For a metrizable convex compact subset A of a locally convex space E, the following two conditions are equivalent (**Choquet's theorem**): (i) A is a **simplex**, i.e., if we put $\tilde{A} = \{(\lambda x, \lambda) \mid x \in A, \lambda > 0\} \subset E \times \mathbf{R}^1$, the vector space $\tilde{A} - \tilde{A}$ becomes a †lattice with positive cone $\tilde{A}$; (ii) for any $x \in A$ there exists a unique positive measure μ on A with $\|\mu\|_1 = 1$ such that $l(x) = \int_A l(y) d\mu(y)$ $(l \in E')$ and the support of μ is contained in the set of extreme points of A.

V. Weakly Compact Set

A subset of a quasicomplete locally convex space is relatively weakly compact if and only if every sequence in the set has a weak accumulation point (**Eberlein's theorem**). If E is a metrizable locally convex space, every weakly compact set of E is weakly sequentially compact (**Shmul'yan's theorem**). If E is a quasicomplete locally convex space, the convex closed hull of any weakly compact subset is weakly compact (**Kreĭn's theorem**). If E is not quasicomplete, this is not necessarily true.

W. Permanence

Each subspace, quotient space, direct product, direct sum, projective limit, and inductive limit (of a family) of locally convex spaces has a unique natural locally convex topology. These spaces, except for quotient spaces and inductive limits, are separated, and a quotient space E/A is separated if and only if the subspace A is closed. The limit of a sequence $E_1 \subset E_2 \subset \ldots$ is said to be a **strictly inductive limit** if E_n has the induced topology as a subspace of E_{n+1}. If E is a strictly inductive limit of a sequence E_n such that E_n is closed in E_{n+1} or if E is the inductive limit of a sequence $E_1 \subset E_2 \subset \ldots$ such that the mapping $E_n \rightarrow E_{n+1}$ maps a neighborhood of 0 to a relatively weakly compact set, then E is separated and each bounded set of E is the image of a bounded set in some E_n. If $E = \bigcup E_n$ is the strictly inductive limit of the sequence $\{E_n\}$, then the topology of E_n coincides with the relative topology of $E_n \subset E$. The strictly inductive limit of a sequence of (F)-spaces is called an **(LF)-space**.

Any complete locally convex space (resp. any locally convex space) is (resp. a dense linear subspace of) the projective limit of Banach spaces. Every (F)-space E is the projective limit of a sequence of Banach spaces $E_1 \leftarrow E_2 \leftarrow \ldots$. In particular, E is said to be a **count-**

ably normed space if the mappings $E \to E_n$ are one-to-one and $\|x\|_n \leqslant \|x\|_{n+1}$ for all $x \in E$ with E considered as a subspace E_n. We call E a **countably Hilbertian space** if, in particular, the E_n are †Hilbert spaces. An (F)-space with at least one continuous norm is a nuclear space if and only if it is a countably Hilbertian space such that the mappings $E_{n+1} \to E_n$ are Hilbert-Schmidt operators or nuclear operators.

A locally convex space is bornological if and only if it is the inductive limit of normed spaces. A locally convex space is said to be **ultrabornological** if it is the inductive limit of Banach spaces, or in particular, if it is quasi-complete and bornological.

Properties of spaces, such as being complete, quasicomplete, semireflexive, or having every bounded closed set compact, are inherited by closed subspaces, direct products, projective limits, direct sums, and strictly inductive limits formed from the original spaces, and properties of spaces, such as being Mackey, quasibarreled, barreled, and bornological, are inherited by quotient spaces, direct sums, inductive limits, and direct products formed from the spaces. (For direct products of high power of bornological spaces, unsolved problems still exist concerning the inheritance of properties.) Quotient spaces of (F)-spaces are (F)-spaces, but quotient spaces of general complete spaces are not necessarily complete. There are examples of a Montel (F)-space whose quotient space is not reflexive and a Montel (DF)-space whose closed subspace is neither a Mackey space nor a (DF)-space. The property of being a Schwartz space or a nuclear space is inherited by the completions, subspaces, quotient spaces of closed subspaces, direct products, projective limits, direct sums of countable families, and inductive limits of countable families formed from such spaces. Tensor products of nuclear spaces are nuclear spaces. Y. Kōmura gave an example of a non-complete space that is quasicomplete, bornological, and nuclear (and hence a Montel space).

X. The Open Mapping Theorem and the Closed Graph Theorem

Let E and F be topological linear spaces. The statement that every continuous linear mapping of E onto F is open is called the **open mapping theorem** (or **homomorphism theorem**), and the statement that every linear mapping of F into E is continuous if its graph is closed in $F \times E$ is called the **closed graph theorem**. These theorems hold if both E and F are complete and metrizable (S. Banach).

A locally convex space is said to be B-**complete** (or **fully complete**) if a subspace C of E' is weak*-closed whenever $C \cap V^\circ$ is weak*-closed for every neighborhood V of 0 in E. (F)-spaces and the dual spaces of reflexive (F)-spaces are B-complete. B-complete spaces are complete, and closed subspaces and quotient spaces by closed subspaces of B-complete spaces are B-complete. If E is B-complete and F is barreled, then the open mapping theorem and the closed graph theorem hold (V. Pták).

Both theorems hold also if F is ultrabornological and E is a locally convex space obtained from a family of (F)-spaces after a finite number of operations of taking closed subspaces, quotient spaces by closed spaces, direct products of countable families, projective limits of countable families, direct sums of countable families, and inductive limits of countable families. This was conjectured by Grothendieck and proved by W. Słowikowski (1961) and D. A. Raikov. Later, L. Schwartz, A. Martineau, M. De Wilde, W. Robertson, and M. Nakamura simplified the proof and enlarged the class of spaces E [15].

(LF)-spaces, the dual spaces of Schwartz (F)-spaces, and the space $\mathscr{D}'$ of distributions are examples of spaces E described in the previous paragraph.

Y. Nonlocally Convex Spaces

The space L_p for $0<p<1$ shows that nonlocally convex spaces are meaningful in functional analysis. Recently, the Banach-Steinhaus theorem, closed graph theorems, etc. have been investigated for nonlocally convex topological linear spaces [13].

Z. Diagram of Topological Linear Spaces

The spaces in Fig. 1 are all locally convex spaces over the real number field or the complex number field and satisfy the separation axiom T_1. The notation $A \to B$ means that spaces with property A have property B. Main properties of dual spaces are listed in Table 1.

References

[1] N. Bourbaki, Eléments de mathématique, Espaces vectoriels topologiques, Actualités Sci. Ind., Hermann, 1189a, 1966; 1229b, 1967; 1230a, 1955.
[2] A. Grothendieck, Espaces vectoriels topologiques, Lecture notes, São Paulo, 1954.
[3] G. Köthe, Topologische lineare Räume I, Springer, second edition, 1966; English translation, Topological vector spaces I, II, Springer, 1969, 1979.

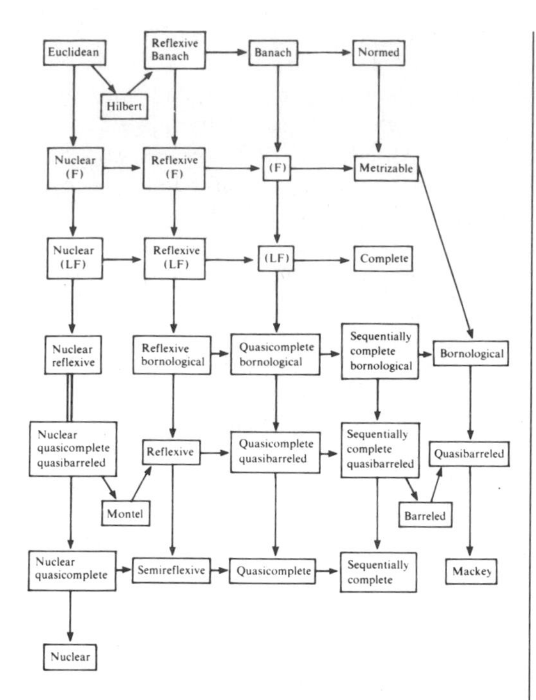

Fig. 1
Topological linear spaces.

Table 1

E	E'
Semireflexive	Barreled
Reflexive	Reflexive
Quasibarreled	Quasicomplete
Bornological	Complete
Reflexive, (F)	Bornological
(F)	(DF)
(DF)	(F)
(M)	(M)
Nuclear, (LF) or (DF)	Nuclear, reflexive
Complete, (S)	Ultrabornological

[4] I. M. Gel'fand et al., Generalized functions. II, Functions and generalized function spaces (with G. E. Shilov), Academic Press, 1968; IV, Applications of harmonic analysis (with N. Ya. Vilenkin), Academic, 1964. (Original in Russian, 1958–1966.)
[5] A. Pietsch, Nukleare lokalkonvexe Räume, Berlin, 1965.
[6] A. P. Robertson and W. Robertson, Topological vector spaces, Cambridge Univ. Press, 1964.
[7] J. Dieudonné and L. Schwartz, La dualité dans les espaces $(\mathfrak{F})$ et $(\mathfrak{LF})$, Ann. Inst. Fourier, 1 (1949), 61–101.
[8] A. Grothendieck, Sur les espaces (F) et (DF), Summa Brasil. Math., 3 (1954), 57–123.
[9] A. Grothendieck, Produits tensoriels topologiques et espaces nucléaires, Mem. Amer. Math. Soc., 1955.
[10] G. Choquet, Le théorème de représentation intégrale dans les ensembles convexes compacts, Ann. Inst. Fourier, 10 (1960), 333–344.
[11] J. L. Kelley, I. Namioka, et al., Linear topological spaces, Van Nostrand, 1963.
[12] L. Schwartz, Théorie des distributions à valeurs vectorielles I, Ann. Inst. Fourier, 7 (1957), 1–141.
[13] N. Adasch, B. Ernst, and D. Keim, Topological vector spaces: The theory without convexity conditions, Lecture notes in math. 639, Springer, 1978.
[14] F. Treves, Topological vector spaces, distributions and kernels, Academic Press, 1967.
[15] M. De Wilde, Closed graph theorems and webbed spaces, Pitman, 1978.

425 (II.16) Topological Spaces

A. Introduction

Convergence and continuity, as well as the algebraic operations on real numbers, are fundamental notions in analysis. In an abstract space too, it is possible to provide an additional structure so that convergence and continuity can be defined and a theory analogous to classical analysis can be developed. Such a structure is called a **topological structure** (for a precise definition, → Section B). There are several ways of giving a topology to a space. One method is to axiomatize the notion of convergence (M. Fréchet [1], 1906; → 87 Convergence). However, defining a topology in terms of either a neighborhood system (due to F. Hausdorff [3], 1914), a closure operation (due to C. Kuratowski, *Fund. Math.*, 3 (1922)), or a family of open sets is more common.

B. Definition of a Topology

Let X be a set. A **neighborhood system** for X is a function $\mathfrak{U}$ that assigns to each point x of X, a family $\mathfrak{U}(x)$ of subsets of X subject to the following axioms (**U**):
(1) $x \in U$ for each U in $\mathfrak{U}(x)$.
(2) If $U_1, U_2 \in \mathfrak{U}(x)$, then $U_1 \cap U_2 \in \mathfrak{U}(x)$.
(3) If $U \in \mathfrak{U}(x)$ and $U \subset V$, then $V \in \mathfrak{U}(x)$.
(4) For each U in $\mathfrak{U}(x)$, there is a member W of $\mathfrak{U}(x)$ such that $U \in \mathfrak{U}(y)$ for each y in W.

A **system of open sets** for a set X is a family $\mathfrak{O}$ of subsets of X satisfying the following axioms (**O**):

(1) $X, \varnothing \in \mathfrak{O}$.
(2) If $O_1, O_2 \in \mathfrak{O}$, then $O_1 \cap O_2 \in \mathfrak{O}$.
(3) If $O_\lambda \in \mathfrak{O}$ $(\lambda \in \Lambda)$, then $\bigcup_{\lambda \in \Lambda} O_\lambda \in \mathfrak{O}$.

A **system of closed sets** for a space X is a family $\mathfrak{F}$ of subsets of X satisfying the following axioms **(F)**:
(1) $X, \varnothing \in \mathfrak{F}$.
(2) If $F_1, F_2 \in \mathfrak{F}$, then $F_1 \cup F_2 \in \mathfrak{F}$.
(3) If $F_\lambda \in \mathfrak{F}$ $(\lambda \in \Lambda)$, then $\bigcap_{\lambda \in \Lambda} F_\lambda \in \mathfrak{F}$.

A **closure operator** for a space X is a function that assigns to each subset A of X, a subset A^a of X satisfying the following axioms **(C)**:
(1) $\varnothing^a = \varnothing$.
(2) $(A \cup B)^a = A^a \cup B^a$.
(3) $A \subset A^a$.
(4) $A^a = A^{aa}$.

An **interior operator** for a space X is a function that assigns to each subset A of X a subset A^i of X satisfying the following axioms **(I)**:
(1) $X^i = X$.
(2) $(A \cap B)^i = A^i \cap B^i$.
(3) $A^i \subset A$.
(4) $A^{ii} = A^i$.

Any one of these five structures for a set X, i.e., a structure satisfying any one of **(U)**, **(O)**, **(F)**, **(C)**, or **(I)**, determines the four other structures in a natural way. For instance, assume that a system of open sets $\mathfrak{O}$ satisfying **(O)** is given. In this case, each member of $\mathfrak{O}$ is called an **open set**. A subset U of X is called a **neighborhood** of a point x in X provided that there is an open set O such that $x \in O \subset U$. If $\mathfrak{U}(x)$ is the family of all neighborhoods of x, the function $x \to \mathfrak{U}(x)$ satisfies **(U)**. The complement of an open set in X is called a **closed set**. The family $\mathfrak{F}$ of all closed sets satisfies **(F)**. Given a subset A of X, the intersection A^a of the family of all closed sets containing A is called the **closure** of A, and each point of A^a is called an **adherent point** of A. The closure A^a is the smallest closed set containing A, and the function $A \to A^a$ satisfies **(A)**. The closure A^a is also denoted by $\bar{A}$ or $\mathrm{Cl}\,A$. Dually, there is a largest open subset A^i of A. The set A^i (also denoted by A° or $\mathrm{Int}\,A$) is called the **interior** of A, and each point of A° is called an **interior point** of A. The closure and interior are related by $A^\circ = X - \overline{(X-A)}$ and $\bar{A} = X - (X-A)^\circ$. The correspondence $A \to A^\circ$ satisfies **(I)**. Conversely, open sets can be characterized variously as follows:

$$\begin{aligned} A \text{ is open} &\Leftrightarrow A \in \mathfrak{U}(x) \text{ for each } x \text{ in } A \\ &\Leftrightarrow X - A \in \mathfrak{F} \\ &\Leftrightarrow \overline{(X-A)} = X - A \\ &\Leftrightarrow A^\circ = A. \end{aligned}$$

When a structure satisfying **(U)**, **(F)**, **(C)**, or **(I)** is given, one of the four characterizations of open sets can be used to define a system of open sets satisfying **(O)** and hence the other structure.

A **topological structure** or simply a **topology** for a space X is any of these five structures for X. If two topologies τ_1 and τ_2 for X give rise to identical systems of open sets, then τ_1 and τ_2 are considered to be identical. For this reason "topology" frequently means simply "system of open sets" in the literature. A **topological space** is a set X provided with a topology τ and is denoted by (X, τ) or simply X when there is no ambiguity.

C. Examples

(1) Discrete Topology. Let X be a set, and let the system $\mathfrak{O}$ of open sets be the family of all subsets of X. The resulting topology is called the **discrete topology**, and X with the discrete topology is a **discrete topological space**. In this space, $\bar{A} = A^\circ = A$ for each subset A, and A is a neighborhood of each of its points.

(2) Trivial Topology. The **trivial** (or **indiscrete**) **topology** for a set X is defined by the system of open sets which consists of X and $\varnothing$ only. If $A \subsetneqq X$, then $A^\circ = \varnothing$, and if $A \neq \varnothing$, then $\bar{A} = X$. Each point of X has only one neighborhood, X itself.

(3) Metric Topology. Let (X, ρ) be a †metric space, i.e., a set X provided with a †metric ρ. For a positive number ε, the ε-neighborhood of a point x is defined to be the set $U_\varepsilon(x) = \{y \mid y \in X, \rho(x, y) < \varepsilon\}$. Let $\mathfrak{U}(x)$ be the family of all sets V such that $U_\varepsilon(x) \subset V$ for some ε; then the assignment $x \to \mathfrak{U}(x)$ satisfies **(U)** and hence defines a topology. This topology is the **metric topology** for the metric space (X, ρ).

(4) Order Topology. Let X be a set †linearly ordered by $\leqslant$. For each point x in X, let $\mathfrak{U}(x)$ be the family of all subsets U such that $x \in \{y \mid a < y < b\} \subset U$ for some a, b. The function $x \to \mathfrak{U}(x)$ satisfies **(U)** and defines the **order topology** for the linearly ordered set X.

(5) Convergence and Topology. We can define the notion of convergence in a topological space, and conversely we can define a topology using convergence as a primitive notion (→ 87 Convergence). In particular, for a metric space, the metric topology can be defined in terms of convergent sequences (→ 273 Metric Spaces).

D. Generalized Topological Spaces

When a space X is equipped with a closure operator that does not satisfy all of **(C)**, the space is called a **generalized topological space** by some authors. Topological implications of each axiom in (**C**) have been investigated for such spaces.

E. Local Bases

Let X be a topological space, and let x be a point of X. A collection $\mathfrak{U}_0(x)$ of neighborhoods of x is called a **base for the neighborhood system (fundamental system of neighborhoods** of a point x or **local base** at x) if each neighborhood of x contains a member of $\mathfrak{U}_0(x)$. Let $\{\mathfrak{U}_0(x) \mid x \in X\}$ be a system of local bases; then the system has the following properties ($\mathbf{U}_0$):
(1) For each V in $\mathfrak{U}_0(x)$, $x \in V \subset X$.
(2) If $V_1, V_2 \in \mathfrak{U}_0(x)$, then there is a V_3 in $\mathfrak{U}_0(x)$ such that $V_3 \subset V_1 \cap V_2$.
(3) For each V in $\mathfrak{U}_0(x)$, there exists a $W \subset V$ in $\mathfrak{U}_0(x)$ such that for each y in W, V contains some member of $\mathfrak{U}_0(y)$.

Conversely, suppose that $\{\mathfrak{U}_0(x) \mid x \in X\}$ is a system satisfying ($\mathbf{U}_0$). For each x in X, let $\mathfrak{U}(x)$ consist of all subsets V of X such that $V \supset U$ for some U in $\mathfrak{U}_0(x)$. Then the system $\{\mathfrak{U}(x) \mid x \in X\}$ satisfies (**U**) and therefore defines a topology for X. This topology is called the topology determined by the system $\{\mathfrak{U}_0(x) \mid x \in X\}$.

For instance, in a metric space X, the set of ε-neighborhoods of $x (\varepsilon > 0)$ is a local base at x with respect to the metric topology. In an arbitrary topological space, the collection of all open sets containing x, i.e., the **open neighborhoods** of x, is a local base at x.

Two systems satisfying ($\mathbf{U}_0$) are called **equivalent** if they determine the same topology. For systems $\{\mathfrak{U}_0(x) \mid x \in X\}$ and $\{\mathfrak{B}_0(x) \mid x \in X\}$ to be equivalent it is necessary and sufficient that for each x in X each member of $\mathfrak{U}_0(x)$ contain a member of $\mathfrak{B}_0(x)$ and each member of $\mathfrak{B}_0(x)$ contain a member of $\mathfrak{U}_0(x)$.

Sometimes the word "neighborhood" stands for a member of a local base or for an open neighborhood. However, this convention is not used here.

F. Bases and Subbases

A family $\mathfrak{O}_0$ of open sets of a topological space X is called a **base for the topology (base for the space**, or **open base**) if each open set is the union of a subfamily of $\mathfrak{O}_0$. A base $\mathfrak{O}_0$ for the topology of a topological space X has the following properties ($\mathbf{O}_0$):
(1) $\bigcup \mathfrak{O}_0 = X$.
(2) If $W_1, W_2 \in \mathfrak{O}_0$ and $x \in W_1 \cap W_2$, then there is a W_3 in $\mathfrak{O}_0$ such that $x \in W_3 \subset W_1 \cap W_2$.

Conversely, if a family $\mathfrak{O}_0$ of subsets of a set X satisfies ($\mathbf{O}_0$), then $\mathfrak{O}_0$ is a base for a unique topology. A member of $\mathfrak{O}_0$ is called a **basic open set**.

A family $\mathfrak{O}_{00}$ of open sets of a topological space X is a **subbase for the topology** (or **subbase for the space**) if the family of all finite intersections of members of $\mathfrak{O}_{00}$ is a base for the topology. If $\mathfrak{O}_{00}$ a subbase for the topology of a topological space X, then $\bigcup \mathfrak{O}_{00} = X$. Conversely, if $\mathfrak{O}_{00}$ is a family of subsets of a set X such that $\bigcup \mathfrak{O}_{00} = X$, then the family of all finite intersections of members of $\mathfrak{O}_{00}$ is a base for a unique topology τ. A subset of X is open for τ if and only if it is the union of a family of finite intersections of members of $\mathfrak{O}_{00}$. The system of open sets relative to τ is said to be **generated** by the family $\mathfrak{O}_{00}$. Thus any family of sets defines a topology for its union.

A set $\mathfrak{F}$ of subsets of a topological space is called a **network** if for each point x and its neighborhood U there is a member $F \in \mathfrak{F}$ such that $x \in F \subset U$ (A. V. Arkhangel'skiĭ, 1959). If all $F \in \mathfrak{F}$ are required to be open, the network $\mathfrak{F}$ is exactly an open base.

G. Continuous Mappings

A mapping f on a topological space X into a topological space Y is called **continuous** at a point a of X if it satisfies one of the following equivalent conditions:
(1) For each neighborhood V of $f(a)$, there is a neighborhood U of a such that $f(U) \subset V$. (1') For each neighborhood V of $f(a)$, the inverse image $f^{-1}(V)$ is a neighborhood of a.
(2) For an arbitrary subset A of X such that $a \in \bar{A}$, $f(a) \in \overline{f(A)}$.

Continuity can also be defined in terms of convergence (→ 87 Convergence).

If f is continuous at each point of X, f is said to be **continuous**. Continuity of f is equivalent to each of the following conditions:
(1) For each open subset O of Y, the inverse image $f^{-1}(O)$ is open in X.
(1') The inverse image under f of each member of a subbase for the topology of Y is open in X.
(2) For each closed subset F of Y, the inverse image $f^{-1}(F)$ is closed.
(3) For each subset A of X, $f(\bar{A}) \subset \overline{f(A)}$.

The image $f(X)$ of X under a continuous mapping f is called a **continuous image** of X. Let X, Y, and Z be topological spaces, and let $f: X \rightarrow Y$ and $g: Y \rightarrow Z$ be mappings. If f is continuous at a point a of X and g is continuous at $f(a)$, then the composite mapping $g \circ f: X \rightarrow Z$ is continuous at the point a. Hence if f and g are continuous, so is $g \circ f$.

When a continuous mapping $f: X \rightarrow Y$ is

†bijective and f^{-1} is continuous, the mapping f is called a **homeomorphism** (named by H. Poincaré, 1895) or **topological mapping**. Two topological spaces X and Y are **homeomorphic**, $X \approx Y$, if there is a homeomorphism $f: X \to Y$.

The relation of being homeomorphic is an †equivalence relation. A property which, when held by a topological space, is also held by each space homeomorphic to it is a **topological property** or **topological invariant**. The problem of deciding whether or not given spaces are homeomorphic is called the **homeomorphism problem**.

A mapping $f: X \to Y$ is called **open** (resp. **closed**) if the image under f of each open (resp. closed) subset of X is open (closed) in Y. A continuous bijection that is either open or closed is a homeomorphism.

A continuous surjection $f: X \to Y$ is called a **quotient mapping** if $U \subset Y$ is open whenever $f^{-1}(U)$ is open (→ Section L). If moreover $f | f^{-1}(S)$ is quotient for each $S \subset Y$ as a mapping from the subspace (→ Section J) $f^{-1}(S)$ onto the subspace S, then f is called a **hereditarily quotient mapping**. Open or closed continuous mappings are hereditarily quotient mappings.

H. Comparison of Topologies

When a set X is provided with two topologies τ_1 and τ_2 and the identity mapping: $(X, \tau_1) \to (X, \tau_2)$ is continuous, the topology τ_1 is said to be **stronger** (**larger** or **finer**) than the topology τ_2, τ_2 is said to be **weaker** (**smaller** or **coarser**) than τ_1, and the notation $\tau_1 \geqslant \tau_2$ or $\tau_2 \leqslant \tau_1$ is used. Let $\mathfrak{O}_i$, $\mathfrak{F}_i$, $\mathfrak{U}_i$, and a_i be the system of open sets, system of closed sets, neighborhood system, and closure operation for X relative to the topology τ_i $(i=1,2)$, respectively. Then each of the following is equivalent to the statement $\tau_1 \geqslant \tau_2$:
(1) $\mathfrak{O}_1 \supset \mathfrak{O}_2$.
(2) $\mathfrak{F}_1 \supset \mathfrak{F}_2$.
(3) For each x in X, $\mathfrak{U}_1(x) \supset \mathfrak{U}_2(x)$.
(4) $A^{a_1} \subset A^{a_2}$ for each subset A of X.

Let S be the family of all topologies for X. Then S is ordered by the relation $\geqslant$. The discrete topology is the strongest topology for X. If $\{\tau_\lambda | \lambda \in \Lambda\}$ is a subfamily of S, then among the topologies stronger than each τ_λ, there is a weakest one $\tau_1 = \sup\{\tau_\lambda | \lambda \in \Lambda\}$. Similarly, among the topologies weaker than each τ_λ, there is a strongest one $\tau_2 = \inf\{\tau_\lambda | \lambda \in \Lambda\}$. In fact, let $\mathfrak{O}_\lambda$ be the family of all open sets relative to τ_λ; then the system of open sets for τ_1 is generated by $\bigcup_{\lambda \in \Lambda} \mathfrak{O}_\lambda$, and the system of open sets for τ_2 is precisely $\bigcap_{\lambda \in \Lambda} \mathfrak{O}_\lambda$. The family S is therefore a †complete lattice.

I. Induced Topology

Let f be a mapping from a set X into a topological space Y. Then the family $\{f^{-1}(O) | O$ is open in $Y\}$ satisfies axioms (**O**) and defines a topology for X. This topology is called the **topology induced by** f (or simply **induced topology**), and it is characterized as the weakest one among the topologies for X relative to which the mapping f is continuous.

J. Subspaces

Let (X, τ) be a topological space and M be a subset of X. The topology for M induced by the inclusion mapping $f: M \to X$, i.e., the mapping f defined by $f(x) = x$ for each x in M, is called the **relativization** of τ to M or the **relative topology**. The set M provided with the relative topology is called a **subspace** of the topological space (X, τ). Topological terms, when applied to a subspace, are frequently preceded by the adjective "relative" to avoid ambiguity. Thus a **relative neighborhood** of a point x in M is a set of the form $M \cap U$, where U is a neighborhood of x in X. A **relatively open** (**relatively closed**) set in M is a set of the form $M \cap A$, where A is open (closed) in X. For a subset A of M, the relative closure of A in M is $M \cap \bar{A}$, where $\bar{A}$ is the closure of A in X. A mapping $f: X \to Y$ is called an **embedding** if f is a homeomorphism from X to the subspace $f(X)$, and in this case X is said to be **embedded** into Y. A property P is said to hold **locally** on a topological space X if each point x of X has a neighborhood U such that the property P holds on the subspace U. A subset A of X is **locally closed** if for each point x of X, there exists a neighborhood V of x such that $V \cap A$ is relatively closed in V. A subset of X is locally closed if and only if it can be represented as $O \cap F$, where O is open and F is closed in X.

K. Product Spaces

Let X be a set, and for each member λ of an index set Λ, let f_λ be a mapping of X into a topological space X_λ. Then there is a weakest topology for X that makes each f_λ continuous. In fact, this topology is $\sup\{\tau_\lambda\}$, where τ_λ is the topology for X induced by f_λ. In particular, let $\{X_\lambda | \lambda \in \Lambda\}$ be a family of topological spaces, and let X be the Cartesian product $\prod_{\lambda \in \Lambda} X_\lambda$. Then the weakest topology for X such that each projection $\mathrm{pr}_\lambda: X \to X_\lambda$ is continuous is called the **product topology** or **weak topology**. The Cartesian product $\prod_{\lambda \in \Lambda} X_\lambda$ equipped with the product topology is called the **product topological space** or simply the **product space** or **direct product** of the family $\{X_\lambda \mid \lambda \in \Lambda\}$ of topological spaces. If $\mathfrak{O}$ is the family of all open subsets of X_λ, the union $\bigcup_\lambda \mathrm{pr}_\lambda^{-1}(\mathfrak{O}_\lambda)$ is a subbase for the product topology. If $x=\{x_\lambda\}$ is a point of X, then sets of the type $\bigcap_{j=1}^{n} \mathrm{pr}_j^{-1}(U_j)=\prod_{\lambda \neq \lambda_1, \ldots, \lambda_n} X_\lambda \times U_1 \times \ldots \times U_n$ form a local base at x for the product topology, where $\lambda_1, \ldots, \lambda_n \in \Lambda$ and U_j is a neighborhood of x_{λ_j}. Each projection $\mathrm{pr}_\lambda: X \rightarrow X_\lambda$ is continuous and open, and a mapping f from a topological space Y into the product space $\prod_\lambda X_\lambda$ is continuous if and only if $\mathrm{pr}_\lambda \circ f: Y \rightarrow X_\lambda$ is continuous for each λ. Given a family $\{f_\lambda\}$ of continuous mappings $f_\lambda: X_\lambda \rightarrow Y_\lambda$, the product mapping $\prod_\lambda f_\lambda: \prod_\lambda X_\lambda \rightarrow \prod_\lambda Y_\lambda$ is continuous with respect to the product topologies.

For the Cartesian product $\prod_\lambda X_\lambda$ of a family $\{X_\lambda \mid \lambda \in \Lambda\}$ of topological spaces, there is another topology called the **box topology** (or **strong topology**). A base for the box topology is the family of all sets $\prod_\lambda O_\lambda$, where O_λ is open in X_λ for each λ. For a point $x=\{x_\lambda\}$, the family of all sets of the form $\prod_\lambda U_\lambda$ is a local base at x relative to the box topology, where U_λ is a neighborhood of x_λ for each λ. With respect to the box topology, each projection $\mathrm{pr}_\lambda: \prod_\lambda X_\lambda \rightarrow X_\lambda$ is continuous and open, and the product mapping $\prod f_\lambda: \prod_\lambda X_\lambda \rightarrow \prod_\lambda Y_\lambda$ of a family $\{f_\lambda\}$ of continuous mappings $f_\lambda: X_\lambda \rightarrow Y_\lambda$ is continuous. For a finite product of topological spaces, the product topology agrees with the box topology, but for an arbitrary product the product topology is weaker than the box topology. For the Cartesian product of topological spaces the usual topology considered is the product topology rather than the box topology.

L. Quotient Spaces

Let f be a mapping of a topological space X onto a set Y. The **quotient topology** for Y (relative to the mapping f) is the strongest topology for Y such that f is continuous. A subset O of Y is open relative to the quotient topology if and only if $f^{-1}(O)$ is open. Given an equivalence relation $\sim$ on a topological space X, the †quotient set $Y=X/\sim$ provided with the quotient topology relative to the projection $\varphi: X \rightarrow Y$ is called the **quotient topological space** (or simply **quotient space**). A mapping f from the quotient space $Y=X/\sim$ into a topological space is continuous if and only if $f \circ \varphi$ is continuous.

A **partition** of a space X is a family $\{A_\lambda \mid \lambda \in \Lambda\}$ of pairwise disjoint subsets of X such that $\bigcup_\lambda A_\lambda=X$. A partition $\{A_\lambda\}$ of a topological space X determines an equivalence relation $\sim$ on X such that the family $\{A_\lambda\}$ is precisely the family of all equivalence classes under $\sim$, and therefore the partition determines the quotient space $Y=X/\sim$. This space is called the **identification space** of X by the given partition. Each member A_λ of the partition can be regarded as a point of Y, and the projection $\varphi: X \rightarrow Y$ satisfies $\varphi(x)=A_\lambda$ whenever $x \in A_\lambda$. A partition $\{A_\lambda \mid \lambda \in \Lambda\}$ of a topological space is called **upper semicontinuous** if for each A_λ and each open set U containing A_λ, there is an open set V such that $A_\lambda \subset V \subset U$, and V is the union of members of $\{A_\lambda \mid \lambda \in \Lambda\}$. A partition $\{A_\lambda \mid \lambda \in \Lambda\}$ is upper semicontinuous if and only if the projection $\varphi: X \rightarrow Y=\{A_\lambda \mid \lambda \in \Lambda\}$ is a closed mapping.

M. Topological Sums

Let X be a set, and for each member λ of an index set Λ, let f_λ be a mapping of a topological space X_λ to X. Then the family $\{O \subset X \mid f_\lambda^{-1}(O)$ is open for any $\lambda\}$ satisfies the axioms of the open sets. This topology τ is characterized as the strongest one for X that makes each f_λ continuous. A mapping g on X with τ to a topological space Y is continuous if and only if $g \circ f_\lambda: X_\lambda \rightarrow Y$ is continuous for each $\lambda \in \Lambda$. The simplest is the case where X is the disjoint union of X_λ and f_λ is the inclusion mapping. Then we call the topological space X the **direct sum** or the **topological sum** of $\{X_\lambda\}$ and denote it by $\oplus X_\lambda$ or $\coprod X_\lambda$. More generally let the set X be the union of topological spaces $\{X_\lambda\}_{\lambda \in \Lambda}$ such that for each λ and $\mu \in \Lambda$ the relative topologies of $X_\lambda \cap X_\mu$ from X_λ and X_μ coincide. Then we call the topology τ the **weak topology** with respect to $\{X_\lambda\}$. If $X_\lambda \cap X_\mu$ is closed (resp. open) in X_μ for any μ, then X_λ is closed (resp. open) in X and the original topology of X_λ coincides with the relative topology. If, moreover, for each subset Γ of Λ, $F=\bigcup_{\lambda \in \Gamma} X_\lambda$ is closed and the weak topology of F with respect to $\{X_\lambda\}_{\lambda \in \Gamma}$ coincides with the relative topology induced by τ, then X with τ is said to have the **hereditarily weak topology** with respect to $\{X_\lambda\}$ (or to be **dominated** by $\{X_\lambda\}$). A topological space has the hereditarily weak topology with respect to any locally finite closed covering, and every CW-complex (→ 70 Complexes) has the hereditarily weak topology with respect to the covering of all finite subcomplexes.

When $\{X_n\}$ is an increasing sequence of topological spaces such that each X_n is a subspace of X_{n+1}, then the union $X=\bigcup X_n$ with the weak topology is called the **inductive limit** of $\{X_n\}$ and is denoted by $\varinjlim X_n$. Each X_n may again be regarded as a subspace of X.

N. Baire Spaces

For a subset A of a topological space X, the set $X-\bar{A}$ is called the **exterior** of A, and the set $\bar{A}\cap\overline{X-A}$ is called the **boundary** of A, denoted by $\mathrm{Bd}\,A$, $\mathrm{Fr}\,A$, or ∂A. A point belonging to the exterior (boundary) of A is an **exterior point** (**boundary point** or **frontier point**) of A. If the closure of A is X, then A is said to be **dense** in X. When $X-A$ is dense in X, i.e., when the interior of A is empty, A is called a **boundary set** (or **border set**), and if the closure $\bar{A}$ is a boundary set, A is said to be **nowhere dense**. The union of a countable family of nowhere dense sets is called a **set of the first category** (or **meager set**). A set that is not of the first category is called a **set of the second category** (or **nonmeager set**). The complement of a set of the first category is called a **residual set**. In the space $\mathbf{R}$ of real numbers, the set $\mathbf{Q}$ of all rational numbers is of the first category, and the set $\mathbf{R}-\mathbf{Q}$ of all irrational numbers is of the second category. Both $\mathbf{Q}$ and $\mathbf{R}-\mathbf{Q}$ are dense in X and hence are boundary sets. The union of a finite family of nowhere dense sets is nowhere dense, and the union of a countable family of sets of the first category is also of the first category. A subset A of X is nowhere dense in X if and only if for each open set O, $O\cap A$ is not dense in O.

A topological space X is called a **Baire space** (Baire, 1899) if each subset of X of the first category has an empty interior. Each of the following conditions is necessary and sufficient for a space X to be a Baire space:
(1) Each nonempty open subset of X is of the second category.
(2) If $F_1, F_2, \ldots$ is a sequence of closed subsets of X such that the union $\bigcup_{n=1}^{\infty} F_n$ has an interior point, then at least one F_n has an interior point.
(3) If $O_1, O_2, \ldots$ is a sequence of dense open subsets of X, then the intersection $\bigcap_{n=1}^{\infty} O_n$ is dense in X.

An open subset of a Baire space is a Baire space for the relative topology. A topological space that is homeomorphic to a complete metric space (→ 436 Uniform Spaces I) is a Baire space (**Baire-Hausdorff theorem**). A locally compact Hausdorff space (→ Section V) is also a Baire space. The class of Čech-complete completely regular spaces (→ Section T) includes both of these spaces, but there are also Baire spaces that are not in the class. A subset A of a topological space is said to satisfy **Baire's condition** or to have the **Baire property** if there exist an open set O and sets P_1, P_2 of the first category such that $A=(O\cup P_1)-P_2$. A †Borel set satisfies Baire's condition.

O. Accumulation Points

A point x is called an **accumulation point**, or a **cluster point** of a subset A of a topological space X if $x\in\overline{A-\{x\}}$. The set of all accumulation points of a set A is called the **derived set** of A and is denoted by A' or A^d. A point x belongs to A' if and only if each neighborhood of x contains a point of A other than x itself. A point belonging to the set $A^s=A-A'$ is called an **isolated point** of A, and a set A consisting of isolated points only, i.e., $A=A^s$, is said to be **discrete**. If each nonempty subset of A contains an isolated point, then A is said to be **scattered**; and if A does not possess an isolated point, i.e., $A\subset A'$, then A is said to be **dense in itself**. The largest subset of A which is dense in itself is called the **kernel** of A. If $A=A'$, then A is called a **perfect set**.

If x is an accumulation point of A, then for each neighborhood U of x, $U\cap(A-\{x\})\neq\varnothing$. Furthermore, it is possible to classify an accumulation point of A according to the †cardinality of $U\cap(A-\{x\})$. A point x is called a **condensation point** of a set A if for each neighborhood U of x, the set $U\cap A$ is uncountable. A point x is a **complete accumulation point** of A if for each neighborhood U of x, the set $U\cap A$ has the same cardinality as A.

P. Countability Axioms

A topological space X satisfies the **first countability axiom** if each point x of X has a countable local base (F. Hausdorff [3]). Metric spaces satisfy the first countability axiom. In fact, the family of $(1/n)$-neighborhoods ($n=1,2,\ldots$) of a point is a local base of the point. The topology of a topological space that satisfies the first countability axiom is completely determined by convergent sequences. For instance, the closure of a subset A of such a space consists of all limits of sequences in A (→ 87 Convergence). A topological space X is said to satisfy the **second countability axiom** or to be **perfectly separable** if there is a countable base for the topology. †Euclidean spaces satisfy the second countability axiom. If X contains a countable dense subset, X is said to be **separable**. A space that satisfies the second countability axiom satisfies the first and is also a separable Lindelöf space (→ Section S). However, the converse is not true. Each of the following properties is independent of the others: separability, the first countability axiom, and the Lindelöf property. If a metric space is separable, then it satisfies the second countability axiom. There are metric spaces that are not separable.

Q. Separation Axioms

Topological spaces that are commonly encountered usually satisfy some of the following separation axioms.

(T_0) **Kolmogorov's axiom**. For each pair of distinct points, there is a neighborhood of one point of the pair that does not contain the other.

(T_1) **The first separation axiom** or **Fréchet's axiom**. For each pair x, y of distinct points, there are neighborhoods U of x and V of y such that $x \notin V$ and $y \notin U$.

Axiom (T_1) can be restated as follows:

(T'_1) For each point x of the space, the singleton $\{x\}$ is closed.

(T_2) **The second separation axiom** or **Hausdorff's axiom** [3]. For each pair x, y of distinct points of the space X, there exist disjoint neighborhoods of x and y.

Axiom (T_2) is equivalent to the following:

(T'_2) In the product space $X \times X$ the diagonal set Δ is closed.

(T_3) **The third separation axiom** or **Vietoris's axiom** (*Monatsh. Math. Phys.*, 31 (1921)). Given a point x and a subset A such that $x \notin \bar{A}$, there exist disjoint open sets O_1 and O_2 such that $x \in O_1$ and $A \subset O_2$. (In this case, the sets $\{x\}$ and A are said to be **separated** by open sets.)

Axiom (T_3) can be restated as (T'_3) or (T''_3):

(T'_3) For each point x of the space, there is a local base at x consisting of closed neighborhoods of x.

(T''_3) An arbitrary closed set and a point not belonging to it can be separated by open sets.

(T_4) **The fourth separation axiom** or **Tietze's first axiom** (*Math. Ann.*, 88 (1923)). Two disjoint closed sets F_1 and F_2 can be separated by open sets, i.e., there exist disjoint open sets O_1 and O_2 such that $F_1 \subset O_1$ and $F_2 \subset O_2$.

(T_5) **Tietze's second axiom**. Whenever two subsets A_1 and A_2 satisfy $A_1 \cap \bar{A}_2 = \bar{A}_1 \cap A_2 = \varnothing$, A_1 and A_2 can be separated by open sets.

It is easily seen that (T_5) $\Rightarrow$ (T_4), (T_0) and (T_3) $\Rightarrow$ (T_2), (T_4) and (T_1) $\Rightarrow$ (T_3). Axiom (T_4) is equivalent to each of (T'_4) and (T''_4):

(T'_4) Whenever F_1 and F_2 are disjoint closed subsets, there exists a continuous function f on the space into the interval $[0, 1]$ such that f is identically 0 on F_1 and 1 on F_2.

(T''_4) Each real-valued continuous function defined on a closed subspace can be extended to a real-valued continuous function on the entire space.

The implications (T_4) $\Rightarrow$ (T'_4) and (T_4) $\Rightarrow$ (T''_4) are known as **Uryson's lemma** (*Math. Ann.*, 94 (1925)) and the **Tietze extension theorem** (*J. Reine Angew. Math.*, 145 (1915)), respectively. In addition, there are two more related axioms:

($T_{3\frac{1}{2}}$) **Tikhonov's separation axiom**. For each closed subset F and each point x not in F, there is a real-valued continuous function f on the space such that $f(x)=0$ and f is identically 1 on F.

(T_6) (N. Vedenisov). For each closed subset F, there is a real-valued continuous function f on the space such that $F = \{x \mid f(x)=0\}$.

Axioms (T_5) and (T_6) are equivalent to the following (T'_5) and (T'_6), respectively:

(T'_5) Each subspace satisfies (T_4)

(T'_6) X satisfies (T_4) and each closed set is a †G_δ-set.

The following implications are valid: ($T_{3\frac{1}{2}}$) $\Rightarrow$ (T_3), (T_6) $\Rightarrow$ (T_5), (T_4) and (T_1) $\Rightarrow$ ($T_{3\frac{1}{2}}$).

Table 1 gives a classification of topological spaces by the separation axioms. Each line represents a special case of the preceding line.

A †metrizable space is perfectly normal, but the converse is false (for metrization theorems → 273 Metric Spaces). Among the spaces satisfying the second countability axiom, regular spaces are normal (**Tikhonov's theorem**, *Math. Ann.*, 95 (1925)) and metrizable (**Tikhonov-Uryson theorem**; P. Uryson, *Math. Ann.*, 94 (1925)).

Table 2 shows whether various topological properties are preserved in subspaces, product spaces, and quotient spaces. The topological properties considered are T_1, T_2 = Hausdorff, T_3 = regular, CR = completely regular, T_4 = normal, T_5 = completely normal, M = metrizable, C_I = first axiom of countability, C_{II} = second axiom of countability, C = compact, S = separable, and L = Lindelöf. Each position is filled with ○ or × according as the property (say, P) listed at the head of the column is preserved or not in the sort of space listed on the left obtained from space(s) all having property P.

R. Coverings

A family $\mathfrak{M} = \{M_\lambda\}_{\lambda \in \Lambda}$ of subsets of a set X is called a **covering** of a subset A of X if $A \subset \bigcup_\lambda M_\lambda$. If $\mathfrak{M}$ is finite (countable), it is called a **finite covering (countable covering)**. An **open (closed) covering** is a covering consisting of open (closed) sets.

A family $\mathfrak{M}$ of subsets of a topological space X is said to be **locally finite** if for each point x of X, there is a neighborhood of x which intersects only a finite number of members of $\mathfrak{M}$. If moreover $\{\bar{M}_\lambda\}_{\lambda \in \Lambda}$ is disjoint, then $\mathfrak{M}$ is called **discrete**. $\mathfrak{M}$ is called **star-finite** if each member of $\mathfrak{M}$ intersects only a finite number of members of $\mathfrak{M}$. A **σ-locally finite** or **σ-discrete** family of subsets of X is respectively the union of a countable number of locally finite or discrete families of subsets of X. A covering $\mathfrak{M}$

Table 1. Separation Axioms

Axioms	Spaces Satisfying the Axioms
(T_0)	**T_0-space (Kolmogorov space)**
(T_1)	**T_1-space (Kuratowski space)**
(T_2)	**T_2-space (Hausdorff space, separated space)**
(T_0) and (T_3)	**T_3-space (regular space)**
(T_1) and $(T_{3\frac{1}{2}})$	**Completely regular space (Tikhonov space)**
(T_1) and (T_4)	**T_4-space (normal space)**
(T_1) and (T_5)	**T_5-space (completely normal space, hereditarily normal space)**
(T_1) and (T_6)	**T_6-space (perfectly normal space)**

Table 2. Topological Properties and Spaces

Space	T_1	T_2	T_3	CR	T_4	T_5	M	C_I	C_{II}	C	S	L
Subspace	○	○	○	○	×	○	○	○	○	×	×	×
Closed subspace	○	○	○	○	○	○	○	○	○	○	×	○
Open subspace	○	○	○	○	×	○	○	○	○	×	○	×
Product	○	○	○	○	×	×	×	×	×	○	×	×
Countable product	○	○	○	○	×	×	○	○	○	○	○	×
Quotient space	×	×	×	×	×	×	×	×	×	○	○	○

is called **point-finite** if each infinite number of members of $\mathfrak{M}$ has an empty intersection. A covering $\mathfrak{M}$ is a **refinement** of a covering $\mathfrak{N}$ (written $\mathfrak{M} \prec \mathfrak{N}$) if each member of $\mathfrak{M}$ is contained in a member of $\mathfrak{N}$. The **order** of the covering $\mathfrak{M}$ is the least integer r such that any subfamily of $\mathfrak{M}$ consisting of $r+1$ members has an empty intersection.

Let $\mathfrak{M}$ be a covering of X, and let A be a subset of X. The **star** of A relative to $\mathfrak{M}$, denoted by $S(A, \mathfrak{M})$, is the union of all members of $\mathfrak{M}$ whose intersection with A is nonempty. Let $\mathfrak{M}^{\Delta}$ denote the family $\{S(\{x\}, \mathfrak{M})\}_{x \in X}$ and $\mathfrak{M}^*$ the family $\{S(M, \mathfrak{M})\}_{M \in \mathfrak{M}}$. Then $\mathfrak{M}^{\Delta}$ and $\mathfrak{M}^*$ are coverings of X, and $\mathfrak{M} \prec \mathfrak{M}^{\Delta} \prec \mathfrak{M}^* \prec \mathfrak{M}^{\Delta\Delta}$. A covering $\mathfrak{M}$ is a **star refinement** of a covering $\mathfrak{N}$ if $\mathfrak{M}^* \prec \mathfrak{N}$, and $\mathfrak{M}$ is a **barycentric refinement** (or **Δ-refinement**) of $\mathfrak{N}$ if $\mathfrak{M}^{\Delta} \prec \mathfrak{N}$.

A sequence $\mathfrak{M}_1, \mathfrak{M}_2, \ldots$ of open coverings of a topological space is called a **normal sequence** if $\mathfrak{M}_{n+1}^{\Delta} \prec \mathfrak{M}_n$ for $n = 1, 2, \ldots$, and an open covering $\mathfrak{M}$ is said to be a **normal covering** if there is a normal sequence $\mathfrak{M}_1, \mathfrak{M}_2, \ldots$ such that $\mathfrak{M}_1 \prec \mathfrak{M}$. The **support** (or **carrier**) of a real-valued function f on a topological space X is defined to be the closure of the set $\{x \mid f(x) \neq 0\}$. Let $\{f_\alpha\}_{\alpha \in A}$ be a family of continuous nonnegative real-valued functions on a topological space X, and for each α in A, let C_α be the support of f_α. The family $\{f_\alpha\}_{\alpha \in A}$ is called a **partition of unity** if the family $\{C_\alpha\}_{\alpha \in A}$ is locally finite and $\Sigma_\alpha f_\alpha(x) = 1$ for each x in X. If the covering $\{C_\alpha\}_{\alpha \in A}$ is a refinement of a covering $\mathfrak{M}$, the family $\{f_\alpha\}_{\alpha \in A}$ is called a **partition of unity subordinate to the covering** $\mathfrak{M}$. A partition of unity subordinate to a covering $\mathfrak{M}$ exists only if $\mathfrak{M}$ is a normal covering (→ Section X). If ρ is a continuous †pseudometric on a T_1-space X, then define a covering M_n for each natural number n by $M_n = \{U(x; 2^{-n})\}_{x \in X}$, where $U(x; \varepsilon) = \{y \mid \rho(x, y) < \varepsilon\}$. Then the sequence $\mathfrak{M}_1, \mathfrak{M}_2, \ldots$ is a normal sequence of open coverings. Conversely, given a normal sequence $\mathfrak{M}_1, \mathfrak{M}_2, \ldots$ of open coverings of X, there exists a continuous pseudometric ρ such that $\rho(x, y) \leqslant 2^{-n}$ whenever $x \in S(y, \mathfrak{M}_n)$, and $\rho(x, y) \geqslant 2^{-n-1}$ whenever $x \notin S(y, \mathfrak{M}_n)$. If in addition for each x the family $\{S(x, \mathfrak{M}_n) \mid n = 1, 2, \ldots\}$ is a local base at x, then the metric topology of ρ agrees with the topology of X.

S. Compactness

If each open covering of a topological space X admits a finite open covering as its refinement, the space X is called **compact**; if each open covering of X admits a countable open refinement, X is said to be a **Lindelöf space** (P. Aleksandrov and P. Uryson, *Verh. Akad. Wetensch.*, Amsterdam, 19 (1929)); if each open covering of X admits a locally finite open refinement, X is called **paracompact** (J. Dieudonné, *J. Math. Pures Appl.*, 23 (1944)); and if each open covering of X admits a star-finite open refinement, X is said to be **strongly paracompact** (C. H. Dowker, *Amer. J. Math.*, 69 (1947)) or to have the **star-finite property** (K. Morita, *Math. Japonicae*, 1 (1948)). The space X is compact (Lindelöf) if for each open covering $\mathfrak{M}$ of X, there is a finite (countable) subfamily of $\mathfrak{M}$ whose union is X.

The following properties for a topological space X are equivalent: (1) The space X is compact. (2) If a family $\{F_\lambda\}_{\lambda \in \Lambda}$ of closed subsets of X has the **finite intersection property**, i.e., each finite subfamily of $\{F_\lambda\}_{\lambda \in \Lambda}$ has nonempty intersection, then $\bigcap_\lambda F_\lambda \neq \varnothing$. (3) Each infinite subset of X has a complete accumulation point. (4) Each †net has a convergent †subnet. (5) Each †universal net and each †ultrafilter converge.

If a subset A of X is compact for the relative topology, A is called a **compact** subset. A subset A of X is said to be **relatively compact** if the closure of A in X is a compact subset. A closed subset of a compact topological space is compact, and a compact subset of a Hausdorff space is closed. A continuous image of a compact space is compact, each continuous mapping of a compact space into a Hausdorff space is a closed mapping, and a continuous bijection of a compact space onto a Hausdorff space is a homeomorphism. The product space of a family $\{X_\lambda\}_{\lambda\in\Lambda}$ of topological spaces is compact if and only if each factor space is compact (**Tikhonov's product theorem**, *Math. Ann.*, 102 (1930)). A compact Hausdorff space is normal. A compact Hausdorff space is metrizable if and only if it satisfies the second countability axiom. A metric space or a †uniform space is compact if and only if it is †totally bounded and †complete. A subset of a Euclidean space is compact if and only if it is closed and bounded. In a discrete space only finite subsets are compact. The cardinality of a compact Hausdorff space with the first countability axiom cannot exceed the power of the continuum (Arkhangel'skiĭ).

There are a number of conditions related to compactness. A topological space is **sequentially compact** if each sequence in X has a convergent subsequence. A space X is **countably compact** (M. Fréchet [1]) if each countable open covering of X contains a finite subfamily that covers X. A space X is **pseudocompact** (E. Hewitt, 1948) if each continuous real-valued function on X is bounded. Some authors use *compact* and *bicompact* for what we call countably compact and compact, respectively. N. Bourbaki [9] uses *compact* and *quasicompact* instead of compact Hausdorff and compact, respectively. A T_1-space is countably compact if and only if each infinite set possesses an accumulation point. If X is countably compact, then X is pseudocompact, and if X is normal, the converse also holds. If a †complete uniform space is pseudocompact, then it is compact. A space satisfying the second countability axiom is compact if and only if it is sequentially compact. If X is sequentially compact, then X is countably compact, and if X satisfies the first countability axiom, the converse is true.

T. Compactification

A **compactification** of a topological space X consists of a compact space Y and a homeomorphism of X onto a dense subspace X_1 of Y. We can always regard X as a dense subspace of a compactification Y. If X is completely regular, then there is a Hausdorff compactification Y such that each bounded real-valued continuous function on X can be extended continuously to Y. Such a compactification is unique up to homeomorphism; it is called the **Stone-Čech compactification** of X (E. Čech, *Ann. Math.*, 38 (1937); M. H. Stone, *Trans. Amer. Math. Soc.*, 41 (1937)) and is denoted by $\beta(X)$. Let $\{f_\lambda\}_{\lambda\in\Lambda}$ be the set of all continuous functions on a completely regular space X into the closed interval $I=[0,1]$. Then a continuous mapping φ of X into a **parallelotope** $I^\Lambda=\prod_\lambda I_\lambda$ $(I_\lambda=I)$ is defined by $\varphi(x)=\{f_\lambda(x)\}_{\lambda\in\Lambda}$, and the mapping φ is a homeomorphism of X onto the subspace $\varphi(X)$ of I^Λ (**Tikhonov's embedding theorem**, *Math. Ann.*, 102 (1930)). The closure $\overline{\varphi(X)}$ of $\varphi(X)$ in I^Λ is the Stone-Čech compactification of X. The natural mapping $\beta(X_1\times X_2)\rightarrow\beta(X_1)\times\beta(X_2)$ is a homeomorphism if and only if $X_1\times X_2$ is pseudocompact (I. Glicksberg, 1959).

For a topological space X, let ∞ be a point not in X, and define a topology on the union $X\cup\{\infty\}$ as follows: A subset U of $X\cup\{\infty\}$ is open if and only if either $\infty\neq U$ and U is open in X, or $\infty\in U$ and $X-U$ is a compact closed subset of X. The topological space $X\cup\{\infty\}$ thus obtained is compact, and if X is not already compact, the space $X\cup\{\infty\}$ is a compactification of X called the **one-point compactification** of X (P. S. Aleksandrov, *C. R. Acad. Sci. Paris*, 178 (1924)). The one-point compactification of a Hausdorff space is not necessarily Hausdorff. The one-point compactification of the n-dimensional Euclidean space $\mathbf{R}^n$ is homeomorphic to the n-dimensional sphere S^n.

A completely regular space X is a †G_δ-set in the Stone-Čech compactification $\beta(X)$ if and only if it is a G_δ-set in any Hausdorff space Y which contains X as a dense subspace. Then X is said to be **Čech-complete**.

U. Absolutely Closed Spaces

A Hausdorff space X is said to be **absolutely closed** (or **H-closed**; P. Aleksandrov and P. Uryson, 1929) if X is closed in each Hausdorff space containing it. A compact Hausdorff space is absolutely closed. A Hausdorff space is absolutely closed if and only if for each open covering $\{N_\lambda\}_{\lambda\in\Lambda}$ of X, there is a finite subfamily of $\{\overline{N}_\lambda\}_{\lambda\in\Lambda}$ that covers X. The product space of a family of absolutely closed spaces is absolutely closed. Each Hausdorff space is a dense subset of an absolutely closed space. Similarly, a regular space X is said to be **r-**

closed if X is closed in each regular space containing it (N. Weinberg, 1941).

V. Locally Compact Spaces

A topological space X is said to be **locally compact** if each point of X has a compact neighborhood (P. Aleksandrov and P. Uryson, 1929). A †uniform space X is said to be **uniformly locally compact** if there is a member U of the †uniformity such that $U(x)$ is compact for each x in X (→ 436 Uniform Spaces). A noncompact space X is locally compact and Hausdorff if and only if the one-point compactification of X is Hausdorff, and this is the case if and only if X is homeomorphic to an open subset of a compact Hausdorff space. A locally compact Hausdorff space is completely regular, and for each point of the space, the family of all of its compact neighborhoods forms a local base at the point. A locally closed, hence open or closed, subset of a locally compact Hausdorff space is also locally compact for the relative topology. If a subspace A of a Hausdorff space X is locally compact, then A is a locally closed subset of X. The Euclidean space $\mathbf{R}^n$ is locally compact, and hence each **locally Euclidean space**, i.e., a space such that each point admits a neighborhood homeomorphic to a Euclidean space, is locally compact. A topological space is called **σ-compact** if it can be expressed as the union of at most countably many compact subsets.

W. Proper (Perfect) Mappings

A mapping f of a topological space X into a topological space Y is said to be **proper** (N. Bourbaki [9]) (or **perfect** [14]) if it is continuous and for each topological space Z, the mapping $f\times 1: X\times Z\rightarrow Y\times Z$ is closed, where $(f\times 1)(x,z)=(f(x),z)$. A continuous mapping $f: X\rightarrow Y$ is proper if and only if it is closed and $f^{-1}(\{y\})$ is compact for each y in Y. Another necessary and sufficient condition is that if $\{x_\alpha\}_{\mathfrak{A}}$ is a †net in X such that its image $\{f(x_\alpha)\}$ converges to $y\in Y$, then a subnet of $\{x_\alpha\}$ converges to an $x\in f^{-1}(y)$ in X. A continuous mapping of a compact space into a Hausdorff space is always proper. For a compact Hausdorff space X, a quotient space Y is Hausdorff if and only if the canonical projection $\varphi: X\rightarrow Y$ is proper.

For a continuous mapping f of a locally compact Hausdorff space X into a locally compact Hausdorff space Y, the following three conditions are equivalent: (1) f is proper. (2) For each compact subset K of Y, the inverse image $f^{-1}(K)$ is compact. (3) If $X\cup\{x_\infty\}$ and $Y\cup\{y_\infty\}$ are the one-point compactifications of X and Y, then the extension f_1 of f such that $f_1(x_\infty)=y_\infty$ is continuous.

The composition of two proper mappings is proper and the direct product of an arbitrary number of proper mappings is proper.

X. Paracompact Hausdorff Spaces

A paracompact Hausdorff space (often called simply a paracompact space) is normal. For a Hausdorff space X, the following five conditions are equivalent: (1) X is paracompact. (2) X is **fully normal** (J. W. Tukey [8]), i.e., each open covering of X admits an open barycentric refinement. (3) Each open covering has a partition of unity subordinate to it. (4) Each open covering is refined by a closed covering $\{F_\alpha\mid\alpha\in A\}$ that is **closure-preserving**, i.e., $\cup\{F_\beta\mid\beta\in B\}$ is closed for each $B\subset A$. (5) Each open covering $\{U_\alpha\mid\alpha\in A\}$ has a **cushioned refinement** $\{V_\alpha\mid\alpha\in A\}$, i.e., $\mathrm{Cl}(\cup\{V_\beta\mid\beta\in B\})\subset\cup\{U_\beta\mid\beta\in B\}$ for each $B\subset A$. The implication (1)–(2) is **Dieudonné's theorem**. The implication (2)→(1) is A. H. **Stone's theorem** (1948), from which it follows that each metric space is paracompact. The implications (5)→(4)→(1) is **Michael's theorem** (1959, 1957).

For normal spaces, the following weaker versions of (2) and (3) hold: A T_1-space X is normal if and only if each finite open covering of X admits a finite open star refinement (or finite open barycentric refinement). For each locally finite open covering of a normal space, there is a partition of unity subordinate to it.

For a regular space X the following three conditions are equivalent: (1) X is paracompact. (2) Each open covering of X is refined by a σ-discrete open covering. (3) Each open covering of X is refined by a σ-locally finite open covering. **Tamano's product theorem**: For a completely regular space X to be paracompact it is necessary and sufficient that $X\times\beta(X)$ be normal (1960).

For a †connected locally compact space X, the following conditions are equivalent: (1) X is paracompact. (2) X is σ-compact. (3) In the one-point compactification $X\cup\{\infty\}$, the point ∞ admits a countable local base. (4) There is a locally finite open covering $\{U_\lambda\}_{\lambda\in\Lambda}$ of X such that $\bar{U}_\lambda$ is compact for each λ. (5) X is the union of a sequence $\{U_n\}$ of open sets such that $\bar{U}_n$ is compact and $\bar{U}_n\subset U_{n+1}$ $(n=1,2,\ldots)$. (6) X is strongly paracompact.

Every †F_σ-set of a paracompact Hausdorff space is paracompact (Michael, 1953). When a T_1-space X has the hereditarily weak topology with respect to a closed covering $\{F_\lambda\}$, then X is paracompact Hausdorff (normal, completely normal or perfectly normal) if and only if each

F_λ is (Morita, 1954; Michael, 1956). In particular, every CW-complex is paracompact and perfectly normal (Morita, 1953).

Y. Normality and Paracompactness of Direct Products

A topological space X is discrete if $X \times Y$ is normal for any normal space Y (M. Atsuji and M. Rudin, 1978). There are a paracompact Lindelöf space X and a separable metric space Y such that the product $X \times Y$ is not normal (Michael, 1963). The following are conditions under which the products are normal or paracompact. Let $\mathfrak{m}$ be an infinite †cardinal number. A topological space X is called $\mathfrak{m}$-paracompact if every open covering consisting of at most $\mathfrak{m}$ open sets admits a locally finite open covering as its refinement. When $\mathfrak{m}$ is countable, it is called **countably paracompact**. If X has an open base of at most $\mathfrak{m}$ members, $\mathfrak{m}$-paracompact means paracompact. The following conditions are equivalent for a topological space X: (1) X is normal and countably paracompact; (2) The product $X \times Y$ is normal and countably paracompact for any compact metric space Y; (3) $X \times I$ is normal, where $I = [0, 1]$ (C. H. Dowker, 1951). Rudin (1971) constructed an example of a collectionwise normal space (→ Section AA) that is not countably paracompact. When $\mathfrak{m}$ is general the following conditions are equivalent: (1) X is normal and $\mathfrak{m}$-paracompact; (2) If Y is a compact Hausdorff space with an open base consisting of at most $\mathfrak{m}$ sets, then $X \times Y$ is normal and $\mathfrak{m}$-paracompact; (3) $X \times I^{\mathfrak{m}}$ is normal; (4) $X \times \{0, 1\}^{\mathfrak{m}}$ is normal (Morita, 1961). In particular, the product $X \times Y$ of a paracompact Hausdorff space X and a compact Hausdorff space Y is paracompact (Dieudonné, 1944).

A topological space X is called a **P-space** if it satisfies the following conditions: Let Ω be an arbitrary set and $\{G(\alpha_1, \ldots, \alpha_i) \mid \alpha_1, \ldots \alpha_i \in \Omega, i = 1, 2, \ldots\}$ be a family of open sets such that $G(\alpha_1, \ldots, \alpha_i) \subset G(\alpha_1, \ldots, \alpha_i, \alpha_{i+1})$. Then there is a family of closed sets $\{F(\alpha_1, \ldots, \alpha_i) \mid \alpha_1, \ldots, \alpha_i \in \Omega, i = 1, 2, \ldots\}$ such that $F(\alpha_1, \ldots, \alpha_i) \subset G(\alpha_1, \ldots, \alpha_i)$ and that if $\bigcup_{i=1}^{\infty} G(\alpha_1, \ldots, \alpha_i) = X$ for a sequence $\{\alpha_i\}$, then $\bigcup_{i=1}^{\infty} F(\alpha_1, \ldots, \alpha_i) = X$. Perfectly normal spaces, countably compact spaces, Čech-complete paracompact spaces and σ-compact regular spaces are P-spaces. Normal P-spaces are countably paracompact. A Hausdorff space X is a normal (resp. paracompact) P-space if and only if the product $X \times Y$ is normal (resp. paracompact) for any metric space Y (Morita, *Math. Ann.*, 154 1964).

The product $X \times Y$ of locally compact Hausdorff spaces X and Y is a locally compact Hausdorff space. If, in this case, X and Y are paracompact, then so is the product. If the direct product space $\prod_\lambda X_\lambda$ of metric spaces is normal, then X_λ are compact except for at most countably many λ, and hence the product space is paracompact (A. H. Stone, 1948).

A class $\mathscr{C}$ of topological spaces is called **countably productive** if for a sequence X_i of members of $\mathscr{C}$ their product $\prod X_i$ is again a member of $\mathscr{C}$. The classes of (complete) (separable) metric spaces form such examples. The class of paracompact and Čech-complete spaces is countably productive (Z. Frolík, 1960). A topological space X is called a **p-space** if it is completely regular and there is a sequence $\mathfrak{M}_i$ of families of open sets in the Stone-Čech compactification $\beta(X)$ such that, for each point $x \in X$, $x \in \bigcap S(x, \mathfrak{M}_i) \subset X$ (Arkhangel'skiĭ, 1963). X is called an **M-space** if there is a normal sequence $\mathfrak{M}_i$ of open coverings of X such that if $K_1 \supset K_2 \supset \ldots$ is a sequence of nonempty closed sets and $K_i \subset S(x, \mathfrak{M}_i)$, $i = 1, 2, \ldots$, for an $x \in X$, then $\bigcap K_i \neq \varnothing$ (Morita, 1963). The class of paracompact p-spaces and that of paracompact Hausdorff M-spaces are the same and are countably productive. For a covering $\mathfrak{F}$ of X and an $x \in X$ we set $C(x, \mathfrak{F}) = \bigcap \{F \mid x \in F \in \mathfrak{F}\}$. X is called a **Σ-space** if X admits a sequence $\mathfrak{F}_i$ of locally finite closed coverings such that if $K_1 \supset K_2 \supset \ldots$ is a sequence of nonempty closed sets and $K_i \subset C(x, \mathfrak{F}_i)$, $i = 1, 2, \ldots$, for an $x \in X$, then $\bigcap K_i \neq \varnothing$ (K. Nagami, 1969). Σ-spaces are P-spaces. The class of all paracompact Σ-spaces is also countably productive. Among the above classes each one is always wider than its predecessors. Yet the product $X \times Y$ of a paracompact Hausdorff P-space X and a paracompact Hausdorff Σ-space Y is paracompact. Other examples of countably productive classes are the Suslin spaces and the Luzin spaces (→ Section CC) introduced by Bourbaki (1958), the **stratifiable spaces** by J. G. Ceder (1961) and C. J. R. Borges (1966), the **$\aleph_0$-spaces** by Michael (1966) and the **σ-spaces** by A. Okuyama (1967).

Z. Strongly Paracompact Spaces

Regular Lindelöf spaces are strongly paracompact. Conversely, if a connected regular space is strongly paracompact, then it is a Lindelöf space (Morita, 1948). Hence a connected nonseparable metric space is not strongly paracompact. Paracompact locally compact Hausdorff spaces and uniformly locally compact Hausdorff spaces are strongly paracompact.

These classes of spaces coincide under suitable †uniform structures.

AA. Collectionwise Normal Spaces

A Hausdorff space X is called a **collectionwise normal space** if for each discrete collection $\{F_\alpha | \alpha \in A\}$ of closed sets of X there exists a disjoint collection $\{U_\alpha | \alpha \in A\}$ of open sets with $F_\alpha \subset U_\alpha$ $(\alpha \in A)$ (R. H. Bing, 1951). If X satisfies an analogous condition for the case where each F_α is a singleton, X is called a **collectionwise Hausdorff space**. Paracompact Hausdorff spaces are collectionwise normal (Bing). Every point-finite open covering of a collectionwise normal space has a locally finite open refinement (Michael, Nagami).

A topological space X is called a **developable space** if it admits a sequence $\mathfrak{U}_i$, $i=1,2,\ldots$, of open coverings such that, for each point $x \in X$, $\{S(x, \mathfrak{U}_i) | i=1,2,\ldots\}$ forms a base for the neighborhood system of x (R. L. Moore, 1916). A regular developable space is called a **Moore space**. The question of whether or not every normal Moore space is metrizable is known as the **normal Moore space problem** (→ 273 Metric Spaces K). Collectionwise normal Moore spaces are metrizable (Bing).

BB. Real-Compact Spaces

A completely regular space X is called **real-compact** if X is complete under the smallest †uniformity such that each continuous real-valued function on X is uniformly continuous (→ 422 Uniform Spaces). This notion was introduced by E. Hewitt (*Trans. Amer. Math. Soc.*, 64 (1948)) under the name of **Q-space**, and independently by L. Nachbin (*Proc. International Congress of Mathematicians*, Cambridge, Mass., 1950).

A Lindelöf space is real-compact. If X_1 and X_2 are real-compact spaces such that the rings $C(X_1)$ and $C(X_2)$ of continuous real-valued functions on X_1 and X_2 are isomorphic, then X_1 and X_2 are homeomorphic (Hewitt). If X is real-compact, then X is homeomorphic to a closed subspace of the product space of copies of the space of real numbers, and conversely.

CC. Images and Inverse Images of Topological Spaces

Each continuous mapping $f: X \to Y$ is decomposed into the product $i \circ h \circ p$ of continuous mappings $p: X \to X/\sim$, $h: X/\sim \to f(X)$ and i: $f(X) \to Y$, where $\sim$ is the equivalence relation such that $x_1 \sim x_2$ if and only if $f(x_1) = f(x_2)$. The mapping f is open (resp. closed) if and only if these mappings are all open (resp. closed). Then h is a homeomorphism. The image of a paracompact Hausdorff space under a closed continuous mapping is paracompact (Michael, 1957).

Let $f: X \to Y$ be a perfect surjection. Then Y is called a **perfect image** of X and X a **perfect inverse image** of Y. If, in this case, one of X and Y satisfies a property such as being compact, locally compact, σ-compact, Lindelöf, or countably compact, then the other also satisfies the property. When X and Y are completely regular, the same is true with regard to Čech completeness. Properties such as regularity, normality, complete normality, perfect normality, and the second countability axiom are preserved in perfect images; but complete regularity and strong paracompactness are not. Perfect images of metric spaces are also metrizable (S. Hanai and Morita, A. H. Stone, 1956). Conversely, perfect inverse images of paracompact spaces are paracompact. If a Hausdorff space is a perfect inverse image of a regular space (resp. k-space; → below), then it is a regular space (resp. k-space). Every paracompact Čech-complete space is a perfect inverse image of a †complete metric space (Z. Frolík, 1961). A completely regular space is a paracompact p-space if and only if it is a perfect inverse image of a metric space (Arkhangel'skiĭ, 1963). A mapping $f: X \to Y$ is called **quasi-perfect** if it is closed and continuous and the inverse image $f^{-1}(y)$ of each point $y \in Y$ is countably compact. A topological space X is an M-space if and only if there is a quasi-perfect mapping from X onto a metric space Y (Morita, 1964). Let $f: X \to Y$ be a quasi-perfect surjection. If one of X and Y is a Σ-space, then the other is also a Σ-space (Nagami, 1969).

A topological space X is called a **Fréchet-Uryson space** (or a **Fréchet space**) if the closure of an arbitrary set $A \subset X$ is the set of all limits of sequences in A (Arkhangel'skiĭ, 1963). X is called a **sequential space** if $A \subset X$ is closed whenever A contains all the limits of sequences in A (S. P. Franklin, 1965). X is called a **k′-space** if the closure of an arbitrary set A is the set of all points adherent to the intersection $A \cap K$ for a compact set K in X (Arkhangel'skiĭ, 1963). X is called a **k-space** if $A \subset X$ is closed whenever $A \cap K$ is closed in K for any compact set K (→ Arkhangel'skiĭ, *Trudy Moskov. Mat. Obshch.*, 13 (1965)). Spaces satisfying the first countability axiom are Fréchet-Uryson spaces. The Fréchet-Uryson spaces (resp. sequential spaces) are characterized as the images under hereditarily quotient (resp. quotient) mappings of metric spaces or locally compact metric spaces. Similarly the k′-spaces (resp. k-spaces) coincide with the images under hereditarily quotient (resp. quotient) mappings of locally compact spaces. The image of a metric space under a closed continuous mapping is called a **Lashnev space**. Any subspace of a Fréchet-Uryson space is a Fréchet-Uryson space. Conversely, a Hausdorff space is a Fréchet-Uryson space if any of its subspaces is a k-space. Čech-complete spaces are k-spaces. A Hausdorff space is called a **Suslin space** (resp. **Luzin space**) if it is the image under a continuous surjection (resp. continuous bijection) of a complete separable metric space (Bourbaki [9]; also → 22 Analytic Sets).

In Figs. 1, 2, and 3, the relationships between the various properties are indicated by the arrows.

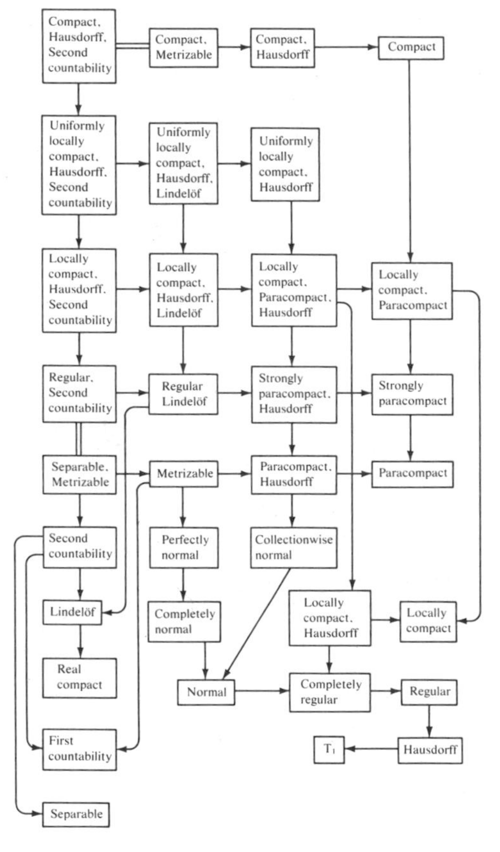

Fig. 1

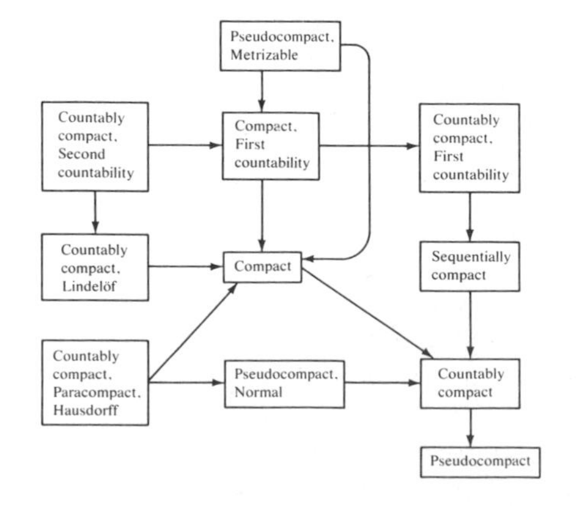

Fig. 2

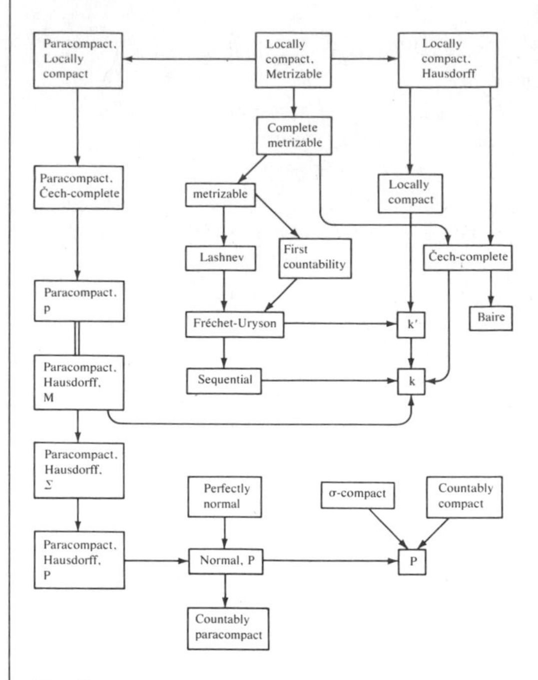

Fig. 3

References

[1] M. Fréchet, Sur quelques points du calcul fonctionnel, Rend. Circ. Mat. Palermo, 22 (1906), 1–74.
[2] F. Riesz, Stetigkeitsbegriff und abstrakte Mengenlehre, Atti del IV Congr. Intern. Mat. Roma 1908, II, 18–24, Rome, 1909.
[3] F. Hausdorff, Grundzüge der Mengenlehre, Teubner, 1914.
[4] M. Fréchet, Les espaces abstraits, Gauthier-Villars, 1928.
[5] C. Kuratowski, Topologie, I, Monograf. Mat., I, 1933, revised edition, 1948; II, 1950.
[6] W. Sierpiński, Introduction to general topology, University of Toronto Press, 1934.
[7] P. S. Aleksandrov (Alexandroff) and H. Hopf, Topologie I, Springer, 1935.
[8] J. W. Tukey, Convergence and uniformity in topology, Ann. Math. Studies, Princeton Univ. Press, 1940.
[9] N. Bourbaki, Eléments de mathématique, III. Topologie générale, 1940–1949; new edition, 1970–1974; Hermann, English translation, General topology, Addison-Wesley, 1966.
[10] J. L. Kelley, General topology, Van Nostrand, 1955.
[11] J. Dugundji, Topology, Allyn and Bacon, 1966.
[12] J. Nagata, Modern general topology, North-Holland, 1968.
[13] R. Engelking, General topology, Polish Scientific Publ., 1977.

[14] P. S. Aleksandrov, Some results in the theory of topological spaces, obtained within the last twenty-five years, Russian Math. Surveys, 16, no. 2 (1960), 23–83.
[15] A. V. Arkhangel'skiĭ, Mappings and spaces, Russian Math. Surveys, 21, no. 4 (1966), 115–162.
[16] Y. Kodama and K. Nagami, Theory of topological spaces (in Japanese), Iwanami, 1974.
[17] L. Steen and J. Seebach, Jr., Counterexamples in topology, Holt, Rinehart and Winston, 1970.
[18] L. Gillman and M. Jerison, Rings of continuous functions, Van Nostrand, 1960.
[19] S. A. Gaal, Point set topology, Academic Press, 1964.
[20] E. Čech, Topological spaces, Interscience, 1966.
[21] R. C. Walker, The Stone-Čech compactification, Springer, 1974.
[22] R. A. Alò and H. L. Shapiro, Normal topological spaces, Cambridge Univ. Press, 1974.
[23] M. D. Weir, Hewitt-Nachbin spaces, North-Holland, 1975.

426 (IX.1) Topology

The term **topology** means a branch of mathematics that deals with topological properties of geometric figures or point sets. A classical result in topology is the Euler relation on polyhedra: Let α_0, α_1, and α_2 be the numbers of vertices, edges, and faces of a polyhedron homeomorphic to the 2-dimensional sphere; then $\alpha_0-\alpha_1+\alpha_2=2$ (†Euler-Poincaré formula for the 2-dimensional case; actually, the formula was known to Descartes). It is one of the earliest results in topology. In 1833, C. F. Gauss used integrals to define the notion of †linking numbers of two closed curves in a space (→ 99 Degree of Mapping). It was in J. B. Listing's classical work *Vorstudien zur Topologie* (1847) that the term *topology* first appeared in print.

In the 19th century, B. Riemann published many works on function theory in which topological methods played an essential role. He solved the homeomorphism problem for compact surfaces (→ 410 Surfaces); his result is basic in the theory of algebraic functions. In the same period, mathematicians began to study topological properties of n-dimensional polyhedra. E. Betti considered the notion of †homology. H. Poincaré, however, was the first to recognize the importance of a topological approach to analysis in general; he defined the †homology groups of a complex [1]. He obtained the famous †Poincaré duality theorem and defined the †fundamental group. He considered †polyhedra as the basic objects in topology, and deduced topological properties utilizing †complexes obtained from polyhedra by †simplicial decompositions. He thus constructed a branch of topology known as **combinatorial topology**.

In its beginning stages combinatorial topology dealt only with polyhedra. In the late 1920s, however, it became possible to apply combinatorial methods to general †compact spaces. P. S. Alexandrov introduced the concept of approximation of a †compact metric space by an inverse sequence of complexes and the definition of homology groups for these spaces. His idea had a precursor in the notion of †simplicial approximations of continuous mappings, which was introduced by L. E. J. Brouwer in 1911. In 1932, E. Čech defined homology groups for arbitrary spaces utilizing the †inductive limit of the homology groups of polyhedra; and †Čech cohomology groups for arbitrary spaces were also defined. S. Eilenberg established †singular (co)homology theory using †singular chain complexes (1944). The axiomatic approach to (co)homology theory is due to Eilenberg and Steenrod, who gave axioms for (co)homology theory in a most comprehensive way and unified various (co)-homology theories (1945) (→ 201 Homology Theory).

The approach using algebraic methods has progressed extensively in connection with the development of homology theory. This branch is called **algebraic topology**. In the 1920s and 1930s, a number of remarkable results in algebraic topology, such as the †Alexander duality theorem, the †Lefschetz fixed-point theorem, and the †Hopf invariant, were obtained. In the late 1930s, W. Hurewicz developed the theory of higher-dimensional †homotopy groups (→ 153 Fixed-Point Theorems, 201 Homology Theory, 202 Homotopy Theory). J. H. C. Whitehead introduced the concept of †CW complexes and proved an algebraic characterization of the homotopy equivalence of CW complexes. N. Steenrod developed †obstruction theory utilizing †squaring operations in the cohomology ring (1947). Subsequently, the theory of †cohomology operations was introduced (→ 64 Cohomology Operations, 305 Obstructions). The theory of †spectral sequences for †fiber spaces was originated by J. Leray (1945) and J.-P. Serre (1951) and was successfully applied to cohomology operations and homotopy theory by H. Cartan and Serre (1954) (→ 148 Fiber Spaces, 200 Homological Algebra). The study of the combinatorial structures of polyhedra and †piecewise linear mappings has flourished since 1940 in the works of Whitehead, S. S. Cairns, and others. S. Smale and, independently, J. Stallings solved the †generalized Poincaré conjecture in 1960. The †Hauptvermutung in combinatorial topology was solved negatively in 1961 by B. Mazur and J. Milnor. E. C. Zeeman proved the unknottedness of codimension 3 (1962). The recent development of the theory in conjunction with progress in †differential topology is notable. The Hauptvermutung for combinatorial manifolds was solved in 1969 by Kirby, Siebenmann, and Wall. In particular, there exist different combinatorial structures on tori of dimension $\geqslant 5$, and there are topological manifolds that do not admit any combinatorial structure (→ 65 Combinatorial Manifolds, 114 Differential Topology, 235 Knot Theory).

The global theory of differentiable manifolds started from the algebraic-topological study of †fiber bundles and †characteristic classes in the 1940s. R. Thom's fundamental theorem of †cobordism (1954) was obtained through extensive use of cohomology operations and homotopy groups. Milnor (1956) showed that the sphere S^7 may have differentiable structures that are essentially distinct from each other by using †Morse theory and the †index theorem of Thom and Hirzebruch. These results led to the creation of a new field, †differential topology (→ 56 Characteristic Classes, 114 Differential Topology).

Since 1959, A. Grothendieck, M. F. Atiyah, F. Hirzebruch, and J. F. Adams have developed †K-theory, which is a generalized cohomology theory constructed using stable classes of †vector bundles (→ 237 K-Theory).

†Knot theory, an interesting branch of topology, was one of the classical branches of topology and is now studied in connection with the theory of low-dimensional manifolds (→ 235 Knot Theory).

On the other hand, G. Cantor established general set theory in the 1870s and introduced such notions as †accumulation points, †open sets, and †closed sets in Euclidean space. The first important generalization of this theory was the concept of †topological space, which was proposed by M. Fréchet and developed by F. Hausdorff at the beginning of the 20th century. The theory subsequently became a new field of study, called **general topology** or **set-theoretic topology**. It deals with the topological properties of point sets in a Euclidean or topological space without reference to polyhedra. There has been a remarkable development of the theory since abount 1920, notably by Polish mathematicians S. Janiszewski, W. Sierpiński, S. Mazurkiewicz, C. Kuratowski, and others. The contributions of R. L. Moore, G. T. Whyburn, and K. Menger are also important (→ 382 Shape Theory, 425 Topological Spaces).

Topology is not only a foundation of various theories, but is also itself one of the most important branches of mathematics. It consists of †homology theory, †homotopy theory, †differential topology, †combinatorial manifolds, †K-theory, †transformation groups, †theory of singularities, †foliations, †dynamical systems, †catastrophe theory, etc. It continues to develop in interaction with other branches of mathematics (→ 51 Catastrophe Theory, 126 Dynamical Systems, 154 Foliations, 418 Theory of Singularities, 431 Transformation Groups).

References

[1] H. Poincaré, Analysis situs, J. Ecole Polytech., (2) 1 (1895), 1–121. (Oeuvres, Gauthier-Villars, 1953, vol. 6, 193–288.)
[2] P. S. Aleksandrov and H. Hopf, Topologie I, Springer, 1935 (Chelsea, 1965).
[3] S. Lefschetz, Algebraic topology, Amer. Math. Soc. Colloq. Publ., 1942.
[4] S. Eilenberg and N. E. Steenrod, Foundations of algebraic topology, Princeton Univ. Press, 1952.

427 (IX.12) Topology of Lie Groups and Homogeneous Spaces

A. General Remarks

Among various topological structures of †Lie groups and †homogeneous spaces, the structures of their †(co)homology groups and †homotopy groups are of special interest. Let G/H be a homogeneous space, where G is a Lie group and H is its closed subgroup. Then $(G, G/H, H)$ is a †fiber bundle, where G/H is the base space and H is the fiber. Thus homology and homotopy theory of fiber bundles (†spectral sequences and †homotopy exact sequences) can be applied. The †cellular decomposition of †Stiefel manifolds, †Grassmann manifolds, and †Kähler homogeneous spaces are known. Concerning †symmetric Riemannian spaces, we have various interesting methods, such as the use of invariant differential forms in connection with real cohomology rings and the use of †Morse theory in order to establish relations between the diagrams of symmetric Riemannian spaces G/H and homological properties of their †loop spaces and some related homogeneous spaces [4, 5]. Lie groups can be regarded as special cases of homogeneous spaces or symmetric spaces, although their group structures are of particular importance. A connected Lie group is homeomorphic to the product of one of its compact subgroups and a Euclidean space (†Cartan-Mal'tsev-Iwasawa theorem). Hence the topological structure of a connected Lie group is essentially determined by the topological structures of its compact subgroups.

B. Homology of Compact Lie Groups

Let G be a connected compact Lie group. Since G is an †H-space whose multiplication is given by its group multiplication h, $H^*(G;k)$ and $H_*(G;k)$ are dual †Hopf algebras for any coefficient field k. Also, $H^*(G;k)$ is isomorphic as a †graded algebra to the tensor product of †elementary Hopf algebras (→ 203 Hopf Algebras), but no factor of the tensor product is isomorphic to a polynomial ring because G is a finite †polyhedron. In particular, if $k=\mathbf{R}$ (the field of real numbers), then $H^*(G;\mathbf{R})\cong \bigwedge_{\mathbf{R}}(x_1,\ldots,x_l)$ (the exterior (Grassmann) algebra over $\mathbf{R}$ with generators $x_1,\ldots,x_l$ of odd degrees). Here we can choose generators x_i such that $h^*(x_i)=1\otimes x_i+x_i\otimes 1$, $1\leqslant i\leqslant l$. The x_i that satisfy this property are said to be **primitive**. Since in this case the †comultiplication h^* is commutative, the multiplication h_* is also commutative and the Hopf algebra $H_*(G;R)$ is an exterior algebra generated by elements y_i having the same degree as x_i $(i=1,\ldots,l)$. When the characteristic of the coefficient field k is nonzero, h_* need not be commutative.

The dimension of a †maximal torus of a connected compact Lie group G is independent of the choice of the maximal torus and is called the **rank** of G. The rank of G coincides with the number l of generators of $H^*(G;\mathbf{R})$. E. Cartan studied $H^*(G;\mathbf{R})$ by utilizing invariant differential forms. The cohomology theory of Lie algebras originated from the method he used in his study. $H^*(G;\mathbf{R})$ is invariant under †local isomorphisms of groups G. For †classical compact simple Lie groups G, R. Brauer calculated $H^*(G;\mathbf{R})$, while C.-T. Yen and C. Chevalley calculated $H^*(G;\mathbf{R})$ for †exceptional compact simple Lie groups (→ Appendix A, Table 6.IV). The degrees of the generators have group-theoretic meaning. Suppose that the degree of the ith generator is $2m_i-1$, $1\leqslant i\leqslant l$, and that $m_1\leqslant m_2\leqslant\ldots\leqslant m_l$. When G is simple, there is a relation $m_i+m_{l-i+1}=\text{constant}$ (Chevalley's duality). We have a proof for this property that does not use classification.

The cohomology groups $H^*(G;\mathbf{Z}_p)$ (where p is a prime and $\mathbf{Z}_p=\mathbf{Z}/p\mathbf{Z}$) have been determined as graded algebras for all compact simple Lie groups by A. Borel, S. Araki, and P. Baum and W. Browder (→ Appendix A, Table 6.IV).

C. Cohomology of Classifying Spaces

Let (E_G, B_G, G) be a †universal bundle of a connected compact Lie group G and p a prime or zero. Suppose that the integral cohomology of G has no p-torsion (no torsion when $p=0$). Then $H^*(G;\mathbf{Z}_p)=\bigwedge_{\mathbf{Z}_p}(x'_1,\ldots,x'_l)$ $(H^*(G;\mathbf{Z})=\bigwedge_{\mathbf{Z}}(x'_1,\ldots,x'_l)$ when $p=0)$, an exterior algebra with $\deg x'_i=2m_i-1$, $1\leqslant i\leqslant l$, and the generators x'_i can be chosen to be †transgressive in the spectral sequence of the universal bundle. Let $y_1,\ldots,y_l$ be their transgression images. Then $\deg y_i=2m_i$, $1\leqslant i\leqslant l$, and the cohomology of the †classifying space B_G over $\mathbf{Z}_p$ (resp. $\mathbf{Z}$) is the polynomial algebra with generators $y_1,\ldots,y_l$. Let T be a maximal torus of G. Then $B_T=E_G/T$ is a classifying space of T, the †Weyl group $W=N(T)/T$ of G with respect to T operates on B_T by †right translations, and $H^*(T;\mathbf{Z})$ has no torsion and is an exterior algebra with l generators of degree 1. Thus $H^*(B_T;\mathbf{Z})=\mathbf{Z}[u_1,\ldots,u_l]$, $\deg u_i=2$. Let I_W be the subalgebra of $H^*(B_T;\mathbf{Z})$ consisting of W-invariant polynomials, and let ρ be the projection of the bundle $(B_T, B_G, G/T)$. Then under the assumption that G has no p-torsion (no torsion), the cohomology mapping ρ^* over $\mathbf{Z}_p(\mathbf{Z})$ is monomorphic, and $\rho^*: H^*(B_G;\mathbf{Z}_p)\cong I_W\otimes\mathbf{Z}_p$ $(H^*(B_G;\mathbf{Z})\cong I_W)$ [1]. In the case of real coefficients, we have $H^*(B_G;\mathbf{R})\cong I_W\otimes\mathbf{R}$ for all G, and $m_1,\ldots,m_l$ are the degrees of generators of the ring I_W of W-invariant polynomials.

Example (1) $G=U(n)$: $l=n$ and G has no torsion. W operates on $H^*(B_T;\mathbf{Z})$ as the group of all permutations of generators $u_1,\ldots,u_n$. Thus generators of I_W are the †elementary symmetric polynomials $\sigma_1,\ldots,\sigma_n$ of $u_1,\ldots,u_n$. Let $c_1,\ldots,c_n$ be the †universal Chern classes; then $\rho^*(c_i)=\sigma_i$ and $H^*(B_{U(n)};\mathbf{Z})=\mathbf{Z}[c_1,\ldots,c_n]$.

Example (2) $G=SO(n)$: $l=[n/2]$ and G has no p-torsion for $p\neq 2$. W operates on $H^*(B_T;\mathbf{Z})$ as the group generated by the permutations of generators $u_1,\ldots,u_l$ and by the transformations $\sigma(u_i)=e_iu_i$, $e_i=\pm 1$, where the number of u_i for which $e_i=-1$ is arbitrary for odd n and even for even n. Thus the generators of I_W are the elementary symmetric polynomials $\sigma'_1,\ldots,\sigma'_l$ of $u_1^2,\ldots,u_l^2$ for odd n and $\sigma'_1,\ldots,\sigma'_{l-1}$ and $u_1\ldots u_l$ for even n. Let $p_1,\ldots,p_l$ be the †universal Pontryagin classes and χ be the †universal Euler-Poincaré class in the case of even n. Then $\rho^*(p_i)=\sigma'_i$ and $\rho^*(\chi)=u_1\ldots u_l$ for integral cohomology. Denote the mod p

reduction of p_i and χ by $\bar{p}_i$ and $\bar{\chi}$, respectively. Then $H^*(B_{SO(2l+1)}; \mathbf{Z}_p) = \mathbf{Z}_p[\bar{p}_1, \ldots, \bar{p}_l]$ and $H^*(B_{SO(2l)}; \mathbf{Z}_p) = \mathbf{Z}_p[\bar{p}_1, \ldots, \bar{p}_{l-1}, \bar{\chi}]$ ($p=0$ or >2).

Example (3) $G = O(n)$: If we use the subgroup Q consisting of all diagonal matrices instead of T, then we can make a similar argument for $\mathbf{Z}_2$-cohomology. Since $Q \cong (\mathbf{Z}_2)^n$, $H^*(B_Q; \mathbf{Z}_2) = \mathbf{Z}_2[v_1, \ldots, v_n]$ ($\mathbf{Z}_2[v_1, \ldots, v_n]$ is a polynomial ring with $\deg v_i = 1$), and $W_2 = N(Q)/Q$ operates on B_Q by right translations and on $H^*(B_Q; \mathbf{Z}_2)$ as the group of all permutations of $v_1, \ldots, v_n$. Let I_{W_2} be the subalgebra of $H^*(B_Q; \mathbf{Z}_2)$ consisting of all W_2-invariant polynomials. Then I_{W_2} is a polynomial ring generated by the elementary symmetric polynomials $\sigma_1'', \ldots, \sigma_n''$ of $v_1, \ldots, v_n$. The projection $\rho_2: B_Q \rightarrow B_{O(n)}$ induces a monomorphic cohomology mapping ρ_2^* over $\mathbf{Z}_2$, and ρ_2^*: $H^*(B_{O(n)}; \mathbf{Z}_2) \cong I_{W_2}$. Let $w_1, \ldots, w_n$ be the †universal Stiefel-Whitney classes. Then $\rho_2^*(w_i) = \sigma_i''$ and $H^*(B_{O(n)}; \mathbf{Z}_2) = \mathbf{Z}_2[w_1, \ldots, w_n]$ [2].

D. Grassmann Manifolds

The following manifolds are called **Grassmann manifolds**: The manifold $M_{n+m,n}(\mathbf{R})$ consisting of all n-subspaces of $\mathbf{R}^{n+m}$; the manifold $\tilde{M}_{n+m,n}(\mathbf{R})$ consisting of all oriented n-subspaces of $\mathbf{R}^{n+m}$; and the manifold $M_{n+m,n}(\mathbf{C})$ consisting of all complex n-subspaces of $\mathbf{C}^{n+m}$. These are expressed as quotient spaces as follows: $M_{n+m,n}(\mathbf{R}) = O(n+m)/O(n) \times O(m)$, $\tilde{M}_{n+m,n}(\mathbf{R}) = SO(n+m)/SO(n) \times SO(m)$, and $M_{n+m,n}(\mathbf{C}) = U(n+m)/U(n) \times U(m)$. They admit cellular decompositions by †Schubert varieties from which their cohomologies can be computed (→ 56 Characteristic Classes). $M_{n+m,n}(\mathbf{R})$ and $\tilde{M}_{n+m,n}(\mathbf{R})$ have no p-torsion for $p \neq 2$, and $M_{n+m,n}(\mathbf{C})$ has no torsion. These spaces are m-, m-, and $(2m+1)$-classifying spaces of $O(n)$, $SO(n)$, and $U(n)$, respectively. Hence their cohomologies are isomorphic to those of B_G ($G = O(n)$, $SO(n)$, $U(n)$) in dimensions $< m$, $< m$, and $\leqslant 2m$, respectively; and they are polynomial rings generated by suitable universal characteristic classes in low dimensions.

E. Cohomologies of Homogeneous Spaces G/U (Rank G = Rank U)

Let G be a compact connected Lie group and U a closed subgroup of G with the same rank as G. Denote the degrees of generators of $H^*(G; \mathbf{R})$ and $H^*(U; \mathbf{R})$ by $2m_1 - 1, \ldots, 2m_l - 1$, and $2n_1 - 1, \ldots, 2n_l - 1$, respectively. Then the real-coefficient †Poincaré polynomial P_0 of the homogeneous space G/U is given by $P_0(G/U, t) = \prod_i (1 - t^{2m_i})/(1 - t^{2n_i})$ (G. Hirsch). When G, U, and G/U have no p-torsion, the same formula is valid for the $\mathbf{Z}_p$-coefficient Poincaré polynomial [1]. When U is the †centralizer of a torus, G/U has a complex analytic cellular decomposition [3]. Hence G/U has no torsion in this case. This was proved by R. Bott and H. Samelson by utilizing Morse theory [5] (→ 279 Morse Theory). The case $U = T$ has also been studied.

F. Homotopy Groups of Compact Lie Groups

The †fundamental group $\pi_1(G)$ of a compact Lie group G is Abelian. Furthermore, $\pi_2(G) = 0$. If we apply Morse theory to G, the variational completeness of G can be utilized to show that the loop space ΩG has no torsion and that its odd-dimensional cohomologies vanish [4]. Consequently, when G is non-Abelian and simple, we have $\pi_3(G) \cong \mathbf{Z}$. A †periodicity theorem on †stable homotopy groups of classical groups proved by Bott is used in K-theory (→ 202 Homotopy Theory; 237 K-Theory). (For explicit forms of homotopy groups → Appendix A, Table 6.VI).

Homotopy groups of Stiefel manifolds are used to define characteristic classes by †obstruction cocycles (→ 147 Fiber Bundles; Appendix A, Table 6.VI).

References

[1] A. Borel, Sur la cohomologie des espaces fibrés principaux et des espaces homogènes de groupes de Lie compacts, Ann. Math., (2) 57 (1953), 115–207.
[2] A. Borel, La cohomologie mod 2 de certains espaces homogènes, Comment. Math. Helv., 27 (1953), 165–197.
[3] A. Borel, Kählerian coset spaces of semisimple Lie groups, Proc. Nat. Acad. Sci. US, 40 (1954), 1147–1151.
[4] R. Bott, An application of the Morse theory to the topology of Lie-groups, Bull. Soc. Math. France, 84 (1956), 251–281.
[5] R. Bott and H. Samelson, Applications of the theory of Morse to symmetric spaces, Amer. J. Math., 80 (1958), 964–1029.

428 (XIII.17) Total Differential Equations

A. Pfaff's Problem

A **total differential equation** is an equation of the form

$$\omega = 0, \qquad (1)$$

where ω is a †differential 1-form $\sum_{i=1}^{n} a_i(x)dx_i$ on a manifold X. A submanifold M of X is called an **integral manifold** of (1) if each vector ξ of the †tangent vector space $T_x(M)$ of M at every point x on M satisfies $\omega(\xi)=0$. We denote the maximal dimension of integral manifolds of (1) by $m(\omega)$. J. F. Pfaff showed that $m(\omega)\geqslant(n-1)/2$ for any ω. The problem of determining $m(\omega)$ for a given form ω is called **Pfaff's problem**. This problem was solved by G. Frobenius, J. G. Darboux, and others as follows: Form an †alternating matrix

$$(a_{ij})_{1\leqslant i,j\leqslant n} \tag{2}$$

from the coefficients of the †exterior derivative of ω,

$$d\omega=\frac{1}{2}\sum_{i=1}^{n}\sum_{j=1}^{n} a_{ij}(x)dx_i\wedge dx_j,$$

where $a_{ij}=\partial a_j/\partial x_i-\partial a_i/\partial x_j$. Suppose that the rank of (2) is $2t$. Then the rank of the matrix

$$\begin{pmatrix} a_{ij} & -a_i \\ a_j & 0 \end{pmatrix}_{1\leqslant i,j\leqslant n}$$

is $2t$ or $2t+2$. In the former case $m(\omega)=n-t$, and ω can be expressed in the form

$$\sum_{i=1}^{t} u_{2i-1}du_{2i}$$

by choosing a suitable coordinate system $(u_1,\ldots,u_n)$. In the latter case $m(\omega)=n-t-1$, and ω can be expressed in the form

$$\sum_{i=1}^{t} u_{2i-1}du_{2i}+du_{2t+1}$$

by choosing a suitable coordinate system $(u_1,\ldots,u_n)$. This theorem is called **Darboux's theorem**.

A 1-form ω is called a **Pfaffian form**, and equation (1) is called a **Pfaffian equation**. A system of equations $\omega_i=0$ $(1\leqslant i\leqslant s)$ for 1-form ω_i is called a **system of Pfaffian equations** or a **system of total differential equations** [6, 12, 26].

B. Systems of Differential Forms and Systems of Partial Differential Equations

Let Ω be a system of differential forms ω_i^p, $0\leqslant p\leqslant n$, $1\leqslant i\leqslant v_p$, on X, where ω_i^p is a p-form on X. A submanifold M of X is called an **integral manifold** of $\Omega=0$ if for each p $(0\leqslant p\leqslant \dim M)$, any p-dimensional subspace E_p of $T_x(M)$ satisfies $\omega_i^p(E_p)=0$ $(1\leqslant i\leqslant v_p)$ at every point x on M. Denote the maximal dimension of integral manifolds of $\Omega=0$ by $m(\Omega)$. The problem of determining $m(\Omega)$ for a given system Ω is called the **generalized Pfaff problem**, and will be explained in later sections. By fixing a local coordinate system of X and dividing it into two systems $(x_1,\ldots,x_r)$ and $(y_1,\ldots,y_m)$ $(m=n-r)$, we can consider the problem of finding an integral manifold of $\Omega=0$ defined by

$$y_\alpha=y_\alpha(x_1,\ldots,x_r),\qquad 1\leqslant\alpha\leqslant m.$$

This problem can be reduced to solving a system of partial differential equations of the first order on the submanifold N with the local coordinate system $(x_1,\ldots,x_r)$.

Consider a system of partial differential equations $\Phi=0$ of order l:

$$\varphi_\lambda(x_i,y_\alpha,p_\beta^{j_1\ldots j_r})=0,\qquad 1\leqslant\lambda\leqslant s, \tag{3}$$

with $1\leqslant i\leqslant r$, $1\leqslant\alpha$, $\beta\leqslant m$, $j_1+\ldots+j_r\leqslant l$, where

$$p_\beta^{j_1\ldots j_r}=\frac{\partial^{j_1+\ldots+j_r}y_\beta}{\partial x_1^{j_1}\ldots\partial x_r^{j_r}}. \tag{4}$$

A submanifold defined by $y_\alpha=y_\alpha(x_1,\ldots,x_r)$, $1\leqslant\alpha\leqslant m$, is called a **solution** of $\Phi=0$ if it satisfies (3) identically. The problem of determining whether a given system $\Phi=0$ has a solution was solved by C. Riquier, who showed that any system can be prolonged either to a passive orthonomic system or to an incompatible system by a finite number of steps. A system of partial differential equations is called a **prolongation** of another system if the former contains the latter and they have the same solution. A **passive orthonomic system** is one whose general solution can be parametrized by an infinite number of arbitrary constants. A solution containing parameters is called a **general solution** if by specifying the parameters we can obtain a solution of the †Cauchy problem for any initial data. A system (3) is said to be **incompatible** if it implies a nontrivial relation $f(x_1,\ldots,x_r)=0$ among the x_i.

The problem of solving a system $\Phi=0$ of partial differential equations can be reduced to that of finding integral manifolds of a system of differential forms Σ as follows: Let J^l be a manifold with the local coordinate system

$$(x_i,y_\alpha,p_\beta^{j_1\ldots j_r};\ 1\leqslant i\leqslant r,\ 1\leqslant\alpha,\ \beta\leqslant m,\ j_1+\ldots+j_r\leqslant l),$$

and Σ be a system of 0-forms φ_λ $(1\leqslant\lambda\leqslant s)$ and 1-forms

$$dy_\alpha-\sum_{i=1}^{r}p_\alpha^i dx_i,$$

$$dp_\beta^{j_1\ldots j_r}-\sum_{k=1}^{r}p_\beta^{j_1\ldots j_k+1\ldots j_r}dx_k$$

$(1\leqslant\alpha,\beta\leqslant m, j_1+\ldots+j_r<l)$. Then an integral manifold of $\Sigma=0$ of the form

$$y_\alpha=y_\alpha(x_1,\ldots,x_r),\qquad 1\leqslant\alpha\leqslant m,$$

$$p_\beta^{j_1\ldots j_r}=p_\beta^{j_1\ldots j_r}(x_1,\ldots,x_r),$$

$$1\leqslant\beta\leqslant m,\qquad j_1+\ldots+j_r\leqslant l,$$

gives a solution $y_\alpha = y_\alpha(x_1, \ldots, x_r)$, $1 \leqslant \alpha \leqslant m$, of $\Phi = 0$, and y_β and $p_\beta^{j_1 \ldots j_r}$ satisfy (4).

Conversely, a solution $y_\alpha = y_\alpha(x_1, \ldots, x_r)$, $1 \leqslant \alpha \leqslant m$, of $\Phi = 0$ gives an integral manifold of $\Sigma = 0$ if we define $p_\beta^{j_1 \ldots j_r}(x_1, \ldots, x_r)$ by (4) [23, 24, 26].

C. Systems of Partial Differential Equations of First Order with One Unknown Function

Consider a system of independent †vector fields on N:

$$L_\lambda = \sum_{i=1}^{r} b_{\lambda i}(x) \frac{\partial}{\partial x_i}, \qquad 1 \leqslant \lambda \leqslant s.$$

We solve a system of inhomogeneous equations

$$L_\lambda y - f_\lambda(x) y - g_\lambda(x) = 0, \qquad 1 \leqslant \lambda \leqslant s, \tag{5}$$

for a given system of $f_\lambda(x)$ and $g_\lambda(x)$. The system (5) is called a **complete system** if each of the expressions

$$[L_\lambda, L_\mu] y - (L_\lambda f_\mu - L_\mu f_\lambda) y - (f_\mu g_\lambda - f_\lambda g_\mu) - (L_\lambda g_\mu - L_\mu g_\lambda), \qquad 1 \leqslant \lambda < \mu \leqslant s, \tag{6}$$

is a linear combination of the left-hand sides of (5), where $[L_\lambda L_\mu]$ means the †commutator of L_λ and L_μ. This condition is called the **complete integrability condition** for (5). Suppose that the homogeneous system

$$L_\lambda y = 0, \qquad 1 \leqslant \lambda \leqslant s, \tag{7}$$

is complete. Then it has a system of †functionally independent solutions $y_1, \ldots, y_{r-s}$, and any solution y of (8) is a function of them: $y = \psi(y_1, \ldots, y_{r-s})$. If the inhomogeneous system (5) is complete, then the homogeneous system (7) is complete. This notion of a complete system is due to Lagrange and was extended to a system of nonlinear equations by Jacobi as follows (→ 324 Partial Differential Equations of First Order C).

Consider a system of nonlinear equations

$$F_\lambda(x_1, \ldots, x_r, y, p_1, \ldots, p_r) = 0, \qquad 1 \leqslant \lambda \leqslant s, \tag{8}$$

where $p_i = \partial y / \partial x_i$. The system (8) is called an **involutory system** if each of $[F_\lambda, F_\mu]$, $1 \leqslant \lambda < \mu \leqslant s$, is a linear combination of $F_1, \ldots, F_s$. Here †Lagrange's bracket $[F, G]$ is defined by

$$[F, G] = \sum_{i=1}^{r} \frac{\partial F}{\partial p_i} \left(\frac{\partial G}{\partial x_i} + p_i \frac{\partial G}{\partial y} \right) - \sum_{i=1}^{r} \frac{\partial G}{\partial p_i} \left(\frac{\partial F}{\partial x_i} + p_i \frac{\partial F}{\partial y} \right).$$

Suppose that the system (8) is involutory and $F_1, \ldots, F_s$ are functionally independent. Then, in general, we can solve the following †Cauchy problem for an $(r-s)$-dimensional submanifold N_{r-s} of N: Given a function f on N_{r-s}, find a solution y of (8) satisfying $y = f$ on N_{r-s}. We can construct a solution by integrating a system of ordinary differential equations called a †characteristic system of differential equations. Hence the solution of these problems may be carried out in the C^∞-category (→ 322 Partial Differential Equations (Methods of Integration) B) [7, 11].

D. Frobenius's Theorem

Let X be a †differentiable manifold of class C^∞ and Ω be a system of independent 1-forms ω_i, $1 \leqslant i \leqslant s$, on X. Then the system of Pfaffian equations $\Omega = 0$ is called a **completely integrable system** if at every point x of X,

$$d\omega_i = \sum_{j=1}^{s} \theta_{ij} \wedge \omega_j, \qquad 1 \leqslant i \leqslant s,$$

for 1-forms θ_{ij} on a neighborhood of x. Suppose that $\Omega = 0$ is completely integrable. Then at every point x of X, there exists a local coordinate system $(f_1, \ldots, f_s, x_{s+1}, \ldots, x_n)$ in a neighborhood U of x for which a tangent vector ξ of X at $z \in U$ satisfies $\omega_i(\xi) = 0$, $1 \leqslant i \leqslant s$, if and only if $\xi f_i = 0$, $1 \leqslant i \leqslant s$. In this case, each of the df_i is a linear combination of $\omega_1, \ldots, \omega_s$, and conversely, each of the ω_i is a linear combination of $df_1, \ldots, df_s$. In general, a function f for which df is a linear combination of $\omega_1, \ldots, \omega_s$ is called a **first integral** of $\Omega = 0$.

The theorem of the previous paragraph is called **Frobenius's theorem**, which can be stated in the dual form as follows: Let $D(X)$ be a †subbundle of the †tangent bundle $T(X)$ over X. The mapping $X \ni x \rightarrow D_x(X)$ is called a **distribution** on X. It is said to be an **involutive distribution** if at every point x of X we can find a system of independent vector fields L_i $(1 \leqslant i \leqslant s)$ on a neighborhood U of x such that the $L_i(z)$ $(1 \leqslant i \leqslant s)$ form a basis of $D_z(X)$ at every $z \in U$ and satisfy $[L_i, L_j] \equiv 0$ $(L_1, \ldots, L_s)$, $1 \leqslant i < j \leqslant s$, on U. A connected submanifold M of X is called an **integral manifold** of $D(X)$ if $T_x(M) = D_x(X)$ at every point x of M. Suppose that $D(X)$ gives an involutive distribution on X. Then every point x of X is in a maximal integral manifold M that contains any integral manifold including x as a submanifold.

E. Cartan-Kähler Existence Theorems

Let X be a †real analytic manifold. Denote the †sheaf of rings of differential forms on X by $\Lambda(X)$ and its subsheaf of $\mathcal{O}(X)$-modules of p-forms on X by $\Lambda_p(X)$, $1 \leqslant p \leqslant n$, where $\mathcal{O}(X)$ is the sheaf of rings of 0-forms on X. A subsheaf of ideals Σ is called a **differential ideal** if it is generated by Σ_p, $0 \leqslant p \leqslant n$, and contains $d\Sigma$, where $\Sigma_p = \Sigma \cap \Lambda_p(X)$. Consider a differential ideal Σ on X. Denote the †Grassmann manifold of p-dimensional subspaces of $T_x(X)$ with origin $x \in X$ by $G_p(x)$, and the Grassmann manifold $\bigcup_{x \in X} G(x)$ over X by $G_p(X)$. An element E_p of $G_p(x)$ is called a p-dimensional **contact element** with **origin** x. An element E_p of $G_p(x)$ is called an **integral element** of Σ_p if $\omega(E_p)=0$ at x for any p-form ω in Σ; furthermore, E_p is called an integral element of Σ if any element E_q contained in E_p, $0 \leqslant q \leqslant p$, is an integral element of Σ_q. In particular, 0-dimensional and 1-dimensional integral elements are called **integral points** and **integral vectors**, respectively. It can be proved that an element E_p is an integral element of Σ if and only if it is an integral element of Σ_p. The **polar element** $H(E_p)$ of an integral element E_p with origin x is defined as the subspace of $T_x(X)$ consisting of all vectors that generate with E_p an integral element of Σ. Let $(\Sigma_p)^0$, $0 \leqslant p \leqslant n$, be the subsheaf of $\mathcal{O}(X)$-modules in $\mathcal{O}(G_p(X))$ consisting of all 0-forms

$$\sum_{1 \leqslant i_1 < \ldots < i_p \leqslant n} a_{i_1 \ldots i_p} z_{i_1 \ldots i_p}$$

on $G_p(X)$ derived from a p-form

$$\sum_{1 \leqslant i_1 < \ldots < i_p \leqslant n} a_{i_1 \ldots i_p} dx_{i_1} \wedge \ldots \wedge dx_{i_p} \in \Sigma_p,$$

where $z_{i_1 \ldots i_p}$ is the †Grassmann coordinate of E_p. An integral element E_p^0 is called a **regular integral element** if the following two conditions are satisfied: (i) $(\Sigma_p)^0$ is a regular local equation of $I\Sigma_p$ at E_p^0, where $I\Sigma_p$ is the set of all integral elements of Σ_p; (ii) $\dim H(E_p) = \text{constant}$ around E_p^0 on $I\Sigma_p$. This definition, due to E. Kähler, is different from that given by E. Cartan [4].

Here, in general, a subsheaf Φ of $\mathcal{O}(X)$ is called a **regular local equation** of $I\Phi$ at an integral point x_0 if there exists a neighborhood U of x_0 and †cross sections $\varphi_1, \ldots, \varphi_s$ of Φ on U that satisfy the following two conditions: (i) $d\varphi_1, \ldots, d\varphi_s$ are linearly independent at every x on U; (ii) a point x of U is an integral point of Φ if and only if $\varphi_1(x) = \ldots = \varphi_s(x) = 0$.

First existence theorem. Suppose that we are given a p-dimensional integral manifold M with a regular integral element $T_x(M)$ at a point x on M. Suppose further that there exists a submanifold F of X containing M such that $\dim F = n - t_{p+1}$, $\dim(T_x(F) \cap H(E_p)) = p+1$, where $E_p = T_x(M)$ and $t_{p+1} = \dim H(E_p) - p - 1$. Then around x there exists a unique integral manifold N such that $\dim N = p+1$ and $F \supset N \supset M$.

This theorem is proved by integrating a system of partial differential equations of Cauchy-Kovalevskaya type. E. Cartan [2–4] also tried to obtain an existence theorem by integrating a system of ordinary differential equations.

A chain of integral elements $E_0 \subset E_1 \subset \ldots \subset E_r$ is called a **regular chain** if each of E_p $(0 \leqslant p < r)$ is a regular integral element. For a regular chain $E_0 \subset E_1 \subset \ldots \subset E_r$, define t_{p+1} by $t_{p+1} = \dim H(E_p) - p - 1$, $0 \leqslant p < r$, and define s_p by $s_p = t_p - t_{p+1} - 1$ $(0 \leqslant p < r)$, $s_r = t_r$, where $t_0 = \dim I\Sigma_0$. Then we have $s_p \geqslant 0$ $(0 \leqslant p \leqslant r)$, $s_0 + \ldots + s_r = t_0 - r$, and we can take a local coordinate system $(x_1, \ldots, x_r, y_1, \ldots, y_m)$, $m = n - r$, around E_0 that satisfies the following four conditions:

(i) $I\Sigma_0$ is defined by $y_{t_0-r+1} = \ldots = y_m = 0$;

(ii) $$H(E_p) = \left\{ \frac{\partial}{\partial x_1}, \ldots, \frac{\partial}{\partial x_r}, \frac{\partial}{\partial y_{s_0+\ldots+s_{p-1}+1}}, \ldots, \frac{\partial}{\partial y_{t_0-r}} \right\}, \quad 0 \leqslant p < r;$$

(iii) $$E_p = \left\{ \frac{\partial}{\partial x_1}, \ldots, \frac{\partial}{\partial x_p} \right\}, \quad 1 \leqslant p \leqslant r;$$

(iv) $E_0 = (0, \ldots, 0, 0, \ldots, 0)$.

The integers $s_0, \ldots, s_r$ are called the **characters** of the regular chain $E_0 \subset \ldots \subset E_r$.

Second existence theorem. Suppose that a chain of integral elements $E_0 \subset \ldots \subset E_r$ is regular, and take a local coordinate system satisfying (i)–(iv). Consider a system of initial data

$$f_1, \ldots, f_{s_0},$$

$$f_{s_0+1}(x_1), \ldots, f_{s_0+s_1}(x_1),$$

$$f_{s_0+s_1+1}(x_1, x_2), \ldots, f_{s_0+s_1+s_2}(x_1, x_2),$$

$$\ldots$$

$$f_{s_0+\ldots+s_{r-1}+1}(x_1, \ldots, x_r), \ldots, f_{t_0-r}(x_1, \ldots, x_r).$$

Then if their values and derivatives of the first order are sufficiently small, there exists a unique integral manifold defined by $y_\alpha = y_\alpha(x_1, \ldots, x_r)$, $y_\beta = 0$, $1 \leqslant \alpha \leqslant t_0 - r < \beta \leqslant m$, such that

$$y_\alpha(x_1, \ldots, x_p, 0, \ldots, 0) = f_\alpha(x_1, \ldots, x_p),$$

$$s_0 + \ldots + s_{p-1} < \alpha \leqslant s_0 + \ldots + s_p, \qquad 0 \leqslant p \leqslant r.$$

This theorem is proved by successive application of the first existence theorem. These two theorems are called the **Cartan-Kähler existence theorems**. Σ is said to be **involutive** at an integral element E_r if there exists a regular chain $E_0 \subset \ldots \subset E_r$. An integral manifold possessing a tangent space at which Σ is involutive is called an **ordinary integral manifold** or **ordinary solution** of Σ. An integral manifold that does not possess such a tangent space is called a **singular integral manifold** or **singular solution** of Σ.

Cartan's definition of ordinary and regular integral elements is as follows: An integral point E_0^0 is an ordinary integral point if Σ_0 is a regular local equation of $I\Sigma_0$ at E_0^0. An ordinary integral point E_0^0 is a regular integral point if $\dim H(E_0)$ is constant on $I\Sigma_0$ around E_0^0. Inductively, an integral element E_p^0 is called an **ordinary integral element** if $(\Sigma_p)^0$ is a regular local equation of $I\Sigma_p$ at E_p^0 and E_p^0 contains a regular integral element E_{p-1}^0. An ordinary integral element E_p^0 is a regular integral element (in the sense of Cartan) if $\dim H(E_p)$ is constant on $I\Sigma_p$ around E_p^0. It can be proved that Σ is involutive at an integral element E_r if and only if E_r is an ordinary integral element of Σ. An integral manifold possessing a tangent space that is a regular integral element of Σ is called a **regular integral manifold** or **regular solution** of Σ. Let m_{p+1} be the minimal dimension of $H(E_p)$, where E_p varies over the set of p-dimensional ordinary integral elements, and g be an integer such that $m_p \geqslant p$ $(1 \leqslant p \leqslant g)$ and $m_{g+1} = p$. Then this integer g is called the **genus** of Σ. It is the maximal dimension of ordinary integral manifolds of Σ. However, in general, it is not the maximal dimension of integral manifolds of Σ.

D. C. Spencer and others have been trying to obtain an existence theorem in the C^∞-category analogous to that of Cartan and Kähler. (For a system of linear partial differential equations → [2, 4, 11, 13, 25, 27].)

F. Involutive Systems of Partial Differential Equations

To give a definition of an involutive system of partial differential equations, we define an involutive subspace of $\mathrm{Hom}(V, W)$, where V and W are finite-dimensional vector spaces over the real number field $\mathbf{R}$. Let A be a subspace of $\mathrm{Hom}(V, W)$. For a system of vectors $v_1, \ldots, v_p$ in V, $A(v_1, \ldots, v_p)$ denotes the subspace of A that annihilates $v_1, \ldots, v_p$. Let g_p be the minimal dimension of $A(v_1, \ldots, v_p)$ as $(v_1, \ldots, v_p)$ varies, where $0 \leqslant p \leqslant r = \dim V$. A basis $(v_1, \ldots, v_r)$ of V is called a generic basis if it satisfies $g_p = \dim A(v_1, \ldots, v_p)$ for each p. There exists a generic basis for any A. Let $W \otimes S^2(V^*)$ be the subspace of $\mathrm{Hom}(V, \mathrm{Hom}(V, W))$ consisting of all elements ξ satisfying $\xi(u)v = \xi(v)u$ for any u and v in V. Then the prolongation pA of A is defined by $pA = \mathrm{Hom}(V, A) \cap W \otimes S^2(V^*)$. For any basis $(v_1, \ldots, v_r)$ of V, we have the inequality

$$\dim pA \leqslant \sum_{p=0}^{r} \dim A(v_1, \ldots, v_p).$$

The subspace A is called an **involutive subspace** of $\mathrm{Hom}(V, W)$ if $\dim pA = \sum_{p=0}^{r} g_p$. This notion of an involutive subspace was obtained by V. W. Guillemin and S. Sternberg [13].

A triple $(X, N; \pi)$ consisting of two manifolds X, N and a projection π from X onto N is called a **fibered manifold** if the †differential π_* is surjective at every point of X. Take the set of all mappings f from a domain in N to X satisfying $\pi \circ f = $ identity for a fibered manifold $(X, N; \pi)$. Then an †l-jet $j_x^l(f)$ is an equivalence class under the equivalence relation defined as follows: $j_x^l(f) = j_u^l(g)$ if and only if $x = u$, $f(x) = g(u)$, and

$$\frac{\partial^{i_1+\ldots+i_r} f}{\partial x_1^{i_1} \ldots \partial x_r^{i_r}}(x) = \frac{\partial^{i_1+\ldots+i_r} g}{\partial x_1^{i_1} \ldots \partial x_r^{i_r}}(u),$$

$i_1 + \ldots + i_r \leqslant l$, where $(x_1, \ldots, x_r)$ is a local coordinate system of N around $x = u$ (→ 105 Differentiable Manifolds X).

Denote the space of all l-jets of a fibered manifold $(X, N; \pi)$ by $J^l(X, N; \pi)$ or simply J^l. Then a subsheaf of ideals Φ in $\mathcal{O}(J^l)$ is called a **system of partial differential equations of order** l on N. A point z of J^l is called an integral point of Φ if $\varphi(z) = 0$ for all $\varphi \in \Phi$. The set of all **integral points** of Φ is denoted by $I\Phi$. Let π^l be the natural projection of J^l onto J^{l-1}. Then at a point z of J^l, we can identify $\mathrm{Ker}\,\pi_*^l$ with $\mathrm{Hom}(T_x(N), \mathrm{Ker}\,\pi_*^l)$, where $x = \pi\pi^1 \ldots \pi^l z$. The principal part $C_z(\Phi)$ of Φ is defined as the subspace of $\mathrm{Ker}\,\pi_*^l$ that annihilates Φ. The **prolongation** $p\Phi$ of Φ is defined as the system of order $l+1$ on N generated by Φ and $\partial_k \Phi$, $1 \leqslant k \leqslant \dim N$, where ∂_k is the formal derivative with respect to a coordinate x_k of N:

$$(\partial_k \varphi)(j_x^{l+1}(f)) = \frac{\partial}{\partial x_k} \varphi(j_x^l(f)), \qquad \varphi \in \mathcal{O}(J^l).$$

Let w be an integral point of $p\Phi$ and z be $\pi^{l+1} w$. Then we have the identity

$$pC_z(\Phi) = C_w(p\Phi).$$

The following definition of an involutive system is due to M. Kuranishi [19]: Φ is involutive at an integral point z if the following two conditions are satisfied: (i) Φ is a regular local equation of $I\Phi$ at z; (ii) there exists a neighborhood U of z in J^l such that $(\pi^{l+1})^{-1} U \cap I(p\Phi)$ forms a fibered manifold with base $U \cap I\Phi$ and projection π^{l+1}.

A system of partial differential equations is said to be **involutive** (or **involutory**) if it has an integral point at which it is involutive. Fix a system of independent variables $(y_1, \ldots, y_N)$ in X. Then a system of differential forms is said to be **involutive** (or **involutory**) if it has an integral element at which it is involutive and $dy_1 \wedge \ldots \wedge dy_N \neq 0$. It can be proved that these two definitions of involutive system are equivalent [19, 25].

G. Prolongation Theorems

Cartan gave a method of prolongation by which we can obtain an involutive system from a given system with two independent variables, if it has a solution. He proposed the following problem: For any $r>2$, construct a method of prolongation by which we can obtain an involutive system from a given system with r independent variables, if it has a solution. To solve this problem, Kuranishi prolonged a given system Φ successively to $p^t\Phi$, $t=1,2,3,\ldots$, and proved the following theorem: Suppose that there exists a sequence of integral points z^t of $p^t\Phi$ with $\pi^{t+l}z^t=z^{t-l}$, $t=1,2,3,\ldots$, that satisfies the following two conditions for each t: (i) $p^t\Phi$ is a regular local equation of $I(p^t\Phi)$ at z^t; (ii) there exists a neighborhood V^t of z^t in $I(p^t\Phi)$ such that $\pi^{t+l}V^t$ contains a neighborhood of z^{t-1} in $I(p^{t-1}\Phi)$ and forms a fibered manifold $(V^t, \pi^{t+l}V^t; \pi^{t+l})$. Then $p^t\Phi$ is involutive at z^t for a sufficiently large integer t.

This prolongation theorem gives a powerful tool to the theory of †infinite Lie groups. However, if we consider a system of partial differential equations of general type, there exist examples of systems that cannot be prolonged to an involutive system by this prolongation, although they have a solution. To improve Kuranishi's prolongation theorem, M. Matsuda [22] defined the prolongation of the same order by $p_0\Phi=p\Phi\cap\mathcal{O}(J^l)$ for a system Φ of order l. This is a generalization of the classical method of completion given by Lagrange and Jacobi. Applying this prolongation successively to a given system Φ, we have $\Psi=\bigcup_{\sigma=1}^{\infty}p_0^{\sigma}\Phi$. Define the p_*-operation by $p_*=\bigcup_{\sigma=1}^{\infty}p_0^{\sigma}p$. Then applying this prolongation successively to Ψ, we have the following theorem: suppose that there exists a sequence of integral points z^t of $p_*^t\Psi$ with $\pi^{l+t}z^t=z^{t-1}$, $t=1,2,3,\ldots$, that satisfies the following two conditions for each t: (i) $p_*^t\Psi$ is a regular local equation of $I(p_*^t\Psi)$ at z^t; (ii) $\dim pC(p_*^t\Psi)$ is constant around z^t on $I(p_*^t\Psi)$. Then $p_*^t\Psi$ is involutive at z^t for a sufficiently large integer t.

To prove this theorem Matsuda applied the following theorem obtained by V. W. Guillemin, S. Sternberg, and J.-P. Serre [25, appendix]: suppose that we are given a subspace A_0 of $\mathrm{Hom}(V,W)$ and subspaces A_t of $\mathrm{Hom}(V,A_{t-1})$ satisfying $A_t\subset pA_{t-1}$, $t=1,2,3,\ldots$. Then A_t is an involutive subspace of $\mathrm{Hom}(V,A_{t-1})$ for a sufficiently large integer t. Thus Cartan's problem was solved affirmatively. To the generalized Pfaff problem these prolongation theorems give another solution, which differs from that obtained by Riquier.

H. Pfaffian Systems in the Complex Domain

Consider a linear system of Pfaffian equations

$$du_i=\sum_{k=1}^{n}\sum_{j=1}^{m}a_{ij}^k(x)u_j\,dx_k,\qquad i=1,\ldots,m,$$

where $x=(x_1,\ldots,x_n)$ is a local coordinate of a complex manifold X and a_{ij}^k are meromorphic functions on X. If we put $u={}^t(u_1,\ldots,u_m)$ and $A^k(x)=(a_{ij}^k(x))$, $k=1,\ldots,n$, the system is written as

$$du=\left(\sum_{k=1}^{n}A^k(x)\,dx_k\right)u. \tag{9}$$

System (9) is completely integrable if and only if

$$\frac{\partial A^j}{\partial x^l}-\frac{\partial A^l}{\partial x^j}=[A^l,A^j],\qquad j,l=1,\ldots,n.$$

Suppose that (9) is completely integrable. If the $A^k(x)$ are holomorphic at $x^0=(x_1^0,\ldots,x_n^0)\in X$, there exists for any $u^0\in\mathbf{C}^m$ one and only one solution of (9) that is holomorphic at x^0 and satisfies $u(x^0)=u^0$. This implies that the solution space of (9) is an m-dimensional vector space; the basis of this space is called a fundamental system of solutions. Therefore any solution is expressible as a linear combination of a fundamental system of solutions and can be continued analytically in a domain where the $A^k(x)$ are holomorphic. A subvariety of X that is the pole set of at least one of the $A^k(x)$ is called a singular locus of (9), and a point on a singular locus is called a singular point.

R. Gérard has given a definition of regular singular points and an analytic expression of a fundamental system of solutions around a regular singular point, and he studied systems of Fuchsian type [8; also 9, 30].

Let $\Omega=\sum_{k=1}^{n}A^k(x)\,dx_k$. Then the system (9) can be rewritten as

$$(d-\Omega)u=0.$$

If we consider a local coordinate (x,u) of a fiber bundle over X, the operator $d-\Omega$ induces a meromorphic linear connection ∇ over X. Starting from this point of view, P. Deligne [5] introduced several important concepts and obtained many results.

The first results for irregular singular points were obtained by Gérard and Y. Sibuya [10], and H. Majima [20] studied irregular singular points of mixed type.

The systems of partial differential equations that are satisfied by the hypergeometric functions of several variables are equivalent to linear systems of Pfaffian equations [1]. This means that such systems of partial differential equations are †holonomic systems. M. Kashiwara and T. Kawai [15] studied holonomic systems with regular singularities from the standpoint of microlocal analysis. Special types of holonomic systems were investigated by T. Terada [28] and M. Yoshida [29].

Consider a system of Pfaffian equations

$$\omega_j=0,\qquad j=1,\ldots,r, \tag{10}$$

where $\omega_j = \sum_{k=1}^n a_{jk}(x) dx_k$ and $x = (x_1, \ldots, x_n)$. Suppose that a_{jk} are holomorphic in a domain D of $\mathbf{C}^n$ and that $d\omega_j \wedge \omega_1 \wedge \ldots \wedge \omega_r = 0$ in D. Denote by S the zero set of $\omega_1 \wedge \ldots \wedge \omega_r = 0$. A point of S is called a singular point of (10). If the codimension of S is $\geqslant 1$, then system (10) is completely integrable in $D - S$. The following theorem was proved by B. Malgrange [21]: Let $x^0 \in S$, and suppose that the codimension of S is $\geqslant 3$ around x^0; then there exist functions $f_j, j = 1, \ldots, r$, and $g_{jk}, j, k = 1, \ldots, r$, that are holomorphic at x^0 and satisfy $\omega_j = \sum_{k=1}^r g_{jk} df_k$ and $\det(g_{jk}(x^0)) \neq 0$.

References

[1] P. Appell and J. Kampé de Fériet, Fonctions hypergéométriques et hypersphériques, Gauthier-Villars, 1926.
[2] E. Cartan, Sur l'intégration des systèmes d'équations aux différentielles totales, Ann. Sci. Ecole Norm. Sup., 18 (1901), 241–311.
[3] E. Cartan, Leçons sur les invariants intégraux, Hermann, 1922.
[4] E. Cartan, Les systèmes différentielles extérieures et leurs applications géométriques, Hermann, Actualités Sci. Ind., 1945.
[5] P. Deligne, Equations différentielles à points singuliers réguliers, Lecture notes in math. 163, Springer, 1970.
[6] A. R. Forsyth, Theory of differential equations, pt. I. Exact equations and Pfaff's problem, Cambridge Univ. Press, 1890.
[7] A. R. Forsyth, Theory of differential equations, pt. IV. Partial differential equations, Cambridge Univ. Press, 1906.
[8] R. Gérard, Théorie de Fuchs sur une variété analytique complexe, J. Math. Pures Appl., 47 (1968), 321–404.
[9] R. Gérard and A. H. M. Levelt, Sur les connexions à singularités régulières dans le cas de plusieurs variables, Funkcial. Ekvac., 19 (1976), 149–173.
[10] R. Gérard and Y. Sibuya, Etude de certains systèms de Pfaff avec singularités, Lecture notes in math. 712, Springer, 1979.
[11] E. Goursat, Leçons sur l'intégration des équations aux dérivées partielles du premier ordre, Hermann, second edition, 1920.
[12] E. Goursat, Leçons sur le problème de Pfaff, Hermann, 1922.
[13] V. W. Guillemin and S. Sternberg, An algebraic model of transitive differential geometry, Bull. Amer. Math. Soc., 70 (1964), 16–47.
[14] E. Kähler, Einführung in die Theorie der Systeme von Differentialgleichungen, Teubner, 1934.
[15] M. Kashiwara and T. Kawai, On holonomic systems of microdifferential equations IV, systems with regular singularities, Publ. Res. Inst. Math. Sci., 17 (1981), 813–979.
[16] M. Kuranishi, On E. Cartan's prolongation theorem of exterior differential systems, Amer. J. Math., 79 (1957), 1–47.
[17] M. Kuranishi, Lectures on exterior differential systems, Lecture notes, Tata Inst., 1962.
[18] M. Kuranishi, On the local theory of continuous infinite pseudo groups I, II, Nagoya Math. J., 15 (1959), 225–260; 19 (1961), 55–91.
[19] M. Kuranishi, Lectures on involutive systems of partial differential equations, Publ. Soc. Math. São Paulo, 1967.
[20] H. Majima, Asymptotic analysis for integrable connections with irregular singular points, Lecture notes in math. 1075, Springer, 1984.
[21] B. Malgrange, Frobenius avec singularités II, Le cas général, Inventiones Math., 39 (1976), 67–89.
[22] M. Matsuda, Cartan-Kuranishi's prolongation of differential systems combined with that of Lagrange and Jacobi, Publ. Res. Inst. Math. Sci., 3 (1967/68), 69–84.
[23] C. Riquier, Les systèmes d'équations aux dérivées partielles, Gauthier-Villars, 1910.
[24] J. F. Ritt, Differential algebra, Amer. Math. Soc. Colloq. Publ., 1950.
[25] I. M. Singer and S. Sternberg, The infinite groups of Lie and Cartan I. The transitive groups, J. Analyse Math., 15 (1965), 1–114.
[26] J. A. Schouten and W. v. d. Kulk, Pfaff's problem and its generalizations, Clarendon Press, 1949.
[27] D. C. Spencer, Overdetermined systems of linear partial differential equations, Bull. Amer. Math. Soc., 75 (1969), 179–239.
[28] T. Terada, Problème de Rieman et fonctions automorphes provenant des fonctions hypergéométriques de plusieurs variables, J. Math. Kyoto Univ., 13 (1973), 557–578.
[29] M. Yoshida, Local theory of Fuchsian systems with certain discrete monodromy groups I, II, Funkcial. Ekvac., 21 (1978), 105–137, 203–221.
[30] M. Yoshida and K. Takano, On a linear system of Pfaffian equations with regular singular point, Funkcial. Ekvac., 19 (1976), 175–189.

429 (XI.6) Transcendental Entire Functions

A. General Remarks

An **entire function** (or **integral function**) $f(z)$ is a complex-valued function of a complex variable

z that is holomorphic in the finite z-plane, $z \neq \infty$. If $f(z)$ has a pole at ∞, then $f(z)$ is a polynomial in z. A polynomial is called a **rational entire function**. If an entire function is bounded, it is constant (†Liouville's theorem). A **transcendental entire function** is an entire function that is not a polynomial, for example, $\exp z$, $\sin z$, $\cos z$. An entire function can be developed in a power series $\sum_{n=0}^{\infty} a_n z^n$ with infinite radius of convergence. If $f(z)$ is a transcendental entire function, this is actually an infinite series.

B. The Order of an Entire Function

If a transcendental entire function $f(z)$ has a zero of order m $(m \geqslant 0)$ at $z=0$ and other zeros at $\alpha_1, \alpha_2, \ldots, \alpha_n, \ldots$ $(0<|\alpha_1| \leqslant |\alpha_2| \leqslant |\alpha_3| \leqslant \ldots \to \infty)$, multiple zeros being repeated, then $f(z)$ can be written in the form

$$f(z)=e^{g(z)} z^m \prod_{k=1}^{\infty}\left(1-\frac{z}{\alpha_k}\right) e^{g_k(z)},$$

where $g(z)$ is an entire function, $g_k(z)=(z/\alpha_k)+(1/2)(z/\alpha_k)^2+(1/3)(z/\alpha_k)^3+\ldots+(1/p_k)(z/\alpha_k)^{p_k}$, and $p_1, p_2, \ldots$ are integers with the property that $\sum_{k=1}^{\infty}|z/\alpha_k|^{p_k+1}$ converges for all z (**Weierstrass's canonical product**).

E. N. Laguerre introduced the concept of the genus of a transcendental entire function $f(z)$. Assume that there exists an integer p for which $\sum_{k=0}^{\infty}|\alpha_k|^{-(p+1)}$ converges, and take the smallest such p. Assume further that in the representation for $f(z)$ in the previous paragraph, when $p_1=p_2=\ldots=p$, the function $g(z)$ reduces to a polynomial of degree q; then $\max(p, q)$ is called the **genus** of $f(z)$. For transcendental entire functions, however, the order is more essential than the genus. The **order** ρ of a transcendental entire function $f(z)$ is defined by

$$\rho=\limsup_{r \to \infty} \log\log M(r)/\log r,$$

where $M(r)$ is the maximum value of $|f(z)|$ on $|z|=r$. By using the coefficients of $f(z)=\sum a_n z^n$, we can write

$$\rho=\limsup_{n \to \infty} n \log n/\log(1/|a_n|).$$

The entire functions of order 0, which were studied by Valiron and others, have properties similar to polynomials, and the entire functions of order less than 1/2 satisfy $\lim_{r_n \to \infty} \min_{|z|=r_n}|f(z)|=\infty$ for some increasing sequence $r_n \uparrow \infty$ (**Wiman's theorem**). Hence entire functions of order less than 1/2 cannot be bounded in any domain extending to infinity. Among the functions of order greater than 1/2 there exist functions bounded in a given angular domain $D: \alpha<\arg z<\alpha+\pi/\mu$. If $|f(z)| <\exp r^\rho$ $(\rho<\mu)$ and $f(z)$ is bounded on the boundary of D, then $f(z)$ is bounded in the angular domain (→ 272 Meromorphic Functions). In particular, if the order ρ of $f(z)$ is an integer p, then it is equal to the genus, and $g(z)$ reduces to a polynomial of degree $\leqslant p$ (J. Hadamard). These theorems originated in the study of the zeros of the †Riemann zeta function and constitute the beginning of the theory of entire functions.

There is some difference between the properties of functions of integral order and those of others. Generally, the point z at which $f(z)=w$ is called a **w-point** of $f(z)$. If $\{z_n\}$ consists of w-points different from the origin, the infimum $\rho_1(w)$ of k for which $\sum 1/|z_n|^k$ converges is called the **exponent of convergence** of $f-w$. If the order ρ of an entire function is integral, then $\rho_1(w)=\rho$ for each value w with one possible exception, and if ρ is not integral, then $\rho_1(w)=\rho$ for all w (É. Borel). Therefore any transcendental entire function has an infinite number of w-points for each value w except for at most one value, called an **exceptional value** of $f(z)$ (**Picard's theorem**). In particular, $f(z)$ has no exceptional values if ρ is not integral. For instance, $\sin z$ and $\cos z$ have no exceptional values, while e^z has 0 as an exceptional value. Since transcendental entire functions have no poles, ∞ can be counted as an exceptional value. Then we must change the statement in Picard's theorem to "except for at most two values." Since the theorem was obtained by E. Picard in 1879, problems of this type have been studied intensively (→ 62 Cluster Sets, 272 Meromorphic Functions).

After Picard proved the theorem by using the inverse of a †modular function, several alternative proofs were given. For instance, there is a proof using the Landau-Schottky theorem and †Bloch's theorem and one using †normal families. Picard's theorem was extended to meromorphic functions and has also been studied for analytic functions defined in more general domains. There are many fully quantitative results, too. For instance, Valiron [3] gave such results by performing some calculations on neighborhoods of points where entire functions attain their maximum absolute values.

Thereafter, the distribution of w-points in a neighborhood of an essential singularity was studied by many people, and in 1925 the Nevanlinna theory of meromorphic functions was established. The core of the theory consists of two fundamental theorems, †Nevanlinna's first and second fundamental theorems (→ 272 Meromorphic Functions). Concerning composite entire functions $F(z)=f(g(z))$, Pólya proved the following fact: The finiteness of the order of F implies that the order of f should be zero unless g is a polynomial. This gives the starting point of the factorization theory, on which several people have been working recently. Several theorems in the theory of meromorphic functions can be applied to the theory. One of the fundamental theorems is the following: Let $F(z)$ be an entire function, which admits the factorizations $F(z)=P_m(f_m(z))$ with a polynomial P_m of degree m and an entire function f_m for all integers m. Then $F(z)=A\cos\sqrt{H(z)}+B$ unless $F(z)=A\exp H(z)+B$. Here, H is a nonconstant entire function and A, B are constant, $A\neq 0$.

C. Julia Directions

Applying the theory of †normal families of holomorphic functions, G. Julia proved the existence of Julia directions as a precise form of Picard's theorem [5]. A transcendental entire function $f(z)$ has at least one direction $\arg z=\theta$ such that for any $\varepsilon>0$, $f(z)$ takes on every (finite) value with one possible exception infinitely often in the angular domain $\theta-\varepsilon<\arg z<\theta+\varepsilon$. This direction $\arg z=\theta$ is called a **Julia direction** of $f(z)$.

D. Asymptotic Values

†Asymptotic values, †asymptotic paths, etc., are defined for entire functions as for meromorphic functions. In relation to †Iversen's theorem and †Gross's theorem for inverse functions and results on †cluster sets, †ordinary singularities of inverse functions hold for entire functions in the same way as for meromorphic functions. Also, as for meromorphic functions, †transcendental singularities of inverse functions are divided into two classes, the †direct and the †indirect transcendental singularities.

The exceptional values in Picard's theorem are asymptotic values of the functions, and ∞ is an asymptotic value of any transcendental entire function. Therefore $f(z)\rightarrow\infty$ along some curve extending to infinity. Between the asymptotic paths corresponding to two distinct asymptotic values, there is always an asymptotic path with asymptotic value ∞. By †Bloch's theorem, A. Bloch showed that the †Riemann surface of the inverse function of a transcendental entire function contains a disk with arbitrarily large radius. Denjoy conjectured in 1907 that $\mu\leqslant 2\rho$, where ρ is the order of an entire function and μ is the number of distinct finite asymptotic values of the function, and L. V. Ahlfors gave the first proof (1929). This result contains Wiman's theorem. There are transcendental entire functions with $\mu=2\rho$. It was shown by W. Gross that among entire functions of infinite order there exists an entire function having every value as its asymptotic value.

References

[1] E. C. Titchmarsh, The theory of functions, Clarendon Press, second edition, 1939.
[2] R. P. Boas, Entire functions, Academic Press, 1954.
[3] G. Valiron, Lectures on the general theory of integral functions, Librairie de l'Université, Deighton, Bell and Co., 1923 (Chelsea, 1949).
[4] L. Bieberbach, Lehrbuch der Funktionentheorie II, Teubner, 1931 (Johnson Reprint Co., 1969).
[5] G. Julia, Sur quelques propriétés nouvelles des fonctions entières ou méromorphes, Ann. Sci. Ecole Norm. Sup., (3) 36 (1919), 93–125.

430 (V.11) Transcendental Numbers

A. History

A complex number α is called a **transcendental number** if α is not †algebraic over the field of rational numbers **Q**. C. Hermite showed in 1873 that e is a transcendental number. Following a similar line of thought as that taken by Hermite, C. L. F. Lindemann showed that π is also transcendental (1882). Among the 23 problems posed by D. Hilbert in 1900 (→ 196 Hilbert), the seventh was the problem of establishing the transcendence of certain numbers (e.g., $2^{\sqrt{2}}$). This stimulated fruitful investigations by A. O. Gel'fond, T. Schneider, C. L. Siegel, and others. The theory of transcendental numbers is, however, far from complete. There is no general criterion that can be utilized to characterize transcendental numbers. For example, neither the transcendence nor even the irrationality of the †Euler constant $C=\lim_{n\rightarrow\infty}(1+1/2+\ldots+1/n-\log n)$ has been established. A survey of the development of the theory of transcendental numbers can be found in [18], in which an extensive list of relevant publications up to 1966 is given.

B. Construction of Transcendental Numbers

Let $\overline{\mathbf{Q}}$ be the field of †algebraic numbers. Suppose that α is an element of $\overline{\mathbf{Q}}$ that satisfies the irreducible equation $f(x)=a_0x^n+a_1x^{n-1}+\ldots+a_n=0$, where the a_i are rational integers, $a_0\neq 0$, and $a_0,a_1,\ldots,a_n$ have no common factors. Then we define $H(\alpha)$ to be the maximum of $|a_i|$ ($i=0, \ldots, n$) and call it the **height** of α. J. Liouville proved the following theorem (1844): Let ξ be a real number ($\xi \notin \mathbf{Q}$). If $\inf\{q^n|\xi - p/q| \mid p/q \in \mathbf{Q}\} = 0$ for any positive integer n, then ξ is transcendental.

Transcendental numbers having this property are called **Liouville numbers**. Examples are: (i) $\xi = \sum_{\nu=1}^{\infty} g^{-\nu!}$, where g is an integer not smaller than 2. (ii) Suppose that we are given a sequence $\{n_k\}$ of positive integers such that $n_k \to \infty$ ($k \to \infty$). Let ξ be the real number expressed as an †infinite simple continued fraction $b_0 + 1/b_1 \dotplus 1/b_2 \dotplus \ldots$. Let B_l be the denominator of the lth †convergent of the continued fraction. If $b_{n_k+1} \geqslant B_{n_k}^{n_k-2}$ for $k \geqslant 1$, then ξ is a Liouville number.

On the other hand, K. Mahler [8, 9] proved the existence of transcendental numbers that are not Liouville numbers. For example, he showed that if $f(x)$ is a nonconstant integral polynomial function mapping the set of positive integers into itself, then a number ξ expressed, e.g., in the decimal system as $0y_1y_2y_3\ldots$ is such a number if we put $y_n = f(n)$, $n = 1, 2, 3, \ldots$. (In particular, from $f(x) = x$ we get the non-Liouville transcendental number $\xi = 0.123456789101112\ldots$.) Mahler proved this result by using †Roth's theorem (1955) (→ 182 Geometry of Numbers). Both Liouville and Mahler utilized the theory of †Diophantine approximation to construct transcendental numbers.

On the other hand, Schneider [10–12] and Siegel [3] constructed transcendental numbers using certain functions. Examples are: $\exp\alpha$ ($\alpha \in \bar{\mathbf{Q}}$, $\alpha \neq 0$); α^β ($\alpha \in \bar{\mathbf{Q}}$, $\alpha \neq 0, 1$; $\beta \in \bar{\mathbf{Q}} - \mathbf{Q}$); $J(\tau)$, where J is the †modular function and τ is an algebraic number that is not contained in any imaginary quadratic number field; $\wp(2\pi i/\alpha)$, where $\wp$ is the Weierstrass $\wp$-function, $\alpha \in \bar{\mathbf{Q}}$, and $\alpha \neq 0$; and $B(p, q)$, where B is the †Beta function and $p, q \in \mathbf{Q} - \mathbf{Z}$.

Since $e = \exp 1$ and $1 = \exp 2\pi i$, the transcendence of e and π is directly implied by the transcendence of $\exp\alpha$ ($\alpha \in \bar{\mathbf{Q}}$, $\alpha \neq 0$).

C. Classification of Transcendental Numbers

(1) Mahler's classification: Given a complex number ξ and positive integers n and H, we consider the following:

$$w_n(H, \xi) = \min\left\{\left|\sum_{\nu=0}^{n} a_\nu \xi^\nu\right| \,\middle|\, a_\nu \in \mathbf{Z},\ |a_\nu| \leqslant H,\ \sum_{\nu=0}^{n} a_\nu \xi^\nu \neq 0\right\},$$

$$w_n(\xi) = w_n = \limsup_{H\to\infty}(-\log w_n(H, \xi)/\log H),$$

$$w(\xi) = w = \limsup_{n\to\infty} w_n(\xi)/n,$$

and let μ = the first number n for which w_n is ∞. Then we have the following four cases: (i) $w=0$, $\mu=\infty$; (ii) $0<w<\infty$, $\mu=\infty$; (iii) $w=\mu=\infty$; (iv) $w=\infty$, $\mu<\infty$, corresponding to which we call ξ an **A-number**, **S-number**, **T-number**, or **U-number**. The set of A-numbers is denoted by **A**, and similarly we have the classes **S**, **T**, and **U**. It is known that $\mathbf{A} = \bar{\mathbf{Q}}$. If two numbers ξ and η are †algebraically dependent over **Q**, then they belong to the same class. If ξ belongs to **S**, the quantity $\theta(\xi) = \sup\{w_n(\xi)/n \mid n = 1, 2, \ldots\}$ is called the **type** of ξ (in the sense of Mahler). Mahler conjectured that almost all transcendental numbers (except a set of Lebesgue measure zero) are S-numbers of the type 1 or 1/2 according as they belong to **R** or not. Various results were obtained concerning this conjecture (W. J. LeVeque, J. F. Koksma, B. Volkmann) until it was proved by V. G. Sprindzhuk in 1965 [14, 15]. The existence of T-numbers was proved by W. M. Schmidt (1968) [16]. All Liouville numbers are U-numbers [7]. On the other hand, $\log\alpha$ ($\alpha \in \mathbf{Q}$, $\alpha > 0$, $\alpha \neq 1$) and π are transcendental numbers that do not belong to **U**.

(2) Koksma's classification: For a given transcendental number ξ and positive numbers n and H, we consider the following:

$$w_n^*(H, \xi) = \min\{|\xi - \alpha| \mid \alpha \in \bar{\mathbf{Q}},\ H(\alpha) \leqslant H,\ [\mathbf{Q}(\alpha):\mathbf{Q}] \leqslant n\},$$

$$w_n^*(\xi) = w_n^* = \limsup_{H\to\infty}(-\log(Hw_n^*(H, \xi))/\log H),$$

$$w^*(\xi) = w^* = \limsup_{n\to\infty} w_n^*(\xi)/n,$$

and let μ^* = the first number n for which w_n^* is ∞. Then we have the following three cases: (i) $w^* < \infty$, $\mu^* = \infty$; (ii) $w^* = \mu^* = \infty$; (iii) $w^* = \infty$, $\mu^* < \infty$. We call ξ an **S*-number**, **T*-number**, or **U*-number** according as (i), (ii), or (iii) holds and denote the set of S*-numbers by **S***, etc. If ξ belongs to **S***, we call $\theta^*(\xi) = \sup\{w_n^*(\xi)/n \mid n = 1, 2, \ldots\}$ the **type** of ξ (in the sense of Koksma). It can be shown that $\mathbf{S} = \mathbf{S}^*$, $\mathbf{T} = \mathbf{T}^*$, and $\mathbf{U} = \mathbf{U}^*$, and that if $\xi \in \mathbf{S}$, then $\theta^*(\xi) \leqslant \theta(\xi) \leqslant \theta^*(\xi) + 1$.

D. Algebraic Independence

Concerning the algebraic relations of transcendental numbers, we have the following three principal theorems:

(1) Let $\alpha_1, \ldots, \alpha_m$ be elements of $\bar{\mathbf{Q}}$ that are linearly independent over **Q**. Then $\exp\alpha_1, \ldots, \exp\alpha_m$ are transcendental and algebraically independent over $\bar{\mathbf{Q}}$ (**Lindemann-Weierstrass theorem**).

(2) Let $J_0(x)$ be the †Bessel function and α a nonzero algebraic number. Then $J_0(\alpha)$ and $J_0'(\alpha)$ are transcendental and algebraically independent over **Q** (Siegel).

(3) Let $\alpha_1, \ldots, \alpha_n$ be nonzero elements of $\bar{\mathbf{Q}}$ such that $\log\alpha_1, \ldots, \log\alpha_n$ are linearly independent over $\mathbf{Q}$. Then $1, \log\alpha_1, \ldots, \log\alpha_n$ are linearly independent over $\bar{\mathbf{Q}}$ (A. Baker).

Besides these theorems, various related results have been obtained by A. B. Shidlovskiĭ, Gel'fond, N. I. Fel'dman, and others. A quantitative extension of theorem (3), also by Baker, will be discussed later.

First we give more detailed descriptions of theorems (1) and (2). Let $\alpha_1, \ldots, \alpha_m$ be as in theorem (1), $s = [\mathbf{Q}(\alpha_1, \ldots, \alpha_m) : \mathbf{Q}]$, $P(X_1, \ldots, X_m)$ be an arbitrary polynomial in $\bar{\mathbf{Q}}[X_1, \ldots, X_m]$ of degree n, and $H(P)$ be the maximum of the absolute values of the coefficients of the polynomial P. Then there exists a positive number C determined only by the numbers $\alpha_1, \ldots, \alpha_m$ and $n(=\deg P)$ such that

$$|P(e^{\alpha_1}, \ldots, e^{\alpha_m})| > CH(P)^{-2s\left(2\binom{2smn+m+n}{m}-1\right)}.$$

In particular, if α is a nonzero algebraic number, then $\exp\alpha$ belongs to $\mathbf{S}$ and $\theta(\exp\alpha) \leqslant 8s^2 + 6s$.

(2′) Let α be a nonzero algebraic number, $s = [\mathbf{Q}(\alpha) : \mathbf{Q}]$, $P \in \mathbf{Q}[X_1, X_2]$, $\deg P = n$. Then there exists a positive number C determined only by α and n such that $|P(J_0(\alpha), J_0'(\alpha))| > CH(P)^{-82s^3n^3}$.

Theorems (1) and (2) are actually special cases of a theorem obtained by Siegel. To state this theorem, the following terminology is used: An entire function $f(z) = \sum_{n=0}^{\infty} C_n \cdot z^n/n!$ is called an ***E*-function** defined over an †algebraic number field K of finite degree if the following three conditions are satisfied: (i) $C_n \in K$ $(n = 0, 1, 2, \ldots)$. (ii) For any positive number ε, $C_n = O(n^{\varepsilon n})$. (iii) Let q_n be the least positive integer such that $C_k q_n$ belongs to the ring $\mathfrak{O}$ of algebraic integers in K $(0 \leqslant n, 0 \leqslant k \leqslant n)$. Then for an arbitrary positive number ε, $q_n = O(n^{\varepsilon n})$.

A system $\{f_1(z), \ldots, f_m(z)\}$ of E-functions defined over K is said to be **normal** if it satisfies the following two conditions: (i) None of the functions $f_i(z)$ is identically zero. (ii) If the functions $w_k = f_k(z)$ $(k = 1, \ldots, m)$ satisfy a system of †homogeneous linear differential equations of the first order, then $w_k' = \sum_{l=1}^{m} Q_{kl}(z) w_l$, where the $Q_{kl}(z)$ are rational functions of z, with coefficients in the ring $\mathfrak{O}$. The matrix (Q_{kl}) can be decomposed by rearranging the order of the indices k, l if necessary into the form

$$\begin{pmatrix} W_1 & & 0 \\ & \ddots & \\ 0 & & W_r \end{pmatrix},$$

where

$$W_t = \begin{bmatrix} Q_{11,t} \cdots Q_{1m_t,t} \\ \cdots \quad \cdots \quad \cdots \\ Q_{m_t1,t} \cdots Q_{m_tm_t,t} \end{bmatrix}, \quad 1 \leqslant t \leqslant r, \quad \sum_{t=1}^{n} m_t = m.$$

The decomposition is unique if we choose r as large as possible, in which case we call $W_1, \ldots, W_r$ the primitive parts of (Q_{kl}). The requirement is that the primitive parts W_i are independent in the following sense: If there are numbers $C_{st} \in K$ and polynomial functions $P_{kt}(z) \in K[z]$ such that

$$\sum_{t=1}^{r} (C_{1t} \ldots C_{m_t t}) W_t \begin{bmatrix} P_{1t}(z) \\ \vdots \\ P_{m_t t}(z) \end{bmatrix} = 0,$$

then $C_{st} = 0$, $P_{kl}(z) = 0$.

Let N be a positive integer. A normal system $\{f_1(z), \ldots, f_m(z)\}$ of E-functions is said to be of degree N if the system $\{F_{n_1, \ldots, n_m}(z) = f_1(z)^{n_1} \ldots f_m(z)^{n_m} \mid n_i \geqslant 0, \sum_{i=1}^{m} n_i \leqslant N\}$ is also a normal system of E-functions. Then the theorem obtained by Siegel [4] is: Let N be an arbitrary positive integer and $\{f_1(z), \ldots, f_m(z)\}$ be a normal system of E-functions of degree N defined over an algebraic number field of finite degree K satisfying the system of differential equations $f_k'(z) = \sum_{l=1}^{m} Q_{kl}(z) f_l(z)$, where $Q_{kl}(z) \in \mathfrak{O}(z)$, $1 \leqslant k \leqslant m$. If α is a nonzero algebraic number that is not a †pole of any one of the functions $Q_{kl}(z)$, then $f_1(\alpha), \ldots, f_m(\alpha)$ are transcendental numbers that are algebraically independent over the field $\bar{\mathbf{Q}}$.

Theorem (3) at the beginning of this section implies, for example, the following: (i) If $\alpha_1, \ldots, \alpha_n$ and $\beta_1, \ldots, \beta_n$ all belong to $\bar{\mathbf{Q}}$ and $\gamma = \alpha_1 \log\beta_1 + \ldots + \alpha_n \log\beta_n \neq 0$, then γ is transcendental. (ii) If $\alpha_1, \ldots, \alpha_n, \beta_0, \beta_1, \ldots, \beta_n$ are nonzero algebraic numbers, then $e^{\beta_0}\alpha_1^{\beta_1} \ldots \alpha_n^{\beta_n}$ is transcendental. (iii) If $\alpha_1, \ldots, \alpha_n$ are algebraic numbers other than 0 and 1, and $\beta_1, \ldots, \beta_n$ also belong to $\bar{\mathbf{Q}}$, with $1, \beta_1, \ldots, \beta_n$ linearly independent over $\mathbf{Q}$, then $\alpha_1^{\beta_1} \ldots \alpha_n^{\beta_n}$ is transcendental.

Baker [17] also obtained a quantitative extension of theorem (3): Suppose that we are given integers $A \geqslant 4$, $d \geqslant 4$ and nonzero algebraic numbers $\alpha_1, \ldots, \alpha_n$ $(n \geqslant 2)$ whose heights and degrees do not exceed A and d, respectively. Suppose further that $0 < \delta \leqslant 1$, and let $\log\alpha_1, \ldots, \log\alpha_n$ be the principal values of the logarithms. If there exist rational integers $b_1, \ldots, b_n$ with absolute value at most H such that

$$0 < |b_1 \log\alpha_1 + \ldots + b_n \log\alpha_n| < e^{-\delta H},$$

then

$$H < (4^{n^2}\delta^{-1} d^{2n} \log A)^{(2n+1)^2}.$$

This theorem has extensive applications in various problems of number theory, including a wide class of †Diophantine problems [19].

A number of new, interesting results on the algebraic independence of values of exponential functions, elliptic functions, and some other special functions have been obtained recently by D. Masser, G. V. Chudnovskiĭ, M. Waldschmidt, and other writers. In particular, Chudnovskiĭ (1975) obtained the remarkable result that $\Gamma(1/3)$ and $\Gamma(1/4)$ are transcendental numbers. See [20–24].

References

[1] A. O. Gel'fond, Transcendental and algebraic numbers, Dover, 1960. (Original in Russian, 1952.)
[2] T. Schneider, Einführung in die transzendenten Zahlen, Springer, 1957.
[3] C. L. Siegel, Über einige Anwendungen diophantischer Approximationen, Abh. Preuss. Akad. Wiss., 1929 (Gesammelte Abhandlungen, Springer, 1966, vol. 1, 209–266).
[4] C. L. Siegel, Transcendental numbers, Ann. Math. Studies, Princeton Univ. Press, 1949.
[5] N. I. Fel'dman, Approximation of certain transcendental numbers I, Amer. Math. Soc. Transl., 59 (1966), 224–245. (Original in Russian, 1951.)
[6] W. J. LeVeque, Note on S-numbers, Proc. Amer. Math. Soc., 4 (1953), 189–190.
[7] W. J. LeVeque, On Mahler's U-numbers, J. London Math. Soc., 28 (1953), 220–229.
[8] K. Mahler, Arithmetische Eigenschaften einer Klasse von Dezimalbrüchen, Proc. Acad. Amsterdam, 40 (1937), 421–428.
[9] K. Mahler, Über die Dezimalbruchentwicklung gewisser Irrationalzahlen, Mathematica B, Zutphen, 6 (1937), 22–36.
[10] T. Schneider, Transzendenzuntersuchungen periodischer Funktionen I, J. Reine Angew. Math., 172 (1935), 65–69.
[11] T. Schneider, Arithmetische Untersuchungen elliptischer Integrale, Math. Ann., 113 (1936), 1–13.
[12] T. Schneider, Zur Theorie der Abelschen Funktionen und Integrale, J. Reine Angew. Math., 183 (1941), 110–128.
[13] S. Lang, Introduction to transcendental numbers, Addison-Wesley, 1966.
[14] V. G. Sprindzhuk, A proof of Mahler's conjecture on the measure of the set of S-numbers, Amer. Math. Soc. Transl., 51 (1966), 215–272. (Original in Russian, 1965.)
[15] V. G. Sprindzhuk, Mahler's problem in metric number theory, Amer. Math. Soc. Transl. Math. Monographs, 1969. (Original in Russian, 1967.)
[16] W. M. Schmidt, T-numbers do exist, Rend. Convegno di Teoria dei Numeri, Roma, 1968.
[17] A. Baker, Linear forms in the logarithms of algebraic numbers I, II, III, IV, Mathematika, 13 (1966), 204–216; 14 (1967), 102–107; 14 (1967), 220–228; 15 (1968), 204–216.
[18] N. I. Fel'dman and A. B. Shidlovskiĭ, The development and present state of the theory of transcendental numbers, Russian Math. Surveys, 22 (1967), 1–79. (Original in Russian, 1967.)
[19] A. Baker, Transcendental number theory, Cambridge Univ. Press, 1975.
[20] D. Masser, Elliptic functions and transcendence, Lecture notes in math. 437, Springer, 1975.
[21] G. V. Chudnovskiĭ, Algebraic independence of values of exponential and elliptic functions, Proc. Int. Congr. Math., Helsinki, 1978.
[22] A. Baker and D. Masser (eds.), Transcendence theory: Advances and applications, Academic Press, 1977.
[23] M. Waldschmidt, Transcendence methods, Queen's papers in pure and appl. math. 52, Queen's Univ., 1979.
[24] M. Waldschmidt, Nombres transcendants et groups algébriques, Astérisque, 69–70 (1980).

431 (IX.19) Transformation Groups

A. Topological Transformation Groups

Let G be a group, M a set, and f a mapping from $G \times M$ into M. Put $f(g,x)=g(x)$ $(g \in G, x \in M)$. Then the group G is said to be a **transformation group** of the set M if the following two conditions are satisfied: (i) $e(x)=x$ $(x \in M)$, where e is the identity element of G; and (ii) $(gh)(x)=g(h(x))$ $(x \in M)$ for any $g, h \in G$. In this case the mapping $x \rightarrow g(x)$ is a one-to-one mapping of M onto itself.

Let G be a transformation group of M. If G is a topological group, M a topological space, and the mapping $(g,x) \rightarrow g(x)$ a continuous mapping from $G \times M$ into M, then G is called a **topological transformation group** of M. In this case $x \rightarrow g(x)$ is a homeomorphism of M onto itself. The mapping $(g,x) \rightarrow g(x)$ is called an **action** of G on M. The space M, together with a given action of G, is called a G-**space**.

For a point x of M, the set $G(x)=\{g(x) \mid g \in G\}$ is called the **orbit** of G passing through the point x. Defining as equivalent two points x and y of M belonging to the same orbit, we get an equivalence relation in M. The quotient space of M by this equivalence relation, denoted by M/G, is called the **orbit space** of G-space M.

If $G(x)=\{x\}$, then x is called a **fixed point**. The set of all fixed points is denoted by M^G. For a point x of M, the set $G_x=\{g \in G \mid g(x)=$

$x\}$ is a subgroup of G called the **isotropy subgroup (stabilizer, stability subgroup)** of G at the point x. A conjugacy class of the subgroup G_x is called an **isotropy type** of the transformation group G on M.

The group G is said to act **nontrivially** (resp. **trivially**) on M if $M \neq M^G$ (resp. $M = M^G$). The group G is said to act **freely** on M if the isotropy subgroup G_x consists only of the identity element for any point x of M.

The group G is said to act **transitively** on M if for any two points x and y of M, there exists an element $g \in G$ such that $g(x) = y$.

Let N be the set of all elements $g \in G$ such that $g(x) = x$ for all points x of M. Then N is a normal subgroup of G. If N consists only of the identity element e, we say that G acts **effectively** on M, and if N is a discrete subgroup of G, we say that G acts **almost effectively** on M. When $N \neq \{e\}$, the quotient topological group G/N acts effectively on M in a natural fashion.

An **equivariant mapping (equivariant map)** (or a G-**mapping**, G-**map**) $h: X \to Y$ between G-spaces is a continuous mapping which commutes with the group actions, that is, $h(g(x)) = g(h(x))$ for all $g \in G$ and $x \in X$. An equivariant mapping which is also a homeomorphism is called an equivalence of G-spaces.

For a G-space M, an equivalence class of the G-spaces $G(x)$, $x \in M$, is called an **orbit type** of the G-space M.

B. Cohomological Properties

We consider only †paracompact G-spaces and †Čech cohomology theory in this section. We shall say that a topological space X is **finitistic** if every open covering has a finite-dimensional refinement. The following theorems are useful [1–3].

(1) If G is finite, X a finitistic paracompact G-space, and K a field of characteristic zero or prime to the order of G, then the induced homomorphism $\pi^*: H^*(X/G; K) \to H^*(X; K)^G$ is an isomorphism. Here, π is a natural projection of X onto X/G. The group G acts naturally on $H^*(X; K)$, and $H^*(X; K)^G$ denotes the fixed-point set of this G-action.

(2) Let X be a finitistic G-space and G cyclic of prime ordor p. Then, with coefficients in $\mathbf{Z}/p\mathbf{Z}$, we have

(a) for each n $\quad \sum_{i=n}^{\infty} \operatorname{rank} H^i(X^G) \leqslant \sum_{i=n}^{\infty} \operatorname{rank} H^i(X)$,

(b) $\quad \chi(X) + (p-1)\chi(X^G) = p\chi(X/G)$.

Here the †Euler-Poincaré characteristics $\chi(\ )$ are defined in terms of mod p cohomology.

(3) **Smith's theorem**: If G is a p-group (p prime) and if x is a finitistic G-space whose mod p cohomology is isomorphic to the n-sphere, then the mod p cohomology of the fixed-point set X^G is isomorphic to that of the r-sphere for some $-1 \leqslant r \leqslant n$, where (-1)-sphere means the empty set.

(4) Let T^k denote the k-dimensional toral group. Let X be a T^k-space whose rational cohomology is isomorphic to the n-sphere, and assume that there are only a finite number of orbit types and that the orbit spaces of all subtori are finitistic. Let H be a subtorus of T^k. Then by the above theorem the rational cohomology of X^H is isomorphic to that of the $r(H)$-sphere for some $-1 \leqslant r(H) \leqslant n$. Assume further that there is no fixed point of the T^k-action. Then, with H ranging over all subtori of dimension $k-1$, we have

$$n+1 = \sum_H (r(H)+1).$$

C. Differentiable Transformation Groups

Suppose that the group G is a transformation group of a †differentiable manifold M, G is a †Lie group, and the mapping $(g, x) \to g(x)$ of $G \times M$ into M is a differentiable mapping. Then G is called a **differentiable transformation group** (or **Lie transformation group**) of M, and M is called a differentiable G-**manifold**.

The following are basic facts about compact differentiable transformation groups [3, 4]:

(5) **Differentiable slice theorem**: Let G be a compact Lie group acting differentiably on a manifold M. Then, by averaging an arbitrary †Riemannian metric on M, we may have a G-invariant Riemannian metric on M. That is, the mapping $x \to g(x)$ is an †isometry of this Riemannian manifold M for each $g \in G$. For each point $x \in M$, the orbit $G(x)$ through x is a compact submanifold of M and the mapping $g \mapsto g(x)$ defines a G-equivariant diffeomorphism $G/G_x \cong G(x)$, where G/G_x is the left quotient space by the isotropy subgroup G_x. G_x acts orthogonally on the †tangent space $T_x M$ at x (resp. the †normal vector space N_x of the orbit $G(x)$); we call it the **isotropy representation** (resp. **slice representation**) of G_x at x. Let E be the †normal vector bundle of the orbit $G(x)$. Since G acts naturally on E as a bundle mapping, the bundle E is equivalent to the bundle $(G \times N_x)/G_x$ over G/G_x as a †G-vector bundle, where G_x acts on N_x by means of the slice representation and G_x acts on G by the right translation. We can choose a small positive real number ε such that the †exponential mapping gives an equivariant †diffeomorphism of the ε-disk bundle of E onto an invariant †tubular neighborhood of $G(x)$.

(6) Assume that a compact Lie group G acts differentiably on M with the orbit space $M^* = M/G$ connected. Then there exists a maximum

orbit type G/H for G on M (i.e., H is an isotropy subgroup and H is conjugate to a subgroup of each isotropy group). The union $M_{(H)}$ of the orbits of type G/H is open and dense in M, and its image $M^*_{(H)}$ in M^* is connected.

The maximum orbit type for orbits in M guaranteed by the above theorem is called the **principal orbit type**, and orbits of this type are called **principal orbits**. The corresponding isotropy groups are called **principal isotropy groups**. Let P be a principal orbit and Q any orbit. If $\dim P > \dim Q$, then Q is called a **singular orbit**. If $\dim P = \dim Q$ but P and Q are not equivalent, then Q is called an **exceptional orbit**.

(7) Let G be a compact Lie group and M a compact G-manifold. Then the orbit types are finite in number.

By applying (5) and (6) we have that an isotropy group is principal if and only if its slice representation is trivial.

The situation is quite different in the case of noncompact transformation groups. For example, there exists an analytic action of $G = SL(4, \mathbf{R})$ on an analytic manifold M such that each orbit of G on M is closed and of codimension one and such that, for $x, y \in M$, G_x is not isomorphic to G_y unless x and y lie on the same G-orbit [5].

D. Compact Differentiable Transformation Groups

Many powerful techniques in †differential topology have been applied to the study of differentiable transformation groups. For example, using the techniques of †surgery, we can show that there are infinitely many free differentiable circle actions on †homotopy $(2n+1)$-spheres $(n \geqslant 3)$ that are differentiably inequivalent and distinguished by the rational †Pontryagin classes of the orbit manifolds (W. C. Hsiang [6]). Also, using †Brieskorn varieties, we can construct many examples of differentiable transformation groups on homotopy spheres [3, 4, 7]. Differentiable actions of compact connected Lie groups on homology spheres have been studied systematically (Hsiang and W. Y. Hsiang [4]).

The Atiyah-Singer †index theorem has many applications in the study of transformation groups. The following are notable applications:

(8) Let M be a compact connected †oriented differentiable manifold of dimension $4k$ with a †spin-structure. If a compact connected Lie group G acts differentiably and nontrivially on M, then the $\hat{A}$-genus $\langle \hat{\mathscr{A}}(M), [M] \rangle$ of M vanishes (where $\hat{\mathscr{A}}(M)$ denotes the †$\hat{A}$-characteristic class of M) (M. F. Atiyah and F. Hirzebruch [8], K. Kawakubo [9]). For further developments, see A. Hattori [10].

(9) Let M be a closed oriented manifold with a differentiable circle action. Then each connected component F_k of the fixed point set can be oriented canonically, and we have

$$I(M) = \sum_k I(F_k),$$

where $I(\ )$ denotes the †Thom-Hirzebruch index [8, 9].

Let G be a compact Lie group and $G \to EG \to BG$ the †universal G-bundle. Then the †singular cohomology $H^*(EG \times_G X)$ is called **equivariant cohomology** for a G-space X and is an $H^*(BG)$-module. Let $G = U(1)$, M a differentiable $U(1)$-manifold, $F = M^G$, and $i: F \to M$ the inclusion mapping. Then the †localization of the induced homomorphism

$$S^{-1} i^* : S^{-1} H^*(EG \times_G M) \to S^{-1} H^*(BG \times F)$$

is an isomorphism, where S^{-1} denotes the localization with respect to the multiplicative set $S = \{at^k\}$ with a, k ranging over all positive integers and t the generator of $H^2(BG)$. Theorems (8) and (9) can be proved by the above localization isomorphism.

Let M be a differentiable manifold. The upper bound $N(M)$ of the dimension of all the compact Lie groups that acts effectively and differentiably on M is called the **degree of symmetry** of M. It measures, in some crude sense, the symmetry of the differentiable manifold M. The number $N(M)$ depends heavily on the differentiable structure. For example, $N(S^m) = m(m+1)/2$ for the standard m-sphere, but $N(\Sigma^m) < (m+1)^2/16 + 5$ for a †homotopy m-sphere $(m \geqslant 300)$ that does not bound a †π-manifold [11]. Also, $N(P_n(\mathbf{C})) = n(n+2)$ for the complex projective n-space $P_n(\mathbf{C})$, but $N(hP_n(\mathbf{C})) < (n+1)(n+2)/2$ for any homotopy complex projective n-space $hP_n(\mathbf{C})$ $(n \geqslant 13)$ other than $P_n(\mathbf{C})$ (T. Watabe [12]).

Let X be a differentiable closed manifold and $h: X \to P_n(\mathbf{C})$ be an orientation-preserving †homotopy equivalence. There is a conjecture about the total $\hat{A}$-classes that states: If X admits a nontrivial differentiable circle action, then $\hat{\mathscr{A}}(X) = h^* \hat{\mathscr{A}}(P_n(\mathbf{C}))$ (T. Petrie [13]). It is known that if the action is free outside the fixed-point set, then the conjecture is true (T. Yoshida [14]).

E. Equivariant Bordism

Fix a compact Lie group G; a compact **oriented G-manifold** (ψ, M) consists of a compact †oriented differentiable manifold M and an orientation-preserving differentiable G-action $\psi: G \times M \to M$ on M.

Given families $F \supset F'$ of subgroups of G, a compact oriented G-manifold (ψ, M) is (F, F')-**free** if the following conditions are satisfied: (i) if $x \in M$, then the isotropy group G_x is conjugate to a member of F; (ii) if $x \in \partial M$, then G_x is conjugate to a member of F'.

If F' is the empty family, then necessarily ∂M is empty and M is closed. In this case we say that (ψ, M) is F-**free**.

Given (ψ, M), define $-(\psi, M)=(\psi, -M)$ with the structure precisely the same as (ψ, M) except for †orientation. Also define $\partial(\psi, M)=(\psi, \partial M)$. Note that if (ψ, M) is (F, F')-free, then $(\psi, \partial M)$ is F'-free. Define (ψ, M) and (ψ', M') to be isomorphic if there exists an equivariant orientation-preserving diffeomorphism of M onto M'.

An (F, F')-free compact oriented n-dimensional G-manifold (ψ, M) is said to **bord** if there exists an (F, F)-free compact oriented $(n+1)$-dimensional G-manifold (Φ, W) together with a regularly embedded compact n-dimensional manifold M_1 in ∂W with M_1 invariant under the G-action Φ such that (Φ, M_1) is isomorphic to (ψ, M) and G_x is conjugate to a member of F' for $x \in \partial W - M_1$. Also, M_1 is required to have its orientation induced by that of W.

We say that (ψ_1, M_1) is **bordant** to (ψ_2, M_2) if the disjoint union $(\psi_1, M_1)+(\psi_2, -M_2)$ bords. Bordism is an equivalence relation on the class of (F, F')-free compact oriented n-dimensional G-manifolds. The bordism classes constitute an Abelian group $\mathbf{O}_n^G(F, F')$ under the operation of disjoint union. If F' is empty, denote the above group by $\mathbf{O}_n^G(F)$. The direct sum

$$\mathbf{O}_*^G(F, F')=\bigoplus_n \mathbf{O}_n^G(F, F')$$

is naturally an Ω-module, where Ω is the †oriented cobordism ring. If F consists of all subgroups of G, then $\mathbf{O}_*^G(F)$ is denoted by $\mathbf{O}_*^G$.

Suppose now that $F \supset F'$ are fixed families of subgroups of G. Every F'-free G-manifold is also F-free, and so this inclusion induces a homomorphism $\alpha: \mathbf{O}_n^G(F') \to \mathbf{O}_n^G(F)$. Similarly every F-free G-manifold is also (F, F')-free, inducing a homomorphism $\beta: \mathbf{O}_n^G(F) \to \mathbf{O}_n^G(F, F')$. Finally, there is a homomorphism $\partial: \mathbf{O}_n^G(F, F') \to \mathbf{O}_{n-1}^G(F')$ given by $\partial(\psi, M)=(\psi, \partial M)$. Then the following sequence is exact [15]:

$$\ldots \xrightarrow{\partial} \mathbf{O}_n^G(F') \xrightarrow{\alpha} \mathbf{O}_n^G(F) \xrightarrow{\beta} \mathbf{O}_n^G(F, F') \xrightarrow{\partial} \mathbf{O}_{n-1}^G(F') \xrightarrow{\alpha} \ldots.$$

A weakly almost complex compact G-manifold (ψ, M) consists of a †weakly almost complex compact manifold M and a differentiable G-action $\psi: G \times M \to M$ that preserves the weakly almost complex structure on M. $\mathbf{U}_*^G(F, F')$, $\mathbf{U}_*^G$ are defined similarly, and they are $\mathbf{U}_*$-modules, where $\mathbf{U}_*$ is the †complex cobordism ring of compact weakly almost complex manifolds.

To study $\mathbf{O}_*^G$ and $\mathbf{U}_*^G$, (co)bordism theory is introduced (P. E. Conner and E. E. Floyd [16]), which is one of the †generalized (co)-homology theories. Miscellaneous results are known, in particular, for G a cyclic group of prime period. By means of the equivariant †Thom spectrum, equivariant cobordism theory can be developed (T. tom Dieck [17]); this is a multiplicative generalized cohomology theory with Thom classes (→ 114 Differential Topology; also → 201 Homology Theory, 56 Characteristic Classes).

F. Equivariant Homotopy

Let G be a compact Lie group. On the category of closed G-manifolds, we say that two objects M, N are χ-**equivalent** if $\chi(M^H)=\chi(N^H)$ for all closed subgroups H of G, where $\chi(\ )$ is the †Euler-Poincaré characteristic. On the set of equivalence classes $\mathbf{A}(G)$, a ring structure is imposed by disjoint union and the Cartesian product. We call $\mathbf{A}(G)$ the **Burnside ring** of G. If G is finite, $\mathbf{A}(G)$ is naturally isomorphic to the classical Burnside ring of G [18].

Denote by $S(V)$ the unit sphere of an orthogonal G-representation space V. Let V, W be orthogonal G-representation spaces. The equivariant stable homotopy group $[[S(V), S(W)]]$, which is defined as the direct limit of the equivariant homotopy sets $[S(V+U), S(W+U)]_G$ taken over orthogonal G-representation spaces U and suspensions, is denoted by ω_α for $\alpha=V-W \in RO(G)$. The †smash product of representatives induces a bilinear pairing $\omega_\alpha \times \omega_\beta \to \omega_{\alpha+\beta}$. Then ω_0 is a ring, and ω_α is an ω_0-module. The ring ω_0 is isomorphic to the Burnside ring of G, and ω_α is a †projective ω_0-module of rank one. The ω_0-module ω_α is free if and only if $S(V)$ and $S(W)$ are stably G-homotopy equivalent [18].

Let E be an orthogonal G-vector bundle over a compact G-space X. Denote by $S(E)$ the sphere bundle associated with E. Let E, F be orthogonal G-vector bundles over X. Then E and F have the **same spherical G-fiber homotopy type** if there exist fiber-preserving G-mappings $f: S(E) \to S(F)$, $f': S(F) \to S(E)$ and fiber-preserving G-homotopies $h_t: S(E) \to S(E)$, $h'_t: S(F) \to S(F)$ such that $h_0=f' \circ f$, $h_1=$identity, $h'_0=f \circ f'$, $h'_1=$identity. Let $KO_G(X)$ be the †equivariant K-group of real G-vector bundles over X. Let $T_G(X)$ be the additive subgroup of $KO_G(X)$ generated by elements of the form $[E]-[F]$, where E and F are orthogonal G-vector bundles having the same spherical G-fiber homotopy type. The factor group $J_G(X)=KO_G(X)/T_G(X)$ and the natural projection

$J_G: KO_G(X) \to J_G(X)$ are called an **equivariant J-group** and an **equivariant J-homomorphism**, respectively (→ 237 K-Theory).

In particular, $J_G(\{x_0\})$ is a factor group of the real representation ring $RO(G)$. †Adams operations on representation rings are the main tools for studying the group $J_G(\{x_0\})$ [18].

G. Infinitesimal Transformations

Let $f: G \times M \to M$ be a differentiable action of a Lie group G on a differentiable manifold M. Let X be a †left invariant vector field on G. Then we can define a differentiable vector field $f^+(X)$ on M as

$$f^+(X)_q h = \lim_{t \to 0} (h(f(\exp(-tX), q)) - h(q))/t$$

for each $q \in M$ and any differentiable function h defined on a neighborhood of q. It is easy to see that $f^+(X)_q = 0$ if and only if q is a fixed point of the one-parameter subgroup $\{\exp(tX)\}$. A vector field $f^+(X)$ is called an **infinitesimal transformation** of the differentiable transformation group G.

The set $\mathfrak{g}$ of all infinitesimal transformations of G forms a finite-dimensional †Lie algebra (the laws of addition and †bracket product are defined from those for the vector fields on M). If G acts effectively on M, $\mathfrak{g}$ is isomorphic to the Lie algebra of the Lie group G (→ 249 Lie Groups). In fact, the correspondence $X \to f^+(X)$ defines a Lie algebra homomorphism f^+ from the Lie algebra of all left invariant vector fields on G into the Lie algebra of all differentiable vector fields on M [19].

The following fact [20] is useful for the study of noncompact real analytic transformation groups. Let $\mathfrak{g}$ be a real †semisimple Lie algebra and $\rho: \mathfrak{g} \to L(M)$ be a Lie algebra homomorphism of $\mathfrak{g}$ into a Lie algebra of real analytic vector fields on a †real analytic manifold M. Let p be a point at which the vector fields in the image $\rho(\mathfrak{g})$ have common zero. Then there exists an analytic system of coordinates $(U; u_1, \ldots, u_m)$ with origin at p in which all the vector fields in $\rho(\mathfrak{g})$ are linear. Namely, there exists $a_{ij} \in \mathfrak{g}^* = \mathrm{Hom}_{\mathbf{R}}(\mathfrak{g}, \mathbf{R})$ such that

$$\rho(X)_q = -\sum_{i,j} a_{ij}(X) u_j(q) \frac{\partial}{\partial u_i}; \qquad X \in \mathfrak{g}, \; q \in U.$$

The correspondence $X \to (a_{ij}(X))$ defines a Lie algebra homomorphism of $\mathfrak{g}$ into $\mathfrak{sl}(m, \mathbf{R})$.

For example, we can show that a real analytic $SL(n, \mathbf{R})$ action on the m-sphere is characterized by a certain real analytic vector field on $(m-n+1)$-sphere $(5 \leqslant n \leqslant m \leqslant 2n-2)$ [21]. In particular, there are infinitely many (at least the cardinality of the real numbers) inequivalent real analytic $SL(n, \mathbf{R})$ actions on the m-sphere $(3 \leqslant n \leqslant m)$.

Conversely, let $\mathfrak{g}$ be a finite-dimensional Lie algebra of vector fields on M. Although there is not always a differentiable transformation group G that admits $\mathfrak{g}$ as its Lie algebra of infinitesimal transformations, the following local result holds. Let $\tilde{G}$ be the †simply connected Lie group corresponding to the Lie algebra $\mathfrak{g}$. Then for each point x of M, there exist a neighborhood $\tilde{U}$ of the identity element e of $\tilde{G}$, neighborhoods V, W $(V \subset W)$ of x, and a differentiable mapping f of $\tilde{U} \times V$ into W with the following properties. Putting $f(g, y) = g(y)$ $(g \in \tilde{U}, y \in V)$, we have: (i) For all $y \in V$, $e(y) = y$. (ii) If $g, h \in \tilde{U}$, $y \in V$, then $(gh)(y) = g(h(y))$, provided that $gh \in \tilde{U}$, $h(y) \in V$. (iii) Let X be an arbitrary element of $\mathfrak{g}$. Put $g_t = \exp(-tX)$, the corresponding one-parameter subgroup of $\tilde{G}$. If $\varepsilon > 0$ is taken small enough, then we have $g_t \in \tilde{U}$ for $|t| < \varepsilon$ so that $g_t(y) (|t| < \varepsilon, y \in V)$ is well defined. Therefore g_t determines a vector field $\tilde{X}$ on V by the formula

$$\tilde{X}_y h = \lim_{t \to 0} (h(g_t(y)) - h(y))/t.$$

The vector field $\tilde{X}$ coincides with the restriction of X to V. This local proposition is often expressed by the statement that $\mathfrak{g}$ generates a **local Lie group of local transformations**, which is called **Lie's fundamental theorem** on local Lie groups of local transformations.

H. Criteria

It is important to know whether a given transformation group is a topological or a Lie transformation group. The following theorems are useful for this purpose [22, 23]:

(10) Let G be a transformation group of a †locally compact Hausdorff space M. If we introduce the †compact-open topology in G, then G is a topological transformation group of M when M is locally connected or M is a †uniform topological space and G acts †equicontinuously on M.

(11) Suppose that M is a †C^1-manifold and G is a topological transformation group of M acting effectively on M. If G is locally compact and the mapping $x \to g(x)$ of M is of class C^1 for each element g of G, then G is a Lie transformation group of M.

(12) Assume that G is a transformation group of a differentiable manifold M and G acts effectively on M. Let $\mathfrak{g}$ be the set of all vector fields on M defined by one-parameter groups of transformations of M contained in G as subgroups. If $\mathfrak{g}$ is a finite-dimensional Lie algebra, then G has a Lie group structure with respect to which G is a Lie transformation group of M, and then $\mathfrak{g}$ coincides with the Lie

algebra formed by the infinitesimal transformations of G.

By applying theorems (10), (11), and (12) we can show that the following groups are Lie transformation groups: the group of all †isometries of a †Riemannian manifold; the group of all affine transformations of a differentiable manifold with a †linear connection (generally, the group of all transformations of a differentiable manifold that leave invariant a given †Cartan connection); the group of all analytic transformations of a compact complex manifold (this group is actually a complex Lie group); and the group of all analytic (holomorphic) transformations of a bounded domain in $\mathbf{C}^n$.

For related topics → 105 Differentiable Manifolds, 114 Differential Topology, 122 Discontinuous Groups, 153 Fixed-Point Theorems, 427 Topology of Lie Groups and Homogeneous Spaces, etc.

References

[1] A. Borel et al., Seminar on transformation groups, Ann. Math. Studies, Princeton Univ. Press, 1960.
[2] W. Y. Hsiang, Cohomology theory of topological transformation groups, Erg. math. 85, Springer, 1975.
[3] G. E. Bredon, Introduction to compact transformation groups, Academic Press, 1972.
[4] W. C. Hsiang and W. Y. Hsiang, Differentiable actions of compact connected classical groups I, II, III, Amer. J. Math. 89 (1967); Ann. Math., (2) 92 (1970); Ann. Math., (2) 99 (1974).
[5] R. W. Richardson, Deformations of Lie subgroups and the variation of isotropy subgroups, Acta Math., 129 (1972).
[6] W. C. Hsiang, A note on free differentiable actions of S^1 and S^3 on homotopy spheres, Ann. Math., (2) 83 (1966).
[7] F. Hirzebruch and K. H. Mayer, $O(n)$-Mannigfaltigkeiten, Exotische Sphären und Singularitäten, Lecture notes in math. 57, Springer, 1968.
[8] M. F. Atiyah and F. Hirzebruch, Spin-manifolds and group actions, Essays on Topology and Related Topics, Memoires dédiés à Georges de Rham, Springer, 1970.
[9] K. Kawakubo, Equivariant Riemann-Roch theorems, localization and formal group law, Osaka J. Math., 17 (1980).
[10] A. Hattori, Spin^c-structures and S^1-actions, Inventiones Math., 48 (1978).
[11] W. C. Hsiang and W. Y. Hsiang, The degree of symmetry of homotopy spheres, Ann. Math., (2) 89 (1969).
[12] T. Watabe, On the degree of symmetry of complex quadric and homotopy complex projective space, Sci. Rep. Niigata Univ., 11 (1974).
[13] T. Petrie, Smooth S^1 actions on homotopy complex projective spaces and related topics, Bull. Amer. Math. Soc., 78 (1972).
[14] T. Yoshida, On smooth semifree S^1 actions on cohomology complex projective spaces, Publ. Res. Inst. Math. Sci., 11 (1976).
[15] P. E. Conner and E. E. Floyd, Maps of odd period, Ann. Math., (2) 84 (1966).
[16] P. E. Conner and E. E. Floyd, Differentiable periodic maps, Erg. Math., 33 (1964).
[17] T. tom Dieck, Bordism of G-manifolds and integrality theorems, Topology, 9 (1970).
[18] T. tom Dieck, Transformation groups and representation theory, Lecture notes in math. 766, Springer, 1979.
[19] R. S. Palais, A global formulation of the Lie theory of transformation groups, Mem. Amer. Math. Soc., 22 (1957).
[20] V. Guillemin and S. Sternberg, Remarks on a paper of Hermann, Trans. Amer. Math. Soc., 130 (1968).
[21] F. Uchida, Real analytic $SL(n, R)$ actions on spheres, Tôhoku Math. J., 33 (1981).
[22] D. Montgomery and L. Zippin, Topological transformation groups, Interscience, 1955.
[23] S. Kobayashi, Transformation groups in differential geometry, Erg. math. 70, Springer, 1972.

432 (VI.8) Trigonometry

A. Plane Trigonometry

Fix an orthogonal frame $O\text{-}XY$ in a plane, and take a point P on the plane such that the angle POX is α. Denote by (x, y) the coordinates of P, and put $OP=r$ (Fig. 1). We call the six ratios $\sin\alpha=y/r$, $\cos\alpha=x/r$, $\tan\alpha=y/x$, $\cot\alpha=x/y$, $\sec\alpha=r/x$, $\operatorname{cosec}\alpha=r/y$ the **sine**, **cosine**, **tangent**, **cotangent**, **secant**, and **cosecant** of α, respectively. These functions of the angle α are called **trigonometric functions** or **circular functions** (→ 131 Elementary Functions). They are periodic functions with the fundamental period π for the tangent and cotangent, and 2π for the others. The relation $\sin^2\alpha+\cos^2\alpha=1$ and the **addition formulas** $\sin(\alpha\pm\beta)=\sin\alpha\cos\beta\pm\cos\alpha\sin\beta$, $\cos(\alpha\pm\beta)=\cos\alpha\cos\beta\mp\sin\alpha\sin\beta$ follow from the definitions (→ Appendix A, Table 2). Given a plane triangle ABC (Fig. 2), we have the following three properties: (i) $a=b\cos C+c\cos B$ (**the first law of cosines**); (ii) $a^2=b^2+c^2-2bc\cos A$ (**the second law of cosines**); (iii) $a/\sin A=b/\sin B=c/\sin C=2R$, where R is the radius of the circle circumscribed about $\triangle ABC$ (**laws of sines**) (→ Appendix A, Table 2). Thus we obtain relations among the six quantities a, b, c, $\angle A$, $\angle B$, and $\angle C$ associated with the triangle ABC. The study of plane figures by means of trigonometric functions is called **plane trigonometry**. For example, if a suitable combination of three of these six quantities (including a side) associated with a triangle is given, then the other three quantities are uniquely determined. The determination of unknown quantities associated with a triangle by means of these laws is called **solving a triangle**.

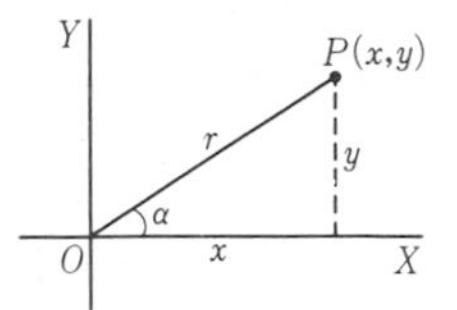

Fig. 1

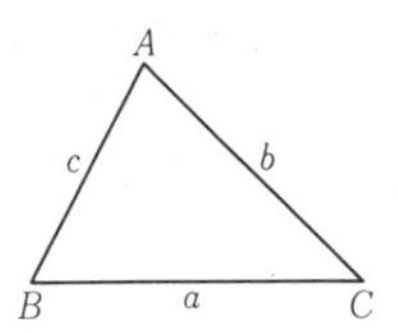

Fig. 2

B. Spherical Trigonometry

The part ABC of a spherical surface bounded by three arcs of great circles is called a **spherical triangle**. Points A, B, C are called the **vertices**; the three arcs a, b, c are called the **sides**; and the angles formed by lines tangent to the sides and intersecting at the vertices are called the **angles** of the spherical triangle (Fig. 3). If we denote the angles by A, B, C, we have the relation $A+B+C-\pi=E>0$, and E is called the **spherical excess**. Spherical triangles have properties similar to those of plane triangles: $\sin a/\sin A=\sin b/\sin B=\sin c/\sin C$ (**laws of sines**), and $\cos a=\cos b\cos c+\sin b\sin c\cos A$ (**law of cosines**). The study of spherical figures by means of trigonometric functions, called **spherical trigonometry**, is widely used in astronomy, geodesy, and navigation (→ Appendix A, Table 2).

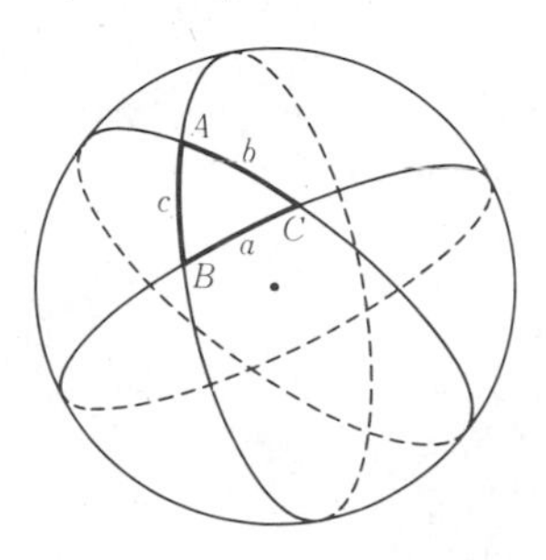

Fig. 3

C. History

Trigonometry originated from practical problems of determining a triangle from three of its elements. The development of spherical trigonometry, which was spurred on by its applications to astronomy, preceded the development of plane trigonometry. In Egypt, Babylon, and China, people had some knowledge of trigonometry, and the founder of trigonometry is believed to have been Hipparchus of Greece (fl. 150 B.C.). In the *Almagest* of Ptolemy (c. 150 A.D.) we find a table for $2\sin\alpha$ for $\alpha=0, 30', 1°, 1°30', \ldots$ that is exact to five decimal places, and the addition formulas. The Greeks calculated $2\sin\alpha$, which is the length of the chord corresponding to the double arc. Indian mathematicians, on the other hand, calculated half of the above quantities, that is, $\sin\alpha$ and $1-\cos\alpha$ for the arc α. In the book by Aryabhatta (c. 500 A.D.) we find laws of cosines. The Arabs, influenced by Indian mathematicians, expressed geometric computations algebraically, a technique also known to the Greeks. Abûl Wafâ (in the latter half of the 10th century A.D.) gave the correct sines of angles for every 30′ to 9 decimal places and studied with Al Battani the projection triangle of the sundial, thereby obtaining the concepts of sine, cosine, secant, and cosecant. Later, a table of sines and cosines for every minute was established by the Arabs. Regiomontanus (d. 1476), a German, elaborated on this table. The form he gave to trigonometry has been maintained nearly intact to the present day. Various theorems in trigonometry were established by G. J. Rhaeticus, J. Napier, J. Kepler, and L. Euler (1748). Euler treated trigonometry as a branch of analysis, generalized it to functions of complex variables, and introduced the abbreviated notations that are still in use (→ 131 Elementary Functions).

References

[1] H. Flanders and J. Price, Trigonometry, Academic Press, 1975.
[2] A. Washington and C. Edmond, Plane trigonometry, Addison-Wesley, 1977.

433 (XX.12) Turbulence and Chaos

Turbulent flow is the irregular motion of fluids, whereas relatively simple types of flows that are either stationary, slowly varying, or periodic in time are called **laminar flow**. When a laminar flow is stable against external disturbances, it remains laminar, but if the flow is unstable, it usually changes into either another type of laminar flow or a turbulent flow.

A. Stability and Bifurcation of Flows

The velocity field $\mathbf{u}(\mathbf{x}, t)$, $\mathbf{x}$ being the space coordinates and t the time, of a flow of an incompressible viscous fluid in a bounded domain G is determined by the †Navier-Stokes equation of motion,

$$\frac{\partial \mathbf{u}}{\partial t}+(\mathbf{u}\cdot\operatorname{grad})\mathbf{u}-\nu\Delta\mathbf{u}+\frac{1}{\rho}\operatorname{grad}p=0, \tag{1}$$

and the equation of continuity,

$$\operatorname{div}\mathbf{u}=0, \tag{2}$$

with the prescribed initial and boundary conditions, where Δ denotes the Laplacian, p the pressure, ρ the density, and ν the kinematic viscosity of the fluid. Suitable extensions must be made in the foregoing system of equations if other field variables, such as the temperature in thermal-convection problems, are to be considered.

The stability of a fluid flow is studied by examining the behavior of the solution of equations (1) and (2) against external disturbances, and, in particular, stability against infinitesimal disturbances constitutes the linear stability problem. The stability characteristics of the solution of equations (1) and (2) depend largely upon the value of the †Reynolds number $R=UL/\nu$, U and L being the representative velocity and length of the flow, respectively. Let a stationary solution of equations (1) and (2) be $\mathbf{u}_0(\mathbf{x}, R)$. If the perturbed flow is given by $\mathbf{u}_0(\mathbf{x}, R)+\mathbf{v}(\mathbf{x}, R)\exp(\sigma t)$, $\mathbf{v}$ being the perturbation velocity, and equation (1) is linearized with respect to $\mathbf{v}$, we obtain a †linear eigenvalue problem for σ. The flow is called linearly stable if $\max(\operatorname{Re}\sigma)$ is negative, and linearly unstable if it is positive. For small values of R, a flow is generally stable, but it becomes unstable if R exceeds a critical value R_c, which is called the critical Reynolds number [1].

The instability of a stationary solution gives rise to the †bifurcation to another solution at a †bifurcation point R_c of the parameter R. If $\operatorname{Im}\sigma=0$ for an eigenvalue σ at $R=R_c$, a stationary solution bifurcates from the solution u_0 at R_c, and if $\operatorname{Im}\sigma\neq 0$, a time-periodic solution bifurcates at R_c. The latter bifurcation is called the Hopf bifurcation. A typical example of stationary bifurcation is the generation of an axially periodic row of Taylor vortices in Couette flow between two rotating coaxial cylinders, which was studied by G. I. Taylor (1923), with excellent agreement between theory and experiment [2]. Hopf bifurcation is exemplified by the generation of Tollmien-Schlichting waves in the laminar †boundary layer along a flat plate, which was predicted theoretically by W. Tollmien (1929) and H. Schlichting (1933) and later confirmed experimentally by G. B. Schubauer and H. K. Skramstad (1947) [3].

In either type of bifurcation ($\operatorname{Im}\sigma=0$ or $\neq 0$) the bifurcation is called supercritical if the bifurcating solution exists only for $R>R_c$, subcritical if it exists only for $R<R_c$, and transcritical if it happens to exist on both sides of R_c. The amplitude of the departure of the bifurcating solution from the unperturbed solution u_0 tends to zero as $R\to R_c$. The behavior of the bifurcating solution around the bifurcation point R_c is dealt with systematically by means of bifurcation analysis. In supercritical bifurcation, the bifurcating solution is stable and represents an equilibrium state to which the perturbed flow approaches just as in the cases of Taylor vortices and Tollmien-Schlichting waves. On the other hand, for subcritical bifurcation the bifurcating solution is unstable and gives a critical amplitude of the disturbance above which the linearly stable basic flow ($R<R_c$) becomes unstable. In this case, the instability of the basic flow gives rise to a sudden change of the flow pattern resulting in either a stationary (or time-periodic) or even turbulent flow. The transition to turbulent flow that takes place in Hagen-Poiseuille flow through a circular tube and is linearly stable at all values of R ($R_c=\infty$) may be attributed to this type of bifurcation.

The concept of bifurcation can be extended to the case where the flow u_0 is nonstationary, but the bifurcation analysis then becomes much more difficult.

B. Onset of Turbulence

The fluctuating flow resulting from an instability does not itself necessarily constitute a turbulent flow. In order that a flow be turbulent, the fluctuations must take on some irregularity. The turbulent flow is usually defined in terms of the long-time behavior of the flow velocity $\mathbf{u}(\mathbf{x}, t)$ at a fixed point $\mathbf{x}$ in space. The flow is expected to be turbulent if the fluctuating velocity

$$\delta\mathbf{u}(\mathbf{x}, t)=\mathbf{u}(\mathbf{x}, t)-\lim_{T\to\infty}\frac{1}{T}\int_0^T \mathbf{u}(\mathbf{x}, t)\,dt \tag{3}$$

satisfies the condition

$$\lim_{\tau\to\infty}\lim_{T\to\infty}\frac{1}{T}\int_0^T \delta u_i(\mathbf{x}, t)\delta u_i(\mathbf{x}, t+\tau)\,dt=0, \tag{4}$$

where the subscripts label the components. Condition (4) implies that the †dynamical system of a fluid has the mixing property. This condition also states that the velocity fluctuation δu_i has a continuous frequency spectrum. In practical situations the frequency spectrum of a turbulent flow may contain both the line and continuous spectra, in which case the flow is said to be partially turbulent.

L. D. Landau (1959) and E. Hopf (1948) proposed a picture of turbulent flow as one composed of a †quasiperiodic motion, $\mathbf{u}(t)=\mathbf{f}(\omega_1 t, \omega_2 t, \ldots, \omega_n t)$, with a large number of rationally independent frequencies $\omega_1, \ldots, \omega_n$ produced by successive supercritical bifurcations of Hopf type. This picture of turbulence is not compatible with the foregoing definition of turbulence, since it does not satisfy the mixing property (4). The fact that the generation of real turbulence is not necessarily preceded by successive supercritical bifurcations casts another limitation on the validity of this picture.

The concept of turbulence is more clearly exhibited with respect to a dynamical system of finite dimension. Although we are without a general proof, it is expected that the Navier-Stokes equation with nonzero viscosity ν can be approximated within any degree of accuracy by a system of finite-dimensional †first-order ordinary differential equations

$$\frac{dX}{dt}=F(X). \tag{5}$$

Thus the onset and some general properties of turbulence are understood in the context of the theory of †dynamical systems. Turbulence is related to those solutions of equation (5) that tend to a †set in the †phase space that is neither a †fixed point, a †closed orbit, nor a †torus. A set of such complicated structure is called a nonperiodic †attractor or a strange attractor. Historically, the strange attractor originates from the strange Axiom A attractor that was found in a certain class of dynamical systems called the Axiom A systems. However, this term has come to be used in a broader sense, and it now represents a variety of nonperiodic motions exhibited by a system that is not necessarily of Axiom A type. The above-mentioned Landau-Hopf picture of turbulence was criticized by D. Ruelle and F. Takens (1971), who proved for the dynamical system (5) that an arbitrary small perturbation on a quasiperiodic †flow on a k-dimensional torus ($k \geqslant 4$) generically (in the sense of residual sets) produces a flow with a strange Axiom A attractor [4].

There exist a number of examples of first-order ordinary differential equations of relatively low dimension whose solutions exhibit nonperiodic behavior. An important model system related to turbulence is the Lorenz model (1963) of thermal convection in a horizontal fluid layer. This model is obtained by taking only three components out of an infinite number of spatial †Fourier components of the velocity and temperature fields. The model is written as

$$\frac{dX}{dt}=-\sigma X+\sigma Y,$$

$$\frac{dY}{dt}=-XZ+rX-Y,$$

$$\frac{dZ}{dt}=XY-bZ, \tag{6}$$

where σ ($>b+1$) and b are positive constants and r is a parameter proportional to the Rayleigh number. Obviously, equations (6) have a fixed point $X=Y=Z=0$ representing the state of thermal convection without fluid flow. For $r<1$, this fixed point is stable, but it becomes unstable for $r>1$, and a pair of new fixed points $X=Y=\pm\sqrt{b(r-1)}$, $Z=r-1$ emerges supercritically. This corresponds to the onset of stationary convection at $r=1$. At a still higher value of $r=\sigma(\sigma+b+3)/(\sigma-b-1)$, a subcritical Hopf bifurcation occurs with respect to this pair of fixed points, and for a certain range of r above this threshold the solutions with almost any initial conditions exhibit nonperiodic behavior. This corresponds to the generation of turbulence. The property

$$\frac{\partial \dot{X}}{\partial X}+\frac{\partial \dot{Y}}{\partial Y}+\frac{\partial \dot{Z}}{\partial Z}=-(\sigma+b+1)<0, \tag{7}$$

where the dots denote time derivatives, shows that each volume element of the phase space shrinks asymptotically to zero as the time increases indefinitely. This property is characteristic of dynamical systems with energy dissipation, in sharp contrast to the †measure-preserving character of †Hamiltonian systems [5].

For a certain class of ordinary differential equations, the bifurcation to nonperiodic motion corresponds neither to the bifurcation of tori, just as in the Ruelle-Takens theory, nor to subcritical bifurcation, as in the Lorenz model. Such a bifurcation takes place when nonperiodic motion emerges as the consequence of an infinite sequence of supercritical bifurcations at each of which a periodic orbit of period T bifurcates into one of period $2T$. If we denote the nth bifurcation point by r_n, the distance $r_{n+1}-r_n$ between two successive bifurcation points decreases exponentially with increasing n, and eventually the bifurcation points accumulate at a point r_c, beyond which nonperiodic motion is expected to emerge. It is not yet clear if any of the above three types of bifurcation leading to nonperiodic behavior is actually responsible for the generation of real turbulence.

Some important properties of a dynamical system with a nonperiodic attractor, which may be either a flow or a †diffeomorphism, can be stated as follows:
(i) The distance between two points in the phase space that are initially close to each other grows exponentially in time, so that the solutions exhibit a sensitive dependence on the initial conditions.
(ii) The nonperiodic attractor has †Lebesgue measure zero, and such a system is expected to have many other †ergodic †invariant measures.

The irregular behavior of a deterministic dynamical system is also called **chaos**, but this concept is more abstract and general than that of turbulence, and covers phenomena exhibited by systems such as nonlinear electric circuits, chemical reactions, and ecological systems.

C. Statistical Theory of Turbulence

The statistical theory of turbulence deals with the statistical behavior of fully developed turbulence. The turbulent field is sometimes idealized for mathematical simplicity to be homogeneous or isotropic. In **homogeneous turbulence** the statistical laws are invariant under all parallel displacements of the coordinates, whereas in **isotropic turbulence** invariance under rotations and reflections of the coordinates is required in addition.

The instantaneous state of the fluid motion is completely determined by specifying the fluid velocity $\mathbf{u}$ at all space points $\mathbf{x}$ and can be expressed as a phase point in the infinite-dimensional †phase space spanned by these velocities. The phase point moves with time along a path uniquely determined by the solution of the Navier-Stokes equation. In the turbulent state the path is unstable to the initial disturbance and describes an irregular line in the phase space. In this situation the deterministic description is no longer useful and should be replaced by a statistical treatment. Abstractly speaking, turbulence is just a view of fluid motion as the random motion of the phase point $\mathbf{u}(\mathbf{x})$ (→ 407 Stochastic Processes). The equation for the †characteristic functional of the random velocity $\mathbf{u}(\mathbf{x})$ was first given by E. Hopf (1952). An exact solution obtained by Hopf represents a †normal distribution associated with a white energy spectrum, but so far no general solution has been obtained [6].

Besides the formulation in terms of the †probability distribution or the characteristic functional, there is another way of describing turbulence by †moments of lower orders. This is the conventional statistical theory originated by G. I. Taylor (1935) and T. von Kármán (1938), which made remarkable progress after World War II. The principal moments in this theory are the **correlation tensor**, whose (i,j)-component is the mean of the product of two velocity components u_i at a point $\mathbf{x}$ and u_j at another point $\mathbf{x}+\mathbf{r}$,

$$B_{ij}(\mathbf{r})=\langle u_i(\mathbf{x})u_j(\mathbf{x}+\mathbf{r})\rangle, \tag{8}$$

and its †Fourier transform, or the **energy spectrum tensor**,

$$\Phi_{ij}(\mathbf{k})=\frac{1}{(2\pi)^3}\int B_{ij}(\mathbf{r})\exp(-\sqrt{-1}\,\mathbf{k}\cdot\mathbf{r})\,d\mathbf{r}. \tag{9}$$

In isotropic turbulence Φ_{ij} is expressed as

$$\Phi_{ij}(\mathbf{k})=\frac{1}{4\pi k^2}E(k)\left(\delta_{ij}-\frac{k_ik_j}{k^2}\right), \qquad k=|\mathbf{k}|, \tag{10}$$

where $E(k)$ is the **energy spectrum function**, representing the amount of energy included in a spherical shell of radius k in the wave number space. The energy of turbulence $\mathscr{E}$ per unit mass is expressed as

$$\mathscr{E}=\frac{1}{2}\langle|\mathbf{u}|^2\rangle=\frac{1}{2}B_{ii}(0)=\frac{1}{2}\int\Phi_{ii}(\mathbf{k})\,d\mathbf{k}$$

$$=\int_0^\infty E(k)\,dk. \tag{11}$$

The state of turbulence is characterized by the Reynolds number $R=E_0^{1/2}/(\nu k_0^{1/2})$, where E_0 and k_0 are representative values of $E(k)$ and k, respectively. For weak turbulence of small R, $E(k)$ is governed by a linear equation with the general solution

$$E(k,t)=E(k,0)\exp(-2\nu k^2t), \tag{12}$$

$E(k,0)$ being an arbitrary function. Thus $E(k)$ decays in time due to the viscous dissipation. For strong turbulence of large R, it is difficult to obtain the precise form of $E(k)$, and this is usually done by way of some assumption that allows us to approximate the nonlinear effects [7].

Some of the similarity laws governing the energy spectrum and other statistical functions can be determined rigorously but not necessarily uniquely. For 3-dimensional incompressible turbulence, the energy spectrum satisfies an inviscid similarity law

$$E(k)/E_0=F_e(k/k_0) \tag{13}$$

in the energy-containing region $k=O(k_0)$ characterized by a wave number k_0, and a viscous similarity law

$$E(k)/E_0=R^{-5/4}F_d(k/(R^{3/4}k_0)), \tag{14}$$

in the energy dissipation region $k=O(R^{3/4}k_0)$, where F_e and F_d denote dimensionless functions generally dependent on the initial condition and the time [6].

If an assumption is made to the effect that the statistical state in the energy-dissipation region depends only upon the energy-dissipation rate $\varepsilon=-d\mathscr{E}/dt$ besides the viscosity ν (or R), then (14) becomes Kolmogorov's equilibrium similarity law (1941):

$$E(k)=\varepsilon^{1/4}\nu^{5/4}F(k/(\varepsilon^{1/4}\nu^{-3/4})), \tag{15}$$

where F is a dimensionless function. For extremely large R (or small ν) there exists an inertial subregion between the energy-containing and energy-dissipation regions such that the viscous effect vanishes and (15) takes the form

$$E(k)=K\varepsilon^{2/3}k^{-5/3}, \tag{16}$$

where K is an absolute constant. **Kolmogorov's spectrum** (16) has been observed experimentally several times, and now its consistency with experimental results at large Reynolds numbers is well established [8].

Kolmogorov (1962) and others modified (16) by taking account of the fluctuation of ε due to the **intermittent structure** of the energy-dissipation region as

$$E(k)=K'\varepsilon^{2/3}k^{-5/3}(Lk)^{-\mu/9}, \tag{17}$$

where ε is now the average of the fluctuating ε, μ is the covariance of the log-normal distribution of ε, and L is the length scale of the spatial domain in which the average of ε is taken [8]. A similar modification, with the exponent $-\mu/3$ in place of $-\mu/9$, is obtained using a fractal model of the energy-cascade process. These corrections to $E(k)$, based upon the experimentally estimated μ of 0.3–0.5, are too small to be detected experimentally, but the deviation is expected to appear more clearly in the higher-order moments [8–10]. It should be noted that Kolmogorov's spectrum (16) itself does not contradict the notion of intermittent turbulence and gives one of the possible asymptotic forms in the limit $R\to\infty$.

The 1-dimensional Burgers model of turbulence satisfies the same similarity laws as (13) and (14), but it has an inviscid spectrum $E(k)\propto k^{-2}$ instead of (16). Two-dimensional incompressible turbulence has no energy-dissipation region, and hence Kolmogorov's theory is not valid for this turbulence. It has an inviscid spectrum $E(k)\propto k^{-3}$, first derived by R. H. Kraichnan (1967), C. E. Leith (1968), and G. K. Batchelor (1969). These inviscid spectra for 1- and 2-dimensional turbulence have been confirmed by numerical simulation [11].

References

[1] C. C. Lin, The theory of hydrodynamical stability, Cambridge Univ. Press, 1955.
[2] S. Chandrasekhar, Hydrodynamic and hydromagnetic stability, Clarendon Press, 1961.
[3] H. L. Dryden, Recent advances in the mechanics of boundary layer flow, Adv. Appl. Mech., 1 (1948), 1–40.
[4] D. Ruelle and F. Takens, On the nature of turbulence, Comm. Math. Phys., 20 (1971), 167–192.
[5] E. N. Lorenz, Deterministic nonperiodic flow, J. Atmospheric Sci., 20 (1963), 130–141.
[6] T. Tatsumi, Theory of homogeneous turbulence, Adv. Appl. Mech., 20 (1980), 39–133.
[7] G. K. Batchelor, The theory of homogeneous turbulence, Cambridge Univ. Press, 1953.
[8] A. S. Monin and A. M. Yaglom, Statistical fluid mechanics II, MIT Press, 1975.
[9] B. Mandelbrot, Fractals and turbulence: Attractors and dispersion, Lecture notes in math. 615, Springer, 1977, 83–93.
[10] B. Mandelbrot, Intermittent turbulence and fractal dimension: Kurtosis and the spectral exponent $5/3+B$, Lecture notes in math. 565, Springer, 1977, 121–145.
[11] T. Tatsumi, Analytical theories and numerical experiments on two-dimensional turbulence, Theor. Appl. Mech., 29 (1979), 375–393.

434 (XX.22) Unified Field Theory

A. History

Unified field theory is a branch of theoretical physics that arose from the success of †general relativity theory. Its purpose is to discuss in a unified way the fields of gravitation, electromagnetism, and nuclear force from the standpoint of the geometric structure of space and time. Studies have continued since 1918, and many theories of mathematical interest have been published without attaining, however, any conclusive physical theory.

A characteristic feature of relativity theory is that it is based on a completely new concept of space and time. That is, in general relativity theory it is considered that when a gravitational field is generated by matter, the structure of space and time changes, and the flat †Minkowski world becomes a 4-dimensional †Riemannian manifold (with signature $(1,3)$) having nonvanishing curvature. The †fundamental tensor g_{ij} of the manifold is interpreted as the gravitational potential, and the basic gravitational equation can be described as a geometric law of the manifold. It is characteristic of general relativity theory that gravitational phenomena are reduced to space-time structure (→ 359 Relativity). The introduction of the Minkowski world in †special relativity theory was a revolutionary advance over the 3-dimensional space of Newtonian mechanics. But the inner structure of the Minkowski world does not reflect gravitational phenomena. The latter shortcoming is overcome by introducing the concept of space-time represented by a Riemannian manifold into general relativity theory.

When a coexisting system of gravitational and electromagnetic fields is discussed in general relativity theory, simultaneous equations (Einstein-Maxwell equations) must be solved for the gravitational potential g_{ij} and the electromagnetic field tensor F_{ij}. Thus the gravitational potential g_{ij} is affected by the existence of an electromagnetic field. As the validity of general relativity began to be accepted, it came to be expected that all physical actions might be attributed to the gravitational and electromagnetic fields. Thus various extensions of general relativity theory have been proposed in order to devise a geometry in which the electromagnetic as well as the gravitational field directly contributes to the space-time structure, and to establish a unified theory of both fields on the basis of the geometry thus obtained. These attempts are illustrated in Fig. 1.

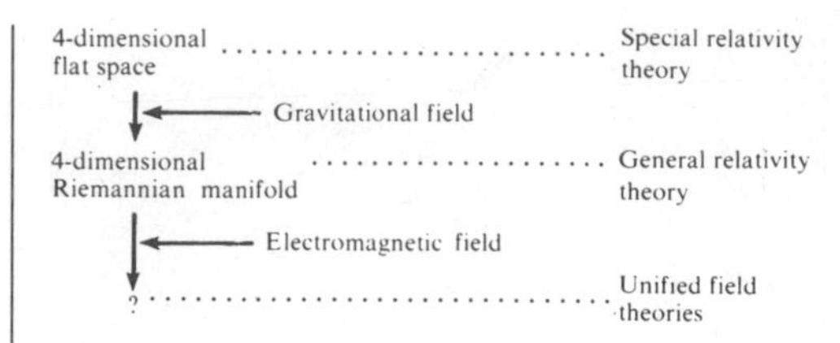

Fig. 1

B. Weyl's Theory

The first unified field theory was proposed by H. Weyl in 1918. In Riemannian geometry, which is the mathematical framework of general relativity theory, the †covariant derivative of the †fundamental tensor g_{ij} vanishes, i.e.,

$$\nabla_i g_{jk} \equiv \partial g_{jk}/\partial x^i - g_{ja}\Gamma^a_{ik} - g_{ak}\Gamma^a_{ij} = 0, \tag{1}$$

where Γ^i_{jk} is the †Christoffel symbol derived from g_{ij}. Conversely, if Γ^i_{jk} is considered as the coefficient of a general †affine connection and (1) is solved with respect to Γ^i_{jk} under the condition $\Gamma^i_{jk} = \Gamma^i_{kj}$, then the Christoffel symbol derived from g_{ij} coincides with Γ^i_{jk}. In this sense, (1) means that the space-time manifold has Riemannian structure. On the other hand, Weyl considered a space whose structure is given by an extension of (1),

$$\nabla_i g_{jk} = 2A_i g_{jk}, \tag{2}$$

and developed a unified field theory by regarding A_i as the electromagnetic potential. This theory has mathematical significance in that it motivated the discovery of Cartan's geometry of connection, but it has some unsatisfactory points concerning the derivation of the field equation and the equation of motion for a charged particle.

The scale transformation given by $\bar{g}_{ij} = \rho^2 g_{ij}$ is important in Weyl's theory. If in addition to this transformation, A_i is changed to

$$\bar{A}_i = A_i - \partial \log \rho/\partial x^i, \tag{3}$$

then (2) is left invariant and the space-time structure in Weyl's theory remains unchanged. We call (3) the **gauge transformation**, corresponding to the fact that the electromagnetic potential A_i is determined by the electromagnetic field tensor F_{ij} up to a gradient vector. In the †field theories known at present, the gauge transformation is generalized to various fields, and the law of charge conservation is derived from the invariance of field equations under generalized gauge transformation.

C. Further Developments

A unified field theory that appeared after Weyl's is **Kaluza's 5-dimensional theory** (Th.

Kaluza, 1921). This theory has been criticized as being artificial, but it is logically consistent, and therefore many of the later unified field theories are improved or generalized versions of it. The underlying space of Kaluza's theory is a 5-dimensional Riemannian manifold with the fundamental form

$$ds^2=(dx^4+A_a dx^a)^2+g_{ab}dx^a dx^b,$$

where A_i and g_{ij} are functions of x^i alone ($a, b, \ldots, i, j=0,1,2,3$). The field equation and the equation of motion of a particle are derived from the variational principle in general relativity theory. The field equation is equivalent to the Einstein-Maxwell equations. The trajectory of a charged particle is given by a geodesic in the manifold, and its equation is reducible to the Lorentz equation in general relativity.

After the introduction of Kaluza's theory, various unifield field theories were proposed, and we give here the underlying manifolds or geometries of some mathematically interesting theories: a manifold with †affine connection admitting absolute parallelism (A. Einstein, 1928); a manifold with †projective connection (O. Veblen, B. Hoffman, 1930 [4]; J. A. Schouten, D. van Dantzig, 1932); **wave geometry** (a theory based on the linearization of the fundamental form; Y. Mimura, 1934 [3]); a nonholonomic geometry (G. Vranceanu, 1936); a manifold with †conformal connection (Hoffman, 1948).

The investigations since 1945 have been motivated by the problem of the representation of matter in general relativity theory. Einstein first represented matter by an energy-momentum tensor T_{ij} of class C^0, which must be determined by information obtained from outside relativity. Afterward he felt that this point was unsatisfactory and tried to develop a theory on the basis of field variables alone, without introducing such a quantity as T_{ij}. This theory is the so-called **unitary field theory**, and a solution without singularities is required from a physical point of view. His first attempt was to remove singularities from an exterior solution in general relativity by changing the topological structure of the space-time manifold. This idea was then extended to a unified field theory by J. A. Wheeler, and an interpretation was given to mass and charge by applying the theory of †harmonic integrals (1957) [2].

Einstein's second attempt was to propose a **nonsymmetric unified field theory** (1945) [1, Appendix II; 6]. The fundamental quantities in this theory are a nonsymmetric tensor g_{ij} and a nonsymmetric affine connection Γ^i_{jk}. The underlying space of the theory can be considered a direct extension of the Riemannian manifold, since (1) is contained in the field equations (notice the order of indices in this equation). E. Schrödinger obtained field equations of almost the same form by taking only Γ^i_{jk} as a fundamental quantity (1947) [5].

References

[1] A. Einstein, The meaning of relativity, Princeton Univ. Press, fifth edition, 1956.
[2] J. A. Wheeler, Geometrodynamics, Academic Press, 1962.
[3] Y. Mimura and H. Takeno, Wave geometry, Sci. Rep. Res. Inst. Theoret. Phys. Hiroshima Univ., no. 2 (1962).
[4] O. Veblen, Projektive Relativitätstheorie, Springer, 1933.
[5] E. Schrödinger, Space-time structure, Cambridge Univ. Press, 1950.
[6] A. Einstein, A generalization of the relativistic theory of gravitation, Ann. Math., (2) 46 (1945), 578–584.
[7] V. Hlavatý, Geometry of Einstein's unified field theory, Noordhoff, 1957.
[8] M. A. Tonnelat, Einstein's unified field theory, Gordon & Breach, 1966. (Original in French, 1955.)

435 (II.23) Uniform Convergence

A. Uniform Convergence of a Sequence of Real-Valued Functions

A sequence of real-valued functions $\{f_n(x)\}$ defined on a set B is said to be **uniformly convergent** (or to **converge uniformly**) to a function $f(x)$ on the set B if it converges with respect to the †norm $\|\varphi\|=\sup\{|\varphi(x)| \mid x\in B\}$, i.e., $\lim_{n\to\infty}\|f_n-f\|=0$ (→ 87 Convergence). In other words, $\{f_n(x)\}$ converges uniformly to $f(x)$ on B if for every positive constant ε we can select a number N independent of the point x such that $|f_n(x)-f(x)|<\varepsilon$ holds for all $n>N$ and $x\in B$. By the †completeness of the real numbers, a sequence of functions $\{f_n(x)\}$ converges uniformly on B if and only if we can select for every positive constant ε a number N independent of the point x such that $|f_m(x)-f_n(x)|<\varepsilon$ holds for all $m, n>N$ and $x\in B$.
The **uniform convergence** of a series $\sum_n f_n(x)$ or of an infinite product $\prod_n f_n(x)$ is defined by the uniform convergence of the sequence of its partial sums or products. If the series of the absolute values $\sum_n |f_n(x)|$ converges uniformly, then the series $\sum_n f_n(x)$ also converges uniformly. In this case the series $\sum_n f_n(x)$ is said to

be **uniformly absolutely convergent**. A sequence of (nonnegative) constants M_n satisfying $|f_n(x)| \leqslant M_n$ is called a **dominant** (or **majorant**) of the sequence of functions $\{f_n(x)\}$. A series of functions $\sum_n f_n(x)$ with converging **majorant series** $\sum_n M_n$ is uniformly absolutely convergent (**Weierstrass's criterion for uniform convergence**).

Let $\{\lambda_n(x)\}$ be another sequence of functions on B. The series $\sum_n \lambda_n(x) f_n(x)$ is uniformly convergent if either of the following conditions holds: (i) The series $\sum_n f_n(x)$ converges uniformly and the partial sums of the series $\sum_n |\lambda_n(x) - \lambda_{n+1}(x)|$ are uniformly bounded, i.e., bounded by a constant independent of $x \in B$ and of the number of terms; or (ii) the series $\sum_n |\lambda_n(x) - \lambda_{n+1}(x)|$ converges uniformly, the sequence $\{\lambda_n(x)\}$ converges uniformly to 0, and the partial sums of $\sum_n |f_n(x)|$ are uniformly bounded.

B. Uniform Convergence and Pointwise Convergence

Let $\{f_n(x)\}$ be a sequence of real-valued functions on B, and let $f(x)$ be a real-valued function also defined on B. If the sequence of numbers $\{f_n(x_0)\}$ converges to $f(x_0)$ for every point $x_0 \in B$, we say that $\{f_n(x)\}$ is **pointwise convergent** (or **simply convergent**) to the function $f(x)$. Pointwise convergence is, of course, weaker than uniform convergence. If we represent the function $f(x)$ by the point $\prod_{x \in B} f(x) = [f]$ of the †Cartesian product $\mathbf{R}^B = \prod_{x \in B} \mathbf{R}$, then the pointwise convergence of $\{f_n(x)\}$ to $f(x)$ is equivalent to the convergence of the sequence of points $\{[f_n]\}$ to $[f]$ in the †product topology of $\mathbf{R}^B$.

When B is a †topological space and every $f_n(x)$ is continuous, the pointwise limit $f(x)$ of the sequence $\{f_n(x)\}$ is not necessarily continuous. However, if the sequence of continuous functions $\{f_n(x)\}$ converges uniformly to $f(x)$, then the limit function $f(x)$ is continuous. On the other hand, the continuity of the limit does not imply that the convergence is uniform. If the set B is †compact and the sequence of continuous functions $\{f_n(x)\}$ is monotone (i.e., $f_n(x) \leqslant f_{n+1}(x)$ for all n or $f_n(x) \geqslant f_{n+1}(x)$ for all n) and pointwise convergent to a continuous function $f(x)$, then the convergence is uniform (**Dini's theorem**).

C. Uniform Convergence on a Family of Sets

Let B be a topological space. We say that a sequence of functions $\{f_n(x)\}$ is **uniformly convergent in the wider sense** to the function $f(x)$, depending on circumstances, in either of the following two cases: (i) Every point $x_0 \in B$ has a neighborhood U on which the sequence $\{f_n(x)\}$ converges uniformly to $f(x)$; or (ii) $\{f_n(x)\}$ converges uniformly to $f(x)$ on every compact subset K in B. If B is †locally compact, the two definitions coincide. The term **uniform convergence on compact sets** is also used for (ii).

In general, given a family $\mathscr{P}$ of subsets in B, we may introduce in the space $\mathscr{F}$ of real-valued functions on B a family of †seminorms $\|f\|_K = \sup\{|f(x)| \mid x \in K\}$ for every set $K \in \mathscr{P}$. Let T be the topology of $\mathscr{F}$ defined by this family of seminorms (→ 424 Topological Linear Spaces). A sequence $\{f_n(x)\}$ is called **uniformly convergent on** $\mathscr{P}$ if it is convergent with respect to T. In particular, when $\mathscr{P}$ coincides with $\{B\}$, $\{\{x\} \mid x \in B\}$, or the family of all compact sets in B, then uniform convergence on $\mathscr{P}$ coincides with the usual uniform convergence, pointwise convergence, or uniform convergence on compact sets, respectively. If $\mathscr{P}$ is a countable set, the topology T is †metrizable. Most of these definitions and results may be extended to the case of functions whose values are in the complex number field, in a †normed space, or in any †uniform space.

D. Topology of the Space of Mappings

Let X, Y be two topological spaces. Denote by $C(X, Y)$ the space of all continuous mappings $f: X \to Y$. This space $C(X, Y)$, or a subspace $\mathscr{F}$ of $C(X, Y)$, is called a **mapping space** (or **function space** or **space of continuous mappings**) from X to Y. A natural mapping $\Phi: \mathscr{F} \times X \to Y$ is defined by $\Phi(f, x) = f(x) (f \in \mathscr{F}, x \in X)$. We define a topology in $\mathscr{F}$ as follows: for a compact set K in X and an open set U in Y, put $W(K, U) = \{f \in \mathscr{F} \mid f(K) \in U\}$, and introduce a topology in $\mathscr{F}$ such that the base for the topology consists of intersections of finite numbers of $W(K_i, U_i)$. This topology is called the **compact-open topology** (R. H. Fox, *Bull. Amer. Math. Soc.*, 51 (1945)). When X is a †locally compact Hausdorff space and Y is a †Hausdorff space, the compact-open topology is the †weakest topology on $\mathscr{F}$ for which the function Φ is continuous. If, in this case, $\mathscr{F}$ is compact with respect to the compact-open topology, then the compact-open topology coincides with the topology of pointwise convergence.

In particular, when Y is a †metric space (or, in general, a †uniform space with the uniformity $\mathfrak{U}$), the compact-open topology in $\mathscr{F}$ coincides with the topology of uniform convergence on compact sets. A family $\mathscr{F}$ is called

equicontinuous at a point $x \in X$ if for every positive number ε (in the case of uniform space, for every $V \in \mathfrak{U}$) there exists a neighborhood U of x such that $\rho(f(x), f(p)) < \varepsilon$ ($(f(x), f(p)) \in V$) for every point $p \in U$ and for every function $f \in \mathcal{F}$ (G. Ascoli, 1883–1884). If X is a †locally compact Hausdorff space, a necessary and sufficient condition for $\mathcal{F}$ to be relatively compact (i.e., for the closure of $\mathcal{F}$ to be compact) with respect to the compact-open topology (i.e., to the topology of uniform convergence on compact sets) is that $\mathcal{F}$ be equicontinuous at every point $x \in X$ and that the set $\{f(x) \mid f \in \mathcal{F}\}$ be relatively compact in Y for every point $x \in X$ (**Ascoli's theorem**). In particular, when X is a σ-compact locally compact Hausdorff space and Y is the space of real numbers, a family of functions $\mathcal{F}$ that are equicontinuous (at every point $x \in X$) and uniformly bounded is relatively compact. Hence, for any sequence of functions $\{f_n\}$ in $\mathcal{F}$, we can select a subsequence $\{f_{n(v)}\}$ which converges uniformly on compact sets (**Ascoli-Arzelà theorem**).

E. Normal Families

P. Montel (1912) gave the name **normal family** to the family of functions that is relatively compact with respect to the topology of uniform convergence on compact sets. This terminology is used mainly for the family of complex analytic functions. In that case, it is customary to compactify the range space and consider Y to be the †Riemann sphere. Using this notion, Montel succeeded in giving a unified treatment of various results in the theory of complex functions.

A family of analytic functions $\mathcal{F}$ on a finite-dimensional †complex manifold X is a normal family if it is uniformly bounded on each compact set (**Montel's theorem**). Another criterion is that there are three values on the Riemann sphere which no function $f \in \mathcal{F}$ takes. More generally, three exceptional values not taken by $f \in \mathcal{F}$ may depend on f, if there is a positive lower bound for the distances between these three values on the Riemann sphere. This gives an easy proof of the †Picard theorem stating that every †transcendental meromorphic function $f(z)$ in $|z| < \infty$ must take all values except possibly two values. In fact the family of functions $f_n(z) = f(z/2^n)$, $n = 1, 2, 3, \ldots$, in $\{1 < |z| < 2\}$ cannot be normal. Using a similar procedure, G. Julia obtained the results on †Julia's direction.

F. Marty introduced the notion of **spherical derivative** $|f'(z)|/(1 + |f(z)|^2)$ for the analytic or meromorphic function $f(z)$ and proved that for a family $\mathcal{F} = \{f(z)\}$ of analytic functions to be normal, it is necessary and sufficient that the spherical derivatives of $f \in \mathcal{F}$ be uniformly bounded. This theorem implies Montel's theorem and its various extensions, including, for example, quantitative results concerning †Borel's direction.

A family $\mathcal{F}$ of analytic functions of one variable defined on X is said to form a **quasi-normal family** if there exists a subset P of X consisting only of isolated points such that from any sequence $\{f_n\}$ ($f_n \in \mathcal{F}$) we can select a subsequence $\{f_{n(v)}\}$ converging uniformly on $X - P$. If P is finite and consists of p points, the family $\mathcal{F}$ is called a quasinormal family of order p. For example, the family of at most †p-valent functions is quasinormal of order p.

The theory of normal families of complex analytic functions is not only applied to †value distribution theory, as above, but also used to show the existence of a function that gives the extremal of functionals. The extremal function is usually obtained as a limit of a subsequence of a sequence in a normal family. A typical example of this method is seen in the proof of the †Riemann mapping theorem. This is perhaps the only general method known today in the study of the iteration of †holomorphic functions. By this method, Julia (1919) made an exhaustive study of the iteration of meromorphic functions; there are several other investigations on the iteration of elementary transcendental functions. On the other hand, A. Wintner (*Comm. Math. Helv.*, 23 (1949)) gave the implicit function theorem for analytic functions in a precise form using the theory of normal families of analytic functions of several variables.

References

[1] N. Bourbaki, Eléments de mathématique, III., Topologie générale, ch. 10, Espaces fonctionnels, Actualités Sci. Ind., 1084b, Hermann, second edition, 1967; English translation, Theory of sets, Addison-Wesley, 1966.
[2] J. L. Kelley, General topology, Van Nostrand, 1955, ch. 7.
[3] J. Dieudonné, Foundations of modern analysis, Academic Press, 1960, enlarged and corrected printing, 1969.
For normal families of complex functions,
[4] P. Montel, Leçons sur les familles normales de fonctions analytiques et leurs applications, Gauthier-Villars, 1927.
[5] G. Valiron, Familles normales et quasinormales de fonctions méromorphes, Gauthier-Villars, 1929.

436 (II.22) Uniform Spaces

A. Introduction

There are certain properties defined on †metric spaces but not on general †topological spaces, for example, †completeness or †uniform continuity of functions. Generalizing metric spaces, A. Weil introduced the notion of uniform spaces. This notion can be defined in several ways [3, 4]. The definition in Section B is that of Weil [1] without the †separation axiom for topology.

We denote by Δ_X the **diagonal** $\{(x, x) | x \in X\}$ of the Cartesian product $X \times X$ of a set X with itself. If U and V are subsets of $X \times X$, then the **composite** $V \circ U$ is defined to be the set of all pairs (x, y) such that for some element z of X, the pair (x, z) is in U and the pair (z, y) is in V. The inverse U^{-1} of U is defined to be the set of all pairs (x, y) such that $(y, x) \in U$.

B. Definitions

Let $\mathscr{U}$ be a nonempty family of subsets of $X \times X$ such that (i) if $U \in \mathscr{U}$ and $U \subset V$, then $V \in \mathscr{U}$; (ii) if U, $V \in \mathscr{U}$, then $U \cap V \in \mathscr{U}$; (iii) if $U \in \mathscr{U}$, then $\Delta_X \subset U$; (iv) if $U \in \mathscr{U}$, then $U^{-1} \in \mathscr{U}$; and (v) if $U \in \mathscr{U}$, then $V \circ V \subset U$ for some $V \in \mathscr{U}$. Then we say that a **uniform structure** (or simply a **uniformity**) is defined on X by $\mathscr{U}$. If a uniformity is defined on X by $\mathscr{U}$, then the pair $(X, \mathscr{U})$ or simply the set X itself is called a **uniform space**, and $\mathscr{U}$ is usually called a **uniformity** for X.

A subfamily $\mathscr{B}$ of the uniformity $\mathscr{U}$ is called a **base for the uniformity** $\mathscr{U}$ if every member of $\mathscr{U}$ contains a member of $\mathscr{B}$. If a family $\mathscr{B}$ of subsets of $X \times X$ is a base for a uniformity $\mathscr{U}$, then the following propositions hold: (ii′) if U, $V \in \mathscr{B}$, then there exists a $W \in \mathscr{B}$ such that $W \subset U \cap V$; (iii′) if $U \in \mathscr{B}$, then $\Delta_X \subset U$; (iv′) if $U \in \mathscr{B}$, then there exists a $V \in \mathscr{B}$ such that $V \subset U^{-1}$; (v′) if $U \in \mathscr{B}$, then there exists a $V \in \mathscr{B}$ such that $V \circ V \subset U$. Conversely, if a family $\mathscr{B}$ of subsets of a Cartesian product $X \times X$ satisfies (ii′)–(v′), then the family $\mathscr{U} = \{U | U \subset X \times X, V \subset U$ for some $V \in \mathscr{B}\}$ defines a uniformity on X and $\mathscr{B}$ is a base for $\mathscr{U}$. Given a uniform space $(X, \mathscr{U})$, a member V of $\mathscr{U}$ is said to be **symmetric** if $V = V^{-1}$. The family of all symmetric members of $\mathscr{U}$ is a base for $\mathscr{U}$.

C. Topology of Uniform Spaces

Given a uniform space $(X, \mathscr{U})$, an element $x \in X$, and $U \in \mathscr{U}$, we put $U(x) = \{y | y \in X, (x, y) \in U\}$. Then the family $\mathscr{U}(x) = \{U(x) | U \in \mathscr{U}\}$ forms a neighborhood system of $x \in X$, which gives rise to a topology of X (→ 425 Topological Spaces). This topology is called the **uniform topology** (or **topology of the uniformity**). When we refer to a topology of a uniform space $(X, \mathscr{U})$, it is understood to be the uniform topology; thus a uniform space is also called a **uniform topological space**. If $\mathscr{B}$ is a base for the uniformity of a uniform space $(X, \mathscr{U})$, then $\mathscr{B}(x) = \{U(x) | U \in \mathfrak{B}\}$ is a base for the neighborhood system at each point $x \in X$. Each member of $\mathscr{U}$ is a subset of the topological space $X \times X$, which is supplied with the product topology. The family of all open (closed) symmetric members of $\mathscr{U}$ forms a base for $\mathscr{U}$. A uniform space $(X, \mathscr{U})$ is a †T_1-topological space if and only if the intersection of all members of $\mathscr{U}$ is the diagonal Δ_X. In this case, the uniformity of $(X, \mathscr{U})$ is called a **T_1-uniformity**, and $(X, \mathscr{U})$ is called a **T_1-uniform space**. A T_1-uniform space is always †regular; a fortiori, it is a T_2-topological space. Hence a T_1-uniform space is also said to be a **Hausdorff uniform space** (or **separated uniform space**). Moreover, a uniform topology satisfies †Tikhonov's separation axiom; in particular, a T_1-uniform space is †completely regular.

D. Examples

(1) Discrete Uniformity. Let X be a nonempty set, and let $\mathscr{U} = \{U | \Delta_X \subset U \subset X \times X\}$. Then $(X, \mathscr{U})$ is a T_1-uniform space and $\mathscr{B} = \{\Delta_X\}$ is a base for $\mathscr{U}$. This uniformity is called the **discrete uniformity** for X.

(2) Uniform Family of Neighborhood System. A family $\{U_\alpha(x)\}_{\alpha \in A} (x \in X)$ of subsets of a set X is called a **uniform neighborhood system** in X if it satisfies the following four requirements: (i) $x \in U_\alpha(x)$ for each $\alpha \in A$ and each $x \in X$; (ii) if x and y are distinct elements of X, then $y \notin U_\alpha(x)$ for some $\alpha \in A$; (iii) if α and β are two elements of A, then there is another element $\gamma \in A$ such that $U_\gamma(x) \subset U_\alpha(x) \cap U_\beta(x)$ for all $x \in X$; (iv) if α is an arbitrary element in A, then there is an element β in A such that $y \in U_\alpha(x)$ whenever x, $y \in U_\beta(z)$ for some z in X. If we denote by $U_\alpha (\alpha \in A)$ the subset of $X \times X$ consisting of all elements (x, y) such that $x \in X$ and $y \in U_\alpha(x)$, then the family $\{U_\alpha | \alpha \in A\}$ satisfies all the conditions for a base for a uniformity. In particular, it follows from (ii) that $\bigcap_{\alpha \in A} U_\alpha = \Delta_X$, which is a stronger condition than (iii′) in Section B. For instance, if $\{U_\alpha | \alpha \in A\}$ is a base for the neighborhood system at the identity element of a T_1-topological group G, then we have two uniform neighborhood systems $\{U^l_\alpha(x)\}$ and $\{U^r_\alpha(x)\}$, where $U^l_\alpha(x) = xU_\alpha$ and

$U_\alpha'(x)=U_\alpha x$. Two uniformities derived from these uniform neighborhood systems are called a †left uniformity and a †right uniformity, respectively. Generally, these two uniformities do not coincide (→ 423 Topological Groups).

(3) Uniform Covering System [4]. A family $\{\mathfrak{U}_\alpha\}_{\alpha\in A}$ of †coverings of a set X is called a **uniform covering system** if the following three conditions are satisfied: (i) if $\mathfrak{U}$ is a covering of X such that $\mathfrak{U}<\mathfrak{U}_\alpha$ for all $\alpha\in A$, then $\mathfrak{U}$ coincides with the covering $\Delta=\{\{x\}\}_{x\in X}$; (ii) if α, $\beta\in A$, then there is a $\gamma\in A$ such that $\mathfrak{U}_\gamma<\mathfrak{U}_\alpha$ and $\mathfrak{U}_\gamma<\mathfrak{U}_\beta$; (iii) if $\alpha\in A$, then there is a $\beta\in A$ such that $\mathfrak{U}_\beta$ is a †Δ-refinement of $\mathfrak{U}_\alpha$ $((\mathfrak{U}_\beta)^\Delta<\mathfrak{U}_\alpha)$. For an example of a uniform covering system of X, suppose that we are given a uniform neighborhood system $\{U_\alpha(x)\}_{\alpha\in A}$ $(x\in X)$. Let $\mathscr{U}_\alpha=\{U_\alpha(x)\}_{\alpha\in X}$ $(\alpha\in A)$. Then $\{\mathscr{U}_\alpha\}_{\alpha\in A}$ is a uniform covering system. On the other hand, for a covering $\mathfrak{U}=\{U_\lambda\}_{\lambda\in\Lambda}$, let $S(x,\mathfrak{U})$ be the union of all members of $\mathfrak{U}$ that contain x. If $\{\mathfrak{U}_\alpha\}_{\alpha\in A}$ is a uniform covering system and $U_\alpha(x)=S(x,\mathfrak{U}_\alpha)$, then $\{U_\alpha(x)\}_{\alpha\in A}$ $(x\in X)$ is a uniform neighborhood system. Hence defining a uniform covering system of X is equivalent to defining a T_1-uniformity on X.

(4). In a metric space (x,d) the subsets $\mathfrak{U}_r=\{(x,y)\mid d(x,Y)<r\}$, $r>0$, form a base of uniformity. The uniform topology defined by this coincides with the topology defined by the metric.

E. Some Notions on Uniform Spaces

Some of the terminology concerning topological spaces can be restated in the language of uniform structures. A mapping f from a uniform space $(X,\mathscr{U})$ into another $(X',\mathscr{U}')$ is said to be **uniformly continuous** if for each member U' in $\mathscr{U}'$ there is a member U in $\mathscr{U}$ such that $(f(x),f(y))\in U'$ for every $(x,y)\in U$. This condition implies that f is continuous with respect to the uniform topologies of the uniform spaces. Equivalently, the mapping is uniformly continuous with respect to the uniform neighborhood system $\{U_\alpha(x)\}_{\alpha\in A}$ if for any index β there is an index α such that $y\in U_\alpha(x)$ implies $f(y)\in U_\beta(f(x))$. If $f:X\rightarrow X'$ and $g:X'\rightarrow X''$ are uniformly continuous, then the composite $g\circ f:X\rightarrow X''$ is also uniformly continuous. A bijection f of a uniform space $(X,\mathscr{U})$ to another $(X',\mathscr{U}')$ is said to be a **uniform isomorphism** if both f and f^{-1} are uniformly continuous; in this case $(X,\mathscr{U})$ and $(X',\mathscr{U}')$ are said to be **uniformly equivalent**. A uniform isomorphism is a homeomorphism with respect to the uniform topologies, and a uniform equivalence defines an equivalence relation between uniform spaces.

If $\mathscr{U}_1$ and $\mathscr{U}_2$ are uniformities for a set X, we say that the uniformity $\mathscr{U}_1$ is **stronger** than the uniformity $\mathscr{U}_2$ and $\mathscr{U}_2$ is **weaker** than $\mathscr{U}_1$ if the identity mapping of $(X,\mathscr{U}_1)$ to $(X,\mathscr{U}_2)$ is uniformly continuous. The discrete uniformity is the strongest among the uniformities for a set X. The weakest uniformity for X is defined by the single member $X\times X$; this uniformity is not a T_1-uniformity unless X is a singleton. Generally, there is no weakest T_1-uniformity. A uniformity $\mathscr{U}_1$ for X is stronger than another $\mathscr{U}_2$ if and only if every member of $\mathscr{U}_2$ is also a member of $\mathscr{U}_1$.

If f is a mapping from a set X into a uniform space $(Y,\mathscr{V})$ and g is the mapping of $X\times X$ into $Y\times Y$ defined by $g(x,y)=(f(x),f(y))$, then $\mathscr{B}=\{g^{-1}(V)\mid V\in\mathscr{V}\}$ satisfies conditions (ii')–(v') in Section B for a base for a uniformity. The uniformity $\mathscr{U}$ for X determined by $\mathscr{B}$ is called the **inverse image** of the uniformity $\mathscr{V}$ for Y by f; $\mathscr{U}$ is the weakest uniformity for X such that f is uniformly continuous. Hence a mapping f from a uniform space $(X,\mathscr{U})$ into another $(Y,\mathscr{V})$ is uniformly continuous if and only if the inverse image of the uniformity $\mathscr{V}$ under f is weaker than the uniformity $\mathscr{U}$. If A is a subset of a uniform space $(X,\mathscr{U})$, then there is a uniformity $\mathscr{V}$ for A determined as the inverse image of $\mathscr{U}$ by the inclusion mapping of A into X. This uniformity $\mathscr{V}$ for A is called the **relative uniformity** for A induced by $\mathscr{U}$, or the **relativization** of $\mathscr{U}$ to A, and the uniform space $(A,\mathscr{V})$ is called a **uniform subspace** of $(X,\mathscr{U})$. The uniform topology for $(A,\mathscr{V})$ is the relative topology for A induced by the uniform topology for $(X,\mathscr{U})$.

If $\{(X_\lambda,\mathscr{U}_\lambda)\}_{\lambda\in\Lambda}$ is a family of uniform spaces, then the **product uniformity** for $X=\prod_{\lambda\in\Lambda}X_\lambda$ is defined to be the weakest uniformity $\mathscr{U}$ such that the projection of X onto each X_λ is uniformly continuous, and $(X,\mathscr{U})$ is called the **product uniform space** of $\{(X_\lambda,\mathscr{U}_\lambda)\}_{\lambda\in\Lambda}$. The topology for $(X,\mathscr{U})$ is the product of the topologies for $(X_\lambda,\mathscr{U}_\lambda)$ $(\lambda\in\Lambda)$.

F. Metrization

Each †pseudometric d for a set X generates a uniformity in the following way. For each positive number r, let $V_{d,r}=\{(x,y)\in X\times X\mid d(x,y)<r\}$. Then the family $\{V_{d,r}\mid r>0\}$ satisfies conditions (ii')–(v') in Section B for a base for a uniformity $\mathscr{U}$. This uniformity is called the **pseudometric uniformity** or **uniformity generated by** d. The uniform topology for $(X,\mathscr{U})$ is the pseudometric topology. A uniform space $(X,\mathscr{U})$ is said to be **pseudometrizable** (**metrizable**) if there is a pseudometric (metric)

d such that the uniformity $\mathscr{U}$ is identical with the uniformity generated by d. A uniform space is pseudometrizable if and only if its uniformity has a countable base. Consequently, a uniform space is metrizable if and only if its uniformity is a T_1-uniformity and has a countable base. For a family P of pseudometrics on a set X, let $V_{d,r}=\{(x,y)\in X\times X \mid d(x,y)<r\}$ for $d\in P$ and positive r. The weakest uniformity containing every $V_{d,r}$ $(d\in P, r>0)$ is called the **uniformity generated by** P. This uniformity may also be described as the weakest one such that each pseudometric in P is uniformly continuous on $X\times X$ with respect to the product uniformity.

Each uniformity $\mathscr{U}$ on a set X coincides with the uniformity generated by the family P_X of all pseudometrics that are uniformly continuous on $X\times X$ with respect to the product uniformity of $\mathscr{U}$ with itself. It follows that each uniform space is uniformly isomorphic to a subspace of a product of pseudometric spaces (in which the number of components is equal to the cardinal number of P_X) and that each T_1-uniform space is uniformly isomorphic to a subspace of a product of metric spaces. A topology τ for a set X is the uniform topology for some uniformity for X if and only if the topological space (X,τ) satisfies †Tikhonov's separation axiom; in particular, the uniformity is a T_1-uniformity if and only if (X,τ) is †completely regular.

G. Completeness

If $(X,\mathscr{U})$ is a uniform space, a subset A of X is called a **small set of order** U $(U\in\mathscr{U})$ if $A\times A\subset U$. A †filter on X is called a **Cauchy filter** (with respect to the uniformity $\mathscr{U}$) if it contains a small set of order U for each U in $\mathscr{U}$. If a filter on X converges to some point in X, then it is a Cauchy filter. If f is a uniformly continuous mapping from a uniform space X into another X', then the image of a base for a Cauchy filter on X under f is a base for a Cauchy filter on X'. A point contained in the closure of every set in a Cauchy filter $\mathfrak{F}$ is the limit point of $\mathfrak{F}$. Hence if a filter converges to x, a Cauchy filter contained in the filter also converges to x.

A †net $x(\mathfrak{A})=\{x_\alpha\}_{\alpha\in\mathfrak{A}}$ (where $\mathfrak{A}$ is a directed set with a preordering $\leqslant$) in a uniform space $(X,\mathscr{U})$ is called a **Cauchy net** if for each U in $\mathscr{U}$ there is a γ in $\mathfrak{A}$ such that $(x_\alpha,x_\beta)\in U$ for every α and β such that $\gamma\leqslant\alpha$, $\gamma\leqslant\beta$. If $\mathfrak{A}$ is the set $\mathbf{N}$ of all natural numbers, a Cauchy net $\{x_n\}_{n\in\mathbf{N}}$ is called a **Cauchy sequence** (or **fundamental sequence**). Given a Cauchy net $\{x_\alpha\}_{\alpha\in\mathfrak{A}}$, let $A_\alpha=\{x_\beta \mid \beta\geqslant\alpha\}$. Then $\mathfrak{B}=\{A_\alpha \mid \alpha\in\mathfrak{A}\}$ is a base for a filter, and the filter is a Cauchy filter. On the other hand, let $\mathfrak{B}$ be a base for a Cauchy filter $\mathfrak{F}$. For $U, V\in\mathfrak{B}$, we put $U\leqslant V$ if and only if $U\supset V$. Then $\mathfrak{B}$ is a directed set with respect to $\leqslant$. The net $\{x_U\}_{U\in\mathfrak{B}}$, where x_U is an arbitrary point in U, is a Cauchy net. A proposition concerning convergence of a Cauchy filter is always equivalent to a proposition concerning convergence of the corresponding Cauchy net.

A Cauchy filter (or Cauchy net) in a uniform space X does not always converge to a point of X. A uniform space is said to be **complete** (with respect to the uniformity) if every Cauchy filter (or Cauchy net) converges to a point of that space. A complete uniform space is called for brevity a **complete space**. A closed subspace of a complete space is complete with respect to the relative uniformity. A pseudometrizable uniform space is complete if and only if every Cauchy sequence in the space converges to a point. Hence in the case of a metric space, our definition of completeness coincides with the usual one (→ 273 Metric Spaces).

A mapping f from a uniform space X to another X' is said to be **uniformly continuous on a subset** A of X if the restriction of f to A is uniformly continuous with respect to the relative uniformity for A. If f is a uniformly continuous mapping from a subset A of a uniform space into a complete T_1-uniform space, then there is a unique uniformly continuous extension $\bar{f}$ of f on the closure $\bar{A}$.

Each T_1-uniform space is uniformly equivalent to a dense subspace of a complete T_1-uniform space; this property is a generalization of the fact that each metric space can be mapped by an isometry onto a dense subset of a complete metric space. **A completion** of a uniform space $(X,\mathscr{U})$ is a pair $(f,(X^*,\mathscr{U}^*))$, where $(X^*,\mathscr{U}^*)$ is a complete space and f is a uniform isomorphism of X onto a dense subspace of X^*. The T_1-completion of a T_1-uniform space is unique up to uniform equivalence.

H. Compact Spaces

A uniformity $\mathscr{U}$ for a topological space (X,τ) is said to be **compatible** with the topology τ if the uniform topology for $(X,\mathscr{U})$ coincides with τ. A topological space (X,τ) is said to be **uniformizable** if there is a uniformity compatible with τ. If (X,τ) is a compact Hausdorff space, then there is a unique uniformity $\mathscr{U}$ compatible with τ; in fact, $\mathscr{U}$ consists of all neighborhoods of the diagonal Δ_X in $X\times X$; and the compact Hausdorff space is complete with this uniformity. Hence every subspace of a compact Hausdorff space is uniformizable, and every †locally compact Hausdorff space is uniformizable. Any continuous mapping from a compact Hausdorff space to a uniform space is uniformly continuous. A uniform space $(X, \mathscr{U})$ is said to be **totally bounded** (or **precompact**) if for each $U \in \mathscr{U}$ there is a finite covering consisting of small sets of order U; a subset of a uniform space is called **totally bounded** if it is totally bounded with respect to the relative uniformity. A uniform space X is said to be **locally totally bounded** if for each point of X there is a base for a neighborhood system consisting of totally bounded open subsets. A uniform space is compact if and only if it is totally bounded and complete. If f is a uniformly continuous mapping from a uniform space X to another, then the image $f(A)$ of a totally bounded subset A of X is totally bounded.

I. Topologically Complete Spaces

A topological space (X, τ) is said to be **topologically complete** (or **Dieudonné complete**) if it admits a uniformity compatible with τ with respect to which X is complete. Each †paracompact Hausdorff space is topologically complete. Actually such a space is complete with respect to its strongest uniformity. A Hausdorff space which is homeomorphic to a †G_δ-set in a compact Hausdorff space is said to be **Čech-complete**; A metric space is homeomorphic to a complete metric space if and only if it is Čech-complete. A Hausdorff space X is paracompact and Čech-complete if and only if there is a †perfect mapping from X onto a complete metric space.

References

[1] A. Weil, Sur les espaces à structure uniforme et sur la topologie générale, Actualités Sci. Ind., Hermann, 1938.
[2] J. W. Tukey, Convergence and uniformity in topology, Ann. Math. Studies, Princeton Univ. Press, 1940.
[3] J. L. Kelley, General topology, Van Nostrand, 1955.
[4] N. Bourbaki, Eléments de mathématique, III. Topologie générale, ch. 2, Actualités Sci. Ind., 1142d, Hermann, fourth edition 1965; English translation, General topology, Addision-Wesley, 1966.
[5] J. R. Isbell, Uniform spaces, Amer. Math. Soc. Math. Surveys, 1964.
[6] H. Nakano, Uniform spaces and transformation groups, Wayne State Univ. Press, 1968.
[7] R. Emgelking, General topology, Polish Scientific Publishers, 1977.
[8] Z. Frolík, Generalization of the G_δ-property of complete metric spaces, Czech. Math. J., 10 (1960), 359–379.

437 (IV.17) Unitary Representations

A. Definitions

A homomorphism U of a †topological group G into the group of †unitary operators on a †Hilbert space $\mathfrak{H}$ $(\neq \{0\})$ is called a **unitary representation** of G if U is **strongly continuous** in the following sense: For any element $x \in \mathfrak{H}$, the mapping $g \rightarrow U_g x$ is a continuous mapping from G into $\mathfrak{H}$. The Hilbert space $\mathfrak{H}$ is called the **representation space** of U and is denoted by $\mathfrak{H}(U)$. Two unitary representations U and U' are said to be **equivalent** (**similar** or **isomorphic**), denoted by $U \cong U'$, if there exists an †isometry T from $\mathfrak{H}(U)$ onto $\mathfrak{H}(U')$ that satisfies the equality $T \circ U_g = U'_g \circ T$ for every g in G. If the representation space $\mathfrak{H}(U)$ contains no closed subspace other than $\mathfrak{H}$ and $\{0\}$ that is invariant under every U_g, the unitary representation U is said to be **irreducible**. An element x in $\mathfrak{H}(U)$ is called a **cyclic vector** if the set of all finite linear combinations of the elements $U_g x (g \in G)$ is dense in $\mathfrak{H}(U)$. A representation U having a cyclic vector is called a **cyclic representation**. Every nonzero element of the representation space of an irreducible representation is a cyclic vector.

Examples. Let G be a †topological transformation group acting on a †locally compact Hausdorff space X from the right. Suppose that there exists a †Radon measure μ that is invariant under the group G. Then a unitary representation R^μ is defined on the Hilbert space $\mathfrak{H} = L^2(X, \mu)$ by the formula $(R^\mu_g f)(x) = f(xg)$ $(f \in \mathfrak{H}, x \in X, g \in G)$. The representation R^μ is called the **regular representation** of G on (X, μ). If G acts on X from the left, then the regular representation L^μ is defined by $(L^\mu_g f)(x) = f(g^{-1}x)$. In particular, when X is the †quotient space $H \backslash G$ of a †locally compact group G by a closed subgroup H, any two invariant measures μ, μ' (if they exist) coincide up to a constant factor. Hence the regular representation R^μ on (X, μ) and the regular representation $R^{\mu'}$ on (X, μ') are equivalent. In this case, the representation R^μ is called the regular representation on X. When $H = \{e\}$, a locally compact group G has a Radon measure $\mu \neq 0$ that is invariant under every right (left) translation $h \rightarrow hg$ $(h \rightarrow gh)$ and is called a right (left) †Haar measure on G. So G has the regular representation R (L) on G. R (L) is called the right (left) regular representation of G.

B. Positive Definite Functions and Existence of Representations

A complex-valued continuous function φ on a topological group G is called **positive definite** if the matrix having $\varphi(g_i^{-1}g_j)$ as the (i,j)-component is a †positive semidefinite Hermitian matrix for any finite number of elements $g_1, \ldots, g_n$ in G. If U is a unitary representation of G, then the function $\varphi(g)=(U_g x, x)$ is positive definite for every element x in $\mathfrak{H}(U)$. Conversely, any positive definite function $\varphi(g)$ on a topological group G can be expressed as $\varphi(g)=(U_g x, x)$ for some unitary representation U and x in $\mathfrak{H}(U)$. Using this fact and the †Kreĭn-Milman theorem, it can be proved that every locally compact group G has **sufficiently many irreducible unitary representations** in the following sense: For every element g in G other than the identity element e, there exists an irreducible unitary representation U, generally depending on g, that satisfies the inequality $U_g \neq 1$. The groups having sufficiently many finite-dimensional (irreducible) unitary representations are called †maximally almost periodic. If a connected locally compact group G is maximally almost periodic, then G is the direct product of a compact group and a vector group $\mathbf{R}^m$. On the other hand, any noncompact connected †simple Lie group has no finite-dimensional irreducible unitary representation other than the unit representation $g \rightarrow 1$ (→ 18 Almost Periodic Functions).

C. Subrepresentations

Let U be a unitary representation of a topological group G. A closed subspace $\mathfrak{N}$ of $\mathfrak{H}(U)$ is called **U-invariant** if $\mathfrak{N}$ is invariant under every U_g $(g \in G)$. Let $\mathfrak{N} \neq \{0\}$ be a closed invariant subspace of $\mathfrak{H}(U)$ and V_g be the restriction of U_g on $\mathfrak{N}$. Then V is a unitary representation of G on the representation space $\mathfrak{N}$ and is called a **subrepresentation** of U. Two unitary representations L and M are called **disjoint** if no subrepresentation of L is equivalent to a subrepresentation of M; they are called **quasi-equivalent** if no subrepresentation of L is disjoint from M and no subrepresentation of M is disjoint from L.

D. Irreducible Representations

Let U be a unitary representation of G, $\mathbf{M}$ be the †von Neumann algebra generated by $\{U_g \mid g \in G\}$, and $\mathbf{M}'$ be the †commutant of $\mathbf{M}$. Then a closed subspace $\mathfrak{N}$ of $\mathfrak{H}(U)$ is invariant under U if and only if the †projection operator P corresponding to $\mathfrak{N}$ belongs to $\mathbf{M}'$. Therefore U is irreducible if and only if $\mathbf{M}'$ consists of scalar operators $\{\alpha 1 \mid \alpha \in \mathbf{C}\}$ **(Schur's lemma)**. A representation space of a cyclic or irreducible representation of a †separable topological group is †separable.

E. Factor Representations

A unitary representation U of G is called a **factor representation** if the von Neumann algebra $\mathbf{M}=\{U_g \mid g \in G\}$ is a †factor, that is, $\mathbf{M} \cap \mathbf{M}'=\{\alpha 1 \mid \alpha \in \mathbf{C}\}$. Two factor representations are quasi-equivalent if and only if they are not disjoint. U is called a **factor representation of type I, II,** or **III** if the von Neumann algebra $\mathbf{M}$ is a factor of †type I, II, or III, respectively (→ 308 Operator Algebras). A topological group G is called a **group of type I** (or **type I group**) if every factor representation of G is of type I. Compact groups, locally compact Abelian groups, connected †nilpotent Lie groups, connected †semisimple Lie groups, and real or complex †linear algebraic groups are examples of groups of type I. There exists a connected solvable Lie group that is not of type I (→ Section U), but a connected solvable Lie group is of type I if the exponential mapping is surjective (O. Takenouchi). A discrete group G with countably many elements is a type I group if and only if G has an Abelian normal subgroup with finite index (E. Thoma).

F. Representation of Direct Products

Let G_1 and G_2 be topological groups, G the †direct product of G_1 and G_2 $(G=G_1 \times G_2)$, and U_i an irreducible unitary representation of G_i $(i=1,2)$. Then the †tensor product representation $U_1 \otimes U_2 : (g_1, g_2) \rightarrow U_{g_1} \otimes U_{g_2}$ is an irreducible unitary representation of G. Conversely, if one of the groups G_1 and G_2 is of type I, then every irreducible unitary representation of G is equivalent to the tensor product $U_1 \otimes U_2$ of some irreducible representations U_i of G_i $(i=1,2)$.

G. Direct Sums

If the representation space $\mathfrak{H}$ of a unitary representation U is the †direct sum $\bigoplus_{\alpha \in I} \mathfrak{H}(\alpha)$ of mutually orthogonal closed invariant subspaces $\{\mathfrak{H}(\alpha)\}_{\alpha \in I}$, then U is called the **direct sum** of the subrepresentations $U(\alpha)$ induced on $\mathfrak{H}(\alpha)$ by U, and is denoted by $U=\bigoplus_{\alpha \in I} U(\alpha)$. Any unitary representation is the direct sum of cyclic representations. A unitary representation U is called a **representation without multiplicity** if U cannot be decomposed as a direct sum $U_1 \oplus U_2$ unless U_1 and U_2 are disjoint. If U is the direct sum of $\{U(\alpha)\}_{\alpha \in I}$ and every $U(\alpha)$ is irreducible, then U is said to be **decomposed into the direct sum of irreducible representations**. Decomposition into direct sums of irreducible representations is essentially unique if it exists; that is, if $U = \bigoplus_{\alpha \in I} U(\alpha) = \bigoplus_{\beta \in J} V(\beta)$ are two decompositions of U into direct sums of irreducible representations, then there exists a bijection φ from I onto J such that $U(\alpha)$ is equivalent to $V(\varphi(\alpha))$ for every α in I. A factor representation U of type I can be decomposed as the direct sum $U = \bigoplus_{\alpha \in I} U(\alpha)$ of equivalent irreducible representations $U(\alpha)$. In general, a unitary representation U cannot be decomposed as the direct sum of irreducible representations even if U is not irreducible. Thus it becomes necessary to use direct integrals to obtain an irreducible decomposition.

H. Direct Integrals

Let U be a unitary representation of a group G and (X, μ) be a †measure space. Assume that the following two conditions are satisfied by U: (i) There exists a unitary representation $U(x)$ of G corresponding to every element x of X, and $\mathfrak{H}(U)$ is a †direct integral (→ 308 Operator Algebras) of $\mathfrak{H}(U(x))$ $(x \in X)$ (written $\mathfrak{H}(U) = \int_X \mathfrak{H}(U(x))\, d\mu(x)$); (ii) for every g in G, the operator U_g is a decomposable operator and can be written as $U_g = \int_X U_g(x)\, d\mu(x)$. Then the unitary representation U is called the **direct integral** of the family $\{U(x)\}_{x \in X}$ of unitary representations and is denoted by $U = \int_X U(x)\, d\mu(x)$. If every point of X has measure 1, then a direct integral is reduced to a direct sum.

I. Decomposition into Factor Representations

We assume that G is a locally compact group satisfying the †second countability axiom, and also that a Hilbert space is separable. Every unitary representation U of G can be decomposed as a direct integral $U = \int_X U(x)\, d\mu(x)$ in such a way that the center $\mathbf{A}$ of the von Neumann algebra $\mathbf{M}'' = \{U_g \mid g \in G\}''$ is the set of all †diagonalizable operators. In this case almost all the $U(x)$ are factor representations. Such a decomposition of U is essentially unique. There exists a †null set N in X such that for every x and x' in $X - N$ $(x \neq x')$, $U(x)$ and $U(x')$ are mutually disjoint factor representations. Hence the space X can be identified with the set G^* of all quasi-equivalence classes of factor representations of G endowed with a suitable structure of a measure space. The space G^* is called the **quasidual** of G. The measure μ is determined by U up to †equivalence of measures.

J. Duals

A topology is introduced on the set $\hat{G}$ of all equivalence classes of irreducible unitary representations of a locally compact group G in the following way. Let H_n be the n-dimensional Hilbert space $l_2(n)$ and I_n the set of all irreducible unitary representations of G realized on H_n $(1 \leqslant n \leqslant \infty)$. We topologize I_n in such a way that a †net $\{U^\lambda\}_{\lambda \in L}$ in I_n converges to U if and only if $(U_g^\lambda x, y)$ converges uniformly to $(U_g x, y)$ on every compact subset of G for any x and y in H_n. Equivalence between representations in I_n is an open relation. Let $\hat{G}_n$ be the set of all equivalence classes of n-dimensional irreducible unitary representations of G with the topology of a quotient space of I_n and $\hat{G} = \bigcup_n \hat{G}_n$ be the direct sum of topological spaces $\hat{G}_n$. Then the topological space $\hat{G}$ is called the **dual** of G. $\hat{G}$ is a locally compact †Baire space with countable open base, but it does not satisfy the †Hausdorff separation axiom in general. If G is a compact Hausdorff topological group, then $\hat{G}$ is discrete. If G is a locally compact Abelian group, then $\hat{G}$ coincides with the †character group of G in the sense of Pontryagin. If G is a type I group, then there exists a dense open subset of $\hat{G}$ that is a locally compact Hausdorff space. The †σ-additive family generated by closed sets in $\hat{G}$ is denoted by $\mathfrak{B}$. In the following sections, a measure on $\hat{G}$ means a measure defined on $\mathfrak{B}$.

K. Irreducible Decompositions

In this section G is assumed to be a locally compact group of type I with countable open base. For any equivalence class x in $\hat{G}$, we choose a representative $U(x) \in x$ with the representation space $H(U(x)) = l_2(n)$ if x is n-dimensional. For any measure μ on $\hat{G}$, the representation $U^\mu = \int_{\hat{G}} U(x) d\mu(x)$ is a unitary representation without multiplicity. Conversely, any unitary representation of G without multiplicity is equivalent to a U^μ for some measure μ on $\hat{G}$. Moreover, U^μ is equivalent to U^ν if and only if the two measures μ and ν are equivalent (that is, μ is absolutely continuous with respect to ν, and vice versa). A unitary representation U with multiplicity on a separable Hilbert space $\mathfrak{H}$ can be decomposed as follows: There exists a countable set of measures $\mu_1, \mu_2, \ldots, \mu_\infty$ whose supports are mutually disjoint such that $U \cong \int_G U(x)d\mu_1(x) \oplus 2\int_G U(x)d\mu_2(x) \oplus \ldots \oplus \infty\int_G U(x)d\mu_\infty(x)$. The measures $\mu_1, \mu_2, \ldots, \mu_\infty$ are uniquely determined by U up to equivalence of measures. Any unitary representation U on a separable Hilbert space $\mathfrak{H}$ of an arbitrary locally compact group with countable open base (even if not of type I) can be decomposed as a direct integral of irreducible representations. In order to obtain such a decomposition, it is sufficient to decompose $\mathfrak{H}$ as a direct integral in such a way that a maximal Abelian von Neumann subalgebra $\mathbf{A}$ of $\mathbf{M}' = \{U_g \mid g \in G\}'$ is the set of all diagonalizable operators. In this case, however, a different choice of $\mathbf{A}$ induces in general an essentially different decomposition, and uniqueness of the decomposition does not hold. For a group of type I, the irreducible representations are the "atoms" of representations, as in the case of compact groups. For a group not of type I, it is more natural to take the factor representations for the irreducible representations, quasi-equivalence for the equivalence, and the quasidual for the dual of G. Therefore the theory of unitary representations for a group not of type I has different features from the one for a type I group. The theory of unitary representation for groups not of type I has not yet been successfully developed, but some important results have been obtained (e.g., L. Pukanszky, *Ann. Sci. Ecole Norm. Sup.*, 4 (1971)).

Tatsuuma [1] proved a duality theorem for general locally compact groups which is an extension of both Pontryagin's and Tannaka's duality theorems considering the direct integral decomposition of tensor product representations.

L. The Plancherel Formula

Let G be a unimodular locally compact group with countable open base, $R(L)$ be the right (left) regular representation of G, and $\mathbf{M}$, $\mathbf{N}$, and $\mathbf{P}$ be the von Neumann algebras generated by $\{R_g\}$, $\{L_g\}$, and $\{R_g, L_g\}$, respectively. Then $\mathbf{M}' = \mathbf{N}$, $\mathbf{N}' = \mathbf{M}$, and $\mathbf{P}' = \mathbf{M} \cap \mathbf{N}$. If we decompose $\mathfrak{H}$ into a direct integral in such a way that $\mathbf{P}'$ is the algebra of all diagonalizable operators, then $\mathbf{M}(x)$ and $\mathbf{N}(x)$ are factors for almost all x. This decomposition of $\mathfrak{H}$ produces a decomposition of the two-sided regular representation $\{R_g, L_g\}$ into irreducible representations and a decomposition of the regular representation $R(L)$ into factor representations. Hence the decomposition is realized as the direct integral over the quasidual G^* of G. Moreover, the factors $\mathbf{M}(x)$ and $\mathbf{N}(x)$ are of type I or II for almost all x in G^*, and there exists a †trace t in the factor $\mathbf{M}(x)$. For any f and g in $L_1(G) \cap L_2(G)$, the **Plancherel formula**

$$\int_G f(s)\overline{g(s)}\,ds = \int_{G^*} t(U_g^*(x)U_f(x))\,d\mu(x) \qquad (1)$$

holds, where $U_f(x) = \int_G f(s)U_s(x)\,ds$ and U^* is the †adjoint of U. The **inversion formula**

$$h(s) = \int_{G^*} t(U_h(x)U_s^*(x))\,d\mu(x) \qquad (2)$$

is derived from (1) for a function $h = f * g$ ($f, g \in L_1(G) \cap L_2(G)$). In (1) and (2), because of the impossibility of normalization of the trace t in a factor of type II_∞, the measure μ cannot in general be determined uniquely. However, if G is a type I group, then (1) and (2) can be rewritten as similar formulas, where the representation $U(x)$ in (1) and (2) is irreducible, the trace t is the usual trace, and the domain of integration is not the quasidual G^* but the dual $\hat{G}$ of G. The revised formula (1) is also called the Plancherel formula. In this case the measure μ on $\hat{G}$ in formulas (1) and (2) is uniquely determined by the given Haar measure on G. The measure μ is called the **Plancherel measure** of G. The support $\hat{G}_r$ of the Plancherel measure μ is called the **reduced dual** of G. The Plancherel formula gives the direct integral decomposition of the regular representation into the irreducible representations belonging to $\hat{G}_r$. Each U in $\hat{G}_r$ is contained in this decomposition, with the multiplicity equal to $\dim \mathfrak{H}(U)$.

M. Square Integrable Representations

An irreducible unitary representation U of a unimodular locally compact group G is said to be **square integrable** when for some element $x \neq 0$, in $\mathfrak{H}(U)$, the function $\varphi(g) = (U_g x, x)$ belongs to $L^2(G, dg)$, where dg is the Haar measure of G. If U is square integrable, then $\varphi_{x,y}(g) = (U_g x, y)$ belongs to $L^2(G, dg)$ for any x and y in $\mathfrak{H}(U)$. Let U and U' be the two square integrable representations of G. Then the following **orthogonality relations** hold:

$$\int_G (U_g x, y)\overline{(U'_g u, v)}\,dg = \begin{cases} 0 & \text{if } U \text{ is not equivalent to } U', \\ d_U^{-1}(x,u)(v,y) & \text{if } U = U'. \end{cases} \qquad (3)$$

When G is compact, every irreducible unitary representation U is square integrable and finite-dimensional. Moreover, the scalar d_U in (3) is the degree of U if the total measure of G is normalized to 1. In the general case, the scalar d_U in (3) is called the **formal degree** of U and is determined uniquely by the given Haar measure dg. Let y be an element in $\mathfrak{H}(U)$ with norm 1 and V be the subspace $\{\varphi_{x,y} \mid x \in \mathfrak{H}(U)\}$ of $L^2(G)$. Then the linear mapping $T: x \rightarrow \sqrt{d_U}\, \varphi_{x,y}$ is an isometry of $\mathfrak{H}(U)$ onto V. Hence U is equivalent to a subrepresentation of the right regular representation R of G. Conversely, every irreducible subrepresentation of R is square integrable. Thus a square integrable representation is an irreducible subrepresentation of R ($\cong L$). Therefore, in the irreducible decomposition of R, the square integrable representations appear as discrete direct summands. Hence every square integrable representation U has a positive Plancherel measure $\mu(U)$ that is equal to the formal degree d_U. There exist noncompact groups that have square integrable representations. An example of such a group is $SL(2, \mathbf{R})$ (→ Section X).

N. Representations of $L_1(G)$

Let G be a locally compact group and $L_1(G)$ be the space of all complex-valued integrable functions on G. Then $L_1(G)$ is an algebra over $\mathbf{C}$, where the convolution

$$(f*g)(s) = \int_G f(st^{-1}) g(t)\, dt$$

is defined to be the product of f and g. Let Δ be the †modular function of G. Then the mapping $f(s) \rightarrow f^*(s) = \Delta(s^{-1})\overline{f(s^{-1})}$ is an †involution of the algebra $L_1(G)$. Let U be a unitary representation of G, and put $U'_f = \int_G U_s f(s)\, ds$. Then the mapping $f \rightarrow U'_f$ gives a **nondegenerate representation** of the Banach algebra $L_1(G)$ with an involution, where nondegenerate means that $\{U'_f x \mid f \in L_1(G), x \in \mathfrak{H}(U)\}^\perp$ reduces to $\{0\}$. The mapping $U \rightarrow U'$ gives a bijection between the set of equivalence classes of unitary representations of G and the set of equivalence classes of nondegenerate representations of the Banach algebra $L_1(G)$ with an involution on Hilbert spaces. U is an irreducible (factor) representation if and only if U' is an irreducible (factor) representation. Therefore the study of unitary representations of G reduces to that of representations of $L_1(G)$. If U'_f is a †compact operator for every f in $L_1(G)$, then U is the discrete direct sum of irreducible representations, and the multiplicity of every irreducible component is finite. (See [2] for Sections A–N.)

O. Induced Representations

Induced representation is the method of constructing a representation of a group G in a canonical way from a representation of a subgroup H of G. It is a fundamental method of obtaining a unitary representation of G. Let G be a locally compact group satisfying the second countability axiom, L be a unitary representation on a separable Hilbert space $\mathfrak{H}(L)$ of a closed subgroup H of G, and m, n, Δ, and δ be the right Haar measures and the modular functions of the groups G and H, respectively. Then there exists a continuous positive function ρ on G satisfying $\rho(hg) = \delta(h)\Delta(h)^{-1}\rho(g)$ for every h in H and g in G. The †quotient measure $\mu = (\rho m)/n$ is a quasi-invariant measure on the coset space $H \backslash G$ (→ 225 Invariant Measures). Let $\mathfrak{H}$ be the vector space of weakly measurable functions f on G with values in $\mathfrak{H}(L)$ satisfying the following two conditions: (i) $f(hg) = L_h f(g)$ for every h in H and g in G; and (ii) $\|f\|^2 = \int_{H \backslash G} \|f(g)\|^2\, d\mu(\dot{g}) < +\infty$, where $\dot{g}$ represents the coset Hg. By condition (i), the norm $\|f(g)\|$ is constant on a coset $Hg = \dot{g}$ and is a function on $H \backslash G$, so the integral in condition (ii) is well defined. Then $\mathfrak{H}$ is a Hilbert space with the norm defined in (ii). A unitary representation U of G on the Hilbert space $\mathfrak{H}$ is defined by the formula

$$(U_s f)(g) = \sqrt{\rho(gs)/\rho(g)}\, f(gs).$$

U is called the **unitary representation induced by the representation** L **of a subgroup** H and is denoted by $U = U^L$ or $\mathrm{Ind}_H^G L$. **Induced representations** have the following properties.

(1) $U^{L_1 \oplus L_2} \cong U^{L_1} \oplus U^{L_2}$ or more generally, $U^{\int U(x) d\mu(x)} \cong \int U^{L(x)} d\mu(x)$. Therefore if U^L is irreducible, L is also irreducible (the converse does not hold in general).

(2) Let H, K be two subgroups of G such that $H \subset K$, L be a unitary representation of H, and M be the representation of K induced by L. Then two unitary representations U^M and U^L of G are equivalent.

An induced representation U^L is the representation on the space of square integrable sections of the †vector bundle with fiber $H(L)$ †associated with the principal bundle $(G, H \backslash G, H)$ (→ G. W. Mackey [3], F. Bruhat [4]).

P. Unitary Representations of Special Groups

In the following sections we describe the fundamental results on the unitary representations of certain special groups.

Q. Compact Groups

Irreducible unitary representations of a compact group are always finite-dimensional. Every unitary representation of a compact group is decomposed into the direct sum of irreducible representations. Irreducible unitary representations of a compact connected Lie group are completely classified. The characters of irreducible representations are calculated in an explicit form (→ 69 Compact Groups; 249 Lie Groups). Every irreducible unitary representation U of a connected compact Lie group G can be extended uniquely to an irreducible holomorphic representation U^C of the complexification G^C of G. U^C is holomorphically induced from a 1-dimensional representation of a Borel subgroup B of G^C (**Borel-Weil theorem**; → R. Bott [5]).

R. Abelian Groups

Every irreducible unitary representation of an Abelian group G is 1-dimensional. †Stone's theorem concerning one-parameter groups of unitary operators, $U_t=\int_{-\infty}^{\infty}e^{i\lambda t}dE_\lambda$, gives irreducible decompositions of unitary representations of the additive group $\mathbf{R}$ of real numbers. †Bochner's theorem on †positive definite functions on $\mathbf{R}$ is a restatement of Stone's theorem in terms of positive definite functions. The theory of the †Fourier transform on $\mathbf{R}$, in particular †Plancherel's theorem, gives the irreducible decomposition of the regular representation of $\mathbf{R}$. The theorems of Stone, Bochner, and Plancherel have been extended to an arbitrary locally compact Abelian group (→ 192 Harmonic Analysis).

S. Representations of Lie Groups and Lie Algebras

Let U be a unitary representation of a Lie group G with the Lie algebra $\mathfrak{g}$. An element x in $\mathfrak{H}(U)$ is called an **analytic vector** with respect to U if the mapping $g\rightarrow U_g x$ is a real analytic function on G with values in $\mathfrak{H}(U)$. The set of all analytic vectors with respect to U forms a dense subspace $\mathfrak{A}=\mathfrak{A}(U)$ of $\mathfrak{H}(U)$. For any elements X in $\mathfrak{g}$ and x in $\mathfrak{A}(U)$, the derivative at $t=0$ of a real analytic function $U_{\exp tX}x$ is denoted by $V(X)x$. Then $V(X)$ is a linear transformation on $\mathfrak{A}$, and the mapping $V:X\rightarrow V(X)$ is a representation of $\mathfrak{g}$ on $\mathfrak{A}$. We call V the **differential representation** of U. The representation V of $\mathfrak{g}$ can be extended uniquely to a representation of the †universal enveloping algebra $\mathfrak{B}$ of $\mathfrak{g}$. Two unitary representations $U^{(1)}$ and $U^{(2)}$ of a connected Lie group G are equivalent if and only if there exists a bijective bounded linear mapping T from $\mathfrak{H}(U^{(1)})$ onto $\mathfrak{H}(U^{(2)})$ such that T maps $\mathfrak{A}(U^{(1)})$ onto $\mathfrak{A}(U^{(2)})$ and satisfies the equality

$$(T\circ V^{(1)}(X))x=(V^{(2)}(X)\circ T)x$$

for all X in $\mathfrak{g}$ and x in $\mathfrak{A}(U^{(1)})$. Let $X_1,\ldots,X_n$ be a basis of $\mathfrak{g}$ and U be a unitary representation of G. Then the element $\Delta=X_1^2+\ldots+X_n^2$ in the universal enveloping algebra $\mathfrak{B}$ of $\mathfrak{g}$ is represented in the differential representation V of U by an †essentially self-adjoint operator $V(\Delta)$. Conversely, if to each element X in $\mathfrak{g}$ there corresponds a (not necessarily bounded) †skew-Hermitian operator $\rho(x)$ that satisfies the following three conditions, then there exists a unique unitary representation U of the simply connected Lie group G with the Lie algebra $\mathfrak{g}$ such that the †closure of $V(X)$ coincides with the closure of $\rho(X)$ for every X in $\mathfrak{g}$: (i) There exists a dense subspace $\mathfrak{D}$ contained in the domain of $\rho(X)\rho(Y)$ for every X and Y in $\mathfrak{g}$; (ii) for each X and Y in $\mathfrak{g}$, a and b in $\mathbf{R}$, and x in $\mathfrak{D}$, $\rho(aX+bY)x=a\rho(X)x+b\rho(Y)x$, $\rho([X,Y])x=(\rho(X)\rho(Y)-\rho(Y)\rho(X))x$; (iii) the restriction of $\rho(X_1)^2+\ldots+\rho(X_n)^2$ to $\mathfrak{D}$ is an essentially self-adjoint operator if $X_1,\ldots,X_n$ is a basis of $\mathfrak{g}$ (E. Nelson [6]).

T. Nilpotent Lie Groups

For every irreducible unitary representation of a connected nilpotent Lie group G, there is some 1-dimensional unitary representation of some subgroup of G that induces it. Let G be a simply connected nilpotent Lie group, $\mathfrak{g}$ be the Lie algebra of G, and ρ be the contragredient representation of the adjoint representation of G. The representation space of ρ is the dual space $\mathfrak{g}^*$ of $\mathfrak{g}$. A subalgebra $\mathfrak{h}$ of $\mathfrak{g}$ is called **subordinate** to an element f in $\mathfrak{g}^*$ if f annihilates each bracket $[X,Y]$ for every X and Y in $\mathfrak{h}$: $(f,[X,Y])=0$. When $\mathfrak{h}$ is subordinate to f, a 1-dimensional unitary representation L of the analytic subgroup H of G with the Lie algebra $\mathfrak{h}$ is defined by the formula $\lambda_f(\exp X)=e^{2\pi i(f,X)}$ $(X\in\mathfrak{h})$. Every 1-dimensional unitary representation λ_f of H is defined as in this formula by an element f in $\mathfrak{g}^*$ to which $\mathfrak{h}$ is subordinate. The unitary representation of G induced by such a λ_f is denoted by $U(f,\mathfrak{h})$. The representation $U(f,\mathfrak{h})$ is irreducible if and only if $\mathfrak{h}$ has maximal dimension among the subalgebras subordinate to f. Two irreducible representations $U(f,\mathfrak{h})$ and $U(f,\mathfrak{h}')$ are equivalent if and only if f and f' are conjugate under the group $\rho(G)$. Therefore there exists a bijection between the set of equivalence classes of the irreducible unitary representations of a simply connected nilpotent Lie group G and the set of orbits of $\rho(G)$ on $\mathfrak{g}^*$ (A. A. Kirillov [7]).

U. Solvable Lie Groups

Let G be a simply connected solvable Lie group. If the exponential mapping is bijective, G is called an **exponential group**. All results stated above for nilpotent Lie groups hold for exponential groups except the irreducibility criterion. In this case the representation $U(f,\mathfrak{h})$ is irreducible if and only if $\mathfrak{h}$ is of maximal dimension among subordinate subalgebras and the orbit $O=\rho(G)f$ contains the affine subspace $f+\mathfrak{h}^{\perp}=f+\{g \mid g(\mathfrak{h})=0\}$ (Pukanszky condition).

The situation is more complicated for general solvable Lie groups. The isotropy subgroup $G_f=\{g\in G \mid \rho(g)f=f\}$ at $f\in\mathfrak{g}^*$ is, in general, not connected. A linear form f is called integral if there exists a unitary character η_f of G_f whose differential is the restriction of $2\pi i f$ to $\mathfrak{g}_f$ (the Lie algebra of G_f). Using the notion of "polarization," an irreducible unitary representation of G is constructed from a pair (f,η_f) of an integral form $f\in\mathfrak{g}^*$ and a character η_f. If G is of type I, then every irreducible unitary representation of G is obtained in this way. A simply connected solvable Lie group G is of type I if and only if (i) every $f\in\mathfrak{g}^*$ is integral and (ii) every G-orbit $\rho(G)f$ in $\mathfrak{g}^*$ is locally closed (Auslander and Kostant [8]).

As an example, let α be an irrational real number. Then the following Lie group G is not of type I: $G=\left\{\begin{pmatrix} e^{it} & 0 & z \\ 0 & e^{i\alpha t} & w \\ 0 & 0 & 1\end{pmatrix} \,\middle|\, t\in\mathbf{R}, z, w\in\mathbf{C}\right\}$.

V. Semisimple Lie Groups

A connected semisimple Lie group is of type I. The character $\chi=\chi_U$ of an irreducible unitary representation U of G is defined as follows: Let $C_0^\infty(G)$ be the set of all complex-valued C^∞-functions with compact support on G. Then for any function f in $C_0^\infty(G)$, the operator $U_f=\int_G U_g f(g)\,dg$ belongs to the †trace class, and the linear form $\chi: f\rightarrow T_r U_f$ is a †distribution in the sense of Schwartz. The distribution χ is called the **character** of an irreducible unitary representation U. A character χ is invariant under any inner automorphism of G and is a simultaneous eigendistribution of the algebra of all two-sided invariant linear differential operators on G. Two irreducible unitary representations of G are equivalent if and only if their characters coincide. The distribution χ is a †locally summable function on G and coincides with a real analytic function on each connected component of the dense open submanifold G' consisting of regular elements in G. In general, χ is not real analytic on all of G (Harish-Chandra [9, III; 10].

W. Complex Semisimple Lie Groups

There are four series of irreducible representations of a complex semisimple Lie group G.

(1) A **principal series** consists of unitary representations of G induced from 1-dimensional unitary representations L of a †Borel subgroup B of G. L is uniquely determined by a unitary character $\nu\in\mathrm{Hom}(A,U(1))=A^*$ of the †Cartan subgroup A of G contained in B. Hence the representations in the principal series are parametrized by the elements in the character group A^* of the Cartan subgroup A. If we denote U^L by U^ν, two representations U^ν and $U^{\nu'}(\nu,\nu'\in A^*)$ are equivalent if and only if ν and ν' are conjugate under the †Weyl group W of G with respect to A.

(2) A **degenerate series** consists of unitary representations induced by 1-dimensional unitary representations of a †parabolic subgroup P of G other than B. (A parabolic subgroup P is any subgroup of G containing a Borel subgroup B.)

(3) A **complementary series** consists of irreducible unitary representations U^L induced by nonunitary 1-dimensional representations L of a Borel subgroup B. In this case, condition (ii) in the definition of U^L (→ Section O) must be changed. When L is a nonunitary representation, then the operator U_g^L is not a unitary operator with respect to the usual L_2-inner product (ii). However, if L satisfies a certain condition, then U_g^L leaves invariant some positive definite Hermitian form on the space of sufficiently nice functions. Completing this space, we get a unitary representation U^L. The representations thus obtained form the complementary series.

(4) A **complementary degenerate series** consists of irreducible unitary representations induced by nonunitary 1-dimensional representations of a parabolic subgroup $P\neq B$.

Representations belonging to different series are never equivalent. It seems certain that any irreducible unitary representation of a connected complex semisimple Lie group is equivalent to a representation belonging to one of the above four series, but this conjecture has not yet been proved. Moreover, E. M. Stein [11] constructed irreducible unitary representations different from any in the list obtained by I. M. Gel'fand and M. A. Naĭmark (Neumark) [12]. These representations belong to the complementary degenerate series. The characters of the representations in these four series are computed in explicit form. For example, the character χ_ν of the representation U^ν in the principal series can be calculated as follows: Let λ be a linear form on a Cartan subalgebra $\mathfrak{a}$ such that $\nu(\exp H)=e^{\lambda(H)}$ for every H in $\mathfrak{a}$, let D be the function on A defined by $D(\exp H)=\prod_\alpha |e^{\alpha(H)/2}-e^{-\alpha(H)/2}|^2$, where α runs over all positive roots. Then the character χ_ν of a representation U^ν in the principal series is given by the formula

$$\chi_\nu(\exp H)=D(\exp H)^{-1}\sum_{s\in W} e^{s\lambda(H)}.$$

In the irreducible decomposition of the regular representation of G, only irreducible representations belonging to the principal series arise. Hence the right-hand side in the Plancherel formula is an integral over the character group A^* of a Cartan subgroup A. Under a suitable normalization of the Haar measures in G and A^*, the Plancherel measure μ of G can be expressed by using the Haar measure dv of A^*:

$$d\mu(v)=w^{-1}\prod_{\alpha}|(\lambda,\alpha)/(\rho,\alpha)|^2 dv,$$

where w is the order of the Weyl group, ρ is the half-sum of all †positive roots, and α runs over all positive roots (Gel'fand and Naĭmark [12]).

X. Real Semisimple Lie Groups

As in the case of a complex semisimple Lie group, a connected real semisimple Lie group G has four series of irreducible unitary representations. However, if G has no parabolic subgroup other than a minimal parabolic subgroup B and G itself, then G has no representation in the degenerate or complementary degenerate series. Examples of such groups are $SL(2,\mathbf{R})$ and higher-dimensional †Lorentz groups. In general, the classification of irreducible unitary representations in the real semisimple case is more complicated than in the complex semisimple case. Irreducible unitary representations arising from the irreducible decomposition of the regular representation are called representations in the **principal series**. The principal series of G are divided into a finite number of subseries corresponding bijectively to the conjugate classes of the †Cartan subgroups of G.

A connected semisimple Lie group G has a square integrable representation if and only if G has a compact Cartan subgroup H. The set of all square integrable representations of G is called the **discrete series** of irreducible unitary representations. The discrete series is the subseries in the principal series corresponding to a compact Cartan subgroup H. The representations in the discrete series were classified by Harish-Chandra. Let $\mathfrak{h}$ be the Lie algebra of H, P the set of all positive roots in $\mathfrak{h}$ for a fixed linear order, π the polynomial $\prod_{\alpha\in P}H_\alpha$, and $\mathscr{F}$ the set of all real-valued linear forms on $\sqrt{-1}\mathfrak{h}$. Moreover, let L be the set of all linear forms λ in $\mathscr{F}$ such that a single-valued character ξ_λ of the group H is defined by the formula $\xi_\lambda(\exp X)=e^{\lambda(X)}$, and let L' be the set of all λ in L such that $\pi(\lambda)\neq 0$. Then for each λ in L', there exists a representation $\omega(\lambda)$ of G in the discrete series, and conversely, every representation in the discrete series is equivalent to $\omega(\lambda)$ for some λ in L'. Two representations $\omega(\lambda_1)$ and $\omega(\lambda_2)$ ($\lambda_1,\lambda_2\in L'$) are equivalent if and only if there exists an element s in $W_G=N(H)/H$ such that $\lambda_2=s\lambda_1$, where $N(H)$ is the normalizer of H in G (W_G can act on $\mathscr{F}$ as a linear transformation group in the natural way). The value of the character χ_λ on the subgroup H of the representation $\omega(\lambda)$ ($\lambda\in L'$) is given as follows: Let $\varepsilon(\lambda)$ be the signature of $\pi(\lambda)=\prod_{\alpha\in P}\lambda(H_\alpha)$, and define q and Δ by $q=(\dim G/K)/2$ and $\Delta(\exp H)=\prod_{\alpha\in P}(e^{\alpha(H)/2}-e^{-\alpha(H)/2})$. Then the character χ_λ of the representation $\omega(\lambda)$ has the value $(-1)^q\varepsilon(\lambda)\chi_\lambda(h)=\Delta(h)^{-1}\sum_{s\in W_G}(\det s)\xi_{s\lambda}(h)$ on a regular element h in H. The formal degree $d(\omega(\lambda))$ of the representation $\omega(\lambda)$ is given by the formula $d(\omega(\lambda))=C^{-1}[W_G]|\pi(\lambda)|$, where C is a positive constant (not depending on λ) and $[W_G]$ is the order of the finite group W_G (Harish-Chandra [13]). A formula expressing the character χ_λ on the whole set of regular elements in G has been given by T. Hirai [14]. The representations in discrete series are realized on L^2-cohomology spaces of homogeneous holomorphic line bundles over G/H (W. Schmid [15]). They are also realized on the spaces of harmonic spinors on the †Riemannian symmetric space G/K (M. Atiyah and Schmid [16]). They are also realized on the eigenspaces of a Casimir operator acting on the sections of vector bundles on G/K (R. Hotta, *J. Math. Soc. Japan*, 23; N. Wallach [17]). An irreducible unitary representation is called **integrable** if at least one of its matrix coefficients belongs to $L^1(G)$. Integrable representations belong to the discrete series. They have been characterized by H. Hecht and Schmid (*Math. Ann.*, 220 (1976)). The theory of the discrete series is easily extended to reductive Lie groups.

The general principal series representations of a connected semisimple Lie group G with finite center are constructed as follows. Let K be a maximal compact subgroup of G. Then there exists a unique involutive automorphism θ of G whose fixed point set coincides with K. θ is called a **Cartan involution** of G. Let H be a θ-stable Cartan subgroup of G. Then H is the direct product of a compact group $T=H\cap K$ and a vector group A. The centralizer $Z(A)$ of A in G is the direct product of a reductive Lie group $M=\theta(M)$ and A. M has a compact Cartan subgroup T. Hence the set $\hat{M}_d$ of the discrete series representations of M is not empty. Let α be an element of the dual space $\mathfrak{a}^*$ of the Lie algebra $\mathfrak{a}$ of A and put $\mathfrak{g}_\alpha=\{X\in\mathfrak{g}\mid[H,X]=\alpha(H)X\ (\forall H\in\mathfrak{a})\}$ and $\Delta=\{\alpha\in\mathfrak{a}^*\mid\mathfrak{g}_\alpha\neq\{0\}\}$. Let Δ^+ be the set of positive elements of Δ in a certain order of $\mathfrak{a}^*$ and put $\mathfrak{n}=\sum_{\alpha\in\Delta^+}\mathfrak{g}_\alpha$ and $N=\exp\mathfrak{n}$. Then $P=MAN$ is a closed subgroup of G. P is called a **cuspidal parabolic subgroup** of G. Let $D\in\hat{M}_d$ and $v\in\mathfrak{a}^*$. Then a unitary representation $D\otimes e^{iv}$ of P

is defined by $(D \otimes e^{i\nu})(man) = D(m)e^{i\nu(\log a)}$ $(m \in M, a \in A, n \in N)$. The unitary representation $\pi_{D,\nu}$ of G induced by $D \otimes e^{i\nu}$ is independent of the choice of Δ^+ up to equivalence. Thus $\pi_{D,\nu}$ depends only on (H, D, ν). The set of representations $\{\pi_{D,\nu} \mid D \in \hat{M}_d, \nu \in \mathfrak{a}^*\}$ is called the **principal H-series**. If ν is regular in $\mathfrak{a}^*$ (i.e., $(\nu, \alpha) \neq 0$ for all $\alpha \in \Delta$), then $\pi_{D,\nu}$ is irreducible. Every $\pi_{D,\nu}$ is a finite sum of irreducible representations. The character $\theta_{D,\nu}$ of $\pi_{D,\nu}$ is a locally summable function which is supported in the closure of $\bigcup_{g \in G} g(MA)g^{-1}$. If two Cartan subgroups H_1 and H_2 are not conjugate in G, then every H_1-series representation is disjoint from every H_2-series representation. Choose a complete system $\{H_1, \ldots, H_r\}$ of conjugacy classes of Cartan subgroups of G. Then every H_i can be chosen as θ-stable. The union of the principal H_i-series $(1 \leqslant i \leqslant r)$ is the principal series of G. The right (or left) regular representation of G is decomposed as the direct integral of the principal series representations. Every complex-valued C^∞-function on G with compact support has an expansion in terms of the matrix coefficients of the principal series representations. Harish-Chandra [18] proved these theorems and determined explicitly the Plancherel measure by studying the asymptotic behavior of the Eisenstein integral [19, 20].

Y. Spherical Functions

Let G be a locally compact †unimodular group and K a compact subgroup of G. The set of all complex-valued continuous functions on G that are invariant under every left translation L_k by elements k in K is denoted by $C(K \backslash G)$. The subset of $C(K \backslash G)$ that consists of all two-sided K-invariant functions is denoted by $C(G, K)$. The subset of $C(G, K)$ consisting of all functions with compact support is denoted by $\mathbf{L} = L(G, K)$. $\mathbf{L}$ is an algebra over $\mathbf{C}$ if the product of two elements f and g in $\mathbf{L}$ is defined by the convolution.

Let λ be an algebra homomorphism from $\mathbf{L}$ into $\mathbf{C}$. Then an element of the eigenspace $F(\lambda) = \{\psi \in C(K, G) \mid f * \psi = \lambda(f)\psi \ (\forall f \in \mathbf{L})\}$ is called a **spherical function** on $K \backslash G$. If $F(\lambda)$ contains a nonzero element, then $F(\lambda)$ contains a unique two-sided K-invariant element ω normalized by $\omega(e) = 1$, where e is the identity element in G. This function ω is called the **zonal spherical function** associated with λ. In this case, the homomorphism λ is defined by $\lambda(f) = \int_G f(g)\omega(g^{-1})\,dg$. Hence the eigenspace $F(\lambda)$ is uniquely determined by the zonal spherical function ω. A function $\omega \neq 0$ in $C(G, K)$ is a zonal spherical function on $K \backslash G$ if and only if ω satisfies either of the following two conditions: (i) The mapping $f \mapsto \int f(g)\omega(g^{-1})\,dg$ is an algebra homomorphism of $\mathbf{L}$ into $\mathbf{C}$; (ii) ω satisfies the functional equation

$$\int_K \omega(gkh)\,dk = \omega(g)\omega(h).$$

When G is a Lie group, every spherical function is a real analytic function on $K \backslash G$.

Z. Expansion by Spherical Functions

In this section, we assume that the algebra $\mathbf{L}$ of two-sided K-invariant functions is commutative. In this case there are sufficiently many spherical functions of $K \backslash G$, and two-sided K-invariant functions are expanded by spherical functions. An irreducible unitary representation U of G is called a **spherical representation** with respect to K if the representation space $\mathfrak{H}(U)$ contains a nonzero vector invariant under every U_k, where k runs over K. By the commutativity of $\mathbf{L}$, the K-invariant vectors in $\mathfrak{H}(U)$ form a 1-dimensional subspace. Let x be a K-invariant vector in $\mathfrak{H}(U)$ with the norm $\|x\| = 1$. Then $\omega(g) = (U_g x, x)$ is a zonal spherical function on $K \backslash G$, and for every y in $\mathfrak{H}(U)$, the function $\varphi_y(g) = (U_g x, y)$ is a spherical function associated with ω. Moreover, in this case the zonal spherical function ω is a positive definite function on G. Conversely, every positive definite zonal spherical function ω can be expressed as $\omega(g) = (U_g x, x)$ for some spherical representation U and some K-invariant vector x in $\mathfrak{H}(U)$.

The set of all positive definite zonal spherical functions becomes a locally compact space Ω by the topology of compact convergence. The **spherical Fourier transform** $\hat{f}$ of a function f in $L_1(K \backslash G)$ is defined by

$$\hat{f}(\omega) = \int_G f(g)\omega(g^{-1})\,dg.$$

There exists a unique †Radon measure μ on Ω such that for every f in $\mathbf{L}$, $\hat{f}$ belongs to $L_2(\Omega, \mu)$. Also, the Plancherel formula

$$\int_G f(s)\overline{g(s)}\,ds = \int_\Omega \hat{f}(\omega)\overline{\hat{g}(\omega)}\,d\mu(\omega) \tag{4}$$

holds for every f and g in $\mathbf{L}$, and an inversion formula $f(s) = \int_\Omega \hat{f}(\omega)\omega(s)\,d\mu(\omega)$ holds for a sufficiently nice two-sided K-invariant function f [21]. Identifying a positive definite zonal spherical function with the corresponding spherical representation, we can regard Ω as a subset of the dual $\hat{G}$ of G. The Plancherel formula for two-sided K-invariant functions is obtained from the general Plancherel formula on G by restricting the domain of the integral from $\hat{G}$ to Ω. When G is a Lie group and $\mathbf{L}$ is commutative, a spherical function on $K \backslash G$ can be characterized as a simultaneous eigenfunction of G-invariant linear differential operators on $K\backslash G$.

AA. Spherical Function on Symmetric Spaces

The most important case where the algebra $\mathbf{L}=L(G,K)$ is commutative is when $K\backslash G$ is a †weakly symmetric Riemannian space or, in particular, a †symmetric Riemannian space When $K\backslash G$ is a compact symmetric Riemannian space, a spherical representation with respect to K is the irreducible component of the regular representation T on $K\backslash G$, and a spherical function on $K\backslash G$ is a function that belongs to the irreducible subspaces in $L_2(K\backslash G)$. In particular, if G is a compact connected semisimple Lie group, the highest weights of spherical representations of G with respect to K are explicitly given by using the **Satake diagram** of $K\backslash G$. The Satake diagram of $K\backslash G$ is the †Satake diagram of the noncompact symmetric Riemannian space $K\backslash G_0$ dual to $K\backslash G$ or the Satake diagram of the Lie algebra of G_0. If a symmetric space is the underlying manifold of a compact Lie group G, then G can be expressed as $G=K\backslash(G\times G)$, where K is the diagonal subgroup of $G\times G$. In this case, a zonal spherical function ω on $G=K\backslash(G\times G)$ is the normalized character of an irreducible unitary representation U of $G:\omega(g)=(\deg U)^{-1}T_rU_g$. The explicit form of ω is given by †Weyl's character formula (→ 249 Lie Groups).

The zonal spherical functions on a symmetric Riemannian space $K\backslash G$ of noncompact type are obtained in the following way: Let G be a connected semisimple Lie group with finite center, K be a maximal compact subgroup of G, and $G=NA_+K$ be an †Iwasawa decomposition. Then for any g in G there exists a unique element $H(g)$ in the Lie algebra $\mathfrak{a}_+$ of A_+ such that g belongs to $N\exp H(g)K$. Let $\mathfrak{a}$ be a Cartan subalgebra containing $\mathfrak{a}_+$, P be the set of all positive roots in $\mathfrak{a}$, and $\rho=(\Sigma_{\alpha\in P}\alpha)/2$. Then for any complex-valued linear form v on $\mathfrak{a}_+$, the function

$$\omega_v(g)=\int_K e^{(iv-\rho)(H(kg))}dk$$

is a zonal spherical function on the symmetric Riemannian space $K\backslash G$. Conversely, every zonal spherical function ω on $K\backslash G$ is equal to ω_v for some v. Two zonal spherical functions ω_v and $\omega_{v'}$ coincide if and only if v and v' are conjugate under the operation of the Weyl group $W_0=N_K(A)/Z_K(A)$ of $K\backslash G$ (Harish-Chandra [22], S. Helgason [23]). If v is real-valued, then ω_v is positive definite. Such a zonal spherical function ω_v is obtained from a spherical representation belonging to the principal A-series. Let Ω_0 be the set of all zonal spherical functions ω_v associated with the real-valued linear form v. Then the support of the Plancherel measure μ on $K\backslash G$ is contained in Ω_0. We can choose v as a parameter on the space Ω_0. Then the right-hand side of the Plancherel formula can be expressed as an integral over the dual space L of $\mathfrak{a}_+$. Moreover, the Plancherel measure μ is absolutely continuous with respect to the Lebesgue measure dv on the Euclidean space L and can be expressed as

$$d\mu(\omega_v)=\omega_0^{-1}|c(v)|^{-2}dv$$

under suitable normalization of μ and dv. The problem of calculating the function $c(v)$ can be reduced to the case of symmetric spaces of rank 1 and can be solved explicitly. Let p_α be the multiplicity of a restricted root α and $I(v)$ be the product

$$I(v)=\prod_\alpha B(\tfrac{1}{2}p_\alpha,\tfrac{1}{4}p_\alpha+(v,\alpha)(\alpha,\alpha)^{-1}),$$

where α runs over all positive restricted roots and B is the †beta function. Then $c(v)=I(iv)/I(\rho)$ [20, 24]. Every spherical function f on $K\backslash G$ is expressed as the Poisson integral of its "boundary values" on the Martin boundary $P\backslash G$ of $K\backslash G$, where $P=MA_+N$ is a minimal parabolic subgroup of G. The boundary values of f form a hyperfunction with values in a line bundle over $P\backslash G$ (K. Okamoto et al. [25]).

BB. Spherical Functions and Special Functions

Some important special functions are obtained as the zonal spherical functions on a certain symmetric Riemannian space $M=K\backslash G$ (G is the motion group of M). In particular when M is of rank 1, then the zonal spherical functions are essentially the functions of a single variable. For example, the zonal spherical functions on an n-dimensional Euclidean space can be expressed as

$$\omega_v(r)=2^m\Gamma(m+1)(vr)^{-m}J_m(vr),$$

where $2m=n-2$ and J_m is the †Bessel function of the mth order. The zonal spherical function on an $(n-1)$-dimensional sphere $S^{n-1}=SO(n-1)\backslash SO(n)$ is given by

$$\omega_v(\theta)=\Gamma(v+1)\Gamma(n-2)\Gamma(v+n-2)^{-1}C_v^m(\cos\theta)\qquad (v=0,1,2,\dots),$$

where $C_v^m(z)$ is the †Gegenbauer polynomial. The zonal spherical functions on an $(n-1)$-dimensional Lobachevskiĭ space can be expressed as

$$\omega_v(t)=2^{m-1/2}\Gamma(m+1/2)\sinh^{-m+1/2}t\times\mathfrak{P}^{1/2-m}_{-1/2-m+v}(\cosh t)$$

using a generalized †associated Legendre function $\mathfrak{P}_\nu^\mu$. Many properties of special functions can be proved from a group-theoretic point of view. For example, the addition theorem is merely the homomorphism property $U_{gh}=U_gU_h$ expressed in terms of the matrix components of U. The differential equation satisfied by these special functions is derived from the fact that a zonal spherical function ω is an eigenfunction of an invariant differential operator. The integral expression of such a special function can be obtained by constructing a spherical representation U in a certain function space and calculating explicitly the inner product in the expression $\omega(g)=(U_gx,x)$ (N. Ya. Vilenkin [26]).

CC. Generalization of the Theory of Spherical Functions

The theory of spherical functions described in Sections Y–BB can be generalized in several ways. First, spherical functions are related to the trivial representation of K. A generalization is obtained if the trivial representation of K is replaced by an irreducible representation of K. The theory of such zonal spherical functions is useful for representation theory [20]. For example, the Plancherel formula for $SL(2,\mathbf{R})$ can be obtained using such spherical functions (R. Takahashi, *Japan. J. Math.*, 31 (1961)). Harish-Chandra's Eisenstein integral is such a spherical function on a general semisimple Lie group G. He used it successfully to obtain the Plancherel measure of G. Another generalization can be obtained by removing the condition that K is compact. In particular, when $K\backslash G$ is a symmetric homogeneous space of a Lie group G, the algebra $\mathscr{D}$ of all G-invariant linear differential operators is commutative if the space $K\backslash G$ has an invariant volume element. In this case, a spherical function on $K\backslash G$ can be defined as a simultaneous eigenfunction of $\mathscr{D}$. The character of a semisimple Lie group is a zonal spherical function (distribution) in this sense. The spherical functions and harmonic analysis on symmetric homogeneous space have been studied by T. Oshima and others. T. Oshima and J. Sekiguchi [27] proved the Poisson integral theorem (→ Section AA) for a certain kind of symmetric homogeneous spaces.

The spherical functions and unitary representations of topological groups that are not locally compact are studied in connection with probability theory and physics. For example, the zonal spherical functions of the rotation group of a real Hilbert space are expressed by Hermite polynomials.

DD. Discontinuous Subgroups and Representations

Let G be a connected semisimple Lie group and Γ be a discrete subgroup of G. Then the regular representation T of G on $\Gamma\backslash G$ is defined by $(T_gf)(x)=f(xg)$ $(f\in L^2(\Gamma\backslash G))$. The problem of decomposing the representation T into irreducible components is important in connection with the theory of †automorphic forms and number theory. First assume that the quotient space $\Gamma\backslash G$ is compact. Then for every function f in $L_1(G)$, the operator $T(f)$ is a compact operator. Hence the regular representation T on $\Gamma\backslash G$ can be decomposed into the discrete sum $T=\sum_{k=1}^{\infty}T^{(k)}$ of irreducible unitary representations $T^{(k)}$, and the multiplicity of every irreducible component is finite. The irreducible unitary representation U of G is related to the automorphic forms of Γ in the following way: Let x be a nonzero element in the representation space $\mathfrak{H}=\mathfrak{H}(U)$ of U. $\mathfrak{H}$ is topologized into a †locally convex topological vector space $\mathfrak{H}_x$ by the set N_x of †seminorms: $N_x=\{P_C(y)=\max_{g\in C}|(U_gx,y)|\}$, where C runs over all compact subsets in G. The topology $\mathscr{T}_x$ of $\mathfrak{H}_x$ is independent of the choice of x provided that $\dim\{T_kx\,|\,k\in K\}<\infty$, where K is a maximal compact subgroup of G. Let $\mathfrak{H}^*$ be the completion of $\mathfrak{H}_x$ with respect to the topology $\mathscr{T}_x$ (the completion is independent of the choice of x). $\mathfrak{H}^*$ contains the original Hilbert space $\mathfrak{H}$ as a subspace. Then the representation U of G on $\mathfrak{H}$ can be extended to a representation U^* of G on the space $\mathfrak{H}^*$. An element f in $\mathfrak{H}^*$ invariant under U_γ^* for every γ in Γ is called an **automorphic form** of Γ of type U. Then the multiplicity of an irreducible representation U in the regular representation T on $\Gamma\backslash G$ is equal to the dimension of the vector space consisting of all automorphic forms of type U. This theorem is called the **Gel'fand–Pyatetskiĭ-Shapiro reciprocity law** [28]. Let $T=\sum_{k=1}^{\infty}T^{(k)}$ be the irreducible decomposition of T and χ_k be the character of the irreducible unitary representation $T^{(k)}$. Then for a suitable function f on G, the integral operator K_f on $\mathfrak{H}(T)=L^2(T\backslash G)$ with kernel $k_f(x,y)=\sum_{\gamma\in\Gamma}f(x^{-1}\gamma y)$ belongs to the trace class. By calculating the trace of K_f in two ways, the following **trace formula** is obtained:

$$\sum_{k=1}^{\infty}\int_G f(g)\chi_k(g)\,dg=\sum_{\{\gamma\}}\int_{\mathfrak{D}_\gamma} f(x^{-1}\gamma x)\,dx,$$

where $\{\gamma\}$ is the conjugate class of γ in Γ and $\mathfrak{D}_\gamma$ is the quotient space of the centralizer G_γ of γ in G by the centralizer Γ_γ of γ in Γ.

When the groups G and Γ are given explicitly, the right-hand side of the trace formula can be expressed in a more explicit form, and the trace formula leads to useful consequences. A similar trace formula holds for the unitary representation U^L induced by a finite-dimensional unitary representation L of Γ instead of the regular representation T on $\Gamma\backslash G$. When the quotient space $\Gamma\backslash G$ is not compact, the irreducible decomposition of the regular representation T on $\Gamma\backslash G$ contains not only the discrete direct sum but also the direct integral (continuous spectrum). A. Selberg showed that even in this case, there are explicit examples for which the trace formula holds for the part with discrete spectrum. Also, the part with continuous spectrum can be described by the †generalized Eisenstein series. Analytic properties and the functional equation of the generalized Eisenstein series have been studied by R. Langlands [30]. Recent developments are surveyed in [31].

EE. History

Finite-dimensional unitary representations of a finite group were studied by Frobenius and Schur (1896–1905). In 1925, †Weyl studied the finite-dimensional unitary representation of compact Lie groups. The theory of infinite-dimensional unitary representation was initiated in 1939 by E. P. Wigner in his work on the inhomogeneous Lorentz group, motivated by problems of quantum mechanics.

In 1943, Gel'fand and D. A. Raikov proved the existence of sufficiently many irreducible unitary representations for an arbitrary locally compact group. The first systematic studies of unitary representations appeared in 1947 in the work of V. Bargmann on $SL(2, \mathbf{R})$ [31] and the work of Gel'fand and Neumark on $SL(2, \mathbf{C})$. Gel'fand and Naĭmark established the theory of unitary representation for complex semisimple Lie groups [12].

Harish-Chandra proved theorems concerning the unitary representations of a general semisimple Lie group; for instance, he proved that a semisimple Lie group G is of type I [7] and defined the character of a unitary representation of G and proved its basic properties [9, III; 10]. Harish-Chandra also determined the discrete series of G and their characters. Harish-Chandra [18] proved the Plancherel formula for an arbitrary connected semisimple Lie group G with finite center. Hence harmonic analysis of square integrable functions on G is established.

Further studies on harmonic analysis on semisimple Lie groups have been carried out. In particular, Paley-Wiener–type theorems, which determine the Fourier transform image of the space $C_C^\infty(G)$ of C^∞-functions with compact support, have been proved for the group $PSL(2, \mathbf{R})$ (L. Ehrenpreis and F. Mautner [33]), complex semisimple Lie groups (Zhelobenko [34]), and two-sided K-invariant functions on general semisimple Lie groups (R. Gangolli [35]). A. W. Knapp and E. M. Stein [36] studied the intertwining operators.

Concerning the construction of irreducible representations, G. W. Mackey [3] and Bruhat [4] developed the theory of induced representations of locally compact groups and Lie groups, respectively. B. Kostant [37] (see Blattner's article in [38]) noticed a relation between homogeneous †symplectic manifolds and unitary representations and proposed a method of constructing irreducible unitary representations of a Lie group. Selberg's research [29] revealed a connection between unitary representations (or spherical functions) and the theory of automorphic forms and number theory. A number of papers along these lines have since appeared [31]. In connection with number-theoretic investigations of an †algebraic group defined over an algebraic number field, unitary representations of the †adele group of G or an algebraic group over a †p-adic number field have been studied (→ [31, 38], Gel'fand, M. I. Grayev, and I. I. Pyatetskiĭ-Shapiro [39], and H. M. Jacquet and R. P. Langlands [40]).

For the algebraic approach to the infinite-dimensional representations of semisimple Lie groups and Lie algebras → [41].

For surveys of the theory of unitary representations → [2, 19, 20, 31, 38].

References

[1] N. Tatsuuma, A duality theorem for locally compact groups, J. Math. Kyoto Univ., 6 (1967), 187–293.
[2] J. Dixmier, Les C^*-algèbres et leurs représentations, Gauthier-Villars, 1964.
[3] G. W. Mackey, Induced representations of locally compact groups I, Ann. Math., 55 (1952), 101–139.
[4] F. Bruhat, Sur les representations induites des groupes de Lie, Bull. Soc. Math. France, 84 (1956), 97–205.
[5] R. Bott, Homogeneous vector bundles, Ann. Math., 66 (1957), 203–248.
[6] E. Nelson, Analytic vectors, Ann. Math., 70 (1959), 572–615.
[7] A. A. Kirillov, Unitary representations of nilpotent Lie groups, Russian Math. Surveys, 17, no. 4 (1962), 53–104. (Original in Russian, 1962.)
[8] L. Auslander and B. Kostant, Polarization

and unitary representations of solvable Lie groups, Inventiones Math., 14 (1971), 255–354.
[9] Harish-Chandra, Representations of a semisimple Lie group on a Banach space I–VI, Trans. Amer. Math. Soc., 75 (1953), 185–243; 76 (1954), 26–65; 76 (1954), 234–253; Amer. J. Math., 77 (1955), 743–777; 78 (1956), 1–41; 78 (1956), 564–628.
[10] Harish-Chandra, Invariant eigendistributions on a semisimple Lie group, Trans. Amer. Math. Soc., 119 (1965), 457–508.
[11] E. M. Stein, Analysis in matrix spaces and some new representations of $SL(N, \mathbf{C})$, Ann. Math., (2) 86 (1967), 461–490.
[12] I. M. Gel'fand and M. A. Naĭmark (Neumark), Unitäre Darstellungen der klassischen Gruppen, Akademische Verlag., 1957. (Original in Russian, 1950.)
[13] Harish-Chandra, Discrete series for semisimple Lie groups I, II, Acta Math., 113 (1965), 241–318; 116 (1966), 1–111.
[14] T. Hirai, The characters of the discrete series for semisimple Lie groups, J. Math. Kyoto Univ., 21 (1981), 417–500.
[15] W. Schmid, L^2-cohomology and the discrete series, Ann. Math., 102 (1976), 375–394.
[16] M. F. Atiyah and W. Schmid, A geometric construction of the discrete series for semisimple Lie groups, Inventiones Math., 42 (1977), 1–62.
[17] N. Wallach, On the Enright-Varadarajan modules. A construction of the discrete series, Ann. Sci. Ecole Norm. Sup., 9 (1976), 81–102.
[18] Harish-Chandra, Harmonic analysis on real reductive groups I–III, J. Functional Anal., 19 (1975), 104–204; Inventiones Math., 36 (1976), 1–55; Ann. Math., 104 (1976), 117–201.
[19] J. A. Wolf, M. Cahen, and M. De Wilde (eds.), Harmonic analysis and representations of semisimple Lie groups, Reidel, 1980.
[20] G. Warner, Harmonic analysis on semisimple Lie groups I, II, Springer, 1972.
[21] H. Yoshizawa, A proof of the Plancherel theorem, Proc. Japan Acad., 30 (1954), 276–281.
[22] Harish-Chandra, Spherical functions on a semisimple Lie group I, II, Amer. J. Math., 80 (1958), 241–310, 553–613.
[23] S. Helgason, Differential geometry and symmetric spaces, Academic Press, 1962.
[24] J. Rosenberg, A quick proof of Harish-Chandra's Plancherel theorem for spherical functions on a semisimple Lie groups, Proc. Amer. Math. Soc., 63 (1977), 143–149.
[25] M. Kashiwara, A. Kowata, K. Minemura, K. Okamoto, T. Oshima, and M. Tanaka, Eigenfunctions of invariant differential operators on a symmetric space, Ann. Math., 107 (1978), 1–39.
[26] N. Ya Vilenkin, Special functions and theory of group representations, Amer. Math. Soc. Transl. of Math. Monographs, 22, 1968. (Original in Russian, 1965.)
[27] T. Oshima and J. Sekiguchi, Eigenspaces of invariant differential operators on an affine symmetric space, Inventiones Math., 57 (1980), 1–81.
[28] I. M. Gel'fand and I. I. Pyatetskiĭ-Shapiro, Theory of representations and theory of automorphic functions, Amer. Math. Soc. Transl., (2) 26 (1963), 173–200. (Original in Russian, 1959.)
[29] A. Selberg, Harmonic analysis and discontinuous groups in weakly symmetric Riemannian spaces with applications to Dirichlet series, J. Indian Math. Soc., 20 (1956), 47–87.
[30] R. P. Langlands, On the functional equations satisfied by Eisenstein series, Lecture notes in math. 544, Springer, 1976.
[31] Automorphic forms, representations and L-functions, Amer. Math. Soc. Proc. Symposia in Pure Math., 33, 1979.
[32] V. Bargmann, Irreducible unitary representations of the Lorentz group, Ann. Math., (2) 48 (1947), 568–640.
[33] L. Ehrenpreis and F. I. Mautner, Some properties of the Fourier transform on semisimple Lie groups I–III, Ann. Math., (2) 61 (1955), 406–439; Trans. Amer. Math. Soc., 84 (1957), 1–55; 90 (1959), 431–484
[34] D. P. Zhelobenko, Harmonic analysis of functions on semisimple Lie groups I, II, Amer. Math. Soc. Transl., (2) 54 (1966), 177–230. (Original in Russian, 1963, 1969.)
[35] R. Gangolli, On the Plancherel formula and the Paley-Wiener theorem for spherical functions on a semisimple Lie group, Ann. Math., 93 (1971), 150–165.
[36] A. W. Knapp and E. M. Stein, Intertwining operators for semisimple groups, Ann. Math., 93 (1971), 489–578.
[37] B. Kostant, Quantization and unitary representations, Lecture notes in math. 170, Springer, 1970, 87–207.
[38] Harmonic analysis on homogeneous spaces, Amer. Math. Soc. Proc. Symposia in Pure Math., 26 (1973).
[39] I. M. Gel'fand, M. I. Grayev, and I. I. Pyatetskiĭ-Shapiro, Representation theory and automorphic functions, Saunders, 1969. (Original in Russian, 1966.)
[40] H. Jacquet and R. P. Langlands, Automorphic forms on $GL(2)$, Lecture notes in math. 114, Springer, 1970.
[41] D. A. Vogan, Representations of real reductive Lie groups, Birkhäuser, 1981.

438 (XI.5) Univalent and Multivalent Functions

A. General Remarks

A single-valued †analytic function $f(z)$ defined in a domain D of the complex plane is said to be **univalent** (or **simple** or **schlicht**) if it is injective, i.e., if $f(z_1) \neq f(z_2)$ for all distinct points z_1, z_2 in D. A multiple-valued function $f(z)$ is also said to be univalent if its distinct function elements always attain distinct values at their centers. The derivative of a univalent function is never zero. The limit function of a †uniformly convergent sequence of univalent functions is univalent unless it reduces to a constant. When $f(z)$ is single-valued, the univalent function $w=f(z)$ gives rise to a one-to-one †conformal mapping between D and its image $f(D)$.

B. Univalent Functions in the Unit Disk

A systematic theory of the family of functions †holomorphic and univalent in the unit disk originates from a **distortion theorem** obtained by P. Koebe (1909) in connection with the uniformization of analytic functions. In general, distortion theorems are theorems for determining bounds of functionals, such as $|f(z)|$, $|f'(z)|$, $\arg f'(z)$, within the family under consideration. In particular, distortion theorems concerning the bounds of the arguments of $f(z)$ and $f'(z)$ are also called **rotation theorems**. Though results were at first qualitative, they were made quantitative subsequently by L. Bieberbach (1916), G. Faber (1916), and others. Any univalent function $f(z)$ holomorphic in the unit disk and normalized by $f(0)=0$ and $f'(0)=1$ satisfies the **distortion inequalities**

$$\frac{|z|}{(1+|z|)^2} \leqslant |f(z)| \leqslant \frac{|z|}{(1-|z|)^2},$$

$$\frac{1-|z|}{(1+|z|)^3} \leqslant |f'(z)| \leqslant \frac{1+|z|}{(1-|z|)^3}.$$

Here the equality holds only if $f(z)$ is of the form $z/(1-\varepsilon z)^2$ ($|\varepsilon|=1$). In deriving these inequalities, Bieberbach centered his attention on the family of †meromorphic functions $g(\zeta) = \zeta + \sum_{\nu=0}^{\infty} b_\nu \zeta^{-\nu}$ univalent in $|\zeta|>1$. He established the **area theorem** $\sum_{\nu=1}^{\infty} \nu |b_\nu|^2 \leqslant 1$, which illustrates the fact that the area of the complementary set of the image domain is nonnegative. Bieberbach, R. Nevanlinna (1919–1920), and others constructed a systematic theory of univalent functions in the unit disk based on this theorem.

After the area theorem, the chief tools in the theory of univalent functions have been Löwner's method, the method of contour integration, the variational method, and the method of the extremal metric. In contrast to the theory of univalent functions based on Bieberbach's area theorem, K. Löwner (1923) introduced a new method. In view of a theorem on the domain kernel (C. Carathéodory, 1912), it suffices to consider an everywhere dense subfamily in order to estimate a continuous functional within the family of univalent functions holomorphic in the unit disk. Löwner used the subfamily of functions mapping the unit disk onto the so-called bounded slit domains. Namely, the range of a member of this subfamily consists of the unit disk slit along a Jordan arc that starts at a periphery point and does not pass through the origin. A mapping function of this nature is determined as the integral $f(z,t_0)$ of **Löwner's differential equation**

$$\frac{\partial f(z,t)}{\partial t} = -f(z,t)\frac{1+\kappa(t)f(z,t)}{1-\kappa(t)f(z,t)}, \qquad 0 \leqslant t \leqslant t_0,$$

with the initial condition $f(z,0)=z$, where $\kappa(t)$ is a continuous function with absolute value equal to 1. Any univalent function $f(z)$ holomorphic in the unit disk and satisfying $f(0)=0$, $f'(0)=1$ has an arbitrarily close approximation by functions of the form $e^{t_0} f(z,t_0)$. By means of this differential equation Löwner proved that $|a_3| \leqslant 3$ for any univalent function $f(z) = z + \sum_{n=2}^{\infty} a_n z^n$ ($|z|<1$) and also derived a decisive estimate concerning a coefficient problem for the inverse function [2].

G. M. Golusin (1935) and I. E. Bazilevich (1936) first noticed that Löwner's method is also a powerful tool for deriving several distortion theorems. They showed that classical distortion theorems can be derived in more detailed form (Golusin, *Mat. Sb.*, 2 (1937), 685); in particular, Golusin (1938) obtained a precise estimate concerning the rotation theorem, i.e.,

$$|\arg f'(z)| \leqslant \begin{cases} 4 \arcsin |z|, & |z| < 1/\sqrt{2}, \\ \pi + \log(|z|^2/(1-|z|^2)), & 1/\sqrt{2} \leqslant |z| < 1. \end{cases}$$

Löwner's method was also investigated by A. C. Schaeffer and D. C. Spencer (1945) [8].

The method of contour integration was introduced by H. Grunsky. It starts with some 2-dimensional integral which can be shown to be positive. Transforming it into a boundary integral and using the †residue theorem, we obtain an appropriate inequality by means of this integral. By this method Grunsky established the following useful inequality (*Math.*

Z., 45 (1939)). For $g(\zeta)=\zeta+\sum_{\nu=0}^{\infty}b_\nu\zeta^{-\nu}$, which is univalent in $|\zeta|>1$, let

$$\log\frac{g(z)-g(\zeta)}{z-\zeta}=\sum_{m=1}^{\infty}\sum_{n=1}^{\infty}c_{mn}z^{-m}\zeta^{-n}\qquad(|z|>1,|\zeta|>1).$$

The coefficients c_{mn} are polynomials in the coefficients b_ν of g. Then **Grunsky's inequality** is: For each integer N and for all complex numbers $\lambda_1,\ldots,\lambda_N$,

$$\left|\sum_{m=1}^{N}\sum_{n=1}^{N}c_{mn}\lambda_m\lambda_n\right|\leqslant\sum_{n=1}^{N}\frac{1}{n}|\lambda_n|^2.$$

It is known that if this inequality holds for an arbitrary integer N and for all complex numbers $\lambda_1,\ldots,\lambda_N$, then $g(\zeta)$ is univalent in $|\zeta|>1$. There are several generalizations of Grunsky's inequality [13].

The variational method was first developed by M. Schiffer for application to the theory of univalent functions. He first used boundary variations (*Proc. London Math. Soc.*, 44 (1938)) and later interior variations (*Amer. J. Math.*, 65 (1943)). The problem of maximizing a given real-valued functional on a family of univalent functions is called an extremal problem, and a function for which the functional attains its maximum is called an extremal function. The **variational method** is used to uncover characteristic properties of an extremal function by comparing it with nearby functions. Typical results are the qualitative information that the extremal function maps the disk $|z|<1$ onto the complement of a system of analytic arcs satisfying a differential equation and that the extremal function satisfies a differential equation. Following Schiffer, Schaeffer and Spencer [8] and Golusin (*Math. Sb.*, 19 (1946)) gave variants of the method of interior variations.

H. Grötzsch (1928–1934) treated the theory of univalent functions in a unified manner by the method of the †extremal metric. The idea of this method is to estimate the length of curves and the area of some region swept out by them together with an application of †Schwarz's inequality (→ 143 Extremal Length). After Grötzsch, the method of the extremal metric has been used by many authors. In particular, O. Teichmüller, in connection with this method, formulated the principle that the solution of a certain type of extremal problem is in general associated with a †quadratic differential, although he did not prove any general result realizing this principle in concrete form. J. A. Jenkins gave a concrete expression of the Teichmüller principle; namely, he established the general coefficient theorem and showed that this theorem contains as special cases a great many of the known results on univalent functions [11].

Univalence criteria have been given by various authors. In particular, Z. Nehari (*Bull. Amer. Math. Soc.*, 55 (1949)) proved that if $|\{f(z),z\}|\leqslant2(1-|z|^2)^{-2}$ in $|z|<1$, then $f(z)$ is univalent in $|z|<1$, and E. Hille (*Bull. Amer. Math. Soc.*, 55 (1949)) proved that 2 is the best possible constant in the above inequality. Here, $\{f(z),z\}$ denotes the †Schwarzian derivative of $f(z)$ with respect to z:

$$\{f(z),z\}=\left(\frac{f''(z)}{f'(z)}\right)'-\frac{1}{2}\left(\frac{f''(z)}{f'(z)}\right)^2.$$

C. Coefficient Problems

In several distortion theorems **Koebe's extremal function** $z/(1-\varepsilon z)^2=\sum_{n=1}^{\infty}n\varepsilon^{n-1}z^n(|\varepsilon|=1)$ is extensively utilized. Concerning this, Bieberbach stated the following conjecture. If $f(z)=z+\sum_{n=2}^{\infty}a_nz^n$ is holomorphic and univalent in $|z|<1$, then $|a_n|\leqslant n$ $(n=2,3,\ldots)$, with equality holding only for Koebe's extremal function $z/(1-\varepsilon z)^2$ $(|\varepsilon|=1)$. This conjecture was solved affirmatively by L. de Branges in 1985 after enormous effort by many mathematicians, as described below.

Bieberbach (1916, [1]) proved $|a_2|\leqslant2$ as a corollary to the area theorem. This result can be proved easily by most of the methods. In 1923 Löwner [2] proved $|a_3|\leqslant3$, introducing his own method. Schaeffer and Spencer gave a proof of $|a_3|\leqslant3$ by the variational method (*Duke Math. J.*, 10 (1943)). Furthermore, Jenkins used the method of the extremal metric to prove a coefficient inequality that implies $|a_3|\leqslant3$ (*Analytic Functions*, Princeton Univ. Press, 1960). The problem of the fourth coefficient remained open until 1955, when P. R. Garabedian and Schiffer [3] proved $|a_4|\leqslant4$ by the variational method. Their proof was extremely complicated. Subsequently, Z. Charzynski and Schiffer gave an alternative brief proof of $|a_4|\leqslant4$ by using the Grunsky inequality (*Arch. Rational Mech. Anal.*, 5 (1960)). M. Ozawa (1969, [4]) and R. N. Pederson (1968, [5]) also used the Grunsky inequality to prove $|a_6|\leqslant6$. In 1972, Pederson and Schiffer [6] proved $|a_5|\leqslant5$. They applied the Garabedian-Schiffer inequality, a generalization of the Grunsky inequality which Garabedian and Schiffer had derived by the variational method.

On the other hand, W. K. Hayman [7] showed that for each fixed $f(z)=z+\sum_{n=2}^{\infty}a_nz^n$,

$$\lim_{n\to\infty}\frac{|a_n|}{n}=\alpha\leqslant1,$$

with the equality holding only for Koebe's extremal function $z/(1-\varepsilon z)^2$ $(|\varepsilon|=1)$. Further, it was shown that Koebe's extremal function $z/(1-z)^2$ gives a local maximum for the nth coefficient in the sense that $\operatorname{Re}\{a_n\} \leqslant n$ whenever $|a_2-2|<\delta_n$ for some $\delta_n>0$ (Garabedian, G. G. Ross, and Schiffer, *J. Math. Mech.*, 14 (1964); E. Bombieri, *Inventiones Math.*, 4 (1967); Garabedian and Schiffer, *Arch. Rational Math. Anal.*, 26 (1967)).

In the most general form, the coefficient problem is to determine the region occupied by the points $(a_2, \ldots, a_n)$ for all functions $f(z)=z+\sum_{n=2}^{\infty} a_n z^n$ univalent in $|z|<1$. Schaeffer and Spencer [8] found explicitly the region for (a_2, a_3).

For the coefficients of functions $g(\zeta)=\zeta+\sum_{v=0}^{\infty} b_v \zeta^{-v}$ univalent in $|\zeta|>1$, the following results are known: $|b_1| \leqslant 1$ (Bieberbach [1]), $|b_2| \leqslant 2/3$ (Schiffer, *Bull. Soc. Math. France*, 66 (1938); Golusin, *Mat. Sb.*, 3 (1938)), $|b_3| \leqslant 1/2+e^{-6}$ (Garabedian and Schiffer, *Ann. Math.*, (2) 61 (1955)).

D. Other Classes of Univalent Functions

We have discussed the general family of functions univalent in the unit disk. There are also several results on distortion theorems and coefficient problems for subfamilies determined by conditions such as that the images are bounded, †starlike with respect to the origin, or †convex. For instance, if $f(z)=z+\sum_{n=2}^{\infty} a_n z^n$ is holomorphic and univalent in $|z|<1$ and its image is starlike with respect to the origin, then $|a_n| \leqslant n$ $(n=2,3,\ldots)$. If the image of $f(z)$ is convex, then $f(z)$ satisfies $|a_n| \leqslant 1$ $(n=2,3,\ldots)$ and the distortion inequalities

$$\frac{|z|}{1+|z|} \leqslant |f(z)| \leqslant \frac{|z|}{1-|z|},$$

$$\frac{1}{(1+|z|)^2} \leqslant |f'(z)| \leqslant \frac{1}{(1-|z|)^2}.$$

Here the equality sign appears at z_0 $(0<|z_0|<1)$ if and only if $f(z)$ is of the form $z/(1+\varepsilon z)$ with $\varepsilon=\pm|z_0|/z_0$.

On the other hand, problems on conformal mappings of multiply connected domains involve essential difficulties in comparison with the simply connected case. Although Bieberbach's method is unsuitable for multiply connected domains, Löwner's method, the method of contour integration, the variational method, and the method of the extremal metric remain useful (→ 77 Conformal Mappings).

E. Multivalent Functions

Multivalent functions are a natural generalization of univalent functions. There are several results that generalize classical results on univalent functions.

A function $f(z)$ that attains every value at most p times and some values exactly p times in a domain D is said to be **p-valent** in D and is called a **multivalent function** provided that $p>1$. In order for $f(z)=\sum_{n=0}^{\infty} a_n z^n$, holomorphic in $|z| \leqslant 1$, to be p-valent there, it is sufficient that it satisfies

$$p-1<\operatorname{Re}(zf'(z)/f(z))<p+1$$

on $|z|=1$. Hence it suffices to have

$$|a_p|-\sum_{n=2}^{p} n|a_{p+1-n}|>\sum_{n=2}^{\infty} n|a_{p-1+n}|.$$

If $f(z)=(1+a_1 z+a_2 z^2+\ldots)/z^p$ is holomorphic and p-valent in $0<|z| \leqslant 1$, then

$$\frac{d}{dr}\int_0^{2\pi} F(|f(re^{i\theta})|)\, d\theta \leqslant 0$$

for any increasing function $F(\rho)$ in $\rho \geqslant 0$. In particular, if $F(\rho)=\rho^2$, this becomes an area theorem from which follow coefficient estimates, etc., for p-valent functions.

Various subfamilies and generalized families of multivalent functions have been considered. Let $f(z)$ be p-valent in D, and $c_0+c_1 z+\ldots+c_{p-1} z^{p-1}+c_p f(z)$ be at most p-valent in D for any constants $c_0, c_1, \ldots, c_p$. Then $f(z)$ is said to be **absolutely p-valent** in D. If a function $f(z)$ holomorphic in a convex domain K satisfies $\operatorname{Re}(e^{i\alpha} f^{(p)}(z))>0$ for a real constant α, then $f(z)$ is absolutely p-valent in K. If $f(z)$ is absolutely p-valent in D, then

$$\left(\sum_{k=0}^{p-1} b_k z^k+b_p f(z)\right) \Bigg/ \left(\sum_{k=0}^{p-1} c_k z^k+c_p f(z)\right)$$

is at most p-value in D for any constants b_k and c_k.

If $f(z)$ is p-valent in the common part of a domain D and the disk centered at each point of D with a fixed radius ρ, then $f(z)$ is said to be **locally p-valent** in D, and ρ is called its **modulus**. A necessary and sufficient condition for $f(z)$, holomorphic in D, to be at most locally p-valent is that $f'(z), \ldots, f^{(p)}(z)$ not vanish simultaneously. In order for $f(z)$, holomorphic in D, to be **locally absolutely p-valent** it is necessary and sufficient that $f^{(p)}(z) \neq 0$. Let the number of $Re^{i\varphi}$-points of $f(z)$ in D be $n(D, Re^{i\varphi})$. If $f(z)$ satisfies

$$\frac{1}{2\pi}\int_0^{2\pi} n(D, Re^{i\varphi})\, d\varphi \leqslant p,$$

for any $R>0$, it is said to be **circumferentially mean p-valent** in D. If $f(z)$ satisfies

$$\int_0^R \int_0^{2\pi} n(D, Re^{i\varphi}) R\, dR\, d\varphi \leqslant p\pi R^2,$$

it is said to be **areally mean ρ-valent** in D. If $f(z)^q$ with $q>1$ is areally mean p-valent in D, then $f(z)$ is areally mean p/q-valent in D. For

$f(z)=(1+a_1 z+a_2 z^2+\ldots)/z^\lambda$ holomorphic and areally mean λ-valent in $0<|z|\leqslant 1$, the following area theorem holds:

$$\sum_{n=1}^{\infty}(n-1)|a_n|^2\leqslant\lambda.$$

Let E be a set containing at least three points. If $f(z)$ in D attains every value of E at most p times and a certain value of E exactly p times (it may attain values outside E more than p times), then $f(z)$ is said to be **quasi-p-valent** in D. If $w=f(z)$ is p-valent in D and $g(w)$ is quasi-q-valent in $f(D)$, then $g(f(z))$ is at most quasi-pq-valent in D.

The first success in obtaining sharp inequalities for multivalent functions was attained by Hayman. In his work, an essential role was played by the method of †symmetrization. For instance, he obtained the following result. If $f(z)=z^p+a_{p+1}z^{p+1}+\ldots$ is holomorphic and circumferentially mean p-valent in $|z|<1$, then $|a_{p+1}|\leqslant 2p$, and for $|z|=r$, $0<r<1$,

$$\frac{r^p}{(1+r)^{2p}}\leqslant|f(z)|\leqslant\frac{r^p}{(1-r)^{2p}},$$

$$|f'(z)|\leqslant\frac{p(1+r)}{r(1-r)}|f(z)|\leqslant\frac{pr^{p-1}(1+r)}{(1-r)^{2p+1}}.$$

References

[1] L. Bieberbach, Über die Koeffizienten derjenigen Potenzreihen, welche eine schlichte Abbildung des Einheitskreises vermitteln, S.-B. Preuss., Akad. Wiss., (1916), 940–955.
[2] K. Löwner, Untersuchungen über schlichte konforme Abbildungen des Einheitskreises I, Math. Ann., 89 (1923), 103–121.
[3] P. R. Garabedian and M. Schiffer, A proof of the Bieberbach conjecture for the fourth coefficient, J. Rational Mech. Anal., 4 (1955), 427–465.
[4] M. Ozawa, On the Bieberbach conjecture for the sixth coefficient, Kôdai Math. Sem. Rep., 21 (1969), 97–128.
[5] R. N. Pederson, A proof of the Bieberbach conjecture for the sixth coefficient, Arch. Rational Mech. Anal., 31 (1968–69), 331–351.
[6] R. N. Pederson and M. Schiffer, A proof of the Bieberbach conjecture for the fifth coefficient, Arch. Rational Mech. Anal., 45 (1972), 161–193.
[7] W. K. Hayman, The asymptotic behavior of p-valent functions, Proc. London Math. Soc., (3) 5 (1955), 257–284.
[8] A. C. Schaeffer and D. C. Spencer, Coefficient regions for schlicht functions, Amer. Math. Soc. Colloq. Publ., 35 (1950).
[9] G. M. Golusin, Geometric theory of functions of a complex variable, Amer. Math. Soc. Transl. of Math. Monographs, 26 (1969). (Original in Russian, 1952.)
[10] W. K. Hayman, Multivalent functions, Cambridge Univ. Press, 1958.
[11] J. A. Jenkins, Univalent functions and conformal mapping, Springer, second edition, 1965.
[12] L. V. Ahlfors, Conformal invariants: Topics in geometric function theory, McGraw-Hill, 1973.
[13] C. Pommerenke, Univalent functions (with a chapter on quadratic differentials by G. Jensen), Vandenhoeck & Ruprecht, 1975.
[14] G. Schober, Univalent functions—Selected topics, Lecture notes in math. 478, Springer, 1975.
[15] P. L. Duren, Coefficients of univalent functions, Bull. Amer. Math. Soc., 83 (1977), 891–911.

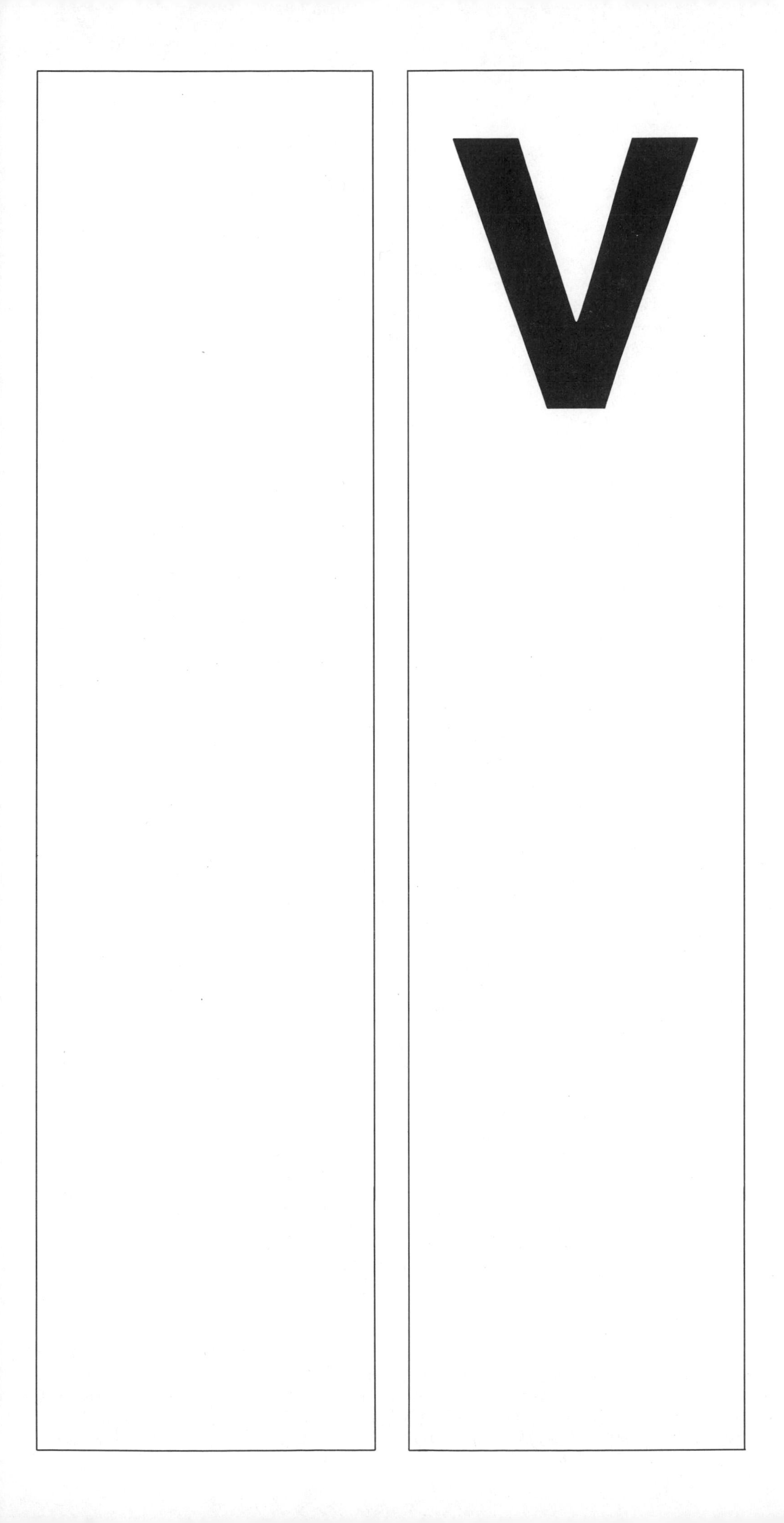
V

439 (III.19) Valuations

A. Introduction

There are two related kinds of valuations, additive (→ Section B) and multiplicative (→ Section C). The notion of valuations, originally defined on (commutative) †fields, has been extended to more general cases (→ Section K); however, we first consider the case of fields.

B. Additive Valuations

In this article, we mean by an **ordered additive group** a **totally ordered additive group**, namely, a commutative group whose operation is addition, which is a †totally ordered set satisfying the condition that $a \geqslant b$ and $c \geqslant d$ imply $a + c \geqslant b + d$ and $-a \leqslant -b$. Suppose that we are given a field K, an ordered additive group G, and an element ∞ defined to be greater than any element of G. Then a mapping $v: K \rightarrow G \cup \{\infty\}$ is called an **additive valuation** (or simply a **valuation**) of the field K if v satisfies the following three conditions: (i) $v(a) = \infty$ if and only if $a = 0$; (ii) $v(ab) = v(a) + v(b)$ for all a, $b \neq 0$; and (iii) $v(a+b) \geqslant \min\{v(a), v(b)\}$.

The set $\{v(a) \mid a \in K - \{0\}\}$ is a submodule of G and is called the **value group** of v, while the set $R_v = \{a \in K \mid v(a) \geqslant 0\}$ is a subring of K and is called the **valuation ring** of v. The ring R_v has only one †maximal ideal $\{a \mid v(a) > 0\}$, called the **valuation ideal** of v (or of R_v), and the †residue class field of R_v modulo the maximal ideal is called the **residue class field** of the valuation v. We have $v(a) \leqslant v(b)$ if and only if $aR_v \supset bR_v$. Two valuations v and v' of the field K are said to be **equivalent** when $v(a) \leqslant v(b)$ if and only if $v'(a) \leqslant v'(b)$; hence v and v' are equivalent if and only if $R_v = R_{v'}$. The **rank** of v is defined to be the †Krull dimension of the valuation ring R_v, and the **rational rank** of v to be the maximum (or supremum) of the numbers of linearly independent elements in the value group. An **extension** (or **prolongation**) of v in a field K' containing K is a valuation v' of K' whose restriction on K is v; such an extension exists for any given v and K'. Sometimes a valuation of rank 1 is called a **special valuation** (or **exponential valuation**), and a valuation of a general rank is called a **generalized valuation**. On the other hand, if k is a subfield of K such that $v(a) = 0$ for every nonzero element a of k, then v is called a **valuation over the subfield** k.

C. Multiplicative Valuations

A **multiplicative valuation** (or **valuation**) of a field K is a mapping $w: K \rightarrow \Gamma \cup \{0\}$ that satisfies the following three conditions, where Γ is the multiplicative group of positive real numbers: (i) $w(a) = 0$ if and only if $a = 0$; (ii) $w(ab) = w(a)w(b)$; and (iii) $w(a+b) \leqslant C(w(a) + w(b))$, where C is a constant (independent of the choice of a and b, but dependent on the choice of w).

The **value group** of w is defined to be $\{w(a) \mid a \in K - \{0\}\}$. Extensions of a valuation and equivalence of valuations are defined as in the case of additive valuations. Thus w' is equivalent to w if and only if there is a positive r such that for all $a \in K$, $w(a) = w'(a)^r$. In each equivalence class of valuations of a field, there exists a valuation for which the constant C in condition (iii) can be taken to be 1. A valuation w is said to be a **valuation over a subfield** k if $w(a) = 1$ for any nonzero element a of k.

We call w an **Archimedean valuation** if for any elements $a, b \in K$, $a \neq 0$, there exists a natural number n such that $w(na) > w(b)$; otherwise, w is said to be a **non-Archimedean valuation**. If w is an Archimedean valuation of a field K, then there is an injection σ from K into the complex number field $\mathbf{C}$ such that w is equivalent to the valuation w' defined by $w'(a) = |\sigma(a)|$. If w is a non-Archimedean valuation of a field K, then $w(a+b) \leqslant \max\{w(a), w(b)\}$. Hence in this case we get an additive valuation v of K when we define $v(a) = -\log w(a)$ $(a \in K)$, and either v is of rank 1 or $v(K) = \{1, 0\}$ (in the latter case, v is called **trivial**). Conversely, every additive valuation of rank 1 of K is equivalent to an additive valuation obtained in this way from a non-Archimedean valuation. (This is why an additive valuation of rank 1 is called an exponential valuation.) Therefore a non-Archimedean valuation determines a valuation ring and valuation ideal in a natural manner. Thus we can identify a non-Archimedean valuation with an additive valuation of rank 1.

D. Topology Defined by a Valuation

Let w be a multiplicative valuation of a field K. When the †distance between two elements a, b of K is defined by $w(a-b)$, K becomes a †topological field. (Although this distance may not make K into a †metric space, there exists a valuation w' equivalent to the valuation w such that K becomes a metric space with respect to the distance $w'(a-b)$ between a and b $(a, b \in K)$.) If K is †complete under the topology, then we say that K is **complete** with respect to w and w is **complete** on K. On the other hand, suppose that w' is an extension of w in a field K' containing K. If w' is complete and K is †dense in K' under the topology defined by w', then we say that the valuation w' is a **completion** of w and that the field K' is a

completion of K with respect to w. For any w, a completion exists and is unique up to isomorphism. When w is a non-Archimedean valuation, the valuation ring of the completion of w is called the **completion** of the valuation ring of w.

When v is an additive valuation of a field K, we can introduce a topology on K by taking the set of all nonzero ideals of the valuation ring R_v of v as a †base for the neighborhood system of zero. Important cases are given by valuations of rank 1, which are the same as those given by non-Archimedean valuations.

If w is a complete non-Archimedean valuation of a field K, then the valuation ring R_w of w is a †Hensel ring, which implies that if K' is a finite algebraic extension of K such that $[K':K]=n$, then w is uniquely extendable to a valuation w' of K' and $w'(a)^n=w(N(a))$, where N is the †norm $N_{K'/K}$.

E. Discrete Valuations

For a non-Archimedean valuation (or an additive valuation of rank 1) w, if the valuation ideal of w is a nonzero †principal ideal generated by an element p, then we say that p is a **prime element** for w, w is a **discrete valuation**, and the valuation ring for w is a **discrete valuation ring**. The condition on the valuation ideal of w holds if and only if the value group of w is a discrete subgroup of the (multiplicative) group Γ of positive real numbers. In the terminology of additive valuations, a valuation w is discrete if and only if it is equivalent to a valuation w' whose value group is the additive group of integers. Such a valuation w' is called a **normalized valuation** (or **normal valuation**). However, we usually mean normalization of a discrete non-Archimedean valuation as in Section H. Sometimes an additive valuation whose value group is isomorphic to the direct sum of a finite number of copies of $\mathbf{Z}$ (the additive group of integers) with a natural †lexicographic order is called a discrete valuation. Concerning a complete discrete valuation w, it is known that if the valuation ring of w contains a field, then it is isomorphic to the ring of †formal power series in one variable over a field (for other cases → 449 Witt Vectors A).

F. Examples

(1) **Trivial valuations** of a field K are the additive valuation v of K such that $v(a)=0$ for all $a\in K-\{0\}$ and the multiplicative valuation w of K such that $w(a)=1$ for all $a\in K-\{0\}$.

(2) If K is isomorphic to a subfield of the complex number field, then we get an Archimedean valuation using the absolute value, and as stated in Section C, every Archimedean valuation of K is equivalent to a valuation obtained in this way.

(3) Let $\mathfrak{p}$ be a †prime ideal of a †Dedekind domain R, $\pi\in\mathfrak{p}$ be such that $\pi\notin\mathfrak{p}^2$, and K be the field of quotients of R. Then each nonzero element α of K can be expressed in the form $\pi^r ab^{-1}$ $(r\in\mathbf{Z}; a, b\in R; a, b\notin\mathfrak{p})$, where r, the **degree** of α with respect to $\mathfrak{p}$, is uniquely determined by α. Hence, letting c be a constant greater than 1, we obtain a non-Archimedean valuation w defined by $w(\alpha)=c^{-r}$. This valuation w is called a **$\mathfrak{p}$-adic valuation**. We also get an additive valuation v defined by $v(\alpha)=r$, called a **$\mathfrak{p}$-adic exponential valuation**. The completion $K_{\mathfrak{p}}$ of K with respect to v is called the **$\mathfrak{p}$-adic extension** of K. If K is a finite †algebraic number field, the $K_{\mathfrak{p}}$ is called a **$\mathfrak{p}$-adic number field**. If $\mathfrak{p}$ is generated by an element p, then "$\mathfrak{p}$-adic" is replaced by "p-adic." For instance, given a rational prime number p, we have a **p-adic valuation** of the rational number field $\mathbf{Q}$, and we obtain the p-adic extension $\mathbf{Q}_p$ of $\mathbf{Q}$, which is called the **p-adic number field**. Every nonzero element α of $\mathbf{Q}_p$ can be written as a uniquely determined expansion $\sum_{n=r}^{\infty} a_n p^n$ $(a_r\neq 0, r\in\mathbf{Z}, a_n\in Z, 0\leqslant a_n<p)$. Then we obtain a valuation v of $\mathbf{Q}_p$ defined by $v(\alpha)=r$. This valuation v is a discrete additive valuation, and $\mathbf{Q}_p$ is complete with respect to v. The valuation ring of v is usually denoted by $\mathbf{Z}_p$, which is called the **ring of p-adic integers**. Each element of $\mathbf{Q}_p$ $(\mathbf{Z}_p)$ is called a **p-adic number** (**p-adic integer**).

(4) Consider the field of †power series $k((t))$ in one variable t over a field k. For $0\neq\alpha\in k((t))$, we define $v(\alpha)=r$ if $\alpha=\sum_{n=r}^{\infty} a_n t^n$ $(a_n\in k, a_r\neq 0)$. Then v is a discrete valuation of $k((t))$, and $k((t))$ is complete with respect to this valuation.

(5) Let v be an additive valuation of a field K with the valuation ring R_v and the valuation ideal $\mathfrak{m}_v$. Let v' be an additive valuation of the field $R_v/\mathfrak{m}_v$ with the valuation ring $R_{v'}$. Then $R''=\{a\in R_v \mid (a \bmod \mathfrak{m}_v)\in R_{v'}\}$ is a valuation ring of K. A valuation v'' whose valuation ring coincides with R'' is called the **composite** of v and v'.

G. The Approximation Theorem and the Independence Theorem

The **approximation theorem** states: Let $w_1,\ldots,w_n$ be mutually nonequivalent and nontrivial multiplicative valuations of a field K. Then for any given n elements $a_1,\ldots,a_n$ of K and a positive number ε, there exists an element a of K such that $w_i(a-a_i)<\varepsilon$ $(i=1,2,\ldots,n)$.

From this follows the **independence theorem**: Let $e_1, \ldots, e_n$ be real numbers, and let w_i and K be as in the approximation theorem. If $\prod_i w_i(a)^{e_i}=1$ for all $a \in K-\{0\}$, then $e_1=\ldots=e_n=0$.

Similar theorems hold for additive valuations. The following independence theorem is basic: Let $v_1, \ldots, v_n$ be additive valuations of a field K, $R_1, \ldots, R_n$ their valuation rings, and $\mathfrak{m}_1, \ldots, \mathfrak{m}_n$ their maximal ideals. Let $D=\bigcap_i R_i$, $\mathfrak{p}_i=\mathfrak{m}_i \cap D$, and consider the rings of quotients $D_{\mathfrak{p}_i}$. Then $D_{\mathfrak{p}_i}=R_i$. If $R_i \not\subset R_j$ (for $i \neq j$), then D has exactly n maximal ideals $\mathfrak{p}_1, \ldots, \mathfrak{p}_n$.

H. Prime Divisors

Let K be an †algebraic number field (algebraic function field of one variable over a field k). An equivalence class of nontrivial multiplicative valuations (over k) is called a **prime divisor** **(prime spot)** of K.

If K is an algebraic number field of degree n, there are exactly n mutually distinct injections $\sigma_1, \ldots, \sigma_n$ of K into the complex number field $\mathbf{C}$. We may assume that $\sigma_i(K)$ is contained in the real number field if and only if $i \leqslant r_1$ and $\sigma_{n-i+1}(a)$ and $\sigma_{r_1+i}(a)$ are conjugate complex numbers $(n-r_1 \geqslant i>0,\ a \in K)$. For $i \leqslant r_1$, let $v_i(a)=|\sigma_i(a)|$, and for $1 \leqslant i \leqslant (n-r_1)/2$, let $v_{r_1+i}(a)=|\sigma_{r_1+i}(a)|^2$. Then $v_1, \ldots, v_{r_1+r_2}$ $(r_2=(n-r_1)/2)$ is a maximal set of mutually nonequivalent Archimedean valuations of K. Equivalence classes of $v_1, \ldots, v_{r_1}$ are called **real (infinite) prime divisors**, and those of $v_{r_1+1}, \ldots, v_{r_1+r_2}$ are called **imaginary (infinite) prime divisors**; all of them are called **infinite prime divisors**. An equivalence class of non-Archimedean valuations of K is called a **finite prime divisor**.

An Archimedean valuation of K is said to be **normal** if it is one of the valuations v_i. If v is non-Archimedean, then v is a $\mathfrak{p}$-adic valuation, where $\mathfrak{p}$ is a prime ideal of the principal order $\mathfrak{o}$ of K (→ Section F, example (3)). Hence if a is an element of K, there exists a constant c $(c>1)$ such that $v(a)=c^{-r}$, where r is the degree of a with respect to $\mathfrak{p}$. In particular, if c is the norm of $\mathfrak{p}$ (i.e., c is the cardinality of the set $\mathfrak{o}/\mathfrak{p}$), then the valuation v is called **normal**. Any finite prime divisor is represented by a normal valuation. Then we have the **product formula** $\prod_w w(a)=1$ for all $a \in K-\{0\}$, where w ranges over all normal valuations of K.

For a function field, a **normal valuation** is defined similarly, using e^f instead of the norm of $\mathfrak{p}$, where e is a fixed real number greater than 1 and f is the degree of the residue class field of the valuation over k. In this case we also have the product formula.

I. Extending Valuations to an Algebraic Extension of Finite Degree

Assume that a field K' is a finite algebraic extension of a field K. Let v be an additive valuation of K and v' be an extension of v to K'. We denote the valuation rings, valuation ideals, and value groups of v and v' by R_v, $R_{v'}$, $\mathfrak{m}_v$, $\mathfrak{m}_{v'}$, and G, G', respectively. Then the degree of the extension $f_{v'}=[R_{v'}/\mathfrak{m}_{v'}:R_v/\mathfrak{m}_v]$ is called the **degree** of v' over v. The group index $e_{v'}=[G':G]$ is called the **ramification index** of v' over v. If v' ranges over all extensions of v in K', then the sum $\sum f_{v'}e_{v'}$ is not greater than $[K':K]$ and the equality holds when v is a discrete valuation and either K' is †separable over K or v is complete.

J. Places

Let k, K, and L be fields, and suppose that $k \subset K$. Let f be a mapping of K onto $L \cup \{\infty\}$ such that $f(ab)=f(a)f(b)$ and $f(a+b)=f(a)+f(b)$, whenever the right member is meaningful, and such that the restriction of f to k is an injection. Here ∞ is an element adjoined to L and satisfying $\infty+a=a+\infty=\infty$, $\infty a=a\infty=\infty$ (for any nonzero element a of K), $1/\infty=0$, and $1/0=\infty$. Then f is called a **place** of K over k. In this case $R=\{x \in K \mid f(x) \neq \infty\}$ is a valuation ring of K containing k. Let $\mathfrak{m}$ be the maximal ideal of R. Then f can be identified with the mapping $g: K \rightarrow R/\mathfrak{m} \cup \{\infty\}$ defined as follows: If $a \in R$, then $g(a)=(a \bmod \mathfrak{m})$; otherwise, $g(a)=\infty$. Places of K over k can be classified in a natural way, and there exists a one-to-one correspondence between the set of classes of places of K over k and the set of equivalence classes of additive valuations over k. When K is an †algebraic function field, we usually consider the case where k is the †ground field. Then if $a_1, \ldots, a_n \in R$, $(a_1, \ldots, a_n) \rightarrow (g(a_1), \ldots, g(a_n))$ gives a †specialization of points over k. Conversely, if $a_i, b_j \in K$ are such that $(a_1, \ldots, a_n) \rightarrow (b_1, \ldots, b_n)$ is a specialization over k, then there is a place f of K over k such that $(b_1, \ldots, b_n)$ is isomorphic to $(f(a_1), \ldots, f(a_n))$ (usually there are infinitely many such f's).

K. Pseudovaluations

A **pseudovaluation** φ of a ring A (not necessarily commutative) is a mapping of A into the set of nonnegative real numbers satisfying the following four conditions: (i) $\varphi(a)=0$ if and only if $a=0$; (ii) $\varphi(ab) \leqslant \varphi(a)\varphi(b)$; (iii) $\varphi(a+b) \leqslant \varphi(a)+\varphi(b)$; and (iv) $\varphi(-a)=\varphi(a)$. These conditions are weaker than those for multiplicative valuations, but with them a topology can be introduced into A as in Section D, with respect to which A becomes a topological ring.

L. History

The theory of valuations was originated by K. Hensel when he introduced p-adic numbers and applied them to number theory [1]. J. Kürschák (*J. Reine Angew. Math.*, 142 (1913)) first treated the theory of multiplicative valuations axiomatically; it was then developed remarkably by A. Ostrowski (*Acta Math.*, 41 (1918)). However, in their theory condition (iii) (→ Section C) was given only in the case $C=1$, thus excluding the normal valuation of an imaginary prime divisor in an algebraic number field. A valuation with general C was introduced by E. Artin [3]. The theory of additive valuations was originated by W. Krull (*J. Reine Angew. Math.*, 167 (1932)), although the concept of exponential valuations existed before. The theory of valuations is used to simplify †class field theory and the theory of algebraic function fields in one variable. For these purposes, the notion of multiplicative valuations is sufficient (→ 9 Algebraic Curves; 59 Class Field Theory). The idea is also used in the theory of normal rings and in algebraic geometry, for both of which the concept of additive valuations is also necessary. Pseudovaluations were used by M. Deuring (*Erg. Math.*, Springer, 1935) in the arithmetic of algebras.

References

[1] K. Hensel, Theorie der algebraischen Zahlen, Teubner, 1908.
[2] O. F. G. Schilling, The theory of valuations, Amer. Math. Soc. Math. Surveys, 1950.
[3] E. Artin, Algebraic numbers and algebraic functions, Gordon & Breach, 1967.
[4] O. Zariski and P. Samuel, Commutative algebra II, Van Nostrand, 1960.
Also → references to 67 Commutative Rings.

440 (X.35) Variational Inequalities

A. Introduction

Variational inequalities arise when we consider extremal problems of functionals under **unilateral constraints**. Some problems in physics and engineering are studied by formulating them as elliptic, parabolic, and hyperbolic variational inequalities [1–8].

B. Stationary Variational Inequality

Let D be a bounded domain in m-dimensional Euclidean space and $f\in L_2(D)$ be a given real-valued function. Consider the variational problem of minimizing the following functional J with the argument function v:

$$J[v]=\int_D |\operatorname{grad} v|^2\,dx-2\int_D fv\,dx.$$

Here, we suppose the set of admissible functions to be the closed convex subset

$$K=\{v\in H_0^1(D)\mid v\leqslant 0 \text{ a.e. in } D\}$$

of the Hilbert space $H_0^1(D)$ (→ 168 Function Spaces). It can be shown by choosing a minimizing sequence that there exists a minimum value of J which is realized by a unique $u\in K$. Since the stationary function u belongs to $H_0^1(D)$, it can be shown that the boundary condition $u|_{\partial D}=0$ is satisfied in the sense that the †trace $\gamma_0 u\in H^{1/2}(\partial D)$ (→ 224 Interpolation of Operators) of u on ∂D vanishes a.e. on ∂D. In view of the fact that $J[u]\leqslant J[v]$ is valid for any $v\in K$, it can be verified that the **stationary variational inequality**

$$\left.\begin{array}{l}-\Delta u-f\leqslant 0\\ u\leqslant 0\\ (-\Delta u-f)\cdot u=0\end{array}\right\} \qquad (1)$$

is satisfied in D in the sense of differentiation of distributions (→ 125 Distributions and Hyperfunctions). The problem (1) is a **Dirichlet problem with obstacle**. Moreover, we can prove the regularity of $u\in H^2(D)$ under an assumption of suitable smoothness for ∂D by establishing the boundedness of the solutions u_ε in $H^2(D)$ of the **penalized problems** associated with (1):

$$-\Delta u_\varepsilon+\frac{1}{\varepsilon}u_\varepsilon^+=f \qquad (\varepsilon>0),$$

$$u_\varepsilon|_{\partial D}=0.$$

Here we note that the u_ε are the stationary functions of the ordinary variational problems of minimization in $H_0^1(D)$ of the functionals

$$J_\varepsilon[v]=\int_D |\operatorname{grad} v|^2\,dx-2\int_D fv\,dx+\frac{1}{\varepsilon}\int |v^+|^2\,dx$$

with the **penalty term** (the third term of the right-hand side of the equality above). We have thus found that the stationary variational inequality (1) is the Euler equation of a conditional problem of variation (→ 46 Calculus of Variations).

C. Variational Inequality of Evolution

Let $\psi\in H^1(D)$ be a given function on D such that $\psi|_{\partial D}\geqslant 0$ and $\Delta\psi\in L_2(D)$. The **variational**

inequality of evolution

$$\frac{\partial u}{\partial t} - \Delta u \leqslant 0,$$

$$u \leqslant \psi,$$

$$\left(\frac{\partial u}{\partial t} - \Delta u\right) \cdot (u - \psi) = 0 \quad (t > 0, x \in D),$$

$$u(0, x) = a(x) \qquad (x \in D),$$

$$u(t, x)|_{\partial D} = 0 \qquad (t > 0)$$

can be formulated as an abstract Cauchy problem (→ 286 Nonlinear Functional Analysis X)

$$\frac{du}{dt} \in Au \qquad (t > 0),$$

$$u(+0) = a$$

in a Hilbert space with a multivalued operator $A = -\partial\varphi$, where $\partial\varphi$ is the subdifferential of the following lower semicontinuous proper convex function on the Hilbert space $L_2(D)$:

$$\varphi(v) = \begin{cases} \dfrac{1}{2}\displaystyle\int_D |\mathrm{grad}\, v|^2 \, dx & \text{if } v \in H_0^1(D) \text{ and } v \leqslant \psi, \\ +\infty & \text{otherwise.} \end{cases}$$

Thus the solution u is given by the vector-valued function

$$u(t) = e^{tA} a.$$

Here e^{tA} is the †nonlinear semigroup generated by A (→ 88 Convex Analysis, 378 Semigroups of Operators and Evolution Equations).

D. Optimal Stopping Time Problem and Variational Inequalities

Let $\{X_t\}_{t \geqslant 0}$ be an m-dimensional Brownian motion (→ 45 Brownian Motion) and consider the problem of finding a †stopping time σ that minimizes

$$J_x[\sigma] = \mathrm{E}_x\left(\int_0^\sigma f(X_t)\, dt\right) \qquad (x \in \mathbf{R}^m)$$

under the restriction that $0 \leqslant \sigma \leqslant \sigma_{\partial D}$, where $\sigma_{\partial D}$ is the †hitting time for the boundary ∂D. Let us define

$$u(x) = \min_\sigma J_x[\sigma].$$

Then the †principle of optimality in dynamic programming gives the stationary variational inequality (1) with Δ replaced by $\frac{1}{2}\Delta$, and we can show by the †Dynkin formula that an optimal stopping time $\hat{\sigma}$ is the hitting time for the set $\{x \in \bar{\Omega} \mid u(x) = 0\}$ (→ 127 Dynamic Programming). We can systematically discuss problems in mathematical programming and operations research by introducing quasivariational inequalities, which are slight generalizations of variational inequalities (→ 227 Inventory Control, 408 Stochastic Programming). The above-mentioned facts are applicable to general †diffusion processes described by †stochastic differential equations (→ 115 Diffusion Processes, 406 Stochastic Differential Equations). We have thus found the relation

free boundary problem ↔ variational inequality
↖↘ optimal stopping time problem ↙↗

(→ 405 Stochastic Control and Stochastic Filtering).

E. Numerical Solution of Variational Inequalities

Since the solution u of the variational inequality (1) is the stationary function for the variational problem, we can apply to the evaluation of the function u numerical methods based on the direct method of the calculus of variations (→ 300 Numerical Methods). The †finite element method, which can be regarded as a type of Ritz-Galerkin method, is extensively employed to calculate numerical solutions. In view of the unilateral constraint $u \leqslant 0$, iteration methods, such as the Gauss-Seidel iteration method, are used with modifications. An algorithm of relaxation with projection is proposed in [3] (→ 304 Numerical Solution of Partial Differential Equations).

References

[1] H. Brézis, Problèmes unilatéraux, J. Math. Pures Appl., 51 (1972), 1–168.
[2] V. Barbu, Nonlinear semigroups and differential equations in Banach spaces, Noordhoff, 1976.
[3] J.-L. Lions, Sur quelques questions d'analyse, de mécanique et de contrôle optimal, Les Presses de l'Université de Montréal, 1976.
[4] J.-L. Lions and G. Stampacchia, Variational inequalities, Comm. Pure Appl. Math., 20 (1967), 493–519.
[5] C. Baiocchi, Su un problema di frontiera libera connesso a questioni di idraulica, Ann. Mat. Pura Appl. 92 (1972), 107–127.
[6] G. Duvaut and J.-L. Lions, Les inéquations en mécanique et en physique, Dunod, 1972; English translation, Springer, 1975.
[7] A. Friedman and D. Kinderlehrer, A one phase Stefan problem, Indiana Univ. Math. J., 24 (1975), 1005–1035.
[8] R. Glowinski, J.-L. Lions, and R. Trémolières, Analyse numérique des inéquations variationelles I, II, Dunod, 1976; English translation, North-Holland, 1981.

441 (XX.3) Variational Principles

A. General Remarks

Among the principles that appear in physics are those expressed not in terms of differential forms but in terms of variational forms. These principles, describing the conditions under which certain quantities attain extremal values, are generally called **variational principles**. Besides Hamilton's principle in classical mechanics (→ Section B) and Fermat's principle in geometric optics (→ Section C), examples are found in †electromagnetism, †relativity theory, †quantum mechanics, †field theory, etc. Independence of the choice of coordinate system is an important characteristic of variational principles. Originally these principles had theological and metaphysical connotations, but a variational principle is now regarded simply as a postulate that precedes a theory and furnishes its foundation. Thus a variational principle is considered to be the supreme form of a law of physics.

B. Mechanics

In 1744 P. L. Maupertuis published an almost theological thesis, dealing with the **principle of least action**. This was the beginning of the search for a single, universal principle of mechanics, contributions to which were made successively by L. Euler, C. F. Gauss, W. R. Hamilton, H. R. Hertz, and others.

Let $\{q_r\}$ be the †generalized coordinates of a system of particles, and consider the integral of a function $L(q_r, \dot{q}_r, t)$ taken from time t_0 to t_1. If we compare the values of the integral taken along any arbitrary path starting from a fixed point P_0 in the coordinate space at time t_0 and arriving at another fixed point P_1 at time t_1, then the actual motion $q_r(t)$ (which obeys the laws of mechanics) is given by the condition that the integral is an †extremum (†stationary value), that is, $\delta\int_{t_0}^{t_1} L\,dt=0$, provided that the function L is properly chosen. This is **Hamilton's principle**, and L is the †Lagrangian function. In †Newtonian mechanics, the †kinetic energy T of a system of particles is expressed as a †quadratic form in $\dot{q}_r$. Furthermore, if the forces acting on the particles can be given by $-\operatorname{grad} V$, where the potential V does not depend explicitly on $\dot{q}_r$, we can choose $L=T-V$. Also, for a charged particle in †special relativity, we can take $L=-m_0c^2(1-v^2/c^2)^{1/2}-e\varphi+e(\mathbf{v}\cdot\mathbf{A})$, where m_0 is the rest mass of the particle, e is the charge, $\mathbf{v}$ is the velocity (with v its magnitude), c is the speed of light in vacuum, and φ and $\mathbf{A}$ are the scalar and vector potentials of the electromagnetic field, respectively.

In general relativity theory, the motion of a particle can be derived from the variational principle $\delta\int ds=0$ (ds is the Riemannian line element). Hence, geometrically, the particle moves along a †geodesic curve in 4-dimensional space-time.

C. Geometric Optics

The path of a light ray between two points P_0 and P_1 (subject to reflection and refraction) is such that the time of transit along the path among all neighboring virtual paths is an extremum (stationary value). This is called **Fermat's principle**. If the index of refraction is n, Fermat's principle can be expressed as $\delta\int_{P_0}^{P_1} n\,ds=0$ (ds is the Euclidean line element). The laws of reflection and refraction of light, as well as the law of rectilinear propagation of light in homogeneous media, can be derived from this principle.

D. Field Theory

Not only the equations of motion of a system of particles, but also various field equations (†Maxwell's equations of the electromagnetic field, †Dirac's equation of the electron field, the meson field equation, the gravitational field equation, etc.) can be derived from variational principles in terms of appropriate Lagrangian functions. In †field theory the essential virtue of the variational principle appears in the fact that the properties of various possible fields as well as conservation laws can be systematically discussed by assuming relativistic invariance and gauge invariance of the Lagrangian functions adopted. In particular, for an electromagnetic field in vacuum, the Lagrangian function density is $L=(\mathbf{H}^2-\mathbf{E}^2)/2$, and the integration is carried out over a certain 4-dimensional domain.

E. Quantum Mechanics

If H is the †Hamiltonian operator for any quantum-mechanical system, the eigenfunction ψ can be determined by the variational principle

$$\delta\int\bar{\psi}H\psi\,d\tau=0,\qquad\text{with}\qquad\int\bar{\psi}\psi\,d\tau=1,$$

where $\bar{\psi}$ is the complex conjugate of ψ and $d\tau$ is the volume element. Based on this variational principle, the †direct method of the calculus of variations is often employed for an approximate numerical calculation of the energy eigenvalues and eigenfunctions. In particular, by restricting the functional form of ψ to the product of one-body wave functions, we can obtain Hartree's equation. A further suitable symmetrization of ψ leads to Fock's equation.

F. Statistical Mechanics

Let φ be a statistical-mechanical state of a system, and let $S(\varphi)$ and $E(\varphi)$ be the state's entropy and energy (mean entropy and mean energy for an infinitely extended system); T is the thermodynamical temperature, and $f(\varphi)=E(\varphi)-TS(\varphi)$ is the free energy. Then the equilibrium state for $T\geqslant 0$ is determined as the state φ that gives the minimum value of $f(\varphi)$ (maximum for $T<0$).

References

[1] R. Courant and D. Hilbert, Methods of mathematical physics, Interscience, I, 1953; II, 1962.
[2] B. L. Moiseiwitsch, Variational principles, Interscience, 1966.
[3] R. Weinstock, Calculus of variations, with applications to physics and engineering, McGraw-Hill, 1952.
Also → references to 46 Calculus of Variations.

442 (VI.12) Vectors

A. Definitions

The **vector** concept originated in physics from such well-known notions as velocity, acceleration, and force. These physical quantities are supplied with length and direction; they can be added or multiplied by scalars. In the Euclidean space E^n (or, in general, an †affine space), a vector **a** is represented by an **oriented segment** $\overrightarrow{pq}$. Two oriented segments $\overrightarrow{p_1q_1}$ and $\overrightarrow{p_2q_2}$ are considered to represent the same vector **a** if and only if the following two conditions are satisfied: (1) The four points p_1, q_1, p_2, q_2 lie in the same plane π. (2) $p_1q_1 // p_2q_2$ and $p_1q_2 // q_1q_2$. Hence a vector in E^n is an equivalence class of oriented segments $\overrightarrow{pq}$, where the equivalence relation $\overrightarrow{p_1q_1}\sim\overrightarrow{p_2q_2}$ is defined by the two conditions just given. Hereafter, we denote the vector by $[\overrightarrow{pq}]$, or simply $\overrightarrow{pq}$. The points p and q are called the **initial point** and **terminal point** of the vector $\overrightarrow{pq}$.

Given a vector $\mathbf{a}=\overrightarrow{pq}$ and a real number λ, we define the **scalar multiple** $\lambda\mathbf{a}$ as the vector $\overrightarrow{pr}$, where r is the point on the straight line containing both p and q such that the ratio $[\overrightarrow{pr}:\overrightarrow{pq}]$ is equal to λ (if $p=q$, then we put $r=p$). The operation $(\lambda,\mathbf{a})\rightarrow\lambda\mathbf{a}$ is called **scalar multiplication**. Given two vectors $\mathbf{a}=\overrightarrow{pq}$ and $\mathbf{b}=\overrightarrow{qs}$, the vector $\mathbf{c}=\overrightarrow{ps}$ is called the **sum** of **a** and **b** and is denoted by $\mathbf{c}=\mathbf{a}+\mathbf{b}$. The vector $\overrightarrow{pp}=\mathbf{0}$ is called the **zero vector**. If $\mathbf{a}=\overrightarrow{pq}$, we put $-\mathbf{a}=\overrightarrow{qp}$.

Scalar multiplication and addition of vectors satisfy the following seven conditions: (1) $\mathbf{a}+\mathbf{b}=\mathbf{b}+\mathbf{a}$ (commutative law); (2) $\mathbf{a}+(\mathbf{b}+\mathbf{c})=(\mathbf{a}+\mathbf{b})+\mathbf{c}$ (associative law); (3) $\mathbf{a}+\mathbf{0}=\mathbf{a}$; (4) for each **a** there is $-\mathbf{a}$ such that $\mathbf{a}+(-\mathbf{a})=\mathbf{0}$; (5) $\lambda(\mathbf{a}+\mathbf{b})=\lambda\mathbf{a}+\lambda\mathbf{b}$, $(\lambda+\mu)\mathbf{a}=\lambda\mathbf{a}+\mu\mathbf{a}$ (distributive laws); (6) $\lambda(\mu\mathbf{a})=(\lambda\mu)\mathbf{a}$ (associative law for scalar multiplication); and (7) $1\mathbf{a}=\mathbf{a}$. Hence the set V of all vectors in E^n forms a †real linear space. Sometimes, a set satisfying (1)–(7), that is, by definition, a linear space, is called a vector space, and its elements are called vectors.

The pair consisting of a vector $\overrightarrow{pq}$ and a specific initial point p of $\overrightarrow{pq}$ is sometimes called a **fixed vector**. An illustration of this is given by the force vector with its initial point being where the force is applied. By contrast, a vector $\overrightarrow{pq}$ is sometimes called a **free vector**. If we fix the origin o in E^n, then for any point p in E^n, the vector $\overrightarrow{op}$ is called the **position vector** of p.

If two vectors $\mathbf{a}=\overrightarrow{op}$ and $\mathbf{b}=\overrightarrow{oq}$ are †linearly dependent, they are sometimes said to be **collinear**. If there vectors $\mathbf{a}=\overrightarrow{op}$, $\mathbf{b}=\overrightarrow{oq}$, and $\mathbf{c}=\overrightarrow{or}$ are linearly dependent, they are sometimes said to be **coplanar**.

If a set of vectors $\mathbf{e}_1,\ldots,\mathbf{e}_n$ forms a †basis of a vector space V, then the vectors $\mathbf{e}_i$ are called **fundamental vectors** in V. Each vector $\mathbf{a}\in V$ is uniquely expressed as $\mathbf{a}=\sum\alpha_i\mathbf{e}_i$ $(\alpha_i\in\mathbf{R})$. We call $(\alpha_1,\ldots,\alpha_n)$ the **components** of the vector **a** with respect to the fundamental vectors $\mathbf{e}_1,\ldots,\mathbf{e}_n$.

B. Inner Product

In the Euclidean space E^n, the length of the line segment $\overline{pq}$ is called the **absolute value** (or **magnitude**) of the vector $\mathbf{a}=\overrightarrow{pq}$ and is denoted by $|\mathbf{a}|$. A vector of length one is called a **unit vector**. For two vectors $\mathbf{a}=\overrightarrow{op}$ and $\mathbf{b}=\overrightarrow{oq}$, the value $(\mathbf{a},\mathbf{b})=|\mathbf{a}|\,|\mathbf{b}|\cos\theta$ is called the **inner product** (or **scalar product**) of **a** and **b**, where θ is the angle $\angle poq$. Instead of $(\mathbf{a},\mathbf{b})$, the notations $\mathbf{a}\cdot\mathbf{b}$, or $\mathbf{ab}$ are also used. If neither vector **a** nor vector **b** is equal to **0**, then $(\mathbf{a},\mathbf{b})=0$ implies $\angle poq=\pi/2$, that is, the orthogonality of the two vectors $\overrightarrow{op}$ and $\overrightarrow{oq}$. If we take an

†orthonormal basis $(\mathbf{e}_1, \ldots, \mathbf{e}_n)$ in E^n (i.e., a set of fundamental vectors with $|\mathbf{e}_i|=1$, $(\mathbf{e}_i, \mathbf{e}_j)=0$ $(i \neq j)$), the inner product of vectors $\mathbf{a}=\sum \alpha_i \mathbf{e}_i$, $\mathbf{b}=\sum \beta_i \mathbf{e}_i$ is equal to $\sum_{i=1}^n \alpha_i \beta_i$. The inner product has the following three properties (i) $(\mathbf{x},\mathbf{x}) \geqslant 0$ and is zero if and only if $\mathbf{x}=\mathbf{0}$; (ii) $(\mathbf{x},\mathbf{y})=(\mathbf{y},\mathbf{x})$; (iii) $(\mathbf{x}_1+\mathbf{x}_2,\mathbf{y})=(\mathbf{x}_1,\mathbf{y})+(\mathbf{x}_2,\mathbf{y})$, $(\alpha\mathbf{x},\mathbf{y})=\alpha(\mathbf{x},\mathbf{y})$ $(\alpha \in \mathbf{R})$. Similar linearity holds for $\mathbf{y}$.

Generally, an $\mathbf{R}$-valued †bilinear form $(\mathbf{x},\mathbf{y})$ on a linear space V satisfying the previous three conditions is also called an inner product. If a linear space V is equipped with an inner product, the space is called an **inner product space** (→ 256 Linear Spaces H; 197 Hilbert Spaces). If V is an inner product space, the absolute value $|\mathbf{x}|$ of $\mathbf{x} \in V$ is defined to be $\sqrt{(\mathbf{x},\mathbf{x})}$.

C. Vector Product

In the 3-dimensional Euclidean space E^3, we take an orthonormal basis $\mathbf{e}_1, \mathbf{e}_2, \mathbf{e}_3$. Let $\mathbf{a}$ and $\mathbf{b}$ be vectors in E^3 whose components with respect to $\mathbf{e}_1, \mathbf{e}_2, \mathbf{e}_3$ are $(\alpha_1, \alpha_2, \alpha_3)$, $(\beta_1, \beta_2, \beta_3)$. The vector

$$\begin{vmatrix}\alpha_2 & \alpha_3\\ \beta_2 & \beta_3\end{vmatrix}\mathbf{e}_1+\begin{vmatrix}\alpha_3 & \alpha_1\\ \beta_3 & \beta_1\end{vmatrix}\mathbf{e}_2+\begin{vmatrix}\alpha_1 & \alpha_2\\ \beta_1 & \beta_2\end{vmatrix}\mathbf{e}_3,$$

which is symbolically written as

$$\begin{vmatrix}\mathbf{e}_1 & \mathbf{e}_2 & \mathbf{e}_3\\ \alpha_1 & \alpha_2 & \alpha_3\\ \beta_1 & \beta_2 & \beta_3\end{vmatrix},$$

is called the **exterior product** or **vector product** of $\mathbf{a}$ and $\mathbf{b}$ and is denoted by $[\mathbf{a},\mathbf{b}]$ or $\mathbf{a} \times \mathbf{b}$. The vector $[\mathbf{a},\mathbf{b}]$ is determined uniquely up to its sign by $\mathbf{a}$ and $\mathbf{b}$ and is independent of the choice of the orthonormal basis.

Suppose that we have $\mathbf{a}=\overrightarrow{op}$, $\mathbf{b}=\overrightarrow{oq}$. Then $|[\mathbf{a},\mathbf{b}]|=|\mathbf{a}|\cdot|\mathbf{b}|\sin\theta$, where $\theta=\angle poq$. Also $|[\mathbf{a},\mathbf{b}]|$ is equal to the area of the parallelogram determined by $\mathbf{a}$ and $\mathbf{b}$. To illustrate the orientation of $[\mathbf{a},\mathbf{b}]$, we sometimes use the idea of a turning screw. That is, the direction of a right-handed screw advancing while turning at o from p to q (within the angle less than 180°) coincides with the direction of $[\mathbf{a},\mathbf{b}]$ (Fig. 1). The exterior product has the following three properties: (1) $[\mathbf{a},\mathbf{b}]=-[\mathbf{b},\mathbf{a}]$ (antisymmetric law); (2) $[\lambda\mathbf{a},\mathbf{b}]=\lambda[\mathbf{a},\mathbf{b}]$ (associative law for scalar multiplication); (3) $[\mathbf{a},\mathbf{b}+\mathbf{c}]=[\mathbf{a},\mathbf{b}]+[\mathbf{a},\mathbf{c}]$ (distributive law). The vector product does not satisfy the associative law, but it does satisfy the †Jacobi identity $[\mathbf{a},[\mathbf{b},\mathbf{c}]]+[\mathbf{b},[\mathbf{c},\mathbf{a}]]+[\mathbf{c},[\mathbf{a},\mathbf{b}]]=\mathbf{0}$. The vector $[\mathbf{a},[\mathbf{b},\mathbf{c}]]$ is sometimes called the **vector triple product**, and for this we have **Lagrange's formula** $[\mathbf{a},[\mathbf{b},\mathbf{c}]]=(\mathbf{a},\mathbf{c})\mathbf{b}-(\mathbf{a},\mathbf{b})\mathbf{c}$.

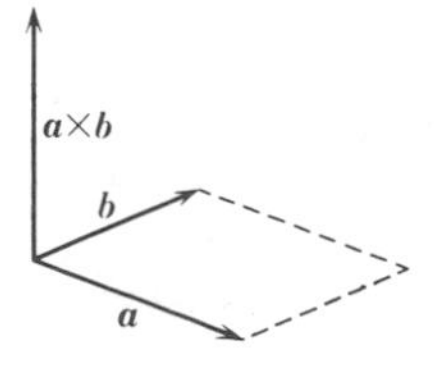

Fig. 1

Let $\mathbf{a}$, $\mathbf{b}$, $\mathbf{c}$ be vectors in E^3 whose components with respect to an orthonormal fundamental basis are $(\alpha_1, \alpha_2, \alpha_3)$, $(\beta_1, \beta_2, \beta_3)$, and $(\gamma_1, \gamma_2, \gamma_3)$. Then $(\mathbf{a},[\mathbf{b},\mathbf{c}])=(\mathbf{b},[\mathbf{c},\mathbf{a}])=(\mathbf{c},[\mathbf{a},\mathbf{b}])=[\mathbf{a},\mathbf{b},\mathbf{c}]$, and the common value is equal to the determinant of the 3×3 matrix

$$\begin{pmatrix}\alpha_1 & \alpha_2 & \alpha_3\\ \beta_1 & \beta_2 & \beta_3\\ \gamma_1 & \gamma_2 & \gamma_3\end{pmatrix}.$$

The value denoted by $[\mathbf{a},\mathbf{b},\mathbf{c}]$ is called the **scalar triple product** of $\mathbf{a}$, $\mathbf{b}$, $\mathbf{c}$ and is equal to the volume of the parallelotope whose three edges are $\mathbf{a}=\overrightarrow{op}$, $\mathbf{b}=\overrightarrow{oq}$, and $\mathbf{c}=\overrightarrow{or}$ with common initial point o. The triple $\mathbf{a}$, $\mathbf{b}$, $\mathbf{c}$ is called a right-hand system or a left-hand system according as $[\mathbf{a},\mathbf{b},\mathbf{c}]$ is positive or negative. We have $[\mathbf{a},\mathbf{b},\mathbf{c}]=0$ if and only if $\mathbf{a}$, $\mathbf{b}$ and $\mathbf{c}$ are coplanar. (For the †exterior product of vectors in E^n and the concept of †p-vectors → 256 Linear Spaces O.)

D. Vector Fields

In this section we consider the case of a 3-dimensional Euclidean space E^3 (for the general case → 105 Differentiable Manifolds). A scalar-valued or a vector-valued function defined on a set D in E^3 is called a **scalar field** or a **vector field**, respectively. The continuity or the differentiability of a vector field is defined by the continuity or the differentiability of its components.

For a differentiable scalar field $f(x,y,z)$, the vector field with the components $(\partial f/\partial x, \partial f/\partial y, \partial f/\partial z)$ is called the **gradient** of f and is denoted by $\mathbf{grad}\, f$. For a differentiable vector field $\mathbf{V}(x,y,z)$ whose components are $(u(x,y,z), v(x,y,z), w(x,y,z))$, the vector field with components

$$\left(\frac{\partial w}{\partial y}-\frac{\partial v}{\partial z}, \frac{\partial u}{\partial z}-\frac{\partial w}{\partial x}, \frac{\partial v}{\partial x}-\frac{\partial u}{\partial y}\right)$$

is called the **rotation** (or **curl**) of $\mathbf{V}$ and is denoted by $\mathbf{rot}\,\mathbf{V}$ (or $\mathbf{curl}\,\mathbf{V}$). Also, for a differentiable vector field $\mathbf{V}$, the scalar field defined by $\partial u/\partial x+\partial v/\partial y+\partial w/\partial z$ is called the **divergence** of $\mathbf{V}$ and is denoted by $\mathbf{div}\,\mathbf{V}$. Utilizing the vector operator ∇ having differential operators $(\partial/\partial x, \partial/\partial y, \partial/\partial z)$ as its components, we may write simply $\operatorname{grad} f=\nabla f$, $\operatorname{div}\mathbf{V}=(\nabla,\mathbf{V})$, $\operatorname{rot}\mathbf{V}=$

$[\nabla, \mathbf{V}]$. The symbol ∇ is called **nabla, atled** (inverse of delta), or **Hamiltonian**.

A vector field $\mathbf{V}$ with $\operatorname{rot} \mathbf{V} = 0$ is said to be **irrotational, (lamellar**, or **without vortex**). A vector field $\mathbf{V}$ with $\operatorname{div} \mathbf{V} = 0$ is said to be **solenoidal** (or **without source**). Thus $\operatorname{grad} f$ is irrotational and $\operatorname{rot} \mathbf{V}$ is solenoidal. In a small neighborhood or in a †simply-connected domain, an irrotational field is a gradient, a solenoidal field is a rotation, and an arbitrary vector field $\mathbf{V}$ is the sum of these two kinds of vector fields: $V = \operatorname{grad} \varphi + \operatorname{rot} \mathbf{u}$ (**Helmholtz theorem**); the function φ is called the **scalar potential** of $\mathbf{V}$, and the vector field $\mathbf{u}$ is called the **vector potential** of $\mathbf{V}$. Furthermore, the operator $\nabla^2 = \nabla\nabla = \operatorname{div} \operatorname{grad} = \partial^2/\partial x^2 + \partial^2/\partial y^2 + \partial^2/\partial z^2$ is called the **Laplace operator** (or **Laplacian**) and is denoted by Δ. A function that satisfies $\Delta\varphi = 0$ is called a †harmonic function. Locally, an irrotational and solenoidal vector field is the gradient of a harmonic function. If $\mathbf{A}$ is a vector field whose components are $(\varphi_1, \varphi_2, \varphi_3)$ (i.e., $\mathbf{A}(\mathbf{v}) = (\varphi_1(\mathbf{v}), \varphi_2(\mathbf{v}), \varphi_3(\mathbf{v}))$), we can let Δ operate on $\mathbf{A}$ by setting $\Delta\mathbf{A} = (\Delta\varphi_1, \Delta\varphi_2, \Delta\varphi_3)$. We then have $\Delta\mathbf{A} = \nabla^2\mathbf{A} = \operatorname{grad} \operatorname{div} \mathbf{A} - \operatorname{rot} \operatorname{rot} \mathbf{A}$.

Suppose that we are given a vector field $\mathbf{V}$ and a curve C such that the vector $\mathbf{V}(p)$ is tangent to the curve at each point $p \in C$. The curve C is the †integral curve of the vector field $\mathbf{V}$ and is called the **vector line** of the vector field $\mathbf{V}$. The set of all vector lines intersecting with a given closed curve C is called a **vector tube**. Given a closed curve C and a vector field $\mathbf{V}$, the †curvilinear integral $\int (\mathbf{V}, ds)$ (where ds is the line element of C) is called the **circulation** (of $\mathbf{V}$) along the closed curve C. A vector field is irrotational if its circulation along every closed curve vanishes; the converse is true in a simply connected domain. Further, let v_n be the †normal component of a vector field $\mathbf{V}$ with respect to a surace S, and let dS be the volume element of the surface. We put $\mathbf{n}dS = d\mathbf{S}$, where $\mathbf{n}$ is the unit normal vector in the positive direction of the surface S. Then the †surface integral $\int v_n dS = \int (\mathbf{V}, d\mathbf{S})$ is called the **vector flux** through the surface S. A vector field whose vector flux vanishes for every closed surface is solenoidal. (For the corresponding formulas → 94 Curvilinear Integrals and Surface Integrals. For generalizations to higher-dimensional manifolds → 105 Differentiable Manifolds; 194 Harmonic Integrals; Appendix A, Table 3.)

References

[1] H. Weyl, Raum, Zeit, Materie, Springer, fifth edition, 1923; English translation, Space, time, matter, Dover, 1952.
[2] S. Banach, Mechanics, Warsaw, 1951.
[3] H. Flanders, Differential forms, with applications to the physical sciences, Academic Press, 1963.
[4] H. K. Nickerson, D. C. Spencer, and N. E. Steenrod, Advanced calculus, Van Nostrand, 1959.

443 (XII.8) Vector-Valued Integrals

A. General Remarks

Integrals whose values are elements (or subsets) of †topological linear spaces are generally called **vector-valued integrals** or **vector integrals**. As in the scalar case, there are vector-valued integrals of Riemann type (→ 37 Banach Spaces K) and of Lebesgue type. In this article we consider only the latter. There are cases where integrands are vector-valued, where measures are vector-valued, and where both are vector-valued. The methods of integration are also divided into the strong type, in which the integrals are defined by means of the original topology of the topological linear space X, and the weak type, in which they are reduced to numerical integrals by applying continuous linear functionals on X. Combining these we can define many kinds of integrals.

Historically, D. Hilbert's †spectral resolution is the first example of vector-valued integrals, but the general theory of vector-valued integrals started only after S. Bochner [1] defined in 1933 an integral of strong type for functions with values in a Banach space with respect to numerical measures. Then G. Birkhoff [2] defined a more general integral by replacing absolutely convergent sums with unconditionally convergent sums. At approximately the same time, N. Dunford introduced integrals equivalent to these. Later, R. S. Phillips (*Trans. Amer. Math. Soc.*, 47 (1940)) generalized the definition to the case where values of functions are in a †locally convex topological linear space, and C. E. Rickart (*Trans. Amer. Math. Soc.*, 52 (1942)) to the case where functions take subsets of a locally convex topological linear space as their values. The theory of integrals of weak type for functions with values in a Banach space and numerical measures was constructed by I. M. Gel'fand [3], Dunford [4], B. J. Pettis [5], and others (1936–1938). N. Bourbaki [6] dealt with the case where integrands take values in a locally convex topological linear space. As for integrals of numerical functions by vector-valued measures, a representative of strong type integrals is the integral of R. G. Bartle, Dunford, and J. T. Schwartz [7] (1955). Weak type integrals have been discussed by Bourbaki [6], D. R. Lewis (*Pacific J. Math.*, 33 (1970)), and I. Kluvánek (*Studia Math.*, 37 (1970)). The bilinear integral of Bartle (*Studia Math.*, 15 (1956)) is typical of integrals in the case where both integrands and measures are vector-valued. For interrelations of these integrals → the papers by Pettis and Bartle cited above and T. H. Hildebrandt's report in the *Bulletin of the American Mathematical Society*, 59 (1953).

Since the earliest investigations [1–3] the main aim of the theory of vector-valued integrals has been to obtain integral representations of vector-valued (set) functions and various linear operators [8]. However, there is the fundamental difficulty of the nonvalidity of the Radon-Nikodým theorem. Whatever definition of integrals we take, the theorem does not hold for vector-valued set functions unconditionally. Many works sought conditions for functions, operators, or spaces such that the conclusion of the theorem would be restored; the works of Dunford and Pettis [9] and Phillips (*Amer. J. Math.*, 65 (1943)) marked a summit of these attempts. Later, after A. Grothendieck's investigations (1953–1956), this problem began to be studied again, beginning in the late 1960s, by many mathematicians (→ J. Diestel and J. J. Uhl, Jr., *Rocky Mountain J. Math.*, 6 (1976); [10]).

Recently, integrals of multivalued vector-valued functions have also been employed in mathematical statistics, economics, control theory, and many other fields. Some contributions are, besides Rickart cited above, G. B. Price (*Trans. Amer. Math. Soc.*, 47 (1940), H. Kudo (*Sci. Rep. Ochanomizu Univ.*, 4 (1953)), H. Richter (*Math. Ann.*, 150 (1963)), R. J. Aumann [11], G. Debreu [12], and M. Hukuhara (*Funkcial. Ekvac.*, 10 (1967)). Furthermore, C. Castaing and M. Varadier [13] have defined weak type integrals of multivalued functions and introduced many results concerning them. In the following we shall give explanations of typical vector-valued integrals with values in a Banach space only.

B. Measurable Vector-Valued Functions

Let $x(s)$ be a function defined on a †σ-finite measure space $(S, \mathfrak{S}, \mu)$ with values in a Banach space X. This is called a **simple function** or **finite-valued function** if there exists a partition of S into a finite number of mutually disjoint measurable sets $A_1, A_2, \ldots, A_n$ in each of which $x(s)$ takes a contant value c_j. Then $x(s)$ can be written as $\sum_{j=1}^{n} c_j \chi_{A_j}(s)$, where $\chi_{A_j}(s)$ is the †characteristic function of A_j. A function $x(s)$ is said to be **measurable** or **strongly measurable** if it is the strong limit of a sequence of simple functions almost everywhere, that is, $\lim_{n\to\infty} \|x_n(s)-x(s)\|=0$ a.e. Then the numerical function $\|x(s)\|$ is measurable. If μ is a †Radon measure on a compact Hausdorff space S, then the measurable functions can be characterized by †Luzin's property (→ 270 Measure Theory I).

A function $x(s)$ is said to be **scalarly measurable** or **weakly measurable** if the numerical function $\langle x(s), x' \rangle$ is measurable for any †continuous linear functional $x' \in X'$. A function $x(s)$ is measurable if and only if it is scalarly measurable and there are a †null set $E_0 \subset S$ and a †separable closed subspace $Y \subset X$ such that $x(s) \in Y$ whenever $s \notin E_0$ (**Pettis measurability theorem**).

C. Bochner Integrals

A measurable vector-valued function $x(s)$ is said to be **Bochner integrable** if the norm $\|x(s)\|$ is †integrable. If $x(s)$ is a Bochner integrable simple function $\sum c_j \chi_{A_j}(s)$, then its Bochner integral is defined by

$$\int_S x(s)\,d\mu = \sum \mu(A_j) c_j.$$

For a general Bochner integrable function $x(s)$ there exists a sequence of simple functions satisfying the following conditions: (i) $\lim_{n\to\infty} \|x_n(s)-x(s)\|=0$ a.e. (ii) $\lim_{n\to\infty} \int_S \|x_n(s)-x(s)\|\,d\mu=0$. Then $\int_S x_n(s)\,d\mu$ converges strongly and its limit does not depend on the choice of the sequence $\{x_n(s)\}$. We call the limit the **Bochner integral** of $x(s)$ and denote it by $\int_S x(s)\,d\mu$ or by (Bn)$\int_S x(s)\,d\mu$ to distinguish it from other kinds of integrals. A Bochner integrable function on S is Bochner integrable on every measurable subset of S. The Bochner integral has the basic properties of Lebesgue integrals, such as linearity, †complete additivity, and †absolute continuity, with absolute values replaced by norms. †Lebesgue's convergence theorem and †Fubini's theorem also hold. However, the Radon-Nikodým theorem does not hold in general (→ Section H). Let T be a †closed linear operator from X to another Banach space Y. If both $x(s)$ and $Tx(s)$ are Bochner integrable, then the integral of $x(s)$ belongs to the domain of T and

$$T\left(\int_S x(s)\,d\mu\right) = \int_S Tx(s)\,d\mu.$$

If, in particular, T is bounded, then the assumption is always satisfied. If μ is the †Lebesgue measure on the Euclidean space $\mathbf{R}^n$, then Lebesgue's differentiability theorem holds for the Bochner integrals regarded as a set function on the regular closed sets (→ 380 Set Functions D).

D. Unconditionally Convergent Series

Let $\sum_{j=1}^{\infty} x_j$ be a series of elements x_j of a Banach space X. It is said to be **absolutely convergent** if $\sum \|x_j\| < \infty$. It is called **unconditionally convergent** if for any rearrangement α the resulting series $\sum x_{\alpha(j)}$ converges strongly. Then the sum does not depend on α. Clearly, an absolutely convergent series is unconditionally convergent. If X is the number space or is finite-dimensional, then the converse holds. However, if X is infinite-dimensional, there is always an unconditionally convergent series which is not absolutely convergent (**Dvoretzky-Rogers theorem**).

A series $\sum x_j$ is unconditionally convergent if and only if each subseries converges weakly (**Orlicz-Pettis theorem**). If $\sum x_j$ is an unconditionally convergent series, then $\sum \langle x_j, x' \rangle$ converges absolutely for any continuous linear functional $x' \in X'$. If X is a Banach space containing no closed linear subspace isomorphic to the †sequence space c_0, then conversely a series $\sum x_j$ converges unconditionally whenever $\sum |\langle x_j, x' \rangle| < \infty$ for any $x' \in X'$ (**Bessaga-Pełczyński theorem**). A Banach space that is †sequentially complete relative to the weak topology, such as a †reflexive Banach space, and a separable Banach space that is the dual of another Banach space, such as l_1 and the †Hardy space $H_1(\mathbf{R}^n)$, satisfy the assumption, while c_0, l_∞, and $L_\infty(\Omega)$ for an infinitely divisible Ω do not. The totality of absolutely convergent series (resp. unconditionally convergent series) in X is identified with the †topological tensor product $l_1 \hat{\otimes} X$ (resp. $l_1 \hat{\hat{\otimes}} X$) (Grothendieck).

E. Birkhoff Integrals

We say that a series $\sum B_j$ of subsets of X converges unconditionally if for any $x_j \in B_j$ the series $\sum x_j$ converges unconditionally. Then $\sum B_j$ denotes the set of such sums. A vector-valued function $x(s)$ is said to be **Birkhoff integrable** if there is a countable partition Δ: $S = \bigcup_{j=1}^{\infty} A_j$ ($A_j \in \mathfrak{S}$, $A_j \cap A_k = \varnothing$ $(j \neq k)$, $\mu(A_j) < \infty$) such that the set $x(A_j)$ of values on A_j are bounded and $\sum \mu(A_j) x(A_j)$ converges unconditionally and if the sum converges to an element of X as the partition is subdivided. The limit is called the **Birkhoff integral** of $x(s)$ and is denoted by $(\mathrm{Bk})\int_S x(s)\,d\mu$ or simply by $\int_S x(s)\,d\mu$. A Birkhoff integrable function is Birkhoff integrable on any measurable set. The Birkhoff integral has, as a set function, complete additivity and absolute continuity in μ. It is linear in the integrand but Fubini's theorem and the Radon-Nikodým theorem do not hold. A Bochner integrable function is Birkhoff integrable, and the integrals coincide. The converse does not hold.

F. Gel'fand-Pettis Integrals

A scalarly measurable function $x(s)$ is said to be **scalarly integrable** or **weakly integrable** if for each $x' \in X'$, $\langle x(s), x' \rangle$ is integrable. Then the linear functional x^* on X' defined by

$$\int_S \langle x(s), x' \rangle \, d\mu = \langle x', x^* \rangle$$

is called the **scalar integral** of $x(s)$. Gel'fand [3] and Dunford [4] proved that x^* belongs to the bidual X''. Hence scalarly integrable functions are often called **Dunford integrable** and the integrals x^* the **Dunford integrals**. More generally, Gel'fand [3] showed that if $x'(s)$ is a function with values in the dual X' of a Banach space X such that $\langle x, x'(s) \rangle$ is integrable for any $x \in X$, then there is an $x' \in X'$ satisfying

$$\int_S \langle x, x'(s) \rangle \, d\mu = \langle x, x' \rangle.$$

This element is sometimes called the **Gel'fand integral** of $x'(s)$. A scalarly integrable function $x(s)$ is scalarly integrable on any measurable subset A. If the scalar integral is always in X, i.e., for each A there is an $x_A \in X$ such that

$$\int_A \langle x, x'(s) \rangle \, d\mu = \langle x_A, x' \rangle, \; x' \in X',$$

then $x(s)$ is said to be **Pettis integrable** or **Gel'fand-Pettis integrable** and x_A is called the **Pettis integral** or **Gel'fand-Pettis integral** on A and is denoted by $(\mathrm{P})\int_A x(s)\,d\mu$ or simply by $\int_A x(s)\,d\mu$. The Pettis integral has complete additivity and absolute continuity as a set function, similarly to the Birkhoff integral. Again, Fubini's theorem and the Radon-Nikodým theorem do not hold. The scalar integral on measurable sets of a scalarly integrable function $x(s)$ is completely additive and absolutely continuous with respect to the †weak* topology of X'' as the dual to X'. It is completely additive or absolutely continuous in the norm topology if and only if $x(s)$ is Pettis integrable (Pettis [5]; [10]). If $x(s)$ is Pettis integrable and $f(s)$ is a numerical function in $L_\infty(S)$, then the product $f(s)x(s)$ is Pettis integrable. Birkhoff integrable functions are Pettis integrable, and the integrals coincide. Conversely, if a measurable function is Pettis integrable, then it is Birkhoff integrable. When X satisfies the Bessaga-Pełczyński condition (→ Section D), a measurable scalarly integrable function is Pettis integrable.

G. Vector Measures

Let Φ be a set function defined on a completely additive class $\mathfrak{S}$ of subsets of the space S and with values in a Banach space X. It is called a **finitely additive vector measure** (resp. a **completely additive vector measure** or simply a **vector measure**) if $\Phi(A_1 \cup A_2)=\Phi(A_1)+\Phi(A_2)$ whenever A_1 and $A_2 \in \mathfrak{S}$ are disjoint (resp. $\Phi(\bigcup_{j=1}^{\infty} A_j)=\sum_{j=1}^{\infty}\Phi(A_j)$ in the norm topology for all $A_j \in \mathfrak{S}$ such that $A_j \cap A_k=\varnothing$ $(j \neq k)$). We remark that the latter sum always converges unconditionally. A set function Φ is completely additive if and only if $\langle \Phi(A), x' \rangle$ is completely additive for all $x' \in X'$ (**Pettis complete additivity theorem**).

Let Φ be a finitely additive vector measure and E be a measurable set. The **total variation** of Φ on E and the **semivariation** of Φ on E are defined by

$$V(\Phi)(E)=\sup \sum_{j=1}^{n} \|\Phi(A_j)\| \tag{1}$$

and

$$\|\Phi\|(E)=\sup \left\| \sum_{j=1}^{n} \alpha_j \Phi(A_j) \right\|, \tag{2}$$

respectively, where the suprema are taken over all finite partitions of E: $E=\bigcup A_j$ $(A_j \in \mathfrak{S}, A_j \cap A_k=\varnothing (j \neq k))$ and all numbers α_j with $|\alpha_j| \leqslant 1$. If $V(\Phi)(S)<\infty$, then Φ is called a measure **of bounded variation.** $\|\Phi\|(S)<\infty$ if and only if $\sup\{\|\Phi(A)\| \mid A \in \mathfrak{S}\}<\infty$. Then Φ is said to be **bounded**. The function $V(\Phi)(E)$ of E is finitely additive but $\|\Phi\|(E)$ is only subadditive: $\|\Phi\|(A \cup B) \leqslant \|\Phi\|(A)+\|\Phi\|(B)$. If Φ is a vector measure of bounded variation, then $V(\Phi)$ is a positive measure. Every vector measure is bounded. A completely additive vector measure on a †finitely additive class $\mathfrak{L}$ can uniquely be extended to a vector measure on the completely additive class $\mathfrak{S}$ generated by $\mathfrak{L}$ (Kluvánek).

Let μ be a positive measure and Φ be a vector measure. Then we have $\Phi(A)\to 0$ as $\mu(A)\to 0$ if and only if Φ vanishes on every A with $\mu(A)=0$. Then Φ is said to be **absolutely continuous** with respect to μ. For every vector measure Φ there is a measure μ such that $\|\Phi\|(A)\to 0$ as $\mu(A)\to 0$ and that $0 \leqslant \mu(A) \leqslant \|\Phi\|(A)$ (Bartle, Dunford, and Schwartz). As a set function, the Bochner integral is a vector measure of bounded variation and the Pettis integral is a bounded vector measure. Both are absolutely continuous with respect to the integrating measure. Let X be $L_p(0,1)$ for $1 \leqslant p \leqslant \infty$, and define $\Phi(E)$ for a Lebesgue measurable set E to be the characteristic function of E. If $p=1$, Φ is a vector measure of bounded variation. If $1<p<\infty$, Φ is a bounded vector measure, but it is not of bounded variation on any set E with $\mu(E)>0$. If $p=\infty$, then Φ is no longer completely additive. These vector measures are absolutely continuous with respect to the Lebesgue measure, but they cannot be represented as the Bochner integral or the Pettis integral.

Let Φ be a vector measure on $\mathfrak{S}$. An $\mathfrak{S}$-measurable numerical function $f(s)$ is said to be Φ-integrable if there exists a sequence of simple functions $f_n(s)$ such that $f_n(s)\to f(s)$ a.e. and that for each $E \in \mathfrak{S}$, $\int_E f_n(s)\,d\Phi$ converges in the norm of X. Then the limit is independent of the choice of f_n. It is called the **Bartle-Dunford-Schwartz integral** and is denoted by $\int_E f(s)\,d\Phi$. Lebesgue's convergence theorem holds for this integral. If Φ is absolutely continuous with respect to the measure μ, then every $f \in L_\infty(\mu)$ is Φ-integrable, and the operator that maps f to $\int_S f\,d\Phi$ is continuous with respect to the weak* topology in $L_\infty(\mu)$ and the weak topology of X. Hence it is a †weakly compact operator. In particular, the range of a vector measure is relatively compact in the weak topology [7]. If Φ is the vector measure of the Pettis integral of a vector-valued function $x(s)$, then the above integral is equal to the Pettis integral of $f(s)x(s)$.

A vector measure Φ is said to be **nonatomic** if for each set A with $\Phi(A)\neq 0$ there is a subset B of A such that $\Phi(B)\neq 0$ and $\Phi(A \setminus B)\neq 0$. If X is finite-dimensional, then the range of a nonatomic vector measure is a compact convex set (**Lyapunov convexity theorem**). This has been generalized to the infinite-dimensional case in many ways, but the conclusion does not hold in the original form (→ Kluvánek and G. Knowles [15]; [10]).

H. The Radon-Nikodým Theorem

As the above examples show, the †Radon-Nikodým theorem does not hold for vector measures in the original form. From 1967 to 1971, M. Metivier, M. A. Rieffel, and S. Moedomo and Uhl improved the classical result of Phillips (1943) and proved the following theorem.

Radon-Nikodým theorem for vector measures. The following conditions are equivalent for μ-absolutely continuous vector measures Φ defined on a finite measure space $(S, \mathfrak{S}, \mu)$: (i) There is a Pettis integrable measurable function $x(s)$ such that

$$\Phi(A)=(\mathrm{P})\int_A x(s)\,d\mu.$$

(ii) For each $\varepsilon>0$ there is an $E\in\mathfrak{S}$ such that $\mu(S\setminus E)<\varepsilon$ and such that $\{\Phi(A)/\mu(A)\mid A\in\mathfrak{S}, A\subset E\}$ is relatively compact. (iii) For each $E\in\mathfrak{S}$ with $\mu(E)>0$ there is a subset F of E with $\mu(F)>0$ such that $\{\Phi(A)/\mu(A)\mid A\in\mathfrak{S}, A\subset F\}$ is relatively weakly compact. Then Φ is of bounded variation if and only if $x(s)$ is Bochner integrable.

On the other hand, since Birkhoff and Gel'fand it has been known that for special Banach spaces X every μ-absolutely continuous vector measure of bounded variation with values in X can be represented as a Bochner integral with respect to μ. Such spaces are said to have the **Radon-Nikodým property**. Separable dual spaces (Gel'fand, Pettis; Dunford and Pettis), reflexive spaces (Gel'fand, Pettis, Phillips), and $l_1(\Omega)$, Ω arbitrary, etc., have the Radon-Nikodým property, while $L_\infty(0,1)$ (Bochner), c_0 (J. A. Clarkson), $L_1(\Omega)$ on a nonatomic Ω (Clarkson, Gel'fand), and $C(\Omega)$ on an infinite compact Hausdorff space Ω, etc., do not. Gel'fand proved that $L_1(0,1)$ (and c_0) is not a dual by means of this fact. From 1967 to 1974, Riefell, H. B. Maynard, R. E. Huff, and W. J. Davis and R. P. Phelps succeeded in characterizing geometrically the Banach spaces with the Radon-Nikodým property. We know today that the following conditions for Banach spaces X are equivalent [10]: (i) X has the Radon-Nikodým property. (ii) Every separable closed linear subspace of X has the Radon-Nikodým property. (iii) Every function $f:[0,1]\to X$ of bounded variation is (strongly or weakly) differentiable a.e. (iv) For any finite measure space $(S,\mathfrak{S},\mu)$ and bounded linear operator $T:L_1(S)\to X$, there is an essentially bounded measurable function $x(s)$ with values in X such that

$$Tf=\int_S f(s)x(s)\,d\mu, \qquad f\in L_1(S).$$

(v) Each nonvoid bounded closed convex set K in X is the †closed convex hull of the set of its strongly exposed points, where a point $x_0\in K$ is called a **strongly exposed point** of K if there is an $x'\in X'$ such that $\langle x_0,x'\rangle>\langle x,x'\rangle$ for all $x\in K\setminus\{x_0\}$ and that any sequence $x_n\in K$ with $\lim\langle x_n,x'\rangle=\langle x_0,x'\rangle$ converges to x_0 strongly.

A Banach space X is said to have the **Kreĭn-Mil'man property** if each bounded closed convex set in X is the closed convex hull of its †extreme points. A Banach space X with the Radon-Nikodým property has the Kreĭn-Mil'man property (J. Lindenstrauss). If X is a dual space, then the converse holds (Huff and P. D. Morris). A Banach space with the Kreĭn-Mil'man property clearly has no closed linear space isomorphic to c_0, but there are Banach spaces that do not contain c_0 and do not have the Kreĭn-Mil'man property. The dual X of a Banach space Y has the Radon-Nikodým property if and only if the dual of every separable closed linear subspace of Y is separable (Uhl, C. Stegall).

I. Integrals of Multivalued Vector Functions

Let $\Gamma(s)$ be a multivalued function defined on a σ-finite complete measure space $(S,\mathfrak{S},\mu)$ with values that are nonempty closed subsets of a separable Banach space X. The inverse image of a subset E of X under $\Gamma(s)$ is, by definition, the set of all s such that $\Gamma(s)\cap E\neq\varnothing$. $\Gamma(s)$ is said to be **measurable** or **strongly measurable** if the inverse image of each open set in X under $\Gamma(s)$ belongs to $\mathfrak{S}$. Let $\mathfrak{B}(X)$ be the †Borel field of X, and $\mathfrak{S}\times\mathfrak{B}(X)$ be the product completely additive class, that is, the smallest completely additive class containing all direct products $A\times B$ of $A\in\mathfrak{S}$ and $B\in\mathfrak{B}(X)$. Then the measurability of $\Gamma(s)$ is equivalent to each of the following: (i) The graph $\{(s,x)\mid x\in\Gamma(s), s\in\mathfrak{S}\}$ of $\Gamma(s)$ belongs to $\mathfrak{S}\times\mathfrak{B}(X)$. (ii) The inverse image of every Borel set in X under $\Gamma(s)$ belongs to $\mathfrak{S}$. (iii) For each $x\in X$, the distance $d(x,\Gamma(s))=\inf\{\|x-y\|\mid y\in\Gamma(s)\}$ between x and $\Gamma(s)$ is measurable as a function on S.

A measurable function $x(s)$ on S with values in X is called a **measurable selection** of $\Gamma(s)$ if $x(s)$ is in $\Gamma(s)$ for all s. (X being separable, we need not discriminate between strong and weak measurability.) The measurability of $\Gamma(s)$ is also equivalent to the following important statement on the existence of measurable selections of $\Gamma(s)$: (iv) There are a countable number of measurable selections $\{x_n(s)\}$ of $\Gamma(s)$ such that the closure of the set $\{x_n(s)\mid n=1,2,\dots\}$ coincides with $\Gamma(s)$ for all $s\in S$. $\Gamma(s)$ is said to be **scalarly measurable** or **weakly measurable** if the support function $\delta'(x',\Gamma(s))=\sup\{\langle x,x'\rangle\mid x\in\Gamma(s)\}$ is measurable on S for all $x'\in X'$. The strong measurability of $\Gamma(s)$ clearly implies the weak one. If the values of $\Gamma(s)$ are nonempty weakly compact convex sets, then the measurabilities are equivalent. Hereafter we shall assume that $\Gamma(s)$ takes the values in the weakly compact convex sets. If the support function $\delta'(x',\Gamma(s))$ is integrable on S for all $x'\in X'$, then $\Gamma(s)$ is said to be **scalarly integrable**. Then the **scalar integral** of $\Gamma(s)$ is defined to be the set in X'' of all scalar integrals of its measurable selections, i.e.,

$$\int_S\Gamma(s)\,d\mu=\left\{\int_S x(s)\,d\mu\;\middle|\;x(s)\text{ is a measurable selection of }\Gamma(s)\right\}.$$

If $\|\Gamma(s)\| = \sup\{\|x\| \mid x \in \Gamma(s)\}$ is integrable, then every measurable selection is Bochner integrable and the integral $\int_S \Gamma(s)\, d\mu$ becomes a nonempty weakly compact convex set in X. When the values of $\Gamma(s)$ are nonempty compact convex sets, there is another method, by G. Debreu, of defining the integral. Let $\mathfrak{L}$ be the class of all nonempty compact convex sets in X and δ be the Hausdorff metric, i.e., for K_1 and $K_2 \in \mathfrak{L}$ define $\delta(K_1, K_2) = \max[\sup\{d(x, K_2) \mid x \in K_1\}, \sup\{d(x, K_1) \mid x \in K_2\}]$. Further, for K_1, $K_2 \in \mathfrak{L}$ and $\alpha \geqslant 0$ define the sum and the nonnegative scalar multiple by $K_1 + K_2 = \{x_1 + x_2 \mid x_1 \in K_1, x_2 \in K_2\}$ and $\alpha \cdot K_1 = \{\alpha x \mid x \in K_1\}$, respectively. Then $\mathfrak{L}$ endowed with the Hausdorff metric and the above addition and scalar multiplication is isometrically embedded in a closed convex cone in a separable Banach space Y by the Rådsröm embedding theorem (*Proc. Amer. Math. Soc.*, 3 (1952)). Let φ be this isometry. Then the **(strong) measurability** and the **(strong) integrability** of $\Gamma(s)$ are defined by the measurability and the Bochner integrability of the Y-valued function $\varphi(\Gamma(s))$, respectively, and its (strong) integral as the inverse image of the Bochner integral of $\varphi(\Gamma(s))$ under φ:

$$\int_S \Gamma(s)\, d\mu = \varphi^{-1}\left(\int_S \varphi(\Gamma(s))\, d\mu\right).$$

This definition of integral for strongly measurable $\Gamma(s)$ is shown to be compatible with that mentioned before. It is clear by the definition that the integral value in this case is a nonempty compact convex set and that most properties of Bochner integrals also hold for this integral.

References

[1] S. Bochner, Integration von Funktionen, deren Werte die Elemente eines Vektorraumes sind, Fund. Math., 20 (1933), 262–276.
[2] G. Birkhoff, Integration of functions with values in a Banach space, Trans. Amer. Math. Soc., 38 (1935), 357–378.
[3] I. Gel'fand, Abstrakte Funktionen und lineare Operatoren, Mat. Sb., 4 (46) (1938), 235–286.
[4] N. Dunford, Uniformity in linear spaces, Trans. Amer. Math. Soc., 44 (1938), 305–356.
[5] B. J. Pettis, On integration in vector spaces, Trans. Amer. Math. Soc., 44 (1938), 277–304.
[6] N. Bourbaki, Eléments de mathématique, Intégration, Hermann, ch. 6, 1959.
[7] R. G. Bartle, N. Dunford, and J. Schwartz, Weak compactness and vector measures, Canad. J. Math., 7 (1955), 289–305.
[8] N. Dunford and J. T. Schwartz, Linear operators I, Interscience, 1958.
[9] N. Dunford and B. J. Pettis, Linear operations on summable functions, Trans. Amer. Math. Soc., 47 (1940), 323–392.
[10] J. Diestel and J. J. Uhl, Jr., Vector measures, Amer. Math. Soc. Math. Surveys 13 (1977).
[11] R. J. Aumann, Integrals of set-valued functions, J. Math. Anal. Appl., 12 (1965), 1–22.
[12] G. Debreu, Integration of correspondences, Proc. Fifth Berkeley Symp. Math. Statist. Probab., II, pt. I (1967), 351–372.
[13] C. Castaing and M. Valadier, Convex analysis and measurable multifunctions, Springer, 1977.
[14] N. Dinculeanu, Vector measures, Pergamon, 1967.
[15] I. Kluvánek and G. Knowles, Vector measures and control systems, North-Holland, 1975.

444 (XXI.42) Viète, François

François Viète (1540–December 13, 1603) was born in Fontenay-le-Comte, Poitou, in western France. He served under Henri IV, first as a lawyer and later as a political advisor. His mathematics was done in his leisure time. He used symbols for known variables for the first time and established the methodology and principles of symbolic algebra. He also systematized the algebra of the time and used it as a method of discovery. He is often called the father of algebra. He improved the methods of solving equations of the third and fourth degrees obtained by G. Cardano and L. Ferrari. Realizing that solving the algebraic equation of the 45th degree proposed by the Belgian mathematician A. van Roomen can be reduced to searching for $\sin(\alpha/45)$ knowing $\sin\alpha$, he was able to solve it almost immediately. However, he would not acknowledge negative roots and refused to add terms of different degrees because of his belief in the Greek principle of homogeneity of magnitudes. He also contributed to trigonometry and represented the number π as an infinite product.

References

[1] Francisci Vietae, Opera mathematica, F. van Schooten (ed.), Leyden, 1646 (Georg Olms, 1970).
[2] Jacob Klain, Die griechische Logistik und die Entstehung der Algebra I, II, Quellen und

Studien zur Gesch. Math., (B) 3 (1934), 18–105; (B) 3 (1936), 122–235.

445 (XXI.43) Von Neumann, John

John von Neumann (December 28, 1903–February 8, 1957) was born in Budapest, Hungary, the son of a banker. By the time he graduated from the university there in 1921, he had already published a paper with M. Fekete. He was later influenced by H. Weyl and E. Schmidt at the universities of Zürich and Berlin, respectively, and he became a lecturer at the universities of Berlin and Hamburg. He moved to the United States in 1930 and in 1933 became professor at the Institute for Advanced Study at Princeton. In 1954 he was appointed a member of the US Atomic Energy Commission. The fields in which he was first interested were †set theory, theory of †functions of real variables, and †foundations of mathematics. He made important contributions to the axiomatization of set theory. At the same time, however, he was deeply interested in theoretical physics, especially in the mathematical foundations of quantum mechanics. From this field, he was led into research on the theory of †Hilbert spaces, and he obtained basic results in the theory of †operator rings of Hilbert spaces. To extend the theory of operator rings, he introduced †continuous geometry. Among his many famous works are the theory of †almost periodic functions on a group and the solving of †Hilbert's fifth problem for compact groups. In his later years, he contributed to †game theory and to the design of computers, thus playing a major role in all fields of applied mathematics.

References

[1] J. von Neumann, Collected works I–VI, Pergamon, 1961–1963.
[2] J. von Neumann, 1903–1957, J. C. Oxtoby, B. J. Pettis, and G. B. Price (eds.), Bull. Amer. Math. Soc., 64 (1958), 1–129.
[3] J. von Neumann, Mathematische Grundlagen der Quantenmechanik, Springer, 1932.
[4] J. von Neumann, Functional operators I, II, Ann. Math. Studies, Princeton Univ. Press, 1950.
[5] J. von Neumann, Continuous geometry, Princeton Univ. Press, 1960.
[6] J. von Neumann and O. Morgenstern, Theory of games and economic behavior, Princeton Univ. Press, third edition, 1953.

W

446 (XX.13) Wave Propagation

A disturbance originating at a point in a medium and propagating at a finite speed in the medium is called a **wave**. For example, a sound wave propagates a change of density or stress in a gas, liquid, or solid. A wave in an elastic solid body is called an elastic wave. **Surface waves** appear near the surface of a medium, such as water or the earth. When electromagnetic disturbances are propagated in a gas, liquid, or solid or in a vacuum, they are called **electromagnetic waves**. Light is a kind of electromagnetic wave. According to †general relativity theory, gravitational action can also be propagated as a wave.

It many cases waves can be described by the **wave equation**:

$$\frac{\partial^2\psi}{\partial t^2}=c^2\left(\frac{\partial^2\psi}{\partial x^2}+\frac{\partial^2\psi}{\partial y^2}+\frac{\partial^2\psi}{\partial z^2}\right).$$

Here t is time, x, y, z are the Cartesian coordinates of points in the space, c is the propagation velocity, and ψ represents the state of the medium.

If we take a closed surface surrounding the origin of the coordinate system, the state $\psi(0,t)$ at the origin at time t can be determined by the state at the points on the closed surface at time $t-r/c$, with r the distance of the point from the origin. More precisely, we have

$$\psi(0,t)=\frac{1}{4\pi}\int\left(\psi\frac{\partial}{\partial n}\left(\frac{1}{r}\right)-\frac{1}{r}\frac{\partial\psi}{\partial n}-\frac{1}{cr}\frac{\partial\psi}{\partial t}\frac{\partial r}{\partial n}\right)_{t-r/c}df.$$

Here n is the inward normal at any point of the closed surface, and the integral is taken over the surface, while the value of the integrand is taken at time $t-r/c$. This relation is a mathematical representation of **Huygens's principle**, which is valid for the 3-dimensional case but does not hold for the 2-dimensional case (→ 325 Partial Differential Equations of Hyperbolic Type).

A **plane wave** propagating in the direction of a unit vector $\mathbf{n}$ can be represented by $\psi=F(t-\mathbf{n}\cdot\mathbf{r}/c)$, where F is an arbitrary function and $\mathbf{r}(x,y,z)$ is the position vector. The simplest case is given by a **sine wave (sinusoidal wave)**: $\psi=A\sin(\omega t-\mathbf{k}\cdot\mathbf{r}+\delta)$. Here A (**amplitude**) and δ (**phase constant**) are arbitrary constants, $\mathbf{k}$ is in the direction of wave propagation and satisfies the relation $|\mathbf{k}|c=\omega$. ω is the **angular frequency**, $\omega/2\pi$ the **frequency**, $\mathbf{k}$ the **wave number vector**, $|\mathbf{k}|$ the **wave number**, $2\pi/\omega$ the **period**, and $2\pi/|\mathbf{k}|$ the **wavelength**. The velocity with which the crest of the wave advances is equal to $\omega/|\mathbf{k}|=c$ and is called the **phase velocity**.

A **spherical wave** radiating from the origin can generally be represented by

$$\psi=\sum_n\varphi_n\left(\frac{d}{r\,dr}\right)^n\frac{1}{r}F\left(t-\frac{r}{c}\right),$$

where φ_n is the †solid harmonic of order n.

Waves are not restricted to those governed by the wave equation. In general, ψ is not a scalar, but has several components (e.g., ψ may be a vector), which satisfy a set of simultaneous differential equations of various kinds. Usually they have solutions in the form of sinusoidal waves, but the phase velocity $c=\omega/|\mathbf{k}|$ is generally a function of the wavelength λ. Such a wave, called a **dispersive wave**, has a propagation velocity (velocity of propagation of the disturbance through the medium) that is not equal to the phase velocity. A disturbance of finite extent that can be approximately represented by a plane wave is propagated with a velocity $c-\lambda\,dc/d\lambda$, called the **group velocity**. Often there exists a definite relationship between the amplitude vector $\mathbf{A}$ (and the corresponding phase constant δ) and wave number vector $\mathbf{k}$, in which case the wave is said to be **polarized**. In particular, when $\mathbf{A}$ and $\mathbf{k}$ are parallel (perpendicular), the wave is called a **longitudinal (transverse) wave**. Usually equations governing the wave are linear, and therefore superposition of two solutions gives a new solution (†principle of superposition). Superposition of two sinusoidal waves traveling in opposite directions gives rise to a wave whose crests do not move (e.g., $\psi=A\sin\omega t\sin\mathbf{k}\cdot\mathbf{r}$). Such a wave is called a **stationary wave**. Since the energy of a wave is proportional to the square of ψ, the energy of the resultant wave formed by superposition of two waves is not equal to the sum of the energies of the component waves. This phenomenon is called **interference**. When a wave reaches an obstacle it propagates into the shadow region of the obstacle, where there is formed a special distribution of energy dependent on the shape and size of the obtacle. This phenomenon is called **diffraction**.

For aerial sound waves and water waves, if the amplitude is so large that the wave equation is no longer valid, we are faced with †nonlinear problems. For instance, **shock waves** appear in the air when surfaces of discontinuity of density and pressure exist. They appear in explosions and for bodies traveling at high speeds. Concerning wave mechanics dealing with atomic phenomena → 351 Quantum Mechanics.

References

[1] H. Lamb, Hydrodynamics, Cambridge Univ. Press, sixth edition, 1932.
[2] Lord Rayleigh, The theory of sound, Macmillan, second revised edition, I, 1937; II, 1929.
[3] M. Born and E. Wolf, Principles of optics, Pergamon, fourth edition, 1970.
[4] F. S. Crawford, Jr., Waves, Berkeley phys. course III, McGraw-Hill, 1968.
[5] C. A. Coulson, Waves; A mathematical theory of the common type of wave motion, Oliver & Boyd, seventh edition, 1955.
[6] L. Brillouin, Wave propagation and group velocity, Academic Press, 1960.
[7] I. Tolstoy, Wave propagation, McGraw-Hill, 1973.
[8] J. D. Achenbach, Wave propagation in elastic solids, North-Holland, 1973.
[9] K. F. Graff, Wave motion in elastic solids, Ohio State Univ. Press, 1975.
[10] J. Lighthill, Waves in fluids, Cambridge Univ. Press, 1978.
[11] R. Courant and D. Hilbert, Methods of mathematical physics II, Interscience, 1962.

447 (XXI.44) Weierstrass, Karl

Karl Weierstrass (October 31, 1815–February 19, 1897) was born into a Catholic family in Ostenfelde, in Westfalen, Germany. From 1834 to 1838 he studied law at the University of Bonn. In 1839 he moved to Münster, where he came under the influence of C. Gudermann, who was then studying the theory of elliptic functions. From this time until 1855, he taught in a parochial junior high school; during this period he published an important paper on the theory of analytic functions. Invited to the University of Berlin in 1856, he worked there with L. Kronecker and E. E. Kummer. In 1864, he was appointed to a full professorship, which he held until his death.

His foundation of the theory of analytic functions of a complex variable at about the same time as Riemann is his most fundamental work. In contrast to Riemann, who utilized geometric and physical intuition, Weierstrass stressed the importance of rigorous analytic formulation. Aside from the theory of analytic functions, he contributed to the theory of functions of real variables by giving examples of continuous functions that were nowhere differentiable. With his theory of †minimal surfaces, he also contributed to geometry. His lectures at the University of Berlin drew many listeners, and in his later years he was a respected authority in the mathematical world.

References

[1] K. Weierstrass, Mathematische Werke I–VII, Mayer & Miller, 1894–1927.
[2] F. Klein, Vorlesungen über die Entwicklung der Mathematik im 19. Jahrhundert I, Springer, 1926 (Chelsea, 1956).

448 (XXI.45) Weyl, Hermann

Hermann Weyl (November 9, 1885–December 8, 1955) was born in Elmshorn in the state of Schleswig-Holstein in Germany. Entering the University of Göttingen in 1904, he also audited courses for a time at the University of Münich. In 1908, he obtained his doctorate from the University of Göttingen with a paper on the theory of integral equations, and by 1910 he was a lecturer at the same university. In 1913, he became a professor at the Federal Technological Institute at Zürich; in 1928–1929, a visiting professor at Princeton University; in 1930, a professor at the University of Göttingen; and in 1933, a professor at the Institute for Advanced Study at Princeton. He retired from his professorship there in 1951, when he became professor emeritus. He died in Zürich in 1955.

Weyl contributed fresh and fundamental works covering all aspects of mathematics and theoretical physics. Among the most notable are results on problems in †integral equations, †Riemann surfaces, the theory of †Diophantine approximation, the representation of groups, in particular compact groups and †semisimple Lie groups (whose structure he elucidated), the space-time problem, the introduction of †affine connections in differential geometry, †quantum mechanics, and the foundations of mathematics. In his later years, with his son Joachim he studied meromorphic functions. In addition to his many mathematical works he left works in philosophy, history, and criticism.

References

[1] H. Weyl, Gesammelte Abhandlungen I–IV, Springer, 1968.
[2] H. Weyl, Die Idee der Riemannschen Fläche, Teubner, 1913, revised edition, 1955; English translation, The concept of a Riemann surface, Addison-Wesley, 1964.

[3] H. Weyl, Raum, Zeit, Materie, Springer, 1918, fifth edition, 1923; English translation, Space, time, matter, Dover, 1952.
[4] H. Weyl, Das Kontinuum, Veit, 1918.
[5] H. Weyl, Gruppentheorie und Quantenmechanik, Hirzel, 1928.
[6] H. Weyl, Classical groups, Princeton Univ. Press, 1939, revised edition, 1946.
[7] H. Weyl and F. J. Weyl, Meromorphic functions and analytic curves, Princeton Univ. Press, 1943.
[8] H. Weyl, Philosophie der Mathematik und Naturwissenschaften, Oldenbourg, 1926; English translation, Philosophy of mathematics and natural science, Princeton Univ. Press, 1949.
[9] H. Weyl, Symmetry, Princeton Univ. Press, 1952.

449 (III.18) Witt Vectors

A. General Remarks

Let Γ be an †integral domain of characteristic 0, and p a fixed prime number. For each infinite-dimensional vector $x=(x_0, x_1, \ldots)$ with components in Γ, we define its **ghost components** $x^{(0)}, x^{(1)}, \ldots$ by $x^{(0)}=x_0, x^{(n)}=x_0^{p^n}+px_1^{p^{n-1}}+\ldots+p^n x_n$. We define the sum of the vectors x and $y=(y_0, y_1, \ldots)$ to be the vector with ghost components $x^{(0)}+y^{(0)}$, $x^{(1)}+y^{(1)}$, $\ldots$, and their product to be the vector with ghost components $x^{(0)}y^{(0)}$, $x^{(1)}y^{(1)}, \ldots$. The sum and product are uniquely determined vectors with components in Γ. Writing their first two terms explicitly, we have

$$x+y$$
$$=\left[x_0+y_0, x_1+y_1-\sum_{\nu=1}^{p-1}\frac{1}{p}\binom{p}{\nu}x_0^{\nu}y_0^{p-\nu}, \ldots\right],$$

$$xy=(x_0y_0, x_1y_0^p+y_1x_0^p+px_1y_1, \ldots).$$

In general, it can be proved that the nth components $\sigma_n(x, y)$ and $\pi_n(x, y)$ of the sum and product are polynomials in $x_0, y_0, x_1, y_1, \ldots, x_n, y_n$ whose coefficients are rational integers. With these operations of addition and multiplication, the set of these vectors forms a †commutative ring, of which the zero element is $(0, 0, \ldots)$ and the unity element is $(1, 0, \ldots)$. Let k be a field of characteristic p. For vectors $(\xi_0, \xi_1, \ldots)$, and $(\eta_0, \eta_1, \ldots)$ with components in k, we define their sum and product by $(\xi_0, \xi_1, \ldots)+(\eta_0, \eta_1, \ldots)=(\ldots, \sigma_n(\xi, \eta), \ldots)$ and $(\xi_0, \xi_1, \ldots)(\eta_0, \eta_1, \ldots)=(\ldots, \pi_n(\xi, \eta), \ldots)$. Since the coefficients of σ_n and π_n are rational integers, these operations are well defined. With these operations, the set of such vectors becomes an integral domain $W(k)$ of characteristic 0. Elements of $W(k)$ are called **Witt vectors** over k.

If we put $V(\xi_0, \xi_1, \ldots)=(0, \xi_0, \xi_1, \ldots)$ and $(\xi_0, \xi_1, \ldots)^p=(\xi_0^p, \xi_1^p, \ldots)$, we get the formula $p\xi=V\xi^p$. (Note that this ξ^p is not the pth power of ξ in $W(k)$ in the usual sense.) Therefore, if we put $|\xi|=p^{-n}$ for a vector ξ whose first nonzero component is ξ_n, then this absolute value $|\ |$ gives a †valuation of $W(k)$. In particular, when k is a †perfect field, denoting the vector $(\xi_0, 0, \ldots)$ by $\{\xi_0\}$ we get $(\xi_0, \xi_1, \ldots)=\sum p^i\{\xi_i^{p^{-i}}\}$, and $W(k)$ is a †complete valuation ring with respect to this valuation. Therefore the †field of quotients of $W(k)$ is a complete valuation field of which p is a prime element and k is the †residue class field. Conversely, let K be a field of characteristic 0 that is complete under a †discrete valuation v, $\mathfrak{o}$ be the valuation ring of v, and k be the residue class field of v. Assume that k is a perfect field of characteristic p. If p is a prime element of v, then $\mathfrak{o}=W(k)$. If $v(p)=v(\pi^e)$ $(e>1)$ with a prime element π of $\mathfrak{o}$, we have $\mathfrak{o}=W(k)[\pi]$, and π is a root of an †Eisenstein polynomial $X^e+a_1X^{e-1}+\ldots+a_e$ $(a_i\in pW(k), a_e\notin p^2W(k))$. In this way we can determine explicitly the structure of a †p-adic number field (→ 257 Local Fields).

B. Applications to Abelian p-Extensions and Cyclic Algebras of Characteristic p

Next we consider $W_n(k)=W(k)/V^nW(k)$. The elements of $W_n(k)$ can be viewed as the n-dimensional vectors $(\xi_0, \ldots, \xi_{n-1})$, but their laws of composition are defined as in the previous section. They are called **Witt vectors of length** n. We define an operator $\wp$ by $\wp\xi=\xi^p-\xi$. Using it, we can generalize the theory of †Artin-Schreier extensions (→ 172 Galois Theory) to the case of Abelian extensions of exponent p^n over a field of characteristic p. Indeed, let k be a field of characteristic p and $\xi=(\xi_0, \ldots, \xi_{n-1})$ an element of $W(k)$. If $\eta=(\eta_0, \ldots, \eta_{n-1})$ is a root of the vector equation $\wp X-\xi=0$, then the other roots are of the form $\eta+\alpha(\alpha=(\alpha_0, \ldots, \alpha_{n-1}), \alpha_i\in\mathbf{F}_p)$. In particular, if $\xi_0\notin\wp k=\{\alpha^p-\alpha\mid\alpha\in k\}$, the field $K=k(\eta_0, \ldots, \eta_{n-1})$ is a cyclic extension of degree p^n over k, and conversely, every cyclic extension of k of degree p^n is obtained in this way. Let $(1/\wp)\xi$ denote the set of all roots of $\wp X-\xi=0$. Then more generally, any finite Abelian extension of exponent p^n of k can be obtained as $K=k((1/\wp)\xi\mid\xi\in H)$ with a suitable finite subgroup $H/\wp W_n(k)$ of $W_n(k)/\wp W_n(k)$, and the Galois group of K/k is isomorphic to $H/\wp W_n(k)$.

Moreover, for a †cyclic extension $K = k((1/\wp)\beta)$ of exponent p^n over k and for $\alpha \in k(\alpha \neq 0)$, we can define a †cyclic algebra $(\alpha, \beta]$ generated by an element u over K by the fundamental relations $u^{p^n} = \alpha$, $\wp\theta = \beta$, $u\theta u^{-1} = \theta + (1, 0, \ldots, 0)$ (where $\theta = (\theta_0, \ldots, \theta_{n-1})$, $u\theta u^{-1} = (u\theta_0 u^{-1}, \ldots, u\theta_{n-1} u^{-1})$), and $(\alpha, \beta]$ is a central simple algebra over k.

Using these results, we can develop the structure theory of the †Brauer group of exponent p^n of a †field of power series in one variable with coefficients in a finite field $\mathbf{F}_q$ (of a †field of algebraic functions in one variable over $\mathbf{F}_q$) exactly as in the case of a $\mathfrak{p}$-adic field (of an algebraic number field) (E. Witt [1]; → 29 Associative Algebras G).

On the other hand, $W_n(k)$ is a commutative †algebraic group over k and is important in the theories of algebraic groups and †formal groups (→ 13 Algebraic Groups).

References

[1] E. Witt, Zyklische Körper und Algebren der Charakteristik p vom Grad p^n, J. Reine Angew. Math., 176 (1937), 126–140.

[2] H. Hasse, Zahlentheorie, Akademie-Verlag., 1949.

[3] N. Jacobson, Lectures in abstract algebra III, Van Nostrand, 1964.

Z

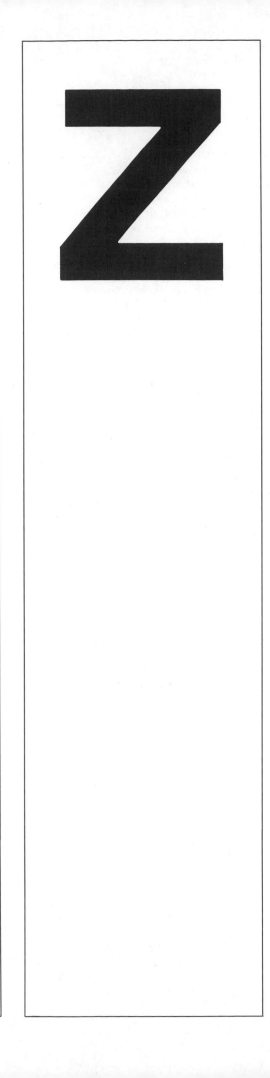

450 (V.19) Zeta Functions

A. Introduction

Since the 19th century, many special functions called ζ-functions (zeta functions) have been defined and investigated. The four main problems concerning ζ-functions are: (1) Methods of defining ζ-functions. (2) Investigation of the properties of ζ-functions. Generally, ζ-functions have the following four properties in common: (i) They are meromorphic on the whole complex plane; (ii) they have †Dirichlet series expansions; (iii) they have Euler product expansions; and (iv) they satisfy certain functional equations. Also, it is an important problem to find the poles, residues, and zeros of ζ-functions. (3) Application to number theory, in particular to the theory of decomposition of prime ideals in finite extensions of algebraic number fields (→ 59 Class Field Theory). (4) Study of the relations between different ζ-functions.

Most of the functions called ζ-functions or L-functions have the four properties of problem (2). The following is a classification of the important types of ζ-functions that are already known, which will be discussed later in this article:

(1) The ζ- and L-functions of algebraic number fields: the Riemann ζ-function, Dirichlet L-functions (study of these functions gave impetus to the theory of ζ-functions), Dedekind ζ-functions, Hecke L-functions, Hecke L-functions with †Grössencharakters, Artin L-functions, and Weil L-functions. (2) The p-adic L-functions related to the works of H. W. Leopoldt, T. Kubota, K. Iwasawa, etc. (3) The ζ-functions of quadratic forms: Epstein ζ-functions, ζ-functions of indefinite quadratic forms (C. L. Siegel), etc. (4) The ζ- and L-functions of algebras: Hey ζ-functions and the ζ-functions given by R. Godement, T. Tamagawa, etc. (5) The ζ-functions associated with Hecke operators, related to the work of E. Hecke, M. Eichler, G. Shimura, H. Jacquet, R. P. Langlands, etc. (6) The congruence ζ- and L-functions attached to algebraic varieties defined over finite fields (E. Artin, A. Weil, A. Grothendieck, P. Deligne), ζ- and L-functions of schemes. (7) Hasse ζ-functions attached to the algebraic varieties defined over algebraic number fields. (8) The ζ-functions attached to discontinuous groups: Selberg ζ-functions, the Eisenstein series defined by A. Selberg, Godement, and I. M. Gel'fand, etc. (9) Y. Ihara's ζ-function related to non-Abelian class field theory over a function field over a finite field. (10) ζ-functions associated with prehomogeneous vector spaces (M. Sato, T. Shintani).

B. The Riemann ζ-Function

Consider the series

$$\zeta(s)=1+\frac{1}{2^s}+\frac{1}{3^s}+\ldots+\frac{1}{n^s}+\ldots,$$

which converges for all real numbers $s>1$. It was already recognized by L. Euler that $\zeta(s)$ can also be expressed by a convergent infinite product $\prod_p(1-p^{-s})^{-1}$, where p runs over all prime numbers (*Werke*, ser. I, vol. VII, ch. XV, § 274). This expansion is called **Euler's infinite product expansion** or simply the **Euler product**. However, Riemann was the first to treat $\zeta(s)$ successfully as a function of a complex variable s (1859) [R1]; for this reason, it is called the **Riemann ζ-function**. As can be seen from its Euler product expansion, $\zeta(s)$ is holomorphic and has no zeros in the domain $\operatorname{Re}s>1$. Riemann proved, moreover, that it has an analytic continuation to the whole complex plane, is meromorphic everywhere, and has a unique pole $s=1$. The functions $(s-1)\zeta(s)$ and $\zeta(s)-1/(s-1)$ are †integral functions of s. This can be seen by considering the integral expression

$$\pi^{-s/2}\Gamma\left(\frac{s}{2}\right)\zeta(s)=\int_0^\infty x^{s/2-1}\left(\sum_{n=1}^\infty e^{-n^2\pi x}\right)dx$$

$$=-\frac{1}{s}-\frac{1}{1-s}+\int_1^\infty (x^{(1-s)/2-1}+x^{(s/2)-1})\times\left(\sum_{n=1}^\infty e^{-n^2\pi x}\right)dx.$$

From this last formula, we also obtain an equality

$$\xi(s)=\xi(1-s),$$

where

$$\xi(s)=\pi^{-s/2}\Gamma(s/2)\zeta(s).$$

This equality is called the **functional equation** for the ζ-function. The residue of $\zeta(s)$ at $s=1$ is 1, and around $s=1$,

$$\zeta(s)=\frac{1}{s-1}+C+O(|s-1|),$$

where C is †Euler's constant. This is called the **Kronecker limit formula** for $\zeta(s)$.

The function $\zeta(s)$ has no zeros in $\operatorname{Re}s\geqslant 1$, and its only zeros in $\operatorname{Re}s\leqslant 0$ are simple zeros at $s=-2,-4,\ldots,-2n,\ldots$. But $\zeta(s)$ has infinitely many zeros in $0<\operatorname{Re}s<1$, which are called the nontrivial zeros. B. Riemann conjectured that all nontrivial zeros lie on the line $\operatorname{Re} s = 1/2$ (1859). This is called the **Riemann hypothesis**, which has been neither proved nor disproved (→ Section I).

If $N(T)$ denotes the number of zeros of $\zeta(s)$ in the rectangle $0 < \operatorname{Re} s < 1$, $0 < \operatorname{Im} s < T$, we have an asymptotic formula

$$N(T) = \frac{1}{2\pi} T \log T - \frac{1 + \log 2\pi}{2\pi} T + O(\log T)$$

(H. von Mangoldt, 1905). Also, $\zeta(s)$ has the following infinite product expansion:

$$(s-1)\zeta(s) = \frac{1}{2} e^{bs} \frac{1}{\Gamma\left(\frac{s}{2}+1\right)} \prod_{\rho} \left(1 - \frac{s}{\rho}\right) e^{s/\rho},$$

where b is a constant and ρ runs over all nontrivial zeros of $\zeta(s)$ (J. Hadamard, 1893). Hadamard and C. de La Vallée-Poussin proved the †prime number theorem, almost simultaneously, by using some properties of $\zeta(s)$ (→ 123 Distribution of Prime Numbers B).

The following **approximate functional equation** is important in investigating the values of $\zeta(s)$:

$$\zeta(s) = \sum_{n < x} \frac{1}{n^s} + \varphi(s) \sum_{n < y} \frac{1}{n^{1-s}} + O(x^{-\sigma}) + O(y^{\sigma-1} |t|^{(1/2)-\sigma}),$$

where φ is †Euler's function, and $\zeta(s) = \varphi(s)\zeta(1-s)$, $s = \sigma + it$, $2\pi xy = |t|$, and the approximation is uniform for $-h \leqslant \sigma \leqslant h$, $x > k$, $y > k$ with h and k positive constants (G. H. Hardy and J. E. Littlewood, 1921).

Euler obtained the values of $\zeta(s)$ for positive even integers s:

$$\zeta(2m) = \frac{2^{2m-1}\pi^{2m} B_{2m}}{(2m)!}$$

($m = 1, 2, 3, \ldots$, and the B_{2m} are †Bernoulli numbers). The values of $\zeta(s)$ for positive odd integers s, however, have not been expressed in such a simple form. The values of $\zeta(s)$ for negative integers s are given by $\zeta(0) = B_1(0) = -\frac{1}{2}$, $\zeta(1-n) = -\frac{B_n(0)}{n}$, $n = 2, 3, \ldots$, where the $B_n(x)$ are †Bernoulli polynomials.

As a slight generalization of $\zeta(s)$, A. Hurwitz (1862) considered

$$\zeta(s, a) = \sum_{n=0}^{\infty} \frac{1}{(n+a)^s}, \quad 0 < a \leqslant 1.$$

This is called the **Hurwitz ζ-function**. Thus $\zeta(s, 1) = \zeta(s)$, and $\zeta(s, 1/2) = (2^s - 1)\zeta(s)$. This function $\zeta(s, a)$ can also be continued analytically to the whole complex plane and satisfies a certain functional equation. But in general it has no Euler product expansion.

C. Dirichlet L-Functions

Let m be a positive integer, and classify all rational integers modulo m. The set of all classes coprime to m forms a multiplicative Abelian group of order $h = \varphi(m)$. Let χ be a †character of this group. Call (n) the residue class of $n \bmod m$, and put $\chi(n) = \chi((n))$ when $(n, m) = 1$ and $\chi(n) = 0$ when $(n, m) \neq 1$. Now, the function of a complex variable s defined by

$$L(s) = L(s, \chi) = \sum_{n=1}^{\infty} \frac{\chi(n)}{n^s}$$

is called a **Dirichlet L-function**. This function converges absolutely for $\operatorname{Re} s > 1$ and has an Euler product expansion

$$L(s, \chi) = \prod_p \frac{1}{1 - \chi(p)p^{-s}}.$$

If there exist a divisor f of m ($f \neq m$) and a character χ^0 modulo f such that $\chi(n) = \chi^0(n)$ for all n with $(n, m) = 1$, we call χ a **nonprimitive character**. Otherwise, χ is called a **primitive character**. If χ is nonprimitive, there exists such a unique primitive χ^0. In this situation, the divisor f of m associated with χ^0 is called the **conductor** of χ (and of χ^0). We have

$$L(s, \chi) = L(s, \chi^0) \prod_{p \mid m} (1 - \chi^0(p)p^{-s}).$$

Let χ be primitive. If the conductor $f = 1$, then χ is a trivial character ($\chi = 1$), and $L(s)$ is equal to the Riemann ζ-function $\zeta(s)$. On the other hand, if $f > 1$, then $L(s)$ is an entire function of s. In particular, if χ is a nontrivial primitive character, $L(1) = L(1, \chi)$ is finite and nonzero. P. G. L. Dirichlet proved the theorem of existence of prime numbers in arithmetic progressions using this fact (→ 123 Distribution of Prime Numbers D).

$L(s, \chi)$ has a functional equation similar to that of $\zeta(s)$; namely, if χ is a primitive character with conductor f and we put

$$\xi(s, \chi) = (f/\pi)^{s/2} \Gamma((s+a)/2) L(s, \chi),$$

where $a = 0$ for $\chi(-1) = 1$ and $a = 1$ for $\chi(-1) = -1$, then we have

$$\xi(s, \chi) = W(\chi) \xi(1-s, \bar{\chi}),$$

where

$$W(\chi) = (-i)^a f^{-1/2} \tau(\chi), \quad \tau(\chi) = \sum_{n \bmod f} \chi(n) \zeta_f^n$$

($\zeta_f = \exp(2\pi i/f)$). The latter sum is called the **Gaussian sum**. Note that $|W(\chi)| = 1$.

The values of $L(s)$ for negative integers s are given by $L(1-m, \chi) = -B_{\chi,m}/m$ ($m = 1, 2, 3, \ldots$), where the $B_{\chi,m}$ are defined by

$$\sum_{\mu=1}^{f} \frac{\chi(\mu) t e^{\mu t}}{e^{ft} - 1} = \sum_{m=0}^{\infty} B_{\chi,m} t^m.$$

Moreover, if $\chi(-1)=-1$, we have

$$L(1,\chi)=\frac{\pi}{i}\frac{\tau(\chi)}{f}\sum_{x=1}^{f}(-\overline{\chi(x)}\cdot x)$$
$$=\pi i\tau(\chi)B_{\bar{\chi},1},$$

and if $\chi(-1)=1$, $\chi\neq 1$, we have

$$L(1,\chi)=2\frac{\tau(\chi)}{f}\sum_{x=1}^{[f/2]}(-\chi(x)\log|1-\zeta_f^x|).$$

In certain cases, the functional equation can be utilized to obtain the values of $L(m,\chi)$ from those of $L(1-m,\chi)$. Actually, if $\chi(-1)=1$, $m=2n=2,4,6,\ldots$, we have

$$L(2n,\chi)=\frac{(-1)^n}{(2n)!}\left(\frac{2\pi}{f}\right)^{2n}\tau(\chi)(-\mathrm{B}_{\chi,2n}),$$

and if $\chi(-1)=-1$, $m=2n+1=3,5,7,\ldots$, we have

$$L(2n+1,\chi)$$
$$=(-i)\frac{(-1)^n}{(2n+1)!}\left(\frac{2\pi}{f}\right)^{2n+1}\tau(\chi)(-B_{\chi,2n+1}).$$

Dirichlet L-functions are important not only in the arithmetic of rational number fields but also in the arithmetic of quadratic or cyclotomic fields.

D. ζ-Functions of Algebraic Number Fields (Dedekind ζ-Functions)

The Riemann ζ-function can be generalized to ζ-functions of algebraic number fields (→ 14 Algebraic Number Fields). Let k be an algebraic number field of degree n, and let $\mathfrak{a}$ run over all integral ideals of k. Consider the sequence $\zeta_k(s)=\sum_{\mathfrak{a}}N(\mathfrak{a})^{-s}$. This sequence converges for $\mathrm{Re}\,s>1$ and has an Euler product expansion $\zeta_k(s)=\prod_{\mathfrak{p}}(1-N(\mathfrak{p})^{-s})^{-1}$, where $\mathfrak{p}$ runs over all prime ideals of k. This function, which is continued analytically to the whole complex plane as a meromorphic function, is called a **Dedekind ζ-function**. Its only pole is a simple pole at $s=1$, with the residue $h_k\kappa_k$. Here h_k is the †class number of k, and $\kappa_k=2^{r_1+r_2}\pi^{r_2}R/(w|d|^{1/2})$, where r_1 $(2r_2)$ is the number of isomorphisms of k into the real (complex) number field, w is the number of roots of unity in k, d is the †discriminant of k, and R is the †regulator of k (R. Dedekind, 1877) [D1].

The function $\zeta_k(s)$ has no zeros in $\mathrm{Re}\,s\geqslant 1$, while in $\mathrm{Re}\,s\leqslant 0$ it has zeros of order r_2 at -1, -3, $-5,\ldots$, zeros of order r_1+r_2 at -2, -4, $-6,\ldots$, and a zero of order r_1+r_2-1 at $s=0$. All other zeros lie in the open strip $0<\mathrm{Re}\,s<1$, which actually contains infinitely many zeros. It is conjectured that all these zeros lie on the line $\mathrm{Re}\,s=1/2$ (the Riemann hypothesis for Dedekind ζ-functions). To obtain a generalization of the functional equation for the Riemann $\zeta(s)$ to the case of $\zeta_k(s)$, we put

$$\Xi_k(s)=\left(\frac{\sqrt{|d|}}{2^{r_2}\pi^{n/2}}\right)^s\Gamma\left(\frac{s}{2}\right)^{r_1}\Gamma(s)^{r_2}\zeta_k(s).$$

Then $\Xi_k(s)=\Xi_k(1-s)$ (Hecke, 1917). If K is a Galois extension of k, then $\zeta_K(s)/\zeta_k(s)$ is an integral function (H. Aramata, 1933; R. Brauer, 1947).

E. Hecke L-Functions

As a generalization of Dirichlet L-functions to algebraic number fields, Hecke (1917) defined the following L-function $L_k(s,\chi)$: Let k be an algebraic number field of finite degree, and let $\tilde{\mathfrak{m}}=\mathfrak{m}\prod\mathfrak{p}_\infty$ be an †integral divisor ($\mathfrak{m}$ the finite part, $\prod\mathfrak{p}_\infty$ the †infinite part). Consider the †ideal class group of k modulo $\tilde{\mathfrak{m}}$ and its character χ (here we put $\chi(\mathfrak{a})=0$ for $(\mathfrak{a},\mathfrak{m})\neq 1$). Then the L-functions are defined by

$$L_k(s,\chi)=\sum_{\mathfrak{a}}\chi(\mathfrak{a})/N(\mathfrak{a})^s$$

[H2], where $\mathfrak{a}$ runs over all integral ideals of k. $L_k(s,\chi)$ is called a **Hecke L-function**. It converges for $\mathrm{Re}\,s>1$ and has an Euler product expansion

$$L_k(s,\chi)=\prod_{\mathfrak{p}}\frac{1}{1-\chi(\mathfrak{p})N(\mathfrak{p})^{-s}}.$$

Here $\mathfrak{p}$ runs over all prime ideals of k. If there is a divisor $\tilde{\mathfrak{f}}|\tilde{\mathfrak{m}}$ ($\tilde{\mathfrak{f}}\neq\tilde{\mathfrak{m}}$) and a character χ^0 modulo $\tilde{\mathfrak{f}}$ such that $\chi^0(\mathfrak{a})=\chi(\mathfrak{a})$ for all $\mathfrak{a}$ with $(\mathfrak{a},\mathfrak{m})=1$, then χ is called **nonprimitive**: otherwise, χ is called a **primitive character**. In general, there exist unique such $\tilde{\mathfrak{f}}$ and χ^0. In this situation, $\tilde{\mathfrak{f}}$ is called the **conductor** of χ. If χ is primitive and the conductor $\tilde{\mathfrak{f}}$ is (1), then χ is a trivial character and $L_k(s,\chi)$ coincides with $\zeta_k(s)$. If χ is primitive and $\chi\neq 1$, then $L_k(s,\chi)$ is an integral function of s, and $L_k(1,\chi)\neq 0$. Utilizing this fact, it can be proved that there exist infinitely many prime ideals in each class of the ideal class group modulo an integral divisor $\tilde{\mathfrak{m}}$ of k.

Let χ be a primitive character with the conductor $\tilde{\mathfrak{f}}$, d be the discriminant of k, $\sigma_1,\ldots,\sigma_{r_1}$ be all distinct isomorphisms of k into the real number field $\mathbf{R}$, and $\mathfrak{f}$ be the finite part of $\tilde{\mathfrak{f}}$. Then if ξ is an integer of k such that $\xi\equiv 1$ (mod $\mathfrak{f}$), we have

$$\chi((\xi))=(\mathrm{sgn}\,\xi^{\sigma_1})^{a_1}\cdot\ldots\cdot(\mathrm{sgn}\,\xi^{\sigma_{r_1}})^{a_{r_1}},$$

where a_m $(m=1,\ldots,r_1)$ is either 0 or 1, depending on χ. By putting

$$\xi_k(s,\chi)=$$
$$\left(\frac{\sqrt{|d|N(\mathfrak{f})}}{2^{r_2}\pi^{n/2}}\right)^s\cdot\prod_{m=1}^{r_1}\Gamma\left(\frac{s+a_m}{2}\right)\Gamma(s)^{r_2}\cdot L_k(s,\chi),$$

we have the following functional equation for the Hecke L-function:

$$\xi_k(s,\chi)=W(\chi)\xi_k(1-s,\bar{\chi}),$$

where $W(\chi)$ is a complex number with absolute value 1 and the exact value of $W(\chi)$ is given as a Gaussian sum. Just as some properties concerning the distribution of prime numbers can be proved using the Riemann ζ-function and Dirichlet L-functions, some properties concerning the distribution of prime ideals can be proved using the Hecke L-functions (→ 123 Distribution of Prime Numbers F).

T. Takagi used Hecke L-functions in founding his †class field theory. In the other direction, this theory implies $L(1,\chi)\neq 0$ $(\chi\neq 1)$.

Let K be a †class field over k that corresponds to an ideal class group H of k with index h. By using class field theory, we obtain $\zeta_K(s)=\prod_\chi L_k(s,\chi)$, where the product is over all characters χ of ideal class groups of k, such that $\chi(H)=1$. This formula can be regarded as an alternative formulation of the decomposition theorem of class field theory (→ 59 Class Field Theory). By taking the residues of both sides of the formula at $s=1$, we obtain $h_K\kappa_K=h_k\kappa_k\prod_{\chi\neq 1}L_k(1,\chi)$.

In particular, if $k=\mathbf{Q}$ (the rational number field) and K is a quadratic number field $\mathbf{Q}(\sqrt{d})$ (d is the discriminant of K), then we have

$$\zeta_K(s)=\zeta(s)\cdot L(s),\quad L(s)=\sum_{n=1}^{\infty}\left(\frac{d}{n}\right)n^{-s},$$

where (d/n) is the †Kronecker symbol, and we put $(d/n)=0$ when $(n,d)\neq 1$. From this, we obtain the class number formula for quadratic number fields (→ 347 Quadratic Fields). A similar method is used for computation of class numbers of cyclotomic fields K (→ 14 Algebraic Number Fields L).

In general, the computation of the relative class number h_K/h_k when K/k is an Abelian extension is reduced to the evaluation of $L(1,\chi)$. This computation has been made successfully for the following cases (besides for the examples in the previous paragraph): k is imaginary quadratic and K is the absolute class field of k or the class field corresponding to †ray $S(\mathfrak{m})$; k is totally real and K is a totally imaginary quadratic extension of k. H. M. Stark and T. Shintani made conjectures about the values of $L(1,\chi)$ [S25, S19].

Let $L(s,\chi)$ be a Hecke L-function for the character χ. Then it follows from the functional equation that the values of $L(s,\chi)$ at $s=0,-1,-2,-3,\ldots$ are zero if k is not totally real. Furthermore, if k is a totally real finite algebraic number field, then these values of $L(s,\chi)$ are algebraic numbers (C. L. Siegel, H. Klingen, T. Shintani).

F. Hecke L-Functions with Grössencharakters

E. Hecke (1918, 1920) extended the notion of characters by introducing the †Grössencharakter χ and defined L-functions with such characters:

$$L_k(s,\chi)=\sum_{\mathfrak{a}}\frac{\chi(\mathfrak{a})}{N(\mathfrak{a})^s}.$$

He also proved the existence of their Euler product expansions and showed that they satisfy certain functional equations [H3]. Moreover, by estimating the sum $\sum_{N(\mathfrak{p})<x}\chi(\mathfrak{p})$, he obtained some results on the distribution of prime ideals.

Later, Iwasawa and J. Tate independently gave clearer definitions of the Grössencharakter χ and $L_k(s,\chi)$ by using harmonic analysis on the adele and idele groups of k (→ 6 Adeles and Ideles) [L3].

Let $\mathbf{J}_k$ be the idele group of k, $\mathbf{P}_k$ be the group of †principal ideles, and $\mathbf{C}_k=\mathbf{J}_k/\mathbf{P}_k$ be the idele class group. Then a Grössencharakter is a continuous character χ of $\mathbf{C}_k$, and χ induces a character of $\mathbf{J}_k$, which is also denoted by χ. Let $\mathbf{J}_k=\mathbf{J}_\infty\times\mathbf{J}_0$ be the decomposition of $\mathbf{J}_k$ into the infinite part $\mathbf{J}_\infty$ and the finite part $\mathbf{J}_0$. Let $\mathbf{U}_0$ be the unit group of $\mathbf{J}_0$, and for each integral ideal $\mathfrak{m}$ of k, put $\mathbf{U}_{\mathfrak{m},0}=\{\mathfrak{u}\in\mathbf{U}_0\mid\mathfrak{u}\equiv 1\ (\mathrm{mod}\,\mathfrak{m})\}$, so that $\{\mathbf{U}_{\mathfrak{m},0}\}$ forms a base for the neighborhood system of 1 in $\mathbf{J}_0$. Put $\mathbf{J}_{\mathfrak{m},0}=\{\mathfrak{a}\in\mathbf{J}_0\mid\mathfrak{a}_\mathfrak{p}=1$ for all $\mathfrak{p}\mid\mathfrak{m}\}$, and with each $\mathfrak{a}\in\mathbf{J}_{\mathfrak{m},0}$, associate an ideal $\tilde{\mathfrak{a}}=\prod_\mathfrak{p}\mathfrak{p}^{\nu_\mathfrak{p}(\mathfrak{a})}$, where $\mathfrak{a}=(\mathfrak{a}_\mathfrak{p})$ and the ideal in $k_\mathfrak{p}$ generated by $\mathfrak{a}_\mathfrak{p}$ is equal to $\mathfrak{p}^{\nu_\mathfrak{p}(\mathfrak{a})}$. Then the mapping $\mathfrak{a}\to\tilde{\mathfrak{a}}$ gives a homomorphism of $\mathbf{J}_{\mathfrak{m},0}$ into the group $G(\mathfrak{m})=\{\tilde{\mathfrak{a}}\mid(\mathfrak{a},\mathfrak{m})=1\}$, and its kernel is contained in $\mathbf{U}_{\mathfrak{m},0}$. Since χ is continuous, $\chi(\mathbf{U}_{\mathfrak{m},0})=1$ for some $\mathfrak{m}$. The greatest common divisor $\mathfrak{f}$ of all such ideals $\mathfrak{m}$ is called the **conductor** of χ. For each $\mathfrak{a}\in\mathbf{J}_{\mathfrak{f},0}$, $\chi(\mathfrak{a})$ depends only on the ideal $\tilde{\mathfrak{a}}$ $(\in G(\mathfrak{f}))$; hence by putting $\chi(\mathfrak{a})=\tilde{\chi}(\tilde{\mathfrak{a}})$, we obtain a character $\tilde{\chi}$ of $G(\mathfrak{f})$. Now put $L_k(s,\chi)=\sum\tilde{\chi}(\tilde{\mathfrak{a}})/N(\tilde{\mathfrak{a}})^s$, where the sum is over all integral ideals $\tilde{\mathfrak{a}}\in G(\mathfrak{f})$. This is called a **Hecke L-function with Grössencharakter** χ. For $\chi\neq 1$, it is an entire function. On the other hand, if we restrict χ to $\mathbf{J}_\infty=\mathbf{R}^{*r_1}\times\mathbf{C}^{*r_2}$, then for $u=(a_1,\ldots,a_{r_1},a_{r_1+1},\ldots,a_{r_1+r_2})\in\mathbf{J}_\infty$, we have

$$\chi(u)=\prod_{j=1}^{r_1+r_2}|a_j|^{\lambda_j\sqrt{-1}}\cdot\prod_{j=1}^{r_1}(\mathrm{sgn}\,a_j)^{e_j}\cdot\prod_{l=r_1+1}^{r_1+r_2}\left(\frac{a_l}{|a_l|}\right)^{e_l},$$

where $e_j=0$ or 1, $e_l\in\mathbf{Z}$, $\lambda_j\in\mathbf{R}$. The numbers e_j, e_l, λ_j are determined uniquely by χ. Putting

$$\xi_k(s,\chi)=\left(\frac{\sqrt{|d|N(\mathfrak{f})}}{2^{r_2}\pi^{n/2}}\right)^s\cdot\prod_{j=1}^{r_1}\Gamma\left(\frac{s+e_j+\lambda_j\sqrt{-1}}{2}\right)\times\prod_{l=r_1+1}^{r_1+r_2}\Gamma\left(s+\frac{|e_l|+\lambda_l\sqrt{-1}}{2}\right)\cdot L(s,\chi),$$

we have a functional equation

$$\xi_k(s,\chi)=W(\chi)\xi_k(1-s,\bar{\chi}),$$

where $W(\chi)$ is a complex number with absolute value 1.

We can express $\xi_k(s,\chi)$ by an integral form on $\mathbf{J}_k$ as

$$\xi_k(s,\chi)=c\int_{\mathbf{J}_k}\varphi(\mathfrak{r})\chi(\mathfrak{r})V(\mathfrak{r})^s d^*\mathfrak{r},$$

where $V(\mathfrak{r})$ is the †total volume of the idele $\mathfrak{r}$, c is a constant that depends on the †Haar measure $d^*\mathfrak{r}$ of $\mathbf{J}_k$, and $\varphi(\mathfrak{r})$ is defined by

$$\varphi(\mathfrak{r})=\prod_{\mathfrak{p}}\varphi_{\mathfrak{p}}(x_{\mathfrak{p}}),\qquad \mathfrak{r}=(\ldots x_{\mathfrak{p}}\ldots),$$

$$\begin{aligned}&\varphi_{\mathfrak{p}_{\infty,i}}(x)\\ &=x^{e_i}e^{-\pi x^2},\quad i\leqslant r_1,\quad k_{\mathfrak{p}_{\infty,i}}=\mathbf{R},\\ &\left.\begin{aligned}&=\frac{1}{2\pi}\bar{x}^{e_i}e^{-2\pi|x|^2},\quad e_i\geqslant 0,\\ &=\frac{1}{2\pi}x^{-e_i}e^{-2\pi|x|^2},\quad e_i<0,\end{aligned}\right\}\ i>r_1,\quad k_{\mathfrak{p}_{\infty,i}}=\mathbf{C},\end{aligned}$$

$$\left.\begin{aligned}\varphi_{\mathfrak{p}}(x)&=e^{2\pi i\lambda(x)},\quad x\in(\mathfrak{b}\mathfrak{f})_{\mathfrak{p}}^{-1},\\ &=0,\qquad x\notin(\mathfrak{b}\mathfrak{f})_{\mathfrak{p}}^{-1}\end{aligned}\right\}\ \mathfrak{p}\text{ finite.}$$

Hence $(\mathfrak{b}\mathfrak{f})_{\mathfrak{p}}^{-1}$ is the $\mathfrak{p}$-component (→ 6 Adeles and Ideles B) of the ideal $(\mathfrak{b}\mathfrak{f})^{-1}$ ($\mathfrak{b}$ is the †different of $k/\mathbf{Q}$) and $\lambda(x)$ is an additive character of $k_{\mathfrak{p}}$ defined as follows. $\mathbf{Q}_p$ is the †p-adic field, $\mathbf{Z}_p$ is the ring of p-adic integers, λ_0 is the mapping $\mathbf{Q}_p\to\mathbf{Q}_p/\mathbf{Z}_p\subset\mathbf{Q}/\mathbf{Z}\subset\mathbf{R}/\mathbf{Z}$, and $\lambda=\lambda_0\circ Tr_{k_{\mathfrak{p}}/\mathbf{Q}_p}$. By putting $\chi(\mathfrak{r})=\prod_{\mathfrak{p}}\chi_{\mathfrak{p}}(x_{\mathfrak{p}})$, $\mathfrak{r}=(\ldots x_{\mathfrak{p}}\ldots)$, we have

$$\xi_k(s,\chi)=c\prod_{\mathfrak{p}}\int_{k_{\mathfrak{p}}}\varphi_{\mathfrak{p}}(x)\chi_{\mathfrak{p}}(x)V_{\mathfrak{p}}(x)^{-s}d^*x,$$

where $\mathfrak{p}$ runs over all prime divisors of k, finite or infinite. Moreover, with a constant $C_{\mathfrak{p}}$, we have

$$\begin{aligned}&\int_{k_{\mathfrak{p}_{\infty,i}}}\varphi_{\mathfrak{p}_{\infty,i}}(x)\chi_{\mathfrak{p}_{\infty,i}}(x)V_{\mathfrak{p}_{\infty,i}}(x)^{-s}d^*x\\ &=C_{\mathfrak{p}_{\infty,i}}\cdot\pi^{-(s+\sqrt{-1}\,\lambda_i+e_i)/2}\\ &\qquad\times\Gamma((s+\sqrt{-1}\,\lambda_i+e_i)/2),\quad k_{\mathfrak{p}_{\infty,i}}=\mathbf{R},\\ &=C_{\mathfrak{p}_{\infty,i}}\cdot(2\pi)^{-(s+(\sqrt{-1}\,\lambda_i+|e_i|)/2)}\\ &\qquad\times\Gamma(s+(\sqrt{-1}\,\lambda_i+|e_i|)/2),\quad k_{\mathfrak{p}_{\infty,i}}=\mathbf{C},\end{aligned}$$

$$\begin{aligned}&\int_{k_{\mathfrak{p}}}\varphi(x)\chi_{\mathfrak{p}}(x)V_{\mathfrak{p}}(x)^{-s}d^*x\\ &=C_{\mathfrak{p}}N(\mathfrak{b}_{\mathfrak{p}})^{s-1/2}\tilde{\chi}(\mathfrak{b}_{\mathfrak{p}}^{-1})\frac{1}{1-\tilde{\chi}(\mathfrak{p})/N(\mathfrak{p})^s},\\ &=C_{\mathfrak{p}}N((\mathfrak{b}\mathfrak{f})_{\mathfrak{p}})^s\tau_{\mathfrak{p}}(\chi_{\mathfrak{p}})\cdot\mu(U_{\mathfrak{f},\mathfrak{p}}),\quad \mathfrak{p}\,|\,\mathfrak{f}.\end{aligned}$$

Here $\tau_{\mathfrak{p}}(\chi_{\mathfrak{p}})$ is a constant called the **local Gaussian sum**, and $\mu(U_{\mathfrak{f},\mathfrak{p}})$ is the volume of $\{u\in k_{\mathfrak{p}}\,|\,u\equiv 1\ (\mathrm{mod}\ \mathfrak{f})\}$. These integrals over $k_{\mathfrak{p}}$ are the Γ-factors and Euler factors of $\xi_k(s,\chi)$, according as $\mathfrak{p}$ is infinite or finite. The functional equation is obtained by applying the †Poisson summation formula for $\varphi(x)$ and its †Fourier transform on the adele group $\mathbf{A}_k$ (→ 6 Adeles and Ideles).

Let $\mathbf{D}_k$ be the connected component of 1 in $\mathbf{C}_k$. If $\chi(\mathbf{D}_k)=1$, the corresponding $\tilde{\chi}$ is a character of an ideal class group of k with a conductor $\mathfrak{f}$. Conversely, all such characters can be obtained in this manner.

As stated in Section E, the Hecke L-functions with characters (of ideal class groups) can be used to describe the decomposition law of prime divisors in class field theory. However, for L-functions with Grössencharakter, such arithmetic implications have not been found yet, except that in the case of Grössencharakters of A_0 type, Y. Taniyama discovered, following the suggestion of A. Weil, that the L-function has a deep connection with the arithmetic of a certain infinite Abelian extension of k [T2, W7]. In particular, when $L(s,\chi)$ is a factor of the †Hasse ζ-function of an Abelian variety A with †complex multiplication, it describes the arithmetic of the field generated by the coordinates of the division points of A.

G. Artin L-Functions

Let K be a finite Galois extension of an algebraic number field k (of degree n), $G=G(K/k)$ be its Galois group, $\sigma\to A(\sigma)$ be a matrix representation (characteristic 0) of G, and χ be its character. Let $\mathfrak{p}$ be a prime ideal of k, and define $L_{\mathfrak{p}}(s,\chi)$ by

$$\log L_{\mathfrak{p}}(s,\chi)=\sum_{m=1}^{\infty}\frac{\chi(\mathfrak{p}^m)}{mN(\mathfrak{p}^m)^s},\qquad \mathrm{Re}\,s>1,$$

with $\chi(\mathfrak{p}^m)=(1/e)\sum_{\tau\in T}\chi(\sigma^m\tau)$, where T is the †inertia group of $\mathfrak{p}$, $|T|=e$, and σ is a †Frobenius automorphism of $\mathfrak{p}$. Then we have

$$L_{\mathfrak{p}}(s,\chi)=\det(E-A_{\mathfrak{p}}\cdot N(\mathfrak{p})^{-s})^{-1},$$

$$A_{\mathfrak{p}}=\frac{1}{e}\sum_{\tau\in T}A(\sigma\tau).$$

In particular, if $T=\{1\}$ (i.e., $\mathfrak{p}$ is †unramified in K/k), then

$$L_{\mathfrak{p}}(s,\chi)=\det(E-A(\sigma)\cdot N(\mathfrak{p})^{-s})^{-1}.$$

Now put

$$L(s,\chi,K/k)=\prod_{\mathfrak{p}}L_{\mathfrak{p}}(s,\chi),\qquad \mathrm{Re}\,s>1,$$

and call $L(s,\chi,K/k)$ an **Artin L-function** [A2].

(1) The most important property of $L(s,\chi,K/k)$ is that if K/k is an Abelian extension and χ is a linear character, it follows from class field theory that $\chi(\mathfrak{p})$ is the character of the ideal class group of k (modulo the †conductor of K/k) and that the Artin L-function equals

a Hecke L-function. This equality is equivalent to Artin's †reciprocity law, and in fact Artin obtained his reciprocity law after he conjectured the equality.

(2) If $K' \supset K \supset k$ and K'/k is a Galois extension, then $L(s, \chi, K/k) = L(s, \chi, K'/k)$.

(3) If $K \supset \Omega \supset k$ and ψ is a character of $G(K/\Omega)$, then $L(s, \psi, K/\Omega) = L(s, \chi_\psi, K/k)$, where χ_ψ is the character of $G(K/k)$ †induced from ψ.

(4) If $\chi_1 = 1$, then $L(s, \chi_1, K/k) = \zeta_k(s)$.

(5) $L(s, \chi_1 + \chi_2, K/k) = L(s, \chi_1, K/k) \cdot L(s, \chi_2, K/k)$.

Conversely, the Artin L-function $L(s, \chi, K/k)$ is characterized by properties (1)–(5).

(6) If χ_R is the †regular representation of G, then $L(s, \chi_R, K/k) = \zeta_K(s)$; hence

$$\zeta_K(s) = \zeta_k(s) \prod_{\chi \neq 1} L(s, \chi, K/k)^{\chi(1)},$$

where χ runs over all irreducible characters $\neq 1$ of G.

(7) Every character of a finite group G can be expressed as $\chi = \sum m_i \chi_{\psi_i}$ $(m_i \in \mathbf{Z})$, where each χ_{ψ_i} is an †induced character from a certain linear character ψ_i of an elementary subgroup of G (**Brauer's theorem**). (Here an elementary subgroup is a subgroup that is the direct product of a cyclic group and a p-group for some prime p.) Hence (3) and (5) imply that an Artin L-function is the product of integral powers (positive or negative) of Hecke L-functions $L_{\Omega_i}(s, \psi_i)$:

$$L(s, \chi, K/k) = \prod_i L_{\Omega_i}(s, \psi_i)^{m_i}.$$

Hence an Artin L-function is a univalent meromorphic function defined over the whole complex plane. Artin made the still open conjecture that if χ is irreducible and $\chi \neq 1$, then $L(s, \chi, K/k)$ is an entire function (**Artin's conjecture**).

This conjecture holds obviously if all m_i are nonnegative. Except for such a case, Artin's conjecture had no affirmative examples until 1974, when Deligne and Serre [D9] proved that each "new cusp form" of weight 1 gives rise to an entire Artin L-function $L(s, \chi, K/k)$ with $\chi(1) = 2$ and $\chi(\rho) = 0$ (ρ is the complex conjugation); by this method, some nontrivial examples were computed by J. Tate and J. Buhler (*Lecture notes in math.* 654 (1978)). Then R. P. Langlands [L5] constructed nontrivial examples of Artin's conjecture for certain 2-dimensional representations

$$\mathrm{Gal}(K/k) \ni \sigma \mapsto A(\sigma) \in GL(2, \mathbf{C})$$

by using ideas of H. Saito and T. Shintani [S1, S20]. This method works for all representations for which the image of the $A(\sigma)$ in $PGL(2, \mathbf{C})$ is the †tetrahedral group. It also works for some †octahedral cases, but a new idea is needed in the †icosahedral case.

(8) Let $\mathfrak{p}_{\infty,i}$ $(i = 1, \ldots, r_1 + r_2)$ be the infinite primes of k. Put

$$\gamma(s, \chi, \mathfrak{p}_{\infty,i}, K/k)$$
$$= (\Gamma(s/2)\Gamma((s+1)/2))^{\chi(1)} \quad \text{for complex } \mathfrak{p}_{\infty,i},$$
$$= \Gamma(s/2)^{(\chi(1)+\chi(\sigma))/2}\Gamma((s+1)/2)^{(\chi(1)-\chi(\sigma))/2} \quad \text{for real } \mathfrak{p}_{\infty,i},$$

where $\sigma \in G$ is the complex conjugation determined by a prime factor of $\mathfrak{p}_{\infty,i}$ in K. Next we introduce the notion of the **conductor** $\mathfrak{f}_\chi$ **with the group character** χ defined by Artin (*J. Reine Angew. Math.*, 164 (1931)). First, for any subset $\mathfrak{m} \subset G$, we put $\chi(\mathfrak{m}) = \sum_{m \in \mathfrak{m}} \chi(m)$; then $\mathfrak{f}_\chi$ is given by

$$\mathfrak{f}_\chi = \mathfrak{f}(\chi, K/k) = \prod_{\mathfrak{p}} \mathfrak{p}^{f(\mathfrak{p})},$$

where

$$f(\mathfrak{p}) = \frac{1}{e}[\{e\chi(1) - \chi(T)\} + \{p^{e_1}\chi(1) - \chi(V_1)\} + \{p^{e_2}\chi(1) - \chi(V_2)\} + \ldots],$$

and where $V_1, V_2, \ldots$, are the higher †ramification groups of prime factors of $\mathfrak{p}$ in K (in lower numbering) and $p^{e_i} = |V_i|$ (→ 14 Algebraic Number Fields I).

Now put

$$\xi(s, \chi, K/k) = \left(\frac{|d|^{\chi(1)} N_k(\mathfrak{f}_\chi)}{\pi^{n\chi(1)}}\right)^{s/2} \times \prod_{\mathfrak{p}_{\infty,i}} \gamma(s, \chi, \mathfrak{p}_{\infty,i}, K/k) \cdot L(s, \chi, K/k).$$

Then the functional equation is written

$$\xi(1-s, \bar{\chi}, K/k) = W(\chi)\xi(s, \chi, K/k), \quad |W(\chi)| = 1.$$

The known proof of this functional equation depends on (7) and the functional equations of Hecke L-functions discussed in Section E. As for the constants $W(\chi)$, there are significant results by B. Dwork, Langlands, and Deligne [D6].

(9) There are some applications to the theory of the distribution of prime ideals.

H. Weil L-Functions

Weil defined a new L-function that is a generalization of both Artin L-functions and Hecke L-functions with Grössencharakter [W5]. Let K be a finite Galois extension of an algebraic number field k, let C_K be the idele class group $K_A^\times / K^\times$ of K, and let $\alpha_{K/k} \in H^2(\mathrm{Gal}(K/k), C_K)$ be the †canonical cohomology class of †class field theory. Then this $\alpha_{K/k}$ determines an extension $W_{K/k}$ of $\mathrm{Gal}(K/k)$ by C_K: $1 \to C_K \to W_{K/k} \to \mathrm{Gal}(K/k) \to 1$ (exact), and the transfer induces an isomorphism $W_{K/k}^{ab} \cong C_k$, where ab denotes the topological commutator quotient. If L is a Galois extension of k containing K, then there is a canonical homomorphism $W_{L/k} \to W_{K/k}$. Hence we define the **Weil group** W_k for $\bar{k}/k$ as the †projective limit group $\mathrm{proj}_K \lim W_{K/k}$ of the $W_{K/k}$. It is obvious that we have a surjective homomorphism $\varphi: W_k \to \mathrm{Gal}(\bar{k}/k)$ and an isomorphism $r_k: C_k \to W_k^{ab}$, where W_k^{ab} is the maximal Abelian Hausdorff quotient of W_k. For $w \in W_k$, let $\|w\|$ be the adelic norm of $r_k^{-1}(w)$.

If k_v is a †local field, then we define the Weil group W_{k_v} for $\bar{k}_v/k_v$ by replacing the idele class group C_K with the multiplicative group $K_w^\times$ in the above definition, where K_w denotes a Galois extension of k_v. If k_v is the completion of a finite algebraic number field k at a place v, then we have natural homomorphisms $k_v^\times \to C_k$ and $\mathrm{Gal}(\bar{k}_v/k_v) \to \mathrm{Gal}(\bar{k}/k)$. Accordingly, we have a homomorphism $W_{k_v} \to W_k$ that commutes with these homomorphisms.

Let W_k be the Weil group of an algebraic number field k, and let $\rho: W_k \to GL(V)$ be a continuous representation of W_k on a complex vector space V. Let $v = \mathfrak{p}$ be a finite prime of k, and let ρ_v be the representation of W_{k_v} induced from ρ. Let Φ be an element of W_{k_v} such that $\varphi(\Phi)$ is the inverse Frobenius element of $\mathfrak{p}$ in $\mathrm{Gal}(\bar{k}_v/k_v)$, and let I be the subgroup of W_{k_v} consisting of elements w such that $\varphi(w)$ belongs to the †inertia group of $\mathfrak{p}$ in $\mathrm{Gal}(\bar{k}_v/k_v)$. Let V^I be the subspace of elements in V fixed by $\rho_v(I)$, let $N\mathfrak{p}$ be the norm of $\mathfrak{p}$, and let

$$L_{\mathfrak{p}}(V,s) = \det(1 - (N\mathfrak{p})^{-s}\rho_v(\Phi) \mid V^I)^{-1}.$$

We can define $L_v(V,s)$ for each Archimedean prime v also, and let

$$L(V,s) = \prod_v L_v(V,s).$$

Then this product converges for s in some right half-plane and defines a function $L(V,s)$. We call $L(V,s)$ the **Weil L-function** for the representation $\rho: W_k \to GL(V)$. This function $L(V,s)$ can be extended to a meromorphic function on the complex plane and satisfies the functional equation

$$L(V,s) = \varepsilon(V,s) L(V^*, 1-s)$$

(T. Tamagawa), where V^* is the dual of V, and $\varepsilon(V,s)$ is an exponential function of s of the form ab^s [T6].

P. Deligne generalized these results in the following manner: Let W_k' be a †group scheme over $\mathbf{Q}$ which is the †semidirect product of W_k by the additive group $\mathbf{G}_a$, on which W_k acts by the rule $wxw^{-1} = \|w\| x$. We can define the notion of representations of W_k' and the L-functions of them in the natural manner [T6].

I. The Riemann Hypothesis

As mentioned in Section B, the Riemann hypothesis asserts that all zeros of the Riemann ζ-function in $0 < \mathrm{Re}\, s < 1$ lie on the line $\mathrm{Re}\, s = 1/2$. In his celebrated paper [R1], Riemann gave six conjectures (including this), and assuming these conjectures, proved the †prime number theorem:

$$\pi(x) \sim \frac{x}{\log x} \sim \mathrm{Li}(x) = \int_2^x \frac{dx}{\log x}, \quad x \to \infty.$$

Here $\pi(x)$ denotes the number of prime numbers smaller than x. Among his six conjectures, all except the Riemann hypothesis have been proved (a detailed discussion is given in [L1]). The prime number theorem was proved independently by Hadamard and de La Vallée-Poussin without using the Riemann hypothesis (→ Section B; 123 Distribution of Prime Numbers B).

R. S. Lehman showed that there are exactly 2,500,000 zeros of $\zeta(\sigma + it)$ for which $0 < t < 170{,}571.35$, all of which lie on the critical line $\sigma = 1/2$ and are simple (*Math. Comp.*, 20 (1966)). Later R. P. Brent extended this computation up to 75,000,000 first zeros (1979).

Hardy proved that there are infinitely many zeros of $\zeta(s)$ on the line $\mathrm{Re}\, s = 1/2$ (1914). Furthermore, A. Selberg [S6] proved that if $N_0(T)$ is the number of zeros of $\zeta(s)$ on the line with $0 < \mathrm{Im}\, s < T$, then $N_0(T) > AT \log T$ (A is a positive constant) (1942). Thus if $N(T)$ is the number of zeros of $\zeta(s)$ in the rectangle $0 < \mathrm{Re}\, s < 1$, $0 < \mathrm{Im}\, s < T$, then $\liminf_{T\to\infty} N_0(T)/N(T) > 0$. N. Levinson proved $\liminf_{T\to\infty} N_0(T)/N(T) > 1/3$ (*Advances in Math.*, 13 (1974)). If $N_\varepsilon(T)$ is the number of zeros of $\zeta(s)$ in $1/2 - \varepsilon < \mathrm{Re}\, s < 1/2 + \varepsilon$, $0 < \mathrm{Im}\, s < T$, then $\lim_{T\to\infty} N_\varepsilon(T)/N(T) = 1$ for any positive number ε (H. Bohr and E. Landau, 1914). Bohr studied the distribution of the values of $\zeta(s)$ in detail and initiated the theory of †almost periodic functions (1925).

D. Hilbert remarked in his lecture at the Paris Congress that the Riemann hypothesis is equivalent to

$$\pi(x) = \mathrm{Li}(x) + O(\sqrt{x} \log x), \quad x \to \infty$$

(H. von Koch, 1901). It is also equivalent to

$$\sum_{n=1}^{N} \mu(n) = O(N^{1/2+\varepsilon}), \quad N \to \infty,$$

for any $\varepsilon > 0$, where $\mu(n)$ is the Möbius function. Assuming the Riemann hypothesis, we get

$$N(T) = \frac{1}{2\pi} T \log T - \frac{1 + \log 2\pi}{2\pi} T + o(\log T)$$

(Littlewood, 1924).

The computation of the zeros of the ζ-functions and the L-functions of general algebraic number fields is more difficult, but conjectures similar to the Riemann hypothesis have been proposed.

Weil showed that a necessary and sufficient condition for the validity of the Riemann hypothesis for all Hecke L-functions $L(s,\chi)$ is that a certain †distribution on the idele group $\mathbf{J}_k$ be positive definite [W1 (1952b)].

It is not known whether the general ζ- and L-functions of algebraic number fields have any zeros in the interval $(0,1)$ on the real axis (see the works of A. Selberg and S. Chowla). Similar problems are considered for the various ζ-functions given in Sections P, Q, and T.

J. p-Adic L-Functions

Let χ be a †primitive Dirichlet character with conductor f, and let $L(s,\chi)$ be the †Dirichlet L-function for χ. Then the values $L(1-n,\chi)$ of $L(s,\chi)$ at nonpositive integers $1-n$ $(n=1,2,\dots)$ are algebraic numbers (→ Section E). Let p be a prime number, let $\mathbf{Q}_p$ be the †p-adic number field, and let $\mathbf{C}_p$ be the completion of the algebraic closure $\bar{\mathbf{Q}}_p$ of $\mathbf{Q}_p$. It is known that $\mathbf{C}_p$ is also algebraically closed. Since $\mathbf{Q}\subset\mathbf{Q}_p$, we fix an embedding $\bar{\mathbf{Q}}\subset\bar{\mathbf{Q}}_p$ and consider $\{L(1-n,\chi)\}_{n=1}^{\infty}$ as a sequence in $\mathbf{C}_p$.

Let $|\ |_p$ be the extension to $\mathbf{C}_p$ of the standard p-adic valuation of $\mathbf{Q}_p$. Let q be p or 4 according as $p\neq 2$ or $p=2$, and let ω be the primitive Dirichlet character with conductor q satisfying $\omega(n)\equiv n \pmod q$ for any integer n prime to p. Then T. Kubota and H. W. Leopoldt proved that there exists a unique function $L_p(s,\chi)$ satisfying the conditions [K5]:

(1) $L_p(s,\chi)=\dfrac{a_{-1}}{s-1}+\sum_{n=0}^{\infty}a_n(s-1)^n \quad (a_n\in\mathbf{C}_p)$;

(2) $a_{-1}=0$ if $\chi\neq 1$ and the series $\Sigma_{n=0}^{\infty}a_n(s-1)^n$ converges for $|s-1|_p<|q^{-1}p^{1/(p-1)}|_p$;

(3) $L_p(1-n,\chi)=(1-\chi\omega^{-n}p^{1-n})L(1-n,\chi\omega^{-n})$ holds for $n=1,2,3,\dots$.

The function $L_p(s,\chi)$ satisfying these three conditions is called the **p-adic L-function** for the character χ. It is easy to see that $L_p(s,\chi)$ is identically zero if $\chi(-1)=-1$, but $L_p(s,\chi)$ is nontrivial if $\chi(-1)=1$.

Let B_n be the Bernoulli number. Then B_n satisfies the conditions: (1) B_n/n is p-integral if $(p-1)\nmid n$ (von Staudt) and (2) $(1/n)B_n\equiv(1/(n+p-1))B_{n+p-1} \pmod p$ holds in this case (Kummer). The generalization of these results for the generalized Bermoulli number $B_{\chi,n}$ was obtained by Leopoldt. Since $L(1-n,\chi)=-(1/n)B_{\chi,n}$, such p-integrabilities and congruences can be naturally interpreted and generalized in terms of the p-adic L-functions $L(s,\chi)$.

We assume $\chi(-1)=1$. Then $L_p(0,\chi)=(1-\chi\omega^{-1}(p))L(0,\chi\omega^{-1})$ and $\chi\omega^{-1}(-1)=-1$. Hence we can express the first factor h_N^- of the class number of a cyclotomic field $\mathbf{Q}(\exp(2\pi i/N))$ as a product of some $L_p(0,\chi)$'s. By using this fact, K. Iwasawa proved [I7] that the p-part $p^{e_n^-}$ of the †first factor $h_{Np^n}^-$ $(N\in\mathbf{N})$ satisfies

$$e_n^-=\lambda n+\mu p^n+\nu \qquad (\lambda,\mu,\nu\in\mathbf{Z};\lambda,\mu\geqslant 0)$$

for any sufficiently large n. Here Iwasawa conjectured $\mu=0$, which was proved by B. Ferrero and L. Washington [F1]. Also, we can obtain some congruences involving the first factor h_N^- of $\mathbf{Q}(\exp(2\pi i/N))$ from this formula.

Let χ be a nontrivial primitive Dirichlet character with conductor f, let

$$\tau(\chi)=\sum_{a=1}^{f}\chi(a)e^{2\pi ia/f}$$

be the †Gaussian sum for χ, and let $\log_p$ be the p-adic logarithmic function. Then Leopoldt [L6] calculated the value $L(1,\chi)$ and obtained

$$L_p(1,\chi)$$
$$=-\left(1-\frac{\chi(p)}{p}\right)\frac{\tau(\chi)}{f}\sum_{a=1}^{f}\chi(a)\log_p(1-e^{\ 2\pi ia/f}).$$

As an application of this formula, Leopoldt obtained a p-adic †class number formula for the maximal real subfield $F=\mathbf{Q}(\cos(2\pi/N))$ of $\mathbf{Q}(\exp(2\pi i/N))$. Let $\zeta_p(s,F)$ be the product of the $L_p(s,\chi)$ for all primitive Dirichlet characters χ such that (1) $\chi(-1)=1$ and (2) the conductor of χ is a divisor of N. We define the **p-adic regulator** R_p by replacing the usual log by the p-adic logarithmic function $\log_p$. Let h be the class number of F, $m=[F:\mathbf{Q}]$, and let d be the discriminant of F. Then the residue of $\zeta_p(s,F)$ at $s=1$ is

$$\prod_{\chi}\left(1-\frac{\chi(p)}{p}\right)\frac{2^{m-1}hR_p}{\sqrt{d}}.$$

Hence $\zeta_p(s,F)$ has a simple pole at $s=1$ if and only if $R_p\neq 0$. In general, for any totally real finite algebraic number field F, Leopoldt conjectured that the p-adic regulator R_p of F is not zero (**Leopoldt's conjecture**). This conjecture was proved by J. Ax and A. Brumer for the case when F is an Abelian extension of $\mathbf{Q}$ [A4, B7].

By making use of the Stickelberger element, Iwasawa gave another proof of the existence of the p-adic L-function [I7]. In particular, he obtained the following result: Let χ be a primitive Dirichlet character with conductor f. Then there exists a primitive Dirichlet character θ such that the p-part of the conductor of θ is either 1 or q and such that the conductor and the order of $\chi\theta^{-1}$ are both powers of p. Let $\mathfrak{o}_\theta$ be the ring generated over the ring $\mathbf{Z}_p$ of p-adic integers by the values of θ. Then there exists a unique element $f(x,\theta)$ of the quotient field of $\mathfrak{o}_\theta[[x]]$ depending only on θ and satisfying

$$L_p(s,\chi)=2f(\zeta(1+q_0)^s-1,\theta),$$

where q_0 is the least common multiple of q and the conductor of θ, and $\zeta=\chi(1+q_0)^{-1}$. Furthermore, Iwasawa proved that $f(x,\theta)$ belongs to $\mathfrak{o}_\theta[[x]]$ if θ is not trivial.

Let $P=\mathbf{Q}(\exp(2\pi i/q))$ and, for any $n\geqslant 1$, let $P_n=\mathbf{Q}(\exp(2\pi i/qp^n))$. Let $P_\infty=\bigcup_{n\geqslant 1}P_n$. Then P_∞ is a Galois extension of $\mathbf{Q}$ satisfying $\mathrm{Gal}(P_\infty/\mathbf{Q})\cong\mathbf{Z}_p^\times$ (the multiplicative group of p-adic units), and P is the subfield of $P_\infty/\mathbf{Q}$ corresponding to the subgroup $1+q\mathbf{Z}_p$ of $\mathbf{Z}_p^\times$.

Let ψ be a $\mathbf{C}_p$-valued primitive Dirichlet character such that (1) $\psi(-1)=-1$ and (2) the p-part of the conductor f_ψ of ψ is either 1 or q. Let K_ψ be the cyclic extension of $\mathbf{Q}$ corresponding to ψ by class field theory. Let $K=K_\psi\cdot P$, $K_n=K\cdot P_n$ and $K_\infty=K\cdot P_\infty$. Let A_n be the p-primary part of the ideal class group of K_n, let $A_n\to A_m$ $(n\geqslant m)$ be the mapping induced by the †relative norm N_{K_n/K_m}, and let $X_K=\varprojlim A_n$. Since each A_n is a finite p-group, X_K is a $\mathbf{Z}_p$-module. Let $V_K=X_k\otimes_{\mathbf{Z}_p}C_p$, and let

$$V_\psi=\{v\in V_K\mid \delta(v)=\psi(\delta)v \text{ for all } \delta\in\mathrm{Gal}(K/\mathbf{Q})\}.$$

Let q_0 be the least common multiple of f_ψ and q, and let γ_0 be the element of $\mathrm{Gal}(K_\infty/K)$ that corresponds to

$$1+q_0\in 1+q\mathbf{Z}_p=\mathrm{Gal}(P_\infty/P)$$

by the restriction mapping $\mathrm{Gal}(K_\infty/K)\hookrightarrow\mathrm{Gal}(P_\infty/P)$. Let $f_\psi(x)$ be the characteristic polynomial of γ_0-1 acting on V_ψ. Hence $f_\psi(x)$ is an element of $\mathfrak{o}_\psi[x]$.

We assume that $\omega\psi^{-1}$ is not trivial. Let $f(x,\omega\psi^{-1})$ be as before. Then $f(x,\omega\psi^{-1})$ is an element of $\mathfrak{o}_\psi[[x]]$. Iwasawa conjectured that $f_\psi(x)$ and $f(x,\omega\psi^{-1})$ coincide up to a unit of $\mathfrak{o}_\psi[[x]]$ (**Iwasawa's main conjecture**). This conjecture was proved recently by B. Mazur and A. Wiles in the case where ψ is a power of ω.

Let F be a totally real finite algebraic number field, let K be a totally real Abelian extension of F, and let χ be a character of $\mathrm{Gal}(K/F)$. Let $L(s,\chi)$ be the †Artin L-function for χ. Then we can construct the p-adic analog $L_p(s,\chi)$ of $L(s,\chi)$ (J.-P. Serre, J. Coates, W. Sinnott, P. Deligne, K. Ribet, P. Cassou-Noguès). But, at present, we have no formula for $L_p(1,\chi)$. Coates generalized Iwasawa's main conjecture to this case, but it has not yet been proven.

P-adic L-functions have been defined in some other cases (e.g. → [K3, M1, M3]).

K. ζ-Functions of Quadratic Forms

Dirichlet defined a Dirichlet series associated with a binary quadratic form and also considered a sum of such Dirichlet series extended over all classes of binary quadratic forms with a given discriminant D, which is actually equivalent to the Dedekind ζ-function of a quadratic field. Dirichlet obtained a formula for the class numbers of binary quadratic forms. The formula is interpreted nowadays as a formula for the class numbers of quadratic fields in the narrow sense.

According as the binary quadratic form is definite or indefinite, we apply different methods to obtain its class number.

Epstein ζ-functions: P. Epstein generalized the definition of the ζ-function of a positive definite binary quadratic form to the case of n variables (*Math. Ann.*, 56 (1903), 63 (1907)). Let V be a real vector space of dimension m with a positive definite quadratic form Q. Let M be a †lattice in V, and put

$$\zeta_Q(s,M)=\sum_{\substack{x\in M\\ x\neq 0}}\frac{1}{Q(x)^s},\qquad \mathrm{Re}\, s>\frac{m}{2}.$$

This series is absolutely convergent in $\mathrm{Re}\, s>m/2$, and

$$\lim_{s\to m/2}\left(s-\frac{m}{2}\right)\zeta_Q(s,M)=D(M)^{-1/2}\pi^{m/2}\Gamma\left(\frac{m}{2}\right)^{-1},$$

$$D(M)=\det|Q(x_i,x_j)|,$$

where $x_1,\ldots,x_m$ is a basis of M and $Q(x,y)=(Q(x+y)-Q(x)-Q(y))/2$. If the $Q(x)$ $(x\in M, x\neq 0)$ are all positive integers, we can write

$$\zeta_Q(s,M)=\sum_{n=1}^{\infty}\frac{a(n)}{n^s},$$

where $a(n)$ is the number of distinct $x\in M$ with $Q(x)=n$. In general, let $x_1,\ldots,x_m$ be a basis of the lattice M and $x_1^*,\ldots,x_m^*$ be its dual basis $(Q(x_i,x_j^*)=\delta_{ij})$. Call $M^*=\sum_i x_i^*\mathbf{Z}$ the **dual lattice** of M. If we consider the ϑ-series (†theta series)

$$\vartheta_Q(u,M)=\sum_{x\in M}\exp(-\pi u Q(x))\quad(\mathrm{Re}\, u>0),$$

then

$$\vartheta_Q(u,M)=(u^{-m/2}D(M)^{-1/2})\vartheta_Q(u^{-1},M^*).$$

With $\xi_Q(s,M)=\pi^{-s}\Gamma(s)\zeta_Q(s,M)$, the displayed equality leads to the functional equation

$$\xi_Q(s,M)=D(M)^{-1/2}\cdot\xi_Q\left(\frac{m}{2}-s,M^*\right).$$

In general, $\zeta_Q(s,M)$ has no Euler product expansion.

Consider the case where $M=\sum\mathbf{Z}x_i$ $(x_i=$

$(0, \ldots, 0, 1, 0, \ldots, 0))$, $Q(x)=\sum_{i=1}^m u_i^2$, for $x=(u_1, \ldots, u_m)$. If we put $\zeta_m(s)=\zeta_Q(s, M)$, $L(s)=\sum_{n=1}^\infty(-4/n)n^{-s}$, then we have

$$\zeta_1(s)=2\zeta(2s),$$

$$\zeta_2(s)=4\zeta(s)\cdot L(s)=4 \times (\text{the Dedekind } \zeta\text{-function of } \mathbf{Q}(\sqrt{-1})),$$

$$\zeta_4(s)=8(1-2^{2-2s})\zeta(s)\zeta(s-1),$$

$$\zeta_6(s)=-4(\zeta(s)L(s-2)-4\zeta(s-2)L(s)),$$

$$\zeta_8(s)=16(1-2^{1-s}+2^{4-2s})\zeta(s)\zeta(s-3),$$

$$\zeta_{10}(s)=(4/5)(\zeta(s)L(s-4)+4^2\zeta(s-4)L(s)) -2\sum_{\substack{\mu\in\mathbf{Z}[\sqrt{-1}]\\ \mu\neq 0}}\frac{\mu^4}{(\mu\bar{\mu})^s},$$

$$\zeta_{12}(s)=c_1 2^{-s}\zeta(s)\zeta(s-5)(2^6-2^{6-s}) +c_2\varphi\{\sqrt{\Delta(\tau)}\},$$

where $\varphi\{\sqrt{\Delta(\tau)}\}$ is the Dirichlet series corresponding to $\sqrt{\Delta(\tau)}$ by the †Mellin transform and $\Delta(\tau)=z\{\prod_{n=1}^\infty(1-z^n)\}^{24}$ with $z=e^{2\pi i\tau}$. $\zeta_m(s)$ has zeros on the line $\operatorname{Re}s=\sigma=m/4$, given explicitly for $m=4$, 8 as follows:

$$m=4:\quad s=1+l\pi i/\log 2,\quad l=1,2,\ldots,$$

$$m=8:\quad s=2+(i/\log 2)(2l\pi\pm\arctan\sqrt{15}),\quad l=0,\pm 1,\ldots.$$

Regarding the Epstein ζ-function of binary quadratic forms

$$\zeta_Q(s)=\sum_{m,n}{}' Q(m,n)^{-s},$$

with

$$Q(x,y)=ax^2+bxy+cy^2,\quad a,b,c\in\mathbf{R}, a>0, c>0, \Delta=4ac-b^2>0,$$

we have the Chowla-Selberg formula (1949):

$$\zeta_Q(s)=\left(2\zeta(2s)a^{-s}+\frac{2^{2s}a^{s-1}\sqrt{\pi}}{\Gamma(s)\Delta^{s-1/2}}\times\zeta(2s-1)\Gamma\left(s-\frac{1}{2}\right)\right)+\left(\frac{\pi^s 2^{s+3/2}}{a^{1/2}\Gamma(s)\Delta^{s/2-1/4}}\times\sum_{n=1}^\infty n^{s-1/2}\sigma_{1-2s}(n)\cos\frac{n\pi b}{a}\right)\times\int_0^\infty\varphi^{s-3/2}\exp\left\{-\frac{\pi n\Delta^{1/2}}{2a}(\varphi+\varphi^{-1})\right\}d\varphi,$$

where $\sigma_k(n)=\sum_{d|n}d^k$ and $\zeta(s)$ is the Riemann ζ-function. By using this formula, we can give another proof of the following result of H. Heilbronn: Let $h(-\Delta)$ be the class number of the imaginary quadratic field with discriminant $-\Delta$. Then $h(-\Delta)\to\infty$ $(\Delta\to\infty)$.

The following generalization of this result was obtained by C. L. Siegel [S22]: Let k be a fixed finite algebraic number field. Let K be a finite Galois extension of k, and let $d=d(K)$, $h=h(K)$, and $R=R(K)$ be the discriminant of K, the class number of K, and the regulator of K, respectively. We assume that K runs over extensions of k such that $[K:k]/\log d\to 0$; then we have

$$\log(hR)\sim\log\sqrt{|d|}.$$

Siegel ζ-functions of indefinite quadratic forms: Siegel defined and investigated some ζ-functions attached to nondegenerate indefinite quadratic forms, which are also meromorphic on the whole complex plane and satisfy certain functional equations [S24].

The case of quadratic forms with irrational algebraic coefficients was treated by Tamagawa and K. G. Ramanathan.

L. ζ-Functions of Algebras

K. Hey defined the ζ-function of a †simple algebra A over the rational number field $\mathbf{Q}$ (M. Deuring [D10]) (→ 27 Arithmetic of Associative Algebras). Consider an arbitrary †maximal order $\mathfrak{o}$ of A, and let

$$\zeta_A(s)=\sum_{\mathfrak{a}}\frac{1}{N(\mathfrak{a})^s},\quad \operatorname{Re}s>1,$$

with the summation taken over all left integral ideals $\mathfrak{a}$ of $\mathfrak{o}$. Then ζ_A is independent of the choice of a maximal order $\mathfrak{o}$. Let k be the †center of A, and put $[A:k]=n^2$. First, ζ_A is decomposed into Euler's infinite product expansion $\zeta_A(s)=\prod_{\mathfrak{p}}\mathbf{Z}_{\mathfrak{p}}(s)$ ($\mathfrak{p}$ runs over the prime ideals of k). For $\mathfrak{p}$ not dividing the discriminant $\mathfrak{d}$ of A, $Z_{\mathfrak{p}}(s)$ coincides with the $\mathfrak{p}$-component of $\prod_{j=0}^{n-1}\zeta_k(ns-j)$. Hence $\zeta_A(s)$ coincides with $\prod_{j=0}^{n-1}\zeta_k(ns-j)$ up to a product of $\mathfrak{p}$-factors for $\mathfrak{p}\,|\,\mathfrak{d}$ which are explicit rational functions of $N(\mathfrak{p})^{-ns}$.

Moreover, if A is the total matrix algebra of degree r over the division algebra $\mathfrak{D}$, then we have $\zeta_A(s)=\prod_{j=0}^{r-1}\zeta_{\mathfrak{D}}(rs-j)$, and $\zeta_{\mathfrak{D}}(s)$ satisfies a functional equation similar to that of $\zeta_k(s)$ (Hey). Also, $\zeta_A(s)$ is meromorphic over the whole complex plane, and at $s=1$, $(n-1)/n, \ldots, 1/n$, it has poles of order 1. Using analytic methods, M. Zorn (1931) showed that the simple algebra A with center k such that $A_{\mathfrak{p}}$ is a matrix algebra over $k_{\mathfrak{p}}$ for every finite or infinite prime divisor $\mathfrak{p}$ of k is itself a matrix algebra over k (→ 27 Arithmetic of Associative Algebras D). A purely algebraic proof of this was given by Brauer, H. Hasse, and E. Noether. G. Fujisaki (1958) gave another proof using the Iwasawa-Tate method. As a direct application of the ζ-function, the computation of the residue at $s=1$ of ζ_A leads to the formula containing the class number of maximal order $\mathfrak{O}$.

Godement defined the ζ-function of fairly general algebras [G1], and Tamagawa investigated in detail the explicit ζ-functions of division algebras, deriving their functional equations [T1].

Let $\tilde{A}=\prod'_{\mathfrak{p}} A_{\mathfrak{p}}$ be the adele ring of A, and let $G=\prod'_{\mathfrak{p}} G_{\mathfrak{p}}$ be the idele group (of A). We take a maximal order $\mathfrak{O}_{\mathfrak{p}}$ of $A_{\mathfrak{p}}$ and a maximal compact subgroup $U_{\mathfrak{p}}$ of $G_{\mathfrak{p}}$. Let $\omega_{\mathfrak{p}}$ be a †zonal spherical function of $G_{\mathfrak{p}}$ with respect to $U_{\mathfrak{p}}$; that is, $\omega_{\mathfrak{p}}$ is a function in $G_{\mathfrak{p}}$ and satisfies

$$\omega_{\mathfrak{p}}(ugv)=\omega_{\mathfrak{p}}(g) \quad (u,v\in U_{\mathfrak{p}}), \quad \omega_{\mathfrak{p}}(1)=1,$$

$$\int_{U_{\mathfrak{p}}}\omega_{\mathfrak{p}}(guh)\,du=\omega_{\mathfrak{p}}(g)\omega_{\mathfrak{p}}(h).$$

In addition, we define the weight function $\varphi_{\mathfrak{p}}$ on $A_{\mathfrak{p}}$ by

$$\varphi_{\mathfrak{p}}(x)=\begin{cases}\text{the characteristic function of } \mathfrak{O}_{\mathfrak{p}} \\ \text{when } \mathfrak{p} \text{ is finite,} \\ \\ \exp(-\pi T_{\mathfrak{p}}(xx^*)) \\ \text{when } \mathfrak{p} \text{ is infinite,}\end{cases}$$

where $T_{\mathfrak{p}}$ is the †reduced trace of $A_{\mathfrak{p}}/\mathbf{R}$ and $*$ is a positive †involution. Tamagawa gave an explicit form of the local ζ-function with the character $\omega_{\mathfrak{p}}$ defined by

$$\zeta_{\mathfrak{p}}(s,\omega_{\mathfrak{p}})=\int_{G_{\mathfrak{p}}}\varphi_{\mathfrak{p}}(g)\omega_{\mathfrak{p}}(g^{-1})|N_{\mathfrak{p}}(g)|_{\mathfrak{p}}^{s}\,dg,$$

where $N_{\mathfrak{p}}$ is the †reduced norm of $A_{\mathfrak{p}}/k_{\mathfrak{p}}$, and $|\ |_{\mathfrak{p}}$ is the valuation of $k_{\mathfrak{p}}$. Then $\omega=\prod_{\mathfrak{p}}\omega_{\mathfrak{p}}$ is the zonal spherical function of G with respect to $\prod U_{\mathfrak{p}}=U$. In particular, if ω is a positive definite zonal spherical function belonging to the spectrum of the discrete subgroup $\Gamma=A^*$ $=\{$all the invertible elements of $A\}$ of G, then the **Tamagawa ζ-function** with character ω is given by

$$\zeta(s,\omega)=\prod_{\mathfrak{p}}\zeta_{\mathfrak{p}}(s,\omega_{\mathfrak{p}})=\int_G\varphi(g)\omega(g^{-1})\|g\|^s\,dg,$$

where $\varphi(g)=\prod\varphi_{\mathfrak{p}}(g_{\mathfrak{p}})$ and $\|\ \|$ is the volume of the element g of G. When A is a division algebra, $\zeta(s,\omega)$ is analytically continued to a meromorphic function over the whole complex plane and satisfies the functional equation. The Tamagawa ζ-function may also be considered as one type of ζ-function of the Hecke operator. When A is an indefinite quaternion algebra over a totally real algebraic number field Φ, the groups of units of various orders of A operate discontinuously on the product of complex upper half-planes. Thus the spaces of holomorphic forms are naturally associated with A. The investigation of ζ-functions associated with these holomorphic automorphic forms was initiated by M. Eichler and extended by G. Shimura, H. Shimizu, and others. Eichler investigated the case $\Phi=\mathbf{Q}$, and Shimura and Shimizu investigated the case for an arbitrary totally real field Φ by defining general holomorphic automorphic forms, Hecke operators, and corresponding ζ-functions. The functional equations of these ζ-functions were proved by Shimizu. Shimizu generalized Eichler's work and found relations among ζ-functions of orders of various quaternion algebras belonging to different discriminants and levels [S10]. For the related results, see, e.g., the work of K. Doi and H. Naganuma [D12].

M. ζ-Functions Defined by Hecke Operators

The ζ-functions of algebraic number fields, algebras, or quadratic forms, and the L-functions are all defined by Dirichlet series, are analytically continued to univalent functions on the complex plane, and satisfy functional equations. One problem is to characterize the functions having such properties.

(1) H. Hamburger (1921–1922) characterized the Riemann ζ-function (up to constant multiples) by the following three properties: (i) It can be expanded as $\zeta(s)=\sum_{n=1}^{\infty}a_n/n^s$ $(\mathrm{Re}\,s\gg 0)$; (ii) it is holomorphic on the complex plane except as $s=1$, and $(s-1)\zeta(s)$ is an entire function of finite †genus; (iii) $G(s)=G(1-s)$, where $G(s)=\pi^{-s/2}\Gamma(s/2)\zeta(s)$.

(2) E. Hecke's theory [H4]: Fixing $\lambda>0$, $k>0$, $\gamma=\pm 1$, and putting

$$R(s)=(2\pi/\lambda)^{-s}\Gamma(s)\varphi(s)$$

for an analytic function $\varphi(s)$, we make the following three assumptions: (i) $(s-k)\varphi(s)$ is an entire function of finite genus; (ii) $R(s)=\gamma R(k-s)$; (iii) $\varphi(s)$ can be expanded as $\varphi(s)=\sum_{n=1}^{\infty}a_n/n^s$ $(\mathrm{Re}\,s>\sigma_0)$. Then we call $\varphi(s)$ a function belonging to the sign (λ,k,γ).

The functions $\zeta(2s)$, $L(2s)$, and $L(2s-1)$ satisfy assumptions (i)–(iii), where L may be either a Dirichlet L-function, an L-function with Grössencharakter of an imaginary quadratic field, or an L-function with class character of a real quadratic form whose Γ-factors are of the form $\Gamma(s/2)\Gamma((s+1)/2)\sim\Gamma(s)$. If $\varphi(s)$ belongs to the sign (λ,k,γ), then $n^{-s}\varphi(s)$ belongs to the sign $(n\lambda,k,\gamma)$. To each Dirichlet series $\varphi(s)=\sum_{n=1}^{\infty}a_n/n^s$ with the sign (λ,k,γ), we attach the series $f(\tau)=a_0+\sum_{n=1}^{\infty}a_n e^{2\pi i n\tau/\lambda}$, where

$$a_0=\gamma(2\pi/\lambda)^{-k}\Gamma(k)\mathrm{Res}_{s=k}(\varphi(s))$$
$$=\gamma\,\mathrm{Res}_{s=k}(R(s)).$$

This correspondence $\varphi(s)\rightarrow f(\tau)$ may also be realized by the †Mellin transform

$$R(s)=\int_0^{\infty}\left(\sum_{n=1}^{\infty}a_ne^{-2\pi ny/\lambda}\right)y^{s-1}\,dy$$

$$=\int_0^{\infty}(f(iy)-a_0)y^{s-1}\,dy,$$

$$f(iy)-a_0=\frac{1}{2\pi i}\int_{\mathrm{Re}\,s=\sigma_0}R(s)y^{-s}ds.$$

In this case, (i) $f(\tau)$ is holomorphic in the upper half-plane and $f(\tau+\lambda)=f(\tau)$, (ii) $f(-1/\tau)/(-i\tau)^k=\gamma f(\tau)$, and (iii) $f(x+iy)=O(y^{\mathrm{const}})$ $(y\to+0)$ uniformly for all x.

Conversely, the Dirichlet series $\varphi(s)=\sum_{n=1}^{\infty}a_nn^{-s}$ formed by the transformation in the previous paragraph from $f(\tau)$ satisfying (i)–(iii) belongs to the sign (λ,k,γ). We also say that the function $f(\tau)$ belongs to the sign (λ,k,γ).

If k is an even integer, then the functions $f(\tau)$ belonging to $(1,k,(-1)^{k/2})$ are the †modular forms of level 1 and weight k. A necessary and sufficient condition for a function $\varphi(s)$ belonging to $(1,k,(-1)^{k/2})$ to have an Euler product is that the corresponding modular form $f(\tau)$ be a simultaneous eigenfunction of the ring formed by the †Hecke operators T_n $(n=1,2,\dots)$. In this case, the coefficient a_n of $\varphi(s)=\sum a_nn^{-s}$ coincides with the eigenvalue of T_n. Namely, if $f\,|\,T_n=t_nf$, we have $\varphi(s)=a_1(\sum_{n=1}^{\infty}t_nn^{-s})$, and this is decomposed into the Euler product $\varphi(s)=a_1\prod_p(1-t_pp^{-s}+p^{k-1-2s})^{-1}$. We call $\varphi(s)/a_1$ a **ζ-function defined by Hecke operators** (Hecke [H5]). For example, $\zeta(s)\cdot\zeta(s-k+1)$ and the Ramanujan function

$$\sum_{n=1}^{\infty}\tau(n)n^{-s}=\prod_p(1-\tau(p)p^{-s}+p^{11-2s})^{-1}$$

are ζ-functions defined by Hecke operators. Hecke applied the theory of Hecke operators to study the group $\Gamma(N)$ [H5]; the situation is more complicated than the case of $\Gamma(1)=SL(2,\mathbf{Z})$. The space of automorphic forms of weight k belonging to the †congruence subgroup

$$\Gamma_0(N)=\left\{\begin{pmatrix}a&b\\c&d\end{pmatrix}\in SL(2,\mathbf{Z})\,\middle|\,c\equiv0\pmod N\right\}$$

is denoted by $\mathfrak{M}_k(\Gamma_0(N))$. The essential part of $\mathfrak{M}_k(\Gamma_0(N))$ is spanned by the functions $f(\tau)=\sum a_ne^{2\pi in\tau}$ satisfying the conditions: (1) $\varphi(s)=\sum a_nn^{-s}$ has the Euler product expansion

$$\varphi(s)=\prod_{p|N}(1-a_pp^{-s})^{-1}$$

$$\times\prod_{p\nmid N}(1-a_pp^{-s}+p^{k-1-2s})^{-1}.$$

(2) The functional equation $R(s)=\gamma R(k-s)$ holds, where $R(s)=(2\pi/\sqrt{N})^{-s}\Gamma(s)\varphi(s)$. (3) When χ is an arbitrary primitive character of $\mathbf{Z}$ such that the conductor f is coprime to N, then

$$R(s,\chi)=(2\pi/\sqrt{N}\,f)^{-s}\Gamma(s)\sum a_n\chi(n)n^{-s}$$

extends to an entire function satisfying the functional equation $R(s,\chi)=\omega R(k-s,\bar{\chi})$ $(|\omega|=1)$ (Shimura). Conversely, (2) and (3) characterize the Dirichlet series $\varphi(s)$ corresponding to $f(\tau)\in\mathfrak{M}_k(\Gamma_0(N))$ (Weil [W1 (1967a)]).

Considering the correspondence $f(\tau)=\sum a_nq^n\to\varphi(s)=\sum a_nn^{-s}$ not as a Mellin transformation but rather as a correspondence effected through Hecke operators, we can derive the ζ-function defined by Hecke operators. When the Hecke operator T_n is defined with respect to a discontinuous group Γ and we have a representation space $\mathfrak{M}$ of the Hecke operator ring $\mathscr{H}$, we denote the matrix of the operation of $T_n\in\mathscr{H}$ on $\mathfrak{M}$ by $(T_n)=(T_n)_{\mathfrak{M}}$ and call the matrix-valued function $\sum_n(T_n)_{\mathfrak{M}}n^{-s}$ the ζ-function defined by Hecke operators. The equation $\varphi(s)=\sum a_nn^{-s}$ is a specific instance of the correspondence in the first sentence, where $\Gamma=\Gamma(N)$, $\mathfrak{M}\subset\mathfrak{M}_k(\Gamma_0(N))$, $\dim\mathfrak{M}=1$. One advantage of this definition is that it may be applied whenever the concept of Hecke operators can be defined with respect to the group Γ (for instance, even for the Fuchsian group without a †cusp). Thus when Γ is a Fuchsian group given by the unit group of a quaternion algebra Φ over the rational number field $\mathbf{Q}$ and $\mathfrak{M}$ is the space of automorphic forms with respect to Γ, the ζ-function $\sum(T_n)n^{-s}$ is defined (Eichler). Moreover, by using its integral expression over the idele group $\mathbf{J}_\Phi$ of Φ, we can obtain its functional equation following the Iwasawa-Tate method (Shimura). Furthermore, by algebrogeometric consideration of T_n, it can be shown that

$$\zeta(s)\zeta(s-1)\det(\sum(T_n)_{\mathfrak{S}_2}n^{-s})$$

$$=\zeta(s)\zeta(s-1)\det\left(\prod_p(1-(T_p)_{\mathfrak{S}_2}p^{-s}+(R_p)_{\mathfrak{S}_2}p^{1-2s})^{-1}\right)$$

coincides (up to a trivial factor) with the Hasse ζ-function of some model of the Riemann surface defined by Γ when $\mathfrak{M}$ is the space $\mathfrak{S}_2$ of all †cusp forms of weight 2 (Eichler [E1], Shimura [S12]).

The algebrogeometric meaning of $\det(\sum(T_n)_{\mathfrak{S}_k}n^{-s})$, when $\mathfrak{M}$ is the space $\mathfrak{S}_k$ of all cusp forms of weight k, has been made clear for the case where Γ is obtained from $\Gamma_0(N)$, $\Gamma(N)$, and the quaternion algebra (M. Kuga, M. Sato, Shimura, Y. Ihara). From these facts, it becomes possible to express $(T_p)_{\mathfrak{S}_k}$, the decomposition of the prime number p in some type of Galois extension (Shimura [S14], Kuga), in terms of Hecke operators. These works gave the first examples of non-Abelian class field theory. Note that this type of ζ-function may be regarded as the analog (or generalization) of L-functions of algebraic number fields, as can be seen from the comparison in Table 1.

Table 1

Algebraic number field	k	Ideal group	Character χ	$\Sigma\chi(n)n^{-s}$
$\updownarrow$	$\updownarrow$	$\updownarrow$	$\updownarrow$	$\updownarrow$
Algebraic group	G	Hecke ring	Representation space $\mathfrak{M}$	$\Sigma(T_n)_{\mathfrak{M}}n^{-s}$

As for special values of ζ-functions defined by Hecke operators, the following fact is known: Let $f(\tau)=\Sigma a_n q^n \in \mathfrak{M}_k(SL(2,\mathbf{Z}))$ be a common eigenfunction of the Hecke operators, and let $\varphi(s)=\Sigma a_n n^{-s}$ be the corresponding Dirichlet series. Let K_f be the field generated over the rational number field $\mathbf{Q}$ by the coefficients a_n of f. Then, for any two integers m and m' satisfying $0<m, m'<k$ and $m\equiv m' \pmod 2$, the ratio $(R(m):R(m'))$ of the special values of

$$R(s)=\frac{\Gamma(s)}{(2\pi)^s}\varphi(s)=\int_0^\infty (f(iy)-a_0)y^{s-1}\,dy$$

at m and m' belongs to the field K_f.

G. Shimura discovered this fact for Ramanujan's function $\Delta(\tau)$ (*J. Math. Soc. Japan*, 11 (1959)), and then Yu. I. Manin generalized it to the above case and, by constructing a p-adic analog of $\varphi(s)$ from it, pointed out the importance of such results [M1]. R. M. Damerell also used such results to study special values of Hecke's L-function with Grössencharakter of an imaginary quadratic field (*Acta Arith.*, 17 (1970), 19 (1971)). Furthermore, Shimura generalized these results to congruence subgroups of $SL(2,\mathbf{Z})$ (*Comm. Pure Appl. Math.*, 29 (1976)), and to Hilbert modular groups (*Ann. Math.*, 102 (1975)). The connection between special values of ζ-functions and the periods of integrals has been studied further by Shimura, Deligne, and others.

In addition, in connection with nonholomorphic automorphic forms H. Maass considered L-functions of real quadratic fields (with class characters) having $\Gamma(s/2)^2$ or $\Gamma((s+1)/2)^2$ as Γ-factors. Furthermore, T. Kubota studied the relation of ζ-functions $\zeta_k(s)$ of an arbitrary algebraic field k or ζ-functions of simple rings to (nonanalytic) automorphic forms of several variables and considered the reciprocity law for the Gaussian sum from a new viewpoint.

N. L-Functions of Automorphic Representations (I)

R. P. Langlands reconstructed the theory of †Hecke operators from the viewpoint of representation theory and defined very general L-functions. He proposed many conjectures about them in [L4], and he and H. Jacquet proved most of them in [J1] for the case $G=GL_2$.

First Langlands defined the L-group ${}^LG$ for any connected reductive algebraic group G defined over a field k in the following manner [B6].

There is a canonical bijection between isomorphism classes of connected †reductive algebraic groups defined over a fixed algebraically closed field $\bar{k}$ and isomorphism classes of †root systems. It is defined by associating to G the root data $\Psi(G)=(X^*(T),\Phi,X_*(T),\Phi^v)$, where T is a †maximal torus of G, $X^*(T)$ ($X_*(T)$) the group of characters (†1-parameter subgroups) of T, Φ (Φ^v) the set of roots (coroots) of G with respect to T.

Since the choice of a †Borel subgroup B of G containing T is equivalent to that of a basis Δ of Φ, the aforementioned bijection yields one between isomorphism classes of triples (G,B,T) and isomorphism classes of based root data $\Psi_0(G)=(X^*(T),\Delta,X_*(T),\Delta^v)$. There is a split exact sequence

$$1\to \operatorname{Int} G\to \operatorname{Aut} G\to \operatorname{Aut}\Psi_0(G)\to 1.$$

and this mapping induces a canonical bijection $\operatorname{Aut}\Psi_0(G)\cong\operatorname{Aut}(G,B,T,\{x_\alpha\}_{\alpha\in\Delta})$ if $x_\alpha\in G_\alpha$ $(\alpha\in\Delta)$ are fixed.

Let G be a connected reductive algebraic group defined over $\bar{k}$. Let T be a maximal torus of G, and let B be a Borel subgroup of G containing T. Let $\Psi_0(G)=(X^*(T),\Delta,X_*(T),\Delta^v)$ be as before. Then there is a connected reductive algebraic group ${}^LG^0$ over $\mathbf{C}$ such that $\Psi_0(G)^\vee=(X_*(T),\Delta^v,X^*(T),\Delta)$ corresponds to the triple $({}^LG^0,{}^LB^0,{}^LT^0)$, where ${}^LB^0$ and ${}^LT^0$ are a Borel subgroup of ${}^LG^0$ and the maximal torus of ${}^LB^0$. For example, (1) if $G=GL_n$, then ${}^LG^0=GL_n(\mathbf{C})$; (2) if $G=Sp_{2n}$, then ${}^LG^0=SO_{2n+1}(\mathbf{C})$.

We assume that $\bar{k}$ is the algebraic closure of k and G is defined over k. Then $\gamma\in\operatorname{Gal}(\bar{k}/k)$ induces an automorphism of the $\bar{k}$-group $G\times_k\bar{k}$. Hence γ defines an element of $\operatorname{Aut}({}^LG^0,{}^LB^0,{}^LT^0)$ because it is a holomorphic image of $\operatorname{Aut}\Psi_0(G\times_k\bar{k})=\operatorname{Aut}\Psi_0(G\times_k\bar{k})^\vee$. Hence we can define the †semidirect product ${}^LG={}^LG^0\rtimes\operatorname{Gal}(\bar{k}/k)$, and call it the ***L*-group** of G.

Let k be a †local field, and let G be a connected reductive algebraic group defined over k. We identify G with the group of its k-rational points. Let W_k' be the Weil-Deligne group of k (→ Section H), and let $\Phi(G)$ be the set of homomorphisms $\varphi: W_k'\to{}^LG$ over $\operatorname{Gal}(\bar{k}/k)$. Let $\Pi(G)$ be the set of infinitesimal equivalence classes of irreducible **admissible** representations of G. If k is a non-Archimedean field, then

$\Pi(G)$ is the set consisting of equivalence classes of irreducible representations $\pi: G \to \mathrm{Aut}\, V$ on complex vector spaces V such that the space V^K of vectors invariant by K is finite dimensional for every compact open subgroup K of G and such that $V = \bigcup V^K$, where K runs over the compact open subgroups of G. If k is an Archimedean field, then $\Pi(G)$ is the set consisting of equivalence classes of representations π of the pair $(\mathfrak{g}, K)$ of the Lie algebra $\mathfrak{g}$ of G and a maximal compact subgroup K satisfying similar conditions [B6]. Then Langlands conjectured that we can parametrize $\Pi(G)$ by $\Phi(G)$ as $\Pi(G) = \bigcup_\varphi \Pi(G)_\varphi$. Let $\pi \in \Pi(G)_\varphi$ $(\varphi \in \Phi(G))$, and let r be a representation of ${}^LG$. Then we can define the **L-function** $L(s, \pi, r)$ and the **ε-factor** $\varepsilon(s, \pi, r)$ of π by

$$L(s, \pi, r) = L(s, r \circ \varphi), \quad \varepsilon(s, \pi, r) = \varepsilon(s, r \circ \varphi, \psi),$$

where the right-hand sides are those of the Weil-Deligne group (→ Section H) and ψ is a nontrivial character of k.

Let G be a connected reductive group over a global field k (i.e., an algebraic number field of finite degree or an algebraic function field of one variable over a finite field), let π be an irreducible admissible representation of $G_{\mathbf{A}}$, where $G_{\mathbf{A}}$ is the group of rational points of G over the †adele ring $k_{\mathbf{A}}$ of k, and let r be a finite-dimensional representation of ${}^LG$. Let ψ be a nontrivial character of $k_{\mathbf{A}}$ which is trivial on k. For any place v of k, let r_v be the representation of the L-group of $G_v = G \times_k k_v$ induced by r, and let ψ_v be the additive character of k_v associated with ψ. It is known that π is decomposed into the tensor product $\bigotimes \pi_v$ of $\pi_v \in \Pi(G(k_v))$ [B6]. Hence we put

$$L(s, \pi, r) = \prod_v L(s, \pi_v, r_v),$$

$$\varepsilon(s, \pi, r) = \prod_v \varepsilon(s, \pi_v, r_v).$$

The local factor $L(s, \pi_v, r_v)$ is in fact defined if v is Archimedean, or G is a †torus, or φ is unramified (i.e., G_v is quasisplit and splits over an unramified extension of k_v, and $G(\mathfrak{o}_v)$ is a special maximal compact subgroup of $G(k_v)$, and π_v is of class one with respect to $G(\mathfrak{o}_v)$, where $\mathfrak{o}_v$ is the integer ring of k_v). It follows that the right-hand side $\prod L(s, \pi_v, r_v)$ is defined up to a finite number of non-Archimedean places v. Furthermore, Langlands proved that $\prod \varepsilon(s, \pi_v, r_v)$ is in fact a finite product, and the infinite product $\prod L(s, \pi_v, r_v)$ converges in some right half-plane if π is automorphic (i.e., if π is a subquotient of the right regular representation of $G_{\mathbf{A}}$ in $G_k \backslash G_{\mathbf{A}}$). It is conjectured that $L(s, \pi, r)$ admits a meromorphic continuation to the whole complex plane and satisfies a functional equation

$$L(s, \pi, r) = \varepsilon(s, \pi, r) L(1-s, \tilde{\pi}, r)$$

if π is automorphic, where $\tilde{\pi}$ is the †contragredient representation of π. Furthermore, if $G = GL_n$ and r is the standard representation of GL_n, then we can construct $L(s, \pi, r)$ and $\varepsilon(s, \pi, r)$ by generalizing the Iwasawa-Tate method. We can also show in this case that $L(s, \pi, r)$ is entire if π is cuspidal. The conjectures are studied in some other cases [B6].

O. L-Functions of Automorphic Representations (II)

A. Weil generalized the theory of †Hecke operators and the corresponding L-functions to the case of †automorphic forms (for holomorphic and nonholomorphic cases together) of GL_2 over a global field [W9]. Then H. Jacquet and Langlands developed a theory from the viewpoint of †representation theory [J1, J2]. They attached L-functions not to automorphic forms but to †automorphic representations of $GL_2(k)$.

Let k be a non-Archimedean local field, and let $\mathfrak{o}_k$ be the maximal order of k. Let $\mathscr{H}_k$ be the space of functions on $G_k = GL_2(k)$ that are locally constant and compactly supported. Then $\mathscr{H}_k$ becomes an algebra with the convolution product

$$(f_1 * f_2)(h) = \int_{G_k} f_1(g) f_2(g^{-1}h)\, dg,$$

where dg is the †Haar measure of G_k that assigns 1 to the maximal compact subgroup $K_k = GL_2(\mathfrak{o}_k)$. Let π be a representation of $\mathscr{H}_k$ on a complex vector space V. Then we say that π is **admissible** if and only if π satisfies the following two conditions: (1) For every v in V, there is an f in $\mathscr{H}_k$ so that $\pi(f)v = v$; (2) Let σ_i $(i = 1, \ldots, r)$ be a family of inequivalent irreducible finite-dimensional representations of K_k, and let

$$\xi(g) = \sum_{i=1}^{r} \dim(\sigma_i)^{-1} \operatorname{tr} \sigma_i(g^{-1}).$$

Then ξ is an idempotent of $\mathscr{H}_k$. We call such a ξ an **elementary idempotent** of $\mathscr{H}_k$. Then for every elementary idempotent ξ of $\mathscr{H}_k$, the operator $\pi(\xi)$ has a finite-dimensional range. If π is an admissible representation of $GL_2(k)$ (→ Section N), then

$$\pi(f) = \int_{G_k} f(g)\pi(g)\, dg \qquad (f \in \mathscr{H}_k)$$

gives an admissible representation of $\mathscr{H}_k$ in this sense. Furthermore, any admissible representation of $\mathscr{H}_k$ can be obtained from an admissible representation of $GL_2(k)$.

Let k be the real number field. Let $\mathscr{H}_1$ be the space of infinitely differentiable compactly supported functions on $G_k(=GL_2(k))$ that are $K_k(=O(2,k))$ finite on both sides, let $\mathscr{H}_2$ be the space of functions on K_k that are finite sums of matrix elements of irreducible representations of K_k, and let $\mathscr{H}_k=\mathscr{H}_1\oplus\mathscr{H}_2$. Then $\mathscr{H}_1$, $\mathscr{H}_2$, and $\mathscr{H}_k$ become algebras with the convolution product. Let π be a representation of $\mathscr{H}_k$ on a complex vector space V. Then π is **admissible** if and only if the following three conditions are satisfied: (1) Every vector v in V is of the form $v=\sum_{i=1}^r \pi(f_i)v_i$ with $f_i\in\mathscr{H}_1$ and $v_i\in V$; (2) for every elementary idempotent $\xi(g)=\sum_{i=1}^r \dim(\sigma_i)^{-1}\operatorname{tr}\sigma_i(g^{-1})$, where the σ_i are a family of inequivalent irreducible representations of K_k, the range of $\pi(\xi)$ is finite-dimensional; (3) for every elementary idempotent ξ of $\mathscr{H}_k$ and for every vector v in $\pi(\xi)V$, the mapping $f\mapsto\pi(f)v$ of $\xi\mathscr{H}_1\xi$ into the finite-dimensional space $\pi(\xi)V$ is continuous. We can define the Hecke algebra $\mathscr{H}_k$ and the notion of admissible representations also in the case $k=\mathbf{C}$. In these cases, an admissible representation of $\mathscr{H}_k$ comes from a representation of the †universal enveloping algebra of $GL_2(k)$ but may not come from a representation of $GL_2(k)$. It is known that for any local field k, the †character of each irreducible representation is a locally integrable function.

Let k be a global field, $G_k=GL_2(k)$, and let $G_{\mathbf{A}}=GL_2(k_{\mathbf{A}})$ be the group of rational points of G_k over the adele ring $k_{\mathbf{A}}$ of k. For any place v of k, let k_v be the completion of k at v, let $G_v=GL_2(k_v)$, and let k_v be the standard maximal compact subgroup of G_v. Let $\mathscr{H}_v$ be the Hecke algebra $\mathscr{H}_{k_v}$ of G_v, and let ε_v be the normalized Haar measure of K_v. Then ε_v is an elementary idempotent of $\mathscr{H}_v$. Let $\mathscr{H}=\bigotimes_{\varepsilon_v}\mathscr{H}_v$ be the restricted tensor product of the local Hecke algebra $\mathscr{H}_v$ with respect to the family $\{\varepsilon_v\}$. We call $\mathscr{H}$ the **global Hecke algebra** of G_A.

Let π be a representation of $\mathscr{H}$ on a complex vector space V. We define the notion of admissibility of π as before. Then we can show that, for any irreducible admissible representation π of $\mathscr{H}$ and for any place v of k, there exists an irreducible admissible representation π_v of $\mathscr{H}_v$ on a complex vector space V_v such that (1) for almost all v, $\dim V_v^{K_v}=1$ and (2) π is equivalent to the restricted tensor product $\bigotimes\pi_v$ of the π_v with respect to a family of nonzero $x_v\in V_v^{K_v}$. Furthermore, the factors $\{\pi_v\}$ are unique up to equivalence.

Let k be a local field, let ψ be a nontrivial character of k, and let $\mathscr{H}_k$ be the Hecke algebra of $G_k=GL_2(k)$. Let π be an infinite-dimensional admissible irreducible representation of $\mathscr{H}_k$. Then there is exactly one space $W(\pi,\psi)$ of continuous functions on G_k with the following three properties: (1) If W is in $W(\pi,\psi)$, then for all g in G_k and for all x in k,

$$W(\begin{pmatrix}1 & x\\ 0 & 1\end{pmatrix}g)=\psi(x)W(g);$$

(2) $W(\pi,\psi)$ is invariant under the right translations of $\mathscr{H}_k$, and the representation on $W(\pi,\psi)$ is equivalent to π; (3) if k is Archimedean and if W is in $W(\pi,\psi)$, then there is a positive number N such that

$$W(\begin{pmatrix}t & 0\\ 0 & 1\end{pmatrix})=O(|t|^N)$$

as $|t|\to\infty$. We call $W(\pi,\psi)$ the **Whittaker model** of π. The Whittaker model exists in the global case if and only if each factor π_v of $\pi=\bigotimes\pi_v$ is infinite-dimensional.

Let k be a local field, and let π be as before. Then the L-function $L(s,\pi)$ and the ε-factor $\varepsilon(s,\pi,\psi)$ are defined in the following manner: Let ω be the quasicharacter of $k^\times$ (i.e., the continuous homomorphism $k^\times\to\mathbf{C}^\times$) defined by

$$\pi(\begin{pmatrix}a & 0\\ 0 & a\end{pmatrix})=\omega(a)\,id_V.$$

Then the †contragredient representation $\tilde{\pi}$ of π is equivalent to $\omega^{-1}\otimes\pi$. For any g in G_k and W in $W(\pi,\psi)$, let

$$\Psi(g,s,W)=\int_{k^\times}W(\begin{pmatrix}a & 0\\ 0 & 1\end{pmatrix}g)|a|^{s-1/2}\,d^\times a,$$

$$\tilde{\Psi}(g,s,W)=\int_{k^\times}W(\begin{pmatrix}a & 0\\ 0 & 1\end{pmatrix}g)|a|^{s-1/2}\omega^{-1}(a)\,d^\times a.$$

Then there is a real number s_0 such that these integrals converge for $\operatorname{Re}(s)>s_0$ for any $g\in G_k$ and $W\in W(\pi,\psi)$. If k is a non-Archimedean local field with $\mathbf{F}_q$ as its residue field, then there is a unique factor $L(s,\pi)$ such that $L(s,\pi)^{-1}$ is a polynomial of q^{-s} with constant term 1,

$$\Phi(g,s,W)=\Psi(g,s,W)/L(s,\pi)$$

is a holomorphic function of s for all g and W, and there is at least one W in $W(\pi,\psi)$ so that $\Phi(e,s,W)=a^s$ with a positive constant a. If k is an Archimedean local field, then we can define the gamma factor $L(s,\pi)$ in the same manner. Furthermore, for any local field k, if

$$\tilde{\Phi}(g,s,W)=\tilde{\Psi}(g,s,W)/L(s,\tilde{\pi}),$$

then there is a unique factor $\varepsilon(s,\psi,\pi)$ which, as a function of s, is an exponential such that

$$\tilde{\Phi}(\begin{pmatrix}0 & 1\\ -1 & 0\end{pmatrix}g,1-s,W)=\varepsilon(s,\pi,\psi)\Phi(g,s,W)$$

for all $g\in G_k$ and $W\in W(\pi,\psi)$.

Let π and π' be two infinite-dimensional irreducible admissible representations of G_k. Then π and π' are equivalent if and only if the quasicharacters ω and ω' are equal and

$$\frac{L(1-s,\chi^{-1}\otimes\tilde{\pi})\varepsilon(s,\chi\otimes\pi,\psi)}{L(s,\chi\otimes\pi)} = \frac{L(1-s,\chi^{-1}\otimes\tilde{\pi}')\varepsilon(s,\chi\otimes\pi',\psi)}{L(s,\chi\otimes\pi')}$$

holds for any quasicharacter χ. In particular, the set $\{L(s,\chi\otimes\pi)$ and $\varepsilon(s,\chi\otimes\pi,\psi)$ for all $\chi\}$ characterizes the representation π.

Let k be a global field, $G_k=GL_2(k)$, $G_{\mathbf{A}}=GL_2(k_{\mathbf{A}})$, and let $K_{\mathbf{A}}=\prod K_v$ be the standard maximal compact subgroup of $G_{\mathbf{A}}$. Then the †global Hecke algebra $\mathscr{H}$ acts on the space of continuous functions on $G_k\backslash G_{\mathbf{A}}$ by the right translations. Let φ be a continuous function on $G_k\backslash G_{\mathbf{A}}$. Then φ is an **automorphic form** if and only if (1) φ is $K_{\mathbf{A}}$-finite on the right, (2) for every †elementary idempotent ξ in $\mathscr{H}$, the space $(\xi\mathscr{H})\varphi$ is finite-dimensional, and (3) φ is slowly increasing if k is an algebraic number field. An automorphic form φ is a **cusp form** if and only if

$$\int_{k\backslash k_{\mathbf{A}}}\varphi\left(\begin{pmatrix}1 & x\\ 0 & 1\end{pmatrix}g\right)dx=0$$

for all g in $G_{\mathbf{A}}$. Let $\mathscr{A}$ be the space of automorphic forms on $G_k\backslash G_{\mathbf{A}}$, and let $\mathscr{A}_0$ be the space of cusp forms on $G_k\backslash G_{\mathbf{A}}$. They are $\mathscr{H}$-modules. Let $\psi=\prod\psi_v$ be a nontrivial character of $k\backslash k_{\mathbf{A}}$, and let π be an irreducible admissible representation $\pi=\bigotimes_v\pi_v$ of the global Hecke algebra $\mathscr{H}=\bigotimes_{\varepsilon_v}\mathscr{H}_v$. If π is a †constituent of the $\mathscr{H}$-module $\mathscr{A}$, then we can define the local factors $L(s,\pi_v)$ and $\varepsilon(s,\pi_v,\psi_v)$ for all v, although π_v may not be infinite-dimensional. Further, the infinite products

$L(s,\pi)=\prod L(s,\pi_v)$ and $L(s,\tilde{\pi})=\prod L(s,\tilde{\pi}_v)$

converge absolutely in a right half-plane, and the functions $L(s,\pi)$ and $L(s,\tilde{\pi})$ can be analytically continued to the whole complex plane as meromorphic functions of s. If π is a constituent of $\mathscr{A}_0$, then all π_v are infinite-dimensional, $L(s,\pi)$ and $L(s,\tilde{\pi})$ are entire functions, and π is contained in $\mathscr{A}_0$ with multiplicity one. If k is an algebraic number field, then they have only a finite number of poles and are bounded at infinity in any vertical strip of finite width. If k is an algebraic function field of one variable with field of constant $\mathbf{F}_q$, then they are rational functions of q^{-s}. In either case, $\varepsilon(s,\pi_v,\psi_v)=1$ for almost all v, and hence

$$\varepsilon(s,\pi)=\prod\varepsilon(s,\pi_v,\psi_v)$$

is well defined. Furthermore, the functional equation

$$L(s,\pi)=\varepsilon(s,\pi)L(1-s,\tilde{\pi})$$

is satisfied.

As for the condition for π being a constituent of $\mathscr{A}_0$, we have the following: Let $\pi=\bigotimes\pi_v$ be an irreducible admissible representation of $\mathscr{H}$. Then π is a constituent of $\mathscr{A}_0$ if and only if (1) for every v, π_v is infinite-dimensional; (2) the quasicharacter η defined by

$$\pi\left(\begin{pmatrix}a & 0\\ 0 & a\end{pmatrix}\right)=\eta(a)id.$$

is trivial on $k^\times$; (3) π satisfies a certain condition so that, for any quasicharacter ω of $k^\times\backslash k_{\mathbf{A}}^\times$, $L(s,\omega\otimes\pi)=\prod L(s,\omega_v\otimes\pi_v)$ and $L(s,\omega^{-1}\otimes\tilde{\pi}_v)=\prod L(s,\omega_v^{-1}\otimes\tilde{\pi}_v)$ converge on a right half-plane; and (4) for any quasicharacter ω of $k^\times\backslash k_{\mathbf{A}}^\times$, $L(s,\omega\otimes\pi)$ and $L(s,\omega^{-1}\otimes\tilde{\pi})$ are entire functions of s which are bounded in vertical strips and satisfy the functional equation

$$L(s,\omega\otimes\pi)=\varepsilon(s,\omega\otimes\pi)L(1-s,\omega^{-1}\otimes\tilde{\pi}).$$

P. Congruence ζ-Functions of Algebraic Function Fields of One Variable or of Algebraic Curves

Let K be an †algebraic function field of one variable over $k=\mathbf{F}_q$ (finite field with q elements). The **ζ-function of the algebraic function field** K/k, denoted by $\zeta_K(s)$, is defined by the infinite sum $\sum_{\mathfrak{A}}N(\mathfrak{A})^{-s}$, where the summation is over all integral divisors $\mathfrak{A}$ of K/k and where the norm $N(\mathfrak{A})$ equals $q^{\deg(\mathfrak{A})}$. Equivalently, $\zeta_K(s)$ is defined by the infinite product $\prod_{\mathfrak{p}}(1-N(\mathfrak{p})^{-s})^{-1}$, where $\mathfrak{p}$ runs over all prime divisors of K/k. By the change of variable $u=q^{-s}$, $\zeta_K(s)=Z_K(u)$ becomes a formal power series in u. $\zeta_K(s)$ and $Z_K(u)$ are sometimes called the **congruence ζ-functions** of K/k.

The fundamental theorem states that (i) (Rationality) $Z_K(u)$ is a rational function of u of the form $Z_K(u)=P(u)/(1-u)(1-qu)$, where $P(u)\in\mathbf{Z}[u]$ is a polynomial of degree $2g$, g being the genus of K; (ii) (Functional equation) $Z_K(u)$ satisfies the functional equation

$$Z_K(1/u)=q^{g-1}u^{2-2g}Z_K(u/q);$$

and (iii) if $P(u)$ is decomposed into linear factors in $\mathbf{C}[u]$: $P(u)=\prod_{i=1}^{2g}(1-\alpha_i u)$, then all the reciprocal roots α_i are complex numbers of absolute value $\sqrt{q}$. Statement (iii) is the analog of the **Riemann hypothesis** because it is equivalent to saying that all the zeros of $\zeta_K(s)=Z_K(q^{-s})$ lie on the line $\mathrm{Re}\,s=1/2$.

The congruence ζ-function was introduced by E. Artin [A1 (1924)] as an analog of the Riemann or Dedekind ζ-functions. Of its fundamental properties, the rationality (i) and the functional equation (ii) were proven by

F. K. Schmidt (1931), using the †Riemann-Roch theorem for the function field K/k. The Riemann hypothesis (iii) was verified first in the elliptic case ($g=1$) by H. Hasse [H1] and then in the general case by A. Weil [W2 (1948)]. For the proof of (iii), it was essential to consider the geometry of algebraic curves that correspond to given function fields.

Let C be a nonsingular complete curve over k with function field K. Then $Z_K(u)$ coincides with the ζ-function of C/k, denoted by $Z(u, C)$, which is defined by the formal power series $\exp(\sum_{m=1}^{\infty} N_m u^m/m)$. Here N_m is the number of rational points of C over the extension k_m of k of degree m. The rationality of $Z_K(u)$ is then equivalent to the formula

$$N_m = 1 + q^m - \sum_{i=1}^{2g} \alpha_i^m \qquad (m \in \mathbf{N}),$$

and the Riemann hypothesis for $Z_K(u)$ is equivalent to the estimate

$$(*) \quad |N_m - 1 - q^m| \leqslant 2g\, q^{m/2} \qquad (m \in \mathbf{N}).$$

Now if F is the qth power morphism of C to itself (the **Frobenius morphism** of C relative to k), then an important observation is that N_m is the number of fixed points of the mth iterate F^m of F. In other words, N_m is equal to the intersection number of the graph of F^m with the diagonal on the surface $C \times C$, and is related to the "trace" of the Frobenius correspondence. Then (*) follows from †Castelnuovo's lemma in the theory of correspondences on a curve. This is Weil's proof of the Riemann hypothesis in [W2]; compare the proof by A. Mattuck and J. Tate (*Abh. Math. Sem. Hamburg* 22 (1958)) and A. Grothendieck (*J. Reine Angew. Math.*, 200 (1958)) using the Riemann-Roch theorem for an algebraic surface.

On the other hand, let J be the †Jacobian variety of C over k. For each prime number l different from the characteristic of k, let $M_l(\alpha)$ denote the †l-adic representation of an endomorphism α of J obtained from its action on points of J of order l^n ($n=1,2,\ldots$). Letting π be the endomorphism of J induced from F (which is the same as the Frobenius morphism of J), we have $P(u) = \det(1 - M_l(\pi)u)$, i.e., the numerator of the ζ-function coincides with the characteristic polynomial of $M_l(\pi)$. In this setting, the Riemann hypothesis is a consequence of the positivity of the Rosati antiautomorphism [E1]. This is the second proof given by Weil [W2], and applies to arbitary Abelian varieties.

Recently E. Bombieri, inspired by Stepanov's idea, gave an elementary proof of (*) using only the Riemann-Roch theorem for a curve (*Sém. Bourbaki*, no. 430 (1973)).

Q. ζ-Functions of Algebraic Varieties over Finite Fields

Let V be an algebraic variety over the finite field with q elements $\mathbf{F}_q$, and let N_m be the number of $\mathbf{F}_q{}^m$-rational points of V. Then the **ζ-function of V over $\mathbf{F}_q$** is the formal power series in $\mathbf{Z}[[u]]$ defined by

$$Z(u, V) = \exp\left(\sum_{m=1}^{\infty} N_m u^m/m\right);$$

alternatively it can be defined by the infinite product $\prod_P (1 - u^{\deg P})^{-1}$, where P runs over the set of prime divisors of V and $\deg P$ is the degree of the residue field of P over $\mathbf{F}_q$ (in other words, P runs over prime rational 0-cycles of V over $\mathbf{F}_q$).

Weil Conjecture. In 1949, the following properties of the ζ-function were conjectured by Weil [W3]. Let V be an n-dimensional complete nonsingular (absolutely irreducible) variety over $\mathbf{F}_q$. Then (1) $Z(u, V)$ is a rational function of u. (2) $Z(u, V)$ satisfies the functional equation

$$Z((q^n u)^{-1}, V) = \pm q^{n\chi/2} u^{\chi} Z(u, V),$$

where the integer χ is the intersection number (the degree of $\Delta_V \cdot \Delta_V$) of the diagonal subvariety Δ_V with itself in the product $V \times V$, which is called the Euler-Poincaré characteristic of V. (3) Moreover, we have

$$Z(u, V) = \frac{P_1(u) \cdot P_3(u) \cdot \ldots \cdot P_{2n-1}(u)}{P_0(u) \cdot P_2(u) \cdot \ldots \cdot P_{2n}(u)},$$

where $P_h(u) = \prod_{j=1}^{B_h} (1 - \alpha_j^{(h)} u)$ is a polynomial with $\mathbf{Z}$-coefficients such that $\alpha_j^{(h)}$ are algebraic integers of absolute value $q^{h/2}$ ($0 \leqslant h \leqslant 2n$); the latter statement is the **Riemann hypothesis for $V/\mathbf{F}_q$**. (4) When V is the reduction mod p of a complete nonsingular variety V^* of characteristic 0, then the degree B_h of $P_h(u)$ is the hth Betti number of V^* considered as a complex manifold.

This conjecture, called the **Weil conjecture**, has been completely proven. To give a brief history, first B. Dwork [D13] proved the rationality of the ζ-function for any (not necessarily complete or nonsingular) variety over $\mathbf{F}_q$. Then A. Grothendieck [A3, G2, G3] developed the l-adic étale cohomology theory with M. Artin and others, and proved the above statements (1)–(4) (except for the Riemann hypothesis) with $P_h(u)$ replaced by some $P_{h,l}(u) \in \mathbf{Q}_l[u]$; and S. Lubkin [L7] obtained similar results for liftable varieties. Finally Deligne [D4] proved the Riemann hypothesis and the independence of l of $P_{h,l}(u)$. More details will be given below. Before the final solution for the general case was obtained, the conjecture had been verified for some special types of varieties. For curves and Abelian varieties, its truth was previously shown by Weil (→ Section P). In the paper [W3] in which he proposed the above conjecture, Weil verified it for Fermat hypersurfaces, i.e., those defined by the equation $a_0x_0^m+\ldots+a_{n+1}x_{n+1}^m=0$ ($a_i\in\mathbf{F}_q^\times$); in this case, the ζ-function is of the form $P(u)^{(-1)^{n+1}}/\prod_{j=0}^n(1-q^ju)$ with a polynomial $P(u)$ that can be explicitly described in terms of Jacobi sums. Dwork [D14] studied by p-adic analysis the case of hypersurfaces in a projective space, verifying the conjecture for them except for the Riemann hypothesis. Further nontrivial examples were provided by †$K3$ surfaces (Deligne [D2], Pyatetskiĭ-Shapiro, Shafarevich [P1]) and cubic 3-folds (E. Bombieri, H. Swinnerton-Dyer [B5]); in these cases the proof of the Riemann hypothesis was reduced to that of certain Abelian varieties naturally attached to these varieties. It can be said that the Weil conjecture has greatly influenced the development of algebraic geometry, as regards both the foundations and the methods of proof of the conjecture itself; see the expositions by N. Katz [K2] or B. Mazur [M2].

Weil Cohomology, *l*-Adic Cohomology. The Weil conjecture suggested the possibility of a good cohomology theory for algebraic varieties over a field of arbitrary characteristic. We first formulate the desired properties of a good cohomology (S. Kleiman [K4]). Let $\bar{k}$ be an algebraically closed field and K a field of characteristic 0, which is called the coefficient field. A contravariant functor $V\to H^*(V)$ from the category of complete connected smooth varieties over $\bar{k}$ to the category of augmented $\mathbf{Z}^+$-graded finite-dimensional anticommutative K-algebras (cup product as multiplication) is called a **Weil cohomology** with coefficients in K if it has the following three properties. (1) **Poincaré duality**: If $n=\dim V$, then a canonical isomorphism $H^{2n}(V)\cong K$ exists and the cup product $H^j(V)\times H^{2n-j}(V)\to H^{2n}(V)\cong K$ induces a perfect pairing. (2) **Künneth formula**: For any V_1 and V_2 the mapping $H^*(V_1)\otimes H^*(V_2)\to H^*(V_1\times V_2)$ defined by $a\otimes b\to \mathrm{Proj}_1^*(a)\cdot\mathrm{Proj}_2^*(b)$ is an isomorphism. (3) Good relation with **algebraic cycles**: Let $C^j(V)$ be the group of algebraic cycles of codimension j on V. There exists a fundamental-class homomorphism FUND: $C^j(V)\to H^{2j}(V)$ for all j, which is functorial in V, compatible with products via Künneth's formula, has compatibility of the intersection with the cup product, and maps 0-cycle$\in C^n(V)$ to its degree as an element of $K\cong H^{2n}(V)$. If a Weil cohomology theory H exists for the V's over $\bar{k}$, we can prove the **Lefschetz fixed-point formula**:

$$((\text{graph of }F)\cdot(\text{diagonal}))_{V\times V}=\sum_{j=0}^{2n}(-1)^j\mathrm{tr}(F^*\,|\,H^j(V))$$

for a morphism $F:V\to V$.

If $\bar{k}=\mathbf{C}$ (the field of complex numbers), the classical cohomology $V\to H^*(V^{an},\mathbf{Q})$, where V^{an} denotes the complex manifold associated with V, gives a Weil cohomology. If $\bar{k}$ is an arbitrary algebraically closed field and if l is a prime number different from the characteristic of $\bar{k}$, then the principal results in the theory of the étale cohomology state that the l-adic cohomology $V\to H^*_{\text{ét}}(V,\mathbf{Q}_l)$ is a Weil cohomology with coefficient field $\mathbf{Q}_l$ (the field of l-adic numbers) [A3, D5, G3, M4]. In defining this, Grothendieck introduced a new concept of topology, which is now called Grothendieck topology. In the étale topology of a variety V, for example, any étale covering of a Zariski open subset is regarded as an "open set." With respect to the étale topology, the cohomology group $H^*(V,\mathbf{Z}/n)$ of V with coefficients in $\mathbf{Z}/n$ is defined in the usual manner and is a finite $\mathbf{Z}/n$-module. If l is a prime number as above, $\varprojlim_\nu H^*(V,\mathbf{Z}/l^\nu)$ is a module over $\mathbf{Z}_l=\varprojlim_\nu\mathbf{Z}/l^\nu$ of finite rank, and

$$H^*_{\text{ét}}(V,\mathbf{Q}_l)=(\varprojlim_\nu H^*(V,\mathbf{Z}/l^\nu))\otimes_{\mathbf{Z}_l}\mathbf{Q}_l$$

defines the l-adic cohomology group, giving rise to a Weil cohomology.

For the characteristic p of k, p-adic étale cohomology does not give Weil cohomology; but the crystalline cohomology (Grothendieck and P. Berthelot [B2, B3]) takes the place of p-adic cohomology and is almost a Weil cohomology: in this theory the fundamental class is defined only for smooth subvarieties.

Now fix a Weil cohomology for $\bar{k}=\bar{\mathbf{F}}_q$, an algebraic closure of a finite field $\mathbf{F}_q$. Given an algebraic variety V over $\mathbf{F}_q$, let $\bar{V}=V\otimes\bar{k}$ denote the base extension of V to $\bar{k}$; then $\mathbf{F}_{q^m}$-rational points of V can be identified with the fixed points of the mth iterate of the Frobenius morphism F of V relative to $\mathbf{F}_q$. Then the Lefschetz fixed-point formula implies the rationality of $Z(u,V)$; more precisely, letting $P_j(u)=\det(1-uF^*\,|\,H^j(\bar{V}))$ be the characteristic polynomial of the automorphism F^* of $H^j(\bar{V})$ induced by F, we have

$$Z(u,V)=\prod_{j=0}^{2n}P_j(u)^{(-1)^{j+1}}.$$

The functional equation of the ζ-function then follows from the Poincaré duality. This proves (1), (2), and a part of (3) in the statement of the Weil conjecture. Further, in the case of l-adic cohomology, (4) means that $\deg P_j(u)=$

$\dim_{\mathbf{Q}_l} H^j(\bar{V}, \mathbf{Q}_l)$ is equal to the jth Betti number of a lifting of V to characteristic 0; this follows from the comparison theorem of M. Artin for the l-adic cohomology and the classical cohomology, combined with the invariance of l-adic cohomology under specialization.

Proof of the Riemann Hypothesis. In 1974, Deligne [D4, I] completed the proof of the Weil conjecture for projective nonsingular varieties by proving that, given such a V over $\mathbf{F}_q$, any eigenvalue of F^* on $H^j_{\text{ét}}(\bar{V}, \mathbf{Q}_l)$ is an algebraic integer, all the conjugates of which are of absolute value $q^{j/2}$. (This implies that $P_j(u) = \det(1 - uF^* \mid H^j_{\text{ét}}(\bar{V}, \mathbf{Q}_l))$ is in $\mathbf{Z}[u]$ and is independent of l.) The proof is done by induction on $n = \dim V$; by the general results in l-adic cohomology (the weak Lefschetz theorem on a hyperplane section, the Poincaré duality, and the Künneth formula), the proof is reduced to the assertion that (*) any eigenvalue α of F^* on $H^n_{\text{ét}}(\bar{V}, \mathbf{Q}_l)$ is an algebraic integer such that $|\alpha'| \leqslant q^{(n+1)/2}$ for all conjugates α' of α. The main ingredients in proving (*) are (1) Grothendieck's theory of L-functions, based on the étale cohomology with compact support and with coefficients in a $\mathbf{Q}_l$-sheaf [G2, G3]; (2) the theory of Lefschetz pencils (Deligne and Katz [D7]), and the Kajdan-Margulis theorem on the monodromy of a Lefschetz pencil (J. L. Verdier, *Sém. Bourbaki*, no. 423 (1972)); and (3) Rankin's methods to estimate the coefficients of modular forms, as adapted to the Grothendieck's L-series. By means of these geometric and arithmetic techniques, Deligne achieved the proof of the Riemann hypothesis for projective nonsingular varieties. For the generalization to complete varieties, see Deligne [D4, II].

Applications of the (Verified) Weil Conjecture. (1) The Ramanujan conjecture (→ 32 Automorphic Functions D): The connection of this conjecture and the Weil conjecture for certain fiber varieties over a modular curve was observed by M. Sato and partially verified by Y. Ihara [I1] and then established by Deligne [D3]. The Weil conjecture as proven above implies the truth of the Ramanujan conjecture and its generalization by H. Petersson.

(2) Estimation of trigonometric sums: Let q be the power of a prime number p. Then

$$\left| \sum_{(x_1, \ldots, x_n) \in \mathbf{F}_q^n} \exp \frac{2\pi i}{p} \operatorname{tr}_{\mathbf{F}_q/\mathbf{F}_p}(F(x_1, \ldots, x_n)) \right| \leqslant (d-1)^n q^{n/2},$$

where $F(X_1, \ldots, X_n) \in \mathbf{F}_q[X_1, \ldots, X_n]$ is a polynomial of degree d that is not divisible by p, and the homogeneous part of the highest degree of F defines a smooth irreducible hypersurface in $\mathbf{P}^{n-1}$. This is a generalization of the Weil estimation of the Kloosterman sum ([D4, W1 (1948c)]; → 4 Additive Number Theory D).

(3) The **hard Lefschetz theorem**: Let $L \in H^2(V)$ be the class of a hyperplane section of an n-dimensional projective nonsingular variety V over an algebraically closed field. Then the cup product by $L^i: H^{n-i}(V) \to H^{n+i}(V)$ is an isomorphism for all $i \leqslant n$. Deligne [D4, II] proved this for l-adic cohomology, from which N. Katz and W. Messing [K1] deduced its validity in any Weil cohomology or in the crystalline cohomology.

Also some geometric properties of an algebraic variety V are reflected in the properties of $Z(u, V)$. The ζ-function $Z(u, A)$ of an Abelian variety A determines the isogeny class of A [T4]. For any algebraic integer α, every conjugate of which has absolute value $q^{1/2}$, there exists an Abelian variety $A/\mathbf{F}_q$ such that α is a root of $\det(1 - uF^* \mid H^1(A)) = 0$ [H6]. J. Tate [T3] conjectured that the rank of the space cohomology classes of algebraic cycles of codimension r is equal to the order of the pole at $u = 1/q^r$ of $Z(u, V)$. This conjecture is still open but has been verified in certain nontrivial cases, e.g., (1) products of curves and Abelian varieties, $r = 1$ (Tate [T4]), (2) Fermat hypersurfaces of dimension $2r$ with some condition on the degree and the characteristic (Tate [T3], T. Shioda, *Proc. Japan Acad.* 55 (1979)), and (3) elliptic $K3$ surfaces, $r = 1$ (M. Artin and Swinnerton-Dyer, *Inventiones Math.* 20 (1973)).

R. ζ- and L-Functions of Schemes

Let X be a †scheme of finite type over $\mathbf{Z}$, and let $|X|$ denote the set of closed points of X; for each $x \in |X|$, the residue field $k(x)$ is finite, and its cardinality is called the norm $N(x)$ of x. The **ζ-function of a scheme** X is defined by the product $\zeta(s, X) = \prod_{x \in |X|} (1 - N(x)^{-s})^{-1}$. This converges absolutely for $\operatorname{Re} s > \dim X$, and it is conjectured to have an analytic continuation in the entire s-plane (Serre [S7]). It reduces to the Riemann (resp. Dedekind) ζ-function if $X = \operatorname{Spec}(\mathbf{Z})$ (resp. $\operatorname{Spec}(\mathfrak{o})$, $\mathfrak{o}$ being the ring of integers of an algebraic number field), and to the ζ-function $Z(q^{-s}, X)$ (→ Section Q) if X is a variety over a finite field $\mathbf{F}_q$. The case of varieties defined over an algebraic number field is discussed in Section S.

Let G be a finite group of automorphisms of a scheme X, and assume that the quotient $Y = X/G$ exists (e.g., X is quasiprojective). For an element x in $|X|$, let y be its image in $|Y|$, and let $D(x) = \{g \in G \mid g(x) = x\}$, the decomposition group of x over y. The natural mapping $D(x) \to \operatorname{Gal}(k(x)/k(y))$ is surjective, and its kernel $I(x)$ is called the inertia group at x. An element of $D(x)$ is called a Frobenius element at x if its image in $\mathrm{Gal}(k(x)/k(y))$ corresponds to the $N(y)$th-power automorphism of $k(x)$. Now let R be a representation of G with character χ. The **Artin L-function** $L(s, X, \chi)$ is defined by

$$L(s,X,\chi)=\exp\left(\sum_{y\in|Y|}\sum_{n=1}^{\infty}\chi(y^n)N(y)^{-ns}/n\right)$$
$$=\prod_{y\in|Y|}\det(1-R(F_y)N(y)^{-s})^{-1},$$

where $\chi(y^n)$ denotes the mean value of χ on the nth power of Frobenius elements F_x at x (x any point of $|X|$ over y), and similarly $R(F_y)$ denotes the mean value of $R(F_x)$; it converges absolutely for $\mathrm{Re}\,s>\dim X$. Again this is reduced to the usual Artin L-function (→ Section G) if X is the spectrum of the ring of integers of an algebraic number field. The Artin L-functions of a scheme have many formal properties analogous to those of Artin L-functions of a number field (Serre [S7]).

Let us consider the case where X is an algebraic variety over a finite field $\mathbf{F}_q$ and elements of G are automorphisms of X over $\mathbf{F}_2$; in this case, $L(s, X, \chi)$ is a formal power series in $u-q^{-s}$, which is called a congruence Artin L-function. For the case where X is a complete nonsingular algebraic curve and χ is an irreducible character of G different from the trivial one, Weil [W2] proved that $L(s, X, \chi)$ is a polynomial in $u=q^{-s}$; thus the analog of †Artin's conjecture holds here. More generally, for any algebraic variety X over $\mathbf{F}_q$, Grothendieck [G2, G3] proved the rationality of L-functions together with the alternating product expression by polynomials in u, as in the case of ζ-functions, by the methods of l-adic cohomology. Actually, Grothendieck treated a more general type of L-function associated with l-adic sheaves on X, which also play an important role in Deligne's proof of the Riemann hypothesis (→ Section Q).

S. Hasse ζ-Functions

For a nonsingular complete algebraic variety V defined over a finite algebraic number field K, let $V_\mathfrak{p}$ be the reduction of V modulo a prime ideal $\mathfrak{p}$ of K, $K_\mathfrak{p}$ be the residue field of $\mathfrak{p}$, and $Z(u, V_\mathfrak{p})$ be the ζ-function of $V_\mathfrak{p}$ over $K_\mathfrak{p}$. The ζ-function $\zeta(s, V)$ of the complex variable s, determined by the infinite product (excluding the finite number of $\mathfrak{p}$'s where $V_\mathfrak{p}$ is not defined),

$$\zeta(s,V)=\prod_{\mathfrak{p}}{}' Z(N(\mathfrak{p})^{-s},V_\mathfrak{p}),$$

is called the **Hasse ζ-function** of V over the algebraic number field K. For this function, we have **Hasse's conjecture** [W4]: $\zeta(s, V)$ is a meromorphic function over the whole complex plane of s and satisfies the functional equation of ordinary type. Sometimes it is more natural to consider

$$\zeta_j(s,V)=\prod_{\mathfrak{p}}{}' P_j(N(\mathfrak{p})^{-s},V_\mathfrak{p})^{-1}\quad(0\leqslant j\leqslant 2\dim V),$$

where $P_j(u, V_\mathfrak{p})$ is the jth factor of $Z(u, V_\mathfrak{p})$, and we have a similar conjecture for them. For the definition of $\zeta_j(s, V)$ taking into account the factors for bad primes and the precise form of the conjectural functional equation, see Serre [S8]. Note that $\zeta_j(s, V)$ converges absolutely for $\mathrm{Re}\,s>j/2+1$ as a consequence of the Weil conjecture.

Hasse's conjecture remains unsolved for the general case, but has been verified when V is one of the following varieties:

($\mathrm{I_a}$) Algebraic curves defined by the equation $y^e=\gamma x^f+\delta$ and Fermat hypersurfaces (Weil [W6]).

($\mathrm{I_b}$) Elliptic curves with complex multiplication (Deuring [D11]).

($\mathrm{I_c}$) Abelian varieties with complex multiplication (Taniyama [T2], Shimura and Taniyama [S11], Shimura, H. Yoshida).

($\mathrm{I_d}$) Singular $K3$ surfaces, i.e., $K3$ surfaces with 20 Picard numbers (Shafarevich and Pyatetskiĭ-Shapiro [P1], Deligne [D2], T. Shioda and H. Inose [S21]).

($\mathrm{II_a}$) Algebraic curves that are suitable models of the elliptic modular function fields (Eichler [E1], Shimura [S12]).

($\mathrm{II_b}$) Algebraic curves that are suitable models of the automorphic function fields obtained from a quaternion algebra (Shimura [S13, S15]).

($\mathrm{II_c}$) Certain fiber varieties of which the base is a curve of type ($\mathrm{II_a}$) or ($\mathrm{II_b}$) and the fibers are Abelian varieties (Kuga and Shimura [K6], Ihara [I1], Deligne [D3]).

($\mathrm{II_d}$) Certain Shimura varieties of higher dimension (Langlands and others; → [B6]).

In these cases, $\zeta(s, V)$ can be expressed by known functions, i.e., by Hecke L-functions with Grössencharakters of algebraic number fields in cases (I) or by Dirichlet series corresponding to modular forms in cases (II). This fact has an essential meaning for the arithmetic properties of these functions. For example, the extended †Ramanujan conjecture concerning the Hecke operator of the automorphic form reduces to Weil's conjecture on varieties related to those in cases II. Moreover, for ($\mathrm{II_a}$)–($\mathrm{II_c}$) the essential point is the congruence relation $\tilde{T}_p=\Pi+\Pi^*$ (Kronecker, Eichler [E1], Shimura). In particular, for ($\mathrm{II_b}$) this formula is related to the problem of constructing class fields over totally imaginary quadratic extensions of a totally real field F utilizing special values of automorphic functions and class fields over F. Actually, the formula is equivalent to the reciprocity law for class fields (Shimura).

One of the facts that makes the Hasse ζ-function important is that it describes the decomposition law of prime ideals of algebraic number fields when V is an algebraic curve or an Abelian variety (Weil, Shimura [S14], Taniyama [T2], T. Honda [H6]). In that case, its Hasse ζ-function has the following arithmetic meaning.

Let C be a complete, nonsingular algebraic curve defined over an algebraic number field K, and let J be the Jacobian variety of C defined over K. For a prime number l, fix an l-adic coordinate system Σ_l on J, and let $K(J,l^\infty)$ be the extension field of K obtained by adjoining to K all the coordinates of the l^νth division points ($\nu=1,2,\ldots$) of J. Then $K(J,l^\infty)/K$ is an infinite Galois extension of K. The corresponding Galois group $\mathfrak{G}(J,l^\infty)$ has the l-adic representation $\sigma\rightarrow M_l^*(\sigma)$ by the l-adic coordinates Σ_l. Almost all prime ideals $\mathfrak{p}$ of K are unramified in $K(J,l^\infty)/K$. Thus when we take an arbitrary prime factor $\mathfrak{P}$ of $\mathfrak{p}$ in $K(J,l^\infty)$, the Frobenius substitution of $\mathfrak{P}$,

$$\sigma_{\mathfrak{P}}=\left[\frac{K(J,l^\infty)/K}{\mathfrak{P}}\right],$$

is uniquely determined. Furthermore, the characteristic polynomial $\det(1-M_l^*(\sigma_{\mathfrak{P}})u)$ is determined only by $\mathfrak{p}$ and does not depend on the choice of the prime factor $\mathfrak{P}$; we denote this polynomial by $P_{\mathfrak{p}}(u,C)$. In this case, for almost all $\mathfrak{p}$, $P_{\mathfrak{p}}(u,C)$ is a polynomial with rational integral coefficients independent of l; namely, the numerator of the ζ-function of the reduction of $C \bmod p$. Thus

$$\zeta_1(s,C)=\prod_{\mathfrak{p}}{}' P_{\mathfrak{p}}(N(\mathfrak{p})^{-s},C)^{-1}$$

$$\sim\prod_{\mathfrak{p}}{}'\det(1-M_l^*(\sigma_{\mathfrak{P}})N(\mathfrak{p})^{-s})^{-1}.$$

Here the product $\prod'\det(1-M_l^*(\sigma_{\mathfrak{P}})N(\mathfrak{p})^{-s})^{-1}$ has the same expression as the Artin L-function if we ignore the fact that M_l^* is the l-adic representation and $K(J,l^\infty)$ is the infinite extension. Thus if we can describe $\zeta(s,C)$ explicitly, then the decomposition process of the prime ideal for intermediate fields between $K(J,l^\infty)$ and K can be made fairly clear. In fact, this is the case for examples (I_a)–(I_c) and (II_a)–(II_c), from which the relations between the arithmetic of the field of division points $K(J,l^\infty)/K$ and the eigenvalues of the Hecke operator have been obtained. Thus for curves and Abelian varieties, $\zeta(s,V)$ is related to the arithmetic of some number fields; but it is not known whether similar arithmetical relations exist for other kinds of varieties except in a few cases.

Tate's Conjecture. For a projective nonsingular variety V over a finite algebraic number field K, let $\mathfrak{A}^r(\bar{V})$ denote the group of algebraic cycles of codimension r on $\bar{V}=V\otimes_K\mathbf{C}$ modulo homological equivalence and let $\mathfrak{A}^r(V)$ be the subgroup of $\mathfrak{A}^r(\bar{V})$ generated by algebraic cycles rational over K. Then Tate [T3] conjectured that the rank of $\mathfrak{A}^r(V)$ is equal to the order of the pole of $\zeta_{2r}(s,V)$ at $s=r+1$. This conjecture is closely connected with **Hodge's conjecture** that the space of rational cohomology classes of type (r,r) on $\bar{V}$ is spanned by $\mathfrak{A}^r(\bar{V})$; in fact, the equivalence of these conjectures is known for Abelian varieties of †CM type (H. Pohlmann, *Ann. Math.*, 88 (1968)) and for Fermat hypersurfaces of dimension $2r$ (Tate [T3], Weil [W6]). Thus, when $r=1$, Tate's conjecture for these varieties holds by Lefschetz's theorem, and when $r>1$, it holds in certain cases where the Hodge conjecture is verified (Shioda, *Math. Ann.*, 245 (1979); Z. Ran, *Compositio Math.*, 42 (1981)). Further examples are given by $K3$ surfaces with large Picard numbers (Shioda and Inose [S21]; T. Oda, *Proc. Japan Acad.*, 56 (1980)).

***L*-Functions of Elliptic Curves.** Let E be an elliptic curve (with a rational point) over the rational number field $\mathbf{Q}$, and let N be its conductor; a prime number p divides N if and only if E has bad reduction $\bmod p$ (Tate [T5]). The L-function of E over $\mathbf{Q}$ is defined as follows:

$$L(s,E)$$
$$=\prod_{p|N}(1-\varepsilon_p p^{-s})^{-1}\prod_{p\nmid N}(1-a_p p^{-s}+p^{1-2s})^{-1},$$

where $\varepsilon_p=0$ or ±1 and $1-a_p u+pu^2=P_1(u,E \bmod p)$. There are many interesting results and conjectures concerning $L(s,E)$ [T5]:

(1) Functional equation. Let

$$\xi(s,E)=N^{s/2}(2\pi)^{-s}\Gamma(s)L(s,E).$$

Then it is conjectured that $\xi(s,E)$ is holomorphic in the entire s-plane and satisfies the functional equation $\xi(s,E)=\pm\xi(2-s,E)$. This is true if E has complex multiplication (Deuring) or E is a certain modular curve (Eichler, Shimura).

(2) **Taniyama-Weil conjecture.** Weil [W1 (1967a)] conjectured that, if $L(s,E)=\sum_{n=1}^{\infty}a_n n^{-s}$, then $f(\tau)=\sum_{n=1}^{\infty}a_n e^{2\pi i n\tau}$ is a cusp form of weight 2 for the congruence subgroup $\Gamma_0(N)$ which is an eigenfunction for Hecke operators; moreover E is isogenous to a factor of the Jacobian variety of the modular curve for $\Gamma_0(N)$ in such a way that $f(\tau)d\tau$ corresponds to the differential of the first kind on E. If this conjecture is true, then the statements in (1) follow.

(3) **Birch–Swinnerton-Dyer conjecture**. Assuming analytic continuation of $L(s,E)$, B. Birch and H. Swinnerton-Dyer [B4] conjectured that the order of the zero of $L(s,E)$ at $s=1$ is equal to the rank r of the group $E(\mathbf{Q})$ of rational points of E which is finitely generated by the Mordell-Weil theorem. They verified this for many examples, especially for curves of the type $y^2=x^3-ax$. J. Coates and A. Wiles (*Inventiones Math.*, 39 (1977)) proved that if E has complex multiplication and if $r>0$ then $L(s,E)$ vanishes at $s=1$. This conjecture has a refinement which extends also to Abelian varieties over a global field (Tate, *Sém. Bourbaki*, no. 306 (1966)).

(4) **Sato's conjecture**. Let

$$1-a_pu+pu^2=(1-\pi_pu)(1-\bar{\pi}_pu),$$

with $\pi_p=\sqrt{p}\,e^{i\theta_p}$ $(0<\theta_p<\pi)$. When E has complex multiplication, the distribution of θ_p for half of p is uniform in the interval $[0,\pi]$, and θ_p is $\pi/2$ for the remaining half of p. Suppose that E does not have complex multiplication. Then Sato conjectured that

$$\lim_{x\to\infty}\frac{\text{(the number of prime numbers } p \text{ less than } x \text{ such that } \theta_p\in[\alpha,\beta])}{\text{(the number of prime numbers less than } x)}$$

$$=\frac{2}{\pi}\int_\alpha^\beta \sin^2\theta\,d\theta \quad (0<\alpha<\beta<\pi)$$

(Tate [T3]).

H. Yoshida [Y1] posed an analog of Sato's conjecture for elliptic curves defined over function fields with finite constant fields and proved it in certain cases.

(5) Formal groups. Letting $L(s,E)=\sum a_nn^{-s}$ as before, set $f(x)=\sum_{n=1}^\infty a_nx^n/n$. Honda [H6] showed that $f^{-1}(f(x)+f(y))$ is a †formal (Lie) group with coefficients in $\mathbf{Z}$ and that this group is isomorphic over $\mathbf{Z}$ to a formal group obtained by power series expansion of the group law of E with respect to suitable †local uniformizing coordinates at the origin. Such an interpretation of the ζ-function also applies to other cases in which ζ-functions of †group varieties may be characterized as Dirichlet series whose coefficients give a normal form of the group law; e.g., the case of algebraic tori (T. Ibukiyama, *J. Fac. Sci. Univ. Tokyo*, (IA) 21 (1974)).

T. Selberg ζ-Functions and ζ-Functions Associated with Discontinuous Groups

Let $\Gamma\subset SL(2,\mathbf{R})$ be a †Fuchsian group operating on the complex upper half-plane $H=\{z=x+iy\mid y>0\}$. When the two eigenvalues of an element $\gamma\in\Gamma$ are distinct real numbers ξ_1, ξ_2 $(\xi_1\xi_2=1,\xi_1<\xi_2)$, we call γ †hyperbolic. Then the number ξ_2^2 is denoted by $N(\gamma)$ and is called the norm of γ. When γ is hyperbolic, γ^n $(n=1,2,3,\dots)$ is also hyperbolic. When $\pm\gamma$ is not a positive power of other hyperbolic elements, γ is called a primitive hyperbolic element. The elements conjugate to primitive hyperbolic elements are also primitive hyperbolic elements and have the same norm as γ. Let P_1, $P_2,\dots$ be the conjugacy classes of primitive hyperbolic elements of Γ, and let $\gamma_i\in P_i$ be their representatives. Suppose that a matrix representation $\gamma\to M(\gamma)$ of Γ is given. Then the analytic function given by

$$Z_\Gamma(s,M)=\prod_i\prod_{n=0}^\infty\det(I-M(\gamma_i)N(\gamma_i)^{-s-n})$$

is called the **Selberg ζ-function** (Selberg [S5]). When $\Gamma\backslash H$ is compact and Γ is torsion-free, then $Z_\Gamma(s,M)$ has the following properties.

(1) It can be analytically continued to the whole complex plane of s and gives an †integral function of genus at most 2.

(2) It has zeros of order $(2n+1)(2g-2)v$ at $-n$ $(n=0,1,2,3,\dots)$. Here g is the genus of the Riemann surface $\Gamma\backslash H$ and v is the degree of the representation M. All other zeros lie on the line $\operatorname{Re}s=1/2$, except for a finite number that lie on the interval $(0,1)$ of the real axis.

(3) It satisfies the functional equation

$$Z_\Gamma(1-s,M)=Z_\Gamma(s,M)\exp\left(-vA(\Gamma\backslash H)\times\int_0^{s-1/2}v\tan(\pi v)\,dv\right),$$

where

$$A(\Gamma\backslash H)=\iint_{\Gamma\backslash H}\frac{dx\,dy}{y^2}=2\pi(2g-2),\qquad x+iy\in H.$$

Property (2) shows that the Riemann hypothesis is almost valid for $Z_\Gamma(s,M)$. The proof is based on the following fact concerning the eigenvalue problem for the variety $\Gamma\backslash H$: The eigenvalue λ of the equation

$$y^2(\partial^2/\partial x^2+\partial^2/\partial y^2)u+\lambda u=0,\quad u\in L^2(\Gamma\backslash H)$$

cannot be a negative number.

Using this function, T. Yamada (1965) investigated the unit distribution of real quadratic fields.

Selberg ζ-functions are defined similarly when $\Gamma\backslash G$ has finite volume but is noncompact. In this case, however, the decomposition of $L_2(\Gamma\backslash G)$ into irreducible representation spaces has a continuous spectrum; hence the properties of the Selberg ζ-function of Γ are quite different from the case when $\Gamma\backslash G$ is compact. Selberg defined the **generalized Eisenstein series** to give the eigenfunctions of this continuous spectrum explicitly. When $\Gamma=$

$SL(2, \mathbf{Z})$, the series is given by $\sum_{(c,d)=1} \frac{y^s}{|c\tau+d|^{2s}}$.

This type of generalized Eisenstein series is also defined for the general semisimple algebraic group G and its arithmetic subgroup. It has been studied by Selberg, Godement, Gel'fand, Harish-Chandra, Langlands, D. Zagier, and others.

U. Ihara ζ-Functions

Let $k_\mathfrak{p}$ be a p-adic field, $\mathfrak{o}_\mathfrak{p}$ the ring of integers in $k_\mathfrak{p}$, and $G = PSL_2(\mathbf{R}) \times PSL_2(k_\mathfrak{p})$. Suppose that Γ is a subgroup of G such that (1) Γ is discrete, (2) $\Gamma \backslash G$ is compact, (3) Γ has no torsion element except the identity, (4) $\Gamma_\mathbf{R}$ (the projection of Γ in $PSL_2(\mathbf{R})$) is dense in $PSL_2(\mathbf{R})$, and (5) $\Gamma_\mathfrak{p}$ (the projection of Γ in $PSL_2(k_\mathfrak{p})$) is dense in $PSL_2(k_\mathfrak{p})$. Then $\Gamma \cong \Gamma_\mathbf{R} \cong \Gamma_\mathfrak{p}$. Let $X = \{x + iy \mid y > 0\}$ be the upper half-plane, and let Γ act on X via $\Gamma_\mathbf{R}$. The action of Γ on X is not discontinuous, but the subgroup $\Gamma_0 = \{\gamma \in \Gamma \mid \text{projection of } \gamma \text{ to } \Gamma_\mathfrak{p} \in PSL_2(\mathfrak{o}_\mathfrak{p})\}$ operates on X properly discontinuously. For each $z \in X$, define $\Gamma_z = \{\gamma \in \Gamma \mid \gamma(z) = z\}$. Then Γ_z is isomorphic to $\mathbf{Z}$ or $\{1\}$. Let $\tilde{\mathbf{P}}(\Gamma) = \{z \in X \mid \Gamma_z \cong \mathbf{Z}\}$. The group Γ acts on $\tilde{\mathbf{P}}(\Gamma)$, since Γ_z and $\Gamma_{\gamma z}$ are conjugate in Γ. Let $\mathbf{P}(\Gamma) = \tilde{\mathbf{P}}(\Gamma)/\Gamma$. Suppose that $P \in \mathbf{P}(\Gamma)$ is represented by $z \in X$. Choose a generator γ of Γ_z and project γ to $\Gamma_\mathfrak{p}$. Then γ is equivalent to a diagonal matrix $\begin{pmatrix} \lambda & 0 \\ 0 & \lambda^{-1} \end{pmatrix}$ with $\lambda \in k_\mathfrak{p}$. We denote the valuation of $k_\mathfrak{p}$ by $\mathrm{ord}_\mathfrak{p}$ and consider $|\mathrm{ord}_\mathfrak{p}(\lambda)|$. This value depends only on P and we denote it by $\deg(P)$. The **Ihara ζ-function** of Γ is defined by

$$Z_\Gamma(u) = \prod_{P \in \mathbf{P}(\Gamma)} (1 - u^{\deg(P)})^{-1}.$$

Ihara proved that

$$Z_\Gamma(u) = \frac{\prod_{i=1}^{g} (1 - \pi_i u)(1 - \pi_i' u)}{(1-u)(1-q^2 u)} (1-u)^H,$$

where q is the number of elements in the residue class field of $\mathfrak{p}$, and g is the genus of the Riemann surface $\Gamma_0 \backslash X$ and $H = (g-1)q(q-1)$. Similar results hold even if Γ has torsion elements and the quotient $\Gamma \backslash G$ is only assumed to have finite volume.

Aside from the factor $(1-u)^H$, this looks like Weil's formula for the congruence ζ-function of an algebraic curve defined over $\mathbf{F}_{q^2}$. Ihara conjectured that the first factor of $Z_\Gamma(u)$ is always the congruence ζ-function of some algebraic curve over $\mathbf{F}_{q^2}$, and furthermore that Γ could be regarded as the fundamental group of a certain Galois covering of this curve which describes the decomposition law of prime divisors in this covering [I2, I3, I4]. He verified the conjecture in the case $\Gamma = PGL_2(\mathbf{Z}[1/p])$ by using the †moduli of elliptic curves. Related results have been obtained by Shimura, Ihara, Y. Morita, and others.

V. ζ-Functions Associated with Prehomogeneous Vector Spaces

M. Sato posed a notion of prehomogeneous vector spaces and defined ζ-functions associated with them. Sato's program has been carried on by himself and T. Shintani [S2, S3, S17, S18]. Let G be a linear algebraic group, V a finite-dimensional linear space of dimension n, and ρ a rational representation $G \to GL(V)$, where G, V, and ρ are defined over $\mathbf{Q}$. The triple (G, ρ, V) is called a **prehomogeneous vector space** if there exists a proper algebraic subset S of $V_\mathbf{C}$ such that $V_\mathbf{C} - S$ is a single $G_\mathbf{C}$-orbit. The algebraic set S is called the set of singular points of V. We also assume that G is reductive and S is an irreducible hypersurface of V. Let V^* be the dual vector space of V, and ρ^* the dual (contragredient) representation of G. Then (G, ρ^*, V^*) is again a prehomogeneous vector space, and we denote its set of singular points by S^*. There are homogeneous polynomials P and Q of the same degree d on V and V^*, respectively, such that $S = \{x \in X \mid P(x) = 0\}$ and $S^* = \{x^* \in V^* \mid Q(x^*) = 0\}$. P and Q are relative invariants of G, i.e., $P(\rho(g)x) = \chi(g)P(x)$ and $Q(\rho^*(g)x^*) = \chi(g)^{-1}Q(x^*)$ (for $g \in G$, $x \in V$, and $x^* \in V^*$) hold with a rational character χ of G. Put $G^1 = \ker \chi = \{g \in G \mid \chi(g) = 1\}$. Denote by $G_\mathbf{R}^+$ the connected component of 1 of the Lie group $G_\mathbf{R}$. Let $V_\mathbf{R} - S = V_1 \cup \ldots \cup V_l$, $V_\mathbf{R}^* - S^* = V_1^* \cup \ldots \cup V_l^*$ be the decompositions of $V_\mathbf{R} - S$ and $V_\mathbf{R}^* - S^*$ into their topologically connected components. Then V_i and V_j^* are $G_\mathbf{R}^+$-orbits. We further assume that $V_\mathbf{R} \cap S$ decomposes into the union of a finite number of $G_\mathbf{R}^1$-orbits. Set $\Gamma = G_\mathbf{R}^+ \cap G_\mathbf{Z}^1$, and take Γ-invariant lattices L and L^* in $V_\mathbf{Q}$ and $V_\mathbf{Q}^*$, respectively. Consider the following functions in s:

$$\Phi_i(f, s) = \int_{V_i} f(x) |P(x)|^s dx,$$

$$\Phi_j^*(f, s) = \int_{V_j^*} f^*(x^*) |Q(x^*)|^s dx^*,$$

and

$$Z_i(f, L, s) = \int_{G_\mathbf{R}^+/\Gamma} \chi(g)^s \sum_{x \in L \cap V_i} f(\rho(g)x)\, dg,$$

$$Z_j^*(f^*, L^*, s)$$

$$= \int_{G_\mathbf{R}^+/\Gamma} \chi(g)^{-s} \sum_{x^* \in L^* \cap V_j^*} f^*(\rho^*(g)x^*)\, dg,$$

where f and f^* are †rapidly decreasing functions on $V_{\mathbf{R}}$ and $V_{\mathbf{R}}^*$, respectively, dx and dx^* are Haar measures of $V_{\mathbf{R}}$ and $V_{\mathbf{R}}^*$, respectively, and dg is a Haar measure of G. Then the ratios

$$\frac{Z_i(f, L, s)}{\Phi_i(f, s-n/d)} = \xi_i(s, L),$$

$$\frac{Z_j^*(f^*, L^*, s)}{\Phi_j^*(f^*, s-n/d)} = \xi_j^*(s, L^*)$$

are independent of the choice of f and f^* and are Dirichlet series in s. These Dirichlet series $\xi_i(s, L)$ and $\xi_j^*(s, L^*)$ are called ξ-**functions associated with the prehomogeneous space**. Considering Fourier transforms of $|P(x)|^s$ and $|Q(x^*)|^s$, we obtain functional equations for ξ_i and ξ_j^* under some additional (but mild) conditions on (G, ρ, V) as follows. The Dirichlet series ξ_i and ξ_j^* are analytically continuable to meromorphic functions on the whole s-plane, and they satisfy

$$v(L^*)\xi_j^*(n/d-s, L^*)$$
$$= \gamma(s-n/d)(2\pi)^{-ds}|b_0|^s \exp(\pi d\sqrt{-1}\, s/2)$$
$$\times \sum_{j=1}^{l} u_{ij}(s)\xi_i(s, L),$$

with a Γ-factor $\gamma(s) = \prod_{i=1}^{d} \Gamma(s-c_i+1)$. Here $u_{ij}(s)$ $(1 \leqslant i, j \leqslant l)$ are polynomials in $\exp(-\pi\sqrt{-1}\, s)$ with degree $\leqslant d$, and b_0 and c_i are constants depending only on (G, ρ, V).

Epstein's ζ-functions and Siegel's Dirichlet series associated with indefinite quadratic forms are examples of the above-defined ζ-functions. Shintani defined such ζ-functions related to integral binary cubic forms and obtained asymptotic formulas concerning the class numbers of irreducible integral binary cubic forms with discriminant n, which are improvements on the results of Davenport [S17].

Recently M. Sato studied ζ-functions of prehomogeneous vector spaces without assuming the conditions that G is reductive and S is irreducible. In this case, ζ-functions of several complex variables are obtained. For examples and classification of prehomogeneous vector spaces → [S4].

References

[A1] E. Artin, Collected papers, S. Lang and J. T. Tate (eds.), Addison-Wesley, 1965.
[A2] E. Artin, Zur Theorie der L-Reihen mit allgemeinen Gruppencharakteren, Abh. Math. Sem. Univ. Hamburg, 8 (1930), 292–306 [A1, 165–179].
[A3] M. Artin, A. Grothendieck, and J. L. Verdier, Théorie des topos et cohomologie étale des schémas (SGA 4), Lecture notes in math. 269, 270, 305, Springer, 1972–1973.
[A4] J. Ax, On the units of an algebraic number fields, Illinois J. Math., 9 (1965), 584–589.
[B1] P. Bayer and J. Neukirch, On values of zeta functions and l-adic Euler characteristics, Inventiones Math., 50 (1978), 35–64.
[B2] P. Berthelot, Cohomogie cristalline des schéma de caracteristique $p > 0$, Lecture notes in math. 407, Springer, 1974.
[B3] P. Berthelot and A. Ogus, Notes on crystalline cohomology, Math. notes 21, Princeton Univ. Press, 1978.
[B4] B. Birch and H. Swinnerton-Dyer, Notes on elliptic curves II, J. Reine Angew. Math., 218 (1965), 79–108.
[B5] E. Bombieri and H. P. F. Swinnerton-Dyer, On the local zeta function of a cubic threefold, Ann. Scuola Norm. Sup. Pisa, (3) 21 (1967), 1–29.
[B6] A. Borel and W. Casselman (eds.), Automorphic forms, representations, and L-functions, Amer. Math. Soc. Proc. Symp. Pure Math., 33 (pts. 1 and 2) (1979).
[B7] A. Brumer, On the units of algebraic number fields, Mathematika, 14 (1967), 121–124.
[C1] K. Chandrasekharan, Lectures on the Riemann zeta-function, Tata Inst., 1953.
[C2] S. Chowla and A. Selberg, On Epstein's zeta function I, Proc. Nat. Acad. Sci. US, 35 (1949), 371–374.
[C3] J. Coates, P-adic L-functions and Iwasawa's theory, Algebraic Number Fields (L-functions and Galois Properties), A. Fröhlich (ed.), Academic Press, 1977.
[D1] R. Dedekind, Über die Theorie der ganzen algebraischen Zahlen, Dirichlet's Vorlesungen über Zahlentheorie, Supplement XI, fourth edition, 1894, secs. 184, 185, 186 (Gesammelte mathematische Werke III, Braunschweig, 1932, 297–314; Chelsea, 1968).
[D2] P. Deligne, La conjecture de Weil pour les surfaces $K3$, Inventiones Math., 75 (1972), 206–226.
[D3] P. Deligne, Formes modulaires et représentations l-adiques, Sém. Bourbaki, exp. 355, Lecture notes in math. 349, Springer, 1973.
[D4] P. Deligne, La conjecture de Weil, I, II, Publ. Math. Inst. HES, 43 (1974), 273–307; 52 (1980), 137–252.
[D5] P. Deligne, Cohomologie étale (SGA $4\frac{1}{2}$), Lecture notes in math. 569, Springer, 1977.
[D6] P. Deligne, Les constantes des équations fonctionnelles des fonctions L, Lecture notes in math. 349, Springer, 1973.
[D7] P. Deligne and N. Katz, Groupes de monodromie en géométrie algébrique (SGA

7II), Lecture notes in math. 340, Springer, 1973.
[D8] P. Deligne and K. Ribet, Values of Abelian L-functions at negative integers over totally real fields, Inventiones Math., 59 (1980), 227–286.
[D9] P. Deligne and J-P. Serre, Formes modulaires de poids 1, Ann. Sci. Ecole Norm. Sup. (4), 7 (1974), 507–530.
[D10] M. Deuring, Algebren, Erg. Math., Springer, second edition, 1968.
[D11] M. Deuring, Die Zetafunktionen einer algebraischen Kurve vom Geschlechte Eins I, II, III, IV, Nachr. Akad. Wiss. Göttingen (1953), 85–94; (1955), 13–42; (1956), 37–76; (1957), 55–80.
[D12] K. Doi and H. Naganuma, On the functional equation of certain Dirichlet series, Inventiones Math., 9 (1969–1970), 1–14.
[D13] B. Dwork, On the rationality of the zeta-function of an algebraic variety, Amer. J. Math., 82 (1960), 631–648.
[D14] B. Dwork, On the zeta function of a hypersurface I, II, III, Publ. Math. Inst. HES, 1962, 5–68; Ann. Math., (2) 80 (1964), 227–299; (2) 83 (1966), 457–519.
[E1] M. Eichler, Einführung in die Theorie der algebraischen Zahlen und Funktionen, Birkhäuser, 1963; English translation, Introduction to the theory of algebraic numbers and functions, Academic Press, 1966.
[F1] B. Ferrero and L. C. Washington, The Iwasawa invariant μ_p vanishes for Abelian number fields, Ann. Math., (2) 109 (1979), 377–395.
[G1] R. Godement, Les fonctions ζ des algèbres simples I, II, Sém. Bourbaki, exp. 171, 176, 1958–1959.
[G2] A. Grothendieck, Formule de Lefschetz et rationalité des fonction L, Sém. Bourbaki, exp. 379, 1964–1965 (Benjamin, 1966).
[G3] A. Grothendieck, Cohomologie l-adique et fonctions L (SGA 5), Lecture notes in math. 589, Springer, 1977.
[H1] H. Hasse, Zur Theorie der abstrakten elliptischen Funktionenkörper I, II, III, J. Reine Angew. Math., 175 (1936), 55–62, 69–88, 193–208.
[H2] E. Hecke, Mathematische Werke, Vandenhoeck & Ruprecht, 1959.
[H3] E. Hecke, Eine neue Art von Zetafunctionen und ihre Beziehung zur Verteilung der Primzahlen I, II, Math. Z., 1 (1918), 357–376; 6 (1920), 11–51 [H2, 215–234, 249–289].
[H4] E. Hecke, Über die Bestimmung Dirichletscher Reihen durch ihre Funktionalgleichung, Math. Ann., 112 (1936), 664–699 (H2, 591–626].
[H5] E. Hecke, Über Modulfunktionen und die Dirichletschen Reihen mit Eulerscher Produktentwicklung I, II, Math. Ann., 114 (1937), 1–28; 316–351 [H2, 644–671, 672–707].
[H6] T. Honda, Formal groups and zeta functions, Osaka J. Math., 5 (1968), 199–213.
[H7] T. Honda, Isogeny classes of Abelian varieties over finite fields, J. Math. Soc. Japan, 20 (1968), 83–95.
[I1] Y. Ihara, Hecke polynomials as congruence ζ functions in elliptic modular case (To validate Sato's identity), Ann. Math., (2) 85 (1967), 267–295.
[I2] Y. Ihara, On congruence monodromy problems, I, II, Lecture notes, Univ. of Tokyo, 1968–1969.
[I3] Y. Ihara, Non-Abelian class fields over function fields in special cases, Actes Congr. Intern. Math., 1970, Nice, Gauthier-Villars, p. 381–389.
[I4] Y. Ihara, Congruence relations and Shimura curves. Amer. Math. Soc. Proc. Symp. Pure Math., 33 (1979), pt. 2, 291–311.
[I5] K. Iwasawa, On Γ-extensions of algebraic number fields, Bull. Amer. Math. Soc., 65 (1959), 183–226.
[I6] K. Iwasawa, On p-adic L-functions, Ann. Math., (2) 89 (1969), 198–205.
[I7] K. Iwasawa, Lectures on p-adic L-functions, Ann. math. studies 74, Princeton Univ. Press, 1972.
[J1] H. Jacquet and R. P. Langlands, Automorphic forms on $GL(2)$, Lecture notes in math. 114, Springer, 1970.
[J2] H. Jacquet, Automorphic forms on $GL(2)$, II, Lecture notes in math. 278, Springer, 1972.
[K1] N. Katz and W. Messing, Some consequences of the Riemann hypothesis for varieties over finite fields, Inventiones Math., 23 (1974), 73–77.
[K2] N. Katz, An overview of Deligne's proof of the Riemann hypothesis for varieties over finite fields, Amer. Math. Soc. Proc. Symp. Pure Math., 28 (1976), 275–305.
[K3] N. Katz, P-adic L-functions for CM-fields, Inventiones Math., 49 (1978), 199–297.
[K4] S. Kleiman, Algebraic cycles and the Weil conjectures, Dix exposés sur la cohomologies des schémas, North-Holland, 1968, 359–386.
[K5] T. Kubota and H. W. Leopoldt, Eine p-adischer Theorie der Zetawerte, J. Reine Angew. Math., 214/215 (1964), 328–339.
[K6] M. Kuga and G. Shimura, On the zeta function of a fiber variety whose fibers are Abelian varieties, Ann. Math., (2) 82 (1965), 478–539.
[L1] E. G. H. Landau, Handbuch der Lehre von der Verteilung der Primzahlen I, II, Teubner, 1909 (Chelsea, 1953).
[L2] S. Lang, Sur les series L d'une variété algébrique, Bull. Soc. Math. France, 84 (1956), 385–407.

[L3] S. Lang, Algebraic number theory, Addison-Wesley, 1970.
[L4] R. P. Langlands, Problems in the theory of automorphic forms, Lecture notes in math. 170, Springer, 1970, 18–86.
[L5] R. P. Langlands, Base change for $GL(2)$, Ann. math. studies 96, Princeton Univ. Press, 1980.
[L6] H. W. Leopoldt, Eine p-adische Theorie der Zetawerte II, J. Reine Angew. Math., 274/275 (1975), 224–239.
[L7] S. Lubkin, A p-adic proof of the Weil conjecture, Ann. Math., (2) 87 (1968), 105–194; (2) (1968), 195–225.
[M1] Yu. I. Manin, Periods of parabolic forms and p-adic Hecke series, Math. USSR-Sb., 21 (1973), 371–393.
[M2] B. Mazur, Eigenvalues of Frobenius acting on algebraic varieties over finite fields, Amer. Math. Soc. Proc. Symp. Pure Math., 29 (1975).
[M3] B. Mazur and H. Swinnerton-Dyer, On the p-adic L-series of an elliptic curve, Inventiones Math., 25 (1974), 1–61.
[M4] J. S. Milne, Etale cohomology, Princeton series in mathematics 33, Princeton Univ. Press, 1980.
[P1] I. I. Pyatetskii-Shapiro and I. R. Shafarevich, The arithmetic of $K3$ surfaces, Proc. Steklov Inst. Math., 132 (1973), 45–57.
[P2] K. Prachar, Primzahlverteilung, Springer, 1957.
[R1] B. Riemann, Über die Anzahl der Primzahlen unter einer gegebenen Grösse, Gesammelte mathematische Werke, 1859, 145–153 (Dover, 1953).
[S1] H. Saito, Automorphic forms and algebraic extensions of number fields, Lectures in math. 8, Kinokuniya, Tokyo, 1975.
[S2] M. Sato, Theory of prehomogeneous vector spaces (notes by T. Shintani) (in Japanese), Sûgaku no Ayumi, 15 (1970), 85–157.
[S3] M. Sato and T. Shintani, On zeta functions associated with prehomogeneous vector spaces, Ann. Math., (2) 100 (1974), 131–170.
[S4] M. Sato and T. Kimura, A classification of irreducible prehomogeneous vector spaces and their invariants, Nagoya Math. J., 65 (1977), 1–155.
[S5] A. Selberg, Harmonic analysis and discontinuous groups on weakly symmetric Riemannian spaces with applications to Dirichlet series, J. Indian Math. Soc., 20 (1956), 47–87.
[S6] A. Selberg, On the zeros of Riemann's zeta-function, Skr. Norske Vid. Akad. Oslo, no. 10 (1942), 1–59.
[S7] J.-P. Serre, Zeta and L functions, Arithmetical Algebraic Geometry, Harper & Row, 1965, 82–92.
[S8] J.-P. Serre, Facteurs locaux des fonctions zêta des variétés algébriques (Définitions et conjectures), Sém. Delange-Pisot-Poitou, 11 (1969/70), no. 19.
[S9] J.-P. Serre, Formes modulaires et fonctions zêta p-adiques, Lecture notes in math. 317, Springer, 1973, 319–338.
[S10] H. Shimizu, On zeta functions of quaternion algebras, Ann. Math., (2) 81 (1965), 166–193.
[S11] G. Shimura and Y. Taniyama, Complex multiplication of Abelian varieties and its applications to number theory, Publ. Math. Soc. Japan, no. 6, 1961.
[S12] G. Shimura, Introduction to the arithmetic theory of automorphic functions, Publ. Math. Soc. Japan, no. 11, 1971.
[S13] G. Shimura, On the zeta-functions of the algebraic curves uniformized by certain automorphic functions, J. Math. Soc. Japan, 13 (1961), 275–331.
[S14] G. Shimura, A reciprocity law in nonsolvable extensions, J. Reine Angew. Math., 221 (1966), 209–220.
[S15] G. Shimura, Construction of class fields and zeta functions of algebraic curves, Ann. Math., (2) 85 (1967), 58–159.
[S16] G. Shimura, On canonical models of arithmetic quotients of bounded symmetric domains, I, II, Ann. Math., (2) 91 (1970) 144–222; 92 (1970), 528–549.
[S17] T. Shintani, On Dirichlet series whose coefficients are class numbers of integral binary cubic forms, J. Math. Soc. Japan, 24 (1972), 132–188.
[S18] T. Shintani, On zeta functions associated with the vector space of quadratic forms, J. Fac. Sci. Univ. Tokyo, (IA) 22 (1975), 25–65.
[S19] T. Shintani, On a Kronecker limit formula for real quadratic fields, J. Fac. Sci. Univ. Tokyo, (IA) 24 (1977), 167–199.
[S20] T. Shintani, On liftings of holomorphic automorphic forms, Amer. Math. Soc. Proc. Symp. Pure Math., 33 (1979), pt. 2, 97–110.
[S21] T. Shioda and H. Inose, On singular $K3$ surfaces, Complex Analysis and Algebraic Geometry, Iwanami and Cambridge Univ. Press, 1977, 119–136.
[S22] C. L. Siegel, Gesammelte Abhandlungen I, II, III, IV, Springer, 1966, 1979.
[S23] C. L. Siegel, Über die Classenzahl quadratischer Zahlkörper, Acta Arith., 1 (1935), 83–86 [S22, I. 406–409].
[S24] C. L. Siegel, Über die Zetafunktionen indefiniter quadratischer Formen I, II, Math. Z., 43 (1938), 682–708; 44 (1939), 398–426 [S22, II, 41–67, 68–96].
[S25] H. M. Stark, L-functions at $s=1$, II, III, Advances in Math., 17 (1975), 60–92; 22 (1976), 64–84.
[T1] T. Tamagawa, On the ζ-functions of a

division algebra, Ann. Math., (2) 77 (1963), 387–405.
[T2] Y. Taniyama, *L*-functions of number fields and zeta functions of Abelian varieties, J. Math. Soc. Japan, 9 (1957), 330–366.
[T3] J. Tate, Algebraic cycles and poles of zeta functions, Arithmetical Algebraic Geometry, Harper & Row, 1965, 93–110.
[T4] J. Tate, Endomorphisms of Abelian varieties over finite fields, Inventiones Math., 2 (1966), 134–144.
[T5] J. Tate, The arithmetic of elliptic curves, Inventiones Math., 23 (1974), 179–206.
[T6] J. Tate, Number theoretic background, Amer. Math. Soc. Proc. Symp. Pure Math., 33 (1979), pt. 2, 3–26.
[T7] T. Tatuzawa, On the Hecke-Landau *L*-series, Nagoya Math. J., 16 (1960), 11–20.
[T8] E. C. Titchmarch, The theory of the Riemann zeta-function, Clarendon Press, 1951.
[W1] A. Weil, Collected papers, I–III, Springer, 1980.
[W2] A. Weil, Courbes algébriques et variétés abéliennes, Hermann, 1971 (orig., 1948).
[W3] A. Weil, Numbers of solutions of equations in finite fields, Bull. Amer. Math. Soc., 55 (1949), 497–508 [W1, I, 399–400].
[W4] A. Weil, Number theory and algebraic geometry, Proc. Intern. Congr. Math., 1950, Cambridge, vol. 2, 90–100 [W1, I, 442–449].
[W5] A. Weil, Sur la théorie du corps de classes, J. Math. Soc. Japan, 3 (1951), 1–35 [W1, I, 483–517].
[W6] A. Weil, Jacobi sums as "Grössencharaktere," Trans. Amer. Math. Soc., 73 (1952), 487–495 [W1, II, 63–71].
[W7] A. Weil, On a certain type of characters of the idele-class group of an algebraic number field, Proc. Intern. Symposium on Algebraic Number Theory, Tokyo-Nikko, 1955, 1–7 [W1, II, 255–261].
[W8] A. Weil, Basic number theory, Springer, 1967.
[W9] A. Weil, Dirichlet series and automorphic forms, Lecture notes in math. 189, Springer, 1971.
[Y1] H. Yoshida, On an analogue of the Sato conjecture, Inventiones Math., 19 (1973), 261–277.

Appendix A
Tables of Formulas

1
Algebraic Equations
2
Trigonometry
3
Vector Analysis and Coordinate Systems
4
Differential Geometry
5
Lie Algebras, Symmetric Riemannian Spaces, and Singularities
6
Algebraic Topology
7
Knot Theory
8
Inequalities
9
Differential and Integral Calculus
10
Series
11
Fourier Analysis
12
Laplace Transforms and Operational Calculus
13
Conformal Mappings
14
Ordinary Differential Equations
15
Total and Partial Differential Equations
16
Elliptic Integrals and Elliptic Functions
17
Gamma Functions and Related Functions
18
Hypergeometric Functions and Spherical Functions
19
Functions of Confluent Type and Bessel Functions
20
Systems of Orthogonal Functions
21
Interpolation
22
Distribution of Typical Random Variables
23
Statistical Estimation and Statistical Hypothesis Testing

1. Algebraic Equations (→ 10 Algebraic Equations)

(I) Quadratic Equation $ax^2+bx+c=0 \qquad (a\neq 0)$

The roots are

$$x=\frac{-b\pm\sqrt{b^2-4ac}}{2a}=\frac{-b'\pm\sqrt{b'^2-ac}}{a} \qquad (b\equiv 2b').$$

The discriminant is b^2-4ac.

(II) Cubic Equation $ax^3+bx^2+cx+d=0 \qquad (a\neq 0)$

By the translation $\xi=x+b/3a$, the equation is transformed into $\xi^3+3p\xi+q=0$, where

$$p\equiv(3ac-b^2)/9a^2, \qquad q\equiv(2b^3-9abc+27a^2d)/27a^3.$$

Its discriminant is $-27(q^2+4p^3)$. The roots of the latter equation are

$$\xi=\sqrt[3]{\alpha}+\sqrt[3]{\beta}, \quad \omega\sqrt[3]{\alpha}+\omega^2\sqrt[3]{\beta}, \quad \omega^2\sqrt[3]{\alpha}+\omega\sqrt[3]{\beta},$$

where

$$\omega=e^{2\pi i/3}=\frac{-1+\sqrt{3}\,i}{2}, \qquad \left.\begin{matrix}\alpha\\ \beta\end{matrix}\right\}=\frac{-q\pm\sqrt{q^2+4p^3}}{2} \qquad \text{(Cardano's formula).}$$

Casus irreducibilis (the case when $q^2+4p^3<0$). Putting $\alpha\equiv re^{i\theta}$ ($\beta=\bar{\alpha}$), the roots are

$$\xi=2\sqrt[3]{r}\cos(\theta/3), \quad 2\sqrt[3]{r}\cos[(\theta+2\pi)/3], \quad 2\sqrt[3]{r}\cos[(\theta+4\pi)/3].$$

(III) Quartic Equation (Biquadratic Equation) $ax^4+bx^3+cx^2+dx+e=0 \qquad (a\neq 0)$

By the translation $\xi=x+b/4a$, the equation is transformed into

$$\xi^4+p\xi^2+q\xi+r=0.$$

The cubic resolvent of the latter is $t^3-pt^2-4rt+(4pr-q^2)=0$. If t_0 is one of the roots of the cubic resolvent, the roots ξ of the above equation are the solutions of two quadratic equations

$$\xi^2\pm\sqrt{t_0-p}\,\left[\xi-q/2(t_0-p)\right]+t_0/2=0 \qquad \text{(Ferrari's formula).}$$

2. Trigonometry

(I) Trigonometric Functions (→ 432 Trigonometry)

(1) In Fig. 1, $OA=OB=OP=1$, and

$$MP=\sin\theta, \qquad OM=\cos\theta, \qquad AT=\tan\theta,$$

$$BL=\cot\theta, \qquad OT=\sec\theta, \qquad OL=\operatorname{cosec}\theta.$$

(2) $\sin^2\theta+\cos^2\theta=1$,

$$\tan\theta=\sin\theta/\cos\theta, \qquad \cot\theta=1/\tan\theta, \qquad \sec\theta=1/\cos\theta,$$

$$\operatorname{cosec}\theta=1/\sin\theta, \qquad 1+\tan^2\theta=\sec^2\theta, \qquad 1+\cot^2\theta=\operatorname{cosec}^2\theta.$$

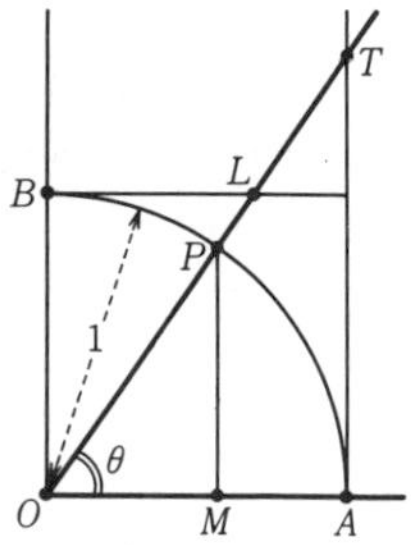

Fig. 1

(3)

	θ	$-\theta$	$\pi/2\pm\theta$	$\pi\pm\theta$	$n\pi\pm\theta$
sin	s	$-s$	c	$\mp s$	$\pm(-1)^n s$
cos	c	c	$\mp s$	$-c$	$(-1)^n c$
tan	t	$-t$	$\mp 1/t$	$\pm t$	$\pm t$

(4)

α	$0°$	$15°$	$18°$	$22.5°$	$30°$	$36°$	$45°$	
↗	0	$\pi/12$	$\pi/10$	$\pi/8$	$\pi/6$	$\pi/5$	$\pi/4$	
$\sin\alpha$	0	$\frac{\sqrt{3}-1}{2\sqrt{2}}$	$\frac{\sqrt{5}-1}{4}$	$\frac{\sqrt{2-\sqrt{2}}}{2}$	$\frac{1}{2}$	$\frac{\sqrt{10-2\sqrt{5}}}{4}$	$\frac{1}{\sqrt{2}}$	$\cos\alpha$
$\cos\alpha$	1	$\frac{\sqrt{3}+1}{2\sqrt{2}}$	$\frac{\sqrt{10+2\sqrt{5}}}{4}$	$\frac{\sqrt{2+\sqrt{2}}}{2}$	$\frac{\sqrt{3}}{2}$	$\frac{\sqrt{5}+1}{4}$	$\frac{1}{\sqrt{2}}$	$\sin\alpha$
	$\pi/2$	$5\pi/12$	$2\pi/5$	$3\pi/8$	$\pi/3$	$3\pi/10$	$\pi/4$	↙
	$90°$	$75°$	$72°$	$67.5°$	$60°$	$54°$	$45°$	α

(5) Addition Formulas

$\sin(\alpha\pm\beta)=\sin\alpha\cos\beta\pm\cos\alpha\sin\beta$, $\cos(\alpha\pm\beta)=\cos\alpha\cos\beta\mp\sin\alpha\sin\beta$,

$\tan(\alpha\pm\beta)=(\tan\alpha\pm\tan\beta)/(1\mp\tan\alpha\tan\beta)$.

(6) $\sin 2\alpha=2\sin\alpha\cos\alpha$, $\cos 2\alpha=\cos^2\alpha-\sin^2\alpha=2\cos^2\alpha-1=1-2\sin^2\alpha$,

$\tan 2\alpha=2\tan\alpha/(1-\tan^2\alpha)$.

$\sin 3\alpha=3\sin\alpha-4\sin^3\alpha$, $\cos 3\alpha=4\cos^3\alpha-3\cos\alpha$,

$\tan 3\alpha=(3\tan\alpha-\tan^3\alpha)/(1-3\tan^2\alpha)$.

$$\sin n\alpha=\sum_{i=0}^{[(n-1)/2]}\binom{n}{2i+1}(-1)^i\sin^{2i+1}\alpha\cos^{n-(2i+1)}\alpha,$$

$$\cos n\alpha=\sum_{i=0}^{[n/2]}\binom{n}{2i}(-1)^i\sin^{2i}\alpha\cos^{n-2i}\alpha.$$

(7) $\sin^2(\alpha/2)=(1-\cos\alpha)/2$, $\cos^2(\alpha/2)=(1+\cos\alpha)/2$,

$\tan^2(\alpha/2)=(1-\cos\alpha)/(1+\cos\alpha)$.

(8) $2\sin\alpha\cos\beta=\sin(\alpha+\beta)+\sin(\alpha-\beta)$, $2\cos\alpha\sin\beta=\sin(\alpha+\beta)-\sin(\alpha-\beta)$,

$2\cos\alpha\cos\beta=\cos(\alpha+\beta)+\cos(\alpha-\beta)$, $-2\sin\alpha\sin\beta=\cos(\alpha+\beta)-\cos(\alpha-\beta)$.

$\sin\alpha+\sin\beta=2\sin[(\alpha+\beta)/2]\cos[(\alpha-\beta)/2]$,

$\sin\alpha-\sin\beta=2\cos[(\alpha+\beta)/2]\sin[(\alpha-\beta)/2]$,

$\cos\alpha+\cos\beta=2\cos[(\alpha+\beta)/2]\cos[(\alpha-\beta)/2]$,

$\cos\alpha-\cos\beta=-2\sin[(\alpha+\beta)/2]\sin[(\alpha-\beta)/2]$.

(II) Plane Triangles

As shown in Fig. 2, we denote the interior angles of a triangle ABC by α, β, γ; the corresponding side lengths by a, b, c; the area by S; the radii of inscribed, circumscribed, and escribed circles by r, R, r_A, respectively; the perpendicular line from the vertex A to the side BC by AH; the midpoint of the side BC by M; bisector of the angle A by AD; and the lengths of AH, AM, AD by h_A, m_A, f_A, respectively. Similar notations are used for B and C. Put $s\equiv(a+b+c)/2$. The symbol ... means similar formulas by the cyclic permutation of the letters A, B, C, and corresponding quantities.

$\dfrac{a}{\sin\alpha}=\dfrac{b}{\sin\beta}=\dfrac{c}{\sin\gamma}=2R$ (law of sines).

$a=b\cos\gamma+c\cos\beta$, ... (the first law of cosines).

$a^2=b^2+c^2-2bc\cos\alpha$, ... (the second law of cosines).

$\sin^2(\alpha/2)=(s-b)(s-c)/bc$, ...; $\cos^2(\alpha/2)=s(s-a)/bc$, ...

$(b+c)\sin(\alpha/2)=a\cos[(\beta-\gamma)/2]$, ...; $(b-c)\cos(\alpha/2)=a\sin[(\beta-\gamma)/2]$, ...

$\dfrac{a+b}{a-b}=\dfrac{\tan[(\alpha+\beta)/2]}{\tan[(\alpha-\beta)/2]}$, ... (Napier's rule).

$$S = ah_A/2 = (1/2)bc\sin\alpha = (1/2)a^2\sin\beta\sin\gamma/\sin\alpha = abc/4R = 2R^2\sin\alpha\sin\beta\sin\gamma$$
$$= rs = r_A(s-a) = \sqrt{rr_Ar_Br_C}$$
$$= \sqrt{s(s-a)(s-b)(s-c)} \qquad \text{(Heron's formula).}$$

$r = (s-a)\tan(\alpha/2) = 4R\sin(\alpha/2)\sin(\beta/2)\sin(\gamma/2)$.

$r_A = s\tan(\alpha/2) = (s-b)\cot(\gamma/2) = 4R\sin(\alpha/2)\cos(\beta/2)\cos(\gamma/2)$.

$1/r = (1/h_A) + (1/h_B) + (1/h_C)$.

$m_A^2 = (2b^2 + 2c^2 - a^2)/4 = (b^2 + c^2 + 2bc\cos\alpha)/4$.

$f_A = 2bc\cos(\alpha/2)/(b+c) = 2\sqrt{bcs(s-a)}\ /(b+c)$.

$f_Af_Bf_C = 8abcrs^2/(b+c)(c+a)(a+b)$.

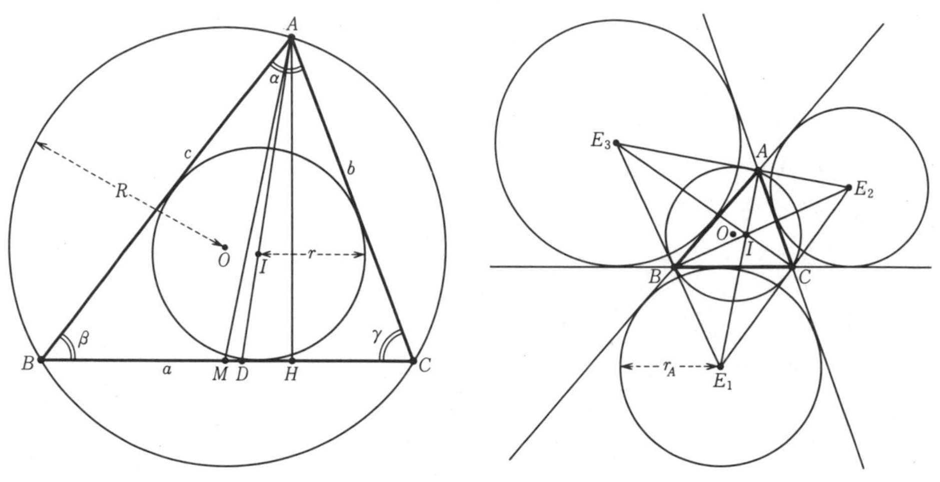

Fig. 2

(III) Spherical Triangles

We denote the interior angles of a spherical triangle by α, β, γ; the corresponding sides by a, b, c; the area by S; and the radius of the supporting sphere by ρ. We have

$\sin a : \sin b : \sin c = \sin\alpha : \sin\beta : \sin\gamma$ (law of sines).

$\cos a = \cos b\cos c + \sin b\sin c\cos\alpha, \quad \dots; \qquad \cos\alpha = -\cos\beta\cos\gamma + \sin\beta\sin\gamma\cos a, \quad \dots$

(law of cosines).

$\sin a\cos\beta = \cos b\sin c - \sin b\cos c\cos\alpha, \quad \dots$ (law of sines and cosines).

$\cot a\sin b = \cos b\cos\gamma + \cot\alpha\sin\gamma, \quad \dots$ (law of cotangents).

$\tan[(a+b)/2]/\tan[(a-b)/2] = \tan[(\alpha+\beta)/2]/\tan[(\alpha-\beta)/2], \quad \dots$ (law of tangents).

$\tan[(\alpha+\beta)/2]\tan(\gamma/2) = \cos[(a-b)/2]/\cos[(a+b)/2], \quad \dots;$

$\tan[(\alpha-\beta)/2]\tan(\gamma/2) = \sin[(a-b)/2]/\sin[(a+b)/2], \quad \dots;$

$\tan[(a+b)/2]\cot(c/2) = \cos[(\alpha-\beta)/2]/\cos[(\alpha+\beta)/2], \quad \dots;$

$\tan[(a-b)/2]\cot(c/2) = \sin[(\alpha-\beta)/2]/\sin[(\alpha+\beta)/2], \quad \dots$ (Napier's analogies).

$$S = (\alpha+\beta+\gamma-\pi)\rho^2 = 2\rho^2\arccos\frac{\cos^2(a/2R) + \cos^2(b/2R) + \cos^2(c/2R)}{2\cos(a/2R)\cos(b/2R)\cos(c/2R)} \qquad \text{(Heron's formula).}$$

For a right triangle ($\gamma = \pi/2$), we have Napier's rule of circular parts: taking the subscripts modulo 5 in Fig. 3,

$$\sin\theta_i = \tan\theta_{i+1}\tan\theta_{i-1} = \cos\theta_{i+2}\cos\theta_{i-2}.$$

For example, we have

$$\cos c = \cos a \cos b = \cot\alpha \cot\beta,$$

$$\cos\beta = \tan a \cot c = \cos b \sin\alpha,$$

$$\sin a = \tan b \cot\beta = \sin c \sin\alpha.$$

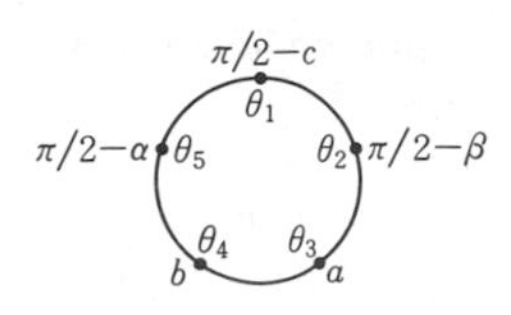

Fig. 3

3. Vector Analysis and Coordinate Systems

We denote a 3-dimensional vector by $\mathbf{A} \equiv (A_x, A_y, A_z) = A_x \boldsymbol{i} + A_y \boldsymbol{j} + A_z \boldsymbol{k}$, $|\mathbf{A}| = \sqrt{A_x^2 + A_y^2 + A_z^2}$.

(I) Vector Algebra (→ 442 Vectors)

Scalar product $\mathbf{A}\cdot\mathbf{B} \equiv \mathbf{AB} \equiv (\mathbf{A},\mathbf{B}) = A_x B_x + A_y B_y + A_z B_z = |\mathbf{A}|\,|\mathbf{B}|\cos\theta$ (where θ is the angle between $\mathbf{A}$ and $\mathbf{B}$).

Vector product

$$\mathbf{A}\times\mathbf{B} \equiv [\mathbf{A},\mathbf{B}] = (A_y B_z - A_z B_y)\mathbf{i} + (A_z B_x - A_x B_z)\mathbf{j} + (A_x B_y - A_y B_x)\mathbf{k} = \begin{vmatrix} \mathbf{i} & \mathbf{j} & \mathbf{k} \\ A_x & A_y & A_z \\ B_x & B_y & B_z \end{vmatrix}.$$

$$|\mathbf{A}\times\mathbf{B}| = |\mathbf{A}|\,|\mathbf{B}|\sin\theta.$$

$$\mathbf{A}\cdot\mathbf{B} = \mathbf{B}\cdot\mathbf{A}. \quad \mathbf{A}\cdot\mathbf{A} \equiv \mathbf{A}^2 = |\mathbf{A}|^2. \quad \mathbf{A}\times\mathbf{A} = 0. \quad \mathbf{A}\cdot(\mathbf{A}\times\mathbf{B}) = 0. \quad (\mathbf{A}\times\mathbf{B})^2 = |\mathbf{A}|^2|\mathbf{B}|^2 - (\mathbf{A}\cdot\mathbf{B})^2.$$

$$\mathbf{A}\times(\mathbf{B}\times\mathbf{C}) = (\mathbf{A}\cdot\mathbf{C})\mathbf{B} - (\mathbf{A}\cdot\mathbf{B})\mathbf{C}. \quad \mathbf{A}\times(\mathbf{B}\times\mathbf{C}) + \mathbf{B}\times(\mathbf{C}\times\mathbf{A}) + \mathbf{C}\times(\mathbf{A}\times\mathbf{B}) = 0.$$

$$(\mathbf{A}\times\mathbf{B})\cdot(\mathbf{C}\times\mathbf{D}) = \mathbf{A}\cdot\{\mathbf{B}\times(\mathbf{C}\times\mathbf{D})\} = (\mathbf{A}\cdot\mathbf{C})(\mathbf{B}\cdot\mathbf{D}) - (\mathbf{B}\cdot\mathbf{C})(\mathbf{A}\cdot\mathbf{D}).$$

Scalar triple product $$[\mathbf{ABC}] \equiv \mathbf{A}\cdot(\mathbf{B}\times\mathbf{C}) = \mathbf{B}\cdot(\mathbf{C}\times\mathbf{A}) = \mathbf{C}\cdot(\mathbf{A}\times\mathbf{B}) = \begin{vmatrix} A_x & A_y & A_z \\ B_x & B_y & B_z \\ C_x & C_y & C_z \end{vmatrix}.$$

$$[\mathbf{BCD}]\mathbf{A} + [\mathbf{ACD}]\mathbf{B} + [\mathbf{ABD}]\mathbf{C} = [\mathbf{ABC}]\mathbf{D}. \quad [\mathbf{ABC}][\mathbf{EFG}] = \begin{vmatrix} \mathbf{A}\cdot\mathbf{E} & \mathbf{A}\cdot\mathbf{F} & \mathbf{A}\cdot\mathbf{G} \\ \mathbf{B}\cdot\mathbf{E} & \mathbf{B}\cdot\mathbf{F} & \mathbf{B}\cdot\mathbf{G} \\ \mathbf{C}\cdot\mathbf{E} & \mathbf{C}\cdot\mathbf{F} & \mathbf{C}\cdot\mathbf{G} \end{vmatrix}.$$

(II) Differentiation of a Vector Field (→ 442 Vectors)

$$\nabla \equiv \mathbf{i}\frac{\partial}{\partial x} + \mathbf{j}\frac{\partial}{\partial y} + \mathbf{k}\frac{\partial}{\partial z} \quad \text{(Nabla)},$$

$$\operatorname{grad}\varphi \equiv \nabla\varphi = \frac{\partial\varphi}{\partial x}\mathbf{i} + \frac{\partial\varphi}{\partial y}\mathbf{j} + \frac{\partial\varphi}{\partial z}\mathbf{k} \quad \text{(gradient of } \varphi),$$

$$\operatorname{rot}\mathbf{A} \equiv \nabla\times\mathbf{A} = \left(\frac{\partial A_z}{\partial y} - \frac{\partial A_y}{\partial z}\right)\mathbf{i} + \left(\frac{\partial A_x}{\partial z} - \frac{\partial A_z}{\partial x}\right)\mathbf{j} + \left(\frac{\partial A_y}{\partial x} - \frac{\partial A_x}{\partial y}\right)\mathbf{k} \quad \text{(rotation of } \mathbf{A}),$$

$$\operatorname{div}\mathbf{A} \equiv \nabla\cdot\mathbf{A} = \frac{\partial A_x}{\partial x} + \frac{\partial A_y}{\partial y} + \frac{\partial A_z}{\partial z} \quad \text{(divergence of } \mathbf{A}),$$

$$\Delta\varphi \equiv \nabla^2\varphi \equiv \operatorname{div}\operatorname{grad}\varphi = \frac{\partial^2\varphi}{\partial x^2} + \frac{\partial^2\varphi}{\partial y^2} + \frac{\partial^2\varphi}{\partial z^2} \quad \text{(Laplacian of } \varphi).$$

$$\operatorname{grad}(\varphi\psi) = \varphi\operatorname{grad}\psi + \psi\operatorname{grad}\varphi,$$

$$\operatorname{grad}(\mathbf{A}\cdot\mathbf{B}) = (\mathbf{B}\cdot\operatorname{grad})\mathbf{A} + (\mathbf{A}\cdot\operatorname{grad})\mathbf{B} + \mathbf{A}\times\operatorname{rot}\mathbf{B} + \mathbf{B}\times\operatorname{rot}\mathbf{A},$$

$$\operatorname{rot}(\varphi\mathbf{A}) = \varphi\operatorname{rot}\mathbf{A} - \mathbf{A}\times\operatorname{grad}\varphi, \quad \operatorname{rot}(\mathbf{A}\times\mathbf{B}) = (\mathbf{B}\cdot\operatorname{grad})\mathbf{A} - (\mathbf{A}\cdot\operatorname{grad})\mathbf{B} + \mathbf{A}\operatorname{div}\mathbf{B} - \mathbf{B}\operatorname{div}\mathbf{A},$$

$$\operatorname{div}(\varphi\mathbf{A}) = \varphi\operatorname{div}\mathbf{A} + \mathbf{A}\cdot\operatorname{grad}\varphi, \quad \operatorname{div}(\mathbf{A}\times\mathbf{B}) = \mathbf{B}\cdot\operatorname{rot}\mathbf{A} - \mathbf{A}\cdot\operatorname{rot}\mathbf{B}.$$

$$\operatorname{rot}\operatorname{grad}\varphi = 0, \quad \operatorname{div}\operatorname{rot}\mathbf{A} = 0. \quad \Delta\mathbf{A} = \operatorname{grad}\operatorname{div}\mathbf{A} - \operatorname{rot}\operatorname{rot}\mathbf{A}.$$

$$\Delta(f\circ\varphi) = (df/d\varphi)\Delta\varphi + (d^2f/d\varphi^2)(\operatorname{grad}\varphi)^2, \quad \Delta(\varphi\psi) = \varphi\Delta\psi + \psi\Delta\varphi + 2(\operatorname{grad}\varphi\cdot\operatorname{grad}\psi).$$

(III) Integration of a Vector Field (→ 94 Curvilinear Integrals and Surface Integrals, 442 Vectors)

Let D be a 3-dimensional domain, B its boundary, dV the volume element of D, dS the surface element of B, and $d\mathbf{S}=\mathbf{n}\,dS$, where $\mathbf{n}$ is the outer normal vector of the surface B. We have

Gauss's formula
$$\iiint_D \operatorname{div}\mathbf{A}\,dV=\iint_B d\mathbf{S}\cdot\mathbf{A}=\iint_B(\mathbf{n}\cdot\mathbf{A})\,dS,$$
$$\iiint_D \operatorname{rot}\mathbf{A}\,dV=\iint_B d\mathbf{S}\times\mathbf{A}=\iint_B(\mathbf{n}\times\mathbf{A})\,dS,$$
$$\iiint_D \operatorname{grad}\varphi\,dV=\iint_B \varphi\,d\mathbf{S};$$

Green's formula
$$\iint_B \varphi\frac{\partial\psi}{\partial n}\,dS=\iiint_D(\varphi\Delta\psi+\operatorname{grad}\varphi\cdot\operatorname{grad}\psi)\,dV,$$
$$\iint_B\left(\varphi\frac{\partial\psi}{\partial n}-\psi\frac{\partial\varphi}{\partial n}\right)dS=\iiint_D(\varphi\Delta\psi-\psi\Delta\varphi)\,dV,$$
$$4\pi\varphi(x_0)=-\iiint_D\frac{\Delta\varphi}{r}\,dV+\iint_B\left\{\frac{1}{r}\frac{\partial\varphi}{\partial n}-\varphi\frac{\partial}{\partial n}\left(\frac{1}{r}\right)\right\}dS,$$

where r is the distance from the point x_0.

Let B be a bordered surface with a boundary curve Γ, ds the line element of Γ, dS the surface element of B, and $d\mathbf{s}=\mathbf{t}\,ds$, $d\mathbf{S}=\mathbf{n}\,dS$, for $\mathbf{t}$ the unit tangent vector of Γ and under the proper choice of the positive direction for the surface normal $\mathbf{n}$. We have

Stokes's formula
$$\iint_B d\mathbf{S}\cdot\operatorname{rot}\mathbf{A}=\oint_\Gamma\mathbf{A}\cdot d\mathbf{s}=\oint_\Gamma(\mathbf{t}\cdot\mathbf{A})\,ds,\quad \iint_B d\mathbf{S}\times\operatorname{grad}\varphi=\oint_\Gamma\varphi\,d\mathbf{s}.$$

If the domain D is simply connected, and the vector field $\mathbf{V}$ tends sufficiently rapidly to 0 near the boundary of D and at infinity, we have

Helmholtz's theorem
$$\mathbf{V}=\operatorname{grad}\varphi+\operatorname{rot}\mathbf{A},\quad \varphi=-\iiint_D\frac{\operatorname{div}\mathbf{V}}{4\pi r}\,dV,\quad \mathbf{A}=\iiint_D\frac{\operatorname{rot}\mathbf{V}}{4\pi r}\,dV.$$

(IV) Moving Coordinate System

Denote differentiation with respect to the rest and the moving systems by d/dt, d^*/dt, respectively. Let the relative velocity of the systems be $\mathbf{v}$. Then we have

$$\frac{d\varphi}{dt}=\frac{d^*\varphi}{dt}-\mathbf{v}\cdot\operatorname{grad}\varphi,\quad \frac{d\mathbf{A}}{dt}=\frac{d^*\mathbf{A}}{dt}-[\mathbf{v}\cdot\operatorname{grad}\mathbf{A}-(\mathbf{A}\cdot\operatorname{grad})\mathbf{v}].$$

With respect to rotating coordinates we have

$$\mathbf{v}=\mathbf{w}\times\mathbf{r}.$$
$$\frac{d\mathbf{A}}{dt}=\frac{d^*\mathbf{A}}{dt}+\left[\mathbf{w}+\mathbf{A}-((\mathbf{w}\times\mathbf{r})\cdot\operatorname{grad})\mathbf{A}\right]$$

When the domain of integration is also a function of t,

$$\frac{d}{dt}\int\mathbf{A}\cdot d\mathbf{s}=\int\left\{\frac{\partial\mathbf{A}}{\partial t}+\operatorname{grad}(\mathbf{v}\cdot\mathbf{A})-\mathbf{v}\times\operatorname{rot}\mathbf{A}\right\}\cdot d\mathbf{s},$$
$$\frac{d}{dt}\iint\mathbf{A}\cdot d\mathbf{S}=\iint\left\{\frac{\partial\mathbf{A}}{\partial t}+\operatorname{rot}(\mathbf{A}\times\mathbf{v})+\mathbf{v}\operatorname{div}\mathbf{A}\right\}\cdot d\mathbf{S},$$
$$\frac{d}{dt}\iiint\varphi\,dV=\iiint\left\{\frac{\partial\varphi}{\partial t}+(\mathbf{v}\cdot\operatorname{grad}\varphi)+\varphi\operatorname{div}\mathbf{v}\right\}dV=\iiint\frac{\partial\varphi}{\partial t}\,dV+\iint\varphi\mathbf{v}\cdot d\mathbf{S}.$$

(V) Curvilinear Coordinates (→ 90 Coordinates)

Let $(x_1,\ldots,x_n)$ be rectangular coordinates in an n-dimensional Euclidean space. If

$$x_j=\varphi_j(u_1,\ldots,u_n)\quad(j=1,\ldots,n),\qquad J\equiv\det(\partial\varphi_j/\partial u_k)\neq0,$$

the system $(u_1,\ldots,u_n)$ may be taken as a coordinate system of an n-dimensional space, and the original space is a Riemannian manifold with the first fundamental form

$$g_{jk}=\sum_{i=1}^{n}\frac{\partial\varphi_i}{\partial u_j}\frac{\partial\varphi_i}{\partial u_k}\qquad (j,k=1,\dots,n),$$

$$g\equiv\det(g_{jk})=J^2.$$

When the metric is of the diagonal form $g_{jk}=g_j^2\delta_{jk}$, the coordinate system $(u_1,\dots,u_n)$ is called an orthogonal curvilinear coordinate system or an isothermal curvilinear coordinate system. In such a case we have $J=g_1\dots g_n$, and the line element is given by $ds^2=\sum_{j=1}^n g_j^2\,du_j^2$.

For a scalar f and a vector $\xi=(\xi_1,\dots,\xi_n)$, we have

$$(\operatorname{grad} f)_j=\frac{1}{g_j}\frac{\partial f}{\partial u_j}\quad (j=1,\dots,n),\qquad \Delta f=\frac{1}{J}\sum_{j=1}^{n}\frac{\partial}{\partial u_j}\left(\frac{J}{g_j^2}\frac{\partial f}{\partial u_j}\right),$$

$$\operatorname{div}\xi=\frac{1}{J}\sum_{j=1}^{n}\frac{\partial}{\partial u_j}\left(\frac{J}{g_j}\xi_j\right),\quad (\operatorname{rot}\xi)_{jk}=\frac{1}{g_jg_k}\left[\frac{\partial(g_k\xi_k)}{\partial u_j}-\frac{\partial(g_j\xi_j)}{\partial u_k}\right]\qquad (j,k=1,\dots,n).$$

When $n=2$, the rot may be considered a scalar, rot $\xi=(\operatorname{rot}\xi)_{12}$, and when $n=3$, the rot may be considered a vector, with components

$$\operatorname{rot}\xi=((\operatorname{rot}\xi)_{23},\ (\operatorname{rot}\xi)_{31},\ (\operatorname{rot}\xi)_{12}).$$

The following are examples of orthogonal coordinates.

(1) Planar Curvilinear Coordinates. In the present Section (1), we put

$$x_1=x,\qquad x_2=y,\qquad u_1=u,\qquad u_2=v,\qquad g_1=p,\qquad g_2=q.$$

$$ds^2=p^2dx^2+q^2dy^2,\qquad J\equiv\partial(x,y)/\partial(u,v)=\sqrt{pq}\ .$$

Planar orthogonal curvilinear coordinates may be represented in the form $x+iy=F(U+iV)$, F being a complex analytic function, by suitable choice of the functions $U=U(u)$, $V=V(v)$.

(i) Polar Coordinates (r,θ) (Fig. 4).

$$x=r\cos\theta,\quad y=r\sin\theta;\quad x+iy=\exp(\log r+i\theta).$$

$$r=\sqrt{x^2+y^2}\ ,\quad \theta=\arctan(y/x).$$

$$p=1,\quad q=r,\quad J=r,\quad ds^2=dr^2+r^2d\theta^2.$$

$$\Delta f=\frac{\partial^2 f}{\partial r^2}+\frac{1}{r}\frac{\partial f}{\partial r}+\frac{1}{r^2}\frac{\partial^2 f}{\partial\theta^2}.$$

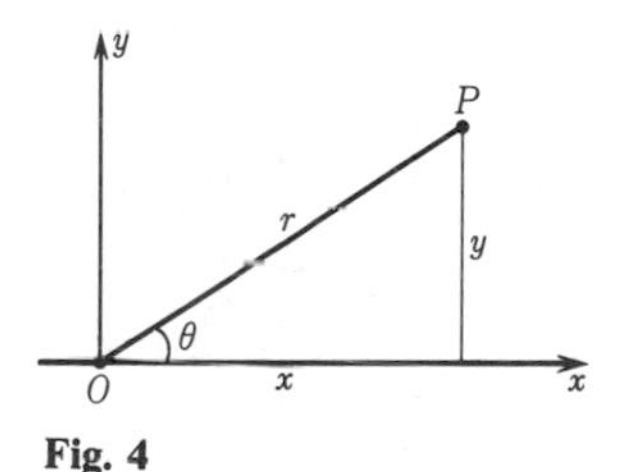

Fig. 4

(ii) Elliptic Coordinates (μ,ν) (Fig. 5). Among the family of confocal conics

$$\frac{x^2}{a^2+\rho}+\frac{y^2}{b^2+\rho}=1\quad (a>b),$$

there are two values of ρ for which the curve passes through a given point $P(x,y)$. Denote the two values of ρ by μ and ν, where $\mu>-b^2>\nu>-a^2$. The curve corresponding to $\rho=\mu$ or $\rho=\nu$ is an ellipse or a hyperbola, respectively. Then we have the relations

$$x^2=(\mu+a^2)(\nu+a^2)/(a^2-b^2),\qquad y^2=(\mu+b^2)(\nu+b^2)/(b^2-a^2).$$

Let the common foci be $(\pm c,0)$ $(c^2=a^2-b^2)$. Then we have

$$r_1=\sqrt{(x-c)^2+y^2}\ ,\qquad r_2=\sqrt{(x+c)^2+y^2}$$

where r_1, r_2 are the distances from the two foci as in Fig. 5, and

$$4(a^2+\mu)=(r_1+r_2)^2,\qquad 4(a^2+\nu)=(r_1-r_2)^2.$$

$$p=\frac{1}{2}\sqrt{\frac{\mu-\nu}{(\mu+a^2)(\mu+b^2)}}\ ,\qquad q=\frac{1}{2}\sqrt{\frac{\nu-\mu}{(\nu+a^2)(\nu+b^2)}}$$

(iii) Parabolic Coordinates (α,β) (Fig. 6). Among the family of parabolas $y^2=4\rho(x+\rho)$ with the focus at the origin and having the x-axis as the principal axis, there are two values of ρ for which the curve passes through a given point $P(x,y)$. Denote the two values of ρ by α, β $(\alpha>0>\beta)$. We have $x=-(\alpha+\beta)$, $y=\sqrt{-4\alpha\beta}$.

(iv) Equilateral (or Rectangular) Hyperbolic Coordinates (u,v) (Fig. 7). This is a system that replaces $x/2$, $y/2$ in (iii) by $-y$ and x, respectively, with $\sqrt{\alpha}=u$, $\sqrt{-\beta}=v$. The relations are

$$x=uv,\quad y=(u^2-v^2)/2;\quad x+iy=i(u-iv)^2/2,\quad u^2, v^2=\sqrt{x^2+y^2}\pm y,\quad p=q=\sqrt{u^2+v^2}.$$

The curves $x=$constant or $y=$constant are equilateral hyperbolas.

(v) Bipolar Coordinates (ξ,η) (Fig. 8). These coordinates represent a point $P(x,y)$ on a plane as the intersection of the family of circles passing through two fixed points $(\pm a,0)$ and the family of loci on which the ratio of distances from the same two fixed points $(\pm a,0)$ is constant. The latter is the set of Apollonius' circles. The relations are

$$x=\frac{a\sinh\xi}{\cosh\xi+\cos\eta},\qquad y=\frac{a\sin\eta}{\cosh\xi+\cos\eta}\qquad(-\infty<\xi<\infty,\ 0\leqslant\eta\leqslant 2\pi).$$

$$p=q=\frac{a}{\cosh\xi+\cos\eta}.$$

(2) Curvilinear Coordinates in 3-Dimensional Space. In the present Section (2), we put $x_1=x$, $x_2=y$, $x_3=z$.

(i) Circular Cylindrical Coordinates (Cylindrical Coordinates) (ρ,φ,z) (Fig. 9).

$$x=\rho\cos\varphi,\quad y=\rho\sin\varphi,\quad z=z.$$

$$ds^2=d\rho^2+\rho^2d\varphi^2+dz^2,\qquad J=\rho.\qquad \Delta f=\frac{1}{\rho}\frac{\partial}{\partial\rho}\left(\rho\frac{\partial f}{\partial\rho}\right)+\frac{1}{\rho^2}\frac{\partial^2 f}{\partial\varphi^2}+\frac{\partial^2 f}{\partial z^2}.$$

(ii) Polar Coordinates (Spherical Coordinates) (Fig. 9).

$$x=r\sin\theta\cos\varphi,\quad y=r\sin\theta\sin\varphi,\quad z=r\cos\theta.$$

$$r=\sqrt{x^2+y^2+z^2},\quad \varphi=\arctan(y/x),\quad \theta=\arctan\left(\sqrt{x^2+y^2}/z\right).$$

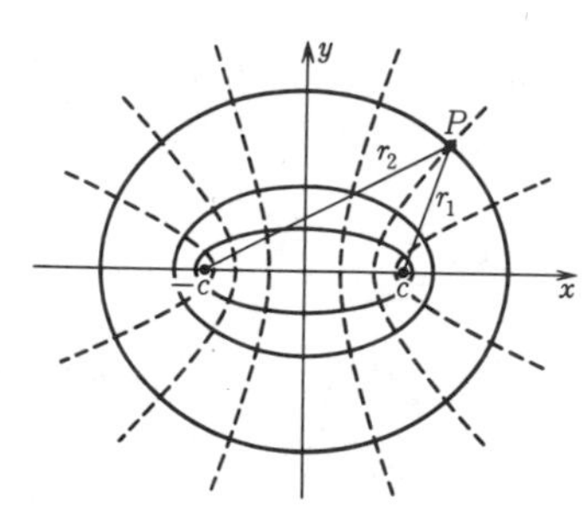

Fig. 5

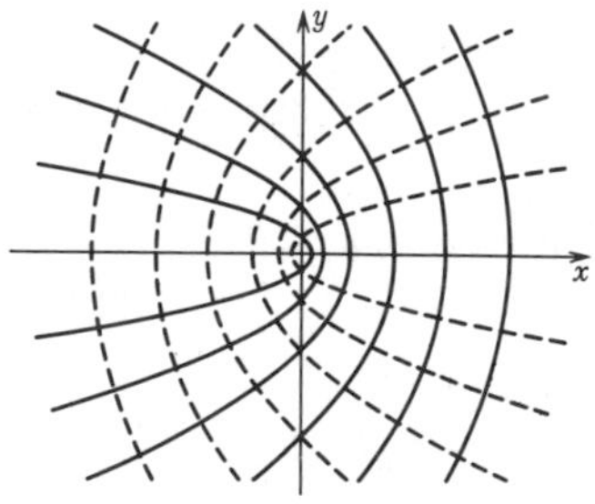

Fig. 6

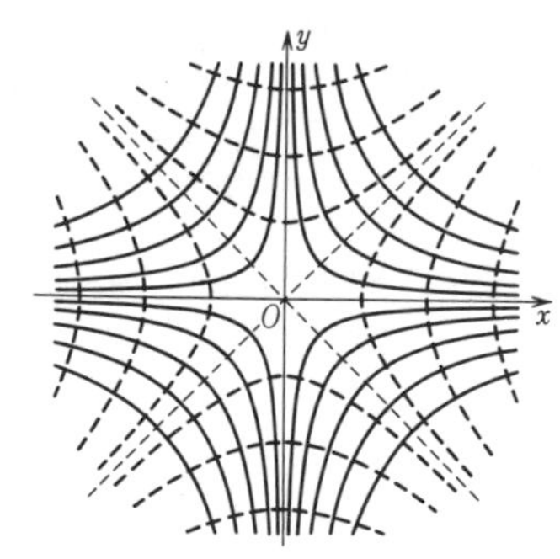

Fig. 7

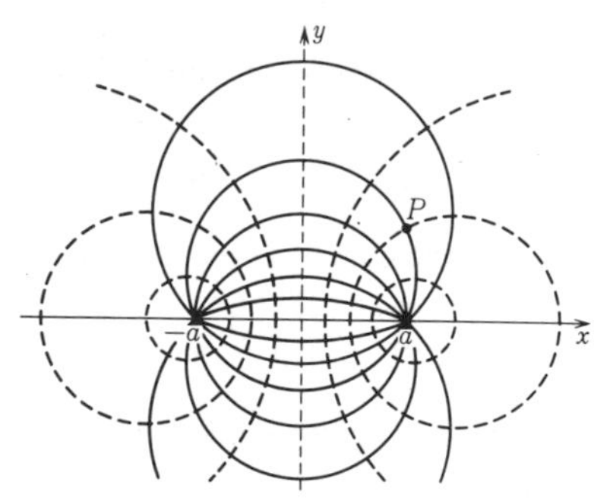

Fig. 8

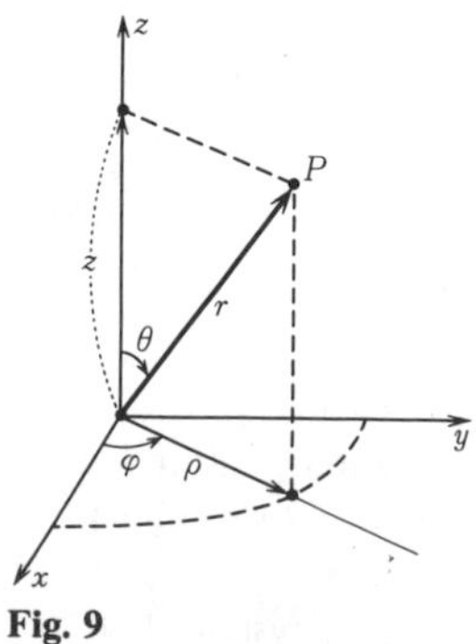

Fig. 9

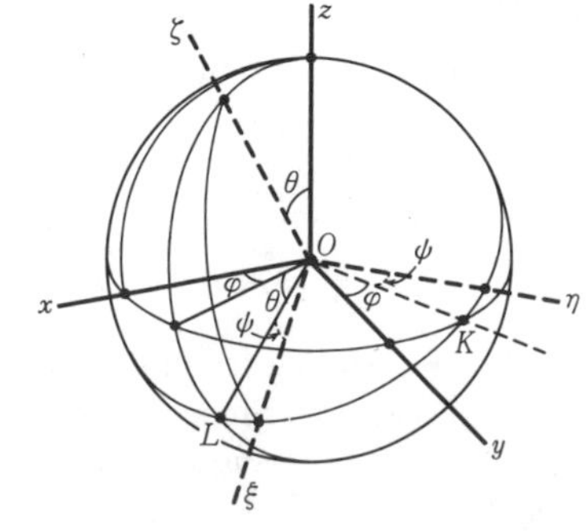

Fig. 10

The angles φ and θ are called azimuth and zenith angle, respectively. We further have

$$ds^2 = dr^2 + r^2 d\theta^2 + r^2 \sin^2\theta \, d\varphi^2, \qquad J = r^2 \sin\theta.$$

$$\Delta f = \frac{1}{r^2}\frac{\partial}{\partial r}\left(r^2 \frac{\partial f}{\partial r}\right) + \frac{1}{r^2 \sin\theta}\frac{\partial}{\partial \theta}\left(\sin\theta \frac{\partial f}{\partial \theta}\right) + \frac{1}{r^2 \sin^2\theta}\frac{\partial^2 f}{\partial \varphi^2}.$$

(iii) Euler's Angles (Fig. 10). Let (x,y,z) and (ξ,η,ζ) be two linear orthogonal coordinate systems with common origin O. Denote the angle between the z-axis and the ζ-axis by θ; the angle between the zx-plane and the $z\zeta$-plane by φ; and the angle between the η-axis and the intersection OK of the xy-plane and the $\xi\eta$-plane (or the angle between the ξ-axis and the intersection OL of the $z\zeta$-plane and the $\xi\eta$-plane) by ψ. The angles θ, φ, and ψ are called Euler's angles. The direction cosines of one coordinate axis with respect to the other coordinate system are as follows:

	x	y	z
ξ	$\cos\varphi\cos\theta\cos\psi - \sin\varphi\sin\psi$	$\sin\varphi\cos\theta\cos\psi + \cos\varphi\sin\psi$	$-\sin\theta\cos\psi$
η	$-\cos\varphi\cos\theta\sin\psi - \sin\varphi\cos\psi$	$-\sin\varphi\cos\theta\sin\psi + \cos\varphi\cos\psi$	$\sin\theta\sin\psi$
ζ	$\cos\varphi\sin\theta$	$\sin\varphi\sin\theta$	$\cos\theta$

(iv) Rotational (or Revolutional) Coordinates (u,v,ρ). Let (u,v) be curvilinear coordinates (Section (1)) on the $z\rho$-plane. The rotational coordinates (u,v,ρ) are given by the combination of $x = \rho\cos\varphi$, $y = \rho\sin\varphi$ with the coordinates on the $z\rho$-plane. We have

$$ds^2 = p^2 du^2 + q^2 dv^2 + \rho^2 d\varphi^2,$$

where p, q are the corresponding values for the coordinates (u,v).

(v) Generalized Cylindrical Coordinates (u,v,z). These are a combination of curvilinear coordinates (u,v) on the xy-plane with z. We have

$$ds^2 = p^2 du^2 + q^2 dv^2 + dz^2.$$

For various selections of (u,v) we have coordinates as follows:

(u,v)	Rotational Coordinate System	Generalized Cylindrical Coordinate System
Linear rectangular coordinates	Circular cylindrical coordinates	Linear rectangular coordinates
Polar coordinates ((1)(i))	Spherical coordinates	Circular cylindrical coordinates
Elliptic coordinates ((1)(ii))	Spheroidal coordinates(1)	Elliptic cylindrical coordinates
Parabolic coordinates ((1)(iii))	Rotational parabolic coordinates(2)	Parabolic cylindrical coordinates
Equilateral hyperbolic coordinates ((1)(iv))	Rotational hyperbolic coordinates	Hyperbolic cylindrical coordinates
Bipolar coordinates ((1)(v))	Toroidal coordinates(3) Bipolar coordinates(4)	Bipolar cylindrical coordinates

Notes

(1) When the ρ-axis is a minor or major axis, we have prolate or oblate spheroidal coordinates, respectively.
(2) We take the z-axis as the common principal axis of the parabolas.
(3) Where the line passing through two fixed points is the ρ-axis.
(4) Where the line passing through two fixed points is the z-axis.

(vi) Ellipsoidal Coordinates (λ,μ,ν) (Fig. 11). Among the family of confocal quadrics

$$\frac{x^2}{a^2+\rho} + \frac{y^2}{b^2+\rho} + \frac{z^2}{c^2+\rho} = 1 \quad (a > b > c > 0),$$

there are three values of ρ for which the surface passes through a given point $P(x,y,z)$. Denote the three values of ρ by λ, μ, ν, where $\lambda>-c^2>\mu>-b^2>\nu>-a^2$. The surfaces corresponding to $\rho=\lambda$, $\rho=\mu$, and $\rho=\nu$ are an ellipsoid, a hyperboloid of one sheet, and a hyperboloid of two sheets, respectively. We have

$$x^2=\frac{h(a)}{(a^2-b^2)(a^2-c^2)},\qquad y^2=\frac{h(b)}{(b^2-c^2)(b^2-a^2)},$$

$$z^2=\frac{h(c)}{(c^2-a^2)(c^2-b^2)};\qquad h(\alpha)\equiv(\lambda+\alpha^2)(\mu+\alpha^2)(\nu+\alpha^2).$$

$$g_1=\frac{\sqrt{(\lambda-\mu)(\lambda-\nu)}}{2\rho(\lambda)},\qquad g_2=\frac{\sqrt{(\mu-\nu)(\mu-\lambda)}}{2\rho(\mu)},$$

$$g_3=\frac{\sqrt{(\nu-\lambda)(\nu-\mu)}}{2\rho(\nu)};\qquad \rho(t)\equiv\sqrt{(t+a^2)(t+b^2)(t+c^2)}\,.$$

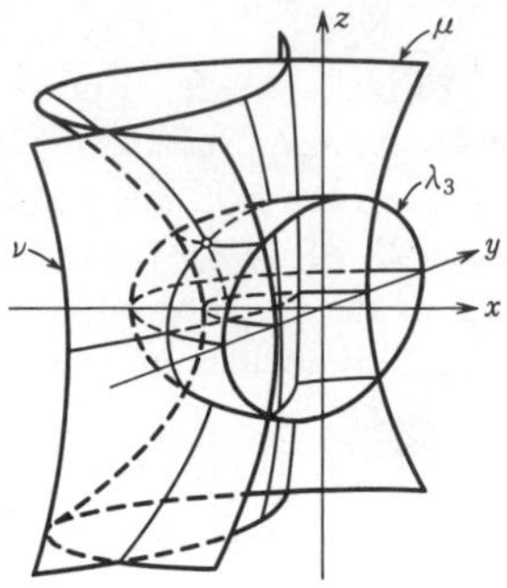

Fig. 11

4. Differential Geometry

(I) Classical Differential Geometry (→ 111 Differential Geometry of Curves and Surfaces)

(1) Plane Curves (Fig. 12). At a point $P(x_0,y_0)$ on a curve $y=f(x)$, the equation of the tangent line is $y-y_0=f'(x_0)(x-x_0)$,

$$PT=\left|y_0\sqrt{1+y_0'^2}\;/y_0'\right|,$$

and the tangential shadow $TM=y_0/y_0'$. The equation of the normal line is $f'(x_0)(y-y_0)+(x-x_0)=0$,

$$PN=\left|y_0\sqrt{1+y_0'^2}\;\right|,$$

and the normal shadow $MN=y_0y_0'$. The slope of the tangent is $\tan\alpha=f'(x_0)=y_0'$. The curvature at P is

$$\kappa=1/PQ=f''(x_0)/\left[1+f'(x_0)^2\right]^{3/2}$$

The coordinates of the center of curvature Q are

$$\left(x_0-f'(x_0)\left[1+f'(x_0)^2\right]/f''(x_0),\;\; f(x_0)+[1+f'(x_0)^2]/f''(x_0)\right).$$

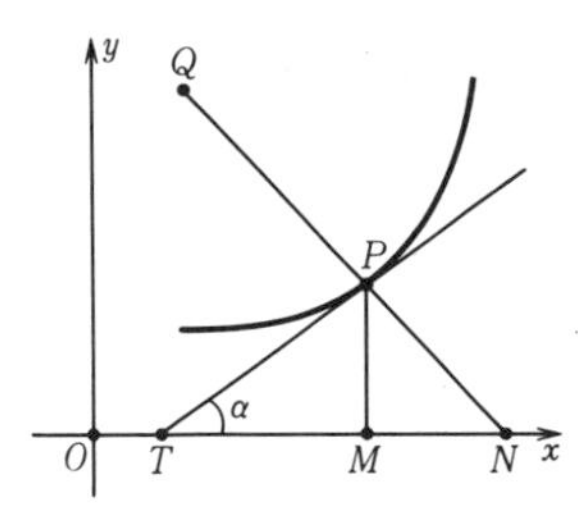

Fig. 12

(2) Space Curves $x_i=x_i(t)$ $(i=1,2,3)$, or $\mathbf{x}=\mathbf{x}(t)$. The line element of a curve $\mathbf{x}=\mathbf{x}(t)$ is

$$ds=\sqrt{(dx_1)^2+(dx_2)^2+(dx_3)^2}=\sqrt{\sum_{\alpha=1}^{3}\dot{x}_\alpha^2}\;dt\qquad\left(\cdot=\frac{d}{dt}\right).$$

The curvature is

$$\kappa=\sqrt{\sum\ddot{x}_\alpha^2-\ddot{s}^2}/\dot{s}^2$$

For $t=s$ (arc length), the curvature is $\kappa=\sqrt{(\sum x_\alpha''^2)}$, and the torsion is $\tau=[\det(x_\alpha',x_\alpha'',x_\alpha''')_{\alpha=1,2,3}]/\kappa^2$, where $'=d/ds$. When we denote Frenet's frame by $(\boldsymbol{\xi},\boldsymbol{\eta},\boldsymbol{\zeta})$, we have $\boldsymbol{\xi}=\mathbf{x}'$, $\boldsymbol{\eta}=\boldsymbol{\xi}'/\kappa$, $\boldsymbol{\zeta}=\boldsymbol{\xi}\times\boldsymbol{\eta}$ (vector product).

The Frenet-Serret formulas are

$$\xi' = \kappa\eta, \quad \eta' = -\kappa\xi + \tau\zeta, \quad \zeta' = -\tau\eta.$$

(3) Surface in 3-Dimensional Space $x_\alpha = x_\alpha(u_1, u_2)$ $(\alpha = 1,2,3)$. The first fundamental form of the surface is

$$g_{jk} = \sum_{\alpha=1}^{3} \frac{\partial x_\alpha}{\partial u_j}\frac{\partial x_\alpha}{\partial u_k} \quad (j,k=1,2). \quad g = \det(g_{jk}) > 0.$$

Let (g^{jk}) be the inverse matrix of (g_{jk}). The tangent plane at the point $x_\alpha^{(0)}$ is given by

$$\det\left(x_\alpha - x_\alpha^{(0)},\ (\partial x_\alpha/\partial u_1)^{(0)},\ (\partial x_\alpha/\partial u_2)^{(0)}\right) = 0.$$

The normal line at the point $x_\alpha^{(0)}$ is given by $x_\alpha - x_\alpha^{(0)} = t\nu_\alpha^{(0)}$, where t is a parameter, and ν_α is the unit normal vector, given by

$$\nu_\alpha = \frac{1}{\sqrt{g}} \frac{\partial(x_\beta, x_\gamma)}{\partial(u_1, u_2)} \quad \left(\delta_{\alpha\beta\gamma}^{123} = +1\right).$$

The second fundamental form is

$$h_{jk} \equiv \sum_{\alpha=1}^{3} \nu_\alpha \frac{\partial^2 x_\alpha}{\partial u_j \partial u_k} = -\sum_{\alpha=1}^{3} \frac{\partial \nu_\alpha}{\partial u_j}\frac{\partial x_\alpha}{\partial u_k}. \quad h \equiv \det(h_{jk}).$$

The principal radii of curvature R_1, R_2 are the roots of the quadratic equation

$$\frac{1}{R^2} - \sum_{j,k} g^{jk} h_{jk} \frac{1}{R} + \frac{h}{g} = 0.$$

The mean curvature (or Germain's curvature) is

$$H \equiv \frac{1}{2}\left(\frac{1}{R_1} + \frac{1}{R_2}\right) = \frac{1}{2}\sum_{j,k} g^{jk} h_{jk},$$

and $H = 0$ is the condition for the given surface to be a minimal surface. The Gaussian curvature (or total curvature) is

$$K \equiv \frac{1}{R_1 R_2} = \frac{h}{g},$$

and $K = 0$ is the condition for the surface to be developable.

We use the notations of Riemannian geometry, with g_{jk} the fundamental tensor:

$$\frac{\partial^2 x_\alpha}{\partial u_j \partial u_k} = \sum_{a=1}^{3} \left\{ {a \atop jk} \right\} \frac{\partial x_\alpha}{\partial u_a} + h_{jk}\nu_\alpha \quad \text{(Gauss's formula).}$$

$$R_{ijkl} = h_{jl}h_{ik} - h_{jk}h_{il} \quad \text{(Gauss's equation).} \qquad h_{jk;l} = h_{jl;k} \quad \text{(Codazzi-Mainardi equation).}$$

$$K = \frac{R_{1212}}{g} = \frac{R}{2} = \frac{1}{\sqrt{g}}\left[\frac{\partial}{\partial u_2}\left(\frac{\sqrt{g}}{g_{11}}\left\{ {2 \atop 11} \right\}\right) - \frac{\partial}{\partial u_1}\left(\frac{\sqrt{g}}{g_{11}}\left\{ {2 \atop 12} \right\}\right)\right]$$

$$= \frac{1}{\sqrt{g}}\left[\frac{\partial}{\partial u_1}\left(\frac{\sqrt{g}}{g_{22}}\left\{ {1 \atop 22} \right\}\right) - \frac{\partial}{\partial u_2}\left(\frac{\sqrt{g}}{g_{22}}\left\{ {1 \atop 12} \right\}\right)\right].$$

$$\frac{\partial \nu_\alpha}{\partial u_j} = -\sum_{k,l} h_{jk} g^{kl} \frac{\partial x_\alpha}{\partial u_l} \quad \text{(Weingarten's formula).}$$

The third fundamental form is given by

$$l_{jk} \equiv \sum_{\alpha=1}^{3} \frac{\partial \nu_\alpha}{\partial u_j}\frac{\partial \nu_\alpha}{\partial u_k} = \sum_{s,t} g^{st} h_{js} h_{kt} = 2Hh_{jk} - Kg_{jk}. \quad \det(l_{jk}) = K^2 g = Kh.$$

(4) Geodesic Curvature. Let $C: u_i = u_i(s)$ be a curve on a surface S and ρ be the curvature of C at a point P. Let θ be the angle between the osculating plane of C and the plane tangent to S. The geodesic curvature ρ_g of C at P is given by

$$\rho_g = \rho\cos\theta = \sqrt{g}\ \det\left(\frac{du_i}{ds},\ \frac{d^2 u_i}{ds^2} + \sum_{j,k} \left\{ {i \atop jk} \right\} \frac{du_j}{ds}\frac{du_k}{ds}\right)_{i=1,2}$$

$\rho_g=0$ is the condition for C to be a geodesic. Let D be a simply connected domain on the surface S, whose boundary Γ consists of n smooth curves. Let θ_α be the outer angle at the intersection of two consecutive curves ($\alpha=1,\ldots,n$). Then we have the Gauss-Bonnet formula:

$$\int_\Gamma \rho_g ds + \iint_D K dS = 2\pi - \sum_{\alpha=1}^{n} \theta_\alpha .$$

(II) Riemannian Geometry, Tensor Calculus (→ 417 Tensor Calculus)

In the present section, we use Einstein's convention (omission of the summation symbol).

(1) Numerical Tensor.

Kronecker's δ $\quad \delta_{jk},\ \delta^{jk},\ \delta^j_k = \begin{cases} 1 & (j=k) \\ 0 & (j\neq k). \end{cases}$

$$\delta^{j_1\ldots j_p}_{k_1\ldots k_p} = \det\left(\delta^{j_\mu}_{k_\nu}\right)_{\mu,\nu=1,\ldots,p} = \begin{cases} 0 & (\{j_\mu\}\neq\{k_\nu\}), \\ +1 & (\{j_\mu\}=\{k_\nu\} \text{ and } (j_\mu) \text{ is an even permutation of } (k_\nu)), \\ -1 & (\{j_\mu\}=\{k_\nu\}) \text{ and } (j_\mu) \text{ is an odd permutation of } (k_\nu). \end{cases}$$

Eddington's ε $\quad \varepsilon_{j_1\ldots j_n} = \delta^{1\ldots n}_{j_1\ldots j_n}, \quad \varepsilon^{j_1\ldots j_n} = \delta^{j_1\ldots j_n}_{1\ldots n}.$

$$\delta^{j_1\ldots j_p j_{p+1}\ldots j_n}_{k_1\ldots k_p j_{p+1}\ldots j_n} = (n-p)!\,\delta^{j_1\ldots j_p}_{k_1\ldots k_p}. \quad \det(a^u_\nu)_{\mu,\nu=1,\ldots,n} = \varepsilon^{j_1\ldots j_n} a^1_{j_1} a^2_{j_2}\ldots a^n_{j_n} = \varepsilon_{j_1\ldots j_n} a^{j_1}_1 a^{j_2}_2 \ldots a^{j_n}_n.$$

(2) Fundamental Objects in Riemannian Geometry. Let g_{jk} be the fundamental tensor, and (g^{jk}) be the inverse matrix of (g_{jk}). We put $g\equiv\det(g_{jk})$.

The Christoffel symbol is

$$\left\{ {i \atop jk} \right\} = \frac{1}{2} g^{ia}\left[\frac{\partial g_{aj}}{\partial x^k} + \frac{\partial g_{ak}}{\partial x^j} - \frac{\partial g_{jk}}{\partial x^a}\right] = \left\{ {i \atop kj} \right\}, \quad \left\{ {a \atop ak} \right\} = \frac{\partial \log \sqrt{g}}{\partial x^k},$$

which has the transformation rule

$$\left\{ {\bar{i} \atop \bar{j}\bar{k}} \right\} = \frac{\partial \bar{x}^{\bar{i}}}{\partial x^i}\left(\frac{\partial x^j}{\partial \bar{x}^{\bar{j}}} \frac{\partial x^k}{\partial \bar{x}^{\bar{k}}} \left\{ {i \atop jk} \right\} + \frac{\partial^2 x^i}{\partial \bar{x}^{\bar{j}} \partial \bar{x}^{\bar{k}}}\right)$$

under a coordinate transformation.

A geometrical object Γ^i_{jk} with a similar transformation rule is called the coefficient of the affine connection. The torsion tensor is

$$S^i_{jk} \equiv \Gamma^i_{jk} - \Gamma^i_{kj}.$$

The equation of a geodesic is

$$\frac{d^2x^i}{ds^2} + \left\{ {i \atop jk} \right\} \frac{dx^j}{ds}\frac{dx^k}{ds} = 0.$$

The covariant derivative of a tensor of weight W with respect to a coefficient of affine connection Γ^i_{jk} is given by

$$T^{j_1\ldots j_p}_{k_1\ldots k_q|l} \equiv \partial T^{j_1\ldots j_p}_{k_1\ldots k_q}/\partial x^l + \sum_{\nu=1}^{p} T^{j_1\ldots j_{\nu-1} a j_{\nu+1}\ldots j_p}_{k_1\ldots\ldots\ldots k_q}\Gamma^{j_\nu}_{al} - \sum_{\mu=1}^{q} T^{j_1\ldots\ldots\ldots j_p}_{k_1\ldots k_{\mu-1} a k_{\mu+1}\ldots k_q}\Gamma^a_{lk_\mu} - W T^{j_1\ldots j_p}_{k_1\ldots k_q}\Gamma^a_{al}.$$

For the Christoffel symbol, we denote the covariant derivative by $;l$. Then we have the following formulas:

$$g_{jk;l}=0, \quad g^{jk}{}_{;l}=0, \quad \delta^j_{k;l}=0, \quad \sqrt{g}\,\varepsilon_{j_1\ldots j_n;l}=0, \quad (1/\sqrt{g})\varepsilon^{j_1\ldots j_n}{}_{;l}=0.$$

For a scalar f $\quad \operatorname{grad} f = (f_{;j})$,

for a covariant vector v_j $\quad \operatorname{rot} v = (v_{j;k} - v_{k;j}) = (\partial v_j/\partial x^k - \partial v_k/\partial x^j)$,

and for a contravariant vector v^j $\quad \operatorname{div} v = v^j{}_{;j} = \dfrac{1}{\sqrt{g}}\dfrac{\partial(\sqrt{g}v^j}{\partial x^j}.$

Beltrami's differential operator of the first kind is

$$\Delta_1 f \equiv g^{jk} f_{;j} f_{;k}.$$

Beltrami's differential operator of the second kind is

$$\Delta_2 f = \operatorname{div}\operatorname{grad} f = \frac{1}{\sqrt{g}} \frac{\partial\left(\sqrt{g}\; g^{jk}(\partial f/\partial x^k)\right)}{\partial x^j}.$$

For a domain D with sufficiently smooth boundary Γ, we denote the directional derivative along the inner normal by $\partial/\partial n$, the volume element by dV, and the surface element on Γ by dS. Then we have Green's formulas,

$$\int_D (\Delta_1(\varphi\psi) + \psi\Delta_2\varphi)\, dV = -\int_\Gamma \psi \frac{\partial\varphi}{\partial n}\, dS, \qquad \int_D (\varphi\Delta_2\psi - \psi\Delta_2\varphi)\, dV = \int_\Gamma \left(\psi\frac{\partial\varphi}{\partial n} - \varphi\frac{\partial\psi}{\partial n}\right) dS.$$

We denote the curvature tensor with respect to the coefficients of a general affine connection Γ^i_{jk} by B^i_{jkl}, and by R^i_{jkl} when $\Gamma^i_{jk} = \left\{ {i \atop jk} \right\}$. We have the following formulas:

$$B^i_{jkl} \equiv \frac{\partial \Gamma^i_{jl}}{\partial x^k} - \frac{\partial \Gamma^i_{jk}}{\partial x^l} + \Gamma^a_{jl}\Gamma^i_{ak} - \Gamma^a_{jk}\Gamma^i_{al};$$

$$R_{ijkl} \equiv g_{ai} R^a_{jkl} = -R_{ijlk} = R_{klij} = \frac{1}{2}\left[\frac{\partial^2 g_{ik}}{\partial x^j \partial x^l} + \frac{\partial^2 g_{jl}}{\partial x^i \partial x^k} - \frac{\partial^2 g_{jk}}{\partial x^i \partial x^l} - \frac{\partial^2 g_{il}}{\partial x^j \partial x^k}\right]$$
$$+ g_{ab}\left(\left\{ {b \atop ik} \right\}\left\{ {a \atop jl} \right\} - \left\{ {b \atop il} \right\}\left\{ {a \atop jk} \right\}\right);$$

Bianchi's first identity $\quad R^i_{jkl} + R^i_{klj} + R^i_{ljk} = 0,$

$$-\left(B^i_{jkl} + B^i_{klj} + B^i_{ljk}\right) = 2\left(S^i_{jk|l} + S^i_{kl|j} + S^i_{lj|k}\right) + 4\left(S^i_{ja}S^a_{kl} + S^i_{ka}S^a_{lj} + S^i_{la}S^a_{jk}\right);$$

Bianchi's second identity $\quad R^i_{jkl;m} + R^i_{jlm;k} + R^i_{jmk;l} = 0,$

$$B^i_{jkl|m} + B^i_{jlm|k} + B^i_{jmk|l} = -2\left(B^i_{jma}S^a_{kl} + B^i_{jka}S^a_{lm} + B^i_{jla}S^a_{mk}\right);$$

Ricci's tensor $\quad R_{jk} \equiv -R^i_{jki} = R_{kj};$

scalar curvature $\quad R \equiv g^{jk} R_{jk};$

Ricci's formula $\quad T^{j_1\cdots j_p}_{k_1\cdots k_q|s|t} - T^{j_1\cdots j_p}_{k_1\cdots k_q|t|s}$

$$= -\sum_{\nu=1}^{p} T^{j_1\cdots j_{\nu-1} a j_{\nu+1}\cdots j_p}_{k_1\cdots\cdots\cdots k_q} B^{j_\nu}_{ast} + \sum_{\mu=1}^{q} T^{j_1\cdots\cdots\cdots j_p}_{k_1\ldots k_{\mu-1} a k_{\mu+1},\ldots,k_q} B^a_{k_\mu st} + 2T^{j_1\cdots j_p}_{k_1\cdots k_q|l} S^l_{st} + WT^{j_1\cdots j_p}_{k_1\cdots k_q} B^a_{ast},$$

where S and B are the torsion and curvature tensors given above, respectively, and W is the weight of the tensor T.

(3) Special Riemannian Spaces (→ 364 Riemannian Manifolds). In the present Section (3), n means the dimension of the space.

(i) Space of Constant Curvature $R^i_{jkl} = \rho(g_{jl}\delta^i_k - g_{jk}\delta^i_l)$; $\quad \rho = R/n(n-1)$,

(ii) Einstein Space $R_{jk} = \rho g_{jk}$, $\rho = R/n$, for $n \geqslant 3$, where R is a constant.

(iii) Locally Symmetric Riemannian Space $R^i_{jkl;m} = 0$.

(iv) Projectively Flat Space. Weyl's projective curvature tensor is defined by

$$W^i_{jkl} \equiv R^i_{jkl} + \frac{1}{n-1}(R_{jk}\delta^i_l - R_{jl}\delta^i_k).$$

The condition for the space to be projectively flat is given by $W^i_{jkl} = 0$, $R_{jk;l} = R_{jl;k}$.

If $n \geqslant 3$, the latter condition follows from the former condition, and the space reduces to a space of constant curvature. If $n = 2$, the former condition $W = 0$ always holds.

(v) Concircularly Flat Space $Z^i_{jkl} \equiv R^i_{jkl} + \dfrac{R}{n(n-1)}(g_{jk}\delta^i_l - g_{jl}\delta^i_k) = 0$. This space reduces to a space of constant curvature.

(vi) Conformally Flat Space. Weyl's conformal curvature tensor is defined by

$$C^i_{jkl} \equiv R^i_{jkl} + \frac{1}{n-2}(R_{jk}\delta^i_l - R_{jl}\delta^i_k + g_{jk}R^i_l - g_{jl}R^i_k) - \frac{R(g_{jk}\delta^i_l - g_{jl}\delta^i_k)}{(n-1)(n-2)},$$

$$\Pi_{jk} \equiv -\frac{R_{jk}}{(n-2)} + \frac{Rg_{jk}}{2(n-1)(n-2)}.$$

The condition for the space to be conformally flat is given by $C^i_{jkl} = 0$, $\Pi_{jk;l} = \Pi_{jl;k}$.

If $n \geqslant 4$, the latter condition follows from the former condition, and if $n = 3$, the former condition $C = 0$ always holds.

5. Lie Algebras, Symmetric Riemannian Spaces, and Singularities

(I) The Classification of Complex Simple Lie Algebras and Compact Real Simple Lie Algebras
(→ 248 Lie Algebras)

(1) Lie Algebra. The unitary restriction of a noncommutative finite-dimensional complex simple Lie algebra $\mathfrak{g}$ is a compact real simple Lie algebra $\mathfrak{g}_u$, and $\mathfrak{g}$ is given by the complexification $\mathfrak{g}_u^c$ of $\mathfrak{g}_u$. There exists a bijective correspondence between the classifications of these two kinds of Lie algebras. Using Dynkin diagrams, the classification is done as in Fig. 14 (→ 248 Lie Algebras). The system of fundamental roots $\{\alpha_1, \dots, \alpha_l\}$ of a simple Lie algebra $\mathfrak{g}$ is in one-to-one correspondence with the vertices of a Dynkin diagram shown by simple circles in Fig. 14. The number of simple circles coincides with the rank l of $\mathfrak{g}$. The double circle in Fig. 14 means -1 times the highest root θ. Sometimes we mean by the term "Dynkin diagram" the diagram without the double circle and the lines issuing from it. Here we call the diagram with double circle representing $-\theta$ the extended Dynkin diagram. Corresponding to the value of the inner product with respect to the Killing form $-2(\alpha_i, \alpha_j)/(\alpha_j, \alpha_j)$ $(i \neq j)$ (which must be 0, 1, 2, or 3), we connect two vertices representing α_i and α_j as in Fig. 13. When the value is 0, we do not connect α_i and α_j. In Fig. 13, the left circle corresponds to α_i and the right circle to α_j.

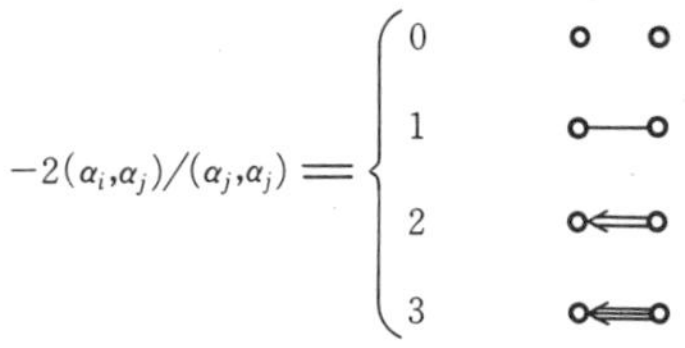

$-2(\alpha_i,\alpha_j)/(\alpha_j,\alpha_j) = \begin{cases} 0 \\ 1 \\ 2 \\ 3 \end{cases}$

Fig. 13

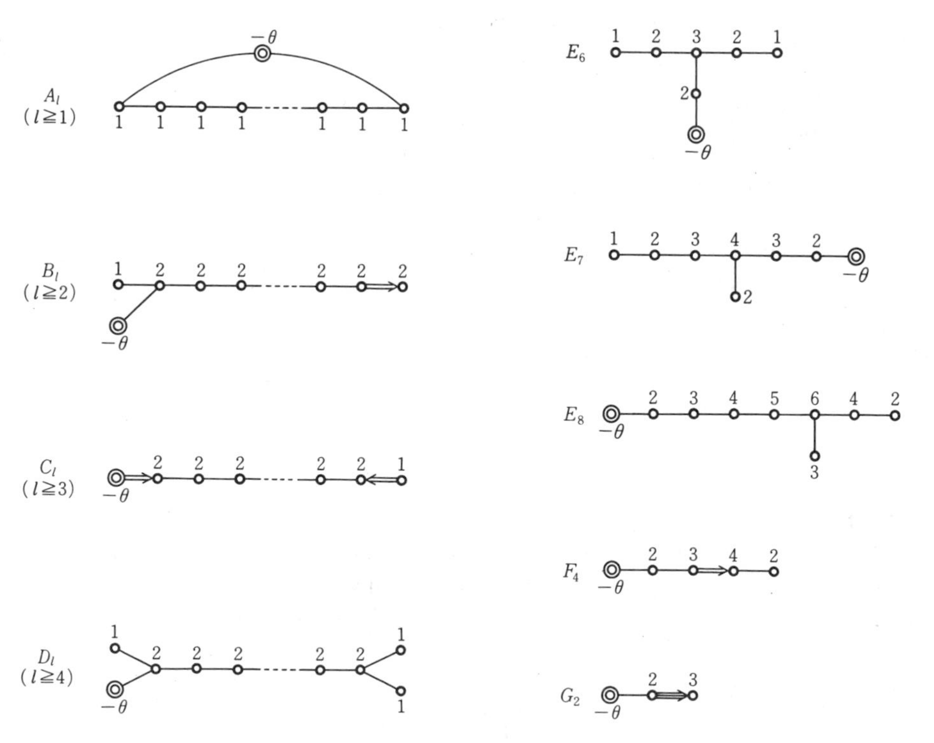

Fig. 14 We have relations $B_1 = C_1 = A_1$, $C_2 = B_2$, and $D_3 = A_3$. ($D_2 = A_1 + A_1$, which is not simple.) In this figure, the number at each vertex means the coefficient m_i in $\theta = \sum m_i \alpha_i$.

From Fig. 14, we have the following information.
(i) The quotient group of the automorphism group $A(\mathfrak{g})$ of $\mathfrak{g}$ with respect to the inner automorphism group $I(\mathfrak{g})$ is isomorphic to the automorphism group of the corresponding Dynkin diagram. The order of the latter group is 2 for A_l $(l \geqslant 2)$ since the diagram is symmetric. It is also 2 for D_l $(l \geqslant 5)$ and for E_6, and it is 6 $(=3!)$ for D_4. For all other cases, the order is 1.

(ii) The order of the center of the simply connected Lie group associated with $\mathfrak{g}$ is equal to the index of the subgroup consisting of elements stabilizing $-\theta$ in the group of automorphisms of the extended Dynkin diagram of $\mathfrak{g}$ (S. Murakami). This index is equal to the order of the fundamental group of the adjoint group of $\mathfrak{g}$ and the number of connected Lie groups, whose Lie algebra is $\mathfrak{g}$.
(iii) Any parabolic Lie subalgebra of $\mathfrak{g}$ is isomorphic to a subalgebra generated by the root vector X_α (and elements of the Cartan subalgebra) such that $\alpha = \sum n_i \alpha_i$, where $\{\alpha_1, \ldots, \alpha_l\}$ is a system of fundamental roots, $n_i \geqslant 0$ $(i=1, \ldots, l)$ or $n_i \leqslant 0$ $(i=1, \ldots, l)$, and $n_j = 0$ for α_j belonging to a fixed subset S of $\{\alpha_1, \ldots, \alpha_l\}$.

Hence, isomorphism classes of parabolic Lie subalgebras are in one-to-one correspondence with the set of subsets S of $\{\alpha_1, \ldots, \alpha_l\}$.
(iv) Maximal Lie subalgebra $\mathfrak{k}$ of $\mathfrak{g}$ with the same rank l as $\mathfrak{g}$. The Lie subalgebra $\mathfrak{k}$ is classified by the following rule. First we remove a vertex α_i from the Dynkin diagram. If the number m_i attached to the vertex is 1, $\mathfrak{k}$ is given by the product of the simple Lie algebra corresponding to the Dynkin diagram after removing the vertex α_i and a one-dimensional Lie subalgebra. If $m_i > 1$, $\mathfrak{k}$ is given by the diagram after removing α_i from the extended Dynkin diagram.

(2) Lie Groups. The classical complex simple Lie groups of rank n represented by A, B, C, D (in Cartan's symbolism) are the complex special linear group $SL(n+1, \mathbf{C})$, the complex special orthogonal group $SO(2n+1, \mathbf{C})$, the complex symplectic group $Sp(n, \mathbf{C})$, and the complex special orthogonal group $SO(2n, \mathbf{C})$, respectively. The classical compact simple Lie groups of rank n represented by A, B, C, D are the special unitary group $SU(n+1)$, the special orthogonal group $SO(2n+1)$, the unitary-symplectic group $Sp(n)$, and the special orthogonal group $SO(2n)$, respectively (→ 60 Classical Groups).

Cartan's Symbol	Complex Form	Compact Form	Dimension	Rank
A_n	$SL(n+1, \mathbf{C})$	$SU(n+1)$	$(n+1)^2-1$	n
B_n	$SO(2n+1, \mathbf{C})$	$SO(2n+1)$	$2n^2+n$	n
C_n	$Sp(n, \mathbf{C})$	$Sp(n)$	$2n^2+n$	n
D_n	$SO(2n, \mathbf{C})$	$SO(2n)$	$2n^2-n$	n
G_2	$\mathrm{Aut}\,\mathfrak{C}^c$	$\mathrm{Aut}\,\mathfrak{C}$	14	2
F_4	$\mathrm{Aut}\,\mathfrak{J}^c$	$\mathrm{Aut}\,\mathfrak{J}$	52	4
E_6			78	6
E_7			133	7
E_8			248	8

Here $\mathfrak{C}$ is the Cayley algebra over $\mathbf{R}$, $\mathfrak{C}^c$ is the complexification of $\mathfrak{C}$, $\mathfrak{J}$ is the Jordan algebra of Hermitian matrices of order 3 over $\mathfrak{C}$, $\mathfrak{J}^c$ is the complexification of $\mathfrak{J}$, and $\mathrm{Aut}\,A$ is the automorphism group of A.

(II) Classification of Noncompact Real Simple Lie Algebras

Classical Cases

Cartan's Symbol	Noncompact Real Simple Lie Algebra $\mathfrak{g}$	Maximal Compact Lie Algebra of $\mathfrak{g}$
AI	$\mathfrak{sl}(p+1; \mathbf{R})$	$\mathfrak{su}(p+1)$
AII	$\mathfrak{sl}(n; \mathbf{H})$	$\mathfrak{sp}(n)$
AIII	$\mathfrak{su}(p,q; \mathbf{C})$	$\mathfrak{su}(p)+\mathfrak{su}(q)$
BI	$\mathfrak{so}(p,q; \mathbf{R})$	$\mathfrak{so}(p)+\mathfrak{so}(q)$ $(p+q=2m+1)$
BII	$\mathfrak{so}(1,n-1; \mathbf{R})$	$\mathfrak{so}(n-1)$ $(n=2m+1)$
CI	$\mathfrak{sp}(p; \mathbf{R})$	$\mathfrak{u}(p)$
CII	$\mathfrak{u}(p,q; \mathbf{H})$	$\mathfrak{sp}(p)+\mathfrak{sp}(q)$
DI	$\mathfrak{so}(p,q; \mathbf{R})$	$\mathfrak{so}(p)+\mathfrak{so}(q)$ $(p+q=2m)$
DII	$\mathfrak{so}(1,n-1; \mathbf{R})$	$\mathfrak{so}(n-1)$ $(n=2m)$
DIII	$\mathfrak{so}(p; \mathbf{H})$	$\mathfrak{u}(2p)$

Here the field F is the real field $\mathbf{R}$, the complex field $\mathbf{C}$, or the quaternion field $\mathbf{H}$ ($\mathbf{R}\subset\mathbf{C}\subset\mathbf{H}$). $\mathbf{H}$ is an algebra over $\mathbf{R}$. For a quaternion $x=x_0+x_1i+x_2j+x_3k\,(x_0,x_1,x_2,x_3\in\mathbf{R})$, we put

$$\bar{x}=x_0-x_1i-x_2j-x_3k,$$
$$x^*=x_0+x_1i-x_2j+x_3k.$$

Then $\mathfrak{gl}(n;F)=\{$set of all square matrices over F of order $n\}$,

$$\mathfrak{sl}(n;F)=\{A\in\mathfrak{gl}(n;F)\,|\,\mathrm{tr}A=0\},$$
$$\mathfrak{so}(p,q;F)=\{A\in\mathfrak{gl}(p+q;F)\,|\,{}^tA^*I_{p,q}+I_{p,q}A=0\},$$

where $I_{p,q}$ is the symmetric transformation of the Euclidean space $\mathbf{R}^{p+q}$ with respect to $\mathbf{R}^p$, i.e.,

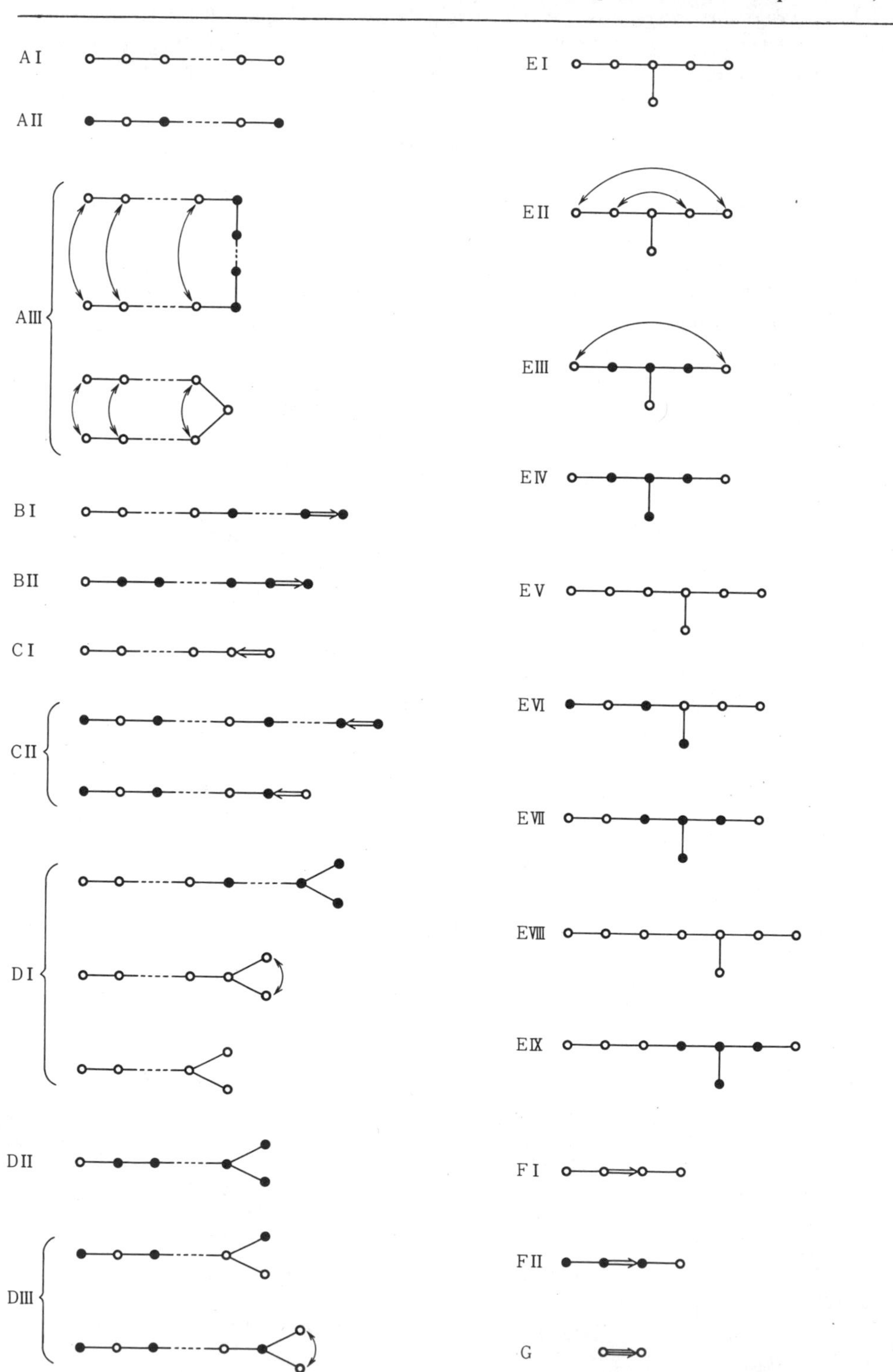

Fig. 15

$I_{p,q}$ is the diagonal sum of the unit matrix I_p of order p and $-I_q$. We have

$\mathfrak{so}(n;F)=\mathfrak{so}(n,0;F)$,

$\mathfrak{u}(p,q;F)=\{A\in\mathfrak{gl}(p+q;F)\mid {}^t\bar{A}I_{p,q}+I_{p,q}A=0\}$,

$\mathfrak{u}(n;F)=\mathfrak{u}(n,0;F)$,

$\mathfrak{sp}(n;F)=\{A\in\mathfrak{gl}(2n;F)\mid {}^t\bar{A}J+JA=0\}$,

where J is the matrix of an alternating form $\sum_{i=1}^{n}(x_i y_{i+n}-x_{i+n}y_i)$ of order $2n$.

A noncompact real simple Lie algebra $\mathfrak{g}$ is classified by the relation of the complex conjugation operator σ with respect to the complexification $\mathfrak{g}^c$ of $\mathfrak{g}$. The results are given by Satake's diagram (Fig. 15).

In the diagram, the fundamental root corresponding to a black circle is multiplied by -1 under σ for a suitable choice of Cartan subalgebra, and the arc with an arrow means that two elements corresponding to both ends of the arc are mutually transformed by a special transformation p such that $\sigma=pw$ $(w\in W)$.

(III) Classification of Irreducible Symmetric Riemannian Spaces (→ 412 Symmetric Riemannian Spaces and Real Forms)

A simply connected irreducible symmetric Riemannian space $M=G/K$ is either a space in the following table or a simply connected compact simple Lie group mentioned in (I). The noncompact forms uniquely corresponding to the compact symmetric Riemannian space are in one-to-one correspondence with the noncompact real simple Lie algebras mentioned in (II).

Cartan's Symbol	$G/K=M$	Dimension	Rank
AI	$SU(n)/SO(n)$ $(n>2)$	$(n-1)(n+2)/2$	$n-1$
AII	$SU(2n)/Sp(n)$ $(n>1)$	$(n-1)(2n+1)$	$n-1$
AIII	$U(p+q)/U(p)\times U(q)$ $(p\geq q\geq 1)$	$2pq$	q
BDI	$SO(p+q)/SO(p)\times SO(q)$ $(p\geq q\geq 2, p+q\neq 4)$	pq	q
BDII	$SO(n+1)/SO(n)$ $(n\geq 2)$	n	1
DIII	$SO(2l)/U(l)$ $(l\geq 4)$	$l(l-1)$	$[l/2]$
CI	$Sp(n)/U(n)$ $(n\geq 3)$	$n(n+1)$	n
CII	$Sp(p+q)/Sp(p)\times Sp(q)$ $(p\geq q\geq 1)$	$4pq$	q
EI	$E_6/Sp(4)$	42	6
EII	$E_6/SU(2)\cdot SU(6)$	40	4
EIII	$E_6/Spin(10)\cdot SO(2)$	32	2
EIV	E_6/F_4	26	2
EV	$E_7/SU(8)$	70	7
EVI	$E_7/Spin(12)\cdot SU(2)$	64	4
EVII	$E_7/E_6\cdot SO(2)$	54	3
EVIII	$E_8/Spin(16)$	128	8
EIX	$E_8/E_7\cdot SU(2)$	112	4
FI	$F_4/Sp(3)\cdot SU(2)$	28	4
FII	$F_4/Spin(9)$	16	1
G	$G_2/SO(4)$	8	2

Notes

The group $G=U(p+q)$ in AIII is not effective, unless it is replaced by $SU(p+q)$. To be precise, $K=Sp(4)$ in EI should be replaced by its quotient group factored by a subgroup of order 2 of its center. K in EII is not a direct product of simple groups; the order of its fundamental group $\pi_1(K)$ is 2. To be precise, K in EV or EVIII should be replaced by its quotient group factored by a subgroup of order 2 of its center. The K's in EII, EIII, EVI, EVII, EIX, and FI are not direct products. The fundamental group $\pi_1(K)$ of K is the infinite cyclic group **Z** for EIII, EVII; for all other cases, the order of $\pi_1(K)$ is 2.

In EIII, EVII, the groups E_6, E_7 are adjoint groups of compact simple Lie algebras. In other cases, E_6 and E_7 (E_8, F_4 and G_2 also) are simply connected Lie groups.

The compact symmetric Riemannian space M is a complex Grassmann manifold for AIII, a real Grassmann manifold for BDI, a sphere for BDII, a quaternion Grassmann manifold for CII, and a Cayley projective plane for FII.

(IV) Isomorphic Relations among Classical Lie Algebras

The isomorphic relations among the classical Lie algebras over **R** or **C** are all given in the following table. In the table, we denote, for example, the real form of type AI of the complex Lie algebra with rank 3 by A_3I in Cartan's symbolism. When there are nonisomorphic real forms of the same type and same rank (e.g., in the case of D_3I) we distinguish them by the rank of the corresponding symmetric Riemannian space and denote them by, e.g., D_3I_p, where p is the index of total isotropy of the sesquilinear form which is invariant under the corresponding Lie algebra.

Cartan's Symbol	Isomorphisms among Classical Lie Algebras
$A_1=B_1=C_1$	$\mathfrak{sl}(2,\mathbf{C})\cong\mathfrak{so}(3,\mathbf{C})\cong\mathfrak{sp}(1,\mathbf{C})$; $\mathfrak{su}(2)\cong\mathfrak{so}(3)\cong\mathfrak{sp}(1)$
$B_2=C_2$	$\mathfrak{so}(5,\mathbf{C})\cong\mathfrak{sp}(2,\mathbf{C})$; $\mathfrak{so}(5)\cong\mathfrak{sp}(2)$
$A_3=D_3$	$\mathfrak{sl}(4,\mathbf{C})\cong\mathfrak{so}(6,\mathbf{C})$; $\mathfrak{su}(4)\cong\mathfrak{so}(6)$
$A_1I=A_1III=B_1I=C_1I$	$\mathfrak{sl}(2,\mathbf{R})\cong\mathfrak{su}(1,1;\mathbf{C})\cong\mathfrak{so}(2,1;\mathbf{R})\cong\mathfrak{sp}(1;\mathbf{R})$
$B_2I_2=C_2I$	$\mathfrak{so}(3,2;\mathbf{R})\cong\mathfrak{sp}(2,\mathbf{R})$
$B_2I_1=C_2II$	$\mathfrak{so}(4,1;\mathbf{R})\cong\mathfrak{u}(1,1;\mathbf{H})$
$A_3I=D_3I_3$	$\mathfrak{sl}(4,\mathbf{R})\cong\mathfrak{so}(3,3;\mathbf{R})$
$A_3II=D_3I_1$	$\mathfrak{sl}(2,\mathbf{H})\cong\mathfrak{so}(5,1;\mathbf{R})$
$A_3III_2=D_3I_2$	$\mathfrak{su}(2,2;\mathbf{C})\cong\mathfrak{so}(4,2;\mathbf{R})$
$A_3III_1=D_3III$	$\mathfrak{su}(3,1;\mathbf{C})\cong\mathfrak{so}(3;\mathbf{H})$
$D_4I_2=D_4III$	$\mathfrak{so}(6,2;\mathbf{R})\cong\mathfrak{so}(4,\mathbf{H})$
$D_2=A_1\times A_1{}^*$	$\mathfrak{so}(4,\mathbf{C})\cong\mathfrak{sl}(2,\mathbf{C})\times\mathfrak{sl}(2,\mathbf{C})$; $\mathfrak{so}(4)\cong\mathfrak{su}(2)\times\mathfrak{su}(2)$
$D_2I_2=A_1I\times A_1I$	$\mathfrak{so}(2,2;\mathbf{R})\cong\mathfrak{sl}(2,\mathbf{R})\times\mathfrak{sl}(2;\mathbf{R})$
$D_2III=A_1\times A_1I^*$	$\mathfrak{so}(2;\mathbf{H})\cong\mathfrak{su}(2)\times\mathfrak{sl}(2;\mathbf{R})$
$D_2I_1=A_1{}^*$	$\mathfrak{so}(3,1;\mathbf{R})\cong\mathfrak{sl}(2,\mathbf{C})$

Note

(*) In these 3 cases, there are isomorphisms given by the replacement of $\mathfrak{sl}(2,\mathbf{C})$ or $\mathfrak{su}(2)$ by isomorphic Lie algebras of type B_1 or type C_1 due to the isomorphism $A_1\cong B_1\cong C_1$.

(V) Lists of Normal Forms of Singularities with Modulus Number $m=0$, 1, and 2 (→ 418 Theory of Singularities)

Letters $A,\ldots,Z$ stand here for stable equivalence classes of function germs (or families of function germs).

(1) Simple Singularities ($m=0$). There are 2 infinite series A, D, and 3 "exceptional" singularities E_6, E_7, E_8:

Notation	Normal form	Restrictions
A_n	$x^{n+1}+y^2+z^2$	$n\geqslant1$
D_n	$x^{n-1}+xy^2+z^2$	$n\geqslant4$
E_6	$x^4+y^3+z^2$	
E_7	$x^3y+y^3+z^2$	
E_8	$x^5+y^3+z^2$	

(2) Unimodular Singularities ($m=1$). There are 3 families of parabolic singularities, one series of hyperbolic singularities (with 3 subscripts), and 14 families of exceptional singularities.

The parabolic singularities

Notation	Normal form	Restrictions
$P_8=\tilde{E}_6$	$x^3+y^3+z^3+axyz$	$a^3+27\neq0$
$X_9=\tilde{E}_7$	$x^4+y^4+z^2+axyz$	$a^4-64\neq0$
$J_{10}=\tilde{E}_8$	$x^6+y^3+z^2+axyz$	$a^6-432\neq0$

The hyperbolic singularities

Notation	Normal form	Restrictions
T_{pqr}	$x^p+y^q+z^r+axyz$	$a\neq 0, \frac{1}{p}+\frac{1}{q}+\frac{1}{r}<1$

The 14 exceptional families

Notation	Normal form	Gabrielov numbers	Dolgachëv numbers	Notation	Normal form	Gabrielov numbers	Dolgachëv numbers
K_{12}	$x^3+y^7+z^2+axy^5$	2 3 7	2 3 7	W_{13}	$x^4+xy^4+z^2+ay^6$	2 5 6	3 4 4
K_{13}	$x^3+xy^5+z^2+ay^8$	2 3 8	2 4 5	Q_{10}	$x^3+y^4+yz^2+axy^3$	3 3 4	2 3 9
K_{14}	$x^3+y^8+z^2+axy^6$	2 3 9	3 3 4	Q_{11}	$x^3+y^2z+xz^3+az^5$	3 3 5	2 4 7
Z_{11}	$x^3y+y^5+z^2+axy^4$	2 4 5	2 3 8	Q_{12}	$x^3+y^5+yz^2+axy^4$	3 3 6	3 3 6
Z_{12}	$x^3y+xy^4+z^2+ax^2y^3$	2 4 6	2 4 6	S_{11}	$x^4+y^2z+xz^2+ax^3z$	3 4 4	2 5 6
Z_{13}	$x^3y+y^6+z^2+axy^5$	2 4 7	3 3 5	S_{12}	$x^2y+y^2z+xz^3+az^5$	3 4 5	3 4 5
W_{12}	$x^4+y^5+z^2+ax^2y^3$	2 5 5	2 5 5	U_{12}	$x^3+y^3+z^4+axyz^2$	4 4 4	4 4 4

(3) Bimodular Singularities ($m=2$). There are 8 infinite series and 14 exceptional families. In all the formulas, $\mathbf{a}=a_0+a_1y$.

The 8 infinite series of bimodular singularities

Notation	Normal form	Restriction	Milnor number
$J_{3,0}$	$x^3+bx^2y^3+y^9+z^2+cxy^7$	$4b^3+27\neq 0$	16
$J_{3,p}$	$x^3+x^2y^3+z^2+ay^{9+p}$	$p>0, a_0\neq 0$	$16+p$
$Z_{1,0}$	$y(x^3+dx^2y^2+cxy^5+y^6)+z^2$	$4d^3+27\neq 0$	15
$Z_{1,p}$	$y(x^3+x^2y^2+ay^{6+p})+z^2$	$p>0, a_0\neq 0$	$15+p$
$W_{1,0}$	$x^4+ax^2y^3+y^6+z^2$	$a_0^2\neq 4$	15
$W_{1,p}$	$x^4+x^2y^3+ay^{6+p}+z^2$	$p>0, a_0\neq 0$	$15+p$
$W^{\#}_{1,2q-1}$	$(x^2+y^3)^2+axy^{4+q}+z^2$	$q>0, a_0\neq 0$	$15+2q-1$
$W^{\#}_{1,2q}$	$(x^2+y^3)^2+ax^2y^{3+q}+z^2$	$q>0, a_0\neq 0$	$15+2q$
$Q_{2,0}$	$x^3+yz^2+ax^2y^2+xy^4$	$a_0^2\neq 4$	14
$Q_{2,p}$	$x^3+yz^2+x^2y^2+az^{6+p}$	$p>0, a_0\neq 0$	$14+p$
$S_{1,0}$	$x^2z+yz^2+y^5+azy^3$	$a_0^2\neq 4$	14
$S_{1,p}$	$x^2z+yz^2+x^2y^2+ay^{5+p}$	$p>0, a_0\neq 0$	$14+p$
$S^{\#}_{1,2q-1}$	$x^2z+yz^2+zy^3+axy^{2+q}$	$q>0, a_0\neq 0$	$14+2q-1$
$S^{\#}_{1,2q}$	$x^2z+yz^2+zy^3+ax^2y^{2+q}$	$q>0, a_0\neq 0$	$14+2q$
$U_{1,0}$	$x^3+xz^2+xy^3+ay^3z$	$a_0(a_0^2+1)\neq 0$	14
$U_{1,2q-1}$	$x^3+xz^2+xy^3+ay^{1+q}z^2$	$q>0, a_0\neq 0$	$14+2q-1$
$U_{1,2q}$	$x^3+xz^2+xy^3+ay^{3+q}z$	$q>0, a_0\neq 0$	$14+2q$

The 14 exceptional families

Notation	Normal form	Notation	Normal form
E_{18}	$x^3+y^{10}+z^2+axy^7$	W_{18}	$x^4+y^7+z^2+ax^2y^4$
E_{19}	$x^3+xy^7+z^2+ay^{11}$	Q_{16}	$x^3+yz^2+y^7+z^2+axy^5$
E_{20}	$x^3+y^{11}+z^2+axy^8$	Q_{17}	$x^3+yz^2+xy^5+z^2+ay^8$
Z_{17}	$x^3y+y^8+z^2+axy^6$	Q_{18}	$x^3+yz^2+y^8+z^2+axy^6$
Z_{18}	$x^3y+xy^6+z^2+ay^9$	S_{16}	$x^2z+yz^2+xy^4+z^2+ay^6$
Z_{19}	$x^3y+y^9+z^2+axy^7$	S_{17}	$x^2z+yz^2+y^6+z^2+azy^4$
W_{17}	$x^4+xy^5+z^2+ay^7$	U_{16}	$x^3+xz^2+y^5+z^2+ax^2y^2$

Adjacency relations between simple and simply elliptic singularities

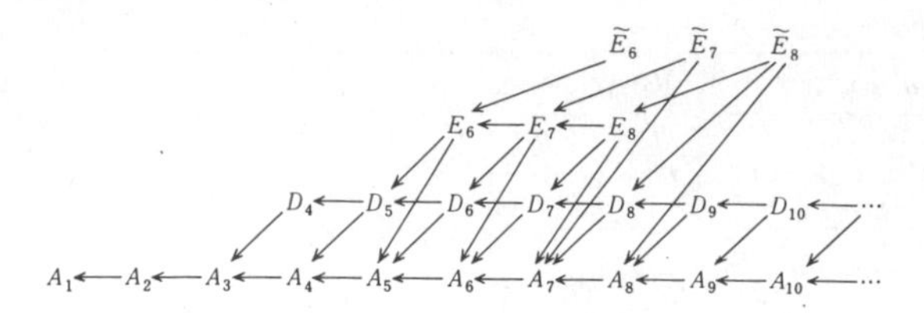

Adjacency relations among unimodular singularities

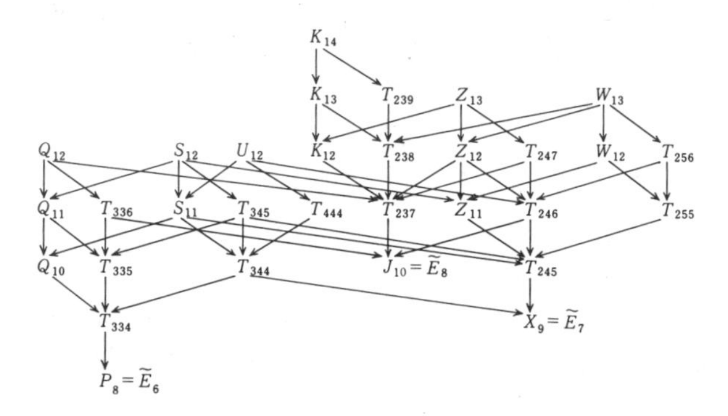

References

[1] V. I. Arnol'd, Singularity theory, Lecture note ser. 53, London Math. Soc., 1981.
[2] E. Breiskorn, Die Hierarchie der 1-modularen Singularitäten, Manuscripta Math., 27 (1979), 183–219.
[3] K. Saito, Einfach elliptische Singularitäten, Inventiones Math., 23 (1974), 289–325.

6. Topology

(I) h-Cobordism Groups of Homotopy Spheres and Groups of Differentiable Structures on Combinatorial Spheres

(1) The Structure of the h-Cobordism Group θ_n of n-Dimensional Homotopy Spheres. In the following table, values of θ_n have the following meanings: 0 means that the group consists only of the identity element, an integer l means that the group is isomorphic to the cyclic group of order l, 2^l means that the group is the direct sum of l groups of order 2, $+$ means the direct sum, and ? means that the structure of the group is unknown.

n	1	2	3	4	5	6	7	8	9	10	11	12	13	14	15	16	17	18
$\theta_n\cong$	0	0	?	0	0	0	28	2	2^3 or $4+2$	6	992	0	3	2	$8128+2$	2	2^4 or $4+2^2$	$8+2$

(2) The Group Γ_n of Differentiable Structures on the n-Dimensional Combinatorial Sphere.

$\Gamma_n \cong \theta_n \quad (n\neq 3), \quad \Gamma_3=0.$

(II) Adem's Formula Concerning Steenrod Operators Sq and $\mathscr{P}$ (→ 64 Cohomology Operations)

For the cohomology operators Sq and $\mathscr{P}$, we have

$$Sq^a Sq^b = \sum_{c=0}^{[a/2]} \binom{b-c-1}{a-2c} Sq^{a+b-c} Sq^c \quad (a<2b).$$

$$\mathcal{P}^a\mathcal{P}^b=\sum_{c=0}^{[a/p]}(-1)^{a+c}\binom{(b-c)(p-1)-1}{a-pc}\mathcal{P}^{a+b-c}\mathcal{P}^c \quad (a<pb),$$

$$\mathcal{P}^a\delta\mathcal{P}^b=\sum_{c=0}^{[a/p]}(-1)^{a+c}\binom{(b-c)(p-1)}{a-pc}\delta\mathcal{P}^{a+b-c}\mathcal{P}^c$$

$$+\sum_{c=0}^{[(a-1)/p]}(-1)^{a+c-1}\binom{(b-c)(p-1)-1}{a-pc-1}\mathcal{P}^{a+b-c}\delta\mathcal{P}^c \quad (a\leqslant pb).$$

Several simple cases of the formula above are as follows.

$Sq^1Sq^{2n}=Sq^{2n+1}$, $\quad Sq^1Sq^{2n+1}=0$,

$Sq^2Sq^{4n}=Sq^{4n+2}+Sq^{4n+1}Sq^1$, $\quad Sq^2Sq^{4n+1}=Sq^{4n+2}Sq^1$,

$Sq^2Sq^{4n+2}=Sq^{4n+3}Sq^1$, $\quad Sq^2Sq^{4n+3}=Sq^{4n+5}+Sq^{4n+4}Sq^1$,

$Sq^4Sq^{8n}=Sq^{8n+4}+Sq^{8n+3}Sq^1+Sq^{8n+2}Sq^2$, $\quad Sq^4Sq^{8n+1}=Sq^{8n+4}Sq^1+Sq^{8n+3}Sq^2$,

$Sq^4Sq^{8n+2}=Sq^{8n+4}Sq^2$, $\quad Sq^4Sq^{8n+3}=Sq^{8n+5}Sq^2$,

$Sq^4Sq^{8n+4}=Sq^{8n+7}Sq^1+Sq^{8n+6}Sq^2$, $\quad Sq^4Sq^{8n+5}=Sq^{8n+9}+Sq^{8n+8}Sq^1+Sq^{8n+7}Sq^2$,

$Sq^4Sq^{8n+6}=Sq^{8n+10}+Sq^{8n+8}Sq^2$, $\quad Sq^4Sq^{8n+7}=Sq^{8n+11}+Sq^{8n+9}Sq^2$,

$\mathcal{P}^1\mathcal{P}^n=(n+1)\mathcal{P}^{n+1}$,

$\mathcal{P}^1\delta\mathcal{P}^n=n\cdot\delta\mathcal{P}^{n+1}+\mathcal{P}^{n+1}\delta$.

(III) Cohomology Ring $H^*(\pi,n;\mathbf{Z}_p)$ of Eilenberg-MacLane Complex (→ 70 Complexes)

$\mathbf{Z}$ means the set of integers, and $\mathbf{Z}_p=\mathbf{Z}/p\mathbf{Z}$, where p is a prime number.

(1) The case $p=2$, $\pi=\mathbf{Z}$ or $\mathbf{Z}_{2^f}$ $(f\geqslant 1)$. The degree of a finite sequence $I=(i_r,i_{r-1},\ldots,i_1)$ of positive integers is defined by $d(I)=i_1+i_2+\ldots+i_r$. If such a sequence satisfies $i_{k+1}\geqslant 2i_k$ $(k=1,\ldots,r-1)$, it is called admissible, and we define its excess by

$$e(I)=(i_r-2i_{r-1})+\ldots+(i_2-2i_1)+i_1=2i_r-d(I).$$

Further, we put $Sq^I=Sq^{i_r}Sq^{i_{r-1}}\ldots Sq^{i_1}$. Then we have $H^*(\mathbf{Z}_{2^f},n;\mathbf{Z}_2)=\mathbf{Z}_2[Sq^Iu_n\,|\,I$ is admissible, $e(I)<n]$, $H^*(\mathbf{Z},n;\mathbf{Z}_2)=\mathbf{Z}_2[Sq^Iu_n\,|\,I$ is admissible, $e(I)<n$, $i_1>1]$.

Here, $u_n\in H^n(\pi,n;\mathbf{Z}_2)$ is the fundamental cohomology class. $I=\varnothing$ (empty) is also admissible, and for this case we put $n(I)=e(I)=0$, $Sq^I=1$. Due to the Künneth theorem, we have

$$H^*(\pi+\pi',n;\mathbf{Z}_p)=H^*(\pi,n;\mathbf{Z}_p)\otimes H^*(\pi',n;\mathbf{Z}_p)$$

if π is finitely generated. In particular, we have

$$H^*(\mathbf{Z}_2,1;\mathbf{Z}_2)=\mathbf{Z}_2[u_1],$$

$$H^*(\mathbf{Z}_2,2;\mathbf{Z}_2)=\mathbf{Z}_2[u_2,Sq^1u_2,Sq^2Sq^1u_2,\ldots,Sq^{2^r}Sq^{2^{r-1}}\ldots Sq^1u_2,\ldots],$$

$$H^*(\mathbf{Z}_2,3;\mathbf{Z}_2)=\mathbf{Z}_2[u_3,Sq^{2^r}Sq^{2^{r-1}}\ldots Sq^1u_3,Sq^{(2^r+1)2^s}Sq^{(2^r+1)2^{s-1}}\ldots Sq^{2^r+1}Sq^{2^{r-1}}\ldots Sq^1u_3\,|\,r\geqslant 0,\quad s\geqslant 0].$$

(2) The case $p\neq 2$, $\pi=\mathbf{Z}$ or $\mathbf{Z}_{p^f}$ $(f\geqslant 1)$. We define the degree of a finite sequence $I=(i_r,i_{r-1},\ldots,i_1,i_0)$ of nonnegative integers by $d(I)=i_r+\ldots+i_1+i_0$. The sequence I is called admissible if it satisfies the following conditions:

$i_k=2\lambda_k(p-1)+\varepsilon_k$ (λ_k is a nonnegative integer, $\varepsilon_k=0$ or 1 $(0\leqslant k\leqslant r)$), and

$i_0=0$ or 1, $\quad i_1\geqslant 2p-2$, $\quad i_{k+1}\geqslant pi_k \quad (1\leqslant k\leqslant r-1)$.

We define its excess by $e(I)=pi_r-(p-1)d(I)$. Further, we put $\mathcal{P}^I=\delta^{\varepsilon_k}\mathcal{P}^{\lambda_k}\ldots\delta^{\varepsilon_1}\mathcal{P}^{\lambda_1}\delta^{\varepsilon_0}$, and assume that $u_n\in H^n(\pi,n;\mathbf{Z})$ is the fundamental cohomology class. Then we have

$$H^*(\mathbf{Z}_{p^f},n;\mathbf{Z}_p)=\mathbf{Z}_p[\mathcal{P}^Iu_n\,|\,I \text{ is admissible, } e(I)<n(p-1),\ n+d(I)\text{ is even}]$$

$$\otimes\wedge_{\mathbf{Z}_p}(\mathcal{P}^Iu_n\,|\,I\text{ is admissible, } e(I)<n(p-1),\ n+d(I)\text{ is odd}).$$

$H^*(\mathbf{Z},n;\mathbf{Z})$ is given by the above formula when the admissible sequence is I with $i_0=0$.

(IV) Cohomology Ring of Compact Connected Lie Groups (→ 427 Topology of Lie Groups and Homogeneous Spaces)

(1) General Remarks. Let G be a compact connected Lie group with rank l and dimension n. We have $H^*(G;\mathbf{R})\cong\bigwedge_{\mathbf{R}}(x_1,\ldots,x_l)$, where $\bigwedge_K(x_1,\ldots,x_l)$ means the exterior algebra over K of a linear space $V=Kx_1+\ldots+Kx_l$ with the basis $\{x_1,\ldots,x_l\}$ over K. We define a new degree in $\bigwedge_K(x_1,\ldots,x_l)$ by putting $\deg x_i=m_i$ (m_i is odd) $(1\leqslant i\leqslant l)$, where $m_1+\ldots+m_l=n$. The $\cong$ means isomorphism as graded rings.

(2) Classical Compact Simple Lie Groups. We set $\deg x_i=i$.

$$H^*(U(n);\mathbf{R})\cong\bigwedge_{\mathbf{R}}(x_1,x_3,\ldots,x_{2n-1}),$$
$$H^*(SU(n);\mathbf{R})\cong\bigwedge_{\mathbf{R}}(x_3,x_5,\ldots,x_{2n-1}),$$
$$H^*(Sp(n);\mathbf{R})\cong\bigwedge_{\mathbf{R}}(x_3,x_7,\ldots,x_{4n-1}),$$
$$H^*(SO(n);\mathbf{Z}_2)\cong(\text{Having } x_1,x_2,\ldots,x_{n-1} \text{ as a simple system of generators})$$
$$\cong\mathbf{Z}_2[x_1,x_3,\ldots,x_{2n'-1}]/(x_i^{2^{s(i)}}\mid i=1,\ldots,n')$$
$$(n'=[n/2],\ s(i) \text{ is the least integer satisfying } 2^{s(i)}(2i-1)\geqslant n)$$
$$H^*(SO(2n);K)\cong\bigwedge_K(x_3,x_7,\ldots,x_{4n-5},x_{2n-1}),$$
$$H^*(SO(2n-1);K)\cong\bigwedge_K(x_3,x_7,\ldots,x_{4n-5}),$$ where K is a commutative field whose characteristic is not 2.

For $SO(n)$, $Sq^a(x_i)=\binom{i}{a}x_{i+a}$. For $SU(n)$, $p^a(x_{2i-1})=\binom{i-1}{a}x_{2i-1+2a(p-1)}$.

For $Sp(n)$, $\mathcal{P}^a(x_{4i-1})=(-1)^{a(p-1)/2}\binom{2i-1}{a}x_{4i-1+2a(p-1)}$.

(3) Exceptional Compact Simple Lie Groups. n and m_i $(1\leqslant i\leqslant l)$ given in (1) are as follows.

G_2:	$n=14$,	$m_i=3,\ 11.$
F_4:	$n=52$,	$m_i=3,\ 11,\ 15,\ 23.$
E_6:	$n=78$,	$m_i=3,\ 9,\ 11,\ 15,\ 17,\ 23.$
E_7:	$n=133$,	$m_i=3,\ 11,\ 15,\ 19,\ 23,\ 27,\ 35.$
E_8:	$n=248$,	$m_i=3,\ 15,\ 23,\ 27,\ 35,\ 39,\ 47,\ 59.$

(4) p-Torsion Groups of Exceptional Groups. The p-torsion groups of exceptional Lie groups are unit groups except when $p=2$ for G_2; $p=2$, 3 for F_4, E_6, E_7; and $p=2$, 3, 5 for E_8. The cohomology ring of $\mathbf{Z}_p$ as a coefficient group in these exceptional cases is given as follows. Here we put $\deg x_i=i$.

$$H^*(G_2;\mathbf{Z}_2)=\mathbf{Z}_2[x_3]/(x_3^4)\otimes\bigwedge_{\mathbf{Z}_2}(Sq^2x_3);$$
$$H^*(F_4;\mathbf{Z}_2)=\mathbf{Z}_2[x_3]/(x_3^4)\otimes\bigwedge_{\mathbf{Z}_2}(Sq^2x_3,x_{15},Sq^8x_{15}),$$
$$H^*(F_4;\mathbf{Z}_3)=\mathbf{Z}_3[\delta\mathcal{P}^1x_3]/((\delta\mathcal{P}^1x_3)^3)\otimes\bigwedge_{\mathbf{Z}_3}(x_3,\mathcal{P}^1x_3,x_{11},\mathcal{P}^1x_{11});$$
$$H^*(E_6;\mathbf{Z}_2)=\mathbf{Z}_2[x_3]/(x_3^4)\otimes\bigwedge_{\mathbf{Z}_2}(Sq^2x_3,Sq^4Sq^2x_3,x_{15},Sq^8Sq^4Sq^2x_3,Sq^8x_{15}),$$
$$H^*(E_6;\mathbf{Z}_3)=\mathbf{Z}_3[\delta\mathcal{P}^1x_3]/((\delta\mathcal{P}^lx_3)^3)\otimes\bigwedge_{\mathbf{Z}_3}(x_3,\mathcal{P}^1x_3,x_9,x_{11},\mathcal{P}^1x_{11},x_{17});$$
$$H^*(E_7;\mathbf{Z}_2)=\mathbf{Z}_2[x_3,Sq^2x_3,Sq^4Sq^2x_3]/(x_3^4,(Sq^2x_3)^4,(Sq^4Sq^2x_3)^4)$$
$$\otimes\bigwedge_{\mathbf{Z}_2}(x_{15},Sq^8Sq^4Sq^2x_3,Sq^8x_{15},Sq^4Sq^8x_{15}),$$
$$H^*(E_7;\mathbf{Z}_3)=\mathbf{Z}_3[\delta\mathcal{P}^1x_3]/((\delta\mathcal{P}^1x_3)^3)\otimes\bigwedge_{\mathbf{Z}_3}(x_3,\mathcal{P}^1x_3,x_{11},\mathcal{P}^1x_{11},\mathcal{P}^3\mathcal{P}^1x_3,x_{27},x_{35});$$
$$H^*(E_8;\mathbf{Z}_2)=\mathbf{Z}_2[x_3,Sq^2x_3,Sq^4Sq^2x_3,x_{15}]/(x_3^{16},(Sq^2x_3)^8,(Sq^4Sq^2x_3)^4,x_{15}^4)$$
$$\otimes\bigwedge_{\mathbf{Z}_2}(Sq^8Sq^4Sq^2x_3,Sq^8x_{15},Sq^4Sq^8x_{15},Sq^2Sq^4Sq^8x_{15}),$$
$$H^*(E_8;\mathbf{Z}_3)=\mathbf{Z}_3[\delta\mathcal{P}^1x_3,\delta\mathcal{P}^3\mathcal{P}^1x_3]/((\delta\mathcal{P}^1x_3)^3,(\delta\mathcal{P}^3\mathcal{P}^1x_3)^3)$$
$$\otimes\bigwedge_{\mathbf{Z}_3}(x_3,\mathcal{P}^1x_3,x_{15},\mathcal{P}^3\mathcal{P}^1x_3,\mathcal{P}^3x_{15},x_{35},x_{39},x_{47}),$$
$$H^*(E_8;\mathbf{Z}_5)=\mathbf{Z}_5[\delta\mathcal{P}^1x_3]/((\delta\mathcal{P}^1x_3)^5)\otimes\bigwedge_{\mathbf{Z}_5}(x_3,\mathcal{P}^1x_3,x_{15},\mathcal{P}^1x_{15},x_{27},x_{35},x_{39},x_{47}).$$

(V) Cohomology Rings of Classifying Spaces (→ 56 Characteristic Classes, 427 Topology of Lie Groups and Homogeneous Spaces C)

(1) Let $H^*(G;K)=\bigwedge_K(x_1,x_2,\ldots,x_n)$. Then the $\deg x_i$ are odd and the x_i may be assumed to be transgressive. y_i being its image, the following formula holds:

$$H^*(BG;K)=K[y_1,y_2,\ldots,y_n] \qquad \text{(Borel's theorem).}$$

(2) $H^*(BU(n))=H^*(BGL(n,\mathbf{C}))=\mathbf{Z}[c_1,c_2,\ldots,c_n]$,

$H^*(BSU(n))=H^*(BSL(n,\mathbf{C}))=\mathbf{Z}[c_2,\ldots,c_n]$,

$H^*(BSp(n))=\mathbf{Z}[q_1,q_2,\ldots,q_n]$,

$H^*(BO(n);K_2)=H^*(BGL(n,\mathbf{R});\mathbf{Z}_2)=K_2[w_1,w_2,\ldots,w_n]$,

$H^*(BSO(n);K_2)=H^*(BSL(n,\mathbf{R});\mathbf{Z}_2)=K_2[w_2,\ldots,w_n]$,

$H^*(BSO(2m+1);K)=K[p_1,p_2,\ldots,p_m]$,

$H^*(BSO(2m);K)=K[p_1,p_2,\ldots,p_{m-1},\chi]$.

Here, K denotes a field of characteristic $\neq 2$, and K_2 is the field of characteristic 2. The c_i denote the ith Chern classes and the q_i the ith symplectic Pontryagin classes, the w_i the ith Stiefel-Whitney classes. Moreover, the p_i denote the ith Pontryagin classes, and χ the Euler class. Their degrees are given as follows: $\deg c_i=2i$, $\deg q_i=\deg p_i=4i$, $\deg w_i=i$, and $\deg\chi=2m$.

(3) Wu's Formula. Let $H^2(BSO(n);\mathbf{Z}_2)=\mathbf{Z}_2[w_2,\ldots,w_n]$ and $H^*(BU(n),\mathbf{Z}_2)=\mathbf{Z}_2[c_1,\ c_2,\ldots,c_n]$. We have

$$Sq^j w_i=\sum_{0\leqslant t\leqslant j}\binom{i-j+t-1}{t}w_{j-t}w_{i+t} \qquad (w_0=1),$$

$$Sq^{2j} c_i=\sum_{0\leqslant t\leqslant j}\binom{i-j+t-1}{t}c_{j-t}c_{i+t} \qquad (c_0=1).$$

Here the symbol $\binom{a}{b}$ denotes the binomial coefficient for $a\geqslant b$; $\binom{a}{0}=1$, and $\binom{a}{b}=0$ otherwise.

(VI) Homotopy Groups of Spheres (→ 202 Homotopy Theory)

Table of the $(n+k)$th Homotopy Group $\pi_{n+k}(S^n)$ of the n-Dimensional Sphere S^n. The table represents Abelian groups. 0 stands for the unit group; integer l the cyclic group of order l; ∞ the infinite cyclic group; 2^l the direct sum of l groups of order 2; and + means the direct sum.

n \ k	<0	0	1	2	3	4	5	6	7	8	9	10	11	12	13
1	0	∞	0	0	0	0	0	0	0	0	0	0	0	0	0
2	0	∞	∞	2	2	12	2	2	3	15	2	2^2	12+2	$84+2^2$	2^2
3	0	∞	2	2	12	2	2	3	15	2	2^2	12+2	$84+2^2$	2^2	6
4	0	∞	2	2	∞+12	2^2	2^2	24+3	15	2	2^3	120+12+2	$84+2^5$	2^6	24+6+2
5	0	∞	2	2	24	2	2	2	30	2	2^3	72+2	$504+2^2$	2^3	6+2
6	0	∞	2	2	24	0	∞	2	60	24+2	2^3	72+2	504+4	240	6
7	0	∞	2	2	24	0	0	2	120	2^3	2^4	24+2	504+2	0	6
8	0	∞	2	2	24	0	0	2	∞+120	2^4	2^5	24+24+2	504+2	0	6+2
9	0	∞	2	2	24	0	0	2	240	2^3	2^4	24+2	504+2	0	6
10	0	∞	2	2	24	0	0	2	240	2^2	$\infty+2^3$	12+2	504	12	6
11	0	∞	2	2	24	0	0	2	240	2^2	2^3	6+2	504	2	6+2
12	0	∞	2	2	24	0	0	2	240	2^2	2^3	6	∞+504	2^2	6+2
13	0	∞	2	2	24	0	0	2	240	2^2	2^3	6	504	2	6
14	0	∞	2	2	24	0	0	2	240	2^2	2^3	6	504	0	∞+3
>15	0	∞	2	2	24	0	0	2	240	2^2	2^3	6	504	0	3

Table of the $(n+k)$th Homotopy Group $\pi_{n+k}(S^n)$ of the n-Dimensional Sphere S^n (Continued)

n \ k	14	15	16	17	18	19	20	21	22
1	0	0	0	0	0	0	0	0	0
2	6	30	30	$6+2$	$12+2^2$	$12+2^2$	$132+2$	2^2	2
3	30	30	$6+2$	$12+2^2$	$12+2^2$	$132+2$	2^2	2	210
4	$2520+6+2$	30	$6+6+2$	$24+12+4+2^2$	$120+12+2^5$	$132+2^5$	2^6	$24+2^2$	$9240+6+2$
5	$6+2$	$30+2$	2^2	$4+2^2$	$24+2^2$	$264+2$	$6+2^2$	$6+2$	$90+2^2$
6	$12+2$	$60+2$	$504+2^2$	2^4	$24+6+2$	$1056+8$	$480+12$	6	$180+2^2$
7	$24+4$	$120+2^3$	2^4	2^4	$24+2$	$264+2$	24	$6+2$	$72+2^3$
8	$240+24+4$	$120+2^5$	2^7	$6+2^4$	$504+24+2$	$264+2$	$24+3$	$12+2^3$	$1440+24+2^4$
9	$16+4$	$240+2^3$	2^4	2^4	$24+2$	$264+2$	24	$6+2^2$	$144+2^3$
10	$16+2$	$240+2^2$	$240+2$	2^3	$24+2^2$	$264+6$	$504+24$	$6+2^2$	$144+6+2$
11	$16+2$	$240+2$	2	2^3	$8+4+2$	$264+2^3$	$24+2^2$	2^4	$48+2^2$
12	$48+4+2$	$240+2$	2	2^4	$480+4+4+2$	$264+2^5$	$24+2^5$	$6+2^4$	$2016+12+2^2$
13	$16+2$	$480+2$	2	2^4	$8+8+2$	$264+2^3$	$24+2^3$	$4+2^3$	$16+2^2$
14	$8+2$	$480+2$	$24+2$	2^4	$8+8+2$	$264+4+2$	$240+24$	$4+2^2$	$16+2^2$
15	$4+2$	$480+2$	2^3	2^5	$8+8+2$	$264+2^2$	2^4	2^3	$16+2^3$
16	$2+2$	$\infty+480+2$	2^4	2^6	$24+8+8+2$	$264+2^2$	2^4	2^4	$240+16+2^3$
17	$2+2$	$480+2$	2^3	2^5	$8+8+2$	$264+2^2$	24	2^3	$16+2^3$
18	$2+2$	$480+2$	2^2	$\infty+2^4$	$8+4+2$	$264+2$	$24+12$	2^3	$16+2^2$
19	$2+2$	$480+2$	2^2	2^4	$8+2^2$	$264+2$	$24+2$	2^4	$16+2^2$
20	$2+2$	$480+2$	2^2	2^4	$8+2$	$\infty+264+2$	$24+2^2$	2^4	$16+2^2$
21	$2+2$	$480+2$	2^2	2^4	$8+2$	$264+2$	$24+2$	2^3	$8+2^2$
22	$2+2$	$480+2$	2^2	2^4	$8+2$	$264+2$	24	$\infty+2^2$	$4+2^2$
23	$2+2$	$480+2$	2^2	2^4	$8+2$	$264+2$	24	2^2	2^3
$\geqslant 24$	$2+2$	$480+2$	2^2	2^4	$8+2$	$264+2$	24	2^2	2^2

Remarks

(1) When $n>k+1$ (below the broken line in the table), $\pi_{n+k}(S^n)$ is independent of n and is isomorphic with the kth stable homotopy group G_k.

(2) Let $\iota_n \in \pi_n(S^n)$ be the identity on S^n; $\eta_2 \in \pi_3(S^2)$, $\nu_4 \in \pi_7(S^4)$, $\sigma_8 \in \pi_{15}(S^8)$ be the Hopf mapping $S^3 \to S^2$, $S^7 \to S^4$, $S^{15} \to S^8$ (induced mapping in the homotopy class), respectively; and $[\iota_{2m}, \iota_{2m}] \in \pi_{4m-1}(S^{2m})$ $(m \neq 1,2,4)$ be the Whitehead product of ι_{2m}. These objects generate infinite cyclic groups which are direct factors of $\pi_{n+k}(S^n)$ corresponding to the original mappings.

(3) $\eta_{n+2}=E^n\eta_2$, $\nu_{n+4}=E^n\nu_4$, $\sigma_{n+8}=E^n\sigma_8$ $(n \geqslant 1)$ (E is the suspension) are the generator for $\pi_{n+k}(S^n)$, which contains the mappings.

(4) The orders of the following compositions are 2:gs$_{n+7}$

$$\eta_n \circ \eta_{n+1} \ (n \geqslant 2), \quad \nu_n \circ \nu_{n+3} \ (n \geqslant 5), \quad \sigma_n \circ \sigma_{n+7} \ (n \geqslant 16), \quad \eta_n \circ \nu_{n+1} \ (n=3,4),$$

$$\nu_n \circ \eta_{n+3} \ (n=4,5), \quad \eta_n \circ \sigma_{n+1} \ (n \geqslant 7), \quad \sigma_n \circ \eta_{n+7} \ (n \geqslant 8), \quad \nu_{10} \circ \sigma_{13}, \quad \sigma_{11} \circ \nu_{18},$$

$$\eta_n \circ \eta_{n+1} \circ \eta_{n+2} \ (n \geqslant 2), \quad \nu_n \circ \nu_{n+3} \circ \nu_{n+6} (n \geqslant 4), \quad \sigma_n \circ \sigma_{n+7} \circ \sigma_{n+14} \ (n \geqslant 9).$$

(VII) The Homotopy Groups $\pi_k(G)$ of Compact Connected Lie Groups G

Here the group G is one of the following:

$$SO(n) \ (n \geqslant 2), \quad Spin(n) \ (n \geqslant 3), \quad U(n) \ (n \geqslant 1), \quad SU(n) \ (n \geqslant 2),$$
$$Sp(n) \ (n \geqslant 1), \quad G_2, \quad F_4, \quad E_6, \quad E_7, \quad E_8.$$

(1) The Fundamental Group $\pi_1(G)$.

$$\pi_1(G) \cong \begin{cases} \infty & (G=U(n)\ (n \geqslant 1), \quad SO(2)), \\ 2 & (G=SO(n)\ (n \geqslant 3)), \\ 0 & (\text{for all other groups } G). \end{cases}$$

(2) Isomorphic Relations $(k \geqslant 2)$.

$\pi_k(U(n)) \cong \pi_k(SU(n))$ $(n \geqslant 2)$,

$\pi_k(U(1)) \cong \pi_k(SO(2)) \cong 0.$

$\pi_k(Spin(n)) \cong \pi_k(SO(n))\ (n \geqslant 3),$

$\pi_k(Spin(3)) \cong \pi_k(Sp(1)) \cong \pi_k(SU(2)) \cong \pi_k(S^3),$

$\pi_k(Spin(4)) \cong \pi_k(Spin(3)) + \pi_k(S^3),$

$\pi_k(Spin(5)) \cong \pi_k(Sp(2)),$

$\pi_k(Spin(6)) \cong \pi_k(SU(4)).$

(3) The Homotopy Group $\pi_k(G)$ $(k \geqslant 2)$.

$\pi_2(G) \cong 0.$

$\pi_3(G) \cong \infty \quad (G \neq SO(2),\ \ U(1),\ \ SO(4),\ \ Spin(4)), \quad \pi_3(SO(4)) \cong \infty + \infty.$

$$\pi_4(G) \cong \begin{cases} 2+2 & (G = SO(4),\ \ Spin(4)), \\ 2 & (G = Sp(n),\ \ SU(2),\ \ SO(3),\ \ SO(5),\ \ Spin(3),\ \ Spin(5)), \\ 0 & (G = SU(n)\ (n \geqslant 3),\ \ SO(n)\ (n \geqslant 6),\ \ G_2,\ \ F_4,\ \ E_6,\ \ E_7,\ \ E_8). \end{cases}$$

$$\pi_5(G) \cong \begin{cases} 2+2 & (G = SO(4),\ \ Spin(4)), \\ 2 & (G = Sp(n),\ \ SU(2),\ \ SO(3),\ \ SO(5)\ \ Spin(3),\ \ Spin(5)), \\ \infty & (G = SU(n)\ (n \geqslant 3),\ \ SO(6),\ \ Spin(6)), \\ 0 & (G = SO(n),\ \ Spin(n)\ (n \geqslant 7),\ \ G_2,\ \ F_4,\ \ E_6,\ \ E_7,\ \ E_8). \end{cases}$$

$\pi_k(G), \quad k \geqslant 6.$

$G \quad k$	6	7	8	9	10	11	12	13	14	15
$Sp(1)$	12	2	2	3	15	2	2^2	$12+2$	$84+2^2$	2^2
$Sp(2)$	0	∞	0	0	120	2	2^2	$4+2$	1680	2
$Sp(3)$	0	∞	0	0	0	∞	2	2	10080	2
$Sp(4)$	0	∞	0	0	0	∞	2	2	0	∞
$SU(2)$	12	2	2	3	15	2	2^2	$12+3$	$84+2^2$	2^2
$SU(3)$	6	0	12	3	30	4	60	6	$84+2$	36
$SU(4)$	0	∞	24	2	$120+2$	4	60	4	$1680+2$	$72+2$
$SU(5)$	0	∞	0	∞	120	0	360	4	1680	6
$SU(6)$	0	∞	0	∞	0	∞	720	2	$5040+2$	6
$SU(7)$	0	∞	0	∞	0	∞	0	∞	5040	0
$SU(8)$	0	∞	0	∞	0	∞	0	∞	0	∞
$SO(5)$	0	∞	0	0	120	2	2^2	$4+2$	1680	2
$SO(6)$	0	∞	24	2	$120+2$	4	60	4	$1680+2$	$72+2$
$SO(7)$	0	∞	2^2	2^2	8	$\infty+2$	0	2	$2520+8+2$	2^4
$SO(8)$	0	$\infty+\infty$	2^3	2^3	$24+8$	$\infty+2$	0	2^2	$2520+120+8+2$	2^7
$SO(9)$	0	∞	2^2	2^2	8	$\infty+2$	0	2	$8+2$	$\infty+2^3$
$SO(10)$	0	∞	2	$\infty+2$	4	∞	12	2	8	$\infty+2^2$
$SO(11)$	0	∞	2	2	2	∞	2	2^2	8	$\infty+2$
$SO(12)$	0	∞	2	2	0	$\infty+\infty$	2^2	2^2	$24+4$	$\infty+2$
$SO(13)$	0	∞	2	2	0	∞	2	2	8	$\infty+2$
$SO(14)$	0	∞	2	2	0	∞	0	∞	4	∞
$SO(15)$	0	∞	2	2	0	∞	0	0	2	∞
$SO(16)$	0	∞	2	2	0	∞	0	0	0	$\infty+\infty$
$SO(17)$	0	∞	2	2	0	∞	0	0	0	∞
G_2	3	0	2	6	0	$\infty+2$	0	0	$168+2$	2
F_4	0	0	2	2	0	$\infty+2$	0	0	2	∞
E_6	0	0	0	∞	0	∞	12	0	0	∞
E_7	0	0	0	0	0	∞	2	2	0	∞
E_8	0	0	0	0	0	0	0	0	0	∞

(4) Stable Homotopy Groups. For sufficiently large n for fixed k, the homotopy groups for classical compact simple Lie groups $G = Sp(n)$, $SU(n)$, $SO(n)$ become stable. We denote them by the following notations. Here we assume $k \geqslant 2$.

$$\pi_k(Sp) = \pi_k(Sp(n)) \qquad (n \geqslant (k-1)/4),$$
$$\pi_k(U) = \pi_k(U(n)) \cong \pi_k(SU(n)) \qquad (n \geqslant (k+1)/2),$$
$$\pi_k(O) = \pi_k(SO(n)) \qquad (n \geqslant k+2).$$

Bott periodicity theorem

$$\pi_k(Sp) \cong \begin{cases} \infty & (k \equiv 3,7 \pmod 8), \\ 2 & (k \equiv 4,5 \pmod 8), \\ 0 & (k \equiv 0,1,2,6 \pmod 8). \end{cases}$$

$$\pi_k(O) \cong \begin{cases} \infty & (k \equiv 3,7 \pmod 8), \\ 2 & (k \equiv 0,1 \pmod 8), \\ 0 & (k \equiv 2,4,5,6 \pmod 8). \end{cases}$$

$$\pi_k(U) \cong \begin{cases} \infty & (k \equiv 1 \pmod 2), \\ 0 & (k \equiv 0 \pmod 2). \end{cases}$$

(5) Metastable Homotopy Groups.

(a,b) means the greatest common divisor of two integers a and b.

$$\pi_{2n}(SU(n)) \cong n!.$$

$$\pi_{2n+1}(SU(n)) \cong \begin{cases} 2 & (n \text{ even}), \\ 0 & (n \text{ odd}). \end{cases}$$

$$\pi_{2n+2}(SU(n)) \cong \begin{cases} (n+1)!+2 & (n \text{ even}, \geqslant 4), \\ (n+1)!/2 & (n \text{ odd}). \end{cases}$$

$$\pi_{2n+3}(SU(n)) \cong \begin{cases} (24,n) & (n \text{ even}), \\ (24,\pi+3)/2 & (n \text{ odd}). \end{cases}$$

$$\pi_{2n+4}(SU(n)) \cong \begin{cases} (n+2)!(24,n)/48 & (n \text{ even}, \geqslant 4), \\ (n+2)!(24,n+3)/24 & (n \text{ odd}). \end{cases}$$

$$\pi_{2n+5}(SU(n)) \cong \pi_{2n+5}(U(n+1)).$$

$$\pi_{2n+6}(SU(n)) \cong \begin{cases} \pi_{2n+6}(U(n+1)) & (n \equiv 2,3 \pmod 4),\ n \geqslant 3), \\ \pi_{2n+6}(U(n+1))+2 & (n \equiv 0,1 \pmod 4)). \end{cases}$$

$$\pi_{4n+2}(Sp(n)) \cong \begin{cases} (2n+1)! & (n \text{ even}), \\ 2(2n+1)! & (n \text{ odd}). \end{cases}$$

$$\pi_{4n+3}(Sp(n)) \cong 2.$$

$$\pi_{4n+4}(Sp(n)) \cong \begin{cases} 2+2 & (n \text{ even}), \\ 2 & (n \text{ odd}). \end{cases}$$

$$\pi_{4n+5}(Sp(n)) \cong \begin{cases} (24, n+2)+2 & (n \text{ even}), \\ (24, n+2) & (n \text{ odd}). \end{cases}$$

$$\pi_{4n+6}(Sp(n)) \cong \begin{cases} (2n+3)!(24, n+2)/12 & (n \text{ even}), \\ (2n+3)!(24, n+2)/24 & (n \text{ odd}). \end{cases}$$

$$\pi_{4n+7}(Sp(n)) \cong 2.$$

$$\pi_{4n+8}(Sp(n)) \cong 2+2.$$

The homotopy groups $\pi_{n+i}(SO(n))$ for $n \geqslant 16$, $3 \geqslant i \geqslant -1$ are determined by the isomorphism

$$\pi_{n+i}(SO(n)) \cong \pi_{n+i}(O) + \pi_{n+i+1}(V_{i+3+n,i+3}(\mathbf{R}))$$

and the homotopy groups of $V_{m+n,m}(\mathbf{R})$ given below.

(6) Homotopy Groups of Real Stiefel Manifolds $V_{m+n,m}(\mathbf{R}) = O(m+n)/I_m \times O(n)$.

$$\pi_{n+k}(V_{n+1,1}) \cong \pi_{n+k}(S^n).$$

$$\pi_{n-k}(V_{m+n,m}) \cong 0 \quad (k \geqslant 1).$$

$$\pi_n(V_{m+n,m}) \cong \begin{cases} 2 & (n=2s-1,\ m \geqslant 2), \\ \infty & (n=2s). \end{cases}$$

$\pi_{n+k}(V_{m+n,m})$ $(k=1,2,3,4,5)$ are given in the following table.

	$n \backslash m$	1	2	3	4	5	6	$8s-1$	$8s$	$8s+1$	$8s+2$	$8s+3$	$8s+4$	$8s+5$	$8s+6$
π_{n+1}	2	0	∞^2	2	$2+\infty$	2	$2+\infty$	2	$2+\infty$	2	$2+\infty$	2	$2+\infty$	2	$2+\infty$
	$\geqslant 3$	0	∞	0	2^2	2	4	0	2^2	2	4	0	2^2	2	4
π_{n+2}	2	∞	2^2	4	2^2	4	2^2	4	2^2	4	2^2	4	2^2	4	2^2
	3	∞^2	2	$2+\infty$	2^2	$4+\infty$	2	$2+\infty$	2^2	$4+\infty$	2	$2+\infty$	2^2	$4+\infty$	2
	$\geqslant 4$	∞	0	2	2^2	8	0	2	2^2	8	0	2	2^2	8	0
π_{n+3}	2	2	2^2	2	$\infty+12+2$	2^2	$24+2$	2^2	$24+2$	2^2	$24+4$	2^2	$24+2$	2^2	$24+2$
	3	2^2	2	2	$\infty+12+4$	2^3	$12+2$	2^2	$24+4$	2^3	$12+2$	2^2	$24+4$	2^3	$12+2$
	4	2	∞	2	∞^2+12+4	2^2	$12+\infty$	2^2	$24+4+\infty$	2^2	$12+\infty$	2^2	$24+4+\infty$	2^2	$12+\infty$
	$\geqslant 5$	0	0	2	$12+4+\infty$	2	12	2	$24+8$	2	12	2^2	$4+48$	2	12
π_{n+4}	2	2	12^2	$\infty+2$	2^2+24	2^2	24	2	24	2	24	2	24	2	24
	3	2^2	0	$\infty+4$	2^4	2^3	2	4	2^2	2^2	2	4	2^2	2^2	2
	4	2	0	$4+\infty$	2^5	2^2	2	8	2^3	2	2	8	2^3	2	2
	5	∞	0	$4+\infty^2$	2^4	$2+\infty$	2	$8+\infty$	2^2	∞	2	$8+\infty$	2^2	∞	2
	$\geqslant 6$	0	0	$4+\infty$	2^3	2	0	8	2	0	2	16	2	0	0
π_{n+5}	2	12	2^2	2	2^3	0	∞	0	0	0	0	0	0	0	0
	3	12^2	∞	$2+24$	2^4	24	$\infty+2$	24	2	24	2	24	2	24	2
	4	0	∞	2^3	2^5	2	$\infty+4$	2^2	2^2	2	4	2^2	2^2	2	4

(VIII) Immersion and Embedding of Projective Spaces (→ 114 Differential Topology)

We denote immersion by $\subset$, and embedding by $\subseteq$. $\mathbf{P}^n(A)$ is an n-dimensional real or complex projective space where $A = \mathbf{R}$ or $\mathbf{C}$, $k\{\mathbf{P}^n(A)\}$ is the integer k such that $\mathbf{P}^n(A) \subset \mathbf{R}^k$ and $\mathbf{P}^n(A) \not\subset \mathbf{R}^{k-1}$, and $\tilde{k}\{\mathbf{P}^n(A)\}$ is the integer k such that $\mathbf{P}^n(A) \subseteq \mathbf{R}^k$ and $\mathbf{P}^n(A) \not\subseteq \mathbf{R}^{k-1}$.

In the table, for example, numbers 9–11 in the row $k\{\mathbf{P}^n(\mathbf{R})\}$ for $n=6$ mean $\mathbf{P}^6(\mathbf{R}) \not\subset \mathbf{R}^8$, $P^6(\mathbf{R}) \subset \mathbf{R}^{11}$.

n	1	2	3	4	5	6	7	8	9	10	11	12
$k\{\mathbf{P}^n(\mathbf{R})\}$	2	4	5	8	9	$9\sim11$	$9\sim12$	16	17	$17\sim19$	...	...
$\tilde{k}\{\mathbf{P}^n(\mathbf{R})\}$	2	3	4	7	7	7	8	15	15	16	16	$17\sim19$
$k\{\mathbf{P}^n(\mathbf{C})\}$	3	7	9	15	17	22	$22\sim25$	31	33	38	$38\sim41$	...
$\tilde{k}\{\mathbf{P}^n(\mathbf{C})\}$	3	7	$8\sim9$	15	$16\sim17$	22	$22\sim25$	31	$32\sim33$	38	$38\sim41$	...

	2^r	2^r+1	2^r+2	2^r+3	2^r+2^s $(r>s>0)$
$k\{\mathbf{P}^n(\mathbf{R})\}$	$2n$	$2n-1$	$2n-3\sim2n-1$	...	...
$\tilde{k}\{\mathbf{P}^n(\mathbf{R})\}$	$2n-1$	$2n-3$	$2n-4$	$2n-6$	...
$k\{\mathbf{P}^n(\mathbf{C})\}$	$4n-1$	$4n-3$	$4n-2$	...	$4n-2$
$\tilde{k}\{\mathbf{P}^n(\mathbf{C})\}$	$4n-1$	$4n-4\sim4n-3$	$4n-2$	...	$4n-2$

7. Knot Theory (→ 235 Knot Theory)

Let k be a projection on a plane of a knot K. We color the domains separated by k, white and black alternatively. The outermost (unbounded) domain determined by k is colored white. In Fig. 16, hatching means black. Take a point (a black point in Fig. 16) in each black domain. The self-intersections of k are represented by white points (Fig. 16). Through each white point we draw a line segment connecting the black points in the black regions meeting at the white point. In Fig. 16, we show this as a broken line. We assign the signature + if the torsion of K at the intersection of k has the orientation of a right-hand screw (as in Fig. 17, left), and the signature − if the orientation is opposite (as in Fig. 17, right). The picture of the line segments with signatures is called the graph corresponding to the projection k of the knot K. Given such a graph, we can reconstruct the original knot K.

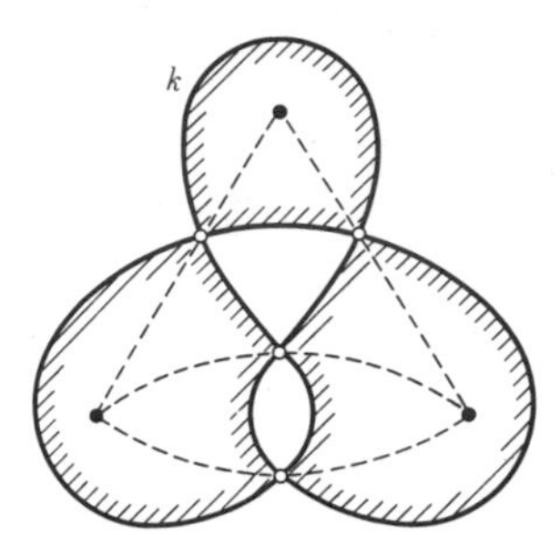

Fig. 16

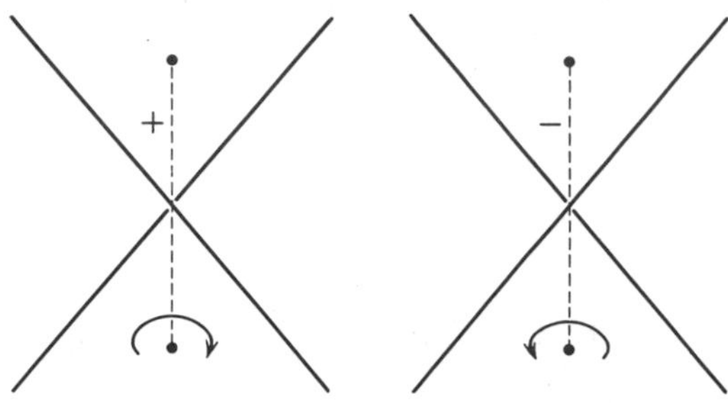

Fig. 17

Fig. 18 shows the classification table of knots for which the numbers of intersections of k are 3 to 8 when we minimize the intersections. The projection of k is described by a solid line, and the graph by broken lines. We omit the signatures since for each graph from 3_1 to 8_{18} they are all + or all −. Such knots are called alternating knots.

8. Inequalities (→ 88 Convex Analysis, 211 Inequalities)

(1) $|a+b| \leqslant |a|+|b|$,

$|a-b| \geqslant ||a|-|b||$.

For real a_ν, we have $\Sigma a_\nu^2 \geqslant 0$, and the equality holds only if all $a_\nu = 0$.

(2) $n! < n^n < (n!)^2 \quad (n \geqslant 3)$.

$e^n \geqslant n^n/n!$.

$n^{1/n} < 3^{1/3} \quad (n \neq 3)$.

(3) $2/\pi < (\sin x)/x < 1 \quad (0 < x < \pi/2)$ (Jordan's inequality).

(4) Denote the elementary symmetric polynomials of positive numbers $a_1, \ldots, a_n > 0$ by S_r $(r = 1, \ldots, n)$. Then

$$S_1 \Big/ \binom{n}{1} \geqslant \left[S_2 \Big/ \binom{n}{2} \right]^{1/2} \geqslant \ldots \geqslant \left[S_r \Big/ \binom{n}{r} \right]^{1/r} \geqslant \ldots \geqslant \left[S_n \Big/ \binom{n}{n} \right]^{1/n}.$$

If at least one equality holds, then $a_1 = \ldots = a_n$. In particular, from the two external terms, we have the following inequalities concerning mean values:

$$\frac{1}{n} \sum_{\nu=1}^{n} a_\nu \geqslant \left(\prod_{\nu=1}^{n} a_\nu \right)^{1/n} \geqslant n \Big/ \sum_{\nu=1}^{n} \frac{1}{a_\nu}.$$

For weighted means, we have

$$\sum_{\nu=1}^{n} \lambda_\nu a_\nu \geqslant \prod_{\nu=1}^{n} a_\nu^{\lambda_\nu} \qquad \left(\Sigma \lambda_\nu = 1, \quad \lambda_\nu > 0 \right).$$

(5) When $a_\nu > 0$, $b_\nu > 0$, $p > 1$, $q > 1$, $(1/p) + (1/q) = 1$,

$$\left[\sum_{\nu=1}^{n} (a_\nu)^p \right]^{1/p} \left[\sum_{\nu=1}^{n} (b_\nu)^q \right]^{1/q} \geqslant \sum_{\nu=1}^{n} a_\nu b_\nu \qquad \text{(Hölder's inequality)}.$$

The equality holds only if $(a_\nu)^p = c(b_\nu)^q$ (c is a constant).

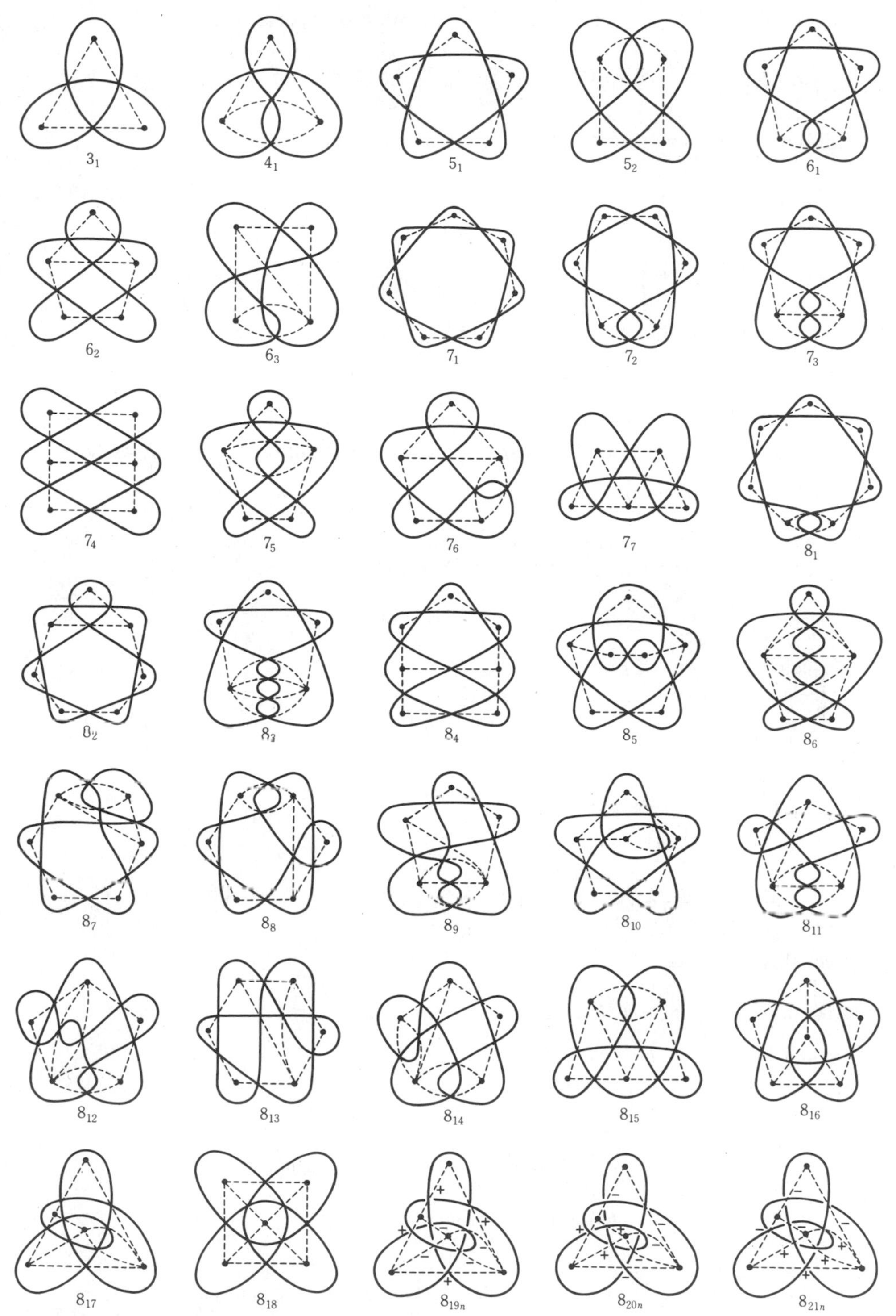

Fig. 18
Classification table of knots. The signatures from 3_1 to 8_{18} are all + or all −.

When $p=q=2$, the inequality is called Cauchy's inequality, the Cauchy-Schwarz inequality, or Bunyakovskiĭ's inequality. As special cases, we have

$$\left(\sum_{\nu=1}^{n} a_\nu\right)\left(\sum_{\nu=1}^{n} \frac{1}{a_\nu}\right) \geqslant n^2 \quad (a_\nu>0),$$

$$\left(\sum_{\nu=1}^{n} a_\nu\right)^2 \leqslant n\left(\sum_{\nu=1}^{n} a_\nu^2\right) \quad (a_\nu>0).$$

When $0<p<1$, we have an inequality by reversing the inequality sign in Hölder's inequality.

(6) When $a_\nu > 0$, $b_\nu > 0$, $p > 0$, and $\{a_\nu\}$ and $\{b_\nu\}$ are not proportional, we have

$$\left[\sum_{\nu=1}^{n}(a_\nu+b_\nu)^p\right]^{1/p} \lesseqgtr \left[\sum_{\nu=1}^{n}(a_\nu)^p\right]^{1/p} + \left[\sum_{\nu=1}^{n}(b_\nu)^p\right]^{1/p} \quad (p \gtreqless 1) \qquad \text{(Minkowski's inequality).}$$

The integral inequality corresponding to (5) or (6) has the same name.

(7) If $a_{\mu\nu} > 0$, $\sum_{\mu=1}^{n} a_{\mu\nu} = \sum_{\nu=1}^{n} a_{\mu\nu} = 1$, $b_\nu \geqslant 0$,

$$\prod_{\nu=1}^{n} b_\nu \leqslant \prod_{\mu=1}^{n}\left(\sum_{\nu=1}^{n} a_{\mu\nu} b_\nu\right).$$

In particular, for the determinant $\Delta = \det(a_{\mu\nu})$,

$$|\Delta|^2 \leqslant \prod_{\nu=1}^{n}\left(\sum_{\mu=1}^{n}|a_{\mu\nu}|^2\right).$$

The equality in this holds only if all rows are mutually orthogonal. If all $|a_{\mu\nu}| \leqslant M$, we have

$$|\Delta| \leqslant n^{n/2} M^n \qquad \text{(Hadamard's estimation).}$$

(8) Suppose that a function $f(x)$ is continuous, strictly monotone increasing in $x \geqslant 0$, and $f(0)=0$. Denote the inverse function of f by f^{-1}. For $a, b > 0$, we have

$$ab \leqslant \int_0^a f(x)\,dx + \int_0^b f^{-1}(x)\,dx \qquad \text{(Young's inequality),}$$

and the equality holds only if $b = f(a)$.

In particular, for $f(x) = x^{p-1}$ $(p > 1)$, we have

$$\frac{a^p}{p} + \frac{b^q}{q} \geqslant ab,$$

where $(1/p)+(1/q)=1$.

(9) If $p, q > 1$, $(1/p)+(1/q)=1$, $a_\mu \geqslant 0$, $b_\nu \geqslant 0$,

$$\sum_{\mu,\nu=0}^{\infty} \frac{a_\mu b_\nu}{\mu+\nu+1} \leqslant \frac{\pi}{\sin(\pi/p)}\left[\sum_{\mu=0}^{\infty}(a_\mu)^p\right]^{1/p}\left[\sum_{\nu=0}^{\infty}(b_\nu)^q\right]^{1/q} \qquad \text{(Hilbert's inequality),}$$

and the equality holds only when the right-hand side vanishes.

(10) For a continuous function $f(x) \geqslant 0$ $(0 \leqslant x < \infty)$, we put

$$F(x) = \int_0^x f(t)\,dt,$$

and assume that $p > 1$. Then

$$\int_0^\infty \left[\frac{F(x)}{x}\right]^p dx \leqslant \left(\frac{p}{p-1}\right)^p \int_0^\infty [f(x)]^p\,dx \qquad \text{(Hardy's inequality),}$$

and the equality holds only if $f(x)$ is identically 0.

Further, if $f(x) > 0$,

$$\int_0^\infty \exp\left[\frac{1}{x}\int_0^x \log f(t)\,dt\right] dx < e\int_0^\infty f(x)\,dx \qquad \text{(Carleman's inequality).}$$

(11) Let $a \leqslant x < \xi \leqslant b$, $p > 1$, and

$$\sup_{\xi} \frac{1}{\xi - x}\int_x^\xi f(t)\,dt = \theta(x).$$

Then

$$\int_a^b [\theta(x)]^p\,dx \leqslant 2\left(\frac{p}{p-1}\right)^p \int_a^b |f(x)|^p\,dx \qquad \text{(Hardy-Littlewood supremum theorem).}$$

(12) If $f(x)$ is piecewise smooth in $0 \leqslant x \leqslant \pi$ and $f(0)=f(\pi)=0$,

$$\int_0^\pi [f'(x)]^2\,dx \geqslant \int_0^\pi [f(x)]^2\,dx \qquad \text{(Wirtinger's inequality),}$$

and the equality holds only if $f(x)$ is a constant multiple of $\sin x$.

9. Differential and Integral Calculus

(I) Derivatives and Primitive Functions (→ 106 Differential Calculus, 216 Integral Calculus)

$F(x)\equiv\int f(x)\,dx$	$f(x)\equiv F'(x)$
$\alpha\varphi+\beta\psi$ (α,β constants)	$\alpha\varphi'+\beta\psi'$
$\varphi\cdot\psi$	$\varphi'\psi+\varphi\psi'$
φ/ψ $(\psi\neq 0)$	$(\varphi'\psi-\varphi\psi')/\psi^2$
$\log\lvert\varphi\rvert$ $(\varphi\neq 0)$	φ'/φ (logarithmic differentiation)
$\Phi(\varphi)$ (composite)	$(d\Phi/d\varphi)\cdot\varphi'$
c (constant)	0
x^n	nx^{n-1}
$x^{n+1}/(n+1)$	x^n $(n\neq -1)$
$\log\lvert x\rvert$	$1/x$
$\log_a\lvert x\rvert$	$(\log_a e)/x$
$x(\log x-1)$	$\log x$
$\exp x=e^x$	$\exp x=e^x$
a^x $(a>0)$	$a^x\log a$
x^x	$x^x(1+\log x)$
$(x-1)e^x$	xe^x
$\sin x$	$\cos x$
$\cos x$	$-\sin x$
$\tan x$	$\sec^2 x$
$\cot x$	$-\operatorname{cosec}^2 x$
$\sec x$	$\sec x\tan x$
$\operatorname{cosec} x$	$-\operatorname{cosec} x\cot x$
$\sinh x=(e^x-e^{-x})/2$	$\cosh x$
$\cosh x=(e^x+e^{-x})/2$	$\sinh x$
$\tanh x=\sinh x/\cosh x$	$\operatorname{sech}^2 x$
$\coth x=\cosh x/\sinh x$	$-\operatorname{cosech}^2 x$
$\operatorname{sech} x=1/\cosh x$	$-\operatorname{sech} x\tanh x$
$\operatorname{cosech} x=1/\sinh x$	$-\operatorname{cosech} x\coth x$
$\arcsin x$ $(\lvert F\rvert<\pi/2)$	$1/\sqrt{1-x^2}$
$\arccos x$ $(0<F<\pi)$	$-1/\sqrt{1-x^2}$
$\arctan x$ $(\lvert F\rvert<\pi/2)$	$1/(1+x^2)$
$\operatorname{arccot} x$ $(\lvert F\rvert<\pi/2)$	$-1/(1+x^2)$
$\operatorname{arcsec} x$ $(0<F<\pi)$	$1/\lvert x\rvert\sqrt{x^2-1}$
$\operatorname{arccosec} x$ $(\lvert F\rvert<\pi/2)$	$-1/\lvert x\rvert\sqrt{x^2-1}$
$\operatorname{arcsinh} x=\log(x+\sqrt{x^2+1})$	$1/\sqrt{x^2+1}$
$\operatorname{arccosh} x=\log(x+\sqrt{x^2-1})$	$1/\sqrt{x^2-1}$
$\frac{1}{2}\log\left\lvert\frac{1+x}{1-x}\right\rvert=\begin{cases}\operatorname{arctanh} x & (\lvert x\rvert<1)\\ \operatorname{arccoth} x & (\lvert x\rvert>1)\end{cases}$	$\frac{1}{1-x^2}$
$\operatorname{arcsech} x$	$-1/x\sqrt{1-x^2}$
$\operatorname{arccosech} x$	$-1/\lvert x\rvert\sqrt{1+x^2}$

$F(x) \equiv \int f(x)\,dx$	$f(x) \equiv F'(x)$
$\frac{1}{2a}\log\left\|\frac{a-x}{a+x}\right\| \quad (a>0)$	$\frac{1}{x^2-a^2}$
$(1/a)\arctan(x/a)$	$1/(x^2+a^2)$
$(x\sqrt{1-x^2}+\arcsin x)/2$	$\sqrt{1-x^2}$
$[x\sqrt{x^2\pm 1}\pm\log(x+\sqrt{x^2\pm 1})]/2$	$\sqrt{x^2\pm 1}$
$-\log\|\cos x\|$	$\tan x$
$\log\|\sin x\|$	$\cot x$
$\log\|\tan x\|$	$1/\sin x\cos x$
$\log\|\tan[(\pi/4)+(x/2)]\|$	$\sec x$
$\log\|\tan(x/2)\|$	$\operatorname{cosec} x$
$(x/2)-(1/4)\sin 2x$	$\sin^2 x$
$\sin x - x\cos x$	$x\sin x$
$\cos x + x\sin x$	$x\cos x$
$\frac{n\sin mx\sin nx + m\cos mx\cos nx}{n^2-m^2} \quad (n^2\neq m^2)$	$\sin mx\cos nx$
$e^{bx}\frac{b\sin ax - a\cos ax}{a^2+b^2}$	$e^{bx}\sin ax$
$e^{bx}\frac{b\cos ax + a\sin ax}{a^2+b^2}$	$e^{bx}\cos ax$
$x\arcsin x+\sqrt{1-x^2}$	$\arcsin x$
$x\arctan x-(1/2)\log(1+x^2)$	$\arctan x$
$\det(\varphi_{jk})_{j,k=1,\ldots,n}$	$\sum\det(\varphi_{j1}\cdots\varphi_{j\nu-1}\varphi'_{j\nu}\varphi_{j\nu+1}\cdots\varphi_{jn})_{j=1,\ldots,n}$

(II) Recurrence Formulas for Indefinite Integrals

(1) $I_m \equiv \int \frac{dx}{(1+x^2)^m}$ (m is a positive integer).

$$I_m = \frac{1}{2m-2}\,\frac{x}{(1+x^2)^{m-1}} + \frac{2m-3}{2m-2}I_{m-1} \quad (m\geqslant 2); \quad I_1=\arctan x.$$

(2) $I_m \equiv \int \frac{x^m}{\sqrt{ax^2+bx+c}}\,dx$ (m is an integer, $a\neq 0$).

The case $m<0$ is reduced to the case $m\geqslant 0$ by the change of variable $1/x=t$.

$$I_m = \frac{1}{ma}x^{m-1}\sqrt{ax^2+bx+c} - \frac{(2m-1)b}{2ma}I_{m-1} - \frac{(m-1)c}{ma}I_{m-2} \quad (m\geqslant 1);$$

$$I_0 = \begin{cases} (1/\sqrt{a})\log\left|2ax+b+2\sqrt{a}\sqrt{ax^2+bx+c}\right| & (a>0), \\ \frac{-1}{\sqrt{-a}}\arcsin\frac{2ax+b}{\sqrt{b^2-4ac}} & (a<0); \end{cases}$$

In this case, for the integrand to be a real function it is necessary that $b^2-4ac>0$.

(3) $I_m \equiv \int x^m e^x\,dx$ (m is an integer).

$$I_m = x^m e^x - mI_{m-1}; \; I_0=e^x, \; I_{-1}=\operatorname{Ei}x,$$

where Ei is the exponential integral function (→ Table 19.II.3, this Appendix).

(4) $I_{m,n}\equiv\int x^m(\log x)^n\,dx$ (m,n are integers, $n\geqslant 0$).

$$I_{m,n}=\frac{x^{m+1}}{m+1}(\log x)^n-\frac{n}{m+1}I_{m,n-1};\ I_{m,0}=\frac{x^{m+1}}{m+1}\quad (m\neq -1),\ I_{-1,n}=(\log x)^{n+1}/(n+1).$$

(5) $I_m\equiv\int x^m\sin x\,dx,\ J_m\equiv\int x^m\cos x\,dx$ (m is a nonnegative integer).

$$I_m=-x^m\cos x+mJ_{m-1}=x^{m-1}(m\sin x-x\cos x)-m(m-1)I_{m-2},$$

$$J_m=x^m\sin x-mI_{m-1}=x^{m-1}(x\sin x+m\cos x)-m(m-1)J_{m-2};$$

$$I_0=-\cos x,$$

$$J_0=\sin x,$$

(6) $I_{m,n}\equiv\int\sin^m x\cdot\cos^n x\,dx$ (m,n are integers).

$$\left.\begin{aligned}I_{m,n}&=\frac{\sin^{m+1}x\cos^{n-1}x}{m+n}+\frac{n-1}{m+n}I_{m,n-2}\\ I_{m,n}&=\frac{-\sin^{m-1}x\cos^{n+1}x}{m+n}+\frac{m-1}{m+n}I_{m-2,n}\end{aligned}\right\}(m+n\neq 0),$$

$$I_{m,n}=\frac{-\sin^{m+1}x\cos^{n+1}x}{n+1}+\frac{m+n+2}{n+1}I_{m,n+2}\ (n\neq -1),$$

$$I_{m,n}=\frac{\sin^{m+1}x\cos^{n+1}x}{m+1}+\frac{m+n+2}{m+1}I_{m+2,n}\ (m\neq -1);$$

$$I_{1,1}=(\sin^2 x)/2,\quad I_{1,0}=-\cos x,\quad I_{1,-1}=-\log|\cos x|,\quad I_{0,1}=\sin x,\quad I_{0,0}=x,$$

$$I_{0,-1}=\log|\tan[(x/2)+(\pi/4)]|,\quad I_{-1,1}=\log|\sin x|,$$

$$I_{-1,0}=\log|\tan(x/2)|,\quad I_{-1,-1}=\log|\tan x|.$$

$$I_{m,-m}\equiv\int\tan^m x\,dx=\frac{\tan^{m-1}x}{m-1}-I_{m-2,-(m-2)}(m\neq 1).$$

(III) Derivatives of Higher Order

$f(x)$	$f^{(n)}(x)$
$\varphi\cdot\psi$	$\sum_{\nu=0}^{n}\binom{n}{\nu}\varphi^{(\nu)}\psi^{(n-\nu)}$ (Leibniz's formula)
x^k	$\prod_{\nu=0}^{n-1}(k-\nu)x^{k-n}$
$(x+a)^n$	$n!$
$\exp x$	$\exp x$
$a^x\ (a>0)$	$a^x(\log a)^n$
$\log x$	$(-1)^{n-1}(n-1)!/x^n$
$\sin x$	$\sin[x+(n\pi/2)]$
$\cos x$	$\cos[x+(n\pi/2)]$
$e^{ax}\cos bx$	$r^n e^{ax}\cos(bx+n\theta)$ (where $a=r\cos\theta,\ b=r\sin\theta$)
$\arcsin x$	$\frac{1}{2^{n-1}}\sum_{\nu=0}^{n-1}(-1)^\nu\binom{n-1}{\nu}(2\nu-1)!!(2n-2\nu-3)!!(1+x)^{-(1/2)-\nu}(1-x)^{(1/2)-n+\nu}$ (where $(2\nu-1)!!\equiv 1\cdot 3\cdot 5\cdot\ldots\cdot(2\nu-1),\ (-1)!!\equiv 1$)
$\arctan x$	$(-1)^{n-1}(n-1)!\sin^n\theta\sin n\theta$ (where $x=\cot\theta$)

$$\left(\frac{1}{f}\right)'=-\frac{f'}{f^2},\quad \left(\frac{1}{f}\right)''=\frac{2f'^2-ff''}{f^3},\quad \left(\frac{1}{f}\right)'''=\frac{6ff'f''-6f'^3-f^2f'''}{f^4}.$$

Higher-order derivatives of a composite function $g(t)\equiv f(x_1(t),\ldots,x_n(t))$

$$\frac{dg}{dt}=\sum_{i=1}^{n}\frac{\partial f}{\partial x_i}\frac{dx_i}{dt},\quad \frac{d^2g}{dt^2}=\sum_{i=1}^{n}\frac{\partial f}{\partial x_i}\frac{d^2x_i}{dt^2}+\sum_{i,j=1}^{n}\frac{\partial^2 f}{\partial x_i\partial x_j}\frac{dx_i}{dt}\frac{dx_j}{dt},$$

$$\frac{d^3g}{dt^3}=\sum_{i=1}^{n}\frac{\partial f}{\partial x_i}\frac{d^3x_i}{dt^3}+2\sum_{i,j=1}^{n}\frac{\partial^2 f}{\partial x_i\partial x_j}\frac{d^2x_i}{dt^2}\frac{dx_j}{dt}+\sum_{i,j,k=1}^{n}\frac{\partial^3 f}{\partial x_i\partial x_j\partial x_k}\frac{dx_i}{dt}\frac{dx_j}{dt}\frac{dx_k}{dt}.$$

For a function $z=z\,(x_1,\ldots,x_n)$ determined implicitly by $F(z;x_1,\ldots,x_n)=0$, we have

$$\frac{\partial z}{\partial x_i}=-\frac{F_{x_i}}{F_z},\quad \frac{\partial^2 z}{\partial x_i\partial x_j}=-\frac{F_{x_ix_j}}{F_z}+\frac{F_{x_i}F_{x_jz}+F_{x_j}F_{x_iz}}{F_z^2}-\frac{F_{x_i}F_{x_j}F_{zz}}{F_z^3}.$$

Schwarzian derivative:

$$\{y;x\}\equiv\left(\frac{d^3y}{dx^3}\right)\Big/\left(\frac{dy}{dx}\right)-\frac{3}{2}\left[\left(\frac{d^2y}{dx^2}\right)\Big/\left(\frac{dy}{dx}\right)\right]^2,\quad \{y;x\}=0\Leftrightarrow y=(ax+b)/(cx+d),$$

$$\{y;x\}=\left(\frac{dz}{dx}\right)^2[\{y;z\}-\{x;z\}]=-\left(\frac{dy}{dx}\right)^2\{x;y\},\ \{(ay+b)/(cy+d);x\}=\{y;x\}.$$

(IV) The Taylor Expansion and Remainder

If $f(x)$ is n times continuously differentiable in the interval $[a,b]$ (i.e., of class C^n),

$$f(b)=\sum_{\nu=0}^{n-1}\frac{(b-a)^\nu}{\nu!}f^{(\nu)}(a)+R_n\qquad\text{(Taylor's formula).}$$

R_n is called the remainder, and is represented as follows:

$$R_n=\frac{1}{(n-1)!}\int_a^b(b-x)^{n-1}f^{(n)}(x)\,dx=\frac{(b-a)^p(b-\xi)^{n-p}}{(n-1)!p}f^{(n)}(\xi)$$

$$=\frac{(b-a)^n}{(n-1)!p}(1-\theta)^{n-p}f^{(n)}(a+\theta(b-a))\quad (n\geqslant p>0,\ 0<\theta<1,\ a<\xi<b,\ \xi=a+\theta(b-a))$$

(Roche-Schlömilch remainder);

$$=\frac{1}{n!}(b-a)^n f^{(n)}(\xi)\qquad\text{(Lagrange's remainder);}$$

$$=\frac{1}{(n-1)!}(b-a)(b-\xi)^{n-1}f^{(n)}(\xi)\qquad\text{(Cauchy's remainder).}$$

If $f(x,y)$ is m times continuously differentiable in a neighborhood of a point (x_0,y_0),

$$f(x_0+h,y_0+k)=\sum_{\lambda=0}^{m-1}\frac{1}{\lambda!}\left(h\frac{\partial}{\partial x}+k\frac{\partial}{\partial y}\right)^\lambda f(x_0,y_0)+R_m$$

$$=\sum_{0\leqslant\mu+\nu\leqslant m-1;\,\mu,\nu\geqslant 0}\frac{1}{\mu!\nu!}h^\mu k^\nu\frac{\partial^{\mu+\nu}f(x_0,y_0)}{\partial x^\mu\partial y^\nu}+R_m,$$

$$R_m=\frac{1}{m!}\left(h\frac{\partial}{\partial x}+k\frac{\partial}{\partial y}\right)^m f(x_0+\theta h,y_0+\theta k)\quad (0<\theta<1).$$

If all partial derivatives up to order $m-1$ are totally differentiable,

$$f(x_1+h_1,\ldots,x_n+h_n)=\sum_{\nu=0}^{m-1}\frac{1}{\nu!}\left(\sum_{\mu=1}^{n}h_\mu\frac{\partial}{\partial x_\mu}\right)^\nu f(x_1,\ldots,x_n)+R_m$$

$$=\sum\frac{1}{\nu_1!\ldots\nu_n!}h_1^{\nu_1}\ldots h_n^{\nu_n}\frac{\partial^{\nu_1+\ldots+\nu_n}f(x_1,\ldots,x_n)}{\partial x_1^{\nu_1}\ldots\partial x_n^{\nu_n}}+R_m,$$

where the Σ means the sum for $\nu_1,\ldots,\nu_n$ in the domain $0\leqslant\nu_1+\ldots+\nu_n\leqslant m-1$; $\nu_1,\ldots,\nu_m\geqslant 0$. The remainder R_m is expressed as

$$R_m=\frac{1}{m!}\left(\sum_{\mu=1}^{n}h_\mu\frac{\partial}{\partial x_\mu}\right)^m f(x_1+\theta h_1,\ldots,x_n+\theta h_n)\quad (0<\theta<1).$$

(V) Definite Integrals [4]

In the following formulas, we assume that m, n are positive integers. δ_{mn} is Kronecker's delta ($\delta_{mn}=0$ or 1 for $m\neq n$ or $m=n$), Γ is the gamma function, B is the beta function, and C is the Euler constant.

For simplicity, we put

$$m!!\equiv\begin{cases}1\cdot3\cdot5\cdot\ldots\cdot(m-2)\cdot m=2^{(m+1)/2}\Gamma[(m/2)+1]/\sqrt{\pi}=m!/2^{(m-1)/2}[(m-1)/2]! & (m\text{ is odd}),\\ 2\cdot4\cdot6\cdot\ldots\cdot(m-2)\cdot m=2^{m/2}\Gamma[(m/2)+1]=2^{m/2}(m/2)! & (m\text{ is even}).\end{cases}$$

In an n-dimensional real space, the volume of the domain

$$|x_1|^p+\ldots+|x_n|^p\leqslant 1\quad(p>0)\qquad\text{is}\qquad \frac{2^n[\Gamma(1/p)]^n}{p^{n-1}n\Gamma(n/p)}.$$

For $p=2$, this is the volume of the unit hypersphere, which is

$$\frac{\pi^{n/2}}{\Gamma[(n/2)+1]}=\begin{cases}(2\pi)^{n/2}/n!! & (n\text{ is even}),\\ 2(2\pi)^{(n-1)/2}/n!! & (n\text{ is odd}).\end{cases}$$

The surface area of the $(n-1)$-dimensional unit hypersphere

$$|x_1|^2+\ldots+|x_n|^2=1\qquad\text{is}\qquad\frac{2\pi^{n/2}}{\Gamma(n/2)}=\begin{cases}(2\pi)^{n/2}/(n-2)!! & (n\text{ is even}),\\ 2(2\pi)^{(n-1)/2}/(n-2)!! & (n\text{ is odd}).\end{cases}$$

$$\int_0^\infty x^{p-1}e^{-x}\,dx=\int_0^1\left(\log\frac{1}{x}\right)^{p-1}dx\equiv\Gamma(p).$$

$$\int_0^1 x^p\left(\log\frac{1}{x}\right)^q dx=\frac{\Gamma(q+1)}{(p+1)^{q+1}}\quad(\mathrm{Re}\,p,\mathrm{Re}\,q>-1).$$

$$\int_0^1 x^{p-1}(1-x)^{q-1}dx=\int_0^\infty\frac{x^{p-1}}{(1+x)^{p+q}}dx\equiv B(p,q)=\frac{\Gamma(p)\Gamma(q)}{\Gamma(p+q)}.$$

$$\int_0^\infty\frac{x^a}{(1+x^c)^{1+b}}dx=\frac{1}{c}\,\frac{\Gamma[(a+1)/c]\Gamma[b-\{(a-c+1)/c\}]}{\Gamma(1+b)}$$
$$\left(\mathrm{Re}\,c>0;\mathrm{Re}\,a,\mathrm{Re}\,b>-1;\mathrm{Re}\,b>\mathrm{Re}\frac{a-c+1}{c}\right).$$

$$\int_0^a\frac{(a-x)^{p-1}(x-b)^{q-1}}{|x-c|^{p+q}}dx=\frac{(a-b)^{p+q-1}}{|a-c|^q|b-c|^p}B(p,q)$$
$$(0<c<b<a\quad\text{or}\quad 0<b<a<c;\mathrm{Re}\,p,\mathrm{Re}\,q>0).$$

$$\int_{-\infty}^\infty\frac{1}{(1+x^2)^{n+1}}dx=\frac{\pi(2n)!}{2^{2n}(n!)^2}=\pi\frac{(2n-1)!!}{(2n)!!}.$$

$$\int_{-\infty}^\infty\frac{x^{2m}}{1+x^{2n}}dx=\frac{\pi}{n\sin[(2m+1)\pi/2n]}\quad(2m+1<2n).$$

$$\int_0^\infty\frac{x^{a-1}}{1+x}dx=\Gamma(a)\Gamma(1-a)=\frac{\pi}{\sin a\pi}\quad(0<a<1).$$

$$\int_0^\infty e^{-a^2x^2}dx=\frac{\sqrt{\pi}}{2|a|}.\quad\int_0^\infty x^pe^{-x^2}dx=\frac{1}{2}\Gamma\left(\frac{p+1}{2}\right)=\begin{cases}(2n-1)!!\sqrt{\pi}/2^{n+1} & (p=2n),\\ n!/2 & (p=2n+1).\end{cases}$$

$$\int_0^\infty(e^{-a^2/x^2}-e^{-b^2/x^2})dx=(b-a)\sqrt{\pi}\quad(a,b\geqslant 0).\quad\int_0^\infty e^{-[x-(1/x)]^2}dx=\frac{\sqrt{\pi}}{2}.$$

$$\int_0^\infty e^{-x^2-(a^2/x^2)}dx=\frac{e^{-2a}\sqrt{\pi}}{2}\quad(a\geqslant 0).\quad\int_0^\infty\frac{1}{e^{ax}+e^{-ax}}dx=\frac{\pi}{4a}.$$

$$\int_0^\infty\frac{x}{e^{ax}-e^{-ax}}dx=\frac{\pi^2}{8a^2}\quad(a>0).$$

$$\int_0^\infty\frac{x}{e^x-e^{-x}}dx=\int_0^1\frac{\log(1/x)}{1-x^2}dx=\frac{\pi^2}{8}.$$

$$\int_0^\infty \frac{x}{e^x+1}\,dx = \int_0^1 \frac{\log(1/x)}{1+x}\,dx = \frac{\pi^2}{12}. \quad \int_0^\infty \frac{x}{e^x-1}\,dx = \int_0^1 \frac{\log(1/x)}{1-x}\,dx = \frac{\pi^2}{6}.$$

$$\int_0^\infty \log\left(\frac{e^x+1}{e^x-1}\right)dx = \int_0^1 \log\left(\frac{1+x}{1-x}\right)\frac{1}{x}\,dx = \frac{\pi^2}{4}.$$

$$\int_0^1 \frac{\log x}{\sqrt{1-x^2}}\,dx = \int_0^{\pi/2} \log\sin x\,dx = -\frac{\pi}{2}\log 2.$$

$$\int_0^1 \frac{\log(1+x)}{1+x^2}\,dx = \frac{\pi}{8}\log 2. \quad \int_0^1 \frac{\log x}{1+x^2}\,dx = -\int_1^\infty \frac{\log x}{1+x^2}\,dx = \sum_{n=0}^\infty \frac{(-1)^{n-1}}{(2n+1)^2} = -0.91596\ldots.$$

$$\int_0^\infty \frac{(\log x)^2}{1+x+x^2}\,dx = \frac{16\pi^3}{81\sqrt{3}}. \quad \int_0^1 \frac{x^p-x^q}{\log x}\,dx = \log\frac{p+1}{q+1} \quad (p,q>-1).$$

$$\int_0^1 \log|\log x|\,dx = \int_0^\infty e^{-t}\log t\,dt = -C = -0.57721\ldots.$$

$$\int_0^\pi \sin mx \sin nx\,dx = \int_0^\pi \cos mx\cos nx\,dx = \delta_{mn}\frac{\pi}{2}.$$

$$\int_0^\pi \sin mx\cos nx\,dx = \begin{cases} \left[1-(-1)^{m+n}\right]\dfrac{m}{m^2-n^2} & (m\neq n);\\ 0 & (m=n).\end{cases}$$

$$\int_0^{\pi/2} \sin^p x\cos^q x\,dx = \frac{1}{2}B\left(\frac{p+1}{2},\frac{q+1}{2}\right) \quad \left(\mathrm{Re}\,p, \mathrm{Re}\,q > -\frac{1}{2}\right);$$

$$= \begin{cases} (\pi/2)(p-1)!!(q-1)!!/(p+q)!! & (p,q \text{ are even positive integers}),\\ (p-1)!!(q-1)!!/(p+q)!! & (p,q \text{ are positive integers not both even}).\end{cases}$$

$$\int_0^{\pi/2} \sin^p x\,dx = \int_0^{\pi/2}\cos^p x\,dx = \frac{\sqrt{\pi}}{2}\frac{\Gamma[(p+1)/2]}{\Gamma[(p/2)+1]} \quad (\mathrm{Re}\,p>-1);$$

$$= \begin{cases} (\pi/2)(2n-1)!!/(2n)!! & (p=2n),\\ (2n)!!/(2n+1)!! & (p=2n+1).\end{cases}$$

$$\int_{-\infty}^\infty \sin(x^2)\,dx = \int_{-\infty}^\infty \cos(x^2)\,dx = \int_0^\infty \frac{\sin x}{\sqrt{x}}\,dx = \int_0^\infty \frac{\cos x}{\sqrt{x}}\,dx = \sqrt{\frac{\pi}{2}} \quad \text{(Fresnel integral).}$$

$$\int_0^\infty \frac{\sin ax}{x}\,dx = \frac{\pi}{2} \quad (a>0). \quad \int_0^\infty \frac{\tan x}{x}\,dx = \frac{\pi}{2}$$

$$\left(\text{take Cauchy's principal value at } x = \left(n+\frac{1}{2}\right)\pi\right).$$

$$\int_0^\infty \frac{\sin^{2n+1}x}{x}\,dx = \frac{\pi}{2}\frac{(2n-1)!!}{(2n)!!}. \quad \int_0^\infty \frac{\sin^2 x}{x^2}\,dx = \frac{\pi}{2}. \quad \int_0^\infty \frac{\sin(x^2)}{x}\,dx = \frac{\pi}{4}.$$

$$\int_0^\infty \frac{\sin qx}{x^p}\,dx = \frac{\pi q^{p-1}}{2\Gamma(p)\sin(p\pi/2)} \quad (0<p<2).$$

$$\int_0^\infty \frac{\cos px}{1+x^2}\,dx = \frac{\pi}{2}e^{-|p|}. \quad \int_0^\infty \frac{\cos^2 ax}{1+x^2}\,dx = \frac{\pi}{4}(1+e^{-2a}) \quad (a>0).$$

$$\int_0^\infty \frac{\sin ax}{x(1+x^2)}\,dx = \frac{\pi}{2}(1-e^{-a}) \quad (a>0). \quad \int_0^\infty \frac{x\sin ax}{1+x^2}\,dx = \frac{\pi}{2}e^{-a} \quad (a>0).$$

$$\int_0^\infty \frac{\sin^{2m+1}x\cos^{2n}x}{x}\,dx = \int_0^\infty \frac{\sin^{2m+1}x\cos^{2n-1}x}{x}\,dx = \frac{\pi}{2}\frac{(2m-1)!!(2n-1)!!}{(2m+2n)!!}.$$

$$\int_0^\infty \frac{\sin ax\cos bx}{x}\,dx = \begin{cases} \pi/2 & (a>b>0),\\ \pi/4 & (a=b>0),\\ 0 & (b>a>0)\end{cases} \quad \text{(Dirichlet's discontinuous factor).}$$

$$\int_0^{2\pi} \frac{1}{1+a\cos x}\,dx = \frac{2\pi}{\sqrt{1-a^2}} \quad (|a|<1). \quad \int_0^{\pi/2} \frac{1}{a^2\cos^2 x + b^2\sin^2 x}\,dx = \frac{\pi}{2ab} \quad (ab\neq 0).$$

$$\int_0^\pi \frac{x\sin x}{1+\cos^2 x}\,dx = \frac{\pi^2}{4}.$$

$$\int_0^{\pi} \frac{\cos nx}{1-2a\cos x + a^2}\, dx = \begin{cases} \pi a^n/(1-a^2) & (|a|<1), \\ \pi/a^n(a^2-1) & (|a|>1). \end{cases}$$

References

[1] B. O. Peirce, A short table of integrals, Ginn, Boston, second revised edition, 1910.
[2] D. Bierens de Haan, Nouvelles tables d'intégrales définies, Leiden, 1867.
There are several mistakes in this table. For the errata, see
[3] C. F. Lindmann, Examen des nouvelles tables de M. Bierens de Haan, Handlingar Svenska Vetenskaps-Akad., 1891.
[4] E. W. Sheldon, Critical revision of de Haan's tables of definite integrals, Amer. J. Math., 34 (1912), 39–114.

10. Series (→ 379 Series)

(I) Finite Series

(1) $S_k \equiv 1^k + 2^k + \ldots + n^k$ (k is an integer). For $k \geqslant 0$, we have

$$S_k = \frac{B_{k+1}(n+1) - B_{k+1}(1)}{k+1} = \sum_{i=0}^{k} (-1)^i \binom{k+1}{i} \frac{B_{2i}(n+1)^{k+1-i}}{k+1},$$

where B_l is a Bernoulli number and $B_l(x)$ is a Bernoulli polynomial. In particular,

$$S_0 = n, \quad S_1 = n(n+1)/2, \quad S_2 = n(n+1)(2n+1)/6, \quad S_3 = n^2(n+1)^2/4,$$
$$S_4 = n(n+1)(2n+1)(3n^2+3n-1)/30.$$

For $k<0$ and $k=-l$,

$$S_{-l} = c_l - [(-1)^l/(l-1)!][d^l \log\Gamma(x)/dx^l]_{x=n+1}$$
$$= c_l - \frac{1}{(l-1)(n+1)^{l-1}} - \frac{1}{2(n+1)^l} + \sum_{i=1}^{\infty} (-1)^i \frac{B_{2(i+1)}}{(i+1)!} \frac{(l+i-1)!}{(l-1)!} \frac{1}{(n+1)^{l+i}}.$$

For $l=1$, the second term in the latter formula is replaced by $\log[1/(n+1)]$. Here Γ is the gamma function, and the constants c_l are

$$c_l = \begin{cases} C & \text{(Euler constant)} \quad (l=1), \\ \zeta(l) & (\zeta \text{ is the Riemann zeta function}) \quad (l \geqslant 2). \end{cases}$$

(2) $$\sum_{i=1}^{n} i(i+1)\ldots(i+m-1) \equiv \sum_{i=1}^{n} \frac{(i+m-1)!}{(i-1)!} = \frac{1}{m+1} \frac{(n+m)!}{(n-1)!},$$

$$\sum_{i=1}^{n} \frac{(i-1)!}{(i+m-1)!} = \frac{1}{m-1}\left[\frac{1}{(m-1)!} - \frac{n!}{(n+m-1)!}\right] \quad (m \geqslant 2),$$

$$\sum_{i=1}^{n} i!i = (n+1)! - 1, \quad \sum_{i=1}^{n} i\binom{n}{i} = n2^{n-1},$$

$$\sum_{i=m}^{n} \binom{i}{m}\binom{n+s-i-1}{n-i} = \binom{n+s}{m+s} \quad (m \leqslant n),$$

$$\sum_{i=0}^{n} \binom{n}{i}\binom{m}{r-i} = \binom{n+m}{r}.$$

$$\sum_{i=1}^{n} a^i = \begin{cases} a(a^n-1)/(a-1) & (a \neq 1) \\ n & (a=1) \end{cases} \quad \text{(geometric progression)}$$

$$\sum_{j=0}^{n} (a+jd) = (n+1)a + \frac{n(n+1)}{2} d = \frac{n+1}{2}(a+a+nd) \quad \text{(arithmetic progression)}$$

$$\sum_{j=0}^{n} \sin(\alpha + j\beta) = \sin\left(\alpha + \frac{n}{2}\beta\right) \sin\frac{(n+1)\beta}{2} \Big/ \sin\frac{\beta}{2},$$

$$\sum_{j=0}^{n} \cos(\alpha + j\beta) = \cos\left(\alpha + \frac{n}{2}\beta\right)\sin\frac{(n+1)\beta}{2} \Big/ \sin\frac{\beta}{2},$$

$$\sum_{j=0}^{n} \operatorname{cosec} 2^j\alpha = \cot(\alpha/2) - \cot 2^n\alpha.$$

(II) Convergence Criteria for Positive Series $\sum a_n$

In the present Section II, we assume that $a_n \geqslant 0$.
Cauchy's criterion: The series converges when $\limsup \sqrt[n]{a_n} < 1$ and it diverges when $\limsup \sqrt[n]{a_n} > 1$.
d'Alembert's criterion: The series converges when $\limsup a_{n+1}/a_n < 1$ and diverges when $\liminf a_{n+1}/a_n > 1$.
Raabe's criterion: The series converges when $\liminf n[(a_n/a_{n+1}) - 1] > 1$ and diverges when $\limsup n[(a_n/a_{n+1}) - 1] < 1$.
Kummer's criterion: For a positive divergent series $\sum(1/b_n)$, the series $\sum a_n$ converges when $\liminf[(b_n a_n/a_{n+1}) - b_{n+1}] > 0$ and diverges when $\limsup[(b_n a_n/a_{n+1}) - b_{n+1}] < 0$ diverges.
Gauss's criterion: Suppose $a_n/a_{n+1} = 1 + (k/n) + (\theta_n/n^\lambda)$, where $\lambda > 1$ and $\{\theta_n\}$ is bounded. Then the series $\sum a_n$ converges when $k > 1$; and diverges when $k \leqslant 1$.
Schlömilch's criterion: For a decreasing positive sequence $a_\nu \downarrow 0$, let n_ν be an increasing sequence of positive integers and suppose that $(n_{\nu+2} - n_{\nu+1})/(n_{\nu+1} - n_\nu)$ is bounded. Then the two series $\sum a_n$ and $\sum(n_{\nu+1} - n_\nu)a_{n_\nu}$ converge or diverge simultaneously.
Logarithmic criterion: For a positive integer k, we put

$$\log_k x \equiv \log(\log_{k-1} x), \quad \log_1 x = \log x.$$

Then for sufficiently large n we have
The first logarithmic criterion: If

$$a_n - 1/(n \log_1 n \ldots \log_{k-1} n (\log_k n)^p) \begin{cases} \leqslant 0, & p > 1 \quad \text{then } \sum a_n \text{ converges,} \\ \geqslant 0, & p \leqslant 1 \quad \text{then } \sum a_n \text{ diverges.} \end{cases}$$

The second logarithmic criterion: If

$$\frac{a_{n+1}}{a_n} - \frac{n}{n+1} \frac{\log_1 n}{\log_1(n+1)} \cdots \frac{\log_{k-1} n}{\log_{k-1}(n+1)} \left(\frac{\log_k n}{\log_k(n+1)}\right)^p$$

$$\begin{cases} \leqslant 0, & p > 1 \quad \text{then } \sum a_n \text{ converges,} \\ \geqslant 0, & p \leqslant 1 \quad \text{then } \sum a_n \text{ diverges.} \end{cases}$$

(III) Infinite Series

$$\sum_{i=1}^{\infty} \frac{(-1)^{i-1}}{i} = \log 2, \quad \sum_{i=1}^{\infty} \frac{(-1)^{i-1}}{2i-1} = \frac{\pi}{4} \quad \text{(Leibniz's formula),}$$

$$\sum_{i=1}^{\infty} \frac{(-1)^{i-1}}{2i(2i+1)(2i+2)} = \frac{\pi - 3}{4},$$

$$\sum_{i=0}^{\infty} \frac{(2i)!}{2^{2i}(i!)^2} \frac{1}{2i+1} = \frac{\pi}{2}, \quad \sum_{i=1}^{\infty} \left(\frac{1}{i} - \log\left(1 + \frac{1}{i}\right)\right) = C \quad (C \text{ is Euler's constant).}$$

Putting

$$\zeta(n) = \sum_{i=1}^{\infty} \frac{1}{i^n}, \quad \alpha(n) = \sum_{i=1}^{\infty} \frac{1}{(2i-1)^n}, \quad \beta(n) = \sum_{i=1}^{\infty} \frac{(-1)^{i-1}}{i^n}, \quad \varepsilon(n) = \sum_{i=1}^{\infty} \frac{(-1)^{i-1}}{(2i-1)^n},$$

we have

$$\zeta(2n)=\frac{(2\pi)^{2n}}{2(2n)!}B_{2n},\quad \alpha(2n)=\frac{(2^{2n}-1)\pi^{2n}}{2(2n)!}B_{2n},$$

$$\beta(2n)=\frac{(2^{2n-1}-1)\pi^{2n}}{(2n)!}B_{2n},\quad \varepsilon(2n+1)=\frac{\pi^{2n+1}}{2^{2n+2}(2n)!}E_{2n}.$$

where B_n is a Bernoulli number, and E_n is an Euler number.

$$\zeta(2)=\pi^2/6,\quad \zeta(4)=\pi^4/90,\quad \zeta(6)=\pi^6/945.$$

$$\alpha(2)=\pi^2/8,\quad \alpha(4)=\pi^4/96,\quad \alpha(6)=\pi^6/960.$$

$$\beta(2)=\pi^2/12,\quad \beta(4)=7\pi^4/720,\quad \beta(6)=31\pi^6/30240.$$

$$\varepsilon(1)=\pi/4,\quad \varepsilon(2)=0.91596\,55941\,77219\,01505\,46035\,14932\ldots \qquad \text{(Catalan's constant)},$$

$$\varepsilon(3)=\pi^3/32,\quad \varepsilon(5)=5\pi^5/1536,\quad \varepsilon(7)=61\pi^7/92160.$$

(IV) Power Series (→ 339 Power Series)

(1) Binomial Series $(1+x)^\alpha=\sum_{i=0}^{\infty}\binom{\alpha}{i}x^i$. This converges always in $|x|<1$. If $\alpha>0$, it converges in $-1\leqslant x\leqslant 1$, and if $-1<\alpha<0$, it converges in $-1<x\leqslant 1$. When α is 0 or a positive integer, it reduces to a polynomial and converges in $|x|<\infty$.

$$\frac{1}{1+x}=\sum_{i=0}^{\infty}(-1)^i x^i\quad(|x|<1),$$

$$\sqrt{1+x}=\sum_{i=0}^{\infty}\frac{(-1)^{i-1}(2i)!}{(2i-1)2^{2i}(i!)^2}x^i\quad(|x|<1),\qquad \frac{1}{\sqrt{1+x}}=\sum_{i=0}^{\infty}\frac{(-1)^i(2i)!}{2^{2i}(i!)^2}x^i\quad(|x|<1).$$

(2) Elementary Transcendental Functions (→ 131 Elementary Functions).

$$e^x\equiv\exp x=\sum_{i=0}^{\infty}\frac{x^i}{i!}=\lim_{n\to\infty}\left(1+\frac{x}{n}\right)^n,\quad a^x=\exp(x\log a)\quad(|x|<\infty).$$

$$\log(1+x)=\sum_{i=1}^{\infty}\frac{(-1)^{i-1}}{i}x^i\quad(-1<x\leqslant 1),\quad \log x=2\sum_{i=0}^{\infty}\frac{1}{2i+1}\left(\frac{x-1}{x+1}\right)^{2i+1}\quad(0<x<\infty).$$

$$\sin x=\sum_{i=0}^{\infty}\frac{(-1)^i}{(2i+1)!}x^{2i+1},\quad \cos x=\sum_{i=0}^{\infty}\frac{(-1)^i}{(2i)!}x^{2i}\quad(|x|<\infty).$$

$$\tan x=\sum_{i=1}^{\infty}\frac{2^{2i}(2^{2i}-1)B_{2i}}{(2i)!}x^{2i-1}\quad\left(|x|<\frac{\pi}{2}\right)\qquad(B_i\text{ is a Bernoulli number}),$$

$$\cot x=\frac{1}{x}-\sum_{i=1}^{\infty}\frac{2^{2i}B_{2i}}{(2i)!}x^{2i-1}\qquad\left(0<|x|<\frac{\pi}{2}\right),$$

$$\sec x=\sum_{i=0}^{\infty}\frac{E_{2i}}{(2i)!}x^{2i}\quad\left(|x|<\frac{\pi}{2}\right)\quad(E_i\text{ is an Euler number}),$$

$$\operatorname{cosec} x=\frac{1}{x}+\sum_{i=1}^{\infty}\frac{(2^{2i}-2)B_{2i}}{(2i)!}x^{2i-1}\quad(0<|x|<\pi).$$

$$\arcsin x=\sum_{i=0}^{\infty}\frac{(2i)!}{2^{2i}(i!)^2}\frac{x^{2i+1}}{2i+1}\quad(|x|\leqslant 1),\quad \arctan x=\sum_{i=0}^{\infty}\frac{(-1)^i}{2i+1}x^{2i+1}\quad(|x|\leqslant 1).$$

(V) Partial Fractions for Elementary Functions

$$\tan x=\sum_{n=0}^{\infty}\frac{8x}{(2n+1)^2\pi^2-4x^2},\quad \cot x=\frac{1}{x}+2x\sum_{n=1}^{\infty}\frac{1}{x^2-n^2\pi^2},$$

$$\sec x=4\sum_{n=1}^{\infty}\frac{(-1)^n(2n-1)\pi}{(2n-1)^2\pi^2-4x^2},\quad \operatorname{cosec} x=\frac{1}{x}+2x\sum_{n=1}^{\infty}\frac{(-1)^n}{x^2-n^2\pi^2},$$

$$\sec^2 x=\sum_{n=-\infty}^{\infty}\frac{1}{\left[x+\{(2n+1)\pi/2\}\right]^2},\quad \operatorname{cosec}^2 x=\sum_{n=-\infty}^{\infty}\frac{1}{(x+n\pi)^2}.$$

(VI) Infinite Products (→ 379 Series F)

$$\prod_{n=1}^{\infty}\frac{4n^2}{4n^2-1}=\frac{\pi}{2}\ \text{(Wallis formula)},\quad \prod_{n=1}^{\infty}\left(1+\frac{x}{a+n}\right)e^{-x/n}=e^{-Cx}\frac{\Gamma(1+a)}{\Gamma(1+a+x)}$$

(C is Euler's constant).

$$\prod_{n=1}^{\infty}\left(1-\frac{x}{2n-1}\right)\left(1+\frac{x}{2n}\right)=\sqrt{\pi}\Big/\Gamma\left(1+\frac{x}{2}\right)\Gamma\left(\frac{1}{2}-\frac{x}{2}\right).$$

$$\prod_{p}1/(1-p^{-s})=\zeta(s)\quad (p \text{ ranges over all prime numbers}, s>1),$$

$$\prod_{n=1}^{\infty}\left(1-\frac{x^2}{n^2\pi^2}\right)=\frac{\sin x}{x},\quad \prod_{n=1}^{\infty}\cos\frac{x}{2^n}=\frac{\sin x}{x},\quad \prod_{n=1}^{\infty}\left(1-\frac{4x^2}{(2n-1)^2\pi^2}\right)=\cos x.$$

For $|q|<1$, putting $q_1\equiv\prod_{n=1}^{\infty}(1+q^{2n})$, $q_2\equiv\prod_{n=1}^{\infty}(1+q^{2n-1})$, $q_3\equiv\prod_{n=1}^{\infty}(1-q^{2n-1})$,

$q_4\equiv\prod_{n=1}^{\infty}(1-q^{2n})$ we have $q_1q_2q_3=1$.

Further, putting $q=e^{i\pi\tau}$, we have the following formulas concerning ϑ-functions (→134 Elliptic Functions):

$$\vartheta_4(0,\tau)=q_4q_3^2,\quad \vartheta_2(0,\tau)=2q^{1/4}q_4q_1^2,\quad \vartheta_3(0,\tau)=q_4q_2^2,\quad \vartheta_1'(0,\tau)=2\pi q^{1/4}q_4^3.$$

11. Fourier Analysis

(I) Fourier Series (→ 159 Fourier Series)

(1) Fourier coefficients

$$a_0=\frac{1}{a}\int_0^a f(x)\,dx,\quad a_n=\frac{2}{a}\int_0^a f(x)\cos\frac{n\pi x}{a}\,dx$$

$$b_n=\frac{2}{a}\int_0^a f(x)\sin\frac{n\pi x}{a}\,dx.$$

Fourier cosinc series

$$a_0+\sum_{n=1}^{\infty}a_n\cos\frac{n\pi x}{a}=\begin{cases}f(x) & (0<x<a),\\ f(-x) & (-a<x<0).\end{cases}$$

Fourier sine series

$$\sum_{n=1}^{\infty}b_n\sin\frac{n\pi x}{a}=\begin{cases}f(x) & (0<x<a),\\ -f(-x) & (-a<x<0).\end{cases}$$

The next table shows the Fourier coefficients of the functions $F(x)$ directly in the following manner from a given function $f(x)$ on the interval $[0,a]$. For x in $[-a,0]$ and when the cosine series $\{a_n\}$ is in question, we set $f(x)=f(-x)$, and when the sine series $\{b_n\}$ is in question we set $f(x)=-f(-x)$. Thus $f(x)$ is extended in two ways to functions on $[-a,a]$. The functions $F(x)$ are the periodic continuations of such functions. We remark that the sum of the Fourier series given by the Fourier coefficients in the right hand side has, in general, some singularities (discontinuity of the function or its higher derivatives, for example) at the points given by the integral multiples of a. We assume that μ is not an integer.

$f(x)$	a_0	$a_n \quad (n=1,2,\ldots)$	$b_n \quad (n=1,2,\ldots)$
1	1	0	$[1+(-1)^{n+1}]2a/n\pi$
x	$\frac{a}{2}$	$[1+(-1)^{n+1}]\frac{-2a}{n^2\pi^2}$	$(-1)^{n+1}\frac{2a}{n\pi}$
x^2	$\frac{a^2}{3}$	$(-1)^n\frac{4a^2}{n^2\pi^2}$	$(-1)^{n-1}\frac{2a^2}{n\pi}-[1+(-1)^{n+1}]\frac{4a^2}{n^3\pi^3}$
e^{kx}	$\frac{e^{ka}-1}{ka}$	$\frac{2ka[(-1)^n e^{ka}-1]}{k^2a^2+n^2\pi^2}$	$\frac{2n\pi[1-(-1)^n]e^{ka}}{k^2a^2+n^2\pi^2}$
$\cos\frac{\mu\pi x}{a}$	$\frac{\sin\mu\pi}{\mu\pi}$	$(-1)^n\frac{2}{\pi}\frac{\mu\sin\mu\pi}{\mu^2-n^2}$	$\frac{2}{\pi}\frac{[(-1)^n\cos\mu\pi-1]}{\mu^2-n^2}$
$\sin\frac{\mu\pi x}{a}$	$\frac{1-\cos\mu\pi}{\mu\pi}$	$\frac{2}{\pi}\frac{\mu[1-(-1)^n\cos\mu\pi]}{\mu^2-n^2}$	$(-1)^n\frac{2}{\pi}\frac{n\sin\mu\pi}{\mu^2-n^2}$
$\frac{1-\lambda^2}{1-2\lambda\cos(\pi x/a)+\lambda^2}$	1	$2\lambda^n \quad (\lvert\lambda\rvert<1)$	
$\frac{\lambda\sin(\pi x/a)}{1-2\lambda\cos(\pi x/a)+\lambda^2}$			$\lambda^n \quad (\lvert\lambda\rvert<1)$
$B_{2m}(x/2a)$	0	$(-1)^{m+1}2(2m)!/(2n\pi)^{2m}$	
$B_{2m+1}(x/a)$			$(-1)^{m+1}2(2m+1)!/(2n\pi)^{2m+1}$
$\log\sin(\pi x/2a)$	$-\log 2$	$-1/n$	
$(1/2)\cot(\pi x/2a)$			$1^{(1)}$

Note

(1)The Fourier series does not converge in the sense of Cauchy, but it is summable, for example, by the Cesàro summation of the first order.

(2) $\sum_{n=1}^{\infty}(-1)^{n-1}\frac{\cos nx}{n}=\log\left(2\cos\frac{x}{2}\right) \quad (-\pi<x<\pi), \qquad \sum_{n=1}^{\infty}\frac{\sin nx}{n}=\frac{1}{2}(\pi-x) \quad (0<x<2\pi).$

$$\sum_{n=1}^{\infty}\frac{\cos(2n-1)x}{2n-1}=\frac{1}{2}\log\left|\cot\frac{x}{2}\right| \quad (0<x<2\pi, \quad x\neq\pi),$$

$$\sum_{n=1}^{\infty}\frac{\sin(2n-1)x}{2n-1}=\begin{cases}\pi/4 & (0<x<\pi),\\ -\pi/4 & (\pi<x<2\pi).\end{cases}$$

$$\sum_{n=1}^{\infty}\frac{\cos nx}{n^2}=\frac{1}{4}(x-\pi)^2-\frac{\pi^2}{12} \quad (0\leqslant x\leqslant 2\pi),$$

$$\sum_{n=1}^{\infty}\frac{\sin nx}{n^2}=-x\log 2-\int_0^x\log\left(\sin\frac{t}{2}\right)dt \quad (0\leqslant x<2\pi).$$

$$\sum_{n=1}^{\infty}\frac{a^n}{n!}\cos nx=e^{a\cos x}\cos(a\sin x)-1, \qquad \sum_{n=1}^{\infty}\frac{a^n}{n!}\sin nx=e^{a\cos x}\sin(a\sin x).$$

$$\sum_{n=1}^{\infty}(-1)^{n-1}\frac{\cos nx}{n^2-a^2}=\frac{\pi\cos ax}{2a\sin a\pi}-\frac{1}{2a^2} \quad (-\pi\leqslant x\leqslant\pi),$$

$$\sum_{n=1}^{\infty}(-1)^{n-1}\frac{n\sin nx}{n^2-a^2}=\frac{\pi\sin ax}{2\sin a\pi} \quad (-\pi<x<\pi).$$

In the final two formulas, we assume that a is not an integer.

(II) Fourier Transforms (→ 160 Fourier Transform)

The Fourier transform $\mathscr{F}[f]$ and the inverse Fourier transform $\bar{\mathscr{F}}[g]$ for integrable functions f and g are defined as

$$\mathcal{F}[f(x)]=\mathcal{F}[f](\xi)=(2\pi)^{-n/2}\int_{\mathbf{R}^n} f(x)e^{-ix\xi}dx,$$

$$\bar{\mathcal{F}}[g(\xi)]=\bar{\mathcal{F}}[g](x)=(2\pi)^{-n/2}\int_{\mathbf{R}^n} g(\xi)e^{ix\xi}d\xi, \qquad x\xi=x_1\xi_1+x_2\xi_2+\ldots+x_n\xi_n.$$

In some textbooks the factor $(2\pi)^{-n/2}$ is deleted or the symbols i and $-i$ are switched when defining $\mathcal{F}$ and $\bar{\mathcal{F}}$. However, conversion of the formulas above to ones due to other definitions is straightforward. These transforms are also defined for some nonintegrable functions, or even more generally for †tempered distributions. The Fourier transform $\mathcal{F}$ and the inverse Fourier transform $\bar{\mathcal{F}}$ defined on the space of tempered distributions $\mathcal{S}'=\mathcal{S}'(\mathbf{R}^n)$ are linear homeomorphic mappings from $\mathcal{S}'$ to itself. Useful formulas of these transforms are given in the table below, where $\alpha=(\alpha_1,\alpha_2,\ldots,\alpha_n)$ $(\alpha_j=0,1,2,\ldots)$, $|\alpha|=\alpha_1+\alpha_2+\ldots+\alpha_n$, C is †Euler's constant, $\lambda\in\mathbf{C}$, and $\mathbf{Z}_+=\{m\in\mathbf{Z}\,|\,m\geqslant 0\}$.

Case 1. $n=1$. First we explain the meaning of the symbols appearing in the table:

$x_+=\max(x,0)$ (the positive part of x),

$x_-=\max(-x,0)$ (the negative part of x),

x_+^λ and x_-^λ are understood in the sense of finite parts (→ 125 Distributions and Hyperfunctions),

$(x+i\varepsilon)^\lambda=\exp[\lambda\,\mathrm{Log}(x+i\varepsilon)]$ ($\varepsilon\neq 0$; Log is the principal value of log)

$\quad=(x^2+\varepsilon^2)^{\lambda/2}\exp[i\lambda\,\mathrm{Arg}(x+i\varepsilon)]$ $(-\pi<\mathrm{Arg}\,z\leqslant\pi)$,

$(x\pm i0)^\lambda=\lim_{\varepsilon\downarrow 0}(x\pm i\varepsilon)^\lambda$ (limit in the sense of distributions).

Then the following formula holds:

$(x\pm i0)^\lambda=x_+^\lambda+e^{\pm i\lambda\pi}x_-^\lambda$

$\mathrm{Pf}\,x^m=x_+^m+(-1)^m x_-^m$ $(m\in Z)$ (Pf is the finite part).

In the special case $m=-1$, $\mathrm{Pf}\,x^{-1}$ coincides with Cauchy's principal value $\mathrm{p.v.}x^{-1}$.

$T\,(\in\mathcal{S}')$	$\mathcal{F}[T]\,(\in\mathcal{S}')$
$\delta(x)$	$\sqrt{2\pi}$
$P(x)$ (polynomial)	$\sqrt{2\pi}\,P(i\,d/d\xi)\delta(\xi)$
p.v. $1/x$	$\sqrt{\pi/2}\;i\,\mathrm{sgn}\,\xi$
$\mathrm{Pf}\,x^{-m}$	$\sqrt{\pi/2}\,[(-i)^m/(m-1)!]\,\xi^{m-1}\,\mathrm{sgn}\,\xi$ $(m\in\mathbf{N},\mathbf{Z})$
x_+^λ	$\dfrac{\Gamma(\lambda+1)}{\sqrt{2\pi}}[e^{-i\pi(\lambda+1)/2}\xi_+^{-\lambda-1}+e^{i\pi(\lambda+1)/2}\xi_-^{-\lambda-1}]$
	$\left[=\dfrac{\Gamma(\lambda+1)}{\sqrt{2\pi}}e^{-i\pi(\lambda+1)/2}(\xi+i0)^{-\lambda-1}\right]$ $(\lambda\notin\mathbf{Z})$
x_+^m	$(i^m/\sqrt{2\pi})[\pi\delta^{(m)}-i(-1)^m m!\,\mathrm{Pf}\,\xi^{-m-1}]$ $(m\in\mathbf{Z}_+)$
x_+^{-m}	$\dfrac{(-i)^{m-1}}{\sqrt{2\pi}(m-1)!}\left[\left(\sum_{j=1}^{m-1}\dfrac{1}{j}-C\right)\xi^{m-1}-\dfrac{i\pi}{2}\xi^{m-1}\,\mathrm{sgn}\,\xi-\xi^{m-1}\log\lvert\xi\rvert\right]$ $(m\in\mathbf{N},\mathbf{Z})$
x_-^λ	$\dfrac{\Gamma(\lambda+1)}{\sqrt{2\pi}}[e^{i\pi(\lambda+1)/2}\xi_+^{-\lambda-1}+e^{-i\pi(\lambda+1)/2}\xi_-^{-\lambda-1}]$
	$\left[=\dfrac{\Gamma(\lambda+1)}{\sqrt{2\pi}}e^{i\pi(\lambda+1)/2}(\xi-i0)^{-\lambda-1}\right]$ $(\lambda\notin\mathbf{Z})$
x_-^m	$\dfrac{(-i)^m}{\sqrt{2\pi}}[\pi\delta^{(m)}+i(-1)^m m!\,\mathrm{Pf}\,\xi^{-m-1}]$ $(m\in\mathbf{Z}_+)$
x_-^{-m}	$\dfrac{i^{m-1}}{\sqrt{2\pi}(m-1)!}\left[\left(\sum_{j=1}^{m-1}\dfrac{1}{j}-C\right)\xi^{m-1}+\dfrac{i\pi}{2}\xi^{m-1}\,\mathrm{sgn}\,\xi-\xi^{m-1}\log\lvert\xi\rvert\right]$ $(m\in\mathbf{N},\mathbf{Z})$
$(x+i0)^\lambda$	$[\sqrt{2\pi}\;e^{i\pi\lambda/2}/\Gamma(-\lambda)]\xi_+^{-\lambda-1}$ $(\lambda\notin\mathbf{Z}_+)$
$(x-i0)^\lambda$	$[\sqrt{2\pi}\;e^{-i\pi\lambda/2}/\Gamma(-\lambda)]\xi_-^{-\lambda-1}$ $(\lambda\notin\mathbf{Z}_+)$
$(x\pm i0)^m=x^m$	$\sqrt{2\pi}\;i^m\delta^{(m)}$ $(m\in\mathbf{Z}_+)$
$x^{-1}\log\lvert x\rvert$	$\sqrt{\pi/2}\;i\,\mathrm{sgn}\,\xi\cdot(C+\log\lvert\xi\rvert)$
e^{-x^2}/α	$\sqrt{a/2}\;e^{-a\xi^2/4}$ $(a>0)$
$\begin{cases}e^{-ax} & (x>0)\\ 0 & (x\leqslant 0)\end{cases}$	$\dfrac{1}{\sqrt{2\pi}}\dfrac{1}{a+i\xi}$ $(a>0)$

$T\ (\in \mathscr{S}')$	$\mathscr{F}[T]\ (\in \mathscr{S}')$
$\dfrac{e^{-a\lvert x\rvert}}{\sqrt{\lvert x\rvert}}$	$\dfrac{\sqrt{a+\sqrt{a^2+\xi^2}}}{\sqrt{a^2+\xi^2}} \quad (a>0)$
$e^{-b\sqrt{x^2+a^2}}$	$\sqrt{\dfrac{2}{\pi}}\dfrac{ab}{\sqrt{\xi^2+b^2}}K_1(a\sqrt{\xi^2+b^2}) \quad (a>0, b>0)$
$\log\dfrac{x^2+a^2}{x^2+b^2}$	$-\dfrac{\sqrt{2\pi}}{\lvert\xi\rvert}(e^{-a\lvert\xi\rvert}-e^{-b\lvert\xi\rvert}) \quad (a\geqslant 0, b\geqslant 0)$
$\arctan(x/a)$	$-\sqrt{\pi/2}\, i\,\mathrm{sgn}\, a\cdot(e^{-\lvert a\rvert\lvert\xi\rvert}/\xi) \quad (a\in\mathbf{R}, a\neq 0)$
$\dfrac{\sin ax}{\lvert x\rvert^{1-\nu}}$	$-\dfrac{i}{\sqrt{2\pi}}\Gamma(\nu)\cos\dfrac{\nu\pi}{2}\left(\dfrac{1}{\lvert\xi-a\rvert^{\nu}}-\dfrac{1}{\lvert\xi+a\rvert^{\nu}}\right) \quad (\nu\notin\mathbf{Z})$
$\dfrac{\cos ax}{\lvert x\rvert^{1-\nu}}$	$\dfrac{1}{\sqrt{2\pi}}\Gamma(\nu)\cos\dfrac{\nu\pi}{2}\left(\dfrac{1}{\lvert\xi-a\rvert^{\nu}}-\dfrac{1}{\lvert\xi+a\rvert^{\nu}}\right) \quad (\nu\notin\mathbf{Z})$
$\dfrac{\sin ax}{x}$	$\begin{cases}\sqrt{\pi/2} & (\lvert\xi\rvert<\lvert a\rvert)\\ 0 & (\lvert\xi\rvert>\lvert a\rvert)\end{cases}$
$\begin{cases}1/(a^2-x^2)^{\nu+(1/2)} & (\lvert x\rvert<a)\\ 0 & (\lvert x\rvert>a)\end{cases}$	$\dfrac{1}{\sqrt{2}}\Gamma((1/2)-\nu)\left(\dfrac{\lvert\xi\rvert}{2a}\right)^{\nu}J_{-\nu}(a\lvert\xi\rvert) \quad (\mathrm{Re}\,\nu<1/2,\ \ a>0)$
$\begin{cases}0 & (\lvert x\rvert<a)\\ 1/(x^2-a^2)^{\nu+(1/2)} & (\lvert x\rvert>a)\end{cases}$	$-\dfrac{1}{\sqrt{2}}\Gamma((1/2)-\nu)\left(\dfrac{\lvert\xi\rvert}{2a}\right)^{\nu}N_{\nu}(a\lvert\xi\rvert) \quad (-1/2<\mathrm{Re}\,\nu<1/2,\ \ a>0)$
$\dfrac{1}{(x^2+a^2)^{\nu+(1/2)}}$	$\dfrac{\sqrt{2}}{\Gamma(\nu+(1/2))}\left(\dfrac{\lvert\xi\rvert}{2a}\right)^{\nu}K_{\nu}(a\lvert\xi\rvert) \quad (\mathrm{Re}\,\nu>-1/2,\ \ a>0)$

Case 2. $n>1$. Let $r=\sqrt{x_1^2+x_2^2+\ldots+x_n^2}$ and $\rho=\sqrt{\xi_1^2+\xi_2^2+\ldots+\xi_n^2}$, where $x=(x_i)\in\mathbf{R}^n$ and $\xi=(\xi_i)\in\mathbf{R}^n$. If $f\in L_1(\mathbf{R}^n)$ depends only on r, then $\mathscr{F}[f]$ depends only on ρ and is expressed as

$$\mathscr{F}[f](\rho)=\rho^{-(n-2)/2}\int_0^{\infty} f(r)r^{n/2}J_{(n-2)/2}(r\rho)\,dr.$$

The constant C in the table stands for Euler's number.

$T\ (\in \mathscr{S}')$	$\mathscr{F}[T]\ (\in \mathscr{S}')$
$\delta(x)$	$(2\pi)^{n/2}$
$P(x)$ (polynomial)	$(2\pi)^{n/2}P(i\partial/\partial\xi)\delta(\xi)$
$\mathrm{Pf}\,r^{\lambda}$	$\dfrac{2^{(n/2)+\lambda}\Gamma((n+\lambda)/2)}{\Gamma(-\lambda/2)}\mathrm{Pf}\,\rho^{-n-\lambda} \quad (\lambda\notin 2\mathbf{Z}_+,\ \ \lambda\notin -n-2\mathbf{Z}_+)$
r^{2m}	$(2\pi)^{n/2}(-\Delta)^m\delta(\xi) \quad (m\in\mathbf{Z}_+)$
$\mathrm{Pf}\,r^{-n-2m}$	$\dfrac{(-1)^m\rho^{2m}}{2^{(n/2)+2m}\Gamma((n/2)+m)m!}\left[2\log\dfrac{2}{\rho}-C+\sum_{j=1}^{m}\dfrac{1}{j}+\dfrac{\Gamma'((n/2)+m)}{\Gamma((n/2)+m)}\right] \quad (m\in\mathbf{Z}_+)$
$(1+r^2)^{\lambda}$	$\dfrac{\rho^{-[(n/2)+\lambda]}K_{(n/2)+\lambda}(\rho)}{2^{-\lambda-1}\Gamma(-\lambda)} \quad (\lambda\notin\mathbf{Z}_+)$
$(1+r^2)^m$	$(2\pi)^{n/2}(1-\Delta)^m\delta(\xi) \quad (m\in\mathbf{Z}_+)$
$\mathrm{Pf}\,r^{\lambda}\log\lambda$	$\dfrac{2^{(n/2)+\lambda}\Gamma((n+\lambda)/2)}{\Gamma(-\lambda/2)}\mathrm{Pf}\,\rho^{-n-\lambda}\left[\log\dfrac{2}{\rho}+\dfrac{1}{2}\dfrac{\Gamma'((n+\lambda)/2)}{\Gamma((n+\lambda)/2)}+\dfrac{1}{2}\dfrac{\Gamma'(-\lambda/2)}{\Gamma(-\lambda/2)}\right]$ $(\lambda\notin 2\mathbf{Z}_+,\ \ \lambda\notin -n-2\mathbf{Z}_+)$
$r^{2m}\log r$	$(-1)^{m-1}2^{(n/2)+2m-1}m!\,\Gamma((n/2)+m)\,\mathrm{Pf}\,\rho^{-n-2m}$ $+(2\pi)^{n/2}\left[\log 2-\dfrac{1}{2}C+\dfrac{1}{2}\sum_{j=1}^{m}\dfrac{1}{j}+\dfrac{\Gamma'((n/2)+m)}{\Gamma((n/2)+m)}\right](-\Delta)^m\delta(\xi) \quad (m\in\mathbf{Z}_+)$
$\mathrm{Pf}\,r^{-n-2m}\log r$	$\dfrac{(-1)^m}{2^{(n/2)+2m}\Gamma((n/2)+m)m!}\rho^{2m}\left[\left\{\log\dfrac{2}{\rho}-\dfrac{1}{2}C+\dfrac{1}{2}\sum_{j=1}^{m}\dfrac{1}{j}+\dfrac{1}{2}\dfrac{\Gamma'((n/2)+m)}{\Gamma((n/2)+m)}\right\}^2\right.$ $\left.+\dfrac{\pi^2}{24}+\dfrac{1}{4}\sum_{j=1}^{m}\dfrac{1}{j^2}-\dfrac{1}{4}\dfrac{\Gamma''((n/2)+m)}{\Gamma((n/2)+m)}+\dfrac{\Gamma'((n/2)+m)^2}{\Gamma((n/2)+m)^2}\right] \quad (m\in\mathbf{Z}_+)$
e^{-ar}	$\dfrac{\sqrt{2^n}}{\sqrt{\pi}}\Gamma\left(\dfrac{n+1}{2}\right)\dfrac{a}{(a^2+\rho^2)^{(n+1)/2}} \quad (a>0)$

The Fourier transform mentioned above is a transformation in the family of complex-valued functions or distributions. Similar transformations in the family of real-valued functions are frequently used in applications:

Fourier cosine transform $f_c(u)=\int_0^\infty F(t)\cos ut\,dt.$

Inverse transform $\frac{2}{\pi}\int_0^\infty f_c(u)\cos ut\,du=\begin{cases}F(t) & (t>0),\\ F(-t) & (t<0).\end{cases}$

Fourier sine transform $f_s(u)=\int_0^\infty F(t)\sin ut\,dt.$

Inverse transform $\frac{2}{\pi}\int_0^\infty f_s(u)\sin ut\,du=\begin{cases}F(t) & (t>0),\\ -F(-t) & (t<0).\end{cases}$

The Fourier transform can be expressed in terms of these transforms. For example (in $\mathbf{R}^1$),

$$\mathscr{F}[f](u)=\frac{1}{\sqrt{2\pi}}\int_0^\infty [f(t)+f(-t)]\cos ut\,dt-\frac{i}{\sqrt{2\pi}}\int_0^\infty [f(t)-f(-t)]\sin ut\,dt.$$

$F(t)$	$f_c(u)$	$f_s(u)$
$\begin{cases}1 & (0<t<a)\\ 0 & (a<t)\end{cases}$	$\frac{\sin au}{u}$	$\frac{1-\cos au}{u}$
t^{-1}	(diverges)	$(\pi/2)\operatorname{sgn} u$
$t^{\alpha-1}$ $(0<\alpha<1)$	$\Gamma(\alpha)\cos(\pi\alpha/2)u^{-\alpha}$	$\Gamma(\alpha)\sin(\pi\alpha/2)u^{-\alpha}$
$1/(a^2+t^2)$	$\pi e^{-a\|u\|}/2a$	$[e^{-au}\operatorname{Ei}(au)-e^{au}\operatorname{Ei}(-au)]/a$ (2)
e^{-at}	$a/(a^2+u^2)$	$u/(a^2+u^2)$
$e^{-\lambda t^2}$ $(\operatorname{Re}\lambda>0)$	$\sqrt{\pi/4\lambda}\,e^{-u^2/4\lambda}$	$e^{-u^2/4\lambda}\varphi(u/2\sqrt{\lambda})/\sqrt{\lambda}$ (3)
$e^{-\lambda t^2}t$		$\sqrt{\pi/4\lambda}\,(u/4\lambda)e^{-u^2/4\lambda}$
$\frac{\sin at}{t}$ $(a>0)$	$\begin{cases}\pi/2 & (0<u<a)\\ 0 & (a<u)\end{cases}$	$\frac{1}{2}\log\left\|\frac{a+u}{a-u}\right\|$
$\tanh(\pi t/2)$		$\operatorname{cosech} u$
$\operatorname{sech}(\pi t/2)$	$\operatorname{sech} u$	
$J_\nu(t)$ $(\operatorname{Re}\nu>-1)$	$\begin{cases}\frac{\cos(\nu\arcsin u)}{\sqrt{1-u^2}}\\ -\frac{(u-\sqrt{u^2-1})^\nu}{\sqrt{u^2-1}}\sin\frac{\nu\pi}{2}\end{cases}$	$\frac{\sin(\nu\arcsin u)}{\sqrt{1-u^2}}$ $(0<u<1)$ $\frac{(u-\sqrt{u^2-1})^\nu}{\sqrt{u^2-1}}\cos\frac{\nu\pi}{2}$ $(1<u)$
$J_0(at)$	$\begin{cases}1/\sqrt{a^2-u^2}\\ 0\end{cases}$	0 $(0\leqslant u<a)$ $1/\sqrt{u^2-a^2}$ $(a<u)$
$N_0(t)$	$\begin{cases}0\\ -\frac{1}{\sqrt{u^2-1}}\end{cases}$	$\frac{2}{\pi}\frac{\arcsin u}{\sqrt{1-u^2}}$ $(0<u<1)$ $\frac{2}{\pi}\frac{\log(u-\sqrt{u^2-1})}{\sqrt{u^2-1}}$ $(1<u)$
$K_0(t)$	$\pi/2\sqrt{1+u^2}$	$(\operatorname{arcsinh} u)/\sqrt{1+u^2}$

Notes

(2) Ei is the exponential integral function (→ Table 19.II.3, this Appendix).

(3) We put $\varphi(x)=\int_0^x e^{t^2}dt$.

12. Laplace Transforms and Operational Calculus

(I) Laplace Transforms (→ 240 Laplace Transform)

Laplace transform $V(p)=\int_0^\infty e^{-pt}F(t)\,dt$ $(\operatorname{Re}p>0)$.

Inverse transform (Bromwich integral) $\frac{1}{2\pi i}\int_{c-i\infty}^{c+i\infty}e^{pt}V(p)\,dp=\begin{cases}F(t) & (t>0),\\ 0 & (t<0).\end{cases}$

$F(t)$	$V(p)$	$F(t)$	$V(p)$
1	$1/p$	$J_\nu(t)$ $(\operatorname{Re}\nu>-1)$	$\dfrac{\left(\sqrt{1+p^2}-p\right)^\nu}{\sqrt{1+p^2}}$
$1(t-a)=\begin{cases}0 & (0\leqslant t<a)\\ 1 & (a\leqslant t)\end{cases}$	e^{-ap}/p	$\frac{1}{t}J_\nu(at)$ $(\operatorname{Re}\nu>0)$	$\dfrac{\left(\sqrt{a^2+p^2}-p\right)^\nu}{\nu a^\nu}$
$[x/a]$ (integral part)	$1/p(e^{ap}-1)$		
$t^{\alpha-1}$ $(\operatorname{Re}\alpha>0)$	$\Gamma(\alpha)/p^\alpha$	$t^\nu J_\nu(at)$ $\left(\operatorname{Re}\nu>-\frac{1}{2}\right)$	$\dfrac{(2a)^\nu\Gamma[\nu+(1/2)]}{\sqrt{\pi}\,(p^2+a^2)^{\nu+(1/2)}}$
e^{-at}	$1/(p+a)$		
$e^{-at}t^{\alpha-1}$ $(\operatorname{Re}\alpha>0,\ a>0)$	$(p+a)^{-\alpha}\Gamma(\alpha)$	$t^{\nu/2}J_\nu(x\sqrt{t})$ $(\operatorname{Re}\nu>-1)$	$\dfrac{x^\nu}{2^\nu p^{\nu+1}}e^{-x^2/4p}$
$e^{-at}F(t)$ $(a>0)$	$V(p+a)$		
$(1-e^{-t})/t$	$\log(1+p^{-1})$	$J_0(t)$	$(1+p^2)^{-1/2}$
$(\pi t)^{-1/2}e^{-x^2/4t}$	$p^{-1/2}e^{-x\sqrt{p}}$ $(x>0)$	$J_0(x\sqrt{t})$	$e^{-x^2/4p}/p$
$\log t$	$-(\log p+C)/p$ [1]	$N_0(t)$	$\dfrac{2}{\pi}\dfrac{\log\left(\sqrt{1+p^2}-p\right)}{\sqrt{1+p^2}}$
$\sin at$	$a/(p^2+a^2)$	$L_n(t)$ [2]	$\dfrac{1}{p}\left(\dfrac{p-1}{p}\right)^n$
$\cos at$	$p/(p^2+a^2)$		
$\sin(x\sqrt{t})$	$\dfrac{\sqrt{\pi}}{2}\dfrac{x}{p^{3/2}}e^{-x^2/4p}$	$t^\alpha L_n^{(\alpha)}(t)$	$\dfrac{\Gamma(\alpha+n+1)}{n!}\dfrac{1}{p^{\alpha+1}}\left(\dfrac{p-1}{p}\right)^n$
$t^{-1/2}\cos(x\sqrt{t})$	$\sqrt{\pi/p}\;e^{-x^2/4p}$	$H_{2n+1}(\sqrt{t})$ [3]	$\sqrt{\dfrac{\pi}{2}}\,(2n+1)!!\dfrac{(1-p)^n}{p^{n+(3/2)}}$ [4]
$t^{-1}\sin xt$	$\arctan(x/p)$		
$x^{-1}(1-\cos ax)$	$\frac{1}{2}\log[1+(a^2/p^2)]$		
$\sinh at$	$a/(p^2-a^2)$	$\dfrac{H_{2n}(\sqrt{t})}{\sqrt{t}}$	$\sqrt{\pi}\,(2n-1)!!\dfrac{(1-p)^n}{p^{n+(1/2)}}$
$\cosh at$	$p/(p^2-a^2)$		

Notes

(1) C is Euler's constant.
(2) $L_n(t)$ is a Laguerre polynomial.
(3) $H_n(t)$ is a Hermite polynomial.
(4) $(2n+1)!!=(2n+1)(2n-1)(2n-3)\ldots 5\cdot 3\cdot 1$.

(II) Operational Calculus (→ 306 Operational Calculus)

Heaviside function (unit function) $\quad \mathbf{1}(t)=\begin{cases}0 & (t<0)\\ 1 & (t\geqslant 0).\end{cases}$

Dirac delta function (unit impulse function) $\quad \delta(t)=\lim_{\varepsilon\to 0}\dfrac{1}{2\varepsilon}[\mathbf{1}(t+\varepsilon)-\mathbf{1}(t-\varepsilon)].$

When an operator $\Omega(p)$ operates on $\mathbf{1}(t)$ and the result is $A(t)$ we write $\Omega(p)\mathbf{1}(t)=A(t)$.
In the following table (i) of general formulas, we assume the relations $\Omega_i(p)\mathbf{1}(t)=A_i(t)$ $(i=1,2)$.

Carson's integral $\quad \Omega(p)=p\displaystyle\int_0^\infty e^{-pt}A(t)\,dt \quad (\operatorname{Re}p>0).$

Laplace transform $\quad V(p)=\dfrac{\Omega(p)}{p}=\displaystyle\int_0^\infty e^{-pt}A(t)\,dt.$

(i) General Formulas		(ii) Examples	
$\Omega(p)$	$A(t)$	$\Omega(p)=pV(p)$	$A(t)$
$\Omega_1(p)+\Omega_2(p)$	$A_1(t)+A_2(t)$	p	$\delta(t)$
$a\Omega_1(p)$	$aA_1(t)$	$1/p^n\ (n=0,1,2,\ldots)$	$(t^n/n!)\mathbf{1}(t)$
$p\Omega_1(p)$	$A_1(0)\delta(t)+A_1'(t)$	$p/(p+a)$	$(e^{-at})\mathbf{1}(t)$
		$p^2/(p^2+a^2)$	$(\cos at)\mathbf{1}(t)$
$\frac{1}{p}\Omega_1(p)$	$\int_0^t A_1(\tau)\,d\tau$	$ap/(p^2+a^2)$	$(\sin at)\mathbf{1}(t)$
$\Omega_1(ap)$	$A_1(t/a)$		
$[p/(p+a)]\Omega_1(p+a)$	$e^{-at}A_1(t)\quad(\mathrm{Re}\,a\geqslant 0)$	$a_0+\frac{a_1}{p}+\frac{a_2}{p^2}+\ldots$	$\left(a_0+a_1\frac{t}{1!}+a_2\frac{t^2}{2!}+\ldots\right)\mathbf{1}(t)$
$\frac{1}{p}\Omega_1(p)\Omega_2(p)$	$\int_0^t A_1(\tau)A_2(t-\tau)\,d\tau$	$\sum_{k=1}^{n}\frac{B_k}{p-p_k}$	$\sum_{k=1}^{n}\frac{B_k}{p_k}(e^{p_kt}-1)\mathbf{1}(t)$
	$=\int_0^t A_1(t-\tau)A_2(\tau)\,d\tau$		$=\Omega(0)\mathbf{1}(t)+\sum_{k=1}^{n}\frac{B_k}{p_k}e^{p_kt}$

13. Conformal Mappings (→ 77 Conformal Mappings)

Original Domain	Image Domain	Mapping Function
$\|z\|<1$ (unit disk)	$\|w\|<1$	$w=\varepsilon\frac{z-z_0}{1-\bar{z}_0z}$, $\|z_0\|<1$, $\|\varepsilon\|=1$ (general form)
$\mathrm{Im}\,z>0$ (upper half-plane)	$\|w\|<1$	$w=\varepsilon\frac{z-z_0}{z-\bar{z}_0}$, $\mathrm{Im}\,z_0>0$, $\|\varepsilon\|=1$ (general form)
$\mathrm{Im}\,z>0$ (upper half-plane)	$\mathrm{Im}\,w>0$	$w=\frac{az+b}{cz+d}$, a,b,c,d are real; $ad-bc>0$ (general form)
$0<\arg z<\alpha$ (angular domain)	$\mathrm{Im}\,w>0$	$w=z^{\pi/\alpha}$
$\|z\|<1, \mathrm{Im}\,z>0$ (upper semidisk)	$\mathrm{Im}\,w>0$	$w=\left(\frac{1+z}{1-z}\right)^2$
$0<\arg z<\alpha$, $\|z\|<1$ (fan shape)	$\mathrm{Im}\,w>0$	$w=\left(\frac{1+z^{\pi/\alpha}}{1-z^{\pi/\alpha}}\right)^2$
$\alpha<\arg\frac{z-p}{z-q}<\beta$ (circular triangle)	$0<\arg w<\gamma$	$w=\left(e^{-i\alpha}\frac{z-p}{z-q}\right)^{\frac{\gamma}{\beta-\alpha}}$
$0<\mathrm{Im}\,z<\eta$ (parallel strip)	$\mathrm{Im}\,w>0$	$w=e^{\pi z/\eta}$
$\mathrm{Re}\,z<0$, $0<\mathrm{Im}\,z<\eta$ (semiparallel strip)	$\mathrm{Im}\,w>0$, $\|w\|<1$	$w=e^{\pi z/\eta}$
$y^2>4c^2(x+c^2)$, $z=x+iy$, $c>0$ (exterior of a parabola)	$\mathrm{Im}\,w>0$	$w=\sqrt{z}-ic$
$y^2<4c^2(x+c^2)$, $z=x+iy$, $c>0$ (interior of a parabola)	$\mathrm{Im}\,w>0$	$w=i\sec\frac{\pi\sqrt{z}}{2ic}$

Original Domain	Image Domain	Mapping Function
$\frac{x^2}{[c+(1/c)]^2}+\frac{y^2}{[c-(1/c)]^2}>1$, $z=x+iy,\quad c>1$ (exterior of an ellipse)	$\lvert w\rvert>c$	$w=\frac{z+\sqrt{z^2-4}}{2},\quad z=w+\frac{1}{w}$
$\frac{x^2}{\cos^2\alpha}-\frac{y^2}{\sin^2\alpha}<4$, $z=x+iy$, $0<\alpha<\pi/2$ (exterior of a hyperbola)	$\operatorname{Im} w>0$	$w=\left(e^{-i\alpha}\frac{z+\sqrt{z^2-4}}{2}\right)^{\pi(\pi-2\alpha)}$
$\frac{x^2}{\cos^2\alpha}-\frac{y^2}{\sin^2\alpha}>4$, $x>0$, $z=x+iy$, $0<\alpha<\pi/2$ (right-hand side interior of a hyperbola)	$\operatorname{Im} w>0$	$w=\frac{1}{2i}\left[\left(\frac{z+\sqrt{z^2-4}}{2}\right)^{\frac{\pi}{2\alpha}}+\left(\frac{z-\sqrt{z^2-4}}{2}\right)^{\frac{\pi}{2\alpha}}\right]$
$\lvert z\rvert<1$	Slit domain with boundary $\lvert\operatorname{Re} w\rvert\leqslant 2$, $\operatorname{Im} w=0$	$w=z+\frac{1}{z}$
$\lvert z\rvert<1$	Slit domain with boundary $\lvert w\rvert\geqslant 1/4$, $\arg w=\lambda$	$w=\frac{z}{(1+e^{-i\lambda}z)^2}$
$\lvert z\rvert<1$	Slit domain with boundary $\lvert w\rvert\geqslant 1/4^{1/p}$ $\arg w=\lambda+(2j\pi/p)$, $j=0,\ldots,p-1$	$w=\frac{z}{(1+e^{-ip\lambda}z^p)^{2/p}}$
$-\pi/2<\operatorname{Re} z<\pi/2$ (parallel strip)	Slit domain with boundary $\lvert\operatorname{Re} w\rvert\geqslant 1$, $\operatorname{Im} w=0$	$w=\sin z$
$-\pi<\operatorname{Im} z<\pi$ (parallel strip)	Slit domain with boundary $\operatorname{Re} w\leqslant -1$, $\operatorname{Im} w=\pm\pi$	$w=z+e^z$
Arbitrary circle or half plane	Interior of an n-gon	$w=c\int^z\prod_{j=1}^{n}(t-z_j)^{\alpha_j-1}dt+c'$ ($c\neq 0$, c' are constants), where the inverse image of the vertex with the inner angle $\alpha_j\pi$ ($j=1,\ldots,n$) is $z=z_j$. When $z_n=\infty$, we omit the factor $(t-z_n)^{\alpha_n-1}$ (Schwarz-Christoffel transformation)
Arbitrary circle or half-plane	Exterior of an n-gon	$w=c\int^z(t-p)^{-2}\prod_{j=1}^{n}(t-z_j)^{1-\alpha_j}dt+c'$ ($c\neq 0$, c' are constants), where the inverse image of the vertex with the inner angle $\alpha_j\pi$ ($j=1,\ldots,n$) is $z=z_j$, and the inverse image of ∞ is $z=p$
$\operatorname{Im} z>0$	Interior of an equilateral triangle	$w=\int_0^z\frac{1}{\sqrt[3]{t^2(1-t)^2}}dt$
$\operatorname{Im} z>0$	Interior of an isosceles right triangle	$w=\int_0^z\frac{1}{\sqrt[5]{t^2(1-t)^3}}dt$

Original Domain	Image Domain	Mapping Function
$\operatorname{Im} z>0$	Interior of a right triangle with one angle $\pi/6$	$w=\int_0^z \frac{1}{\sqrt[6]{t^3(1-t)^4}}\,dt$
$\lvert z\rvert<1$	Interior of a regular n-gon	$w=\int_0^z (1-t^n)^{-2/n}\,dt$
$0<\operatorname{Re} z<\omega_1$, $0<\operatorname{Im} z<\omega_3/i$ (rectangle)	$\operatorname{Im} w>0$	$w=\wp(z\lvert 2\omega_1,2\omega_3)$ ($\wp$ is the Weierstrass $\wp$-function)
$-K<\operatorname{Re} z<K$, $0<\operatorname{Im} z<K'$ (rectangle)(1)	$\operatorname{Im} w>0$	$w=\operatorname{sn}(z,k)$, $z=\int_0^w \frac{1}{\sqrt{(1-t^2)(1-k^2t^2)}}\,dt$ (sn is Jacobi's sn function)
$v<\lvert z\rvert<1$ $\operatorname{Im} z<0$ (upper half-ring domain)	$\log q<\operatorname{Re} w<0$, $0<\operatorname{Im} w<\pi$ (rectangle)	$w=\log z$
$\lvert z\rvert<1$	$\frac{u^2}{A^2}+\frac{v^2}{B^2}<1$, $w=u+v$, $A>B>0$; (interior of an ellipse)	$w=\sqrt{A^2-B^2}\,\sin\left(\frac{\pi}{2K}\int_0^{2z/k(1+z^2)}\frac{dt}{\sqrt{(1-t^2)(1-k^2t^2)}}\right)$, $\frac{1}{\pi}\log\frac{A+B}{A-B}=\frac{K'}{K}$ (1)
$\operatorname{Im} z>0$	interior of a circular polygon	$\{w;z\}\equiv\frac{w'''}{w'}-\frac{3}{2}\left(\frac{w''}{w'}\right)^2=R(z)$ ($R(z)$ is a rational function)
$\lvert z\rvert<1$	Interior of an equilateral circular triangle with inner angle π/k, $1<k\leqslant\infty$	$w=\dfrac{z\int_0^1 t^{-\frac{1}{2}-\frac{1}{2k}}(1-t)^{-\frac{1}{6}+\frac{1}{2k}}(1-z^3t)^{-\frac{5}{6}+\frac{1}{2k}}\,dt}{\int_0^1 t^{-\frac{1}{2}-\frac{1}{2k}}(1-t)^{-\frac{5}{6}+\frac{1}{2k}}(1-z^3t)^{-\frac{1}{6}+\frac{1}{2k}}\,dt}$ The vertices are the images of $z=1$, $e^{2\pi i/3}$, and $e^{4\pi i/3}$; and $\left[\frac{dw}{dz}\right]_{z=0}=\frac{\Gamma[(5/6)+(1/2k)]\Gamma(2/3)}{\Gamma[(1/6)+(1/2k)]\Gamma(4/3)}$
$\operatorname{Im} z>0$	Interior of a circular triangle with inner angles $\pi\alpha$, $\pi\beta$, $\pi\gamma$, $\alpha+\beta+\gamma<1$(2)	$w=\dfrac{\int_0^1 t^{-\frac{1+\alpha+\beta+\gamma}{2}}(1-t)^{-\frac{1+\alpha-\beta-\gamma}{2}}(1-zt)^{-\frac{1-\alpha+\beta-\gamma}{2}}\,dt}{\int_0^1 t^{-\frac{1+\alpha+\beta+\gamma}{2}}(1-t)^{-\frac{1-\alpha-\beta+\gamma}{2}}(1-t+zt)^{-\frac{1-\alpha+\beta-\gamma}{2}}\,dt}$
$\lvert\tau\rvert>1$, $-1/2<\operatorname{Re}\tau<0$	$\operatorname{Im} J>0$	$J=J(\tau)$, $\tau=\omega_3/\omega_1$, $J=g_2^3/(g_2^3-27g_3^2)$ (the absolute invariant of the elliptic modular function); $J(e^{2\pi i/3})=0$, $J(i)=1$, $J(\infty)=\infty$
$\lvert\tau+1/2\rvert<1/2$, $-1<\operatorname{Re}\tau<0$	$\operatorname{Im}\lambda<0$	$\lambda=\lambda(\tau)$, $\tau=\omega_3/\omega_1$, $\lambda=(e_2-e_3)/(e_1-e_3)$; $J(\tau)\equiv\frac{4}{27}\frac{\left[\lambda(\tau)^2-\lambda(\tau)+1\right]^2}{\lambda(\tau)^2[\lambda(\tau)-1]^2}$, $\lambda(-1)=\infty$, $\lambda(0)=1$, $\lambda(\infty)=0$

Notes

(1) K, K', k' are the usual notations in the in the theory of elliptic integrals:

$$K\equiv\int_0^1\frac{1}{\sqrt{(1-t^2)(1-k^2t^2)}}\,dt,\qquad K'=\int_0^1\frac{1}{\sqrt{(1-t^2)(1-k'^2t^2)}}\,dt,\qquad k^2+k'^2=1.$$

(2) When $\alpha+\beta+\gamma=1$, the circular triangle is mapped into the ordinary linear triangle by a suitable linear transformation, and we can apply the Schwarz-Christoffel transformation. When $\alpha+\beta+\gamma>1$, we have a similar mapping function replacing the integral representations of hypergeometric functions in the formula by the corresponding integral representations of the hypergeometric functions converging at α, β, and γ.

14. Ordinary Differential Equations

(I) Solution by Quadrature

$a,b,c,\ldots$ are integral constants.

(1) Solution of the First-Order Differential Equations (→ 313 Ordinary Differential Equations).
(i) Separated type $dy/dx=X(x)Y(y)$. The general solution is

$$\int^y \frac{dy}{Y(y)}=\int^x X(x)\,dx+c.$$

(ii) Homogeneous ordinary differential equation $dy/dx=f(y/x)$. Putting $y=ux$, we have $du/dx=[f(u)-u]x$, and the equation reduces to type (i). The general solution is

$$x=c\exp\left[\int^u\frac{du}{f(u)-u}\right]\quad\left(u=\frac{y}{x}\right).$$

(iii) Linear ordinary differential equation of the first order. $dy/dx+p(x)y+q(x)=0$. The general solution is

$$y=\left[c-\int q(x)P(x)\,dx\right]\Big/P(x),$$

where

$$P(x)\equiv\exp\left[\int p(x)\,dx\right].$$

(iv) Bernoulli's differential equation $dy/dx+p(x)y+q(x)y^\alpha=0$ $(\alpha\neq 0,1)$. Putting $z=y^{1-\alpha}$, the equation is transformed into

$$dz/dx+(1-\alpha)p(x)z+(1-\alpha)q(x)=0,$$

which reduces to (iii).
(v) Riccati's differential equation $dy/dx+ay^2=bx^m$. If $m=-2,4k/(1-2k)$ (k an integer), this is solved by quadrature. In general, it is reduced to Bessel's differential equation by $ay=u'/u$.
(vi) Generalized Riccati differential equation $dy/dx+p(x)y^2+q(x)y+r(x)=0$. If we know one, two, or three special solutions $y=y_i(x)$, the general solution is represented as follows. When $y_1(x)$ is one known special solution,

$$y=y_1(x)+P(x)\Big/\left[\int p(x)P(x)\,dx+c\right],$$

where

$$P(x)\equiv\exp\left[-\int\{q(x)+2p(x)y_1(x)\}\,dx\right].$$

When $y_1(x)$, $y_2(x)$ are the known solutions,

$$\frac{y-y_1(x)}{y-y_2(x)}=c\exp\left[\int p(x)\{y_2(x)-y_1(x)\}\,dx\right].$$

When $y_1(x)$, $y_2(x)$, $y_3(x)$ are known solutions,

$$\frac{y-y_1(x)}{y-y_2(x)}=c\frac{y_3(x)-y_1(x)}{y_3(x)-y_2(x)}.$$

(vii) Exact differential equation $P(x,y)dx+Q(x,y)dy=0$. If the left-hand side is an exact differential form, the condition is $\partial P/\partial y=\partial Q/\partial x$. The general solution is

$$\int P\,dx+\int\left(Q-\frac{\partial}{\partial y}\int P\,dx\right)dy=c.$$

(viii) Integrating factors. A function $M(x,y)$ is called an integrating factor of a differential equation $P(x,y)dx+Q(x,y)dy=0$, if $M(x,y)[P(x,y)dx+Q(x,y)dy]$ is an exact differential form $d\varphi(x,y)$. If we know an integrating factor, the general solution is given by $\varphi(x,y)=c$. If we know two independent integrating factors M and N, the general solution is given by $M/N=c$.
(ix) Clairaut's differential equation $y=xp+f(p)$ $(p=dy/dx)$. The general solution is the family of straight lines $y=cx+f(c)$, and the singular solution is the envelope of this family, which is given by eliminating p from the original equation and $x+f'(p)=0$.
(x) Lagrange's differential equation $y=x\varphi(p)+\psi(p)$ $(p\equiv dy/dx)$. Differentiation with respect to x reduces the equation to a linear differential equation $[\varphi(p)-p](dx/dp)+\varphi'(p)x+\psi'(p)=0$

with respect to x, p (see (iii)). The general solution of the original equation is given by eliminating p from the original equation and the solution of the latter linear equation. The parameter p may represent the solution. If the equation $p=\varphi(p)$ has a solution $p=p_0$, we have a solution $y=p_0x+\psi(p_0)$ (straight line). This solution is sometimes the singular solution.

(xi) Singular solutions. The singular solution of $f(x,y,p)=0$ is included in the equation resulting from eliminating p from $f=0$ and $\partial f/\partial p=0$, though the eliminant may contain various curves that are not the singular solutions.

(xii) System of differential equations.

eq. (1) $\quad dx:dy:dz=P:Q:R.$

A function $M(x,y,z)$ is called a Jacobi's last multiplier for eq. (1) if M is a solution of a partial differential equation $(\partial MP/\partial x)+(\partial MQ/\partial y)+(\partial MR/\partial z)=0$. If we know two independent last multipliers M and N, then $M/N=c$ is a solution of eq. (1). If we know a last multiplier M and a solution $f=a$ of eq. (1), we may find another solution of (1) as follows: solving $f=a$ with respect to z and inserting the solution into eq. (1), we see that $M(Qdx-Pdy)/f_z$ is an exact differential form $dG(x,y,a)$ in three variables x, y, and a. Then $G(x,y,f(x,y,z))=b$ is another solution of eq. (1).

(2) Solutions of Higher-Order Ordinary Differential Equations. The following (i)–(iv) are several examples of depression.

(i) $f(x,y^{(k)},y^{(k+1)},\ldots,y^{(n)})=0$ $(0<k\leqslant n)$. Set $y^{(k)}=z$; the equation reduces to one of the $(n-k)$th order in z.

(ii) $f(y,y',y'',\ldots,y^{(n)})=0$. This is reduced to $(n-1)$st order if we consider $y'=p$ as a variable dependent on y.

(iii) $y''=f(y)$. The general solution is given by

$$x=a\pm\int\left[2\int f(y)\,dy+b\right]^{-1/2}dy.$$

We have a similar formula for $y^{(n)}=f(y^{(n-2)})$.

(iv) Homogeneous ordinary differential equation of higher order. If the left-hand side of $F(x,y,y',\ldots,y^{(n)})=0$ satisfies the homogeneity relation $F(x,\rho y,\rho y',\ldots,\rho y^{(n)})=\rho^{\alpha}F(x,y,y',\ldots,y^{(n)})$, the equation is reduced to one of the $(n-1)$st order in u by $u=y'/y$.

If F satisfies $F(\rho x,\rho^{l}y,\rho^{l-1}y',\ldots,\rho^{l-n}y^{(n)})=\rho^{\alpha}F(x,y,y',\ldots,y^{(n)})$, then $u=y/x^{l}$, $t=\log x$ reduces the equation to one of type (ii) not containing t.

(v) Euler's linear ordinary differential equation.

$$p_n(x)x^n y^{(n)}+p_{n-1}(x)x^{n-1}y^{(n-1)}+\ldots+p_1(x)xy'+p_0(x)y=q(x)$$

is reduced to a linear equation by $t=\log x$.

(vi) Linear ordinary differential equations of higher order (exact equations). A necessary and sufficient condition that $L[y]\equiv\sum_{j=0}^{n}p_j(x)y^{(j)}=X(x)$ is an exact differential form is $\sum_{j=0}^{n}(-1)^j p_j^{(j)}=0$, and then the first integral of the equation is given by

$$\sum_{j=0}^{n-1}\sum_{k=0}^{n-j-1}(-1)^k p_{k+j+1}^{(k)}y^{(j)}=\int X(x)\,dx+c.$$

(vii) Linear ordinary differential equation of higher order (depression).

$$L[y]\equiv\sum_{j=0}^{n}p_j(x)y^{(j)}=X(x).$$

If we know mutually independent special solutions $y_1(x),\ldots,y_m(x)$ for the homogeneous linear ordinary differential equation $L[y]=0$, the equation is reduced to the $(n-m)$th linear ordinary differential equation with respect to z by a transformation $z=A(y)$, where $A(y)=0$ is the mth linear ordinary differential equation with solutions $y_1(x),\ldots,y_m(x)$. For example, if $m=1$, the equation is reduced to the $(n-1)$st linear ordinary differential equation with respect to z by the transformation

$$y(x)=y_1(x)\int z(x)\,dx.$$

Also, if $n=m=2$, the general solution is

$$y=c_1y_1+c_2y_2-y_1\int Ty_2\,dx+y_2\int Ty_1\,dx,$$

where $T(x)\equiv X(x)/[y_1(x)y_2'(x)-y_2(x)y_1'(x)]$. The denominator of the last expression is the Wronskian of y_1 and y_2.

(viii) Regular singularity. For a linear ordinary differential equation of higher order,

eq. (1) $x^n y^{(n)} + x^{n-1} p_1(x) y^{(n-1)} + \dots + p_n(x) y = 0,$

the point $x=0$ is its regular singularity if $p_1(x), \dots, p_n(x)$ are analytic at $x=0$.

We put $p_0 = 1$ and

$$\sum_{\nu=0}^{\infty} f_\nu(\rho) x^\nu \equiv \sum_{j=0}^{n} p_{n-j}(x) \rho(\rho-1)\dots(\rho-j+1).$$

If ρ is a root of the characteristic equation $f_0(\rho)=0$ and $\rho+1, \rho+2, \dots$ are not roots, we can determine the coefficients c_ν uniquely from

eq. (2) $\sum_{\nu=0}^{m} c_\nu f_{m-\nu}(\rho+\nu) = 0 \qquad (m=1,2,\dots),$

starting from a fixed value c_0 ($\neq 0$), and the series $y = x^\rho \sum_{\nu=0}^{\infty} c_\nu x^\nu$ converges and represents a solution of eq. (1). If the differences of all pairs of roots of the determining equation are not integers, we have n linearly independent solutions of eq. (1) applying the process for each characteristic root.

If there are roots whose differences are integers (including multiple roots), we denote such a system of roots by $\rho_1, \dots, \rho_l$. We arrange them in increasing order, and denote the multiplicities of the roots by $e_1, \dots, e_l$, respectively. Put $q_k = \rho_k - \rho_1$ $(k=1,2,\dots,l; 0=q_1<q_2<\dots<q_l)$. Take $N \geqslant q_l$ and a constant c ($\neq 0$). Let λ be a parameter, and starting from $c_0 = c_0(\lambda) \equiv c\prod_{k=1}^{N} f_0(\lambda+k)$, we determine $c_\nu = c_\nu(\lambda)$ uniquely by the relation (2). Putting

$m_k \equiv e_k + e_{k+1} + \dots + e_l \quad (k=1,\dots,l) \qquad$ for h in $\quad m_{k+1} \leqslant h \leqslant m_k - 1,$

the series

eq. (3) $y = \left[\frac{\partial^h}{\partial \lambda^h} x^\lambda \sum_{\nu=0}^{\infty} c_\nu(\lambda) x^\nu \right]_{\lambda=\rho_k} = x^{\rho_k} \sum_{\nu=0}^{\infty} x^\nu \left[\sum_{j=0}^{h} \binom{h}{j} c_\nu^{(j)}(\rho_k) (\log x)^{h-j} \right]$

converges and gives e_k independent solutions of eq. (1). Hence for $k=1,\dots,l$, we may have $\sum_{k=1}^{l} e_k = m_1$ mutually independent solutions of (1). Applying this process to every characteristic root, we have finally n independent solutions of (1) (Frobenius method).

In the practical computation of the solution, since it is known to have the expression (3), we often determine its coefficients successively by the method of undetermined coefficients.

(3) Solution of Linear Ordinary Differential Equations with Constant Coefficient (→ 252 Linear Ordinary Differential Equations). Let α_i, α_{jk} be constants. We consider the following linear ordinary differential equation of higher order (eq. (1)) and system of linear ordinary differential equations (eq. (2)).

eq. (1) $\sum_{i=0}^{n} \alpha_i y^{(i)} = X(x).$

eq. (2) $y_j' = \sum_{k=1}^{n} \alpha_{jk} y_k + X_j(x) \quad (j=1,\dots,n).$

(i) The general solution of the homogeneous equation (cofactor) is given by the following formulas:

for eq. (1) $y = x^j \exp \lambda_k x \quad (j=0,1,,\dots,e_k-1; k=1,\dots,m),$

for eq. (2) $y_j(x) = \sum_{k=1}^{m} p_{jk}(x) \exp \lambda_k x \quad (j=1,\dots,n),$

where $\lambda_1, \dots, \lambda_m$ are the roots of the characteristic equation of eq. (1) or eq. (2) given by

eq. (1′) $\sum_{i=0}^{n} \alpha_i \lambda^i = 0,$

eq. (2′) $\det(\alpha_{jk} - \lambda \delta_{jk}) = 0,$

respectively. We denote the multiplicities of the roots by $e_1, \dots, e_m$ $(e_1 + \dots + e_m = n)$; $p_{jk}(x)$ is a polynomial of degree at most $e_k - 1$ containing e_k arbitrary constants.

If all the coefficients in the original equation are real, and the root $\lambda_k = \mu_k + i\nu_k$ is imaginary, then $\bar{\lambda}_k = \mu_k - i\nu_k$ is also a root with the same multiplicity. Then we may replace $\exp \lambda_k x$ and $\exp \bar{\lambda}_k x$ by $\exp \mu_k x \cos \nu_k x$ and $\exp \mu_k x \sin \nu_k x$, and in this way we can represent the solution using real functions.

(ii) Inhomogenous equation. The solution of an inhomogeneous linear ordinary differential equation is given by the method of variation of parameters or by the method described in Section (2)(vii).

We explain the method of variation of parameters for eq. (2). First we use (i) to find a fundamental system of n independent solutions $y_j=\varphi_{jk}(x)$ $(k=1,\ldots,n)$ by (i)). Inserting $y_j=\sum_{k=1}^{n}c_k(x)\varphi_{jk}(x)$ into eq. (2), we have a system of linear equations in the $c_k'(x)$. Solving for the $c_k'(x)$ and integrating, we have $c_k(x)$.

Special forms of $X(x)$ or $X_j(x)$ determine the form of the solutions, and the parameters may be found by the method of undetermined coefficients. The following table shows some examples of special solutions for eq. (1). In the table, α, k, a, b, c, are constants, p_r, q_r are polynomials of degree r, and I_a is the operator defined by

$$I_a\cdot F=\frac{1}{a}\left[\sin ax\int\cos ax\cdot F(x)\,dx-\cos ax\int\sin ax\cdot F(x)\,dx\right]\quad(a\neq 0).$$

$X(x)$	Condition	Special Solution
$p_r(x)$	$\lambda=0$ is an m-tuple root of(1′)	$x^m q_r(x)$
$ke^{\alpha x}$	$\lambda=\alpha$ is an m-tuple root of (1′)	$cx^m e^{\alpha x}$
$e^{\alpha x}p_r(x)$	$\lambda=\alpha$ is an m-tuple root of (1′)	$x^m q_r(x)e^{\alpha x}$
$\left.\begin{matrix}\cos(ax+b)\\ \sin(ax+b)\end{matrix}\right\}$	$\left\{\begin{matrix}(1')\equiv\varphi(\lambda^2)+\lambda\psi(\lambda^2),\text{ and}\\ \varphi(-a^2)+a^2\psi(-a^2)\neq 0\end{matrix}\right\}$	$c_1\cos(ax+b)+c_2\sin(ax+b)$
$\left.\begin{matrix}\cos(ax+b)\\ \sin(ax+b)\end{matrix}\right\}$	$\left\{\begin{matrix}(1')=g(\lambda)/f(\lambda^2)\text{ and }f(\lambda^2)\\ \text{is divisible by }(\lambda^2+a^2)^m\\ (\text{but not by }(\lambda^2+a^2)^{m+1})\end{matrix}\right\}$	$c(I_a)^m\left\{\begin{matrix}\cos(ax+b)\\ \sin(ax+b)\end{matrix}\right.$

(II) Riemann's P-Function and Special Functions (→ 253 Linear Ordinary Differential Equations (Global Theory))

(1) Some Examples Expressed by Elementary Functions. A,B are integral constants.

$$P\begin{Bmatrix} a & b & c & \\ 0 & \mu & -\mu & x \\ 1 & \mu' & -\mu' & \end{Bmatrix}=\begin{cases} A\left(\dfrac{x-b}{x-c}\right)^{\mu}+B\left(\dfrac{x-b}{x-c}\right)^{\mu'} & (\mu\neq\mu'),\\ \left(\dfrac{x-b}{x-c}\right)^{\mu}\left[A+B\log\left(\dfrac{x-b}{x-c}\right)\right] & (\mu=\mu'). \end{cases}$$

$$P\begin{Bmatrix} a & b & c & \\ \lambda & \mu & \nu & x \\ 0 & 0 & 0 & \end{Bmatrix}=A+B\int(x-a)^{\lambda-1}(x-b)^{\mu-1}(x-c)^{\nu-1}dx\quad(\lambda+\mu+\nu=1).$$

These are for finite a, b, c. If $c=\infty$, $x-c$ should be replaced by 1.

$$P\begin{Bmatrix} \overbrace{\quad\infty\quad} & 0 & \\ \alpha\quad 0 & 0 & x \\ \alpha'\quad 0 & 1 & \end{Bmatrix}=\begin{cases} Ae^{\alpha x}+Be^{\alpha' x} & (\alpha\neq\alpha'),\\ e^{\alpha x}(Ax+B) & (\alpha=\alpha'). \end{cases}$$

$$P\begin{Bmatrix} \overbrace{\quad\infty\quad} & 0 & \\ \alpha\quad 1-\sigma & \sigma & x \\ 0\quad 0 & 0 & \end{Bmatrix}=A+B\int e^{\alpha x}x^{\sigma-1}dx\quad(\alpha\neq 0).$$

$$P\begin{Bmatrix} \overbrace{\quad\infty\quad} & 0 & \\ \alpha\quad -\sigma & \sigma & x \\ \alpha'\quad 1-\sigma' & \sigma' & \end{Bmatrix}=x^{\sigma}e^{\alpha x}\left[A+B\int e^{(\alpha'-\alpha)x}x^{\sigma'-\sigma-1}dx\right]\quad(\alpha\neq\alpha').$$

Riemann's P-function is reduced to Gauss's hypergeometric function with parameters $\alpha=\lambda+\mu+\nu$, $\beta=\lambda+\mu+\nu'$, $\gamma=1+\lambda+\lambda'$ by transforming a, b, c to 0, 1, ∞ by a suitable linear transformation and by putting $z=x^{-\lambda}(x-1)^{-\mu}y$.

(2) Representation of Special Functions by Riemann's P-function.

(i) Gauss's hypergeometric differential equation $x(1-x)y''+[\gamma-(\alpha+\beta+1)x]y'-\alpha\beta y=0$.

$$y=P\begin{Bmatrix} 0 & 1 & \infty & \\ 0 & 0 & \alpha & x \\ 1-\gamma & \gamma-\alpha-\beta & \beta & \end{Bmatrix}.$$

A special solution is $F(\alpha,\beta,\gamma;x)$ (→ 206 Hypergeometric Functions).

(ii) Confluent hypergeometric differential equation $xy''-(x-\mu)y'-\lambda y=0$.

$$y=P\left\{\begin{matrix}\overbrace{\qquad\qquad}^{\infty} & 0 & \\ 0 \quad \lambda & 0 & x \\ 1 \quad \mu-\lambda & 1-\mu & \end{matrix}\right\}.$$

A special solution is

$${}_1F_1(\lambda,\mu;x)\equiv\sum_{k=0}^{\infty}\frac{\Gamma(\lambda+k)}{\Gamma(\lambda)}\frac{\Gamma(\mu)}{\Gamma(\mu+k)}\frac{x^k}{k!}.$$

(iii) Whittaker's differential equation $y''+\left[-\frac{1}{4}+\frac{k}{x}+\frac{(1/4)-n^2}{x^2}\right]y=0$.

$$y=P\left\{\begin{matrix}\overbrace{\qquad\qquad}^{\infty} & 0 & \\ 1/2 \quad k & 1/2+n & x \\ -1/2 \quad -k & 1/2-n & \end{matrix}\right\}.$$

Special solutions are $M_{k,n}(x), W_{k,n}(x)$.

(iv) Bessel's differential equation $x^2y''+xy'+(x^2-\nu^2)y=0$.

$$y=P\left\{\begin{matrix}\overbrace{\qquad\qquad}^{\infty} & 0 & \\ i \quad 1/2 & \nu & x \\ -i \quad 1/2 & -\nu & \end{matrix}\right\}.$$

Special solutions are $J_\nu(x)$, $N_\nu(x)$ (→ 39 Bessel Functions). When $m=0,1,2,\ldots$, $J_{m-1/2}(x)=(-1)^m2^{m+1/2}\pi^{-1/2}x^{m-1/2}d^m(\cos x)/d(x^2)^m$.

(v) Hermite's different equation (parabolic cylindrical equation) $y''-2xy'+2ny=0$.

$$y=P\left\{\begin{matrix}\overbrace{\qquad\qquad}^{\infty} & 0 & \\ 0 \quad -n/2 & 0 & x^2 \\ 1 \quad (n+1)/2 & 1/2 & \end{matrix}\right\}.$$

When $n=0,1,2,\ldots$, the Hermite polynomial $H_n(x)=(-1)^n2^{-n/2}e^{x^2}d^n(e^{-x^2})/dx^n$ is the solution.

(vi) Laguerre's differential equation $xy''+(l-x+1)y'+ny=0$.

$$y=P\left\{\begin{matrix}\overbrace{\qquad\qquad}^{\infty} & 0 & \\ 0 \quad -n & 0 & x \\ 1 \quad l+n+1 & -l & \end{matrix}\right\}.$$

When $n=0,1,2,\ldots$, the Laguerre polynomial $L_n^l(x)=(1/n!)x^{-l}e^xd^n(x^{n+l}e^{-x})/dx^n$ is the solution.

(vii) Jacobi's differential equation $x(1-x)y''+[q-(p+1)x]y'+n(n+p)y=0$.

$$y=P\left\{\begin{matrix}0 & 1 & \infty & \\ 1-q & q-p & p+n & x \\ 0 & 0 & -n & \end{matrix}\right\}.$$

When $n=0,1,2,\ldots$, the Jacobi polynomial

$$G_n(p,q;x)=\frac{\Gamma(q)x^{1-q}(1-x)^{q-p}}{\Gamma(n+q)}\frac{d^n\left[x^{q+n-1}(1-x)^{p+n-q}\right]}{dx^n}$$

is the solution.

(viii) Legendre's differential equation $(1-x^2)y''-2xy'+n(n+1)y=0$.

$$y=P\left\{\begin{matrix}1 & -1 & \infty & \\ 0 & 0 & n+1 & x \\ 0 & 0 & -n & \end{matrix}\right\}.$$

When $n=0,1,2,\ldots$, the general solution is

$$\frac{d^n}{dx^n}\left[A(x^2-1)^n+B(x^2-1)^n\int\frac{dx}{(x^2-1)^{n+1}}\right].$$

The Legendre polynomial $P_n(x)=[d^n\{(x^2-1)^n\}/dx^n]/2^nn!$ is a special solution.

(3) Solution by Cylindrical Functions of Ordinary Linear Differential Equations of the Second Order. We denote cylindrical functions by $C_\nu(x)$ (→ 39 Bessel Functions).

Equation	Solution
$y''+\dfrac{1-2\alpha}{x}y'+\left[(\beta\gamma x^{\gamma-1})^2+\dfrac{\alpha^2-\nu^2\gamma^2}{x^2}\right]y=0$	$y=x^\alpha C_\nu(\beta x^\gamma)$
$y''+\left[\dfrac{1-2\alpha}{x}-2\beta\gamma i x^{\gamma-1}\right]y'+\left[\dfrac{\alpha^2-\nu^2\gamma^2}{x^2}-\beta\gamma(\gamma-2\alpha)ix^{\gamma-2}\right]y=0$	$y=x^\alpha\exp(i\beta x^\gamma)C_\nu(\beta x^\gamma)$
$y''+\left[\dfrac{1}{x}-2u(x)\right]y'+\left[1-\dfrac{\nu^2}{x^2}+u(x)^2-u'(x)-\dfrac{u(x)}{x}\right]y=0$	$y=\exp\left[\int u(x)\,dx\right]C_\nu(x)$
$y''+\alpha^2\nu^2x^{2\nu-2}y=0$	$y=\sqrt{x}\,C_{1/2\nu}(\alpha x^\nu)$
$y''+(e^{2x}-\nu^2)y=0$	$y=C_\nu(e^x)$
$x^2y''+xy'+(\beta^2x^2-\nu^2)y=0$	$y=C_\nu(\beta x)$
$x^2y''+xy'-(x^2+\nu^2)y=0$	$y=C_\nu(ix)$ (modified Bessel function)

(III) Transformation Groups and Invariants

Let $U\equiv\xi\dfrac{\partial}{\partial x}+\eta\dfrac{\partial}{\partial y}$ be the infinitesimal transformation of a given continuous transformation group of two variables, and $U'\equiv\xi\dfrac{\partial}{\partial x}+\eta\dfrac{\partial}{\partial y}+\zeta\dfrac{\partial}{\partial p}$ be that of its extended group.

We have

$$\zeta=\frac{\partial\eta}{\partial x}+p\left(\frac{\partial\eta}{\partial y}-\frac{\partial\xi}{\partial x}\right)-p^2\frac{d\xi}{\partial x}.$$

We put

$$p\equiv\frac{dy}{dx},\qquad r\equiv\frac{d^2y}{dx^2}.$$

Let α, β and γ be invariants of the 0th, first, and second order, respectively. The general form of the diffferential equation of the first or of the second order invariant under U is given by $\Phi(\alpha,\beta)=0$ (or $\beta=F(\alpha)$), and $\Psi(\alpha,\beta,\gamma)=0$ (or $\gamma=G(\alpha,\beta)$), respectively, where F, Φ, Ψ, G denote arbitrary functions of the corresponding variables.

Group With Infinitesimal Transformation U			Invariants			Note
ξ	η	ζ	0th	1st	2nd	
0	1	0	x	p	r	(1)
1	0	0	y	p	r	(1)
$-y$	x	$1+p^2$	x^2+y^2	$(y-xp)/(x+yp)$	$r/(1+p^2)^{3/2}$	(2)
0	y	p	x	p/y	r/y	(3)
x	0	$-p$	y	xp	x^2r	(3)
x	y	0	y/x	p	xr	
x	$-y$	$-2p$	xy	x^2p	x^3r	
μx	νy	$(\nu-\mu)p$	y^μ/x^ν	$x^{1-\nu/\mu}p$ or px/y	$r\mu/x^{\nu-\mu-1}$	(4)
μ	ν	0	$\nu x-\mu y$	p	r	
0	$h(x)$	$h'(x)$	x	$h(x)p-h'(x)y$	$h(x)r-h''(x)y$	
$k(y)$	0	$-k'(y)p^2$	y	$\dfrac{1}{p}-\dfrac{k'(y)}{k(y)}x$	$\dfrac{r}{p^3}+\dfrac{k''(y)}{k'(y)p}$	(5)
0	$k(y)$	$k'(y)p$	x	$\dfrac{p}{k(y)}$	$\dfrac{r}{k(y)}-\dfrac{k'(y)p^2}{[k(y)]^2}$	(6)
$h(x)$	0	$-h'(x)p$	y	$h(x)p$	$(h(x))^2r+h(x)h'(x)p$	

Group With Infinitesimal Transformation U			Invariants			Note
ξ	η	ζ	0th	1st	2nd	
0	$h(x)k(y)$	$h'(x)k(y) + h(x)k'(y)p$	x	$\frac{p}{k(y)} - \frac{h'(x)}{h(x)} \int^y \frac{dy}{k(y)}$	—	(7)
$xh(x)$	$yh(x)$	$h'(x)(y-xp)$	$\frac{y}{x}$	$\left(p-\frac{y}{x}\right)h(x)$	$\left(\frac{x^2r}{xp-y}-1\right)h(x) + h'(x)$	
y	x	$1-p^2$	x^2-y^2	$\frac{1-p}{1+p}\frac{x+y}{x-y}$ or $(x-yp)/(1+p)(x-y)$	$\frac{r}{(1-p^2)^{3/2}}$	

Notes

(1) Parallel translation.
(2) Rotation.
(3) Affine transformation.
(4) Similar transformation; the equation is a homogeneous differential equation.
(5) Linear differential equation.
(6) Separated variable type.
(7) When $k(y)=y^n$, the equation is Bernoulli's differential equation.

Reference

[1] A. R. Forsyth, A treatise on differential equations, Macmillan, fourth edition, 1914.

15. Total and Partial Differential Equations

(I) Total Differential Equations (→ 428 Total Differential Equations)

Suppose we are given a system of total differential equations

$$dz_j = \sum_{k=1}^{n} P_{jk}(x;z)\,dx_k \quad (j=1,2,\ldots,m).$$

A condition for complete integrability is given by

$$\frac{\partial P_{jk}(x;z)}{\partial x_l} + \sum_i \frac{\partial P_{jk}(x;z)}{\partial z_i} P_{il}(x;z) = \frac{\partial P_{jl}(x;z)}{\partial x_k} + \sum_i \frac{\partial P_{jl}(x;z)}{\partial z_i} P_{ik}(x;z).$$

Under this condition, the solution with the initial condition $(x_1^0,\ldots,x_n^0;z_1^0,\ldots,z_m^0)$ is obtained as follows: First, solve the system of differential equations $dz_j/dx_1 = P_{j1}(x_1,x_2^0,\ldots,x_n^0;z)$ in x_1 with the initial condition $z_j(x_1^0)=z_j^0$, and denote the solution by $z_j=\varphi_j(x_1)$. Next, considering x_1 as a parameter, solve the system of differential equations $dz_j/dx_2 = P_{j2}(x_1,x_2,x_3^0,\ldots,x_n^0;z)$ in x_2 with the initial condition $z_j(x_2^0)=\varphi_j(x_1)$, and denote the solution by $z_j=\varphi_j(x_1,x_2)$. Repeat the process, until we finally have $z_j=\varphi_j(x_1,\ldots,x_n)$, which is the solution of the original equation. Or, if we have m independent first integrals $f_j(x;z)=c_j$ of the equation $dz_j/dx_1=P_{j1}(x;z)$, we may transform the equation into $du_j=\sum_{k=1}^n Q_{jk}(x;u)\,dx_k$ by the transformation $u_j=f_j(x;z)$. Since the $Q_{jk}(x;u)$ do not involve x_1 and the equation is a completely integrable total differential equation, we have reduced the number of variables. We obtain the general solution by repeating this process n times.

For

$$P(x,y,z)\,dx + Q(x,y,z)\,dy + R(x,y,z)\,dz = 0$$

$(n=3, m=1)$, the complete integrability condition is

$$P\left(\frac{\partial Q}{\partial z}-\frac{\partial R}{\partial y}\right)+Q\left(\frac{\partial R}{\partial x}-\frac{\partial P}{\partial z}\right)+R\left(\frac{\partial P}{\partial y}-\frac{\partial Q}{\partial x}\right)=0.$$

(II) Solution of Partial Differential Equations of First Order (→ 322 Partial Differential Equations (Methods of Integration), 324 Partial Differential Equations of First Order)

Let z be a function of x and y, and

$$p\equiv\partial z/\partial x,\quad q\equiv\partial z/\partial y,\quad r\equiv\partial^2 z/\partial x^2,\quad s\equiv\partial^2 z/\partial x\partial y,\quad t\equiv\partial^2 z/\partial y^2.$$

We consider a partial differential equation of the first order $F(x,y,z,p,q)=0$.

(1) The Lagrange-Charpit Method. We consider the auxiliary equation

$$\frac{dx}{F_p}=\frac{dy}{F_q}=\frac{dz}{pF_p+qF_q}=\frac{-dp}{F_x+pF_z}=\frac{-dq}{F_y+qF_z},$$

which is a system of ordinary differential equations. Let $G(x,y,z,p,q)=a$ be the solution of the auxiliary equation. Using this together with the original equation $F=0$, we obtain $p=P(x,y,z,a)$, $q=Q(x,y,z,a)$, and the complete solution by integrating $dz=P\,dx+Q\,dy$. If we know another solution of the auxiliary equation $H(x,y,z,p,q)=b$ independent of $G=a$, we have the complete solution $z=\Phi(x,y,a,b)$ by eliminating p and q from $F=0$, $G=a$, and $H=b$.

(2) Solution of Various Standard Forms of Partial Differential Equations of the First Order. The integration constants are a, b.
(i) $f(p,q)=0$. The complete solution is $z=ax+\varphi(a)y+b$, where the function $t=\varphi(a)$ is defined by $f(t,a)=0$.
(ii) $f(px,q)=0$, $f(x,qy)=0$, $f(p/z,q/z)=0$. These equations reduce to (i) if $x=e^X$, $y=e^Y$, $z=e^Z$, respectively.
(iii) $f(x,p,q)=0$. If we can solve for $p=F(x,q)$, the complete solution is $z=\int F(x,a)\,dx+ay+b$. A similar procedure applies to $f(y,p,q)=0$.
(iv) $f(z,p,q)=0$. Solve $f(z,t,at)=0$ for $t=F(z,a)$. The complete solution is then given by $x+ay+b=\int dz/F(z,a)$. If we eliminate a and b from the complete solution $\Phi(x,y,z,a,b)=0$ and $\partial\Phi/\partial a=\partial\Phi/\partial b=0$, we have the singular solution of the original equation.
(v) Separated variable type $f(x,p)=g(y,q)$. Solve the two ordinary differential equations $f(x,p)=a$ and $g(y,q)=a$ for the solutions $p=P(x,a)$ and $q=Q(y,a)$, respectively. Then the complete solution is $z=\int P(x,a)\,dx+\int Q(y,a)\,dy+b$.
(vi) Lagrange's partial differential equation $Pp+Qq=R$. Here P, Q, R are functions of x, y, and z. Denote the solutions of the system of differential equations $dx:dy:dz=P:Q:R$ by $u(x,y,z)=a$, $v(x,y,z)=b$. Then the general solution is $\Phi(u,v)=0$, where Φ is an arbitrary function. A similar method is applicable to

$$\sum_{j=1}^{n}P_j(x_1,\ldots,x_n)\frac{\partial z}{\partial x_j}=R(x_1,\ldots,x_n).$$

If we have n independent solutions $u_j(x)=a_j$ of a system of n differential equations $dx_j/P_j=dz/R$ $(j=1,\ldots,n)$, the general solution is given by $\Phi(u_1,\ldots,u_n)=0$.
(vii) Clairaut's partial differential equation $z=px+qy+f(p,q)$. The complete solution is given by the family of planes $z=ax+by+f(a,b)$. The singular solution as the envelope of the family of planes is given by eliminating p and q from the original equation and $x=-\partial f/\partial p$ and $y=-\partial f/\partial q$.

(III) Solutions of Partial Differential Equations of Second Order (→ 322 Partial Differential Equations (Methods of Integration))

(1) Quadrature. Here φ and ψ are arbitrary functions.
(i) $r=f(x)$. The general solution is $z=\iint f(x)\,dx\,dx+\varphi(y)x+\psi(y)$. A similar rule applies to $t=f(y)$.
(ii) $s=f(x,y)$. The general solution is $z=\iint f(x,y)\,dx\,dy+\varphi(x)+\psi(y)$.
(iii) Wave equation. $r-t=0$. The general solution is $z=\varphi(x+y)+\psi(x-y)$.
(iv) Laplace's differential equation. $r+t=0$. Let $x+iy=\zeta$ and φ, ψ be complex analytic functions of ζ. The general solution is $z=\varphi(\zeta)+\psi(\bar{\zeta})$, and a real solution is $z=\varphi(\zeta)+\bar{\varphi}(\bar{\zeta})$.
(v) $r+Mp=N$, where M and N are functions of x and y. The general solution is given by $z=\int[\int L(x,y)N(x,y)\,dx+\varphi(y)]/L(x,y)\,dx+\psi(y)$, $L(x,y)=\exp\int[M(x,y)\,dx]$. In the integration, y is considered a constant.

A similar method is applicable to $s+Mp=N$, $s+Mq=N$, and $t+Mq=N$.
(vi) Monge-Ampère partial differential equation. $Rr+Ss+Tt+U(rt-s^2)=V$, where R, S, T, U, V are functions of x, y, z, p, q.

First, in the case $U=0$, we take auxiliary equations

eq. (1) $\quad R\,dy^2+T\,dx^2-S\,dx\,dy=0,$

eq. (2) $\quad R\,dp\,dy+T\,dq\,dx=V\,dx\,dy.$

Equation (1) is decomposed into two linear differential forms $X_i\,dx+Y_i\,dy=0$ $(i=1,2)$. The combination with (2) gives a solution $u_i(x,y,z,p,q)=a_i$, $v_i(x,y,z,p,q)=b_i$ $(i=1,2)$, and we have intermediate integrals $F_i(u_i,v_i)=0$ $(i=1,2)$ for an arbitrary function F_i. We have the solution of the original equation by solving the intermediate integrals. If $S^2\neq 4RT$, two intermediate integrals are distinct, and hence we can solve them in the form $p=P(x,y,z)$, $q=Q(x,y,z)$, and then we may integrate $dz=P\,dx+Q\,dy$.

Next, in the case $U\neq 0$, let λ_1 and λ_2 be the solutions of $U^2\lambda^2+US\lambda+TR+UV=0$. We have two auxiliary equations

$$\begin{cases}\lambda_1 U\,dy+T\,dx+U\,dp=0,\\ \lambda_2 U\,dx+R\,dy+U\,dq=0,\end{cases}\quad\text{or}\quad\begin{cases}\lambda_2 U\,dy+T\,dx+U\,dp=0,\\ \lambda_1 U\,dx+R\,dy+U\,dq=0,\end{cases}$$

and from the solutions $u_i=a_i$, $v_i=b_i$ $(i=1,2)$, we have intermediate integrals $F_i(u_i,v_i)=0$ $(i=1,2)$. If $4(TR+UV)\neq S^2$, $\lambda_1\neq\lambda_2$, we have two different intermediate integrals $F_i=0$. Solving the simultaneous equations $F_i=0$ in $p=P(x,y,z)$, $q=Q(x,y,z)$, we may also find the solution by integrating $dz=P\,dx+Q\,dy$.

(vii) Poisson's differential equation. $P=(rt-s^2)^nQ$, where $P=P(p,q,r,s,t)$ is homogeneous with respect to r, s, t and we assume that $Q=Q(x,y,z)$ satisfies $\partial Q/\partial z\neq\infty$ for x, y, z when $rt=s^2$. The equation $P(p,\varphi(p),r,r\varphi'(p),r\{\varphi'(p)\}^2)=0$ is then an ordinary differential equation in φ as a function of p. We first solve this for φ, and then solve a partial differential equation of the first order $q=\varphi(p)$ by the method (II)(2)(i).

(2) Intermediate Integrals. Let $f(x,y,z,p,q,r,s,t)$ be polynomials with respect to r, s, t. Suppose that $f(x,y,z,p,q,r,s,t)=0$ has the first integral $u(x,y,z,p,q)=0$. We insert

$$r=-\left(\frac{\partial u}{\partial x}+p\frac{\partial u}{\partial z}+s\frac{\partial u}{\partial q}\right)\Big/\frac{\partial u}{\partial p},\qquad t=-\left(\frac{\partial u}{\partial y}+q\frac{\partial u}{\partial z}+s\frac{\partial u}{\partial p}\right)\Big/\frac{\partial u}{\partial q}$$

into the original equation, and replace all the coefficients that are polynomials of s by 0. We thus obtain a system of differential equations in u. If u and v are two independent solutions of this system, an intermediate integral of the original equation is given in the form $\Phi(u,v)=0$.

(3) Initial Value Problem for a Hyperbolic Partial Differential Equation $L[u]\equiv u_{xy}+au_x+bu_y+cu=h$.

$$u(\xi,\eta)=[(uR)_A+(uR)_B]/2+\iint_\Delta R(x,y;\xi,\eta)h(x,y)\,dx\,dy$$
$$+\int_A^B\left[\frac{1}{2}\left(u\frac{\partial R}{\partial n'}-R\frac{\partial u}{\partial n'}\right)-\{a\cos(n,x)+b\cos(n,y)\}uR\right]ds,$$

where Δ is the hatched region in Fig. 19, and the conormal n' is the mirror image of the normal n with respect to $x=y$.

$$u(\xi,\eta)=(uR)_C+\int_C^A R(u_y+au)\,dy+\int_C^B R(u_x+bu)\,dx+\iint_\Box R(x,y;\xi,\eta)h(x,y)\,dx\,dy$$

(characteristic initial value problem).

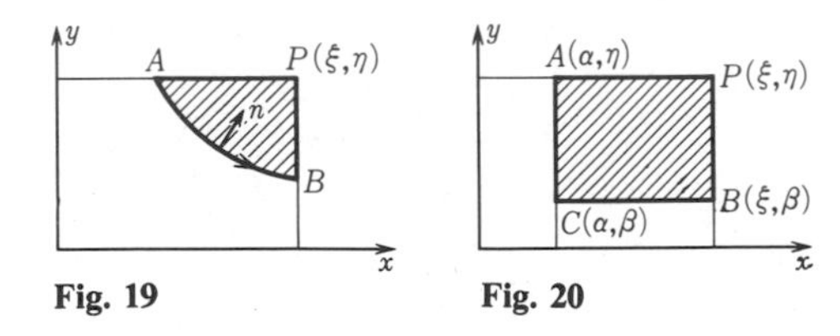

Fig. 19 Fig. 20

Here $\Box$ is the hatched rectangular region in Fig. 20. $R(x,y;\xi,\eta)$ is the Riemann function; it satisfies

$$M[R(x,y;\xi,\eta)]=0,$$

$$R_x-bR=0\ (\text{on } x=\xi),$$

$$R_y-aR=0\ (\text{on } y=\eta),$$

$$R(\xi,\eta;\xi,\eta)=1.$$

Example (i). $u_{xy}=h(x,y)$. $R(x,y;\xi,\eta)=1$.

$$u(\xi,\eta)=\frac{1}{2}[u_A+u_B]+\frac{1}{2}\int_A^B\left[u_y\cos(n,x)+u_x\cos(n,y)\right]ds+\iint_\Delta h(x,y)\,dx\,dy.$$

Example (ii). Telegraph equation $u_{xy}+cu=0$ $(c>0)$. $R(x,y;\xi,\eta)=J_0\left(2\sqrt{c(x-\xi)(y-\eta)}\right)$.
Example (iii). $u_{xy}+\dfrac{n}{x+y}(u_x+u_y)=0$ ($n=$ a constant >0).

$$R(x,y;\xi,\eta)=\left(\frac{x+y}{\xi+\eta}\right)^n F\left(1-n,n;1;-\frac{(x-\xi)(y-\eta)}{(x+y)(\xi+\eta)}\right).$$

(IV) Contact Transformations (→ 82 Contact Transformations)

We consider a transformation $(x_1,\ldots,x_n;z)\to(X_1,\ldots,X_n;Z)$. We put $p_j\equiv\partial z/\partial x_j$, $P_j\equiv\partial Z/\partial X_j$ $(j=1,\ldots,n)$. The transformation is called a contact transformation if there exists a function $\rho(x,z,p)\neq 0$ satisfying $dZ-\sum P_j dX_j=\rho(x,z,p)(dz-\sum p_j dx_j)$.
A transformation given by $(2n+1)$ equations $\Omega=0$, $\partial\Omega/\partial X_j+P_j\partial\Omega/\partial Z=0$, $\partial\Omega/\partial x_j+p_j\partial\Omega/\partial z=0$ generated by a generating function $\Omega(x,z,X,Z)$ is a contact transformation.

Generating Function	ρ	Transformation	Name
$\sum x_jX_j+z+Z$	-1	$X_j=-p_j,\quad P_j=-x_j,$ $Z=\sum p_jx_j-z$	Legendre's transformation
$\sum X_j^2+Z^2-\sum x_jX_j-zZ$	$Z/(2Z-z)$	$X_j=-p_jZ,$ $p_j=-(2X_j-x_j)/(2Z-z)$	Pedal transformation
$\sum(X_j-x_j)^2+(Z-z)^2-a^2$	1	$X_j=x_j-ap_j(1+\sum p_j^2)^{-1/2},$ $P_j=p_j,$ $Z=z_j+a(1+\sum p_j^2)^{-1/2}$	Similarity
$\sum(X_j-x_j)^2-Z^2-z^2$	$-\dfrac{1}{\sqrt{\sum p_j^2-1}}$	$X_j=x_j-p_jz,$ $P_j=-p_j(\sum p_j^2-1)^{-1/2},$ $Z=z(\sum p_j^2-1)^{1/2}$	

(V) Fundamental Solutions (→ 320 Partial Differential Equations H)

A function (or a generalized function such as a distribution) T satisfying $LT=\delta$ (δ is the Dirac delta function) for a linear differential operator L is called the fundamental (or elementary) solution of L. In the following table, we put

$$\Delta\equiv\sum_{i=1}^{n}\frac{\partial^2}{\partial x_i^2},\quad \Box\equiv\frac{\partial^2}{\partial x_n^2}-\sum_{i=1}^{n-1}\frac{\partial^2}{\partial x_i^2},\quad r^2\equiv\sum_{i=1}^{n}x_i^2,\quad \mathbf{1}(x)=\begin{cases}1 & (x>0)\\ 0 & (x\leqslant 0)\end{cases}\quad \text{(Heaviside function).}$$

J_ν is the Bessel function of the first kind; K_ν and I_ν are the modified Bessel functions. (→Table 19.IV, this Appendix.)

$$s=\begin{cases}\sqrt{x_n^2-x_1^2-\ldots-x_{n-1}^2} & \text{(if } x_n>0 \text{ and the quantity under the radical sign is positive),}\\ 0 & \text{(otherwise).}\end{cases}$$

(For Pf (finite part) → 125 Distributions and Hyperfunctions.)

Operator	Fundamental Solution
d/dx	$\mathbf{1}(x)$
$\dfrac{d^m}{dx^m}$	$\begin{cases}x^{m-1}/(m-1)! & (x>0)\\ 0 & (x\leqslant 0)\end{cases}$
$\partial^n/\partial x_1\partial x_2\ldots\partial x_n$	$\mathbf{1}(x_1)\mathbf{1}(x_2)\ldots\mathbf{1}(x_n)$
$\dfrac{\partial}{\partial x}+i\dfrac{\partial}{\partial y}$	$\dfrac{1}{2\pi}\dfrac{1}{x+iy}\quad(i\equiv\sqrt{-1})$

Operator	Fundamental Solution
Δ	$\begin{cases} -\left[\Gamma\left(\frac{n}{2}\right)/2(n-2)\pi^{n/2}\right]\frac{1}{r^{n-2}} & (n\geqslant 3) \\ (1/2\pi)\log r & (n=2) \end{cases}$
Δ^m	$\begin{cases} \left[\Gamma\left(\frac{n}{2}\right)/2^m(m-1)!\pi^{n/2}\prod_{k=1}^{m}(2k-n)\right]\frac{1}{r^{n-2m}} & \left(\begin{matrix} n-2m \text{ is a positive integer} \\ \text{or a negative odd integer}\end{matrix}\right) \\ \left[\Gamma\left(\frac{n}{2}\right)/2^m(m-1)!\pi^{n/2}\prod_{\substack{k=1\\k\neq m-h}}^{m}(2k-n)\right]\frac{\log r}{r^{n-2m}} & \left(\begin{matrix} n-2m=-2h, \\ h=0,1,2,\ldots \end{matrix}\right) \end{cases}$
$\left(1-\frac{\Delta}{4\pi^2}\right)^m$	$\frac{2\pi^m}{(m-1)!}r^{m-(n/2)}K_{(n/2)-m}(2\pi r)$
$\Box^m$	$(\mathrm{Pf}\,s^{2m-n})/\pi^{(n/2)-1}2^{2m-1}(m-1)!\Gamma[m+1-(n/2)]$
$(\Box-\lambda)^m$ (λ is real and $\neq 0$)	$\frac{\lvert\lambda\rvert^{(n/4)-(m/2)}}{\pi^{(n/2)-1}(m-1)!2^{m-1+(n/2)}}\mathrm{Pf}\,s^{m-(n/2)}\begin{cases} I_{m-(n/2)}(\sqrt{\lambda s}) & (\lambda>0) \\ J_{m-(n/2)}(\sqrt{\lvert\lambda\rvert s}) & (\lambda<0) \end{cases}$
$\left(\frac{\partial}{\partial x_n}-\sum_{i=1}^{n-1}\frac{\partial^2}{\partial x_i^2}\right)^m$	$\begin{cases} \frac{x_n^{m-1}}{(m-1)!}\left(\frac{1}{2\sqrt{\pi x_n}}\right)^{n-1}\exp\left(-\sum_{i=1}^{n-1}x_i^2/4x_n\right) & (x_n>0) \\ 0 & (x_n\leqslant 0) \end{cases}$

(VI) Solution of Boundary Value Problems (→ 188 Green's Functions, 323 Partial Differential Equations of Elliptic Type, 327 Partial Differential Equations of Parabolic Type)

$$L[u]=Au_{xx}+2Bu_{xy}+Cu_{yy}+Du_x+Eu_y+Fu,$$

$$M[v]=(Av)_{xx}+2(Bv)_{xy}+(Cv)_{yy}-(Dv)_x-(Ev)_y+Fv.$$

Green's formula $\displaystyle\int\int_D\{vL[u]-uM[v]\}\,dx\,dy=\int_C\left\{P\left(u\frac{\partial v}{\partial n'}-v\frac{\partial u}{\partial n'}\right)+Quv\right\}ds.$

eq. (1) $\begin{cases} A\cos(n,x)+B\cos(n,y)=P\cos(n',x), \\ B\cos(n,x)+C\cos(n,y)=P\cos(n',y). \end{cases}$

$$Q=(A_x+B_y-D)\cos(n,x)+(B_x+C_y-E)\cos(n,y).$$

The integration contour C is the boundary of the domain D (Fig. 21), n is the inner normal of C, and n', called the conormal, is given by (1).

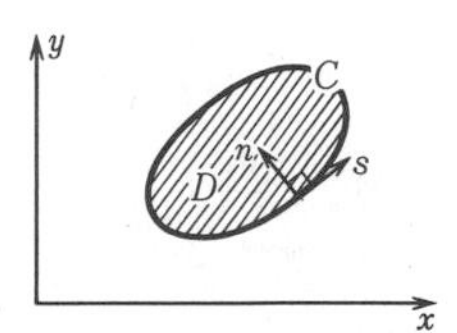

Fig. 21

(1) Elliptic Partial Differential Equation $L[u]\equiv u_{xx}+u_{yy}+au_x+bu_y+cu=h$.

$$u(\xi,\eta)=-\int_C u(s)\frac{\partial G}{\partial n}\,ds+\int\int_D G(x,y;\xi,\eta)h(x,y)\,dx\,dy.$$

Here $G(x,y;\xi,\eta)$ is Green's function, which satisfies $M(G(x,y;\xi,\eta))=0$ in the interior of D except at $(x,y)\neq(\xi,\eta)$, and

$$G(x,y;\xi,\eta)=-(1/2\pi)\log\sqrt{(x-\xi)^2+(y-\eta)^2}\ +\text{ a regular function},$$

$$G(x,y;\xi,\eta)=0\quad((x,y)\in C).$$

(2) Laplace's Differential Equation in the 2-Dimensional Case $\dfrac{\partial^2 u}{\partial x^2}+\dfrac{\partial^2 u}{\partial y^2}=0$.

$u(x,y)\equiv \tilde{u}(r,\varphi)=\operatorname{Re} f(z) \quad (z\equiv x+iy=re^{i\theta})$.

(i) Interior of a disk ($r\leqslant 1$).

$$f(z)=\frac{1}{2\pi}\int_0^{2\pi}\tilde{u}(1,\varphi)\frac{e^{i\varphi}+z}{e^{i\varphi}-z}\,d\varphi, \qquad \tilde{u}(r,\theta)=\frac{1}{2\pi}\int_0^{2\pi}\tilde{u}(1,\varphi)\frac{1-r^2}{1-2r\cos(\theta-\varphi)+r^2}\,d\varphi$$

(Poisson's integration formula).

(ii) Annulus ($0<q\leqslant r\leqslant 1$).

$$f(z)=\frac{\omega_1}{\pi^2 i}\left\{\int_0^{2\pi}\tilde{u}(1,\varphi)\zeta_1(w)\,d\varphi-\int_0^{2\pi}\tilde{u}(q,\varphi)\zeta_3(w)\,d\varphi-a\log z\right\}$$

(Villat's integration formula).

$$w=\frac{\omega_1}{\pi}(i\log z+\varphi), \quad a=\left(\frac{1}{2\omega_3}-\frac{\eta_1}{\pi i}\right)\int_0^{2\pi}\{\tilde{u}(1,\varphi)-\tilde{u}(q,\varphi)\}\,d\varphi, \quad \frac{\omega_3}{\omega_1}=-\frac{i}{\pi}\log q.$$

Here ζ_1 and ζ_3 are the Weierstrass ζ-functions (→ 134 Elliptic Functions) with the fundamental periods $2\omega_1$ and $2\omega_3$.

(iii) Half-plane ($y\geqslant 0$). $f(z)=\dfrac{i}{\pi}\displaystyle\int_{-\infty}^{\infty}\frac{u(t,0)}{z-t}\,dt, \quad u(x,y)=\frac{1}{\pi}\int_{-\infty}^{\infty}\frac{u(t,0)y}{(x-t)^2+y^2}\,dt.$

(3) Laplace's Differential Equation in the 3-Dimensional Case.

(i) Interior of a sphere ($r\leqslant 1$).

$$\tilde{u}(r,\varphi,\theta)=\frac{1}{4\pi}\int_0^{\pi}\int_0^{2\pi}\tilde{u}(1,\Phi,\Theta)\frac{1-r^2}{(1-2r\cos\gamma+r^2)^{3/2}}\sin\Theta\,d\Theta\,d\Phi,$$

where

$$\cos\gamma=\cos\Theta\cos\theta+\sin\Theta\sin\theta\cos(\Phi-\varphi).$$

(ii) Half-space ($z\geqslant 0$).

$$u(x,y,z)=\frac{z}{2\pi}\int\int_{-\infty}^{\infty}\frac{u(\xi,\eta,0)}{\{(x-\xi)^2+(y-\eta)^2+z^2\}^{3/2}}\,d\xi\,d\eta.$$

(4) Equation of Oscillation (Helmholtz Differential Equation) $\Delta u+k^2u=0$. Let u_n be the normalized eigenfunction with the same boundary condition for the eigenvalue k_n. Green's function is

$$G(P,Q)=\sum\frac{u_n(P)u_n^*(Q)}{k^2-k_n^2}$$

Domain	Boundary Condition	Eigenvalue	Eigenfunction
rectangle $0\leqslant x\leqslant a,\ 0\leqslant y\leqslant b$	$u=0$	$k_{nm}=\pi\sqrt{\dfrac{n^2}{a^2}+\dfrac{m^2}{b^2}}$ $(n,\ m=1,2,\ldots)$	$\sin n\pi\dfrac{x}{a}\sin m\pi\dfrac{y}{b}$
circle $0\leqslant r\leqslant a$	$u=0$	k_{nm} is the root of $J_m(kx)=0$	$J_m(k_{nm}r)e^{\pm im\varphi}$
annulus $b\leqslant r\leqslant a$	$u=0$	k_{nm} is the root of $J_m(ka)N_m(kb)-J_m(kb)N_m(ka)=0$	$\left\{\dfrac{J_m(k_{nm}r)}{J_m(k_{nm}a)}-\dfrac{N_m(k_{nm}r)}{N_m(k_{nm}a)}\right\}e^{\pm im\varphi}$
fan shape $0\leqslant r\leqslant a,\ 0\leqslant\varphi\leqslant\alpha$	$u=0$	k_{nm} is the root of $J_\mu(ka)=0$ $(\mu=m\pi/\alpha)$	$J_\mu(k_{nm}r)\sin\mu\varphi$
rectangular parallelepiped $0\leqslant x\leqslant a,\ 0\leqslant y\leqslant b,\ 0\leqslant z\leqslant c$	$\dfrac{\partial u}{\partial n}=0$	$k_{nml}=\pi\sqrt{\dfrac{n^2}{a^2}+\dfrac{m^2}{b^2}+\dfrac{l^2}{c^2}}$	$\cos n\pi\dfrac{x}{a}\cos m\pi\dfrac{y}{b}\cos l\pi\dfrac{z}{c}$
sphere $0\leqslant r\leqslant a$	$\dfrac{\partial u}{\partial n}=0$	k_{nl} is the root of $\psi_n'(ka)=0$, where $\psi_n(\rho)\equiv\sqrt{\pi/2}\,J_{n+(1/2)}(\rho)$	$\psi_n(k_{nl}r)P_n^m(\cos\theta)^{\pm im\varphi}$

(5) Heat Equation. $\dfrac{\partial u}{\partial t}=\kappa\Delta u\left(\Delta=\dfrac{\partial^2}{\partial x_1^2}+\ldots+\dfrac{\partial^2}{\partial x_m^2};\ \kappa \text{ is a positive constant}\right)$. Boundary condi-

tion: $hu-k\,\partial u/\partial n=\varphi$, where h and k are nonnegative constants with $h+k=1$, and φ is a given function.

$$u(P,t)=\int_V G(P,Q,t)u(Q,0)\,dV_Q+\kappa\int_0^t d\tau\int_S\left\{\frac{\partial G(P,Q,t-\tau)}{\partial n_Q}+G(P,Q,t-\tau)\right\}\varphi(Q,\tau)\,dS_Q.$$

Here V is the domain, and S is its boundary. $G(P,Q,t)$ is the elementary solution that satisfies $\partial G/\partial t=\kappa\Delta G$ in V and $k\partial G/\partial n=hG$ on S, and further in the neighborhood of $P=Q$, $t=0$, it has the form $G(P,Q,t)=(4\pi\kappa t)^{-m/2}e^{-R^2/4\kappa t}$ + terms of lower degree ($R=\overline{PQ}$).

(i) $-\infty<x<\infty$, $G=U(x-\xi,t)$, where $U(x,t)=e^{-x^2/4\kappa t}/\sqrt{4\pi\kappa t}$ (similar in the following case (ii)).

(ii) $0\leqslant x<\infty$. $\quad u(0,t)=0$: $\quad G=U(x-\xi,t)-U(x+\xi,t)$.

$$\frac{\partial u}{\partial x}=hu:\qquad G=U(x-\xi,t)+U(x+\xi,t)-2he^{h\xi}\int_{-\infty}^{-\xi}e^{h\eta}U(x-\eta,t)\,d\eta.$$

(iii) $0\leqslant x\leqslant l$.

$$u(0,t)=u(l,t)=0:\qquad G=\vartheta\left(\frac{x-\xi}{2l}\middle|\tau\right)-\vartheta\left(\frac{x+\xi}{2l}\middle|\tau\right).$$

$$\frac{\partial u}{\partial x}(0,t)=\frac{\partial u}{\partial x}(l,t)=0:\qquad G=\vartheta\left(\frac{x-\xi}{2l}\middle|\tau\right)+\vartheta\left(\frac{x+\xi}{2l}\middle|\tau\right).$$

$$u(0,t)=\frac{\partial u}{\partial x}(l,t)=0:\qquad G=\vartheta\left(\frac{x-\xi}{4l}\middle|\tau\right)-\vartheta\left(\frac{x+\xi}{4l}\middle|\tau\right)+\vartheta\left(\frac{x+\xi-2l}{4l}\middle|\tau\right)-\vartheta\left(\frac{x-\xi-2l}{4l}\middle|\tau\right).$$

Here ϑ is the elliptic theta function: $\vartheta(x|\tau)=\vartheta_3(x,\tau)\equiv 1+2\sum e^{i\pi\tau n^2}\cos 2n\pi x$.

(iv) $0\leqslant x<\infty$, $\quad 0\leqslant y<\infty$. $\quad u(x,0,t)=u(0,y,t)=0$:

$$G=\left(e^{-(x-\xi)^2/4\kappa t}-e^{-(x+\xi)^2/4\kappa t}\right)\left(e^{-(y-\eta)^2/4\kappa t}-e^{-(y+\eta)^2/4\kappa t}\right)/4\pi\kappa t.$$

(v) $0\leqslant x\leqslant a$, $\quad 0\leqslant y\leqslant b$. $\quad u=0$ on the boundary:

$$G=\frac{4}{ab}\sum_{m=1}^{\infty}\sum_{n=1}^{\infty}\exp\left\{-\kappa\pi^2t\left(\frac{m^2}{a^2}+\frac{n^2}{b^2}\right)\right\}\sin\frac{m\pi x}{a}\sin\frac{m\pi\xi}{a}\sin\frac{n\pi y}{b}\sin\frac{n\pi\eta}{b}.$$

(vi) $0\leqslant x\leqslant a$, $0\leqslant y\leqslant b$, $0\leqslant z\leqslant c$. $u=0$ on the boundary:

$$G=\frac{8}{abc}\sum_{l=1}^{\infty}\sum_{m=1}^{\infty}\sum_{n=1}^{\infty}\exp\left\{-\kappa\pi^2t\left(\frac{l^2}{a^2}+\frac{m^2}{b^2}+\frac{n^2}{c^2}\right)\right\}\times\sin\frac{l\pi x}{a}\sin\frac{l\pi\xi}{a}\sin\frac{m\pi y}{b}\sin\frac{m\pi\eta}{b}\sin\frac{n\pi z}{c}\sin\frac{n\pi\zeta}{c}.$$

(vii) $0\leqslant r<\infty$. Spherically symmetric. $\quad |x|=r$, $|\xi|=r'$:

$$G=\left(e^{-(r-r')^2/4\kappa t}-e^{-(r+r')^2/4\kappa t}\right)/8\pi rr'(\pi\kappa t)^{1/2}.$$

(viii) $0\leqslant r\leqslant a$. Spherically symmetric. $\quad u=0$ on the boundary:

$$G=\frac{1}{2\pi arr'}\sum_{n=1}^{\infty}e^{-\kappa n^2\pi^2t/a^2}\sin\frac{n\pi r}{a}\sin\frac{n\pi r'}{a}.$$

(ix) $a\leqslant r<\infty$. Spherically symmetric. $\quad k\partial u/\partial r-hu=0$ on the boundary:

$$G=\frac{1}{8\pi rr'(\pi\kappa t)^{1/2}}\left[e^{-(r-r')^2/4\kappa t}+e^{-(r+r'-2a)^2/4\kappa t}-\frac{ah+k}{ak}(4\pi\kappa t)^{1/2}\times\exp\left\{\kappa t\left(\frac{ah+k}{ak}\right)^2+(r+r'-2a)\frac{ah+k}{ak}\right\}\times\operatorname{erfc}\left\{\frac{r+r'-2a}{2\sqrt{\kappa t}}+\frac{ah+k}{ak}\sqrt{\kappa t}\right\}\right]\qquad\left(\operatorname{erfc}x\equiv\int_x^{\infty}e^{-t^2}dt\right).$$

(x) $0\leqslant r<\infty$. Axially symmetric: $\quad G=e^{-(r^2+r'^2)/4\kappa t}I_0(rr'/2\kappa t)/4\pi\kappa t$.

(xi) $0\leqslant r\leqslant a$. Axially symmetric. $\quad k\dfrac{\partial u}{\partial r}-hu=0$ on the boundary:

$$G=\frac{1}{\pi a^2}\sum_{n=1}^{\infty}\frac{J_0(r\alpha_n)J_0(r'\alpha_n)}{\{J_0(a\alpha_n)\}^2+\{J_1(a\alpha_n)\}^2}e^{-\kappa\alpha_n t/a},$$

where α_n is given by $k\alpha_nJ_1(a\alpha_n)-hJ_0(a\alpha_n)=0$.

16. Elliptic Integrals and Elliptic Functions

(I) Elliptic Integrals (→ 134 Elliptic Functions)

(1) Legendre-Jacobi Standard Form.

Elliptic integral of the first kind

$$F(k,\varphi)\equiv\int_0^{\varphi}\frac{d\psi}{\sqrt{1-k^2\sin^2\psi}}=\int_0^{\sin\varphi}\frac{dt}{\sqrt{(1-t^2)(1-k^2t^2)}}\qquad (k \text{ is the modulus}).$$

Elliptic integral of the second kind

$$E(k,\varphi)\equiv\int_0^{\varphi}\sqrt{1-k^2\sin^2\psi}\,d\psi=\int_0^{\sin\varphi}\sqrt{\frac{1-k^2t^2}{1-t^2}}\,dt.$$

Elliptic integral of the third kind

$$\Pi(\varphi,n,k)=\int_0^{\varphi}\frac{d\psi}{(1+n\sin^2\psi)\sqrt{1-k^2\sin^2\psi}}=\int_0^{\sin\varphi}\frac{dt}{(1+nt^2)\sqrt{(1-t^2)(1-k^2t^2)}}.$$

When $\varphi=\pi/2$, elliptic integrals of the first and the second kinds are called complete elliptic integrals:

$$K(k)\equiv F\left(k,\frac{\pi}{2}\right)=\int_0^{\pi/2}\frac{d\psi}{\sqrt{1-k^2\sin^2\psi}}=\int_0^1\frac{dt}{\sqrt{(1-t^2)(1-k^2t^2)}}=\frac{\pi}{2}F\left(\frac{1}{2},\frac{1}{2};1;k^2\right),$$

$$E(k)\equiv E\left(k,\frac{\pi}{2}\right)=\int_0^{\pi/2}\sqrt{1-k^2\sin^2\psi}\,d\psi=\int_0^1\sqrt{\frac{1-k^2t^2}{1-t^2}}\,dt=\frac{\pi}{2}F\left(-\frac{1}{2},\frac{1}{2};1;k^2\right),$$

where F is the hypergeometric function.

$$K(k')=K(\sqrt{1-k^2})\equiv K'(k),\qquad E(k')=E(\sqrt{1-k^2})\equiv E'(k)\qquad (k'^2=1-k^2;\ k' \text{ is the complementary modulus}).$$

$$EK'+E'K-KK'=\frac{\pi}{2}\ \text{(Legendre's relation)}.\qquad K\left(\frac{1}{\sqrt{2}}\right)=\frac{\Gamma(1/4)^2}{4\sqrt{\pi}}.$$

$$\frac{\partial F}{\partial k}=\frac{1}{k'^2}\left(\frac{E-k'^2F}{k}-\frac{\sin\varphi\cos\varphi}{\sqrt{1-k^2\sin^2\varphi}}\right),\qquad \frac{\partial E}{\partial k}=\frac{E-F}{k}.$$

(2) Change of Variables.

$$\tan(\psi-\varphi)=k'\tan\varphi:\quad F\left(\frac{1-k'}{1+k'},\psi\right)=(1+k')F(k,\varphi),$$

$$E\left(\frac{1-k'}{1+k'},\psi\right)=\frac{2}{1+k'}[E(k,\varphi)+k'F(k,\varphi)]-\frac{1-k'}{1+k'}\sin\psi.$$

$$\sin\chi=\frac{(1+k)\sin\varphi}{1+k\sin^2\varphi}:\quad F\left(\frac{2\sqrt{k}}{1-k},\chi\right)=(1+k)F(k,\varphi),$$

$$E\left(\frac{2\sqrt{k}}{1+k},\chi\right)=\frac{1}{1+k}\left[2E(k,\varphi)-k'^2F(k,\varphi)+2k\frac{\sin\varphi\cos\varphi}{1+k\sin^2\varphi}\sqrt{1-k^2\sin^2\varphi}\right].$$

k_1	$\sin\varphi_1$	$\cos\varphi_1$	$F(k_1,\varphi_1)$	$E(k_1,\varphi_1)$
$i\frac{k}{k'}$	$k'\frac{\sin\varphi}{\sqrt{1-k^2\sin^2\varphi}}$	$\frac{\cos\varphi}{\sqrt{1-k^2\sin^2\varphi}}$	$k'F(k,\varphi)$	$\frac{1}{k'}\left[E(k,\varphi)-k^2\frac{\sin\varphi\cos\varphi}{\sqrt{1-k^2\sin^2\varphi}}\right]$
k'	$-i\tan\varphi$	$\frac{1}{\cos\varphi}$	$-iF(k,\varphi)$	$i[E(k,\varphi)-F(k,\varphi)-\sqrt{1-k^2\sin^2\varphi}\tan\varphi]$
$\frac{1}{k}$	$k\sin\varphi$	$\sqrt{1-k^2\sin^2\varphi}$	$kF(k,\varphi)$	$\frac{1}{k}[E(k,\varphi)-k'^2F(k,\varphi)]$

(3) Transformation into Standard Form.
(i) The following are reducible to elliptic integrals of the first kind (we assume $a>b>0$ for parameters).

$AF(k,\varphi)$	A	k	φ
$\int_1^x \frac{dt}{\sqrt{t^3-1}}$	$\frac{1}{\sqrt[4]{3}}$	$\frac{\sqrt{3}-1}{2\sqrt{2}}$	$\arccos\frac{\sqrt{3}+1-x}{\sqrt{3}-1+x}$
$\int_1^x \frac{dt}{\sqrt{1-t^3}}$	$\frac{1}{\sqrt[4]{3}}$	$\frac{\sqrt{3}+1}{2\sqrt{2}}$	$\arccos\frac{\sqrt{3}-1+x}{\sqrt{3}+1-x}$
$\int_0^x \frac{dt}{\sqrt{1+t^4}}$	$\frac{1}{2}$	$\frac{1}{\sqrt{2}}$	$\arccos\frac{1-x^2}{1+x^2}$
$\int_0^x \frac{dt}{\sqrt{(a^2-t^2)(b^2-t^2)}}$	$\frac{1}{a}$	$\frac{b}{a}$	$\arcsin\frac{x}{b}$
$\int_b^x \frac{dt}{\sqrt{(a^2-t^2)(t^2-b^2)}}$	$\frac{1}{a}$	$\sqrt{1-(b/a)^2}$	$\arcsin\sqrt{\frac{1-(b/x)^2}{1-(b/a)^2}}$
$\int_a^x \frac{dt}{\sqrt{(t^2-a^2)(t^2-b^2)}}$	$\frac{1}{a}$	$\frac{b}{a}$	$\arcsin\sqrt{\frac{1-(a/x)^2}{1-(b/x)^2}}$
$\int_0^x \frac{dt}{\sqrt{(a^2+t^2)(b^2+t^2)}}$	$\frac{1}{a}$	$\sqrt{1-(b/a)^2}$	$\arctan\frac{x}{b}$
$\int_0^x \frac{dt}{\sqrt{(a^2-t^2)(b^2+t^2)}}$	$\frac{1}{\sqrt{a^2+b^2}}$	$\frac{a}{\sqrt{a^2+b^2}}$	$\arcsin\sqrt{\frac{1+(b/a)^2}{1+(b/x)^2}}$
$\int_b^x \frac{dt}{\sqrt{(a^2+t^2)(t^2-b^2)}}$	$\frac{1}{\sqrt{a^2+b^2}}$	$\frac{a}{\sqrt{a^2+b^2}}$	$\arccos\frac{b}{x}$

(ii) The following are reducible to elliptic integrals of the second kind (we assume $a>b>0$ for parameters).

$AE(k,\varphi)$	A	k	φ
$\int_0^x \sqrt{\frac{a^2-t^2}{b^2-t^2}}\,dt$	a	$\frac{b}{a}$	$\arcsin\frac{x}{b}$
$\int_x^a \sqrt{\frac{b^2+t^2}{a^2-t^2}}\,dt$	$\sqrt{a^2+b^2}$	$\frac{a}{\sqrt{a^2+b^2}}$	$\arccos\frac{x}{a}$
$\int_b^x \frac{1}{t^2}\sqrt{\frac{t^2+a^2}{t^2-b^2}}\,dt$	$\frac{\sqrt{a^2+b^2}}{b^2}$	$\frac{a}{\sqrt{a^2+b^2}}$	$\arccos\frac{b}{x}$
$\int_\infty^x t^2\sqrt{\frac{t^2+a^2}{t^2-b^2}}\,dt$	$\sqrt{a^2+b^2}$	$\frac{a}{\sqrt{a^2+b^2}}$	$\arcsin\sqrt{\frac{1+(b/a)^2}{1+(x/a)^2}}$
$\int_0^x \sqrt{\frac{a^2+t^2}{(b^2+t^2)^3}}\,dt$	$\frac{a}{b^2}$	$\frac{\sqrt{a^2-b^2}}{a}$	$\arctan\frac{x}{b}$
$\int_b^x \frac{dt}{t^2\sqrt{(t^2-b^2)(a^2-t^2)}}$	$\frac{1}{ab^2}$	$\frac{\sqrt{a^2-b^2}}{a}$	$\arcsin\sqrt{\frac{1-(b/x)^2}{1-(b/a)^2}}$

(II) Elliptic Theta Functions

(1) For $\operatorname{Im}\tau>0$, we put $q\equiv e^{i\pi\tau}$ and define

$$\vartheta_0(u,\tau)\equiv\vartheta_4(u,\tau)\equiv 1+2\sum_{n=1}^{\infty}(-1)^n q^{n^2}\cos 2n\pi u,$$

$$\vartheta_1(u,\tau)\equiv 2\sum_{n=0}^{\infty}(-1)^n q^{[n+(1/2)]^2}\sin(2n+1)\pi u,$$

$$\vartheta_2(u,\tau)\equiv 2\sum_{n=0}^{\infty} q^{[n+(1/2)]^2}\cos(2n+1)\pi u,$$

$$\vartheta_3(u,\tau)\equiv 1+2\sum_{n=1}^{\infty} q^{n^2}\cos 2n\pi u.$$

Each of the four functions ϑ_j $(j=0,1,2,3)$ as a function of two variables u and τ satisfies the following partial differential equation

$$\frac{\partial^2\vartheta(u,\tau)}{\partial u^2}=4\pi i\frac{\partial\vartheta(u,\tau)}{\partial\tau}.$$

(2) Mutual Relations.

$$\vartheta_0^4(u)+\vartheta_2^4(u)=\vartheta_1^4(u)+\vartheta_3^4(u),\qquad \vartheta_0^2(u)=k\vartheta_1^2(u)+k'\vartheta_3^2(u),$$

$$\vartheta_2^2(u)=-k'\vartheta_1^2(u)+k\vartheta_3^2(u),\qquad \vartheta_1^2(u)=k\vartheta_0^2(u)-k'\vartheta_2^2(u),$$

where k is the modulus such that $iK'(k)/K(k)=\tau$, and k' is the corresponding complementary modulus.

$$k=\vartheta_2^2(0)/\vartheta_3^2(0),\qquad k'=\vartheta_0^2(0)/\vartheta_3^2(0).$$

$$\vartheta_1'(0)=\pi\vartheta_2(0)\vartheta_3(0)\vartheta_0(0),\qquad \frac{\vartheta_1'''(0)}{\vartheta_1'(0)}=\frac{\vartheta_2''(0)}{\vartheta_0(0)}+\frac{\vartheta_3''(0)}{\vartheta_3(0)}+\frac{\vartheta_0''(0)}{\vartheta_0(0)}.$$

(3) Pseudoperiodicity. In the following table, the only variables in ϑ are u and τ. m and n are integers.

Increment of u	ϑ_0	ϑ_1	ϑ_2	ϑ_3	Exponential Factor
$m+n\tau$	$(-1)^n\vartheta_0$	$(-1)^{m+n}\vartheta_1$	$(-1)^m\vartheta_2$	ϑ_3	$\exp[-n\pi i \times(2u+n\tau)]$
$m-\frac{1}{2}+n\tau$	ϑ_3	$(-1)^{m+1}\vartheta_2$	$(-1)^{m+n}\vartheta_1$	$(-1)^n\vartheta_0$	
$m+\left(n+\frac{1}{2}\right)\tau$	$(-1)^n i\vartheta_1$	$(-1)^{m+n}i\vartheta_0$	$(-1)^m\vartheta_3$	ϑ_2	$\exp\left[-\left(n+\frac{1}{2}\right)\pi i \times\left\{2u+\left(n+\frac{1}{2}\right)\tau\right\}\right]$
$m-\frac{1}{2}+\left(n+\frac{1}{2}\right)\tau$	ϑ_2	ϑ_3	$(-1)^{m+n}i\vartheta_0$	$(-1)^n i\vartheta_1$	
Zeros $u=$	$m+\left(n+\frac{1}{2}\right)\tau$	$m+n\tau$	$m+\frac{1}{2}+n\tau$	$m+\frac{1}{2}+\left(n+\frac{1}{2}\right)\tau$	

(4) Expansion into Infinite Products. We put $Q_0\equiv\prod_{n=1}^{\infty}(1-q^{2n})$. Then we have

$$\vartheta_0(u)=Q_0\prod_{n=1}^{\infty}(1-2q^{2n-1}\cos 2\pi u+q^{4n-2}),$$

$$\vartheta_1(u)=2Q_0q^{1/4}\sin\pi u\prod_{n=1}^{\infty}(1-2q^{2n}\cos 2\pi u+q^{4n}),$$

$$\vartheta_2(u)=2Q_0q^{1/4}\cos\pi u\prod_{n=1}^{\infty}(1+2q^{2n}\cos 2\pi u+q^{4n}),$$

$$\vartheta_3(u)=Q_0\prod_{n=1}^{\infty}(1+2q^{2n-1}\cos 2\pi u+q^{4n-2}).$$

(III) Jacobi's Elliptic Functions

(1) We express the modulus k and the complementary modulus as follows.

$$k=\frac{\vartheta_2^2(0)}{\vartheta_3^2(0)},\qquad k'=\frac{\vartheta_0^2(0)}{\vartheta_3^2(0)},\qquad k^2+k'^2=1.$$

Then we have

$$K(k)=K=\frac{\pi}{2}\vartheta_3^2(0),\qquad K'(k)=K'=-i\tau K.$$

The relation between q and k is

$$q=e^{i\pi\tau}=e^{-\pi(K'/K)},$$

$$q^{1/4}=\left(\frac{k}{4}\right)^{1/2}\left[1+2\left(\frac{k}{4}\right)^2+15\left(\frac{k}{4}\right)^4+150\left(\frac{k}{4}\right)^6+1707\left(\frac{k}{4}\right)^8+\dots\right],$$

$$q=\frac{1}{2}L+\frac{2}{2^5}L^5+\frac{15}{2^9}L^9+\frac{150}{2^{13}}L^{13}+\frac{1707}{2^{17}}L^{17}+\dots,\quad\text{where}\quad L=\frac{1-\sqrt[4]{1-k^2}}{1+\sqrt[4]{1-k^2}}.$$

(2) Functions sn, cn, dn; Addition Theorem.

$$\operatorname{sn}(u,k)\equiv\frac{1}{\sqrt{k}}\frac{\vartheta_1(u/2K)}{\vartheta_0(u/2K)},\qquad \operatorname{cn}(u,k)\equiv\sqrt{\frac{k'}{k}}\frac{\vartheta_2(u/2K)}{\vartheta_0(u/2K)},\qquad \operatorname{dn}(u,k)\equiv\sqrt{k'}\frac{\vartheta_3(u/2K)}{\vartheta_0(u/2K)}.$$

$$\operatorname{sn}^2u+\operatorname{cn}^2u=1,\qquad \operatorname{dn}^2u+k^2\operatorname{sn}^2u=1.$$

$$\operatorname{sn}(u+v)=\frac{\operatorname{sn}u\operatorname{cn}v\operatorname{dn}v+\operatorname{sn}v\operatorname{cn}u\operatorname{dn}u}{1-k^2\operatorname{sn}^2u\operatorname{sn}^2v},\qquad \operatorname{cn}(u+v)=\frac{\operatorname{cn}u\operatorname{cn}v-\operatorname{sn}u\operatorname{dn}u\operatorname{sn}v\operatorname{dn}v}{1-k^2\operatorname{sn}^2u\operatorname{sn}^2v},$$

$$\operatorname{dn}(u+v)=\frac{\operatorname{dn}u\operatorname{dn}v-k^2\operatorname{sn}u\operatorname{cn}u\operatorname{sn}v\operatorname{cn}v}{1-k^2\operatorname{sn}^2u\operatorname{sn}^2v}.$$

$$\frac{d\operatorname{sn}u}{du}=\operatorname{cn}u\operatorname{dn}u,\qquad \frac{d\operatorname{cn}u}{du}=-\operatorname{sn}u\operatorname{dn}u,\qquad \frac{d\operatorname{dn}u}{du}=-k^2\operatorname{sn}u\operatorname{cn}u.$$

(3) Periodicity. In the next table, m and n are integers.

Increment of u	$\operatorname{sn} u$	$\operatorname{cn} u$	$\operatorname{dn} u$
$2mK+2niK'$	$(-1)^m \operatorname{sn} u$	$(-1)^{m+n} \operatorname{cn} u$	$(-1)^n \operatorname{dn} u$
$(2m-1)K+2niK'$	$(-1)^{m+1} \dfrac{\operatorname{cn} u}{\operatorname{dn} u}$	$(-1)^{m+n} k' \dfrac{\operatorname{sn} u}{\operatorname{dn} u}$	$(-1)^n k' \dfrac{1}{\operatorname{dn} u}$
$2mK+(2n+1)iK'$	$(-1)^m k^{-1} \dfrac{1}{\operatorname{sn} u}$	$(-1)^{m+n+1} i k^{-1} \dfrac{\operatorname{dn} u}{\operatorname{sn} u}$	$i(-1)^{n+1} \dfrac{\operatorname{cn} u}{\operatorname{sn} u}$
$(2m-1)K+(2n+1)iK'$	$(-1)^{m+1} k^{-1} \dfrac{\operatorname{dn} u}{\operatorname{cn} u}$	$(-1)^{m+n} i k' k^{-1} \dfrac{1}{\operatorname{cn} u}$	$(-1)^n i k' \dfrac{\operatorname{sn} u}{\operatorname{cn} u}$
Zeros $u=$	$2nK+2miK'$	$(2n+1)K+2miK'$	$(2n+1)K+(2m+1)iK'$
Poles $u=$	$2nK+(2m+1)iK'$	$2nK+(2m+1)iK'$	$2nK+(2m+1)iK'$
Fundamental periods	$4K, \quad 2iK'$	$4K, \quad 2K+2iK'$	$2K, \quad 4iK'$

(4) Change of Variables. In the next table, the second column, for example, means the relation $\operatorname{sn}(ku, 1/k) = k \operatorname{sn}(u,k)$.

u	k	sn	cn	dn
ku	$1/k$	$k \operatorname{sn}$	dn	cn
iu	k'	$i \dfrac{\operatorname{sn}}{\operatorname{cn}}$	$\dfrac{1}{\operatorname{cn}}$	$\dfrac{\operatorname{dn}}{\operatorname{cn}}$
$k'u$	$i\dfrac{k}{k'}$	$k' \dfrac{\operatorname{sn}}{\operatorname{dn}}$	$\dfrac{\operatorname{cn}}{\operatorname{dn}}$	$\dfrac{1}{\operatorname{dn}}$
iku	$i\dfrac{k'}{k}$	$ik \dfrac{\operatorname{sn}}{\operatorname{dn}}$	$\dfrac{1}{\operatorname{dn}}$	$\dfrac{\operatorname{cn}}{\operatorname{dn}}$
$ik'u$	$\dfrac{1}{k'}$	$ik' \dfrac{\operatorname{sn}}{\operatorname{cn}}$	$\dfrac{\operatorname{dn}}{\operatorname{cn}}$	$\dfrac{1}{\operatorname{cn}}$
$(1+k)u$	$\dfrac{2\sqrt{k}}{1+k}$	$\dfrac{(1+k)\operatorname{sn}}{1+k\operatorname{sn}^2}$	$\dfrac{\operatorname{cn}\operatorname{dn}}{1+k\operatorname{sn}^2}$	$\dfrac{1-k\operatorname{sn}^2}{1+k\operatorname{sn}^2}$ (Gauss's transformation)
$(1+k')u$	$\dfrac{1-k'}{1+k'}$	$(1+k')\dfrac{\operatorname{sn}\operatorname{cn}}{\operatorname{dn}}$	$\dfrac{1-(1+k')\operatorname{sn}^2}{\operatorname{dn}}$	$\dfrac{1-(1-k')\operatorname{sn}^2}{\operatorname{dn}}$ (Landen's transformation)
$\dfrac{(1+k')^2 u}{2}$	$\left(\dfrac{1-\sqrt{k'}}{1+\sqrt{k'}}\right)^2$	$\dfrac{1+\sqrt{k'}}{1-\sqrt{k'}} \dfrac{k^2 \operatorname{sn}\operatorname{cn}}{(1+\operatorname{dn})(k'+\operatorname{dn})}$	$\dfrac{\operatorname{dn}-\sqrt{k'}}{1-\sqrt{k'}} \times \sqrt{\dfrac{2(1+k')}{(1+\operatorname{dn})(k'+\operatorname{dn})}}$	$\dfrac{\sqrt{2(1+k')}}{1+\sqrt{k'}} \dfrac{\operatorname{dn}+\sqrt{k'}}{\sqrt{(1+\operatorname{dn})(k'+\operatorname{dn})}}$

Jacobi's transformation. $\operatorname{sn}(iu,k) = i\dfrac{\operatorname{sn}(u,k')}{\operatorname{cn}(u,k')}$, $\quad \operatorname{cn}(iu,k) = \dfrac{1}{\operatorname{cn}(u,k')}$,

$\operatorname{dn}(iu,k) = \dfrac{\operatorname{dn}(u,k')}{\operatorname{cn}(u,k')}$.

(5) Amplitude.
The inverse function $\varphi = \operatorname{am}(u,k)$ of $u(k,\varphi) \equiv F(k,\varphi) = \displaystyle\int_0^{\varphi} \frac{d\psi}{\sqrt{1-k^2\sin^2\psi}}$

is called the amplitude.

$$\mathrm{sn}(u,k)=\sin\varphi=\sin\mathrm{am}(u,k),\qquad \mathrm{cn}(u,k)=\cos\varphi=\cos\mathrm{am}(u,k),$$

$$\mathrm{dn}(u,k)=\sqrt{1-k^2\sin^2\varphi}=\sqrt{1-k^2\,\mathrm{sn}^2(u,k)}\,.$$

$$u(k,x)=\int_0^x\frac{dt}{\sqrt{(1-t^2)(1-k^2t^2)}},\qquad x=\mathrm{sn}(u,k).$$

$$\mathrm{am}(u,k)=\frac{\pi u}{2K}+\sum_{n=1}^{\infty}\frac{2q^n}{n(1+q^{2n})}\sin\left(n\pi\frac{u}{2K}\right)\qquad (q=e^{i\pi\tau}=e^{-\pi(K'/K)}).$$

$\mathrm{am}(\theta,1)=\mathrm{gd}\,\theta$ (Gudermann function).

(IV) Weierstrass's Elliptic Functions

(1) Weierstrass's $\wp$-function. For the fundamental periods $2\omega_1$, $2\omega_3$, we have

$$\wp(u)\equiv\frac{1}{u^2}+\sum_{n,m}{}'\left[\frac{1}{(u-2n\omega_1-2m\omega_3)^2}-\frac{1}{(2n\omega_1+2m\omega_3)^2}\right]$$

$$=\frac{1}{u^2}+\frac{g_2}{20}u^2+\frac{g_3}{28}u^4+\frac{g_2^2}{1200}u^6+\frac{3g_2g_3}{6160}u^8+\dots.$$

$$g_2\equiv 60\sum_{n,m}{}'\frac{1}{(2n\omega_1+2m\omega_3)^4},\qquad g_3\equiv 140\sum_{n,m}{}'\frac{1}{(2n\omega_1+12m\omega_3)^6},$$

where Σ' means the sum over all integers except $m=n=0$.

$\wp(-u)=\wp(u)$. Putting $\omega_2\equiv-(\omega_1+\omega_3)$, $e_j\equiv\wp(\omega_j)$ $(j=1,2,3)$ we have

$$e_1+e_2+e_3=0,\qquad e_1e_2+e_2e_3+e_3e_1=-g_2/4,\qquad e_1e_2e_3=g_3/4.$$

$$\wp'(u)\equiv d\wp/du=-2\sum_{m,n}\frac{1}{(u-2n\omega_1-2m\omega_3)^3}.$$

$$\wp'^2(u)=4[\wp(u)-e_1][\wp(u)-e_2][\wp(u)-e_3]=4\wp^3(u)-g_2\wp(u)-g_3.$$

Addition theorem

$$\wp(u+v)=-\wp(u)-\wp(v)+\frac{1}{4}\left[\frac{\wp'(u)-\wp'(v)}{\wp(u)-\wp(v)}\right]^2,$$

$$\wp(u+\omega_j)=e_j+\frac{(e_j-e_k)(e_j-e_l)}{\wp(u)-e_j}\qquad (j,k,l)=(1,2,3).$$

Using theta functions corresponding to $\tau=\omega_3/\omega_1$,

$$\wp(u)=-\frac{\eta_1}{\omega_1}-\frac{d^2\log\vartheta_1(u/2\omega_1)}{du^2}\qquad\left(\eta_1=\zeta(\omega_1)=-\frac{1}{12\omega_1}\frac{\vartheta_1'''(0)}{\vartheta_1'(0)}\right),$$

$$\wp'(u)=-\frac{1}{4\omega_1^3}\frac{\vartheta_1'^3(0)\vartheta_2(u/2\omega_1)\vartheta_3(u/2\omega_1)\vartheta_0(u/2\omega_1)}{\vartheta_2(0)\vartheta_0(0)\vartheta_1^3(u/2\omega_1)}.$$

The relations to Jacobi's elliptic functions are

$$q\equiv\exp(i\pi\omega_3/\omega_1).$$

$$\wp\left(\frac{u}{\sqrt{e_1-e_3}}\right)=e_1+(e_1-e_3)\frac{\mathrm{cn}^2u}{\mathrm{sn}^2u}=e_2+(e_1-e_3)\frac{\mathrm{dn}^2u}{\mathrm{sn}^2u}=e_3+(e_1-e_3)\frac{1}{\mathrm{sn}^2u},$$

where the modulus is $k=\sqrt{\dfrac{e_2-e_3}{e_1-e_3}}$, $\quad K(k)=\omega_1\sqrt{e_1-e_3}$.

(2) ζ-function.

$$\zeta(u)\equiv\frac{1}{u}+\sum_{n,m}{}'\left[\frac{1}{u-2n\omega_1-2m\omega_3}+\frac{u}{(2n\omega_1+2m\omega_3)^2}+\frac{1}{2n\omega_1+2m\omega_3}\right]$$

$$=\frac{1}{u}-\frac{g_2}{60}u^3-\frac{g_3}{140}u^5-\frac{g_2^2}{8400}u^7-\frac{g_2g_3}{18480}u^9-\dots$$

$$=(\zeta_1/\omega_1)u+d\log\vartheta_1(u/2\omega_1)/du.$$

$$\zeta'(u)=-\wp(u).$$

Pseudoperiodicity. Putting $\eta_j \equiv \zeta(\omega_j)$ $(j=1,2,3)$ we have

$$\zeta(u+2n\omega_1+2m\omega_3)=\zeta(u)+2n\eta_1+2m\eta_3 \qquad (n,m=0,\pm 1,\pm 2,\ldots),$$

$$\eta_1=-\frac{1}{12\omega_1}\frac{\vartheta_1'''(0)}{\vartheta_1'(0)}, \qquad \eta_1+\eta_2+\eta_3=0,$$

$$\eta_1\omega_3-\eta_3\omega_1=\eta_2\omega_1-\eta_1\omega_2=\eta_3\omega_2-\eta_2\omega_3=\pi i/2 \text{ (Legendre's relation)}.$$

Addition theorem $\zeta(u+v)=\zeta(u)+\zeta(v)+\dfrac{1}{2}\dfrac{\zeta''(u)-\zeta''(v)}{\zeta'(u)-\zeta'(v)}$.

(3) σ-function.

$$\sigma(u)\equiv u\prod_{n,m}{}'\left(1-\frac{u}{2n\omega_1+2m\omega_3}\right)\exp\left[\frac{u}{2n\omega_1+2m\omega_3}+\frac{1}{2}\left(\frac{u}{2n\omega_1+2m\omega_3}\right)^2\right] \qquad \left(\begin{array}{c} n,m=0,\pm 1,\\ \pm 2,\ldots,\\ (n,m)\neq(0,0)\end{array}\right)$$

$$=u-\frac{g_2}{2^4\cdot 3\cdot 5}u^5-\frac{g_3}{2^3\cdot 3\cdot 5\cdot 7}u^7-\frac{g_2^2}{2^9\cdot 3^2\cdot 5\cdot 7}u^9-\ldots$$

$$=2\omega_1\left(\exp\frac{\eta_1 u^2}{2\omega_1}\right)\frac{\vartheta_1(u/2\omega_1)}{\vartheta_1'(0)}.$$

$\zeta(u)=\sigma'(u)/\sigma(u)$. $\quad\sigma(-u)=-\sigma(u)$.

Pseudoperiodicity. $\sigma(u+2n\omega_1+2m\omega_3)=(-1)^{n+m+mn}[\exp(2n\eta_1+2m\eta_3)(u+n\omega_1+m\omega_3)]\sigma(u)$.

(4) Cosigma functions σ_1, σ_2, σ_3.

$$\sigma_j(u)\equiv -e^{\eta_j u}\frac{\sigma(u+\omega_j)}{\sigma(\omega_j)}=\left(\exp\frac{\eta_1 u^2}{2\omega_1}\right)\frac{\vartheta_{j+1}(u/2\omega_1)}{\vartheta_{j+1}(0)} \qquad (j=1,2,3;\ \vartheta_4\equiv\vartheta_0).$$

$$\wp(u)-e_j=\left[\frac{\sigma_j(u)}{\sigma(u)}\right]^2, \qquad \wp'(2u)=-\frac{2\sigma_1(u)\sigma_2(u)\sigma_3(u)}{\sigma^3(u)}=-\frac{\sigma(2u)}{\sigma^4(u)}.$$

$$\operatorname{sn}u=\alpha\frac{\sigma(u/\alpha)}{\sigma_3(u/\sigma)}, \qquad \operatorname{cn}u=\frac{\sigma_1(u/\alpha)}{\sigma_3(u/\alpha)}, \qquad \operatorname{dn}u=\frac{\sigma_2(u/\alpha)}{\sigma_3(u/\alpha)}, \qquad \text{where } \alpha=\sqrt{e_1-e_3}=\frac{K}{\omega_1}.$$

References

[1] W. F. Magnus, F. Oberhettinger, and R. P. Soni, Formulas and theorems for the special functions of mathematical physics, Springer, third enlarged edition, 1966.
[2] Y. L. Luke, The special functions and their approximations I, II, Academic Press, 1969.
[3] A. Erdelyi, Higher transcendental functions I, II, III (Bateman manuscript project) McGraw-Hill, 1953, 1955.
In particular, for hypergeometric functions of two variables see
[4] P. Appell, Sur les fonctions hypergéométriques de plusieurs variables, Mémor. Sci. Math., Gauthier-Villars, 1925.
Also → references to 39 Bessel Functions, 134 Elliptic Functions, 174 Gamma Function, 389 Special Functions.

17. Gamma Functions and Related Functions

(I) Gamma Functions and Beta Functions (→ 174 Gamma Function)

In this Section (I), C means Euler's constant, B_n means a Bernoulli number, ζ means the Riemann zeta function.

(1) Gamma function. $\Gamma(z)\equiv\displaystyle\int_0^\infty e^{-t}t^{z-1}\,dt \qquad (\operatorname{Re}z>0)$

$$=\frac{1}{e^{2\pi iz}-1}\int_\infty^{(0+)} e^{-t}t^{z-1}\,dt.$$

In the last integral, the integration contour goes once around the positive real axis in the positive direction.

$$\Gamma(n+1)=n! \qquad (n=0,1,2,\ldots), \qquad \Gamma(1/2)=\sqrt{\pi}.$$

$$\Gamma(z+1)=z\Gamma(z), \qquad \Gamma(z)\Gamma(1-z)=\frac{\pi}{\sin \pi z}, \qquad \prod_{j=0}^{n-1}\Gamma\left(z+\frac{j}{n}\right)=(2\pi)^{(n-1)/2}n^{(1/2)-nz}\Gamma(nz),$$

$$\frac{1}{\Gamma(z)}=ze^{Cz}\prod_{n=1}^{\infty}\left(1+\frac{z}{n}\right)e^{-z/n}, \qquad \log\Gamma(1+z)=-\frac{1}{2}\log\frac{\sin \pi z}{\pi z}-Cz-\sum_{n=1}^{\infty}\frac{\zeta(2n+1)}{2n+1}z^{2n+1}$$

$$(|z|<1).$$

$$\left|\frac{\Gamma(x+iy)}{\Gamma(x)}\right|^2=\prod_{n=0}^{\infty}1\Big/\left(1+\frac{y^2}{(x+n)^2}\right) \qquad (x,y \text{ are real and } x>0).$$

Asymptotic expansion (Stirling formula).

$$\Gamma(z)\approx e^{-z}z^{z-1}\sqrt{2\pi z}\,\exp\left[\sum_{n=1}^{\infty}\frac{(-1)^{n-1}B_{2n}z^{1-2n}}{2n(2n-1)}\right] \qquad (|\arg z|<\pi)$$

$$=e^{-z}z^{z-(1/2)}\sqrt{2\pi}\left[1+\frac{1}{12z}+\frac{1}{288z^2}-\frac{139}{51840z^3}-\frac{571}{2488320z^4}+O(z^{-5})\right].$$

(2) Beta Function. $B(x,y)\equiv\int_0^1 t^{x-1}(1-t)^{y-1}dt \qquad (\operatorname{Re}x, \quad \operatorname{Re}y>0)$

$$=\Gamma(x)\Gamma(y)/\Gamma(x+y).$$

(3) Incomplete Gamma Function.

$$\gamma(\nu,x)\equiv\int_0^x t^{\nu-1}e^{-t}dt=\Gamma(\nu)-x^{(\nu-1)/2}e^{-x/2}W_{(\nu-1)/2,\nu/2}(x) \qquad (\operatorname{Re}\nu>0).$$

(4) Incomplete Beta Function. $B_\alpha(x,y)\equiv\int_0^\alpha t^{x-1}(1-t)^{y-1}dt \qquad (0<\alpha\leqslant 1).$

(5) Polygamma Functions. $\psi(z)\equiv\dfrac{d}{dz}\log\Gamma(z)$

$$=\frac{\Gamma'(z)}{\Gamma(z)}=\int_0^\infty\left[\frac{e^{-t}}{t}-\frac{e^{-zt}}{1-e^{-t}}\right]dt=-C+\sum_{n=0}^{\infty}\left(\frac{1}{n+1}-\frac{1}{z+n}\right).$$

$$\psi'(z)=\sum_{n=0}^{\infty}\frac{1}{(z+n)^2}, \qquad \psi^{(k)}(z)=\sum_{n=0}^{\infty}\frac{(-1)^{k+1}k!}{(z+n)^{k+1}} \qquad (k=1,2,\ldots).$$

(II) Combinatorial Problems (→ 330 Permutations and Combinations)

Factorial $\quad n!=n(n-1)(n-2)\ldots3\cdot2\cdot1. \qquad 0!=1.$

Binomial coefficient $\quad \binom{\alpha}{r}=\dfrac{\alpha(\alpha-1)\ldots(\alpha-r+1)}{r!}.$

(1) Number of Permutations of n Elements Taken r at a Time.

$${}_nP_r=n(n-1)\ldots(n-r+1)=n!/(n-r)!.$$

Number of combinations of n elements taken r at a time

$${}_nC_r=\frac{{}_nP_r}{r!}=\frac{n!}{r!(n-r)!}=\binom{n}{r}.$$

$${}_nC_r={}_nC_{n-r}, \qquad {}_nC_r={}_{n-1}C_r+{}_{n-1}C_{r-1}.$$

Number of multiple permutations $\quad {}_n\Pi_r=n^r.$

Number of multiple combinations $\quad {}_nH_r={}_{n+r-1}C_r=\dfrac{(n+r-1)!}{r!(n-1)!}.$

Number of circular permutations $\quad {}_nP_r/r.$

(2) Binomial Theorem. $(a+b)^n = \sum_{r=0}^{n} \binom{n}{r} a^{n-r} b^r$.

Multinomial theorem $(a_1 + \ldots + a_m)^n = \sum \frac{n!}{p_1! \ldots p_m!} a_1^{p_1} \ldots a_m^{p_m}$.

The latter summation runs over all nonnegative integers satisfying $p_1 + \ldots + p_m = n$.

References

See references to Table 16, this Appendix.

18. Hypergeometric Functions and Spherical Functions

(I) Hypergeometric Function (→ 206 Hypergeometric Functions)

(1) Hypergeometric Function. $F(a,b;c;z) \equiv \sum_{n=0}^{\infty} \frac{\Gamma(a+n)}{\Gamma(a)} \frac{\Gamma(b+n)}{\Gamma(b)} \frac{\Gamma(c)}{\Gamma(c+n)} \frac{z^n}{n!}$.

The fundamental system of solutions of the hypergeometric differential equation $z(1-z)\frac{d^2u}{dz^2} + [c-(a+b+1)z]\frac{du}{dz} - abu = 0$ at $z=0$ is given by

$$u_1 = F(a,b;c;z), \qquad u_2 = z^{1-c}F(a-c+1,\ b-c+1;\ 2-c;\ z) \qquad (c \neq 0, -1, -2, \ldots).$$

$$F(a,b;c;z) = F(b,a;c;z). \qquad dF/dz = (ab/c)F(a+1,\ b+1;\ c+1;\ z).$$

$$F(a,b;c;1) = \frac{\Gamma(c)\Gamma(c-a-b)}{\Gamma(c-a)\Gamma(c-b)} \qquad (\mathrm{Re}(a+b-c)<0).$$

$$F(a,b;c;z) = \frac{\Gamma(c)}{\Gamma(b)\Gamma(c-b)} \int_0^1 t^{b-1}(1-t)^{c-b-1}(1-tz)^{-a}\,dt \qquad (\mathrm{Re}\,c > \mathrm{Re}\,b > 0, \quad |z|<1),$$

$$F(a,b;c;z) = \frac{1}{2\pi i} \frac{\Gamma(c)}{\Gamma(a)\Gamma(b)} \int_{-i\infty}^{i\infty} \frac{\Gamma(a+s)\Gamma(b+s)\Gamma(-s)}{\Gamma(c+s)} (-z)^s\,ds.$$

(2) Transformations of the Hypergeometric Function.

$$F(a,b;\ c;\ z) = (1-z)^{-a}F\left(a, c-b;\ c;\ \frac{z}{z-1}\right)$$

$$= (1-z)^{c-a-b}F(c-a, c-b;\ c;\ z)$$

$$= (1-z)^{-a}\frac{\Gamma(c)\Gamma(b-a)}{\Gamma(b)\Gamma(c-a)} F\left(a, c-b;\ a-b+1;\ \frac{1}{1-z}\right)$$

$$+ (1-z)^{-b}\frac{\Gamma(c)\Gamma(a-b)}{\Gamma(a)\Gamma(c-b)} F\left(b, c-a;\ b-a+1;\ \frac{1}{1-z}\right)$$

$$= \frac{\Gamma(c)\Gamma(c-a-b)}{\Gamma(c-a)\Gamma(c-b)} F(a,b;\ a+b-c+1;\ 1-z)$$

$$+ (1-z)^{c-a-b}\frac{\Gamma(c)\Gamma(a+b-c)}{\Gamma(a)\Gamma(b)} F(c-a, c-b;\ c-a-b+1;\ 1-z)$$

$$= \frac{\Gamma(c)\Gamma(b-a)}{\Gamma(b)\Gamma(c-a)}(-z)^{-a}F\left(a, 1-c+a;\ 1-b+a;\ \frac{1}{z}\right)$$

$$+ \frac{\Gamma(c)\Gamma(a-b)}{\Gamma(a)\Gamma(c-b)}(-z)^{-b}F\left(b, 1-c+b;\ 1-a+b;\ \frac{1}{z}\right).$$

(3) Riemann's Differential Equation (→ Table 14.II, this Appendix).

$$\frac{d^2u}{dz^2}+\left[\frac{1-\alpha-\alpha'}{z-a}+\frac{1-\beta-\beta'}{z-b}+\frac{1-\gamma-\gamma'}{z-c}\right]\frac{du}{dz}$$

$$+\left[\frac{\alpha\alpha'(a-b)(a-c)}{z-a}+\frac{\beta\beta'(b-c)(b-a)}{z-b}+\frac{\gamma\gamma'(c-a)(c-b)}{z-c}\right]\frac{u}{(z-a)(z-b)(z-c)}=0.$$

Here we have $\alpha+\alpha'+\beta+\beta'+\gamma+\gamma'=1$ (Fuchsian relation). The solution of this equation is given by Riemann's P-function

$$u=P\begin{Bmatrix} a & b & c & \\ \alpha & \beta & \gamma & z \\ \alpha' & \beta' & \gamma' & \end{Bmatrix}$$

$$=\left(\frac{z-a}{z-b}\right)^{\alpha}\left(\frac{z-c}{z-b}\right)^{\gamma}F\left(\alpha+\beta+\gamma,\alpha+\beta'+\gamma;\ 1+\alpha-\alpha';\ \frac{(c-b)(z-a)}{(c-a)(z-b)}\right)$$

$(\alpha-\alpha',\beta-\beta',\gamma-\gamma'\neq\text{integer})$.

We have 24 representations of the above function by interchanging the parameters a, b, c; α, α'; β, β'; γ, γ' in the right-hand side.

(4) Barnes's Extended Hypergeometric Function.

$${}_pF_q(\alpha_1,\ldots,\alpha_p;\beta_1,\ldots,\beta_q;z)\equiv\sum_{n=0}^{\infty}\frac{(\alpha_1)_n\cdots(\alpha_p)_n}{(\beta_1)_n\cdots(\beta_q)_n}\frac{z^n}{n!},$$ where $(\alpha)_n=\alpha(\alpha+1)\ldots(\alpha+n-1)$ $=\Gamma(\alpha+n)/\Gamma(\alpha)$. $F(a,b;c;z)={}_2F_1(a,b;c;z)$. ${}_0F_0(x)=e^x$, ${}_1F_0(\alpha;x)=(1-x)^{-\alpha}$.

(5) Appell's Hypergeometric Functions of Two Variables.

$$F_1(\alpha;\beta,\beta';\gamma;x,y)=\sum_{m=0}^{\infty}\sum_{n=0}^{\infty}\frac{(\alpha)_{m+n}(\beta)_m(\beta')_n}{m!n!(\gamma)_{m+n}}x^m y^n,$$

$$F_2(\alpha;\beta,\beta';\gamma,\gamma';x,y)=\sum_{m=0}^{\infty}\sum_{n=0}^{\infty}\frac{(\alpha)_{m+n}(\beta)_m(\beta')_n}{m!n!(\gamma)_m(\gamma')_n}x^m y^n,$$

$$F_3(\alpha,\alpha';\beta,\beta';\gamma;x,y)=\sum_{m=0}^{\infty}\sum_{n=0}^{\infty}\frac{(\alpha)_m(\alpha')_n(\beta)_m(\beta')_n}{m!n!(\gamma)_{m+n}}x^m y^n,$$

$$F_4(\alpha;\beta;\gamma,\gamma';x,y)=\sum_{m=0}^{\infty}\sum_{n=0}^{\infty}\frac{(\alpha)_{m+n}(\beta)_{m+n}}{m!n!(\gamma)_m(\gamma')_n}x^m y^n.$$

(6) Representation of Various Special Functions by Hypergeometric Functions.

$$(1-x)^{\nu}=F(-\nu,b;b;x),\qquad e^{-nx}=\left(\frac{\operatorname{sech}x}{2}\right)^n(\tanh x)F\left(1+\frac{n}{2},\frac{1+n}{2};\ 1+n;\ \operatorname{sech}^2x\right).$$

$$\log(1+x)=xF(1,1;\ 2;\ -x),\qquad \frac{1}{2}\log\frac{1+x}{1-x}=xF\left(\frac{1}{2},1;\ \frac{3}{2};\ x^2\right),$$

$$\sin nx=n(\sin x)F\left(\frac{1+n}{2},\frac{1-n}{2};\frac{3}{2};\sin^2x\right),$$

$$\cos nx=F\left(\frac{n}{2},-\frac{n}{2};\ \frac{1}{2};\ \sin^2x\right)=(\cos x)F\left(\frac{1+n}{2},\frac{1-n}{2};\ \frac{1}{2};\ \sin^2x\right),$$

$$\arcsin x=xF\left(\frac{1}{2},\frac{1}{2};\ \frac{3}{2};\ x^2\right),\qquad \arctan x=xF\left(\frac{1}{2},1;\ \frac{3}{2};\ -x^2\right).$$

$$P_{2n}(x)=(-1)^n\frac{(2n-1)!!}{(2n)!!}F\left(-n,n+\frac{1}{2};\ \frac{1}{2};\ x^2\right),$$

$$P_{2n+1}(x)=(-1)^n\frac{(2n+1)!!}{(2n)!!}xF\left(-n,n+\frac{3}{2};\ \frac{3}{2};\ x^2\right)\qquad\text{(spherical function)},$$

where

$$n=0,1,2,\ldots; \qquad m!!=\begin{cases} m(m-2)\ldots 4\cdot 2 & (m \text{ even}),\\ m(m-2)\ldots 3\cdot 1 & (m \text{ odd}),\end{cases} \qquad 0!!=(-1)!!=1.$$

$$K(x)=\frac{\pi}{2}F\left(\frac{1}{2},\frac{1}{2};1;x^2\right), \qquad E(x)=\frac{\pi}{2}F\left(-\frac{1}{2},\frac{1}{2};1;x^2\right) \quad \text{(complete elliptic integral).}$$

$$J_\nu(x)=\frac{x}{2\Gamma(\nu+1)}{}_0F_1\left(\nu+1;-\frac{x^2}{4}\right)=\frac{x^\nu e^{-ix}}{2^\nu\Gamma(\nu+1)}{}_1F_1\left(\nu+\frac{1}{2};2\nu+1;2ix\right).$$

$$e^x=\lim_{b\to\infty}F(a,b;a;x/b)={}_1F_1(a;a;x)={}_0F_0(x).$$

(II) Legendre Function (→ 393 Spherical Functions)

(1) Legendre Functions. The generalized spherical function corresponding to the rotation group of 3-dimensional space is the solution of the following differential equation.

$$(1-z^2)\frac{d^2u}{dz^2}-2z\frac{du}{dz}+\left[\nu(\nu+1)-\frac{\mu^2}{1-z^2}\right]u=0.$$

When $\mu=0$, the equation is Legendre's differential equation, and the fundamental system of solutions is given by the following two kind of functions.

Legendre function of the first kind $\mathfrak{P}_\nu(z)\equiv P_\nu(z)\equiv {}_2F_1\left(-\nu,\ \nu+1;\ 1;\ \frac{1-z}{2}\right)$.

Legendre function of the second kind

$$\mathfrak{Q}_\nu(z)\equiv\frac{\Gamma(\nu+1)\sqrt{\pi}}{2^{\nu+1}\Gamma[\nu+(3/2)]}z^{-\nu-1}{}_2F_1\left(\frac{\nu+2}{2},\ \frac{\nu+1}{2};\ \nu+\frac{3}{2};\ \frac{1}{z^2}\right).$$

$$Q_\nu(x)\equiv\frac{1}{2}[\mathfrak{Q}_\nu(x+i0)+\mathfrak{Q}_\nu(x-i0)]$$

$$=\pi\frac{(\cos\nu\pi)P_\nu(x)-P_\nu(-x)}{2\sin\nu\pi} \qquad (\nu\neq\text{integer};\ -1<x<1).$$

Recurrence formulas:

$$\mathfrak{P}_\nu(z)=\mathfrak{P}_{-\nu-1}(z). \qquad \mathfrak{Q}_\nu(z)-\mathfrak{Q}_{-\nu-1}(z)=\pi(\cot\nu\pi)\mathfrak{P}_\nu(z) \qquad (\nu\neq\text{integer}).$$

$$\mathfrak{P}_\nu(-z)=e^{\pm\nu\pi i}\mathfrak{P}_\nu(z)-(2/\pi)(\sin\nu\pi)\mathfrak{Q}_\nu(z), \qquad \mathfrak{Q}_\nu(-z)=-e^{\pm\nu\pi i}\mathfrak{Q}_\nu(z) \qquad (\pm=\operatorname{sgn}(\operatorname{Im}z)).$$

$$(z^2-1)d\mathfrak{P}_\nu(z)/dz=(\nu+1)[\mathfrak{P}_{\nu+1}(z)-z\mathfrak{P}_\nu(z)],$$

$$(2\nu+1)z\mathfrak{P}_\nu(z)=(\nu+1)\mathfrak{P}_{\nu+1}(z)+\nu\mathfrak{P}_{\nu-1}(z),$$

$$(z^2-1)d\mathfrak{Q}_\nu(z)/dz=(\nu+1)[\mathfrak{Q}_{\nu+1}(z)-z\mathfrak{Q}_\nu(z)],$$

$$(2\nu+1)z\mathfrak{Q}_\nu(z)=(\nu+1)\mathfrak{Q}_{\nu+1}(z)+\nu\mathfrak{Q}_{\nu-1}(z).$$

$$\mathfrak{P}_\nu(z)=\pi^{-1/2}2^{-\nu-1}\tan\nu\pi\frac{\Gamma(\nu+1)}{\Gamma[\nu+(3/2)]}z^{-\nu-1}{}_2F_1\left(\frac{\nu}{2}+1,\ \frac{\nu+1}{2};\ \nu+\frac{3}{2};\ \frac{1}{z^2}\right)$$

$$+\pi^{-1/2}2^\nu\frac{\Gamma[\nu+(1/2)]}{\Gamma(\nu+1)}z^\nu{}_2F_1\left(\frac{1-\nu}{2},\ \frac{-\nu}{2};\ \frac{1}{2}-\nu;\ \frac{1}{z^2}\right).$$

$$P_\nu(\cos\theta)=\frac{\sin\nu\pi}{\pi}\sum_{n=0}^{\infty}(-1)^n\left(\frac{1}{\nu-n}-\frac{1}{\nu+n+1}\right)P_n(\cos\theta) \qquad (\nu\neq\text{integer};\ 0\leqslant\theta<\pi).$$

Estimation: $|P_\nu(\cos\theta)|\leqslant\dfrac{2}{\sqrt{\nu\pi}\sin\theta}, \qquad |Q_\nu(\cos\theta)|\leqslant\dfrac{\sqrt{\pi}}{\sqrt{\nu}\sin\theta} \qquad (0<\theta<\pi;\ \nu>1).$

$$P_\nu(1)=1, \qquad P_\nu(0)=-\frac{\sin\nu\pi}{2\pi^{3/2}}\Gamma\left(\frac{\nu+1}{2}\right)\Gamma\left(\frac{-\nu}{2}\right),$$

$$Q_\nu(0)=\frac{1}{4\sqrt{\pi}}(1-\cos\nu\pi)\Gamma\left(\frac{\nu+1}{2}\right)\Gamma\left(\frac{-\nu}{2}\right).$$

(2) The Case $\nu=n$ $(=0,1,2,\ldots)$. In the following, the symbol !! means

$$m!!\equiv\begin{cases} m(m-2)\ldots 4\cdot 2 & (m \text{ even}),\\ m(m-2)\ldots 5\cdot 3\cdot 1 & (m \text{ odd}).\end{cases}$$

The function P_n is a polynomial of degree n (Legendre polynomial) and is represented as follows:

$$P_n(z) \equiv \frac{1}{2^n n!}\frac{d^n}{dz^n}(z^2-1)^n$$

$$= \frac{(2n)!}{2^n (n!)^2} z^n {}_2F_1\left(-\frac{n}{2}, \frac{1-n}{2}; \frac{1}{2}-n; \frac{1}{z^2}\right)$$

$$= \frac{(2n-1)!!}{n!}\left[z^n - \frac{n(n-1)}{(2n-1)}z^{n-2} + \frac{n(n-1)(n-2)(n-3)}{2\cdot 4\cdot(2n-1)(2n-3)}z^{n-4} \mp \ldots\right].$$

$$P_{2m}(z) = \sum_{j=0}^{m}(-1)^{m-j}\frac{(2m+2j-1)!!}{(2j)!(2m-2j)!!}z^{2j},$$

$$P_{2m+1}(z) = \sum_{j=0}^{m}(-1)^{m-j}\frac{(2m+2j+1)!!}{(2j+1)!(2m-2j)!!}z^{2j+1}.$$

$$P_n(\cos\theta) = \frac{(2n)!}{2^{2n}(n!)^2}e^{\mp in\theta}{}_2F_1\left(\frac{1}{2}, -n; \frac{1}{2}-n; e^{\pm 2i\theta}\right)$$

$$= \frac{2(2n-1)!!}{(2n)!!}\left[\cos n\theta + \frac{1}{1}\frac{n}{(2n-1)}\cos(n-2)\theta + \frac{1\cdot 3}{1\cdot 2}\frac{n(n-1)}{(2n-1)(2n-3)}\cos(n-4)\theta\right.$$

$$\left. + \frac{1\cdot 3\cdot 5}{1\cdot 2\cdot 3}\frac{n(n-1)(n-2)}{(2n-1)(2n-3)(2n-5)}\cos(n-6)\theta + \ldots\right]$$

$$+\begin{cases}\left[\dfrac{(n-1)!!}{n!!}\right]^2 & (n \text{ even}),\\ 0 & (n \text{ odd}).\end{cases}$$

$$= \frac{4}{\pi}\frac{(2n)!!}{(2n+1)!!}\left[\sin(n+1)\theta + \frac{1\cdot(n+1)}{1\cdot(2n+3)}\sin(n+3)\theta\right.$$

$$\left. + \frac{1\cdot 3\cdot(n+1)(n+2)}{1\cdot 2\cdot(2n+3)(2n+5)}\sin(n+5)\theta + \ldots(\text{ad infinitum})\right] \qquad (0<\theta<\pi).$$

Laplace-Mehler integral representation

$$P_n(\cos\theta) = \frac{1}{\pi}\int_0^\pi(\cos\theta + i\sin\theta\cos\varphi)^n\,d\varphi$$

$$= \frac{\sqrt{2}}{\pi}\int_0^\theta\frac{\cos[n+(1/2)]\varphi}{\sqrt{\cos\varphi-\cos\theta}}\,d\varphi = \frac{\sqrt{2}}{\pi}\int_\theta^\pi\frac{\sin[n+(1/2)]\varphi}{\sqrt{\cos\theta-\cos\varphi}}\,d\varphi.$$

$$P_n(x) = \frac{(-1)^n}{n!}r^{n+1}\frac{\partial^n}{\partial z^n}\left(\frac{1}{r}\right) \qquad \left(x=\frac{z}{r}, r=\sqrt{z^2+\rho^2}\right).$$

$$P_n(1)=1, \qquad P_n(-1)=(-1)^n, \qquad P_{2n+1}(0)=0,$$

$$P_{2n}(0) = (-1)^n\frac{(2n)!}{2^{2n}(n!)^2} = \frac{(-1)^n(2n-1)!!}{(2n)!!}.$$

Recurrence formulas: $\quad nP_n(z)-(2n-1)zP_{n-1}(z)+(n-1)P_{n-2}(z)=0,$

$$(z^2-1)\frac{dP_n}{dz} = n(zP_n - P_{n-1}) = \frac{n(n+1)}{2n+1}(P_{n+1}-P_{n-1}) = (n+1)(P_{n+1}-zP_n).$$

$$\mathfrak{Q}_n(z) = \frac{1}{2^n n!}\frac{d^n}{dz^n}\left[(z^2-1)^n\log\frac{z+1}{z-1}\right] - \frac{1}{2}P_n(z)\log\frac{z+1}{z-1}$$

$$= 2^n n!\int_z^\infty\cdots\int_z^\infty\frac{(dz)^{n+1}}{(z^2-1)^{n+1}}$$

$$= 2^n\int_z^\infty\frac{(t-z)^n}{(t^2-1)^{n+1}}\,dt$$

$$= (-1)^n\frac{1}{(2n-1)!!}\frac{d^n}{dz^n}\left[(z^2-1)^n\int_z^\infty\frac{dt}{(t^2-1)^{n+1}}\right] \qquad (\operatorname{Re} z > 1).$$

$$Q_n(\cos\theta)=\frac{2\cdot(2n)!!}{(2n+1)!!}\left[\cos(n+1)\theta+\frac{1\cdot(n+1)}{1\cdot(2n+3)}\cos(n+3)\theta\right.$$

$$\left.+\frac{1\cdot 3\cdot(n+1)(n+2)}{1\cdot 2\cdot(2n+3)(2n+5)}\cos(n+5)\theta+\dots\right]\qquad(0<\theta<\pi).$$

$$Q_n(x)=\frac{1}{2^n n!}\frac{d^n}{dx^n}\left[(x^2-1)^n\log\frac{1+x}{1-x}\right]-\frac{1}{2}P_n(x)\log\frac{1+x}{1-x}$$

$$=\frac{1}{2}P_n(x)\log\frac{1+x}{1-x}-\sum_{j=1}^{n}\frac{1}{j}P_{j-1}(x)P_{n-j}(x).$$

$$Q_0(x)=\frac{1}{2}\log\frac{1+x}{1-x},\qquad Q_1(x)=\frac{x}{2}\log\frac{1+x}{1-x}-1,\qquad Q_2(x)=\frac{1}{4}(3x^2-1)\log\frac{1+x}{1-x}-\frac{3}{2}x.$$

(3) Generating Functions.

$$\frac{1}{\sqrt{1-2hz+h^2}}=\begin{cases}\sum_{n=0}^{\infty}h^nP_n(z) & (|h|<\min|z\pm\sqrt{z^2-1}\,|),\\ \sum_{n=0}^{\infty}\frac{1}{h^{n+1}}P_n(z) & (|h|>\max|z\pm\sqrt{z^2-1}\,|).\end{cases}$$

(If $-1\leqslant z\leqslant 1$, the right-hand side is equal to 1.)

$$\frac{1}{z-t}=\sum_{n=0}^{\infty}(2n+1)\mathfrak{P}_n(t)\mathfrak{Q}_n(z)\qquad(|t+\sqrt{t^2-1}\,|<|z+\sqrt{z^2-1}\,|).$$

$$\frac{1}{\sqrt{1-2tz+z^2}}\log\frac{z-t+\sqrt{1-2tz+z^2}}{\sqrt{z^2-1}}=\sum_{n=0}^{\infty}t^n\mathfrak{Q}_n(z)\qquad(\mathrm{Re}\,z>1,\quad |t|<1).$$

$$r=\sqrt{x^2+y^2+z^2},\qquad \cos\theta=z/r,\qquad x,y\text{ real},$$

$$\frac{1}{r}+\frac{1}{2}+\sum_{m=1}^{\infty}\left[\frac{1}{\sqrt{(2mi\pi+z)^2+x^2+y^2}}+\frac{1}{\sqrt{(2mi\pi-z)^2+x^2+y^2}}\right]$$

(Here the square root of a complex number is taken so that its real part is positive.)

$$=\begin{cases}1+\sum_{n=1}^{\infty}e^{-nz}J_0\left(n\sqrt{x^2+y^2}\,\right) & (\mathrm{Re}\,z>0),\\ \frac{1}{r}+\frac{1}{2}+\sum_{n=1}^{\infty}\frac{(-1)^nB_{2n}}{(2n)!}r^{2n-1}P_{2n-1}(\cos\theta) & (0<\theta<2\pi;\,z\text{ real}).\end{cases}$$

(4) Integrals of Legendre Polynomials.

Orthogonal relations: $\int_{-1}^{+1}P_n(z)P_m(z)\,dz=\delta_{nm}\frac{2}{2n+1}$.

$$\int_{-1}^{+1}z^kP_n(z)\,dz=0\qquad(k=0,1,\dots,n-1).$$

$$\int_0^1 z^\lambda P_n(z)\,dz=\begin{cases}\frac{\lambda(\lambda-2)\dots(\lambda-n+2)}{(\lambda+n+1)(\lambda+n-1)\dots(\lambda+1)} & (n\text{ even}),\\ \frac{(\lambda-1)(\lambda-3)\dots(\lambda-n+2)}{(\lambda+n+1)(\lambda+n-1)\dots(\lambda+2)} & (n\text{ odd})\qquad(\mathrm{Re}\,\lambda>-1).\end{cases}$$

$$\int_0^\pi P_n(\cos\theta)\sin m\theta\,d\theta=\begin{cases}\frac{2(m-n+1)(m-n+3)\dots(m+n-1)}{(m-n)(m-n+2)\dots(m+n)} & (m>n;\ m+n\text{ is odd}),\\ 0 & (\text{otherwise}).\end{cases}$$

(5) Conical Function (Kegelfunktion). This is the Legendre function corresponding to the case $\nu=-(1/2)+i\lambda$ (λ is a real parameter),

$$P_{-(1/2)+i\lambda}(\cos\theta)=1+\frac{4\lambda^2+1^2}{2^2}\sin^2\frac{\theta}{2}+\frac{(4\lambda^2+1^2)(4\lambda^2+3^2)}{2^2\cdot 4^2}\sin^4\frac{\theta}{2}+\dots.$$

$$P_{-(1/2)+i\lambda}(x)\equiv P_{-(1/2)-i\lambda}(x).$$

(III) Associated Legendre Functions (→ 393 Spherical Functions)

(1) Associated Legendre Functions. The fundamental system of solutions of the differential equation in (II) (1) is given by the following two kind of functions when $\mu \neq 0$.

Associated Legendre function of the first kind:

$$\mathfrak{P}_\nu^\mu(z) \equiv \frac{1}{\Gamma(1-\mu)}\left(\frac{z+1}{z-1}\right)^{\mu/2} {}_2F_1\left(-\nu, \nu+1; 1-\mu; \frac{1-z}{2}\right),$$

where we take the branch satisfying $\arg[(z+1)/(z-1)]^{\mu/2}=0$ for $z>1$ in the expression raised to the $(\mu/2)$th power.

Associated Legendre function of the second kind:

$$\mathfrak{Q}_\nu^\mu(z) \equiv \frac{e^{i\mu\pi}}{2^{\nu+1}} \frac{\Gamma(\nu+\mu+1)\sqrt{\pi}}{\Gamma[\nu+(3/2)]}(z^2-1)^{\mu/2} z^{-\nu-\mu-1} {}_2F_1\left(\frac{\nu+\mu+2}{2}, \frac{\nu+\mu+1}{2}; \nu+\frac{3}{2}; \frac{1}{z^2}\right),$$

where we take the branch satisfying $\arg(z^2-1)^{\mu/2}=0$ for $z>1$ in $(z^2-1)^{\mu/2}$, and $\arg z^{-\nu-\mu-1}=0$ for $z>0$ in $z^{-\nu-\mu-1}$, respectively.

$$P_\nu^\mu(x) \equiv e^{i\mu\pi/2}\mathfrak{P}_\nu^\mu(x+i0) = e^{i\mu\pi/2}\mathfrak{P}_\nu^\mu(x-i0)$$

$$= \frac{1}{\Gamma(1-\mu)}\left(\frac{1+x}{1-x}\right)^{\mu/2} {}_2F_1\left(-\nu, \nu+1; 1-\mu; \frac{1-x}{2}\right) \qquad (-1 \leqslant x \leqslant 1).$$

$$Q_\nu^\mu(x) \equiv e^{-i\mu\pi}\left[e^{-i\mu\pi/2}\mathfrak{Q}_\nu^\mu(x+i0) + e^{i\mu\pi/2}\mathfrak{Q}_\nu^\mu(x-i0)\right]/2$$

$$= \frac{\pi}{2\sin\mu\pi}\left[P_\nu^\mu(x)\cos\mu\pi - \frac{\Gamma(\nu+\mu+1)}{\Gamma(\nu-\mu+1)}P_\nu^{-\mu}(x)\right] \qquad (-1 \leqslant x \leqslant 1).$$

Integral representations:

$$\mathfrak{P}_\nu^{-\mu}(z) = \frac{(z^2-1)^{\mu/2}}{2\mu\sqrt{\pi}\,\Gamma[\mu+(1/2)]}\int_{-1}^{+1}\frac{(1-t^2)^{\mu-1/2}}{(z+t\sqrt{z^2-1}\,)^{\mu-\nu}}dt$$

$$\left(\operatorname{Re}\mu > -\frac{1}{2}, |\arg(z\pm 1)| < \pi\right).$$

$$\mathfrak{P}_\nu^{-\mu}(z) = \frac{(z^2-1)^{\mu-2}}{2^\nu\Gamma(\mu-\nu)\Gamma(\nu+1)}\int_0^\infty\frac{(\sinh t)^{2\nu+1}}{(z+\cosh t)^{\mu+\nu+1}}dt$$

$$(\operatorname{Re}z > -1, |\arg(z\pm 1)| < \pi, \operatorname{Re}\nu > -1, \operatorname{Re}(\mu-\nu) > 0).$$

$$\mathfrak{P}_\nu^{-\mu}(z) = \sqrt{\frac{2}{\pi}}\frac{\Gamma[\mu+(1/2)](z^2-1)^{\mu-2}}{\Gamma(\mu+\nu+1)\Gamma(\mu-\nu)}\int_0^\infty\frac{\cosh[\nu+(1/2)]t}{(z+\cosh t)^{\mu+(1/2)}}dt$$

$$(\operatorname{Re}z > -1, |\arg(z\pm 1)| < \pi, \operatorname{Re}(\mu+\nu) > -1, \operatorname{Re}(\mu-\nu) > 0).$$

$$\mathfrak{P}_\nu^\mu(\cosh\alpha) = \sqrt{\frac{2}{\pi}}\frac{(\sinh\alpha)^\mu}{\Gamma[-\mu+(1/2)]}\int_0^\alpha\frac{\cosh[\{\nu+(1/2)\}t]\,dt}{(\cosh\alpha-\cosh t)^{\mu+(1/2)}} \qquad \left(\alpha>0, \operatorname{Re}\mu < \frac{1}{2}\right).$$

$$P_\nu^\mu(\cos\theta) = \sqrt{\frac{2}{\pi}}\frac{(\sin\theta)^\mu}{\Gamma[-\mu+(1/2)]}\int_0^\theta\frac{\cos[\{\nu+(1/2)\}\varphi]\,d\varphi}{(\cos\varphi-\cos\theta)^{\mu+(1/2)}} \qquad \left(0<\theta<\pi, \operatorname{Re}\mu < \frac{1}{2}\right).$$

$$P_\nu^{-\mu}(\cos\theta) = \frac{\Gamma(2\mu+1)2^{-\mu}(\sin\theta)^\mu}{\Gamma(\mu+1)\Gamma(\mu+\nu+1)\Gamma(\mu-\nu)}\int_0^\infty\frac{t^{\mu+\nu}\,dt}{(1+2t\cos\theta+t^2)^{\mu+(1/2)}}$$

$$(\operatorname{Re}(\mu+\nu) > -1, \operatorname{Re}(\mu-\nu) > 0).$$

$$P_\nu^{-\mu}(\cos\theta) = \frac{1}{\Gamma(\nu+\mu+1)}\int_0^\infty e^{-t\cos\theta}J_\mu(t\sin\theta)t^\nu\,dt \qquad \left(0<\theta<\frac{\pi}{2}, \operatorname{Re}(\mu+\nu) > -1\right).$$

$$\mathfrak{Q}_\nu^\mu(z) = \frac{e^{i\mu\pi}}{2^{\nu+1}}\frac{\Gamma(\nu+\mu+1)}{\Gamma(\nu+1)}(z^2-1)^{\mu/2}\int_{-1}^{+1}(1-t)^\nu(z-1)^{-\nu-\mu-1}dt$$

$$(\operatorname{Re}(\nu+\mu) > -1, \operatorname{Re}\nu > -1, |\arg(z\pm 1)| < \pi).$$

$$\mathfrak{Q}_\nu^\mu(\cosh\alpha) = \sqrt{\frac{\pi}{2}}\frac{e^{i\mu\pi}(\sinh\alpha)^\mu}{\Gamma[-\mu+(1/2)]}\int_\alpha^\infty\frac{e^{-[\nu+(1/2)]t}\,dt}{(\cosh t-\cosh\alpha)^{\mu+(1/2)}}$$

$$(\alpha > 0, \operatorname{Re}\mu < 1/2, \operatorname{Re}(\nu+\mu) > -1).$$

Recurrence formulas:

$$(z^2-1)d\mathfrak{P}_\nu^\mu(z)/dz=(\nu-\mu+1)\mathfrak{P}_{\nu+1}^\mu(z)-(\nu+1)z\mathfrak{P}_\nu^\mu(z),$$

$$(2\nu+1)z\mathfrak{P}_\nu^\mu(z)=(\nu-\mu+1)\mathfrak{P}_{\nu+1}^\mu(z)+(\nu+\mu)\mathfrak{P}_{\nu-1}^\mu(z),$$

$$\mathfrak{P}_{-\nu-1}^\mu(z)=\mathfrak{P}_\nu^\mu(z),$$

$$\mathfrak{Q}_\nu^{-\mu}(z)=e^{-2i\mu\pi}\frac{\Gamma(\nu-\mu+1)}{\Gamma(\nu+\mu+1)}\mathfrak{Q}_\nu^\mu(z),$$

$$(1-x^2)dP_\nu^\mu(x)/dx=(\nu+1)xP_\nu^\mu(x)-(\nu-\mu+1)P_{\nu+1}^\mu(x).$$

The case when μ is an integer $m(m=0,1,2,\ldots)$ and ν is also an integer n:

$$P_n^{m+2}(x)+2(m+1)x(1-x^2)^{-1/2}P_n^{m+1}(x)+(n-m)(n+m+1)P_n^m(x)=0.$$

$$(2n+1)xP_n^m(x)-(n-m+1)P_{n+1}^m(x)-(n+m)P_{n-1}^m(x)=0 \qquad (0\leqslant m\leqslant n-2),$$

$$(x^2-1)dP_n^m(x)/dx-(n-m+1)P_{n+1}^m(x)+(n+1)xP_n^m(x)=0,$$

$$P_{n-1}^m(x)-P_{n+1}^m(x)=(2n+1)\sqrt{1-x^2}\;P_n^{m-1}(x).$$

$$\mathfrak{P}_\nu^{-\mu}(z)=\frac{\Gamma(\nu-\mu+1)}{\Gamma(\nu+\mu+1)}\left[\mathfrak{P}_\nu^\mu(z)-\frac{2}{\pi}e^{-i\mu\pi}(\sin\mu\pi)\mathfrak{Q}_\nu^\mu(z)\right],$$

$$\mathfrak{Q}_\nu^\mu(z)\sin[(\nu+\mu)\pi]-\mathfrak{Q}_{-\nu-1}^\mu(z)\sin[(\nu-\mu)\pi]=\pi e^{i\mu\pi}(\cos\nu\pi)\mathfrak{P}_\nu^\mu(z),$$

$$\mathfrak{P}_\nu^\mu(-z)=e^{\mp i\nu\pi}\mathfrak{P}_\nu^\mu(z)-(2/\pi)[\sin(\nu+\mu)\pi]e^{-i\mu\pi}\mathfrak{Q}_\nu(z) \qquad (\mp=-\operatorname{sgn}(\operatorname{Im}z)),$$

$$\mathfrak{Q}_\nu^\mu(-z)=-e^{\pm i\nu\pi}\mathfrak{Q}_\nu^\mu(z) \qquad (\pm=\operatorname{sgn}(\operatorname{Im}z)).$$

$$e^{-i\mu\pi}\mathfrak{Q}_\nu^\mu(\cosh\alpha)=\frac{\pi\Gamma(1+\mu+\nu)}{\sqrt{2\pi\sinh\alpha}}\mathfrak{P}_{-\mu-1/2}^{-\nu-1/2}(\coth\alpha) \qquad (\operatorname{Re}\cosh\alpha>0).$$

$$e^{-i\mu\pi}\mathfrak{Q}_\nu^\mu(x\pm i0)=e^{\pm i\mu\pi/2}[Q_\nu^\mu(x)\mp(i\pi/2)P_\nu^\mu(x)].$$

$$Q_{-\nu-1}^\mu(x)=\frac{\sin(\nu+\mu)\pi}{\sin(\nu-\mu)\pi}Q_\nu^\mu(x)-\frac{\pi\cos\nu\pi\cos\mu\pi}{\sin(\nu-\mu)\pi}P_\nu^\mu(x),$$

$$P_\nu^{-\mu}(x)=\frac{\Gamma(\nu-\mu+1)}{\Gamma(\nu+\mu+1)}\left[\cos\mu\pi P_\nu^\mu(x)-\frac{2}{\pi}\sin\mu\pi Q_\nu^\mu(x)\right],$$

$$P_\nu^\mu(-x)=[\cos(\nu+\mu)\pi]P_\nu^\mu(x)-(2/\pi)[\sin(\nu+\mu)\pi]Q_\nu^\mu(x),$$

$$Q_\nu^\mu(-x)=-[\cos(\nu+\mu)\pi]Q_\nu^\mu(x)+(\pi/2)[\sin(\nu+\mu)\pi]P_\nu^\mu(x).$$

$$\mathfrak{P}_\nu^m(z)=\frac{\Gamma(1+\nu+m)(z^2-1)^{m/2}}{\Gamma(1+\nu-m)m!2^m}\,{}_2F_1\left(m-\nu,\,m+\nu+1;\,m+1;\,\frac{1-z}{2}\right)=(z^2-1)^{m/2}\frac{d^m\mathfrak{P}_\nu(z)}{dz^m},$$

$$\mathfrak{P}_\nu^{-m}(z)=(z^2-1)^{-m/2}\int_1^z\cdots\int_1^z P_\nu(z)(dz)^m.$$

$$\mathfrak{Q}_\nu^m(z)=(z^2-1)^{m/2}\frac{d^m\mathfrak{Q}_\nu(z)}{dx^m}, \qquad \mathfrak{Q}_\nu^{-m}(z)=(-1)^m(z^2-1)^{-m/2}\int_z^\infty\cdots\int_z^\infty\mathfrak{Q}_\nu(z)(dz)^m.$$

$$P_\nu^m(x)=(-1)^m\frac{\Gamma(1+\nu+m)(1-x^2)^{m/2}}{\Gamma(1+\nu-m)m!2^m}\,{}_2F_1\left(m-\nu,\,m+\nu+1;\,m+1;\,\frac{1-x}{2}\right)$$

$$=(-1)^m(1-x^2)^{m/2}\frac{d^mP_\nu(x)}{dx^m},$$

$$P_\nu^{-m}(x)=(1-x^2)^{-m/2}\int_x^1\cdots\int_x^1 P_\nu(x)(dx)^m=(-1)^m\frac{\Gamma(\nu-m+1)}{\Gamma(\nu+m+1)}P_\nu^m(x).$$

$$Q_\nu^m(x)=(-1)^m(1-x^2)^{m/2}\frac{d^mQ_\nu(x)}{dx^m}, \qquad Q_\nu^{-m}(x)=(-1)^m\frac{\Gamma(\nu-m+1)}{\Gamma(\nu+m+1)}Q_\nu^m(x).$$

The values at the origin are

$$P_\nu^\mu(0)=\frac{\sqrt{\pi}\,2^\mu}{\Gamma[(\nu-\mu)/2+1]\Gamma[(-\nu-\mu+1)/2]},$$

$$\frac{dP_\nu^\mu(0)}{dx}=\frac{2^{\mu+1}\sin[\pi(\nu+\mu)/2]\Gamma[(\nu+\mu+2)/2]}{\Gamma[(\nu-\mu+1)/2]\sqrt{\pi}}=\frac{\sqrt{\pi}\,2^{\mu+1}}{\Gamma[(\nu-\mu+1)/2]\Gamma[(-\nu-\mu)/2]},$$

$$Q_\nu^\mu(0)=-2^{\mu-1}\sqrt{\pi}\,\sin\left(\frac{\nu+\mu}{2}\pi\right)\frac{\Gamma[(\nu+\mu+1)/2]}{\Gamma[(\nu-\mu+2)/2]},$$

$$\frac{dQ_\nu^\mu(0)}{dx}=2^\mu\sqrt{\pi}\,\cos\left(\frac{\nu+\mu}{2}\pi\right)\frac{\Gamma[(\nu+\mu+2)/2]}{\Gamma[(\nu-\mu+1)/2]}.$$

(2) Generating Functions.

$$(\cos\theta+i\sin\theta\sin\varphi)^n=P_n(\cos\theta)+2\sum_{m=1}^{n}(-i)^m\frac{n!}{(n+m)!}(\cos m\varphi)P_n^m(\cos\theta).$$

$$P_\nu^{-\mu}(\cos\theta)=\frac{\sin\nu\pi}{\pi}\sum_{n=0}^{\infty}(-1)^n\left(\frac{1}{\nu-n}-\frac{1}{\nu+n+1}\right)P_n^{-\mu}(\cos\theta)\qquad(0<\theta<\pi,\ \mu\geqslant 0).$$

(3) Orthogonal Relations.

$$\int_{-1}^{+1}P_n^m(x)P_{n'}^m(x)\,dx=\frac{2}{2n+1}\frac{(n+m)!}{(n-m)!}\delta_{nn'}.$$

$$\int_0^{2\pi}d\varphi\int_0^{\pi}\sin\theta e^{\pm i(m-m')\varphi}P_n^m(\cos\theta)P_{n'}^{m'}(\cos\theta)\,d\theta=\frac{4\pi}{2n+1}\frac{(n+m)!}{(n-m)!}\delta_{nn'}\delta_{mm'}.$$

(4) Addition Theorems.

$$\mathfrak{P}_\nu\left(z\zeta-\sqrt{z^2-1}\,\sqrt{\zeta^2-1}\,\cos\varphi\right)=\mathfrak{P}_\nu(z)\mathfrak{P}_\nu(\zeta)+2\sum_{m=1}^{\infty}(-1)^m\mathfrak{P}_\nu^m(z)\mathfrak{P}_\nu^{-m}(\zeta)\cos m\varphi$$

$$(\mathrm{Re}\,z>0,\ \mathrm{Re}\,\zeta>0,\ |\arg(z-1)|<\pi,\ |\arg(\zeta-1)|<\pi).$$

$$\mathfrak{Q}_\nu\left(tt'-\sqrt{t^2-1}\,\sqrt{t'^2-1}\,\cos\varphi\right)=\mathfrak{Q}_\nu(t)\mathfrak{P}_\nu(t')+2\sum_{m-1}^{\infty}(-1)^m\mathfrak{Q}_\nu^m(t)\mathfrak{P}_\nu^{-m}(t')\cos m\varphi$$

$$(t,t'\ \text{real},\ 1<t'<t,\ \nu\neq\text{negative integer},\ \varphi\ \text{real}).$$

$$P_\nu(\cos\theta\cos\theta'+\sin\theta\sin\theta'\cos\varphi)=P_\nu(\cos\theta)P_\nu(\cos\theta')+2\sum_{m=1}^{\infty}(-1)^mP_\nu^{-m}(\cos\theta')P_\nu^m(\cos\theta')\cos m\varphi$$

$$=P_\nu(\cos\theta)P_\nu(\cos\theta')+2\sum_{m=1}^{\infty}\frac{\Gamma(\nu-m+1)}{\Gamma(\nu+m+1)}P_\nu^m(\cos\theta)P_\nu^m(\cos\theta')\cos m\varphi$$

$$(0\leqslant\theta<\pi,\ 0\leqslant\theta'<\pi,\ \theta+\theta'<\pi,\ \varphi\ \text{real}).$$

$$Q_\nu(\cos\theta\cos\theta'+\sin\theta\sin\theta'\cos\varphi)=P_\nu(\cos\theta')Q_\nu(\cos\theta)+2\sum_{m=1}^{\infty}(-1)^mP_\nu^{-m}(\cos\theta')Q_\nu^m(\cos\theta)\cos m\varphi$$

$$(0<\theta'<\pi/2,\ 0<\theta<\pi,\ \theta+\theta'<\pi,\ \varphi\ \text{real}).$$

$$\mathfrak{Q}_n\left(\tau\tau'+\sqrt{\tau^2+1}\,\sqrt{\tau'^2+1}\,\cosh\alpha\right)=\sum_{m=n+1}^{\infty}\frac{1}{(m-n-1)!(m+n)!}\mathfrak{Q}_n^m(i\tau)\mathfrak{Q}_n^m(i\tau')e^{-m\alpha}$$

$$(\tau,\ \tau',\ \alpha>0).$$

(5) Asymptotic Expansions.

$$\mathfrak{P}_\nu^\mu(z)=\left[\frac{2^\nu\Gamma[\nu+(1/2)]}{\sqrt{\pi}\,\Gamma(\nu-\mu+1)}z^\nu+\frac{2^{-\nu-1}\Gamma[-\nu-(1/2)]}{\sqrt{\pi}\,\Gamma(-\mu-\nu)}z^{-\nu-1}\right][1+O(z^{-2})]$$

$$(\nu+(1/2)\neq\text{integer},\ |\arg z|<\pi,\ |z|\gg 1).$$

$$\mathfrak{Q}_\nu^\mu(z)=\frac{\sqrt{\pi}\,e^{i\mu\pi}}{2^{\nu+1}}\frac{\Gamma(\nu+\mu+1)}{\Gamma[\nu+(3/2)]}z^{-\nu-1}[1+O(z^{-2})]$$

$$(\nu+(1/2)\neq\text{negative integer},\ |\arg z|<\pi,\ |z|\gg 1).$$

$$P_\nu^\mu(\cos\theta)=\frac{2}{\sqrt{\pi}}\frac{\Gamma(\nu+\mu+1)}{\Gamma[\nu+(3/2)]}\frac{\cos[\{\nu+(1/2)\}\theta-(\pi/4)+(\mu\pi/2)]}{\sqrt{2\sin\theta}}[1+O(\nu^{-1})]$$

$$(\varepsilon\leqslant\theta\leqslant\pi-\varepsilon,\ \varepsilon>0,\ |\nu|\gg 1/\varepsilon).$$

$$P_\nu^\mu(\cos\theta)=\frac{2\Gamma(\nu+\mu+1)}{\sqrt{\pi}\,\Gamma[\nu+(3/2)]}$$
$$\times\sum_{l=0}^{\infty}\frac{\Gamma\left(\frac{1}{2}+\mu+l\right)\Gamma\left(\frac{1}{2}-\mu+l\right)\Gamma\left(\nu+\frac{3}{2}\right)}{\Gamma\left(\frac{1}{2}+\mu\right)\Gamma\left(\frac{1}{2}-\mu\right)\Gamma\left(\nu+l+\frac{3}{2}\right)l!}\frac{\cos\left[\left(\nu+\frac{2l+1}{2}\right)\theta-\frac{(2l+1)\pi}{4}+\frac{\mu\pi}{2}\right]}{(2\sin\theta)^{l+(1/2)}},$$

$$Q_\nu^\mu(\cos\theta)=\sqrt{\pi}\,\frac{\Gamma(\nu+\mu+1)}{\Gamma[\nu+(3/2)]}$$
$$\times\sum_{l=0}^{\infty}(-1)^l\frac{\Gamma\left(\frac{1}{2}+\mu+l\right)\Gamma\left(\frac{1}{2}-\mu+l\right)\Gamma\left(\nu+\frac{3}{2}\right)}{\Gamma\left(\frac{1}{2}+\mu\right)\Gamma\left(\frac{1}{2}-\mu\right)\Gamma\left(\nu+l+\frac{3}{2}\right)l!}\frac{\cos\left[\left(\nu+\frac{2l+1}{2}\right)\theta+\frac{(2l+1)\pi}{4}+\frac{\mu\pi}{2}\right]}{(2\sin\theta)^{l+(1/2)}}.$$

(In the final two formulas the series converges when $\nu+\mu\neq$negative integer, $\nu+(1/2)\neq$negative integer, $\pi/6<\theta<5\pi/6$.)

$$\left[\left(\nu+\frac{1}{2}\right)\cos\frac{\theta}{2}\right]^\mu P_\nu^{-\mu}(\cos\theta)=J_\mu(\eta)+\sin^2\frac{\theta}{2}\left[\frac{J_{\mu+1}(\eta)}{2\eta}-J_{\mu+2}(\eta)+\frac{\eta}{6}J_{\mu+3}(\eta)\right]+O\left(\sin^4\frac{\theta}{2}\right)$$
$$(\eta=(2\nu+1)\sin(\theta/2)).$$

(6) Estimation. When $\nu\geqslant 1,\quad \nu-\mu+1>0,\quad \mu\geqslant 0,$

$$|P_\nu^{\pm\mu}(\cos\theta)|<\frac{\Gamma(\nu\pm\mu+1)}{\Gamma(\nu+1)}\left(\frac{8}{\nu\pi\sin\theta}\right)^{1/2}\frac{1}{(\sin\theta)^\mu},$$

$$|Q_\nu^{\pm\mu}(\cos\theta)|<\frac{\Gamma(\nu\pm\mu+1)}{\Gamma(\nu+1)}\left(\frac{2\pi}{\nu\sin\theta}\right)^{1/2}\frac{1}{(\sin\theta)^\mu},$$

$$|P_\nu^{\pm m}(\cos\theta)|<\frac{\Gamma(\nu\pm m+1)}{\Gamma(\nu+1)}\left(\frac{4}{\nu\pi\sin\theta}\right)^{1/2}\frac{1}{(\sin\theta)^m},$$

$$|Q_\nu^{\pm m}(\cos\theta)|\leqslant\frac{\Gamma(\nu\pm m+1)}{\Gamma(\nu+1)}\left(\frac{4}{\nu\sin\theta}\right)^{1/2}\frac{1}{(\sin\theta)^m}.$$

(7) Torus Functions. These are solutions of the differential equation

$$\frac{d^2u}{d\eta^2}+\coth\eta\frac{du}{d\eta}-\left(n^2-\frac{1}{4}+\frac{m^2}{\sinh^2\eta}\right)u=0.$$

The fundamental system of solutions is given by

$\mathfrak{P}^m_{n-(1/2)}(\cosh\eta),\ \mathfrak{Q}^m_{n-(1/2)}(\cosh\eta).$

The asymptotic expansion when $m=0$ is

$$\mathfrak{P}_{n-(1/2)}(\cosh\eta)$$
$$=\frac{(n-1)!e^{n-(1/2)\eta}}{\Gamma[n+(1/2)]\sqrt{\pi}}\left[\frac{2\Gamma^2[n+(1/2)]}{\pi n!(n-1)!}(\log 4+\eta)e^{-2m\eta}{}_2F_1\left(\frac{1}{2},n+\frac{1}{2};n+1;e^{-2\eta}\right)+A+B\right].$$

Here

$$A=1+\frac{(1/2)[n-(1/2)]}{1\cdot(n-1)}e^{-2\eta}+\frac{(1/2)(3/2)[n-(1/2)][n-(3/2)]}{1\cdot 2\cdot(n-1)(n-2)}e^{-4\eta}+\ldots$$
$$+\frac{(2n-3)!!(2n-1)!!}{[(2n-2)!!]^2}e^{-2(n-1)\eta},$$

$$B=\frac{\Gamma[n+(1/2)]}{\pi^{3/2}(n-1)!}\sum_{l=1}^{\infty}\frac{\Gamma[l+(1/2)]\Gamma[n+l+(1/2)]}{(n+l)!l!}(u_{n+l}+u_l-v_{l-(1/2)}-v_{n+l-(1/2)})e^{-2(l+n)\eta},$$

where

$$u_r\equiv 1+\frac{1}{2}+\ldots+\frac{1}{r},\qquad v_{r-(1/2)}\equiv\frac{2}{1}+\frac{2}{3}+\frac{2}{5}+\ldots+\frac{2}{2r-1}=2u_{2r}-u_r.$$

References

See references to Table 16, this Appendix.

19. Functions of Confluent Type and Bessel Functions

(I) Hypergeometric Function of Confluent Type (→ 167 Functions of Confluent Type)

(1) Kummer Functions.

$$v(z)={}_1F_1(a;\ c;\ z)\equiv\sum_{n=0}^{\infty}\frac{\Gamma(a+n)}{\Gamma(a)}\frac{\Gamma(c)}{\Gamma(c+n)}\frac{z^n}{n!}$$

$$=\frac{\Gamma(c)}{\Gamma(a)\Gamma(c-a)}z^{1-c}\int_0^z e^t t^{a-1}(z-t)^{c-a-1}dt \qquad (0<\mathrm{Re}\,a<\mathrm{Re}\,c)$$

$$=\frac{\Gamma(c)2^{1-c}}{\Gamma(a)\Gamma(c-a)}e^{z/2}\int_{-1}^{+1}e^{zt/2}(1-t)^{c-a-1}(1+t)^{a-1}dt \qquad (0<\mathrm{Re}\,a<\mathrm{Re}\,c).$$

The fundamental system of solutions of the confluent hypergeometric differential equation (Kummer's differential equation)

$$z\frac{d^2v}{dz^2}+(c-z)\frac{dv}{dz}-av=0,$$

when $c\neq 0,-1,-2,\ldots$, is given by

$$\mathring{v}_1(z)\equiv{}_1F_1(a;c;z), \qquad \mathring{v}_2(z)\equiv z^{1-c}{}_1F_1(a-c+1;\ 2-c;\ z).$$

$$d\,{}_1F_1(a;\ c;\ z)/dz=(a/c){}_1F_1(a+1;c+1;z),$$

$${}_1F_1(a;c;z)=e^z{}_1F_1(c-a;c;-z),$$

$$a\,{}_1F_1(a+1;c+1;z)=(a-c){}_1F_1(a;c+1;z)+c\,{}_1F_1(a;c;z),$$

$$a\,{}_1F_1(a+1;c;z)=(z+2a-c){}_1F_1(a;c;z)+(c-a){}_1F_1(a-1;c;z).$$

Putting $(a)_n=a(a+1)\ldots(a+n-1)=\Gamma(a+n)/\Gamma(a)$ we have

$$\lim_{c\to -n}\frac{1}{\Gamma(c)}{}_1F_1(a;c;z)=\frac{z^{n+1}(a)_{n+1}}{(n+1)!}{}_1F_1(a+n+1;n+2;z) \qquad (n=0,1,2,\ldots).$$

Asymptotic expansion:

$$\mathring{v}_1\approx A_1z^{-a}\sum_{n=0}^{\infty}\frac{(a)_n(a-c+1)_n}{n!}(-z)^{-n}+B_1e^zz^{a-c}\sum_{n=0}^{\infty}\frac{(c-a)_n(1-a)_n}{n!}z^n,$$

$$\mathring{v}_2\approx A_2z^{-a}\sum_{n=0}^{\infty}\frac{(a)_n(a-c+1)_n}{n!}(-z)^{-n}+B_2e^zz^{a-c}\sum_{n=0}^{\infty}\frac{(c-a)_n(1-a)_n}{n!}z^n$$

$$(|z|\gg|a|,\ |z|\gg|c|,\ -3\pi/2<\arg z<\pi/2,\ c\neq\text{integer}),$$

where

$$A_1=e^{-i\pi a}\Gamma(c)/\Gamma(c-a), \qquad B_1=\Gamma(c)/\Gamma(a),$$

$$A_2=e^{-i\pi(a-c+1)}\Gamma(2-c)/\Gamma(1-a), \qquad B_2=\Gamma(2-c)/\Gamma(a-c+1).$$

(2) The fundamental system of solutions at $z=0$ of the hypergeometric differential equation of confluent type

$$\frac{d^2u}{dz^2}+\frac{du}{dz}+\left[\frac{\kappa}{z}+\frac{(1/4)-\mu^2}{z^2}\right]u=0$$

is given by

$$z^{(1/2)\pm\mu}e^{-z}{}_1F_1[(1/2)\pm\mu-\kappa;\ \pm2\mu+1;z].$$

(II) Whittaker Functions (→ 167 Functions of Confluent Type)

(1) A pair of linearly independent solutions of Whittaker's differential equation

$$\frac{d^2W}{dz^2}+\left[-\frac{1}{4}+\frac{\kappa}{z}+\frac{(1/4)-\mu^2}{z^2}\right]W=0$$

is given by $M_{\kappa,\pm\mu}(z)=z^{\pm\mu+(1/2)}e^{-z/2}{}_1F_1[\pm\mu-\kappa+(1/2);\pm 2\mu+1;z]$.

Whittaker functions:

$$W_{\kappa,\mu}(z)\equiv\frac{\Gamma(-2\mu)}{\Gamma[(1/2)-\mu-\kappa]}M_{\kappa,\mu}(z)+\frac{\Gamma(2\mu)}{\Gamma[(1/2)+\mu-\kappa]}M_{\kappa,-\mu}(z)=W_{\kappa,-\mu}(z).$$

When 2μ is an integer, the above definition of $W_{\kappa,\mu}(z)$ loses meaning, but by taking the limit with respect to μ we can define it in terms of the following integrals.

$$W_{\kappa,\mu}(z)=\frac{z^{\mu+(1/2)}e^{-z/2}}{\Gamma[\mu+(1/2)-\kappa]}\int_0^\infty e^{-z\tau}\tau^{\mu-\kappa-(1/2)}(1+\tau)^{\mu+\kappa-(1/2)}d\tau$$

$$=\frac{z^{\kappa}e^{-z/2}}{\Gamma[\mu+(1/2)-\kappa]}\int_0^\infty t^{\mu-\kappa-(1/2)}e^{-t}\left(1+\frac{t}{z}\right)^{\mu+\kappa-(1/2)}dt$$

$(\mathrm{Re}[\mu+(1/2)-\kappa]>0,\quad |\arg z|<\pi)$.

$$W_{\kappa,\mu}(z)=\frac{e^{-z/2}}{2\pi i}\int_{-i\infty}^{+i\infty}\frac{\Gamma(s-\kappa)\Gamma[-s-\mu+(1/2)]\Gamma[-s+\mu+(1/2)]}{\Gamma[-\kappa+\mu+(1/2)]\Gamma[-\kappa-\mu+(1/2)]}z^s\,ds.$$

$$M_{l+\mu+(1/2),\mu}(z)=(-1)^l z^{\mu+(1/2)}e^{-z/2}(2\mu+1)_{l\,1}F_1(-l;2\mu+1;z)\qquad(l=0,1,2,\ldots).$$

$$M_{\kappa,\mu}(z)=e^{-i\pi[\mu+(1/2)]}M_{-\kappa,\mu}(e^{i\pi}z).$$

$$M_{\kappa,\mu}(z)=\frac{\Gamma(2\mu+1)}{\Gamma[\mu+(1/2)-\kappa]}e^{i\pi\kappa}W_{-\kappa,\mu}(e^{i\pi}z)+\frac{\Gamma(2\mu+1)}{\Gamma[\mu+(1/2)+\kappa]}e^{i\pi[\kappa-\mu-(1/2)]}W_{\kappa,\mu}(z)$$

$(-3\pi/2<\arg z<\pi/2,\quad 2\mu\neq-1,-2,\ldots)$.

$$M_{\kappa,\mu}(z)=\frac{\Gamma(2\mu+1)}{\Gamma[\mu+(1/2)-\kappa]}e^{-i\pi\kappa}W_{-\kappa,\mu}(e^{-i\pi}z)+\frac{\Gamma(2\mu+1)}{\Gamma[\mu+(1/2)+\kappa]}e^{-i\pi[\kappa-\mu-(1/2)]}W_{\kappa,\mu}(z)$$

$(-\pi/2<\arg z<3\pi/2,\quad 2\mu\neq-1,-2,\ldots)$.

$$W_{\kappa,\mu}(z)=z^{1/2}W_{\kappa-(1/2),\mu-(1/2)}(z)+[(1/2)-\kappa+\mu]W_{\kappa-1,\mu}(z)$$

$$=z^{1/2}W_{\kappa-(1/2),\mu+(1/2)}(z)+[(1/2)-\kappa-\mu]W_{\kappa-1,\mu}(z).$$

$$z\,dW_{\kappa,\mu}(z)/dz=[\kappa-(z/2)]W_{\kappa,\mu}(z)-\left[\mu^2-\{\kappa-(1/2)\}^2\right]W_{\kappa-1,\mu}(z).$$

When κ is sufficiently large we have

$$M_{\kappa,\mu}(z)\sim\pi^{-1/2}\Gamma(2\mu+1)\kappa^{-\mu-(1/4)}z^{1/4}\cos\left[2(z\kappa)^{1/2}-\mu\pi-(\pi/4)\right],$$

$$W_{\kappa,\mu}(z)\sim-(4z/\kappa)^{1/4}\exp(-\kappa+\kappa\log\kappa)\sin\left[2(z\kappa)^{1/2}-\pi\kappa-(\pi/4)\right],$$

$$W_{-\kappa,\mu}(z)\sim(z/4\kappa)^{1/4}\exp\left(\kappa-\kappa\log\kappa-2(z\kappa)^{1/2}\right).$$

Asymptotic expansion:

$$W_{\kappa,\mu}(z)\approx e^{-z/2}z^{\kappa}$$

$$\times\left(1+\sum_{n=1}^{\infty}\frac{[\mu^2-\{\kappa-(1/2)\}^2][\mu^2-\{\kappa-(3/2)\}^2]\ldots[\mu^2-\{\kappa-n+(1/2)\}^2]}{n!z^n}\right).$$

(2) Representation of Various Special Functions by Whittaker Functions.

(i) Probability integral (error function) $\quad\mathrm{erf}\,x\equiv\Phi(x)\equiv\dfrac{2}{\sqrt{\pi}}\displaystyle\int_0^x e^{-t^2}dt$

$$=1-\pi^{-1/2}x^{-1/2}e^{-x^2/2}W_{-1/4,1/4}(x^2)$$

$$=\frac{2x}{\sqrt{\pi}}{}_1F_1\left(\frac{1}{2};\frac{3}{2};-x^2\right)=\frac{2}{\sqrt{\pi}}\left(x-\frac{x^3}{1!3}+\frac{x^5}{2!5}-\frac{x^7}{3!7}\pm\ldots\right).$$

Asymptotic expansion:

$$\frac{\sqrt{\pi}}{2}[1-\Phi(x)]\approx\frac{e^{-x^2}}{2x}\left(1-\frac{1}{2x^2}+\frac{1\cdot3}{(2x^2)^2}-\frac{1\cdot3\cdot5}{(2x^2)^3}\pm\ldots\right).$$

$$\frac{1}{1+i}\Phi\left(x\frac{1+i}{2}\sqrt{\pi}\right)=C(x)-iS(x),$$

where $C(x)$, $S(x)$ are the following Fresnel integrals.

$$C(x)\equiv\int_0^x\cos\frac{\pi}{2}t^2dt=\frac{1}{2}+\frac{1}{\pi x}\sin\frac{\pi}{2}x^2+O\left(\frac{1}{x^2}\right),$$

$$S(x)\equiv\int_0^x\sin\frac{\pi}{2}t^2dt=\frac{1}{2}-\frac{1}{\pi x}\cos\frac{\pi}{2}x^2+O\left(\frac{1}{x^2}\right).$$

(ii) Logarithmic integral

$$\mathrm{Li}\,z\equiv\int_0^z\frac{dt}{\log t}\qquad(\text{When } z>1, \text{ take Cauchy's principal value at } t=1.)$$

$$=-(\log 1/z)^{-1/2}z^{1/2}W_{-1/2,0}(-\log z).$$

$\mathrm{Li}\,z$ is sometimes written as $\mathrm{li}\,z$.

(3) Exponential Integral

$$\mathrm{Ei}\,x\equiv\int_{-\infty}^x\frac{e^t}{t}dt\quad(\text{When } x>0, \text{ take the Cauchy's principal value at } t=0 \text{ while integrating.})$$

$$=C+\log|x|+\sum_{n=1}^{\infty}\frac{x^n}{n\cdot n!}\qquad(x \text{ real}, \neq 0)$$

$$=e^x\sum_{n=1}^{N}\frac{(n-1)!}{t^n}+N!\sum_{\substack{n=0\\ n\neq N}}^{\infty}\frac{t^{n-N}}{n!(n-N)}-\left(1+\frac{1}{2}+\cdots+\frac{1}{N}\right)+C+\log|x|.$$

Cosine integral $\quad \mathrm{Ci}\,x\equiv-\int_x^\infty\frac{\cos t}{t}dt=C+\log x-\int_0^x\frac{1-\cos t}{t}dt.$

Sine integral $\quad \mathrm{Si}\,x\equiv\int_0^x\frac{\sin t}{t}dt,$

$$\mathrm{si}\,x\equiv-\int_x^\infty\frac{\sin t}{t}dt=\mathrm{Si}\,x-\frac{\pi}{2}.$$

Asymptotic expansion $\quad \mathrm{Ei}\,ix=\mathrm{Ci}\,x+i\,\mathrm{si}\,x\approx e^{ix}\left(\frac{1}{ix}+\frac{1!}{(ix)^2}+\frac{2!}{(ix)^3}+\frac{3!}{(ix)^4}+\cdots\right).$

(III) Bessel Functions (→ 39 Bessel Functions)

(1) Cylindrical Functions. A cylindrical function Z_ν is a solution of Bessel's differential equation

$$\frac{d^2Z_\nu}{dz^2}+\frac{1}{z}\frac{dZ_\nu}{dz}+\left(1-\frac{\nu^2}{z^2}\right)Z_\nu=0.$$

Recurrence formulas:

$$Z_{\nu-1}(z)+Z_{\nu+1}(z)=(2\nu/z)Z_\nu(z),\qquad Z_{\nu-1}(z)-Z_{\nu+1}(z)=2dZ_\nu(z)/dz.$$

$$\int z^{\nu+1}Z_\nu(z)dz=z^{\nu+1}Z_{\nu+1}(z),\qquad \int z^{-\nu}Z_{\nu+1}(z)dz=-z^{-\nu}Z_\nu(z).$$

As special solutions, we have the following three kinds of functions.

(i) Bessel function (Bessel function of the first kind).

$$J_\nu(z)\equiv\left(\frac{z}{2}\right)^\nu\sum_{l=0}^{\infty}\frac{(-1)^l}{l!\Gamma(\nu+l+1)}\left(\frac{z}{2}\right)^{2l}=\frac{M_{0,\nu}(2iz)}{(2iz)^{1/2}2^{2\nu}i^\nu\Gamma(\nu+1)}\qquad(|\arg z|<\pi).$$

$$J_\nu(e^{im\pi}z)=e^{im\nu\pi}J_\nu(z).$$

$$J_{-n}(z)=(-1)^nJ_n(z).$$

$$J_{n+(1/2)}(z)=\sqrt{\frac{2}{\pi}}\,z^{n+(1/2)}\left(-\frac{1}{z}\frac{d}{dz}\right)^n\left(\frac{\sin z}{z}\right)\qquad(n=0,1,2,\ldots).$$

(ii) Neumann function (Bessel function of the second kind).

$$N_\nu(z)\equiv\frac{1}{\sin\nu\pi}\left[(\cos\nu\pi)J_\nu(z)-J_{-\nu}(z)\right]\qquad(\nu\neq\text{integer};\ \ |\arg z|<\pi),$$

$$N_n(z)\equiv\frac{2}{\pi}J_n(z)\left(C+\log\frac{z}{2}\right)-\frac{1}{\pi}\left(\frac{z}{2}\right)^n\sum_{l=0}^{\infty}\frac{(-1)^l}{l!(n+l)!}\left(\frac{z}{2}\right)^{2l}[\varphi(l)+\varphi(l+n)]$$

$$-\frac{1}{\pi}\left(\frac{z}{2}\right)^{-n}\sum_{l=0}^{n-1}\frac{(n-l-1)!}{l!}\left(\frac{z}{2}\right)^{2l}\qquad\left(\varphi(l)\equiv\sum_{m=1}^{l}\frac{1}{m}\right),$$

$$N_{-n}(z) \equiv (-1)^n N_n(z) \qquad (n=0,1,2,\ldots; \qquad |\arg z|<\pi).$$
$$N_\nu(e^{im\pi}z) = e^{-im\nu\pi}N_\nu(z) + 2i(\sin m\nu\pi \cot \nu\pi)J_\nu(z).$$
$$N_{n+(1/2)}(z) = (-1)^{n+1}J_{-[n+(1/2)]}(z).$$

(iii) Hankel function (Bessel function of the third kind).

$$H_\nu^{(1)}(z) \equiv J_\nu(z) + iN_\nu(z),$$
$$H_\nu^{(2)}(z) \equiv J_\nu(z) - iN_\nu(z).$$
$$H_\nu^{(1)}(iz/2) = -2ie^{-i\nu\pi/2}(\pi z)^{-1/2}W_{0,\nu}(z).$$
$$H_{-\nu}^{(1)}(z) = e^{i\nu\pi}H_\nu^{(1)}(z), \quad H_{-\nu}^{(2)}(z) = e^{-i\nu\pi}H_\nu^{(2)}(z). \quad \overline{H_\nu^{(2)}(x)} = H_\nu^{(1)}(x) \quad (x,\nu \text{ real}).$$

(2) Integral Representation.

Hansen-Bessel formula
$$J_n(z) = \frac{1}{2\pi}\int_{-\pi}^{+\pi} e^{iz\cos t}e^{in[t-(\pi/2)]}\,dt$$
$$= \frac{i^{-n}}{\pi}\int_0^\pi e^{iz\cos t}\cos nt\,dt$$
$$= \frac{1}{\pi}\int_0^\pi \cos(z\sin t - nt)\,dt \qquad (n=0,1,2,\ldots).$$

Mehler's formula
$$J_0(x) = \frac{2}{\pi}\int_0^\infty \sin(x\cosh t)\,dt,$$
$$N_0(x) = -\frac{2}{\pi}\int_0^\infty \cos(x\cosh t)\,dt \qquad (x>0).$$

Poisson's formula
$$J_\nu(z) = \frac{2(z/2)^\nu}{\sqrt{\pi}\,\Gamma[\nu+(1/2)]}\int_0^{\pi/2}\cos(z\cos t)\sin^{2\nu}t\,dt \qquad \left(\mathrm{Re}\,\nu > -\frac{1}{2}\right),$$
$$N_\nu(z) = \frac{2(z/2)^\nu}{\sqrt{\pi}\,\Gamma[\nu+(1/2)]}\left[\int_0^{\pi/2}\sin(z\sin t)\cos^{2\nu}t\,dt - \int_0^\infty e^{-z\sinh t}\cosh^{2\nu}t\,dt\right]$$
$$(\mathrm{Re}\,z>0, \quad \mathrm{Re}\,\nu > -1/2).$$

Schläfli's formula
$$J_\nu(z) = \frac{1}{\pi}\int_0^\pi \cos(z\sin t - \nu t)\,dt - \frac{\sin\nu\pi}{\pi}\int_0^\infty e^{-z\sinh t}e^{-\nu t}\,dt \qquad (\mathrm{Re}\,z>0),$$
$$N_\nu(z) = \frac{1}{\pi}\int_0^\pi \sin(z\sin t - \nu t)\,dt - \frac{1}{\pi}\int_0^\infty e^{-z\sinh t}[e^{\nu t} + (\cos\nu\pi)e^{-\nu t}]\,dt \qquad (\mathrm{Re}\,z>0).$$

$$J_\nu(z) = \frac{z^\nu}{2\pi i}\int_{c-i\infty}^{c+i\infty}\exp\left[\frac{1}{2}\left(t - \frac{z^2}{t}\right)\right]t^{-\nu-1}\,dt \qquad (c>0, \quad |\arg z|<\pi, \quad \mathrm{Re}\,\nu>-1).$$

$$J_\nu(x) = \frac{2(x/2)^{-\nu}}{\sqrt{\pi}\,\Gamma[(1/2)-\nu]}\int_1^\infty \frac{\sin xt}{(t^2-1)^{\nu+(1/2)}}\,dt,$$

$$N_\nu(x) = -\frac{2(x/2)^{-\nu}}{\sqrt{\pi}\,\Gamma[(1/2)-\nu]}\int_1^\infty \frac{\cos xt}{(t^2-1)^{\nu+(1/2)}}\,dt \qquad \left(x>0, \quad -\frac{1}{2}<\mathrm{Re}\,\nu<\frac{1}{2}\right).$$

$$J_\nu(z) = \frac{1}{2\pi i}\int_{-\infty}^{(0+)} e^{z[t-(1/t)]/2}t^{-\nu-1}\,dt \qquad (\mathrm{Re}\,z>0).$$

(The contour goes once around the negative real axis in the positive direction.)

Sommerfeld's formula
$$J_\nu(z) = \frac{1}{2\pi}\int_{-\eta+i\infty}^{2\pi-\eta+i\infty} e^{iz\cos t}e^{i\nu[t-(\pi/2)]}\,dt,$$

$$H_\nu^{(1)}(z) = \frac{1}{\pi}\int_{-\eta+i\infty}^{\eta-i\infty} e^{iz\cos t}e^{i\nu[t-(\pi/2)]}\,dt,$$

$$H_\nu^{(2)}(z) = \frac{1}{\pi}\int_{\eta-i\infty}^{2\pi-\eta+i\infty} e^{iz\cos t}e^{i\nu[t-(\pi/2)]}\,dt \qquad (-\eta<\arg z<\pi-\eta, \qquad 0<\eta<\pi).$$

$$H_\nu^{(1)}(z) = -\frac{2i}{\pi}e^{-i\nu\pi/2}\int_0^\infty e^{iz\cosh t}\cosh\nu t\,dt \qquad (0<\arg z<\pi;\ \text{when } \nu=0, \text{ it holds also at } z=0).$$

$$H_\nu^{(1)}(z) = -\frac{2ie^{-i\nu\pi}(z/2)^\nu}{\sqrt{\pi}\,\Gamma[\nu+(1/2)]}\int_0^\infty e^{iz\cosh t}\sinh^{2\nu}t\,dt$$

$$(0<\arg z<\pi, \quad \mathrm{Re}\,\nu>-1/2;\ \text{when } z=0,\ -1/2<\mathrm{Re}\,\nu<1/2).$$

$$H_\nu^{(1)}(z) = -i\frac{e^{-i\nu\pi/2}}{\pi}\int_0^\infty e^{iz[t-(1/t)]/2}t^{-\nu-1}\,dt \qquad (0<\arg z<\pi;\ \text{when } \arg z=0,\ -1<\operatorname{Re}\nu<1).$$

(3) Generating Function.

$$\exp\left[\frac{z(t-t^{-1})}{2}\right] = J_0(z) + \sum_{n=1}^{\infty}[t^n+(-t)^{-n}]J_n(z),$$

$$\exp(iz\cos\theta) = \sum_{n=-\infty}^{\infty} i^n J_n(z)e^{in\theta} = J_0(z) + 2\sum_{n=1}^{\infty} i^n J_n(z)\cos n\theta.$$

$$\int J_\nu(z)\,dz = 2\sum_{n=0}^{\infty} J_{\nu+2n+1}(z).$$

Kapteyn's series $\quad \dfrac{1}{1-z} = 1 + 2\displaystyle\sum_{n=1}^{\infty} J_n(nz),$

$$\frac{1}{2}\frac{z^2}{1-z^2} = \sum_{n=1}^{\infty} J_{2n}(2nz) \qquad \left(\left|\frac{z\exp\sqrt{1-z^2}}{1+\sqrt{1-z^2}}\right|<1\right).$$

Schlömilch's series. Supposing that $f(x)$ is twice continuously differentiable with respect to the real variable x in $0 \leqslant x \leqslant \pi$, we have

$$f(x) = \frac{1}{2}a_0 + \sum_{n=1}^{\infty} a_n J_0(nx) \quad (0<x<\pi),$$

where $\quad a_0 \equiv 2f(0) + \dfrac{2}{\pi}\displaystyle\int_0^\pi du \int_0^{\pi/2} f'(u\sin\varphi)\,d\varphi,$

$$a_n \equiv \frac{2}{\pi}\int_0^\pi du\int_0^{\pi/2} uf'(u\sin\varphi)\cos n\varphi\,d\varphi.$$

$$1 = J_0(z) + 2\sum_{n=1}^{\infty} J_{2n}(z) = [J_0(z)]^2 + 2\sum_{n=1}^{\infty}[J_n(z)]^2.$$

(4) Addition Theorem. For the cylindrical function Z_ν, we have

$$e^{i\nu\psi}Z_\nu(kR) = \sum_{n=-\infty}^{\infty} J_n(k\rho)Z_{\nu+n}(kr)e^{in\varphi}$$

$$\left(R=\sqrt{r^2+\rho^2-2r\rho\cos\varphi}\,,\quad 0<\psi<\frac{\pi}{2},\qquad e^{2i\psi}=\frac{r-\rho e^{-i\varphi}}{r-\rho e^{i\varphi}},\quad 0<\rho<r,\right.$$

k is an arbitrary complex number),

$$\frac{Z_\nu(kR)}{R^\nu} = 2^\nu k^{-\nu}\Gamma(\nu)\sum_{m=0}^{\infty}(\nu+m)\frac{J_{\nu+m}(k\rho)}{\rho^\nu}\frac{Z_{\nu+m}(kr)}{r^\nu}C_m^{(\nu)}(\cos\varphi)$$

($\nu \neq$ negative integer).

$$\frac{\exp\left[(-1)^{\iota+1}ikR\right]}{R} = \frac{\pi}{2}\frac{(-1)^{\iota+1}i}{\sqrt{r\rho}}\sum_{m=0}^{\infty}(2m+1)^{\iota}J_{m+(1/2)}(k\rho)H_{m+(1/2)}^{(\iota)}(kr)P_m(\cos\varphi)$$

($\iota = 1,2$).

$$e^{ik\rho\cos\varphi} = \left(\frac{\pi}{2k\rho}\right)^{1/2}\sum_{m=0}^{\infty} i^m(2m+1)J_{m+(1/2)}(k\rho)P_m(\cos\varphi)$$

$$= 2^\nu\Gamma(\nu)\sum_{m=0}^{\infty}(\nu+m)i^m J_{\nu+m}(k\rho)(k\rho)^{-\nu}C_m^{(\nu)}(\cos\varphi) \qquad (\nu\neq 0,-1,-2,\ldots),$$

where P_m is a Legendre polynomial, and $C_m^{(\nu)}$ is a Gegenbauer polynomial.

(5) Infinite Products and Partial Fractions. Let $j_{\nu,n}$ be the zeros of $z^{-\nu}J_\nu(z)$ in ascending order with respect to the real part. We have

$$J_\nu(z) = \frac{(z/2)^\nu}{\Gamma(\nu+1)}\prod_{n=1}^{\infty}\left(1-\frac{z^2}{j_{\nu,n}^2}\right) \qquad (\nu\neq -1,-2,-3,\ldots).$$

Note that if ν is real and greater than -1, all zeros are real.

Kneser-Sommerfeld formula

$$\frac{\pi J_\nu(xz)}{4J_\nu(z)}[J_\nu(z)N_\nu(Xz)-N_\nu(z)J_\nu(Xz)]=\sum_{n=1}^{\infty}\frac{J_\nu(j_{\nu,n}x)J_\nu(j_{\nu,n}X)}{(z^2-j_{\nu,n}^2)J_{\nu,n}'^2(j_{\nu,n})}$$

$(0<x<X<1,\quad \mathrm{Re}\, z>0).$

(6) Definite Integrals.

$$\int_0^{\pi/2}J_\nu(z\cos\theta)\cos\theta\,d\theta=\frac{1}{2z}\int_0^{2z}J_{2\nu}(t)\,dt,\qquad \int_0^z J_\mu(t)\,dt=\frac{1}{\pi}\int_0^\pi\frac{\sin(z\sin\theta)}{\sin\theta}\cos\mu\theta\,d\theta.$$

$$\int_0^\infty e^{-at}J_\nu(bt)t^{\mu-1}\,dt=\frac{(b/2a)^\nu\Gamma(\mu+\nu)}{a^\mu\Gamma(\nu+1)}{}_2F_1\left(\frac{\mu+\nu}{2},\ \frac{\mu+\nu+1}{2};\ \nu+1;\ \frac{-b^2}{a^2}\right)$$

$(\mathrm{Re}(a+ib)>0,\quad \mathrm{Re}(a-ib)>0,\quad \mathrm{Re}(\mu+\nu)>0).$

$$\int_0^\infty e^{-at}J_\nu(bt)t^\nu\,dt=\frac{(2x)^\nu\Gamma[\nu+(1/2)]}{(a^2+b^2)^{\nu+(1/2)}\sqrt{\pi}}\qquad\left(\mathrm{Re}\,\nu>-\frac{1}{2},\quad \mathrm{Re}\,a>|\mathrm{Im}\,b|\right).$$

$$\int_0^\infty e^{-at}J_\nu(bt)\frac{dt}{t}=\frac{(\sqrt{a^2+b^2}-a)^\nu}{\nu b^\nu}\qquad(\mathrm{Re}\,\nu>0,\quad \mathrm{Re}\,a>|\mathrm{Im}\,b|).$$

Sommerfeld's formula $\displaystyle\int_0^\infty J_0(\tau r)e^{-|x|\sqrt{\tau^2-k^2}}\frac{\tau\,d\tau}{\sqrt{\tau^2-k^2}}=\frac{e^{ik\sqrt{r^2+k^2}}}{\sqrt{r^2+x^2}}$

$(r,x\text{ real};\quad -\pi/2\leqslant\arg\sqrt{\tau^2-k^2}<\pi/2,\quad 0\leqslant\arg k<\pi).$

Weyrich's formula $\displaystyle\frac{i}{2}\int_{-\infty}^{+\infty}e^{i\tau x}H_0^{(1)}(r\sqrt{k^2-\tau^2})\,d\tau=\frac{e^{ik\sqrt{r^2+x^2}}}{\sqrt{r^2+x^2}}$

$(r,x\text{ real};\quad 0\leqslant\arg\sqrt{k^2-\tau^2}<\pi,\quad 0\leqslant\arg k<\pi).$

Weber-Sonine formula

$$\int_0^\infty J_\nu(at)e^{-p^2t^2}t^{\mu-1}\,dt=\frac{(a/2p)^\nu\Gamma[(\nu+\mu)/2]}{2p^\mu\Gamma(\nu+1)}{}_1F_1\left(\frac{\nu+\mu}{2};\ \nu+1;\ \frac{-a^2}{4p^2}\right)$$

$(\mathrm{Re}(\mu+\nu)>0,\quad |\arg p|<\pi/4,\quad a>0),$

$$\int_0^\infty J_\nu(at)e^{-p^2t^2}t^{\nu+1}\,dt=\frac{a^\nu}{(2p^2)^{\nu+1}}e^{-a^2/4p^2}\qquad(\mathrm{Re}\,\nu>-1,\quad |\arg p|<\pi/4).$$

Sonine-Schafheitlin formula

$$\int_0^\infty J_\mu(at)J_\nu(bt)t^{-\lambda}\,dt=\frac{a^\mu\Gamma[(\mu+\nu-\lambda+1)/2]}{2^\lambda b^{\mu-\lambda+1}\Gamma[(-\mu+\nu+\lambda+1)/2]\Gamma(\mu+1)}$$

$$\times{}_2F_1\left(\frac{\mu+\nu-\lambda+1}{2},\ \frac{\mu-\nu-\lambda+1}{2};\ \mu+1;\ \frac{a^2}{b^2}\right)$$

$(\mathrm{Re}(\mu+\nu-\lambda+1)>0,\quad \mathrm{Re}\,\lambda>-1,\quad 0<a<b).$

(7) Asymptotic Expansion.

(i) Hankel's asymptotic representation. We put

$$(\nu,m)\equiv\frac{[4\nu^2-1^2][4\nu^2-3^2]\dots[4\nu^2-(2m-1)^2]}{2^{2m}m!}\qquad(m=1,2,3,\dots);\qquad(\nu,0)\equiv1.$$

For $|z|\gg|\nu|,\quad |z|\gg1,$

$$J_\nu(z)=\sqrt{\frac{2}{\pi z}}\cos\left(z-\frac{\nu\pi}{2}-\frac{\pi}{4}\right)\left[\sum_{m=0}^{M-1}(-1)^m\frac{(\nu,2m)}{(2z)^{2m}}+O(|z|^{-2M})\right]$$

$$-\sqrt{\frac{2}{\pi z}}\sin\left(z-\frac{\nu\pi}{2}-\frac{\pi}{4}\right)\left[\sum_{m=0}^{M-1}(-1)^m\frac{(\nu,2m+1)}{(2z)^{2m+1}}+O(|z|^{-2M-1})\right]$$

$(-\pi<\arg z<\pi),$

$$N_\nu(z)=\sqrt{\frac{2}{\pi z}}\sin\left(z-\frac{\nu\pi}{2}-\frac{\pi}{4}\right)\left[\sum_{m=0}^{M-1}(-1)^m\frac{(\nu,2m)}{(2z)^{2m}}+O(|z|^{-2M})\right]$$

$$+\sqrt{\frac{2}{\pi z}}\cos\left(z-\frac{\nu\pi}{2}-\frac{\pi}{4}\right)\left[\sum_{m=0}^{M-1}(-1)^m\frac{(\nu,2m+1)}{(2z)^{2m+1}}+O(|z|^{-2M-1})\right]$$

$$(-\pi<\arg z<\pi),$$

$$H_\nu^{(1)}(z)=\sqrt{\frac{2}{\pi z}}\exp\left[i\left(z-\frac{\nu\pi}{2}-\frac{\pi}{4}\right)\right]\left[\sum_{m=0}^{M-1}\frac{(\nu,m)}{(-2iz)^m}+O(|z|^{-M})\right]\qquad(-\pi<\arg z<2\pi),$$

$$H_\nu^{(2)}(z)=\sqrt{\frac{2}{\pi z}}\exp\left[-i\left(z-\frac{\nu\pi}{2}-\frac{\pi}{4}\right)\right]\left[\sum_{m=0}^{M-1}\frac{(\nu,m)}{(2iz)^m}+O(|z|^{-M})\right]\qquad(-2\pi<\arg z<\pi).$$

(ii) Debye's asymptotic representation.

$\nu \fallingdotseq x$, $1-(\nu/x)>\varepsilon$, $\nu/x=\sin\alpha$, when $1-(\nu/x)>(3/x)\nu^{1/2}$,

$$H_\nu^{(1)}(x)\sim\frac{1}{\sqrt{\pi}}\exp\left[ix\left\{\cos\alpha+\left(\alpha-\frac{\pi}{2}\right)\sin\alpha\right\}\right]$$

$$\times\left[\frac{e^{i\pi/4}}{X}+\left(\frac{1}{8}+\frac{5}{24}\tan^2\alpha\right)\frac{3e^{3\pi i/4}}{2X^3}\right.$$

$$\left.+\left(\frac{3}{128}+\frac{77}{576}\tan^2\alpha+\frac{385}{3456}\tan^4\alpha\right)\frac{3\cdot 5e^{5\pi i/4}}{2^2X^5}+\ldots\right]$$

$$(X=[-x\cos(\alpha/2)]^{1/2}).$$

$\nu \fallingdotseq x$, $(\nu/x)-1>\varepsilon$, $\nu/x=\cosh\sigma$, when $|\nu^2-x^2|^{1/2}\gg 1$, $|\nu^2-x^2|^{3/2}\nu^{-2}\gg 1$

$$H_\nu^{(1)}(x)\sim\frac{1}{\sqrt{\pi}}\exp[x(\sigma\cosh\sigma-\sinh\sigma)]$$

$$\times\left[\frac{1}{X}+\left(\frac{1}{8}-\frac{5}{24}\coth^2\sigma\right)\frac{3}{2X^3}+\left(\frac{3}{128}-\frac{77}{576}\coth^2\sigma+\frac{385}{3456}\coth^4\sigma\right)\frac{3\cdot 5}{2^2X^5}+\ldots\right]$$

$$(X=[-x\sinh(\sigma/2)]^{1/2}).$$

When $\nu \fallingdotseq x$, $|x-\nu|\ll x^{1/3}$, $x\gg 1$, $x-\nu=\delta$,

$$H_\nu^{(2)}(x)\sim\frac{6^{1/3}e^{i\pi/3}}{3^{1/2}\pi}\left[\frac{\Gamma(1/3)}{x^{1/3}}-6^{1/3}e^{i\pi/3}\delta\frac{\Gamma(2/3)}{x^{2/3}}+\left(\frac{2}{5}\delta-\delta^3\right)\frac{\Gamma(4/3)}{x^{4/3}}\right.$$

$$\left.+\left(\frac{3}{140}-\frac{\delta^2}{4}+\frac{\delta^4}{4}\right)6^{1/3}e^{i\pi/3}\frac{\Gamma(5/3)}{x^{5/3}}+\ldots\right]$$

(iii) Watson-Nicholson formula. When $x,\nu>0$, $w=[(x/\nu)^2-1]^{1/2}$,

$$H_\nu^{(\iota)}(x)=3^{-1/2}w\exp[(-1)^{\iota+1}i\{(\pi/6)+\nu(w-(w^3/3)-\arctan w)\}]H_{1/3}^{(\iota)}(\nu w^3/3)+O|\nu^{-1}|$$

$$(\iota=1,2).$$

(IV) Functions Related to Bessel Functions

(1) Modified Bessel Functions.

$$I_\nu(z)\equiv e^{-i\nu\pi/2}J_\nu(e^{i\pi/2}z)$$

$$=\sum_{n=0}^{\infty}\frac{(z/2)^{\nu+2n}}{n!\Gamma(\nu+n+1)},$$

$$K_\nu(z)\equiv\frac{i\pi}{2}e^{i\nu\pi/2}H_\nu^{(1)}(e^{i\pi/2}z)=-\frac{i\pi}{2}e^{-i\nu\pi/2}H_\nu^{(2)}(e^{-i\pi/2}z)$$

$$=\frac{\pi}{2}\frac{I_{-\nu}(z)-I_\nu(z)}{\sin\nu\pi}=\left(\frac{\pi}{2z}\right)^{1/2}W_{0,\nu}(2z).$$

Recurrence formulas:

$$I_{\nu-1}(z)-I_{\nu+1}(z)=(2\nu/z)I_\nu(z),$$
$$I_{\nu-1}(z)+I_{\nu+1}(z)=2I_\nu'(z).$$
$$K_{\nu-1}(z)-K_{\nu+1}(z)=-(2\nu/z)K_\nu(z),$$
$$K_{\nu-1}(z)+K_{\nu+1}(z)=-2K_\nu'(z),$$
$$K_{-\nu}(z)=K_\nu(z).$$

Airy's integral: $$\int_0^\infty \cos(t^3-tx)\,dt=\frac{\pi}{3}\sqrt{\frac{x}{3}}\left[J_{1/3}\left(\frac{2x\sqrt{x}}{3\sqrt{3}}\right)+J_{-1/3}\left(\frac{2x\sqrt{x}}{3\sqrt{3}}\right)\right],$$

$$\int_0^\infty \cos(t^3+tx)\,dt=\tfrac{1}{3}\sqrt{x}\,K_{1/3}\left(\frac{2x\sqrt{x}}{3\sqrt{3}}\right)\qquad(x>0).$$

H. Weber's formula: $$\frac{1}{2p^2}e^{-(a^2+b^2)/4p^2}I_\nu\left(\frac{ab}{2p^2}\right)=\int_0^\infty e^{-p^2t^2}J_\nu(at)J_\nu(bt)\,t\,dt$$

$$(\mathrm{Re}\,\nu>-1,\quad |\arg p|<\pi/4;\quad a,b>0).$$

Watson's formula: $$J_\mu(z)N_\nu(z)-J_\nu(z)N_\mu(z)=\frac{4\sin(\mu-\nu)\pi}{\pi^2}\int_0^\infty K_{\nu-\mu}(2z\sinh t)e^{(\mu+\nu)t}\,dt$$

$$(\mathrm{Re}\,z>0,\quad \mathrm{Re}(\mu-\nu)<1),$$

$$J_\nu(z)\frac{\partial N_\nu(z)}{\partial\nu}-N_\nu(z)\frac{\partial J_\nu(z)}{\partial\nu}=-\frac{4}{\pi}\int_0^\infty K_0(2z\sinh t)e^{-2\nu t}\,dt\qquad(\mathrm{Re}\,z>0).$$

Nicholson's formula: $$J_\nu^2(z)+N_\nu^2(z)=\frac{8}{\pi^2}\int_0^\infty K_0(2z\sinh t)\cosh 2\nu t\,dt\qquad(\mathrm{Re}\,z>0).$$

Dixon-Ferrar formula: $$J_\nu^2(z)+N_\nu^2(z)=\frac{8\cos\nu\pi}{\pi^2}\int_0^\infty K_{2\nu}(2z\sinh t)\,dt$$

$$\left(\mathrm{Re}\,z>0;\qquad -\frac{1}{2}<\mathrm{Re}\,\nu<\frac{1}{2}\right).$$

(2) Kelvin Functions.

$$\mathrm{ber}_\nu(z)\pm i\,\mathrm{bei}_\nu(z)\equiv J_\nu(e^{\pm 3\pi i/4}z),$$
$$\mathrm{her}_\nu(z)\pm i\,\mathrm{hei}_\nu(z)\equiv H_\nu^{(1)}(e^{\pm 3\pi i/4}z),$$
$$\mathrm{ker}_\nu(z)\equiv-(\pi/2)\mathrm{hei}_\nu(z),$$
$$\mathrm{kei}_\nu(z)\equiv(\pi/2)\mathrm{her}_\nu(z).$$

When ν is an integer n,

$$\mathrm{ber}_n(x)-i\,\mathrm{bei}_n(x)=(-1)^nJ_n(\sqrt{i}\,x),$$
$$\mathrm{her}_n(x)-i\,\mathrm{hei}_n(x)=(-1)^{n+1}H_n^{(1)}(\sqrt{i}\,x)\qquad(x\text{ real}).$$

(3) Struve Function.

$$\boldsymbol{H}_\nu(x)\equiv\frac{2(z/2)^\nu}{\Gamma[\nu+(1/2)]\sqrt{\pi}}\int_0^{\pi/2}\sin(z\cos\theta)\sin^{2\nu}\theta\,d\theta$$

$$=\sum_{m=0}^{\infty}\frac{(-1)^m(z/2)^{\nu+2m+1}}{\Gamma[m+(3/2)]\Gamma[\nu+m+(3/2)]}.$$

Anger function: $$\boldsymbol{J}_\nu(z)\equiv\frac{1}{\pi}\int_0^\pi\cos(\nu\theta-z\sin\theta)\,d\theta.$$

H. F. Weber function: $$\boldsymbol{E}_\nu(z)\equiv\frac{1}{\pi}\int_0^\pi\sin(\nu\theta-z\sin\theta)\,d\theta.$$

Putting $\nabla_\nu\equiv z^2\dfrac{d^2}{dz^2}+z\dfrac{d}{dz}+z^2-\nu^2$,

$$\nabla_\nu\boldsymbol{H}_\nu(z)=\frac{4(z/2)^{\nu+1}}{\Gamma[\nu+(1/2)]\sqrt{\pi}},\qquad \nabla_\nu\boldsymbol{J}_\nu(z)=\frac{(z-\nu)\sin\nu\pi}{\pi},$$

$$\nabla_\nu\boldsymbol{E}_\nu(z)=-\frac{z+\nu}{\pi}-\frac{(z-\nu)\cos\nu\pi}{\pi}.$$

When ν is an integer n, $\mathbf{J}_n(z)=J_n(z)$.

$$\int_0^z J_0(t)\,dt = zJ_0(z)+\frac{\pi z}{2}[J_1(z)\mathbf{H}_0(z)-J_0(z)\mathbf{H}_1(z)],$$

$$\int_0^z N_0(t)\,dt = zN_0(z)+\frac{\pi z}{2}[N_1(z)\mathbf{H}_0(z)-N_0(z)\mathbf{H}_1(z)].$$

(4) Neumann Polynomials. $O_n(t)=\sum_{j=0}^{[n/2]}\frac{n(n-j-1)!}{j!(t/2)^{n-2j+1}}$ (n is a positive integer),

$$O_0(t)=1/t.$$

$$\frac{1}{t-z}\equiv 1+2\sum_{n=1}^{\infty}O_n(t)J_n(z) \qquad (|t|>|z|).$$

Schläfli polynomials: $S_n(t)\equiv\frac{2}{n}\left[tO_n(t)-\cos^2\frac{n\pi}{2}\right]$ (n is a positive integer),

$$S_0(t)\equiv 0.$$

$\nabla_n S_n(x)=2n+2(x-n)\sin^2(n\pi/2)$ (∇_n is the same operator defined in (3)).

Lommel polynomials: $$R_{m,\nu}(z)\equiv\frac{\Gamma(\nu+m)}{\Gamma(\nu)(z/2)^m}{}_2F_3\left(\frac{1-m}{2},\ \frac{-m}{2};\ \nu,\ -m,\ 1-\nu-m;\ -z^2\right)$$

$$=(\pi z/2\sin\nu\pi)[J_{\nu+m}(z)J_{-\nu+1}(z)+(-1)^mJ_{-\nu-m}(z)J_{\nu-1}(z)]$$

(m is a nonnegative integer).

References

See references to Table 16, this Appendix.

20. Systems of Orthogonal Functions (→ 317 Orthogonal Functions)

$$\int_a^b p_n(x)p_m(x)\varphi(x)\,dx=\delta_{nm}A_n$$

Name	Notation $p_n(x)$	Interval (a,b)	Weight $\varphi(x)$	Norm A_n
Legendre	$P_n(x)$	$(-1,+1)$	1	$2/(2n+1)$
Gegenbauer	$C_n^\nu(x)$	$(-1,+1)$	$(1-x^2)^{\nu-(1/2)}$	$2\pi\Gamma(2\nu+n)/2^{2\nu}(n+\nu)n![\Gamma(\nu)]^2$
Chebyshev	$T_n(x)$	$(-1,+1)$	$(1-x^2)^{-1/2}$	$\pi(n=0);\ \pi/2\ (n\geqslant 1)$
Hermite	$H_n(x)$	$(-\infty,+\infty)$	e^{-x^2}	$\sqrt{\pi}\cdot n!$
Jacobi	$G_n(\alpha,\gamma;x)$	$(0,1)$	$x^{\gamma-1}(1-x)^{\alpha-\gamma}$	$\dfrac{n![\Gamma(\gamma)]^2\Gamma(\alpha+n-\gamma+1)}{(\alpha+2n)\Gamma(\alpha+n)\Gamma(\gamma+n)}$
Laguerre	$L_n^\alpha(x)$	$(0,\infty)$	$x^\alpha e^{-x}$	$\Gamma(\alpha+n+1)/n!$

For Legendre polynomials $P_n(x)$ → Table 18.II, this Appendix.

(I) Gegenbauer Polynomials (Gegenbauer Functions)

$$C_n^\nu(t)\equiv\frac{\Gamma(n+2\nu)}{n!\Gamma(2\nu)}{}_2F_1\left(n+2\nu,\ -n;\ \nu+\frac{1}{2};\ \frac{1-t}{2}\right)$$

$$=\frac{\Gamma(2\nu+n)\Gamma[\nu+(1/2)]}{\Gamma(2\nu)n!}\left[\frac{1}{4}(t^2-1)\right]^{(1/4)-(\nu/2)}\mathfrak{P}^{(1/2)-\nu}_{n+\nu-(1/2)}(t).$$

Generating function $\quad (1-2\alpha t+\alpha^2)^{-\nu}\equiv\sum_{n=0}^{\infty}C_n^\nu(t)\alpha^n.\quad C_{n-l}^{l+(1/2)}(x)=\frac{1}{(2l-1)!!}\frac{d^lP_n(t)}{dt^l}.$

Orthogonal relation $\quad \int_0^\pi(\sin^{2\nu}\theta)C_m^\nu(\cos\theta)C_n^\nu(\cos\theta)d\theta=\frac{\pi\Gamma(2\nu+n)}{2^{2\nu-1}(\nu+n)n![\Gamma(\nu)]^2}\delta_{nm}.$

(II) Chebyshev (Tschebyscheff) Polynomials

(1) Chebyshev Polynomial (Chebyshev Function of the First Kind)

$$\begin{aligned}T_n(x)&\equiv\cos(n\arccos x)\\&=(1/2)\left[\left(x+i\sqrt{1-x^2}\right)^n+\left(x-i\sqrt{1-x^2}\right)^n\right]\\&=F(n,-n;1/2;(1-x)/2)\\&=\sum_{j=0}^{[n/2]}(-1)^j\binom{n}{2j}x^{n-2j}(1-x^2)^j\\&=\frac{(-1)^n(1-x^2)^{1/2}}{(2n-1)!!}\frac{d^n(1-x^2)^{n-(1/2)}}{dx^n},\end{aligned}$$

Chebyshev function of the second kind

$$\begin{aligned}U_n(x)&\equiv\sin(n\arccos x)\\&=(1/2i)\left[\left(x+i\sqrt{1-x^2}\right)^n-\left(x-i\sqrt{1-x^2}\right)^n\right]\\&=\frac{(-1)^{n-1}n}{(2n-1)!!}\frac{d^{n-1}(1-x^2)^{n-(1/2)}}{dx^{n-1}}.\end{aligned}$$

$T_n(x)$, $U_n(x)$ are mutually linearly independent solutions of Chebyshev's differential equation $(1-x^2)y''-xy'+n^2y=0$. Recurrence relations are

$$T_{n+1}(x)-2xT_n(x)+T_{n-1}(x)=0,\quad U_{n+1}(x)-2xU_n(x)+U_{n-1}(x)=0.$$

Generating function:

$$\frac{1-t^2}{1-2tx+t^2}=T_0(x)+2\sum_{n=0}^{\infty}T_n(x)t^n,\quad \frac{1}{1-2tx+t^2}=\frac{1}{\sqrt{1-x^2}}\sum_{n=0}^{\infty}U_{n+1}(x)t^n.$$

Orthogonal relation:

$$\int_{-1}^{+1}\frac{T_m(x)T_n(x)}{\sqrt{1-x^2}}dx=\begin{cases}0 & (m\neq n),\\ \pi/2 & (m=n\neq0),\\ \pi & (m=n=0);\end{cases}\quad \int_{-1}^{+1}\frac{U_m(x)U_n(x)}{\sqrt{1-x^2}}dx=\begin{cases}\frac{\pi}{2} & (m=n\neq0),\\ 0 & (\text{otherwise}).\end{cases}$$

Orthogonality in finite sums. Let $u_0, u_1, \dots, u_k$ be the zeros of $T_{k+1}(x)$. All zeros are real and situated in the interval $(-1,1)$. Then we have

$$\sum_{i=0}^{k}T_m(u_i)T_n(u_i)=\begin{cases}0 & (m\neq n,\text{ or } m=n=k+1),\\ (k+1)/2 & (1\leqslant m=n\leqslant k),\\ k+1 & (m=n=0).\end{cases}$$

Let $p_n(x)$ be the best approximation of x^n in $-1\leqslant x\leqslant 1$ by polynomials of degree at most $n-1$. Then we have $x^n-p_n(x)=2^{-n+1}T_n(x)$.

(2) Expansions by $T_n(x)$.

$$e^{ax}=I_0(a)+2\sum_{n=1}^{\infty}I_n(a)T_n(x),$$

$$\sin ax=2\sum_{n=0}^{\infty}(-1)^nJ_{2n+1}(a)T_{2n+1}(x),$$

$$\cos ax=J_0(a)+2\sum_{n=1}^{\infty}(-1)^nJ_{2n}(a)T_{2n}(x),$$

$$\log(1+x\sin 2\alpha)=2\log\cos\alpha-2\sum_{n=1}^{\infty}\frac{1}{n}(-\tan\alpha)^n T_n(x),$$

$$\arctan x=2\sum_{n=1}^{\infty}(-1)^n\frac{(\sqrt{2}-1)^{2n+1}}{2n+1}T_{2n+1}(x).$$

(III) Parabolic Cylinder Functions (Weber Functions) (→ 167 Functions of Confluent Type)

Parabolic cylinder functions:

$$D_\nu(z)\equiv 2^{(1/4)+(\nu/2)}z^{-1/2}W_{(1/4)+(\nu/2),\,-1/4}(z^2/2)$$

$$=\sqrt{\pi}\,2^{(1/4)+(\nu/2)}z^{-1/2}\left[\frac{M_{(1/4)+(\nu/2),\,-1/4}(z^2/2)}{\Gamma[(1-\nu)/2]}+\frac{M_{(1/4)-(\nu/2),\,-1/4}(z^2/2)}{\Gamma(-\nu/2)}\right]$$

$$=2^{\nu/2}e^{-z^2/4}\sqrt{\pi}\left[\frac{1}{\Gamma[(1-\nu)/2]}\,{}_1F_1\left(\frac{-\nu}{2};\frac{1}{2};\frac{z^2}{2}\right)-\frac{\sqrt{2}\,z}{\Gamma(-\nu/2)}\,{}_1F_1\left(\frac{1-\nu}{2};\frac{3}{2};\frac{z^2}{2}\right)\right].$$

The solutions of Weber's differential equation

$$\frac{d^2u}{dz^2}+\left(\nu+\frac{1}{2}-\frac{z^2}{4}\right)u=0$$

are given by

$$D_\nu(z),\ D_\nu(-z),\ D_{-\nu-1}(iz),\ D_{-\nu-1}(-iz),$$

and the following relations hold among them.

$$D_\nu(z)=\left[\Gamma(\nu+1)/\sqrt{2\pi}\right]\left[e^{i\nu\pi/2}D_{-\nu-1}(iz)+e^{-i\nu\pi/2}D_{-\nu-1}(-iz)\right]$$

$$=e^{-i\nu\pi}D_\nu(-z)+\left[\sqrt{2\pi}/\Gamma(-\nu)\right]e^{-i(\nu+1)\pi/2}D_{-\nu-1}(iz)$$

$$=e^{i\nu\pi}D_\nu(-z)+\left[\sqrt{2\pi}/\Gamma(-\nu)\right]e^{i(\nu+1)\pi/2}D_{-\nu-1}(-iz).$$

Integral representation:

$$D_\nu(z)=\frac{e^{-z^2/4}}{\Gamma(-\nu)}\int_0^\infty e^{-zt-(t^2/2)}t^{-\nu-1}dt\quad(\mathrm{Re}\,\nu<0).$$

$$e^{-(z^2/4)-zt-(t^2/2)}=\sum_{n=0}^{\infty}\frac{(-t)^n}{n!}D_n(z)=\frac{1}{2\pi i}\int_{c-i\infty}^{c+i\infty}t^\nu\Gamma(-\nu)D_\nu(z)d\nu\quad(c<0,\ |\arg t|<\pi/4).$$

Recurrence formula:

$$D_{\nu+1}(z)-zD_\nu(z)+\nu D_{\nu-1}(z)=0,\quad dD_\nu(z)/dz+(1/2)zD_\nu(z)-\nu D_{\nu-1}(z)=0.$$

$$D_\nu(0)=\frac{2^{\nu/2}\sqrt{\pi}}{\Gamma[(1-\nu)/2]},\quad D_\nu'(0)=-\frac{2^{(\nu+1)/2}\sqrt{\pi}}{\Gamma(-\nu/2)}.$$

Asymptotic expansion:

$$D_\nu(z)\approx e^{-z^2/4}z^\nu\left(1-\frac{\nu(\nu-1)}{2z^2}+\frac{\nu(\nu-1)(\nu-2)(\nu-3)}{2\cdot 4z^4}\mp\ldots\right)\quad\left(|\arg z|<\frac{3}{4}\pi\right).$$

$$D_{-1}(z)=e^{z^2/4}\sqrt{\frac{\pi}{2}}\left[1-\mathrm{erf}\left(\frac{z}{\sqrt{2}}\right)\right],\quad \mathrm{erf}(x)\equiv\frac{2}{\sqrt{\pi}}\int_0^x e^{-t^2}dt\quad\text{(error function)}.$$

(IV) Hermite Polynomials

For the parabolic cylinder functions, when ν is an integer n, we have

$$D_n(z)=(-1)^n e^{z^2/4}d^n(e^{-z^2/2})/dz^n=e^{-z^2/4}H_n(z/\sqrt{2}),$$

where $H_n(x)$ is the Hermite polynomial

$$H_n(x) \equiv 2^{-n/2}(-1)^n e^{x^2} d^n (e^{-x^2})/dx^n = e^{x^2/2} D_n(\sqrt{2}\,x).$$

A Hermite polynomial is more often defined by the following function $\mathrm{He}_n(x)$ (e.g., in W.F. Magnus, F. Oberhettinger, and R. P. Soni [1]).

$$\mathrm{He}_n(x) \equiv (-1)^n e^{x^2/2} d^n (e^{-x^2/2})/dx^n = e^{x^2/4} D_n(x) = H_n(x/\sqrt{2}).$$

The function $y = H_n(x)$ is a solution of Hermite's differential equation

$$y'' - 2xy' + 2ny = 0.$$

$H_n(x)$ is a polynomial in x of degree n, and is an even or odd function according to whether n is even or odd.

$$H_{2n}(x) = (-1)^n (2n-1)!!\,{}_1F_1(-n; 1/2; x^2),$$

$$H_{2n+1}(x) = (-1)^n (2n+1)!!\sqrt{2}\,x\,{}_1F_1(-n; 3/2; x^2).$$

Recurrence formula:

$$H_{n+1}(x) = \sqrt{2}\,xH_n(x) - nH_{n-1}(x) = \sqrt{2}\,xH_n(x) - H_n'(x)/\sqrt{2},$$

$$H_n'(x) = \sqrt{2}\,nH_{n-1}(x).$$

$$H_{2n}(0) = \frac{(-1)^n (2n)!}{2^n n!} = (-1)^n (2n-1)!!, \quad H_{2n+1}(0) = 0.$$

Generating function:

$$e^{\sqrt{2}\,tx - (t^2/2)} = \sum_{n=0}^{\infty} H_n(x) t^n / n!.$$

Orthogonal relation:

$$\int_{-\infty}^{+\infty} H_n(x) H_m(x) e^{-x^2} dx = \delta_{nm} n! \sqrt{\pi}.$$

(V) Jacobi Polynomials

$$G_n(\alpha, \gamma; x) \equiv F(-n, \alpha+n; \gamma; x)$$

$$= x^{1-\gamma}(1-x)^{\gamma-\alpha} \frac{\Gamma(\gamma+n)}{\Gamma(\gamma)} \frac{d^n}{dx^n}\left[x^{\gamma+n-1}(1-x)^{\alpha+n-\gamma}\right].$$

These satisfy Jacobi's differential equation $x(1-x)y'' + [\gamma - (\alpha+1)x]y' + n(\alpha+n)y = 0.$

Orthogonal relation:

$$\int_0^1 x^{\gamma-1}(1-x)^{\alpha-\gamma} G_m(\alpha, \gamma; x) G_n(\alpha, \gamma; x)\, dx = \frac{n!\Gamma(\alpha+n-\gamma+1)\Gamma(\gamma)^2}{(\alpha+2n)\Gamma(\alpha+n)\Gamma(\gamma+n)} \delta_{mn}$$

$$(\mathrm{Re}\gamma > 0, \quad \mathrm{Re}(\alpha-\gamma) > -1).$$

Representation of other functions:

$$P_n(x) = G\left(1, 1; \frac{1-x}{2}\right), \quad T_n(x) = G\left(0, \frac{1}{2}; \frac{1-x}{2}\right),$$

$$C_n^{\nu}(x) = (-1)^n \frac{\Gamma(2\nu+n)}{\Gamma(2\nu)\cdot n!} G_n\left(2\nu, \nu+\frac{1}{2}; \frac{1+x}{2}\right).$$

(VI) Laguerre Functions

(1) Laguerre Functions.

$$L_\nu^{(\alpha)}(z) \equiv \frac{\Gamma(\alpha+\nu+1)}{\Gamma(\alpha+1)\Gamma(\nu+1)} z^{-(\alpha+1)/2} e^{z/2} M_{[(\alpha+1)/2]+\nu,\alpha/2}(z)$$

$$= \frac{\Gamma(\alpha+\nu+1)}{\Gamma(\alpha+1)\Gamma(\nu+1)} {}_1F_1(-\nu;\alpha+1;z),$$

These satisfy Laguerre's differential equation

$$zd^2[L_\nu^{(\alpha)}(z)]/dz^2 + (\alpha+1-z)d[L_\nu^{(\alpha)}(z)]/dz + \nu L_\nu^{(\alpha)}(z) = 0.$$

(2) Laguerre Polynomials. When ν is an integer n ($n=0,1,2,\ldots$), the function $L_n^{(\alpha)}(x)$ reduces to a polynomial of degree n as follows.

Laguerre polynomials:

$$L_n^{(\alpha)}(x) = \frac{e^x x^{-\alpha}}{n!}\frac{d^n}{dx^n}(e^{-x}x^{n+\alpha}) = \sum_{j=0}^{n}\binom{n+\alpha}{n-j}\frac{(-x)^j}{j!}.$$

$$L_n^{(0)}(x)=1, \quad L_0^{(m)}(x)=1, \quad L_{n+m}^{(-m)}(x) = \frac{(-1)^m n!}{(n+m)!} x^m L_n^{(m)}(x) \quad (m=0,1,2,\ldots).$$

Recurrence formulas:

$$nL_n^{(\alpha)}(x) = (-x+2n+\alpha-1)L_{n-1}^{(\alpha)}(x) - (n+\alpha-1)L_{n-2}^{(\alpha)}(x),$$

$$xd[L_n^{(\alpha)}(x)]/dx = nL_n^{(\alpha)}(x) - (n+\alpha)L_{n-1}^{(\alpha)}(x) \quad (n=2,3,\ldots).$$

Generating function:

$$\frac{e^{-xt/(1-t)}}{(1-t)^{\alpha+1}} = \sum_{n=0}^{\infty} L_n^{(\alpha)}(x)t^n \quad (|t|<1).$$

Orthogonal relations:

$$\int_0^\infty e^{-x}x^\alpha L_m^{(\alpha)}(x)L_n^{(\alpha)}dx = \delta_{mn}\Gamma(\alpha+n+1)/n! = \delta_{mn}\Gamma(1+\alpha)\binom{n+\alpha}{n}.$$

$$H_{2n}(x) = (-2)^n n! L_n^{(-1/2)}(x^2), \quad H_{2n+1}(x) = (-2)^n n!\sqrt{2}\, xL_n^{(1/2)}(x^2).$$

(3) Sonine Polynomials.

$$S_n^{(\alpha)}(x) \equiv \frac{(-1)^n}{\Gamma(\alpha+n+1)} L_n^{(\alpha)}(x).$$

(VII) Orthogonal Polynomials

$$P_{n,m}(x) = \sum_{k=0}^{n}(-1)^k\binom{n}{k}\binom{n+k}{k}\frac{x(x-1)\ldots(x-k+1)}{m(m-1)\ldots(m-k+1)}$$

(where n, m are positive integers and $n \leqslant m$).

We have the same polynomials if we replace x^k in $P_n(1-2x)$ by

$$x(x-1)\ldots(x-k+1)/m(m-1)\ldots(m-k+1) \quad (k=0,1,\ldots,n).$$

Orthogonality in finite sums:

$$\sum_{k=0}^{m} P_{n,m}(k)P_{l,m}(k) = \delta_{nl}\frac{(m+n+1)!(m-n)!}{(2n+1)(m!)^2}.$$

Chebyshev's q functions:

$$q_n(m,x) = \frac{(-1)^n(m-1)!}{2^n(m-n-1)!}P_{n,m-1}(x), \quad \xi_{n,m}(x) = \left[2^n(n!)^2/(2n)!\right]q_n(m,x-1).$$

For given data y_k at m points $x_k = x_1+(k-1)h$ ($k=1,\ldots,m$) that are equally spaced with step h, the least square approximation among the polynomials $Q(x)$ of degree $n(<m)$, i.e., the polynomial that minimizes the square sum of the residues $S=\sum_{k=1}^{m}[y_k-Q(x_k)]^2$ is given by the

following formula (→ 19 Analog Computation):

$$Q(x)=\sum_{k=0}^{n}\frac{B_k}{S_k}\xi_{k,m}\left(\frac{x-x_1}{h}+1\right),\quad S=\sum_{k=1}^{m}(y_k)^2-\sum_{k=0}^{m}\frac{B_k^2}{S_k},$$

$$B_k=\sum_{i=1}^{m}y_i\xi_{k,m}(i),\quad S_k=\sum_{i=1}^{m}[\xi_{k,m}(i)]^2.$$

References

See references to Table 16, this Appendix.

21. Interpolation (→ 223 Interpolation)

(1) Lagrange's Interpolation Polynomial.

$$f(x)=\sum_{s=0}^{n}f(x_s)\frac{(x-x_0)(x-x_1)\dots(x-x_{s-1})(x-x_{s+1})\dots(x-x_n)}{(x_s-x_0)(x_s-x_1)\dots(x_s-x_{s-1})(x_s-x_{s+1})\dots(x_s-x_n)}.$$

Aitken's interpolation scheme. The interpolation polynomial $f(x)$ corresponding to the value $y_s=f(x_s)$ $(s=0,1,\dots,n)$ is given inductively by the following procedure. The order of $x_0,x_1,\dots,x_s$ is quite arbitrary.

$$p_{s,0}(x)=y_s\quad(s=0,1,\dots,n),$$
$$p_{s,k+1}(x)=\left[(x_s-x)p_{k,k}(x)-(x_k-x)p_{s,k}(x)\right]/(x_s-x_k)\quad(s=k+1,k+2,\dots,n),$$
$$f(x)\equiv p_{n,n}(x).$$

(2) Interpolation for Equally Spaced Points. When the points x_k lie in the order of their subscripts at a uniform distance h $(x_s=x_0+sh)$, we make the following difference table $(\Delta x=h)$. Forward difference:

$$\Delta_i\equiv\Delta_i^1\equiv f_{i+1}-f_i=f(x_{i+1})-f(x_i),\quad \Delta_i^s\equiv\Delta_{i+1}^{s-1}-\Delta_i^{s-1}.$$

Variable	Value of Function	Difference (1st)	(2nd)	(3rd)	(4th)	…
…	…					
$x_0-2\Delta x$	f_{-2}	…	…			
$x_0-\Delta x$	f_{-1}	Δ_{-2}	Δ_{-2}^2	…	…	
x_0	f_0	Δ_{-1}	Δ_{-1}^2	Δ_{-2}^3	Δ_{-2}^4	…
$x_0+\Delta x$	f_1	Δ_0	Δ_0^2	Δ_{-1}^3	Δ_{-1}^4	…
$x_0+2\Delta x$	f_2	Δ_1	Δ_1^2	Δ_0^3	…	
$x_0+3\Delta x$	f_3	Δ_2	…	…		
…	…	…				

Backward difference:

$$\bar{\Delta}_i^s\equiv\bar{\Delta}_i^{s-1}-\bar{\Delta}_{i-1}^{s-1}=\Delta_{s-i}^s.$$

Central difference:

$$\delta_i^s=\delta_{i+(1/2)}^{s-1}-\delta_{i-(1/2)}^{s-1},\quad \delta_{i+(s/2)}^s=\Delta_i^s.$$

Newton interpolation formula (forward type):

$$f(x_0+u\Delta x)=f(x_0)+\frac{u}{1!}\Delta_0+\frac{u(u-1)}{2!}\Delta_0^2+\frac{u(u-1)(u-2)}{3!}\Delta_0^3+\frac{u(u-1)(u-2)(u-3)}{4!}\Delta_0^4+\dots.$$

Gauss's interpolation formula (forward type):

$$f(x_0+u\Delta x)=f(x_0)+\frac{u}{1!}\Delta_0+\frac{u(u-1)}{2!}\Delta_{-1}^2+\frac{u(u-1)(u+1)}{3!}\Delta_{-1}^3+\frac{u(u-1)(u+1)(u-2)}{4!}\Delta_{-2}^4+\dots.$$

Stirling's interpolation formula:

$$f(x_0+u\Delta x)=f(x_0)+\frac{u}{1!}\frac{\Delta_{-1}+\Delta_0}{2}+\frac{u^2}{2!}\Delta^2_{-1}+\frac{u(u^2-1)}{3!}\frac{\Delta^3_{-2}+\Delta^3_{-1}}{2}+\frac{u^2(u^2-1)}{4!}\Delta^4_{-2}+\ldots.$$

Bessel's interpolation formula:

$$f\left(\frac{x_0+x_1}{2}+v\Delta x\right)=\frac{f(x_0)+f(x_1)}{2}+\frac{v}{1!}\Delta_0+\frac{1}{2!}\left(v^2-\frac{1}{4}\right)\frac{\Delta^2_{-1}+\Delta^2_0}{2}+\frac{v}{3!}\left(v^2-\frac{1}{4}\right)\Delta^3_{-1}$$

$$+\frac{1}{4!}\left(v^2-\frac{1}{4}\right)\left(v^2-\frac{9}{4}\right)\frac{\Delta^4_{-2}+\Delta^4_{-1}}{2}+\ldots.$$

Everett's interpolation formula:

$$f(x_0+u\Delta x)=f(x_1-\xi\Delta x)=\xi f(x_0)+\frac{\xi(\xi^2-1)}{3!}\Delta^2_{-1}+\frac{\xi(\xi^2-1)(\xi^2-4)}{5!}\Delta^4_{-2}+\ldots$$

$$+uf(x_1)+\frac{u(u^2-1)}{3!}\Delta^2_0+\frac{u(u^2-1)(u^2-4)}{5!}\Delta^4_{-1}+\ldots.\quad(\xi=1-u)$$

(3) Interpolation for Functions of Two Variables. Let $x_m=x_0+m\Delta x$, $y_n\equiv y_0+n\Delta y$ (m and n are integers). We define the finite differences as follows:

$$\Delta_x(x_0,y_0)\equiv f(x_1,y_0)-f(x_0,y_0),$$

$$\Delta_y(x_0,y_0)\equiv f(x_0,y_1)-f(x_0,y_0),$$

$$\Delta^2_x(x_0,y_0)\equiv\Delta_x(x_1,y_0)-\Delta_x(x_0,y_0)\equiv\delta^2_x(x_1,y_0),$$

$$\Delta_{xy}(x_0,y_0)\equiv\Delta_y(x_1,y_0)-\Delta_y(x_0,y_0)=\Delta_x(x_0,y_1)-\Delta_x(x_0,y_0),$$

$$\Delta^2_y(x_0,y_0)\equiv\Delta_y(x_0,y_1)-\Delta_y(x_0,y_0)\equiv\delta^2_y(x_0,y_1),\quad\ldots.$$

Newton's formula:

$$f(x_0+u\Delta x,y_0+v\Delta y)=f(x_0,y_0)+(u\Delta_x+v\Delta_y)(x_0,y_0)$$

$$+(1/2!)\left[u(u-1)\Delta^2_x+2uv\Delta_{xy}+v(v-1)\Delta^2_y\right](x_0,y_0)+\ldots.$$

Everett's formula. Putting $s\equiv1-u,t\equiv1-v$ we have

$$f(x_0+u\Delta x,y_0+v\Delta y)=stf(x_0,y_0)+svf(x_0,y_1)+utf(x_1,y_0)+uvf(x_1,y_1)$$

$$-(1/6)\left[us(1+s)\left\{t\delta^2_x(x_0,y_0)+v\delta^2_x(x_0,y_1)\right\}+us(1+u)\left\{t\delta^2_x(x_1,y_0)+v\delta^2_x(x_1,y_1)\right\}\right.$$

$$\left.+vt(1+t)\left\{s\delta^2_y(x_0,y_0)+u\delta^2_y(x_1,y_0)\right\}+vt(1+v)\left\{s\delta^2_y(x_0,y_1)+u\delta^2_y(x_1,y_1)\right\}\right]+\ldots.$$

References

[1] F. J. Thompson, Table of the coefficients of Everett's central-difference interpolation formula, Tracts for computers, no. V, Cambridge Univ. Press, 1921.
[2] M. Lindow, Numerische Infinitesimalrechnung, Dummler, Berlin, 1928.
[3] H. T. Davis, Tables of the higher mathematical functions I, Principia Press, Bloomington, 1933.
[4] K. Hayashi and S. Moriguti, Table of higher transcendental functions (in Japanese), Iwanami, second revised edition, 1967.

22. Distribution of Typical Random Variables

(→ 341 Probability Measures, 374 Sampling Distributions)

In the following table, Nos. 1–13 are 1-dimensional continuous distributions, and Nos. 20–21 are k-dimensional continuous distributions, for which the distribution density is the one with respect to Lebesgue measure. Nos. 14–19 are 1-dimensional discrete distributions, and Nos. 22–24 are k-dimensional discrete distributions, where the density function $P(x)$ means the probability at the point x.

The characteristic function, average, and variance are given only for those represented in a simple form.

No.	Name	Symbol	Density Function	Domains
1	Normal	$N(\mu,\sigma^2)$	$\frac{1}{(2\pi\sigma^2)^{1/2}}\exp\left[-\frac{(x-\mu)^2}{2\sigma^2}\right]$	$-\infty<x<\infty$
2	Logarithmic normal		$\frac{1}{(2\pi\sigma^2)^{1/2}}\frac{1}{y}\exp\left[-\frac{(\log y-\mu)^2}{2\sigma^2}\right]$	$0<y<\infty$
3	Gamma	$\Gamma(p,\sigma)$	$[\Gamma(p)]^{-1}\sigma^{-p}x^{p-1}e^{-x/\sigma}$	$0<x<\infty$
4	Exponential	$e(\mu,\sigma)$	$(1/\sigma)\exp(-(x-\mu)/\sigma)$	$\mu<x<\infty$
5	Two-sided exponential		$(1/2\sigma)e^{-\lvert x\rvert/\sigma}$	$-\infty<x<\infty$
6	Chi square	$\chi^2(n)$	$2^{-n/2}[\Gamma(n/2)]^{-1}x^{(n/2)-1}e^{-x/2}$	$0<x<\infty$
7	Beta	$B(p,q)$	$[B(p,q)]^{-1}x^{p-1}(1-x)^{q-1}$	$0<x<1$
8	F	$F(m,n)$	$2K_F x^{(m/2)-1}[1+(mx/n)]^{-(m+n)/2}$, $K_F\equiv[B(m/2,n/2)]^{-1}(m/n)^{m/2}$	$0<x<\infty$
9	z	$z(m,n)$	$K_F e^{mz}[1+(me^{2z}/n)]^{-(m+n)/2}$, $K_F\equiv[B(m/2,n/2)]^{-1}(m/n)^{m/2}$	$-\infty<z<\infty$
10	t	$t(n)$	$[\sqrt{n}\,B(n/2,1/2)]^{-1}[1+(t^2/n)]^{-(n+1)/2}$	$-\infty<t<\infty$
11	Cauchy	$C(\mu,\sigma)$	$(\pi\sigma)^{-1}\left[1+\frac{(x-\mu)^2}{\sigma^2}\right]^{-1}$	$-\infty<x<\infty$
12	One-side stable for exponent 1/2		$c(2\pi)^{-1/2}x^{-3/2}\exp(-c^2/2x)$	$0<x<\infty$
13	Uniform rectangular	$U(\alpha,\beta)$	$1/(\beta-\alpha)$	$\alpha<x<\beta$
14	Binomial	$Bin(n,p)$	$\binom{n}{x}p^x q^{n-x}$	$x=0,1,2,\ldots,n$
15	Poisson	$P(\lambda)$	$e^{-\lambda}\lambda^x/x!$	$x=0,1,2,\ldots$
16	Hypergeometric	$H(N,n,p)$	$\binom{Np}{x}\binom{Nq}{n-x}\Big/\binom{N}{n}$	x integer $0\leqslant x\leqslant Np$, $0\leqslant n-x\leqslant Nq$
17	Negative binomial	$NB(m,p)$	$\Gamma(m+x)[\Gamma(m)x!]^{-1}p^m q^x$	$x=0,1,2,\ldots$
18	Geometric	$G(p)$	pq^x	$x=0,1,2,\ldots$
19	Logarithmic		$K_L q^x/x,\ K_L\equiv-1/\log p$	$x=1,2,3,\ldots$
20	Multidimensional normal	$N(\mu,\Sigma)$	$(2\pi)^{-k/2}\lvert\Sigma\rvert^{-1/2}$ $\times\exp[-(x-\mu)\Sigma^{-1}(x-\mu)'/2]$, $x=(x_1,\ldots,x_k),\ \mu=(\mu_1,\ldots,\mu_k),\ \Sigma=(\sigma_{ij})$	$-\infty<x_1,\ldots,x_k<\infty$
21	Dirichlet		$\frac{\Gamma(\nu_1+\ldots+\nu_{k+1})}{\Gamma(\nu_1)\ldots\Gamma(\nu_{k+1})}x_1^{\nu_1-1}\ldots x_{k+1}^{\nu_{k+1}-1}$, $x_{k+1}=1-(x_1+\ldots+x_k)$	$x_1,\ldots,x_k>0$, $x_1+\ldots+x_k<1$
22	Multinomial	$M(n,(p_i))$	$n!(x_1!\ldots x_{k+1}!)^{-1}p_1^{x_1}\ldots p_{k+1}^{x_{k+1}}$, $x_{k+1}=n-(x_1+\ldots+x_k)$	$x_1,\ldots,x_k=0,1,\ldots,n$, $x_1+\ldots+x_k\leqslant n$
23	Multidimensional hypergeometric	$H(N,n,(p_i))$	$\binom{Np_1}{x_1}\cdots\binom{Np_{k+1}}{x_{k+1}}\Big/\binom{N}{n}$, $x_{k+1}=n-(x_1+\ldots+x_k)$	$x_1,\ldots,x_k$ integers $0\leqslant x_i\leqslant Np_i$ $(i=1,\ldots,k+1)$
24	Negative polynomial		$\frac{\Gamma(m+x_1+\ldots+x_k)}{\Gamma(m)x_1!\ldots x_k!}p_0^m p_1^{x_1}\ldots p_k^{x_k}$,	$x_1,\ldots,x_k=0,1,2,\ldots$

Conditions for Parameters	Characteristic Function	Mean	Variance	No.
$-\infty<\mu<\infty$, $\sigma>0$	$\exp\left(i\mu t-\frac{\sigma^2 t^2}{2}\right)$	μ	σ^2	1
$-\infty<\mu<\infty$, $\sigma>0$		$e^{\mu+(\sigma^2/2)}$	$e^{2\mu}(e^{2\sigma^2}-e^{\sigma^2})$	2
$p,\sigma>0$	$(1-i\sigma t)^{-p}$	σp	$\sigma^2 p$	3
$-\infty<\mu<\infty, \sigma>0$	$e^{i\mu t}(1-i\sigma t)^{-1}$	$\mu+\sigma$	σ^2	4
$\sigma>0$	$(1+\sigma^2 t^2)^{-1}$	0	$2\sigma^2$	5
n positive integer	$(1-2it)^{-n/2}$	n	$2n$	6
$p,q>0$		$\frac{p}{p+q}$	$\frac{pq}{(p+q)^2(p+q+1)}$	7
m,n positive integers		$\frac{n}{n-2}$ $(n>2)$	$\frac{2n^2(m+n-2)}{m(n-2)^2(n-4)}$ $(n>4)$	8
m, n positive integers				9
n positive integer		0 $(n>1)$	$n/(n-2)$ $(n>2)$	10
$-\infty<\mu<\infty$, $\sigma>0$	$\exp(i\mu t-\sigma\|t\|)$	none	none	11
$0<c<\infty$	$\exp[-c\|t\|^{1/2}(1-it/\|t\|)]$	none	none	12
$-\infty<\alpha<\beta<\infty$	$(e^{i\beta t}-e^{i\alpha t})/it(\beta-\alpha)$	$(\alpha+\beta)/2$	$(\beta-\alpha)^2/12$	13
$p+q=1,p,q>0$, n positive integer	$(pe^{it}+q)^n$	np	npq	14
$\lambda>0$	$\exp[-\lambda(1-e^{it})]$	λ	λ	15
$p+q=1,p,q>0$, N,Np,n positive integers $N\geqslant n$	$(Nq)^{[n]}(N^{[n]})^{-1}$ $\times F(-n,-Np;Nq-n+1;e^{it})$, $m^{[n]}\equiv m!/(m-n)!$	np	$\frac{npq(N-n)}{N-1}$	16
$p+q=1,p,q>0$, $m>0$	$\frac{p^m}{(1-qe^{it})^m}$	$\frac{mq}{p}$	$\frac{mq}{p^2}$	17
$p+q=1, p,q>0$	$\frac{p}{1-qe^{it}}$	$\frac{q}{p}$	$\frac{q}{p^2}$	18
$p+q=1, p,q>0$	$-K_L\log(1-qe^{it})$	$K_L q/p$	$K_L q(1-K_L q)/p^2$	19
$-\infty<\mu_1,\ldots,\mu_k<\infty$, Σ symmetric positive definite quadratic form	$\exp\left(i\mu t'-\frac{t\Sigma t'}{2}\right)$, $t=(t_1,\ldots,t_k)$	$E(x_i)=\mu_i$	$V(x_i)=\sigma_{ii}$, $\mathrm{Cov}(x_i,x_j)=\sigma_{ij}$	20
$\nu_1,\ldots,\nu_{k+1}>0$		$E(x_i)=\frac{\nu_i}{\nu_1+\ldots+\nu_{k+1}}$	$V(x_i)=C\nu_i(\nu_1+\ldots+\nu_{k+1}-\nu_i)$, $\mathrm{Cov}(x_i,x_j)=-C\nu_i\nu_j$, $C\equiv(\nu_1+\ldots+\nu_{k+1})^{-2}\times(\nu_1+\ldots+\nu_{k+1}+1)^{-1}$	21
$p_1+\ldots+p_{k+1}=1$, $p_1,\ldots,p_{k+1}>0$ n positive integer	$(p_1e^{it_1}+\ldots+p_ke^{it_k}+p_{k+1})^n$	$E(x_i)=np_i$	$V(x_i)=np_i(1-p_i)$, $\mathrm{Cov}(x_i,x_j)=-np_ip_j$	22
$p_1+\ldots+p_{k+1}=1$, $p_1,\ldots,p_{k+1}>0$, $N,Np_1,\ldots,Np_k,n$ positive integers		$E(x_i)=np_i$	$V(x_i)=Cnp_i(1-p_i)$, $\mathrm{Cov}(x_i,x_j)=-Cnp_ip_j$, $C\equiv\frac{N-n}{N-1}$	23
$p_0+p_1+\ldots+p_k=1$, $p_0,p_1,\ldots,p_k>0$, $m>0$	$p_0^m(1-p_1e^{it_1}-\ldots-p_ke^{it_k})^m$	$E(x_i)=\frac{mp_i}{p_0}$	$V(x_i)=mp_i(p_0+p_i)/p_0^2$, $\mathrm{Cov}(x_i,x_j)=mp_ip_j/p_0^2$	24

Remarks
1. Reproducing property with respect to μ, σ^2.
2. $X = \log Y : N(\mu, \sigma^2)$.
3. Reproducing property with respect to p.
4. $e(0,\sigma) = \Gamma(1,\sigma)$.
6. n is the number of degrees of freedom; reproducing property with respect to n.
8. m and n are the numbers of degrees of freedom.
9. $e^{2z} = F(m,n)$.
10. n is the number of degrees of freedom.
11. $C(0,1) = t(1)$; reproducing property with respect to μ and σ.
14. Reproducing property with respect to n.
15. Reproducing property with respect to λ.
17. Reproducing property with respect to m.
18. $G(p) = NB(1,p)$.
20. Generalization of normal distribution; reproducing property with respect to μ and Σ.
22. Generalization of binomial distribution; reproducing property with respect to n.
23. Generalization of hypergeometric distribution.
24. Generalization of negative binomial distribution; reproducing property with respect to m.

23. Statistical Estimation and Statistical Hypothesis Testing

Listed below are some frequently used and well-investigated statistical procedures. (Concerning main probability distributions → 398 Statistical Decision Functions, 399 Statistical Estimation, 400 Statistical Hypothesis Testing). The following notations and conventions are adopted, unless otherwise stated.

Immediately after the heading number, the distribution is indicated by the symbol as defined in Table 22, this Appendix. It is to be understood that a random sample $(x_1, x_2, \ldots, x_n)$ is observed from this distribution. Where two distributions are involved, samples $(x_1, \ldots, x_{n_1})$ and $(y_1, \ldots, y_{n_2})$ are understood to be observed from the respective distributions.

Next, a necessary and sufficient statistic based on the sample is marked with * when it is complete, and # otherwise. Then appears the sampling distribution of this statistic. For those statistics consisting of several independent components, the distribution of these are shown. Greek lowercase letters except α and χ denote unknown parameters. Italic lowercase letters denote constants, each taking arbitrary real values. Italic capital letters denote constants whose values are specified in each procedure; repeated occurrences of the same letter under the same heading number specify a certain common real value.

Problems of point estimation, interval estimation, and hypothesis testing are presented, with corresponding estimators, confidence intervals, and tests (critical regions) as their solutions. All the confidence intervals here are those constructed from UMP unbiased tests, having $1-\alpha$ as confidence levels. Alternative hypotheses are understood to be the negations of corresponding null hypotheses. Significance levels of all the tests are α. The following symbols are attached to each procedure to describe its properties.

For estimators:
UMV: uniformly minimum variance unbiased.
ML: maximum likelihood.
AD: admissibility with respect to quadratic loss function.
IAD: inadmissibility with respect to quadratic loss function.

For tests:
UMP: uniformly most powerful.
UMPU: uniformly most powerful unbiased.
UMPI( ): uniformly most powerful invariant with respect to the product of transformation groups shown in ( ).
LR: likelihood ratio.
O: group of orthogonal transformations.
L: group of shift transformations.
S: group of change of scales.
AD: admissibility with respect to simple loss function.

IAD: inadmissibility with respect to simple loss function. (Note that UMPU implies AD.)

The following symbols denote $100(1-\alpha)\%$ points of respective distributions, α being sufficiently small.

$u(\alpha)$: standard normal distribution.
$t_f(\alpha)$: t-distribution with f degrees of freedom.
$\chi_f^2(\alpha)$: χ^2 distribution with f degrees of freedom.
$F_{f_2}^{f_1}(\alpha)$: F-distribution with (f_1,f_2) degrees of freedom.

(1) $N(\mu,b^2)$. Σx_i*. $N(n\mu,nb^2)$.
Point estimation of μ. $\bar{x}=\frac{1}{2}\Sigma x_i$: UMV, ML, AD.
Interval estimation of μ. $\left(\bar{x}\pm u(\alpha/2)\frac{b}{\sqrt{n}}\right)$.
Hypothesis $[\mu\leqslant k]$. $\bar{x}>k+u(\alpha)\frac{b}{\sqrt{n}}$: UMP, LR.
Hypothesis $[h\leqslant\mu\leqslant l]$. $\bar{x}<h-C$ or $\bar{x}>l+C$: UMPU, LR.

(2) $N(a,\sigma^2)$. $\Sigma(x_i-a)^2$*. $\sigma^2\chi_n^2$. ($\sigma^2\chi_n^2$ is the σ^2-multiplication of a random variable obeying the $\chi^2(n)$ distribution. We use similar notations in the following.)

Point estimation of σ^2. $\frac{\Sigma(x_i-a)^2}{n}$: UMV, ML, IAD.
Interval estimation of σ^2. $(A\Sigma(x_i-a)^2,\ B\Sigma(x_i-a)^2)$.
Hypothesis $[\sigma^2\leqslant k]$. $\Sigma(x_i-a)^2>\chi_n^2(\alpha)k$: UMP, LR.
Hypothesis $[\sigma^2=k]$. $\Sigma(x_i-a)^2<Ak$ or $\Sigma(x_i-a)^2>Bk$: UMPU.

(3) $N(\mu,\sigma^2)$. $\begin{pmatrix}\Sigma x_i\\ \Sigma(x_i-\bar{x})^2\end{pmatrix}^*$. $\begin{pmatrix}N(n\mu,n\sigma^2)\\ \sigma^2\chi_{n-1}^2\end{pmatrix}$.
Point estimation of μ. $\bar{x}$: UMV, ML, AD.
Interval estimation of μ. $\left[\bar{x}\pm t_{n-1}(\alpha/2)\frac{\sqrt{\Sigma(x_i-\bar{x})^2}}{\sqrt{n(n-1)}}\right]$.
Hypothesis $[\mu\leqslant k]$. $\frac{\bar{x}-k}{\sqrt{\Sigma(x_i-\bar{x})^2}}>\frac{t_{n-1}(\alpha)}{\sqrt{n(n-1)}}$: UMPU, LR.
Hypothesis $[\mu=k]$. $\frac{|\bar{x}-k|}{\sqrt{\Sigma(x_i-\bar{x})^2}}>\frac{t_{n-1}(\alpha/2)}{\sqrt{n(n-1)}}$: UMPU, LR, UMPI(S, O) for $k=0$.
Point estimation of σ^2. $\frac{\Sigma(x_i-\bar{x})^2}{n-1}$: UMV, IAD. $\frac{\Sigma(x_i-\bar{x})^2}{n}$: ML, IAD.
Point estimation of σ. $\frac{\Gamma[(n-1)/2]}{\sqrt{2}\,\Gamma(n/2)}\sqrt{\frac{\Sigma(x_i-\bar{x})^2}{n-1}}$: UMV, IAD.
Interval estimation of σ^2. $(A\Sigma(x_i-\bar{x})^2,\ B\Sigma(x_i-\bar{x})^2)$.
Hypothesis $[\sigma^2\leqslant k]$. $\Sigma(x_i-\bar{x})^2>\chi_{n-1}^2(\alpha)k$: UMP, LR.
Hypothesis $[\sigma^2=k]$. $\Sigma(x_i-\bar{x})^2<Ak$ or $\Sigma(x_i-\bar{x})^2>Bk$: UMPU.
Hypothesis $[\sigma^2\geqslant k]$. $\Sigma(x_i-\bar{x})^2<\chi_{n-1}^2(1-\alpha)k$: UMPU, UMPI(L).
Hypothesis $\left[\frac{\mu}{\sigma}\leqslant k\right]$. $\frac{\bar{x}}{\sqrt{\Sigma(x_i-\bar{x})^2}}>E$: UMPI(S), AD.

(4) $Bin(N,\theta)$. Σx_i*. $Bin(Nn,\theta)$.
Point estimation of θ. $\frac{\bar{x}}{N}$: UMV, ML, AD.
Hypothesis $[\theta\leqslant k]$. $\bar{x}>A$: UMP.
Hypothesis $[h\leqslant\theta\leqslant l]$. $\bar{x}<B$ or $\bar{x}>C$: UMPU.

(5) $H(N,m,\theta)$ $(n=1)$. x*.
Point estimation of θ. $\frac{Nx}{m}$: UMV, AD.
Hypothesis $[\theta\leqslant k]$. $x>A$: UMP.

(6) $NB(N,\theta)$. $\Sigma x_i{}^*$. $NB(Nn,\theta)$.
Point estimation of θ. $\frac{Nn-1}{Nn+\Sigma x_i-1}$ (1 when the denominator is 0): UMV, AD.

$\frac{Nn}{Nn+\Sigma x_i}$: ML.

Hypothesis $[\theta \leqslant k]$. $\Sigma x_i < A$: UMP.
Hypothesis $[h \leqslant \theta \leqslant l]$. $\Sigma x_i < B$ or $\Sigma x_i > C$: UMPU.

(7) $P(\lambda)$. $\Sigma x_i{}^*$. $P(n\lambda)$.
Point estimation of λ. $\bar{x}$: UMV, ML, AD.
Hypothesis $[\lambda \leqslant k]$. $\bar{x} > A$: UMP.
Hypothesis $[h \leqslant \lambda \leqslant l]$. $\bar{x} < B$ or $\bar{x} > C$: UMPU.

(8) $G(\theta)$. $\Sigma x_i{}^*$. $NB(n,\theta)$.
For the point estimation of θ and hypothesis testing $\rightharpoonup$ (6).

(9) $U[0,\theta]$. $\max x_i{}^*$.
Point estimation of θ. $\max x_i$: ML, IAD. $\frac{n+1}{n}\max x_i$: UMV, IAD.
Hypothesis $[\theta \leqslant k]$. $\max x_i > (1-\alpha)^{1/n}k$: UMP.
Hypothesis $[\theta = k]$. $\max x_i < k\alpha^{1/n}$ or $\max x_i > k$: UMP.

(10) $U[\xi,\eta]$. $(\min x_i, \max x_i)^*$.
Point estimation of ξ. $\frac{n\min x_i - \max x_i}{n-1}$: UMV, IAD. $\min x_i$: ML, IAD.
Point estimation of $\frac{\xi+\eta}{2}$. $\frac{\min x_i + \max x_i}{2}$: UMV, AD.
Hypothesis $[\eta - \xi \leqslant k]$. $\max x_i - \min x_i > k\alpha^{1/n}$: UMP.

(11) $U\left[\theta - \frac{1}{2}, \theta + \frac{1}{2}\right]$. $(\min x_i, \max x_i)^\#$.
Point estimation of θ. $\frac{\min x_i + \max x_i}{2}$: ML, AD.
Hypothesis $[\theta \leqslant k]$. $\min x_i > k + \frac{1}{2} - \alpha^{1/n}$ or $\max x_i > k + \frac{1}{2}$: UMP.

(12) $e(\mu,\sigma)$. $\begin{pmatrix}\Sigma x_i \\ \min x_i\end{pmatrix}^*$. $\begin{pmatrix}\Gamma(n,\sigma)+n\mu \\ e(\mu,\sigma/n)\end{pmatrix}$.

Point estimation of σ. $\frac{\Sigma x_i - n\min x_i}{n-1}$: UMV, IAD. $\bar{x} - \min x_i$: ML, IAD.
Point estimation of μ. $\frac{n}{n-1}\min x_i - \frac{1}{n-1}\bar{x}$: UMV, IAD. $\min x_i$: ML, IAD.
Hypothesis $[\sigma \leqslant k, \mu = h]$. $\Sigma x_i < h$ or $\Sigma x_i > k\log\alpha^{-1/n} + h$: UMP.
Hypothesis $[h \leqslant \sigma \leqslant l]$. $\Sigma x_i - n\min x_i < A$ or $\Sigma x_i - n\min x_i > B$: UMPU.
Hypothesis $[\mu = k]$. $\frac{n\min x_i - k}{\Sigma x_i - n\min x_i} < 0$ or $\frac{n\min x_i - k}{\Sigma x_i - n\min x_i} > C$: UMPU.

(13) $\Gamma(p,\sigma)$. $\Sigma x_i{}^*$. $\Gamma(np,\sigma)$.
Point estimation of σ. $\frac{\bar{x}}{p}$: UMV, ML, IAD.
Interval estimation of σ. $(C\Sigma x_i, D\Sigma x_i)$.
Hypothesis $[\sigma \leqslant k]$. $\Sigma x_i > A$: UMP.
Hypothesis $[\sigma = k]$. $\Sigma x_i < Ck$ or $\Sigma x_i > Dk$: UMPU.

(14) $\begin{matrix} N(\mu_1, a^2) \\ N(\mu_2, b^2).\end{matrix}$ $\begin{pmatrix}\Sigma x_i \\ \Sigma y_i\end{pmatrix}^*$. $\begin{pmatrix}N(n_1\mu_1, n_1 a^2) \\ N(n_2\mu_2, n_2 b^2)\end{pmatrix}$.

Point estimation of $\mu_1 - \mu_2$. $\bar{x} - \bar{y}$: UMV, ML, AD.

Interval estimation of $\mu_1 - \mu_2$. $\left(\bar{x} - \bar{y} \pm u(\alpha/2)\sqrt{\frac{a^2}{n_1} + \frac{b^2}{n_2}}\right)$.

Hypothesis $[\mu_1 - \mu_2 \leqslant k]$. $\bar{x} - \bar{y} > k + u(\alpha)\sqrt{\frac{a^2}{n_1} + \frac{b^2}{n_2}}$: UMP, LR.

Hypothesis $[\mu_1-\mu_2=k]$. $|\bar{x}-\bar{y}-k|>u(\alpha/2)\sqrt{\dfrac{a^2}{n_1}+\dfrac{b^2}{n_2}}$: UMPU, UMPI(L), LR.

(15) $N(\mu_1,\sigma^2)$, $N(\mu_2,\sigma^2)$. $\begin{bmatrix}\Sigma x_i\\ \Sigma y_i\\ s^2=\Sigma(x_i-\bar{x})^2+\Sigma(y_i-\bar{y})^2\end{bmatrix}^*$. $\begin{bmatrix}N(n_1\mu_1,n_1\sigma^2)\\ N(n_2\mu_2,n_2\sigma^2)\\ \sigma^2\chi^2_{n_1+n_2-2}\end{bmatrix}$.

Point estimation of $\mu_1-\mu_2$. $\bar{x}-\bar{y}$: UMV, ML, AD.

Interval estimation of $\mu_1-\mu_2$. $\left(\bar{x}-\bar{y}\pm t_{n_1+n_2-2}(\alpha/2)\sqrt{\dfrac{1}{n_1}+\dfrac{1}{n_2}}\sqrt{\dfrac{s^2}{n_1+n_2-2}}\right)$.

Hypothesis $[\mu_1-\mu_2\leqslant k]$. $t=\dfrac{(\bar{x}-\bar{y}-k)\sqrt{n_1n_2}\sqrt{n_1+n_2-2}}{\sqrt{n_1+n_2}\sqrt{s^2}}>t_{n_1+n_2-2}(\alpha)$: UMPU, UMPI(L), LR.

Hypothesis $[\mu_1-\mu_2=k]$. $|t|>t_{n_1+n_2-2}(\alpha)$: UMPU, UMPI(L), LR.

Point estimation of σ^2. $\dfrac{s^2}{n_1+n_2-2}$: UMV, IAD. $\dfrac{s^2}{n_1+n_2}$: ML, IAD.

Interval estimation of σ^2. (As^2, Bs^2).

Hypothesis $[\sigma^2\leqslant k]$. $s^2>\chi^2_{n_1+n_2-2}(\alpha)k$: UMP, LR.

Hypothesis $[\sigma^2=k]$. $s^2<Ak$ or $s^2>Bk$: UMPU.

Hypothesis $[\sigma^2\geqslant k]$. $s^2>\chi^2_{n_1+n_2-2}(1-\alpha)k$: UMPU, UMPI(L), LR.

(16) $N(\mu_1,\sigma_1^2)$, $N(\mu_2,\sigma_2^2)$. $\begin{bmatrix}\Sigma x_i, & \Sigma(x_i-\bar{x})^2\\ \Sigma y_i, & \Sigma(y_i-\bar{y})^2\end{bmatrix}^*$.

Interval estimation of $\dfrac{\sigma_1^2}{\sigma_2^2}$. $\left[A\dfrac{\Sigma(x_i-\bar{x})^2}{\Sigma(y_i-\bar{y})^2}, B\dfrac{\Sigma(x_i-\bar{x})^2}{\Sigma(y_i-\bar{y})^2}\right]$.

Hypothesis $\left[\dfrac{\sigma_1^2}{\sigma_2^2}\leqslant k\right]$. $\dfrac{(n_2-1)}{(n_1-1)}\dfrac{\Sigma(x_i-\bar{x})^2}{\Sigma(y_i-\bar{y})^2}>F^{n_1-1}_{n_2-1}(\alpha)k$: UMPU, UMPI(L, S), LR.

(17) $N(\mu_1,\mu_2,\sigma_1^2,\sigma_2^2,\rho)$. $\begin{bmatrix}\Sigma x_i, & \Sigma(x_i-\bar{x})^2, & \Sigma(x_i-\bar{x})(y_i-\bar{y})\\ \Sigma y_i, & \Sigma(y_i-\bar{y})^2, & \end{bmatrix}^*$.

Point estimation of ρ. $r=\dfrac{\Sigma(x_i-\bar{x})(y_i-\bar{y})}{\sqrt{\Sigma(x_i-\bar{x})^2\Sigma(y_i-\bar{y})^2}}$: ML.

Hypothesis $[\rho=0]$. $|r|>\dfrac{t_{n-1}(\alpha/2)}{\sqrt{t_{n-1}(\alpha/2)^2+n-2}}$: UMPU, LR.

Appendix B
Numerical Tables

1. Prime Numbers and Primitive Roots (→ 297 Number Theory, Elementary

In the following table, p is a prime number and r is a corresponding primitive root.

p	r	p	r	p	r	p	r	p	r	p	r	p	r	p	r
2		79	3	191	19	311	17	439	17	577	5	709	2	857	3
3	2	83	2	193	5	313	17	443	2	587	2	719	11	859	2
5	2	89	3	197	2	317	2	449	3	593	3	727	5	863	5
7	3	97	5	199	3	331	3	457	13	599	7	733	7	877	2
11	2	101	2	211	2	337	19	461	2	601	7	739	3	881	3
13	2	103	5	223	3	347	2	463	3	607	3	743	5	883	2
17	3	107	2	227	2	349	2	467	2	613	2	751	3	887	5
19	2	109	11	229	7	353	3	479	13	617	3	757	2	907	2
23	5	113	3	233	3	359	7	487	3	619	2	761	7	911	17
29	2	127	3	239	7	367	11	491	2	631	3	769	11	919	7
31	3	131	2	241	7	373	2	499	7	641	3	773	2	929	3
37	2	137	3	251	11	379	2	503	5	643	11	787	2	937	5
41	7	139	2	257	3	383	5	509	2	647	5	797	2	941	2
43	3	149	2	263	5	389	2	521	3	653	2	809	3	947	2
47	5	151	7	269	2	397	5	523	2	659	2	811	3	953	3
53	2	157	5	271	43	401	3	541	2	661	2	821	2	967	5
59	2	163	2	277	5	409	29	547	2	673	5	823	3	971	11
61	2	167	5	281	3	419	2	557	2	677	2	827	2	977	3
67	2	173	2	283	3	421	2	563	2	683	5	829	2	983	5
71	7	179	2	293	2	431	7	569	3	691	3	839	11	991	7
73	5	181	2	307	5	433	5	571	3	701	2	853	2	997	7

†Mersenne numbers. A prime number of the form 2^p-1 is called a Mersenne number. There exist 27 such p's less than 44500: $p=2$, 3, 5, 7, 13, 17, 19, 31, 61, 89, 107, 127, 521, 607, 1279, 2203, 2281, 3217, 4253, 4423, 9689, 9941, 11213, 19937, 21701, 23209, 44497. The even perfect numbers are the numbers of the form $2^{p-1}(2^p-1)$, where 2^p-1 is a Mersenne number.

2. Indices Modulo p (→ 297 Number Theory, Elementary)

Let r be a primitive root corresponding to a prime number p. The index $l=\mathrm{Ind}_r a$ of a with respect to the basis r is the integer l in $0\leqslant l<p-1$ satisfying $r^l\equiv a(\mathrm{mod}\,p)$. $a\equiv b(\mathrm{mod}\,p)$ is equivalent to $\mathrm{Ind}_r a\equiv\mathrm{Ind}_r b(\mathrm{mod}(p-1))$. The index satisfies the following congruence relations with respect to $\mathrm{mod}(p-1)$: $\mathrm{Ind}_r ab\equiv\mathrm{Ind}_r a+\mathrm{Ind}_r b$, $\mathrm{Ind}_r a^n\equiv n\,\mathrm{Ind}_r a$, $\mathrm{Ind}_s a\equiv\mathrm{Ind}_s r\,\mathrm{Ind}_r a$.

We can solve congruence equations using these relations. The following is a table of indices.

p	$p-1$	r	2	3	5	7	11	13	17	19	23	29	31	37	41	43	47
2	1		—														
3	2	2	1	—													
5	4	2	1	3	—												
7	$2\cdot 3$	3	2	1	5	—											
11	$2\cdot 5$	2	1	8	4	7	—										
13	$2^2\cdot 3$	2	1	4	9	11	7	—									
17	2^4	3	14	1	5	11	7	4	—								
19	$2\cdot 3^2$	2	1	13	16	6	12	5	10	—							
23	$2\cdot 11$	5	2	16	1	19	9	14	7	15	—						
29	$2^2\cdot 7$	2	1	5	22	12	25	18	21	9	20	—					
31	$2\cdot 3\cdot 5$	3	24	1	20	28	23	11	7	4	27	9	—				
37	$2^2\cdot 3^2$	2	1	26	23	32	30	11	7	35	15	21	9	—			
41	$2^3\cdot 5$	7	14	25	18	1	37	9	7	31	4	33	12	8	—		
43	$2\cdot 3\cdot 7$	3	27	1	25	35	30	32	38	19	16	41	34	7	6	—	
47	$2\cdot 23$	5	18	20	1	32	7	11	16	45	5	35	3	42	15	13	—
53	$2^2\cdot 13$	2	1	17	47	14	6	24	10	37	39	46	33	30	45	22	44
59	$2\cdot 29$	2	1	50	6	18	25	45	40	38	15	28	49	55	14	33	23
61	$2^2\cdot 3\cdot 5$	2	1	6	22	49	15	40	47	26	57	35	59	39	54	43	20
67	$2\cdot 3\cdot 11$	2	1	39	15	23	59	19	64	10	28	44	47	22	53	9	50
71	$2\cdot 5\cdot 7$	7	6	26	28	1	31	39	49	16	15	68	11	20	25	48	9
73	$2^3\cdot 3^2$	5	8	6	1	33	55	59	21	62	46	35	11	64	4	51	31
79	$2\cdot 3\cdot 13$	3	4	1	62	53	68	34	21	32	26	11	56	19	75	49	59
83	$2\cdot 41$	2	1	72	27	8	24	77	56	47	60	12	38	20	40	71	23
89	$2^3\cdot 11$	3	16	1	70	81	84	23	6	35	57	59	31	11	21	29	54
97	$2^5\cdot 3$	5	34	70	1	31	86	25	89	81	77	13	46	91	85	4	84
101	$2^2\cdot 5^2$	2	1	69	24	9	13	66	30	96	86	91	84	56	45	42	58
103	$2\cdot 3\cdot 17$	5	44	39	1	4	61	72	70	80	24	86	57	93	50	77	85
107	$2\cdot 53$	2	1	70	47	43	22	14	29	78	62	32	27	38	40	59	66
109	$2^2\cdot 3^3$	11	15	80	92	20	1	101	87	105	3	98	34	43	63	42	103
113	$2^4\cdot 7$	3	12	1	83	8	74	22	5	99	41	89	50	67	94	47	31
127	$2\cdot 3^2\cdot 7$	3	72	1	87	115	68	94	38	84	121	113	46	98	80	71	60
131	$2\cdot 5\cdot 13$	2	1	72	46	96	56	18	43	35	23	51	29	41	126	124	105
137	$2^3\cdot 17$	3	10	1	75	42	122	25	38	46	125	91	73	102	119	97	19
139	$2\cdot 3\cdot 23$	2	1	41	86	50	76	64	107	61	27	94	56	80	32	115	98
149	$2^2\cdot 37$	2	1	87	104	142	109	53	124	84	95	120	132	72	41	93	138
151	$2\cdot 3\cdot 5^2$	7	10	93	136	1	82	23	124	120	145	42	34	148	3	74	128
157	$2^2\cdot 3\cdot 13$	5	141	82	1	147	28	26	40	124	135	129	62	116	21	113	92
163	$2\cdot 3^4$	2	1	101	15	73	47	51	57	125	9	107	69	33	160	38	28
167	$2\cdot 83$	5	40	94	1	118	28	103	53	58	99	150	90	61	97	87	132
173	$2^2\cdot 43$	2	1	27	39	95	23	130	73	33	20	144	102	162	138	84	64
179	$2\cdot 89$	2	1	108	138	171	15	114	166	54	135	118	62	149	155	80	36
181	$2^2\cdot 3^2\cdot 5$	2	1	56	156	15	62	164	175	135	53	48	99	26	83	20	13
191	$2\cdot 5\cdot 19$	19	44	116	50	171	85	112	98	1	134	33	175	15	165	8	123
193	$2^6\cdot 3$	5	34	84	1	104	183	141	31	145	162	123	82	5	151	24	29
197	$2^2\cdot 7^2$	2	1	181	89	146	29	25	159	154	120	36	141	192	110	78	66
199	$2\cdot 3^2\cdot 11$	3	106	1	138	142	189	172	123	55	118	70	164	11	167	88	76
211	$2\cdot 3\cdot 5\cdot 7$	2	1	43	132	139	162	144	199	154	21	179	115	118	17	80	124
223	$2\cdot 3\cdot 37$	3	180	1	89	210	107	147	144	172	163	128	82	152	204	118	50
227	$2\cdot 113$	2	1	46	11	154	28	61	99	178	34	8	197	77	131	150	218
229	$2^2\cdot 3\cdot 19$	7	111	68	214	1	42	195	24	52	131	191	175	164	73	12	193

(table continued on following page)

p	$p-1$	r	2	3	5	7	11	13	17	19	23	29	31	37	41	43	47
233	$2^3\cdot 29$	3	72	1	165	222	197	158	103	136	112	132	182	8	85	25	139
239	$2\cdot 7\cdot 17$	7	66	74	138	1	4	43	52	155	63	160	188	31	99	15	113
241	$2^4\cdot 3\cdot 5$	7	190	182	138	1	25	47	111	85	57	154	151	73	6	219	114
251	$2\cdot 5^3$	11	135	6	80	218	1	162	184	233	134	203	226	187	64	77	85
257	2^8	3	48	1	55	85	196	106	120	125	28	94	242	219	19	207	61
263	$2\cdot 131$	5	190	50	1	79	166	62	126	43	156	221	136	170	17	154	65
269	$2^2\cdot 67$	2	1	109	208	19	230	142	105	223	176	187	259	56	200	254	32
271	$2\cdot 3^3\cdot 5$	43	266	153	220	98	92	15	16	261	75	45	222	182	156	1	213
277	$2^2\cdot 3\cdot 23$	5	147	188	1	22	7	222	103	252	208	74	47	87	126	55	218
281	$2^3\cdot 5\cdot 7$	3	204	1	186	182	253	9	166	221	197	172	62	135	23	132	75

3. Bernoulli Numbers and Euler Numbers (→ 177 Generating Functions)

B_n are Bernoulli numbers; E_n are Euler numbers.

n	Numerator of B_n	Denominator of B_n	B_n	E_n
2	1	6	0.16667	1
4	1	30	0.03333	5
6	1	42	0.02381	61
8	1	30	0.03333	1385
10	5	66	0.07576	50521
12	691	2730	0.25311	2702765
14	7	6	1.16667	199360981
16	3617	510	7.09216	19391512145
18	43867	798	54.97118	2404879675441
20	174611	330	529.12424	370371188237525
22	854513	138	6192.12319	6.934887×10^{16}
24	236364091	2730	86580.25311	1.551453×10^{19}
26	8553103	6	1425517.16667	4.087073×10^{21}
28	23749461029	870	27298231.06782	1.252260×10^{24}
30	8615841276005	14322	601580873.90064	4.415439×10^{26}

4. Class Numbers of Algebraic Number Fields

(I) Class Numbers of Real Quadratic Field (→ 347 Quadratic Fields)

Let $k=\mathbf{Q}(\sqrt{m})$, where m is a positive integer without square factor ($1<m\leqslant 501$). h is the class number (in the wider sense) of k. The $-$sign in the row of $N(\varepsilon)$ means that the norm $N(\varepsilon)$ of the fundamental unit is -1. When $N(\varepsilon)=+1$, the class number in the narrow sense is $2h$, and when $N(\varepsilon)=-1$, the class number in the narrow sense is also h.

m	h	$N(\varepsilon)$	m	h	$N(\varepsilon)$	m	h	$N(\varepsilon)$	m	h	$N(\varepsilon)$	m	h	$N(\varepsilon)$	m	h	$N(\varepsilon)$
2	1	−	85	2	−	170	4	−	253	1		335	2		421	1	−
3	1		86	1		173	1	−	254	3		337	1	−	422	1	
5	1	−	87	2		174	2		255	4		339	2		426	2	
6	1		89	1	−	177	1		257	3	−	341	1		427	6	
7	1		91	2		178	2		258	2		345	2		429	2	
10	2	−	93	1		179	1		259	2		346	6	−	430	2	
11	1		94	1		181	1	−	262	1		347	1		431	1	
13	1	−	95	2		182	2		263	1		349	1	−	433	1	−
14	1		97	1	−	183	2		265	2	−	353	1	−	434	4	

m	h	$N(\varepsilon)$	m	h	$N(\varepsilon)$	m	h	$N(\varepsilon)$	m	h	$N(\varepsilon)$	m	h	$N(\varepsilon)$	m	h	$N(\varepsilon)$
15	2		101	1	–	185	2	–	266	2		354	2		435	4	
17	1	–	102	2		186	2		267	2		355	2		437	1	
19	1		103	1		187	2		269	1	–	357	2		438	4	
21	1		105	2		190	2		271	1		358	1		439	5	
22	1		106	2	–	191	1		273	2		359	3		442	8	–
23	1		107	1		193	1	–	274	4	–	362	2	–	443	3	
26	2	–	109	1	–	194	2		277	1	–	365	2	–	445	4	–
29	1	–	110	2		195	4		278	1		366	2		446	1	
30	2		111	2		197	1	–	281	1	–	367	1		447	2	
31	1		113	1	–	199	1		282	2		370	4	–	449	1	–
33	1		114	2		201	1		283	1		371	2		451	2	
34	2		115	2		202	2	–	285	2		373	1	–	453	1	
35	2		118	1		203	2		286	2		374	2		454	1	
37	1	–	119	2		205	2		287	2		377	2		455	4	
38	1		122	2	–	206	1		290	4	–	379	1		457	1	–
39	2		123	2		209	1		291	4		381	1		458	2	–
41	1	–	127	1		210	4		293	1	–	382	1		461	1	–
42	2		129	1		211	1		295	2		383	1		462	4	
43	1		130	4	–	213	1		298	2	–	385	2		463	1	
46	1		131	1		214	1		299	2		386	2		465	2	
47	1		133	1		215	2		301	1		389	1	–	466	2	
51	2		134	1		217	1		302	1		390	4		467	1	
53	1	–	137	1	–	218	2	–	303	2		391	2		469	3	
55	2		138	2		219	4		305	2		393	1		470	2	
57	1		139	1		221	2		307	1		394	2	–	471	2	
58	2	–	141	1		222	2		309	1		395	2		473	3	
59	1		142	3		223	3		310	2		397	1	–	474	2	
61	1	–	143	2		226	8	–	311	1		398	1		478	1	
62	1		145	4	–	227	1		313	1	–	399	8		479	1	
65	2	–	146	2		229	3	–	314	2	–	401	5	–	481	2	–
66	2		149	1	–	230	2		317	1	–	402	2		482	2	
67	1		151	1		231	4		318	2		403	2		483	4	
69	1		154	2		233	1	–	319	2		406	2		485	2	–
70	2		155	2		235	6		321	3		407	2		487	1	
71	1		157	1	–	237	1		322	4		409	1	–	489	1	
73	1	–	158	1		238	2		323	4		410	4		491	1	
74	2	–	159	2		239	1		326	3		411	2		493	2	–
77	1		161	1		241	1	–	327	2		413	1		494	2	
78	2		163	1		246	2		329	1		415	2		497	1	
79	3		165	2		247	2		330	4		417	1		498	2	
82	4	–	166	1		249	1		331	1		418	2		499	5	
83	1		167	1		251	1		334	1		419	1		501	1	

One can find a table of fundamental units and representatives of ideal classes for $0<m<2025$ in E. L. Ince, *Cycles of reduced ideals in quadratic fields*, Royal Society, London, 1968.

(II) Class Numbers of Imaginary Quadratic Fields (→ 347 Quadratic Fields)

Let $k=\mathbf{Q}(\sqrt{-m}\,)$, where m is a positive integer without square factor ($1\leqslant m\leqslant 509$). h is the class number of k. In the present case, there is no distinction between the class numbers in the wider and narrow senses.

m	h	m	h	m	h	m	h	m	h	m	h	m	h	m	h
1	1	65	8	129	12	193	4	255	12	319	10	389	22	447	14
2	1	66	8	130	4	194	20	257	16	321	20	390	16	449	20
3	1	67	1	131	5	195	4	258	8	322	8	391	14	451	6
5	2	69	8	133	4	197	10	259	4	323	4	393	12	453	12
6	2	70	4	134	14	199	9	262	6	326	22	394	10	454	14
7	1	71	7	137	8	201	12	263	13	327	12	395	8	455	20
10	2	73	4	138	8	202	6	265	8	329	24	397	6	457	8
11	1	74	10	139	3	203	4	266	20	330	8	398	20	458	26

m	h	m	h	m	h	m	h	m	h	m	h	m	h	m	h
13	2	77	8	141	8	205	8	267	2	331	3	399	16	461	30
14	4	78	4	142	4	206	20	269	22	334	12	401	20	462	8
15	2	79	5	143	10	209	20	271	11	335	18	402	16	463	7
17	4	82	4	145	8	210	8	273	8	337	8	403	2	465	16
19	1	83	3	146	16	211	3	274	12	339	6	406	16	466	8
21	4	85	4	149	14	213	8	277	6	341	28	407	16	467	7
22	2	86	10	151	7	214	6	278	14	345	8	409	16	469	16
23	3	87	6	154	8	215	14	281	20	346	10	410	16	470	20
26	6	89	12	155	4	217	8	282	8	347	5	411	6	471	16
29	6	91	2	157	6	218	10	283	3	349	14	413	20	473	12
30	4	93	4	158	8	219	4	285	16	353	16	415	10	474	20
31	3	94	8	159	10	221	16	286	12	354	16	417	12	478	8
33	4	95	8	161	16	222	12	287	14	355	4	418	8	479	25
34	4	97	4	163	1	223	7	290	20	357	8	419	9	481	16
35	2	101	14	165	8	226	8	291	4	358	6	421	10	482	20
37	2	102	4	166	10	227	5	293	18	359	19	422	10	483	4
38	6	103	5	167	11	229	10	295	8	362	18	426	24	485	20
39	4	105	8	170	12	230	20	298	6	365	20	427	2	487	7
41	8	106	6	173	14	231	12	299	8	366	12	429	16	489	20
42	4	107	3	174	12	233	12	301	8	367	9	430	12	491	9
43	1	109	6	177	4	235	2	302	12	370	12	431	21	493	12
46	4	110	12	178	8	237	12	303	10	371	8	433	12	494	28
47	5	111	8	179	5	238	8	305	16	373	10	434	24	497	24
51	2	113	8	181	10	239	15	307	3	374	28	435	4	498	8
53	6	114	8	182	12	241	12	309	12	377	16	437	20	499	3
55	4	115	2	183	8	246	12	310	8	379	3	438	8	501	16
57	4	118	6	185	16	247	6	311	19	381	20	439	15	502	14
58	2	119	10	186	12	249	12	313	8	382	8	442	8	503	21
59	3	122	10	187	2	251	7	314	26	383	17	443	5	505	8
61	6	123	2	190	4	253	4	317	10	385	8	445	8	506	28
62	8	127	5	191	13	254	16	318	12	386	20	446	32	509	30

There are only 9 instances of m for which $h=1$, and only 18 instances of m for which $h=2$ (Baker, Stark). All these cases are in this table.

One can find a table of structures of the ideal class groups and representatives of ideal classes for $m<24000$ in H. Wada, *A table of ideal class groups of imaginary quadratic fields*, Proc. Japan Acad., 46 (1970), 401–403.

(III) Class Numbers of Cyclotomic Fields

Cyclotomic field $k=\mathbf{Q}(e^{2\pi i/l})$ $(1<l<100$; l prime$)$. h_1 is the first factor of the class number of k (→ 14 Algebraic Number Fields).

l	h_1	l	h_1	l	h_1	l	h_1	l	h_1	l	h_1
3	1	13	1	29	2^3	43	211	61	$41\cdot 1861$	79	$5\cdot 53\cdot 377911$
5	1	17	1	31	3^2	47	$5\cdot 139$	67	$67\cdot 12739$	83	$3\cdot 279405653$
7	1	19	1	37	37	53	4889	71	$7^2\cdot 79241$	89	$113\cdot 118401449$
11	1	23	3	41	11^2	59	$3\cdot 59\cdot 233$	73	$89\cdot 134353$	97	$577\cdot 3457\cdot 206209$

$h_1>1$ for $l>19$ (Uchida).

5. Characters of Finite Groups; Crystallographic Groups

(I) Symmetric Groups S_n, Alternating Groups A_n $(3\leqslant n\leqslant 7)$, and Mathieu Groups M_n $(n=11,12,22,23,24)$

(1) In each table, the first column gives the representation of the conjugate class as we represent a permutation by the product of cyclic permutations. For example, $(3)(2)^2$ means the conjugate class containing (123)(45)(67).

(2) The second column gives the order of the centralizer of the elements of the conjugate class.
(3) In the table of S_n, the first row gives the type of Young diagram corresponding to each irreducible character. For example, $[3, 2^2, 1]$ means $T(3, 2, 2, 1)$.
(4) In the table of A_n, when we restrict the self-conjugate character of S_n (the character with *) to A_n, it is decomposed into two mutually algebraically conjugate irreducible characters, and therefore we show only one of them. The other irreducible character of A_n is given by the restriction to A_n of the character of S_n that is not self-conjugate.
(5) In the table of M_n, each character with a bar over the degree is one of the two mutually algebraically conjugate characters.

$$\varepsilon_1^{\pm} = (-1 \pm \sqrt{-3})/2, \quad \varepsilon_2^{\pm} = (1 \pm \sqrt{5})/2, \quad \varepsilon_3^{\pm} = (-1 \pm \sqrt{-7})/2$$
$$\varepsilon_4^{\pm} = (-1 \pm \sqrt{-11})/2, \quad \varepsilon_5^{\pm} = (-1 \pm \sqrt{-15})/2, \quad \varepsilon_6^{\pm} = (-1 \pm \sqrt{-23})/2$$

S_3		[3]	[2, 1]*	[1³]
(1)	6	1	2	1
(2)	2	1	0	−1
(3)	3	1	−1	1

A_3		[2, 1]*
(1)	3	1
(3)	3	ε_1^+
(3)	3	ε_1^-

S_4		[4]	[3, 1]	[2²]*	[2, 1²]	[1⁴]
(1)	24	1	3	2	3	1
(2)	4	1	1	0	−1	−1
(3)	3	1	0	−1	0	1
(4)	4	1	−1	0	1	−1
(2)²	8	1	−1	2	−1	1

A_4		[2²]*
(1)	12	1
(3)	3	ε_1^+
(3)	3	ε_1^-
(2)²	4	1

S_5		[5]	[4, 1]	[3, 2]	[3, 1²]*	[2², 1]	[2, 1³]	[1⁵]
(1)	120	1	4	5	6	5	4	1
(2)	12	1	2	1	0	−1	−2	−1
(3)	6	1	1	−1	0	−1	1	1
(4)	4	1	0	−1	0	1	0	−1
(2)²	8	1	0	1	−2	1	0	1
(3)(2)	6	1	−1	1	0	−1	1	−1
(5)	5	1	−1	0	1	0	−1	1

A_5		[3, 1²]*
(1)	60	3
(3)	3	0
(2)²	4	−1
(5)	5	ε_2^+
(5)	5	ε_2^-

S_6		[6]	[5, 1]	[4, 2]	[4, 1²]	[3²]	[3, 2, 1]*	[2³]	[3, 1³]	[2², 1²]	[2, 1⁴]	[1⁶]
(1)	720	1	5	9	10	5	16	5	10	9	5	1
(2)	48	1	3	3	2	1	0	−1	−2	−3	−3	−1
(3)	18	1	2	0	1	−1	−2	−1	1	0	2	1
(4)	8	1	1	−1	0	−1	0	1	0	1	−1	−1
(2)²	16	1	1	1	−2	1	0	1	−2	1	1	1
(3)(2)	6	1	0	0	−1	1	0	−1	1	0	0	−1
(5)	5	1	0	−1	0	0	1	0	0	−1	0	1
(6)	6	1	−1	0	1	0	0	0	−1	0	1	−1
(4)(2)	8	1	−1	1	0	−1	0	−1	0	1	−1	1
(2)³	48	1	−1	3	−2	−3	0	3	2	−3	1	−1
(3)²	18	1	−1	0	1	2	−2	2	1	0	−1	1

A_6		$[3,2,1]^*$
(1)	360	8
(3)	9	−1
$(2)^2$	8	0
(5)	5	ε_2^+
(5)	5	ε_2^-
(4)(2)	4	0
$(3)^2$	9	−1

S_7		[7]	[6, 1]	[5, 2]	$[5, 1^2]$	[4, 3]	[4, 2, 1]	$[3^2, 1]$	$[4, 1^3]^*$	$[3, 2^2]$	$[3, 2, 1^2]$	$[2^3, 1]$	$[3, 1^4]$	$[2^2, 1^3]$	$[2, 1^5]$	$[1^7]$
(1)	5040	1	6	14	15	14	35	21	20	21	35	14	15	14	6	1
(2)	240	1	4	6	5	4	5	1	0	−1	−5	−4	−5	−6	−4	−1
(3)	72	1	3	2	3	−1	−1	−3	2	−3	−1	−1	3	2	3	1
(4)	24	1	2	0	1	−2	−1	−1	0	1	1	2	−1	0	−2	−1
$(2)^2$	48	1	2	2	−1	2	−1	1	−4	1	−1	2	−1	2	2	1
(3)(2)	12	1	1	0	−1	1	−1	1	0	−1	1	−1	1	0	−1	−1
(5)	10	1	1	−1	0	−1	0	1	0	1	0	−1	0	−1	1	1
(6)	6	1	0	−1	0	0	1	0	0	0	−1	0	0	1	0	−1
(4)(2)	8	1	0	0	−1	0	1	−1	0	−1	1	0	−1	0	0	1
$(2)^3$	48	1	0	2	−3	0	1	−3	0	3	−1	0	3	−2	0	−1
$(3)^2$	18	1	0	−1	0	2	−1	0	2	0	−1	2	0	−1	0	1
(5)(2)	10	1	−1	1	0	−1	0	1	0	−1	0	1	0	−1	1	−1
$(3)(2)^2$	24	1	−1	2	−1	−1	−1	1	2	1	−1	−1	−1	2	−1	1
(4)(3)	12	1	−1	0	1	1	−1	−1	0	1	1	−1	−1	0	1	−1
(7)	7	1	−1	0	1	0	0	0	−1	0	0	0	1	0	−1	1

A_7		$[4,1^3]^*$
(1)	2520	10
(3)	36	1
$(2)^2$	24	−2
(5)	5	0
(4)(2)	4	0
$(3)^2$	9	1
$(3)(2)^2$	12	1
(7)	7	ε_3^+
(7)	7	ε_3^-

M_{11}									
(1)	g	1	10	11	55	45	44	$\overline{16}$	$\overline{10}$
$(2)^4$	48	1	2	3	−1	−3	4	0	−2
$(4)^2$	8	1	2	−1	−1	1	0	0	0
$(3)^3$	18	1	1	2	1	0	−1	−2	1
$(5)^2$	5	1	0	1	0	0	−1	1	0
(8)(2)	8	1	0	−1	1	−1	0	0	$\pm i\sqrt{2}$
(8)(2)	8	1	0	−1	1	−1	0	0	$\mp i\sqrt{2}$
(6)(3)(2)	6	1	−1	0	−1	0	1	0	1
(11)	11	1	−1	0	0	1	0	ε_4^+	−1
(11)	11	1	−1	0	0	1	0	ε_4^-	−1

$g = 11 \cdot 10 \cdot 9 \cdot 8 = 7920.$

M_{12}															
(1)	g	1	11	11	55	55	55	45	54	66	99	120	144	176	$\overline{16}$
$(2)^4$	192	1	3	3	−1	−1	7	−3	6	2	3	−8	0	0	0
$(4)^2$	32	1	3	−1	3	−1	−1	1	2	−2	−1	0	0	0	0
$(3)^3$	54	1	2	2	1	1	1	0	0	3	0	3	0	−4	−2
$(5)^2$	10	1	1	1	0	0	0	0	−1	1	−1	0	−1	1	1
(8)(2)	8	1	1	−1	−1	1	−1	−1	0	0	1	0	0	0	0
(6)(3)(2)	6	1	0	0	−1	−1	1	0	0	−1	0	1	0	0	0
(11)	11	1	0	0	0	0	0	1	−1	0	0	−1	1	0	ε_4^+
(11)	11	1	0	0	0	0	0	1	−1	0	0	−1	1	0	ε_4^-
$(2)^6$	240	1	−1	−1	−5	−5	−5	5	6	6	−1	0	4	−4	4
(10)(2)	10	1	−1	−1	0	0	0	0	1	1	−1	0	−1	1	−1
$(4)^2(2)^2$	32	1	−1	3	−1	3	−1	1	2	−2	−1	0	0	0	0
$(3)^4$	36	1	−1	−1	1	1	1	3	0	0	3	0	−3	−1	1
$(6)^2$	12	1	−1	−1	1	1	1	−1	0	0	−1	0	1	−1	1
(8)(4)	8	1	−1	1	1	−1	−1	−1	0	0	1	0	0	0	0

$g = 12 \cdot 11 \cdot 10 \cdot 9 \cdot 8 = 95040.$

M_{22}

(1)	g	1	21	55	154	210	$\overline{280}$	231	385	99	$\overline{45}$
$(2)^8$	384	1	5	7	10	2	−8	7	1	3	−3
$(3)^6$	36	1	3	1	1	3	1	−3	−2	0	0
$(5)^4$	5	1	1	0	−1	0	0	1	0	−1	0
$(4)^4(2)^2$	16	1	1	−1	2	−2	0	−1	1	−1	1
$(4)^4(2)^2$	32	1	1	3	−2	−2	0	−1	1	3	1
$(7)^3$	7	1	0	−1	0	0	0	0	0	1	ε_3^+
$(7)^3$	7	1	0	−1	0	0	0	0	0	1	ε_3^-
$(8)^2(4)(2)$	8	1	−1	1	0	0	0	−1	1	−1	−1
$(6)^2(3)^2(2)^2$	12	1	−1	1	1	−1	1	1	−2	0	0
$(11)^2$	11	1	−1	0	0	1	ε_4^+	0	0	0	1
$(11)^2$	11	1	−1	0	0	1	ε_4^-	0	0	0	1

$g = 22 \cdot 21 \cdot 20 \cdot 48 = 443520.$

M_{23}

(1)	g	1	22	230	231	$\overline{770}$	1035	2024	$\overline{45}$	$\overline{990}$	$\overline{231}$	253	$\overline{896}$
$(2)^8$	2688	1	6	22	7	−14	27	8	−3	−18	7	13	0
$(3)^6$	180	1	4	5	6	5	0	−1	0	0	−3	1	−4
$(5)^4$	15	1	2	0	1	0	0	−1	0	0	1	−2	1
$(4)^4(2)^2$	32	1	2	2	−1	−2	−1	0	1	2	−1	1	0
$(7)^3$	14	1	1	−1	0	0	−1	1	ε_3^+	ε_3^+	0	1	0
$(7)^3$	14	1	1	−1	0	0	−1	1	ε_3^-	ε_3^-	0	1	0
$(8)^2(4)(2)$	8	1	0	0	−1	0	1	0	−1	0	−1	−1	0
$(6)^2(3)^2(2)^2$	12	1	0	1	−2	1	0	−1	0	0	1	1	0
$(11)^2$	11	1	0	−1	0	0	1	0	1	0	0	0	ε_4^+
$(11)^2$	11	1	0	−1	0	0	1	0	1	0	0	0	ε_4^-
(15)(5)(3)	15	1	−1	0	1	0	0	−1	0	0	ε_5^+	1	1
(15)(5)(3)	15	1	−1	0	1	0	0	−1	0	0	ε_5^-	1	1
(14)(7)(2)	14	1	−1	1	0	0	−1	1	$-\varepsilon_3^+$	ε_3^+	0	−1	0
(14)(7)(2)	14	1	−1	1	0	0	1	1	ε_3^-	ε_3^-	0	−1	0
(23)	23	1	−1	0	1	ε_6^+	0	0	−1	1	1	0	−1
(23)	23	1	−1	0	1	ε_6^-	0	0	−1	1	1	0	−1

$g = 23 \cdot 22 \cdot 21 \cdot 20 \cdot 48 = 10200960.$

M_{24}

$(1)^{24}$	g	1	23	$7 \cdot 36$	$23 \cdot 11$	$23 \cdot 77$	$55 \cdot 64$	$\overline{45}$	$\overline{22 \cdot 45}$	$\overline{23 \cdot 45}$	$23 \cdot 45$	$\overline{11 \cdot 21}$	$\overline{770}$
$(2)^8$	$21 \cdot 2^{10}$	1	7	28	13	−21	64	−3	−18	−21	27	7	−14
$(3)^6$	$27 \cdot 40$	1	5	9	10	16	10	0	0	0	0	−3	5
$(5)^4$	60	1	3	2	3	1	0	0	0	0	0	1	0
$(4)^4(2)^2$	128	1	3	4	1	−5	0	1	2	3	−1	−1	−2
$(7)^3$	42	1	2	0	1	0	−1	ε_3^+	ε_3^+	$2\varepsilon_3^+$	−1	0	0
$(7)^3$	42	1	2	0	1	0	−1	ε_3^-	ε_3^-	$2\varepsilon_3^-$	−1	0	0
$(8)^2(4)(2)$	16	1	1	0	−1	−1	0	−1	0	−1	1	−1	0
$(6)^2(3)^2(2)^2$	24	1	1	1	−2	0	−2	0	0	0	0	1	1
$(11)^2$	11	1	1	−1	0	0	0	1	0	1	1	0	0
(15)(5)(3)	15	1	0	−1	0	1	0	0	0	0	0	ε_5^+	0
(15)(5)(3)	15	1	0	−1	0	1	0	0	0	0	0	ε_5^-	0
(14)(7)(2)	14	1	0	0	−1	0	1	$-\varepsilon_3^-$	ε_3^+	0	−1	0	0
(14)(7)(2)	14	1	0	0	−1	0	1	$-\varepsilon_3^+$	ε_3^-	0	−1	0	0
(23)	23	1	0	−1	0	0	1	−1	1	0	0	1	ε_6^+
(23)	23	1	0	−1	0	0	1	−1	1	0	0	1	ε_6^-
$(12)^2$	12	1	−1	0	1	−1	0	1	1	−1	0	0	1
$(6)^4$	24	1	−1	0	1	−1	0	−1	−1	1	2	0	1
$(4)^6$	96	1	−1	0	1	−1	0	1	−2	−1	3	3	−2
$(3)^8$	$7 \cdot 72$	1	−1	0	1	7	−8	3	3	−3	6	0	−7
$(2)^{12}$	$15 \cdot 2^9$	1	−1	12	−11	11	0	5	−10	−5	35	−9	10
$(10)^2(2)^2$	20	1	−1	2	−1	1	0	0	0	0	0	1	0
(21)(3)	21	1	−1	0	1	0	−1	ε_3^-	ε_3^-	$-\varepsilon_3^+$	−1	0	0
(21)(3)	21	1	−1	0	1	0	−1	ε_3^+	ε_3^+	$-\varepsilon_3^-$	−1	0	0
$(4)^4(2)^4$	$3 \cdot 2^7$	1	−1	4	−3	3	0	−3	6	3	3	−1	2
(12)(6)(4)(2)	12	1	−1	1	0	0	0	0	0	0	0	−1	−1

$(1)^{24}$	g	$23\cdot 21$	$23\cdot 55$	$23\cdot 88$	$23\cdot 99$	$23\cdot 144$	$23\cdot 11\cdot 21$	$23\cdot 7\cdot 36$	$77\cdot 72$	$11\cdot 35\cdot 27$
$(2)^8$	$21\cdot 2^{10}$	35	49	8	21	48	49	−28	−56	−21
$(3)^6$	$27\cdot 40$	6	5	−1	0	0	−15	−9	9	0
$(5)^4$	60	−2	0	−1	−3	−3	3	1	−1	0
$(4)^4(2)^2$	128	3	1	0	1	0	−3	4	0	−1
$(7)^3$	42	0	−2	1	2	1	0	0	0	0
$(7)^3$	42	0	−2	1	2	1	0	0	0	0
$(8)^2(4)(2)$	16	−1	1	0	−1	0	−1	0	0	1
$(6)^2(3)^2(2)^2$	24	2	1	−1	0	0	1	−1	1	0
$(11)^2$	11	−1	0	0	0	1	0	−1	0	0
$(15)(5)(3)$	15	1	0	−1	0	0	0	1	−1	0
$(15)(5)(3)$	15	1	0	−1	0	0	0	1	−1	0
$(14)(7)(2)$	14	0	0	1	0	−1	0	0	0	0
$(14)(7)(2)$	14	0	0	1	0	−1	0	0	0	0
(23)	23	0	0	0	0	0	0	0	1	−1
(23)	23	0	0	0	0	0	0	0	1	−1
$(12)^2$	12	0	0	0	0	0	0	0	0	0
$(6)^4$	24	0	0	0	2	−2	0	0	0	0
$(4)^6$	96	3	−3	0	−3	0	−3	0	0	3
$(3)^8$	$7\cdot 72$	0	8	8	6	−6	0	0	0	0
$(2)^{12}$	$15\cdot 2^9$	3	−15	24	−19	16	9	36	24	−45
$(10)^2(2)^2$	20	−2	0	−1	1	1	−1	1	−1	0
$(21)(3)$	21	0	1	1	−1	1	0	0	0	0
$(21)(3)$	21	0	1	1	−1	1	0	0	0	0
$(4)^4(2)^4$	$3\cdot 2^7$	3	−7	8	−3	0	1	−4	−8	3
$(12)(6)(4)(2)$	12	0	−1	−1	0	0	1	−1	1	0

$g=24\cdot 23\cdot 22\cdot 21\cdot 20\cdot 48=244823040.$

(II) General Linear Groups $GL(2,q)$, Unitary Groups $U(2,q)$, and Special Linear Groups $SL(2,q)$ (q is a power of a prime) (→ 151 Finite Groups I)

(1) The notations are as follows. $\varepsilon=\exp[2\pi\sqrt{-1}/(q-1)]$, $\eta=\exp[2\pi\sqrt{-1}/(q^2-1)]$, $\sigma=\exp[2\pi\sqrt{-1}/(q+1)]$, ρ is the generator of the multiplicative group of $GF(q)-\{0\}$, ω is the generator of the multiplicative group of $GF(q^2)-\{0\}$, $\omega^{q-1}=\alpha$, B is an element of $GL(2,q)$ with order q^2-1, and $B_1=B^{q-1}$.
(2) The first column gives a representative of the conjugate class.

General Linear Group $GL(2,q)$.

	$X_n(1)$	$X_n(q)$	$Y_{m,n}$	Z_n
$\begin{pmatrix}\rho^a & \\ & \rho^a\end{pmatrix}$	ε^{2na}	$q\varepsilon^{2na}$	$(q+1)\varepsilon^{(m+n)a}$	$(q-1)\eta^{na(q+1)}$
$\begin{pmatrix}\rho^a & \\ 1 & \rho^a\end{pmatrix}$	ε^{2na}	0	$\varepsilon^{(m+n)a}$	$-\eta^{na(q+1)}$
$\begin{pmatrix}\rho^a & \\ & \rho^b\end{pmatrix}$	$\varepsilon^{n(a+b)}$	$\varepsilon^{n(a+b)}$	$\varepsilon^{ma+nb}+\varepsilon^{mb+na}$	0
B^c	ε^{nc}	$-\varepsilon^{nc}$	0	$-(\eta^{nc}+\eta^{ncq})$

(1) $1\leqslant a\leqslant q-1$, $1\leqslant b\leqslant q-1$, $a\not\equiv b \pmod{q-1}$, $1\leqslant c<q^2-1$, $c\not\equiv 0 \pmod{q+1}$.
(2) We assume that $1\leqslant n\leqslant q-1$, for $X_n(1), X_n(q)$, $1\leqslant m<n\leqslant q-1$, for $Y_{m,n}$, $1\leqslant n<q^2-1$ for Z_n, $n\not\equiv 0 \pmod{q+1}$. Here, $Z_n=Z_{n'}$ when $n\equiv n'q \pmod{q^2-1}$.

Unitary Group $U(2,q)$.

	$X'_n(1)$	$X'_n(q)$	$Y'_{m,n}$	Z'_n
$\begin{pmatrix}\alpha^s & \\ & \alpha^s\end{pmatrix}$	σ^{2ns}	$q\sigma^{2ns}$	$(q-1)\sigma^{(m+n)s}$	$(q+1)\sigma^{ns}$
$\begin{pmatrix}\alpha^s & \\ 1 & \alpha^s\end{pmatrix}$	σ^{2ns}	0	$-\sigma^{(m+n)s}$	σ^{ns}
$\begin{pmatrix}\alpha^s & \\ & \alpha^t\end{pmatrix}$	$\sigma^{n(s+t)}$	$-\sigma^{n(s+t)}$	$-(\sigma^{ms+nt}+\sigma^{mt+ns})$	0
$\begin{pmatrix}\omega^u & \\ & \omega^{-uq}\end{pmatrix}$	σ^{-nu}	σ^{-nu}	0	$\eta^{nu}+\eta^{-nuq}$

(1) $\begin{pmatrix} \alpha^s & \\ 1 & \alpha^s \end{pmatrix}$, $\begin{pmatrix} \omega^u & \\ & \omega^{-uq} \end{pmatrix}$ are the canonical forms of an element of $U(2,q)$ in $GL(2,q^2)$.
(2) $1 \leqslant s \leqslant q+1$, $1 \leqslant t \leqslant q+1$, $s \not\equiv t \pmod{q+1}$, $1 \leqslant u < q^2-1$, $u \not\equiv 0 \pmod{q-1}$. When $u \equiv -u'q \pmod{q^2-1}$ u, u' gives the same conjugate class.
(3) The ranges are $1 \leqslant n \leqslant q+1$ for $X_n'(1)$, $X_n'(q)$, $1 \leqslant m < n \leqslant q+1$ for $Y_{m,n}'$, $1 \leqslant n < q^2-1$ for Z_n', $n \not\equiv 0 \pmod{q-1}$. When $n' \equiv -nq \pmod{q^2-1}$, we have $Z_n' = Z_{n'}'$.

Special Linear Group $SL(2,2^n)$ (the case when $q=2^n$).

			Y_n	Z_m
$\begin{pmatrix} 1 & \\ & 1 \end{pmatrix}$	1	q	$q+1$	$q-1$
$\begin{pmatrix} 1 & \\ 1 & 1 \end{pmatrix}$	1	0	1	-1
$\begin{pmatrix} \rho^a & \\ & \rho^{-a} \end{pmatrix}$	1	1	$\varepsilon^{na}+\varepsilon^{-na}$	0
B_1^c	1	-1	0	$-(\sigma^{mc}+\sigma^{-mc})$

(1) $1 \leqslant a \leqslant (q-2)/2$, $1 \leqslant c \leqslant q/2$.
(2) $1 \leqslant n \leqslant (q-2)/2$, $1 \leqslant m \leqslant q/2$.

Special Linear Group $SL(2,q)$ ($q=$ power of an odd prime number, $e=(q-1)/2$, $e'=(q+1)/2$).

			Y_n	Z_m		
$\begin{pmatrix} 1 & \\ & 1 \end{pmatrix}$	1	q	$q+1$	$q-1$	$\frac{q+1}{2}$	$\frac{q-1}{2}$
$Z=\begin{pmatrix} -1 & \\ & -1 \end{pmatrix}$	1	q	$(-1)^n(q+1)$	$(-1)^m(q-1)$	$(-1)^e\frac{q+1}{2}$	$(-1)^{e'}\frac{q-1}{2}$
$P_1=\begin{pmatrix} 1 & \\ 1 & 1 \end{pmatrix}$	1	0	1	-1	$\mu^{\pm}$	$\lambda^{\pm}$
$P_2=\begin{pmatrix} 1 & \\ \beta & 1 \end{pmatrix}$	1	0	1	-1	$\mu^{\mp}$	$\lambda^{\mp}$
P_1Z	1	0	$(-1)^n$	$-(-1)^m$	$(-1)^e\mu^{\pm}$	$(-1)^{e'}\lambda^{\pm}$
P_2Z	1	0	$(-1)^n$	$-(-1)^m$	$(-1)^e\mu^{\mp}$	$(-1)^{e'}\lambda^{\mp}$
$\begin{pmatrix} \rho^a & \\ & \rho^{-a} \end{pmatrix}$	1	1	$\varepsilon^{na}+\varepsilon^{-na}$	0	$(-1)^a$	0
B_1^c	1	-1	0	$-(\sigma^{mc}+\sigma^{-mc})$	0	$-(-1)^c$

(1) $1 \leqslant a \leqslant (q-3)/2$, $1 \leqslant c \leqslant (q-1)/2$, $1 \leqslant n \leqslant (q-3)/2$, $1 \leqslant m \leqslant (q-1)/2$,
$\lambda^{\pm} = \{-1 \pm [(-1)^e q]^{1/2}\}/2$, $\mu^{\pm} = \{1 \pm [(-1)^e q]^{1/2}\}/2$.
(2) The last two columns mean two characters (with the same signs), respectively.

(III) Ree group $Re(q)$, **Suzuki Group** $Sz(q)$, **and Janko Group** J.

Ree group $Re(q)$ ($q=3^{2n+1}=3m^2$).
The order of $Re(q)$ is $q^3(q^3+1)(q-1)$, $q_0=q^2-q+1$, $m_+=q+3m+1$, $m_-=q-3m+1$.

						A	B	C	X_μ
1	1	1	q_0	q^3	qq_0	$(q-1)mm_+/2$	$(q-1)mm_-/2$	$m(q^2-1)$	q^3+1
J	2	1	-1	q	$-q$	$-(q-1)/2$	$(q-1)/2$	0	$q+1$
X	3	1	$-(q-1)$	0	q	$-(q+m)/2$	$(q-m)/2$	$-m$	1
Y	9	1	1	0	0	m	m	$-m$	1
T	3	1	1	0	0	α	α	2α	1
T^{-1}	3	1	1	0	0	$\bar{\alpha}$	$\bar{\alpha}$	$2\bar{\alpha}$	1
YT	9	1	1	0	0	β	β	$-\beta$	1
YT^{-1}	9	1	1	0	0	$\bar{\beta}$	$\bar{\beta}$	$-\bar{\beta}$	1
JT	6	1	-1	0	0	γ	$-\gamma$	0	1
JT^{-1}	6	1	-1	0	0	$\bar{\gamma}$	$-\bar{\gamma}$	0	1
R^a		1	1	1	1	0	0	0	$\rho^{\mu a}+\rho^{-\mu a}$
S^b		1	3	-1	-3	1	-1	0	0
JR^a		1	-1	1	-1	0	0	0	$\rho^{\mu a}+\rho^{-\mu a}$
JS^b		1	-1	-1	1	1	-1	0	0
V^s		1	0	-1	0	-1	0	-1	0
W^t		1	0	-1	0	0	1	1	0

		X'_μ	Y_ν	Y'_λ	Z_κ	Z'_τ
1	1	q^3+1	$(q-1)q_0$	$(q-1)q_0$	$(q^2-1)m_+$	$(q^2-1)m_-$
J	2	$-(q+1)$	$3(q-1)$	$-(q-1)$	0	0
X	3	1	$2q-1$	$2q-1$	$-m_+$	$-m_-$
Y	9	1	-1	-1	-1	-1
T	3	1	-1	-1	$-3m-1$	$3m-1$
T^{-1}	3	1	-1	-1	$-3m-1$	$3m-1$
YT	9	1	-1	-1	-1	-1
YT^{-1}	9	1	-1	-1	-1	-1
JT	6	-1	-3	1	0	0
JT^{-1}	6	-1	-3	1	0	0
R^a		$\rho^{\mu a}+\rho^{-\mu a}$	0	0	0	0
S^b		0	$\sigma(\nu b)$	$\sigma'(\lambda b)$	0	0
JR^a		$-(\rho^{\mu a}+\rho^{-\mu a})$	0	0	0	0
JS^b		0	$\sigma(\nu b)$	$\sigma'(\lambda b)$	0	0
V^s		0	0	0	$-\sum_{i=0}^{2}(v^{\kappa s q^i}+v^{-\kappa s q^i})$	0
W^t		0	0	0	0	$-\sum_{i=0}^{2}(w^{\tau t q^i}+w^{-\tau t q^i})$

(1) The first column gives a representative of conjugate class, and the second column gives its order. The orders of R, S, V, W are $(q-1)/2$, $(q+1)/4$, m_-, m_+, respectively. R, S, T are commutative with J.
(2) $R^a \sim R^{-a}$, $V^s \sim V^{sq} \sim V^{sq^2} \sim V^{-s} \sim V^{-sq} \sim V^{-sq^2}$, $W^t \sim W^{tq} \sim W^{tq^2} \sim W^{-t} \sim W^{-tq} \sim W^{-tq^2}$. Here we fix an integer δ satisfying $\delta^3 \equiv 1\ [\mathrm{mod}(q+1)/4]$, $(\delta-1,\ (q+1)/4)=1$.

$$S^b \sim S^{b\delta} \sim S^{b\delta^2} \sim S^{-b} \sim S^{-b\delta} \sim S^{-b\delta^2}, \qquad JR^a \sim JR^{-a}, \qquad JS^b \sim JS^{-b},$$

where $A \sim B$ means that A and B are mutually conjugate.
(3) $\rho=\exp[4\pi\sqrt{-1}\,/(q-1)]$, $v=\exp(2\pi\sqrt{-1}\,/m_-)$, $w=\exp(2\pi\sqrt{-1}\,/m_+)$, $\sigma=\exp[8\pi\sqrt{-1}\,/(q+1)]$.
(4) $1 \leqslant \mu \leqslant (q-3)/4$, $1 \leqslant \lambda \leqslant (q-3)/8$.
Here ν is considered $\mathrm{mod}(q+1)/4$ and

$$Y_\nu = Y_{\nu\delta} = Y_{\nu\delta^2} = Y_{-\nu} = Y_{-\nu\delta} = Y_{-\nu\delta^2},$$

κ is considered $\mathrm{mod}\, m_-$ and

$$Z_\kappa = Z_{\kappa q} = Z_{\kappa q^2} = Z_{-\kappa} = Z_{-\kappa q} = Z_{-\kappa q^2},$$

τ is considered $\mathrm{mod}\, m_+$ and

$$Z'_\tau = Z'_{\tau q} = Z'_{\tau q^2} = Z'_{-\tau} = Z'_{-\tau q} = Z'_{-\tau q^2}.$$

(5) $\sigma(\nu b) = -\sum_{i=0}^{2}(\sigma^{\nu b\delta^i}+\sigma^{-\nu b\delta^i})$, $\sigma'(\lambda b)=\sum_{i=0}^{1}(\sigma^{\lambda b\delta^i}+\sigma^{-\lambda b\delta^i})-(\sigma^{\lambda b\delta^2}+\sigma^{-\lambda b\delta^2})$.

(6) $\alpha=\dfrac{-m+m\sqrt{-q}}{2}$, $\beta=\dfrac{-m-\sqrt{-q}}{2}$, $\gamma=\dfrac{1-\sqrt{-q}}{2}$. We show one of the two mutually complex conjugate characters, for the characters A, B, C.

Suzuki group $Sz(q)$. The order of $Sz(q)$ is $q^2(q^2+1)(q-1)$ $(q=2^{2n+1}, 2q=r^2)$.

			X_α	Y_β	Z_γ		
1	1	q^2	q^2+1	$(q-r+1)(q-1)$	$(q+r+1)(q-1)$	$r(q-1)/2$	$r(q-1)/2$
σ	1	0	1	$r-1$	$-r-1$	$-r/2$	$-r/2$
ρ	1	0	1	-1	-1	$r\sqrt{-1}\,/2$	$-r\sqrt{-1}\,/2$
ρ^{-1}	1	0	1	-1	-1	$-r\sqrt{-1}\,/2$	$r\sqrt{-1}\,/2$
π_0^i	1	1	$\varepsilon_0^{\alpha i}+\varepsilon_0^{-\alpha i}$	0	0	0	0
π_1^j	1	-1	0	$-(\varepsilon_1^{\beta j}+\varepsilon_1^{\beta jq}+\varepsilon_1^{-\beta j}+\varepsilon_1^{-\beta jq})$	0	1	1
π_2^k	1	-1	0	0	$-(\varepsilon_2^{\gamma k}+\varepsilon_2^{\gamma kq}+\varepsilon_2^{-\gamma k}+\varepsilon_2^{-\gamma kq})$	-1	-1

(1) The first column gives a representative of the conjugate class.
(2) π_0, π_1, π_2 are the elements of order $q-1$ $q+r+1$, $q-r+1$, respectively.
(3) ε_0, ε_1, ε_2 are the primitive $q-1$, $q+r+1$, $q-r+1$ roots of 1, respectively.
(4) π_0^i and π_0^{-i} are mutually conjugate elements, and hence X_α and $X_{-\alpha}$ give the same character. i, α run over the representatives of mod $q-1$, and $i,\alpha \not\equiv 0$ $(\mathrm{mod}\, q-1)$.

(5) π_1^j, π_1^{-j}, π_1^{jq}, π_1^{-jq} are mutually conjugate, and hence Y_β, $Y_{-\beta}$, $Y_{\beta q}$, $Y_{-\beta q}$ give the same character. j, β run over the representatives of $\bmod q+r+1$, and j, $\beta \not\equiv 0 \pmod{q+r+1}$.
(6) π_2^k, π_2^{-k}, π_2^{kq}, π_2^{-kq} are mutually conjugate, and hence Z_γ, $Z_{-\gamma}$, $Z_{\gamma q}$, $Z_{-\gamma q}$ give the same character. k, γ run over the representatives of mod $q-r+1$, and k, $\gamma \not\equiv 0$ (mod $q-r+1$).

Janko Group J.

1	1	77	133	209	133	77	77	133	76	76	56	56	120	120	120
2	1	5	5	1	-3	-3	-3	-3	4	-4	0	0	0	0	0
3	1	-1	1	-1	-2	2	2	-2	1	1	2	2	0	0	0
5	1	2	-2	-1	ε^+	$-\varepsilon^+$	$-\varepsilon^-$	ε^-	1	1	$2\varepsilon^-$	$2\varepsilon^+$	0	0	0
5	1	2	-2	-1	ε^-	$-\varepsilon^-$	$-\varepsilon^+$	ε^+	1	1	$2\varepsilon^+$	$2\varepsilon^-$	0	0	0
6	1	-1	-1	1	0	0	0	0	1	-1	0	0	0	0	0
7	1	0	0	-1	0	0	0	0	-1	-1	0	0	1	1	1
10	1	0	0	1	$-\varepsilon^+$	$-\varepsilon^+$	$-\varepsilon^-$	$-\varepsilon^-$	-1	1	0	0	0	0	0
10	1	0	0	1	$-\varepsilon^-$	$-\varepsilon^-$	$-\varepsilon^+$	$-\varepsilon^+$	-1	1	0	0	0	0	0
11	1	0	1	0	1	0	0	1	-1	-1	1	1	-1	-1	-1
15	1	-1	1	-1	ε^+	$-\varepsilon^+$	$-\varepsilon^-$	ε^-	1	1	$-\varepsilon^-$	$-\varepsilon^+$	0	0	0
15	1	-1	1	-1	ε^-	$-\varepsilon^-$	$-\varepsilon^+$	ε^+	1	1	$-\varepsilon^+$	$-\varepsilon^-$	0	0	0
19	1	1	0	0	0	1	1	0	0	0	-1	-1	λ_1	λ_2	λ_3
19	1	1	0	0	0	1	1	0	0	0	-1	-1	λ_2	λ_3	λ_1
19	1	1	0	0	0	1	1	0	0	0	-1	-1	λ_3	λ_1	λ_2

(1) The order of J is $8\cdot 3\cdot 5\cdot 7\cdot 11\cdot 19=175560$.
(2) The first column gives the order of the elements of each conjugate class.
(3) $\rho=\exp(2\pi\sqrt{-1}/19)$, $\lambda_1=\rho+\rho^7+\rho^8+\rho^{11}+\rho^{12}+\rho^{18}$, $\lambda_2=\rho^2+\rho^{14}+\rho^{16}+\rho^3+\rho^5+\rho^{17}$, $\lambda_3=\rho^4+\rho^9+\rho^{13}+\rho^6+\rho^{10}+\rho^{15}$, $\varepsilon^\pm=(1\pm\sqrt{5})/2$.

References

For $S_n (2\leqslant n\leqslant 10)$, $A_n (3\leqslant n\leqslant 9)$:
[1] D. E. Littlewood, The theory of group characters, second edition, Oxford Univ. Press, 1950.
For $S_n (n=11,12,13)$:
[2] M. Zia-ud-Din, Proc. London Math. Soc., 39(1935), 200–204, 42 (1937), 340–355.
For S_{14}:
[3] K. Kondô, Table of characters of the symmetric group of degree 14, Proc. Phys. Math. Soc. Japan, 22 (1940), 585–593.
For M_n $(n=12,24)$:
[4] G. Frobenius, Über die Charaktere der mehrfach transitive Gruppen, S. B. Preuss. Akad. Wiss., 1904, 558–571.
For M_n $(n=11,22,23)$:
[5] N. Burgoyne and P. Fong, The Schur multipliers of the Mathieu groups, Nagoya Math. J., 27 (1966), 733–745.
For $GL(2,q)$, $SL(2,q)$:
[6] I. Schur, Untersuchungen über die Darstellung der endlichen Gruppen durch gebrochene lineare Substitutionen, J. Reine Angew. Math., 132 (1907), 85–137.
[7] H. E. Jordan, Group characters of various types of linear groups, Amer. J. Math., 29 (1907), 387–405.
For $GL(3,q)$, $GL(4,q)$, $GL(n,q)$:
[8] R. Steinberg, The representations of $GL(3,q)$, $GL(4,q)$, $PGL(3,q)$ and $PGL(4,q)$, Canad. J. Math. 3 (1951), 225–235.
[9] J. A. Green, The characters of the finite general linear groups, Trans. Amer. Math. Soc., 80 (1955), 402–447.
For $U(2,q)$, $U(3,q)$:
[10] V. Ennola, On the characters of the finite unitary groups, Ann. Acad. Sci. Fenn., 323 (1963), 1–34.
For $Sz(q)$, $Re(q)$, J:
[11] M. Suzuki, On a class of doubly transitive groups, Ann. of Math., (2) 75 (1962), 105–145.
[12] H. N. Ward, On Ree's series of simple groups, Trans. Amer. Math. Soc., 121 (1966), 62–89.
[13] Z. Janko, A new finite simple group with Abelian Sylow 2-subgroups and its characterization, J. Algebra, 3 (1966), 147–186.

For other Lie-type groups:
[14] B. Srinivasan, The characters of the finite symplectic group $Sp(4,q)$, Trans. Amer. Math. Soc., 131 (1968), 488–525.
[15] H. Enomoto, The characters of the finite symplectic group $Sp(4,q)$, $q=2^f$, Osaka J. Math., 9 (1972), 75–94.
[16] G. I. Lehrer, The characters of the finite special linear groups, J. Algebra, 26 (1973), 564–583.
[17] P. Deligne and G. Lusztig, Representations of reductive groups over finite fields, Ann. of Math., (2) 103 (1976), 103–161.
For sporadic groups:
[18] M. Hall, Jr. and D. Wales, The simple groups of order 604, 800, J. Algebra, 9 (1968), 417–450.
[19] J. S. Frame, Computation of characters of the Higman-Sims group and its automorphism group, J. Algebra, 20 (1972), 320–349.
[20] D. Fendel, A characterization of Conway's group 3, J. Algebra, 24 (1973), 159–196.
[21] D. Wright, The irreducible characters of the simple group of M. Suzuki of order 448, 345, 497, 600, J. Algebra, 29 (1974), 303–323.

(IV) Three-Dimensional Crystal Classes (→ 92 Crystallographic Groups)

Crystal System Bravais Types	Geometric Crystal Classes			Arithmetic Crystal Classes	Number of Space Groups[2]
	Schoenflies Notation	International Notation[1]			
		Short	Full		
Triclinic P	C_1	1	1	$(P,1)$	1
	$S_2(C_i)$	$\bar{1}$	$\bar{1}$	$(P,\bar{1})$	2
Monoclinic P, C	C_2	2	2	$(P,2)\,(C,2)$	3–5
	C_{1h}	m	m	$(P,m)\,(C,m)$	6–9
	C_{2h}	$2/m$	$\frac{2}{m}$	$(P,2/m)\,(C,2/m)$	10–15
Orthorhombic P, C, F, I	$D_2(V)$	222	222	$(P,222)\,(C,222)\,(F,222)\,(I,222)$	16–24
	C_{2v}	$mm2$	$mm2$	$(P,mm2)\,(C,mm2)\,(A,mm2)\,(F,mm2)\,(I,mm2)$	25–46
	$D_{2h}(V_h)$	mmm	$\frac{2}{m}\frac{2}{m}\frac{2}{m}$	$(P,mmm)\,(C,mmm)\,(F,mmm)\,(I,mmm)$	47–74
Tetragonal P, I	C_4	4	4	$(P,4)$[3] $(I,4)$	75–80
	S_4	$\bar{4}$	$\bar{4}$	$(P,\bar{4})\,(I,\bar{4})$	81–82
	C_{4h}	$4/m$	$\frac{4}{m}$	$(P,4/m)\,(I,4/m)$	83–88
	D_4	422	422	$(P,422)$[4] $(I,422)$	89–98
	C_{4v}	$4mm$	$4mm$	$(P,4mm)\,(I,4mm)$	99–110
	$D_{2d}(V_d)$	$\bar{4}2m$	$\bar{4}2m$	$(P,\bar{4}2m)\,(P,\bar{4}m2)\,(I,\bar{4}m2)\,(I,\bar{4}2m)$	111–122
	D_{4h}	$4/mmm$	$\frac{4}{m}\frac{2}{m}\frac{2}{m}$	$(P,4/mmm)\,(I,4/mmm)$	123–142
Trigonal P, R	C_3	3	3	$(P,3)$[5] $(R,3)$	143–146
	$S_6(C_{3i})$	$\bar{3}$	$\bar{3}$	$(P,\bar{3})\,(R,\bar{3})$	147–148
	D_3	32	32	$(P,312)$[6] $(P,321)$[7] $(R,32)$	149–155
	C_{3v}	$3m$	$3m$	$(P,3m1)\,(P,31m)\,(R,3m)$	156–161
	D_{3d}	$\bar{3}m$	$\bar{3}\frac{2}{m}$	$(P,\bar{3}1m)\,(P,\bar{3}m1)\,(R,\bar{3}m)$	162–167
Hexagonal P	C_6	6	6	$(P,6)$[8]	168–173
	C_{3h}	$\bar{6}$	$\bar{6}$	$(P,\bar{6})$	–174
	C_{6h}	$6/m$	$\frac{6}{m}$	$(P,6/m)$	175–176
	D_6	622	622	$(P,622)$[9]	177–182

Crystal System Bravais Types	Geometric Crystal Classes			Arithmetic Crystal Classes	Number of Space Groups[2]
	Schoenflies Notation	International Notation[1]			
		Short	Full		
Hexagonal	C_{6v}	$6mm$	$6mm$	$(P, 6mm)$	183–186
P	D_{3h}	$\bar{6}m2$	$\bar{6}m2$	$(P, \bar{6}m2)\ (P, \bar{6}2m)$	187–190
(cont.)	D_{6h}	$6/mmm$	$\frac{6}{m}\frac{2}{m}\frac{2}{m}$	$(P, 6/mmm)$	191–194
Cubic	T	23	23	$(P, 23)\ (F, 23)\ (I, 23)$	195–199
P, F, I	T_h	$m3$	$\frac{2}{m}\bar{3}$	$(P, m3)\ (F, m3)\ (I, m3)$	200–206
	O	432	432	$(P, 432)$[10] $(F, 432)\ (I, 432)$	207–214
	T_d	$\bar{4}3m$	$\bar{4}3m$	$(P, \bar{4}3m)\ (F, \bar{4}3m)\ (I, \bar{4}3m)$	215–220
	O_h	$m3m$	$\frac{4}{m}\bar{3}\frac{2}{m}$	$(P, m3m)\ (F, m3m)\ (I, m3m)$	221–230

Notes
(1) The notation is based upon *International tables for X-ray crystallography* I, Kynoch, 1969. In each crystal system, the lowest class is a holohedry.
(2) These correspond to the consecutive numbers of space groups in the book cited in (1).
(3)–(10) Enantiomorphic pairs arise from these classes: two pairs for (4), (8), (9), and one pair for the others.
For the shapes of Bravais lattices → 92 Crystallographic Groups E, Fig. 3.

6. Miscellaneous Constants

$\sqrt{2} = 1.41421\ 35623\ 73095$, $\sqrt{10} = 3.16227\ 76601\ 68379$.
$\sqrt[3]{2} = 1.25992\ 10498\ 94873$, $\sqrt[3]{100} = 4.64158\ 88336\ 12779$.
$\log_{10} 2 = 0.30102\ 99956\ 63981 = 1/3.32192\ 80948\ 87364$.

(I) Base of Natural Logarithm e (1000 decimals)

e=2.71828 18284 59045 23536 02874 71352 66249 77572 47093 69995 95749 66967 62772 40766 30353 54759
45713 82178 52516 64274 27466 39193 20030 59921 81741 35966 29043 57290 03342 95260 59563 07381
32328 62794 34907 63233 82988 07531 95251 01901 15738 34187 93070 21540 89149 93488 41675 09244
76146 06680 82264 80016 84774 11853 74234 54424 37107 53907 77449 92069 55170 27618 38606 26133
13845 83000 75204 49338 26560 29760 67371 13200 70932 87091 27443 74704 72306 96977 20931 01416
92836 81902 55151 08657 46377 21112 52389 78442 50569 53696 77078 54499 69967 94686 44549 05987
93163 68892 30098 79312 77361 78215 42499 92295 76351 48220 82698 95193 66803 31825 28869 39849
64651 05820 93923 98294 88793 32036 25094 43117 30123 81970 68416 14039 70198 37679 32068 32823
76464 80429 53118 02328 78250 98194 55815 30175 67173 61332 06981 12509 96181 88159 30416 90351
59888 85193 45807 27386 67385 89422 87922 84998 92086 80582 57492 79610 48419 84443 63463 24496
84875 60233 62482 70419 78623 20900 21609 90235 30436 99418 49164 31409 34317 38143 64054 62531
52096 18369 08887 07016 76839 64243 78140 59271 45635 49061 30310 72085 10383 75051 01157 47704
17189 86106 87396 96552 12671 54688 95703 50354.

e (in octal) = 2.55760 52130 50535 5.
$1/e = 0.36787\ 94411\ 71442$, $e^2 = 7.38905\ 60989\ 30650 = 1/0.13533\ 52832\ 36613$,
$\sqrt{e} = 1.64872\ 12707\ 00128 = 1/0.60653\ 06597\ 12633$.
$\log_e 10 = 2.30258\ 50929\ 94046 = 1/0.43429\ 44819\ 03252$,
$\log_e 2 = 0.69314\ 71805\ 59945 = 1/1.44269\ 50408\ 88964$.

(II) The Number π (1000 decimals) (→ 328 Pi(π))

π=3.14159 26535 89793 23846 26433 83279 50288 41971 69399 37510 58209 74944 59230 78164 06286 20899
86280 34825 34211 70679 82148 08651 32823 06647 09384 46095 50582 23172 53594 08128 48111 74502
84102 70193 85211 05559 64462 29489 54930 38196 44288 10975 66593 34461 28475 64823 37867 83165
27120 19091 45648 56692 34603 48610 45432 66482 13393 60726 02491 41273 72458 70066 06315 58817
48815 20920 96282 92540 91715 36436 78925 90360 01133 05305 48820 46652 13841 46951 94151 16094
33057 27036 57595 91953 09218 61173 81932 61179 31051 18548 07446 23799 62749 56735 18857 52724
89122 79381 83011 94912 98336 73362 44065 66430 86021 39494 63952 24737 19070 21798 60943 70277
05392 17176 29317 67523 84674 81846 76694 05132 00056 81271 45263 56082 77857 71342 75778 96091
73637 17872 14684 40901 22495 34301 46549 58537 10507 92279 68925 89235 42019 95611 21290 21960
86403 44181 59813 62977 47713 09960 51870 72113 49999 99837 29780 49951 05973 17328 16096 31859
50244 59455 34690 83026 42522 30825 33446 85035 26193 11881 71010 00313 78387 52886 58753 32083
81420 61717 76691 47303 59825 34904 28755 46873 11595 62863 88235 37875 93751 95778 18577 80532
17122 68066 13001 92787 66111 95909 21642 01989.

π (in octal)=3.11037 55242 10264 3.
$1/\pi$=0.31830 98861 83791, π^2=9.86960 44010 89359=1/0.10132 11836 42338,
$\sqrt{\pi}$ =1.77245 38509 05516=1/0.56418 95835 47756,
$\sqrt{2\pi}$ =2.50662 82746 31001=1/0.39894 22804 01433,
$\sqrt{\pi/2}$ =1.25331 41373 15500=1/0.79788 45608 02865,
$\sqrt[3]{\pi}$ =1.46459 18875 61523=1/0.68278 40632 55296.
$\log_{10}\pi$=0.49714 98726 94134, $\log_e \pi$=1.14472 98858 49400.

(III) Radian rad

1 rad=57°.29577 95130 82321=3437′.74677 07849 393=20626 4″.80624 70964.
1°=0.01745 32925 19943 rad, 1′=0.00029 08882 08666 rad, 1″=0.00000 48481 36811 rad.

(IV) Euler's Constant C (100 decimals) (→ 174 Gamma Function)

C=0.57721 56649 01532 86060 65120 90082 40243 10421 59335 93992
35988 05767 23488 48677 26777 66467 09369 47063 29174 67495.

e^C=1.78107 24179 90197 98522.

$$S_n=\sum_{i=1}^{n}\frac{1}{i}.$$

n	S_n	n	S_n	n	S_n	n	S_n
3	1.83333 333	6	2.45000 000	15	3.31822 899	100	5.18737 752
4	2.08333 333	8	2.71785 714	20	3.59773 966	500	6.79282 343
5	2.28333 333	10	2.92896 825	50	4.79920 534	1000	7.48547 086

7. Coefficients of Polynomial Approximations

In this table, we give some typical examples of approximation formulas for computation of functions on a digital computer (→ 19 Analog Computation, 336 Polynomial Approximation).

(I) Exponential Function

(1) Putting $\frac{x}{\log 2}+1=q+y+\frac{1}{2}\left(q \text{ is an integer}, -\frac{1}{2}\leqslant y<\frac{1}{2}\right)$, we have $e^x=2^q v(y), v(y)\fallingdotseq\sum a_i y^i$, which gives an approximation by a polynomial of the 7th degree, where the maximal error is 3×10^{-11}.
a_0=0.70710 67811 6, a_1=0.49012 90717 2, a_2=0.16986 57957 2, a_3=0.03924 73321 5,
a_4=0.00680 09712, a_5=0.00094 28173, a_6=0.00010 93869, a_7=0.00001 0826.

(2) An approximation by a polynomial of the 11th degree: $e^x\fallingdotseq\sum a_i x^i$ $(-1\leqslant x\leqslant 0)$.
Maximal error 1×10^{-12}.
a_0=0.99999 99999 990, a_1=0.99999 99999 995, a_2=0.50000 00000 747,
a_3=0.16666 66666 812, a_4=0.04166 66657 960, a_5=0.00833 33332 174,
a_6=0.00138 88925 998, a_7=0.00019 84130 955, a_8=0.00002 47944 428,
a_9=0.00000 27550 711, a_{10}=0.00000 02819 019, a_{11}=0.00000 00255 791.

(3) $e^x \doteqdot 1+\cfrac{x}{-\frac{x}{2}+\cfrac{k_0+k_1x^2+k_2x^4}{1+k_3x^2}}$ $(-\log\sqrt{2} \leqslant x \leqslant \log\sqrt{2})$.

Maximal error 1.4×10^{-14}.
$k_0=1.00000\ 00000\ 00327\ 1$, $k_1=0.10713\ 50664\ 56464\ 2$,
$k_2=0.00059\ 45898\ 69018\ 8$, $k_3=0.02380\ 17331\ 57418\ 6$.

(II) Logarithmic Function

(1) An approximation by a polynomial of the 11th degree: $\log(1+x)\doteqdot\Sigma a_i x^i$ $(0\leqslant x\leqslant 1)$.
Maximal error 1.1×10^{-10}.
$a_0=0.00000\ 00001\ 10$, $a_1=0.99999\ 99654\ 98$, $a_2=-0.49999\ 82537\ 98$,
$a_3=0.33329\ 85059\ 64$, $a_4=-0.24963\ 72428\ 65$, $a_5=0.19773\ 31015\ 60$,
$a_6=-0.15744\ 88954\ 13$, $a_7=0.11712\ 91156\ 18$, $a_8=-0.07364\ 03719\ 14$,
$a_9=0.03469\ 74937\ 56$, $a_{10}=-0.01046\ 82295\ 69$, $a_{11}=0.00148\ 19917\ 22$.

(2) For $1\leqslant x\leqslant 2$, and putting $y=\frac{x-\sqrt{2}}{x+\sqrt{2}}(3+2\sqrt{2})$ $(-1\leqslant y\leqslant 1)$, then $\log x\doteqdot\log\sqrt{2}+\sum a_i y^{2i+1}$ gives an approximation by a polynomial of the 11th degree $(0\leqslant i\leqslant 5)$, where the maximal error is 9.2×10^{-15}.
$a_0=0.34314\ 57505\ 07610\ 6$, $a_1=0.00336\ 70892\ 56222\ 5$, $a_2=0.00005\ 94707\ 04347\ 4$,
$a_3=0.00000\ 12504\ 99776\ 2$, $a_4=0.00000\ 00285\ 68292\ 8$, $a_5=0.00000\ 00007\ 43713\ 9$.

(III) Trigonometric Functions

(1) We put $\frac{x}{2\pi}=p+\frac{q}{2}+\frac{r}{4}+\frac{z}{8}$ (p is an integer; $q=0,1$; $r=0,1$; $-1\leqslant z<1$), and $s=\sin\frac{\pi z}{4}$, $c=\cos\frac{\pi z}{4}$.
If $r=0$, $\sin x=(-1)^q s$, $\cos x=(-1)^q c$,
If $r=1$, $\sin x=(-1)^q c$, $\cos x=-(-1)^q s$.
Here s and c are computed by the following approximation formulas. Putting $-z^2/2=y$, $s(y)=\sin(\pi z/4)\doteqdot z\Sigma a_i y^i$, $c(y)=\cos(\pi z/4)\doteqdot\Sigma b_i y^i$ gives an approximation by a polynomial of the 5th degree, where the maximal errors are s: 2×10^{-15}, c: 2×10^{-13}.
$a_0=0.78539\ 81633\ 97426$, $a_1=0.16149\ 10243\ 75338$, $a_2=0.00996\ 15782\ 61200$,
$a_3=0.00029\ 26094\ 99152$, $a_4=0.00000\ 50133\ 389$, $a_5=0.00000\ 00555\ 1357$.
$b_0=0.99999\ 99999\ 999$, $b_1=0.61685\ 02750\ 601$, $b_2=0.06341\ 73767\ 885$,
$b_3=0.00260\ 79335\ 007$, $b_4=0.00005\ 74476\ 09$, $b_5=0.00000\ 07765\ 93$.

(2) $\frac{\sin(\pi x/2)}{x}\doteqdot\Sigma(-1)^i a_i x^{2i}$ $(-1\leqslant x\leqslant 1)$. This gives an approximation by a polynomial of 10th degree $(0\leqslant i\leqslant 5)$, where the maximal error is 2.67×10^{-11}.
$a_0=1.57079\ 63267\ 682$, $a_1=0.64596\ 40955\ 820$, $a_2=0.07969\ 26037\ 435$,
$a_3=0.00468\ 16578\ 837$, $a_4=0.00016\ 02547\ 767$, $a_5=0.00000\ 34318\ 696$.

(3) $\tan\frac{\pi x}{4}\doteqdot x\left(k_0+\frac{x^2|}{|k_1}+\ldots+\frac{x^2|}{|k_4}\right)$ (continued fraction) $(-1\leqslant x\leqslant 1)$.

Maximal error 9.8×10^{-12}.
$k_0=0.78539\ 81634\ 9907$, $k_1=6.19229\ 46807\ 1350$, $k_2=-0.65449\ 83095\ 2316$,
$k_3=520.24599\ 06398\ 9939$, $k_4=-0.07797\ 95098\ 7751$.

(IV) Inverse Trigonometric Functions

(1) An approximation by a polynomial of the 21st degree $(0\leqslant i\leqslant 10)$:
$\arcsin x\doteqdot\Sigma a_i x^{2i+1}$ $(|x|\leqslant 1/\sqrt{2})$.
Maximal error 10^{-10}.
$a_0=1.00000\ 00005\ 3$, $a_1=0.16666\ 65754\ 5$, $a_2=0.07500\ 46066\ 5$, $a_3=0.04453\ 58425\ 7$,
$a_4=0.03175\ 26509\ 6$, $a_5=0.01176\ 58281\ 9$, $a_6=0.06921\ 26185\ 7$, $a_7=-0.14821\ 09628\ 8$,
$a_8=0.32889\ 76635\ 2$, $a_9=-0.35020\ 41201\ 5$, $a_{10}=0.19740\ 50325\ 0$.

(2) Putting $x = w + u$ ($w = \frac{1}{8}, \frac{3}{8}, \frac{5}{8}, \frac{7}{8}$; $-\frac{1}{8} \leqslant u \leqslant \frac{1}{8}$), $v = \dfrac{x-w}{1+xw}$ ($|v| \leqslant \frac{1}{2}$)

$\arctan x = \arctan w + t(v)$, $t(v) = \arctan v$.

The values of $\arctan w$:

$\arctan(1/8) = 0.12435\,49945\,46711$, $\arctan(3/8) = 0.35877\,06702\,70611$,
$\arctan(5/8) = 0.55859\,93153\,43560$, $\arctan(7/8) = 0.71882\,99996\,21623$.

$t(v)$ is computed by an approximation by a polynomial of the 9th degree ($0 \leqslant i \leqslant 4$), where $t(v) = \arctan v \fallingdotseq \Sigma(-1)^i a_i v^{2i+1}$.

Maximal error 1.6×10^{-13}.

$a_0 = 0.99999\,99999\,9992$, $a_1 = 0.33333\,33328\,220$, $a_2 = 0.19999\,97377\,6$,
$a_3 = 0.14280\,9976$, $a_4 = 0.10763\,60$.

(3) $\arctan x \fallingdotseq x\left(k_0 + \dfrac{x^2|}{|k_1} + \ldots + \dfrac{x^2|}{|k_6}\right)$ (continued fraction) ($-1 \leqslant x \leqslant 1$).

Maximal error 3.6×10^{-10}.

$k_0 = 0.99999\,99936\,2$, $k_1 = -3.00000\,30869\,4$, $k_2 = -0.55556\,97728\,4$,
$k_3 = -15.77401\,81127\,3$,
$k_4 = -0.16190\,80978\,0$, $k_5 = -44.57191\,79508\,8$, $k_6 = -0.10810\,67493\,1$.

(V) Gamma Function

An approximation by a polynomial of the 8th degree:

$\Gamma(2+x) \fallingdotseq \sum a_i x^i$ ($-1/2 \leqslant x \leqslant 1/2$).

Maximal error 7.6×10^{-8}.

$a_0 = 0.99999\,9926$, $a_1 = 0.42278\,4604$, $a_2 = 0.41184\,9671$, $a_3 = 0.08156\,52323$,
$a_4 = 0.07406\,48982$. $a_5 = -0.00012\,51376\,7$, $a_6 = 0.01229\,95771$, $a_7 = -0.00349\,61289$,
$a_8 = 0.00213\,85778$.

(VI) Normal Distribution

(1) $\dfrac{2}{\sqrt{\pi}} \displaystyle\int_x^\infty e^{-t^2}\,dt \fallingdotseq \dfrac{1}{(1+\Sigma a_i x^{i+1})^{16}}$ ($0 \leqslant x < \infty$). This gives an approximation by a polynomial of the 6th degree.

Maximal error 2.8×10^{-7}.

$a_0 = 0.07052\,30784$, $a_1 = 0.04228\,20123$, $a_2 = 0.00927\,05272$,
$a_3 = 0.00015\,20143$, $a_4 = 0.00027\,65672$, $a_5 = 0.00004\,30638$.

(2) $P(x) \equiv \dfrac{1}{\sqrt{2\pi}} \displaystyle\int_x^\infty e^{-t^2/2}\,dt$,

$$4P(x)(1-P(x)) \fallingdotseq \left[\exp\left(-\frac{2x^2}{\pi}\right)\right]\left[1 + x^4\left(a_0 + \frac{a_1}{x^2+a_2}\right)\right] \qquad (0 \leqslant x < \infty).$$

Maximal error 2×10^{-5}.

$a_0 = 0.0055$, $a_1 = 0.0551$, $a_2 = 14.4$.

(3) The inverse function of (2)

$$x \fallingdotseq \left[y\left(a_0 + \frac{a_1}{y+a_2}\right)\right]^{1/2}, \quad y = -\log[4P(x)(1-P(x))] \qquad (0 \leqslant y < \infty).$$

Maximal error 4.9×10^{-4}.

$a_0 = 2.06117\,86$, $a_1 = -5.72622\,04$, $a_2 = 11.64059\,5$.

Statistical Tables for Reference

Statistical Tables

[1] J. A. Greenwood and H. O. Hartley, Guide to tables in mathematical statistics, Princeton Univ. Press, 1962.
[2] Research Group for Statistical Sciences (T. Kitagawa and M. Masuyama, eds.) New statistical tables (Japanese), explanation p. 264, table p. 214, Kawade, 1952.
[3] R. A. Fisher and F. Yates, Statistical tables for biological, agricultural and medical research, explanation p. 30, table p. 137, Oliver & Boyd, third edition, 1948.
[4] E. S. Pearson and H. O. Hartley, Biometrika tables for statisticians, explanation p. 104, table p. 154, Cambridge Univ. Press, third edition, 1970.
[5] K. Pearson, Tables for statisticians and biometricians I, 1930, explanation p. 83, table p. 143; II, 1931, explanation p. 250, table p. 262, Cambridge Univ. Press.
[6] Statistical tables JSA-1972 (Japanese), table p. 454, explanation p. 260, Japanese Standards Association, 1972.

Tables of Special Statistical Values

[8] Harvard Univ., Tables of the cumulative binomial probability distribution, Harvard, 1955,

$$\sum_{i=r}^{n}\binom{n}{i}p^iq^{n-i}\text{: 5 dec.,}$$

$p=0.01(0.01)0.50$,
$n=1(1)50(2)100(10)200(20)500(50)1000$.
[9] National Bureau of Standards, NBS applied mathematical series, no. 6, Tables of the binomial probability distribution, 1950, $\binom{n}{i}p^iq^{n-i}$ and the partial sum: 7 dec., $p=0.01(0.01)0.50$, $n=2(1)49$.
[10] T. Kitagawa, Table of Poisson distribution (Japanese), Baihûkan, 1951, $e^{-m}m^i/i!$: 7-8 dec., $m=0.001(0.001)1.000(0.01)10.00$.
[11] G. J. Lieberman and D. B. Owen, Tables of the hypergeometric probability distribution, Stanford, 1961, $\binom{k}{x}\binom{N-k}{n-x}\Big/\binom{N}{n}$: 6 dec., $N=2(1)50(10)100(100)2000$, $n=1(1)(N/2)$, $k=1(1)n$.
[12] National Bureau of Standards, NBS no. 23, Tables of normal probability functions, 1942,
$\varphi(x)=(1/\sqrt{2\pi})\exp(-\frac{1}{2}x^2)$,
$\Phi(x)=\int_{-x}^{x}\varphi(x)\,dx$: 15 dec.,
$x=0(0.00001)1.0000(0.001)8.285$.
[13] K. Pearson, Tables of the incomplete beta-functions, Cambridge, second edition, 1968, $I_x(p,q)$: 8 dec., p, $q=0.5(0.5)\,11(1)50$.
[14] K. Pearson, Tables of the incomplete gamma-function, Cambridge, 1922, revised edition, 1951,

$$I(u,p)=\int_0^{u\sqrt{p+1}}(1/e^v)(v^p/\Gamma(p+1))\,dv:$$

7 dec., $p=0.0(0.1)5.0(0.2)50.0$, $u=0.1(0.1)20.0$; $p=-1.0(0.05)0.0$, $u=0.1(0.1)51.3$.
[15] N. V. Smirnov, Tables for the distribution and density function of the t-distribution, Pergamon, 1961, 6 dec., $f=1(1)35$, $t=0(0.01)3.00(0.02)4.50(0.05)6.50$.
[16] G. J. Resnikoff and G. J. Lieberman, Tables of the non-central t-distribution, Stanford, 1957.
[17] F. N. David, Tables of the ordinates and probability integral of the distribution of the correlation coefficient in small samples, Cambridge, 1938.
[18] D. B. Owen, The bivariate normal probability distribution, Sandia Corp., 1957, $T(h,a)$: 6 dec., $a=0.000(0.025)1.000, \infty$, $h=0.00(0.01)3.50(0.05)4.75$, $T(h,a)$

$$=\int_0^h\int_0^{ax}\frac{1}{2\pi}\exp\left(-\frac{x^2+y^2}{2}\right)dy\,dx.$$

[19] National Bureau of Standards, NBS no. 50, Tables of the bivariate normal distribution function and related functions, 1959, $L(h,k,r)$

$$=\int_h^\infty\int_k^\infty\frac{1}{2\pi\sqrt{1-r^2}}\times\exp\left[-\frac{x^2+y^2-2rxy}{2(1-r^2)}\right]dy\,dx\text{: 6 dec.,}$$

$r=\pm 0.00(0.05)0.95(0.01)0.99$,
$h,\ k=0.0(0.1)4.0$.

Tables of Allocation

[20] T. Kitagawa and M. Midome, Table of allocation of elements for experimental design (Japanese), Baihûkan, 1953.
[21] R. C. Bose, W. H. Clatworthy, and S. S. Shrikhande, Tables of partially balanced designs with two associate classes, North Carolina Agric. Expt. Station Tech. Bull., 1954 (table of PBIBD).

Numerical Tables for Reference

General Tables

[1] M. Boll, Tables numériques universelles, Dunod, 1947.
[2] P. Barlow, Barlow's tables, Robinson, 1814, third edition, 1930.
[3] W. Shibagaki, 0.01% table of elementary functions (Japanese), Kyôritu, 1952.
[4] K. Hayashi, Table of higher functions (Japanese), Iwanami, second edition, 1967.
[5] E. Jahnke and F. Emde, Funktionentafeln mit Formelin und Kurven, Teubner, second edition 1933 (English translation: Tables of functions with formulae and curves, Dover, fourth edition, 1945).
[6] Y. Yoshida and M. Yoshida, Mathematical tables (Japanese), Baihûkan, 1958.
[7] M. Abramowitz and I. A. Stegun (eds.), Handbook of mathematical functions with formulas, graphs and mathematical tables, National Bureau of Standards, 1964 (Dover, 1965).
[8] A. Fletcher et al. (eds.), Index of mathematical tables I, II, Scientific Computing Service, Addison–Wesley, second edition, 1962.

Multiplication Table

[9] A. L. Crelle, Rechentafeln welche alles Mutiplicieren und Dividiren mit Zahlen unter 1000 ersparen, bei grösseren Zahlen aber die Rechung erleichtern und sicherer machen, W. de Gruyter, new edition 1944.

Table of Prime Numbers

[10] D. N. Lehmer, List of prime numbers from 1 to 10,006,721, Carnegie Institution of Washington, 1914.

Series of Tables of Functions

[11] British Association for the Advancement of Science, Mathematical tables, vol. 2, Emden functions, 1932; vol. 6, Bessel functions, pt. 1, 1937; vol. 8, Number-divisor tables, 1940; vol. 9, Tables of powers giving integral powers of integrals, 1940; vol. 10, Bessel functions, pt. 2, 1952.
[12] Harvard University, Computation Laboratory, Annals, vol. 2, Tables of the modified Hankel functions of order one-third and their derivatives, 1945; vol. 3, Tables of the Bessel functions of the first kind of orders zero and one, 1947; vol. 14, Orders seventy-nine through one hundred thirty-five, 1951; vol. 18, Tables of generalized sine- and cosine-integral functions, pt. 1, 1949; vol. 19, pt. 2, 1949; vol. 20, Tables of inverse hyperbolic functions, 1949; vol. 21, Tables of the generalized exponential-integral functions, 1949.
[13] National Bureau of Standards, Applied Mathematics Series (AMS), AMS 1, Tables of the Bessel functions $Y_0(x)$, $Y_1(x)$, $K_0(x)$, $K_1(x)$, $0 \leqslant x \leqslant 1$, 1948; AMS 5, Tables of sines and cosines to fifteen decimal places at hundredths of a degree, 1949; AMS 11, Table of arctangents of rational numbers, 1951; AMS 14, Tables of the exponential function e^x (including e^{-x}), 1951; AMS 16, Tables of $n!$ and $\Gamma(n+\frac{1}{2})$ for the first thousand values of n, 1951; AMS 23, Tables of normal probability functions, 1953; AMS 25, Tables of the Bessel functions $Y_0(x)$, $Y_1(x)$, $K_0(x)$, $K_1(x)$, $0 \leqslant x \leqslant 1$, 1952; AMS 26, Tables of $\mathrm{Arctan}\, x$, 1953; Tables of 10^x, 1953; AMS 32, Table of sine and cosine integrals for arguments from 10 to 100, 1954; AMS 34, Table of the gamma function for complex arguments, 1954; AMS 36, Tables of circular and hyperbolic sines and cosines for radian arguments, 1953; AMS 40, Table of secants and cosecants to nine significant figures at hundredths of a degree, 1954; AMS 41, Tables of the error function and its derivative, 1954; AMS 43, Tables of sines and cosines for radian arguments, 1955; AMS 45, Table of hyperbolic sines and cosines, 1955; AMS 46, Table of the descending exponential, 1955.

Tables of Special Functions

[14] Akademiya Nauk SSSR, Tables of the exponential integral functions, 1954.
[15] J. Brownlee, Table of $\log \Gamma(x)$, Tracts for computers, no. 9, Cambridge Univ. Press, 1923.
[16] L. Dolansky and M. P. Dolansky, Table of $\log_2(1/p)$, $p \cdot \log_2(1/p)$ and $p \cdot \log_2(1/p) + (1-p)\log_2(1/(1-p))$, M.I.T. Research Lab. of Electronics tech. report 227, 1952.
[17] L. M. Milne-Thomson, Die elliptischen Funktionen von Jacobi, Springer, 1931; English translation: Jacobian elliptic function tables, Dover, 1950.
[18] H. T. Davis, Tables of the higher mathematical functions, Principia Press, vol. 1, Gamma function, 1933; vol. 2, Polygamma functions, 1935; vol. 3, Arithmetical tables, 1962.
[19] R. G. Selfridge and J. E. Maxfield, A table of the incomplete elliptic integral of the third kind, Dover, 1958.

[20] W. Shibagaki, 0.01% table of modified Bessel functions and the method of numerical computation for them (Japanese), Baihûkan, 1955.
[21] W. Shibagaki, Theory and application of gamma function, Appendix. Table of gamma function of complex variable effective up to 6 decimals (Japanese), Iwanami, 1952.
[22] L. J. Slater, A short table of the Laguerre polynomials, Proc. IEE, monograph no. 103c, 1955, 46–50.
[23] G. N. Watson, A treatise on the theory of Bessel functions, Appendix, Cambridge Univ. Press, 1922, second edition, 1958.
See also Statistical Tables for Reference.

Tables of Approximation Formulas of Functions

[24] S. Hitotumatu, Approximation formula (Japanese) Takeuti, 1963.
[25] T. Uno (ed.), Approximation formulas of functions for computers 1–3, Joint research work for mathematical sciences, pt. IV, section 5, 1961–1963 (Japanese, mimeographed).
[26] Z. Yamauti, S. Moriguti, and S. Hitotumatu (eds.), Numerical methods for electronic computers (Japanese), Baihûkan, I, 1965; II, 1967.
[27] C. Hastings (ed.), Approximations for digital computers, Princeton, 1955.
[28] A. J. W. Duijvestijn and A. J. Dekkers, Chebyshev approximations of some transcendental functions for use in digital computing, Philips Research Reports, 16 (1961), 145–174.
[29] L. A. Lyusternik, O. A. Chervonenkis, and A. P. Yanpol'skiĭ, Mathematical analysis: functions, limits, series, continued fractions, Pergamon, 1965. (Original in Russian, 1963.)
[30] J. F. Hart et al. (eds.), Computer approximations, Wiley, 1968.

Journals

Abh. Akad. Wiss. Göttingen
Abhandlungen der Akademie der Wissenschaften in Göttingen (Göttingen)

Abh. Bayer. Akad. Wiss. Math.-Nat. Kl.
Abhandlungen der Bayerschen Akademie der Wissenschaften. Mathematisch-Naturwissenschaftliche Klasse (Munich)

Abh. Math. Sem. Univ. Hamburg
Abhandlungen aus dem Mathematischen Seminar der Universität Hamburg (Hamburg)

Acta Arith.
Acta Arithmetica, Polska Akademia Nauk, Instytut Matematyczny (Warsaw)

Acta Informat.
Acta Informatica (Berlin)

Acta Math.
Acta Mathematica (Uppsala)

Acta Math. Acad. Sci. Hungar.
Acta Mathematica Academiae Scientiarum Hungaricae (Budapest)

Acta Math. Sinica
Acta Mathematica Sinica (Peking). Chinese Math. Acta is its English translation

Acta Sci. Math. Szeged.
Acta Universitatis Szegediensis, Acta Scientiarum Mathematicarum (Szeged)

Actualités Sci. Ind.
Actualités Scientifiques et Industrielles (Paris)

Advances in Math.
Advances in Mathematics (New York)

Aequationes Math.
Aequationes Mathematicae (Basel-Waterloo)

Algebra and Logic
Algebra and Logic (New York). Translation of Algebra i Logika

Algebra i Logika
Akademiya Nauk SSSR Sibirskoe Otdelenie. Institut Matematiki. Algebra i Logika (Novosibirsk). Translated as Algebra and Logic

Algebra Universalis
Algebra Universalis (Basel)

Amer. J. Math.
American Journal of Mathematics (Baltimore)

Amer. Math. Monthly
The American Mathematical Monthly (Menasha)

Amer. Math. Soc. Colloq. Publ.
American Mathematical Society Colloquium Publications (Providence)

Amer. Math. Soc. Math. Surveys
American Mathematical Society Mathematical Surveys (Providence)

Amer. Math. Soc. Proc. Symp. Pure Math.
American Mathematical Society Proceedings of Symposia in Pure Mathematics (Providence)

Amer. Math. Soc. Transl.
American Mathematical Society Translations (Providence)

Amer. Math. Soc. Transl. Math. Monographs
American Mathematical Society Translations of Mathematical Monographs (Providence)

Ann. Acad. Sci. Fenn.
Suomalaisen Tiedeakatemian Toimituksia, Annales Academiae Scientiarum Fennicae. Series A. I. Mathematica (Helsinki)

Ann. der Phys.
Annalen der Physik (Leipzig)

Ann. Fac. Sci. Univ. Toulouse
Annales de la Faculté des Sciences de l'Université de Toulouse pour les Sciences Mathématiques et les Sciences Physiques (Toulouse)

Ann. Inst. Fourier
Annales de l'Institut Fourier, Université de Grenoble (Grenoble)

Ann. Inst. H. Poincaré
Annales de l'Institut Henri Poincaré (Paris)

Ann. Inst. Statist. Math.
Annals of the Institute of Statistical Mathematics (Tokyo)

Ann. Mat. Pura Appl.
Annali di Matematica Pura ed Applicata (Bologna)

Ann. Math. Statist.
The Annals of Mathematical Statistics (Baltimore)

Ann. Math.
Annals of Mathematics (Princeton)

Ann. Physique
Annales de Physique (Paris)

Ann. Probability
The Annals of Probability (San Francisco)

Ann. Polon. Math.
Annales Polonici Mathematici. Polska Akademia Nauk (Warsaw)

Ann. Roumaines Math.
Annales Roumaines de Mathématiques. Journal de l'Institut Mathématique Roumain (Bucharest)

Ann. Sci. Ecole Norm. Sup.
Annales Scientifiques de l'Ecole Normale Supérieure (Paris)

Ann. Scuola Norm. Sup. Pisa
Annali della Scuola Normale Superiore de Pisa, Scienze Fisiche e Matematiche (Pisa)

Ann. Statist.
The Annals of Statistics (San Francisco)

Arch. History Exact Sci.
Archive for History of Exact Sciences (Berlin-New York)

Arch. Math.
Archiv der Mathematik (Basel-Stuttgart)

Arch. Rational Mech. Anal.
Archive for Rational Mechanics and Analysis (Berlin)

Ark. Mat.
Arkiv för Matematik (Stockholm)

Ark. Mat. Astr. Fys.
Arkiv för Matematik, Astronomi och Fysik (Uppsala)

Astérique
Astérique. La Société Mathématique de France (Paris)

Atti Accad. Naz. Lincei, Mem.
Atti della Accademia Nazionale dei Lincei, Memorie (Rome)

Atti Accad. Naz. Lincei, Rend.
Atti della Accademia Nazionale dei Lincei, Rendiconti (Rome)

Automat. Control Comput. Sci.
Automatic Control and Computer Sciences (New York). Translation of Avtomatika i Vychislitel'naya Tekhnik. Akademiya Nauk Latviĭskoĭ SSR (Riga)

Bell System Tech. J.
The Bell System Technical Journal (New York)

Biometrika
Biometrika, A Journal for the Statistical Study of Biological Problems (London)

Bol. Soc. Mat. São Paulo
Boletim da Sociedade de Matemática de São Paulo (São Paulo)

Bull. Acad. Pol. Sci.
Bulletin de l'Academie Polonaise des Sciences (Warsaw)

Bull. Amer. Math. Soc.
Bulletin of the American Mathematical Society (Providence)

Bull. Calcutta Math. Soc.
Bulletin of the Calcutta Mathematical Society (Calcutta)

Bull. Math. Statist.
Bulletin of Mathematical Statistics (Fukuoka, Japan)

Bull. Nat. Res. Council
Bulletin of the National Research Council (Washington)

Bull. Sci. Math.
Bulletin des Sciences Mathématiques (Paris)

Bull. Soc. Math. Belg.
Bulletin de la Société Mathématique de Belgique (Brussels)

Bull. Soc. Math. France
Bulletin de la Société Mathématique de France (Paris)

Bull. Soc. Roy. Sci. Liège
Bulletin de la Société Royale des Sciences de Liège (Liège)

C. R. Acad. Sci. Paris
Comptes Rendus Hebdomadaires des Séances de l'Académie des Sciences (Paris)

Canad. J. Math.
Canadian Journal of Mathematics (Toronto)

Časopis Pěst. Mat.
Časopis pro Pěstování Matematiky, Československá Akademie Věd (Prague)

Colloq. Math.
Colloquium Mathematicum (Warsaw)

Comm. ACM
Communications of the Association for Computing Machinery (New York)

Comm. Math. Phys.
Communications in Mathematical Physics (Berlin)

Comm. Pure Appl. Math.
Communications on Pure and Applied Mathematics (New York)

Comment. Math. Helv.
Commentarii Mathematici Helvetici (Zurich)

Compositio Math.
Compositio Mathematica (Groningen)

Comput. J.
The Computer Journal (London)

Crelles J.
= J. Reine Angew. Math.

CWI Newslett.
Centrum voor Wiskunde en Informatica. Newsletter (Amsterdam)

Cybernetics
Cybernetics (New York). Translation of Kibernetika (Kiev)

Czech. Math. J.
Czechoslovak Mathematical Journal (Prague)

Deutsche Math.
Deutsche Mathematik (Berlin)

Differentsial'nye Uravneniya
Differentsial'nye Uravneniya (Minsk). Translated as Differential Equations

Differential Equations
Differential Equations (New York). Translation of Differentsial'nye Uravneniya

Dokl. Akad. Nauk SSSR
Doklady Akademii Nauk SSSR (Moscow). Soviet Math. Dokl. is the English translation of its mathematics section

Duke Math. J.
Duke Mathematical Journal (Durham)

Econometrica
Econometrica, Journal of the Econometric Society (Chicago)

Edinburgh Math. Notes
The Edinburgh Mathematical Notes (Edinburgh)

Enseignement Math.
L'Enseignement Mathématique (Geneva)

Enzykl. Math.
Enzyklopädie der Mathematischen Wissenschaften mit Einschluss ihrer Anwendungen (Berlin)

Erg. Angew. Math.
Ergebnisse der Angewandte Mathematik (Berlin-New York)

Erg. Math.
Ergebnisse der Mathematik und ihrer Grenzgebiete (Berlin-New York)

Fund. Math.
Fundamenta Mathematicae (Warsaw)

Funkcial. Ekvac.
Fako de l'Funkcialaj Ekvacioj Japana Matematika Societo. Funkcialaj Ekvacioj (Serio Internacia) (Kobe, Japan)

Functional Anal. Appl.
Functional Analysis and its Applications (New York). Translation of Funktsional. Anal. Prilozhen.

Funktsional. Anal. Prilozhen.
Funktsional'nyi Analiz i ego Prilozheniya. Akademiya Nauk SSSR (Moscow). Translated as Functional Anal. Appl.

General Topology and Appl.
General Topology and its Applications (Amsterdam)

Hiroshima Math. J.
Hiroshima Mathematical Journal. Hiroshima Univ. (Hiroshima, Japan)

Hokkaido Math. J.
Hokkaido Mathematical Journal. Hokkaido Univ. (Sapporo, Japan)

IBM J. Res. Develop.
IBM Journal of Research and Development (Armonk, N.Y.)

Illinois J. Math.
Illinois Journal of Mathematics (Urbana)

Indag. Math.
Indagationes Mathematicae = Nederl. Akad. Wetensch. Proc.

Indian J. Math.
Indian Journal of Mathematics (Allahabad)

Indiana Univ. Math. J.
Indiana University Mathematics Journal (Bloomington)

Information and Control
Information and Control (New York)

Inventiones Math.
Inventiones Mathematicae (Berlin)

Izv. Akad. Nauk SSSR
Izvestiya Akademii Nauk SSSR (Moscow). Math. USSR-Izv. is the English translation of its mathematics section

J. Algebra
Journal of Algebra (New York)

J. Analyse Math.
Journal d'Analyse Mathématiques (Jerusalem)

J. Appl. Math. Mech.
Journal of Applied Mathematics and Mechanics (New York). Translation of Prikl. Mat. Mekh.

J. Approximation Theory
Journal of Approximation Theory (New York)

J. Assoc. Comput. Mach. (J. ACM)
Journal of the Association for Computing Machinery (New York)

J. Austral. Math. Soc.
The Journal of the Australian Mathematical Society (Sydney)

J. Combinatorial Theory
Journal of Combinatorial Theory. Series A and Series B (New York)

J. Comput. System Sci.
Journal of Computer and System Sciences (New York)

J. Differential Equations
Journal of Differential Equations (New York)

J. Differential Geometry
Journal of Differential Geometry (Bethlehem, Pa.)

J. Ecole Polytech.
Journal de l'Ecole Polytechnique (Paris)

J. Fac. Sci. Hokkaido Univ.
Journal of the Faculty of Science, Hokkaido University. Series I. Mathematics (Sapporo, Japan)

J. Fac. Sci. Univ. Tokyo
Journal of the Faculty of Science, University of Tokyo. Section I. (Tokyo)

J. Franklin Inst.
Journal of the Franklin Institute (Philadelphia)

J. Functional Anal.
Journal of Functional Analysis (New York)

J. Indian Math. Soc.
The Journal of the Indian Mathematical Society (Madras)

J. Inst. Elec. Engrs.
Journal of the Institution of Electrical Engineers (London)

J. Inst. Polytech. Osaka City Univ.
Journal of the Institute of Polytechnics, Osaka City University. Series A. Mathematics (Osaka)

J. London Math. Soc.
The Journal of the London Mathematical Society (London)

J. Math. Anal. Appl.
Journal of Mathematical Analysis and Applications (New York)

J. Math. and Phys.
Journal of Mathematics and Physics (Cambridge, Massachusetts, for issues prior to 1975; for 1975 and later, New York)

J. Math. Econom.
Journal of Mathematical Economics (Amsterdam)

J. Math. Kyoto Univ.
Journal of Mathematics of Kyoto University (Kyoto)

J. Math. Mech.
Journal of Mathematics and Mechanics (Bloomington)

J. Math. Pures Appl.
Journal de Mathématiques Pures et Appliquées (Paris)

J. Math. Soc. Japan
Journal of the Mathematical Society of Japan (Tokyo)

J. Mathematical Phys.
Journal of Mathematical Physics (New York)

J. Multivariate Anal.
Journal of Multivariate Analysis (New York)

J. Number Theory
Journal of Number Theory (New York)

J. Operations Res. Soc. Japan
Journal of the Operations Research Society of Japan (Tokyo)

J. Optimization Theory Appl.
Journal of Optimization Theory and Applications (New York)

J. Phys. Soc. Japan
Journal of the Physical Society of Japan (Tokyo)

J. Pure Appl. Algebra
Journal of Pure and Applied Algebra (Amsterdam)

J. Rational Mech. Anal.
Journal of Rational Mechanics and Analysis (Bloomington)

J. Reine Angew. Math.
Journal für die Reine und Angewandte Mathematik (Berlin). = Crelles J.

J. Res. Nat. Bur. Standards
Journal of Research of the National Bureau of Standards. Section B. Mathematics and Mathematical Physics (Washington)

J. Sci. Hiroshima Univ.
Journal of Science of Hiroshima University. Series A (Mathematics, Physics, Chemistry); Series A-I. (Mathematics) (Hiroshima)

J. Soviet Math.
Journal of Soviet Mathematics (New York). Translation of (1) Itogi Nauki—Seriya Matematika (Progress in Science—Mathematical Series); (2) Problemy Matematicheskogo Analiza (Problems in Mathematical Analysis); (3) Zap. Nauchn. Sem. Leningrad. Otdel. Mat. Inst. Steklov.

J. Symbolic Logic
The Journal of Symbolic Logic (New Brunswick)

Japan. J. Math.
Japanese Journal of Mathematics (Tokyo)

Jber. Deutsch. Math. Verein. (Jber. D.M.V.)
Jahresbericht der Deutschen Mathematiker Vereinigung (Stuttgart)

Kibernetika (Kiev)
Otdelenie Matematiki, Mekhaniki i Kibernetiki Akademii Nauk Ukrainskoĭ SSR. Kibernetika (Kiev). Translated as Cybernetics

Kôdai Math. Sem. Rep.
Kôdai Mathematical Seminar Reports (Tokyo)

Linear Algebra and Appl.
Linear Algebra and Its Applications (New York)

Linear and Multilinear Algebra.
Linear and Multilinear Algebra (New York)

Mat. Sb.
Matematicheskiĭ Sbornik (Recueil Mathématique). Akademiya Nauk SSSR (Moscow). Translated as Math. USSR-Sb.

Mat. Tidsskr. A
Matematisk Tidsskrift. A (Copenhagen)

Mat. Zametki
Matematicheskiĭ Zametki.
Akademiya Nauk SSSR (Moscow). Translated as Math. Notes

Math. Ann.
Mathematische Annalen (Berlin-Göttingen-Heidelberg)

Math. Comp.
Mathematics of Computation (Providence). Formerly Math. Tables Aids Comput.

Math. J. Okayama Univ.
Mathematical Journal of Okayama University (Okayama, Japan)

Math. Japonicae
Mathematica Japonicae (Osaka)

Math. Nachr.
Mathematische Nachrichten (Berlin)

Math. Notes
Mathematical Notes of the Academy of Sciences of the USSR (New York). Translation of Mat. Zametki

Math. Rev.
Mathematical Reviews (Ann Arbor)

Math. Scand.
Mathematica Scandinavica (Copenhagen)

Math. Student
The Mathematical Student (Madras)

Math. Tables Aids Comput. (MTAC)
Mathematical Tables and Other Aids to Computation (Washington). Name changed to Mathematics of Computation in 1960 (vol. 14ff.)

Math. USSR-Izv.
Mathematics of the USSR-Izvestiya (Providence). Translation of Izv. Akad. Nauk SSSR

Math. USSR-Sb.
Mathematics of the USSR-Sbornik (Providence). Translation of Mat. Sb.

Math. Z.
Mathematische Zeitschrift (Berlin-Göttingen-Heidelberg)

Mathematika
Mathematika, A Journal of Pure and Applied Mathematics (London)

Meed. Lunds Univ. Mat. Sem.
Meddelanden från Lunds Universitets Matematiska Seminarium = Communications du Séminaire Mathématique de l'Université de Lund (Lund)

Mem. Amer. Math. Soc.
Memoirs of the American Mathematical Society (Providence)

Mem. Coll. Sci. Univ. Kyôto
Memoirs of the College of Science, University of Kyôto. Series A (Kyoto)

Mem. Fac. Sci. Kyushu Univ.
Memoirs of the Faculty of Science, Kyushu University. Series A. Mathematics (Fukuoka, Japan)

Mémor. Sci. Math.
Mémorial des Sciences Mathématiques (Paris)

Michigan Math. J.
The Michigan Mathematical Journal (Ann Arbor)

Mitt. Math. Ges. Hamburg
Mitteilungen der Mathematischen Gesellschaft in Hamburg (Hamburg)

Monatsh. Math. Phys.
Monatschefte für Mathematik und Physik (Vienna)

Monatsh. Math.
Monatshefte für Mathematik (Vienna)

Monograf. Mat.
Monografje Matematyczne (Warsaw)

Moscow Univ. Math. Bull.
Moscow University Mathematics Bulletin (New York). Translation of the mathematics section of Vestnik Moskov. Univ., Ser. I, Mat. Mekh.

Nachr. Akad. Wiss. Göttingen
Nachrichten der Akademie der Wissenschaften in Göttingen. Math.-Phys. Kl. (Göttingen)

Nagoya Math. J.
Nagoya Mathematical Journal (Nagoya)

Nederl. Akad. Wetensch. Proc.
Koninklijke Nederlandse Akademie van Wetenschappen, Proceedings. Series A. Mathematical Sciences (Amsterdam) = Indag. Math., Proc. Acad. Amsterdam

Nieuw Arch. Wisk.
Nieuw Archief voor Wiskunde (Groningen)

Numerische Math.
Numerische Mathematik (Berlin-Göttingen-Heidelberg)

Nuovo Cimento
Il Nuovo Cimento (Bologna)

Osaka J. Math.
Osaka Journal of Mathematics (Osaka)

Osaka Math. J.
Osaka Mathematical Journal (Osaka)

Pacific J. Math.
Pacific Journal of Mathematics (Berkeley)

Philos. Trans. Roy. Soc. London
Philosophical Transactions of the Royal Society of London. Series A (London)

Phys. Rev.
The Physical Review (New York)

Portugal. Math.
Portugaliae Mathematica (Lisbon)

Prikl. Mat. Mekh.
Adademiya Nauk SSSR. Otdelenie Tekhnicheskikh Nauk. Institut Mekhaniki Prikladnaya Matematika i Mekhanika (Moscow). Translated as J. Appl. Mat. Mech.

Proc. Acad. Amsterdam
= Nederl. Akad. Wetensch. Proc.

Proc. Amer. Math. Soc.
Proceedings of the American Mathematical Society (Providence)

Proc. Cambridge Philos. Soc.
Proceedings of the Cambridge Philosophical Society (Cambridge)

Proc. Imp. Acad. Tokyo
Proceedings of the Imperial Academy (Tokyo)

Proc. Japan Acad.
Proceedings of the Japan Academy (Tokyo)

Proc. London Math. Soc.
Proceedings of the London Mathematical Society (London)

Proc. Nat. Acad. Sci. US
Proceedings of the National Academy of Sciences of the United States of America (Washington)

Proc. Phys.-Math. Soc. Japan
Proceedings of the Physico-Mathematical Society of Japan (Tokyo)

Proc. Roy. Soc. London
Proceedings of the Royal Society of London. Series A (London)

Proc. Steklov Inst. Math.
Proceedings of the Steklov Institute of Mathematics (Providence). Translation of Trudy Mat. Inst. Steklov.

Prog. Theoret. Phys.
Progress of Theoretical Physics (Kyoto)

Publ. Inst. Math.
Publications de l'Institut Mathématique (Belgrade)

Publ. Inst. Math. Univ. Strasbourg
Publications de l'Institut de Mathématiques de l'Université de Strasbourg (Strasbourg)

Publ. Math. Inst. HES
Publications Mathématiques de l'Institut des Hautes Etudes Scientifiques (Paris)

Publ. Res. Inst. Math. Sci.
Publications of the Research Institute for Mathematical Sciences (Kyoto)

Quart. Appl. Math.
Quarterly of Applied Mathematics (Providence)

Quart. J. Math.
The Quarterly Journal of Mathematics, Oxford. Second Series (Oxford)

Quart. J. Mech. Appl. Math.
The Quarterly Journal of Mechanics and Applied Mathematics (Oxford)

Rend. Circ. Mat. Palermo
Rendiconti del Circolo Matematico de Palermo (Palermo)

Rend. Sem. Mat. Univ. Padova
Rendiconti del Seminario Matematico dell'Universitá di Padova (Padua)

Rev. Mat. Hisp. Amer.
Revista Matemática Hispaño-Americana (Madrid)

Rev. Mod. Phys.
Reviews of Modern Physics (New York)

Rev. Un. Mat. Argentina
Revista de la Unión Matemática Argentina (Buenos Aires)

Rev. Univ. Tucumán
Revista Universidad Nacional de Tucumán, Facultad de Ciencias Exactas y Tecnologia. Serie A. Matemáticas y Fisica Teorica (Tucumán)

Roczniki Polsk. Towar. Mat.
Roczniki Polskiego Towarzystwa Matematycznego. Serja I. Prace Matematyczne (Krakow)

Rozprawy Mat.
Rozprawy Matematyczne, Polska Akademia Nauk, Instytut Matematyczny (Warsaw)

Russian Math. Surveys.
Russian Mathematical Surveys (London). Translation of Uspekhi Mat. Nauk

Sammul. Göschen
Sammulung Göschen (Leipzig)

Sankhyā
Sankhyā, The Indian Journal of Statistics. Series A and Series B (Calcutta)

S.-B. Berlin. Math. Ges.
Sitzungsberichte der Berliner Mathematischen Gesellschaft (Berlin)

S.-B. Deutsch. Akad. Wiss. Berlin
Sitzungsberichte der Deutschen Akademie der Wissenschaften zu Berlin, Mathematisch-Naturwissenschaftliche Klasse (Berlin)

S.-B. Heidelberger Akad. Wiss.
Sitzungsberichte der Heidelberger Akademie der Wissenschaften, Mathematisch-Naturwissenschaftliche Klasse (Heidelberg)

S.-B. Math.-Nat. Kl. Bayer. Akad. Wiss.
Sitzungsberichte der Mathematisch-Naturwissenschaflichen Klasse der Bayerischen Akademie der Wissenschaften zu München (Munich)

S.-B. Öster. Akad. Wiss.
Sitzungsberichte der Österreichische Akademie der Wissenschaften (Vienna)

S.-B. Phys.-Med. Soz. Erlangen
Sitzungsberichte der Physikalisch-Medizinischen Sozietät zu Erlangen (Erlangen)

S.-B. Preuss. Akad. Wiss.
Sitzungsberichte der Preussischen Akademie der Wissenschaften. Physikalisch-Mathematische Klasse (Berlin)

Schr. Math. Inst. u. Inst. Angew. Math. Univ. Berlin
Schriften des Mathematischen Instituts und des Instituts für Angewandte Mathematik der Universität Berlin (Berlin)

Schr. Math. Inst. Univ. Münster
Schriftenreihe des Mathematischen Instituts der Universität Münster (Münster)

Sci. Papers Coll. Gen. Ed. Univ. Tokyo
Scientific Papers of the College of General Education, University of Tokyo (Tokyo)

Sci. Rep. Tokyo Kyoiku Daigaku
Science Reports of the Tokyo Kyoiku Daigaku. Section A (Tokyo)

Scripta Math.
Scripta Mathematica. A Quarterly Journal devoted to the Philosophy, History, and Expository Treatment of Mathematics (New York)

Sém. Bourbaki
Séminaire Bourbaki (Paris)

SIAM J. Appl. Math.
SIAM Journal of Applied Mathematics. A Publication of the Society for Industrial and Applied Mathematics (Philadelphia)

SIAM J. Comput.
SIAM Journal on Computing (Philadelphia)

SIAM J. Control
SIAM Journal on Control (Philadelphia)

SIAM J. Math. Anal.
SIAM Journal on Mathematical Analysis (Philadelphia)

SIAM J. Numer. Anal.
SIAM Journal on Numerical Analysis (Philadelphia)

SIAM Rev.
SIAM Review (Philadelphia)

Siberian Math. J.
Siberian Mathematical Journal (New York). Translation of Sibirsk. Mat. Zh.

Sibirsk. Mat. Zh.
Akademiya Nauk SSSR. Sibirskoe Otdelenie. Sibirskiĭ Matematicheskiĭ Zhurnal (Moscow). Translated as Siberian Math. J.

Skr. Norske Vid. Akad. Oslo
Skrifter Utgitt av det Norske Videnskaps-Akademii Oslo. Matematisk-Naturvidenskapelig Klasse (Oslo)

Soviet Math. Dokl.
Soviet Mathematics, Doklady (Providence). Translation of mathematical section of Dokl. Akad. Nauk SSSR

SRI J.
Stanford Research Institute Journal (Menlo Park)

Studia Math.
Studia Mathematica. (Wrocław)

Sûbutu-kaisi
Nihon Sûgaku-buturi-gakkai Kaisi (Tokyo)

Sûgaku
Sûgaku, Mathematical Society of Japan (Tokyo)

Summa Brasil. Math.
Summa Brasiliensis Mathematicae (Rio de Janeiro)

Tensor
Tensor (Chigasaki, Japan)

Teor. Veroyatnost. i Primenen.
Teoriya Veroyatnosteĭ i ee Primenenie. Akademiya Nauk SSSR (Moscow). Translated as Theor. Prob. Appl.

Theor. Prob. Appl.
Theory of Probability and Its Applications. Society for Industrial and Applied Mathematics. English translation of Teor. Veroyatnost. i Primenen. (Philadelphia)

Tôhoku Math. J.
The Tôhoku Mathematical Journal (Sendai, Japan)

Tôhoku-rihô
Tôhoku Teikokudaigaku Rikahôkoku (Sendai, Japan)

Topology
Topology. An International Journal of Mathematics (Oxford)

Trans. Amer. Math. Soc.
Transactions of the American Mathematical Society (Providence)

Trans. Moscow Math. Soc.
Transactions of the Moscow Mathematical Society (Providence). Translation of Trudy Moskov. Mat. Obshch.

Trudy Mat. Inst. Steklov.
Trudy Matematicheskogo Instituta im. V. A.

Steklova. Akademiya Nauk SSSR (Moscow-Leningrad). Translated as Proc. Steklov Inst. Math.

Trudy Moskov. Obshch.
Trudy Moskovskogo Matematicheskogo Obshchestva (Moscow). Translated as Trans. Moscow Math. Soc.

Tsukuba J. Math.
Tsukuba Journal of Mathematics. Univ. Tsukuba (Ibaraki, Japan)

Ukrain. Mat. Zh.
Akademiya Nauk Ukrainskoĭ SSR. Institut Matematiki. Ukrainskiĭ Matematicheskiĭ Zhurnal (Kiev). Translated as Ukrainian Math. J.

Ukrainian Math. J.
Ukrainian Mathematical Journal (New York). Translation of Ukrain. Mat. Zh.

Uspekhi Mat. Nauk
Uspekhi Matematicheskikh Nauk (Moscow-Leningrad). Translated as Russian Math. Surveys

Vestnik Moskov. Univ.
Vestnik Moskovskogo Universiteta. I, Matematika i Mekhanika (Moscow). Mathematical section translated as Moscow Univ. Math. Bull.

Vierteljschr. Naturf. Ges. Zürich
Vierteljahrsschrifte der Naturforschenden Gesellschaft in Zürich (Zurich)

Z. Angew. Math. Mech. (Z.A.M.M.)
Zeitschrift für Angewandte Mathematik und Mechanik, Ingenieurwissenschaftliche Forschungsarbeiten (Berlin)

Z. Angew. Math. Phys. (Z.A.M.P.)
Zeitschrift für Angewandte Mathematik und Physik (Basel)

Z. Wahrscheinlichkeitstheorie
Zeitschrift für Wahrscheinlichkeitstheorie und Verwandte Gebiete (Berlin)

Zbl. Angew. Math.
Zentralblatt für Angewandte Mathematik (Berlin)

Zbl. Math.
Zentralblatt für Mathematik und ihre Grenzgebiete (Berlin-Göttingen-Heidelberg)

Zh. Èksper. Teoret. Fiz.
Zhurnal Èksperimental'noĭ i Teoreticheskoĭ Fiziki (Moscow)

Publishers

Academic Press
Academic Press Inc., New York-London

Addison-Wesley
Addison-Wesley Publishing Company, Inc., Reading (Massachusetts)-Menlo Park (California)-London-Don Mills (Ontario)

Akadémiai Kiadó
A kiadásért felös: az Adadémiai Kiadó igazatója (Publishing House of the Hungarian Academy of Sciences), Budapest

Akademie-Verlag
Berlin

Akademische Verlag.
Akademische Verlagsgesellschaft, Leipzig

Allen
W. H. Allen & Co. Ltd., London

Allen & Unwin
Allen & Unwin, Inc., Winchester (Massachusetts)

Allyn & Bacon
Allyn & Bacon, Inc., Newton (Massachusetts)

Almqvist and Wiksell
Almqvist och Wiksell Förlag, Stockholm

Asakura
Asakura-syoten, Tokyo

Aschelhoug
H. Aschelhoug and Company, Oslo

Baihûkan
Tokyo

Benjamin
W. A. Benjamin, Inc., New York-London

Birkhäuser
Birkhäuser Verlag, Basel-Stuttgart

Blackie
Blackie & Son Ltd., London-Glasgow

Cambridge Univ. Press
Cambridge University Press, London-New York

Chapman & Hall
Chapman & Hall Ltd., London

Chelsea
Chelsea Publishing Company, New York

Clarendon Press
Oxford University Press, Oxford

Cremona
Edizioni Cremonese, Rome

de Gruyter
Walter de Gruyter and Company, Berlin

Dekker
Marcel Dekker, Inc., New York

Deutscher Verlag der Wiss.
Deutscher Verlag der Wissenschaften, Berlin

Dover
Dover Publications, Inc., New York

Dunod
Dunod, Editeur, Paris

Elsevier
Elsevier Publishing Company, Amsterdam-London-New York

Fizmatgiz
Gosudarstvennoe Izdatel'stvo Fiziko-Matematicheskoĭ Literatury, Moscow

Gauthier-Villars
Gauthier-Villars & Cie, Editeur, Paris

Ginn
Ginn and Company, Waltham (Massachusetts)-Toronto-London

Gordon & Breach
Gordon & Breach, Science Publishers Ltd., London

Goztekhizdat
Gosudarstvennoe Izdatel'stvo Tekhniko-Teoreticheskoĭ Literatury, Moscow

Griffin
Charles Griffin and Company Ltd., London

Hafner
Hafner Publishing Company, New York

Harper & Row
Harper & Row Publishers, New York-Evanston-London

Hermann
Hermann & Cie, Paris

Hirokawa
Hirokawa-syoten, Tokyo

Hirzel
Verlag von S. Hirzel, Leipzig

Holden-Day
Holden-Day, Inc., San Francisco-London-Amsterdam

Holt, Rinehart and Winston
Holt, Rinehart and Winston, Inc., New York-Chicago-San Francisco-Toronto-London

Interscience
Interscience Publishers, Inc., New York-London

Iwanami
Iwanami Shoten, Tokyo

Kawade
Kawade-syobô, Tokyo

Kinokoniya
Kinokoniya Company, Tokyo

Kyôritu
Kyôritu-syuppan, Tokyo

Lippincott
J. B. Lippincott Company, Philadelphia

Longman
Longman Group, Ltd., Harlow (Essex)

Longmans, Green
Longmans, Green and Company, Ltd., London-New York-Toronto-Bombay-Calcutta-Madras

Macmillan
The Macmillan Company, New York-London

Maki
Maki-syoten, Tokyo

Maruzen
Maruzen Company Ltd., Tokyo

Masson
Masson et Cie, Paris

Math-Sci Press
Math-Sci Press, Brookline (Massachusetts)

McGraw-Hill
McGraw-Hill Book Company, Inc., New York-London-Toronto

Methuen
Methuen and Company Ltd., London

MIT Press
The MIT Press, Cambridge (Massachusetts)-London

Nauka
Izdatel'stvo Nauka, Moscow

Noordhoff
P. Noordhoff Ltd., Groningen

North-Holland
North-Holland Publishing Company, Amsterdam

Oldenbourg
Verlag von R. Oldenbourg, Munich-Vienna

Oliver & Boyd
Oliver & Boyd Ltd., Edinburgh-London

Oxford Univ. Press
Oxford University Press, London-New York

Pergamon
Pergamon Press, Oxford-London-Edinburgh-New York-Paris-Frankfurt

Polish Scientific Publishers
Państwowe Wydawnictwo Naukowe, Warsaw

Prentice-Hall
Prentice-Hall, Inc., Englewood Cliffs (New Jersey)

Princeton Univ. Press
Princeton University Press, Princeton

Random House
Random House, Inc., New York

Sibundô
Tokyo

Springer
Springer-Verlag, Berlin (-Göttingen)-Heidelberg-New York

Teubner
B. G. Teubner Verlagsgesellschaft, Leipzig-Stuttgart

Tôkai
Tôkai-syobô, Tokyo

Tokyo-tosyo
Tokyo

Tokyo Univ. Press
Tokyo University Press, Tokyo

Ungar
Frederick Ungar Publishing Company, New York

Univ. of Tokyo Press
University of Tokyo Press, Tokyo

Utida-rôkakuho
Tokyo

Van Nostrand
D. Van Nostrand Company, Inc., Toronto-New York-London

Vandenhoeck & Ruprecht
Göttingen

Veit
Verlag von Veit & Company, Leipzig

Vieweg
Friedr Vieweg und Sohn Verlagsgesellschaft mbH, Wiesbaden

Wiley
Wiley & Sons, Inc., New York-London

Wiley-Interscience
Wiley & Sons, Inc., New York-London

Zanichelli
Nicola Zanichelli Editore, Bologna

Special Notation

This list contains the notation commonly and frequently used throughout this work. The symbol * means that the same notation is used with more than one meaning. For more detailed definitions or further properties of the notation, see the articles cited.

Notation	Example	Definition	Article and Section
I. Logic			
$\forall$	$\forall x F(x)$	Universal quantifier (for all x, $F(x)$ holds)	411B, C
$\exists$	$\exists x F(x)$	Existential quantifier (there exists an x such that $F(x)$ holds)	411B, C
$\wedge$, &	$A \wedge B$, $A \& B$	Conjunction, logical product (A and B)	411B*
$\vee$	$A \vee B$	Disjunction, logical sum (A or B)	411B*
$\neg$	$\neg A$	Negation (not A)	411B
$\rightarrow$, $\supset$, $\Rightarrow$	$A \rightarrow B$, $A \Rightarrow B$	Implication (A implies B)	411B*
$\leftrightarrow$, $\Leftrightarrow$, $\rightleftarrows$	$A \Leftrightarrow B$	Equivalence (A and B are logically equivalent)	411B
II. Sets			
$\in$	$x \in X$	Membership (element x is a member of the set X)	381A
$\notin$	$x \notin X$	Nonmembership (element x is not a member of the set X)	381A
$\subset$	$A \subset B$	Inclusion (A is a subset of B)	381A
$\not\subset$	$A \not\subset B$	Noninclusion (A is not a subset of B)	381A
$\subsetneqq$	$A \subsetneqq B$	Proper inclusion (A is a proper subset of B)	381A
$\varnothing$		Empty set	381A
$\cup$, $\bigcup$	$A \cup B$, $\bigcup A_\lambda$	Union, join	381B, D*
$\cap$, $\bigcap$	$A \cap B$, $\bigcap A_\lambda$	Intersection, meet	381B, D*
^{c}, C	A^c, $C(A)$	Complement (of a set A)	381B
$-$, $\setminus$	$A-B$, $A \setminus B$	Difference ($A-B=A \cap B^c$)	381B
$\times$	$A \times B$	Cartesian product (of A and B)	381B*
R, $\sim$	xRy, $x \sim y$	Equivalence relation (for two elements x, y)	135A*
$/$	A/R	Quotient set (set of equivalence classes of A with respect to an equivalence relation R)	135B*
Π	$\Pi_\lambda A_\lambda$	Cartesian product (of the A_λ)	381E
Σ, $\amalg$	ΣA_λ, $\amalg A_\lambda$	Direct sum (of the A_λ)	381E
$\mathfrak{B}$	$\mathfrak{B}(A)$	Power set (set of all subsets of A)	381E
	B^A	Set of all mappings from A to B	381C
$\{\mid\}$	$\{x \mid P(x)\}$	Set of all elements x with the property $P(x)$	381A

Notation	Example	Definition	Article and Section
{ }	$\{a_\lambda\}_{\lambda\in\Lambda}$	Family with index set Λ	165D
	$\{a_n\}$	Sequence (of numbers, points, functions, or sets)	165D
$\bar{\bar{}}$, | |, #	$\bar{\bar{X}}$, $\|X\|$, $\#X$	Cardinal number (of the set X)	49A*
$\aleph$	$\aleph_\beta$	Aleph (transfinite cardinal)	49E
$\rightarrow$	$f:X\rightarrow Y$	Mapping (f from X to Y)	381C*
$\mapsto$	$f:X\mapsto Y$	Mapping (where $f(X)=Y$, but not in the present volumes)	381C
1, id	1_A, id_A	Identity mapping (identity function)	381C
c, χ	$c_X(x)$, $\chi_X(x)$	Characteristic function (representing function)	381C
|	$f\|A$	Restriction (of a mapping f to A)	381C*
$\circ$	$g\circ f$	Composite (of mappings f and g)	381C
lim sup, $\overline{\lim}$	$\limsup A_n$	Superior limit (of the sequence of sets A_n)	270C*
lim inf, $\underline{\lim}$	$\liminf A_n$	Inferior limit (of the sequence of sets A_n)	270C*
lim	$\lim A_n$	Limit (of the sequence of scts A_n)	270C*
$\varinjlim$	$\varinjlim A_\lambda$	Inductive limit (of A_λ)	210B
$\varprojlim$	$\varprojlim A_\lambda$	Projective limit (of A_λ)	210B
III. Order			
(,)	(a,b)	Open interval $\{x\|a<x<b\}$	355C*
[,]	$[a,b]$	Closed interval $\{x\|a\leqslant x\leqslant b\}$	355C*
(,]	$(a,b]$	Half-open-interval $\{x\|a<x\leqslant b\}$	355C
[,)	$[a,b)$	Half-open interval $\{x\|a\leqslant x<b\}$	355C
max	$\max A$	Maximum (of A)	311B
min	$\min A$	Minimum (of A)	311B
sup	$\sup A$	Supremum, least upper bound (of A)	311B
inf	$\inf A$	Infimum, greatest lower bound (of A)	311B
$\ll$	$a\ll b$	Very large (b is very large compared to a)	
$\cup$, $\vee$	$a\cup b$, $a\vee b$	Join of a, b in an ordered set	243A*
$\cap$, $\wedge$	$a\cap b$, $a\wedge b$	Meet of a, b in an ordered set	243A*
IV. Algebra			
mod	$a\equiv b \pmod n$	Modulo (a and b are congruent modulo n)	297G
|	$a\|b$	Divisibility (a divides b)	297A*
$\nmid$	$a\nmid b$	Nondivisibility (a does not divide b)	297A
det, | |	$\det A$, $\|A\|$	Determinant (of a square matrix A)	103A*

Notation	Example	Definition	Article and Section
tr, Sp	$\operatorname{tr} A$, $\operatorname{Sp} A$	Trace (of a square matrix A)	269F
^{t}, ^{T}, $'$	${}^{t}A; A^{t}, A^{T}, A'$	Transpose (of a matrix A)	269B
I	I_n	Unit matrix (of degree n)	269A
E_{ij}		Matrix unit (matrix whose (i,j)-component is 1 and all others are 0)	269B
$\otimes$	$A \otimes B$	Kronecker product (of two matrices A and B)	269C*
$\cong$	$M \cong N$	Isomorphism (of two algebraic systems M and N)	256B
$/$	M/N	Quotient space (of an algebraic system M by N)	256F*
dim	$\dim M$	Dimension (of a linear space, etc.)	256C
Im	$\operatorname{Im} f$	Image (of a mapping f)	277E*
Ker	$\operatorname{Ker} f$	Kernel (of a mapping f)	277E
Coim	$\operatorname{Coim} f$	Coimage (of a mapping f)	277E
Coker	$\operatorname{Coker} f$	Cokernel (of a mapping f)	277E
δ_{ij}, δ_i^j		Kronecker delta ($\delta_{ii}=1$ and $\delta_{ij}=0$ for $i \neq j$)	269A
$(\; , \;)$, $\cdot$	$(\mathbf{a}, \mathbf{b})$, $\mathbf{a} \cdot \mathbf{b}$	Inner product (of two vectors **a** and **b**)	442B*
$[\; , \;]$, $\times$	$[\mathbf{a}, \mathbf{b}]$, $\mathbf{a} \times \mathbf{b}$	Vector product (of two 3-dimensional vectors **a** and **b**)	442C*
$\otimes$	$M \otimes N$	Tensor product (of two modules M and N)	277J, 256I*
Hom	$\operatorname{Hom}(M, N)$	Set of all homomorphisms (from M to N)	277B
Hom_A	$\operatorname{Hom}_A(M, N)$	Set of all A-homomorphisms (of an A-module M to an A-module N)	277E
Tor	$\operatorname{Tor}_n(M, N)$	Torsion product (of M, N)	200D
Ext	$\operatorname{Ext}^n(M, N)$	Extension (of M, N)	200G
$\wedge$, $\wedge^p$	$\wedge M$, $\wedge^p M$	Exterior algebra (of a linear space M), pth exterior product (of M)	256O
V. Algebraic Systems			
$\mathbf{N}$		Set of all natural numbers	294A
$\mathbf{Z}$		Set of all rational integers	294A
$\mathbf{Z}_m$		$\mathbf{Z}/m\mathbf{Z}$ (set of all residue classes modulo m)	297G*
$\mathbf{Q}$		Set of all rational numbers	294A
$\mathbf{R}$		Set of all real numbers	294A
$\mathbf{C}$		Set of all complex numbers	294A
$\mathbf{H}$		Set of all quaternions	29B
$GF(q)$, $\mathbf{F}_q$		Finite field (with q elements)	149M

Notation	Example	Definition	Article and Section
$\mathbf{Q}_p$		p-adic number field (p is a prime)	439F
$\mathbf{Z}_p$		Ring of p-adic integers	439F
[ ]	$k[x_1,\ldots,x_n]$	Polynomial ring (of variables $x_1,\ldots,x_n$ with coefficients in k)	369A
( )	$k(x_1,\ldots,x_n)$	Field extension (of k by $x_1,\ldots,x_n$)	149D
[[ ]], { }	$k[[x_1,\ldots,x_n]]$	Formal power series ring (with coefficients in k).	370A
		Note: The symbols **N**, **Z**, **Q**, **R**, **C**, and **H** stand for sets, each with its own natural mathematical structure	

VI. Groups

Notation	Example	Definition	Article and Section
GL	$GL(V)$, $GL(n, K)$	General linear group (over V, or over K of degree n)	60B
SL	$SL(n, K)$	Special linear group (over K of degree n)	60B
PSL	$PSL(n, K)$	Projective special linear group (over K of degree n)	60B
U	$U(n)$	Unitary group (of degree n)	60F
SU	$SU(n)$	Special unitary group (of degree n)	60F
O	$O(n)$	Orthogonal group (of degree n)	60I
SO	$SO(n)$	Special orthogonal group, rotation group (of degree n)	60I
$Spin$	$Spin(n)$	Spinor group (of degree n)	61D
Sp	$Sp(n)$	Symplectic group (of degree n)	60L

[For $PGL(n, K)$, $LF(n, K)$, $PU(n)$, $Sp(n)$, $PSp(n, K)$ → 60 Classical Groups]

VII. Topology (Convergence)

Notation	Example	Definition	Article and Section
$\rightarrow$	$a_n \rightarrow a$	Convergence (sequence a_n converges to a)	87B, E*
$\downarrow$, $\searrow$	$a_n \downarrow a$, $a_n \searrow a$	Convergence monotonically decreasing	87B
$\uparrow$, $\nearrow$	$a_n \uparrow a$, $a_n \nearrow a$	Convergence monotonically increasing	87B
lim	$\lim a_n$	Limit (of a sequence a_n)	87B, E*
lim sup, $\overline{\lim}$	$\limsup a_n$, $\overline{\lim}\, a_n$	Superior limit (of a sequence a_n)	87C*
lim inf, $\underline{\lim}$	$\liminf a_n$, $\underline{\lim}\, a_n$	Inferior limit (of a sequence a_n)	87C*
^{a}, $^-$, Cl	E^a, $\bar{E}$, $\mathrm{Cl}\, E$	Closure (of a set E)	425B
^{i}, $^\circ$, Int	E^i, E°, $\mathrm{Int}\, E$	Interior (of a set E)	425B
ρ, d	$\rho(x, y)$, $d(x, y)$	Distance (between two points x and y)	273B*
$\Vert\ \Vert$	$\Vert x \Vert$	Norm (of x)	37B
l.i.m.	$\mathrm{l.i.m.}\, f_n$	Limit in the mean (of a sequence f_n)	168B

Notation	Example	Definition	Article and Section
s-lim	s-lim x_n	Strong limit (of a sequence x_n)	37B
w-lim	w-lim x_n	Weak limit (of a sequence x_n)	37E
$\simeq$	$f \simeq g$	Homotopy (of two mappings f and g)	202B
$\approx$	$X \approx Y$	Homeomorphism (of two topological spaces X and Y)	425G

VIII. Geometry and Algebraic Topology

Notation	Example	Definition	Article and Section
E^n		Euclidean space (of dimension n)	140
P^n		Projective space (of dimension n)	343B
S^n		Spherical surface (of dimension n)	140
$\mathbf{T}^n$		Torus (of dimension n)	422E
H^n	$H^n(X, A)$	n-dimensional cohomology group (of X with coefficients in A)	201H
H_n	$H_n(X, A)$	n-dimensional homology group (of X with coefficients in A)	201G
	$H_n(C)$	(of chain complex C)	201B
π_n	$\pi_n(X)$	n-dimensional homotopy group (of X)	202J, 170
∂	∂C	Boundary (of C)	201B
δ	δf	Coboundary (of f)	201H*
Sq	$Sq^i x$	Streenrod square (of x)	64B
$\mathscr{P}$	$\mathscr{P}_p^r(x)$	Steenrod pth power (of x)	64B
$\smile$	$z_1 \smile z_2$	Cup product (of z_1 and z_2)	201I
$\frown$	$z_1 \frown z_2$	Cap product (of z_1 and z_2)	201K
$\wedge$	$\omega \wedge \eta$	Exterior product (of two differential forms ω and η)	105Q*
d	$d\omega$	Exterior derivative (of a differential form ω)	105Q
grad	grad φ	Gradient (of a function φ)	442D
rot	rot $\mathbf{u}$	Rotation (of a vector $\mathbf{u}$)	442D
div	div $\mathbf{u}$	Divergence (of a vector $\mathbf{u}$)	442D
Δ	$\Delta\varphi$	Laplacian (of a function φ)	323A
$\square$	$\square\varphi$	d'Alembertian (of a function φ)	130A
D	$D\varphi$	Differential operator	112A*
$\dfrac{D(u_1, \ldots, u_n)}{D(x_1, \ldots, x_n)}, \left\lvert\dfrac{\partial u_i}{\partial x_j}\right\rvert, \det\left(\dfrac{\partial u_i}{\partial x_j}\right)$		Jacobian determinant (of $(u_1, \ldots, u_n)$ with respect to $(x_1, \ldots, x_n)$)	208B
$\dfrac{\partial(u_1, \ldots, u_n)}{\partial(x_1, \ldots, x_n)}, \left(\dfrac{\partial u_i}{\partial x_j}\right)$		Jacobian matrix (of $(u_1, \ldots, u_n)$ with respect to $(x_1, \ldots, x_n)$)	208B

IX. Function Spaces

Notation	Example	Definition	Article and Section
C	$C(\Omega)$	Space of continuous functions (on Ω)	168B(1)

Notation	Example	Definition	Article and Section
L_p	$L_p(\Omega)$, $L_p(a, b)$	Space of functions such that $\vert f\vert^p$ is integrable on Ω	168B(2)
C^l	$C^l(L)(1 \leqslant l \leqslant \infty)$	Space of functions of class C^l	168B(9)
$\mathscr{D}$	$\mathscr{D}(\Omega)$	Space of C^∞ functions with compact support	168B(13)
$\mathscr{E}$	$\mathscr{E}(\Omega)$	Space of C^∞ functions	168B(13)

[For $\mathscr{A}(\Omega)$, $A(\Omega)$, $A_p(\Omega)$, $\mathscr{B}(\Omega)(=\mathrm{D}_{L^\infty}(\Omega))$, $BMO(\mathbf{R}^n)$, $BV(\Omega)$, c, C, $C_0(\Omega)$, $C_\infty(\Omega)$, $C_0^l(\Omega)$, $\mathscr{D}_{L^p}(\Omega)$, $\mathscr{D}_{\{M^p\}}(\Omega)$, $\mathscr{D}_{(M^p)}(\Omega)$, $\mathscr{E}_{\{M^p\}}(\Omega)$, $\mathscr{E}_{(M^p)}(\Omega)$, $H_p(\mathbf{R}^n)$, $H^l(\Omega)$, $H_0^l(\Omega)$, $\Lambda^S(\mathbf{R}^n)$, $\bigcap \lambda(\alpha^{(k)})$, $\sum \lambda^\times(\alpha^{(k)})$, l_p, $L_{(p,q)}(\Omega)$, m, $M(\Omega)$, $\mathscr{O}(\Omega)$, $\mathscr{O}_p(\Omega)$, $\mathscr{S}$, s, $S(\Omega)$, $W_p^l(\Omega)$ → 168 Function Spaces. For $\mathscr{B}(\Omega)$ (Space of Sato hyperfunctions), $\mathscr{D}'(\Omega)$, $\mathscr{E}'(\Omega)$, $\mathscr{O}_c$, $\mathscr{O}_M$, $\mathscr{S}'(\mathbf{R}^n)$ → 125 Distributions and Hyperfunctions]

X. Functions

Notation	Example	Definition	Article and Section
$\vert\ \vert$	$\vert z\vert$	Absolute value (of a complex number z)	74B*
Re	$\mathrm{Re}\, z$	Real part (of a complex number z)	74A
Im	$\mathrm{Im}\, z$	Imaginary part (of a complex number z)	74A*
$\bar{\ }$	$\bar{z}$	Complex conjugate (of a complex number z)	74A
arg	$\arg z$	Argument (of a complex number z)	74C
[ ]	$[\alpha]$	Gauss symbol (greatest integer not exceeding a real number α)	83A
O	$f(x)=O(g(x))$	Landau's notation ($f(x)/g(x)$ is bounded for $x \to \alpha$)	87G
o	$f(x)=o(g(x))$	Landau's notation ($f(x)/g(x)$ tends to 0 for $x \to \alpha$)	87G
$\sim$	$f(x) \sim g(x)$	Infinite or infinitesimal of the same order (for $x \to \alpha$)	87G*
D	$D(T)$	Domain (of an operator T)	251A
R	$R(T)$	Range (of an operator T)	251A
supp	$\mathrm{supp}\, f$	Support (of a function f)	168B(1)
p.v.	$\mathrm{p.v.} \int_a^b f(x)dx$	Cauchy's principal value (of an integral)	216D
Pf	$\mathrm{Pf} \int f(x)dx$	Finite part (of an integral)	125C
δ	$\delta(x)$, δ_x	Dirac's delta function (measure or distribution)	125C*
exp	$\exp x$	Exponential function ($\exp x = e^x$)	113D, 269H
log, Log	$\log x$, $\mathrm{Log}\, x$	Natural logarithmic function and its principal value, respectively	131D, G
$\sin x$, $\cos x$, $\tan x$, $\sec x$, $\mathrm{cosec}\, x$, $\mathrm{cotan}\, x$		Trigonometric functions	131E
$\arcsin x$, $\arccos x$, $\arctan x$		Inverse trigonometric functions	131E
$\mathrm{Arc}\sin x$, $\mathrm{Arc}\cos x$, $\mathrm{Arc}\tan x$		Principal value of inverse trigonometric functions	131E

Notation	Example	Definition	Article and Section
$\sinh x$, $\cosh x$, $\tanh x$		Hyperbolic functions	131F
$\binom{\ }{\ }$, C	$\binom{n}{r}$, ${}_nC_r$	Binomial coefficient, combination	330
P	${}_nP_r$	Permutation	330
!	$n!$	Factorial (of n)	330
φ	$\varphi(n)$	Euler function	295C*
μ	$\mu(n)$	Möbius function	295C
ζ	$\zeta(z)$	Riemann zeta function	450B*
J_v	$J_v(z)$	Bessel function of the first kind	39B
Γ	$\Gamma(x)$	Gamma function	174A
B	$B(x, y)$	Beta function	174C
F	$F(\alpha, \beta, \gamma; z)$	Gauss's hypergeometric function	206A
P	$P\begin{Bmatrix} a & b & c \\ \lambda & \mu & \nu \\ \lambda' & \mu' & \nu' \end{Bmatrix} x$	Riemann's P function	253B
Li	$\mathrm{Li}(x)$	Logarithmic integral	167D
XI. Probability			
P, Pr	$P(E)$, $\Pr(\varepsilon)$	Probability (of an event)	342B*
E	$E(X)$	Mean or expectation (of a random variable X)	342C
V, σ^2	$V(X)$, $\sigma^2(X)$	Variance (of a random variable X)	342C
ρ	$\rho(X, Y)$	Correlation coefficient (of two random variables X and Y)	342C*
$P(\mid)$	$P(E\mid F)$	Conditional probability (of an event E under the condition F)	342E
$E(\mid)$	$E(X\mid Y)$	Conditional mean (of a random variable X under the condition Y)	342E
N	$N(m, \sigma^2)$	One-dimensional normal distribution (with mean m and variance σ^2)	Appendix A, Table 22
	$N(\boldsymbol{\mu}, \Sigma)$	Multidimensional normal distribution (with mean vector $\boldsymbol{\mu}$ and variance matrix Σ)	Appendix A, Table 22
P	$P(\lambda)$	Poisson distribution (with parameter λ)	Appendix A, Table 22*

Systematic List of Articles

XI
Complex Analysis

XII
Functional Analysis

XIII
Differential, Integral, and Functional Equations

XIV
Special Functions

XV
Numerical Analysis

XVI
Computer Science and Combinatorics

XVII
Probability Theory

XVIII
Statistics

XIX
Mathematical Programming and Operations Research

XX
Mechanics and Theoretical Physics

XXI
History of Mathematics

Alphabetical List of Articles

Contributors to the Second Edition

Abe Eiichi
Akahira Masafumi
Akaike Hirotugu
Akaza Touru
Amari Shun-ichi
Ando Tsuyoshi
Araki Huzihiro
Araki Shôrô
Arimoto Suguru
Asai Akira
Fuji'i'e Tatuo
Fujiki Akira
Fujisaki Genjiro
Fujita Hiroshi
Fujiwara Daisuke
Fujiwara Masahiko
Fukushima Masatoshi
Furuya Shigeru
Hamada Yûsaku
Hasumi Morisuke
Hattori Akio
Hayashi Kiyoshi
Hida Takeyuki
Hijikata Hiroaki
Hirose Ken
Hitotumatu Sin
Huzii Mituaki
Ibaraki Toshihide
Ihara Shin-ichiro
Iitaka Shigeru
Ikawa Mitsuru
Ikebe Teruo
Ikebe Yasuhiko
Ikeda Nobuyuki
Inagaki Nobuo
Inoue Atsushi
Iri Masao
Itaya Nobutoshi
Itô Kiyosi
Itô Seizô
Ito Yuji
Itoh Mitsuhiro
Iwahori Nobuyoshi
Iyanaga Shôkichi
Kadoya Norihiko
Kageyama Sampei
Kamae Teturo
Kaneko Akira
Kaneyuki Soji
Kasahara Kōji
Kataoka Shinji
Kato Junji
Kato Mitsuyoshi
Kawai Takahiro
Kawasaki Tetsuro
Kimura Tosihusa
Kishi Masanori
Kobayashi Shoshichi
Kodama Yukihiro
Komatsu Hikosaburo
Kômura Yukio
Kondo Takeshi
Konishi Yoshio
Konno Hiroshi
Kotani Shinichi
Kubota Yoshihisa
Kumano-go Hajime
Kunita Hiroshi
Kuramoto Yuki
Kuroda Shige-Toshi
Kuroda Tadashi
Kusama Tokitake
Kusano Takaŝi
Kusunoki Yukio
Mabuchi Toshiki
Maeda Shūichiro
Maeda Yoshiaki
Maehara Shôji
Makabe Hajime
Maruyama Masaki
Matsumoto Kikuji
Matsumoto Yukio
Matsumura Mutsuhide
Mitsui Takayoshi
Mizohata Sigeru
Mizutani Tadayoshi
Mori Masatake
Morimoto Akihiko
Morimura Hidenori
Morita Shigeyuki
Morita Yasuo
Motohashi Nobuyoshi
Murasugi Kunio
Nagami Keio
Nagata Masayoshi
Nakagawa Hisao
Nakamura Tokushi
Nakaoka Minoru
Namba Kanji
Namba Makoto
Namikawa Yukihiko
Naruki Isao
Niiro Fumio
Nishikawa Seiki
Nishimura Junichi
Nishitani Kensaku
Nisio Makiko
Noguchi Hiroshi
Nozaki Akihiro
Ôaku Toshinori
Obata Morio
Ochiai Takushiro
Ogiue Koichi
Ohta Masami
Ohya Yujiro
Oikawa Kôtaro
Oka Mutsuo
Okabe Yasunori

Okuno Tadakazu
Omori Hideki
Oshima Toshio
Ozawa Mitsuru
Saito Kyoji
Saito Masahiko
Saito Tosiya
Sakai Fumio
Sakawa Yoshiyuki
Sato Fumitaka
Sato Ken-iti
Sawashima Ikuko
Shibata Katsuyuki
Shibata Keiichi
Shiga Koji
Shiga Tokuzo
Shikata Yoshihiro
Shimada Nobuo
Shimakura Norio
Shimizu Ryoichi
Shioda Tetsuji
Shiohama Katsuhiro
Shiraiwa Kenichi
Shiratani Katsumi
Sibuya Masaaki
Sugie Toru
Sugiura Mitsuo
Sugiura Shigeaki
Suita Nobuyuki
Sumihiro Hideyasu
Suzuki Masuo
Suzuki Mitsuo
Takahashi Tsuneo
Takakuwa Shoichiro
Takami Hideo
Takasu Toru
Takeuchi Kei
Takeuchi Masaru
Takeuti Gaishi
Tamura Itiro
Tanabe Kunio
Tanaka Hiroshi
Tanaka Masatsugu
Tanaka Shunichi
Tanno Shukichi
Tatsumi Tomomasa
Terada Toshiaki
Toda Hideo
Toda Hirosi
Toda Morikazu
Toda Nobushige
Tomiyama Jun
Totoki Haruo
Tugé Tosiyuki
Uchida Fuichi
Uchiyama Saburô
Ueno Kenji
Ukai Seiji
Ura Shoji
Ushijima Teruo
Wakabayashi Isao
Washizu Kyuichiro
Watanabe Kimio
Watanabe Shinzo
Watanabe Takesi
Yamaguchi Masaya
Yamamoto Koichi
Yamamoto Sumiyasu
Yamamoto Tetsuro
Yamazaki Masao
Yamazaki Tadashi
Yanagawa Takashi
Yanagiwara Hiroshi
Yanai Haruo
Yano Tamaki
Yokonuma Takeo
Yoshikawa Atsushi
Yoshida Junzo
Yosida Kôsaku

Contributors to the First Edition

Aizawa Sadakazu
Akao Kazuo
Akizuki Yasuo
Amemiya Ayao
Anzai Hirotada
Aoki Kiyoshi
Araki Huzihiro
Araki Shôrô
Arima Akito
Asaka Tetsuichi
Asano Keizo
Asatani Teruo
Azumaya Gorô
Fujinaka Megumu
Fujita Hiroshi
Fujiwara Daisuke
Fukuda Takeo
Fukutomi Setsuo
Furuya Shigeru
Goto Morikuni
Gotô Motinori
Gotō Yūzo
Hagihara Yusuke
Harada Manabu
Hasimoto Hidenori
Hattori Akio
Hattori Akira
Hayashi Chikio
Hayashi Keiichi
Hayashi Tsuyoshi
Hida Takeyuki
Hidaka Koji
Hirakawa Junkô
Hirano Tomoharu
Hirasawa Yoshikazu
Hirayama Akira
Hirose Ken
Hisatake Masao
Hitotumatu Sin
Hokari Shisanji
Homma Kiyomi
Homma Tatsuo
Honbu Hitoshi
Hong Imsik
Hosokawa Fujitsugu
Hukuhara Masuo
Husimi Kozi
Huzii Mituaki
Ichida Asajiro
Ihara Shin-ichiro
Ihara Yasutaka
Iitaka Shigeru
Ikawa Mitsuru
Ikeda Mineo
Ikeda Nobuyuki
Ikehara Shikao
Imai Isao
Inaba Eiji
Inagaki Takeshi
Inoue Masao
Inui Teturo
Iri Masao
Irie Shoji
Ise Mikio
Iseki Kanesiroo
Ishida Yasushi
Ishii Goro
Isizu Takehiko
Itô Juniti
Itô Kiyosi
Itô Noboru
Itô Seizô
Itô Teiiti
Ito Tsutomu
Itoh Makoto
Iwahashi Ryōsuke
Iwahori Nagayoshi
Iwamura Tsurane
Iwano Masahiro
Iwasawa Kenkichi
Iwata Giiti
Iyanaga Shôkichi
Izumi Shin-ichi
Kaburaki Masaki
Kamae Keiko
Kametani Shunji
Kanbe Tsunekazu
Kanitani Jōyō
Kasahara Koji
Katase Kiyoshi
Kato Tosio
Katuura Sutezo
Kawada Yukiyosi
Kawaguchi Akitsugu
Kawai Saburo
Kawakami Riiti
Kawata Tatuo
Kihara Taro
Kimura Motoo
Kimura Tosihusa
Kinoshita Shin'ichi
Kitagawa Toshio
Kiyasu Zen'ichi
Koba Ziro
Kobayashi Mikio
Kobori Akira
Koda Akira
Kodaira Kunihiko
Koga Yukiyoshi
Komatsu Hikosaburo
Komatu Atuo
Komatu Yûsaku
Kondo Jiro
Kondo Kazuo
Kondô Motokiti
Kondō Ryōji
Kondo Takeshi
Kondo Yōitsu

Kôta Osamu
Kotake Takesi
Kotani Masao
Kozai Yoshihide
Kubo Izumi
Kubo Ryogo
Kubota Tadahiko
Kudo Akio
Kudō Hirokichi
Kudo Tatsuji
Kuga Michio
Kunisawa Kiyonori
Kunita Hiroshi
Kunugi Kinjiro
Kuranishi Masatake
Kuroda Shige-Toshi
Kuroda Sigekatu
Kuroda Sige-Nobu
Kusama Tokitake
Kusano Takasi
Kusunoki Yukio
Maehara Shôji
Maruyama Gisiro
Mashio Shinji
Masuda Kyūya
Masuyama Motosaburo
Matsukuma Ryozai
Matsumoto Kikuji
Matsumoto Makoto
Matsumoto Toshizô
Matsumura Hideyuki
Matsumura Mutsuhide
Matsushima Yozo
Matsuyama Noboru
Matuda Tizuko
Matusita Kameo
Matuzaka Kazuo
Mikami Masao
Mimura Yukio
Minagawa Takizo
Mita Hiroo
Mitome Michio
Mitsui Takayoshi
Miyake Kazuo
Miyasawa Koichi
Miyashita Totaro
Miyazawa Hironari
Miyazima Tatuoki
Mizohata Sigeru
Mizuno Katsuhiko
Mogi Isamu
Mori Shigeo
Moriguti Sigeiti
Morimoto Haruki
Morimoto Seigo
Morimura Hidenori
Morishima Taro
Morita Kiiti
Morita Yuzo
Moriya Mikao
Moriya Tomijiro
Motohashi Nobuyoshi
Motoo Minoru
Murakami Shingo
Muramatu Tosinobu
Murata Tamotsu
Nabeya Seiji
Nagano Tadashi
Nagao Hirosi
Nagashima Takashi
Nagata Jun-iti
Nagata Masayoshi
Nagumo Mitio
Nakai Mitsuru
Nakai Yoshikazu
Nakamori Kanji
Nakamura Koshiro
Nakamura Masahiro
Nakamura Tokushi
Nakanishi Shizu
Nakano Shigeo
Nakaoka Minoru
Nakayama Tadasi
Nakayama Takashi
Namikawa Yukihiko
Nanba Kanji
Narita Masao
Narumi Seimatu
Niiro Fumio
Nikaido Hukukane
Nishi Mieo
Nishimura Toshio
Noguchi Hiroshi
Nomizu Katsumi
Nomoto Hisao
Nomura Yukichi
Nonaka Toshio
Noshiro Kiyoshi
Nozaki Akihiro
Nozaki Yasuo
Ochiai Kiichiro
Ogasawara Tôjirō
Ogawa Junjiro
Ogawara Masami
Ogino Shusaku
Ōhasi Saburo
Ohtsuka Makoto
Oikawa Kotaro
Ôishi Kiyoshi
Oka Syoten
Okamoto Kiyosato
Okamoto Masashi
Okubo Kenjiro
Okubo Tanjiro
Okugawa Kōtaro
Ōnari Setsuo
Ono Akimasa
Ono Katuzi
Ono Shigeru
Ono Takashi
Onoyama Takuji
Oshida Isao

Oshima Nobunori
Osima Masaru
Otsuki Tominosuke
Ozaki Shigeo
Ozawa Mitsuru
Ozeki Hideki
Saito Kinichiro
Saito Tosiya
Saito Yoshihiro
Sakai Eiichi
Sakai Takuzo
Sakamoto Heihachi
Sakata Shôichi
Sasaki Shigeo
Satake Ichiro
Sato Ken-iti
Sato Ryoichiro
Sato Tokui
Sato Yumiko
Sawashima Ikuko
Seimiya Toshio
Seki Setsuya
Shibagaki Wasao
Shiga Kôji
Shimada Nobuo
Shimizu Hideo
Shimizu Tatsujiro
Shimura Goro
Shintani Takuro
Shizuma Ryoji
Shoda Kenjiro
Shono Shigekata
Sibuya Masaaki
Simauti Takakazu
Sirao Tunekiti
Sueoka Seiichi
Suetuna Zyoiti
Sugawara Masahiro
Sugawara Masao
Sugiura Mitsuo
Sugiura Nariaki
Sugiyama Shohei
Suita Nobuyuki
Sumitomo Takeshi
Sunouchi Genichiro
Suzuki Michio
Suzuki Yoshindo
Takada Masaru
Takagi Teiji
Takagi Yutaka
Takahashi Hidetoshi
Takahashi Tsunero
Takenouchi Osamu
Takenouchi Tadao
Takeuchi Kei
Takeuti Gaisi
Takizawa Seizi
Tamagawa Tsuneo
Tamura Itiro
Tamura Jirô
Tanaka Chuji
Tanaka Hiroshi
Tanaka Hisao
Tanaka Minoru
Tanaka Sen-ichiro
Tannaka Tadao
Tatsumi Tomomasa
Tatuzawa Tikao
Terasaka Hidetaka
Toda Hideo
Toda Hirosi
Toda Morikazu
Tôki Yukinari
Tomotika Susumu
Totoki Haruo
Toyama Hiraku
Tsuboi Chuji
Tsuchikura Tamotsu
Tsuji Masatsugu
Tsujioka Kunio
Tsukamoto Yôtarô
Tsuzuku Toshiro
Tugué Tosiyuki
Tumura Yosiro
Uchiyama Saburo
Uchiyama Tatsuo
Udagawa Masatomo
Ueno Kenji
Ueno Tadashi
Ugaeri Tadashi
Umegaki Hisaharu
Umezawa Hiroomi
Umezawa Toshio
Uno Toshio
Ura Shoji
Ura Taro
Urabe Minoru
Wada Yasaku
Washimi Shinichi
Washio Yasutoshi
Washizu Kyuichiro
Watanabe Hiroshi
Watanabe Hisao
Watanabe Masaru
Watanabe Shinzo
Yamada Isamu
Yamaguti Masaya
Yamamoto Koichi
Yamanaka Takesi
Yamanoshita Tsuneyo
Yamanouti Takahiko
Yamashita Hideo
Yamauti Ziro
Yamazaki Keijiro
Yanagihara Kichiji
Yano Kentaro
Yano Shigeki
Yoneda Nobuo
Yoshizawa Hisaaki
Yosida Kôsaku
Yosida Setuzô
Yosida Yôiti

Translators (for the First Edition)

Adachi Masahisa
Aizawa Sadakazu
Akizuki Yasuo
Akô Kiyoshi
Araki Shôrô
Arima Akito
Asano Keizo
Azumaya Goro
Chiba Katsuhiro
Fujisawa Takehisa
Fujita Hiroshi
Fukada Ichiro
Fukuyama Masaru
Furuya Shigeru
Hashimoto Yasuko
Hasimoto Hidenori
Hasumi Morisuke
Hattori Akio
Hattori Akira
Hayashi Kazumichi
Hida Takeyuki
Hinata Shigeru
Hirasawa Yoshikazu
Hirose Ken
Hitotumatu Sin
Honda Taira
Hong Imsik
Hosoi Tsutomu
Hukuhara Masuo
Ihara Shin-ichiro
Ihara Yasutaka
Ikeda Mineo
Ikeda Nobuyuki
Imai Isao
Inoue Masao
Iri Masao
Ishii Goro
Itô Seizô
Ito Yuji
Iwahashi Ryōsuke
Iwahori Nagayoshi
Iwarmura Tsurane
Iwano Masahiro
Iwata Giichi
Iyanaga Kenichi
Iyanaga Shôkichi
Kametani Shunji
Kasahara Koji
Kato Junji
Kawada Yukiyosi
Kawano Sanehiko
Kimura Tosihusa
Kitagawa Toshio
Kobori Akira
Kodaira Kunihiko
Komatsu Hikosaburo
Komatu Atuo
Komatu Yûsaku
Kondo Ryōji
Kondo Takeshi
Kozai Yoshihide
Kubo Ryogo
Kubota Tomio
Kudo Akio
Kudo Hirokichi
Kumano-go Hitoshi
Kunisawa Kiyonori
Kunita Hiroshi
Kurita Minoru
Kuroda Shige-Toshi
Kusama Tokitake
Kusano Takaŝi
Maruyama Gisiro
Matsuda Michihiko
Matsumoto Kikuji
Matsumoto Makoto
Matsumura Hideyuki
Matsushima Yozo
Mimura Yukio
Miyazawa Hironari
Mizohata Sigeru
Mori Toshio
Morimoto Haruki
Morimura Hidenori
Morita Kiiti
Morita Yasuo
Motoo Minoru
Murakami Shingo
Muramatu Tosinobu
Murase Ichiro
Nagao Hirosi
Nagasawa Masao
Nagata Masayoshi
Nakai Mitsuru
Nakai Yoshikazu
Nakamura Tokushi
Nakanishi Shizu
Nakano Shigeo
Nakaoka Minoru
Namikawa Yukihiko
Namioka Isaac
Nanba Kanji
Niiro Fumio
Nikaido Hukukane
Nishi Mieo
Nishijima Kazuhiko
Nishimiya Han
Noguchi Hiroshi
Nonaka Toshio
Noshiro Kiyoshi
Nozaki Akihiro
Ôbayashi Tadao
Ogino Shusaku
Ohtsuka Makoto
Oikawa Kotaro
Okamoto Masashi
Okubo Kenjiro
Okugawa Kōtaro

Ono Katuzi
Ono Shigeru
Osima Masaru
Ozawa Mitsuru
Ozeki Hideki
Saito Tosiya
Sakai Shoichiro
Sakamoto Minoru
Sasaki Shigeo
Satake Ichiro
Sato Kenkichi
Sawashima Ikuko
Shibuya Masaaki
Shiga Kôji
Shimada Nobuo
Shintani Takuro
Shioda Tetsuji
Shiraiwa Kenichi
Shiratani Katsumi
Shizuma Ryoji
Simauti Takakazu
Sugawara Masahiro
Sugiura Mitsuo
Sugiyama Shohei
Suita Nobuyuki
Sumitomo Takeshi
Sunouchi Genichiro
Suzuki Michio
Takahashi Moto-o
Takahashi Reiji
Takahashi Tsunero
Takami Hideo
Takaoka Seiki
Takenouchi Osamu
Takeuchi Kei
Takeuti Gaisi
Tamura Itiro
Tanaka Hiroshi
Tanaka Hisao
Tanaka Minoru
Tatsumi Tomomasa
Tatuzawa Tikao
Terasaka Hidetaka
Toda Hideo
Toda Hirosi
Tsuka-da Nobutaka
Tsukamoto Yôtarô
Tugué Tosiyuki
Uchiyama Saburo
Uesu Tadahiro
Umegaki Hisaharu
Umezawa Toshio
Uno Toshio
Ura Shoji
Ura Taro
Urabe Minoru
Uzawa Hirobumi
Wada Junzo
Watanabe Hiroshi
Watanabe Hisao
Watanabe Shinzo
Watanabe Takesi
Yamaguti Masaya
Yamamoto Koichi
Yamanaka Takesi
Yamazaki Keijiro
Yano Kentaro
Yano Shigeki
Yoshizawa Taro
Yosida Kôsaku
Yosida Setuzô

Name Index

Note. Citation is to article and section, not to page. r = references; STR = Statistical Tables for Reference (to be found after App. B); NTR = Numerical Tables for Reference. * means that the article is not divided into sections.

A

Abadie, Jean M. (1919–) 292.r
Abe, Eiichi (1927–) 13.R
Abe, Kinetsu (1941–) 365.L
Abel, Niels Henrik (1802–29) 1 2.A, B, C 3.A, B, G, J, L, M 8 10.D 11.B, C, E 20 21.B 52.B, N 60.L 73.A 121.D 136.B 172.A, B, G, H 190.A, H, Q 201.R 202.N 217.A, L 240.G 267 308.E 339.B 367.H 379.D, F, K, N 383.B 388.D 422.A, E
Abers, Ernest S. 132.r
Aberth, Oliver George (1929–) 301.F
Abhyankar, Shreeram Shankar (1930–) 15.B 16.L 23.r
Abikoff, William (1944–) 234.E
Abraham, C. T. 96.F
Abraham, Ralph H. (1936–) 126.J, r 183 271.r 316.r 420.r
Abramov, Leonid Mikhaĭlovich (1931–) 136.E
Abramowitz, Milton J. NTR
Abrikosov, Alekseĭ Alekseevich (1928–) 402.r
Abûl Wafâ (940–98) 432.C
Achenbach, J. D. 446.r
Ackermann, Wilhelm (1896–1962) 97.*, r 156.E, r 356.B 411.J, r
Ackoff, Russell Lincoln (1919–) 307.A
Aczél, Janós D. (1924–) 388.r
Adams, Douglas Payne 19.r
Adams, John Couch (1819–92) 303.E
Adams, John Frank (1930–) 64.C 200.r 202.S, T, V 237.A, E, I 249.r 426
Adamson, Iain Thomas 200.M
Addison, John West, Jr. (1930–) 22.D, F, G, H 81.r 356.H, r
Adem, José (1921–) 64.B, C 305.A App. A, Table 6.II
Adler, Mark 387.C
Adler, Roy L. (1931–) 126.K 136.E, H
Adler, Stephen L. (1939–) 132.C, r
Ado, Igor Dmitrievich 248.F, r
Adriaan, Anthonisz (c. 1543–1620) 332
Agmon, Shmuel (1922–) 112.F, H, Q 323.H, r 375.B, C
Agnesi, Maria Gaetana (1718–99) 93.H
Aguilar, Joseph 331.F
Ahern, Patrick Robert (1936–) 164.K
Ahlberg, John Harold (1927–) 223.r
Ahlfors, Lars Valerian (1907–) 17.D 21.N 43.G, K 74.r 77.E, F, r 122.I 124.B, r 143.A 169.E 198.r 234.D, E, r 272.I, J, L 352.A, B, C, F 367.B, G, I, r 416 429.D 438.r
Aho, Alfred V. 31.r 71.r 75.r 186.r
Aida, Yasuaki (1747–1817) 230
Airy, Sir George Biddell (1801–92) 325.L App. A, Table 19.IV
Aitken, Alexander Craig (1895–1967) 223.B App. A, Table 21
Aizawa, Sadakazu (1934–) 286.X
Aizenman, Michael (1945–) 136.G 340.r 402.G
Akahira, Masafumi (1945–) 128.r 399.K, O 400.r
Akahori, Takao (1949–) 72.r
Akaike, Hirotugu (1927–) 421.D
Akaza, Tohru (1927–83) 234.r
Akbar-Zadeh, Hassan (1927–) 152.C
Akcoglu, Mustafa A. (1934–) 136.B
Akemann, Charles A. (1941–) 36.K
Akhiezer, Naum Il'ich (1901–80) 197.r 251.r 336.r 390.r
Akizuki, Yasuo (1902–84) 8 59.H 284.F, G 368.F
Alaoglu, Leonidas (1914–) 37.E 424.H
Albanese, Giacomo (1890–1947) 16.P 232.C
al-Battānī, Mohamed ibn Gabis ibn Sinan, Abu Abdallah (858?–929) 26 432.C
Albert, Abraham Adrian (1905–72) 29.F, r 149.r 231.r
Albertus Magnus (1193?–1280) 372
Alcuin (735–804) 372
Aleksandrov, Aleksandr Danilovich (1912–) 111.r 178.A 255.D 365.H 425.r
Aleksandrov (Alexandroff), Pavel Sergeevich (1896–1982) 22.I 65.r 93.r 99.r 117.A, E, F, r 201.A, r 207.C 273.K 425.S–V, r 426. *, r
Alekseev, Vladimir Mikhaĭlovich (1932–80) 420.r
Alekseevskiĭ, Dmitriĭ Vladimirovich (1940–) 364.r
Alexander, Herbert James (1940–) 344.F
Alexander, James Waddell (1888–1971) 65.G 201.A, J, M, O, P 235.A, C, D, E 426
Alexits, György (1899–1978) 317.r
Alfsen, Erik Magnus (1930–) 351.L
Alfvén, Hannes (1908–) 259.*, r
Alinhac, Serge (1948–) 345.A
al-Khwarizmi (Alkwarizmi), Mohammed ibn Musa (c. 780–c. 850) 26
Allard, William Kenneth (1941–) 275.G
Allendoerfer, Carl Bennett (1911–74) 109 365.E
Almgren, Frederick Justin, Jr. (1933–) 275.A, F, G, r 334.F
Altman, Allen B. 16.r
Amari, Shun-ichi (1936–) 399.O
Ambrose, Warren (1914–) 136.D
Amemiya, Ichiro (1923–) 72.r
Amitsur, Shimshon A. (1921–) 200.P
Ampère, André-Marie (1775–1836) 82.A 107.B 278.A
Amrein, Werner O. 375.B, r
Ananda-Rau, K. 121.D
Andersen, Erik Sparre (1919–) 260.J
Anderson, Brian D. O. 86.r
Anderson, Joel H. (1935–) 36.J
Anderson, Richard Louis (1915–) 19.r
Anderson, Robert Murdoch (1951–) 293.D, r
Anderson, Theodore Wilbur (1918–) 280.r 374.r 421.D
Andersson, Karl Gustav (1943–) 274.I
Ando, Tsuyoshi (1932–) 310.H
Ando, Y. 304.r
Andreotti, Aldo (1924–80) 32.F 72.r
Andrews, David F. 371.H
Andrews, Frank C. 419.r
Andrianov, Anatoliĭ N. (1936–) 32.F
Andronov, Aleksandr Aleksandrovich (1901–52) 126.A, I, r 290.r 318.r
Anger, Carl Theodor (1803–58) 39.G App. A, Table 19.IV
Anikin, S. A. 146.A
Anosov, Dmitriĭ Viktorovich (1936–) 126.A, J 136.G
Antiphon (fl. 430? B.C.) 187
Antoine, Louis August (1888–1971) 65.G
Anzai, Hirotada (1919–55) 136.E
Apéry, Roger (1916–) 182.G

B

92.F, r 107.r 179.B, r 198.r 254.r 288.r 339. 429.r 438.B, C
Biedenharn, Lawrence C. (1922–) 353.r
Biezeno, Cornelis Benjamin (1888–1975) 19.r
Biggeri, Carlos 121.C
Biggs, Norman Linstead (1941–) 157.r
Billera, Louis J. (1943–) 173.E
Billingsley, Patrick P. (1925–) 45.r 136.r 250.r 341.r 374.r
Binet, Jacques Philippe Marie (1786–1856) 174.A 295.A
Bing, Rudolf H. (1914–86) 65.F, G 79.D 273.K 382.D 425.AA
Birch, Bryan John (1931–) 4.E 118.C–E 450.S
Birkeland, R. 206.D
Birkhoff, Garrett (1911–) 8.r 87.r 103.r 183.r 243.r 248.J 310.A 311.r 343.r 443.A, E, H
Birkhoff, George David (1884–1944) 30.r 107.A 109 111.I 126.A, E 136.A, B 139.r 153.B, D 157.A 162 253.C 254.D 279.A 286.D 420.F
Birman, Joan S. (1922–) 235.r
Birman, Mikhail Shlemovich (1928–) 331.E
Birnbaum, Allan (1923–76) 399.C, r 400.r
Birtel, Frank Thomas (1932–) 164.r
Bisconcini, Giulio 420.C
Bishop, Errett A. (1928–83) 164.D–F, J, K 367.r
Bishop, Richard L. (1931–) 105.r 178.r 417.r
Bishop, Yvonne M. M. 280.r 403.r
Bitsadze, Andreĭ Vasil'evich (1916–) 326.r
Bjerknes, Carl Anton (1825–1903) 1.r
Björck, Åke (1934–) 302.r
Björck, Göran (1930–) 125.r
Björk, Jan-Erik 112.r 125.EE 274.r
Bjorken, James Daniel (1934–) 132.r 146.A, C 150.r
Blackman, R. B. 421.r
Blackwell, David (Harold) (1919–) 22.H 398.r 399.C
Blahut, Richard E. (1937–) 213.r
Blair, David E. (1940–) 110.E 364.G
Blakers, Albert Laurence (1917–) 202.M
Blanchard, André (1928–) 72.r
Blanc-Lapierre, André Joseph (1915–) 395.r
Bland, Robert G. (1948–) 255.C
Blaschke, Wilhelm (1885–1962) 43.F 76.r 89.C, r 109.*, r 110.C, r 111.r 178.G 218.A, C, H 228.r
Blatt, John Markus 353.r
Blattner, Robert James (1931–) 437.W, EE
Bleaney, B. I. 130.r
Bleuler, Konrad (1912–) 150.G
Bloch, André (1893–1948) 21.N, O 77.F 272.L 429.D
Bloch, Felix (1905–) 353.r 402.H
Bloch, Spencer 16.R
Block, Henry David (1920–78) 420.C
Bloxham, M. J. D. 386.C
Blum, Julius Rubin (1922–82) 136.E
Blum, Manuel 71.D, r
Blumenthal, Ludwig Otto (1876–1944) 32.G 122.E
Blumenthal, Robert McCallum (1931–) 5.r 261.B, r
Boas, Ralph Philip, Jr. (1912–) 58.r 220.D 240.K 429.r
Bôcher, Maxime (1867–1918) 107.A 167.E 193.D
Bochner, Salomon (1899–1982) 5.r 18.A, r 21.Q, r 36.L 80.r 109.*, r 125.A 160.C, r 164.G 192.B, O 194.G 261.F 327.r 341.C, J, r 367.F 378.D 443.A, C, H
Bodewig, Ewald 298.r
Boerner, Hermann (1906–82) 362.r
Boetius, Anicius Manlius Torquatus Severinus (c. 480–524) 372
Bogolyubov, Nikolaĭ Nikolaevich (1909–) 125.W 136.H 146.A 150.r 212.B 290.A, D 361.r 402.J
Böhme, Reinhold (1944–) 275.C
Bohnenblust, (Henri) Frederic (1906–) 28 310.A, G
Bohr, Harald (1887–1951) 18.A, B, H, r 69.B 121.B, C 123.r 450.I
Bohr, Niels Henrik David (1885–1962) 351.A
Bokshteĭn (Bockstein), Meer Feliksovich (1913–) 64.B 117.F
Bol, Gerrit (1906–) 110.r
Boll, Marcel (1886–) NTR
Bolley, Pierre (1943–) 323.N
Boltyanskiĭ, Vladimir Grigor'evich (1925–) 86.r 89.r 117.F 127.G
Boltzmann, Ludwig (1844–1906) 41.A, B, r 136.A 402.B, H, r 403.B, r
Bolyai, János (Johann) (1802–60) 35.A 181 267 285.A
Bolza, Oskar (1857–1942) 46.r
Bolzano, Bernard (1781–1848) 140 273.F
Bombieri, Enrico (1940–) 15.r 72.K 118.B 123.D, E, r 151.J 275.F 438.C 450.P, Q
Bompiani, Enrico (1889–1975) 110.B
Bonnesen, Tommy (1873–) 89.r 228.A
Bonnet, Ossian Pierre (1819–92) 109 111.H 275.A, C 364.D App. A, Table 4.I
Bonsall, Frank Featherstone (1920–) 310.H
Bony, Jean-Michel (1942–) 274.r
Book, D. L. 304.r
Boole, George (1815–64) 33.E 42.A–D, r 104.r 156.B 243.E 267 379.J 411.A, r
Boone, William Werner (1920–83) 97.*, r 161.B
Boothby, William M. (1918–) 110.E
Borchardt, Carl Wilhelm (1817–80) 229.r
Borchers, Hans-Jürgen (1926–) 150.E
Borel, Armand (1923–) 12.B 13.A, G, r 16.Z 32.H, r 56.r 73.r 122.F, G, r 147.K 148.E, r 199.r 203.A 248.O 249.J, V, r 366.D 383.r 384.D 427.B, r 431.r 437.Q 450.r App. A, Table 6.V
Borel, Emile (1871–1956) 20 21.O 22.A, G 58.D 83.B 124.B 156.C 198.Q, r 261.D 270.B, C, G, J 272.E, F 273.F 339.D 342.A, B 379.O 429.B
Borevich, Zenon Ivanovich (1922–) 14.r 297.r 347.r
Borges, Carlos J. Rego (1939–) 273.K 425.Y
Borisovich, Yuriĭ Grigor'evich (1930–) 286.r
Born, Max (1882–1970) 402.J 446.r
Borovkov, Aleksandr Alekseevich (1931–) 260.H
Borsuk, Karol (1905–82) 79.C, r 153.B 202.B, I 382. A, C
Bortolotti, Ettore (1866–1947) 417.E
Bose, Raj Chandra (1901–) 63.D 241.B STR
Bose, Satyendra Nath (1894–1974) 132.A, C 351.H 377.B 402.E
Bott, Raoul (1923–) 105.r 109 153.C 154.F–H 178.G 202.V, r 237.D, H, r 248.r 272.L 279.D 325.J 345.A 366.r 391.N, r 413.r 427.E, r 437.Q App. A, Table 6.VII
Bouligand, Georges (1889–?) 120.D
Bouquet, Jean-Claude (1819–85) 107.A 111.F 288.B 289.B
Bourbaki, Nicolas 8 13.r 20.r 22.r 34.r 35.r 60.r 61.r 67.r 74.r 84.r 87.r 88.r 103.r 105.r 106.r 122.r 131.r 135.r 149.r 162 168.C 172.r 187.r 216.r 221.r 225.r 248.r 249.r 256.r 265.r 266.r 267.r 270.r 277.r 284.r 310.I 311.r 312.r 337.r 348.r 355.r 360.r 362.r 368.r 379.r 381.r 409.r 423.r 424.I, r 425.S, W, Y, CC, r 435.r 436.r 443.A
Bourgin, David G. 201.r
Bourgne, Robert 171.r
Bourguignon, Jean-Pierre (1947–) 80.r 364.r

C

E

F

G

H

K

Q

R

S

T

V

W

X

Y

Z

Subject Index

Note: Citation is to article and section or to a table in an appendix, not to page.

A

C

D

E

G

H

I

K

L

M

N

O

P

Q

R

S

U

W

X

Y